www.mathprojects.com
MPJ
UH math lessons
the futures channel

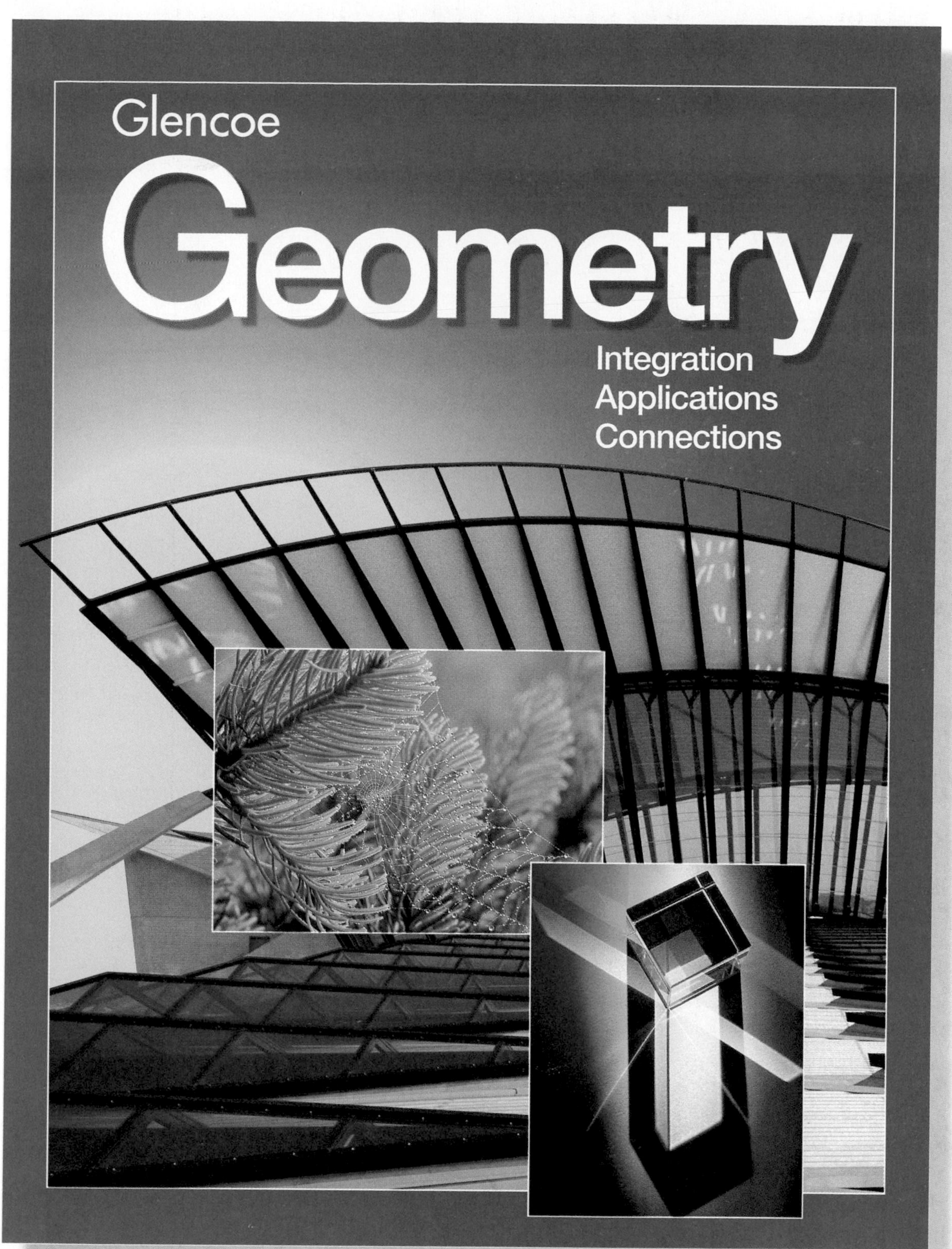

GLENCOE

McGraw-Hill

New York, New York Columbus, Ohio Woodland Hills, California Peoria, Illinois

Glencoe/McGraw-Hill

A Division of The McGraw-Hill Companies

Send all inquiries to:
Glencoe/McGraw-Hill
8787 Orion Place
Columbus, OH 43240

ISBN: 0-07-822880-8

6 7 8 9 10 071/055 05 04 03

WHY IS GEOMETRY IMPORTANT?

Why do I need to study geometry? When am I ever going to have to use geometry in the real world?

Many people, not just geometry students, wonder why mathematics is important. ***Geometry*** is designed to answer those questions through **integration**, **applications**, and **connections**.

Algebra

Did you know that geometry and algebra are closely related? Topics from all branches of mathematics, like algebra and statistics, are integrated throughout the text.

You'll apply algebra to the midpoint formula. (Lesson 1–5, page 38)

Physical Therapy

Would you believe that medical personnel use angles and line segments in their work? Real-world uses of geometry are presented.

Physical therapists apply inequalities in two triangles to help people regain strength after injury by using special exercises. (Lesson 5–6, page 273)

Literature

What does literature have to do with mathematics? Mathematical topics are connected to other subjects you study.

Lewis Carroll, author of *Alice's Adventures in Wonderland,* frequently used the logic of mathematics in his writing. (Lesson 2–2, page 76)

Authors

CINDY J. BOYD teaches mathematics at Abilene High School in Abilene, Texas. She received her B.S.Ed. and M.Ed. degrees from Abilene Christian University. She has received numerous awards and is the first teacher to receive the Texas Presidential State Award for four consecutive years (1994–1997). Ms. Boyd was selected as the 1995–1996 Mathematics Teacher of the Year by Disney and McDonald's American Teacher Awards. Ms. Boyd is an active member of the National Council of Teachers of Mathematics, presenting *Skit-So-Phrenia!* at numerous conventions each year, and she serves on the regional Services Committee as the Southern-Region 2 representative. She is also a consultant on Glencoe's *Algebra 1* and *Algebra 2*.

"Glencoe Geometry is well suited for teachers and students because concepts are developed and built for deeper understanding. A strong problem-solving base can be developed and maintained by all students so they enter the workforce with the tools needed for whatever the twenty-first century holds."

GAIL F. BURRILL taught mathematics at Whitnall High School in Greenfield, Wisconsin. Ms. Burrill obtained her B.S. in Mathematics from Marquette University and her M.S. degree in Mathematics from Loyola University. Ms. Burrill received a Presidential Award for Excellence in Teaching Mathematics and Science in 1985. She is a past president of the Wisconsin Mathematics Council and received a Wisconsin Distinguished Mathematics Educator Award. Ms. Burrill is currently the president of the National Council of Teachers of Mathematics. She is a member of the National Board for Professional Teaching Standards and was Chair of the Mathematics Committee for Initial Teacher Certification for the Council of Chief State School Officers. She is a Fellow of the American Statistical Association, has spoken internationally, and has authored many articles on teaching math and statistics.

JERRY J. CUMMINS is a Staff Development Specialist for the Bureau of Education and Research. He is also a consultant on Systemic Leadership Initiatives in Mathematics and Science in the State of Illinois. He was the Chair of the Mathematics/Science Division at Lyons Township High School, in LaGrange, Illinois. Mr. Cummins obtained his B.S. degree in mathematics education and M.S. degree in educational administration and supervision from Southern Illinois University at Carbondale. He also holds an M.S. degree in mathematics education from the University of Oregon. Mr. Cummins has spoken at many local, state, and national mathematics conferences, and is a member of the Regional Services Committee of the National Council of Teachers of Mathematics. He is currently the First Vice-President of the National Council of Supervisors of Mathematics. Mr. Cummins received an Illinois State Presidential Award for Excellence in Teaching of Mathematics.

"Glencoe Geometry places the highest priority possible on student achievement. Motivation is one of the primary factors in learning. Students are able to see or experience reasons for studying each topic."

TIMOTHY D. KANOLD is the Director of Mathematics and a mathematics teacher at Adlai Stevenson High School in Lincolnshire, Illinois. Mr. Kanold obtained his B.S. degree in mathematics education and M.S. degree in mathematics from Illinois State University. He also holds a C.A.S. degree from the University of Illinois. In 1995, he received the Award of Excellence from the Illinois State Board of Education for outstanding contributions to Illinois education. A member of the National Council of Teachers of Mathematics, he served on NCTM's "Professional Standards for Teaching Mathematics" Commission, was a member of the Regional Services Committee, and served as a speaker for "New Dimensions in Leadership." He is a past president of the Council for Presidential Awardees of Mathematics. He is an author of numerous articles on effective classroom teaching practices including a chapter in NCTM's 1990 Yearbook. He is a nationally known, well-respected, motivational speaker.

"The text provides numerous formats and opportunities for students to clarify their thinking and formulate generalizations through guided discovery and investigations while maintaining a formal sense of rigor necessary in a contemporary geometry curriculum."

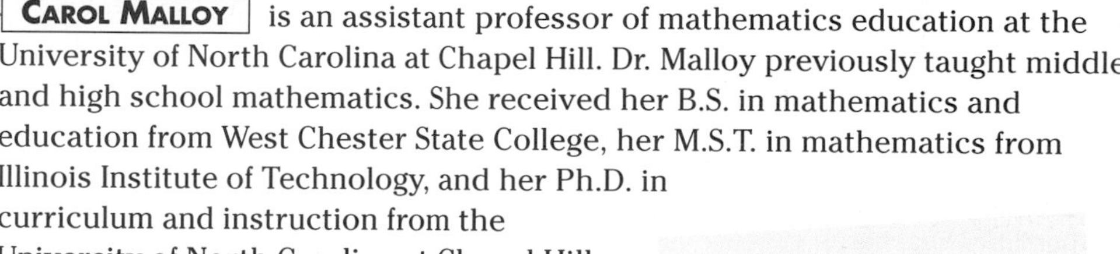

CAROL MALLOY is an assistant professor of mathematics education at the University of North Carolina at Chapel Hill. Dr. Malloy previously taught middle and high school mathematics. She received her B.S. in mathematics and education from West Chester State College, her M.S.T. in mathematics from Illinois Institute of Technology, and her Ph.D. in curriculum and instruction from the University of North Carolina at Chapel Hill. Dr. Malloy is the president of the Benjamin Banneker Association, Inc. Dr. Malloy is an active member of numerous local, state, and national professional organizations and boards, including the National Council of Teachers of Mathematics (NCTM) and Association for Supervision and Curriculum Development (ASCD). She frequently speaks at conferences and schools on the topics of mathematical problem solving, influence of culture on mathematics learning, and mathematics for all students.

"Geometry often has been an exciting course for many teachers and students. This textbook captures the excitement of learning and using geometry for students from all levels and experiences. Diverse approaches to learning and teaching from hands-on activities and technology provide a strong axiomatic structure in the development of deductive and inductive reasoning."

LEE E. YUNKER (1941–1994)

This edition of *Geometry: Integration, Applications, and Connections* is dedicated to the memory of Lee E. Yunker. For 31 years, he educated students and teachers alike. He was personally committed to life-long learning and strove to help all teachers best serve the needs of their students. His many contributions to mathematics education continue in this and the 15 other mathematics books he authored.

Consultants, Writers, and Reviewers

Consultants

Pamela Ann Chandler, Ed. D.
Mathematics Coordinator
Fort Bend ISD
Sugar Land, TX 77478

Eva Gates
Independent Mathematics Consultant
Pearland, Texas

Dr. Luis Ortiz-Franco
Consultant, Diversity
Associate Professor of Mathematics
Chapman University
Orange, California

Gilbert J. Cuevas
Professor of Mathematics Education
University of Miami
Coral Gables, Florida

Deborah Ann Haver, Ed. D.
Mathematics Consultant
Chesapeake, Virginia

Joan Estlow
Mathematics Teacher
Egg Harbor Township High School
Pleasantville, New Jersey

Melissa McClure
Mathematics Consultant
Teaching for Tomorrow
Fort Worth, Texas

Writers

David Foster
Writer, Investigations
Glencoe Author and Mathematics Consultant
Morgan Hill, California

Reviewers

Rhea Baldino
Mathematics Teacher
Charlottesville High School
Charlottesville, Virginia

Paul J. Bohney
Supervisor of Math/Science-Retired
Gary Community School Corporation
Gary, Indiana

Judith L. Conley
Geometry/Algebra II Teacher
Shades Valley High School
Birmingham, Alabama

Sandra Argüelles Daire
Mathematics Teacher
Miami Senior High School
Miami, Florida

Pamela M. Fey
Mathematics Teacher
Northside ISD
San Antonio, Texas

Linda M. Gordius
Mathematics Department Chair
Portland High School
Portland, Maine

Joy E. Hine
Mathematics Teacher
Carmel High School
Carmel, Indiana

Brenda Berg
Math Department Chair
Carbondale Community High School
Carbondale, Illinois

Marcia S. Chumas
Mathematics Teacher
East Mecklenburg High School
Charlotte, North Carolina

Amy L. Craig
Mathematics Teacher
Princeton High School
Cincinnati, Ohio

Walter S. Dewar
Mathematics Department Chair
Rowlett High School
Rowlett, Texas

John H. Geiger
Supervisor of Mathematics
Pasco County School District
Land O' Lakes, Florida

Barbara R. Gray
Mathematics Department Chair
West Forsyth High School
Clemmons, North Carolina

Craig Hochhaus
Mathematics Department Chair
Agoura High School
Agoura Hills, California

Mark Bleiler
Geometry Teacher
Breckenridge High School
Breckenridge, Michigan

Judy Cline
Math Teacher/Department Chair
North Mecklenburg High School
Huntersville, North Carolina

David Crowe
Mathematics Teacher
Thomas W. Harvey High School
Painesville, Ohio

Dollie Driver
Mathematics Teacher
Coronado High School
Lubbock, Texas

Norma L. Goldner
Mathematics Teacher
Benton Harbor High School
Benton Harbor, Michigan

Karen Sparks Gray
Teacher
Clay-Chalkville High School
Clay, Alabama

Nahid Mazdeh Huff
Mathematics Teacher
Colonel White High School
Dayton, Ohio

Table of Contents

Jeff Killion
Math Challenge Tutor
Lower Merion School District
Ardmore, Pennsylvania

Bruce Linn
Mathematics Teacher
Fort Zumwalt North High School
O'Fallon, Missouri

David F. McReynolds
Mathematics Teacher
Boise City High School
Boise City, Oklahoma

Kathleen Walker Murrell
Mathematics Teacher
J. Frank Dobie High School
Pasadena ISD
Houston, Texas

Virginia A. Palmer
Mathematics Teacher
East Lyme High School
East Lyme, Connecticut

Nancy Puhlmann
Mathematics Teacher
Logan High School
Logan, Utah

Linda C. Ruppenthal
Mathematics Department Chairperson
Hancock Middle-Senior High School
Hancock, Maryland

John Sico, Jr.
Vice Principal
Eastside High School
Paterson, New Jersey

Erik L. Thompson
Chemistry Teacher
Thomas Worthington High School
Worthington, Ohio

Peggy Turman
Mathematics Teacher
Cabell Midland High School
Ona, West Virginia

Susan K. White
Math Supervisor
West Chester Area School District
West Chester, Pennsylvania

Lynda B. (Lucy) Kay
Mathematics Department Chair
Martin Middle School
Raleigh, North Carolina

Nancy W. Kinard
Mathematics Teacher
Palm Beach Gardens High School
Palm Beach Gardens, Florida

Kimberly K. Loomis
Mathematics Teacher
Southern Nevada Vocational-Technical Center
Las Vegas, Nevada

Patrick Miller
Instructional Support Team
Fort Worth ISD
Fort Worth, Texas

Maxine Nesbitt
Mathematics Teacher
Carmel High School
Carmel, Indiana

Joseph P. Pimental
Mathematics Teacher
Taunton High School
Taunton, Massachusetts

Scott Reed
Mathematics Teacher
Bogle Junior High School
Chandler, Arizona

Lee Ann Russell
Mathematics Teacher
George Washington High School
Danville, Virginia

Edith R. Simon
Teacher
McMain Magnet Secondary School
New Orleans, Louisiana

Margaretta L. Thornton
Mathematics Teacher
Sam Houston High School
Moss Bluff, Louisiana

Connie G. Turner
Mathematics Teacher
East Rutherford High School
Forest City, North Carolina

Nancy Keen
Geometry Teacher
Martinsville High School
Martinsville, Indiana

Susan Rypkema Lambert
Mathematics Teacher
Plano Senior High School
Plano, Texas

Gerald Martau
Deputy Superintendent-Retired
Lakewood City Schools
Lakewood, Ohio

Melleretha Moses-Johnson
Mathematics Coordinator
Saginaw Public Schools
Saginaw, Michigan

Roger O'Brien
Mathematics Supervisor
Polk County Schools
Bartow, Florida

Joseph "Joe" Ruffino
Mathematics Teacher
Covington High School
Covington, Louisiana

William Scott
Mathematics Supervisor (K–12)
Colorado Springs School District 11
Colorado Springs, Colorado

Mary Beth Smith
Mathematics Teacher
Dublin Coffman High School
Dublin, Ohio

John Tithof
Mathematics Teacher
Gaylord High School
Gaylord, Michigan

Simonette Urbain
Mathematics Teacher
J. Sterling Morton High School, West Campus
Berwyn, Illinois

CHAPTER 2 Connecting Reasoning and Proof 68

CHAPTER 3 Using Perpendicular and Parallel Lines 122

interNET CONNECTION

Content Integration

What does algebra have to do with geometry? Believe it or not, you can study most math topics from more than one point of view. Here are some examples.

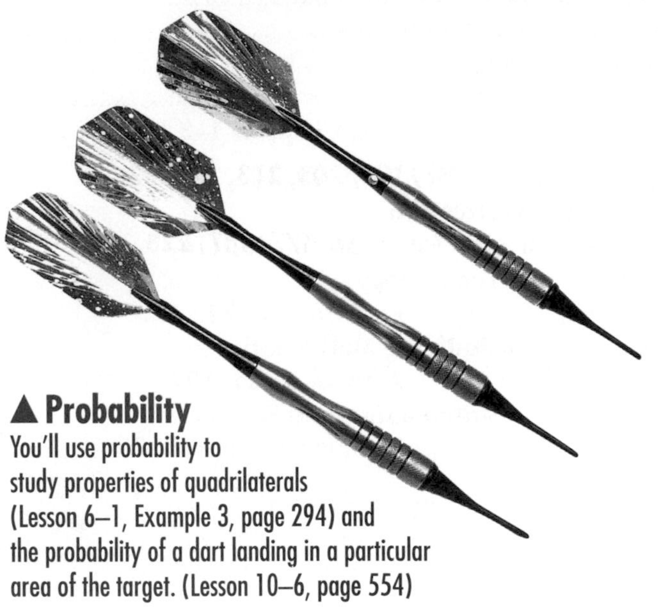

▲ Probability
You'll use probability to study properties of quadrilaterals (Lesson 6–1, Example 3, page 294) and the probability of a dart landing in a particular area of the target. (Lesson 10–6, page 554)

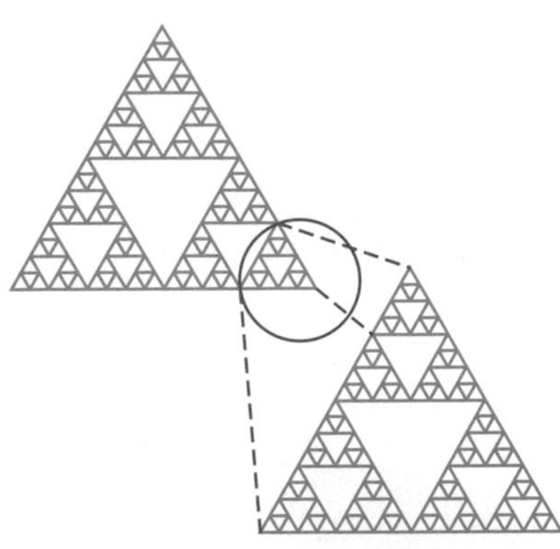

▲ Problem Solving You'll solve a simpler problem to find values in Pascal's triangle and look for a pattern in Sierpenski's triangle. (Lesson 7–6, pages 379 and 380)

LOOK BACK

You can refer to Lesson 1-4 for information on finding the distance between points.

Look Back refers you to skills and concepts that have been taught earlier in the book. (Lesson 3–5, page 156)

Discrete Mathematics ▶
You'll investigate sequences as you study geometric means and Pythagorean triples. (Lesson 8–1, pages 397 and 401)

Statistics You'll learn ▶ how to display data on a circle graph. (Lesson 9–2, page 452)

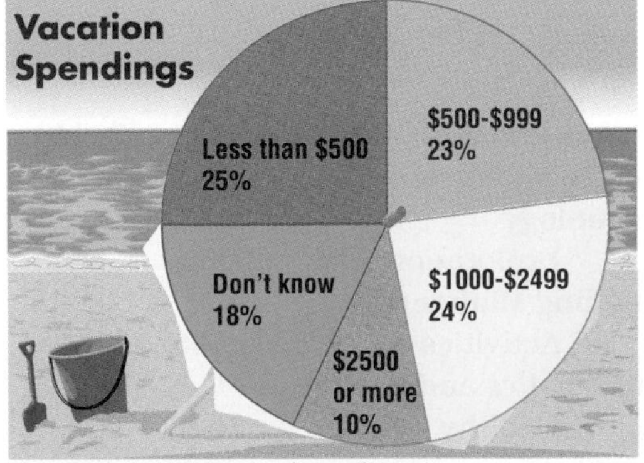

Source: *Economic Analysis of North American Ski Areas*

Algebra You'll use your algebra skills to prove that two triangles are congruent. (Lesson 4–4, page 206)

CHAPTER 4 Identifying Congruent Triangles 178

CHAPTER 5 Applying Congruent Triangles 236

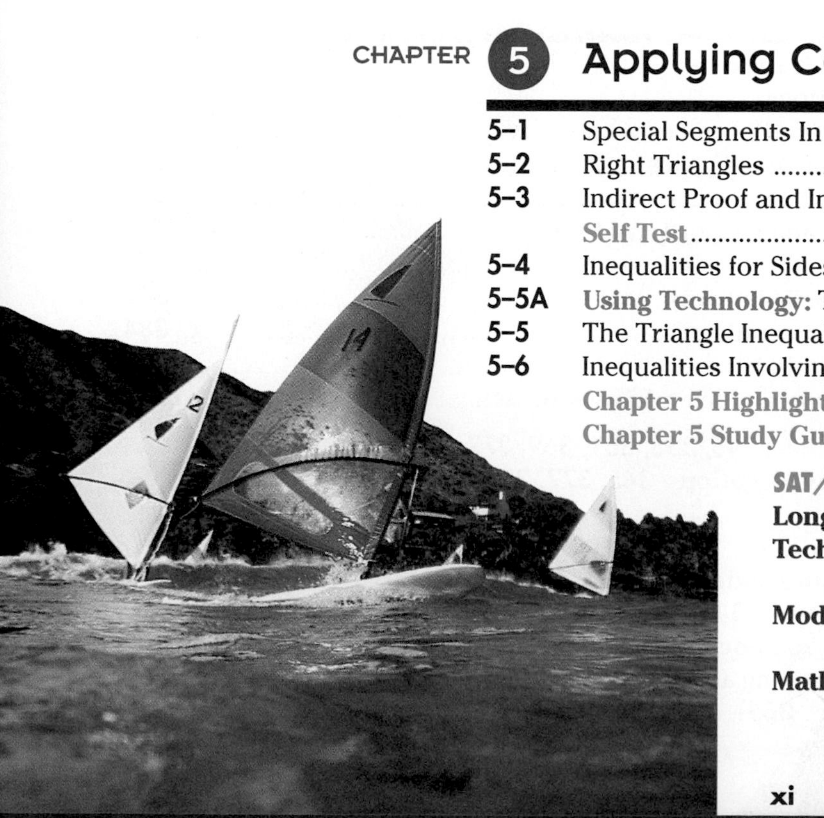

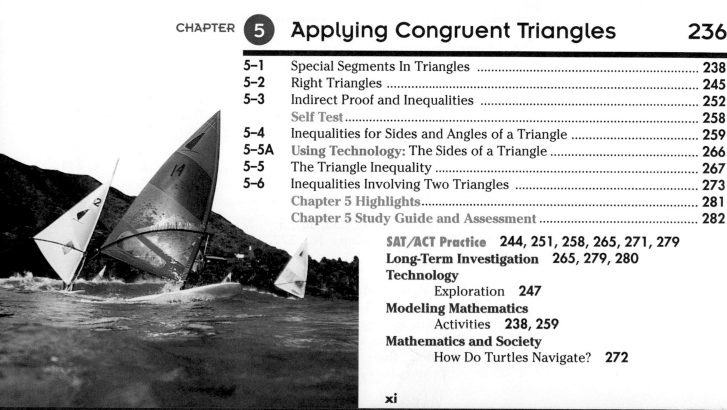

CHAPTER 6 Exploring Quadrilaterals 288

CHAPTER 7 Connecting Proportion and Similarity 336

interNET CONNECTION

Real World APPLICATIONS

Have you ever wondered if you'll ever actually use math? Every lesson in this book is designed to help show you where and when math is used in the real world. Since you'll explore many interesting topics, we believe you'll discover that math is relevant and exciting. Here are some examples.

Top Five List, FYI, and **Fabulous Firsts** contain interesting facts that enhance the applications.

fabulous FIRSTS

Norma Merrick Sklarek (1928–)

Norma Merrick Sklarek was the first African-American woman to become a registered architect.

Source: Lesson 6–5, page 327

FYI

During the 20th century, 78 total solar eclipses occurred. Only 15 of these eclipses were visible from some part of the United States. The next total eclipse visible from the United States will be in 2017.

Source: Lesson 9–5, page 475

Construction You'll study an industrial technology application that involves congruent triangles formed by a power line and tower. (Lesson 5–2, page 246)

SHOE

▲ **Comics** You'll discuss a cartoon and whether Skyler's conjecture is true. (Lesson 2–1, page 70)

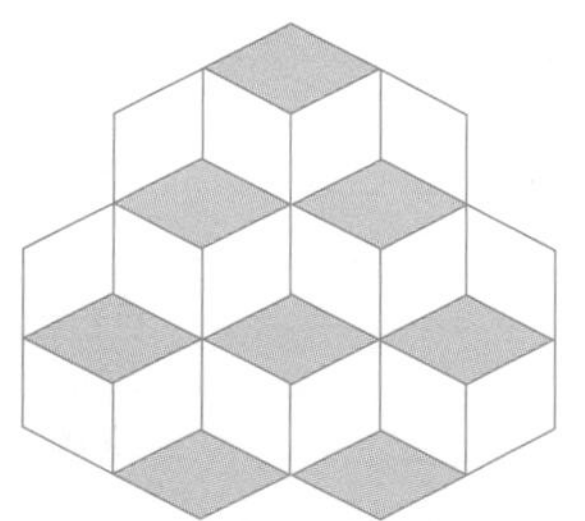

▲ **Optical Illusions** Rombi can be used to create an optical illusion. (Lesson 6–4, page 313)

Stealthy Geometry

What do radar waves and building design have to do with mathematics? Actual reprinted articles illustrate how mathematics is a part of our society. (Lesson 3–5, page 161)

CHAPTER 8 Applying Right Triangles and Trigonometry 394

interNET CONNECTION

CHAPTER 9 Analyzing Circles 444

CHAPTER 10 Exploring Polygons And Area 512

CHAPTER 11 Investigating Surface Area And Volume 572

Interdisciplinary Connections

Did you realize that mathematics is used in botany? in history? in geography? Yes, it may be hard to believe, but mathematics is frequently connected to other subjects that you are studying.

◀ **Botany** You'll study angles made by tree branches (Lesson 2–6, page 111)

▲ **Geography** You'll apply spherical geometry to find the flying distance between two cities. (Lesson 3–6, page 165)

▼ **Career Choices** include information on interesting careers.

CAREER CHOICES

Architects have played an important role in history from the earliest cities to our modern skyscrapers. The word *architect* is from the Greek meaning "master builder." A licensed architect requires a bachelor's degree in architecture, participation in an internship of 3 years, and passing an exam.

For more information, contact:

Society of American Registered Architects
1245 S. Highland Ave.
Lombard, IL 60148

Source: Lesson 4–1, page 181

Evidence of Nine men's morris has been found in Ceylon, now Sri Lanka. It was also known to inhabitants of Troy and to the Vikings.

Source: Lesson 6–4, page 317

◀ **Global Connections** introduce you to a variety of world cultures.

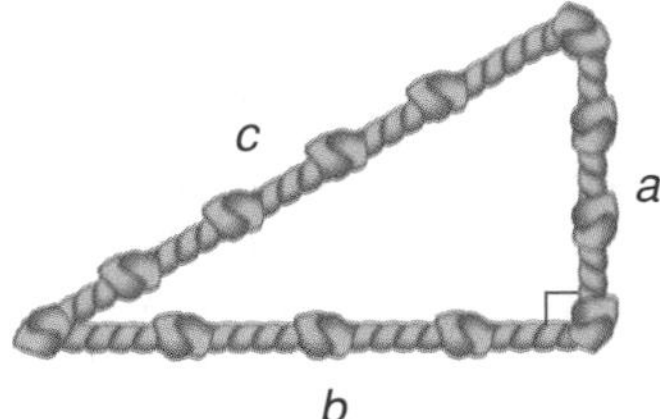

◀ **History** You'll use Egyptian techniques to sketch and classify triangles. (Lesson 5–5, pages 268–269)

Math Journal exercises give you the opportunity to assess yourself and write about your understanding of key math concepts. (Lesson 2–3, page 88)

5. Make up several logic puzzles. Have your friends try to solve them. Write the puzzles and solutions in your journal.

CHAPTER 12 Continuing Coordinate Geometry 644

CHAPTER 13 Investigating Loci and Coordinate Transformations 694

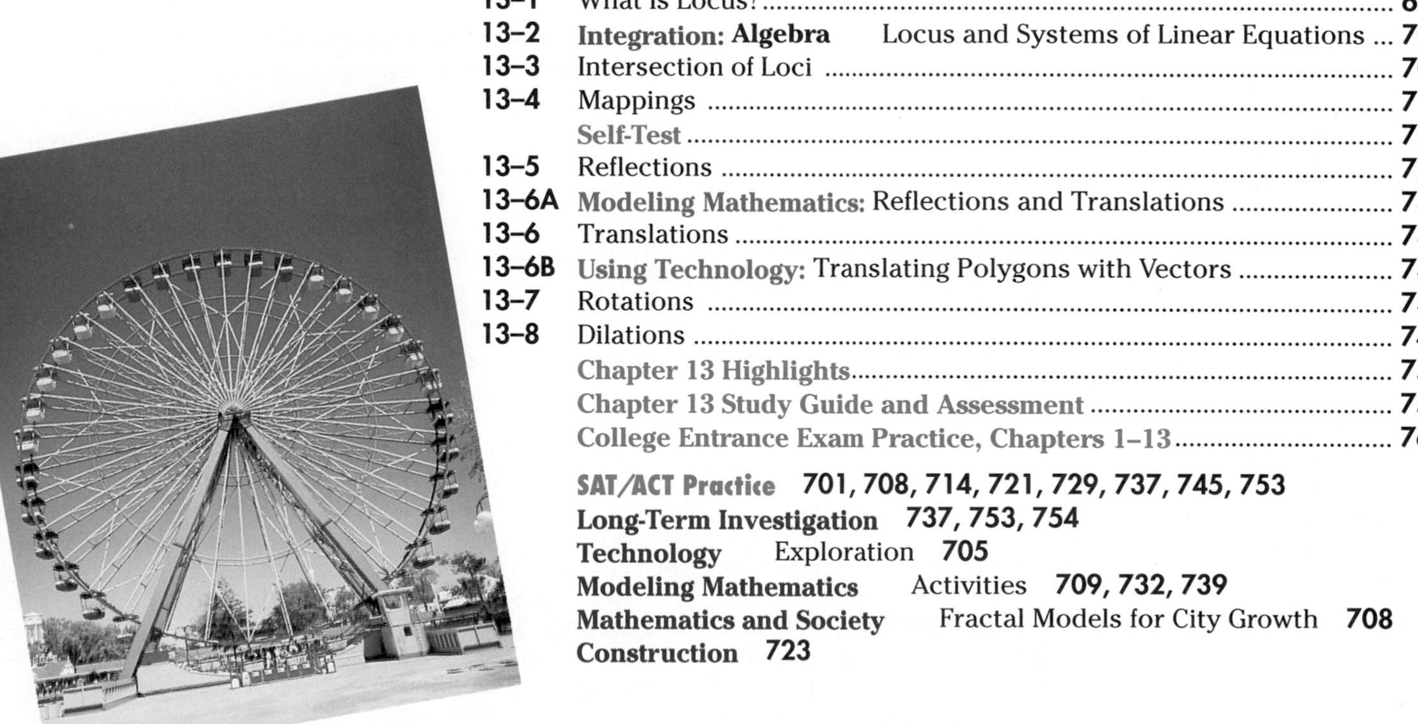

TECHNOLOGY

Do you know how to use computers and graphing calculators? If you do, you'll have a much better chance of being successful in today's high-tech society and workplace.

GRAPHING CALCULATORS

There are several ways in which graphing calculators are integrated.

- **Getting Acquainted with the Graphing Calculator** On pages 2–3, you'll get acquainted with the basic features and functions of a graphing calculator.
- **Graphing Calculator Explorations** You'll learn how to use a graphing calculator to explore slope-intercept form in Lesson 3–3.
- **Graphing Calculator Programs** In Lesson 2–2, Exercise 51 includes a graphing calculator program that can be used to find the volume of a cylinder.
- **Graphing Calculator Exercises** Many exercises are designed to be solved using a graphing calculator. For example, see Exercise 37 in Lesson 9–8.

- **Using Technology Lessons** On pages 26–27, you'll learn how to use a calculator or Cabri Geometry to measure segments.

COMPUTER SOFTWARE

- **Spreadsheets** On page 21 of Lesson 1–3, a spreadsheet is used to calculate quantities determined by a formula.
- **Geometry Software** Software on a computer or on a calculator is used in Exercise 25 on page 526 to create a tessellation by reflecting a polygon.

Technology Tips, such as this one on page 516 of Lesson 10–1, are designed to help you make more efficient use of technology through practical hints and suggestions.

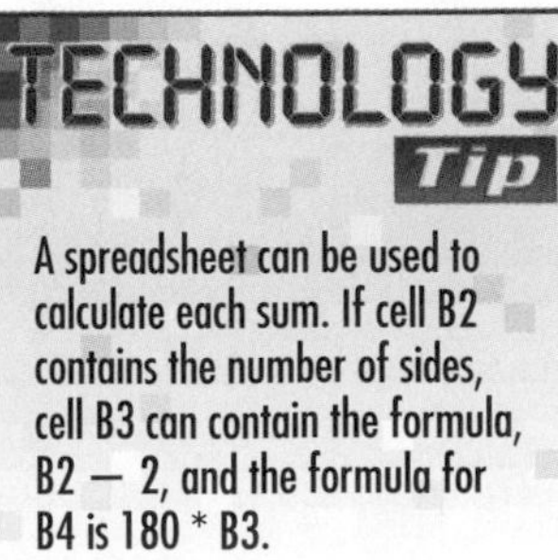

TECHNOLOGY
Tip
A spreadsheet can be used to calculate each sum. If cell B2 contains the number of sides, cell B3 can contain the formula, B2 − 2, and the formula for B4 is 180 * B3.

SYMBOLS

h	altitude*
$\angle$	angle
$\measuredangle s$	angles
a	apothem*
$\approx$	is approximately equal to, approximately
$\overset{\frown}{AB}$	minor arc with endpoints A and B
$\overset{\frown}{ACB}$	major arc with endpoints A and B
A	area of a polygon or circle* surface area of a sphere*
B	area of the base of a prism, cylinder, pyramid, or cone*
b	base of a triangle, parallelogram, or trapezoid*
$\odot P$	circle with center P
C	circumference*
$\cong$	is congruent to, congruent
$\leftrightarrow$	corresponds to
cos	cosine
$^\circ$	degree
d	diameter of a circle* distance*
AB	distance between points A and B*
$=$	equals, is equal to
$\neq$	is not equal to
$>$	is greater than
A'	the image of preimage A
$<$	is less than
L	lateral area*
$\overleftrightarrow{DE}$	line containing points D and E

$\rightarrow$	is mapped onto
$m\angle A$	degree measure of $\angle A$
$m\overset{\frown}{AB}$	degree measure of arc AB
$\sqrt{}$	nonnegative square root
(x, y)	ordered pair
(x, y, z)	ordered triple
$\parallel$	is parallel to, parallel
$\nparallel$	is not parallel to
$\square$	parallelogram
P	perimeter*
$\perp$	is perpendicular to, perpendicular
π	pi
n-gon	polygon with n sides
r	radius of a circle*
$\overrightarrow{PQ}$	ray with endpoint P passing through Q
$\overline{RS}$	segment with endpoints R and S
s	side of a regular polygon*
$\sim$	is similar to, similar
sin	sine
ℓ	line ℓ length of a rectangle slant height
m	slope
tan	tangent
T	total surface area
$\triangle$	triangle
$\overrightarrow{AB}$	vector from A to B
V	volume*

* indicates that this is the symbol for the measure of the item listed.

GETTING ACQUAINTED WITH THE GRAPHING CALCULATOR

What is it?
What does it do?
How is it going to help me learn math?

These are just a few of the questions many students ask themselves when they first see a graphing calculator. Some students may think, "Oh, no! Do we ***have*** to use one?", while others may think, "All right! We get to use these neat calculators!" There are as many thoughts and feelings about graphing calculators as there are students, but one thing is for sure: a graphing calculator can help you learn mathematics.

So what is a graphing calculator? Very simply, it is a calculator that draws graphs. This means that it will do all of the things that a "regular" scientific calculator will do, *plus* it will draw graphs of equations. In geometry, this capability is helpful as you analyze the graphs of equations.

But a graphing calculator can do more than just calculate and draw graphs. For example, you can program it or work with data to make statistical graphs and perform computations. If you need to generate random numbers, you can do that on the graphing calculator. You can even draw and manipulate geometric figures on some graphing calculators. It's really a very powerful tool—so powerful that it is often called a pocket computer.

As you may have noticed, graphing calculators have some keys that other calculators do not. The keys on the bottom half of the graphing calculator are those found on scientific calculators. The keys located just below the screen are the graphing keys. You will also notice the up, down, left, and right arrow keys. These allow you to move the cursor around on the screen, to "trace" graphs that have been plotted, and to choose items from the menus. The other keys located on the top half of the calculator access the special features such as statistical computations and programming features.

A few of the keystrokes that can save you time when using the graphing calculator are listed below.

- The commands above the calculator keys are accessed with the 2nd or ALPHA key. On some calculators, the 2nd key and the commands accessed by it are blue, and the ALPHA key and its commands are gray. On other calculators, the 2nd key and its commands are yellow, and the ALPHA and its commands are green.
- 2nd [ENTRY] copies the previous calculation so you can edit and use it again.
- Pressing ON while the calculator is graphing stops the calculator from completing the graph.
- 2nd [QUIT] will return you to the home (or text) screen.
- 2nd [A-LOCK] locks the ALPHA key, which is like pressing "shift lock" or "caps locks" on a typewriter or computer. The result is that all letters will be typed and you do not have to repeatedly press the ALPHA key. (This is handy for programming.) Stop typing letters by pressing ALPHA again.
- 2nd [OFF] turns the calculator off.

Some commonly used mathematical functions are shown in the table below. As with any scientific calculator, the graphing calculator observes the order of operations.

Mathematical Operation	Examples	Keys	Display
evaluate expressions	Find 2 + 5.	2 + 5 ENTER	2+5 7
exponents	Find 3^5.	3 ^ 5 ENTER	3^5 243
multiplication	Evaluate 3(9.1 + 0.8).	3 (9.1 + .8) ENTER	3(9.1+.8) 29.7
roots	Find $\sqrt{14}$.	2nd [√‾] 14 ENTER	√‾ 14 3.741657387
opposites	Enter −3.	(−) 3	−3

Graphing

Before graphing, we must instruct the calculator how to set up the axes in the coordinate plane. To do this, we define a **viewing window**. The viewing window for a graph is the portion of the coordinate grid that is displayed on the graphics screen of the calculator. The viewing window is written as [left, right] by [bottom, top] or [Xmin, Xmax] by [Ymin, Ymax]. A viewing window of [−10, 10] by [−10, 10] is called the **standard viewing window** and is a good viewing window to start with to graph an equation. The standard viewing window can be easily obtained by pressing ZOOM 6. Try this.

Move the arrow keys around and observe what happens. You are seeing a portion of the coordinate plane that includes the region from -10 to 10 on the x-axis and from -10 to 10 on the y-axis. Move the cursor, and you can see the coordinates of the point for the position of the cursor.

Any viewing window can be set manually by pressing the WINDOW key. The window screen will appear and display the current settings for your viewing window. Using the arrow and ENTER keys, move the cursor to edit the window settings. Xscl and Yscl refer to the x-scale and y-scale. This is the number of units between tick marks on the x- and y-axes. Xscl=1 means that there will be a tick mark for every unit of one along the x-axis.

Programming

Graphing calculators have programming features that allow us to write and execute a series of commands to perform tasks that may be too complex or cumbersome to perform otherwise. Each program is given a name. Commands begin with a colon (:), which the calculator enters automatically, followed by an expression or an instruction. Most of the features of the calculator are accessible from program mode.

When you press PRGM, you see three menus: EXEC, EDIT, and NEW. EXEC allows you to execute a stored program, EDIT allows you to edit or change a program, and NEW allows you to create a program. The following tips will help you as you enter and run programs.

- To begin entering a new program, press PRGM ▶ ▶ ENTER.
- After a program is entered, press 2nd [QUIT] to exit the program mode and return to the home screen.
- To execute a program, press PRGM. Then use the down arrow key to locate the program name and press ENTER, or press the number or letter next to the program name.
- If you wish to edit a program, press PRGM ▶ and choose the program from the menu.
- To immediately re-execute a program after it is run, press ENTER when Done appears on the screen.
- To stop a program during execution, press ON.

Geometry on the Calculator

A certain calculator includes a geometry program that allows you to perform operations on geometric figures. You can draw, move, measure, graph, and alter figures. To access the geometry applications on your calculator, press APPS and choose 8: Geometry. If you are starting a new session, choose 3: New, press down on the cursor pad, enter a name for the construction (up to eight letters), and press ENTER.

The toolbar is organized into eight menus which are accessed by pressing the function keys F1 through F8. The list below describes the types of functions on the menu for each key.

Key	Menu Functions
F1	Freehand transformations
F2	Construction of points or linear objects
F3	Construction of curves and polygons
F4	Euclidean constructions and creating macros
F5	Transformational geometry constructions
F6	Measurements and calculations
F7	Labels and animation
F8	File operations and editing

All figures are drawn by choosing one or more points on the screen. To draw a figure, first choose the object you would like to draw from a menu, then create or select the points to define the object. For example to draw a triangle, press F3 3 to choose triangle. Then move the cursor to where you would like the first vertex and press ENTER. Move the cursor to locate the second vertex and press ENTER. Finally move the cursor to the third vertex and press ENTER to complete the triangle.

The procedures for some common operations you will use in the geometry application are listed below.

- To select an object that you have drawn, move the cursor toward the object until the object's name appears next to the cursor. Then press ENTER.
- If you would like to delete an object you have drawn, select the object and press ← or F8 7.
- Move objects on the screen by choosing the pointer from the F1 menu. Then select the object, press and hold the grasping hand key, and use the arrow keys to move the object.

While a calculator cannot do everything, it can make some things easier. To prepare for whatever lies ahead, you should try to learn as much as you can. The future will definitely involve technology, and using a graphing calculator is a good start toward becoming familiar with technology. Who knows? Maybe one day you will be designing the next satellite, building the next skyscraper, or helping students learn mathematics with the aid of a graphing calculator!

Discovering Points, Lines, Planes, and Angles

What You'll Learn

In this chapter, you will:

- identify and model points, lines, and planes in space and on a coordinate plane,
- identify collinear points and coplanar points and lines,
- solve problems by listing the possibilities and by using formulas, finding maximum and minimum values for a given perimeter, and
- find the measures of segments and angles and the relationships that exist among them.

Why It's Important

Real World

Cartography Algebra and geometry can be integrated to help map makers mark locations on maps. In the 19th century, seafaring peoples in the South Pacific developed maps made of stiff palm fibers. Today, advances in technology have enabled us to make highly accurate maps, many of which are computerized. Maps showing latitude and longitude are very similar to the coordinate system you used in algebra. You will learn more about the coordinate plane in Lesson 1–1.

PREREQUISITE SKILLS

To be successful in this chapter, you'll need to understand these concepts and be able to apply them. Refer to the examples to help you complete these problems.

Evaluate expressions.

Example Evaluate $3x + 2$ if $x = -4$

$3x + 2 = 3(-4) + 2$ *Replace x with −4.*
$= -12 + 2$ *Follow the order of operations.*
$= -10$ *Simplify.*

Evaluate each expression if $x = 3$, $y = -4$, and $z = -1$.

1. $-4x - 5$
2. $-6 + 2z$
3. $3 - y$
4. $\frac{x+y}{z}$
5. $xy - z$
6. $2(x + z)$

Solve equations involving more than one operation.

Example Solve $6a - 2 - a + 5 = 18$.

$6a - 2 - a + 5 = 18$
$5a + 3 = 18$ *Combine like terms.*
$5a = 15$ *Subtract 3 from each side.*
$a = 3$ *Divide each side by 5.*

Solve each equation.

7. $2x + 3 + 5x - 2 = 13$
8. $7b + 5 - 8b + 9 = 10$
9. $-s - 4 + 6s + 4 = 5$
10. $9a - 6 - 13a - 12 = -2$

Focus On

READING SKILLS

In this chapter, you'll be learning about basic terms in geometry. To be successful in geometry, you must understand the language of geometry. Many terms are easy to understand such as point, line, plane, and angle. The meaning of other terms can usually be determined by examining the word. For example, the word *collinear* means points all lying on one line. You can see that the word collinear contains the word *line*. You might also know that the prefix *co* means together, as in coworker. Therefore, collinear means points lying together.

1-1 Integration: Algebra The Coordinate Plane

What **YOU'LL LEARN**

- To graph ordered pairs on a coordinate plane, and
- to identify collinear points.

Why **IT'S IMPORTANT**

Coordinate systems are used to locate items in publishing, archeology, and aquatic explorations.

Real World APPLICATION

Games

Have you ever played the game in which you try to guess the location of your opponent's ships? Your goal is to sink your opponent's ships by calling out row and column coordinates. Then, by using logic and your opponent's responses, you guess more wisely with each turn, trying to sink all of your opponent's ships before he or she sinks yours.

The grid system used in playing the game is an example of plotting points in a coordinate system. In algebra, you used a **coordinate plane** in which a horizontal number line called the ***x*-axis** and a vertical number line called the ***y*-axis** intersect at their zero points. Their intersection is called the **origin**. The axes divide the coordinate plane into four **quadrants**, as shown in the figure at the right.

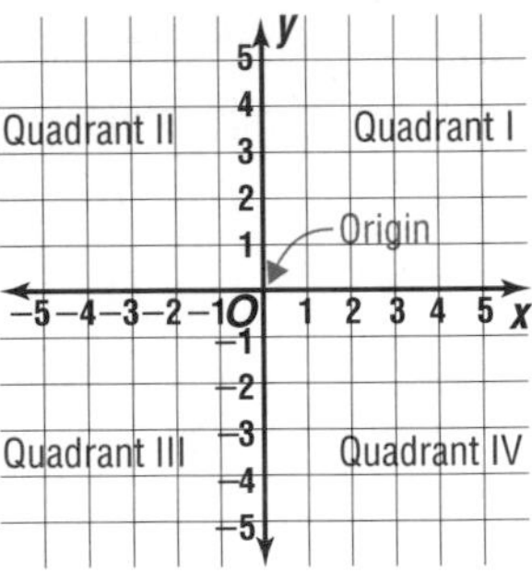

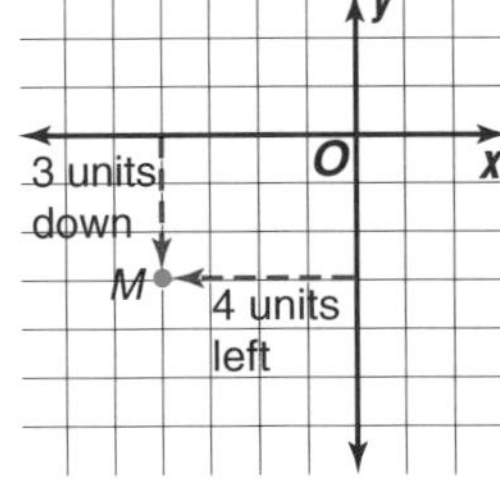

The notation M(−4, −3) indicates that point M is named by the ordered pair (−4, −3).

There are an infinite number of points on a coordinate plane. Each point can be named by an **ordered pair** of coordinates. The first coordinate of an ordered pair, usually called the ***x*-coordinate**, tells how far left or right of the *y*-axis the point is located. The second coordinate, the ***y*-coordinate**, tells how far up or down from the *x*-axis the point is located. Point *M* is located in Quadrant III and has coordinates $(-4, -3)$. This means that the point is located 4 units left and 3 units down from the origin.

If the axes on a coordinate plane do not have numbers on them, you can assume that each square on the grid represents 1 unit.

Example  **Write the coordinates for points P, Q, and R.**

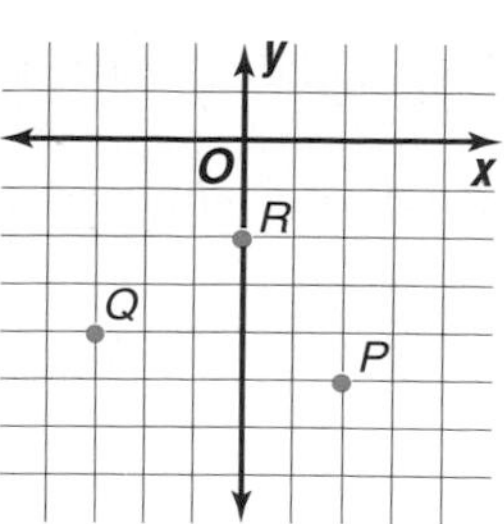

- Point P is 2 units right and 5 units down from the origin. Its coordinates are $(2, -5)$.
- Point Q is 3 units left and 4 units down from the origin. Its coordinates are $(-3, -4)$.
- Point R is at -2 on the y-axis. Its coordinates are $(0, -2)$.

There are many types of coordinate systems. In publishing, computers are often used to arrange type and pictures on the page. Photographs, artwork, and special text features are positioned by putting each item in a special box that can be moved around the page.

Example **2**

Real World APPLICATION

Publishing

FYI

A *pica* is a unit of measure used in publishing that is equal to $\frac{1}{6}$ of an inch.

In a certain page layout software program, the position of the upper left corner of a box is expressed by X (how many picas right) and Y (how many picas down) it is from the upper left-hand corner of the page designated as 0. The drawing below shows a layout screen. Identify which box is described by each location.

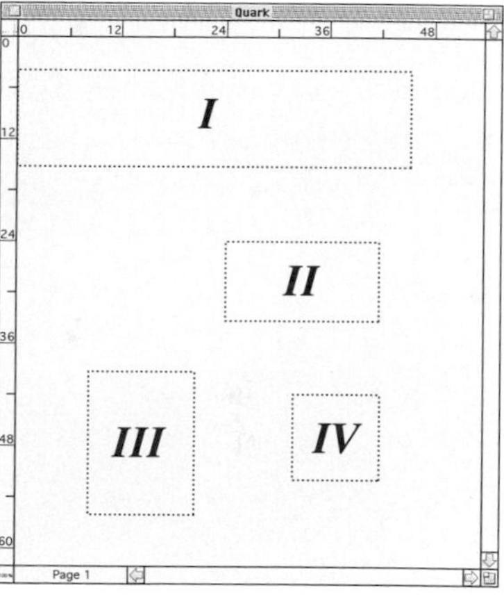

a. [X: 24, Y: 24]

This means the upper left corner of the box is located 24 picas right of 0 and 24 picas down. This is the location of box II.

b. [X: 8, Y: 39]

The box that is located 8 picas right and 39 picas down is box III.

c. [X: 0, Y: 4]

The box located 0 units right and 4 units down is box I.

An algebraic relation can be represented in several ways.

In algebra, you used a table to find values that satisfy an equation. Then you graphed those values and connected them with a line to draw the graph of the equation. For example, the table below shows five pairs of values that satisfy the equation $y = 2x - 1$. The coordinate plane shows the graphs of those points and the graph of the equation, which is a line.

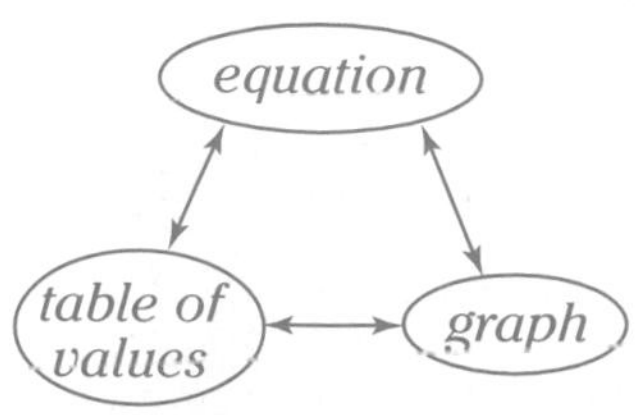

$y = 2x - 1$		
x	y	(x, y)
−2	−5	$A(-2, -5)$
−1	−3	$B(-1, -3)$
0	−1	$C(0, -1)$
2	3	$D(2, 3)$
3	5	$E(3, 5)$

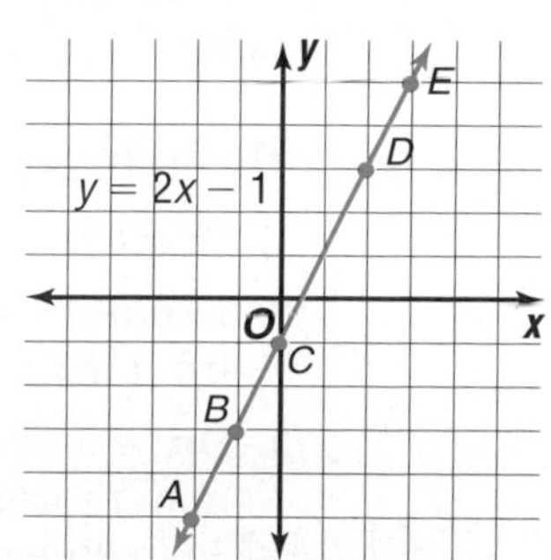

The notation ABCDE is sometimes used to show that points are collinear.

The coordinates of points A, B, C, D, and E satisfy the equation $y = 2x - 1$. The graph of $y = 2x - 1$ is a line, and points A, B, C, D, and E lie on that line. Thus, they are **collinear**. Is the point $M(4, 8)$ collinear with points A, B, C, D, and E? To determine if M lies on the same line with A, B, C, D, and E, see whether the coordinates of M satisfy the equation $y = 2x - 1$.

$$y = 2x - 1$$
$$8 \stackrel{?}{=} 2(4) - 1 \quad (x, y) = (4, 8)$$
$$8 \neq 7$$

If you graphed M(4, 8), you could also see that it does not lie on the line containing A, B, C, D, and E.

The coordinates of M do not satisfy the equation. Therefore, M does not lie on the line representing $y = 2x - 1$. Points A, B, C, D, E, and M are **noncollinear** points.

Example 3

Algebra

a. Find the coordinates of three points that lie on the graph of $y = -3x + 3$.
b. Graph the points and draw the line representing $y = -3x + 3$.
c. Name the coordinates of one point not on the line.

a. Choose three values for x and make a table of values.

$y = -3x + 3$		
x	y	(x, y)
−1	6	(−1, 6)
0	3	(0, 3)
3	−6	(3, −6)

b. Graph the points from the table and draw the line representing the equation.

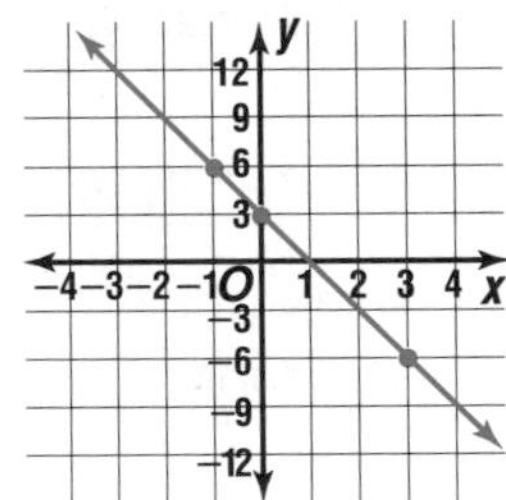

c. There are an infinite number of points in the coordinate plane that do not lie on the line representing $y = -3x + 3$. Visually, select any point that does not appear to lie on the line, for example, (1, 2). You can check your selection by testing its coordinates to make sure that they do not satisfy the equation.

$$y = -3x + 3$$
$$2 \stackrel{?}{=} -3(1) + 3$$
$$2 \neq 0$$

The point named by (1, 2) does not lie on the graph of $y = -3x + 3$.

CHECK FOR UNDERSTANDING

Communicating Mathematics

Study the lesson. Then complete the following.

1. **Describe** how to graph a point if you know the coordinates of the point.
2. **Explain** some of the ways that the coordinate system in Example 2 differs from the coordinate plane used in mathematics.
3. **Draw** a diagram showing how you might determine whether the points on a map representing Cairo, Egypt; Mumbia (formerly Bombay), India; and San Juan, Puerto Rico, are noncollinear.

Guided Practice

Write the ordered pair for each point shown at the right.

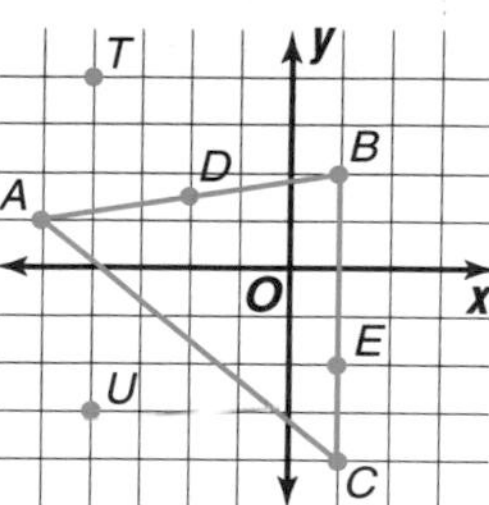

4. T
5. U

Graph each point on the same coordinate plane.

6. $R(3, -6)$
7. $S(0, 2)$

8. Use triangle ABC ($\triangle ABC$) shown in the coordinate plane above to answer each question.
   a. In which quadrant is point A located?
   b. In which quadrant is point B located?
   c. Which axis intersects the side connecting A and B?
   d. Which point, A, B, or D, has the greatest x-coordinate?
   e. Which point, B, C, or E, has the greatest y-coordinate?

9. Points $A(5, 7)$ and $B(-1, 1)$ lie on the graph of $y = x + 2$. Determine whether $Z(-3, -2)$ is collinear with A and B.

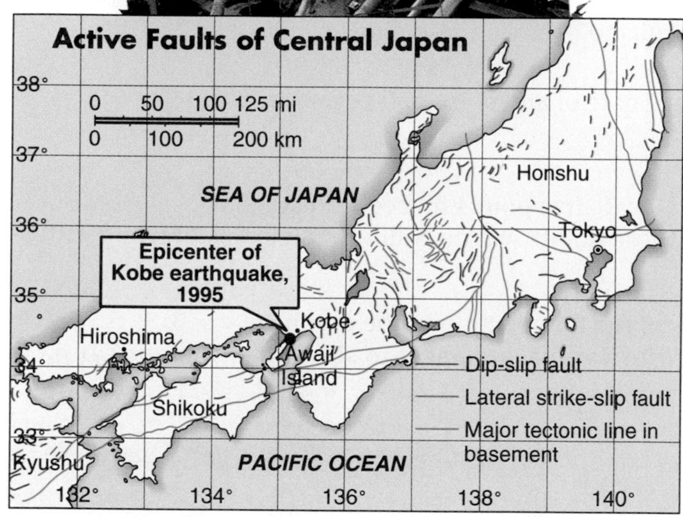

10. **Seismology** Mapmakers use letters and numbers to define sectors on a map. In January, 1995, a powerful earthquake struck the Japanese port of Kobe, killing more than 5100 people and destroying more than 45,000 buildings. The map at the left shows the active faults of central Japan.
    a. Use the longitude and latitude degree measures to write an ordered pair to approximate the location of Kobe earthquake on the map.
    b. Name the town located near the coordinates (139.5°, 35.5°).

EXERCISES

Practice

Write the ordered pair for each point shown at the right.

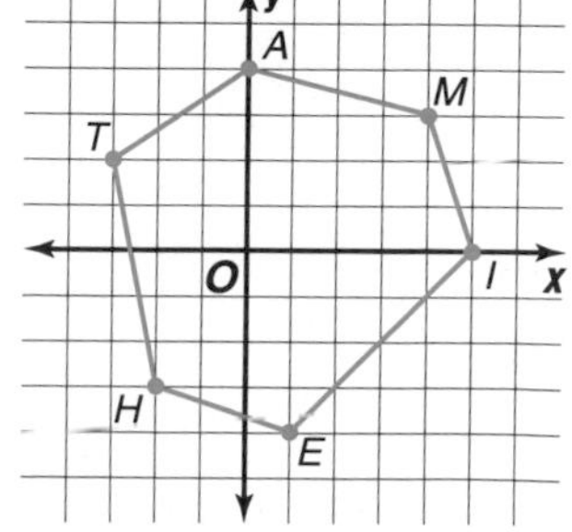

11. M
12. A
13. T
14. H
15. E
16. I

Graph each point on the same coordinate plane.

17. $C(0, 5)$
18. $D(-3, -2)$
19. $E(4, -5)$
20. $F(-1, 4)$
21. $G(2.5, -3)$
22. $H(-4.5, 0)$

Points $M(1, -3)$ and $N(-2, -15)$ lie on the graph of $y = 4x - 7$. Determine whether each point is collinear with M and N.

23. $Q(0, -7)$
24. $R(0.5, -5)$
25. $S(-1, 11)$

In the figure at the right, the blue segments forming the rectangle lie on the gridlines of a coordinate plane. Determine the ordered pair that represents each point.

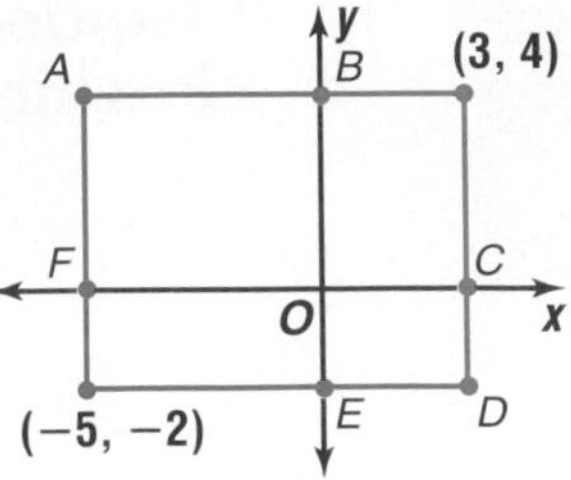

26. A 27. B

28. C 29. D

30. E 31. F

Find the coordinates of three points that lie on the graph of each equation. Graph the points and the line representing the equation. Then name the coordinates of one point not on the line.

32. $y = x + 5$ 33. $y = 6x - 9$ 34. $3x = 2y - 7$

35. Refer to square $PQRS$ graphed at the right.
   a. In how many quadrants does $PQRS$ lie? Name them.
   b. Name the coordinates of the points where segments PS and QR intersect the x-axis.
   c. Name the coordinates of a point that would be collinear with S and R.
   d. What is the y-coordinate of any point collinear with S and R?
   e. What is the x-coordinate of any point collinear with P and S?

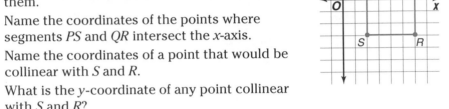

Critical Thinking

36. Explain how you can locate five points in a coordinate plane so that you are sure no three of these points will be collinear. Include a drawing with your explanation.

37. Find the coordinates of the intersection of the graphs of $y = 3x + 5$ and $y = -2x - 10$. Write a verbal argument to show that your answer is correct.

Applications and Problem Solving

38. **Aquatic Engineering** The Dutch Delta Plan controls the flow of the Atlantic Ocean on the southwest coast of the Netherlands. The system was built using a grid of nylon mattresses with graded gravel and rocks to support concrete piers and steel gates. The horizontal axis is labeled with letters and the vertical axis with numbers. Suppose the first five piers were placed at 5B, 2K, 8D, 7A, and 12E. Draw a graph showing the positions of the first five piers.

39. **Archaeology** Divers investigating the remains of the seaport city of Caesarea beneath the Mediterranean Sea, found statues, lamps, pieces of pottery, several ancient shipwrecks, and submerged portions of a port. The divers laid out ropes to form a grid system for mapping the ocean floor. As items were found, they were recorded on paper using a smaller grid. Suppose pieces of pottery were found at (4, 1), (2, 9), (8, 0), and (11, 6). Draw a graph showing the locations of the finds.

Mixed Review

Simplify.

40. $-2 + 3$ 41. $-4 - (-5)$ 42. $7(-2)$

43. $-16 \div (-4)$ 44. $3(-5) + 2(7)$ 45. $-2(-7 + 4)$

46. **ACT Practice** After $\frac{12\frac{1}{4}}{5\frac{4}{9}}$ has been simplified to a single fraction in lowest terms, what is the denominator?

A 2 **B** 4 **C** 9 **D** 16 **E** 36

47. **Statistics** The graph shows how much time parents say they spend per week helping their children with schoolwork.

Source: 20/20 Research for Shoney's Restaurants

a. Which response was most frequently given by parents?

b. If 8000 parents were surveyed, how many said they do not help their children with their schoolwork?

c. Which category contains about one third of the parents polled in this survey?

d. Keep track of how much time you get help with your schoolwork at home. In which category would your household fit?

For **Extra Practice**, see page 764.

1-2 Points, Lines, and Planes

What YOU'LL LEARN

- To identify and model points, lines, and planes,
- to identify coplanar points and intersecting lines and planes, and
- to solve problems by listing the possibilities.

Why IT'S IMPORTANT

You can use points, lines, and planes to represent real-life objects.

Real World APPLICATION

Art Design

Art can be functional as well as interesting from a geometric point of view. The chair shown at the right is a drawing of a design by Gerrit Rietveld in Holland in 1917. The artist used red and blue rectangles to represent **planes** extending into space. The black bars represented **lines** through space. Rietveld used yellow squares to represent **points** in space. Rietveld's chair demonstrates the effect of the cubism art movement.

You are already familiar with the terms *plane, line,* and *point* from your experiences with graphing in algebra. The terms have similar meaning in geometry.

Unlike the squares used to represent points in the chair design, in geometry, points do not have any actual size. They can, however, represent objects that do have size. A point is usually named by a capital letter, just like the graphs of ordered pairs in algebra. *All* geometric figures consist of points. **Space** is a boundless, three-dimensional set of all points.

In algebra, you found the equation for a line that passes through two points, and you know there are many points on that line. The same is true in geometry.

*A line **segment** is a section of a line. It has endpoints and does not extend indefinitely.*

A line has no thickness or width, although a picture of a line does. Arrows on each end of the line symbolize that the line extends indefinitely in both directions. A line is often named by a lowercase script letter. The line at the right is line n. If the names of two points on a line are known, then the line can be named by those points. In this example, points A and B lie on line n, so line n can be referred to in the following ways.

words	line AB	line BA
symbols	$\overleftrightarrow{AB}$	$\overleftrightarrow{BA}$

In algebra, you used a coordinate plane. This plane contained points and lines. In geometry, a plane is a flat surface that extends indefinitely in all directions. We use four-sided figures like the one at the right to model a plane. A plane can be named by a capital script letter or by three *noncollinear* points in the plane. The figure at the right can be named as plane $\mathcal{R}$ or plane PQS.

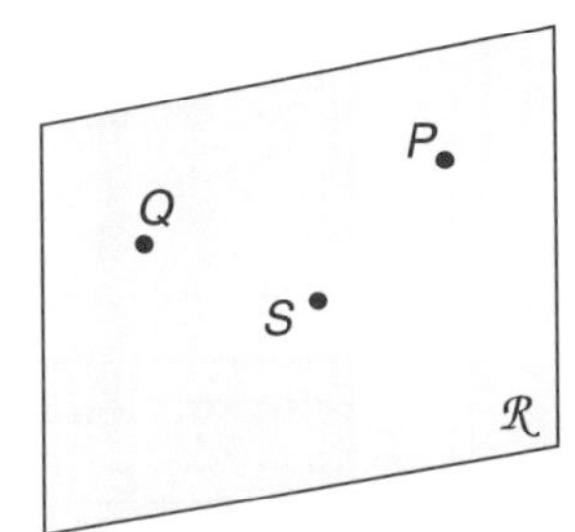

There are often many names for the same line or plane, all of which are correct. One way to recognize all the names for a given figure is to **list the possibilities**.

Example 1

PROBLEM SOLVING

List the Possibilities

List all of the possible names for each figure.

a. Line *RS*

Points T and U also lie on $\overleftrightarrow{RS}$. Choose two letters from the four named in the figure to name this line.

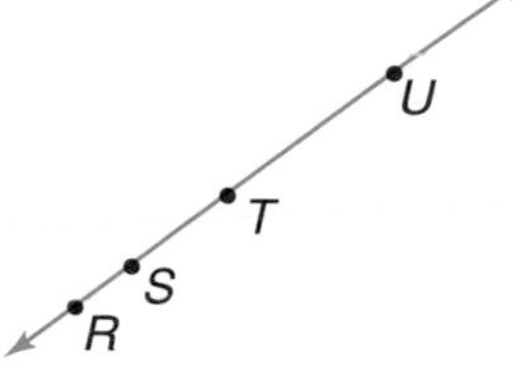

$\overleftrightarrow{RS}$	$\overleftrightarrow{SR}$	$\overleftrightarrow{RT}$	$\overleftrightarrow{TR}$	$\overleftrightarrow{RU}$	$\overleftrightarrow{UR}$
$\overleftrightarrow{ST}$	$\overleftrightarrow{TS}$	$\overleftrightarrow{SU}$	$\overleftrightarrow{US}$	$\overleftrightarrow{TU}$	$\overleftrightarrow{UT}$

b. Plane $\mathcal{M}$

Points A, B, and C lie on plane $\mathcal{M}$. Use different orders of these letters to name the plane.

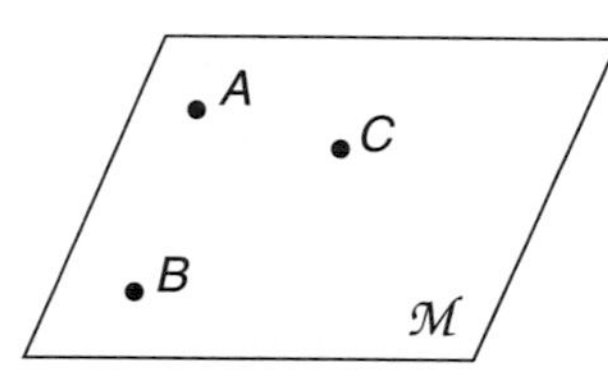

plane *ABC*	plane *ACB*	plane *BCA*
plane *BAC*	plane *CAB*	plane *CBA*

In geometry, the terms *point, line,* and *plane* are considered *undefined terms* because they have only been explained using examples and descriptions. Even though they are undefined, these terms can still be used to define other geometric terms and properties.

You may recall that *collinear* refers to points that lie on the same line. If points are **coplanar**, they lie on the same plane. All the points in a coordinate plane are coplanar. Sometimes it is difficult to identify coplanar points in space unless you understand what a drawing represents. In the figure in Example 2 shown below, dashed segments and lines are used to represent parts of the three-dimensional figure that are hidden from view. Often an artist will use different shades of color to denote different planes.

Example  2

The figure shows a pyramid sitting on plane $\mathcal{N}$.

Refer to the figure at the right to answer each question.

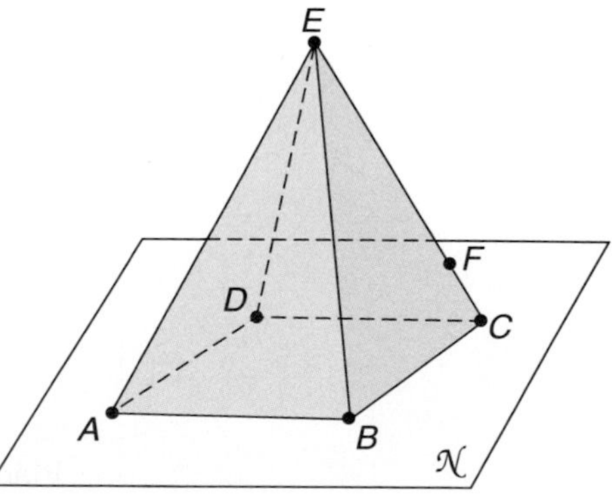

a. Are points *E, F,* and *C* collinear?

Since points E, F, and C lie on segment EC which is part of line EC, they are collinear.

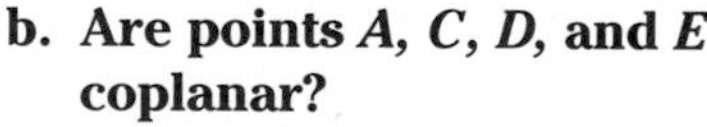

b. Are points *A, C, D,* and *E* coplanar?

Points A, C, and D lie in plane $\mathcal{N}$, but point E does not lie in plane $\mathcal{N}$. Thus, the four points are not coplanar.

Points C, D, and E lie in plane CDE, but point A does not.

c. How many planes appear in this figure?

There are five planes: plane $\mathcal{N}$, plane *ABE*, plane *EBC*, plane *EDC*, and plane *ADE*. *Plane $\mathcal{N}$ can also be named by using any three of the four points named in the plane.*

Note that information about measurement and equality cannot be determined by looking at a figure.

Figures play an important role in understanding geometric concepts. Drawing and labeling figures can help you model and visualize various geometric relationships. For example, the figures and descriptions below can help you visualize and write about some important relationships among points, lines, and planes.

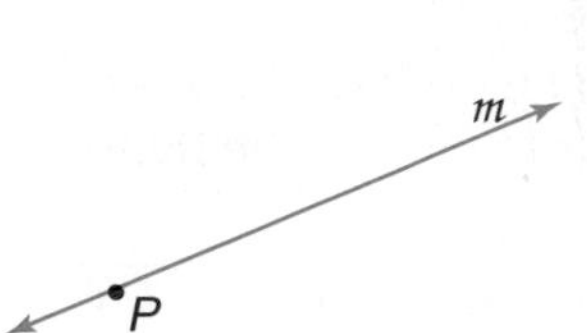

P is on *m*.

m contains *P*.

m passes through *P*.

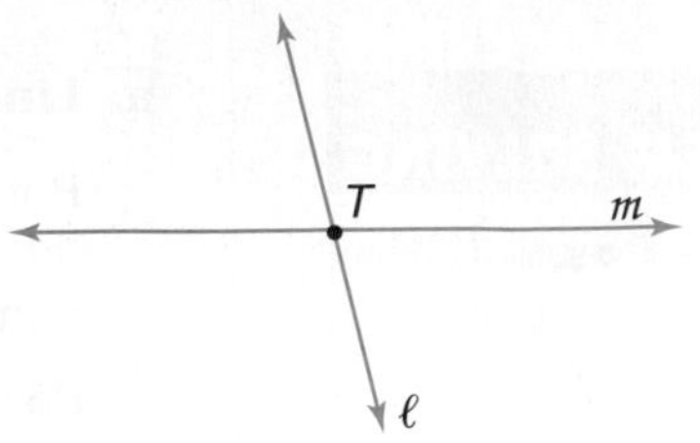

ℓ and *m* intersect in *T*.

ℓ and *m* both contain *T*.

T is the intersection of ℓ and *m*.

*The **intersection** of two figures is the set of points that are contained in both figures.*

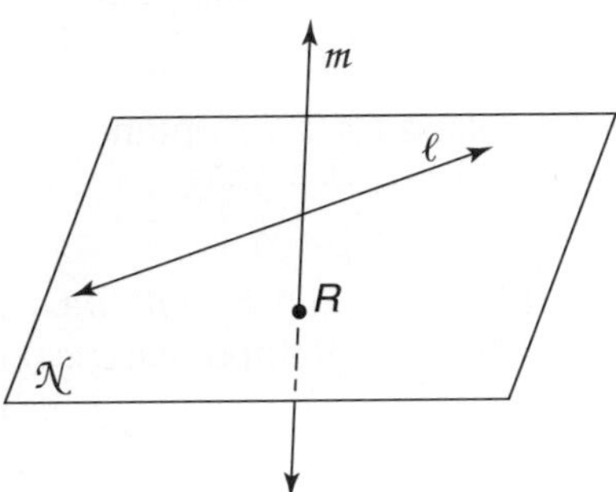

ℓ and *R* are in 𝒩.

𝒩 contains *R* and ℓ.

m intersects 𝒩 at *R*.

R is the intersection of *m* with 𝒩.

The symbol ∩ is often used to denote intersection as in $m \cap \mathcal{N} = R$.

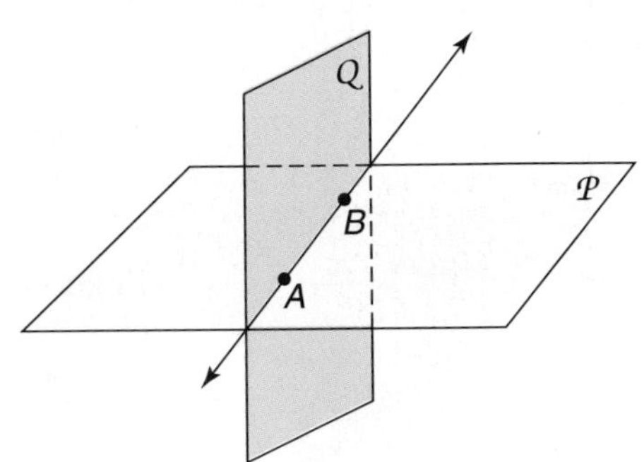

$\overleftrightarrow{AB}$ is in 𝒫 and is in 𝒬.

𝒫 and 𝒬 both contain $\overleftrightarrow{AB}$.

𝒫 and 𝒬 intersect in $\overleftrightarrow{AB}$.

$\overleftrightarrow{AB}$ is the intersection of 𝒫 and 𝒬.

Example 3 **Suppose four points on a coordinate plane are *A*(3, −4), *B*(−2, 3), *C*(4, 4), and *D*(0, −5). Draw and label a figure showing lines *AB* and *CD* intersecting at *F* and a point *G* that is coplanar with *A, B, C, D,* and *F,* but is not contained in either line.**

First graph the four given points on a coordinate plane.

Then draw line *AB* and line *CD*.

Label the point where the two lines intersect as *F*.

Place point *G* anywhere on the graph so that it does not lie on $\overleftrightarrow{AB}$ or $\overleftrightarrow{CD}$.

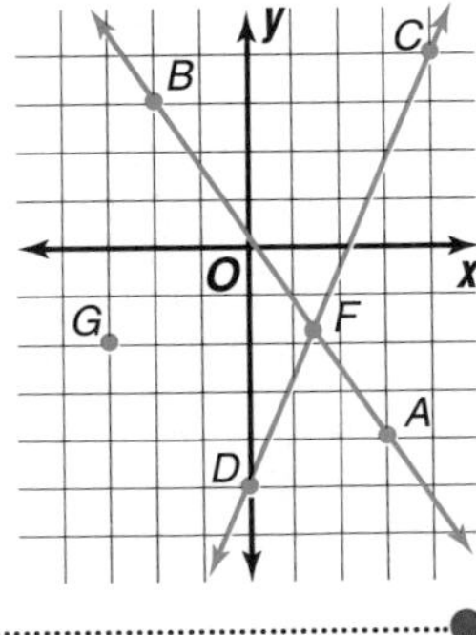

Sometimes it is helpful to make a model of a geometric situation in order to better visualize the information being presented. This is especially true of **three-dimensional figures**. You can use sheets of paper to model planes. The following activity shows you how to model two intersecting planes.

MODELING MATHEMATICS

Modeling Intersecting Planes

Materials:  two sheets of different-colored paper  scissors  tape

Make a model of planes $\mathcal{M}$ and $\mathcal{N}$ that intersect in $\overleftrightarrow{AB}$. Point C lies in $\mathcal{M}$, but not in $\mathcal{N}$. Point D lies in $\mathcal{N}$, but not in $\mathcal{M}$. Point E lies in both $\mathcal{M}$ and $\mathcal{N}$.

- Label one sheet of paper as $\mathcal{M}$ and the other as $\mathcal{N}$. Hold the two sheets of paper together and cut a slit halfway through both sheets.

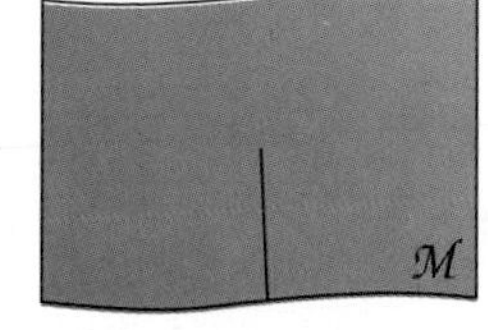

- Turn the papers so that the two slits meet and insert one sheet into the slit on the other sheet. Use tape to hold the two sheets together.

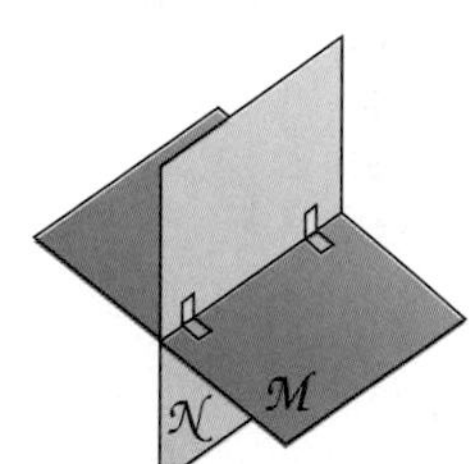

- The line where the two sheets meet is line AB. Draw the line and label points A and B.

Your Turn

a. Now draw point C so that it lies on $\mathcal{M}$ but not on $\mathcal{N}$. Can C lie on $\overleftrightarrow{AB}$?

b. Draw point D so that it lies on $\mathcal{N}$ but not on $\mathcal{M}$. Can D lie on $\overleftrightarrow{AB}$?

c. If point E lies in both $\mathcal{M}$ and $\mathcal{N}$, where would it lie? Draw point E.

d. Look at your model. Now draw a sketch of your model on your paper, labeling each point, line, and plane appropriately.

CHECK FOR UNDERSTANDING

Communicating Mathematics

Study the lesson. Then complete the following.

1. **Describe** how the walls in your classroom can represent planes, lines, and points.
2. **Name** three undefined terms listed in this lesson.
3. **Draw** plane Q with a line m intersecting Q at point E.
4. **You Decide** Malia told Lucia that she had figured out a pattern for computing the number of 3-letter names for any plane if she knew how many letters were named on the plane. For example, if there were four points, there were $4 \cdot 3 \cdot 2 \cdot 1$ names possible. Lucia said this pattern wouldn't work for all sets of letters. Who was right? Is there a pattern for n number of points? If so, describe it.
5. **List the possibilities** for naming a plane that contains points P, Q, and R.

6. Fold a sheet of paper. Open the paper and fold it again in a different way. Open the paper and label the creases as lines m and n.
   a. Do the two creases intersect? If so, in how many points?
   b. Suppose the creases did not intersect on the paper. Would the lines they represent intersect?
   c. Do you think it is possible that the two lines represented by the creases would not intersect?

Guided Practice

State whether each is best modeled by a *point, line, or plane.*

7. a star in the sky
8. an ice skating rink
9. a telephone wire strung between two poles

Refer to the figure at the right to name each of the following.

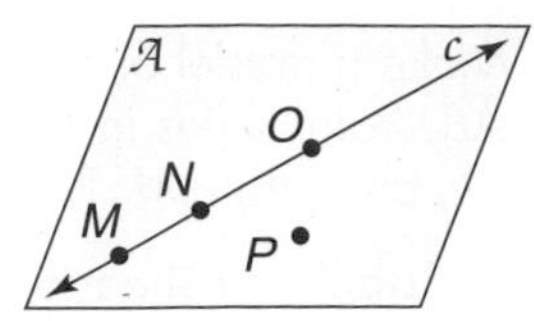

10. a line containing point N
11. a plane containing points P and M

Draw and label a figure for each relationship.

12. A line passes through $C(-3, -4)$, $R(-1, -3)$, and $S(3, -1)$, but point D does not lie on $\overleftrightarrow{RS}$.
13. Plane Q contains lines r and s that intersect in P.
14. $\overleftrightarrow{AB}$ intersects plane $\mathcal{P}$ in W.

Refer to the figure at the right to answer each question.

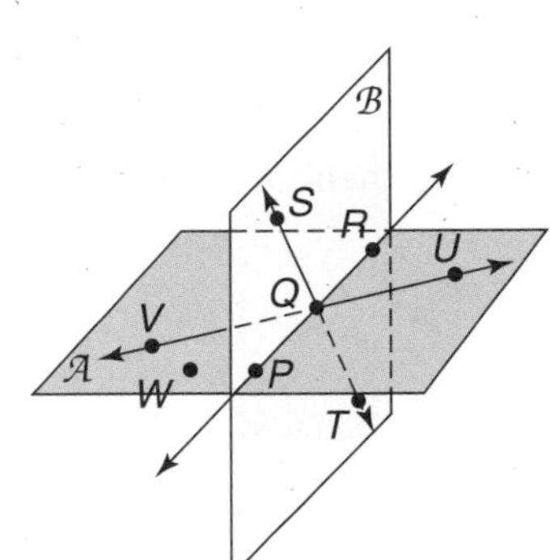

15. Name the intersection of planes $\mathcal{A}$ and $\mathcal{B}$.
16. Name another point that is collinear with points S and Q.
17. Name a line that is coplanar with $\overleftrightarrow{VU}$ and point W.
18. **List the Possibilities** The supermarket has a soft-drink machine that dispenses cans for 50¢ a can. The machine will only accept quarters, dimes, and nickels. The order in which the coins are placed in the machine does not matter. How many different combinations of coins must the machine be programmed to accept?

EXERCISES

Practice

State whether each is best modeled by a *point, line, or plane.*

19. a taut piece of thread
20. a knot in a piece of thread
21. a piece of cloth
22. the corner of a room
23. the rules on your notebook paper
24. your desktop
25. each color dot, or pixel, on a video-game screen
26. a telecommunications beam to a satellite in space
27. the crease in a folded sheet of wrapping paper

Refer to the figure at the right to name each of the following.

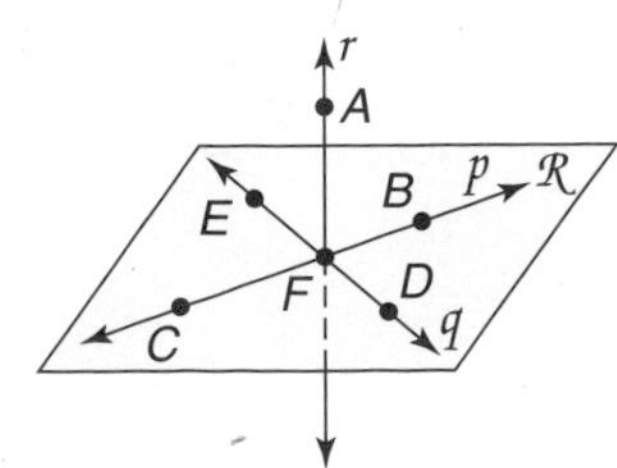

28. a line containing point A
29. a line passing through B
30. two points collinear with point D
31. two points coplanar with point B
32. a plane containing points B, C, and E
33. a plane containing lines p and q

Draw and label a figure for each relationship.

34. Point S lies on $\overrightarrow{PR}$.

35. Points $A(2, 4)$, $B(2, -4)$, and C are collinear, but points F, A, B, and C are noncollinear.

36. $\overleftrightarrow{TU}$ lies in plane Q and contains point R.

37. $\overleftrightarrow{CD}$ and $\overleftrightarrow{RS}$ intersect at $P(3, 2)$ for $C(-1, 4)$ and $R(6, 4)$.

38. Line m contains A and B, but does not contain C.

39. Lines a, b, and c are coplanar, but do not intersect.

40. Planes $\mathcal{P}$ and $\mathcal{R}$ intersect in ℓ.

41. Point C and line m lie in Q. Line m intersects line n at T. Line m and C are coplanar but m, n, and C are not.

42. Lines a, b, and c are coplanar and meet at point Z.

Refer to the figure below to answer each question. The figure is a rectangular prism formed by six planes. Only a portion of each plane is shown.

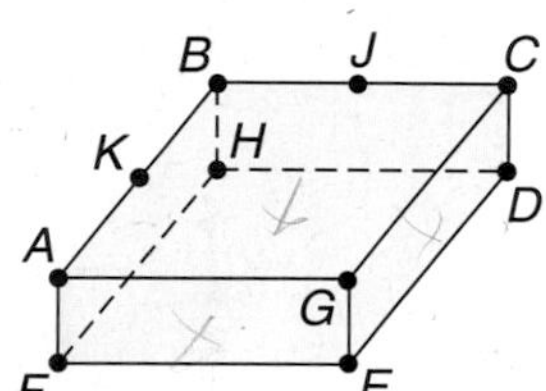

43. Name the six planes that form the rectangular prism.

44. Name two points that are coplanar with points B, H, and K.

45. Name the lines that intersect at E.

46. Which two planes intersect in $\overleftrightarrow{CD}$?

47. Name two lines that lie in plane GEC.

48. Name a plane and a line that intersect at A.

49. Which point(s) do planes ABC, CDE, and AGE have in common?

50. What do the dashed lines in the figure represent?

51. Are points B, H, E, and G coplanar? Explain your answer.

Draw and label a figure for each relationship.

52. Planes $\mathcal{A}$ and $\mathcal{B}$ intersect, and planes $\mathcal{B}$ and $\mathcal{C}$ intersect, but planes $\mathcal{A}$ and $\mathcal{C}$ do not intersect.

53. Line t lies in planes $\mathcal{P}$, $\mathcal{Q}$, and $\mathcal{R}$.

54. Planes $\mathcal{P}$ and $\mathcal{Q}$ intersect each other. They both intersect plane $\mathcal{R}$.

55. The intersection of planes $\mathcal{A}$, $\mathcal{B}$, $\mathcal{C}$, and $\mathcal{D}$ is point E.

Critical Thinking

56. Think of your classroom as a model of six planes. The floor and ceiling are models of horizontal planes $\mathcal{A}$ and $\mathcal{B}$, and the four walls are models of vertical planes $\mathcal{C}$, $\mathcal{D}$, $\mathcal{E}$, and $\mathcal{F}$. Use your classroom as a model to explain your answers to the following questions.
    a. Can planes $\mathcal{A}$ and $\mathcal{B}$ intersect?
    b. Is it possible for any of the vertical planes to intersect?
    c. Are all lines in plane $\mathcal{C}$ vertical lines?
    d. What are the possiblities for the intersection of two of the vertical planes and plane $\mathcal{B}$?

57. Four lines are coplanar. Draw figures to show the maximum number of intersection points and the minimum number of intersection points using the four lines.

Applications and Problem Solving

58. **List the Possibilities** The Hawaiian game of lu-lu is played with four disks of volcanic stone. The face of each stone is marked with a series of dots.

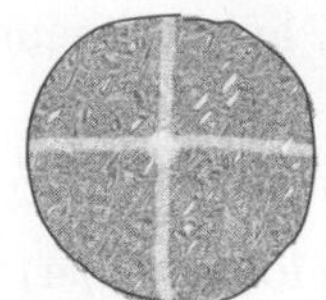
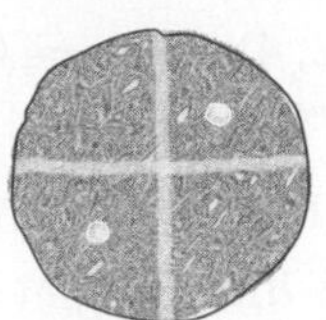
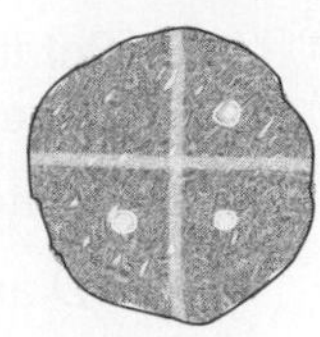
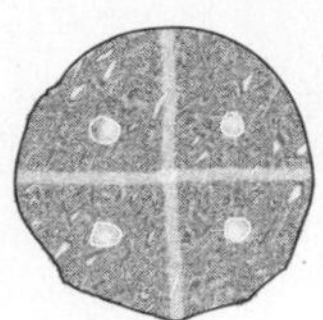

A player tosses the four disks and if they land all faceup, 10 points are scored, and the player tosses again. If any of the disks land facedown on the first toss, the player gets to toss those pieces again. The score is the total number of dots showing after the second toss. List the possible outcomes after the first toss.

59. **Anatomy** Has anyone ever told you to stand up straight? If your posture is perfect, you should be able to draw a straight line from your ear to your ankle, running through your shoulder, hip, and knee.
    a. Study the posture of five of your friends or relatives. How many of them seem to have good posture according to the straight line rule?
    b. What percent of the people you observed have good posture?

60. **Art** In the 1950s, a movement called *op art* stressed pure abstract images rather than real-life images. Op art consists of carefully arranged colors and geometric patterns that create optical illusions, sometimes even movement, on a painted surface. The image at the right is an optical illusion. Use your knowledge of planes to describe what you see. Then look again to see if you can see the objects in a different perspective. Write about your observations.

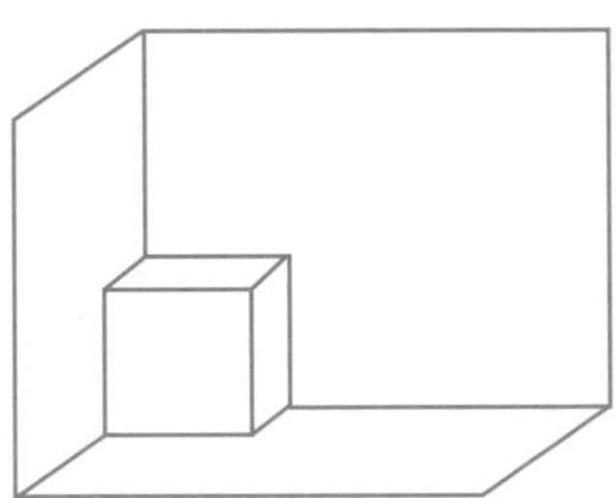

Mixed Review

61. Graph points $A(-2, 5)$, $B(3, -4)$, $C(0, -3)$, and $D(5, 5)$ on the same coordinate plane. (Lesson 1–1)

62. Points $F(2, 4)$ and $G(4, 8)$ lie on the graph of $y = 2x$. Determine whether $H(-3, -6)$ is collinear with F and G. (Lesson 1–1)

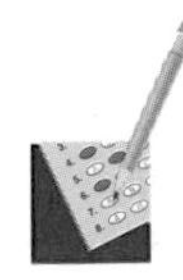

63. **SAT Practice** If n is an odd integer, which of the following must be an even integer?

    A $3(n + 2)$ **B** $3(n + 1)$ **C** $n - 2$ **D** $3n$ **E** n^2

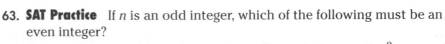

Algebra

Solve each equation.

64. $x + 3 = 8$
65. $4x = -44$
66. $-y + 8 = -2$
67. $3p + 4 = 2p$
68. $-4c + 12 = 15$
69. $3(x + 2) = -6$

Evaluate each expression if $a = 3$, $b = -2$, and $c = 0$.

70. $ab + 6c$
71. $a^2 + b$
72. $bc + ac - c^2$

For **Extra Practice**, see page 764.

1-3

Integration: Algebra Using Formulas

What YOU'LL LEARN

- To solve problems by using formulas, and
- to find maximum area of a rectangle for a given perimeter.

Why IT'S IMPORTANT

You can use formulas to solve problems in science, social studies, and business as well as mathematics.

Real World APPLICATION

Photography

One of the easiest ways to "capture the moment" these days without investing in expensive equipment is with a single-use camera. A new type of single-use camera uses a panoramic lens. The diagram below shows what portion of a scene is captured by a 35-mm camera that has a panoramic mode, a standard single-use camera, and a panoramic single-use camera. How does the area of the view of a panoramic single-use camera compare with the area of the view of a 35-mm camera that has a panoramic mode? *This problem will be solved in Example 1.*

To be a good problem solver, you need to develop a plan for finding a solution. Four steps that can be used to solve any problem are listed below.

Problem-Solving Plan	**1. Explore the problem.** **2. Plan the solution.** **3. Solve the problem.** **4. Examine the solution.**

To solve the problem presented above, we use the formula for the **area** of a rectangle since each photographic image is a rectangle.

Area of a Rectangle	**The formula for the area of a rectangle is $A = \ell w$, where A represents the area expressed in square units, ℓ represents the length, and w represents the width.**

Example 1

Real World APPLICATION

Photography

Refer to the application at the beginning of the lesson. Compare the area of the view of a panoramic single-use camera to the area of the view of a 35-mm camera that has a panoramic mode.

Explore The view of the panoramic single-use camera is much wider and taller than that of the 35-mm camera. To compare areas, we need to determine the dimensions of each view.

(continued on the next page)

Moneta J. Sleet, Jr. (1926–1996)

The first African-American journalist to win a Pulitzer Prize for feature photography was Moneta Sleet Jr. in 1969. He won the award for his photograph of Coretta Scott King comforting her young daughter, Bernice, at the funeral of Martin Luther King, Jr.

Plan Measure each rectangle using a millimeter ruler. Then use the measurements in the formula $A = \ell w$. Finally, compare the areas.

Solve

panoramic camera	*35-mm camera*
$A = \ell w$	$A = \ell w$
$= 80 \cdot 28$ or 2240	$= 38 \cdot 14$ or 532
The area is 2240 mm^2.	The area is 532 mm^2.

The view in the panoramic single-use camera has an area about 4 times that of the view in the 35-mm camera with panoramic mode.

Examine Suppose you draw a rectangle to model the view of the panoramic single-use camera. Then draw and cut out a rectangle to model the view of the 35-mm camera. See how many of the smaller rectangles will fit on the larger one.

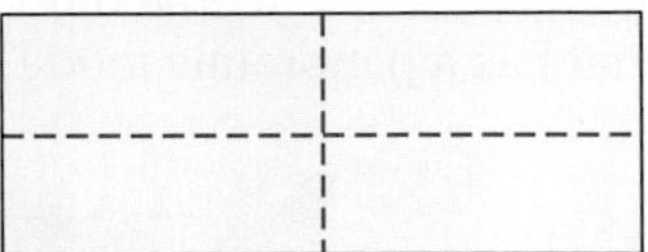

Four of the smaller rectangles are slightly larger than the area of the larger rectangle. Our estimate is correct.

There are many formulas that we will use in geometry. Another formula used with rectangles that you may remember is the **perimeter** formula. The perimeter is the distance around a figure. The formula for the perimeter P of a rectangle is $P = 2\ell + 2w$.

Perimeter of a Rectangle	**The formula for the perimeter of a rectangle is $P = 2\ell + 2w$, where P represents the perimeter, ℓ represents the length of the rectangle, and w represents the width of the rectangle.**

If you know the perimeter of a rectangle, you can find the maximum area that a rectangle with that perimeter can have.

Example 2

Pets

One type of pet enclosure is made up of 12 sections of fencing, connected by hinges. This enclosure folds for easy storage and is flexible for making different-sized areas in which your pet can play. Suppose the sections are positioned to make a rectangle. What are the dimensions of the rectangle that provide the maximum area for your pet?

Explore Since the perimeter will always be 12 units and fractions of a unit do not apply in this situation, we can draw all of the possibilities for rectangular shapes.

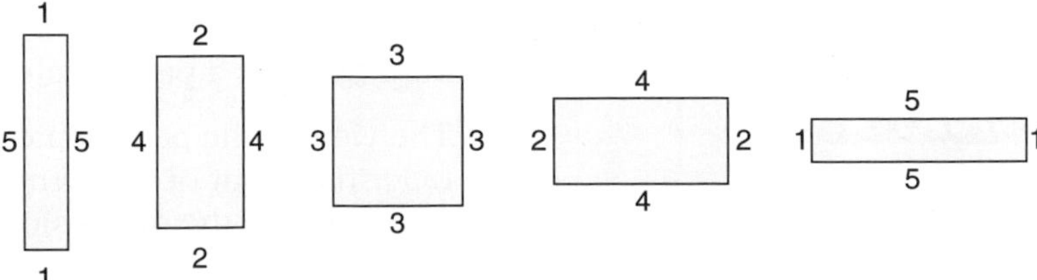

Plan Use a table to find the area of each of the possible rectangles. Then select the one with the greatest area.

Solve Make a table that has columns for width w, length ℓ, and area A. It appears that the greatest area for the pet enclosure is 9 square units. This area occurs when the width is 3 units and the length is 3 units.

w	ℓ	$A = \ell w$
1	5	5
2	4	8
3	3	9
4	2	8
5	1	5

Why can't w be greater than or equal to 6?

The shape of the enclosure is a square.

Examine We can graph all ordered pairs (w, A) to look for a maximum.

w	A
1	5
2	8
3	9
4	8
5	5

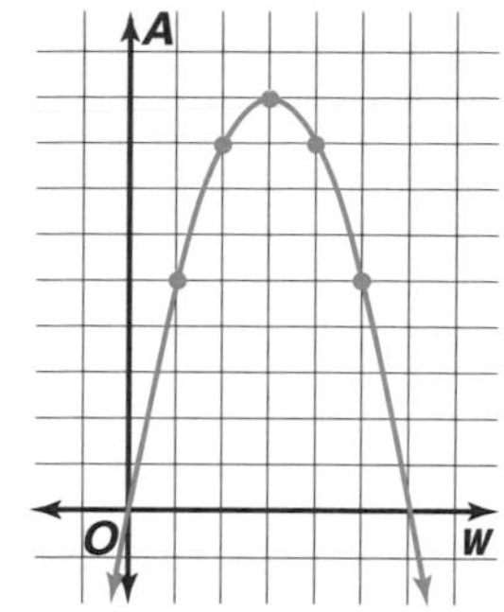

*The graph shows **all** the possible widths and areas, including fractional values.*

The graph is U-shaped and the greatest value of A appears to be when $w = 3$.

A spreadsheet can be used to quickly calculate quantities determined by a formula. This tool can help you look for patterns in the calculated data.

EXPLORATION SPREADSHEETS

A spreadsheet is a table of cells that can contain text (labels) or numbers and formulas. Each cell is named by the column and row in which it is located. In the spreadsheet below, cells A1, B1, C1, and D1 contain labels.

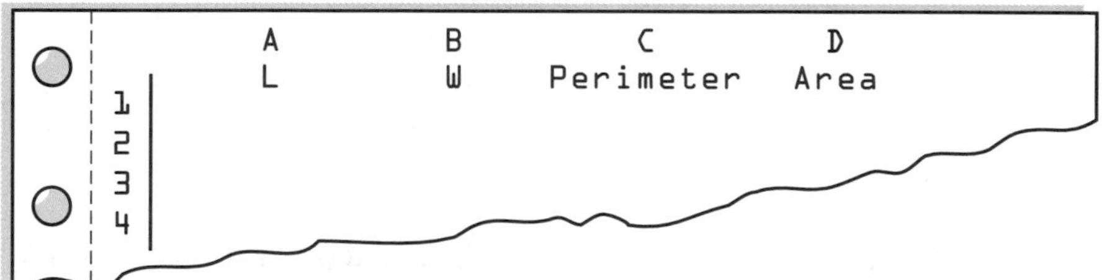

- In cell C2, enter the formula 2*A2 +2*B2, which represents the formula for the perimeter of a rectangle.
- Copy the formula to all of the cells of column C. When you do, the spreadsheet automatically uses the formula with the values in that row. For example, in cell D3, the formula becomes 2*A3 +2*B3 and in cell D4, it becomes 2*A4 +2*B4.

Your Turn

a. Enter a formula for the area of a rectangle in cell D2 and copy it to the other cells.

b. Enter values in columns A and B and look for patterns in the values in columns C and D.

Many problems in mathematics use formulas. Sometimes it is necessary to solve a formula for a certain variable to find the answer to the problem. This requires some of the skills you learned in algebra.

Example 3

Money

A formula for computing simple interest is $I = prt$, where I is the interest earned, p is the principal (or the amount of money invested), r is the rate of interest (expressed as a decimal value), and t is the length of time in years it is invested. Soledad offered to lend her younger sister Lola \$200 at a 3.5% interest rate. She told Lola that her loan would cost Lola \$217.50 at the end of the loan period. How long is Soledad giving Lola to pay back the loan?

The interest on her loan is \$217.50 − \$200 or \$17.50. There are two ways to determine the time of the loan.

Method 1

Solve the formula for t.

$I = prt$

$\frac{I}{pr} = t$ *Divide each side by pr.*

Now substitute 17.50 for I, 200 for p, and 0.035 for r. *3.5% = 0.035*

$t = \frac{I}{pr}$

$= \frac{17.50}{200 \cdot 0.035}$ or 2.5 *Use a calculator.*

Method 2

Substitute the known values into the formula and solve for t.

$I = prt$

$17.50 = 200 \cdot 0.035 \cdot t$

$17.50 = 7t$ *Simplify.*

$2.5 = t$ *Divide each side by 7.*

Regardless of the method used, Lola has 2.5 years to pay back the loan.

CHECK FOR UNDERSTANDING

Communicating Mathematics

Study the lesson. Then complete the following.

1. **Describe** what each variable in the formula $P = 2\ell + 2w$ means.
2. **Demonstrate** how you could find the formulas for the perimeter and area of a square by using the formulas for a rectangle.
3. **Explain** the four-step plan for problem solving.

4. **Make a list** of other mathematical formulas you recall from previous mathematics or science courses. Explain what each variable represents.

Guided Practice

Find the perimeter and area of each rectangle.

5. Rectangle: 10 cm by 4 cm

6. Rectangle: 8.5 in. by 7 in.

Find the missing measure in each formula.

7. $A = \ell w$; $\ell = 4$, $w = 7$, $A = ?$
8. $P = 2\ell + 2w$; $\ell = 3$, $w = 5$, $P = ?$
9. $I = prt$; $p = 350$, $r = 6\%$, $I = 42$, $t = ?$

10. **Distance** Solve the formula $d = rt$ for t. In the formula, d represents the distance, r the rate of speed, and t the time traveled.

11. **Meteorology** In the formula $\frac{T - t}{5} = d$, T represents the time when the flash of lightning occurs, t, the time when the sound of the thunder begins; and d, the distance in miles the lightning is from you. Solve the formula for T.

Find the maximum area for the given perimeter of a rectangle. State the length and width of the rectangle.

12. 24 millimeters
13. 36 miles

Data Update For more information on weather around the world, visit: **www.geometry.glencoe.com**

14. **Travel** Marguerite is going to Spain during spring break with the foreign language club. She receives an information sheet about the trip that says the average temperature during April is 18°C and the average rainfall is 8.6 centimeters. However, Marguerite is not familiar with metric units of measure.
   a. The formula for converting Celsius (C) temperatures to Fahrenheit (F) temperatures is $\frac{9}{5}C + 32 = F$. What is the average temperature in Spain in April in degrees Fahrenheit?
   b. The formula for estimating the number of inches in c centimeters is $\ell = \frac{c}{2.54}$. Find the average rainfall in inches.

EXERCISES

Practice

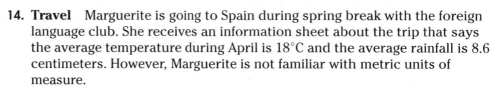

Find the perimeter and area of each rectangle.

15. 10 m, 3 m

16.

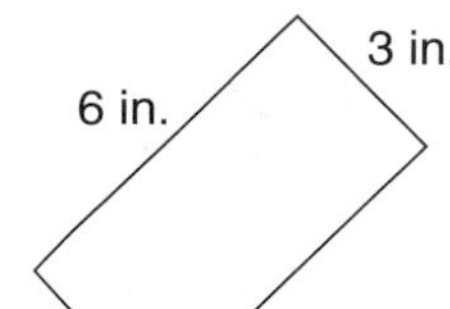

17. 2.5 cm, 2.5 cm

18. 1 yd, 12 yd

19.

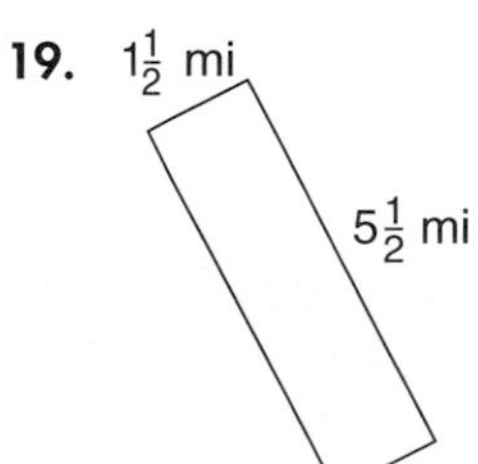

20.

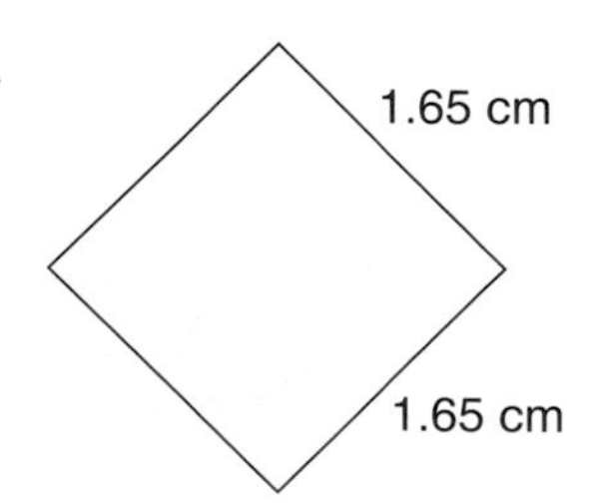

Find the missing measure in each formula if $P = 2\ell + 2w$ and $A = \ell w$.

21. $\ell = 7, w = 3, P = ?$
22. $\ell = 4.5, w = 1.5, P = ?$
23. $\ell = 8, w = 4, A = ?$
24. $\ell = 2.2, w = 1.1, A = ?$
25. $\ell = 6, A = 36, w = ?$
26. $\ell = 12, A = 30, w = ?$
27. $A = 34, w = 2, \ell = ?$
28. $A = 3\frac{1}{2}, w = \frac{1}{2}, \ell = ?$
29. $P = 84, \ell = 12, w = ?$
30. $P = 13, w = 2.5, \ell = ?$

Find the maximum area for the given perimeter of a rectangle. State the length and width of the rectangle.

31. 28 inches **32.** 44 centimeters **33.** 32 feet

34. 26 meters **35.** 15 yards **36.** 5 millimeters

Evaluate each formula for the values given.

37. Area of a triangle: $A = \frac{1}{2}bh$, if $b = 10$ and $h = 12$

38. Changing °F to °C: $C = \frac{5}{9}(F - 32)$ for 212°F

39. Volume of a rectangular solid: $V = \ell wh$, if $\ell = 3.5$, $w = 4$, and $h = 1.25$

Critical Thinking

40. Suppose the area of a rectangle is 36 square inches. Find the *minimum* perimeter for a rectangle with this area. Explain your procedure for finding the perimeter.

Applications and Problem Solving

41. Ranching A rancher is adding a corral to his barn so that the barn opens directly into the corral. He has 195 feet of fencing left over from another project. Find the greatest possible area for his corral using this length of fence. (*Hint*: The corral with maximum area is not a square.)

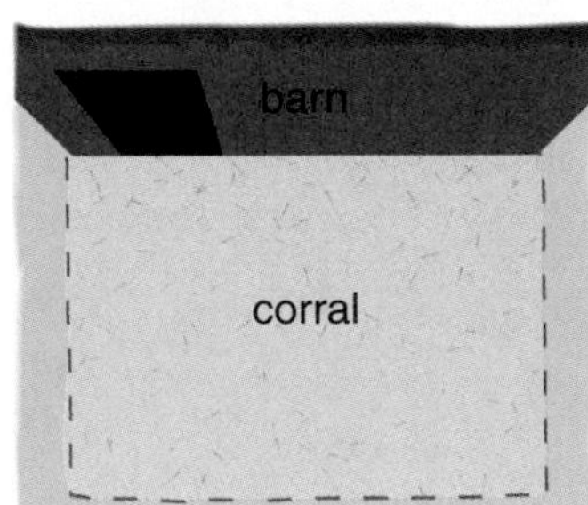

42. Printing Dots of four colors are used to break a photo down into its component parts for printing. Each photo is separated into four primary colors.

The photos in this book are printed at 1200 dpi (dots per square inch). What is the maximum number of different colored dots that could be used to print a 5-inch by 8-inch rectangular color photo?

43. Energy Conservation A house that has less exterior wall area loses less energy. You can compare the amount of exterior wall area on houses of the same height by comparing the possible perimeters of the house. A house with a greater perimeter will have a greater amount of exterior wall area.

a. If the houses below have about the same amount of living area, which house will lose the least amount of energy through the exterior walls?

b. Which house actually has the greatest amount of living area?

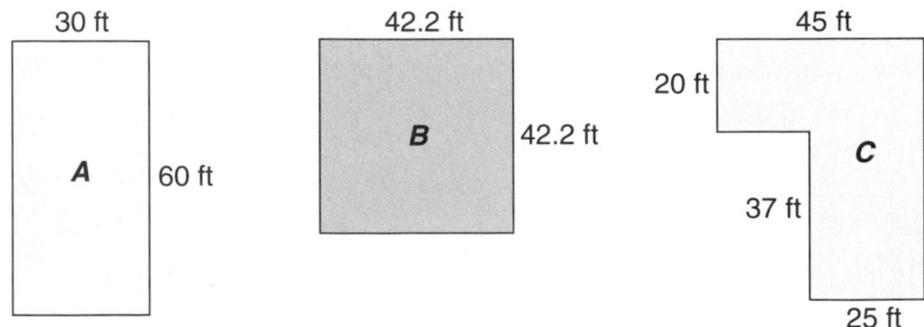

Mixed Review

Refer to the figure at the right to answer each question. (Lesson 1–2)

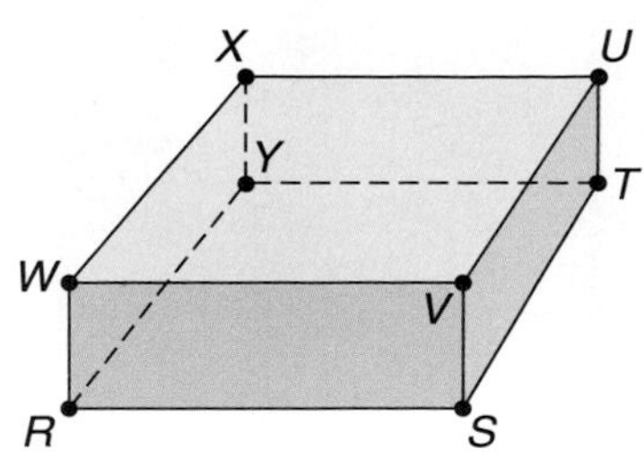

44. Name a point not coplanar with R, S, and T.

45. Name three lines that contain S.

46. Name the intersection of planes RWX and UTY.

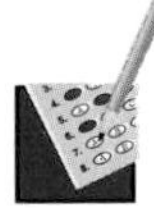

47. **SAT Practice** In the figure, which of the following points lies within the shaded region?

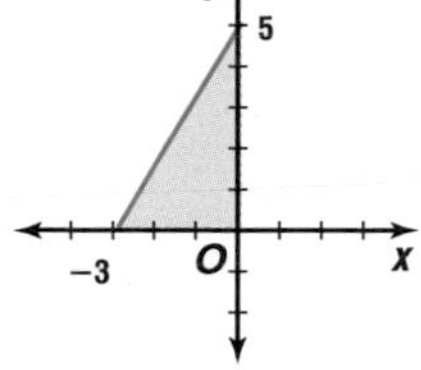

A $(-2, 4)$ B $(-1, 3)$
C $(1, -3)$ D $(2, -4)$
E $(4, -2)$

48. Graph $A(-3, 4)$, $B(3, 4)$, $C(3, -4)$, and $D(-3, -4)$ on the same coordinate plane. (Lesson 1–1)

Algebra

Simplify each expression.

For **Extra Practice**, see page 764.

49. $4x(x + 3) + 3x^2 - 5x$

50. $3x + 4x - 9x + 2$

51. $6c + 7d + 2c - 10d$

52. $a^2b + 3a^2b - b + a^2$

SELF TEST

1. Name the ordered pair for each point graphed in the coordinate plane at the right. (Lesson 1–1)
   a. P b. Q c. R

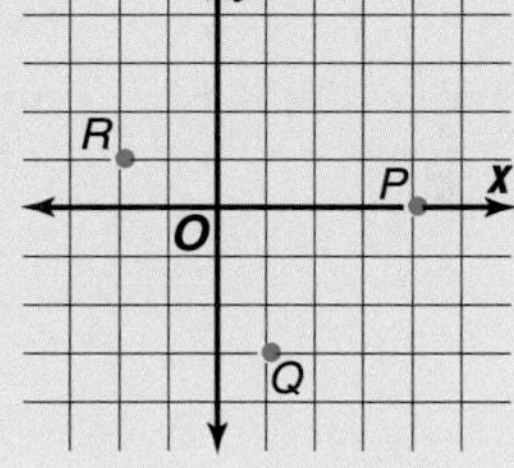

2. Graph each point on the same coordinate plane. (Lesson 1–1)
   a. $A(3, 3)$ b. $B(0, -5)$ c. $C(-3, -6)$
3. List all the possible names for the figure below. (Lesson 1–2)

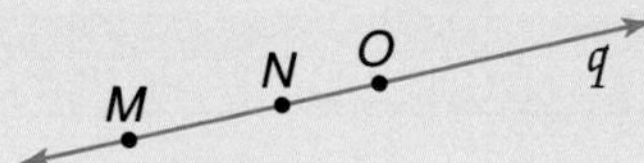

Refer to the figure at the right to answer each question. (Lesson 1–2)

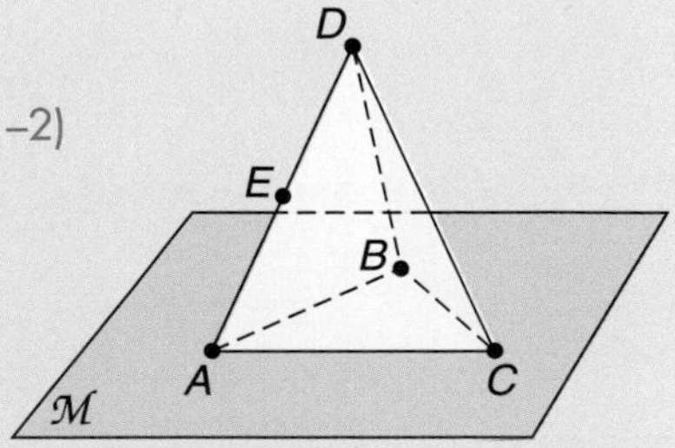

4. Are points A, E, and D collinear? Explain.
5. Are points A, B, C, and D coplanar? Explain.
6. How many planes appear in this figure? Name them.
7. Draw and label a figure showing lines a and b intersecting at T with line a in plane Q, but line b not in Q. (Lesson 1–2)

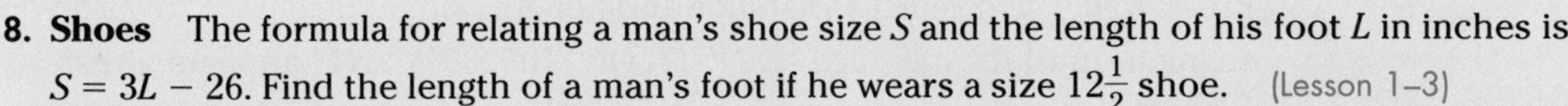

8. **Shoes** The formula for relating a man's shoe size S and the length of his foot L in inches is $S = 3L - 26$. Find the length of a man's foot if he wears a size $12\frac{1}{2}$ shoe. (Lesson 1–3)
9. Find the perimeter of a rectangle that is 8 kilometers wide and 22.5 kilometers long. (Lesson 1–3)
10. What is the maximum area for a rectangle whose perimeter is 48 meters? (Lesson 1–3)

1–4A Using Technology Measuring Segments

A Preview of Lesson 1–4

This lesson can be done using software on a computer. Instead of keys, you use similar pulldown menus to perform the functions.

One kind of calculator is actually a small computer. It works very much like a graphing calculator in some respects, but has many more features, one of which is a keyboard like you find on a computer. The calculator allows you to construct geometric figures, measure them, and reposition them to study their characteristics. Many of the graphing calculator lessons in this book will use these geometry features.

To select an item on a menu, press down on the cursor pad until the selection is highlighted and then press ENTER. You can also select the item by entering the item's number on the menu.

To **access the geometry applications**, press the applications button APPS, select 8:Geometry, and then 3:New. Press down on the cursor pad and type in any name for your work session by using the letter keys. Press ENTER twice. A screen like the one at the right appears. The cursor is a small + sign.

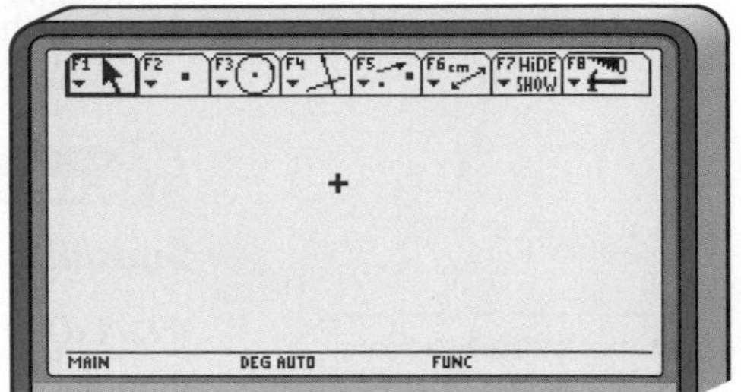

To **construct a segment**, press F2 (the drawing menu) and select 5:Segment. The cursor becomes a small pencil. Use the cursor pad to move the cursor to any location on the screen to start your drawing. Press ENTER to locate your first endpoint. If you wish to name your endpoint easily, do it now by pressing the shift key ⬆ while typing in your letter to name the point. *If you move the cursor before naming the point, the automatic label function will not work.*

Now move your cursor to wherever you wish your second endpoint to be. Press ENTER to locate the point. Name the other endpoint. The screen at the right shows a sample segment *AB*.

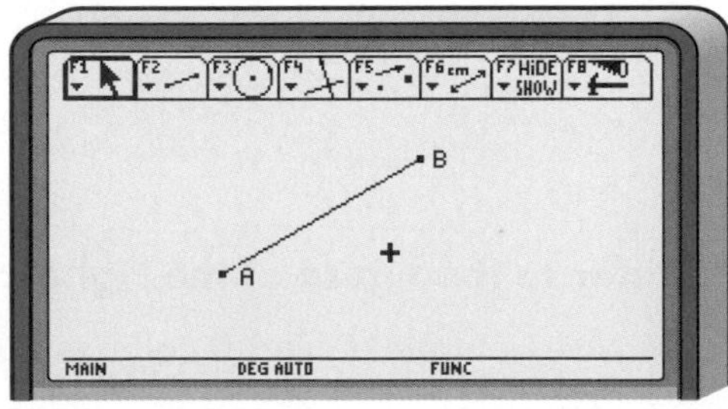

To select a different unit of measure, press F8 and select 9:Format. Select Length & Area on the menu. Press the cursor pad right. Select your preferred unit of measure.

The geometry applications allow you to measure each segment you draw. Press F6 and select 1:Distance and Length. Move the cursor toward your segment until the message "LENGTH OF THIS SEGMENT" appears. Press ENTER and the measurement appears. You can **relocate the label** by moving the cursor toward the label until the message "THIS NUMBER" appears. Hold down the hand key and use the position pad to move the label to wherever you wish. Release the hand key to finalize its position.

All of the commands referenced are on the F2 *menu.*

You can select the items from the F2 menu to **draw additional points, segments, and lines** on the screen.

- Draw other segments by using the same method you used to draw $\overline{AB}$.
- Name a point of intersection by using 3:Intersection Point command. Once you select the command, move your cursor to where you think the intersection should be. When you are close enough to the point, the message "POINT AT THIS INTERSECTION" will appear. Press ENTER and name the point *E*.
- To name other points on a segment, select 2:Point on Object. Move the cursor to where you wish to place the point. The message "ON THIS SEGMENT" will appear. Press ENTER and name the point. Because you are in "draw a point" mode, you can put additional points on the segment without returning to the F2 menu. Just press ENTER at each location and name the point.
- To draw points not on a segment, select 1:Point, press ENTER, and name the point.
- To draw a line through two points, select 4:Line and move the cursor to one of the points through which the line will pass. When the message "THRU THIS POINT" appears, press ENTER and move the cursor to the second point. Press ENTER at the message and a line will appear.

You can **clear your screen** or **erase a figure** by using 7:Delete or 8:Clear All on the F8 menu.

EXERCISES

Analyze your drawing.

1. How many points could you put on a segment?
2. Which unit of measure seems more reasonable when drawing segments?
3. Try to measure the distance from *E* to *A*. What happens?
4. Move the cursor to the position of one of the endpoints of $\overline{AB}$. Hold down the hand key and use the cursor pad to move the endpoint to a new location. What happens to the measure of the segment?

Construct each of the following. Sketch each drawing on your paper.

5. Lines *PQ* and *RS* meet at *T*.
6. Segment *WT* intersects $\overline{TR}$.
7. Point *O* lies between *N* and *M* on $\overleftrightarrow{PQ}$.
8. Segments *AB* and *CD* do not intersect.
9. segment *XY* and its length
10. Experiment with the 4:Line command on the F2 menu when you do not draw two points. Write about your findings.

If you have linking software and a cable for the calculator, you can print out your sketches from a computer.

1–4 Measuring Segments

What YOU'LL LEARN

- To find the distance between two points on a number line and between two points in a coordinate plane, and
- to use the Pythagorean Theorem to find the length of the hypotenuse of a right triangle.

Why IT'S IMPORTANT

Segment measures are used in discovering characteristics of many geometric shapes.

Real World APPLICATION

Textiles

The woman in Guatemala shown in the photo at the right is weaving a colorful fabric by an age-old method. However, most fabric for your clothing was produced on an electronic high-speed loom.

Fabric is measured in bolts. A bolt, which is 120 feet long, is only one of many units of linear measure used in the world. A piece of thread running the length of the fabric could be a model for a line segment. The ends of the thread could be thought of as the endpoints of the segment. In geometry, the length of a segment is the distance between its two endpoints. Segments can be defined by using the idea of *betweenness* of points.

THE FAR SIDE By GARY LARSON

"Well, lemme think. ... You've stumped me, son. Most folks only wanna know how to go the other way."

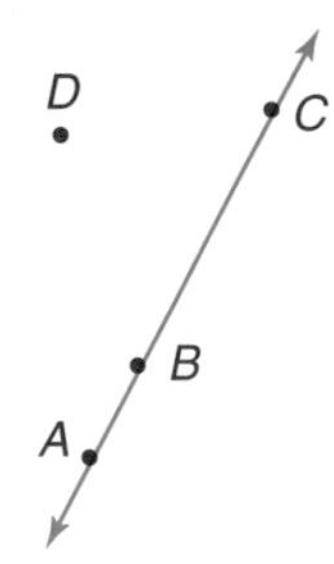

In the figure at the right, point B is **between** points A and C, while point D is *not* between A and C. For B to be between A and C, all three points must be collinear. Segment AB, written $\overline{AB}$, consist of points A and B and all points between A and B. The **measure** of $\overline{AB}$, written AB (no bar over the letters), is the distance between A and B. Thus, the measure of a segment is the same as the distance between its two endpoints.

You have probably measured segments by using a centimeter- or inch-ruler. Centimeters and inches are just two examples of *units of measure.*

Whether you are trying to measure the distance from A to B or from B to A as in the cartoon at the left, the measure is the same. In fact, two points on any line can always be paired with real numbers so that one point is paired with zero and the other paired with a positive number. This correspondence suggests the **Ruler Postulate**.

A ***postulate*** *is a statement that is assumed to be true.*

Postulate 1–1 **Ruler Postulate**	**The points on any line can be paired with real numbers so that, given any two points *P* and *Q* on the line, *P* corresponds to zero, and *Q* corresponds to a positive number.**

One way to measure the distance between two points is to use a number line. The numbers on a ruler are a real-life example of a number line.

To find the measure of $\overline{XY}$, you first need to identify the coordinates of X and Y. The coordinate of X is 2, and the coordinate of Y is 8. One way to find the distance between the two points is to count the number of units between X and Y, which is 6. You could also use your algebra skills. Since measure is always a positive number, you can find the *absolute value* of the difference between the two coordinates. When you use absolute value, the order in which you subtract the coordinates does not matter.

distance from X to Y	*distance from X to Y*
$\lvert 8 - 2 \rvert = \lvert 6 \rvert$ or 6	$\lvert 2 - 8 \rvert = \lvert -6 \rvert$ or 6

The measure of $\overline{XY}$ is 6, or $XY = 6$.

Example  **Find *PQ*, *QR*, and *PR* on the number line shown below.**

$PQ = \lvert -3.5 - (-2) \rvert$
$= \lvert -3.5 + 2 \rvert$
$= \lvert -1.5 \rvert$ or 1.5

$QR = \lvert -2 - 2.5 \rvert$
$= \lvert -4.5 \rvert$ or 4.5

$PR = \lvert -3.5 - 2.5 \rvert$
$= \lvert -6 \rvert$ or 6

Examine the measures PQ, QR, and PR in Example 1. Notice that $1.5 + 4.5 = 6$. So $PQ + QR = PR$. This suggests the following postulate.

Postulate 1–2 **Segment Addition Postulate**	**If *Q* is between *P* and *R*, then *PQ* + *QR* = *PR*.** **If *PQ* + *QR* = *PR*, then *Q* is between *P* and *R*.**

Example  **Find *LM* if *L* is between *N* and *M*, $NL = 6x - 5$, $LM = 2x + 3$, and $NM = 30$.**

Algebra

A good way to begin solving this problem is to make a drawing of the given information. Draw L between N and M and label the measures of the segments.

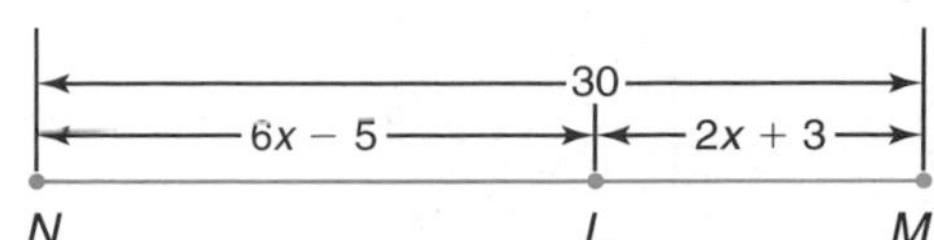

(continued on the next page)

Since L is between N and M, $NL + LM = NM$. Use this equation and the values we know to write a new equation.

$$NL + LM = NM$$

$6x - 5 + 2x + 3 = 30$ *$NL = 6x - 5$, $LM = 2x + 3$, $NM = 30$*

$8x - 2 = 30$ *Combine like terms.*

$8x = 32$ *Add 2 to each side.*

$x = 4$ *Divide each side by 8.*

Now use the value of x to find LM.

$LM = 2x + 3$

$= 2(4) + 3$ or 11

FYI

In 2000 B.C., Egyptians used a loop of rope knotted at 12 equal intervals to determine the boundaries of their properties each year after the Nile flooded. They stretched the rope around three stakes so that the measures of the sides of the triangle were 3, 4, and 5. This right triangle was used to establish the corners of their fields.

You have probably encountered the **Pythagorean Theorem** in other math classes. The theorem is often expressed as $a^2 + b^2 = c^2$, where a and b are the measures of the shorter sides (legs) of the right triangle and c is the measure of the longest side (hypotenuse).

You will prove the Pythagorean Theorem in Lesson 8–1.

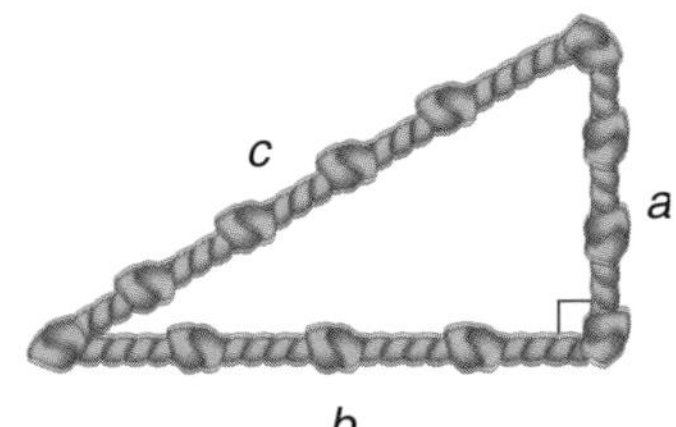

Pythagorean Theorem

In a right triangle, the sum of the squares of the measures of the legs equals the square of the measure of the hypotenuse.

If a and b represent the measures of the legs of a right triangle, and c represents the measure of the hypotenuse, then $a^2 + b^2 = c^2$.

Example 

Find the distance from $A(1, 2)$ to $B(6, 14)$ using the Pythagorean Theorem.

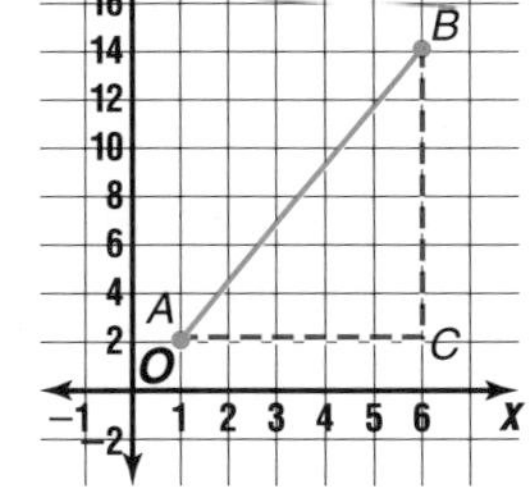

The grid lines on a coordinate plane meet at right angles, so we can draw segments along the grid lines to form a right triangle that has $\overline{AB}$ as its longest side. The coordinates of C are (6, 2).

Since the y-coordinates of A and C are the same, we can use the x-axis as the number line to find AC.

$$AC = |1 - 6| \text{ or } 5$$

Likewise, since the x-coordinates of B and C are the same, we can use the y-axis as the number line to measure the length of $\overline{BC}$.

$$BC = |14 - 2| \text{ or } 12$$

Now use the Pythagorean Theorem to find the length of $\overline{AB}$ if $AB = x$.

$a^2 + b^2 = c^2$

$5^2 + 12^2 = x^2$ *$a = 5$, $b = 12$, $c = x$*

$25 + 144 = x^2$

$169 = x^2$

$\pm\sqrt{169} = x$ *Take the square root of each side.*

$13 = x$ *Ignore the negative value of x.*

The distance from A to B is 13.

You will derive the distance formula in Chapter 8.

The process in Example 3 could become very tedious if you had to find a lot of distances by using the Pythagorean Theorem. A formula has been developed from the Pythagorean Theorem to find the distance between two points in a coordinate plane. This formula is called the **distance formula**.

Distance Formula	**The distance d between any two points with coordinates (x_1, y_1) and (x_2, y_2) is given by the formula** $d = \sqrt{(x_2 - x_1)^2 + (y_2 - y_1)^2}$.

The distance between two points is the same no matter which point is used as (x_1, y_1).

Example 4

Find PQ for $P(-3, -5)$ and $Q(4, -6)$.

Let (x_1, y_1) be $(-3, -5)$ and (x_2, y_2) be $(4, -6)$.

$$d = \sqrt{(x_2 - x_1)^2 + (y_2 - y_1)^2} \quad \textit{Distance formula}$$

$$PQ = \sqrt{[4 - (-3)]^2 + [-6 - (-5)]^2}$$

$$= \sqrt{7^2 + (-1)^2}$$

$$= \sqrt{49 + 1}$$

$$= \sqrt{50} \text{ or about } 7.07 \quad \textit{Use a calculator.}$$

When two segments have the same length, they are said to be **congruent** segments. For example, if $AB = CD$, then we write $\overline{AB} \cong \overline{CD}$, which is read "segment AB is congruent to segment CD."

We can use a compass and straightedge to construct a segment that is congruent to another segment.

CONSTRUCTION

Congruent Segments

A straightedge is any object that can be used to draw a straight line, such as a ruler or the edge of a protractor. A straightedge is not used to measure length.

Use a compass and a straightedge to construct a segment congruent to another segment.

1. Draw a segment XY.

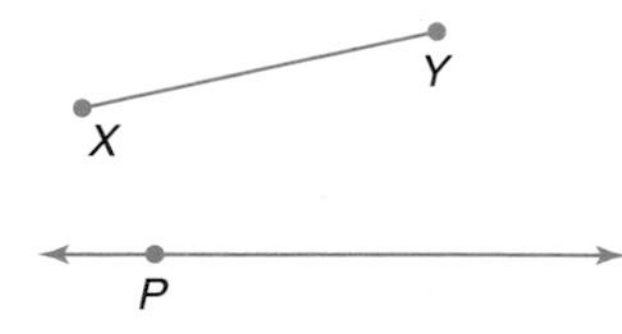

2. Elsewhere on your paper, draw a line and a point on the line. Label the point P.

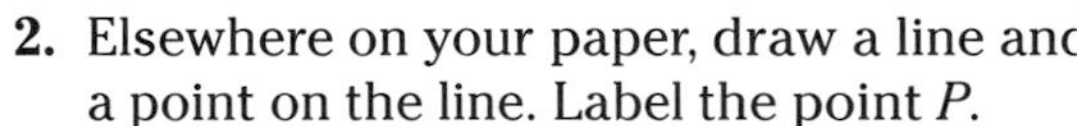

3. Place the compass at point X and adjust the compass setting so that the pencil is at point Y.

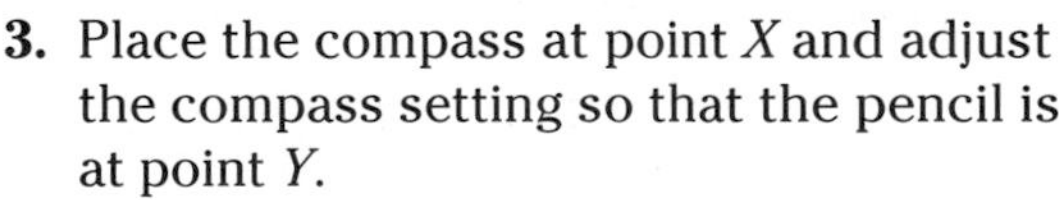

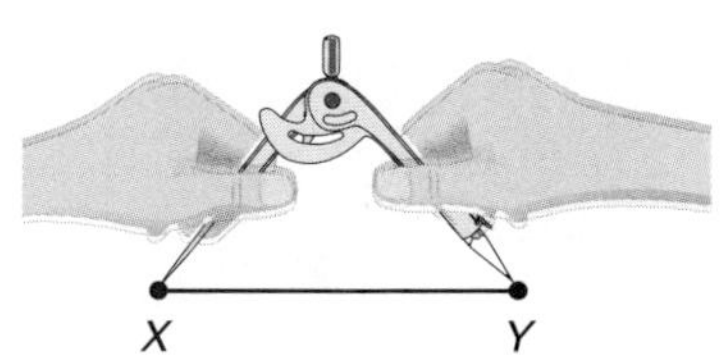

4. Using that setting, place the compass at point P and draw an arc that intersects the line. Label the point of intersection Q.

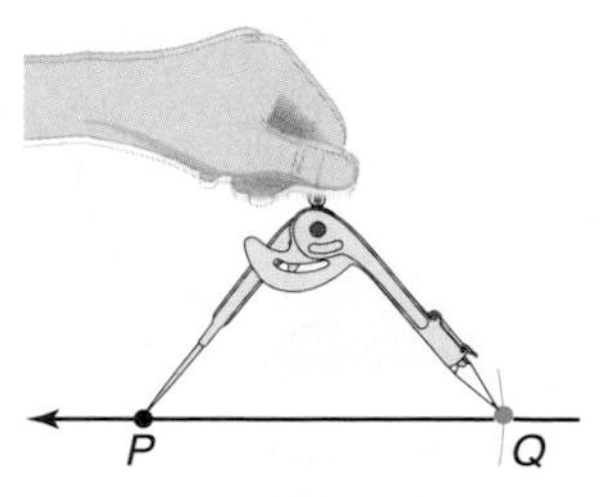

Conclusion: Since the compass setting used to construct $\overline{PQ}$ is the same as the distance from X to Y, $PQ = XY$. Thus, $\overline{PQ} \cong \overline{XY}$.

When comparing measures of segments with no unit of measure given, assume that the units are the same.

Because the measures of segments are real numbers, they can be compared. To compare the measure of $\overline{AB}$ to the measures of $\overline{CD}$, $\overline{EF}$, and $\overline{GH}$ shown below, you can set your compass width to match the measure of $\overline{AB}$ and then compare this to the measure of each of the segments.

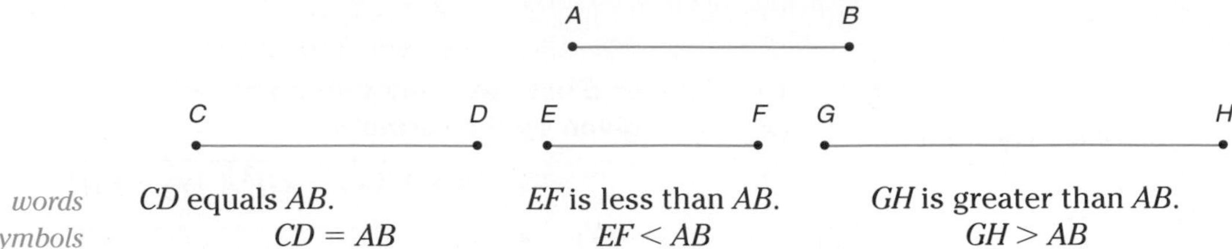

words	*CD* equals *AB*.	*EF* is less than *AB*.	*GH* is greater than *AB*.
symbols	$CD = AB$	$EF < AB$	$GH > AB$

CHECK FOR UNDERSTANDING

Communicating Mathematics

Study the lesson. Then complete the following.

1. **Write** an equation that relates the measures of $\overline{XY}$, $\overline{XZ}$, and $\overline{YZ}$, if point Y is between points X and Z.
2. **Describe** how you know which sides of a right triangle are represented by a, b, and c in the Pythagorean Theorem.
3. **Explain** the difference between $\overline{AB}$ and AB.
4. If $DE + EF = DF$, explain what must be true of point E.

Guided Practice

Refer to the number line at the right to find each measure.

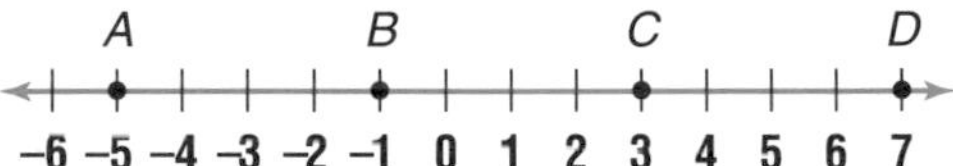

5. AC
6. BD
7. Find HK if J is between H and K, $HJ = 17$, and $JK = 6$.

Refer to the coordinate plane at the right to find each measure. Round your answers to the nearest hundredth.

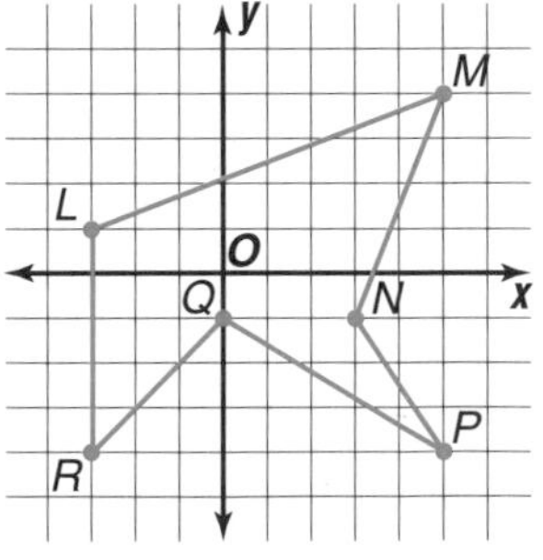

8. MP
9. NP
10. QR

11. Write a mathematical sentence to compare NP and QR.

12. Use the Pythagorean Theorem to find the missing side measure of $\triangle XYZ$.

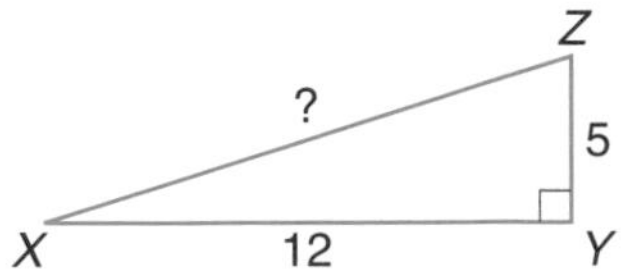

13. If W is between R and S, $RS = 7n + 8$, $RW = 4n - 3$, and $WS = 6n + 2$, find the value of n and WS.

14. Use a compass to compare the length of the segments shown below. Then list the measures of the segments in order from least to greatest.

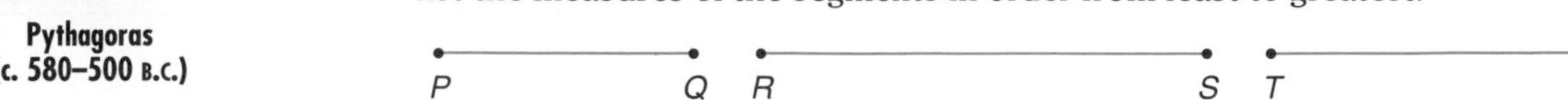

Pythagoras (c. 580–500 B.C.)

15. **World Records** On September 8, 1989, the British Royal Marines stretched a rope from the top of Blackpool Tower (416 feet high) in Lancashire, Great Britain, to a fixed point on the ground 1128 feet from the base of the tower. Then Sgt. Alan Heward and Cpl. Mick Heap of the Royal Marines, John Herbert of Blackpool Tower, and TV show hosts Cheryl Baker and Roy Castle slid down the rope establishing the greatest distance recorded in a rope slide.
    a. Draw a right triangle to represent this event.
    b. How far did they slide?

EXERCISES

Practice

Refer to the number line below to find each measure.

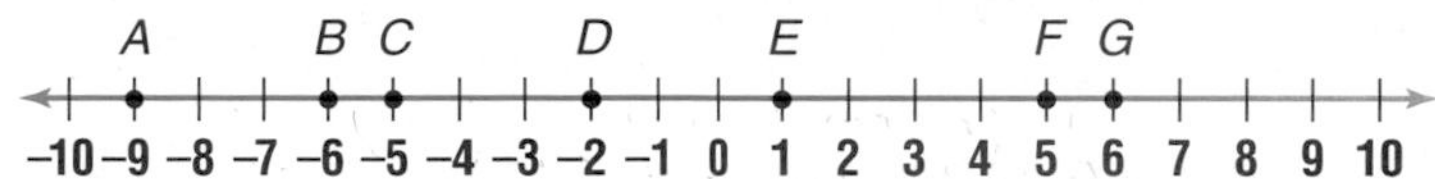

16. AE 17. BD 18. EC 19. EG 20. FC 21. CA

Given that R is between S and T, find each missing measure.

22. $RS = 6$, $TR = 4.5$, $TS = \underline{?}$
23. $SR = 3\frac{2}{3}$, $RT = 1\frac{2}{3}$, $ST = \underline{?}$
24. $ST = 15$, $SR = 6$, $RT = \underline{?}$
25. $TS = 11.75$, $TR = 3.4$, $RS = \underline{?}$

Refer to the coordinate plane at the right to find each measure. Round your answers to the nearest hundredth.

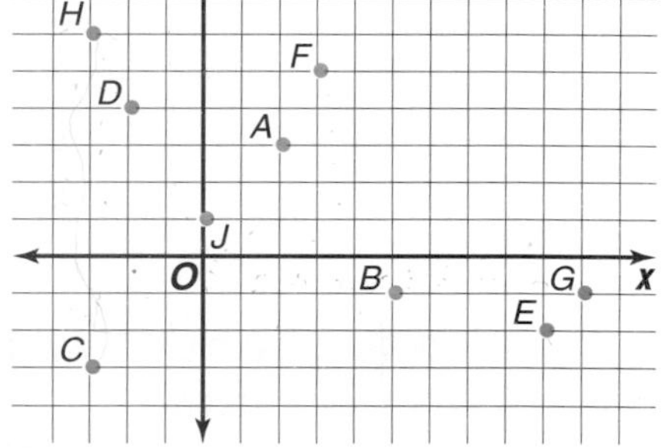

26. BG 27. HC 28. GH
29. EG 30. FJ 31. JC

Write a mathematical sentence to compare each pair of measures.

32. GB and GF
33. FJ and JC
34. AC and AD

Use the Pythagorean Theorem to find the missing length x in each right triangle.

35.
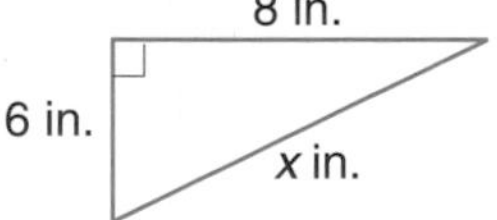

36. 37.
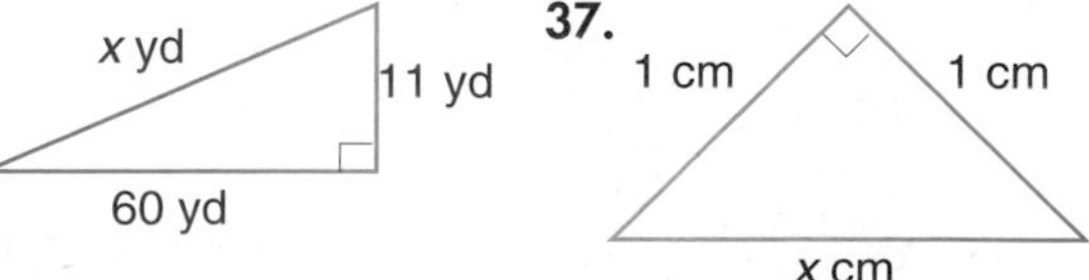

If U is between T and B, find the value of x and the measure of $\overline{TU}$.

38. $TU = 2x$, $UB = 3x + 1$, $TB = 21$
39. $TU = 4x - 1$, $UB = 2x - 1$, $TB = 5x$
40. $TU = 1 - x$, $UB = 4x + 17$, $TB = -3x$

Use $\overline{RS}$, $\overline{PQ}$, a compass, and a straightedge to construct $\overline{XY}$ for each set of measures.

R S

41. $XY = PQ + RS$
42. $XY = 4(RS)$
43. $XY = PQ - RS$
44. $XY = 3(PQ) - RS$

45. A rectangle has vertices $A(-1, 1)$, $B(3, 4)$, $C(6, 0)$, and $D(2, -3)$.
   a. Graph the rectangle.
   b. Find the area and perimeter of the rectangle.

Programming

46. The TI-82/83 graphing calculator program at the right calculates the distance between two points given their coordinates as expressed in the distance formula.

```
PROGRAM:DISTANCE
: Input "ENTER X1=",A
: Input "ENTER Y1=",B
: Input "Enter X2=",C
: Input "Enter Y2=",D
: (C−A)²+(D−B)²→Y
: Disp "DISTANCE=√‾",
    Y,√‾Y
: Stop
```

Use the program to find the distance between each pair of points.

   a. (7, 11), (−1, 5)
   b. (−3, 5), (12, −2)
   c. (3, −2), (0.67, −4)

Critical Thinking

47. In the figure, $BE = EC$ and $AF = FB$. $\overline{FG}$ and $\overline{GE}$ lie on grid lines of the coordinate plane.

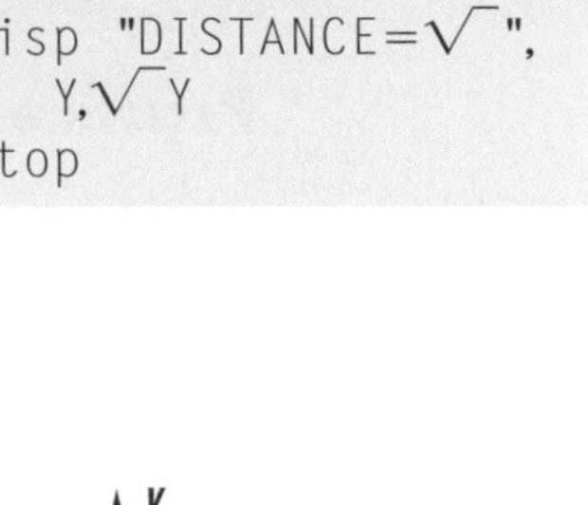

   a. Find the coordinates of E and F.
   b. How might you find the coordinates of G?
   c. How does DG relate to GB? Explain.

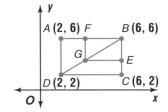

48. Draw a figure that satisfies all of the following conditions.
   - Points A, B, C, D, and E are collinear.
   - Point A lies between points D and E.
   - Point C is next to point A, and $BD = DC$.

Applications and Problem Solving

49. **Comics** Study the *Peanuts* comic strip below. Explain how Sally knew that the length of Snoopy's mouth was "lip to lip, three inches."

Pack	Durability	Comfort
Super Dealer Briefcase	5	2
Day Pak' R	3	4
X-Country Pak' R II	2	2
Big Bag	1	1
Collegian	3	3
Santa Fe	5	3
Sport	2	3
Bookpack	2	5
Book Pack	3	4
Elite Gear Bag	2	3

Source: *Zillions*

50. **Consumer Awareness** A consumer magazine rated backpacks based on durability and comfort. Packs were given a score of 1 to 5 stars for each category. One way to compare the rating to find the best backpack is to graph a point for each pack using the durability score as the x-coordinate and the comfort score as the y-coordinate. Suppose the point farthest from the origin represents the best-rated pack. Of the packs listed in the table at the left, which one was rated the best?

Mixed Review

51. Find the length of a rectangle whose area is 15 square centimeters and whose width is 2.5 centimeters. (Lesson 1–3)

52. Find the maximum area of a rectangle whose perimeter is 8 miles. (Lesson 1–3)

53. Draw a plane $\mathcal{B}$ containing line p. Points R, S, and T lie in $\mathcal{B}$, but only points R and S lie on p. (Lesson 1–2)

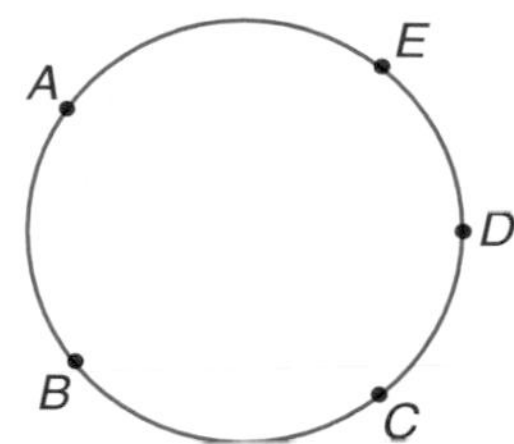

54. **List the Possibilities** Points A, B, C, D, and E lie on a circle. List all of the lines that contain exactly two of these five points. (Lesson 1–2)

55. A five-sided figure is composed of segments that meet at $F(-2, 2)$, $G(0, 4)$, $H(3, 0)$, $I(2, -6)$, and $J(-3, -3)$. Graph these points and draw the figure. (Lesson 1–1)

56. **SAT Practice** Benito has a collection of 60 video tapes. If 30 percent of his videos are TV shows and the rest are movies, how many movies does he have on video tape?

A 18 **B** 30 **C** 35 **D** 42 **E** 48

Algebra

Multiply.

For **Extra Practice**, see page 765.

57. $3x(x + 6)$
58. $(x + 2)(x + 2)$
59. $(x + 3)(x - 4)$
60. $(x - 4)(x + 4)$

Money Lines and Angles

The excerpt below appeared in an article in *Science News* on January 27, 1996.

IN 1994, THE SECRET SERVICE SEIZED $45.7 million in counterfeit bills before they entered U.S. circulation—though forgers sneaked $25.3 million in fakes into the economy.... The potential of desktop counterfeiting has guided the design, materials, and production decisions of the next generation of U.S. currency.... The new currency sports several security-enhancing features.... An enlarged, off-center portrait...increases recognition, reduces wear, and opens space for a watermark; extra detail stymies duplication.... Concentric, fine lines behind the portrait...frustrate replication.... A watermark...becomes visible only in transmitted light.... Numerals printed in color-shifting ink appear green when viewed straight on and black when seen from an angle.... The Secret Service will monitor the new features to see how well they deter counterfeiters and increase the capture of fakes. ■

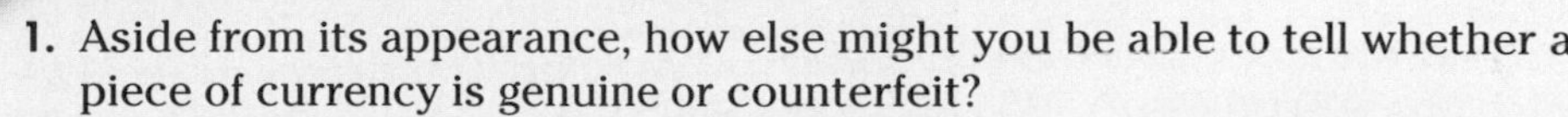

1. Aside from its appearance, how else might you be able to tell whether a piece of currency is genuine or counterfeit?
2. In addition to the fine, concentric lines, the new currency has tiny, microprinted words that cannot be read with the naked eye. What do you think would be the result if you tried to counterfeit these bills by photocopying them?
3. If you were designing currency, what other types of features could you use so it would be harder to counterfeit?

1–5 Midpoints and Segment Congruence

What YOU'LL LEARN

- To find the midpoint of a segment, and
- to complete proofs involving segment theorems.

Why IT'S IMPORTANT

Finding the midpoint is often used to connect algebra to geometry.

In this lesson, you will study several methods for finding the midpoint of a line segment, including paper folding.

MODELING MATHEMATICS

Locating the Midpoint of a Segment

Materials: 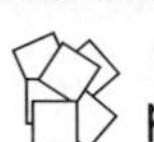 patty paper  ruler

You can locate the midpoint of any segment by using paper folding.

- Draw points A and B anywhere on a sheet of patty paper.
- Connect the points to form segment AB.

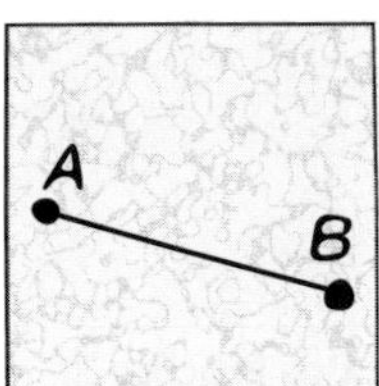

- Fold the paper so that the endpoints A and B lie on top of each other. Pinch the paper to make a crease on the segment.

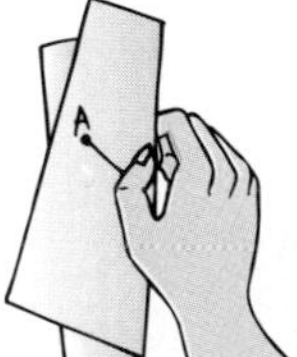

- Open the paper and label the point where the crease intersects $\overline{AB}$ as C. Point C is the midpoint of $\overline{AB}$.

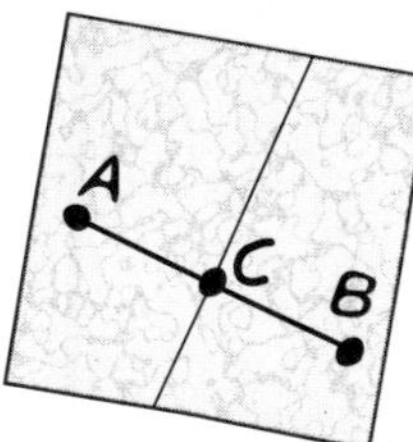

Your Turn

a. Use a ruler to measure $\overline{AC}$ and $\overline{CB}$.

b. Repeat the activity with two other segments.

c. Write a sentence to summarize your observations.

In the Modeling Mathematics activity above, you discovered that the midpoint of a segment separates the segment into two segments that have equal measures.

Definition of Midpoint	**The midpoint M of $\overline{PQ}$ is the point between P and Q such that $PM = MQ$.**

You may remember from algebra that if a segment is graphed on a number line or coordinate plane, there is a formula for finding the coordinate(s) of the midpoint.

Midpoint Formulas	**1. On a number line, the coordinate of the midpoint of a segment whose endpoints have coordinates a and b is $\frac{a+b}{2}$.** **2. In a coordinate plane, the coordinates of the midpoint of a segment whose endpoints have coordinates (x_1, y_1) and (x_2, y_2) are $\left(\frac{x_1+x_2}{2}, \frac{y_1+y_2}{2}\right)$.**

Example 1 **a. Use the number line to find the coordinate of the midpoint of $\overline{FG}$.**

F G

−4 −3 −2 −1 0 1 2 3 4 5 6 7 8

The coordinate of F is −2, and the coordinate of G is 6.

Use these coordinates with the midpoint formula for a number line.

$$\frac{a+b}{2} = \frac{-2+6}{2}$$
$$= \frac{4}{2} \text{ or } 2$$

The coordinate of the midpoint of $\overline{FG}$ is 2.

b. Find the coordinates of Q, the midpoint of $\overline{RS}$, if the endpoints of $\overline{RS}$ are $R(-3, -4)$ and $S(5, 7)$.

Graph points R and S and connect them.

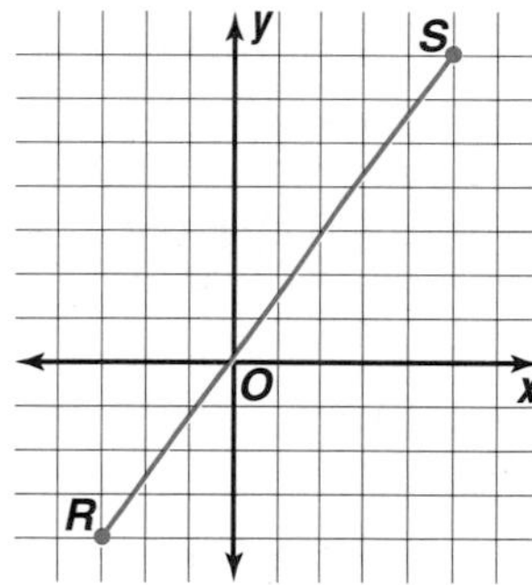

Use the midpoint formula to find the coordinates of Q.

$$\left(\frac{x_1+x_2}{2}, \frac{y_1+y_2}{2}\right) = \left(\frac{-3+5}{2}, \frac{-4+7}{2}\right) \quad (x_1, y_1) = (-3, -4) \quad (x_2, y_2) = (5, 7)$$
$$= \left(1, \frac{3}{2}\right)$$

The coordinates of midpoint Q are $\left(1, \frac{3}{2}\right)$.

You can use the midpoint formula to help you find one of the endpoints of a segment if you know the coordinates of one of its endpoints and its midpoint.

Example 2

a. Find the coordinates of point Q if $L(4, -6)$ is the midpoint of $\overline{NQ}$ and the coordinates of N are $(8, -9)$.

Graph L and N and connect them. Q will be located to the left of L. Use the midpoint formula. Let (x_1, y_1) be $(8, -9)$ and (x_2, y_2) be the coordinates of Q.

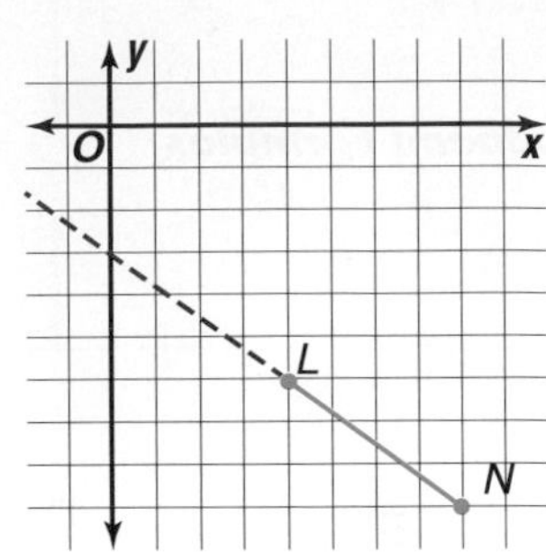

$$\left(\frac{x_1 + x_2}{2}, \frac{y_1 + y_2}{2}\right) = (4, -6)$$

$$\frac{x_1 + x_2}{2} = 4 \qquad \frac{y_1 + y_2}{2} = -6$$

$$\frac{8 + x_2}{2} = 4 \qquad \frac{-9 + y_2}{2} = -6$$

$$8 + x_2 = 8 \qquad -9 + y_2 = -12$$

$$x_2 = 0 \qquad y_2 = -3$$

The coordinates of Q are $(0, -3)$.

b. If Y is the midpoint of $\overline{XZ}$, $XY = 2a + 11$, and $YZ = 4a - 5$, find the value of a and the measure of $\overline{XZ}$.

Y is the midpoint of $\overline{XZ}$, so $XY = YZ$. Write an equation and solve for a.

$$XY = YZ$$

$$2a + 11 = 4a - 5$$

$11 = 2a - 5$ *Subtract 2a from each side.*

$16 = 2a$ *Add 5 to each side.*

$8 = a$ *Divide each side by 2.*

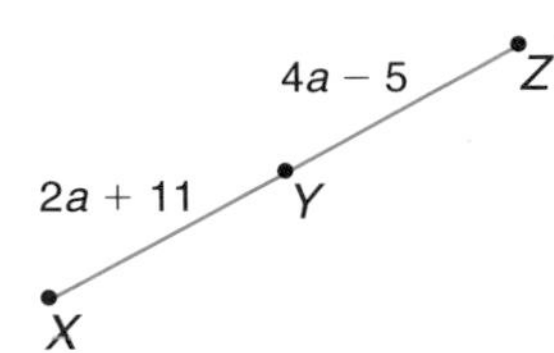

Now use the value of a to find XZ.

$XZ = XY + YZ$ *Segment Addition Postulate*

$= (2a + 11) + (4a - 5)$ *Substitution*

$= 6a + 6$

$= 6(8) + 6$ or 54

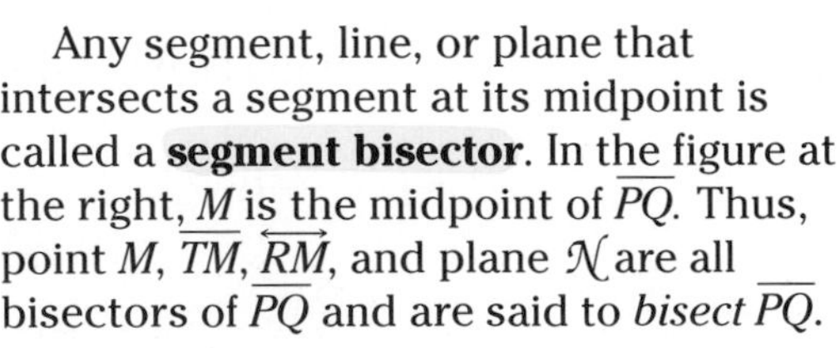

Any segment, line, or plane that intersects a segment at its midpoint is called a **segment bisector**. In the figure at the right, M is the midpoint of $\overline{PQ}$. Thus, point M, $\overline{TM}$, $\overleftrightarrow{RM}$, and plane $\mathcal{N}$ are all bisectors of $\overline{PQ}$ and are said to *bisect* $\overline{PQ}$.

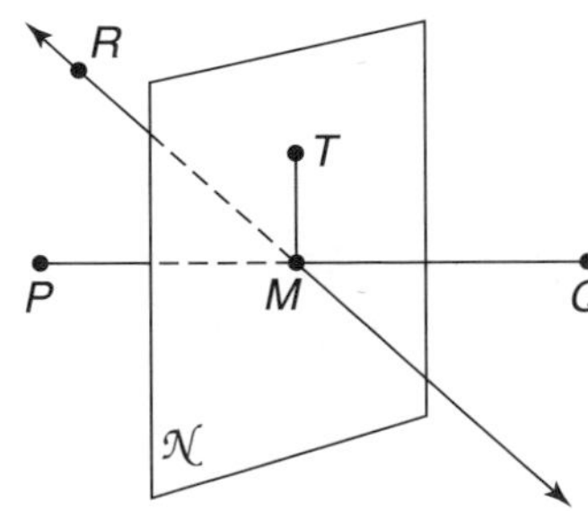

In the Modeling Mathematics activity at the beginning of this lesson, you used paper folding to find the midpoint of a segment. In Exercise 44, you will use a TI-92 calculator to construct a segment and find its midpoint. You can also use a compass and straightedge to find the midpoint of any segment, and thus, bisect the segment.

CONSTRUCTION

Bisecting a Segment

Use a compass and a straightedge to bisect a segment.

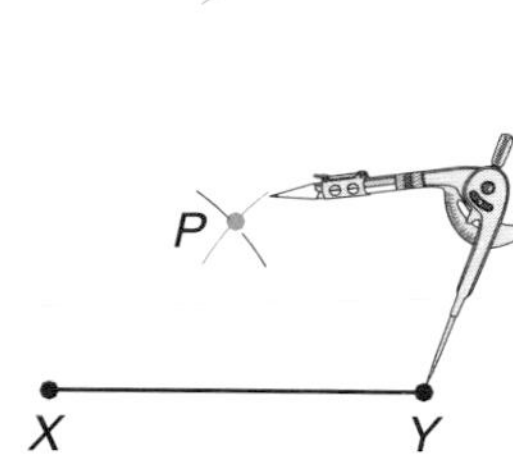

1. Draw a segment and name it $\overline{XY}$.
2. Place the compass at point X. Adjust the compass so that its width is greater than $\frac{1}{2}XY$.
3. Draw arcs above and below $\overline{XY}$.
4. Using the same compass setting, place the compass at point Y and draw arcs above and below $\overline{XY}$ so that they intersect the two arcs previously drawn. Label the points of the intersection of the arcs as P and Q.

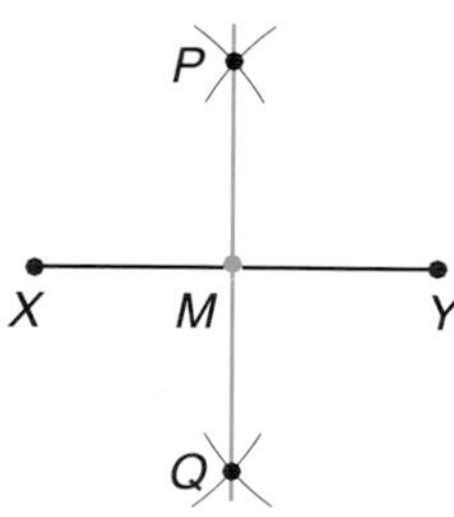

5. Use a straightedge to draw $\overleftrightarrow{PQ}$. Label the point where it intersects $\overline{XY}$ as M.

Conclusion: Point M is the midpoint of $\overline{XY}$, and $\overleftrightarrow{PQ}$ is a bisector of $\overline{XY}$. Also $XM = MY = \frac{1}{2}XY$.

In the study of geometry, definitions, postulates, and undefined terms are accepted as true without verification or proof. These three types of statements can be used to prove that other statements called **theorems** are true. The following theorem states that a congruence relationship exists between the segments formed by the midpoint of a segment.

Theorem 1–1 ***Midpoint Theorem***	**If *M* is the midpoint of $\overline{AB}$, then $\overline{AM} \cong \overline{MB}$.**

You can use definitions, postulates, and, previously proven theorems to justify a statement.

In this book, we will study and use various types of proof. A **proof** is a logical argument in which each statement you make is backed up by a statement that is accepted as true. One type of proof is called a **paragraph** or **informal proof**. In this type of proof, you write a paragraph to explain why a conjecture for a given situation is true. A proof usually is accompanied by a figure for clarification.

Proof of Theorem 1–1

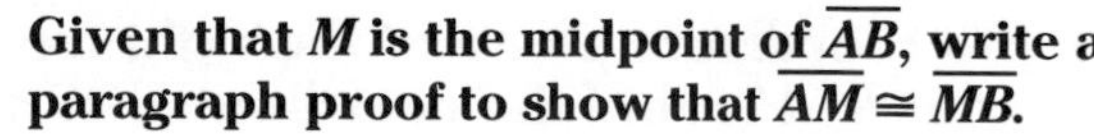

Given that *M* is the midpoint of $\overline{AB}$, write a paragraph proof to show that $\overline{AM} \cong \overline{MB}$.

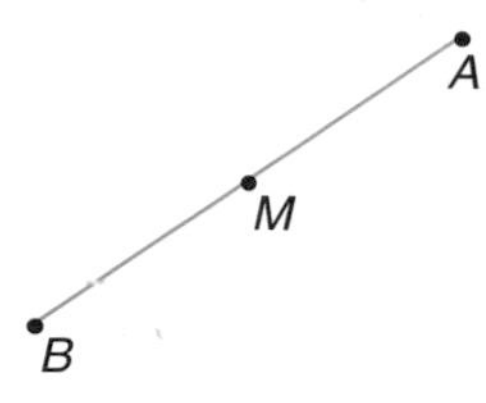

From the definition of the midpoint of a segment, we know that $AM = MB$. That means that $\overline{AM}$ and $\overline{MB}$ have the same measures. By the definition of congruence, if AM and MB have the same measure, they are congruent segments. Thus, $\overline{AM} \cong \overline{MB}$.

A paragraph proof can be used to show that other statements are also true.

Example 3 **In the figure at the right, T is the midpoint of $\overline{SE}$, and E is the midpoint of $\overline{TP}$. Show that $\overline{ST} \cong \overline{EP}$.**

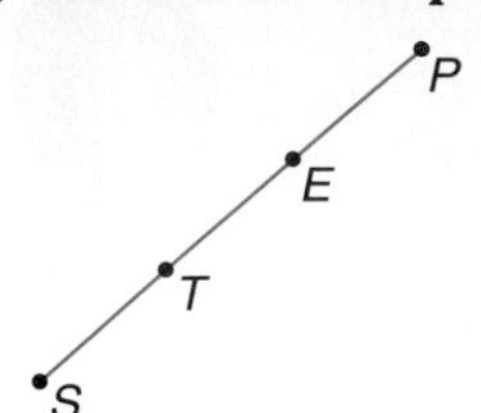

Since T is the midpoint of $\overline{SE}$, then $ST = TE$. Since E is the midpoint of TP, then $TE = EP$. Substitute EP for TE in the statement $ST = TE$ to get $ST = EP$. Then, by the definition of congruence, $\overline{ST} \cong \overline{EP}$.

CHECK FOR UNDERSTANDING

Communicating Mathematics

Study the lesson. Then complete the following.

1. **Describe** three ways to locate the midpoint of a segment.
2. **You Decide** Jadine says that if $DE = EF$, then E is the midpoint of $\overline{DF}$. Misha says this is only true in a special case. Who is correct? Include drawings to demonstrate your argument.
3. **Write** a convincing argument to show that the number of bisectors a segment has is greater than the number of its midpoints.
4. **Explain** how you would find the value of x if R is the midpoint of $\overline{QS}$, $QR = 8 - x$, and $RS = 5x - 10$.

5. **Draw** a large rectangle on your paper. Determine how you could find the midpoint of each side with a minimum number of paper folds.

Guided Practice

Use the number line below for Exercises 6–8.

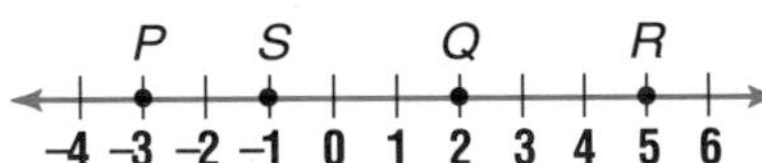

6. Find the coordinate of the midpoint of $\overline{PQ}$.

Determine whether each statement is *true* or *false*. If false, state why.

7. Q is the midpoint of $\overline{SR}$
8. $PS = QR$

Use the coordinate plane at the right for Exercises 9–11. Find the coordinates of the midpoint of each segment.

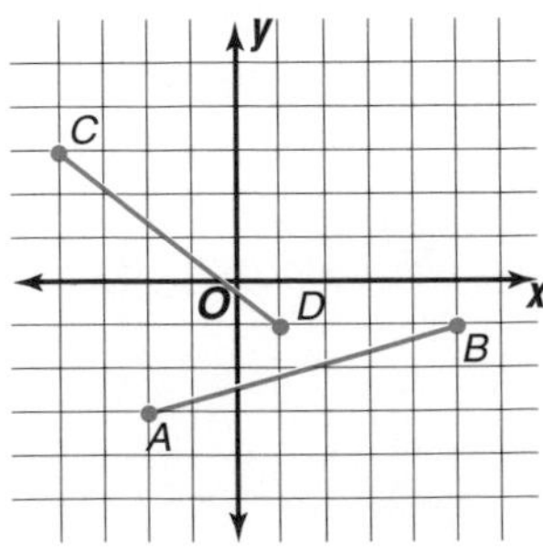

9. $\overline{AB}$
10. $\overline{CD}$
11. Copy $\overline{AB}$ on a piece of grid paper. Then use a compass and straightedge to find the midpoint of $\overline{AB}$. How does your midpoint compare to the result of Exercise 9?

Algebra

12. If $R(2, 5)$ is the midpoint of $\overline{ST}$ and the coordinates of T are $(-1, 8)$, find the coordinates of S.

13. In the figure at the right, $\overline{AC}$ and $\overline{BD}$ bisect each other at E. For each of the following, find the value of x and the measure of the indicated segment.
   a. $DE = 2x + 5$; $EB = 13 - 2x$; $\overline{DB}$
   b. $AC = x + 3$, $EC = 3x - 1$; $\overline{EA}$

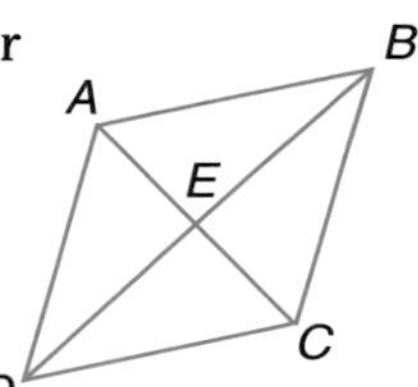

14. **Logical Reasoning** Refer to the figure for Exercise 13. Suppose that $\overline{DB}$ is extended to point F so that B is the midpoint of $\overline{DF}$. Write a paragraph proof to show that ED is one-fourth the length of $\overline{DF}$.

EXERCISES

Practice

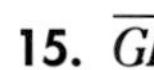

Use the figure at the right for Exercises 15–23.

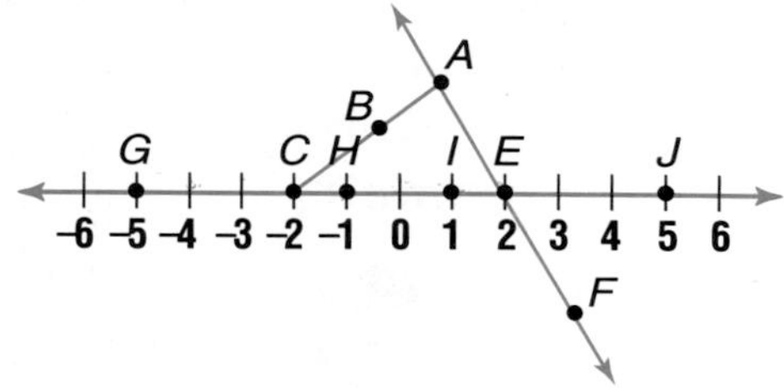

Find the coordinates of the midpoint of each segment.

15. $\overline{GI}$
16. $\overline{HJ}$
17. $\overline{GC}$

Determine whether each statement is *true* or *false*. If false, state why.

18. $\overline{GI} \cong \overline{JH}$
19. E is the midpoint of $\overline{CJ}$.
20. $\overline{AC}$ bisects $\overline{GE}$.
21. $\overline{EJ} \cong \overline{CI}$
22. $\overleftrightarrow{AE}$ bisects $\overline{HJ}$.
23. $CH = \frac{1}{2}HI$

W, R, and S are points on a number line, and W is the midpoint of $\overline{RS}$. For each pair of coordinates given, find the coordinate of the third point.

24. $R = 4, S = -6$
25. $R = 12, W = -3$
26. $W = -4, S = 2$

Y is the midpoint of $\overline{XZ}$. For each pair of points given, find the coordinates of the third point.

27. $X(5, 5), Z(-1, 5)$
28. $X(-4, 3), Y(-1, 5)$
29. $Z(2, 8), Y(-2, 2)$
30. $Z(-3, 6), Y(0, 5.5)$
31. $X\left(\frac{2}{3}, -5\right), Y\left(\frac{5}{3}, 3\right)$
32. $X(2, -10), Y\left(\frac{1}{2}, -6\frac{1}{2}\right)$

In the figure below, $\overline{EC}$ bisects $\overline{AD}$ at C, and $\overline{EF}$ bisects $\overline{AC}$ at B. For each of the following, find the value of x and the measure of the indicated segment.

33. $AB = 3x + 6$, $BC = 2x + 14$; $\overline{AC}$
34. $AC = 5x - 8$, $CD = 16 - 3x$; $\overline{AD}$
35. $AD = 6x - 4$, $AC = 4x - 3$; $\overline{CD}$
36. $AC = 3x - 1$, $BC = 12 - x$; $\overline{AB}$
37. $AD = 5x + 2$, $BC = 7 - 2x$; $\overline{CD}$
38. $AB = 4x + 17$, $CD = 25 + 5x$; $\overline{BC}$

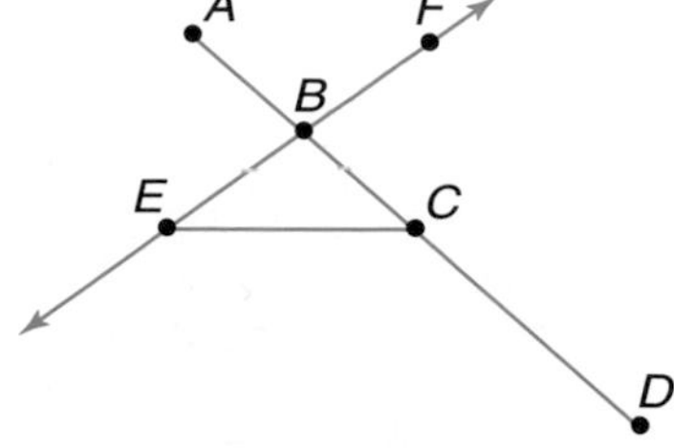

39. Draw a line segment about 5 inches long. Use a compass, a straightedge, and the segment bisector construction to find a segment that is $\frac{1}{8}$ the length of the original segment.

40. Point C lies on $\overline{AB}$ such that $AC = \frac{1}{4}AB$. If the endpoints of $\overline{AB}$ are $A(8, 12)$ and $B(-4, 0)$, find the coordinates of C.

41. Suppose $\overline{PQ}$ has endpoints $P(2, 3)$ and $Q(8, -9)$. Find the coordinates of R and S so that R lies between P and S and $\overline{PR} \cong \overline{RS} \cong \overline{SQ}$.

42. In the figure at the right, E is the midpoint of $\overline{FG}$ and $\overline{HI}$, and $\overline{FG} \cong \overline{HI}$. Write a paragraph proof to show that $FE = HE$.

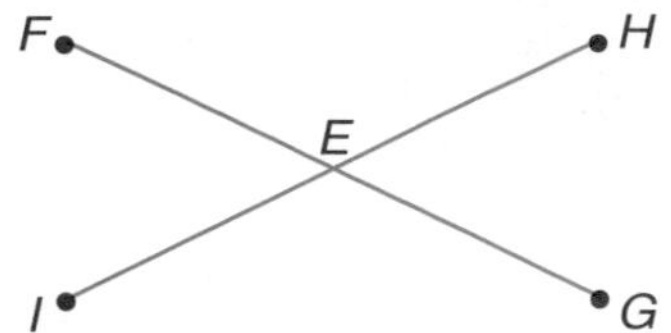

43. The notation $\overline{STEP}$ means that S, T, E, and P are collinear points in that order. Write a paragraph proof to show that $SP = ST + TE + EP$ for $\overline{STEP}$. Be sure to include a drawing with your proof.

Cabri Geometry

44. Use a calculator to construct a segment AB.
    a. Locate midpoint M on $\overline{AB}$ by selecting 3:Midpoint from the F4 menu.
    b. Use the 1:Distance and Length command from the F6 menu to measure the distance from A to M and from M to B. What do you find?
    c. Use the hand key to move the endpoints of the segment to new locations. What do you notice about the measures of the two segments formed by the midpoint?

Critical Thinking

45. Draw a large triangle similar in shape to the one shown at the right.
    a. Find the midpoints of each side of the triangle. Label them D, E, and F.
    b. Connect the points to form triangle DEF.
    c. Measure the sides of triangles DEF and ABC.
    d. How do the perimeters of the two triangles compare?
    e. Make a conjecture about the areas of the two triangles. Explain how you could show that your conjecture is true.

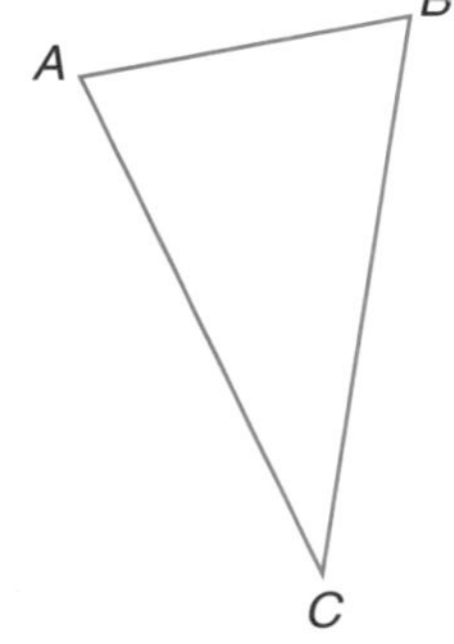

Applications and Problem Solving

46. **Music** A compact disc is actually a metallic disc covered with a plastic protective coating. The metal disc is pitted by an encoder that stores a small bit of information in each pit. A finished CD is 12 centimeters wide at its diameter. The hole in its center is 1.5 centimeters wide. Find the value of x in the picture.

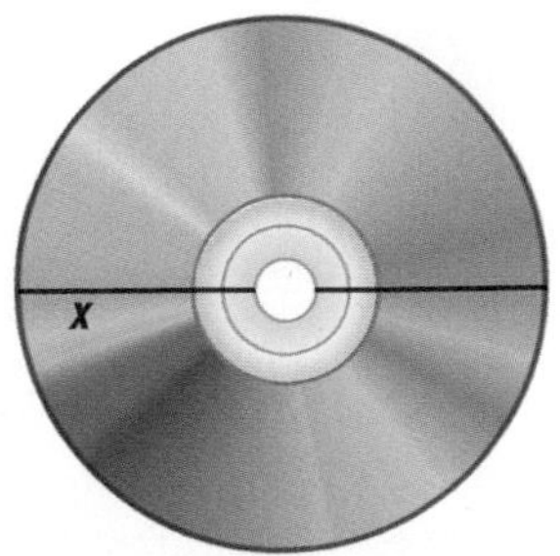

47. **Transportation** Interstate 70 passes through Kansas. Mile markers are used to name many of the exits. The exit for U.S. Route 283 North is Exit 128, and the exit to use U.S. Route 281 to Russell is Exit 184. The exit for Hays on U.S. Route 183 is 3 miles farther than halfway between Exits 128 and 184. What is the exit number for the Hays exit?

Mixed Review

48. Draw a segment with points A, B, C, and X that satisfies the following requirements. (Lesson 1–4)
    - $AX = 10$
    - $AB = 12$
    - $XC = 6$

49. The coordinate grids below show three paths to get from point $A(-3, -1)$ to point $B(2, 2)$. (Lesson 1–4)

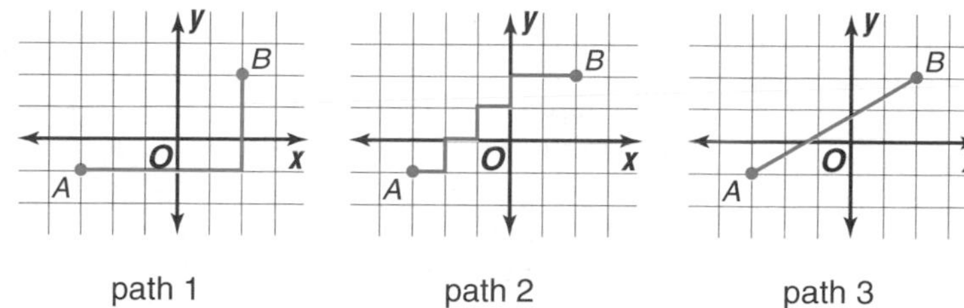

a. Find the distance along each path.

b. If the grid represents streets in a city, which path would a taxicab be least likely to take? Explain your reasoning.

50. **SAT Practice** For all integers n,

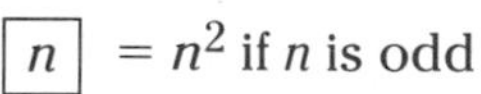

$\boxed{n} = n^2$ if n is odd

$\boxed{n} = \sqrt{n}$ if n is even

What is the value of $\boxed{16} + \boxed{9}$?

A 7 **B** 25 **C** 85 **D** 97 **E** 337

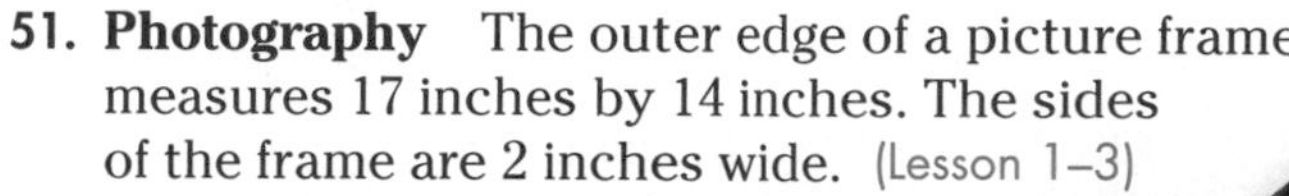

51. **Photography** The outer edge of a picture frame measures 17 inches by 14 inches. The sides of the frame are 2 inches wide. (Lesson 1–3)

a. Draw a picture to represent the frame, labeling all the information presented in the problem.

b. Find the area of a picture that will show in this frame.

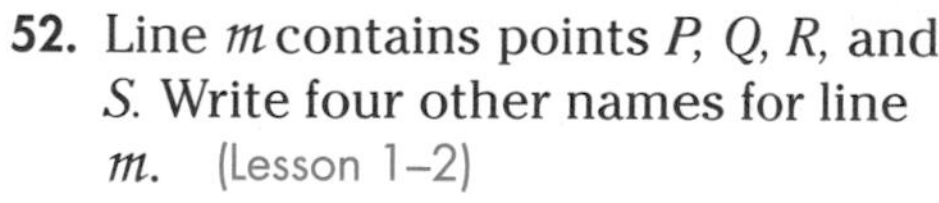

52. Line m contains points P, Q, R, and S. Write four other names for line m. (Lesson 1–2)

53. Draw and label figures to show the different ways in which planes $\mathcal{A}$, $\mathcal{B}$, and $\mathcal{C}$ can intersect. (Lesson 1–2)

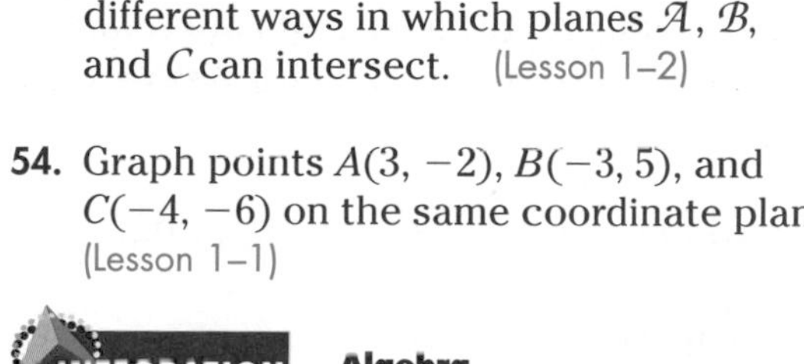

54. Graph points $A(3, -2)$, $B(-3, 5)$, and $C(-4, -6)$ on the same coordinate plane. (Lesson 1–1)

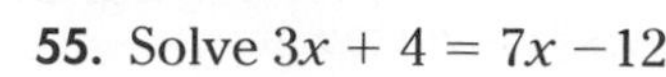

Algebra

55. Solve $3x + 4 = 7x - 12$.

56. Simplify $3(2x - 8) - 4(x + 2)$.

For **Extra Practice**, see page 765.

1–6 Exploring Angles

What YOU'LL LEARN

- To identify angles and classify angles,
- to use the Angle Addition Postulate to find the measures of angles, and
- to identify and use congruent angles and the bisector of an angle.

Why IT'S IMPORTANT

Angles and their measures play a major part in many aspects of your study of geometry.

CONNECTION

Art

The Japanese have long appreciated the beauty of nature through the art of *ikebana*. In ikebana, flowers and branches are arranged in a simple way that allows the beauty of each piece to be seen clearly. When creating an arrangement, the placement of each branch or flower is determined by forming an **angle** of a specific size.

An angle is defined in terms of the two **rays** that form the angle. A ray extends indefinitely in *one* direction. A ray has exactly one endpoint and that point is always named first when naming the ray. Like segments, rays can be defined using betweenness of points.

A beam of light shooting into space is a model for a ray.

Ray DE, written $\overrightarrow{DE}$, consists of the points on $\overline{DE}$ and all points F on $\overleftrightarrow{DF}$ such that E is between D and F.

If you choose any point on a line, that point determines exactly two rays, called **opposite rays**. The point is the common endpoint of the opposite rays. In the figure below, $\overrightarrow{GH}$ and $\overrightarrow{GI}$ are opposite rays, and G is the common endpoint. Opposite rays are collinear.

*The figure formed by opposite rays is sometimes referred to as a **straight angle**. Unless otherwise specified, the term "angle" in this book means a nonstraight angle.*

An *angle* is a figure formed by two *noncollinear* rays with a common endpoint. The two rays are called the **sides** of the angle. The common endpoint is called the **vertex**.

The sides of the angle at the right are $\overrightarrow{BA}$ and $\overrightarrow{BC}$, and the vertex is B. You could name this angle as $\angle B$, $\angle ABC$, $\angle CBA$, or $\angle 1$. When letters are used to name an angle, the letter that names the vertex is used either as the only letter or as the middle of three letters.

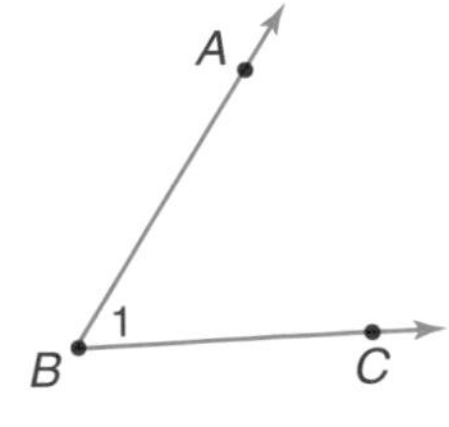

You should name an angle by a single letter only when there is no chance of confusion. For example, it is not obvious which angle shown at the right is $\angle P$ since there are three different angles that have P as a vertex. *Can you name the three angles?*

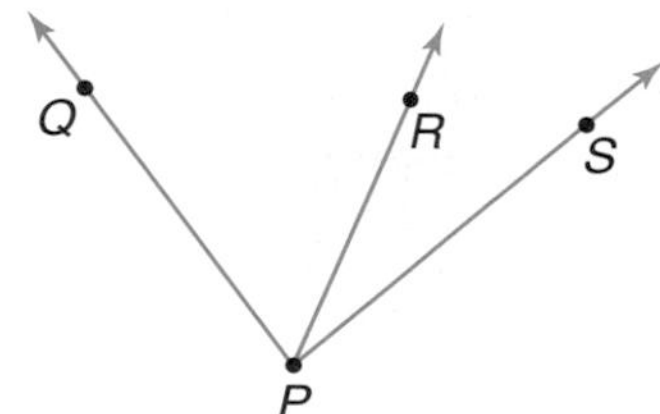

Example 1 **Refer to the figure at the right to answer each question.**

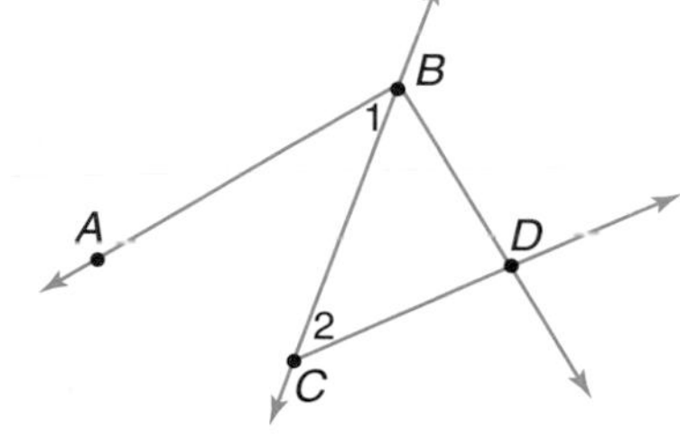

a. What other names could be used to identify $\angle BCD$?
$\angle 2$, $\angle C$, and $\angle DCB$ can be used to identify $\angle BCD$.

b. Name the vertex of $\angle 1$.
B is the vertex of $\angle 1$.

c. What are the sides of $\angle 2$?
$\overrightarrow{CB}$ and $\overrightarrow{CD}$ are the sides of $\angle 2$.

An angle separates a plane into three distinct parts, the **interior** of the angle, the **exterior** of the angle, and the angle itself. If a point does not lie on an angle and it lies on a segment whose endpoints are on each side of the angle, the point is in the interior of the angle. Neither of the endpoints of the segment can be the vertex of the angle.

In the figure at the right, point Z and all other points in the *blue* region are in the interior of $\angle X$. Any point that is not on the angle or in the interior of the angle is in the exterior of the angle. The *yellow* region is the exterior of $\angle X$.

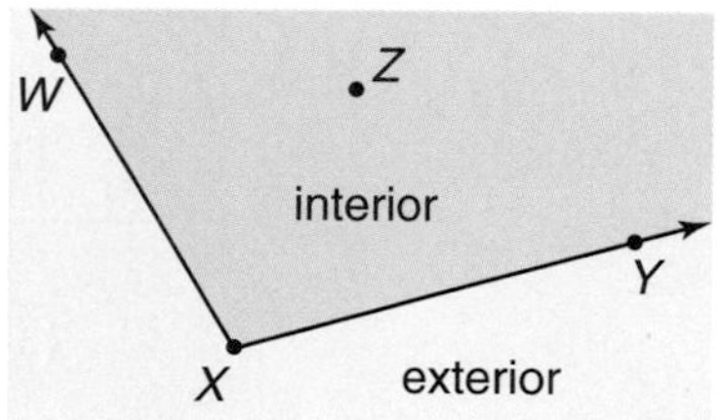

Angles can be measured in units called **degrees**. A *protractor* can be used to find the **measure** of an angle just as a ruler is used to find the measure of a segment. To find the measure of an angle, place the center point of the protractor over the vertex of the angle. Then align the mark labeled 0 on either side of the scale with one side of the angle.

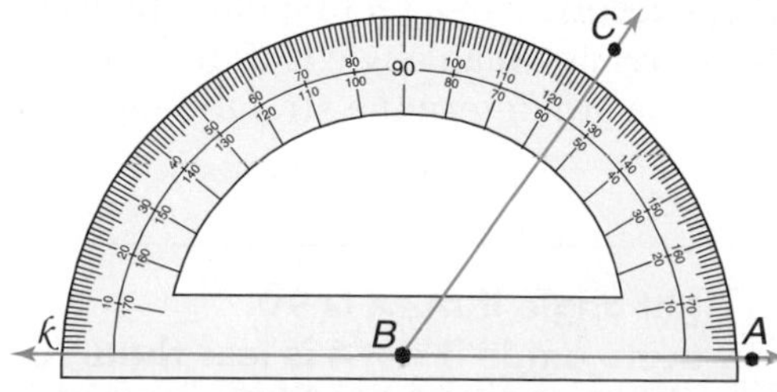

Using the inner scale of the protractor, you can see that $\angle B$ is a 55-degree (55°) angle. Thus, we say that the degree measure of $\angle ABC$ is 55 or $m\angle ABC = 55$.

The **Protractor Postulate** states that there can only be one 55° angle on either side of a given ray.

Postulate 1–3 Protractor Postulate	**Given $\overrightarrow{AB}$ and a number *r* between 0 and 180, there is exactly one ray with endpoint *A*, extending on either side of $\overrightarrow{AB}$, such that the measure of the angle formed is *r*.**

In Lesson 1–4, a measurement relationship between the lengths of segments, called the *Segment Addition Postulate*, was introduced. A similar relationship exists between the measure of angles.

Postulate 1–4 Angle Addition Postulate	**If *R* is in the interior of $\angle PQS$, then $m\angle PQR + m\angle RQS = m\angle PQS$. If $m\angle PQR + m\angle RQS = m\angle PQS$, then *R* is in the interior of $\angle PQS$.**

Example 2

Art

Refer to the application at the beginning of the lesson. The first branch placed in an ikebana arrangement is to be placed so that it forms a 10° angle with an imaginary vertical ray. The second branch is placed on the same side of the vertical as the first branch so that it forms an angle of 45° with the vertical. What is the measure of the angle formed by the first and second branches?

$\overrightarrow{AB}$ represents the imaginary vertical ray. The first branch forms a 10° angle. It is represented by $\overrightarrow{AC}$. The second branch, represented by $\overrightarrow{AD}$, forms an angle of 45° with the vertical. We can use the Angle Addition Postulate to find $m\angle DAC$.

$$m\angle DAB = m\angle DAC + \angle CAB$$
$$45 = m\angle DAC + 10$$
$$35 = m\angle DAC$$

Therefore, the first and second branches form an angle of 35°.

Angles can be classified by their measures.

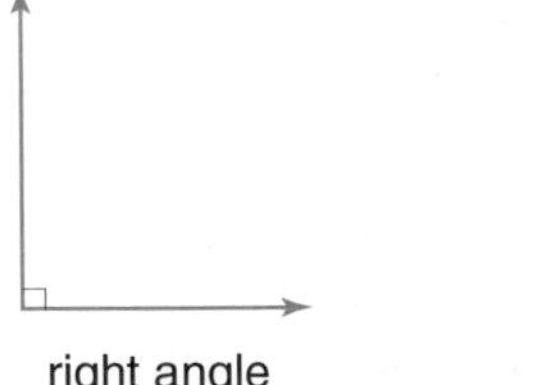

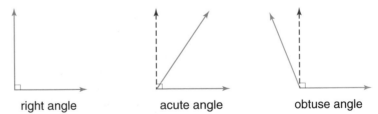

right angle

acute angle

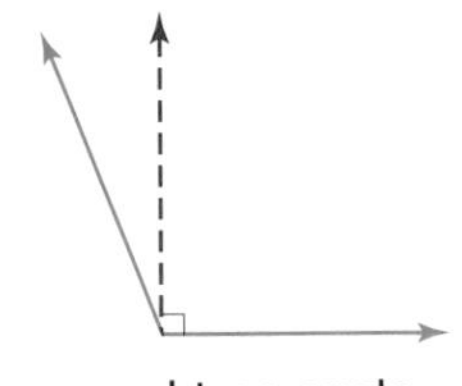

obtuse angle

The symbol ⌝ indicates a right angle.

You can use a corner of a rectangular sheet of paper, which is a right angle, to help you determine if an angle is right, acute, or obtuse. The edge of the paper can be used to determine straight angles.

Definition of Right, Acute, and Obtuse Angles	**$\angle A$ is a right angle if $m\angle A$ is 90. $\angle A$ is an acute angle if $m\angle A$ is less than 90. $\angle A$ is an obtuse angle if $m\angle A$ is greater than 90 and less than 180.**

A straight angle has a measure of 180.

You know that congruent segments are segments that have the same length. Similarly, **congruent** angles are angles that have the same measure.

The definition of congruent angles tells us that statements such as $m\angle A = m\angle B$ and $\angle A \cong \angle B$ are equivalent. Therefore, we will use them interchangeably throughout this book.

Angle *NMP* and angle *RMQ* in the figure at the right each measure 70° Therefore, the two angles are congruent angles. You can write this as $\angle NMP \cong \angle RMQ$. Small arcs, like the ones in red shown at the right, are used to indicate congruent angles in a figure.

You can use a compass and straightedge to construct an angle that is congruent to a given angle without knowing the degree measure of the angle.

CONSTRUCTION

Congruent Angles

Construct an angle congruent to a given angle.

1. Draw an angle like $\angle P$ on your paper.

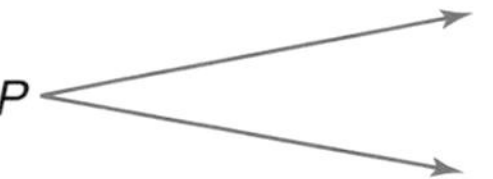

2. Use a straightedge to draw a ray on your paper. Label its endpoint *T*.

3. Place the tip of the compass at point *P* and draw a large arc that intersects both sides of $\angle P$. Label the points of intersection *Q* and *R*.

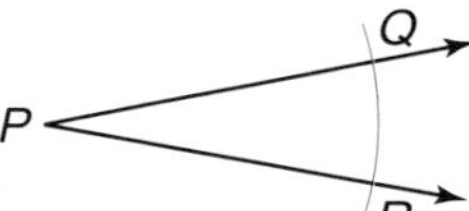

4. Using the same compass setting, put the compass at point *T* and draw a large arc that starts above the ray and intersects the ray. Label the point of intersection *S*.

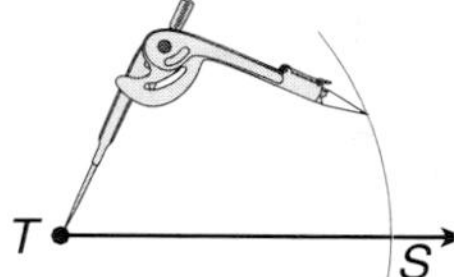

5. Place the point of your compass on *R* and adjust so that the pencil tip is on *Q*.

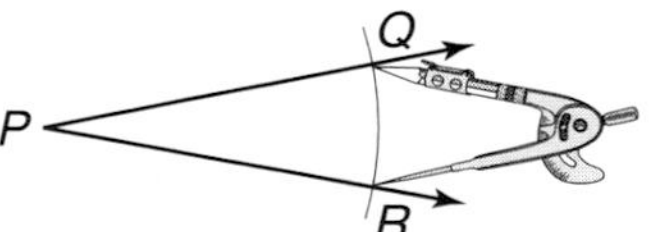

6. Without changing the setting, place the compass at point *S* and draw an arc to intersect the larger arc you drew in Step 4. Label the point of intersection *U*.

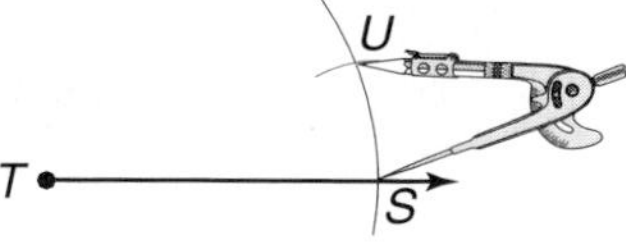

7. Use a straightedge to draw $\overrightarrow{TU}$.

Conclusion: $m\angle QPR = m\angle UTS$ by construction. Thus, $\angle QPR \cong \angle UTS$ by the definition of congruent angles.

A segment that divides another segment into two congruent segments is called a *segment bisector*. A segment bisector intersects the midpoint of the segment that it bisects. Similarly, an **angle bisector** divides an angle into two congruent angles.

If $\overrightarrow{XZ}$ is the angle bisector of $\angle WXY$, point Z must be in the interior of $\angle WXY$ and $\angle WXZ \cong \angle ZXY$, as shown at the right.

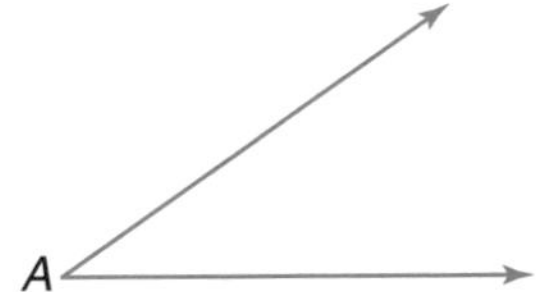

The angle bisector of a given angle can be constructed using the following procedure.

CONSTRUCTION

Angle Bisector

Construct the bisector of a given angle.

1. Draw an angle like $\angle A$ on your paper.

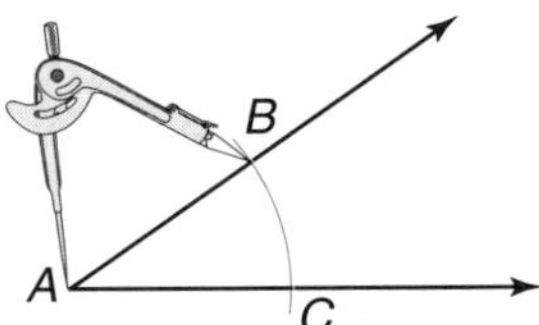

2. Put your compass at point A and draw a large arc that intersects both sides of $\angle A$. Label the points of intersection B and C.

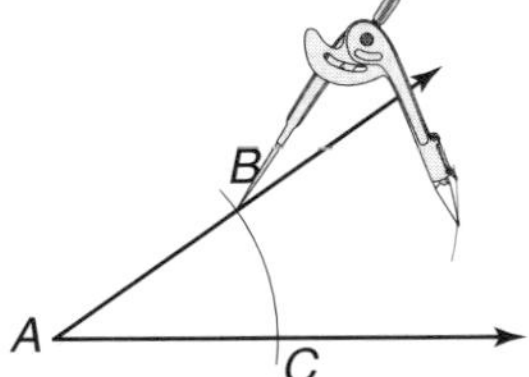

3. With the compass at point B, draw an arc in the interior of the angle.

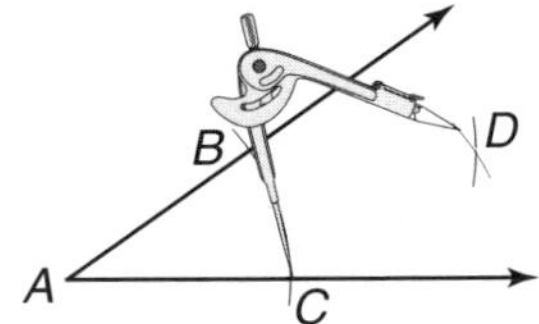

4. Keeping the same compass setting, place the compass at point C and draw an arc that intersects the arc drawn in Step 3. Label the point of intersection D.

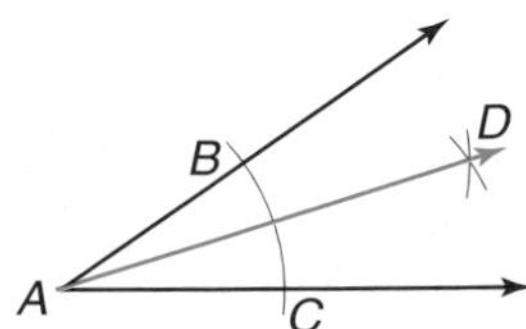

5. Draw $\overrightarrow{AD}$.

B
D
A
C

Conclusion: By construction, $\overrightarrow{AD}$ is the bisector of $\angle BAC$. $m\angle BAD = m\angle DAC$, and $\angle BAD \cong \angle DAC$.

CHECK FOR UNDERSTANDING

Communicating Mathematics

Study the lesson. Then complete the following.

1. **State** how many opposite rays are determined by a given point on a line.
2. **Write in your own words** why a straight angle does not fit our definition of an angle.
3. *True* or *false*: All right angles are congruent. Explain.
4. **Draw and label** a figure to show $\angle MNO$ with angle bisector $\overrightarrow{NQ}$ and $\angle MNQ$ with angle bisector $\overrightarrow{NR}$. Indicate any pairs of congruent angles.
5. **You Decide** Angeni says that for any three angles $\angle ABD$, $\angle DBC$, and $\angle ABC$ that share vertex B, it must be true that $m\angle ABD + m\angle DBC = m\angle ABC$. Reta says that the relationship is not necessarily true. Who is correct and why?

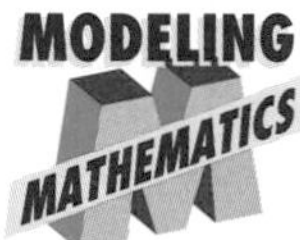

6. Draw an angle on a piece of patty paper. Fold the paper through the vertex of the angle so that the sides of the angle match. Make a crease. Open the paper. What does the crease represent? Explain.

Guided Practice

Refer to the figure below for Exercises 7–12.

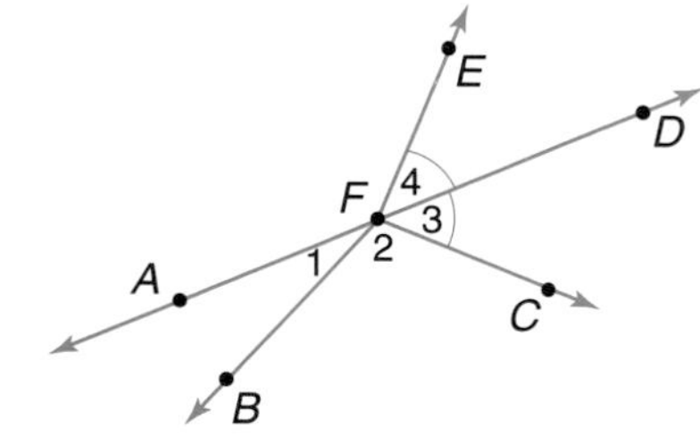

7. State two other names for $\angle 1$.
8. Does $\angle EFC$ appear to be acute, right, obtuse, or straight?
9. Name a ray that appears to bisect an angle and the angle that it appears to bisect.
10. Name all of the angles that have $\overrightarrow{FC}$ for a side.
11. Complete: $m\angle DFB = m\angle 2 + \underline{\ ?\ }$
12. Name a point in the exterior of $\angle AFB$.
13. Draw two angles that intersect in exactly two points.
14. If $\angle G$ is an acute angle and $m\angle G = 4x + 16$, write a compound inequality to describe all the possible values for x.
15. Suppose $\angle T \cong \angle S$, $m\angle T = 12n - 6$, and $m\angle S = 4n + 18$. Is $\angle T$ acute, right, obtuse, or straight?
16. **Aviation** Compass headings are used to indicate the directions of airplanes. A heading is stated as the measure of the angle formed by the flight path of the airplane and an imaginary path in the direction of due north. A pilot is on approach to land at a compass heading of 68° NW (northwest). The tower has instructed her to land on runway 9, which has a heading of 90° NW. How many degrees must the pilot turn her plane in order to land?

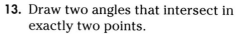

U.S. Navy Blue Angels

EXERCISES

Practice

Refer to the figure below for Exercises 17–28.

17. Name two angles that have N as a vertex.
18. If $\overrightarrow{MQ}$ bisects $\angle PMN$, name two congruent angles.
19. Name a point on the interior of $\angle JMQ$.
20. List all the angles that have O as the vertex.
21. Does $\angle QMJ$ appear to be acute, obtuse, right, or straight?
22. Name a pair of opposite rays.
23. List all the angles that have $\overrightarrow{MN}$ as a side.
24. Name an angle that appears to be acute.
25. If $m\angle PMO = 55$ and $m\angle OMN = 65$, what is the measure of $\angle PMN$?
26. Is $\angle P$ a valid name for one of the angles? Explain.
27. Name a pair of angles that share exactly one point.
28. Is $\overrightarrow{PN}$ the same as $\overrightarrow{NP}$? Explain.

Draw two angles that satisfy the following conditions.

29. The angles intersect in one point.
30. The angles intersect in four points.
31. The angles intersect in a segment.

32. Use a protractor to draw $\angle ABC$ so that it measures 80°.
    a. Use a compass and a straightedge to construct $\overrightarrow{BD}$, the bisector of the angle.
    b. Measure $\angle ABD$ and $\angle CBD$. What do you observe?

In the figure, $\overrightarrow{BA}$ and $\overrightarrow{BC}$ are opposite rays, and $\overrightarrow{BE}$ bisects $\angle ABD$.

33. If $m\angle ABE = 6x + 2$ and $m\angle DBE = 8x - 14$, find $m\angle ABE$.
34. Given that $m\angle ABD = 2y$ and $m\angle DBC = 6y - 12$, find $m\angle DBC$.
35. Find $m\angle EBD$ if $m\angle ABE = 12n - 8$ and $m\angle ABD = 22n - 11$.
36. If $m\angle ABE = 9x - 1$ and $m\angle DBC = 24x + 14$, find $m\angle EBD$.
37. Suppose $m\angle ABE = 15y$ and $m\angle DBC = 45y - 30$. Find the value of y.
38. If $\angle ABD$ is a right angle and $m\angle ABE = 13x - 7$, what is the value of x?

Critical Thinking

39. Each figure below shows noncollinear rays with a common endpoint.

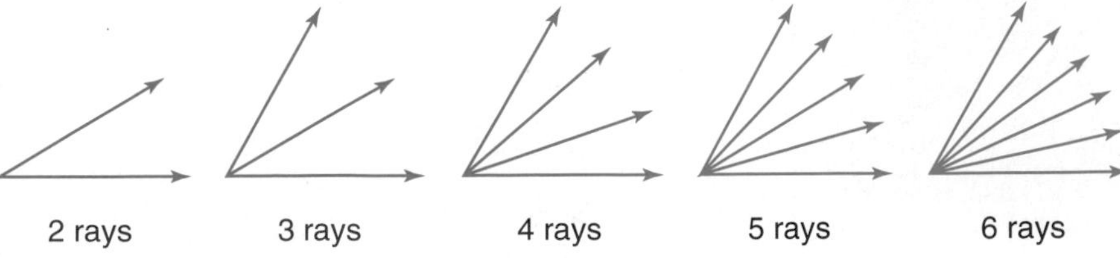

a. Count the number of angles in each figure.

b. Describe the pattern that exists between the number of rays and the number of angles in each figure.

c. Predict the number of angles that are formed by seven rays and by 10 rays.

d. Write a formula for the number of angles formed by n noncollinear rays with a common endpoint.

Applications and Problem Solving

40. Physics A ripple tank can be used to study the behavior of waves in two dimensions. As a wave strikes a barrier, it is reflected. The angle of incidence and the angle of reflection are congruent. In the diagram at the right, if $m\angle IBR = 78$, find the measure of the angle of reflection and $m\angle IBA$.

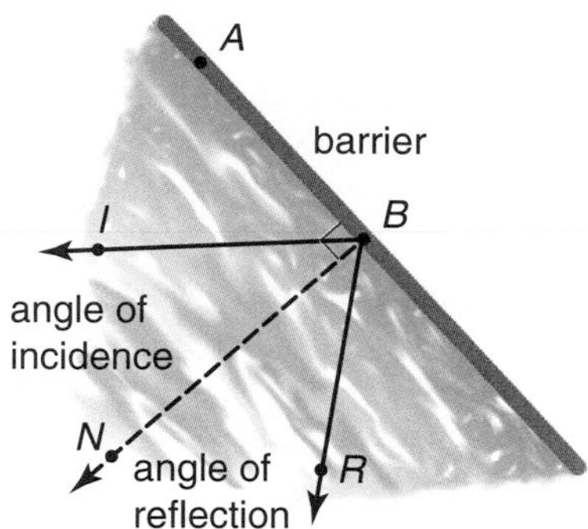

41. Entertainment The angle that a pinball machine tilts, or the pitch of the machine, determines the difficulty of the game. The angles at which the flipper will hit the ball are considered when the ramps, loops, and targets of the game are placed.

a. A pinball machine designer recommends that a machine be installed with a pitch of 6° to 7°. Is this an acute, right, straight, or obtuse angle?

b. Suppose a line is drawn down the center of a pinball machine. If the path of a ball hit by the flipper intersects that center line, an angle of between 5° and 20° is formed. If you drew an angle whose interior contained most of the paths of these balls, what would be the measure of that angle?

Mixed Review

42. Find B, if A is at $(5, -3)$ and $M(-4, 2)$ is the midpoint of $\overline{AB}$. (Lesson 1–5)

43. G is between F and H. If $GH = 6$, $FG = 8x + 2$, and $FH = 16$, find FG. (Lesson 1–4)

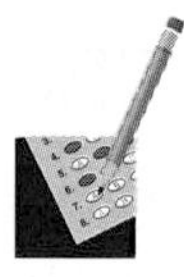

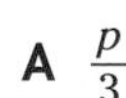

44. SAT Practice If the perimeter of the rectangle $ABCD$ is equal to p, and $x = \frac{1}{5}y$, what is the value of y in terms of p?

A $\frac{p}{3}$ B $\frac{5p}{12}$ C $\frac{5p}{8}$

D $\frac{5p}{6}$ E $\frac{p}{12}$

A y B
x x
D y C

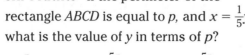

Note: Figure is not drawn to scale.

45. The perimeter of a rectangle is 27 millimeters. If the length is 8 millimeters, what is the width? (Lesson 1–3)

46. Draw a figure for plane $\mathcal{C}$ containing line ℓ and point Y, not on ℓ. (Lesson 1–2)

47. Graph $R(-4, 7)$ in a coordinate plane. (Lesson 1–1)

Algebra

48. Find $\sqrt{49}$.

49. Solve $x^2 = 25$.

For **Extra Practice**, see page 765.

1–7A Using Technology Angle Relationships

A Preview of Lesson 1–7

LOOK BACK

Refer to Lesson 1-4A to review how to start a new drawing screen and draw a segment.

In Lesson 1–4A, you learned to use a calculator to draw and measure segments. You can also draw and measure angles on the calculator.

- Draw $\overleftrightarrow{AD}$.
- Draw $\overleftrightarrow{CE}$ so that it intersects $\overleftrightarrow{AD}$. Label the intersection point *B*.
- To find the measures of the angles created by the intersecting lines, press F6 and select 3:Angle. You will define the angle to be measured by selecting three points on the angle in the same order you would name the angle. Move the pointer to one of the points on the angle and press ENTER. Move to the next and press ENTER. Then move to the last and press ENTER. The measure of the angle to the nearest hundredth degree appears.

This process can be used to measure any angle, regardless of how it was created.

- Repeat this process to measure the other three angles.

The screen shows a sample result of this activity.

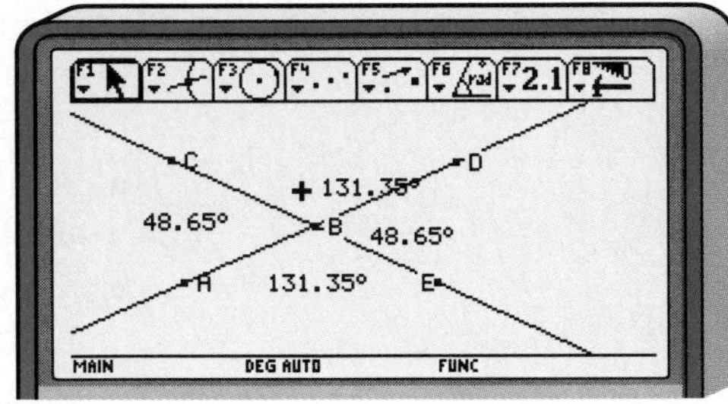

EXERCISES

Analyze your drawing.

1. Angles that are formed by intersecting lines and share a common vertex are called *vertical angles.*
   a. Name two pairs of vertical angles from your screen.
   b. Compare the measures of the vertical angles. What do you notice?
2. Angles that have a common side and vertex and whose noncommon sides are opposite rays are called a *linear pair.*
   a. Name the linear pairs of angles from your screen.
   b. Compare the measures of each linear pair. What do you notice?

Test your conclusion.

3. Move the cursor to one of the lines. Hold down the hand key and move the line. Record the angle measures after moving one of the lines.
4. Use the hand key to move the other line. Record the angle measures after this move.
5. Write a sentence to summarize the relationship you notice among the measures of each pair of vertical angles.
6. Write a sentence to summarize the relationship you notice among the measures of each linear pair of angles.

1–7 Angle Relationships

What YOU'LL LEARN

- To identify and use adjacent, vertical, complementary, supplementary, and linear pairs of angles, and perpendicular lines, and
- to determine what information can and cannot be assumed from a diagram.

Why IT'S IMPORTANT

You can use the measures of angles to solve problems in geology, sports, and construction.

CONNECTION
Geology

Geophysicists at the U.S. Geological Survey are studying the way that the continents and seas have been formed. In order to describe a section of Earth's crust accurately, they measure the strike and dip of the area.

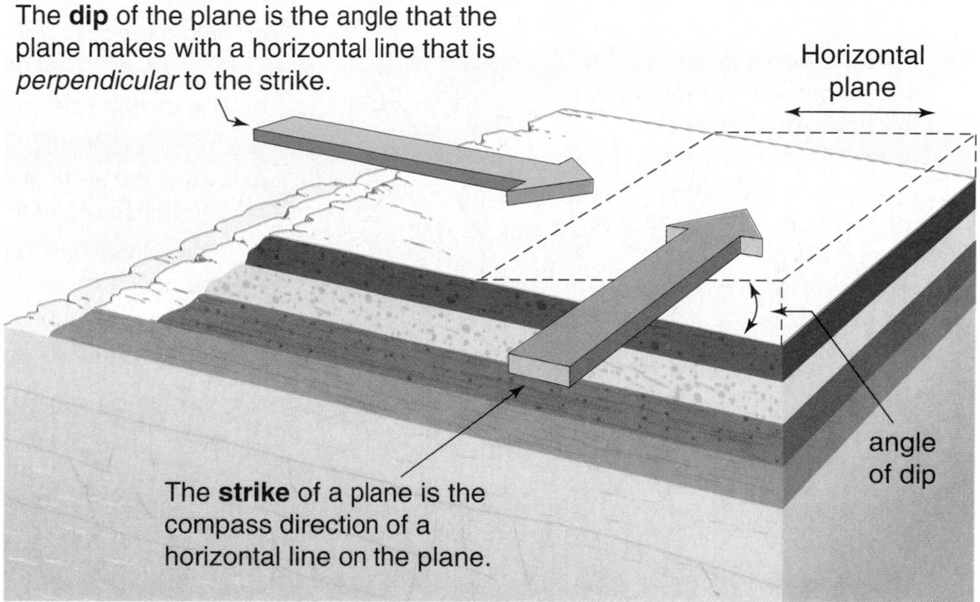
The **dip** of the plane is the angle that the plane makes with a horizontal line that is *perpendicular* to the strike.
Horizontal plane
angle of dip
The **strike** of a plane is the compass direction of a horizontal line on the plane.

CAREER CHOICES

A **geophysicist** combines geology and physics to study the physical properties and structure of Earth. They study electrical, thermal, gravitational, seismic, and magnetic phenomena.

For more information, contact:

American Geophysical Union
2000 Florida Avenue, NW
Washington, DC 20009

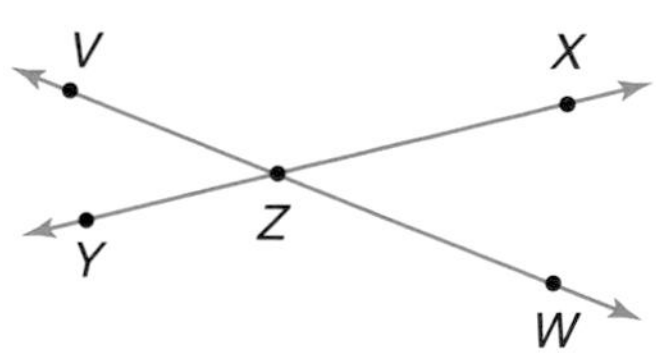
V
X
Z
Y
W

Perpendicular lines are special intersecting lines that form right angles. Not all lines that intersect are perpendicular lines. When two lines intersect, they form four angles. These angles are not necessarily right angles. In the figure at the left, $\overleftrightarrow{VW}$ intersects $\overleftrightarrow{XY}$ at Z forming $\angle VZX$, $\angle XZW$, $\angle WZY$, and $\angle YZV$. Certain pairs of angles formed by intersecting lines have special names that are used to describe the relationship between the angles. The chart below summarizes the names for special pairs of angles formed by these two lines.

Special Name	Definition	Examples
adjacent angles	angles in the same plane that have a common vertex and a common side, but no common interior points	$\angle VZX$ and $\angle XZW$ $\angle XZW$ and $\angle WZY$ $\angle WZY$ and $\angle YZV$ $\angle YZV$ and $\angle VZX$
vertical angles	two nonadjacent angles formed by two intersecting lines	$\angle VZY$ and $\angle XZW$ $\angle VZX$ and $\angle YZW$
linear pair	adjacent angles whose noncommon sides are opposite rays	$\angle VZX$ and $\angle XZW$ $\angle XZW$ and $\angle WZY$ $\angle WZY$ and $\angle YZV$ $\angle YZV$ and $\angle VZX$

The measures of vertical angles and linear pairs of angles have special relationships. The following activity will explore those relationships.

MODELING MATHEMATICS

Angle Relationships

Materials:  patty paper  protractor

You can create angles formed by intersecting lines by using paper folding.

- Fold a piece of patty paper so that it makes a crease across the paper. Open the paper and trace the crease with a pencil. Place two points, *A* and *B*, on the line.

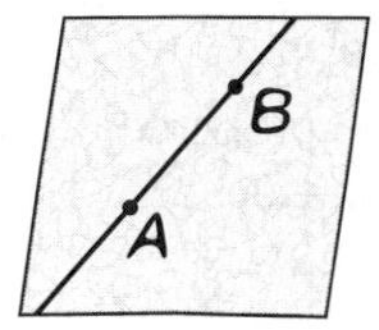

- Fold the paper again so that the second crease intersects the first crease between points *A* and *B*. Trace this crease and label the intersection as *C*. Label two points, *D* and *E*, on the second line so that *C* is between *D* and *E*.

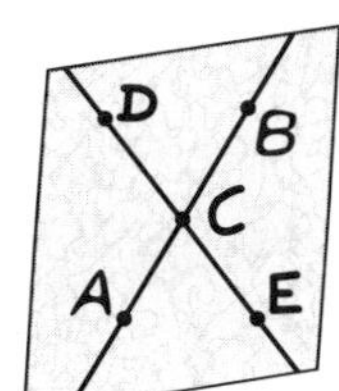

- Fold the paper again through point *C* so that $\angle ACD$ lies on top of $\angle ECB$. What do you notice?

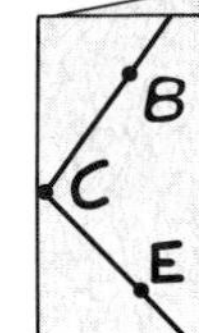

Your Turn

a. Fold the paper again through *C* so that $\angle ACE$ lies on $\angle DCB$. What do you notice?

b. Use a protractor to measure each angle. Write the measures on your drawing.

c. Name the vertical angles in this activity. What do you notice about their measures?

d. Name the linear pairs in this activity. What do you notice about their measures?

e. Repeat this activity with another piece of patty paper. What do you notice?

In the Modeling Mathematics activity, the following relationships among vertical angles and linear pairs are suggested. *You will be asked to prove these in Chapter 2.*

Vertical angles are congruent.
The sum of the measures of the angles in a linear pair is 180.

These relationships can help you solve algebraic problems related to geometry.

Example 

Algebra

In the figure, $\overleftrightarrow{GH}$ and $\overleftrightarrow{JK}$ intersect at *I*. Find the value of *x* and the measure of $\angle JIH$.

$\angle GIJ$ and $\angle KIH$ are vertical angles, so $m\angle GIJ = m\angle KIH$.

$$m\angle GIJ = m\angle KIH$$
$$16x - 20 = 13x + 7$$
$$3x = 27$$
$$x = 9$$

$$m\angle GIJ = 16x - 20$$
$$= 16 \cdot 9 - 20 \text{ or } 124$$

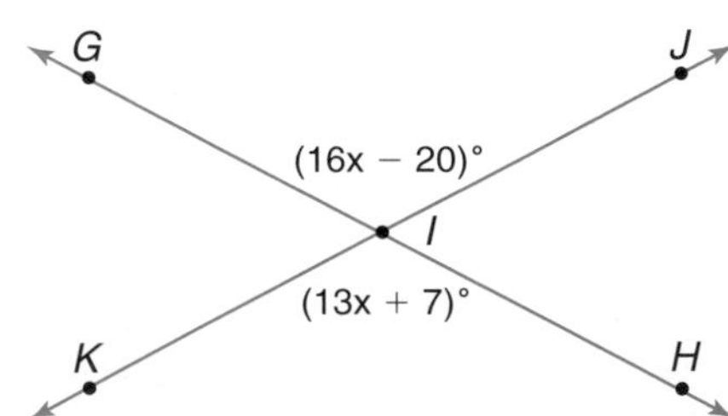

Since $\angle GIJ$ and $\angle JIH$ are a linear pair, $m\angle GIJ + m\angle JIH = 180$.

$$m\angle GIJ + m\angle JIH = 180$$
$$124 + m\angle JIH = 180$$
$$m\angle JIH = 56$$

The measure of $\angle JIH$ is 56.

The sum of the measures of $\angle GIJ$ and $\angle JIH$ in Example 1 is 180. Two angles whose measures have a sum of 180 are called **supplementary angles**. When two angles are supplementary, each angle is said to be a *supplement* of the other angle. Similarly, if the sum of the measures of two angles is 90, the angles are called **complementary angles**. Whenever two angles are complementary, each angle is said to be the *complement* of the other angle.

Since we previously saw that the sum of the measures of a linear pair of angles is 180, we can now say that *any two angles that form a linear pair must be supplementary angles.*

Example 2

Algebra

The measure of the supplement of an angle is 60 less than three times the measure of the complement of the angle. Find the measure of the angle.

Explore Let x represent the measure of the angle. Then $180 - x$ is the measure of its supplement, and $90 - x$ is the measure of its complement.

Plan Write an equation.

$$\underbrace{\text{The supplement}}_{180 - x} \quad \underbrace{\text{is}}_{=} \quad \underbrace{\text{60 less than three times the complement.}}_{3(90 - x) - 60}$$

Solve

$180 - x = 3(90 - x) - 60$	
$180 - x = 270 - 3x - 60$	*Use the distributive property.*
$180 - x = 210 - 3x$	*Combine like terms.*
$180 + 2x = 210$	*Add 3x to each side*
$2x = 30$	*Subtract 180 from each side.*
$x = 15$	*Divide each side by 2.*

The measure of the angle is 15.

Examine Is the measure of the supplement of a 15° angle 60 less than 3 times the measure of its complement?

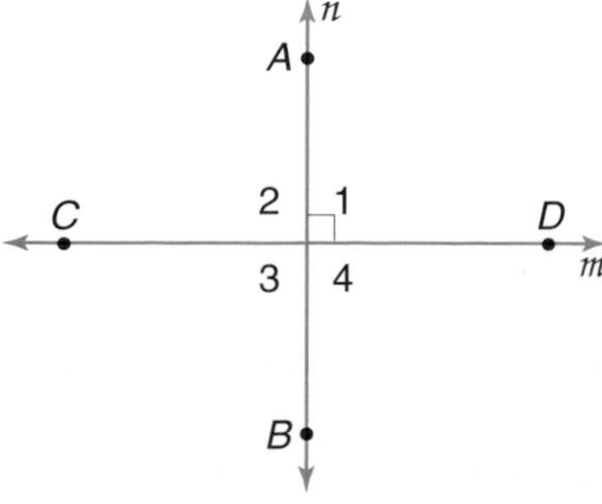

Perpendicular lines are two lines that intersect to form a right angle. In the figure at the left, lines m and n are perpendicular. To indicate this, we write $m \perp n$, which is read, "m is perpendicular to n." Line segments and rays can be perpendicular to lines or other line segments and rays if they intersect to form a right angle. For example, in the figure, $\overline{AB} \perp \overleftrightarrow{CD}$, $\overline{DC} \perp \overrightarrow{BA}$ and $\overline{AB} \perp \overline{DC}$.

Let's examine the relationships among the measures of the angles formed by these two perpendicular lines.

- $\angle 1$ is a right angle. $\angle 1$ and $\angle 3$ are vertical angles.
  What can you conclude about $\angle 3$? *It is also a right angle.*
- Consider $\angle 2$ and $\angle 4$. $\angle 1$ forms a linear pair with $\angle 2$ and also with $\angle 4$.
  What can you conclude about $\angle 2$ and $\angle 4$? *They are both right angles.*

Thus, $\angle 1$, $\angle 2$, $\angle 3$, and $\angle 4$ are all right angles.

If you draw two different lines, ℓ and p, that are perpendicular, do you think the relationships between the four angles formed will be the same? Based on this example, we could make the following conclusion, which will be proved in Chapter 2.

Perpendicular lines intersect to form *four* right angles.

You can use a compass and a straightedge to construct a line perpendicular to a given line through a point on the line, *or* through a point *not* on the line.

CONSTRUCTION

Perpendicular Lines Through a Point on the Line

Construct a line perpendicular to line n and passing through point C on n.

1. Place the compass at point C. Using the same compass setting, draw arcs to the right and left of C, intersecting line n. Label the points of intersection A and B.

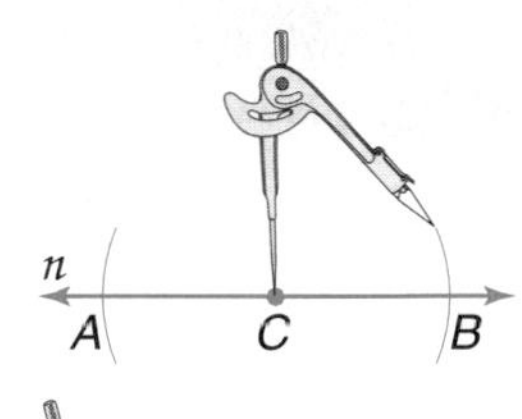

2. Open the compass to a setting greater than AC. Put the compass at point A and draw an arc above line n.

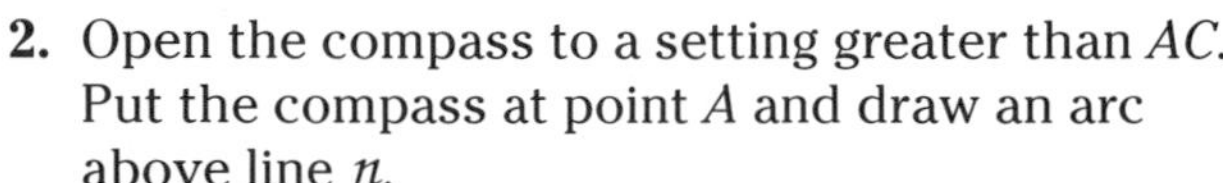

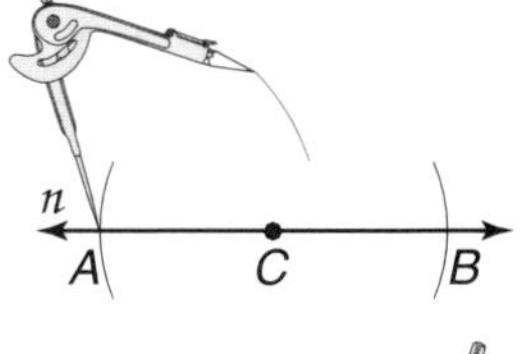

3. Using the same compass setting as in Step 2, place the compass at point B and draw an arc intersecting the arc previously drawn. Label the point of intersection D.

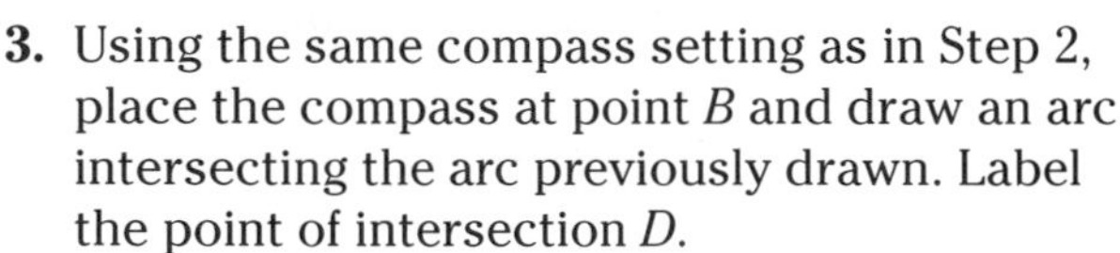

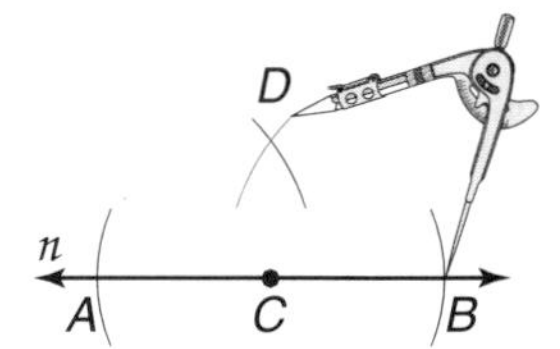

4. Use a straightedge to draw $\overleftrightarrow{CD}$.

Conclusion: By construction, $\overleftrightarrow{CD}$ is perpendicular to n at C.

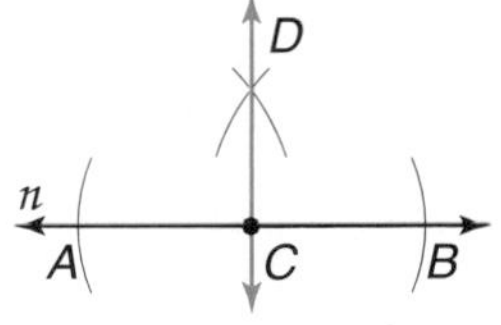

CONSTRUCTION

Perpendicular Lines Through a Point Not on the Line

Construct a line perpendicular to line m and passing through point Z *not* on m.

1. Place the compass at point Z. Draw an arc that intersects line m in two different places. Label the points of intersection X and Y.

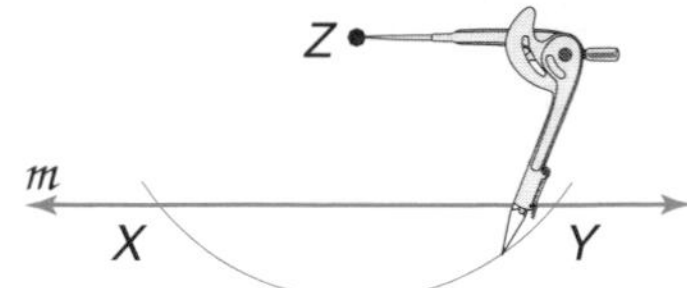

2. Open the compass to a setting greater than $\frac{1}{2}XY$. Put the compass at point X and draw an arc below line m.

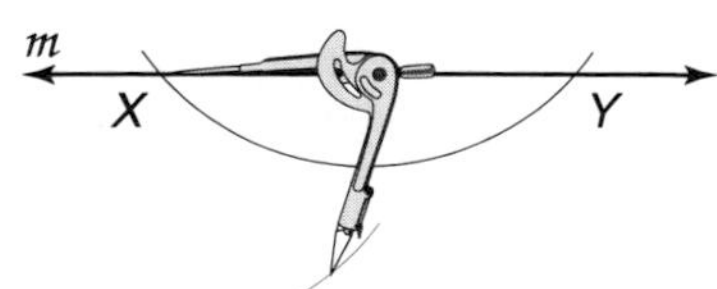

3. Using the same compass setting, place the compass at point Y and draw an arc intersecting the arc drawn in Step 2. Label the point of intersection A.

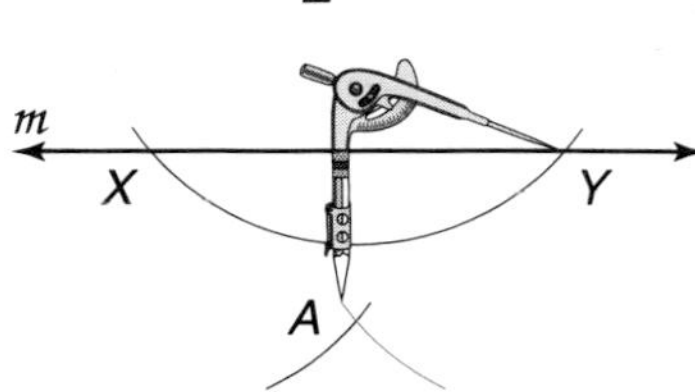

4. Use a straightedge to draw $\overleftrightarrow{ZA}$.

Conclusion: By construction, $\overleftrightarrow{ZA}$ is perpendicular to m.

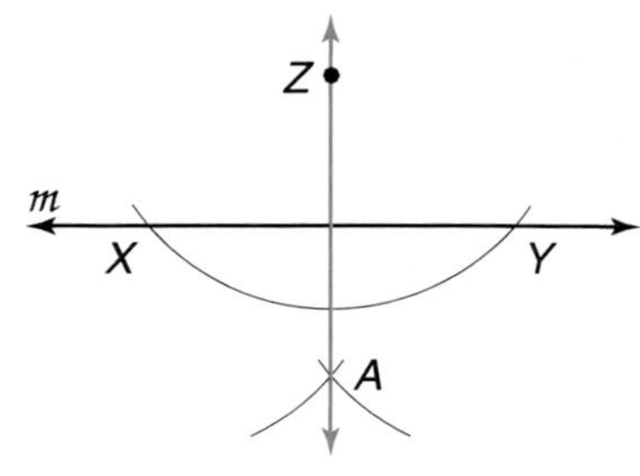

The concept of perpendicularity can be extended to include planes. If a line is perpendicular to a plane, then the line must be perpendicular to every line in the plane that intersects it. Thus, in the figure at the right, $\overleftrightarrow{AB} \perp \mathcal{E}$ and $\overleftrightarrow{AB}$ must be perpendicular to every line in $\mathcal{E}$ that intersects it.

We can also refer to line segments and rays as being perpendicular to planes. For example, in the figure, $\overline{AB} \perp \mathcal{E}$ and $\overrightarrow{AB} \perp \mathcal{E}$.

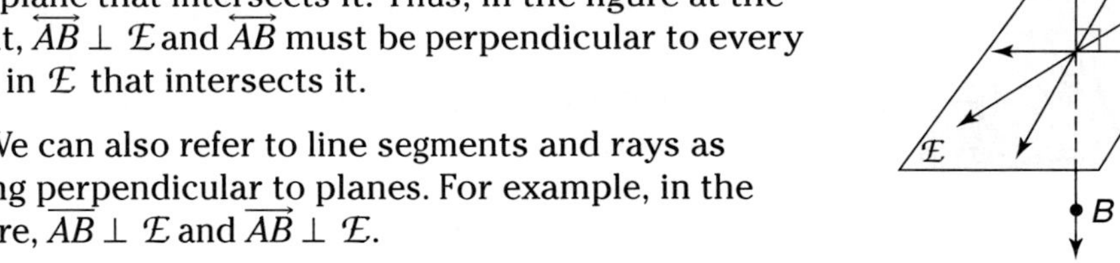

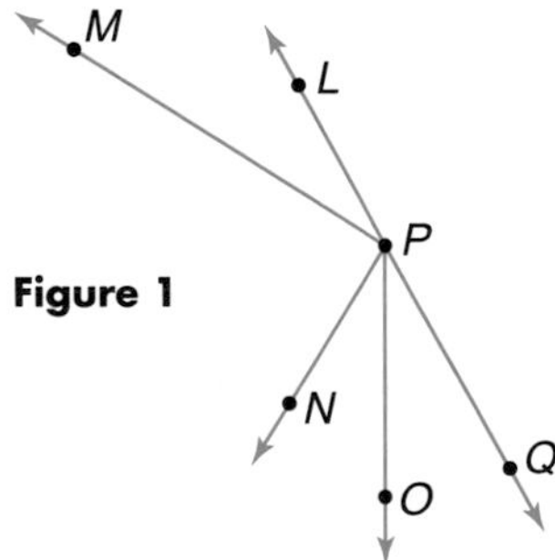

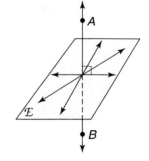

Figure 1

In this chapter, you have used figures to help describe or demonstrate different relationships among points, segments, lines, rays, and angles. Whenever you draw a figure, there are certain relationships that can be assumed from the figure and others that cannot be assumed. Consider the figure shown at the left.

Can Be Assumed from Figure 1	Cannot Be Assumed from Figure 1
All points shown are coplanar.	$\overrightarrow{PN} \perp \overrightarrow{PM}$
L, P and Q are collinear.	$\angle QPO \cong \angle LPM$
$\overrightarrow{PM}$, $\overrightarrow{PN}$, $\overrightarrow{PO}$, and $\overleftrightarrow{LQ}$ intersect at P.	$\overline{LP} \cong \overline{PQ}$
P is between L and Q.	$\overline{PQ} \cong \overline{PO}$
N is in the interior of $\angle MPO$.	$\angle QPO \cong \angle OPN$
$\angle LPQ$ is a straight angle.	$\angle OPN \cong \angle LPM$
$\angle LPM$ and $\angle MPN$ are adjacent angles.	$\overline{PO} \cong \overline{PN}$
$\angle LPN$ and $\angle NPQ$ are a linear pair.	$\overline{PN} \cong \overline{PL}$
$\angle QPO$ and $\angle OPL$ are supplementary.	

Compare Figure 2, shown at the right, with Figure 1 on the previous page. With these additional markings, some of the relationships that could *not* be assumed from Figure 1 *can* now be assumed from Figure 2.

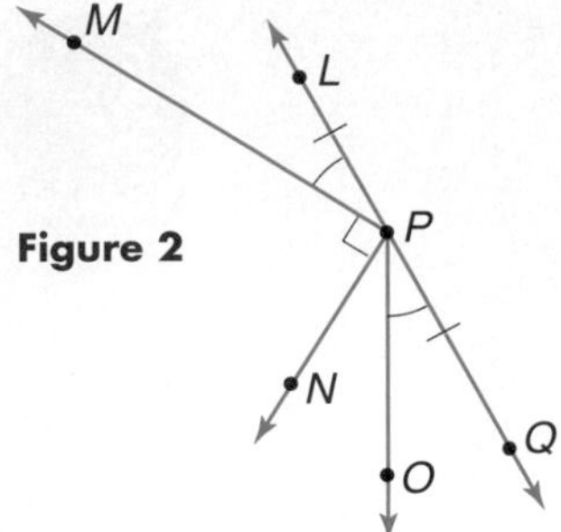

Figure 2

$\overrightarrow{PN} \perp \overrightarrow{PM}$
$\angle QPO \cong \angle LPM$
$\overline{LP} \cong \overline{PQ}$

CHECK FOR UNDERSTANDING

Communicating Mathematics

Study the lesson. Then complete the following.

1. **Compare and contrast** supplementary and complementary angles.
2. If two adjacent angles are supplementary, must they be a linear pair? Explain.
3. **Complete:** Line ℓ intersects plane C at point P and $\ell \perp C$. Thus, for any line n in C that passes through P, it must be true that __?__.

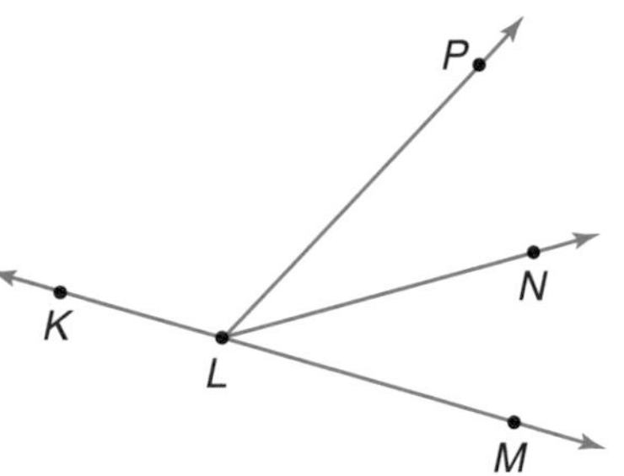

4. Can you assume that K, L, and M are collinear from the figure at the right? Can you assume that $\overrightarrow{LN}$ bisects $\angle PLM$? that $m\angle KLP + m\angle PLN = m\angle KLN$? Explain your answers.

5. Write an explanation as to why you think a linear pair of angles is called "linear."

Guided Practice

Refer to the figure at the right to answer Exercises 6–11.

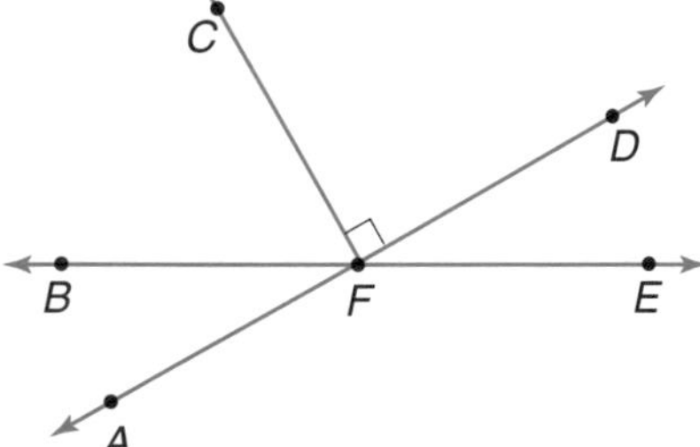

6. Identify a pair of obtuse vertical angles.
7. Name a segment that is perpendicular to $\overrightarrow{FC}$.
8. Which angle forms a linear pair with $\angle DFE$?
9. Identify a pair of adjacent angles that *do not* form a linear pair.
10. Which angle is complementary to $\angle CFB$?
11. Can you assume that $\overrightarrow{FC}$ bisects $\overline{AD}$ from the figure? Explain.
12. $\angle N$ is a complement of $\angle M$, $m\angle N = 8x - 6$ and $m\angle M = 14x + 8$. Find the value of x and $m\angle M$.
13. Find $m\angle S$ if $m\angle S$ is 20 more than four times its supplement.

14. **Sports** In the 1998 Winter Olympic Games, Jani Soininen of Finland claimed the gold medal in the 90-meter ski jump. When a skier completes a jump, he or she tries to make the angle between his or her body and the front of his or her skis as small as possible. If Jani is aligned so that the front of his skis make a 15° angle with his body, what angle is formed by the tail of the skis and his body?

EXERCISES

Practice **Refer to the figure at the right to answer Exercises 15–22.**

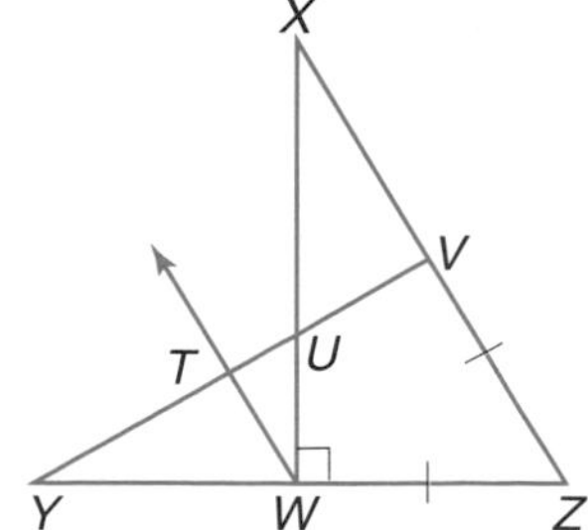

15. Name a pair of vertical angles.

16. Can you assume that $\angle YUW$ and $\angle WUV$ are supplementary from the figure? Explain.

17. Which angle is complementary to $\angle YWT$?

18. Identify a pair of angles that are congruent.

19. Can you assume that $\overrightarrow{WT}$ is perpendicular to $\overline{YV}$? Explain.

20. Name a pair of angles that are noncongruent and supplementary.

21. Identify $\angle YWU$ and $\angle UWZ$ as congruent, adjacent, vertical, complementary, supplementary, and/or a linear pair.

22. Can you assume that $\overline{VZ} \cong \overline{WZ}$? Explain.

Refer to the figure at the right to answer Exercises 23–26.

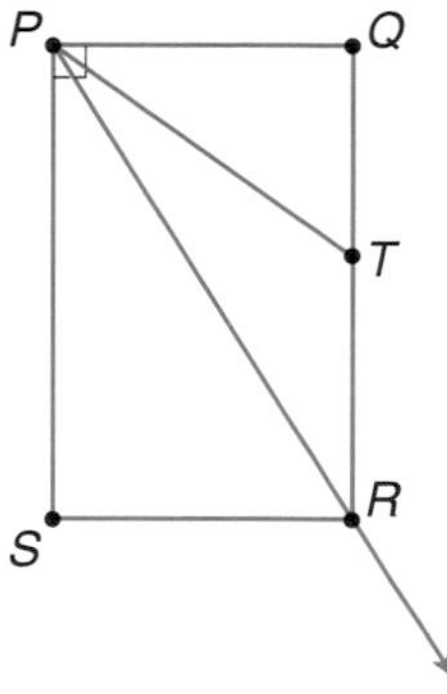

23. Can you determine that $\angle SRP$ and $\angle PRT$ are complementary from the figure? Explain.

24. Name two angles that are adjacent, but not complementary or supplementary.

25. Given that$\angle Q$ and $\angle QPS$ are supplementary, is $\angle Q$ obtuse, acute, or right?

26. List four things that you *cannot* assume from the figure.

27. Fold a piece of patty paper to make a crease. Open the paper and label this as line a. Fold the paper again so that the two parts of line a lie on top of each other. Open the paper and label the second crease as line b.

- **a.** What types of lines are a and b?
- **b.** Fold the paper to find the angle bisectors of two adjacent angles.
- **c.** What is the relationship of the lines containing the angle bisectors? Explain.

28. Find the measures of two complementary angles, $\angle A$ and $\angle B$, if $m\angle A = 7x + 4$ and $m\angle B = 4x + 9$.

29. The measure of an angle is 44 more than the measure of its supplement. Find the measures of the angles.

30. What are the measures of two complementary angles if the difference in the measures of the two angles is 12?

31. Suppose $\angle T$ and $\angle U$ are complementary. Find $m\angle T$ and $m\angle U$ if $m\angle T = 16x - 9$ and $m\angle U = 4x + 3$.

32. Find the measures of two supplementary angles, $\angle N$ and $\angle M$, if the measure of angle N is 5 less than 4 times the measure of angle M.

33. $\angle R$ and $\angle S$ are complementary angles, and $\angle U$ and $\angle V$ are also complementary angles. If $m\angle R = y - 2$, $m\angle S = 2x + 3$, $m\angle U = 2x - y$, and $m\angle V = x - 1$, find the values of x, y, $m\angle R$, $m\angle S$, $m\angle U$, and $m\angle V$.

Proof

34. Suppose $\angle ABC$ and $\angle CBD$ form a linear pair, $\angle CBD$ and $\angle DBE$ form a linear pair, and $m\angle CBD = 30$.
    a. Draw a figure representing these angles.
    b. Write a paragraph proof showing that $\angle ABC \cong \angle DBE$.

35. $\angle PQR$ and $\angle RQS$ are complementary angles, and $m\angle PQR = 45$. Write a paragraph proof showing that $\angle PQR \cong \angle RQS$.

Critical Thinking

36. Line ℓ in plane C contains point X. How many lines in the plane are perpendicular to line ℓ and pass through X? Justify your answer.

Applications and Problem Solving

37. **Language** Look up the words *complementary* and *supplementary* in a dictionary. How do their everyday meanings relate to their mathematical meanings?

38. **Construction** A framer is installing a cathedral ceiling in a newly-built home. A protractor and a plumb bob are used to check the angle at the joint between the ceiling and wall. The wall is vertical, so the angle between the vertical plumb line and the ceiling is the same as the angle between the wall and the ceiling.
    a. How are $\angle ABC$ and $\angle CBD$ related?
    b. If $m\angle ABC = 110$, what is $m\angle CBD$?

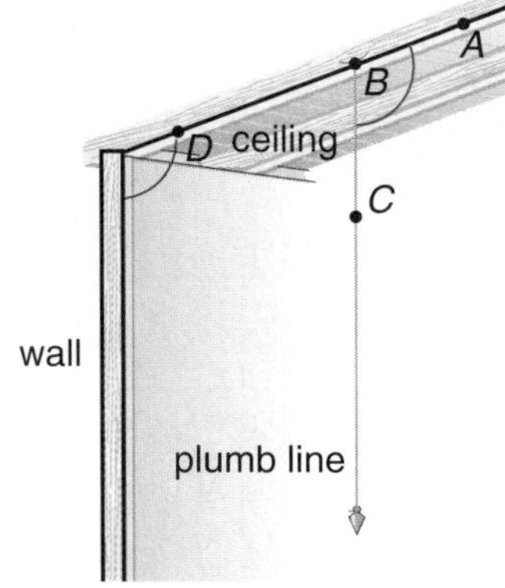

Mixed Review

39. If $\angle Q \cong \angle R$, $m\angle Q = 8x - 17$, and $m\angle R = 7x - 3$, is $\angle Q$ acute, right, obtuse, or straight? (Lesson 1–6)

40. If $FG = 12x - 11$, $GH = 5x + 10$, and $FH = 36$, for what value of x are $\overline{FG}$ and $\overline{GH}$ congruent? (Lesson 1–5)

41. **Safety** The National Safety Council recommends placing the base of a 15-foot ladder 5 feet from the wall. How high can the ladder safely reach? (Lesson 1–4)

15 ft

5 ft

42. **SAT Practice** For which of the following values of x is $\frac{x^3}{x^4}$ the LEAST?

    A -4 B -3 C -2
    D -1 E 1

43. Draw and label a figure that shows line k that intersects plane C at point P. (Lesson 1–2)

44. Points $U(-1, 3)$ and $V(5, 9)$ lie on the graph of $y = x + 4$. Determine whether $W(-2, 3)$ is collinear with U and V. (Lesson 1–1)

Algebra

45. Find the greatest common factor of $8ab^2$, $16a^2b$, and $12ab$.

46. Factor $x^2 - 4x + 4$.

For **Extra Practice**, see page 766.

HIGHLIGHTS

Chapter Review For additional lesson-by-lesson review, visit: **www.geometry.glencoe.com**

VOCABULARY

After completing this chapter, you should be able to define each term, property, or phrase and give an example or two of each.

Geometry

acute angle (p. 46)
adjacent angles (p. 53)
angle (p. 44)
angle bisector (p. 48)
area (p. 19)
between (p. 28)
collinear (p. 8)
complementary angles (p. 55)
congruent (pp. 31, 47)
coplanar (p. 13)
degree (p. 45)
exterior of an angle (p. 45)
informal proof (p. 39)
interior of an angle (p. 45)
intersection (p. 14)
line (p. 12)
linear pair (p. 53)
measure (pp. 28, 45)
midpoint (p. 36)
noncollinear (p. 8)
obtuse angle (p. 46)
opposite rays (p. 44)
paragraph proof (p. 39)
perimeter (p. 20)
perpendicular lines (p. 53)
plane (p. 12)
point (p. 12)
postulate (p. 28)
proof (p. 39)
Protractor Postulate (p. 45)
Pythagorean Theorem (p. 30)
ray (p. 44)
right angle (p. 46)
Ruler Postulate (p. 28)
segment (p. 12)
segment bisector (p. 38)
sides of an angle (p. 44)
space (p. 12)
straight angle (p. 44)
supplementary angles (p. 55)
theorem (p. 39)
three-dimensional figure (p. 14)
vertex of an angle (p. 44)
vertical angles (p. 53)

Algebra

coordinate plane (p. 6)
distance formula (p. 31)
midpoint formulas (p. 37)
ordered pair (p. 6)
origin (p. 6)
quadrants (p. 6)
x-axis (p. 6)
x-coordinate (p. 6)
y-axis (p. 6)
y-coordinate (p. 6)

Problem Solving

list the possibilities (p. 13)
problem-solving plan (p. 19)

UNDERSTANDING AND USING THE VOCABULARY

Choose the correct term to complete each sentence.

1. The x- and y-axes divide the coordinate plane into (*quadrants, perpendicular lines*).
2. An angle that measures 120° is an (*acute, obtuse*) angle.
3. Two angles whose measures have a sum of 180 are (*complementary, supplementary*).
4. Opposite rays form a (*plane, line*).
5. The intersection of two planes is a (*point, line*).
6. Adjacent angles share a (*ray, point*).
7. The intersection of the x- and y-axes is the (*origin, midpoint*).
8. A segment bisector divides a segment into (*two congruent segments, a linear pair*).
9. (*Adjacent, Vertical*) angles are congruent.
10. Statements that are assumed to be true are (*postulates, theorems*).

CHAPTER 1 STUDY GUIDE AND ASSESSMENT

SKILLS AND CONCEPTS

OBJECTIVES AND EXAMPLES

Upon completing this chapter, you should be able to:

- graph ordered pairs on a coordinate plane (Lesson 1–1)

Graph $P(2, -1)$, $Q(5, 0)$, and $R(-4, -3)$.

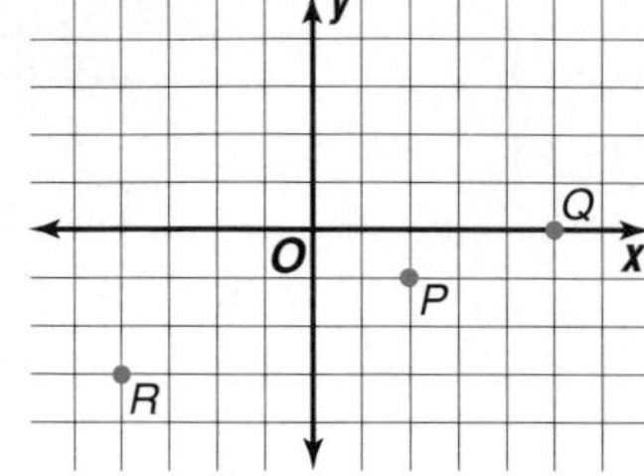

REVIEW EXERCISES

Use these exercises to review and prepare for the chapter test.

Graph each point on the same coordinate plane.

11. $S(1, 7)$
12. $T(-4, -1)$
13. $U(-8, -2)$
14. $V(0, 6)$
15. In which quadrant is the point $W(x, y)$ located if $x < 0$ and $y < 0$?

- identify collinear and coplanar points and intersecting lines and planes (Lesson 1–2)

Name the intersection of plane *ACD* and plane $\mathcal{Q}$.

Line *CD* is the intersection of *ACD* and $\mathcal{Q}$.

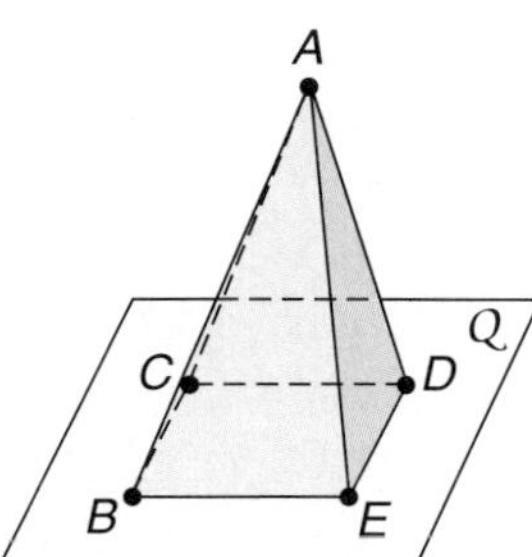

Refer to the figure at the left to answer each question.

16. Are $\overline{BE}$ and $\overline{AD}$ coplanar?
17. What points do planes *ACB* and *ADE* have in common?
18. Name two lines that intersect at point *A*.

- solve problems using formulas (Lesson 1–3)

The formula $W = I \times V$ relates the electrical wattage (W), amperage (I), and voltage (V) of a circuit. If current flow through a lightbulb is 0.5 amps and it uses 150 volts, what is the wattage of the bulb?

The wattage of the bulb is 75 watts.

19. Use the formula $d = rt$ to find the rate r if the distance d is 135 miles and the time t is 3 hours.
20. Find the width of a rectangle if its perimeter is 45 inches and its length is 17 inches.
21. What is the greatest possible area of a rectangle if its perimeter is 30 centimeters?

- find the distance between points on a number line (Lesson 1–4)

Find *UW* and *VT* on the number line shown below.

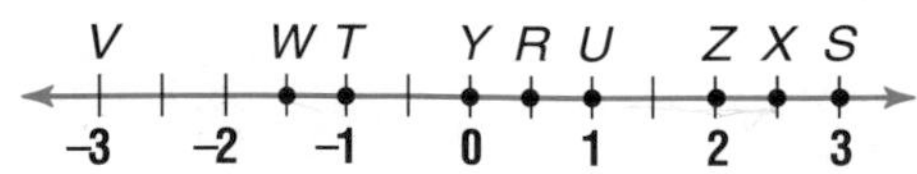

$$UW = \left|1 - \left(-1\tfrac{1}{2}\right)\right| = \left|1 + 1\tfrac{1}{2}\right| = \left|2\tfrac{1}{2}\right| \text{ or } 2\tfrac{1}{2}$$

$$VT = |-3 - (-1)| = |-3 + 1| = |-2| \text{ or } 2$$

Refer to the number line at the left to answer each question.

22. Find the distance between *X* and *T*.
23. What is the distance between *Z* and *W*?
24. Which point(s) is 2 units from point *U*?

OBJECTIVES AND EXAMPLES

- find the distance between points on a coordinate plane (Lesson 1–4)

What is the distance between $F(-1, 4)$ and $G(6, -2)$?

$$d = \sqrt{(x_2 - x_1)^2 + (y_2 - y_1)^2}$$
$$= \sqrt{(6 - (-1))^2 + (-2 - 4)^2}$$
$$= \sqrt{7^2 + (-6)^2}$$
$$= \sqrt{49 + 36}$$
$$= \sqrt{85} \text{ or about } 9.22$$

REVIEW EXERCISES

Refer to the graph below to find each measure. If the measure is not a whole number, round the result to the nearest hundredth.

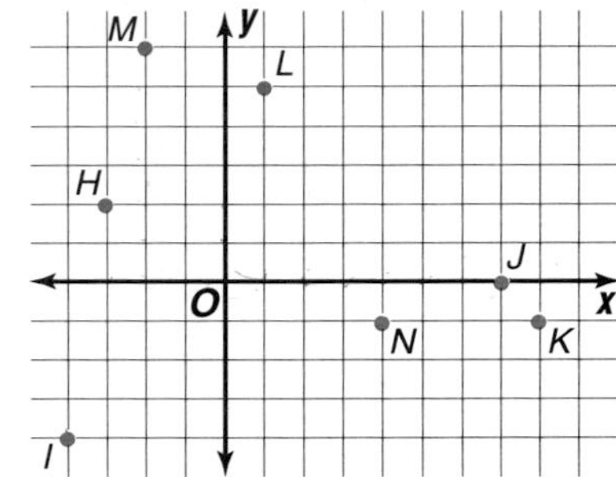

25. Find the distance between H and L.

26. What is the distance between J and H?

27. Find the distance between K and M.

28. What is the measure of segment IN?

- find the midpoint of a segment (Lesson 1–5)

For the graph above at the right, what is the midpoint of $\overline{HK}$?

Let $H(-3, 2)$ be (x_1, y_1).

Let $K(8, -1)$ be (x_2, y_2).

$$\left(\frac{x_1 + x_2}{2}, \frac{y_1 + y_2}{2}\right) = \left(\frac{-3 + 8}{2}, \frac{2 + (-1)}{2}\right)$$
$$= \left(\frac{5}{2}, \frac{1}{2}\right) \text{ or } \left(2\frac{1}{2}, \frac{1}{2}\right)$$

29. Refer to the graph above to find the midpoint of $\overline{NJ}$.

30. If the coordinate of point B on a number line is 3, name the coordinates of two points A and C such that B is the midpoint of $\overline{AC}$.

31. Algebra F is the midpoint of $\overline{EG}$. If $EF = 2x + 3$ and $EG = 6x - 3$, find FG.

- identify and use congruent segments (Lesson 1–5)

For the number line below, which segments are congruent to $\overline{AD}$?

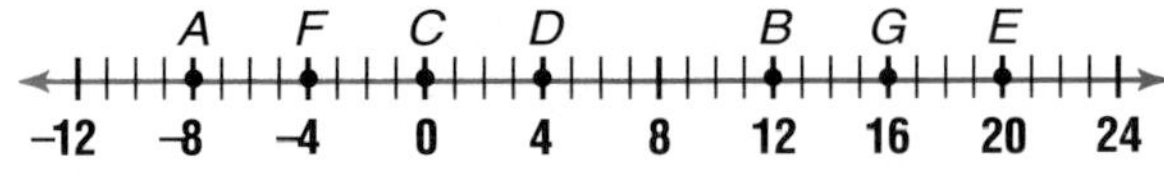

$AD = |-8 - 4|$

$= |-12|$ or 12

Since $BC = |12 - 0|$ or 12, $\overline{BC} \cong \overline{AD}$.

Since $DG = |4 - 16|$ or 12, $\overline{DG} \cong \overline{AD}$.

32. The coordinate of P on a number line is 6. If $QR = 18$, find the coordinate of S if $\overline{PS} \cong \overline{QR}$.

33. N, M, and O are collinear and $\overline{NO} \cong \overline{MO}$. Is O the midpoint of $\overline{NM}$?

34. Algebra $UX = 4x - 2$ and $WV = 2x + 8$. If $\overline{UX} \cong \overline{WV}$, what is the measure of $\overline{UX}$?

OBJECTIVES AND EXAMPLES

- use angle addition to find the measures of angles (Lesson 1–6)

If $m\angle AEG = 75$, $m\angle 1 = 25 - x$, and $m\angle 4 = 5x + 20$, find the value of x.

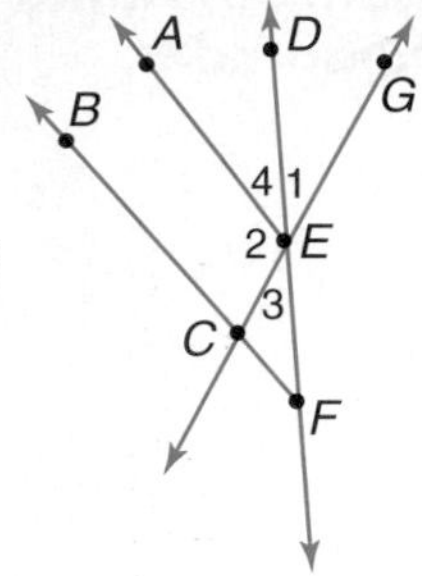

$$m\angle 1 + m\angle 4 = m\angle AEG$$
$$(25 - x) + (5x + 20) = 75$$
$$4x + 45 = 75$$
$$4x = 30$$
$$x = 7.5$$

REVIEW EXERCISES

Refer to the figure at the left for Exercises 35–37.

35. If $m\angle 3 = 32$, find $m\angle CED$.

36. If $m\angle 2 = 6x - 20$, $m\angle 4 = 3x + 18$, and $m\angle CED = 151$, find the value of x.

37. If $m\angle 1 = 49 - 2x$, $m\angle 4 = 4x + 12$, and $m\angle 2 = 15x$, find $m\angle 4$.

- identify and use adjacent angles, vertical angles, complementary angles, supplementary angles, linear pairs of angles, and perpendicular lines (Lesson 1–7)

Name a pair of adjacent angles and a pair of perpendicular lines.

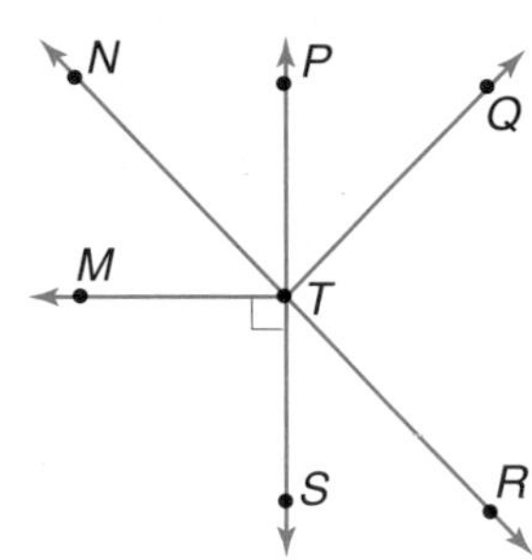

$\angle NTP$ and $\angle PTQ$ are one pair of adjacent angles.

$\overrightarrow{TM}$ is perpendicular to $\overleftrightarrow{PS}$.

Refer to the figure at the left for Exercises 38–41.

38. Name a pair of adjacent complementary angles.

39. Which angle is a vertical angle to $\angle STR$?

40. If $\overline{QT}$ is perpendicular to $\overleftrightarrow{NR}$, what can you conclude about $\angle PTQ$ and $\angle NTM$?

41. Algebra The measure of $\angle PTR$ is twice the measure of $\angle NTP$. Find the measure of both angles.

42. Algebra An angle measures 43° less than six times the measure of its complement. Find the measure of both angles.

APPLICATIONS AND PROBLEM SOLVING

43. Technology Digital displays are made up of points turned on or off in a grid system. Draw a coordinate grid to represent points turned on at (0, 3), (4, 12), (−3, −16), (−3, −5), and (10, −4). (Lesson 1–1)

44. Boat Safety The Coast Guard recommends that the number of people that can safely occupy a boat equals the length of the boat multiplied by its width and then divided by 15. How many people can safely occupy a sailboat that is 25 feet long and 12 feet wide? (Lesson 1–3)

A practice test for Chapter 1 is provided on page 793.

45. Sports The face of a golf club is angled so that a ball will travel in different ways. The measure of the angle that the head of the club would form with a vertical ray is called the *loft*. Wood and iron clubs are numbered so that greater numbers indicate greater amounts of loft. For example, an 8-iron has more loft than a 4-iron. (Lesson 1–6)

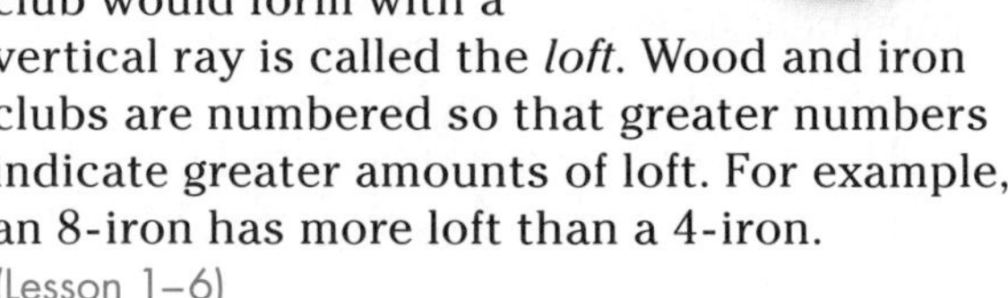

a. How do you think the loft of a club affects the path of a shot?

b. Draw a picture to show how the path of a shot hit with an 8-iron might differ from the path of a shot hit with a 4-iron.

ALTERNATIVE ASSESSMENT

COOPERATIVE LEARNING PROJECT

Astronomy In this project, you will use one of the constellations in the night sky to apply the concepts you have learned in this chapter.

A star's position in the sky is given by two numbers, its *right ascension* and its *declination*. The right ascension is given in hours (h), minutes (m), and seconds (s). Each of the 24 hours is subdivided into 60 minutes, and each minute is subdivided into 60 seconds. Each hour is equivalent to 15°. For example, the star alpha Andromeda's right ascension is 0 h 8 m 23.2 s or about 2.09°.

A star's declination is given in degrees (°), minutes (′), and seconds (″). Each of the 360 degrees is subdivided into 60 minutes, and each minute is subdivided into 60 seconds. Alpha Andromeda's declination is +29° 5′ 26″ or about 29.09°.

Follow these steps to analyze the apparent geometry of the constellation Andromeda.

- Go to the library to find a book with detailed star charts. Good sources are *Starlist 2000* by Richard Dibon-Smith, *Cambridge Star Atlas 2000.0* by Wil Tirion, *The Star Guide* by Robin Kerrod, or a world atlas.
- Find the star chart for the constellation Andromeda, and locate the stars α (alpha), δ (delta), π (pi), β (beta), γ (gamma), ν (nu), and υ (upsilon). Draw a diagram of the seven stars with straight line segments connecting each pair of stars.
- Look up the right ascension and declination of each of the seven stars. Write this information as an ordered pair for each star, (right ascension, declination). Convert each measurement to the nearest hundredth degree.
- Use the Pythagorean Theorem to find the lengths of the line segments joining each pair of stars. This is known as the *angular distance* between stars.
- Use your eye to estimate whether each angle formed by each triple of stars is acute, right, or obtuse.

Follow these steps to analyze the real geometry of the constellation Andromeda.

- Look up the distances, in light years, to each of the seven stars.
- Since any three noncollinear points define a plane, any three stars define a plane in astronomical space. How many different planes are there in Andromeda? Use the strategy of listing all possibilities.

THINKING CRITICALLY

In which quadrant is the point (1, 0)? What about the points, (0, 1), (−1, 0), (0, −1), and (0, 0)? What can you conclude about any ordered pair that has 0 as one of its coordinates. Explain your answer.

PORTFOLIO

Find a photograph of a building with a large number of sides in different planes. Count the number of such planes. Keep the photograph in your portfolio.

LONG-TERM PROJECT

In·ves·ti·ga·tion

art FOR ART'S SAKE

MATERIALS NEEDED

- markers
- sheet of paper 8½″×11″
- protractor
- ruler

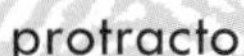

Cygnus Software, Inc. is planning to renovate a floor of a high-rise office building to house their new corporate headquarters. The CEO (Chief Executive Officer) wants to design the office around a mural that will be placed on the wall of the lobby. Your company has been hired to create the mural and choose the furnishings, flooring, and wall coverings to complete the lobby.

FUNCTION AND FEELING

A good room design is determined by the way that the room will be used and the feeling that the designer wants to create. Discuss the image that you think Cygnus Software, Inc. would like to present to the people who visit their lobby. Use the following questions to guide your discussion.

- Should the company appear friendly, small, and warm?
- Would a professional, high-tech image be more appropriate?
- Will there be a reception desk or other furniture in the lobby?
- How much time will people spend in the lobby? Will people need to be comfortable waiting there for long periods of time?
- Will there be art other than the mural, such as smaller paintings or sculptures, in the lobby?

CHOOSING A STYLE

Art is a means of communicating. The way in which works of art communicate that message can be vastly different. Scholars classify art in

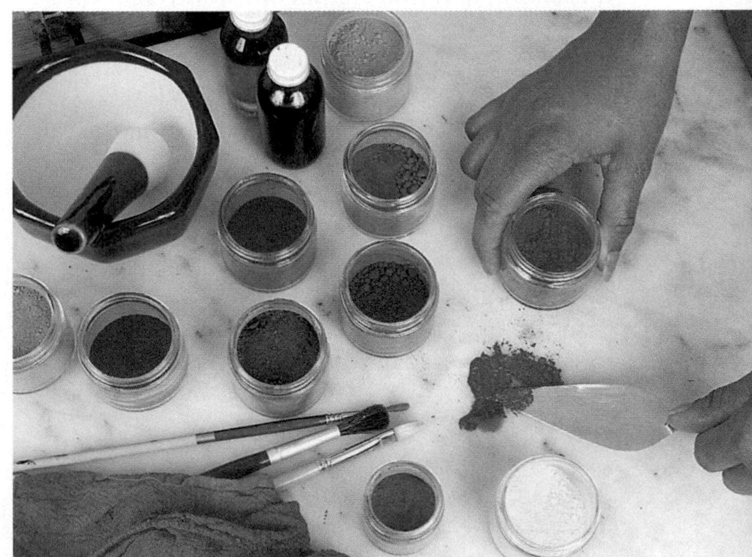

different styles. Classical art, art deco, rococo, avant garde, Byzantine, Islamic, realism, and surrealism are a few examples of art styles. Classical art emphasizes order, balance and simplicity. Art deco became popular in the 1920s and 1930s and is characterized by geometric shapes. Repeated patterns of geometric shapes are common in Islamic art.

The style of art that you choose for your mural will dictate the look of the lobby. Research several different styles of art and architecture. You may wish to consult a reference book in a library or do some research on the Internet. Study the characteristics of the style and the works of some prominent artists and architects who worked in that style. You may wish to contact an art instructor or an interior designer to help you in your research. Compare the artistic styles you are considering to the image that you believe the company would like to present. Which style communicates that message best?

Begin an Investigation folder to keep your ideas and materials organized throughout the artistic process.

You will continue working on this Investigation throughout Chapters 2 and 3.

Be sure to keep your research, discussion notes, and other materials in your Investigation folder.

Art for Art's Sake Investigation

Working on the Investigation
Lesson 2–2, p. 83

Working on the Investigation
Lesson 2–6, p. 114

Working on the Investigation
Lesson 3–4, p. 153

Closing the Investigation
End of Chapter 3, p. 170

CHAPTER 2

Connecting Reasoning and Proof

What You'll Learn

In this chapter, you will:

- make conjectures,
- use the laws of logic to make conclusions,
- solve problems by looking for a pattern,
- write algebraic proofs, and
- write proofs involving segment and angle theorems.

Why It's Important

Real World

Law The job of a lawyer is to present the client's case for guilt or innocence so that jurors can use logical reasoning to determine whether the client is guilty or not. In this chapter, you will learn about two basic types of logic that can be used to help a person determine whether something is true or false. Inductive reasoning relies on patterns in past occurrences to reach a conclusion. Deductive reasoning uses a rule to reach a conclusion. Lawyers may use both types of logic as they present their cases to juries.

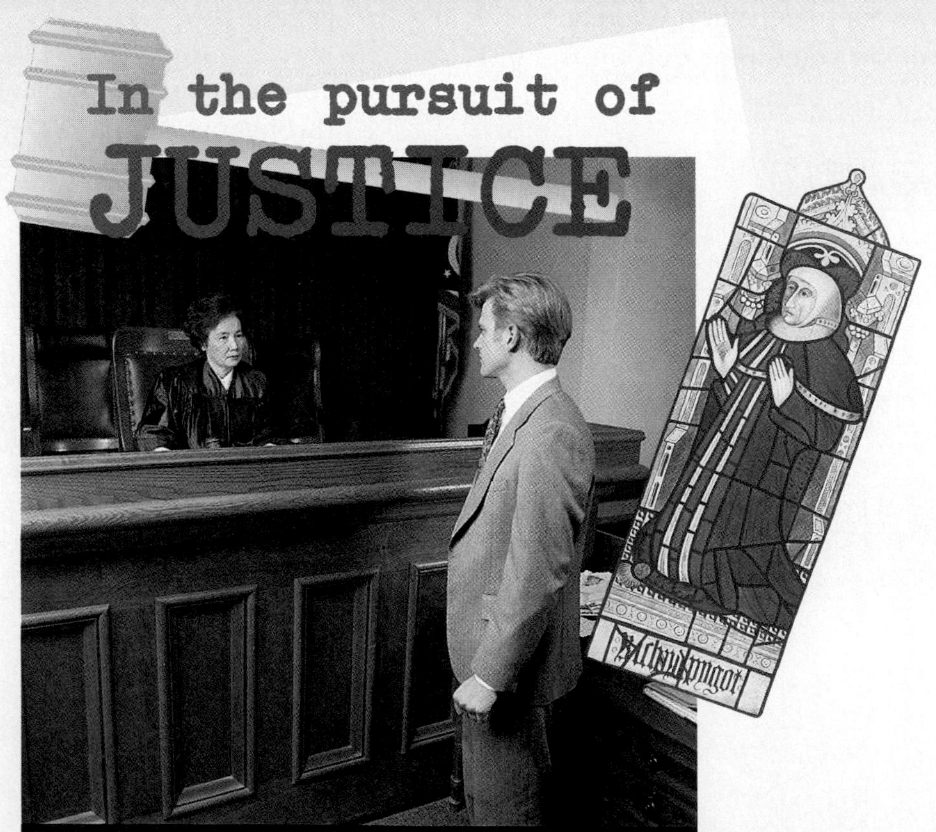

Focus On
PREREQUISITE SKILLS

To be successful in this chapter, you'll need to understand these concepts and be able to apply them. Refer to the example or to the lesson in parentheses if you need more review before beginning the chapter.

Identify and use adjacent and vertical pairs of angles. (Lesson 1–7)

Refer to the figure at the right.

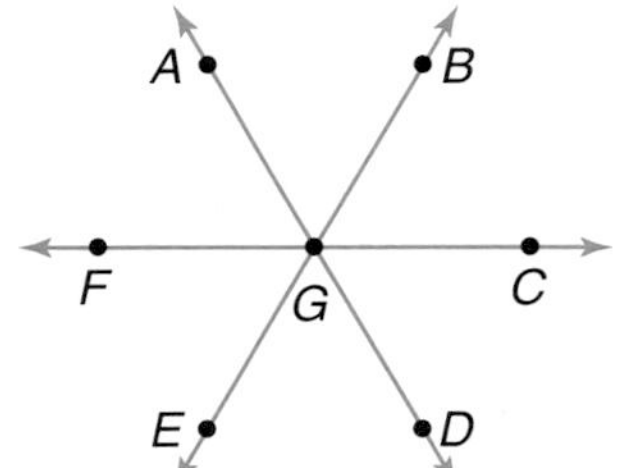

1. Identify a pair of vertical angles that appear to be acute.
2. Identify a pair of adjacent angles that appear to be obtuse.
3. If $\angle AGB = (4x + 7)^\circ$ and $\angle EGD = 71^\circ$, find the value of x.
4. If $\angle BGC = 45^\circ$, $\angle CGD = (8x + 4)^\circ$, and $\angle DGE = (15x - 7)^\circ$, find the value of x.

Solve equations with the variable on both sides.

Example Solve $5x + 1 = 3x - 4$.

$5x + 1 = 3x - 4$

$2x + 1 = -4$ *Subtract 3x from each side.*

$2x = -5$ *Subtract 1 from each side.*

$x = -2.5$ *Divide each side by 2.*

Solve each equation.

5. $6x - 42 = 4x$
6. $8 - 3n = -2 + 2n$
7. $3(y + 2) = -12 + y$
8. $12 + 7x = x - 18$
9. $3x + 4 = \frac{1}{2}x - 5$
10. $2 - 2x = \frac{2}{3}x - 2$

Focus On
READING SKILLS

In this chapter, you'll be learning about two forms of logical reasoning used in geometry. When you observe the same thing happening again and again and form a conclusion from those observations, you are using ***inductive reasoning.*** When you use laws of logic and statements that are known to be true to reach a conclusion, you are using ***deductive reasoning.***

2–1

Inductive Reasoning and Conjecturing

What YOU'LL LEARN

- To make conjectures based on inductive reasoning.

Why IT'S IMPORTANT

Inductive reasoning and conjecturing help you arrive at valid conclusions.

Comics

In the cartoon *Shoe* shown below, Skyler observed two specific situations in order to make the **conjecture**, "I'll be the next president of Mexico."

SHOE

A conjecture is an educated guess. Looking at several specific situations to arrive at a conjecture is called **inductive reasoning**. For centuries, mathematicians have used inductive reasoning to develop the geometry that we study today.

Developing skills in games or sports usually involves some experimentation followed by conjectures about how to effectively participate in the activity. A game that involves striking a ball to make it go a specific direction is *carom billiards.* A carom billiard table is like a pool table without pockets. It is twice as long as it is wide, usually 10 feet by 5 feet.

Practice may improve your game. Knowledge of geometry could also be helpful in becoming a better player.

Example 1

Billiards

This example does not take into account other factors that may affect the path of the ball, such as friction, spin, and the various materials of which the ball and table are made.

If you place the ball in one corner of a 10-foot by 5-foot carom billiard table and shoot it at a 45° angle with respect to the sides of the table, where will it go, assuming there are no other balls on the table? Use graph paper to draw the path of the ball. Make a conjecture about the path of the ball if it is shot at a 45° angle from any corner of the table.

Start

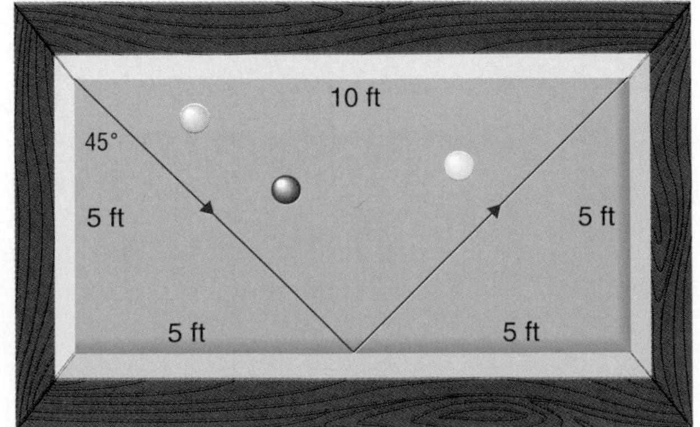

Some conjectures are:

- The ball will strike the long side of the table at its midpoint.
- The ball will then bounce off the rail at the same angle.
- The ball will continue on a path and touch the opposite corner.

What happens to the path of the ball if we change the dimensions of the table?

In Chapter 1, you had some experience with basic geometric ideas. You can use some of these ideas to make conjectures in geometry.

Example 2 **For points A, B, and C, $AB = 10$, $BC = 8$, and $AC = 5$. Make a conjecture and draw a figure to illustrate your conjecture.**

Some conjectures can be made from a figure. However, you should not judge measurements from the appearance of a figure.

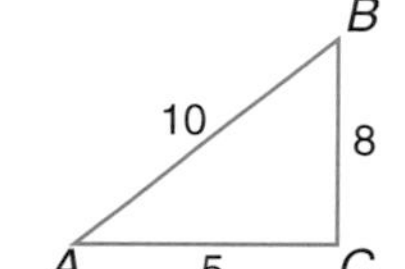

Given: Points A, B, and C, $AB = 10$, $BC = 8$, and $AC = 5$.

Conjecture: A, B, and C are noncollinear.

Conjectures are made based on observations of a particular situation. You can use a calculator to make conjectures about the properties of a rectangle.

EXPLORATION CABRI GEOMETRY

Use 4:Polygon on the F3 menu to draw a rectangle named $ABCD$. Press ENTER when you want to locate each corner of the rectangle. Your conjecture is that you have drawn a rectangle with opposite sides that are equal in length and angles that are right angles. How could you test your conjecture?

Your Turn

a. Use the Distance & Length option on the F6 menu to find the measures of the sides of $ABCD$. Then measure the angle at each corner. Is your drawing a rectangle? Explain.

b. Use the hand key to adjust the appearance of your figure so that it is a rectangle. Then draw $\overline{AC}$ and $\overline{BD}$. Make a conjecture about the relationship of the lengths $\overline{AC}$ and $\overline{BD}$. How could you test your conjecture? Is your conjecture true?

A conjecture based on several observations may be true or false.

Example 3 **Eric Pham was driving his friends to school when his car suddenly stopped two blocks away from school. Make a list of conjectures that Eric can make and investigate as to why his car stopped.**

Transportation

Some conjectures are:

- The car ran out of gas.
- The battery cable lost its contact with the battery.

Can you make some more conjectures about what might have happened to Eric's car?

Like the character in the cartoon *Shoe*, we sometimes make a conjecture and later determine that the conjecture is false. It takes only one false example to show that a conjecture is not true. The false example is called a **counterexample**.

Example 4 **Given that points P, Q, and R are collinear, Joel made a conjecture that Q is between P and R. Determine if his conjecture is *true* or *false*. Explain your answer.**

Given: Points P, Q, and R are collinear.

Conjecture: Q is between P and R.

The figure at the right can be used to disprove the conjecture. In this case, P, Q, and R are collinear and R is between Q and P. Since we can find a counterexample for the conjecture, the conjecture is false.

CHECK FOR UNDERSTANDING

Communicating Mathematics

Study the lesson. Then complete the following.

1. **Explain** the meaning of *conjecture* in your own words.
2. **Describe** why three points on a circle could never be collinear.
3. **Explain** how Skyler's reasoning about Churchhill and Einstein in the application at the beginning of the lesson led him to a false conjecture. What could he have done differently to show that the conjecture was false?
4. **Explain** how you can prove that a conjecture is false.

5. **Assess Yourself** Describe a situation in which you had several experiences that led you to make a true conjecture. Then describe a situation where you had several experiences that led you to make a false conjecture.

Guided Practice

6. Determine if the conjecture is *true* or *false* based on the given information. Explain your answer.
   Given: $\angle 1$ and $\angle 2$ are supplementary angles.
   $\angle 1$ and $\angle 3$ are supplementary angles.
   Conjecture: $\angle 2 \cong \angle 3$

Write a conjecture based on the given information. If appropriate, draw a figure to illustrate your conjecture.

7. **Given:** Lines ℓ and m are perpendicular.
8. **Given:** $A(-1, 0)$, $B(0, 2)$, $C(1, 4)$
9. Points H, I, and J are each located on different sides of a triangle. Make a conjecture about points H, I, and J.
10. Determine if the conjecture is *true* or *false*. Explain your answer and give a counterexample if the conjecture is false.
    Given: $\overline{FG} \cong \overline{GH}$
    Conjecture: G is the midpoint of $\overline{FH}$.

11. **Astronomy** For thousands of years, astronomers believed that Earth, Mercury, Venus, Mars, Jupiter, and Saturn were the only planets in the Solar System. That conjecture was proved to be false when Uranus, Neptune, and Pluto were discovered during the last couple of centuries. Name some other famous conjectures that proved to be false.

EXERCISES

Practice

Determine if each conjecture is *true* or *false* based on the given information. Explain your answer.

12. **Given:** points D, E, F, and G
    Conjecture: D, E, F, and G are noncollinear.
13. **Given:** collinear points X, Y, and Z; Z is between X and Y.
    Conjecture: $XY + YZ = XZ$
14. **Given:** noncollinear points L, M, and N
    Conjecture: $\overline{LM}$, $\overline{MN}$, and $\overline{LN}$ form a triangle.

Write a conjecture based on the given information. If appropriate, draw a figure to illustrate your conjecture.

15. **Given:** Points A, B, and C are noncollinear.
16. **Given:** $AB = CD$ and $CD = EF$.
17. **Given:** $\overleftrightarrow{XY}$ and $\overleftrightarrow{ZW}$ intersect at A.
18. **Given:** $\angle 1$ and $\angle 2$ are adjacent.
19. **Given:** $R(3, -4)$, $S(-2, -4)$, $T(0, -4)$

Algebra

20. **Given:** $x > y, y > 5$

Make a conjecture about points *A, B, C,* and *D* based on the given information.

21. $\overrightarrow{DB}$ is an angle bisector of $\angle ABC$.

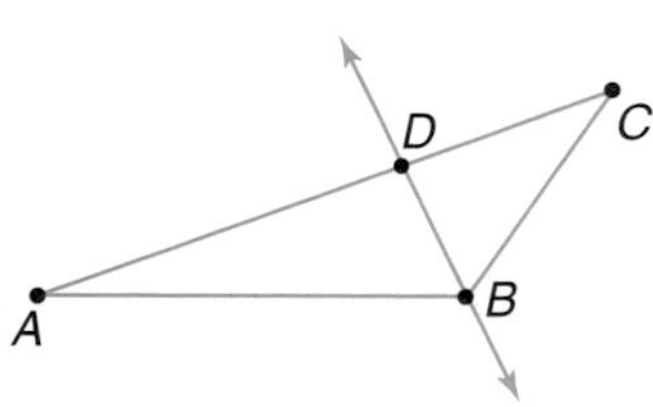

22. $ABCD$ is a square.

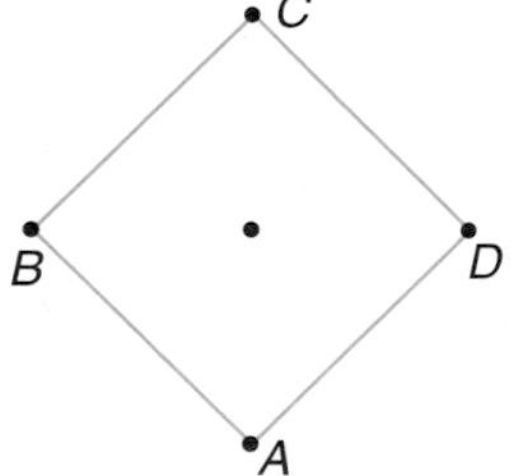

Determine if each conjecture is *true* or *false*. Explain your answer and give a counterexample for any false conjecture.

23. **Given:** $WX = XY$
    Conjecture: W, X, and Y are collinear.
24. **Given:** $PQRS$ is a rectangle.
    Conjecture: $PQ = RS$ and $QR = SP$
25. **Given:** $K(-1, 0)$, $L(1, 1)$, $M(5, 3)$
    Conjecture: Points K, L, and M form a triangle.
26. **Given:** x is an integer.
    Conjecture: $-x$ is negative.

Write a conjecture based on the given information. Draw a figure to illustrate your conjecture. Write a sentence or two to explain why you think your conjecture is true.

27. points P, Q, R, S, and T with no three collinear

28. $\overline{JK}$, $\overline{KL}$, $\overline{LM}$, $\overline{MN}$, and $\overline{NJ}$ with only J, K, and L collinear

Cabri Geometry

29. Use a graphing calculator to draw a square named $PQRS$. Draw $\overline{PR}$ and $\overline{QS}$.
    a. Make a conjecture about $\angle PQS$ and $\angle QSR$.
    b. How could you test your conjecture?
    c. Make another conjecture about the figure and test it. Write about your findings.

Critical Thinking

30. **Science** Eratosthenes, born around 274 B.C., made conjectures based on the position of the sun. At noon on June 21, the sun cast no shadows in the town of Syene. Eratosthenes concluded that the sun must be directly overhead. He used this conjecture and his knowledge of geometry to calculate the distance around Earth. Research various fields of science and give some examples where conjectures were helpful in proving scientific facts.

Applications and Problem Solving

31. **Billiards** Consider a carom billiard table with a length of 6 feet and a width of 3 feet. Suppose you start in the upper left-hand corner and shoot the ball at a 45° angle. Use graph paper to trace the path of the ball. Make a conjecture about shooting the ball from any corner of a table this size at a 45° angle.

32. **Recycling** The graph at the right shows the percent of aluminum cans in the U.S. recycled from 1974 to 1998. Make a list of conjectures that could be made from the data shown in the graph. What additional information might you need to know to determine the truth of your conjectures?

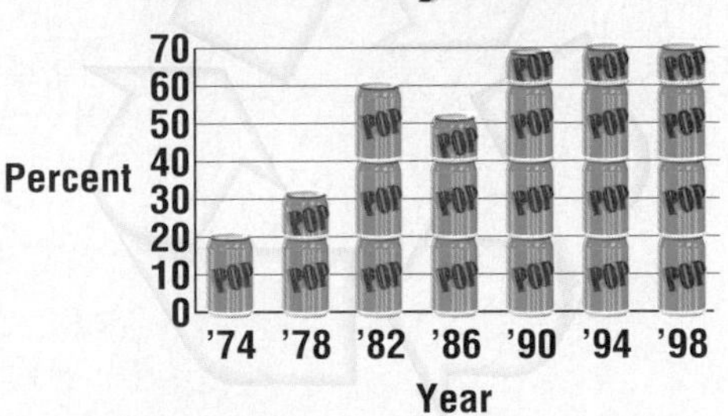

Mixed Review

33. Refer to the figure at the right. Find the value of p and determine if $\overline{DE} \perp \overline{FG}$. (Lesson 1–7)

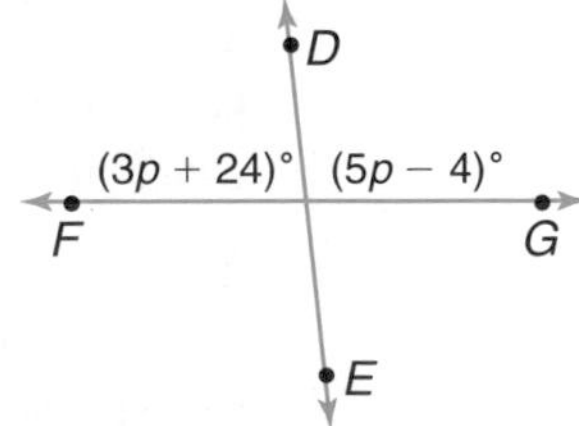

34. Suppose $\angle MON$ is a right angle and L is in the interior of $\angle MON$. If $m\angle MOL$ is five times $m\angle LON$, find $m\angle LON$. (Lesson 1–6)

35. **Cartography** On a map of Ohio, Cincinnati is located at (3, 5), and Massillon is located at (17, 19). If Columbus is halfway between the two cities, what ordered pair describes its position? (Lesson 1–5)

36. Use the distance formula to find the measure of $\overline{AB}$ with endpoints $A(5, -3)$ and $B(0, -5)$. (Lesson 1–4)

37. If the perimeter of the rectangle is 44 centimeters, find x and the dimensions of the rectangle. (Lesson 1–3)

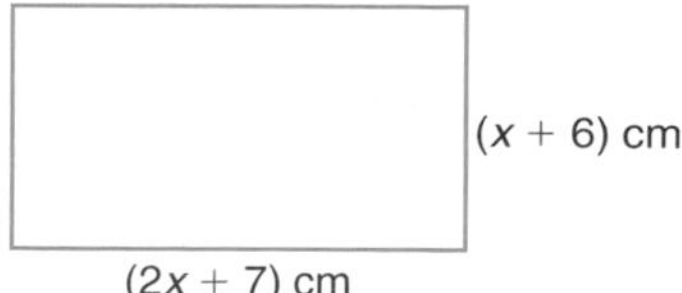

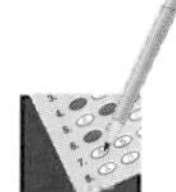

38. **ACT Practice** Which of the following expresses the prime factorization of 72?

A 9×8
B $3 \times 3 \times 8$
C $4 \times 3 \times 3 \times 3 \times 2$
D $3 \times 3 \times 2 \times 2 \times 2$
E $4 \times 3 \times 3 \times 2$

Algebra

39. Find $-\frac{1}{4} + \frac{2}{5}$.

40. **Entertainment** Refer to the graph at the right.

a. Describe the information that the graph shows.

b. Explain what the ordered pair (7, 11.8) represents.

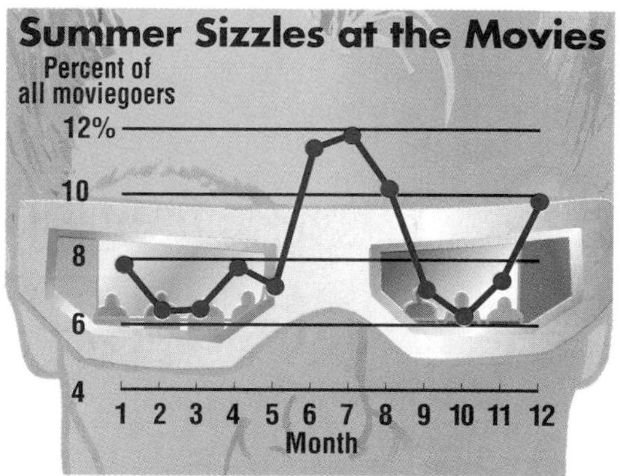

Source: *Frequent Moviegoer*

For **Extra Practice**, see page 766.

Mathematics and SOCIETY

Finding a Better Fit

The excerpt below appeared in an article in *Science News* on May 18, 1996.

Researchers are designing systems that can scan the surface of a person's body and produce an accurate three-dimensional image of it. These scanners offer the possibility of crafting quick, computerized representations that can be used to design a wide variety of items, from custom-fit clothing and shoes to automobile passenger compartments and crash helmets....To obtain a 3-D image for fitting clothes, a customer puts on a stretchy, tight-fitting garment and steps onto a platform surrounded by an array of lights and mirrors. The system passes harmless laser beams over the person's body from head to toe, while cameras record the patterns they make. The system can then put together a detailed surface map of the individual's shape in a process that takes only 17 seconds. ■

1. Make a conjecture about how the 3-D scanning process might be used to create an artificial limb for a person.
2. In the 3-D scanning process, why would you want to use cameras that view the person from several different angles and perspectives?

2–2 If-Then Statements and Postulates

What YOU'LL LEARN

- To write a statement in if-then form,
- to write the converse, inverse, and contrapositive of an if-then statement, and
- to identify and use basic postulates about points, lines, and planes.

Why IT'S IMPORTANT

Understanding if-then statements helps determine the validity of conclusions.

CONNECTION

Literature

Lewis Carroll, author of *Alice's Adventures in Wonderland* and *Through the Looking Glass,* was a mathematician as well as a writer. He was a master at creating puzzles and making connections between mathematics and literature. Following is an example of some of his statements.

Babies are illogical.
Nobody is despised who can manage a crocodile.
Illogical persons are despised.

The conclusion in the last statement may seem confusing. Rewriting each statement in "if-then" form helps to make the progression of the logic easier to understand. *A conclusion will be written in Lesson 2–3, Exercise 37.*

Example  **Write these three Lewis Carroll statements in if-then form.**

CONNECTION **Literature**

Statement	If-Then Form
Babies are illogical.	If a person is a baby, then the person is not logical.
Nobody is despised who can manage a crocodile.	If a person can manage a crocodile, then that person is not despised.
Illogical persons are despised.	If a person is not logical, then the person is despised.

If-then statements can be used to clarify statements that may seem confusing as in Example 1 above.

If-then statements are called **conditional statements** or *conditionals*. The portion of the sentence following *if* is called the **hypothesis**, and the part immediately following *then* is called the **conclusion.** *$p \rightarrow q$ represents the conditional statement "if p, then q."*

Sometimes the word then is left out of a conditional statement

Example 2 **The following statement is part of a message frequently played by radio stations across the nation. "If this had been an actual emergency, (then) the attention signal you just heard would have been followed by official information, news, or instruction." Identify the hypothesis and conclusion of the conditional.**

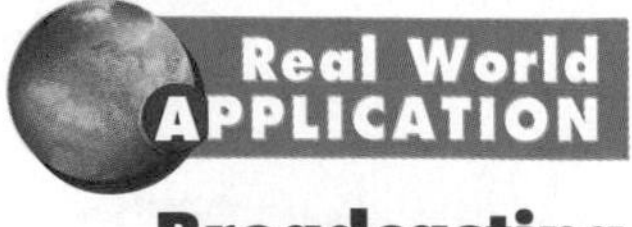

Real World APPLICATION

Broadcasting

Hypothesis: this had been an actual emergency

Conclusion: the attention signal you just heard would have been followed by official information, news, or instruction

Note that "if" is not used when you write the hypothesis and "then" is not used when you write the conclusion.

Sometimes a conditional statement is written without using the "if" and "then" such as the Lewis Carroll statements in Example 1.

Sometimes you must add information to a statement when you write it in if-then form. For example, the statement *Perpendicular lines intersect* written in if-then form is *If two lines are perpendicular, then the lines intersect.* It was necessary to know that perpendicular lines come in pairs in order for the if-then statement to be clear.

You can form another if-then statement by exchanging the hypothesis and conclusion of a conditional. This new statement is called the **converse** of the conditional. The converse of *If two lines are perpendicular, then they intersect* is *If two lines intersect, then they are perpendicular.*

Lewis Carroll

The converse of $p \rightarrow q$ is $q \rightarrow p$.

It may be easier to write a conditional in if-then form first before writing the converse. The converse of a true statement is not necessarily true.

LOOK BACK

Refer to Lesson 1-7 to review adjacent and vertical angles.

Example 3

Write the converse of the true conditional *Adjacent angles have a common side.* Determine if the converse is *true* or *false*. If false, give a counterexample.

Explore The problem has three parts.

1. Write the converse.
2. Determine if it is true or false.
3. Give a counterexample if it is not true.

Plan First, write the conditional in if-then form. Then switch the hypothesis and conclusion to write the converse.

Solve

1. The statement in if-then form is *If two angles are adjacent, then they have a common side.*
2. **Converse:** If two angles have a common side, then they are adjacent.
3. The converse of the true conditional is false since two angles with a common side are not necessarily adjacent.

Examine Draw a figure to show that the converse is false.

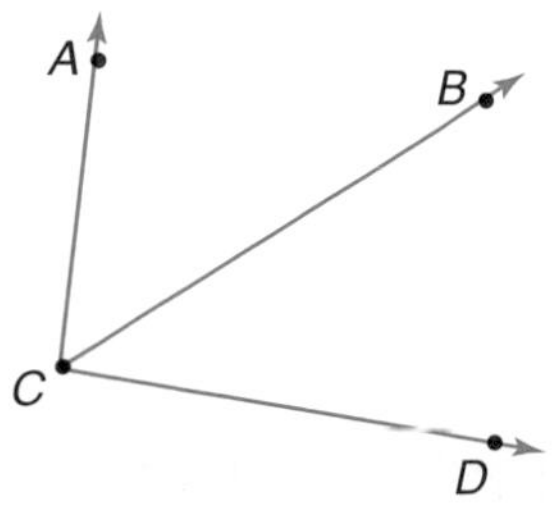

$\angle BCD$ and $\angle ACD$ have side $\overrightarrow{CD}$ in common. Yet, the two angles are not adjacent.

The denial of a statement is called a **negation**. For example, the negation of *An angle is obtuse* is *An angle is not obtuse*. If a statement is true, then its negation is false. If a statement is false, then its negation is true.

$\sim p$ represents "not p" or the negation of p.

Given a conditional statement, its **inverse** can be formed by negating both the hypothesis and conclusion. The inverse of a true statement is not necessarily true. *The inverse of $p \to q$ is $\sim p \to \sim q$.*

Example  **Write the inverse of the true conditional *Vertical angles are congruent.* Determine if the inverse is *true* or *false*. If false, give a counterexample.**

First, write the conditional in if-then form.

The hypothesis is *two angles are vertical,* and the conclusion is *the angles are congruent.* So, the if-then form of the conditional is *If two angles are vertical, then they are congruent.*

Now, negate both the hypothesis and the conclusion to form the inverse of the conditional.

Inverse: If two angles are not vertical, then they are not congruent.

The inverse of the true conditional is false since two angles may be congruent without being vertical. A counterexample is shown below.

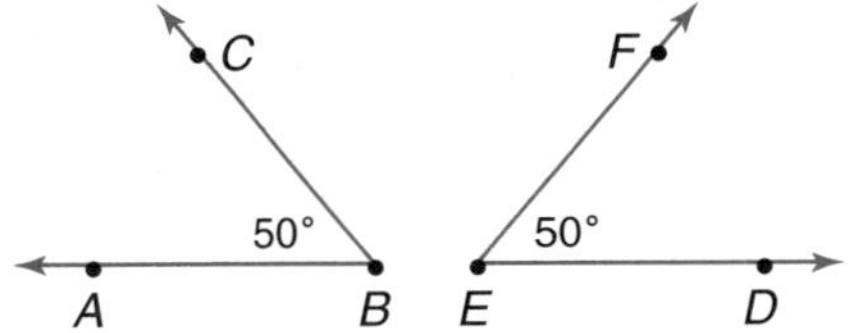

$\angle ABC \cong \angle DEF$, yet the two angles are not vertical.

Given a conditional statement, its **contrapositive** can be formed by negating the hypothesis and conclusion of the converse of the given conditional. When forming a contrapositive of a conditional it may be easier to write the converse first. *The contrapositive of $p \to q$ is $\sim q \to \sim p$.*

Example  **Write the contrapositive of the true conditional *If two angles are vertical, then they are congruent.* Determine if the contrapositive is *true* or *false*.**

Statement: If two angles are vertical, then they are congruent.

Converse: If two angles are congruent, then they are vertical.

Contrapositive: If two angles are not congruent, then they are not vertical.

The contrapositive of the conditional is true.

The contrapositive of a true conditional is always true, and the contrapositive of a false conditional is always false.

Remember that postulates are principles that are accepted to be true without proof. The following postulates describe the ways that points, lines, and planes are related.

Postulate 2–1	**Through any two points, there is exactly one line.**

Through any 2 pts., there is 1 line.

Postulate 2–2	**Through any three points not on the same line, there is exactly one plane.**

Through any 3 noncollinear pts., there is 1 plane.

The relationships between points, lines, and planes can be used to solve problems.

Example 6 **Four people meet each other for the first time. Their first names are Kamaria (*K*), Juan (*J*), Colleen (*C*), and Mara (*M*). They each shake hands with each other once. How many handshakes will there be? Draw a diagram to illustrate the solution.**

PROBLEM SOLVING

Draw a Diagram

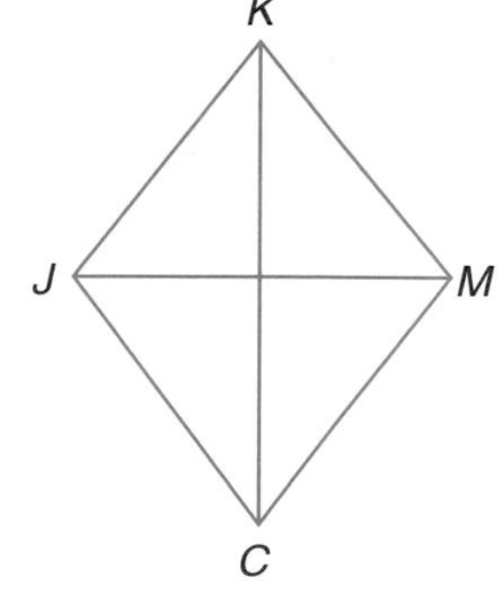

Let noncollinear points *K*, *J*, *C*, and *M* represent the people. For every two points there is exactly one line (or line segment). So for four points, there are six segments that can be drawn connecting the points.

In the figure, $\overline{KJ}$, $\overline{JC}$, $\overline{CM}$, $\overline{MK}$, $\overline{KC}$, and $\overline{JM}$ represent handshakes. Six handshakes will take place between the four people.

The next four postulates state more relationships among points, lines, and planes.

Postulate 2–3	**A line contains at least two points.**
Postulate 2–4	**A plane contains at least three points not on the same line.**
Postulate 2–5	**If two points lie in a plane, then the entire line containing those two points lies in that plane.**
Postulate 2–6	**If two planes intersect, then their intersection is a line.**

Binti and baby, Brookfield Zoo (IL)

Another type of diagram called a **Venn diagram** can be used to illustrate a conditional. A Venn diagram shows how different sets of data are related. The Venn diagram at the right shows that gorillas are a subset of the set of primates. Therefore, any animal that is a gorilla is also a primate. This diagram can also be an illustration of the conditional statement *If Binti is a gorilla, then she is a primate.*

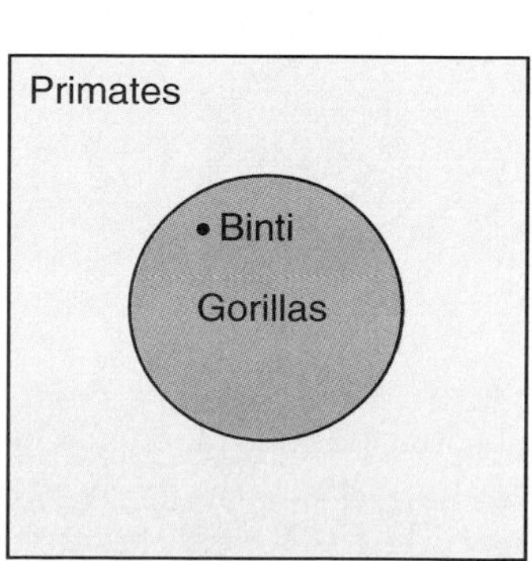

CHECK FOR UNDERSTANDING

Communicating Mathematics

Study the lesson. Then complete the following.

1. **Explain** why writing statements in if-then form is helpful.
2. **Define** *conditional statement* in your own words and give an example of a conditional.
3. **You Decide** Omar says that all squares are rectangles. Julia argues that all rectangles are squares. Minaku says that some rectangles are squares. Who is correct? Draw a diagram to help explain your answer.
4. **Discuss** similarities and differences between the inverse and the contrapositive of a conditional.

5. Find some ads in magazines or newspapers that contain if-then statements. Tape the ads in your journal. Write the converse, inverse, and contrapositive of each statement in your journal. Do you think the messages these ads convey are valid or necessarily true? Explain.

Guided Practice

Identify the hypothesis and conclusion of each conditional statement.

6. "If you can sell green toothpaste in this country, you can sell opera." *(Sarah Caldwell, 1975)*
7. If three points lie on a line, then they are collinear.

Write each conditional statement in if-then form.

8. A piranha eats other fish.
9. Angles with the same measure are congruent.

Write the negation of each statement.

10. Three points are collinear
11. Four points are noncoplanar.

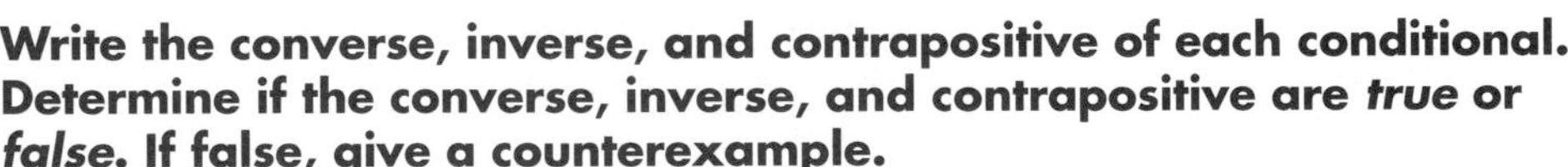

Write the converse, inverse, and contrapositive of each conditional. Determine if the converse, inverse, and contrapositive are *true* or *false*. If false, give a counterexample.

12. If you are 13 years old, then you are a teenager.
13. If an angle measures 90°, then it is a right angle.

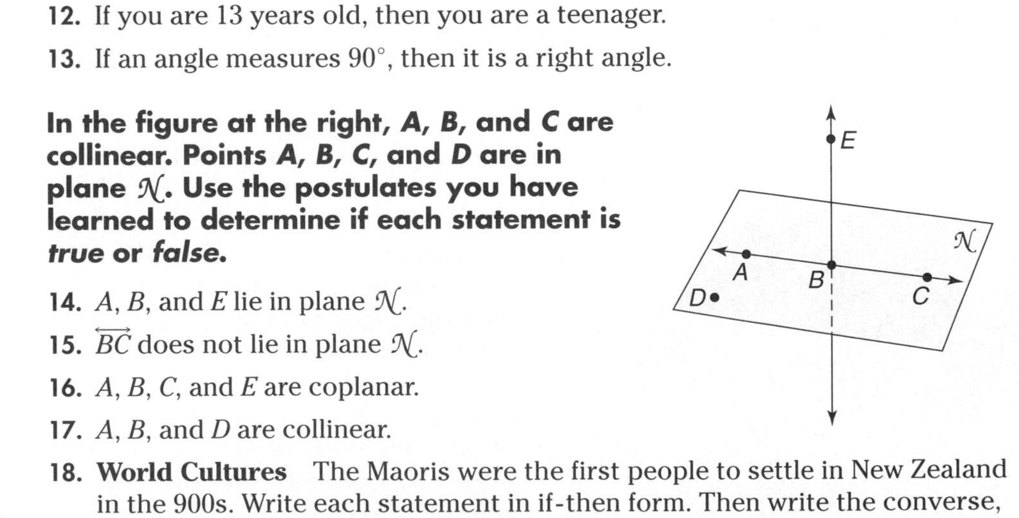

In the figure at the right, *A*, *B*, and *C* are collinear. Points *A*, *B*, *C*, and *D* are in plane $\mathcal{N}$. Use the postulates you have learned to determine if each statement is *true* or *false*.

14. A, B, and E lie in plane $\mathcal{N}$.
15. $\overleftrightarrow{BC}$ does not lie in plane $\mathcal{N}$.
16. A, B, C, and E are coplanar.
17. A, B, and D are collinear.
18. **World Cultures** The Maoris were the first people to settle in New Zealand in the 900s. Write each statement in if-then form. Then write the converse, inverse, and contrapositive of each conditional.
    a. The Maoris were outstanding sculptors.
    b. The Maori treaty house has pillars of carved wood.

EXERCISES

Practice

Identify the hypothesis and conclusion of each conditional statement.

19. "If a man hasn't discovered something that he will die for, he isn't fit to live." (*Martin Luther King, Jr., 1963*)
20. "If you don't know where you are going, you will probably end up somewhere else." (*Laurence Peters, 1969*)
21. "If we would have new knowledge, we must get a whole world of new questions." (*Susanne K. Langer, 1957*)
22. If you are an NBA basketball player, then you are at least 5′ 2″ tall.
23. If $3x - 5 = -11$, then $x = -2$.
24. If you are an adult, then you are at least 21 years old.

Write each conditional in if-then form.

25. "Happy people rarely correct their faults." (*La Rochefoucauld, 1678*)
26. "A champion is afraid of losing." (*Billie Jean King, 1970s*)
27. Adjacent angles have a common vertex.
28. Equiangular triangles are equilateral.
29. Angles whose measures are between 90 and 180 are obtuse angles.
30. Perpendicular lines form right angles.

Write the negation of each statement.

31. A book is a mirror.
32. Right angles are not acute angles.
33. Rectangles are not squares.
34. A cardinal is not a dog.
35. You live in Dallas.

Write the converse, inverse, and contrapositive of each conditional. Determine if the converse, inverse, and contrapositive are *true* or *false*. If false, give a counterexample.

36. All squares are quadrilaterals.
37. Three points not on the same line are noncollinear.
38. If a ray bisects an angle, then the two angles formed are congruent.
39. Acute angles have measures less than 90.
40. Vertical angles are congruent.
41. If you don't live in Chicago, then you don't live in Illinois.

In the figure at the right, *A*, *B*, and *C* are collinear. Points *A* and *X* lie in plane $\mathcal{M}$. Points *B* and *Z* lie in plane $\mathcal{N}$. Determine whether each statement is *true* or *false*.

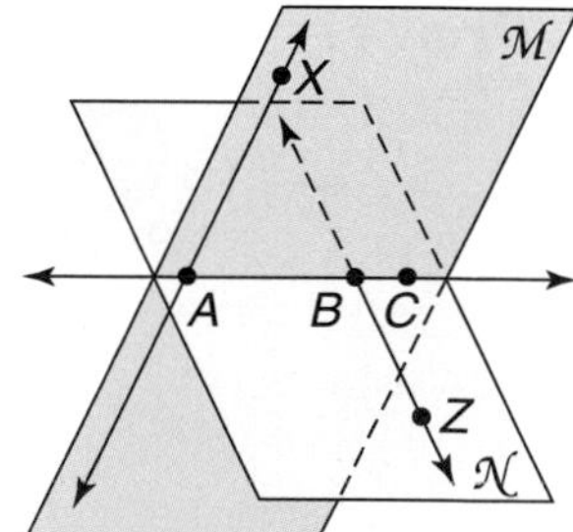

42. B lies in plane $\mathcal{M}$.
43. A, B, and C lie in plane $\mathcal{M}$.
44. A, B, X, and Z are coplanar.
45. $\overleftrightarrow{BZ}$ lies in plane $\mathcal{N}$.

State the number of lines that can be drawn containing each set of points taken two at a time. Draw a figure for each.

46. three collinear points
47. three noncollinear points
48. four points, no three of which are collinear

State the number of planes that can be drawn that contain the given set of points.

49. three noncollinear points

50. a line and a point not on the line

Programming

51. The TI-82/83 graphing calculator program at the right finds the volume of a cylinder, given the height and radius.

```
PROGRAM: CYLINDER
:Input "HEIGHT?",H
:Input "RADIUS?",R
:πR²H→V
:Disp "VOLUME IS",V
```

Suppose the height of a cylinder is 8 centimeters and the radius is 3 centimeters. Fill in the blanks by making a conjecture. Then use the program to test your conjectures.

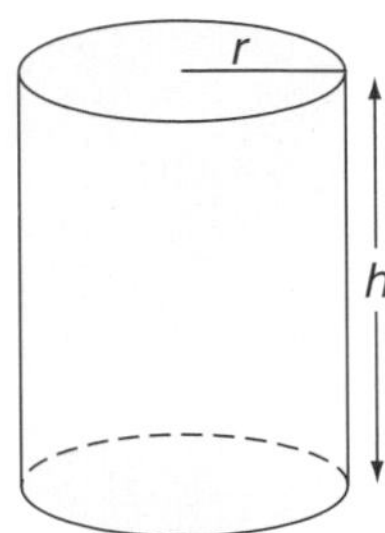

a. If the height of the cylinder is doubled and the radius remains the same, then the volume ______.

b. If the radius of the cylinder is doubled and the height remains the same, then the volume ______.

c. If the radius and the height of the cylinder are both doubled, then the volume ______.

Critical Thinking

52. Consider the conditional *If two angles are adjacent, they are not both acute.* Write the converse of the contrapositive of the inverse of the conditional. Explain how the result is related to the original conditional.

53. **Literature** The following quote is from Lewis Carroll's *Alice's Adventures in Wonderland.*

"Then you should say what you mean," the March Hare went on.

"I do," Alice hastily replied; "at least—at least I mean what I say—that's the same thing, you know."

"Not the same thing a bit!" said the Hatter. "Why, you might just as well say that 'I see what I eat' is the same thing as 'I eat what I see'!"

a. Who is right, Alice or the Hatter? Explain your reasoning.

b. How are the statements *say what you mean* and *mean what you say* related to each other?

Applications and Problem Solving

54. **Biology** Use a Venn diagram to illustrate the following conditional about the animal kingdom. *If an animal is a butterfly, then it is an arthropod.*

55. **Advertising** Advertising writers frequently use if-then statements to relay a message and promote their product. An ad for a type of Mexican food reads, *If you're looking for a fast, easy way to add some fun to your family's menu, try Casa Fiesta.*

a. Write the converse of the conditional.

b. What do you think the advertiser wants people to conclude about Casa Fiesta products?

c. Does the advertisement say that Casa Fiesta adds fun to your family's menu?

56. **Draw a Diagram** Eleven students are riding bumper cars at a carnival. They are playing a game in which if one car bumps another car in the back, the car bumped in the back is out of the game. How many collisions will there be before there is a winner? Name the bumper cars with points *A* through *K* for your drawing.

Mixed Review

57. Write a conjecture given that $\angle C$ and $\angle D$ are right angles. (Lesson 2–1)

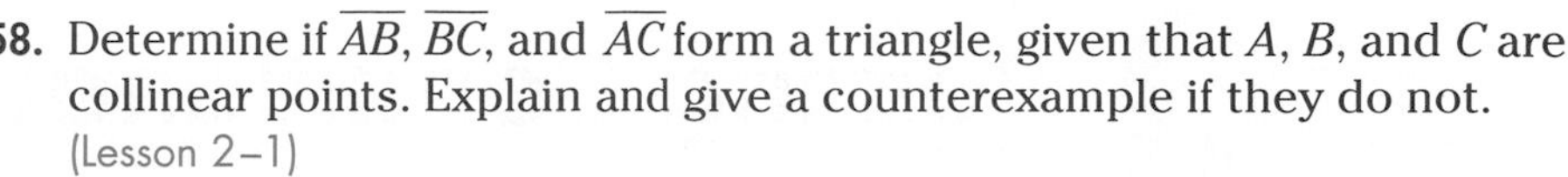

58. Determine if $\overline{AB}$, $\overline{BC}$, and $\overline{AC}$ form a triangle, given that A, B, and C are collinear points. Explain and give a counterexample if they do not. (Lesson 2–1)

59. **SAT Practice** If 6 and 9 are factors of P, what is the value of P?
A 18 **B** 27 **C** 36 **D** 54
E It cannot be determined from the information given.

60. Angles AND and NOR are complementary. If $m\angle AND = 4m\angle NOR$, find the measures of the angles. (Lesson 1–7)

61. Is a 67° angle acute, obtuse, right, or straight? (Lesson 1–6)

62. Find the coordinates of the midpoint of $\overline{QP}$ with endpoints $Q(4, 8)$ and $P(-3, 0)$. (Lesson 1–5)

63. **Manufacturing** *Stars and Stripes Flag Company* manufactures small flags. The length of a flag must be 1.5 times the width, and the area of the flag must be between 60 square inches and 300 square inches. What are the possible whole-number dimensions for the flags? (Lesson 1–4)

64. Find the maximum area for a rectangle whose perimeter is 44 meters. (Lesson 1–3)

65. Points $R(-2, 3)$ and $S(4, 9)$ lie on the graph of $y = x + 5$. Determine whether point $T(-1, -4)$ is collinear with R and S. (Lesson 1–1)

Algebra

For **Extra Practice**, see page 766.

66. Write an algebraic expression for *15 increased by three times a number n.*

67. State whether $t^2 - 5 = 6$ is *true* or *false* for $t = 3$.

Refer to the Investigation on pages 66–67.

One popular style of geometric art is called op art, or optical art. Op art is abstract and uses straight lines or geometric patterns to create a special visual effect. Use the following steps to create an op art sample.

1 Draw three evenly-spaced points at the top, bottom, and sides of a piece of unlined paper. Choose a method for connecting some of the points. For example, connect the top corners of the page to each point on the bottom and opposite side.

2 Choose two markers and shade the regions bounded by the lines. No two regions that share a side should be the same color.

3 Create two other op art designs using different grid systems or coloring schemes.

4 Refer to your discussion notes about the mural project. Is op art a style that would be appropriate for the mural?

Add the results of your work to your Investigation Folder.

2–2B Using Technology
Testing Conditional Statements

An Extension of Lesson 2–2

A graphing calculator can be helpful in testing the validity of some conditional statements that involve mathematical sentences in one variable.

Example **Given the conditional statement *If $3x + 6 > 4x + 9$, then $x > -3$*, use a graphing calculator to verify whether the conclusion is true.**

You can use the symbols on the TEST menu.

- Use the standard viewing window by pressing ZOOM 6.
- Clear the calculator's Y= list.
- Enter $3x + 6 > 4x + 9$ as Y1. To enter the $>$ symbol, press 2nd [TEST] 3.
- Press GRAPH. A screen like the one shown at the right should appear.
- Use the TRACE function to scan the Y values along the graph.

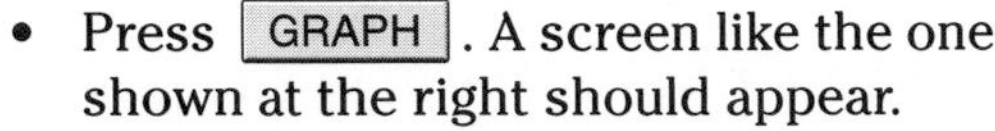

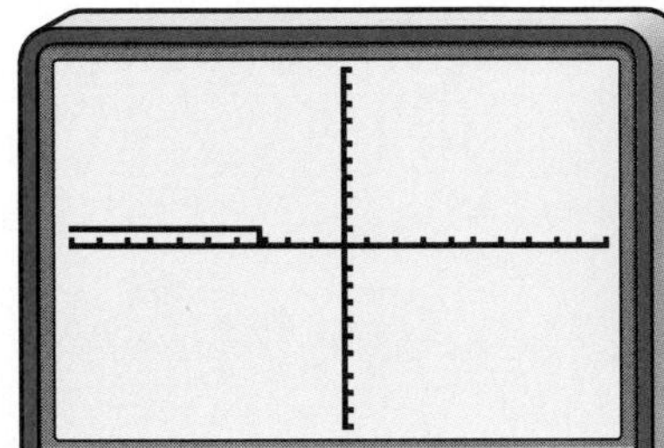

TECHNOLOGY *Tip* The TEST menu is the second function of MATH.

Notice that the value of Y is either 1 or 0. When the statement $3x + 6 > 4x + 9$ is true, the value of Y is 1. When it is false, the value of Y is 0. So, the hypothesis is true only when $x < -3$. Assuming the hypothesis $3x + 6 > 4x + 9$ is true, then the conclusion $x > -3$ is false. The correct conditional would be *If $3x + 6 > 4x + 9$, then $x < -3$.*

EXERCISES

Analyze the activity.

1. Solve the inequality in the hypothesis in the Example algebraically. How does it compare with the result from the calculator test?
2. Write a paragraph to explain how you could use the test menu to determine the validity of a solution of an equation.

Use a graphing calculator to test the validity of the conclusion in each conditional statement. Rewrite the conditional if necessary to make a true statement.

3. If $x - 9 < 5x - 1$, then $x > -2$.
4. If $4x + 19 \leq -8 + 7x$, then $x \leq 9$.
5. If $12 - 3x > 23 - 14x$, then $x < 1$.

2-3 Deductive Reasoning

What YOU'LL LEARN
- To use the Law of Detachment and the Law of Syllogism in deductive reasoning, and
- to solve problems by looking for a pattern.

Why IT'S IMPORTANT

You can use deductive reasoning to reach logical conclusions.

You can use paper folding to explore the measures of vertical angles.

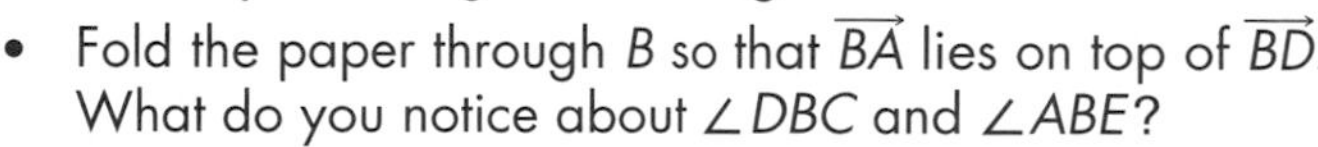

MODELING MATHEMATICS

Vertical Angles

Materials: 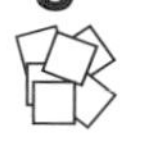 patty paper  straightedge

- Fold the paper and make a crease. Fold the paper again so that the second fold intersects the first.
- Open the paper and use a straightedge to draw the two lines formed by the creases. Label your paper similarly to the figure at the right.
- Fold the paper through B so that $\overrightarrow{BA}$ lies on top of $\overrightarrow{BD}$. What do you notice about $\angle DBC$ and $\angle ABE$?
- Refold the paper through B so that $\overrightarrow{BA}$ lies on $\overrightarrow{BE}$. What do you notice about $\angle ABD$ and $\angle EBC$?

Your Turn

Make a conjecture about the opposite pairs of angles formed by any two intersecting lines.

In the Modeling Mathematics activity above, you discovered a pattern with vertical angles and made a conclusion based on the pattern. **Looking for a pattern** is a good way to help you make a conjecture.

Example 1 **Find the number of angles formed by 10 distinct rays with a common endpoint.**

Look for a Pattern

Explore A figure drawn with 10 rays may look confusing. Instead, we can draw a series of figures with fewer rays.

no angle

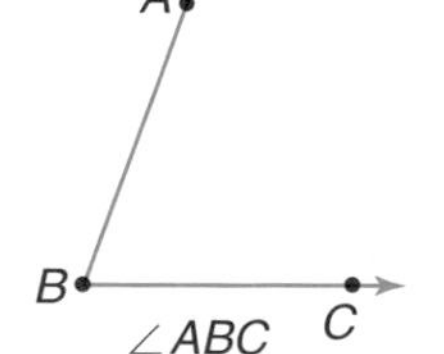

$\angle ABC$

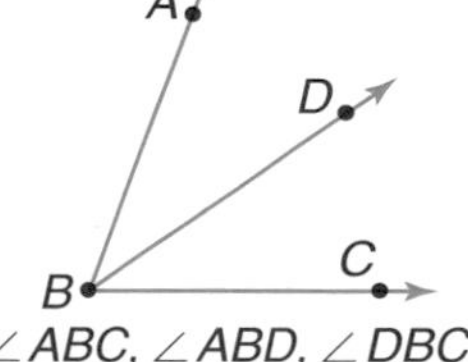

$\angle ABC$, $\angle ABD$, $\angle DBC$

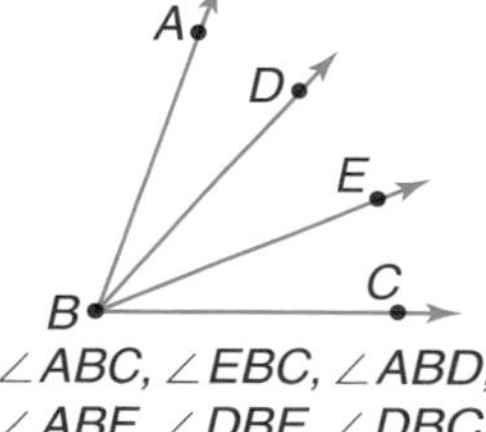

$\angle ABC$, $\angle EBC$, $\angle ABD$, $\angle ABE$, $\angle DBE$, $\angle DBC$

Plan Make a table. Look for a pattern in the number of angles formed.

Solve Record the number of angles in each figure.

number of rays	1	2	3	4	5	6
number of angles formed	0	1	3	6	10	?

+1 +2 +3 +4 +?

(continued on the next page)

After drawing a figure with 5 rays, you may conclude that a figure with 6 rays forms 10 + 5 or 15 angles, a figure with 7 rays forms 15 + 6 or 21 angles, and so on until a figure with 10 rays forms 36 + 9 or 45 angles.

Examine To check your conjecture that a figure with 10 rays forms 45 angles, you could draw a figure with 10 rays and count them all. Although this may be confusing and tedious, you would find that your conjecture is true. You would also discover that looking for a pattern is an easier way to solve the problem.

Looking for and finding a pattern does not guarantee that a statement is always true. It is necessary to prove a statement in order to assume that it is absolutely true for all cases. Even if we listed 1000 examples in which vertical angles were congruent, it would not be enough to assume that it is true for all cases. We will prove the theorem in Lesson 2–6. In order to prove statements, we need some "tools" of the trade.

In the June 3, 1996 issue of a weekly magazine, Della Reese made the following statement about Roma Downey who stars in a TV drama.

If-Then Statement:
"If you can't look at Roma and see that she is sweet, caring, and tender, then I don't know what I can tell you."

True statement about the hypothesis:
A person can't look at Roma and see that she is sweet, caring, and tender.

True Conclusion: Della Reese doesn't know what she can tell that person.

This form of reasoning is used in proof and is called the **Law of Detachment**. The Law of Detachment offers us a way to draw conclusions from if-then statements. It states that whenever a conditional is true and its hypothesis is true, we can assume that its conclusion is true.

Law of Detachment	**If $p \rightarrow q$ is a true conditional and p is true, then q is true.**

Deductive reasoning uses a rule to make a conclusion. Inductive reasoning uses examples to make a conjecture or rule.

The Law of Detachment and other laws of logic can be used to provide a system for reaching logical conclusions. This system is called **deductive reasoning**. Deductive reasoning is one of the cornerstones of the study of geometry.

Example  ***If two numbers are odd, then their sum is even*** **is a true conditional, and 3 and 5 are odd numbers. Use the Law of Detachment to reach a logical conclusion.**

The hypothesis is *two numbers are odd*. 3 and 5 are indeed two odd numbers. Since the conditional is true and the given statement satisfies the hypothesis, the conclusion is true. So, the sum of 3 and 5 must be even.

Knowing that a conditional is true and that its conclusion is true does not allow us to say the hypothesis is true. Consider a counterexample from Example 2. The sum of 8 and 12 is 20, which is an even number, but 8 and 12 are not two odd numbers.

Example 3

Sports

Determine if statement (3) follows from statements (1) and (2) by the Law of Detachment. *Assume that statements (1) and (2) are true.*

(1) If you played baseball in the '96 Summer Olympics, then you played baseball in Atlanta in the summer of '96.

(2) Greg Maddux (Atlanta Braves) played baseball in Atlanta in the summer of '96.

(3) Greg Maddux played baseball in the '96 Summer Olympics.

Hypothesis: you played baseball in the '96 Summer Olympics

Conclusion: you played baseball in Atlanta in the summer of '96

Given: Greg Maddux played baseball in Atlanta in the summer of '96.

A tendency may be to conclude that Greg Maddux played baseball in the '96 Summer Olympics. In fact, all of the Atlanta Braves players played baseball in Atlanta in the summer of '96, but none of them played in the Olympics. The given information satisfied the conclusion rather than the hypothesis. This led to the invalid conclusion even though the conditional was true and the given was true.

A second law of logic is the **Law of Syllogism**. It is similar to the Transitive Property of Equality that you should remember from your studies in algebra.

Law of Syllogism	**If $p \rightarrow q$ and $q \rightarrow r$ are true conditionals, then $p \rightarrow r$ is also true.**

Example 4

Ancient Artifacts

The Noks were one of Africa's earliest civilizations, dating back to 500 B.C. Determine if a valid conclusion can be reached from the two true statements *If it is a Nok sculpture, then it has hollowed-out eyes and mouth* and *If a sculpture has hollowed out eyes and mouth, then it has air vents that prevented cracking.*

Let p, q, and r represent the parts of the statements.

p: it is a Nok sculpture

q: it has hollowed-out eyes and mouth

r: it has air vents that prevented cracking

Using these letters, the given statements can be represented as $p \rightarrow q$ and $q \rightarrow r$. Since the given statements are true, we can use the Law of Syllogism to conclude $p \rightarrow r$. That is, *If it is a Nok sculpture, then it has air vents that prevented cracking.*

CHECK FOR UNDERSTANDING

Communicating Mathematics

Study the lesson. Then complete the following.

1. **State** which property in algebra is similar to the Law of Syllogism. Explain.
2. "Those who choose Tint-and-Trim Hair Salon have impeccable taste; and you have impeccable taste" is an example of how an advertiser can misuse the Law of Detachment to make you come to an invalid conclusion.
   a. What conclusion do they want you to make?
   b. Write another example that illustrates incorrect logic.

3. **Write** your own example to illustrate the correct use of the Law of Detachment.
4. **You Decide** Marlene said, "The volleyball coach said that if I wanted to be on the varsity team, then I had to practice more on the weekends. Since I practice every weekend now, I'm going to make the team." Nina pointed out, "But you might not make the varsity team this season. You might have a better chance next year." Whose reasoning is correct? Explain.

5. Make up several logic puzzles. Have your friends try to solve them. Write the puzzles and solutions in your journal.

Guided Practice

Determine if statement (3) follows from statements (1) and (2) by the Law of Detachment or the Law of Syllogism. If it does, state which law was used. If it does not, write *invalid*.

6. (1) If you plan to attend the University of Notre Dame, then you need to be in the top 10% of your class.
   (2) Rosita Nathan plans to attend Notre Dame.
   (3) Rosita Nathan needs to be in the top 10% of her class.
7. (1) If an angle has a measure less than 90, it is acute.
   (2) If an angle is acute, then its supplement is obtuse.
   (3) If an angle has a measure less than 90, then its supplement is obtuse.
8. (1) Vertical angles are congruent.
   (2) $\angle 1 \cong \angle 2$
   (3) $\angle 1$ and $\angle 2$ are vertical.

Determine if a valid conclusion can be reached from the two true statements using the Law of Detachment or the Law of Syllogism. If a valid conclusion is possible, state it and the law that is used. If a valid conclusion does not follow, write *no conclusion*.

9. (1) If you want good health, then you should get 8 hours of sleep each day.
   (2) Patricia Gorman wants good health.
10. (1) If $AB = BC$ and $BC = CD$, then $AB = CD$.
    (2) $AB = CD$
11. (1) If the measure of an angle is less than 90, then it is acute.
    (2) If an angle is acute, then it is not obtuse.
12. (1) If there are two points, then there is exactly one line that contains them.
    (2) There exist two points A and B.

13. **Music** The October/November, 1995 issue of *Zillions* magazine featured a music review of a Beatles album. It stated, "If you're a true Beatles fan, you'll love this, and if not, you'll become one." Determine if a possible conclusion can be reached, given each true statement. If a conclusion is possible, state it.
    a. You're not a true Beatles fan.
    b. You'll love this album.
    c. You'll become a true Beatles fan.
    d. You're a true Beatles fan.

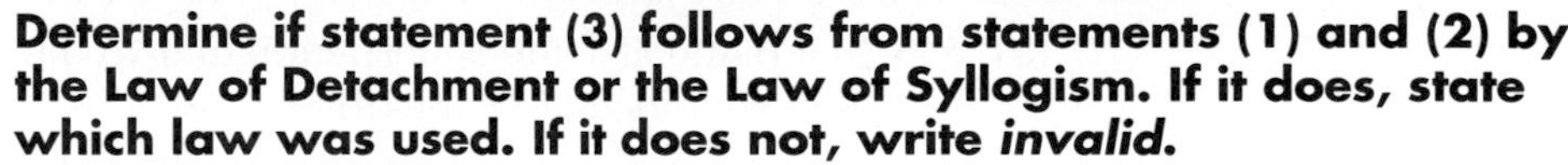

EXERCISES

Practice

Determine if statement (3) follows from statements (1) and (2) by the Law of Detachment or the Law of Syllogism. If it does, state which law was used. If it does not, write *invalid*.

14. (1) In-line skaters live dangerously.
(2) If you live dangerously, then you like to dance.
(3) If you are an in-line skater, then you like to dance.

15. (1) If you drive safely, the life you save may be your own.
(2) Shani drives safely.
(3) The life she saves may be her own.

16. (1) If a figure is a rectangle, then its opposite sides are congruent.
(2) $\overline{AB} \cong \overline{DC}$ and $\overline{AD} \cong \overline{BC}$
(3) $ABCD$ is a rectangle.

17. (1) Right angles are congruent.
(2) $\angle A \cong \angle B$
(3) $\angle A$ and $\angle B$ are right angles.

18. (1) If an angle is obtuse, then it is not acute.
(2) $\angle 1$ is obtuse.
(3) $\angle 1$ is not acute.

19. (1) If you are a customer, then you are always right.
(2) If you are a teenager, then you are always right.
(3) If you are a teenager, then you are a customer.

20. (1) Vertical angles are congruent.
(2) If two angles are congruent, then their measures are equal.
(3) If two angles are vertical, then their measures are equal.

21. (1) If you do well, then you like criticism.
(2) If you resent criticism, then you do not do well.
(3) If you like criticism, then you do not do well.

Determine if a valid conclusion can be reached from the two true statements using the Law of Detachment or the Law of Syllogism. If a valid conclusion is possible, state it and the law that is used. If a valid conclusion does not follow, write *no conclusion*.

22. (1) If two angles are vertical, then they do not form a linear pair.
(2) If two angles are vertical, then they are congruent.

23. (1) If you eat to live, then you live to eat.
(2) Odina eats to live.

24. (1) If a plane exists, then it contains at least three points not on the same line.
(2) Plane $\mathcal{N}$ contains points A, B, and C, which are not on the same line.

25. (1) If M is the midpoint of $\overline{AB}$, then $AM = MB$.
(2) If the measures of two segments are equal, then they are congruent.

26. (1) If an angle is obtuse, then its measure is greater than 90.
(2) $\angle 2$ is obtuse.

27. (1) If you spend money on it, then it's a business.
(2) If you spend money on it, then it's fun.

28. (1) If two lines intersect to form a right angle, then they are perpendicular.
(2) ℓ and m are perpendicular.

29. (1) If two planes intersect, then their intersection is a line.
(2) Planes $\mathcal{M}$ and $\mathcal{N}$ intersect.

30. (1) If two angles of a triangle are congruent, then the sides opposite these angles are congruent.
(2) If two sides of a triangle are congruent, then the triangle is isosceles.

31. (1) In a plane, if a line is perpendicular to one of two parallel lines, then it is perpendicular to the other.
(2) Line t is perpendicular to line p, which is parallel to line q in the same plane as t.

32. (1) Cars are useful.
(2) Useful cars are practical.

33. (1) If two points lie in a plane, then the entire line containing those two points lie in that plane.
(2) Points X and Y lie in plane $\mathcal{P}$.

Using the given statement, create a second statement and a valid conclusion that illustrates the correct use of the Law of Detachment. Then write a statement and a conclusion that illustrate the correct use of the Law of Syllogism.

34. If you're a careful bicycle rider, then you wear a helmet.
35. If you like pizza with everything, then you'll like Jimmy's Pizza.
36. If two angles form a linear pair, then they share a common ray.

Critical Thinking

37. The following if-then statements were used in the application at the beginning of Lesson 2–2. Assume all are true. Use deductive reasoning laws to write a true conclusion. You must use all three statements. Explain all steps used to arrive at your conclusion. (*Hint:* Remember if a conditional is true, then its contrapositive is true.)
If a person is a baby, then the person is not logical.
If a person can manage a crocodile, then that person is not despised.
If a person is not logical, then the person is despised.

Applications and Problem Solving

38. **Airline Safety** The sign at the the right is posted in airports throughout the U.S. Provide information necessary to illustrate logical reasoning using the Law of Detachment with this if-then statement.

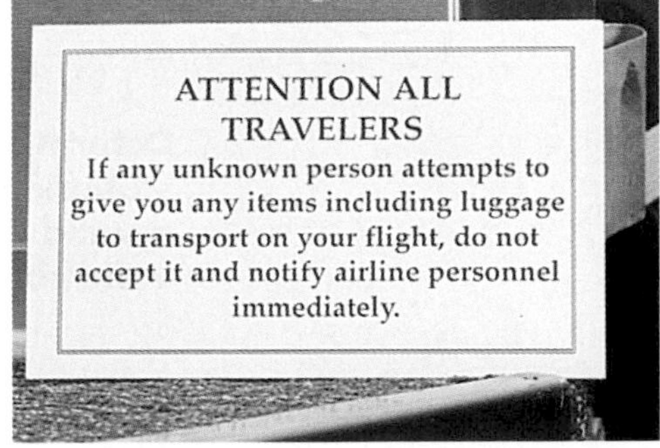

ATTENTION ALL TRAVELERS
If any unknown person attempts to give you any items including luggage to transport on your flight, do not accept it and notify airline personnel immediately.

39. **Look for a Pattern** What is the ones digit of 9^{46}?

If you can answer "YES" to these three questions, you probably qualify for savings of up to 70% on your life insurance.	1. Are you a non-smoker?........... ☐YES ☐NO 2. Do you watch your weight and your fat intake?........... ☐YES ☐NO 3. Are you free of any major health problems, such as heart disease?............ ☐YES ☐NO

40. **Advertising** The magazine ad at the left appeared in *Kiplinger's Personal Finance Magazine*. Describe how the Law of Detachment could be used to arrive at a logical conclusion, assuming the statements in the ad are true.

Mixed Review

41. Write the converse of the conditional *If two lines are perpendicular, then they intersect.* Determine if the converse is *true* or *false*. If false, give a counterexample. (Lesson 2–2)

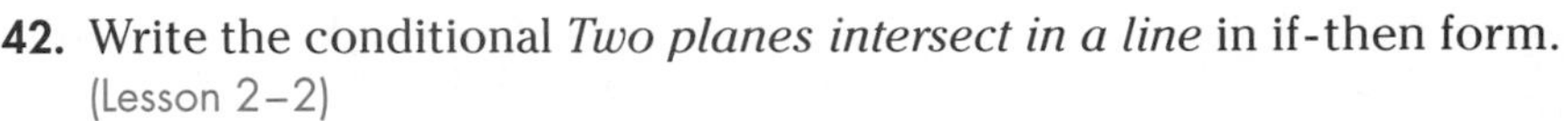

42. Write the conditional *Two planes intersect in a line* in if-then form. (Lesson 2–2)

43. Given $\overline{AB}$, $\overline{BC}$, and $\overline{CD}$, write a conjecture. Draw a figure to illustrate your conjecture. (Lesson 2–1)

44. **SAT Practice** **Grid-in** What is the sum of the positive odd factors of 30?

45. Use a protractor to draw a 65° angle. (Lesson 1–6)

46. Q is between R and S, $RQ = 6x - 1$, $QS = 2x + 4$, and $RS = 9x - 3$. Find RS. (Lesson 1–4)

47. Find the length of the segment with endpoints $B(4, -1)$ and $C(2, 5)$. (Lesson 1–4)

Algebra

48. Find $3.8 + (-4.7)$.

49. **Sales** Carmen bought a pair of shoes for \$36.78 including tax. If she gave the clerk 2 twenty-dollar bills, how much should Carmen receive in change?

For **Extra Practice**, see page 767.

SELF TEST

Determine if the conjecture is *true* or *false* based on the given information. Explain your answer. (Lesson 2–1)

1. **Given:** points A, B, and C
   Conjecture: A, B, and C are collinear.

2. **Given:** $\angle A$ and $\angle C$ are complementary angles.
   $\angle B$ and $\angle C$ are complementary angles.
   Conjecture: $\angle A \cong \angle B$

3. **Botany** In the 1920s, some Japanese farmers observed that certain rice plants were growing taller and thinner than normal rice plants and then drooping over, making them impossible to harvest. (Lesson 2–1)
   a. Make some conjectures about why the plants were drooping.
   b. The scientists who researched the problem discovered that fungus was growing on the drooping rice plants, while the healthy plants had no fungus. Does this observation prompt a new conjecture?
   c. How might the scientists have tested the conjecture they made?

4. Identify the hypothesis and conclusion of the statement *If you're there before it's over, then you're on time.* (Lesson 2–2)

5. Write the converse of the statement *If it is raining, then there are clouds.* (Lesson 2–2)

6. Write the inverse of the statement *If two lines are parallel, then they do not intersect.* (Lesson 2–2)

7. Write the contrapositive of the statement *If a figure is a square, then it has four sides.* (Lesson 2–2)

Determine if statement (3) follows from statements (1) and (2). If it does, state which law was used. If it does not, write *invalid*. (Lesson 2–3)

8. (1) If you are not satisfied with a tape, then you can return it within a week for a full refund.
   (2) Yong is not satisfied with a tape.
   (3) Yong can return the tape within a week for a full refund.

9. (1) If x is a real number, then x is an integer.
   (2) x is an integer.
   (3) x is a real number.

10. (1) If fossil fuels are burned, then acid rain is produced.
    (2) If acid rain is produced, wildlife suffers.
    (3) If fossil fuels are burned, then wildlife suffers.

2-4

Integration: Algebra
Using Proof in Algebra

What YOU'LL LEARN

- To use properties of equality in algebraic and geometric proofs.

Why IT'S IMPORTANT

You can use proof to solve problems in algebra.

Exercise

A recent study by the University of Massachusetts reveals that in-line skating at moderate speeds burns as many Calories as running. The chart below shows the Calories burned per minute at various skating speeds for various body weights.

Calories Burned Per Minute

Weight (lb)	Skating Speed				
	8 mph	9 mph	10 mph	11 mph	12 mph
120	4.2	5.8	7.4	8.9	10.5
140	5.1	6.7	8.3	9.9	11.4
160	6.1	7.7	9.2	10.8	12.4
180	7.0	8.6	10.2	11.7	13.3
200	7.9	9.5	11.2	12.6	14.2

You are familiar with many of the properties of algebra. These properties, along with various defined operations and sets of numbers, form a mathematical system. Working within the properties of the system allows you to perform algebraic operations.

For example, you can use the Multiplication Property of Equality to determine your skating speed in miles per hour. Use the formula $\frac{s}{d} = \frac{60}{t}$, where s represents the average speed (mph), d represents the distance (mi), 60 represents 60 minutes in an hour, and t represents time skating (min).

The Multiplication Property of Equality allows you to multiply each side of the equation by d to solve the equation in terms of s.

$$\frac{s}{d} \cdot d = \frac{60}{t} \cdot d$$

$$s = \frac{60}{t} \cdot d$$

So, if you skated 3 miles in 20 minutes, then the equation would become $s = \frac{60}{20} \cdot 3$ or $s = 9$ mph. You skated 9 mph for 20 minutes. Now use the table to determine the Calories burned. If you weighed 120 pounds, you would have burned $20 \cdot 5.8$ or 116 Calories.

Another example of a mathematical system can be found in geometry. Since geometry also deals with variables, numbers, and operations, many of the properties of algebra are also true in geometry. Some of the other important properties of algebra are listed in the following table.

Properties of Equality for Real Numbers	
Reflexive Property	For every number a, $a = a$.
Symmetric Property	For all numbers a and b, if $a = b$, then $b = a$.
Transitive Property	For all numbers a, b, and c, if $a = b$ and $b = c$, then $a = c$.
Addition and Subtraction Properties	For all numbers a, b, and c, if $a = b$, then $a + c = b + c$ and $a - c = b - c$.
Multiplication and Division Properties	For all numbers a, b, and c, if $a = b$, then $a \cdot c = b \cdot c$, and if $c \neq 0$, $\frac{a}{c} = \frac{b}{c}$.
Substitution Property	For all numbers a and b, if $a = b$, then a may be replaced by b in any equation or expression.
Distributive Property	For all numbers a, b, and c, $a(b + c) = ab + ac$.

Since segment measures and angle measures are real numbers, these properties from algebra can be used to discuss their relationships. Some examples of these applications are shown below.

Property	Segments	Angles
Reflexive	$PQ = PQ$	$m\angle 1 = m\angle 1$
Symmetric	If $AB = CD$, then $CD = AB$.	If $m\angle A = m\angle B$, then $m\angle B = m\angle A$.
Transitive	If $GH = JK$ and $JK = LM$, then $GH = LM$.	If $m\angle 1 = m\angle 2$ and $m\angle 2 = m\angle 3$, then $m\angle 1 = m\angle 3$.

Example  **Name the property of equality that justifies each statement.**

Algebra

Statements	Reasons
a. If $AB + BC = DE + BC$, then $AB = DE$.	**a.** Subtraction Property (=)
b. $m\angle ABC = m\angle ABC$	**b.** Reflexive Property (=)
c. If $XY = PQ$ and $XY = RS$, then $PQ = RS$.	**c.** Substitution Property (=)
d. If $\frac{1}{3}x = 5$, then $x = 15$.	**d.** Multiplication Property (=)
e. If $2x = 9$, then $x = \frac{9}{2}$.	**e.** Division Property (=)

You can use these properties as reasons for the step-by-step solution of an equation.

Example  **Justify each step in solving $\frac{3x + 5}{2} = 7$.**

Algebra

Statements	Reasons
1. $\frac{3x + 5}{2} = 7$	**1.** Given
2. $2\left(\frac{3x + 5}{2}\right) = 2(7)$	**2.** Multiplication Property (=)
3. $3x + 5 = 14$	**3.** Distributive Property (=)
4. $3x = 9$	**4.** Subtraction Property (=)
5. $x = 3$	**5.** Division Property (=)

Example 2 is a proof of the conditional *If* $\frac{3x + 5}{2} = 7$, *then* $x = 3$. The given information comes from the hypothesis of the conditional. It is the starting point of the proof. The conclusion, $x = 3$, is the end of the proof. Reasons (properties) listed for each step leading to the conclusion make this sequence a **proof**. This type of proof is called a **two-column proof**.

Proofs in geometry can be organized in the same manner. Algebra properties, definitions, postulates, and previously-proven theorems can be used for reasons. Proofs in geometry are usually written in two-column or in paragraph form. *Two-column proofs are sometimes called formal proofs.*

Example 3 **Justify the steps for the proof of the conditional *If PR = QS, then PQ = RS.*** *Remember that PR, QS, PQ, and RS represent real numbers.*

Given: $PR = QS$

Prove: $PQ = RS$

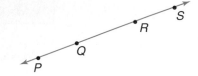

Proof:

Statements	Reasons
1. $PR = QS$	1. ?
2. $PQ + QR = PR$ $QR + RS = QS$	2. ?
3. $PQ + QR = QR + RS$	3. ? *Step 3 uses information from steps 1 and 2.*
4. $PQ = RS$	4. ?

Reason 1: Given (since it follows from the hypothesis)

Reason 2: Segment Addition Postulate

Reason 3: Substitution Property (=)

Reason 4: Subtraction Property (=)

CHECK FOR UNDERSTANDING

Communicating Mathematics

Study the lesson. Then complete the following.

1. **Describe** the parts of a two-column proof.
2. **State** the part of the conditional that is related to the *Given* statement of a proof. What part is related to the *Prove* statement?
3. **Write** a statement that illustrates the Substitution Property of Equality.
4. **List** the types of reasons that can be used to justify a statement in a geometric proof.
5. **Choose** the number of the reason in the right column that best matches each statement in the left column.

Statements	Reasons
a. If $x - 7 = 12$, then $x = 19$.	(1) Distributive Property
b. If $MK = NJ$ and $BG = NJ$, then $MK = BG$.	(2) Addition Property (=)
c. If $m\angle 4 = m\angle 5$ and $m\angle 5 = m\angle 6$, then $m\angle 4 = m\angle 6$.	(3) Symmetric Property (=)
d. If $ST = UV$, then $UV = ST$.	(4) Substitution Property (=)
e. If $x = -3(2x - 4)$, then $x = -6x + 12$.	(5) Transitive Property (=)

Guided Practice

Name the property of equality that justifies each statement.

6. If $2x = 3$, then $x = \frac{3}{2}$.
7. If $XY - AB = WZ - AB$, then $XY = WZ$.
8. If $m\angle 1 + m\angle 2 = 90$ and $m\angle 2 = m\angle 3$, then $m\angle 1 + m\angle 3 = 90$.
9. For the proof below, the reasons in the right column are not in the proper order. Reorder the reasons to properly match the statements in the left column.

Prove that if $m\angle AXC = m\angle DYF$ and $m\angle 1 = m\angle 3$, then $m\angle 2 = m\angle 4$.

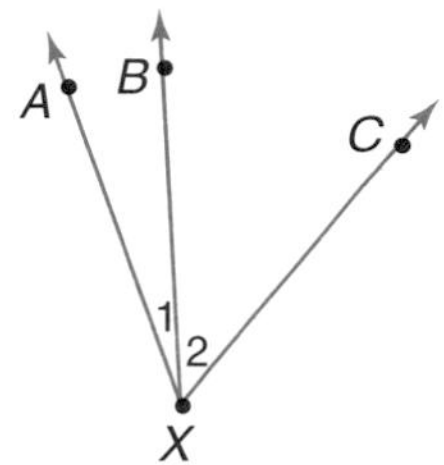

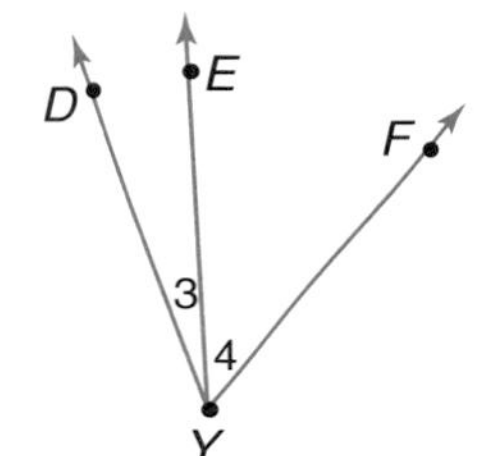

Given: $m\angle AXC = m\angle DYF$
$m\angle 1 = m\angle 3$

Prove: $m\angle 2 = m\angle 4$

Proof:

Statements	Reasons
a. $m\angle AXC = m\angle DYF$ $m\angle 1 = m\angle 3$	(1) Given
b. $m\angle AXC = m\angle 1 + m\angle 2$ $m\angle DYF = m\angle 3 + m\angle 4$	(2) Subtraction Property (=)
c. $m\angle 1 + m\angle 2 = m\angle 3 + m\angle 4$	(3) Angle Addition Postulate
d. $m\angle 3 + m\angle 2 = m\angle 3 + m\angle 4$	(4) Substitution Property (=)
e. $m\angle 2 = m\angle 4$	(5) Substitution Property (=)

10. Copy the proof. Then name the property that justifies each statement.
Prove that if $-2x + \frac{3}{2} = 8$, then $x = -\frac{13}{4}$.

Given: $-2x + \frac{3}{2} = 8$

Prove: $x = -\frac{13}{4}$

Proof:

Statements	Reasons
a. $-2x + \frac{3}{2} = 8$	**a.** ?
b. $2\left(-2x + \frac{3}{2}\right) = 2(8)$	**b.** ?
c. $-4x + 3 = 16$	**c.** ?
d. $-4x = 13$	**d.** ?
e. $x = -\frac{13}{4}$	**e.** ?

11. **Construction** When bridges, buildings, or roads are constructed, engineers must consider the expansion and contraction of materials during temperature changes. The coefficient of linear expansion k is different for every substance, and physicists find k by using the formula $k = \frac{\Delta\ell}{\ell(T - t)}$, where ℓ represents original length, $\Delta\ell$ represents change in length, T represents highest temperature, and t represents lowest temperature. Solve this formula for T and justify each step.

12. Copy and complete the proof.

Prove that if $m\angle 1 = m\angle 2$, then $m\angle PXR = m\angle SXQ$.

Given: $m\angle 1 = m\angle 2$

Prove: $m\angle PXR = m\angle SXQ$

Proof:

Statements	Reasons
a. $m\angle 1 = m\angle 2$	a. ?
b. $m\angle 3 = m\angle 3$	b. ?
c. ?	c. Addition Property (=)
d. $m\angle PXR = m\angle 2 + m\angle 3$ $m\angle SXQ = m\angle 1 + m\angle 3$	d. ?
e. ?	e. Substitution Property (=)

EXERCISES

Practice

Name the property of equality that justifies each statement.

13. If $5 = 3x - 4$, then $3x - 4 = 5$.

14. If $3\left(x - \frac{5}{3}\right) = 1$, then $3x - 5 = 1$.

15. If $0.5AB = 0.5CD$, then $AB = CD$.

16. If $m\angle 1 = 90$ and $m\angle 2 = 90$, then $m\angle 1 = m\angle 2$.

17. If $2m\angle ABC = 180$, then $m\angle ABC = 90$.

18. For XY, $XY = XY$.

19. If $EF = GH$ and $GH = JK$, then $EF = JK$.

20. If $m\angle 1 + 30 = 90$, then $m\angle 1 = 60$.

21. If $AB + IJ = MX + IJ$, then $AB = MX$.

For each proof, the reasons in the right column are not in the proper order. Reorder the reasons to properly match the statements in the left column.

Proof

22. Prove that if $2x - 7 = \frac{1}{3}x - 2$, then $x = 3$.

Given: $2x - 7 = \frac{1}{3}x - 2$

Prove: $x = 3$

Proof:

Statements	Reasons
a. $2x - 7 = \frac{1}{3}x - 2$	(1) Given
b. $3(2x - 7) = 3\left(\frac{1}{3}x - 2\right)$	(2) Distributive Property
c. $6x - 21 = x - 6$	(3) Addition Property (=)
d. $5x - 21 = -6$	(4) Multiplication Property (=)
e. $5x = 15$	(5) Division Property (=)
f. $x = 3$	(6) Subtraction Property (=)

23. Prove that if $\angle XWY \cong \angle XYW$, then $\angle AWX \cong \angle BYX$.

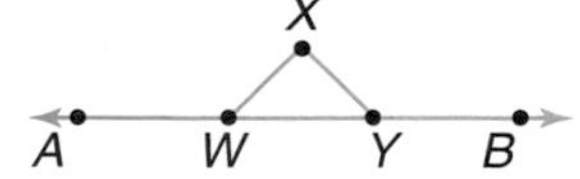

Given: $\angle XWY \cong \angle XYW$

Prove: $\angle AWX \cong \angle BYX$

Proof:

Statements	Reasons
a. $\angle XWY \cong \angle XYW$	(1) Def. supplementary ≰
b. $m\angle AWX + m\angle XWY = 180$ $m\angle BYX + m\angle XYW = 180$	(2) Given
c. $m\angle AWX + m\angle XWY =$ $m\angle BYX + m\angle XYW$	(3) Substitution Property (=)
d. $m\angle AWX + m\angle XWY =$ $m\angle BYX + m\angle XWY$	(4) Substitution Property (=)
e. $m\angle AWX = m\angle BYX$	(5) Subtraction Property (=)
f. $\angle AWX \cong \angle BYX$	(6) Def. $\cong$ ≰

Copy each proof. Then name the property that justifies each statement.

24. Prove that if $-\frac{1}{2}x = 9$, then $x = -18$.

Given: $-\frac{1}{2}x = 9$

Prove: $x = -18$

Proof:

Statements	Reasons
a. $-\frac{1}{2}x = 9$	a. ?
b. $-1x = 18$	b. ?
c. $x = -18$	c. ?

25. Prove that if $5 - \frac{2}{3}x = 1$, then $x = 6$.

Given: $5 - \frac{2}{3}x = 1$

Prove: $x = 6$

Proof:

Statements	Reasons
a. $5 - \frac{2}{3}x = 1$	a. ?
b. $3\left(5 - \frac{2}{3}x\right) = 3(1)$	b. ?
c. $15 - 2x = 3$	c. ?
d. $-2x = -12$	d. ?
e. $x = 6$	e. ?

26. Prove that if $AC = DF$ and $AB = DE$, then $BC = EF$.

A B C
D E F

Given: $AC = DF$ and $AB = DE$

Prove: $BC = EF$

Proof:

Statements	**Reasons**
a. $AC = DF$	**a.** ?
b. $AC = AB + BC$ $DF = DE + EF$	**b.** ?
c. $AB + BC = DE + EF$	**c.** ?
d. $AB = DE$	**d.** ?
e. $BC = EF$	**e.** ?

Copy and complete the following proofs.

27. If $m\angle TUV = 90$, $m\angle XWV = 90$, and $m\angle 1 = m\angle 3$, then $m\angle 2 = m\angle 4$.

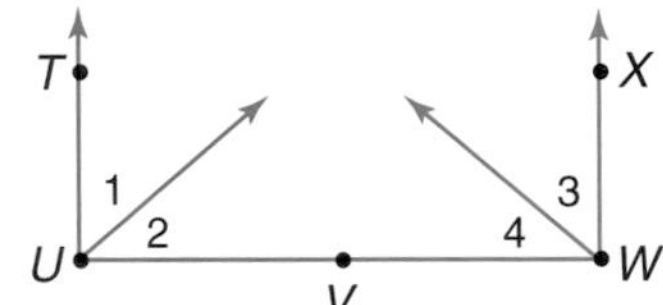

Given: $m\angle TUV = 90$, $m\angle XWV = 90$, and $m\angle 1 = m\angle 3$

Prove: $m\angle 2 = m\angle 4$

Proof:

Statements	**Reasons**
a. ?	**a.** Given
b. $m\angle TUV = m\angle XWV$	**b.** ?
c. $m\angle TUV = m\angle 1 + m\angle 2$ $m\angle XWV = m\angle 3 + m\angle 4$	**c.** ?
d. ?	**d.** Substitution Property (=)
e. $m\angle 1 + m\angle 2 = m\angle 1 + m\angle 4$	**e.** ?
f. ?	**f.** Subtraction Property (=)

28. If $4 - \frac{1}{2}x = \frac{7}{2} - x$, then $x = -1$.

Given: $4 - \frac{1}{2}x = \frac{7}{2} - x$

Prove: $x = -1$

Proof:

Statements	**Reasons**
a. $4 - \frac{1}{2}x = \frac{7}{2} - x$	**a.** ?
b. $2\left(4 - \frac{1}{2}x\right) = 2\left(\frac{7}{2} - x\right)$	**b.** ?
c. ?	**c.** Distributive Property
d. $1 - x = -2x$	**d.** ?
e. ?	**e.** Addition Property (=)
f. $-1 = x$	**f.** ?
g. $x = -1$	**g.** ?

Write a complete proof for each of the following.

29. If $2x + 6 = 3 + \frac{5}{3}x$, then $x = -9$.
30. If $AC = AB$, $AC = 4x + 1$, and $AB = 6x - 13$, then $x = 7$.

Critical Thinking

31. Discuss similarities and differences between the Transitive Property of Equality and the Transitive Property of Congruent Segments. Give an example of each property using segments and angles.

Applications and Problem Solving

32. **Physics** Kinetic energy is the energy of motion. The formula for kinetic energy is $E_k = h \cdot f + W$, where h represents Planck's Constant, f represents the frequency of its photon, and W represents the work function of the material being used. Solve this formula for f and justify each step.
33. **Photography** Film in a camera is fed through the camera by gears that peel the perforation in the film. The distance from the left edge of the film, A, to the right edge of the image, C, is the same as the distance from the left edge of the image, B, to the right edge of the film, D. Show that the two perforated strips are the same width.

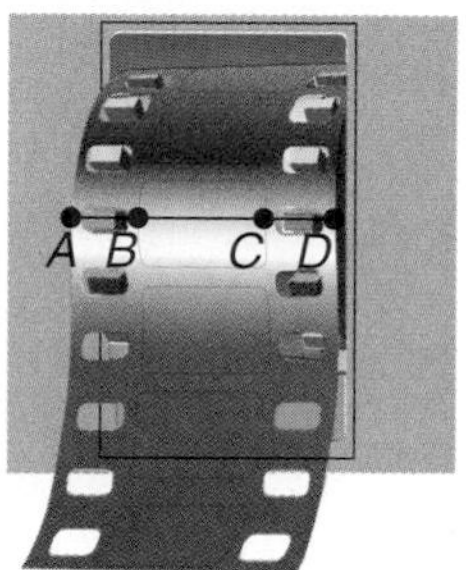

Mixed Review

34. **Ecology** Determine if a valid conclusion can be reached from the two true statements using the Law of Detachment or the Law of Syllogism. If a valid conclusion does not follow, write *no conclusion*. (Lesson 2–3)
    (1) Sponges belong to the phylum porifera.
    (2) Sponges are animals.
35. Write the inverse of the statement *If* $m\angle 1 = 27$*, then* $\angle 1$ *is acute*. Determine if the inverse is true or false. If false, give a counterexample. (Lesson 2–2)
36. Write a conjecture if $\overleftrightarrow{XY}$ and $\overleftrightarrow{ST}$ intersect at P. Draw a diagram to illustrate your conjecture. (Lesson 2–1)
37. Find the measure of the complement and the supplement of an angle that measures 159°. (Lesson 1–7)
38. **SAT Practice** **Quantitative Comparison**

Column A	Column B
x^2	$(x - 1)^2$

A if the quantity in Column A is greater
B if the quantity in Column B is greater
C if the two quantities are equal
D if the relationship cannot be determined from the information given

39. Draw and label a figure that shows lines ℓ and m intersecting at point P. (Lesson 1–2)

Algebra

Simplify each expression.

40. $6 + 3(bc - 4a) + 2bc$
41. $3(m + n) + 4m - 2(m + n)$

For **Extra Practice**, see page 767.

2-5

Verifying Segment Relationships

What YOU'LL LEARN

- To complete proofs involving segment theorems.

Why IT'S IMPORTANT

You can use segment relationships to solve problems in geography.

Geography

The U.S. mileage map below contains many line segments.

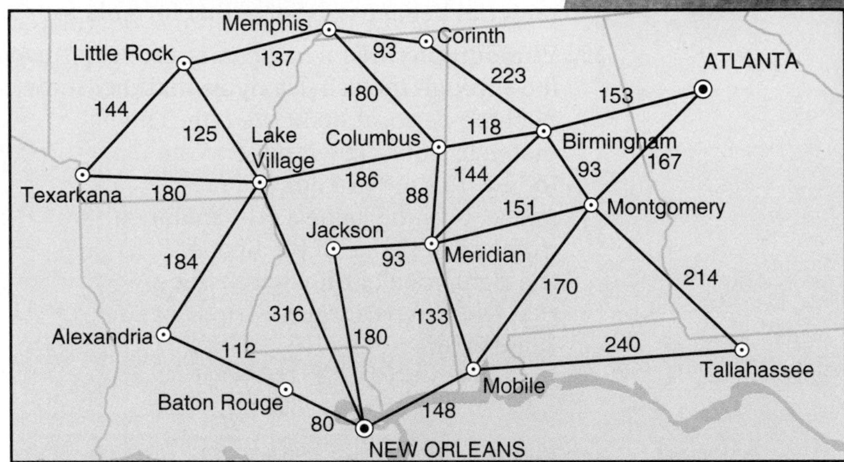

It is possible to use the mileage from city to city to illustrate some properties of equality and other facts about segments.

Example 1

Geography

a. Find two pairs of cities that illustrate the definition of congruent segments. Describe that relationship.

The distance from Jackson, Mississippi, to New Orleans, Louisiana, is 180 miles. The distance from Texarkana, Arkansas, to Lake Village, Arkansas, is 180 miles. Therefore, the segment connecting Jackson to New Orleans is congruent to the segment connecting Texarkana to Lake Village.

b. Find three pairs of cities that illustrate the Transitive Property of Equality. Describe that relationship.

City	City	Distance between Cities
Memphis, Tennessee	Corinth, Tennessee	93 miles
Birmingham, Alabama	Montgomery, Alabama	93 miles
Jackson, Mississippi	Meridian, Mississippi	93 miles

The Transitive Property of Equality can be illustrated by saying, *If the distance from Memphis to Corinth is the same as the distance from Birmingham to Montgomery, and the distance from Birmingham to Montgomery is the same as the distance from Jackson to Meridian, then the distance from Memphis to Corinth is the same as the distance from Jackson to Meridian.*

There are five essential parts in a good proof.

- State the theorem to be proved.
- List the given information.
- If possible, draw a diagram to illustrate the given information.
- State what is to be proved.
- Develop a system of deductive reasoning.

In order to use deductive reasoning to construct a valid proof, we must rely on statements that are accepted to be true. In geometry, those statements are definitions and postulates. We also depend on a list of undefined terms.

In Chapter 1, you learned that statements that are proved through deductive reasoning using definitions, postulates, and undefined terms are called *theorems.* Once a theorem is proved, it becomes another tool that we can use in the system. That is, proved theorems can be used in the proof of new theorems.

The first theorem we will prove is similar to properties of equality from algebra.

Theorem 2–1	**Congruence of segments is reflexive, symmetric, and transitive.**

This theorem can be abbreviated as "Reflexive Prop. of ≅ Segments," "Symmetric Prop. of ≅ Segments," or "Transitive Prop. of ≅ Segments."

Theorem 2–1 can be written in symbols as follows.

Reflexive Property $\overline{AB} \cong \overline{AB}$

Symmetric Property If $\overline{AB} \cong \overline{CD}$, then $\overline{CD} \cong \overline{AB}$.

Transitive Property If $\overline{AB} \cong \overline{CD}$ and $\overline{CD} \cong \overline{EF}$, then $\overline{AB} \cong \overline{EF}$.

The symmetric part of Theorem 2–1 is proved below. *You will be asked to prove the transitive and reflexive parts in Exercises 11 and 32, respectively.*

You can use the properties of algebra in geometric proofs. Notice that the symmetric property of equality is used in the proof of Theorem 2–1.

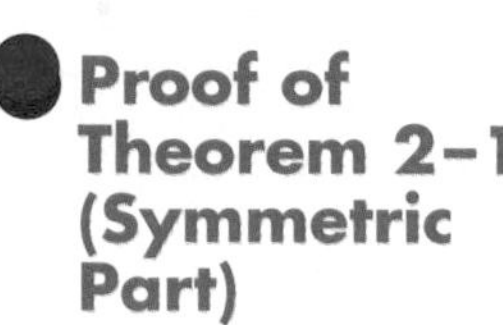

Proof of Theorem 2–1 (Symmetric Part)

Since measures of segments are real numbers, properties of algebra can be used to prove relationships with segment measures.

Given: $\overline{PQ} \cong \overline{RS}$

Prove: $\overline{RS} \cong \overline{PQ}$

Proof:

Statements	**Reasons**
1. $\overline{PQ} \cong \overline{RS}$	**1.** Given
2. $PQ = RS$	**2.** Definition of congruent segments
3. $RS = PQ$	**3.** Symmetric Property (=)
4. $\overline{RS} \cong \overline{PQ}$	**4.** Definition of congruent segments

Example 2 **Justify each step in the proof.**

$\overline{ABCD}$ means A, B, C, and D are collinear.

Given: $\overline{ABCD}$

Prove: $AD = AB + BC + CD$

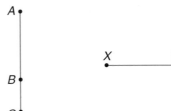

Proof:

Statements	Reasons
1. $\overline{ABCD}$	1. ?
2. $AD = AB + BD$	2. ?
3. $BD = BC + CD$	3. ?
4. $AD = AB + BC + CD$	4. ?

Reason 1: Given
Reason 2: Segment Addition Postulate
Reason 3: Segment Addition Postulate
Reason 4: Substitution Property (=)

Proofs can be written in paragraph format or in two-column format.

Example 3 **Prove the following.**

Given: $\overline{AB} \cong \overline{XY}$
$\overline{BC} \cong \overline{YZ}$

Prove: $\overline{AC} \cong \overline{XZ}$

Method 1: Paragraph Proof

Since $\overline{AB} \cong \overline{XY}$ and $\overline{BC} \cong \overline{YZ}$, then by the definition of congruence, $AB = XY$ and $BC = YZ$. By the Addition Property of Equality, $AB + BC = XY + YZ$. From the Segment Addition Postulate, $AB + BC = AC$ and $XY + YZ = XZ$. Substituting AC for $AB + BC$ and substituting XZ for $XY + YZ$ in the statement $AB + BC = XY + YZ$, we get $AC = XZ$. Then, by the definition of congruence, $\overline{AC} \cong \overline{XZ}$.

Method 2: Two-Column Proof

Statements	Reasons
1. $\overline{AB} \cong \overline{XY}$, $\overline{BC} \cong \overline{YZ}$	1. Given
2. $AB = XY$, $BC = YZ$	2. Definition of congruent segments
3. $AB + BC = XY + YZ$	3. Addition Property (=)
4. $AB + BC = AC$ $XY + YZ = XZ$	4. Segment Addition Postulate
5. $AC = XZ$	5. Substitution Property (=) *twice*
6. $\overline{AC} \cong \overline{XZ}$	6. Definition of congruent segments

CHECK FOR UNDERSTANDING

Communicating Mathematics

Study the lesson. Then complete the following.

1. **Explain** the five essential parts needed to construct a good proof.
2. **Describe** what a theorem is and explain why theorems are useful.

3. **Describe** in your own words what is meant by a system of deductive reasoning.
4. **Select** two cities from the U.S. mileage map. Describe the Reflexive Property of Equality using those two cities.

Guided Practice

Justify each statement with a property from algebra or a property of congruent segments.

5. If $\overline{KL} \cong \overline{MN}$, then $\overline{MN} \cong \overline{KL}$.
6. If $AB = 10$ and $CD = 10$, then $AB = CD$.
7. If $SY - 5 = RT - 5$, then $SY = RT$.

Write the given and prove statements you would use to prove each conjecture. Draw a figure if applicable.

8. If a figure is a square, then it has four right angles.
9. All whole numbers are integers.
10. Acute angles are angles with measures less than 90.
11. The reasons in the right column are not in the proper order. Reorder the reasons to properly match the statements in the left column for the proof of the transitive part of Theorem 2–1.

Given: $\overline{GH} \cong \overline{IJ}$
$\overline{IJ} \cong \overline{KL}$

Prove: $\overline{GH} \cong \overline{KL}$

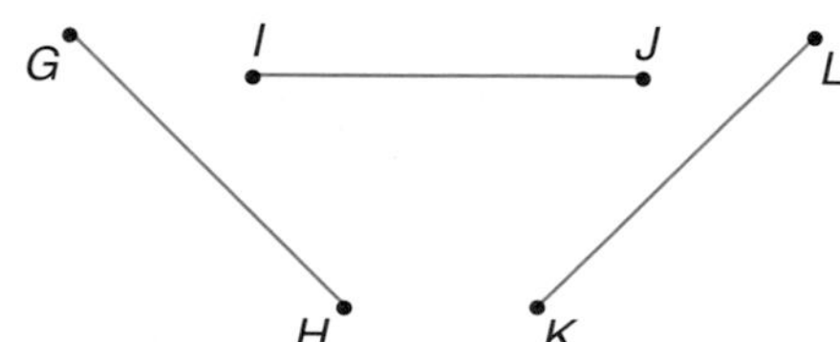

Proof:

Statements	Reasons
a. $\overline{GH} \cong \overline{IJ}$ $\overline{IJ} \cong \overline{KL}$	(1) Definition of congruent segments
b. $GH = IJ$ $IJ = KL$	(2) Transitive Property (=)
c. $GH = KL$	(3) Given
d. $\overline{GH} \cong \overline{KL}$	(4) Definition of congruent segments

12. Copy and complete the proof.

Given: $\overline{AB} \cong \overline{CD}$
$\overline{BD} \cong \overline{DE}$

Prove: $\overline{AD} \cong \overline{CE}$

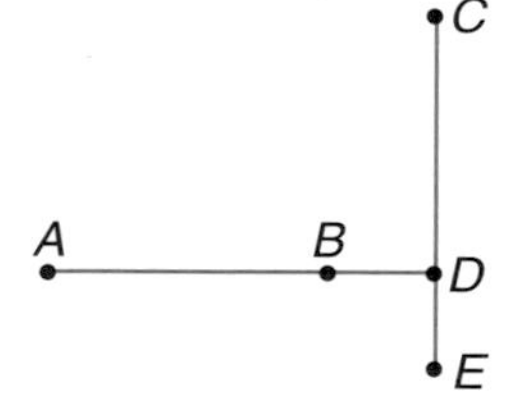

Proof:

Statements	Reasons
a. ? ?	a. Given
b. $AB = CD$ $BD = DE$	b. ?
c. $AD = AB + BD$ $CE = CD + DE$	c. ?
d. $AB + BD = CD + DE$	d. Addition Property (=)
e. ?	e. Substitution Property (=)
f. $\overline{AD} \cong \overline{CE}$	f. ?

13. Write a two-column proof.

Given: $WX = XY$

Prove: $WY = 2XY$

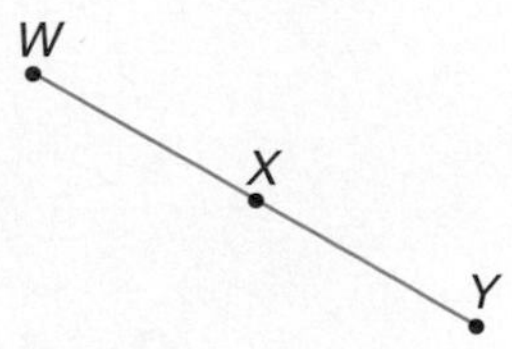

Data Update For more information on mileage between U.S. cities, visit: **www.geometry.glencoe.com**

14. **Geography** Refer to the connection at the beginning of the lesson.
    a. Select two pairs of cities from a U.S. mileage map in which the distance between one pair is twice the distance between the other pair.
    b. Describe this relationship as an equality.
    c. Divide each side of the equality by 2 and describe it again.

EXERCISES

Practice

Justify each statement with a property from algebra or a property of congruent segments.

15. $\overline{TJ} \cong \overline{TJ}$
16. If $2AB = 2WV$, then $AB = WV$.
17. If $AB = XY$ and $DF = MN$, then $AB + DF = XY + MN$.
18. $\overline{PQ} \cong \overline{PQ}$
19. If $\overline{PO} \cong \overline{WE}$ and $\overline{WE} \cong \overline{QR}$, then $\overline{PO} \cong \overline{QR}$.
20. If $AB = AC + CB$, then $AB - AC = CB$.
21. $AB \cdot BC - PQ \cdot BC = (AB - PQ) \cdot BC$.

Write the given and prove statements you would use to prove each conjecture. Draw a figure if applicable.

22. If two segments have equal measures, then they are congruent.
23. If a number greater than 2 is prime, then it is odd.
24. If two segments intersect at a point between the endpoints of the segments, then four angles are formed.
25. If two segments have measures equal to a third segment, then the two segments have equal measures.
26. If two segments are perpendicular, then they form a right angle.
27. If two points of a segment are on a line, then every point on that segment is on the line.
28. If a segment joins the midpoints of two sides of a triangle, then its measure is half the measure of the third side.
29. All rational numbers are real numbers.
30. Rewrite the proof as a two-column proof.

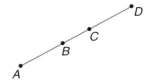

    Given: $AC = BD$

    Prove: $AB = CD$

    Paragraph Proof:

    We are given that $AC = BD$. From the Segment Addition Postulate, $AC = AB + BC$ and $BD = BC + CD$. Substituting $AB + BC$ for AC and $BC + CD$ for BD in the statement $AC = BD$, we get $AB + BC = BC + CD$. By the Subtraction Property of Equality, we can subtract BC from each side to get $AB = CD$.

31. Copy and complete the proof.

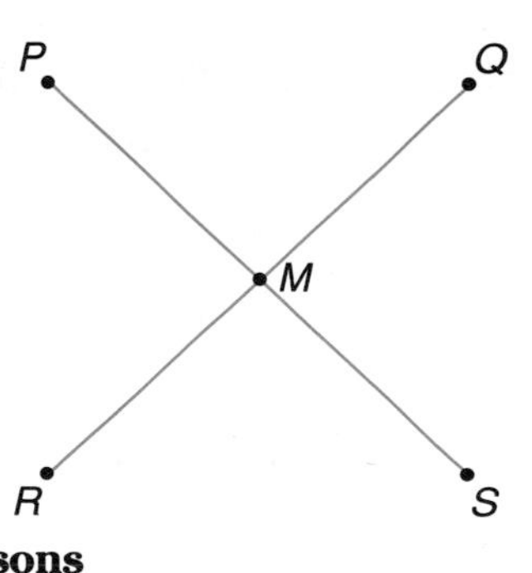

Given: $\overline{PS} \cong \overline{RQ}$
M is the midpoint of $\overline{PS}$.
M is the midpoint of $\overline{RQ}$.

Prove: $\overline{PM} \cong \overline{RM}$

Proof:

Statements	Reasons
a. $\overline{PS} \cong \overline{RQ}$ M is the midpoint of $\overline{PS}$. M is the midpoint of $\overline{RQ}$.	a. ?
b. $PS = RQ$	b. ?
c. ? ?	c. Definition of midpoint
d. $PS = PM + MS$ $RQ = RM + MQ$	d. ?
e. $PM + MS = RM + MQ$	e. ?
f. $PM + PM = RM + RM$	f. ?
g. $2PM = 2RM$	g. Substitution Property (=)
h. $PM = RM$	h. ?
i. $\overline{PM} \cong \overline{RM}$	i. ?

32. Write a two-column proof of the reflexive part of Theorem 2–1.

Write a two-column proof.

33. Given: $NL = NM$
$AL = BM$

Prove: $NA = NB$

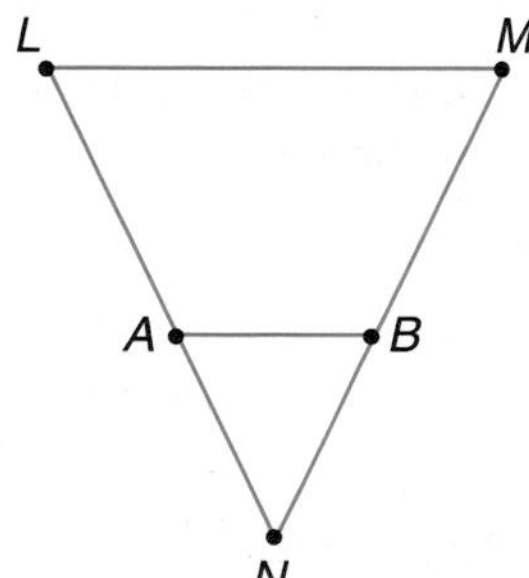

34. Given: $\overline{GR} \cong \overline{IL}$
$\overline{SR} \cong \overline{SL}$

Prove: $\overline{GS} \cong \overline{IS}$

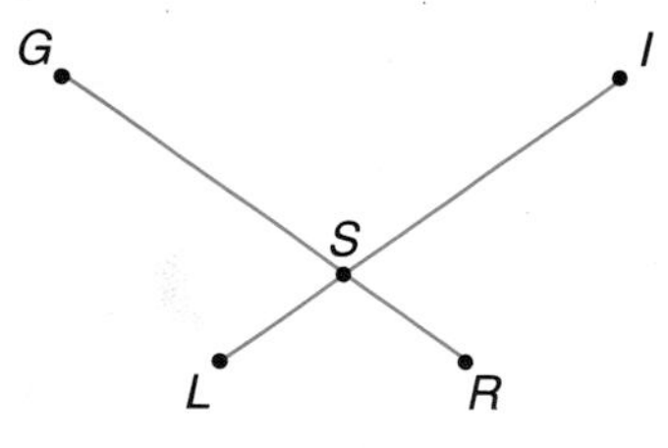

Critical Thinking

35. Given that $\overline{LN} \cong \overline{RT}$, $\overline{RT} \cong \overline{QO}$, $\overline{LQ} \cong \overline{NO}$, $\overline{MP} \cong \overline{NO}$, M is the midpoint of $\overline{LN}$, S is the midpoint of $\overline{RT}$, and P is the midpoint of $\overline{QO}$, list three statements that you could prove using the postulates, theorems, and definitions that you have learned.

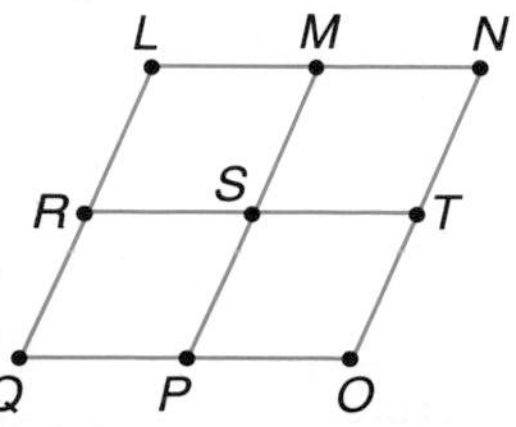

Applications and Problem Solving

36. **Geography** On a U.S. mileage map, there is a scale that lists kilometers on the top and miles on the bottom. It appears that a segment can have two lengths, but the difference is the units.

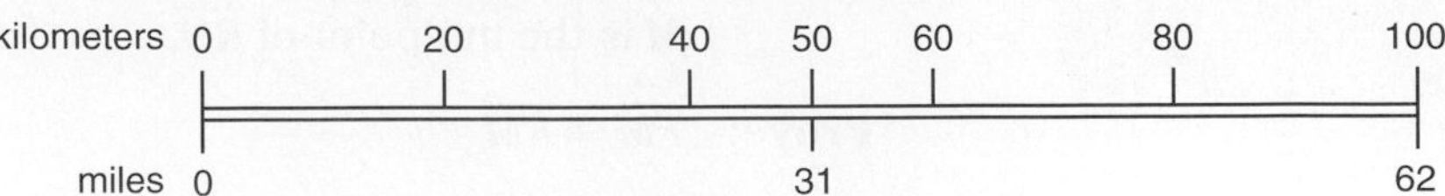

Suppose $\overline{AB}$ and $\overline{CD}$ are segments on the map with the scale above.

a. If $\overline{AB}$ is 100 kilometers long and $\overline{CD}$ is 62 miles long, is $\overline{AB} \cong \overline{CD}$?

b. It is common for charity fund-raising walks to be 5k or 10k (kilometers). How far is each distance in miles?

37. **Measure** Some rulers have centimeters on one edge and inches on the other edge.

a. About how long in centimeters is a segment that is 6 inches long?

b. Are the two segments congruent? Explain.

Mixed Review

38. Name the property of equality that justifies the statement *If* $3(x - 1) = 10$, *then* $3x - 3 = 10$. (Lesson 2–4)

39. Determine if a valid conclusion can be reached from the two true statements using the Law of Detachment or the Law of Syllogism. If a valid conclusion is possible, state it. (Lesson 2–3)

(1) Parallel lines do not intersect.

(2) If lines do not intersect, then they have no points in common.

40. **Advertising** A billboard reads *If you want a fabulous vacation, try Georgia.* (Lesson 2–2)

a. Write the converse of the conditional.

b. What do you think the advertiser wants you to conclude about vacations in Georgia?

c. Does the advertisement say that vacations in Georgia are fabulous?

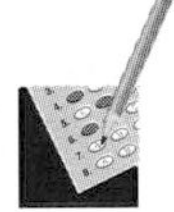

41. **SAT Practice** A breakfast cereal contains wheat, rice, and oats in the ratio 3:1:2. If the manufacturer makes a mixture using 120 pounds of oats, how many pounds of wheat will be used?

A 60 **B** 80 **C** 120 **D** 180 **E** 360

42. Use the Pythagorean Theorem to find the missing side measure in the triangle at the right. (Lesson 1–4)

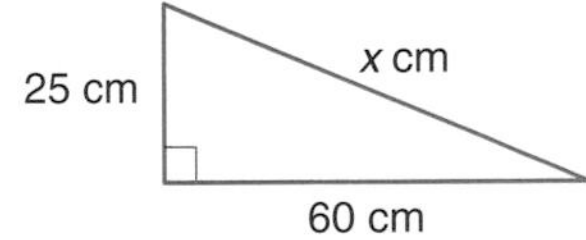

43. Find the missing measure in the formula $A = \ell w$ if $\ell = 17$ and $w = 8$. (Lesson 1–3)

44. Graph points $G(4, -3)$, $H(4, -1)$, $I(4, 6)$, and $K(4, 2)$ on the same coordinate plane. What pattern do you notice? (Lesson 1–1)

Algebra

45. **Basketball** According to NBA standards, a basketball should bounce back $\frac{2}{3}$ of the distance from which it is dropped. How high should a basketball bounce when it is dropped from a height of $3\frac{3}{4}$ yards?

46. Name the property illustrated by $(a + b)c = ac + bc$.

For **Extra Practice**, see page 767.

2–6 Verifying Angle Relationships

What **YOU'LL LEARN**

- To complete proofs involving angle theorems.

Why **IT'S IMPORTANT**

Understanding angle relationships is useful in solving problems involving art, nature, and architecture.

Illusions

The following excerpt is from *Can You Believe Your Eyes*, by J. Richard Block and Harold Yuker.

> Illusions are misperceptions. Illusions sometimes involve angles. For example, the following two lines intersect to form two pairs of vertical angles. Hold the book parallel to the floor at eye level so the intersection point of the two lines is directly in front of you. A third line should appear. More angles are formed if you move the book so the third line goes through the intersection point. How many more angles do you see?

Two adjacent angles in the figure above form a linear pair. Theorem 2–2 states that if two angles form a linear pair, the angles are supplementary.

Theorem 2–2 Supplement Theorem	**If two angles form a linear pair, then they are supplementary angles.**

You will be asked to prove this theorem in Exercise 13.

Example  **If $\angle 1$ and $\angle 2$ form a linear pair and $m\angle 1 = 72$, find $m\angle 2$.**

$$m\angle 1 + m\angle 2 = 180$$
$$72 + m\angle 2 = 180$$
$$m\angle 2 = 108$$

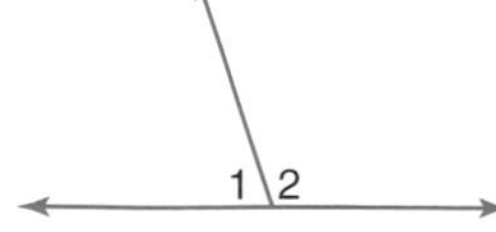

The properties of algebra that applied to the congruence of segments and the equality of their measures also hold true for the congruence of angles and the equality of their measures. The congruence relationships are stated in Theorem 2–3.

Theorem 2–3	**Congruence of angles is reflexive, symmetric, and transitive.**

The parts of this theorem can be abbreviated as "Reflexive Prop. of ≅ ∡s," "Symmetric Prop. of ≅ ∡s," or "Transitive Prop. of ≅ ∡s."

The proof of the transitive portion of Theorem 2–3 is given below. *You will be asked to prove the reflexive and symmetric portions in Exercises 36 and 37, respectively.*

Proof of Theorem 2–3 (Transitive Part)

Given: $\angle 1 \cong \angle 2$
$\angle 2 \cong \angle 3$

Prove: $\angle 1 \cong \angle 3$

Proof:

Statements	**Reasons**
1. $\angle 1 \cong \angle 2$, $\angle 2 \cong \angle 3$	**1.** Given
2. $m\angle 1 = m\angle 2$, $m\angle 2 = m\angle 3$	**2.** Definition of congruent angles
3. $m\angle 1 = m\angle 3$	**3.** Transitive Property (=)
4. $\angle 1 \cong \angle 3$	**4.** Definition of congruent angles

Example 2

Art

The gold disk shown at the right once belonged to the high rulers of the West African kingdom of Ghana, which was founded about A.D. 200. If $m\angle 1 = 40$ and $\angle 1 \cong \angle 2$ and $\angle 2 \cong \angle 3$, find $m\angle 3$.

Since congruence of angles is transitive, $\angle 1 \cong \angle 3$. Therefore, $m\angle 3 = 40$.

If two angles are supplementary to the same angle, what do you think is true about the angles? Draw several examples and make a conjecture.

Theorem 2–4	**Angles supplementary to the same angle or to congruent angles are congruent.**

This theorem can be abbreviated as "$\measuredangle$s supp. to same $\angle$ or $\cong$ $\measuredangle$s are $\cong$."

Proof of Theorem 2–4

Given: $\angle 1$ and $\angle 3$ are supplementary.
$\angle 2$ and $\angle 3$ are supplementary.

Prove: $\angle 1 \cong \angle 2$

Proof:

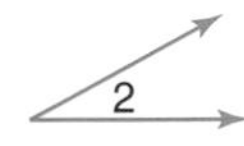

Statements	**Reasons**
1. $\angle 1$ and $\angle 3$ are supplementary. $\angle 2$ and $\angle 3$ are supplementary.	**1.** Given
2. $m\angle 1 + m\angle 3 = 180$ $m\angle 2 + m\angle 3 = 180$	**2.** Definition of supplementary
3. $m\angle 1 + m\angle 3 = m\angle 2 + m\angle 3$	**3.** Substitution Property (=)
4. $m\angle 1 = m\angle 2$	**4.** Subtraction Property (=)
5. $\angle 1 \cong \angle 2$	**5.** Definition of congruent angles

Theorem 2–5 is a similar theorem for complementary angles. *You will be asked to prove Theorem 2–5 in Exercise 5.*

Theorem 2–5	Angles complementary to the same angle or to congruent angles are congruent.

This theorem can be abbreviated as "∡ compl. to same ∠ or ≅ ∡ are ≅."

You can create right angles and investigate congruent angles by paperfolding.

MODELING MATHEMATICS

Right Angles

Materials:  patty paper  straightedge

- Fold the paper so that one corner is bent downward.
- Fold along the crease so that the top edge meets the side edge.
- Unfold the paper.

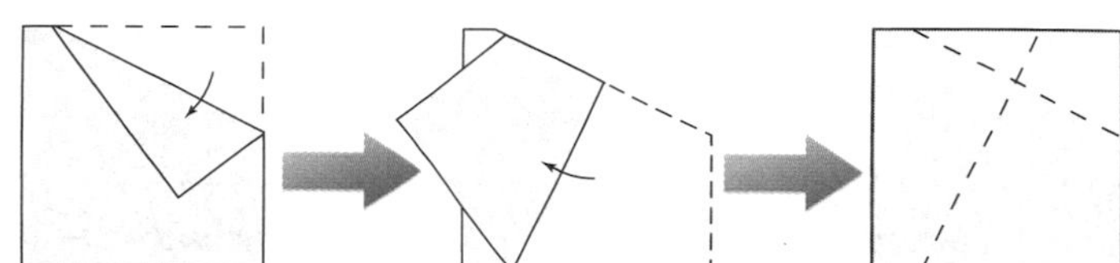

Your Turn

a. Measure each of the angles formed.
b. Repeat the activity three more times.
c. Make a conjecture about what is true of all right angles.

The following theorem supports the conjecture you formed in the Modeling Mathematics activity. *You will be asked to prove Theorem 2–6 in Exercise 38.*

Theorem 2–6	All right angles are congruent.

This theorem can be abbreviated as "All rt. ∡ are ≅."

In Lesson 2–3, you discovered that vertical angles are congruent. We stated that even 1000 cases would not represent a proof of the fact that vertical angles are congruent for every case. When we prove this as a theorem, it will be true for all cases. That is why theorems are so important.

Theorem 2–7	Vertical angles are congruent.

This theorem can be abbreviated as "Vert. ∡ are ≅."

Proof of Theorem 2–7

Given: $\angle 1$ and $\angle 3$ are vertical angles.

Prove: $\angle 1 \cong \angle 3$

Proof:

Statements	Reasons
1. $\angle 1$ and $\angle 3$ are vertical angles.	1. Given
2. $\angle 1$ and $\angle 2$ form a linear pair. $\angle 2$ and $\angle 3$ form a linear pair.	2. Definition of linear pair
3. $\angle 1$ and $\angle 2$ are supplementary. $\angle 2$ and $\angle 3$ are supplementary.	3. If 2 ∡ form a linear pair, then they are supp.
4. $\angle 1 \cong \angle 3$	4. ∡ supp. to same ∠ are ≅.

Example 3 **If $\angle 7$ and $\angle 8$ are vertical angles and $m\angle 7 = 3x + 6$ and $m\angle 8 = x + 26$, find $m\angle 7$ and $m\angle 8$.**

$m\angle 7 = m\angle 8$	*Vertical angles are congruent.*
$3x + 6 = x + 26$	*Substitution Property (=)*
$2x + 6 = 26$	*Substitution Property (=)*
$2x = 20$	*Substitution Property (=)*
$x = 10$	*Division Property (=)*

$m\angle 7 = 3x + 6$
$= 3(10) + 6$ or 36

$m\angle 8 = x + 26$
$= 10 + 26$ or 36

A theorem that follows from Theorem 2–7 is stated below.

Theorem 2–8	**Perpendicular lines intersect to form four right angles.**

This theorem can be abbreviated as "⊥ lines form 4 rt. ⦞."

CHECK FOR UNDERSTANDING

Communicating Mathematics

Study the lesson. Then complete the following.

1. **Explain** in your own words what an illusion is.
2. **Compare and contrast** the Transitive Property of Equality, the Transitive Property of Congruent Segments, and the Transitive Property of Congruent Angles.
3. The definition of congruent angles involves a statement and its converse. Write each one.
4. **Explain** why many true examples of a statement are not sufficient evidence to conclude that a statement is true for all cases. Give an illustration.
5. The reasons in the right column of the proof are not in the proper order. Reorder the reasons to properly match the statements in the left column.

Given: $\angle 1$ and $\angle 2$ are complementary.
$\angle 3$ and $\angle 2$ are complementary.

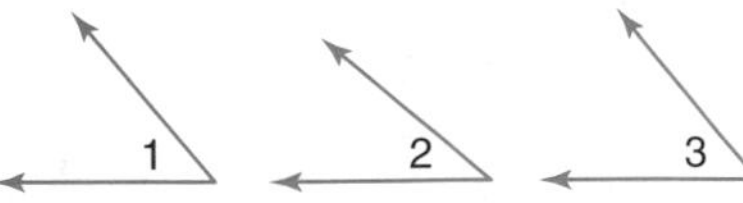

Prove:  $\angle 1 \cong \angle 3$

Proof:

Statements	Reasons
a. $\angle 1$ and $\angle 2$ are complementary. $\angle 3$ and $\angle 2$ are complementary.	(1) Subtraction Property (=)
b. $m\angle 1 + m\angle 2 = 90$ $m\angle 3 + m\angle 2 = 90$	(2) Given
c. $m\angle 1 + m\angle 2 = m\angle 3 + m\angle 2$	(3) Definition of complementary ⦞
d. $m\angle 1 = m\angle 3$	(4) Definition of congruent ⦞
e. $\angle 1 \cong \angle 3$	(5) Substitution Property (=)

6. Fold a piece of patty paper so that all four corners lie on top of each other. Write a conjecture about the angles of a square.

Complete each statement with *always, sometimes,* or *never.*

7. Two angles that are supplementary __?__ form a linear pair.
8. Two angles that form a linear pair are __?__ supplementary.
9. Two angles that are congruent are __?__ right.
10. Two angles that are right are __?__ congruent.

Guided Practice

Find the measure of each numbered angle.

11. $m\angle 1 = 2x - 5$ and $m\angle 2 = x - 4$

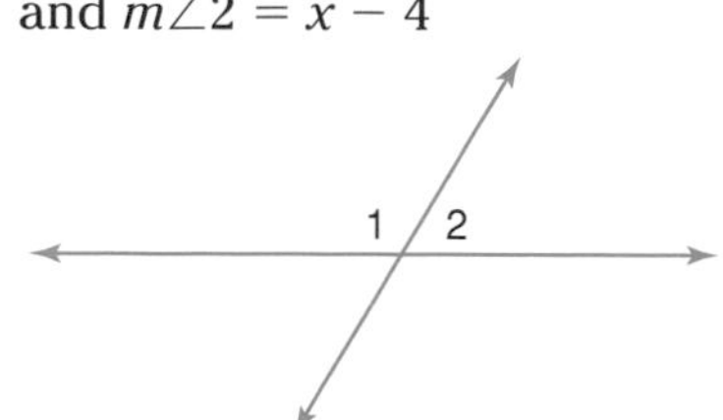

12. $m\angle 3 = 228 - 3x$ and $m\angle 4 = x$

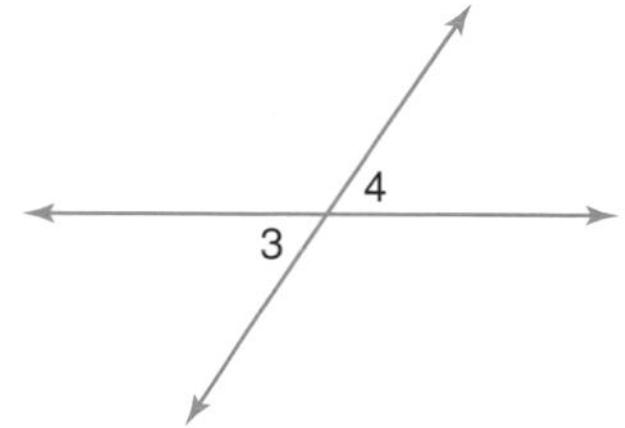

13. Copy and complete the proof of the Supplement Theorem (Theorem 2–2).

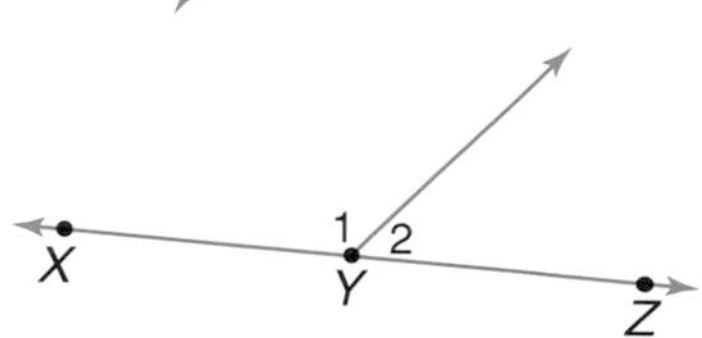

Given: $\angle 1$ and $\angle 2$ form a linear pair.

Prove: $\angle 1$ and $\angle 2$ are supplementary.

Proof:

Statements	Reasons
a. $\angle 1$ and $\angle 2$ form a linear pair.	a. Given
b. $\overrightarrow{YX}$ and $\overrightarrow{YZ}$ are opposite rays.	b. Definition of opposite rays
c. $m\angle XYZ = 180$	c. Definition of straight angle
d. __?__	d. Angle Addition Postulate
e. $m\angle 1 + m\angle 2 = 180$	e. __?__
f. $\angle 1$ and $\angle 2$ are supplementary.	f. __?__

CAREER CHOICES

A **botanist** studies plants, including their identification and classification, their geological record, and the structure and function of plant parts. The Ph.D. degree is generally required for biological scientists.

For more information, contact:

Business Office, Botanical Society of America
1725 Neil Ave.
Columbus, OH 43210

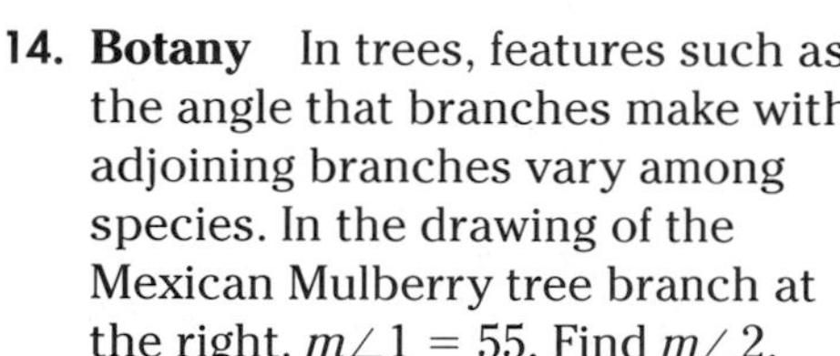

14. **Botany** In trees, features such as the angle that branches make with adjoining branches vary among species. In the drawing of the Mexican Mulberry tree branch at the right, $m\angle 1 = 55$. Find $m\angle 2$.

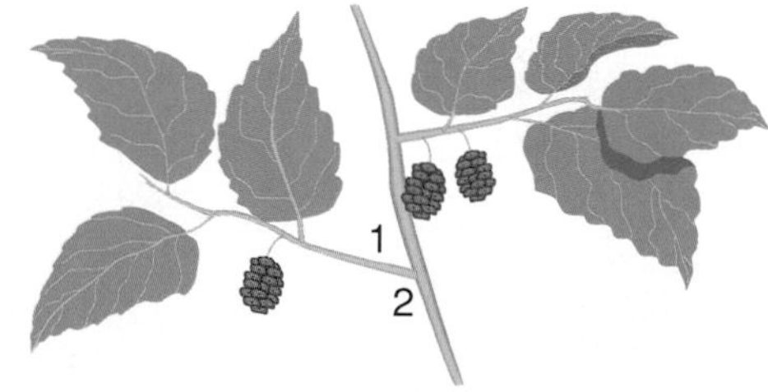

EXERCISES

Practice

Complete each statement in Exercises 15–26 with *always, sometimes,* or *never.*

15. Two angles that are congruent are __?__ complementary to the same angle.
16. Two angles that are complementary to the same angle are __?__ congruent.
17. Two angles that are vertical are __?__ nonadjacent.
18. Two angles that are nonadjacent are __?__ vertical.
19. Two angles that are complementary __?__ form a right angle.
20. Two angles that form a right angle are __?__ complementary.
21. Two angles that form a linear pair are __?__ congruent.
22. Two angles that are supplementary are __?__ congruent.

23. Two angles that are supplementary are _?_ complementary.

24. Two right angles are _?_ supplementary.

25. Vertical angles are _?_ complementary.

26. Angles with a common side and a common vertex _?_ form a linear pair.

Find the measure of each numbered angle.

27. $m\angle 5 = x$ and
$m\angle 6 = 6x - 290$

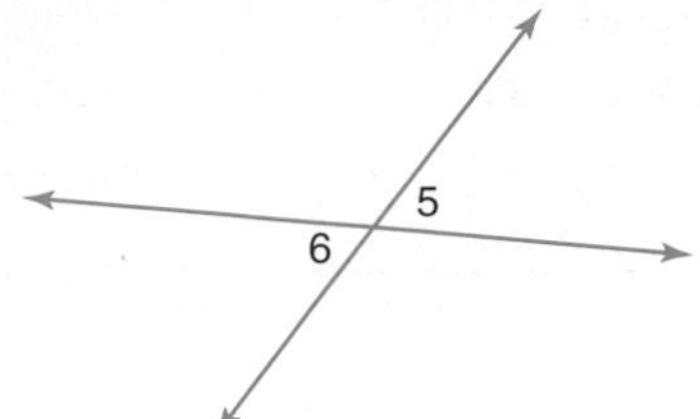

28. $m\angle 7 = 2x - 4$ and
$m\angle 8 = 2x + 4$

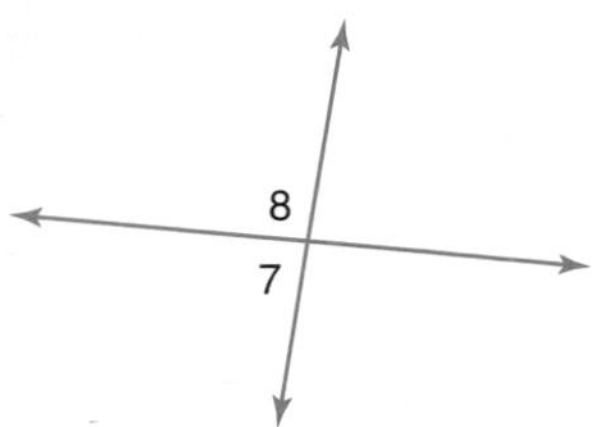

29. $m\angle 1 = 4x$ and
$m\angle 2 = 2x - 6$

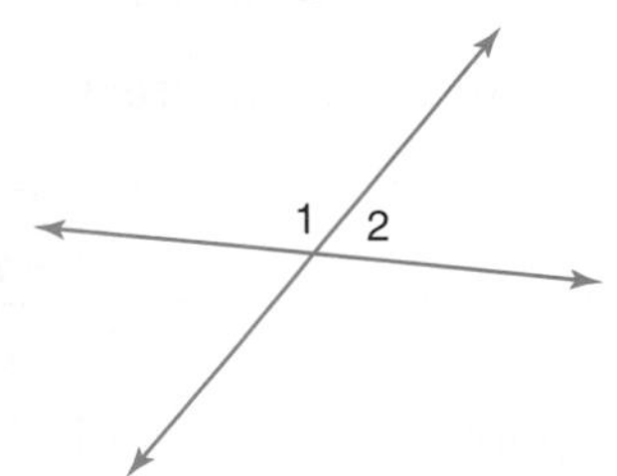

30. $m\angle 3 = 2x + 7$ and
$m\angle 4 = x + 30$

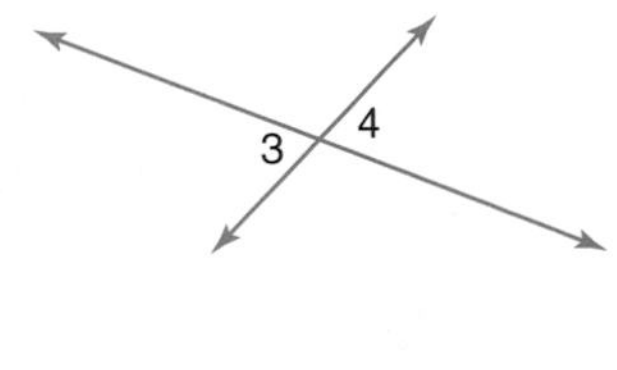

31. $\angle 5$ and $\angle A$ are complementary.
$\angle 6$ and $\angle A$ are complementary.
$m\angle 5 = 2x + 2$ and
$m\angle 6 = x + 32$

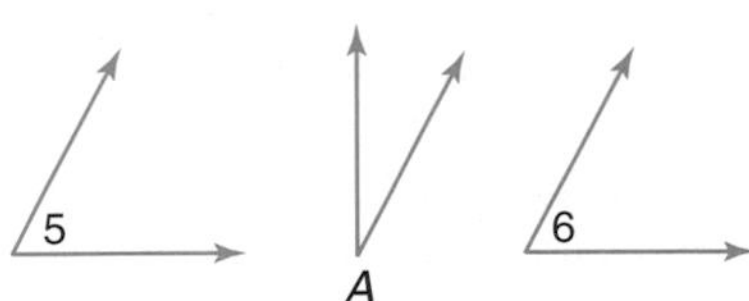

32. $m\angle 7 = x + 20$, $m\angle 8 = x + 40$, and $m\angle 9 = x + 30$

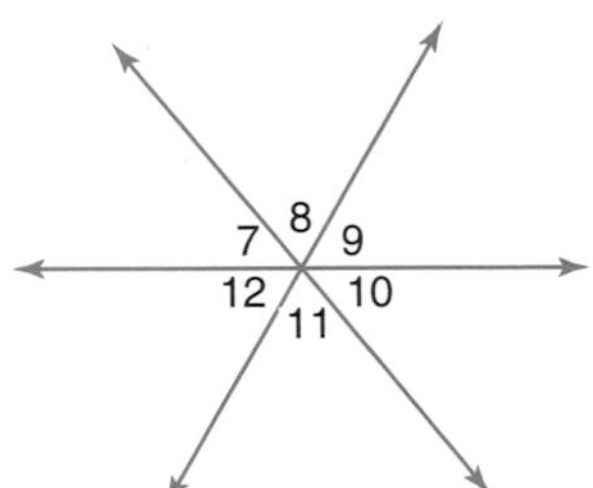

33. Refer to the figure at the right.
    a. Use numbers to name four pairs of vertical angles.
    b. Use letters to name eight different linear pairs of angles.

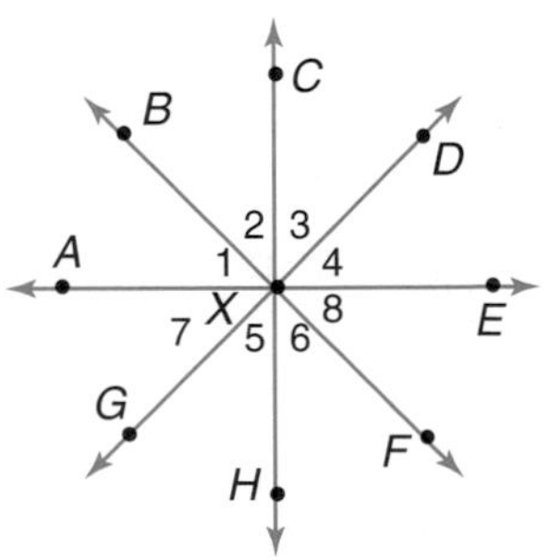

Write a two-column proof.

34. **Given:** $m\angle ABC = m\angle DFE$
$m\angle 1 = m\angle 4$

Prove: $m\angle 2 = m\angle 3$

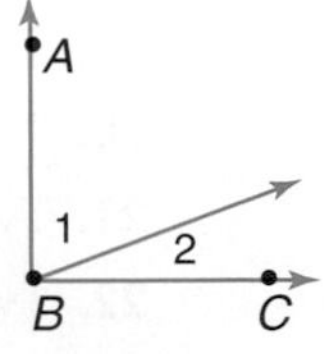

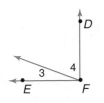

35. **Given:** $\angle ABD$ and $\angle CBD$ form a linear pair.
$\angle YXZ$ and $\angle WXZ$ form a linear pair.
$\angle ABD \cong \angle YXZ$

Prove: $\angle CBD \cong \angle WXZ$

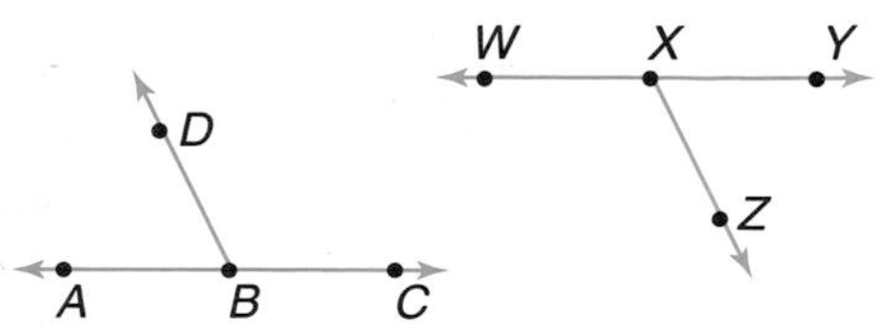

For each theorem, name the given and prove statements and draw a figure. Then write a two-column proof.

36. Congruence of angles is reflexive. (Theorem 2–3)
37. Congruence of angles is symmetric. (Theorem 2–3)
38. All right angles are congruent. (Theorem 2–6)
39. If one angle in a linear pair is a right angle, then the other angle is a right angle also.

Critical Thinking

40. What conclusion can you draw about the sum of $m\angle 1$ and $m\angle 4$ if $m\angle 1 = m\angle 2$ and $m\angle 3 = m\angle 4$? Explain.

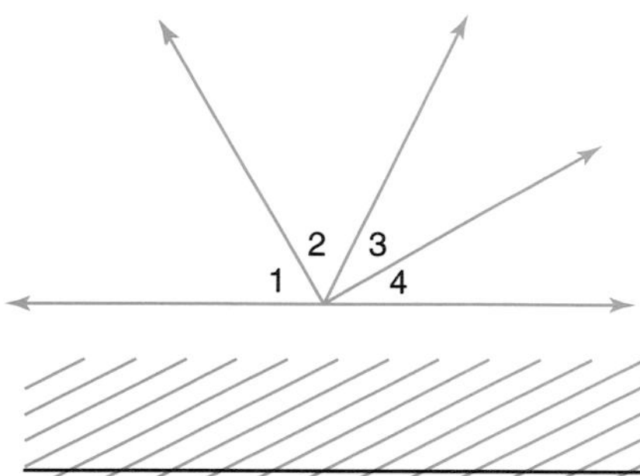

Applications and Problem Solving

41. **Illusions** At the right is an illusion created by lines and angles. If extended, will the long lines meet? Explain. Create your own illusion using the idea of angles.

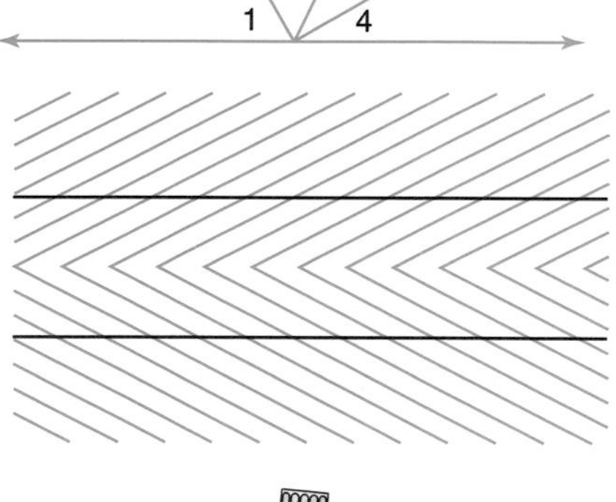

42. **Architecture** The Leaning Tower of Pisa in Italy makes an angle with the ground of about 84° on one side. If you look at the building as a ray and the ground as a line, then the angles that the tower forms with the ground form a linear pair. Find the measure x of the other angle that the tower makes with the ground.

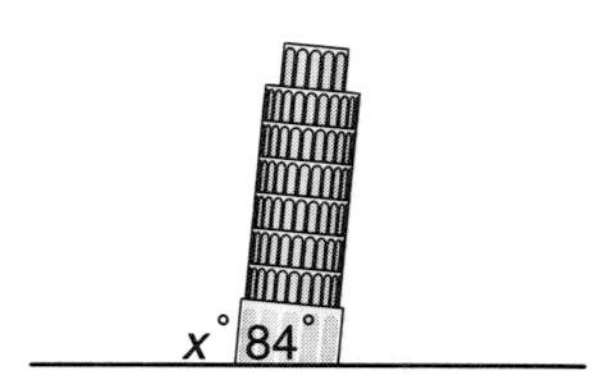

Wilhelm Wundt

43. **Illusions** In 1879, psychologist Wilhelm Wundt of Germany used angular lines on the outside of two vertical lines to create the illusions at the right.
    a. Describe what appears to be a relationship between the two vertical lines in each case.
    b. Tilt the page at an angle of about 45° so the vertical lines are pointing toward you. Redescribe what now appears to be a relationship between the two vertical lines.
    c. A close-up of the angular lines is shown at the right. If $\angle 4 \cong \angle 2$, prove that $\angle 3 \cong \angle 1$.

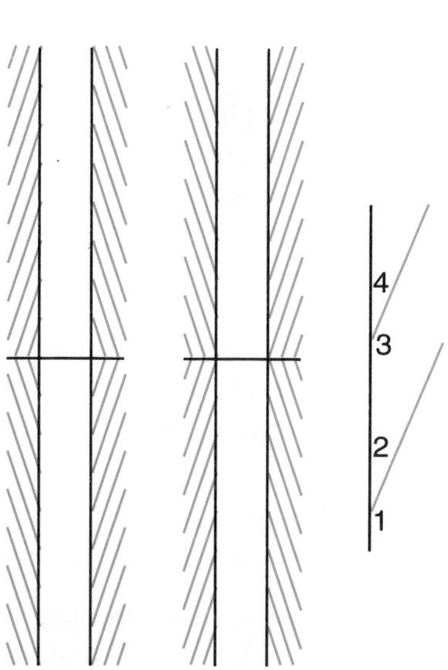

Mixed Review

44. Write the given and prove statements you would use to prove the statement *If an angle is a right angle, then its measure is 90.* (Lesson 2–5)

45. Name the property of equality that justifies the statement *If* $y = 4x + 9$ *and* $x = 2$, *then* $y = 17$. (Lesson 2–4)

46. **ACT Practice** $-|-8| - |-3| + 2|-5| = ?$

    A -21 **B** -14 **C** -1 **D** 1 **E** 21

47. If K is between J and L, $JK = 3x$, $KL = 2x - 1$, and $JL = 24$, find the value of x and the measure of $\overline{JK}$. (Lesson 1–4)

48. State whether a page of your geometry book is best modeled by a *point*, *line*, or *plane*. (Lesson 1–2)

49. **List the Possibilities** Emilio is planting a garden along the back of his house. The garden must have an area of at least 70 square feet. One side of the garden will be bounded by the house, but the other three sides will be bounded by 25 feet of fencing. What are the possible dimensions of Emilio's garden if he plans to use all the fencing and each side has a whole number length? (Lesson 1–2)

Algebra

50. Evaluate $6(5) - 1 + 8 \cdot 4$.

51. The graph shows how many people watched prime-time TV on average each day of the week in a recent year. If you were buying commercial time on one night of the week for your new product, which night would you choose? Why?

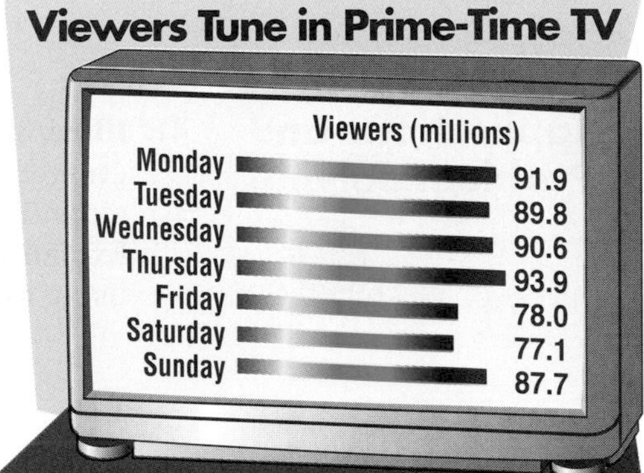

Source: *Nielsen Media Research*

For **Extra Practice**, see page 768.

WORKING ON THE

In·ves·ti·ga·tion

Refer to the Investigation on pages 66–67.

Although you may not realize it, your eyes respond in different ways to the shapes within a painting. Artists make use of these patterns when they create their works. In general, diagonal lines give a painting energy while circles draw the eye toward the center.

1 Find several examples of paintings of different styles. You may wish to consult a reference book in a library or the Internet home page for a large museum.

2 Look at the paintings with your group members. Take turns looking at a painting and observing the ways that the observers' eyes move around the painting.

3 Discuss the theme of your mural. Do you want the mural to appear lively or comforting? Which geometric shape is best for arranging your mural to accomplish your goal?

Add the results of your work to your Investigation folder.

CHAPTER 2 HIGHLIGHTS

Chapter Review For additional lesson-by-lesson review, visit: **www.geometry.glencoe.com**

VOCABULARY

After completing this chapter, you should be able to define each term, property, or phrase and give an example or two of each.

Geometry

conclusion (p. 76)
conditional statement (p. 76)
conjecture (p. 70)
contrapositive (p. 78)
converse (p. 77)
counterexample (p. 72)
deductive reasoning (p. 86)
hypothesis (p. 76)
if-then statement (p. 76)
inductive reasoning (p. 70)
inverse (p. 78)
Law of Detachment (p. 86)
Law of Syllogism (p. 87)
negation (p. 78)
proof (p. 94)
two-column proof (p. 94)
Venn diagram (p. 79)

Algebra

Addition and Subtraction Properties (p. 93)
Distributive Property (p. 93)
Multiplication and Division Properties (p. 93)
Reflexive Property (p. 93)
Substitution Property (p. 93)
Symmetric Property (p. 93)
Transitive Property (p. 93)

Problem Solving

draw a diagram (p. 79)
look for a pattern (p. 85)

UNDERSTANDING AND USING THE VOCABULARY

Choose the letter of the term that best matches each statement or phrase.

1. the portion of a conditional statement immediately following *if*
2. using several examples to make a conjecture
3. a statement formed by interchanging the hypothesis and conclusion of a conditional statement
4. When a conditional is true and its hypothesis is true, then its conclusion is true.
5. a false example that shows a conjecture is not true
6. a statement formed by negating the hypothesis and conclusion of the converse of the given conditional
7. an educated guess about what will always be true
8. the denial of a statement
9. If $p \rightarrow q$ and $q \rightarrow r$ are true conditionals, then $p \rightarrow r$ is also true.
10. a statement formed by negating both the hypothesis and conclusion
11. a method of constructing a valid argument by listing reasons for each step
12. using rules of logic to reach a conclusion

a. conjecture
b. converse
c. counterexample
d. deductive reasoning
e. inductive reasoning
f. inverse
g. Law of Detachment
h. Law of Syllogism
i. negation
j. proof
k. hypothesis
l. contrapositive

CHAPTER 2 STUDY GUIDE AND ASSESSMENT

SKILLS AND CONCEPTS

OBJECTIVES AND EXAMPLES

Upon completing this chapter, you should be able to:

- make conjectures based on inductive reasoning (Lesson 2–1)

To determine if a conjecture made from inductive reasoning is false, look at situations where the given information is true. Determine if there are situations where the given is true and the conjecture is false. A proof is required to determine if a conjecture is true.

REVIEW EXERCISES

Use these exercises to review and prepare for the chapter test.

Determine if each conjecture is *true* or *false* based on the given information. Explain your answer.

13. **Given:** X, Y, and Z are collinear and $XY = YZ$.
    Conjecture: Y is the midpoint of $\overline{XZ}$.
14. **Given:** $\angle 1$ and $\angle 2$ are supplementary.
    Conjecture: $\angle 1 \cong \angle 2$

- write a conditional in if-then form (Lesson 2–2)

Write the statement *Adjacent angles have a common ray* in if-then form.

If angles are adjacent, then they have a common ray.

Write each conditional in if-then form.

15. Every cloud has a silver lining.
16. A rectangle has four right angles.
17. Obsidian is a glassy rock produced by a volcano.
18. The intersection of two planes is a line.

- write the converse, inverse, and contrapositive of an if-then statement (Lesson 2–2)

Given: If an angle measures 120°, then it is obtuse.

Converse: If an angle is obtuse, then it measures 120°.

Inverse: If an angle does not measure 120°, then it is not obtuse.

Contrapositive: If an angle is not obtuse, then it does not measure 120°.

Write the converse, inverse, and contrapositive of each statement.

19. If a rectangle has four congruent sides, then it is a square.
20. If three points are collinear, then they lie on a straight line.
21. If the month is January, then it has 31 days.
22. If an ordered pair for a point has 0 as its y-coordinate, then the point lies on the x-axis.

OBJECTIVES AND EXAMPLES

- use the Law of Detachment and the Law of Syllogism in deductive reasoning (Lesson 2–3)

The Law of Detachment states that if $p \rightarrow q$ is a true conditional and p is true, then q is true.

The Law of Syllogism states that if $p \rightarrow q$ and $q \rightarrow r$ are true conditionals, then $p \rightarrow r$ is also true.

REVIEW EXERCISES

Determine if a valid conclusion can be reached from the two true statements using the Law of Detachment or the Law of Syllogism. If a valid conclusion is possible, state it and the law that is used. If a valid conclusion does not follow, write *no conclusion*.

23. (1) Angles that are supplementary have measures with a sum of 180.
(2) $\angle A$ and $\angle B$ are supplementary.

24. (1) Well-known athletes appear on Wheaties™ boxes.
(2) Michael Jordan appeared on a Wheaties™ box.

25. (1) The sun is a star.
(2) Stars are in constant motion.

- use properties of equality in algebraic and geometric proofs (Lesson 2–4)

Prove that if $ST = UV$, then $SU = TV$.

Given: $ST = UV$

Prove: $SU = TV$

S T U V

Statements	Reasons
1. $ST = UV$	1. Given
2. $TU = TU$	2. Reflexive Property (=)
3. $ST + TU = TU + UV$	3. Addition Property (=)
4. $SU = ST + TU$ $TV = TU + UV$	4. Segment Addition Postulate
5. $SU = TV$	5. Substitution Property (=)

Name the property of equality that justifies each statement.

26. If $12x = 24$, then $x = 2$.

27. For MN, $MN = MN$.

28. Copy the proof. Then name the property that justifies each statement.

Given: $MN = PN$
$NL = NO$

Prove: $ML = PO$

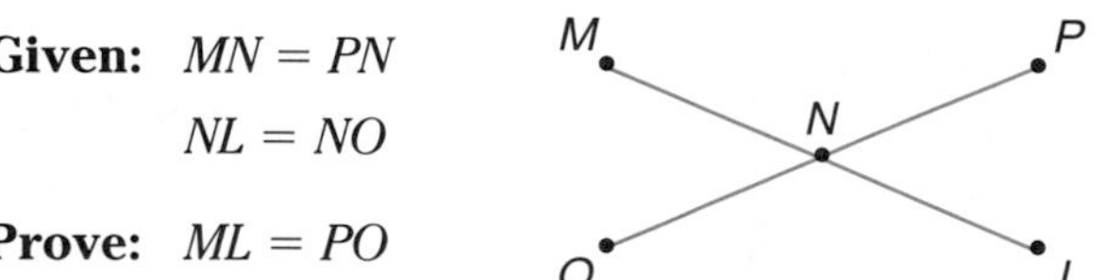

Proof:

Statements	Reasons
a. $MN = PN$, $NL = NO$	a. ?
b. $MN + NL = PN + NO$	b. ?
c. $ML = MN + NL$ $PO = PN + NO$	c. ?
d. $ML = PO$	d. ?

- complete proofs involving segment theorems (Lesson 2–5)

Theorem 2–1 states that congruence of segments is reflexive, symmetric, and transitive.

Justify each statement with a property from algebra or a property of congruence.

29. If $AB + BC = BC + CD$, then $AB = CD$.

30. If $\overline{XY} \cong \overline{OP}$, then $\overline{OP} \cong \overline{XY}$.

31. If $GH = 12$ and $GH + HI = GI$, then $12 + HI = GI$.

OBJECTIVES AND EXAMPLES

- complete proofs involving angle theorems (Lesson 2–6)

Given: $\angle 1 \cong \angle 2$

Prove: $\angle 3 \cong \angle 4$

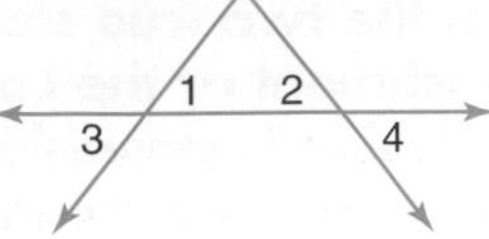

Proof:

Statements	Reasons
1. $\angle 1 \cong \angle 2$	1. Given
2. $\angle 3 \cong \angle 1$ $\angle 2 \cong \angle 4$	2. Vertical $\angle$s are $\cong$.
3. $\angle 3 \cong \angle 4$	3. Congruence of angles is transitive. *twice*

REVIEW EXERCISES

Complete each statement with *always, sometimes,* or *never.*

32. If two angles are right angles, they are __?__ adjacent.

33. If two angles are complementary, they are __?__ right angles.

34. An angle is __?__ congruent to itself.

35. Vertical angles are __?__ adjacent angles.

APPLICATIONS AND PROBLEM SOLVING

36. Advertising Write the conditional *Hard-working people deserve a great vacation* in if-then form. Identify the hypothesis and the conclusion of the conditional. Then write the converse, inverse, and contrapositive. (Lesson 2–2)

37. Look for a Pattern Find the number of pairs of vertical angles determined by eight distinct lines passing through a point. (Lesson 2–3)

38. Biology If possible, write a valid conclusion that can be reached from the two true statements. State the law of logic that you used. (Lesson 2–3)

(1) A sponge is a sessile animal.

(2) A sessile animal is one that remains permanently attached to a surface for all of its adult life.

39. Algebra Name the property of equality that justifies the statement *If* $x + y = 3$ *and* $3 = w + v$, *then* $x + y = w + v$. (Lesson 2–4)

40. Geology The underground temperature of rocks varies with their depth below Earth's surface. The deeper that a rock is, the hotter it is. The temperature t in degrees Celsius is estimated by the equation $t = 35d + 20$, where d is the depth in kilometers. Solve the formula for d and justify each step. (Lesson 2–4)

A practice test for Chapter 2 is provided on page 794.

ALTERNATIVE ASSESSMENT

COOPERATIVE LEARNING PROJECT

Anatomy The human brain consists of two hemispheres, the left and the right. Mental activities appear to be split between the two sides. The left hemisphere is responsible for the ways we think. These can be described as analytical, linear, sequential, verbal, and concrete. The right side is responsible for very different types of mental skills. These are often described as patterning, holistic, visual, spatial, intuitive, creative, symbolic, nonverbal, artistic, and spontaneous.

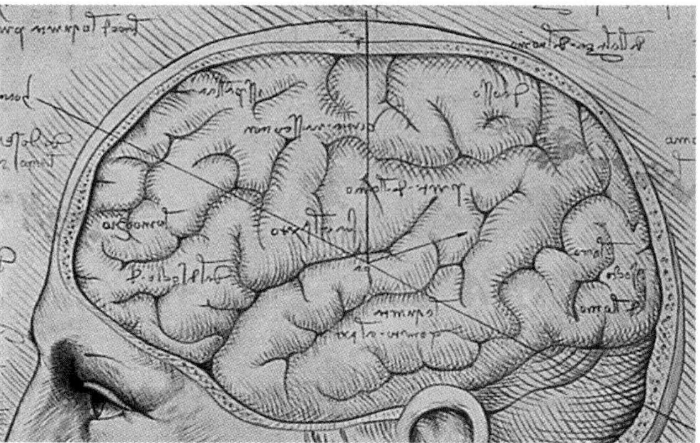

Follow these steps to complete your task.

Select three problems from this chapter.

- Choose one where you came up with the conjecture. Recall and write the steps you went through in your mind to arrive at your conjecture.
- Choose one where the conjecture was provided by the text. Recall and write the steps you went through in your mind to arrive at the proof when you were given the conjecture.
- Choose one where the entire proof was provided. Recall and write the steps you went through in your mind to understand the proof given in the text.

Complete the following table recording the steps you went through for each of the three problems.

Step	Left-Brain	Right-Brain
1		
2		
3		
4		

List each mental activity you performed in the order in which you performed it. Decide whether the activity is left-brain or right-brain. As a group, discuss which activities in mathematics are left-brain and which are right-brain.

THINKING CRITICALLY

Mathematical sentences use very few words. Many words can be replaced by symbols. *If A, then B* can be written as $A \Rightarrow B$. *A and B* is often replaced by $A \wedge B$, and *A or B* is often written as $A \vee B$.

- Do you think it is possible to replace all words in mathematics with symbols?
- What would be some advantages and disadvantages of using all symbols?

PORTFOLIO

Select a proof from the chapter that you completed. Draw a diagram of the proof by putting each step of the proof in a circle. Then connect circles that are logically related. If, in the proof, the step in circle B used the step in circle A, draw a directed arrow from circle A to circle B. Include the diagram in your portfolio.

STANDARDIZED TEST PRACTICE

CHAPTERS 1–2

Section One: MULTIPLE CHOICE

There are ten multiple-choice questions in this section. After working each problem, write the letter of the correct answer on your paper.

1. If $5n + 5 = 10$, find the value of $11 - n$.

A. -10

B. 0

C. 5

D. 10

2. Find the result when 14^{12} is divided by 14^{10}.

A. 14^{22}

B. 14^{-2}

C. 14^{-22}

D. 14^{2}

3. Find the distance between X and Z.

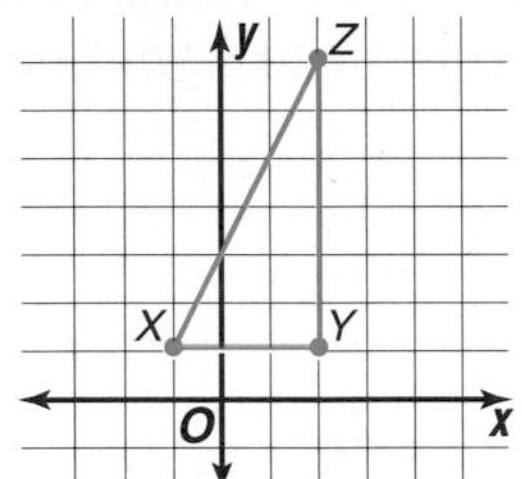

A. $3\sqrt{5}$ units

B. 12 units

C. $3\sqrt{3}$ units

D. 6 units

4. Find x if $0.08x = 32$.

A. 800

B. 400

C. 40

D. 0.04

5. Find the perimeter of a square whose side is 8 meters long.

A. 64 m

B. $64\ m^2$

C. 32 m

D. $32\ m^2$

6. If two planes intersect, their intersection can be

I. a line.

II. three noncollinear points.

III. two intersecting lines.

A. I only

B. II only

C. III only

D. I and II only

7. If $\overrightarrow{BX}$ bisects $\angle ABC$, which of the following is true?

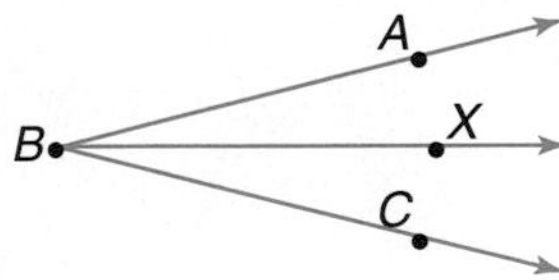

A. $m\angle ABX = m\angle XBC$

B. $m\angle ABX = \frac{1}{2}m\angle ABC$

C. $\frac{1}{2}m\angle ABC = m\angle XBC$

D. all of the above

8. Find x if $2x + 3 = \frac{1}{5}(10x + 15)$.

A. 3

B. 2

C. no real number

D. any real number

9. 30% of 120 is one-third of what number?

A. 10.8

B. 1200

C. 108

D. 12

10. Solve $y = mx + b$ for x.

A. $x = \frac{y}{m} - b$

B. $x = \frac{y - b}{m}$

C. $x = \frac{b - y}{m}$

D. $x = \frac{y}{b} - m$

Test-Taking Tip Sketching a picture for a problem for which no diagram is provided can help you understand and solve many problems.

Section Two: SHORT ANSWER

This section contains seven questions for which you will provide short answers. Write your answer on your paper.

11. In the graph below, the axes and the origin are not shown. If point M has coordinates $(-3,-4)$, what are the coordinates of point N?

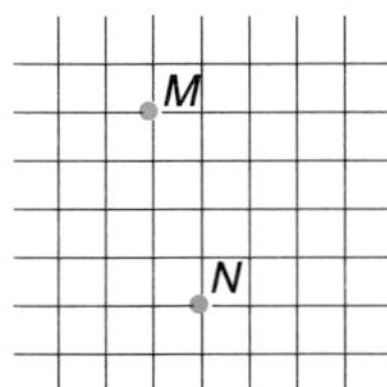

12. Find the expanded form of $(5x - 2)^2$.

13. There are 120 Calories in one cup of grapefruit juice. Write an equation to represent the number of Calories in four cups of grapefruit juice.

14. A farmer fenced all but one side of a square field. If he has already used $3x$ meters of fencing, how many meters will he need for the last side?

15. In the figure below, the hand pointing to 8 moves clockwise to the next numeral every hour. At the end of 24 hours, to which numeral will it point?

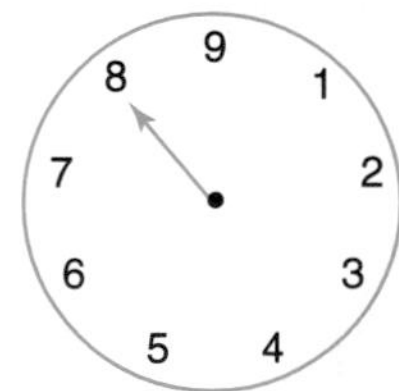

16. What is the greatest integer x such that $9x + 5 < 100$?

17. Westlake has 1600 public payphones. Three-fourths of the phones have dials. If one-third of the dial phones are replaced by push-button phones, how many dial phones remain?

Section Three: COMPARISON

This section contains five comparison problems that involve comparing two quantities, one in column A and one in column B. In certain questions, information related to one or both quantities is centered above them. All variables used represent real numbers.

Compare quantities A and B below.

- Write A if quantity A is greater.
- Write B if quantity B is greater.
- Write C if the two quantities are equal.
- Write D if there is not enough information to determine the relationship.

Column A		Column B
	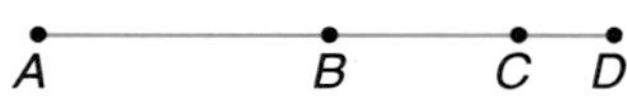	
18.	$AC = 60$, $CD = 12$, B is the midpoint of $\overline{AD}$.	
BC		CD
19. $\dfrac{1}{\frac{3}{2}}$		$\left(\frac{3}{2}\right)^2$
	$\frac{3}{4}x = -24$	
20. $\frac{1}{2}x$		$\frac{3}{2}x$
	$a = -4$	
21. $a^2 + a$		12
	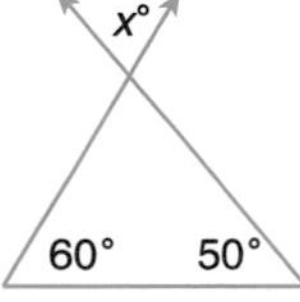	
22. x		60

Test Practice For additional test practice questions, visit: **www.geometry.glencoe.com**

CHAPTER

3 Using Perpendicular and Parallel Lines

What You'll Learn

In this chapter, you will:

- solve problems by drawing a diagram,
- use properties of parallel lines,
- use slope to identify parallel and perpendicular lines,
- prove lines parallel,
- apply distance relationships among points, lines, and planes, and
- analyze properties of spherical geometry.

Why It's Important

Real World

Sports Do you like to ski? One of the keys to the basic downhill movement, the schuss, is keeping the skis parallel. The concept of parallel is often used in geometry. Another geometry concept that you will learn in this chapter is slope. Slope is also commonly used in algebra. Just as a hill or mountain that you ski has a slope or steepness, so does a line. The steeper the line, the larger the absolute value of the number representing the slope. You will learn more about slope in Lesson 3–3.

PREREQUISITE SKILLS

To be successful in this chapter, you'll need to understand these concepts and be able to apply them. Refer to the examples or to the lesson in parentheses if you need more review before beginning the chapter.

Simplify fractions.

Example Write $\frac{18}{45}$ in simplest form.

$\frac{18}{45} = \frac{2 \cdot 3 \cdot 3}{3 \cdot 3 \cdot 5}$ *Write the numerator and denominator in factored form.*

$= \frac{2}{5}$ *Divide the numerator and denominator by $3 \cdot 3$.*

Write each fraction in simplest form.

1. $\frac{8}{10}$ **2.** $-\frac{12}{9}$ **3.** $\frac{54}{18}$ **4.** $-\frac{12}{72}$

Solve proportions using cross products.

Example Solve $-\frac{1}{3} = \frac{a}{15}$.

$(-1)(15) = (a)(3)$ *The cross products of a proportion are equal.*

$-15 = 3a$ *Multiply.*

$-5 = a$ *Divide each side by 3.*

Solve each proportion.

5. $\frac{x}{-2} = \frac{4}{8}$ **6.** $\frac{-5}{6} = \frac{15}{z}$ **7.** $\frac{3}{-b} = \frac{-4}{12}$ **8.** $\frac{17}{51} = \frac{y}{-6}$

Find the distance between two points in a coordinate plane.

(Lesson 1–4)

Find the distance between each pair of points. Round to the nearest hundredth.

9. (0, 0), (2, −2) **10.** (−5, 1), (4, 3)

11. (3, −1), (2, 5) **12.** (−1, −1), (3, −1)

READING SKILLS

In this chapter, you'll be learning about ***parallel*** and ***perpendicular lines***. When other lines intersect with parallel lines, some angles with special relationships are formed. You will need to remember the names of these types of angles. The words *interior* and *exterior* are used to describe some of these special angles. Remember that interior means "inside" and exterior means "outside."

3-1 Parallel Lines and Transversals

What YOU'LL LEARN

- To solve problems by drawing a diagram,
- to identify the relationships between two lines or between two planes, and
- to name angles formed by a pair of lines and a transversal.

Why IT'S IMPORTANT

You can identify the relationship of lines and planes in architecture, agriculture, and air travel.

Business

A corn harvest in Kansas and the multilevel shopping center in Hong Kong show examples of *parallel* planes, *parallel* lines, and *skew* lines.

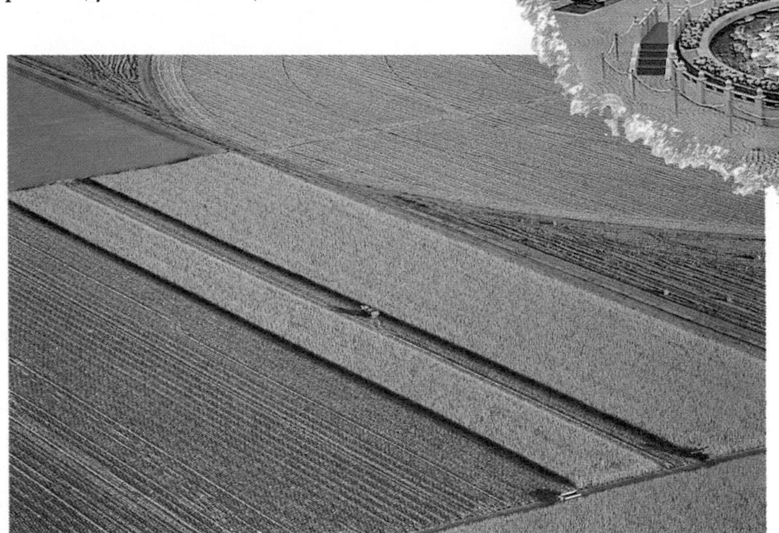

In the picture of the corn field, the rows are considered to be coplanar since they lie in the same plane. The rows represent lines in the plane that never meet. Two lines in a plane that never meet are called **parallel lines**.

In geometry, the symbol $\parallel$ means *is parallel to*. Red arrows are used in diagrams to indicate that lines are parallel. In the figure at the right, the arrows indicate that $\overleftrightarrow{BC}$ is parallel to $\overleftrightarrow{AD}$.

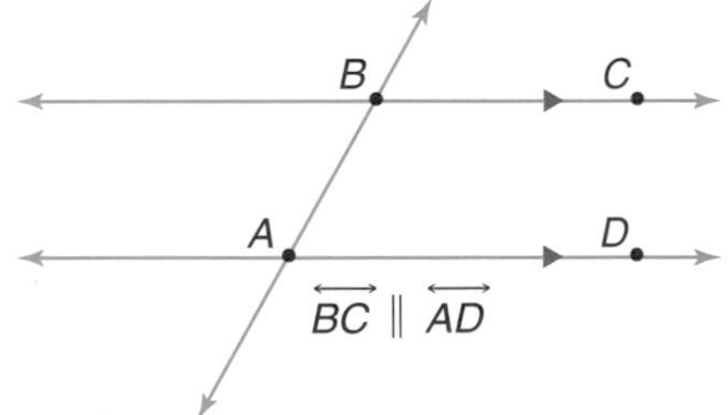

The term parallel and the notation $\parallel$ are used for lines, segments, rays, and planes. The symbol $\nparallel$ means "is not parallel to."

Similarly, two planes can intersect or be parallel. In the photograph of the shopping center in Hong Kong, the floors of each level of shops are contained in parallel planes. The walls and the floor of each level are contained in intersecting planes.

When dealing with geometric terms, it is often helpful to **draw a diagram**. The activity on the next page explains how to draw a three-dimensional figure involving parallel planes.

Drawing a Rectangular Prism

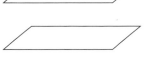

Materials:  plain paper  straightedge  colored pencils

A rectangular prism resembles a shoe box or a cereal box. You can draw a diagram of a rectangular prism by drawing parallel planes.

- Draw two figures to represent the top and bottom of the prism. *These are similar to the figures used to represent parallel planes.*

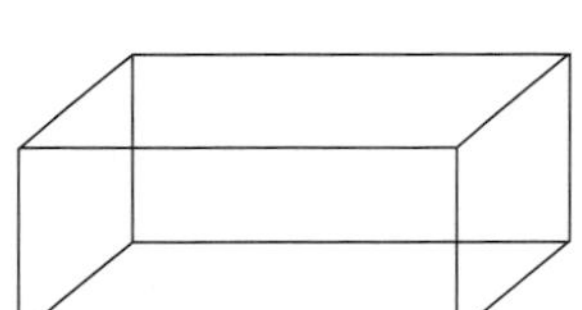

- Draw the vertical edges.

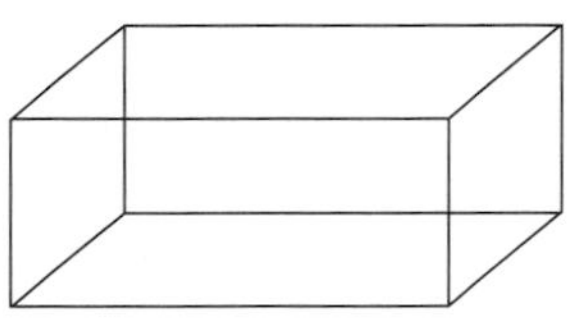

- Make the hidden edges of the prism dashed.

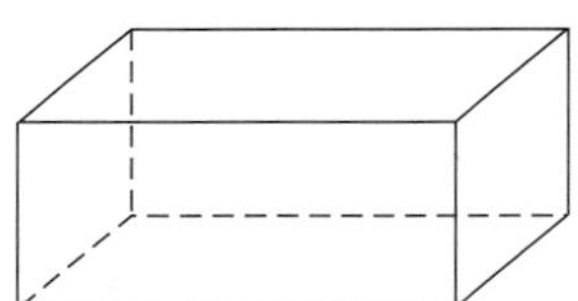

- Label the vertices of the prism.

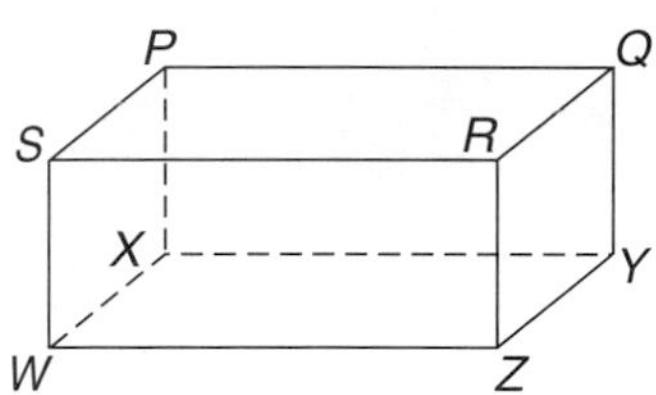

Your Turn

a. Name the parallel planes.

b. Name the planes that intersect plane *PQR* and name their intersection.

In the Modeling Mathematics activity, notice that $\overleftrightarrow{QR}$ and $\overleftrightarrow{WZ}$ do not intersect. Yet they are not parallel since they are not in the same plane. These lines are **skew lines**. The photograph at the right shows a cotton gin in Lubbock, Texas. Walkways $\overline{AB}$ and $\overline{CD}$ in the photograph are an example of skew lines.

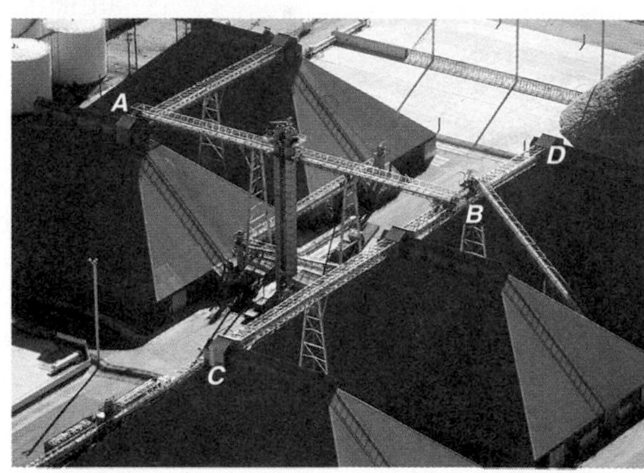

Definition of Skew Lines	**Two lines are skew if they do not intersect and are not in the same plane.**

Example 1 **Refer to the labeled planes and segments in the figure at the right.**

a. Name all planes that are parallel to plane *ABC*.

plane *TKG*

b. Name all segments that intersect $\overline{AB}$.

$\overline{BC}$, $\overline{AC}$, $\overline{AD}$, $\overline{AT}$, and $\overline{BK}$

c. Name all segments that are parallel to $\overline{KG}$.

$\overline{BC}$, $\overline{AD}$, and $\overline{TH}$

d. Name all segments that are skew to $\overline{TK}$.

$\overline{CG}$, $\overline{DH}$, $\overline{AD}$, $\overline{AC}$, and $\overline{BC}$

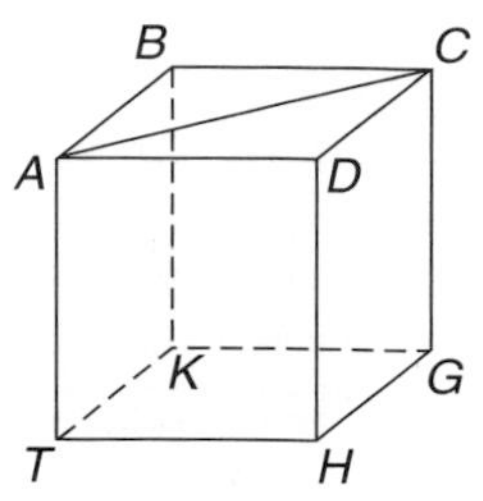

A transversal can also be a ray or line segment.

In the photograph of the irrigated field, notice that line t intersects the rows of plants represented by lines ℓ and m. A line that intersects two or more lines in a plane at different points is called a **transversal**. The lines the transversal intersects need not be parallel.

When the transversal t intersects lines ℓ and m, eight angles are formed. These angles are given special names.

- **exterior angles** $\angle 1, \angle 2, \angle 7, \angle 8$
- **interior angles** $\angle 3, \angle 4, \angle 5, \angle 6$
- **consecutive interior angles** $\angle 3$ and $\angle 5$, $\angle 4$ and $\angle 6$
- **alternate exterior angles** $\angle 1$ and $\angle 8$, $\angle 2$ and $\angle 7$
- **alternate interior angles** $\angle 3$ and $\angle 6$, $\angle 4$ and $\angle 5$
- **corresponding angles** $\angle 1$ and $\angle 5$, $\angle 2$ and $\angle 6$, $\angle 3$ and $\angle 7$, $\angle 4$ and $\angle 8$

Consecutive interior angles are sometimes referred to as same-side interior angles.

Example 2 **Refer to the figure below. Identify each pair of angles as *alternate interior*, *alternate exterior*, *corresponding*, or *consecutive interior* angles.**

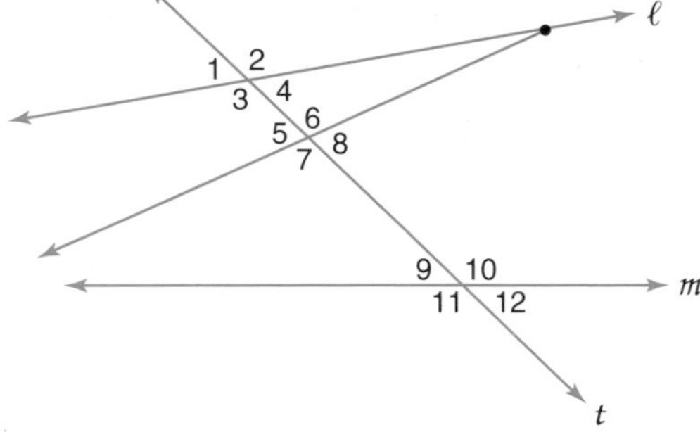

a. **$\angle 1$ and $\angle 8$**
alternate exterior

b. **$\angle 7$ and $\angle 10$**
alternate interior

c. **$\angle 8$ and $\angle 12$**
corresponding

d. **$\angle 1$ and $\angle 5$**
corresponding

e. **$\angle 4$ and $\angle 6$**
consecutive interior

f. **$\angle 8$ and $\angle 9$**
alternate interior

Example 3

Agribusiness

Texas is the leading producer of livestock in the United States. The photograph shows a feed lot in Trent, Texas.

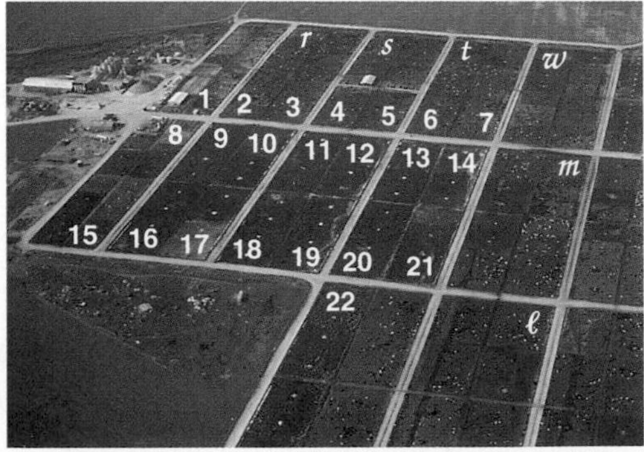

a. **Identify all transversals to lines ℓ and m in the photo.**
r, s, t, w

b. **Identify the special name given to each pair of angles.**

$\angle 2$ and $\angle 10$	alternate interior angles
$\angle 19$ and $\angle 12$	consecutive interior angles
$\angle 15$ and $\angle 17$	corresponding angles

CHECK FOR UNDERSTANDING

Communicating Mathematics

Study the lesson. Then complete the following.

1. The word *parallel* comes from the Greek *parallelos* derived from *para*, beside, and *allelon*, of one another. Do you think these words describe parallel lines? Explain your reasoning.

2. **You Decide** Refer to the figure at the right. Jessica claims that ∠2 and ∠9 are corresponding angles. Enrique disagrees, stating that ∠2 and ∠6 are corresponding angles. Who is correct? Explain your reasoning.

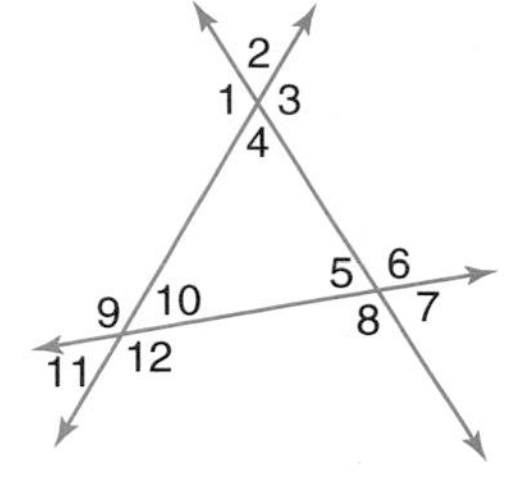

3. **Write** a description of each figure using terms described in this lesson.

a. 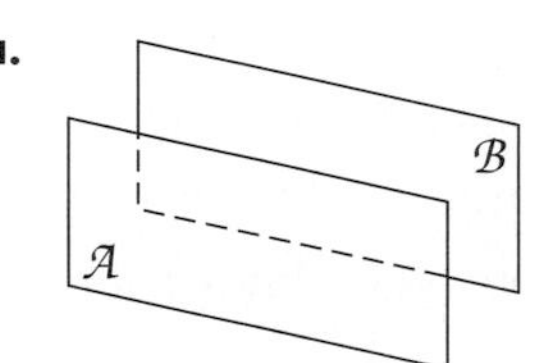

b. 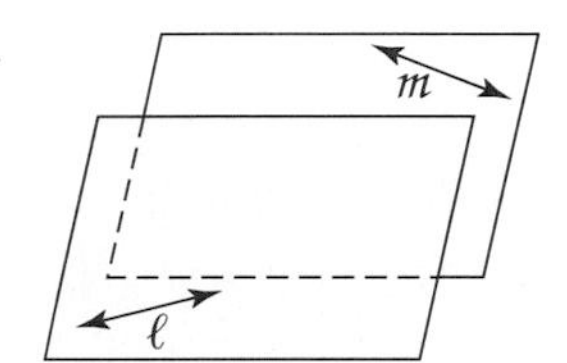

c. 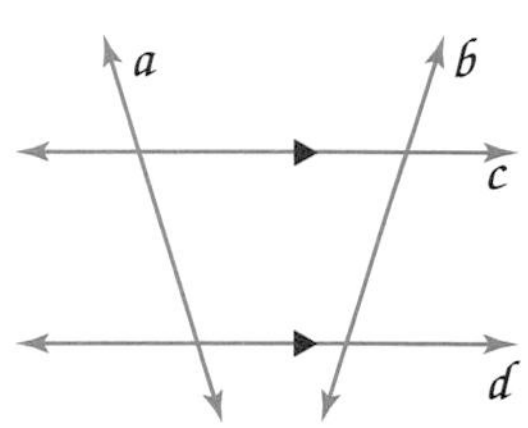

4. Refer to the drawing technique illustrated in the Modeling Mathematics activity on page 125. Draw a prism with triangular bases.

Guided Practice

Describe each of the following as *intersecting, parallel,* or *skew.*

5. railroad crossing sign
6. opposite sides of a cereal box

Determine whether each statement is *true* or *false*. Explain your reasoning.

7. A line that intersects two skew lines is a transversal.
8. If two lines do not intersect, they are parallel.

State the transversal that forms each pair of angles. Then identify the special name for the angle pair.

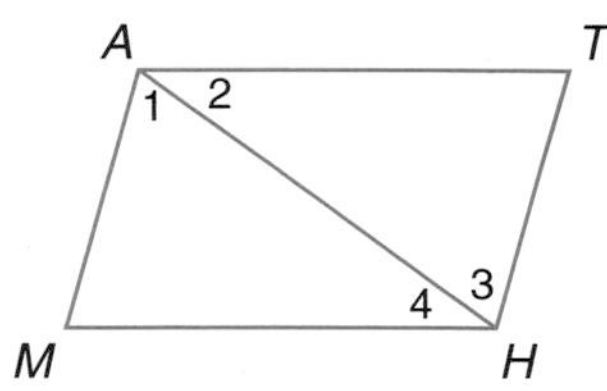

9. ∠1 and ∠3
10. ∠ATH and ∠MAT

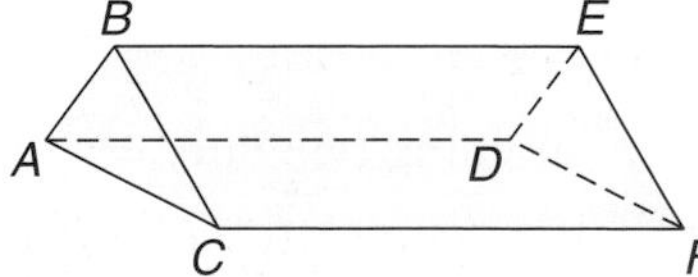

Name each of the following from the figure at the left.

11. all pairs of parallel planes
12. all pairs of skew lines

Draw a diagram to illustrate each of the following.

13. two parallel lines with two nonparallel transversals
14. two intersecting planes
15. **Air Traffic Control** For safety, airplanes heading eastbound are assigned an altitude level that is an odd number of thousands of feet and airplanes heading westbound are assigned an altitude level that is an even number of thousands of feet. If you are in an airplane flying northwest at 32,000 feet and your best friend is in an airplane flying east at 23,000 feet, describe the type of lines formed by the paths of the airplanes. Explain your reasoning.

EXERCISES

Practice

Describe each of the following as *intersecting, parallel,* or *skew.*

16. yard lines on a football field

17. the sides of the Great Pyramid

18. ceiling and wall of a room

19. shelves of a bookcase

20. a flag pole in a park and a road that runs along the edge of the park

21. the service lines on a tennis court

Determine whether each statement is *true* or *false*. Explain your reasoning.

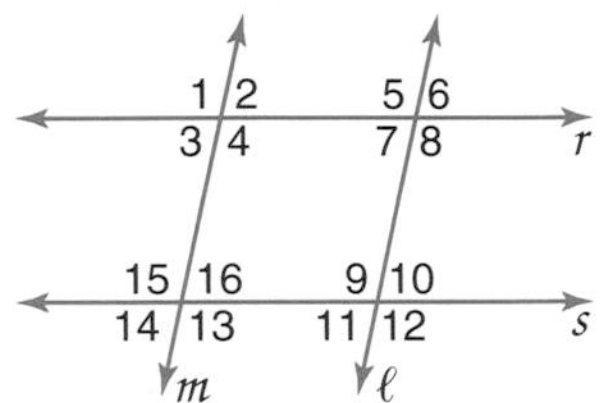

22. Line m is a transversal for lines r and s.

23. $\angle 4$ and $\angle 9$ are consecutive interior angles.

24. $\angle 14$ and $\angle 10$ are alternate exterior angles.

25. $\angle 2$ and $\angle 16$ are corresponding angles.

26. $\angle 7$ and $\angle 10$ are alternate interior angles.

27. $\angle 13$ and $\angle 11$ are formed by lines ℓ and m and transversal r.

State the transversal that forms each pair of angles. Then identify the special name for the angle pair.

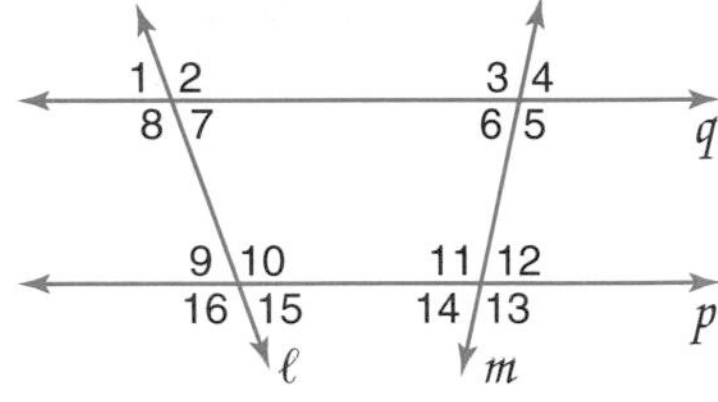

28. $\angle 6$ and $\angle 7$

29. $\angle 16$ and $\angle 2$

30. $\angle 13$ and $\angle 5$

31. $\angle 8$ and $\angle 10$

32. $\angle 11$ and $\angle 15$

33. $\angle 4$ and $\angle 8$

Name each of the following from the figure at the right.

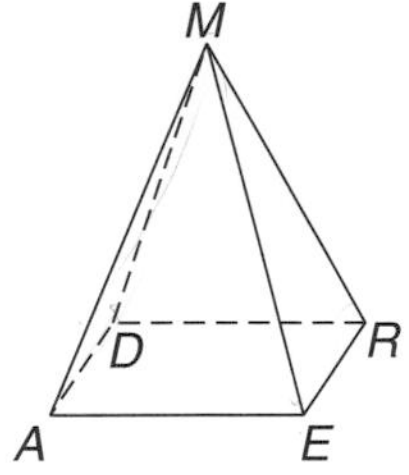

34. all pairs of intersecting planes

35. all pairs of parallel segments

36. all pairs of skew segments

37. all pairs of parallel planes

38. all points contained in four lines

39. all planes intersecting with plane ADM

Draw a diagram to illustrate each of the following.

40. two parallel planes

41. two parallel planes containing two lines that are skew

42. three parallel planes with a line intersecting the planes

43. two parallel lines with a plane intersecting the lines

44. Draw a Diagram Use the technique in the Modeling Mathematics activity on page 125 to draw a prism with bases that are hexagons (6-sided figures).

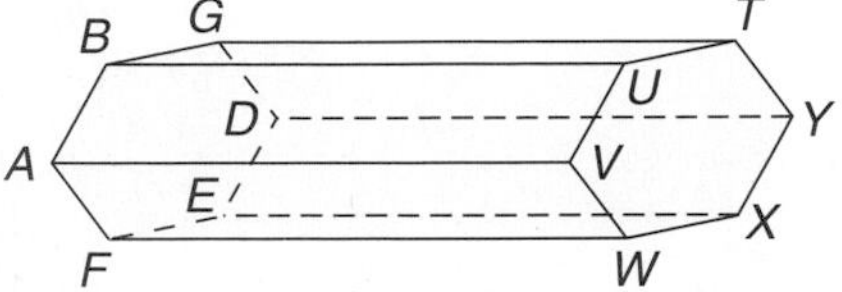

a. The figure at the right may show a different perspective than the prism you drew. Explain the difference between the "views" of the two prisms.

b. Identify all segments skew to $\overline{TU}$.

c. Identify all planes parallel to plane AFW.

Critical Thinking

45. **Optical Illusion** In the artist's drawing, lines ℓ and m are parallel, but appear to be bowed, due to the many transversals drawn through ℓ and m.
    a. Use a protractor to measure $\angle 1$ and $\angle 2$.
    b. Notice that $\angle 1$ and $\angle 2$ are a pair of alternate interior angles and make a conjecture about the relationship between the alternate interior angles if lines ℓ and m are parallel.

46. Suppose there is a line m and a point A not on the line. In space, how many lines can be drawn through A that do not intersect m? In space, how many lines can be drawn through A that are parallel to m?

Applications and Problem Solving

47. **Music** The word *parallel* is used in music to describe songs moving consistently by the same intervals such as harmony with parallel voices. Find at least two additional uses of the word *parallel* in other school subjects such as history, electronics, computer science, or English.

48. **Graphics** Locate an example of parallel lines with a transversal in a newspaper or magazine. Make a copy of the picture or diagram and label the lines and angles.

49. **Draw a Diagram** Square dancing involves four couples. If each member of the square shakes hands with every other member of the square except his or her partner before the dance begins, what is the total number of handshakes?

Mixed Review

50. Write a two-column proof. (Lesson 2–5)

    Given: $\overline{AB} \cong \overline{FE}$
    $\overline{BC} \cong \overline{ED}$

    Prove: $\overline{AC} \cong \overline{FD}$

    A B C

    D E F

51. Write the given and prove statements that you would use to prove the statement *Opposite angles of a parallelogram are congruent.* Draw a figure. (Lesson 2–5)

52. Name the property of equality that justifies the statement *If $m\angle A = m\angle B$, then $m\angle B = m\angle A$.* (Lesson 2–4)

53. **Science** Write *A cloud is composed of millions of water droplets* in if-then form. (Lesson 2–2)

54. The measure of an angle is $9x + 14$, and the measure of its supplement is $12x + 19$. Find the value of x. (Lesson 1–7)

55. **SAT Practice** For a positive integer x, 1 percent of x percent of 10,000 equals

    A $0x$ **B** x **C** $10x$ **D** $100x$ **E** $1000x$

56. T is between R and S. If $TS = 7$ and $RS = 20$, find RT. (Lesson 1–4)

57. Explain the four steps for solving any problem. (Lesson 1–3)

58. Graph points $A(-4, 7)$ and $B(3, -5)$ on the same coordinate plane. Name the quadrants in which each point lies. (Lesson 1–1)

Algebra

59. Name the coefficient of p^2 in $\frac{2p^2}{5}$.

60. State the additive inverse and absolute value of -16.

For **Extra Practice**, see page 768.

3–2A Using Technology Angles and Parallel Lines

A Preview of Lesson 3–2

You can use a calculator to investigate the measures of the angles formed by two parallel lines and a transversal.

- First clear your screen by using the 8:Clear All on the F8 menu.
- Draw any line on your screen. Label two points A and B on the line.
- You can draw a line parallel to $\overleftrightarrow{AB}$ by selecting 2:Parallel Line from the F4 menu. Move the cursor onto AB and the message "PARALLEL TO THIS LINE" appears. Press ENTER. Move the cursor to a new location and press ENTER. The parallel line will appear. Label two points C and D on the second line.
- Draw any line GH intersecting the two parallel lines and label the intersection points as E and F.

The screen at the right is a sample of two parallel lines cut by a transversal.

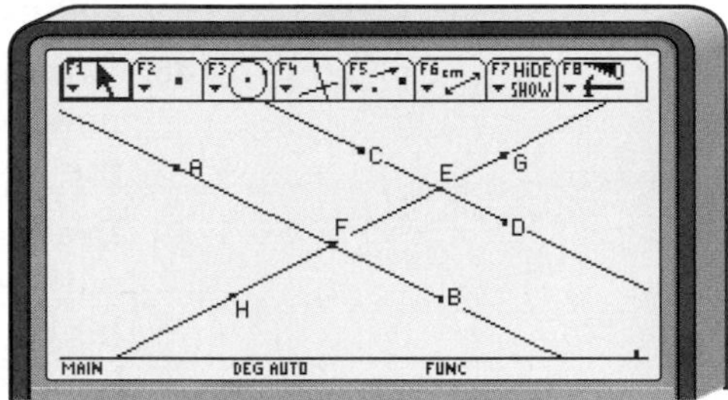

- Measure each of the angles by selecting 3:Angle on the F6 menu. Remember that you need to select three points on the angle to define it in order for the calculator to measure the angle.

EXERCISES

Analyze your drawing.

1. List pairs of angles by the special names you learned in Lesson 3–1.
2. Which pairs of angles listed in Exercise 1 are congruent?
3. Make a conjecture about the following pairs of angles formed by two parallel lines and a transversal.
   a. corresponding angles
   b. alternate interior angles
   c. alternate exterior angles

Test your conjectures.

4. Rotate the transversal. Are the angles with equal measures in the same relative location as the angles with equal measures in the original drawing?
5. Rotate the transversal so that the measure of one of the angles is 90.
   a. What do you notice about the measures of the other angles?
   b. Make a conjecture about a transversal that is perpendicular to one of two parallel lines.

3-2 Angles and Parallel Lines

What YOU'LL LEARN

- To use the properties of parallel lines to determine angle measures.

Why IT'S IMPORTANT

You can apply properties of parallel lines in construction and interior decorating.

Real World APPLICATION

Graphic Art

In the graph at the right, examine the method used by the artist to illustrate the number of teams making the playoffs. The parallel line segments illustrate the total number of teams, and a *transversal* is used to indicate the number of teams missing the playoffs.

Teams Making the Playoff

Source: *USA TODAY* research

In the Major League Baseball display, $\angle 1$ and $\angle 2$ are a pair of *corresponding angles*. What is the relationship between a pair of corresponding angles formed by two parallel lines cut by a transversal?

MODELING MATHEMATICS

Corresponding Angle Measures

Materials: notebook paper  protractor 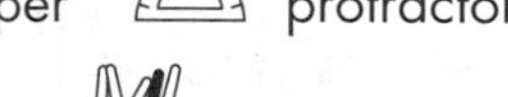 straightedge colored pencils

You can model parallel lines by using the lines printed on notebook paper.

- Use a pencil and straightedge to darken two lines on a piece of notebook paper. Use your straightedge to draw transversal *t*.

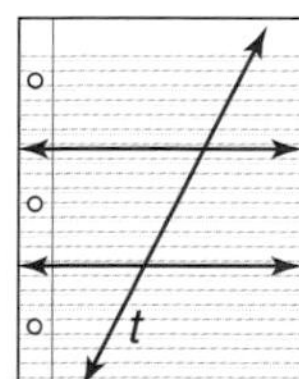

- Label each angle.

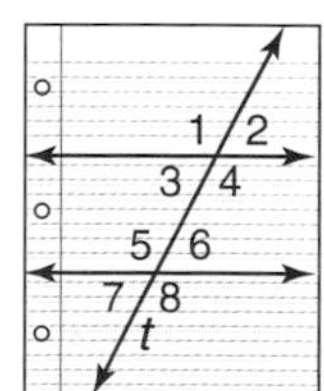

Your Turn

a. Use your protractor to measure each of the four pairs of corresponding angles.

b. Make a conjecture about the corresponding angles formed by two parallel lines cut by a transversal.

c. What appears to be true about alternate interior angles? consecutive interior angles? alternate exterior angles?

The property relating corresponding angles shown in the Modeling Mathematics activity is accepted as a postulate.

Postulate 3–1 Corresponding Angles Postulate	**If two parallel lines are cut by a transversal, then each pair of corresponding angles is congruent.**

The postulate, combined with linear pair and vertical angle properties, helps to establish several angle relationships.

Example 1

Agriculture

The road displayed in the diagram divides a farm into two parts. The opposite edges of the field are parallel, and the road is a transversal. If $m\angle 6 = 115$, find $m\angle 3$.

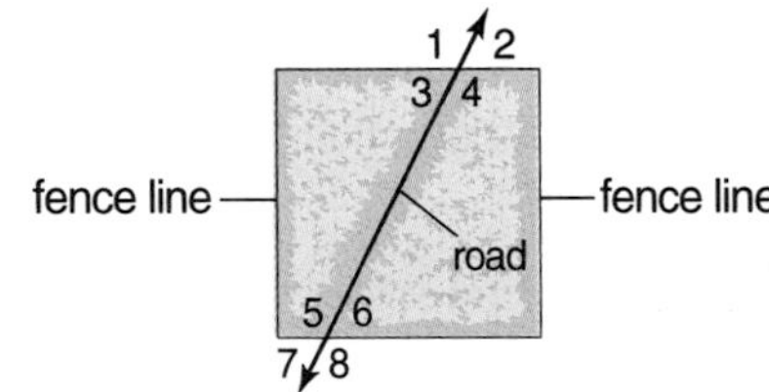

$m\angle 6 = m\angle 2$ *Corresponding Angle Postulate*

$m\angle 2 = m\angle 3$ *Vertical angles have equal measures.*

$m\angle 6 = m\angle 3$ *Transitive Property (=)*

$115 = m\angle 3$ *Substitution Property (=)*

Notice that in Example 1, $\angle 3$ and $\angle 6$ form a pair of alternate interior angles and were shown to have equal measures. This is an application of another of the special relationships between the angles formed by two parallel lines and a transversal. These relationships are summarized in Theorems 3–1, 3–2, and 3–3. *You will be asked to prove Theorems 3–1 and 3–2 in Exercises 44 and 45, respectively.*

Theorem 3–1 Alternate Interior Angles Theorem	**If two parallel lines are cut by a transversal, then each pair of alternate interior angles is congruent.**

Theorem 3–2 Consecutive Interior Angles Theorem	**If two parallel lines are cut by a transversal, then each pair of consecutive interior angles is supplementary.**

Theorem 3–3 Alternate Exterior Angles Theorem	**If two parallel lines are cut by a transversal, then each pair of alternate exterior angles is congruent.**

Proof of Theorem 3–3

Given: $p \parallel q$

ℓ is a transversal of p and q.

Prove: $\angle 1 \cong \angle 8$, $\angle 2 \cong \angle 7$

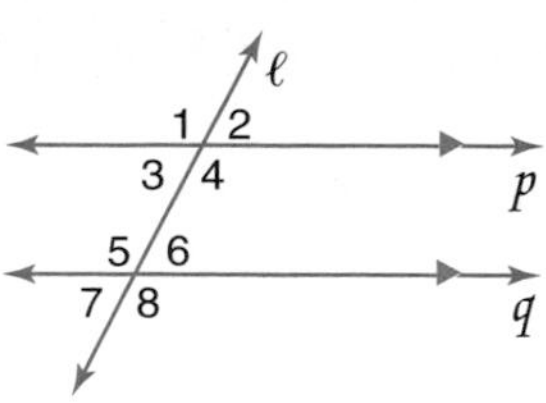

Paragraph Proof:

We are given that $p \parallel q$. If two parallel lines are cut by a transversal, corresponding angles are congruent. So, $\angle 1 \cong \angle 5$ and $\angle 2 \cong \angle 6$. $\angle 5 \cong \angle 8$ and $\angle 6 \cong \angle 7$ because vertical angles are congruent. Therefore, $\angle 1 \cong \angle 8$ and $\angle 2 \cong \angle 7$ since congruence of angles is transitive.

The special angle relationships between the angles formed by two parallel lines and a transversal can be used to solve for unknown values.

Example 2

Algebra

In the figure at the right, $\overleftrightarrow{MA} \parallel \overleftrightarrow{HT}$ and $\overleftrightarrow{NG} \parallel \overleftrightarrow{EL}$. Find the values of x, y, and z.

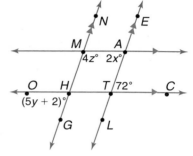

Find x. Since $\overleftrightarrow{MA} \parallel \overleftrightarrow{HT}$, $\angle MAT \cong \angle ATC$ by the Alternate Interior Angles Theorem.

$m\angle MAT = m\angle ATC$ *Def. ≅ ∠s*

$2x = 72$ *$m\angle MAT = 2x$, $m\angle ATC = 72$*

$x = 36$

Find z. Since $\overleftrightarrow{NG} \parallel \overleftrightarrow{EL}$, $\angle GMA$ and $\angle TAM$ are supplementary by the Consecutive Interior Angles Theorem.

$m\angle GMA + m\angle TAM = 180$ *Def. supplementary ∠s*

$4z + 2x = 180$ *$m\angle GMA = 4z$, $m\angle TAM = 2x$*

$4z + 2(36) = 180$ *Substitute 36 for x.*

$4z + 72 = 180$

$4z = 108$

$z = 27$

Find y. Since $\overleftrightarrow{NG} \parallel \overleftrightarrow{EL}$, $\angle GHO \cong \angle ATC$ by the Alternate Exterior Angles Theorem.

$m\angle GHO = m\angle ATC$ *Def. ≅ ∠s*

$5y + 2 = 72$ *$m\angle GHO = 5y + 2$, $m\angle ATC = 72$*

$5y = 70$

$y = 14$

Therefore, $x = 36$, $y = 14$, and $z = 27$.

There is a special relationship that occurs when one of two parallel lines is cut by a perpendicular line.

Theorem 3–4 Perpendicular Transversal Theorem	**In a plane, if a line is perpendicular to one of two parallel lines, then it is perpendicular to the other.**

You will be asked to prove this theorem in Exercise 43.

CHECK FOR UNDERSTANDING

Communicating Mathematics

Study the lesson. Then complete the following.

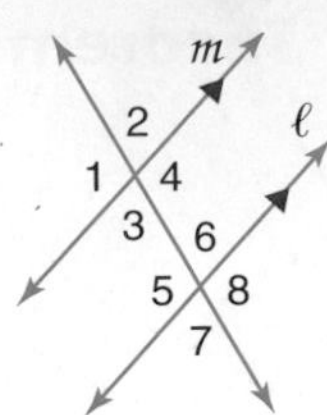

1. **Explain** why $\angle 3$ and $\angle 5$ must be supplementary.
2. **Describe** two different methods you could use to find $m\angle 3$ if $m\angle 8 = 130$.

Refer to the figure at the right for Exercises 3–5.

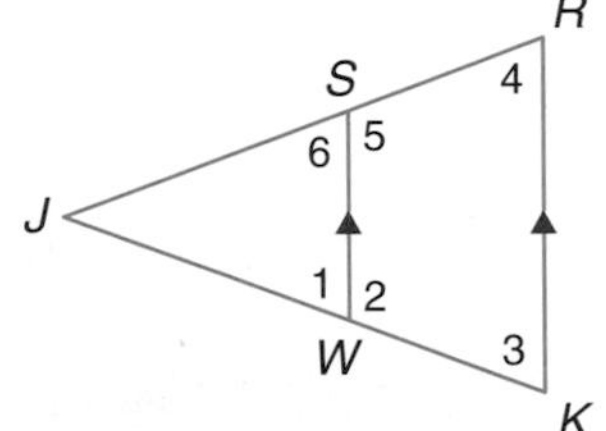

3. **Explain** what the arrowheads on the lines in the diagrams at the right indicate.
4. **Explain** why $\angle 1 \cong \angle 3$ and $\angle 6 \cong \angle 4$ if $\overline{SW} \parallel \overline{RK}$.
5. **Describe** how you could find $m\angle 4$ if $\overline{SW} \parallel \overline{RK}$ and $m\angle 5 = 110$.

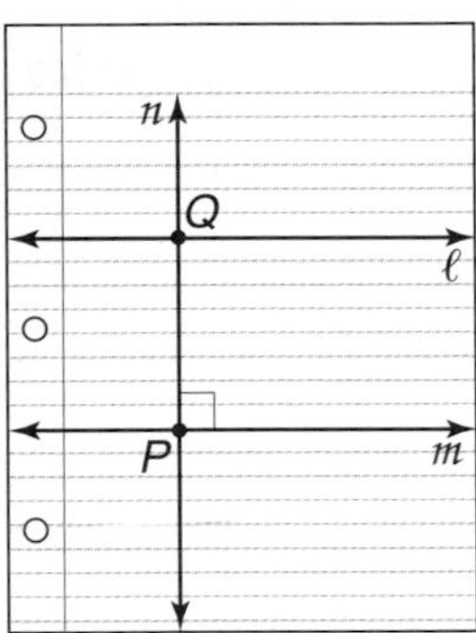

6. Draw and label parallel lines ℓ and m on notebook paper with point P on line m. Use a protractor to draw line n through P, perpendicular to line m. Label the point of intersection Q.
   a. Use your protractor to measure one of the four angles at Q.
   b. Does the result verify the conclusion stated in the Perpendicular Transversal Theorem (Theorem 3–4)? Explain your reasoning.

Guided Practice

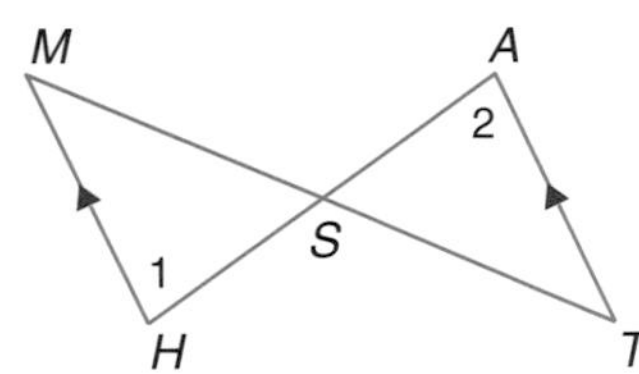

7. State the postulate or theorem that allows you to conclude that $\angle 1 \cong \angle 2$.

In the figure at the right, $p \parallel q$, $m\angle 1 = 107$, and $m\angle 11 = 48$. Find the measure of each angle.

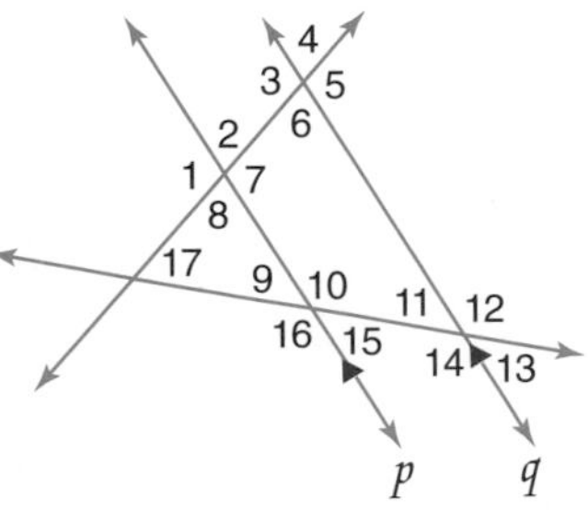

8. $\angle 3$
9. $\angle 5$
10. $\angle 13$
11. $\angle 9$
12. $\angle 15$
13. $\angle 17$

Find the values of x and y in each figure.

14.

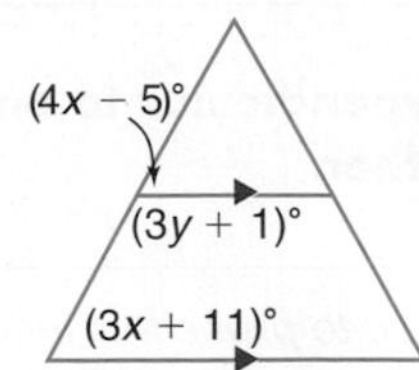

15.

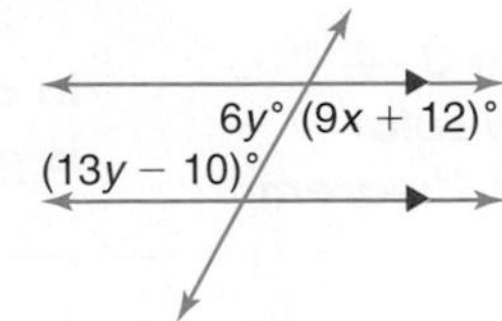

16. **Construction** A carpenter is building a flight of stairs. If the tops of the stairs are parallel to the floor and the stringer makes a 25° angle with the floor, find the measure of the angle formed by the tops of the steps and the stringer.

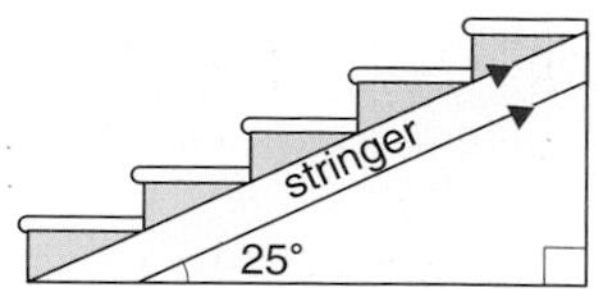

EXERCISES

Practice

In Exercises 17–19, state the postulate or theorem that allows you to conclude that $\angle 1 \cong \angle 2$.

17. 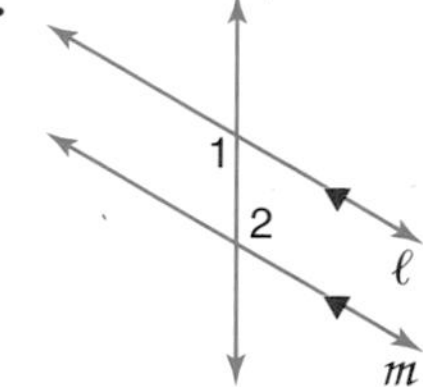

18. 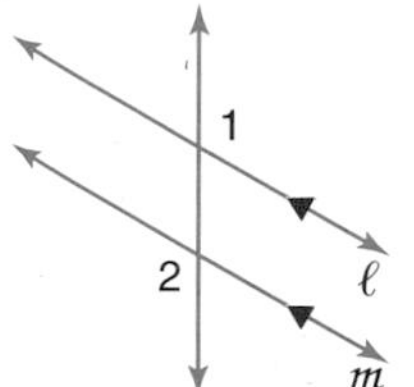

19. 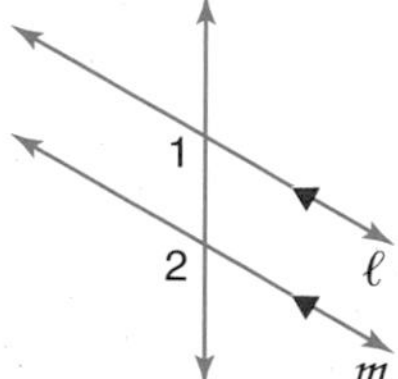

In the figure, $x \parallel y$, $\overline{ST} \parallel \overline{RQ}$, and $m\angle 1 = 131$. Find the measure of each angle.

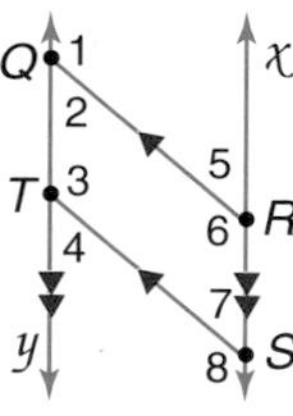

20. $\angle 6$
21. $\angle 7$
22. $\angle 4$
23. $\angle 2$
24. $\angle 5$
25. $\angle 8$

In the figure, $\overline{AB} \parallel \overline{EC}$, $m\angle 1 = 58$, $m\angle 2 = 47$, and $m\angle 3 = 26$. Find the measure of each angle.

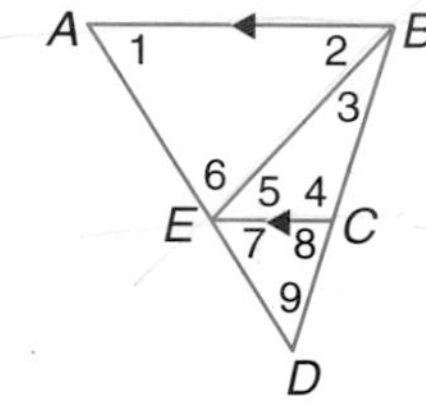

26. $\angle 7$
27. $\angle 5$
28. $\angle 6$
29. $\angle 4$
30. $\angle 8$
31. $\angle 9$

In the figure $\overline{BG} \parallel \overline{CE}$, $\overline{BE} \parallel \overline{CD}$, $\overrightarrow{BG}$ bisects $\angle EBA$, $m\angle 8 = 42$, and $m\angle 3 = 18$. Find the measure of each angle.

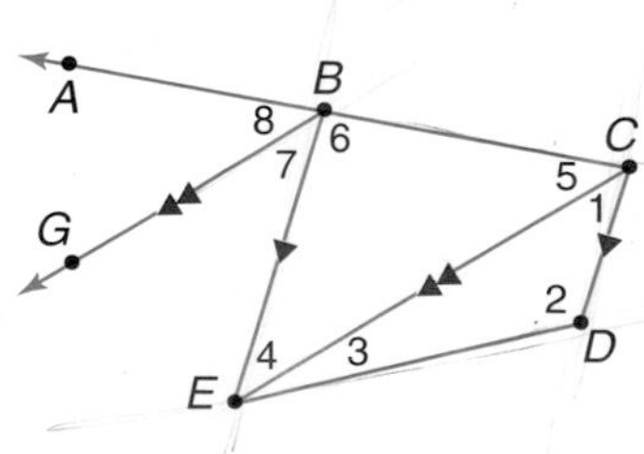

32. $\angle 7$
33. $\angle 5$
34. $\angle 1$
35. $\angle 4$
36. $\angle 6$
37. $\angle 2$

Find the values of *x*, *y*, and *z* in each figure.

38. 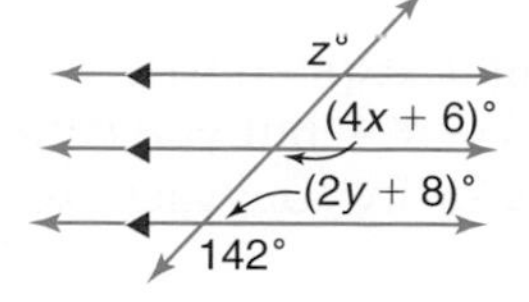

39. 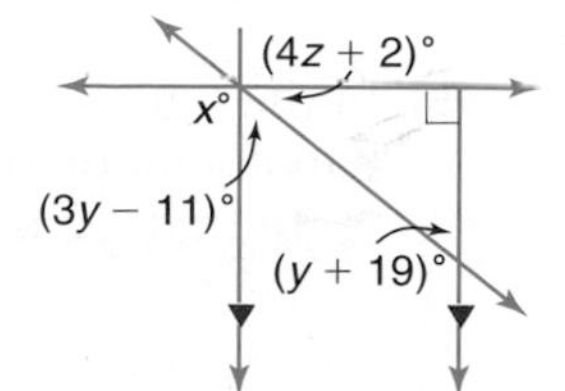

40. 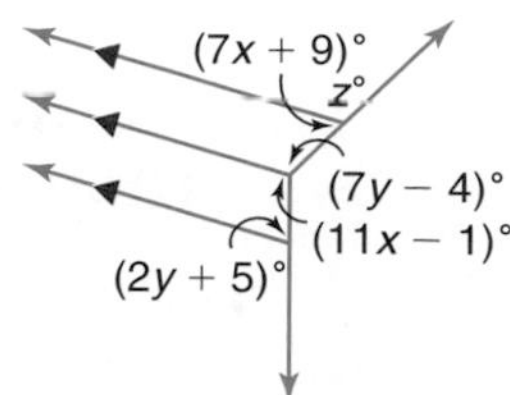

Refer to the figure at the right for Exercises 41–42.

41. If $m\angle 4 = 2x - 25$ and $m\angle 8 = x + 26$, find $m\angle 2$. Explain your reasoning.

42. If $m\angle 6 = 2x + 43$ and $m\angle 7 = 5x + 11$, find $m\angle 5$. Explain your reasoning.

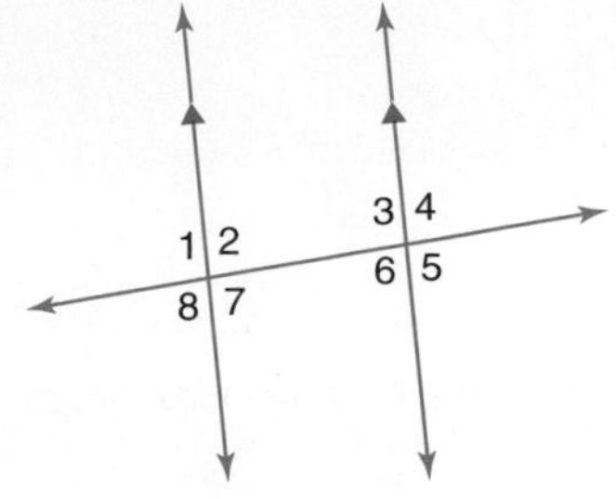

43. Copy and complete the proof of Theorem 3–4.

Given: $m \perp \ell$

$\ell \parallel p$

Prove: $m \perp p$

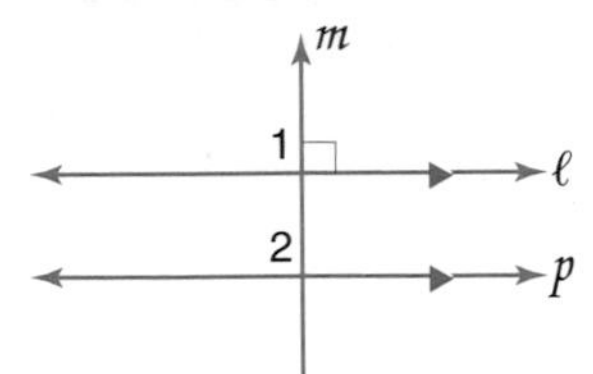

Proof:

Statements	Reasons
1. $m \perp \ell$, $\ell \parallel p$	1. ?
2. $\angle 1$ is a right angle.	2. ?
3. $m\angle 1 = 90$	3. ?
4. $\angle 1 \cong \angle 2$	4. ?
5. $m\angle 1 = m\angle 2$	5. ?
6. $m\angle 2 = 90$	6. ?
7. $\angle 2$ is a right angle.	7. ?
8. $m \perp p$	8. ?

44. Write a two-column proof of Theorem 3–1.

45. Write a paragraph proof of Theorem 3–2.

46. Find the measures x and y in the figure at the right, if $\overline{AF} \parallel \overline{BC}$ and $\overline{AE} \parallel \overline{CD}$.

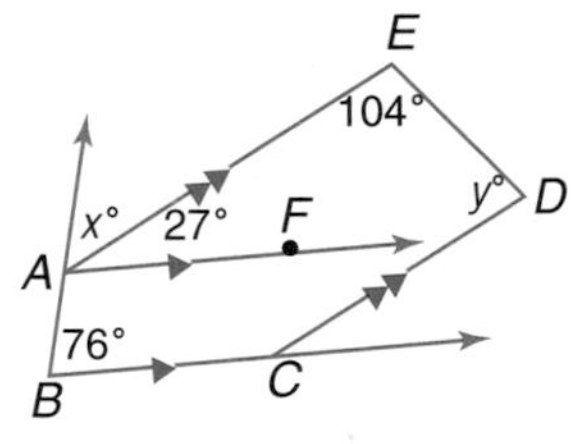

47. Given: $\overline{MQ} \parallel \overline{NP}$

$\angle 4 \cong \angle 3$

Prove: $\angle 1 \cong \angle 5$

48. Find $m\angle TDK$ in the figure at the right.

Critical Thinking

49. In the figure at the right, explain why you can conclude that $\angle 1 \cong \angle 4$, but you cannot state that $\angle 3$ is necessarily congruent to $\angle 2$.

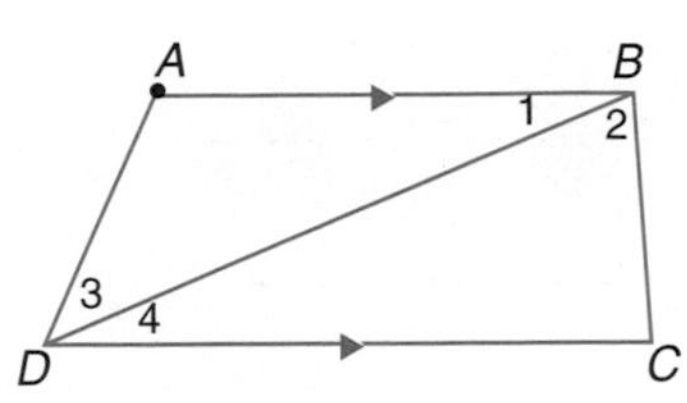

50. In the figure at the right, explain why you can conclude that $\angle 2$ and $\angle 6$ are supplementary, but you cannot state that $\angle 4$ and $\angle 6$ are necessarily supplementary.

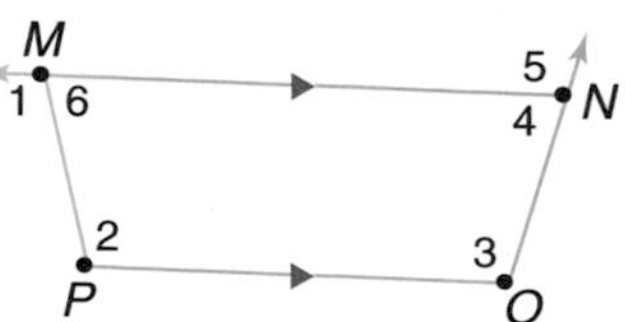

Applications and Problem Solving

51. **Statistics** Refer to the application at the beginning of the lesson.
    a. Did the artist use the transversal segment to include or exclude the number of playoff teams? Explain your reasoning.
    b. Why do you think professional sports allow so many teams into the playoffs?

52. **Interior Decorating** Walls in houses are never perfectly vertical. To hang wallpaper, a true vertical line must be established so the pattern looks nice. The paperhanger uses a plumb line, which is a piece of string with a weight at the bottom, to make the vertical line for the first piece of wallpaper. How can she be sure that all of the pieces of wallpaper are vertical if she does not use the plumb line again?

Mixed Review

53. If $m \parallel n$ and $n \parallel p$, is $m \parallel p$? Explain. (Lesson 3–1)

54. **Draw a Diagram** Halfway through her bus trip from Savannah to Jacksonville, Niabi fell asleep. When she awoke, she still had to travel half of the distance she traveled when asleep. For what fraction of the trip was Niabi asleep? (Lesson 3–1)

55. Find the value of x. (Lesson 2–6)

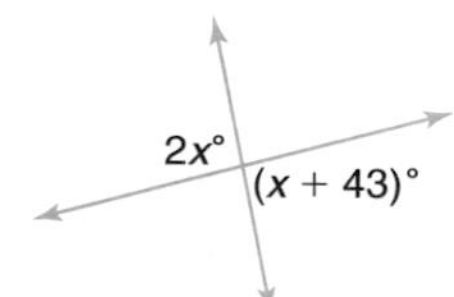

56. Write a complete proof of *If* $5x - 7 = x + 1$, *then* $x = 2$. (Lesson 2–4)

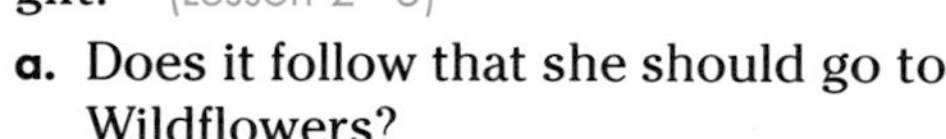

57. **Advertising** An ad for Wildflowers Gift Boutique says *When it has to be special, it has to be Wildflowers.* Catalina needs a special gift. (Lesson 2–3)
    a. Does it follow that she should go to Wildflowers?
    b. Which law was used to make this conclusion?

58. Write the converse of the conditional *If two lines are parallel, then they lie in the same plane and do not intersect.* (Lesson 2–2)

59. What type of angles are $\angle M$ and $\angle N$? $\angle M$ and $\angle N$ are vertical angles and $m\angle M = 4x + 14$ and $m\angle N = 6x - 24$. (Lesson 1–7)

60. **ACT Practice** $(-2)^4 + 4^{-2} - \frac{9}{16} =$

    A $-\frac{9}{16}$ B $-\frac{7}{16}$ C $17\frac{7}{16}$ D $15\frac{1}{2}$ E $16\frac{1}{2}$

61. Points W, X, Y, and Z are collinear. How many ways can you use two letters to name the line that contains these four points? (Lesson 1–2)

For Extra Practice, see page 768.

Algebra

62. Find $\sqrt{576}$.

63. Solve $\frac{r}{28} = \frac{5}{7}$.

3–3 Integration: Algebra Slopes of Lines

What YOU'LL LEARN

- To find the slopes of lines, and
- to use slope to identify parallel and perpendicular lines.

Why IT'S IMPORTANT

The slopes of parallel and perpendicular lines will help you to investigate ways to prove lines are parallel and to find the distance between points and lines.

Real World APPLICATION

Skiing

Snowbird Ski Resort is located in the Wasatch-Cache National Forest outside Salt Lake City, Utah. The base of the resort is at an elevation of 7900 feet. The highest ski run starts on Hidden Peak at an elevation of 11,000 feet. A brochure for the Snowbird Resort claims that the vertical rise for the mountain at Snowbird is 11,000 − 7900 or 3100 feet. Skiers sometimes refer to this as the *slope* of the mountain.

In a coordinate plane, the **slope** of a line is the ratio of its vertical rise to its horizontal run.

$$\text{slope} = \frac{\text{vertical rise}}{\text{horizontal run}}$$

Definition of Slope	**The slope m of a line containing two points with coordinates (x_1, y_1) and (x_2, y_2) is given by the formula** $m = \frac{y_2 - y_1}{x_2 - x_1}$, **where $x_1 \neq x_2$.**

The slope of a vertical line, where $x_1 = x_2$, is undefined.

The slope of a line indicates whether the line rises to the right, falls to the right, or is horizontal.

Example  **Find the slope of each line.**

a.

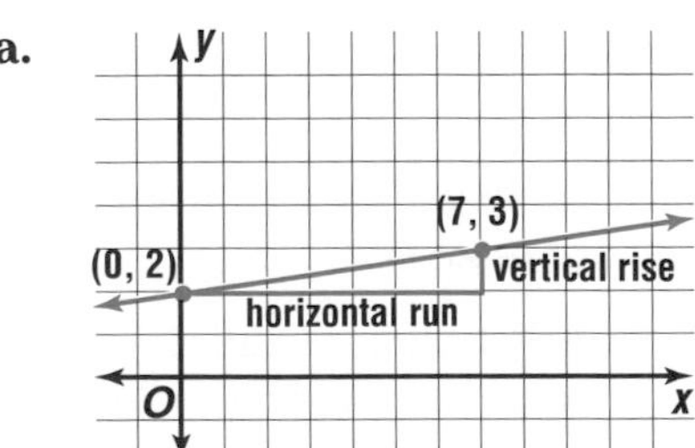

Let (0, 2) be (x_1, y_1) and (7, 3) be (x_2, y_2).

$$m = \frac{y_2 - y_1}{x_2 - x_1} = \frac{3 - 2}{7 - 0} \text{ or } \frac{1}{7}$$

Lines with positive slope rise as you move from left to right.

b.

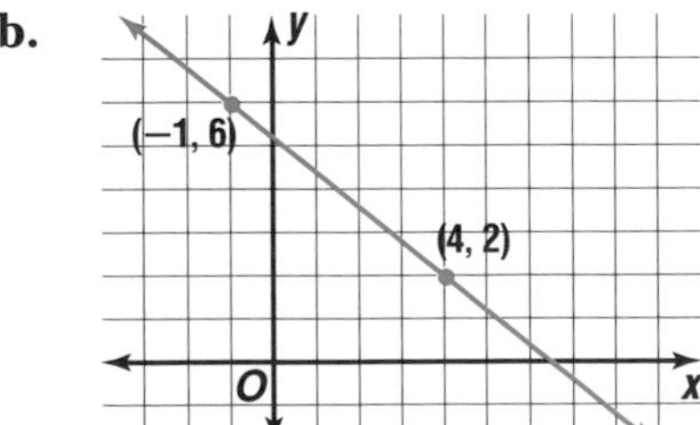

Let (−1, 6) be (x_1, y_1) and (4, 2) be (x_2, y_2).

$$m = \frac{y_2 - y_1}{x_2 - x_1} = \frac{2 - 6}{4 - (-1)} \text{ or } -\frac{4}{5}$$

Lines with negative slope fall as you move from left to right.

When finding slope, either point can be (x_1, y_1). Just be sure to subtract the coordinates in the same order.

c.

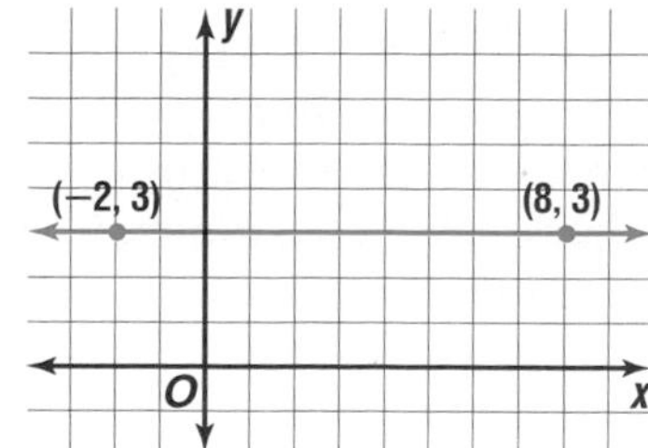

$$m = \frac{y_2 - y_1}{x_2 - x_1}$$
$$= \frac{3 - 3}{8 - (-2)}$$
$$= \frac{0}{10} \text{ or } 0$$

Lines with a slope of 0 are horizontal.

d.

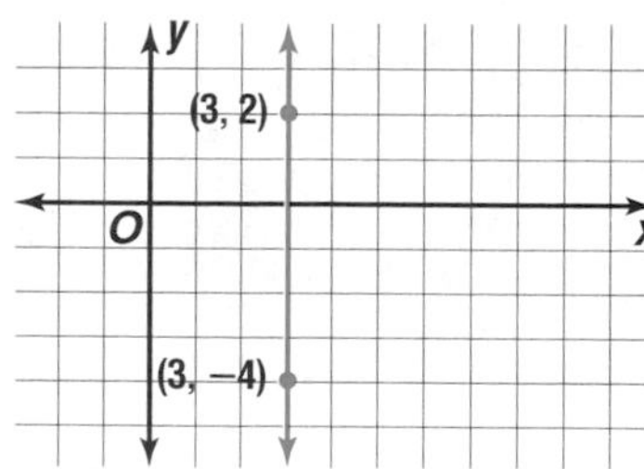

$$m = \frac{y_2 - y_1}{x_2 - x_1}$$
$$= \frac{2 - (-4)}{3 - 3}$$
$$= \frac{6}{0} \text{ or } \textit{undefined}$$

Lines with an undefined slope are vertical.

The graphs of lines r, s, and t are shown at the right. Lines r and s are parallel, and t is perpendicular to r and s. Let's investigate the slopes of these lines.

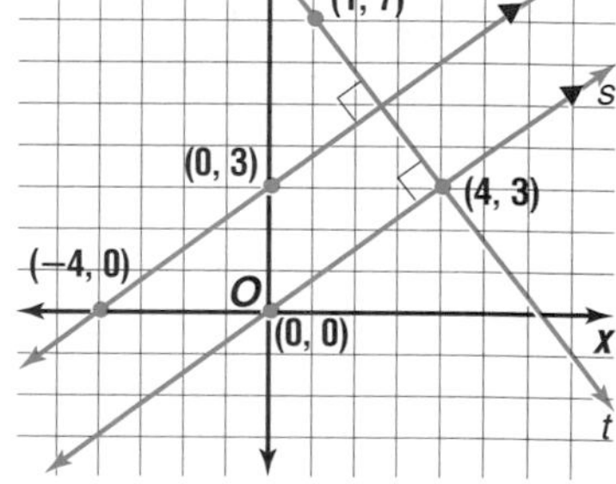

slope of r	slope of s	slope of t
$m = \frac{0 - 3}{-4 - 0}$	$m = \frac{0 - 3}{0 - 4}$	$m = \frac{3 - 7}{4 - 1}$
$= \frac{3}{4}$	$= \frac{3}{4}$	$= -\frac{4}{3}$

Lines r and s are parallel, and their slopes are the same. Line t is perpendicular to lines r and s, and its slope is the negative reciprocal of the slopes of r and s; that is, $-\frac{4}{3} \cdot \frac{3}{4} = -1$. These results suggest two important algebraic properties of parallel and perpendicular lines.

Postulate 3–2	**Two nonvertical lines have the same slope if and only if they are parallel.**
Postulate 3–3	**Two nonvertical lines are perpendicular if and only if the product of their slopes is −1.**

Note that Postulates 3–2 and 3–3 are written in **if and only if** form. If both a conditional and its converse are true, it can be written in *if and only if* form.

Example 2 **Given $A(-3, -2)$, $B(9, 1)$, $C(3, 6)$, and $D(5, -2)$, determine if $\overleftrightarrow{AB}$ is parallel or perpendicular to $\overleftrightarrow{CD}$.**

First find the slopes of $\overleftrightarrow{AB}$ and $\overleftrightarrow{CD}$.

slope of $\overleftrightarrow{AB} = \frac{1 - (-2)}{9 - (-3)}$
$= \frac{3}{12}$ or $\frac{1}{4}$

slope of $\overleftrightarrow{CD} = \frac{-2 - 6}{5 - 3}$
$= \frac{-8}{2}$ or -4

The product of the slopes for $\overleftrightarrow{AB}$ and $\overleftrightarrow{CD}$ is $\left(\frac{1}{4}\right)(-4)$ or -1. So, $\overleftrightarrow{AB} \perp \overleftrightarrow{CD}$.

Example 3

Transportation

You will learn to write equations of lines that are parallel or perpendicular to a given line in Lesson 12–2.

A maintenance crew is marking lines for parking spaces in a parking lot. Assume the parking lot is on an imaginary coordinate plane with each grid segment representing one yard. How can the maintenance crew draw $\overline{CD}$ parallel to $\overline{AB}$ for $C(0, 3)$, $A(0, 0)$, and $B(5, 2)$?

First, find the slope of $\overline{AB}$.

$$m = \frac{2-0}{5-0} \text{ or } \frac{2}{5}$$

Next, find point D so the slope of $\overline{CD}$ is the same as the slope of $\overline{AB}$ or $\frac{2}{5}$.

Since slope is $\frac{\text{vertical rise}}{\text{horizontal run}}$, start at point C.

Move vertically 2 yards, and then horizontally 5 yards.

Locate point D. Draw $\overline{CD}$.

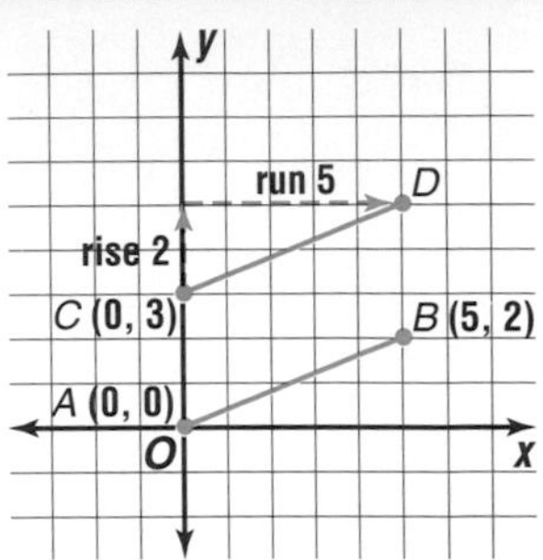

You can use what you know about slope to find missing values.

Example 4

Algebra

Find the value of x so the line that passes through $(x, 5)$ and $(6, -1)$ is perpendicular to the line that passes through $(2, 3)$ and $(-3, -7)$.

First, find the slope of the line that passes through $(2, 3)$ and $(-3, -7)$.

$$m = \frac{y_2 - y_1}{x_2 - x_1} \quad \textit{Definition of slope}$$

$$= \frac{3-(-7)}{2-(-3)} \quad \textit{Let } (x_1, y_1) = (-3, -7) \textit{ and } (x_2, y_2) = (2, 3).$$

$$= \frac{10}{5} \text{ or } 2$$

The product of the slopes of the two perpendicular lines is -1. Since $2\left(-\frac{1}{2}\right) = -1$, the slope of the line through $(x, 5)$ and $(6, -1)$ is $-\frac{1}{2}$. Now use the formula for the slope of a line to find the value of x.

$$m = \frac{y_2 - y_1}{x_2 - x_1}$$

$$-\frac{1}{2} = \frac{5-(-1)}{x-6} \quad \textit{slope} = -\tfrac{1}{2}, (x_1, y_1) = (6, -1), \textit{and } (x_2, y_2) = (x, 5)$$

$$\frac{1}{-2} = \frac{6}{x-6}$$

$$x - 6 = -12 \quad \textit{Cross multiply.}$$

$$x = -6$$

The line that passes through $(-6, 5)$ and $(6, -1)$ is perpendicular to the line that passes through $(2, 3)$ and $(-3, -7)$. To check your answer, graph the two lines.

An equation such as $y = x + 3$ is said to be in slope-intercept form. An equation can be graphed easily if it is in slope-intercept form.

EXPLORATION FAMILY OF LINES

Use a graphing calculator to graph $y = x$, $y = x + 3$, and $y = x - 2$.

Set the calculator on the standard viewing screen, $[-10, 10]$ by $[-10, 10]$. Enter each equation in the Y= list and press GRAPH .

a. Compare the slopes of the lines.

b. Compare the points where the lines intersect the y-axis.

c. Predict what the graph of $y = x + 2$ will look like. Then graph the equation to check your answer.

d. Graph $y = -x$, $y = -x + 2$, and $y = -x - 3$. Describe the similarities and differences among the graphs.

e. Predict what the graph of $y = -x - 2$ will look like. Check your answer.

f. The slope-intercept form of an equation of the line is $y = mx + b$, where m is the slope and b is the y-intercept. Write a paragraph explaining how the values of m and b affect the graph of the equation.

CHECK FOR UNDERSTANDING

Communicating Mathematics

Study the lesson. Then complete the following.

1. **Describe** a line whose slope is 0 and a line whose slope is undefined.
2. **Explain** why Postulate 3–2 excludes vertical lines. Are vertical lines parallel?
3. **Draw** a vertical line and a line perpendicular to it. Describe the line perpendicular to the vertical line.
4. **Refer** to the application at the beginning of the lesson. Explain why there is a vertical rise of 3100 feet.
5. **Explain** how to determine if the lines in the graph at the right are parallel, perpendicular, or neither. Describe the relationship of these two lines.

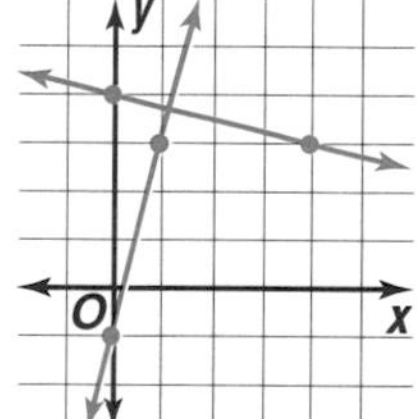

MATH JOURNAL

6. **Compare and contrast** each set of lines. Determine which line is steeper and make a drawing to support your reasoning.
   a. a line with a slope of 2 and a line with a slope of $\frac{1}{3}$
   b. a line with a slope of $\frac{3}{4}$ and a line with a slope of $\frac{2}{3}$
   c. a line with a slope of -3 and a line with a slope of $\frac{1}{4}$

Guided Practice

Find the slope of the line passing through the given points. Then describe each line as you move from left to right as *rising, falling, horizontal,* or *vertical.*

7. $A(3, 2)$, $B(4, -3)$
8. $C(-3, 6)$, $D(-3, -2)$

Determine the slope of each line named below.

9. q
10. any line perpendicular to s
11. any line parallel to r

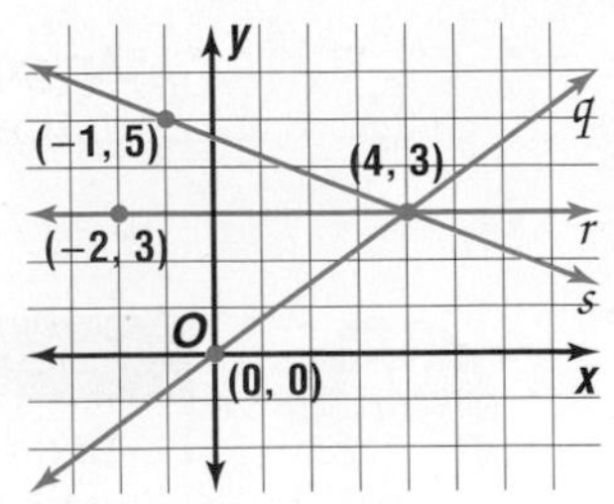

Graph the line that satisfies each description.

12. slope = −4, passes through $P(-2, 1)$
13. passes through $P(4, 1)$, perpendicular to $\overleftrightarrow{CD}$ with $C(0, 3)$ and $D(-3, 0)$
14. Given $A(16, 7)$, $B(4, -3)$, $C(1, 6)$, and $D(7, -2)$, determine if the intersection of $\overleftrightarrow{AB}$ and $\overleftrightarrow{CD}$ forms a right angle. Explain your reasoning.
15. Find the value of x so that the line passing through $(x, 6)$ and $(2, -3)$ is perpendicular to the line passing through $(1, 6)$ and $(7, 2)$.
16. **Aviation** An airplane passing over Richmond at an elevation of 33,000 feet begins its descent to land at Washington, D.C., 107 miles away. How many feet should the airplane descend per mile to land in Washington, which has an elevation of 25 feet?

EXERCISES

Practice

Find the slope of the line passing through the given points. Then, describe each line as you move from left to right as *rising, falling, horizontal,* or *vertical.*

17. $A(0, 6)$, $B(4, 0)$
18. $C(-3, 8)$, $D(4, 2)$
19. $E(6, 3)$, $F(-6, 3)$
20. $G(8, 1)$, $H(8, -6)$
21. $I(-2, -3)$, $J(-6, -5)$
22. $K(-2, 5)$, $L(4, 9)$

Determine the slope of each line named below.

23. $\overleftrightarrow{BD}$
24. $\overleftrightarrow{CD}$
25. $\overleftrightarrow{AB}$
26. $\overleftrightarrow{EO}$
27. any line parallel to $\overleftrightarrow{DE}$
28. any line parallel to $\overleftrightarrow{EO}$
29. any line perpendicular to $\overleftrightarrow{BD}$
30. any line perpendicular to $\overleftrightarrow{CD}$
31. any line perpendicular to $\overleftrightarrow{DE}$

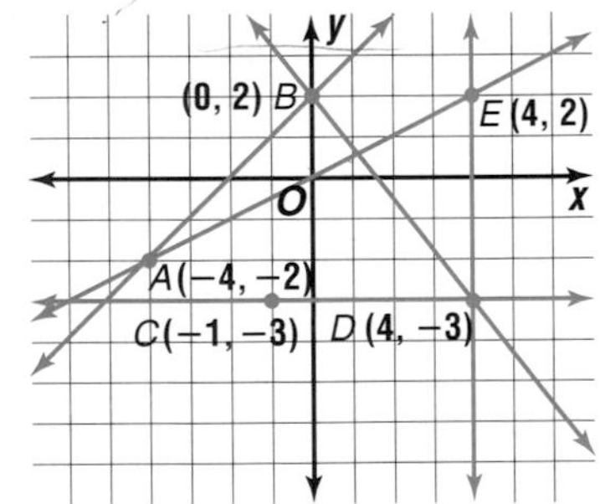

Graph the line that satisfies each description.

32. undefined slope, passes through $P(-3, -4)$
33. slope = $\frac{3}{5}$, passes through $P(1, 2)$
34. slope = −2, passes through $P(8, 1)$
35. slope = 0, passes through $P(-7, 3)$
36. passes through $P(6, 4)$ and is perpendicular to $\overleftrightarrow{TK}$ with $T(0, 2)$ and $K(5, 0)$
37. passes through $P(-1, -3)$ and is parallel to $\overleftrightarrow{CR}$ with $C(-1, 7)$ and $R(5, 1)$

Given each set of points, determine if $\overleftrightarrow{MA} \parallel \overleftrightarrow{TH}$. Explain your reasoning.

38. $M(-6, 1), A(2, 8), T(1, 1), H(9, 8)$
39. $M(-3, -4), A(7, -10), T(-1, 6), H(4, 3)$

Determine if the intersection of $\overleftrightarrow{PQ}$ and $\overleftrightarrow{RS}$ forms a right angle. Explain your reasoning.

40. $P(-9, 2), Q(0, 1), R(-1, 8), S(-2, -1)$
41. $P(3, 6), Q(-1, 4), R(4, 0), S(0, 8)$

Determine the value of *x*, so that a line through points with the given coordinates has the given slope. Draw a sketch of each situation.

42. $(x, 2), (-4, -6)$, slope $= \frac{4}{5}$
43. $(6, 2), (x, -1)$, slope $= -\frac{3}{7}$
44. Find the value of x so that the line passing through the points at $(x, 2)$ and $(-4, 5)$ is perpendicular to the line that passes through the points at $(4, 8)$ and $(2, -1)$.
45. The vertices of figure *PQRS* are $P(5, 2)$, $Q(1, 6)$, $R(-3, 2)$, and $S(1, -2)$.
    a. Show that the opposite sides are parallel.
    b. Show that the adjacent sides are perpendicular.
    c. Show that all sides are congruent.
    d. What type of figure is *PQRS*?
46. A parallelogram is a four-sided figure whose opposite sides are parallel. Given $E(2, 3)$, $F(1, -6)$, and $G(-2, 5)$, find the coordinates of point H so that the points are the vertices of a parallelogram. (*Hint:* There is more than one location.)

Programming

47. The graphing calculator program at the right finds the slope of a line containing two points at (x_1, y_1) and (x_2, y_2).
    Use the program to find the slope of each line.
    a. $\overleftrightarrow{AB}$ if $A(5, 3)$ and $B(-1, 8)$
    b. $\overleftrightarrow{GH}$ if $G(-3, -7)$ and $H(-5, -2)$
    c. $\overleftrightarrow{TS}$ if $T(3, -1)$ and $S(5, 4)$
    d. $\overleftrightarrow{VW}$ if $V(-5, 7)$ and $W(3, 1)$

```
PROGRAM: SLOPE
:Disp "ENTER THE","COORDINATES"
:Input "X1=", A
:Input "Y1=", B
:Input "X2=", C
:Input "Y2=", D
:(D-B)/(C-A)→M
:Disp "SLOPE=",M
:Stop
```

Critical Thinking

48. A line contains the points at $(-3, 6)$ and $(1, 2)$. Using slope, write a convincing argument that the line intersects the x-axis at $(3, 0)$. Graph the points to verify your conclusion.

Applications and Problem Solving

49. **Construction** According to the building code in Crystal Lake, Illinois, the slope of a stairway cannot be steeper than 0.88. The stairs in Li-Chih's home measure 11 inches deep and 7 inches high. Do the stairs in his home meet the code requirements?

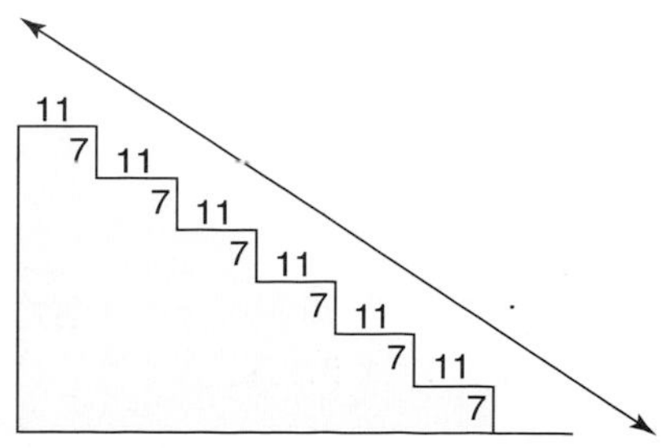

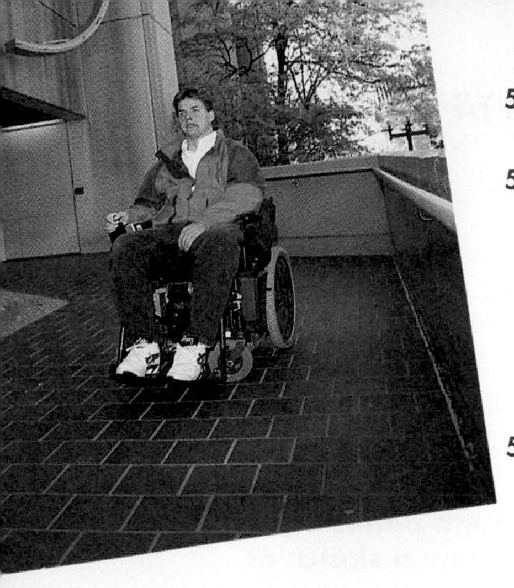

50. **Construction** Measure the height and width of stairs in your school or home and determine the slope of those stairs.

51. **Architecture** The *Americans with Disabilities Act* (ADA) of 1990 requires that ramps should have at least a 12-inch run for each rise of 1 inch, with a maximum rise of 30 inches. Your school needs to design a ramp 45 feet long.
   a. Determine the appropriate slope of the ramp in order to conform with the ADA.
   b. Find the maximum run in feet for each 1-inch rise.

52. **Computer Software** Multimedia software on CD-ROM has been growing since 1992. Five million software packages were shipped in 1992, and 39 million packages were shipped in 1995.
   a. What is the average rate of change in software packages per year during this time period?
   b. How does the rate of change relate to the slope of a line?
   c. Predict how many CD-ROM software packages will be shipped in the year 2001.

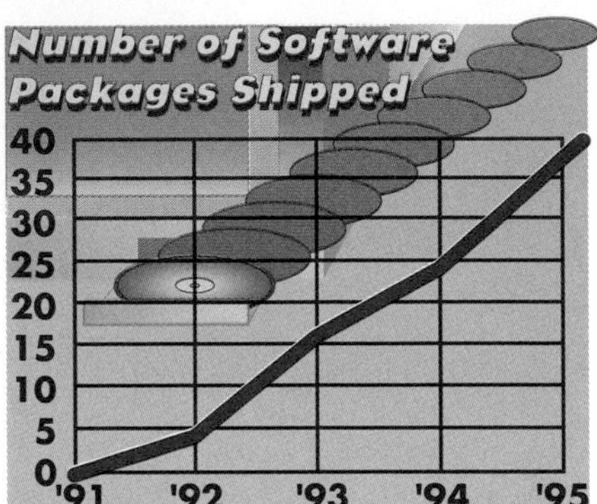

Source: *Information and Interactive Services Report,* Dataquest

Data Update For more information on software sales, visit: **www.geometry.glencoe.com**

Mixed Review

53. Find the values of x and y. (Lesson 3–2)

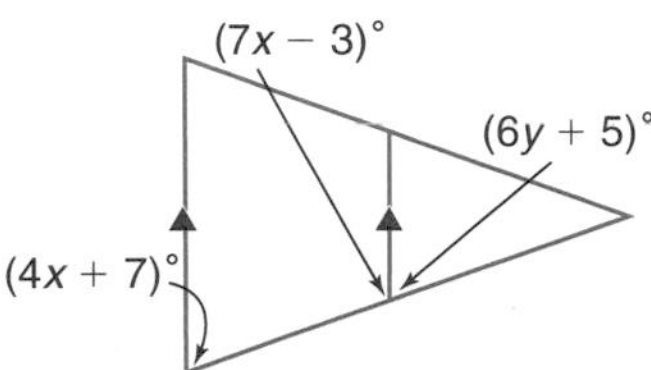

54. Identify each pair as *intersecting, parallel,* or *skew.* (Lesson 3–1)
   a. $\overline{BA}$ and $\overline{GH}$
   b. $\overline{EH}$ and $\overline{CD}$
   c. plane EAB and plane GCB
   d. $\overline{HG}$ and plane EAB
   e. plane HGC and $\overline{BC}$

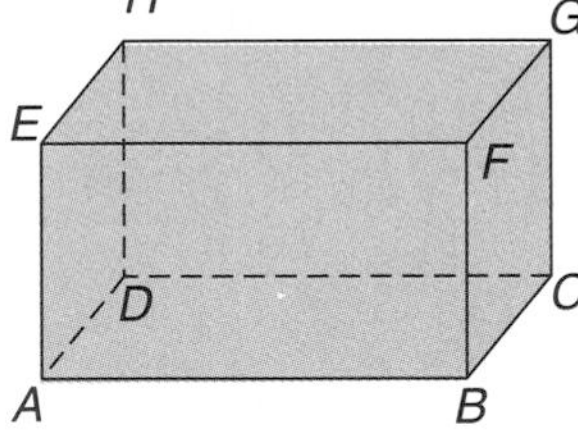

55. Write a two-column proof. (Lesson 2–6)
   Given: $\angle 1 \cong \angle 2$
   Prove: $\angle 1 \cong \angle 3$

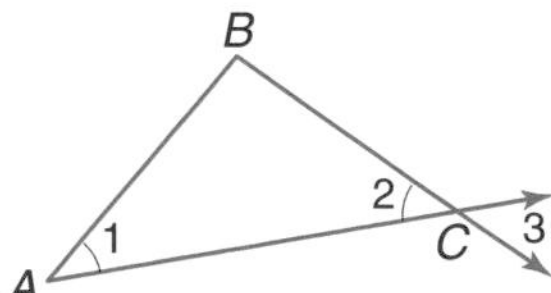

56. Justify each statement with a property from algebra or a property of congruent segments. (Lesson 2–5)
   a. If $JL + 5 = KM + 5$, then $JL = KM$.
   b. If $\overline{RS} \cong \overline{TU}$, then $\overline{TU} \cong \overline{RS}$.
   c. If $\frac{2}{3}g = 12$, then $g = 18$.

57. The formula for finding the total surface area of a cylinder is $A = 2\pi r^2 + 2\pi rh$, where r is the radius of the base and h is the height. Solve the formula for h and justify each step. (Lesson 2–4)

58. **Recreation** The sign in front of the Screaming Eagle Roller Coaster states *If you are over 48 inches tall, then you may ride the Screaming Eagle*. Rafael is 54 inches tall. Can he ride the Screaming Eagle? Which law of logic leads you to this conclusion? (Lesson 2–3)

59. Beng observed that 2^2 is greater than 2 and 5^2 is greater than 5, and made a conjecture that *The square of any real number is greater than the number.* Give a counterexample to this conjecture. (Lesson 2–1)

60. Is $\overrightarrow{DB}$ the bisector of $\angle ADC$ if $m\angle 1 = 2x$, $m\angle 2 = 28$, and $m\angle ADC = 5x - 14$? Justify your answer. (Lesson 1–6)

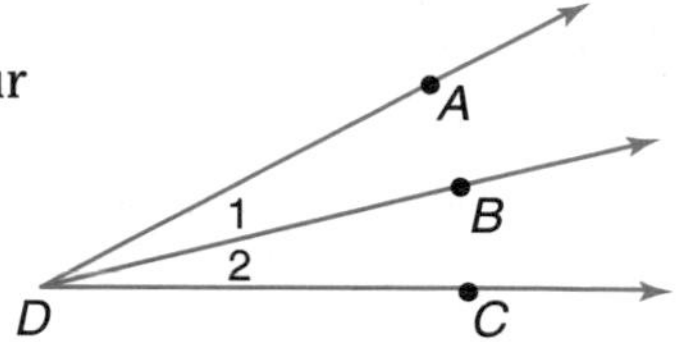

61. **SAT Practice** The winning sailboat completed a 24-mile race at an average speed of 9 miles per hour. The second-place boat in the same race had an average speed of 8 miles per hour. How many minutes longer than the winner did the second-place boat take to finish the race?

A 20 **B** 33 **C** 60 **D** 120 **E** 180

62. If $x < 0$ and $y < 0$, in which quadrant is the point $Q(x, y)$ located? (Lesson 1–1)

For **Extra Practice**, see page 769.

Algebra

63. Evaluate $4x^2 - 3x$ if $x = -2$.

64. Solve $4n + 23 = 3 - 6n$.

SELF TEST

The three-dimensional figure shown at the right is called a *right-triangular prism*. (Lesson 3–1)

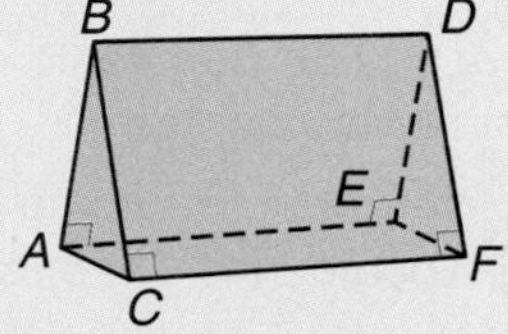
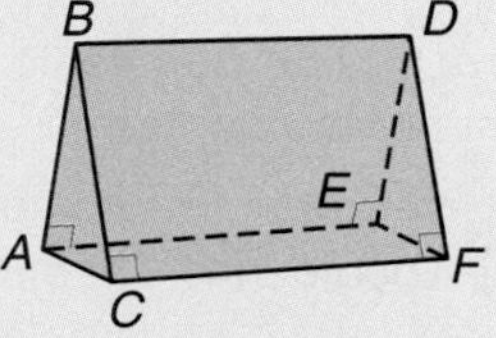

1. Name all segments parallel to $\overline{CF}$.
2. Name all segments skew to $\overline{AC}$.
3. Name all pairs of parallel planes.

Given $\ell \parallel m$, $m\angle 1 = 98$, and $m\angle 2 = 40$, find the measure of each angle. (Lesson 3–2)

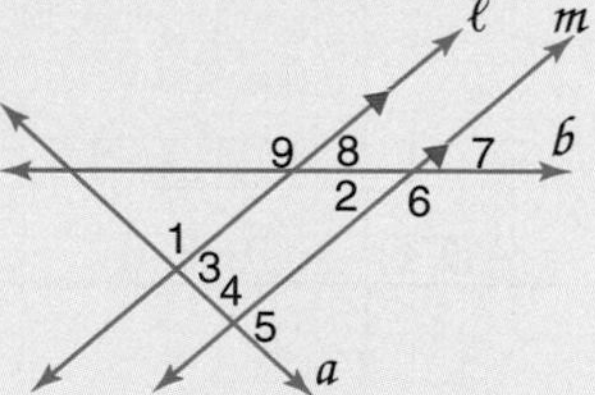

4. $\angle 4$
5. $\angle 8$
6. $\angle 9$

Determine the slope of each line named below. (Lesson 3–3)

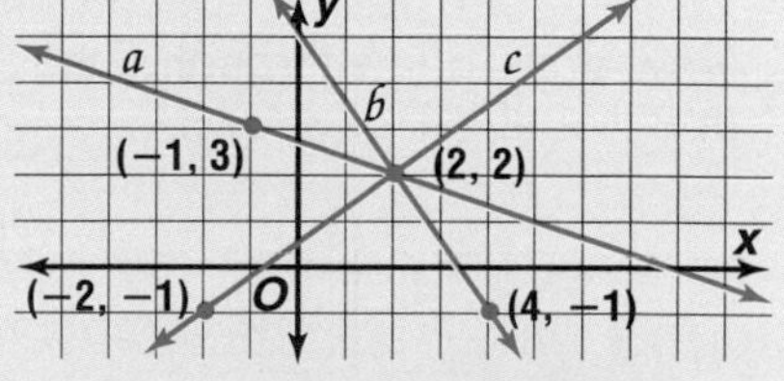

7. a
8. any line parallel to b
9. any line perpendicular to c

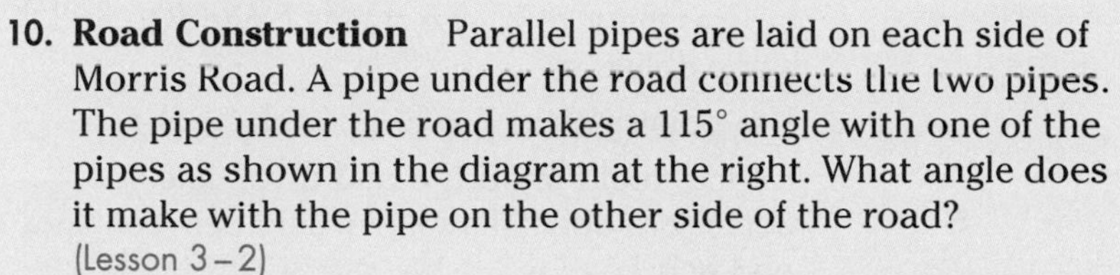

10. **Road Construction** Parallel pipes are laid on each side of Morris Road. A pipe under the road connects the two pipes. The pipe under the road makes a 115° angle with one of the pipes as shown in the diagram at the right. What angle does it make with the pipe on the other side of the road? (Lesson 3–2)

3–4 Proving Lines Parallel

- To recognize angle conditions that produce parallel lines, and
- to prove two lines are parallel based on given angle relationships.

Why IT'S IMPORTANT

You can use the various ways to prove lines are parallel in construction and physics.

Farming

The orange trees on a farm in Polk County, Florida, appear to be planted in parallel lines. Using geometry, how could we prove the lines are parallel? The following Modeling Mathematics activity suggests one way to prove the lines are parallel.

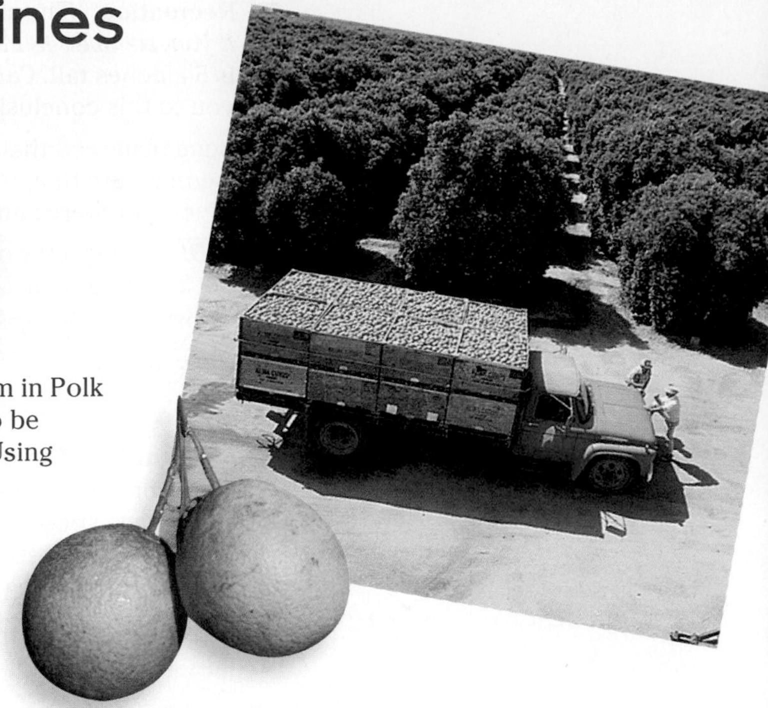

MODELING MATHEMATICS

Construction of Parallel Lines

Materials:  graph paper  straightedge  compass  colored pencils

You can construct a line parallel to a given line through a point not on the line.

- Use a straightedge to draw line ℓ through $A(0, 0)$ and $R(2, 1)$. Locate $P(4, 5)$ not on ℓ.

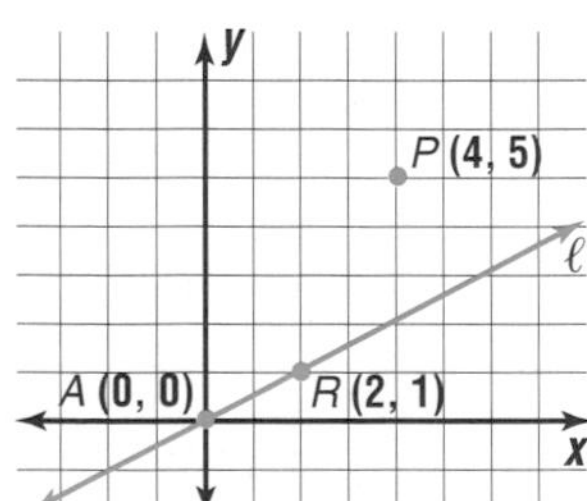

- Now draw a line through P that intersects line ℓ at R. Label $\angle 1$ as shown.

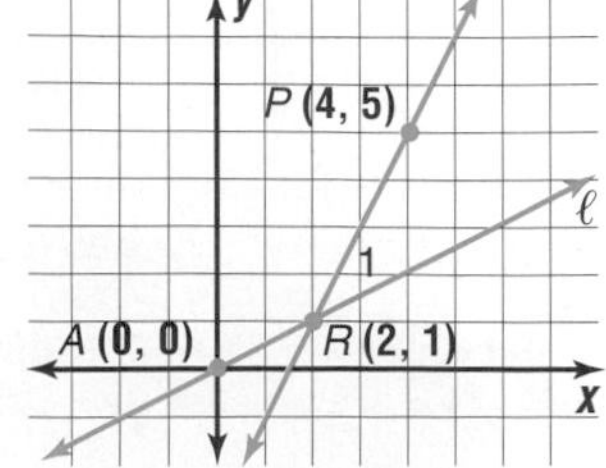

- Construct an angle congruent to $\angle 1$ using P as a vertex and $\overleftrightarrow{PR}$ as one side. Draw a line through P to form an angle congruent to $\angle 1$. Label the line m, the angle 2, and the point where m intersects the y-axis B.

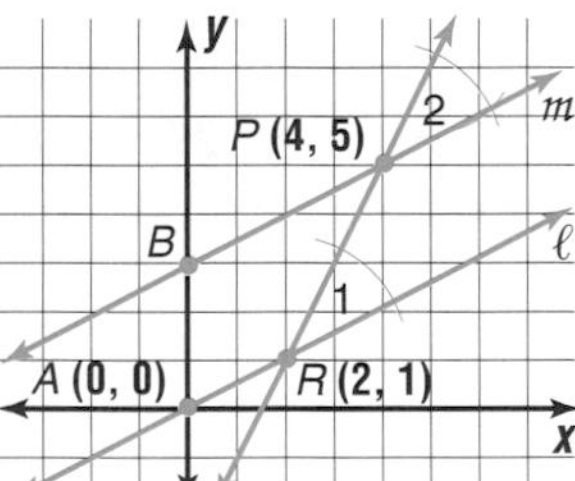

Your Turn

a. Identify the special angle pair name for $\angle 1$ and $\angle 2$.

b. Recall that $\angle 2$ was constructed congruent to $\angle 1$ and make a conjecture about ℓ and m.

c. Find the coordinates of point B where line m crosses the y-axis. Use points P and B to find the slope of line m.

d. Use points A and R to find the slope of ℓ.

e. Do the results in parts c and d validate your conjecture about the relationship between lines ℓ and m?

The Modeling Mathematics activity illustrates a postulate that helps to prove two lines are parallel. Notice that this postulate is the converse of Postulate 3–1.

Postulate 3–4	**If two lines in a plane are cut by a transversal so that corresponding angles are congruent, then the lines are parallel.**

This can be abbreviated as "If ⇄ and corr. ∠s are ≅, then the lines are ∥."

The Modeling Mathematics activity establishes *at least* one line through P parallel to ℓ. In 1795, Scottish physicist and mathematician John Playfair provided the modern version of Euclid's famous **Parallel Postulate**, which states there is *exactly* one line parallel to a line through a given point not on the line.

Postulate 3–5 Parallel Postulate	**If there is a line and a point not on the line, then there exists exactly one line through the point that is parallel to the given line.**

There are sets of conditions other than those in Postulate 3–4 that prove that two lines are parallel. One of them is stated in Theorem 3–5.

Theorem 3–5	**If two lines in a plane are cut by a transversal so that a pair of alternate exterior angles are congruent, then the two lines are parallel.**

Proof of Theorem 3–5

Given: $\angle 1 \cong \angle 2$

Prove: $\ell \parallel m$

Proof:

Statements	Reasons
1. $\angle 1 \cong \angle 2$	**1.** Given
2. $\angle 2 \cong \angle 3$	**2.** Vertical angles are congruent.
3. $\angle 1 \cong \angle 3$	**3.** Congruence of angles is transitive.
4. $\ell \parallel m$	**4.** If ⇄ and corr. ∠s are ≅, then the lines are ∥.

Theorems 3–6, 3–7, and 3–8 state three more ways to prove that two lines are parallel. *You will be asked to prove these theorems in Exercises 33, 13, and 34, respectively.*

Theorem 3–6	**If two lines in a plane are cut by a transversal so that a pair of consecutive interior angles is supplementary, then the lines are parallel.**
Theorem 3–7	**If two lines in a plane are cut by a transversal so that a pair of alternate interior angles is congruent, then the lines are parallel.**
Theorem 3–8	**In a plane, if two lines are perpendicular to the same line, then they are parallel.**

Algebra

Example 1 **Find the value of x and $m\angle ABC$ so that $p \parallel q$.**

Explore From the figure, you know that $m\angle ABC = 5x + 90$ and $m\angle BEF = 14x + 9$. You want to find the value of x and $m\angle ABC$ so that $p \parallel q$.

Plan If two lines are cut by a transversal so that corresponding angles are congruent, then the lines are parallel. So if $m\angle ABC = m\angle BEF$, then $p \parallel q$.

Solve $m\angle ABC = m\angle BEF$

$5x + 90 = 14x + 9$ *Substitution*

$81 = 9x$ *Subtract 9 and 5x from each side.*

$9 = x$ *Divide each side by 9.*

Use the value of x to find $m\angle ABC$.

$m\angle ABC = 5x + 90$

$= 5(9) + 90$ or 135

Examine Since $m\angle BEF = 14x + 9$ and $x = 9$, $m\angle BEF = 14(9) + 9$ or 135. Therefore, $\angle ABC \cong \angle BEF$ and $p \parallel q$.

You can use relationships between angles to determine if lines are parallel.

Example  **If $\angle 1 \cong \angle 2$ and $\angle RAB \cong \angle CBS$, which lines must be parallel? Explain.**

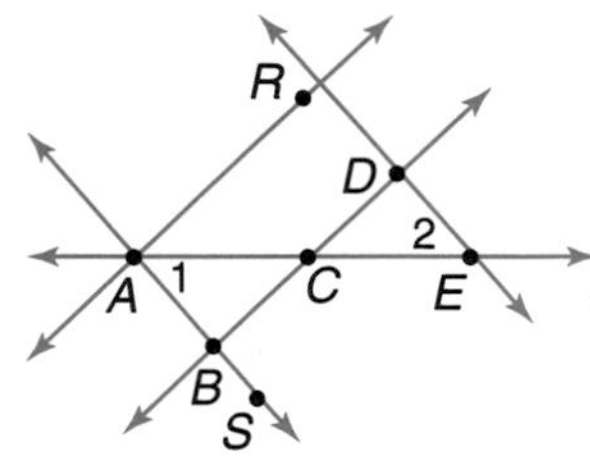

$\overleftrightarrow{AE}$ is a transversal for $\overleftrightarrow{AB}$ and $\overleftrightarrow{DE}$. Since $\angle 1$ and $\angle 2$ are congruent alternate interior angles, $\overleftrightarrow{AB} \parallel \overleftrightarrow{DE}$.

$\overleftrightarrow{AS}$ is a transversal for $\overleftrightarrow{AR}$ and $\overleftrightarrow{BD}$. Since $\angle RAB$ and $\angle CBS$ are congruent corresponding angles, $\overleftrightarrow{AR} \parallel \overleftrightarrow{BD}$.

Some problems can be solved using either a geometric method or an algebraic method.

Example 3 **Prove $\overleftrightarrow{AB} \parallel \overleftrightarrow{CD}$ using two methods.**

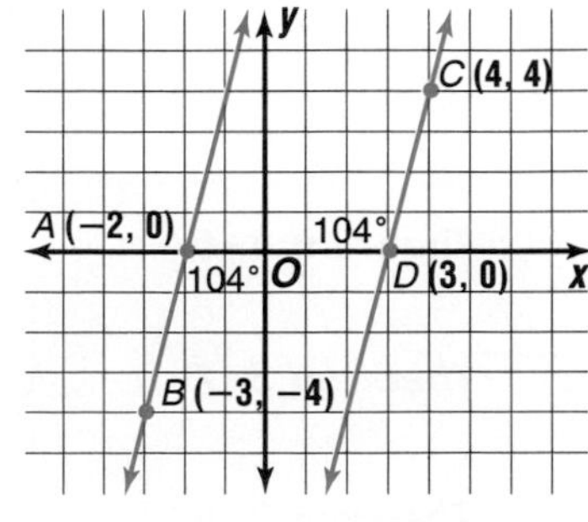

Method 1: Using Geometric Theorems

$\angle BAD$ and $\angle CDA$ are congruent alternate interior angles, so $\overleftrightarrow{AB} \parallel \overleftrightarrow{CD}$ by Theorem 3–7.

Method 2: Using Slope

slope of $\overleftrightarrow{CD}$: $\frac{4-0}{4-3} = \frac{4}{1}$ or 4

slope of $\overleftrightarrow{AB}$: $\frac{0-(-4)}{-2-(-3)} = \frac{4}{1}$ or 4

Since the slopes are the same, $\overleftrightarrow{AB} \parallel \overleftrightarrow{CD}$.

FYI

Euclid's great book on geometry, *Elements*, has no reference to the word *geometry*.

Communicating Mathematics

Study the lesson. Then complete the following.

1. **Summarize** five different methods you can use to prove two lines are parallel.
2. **Explain** the meaning of *exactly* in the Parallel Postulate. Try to draw two lines parallel to a given line through a given point not on the line.
3. **You Decide** Using the figure at the right, Nancy claims $\overline{QU} \parallel \overline{AD}$, but Jasmine claims they definitely are not parallel. Who is right? Explain your answer.

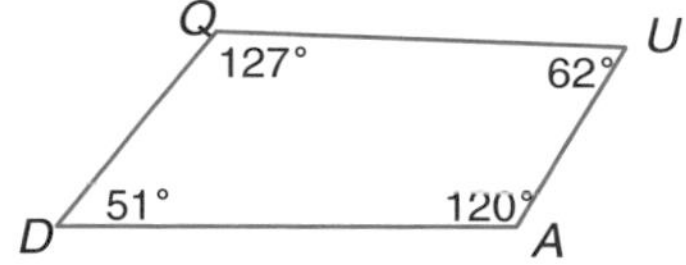

4. **Describe** two situations in your own life in which you encounter parallel lines. How could you guarantee the lines are parallel?

5. **Rearrange** the statements and their corresponding reasons in the following proof to write the seven steps of a correct proof.

Given: $r \parallel s$; $\angle 5 \cong \angle 6$

Prove: $\ell \parallel m$

Proof:

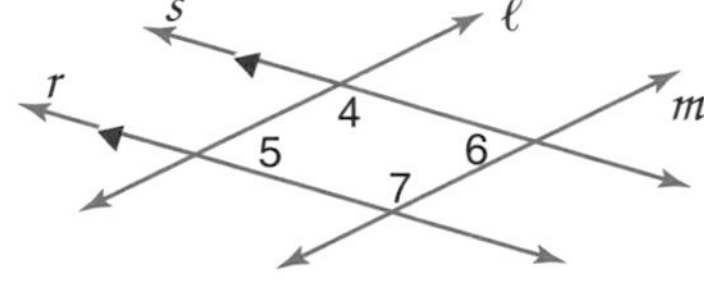

Statements	Reasons
1. $m\angle 4 + m\angle 6 = 180$	a. Substitution Property (=)
2. $r \parallel s$, $\angle 5 \cong \angle 6$	b. Given
3. $m\angle 4 + m\angle 5 = 180$	c. Definition of supplementary
4. $\ell \parallel m$	d. If ⇸ and a pair of consecutive angles are supplementary, then the lines are parallel.
5. $\angle 4$ and $\angle 6$ are supplementary.	e. Definition of supplementary angles
6. $\angle 4$ and $\angle 5$ are supplementary.	f. Consecutive Interior Angle Theorem.
7. $m\angle 5 = m\angle 6$	g. Definition of congruent angles

6. **Assess Yourself** Explain how you would construct parallel lines as described in Theorem 3–8.

Guided Practice

Given the following information, determine which lines, if any, are parallel. State the postulate or theorem that justifies your answer.

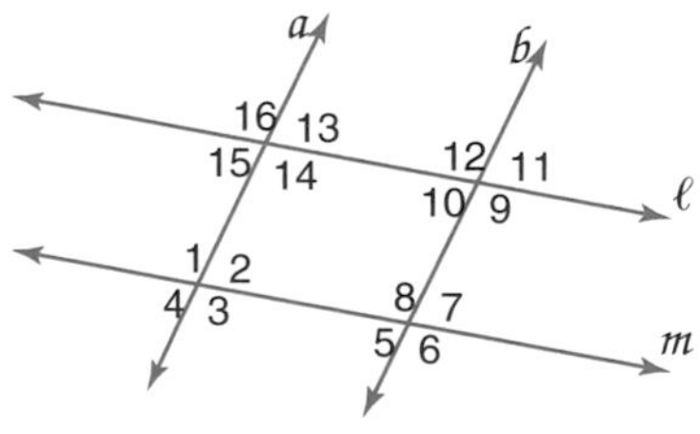

7. $\angle 10 \cong \angle 13$

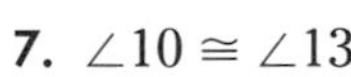

8. $\angle 1 \cong \angle 6$
9. $\angle 2 \cong \angle 13$

Find the value of x so that $\ell \parallel m$.

10.

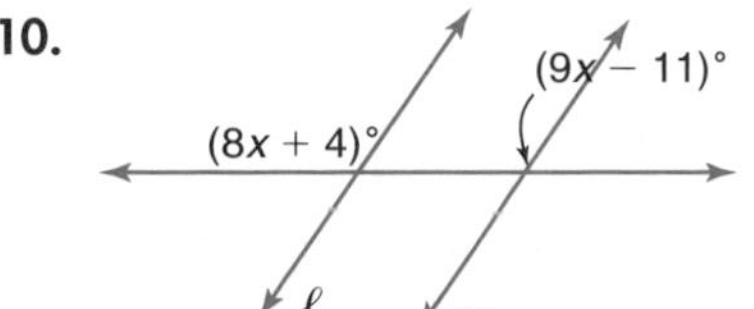

11.

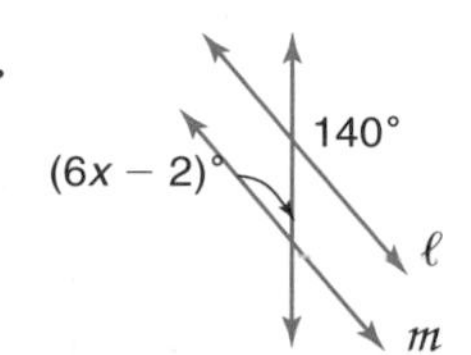

12. State which lines, if any, are parallel. State the postulate or theorem that justifies your answer.

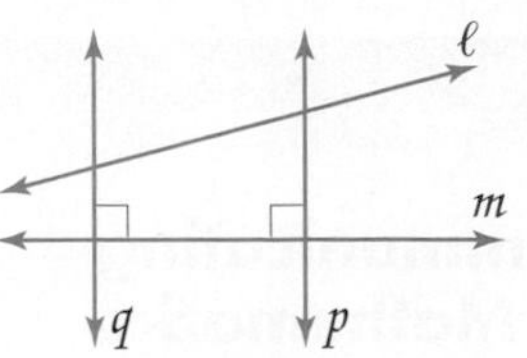

Proof

13. Copy and provide a reason for each step in the proof of Theorem 3–7.

Given: $\angle 4 \cong \angle 6$

Prove: $\ell \parallel m$

Proof:

Statements	Reasons
1. $\angle 4 \cong \angle 6$	**1.** ?
2. $\angle 6 \cong \angle 7$	**2.** ?
3. $\angle 4 \cong \angle 7$	**3.** ?
4. $\ell \parallel m$	**4.** ?

14. **Construction** A carpenter uses a special instrument called an *engineer and carpenter square* to draw parallel line segments. Carmen wants to make two parallel cuts at an angle of 120° though points *D* and *P*. Explain why these lines will be parallel.

EXERCISES

Practice

Given the following information, determine which lines, if any, are parallel. State the postulate or theorem that justifies your answer.

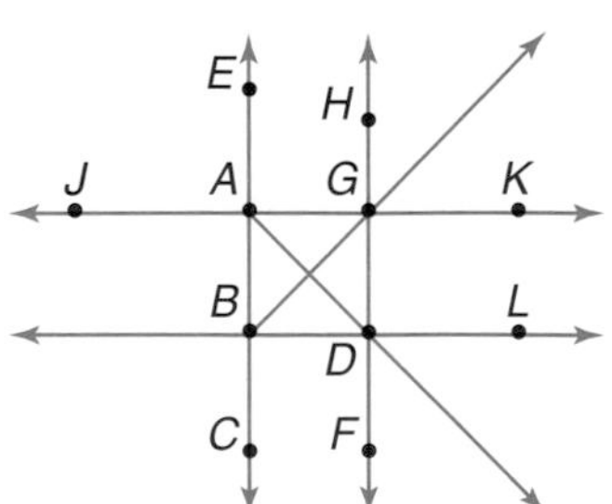

15. $\angle EAJ \cong \angle HGA$
16. $\angle BAD \cong \angle GDA$
17. $m\angle GAB + m\angle LBA = 180$
18. $\overleftrightarrow{EC} \perp \overleftrightarrow{BL}, \overleftrightarrow{FH} \perp \overleftrightarrow{BL}$
19. $\angle 1 \cong \angle 7$
20. $\angle 16 \cong \angle 3$
21. $m\angle 14 + m\angle 10 = 180$
22. $\angle 4 \cong \angle 13$
23. $m\angle 8 + m\angle 10 = 180$

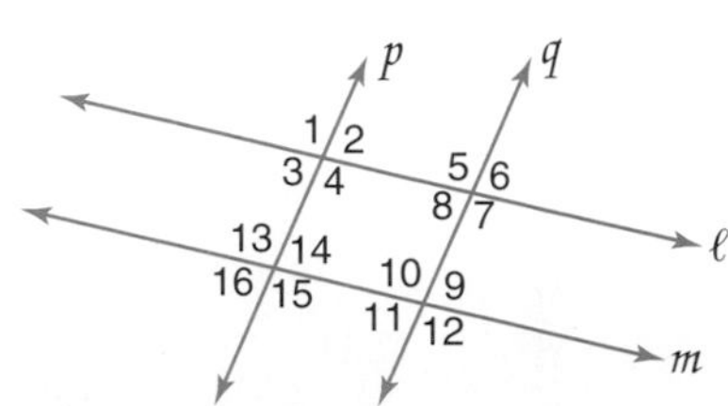

24. Use slope to determine if $\ell \parallel m$. Verify your results by measuring $\angle 1$ and $\angle 2$. Which method do you prefer for showing whether ℓ is parallel to m?

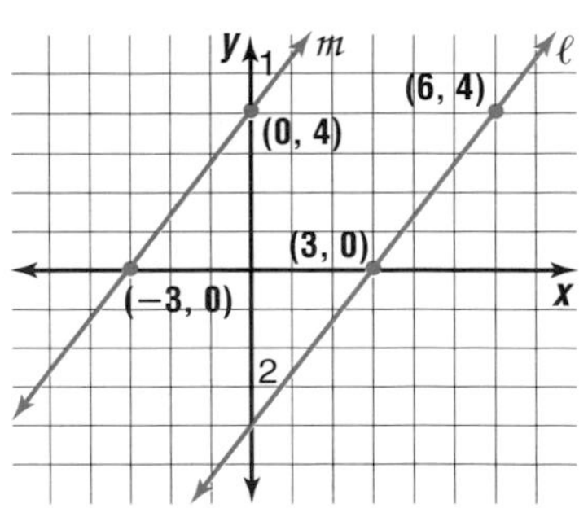

Find the value of x so that $\ell \parallel m$.

25.

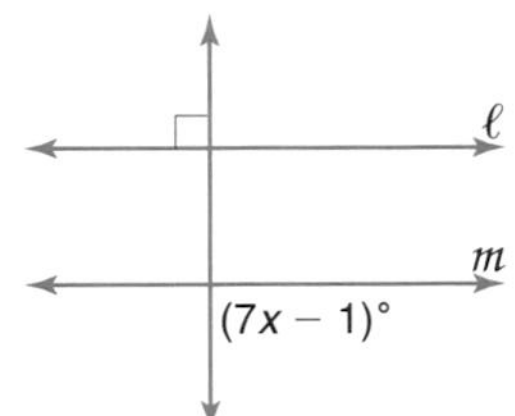

26.

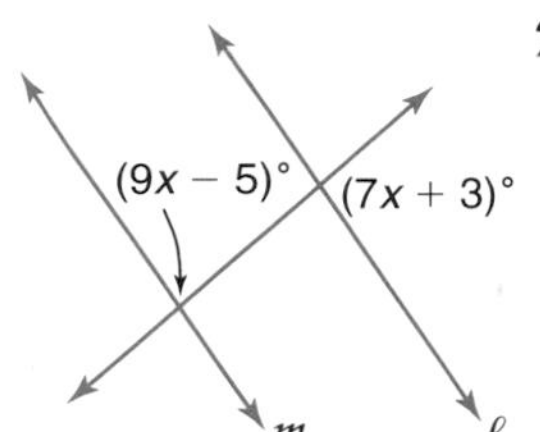

27.

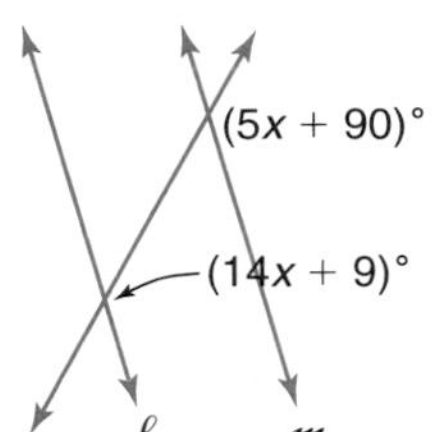

Find the values of x and y that make the blue lines parallel and the red lines parallel.

28.

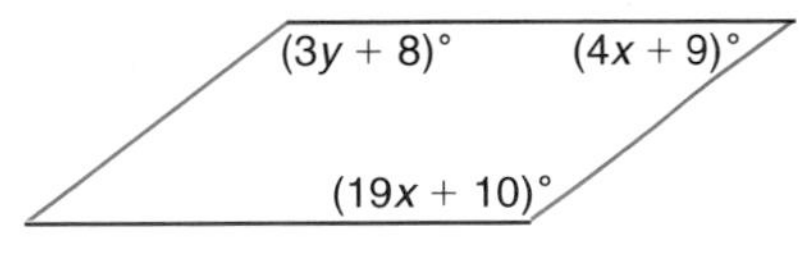

29.

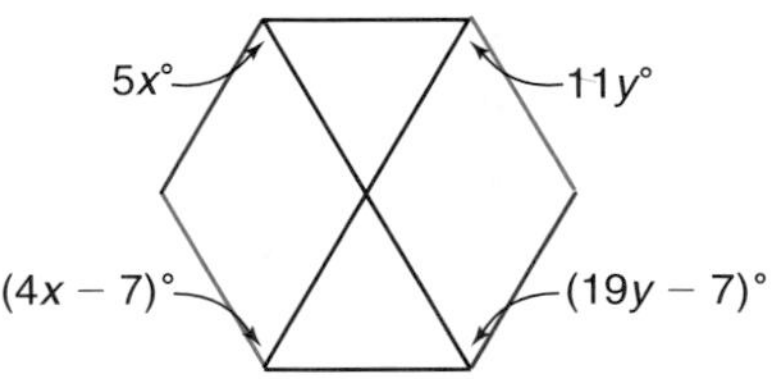

State which lines, if any, are parallel. State the postulate or theorem that justifies your answer.

Proof

30.

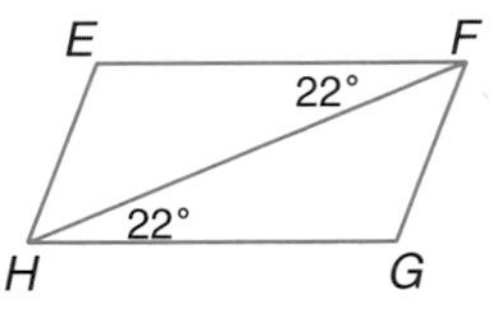

31.

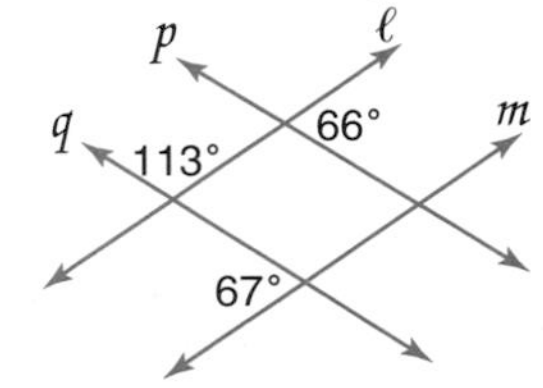

32.

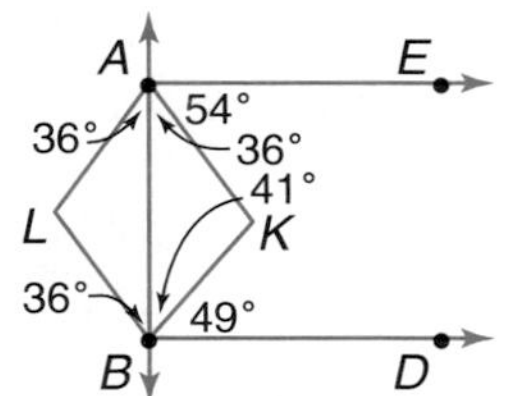

33. Copy and complete the proof of Theorem 3–6.

Given: $\angle 1$ and $\angle 2$ are supplementary.

Prove: $\ell \parallel m$

Proof:

Statements	Reasons
1. $\angle 1$ and $\angle 2$ are supplementary.	1. ?
2. $\angle 2$ and $\angle 3$ form a linear pair	2. ?
3. ?	3. If 2 angles form a linear pair, they are supplementary.
4. $\angle 1 \cong \angle 3$	4. ?
5. $\ell \parallel m$	5. ?

34. Write a paragraph proof of Theorem 3–8.

Write a two-column proof.

35. **Given:** $\angle 2 \cong \angle 1$
$\angle 1 \cong \angle 3$
Prove: $\overline{ST} \parallel \overline{YZ}$

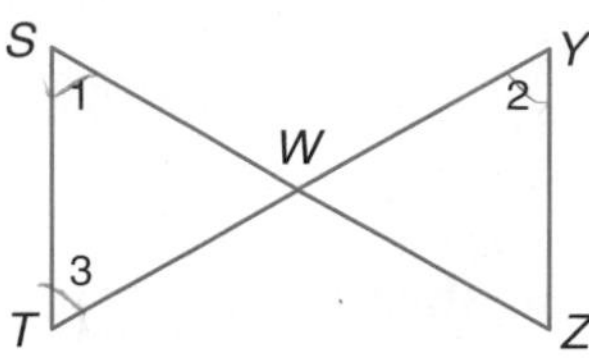

36. **Given:** $\overline{JO} \parallel \overline{KN}$
$\angle 1 \cong \angle 2$
$\angle 3 \cong \angle 4$
Prove: $\overline{KO} \parallel \overline{AN}$

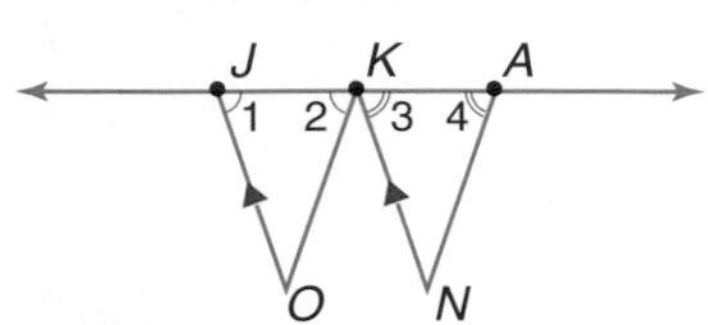

37. **Given:** $\overline{AU} \perp \overline{QU}$
$\angle 1 \cong \angle 2$
Prove: $\overline{DQ} \perp \overline{QU}$

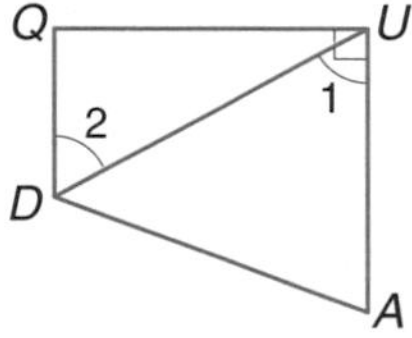

38. Use the figure below to determine the relationship between a and b so that $\ell \parallel m$.

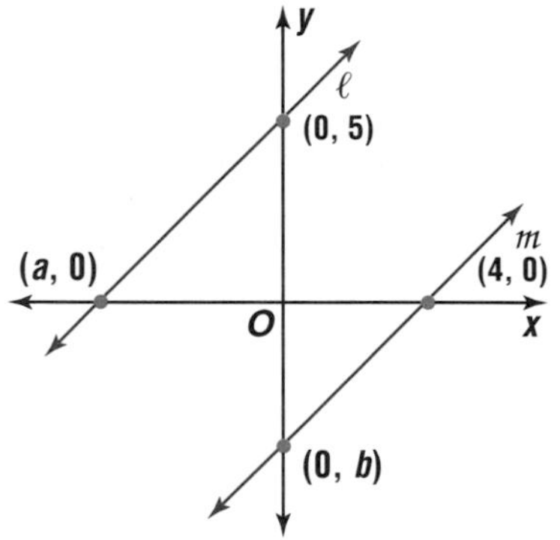

Critical Thinking

39. Suppose lines a, b, and c lie in the same plane and $a \parallel b$ and $a \parallel c$.
   a. Draw a figure showing lines a, b, and c.
   b. Explain how you would prove that $b \parallel c$.

Applications and Problem Solving

40. **Construction** Carpenters use parallel lines in creating walls for construction projects. Copy the drawing at the right. Label your drawing and describe three different ways to guarantee that the wall studs are parallel.

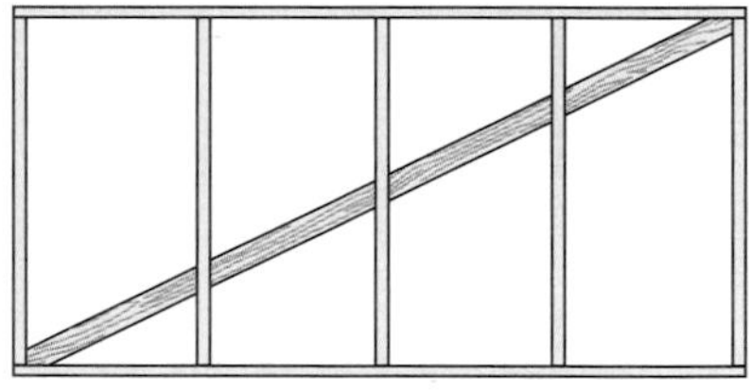

FYI

The Hubble Telescope orbits 380 miles above Earth allowing it to obtain images before they are distorted by Earth's atmosphere. It can observe objects that are only 2% as bright as those barely detected by telescopes on Earth.

41. **Physics** The diagram represents a parabolic concave mirror as used in the Hubble Telescope. The Hubble Telescope gathers parallel light rays and directs them to a central focal point. Use a protractor to measure several of the angles shown in the diagram. Are the lines parallel?

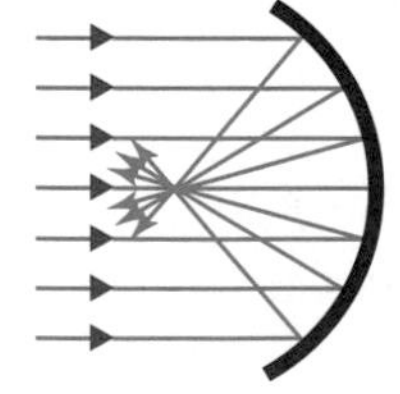

42. **Architecture** Check with a classmate or teacher who participates in your school drafting classes. How do they draw parallel lines? If your school uses computer-aided design (CAD), determine the applications of parallel lines used in CAD.

Mixed Review

43. Find the slope of the line that passes through points at $(-2, 3)$ and $(-7, -1)$. Then describe the line as you move from left to right as *rising*, *falling*, *horizontal*, or *vertical*. (Lesson 3–3)

44. True or false: *If two parallel lines are cut by a transversal, the alternate interior angles are supplementary.* If false, explain why. (Lesson 3–2)

45. **Recreation** Are the spokes on a wheel a model of parallel, intersecting, or skew lines? Explain. (Lesson 3–1)

46. **SAT Practice** If the average of six numbers is 18 and the average of three of the numbers is 15, then what is the *sum* of the remaining three numbers?

A 21 **B** 45 **C** 53 **D** 63 **E** 108

47. If possible, write a valid conclusion from the two true statements *If two lines are parallel, then they never meet* and *Lines p and m are parallel.* State the law of logic that you used. (Lesson 2–3)

48. Find the measure of the supplement of an angle that measures 159. (Lesson 1–7)

49. Find the midpoint of the segment whose endpoints have coordinates (8, 11) and (−4, 7). (Lesson 1–5)

50. **Education** James forgot the combination to his gym locker. He remembers that the numbers are 18, 37, and 12, but doesn't remember the order. How many different combinations are possible? (Lesson 1–2)

For **Extra Practice**, see page 769.

Algebra

51. Factor $16x^2 - 22xy - 3y^2$.

52. Solve $f = \frac{W}{g} \cdot \frac{V^2}{R}$ for R.

Refer to the Investigation on pages 66–67.

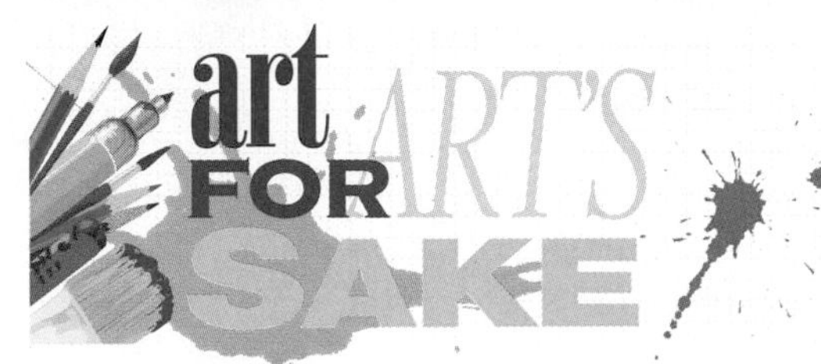

You may want to use a system for making the three-dimensional objects appear realistic in your mural by using linear perspective. Use the following steps to draw a room with a tile floor using linear perspective.

1. Draw a square with sides 15 centimeters long and separate the bottom edge into ten congruent segments.
2. Select a vanishing point V in the upper portion of the square, along the perpendicular bisector of the bottom edge of the square.
3. Mark points Z_1 and Z_2 ten centimeters to the left and right of V so that V, Z_1 and Z_2 are collinear and $\overline{Z_1Z_2}$ is parallel to the bottom edge of the square.
4. Draw lines connecting each point on the bottom edge with V. Lightly draw lines connecting the three rightmost points with Z_1 and lines connecting the three leftmost points with Z_2.
5. Start at the bottom of the square and draw ten horizontal lines between the leftmost and rightmost lines to V. The lines should pass through the points of intersection of the lines to V and the lines to Z_1 and Z_2.
6. Draw a line from the upper corners of the square to point V. Then draw two vertical lines from these lines to the leftmost and rightmost intersection points on the top horizontal line.
7. Finally, draw a horizontal line between the two verticals. Erase the lines to Z_1 and Z_2 and the lines within the small square that represents the back wall of your room to complete the picture.
8. You may wish to experiment with different methods of perspective, such as moving the vanishing point or changing the distance between V and Z_1 and Z_2.
9. Work with your group to begin creating your mural.

Add the results of your work to your Investigation folder.

3-5 Parallels and Distance

What YOU'LL LEARN

- To recognize and use distance relationships among points, lines, and planes.

Why IT'S IMPORTANT

You can find the correct distance between points and lines and between parallel lines and parallel planes.

The following Modeling Mathematics activity investigates the shortest distance between a line and a point not on the line.

MODELING MATHEMATICS

Distance Between a Line and a Point

Materials: lined notebook paper

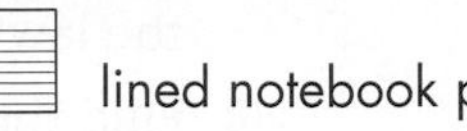

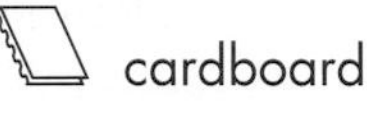 cardboard 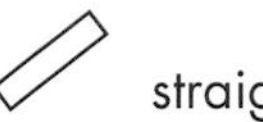 straightedge

scissors thumbtack

You can draw many line segments from a line to a point not on the line. But which one of the segments is the shortest one?

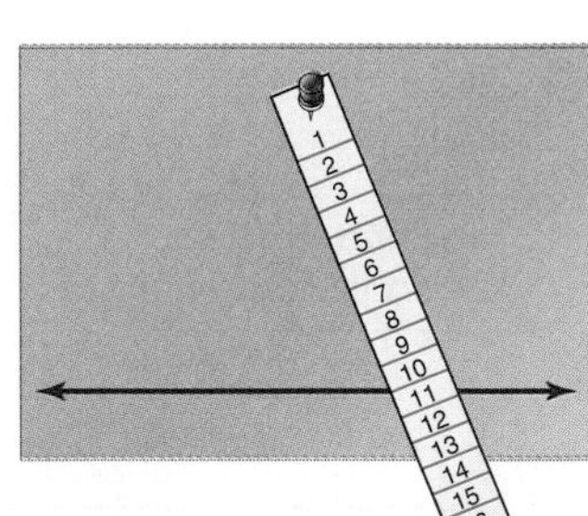

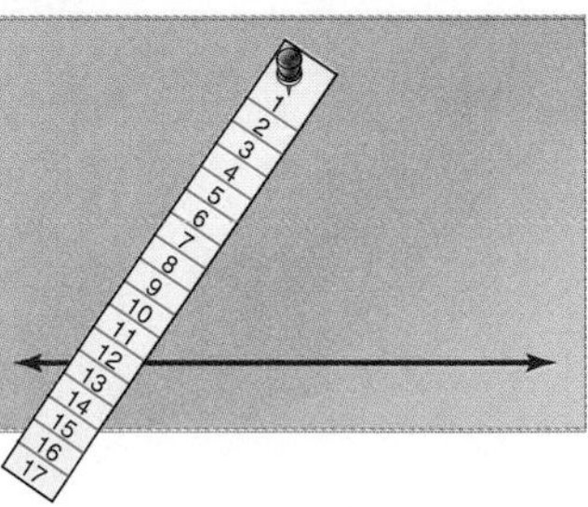

- Cut a vertical strip from a sheet of lined notebook paper, then number the lines from top to bottom.
- Draw a horizontal line on a piece of cardboard. Use a thumbtack to attach the numbered paper strip at a point above the horizontal line.
- Move the strip to the far right. Using the numbers on the strip as a scale, record the measure from the thumbtack to the horizontal line. Move the strip a little to the left and record the new distance. Continue moving the strip to the left and recording the measure until the strip is at the far left. You should record at least five measures.

Your Turn

a. Move the strip to the position of your smallest measure. Can you locate any other position with a smaller measure? If so, move the strip to that position.

b. The last position of the strip of paper in part a represents the shortest segment from the line to the point. What seems true about the strip of paper and the horizontal line?

c. Make a conjecture about the shortest segment from a line to a point not on the line.

The shortest segment from a point to a line is the perpendicular segment from the point to the line.

Definition of the Distance Between a Point and a Line	**The distance from a line to a point not on the line is the length of the segment perpendicular to the line from the point.**

The measure of the distance between a line and a point on the line is zero.

Example **Draw the segment that represents the distance from R to $\overleftrightarrow{AB}$.**

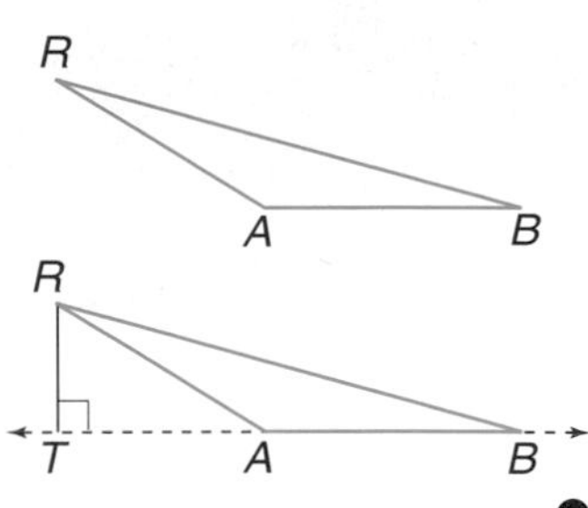

Since the distance from a line to a point not on the line is the length of the segment perpendicular to the line from the point, extend $\overline{AB}$ and draw $\overline{RT}$ so that $\overline{RT} \perp \overleftrightarrow{AB}$.

In Example 1, you can guarantee $\overline{RT} \perp \overleftrightarrow{AB}$ when you draw the segment by using the construction of a line perpendicular to a line through a point not on the line.

Example **Construct a line perpendicular to line r through $A(-4, 0)$, not on r. Then find the distance from A to r.**

Copy line r and point A. Place the compass point at point A. Make the setting wide enough so that when an arc is drawn, it intersects r in two places. Label these points of intersection B and C.

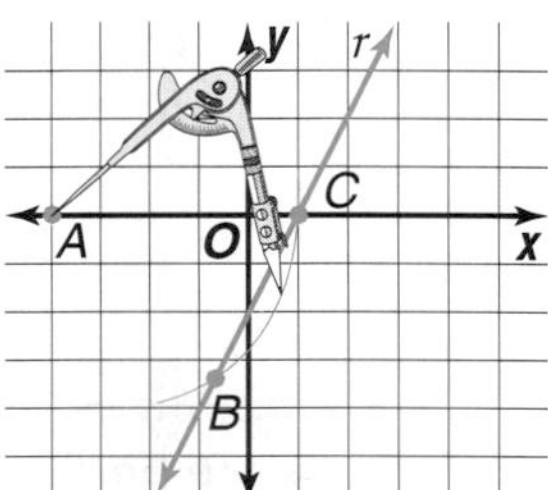

Put the compass at point B and draw an arc below line r. (*Hint:* Any compass setting greater than $\frac{1}{2}BC$ will work.)

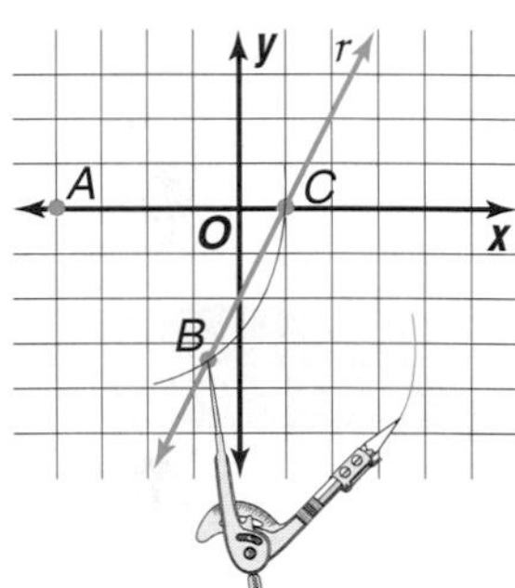

Using the same compass setting, put the compass at point C and draw an arc to intersect the one drawn in the previous step. Label the point of intersection D.

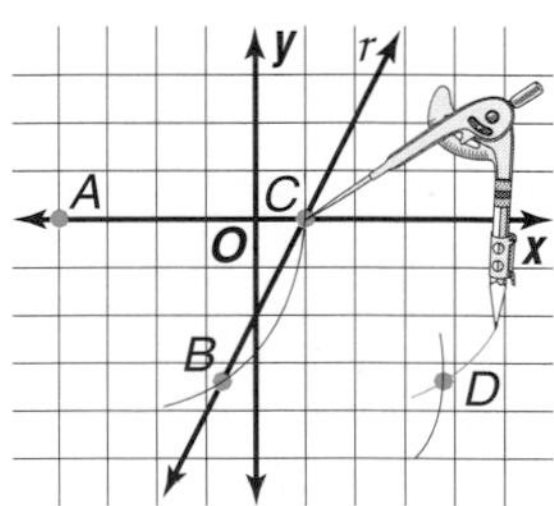

Draw $\overleftrightarrow{AD}$. $\overleftrightarrow{AD} \perp r$. Label point E at the intersection of $\overleftrightarrow{AD}$ and r. *Use the slopes of $\overleftrightarrow{AD}$ and r to verify that the lines are perpendicular.*

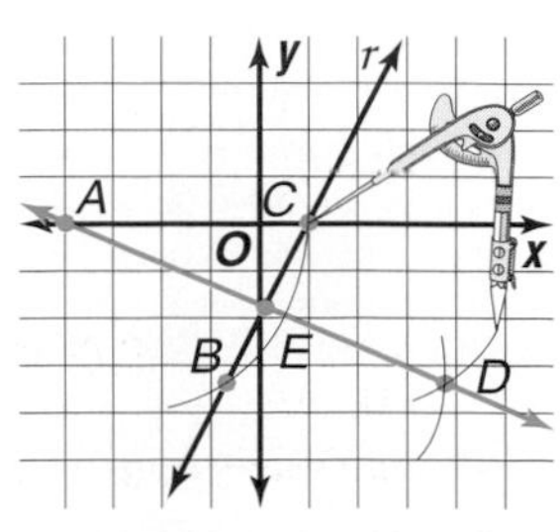

(continued on the next page)

The line segment constructed from point $A(-4, 0)$ perpendicular to the line r intersects line r at $E(0, -2)$. We can use the distance formula to find the distance between point A and line r.

LOOK BACK

You can refer to Lesson 1-4 for information on finding the distance between points.

$$\begin{aligned} d &= \sqrt{(x_2 - x_1)^2 + (y_2 - y_1)^2} \\ &= \sqrt{(-4 - 0)^2 + (0 - (-2))^2} \\ &= \sqrt{20} \end{aligned}$$

The distance between A and r is $\sqrt{20}$ or about 4.47 units.

Remember that the distance from a point to a line is the perpendicular distance from the point to the line.

In this chapter, we have learned how to prove two lines parallel using slopes in the coordinate plane or by using properties of angle relationships formed by a transversal to the parallel lines. Yet, how could you show two lines parallel if there were no coordinate plane and no transversal to the lines?

According to the definition, two parallel lines do not intersect. An alternate definition states that two lines in a plane are parallel if they are everywhere **equidistant**. Equidistant means that the distance between two lines measured along a perpendicular line to the line is always the same. To measure the distance between two parallel lines, we can measure the distance between one of the lines and any point on the other line.

Definition of the Distance Between Parallel Lines	The distance between two parallel lines is the distance between one of the lines and any point on the other line.

Example 3 INTEGRATION Algebra

Find the distance between the parallel lines ℓ and m whose equations are $y = 2x + 2$ and $y = 2x - 3$, respectively.

Explore You know the equations of two parallel lines and wish to find the distance between the lines.

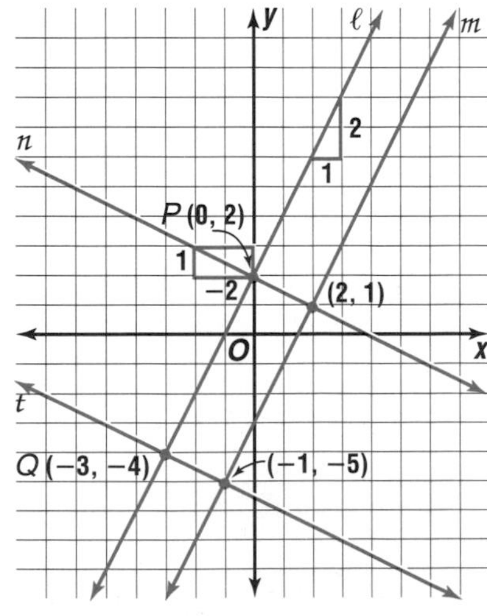

Plan Graph the lines on a coordinate plane. Pick a point such as $P(0, 2)$ on line ℓ. Draw a line perpendicular to ℓ through P and use the distance formula to find the distance between ℓ and P.

Recall that the product of the slopes of perpendicular lines is -1.

Solve Since line ℓ has a vertical change of 2 units for a horizontal change of 1 unit, its slope is 2. The slope of a line perpendicular to ℓ is $-\frac{1}{2}$. From point P, go left 2 units and up 1 unit to $(-2, 3)$. Draw line n through $(0, 2)$ and $(-2, 3)$. Line n is perpendicular to lines ℓ and m and appears to intersect line m at $(2, 1)$. Find the distance between $(0, 2)$ and $(2, 1)$.

$$d = \sqrt{(x_2 - x_1)^2 + (y_2 - y_1)^2}$$
$$= \sqrt{(0 - 2)^2 + (2 - 1)^2}$$
$$= \sqrt{5}$$

The distance between the lines is $\sqrt{5}$ or about 2.24 units.

Examine Pick another point $Q(-3, -4)$ on line ℓ. Since $\ell \parallel m$, the distance from Q to m should be the same, $\sqrt{5}$. Draw a line perpendicular to line m through Q. This line intersects m at $(-1, -5)$. Use the distance formula to find the distance between $Q(-3, -4)$ and $(-1, -5)$.

$$d = \sqrt{(x_2 - x_1)^2 + (y_2 - y_1)^2}$$
$$= \sqrt{(-3 - (-1))^2 + (-4 - (-5))^2}$$
$$= \sqrt{5}$$

In both cases, the distance between the lines is $\sqrt{5}$, so the answer checks.

For figures not on a coordinate plane, you can actually measure the distance between lines at several different points to determine if they are parallel.

Example 4

Art

Use a ruler to determine whether the lines shown in black by the artist are parallel.

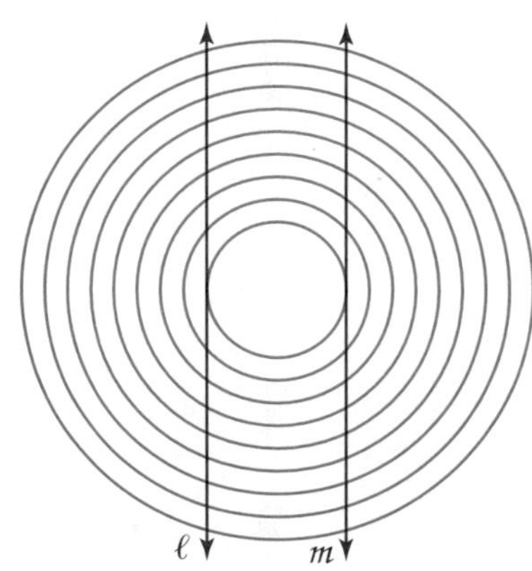

Choose several points on either line, such as line ℓ, and measure the perpendicular distance from ℓ to m with a ruler. The measurements will verify that ℓ and m are everywhere equidistant.

CHECK FOR UNDERSTANDING

Communicating Mathematics

Study the lesson. Then complete the following.

1. **Compare** the following definition for *parallel* from the *American Heritage Dictionary* and the alternate definition given in this lesson. How could you verify the dancers are in "parallel rows"?

   parallel being equal distance apart everywhere: *dancers in two parallel rows*

2. **Examine** the linear equations used for lines ℓ and m in Example 3. Describe the coefficient of the x term in each equation. What information does the coefficient provide regarding the slope of the lines?
3. **Make up a problem** involving an everyday situation where you need to find the distance between a point to a line or the distance between two lines. For example, find the shortest path from the garage of a house to the street to minimize the length of a driveway and material used in its construction.

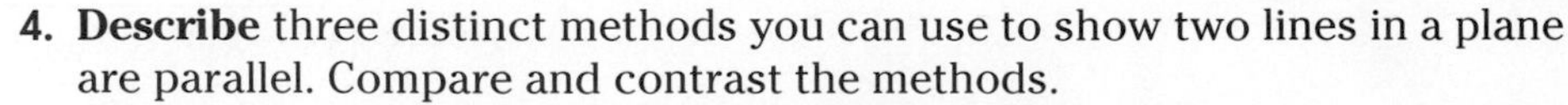

4. **Describe** three distinct methods you can use to show two lines in a plane are parallel. Compare and contrast the methods.

5. **Explain** how to construct a segment that represents the distance between two parallel lines, not on a coordinate plane.

Guided Practice

Copy each figure and draw the segment that represents the distance indicated.

6. Q to $\overleftrightarrow{AD}$

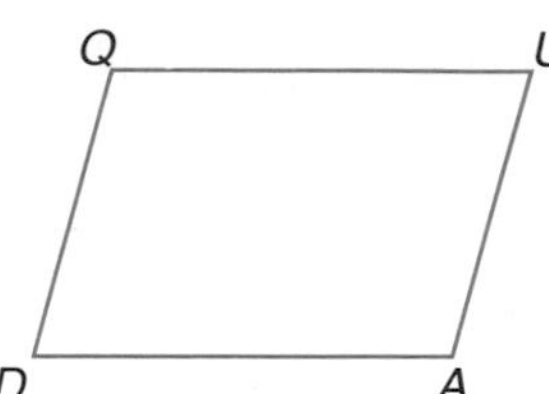

7. s to t

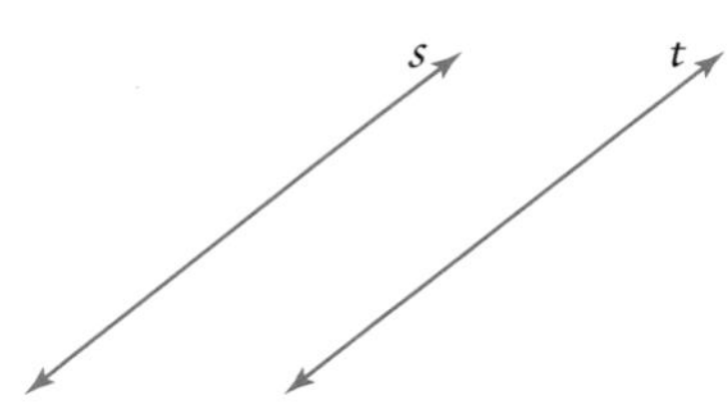

8. **Art** Use a ruler to determine whether the black lines are parallel. Write *yes* or *no*. Explain your answer.

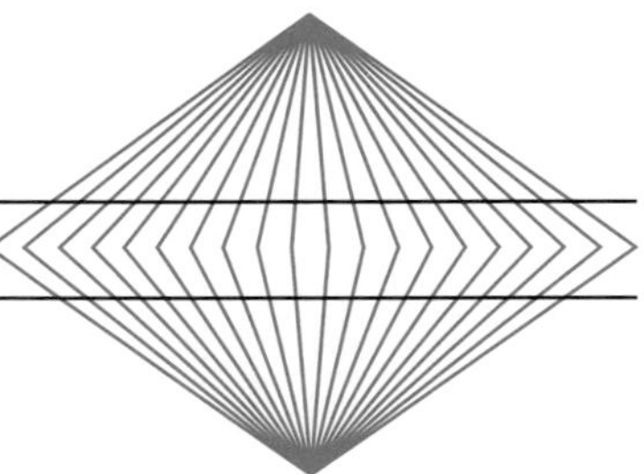

Graph each equation and plot the given ordered pair. Then construct a perpendicular segment and find the distance from the point to the line.

9. $y = 5, (-2, 4)$

10. $x + 3y = 6, (-5, -3)$

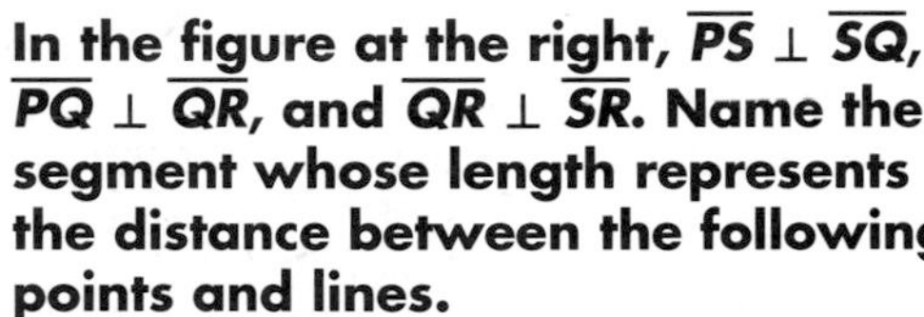

In the figure at the right, $\overline{PS} \perp \overline{SQ}$, $\overline{PQ} \perp \overline{QR}$, and $\overline{QR} \perp \overline{SR}$. Name the segment whose length represents the distance between the following points and lines.

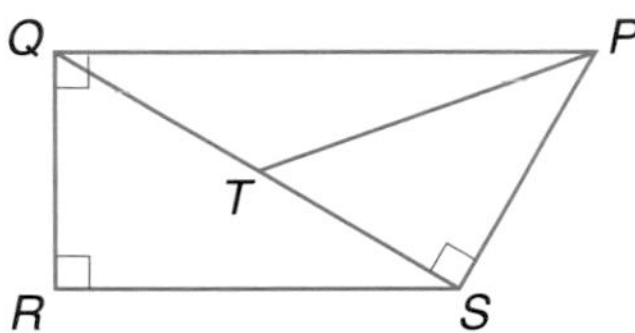

11. P to $\overleftrightarrow{SQ}$

12. S to $\overleftrightarrow{QR}$

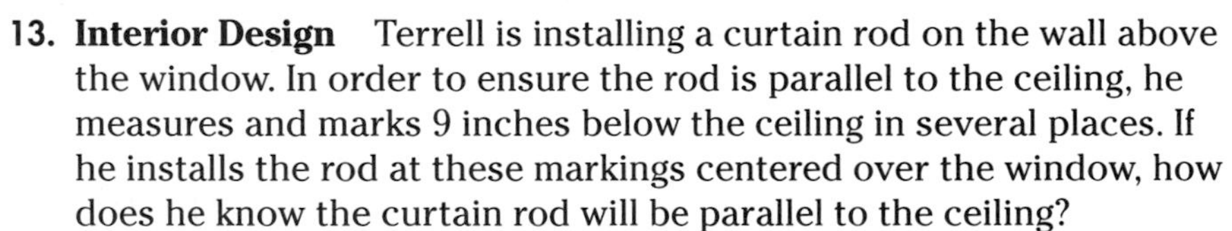

13. **Interior Design** Terrell is installing a curtain rod on the wall above the window. In order to ensure the rod is parallel to the ceiling, he measures and marks 9 inches below the ceiling in several places. If he installs the rod at these markings centered over the window, how does he know the curtain rod will be parallel to the ceiling?

EXERCISES

Practice

Copy each figure and draw the segment that represents the distance indicated.

14. G to $\overleftrightarrow{EA}$

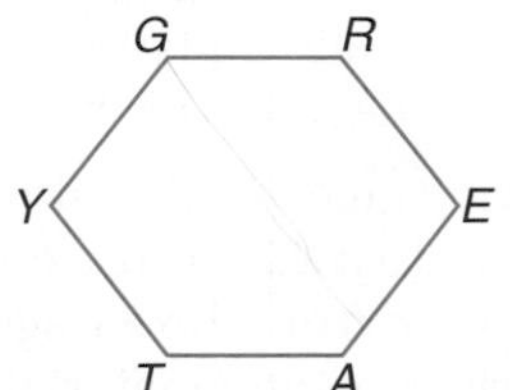

15. R to $\overleftrightarrow{LM}$

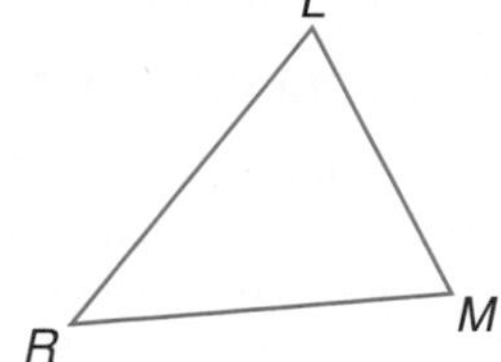

16. A to $\overleftrightarrow{HT}$

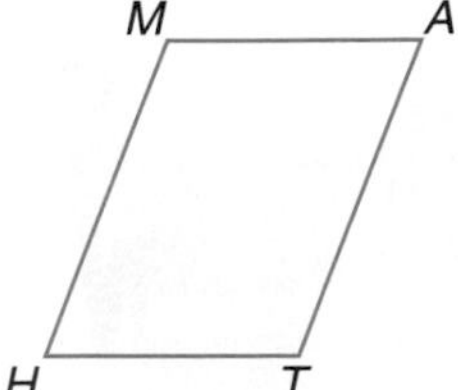

17. T to $\overleftrightarrow{AP}$

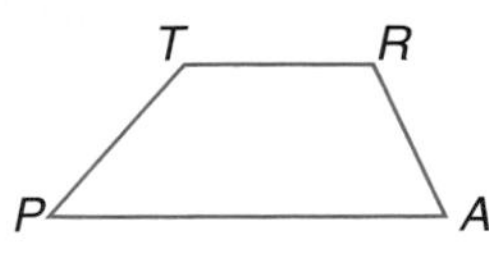

18. S to $\overleftrightarrow{PT}$

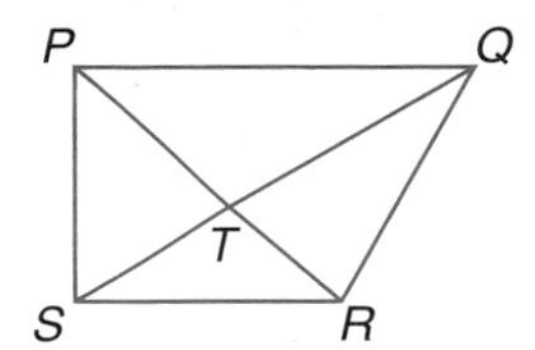

19. G to $\overleftrightarrow{AD}$

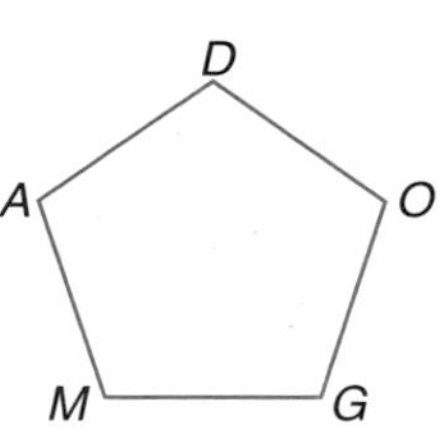

Use a ruler to determine whether the black lines are parallel. Explain your reasoning.

20.

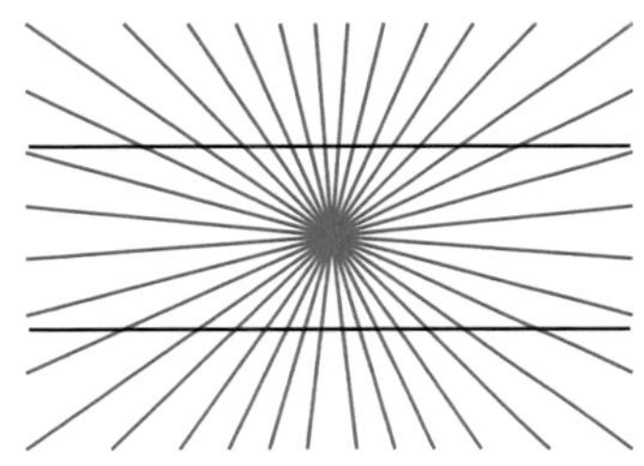

21.

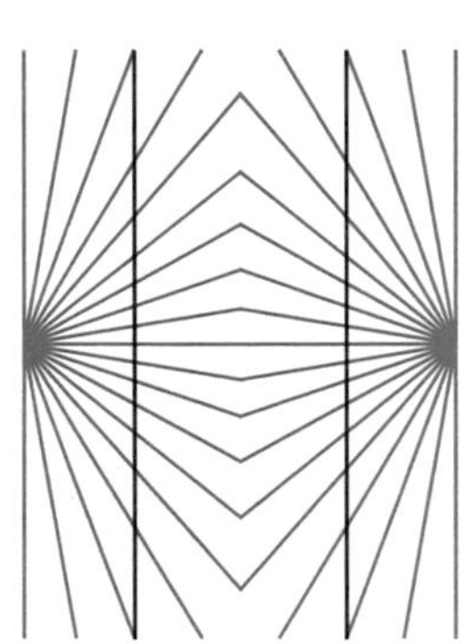

Using graph paper, graph each equation and plot the given ordered pair. Then construct a perpendicular segment and find the distance from the point to the line.

22. $y = 3$, $(4, -1)$
23. $x = -2$, $(3, 2)$
24. $y = 4x$, $(5, 3)$
25. $2x - y = 3$, $(2, 6)$
26. $3x + 4y = 1$, $(2, 5)$
27. $2x - 3y = -9$, $(2, 0)$

In the figure at the right, $\overline{MT} \perp \overline{HM}$, $\overline{AT} \perp \overline{HT}$, and $\overline{AT} \perp \overline{AM}$. Name the segment whose length represents the distance between the following points and lines.

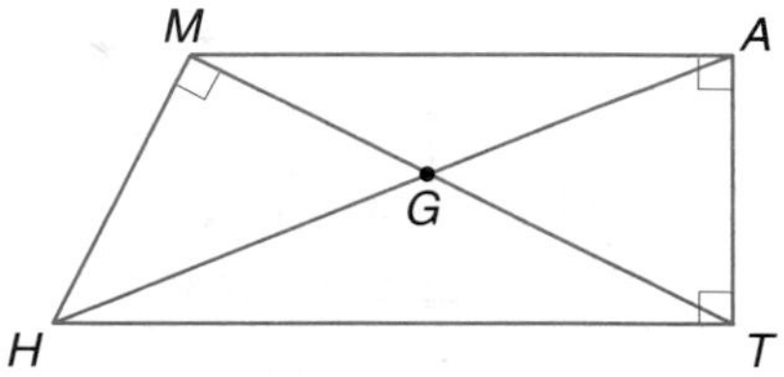

M A G H T

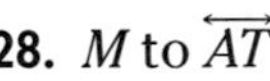

28. M to $\overleftrightarrow{AT}$
29. T to $\overleftrightarrow{HM}$
30. H to $\overleftrightarrow{AT}$
31. G to $\overleftrightarrow{HM}$
32. T to $\overleftrightarrow{AM}$
33. H to $\overleftrightarrow{MT}$

Proof

34. Provide the reason for each statement.

Given: $\overline{XW} \parallel \overline{YZ}$
$\overline{XW} \perp \overline{WZ}$
$\overline{XY} \perp \overline{YZ}$

Prove: $\overline{XY} \parallel \overline{WZ}$

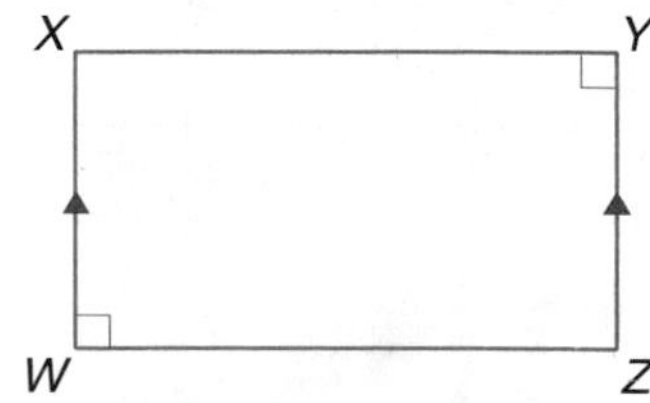

Proof:

Statements	**Reasons**
1. $\overline{XW} \parallel \overline{YZ}$ $\overline{XW} \perp \overline{WZ}$ $\overline{XY} \perp \overline{YZ}$	1. ?
2. $\overline{WZ} \perp \overline{YZ}$	2. ?
3. $\overline{XY} \parallel \overline{WZ}$	3. ?

35. Find the distance between the two parallel lines shown at the right.

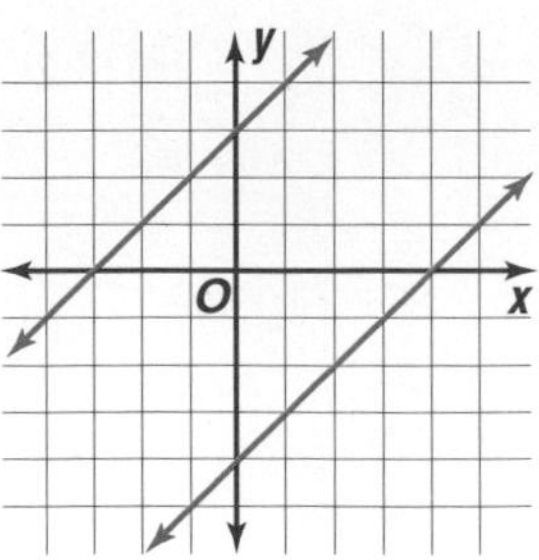

36. In the figure at the right, plane *ABC* is parallel to plane *EFG*. Explain how you could find the distance between two parallel planes.

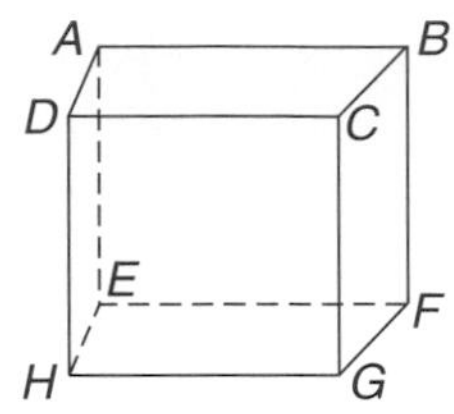

37. The distance from a point $P(x_1, y_1)$ to a line ℓ with equation $Ax + By + C = 0$ can be found by the formula $d = \frac{|Ax_1 + By_1 + C|}{\sqrt{A^2 + B^2}}$. Use the formula to find the distance between the line $3x + 4y = 1$ and the point $P(2, 5)$. Check your answer against your solution to Exercise 26. Did the formula verify your answer? Explain.

Critical Thinking

38. In the figure at the right, line ℓ is perpendicular to plane $\mathcal{P}$. Planes $\mathcal{P}$ and Q are parallel.
    a. Is line ℓ perpendicular to plane Q? Explain your reasoning.
    b. Create a three-dimensional model that verifies your conjecture.

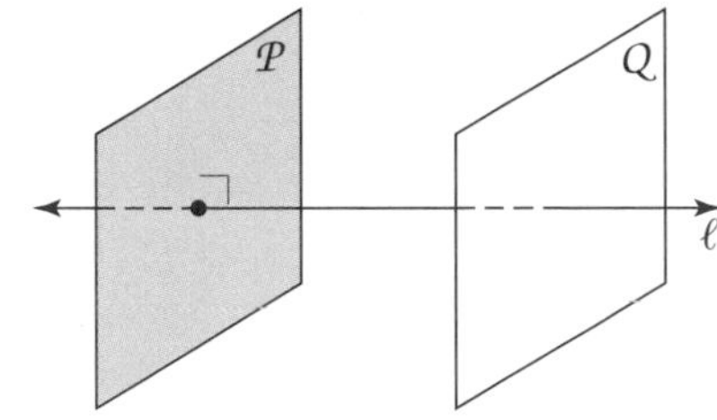

Applications and Problem Solving

39. **Graphic Arts** A graphic artist used the bar graph to illustrate statistics that compare the number of words used. Use a ruler to measure the distances between the bars.

Read All About It!

	Number of Words
Gettysburg Address	272
Back of Potato Chips bag	401
IRS Form 1040 EZ	418
Average *USA TODAY* cover story	1200

Source: *USA TODAY* research

a. Are the bars parallel?
b. Are the bars the same distance apart?
c. Do the statistical results surprise you? Explain.

40. **Construction** Dominique wants to install vertical boards to strengthen the handrail structure on her deck. How can she guarantee the vertical boards will be parallel?

Mixed Review

Refer to the figure at the right for Exercises 41–43.

41. If $m\angle 6 = 50$ and $m\angle 7 = 120$, what can you say about lines p and q? Explain. (Lesson 3–4)

42. If p and q are parallel, what can you say about their slopes? (Lesson 3–3)

43. If p and q are parallel, what can you say about $m\angle 2$ and $m\angle 4$? Explain. (Lesson 3–2)

44. Sports During the winter, Alma is a ski instructor. When helping a novice skier, should she describe the correct position for holding the skis as parallel, intersecting, or skew lines? (Lesson 3–1)

45. Two angles are congruent and supplementary. What else do you know about these angles? (Lesson 2–6)

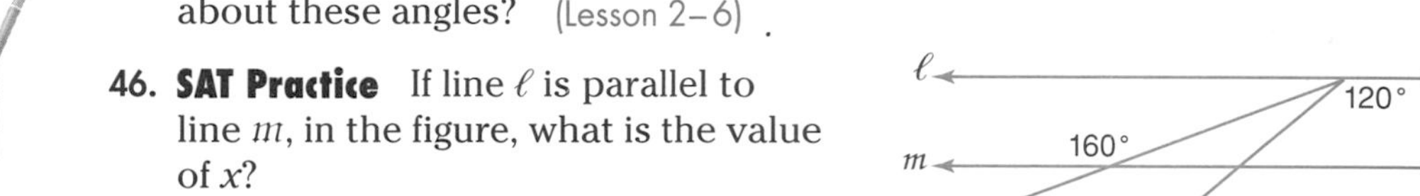
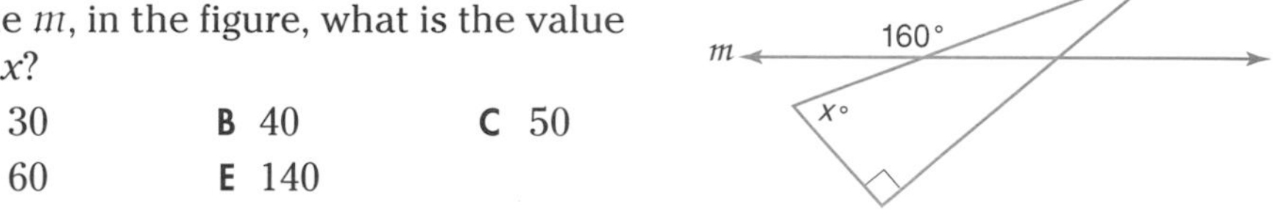

46. SAT Practice If line ℓ is parallel to line m, in the figure, what is the value of x?

A 30 B 40 C 50
D 60 E 140

47. Modeling Draw and label a figure to show plane $\mathcal{R}$ contains point A and line ℓ. (Lesson 1–2)

Algebra

For **Extra Practice**, see page 769.

48. Find $68 \div (-17)$.

49. Solve $15 - 2.7a = 22.29 - 3.6a$.

Stealthy Geometry

The excerpt below appeared in an article in *Popular Mechanics* in September, 1995.

A BRAND-NEW BUILDING AT LONDON'S Heathrow Airport sports unique architecture that renders it virtually invisible to radar. Airport authorities were worried that traffic-control radar would ricochet off the new... nerve center and interfere with runway operation. To squelch the problem, architect Norman Grimshaw incorporated distinctive geometry. The building's glass face looms outward as it rises, reaching an angle of 21° at the top. This curvature, combined with a series of external baffles and ribs, scatters radio energy down into an adjoining parking lot, which is paved with blocks rather than radar-reflecting asphalt. These features have shaved the building's radar signature by 99%. ■

1. Two things can happen to a radar wave when it strikes an object. What do you think they are?
2. What did designers need to know about the reflection of the radar waves off the new building in order to properly design it?
3. The article refers to the building's *radar signature.* What do you think is meant by this term?

3–5B Using Technology Distance Between a Point and a Line

An Extension of Lesson 3–5

You can use a graphing calculator to find the distance between a point and a line.

The standard viewing window can be selected in the ZOOM menu.

Notice the messages that apppear on the screen before each ENTER. By pressing ENTER, you answer yes to the prompt.

Example **Find the distance between the line whose equation is $y = 2x - 5$ and the point at $(7, -1)$. The equation of the line perpendicular to the first line through $(7, -1)$ is $y = -\frac{1}{2}x + \frac{5}{2}$.**

Clear the graphics screen and select the standard viewing window. Graph the lines by entering the equations in the Y= list.

Enter: [Y=] 2 [X,T,θ] [—] 5

[ENTER]

[(–)] .5 [X,T,θ] [+] 2.5

[GRAPH]

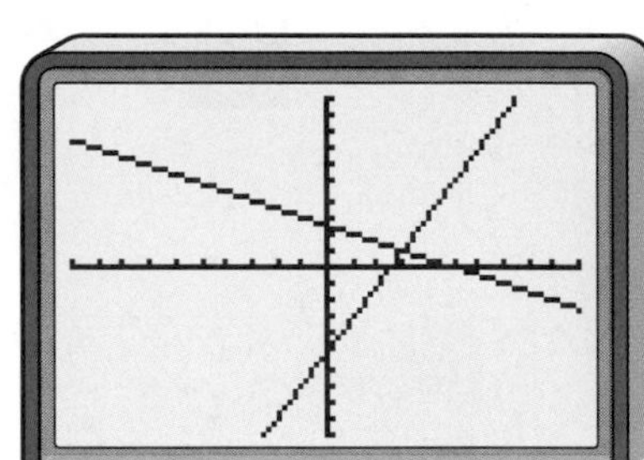

You can select the 5:Intersect command in the CALC menu to find the coordinates of the intersection of two graphed lines.

Enter: [2nd] [CALC] 5 [ENTER] [ENTER] [ENTER].

The coordinates at the bottom of the screen are X=3 and Y=1, or (3, 1).

Use your calculator to find the distance between the intersection point and the given point.

$$d = \sqrt{(x_2 - x_1)^2 + (y_2 - y_1)^2} \text{ or } \sqrt{(7 - 3)^2 + (-1 - 1)^2}$$

Enter: [2nd] [$\sqrt{\ }$] [(] [(] 7 [—] 3 [)] [x^2] [+]

[(] [(–)] 1 [—] 1 [)] [x^2] [)] [ENTER]

4.472135955

The distance between the point and the line is about 4.47.

EXERCISES

Use a graphing calculator to find the distance between the graph of each equation and the point, rounded to the nearest hundredth. The equation of the line perpendicular to the first line through the point is given.

1. $y = x + 7$, $A(2, 7)$
perpendicular: $y = -x + 9$

2. $y = \frac{2}{5}x + 3$, $B(-8, -6)$
perpendicular: $y = -\frac{5}{2}x - 26$

3. $y = 3x - 13$, $C(1, 0)$
perpendicular: $y = -\frac{1}{3}x + \frac{1}{3}$

4. $y = \frac{3}{2}x$, $D(-4, 7)$
perpendicular: $y = -\frac{2}{3}x + \frac{13}{3}$

3-6 Integration: Non-Euclidean Geometry Spherical Geometry

What YOU'LL LEARN

- To identify points, lines, and planes in spherical geometry, and
- to compare and contrast basic properties of plane and spherical geometry.

Why IT'S IMPORTANT

You can understand the relationship of locations on the surface of Earth.

CONNECTION

Geography

This world map uses a system of coordinates to locate places on the surface of Earth.

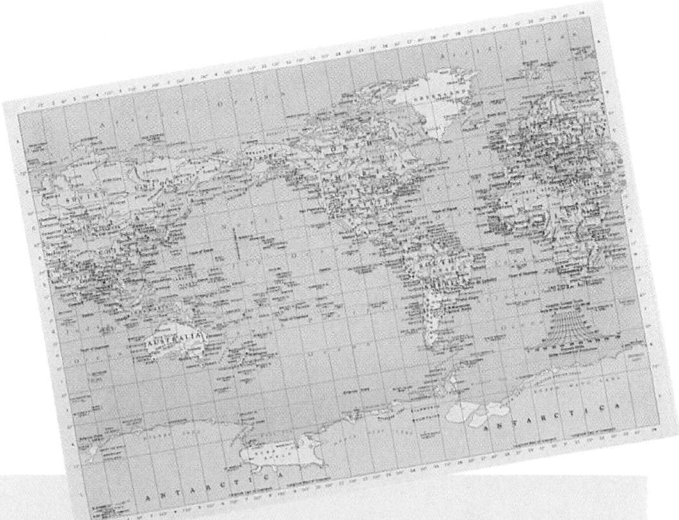

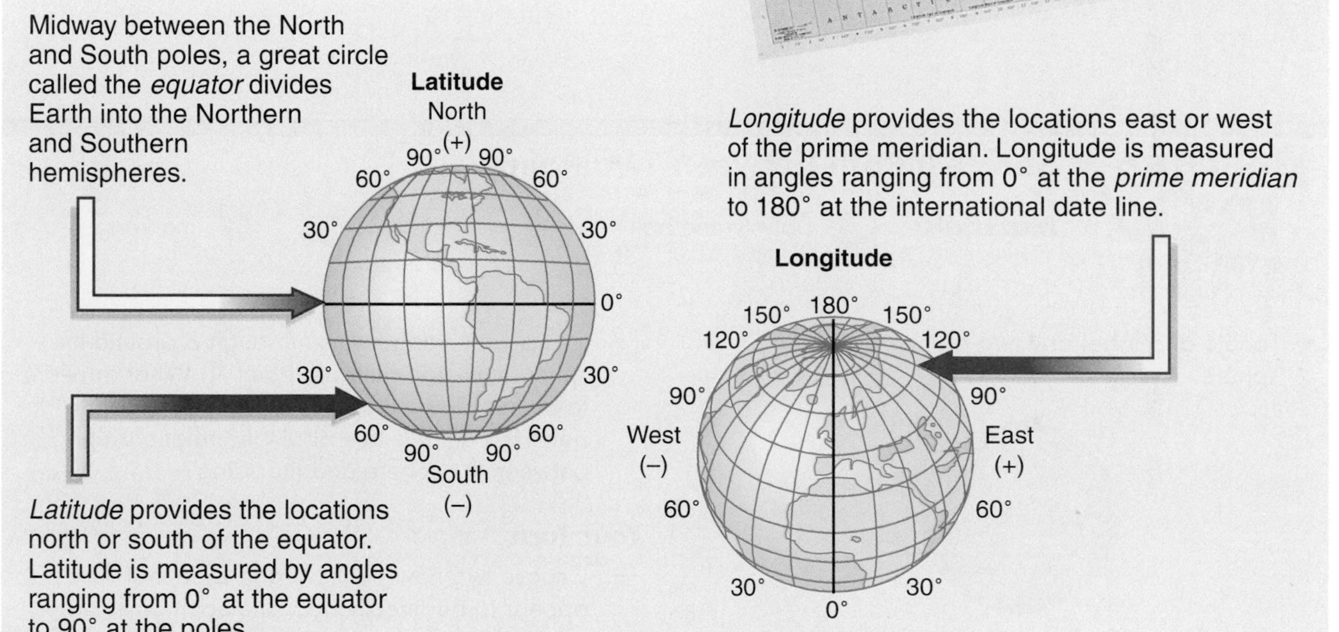

Example 1 Find a major city in the United States whose location is near the point with coordinates of 90° W and 35° N.

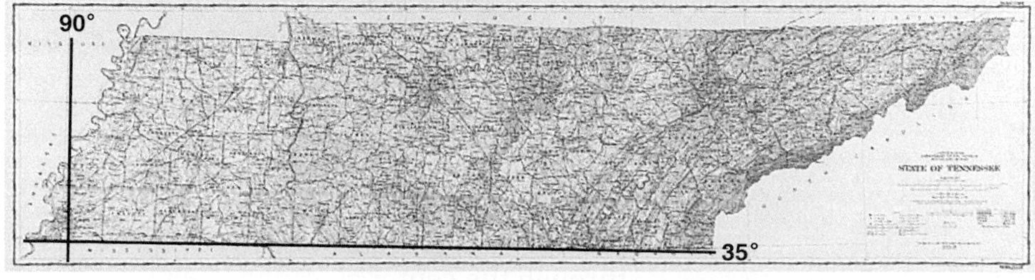

Memphis, Tennessee, is located near the intersection of 90° W and 35° N.

The equator is called a *great circle* of Earth because it divides Earth into two equal halves. Similarly in spherical geometry, a line is defined as a great circle of a sphere. A line is a circle that divides the sphere into two equal *half-spheres*. A plane is the sphere itself.

So far in our studies of **plane Euclidean geometry**, we studied a system of points, lines, and planes. In **spherical geometry**, we study a system of points, great circles (lines), and spheres (planes). Spherical geometry is one type of **non-Euclidean geometry**.

Plane Euclidean Geometry

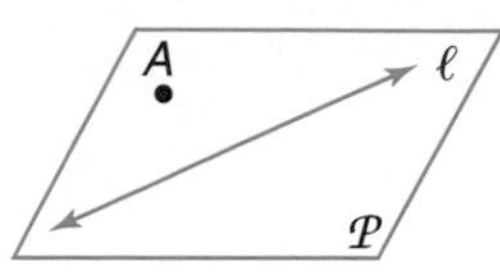

Plane 𝒫 contains line ℓ and point A not on ℓ.

Spherical Geometry

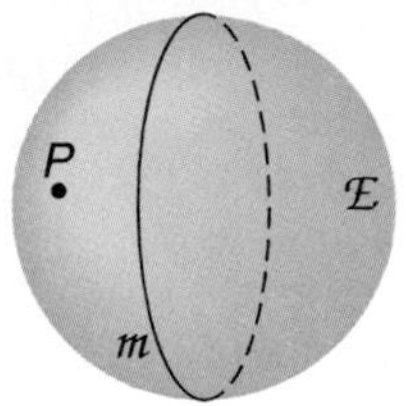

Sphere ℰ contains great circle m and point P not on m. m is a line on sphere ℰ.

MODELING MATHEMATICS Shortest Path between Two Points

Materials:  polystyrene ball  ruler  string markers

- Locate and label any two points A and B on your sphere.

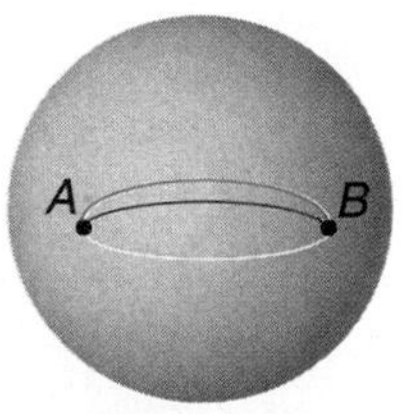

- How many curved paths are there from A to B?
- Use string to show several curved paths from A to B. Measure the length of each string to the nearest millimeter. Make a conjecture about the shortest distance between two points in spherical geometry.
- Wrap your string from A through B around the sphere until you return to point A. What appears to be true about the shortest path from point A to point B? What is true about the relationship between the sphere and the string?

Your Turn

a. Choose two new points C and D so that they appear to be directly opposite each other on your sphere. These are called *polar points* of the sphere. Draw a figure showing the sphere and points C and D.

b. In how many different ways can you measure the shortest path from C to D?

The polar points discussed in the Modeling Mathematics activity are points of intersection of a line through the center of a sphere with the sphere.

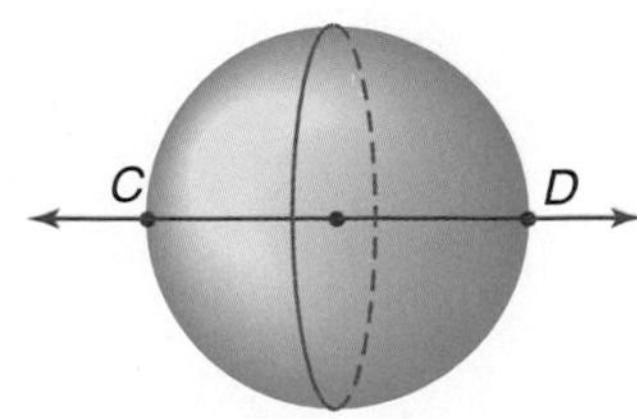

The table below compares and contrasts lines in the system of plane Euclidean geometry and lines (great circles) in spherical geometry.

Plane Euclidean Geometry Lines on the Plane	Spherical Geometry Great Circles (Lines) on the Sphere
1. A line segment is the shortest path between two points.	**1.** An arc of a great circle is the shortest path between two points.
2. There is a unique straight line passing through any two points.	**2.** There is a unique great circle passing through any pair of nonpolar points.
3. A straight line is infinite.	**3.** A great circle is finite and returns to its original starting point.
4. If three points are collinear, exactly one is between the other two. *A* *B* *C* *B* is between *A* and *C*.	**4.** If three points are collinear, any one of the three points is between the other two. *A* is between *B* and *C*. *B* is between *A* and *C*. *C* is between *A* and *B*. *C* *A* *B*

The Segment Addition Postulate holds true in spherical geometry as well as plane Euclidean geometry.

Example 2

Geography

The cities of London (England), Castellón de la Plana (Spain), and East Cape (New Zealand) lie approximately on the great circle containing the prime meridian. The distance between London and Castellón de la Plana is 820 miles. The distance from Castellón de la Plana to East Cape is 5300 miles. Find the distance from London to East Cape through Castellón de la Plana.

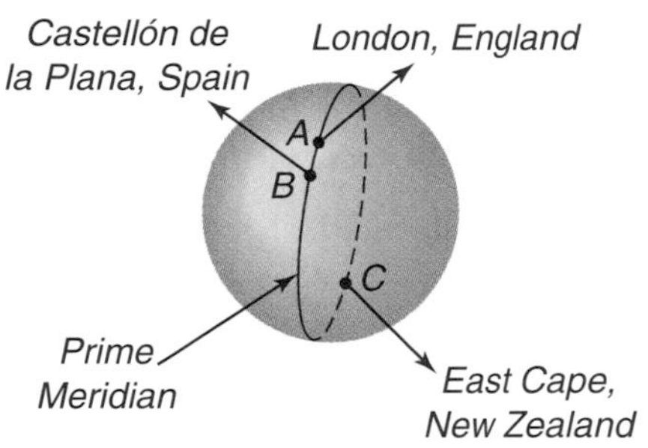

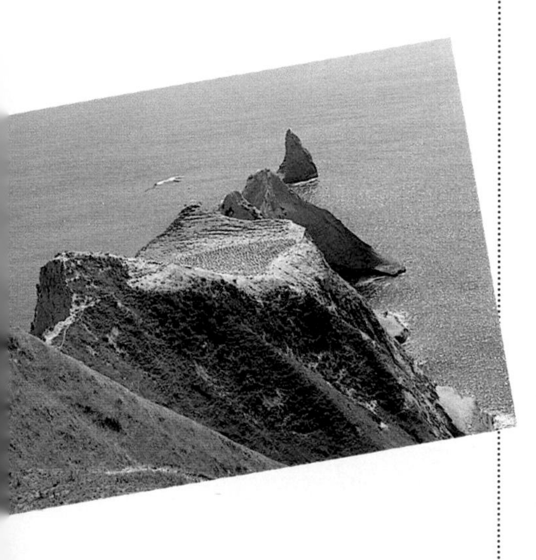

Since the three cities lie on the same great circle, you can state that Castellón de la Plana (B) lies between London (A) and East Cape (C). By the segment addition postulate, $AC = AB + BC$.

$$\begin{aligned} AC &= AB + BC \\ &= 820 + 5300 \\ &= 6120 \end{aligned}$$

Therefore, the distance from London to East Cape through Castellón de la Plana is approximately 6120 miles.

FYI

Spherical Geometry, also known as *Riemannian geometry*, was discovered by German mathematician G.F.B. Reimann about 1860.

In spherical geometry, Euclid's first four postulates (Postulates 2–1 through 2–4) and their related theorems hold true. However, theorems that depend on the parallel postulate may not be true.

The definition of parallel lines states that they lie in the same plane and never intersect. In spherical geometry, the sphere is the plane, and a line must be a great circle. Every great circle containing A intersects ℓ. Thus, there exists no line through point A that is parallel to ℓ.

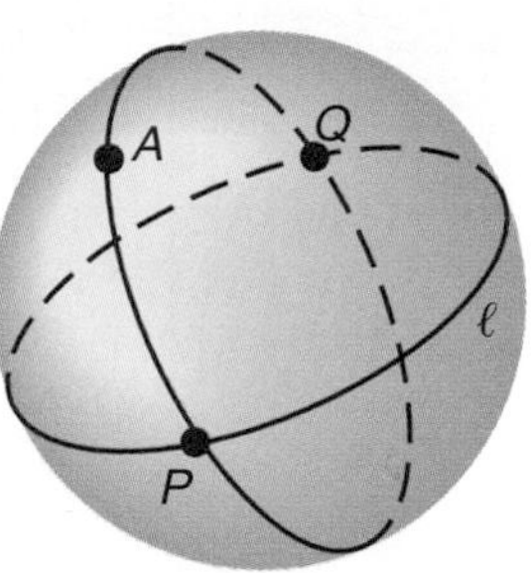

Every great circle of a sphere intersects all other great circles of a sphere in exactly two points. In the figure at the left, one possible line through point A intersects line ℓ at P and Q.

If two great circles divide a sphere into four congruent regions, the lines are perpendicular to each other.

Example 3 **For each property listed from plane Euclidean geometry, write a corresponding statement for non-Euclidean spherical geometry.**

a. Perpendicular lines intersect at one point.

Perpendicular great circles intersect at two points.

b. Perpendicular lines form four right angles.

Perpendicular great circles form eight right angles.

c. Perpendicular lines divide a plane into four infinite regions.

Perpendicular great circles divide the sphere into four finite regions.

Example 4 **In the figure at the right, both lines ℓ and m are perpendicular to line s.**

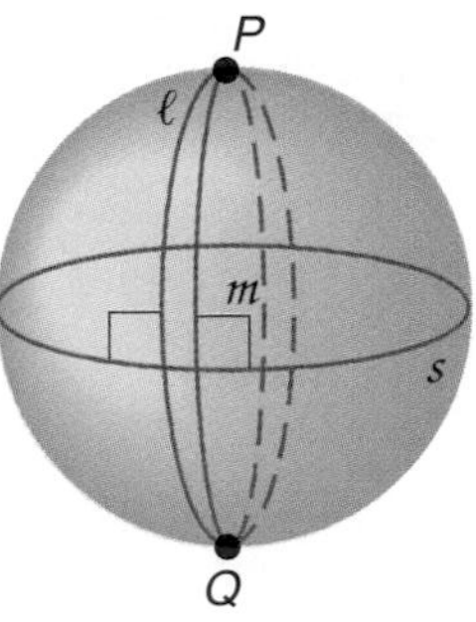

a. In plane Euclidean geometry, what relationship must exist between lines ℓ and m if they are both perpendicular to line s?

In plane Euclidean geometry, Theorem 3–8 states that two lines perpendicular to the same line are parallel.

b. Is this true in spherical non-Euclidean geometry? Why or why not?

In spherical geometry, this cannot be true since parallel lines (great circles) fail to exist. In this case, although $\ell \perp s$ and $m \perp s$, ℓ and m do intersect at points P and Q.

CHECK FOR UNDERSTANDING

Communicating Mathematics

Study the lesson. Then complete the following.

1. In spherical geometry, there is a unique great circle passing through any pair of nonpolar points. Explain why the points must be nonpolar.
2. **Compare** lines in plane Euclidean geometry with lines in spherical geometry.
   a. Which lines divide the plane into two finite congruent regions?
   b. Which lines have one point of intersection instead of two?
3. **Refer** to Example 2 on page 165. If the approximate distance around Earth is 24,900 miles, explain whether or not it would be shorter to go from London to East Cape without going through Castellón de la Plana.

4. a. **Draw** a great circle on a ball. Also, draw point P not on ℓ. Draw a circle r on the sphere that passes through P and appears to be parallel to ℓ. Why is the circle r not a great circle of the sphere?

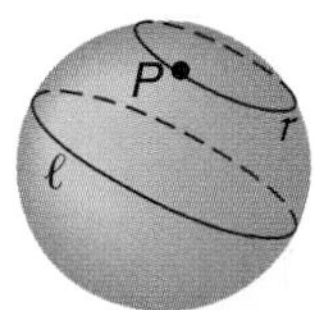

   b. **Draw** a line (great circle) m perpendicular to ℓ passing through P. Can any lines other than m be drawn through point P and be perpendicular to ℓ? Explain.

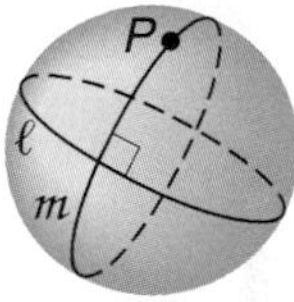

Guided Practice

If spherical points are restricted to be nonpolar points, determine if each statement from Euclidean geometry is also *true* in spherical geometry. If *false*, explain your reasoning.

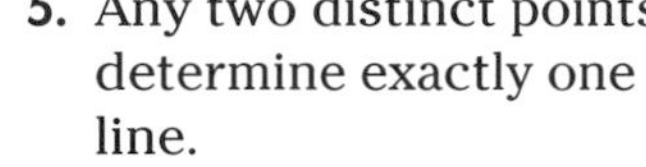

CAREER CHOICES

Cartographers prepare maps using information provided by aerial photographs, geodetic surveys, and satellite data. A bachelor's degree in engineering or physical science is necessary for the field, with mathematics, geography, and computer science training very helpful.

For more information, contact:

American Congress on Surveying and Mapping
5410 Grosvenor Lane
Bethesda, MD 20814-2122

5. Any two distinct points determine exactly one line.
6. Two perpendicular lines create four right angles.
7. Use a map of southwestern Asia to name the latitude and longitude of Istanbul, Turkey.
8. Use a map to name the city and country located near 25° N, 55° E.

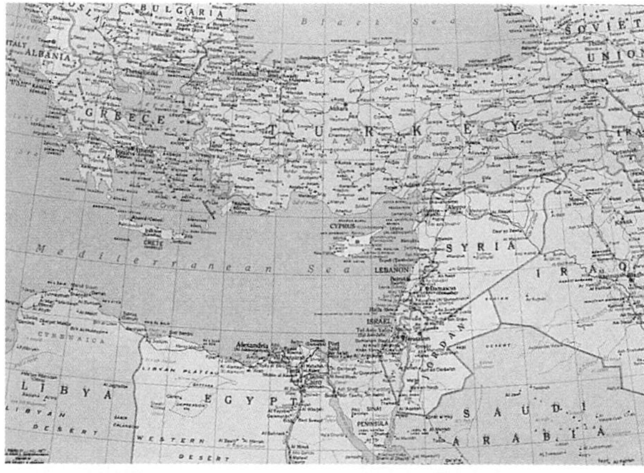

For each property from plane Euclidean geometry, write a corresponding statement for non-Euclidean spherical geometry.

9. The straight line is infinite.
10. A straight line segment is the shortest path between two points.
11. **Geography** Suppose the town where you live is a pole of Earth.
    a. Use a globe to state the latitude and longitude of your home town.
    b. State the latitude and longitude of the location of your opposite pole. What city is near it?

EXERCISES

Practice **If spherical points are restricted to be nonpolar points, decide which statements from Euclidean geometry are *true* in spherical geometry. If *false*, explain your reasoning.**

12. If three points are collinear, exactly one point is between the other two.
13. A line has infinite length.
14. Given a line ℓ and point P not on ℓ, there exists exactly one line parallel to ℓ passing through P.
15. Given a line ℓ and a point P not on ℓ, there exists exactly one line perpendicular to ℓ passing through P.
16. Two lines perpendicular to the same line are parallel to each other.
17. Two intersecting lines divide the plane into four regions.

Use a globe or world map to name the latitude and longitude of each city.

18. Houston, Texas
19. Panama City, Florida
20. Johannesburg, South Africa
21. Mexico City, Mexico

Use a globe or world map to name the city located near each set of coordinates.

22. 30° N, 90° W
23. 40° N, 120° W
24. 32° S, 116° E

For each property listed from plane Euclidean geometry, write a corresponding statement for non-Euclidean spherical geometry.

25. Two distinct lines with no point of intersection are parallel.
26. Two distinct intersecting lines intersect in exactly one point.
27. A pair of perpendicular straight lines divides the plane into four infinite regions.
28. A pair of perpendicular straight lines intersects once and creates four right angles.
29. Parallel lines have infinitely many common perpendicular lines.
30. There is only one distance that can be measured between two points.
31. Draw two great circles, ℓ and m, on a sphere as shown. Draw as many great circles as possible that are perpendicular to both ℓ and m. How many perpendiculars were you able to draw?

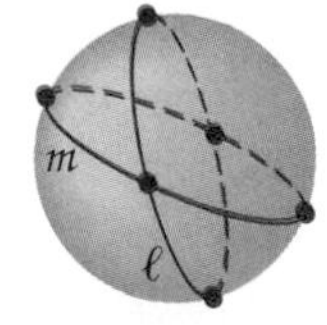

32. Compare each of the following lengths to the length of a great circle of a sphere.
    a. distance between any pair of pole points
    b. distance between any pole point and its equator
33. Draw two perpendicular great circles and then draw the common perpendicular.
    a. What relationship exists between the great circles?
    b. Describe the location of polar points for each great circle.

34. Consider another version of the Parallel Postulate: *Given a line and a point not on the line, there are many lines that pass through the point and that are parallel to the given line.*
   a. Draw a model that would show how this might be true in space.
   b. Do some research to find the name of this type of non-Euclidean geometry.

Critical Thinking

35. a. Explain why $\triangle ABC$ could not exist in plane Euclidean geometry.
   b. Could $\triangle ABC$ exist in non-Euclidean spherical geometry? Explain your reasoning. Include a drawing.

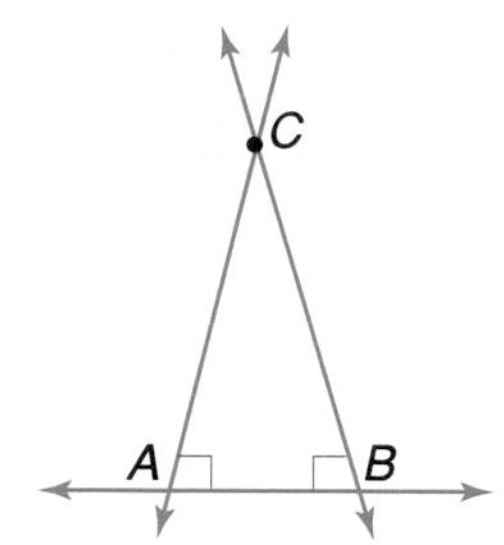

Applications and Problem Solving

36. **Climate** Locate Philadelphia, Pennsylvania, in an atlas.
   a. Determine its approximate latitude above the equator.
   b. Use a globe to locate two other cities in the Northern Hemisphere that lie on the same latitude.
   c. Is there a connection between latitude and climate? Explain your reasoning.

37. **Space Travel** According to *Einstein's Spherical Universe*, geometric properties of space are similar to those on the surface of a sphere. Based on this model, what conclusion follows about the path of a spaceship along a straight line? Explain your reasoning.

Albert Einstein

Mixed Review

38. Find the distance between point $P(6, -2)$ and the graph of line ℓ whose equation is $y = 7$. (Lesson 3–5)

39. Two lines are cut by a transversal and the consecutive interior angles are congruent. What must be true for the lines to be parallel? (Lesson 3–4)

40. Find the slope of the line that passes through (3, 0) and (8, −2). (Lesson 3–3)

41. **Draw a Diagram** Hector and Shantel are building steps to their new shed. It takes one concrete block for one step, three blocks for two steps, and six blocks for three steps. How many concrete blocks will it take to build six steps? (Lesson 3–1)

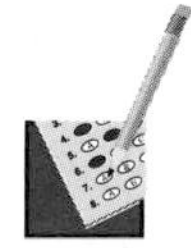

42. **SAT Practice Grid-in** What is the arithmetic mean of the numbers below?
$$-60, -48, -24, 0, 2, 4, 6, 24, 48, 60$$

43. Write a conditional statement about angle measures. Identify the hypothesis and conclusion. (Lesson 2–2)

44. Find AB, given $A(-1, 4)$ and $B(6, -20)$. (Lesson 1–4)

Algebra

For **Extra Practice**, see page 770.

45. Name the property illustrated by $15 + 41 = 41 + 15$.

46. Solve $-8 - m = 13$.

CLOSING THE

In·ves·ti·ga·tion

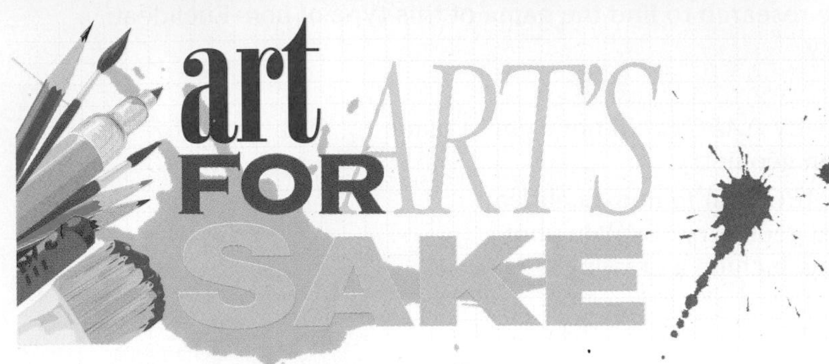

Refer to the Investigation on pages 66–67.

Commercial artists create artwork for specific uses for their clients. Many art pieces are used in advertising, as logos, or in decorating. Creating art for a client is different than working for yourself, since the client's tastes, goals, and budget must be considered.

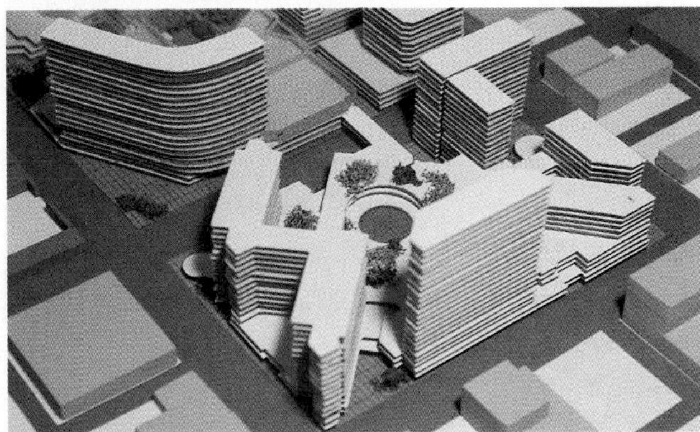

Analyze

You have chosen a style for the lobby and created your mural around a chosen theme. It is now time to analyze your work and finalize the project.

PORTFOLIO ASSESSMENT

You may want to keep your work on this Investigation in your portfolio.

1. Look over your mural and the style you have chosen for the lobby furnishings. Do they harmonize well? Do they communicate the client's image?
2. Refer to your notes about the uses for the lobby. Compile a list of suggested furniture, flooring, wallcovering, and any additional artwork for the lobby.
3. Acquire samples of wallcoverings and photographs of furniture and flooring materials for the lobby. Place these and the mural design in a portfolio for the client.
4. You now need to do a cost analysis and determine what to charge the client. First determine the total number of hours your team spent on the project, including the time you would spend installing the mural and furniture in the lobby. Consult a reference book or a working artist to determine a reasonable hourly rate. Add the cost of materials and furnishings to create a final bill.

Present

It is time to unveil the lobby. Choose one or more members of your team to present the lobby to the CEO and the board members of Cygnus Software, Inc.

5. Give a short speech describing the process of the design, the goals, the image that the lobby is to present, and the reasons for your choices.
6. Present the bill for your work. Be sure to justify your pricing method.

Data Collection and Comparison To share and compare your data with other students, visit:

www.geometry.glencoe.com

CHAPTER 3 HIGHLIGHTS

Chapter Review For additional lesson-by-lesson review, visit: **www.geometry.glencoe.com**

VOCABULARY

After completing this chapter, you should be able to define each term, property, or phrase and give an example or two of each.

Geometry

alternate exterior angles (p. 126)
alternate interior angles (p. 126)
consecutive interior angles (p. 126)
corresponding angles (p. 126)
distance between a point and a line (p. 154)
distance between parallel lines (p. 156)
equidistant (p. 156)
exterior angle (p. 126)
if and only if (p. 139)
interior angle (p. 126)
non-Euclidean Geometry (p. 164)
parallel lines (p. 124)
Parallel Postulate (p. 147)
plane Euclidean geometry (p. 164)
skew lines (p. 125)
spherical geometry (p. 164)
transversal (p. 126)

Algebra

slope (p. 138)

Problem Solving

draw a diagram (p. 124)

UNDERSTANDING AND USING THE VOCABULARY

Refer to the figure below and choose the letter of the term that best matches each statement or phrase.

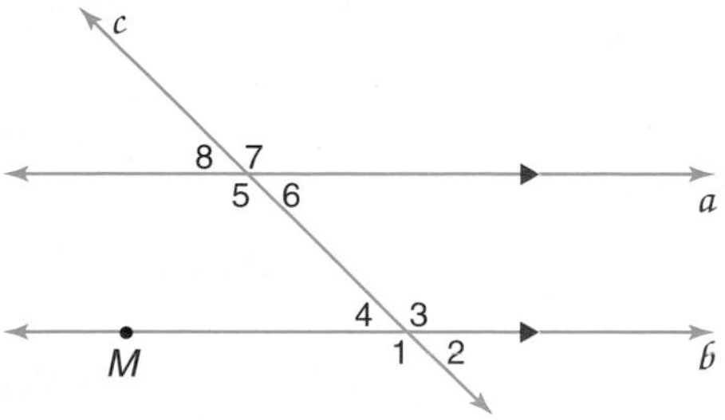

1. ∠3, ∠4, ∠5, and ∠6
2. line c
3. ∠3 and ∠7
4. ∠2 and ∠8
5. lines a and b
6. ∠1, ∠2, ∠7, and ∠8
7. ∠5 and ∠4
8. There is exactly one line parallel to line a through point M.
9. ∠5 and ∠3
10. line c and a line perpendicular to the plane of this page passing through point M
11. the ratio of the vertical rise to the horizontal run of line b

a. alternate exterior angles
b. alternate interior angles
c. consecutive interior angles
d. corresponding angles
e. exterior angles
f. interior angles
g. parallel lines
h. parallel postulate
i. skew lines
j. slope
k. spherical geometry
l. transversal

CHAPTER 3 STUDY GUIDE AND ASSESSMENT

SKILLS AND CONCEPTS

OBJECTIVES AND EXAMPLES

Upon completing this chapter, you should be able to:

- describe the relationships between two lines, between two planes, and between angles formed by two lines and a transversal (Lesson 3–1)

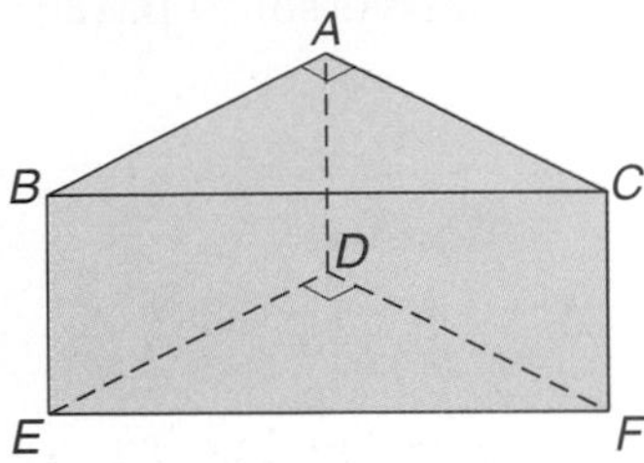

Plane *ABC* and plane *DEF* are parallel.
Segments *BC* and *EF* are parallel.
Segments *AB* and *DF* are skew.

REVIEW EXERCISES

Use these exercises to review and prepare for the chapter test.

Refer to the figure at the right to find an example of each of the following.

12. two parallel lines
13. two skew lines
14. two parallel lines and a transversal
15. two intersecting lines
16. two noncoplanar lines

- use the properties of parallel lines to determine angle measures (Lesson 3–2)

List the conclusions that can be drawn if $\ell \parallel m$.

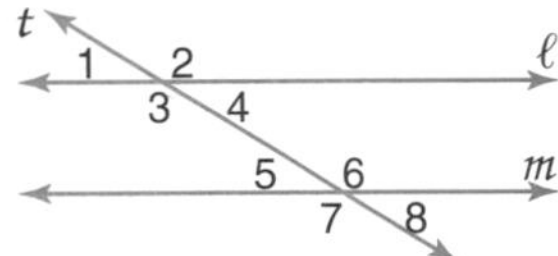

corresponding angles $\angle 1 \cong \angle 5$, $\angle 2 \cong \angle 6$, $\angle 3 \cong \angle 7$, and $\angle 4 \cong \angle 8$

alternate interior angles $\angle 3 \cong \angle 6$, and $\angle 4 \cong \angle 5$

consecutive interior angles $\angle 3$ and $\angle 5$, and $\angle 4$ and $\angle 6$ are supplementary.

alternate exterior angles $\angle 1 \cong \angle 8$, and $\angle 2 \cong \angle 7$

Refer to the figure below for Exercises 17–20.

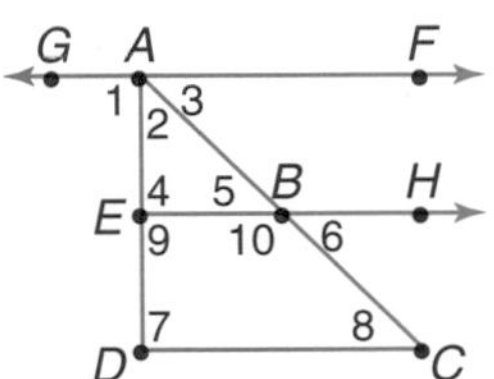

17. If $\overline{GF} \parallel \overline{EH}$, which angles are congruent?
18. If $\overline{GF} \parallel \overline{DC}$, which angles are congruent?
19. If $\overline{GF} \parallel \overline{DC}$, which angles are supplementary to $\angle CAG$?
20. If $\overline{EH} \parallel \overline{DC}$, which angles are supplementary to $\angle BCD$?

- find the slopes of lines (Lesson 3–3)

Find the slope of the line passing through (5, −2) and (−1, −5). Describe the line.

$$m = \frac{y_2 - y_1}{x_2 - x_1} = \frac{-2 - (-5)}{5 - (-1)} \text{ or } \frac{1}{2}$$

Since the slope is positive, the line rises from left to right.

Find the slope of the line passing through points having the given coordinates. Then describe the line as you move from left to right as *rising, falling, horizontal,* or *vertical.*

21. (0, 4), (−1, −2)
22. (2, 0), (2, −6)
23. (11, 2), (5, 4)
24. (−1, −5), (3, −7)

OBJECTIVES AND EXAMPLES	REVIEW EXERCISES

- Use slope to identify parallel and perpendicular lines (Lesson 3–3)

Find the slopes of the lines parallel to and perpendicular to a line passing through (4, −2) and (5, 3).

The slope of the line passing through (4, −2) and (5, 3) is $\frac{-2-3}{4-5}$ or 5.

A parallel line has slope 5, and a perpendicular line has slope $-\frac{1}{5}$.

Find the slopes of the lines parallel and perpendicular to the line through points having the given coordinates.

25. (−3, 7), (4, −2)

26. (0, 6), (3, 6)

27. (7, −2), (1, −3)

28. (9, −2), (−1, 4)

- prove two lines are parallel based on given angle relationships (Lesson 3–4)

Ways to Show Two Lines Parallel

- Corresponding angles are congruent.
- Alternate exterior angles are congruent.
- Consecutive interior angles are supplementary.
- Alternate interior angles are congruent.
- Two lines are perpendicular to a third line.

Refer to the figure below for Exercises 29–31.

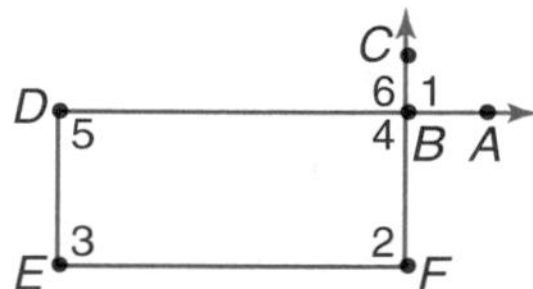

29. Given ∠1 and ∠2 are supplementary, which lines are parallel and why?

30. Given ∠5 ≅ ∠6, which lines are parallel and why?

31. Given ∠6 ≅ ∠2, which lines are parallel and why?

- recognize and use distance relationships among points, lines, and planes (Lesson 3–5)

The distance from B to $\overline{AD}$ is BD.

The distance from $\overline{BA}$ to $\overline{CD}$ is BC.

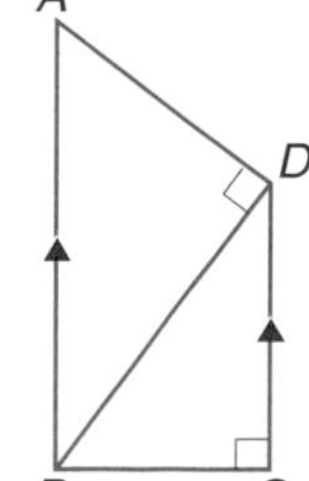

In the figure below, $\overline{PS} \perp \overline{SQ}$, $\overline{PQ} \perp \overline{QR}$, $\overline{PM} \perp \overline{RM}$, and $\overline{QR} \perp \overline{SR}$. Name the segment whose length represents the distance between the following points and lines.

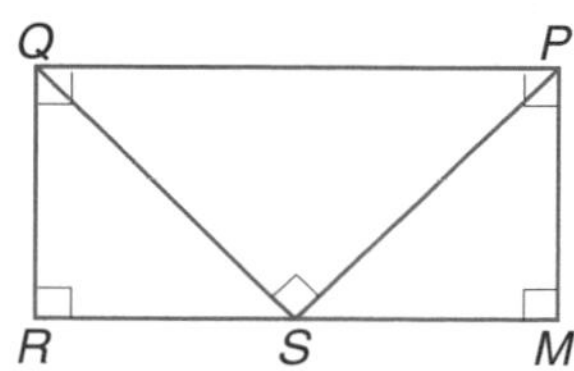

32. from P to $\overleftrightarrow{SQ}$

33. from R to $\overleftrightarrow{PQ}$

34. from $\overleftrightarrow{RM}$ to $\overleftrightarrow{QP}$

35. from $\overleftrightarrow{RQ}$ to $\overleftrightarrow{MP}$

OBJECTIVES AND EXAMPLES

- compare and contrast basic properties of plane and spherical geometry (Lesson 3–6)

Plane Euclidean geometry studies a system of points, lines, and planes.

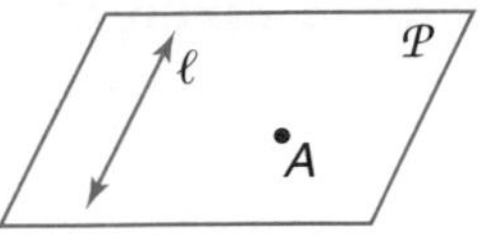

Spherical geometry studies a system of points, great circles, and spheres.

REVIEW EXERCISES

For each property listed from plane Euclidean geometry, write a corresponding statement for spherical geometry.

36. If three points are collinear, exactly one point is between the other two.

37. The intersection of two lines creates four angles.

38. If two lines are parallel to a given line, the lines are parallel to each other.

39. The shortest path between two points is a straight line segment.

40. There is exactly one line passing through two points.

APPLICATIONS AND PROBLEM SOLVING

41. Draw a Diagram The students in a gym class are standing in a circle. When they count off, the students with numbers 5 and 23 are standing exactly opposite one another. Assuming the students are evenly spaced around the circle, how many students are in the class? (Lesson 3–1)

42. Construction A ramp was installed to give handicapped people access to the new public library. The top of the ramp is two feet higher than the bottom. The lower end of the ramp is 24 feet from the door of the library. (Lesson 3–3)

a. Draw a diagram of the ramp.

b. Find the slope of the ramp.

43. Travel At 10:00 A.M., Luisa had completed 195 miles of her cross-country trip. By 2:00 P.M., she had traveled a total of 455 miles. Use slope to determine Luisa's average rate of travel. (Lesson 3–3)

44. Nature Studies Hong is a park ranger and she estimates that there are 6000 deer in the Blendon Woods Park. She also estimates that one year ago there were 6100 deer in the park. (Lesson 3–3)

a. What is the rate of change for the number of deer in Blendon Woods Park?

b. At the same rate, how many deer will there be in the park in ten years?

c. Without using a graph, determine if the graph of this rate would rise or fall. Explain.

A practice test for Chapter 3 is available on page 795.

ALTERNATIVE ASSESSMENT

COOPERATIVE LEARNING PROJECT

Construction In this project, you will explore parallel lines in the real world. Read the project completely before beginning the first task.

Follow these steps to accomplish your task.

- Look around you to locate five examples of parallel lines.
- Measure the shortest distance between the parallel segments at their endpoints and at their midpoints.
- Record your measurements in a table.
- Are the lines parallel, as defined in this chapter?
- Discuss the following questions with your group. Why may the lines not be exactly parallel? Is it due to poor workmanship? Could the builder have made the lines parallel?

Follow these steps to accomplish your next task.

- Select three pairs of parallel lines you measured in the activity. Stand between the lines near one of the endpoints and photograph them so that all of both lines are included in the image. Measure the distance from your standpoint to the nearest endpoint. (Instead of taking a picture, you could sketch the lines from this perspective.)
- On the photograph or drawing, measure the shortest distance between the lines at each endpoint and at the midpoint.
- Record the measurements in a table.

- Discuss the following questions with your group. Do you think you could find a mathematical function to model the decreasing distance between the lines as they are farther away? Do you think you could find a mathematical function relating your distance from the nearest endpoint to the mathematical function mentioned in the previous question? Could you use this relationship to determine whether two converging lines in a photograph or drawing are parallel or not?

THINKING CRITICALLY

- If $\mathcal{A}$ and $\mathcal{B}$ are parallel planes, ℓ is a line on $\mathcal{A}$, and m is a line on $\mathcal{B}$, must it be true that ℓ and m are parallel?
- If ℓ is a line parallel to the plane $\mathcal{A}$ and $\mathcal{B}$ is a plane containing the line ℓ, must it be true that $\mathcal{A}$ and $\mathcal{B}$ are parallel?

PORTFOLIO

Use word processing software to write a one-page report on the history of projective geometry and its relationship to perspective drawing. The Internet may help you to gather information. Try to answer these questions.

- Who developed the ideas first, mathematicians or artists?
- How did artists handle the problem of perspective before this development?

Place your report in your portfolio.

LONG-TERM PROJECT

In•ves•ti•ga•tion

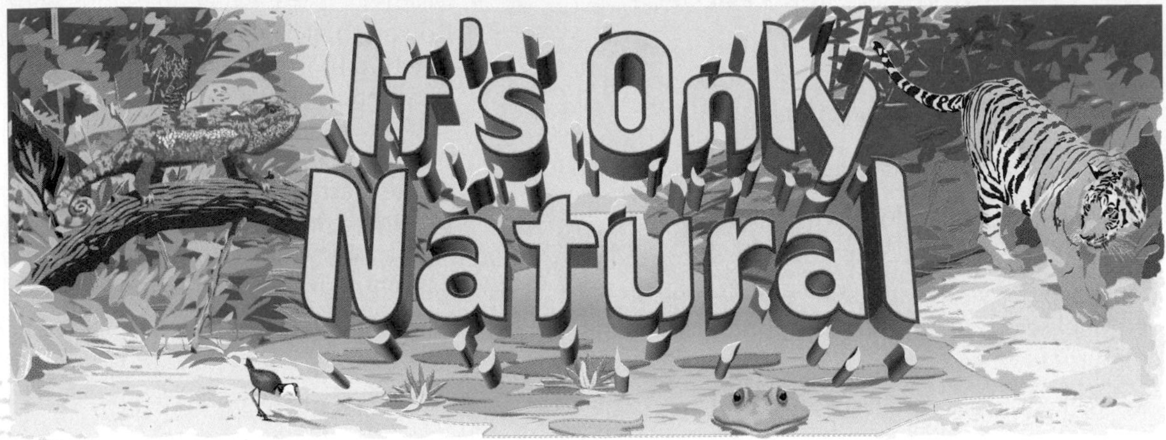

MATERIALS NEEDED

camera 

slide film 

posterboard 

markers 

scissors 

glue 

video camera 

videotape 

soft pencil 

clear tape 

The universe is filled with patterns. If you could speed up time, you would see the stars gliding across the sky in circles. As the poets say, every snowflake is different. Yet almost all snowflakes have six sides. As the wind blows across the water, waves form in parallel lines. If you set your mind on looking for patterns, you will be surprised by the new things that you will see.

In this Investigation, you will work with your group to investigate a pattern found in nature. You will describe where the pattern occurs and any mathematical formulas that can be applied to the pattern. You will complete the project by presenting your findings in a poster display, slide show, or video.

PATTERNS TO FOLLOW

Begin your project by choosing a pattern to investigate. Use reference books from the library or do some research on the Internet to read about the patterns that have been observed. Some areas you may decide to study in this Investigation are described below.

Symmetry Symmetric objects show a one-to-one correspondence among their parts. For example, things like butterflies that could be folded so that the halves match are symmetric. Another type of symmetry involves rotating an object so that its parts correspond. An apple blossom has this type of rotational symmetry.

Packing Natural objects are almost always formed so that space is used efficiently. For example, the cells of a honeycomb are arranged to use the available space most efficiently. How do you think three-dimensional space is used most efficiently?

Cracking Have you ever noticed the way that dry soil or concrete cracks? The same patterns are repeated many times. What are the measures of the angles formed when two, three, or more cracks meet?

Spirals The horns of a ram, a nautilus shell, hurricanes, and galaxies all form spirals. You can see a spiral by filling a sink with water, then releasing it and observing the path the water takes as it drains.

Motion When you pluck a guitar string, the string moves back and forth in a regular pattern called an *oscillation.* You can also see a pattern in the way that an animal's limbs move when it walks. For example, when a horse trots its front left and back right legs hit the ground together. On the next step, the front right and back left legs hit the ground together.

Fibonacci Numbers and the Golden Section The sequence 1, 1, 2, 3, 5, 8, 13, 21, ... is called the *Fibonacci sequence* in honor of its discoverer Leonardo Fibonacci. The numbers in this sequence, called *Fibonacci numbers*, are found in many natural objects such as the rows of seeds in a sunflower and the numbers of leaves on plant stems. The ratios between successive Fibonacci numbers approach the *Golden Section.* This special irrational number is also associated with many natural objects.

Spheres and Near-Spheres A beachball is a good model for a sphere. As raindrops fall, they form spheres. Bubbles are also spherical. The 1996 Nobel Prize for Chemistry was awarded to Harold Kroto, Robert Curl, and Richard Smalley for their work on an unusual carbon molecule that is nearly spherical. The molecule is roughly shaped like a soccer ball and is called *buckminsterfullerene.*

There are many other patterns displayed in nature. Also, more than one of the patterns can often be found in any one object. After you have chosen the pattern you will study, begin looking for information in books or on the Internet. Gather photographs for a poster display or slide presentation. If the equipment is available, you may wish to make a video presentation of your findings.

You will continue working on this Investigation throughout Chapters 4 and 5.

Be sure to keep your research, designs, and other materials in your Investigation Folder.

It's Only Natural Investigation

Working on the Investigation
Lesson 4–1, p. 187

Working on the Investigation
Lesson 4–6, p. 228

Working on the Investigation
Lesson 5–4, p. 265

Working on the Investigation
Lesson 5–6, p. 279

Closing the Investigation
End of Chapter 5, p. 280

CHAPTER 4

Identifying Congruent Triangles

What You'll Learn

In this chapter, you will:

- classify triangles by their parts,
- apply the Angle Sum Theorem and the Exterior Angle Theorem,
- use CPCTC, SSS, SAS, ASA, and AAS to test triangle congruence,
- solve problems by eliminating the possibilities, and
- use properties of isosceles and equilateral triangles.

Why It's Important

Real World

Architecture and Art Geometric shapes are often used in architecture and art. In the picture, the building appears to be a pyramid with a triangular face. That face is also adorned with many triangles of the same size and shape. Look closely at the monorail train. You will see that some of the windows are shaped like triangles. The picture of ancient cave art also contains triangular shapes. In this chapter, you will learn about special types of triangles, measuring angles of triangles, and congruent triangles.

Focus On

PREREQUISITE SKILLS

To be successful in this chapter, you'll need to understand these concepts and be able to apply them. Refer to the lesson in parentheses if you need more review before beginning the chapter.

Name properties of equality to justify statements. (Lesson 2–4)

Name the property of equality that justifies each statement.

1. If $AB = 3x$ and $CD = 10$, then $AB + CD = 3x + 10$.

2. For EF, $EF = EF$.

3. If $0.5x = 10$, then $x = 20$.

4. If $3a + 5 = 20$, then $3a = 15$.

Use the Pythagorean Theorem to find the length of the hypotenuse of a right triangle. (Lesson 1–4)

Find the measure of the hypotenuse of each right triangle with legs having the given measures. Round to the nearest hundredth.

5. 6 and 8 **6.** 4 and 9

7. 2 and 3 **8.** 30 and 40

Use the properties of parallel lines to determine angle measures. (Lesson 3–2)

In the figure, $a \parallel b$, $c \parallel d$, and $m\angle 2 = 55°$. Find the measure of each angle.

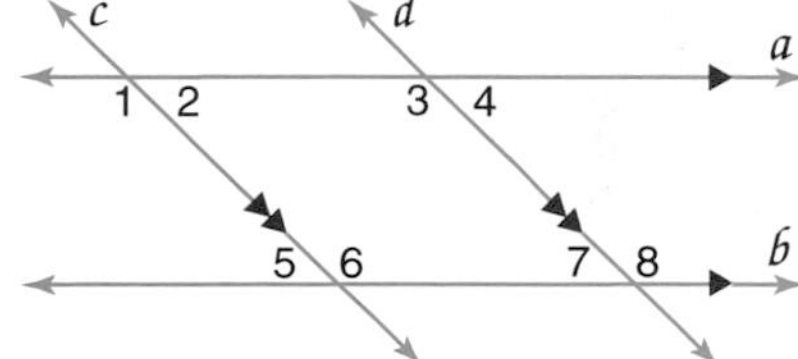

9. $\angle 1$ **10.** $\angle 3$

11. $\angle 4$ **12.** $\angle 5$

13. $\angle 6$ **14.** $\angle 7$

Focus On

READING SKILLS

In this chapter, you'll be learning about triangles. Triangles are classified by the measures of their angles and by the number of sides they have of equal length. The classifications by angles are ***acute, right,*** and ***obtuse.*** These names correspond to the names of angles. For example, a right angle measures 90°, and a right triangle has one right angle. The names for triangles classified by number of congruent sides are more difficult to remember. They are ***scalene, isosceles,*** and ***equilateral.***

4–1 Classifying Triangles

What YOU'LL LEARN

- To identify the parts of triangles and classify triangles by their parts.

Why IT'S IMPORTANT

You can classify triangles found in architecture and crafts.

Architecture

EPCOT Center, near Orlando, Florida, includes one of the world's most recognized structures. It looks like a giant golf ball and is the first completely spherical *geodesic dome* ever built. The framework of the geodesic dome consists of 1450 steel beams covered with waterproof neoprene sheeting. The external cladding of nearly 1000 *triangular* aluminum panels is bolted to the steel framework. The basic structure is drawn below.

There are two types of triangles in a geodesic dome. A **triangle** is a three-sided polygon. A **polygon** is a closed figure in a plane that is made up of segments, called **sides**, that intersect only at their endpoints, called **vertices**.

The vertices of the triangle can be named in any order.

Triangle *CDE*, written $\triangle CDE$, has the following parts.

sides: $\overline{CD}$, $\overline{DE}$, $\overline{CE}$

vertices: *C*, *D*, *E*

angles: $\angle CDE$ or $\angle D$, $\angle CED$ or $\angle E$, $\angle DCE$ or $\angle C$

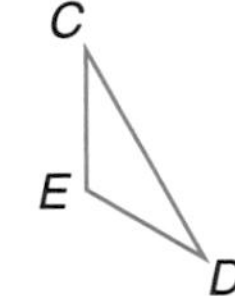

The side *opposite* $\angle C$ is $\overline{DE}$. The angle *opposite* $\overline{CE}$ is $\angle D$. $\angle E$ is opposite $\overline{CD}$.

FYI

Inventor and designer R. Buckminster (Bucky) Fuller (1895–1983) used geometry to create the geodesic dome. Recently, a new molecule, C_{60}, was discovered and named *buckminsterfullerene*, or *buckyball* for short, after the famous designer.

One way of classifying triangles is by their angles. All triangles have at least two acute angles, but the third angle, which is used to classify the triangle, can be acute, right, or obtuse. In the basic structure of the geodesic dome, $\triangle CED$ is an obtuse triangle because $\angle CED$ is obtuse.

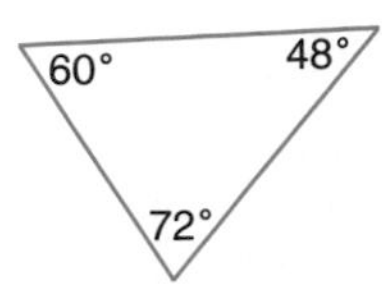

In an **acute triangle**, all the angles are acute.

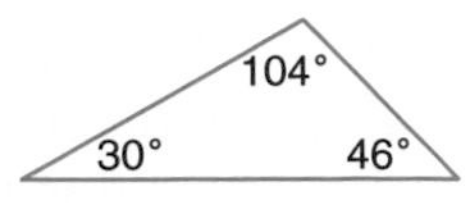

In an **obtuse triangle**, one angle is obtuse.

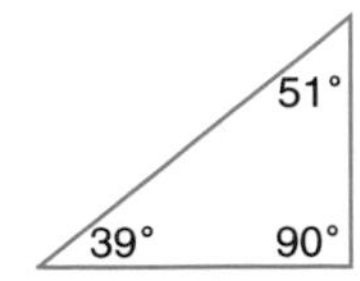

In a **right triangle**, one angle is right.

An **equiangular triangle** is an acute triangle in which all angles are congruent. In the figure showing the basic structure of the geodesic dome, $\triangle ABC$ is an equiangular triangle.

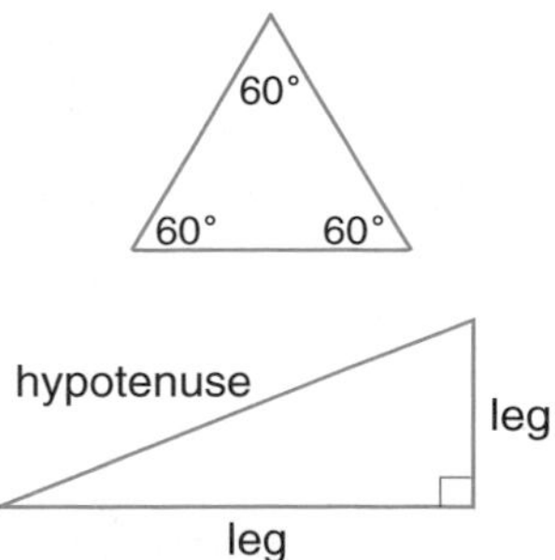

In Chapter 1, you learned that the sides of a right triangle have special names. The side opposite the right angle is called the *hypotenuse*. The two sides that form the right angle are called the *legs*.

Example 1

Architecture

The Alcoa Office Building shown at the right is located in San Francisco, California. Triangular bracings help to secure the building in the event of high winds or an earthquake. Classify $\triangle ABC$, $\triangle BCD$, and $\triangle BCE$ as acute, obtuse, right, or equiangular.

$\triangle ABC$ appears to be an equiangular triangle.

$\triangle BCD$ appears to be a right triangle.

$\triangle BCE$ appears to be an obtuse triangle.

CAREER CHOICES

Architects have played an important role in history from the earliest cities to our modern skyscrapers. The word *architect* is from the Greek meaning "master builder." A licensed architect requires a bachelor's degree in architecture, participation in an internship of 3 years, and passing an exam.

For more information, contact:

Society of American Registered Architects
1245 S. Highland Ave.
Lombard, IL 60148

Triangles can also be classified according to the number of congruent sides they have. An equal number of slashes on sides of a triangle indicate that those sides are congruent.

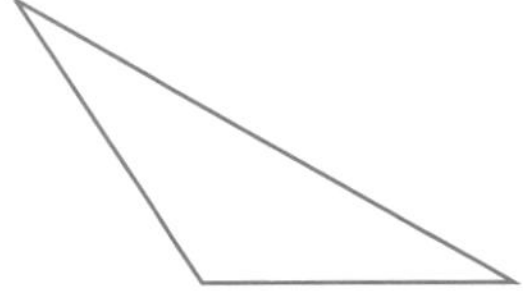

No two sides of a **scalene triangle** are congruent.

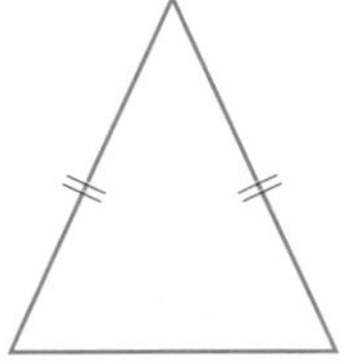

At least two sides of an **isosceles triangle** are congruent.

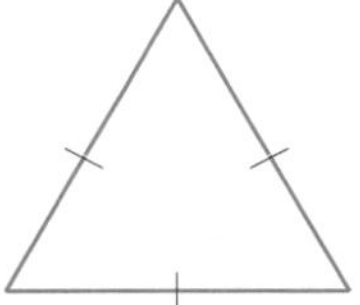

All the sides of an **equilateral triangle** are congruent.

MODELING MATHEMATICS

Making Triangles

Materials:  patty paper

Make an equilateral triangle.

- Take three pieces of patty paper and align them as indicated below. Make a dot at *A* with your pencil.

- Make a fold from *C* through *A* and from *B* through *A*. $\triangle ABC$ is equilateral.

Your Turn

a. How do you know that $\triangle ABC$ is equilateral?

b. Make a triangle that is isosceles but not equilateral.

c. Make a triangle that is scalene.

Like the right triangle, the parts of an isosceles triangle have special names.

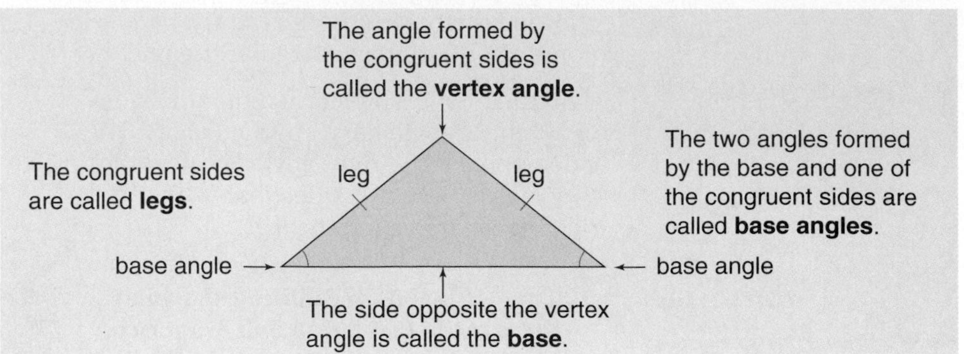

The angle formed by the congruent sides is called the **vertex angle**.
The congruent sides are called **legs**.
leg
leg
The two angles formed by the base and one of the congruent sides are called **base angles**.
base angle
base angle
The side opposite the vertex angle is called the **base**.

Example 2

INTEGRATION
Algebra

Triangle *RST* is an isosceles triangle. $\angle R$ is the vertex angle, $RS = x + 7$, $ST = x - 1$, and $RT = 3x - 5$. Find x, RS, ST, and RT.

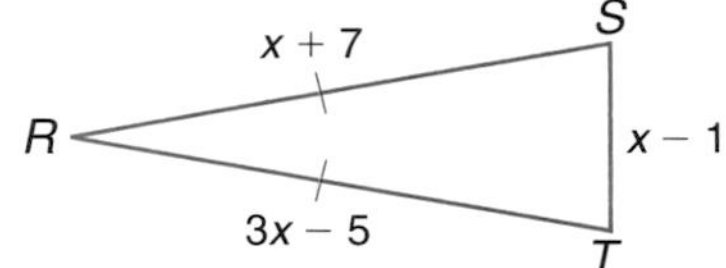

S
x + 7
R
x − 1
3x − 5
T

Since $\angle R$ is the vertex angle, the side opposite $\angle R$, $\overline{ST}$, is the base of the triangle. The congruent legs are $\overline{RS}$ and $\overline{RT}$. So, $RS = RT$.

$$RS = RT$$

$$x + 7 = 3x - 5 \quad \textit{Substitution Property (=)}$$

$$12 = 2x \quad \textit{Addition Property (=)}$$

$$6 = x \quad \textit{Division Property (=)}$$

If $x = 6$, then $RS = 6 + 7$ or 13. Since $RS = RT$, $RT = 13$. Since $ST = x - 1$, $ST = 6 - 1$ or 5. The legs of the isosceles triangle are each 13 units long, and the base is 5 units long.

Triangles can be graphed on the coordinate plane.

Example 3

INTEGRATION
Algebra

Given $\triangle DAR$ with vertices $D(2, 6)$, $A(4, -5)$, and $R(-3, 0)$, use the distance formula to show that $\triangle DAR$ is scalene.

According to the distance formula, the distance between the points at (x_1, y_1) and (x_2, y_2) is $\sqrt{(x_2 - x_1)^2 + (y_2 - y_1)^2}$.

$$DR = \sqrt{(-3 - 2)^2 + (0 - 6)^2} = \sqrt{25 + 36} \text{ or } \sqrt{61}$$

$$AD = \sqrt{(2 - 4)^2 + (6 - (-5))^2} = \sqrt{4 + 121} \text{ or } \sqrt{125}$$

$$RA = \sqrt{(4 - (-3))^2 + (-5 - 0)^2} = \sqrt{49 + 25} \text{ or } \sqrt{74}$$

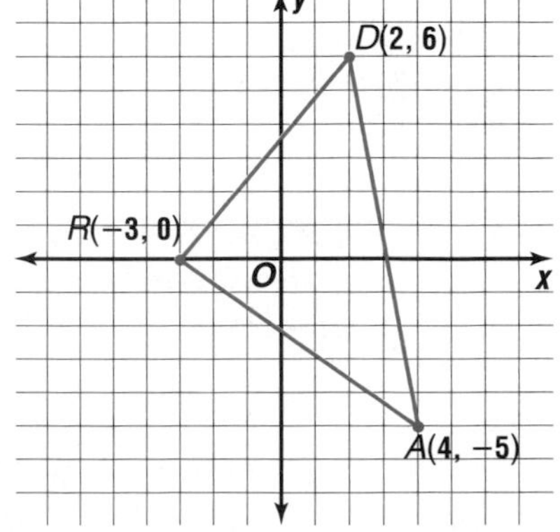

y
D(2, 6)
R(−3, 0)
O
x
A(4, −5)

Since no two sides have the same length, the triangle is scalene.

LOOK BACK

You can refer to Lesson 1-4 for information on finding the distance between points.

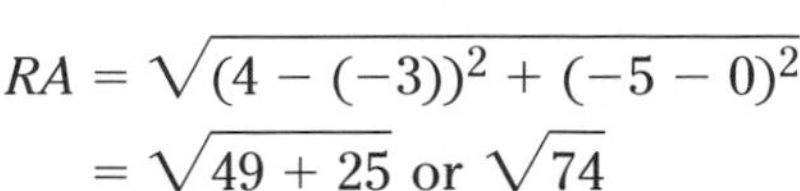

You can draw a triangle on a graphing calculator by plotting its vertices and connecting them.

EXPLORATION GRAPHING CALCULATOR

Use a graphing calculator to graph $\triangle ABC$ whose vertices are $A(0, 2)$, $B(6, 2)$, and $C(5, 4)$.

Set your calculator to the standard viewing screen and make sure the Y= list is cleared. Use the 2:Line command on the Draw menu to draw the segment connecting two of the points. Enter the points using Line (x_1, y_1, x_2, y_2) followed by ENTER. Select the Line command again and use the arrow keys to move the cursor to one endpoint of the segment. Press ENTER. Then move the cursor to the third vertex, watching the coordinates given at the bottom of the screen to approximate the position of the point. Press ENTER twice. Finally, move the cursor to the other endpoint of the segment, and press ENTER to complete the triangle.

Your Turn

a. Clear the screen. Select the standard viewing window and graph $\triangle ABC$ with vertices $A(-2, -2)$, $B(4, -2)$, and $C(1, 3.2)$. Classify the triangle by its appearance.

b. Regraph $\triangle ABC$ using the square viewing window. What type of triangle does it appear to be?

c. Why is it unwise to use the appearance of a triangle on a graphing calculator screen to classify a triangle?

CHECK FOR UNDERSTANDING

Communicating Mathematics

Study the lesson. Then complete the following.

1. **Name** any terms at the right that describe each triangle whose angle measures are given.

   a. 30, 20, 130
   b. 45, 45, 90
   c. 60, 60, 60
   d. 85, 45, 50

   acute
   equiangular
   obtuse
   right

2. **Describe** three types of triangles that are classified by their sides.
3. **You Decide** Kanisha says the figure at the right is an isosceles triangle, but Juana says it is a right triangle. Who is correct? Explain.

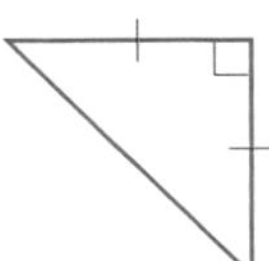

4. **Draw** a scalene right triangle and label the hypotenuse and legs.

5. Use three pieces of patty paper to create an equilateral triangle. Use your protractor to measure each of the angles. Compare your results with a partner. What can you conclude?

Guided Practice

Refer to the figure at the right for Exercises 6–8.

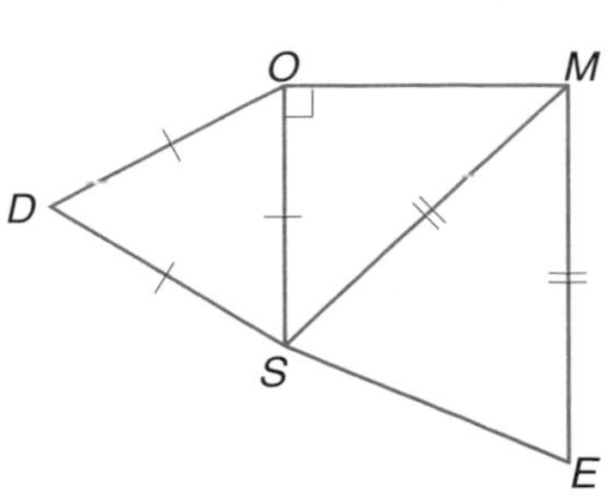

6. Identify an equilateral triangle.
7. Name the hypotenuse of the right triangle.
8. Name the legs of the isosceles triangle that is not equilateral.

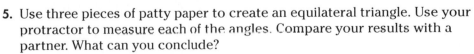

Use a protractor and ruler to draw $\triangle ALT$ using the given conditions. If possible, classify each triangle by the measures of its angles and sides.

9. $m\angle A = 90$ and $m\angle L = m\angle T$ 10. $AL < LT < AT$

11. $\triangle EQU$ is an equilateral triangle. Find x and the measure of each side if $EQ = 4x - 3$ and $QU = 3x + 4$.

12. Given $\triangle ABC$ with vertices $A(0, \sqrt{27})$, $B(3, 0)$, and $C(-3, 0)$, use the distance formula to classify the triangle by the lengths of its sides.

Copy each sentence. Fill in the blank with *sometimes, always,* or *never.*

13. Isosceles triangles are __?__ equilateral.

14. Right triangles are __?__ obtuse.

15. **Crafts** The Zuni Indians are a branch of the Pueblo Indians who live in the western part of New Mexico. They are known for their beautiful pottery, basketry, jewelry, and weavings. The rain-bird is frequently found on the water jars they make. Two examples of rain-birds are shown at the right.

   a. Describe the triangles used to draw these two rain-birds.
   b. Draw a rain-bird using an equilateral triangle.
   c. Draw a rain-bird using an obtuse, isosceles triangle.

EXERCISES

Practice

In the figure at the right, $\triangle BLM$ is isosceles with base $\overline{ML}$. Refer to the figure for Exercises 16–23.

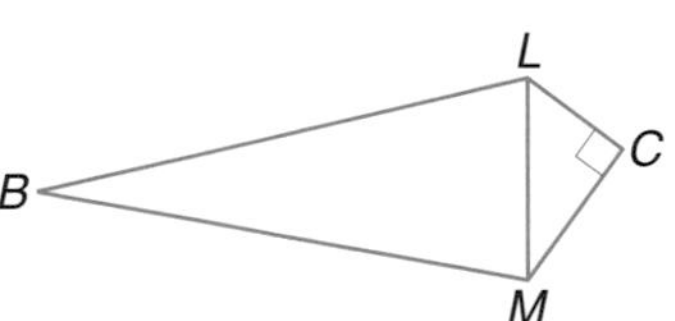

16. Identify an acute triangle.
17. Name the hypotenuse.
18. Name the vertex angle.
19. Name the side opposite $\angle C$.
20. Name the angle opposite $\overline{MB}$.
21. Name the base angles.
22. Name the vertices of the right triangle.
23. Name the legs of the isosceles triangle.

Use a protractor and ruler to draw $\triangle BQS$ using the given conditions. If possible, classify each triangle by the measures of its angles and sides.

24. $m\angle B < 90$ and $\overline{BQ}$ is the hypotenuse.
25. $SB = SQ$ and $m\angle S = 90$.
26. $m\angle S > 90$ and $SQ < BQ$.
27. $SB > SQ > QB$
28. $SB = SQ = QB$
29. $\angle S$ is obtuse and $\angle BQS$ is isosceles.

30. $\triangle BCD$ is isosceles with $\angle C$ as the vertex angle. Find x and the measure of each side if $BC = 2x + 4$, $BD = x + 2$, and $CD = 10$. (*Hint:* Draw a diagram.)

31. $\triangle HKT$ is equilateral. Find x and the measure of each side if $HK = x + 7$ and $HT = 4x - 8$.

32. $\triangle ABC$ is isosceles with $\angle A$ as the vertex angle. AC is five less than two times a number. AB is three more than the number. BC is one less than the number. Find the measure of each side.

Use the distance formula to classify each triangle by the measures of its sides.

33. $\triangle PQR$ with vertices $P(0, 6)$, $Q(3, 6)$, and $R(3, 0)$

34. $\triangle SUV$ with vertices $S(-3, -1)$, $U(2, 1)$, and $V(2, -3)$

35. $\triangle KLM$ with vertices $K(4, 0)$, $L(-2, 0)$, and $M(1, 5)$

Copy each sentence. Fill in the blank with *sometimes, always,* or *never.*

36. Equilateral triangles are __?__ isosceles.

37. Scalene triangles are __?__ isosceles.

38. Right triangles are __?__ acute.

39. Acute triangles are __?__ equilateral.

40. Obtuse triangles are __?__ scalene.

41. Equiangular triangles are __?__ acute.

42. $\triangle RST$ is equilateral, and V lies on $\overline{RS}$ so that $\overline{TV} \perp \overline{RS}$. Classify $\triangle TVS$ by the measures of its angles and its sides.

Proof

43. Copy and complete the proof.

Given: $\angle RGH$ is a right angle.
$\overline{TS} \parallel \overline{HG}$

Prove: $\triangle RST$ is a right triangle.

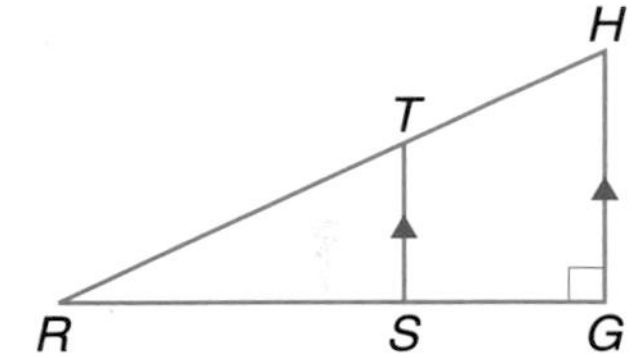

Proof:

Statements	Reasons
1. $\angle RGH$ is a right angle. $TS \parallel HG$	1. Given
2. $\overline{RG} \perp \overline{GH}$	2. __?__
3. $\overline{RS} \perp \overline{ST}$	3. __?__
4. __?__	4. Definition of perpendicular lines
5. __?__	5. Definition of right triangle

44. Write a two-column proof.

Given: $m\angle LGM = 25$

Prove: $\triangle LGS$ is an obtuse triangle.

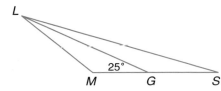

45. $\triangle DEF$ is isosceles with a perimeter between 23 and 32 units. Which angle is the vertex angle? Explain your answer.

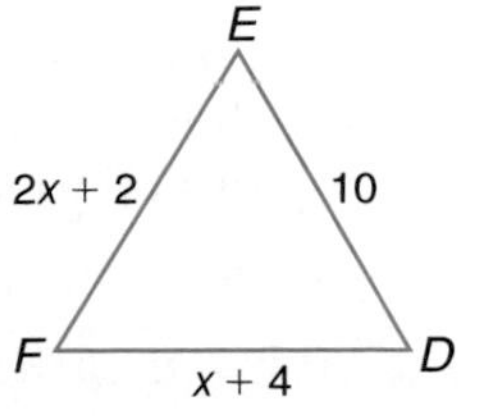

46. In the figure at the right, $m \parallel n$, $\angle 4 \cong \angle 6$, and $m\angle 1 = 120$. Classify $\triangle RST$ by the measure of its angles. Justify your answer.

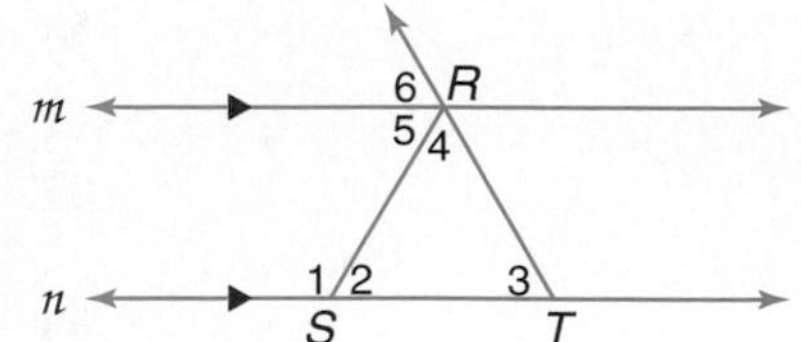

Critical Thinking

47. The Pythagorean Theorem states that in a right triangle, the square of the measure of the hypotenuse is equal to the sum of the squares of the measures of the legs. Use patty paper to create several obtuse and acute triangles. Measure the sides of each. Square the measure of each side. Compare the square of the measure of the longest side to the sum of the squares of the measures of the other two sides. Make a conjecture about the relationship of the square of the measure of the longest side compared to the sum of the squares of the measures of the two shorter sides.

Applications and Problem Solving

48. **Architecture** Consider a figure of the basic structure of the geodesic dome. How many equilateral triangles are in the figure. How many obtuse triangles are in the figure?

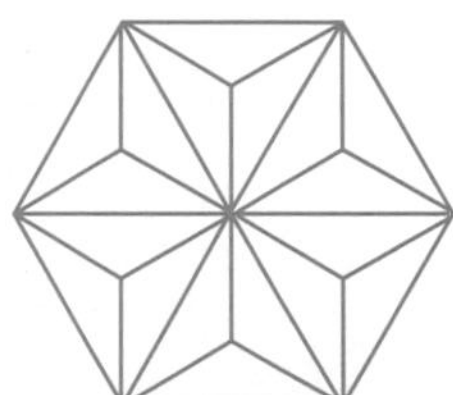

49. **Architecture** At the right is a diagram of an A-frame house.
    a. Identify the triangles that appear to be equilateral.
    b. Identify the triangles that appear to be isosceles, but not equilateral.
    c. Identify the triangles that appear to be right triangles.

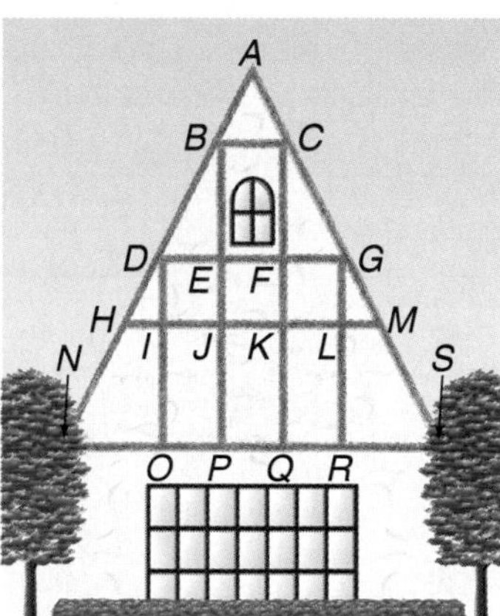

50. **Look for a Pattern** Consider the pattern formed by the dots.

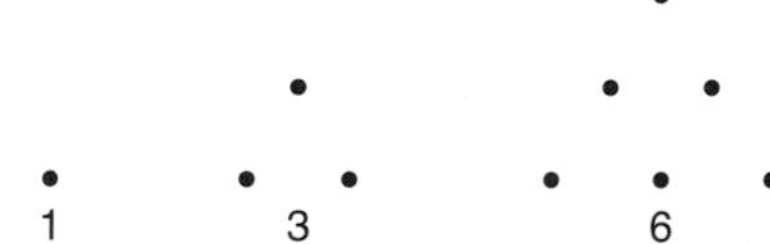

The numbers used to describe each array of dots are called *triangular numbers*. The third triangular number is 6.

a. Draw the array for the fourth triangular number.
b. How many dots will be in the array for the eighth triangular number?

Mixed Review

51. In spherical geometry, if points M and N are polar points, how many great circles can be drawn through M and N? (Lesson 3–6)

52. Copy the figure and draw a line segment that represents the shortest distance from C to $\overleftrightarrow{ED}$. (Lesson 3–5)

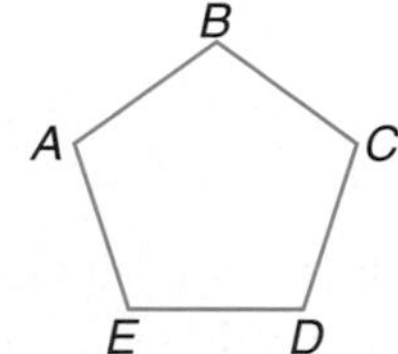

53. Name five ways to prove that two lines are parallel. (Lesson 3–4)

54. **Algebra** Determine the value of r so that a line through the points at $(r, 2)$ and $(4, -6)$ has a slope of $-\frac{8}{3}$. (Lesson 3–3)

55. Give a real-world example of two skew lines. (Lesson 3–1)

56. Name the congruence property that justifies the statement $\angle A \cong \angle A$. (Lesson 2–6)

57. **Algebra** Which property of equality is similar to the Law of Syllogism? Explain. (Lesson 2–3)

58. **ACT Practice** $\sqrt{16 + 9} = ?$

A 5 B 7 C 12 D 14 E 25

59. The measure of an angle is one-third the measure of its supplement. Find the measure of the angle. (Lesson 1–7)

60. Find the value of a so that the distance between $M(5, 6)$ and $N(a, 10)$ is 5 units. (Lesson 1–4)

Algebra

61. Solve $-6 = 5u + 9$.

62. If y varies inversely as x and $y = 27$ when $x = 40$, find x when $y = 10$.

For **Extra Practice**, see page 770.

WORKING ON THE In·ves·ti·ga·tion

Refer to the Investigation on pages 176–177.

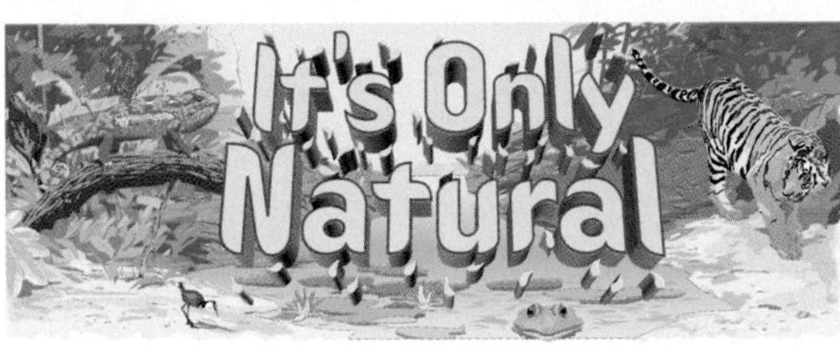

Many people are afraid of spiders. But of the more than 30,000 different species of spiders, only a few are actually dangerous to humans. Spiders feed on insects. And while all spiders spin silk, not all of them make webs to trap their food. All web-spinning spiders that are of the same species spin the same type of web.

1 The triangle spider spins its web between two twigs. Identify the type of triangle formed by the web shown at the right.

2 It is possible for a triangle spider to spin a web that is a different type of triangle than the one shown? If so, when would this occur?

3 Investigate the different types of webs spun by different spiders. Do any of the web types fit a pattern described on pages 176–177 of the Investigation?

4 If a type of spider web fits the pattern you have chosen to research, add a photograph of the web to your presentation. Also include an explanation of the pattern in the web.

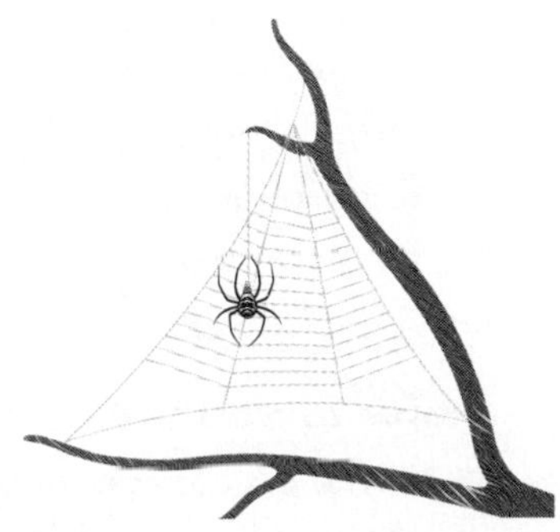

Add the results of your work to your Investigation Folder.

4–2A Angles of Triangles

Materials:  straightedge  scissors

There are special relationships among the angles of a triangle.

A Preview of Lesson 4–2

Activity 1 Find the relationships among the measures of the angles of a triangle.

- Use a straightedge to draw an acute triangle.
- Cut out your acute triangle and label the angles 1, 2, and 3.
- Tear off each angle and arrange the angles to form three adjacent angles as shown.

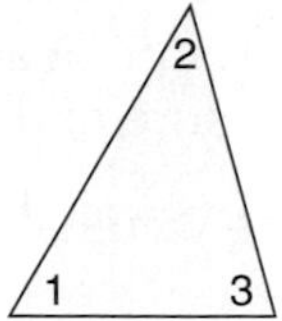

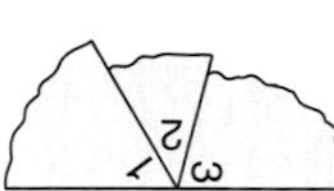

What seems to be true about the sum of the measures of the angles of a triangle?

In the figure at the right, $\angle 4$ is called an **exterior angle** of the triangle, and $\angle 2$ and $\angle 3$ are called its **remote interior angles**.

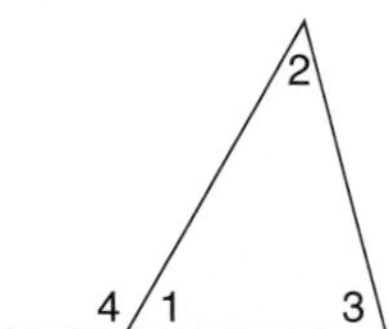

Activity 2 Find the relationship among $m\angle 4$, $m\angle 2$, and $m\angle 3$.

- Arrange the angles from Activity 1 back into a triangle.
- Use a straightedge to draw an exterior angle of your triangle at the vertex of $\angle 1$.
- Place $\angle 2$ and $\angle 3$, the remote interior angles, over the exterior angle you drew. What do you observe?
- Draw exterior angles for angles 2 and 3.

Make a conjecture about the relationship of the remote interior angles of a triangle and their exterior angles.

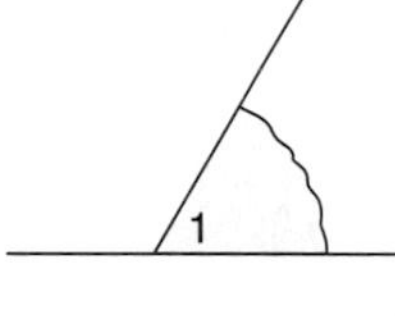

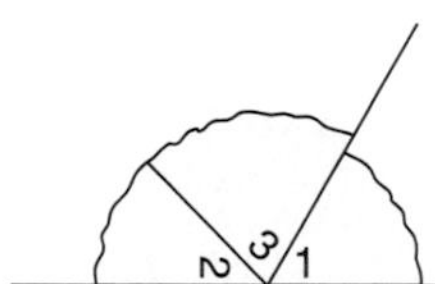

Model

1. Draw an obtuse triangle. Cut out the triangle and tear off each angle.
   a. Arrange the angles to form three adjacent angles.
   b. Draw three exterior angles of the triangle. Compare each exterior angle with its remote interior angles.
2. Draw a right triangle. Cut out the triangle and tear off each angle.
   a. Arrange the angles to form three adjacent angles.
   b. Draw three exterior angles of the triangle. Compare each exterior angle with its remote interior angles.

Write

Refer to your observations in Activities 1 and 2.

3. Write a statement about the sum of the measures of the angles of a triangle.
4. Write a statement about the measure of an exterior angle and the sum of the measures of the two remote interior angles.

4–2 Measuring Angles in Triangles

What YOU'LL LEARN

- To apply the Angle Sum Theorem, and
- to apply the Exterior Angle Theorem.

Why IT'S IMPORTANT

You can determine the measures of angles in triangles involved in countruction and designs.

Real World APPLICATION

Astronomy

The Summer Triangle can be viewed during the month of June. It consists of three bright stars from different constellations. They are Vega in the constellation Lyra, Altair in the constellation Aquile, and Deneb in the constellation Cygnus. If the measure of the angle at Vega is 74 and the measure of the angle at Altair is 41, what is the measure of the angle at Deneb? To answer this question, you need to use the Angle Sum Theorem. *This problem will be solved in Example 1.*

Theorem 4–1 Angle Sum Theorem	**The sum of the measures of the angles of a triangle is 180.**

In order to prove the Angle Sum Theorem, we will need to draw an **auxiliary line**. An auxiliary line is a line or line segment added to a diagram to help in a proof. These are shown as dashed lines in the diagram. Be sure it is possible to draw any auxiliary lines that you use.

Proof of Theorem 4–1

Given: $\triangle PQR$

Prove: $m\angle R + m\angle 2 + m\angle Q = 180$

Proof:

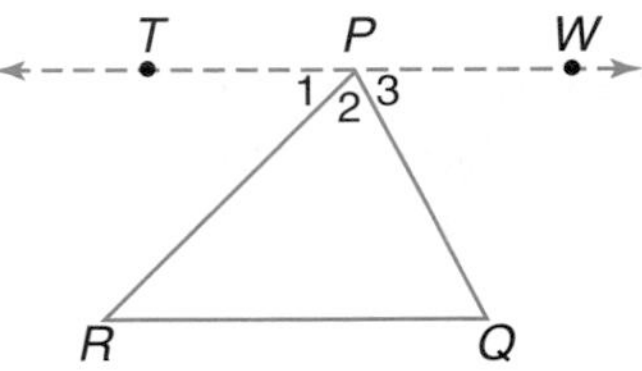

Statements	Reasons
1. $\triangle PQR$	1. Given
2. Draw $\overleftrightarrow{TW}$ through P parallel to $\overleftrightarrow{RQ}$.	2. Parallel Postulate
3. $\angle 1$ and $\angle RPW$ form a linear pair.	3. Definition of a linear pair
4. $\angle 1$ and $\angle RPW$ are supplementary.	4. If two angles form a linear pair, they are supplementary.
5. $m\angle 1 + m\angle RPW = 180$	5. Definition of supplementary angles
6. $m\angle RPW = m\angle 2 + m\angle 3$	6. Angle Addition Postulate
7. $m\angle 1 + m\angle 2 + m\angle 3 = 180$	7. Substitution Property (=)
8. $\angle 1 \cong \angle R$ $\angle 3 \cong \angle Q$	8. Alternate Interior Angles Theorem
9. $m\angle 1 = m\angle R$ $m\angle 3 = m\angle Q$	9. Definition of congruent angles
10. $m\angle R + m\angle 2 + m\angle Q = 180$	10. Substitution Property (=)

fabulous FIRSTS

Carolyn Shoemaker (1929–)

Carolyn Shoemaker is the first person to discover more than 30 comets. So far, she has found 32 comets, all of which bear her name. Her most famous comet, Shoemaker-Levy 9, was discovered with her husband Eugene Shoemaker and her coworker, David Levy, in 1993. This comet looked like a string of pearls when it collided with Jupiter in 1994.

If you know the measures of two angles of a triangle, you can find the measure of the third angle.

Example 1 Refer to the application at the beginning of the lesson. Find the measure of the angle at Deneb.

Astronomy

Explore Let V, A, and D be the vertices of the triangle. We are given that $m\angle V = 74$ and $m\angle A = 41$. We are asked to find the $m\angle D$.

Plan Use the Angle Sum Theorem to write an equation. Then solve the equation for the measure of the missing angle.

Solve

$m\angle V + m\angle A + m\angle D = 180$ *Angle Sum Theorem*

$74 + 41 + m\angle D = 180$ *Substitution Property (=)*

$m\angle D = 65$ *Subtraction Property (=)*

The measure of the angle at Deneb is 65.

Examine Find the sum of the measures of the three angles of the triangle.

$$74 + 41 + 65 = 180$$

Since the sum is 180, the answer is correct.

The Angle Sum Theorem leads to a useful theorem about the angles in two triangles. *You will be asked to prove this theorem in Exercise 36.*

Theorem 4–2 ***Third Angle Theorem***	**If two angles of one triangle are congruent to two angles of a second triangle, then the third angles of the triangles are congruent.**

In the figure at the right, $\angle CBD$ is an **exterior angle** of $\triangle ABC$. An exterior angle is formed by one side of a triangle and the extension of another side. The interior angles of the triangle not adjacent to a given exterior angle are called **remote interior angles** of the exterior angle. In the figure, $\angle CBD$ is an exterior angle with $\angle A$ and $\angle C$ as its remote interior angles. Theorem 4–3 relates the measure of an exterior angle of a triangle to the sum of the measures of its two remote interior angles.

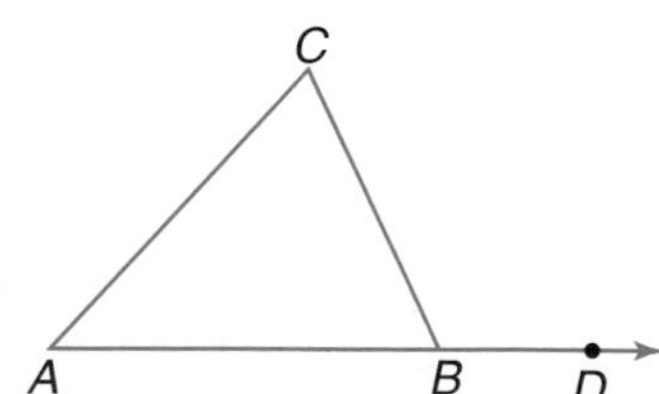

Theorem 4–3 ***Exterior Angle Theorem***	**The measure of an exterior angle of a triangle is equal to the sum of the measures of the two remote interior angles.**

To prove this theorem, we will use a **flow proof**. A flow proof organizes a series of statements in logical order, starting with the given statements. Each statement along with its reason is written in a box, and arrows are used to show how each statement leads to another.

Proof of Theorem 4–3

A flow proof can be drawn in a horizontal or a vertical direction.

Given: $\triangle ABC$

Prove: $m\angle CBD = m\angle A + m\angle C$

Flow Proof:

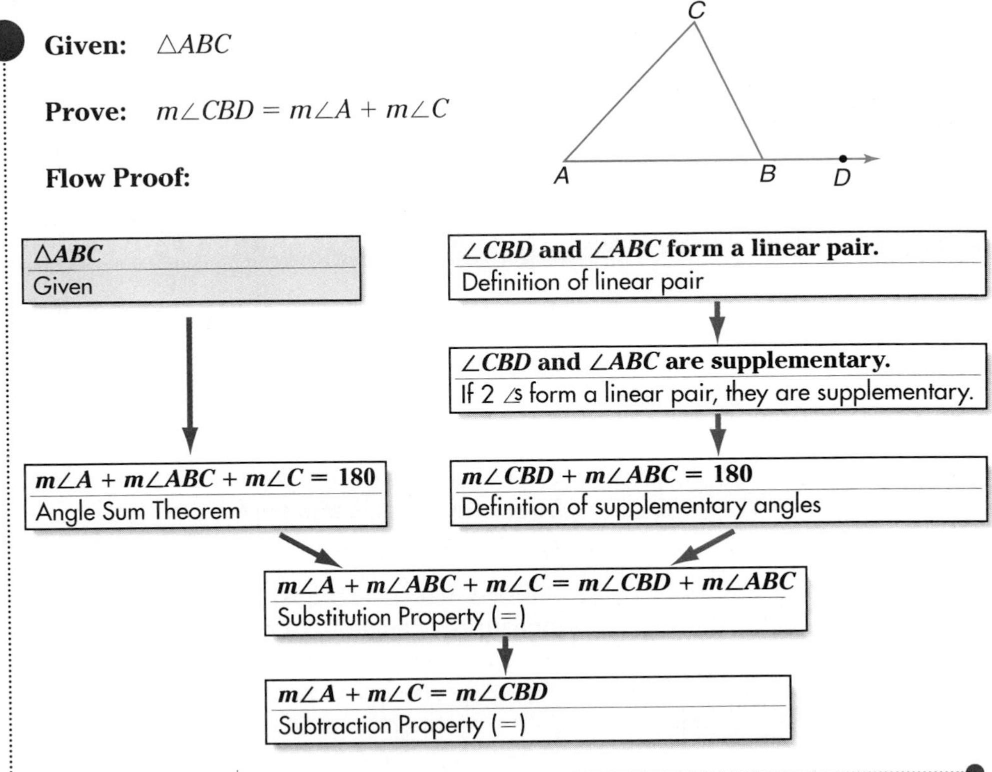

C
A
B
D
△ABC
Given
∠CBD and ∠ABC form a linear pair.
Definition of linear pair
∠CBD and ∠ABC are supplementary.
If 2 ∠s form a linear pair, they are supplementary.
m∠A + m∠ABC + m∠C = 180
Angle Sum Theorem
m∠CBD + m∠ABC = 180
Definition of supplementary angles
m∠A + m∠ABC + m∠C = m∠CBD + m∠ABC
Substitution Property (=)
m∠A + m∠C = m∠CBD
Subtraction Property (=)

Sometimes the measures of angles can be determined by using a combination of theorems.

Example 2 **Find the measure of each numbered angle in the figure at the right.**

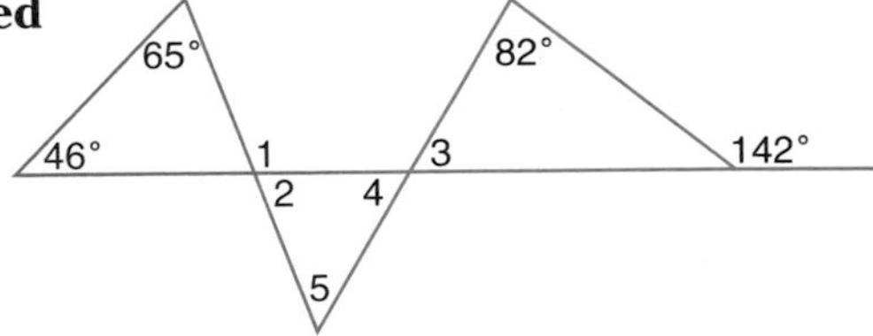

$m\angle 1 = 46 + 65$ *Exterior Angle Theorem*

$= 111$

$m\angle 1 + m\angle 2 = 180$ *If 2 ∠s form a linear pair, they are supplementary.*

$111 + m\angle 2 = 180$ *Substitution Property (=)*

$m\angle 2 = 69$ *Subtraction Property (=)*

$m\angle 3 + 82 = 142$ *Exterior Angle Theorem*

$m\angle 3 = 60$ *Subtraction Property (=)*

Since $\angle 3$ and $\angle 4$ are vertical angles, $m\angle 4 = 60$.

$m\angle 2 + m\angle 4 + m\angle 5 = 180$ *Angle Sum Theorem*

$69 + 60 + m\angle 5 = 180$ *Substitution Property (=)*

$m\angle 5 = 51$ *Subtraction Property (=)*

Therefore, $m\angle 1 = 111$, $m\angle 2 = 69$, $m\angle 3 = 60$, $m\angle 4 = 60$, and $m\angle 5 = 51$.

A statement that can be easily proved using a theorem is often called a **corollary** of that theorem. A corollary, just like a theorem, can be used as a reason in a proof. *You will be asked to prove Corollaries 4–1 and 4–2 in Exercises 37 and 38, respectively.*

Corollary 4–1	**The acute angles of a right triangle are complementary.**
Corollary 4–2	**There can be at most one right or obtuse angle in a triangle.**

CHECK FOR UNDERSTANDING

Communicating Mathematics

Study the lesson. Then complete the following.

1. **Explain** in your own words how to find the third angle of a triangle if you know the measure of the other two angles.
2. **Draw** a triangle and extend all of its sides. There are four angles at each vertex of the triangle. Describe how you can distinguish exterior angles from other angles at a vertex.
3. **State** why it is impossible to have two obtuse angles in a triangle.
4. **Choose** two angle measures other than the right angle that could be measures of angles of a right triangle. Explain why you know these two angles along with the right angle are angles of a right triangle.

5. Fold a rectangular piece of paper along a diagonal from *A* to *C*. Then cut along the fold to form a right triangle *ABC*. Write the name of each angle on the inside of the triangle. Tear off angles *A* and *C*. Place the torn angles at ∠*B*.
   a. What seems to be true about the sum of the measures of the acute angles of a right triangle?
   b. Do your results seem to confirm or deny Corollary 4–1?

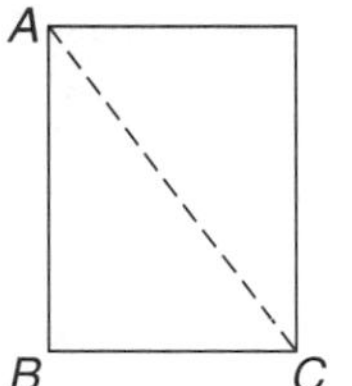

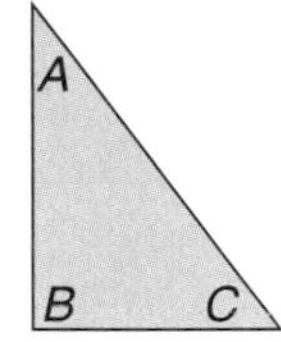

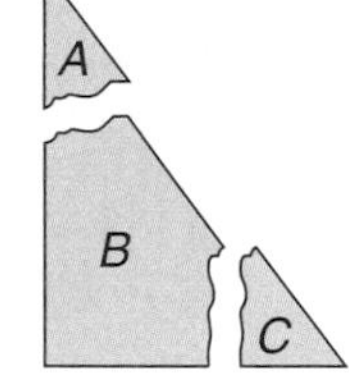

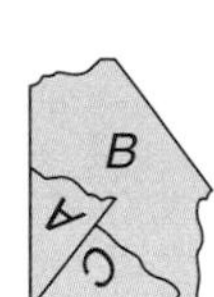

Guided Practice

Find the value of *x*.

6. 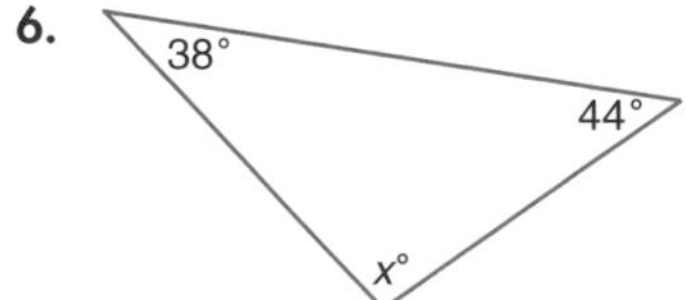

7. 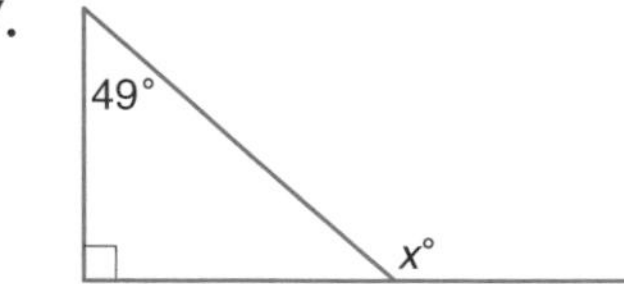

If $\overline{KH}$ is parallel to $\overline{JI}$, find the measure of each angle in the figure at the right.

8. ∠1
9. ∠2
10. ∠3

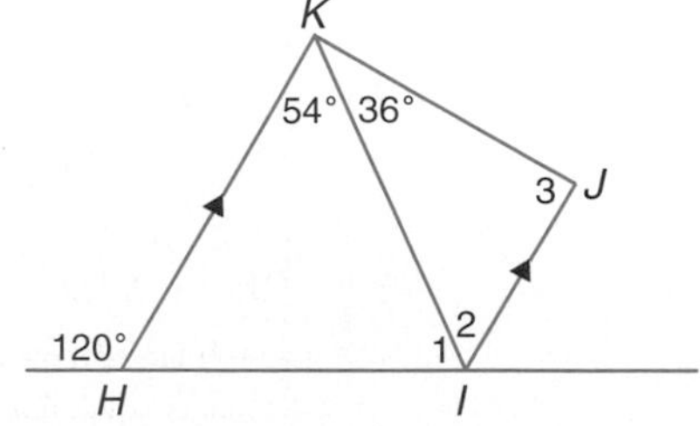

11. Refer to the figure at the right.
    a. Find x.
    b. Find $m\angle A$.
    c. Classify the triangle by the measure of its angles.

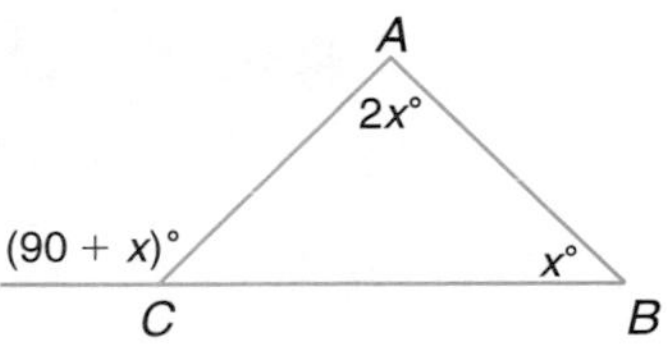

12. Is it possible to have two right angles as exterior angles of a triangle? Explain your reasoning.

13. **Astronomy** The Big Dipper is a part of the larger constellation Ursa Major, which means *great bear*. Three of the brighter stars in the constellation form a triangle SRA. If $m\angle S = 109$ and $m\angle R = 41$, find $m\angle A$.

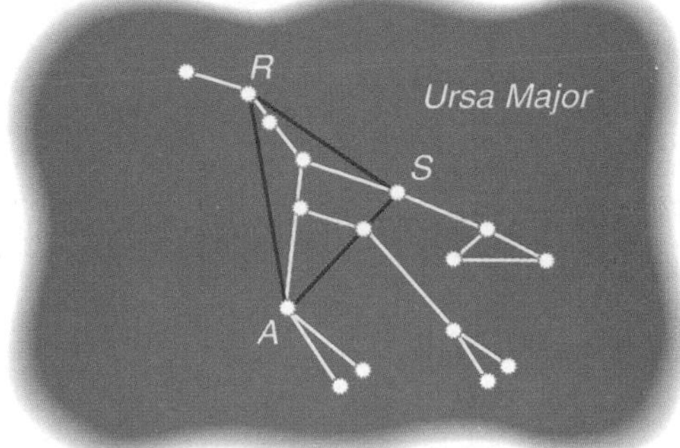

EXERCISES

Practice

Find the value of x.

14. 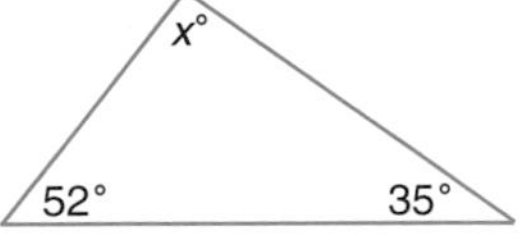

15. 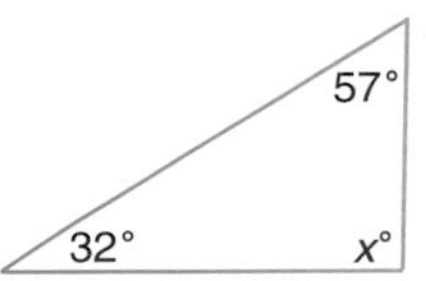

16. 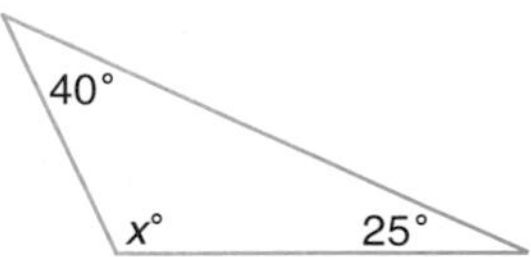

17. 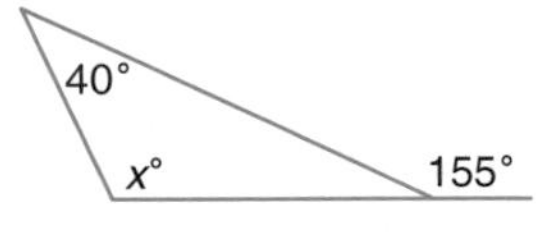

18. 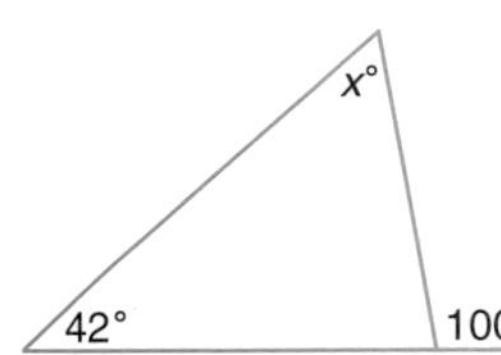

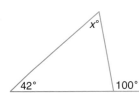

19. 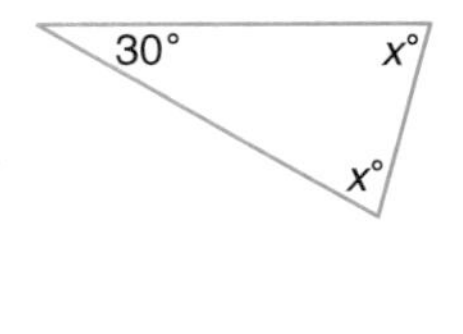

If $\overline{AB}$ is perpendicular to $\overline{BC}$, find the measure of each angle in the figure below.

20. $\angle 1$ 21. $\angle 2$
22. $\angle 3$ 23. $\angle 4$
24. $\angle 5$ 25. $\angle 6$
26. $\angle 7$ 27. $\angle 8$

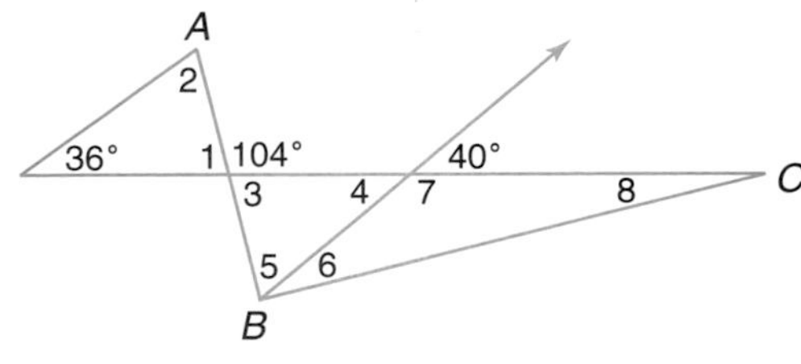

For each triangle, find $m\angle A$.

28. 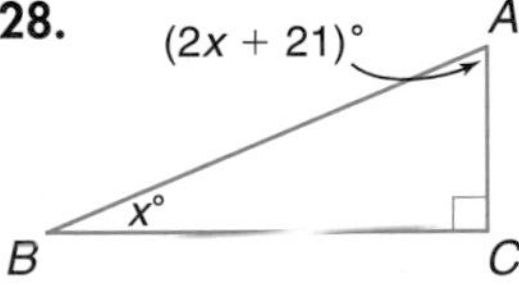

29. 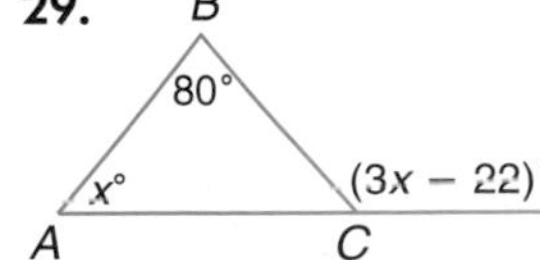

30. 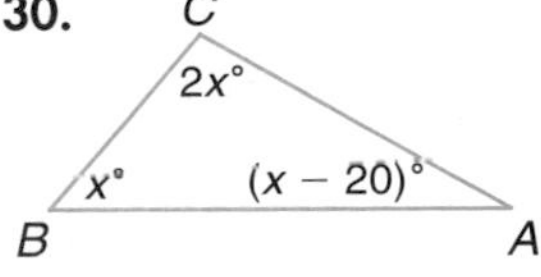

31. Refer to $\triangle ABC$ in Exercise 29. Is this triangle scalene, equilateral, isosceles, or right? Explain.

32. In $\triangle ABC$, $m\angle A$ is 16 more than $m\angle B$, and $m\angle C$ is 29 more than $m\angle B$.
    a. Write an equation relating the measures.
    b. Find the measure of each angle.

33. The statements and reasons used in a flow proof to prove $\angle V \cong \angle W$ are listed below. Put the boxes in the correct order and add the arrows.

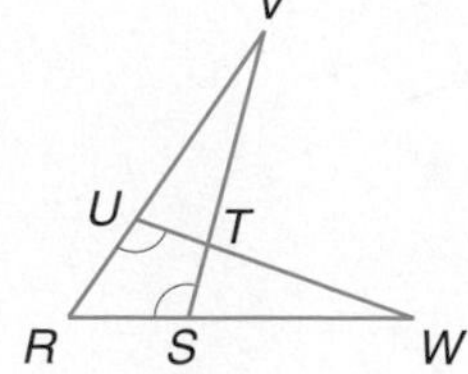

Given: $\angle RUW \cong \angle VSR$

Prove: $\angle W \cong \angle V$

Proof:

$\angle W \cong \angle V$ Third Angle Theorem	$\angle RUW \cong \angle VSR$ Given	$\angle URW \cong \angle VRS$ Congruence of angles is reflexive.

34. The measures of two angles of a triangle are 45 and 70. What is the largest exterior angle of the triangle? Explain your reasoning.

35. Write a two-column proof to show that if a triangle is equiangular, the measure of each angle is 60.

36. Write a two-column proof for the Third Angle Theorem. (Theorem 4–2).

37. Write a flow proof to show that the acute angles of a right triangle are complementary. (Corollary 4–1).

38. Write a paragraph proof to show there can be at most one right or obtuse angle in a triangle. (Corollary 4–2).

39. What is the sum of the interior angles of the quadrilateral at the right? (*Hint*: Think in terms of triangles.)

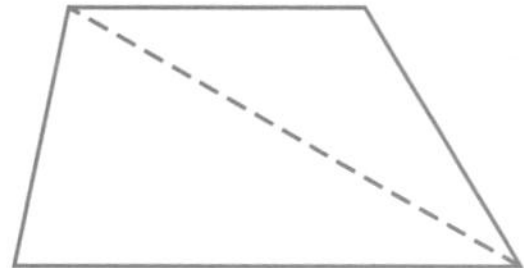

Cabri Geometry

40. Use a calculator to draw a triangle and a line through a vertex of the triangle that is parallel to the side opposite that vertex.
   a. Calculate all the angle measures in the figure.
   b. Drag a vertex around on the screen. Describe how the measure changes as you move the location of the vertex.

Critical Thinking

41. In spherical geometry, triangles can be drawn on a sphere. Sides of a triangle can consist of part of the equator and any great circle that passes through the two poles. In spherical geometry, any great circle passing through the two poles is perpendicular to the equator. Three such triangles are drawn below.

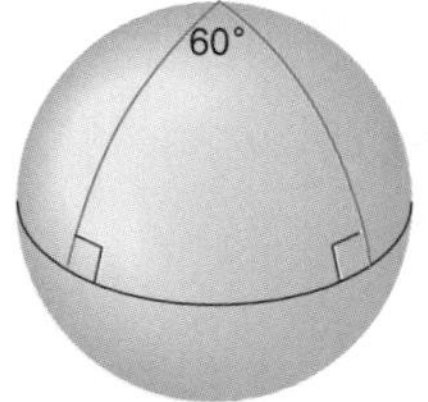

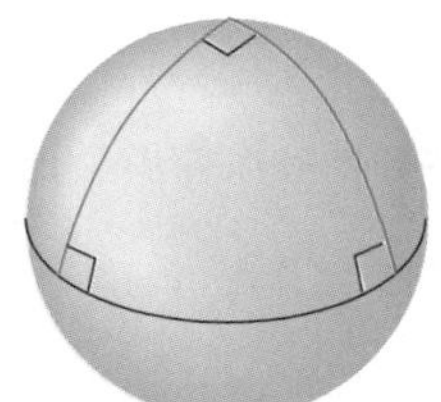

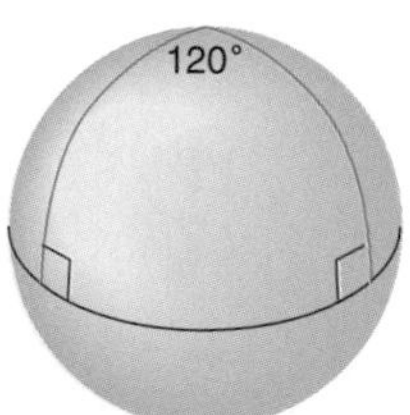

LOOK BACK

You can refer to Lesson 3-6 for information on spherical geometry.

   a. Is it possible for the sum of the measures of the angles of a triangle in this system to be 180? Explain your reasoning.
   b. What is the possible range of the sum of the measures of angles of a triangle in this system? Explain your reasoning.
   c. How would you define an acute triangle in this system?
   d. How would you define an obtuse triangle in this system?
   e. How would you define a right triangle in this system?

Applications and Problem Solving

42. **Astronomy** Leo is a constellation that represents a lion. Three of the brighter stars in the constellation form a triangle *LEO*. If the angles have measures indicated in the figure at the right, find $m\angle L$.

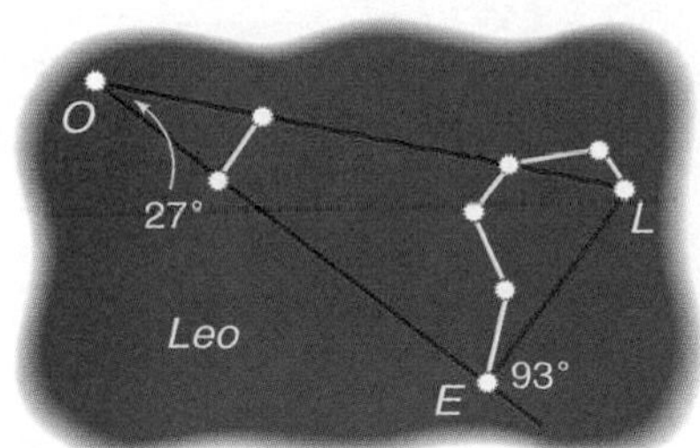

43. **Construction** The roof support at the right is shaped like a triangle. Two angles each have a measure of 25. Find $m\angle R$.

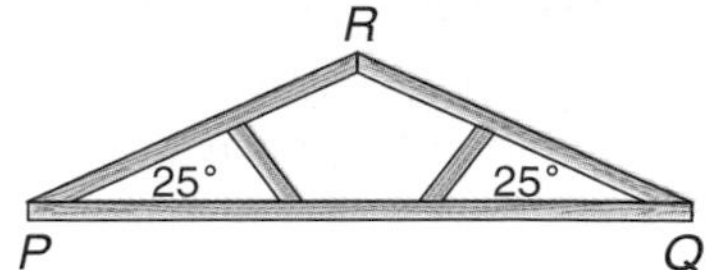

Mixed Review

44. The measures of the angles of a triangle are 65, 95, and 20. Classify the triangle by its angles. (Lesson 4–1)

45. Determine if $\overleftrightarrow{RS} \parallel \overleftrightarrow{LM}$ given that $m\angle 1 = 42$ and $m\angle 5 = 48$. Justify your answer. (Lesson 3–4)

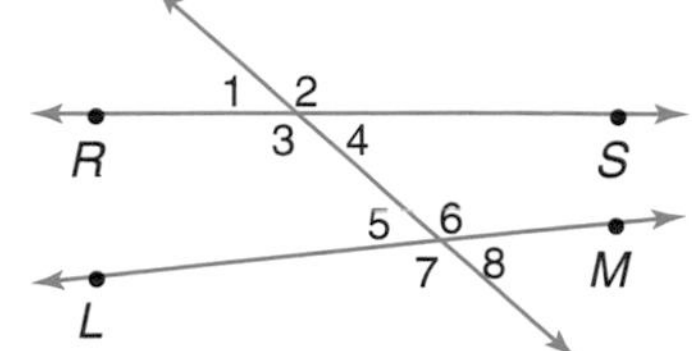

Data Update For more information on population of U.S. cities and their growth rates, visit: **www.geometry.glencoe.com**

46. **Demographics** The population of Mississippi was 2,575,000 in 1990 and 2,752,000 in 1998. What was the average annual rate of change for the population of Mississippi from 1990 to 1998? (Lesson 3–3)

47. **Draw a Diagram** The Cedarville Club is holding its annual women's tennis tournament. If 8 women are registered for the tournament, and each woman plays every other woman exactly once, how many games will be played? (Lesson 3–1)

48. Write the given and prove statements that you would use to prove the theorem *The acute angles of a right triangle are complementary.* Then draw a figure. (Lesson 2–5)

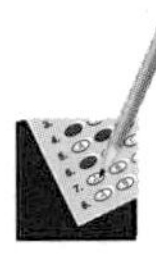

49. **SAT Practice** If $x \otimes y = \frac{1}{x - y}$, what is the value of $\frac{1}{2} \otimes \frac{3}{4}$?

A -4 **B** $-\frac{1}{4}$ **C** $\frac{4}{5}$ **D** $\frac{5}{4}$ **E** 4

50. Copy segment *LN*. Then use a compass and straightedge to bisect the segment. (Lesson 1–5)

L ——————————— N

51. In which quadrant is $P(-4, -6)$ located? (Lesson 1–1)

Algebra

52. Solve $\frac{6}{t} = \frac{2}{7}$.

53. **Mechanics** A nationwide poll of 500 women was taken to determine how many women know how to take care of their cars. The poll results are shown at the right.

    a. How many of the women surveyed knew how to charge a car battery?

    b. How many of the women surveyed knew how to check the oil?

Women Care for Their Cars

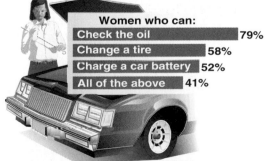

Women who can:

Check the oil	79%
Change a tire	58%
Charge a car battery	52%
All of the above	41%

Source: *EDK Forecast*

For **Extra Practice**, see page 770.

4-3

Exploring Congruent Triangles

What YOU'LL LEARN

- To name and label corresponding parts of congruent triangles.

Why IT'S IMPORTANT

You can identify congruent triangles and their corresponding parts involved in crafts, arts, and construction.

Crafts

Diane Leighton of Yuba City, California, learned the craft of quilting by reading books on the subject from her local library. She teaches quilting while maintaining her career as a registered nurse. One of Ms. Leighton's quilts, which includes several isosceles right triangles, is pictured.

The numbered triangles are all the same size and same shape. Triangles that are the same size and same shape are **congruent triangles**. Each triangle has six parts, three angles and three sides. If the corresponding six parts of one triangle are congruent to the six parts of another triangle, then the triangles are congruent.

FYI

To celebrate the 1989 centennial of North Dakota, 7000 citizens made the largest quilt in the world. It is 85 feet by 134 feet.

In the figure below, $\triangle DEF$ is congruent to $\triangle ABC$. If you *slide* $\triangle DEF$ up and to the right, $\triangle DEF$ is still congruent to $\triangle ABC$.

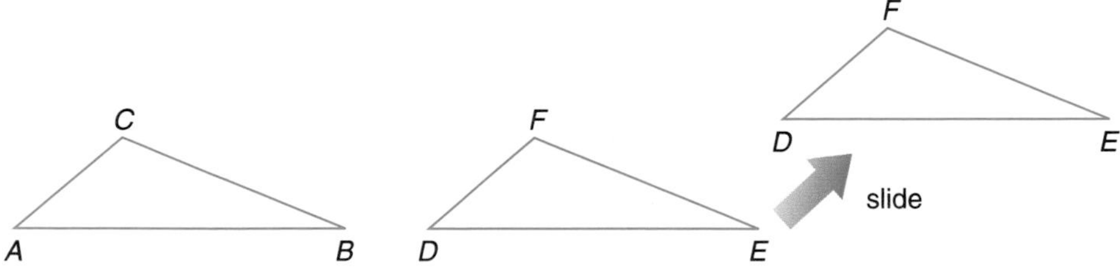

Also, if you *rotate* $\triangle DEF$, $\triangle DEF$ remains congruent to $\triangle ABC$. If you *flip* $\triangle DEF$, $\triangle DEF$ remains congruent.

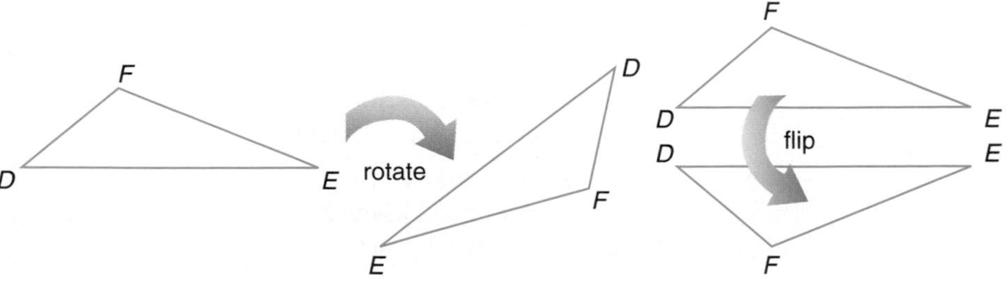

If you slide, rotate, or flip a figure, congruence will not change. These three transformations are called **congruence transformations**.

Example 1

Crafts

A design for a quilt has been drawn on graph paper. Identify the triangles that appear to be congruent.

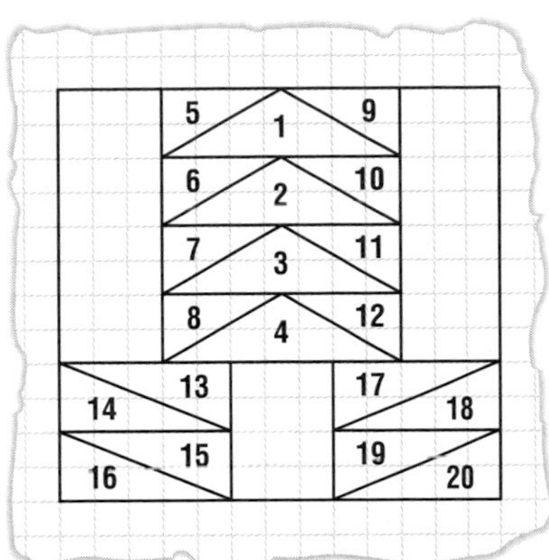

Triangles 1 through 4 appear to be congruent.

Triangles 5 through 12 appear to be congruent.

Triangles 13 through 20 appear to be congruent.

The symbol ↔ means "corresponds to."

If $\triangle LMN$ is congruent to $\triangle PQR$ ($\triangle LMN \cong \triangle PQR$), the vertices of the two triangles correspond in the same order as the letters naming the triangles.

$L \leftrightarrow P$ $\qquad$ $M \leftrightarrow Q$ $\qquad$ $N \leftrightarrow R$

This correspondence of vertices can be used to name the corresponding congruent sides and angles of the two triangles.

$\angle L \cong \angle P$ $\qquad$ $\angle M \cong \angle Q$ $\qquad$ $\angle N \cong \angle R$

$\overline{LM} \cong \overline{PQ}$ $\qquad$ $\overline{MN} \cong \overline{QR}$ $\qquad$ $\overline{LN} \cong \overline{PR}$

The corresponding sides and angles can be determined from any congruence statement by following the order of the letters. For example, $\triangle ABC \cong \triangle FGH$ indicates the following congruences.

$\angle A \cong \angle F$ $\qquad$ $\angle B \cong \angle G$ $\qquad$ $\angle C \cong \angle H$

$\overline{AB} \cong \overline{FG}$ $\qquad$ $\overline{BC} \cong \overline{GH}$ $\qquad$ $\overline{AC} \cong \overline{FH}$

It is important that you list the letters of the vertices in the correct order whenever you write a congruence statement.

Definition of Congruent Triangles (CPCTC)	Two triangles are congruent if and only if their corresponding parts are congruent.

The abbreviation CPCTC means Corresponding Parts of Congruent Triangles are Congruent. "If and only if" is used to show that the conditional and its converse are both true.

Example 2 **The bridge in the picture below uses a simple triangular truss design. The vertices of two triangles are labeled and $\triangle TRU \cong \triangle TSU$. Name the corresponding congruent angles and sides.**

Construction

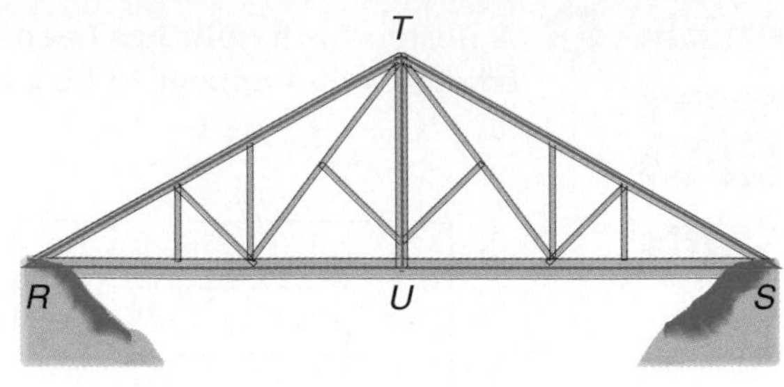

$\angle RTU \cong \angle STU$ $\quad \angle R \cong \angle S$ $\quad \angle TUR \cong \angle TUS$

$\overline{TR} \cong \overline{TS}$ $\quad \overline{RU} \cong \overline{SU}$ $\quad \overline{TU} \cong \overline{TU}$

Congruence of triangles, like congruence of segments and angles, is reflexive, symmetric, and transitive. This is stated in Theorem 4–4. The proof of the transitive part of the theorem is given. *You will be asked to prove the reflexive and symmetric parts of the theorem in Exercises 9 and 22, respectively.*

Theorem 4–4	**Congruence of triangles is reflexive, symmetric, and transitive.**

Proof of Theorem 4–4 (Transitive Part)

Given: $\triangle ABC \cong \triangle DEF$
$\triangle DEF \cong \triangle GHI$

Prove: $\triangle ABC \cong \triangle GHI$

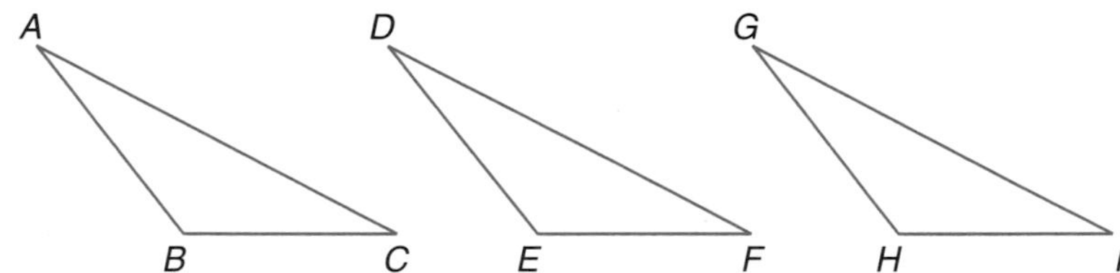

Proof:

Statements	**Reasons**
1. $\triangle ABC \cong \triangle DEF$	**1.** Given
2. $\angle A \cong \angle D, \angle B \cong \angle E, \angle C \cong \angle F$ $\overline{AB} \cong \overline{DE}, \overline{BC} \cong \overline{EF}, \overline{AC} \cong \overline{DF}$	**2.** CPCTC
3. $\triangle DEF \cong \triangle GHI$	**3.** Given
4. $\angle D \cong \angle G, \angle E \cong \angle H, \angle F \cong \angle I$ $\overline{DE} \cong \overline{GH}, \overline{EF} \cong \overline{HI}, \overline{DF} \cong \overline{GI}$	**4.** CPCTC
5. $\angle A \cong \angle G, \angle B \cong \angle H, \angle C \cong \angle I$	**5.** Congruence of angles is transitive.
6. $\overline{AB} \cong \overline{GH}, \overline{BC} \cong \overline{HI}, \overline{AC} \cong \overline{GI}$	**6.** Congruence of segments is transitive.
7. $\triangle ABC \cong \triangle GHI$	**7.** Definition of congruent triangles

CHECK FOR UNDERSTANDING

Communicating Mathematics

Study the lesson. Then complete the following.

1. **Draw** two triangles that are the same size and shape. What is the name given to these two triangles?
2. **Explain** in your own words the meaning of *congruence of triangles is reflexive*.
3. **Name** the corresponding angles if $\triangle BCD \cong \triangle EFG$.
4. **Describe** how a congruence transformation would affect the perimeter of a triangle. Explain your reasoning.

5. **Look** around your school or community. Describe where congruent triangles are used as part of the architecture. Identify each set of congruent triangles as acute, obtuse, right, equiangular, scalene, isosceles, and/or equilateral.

Guided Practice

6. In the quilt design at the right, indicate which triangles appear to be congruent.

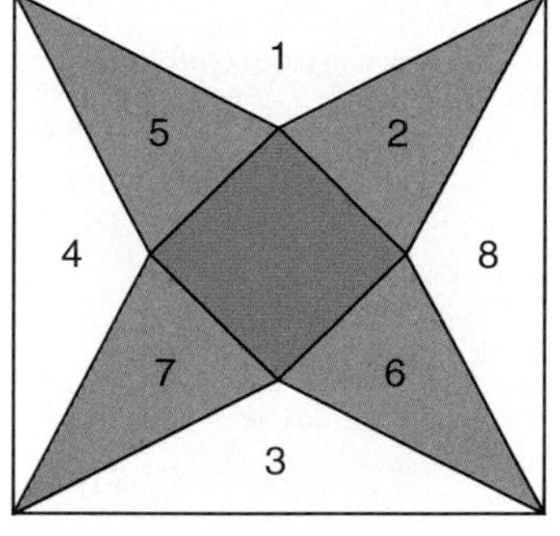

7. If $\triangle ABC \cong \triangle XYZ$, name all the corresponding angles and corresponding sides. Use $\leftrightarrow$ to indicate each correspondence. Draw a figure showing the two triangles, and mark the corresponding parts.
8. Refer to the figure at the right. Complete the congruence statement.

   $\triangle LMN \cong \triangle$ <u>?</u>

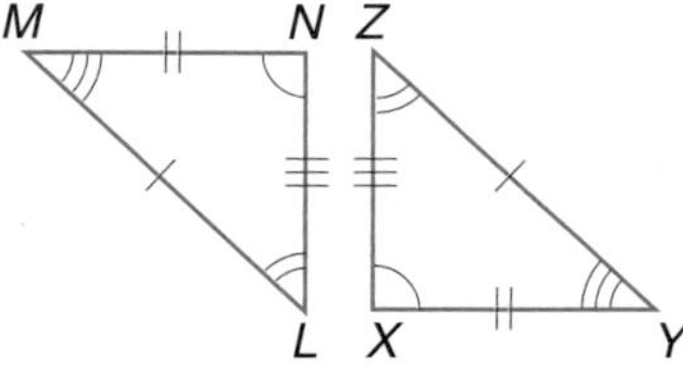

9. Copy and complete the proof that congruence of triangles is reflexive.

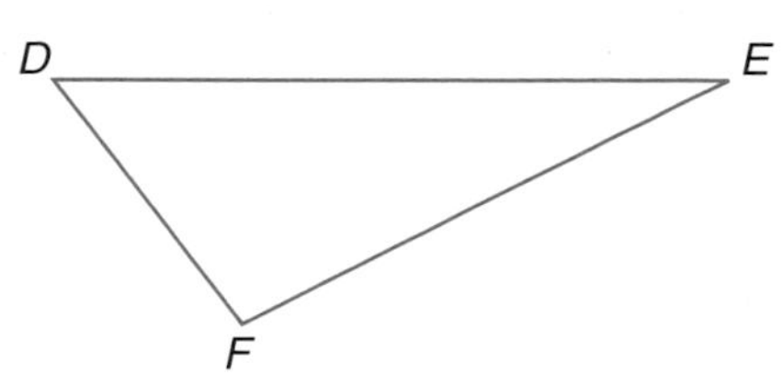

Given: $\triangle DEF$

Prove: $\triangle DEF \cong \triangle DEF$

Proof:

Statements	Reasons
1. $\triangle DEF$	1. ?
2. $\overline{DE} \cong \overline{DE}$, $\overline{EF} \cong \overline{EF}$, $\overline{FD} \cong \overline{FD}$	2. ?
3. $\angle D \cong \angle D$, $\angle E \cong \angle E$, $\angle F \cong \angle F$	3. ?
4. $\triangle DEF \cong \triangle DEF$	4. ?

10. Write a flow proof that congruence of triangles is reflexive.
11. Given $\triangle CAT \cong \triangle DOG$, $CA = 14$, $AT = 18$, $TC = 21$, and $DG - 2x + 7$.
    a. Draw and label a figure showing the congruent triangles.
    b. Find the value of x.
12. Draw two noncongruent triangles that have the same perimeter.

Proof

13. Write either a two-column or a flow proof for the following.

Given: $\triangle MXR$ is a right isosceles triangle with $\angle X$ the vertex angle.
$\overline{XY} \perp \overline{MR}$
Y is the midpoint of $\overline{MR}$.
$\angle M \cong \angle R$
$\overline{YX}$ bisects $\angle MXR$.

Prove: $\triangle MXY \cong \triangle RXY$

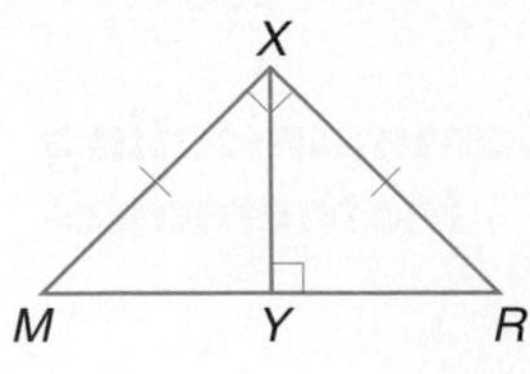

FYI

Robert Mangold (1937–) became interested in geometric art while he was working as a guard at the Museum of Modern Art in New York City.

14. **Art** Draw a sketch of the painting by Robert Mangold shown at the right. Then label the intersections of lines and name the triangles that appear to be congruent.

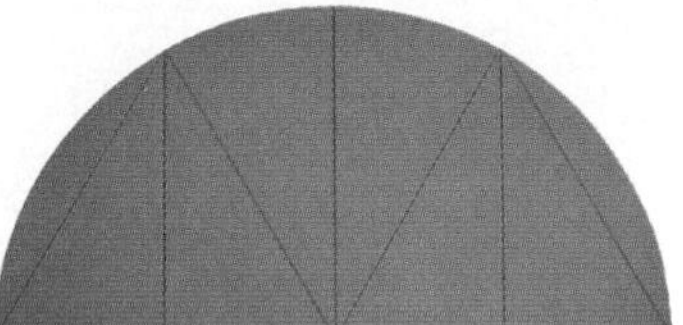

Half–W Series

EXERCISES

Practice

15. In the quilt design at the right, indicate which triangles appear to be congruent.

For each pair of congruent triangles, name all the corresponding sides and angles. Use ↔ to indicate each correspondence. Draw a figure for each pair of triangles, and mark the corresponding parts.

16. $\triangle BIG \cong \triangle DEN$
17. $\triangle PQR \cong \triangle RST$
18. $\triangle EGO \cong \triangle PGO$

Complete each congruence statement.

19. $\triangle YZW \cong \triangle \underline{\ ?\ }$
20. $\triangle MQN \cong \triangle \underline{\ ?\ }$
21. $\triangle ARZ \cong \triangle \underline{\ ?\ }$

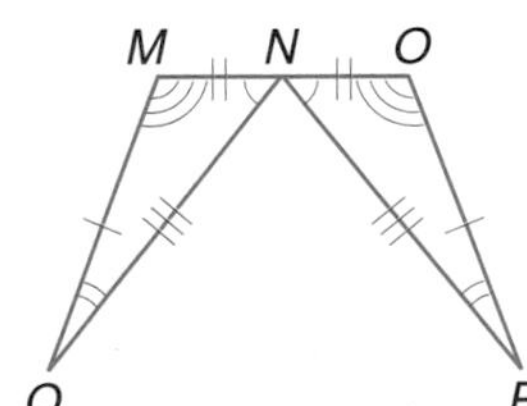

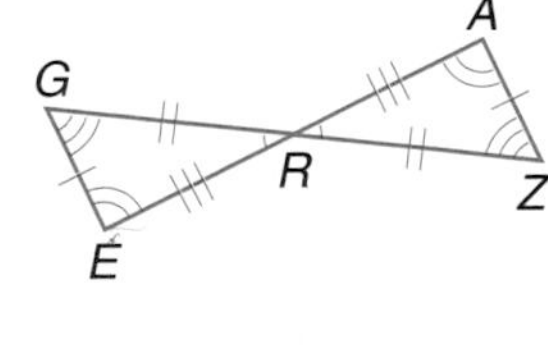

Proof

22. Copy the flow proof of *Congruence of triangles is symmetric*. Provide the reasons for each statement.

Given: $\triangle RST \cong \triangle XYZ$

Prove: $\triangle XYZ \cong \triangle RST$

Proof:

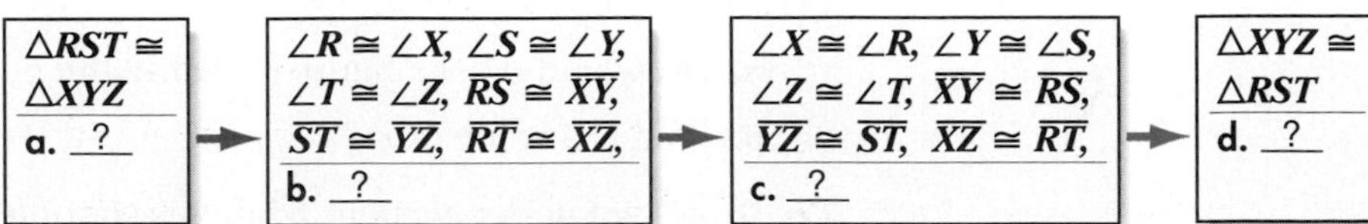

23. Copy the flow proof and provide the reasons for each statement.

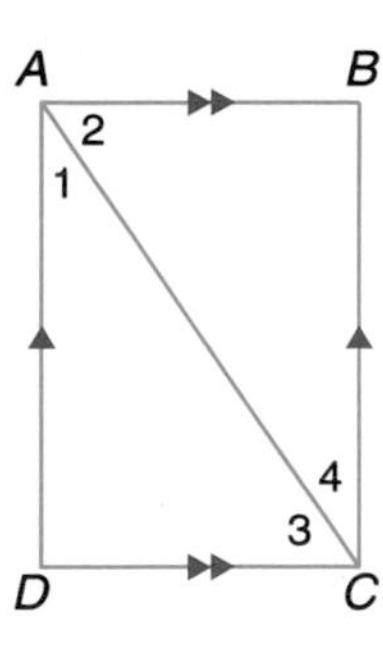

Given: $\overline{AB} \cong \overline{CD}$, $\overline{AD} \cong \overline{CB}$, $\overline{AD} \perp \overline{DC}$, $\overline{AB} \perp \overline{BC}$, $\overline{AD} \parallel \overline{BC}$, $\overline{AB} \parallel \overline{CD}$

Prove: $\triangle ACD \cong \triangle CAB$

Proof:

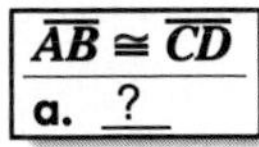

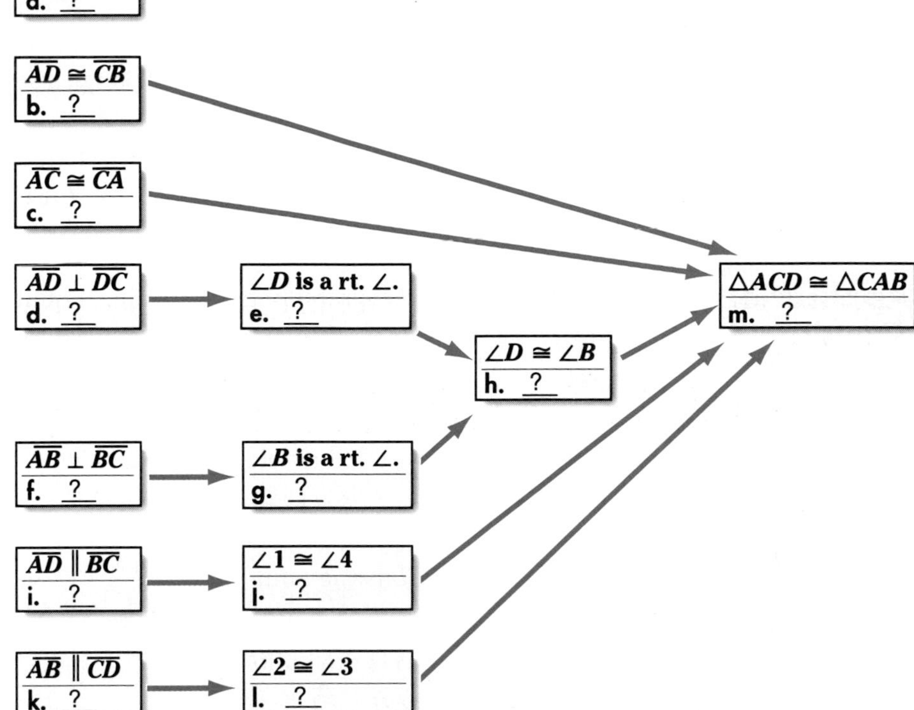

24. Given $\triangle ABC \cong \triangle DEF$, $BC = 12$, $AB = 10$, $AC = 6$, and $EF = 2x - 4$.
    a. Draw and label a figure to show the congruent triangles.
    b. Find the value of x.

25. Given $\triangle JKL \cong \triangle ABC$, $m\angle J = 36$, $m\angle B = 64$, and $m\angle C = 3x + 52$.
    a. Draw and label a figure to show the congruent triangles.
    b. Find the value of x.

26. If $\triangle PQR \cong \triangle CDE$, PQ is 10 less than 3 times a number, PR is 2 less than twice the number, CE is 5 more than the number, CD is 4 more than the number, find PQ and CE.

Draw two triangles that have the following properties.

27. equal areas and are congruent
28. equal perimeters and are congruent
29. equal areas but are not congruent

If $\triangle DNB \cong \triangle ANC$, determine if each statement is *true* or *not necessarily true*. Explain your reasoning.

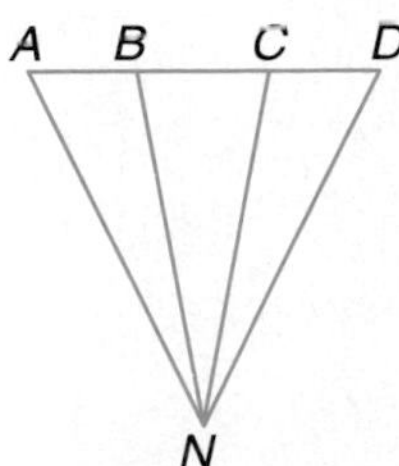

30. $\overline{DB} \cong \overline{AC}$
31. $\overline{DC} \cong \overline{AB}$
32. $\overline{DN} \cong \overline{CN}$
33. $\overline{CN} \cong \overline{BN}$
34. $\angle NAB \cong \angle NDC$
35. $\angle ANB \cong \angle DNC$

36. $\triangle MNO$ is both equilateral and equiangular. $\triangle PQR$ has three congruent sides, $m\angle P = 60$, and $m\angle Q = 60$. Is $\triangle MNO \cong \triangle PQR$? Explain your reasoning.

37. If $\triangle SIO \cong \triangle OIS$, prove $\triangle SIO$ is isosceles by using a flow proof.

38. If $\triangle ANG \cong \triangle NGA$ and $\triangle NGA \cong \triangle GAN$, prove $\triangle AGN$ is both equilateral and equilangular by using a two-column proof.

39. On graph paper, draw six separate congruent right scalene triangles. Cut out the triangles. Arrange the triangles so that congruent sides fit together. Try several different arrangements.

a. How many different shapes with four sides can you make?

b. How many different shapes with three sides can you make?

Applications and Problem Solving

40. **Crafts** Use graph paper to design a quilt using congruent triangles. Classify the triangles used by name and identify those that can be proved congruent.

41. **Construction** The figure below is a drawing of a truss roof. Using the given information below, name all triangles that are congruent.

Given: $\triangle RPT$ is isosceles with vertex $\angle R$.
U is the midpoint of $\overline{PT}$.
Q is the midpoint of $\overline{PR}$.
S is the midpoint of $\overline{RT}$.
$\overline{RU} \perp \overline{PT}$
$\angle PQU \cong \angle TSU$
$\overline{RU}$ bisects $\angle PRT$.
$\angle RPT \cong \angle RTP$

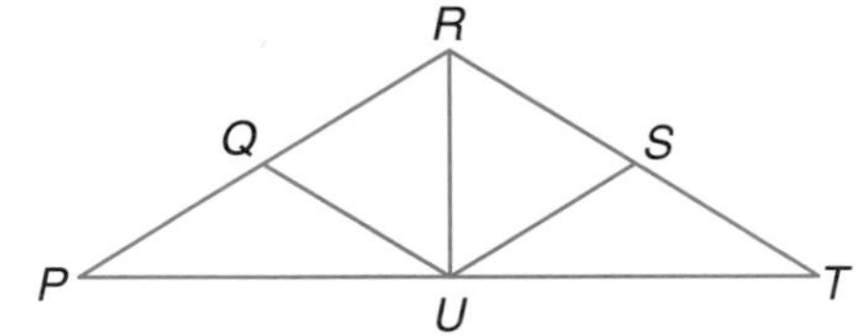

Mixed Review

42. **Construction** A brace shown at the right is used to keep a shelf perpendicular to the wall. If $m\angle CBA = 35$, find $m\angle BCD$. (Lesson 4–2)

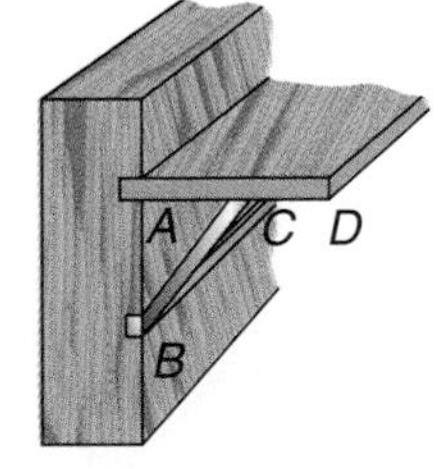

43. Find the perimeter of equilateral $\triangle JLK$ if $JK = x + 3$ and $KJ = 2x - 5$. (Lesson 4–1)

44. Copy the figure and draw the segment that represents the distance from R to $\overline{PQ}$. (Lesson 3–5)

45. List the conclusions that can be drawn from the figure at the right. (Lesson 3–2)

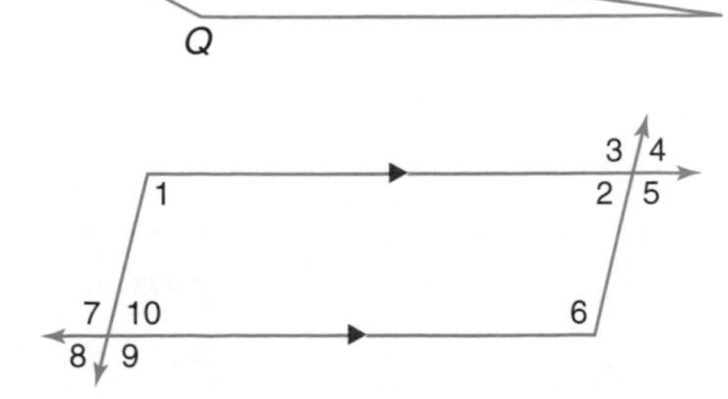

46. Draw a figure to illustrate two intersecting lines with a plane parallel to both lines. (Lesson 3–1)

47. **Algebra** Name the algebraic property that justifies the statement *If* $a = x$ *and* $a = y$, *then* $x = y$. (Lesson 2–4)

48. **SAT Practice** How many even integers are there between 1 and 50, not including 50?
A 20 **B** 24 **C** 25 **D** 48 **E** 49

49. $\overrightarrow{QT}$ and $\overrightarrow{QS}$ are opposite rays. Describe $\angle TQS$. (Lesson 1–6)

Algebra

50. **Government** The graphic shows the number of countries in the world listed in various sources.
   a. Write a conclusion from the data given in the table. Justify your statement.
   b. Would the average of these numbers have any meaning? Explain.

51. Solve $-10 < 3x - 1 \leq 5$.

The Number of Countries

World Bank 166
Int'l Monetary Fund 178
U.N. members 185
U.S. State Dept. 189
World Book encyclopedia 235
Guiness Book of World Records 256

Source: *USA TODAY* research

For **Extra Practice**, see page 771.

SELF TEST

In figure *ACDE*, $\angle E$ and $\angle ADC$ are right angles and the congruent parts are indicated. (Lesson 4–1)

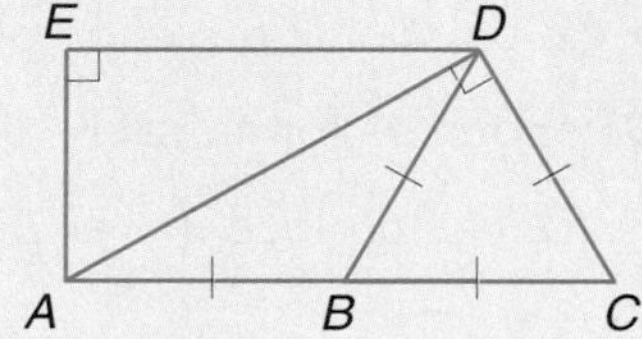

1. Identify any isosceles triangles.
2. Identify any obtuse triangles.
3. Identify any segments that are hypotenuses.
4. Which segments are opposite $\angle C$?

Find the value of x. (Lesson 4–2)

5.
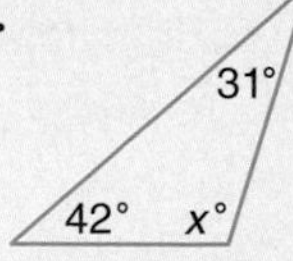

6.
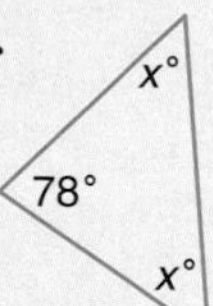

7.
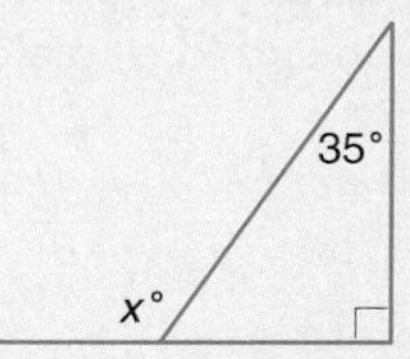

Complete each congruence statement. (Lesson 4–3)

8. $\triangle CDB \cong$ __?__

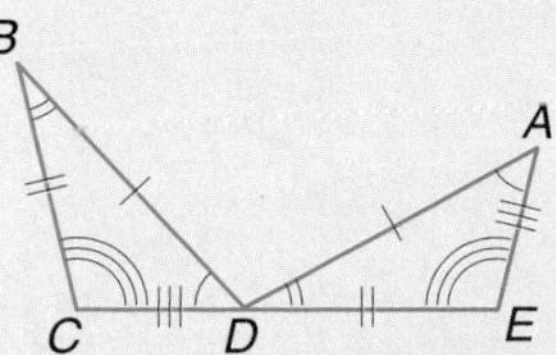

9. $\triangle PTL \cong$ __?__

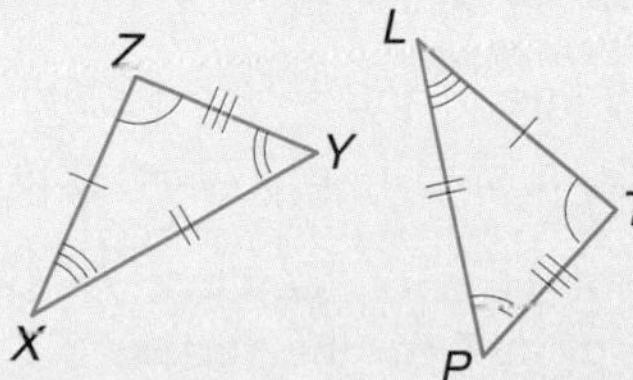

10. Given $\triangle BCD \cong \triangle RST$, $m\angle C = 57$, $m\angle R = 64$, and $m\angle D = 5x + 4$, find the value of x. (Lesson 4–3)

MODELING MATHEMATICS

4–4A Constructing Congruent Triangles

Materials: 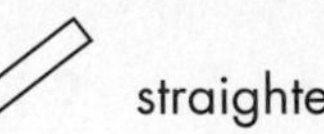 straightedge protractor

 compass  scissors

A Preview of Lesson 4–4

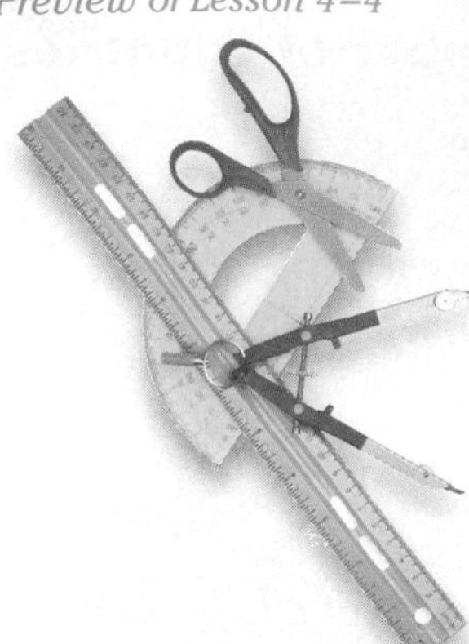

How can you draw congruent triangles? To investigate this question, use your protractor to draw a triangle with angle measures of 20, 40, and 120. Label your triangle ABC as shown at the right. How does your triangle compare to triangles drawn by other students?

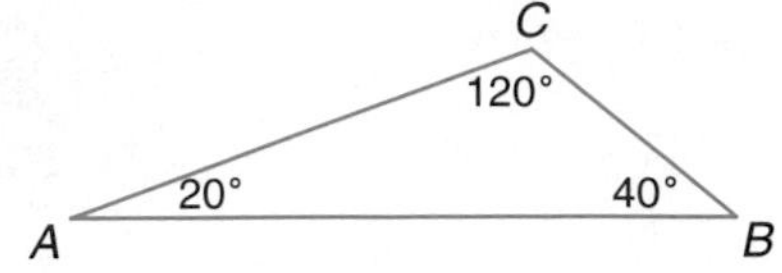

Use your $\triangle ABC$ to complete the activities.

Activity 1 Construct a triangle in which the three sides are congruent to the three sides of △*ABC* and compare it to △*ABC*.

- Use a straightedge to draw any line ℓ and select a point D.
- Use a compass to construct $\overline{DE}$ on ℓ such that $\overline{DE} \cong \overline{AB}$.
- Using D as the center, draw an arc with radius equal to AC.
- Using E as the center, draw an arc with radius equal to BC.
- Let F be the point of intersection of the two arcs.
- Draw $\overline{DF}$ and $\overline{EF}$ to form $\triangle DEF$.
- Cut out $\triangle DEF$ and place it over $\triangle ABC$. How does $\triangle DEF$ compare to $\triangle ABC$?

In a triangle, the angle formed by two given sides is called the **included angle** of the sides. In $\triangle ABC$, $\angle A$ is the included angle of sides $\overline{AB}$ and $\overline{AC}$. *$\angle B$ is also the included angle of sides $\overline{AB}$ and $\overline{CB}$, and $\angle C$ is the included angle of sides $\overline{AC}$ and $\overline{BC}$.*

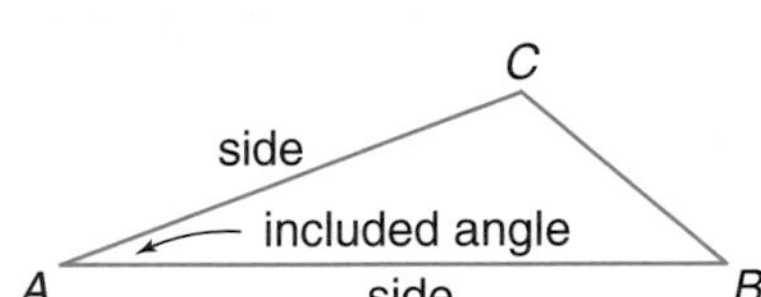

Activity 2 Construct a triangle in which two sides are congruent to two sides of △*ABC* and the included angle is congruent to the included angle in △*ABC*. Compare the triangle to △*ABC*.

- Use a straightedge to draw any line m and select a point G.
- Use a compass to construct $\overline{GH}$ on m such that $\overline{GH} \cong \overline{AB}$.
- Use a compass to construct an angle congruent to $\angle A$ using $\overline{GH}$ as a side of the angle and point G as the vertex.
- Use a compass to construct $\overline{GI}$ on the new side of the angle such that $\overline{GI} \cong \overline{AC}$.

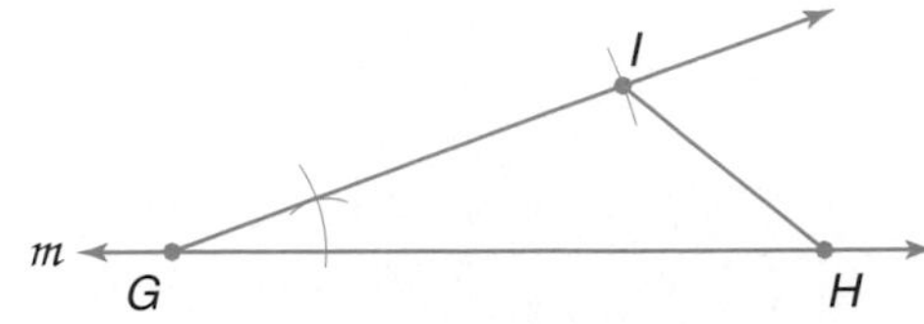

- Draw $\overline{HI}$ to complete $\triangle GHI$.
- Cut out $\triangle GHI$ and place it over $\triangle ABC$. How does $\triangle GHI$ compare to $\triangle ABC$?

The side of a triangle that forms a side of two of its angles is called the **included side** of the angles. In $\triangle ABC$, $\overline{AB}$ is the included side of $\angle A$ and $\angle B$.
$\overline{AC}$ is the included side of $\angle A$ and $\angle C$.
$\overline{BC}$ is the included side of $\angle B$ and $\angle C$.

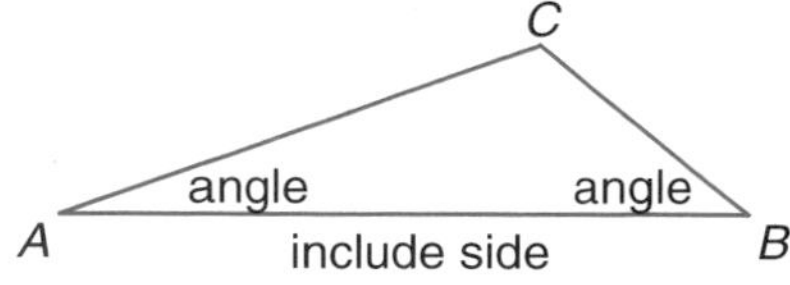

Activity 3 Construct a triangle in which two angles are congruent to two angles of $\triangle ABC$ and the included side is congruent to the included side of $\triangle ABC$.

- Use a straightedge to draw any line m and select a point J.
- Use a compass to draw $\overline{JK}$ such that $\overline{JK} \cong \overline{AB}$.
- Use a compass to construct an angle congruent to $\angle A$ using $\overrightarrow{JK}$ as a side of the angle and point J as the vertex.
- Use a compass to construct an angle congruent to $\angle B$ using $\overrightarrow{KJ}$ as a side of the angle and point K as the vertex.
- Label the point where the new sides of the angles meet L.
- Cut out $\triangle JKL$ and place it over $\triangle ABC$. How does $\triangle JKL$ compare to $\triangle ABC$?

Model **Use $\triangle$ABC from the beginning of the lesson.**

1. Construct a triangle in which two sides are congruent to $\overline{AC}$ and $\overline{BC}$ and the included angle is congruent to $\angle C$. How does this triangle compare to $\triangle ABC$?
2. Construct a triangle in which two sides are congruent to $\overline{AB}$ and $\overline{BC}$ and the included angle is congruent to $\angle B$. How does this triangle compare to $\triangle ABC$?
3. Construct a triangle in which two angles are congruent to $\angle A$ and $\angle C$ and the included side is congruent to $\overline{AC}$. How does this triangle compare to $\triangle ABC$?
4. Construct a triangle in which two angles are congruent to $\angle B$ and $\angle C$ and the included side is congruent to $\overline{BC}$. How does this triangle compare to $\triangle ABC$?

Write

5. Write a conjecture about two triangles in which the sides of one triangle are congruent to the sides of the other triangle.
6. Write a conjecture about two triangles in which two sides and the included angle of one triangle are congruent to two sides and the included angle of the other triangle.
7. Write a conjecture about two triangles in which two angles and the included side of one triangle are congruent to two angles and the included side of the other triangle.
8. What can you say about two triangles in which the angles of one triangle are congruent to the angles of the other triangle?

4-4 Proving Triangles Congruent

What YOU'LL LEARN

- To use SSS, SAS, and ASA Postulates to test for triangle congruence.

Why IT'S IMPORTANT

You can use postulates to determine if two triangles are congruent.

Construction

Is it always necessary to show that all of the corresponding parts of two triangles are congruent to be sure that the two triangles are congruent? The design of a deck at the right offers a carpenter many opportunities to use congruent triangles.

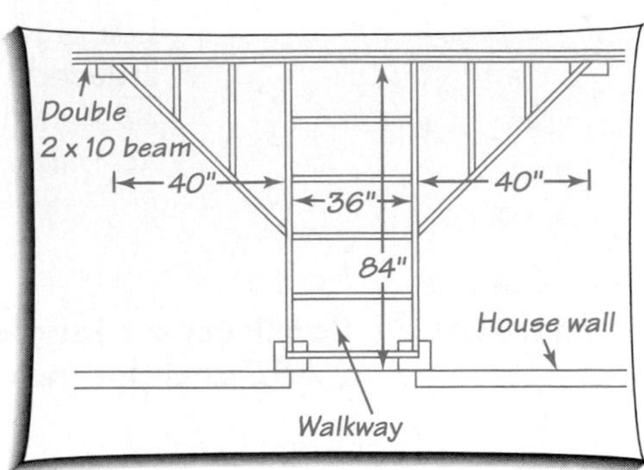

One method of guaranteeing a square corner in construction is called the 3-4-5 method. The 3-4-5 method lets the carpenter check a right angle using a tape measure instead of a framing square. The carpenter marks a point 3 feet from a corner along one side. Then he marks another point 4 feet from the corner along the other side. If these points are 5 feet apart, the corner forms a right angle. The 3-4-5 method is most useful when a carpenter is doing a large-scale layout that is beyond the accuracy of a framing square.

The carpenter measures several 3-4-5 triangles in the deck. All of these triangles are congruent. Only the measures of the sides were necessary to have congruent triangles.

Postulate 4–1 SSS Postulate (Side-Side-Side)

If the sides of one triangle are congruent to the sides of a second triangle, then the triangles are congruent.

The SSS Postulate can be used to prove that two triangles are congruent.

Example 1

Algebra

Given $\triangle STU$ with vertices $S(0, 5)$, $T(0, 0)$, and $U(-2, 0)$ and $\triangle XYZ$ with vertices $X(4, 8)$, $Y(4, 3)$, and $Z(6, 3)$, determine if $\triangle STU \cong \triangle XYZ$.

Use the distance formula to show that the corresponding sides are congruent.

$ST = \sqrt{(0-0)^2 + (0-5)^2} = \sqrt{25}$ or 5

$TU = \sqrt{(-2-0)^2 + (0-0)^2} = \sqrt{4}$ or 2

$SU = \sqrt{(-2-0)^2 + (0-5)^2} = \sqrt{29}$

Notice that $\triangle XYZ$ is a flip and slide of $\triangle STU$.

$XY = \sqrt{(4-4)^2 + (3-8)^2} = \sqrt{25}$ or 5

$YZ = \sqrt{(6-4)^2 + (3-3)^2} = \sqrt{4}$ or 2

$XZ = \sqrt{(6-4)^2 + (3-8)^2} = \sqrt{29}$

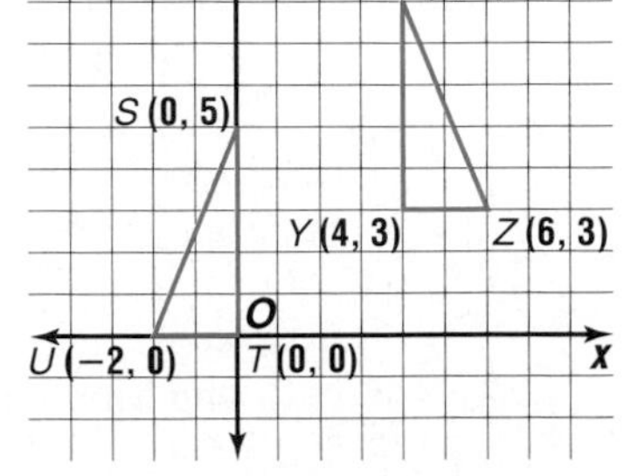

$ST = XY$, $TU = YZ$, and $SU = XZ$, or by definition of congruent segments, all corresponding segments are congruent. Therefore, $\triangle STU \cong \triangle XYZ$ by SSS.

Will any other combinations of corresponding congruent sides and angles determine a unique triangle? Suppose you were given the measures of two sides and the angle that they form, which is called the **included angle**. According to the SAS Postulate, only one triangle can be formed with these given conditions.

Postulate 4–2 SAS Postulate (Side-Angle-Side)	**If two sides and the included angle of one triangle are congruent to two sides and the included angle of another triangle, then the triangles are congruent.**

The following proof uses the SAS Postulate.

Example 2

Write a proof for the following.

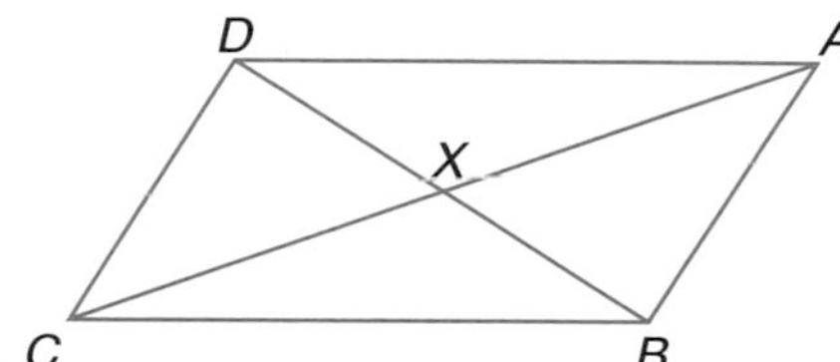

Given: X is the midpoint of $\overline{BD}$.
X is the midpoint of $\overline{AC}$.

Prove: $\triangle DXC \cong \triangle BXA$

Method 1: Two-Column Proof

Statements	**Reasons**
1. X is the midpoint of $\overline{BD}$.	**1.** Given
2. $\overline{DX} \cong \overline{BX}$	**2.** Definition of midpoint
3. X is the midpoint of $\overline{AC}$.	**3.** Given
4. $\overline{CX} \cong \overline{AX}$	**4.** Definition of midpoint
5. $\angle DXC \cong \angle BXA$	**5.** Vertical angles are congruent.
6. $\triangle DXC \cong \triangle BXA$	**6.** SAS

Method 2: Flow Proof

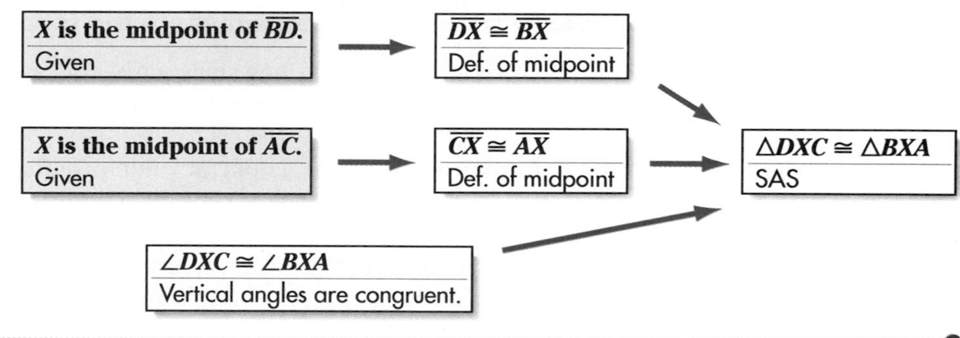

Suppose you were given the measures of two angles and the side of the triangle that forms a side of the two angles called the **included side**. According to the ASA Postulate, only one triangle can be formed with these given conditions.

Postulate 4–3 ASA Postulate (Angle-Side-Angle)	**If two angles and the included side of one triangle are congruent to two angles and the included side of another triangle, the triangles are congruent.**

Some proofs ask you to show that a pair of corresponding parts of two triangles are congruent. Often you can do this by first proving that the two triangles are congruent. Then the definition of congruent triangles (CPCTC) can be used to show that the corresponding parts are congruent.

Example 

Write a two-column proof.

Given: $\overline{VR} \perp \overline{RS}$
$\overline{UT} \perp \overline{SU}$
$\overline{RS} \cong \overline{US}$

Prove: $\overline{VR} \cong \overline{TU}$

Proof:

Statements	Reasons
1. $\overline{VR} \perp \overline{RS}$, $\overline{UT} \perp \overline{SU}$	1. Given
2. $\angle R$ is a right angle. $\angle U$ is a right angle.	2. Perpendicular lines form four right angles.
3. $\angle R \cong \angle U$	3. All right angles are congruent.
4. $\overline{RS} \cong \overline{US}$	4. Given
5. $\angle RSV \cong \angle UST$	5. Vertical angles are congruent.
6. $\triangle VRS \cong \triangle TUS$	6. ASA
7. $\overline{VR} \cong \overline{TU}$	7. CPCTC

CHECK FOR UNDERSTANDING

Communicating Mathematics

Study the lesson. Then complete the following.

1. **Draw** a triangle that has sides with lengths 3 inches, 4 inches, and 5 inches. Identify the triangle according to its angles.

2. **Explain** how CPCTC is useful when proving that parts of two triangles are congruent.

3. **You Decide** Patty observes that the two triangles at the right have two pairs of congruent sides and a pair of congruent angles. She says the triangles are congruent by the SAS Postulate. Mirna says the triangles are not necessarily congruent. Who is correct? Explain your reasoning.

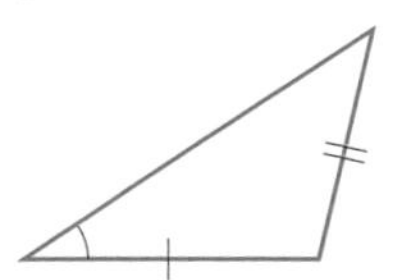

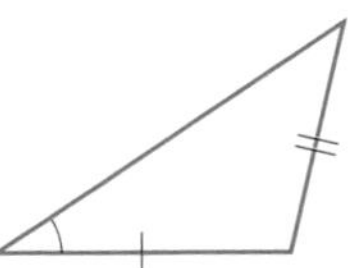

4. Refer to Example 3. Describe $\triangle TUS$ as a congruence transformation of $\triangle VRS$.

5. **Assess Yourself** Compare and contrast the SSS Postulate, the SAS Postulate, and the ASA Postulate. Give an example of each.

Guided Practice

Determine which postulate can be used to prove the triangles are congruent. If it is not possible to prove that they are congruent, write *not possible.*

6.

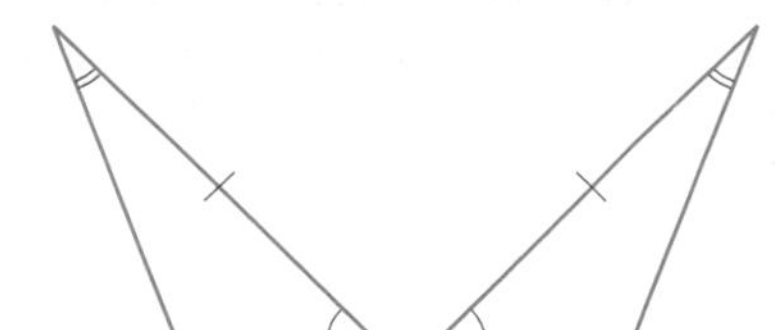

7.

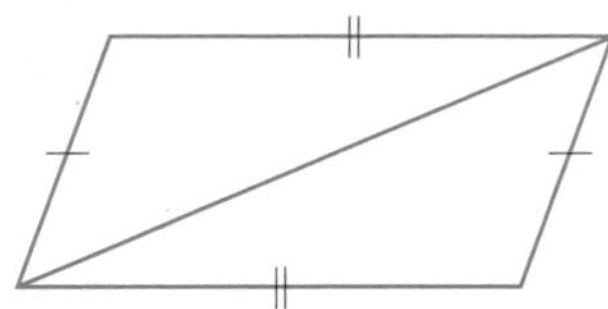

8. If the vertices of $\triangle ACM$ are $A(0, 4)$, $C(0, 0)$, and $M(3, 0)$ and the vertices of $\triangle SCR$ are $S(-4, 0)$, $C(0, 0)$, and $R(0, -4)$, determine if $\triangle ACM \cong \triangle SCR$. Justify your answer.

9. Supply the reason for each step in the two-column proof.

Given: $\overline{AD}$ bisects $\overline{BE}$.
$\overline{AB} \parallel \overline{DE}$

Prove: $\triangle ABC \cong \triangle DEC$

Proof:

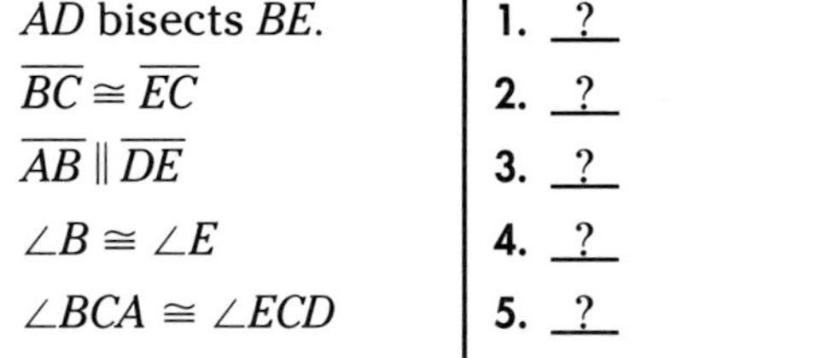

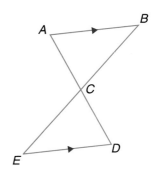

Statements	Reasons
1. $\overline{AD}$ bisects $\overline{BE}$.	1. ?
2. $\overline{BC} \cong \overline{EC}$	2. ?
3. $\overline{AB} \parallel \overline{DE}$	3. ?
4. $\angle B \cong \angle E$	4. ?
5. $\angle BCA \cong \angle ECD$	5. ?
6. $\triangle ABC \cong \triangle DEC$	6. ?

10. Write a flow proof.

Given: $\overline{OM}$ bisects $\angle LMN$.
$\overline{LM} \cong \overline{NM}$

Prove: $\triangle MOL \cong \triangle MON$

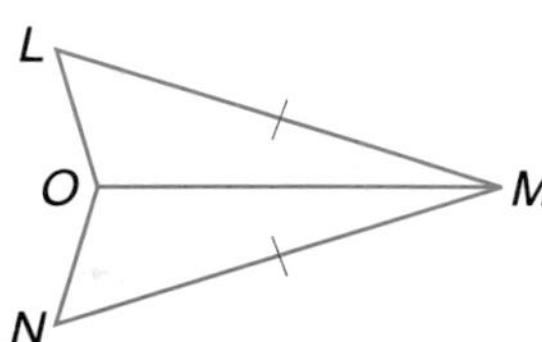

Use the figure at the right for Exercises 11–12.

11. Given: $\overline{MO} \cong \overline{PO}$
$\overline{NO}$ bisects $\overline{MP}$.

Prove: $\triangle MNO \cong \triangle PNO$

12. Given: $\overline{NO}$ bisects $\angle POM$.
$\overline{NO} \perp \overline{MP}$

Prove: $\triangle MNO \cong \triangle PNO$

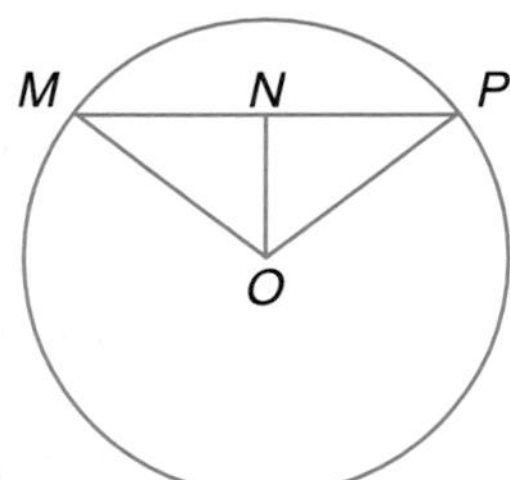

13. Construction According to the SSS Postulate, only one triangle can be formed with three fixed segments lengths. For this reason, a triangle is rigid and is frequently used to reinforce structures.

a. Which of the following frameworks is rigid?

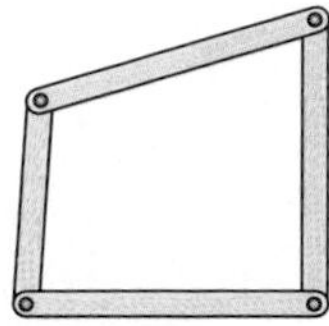

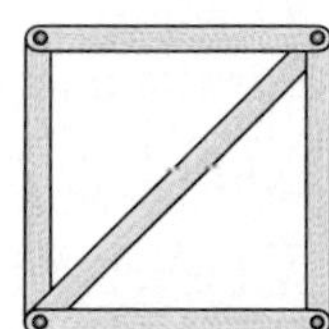

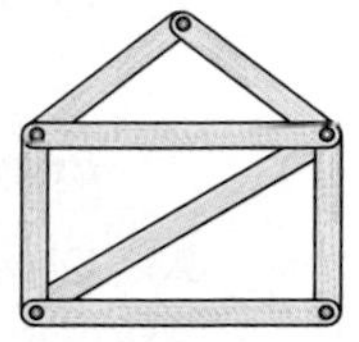

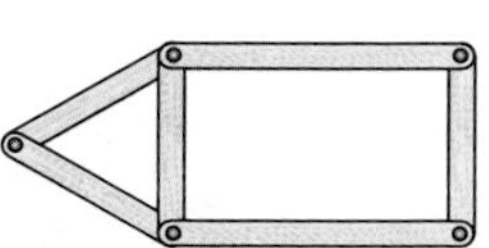

b. Find three examples in which triangles are used to reinforce a structure.

EXERCISES

Practice **Determine which postulate can be used to prove the triangles are congruent. If it is not possible to prove that they are congruent, write *not possible.***

14. 

15. 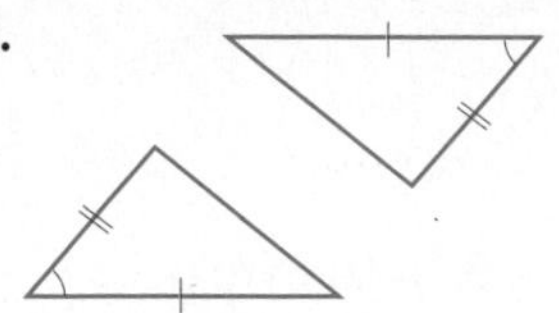

16. 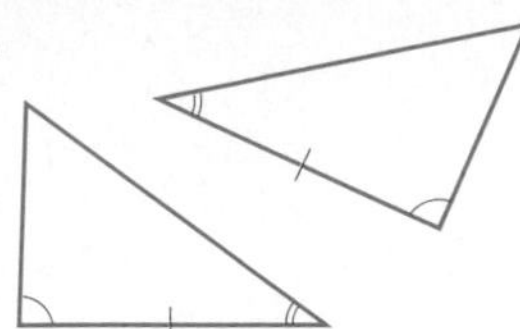

17. 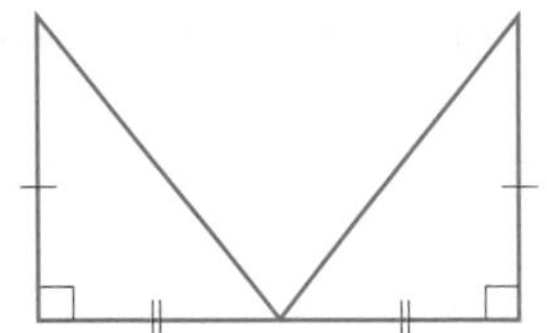

18. 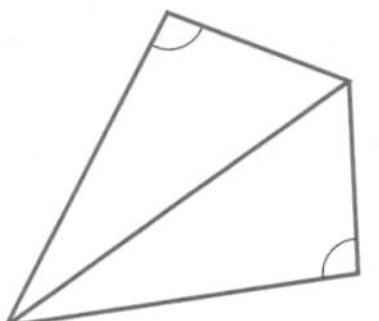

19. 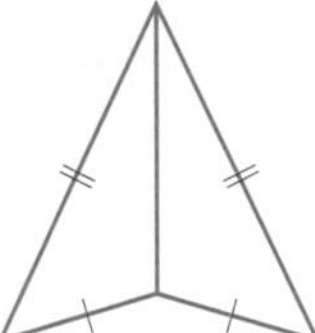

Determine if each pair of triangles is congruent. Justify your answer.

20. The vertices of $\triangle RTY$ are $R(2, 5)$, $T(5, 2)$, and $Y(1, 1)$, and the vertices of $\triangle MGE$ are $M(-4, 4)$, $G(-7, 1)$, and $E(-3, 0)$.

21. The vertices of $\triangle PQR$ are $P(-1, -1)$, $Q(0, 6)$, and $R(2, 3)$, and the vertices of $\triangle XYZ$ are $X(3, 1)$, $Y(5, 3)$, and $Z(8, 1)$.

22. The vertices of $\triangle TSR$ are $T(-1, -1)$, $S(-2, -2)$, and $R(-5, -1)$, and the vertices of $\triangle HND$ are $H(2, -1)$, $N(3, -2)$, and $D(2, -5)$.

23. In the figure at the right, $\angle SUR \cong \angle TUR$. State one other fact that you would need to know to prove $\triangle SRU \cong \triangle TRU$ by ASA.

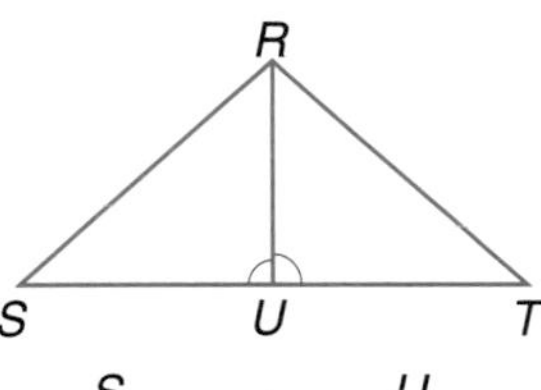

24. Justify each step in the flow proof.

Given: $\overline{SU} \parallel \overline{AC}$
$\overline{SU} \cong \overline{AC}$

Prove: $\triangle SUA \cong \triangle ACS$

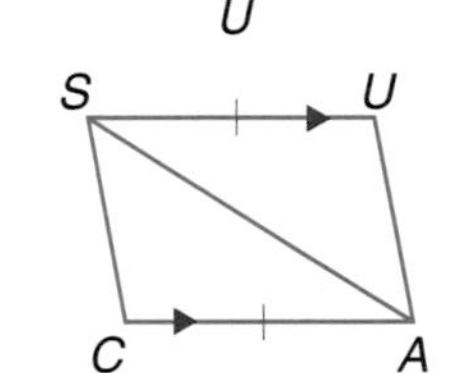

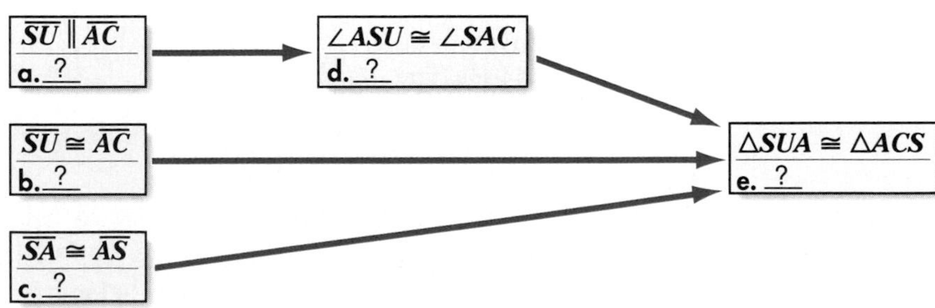

Write a proof.

25. Given: $\angle J \cong \angle L$
B is the midpoint of $\overline{JL}$.

Prove: $\triangle JHB \cong \triangle LCB$

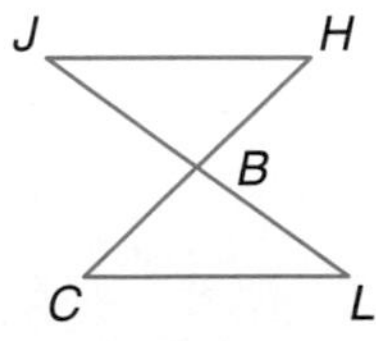

26. Given: $\overline{JL}$ bisects $\overline{HC}$.
$\overline{HC}$ bisects $\overline{JL}$.

Prove: $\triangle JHB \cong \triangle LCB$

Write a two-column proof.

27. Given: $\triangle MGR$ is an isosceles triangle with vertex $\angle MGR$. K is the midpoint of $\overline{MR}$.

Prove: $\triangle MGK \cong \triangle RGK$

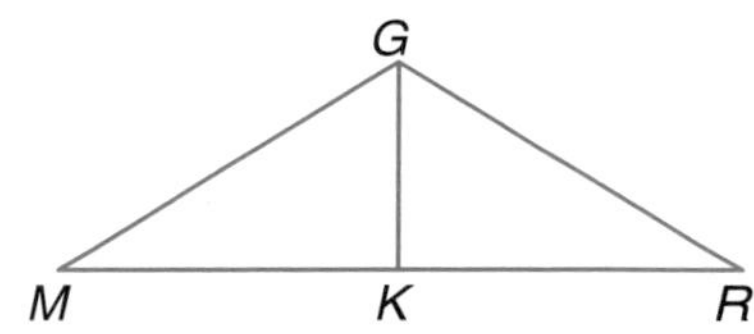

28. Given: $\overline{GK} \perp \overline{MR}$
$\overline{GK}$ bisects $\overline{MR}$.

Prove: $\triangle MGK \cong \triangle RGK$

Use quadrilateral *CDRL* for Exercises 29–30.

29. Given: $\overline{RL} \parallel \overline{DC}$
$\overline{LC} \parallel \overline{RD}$

Prove: $\angle R \cong \angle C$

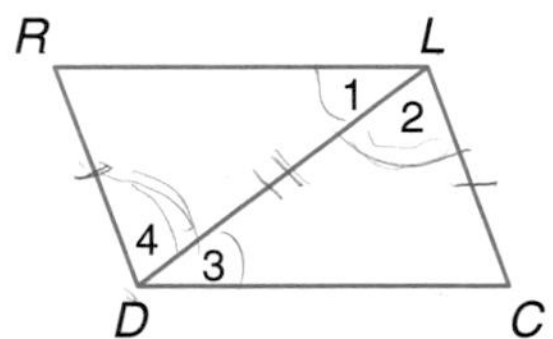

30. Given: $\angle 4 \cong \angle 2$
$\overline{DR} \cong \overline{LC}$

Prove: $\overline{RL} \cong \overline{CD}$

Use $\triangle$*LRW* for Exercises 31–34.

31. Given: $\angle 1 \cong \angle 2$
$\angle 3 \cong \angle 4$
$\overline{LA} \cong \overline{RU}$

Prove: $\triangle WLU \cong \triangle WRA$

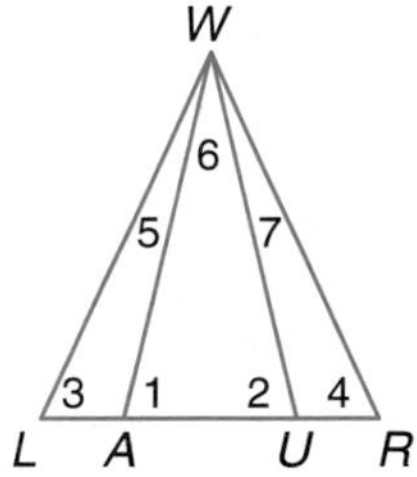

32. Given: $\angle LWU \cong \angle RWA$
$\angle 3 \cong \angle 4$
$\overline{LW} \cong \overline{RW}$

Prove: $\triangle LWA \cong \triangle RWU$

33. Given: $\angle 5 \cong \angle 7$
$\angle 3 \cong \angle 4$
$\overline{LW} \cong \overline{RW}$

Prove: $\overline{LU} \cong \overline{RA}$

34. Given: $\overline{LU} \cong \overline{RA}$
$\angle 3 \cong \angle 4$
$\overline{LW} \cong \overline{RW}$

Prove: $\overline{AW} \cong \overline{UW}$

Critical Thinking

35. Explain why the following proof is not valid.

Given: $\overline{TI} \cong \overline{TN}$

Prove: $\triangle ATI \cong \triangle ATN$

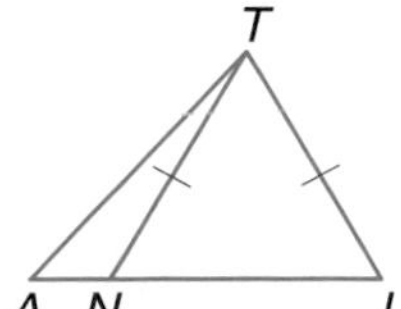

Proof:

Statements	Reasons
1. $\overline{TI} \cong \overline{TN}$	1. Given
2. $\angle A \cong \angle A$	2. Congruence of angles is reflexive.
3. $\overline{AT} \cong \overline{AT}$	3. Congruence of segments is reflexive.
4. $\triangle ATI \cong \triangle ATN$	4. SAS

Applications and Problem Solving

36. Recreation Tapatan is a game played in the Philippines on a square board as shown at the right. The players take turns placing each of their three pieces on a different point of intersection. After all the pieces have been played, the players take turns moving a piece along a line to another intersection. A piece cannot jump over another piece. A player who gets all his or her pieces in a straight line wins the game. Point E bisects all four line segments that pass through it.

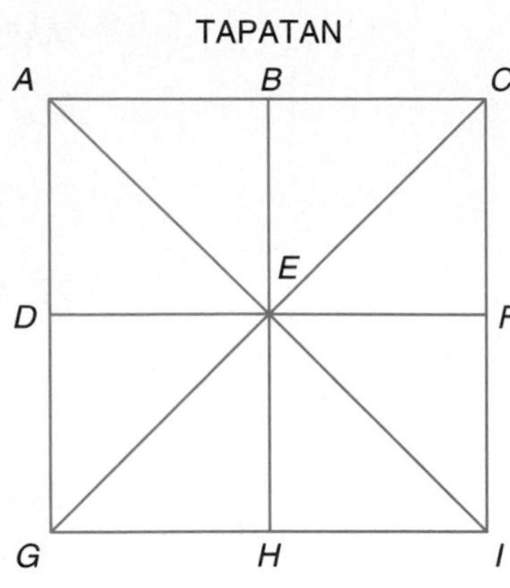

a. What can you say about $\triangle GHE$ and $\triangle CBE$? Explain your reasoning.

b. What can you say about $\triangle AEG$ and $\triangle IEG$? Explain your reasoning.

c. What can you say about $\triangle ACI$ and $\triangle CAG$? Explain your reasoning.

37. Estimation Jamal observes a duckling on the other side of the river. He wishes to estimate his distance from the duckling. Jamal adjusts the visor of his cap so that it is in line with his line of sight to the duckling. He keeps his neck stiff and turns his body. Using the visor of his cap, he uses the same line of sight to observe another point on the ground. He then paces out the distance to the new point. Jamal concludes that his distance to the duckling was the same as the distance he just paced out. Do you agree with Jamal's method of estimation? Explain your reasoning.

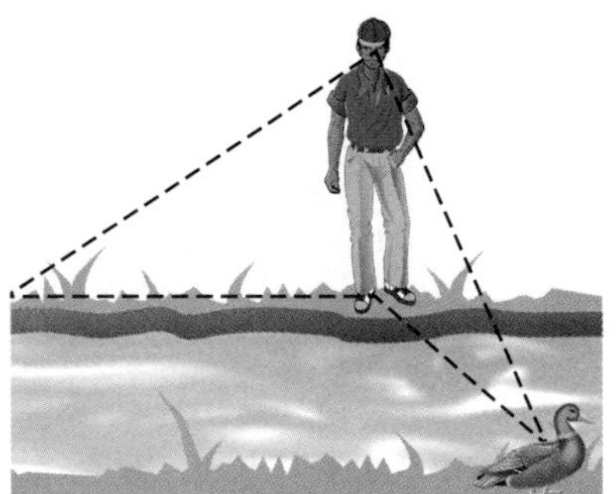

Mixed Review

38. Given $\triangle ABC \cong \triangle RST$, $m\angle A = 2x + 5$, $m\angle S = x - 15$, and $m\angle R = 55$. (Lesson 4–3)

a. Draw and label a figure to show the congruent triangles.

b. Find the value of x.

c. Find the measure of each of the angles in the triangles.

39. Refer to the figure at the right. Find the value of x. (Lesson 4–2)

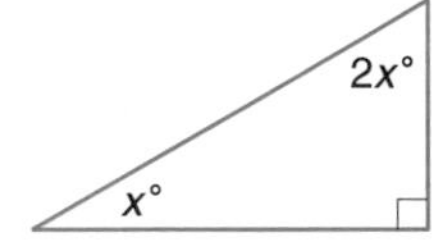

40. Triangle ROM is an isosceles triangle with the congruent sides as marked. Name each of the following. (Lesson 4–1)

a. sides

b. angles

c. vertex angle

d. base angles

e. side opposite $\angle R$

f. angle opposite $\overline{OR}$

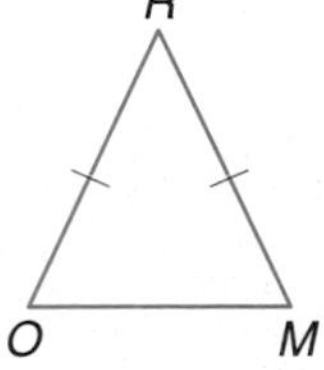

41. **Algebra** State the slope of the line parallel to the line passing through (3, 9) and (−7, 8). (Lesson 3–3)

42. Write the converse, inverse, and contrapositive of *If the measure of an angle is 60, then it is acute*. Then determine if each is true or false. If false, give a counterexample. (Lesson 2–2)

43. **Botany** Delaney noticed that several potted plants on her porch were dying. All of these plants were supposed to receive partial sun. Make a conjecture about why the plants were dying and describe a method for testing this conjecture. (Lesson 2–1)

44. **ACT Practice** In ΔABC, $\angle A \cong \angle B$, and $m\angle C$ is three times the measure of $\angle B$. What is the measure, in degrees, of $\angle A$?

A 30 **B** 36 **C** 45
D 72 **E** 108

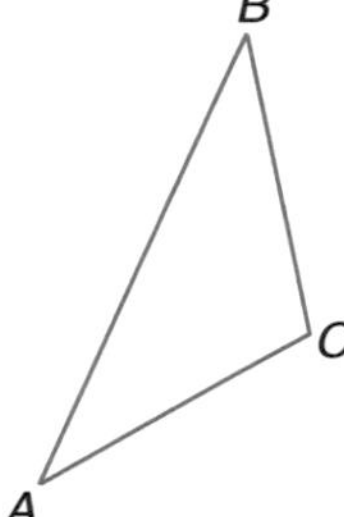

Algebra

45. **Travel** Rachel and Mike leave their home at the same time, traveling in opposite directions. Rachel travels at 55 miles per hour, and Mike travels at 45 miles per hour. In how many hours will they be 150 miles apart?

46. Use substitution to solve the system of equations. If the system does not have exactly one solution, state whether it has *no* solution, or *infinitely many* solutions.
$c = 4d + 2$ and $3c - 12d = 6$

For **Extra Practice**, see page 771.

Light Shows in the Sky

The excerpt below appeared in an article in *Earth* in October, 1996.

The interplay of sunlight and ice crystals in the sky can produce spectacular displays of halos and other phenomena. Many people have seen a ring of light around the Sun or Moon—a halo that forms when light passes through ice crystals in the air. But on occasion, halos, arcs and pillars of light can fill the sky, creating bright geometrical patterns....What are these beautiful apparitions? How do they form?....The location of the Sun and the geometry of the ice crystals determine which halos, arcs or pillars form in the sky ....Scientists now know that the ice crystals that play a crucial role in halo formation are not intricate patterns like snowflakes, but rather simple six-sided crystals shaped like plates and columns....there are dozens of possible light paths through the crystals, and these explain the enormous variety of displays. ■

1. As the light passes through a prism and is reflected, its path outlines a triangle. What characteristics might two crystals have so they produce congruent triangular light paths?
2. Would you expect to see more of these light displays where you live or near the South Pole? Why?
3. Which would be more likely to produce one of these displays, ice crystals tumbling at random or crystals falling so that one of their sides remains horizontal as they fall?

4–5 More Congruent Triangles

What YOU'LL LEARN

- To use the AAS Theorem to test triangle congruence, and
- to solve problems by eliminating the possibilities.

Why IT'S IMPORTANT

You can determine if two triangles are congruent by eliminating the possibilities.

You have studied different methods for proving that two triangles are congruent. Is proving that two angles and a nonincluded side of two triangles are congruent, or AAS, sufficient for proving the triangles congruent?

MODELING MATHEMATICS

Angle-Angle-Side

Materials:  patty paper  straightedge

You can draw a triangle given two angles and a nonincluded side.

- Draw a triangle on a piece of patty paper. Label the vertices *A*, *B*, and *C*.

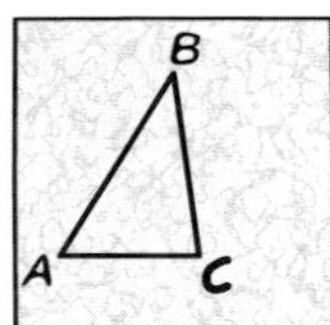

- Copy $\overline{AB}$, $\angle B$, and $\angle C$ on another piece of patty paper.
- Cut out the side and two angles.

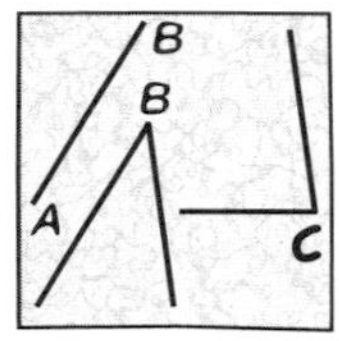

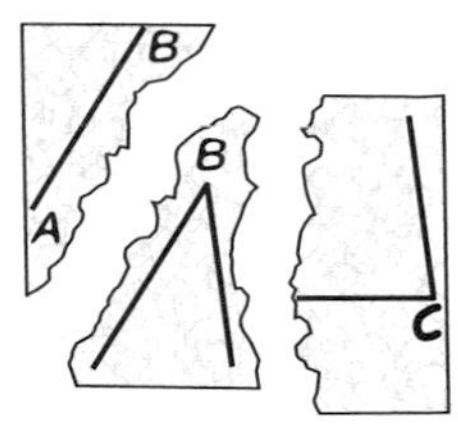

- Place them together to form a triangle in which the side is not the included side of the angles.
- Place another piece of patty paper over the triangle and trace it.
- Place this patty paper over $\triangle ABC$. How does the new triangle compare to $\triangle ABC$?

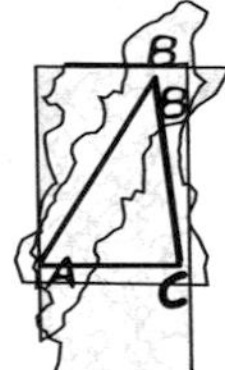

Your Turn

a. Copy $\overline{BC}$, $\angle A$, and $\angle B$. Use these to form a triangle in which the side is not the included side of the angles. How does this triangle compare to $\triangle ABC$?

b. Write a conjecture about two triangles in which two angles and the nonincluded side of one triangle are congruent to two angles and the nonincluded side of the other triangle.

This Modeling Mathematics activity suggests Theorem 4–5.

Theorem 4–5 AAS (Angle-Angle-Side)

If two angles and a nonincluded side of one triangle are congruent to the corresponding two angles and side of a second triangle, the two triangles are congruent.

Proof of Theorem 4–5

Given: $\angle N \cong \angle D$
$\angle G \cong \angle I$
$\overline{AN} \cong \overline{SD}$

Prove: $\triangle ANG \cong \triangle SDI$

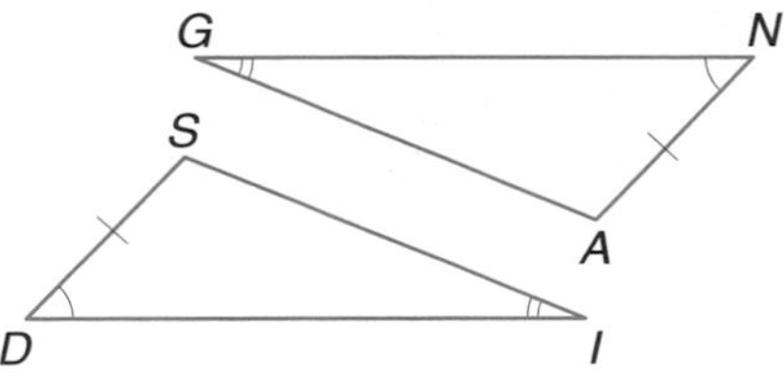

Proof:

Statements	Reasons
1. $\angle N \cong \angle D$ $\angle G \cong \angle I$ $\overline{AN} \cong \overline{SD}$	**1.** Given
2. $\angle A \cong \angle S$	**2.** Third Angle Theorem
3. $\triangle ANG \cong \triangle SDI$	**3.** ASA

Sometimes the triangles we would like to prove congruent are overlapping. Then it is helpful to slide the two triangles apart or to use different colors to distinguish each triangle.

Example 1

Proof

Write a paragraph proof.

Given: $\angle PSU \cong \angle PTR$
$\overline{SU} \cong \overline{TR}$

Prove: $\overline{SP} \cong \overline{TP}$

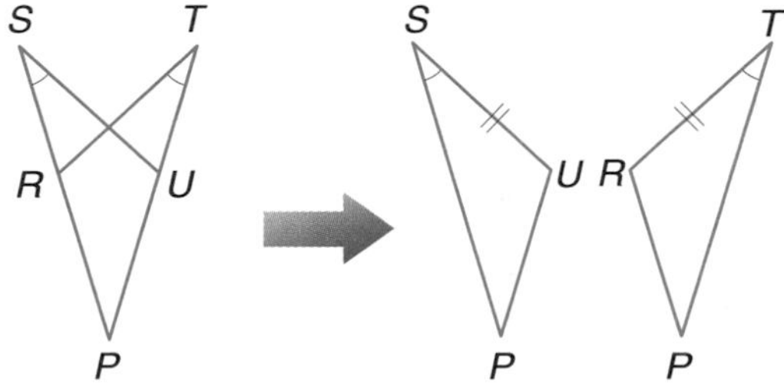

Plan for Proof:

Slide $\triangle SUP$ and $\triangle TRP$ to show two separate triangles. Prove that these two triangles are congruent, and then use CPCTC to prove that $\overline{SP} \cong \overline{TP}$.

Paragraph Proof:

We are given that $\angle PSU \cong \angle PTR$ and $\overline{SU} \cong \overline{TR}$. By the Reflexive Property of Congruent Angles, $\angle P \cong \angle P$. Then $\triangle SUP \cong \triangle TRP$ by AAS and $\overline{SP} \cong \overline{TP}$ by CPCTC.

One important problem-solving strategy is to **eliminate the possibilities**.

Example 2

PROBLEM SOLVING

Eliminate the Possibilities

Some of the measurements of $\triangle ABC$ and $\triangle DEF$ are given at the right. Can you determine if the two triangles are congruent from this information?

Explore We are given three measurements of each triangle. We are asked if we can determine that the triangles are congruent from the three given measurements.

Plan Since $m\angle A = 30$ and $m\angle D = 30$, by definition $\angle A \cong \angle D$. Since $AB = DE$ and $BC = EF$, by definition $\overline{AB} \cong \overline{DE}$ and $\overline{BC} \cong \overline{EF}$. We know five methods to prove that two triangles are congruent: the definition of congruent triangles, SSS, SAS, ASA, and AAS. We can check each of these possibilities.

(continued on the next page)

Solve

Method	Qualifications	Does It Apply?
Definition of Congruent Triangles	All six parts of one triangle must be congruent with all six parts of the other triangle.	No, we only know about three parts of each triangle.
SSS	The three sides of one triangle must be congruent to the three sides of the other triangle.	No, we only know about two of the sides of the triangles.
SAS	Two sides and the included angle of one triangle must be congruent to two sides and the included angle of the other triangle.	No, we know that two sides are congruent, but we do not know about the included angle.
ASA	Two angles and the included side of one triangle must be congruent to two angles and the included side of the other triangle.	No, we only know about one angle of each triangle.
AAS	Two angles and a nonincluded side of one triangle must be congruent to the corresponding two angles and side of the other triangle.	No, we only know about one angle of each triangle.

The triangles cannot be proven to be congruent by just the information we are given and the methods we know at this time.

Examine Use a ruler, protractor, and a compass to try to draw a triangle as described in this example. First draw a segment 4 centimeters long. At one end of the segment, draw an angle that is 30°. Open a compass so that it has a radius of 2.5 centimeters.

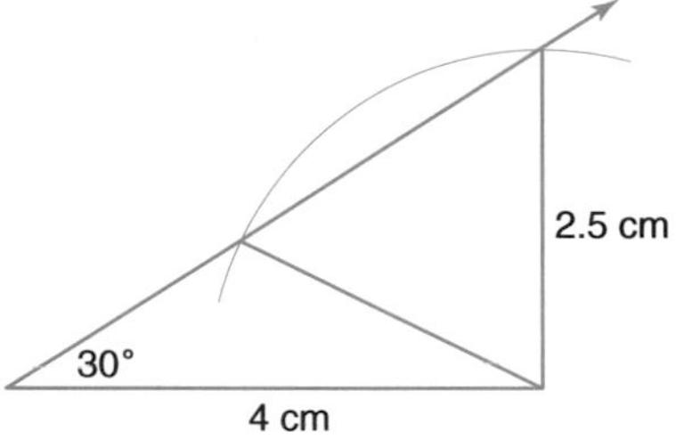

Place the point of the compass on the other end of the 4-centimeter segment and draw an arc to find the other vertex of the triangle. Notice that there are two different vertices that could satisfy the conditions. With the given information, the triangles are not necessarily congruent.

Example 2 shows that more than one triangle can be drawn with two given sides and a nonincluded angle. By this counterexample, we know that side-side-angle (SSA) cannot be used as a proof of congruent triangles.

CHECK FOR UNDERSTANDING

Communicating Mathematics

Study the lesson. Then complete the following.

1. **Explain** why AAS is different than ASA.
2. **Compare and contrast** AAS and SSA.
3. **Describe** how a congruence transformation might help you to write a proof involving overlapping triangles.

4. **List** the methods used to show that two triangles are congruent. For each method, draw a diagram showing the congruent parts needed to show the congruence.

5. **Draw** a horizontal line using a straightedge.
   a. Connect two straws of unequal length with a pin, then position the straws so that the two straws form a triangle with the segment. Anchor one of the straws with tape, and label the vertices of the triangle *A*, *B*, and *C*.
   b. Now swing the straw representing $\overline{BC}$ so that the end of the straw touches the segment again. Label this point *D*.
   c. Is $\triangle ABC$ congruent to $\triangle ABD$?
   d. Is SSA a valid test for determining triangle congruence? Explain your reasoning.

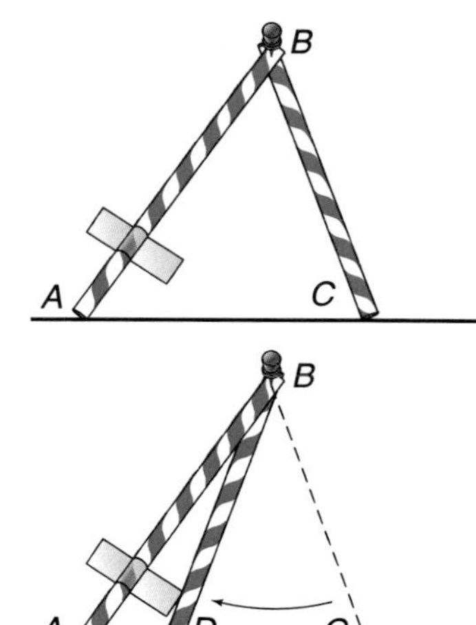

Guided Practice

Refer to the figure at the right for Exercises 6–9.

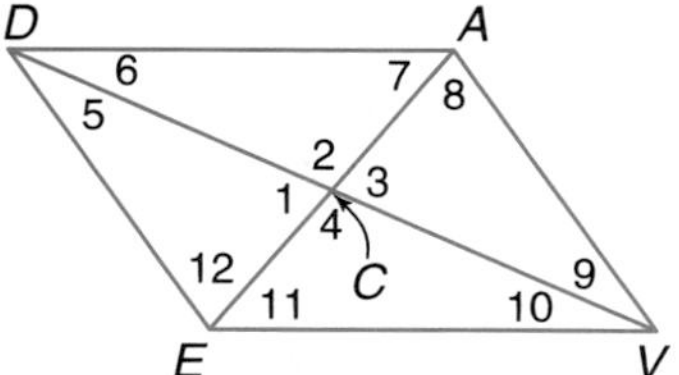

6. Name the included side for $\angle 1$ and $\angle 5$.
7. Name a pair of angles in which $\overline{DE}$ is not included.
8. If $\angle 6 \cong \angle 10$, and $\overline{DC} \cong \overline{VC}$, then $\triangle DCA \cong \triangle$ __?__ by __?__.
9. Given $\angle 7 \cong \angle 11$, $\overline{AD} \cong \overline{EV}$, and $\overline{DC} \cong \overline{VC}$, can $\triangle ADC$ and $\triangle EVC$ be congruent? Explain.

10. The statements and reasons used in a flow proof to prove $\triangle CTJ \cong \triangle HTN$ are listed below. Rearrange the boxes and add arrows to complete the flow proof.

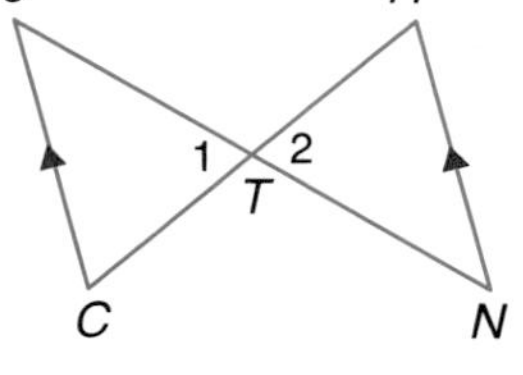

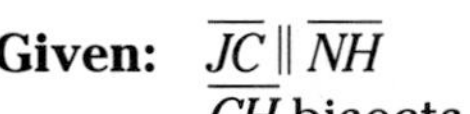

Given: $\overline{JC} \parallel \overline{NH}$
$\overline{CH}$ bisects $\overline{JN}$.

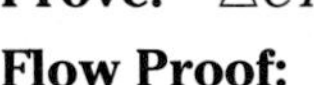

Prove: $\triangle CTJ \cong \triangle HTN$

Flow Proof:

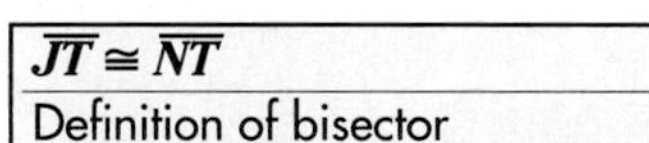
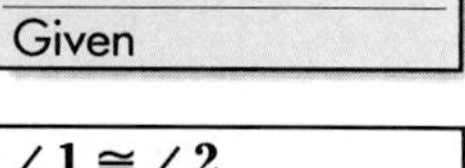
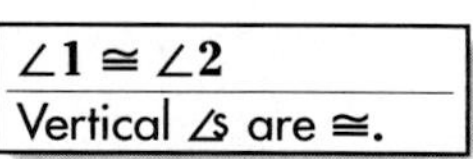

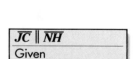

$\overline{JT} \cong \overline{NT}$ Definition of bisector	$\triangle CTJ \cong \triangle HTN$ ASA	$\overline{JC} \parallel \overline{NH}$ Given
$\angle CJT \cong \angle HNT$ Alternate Interior Angle Theorem	$\overline{CH}$ bisects $\overline{JN}$. Given	$\angle 1 \cong \angle 2$ Vertical $\angle$s are $\cong$.

11. Supply the reason for each step in the two-column proof.

Given: $\angle E \cong \angle C$
$\overline{AE} \cong \overline{DC}$

Prove: $\overline{EB} \cong \overline{CB}$

Proof:

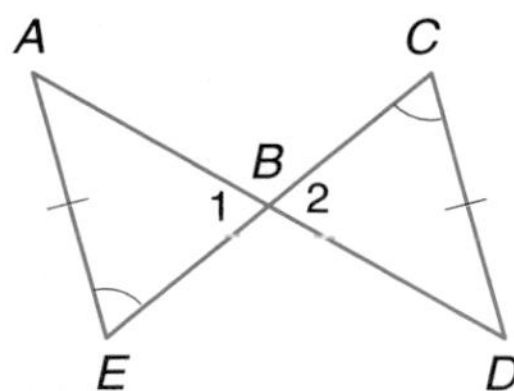

Statements	Reasons
1. $\angle E \cong \angle C$ $\overline{AE} \cong \overline{DC}$	1. __?__
2. $\angle 1 \cong \angle 2$	2. __?__
3. $\triangle EBA \cong \triangle CBD$	3. __?__
4. $\overline{EB} \cong \overline{CB}$	4. __?__

Refer to the figure at the right for Exercises 12–13.

12. **Given:** X is the midpoint of $\overline{RD}$.
$\overline{RM} \parallel \overline{DN}$
Prove: $\overline{MX} \cong \overline{NX}$

13. **Given:** $\overline{NM}$ bisects $\overline{RD}$.
$\angle 7 \cong \angle 8$
Prove: $\overline{MD} \cong \overline{NR}$

14. $\triangle RTE$ is a right triangle with right angle T. $\triangle ANG$ is a right triangle with right angle N. Suppose $m\angle A$ is 26, $m\angle R$ is 14 less than three times $m\angle E$, and $\overline{RE} \cong \overline{AG}$.
   a. Find $m\angle R$ and $m\angle E$.
   b. Are the triangles congruent? Explain your reasoning.

EXERCISES

Practice

Refer to the figure at the right for Exercises 15–26.

15. Name the included side for $\angle 1$ and $\angle 4$.
16. Name the included side for $\angle 7$ and $\angle 8$.
17. Name a nonincluded side for $\angle 5$ and $\angle 6$.
18. Name a nonincluded side for $\angle 9$ and $\angle 10$.
19. $\overline{CT}$ is included between what two angles?
20. $\overline{SR}$ is included between what two angles?
21. In $\triangle FDR$, name a pair of angles so that $\overline{FR}$ is not included.
22. In $\triangle SDT$, name a pair of angles so that $\overline{ST}$ is not included.
23. If $\angle 1 \cong \angle 6$, $\angle 4 \cong \angle 3$, and $\overline{FR} \cong \overline{DS}$, then $\triangle FDR \cong \triangle$ __?__ by __?__.
24. If $\angle 5 \cong \angle 7$, $\angle 6 \cong \angle 8$, and $\overline{DS} \cong \overline{DS}$, then $\triangle TDS \cong \triangle$ __?__ by __?__.
25. If $\angle 4 \cong \angle 9$, what sides would need to be congruent to show $\triangle FDR \cong \triangle CDT$?
26. If $\overline{RS} \cong \overline{TS}$ and $\overline{DR} \cong \overline{DT}$, name a pair of angles that would create an SSA relationship.

27. Justify each step.

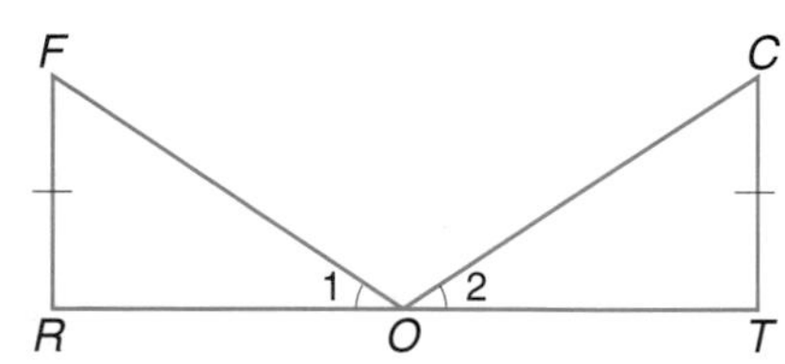

Given: $\angle 1 \cong \angle 2$
$\overline{FR} \perp \overline{RT}$
$\overline{CT} \perp \overline{RT}$
$\overline{FR} \cong \overline{CT}$

Prove: $\overline{RO} \cong \overline{TO}$

Flow Proof:

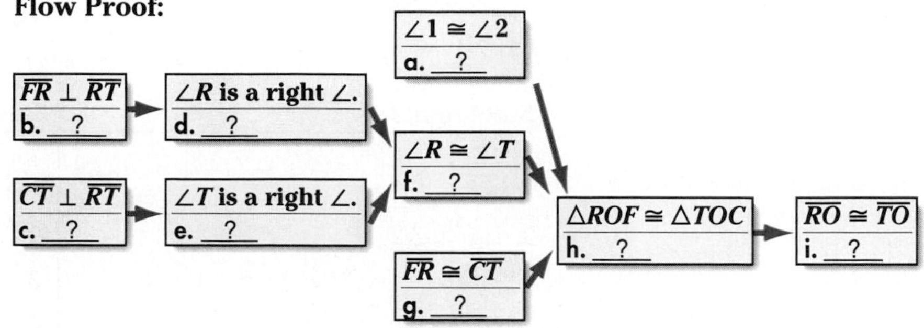

28. Supply the reason for each step in the two-column proof.

Given: $\overline{VZ} \parallel \overline{WX}$

$\overline{VY} \cong \overline{WY}$

$\angle V \cong \angle W$

$\angle X$ is a right angle.

Prove: $\overline{VZ} \cong \overline{WX}$

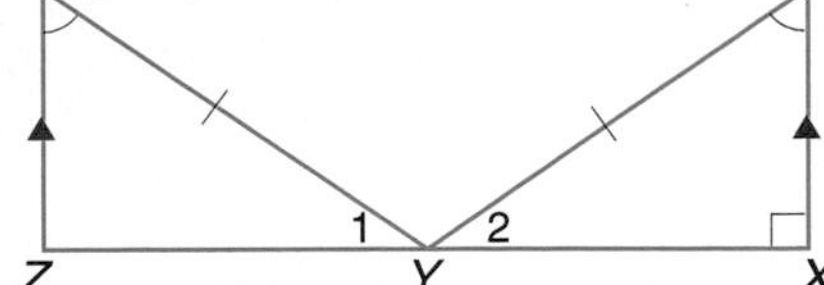

Proof:

Statements	Reasons
1. $\angle X$ is a right angle.	1. ?
2. $\overline{WX} \perp \overline{ZX}$	2. ?
3. $\overline{VZ} \parallel \overline{WX}$	3. ?
4. $\overline{VZ} \perp \overline{ZX}$	4. ?
5. $\angle Z$ is a right angle.	5. ?
6. $\angle X \cong \angle Z$	6. ?
7. $\overline{VY} \cong \overline{WY}$	7. ?
8. $\angle V \cong \angle W$	8. ?
9. $\triangle ZYV \cong \triangle XYW$	9. ?
10. $\overline{VZ} \cong \overline{WX}$	10. ?

29. Write a paragraph proof for the following.

Given: $\angle A \cong \angle D$

$\angle EBC \cong \angle ECB$

$\overline{AE} \cong \overline{DE}$

Prove: $\triangle ABE \cong \triangle DCE$

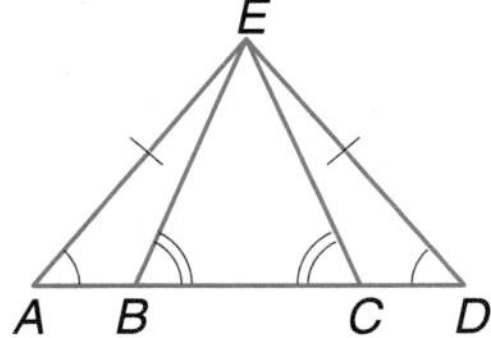

Refer to △*TEN* for Exercises 30–31.

30. Given: $\triangle TEN$ is an isosceles triangle with base $\overline{TN}$.

$\angle 2 \cong \angle 3$

$\angle T \cong \angle N$

Prove: $\triangle TEA \cong \triangle NEC$

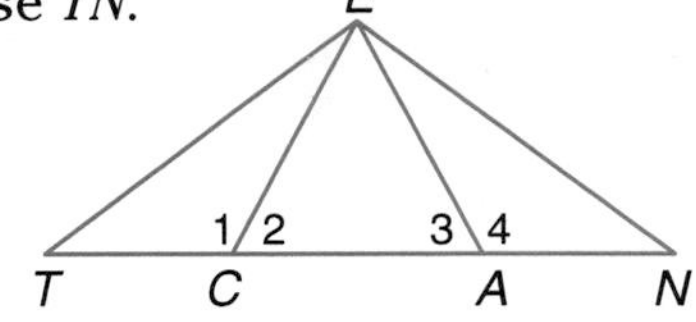

31. Given: $\angle 3 \cong \angle 2$

$\angle T \cong \angle N$

$\overline{TC} \cong \overline{NA}$

Prove: $\triangle TEA \cong \triangle NEC$

Refer to quadrilateral *FLMP* for Exercises 32–33.

32. Given: $\angle F \cong \angle M$

$\angle 1 \cong \angle 2$

Prove: $\overline{FP} \cong \overline{ML}$

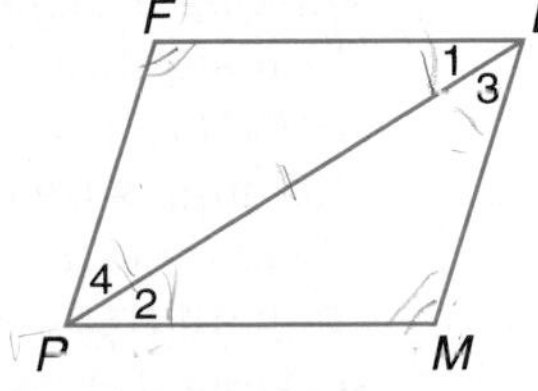

33. Given: $\overline{FP} \parallel \overline{ML}$

$\overline{FL} \parallel \overline{MP}$

Prove: $\overline{PM} \cong \overline{LF}$

Refer to pentagon *GRSTV* for Exercises 34–35.

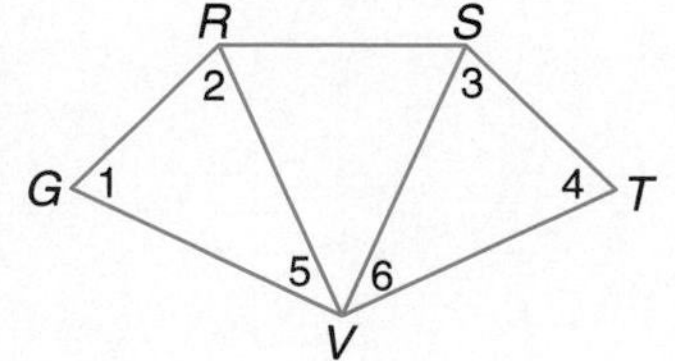

34. **Given:** $\angle 1 \cong \angle 2$
$\angle 3 \cong \angle 4$
$\angle 1 \cong \angle 4$
$\overline{GV} \cong \overline{TV}$

Prove: $\triangle RVS$ is an isosceles triangle.

35. **Given:** $\triangle GVR$ is an isosceles triangle with base $\overline{GR}$.
$\triangle TVS$ is an isosceles triangle with base $\overline{TS}$.
$\overline{GV} \cong \overline{TV}$
$\angle 5 \cong \angle 6$

Prove: $\overline{GR} \cong \overline{TS}$

Refer to quadrilateral *PLXT* for Exercises 36–37.

36. **Given:** $\angle 1 \cong \angle 2$
$\angle 3 \cong \angle 4$

Prove: $\overline{PT} \cong \overline{LX}$

37. **Given:** $\overline{PX} \cong \overline{LT}$
$\triangle PRL$ is an isosceles triangle with base $\overline{PL}$.

Prove: $\triangle TRX$ is an isosceles triangle.

Cabri Geometry

38. Draw an obtuse, scalene triangle. Label it $\triangle ABC$ with $\angle A$ as the obtuse angle and $AC < AB$.
    a. Find the measures of the sides and angles of the triangle.
    b. Draw a circle with center at A and radius equal to AC. The circle intersects $\overline{BC}$ at two points. Label the new point D.
    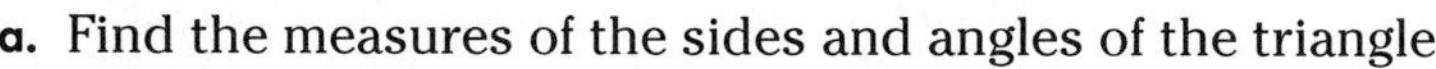
    c. Find BD, AD, $m\angle BDA$, and $m\angle BAD$.
    d. Consider $\triangle ABC$ and $\triangle ABD$. How many sides of $\triangle ABD$ have the same length as a side of $\triangle ABC$? How many angles of $\triangle ABD$ are the same measure as an angle of $\triangle ABC$?
    e. Is $\triangle ABD$ congruent to $\triangle ABC$? What does this say about SSA as a proof of congruent triangles?

Critical Thinking

39. Can two triangles be proved congruent by AAA (Angle-Angle-Angle)? Justify your answer completely.

Applications and Problem Solving

40. **History** It is said that Thales determined the distance from the shore to enemy Greek ships during an early war by sighting the angle to the ship from a point P on the shore, walking a distance to point Q, and then sighting the angle to the ship from that point. He then reproduced the angles on the other side of line PQ and continued these lines until they intersected.

    a. How did he determine the distance to the ship in this way?
    b. Why does this method work?

41. **Broadcasting** In order to stabilize a television antenna so that it remains perpendicular to the ground, three guy wires are attached as shown at the right. If the guy wires are attached to the antenna at the same height from the ground and the measures of $\angle A$, $\angle B$, and $\angle C$ are the same, what can you say about the length of the guy wires? Explain your reasoning.

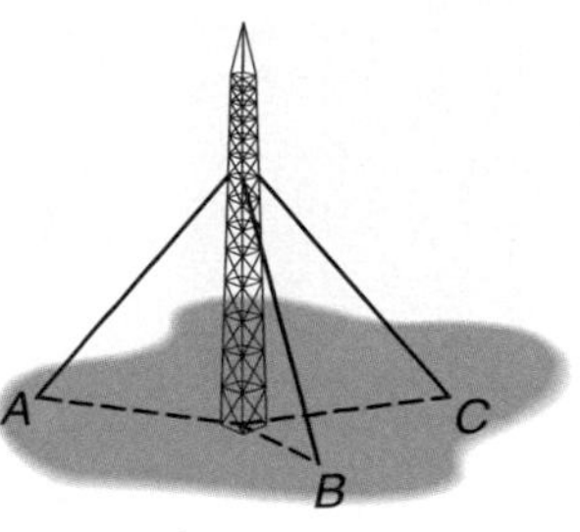

42. **Eliminate the Possibilities** On a recent geology test, Raul was given five different mineral samples to identify using the portion of a mineral characteristics chart shown at the right. Raul made the following observations about the samples.

- Sample C is softer than glass.
- Samples D and E are red and Sample C is brown.
- Samples B and E are harder than glass.

Identify each of the samples.

Mineral	Color	Hardness
Biotite	brown or black	softer than glass
Halite	white	softer than glass
Hematite	red	softer than glass
Feldspar	white, pink, or green	harder than glass
Jasper	red	harder than glass

Mixed Review

43. **Hockey** The width of the opening of a hockey net is 2 meters. A player shoots from a point 9 meters from point A and 8 meters from point B. Another player shoots from a point 8 meters from point A and 9 meters from point B. Which player has a greater angle of possible shots and therefore the better chance of making a score? Explain your reasoning. (Lesson 4–4)

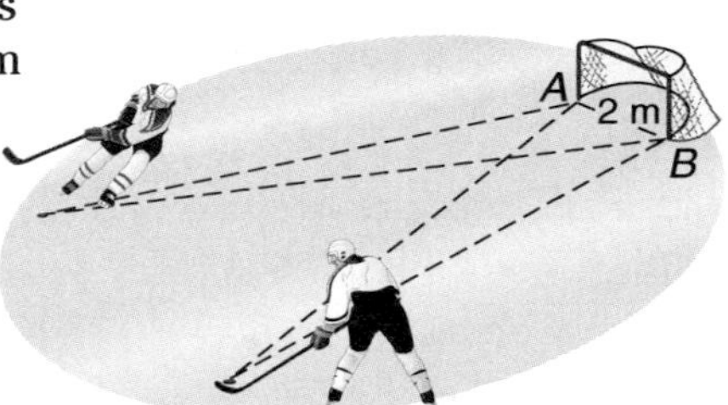

44. If $\triangle HRT \cong \triangle MNP$, complete each statement. (Lesson 4–3)
   a. $\angle R \cong$ _?_ **b.** $\overline{HT} \cong$ _?_ **c.** $\angle P \cong$ _?_

45. Decide whether it is possible to have a triangle with no acute angles. (Lesson 4–2)

46. Draw an acute, scalene triangle. (Lesson 4–1)

47. **Algebra** The slope of line n is 9. What is the slope of any line perpendicular to n? (Lesson 3–3)

48. If $\angle A$ is complementary to $\angle B$, $\angle C$ is complementary to $\angle B$, and $m\angle A = 34$, find $m\angle B$ and $m\angle C$. (Lesson 2–6)

49. $\overrightarrow{QT}$ and $\overrightarrow{QS}$ are opposite rays. Describe $\angle TQS$. (Lesson 1–7)

50. **ACT Practice** Property tax is 7% of the assessed value of a house. How much would the property tax be on a house with an assessed value of $140,000?
   A $980 **B** $2,000 **C** $9,800 **D** $20,000 **E** $98,000

Algebra

51. State whether $18 = rs$ represents an inverse variation or a direct variation. Then find the constant of variation.

52. Which ordered pairs are solutions for the equation $4x - y = 7$?
   a. (2, 6) **b.** (3, 0) **c.** (2, 1) **d.** (0, −7)

For **Extra Practice**, see page 771.

4–6

Analyzing Isosceles Triangles

What YOU'LL LEARN

- To use properties of isosceles and equilateral triangles.

Why IT'S IMPORTANT

You can use the properties of isosceles and equilateral triangles involved in designs, carpentry, and navigation.

Art and Design

Many teens like to collect pennants displaying the names of their school or favorite sports teams. Pennants are designed by artists and usually are in the shape of isosceles triangles.

MODELING MATHEMATICS Isosceles Triangles

Materials:  patty paper  straight edge  compass

Make an isosceles triangle and compare its base angles.

- Draw an acute angle with vertex C on a piece of patty paper.

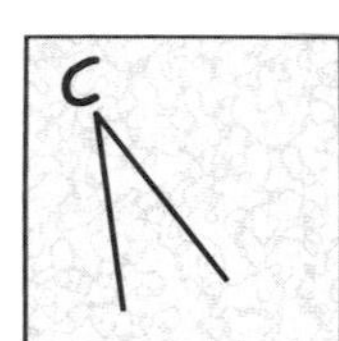

- Mark equal lengths on the sides of $\angle C$. Name the points A and B. Draw $\overline{AB}$. $\triangle ABC$ is isosceles with base $\overline{AB}$.

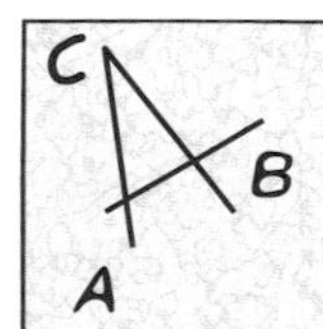

- Fold the patty paper through C so that A coincides with point B. What do you notice about $\angle A$ and $\angle B$?

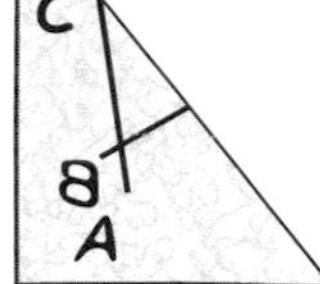

Your Turn

a. Draw an obtuse, isosceles triangle. Compare the base angles.

b. Draw a right, isosceles triangle. Compare the base angles.

c. Use three pieces of patty paper to form an equilateral triangle. Compare the three angles of the triangle.

The Modeling Mathematics activity suggests Theorem 4–6.

Theorem 4–6 Isosceles Triangle Theorem	**If two sides of a triangle are congruent, then the angles opposite those sides are congruent.**

Proof of Theorem 4–6

Given: $\triangle PQR$, $\overline{PQ} \cong \overline{RQ}$

Prove: $\angle P \cong \angle R$

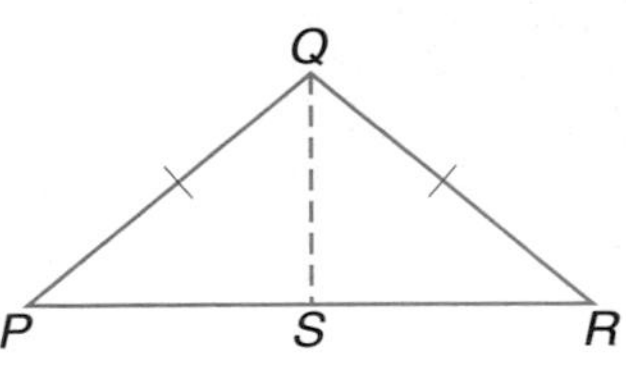

Proof:

Statements	Reasons
1. Let S be the midpoint of $\overline{PR}$.	**1.** Every segment has exactly one midpoint.
2. Draw auxiliary segment $\overline{QS}$.	**2.** Through any two points there is one line.
3. $\overline{PS} \cong \overline{RS}$	**3.** Definition of midpoint
4. $\overline{QS} \cong \overline{QS}$	**4.** Congruence of segments is reflexive.
5. $\overline{PQ} \cong \overline{RQ}$	**5.** Given
6. $\triangle PQS \cong \triangle RQS$	**6.** SSS
7. $\angle P \cong \angle R$	**7.** CPCTC

You can use Theorem 4–6 and algebra to find missing measures of an isosceles triangle.

Example 1

Algebra

In isosceles $\triangle ISO$ with base $\overline{SO}$, $m\angle S = 5x - 18$ and $m\angle O = 2x + 21$. Find the measure of each angle of the triangle.

$m\angle S = m\angle O$ *Isosceles Triangle Theorem*

$5x - 18 = 2x + 21$

$3x = 39$

$x = 13$

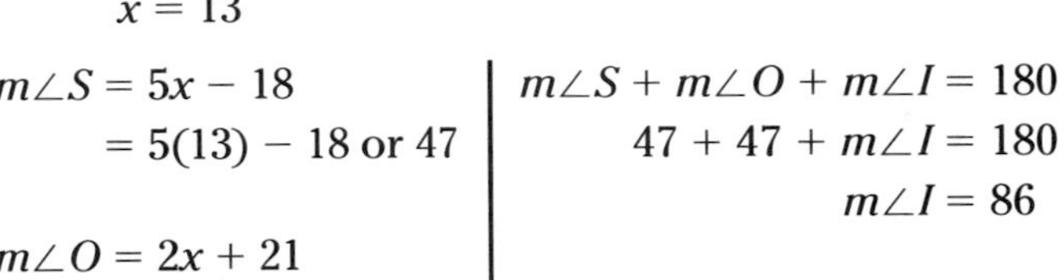

$m\angle S = 5x - 18$
$= 5(13) - 18$ or 47

$m\angle O = 2x + 21$
$= 2(13) + 21$ or 47

$m\angle S + m\angle O + m\angle I = 180$
$47 + 47 + m\angle I = 180$
$m\angle I = 86$

The measures of $\angle S$, $\angle O$, and $\angle I$ are 47, 47, and 86, respectively.

The converse of the Isosceles Triangle Theorem is also true.

Theorem 4–7	**If two angles of a triangle are congruent, then the sides opposite those angles are congruent.**

Example 2

Write a plan for a two-column proof for Theorem 4–7.

Draw an auxiliary ray that is the bisector of $\angle ABC$ and let D be the point where the bisector intersects $\overline{AC}$. Show that $\triangle ABD \cong \triangle CBD$ by AAS using the given information and the auxiliary ray. Then, $\overline{AB} \cong \overline{CB}$ by CPCTC.

You will be asked to prove this theorem in Exercise 13.

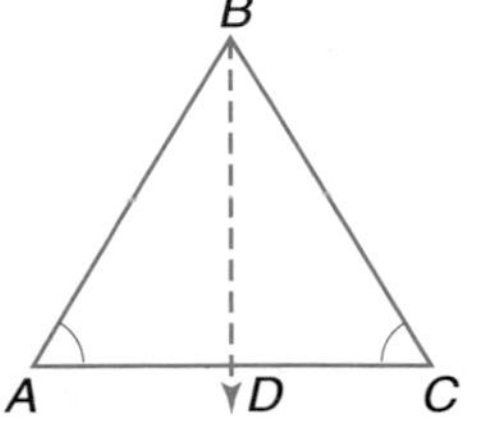

The Isosceles Triangle Theorem leads us to some interesting corollaries. *You will be asked to prove Corollaries 4–3 and 4–4 in Exercises 42 and 43, respectively.*

Corollary 4–3	**A triangle is equilateral if and only if it is equiangular.**
Corollary 4–4	**Each angle of an equilateral triangle measures 60°.**

Recall that corollaries can be used as reasons in proofs just like theorems.

Example 3 Proof

Write a paragraph proof.

Given: $\overline{AB} \cong \overline{EB}$
$\angle DEC \cong \angle B$

Prove: $\triangle ABE$ is equilateral.

Proof:

Since $\overline{AB} \cong \overline{EB}$, $\angle A \cong \angle AEB$ by the Isosceles Triangle Theorem. Since vertical angles are congruent, $\angle AEB \cong \angle DEC$. But $\angle DEC \cong \angle B$, so $\angle AEB \cong \angle B$ because congruence of angles is transitive. Also, $\angle A \cong \angle AEB \cong \angle B$ for the same reason. By definition of equiangular triangles, $\triangle AEB$ is equiangular. However, a triangle that is equiangular is also equilateral, so $\triangle AEB$ is an equilateral triangle.

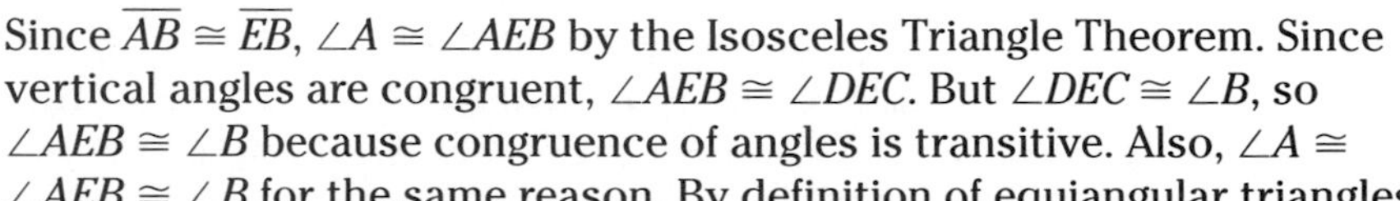

CHECK FOR UNDERSTANDING

Communicating Mathematics

Study the lesson. Then complete the following.

1. **Combine** Theorems 4–6 and 4–7 to form one *if and only if* statement.
2. **Name** the congruent sides and angles of isosceles $\triangle RST$ with base $\overline{RS}$.
3. $\triangle ABC$ with right angle C is flipped over $\overleftrightarrow{BC}$ to form $\triangle A'BC$ as shown at the right.
   a. Since a flip is a congruence transformation, what is true about $\overline{AB}$ and $\overline{A'B}$?
   b. What is true about $\angle A$ and $\angle A'$?
   c. Identify $\triangle ABA'$.

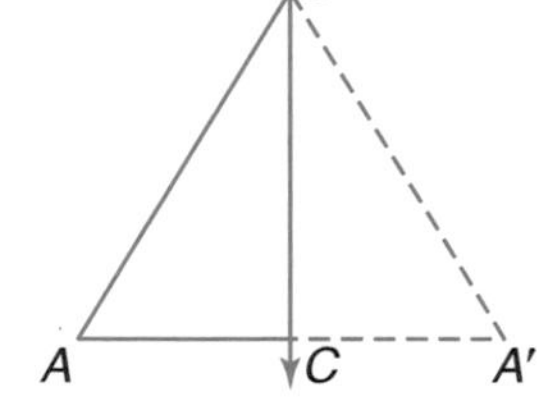

4. Use a compass and a straightedge to construct an equilateral triangle, an isosceles triangle that is not equilateral, and a scalene triangle. Cut out each triangle.
   a. How many ways can an equilateral triangle be folded to form two congruent triangles?
   b. How many ways can an isosceles triangle that is not equilateral be folded to form two congruent triangles?
   c. How many ways can a scalene triangle be folded to form two congruent triangles?

Guided Practice

Refer to the figure at the right for Exercises 5–7.

5. If $\overline{NL} \cong \overline{SL}$, name two congruent angles.
6. If $\angle 1 \cong \angle 4$, name two congruent segments.
7. If $\angle 9 \cong \angle 10$, name two congruent segments.
8. Draw $\triangle ABC$ with point D on $\overline{BC}$ such that $\overline{AD} \perp \overline{BC}$, but $\overline{BD}$ and $\overline{DC}$ are not congruent.

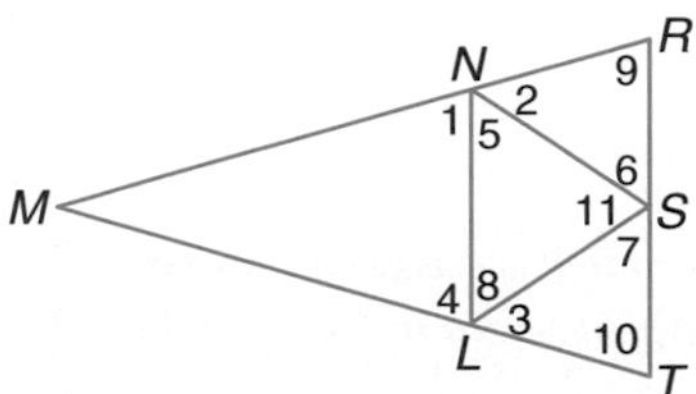

Find the value of x.

9. 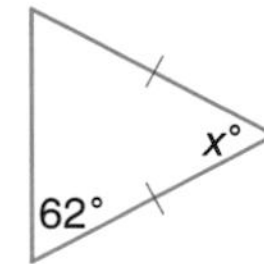

10.

11. $\triangle ABC$ has vertices $A(2, 5)$, $B(5, 2)$, and $C(2, -1)$.
 a. Use the distance formula to show that $\triangle ABC$ is isosceles.
 b. Name the pair of congruent angles.

Proof

12. **Given:** $\overline{HT} \cong \overline{HA}$

 Prove: $\angle 1 \cong \angle 3$

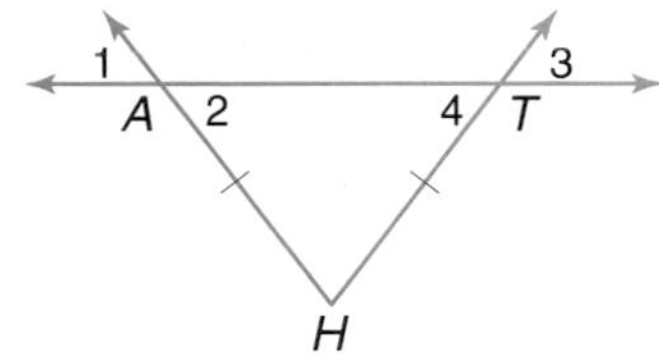

13. Write a two-column proof for Theorem 4–7. If two angles of a triangle are congruent, then the sides opposite those angles are congruent.

14. **Construction** A frame for a roof is in the form of an isosceles triangle. If the measure of the vertex angle is 120, find the measures of the other two angles.

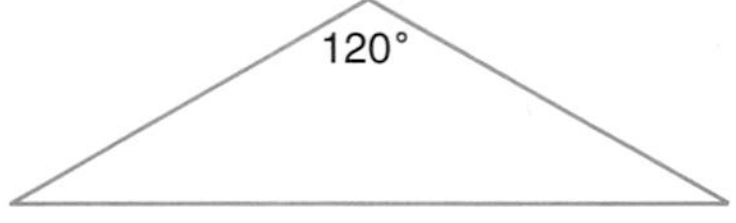

EXERCISES

Practice

Refer to the figure at the right for Exercises 15–23.

15. If $\overline{FL} \cong \overline{FD}$, name two congruent angles.
16. If $\overline{CF} \cong \overline{KF}$, name two congruent angles.
17. If $\overline{FA} \cong \overline{FB}$, name two congruent angles.
18. If $\overline{OF} \cong \overline{EF}$, name two congruent angles.
19. If $\overline{FO} \cong \overline{OE}$, name two congruent angles.
20. If $\angle 10 \cong \angle DFL$, name two congruent segments.
21. If $\angle EOF \cong \angle EFO$, name two congruent segments.
22. If $\angle 6 \cong \angle LFD$, name two congruent segments.
23. If $\angle 12 \cong \angle 8$, name two congruent segments.

For Exercise 24–26, draw $\triangle MTN$ with point X on $\overline{TN}$ such that the following conditions are satisfied.

24. $\overline{XM}$ bisects $\angle TMN$, but $\overline{MT}$ is not congruent to $\overline{MN}$.
25. $\overline{TX} \cong \overline{XN}$, but $\overline{XM}$ does not bisect $\angle NMT$.
26. $\overline{TX} \cong \overline{XN}$, but $\overline{MX}$ is not perpendicular to $\overline{TN}$.

Find the value of x.

27. 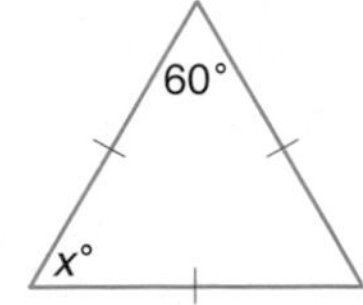

28. 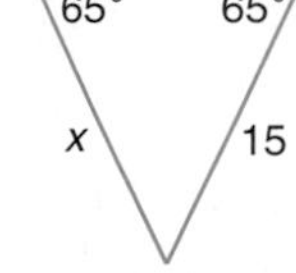

29.

Find the value of x.

30.

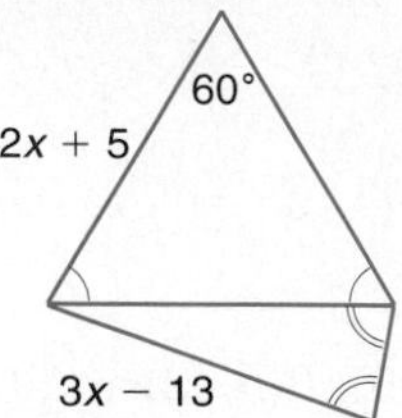

31.

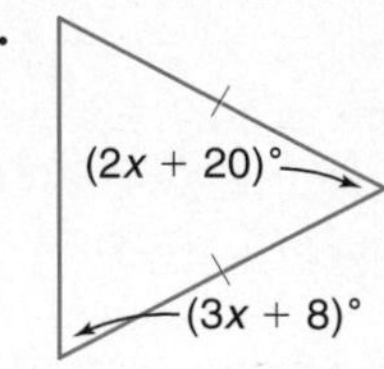

32.

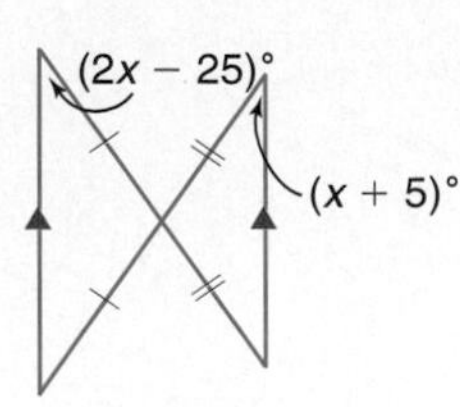

33. If $\triangle DEF$ has vertices $D(-4, -3)$, $E(-2, -1)$, and $F(0, -3)$, use the distance formula and slopes of the congruent sides to show that $\triangle DEF$ is a right isosceles triangle.

34. In $\triangle ABD$, $\overline{AB} \cong \overline{BD}$, $m\angle A$ is 12 less than 3 times a number, and $m\angle D$ is 13 more than twice the same number. Find $m\angle B$.

35. Find each angle measure if $m\angle 1 = 30$.

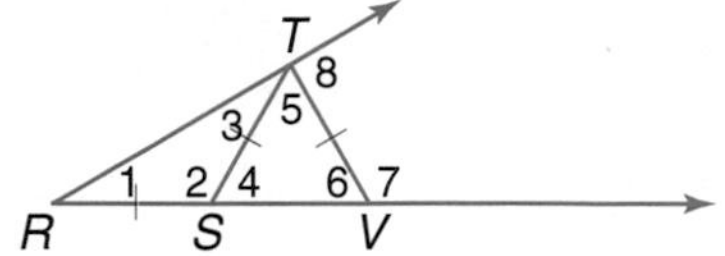

Write a proof for each of the following.

36. Given: $\angle 3 \cong \angle 4$

Prove: $\overline{MA} \cong \overline{MC}$

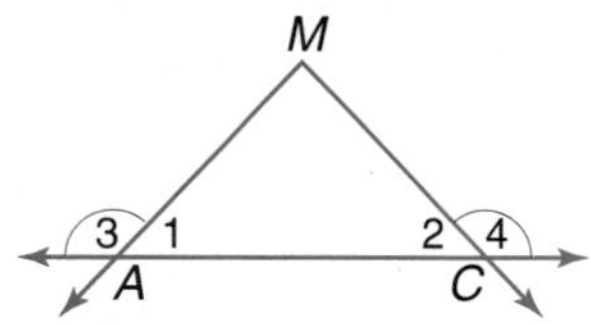

37. Given: $\angle 5 \cong \angle 6$

$\overline{FR} \cong \overline{GS}$

Prove: $\angle 4 \cong \angle 3$

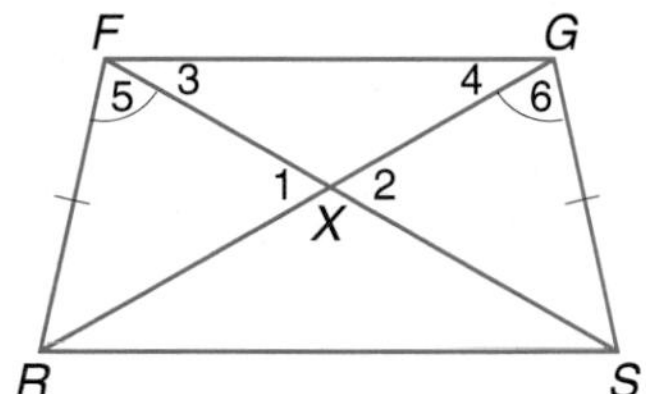

38. Given: $\angle 1 \cong \angle 4$

$\overline{NA} \cong \overline{TC}$

Prove: $\angle 3 \cong \angle 2$

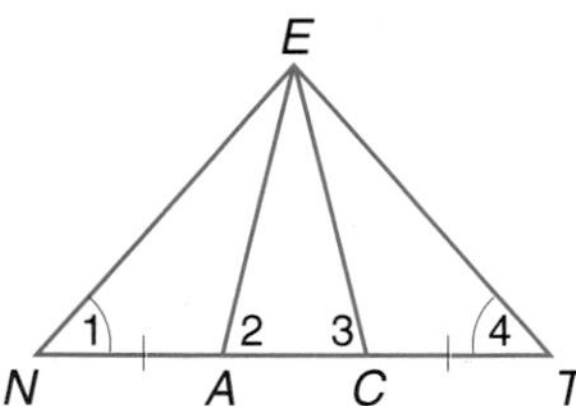

39. Given: $\triangle CAN$ is an isosceles triangle with vertex $\angle N$.

$\overline{CA} \parallel \overline{BE}$

Prove: $\triangle NEB$ is an isosceles triangle.

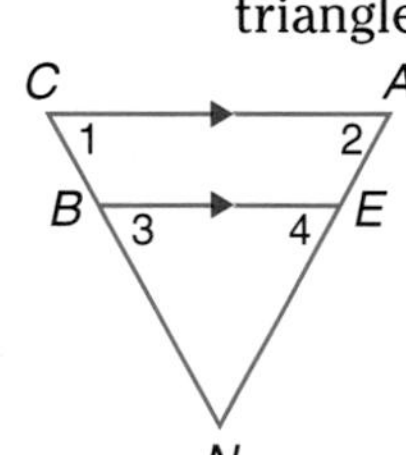

40. Given: $\overline{PH}$ bisects $\angle YHX$.

$\overline{HP} \perp \overline{YX}$

Prove: $\triangle YHX$ is an isosceles triangle.

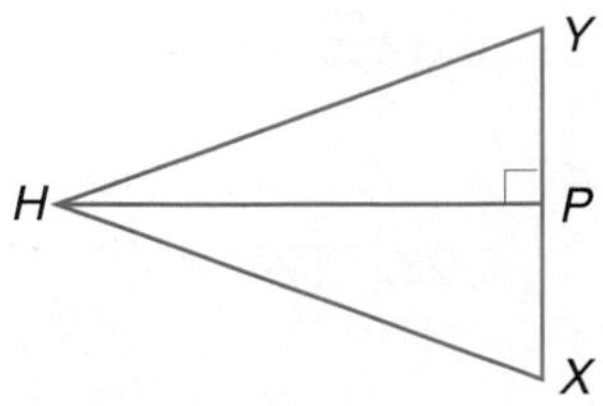

41. Given: $\triangle IOE$ is an isosceles triangle with base $\overline{OE}$.

$\overline{AO}$ bisects $\angle IOE$.

$\overline{AE}$ bisects $\angle IEO$.

Prove: $\triangle AEO$ is an isosceles triangle.

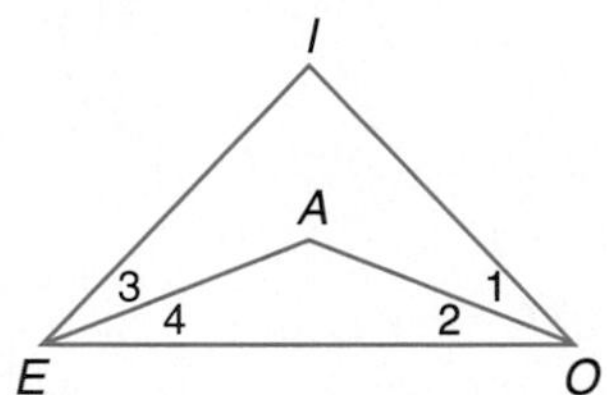

42. A triangle is equilateral if and only if it is equiangular. (Corollary 4–3) (*Hint*: Since the statement contains if and only if, the proof must show an equilateral triangle is equiangular and that an equiangular triangle is equilateral.)

43. Each angle of an equilateral triangle measures 60°. (Corollary 4–4)

Programming

44. Use the graphing calculator program to find the measures of the base angles of an isosceles triangle that has a vertex angle of the given measure.

a. 26 **b.** 120

c. 78 **d.** 101

```
PROGRAM:ISOTRI
: Input "VERTEX ANGLE =", A
: (180-A)/2→B
: Disp "BASE ANGLE =", B
: Stop
```

Critical Thinking

45. Draw an isosceles triangle ABC with vertex angle at A. Find the midpoints of each side. Label the midpoint of $\overline{AB}$ point D, the midpoint of $\overline{BC}$ point E, and the midpoint of $\overline{AC}$ point F. Draw $\triangle DEF$.

a. Name a pair of congruent triangles. Explain your reasoning.

b. Name three isosceles triangles. Explain your reasoning.

Applications and Problem Solving

46. Design Tiled patterns formed by repeating figures to fill a plane without gaps or overlaps are called *tessellations*. Tessellations are found in woven rugs, pottery, and flooring. An example of a tessellation using octagons and squares is given at the right.

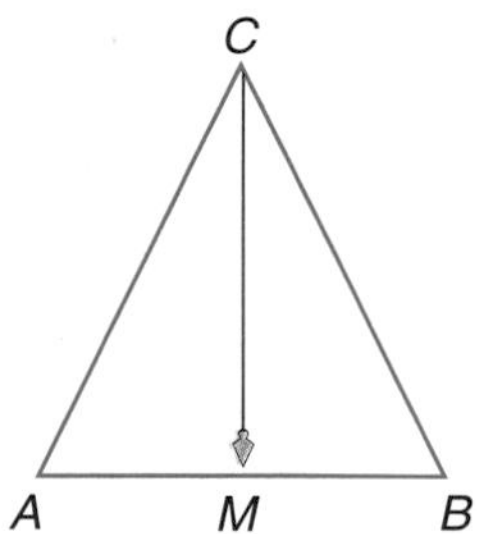

a. Draw a tessellation using only isosceles triangles.

b. Draw a tessellation using squares and equilateral triangles.

c. Draw a tessellation using other types of figures.

47. Carpentry Before the invention of the bubble level, carpenters used a device called a *plumb level* to verify that a surface was level. This level consisted of a frame in the shape of an isosceles triangle with the midpoint of the base (point M in the figure at the right) marked. A plumb line was suspended from the vertex angle. To use this instrument, the carpenter would hold it upright with the base resting on the surface to be leveled. If the surface were level, over what point on the base do you think the plumb line would hang? Explain your reasoning.

48. Navigation Sau-Lim is the captain of a ship and he uses an instrument called a *pelorus* to measure the angle between the ship's path and the line from the ship to a lighthouse. Sau-Lim finds the distance that the ship travels and the change in the measure of the angle with the lighthouse as the ship sails. When the angle with the lighthouse is twice that of the original angle, Sau-Lim knows that the ship is as far from the lighthouse as the ship has traveled since the lighthouse was first sighted. Why?

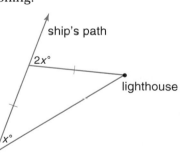

Mixed Review

49. Explain the difference between SSA and SAS. (Lesson 4–5)
50. **Eliminate the Possibilities** Four friends have birthdays in January, February, August, and September. Amy was not born in the winter. Kiana celebrates her birthday during the summer vacation. Timothy's birthday is the month after Pablo's. Give the month of each person's birthday. (Lesson 4–5)
51. Suppose $\triangle CDE \cong \triangle PQR$. List all the pairs of corresponding parts. (Lesson 4–3)
52. The measures of two interior angles of a triangle are 54 and 79. What is the measure of the exterior angle opposite these angles? (Lesson 4–2)
53. **Algebra** The legs of an isosceles triangle are $(6x - 6)$ units and $(x + 9)$ units long. Find the length of the legs. (Lesson 4–1)
54. Find the slope of the line passing through $A(7, -3)$ and $B(6, -1)$. (Lesson 3–3)
55. **ACT Practice** A repair technician charges $80 for the first thirty minutes of each house call plus $2 for each additional minute. The repair technician charged a total of $170 for a job. How many minutes did the repair technician work?
    A 45 **B** 55 **C** 75 **D** 85 **E** 95
56. Write an if-then statement for *Every whole number is an integer.* (Lesson 2–2)
57. Find the length and midpoint of the segment with endpoints $C(7, 4)$ and $D(-3, 8)$. (Lesson 1–4)

INTEGRATION **Algebra**

For **Extra Practice**, see page 772.

58. State the domain and range of the relation $\{(7, -2), (4, 0), (-19, -1), (5, 5)\}$.
59. Find the x- and y-intercepts of the graph of $9x - 2y = 18$.

WORKING ON THE In•ves•ti•ga•tion

Refer to the Investigation on pages 176–177.

Because each person has unique fingerprints, they are often used to identify people. Even though each fingerprint is different, there are patterns that can be observed in fingerprints.

1. Rub a pencil with soft lead on a piece of paper.
2. Roll your thumb through the pencil mark a few times to coat the thumb with dust.
3. Press the thumb on a piece of clear tape to create a thumbprint.
4. Place the tape on a recording sheet and note your name.
5. Create a thumbprint for each person in your group.
6. Look for similarities and differences among the thumbprints. Are any of the natural patterns described on pages 176–177 displayed in any of the thumbprints?
7. Research fingerprinting techniques. What patterns are used?

Add the results of your work to your Investigation Folder.

CHAPTER 4 HIGHLIGHTS

Chapter Review For additional lesson-by-lesson review, visit: **www.geometry.glencoe.com**

VOCABULARY

After completing this chapter, you should be able to define each term, property, or phrase and give an example or two of each.

Geometry

acute triangle (p. 180)
auxiliary line (p. 189)
base (p. 182)
base angles (p. 182)
congruence transformation (p. 197)
congruent triangles (p. 196)
corollary (p. 192)
equiangular triangle (p. 181)
equilateral triangle (p. 181)
exterior angle (pp. 188, 190)
flow proof (p. 191)
included angle (pp. 204, 207)
included side (pp. 205, 207)
isosceles triangle (p. 181)
legs (p.182)
obtuse triangle (p. 180)
polygon (p. 180)
remote interior angles (pp. 188, 190)
right triangle (p. 180)
scalene triangle (p. 181)
sides (p. 180)
triangle (p. 180)
vertex angle (p. 182)
vertices (p. 180)

Problem Solving

eliminate the possibilities (p. 215)

UNDERSTANDING AND USING THE VOCABULARY

State whether each sentence is *true* or *false*. If false, replace the underlined word(s) to make a true statement.

1. A triangle is a polygon with three sides.
2. The side opposite the right angle of a triangle is called the leg.
3. An isosceles triangle has two vertex angles.
4. A triangle in which no two sides are congruent is called a scalene triangle.
5. AAS refers to two angles and their included side.
6. The endpoints of the sides of a polygon are called vertices.
7. SSS, SAS, ASA, and AAS are ways to prove that two triangles are congruent.
8. A triangle with two acute angles is called acute.
9. A paragraph proof uses boxes to show the steps of a proof and arrows to show the order of the steps.
10. An equiangular triangle is defined as a triangle with three congruent sides.
11. The sum of the measures of two remote interior angles is equal to the measure of the exterior angle.
12. An equilateral triangle is also an isosceles triangle.
13. In an isosceles triangle, the side opposite the vertex angle is the hypotenuse.
14. An auxiliary line is a statement that can be easily proved using a theorem.
15. A triangle with angle measures 95, 35, and 50 is an obtuse triangle.

SKILLS AND CONCEPTS

OBJECTIVES AND EXAMPLES

Upon completing this chapter, you should be able to:

- identify parts of triangles and classify triangles by their parts (Lesson 4–1)

Types of Triangles

Classification by Angles	
acute	three acute angles
obtuse	one obtuse angle
right	one right angle
equiangular	three congruent angles
Classification by Sides	
scalene	no two sides congruent
isosceles	at least two sides congruent
equilateral	three sides congruent

REVIEW EXERCISES

Use these exercises to review and prepare for the chapter test.

In the figure, ∠*STU* and ∠*SVU* are right angles, and the congruent parts are indicated.

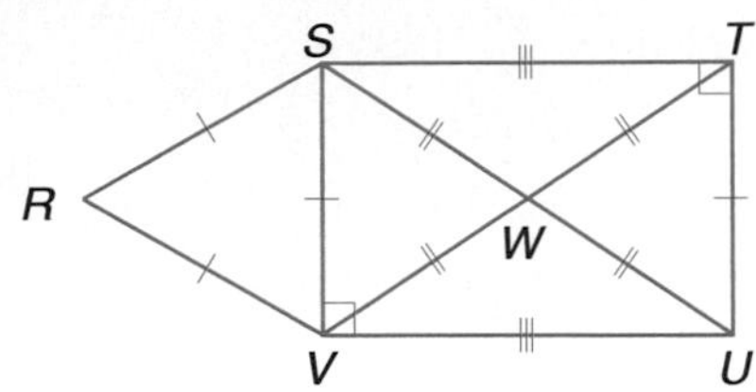

16. Identify the equilateral triangle(s).
17. Identify the right triangle(s).
18. Identify the acute triangle(s).
19. Identify the obtuse triangle(s).
20. Identify the triangles that are isosceles but not equilateral.
21. Name the side opposite $\angle R$.
22. Name the angle opposite $\overline{RV}$.

- apply the Angle Sum Theorem and Exterior Angle Theorem (Lesson 4–2)

Find the values of x and y.

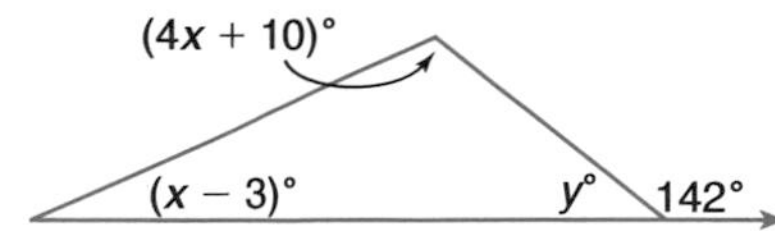

By the Exterior Angle Theorem,

$$(x - 3) + (4x + 10) = 142$$
$$5x = 135$$
$$x = 27$$

By the Angle Sum Theorem,

$$(x - 3) + (4x + 10) + y = 180$$
$$(27 - 3) + [4(27) + 10] + y = 180$$
$$142 + y = 180$$
$$y = 38$$

In the figure, $\overline{RK} \parallel \overline{SL}$ and $\overline{RS} \perp \overline{SL}$. Find the measure of each angle.

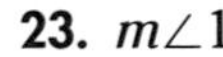

23. $m\angle 1$
24. $m\angle 2$
25. $m\angle 3$
26. $m\angle 4$
27. $m\angle 5$

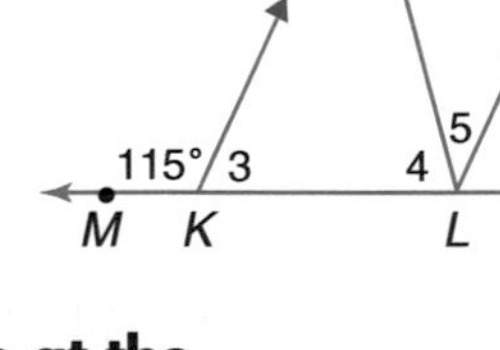

Refer to the figure at the right to find each value.

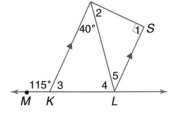

28. x
29. r
30. w
31. y
32. z
33. s
34. v
35. t

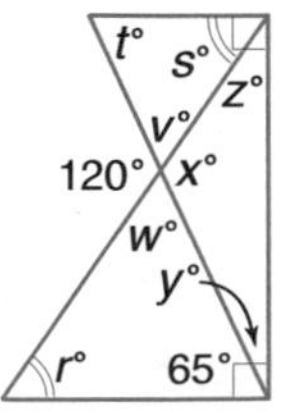

OBJECTIVES AND EXAMPLES

- name and label corresponding parts of congruent triangles (Lesson 4–3)

Definition of Congruent Triangles

Two triangles are congruent if and only if their corresponding parts are congruent.

REVIEW EXERCISES

Draw $\triangle MNO$ and $\triangle RST$. Label the corresponding parts if $\triangle MNO \cong \triangle RST$. Use the figures to complete each statement.

36. $\angle T \cong$ __?__
37. $\overline{MO} \cong$ __?__
38. $\overline{SR} \cong$ __?__
39. $\angle TSR \cong$ __?__
40. $\overline{TR} \cong$ __?__
41. $\angle NOM \cong$ __?__

- use SSS, SAS, and ASA Postulates to test for triangle congruence (Lesson 4–4)

Given: $\overline{LN}$ and $\overline{OP}$ bisect each other at M.

Prove: $\angle O \cong \angle P$

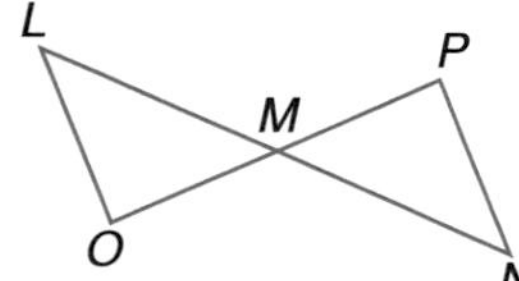

Proof:

Statements	Reasons
1. $\overline{LN}$ and $\overline{OP}$ bisect each other at M.	1. Given
2. $\overline{LM} \cong \overline{MN}$, $\overline{PM} \cong \overline{MO}$	2. Definition of bisector
3. $\angle LMO \cong \angle NMP$	3. Vertical ∠s are ≅.
4. $\triangle LMO \cong \triangle NMP$	4. SAS
5. $\angle O \cong \angle P$	5. CPCTC

Write a flow proof.

42. **Given:** $\overline{EF} \cong \overline{GH}$

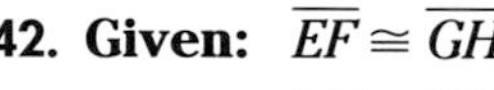

$\overline{EH} \cong \overline{GF}$

Prove: $\triangle EFH \cong \triangle GHF$

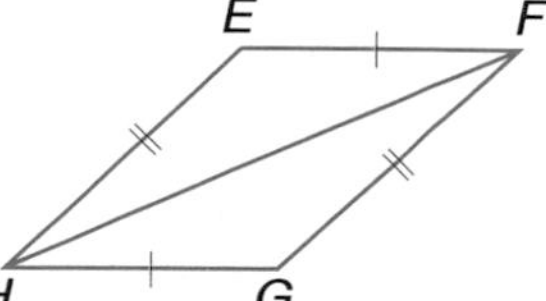

Write a two-column proof.

43. **Given:** $\overline{AM} \parallel \overline{CR}$

B is the midpoint of $\overline{AR}$.

Prove: $\overline{AM} \cong \overline{RC}$

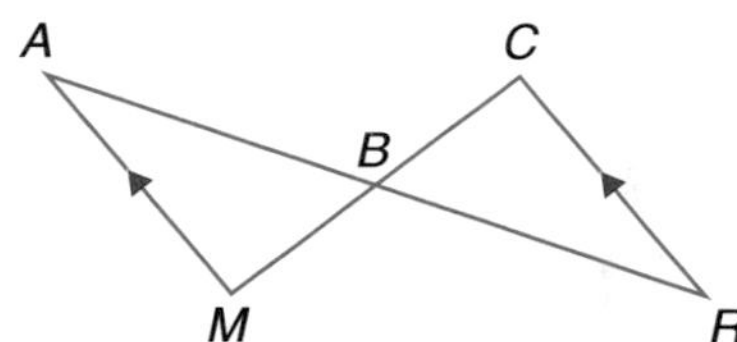

- use AAS Theorem to test for triangle congruence (Lesson 4–5)

AAS

If two angles and a nonincluded side of one triangle are congruent to the corresponding two angles and side of a second triangle, the two triangles are congruent.

Write a two-column proof.

44. **Given:** $\overline{BC} \cong \overline{DC}$

$\angle A \cong \angle E$

$\angle 1 \cong \angle 2$

Prove: $\overline{AB} \cong \overline{ED}$

45. **Given:** $\overline{AC} \cong \overline{EC}$

$\angle 1 \cong \angle 2$

$\overline{BC} \cong \overline{DC}$

Prove: $\angle B \cong \angle D$

OBJECTIVES AND EXAMPLES

- use the properties of isosceles and equilateral triangles (Lesson 4–6)

In isosceles triangle XYZ, $\angle Z$ is the vertex angle. If $m\angle X = 4a + 5$ and $m\angle Y = 2a + 27$, find the measure of each angle of the triangle.

Since $\angle XYZ$ is isosceles and $\angle Z$ is the vertex angle, $m\angle X = m\angle Y$.

$$4a + 5 = 2a + 27 \qquad m\angle X = 4a + 5$$
$$2a = 22 \qquad = 4(11) + 5$$
$$a = 11 \qquad = 49$$

$$m\angle X + m\angle Y + m\angle Z = 180$$
$$49 + 49 + m\angle Z = 180$$
$$m\angle Z = 82$$

The measures of the angles are 49, 49, and 82.

REVIEW EXERCISES

Find the value of x.

46.

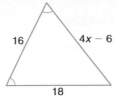

16
4x − 6
18

47.

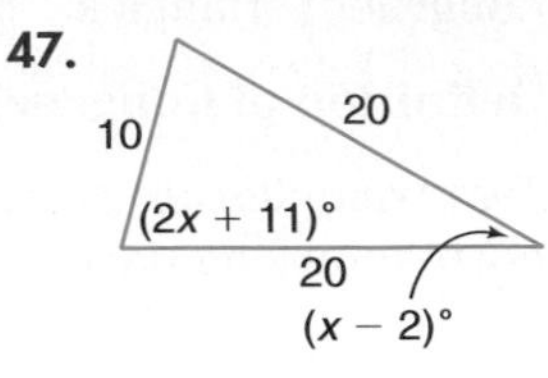

Write a proof.

48. Given: $\angle 2 \cong \angle 1$
$\angle 4 \cong \angle 3$

Prove: $\overline{AM} \cong \overline{AO}$

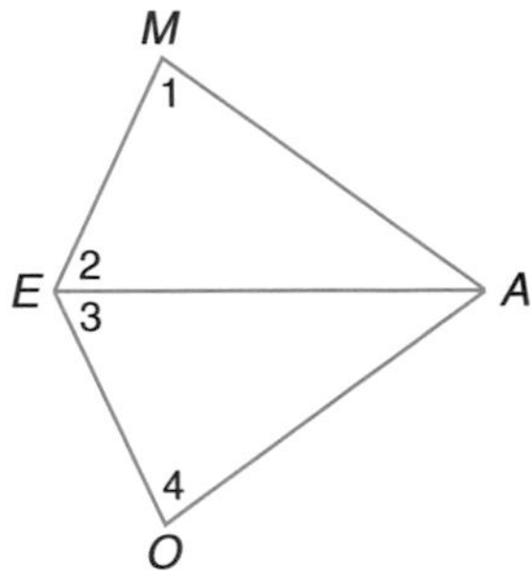

APPLICATIONS AND PROBLEM SOLVING

49. Sports The sail for a sailboat resembles a right triangle. If the angle at the top of the sail measures 54°, what is the measure of the acute angle at the bottom? (Lesson 4–2)

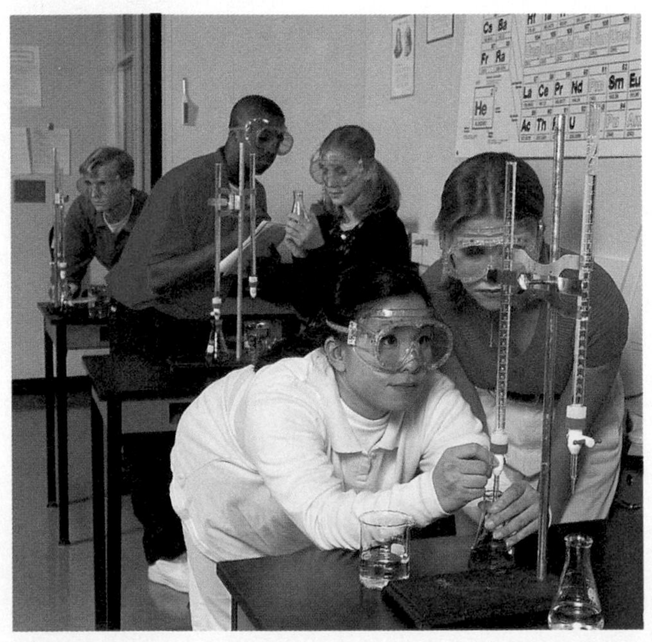

50. Eliminate the Possibilities In chemistry lab, Ebony was given four different metal samples to identify. They were gallium, a metal that melts when held in the hand; mercury, which is liquid at room temperature; lithium, which will float in water; and calcium, which bubbles slowly when placed in water. When Ebony held Sample 1, it was a liquid. Sample 2 was a solid and sank to the bottom when dropped in a beaker of water. Bubbling action occurred when Ebony dropped Sample 4 in the water. Identify each sample. (Lesson 4–5)

51. Hang Gliding The sail of a certain type of hang glider consists of two congruent isosceles triangles joined along a keel so that a 90° angle is formed at the nose of the sail. In order to construct such a sail, what must be the measure of $\angle BCD$? (Lesson 4–6)

A
keel
B
D
C

A practice test for Chapter 4 is available on page 796.

ALTERNATIVE ASSESSMENT

COOPERATIVE LEARNING PROJECT

Three-Dimensional Design In this project, you will apply your knowledge of congruent triangles to construct a variety of geometric sculptures.

Follow these steps to explore geometric sculptures.

- You will need several pieces of heavy construction paper and a roll of transparent tape.
- Each group member will cut out ten congruent equilateral triangles. (Sides measuring 4 inches are a convenient size.)
- See how many different solids you can construct using various numbers of triangles. Be creative!
- For each solid you construct, record the number of triangles you used, the number of edges, and the number of sides.
- Discuss the following questions with your group. Are there any numerical relationships between the number of triangles, edges, and sides?

Follow these steps to explore more geometric sculptures.

- Cut out ten right triangles, with sides 3 inches, 4 inches, and 5 inches.
- Construct a variety of solids out of these triangles. Be creative!
- For each solid you construct, record the number of triangles, the number of edges, and the number of sides.
- Discuss the following questions with your group. Are there any numerical relationships between the number of triangles, edges, and sides? Was it harder or easier to construct solids from the nonequilateral triangles than from the equilateral triangles? Why?

THINKING CRITICALLY

Imagine intersecting a sphere with three planes, each of which passes through the center of the sphere. Note that each forms a circle on the surface of the sphere. Adjust the planes so that the three circles intersect to form a three-sided figure on the surface of the sphere. Is it possible to create a two-sided figure on the surface using two circles?

PORTFOLIO

Travel around your school or neighborhood to find examples of triangles. Make a particular effort to find objects that contain congruent triangles. Compile photographs or drawings. Organize the images by triangle type; obtuse, right, or acute. Place your collection in your portfolio.

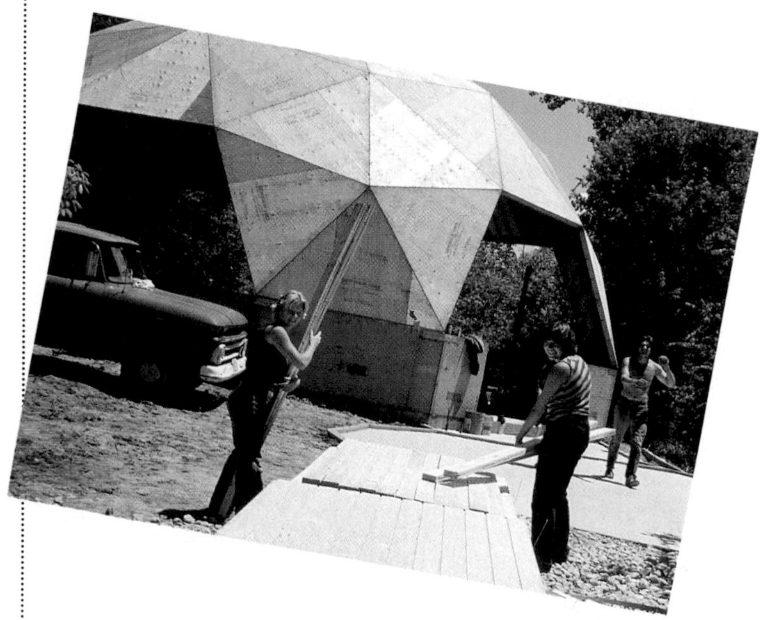

STANDARDIZED TEST PRACTICE

CHAPTERS 1–4

Section One: MULTIPLE CHOICE

There are eight multiple-choice questions in this section. After working each problem, write the letter of the correct answer on your paper.

1. What is the slope of a line that passes through the points at (3, −2) and (−6, 4)?

A. $-\frac{3}{2}$ **C.** $\frac{3}{2}$

B. $-\frac{2}{3}$ **D.** $\frac{2}{3}$

2. If $0 < x < \frac{1}{2}$, which of the following could not be be x?

A. $\frac{1}{4}$ **C.** $\frac{5}{8}$

B. $\frac{3}{10}$ **D.** $\frac{7}{16}$

3. In the figure below, if the three equilateral triangles have a common vertex, then find the value of $a + b + c$.

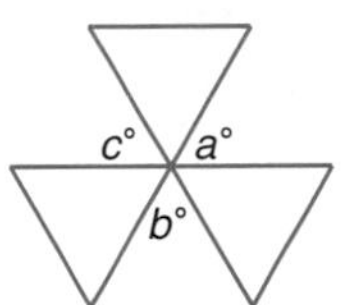

A. 120 **C.** 180

B. 90 **D.** 360

4. In the figure below, if $\ell_1 \parallel \ell_2$, $\ell_2 \parallel \ell_3$, and $\ell_1 \perp \ell_4$, which of the following statements must be true?

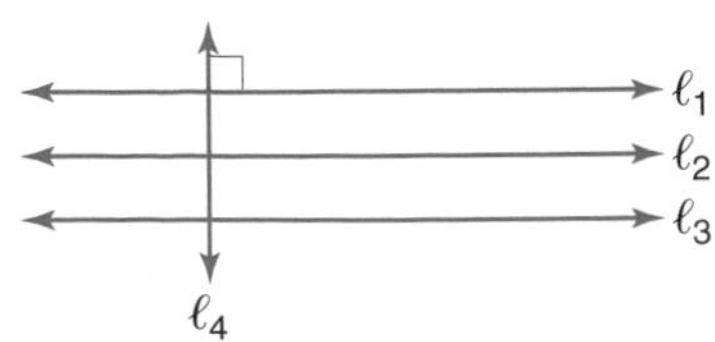

I. $\ell_1 \parallel \ell_3$

II. $\ell_2 \perp \ell_4$

III. $\ell_3 \perp \ell_4$

A. I only **C.** I, II, and III

B. II and III only **D.** none

5. 0.08 is the ratio of 8 to which number?

A. 1000 **C.** 10

B. 100 **D.** $\frac{1}{100}$

6. If $\frac{6}{n}$ is an odd integer, which of the following could be a value of n?

A. $\frac{6}{5}$ **C.** $\frac{3}{4}$

B. $\frac{1}{3}$ **D.** $\frac{4}{3}$

7. In the figure below, find x.

A. 60

B. 120

C. 45

D. 30

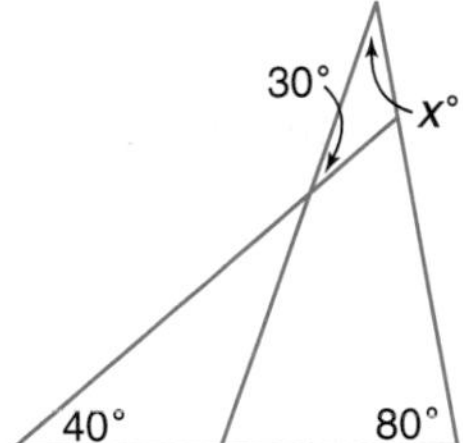

8. In the figure below, which of the following must be equal to 180?

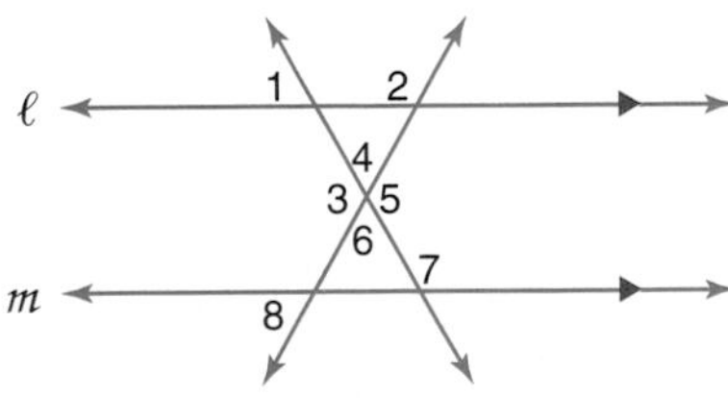

I. $m\angle 3 + m\angle 5$

II. $m\angle 4 + m\angle 6$

III. $m\angle 1 + m\angle 7$

IV. $m\angle 2 + m\angle 8$

V. $m\angle 7 + m\angle 8$

A. I and II only **C.** V only

B. III and IV only **D.** I, II, III, and IV only

Test-Taking Tip On multiple-choice problems that ask you to find the value of a variable, you can use a strategy called *working backward.* Replace the variable with each answer choice and see if the statement is true. Try the other choices, as needed, to find the correct answer.

Section Two: SHORT ANSWER

This section contains seven questions for which you will provide short answers. Write your answer on your paper.

9. What is the slope of a line parallel to $5x - y = 10$?

10. The product of 4, 5, and 6 is equal to twice the sum of 10 and what number?

11. If $\frac{x}{2} + \frac{x}{3} + \frac{x}{6} = kx$, then find k if $x \neq 0$.

12. Find $10x - 5$ if $10x - 4 = 8$.

13. In the figure below, if $\triangle ABC$ is equilateral, what is the ratio of AC to CD?

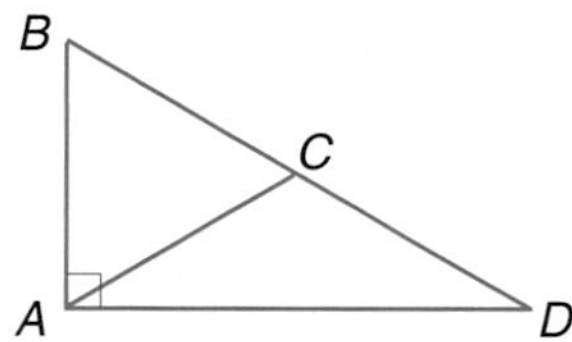

14. If $x = 1$, $y = -1$, and $z = -2$, then find $\frac{x^2y}{(x - z)^2}$.

15. In the figure below, lines r and s are parallel. List three angles all having the same measure.

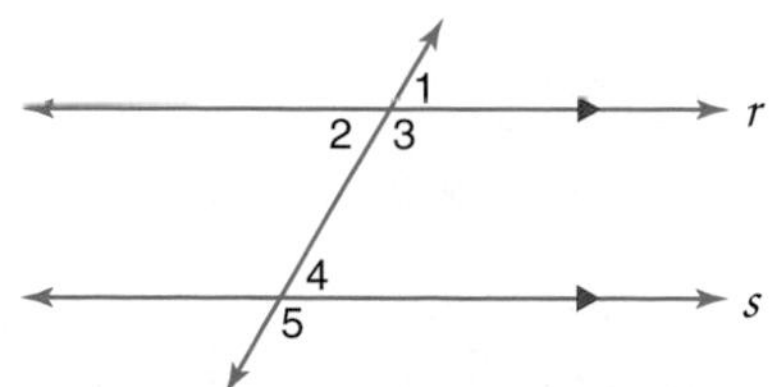

Section Three: COMPARISON

This section contains five comparison problems that involve comparing two quantities, one in column A and one in column B. In certain questions, information related to one or both quantities is centered above them. All variables used represent real numbers.

Compare quantities A and B below.

- Write A if quantity A is greater.
- Write B if quantity B is greater.
- Write C if the two quantities are equal.
- Write D if there is not enough information to determine the relationship.

Column A	Column B
A jacket that was priced at \$36.50 is sold at a 30% discount.	
16. the price of the coat after the discount	\$25.50
$x \neq 0$	
17. $3x^2$	$(3x)^2$
$\frac{3}{5} = \frac{x}{20}$; $\frac{4}{8} = \frac{y}{24}$	
18. x	y
19. $2(9 + 6) - 4(10 \div 2)$	$2 \times 9 + 6 - 4 \times 10 \div 2$

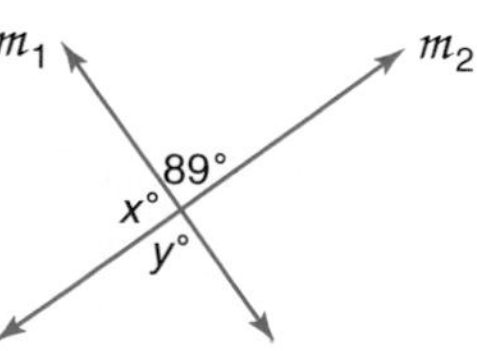

20. x — y

Test Practice For additional test practice questions, visit: **www.geometry.glencoe.com**

CHAPTER

5 Applying Congruent Triangles

What You'll Learn

In this chapter, you will:

- identify and use the special segments in triangles,
- prove right triangles congruent, and
- recognize and apply relationships between the sides and angles in a triangle.

Why It's Important

Real World

Art The Alhambra, a palace built between 1248 and 1254 in Granada, Spain, has many triangular patterns based upon congruent triangles. The pattern of tiles in the upper right was designed by Dutch artist M.C. Escher, who was inspired by the patterns of tiles in the Alhambra. You can see that Escher's design is also based upon congruent triangles. Escher is famous for his tessellations. Triangles and other polygons can be used to produce tessellations. In order for a regular polygon to tessellate the plane, the measure of an interior angle of the polygon must divide evenly into 360°. In this chapter, you will learn more about congruent triangles and the relationships of angle measures and side lengths in triangles.

PREREQUISITE SKILLS

To be successful in this chapter, you'll need to understand these concepts and be able to apply them. Refer to the lesson in parentheses if you need more review before beginning the chapter.

Find the midpoint of a segment. (Lesson 1–5)

B is the midpoint of $\overline{AC}$. For each pair of points given, find the coordinates of the third point.

1. $A(0, 0)$, $C(6, 2)$ **2.** $A(-1, 1)$, $B(3, 5)$ **3.** $B(2, -4)$, $C(-5, 1)$

Find the slopes of lines. (Lesson 3–3)

Find the slope of the line passing through the given points.

4. $A(-2, -2)$, $B(2, 2)$ **5.** $X(0, 0)$, $Z(-7, 9)$

6. $N(3, -4)$, $O(4, -3)$ **7.** $R(-8, 5)$, $W(1, 2)$

Use the Law of Detachment in deductive reasoning. (Lesson 2–3)

Determine if a valid conclusion can be reached from the two true statements using the Law of Detachment. If a valid conclusion is possible, state it. If a valid conclusion does not follow, write *no conclusion*.

8. (1) If you are a skier, then you like snow.
(2) You like snow.

9. (1) If the three sides of one triangle are congruent to the three sides of a second triangle, then the triangles are congruent.
(2) ΔABC and ΔXYZ are congruent.

10. (1) The sum of the measures of a triangle is 180°.
(2) The figure is a triangle.

Focus On

READING SKILLS

In this chapter, you'll be learning about ***inequalities*** for the measures of angles and sides in triangles. Remember that an inequality is a mathematical sentence that contains $\leq$, $<$, $\geq$, or $>$. When the sides of a triangle are not equal, you can write certain inequalities that are always true for the angles of the triangle. You can also write inequalities about the relationship between the angles or sides of two triangles that are not congruent.

5-1 Special Segments in Triangles

What **YOU'LL LEARN**

- To identify and use medians, altitudes, angle bisectors, and perpendicular bisectors in a triangle.

Why **IT'S IMPORTANT**

You can use the special segments in triangles to solve problems involving engineering, sports, and physics.

Triangles have four types of special segments. You can use paper folding to model some of these segments.

Special Segments of a Triangle

Materials:  patty paper  ruler

- Draw a triangle like $\triangle ABC$ on a piece of patty paper. Fold the paper so that point A falls on point B and make a crease through $\overline{AB}$. Unfold the paper and label the point where the crease intersects $\overline{AB}$ as M and the point where it intersects $\overline{AC}$ as P. $\overleftrightarrow{MP}$ is a **perpendicular bisector**.

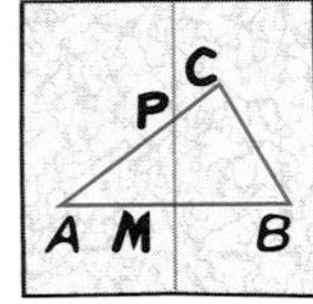

- Fold the paper and make a crease from point C to M. $\overline{CM}$ is a **median**.

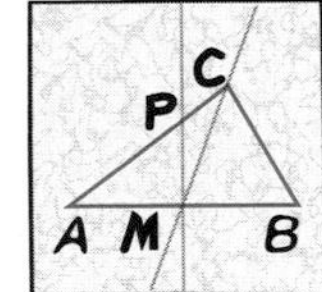

- Next, fold the paper so that B lies on $\overline{AB}$ and the fold passes through point C. Make a crease. Label the point where the crease intersects $\overline{AB}$ as T. $\overline{CT}$ is an **altitude**.

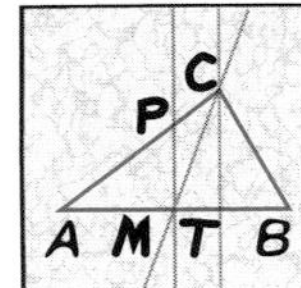

Your Turn

Write a list of observations about each of the three special segments you have folded.

The first special segment you found in the Modeling Mathematics activity was the *perpendicular bisector* of the side of a triangle. A line or line segment that passes through the midpoint of a side of a triangle *and* is perpendicular to that side is the perpendicular bisector of the side of the triangle. The red lines in the figure at the right are the perpendicular bisectors of the sides of $\triangle GHI$. Since there are three sides in every triangle, every triangle has three perpendicular bisectors.

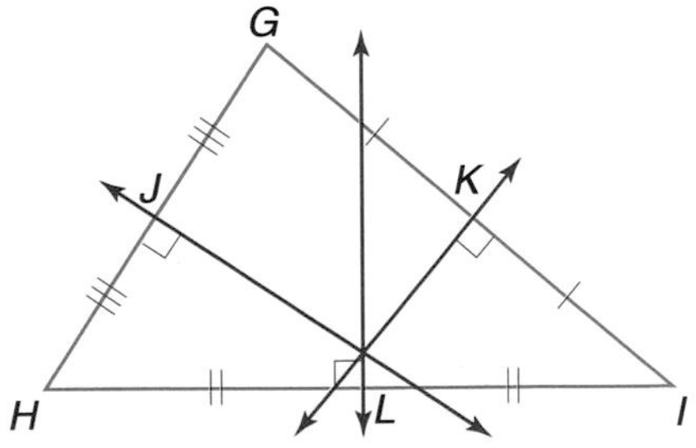

Perpendicular bisectors of segments have some special properties. These properties are described in Theorems 5–1 and 5–2. *You will be asked to prove these theorems in Exercises 35 and 36, respectively.*

Theorem 5–1	**Any point on the perpendicular bisector of a segment is equidistant from the endpoints of the segment.**
Theorem 5–2	**Any point equidistant from the endpoints of a segment lies on the perpendicular bisector of the segment.**

You also folded a median and an altitude of $\triangle ABC$. A *median* is a segment that connects a vertex of a triangle to the midpoint of the side opposite that vertex. For $\triangle JKL$, $\overline{JM}$, $\overline{KN}$, and $\overline{LD}$ are the medians. Every triangle has three medians.

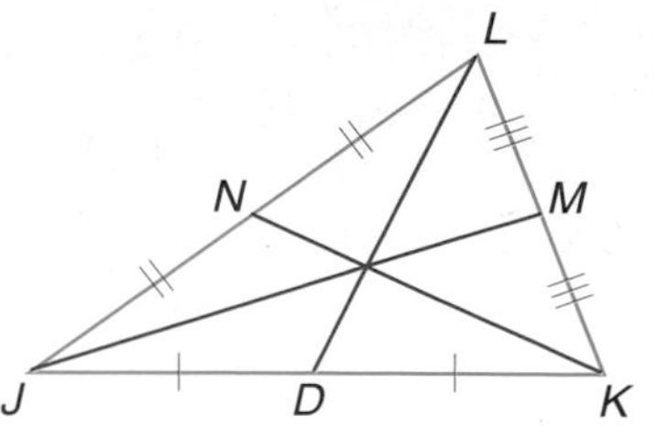

An *altitude* has one endpoint at a vertex of a triangle and the other on the line that contains the side opposite that vertex so that the segment is perpendicular to this line. For acute triangle PQR, $\overline{PS}$, $\overline{QT}$, and $\overline{RV}$ are the altitudes. Every triangle has three altitudes.

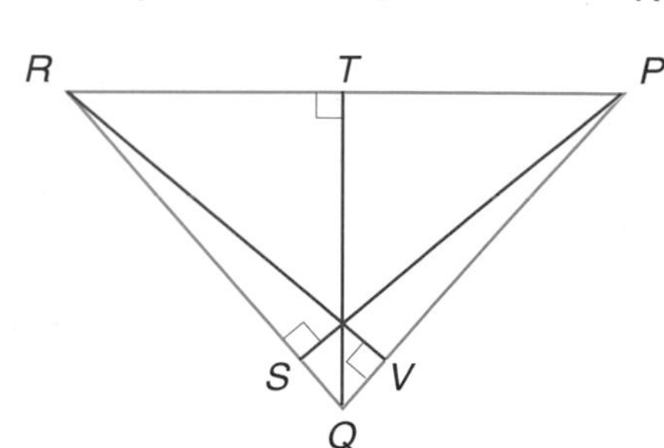

The diagrams below show the altitudes for a right triangle and an obtuse triangle. The legs of a right triangle are two of the altitudes. For an obtuse triangle, two of the altitudes lie outside the triangle.

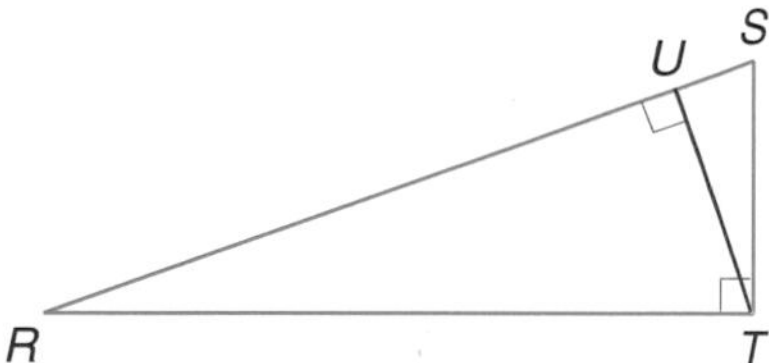

In right triangle TRS, $\overline{ST}$ is the altitude from S to $\overline{RT}$, $\overline{RT}$ is the altitude from R to $\overline{TS}$, and $\overline{TU}$ is the altitude from T to $\overline{RS}$.

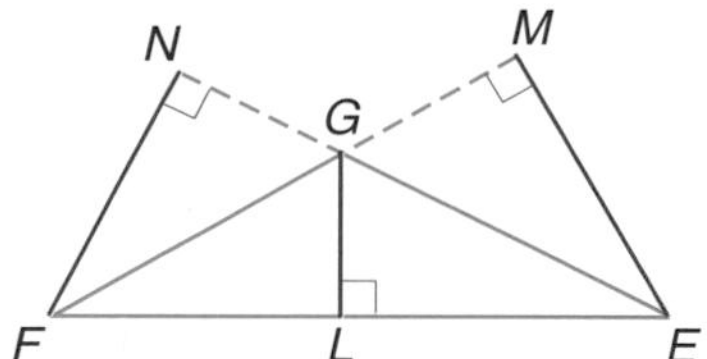

In obtuse triangle EFG, $\overline{FN}$, $\overline{GL}$, and $\overline{EM}$ are the altitudes. Notice that $\overline{FN}$ and $\overline{EM}$ are outside $\triangle EFG$.

Example 1

Algebra

$\triangle ABC$ has vertices $A(-3, 10)$, $B(9, 2)$, and $C(9, 15)$.

a. Determine the coordinates of point P on $\overline{AB}$ so that $\overline{CP}$ is a median of $\triangle ABC$.

b. Determine if $\overline{CP}$ is an altitude of $\triangle ABC$.

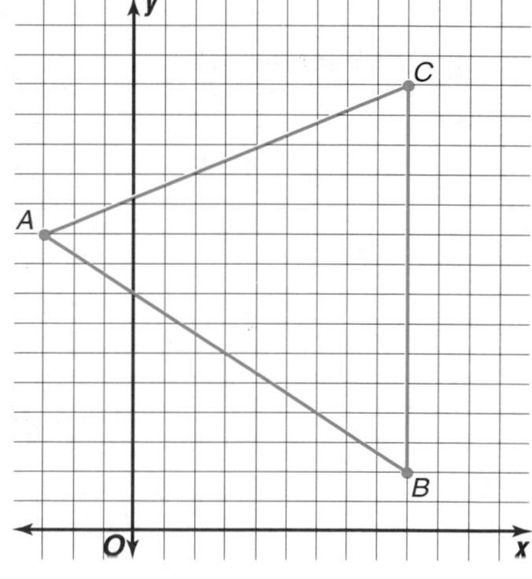

LOOK BACK

You can review finding the midpoint in a coordinate plane in Lesson 1-5.

a. By the definition of median, $\overline{CP}$ is a median of $\triangle ABC$ if P is the midpoint of $\overline{AB}$.

$$P\left(\frac{x_1 + x_2}{2}, \frac{y_1 + y_2}{2}\right) = \left(\frac{-3 + 9}{2}, \frac{10 + 2}{2}\right)$$ *Substitute $A(-3, 10)$ for (x_1, y_1) and $B(9, 2)$ for (x_2, y_2).*

$$= (3, 6)$$

The coordinates of P are (3, 6).

(continued on the next page)

b. For $\overline{CP}$ to be an altitude of $\triangle ABC$, $\overline{CP}$ must be perpendicular to $\overline{AB}$. This means that the product of the slopes of $\overline{CP}$ and $\overline{AB}$ must equal -1.

$$\text{slope of } \overline{CP} = \frac{15-6}{9-3} = \frac{9}{6} \text{ or } \frac{3}{2}$$

$$\text{slope of } \overline{AB} = \frac{10-2}{-3-9} = \frac{8}{-12} \text{ or } -\frac{2}{3}$$

$$\text{product of slopes} = \frac{3}{2} \cdot -\frac{2}{3} \text{ or } -1$$

Since the product of these slopes is -1, $\overline{CP}$ is perpendicular to $\overline{AB}$. Thus, $\overline{CP}$ is an altitude of $\triangle ABC$.

The fourth special segment of a triangle is an **angle bisector**. In Lesson 1–6, you learned that an angle bisector is a ray that divides an angle into two congruent angles. An angle bisector of a triangle is a segment that bisects an angle of the triangle and has one endpoint at a vertex of the triangle and the other endpoint at another point on the triangle. For $\triangle LMN$, $\overline{LA}$, $\overline{MB}$, and $\overline{NC}$ are the angle bisectors. Since there are three angles in every triangle, every triangle has three angle bisectors.

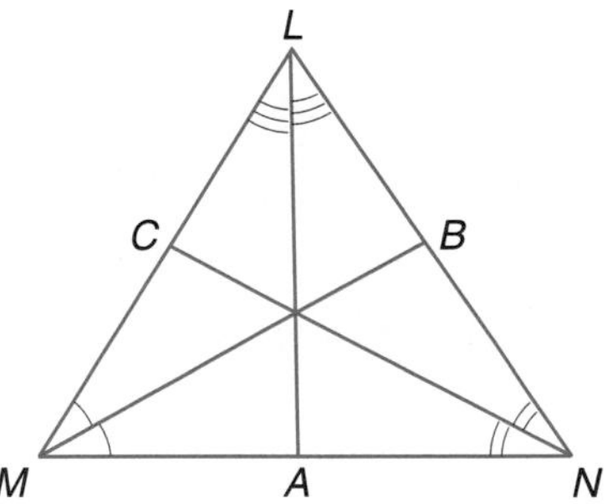

Theorems 5–3 and 5–4 involve special properties related to angle bisectors.

Theorem 5–3	**Any point on the bisector of an angle is equidistant from the sides of the angle.**
Theorem 5–4	**Any point on or in the interior of an angle and equidistant from the sides of an angle lies on the bisector of the angle.**

You will be asked to prove Theorem 5–3 in Exercise 37.

The special segments of different types of triangles have special properties.

Example  **Prove that if a triangle is isosceles, then the bisector of the vertex angle of the triangle is also a median.**

Begin by drawing and labeling a diagram of the situation.

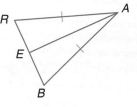

Given: $\triangle RAB$ is isosceles with vertex angle RAB. $\overline{EA}$ is the bisector of $\angle RAB$.

Prove: $\overline{EA}$ is a median.

Proof:

Statements	Reasons
1. $\triangle RAB$ is isosceles. $\overline{EA}$ is the bisector of $\angle RAB$.	1. Given
2. $\overline{RA} \cong \overline{BA}$	2. Def. isosceles $\triangle$
3. $\angle RAE \cong \angle BAE$	3. Def. $\angle$ bisector
4. $\overline{AE} \cong \overline{AE}$	4. Congruence of segments is reflexive.
5. $\triangle AER \cong \triangle AEB$	5. SAS
6. $\overline{ER} \cong \overline{EB}$	6. CPCTC
7. $\overline{AE}$ is a median.	7. Def. median

The special segments of a triangle are often used to make patterns involving triangles more interesting. Triangles are also used in designing bridges and buildings because they are the most rigid shape.

Example 3

Civil Engineering

Study the picture of the Waddell "A" Truss Bridge carefully. Write at least one conclusion that you can make from each statement.

a. $\overline{LO}$ is an altitude of $\triangle JKL$.

By the definition of an altitude, $\overline{LO} \perp \overline{JK}$. $\angle LOJ \cong \angle LOK$ because perpendicular lines form right angles and all right angles are congruent.

b. $\overline{JM} \cong \overline{ML}$

M is the midpoint of $\overline{JL}$. Thus, $\overline{OM}$ is a median of $\triangle JOL$.

c. $\overline{NQ}$ is an angle bisector and an altitude of $\triangle ONK$.

Since $\overline{NQ}$ is an angle bisector, $\angle ONQ \cong \angle KNQ$. $\overline{NQ} \perp \overline{OK}$ because $\overline{NQ}$ is an altitude. $\angle OQN \cong \angle KQN$ because perpendicular lines form right angles and all right angles are congruent. Congruence of segments is reflexive, so $\overline{NQ} \cong \overline{NQ}$. Therefore, by ASA, $\triangle ONQ \cong \triangle KNQ$. By CPCTC, $\overline{ON} \cong \overline{KN}$. Thus, $\triangle ONK$ is isosceles.

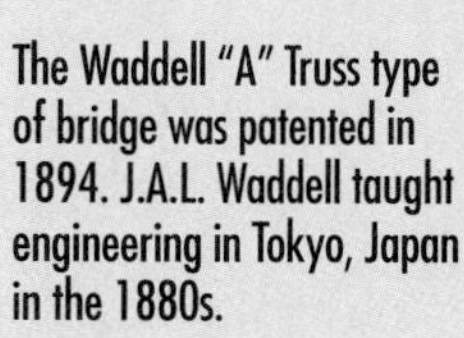

The Waddell "A" Truss type of bridge was patented in 1894. J.A.L. Waddell taught engineering in Tokyo, Japan in the 1880s.

CHECK FOR UNDERSTANDING

Communicating Mathematics

Study the lesson. Then complete the following.

1. **Compare and contrast** a perpendicular bisector and an altitude of a triangle.
2. **Determine** which of the four special segments might occur outside of a triangle. Draw an example of each.
3. **Find a counterexample** to the statement *An altitude and an angle bisector of a triangle are never the same segment.*
4. Draw $\triangle EFG$ with altitude $\overline{FH}$. What type of triangle is $\triangle EFH$?

5. Draw a right triangle ABC on patty paper with the right angle at B. Fold the paper to create the median, altitude, and the angle bisector from vertex B, and the perpendicular bisector of $\overline{AC}$. Are any of the segments parallel? Justify your answer by making one or more additional folds in the paper.

Guided Practice

Draw and label a figure to illustrate each situation.

6. $\overline{AB}$ is a median of $\triangle BOC$, and A is between O and C.
7. $\overline{RA}$ is an altitude and a median in $\triangle RST$. A is between S and T.
8. $\overline{TU}$ is an altitude in $\triangle TUL$.
9. $\overline{QL}$ is an angle bisector in $\triangle PQR$, and L is between P and R.

Refer to $\triangle RES$. Write at least one conclusion you can make from each statement.

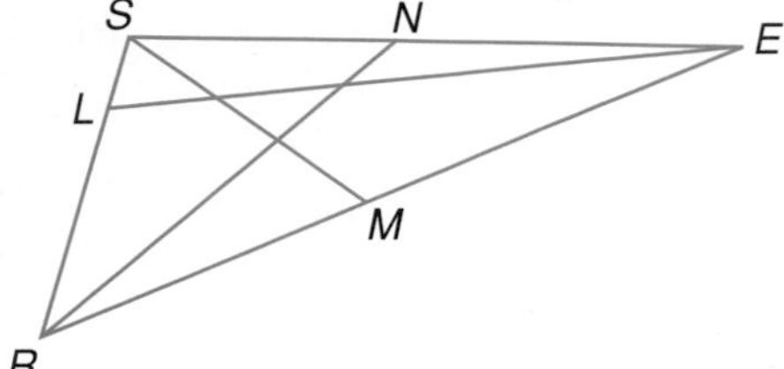

10. $\overline{SM}$ is an altitude to $\overline{RE}$.
11. $\overline{SN} \cong \overline{NE}$
12. M is equidistant from R and E, and $\angle RMS$ is a right angle.
13. $\angle ERN \cong \angle SRN$
14. $\overline{EL} \perp \overline{SR}$

Algebra

15. $\triangle ABC$ has vertices $A(-3, -9)$, $B(5, 11)$, and $C(9, -1)$. $\overline{AT}$ is a median of $\triangle ABC$ with T on $\overline{BC}$.
    a. What are the coordinates of T?
    b. Find the slope of $\overline{AT}$.
    c. Is $\overline{AT}$ an altitude of $\triangle ABC$? Explain.
    d. Determine if $\overline{AT}$ is longer than $\overline{AB}$. Explain your findings.

Proof

16. **Prove** that if $\overline{EG} \cong \overline{EH}$ in $\triangle EGH$, the altitude to $\overline{GH}$ is also a median of $\triangle EGH$.

17. **Sports** Participants in a trail contest try to be the first to find the flag. Suppose each person in a contest is given the map shown at the right and the following instructions.
    - The flag is as far from the Lookout Tower as it is from the park entrance.
    - If you walk from West Road to the flag or from Shore Road to the flag, you would walk the same distance.

    Describe the position of the flag.

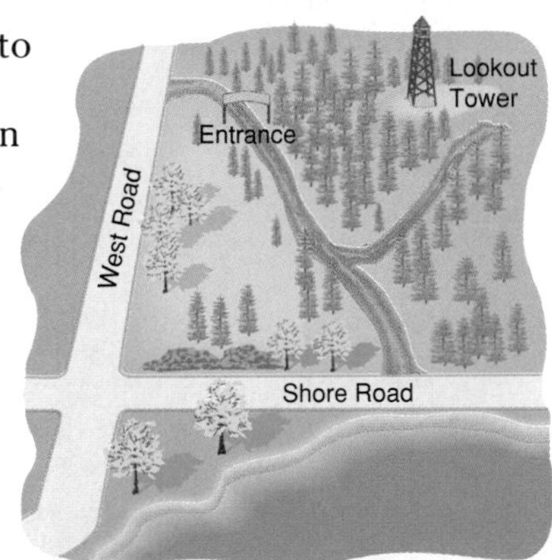

EXERCISES

Practice

Draw and label a figure to illustrate each situation.

18. $\overline{RS}$ is an angle bisector and an altitude of $\triangle RQP$.
19. $\overline{AB}$ and $\overline{CD}$ are angle bisectors of $\triangle ACT$ and intersect at X.
20. $\overrightarrow{LT}$ and $\overrightarrow{OS}$ contain altitudes of $\triangle LOC$ and intersect at M outside of $\triangle LOC$.
21. $\triangle EFG$ is a right triangle with right angle F. $\overline{FT}$ is both an altitude and a median of $\triangle EFG$.
22. $\overline{PM}$ is a median and an angle bisector of $\triangle PQR$, and $\overline{RP}$ is an altitude.
23. $\triangle JKL$ is a right triangle with right angle at L. $\overline{LM}$ is a median of $\triangle JKL$, and $\overline{MN}$ is the perpendicular bisector of $\overline{JK}$.

Refer to $\triangle ABC$. Write at least one conclusion you can make from each statement.

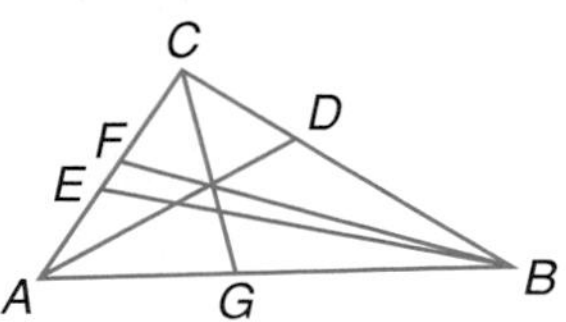

24. $\overline{AD}$ is an altitude.
25. $\overline{BE}$ is an altitude and a median.
26. $\overline{AF} \cong \overline{FC}$
27. $\overline{AD}$ is an angle bisector.

If possible, describe a triangle for which each statement is true. If no triangle exists, write *no such triangle*.

28. The three angle bisectors of a triangle intersect at a point inside the triangle.
29. The three altitudes of a triangle intersect on the triangle.
30. The three angle bisectors of a triangle will not intersect.
31. The perpendicular bisectors intersect outside of a triangle.

Algebra

32. $\overline{RT}$ is a median in $\triangle RLB$ with points $R(3, 8)$, $T(12, 3)$, and $B(9, 12)$.
    a. What are the coordinates of L?
    b. Is $\overline{RT}$ an altitude of $\triangle RLB$?
    c. The graph of point S is at $(4, 13)$. $\overline{SC}$ intersects $\overline{RB}$ at C. If C is at $(6, 10)$, is $\overline{SC}$ a perpendicular bisector of $\overline{RB}$?

33. In $\triangle RTE$, $AE = 3x - 11$, $AR = x + 5$, $RY = 2z - 1$, $YT = 4z - 11$, $m\angle RTA = 4y - 17$, $m\angle ATE = 3y - 4$, and $m\angle RST = 2x + 10$.

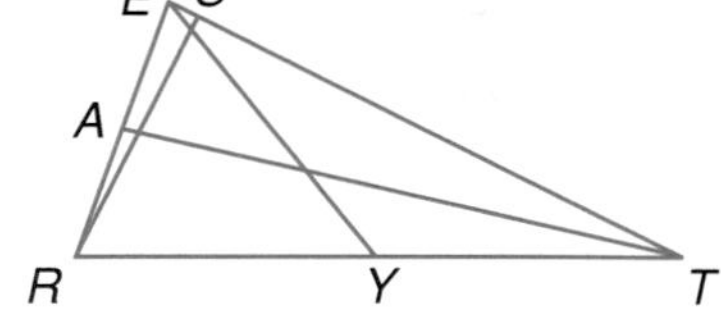

    a. $\overline{RS}$ is an altitude of $\triangle RTE$. Find the value of x.
    b. If $\overline{TA}$ is an angle bisector of $\angle RTE$, find $m\angle RTA$.
    c. $\overline{EY}$ is a median of $\triangle RTE$. Find RT.

Draw and label a figure for each statement. List the information that is given and the statement to be proved in terms of your figure. Then write a proof of the statement.

34. If an angle bisector of a triangle is also an altitude, then the triangle is isosceles.
35. A point on the perpendicular bisector of a segment is equidistant from the endpoints of the segment. (Theorem 5–1)
36. A point equidistant from the endpoints of a segment lies on the perpendicular bisector of the segment. (Theorem 5–2)
37. A point on the bisector of an angle is equidistant from the sides of the angle. (Theorem 5–3)
38. The medians drawn to the congruent sides of an isosceles triangle are congruent.
39. The median to the base of an isosceles triangle bisects the vertex angle.
40. Corresponding angle bisectors of congruent triangles are congruent.
41. Corresponding medians of two congruent triangles are congruent.

Algebra

42. $\triangle ABC$ has vertices $A(-2, 3)$, $B(10, 13)$, and $C(4, 17)$. Draw $\triangle ABC$.
    a. Draw the medians of $\triangle ABC$. Label the point of intersection of the medians D. What are the coordinates of D?
    b. Show that the distance from D to the midpoint of $\overline{AB}$ is one-third the distance from C to the midpoint of $\overline{AB}$.

Cabri Geometry

43. Use a calculator to draw different kinds of triangles and the special segments in each triangle. Determine where the special types of segments for each kind of triangle meet.

Critical Thinking

44. Draw any $\triangle ABC$ with median $\overline{AD}$ and altitude $\overline{AE}$. Recall that the area of a triangle is one-half the product of the measures of the base and the altitude. What conclusion can you make about the relationship between the areas of $\triangle ABD$ and $\triangle ACD$?

Applications and Problem Solving

Real World

45. **Statistics** $\triangle TSR$ has vertices $T(-6, 12)$, $S(2, 4)$, and $R(16, 8)$. Graph $\triangle TSR$.

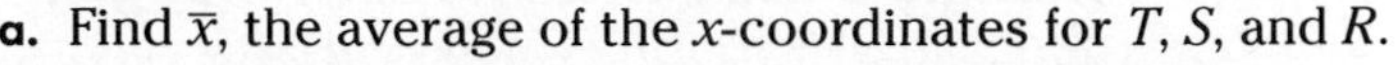

a. Find $\bar{x}$, the average of the x-coordinates for T, S, and R.

b. Find $\bar{y}$, the average of the y-coordinates for T, S, and R.

c. Plot $M(\bar{x}, \bar{y})$ on the graph of $\triangle TSR$.

d. Draw the medians of $\triangle TSR$. What do you observe?

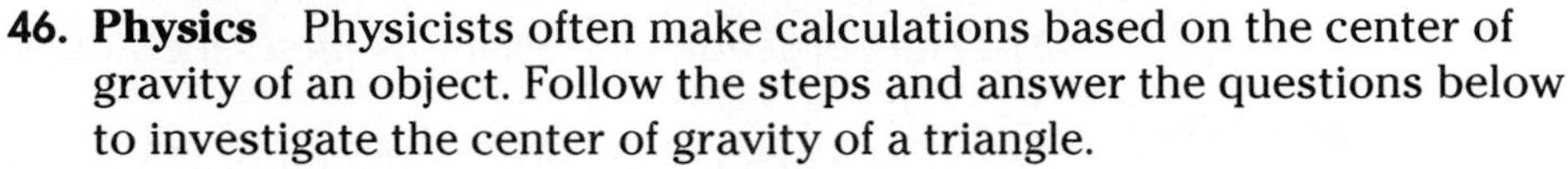

46. **Physics** Physicists often make calculations based on the center of gravity of an object. Follow the steps and answer the questions below to investigate the center of gravity of a triangle.

a. Draw a large acute triangle that is not isosceles.

b. Construct the three medians of the triangle. What do you notice?

c. Construct the three angle bisectors of the triangle. What do you notice?

d. Construct the three altitudes of the triangle. What do you notice?

e. Construct the three perpendicular bisectors of the sides of the triangle. What do you notice?

f. Cut out the triangles you made for parts a–e. Place the point where the segments intersect on the flat end of a pencil for each of the triangles. What do you observe?

g. What changes would occur in the construction in parts b–e if the triangle were right or obtuse instead of acute?

Mixed Review

47. Graph $\triangle BAY$ with vertices $B(-1, 1)$, $A(-5, 4)$, and $Y(3, 4)$. Prove that $\triangle BAY$ is isosceles. (Lesson 4–6)

48. **Eliminate the Possibilities** Raul, Molesha, and Jennifer just began after-school jobs. One works at a pet store, one at a day camp, and one at a fast-food restaurant. Raul takes his sister to the day camp on his way to work. Jennifer is allergic to pet hair. Molesha receives free meals from her job. Who works at which job? (Lesson 4–5)

49. Without graphing, determine if $\overleftrightarrow{PQ} \parallel \overleftrightarrow{RS}$ given points $P(0, 0)$, $Q(3, -2)$, $R(1, -7)$, and $S(-5, -3)$. Explain. (Lesson 3–4)

50. **ACT Practice** What is the length of the line segment whose endpoints are represented on the coordinate plane by points $(1, -2)$ and $(-3, 1)$?

A 3 B 4 C 5 D 6 E 7

51. Write a two-column proof for *If* $\overline{PQ}$ *bisects* $\overline{AB}$ *at point M, then* $\overline{AM} \cong \overline{MB}$. (Lesson 2–5)

52. **Algebra** Write a conjecture based on $12x - 42 = 54$. (Lesson 2–1)

53. If S is between R and T, $RS = 13$, $ST = 2x + 7$, and $RT = 3x + 8$, find the value of x and RT. (Lesson 1–4)

54. Draw and label a figure to show points A, B, C, and D so that no more than two points are collinear. (Lesson 1–2)

INTEGRATION **Algebra**

55. Simplify $2x - 5x$.

56. Simplify $12r - (-10r)$.

For **Extra Practice**, see page 772.

5–2 Right Triangles

What YOU'LL LEARN

- To recognize and use tests for congruence of right triangles.

Why IT'S IMPORTANT

Right triangles are used in trigonometry, a branch of mathematics you will study later in this course.

Real World APPLICATION

Windsurfing

Harnessing the wind may seem like it would take a great deal of strength. But Rhonda Smith, one of the pioneers in women's windsurfing, says that it's more a matter of proper technique than strength. She should know—Rhonda has won five world windsurfing championships! She now runs a windsurfing school in Hood River, Oregon.

A light-wind sail for a sailboard is a right triangle. Suppose right triangles *DEF* and *RST* model the sails for two of Rhonda's boards. What would be required to prove the triangles congruent by SAS? The conditions necessary for this congruence are stated in Theorem 5–5.

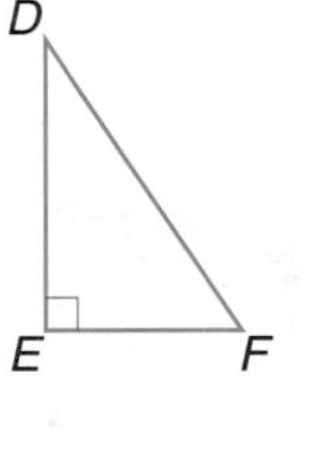

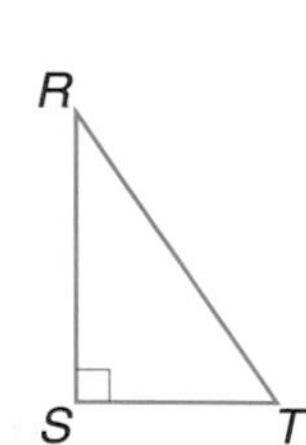

Theorem 5–5 ***LL***	**If the legs of one right triangle are congruent to the corresponding legs of another right triangle, then the triangles are congruent.**

Proof of Theorem 5–5

Given: $\triangle DEF$ and $\triangle RST$ are right triangles.
$\angle E$ and $\angle S$ are right angles.
$\overline{EF} \cong \overline{ST}$
$\overline{ED} \cong \overline{SR}$

Prove: $\triangle DEF \cong \triangle RST$

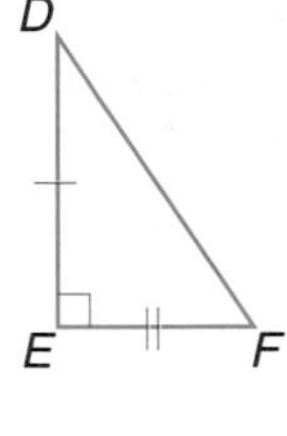

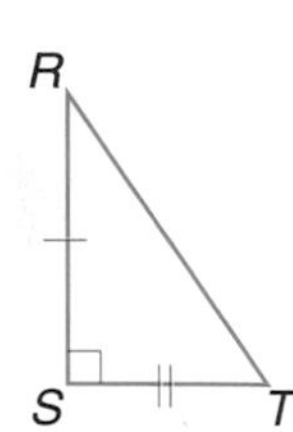

Paragraph Proof:

We are given that $\overline{EF} \cong \overline{ST}$, $\overline{ED} \cong \overline{SR}$, and $\angle E$ and $\angle S$ are right angles. Since all right angles are congruent, $\angle E \cong \angle S$. Therefore, by SAS, $\triangle DEF \cong \triangle RST$.

Suppose the hypotenuse and an acute angle of one sail are congruent to the hypotenuse and corresponding acute angle of the other sail. Is that enough information to prove that the sails are congruent?

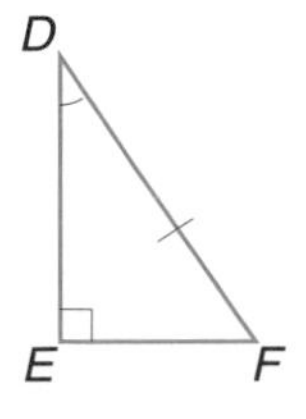

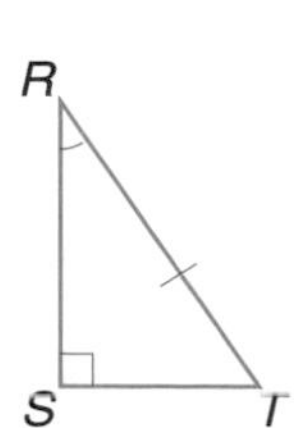

The right angles in each triangle are congruent. Knowing that another angle and the hypotenuse are congruent indicates that two angles and a non-included side are congruent. Thus, the triangles are congruent by AAS. The two sails would be the same size. This leads us to Theorem 5–6. *You will be asked to prove Theorem 5–6 in Exercise 28.*

Theorem 5–6 HA	If the hypotenuse and an acute angle of one right triangle are congruent to the hypotenuse and an acute angle of another right triangle, then the two triangles are congruent.

You can use the HA theorem to complete proofs involving right triangles.

Example 1

Construction

Part of the brace used to support electrical power lines is shaped like an isosceles triangle with an altitude from the vertex to the base. Prove that the two right triangles formed by the altitude are congruent.

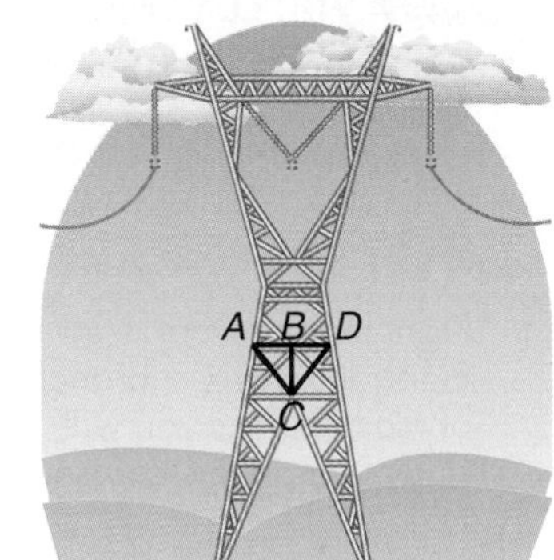

Given: $\overline{CB}$ is an altitude of $\triangle ACD$.
$\triangle ACD$ is an isosceles triangle with legs $\overline{CA}$ and $\overline{CD}$.

Prove: $\triangle ABC \cong \triangle DBC$

Two-column Proof:

Statements	Reasons
1. $\triangle ACD$ is an isosceles triangle.	1. Given
2. $\overline{CA} \cong \overline{CD}$	2. Def. isos. $\triangle$
3. $\angle BAC \cong \angle BDC$	3. Isosceles Triangle Theorem
4. $\overline{CB}$ is an altitude of $\triangle ACD$.	4. Given
5. $\overline{BC} \perp \overline{AD}$	5. Def. altitude
6. $\angle ABC$ and $\angle DBC$ are right angles.	6. $\perp$ lines form 4 rt. $\angle$s.
7. $\triangle ABC$ and $\triangle DBC$ are right triangles.	7. Def. rt. $\triangle$
8. $\triangle ABC \cong \triangle DBC$	8. HA

We have already determined that right triangles are congruent if corresponding legs are congruent (LL) or if the hypotenuses and corresponding acute angles are congruent (HA). As a third possibility, suppose a leg and an acute angle of one sail are congruent to the corresponding leg and acute angle of the second sail. Will the two sails be congruent under these conditions? There are two different cases to consider.

Case 1

The leg is included between the acute angle and the right angle.

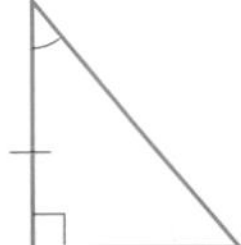
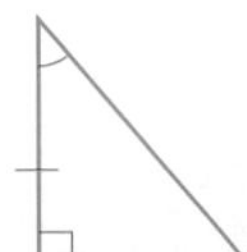

Case 2

The leg is not included between the acute angle and the right angle.

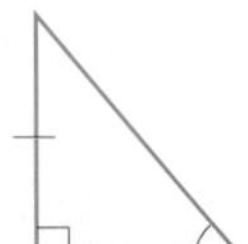
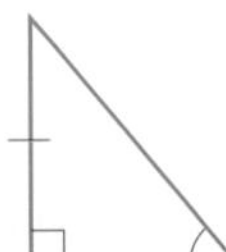

The triangles are congruent in both cases. For Case 1, the triangles are congruent by ASA. For Case 2, the triangles are congruent by AAS. This observation suggests Theorem 5–7. *You will be asked to prove Theorem 5–7 in Exercise 29.*

Theorem 5–7
LA

If one leg and an acute angle of one right triangle are congruent to the corresponding leg and acute angle of another right triangle, then the triangles are congruent.

Example 

INTEGRATION Algebra

Find the values of x and y so that $\triangle MNO$ is congruent to $\triangle RST$.

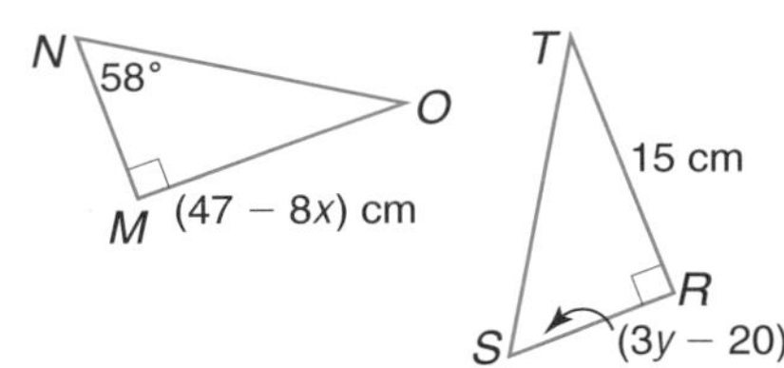

Assume $\triangle MNO \cong \triangle RST$. Then $\angle N \cong \angle S$ and $\overline{MO} \cong \overline{RT}$.

$m\angle N = m\angle S$	$MO = RT$
$58 = 3y - 20$	$47 - 8x = 15$
$78 = 3y$	$-8x = -32$
$26 = y$	$x = 4$

By LA, $\triangle MNO \cong \triangle RST$ for $x = 4$ and $y = 26$.

You can use a graphing calculator to investigate the following question.

EXPLORATION — CABRI GEOMETRY

If the hypotenuse and one leg of one right triangle are congruent to the hypotenuse and corresponding leg of a second right triangle, are the triangles congruent?

- Use a calculator to draw a right triangle.
- Draw two perpendicular rays in another place on the screen.
- Measure one of the legs of the right triangle. Then, use 9:Measurement Transfer on F4 to transfer the measurement to one of the perpendicular rays to create a congruent leg.
- Measure the hypotenuse of the right triangle. Then transfer the measurement to the endpoint of the segment created above that is not the vertex of the right angle. Position the segment being drawn so that the vertex lies on the perpendicular ray in order to form a right triangle.
- Measure the angles and sides of both triangles.

Your Turn

a. How do the measurements of the two right triangles compare?

b. Write a conjecture about two right triangles that have one pair of corresponding legs congruent and congruent hypotenuses.

The results of the Exploration lead us to Postulate 5–1.

Postulate 5–1
HL

If the hypotenuse and a leg of one right triangle are congruent to the hypotenuse and corresponding leg of another right triangle, then the triangles are congruent.

Since this is a postulate, it need not be proved.

Example 3 **In the figure at the right, $\overline{PR} \perp \overline{QR}$, $\overline{PS} \perp \overline{QS}$, and $\overline{QR} \cong \overline{QS}$. Write a flow proof to show that $\overline{QP}$ bisects $\angle SQR$.**

Proof

Given: $\overline{PR} \perp \overline{QR}$, $\overline{PS} \perp \overline{QS}$, $\overline{QR} \cong \overline{QS}$

Prove: $\overline{QP}$ bisects $\angle SQR$

Flow Proof:

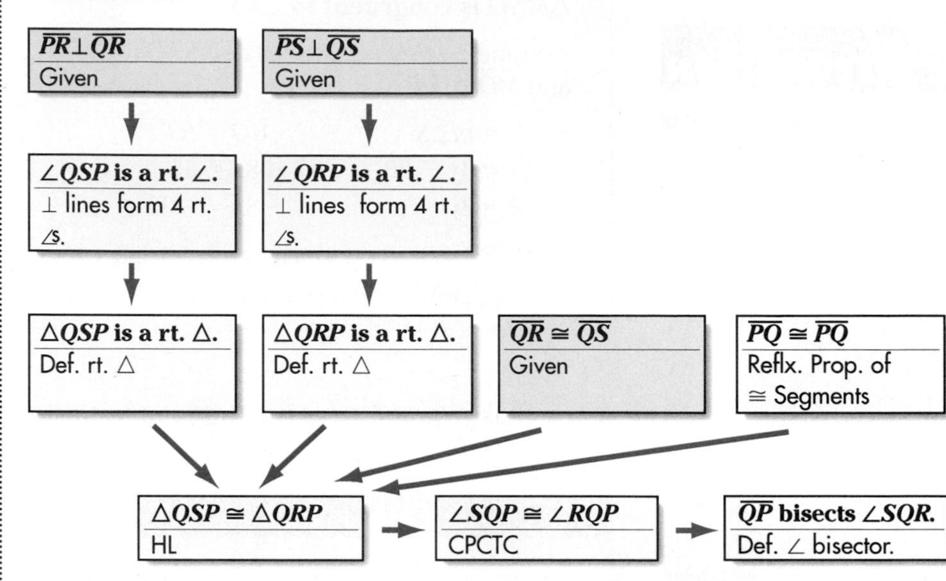

$\overline{PR} \perp \overline{QR}$
Given
$\overline{PS} \perp \overline{QS}$
Given
$\angle QSP$ is a rt. $\angle$.
$\perp$ lines form 4 rt. $\angle$s.
$\angle QRP$ is a rt. $\angle$.
$\perp$ lines form 4 rt. $\angle$s.
ΔQSP is a rt. Δ.
Def. rt. Δ
ΔQRP is a rt. Δ.
Def. rt. Δ
$\overline{QR} \cong \overline{QS}$
Given
$\overline{PQ} \cong \overline{PQ}$
Reflx. Prop. of $\cong$ Segments
$\Delta QSP \cong \Delta QRP$
HL
$\angle SQP \cong \angle RQP$
CPCTC
$\overline{QP}$ bisects $\angle SQR$.
Def. $\angle$ bisector.

CHECK FOR UNDERSTANDING

Communicating Mathematics

Study the lesson. Then complete the following.

1. **Explain** why there are only two requirements to prove two right triangles congruent while tests for other triangles have three.
2. **Compare and contrast** the tests for triangle congruence (SAS, AAS, ASA) and the tests for congruence of right triangles (LL, HA, and LA).
3. **Show** why two cases must be considered when proving LA.
4. **You Decide** Carlos said that you could justify the Leg-Leg test for congruent right triangles by SSS. Jeannie disagrees. Who is correct and why?

MATH JOURNAL

5. Write a short paragraph describing each of the tests for congruence of right triangles presented in this lesson. Use word processing software if it is available.

Guided Practice

State the additional information needed to prove each pair of triangles congruent by the given theorem or postulate.

6. HL

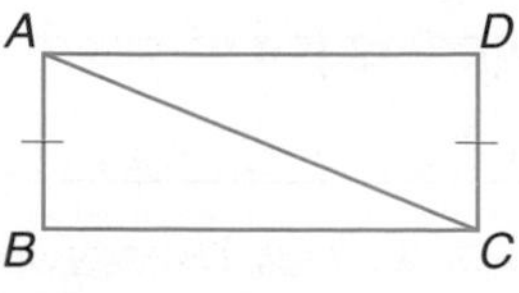

A
D
B
C

7. LL

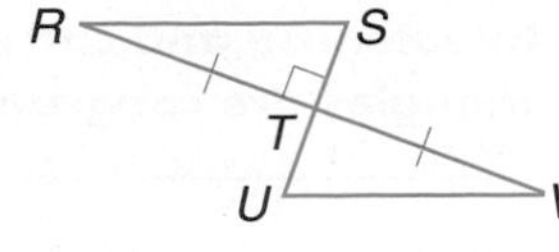

R
S
T
U
V

8. LA

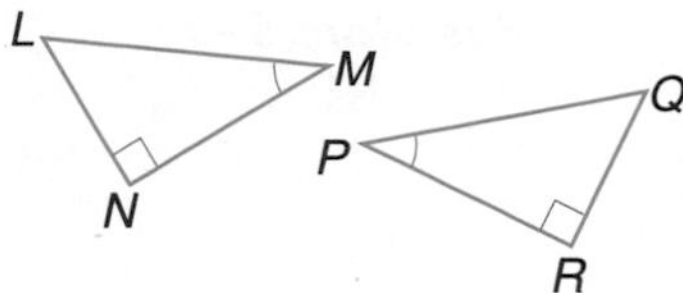

L
M
N
P
Q
R

INTEGRATION Algebra

For each figure, find the value of *x* so that each pair of triangles is congruent by the indicated theorem or postulate.

9. HA

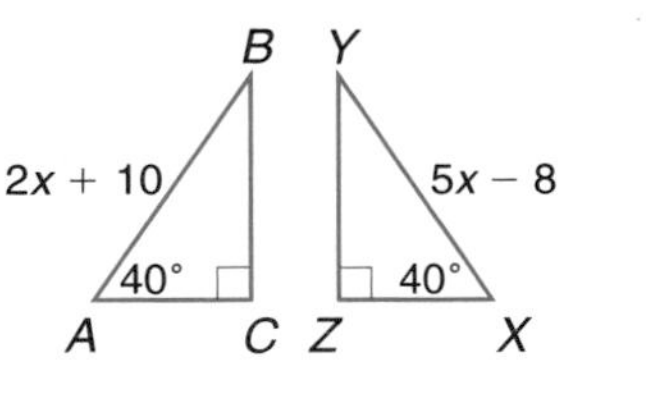

10. LA

11. HL

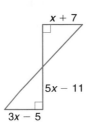

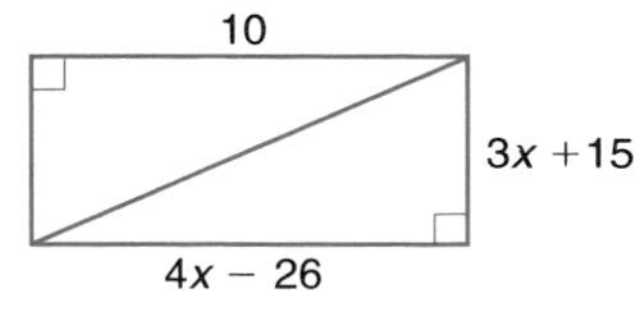

The Sculptured Space

Proof

12. Use the figure at the right to write a two-column proof.

Given: $\angle RLT$ and $\angle SLT$ are right angles.
$\angle LTR \cong \angle LTS$

Prove: $\overline{LR} \cong \overline{LS}$

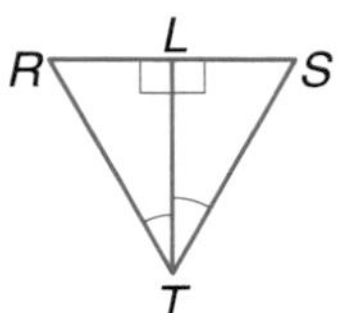

13. Art *The Sculptured Space* is a sculpture built by a team of sculptors headed by Helen Escobedo. The sculpture is made up of triangular pieces of concrete surrounding a circle of petrified lava. One leg of each right triangle is 5 meters long, and the other leg is 7.31 meters long. Can you conclude that all the triangles are congruent? Explain.

EXERCISES

Practice

State the additional information needed to prove each pair of triangles congruent by the given theorem or postulate.

14. LL

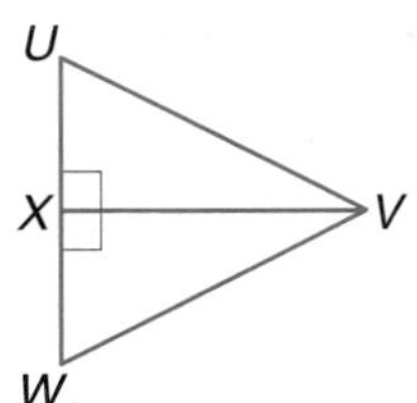

15. HA

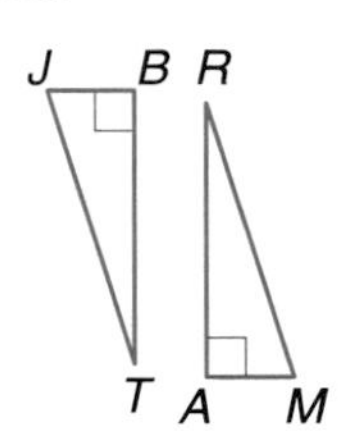

16. HL

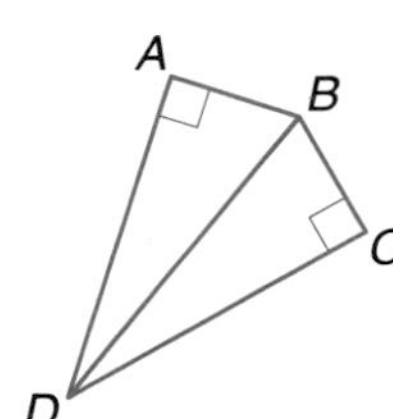

17. LL

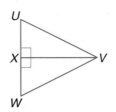

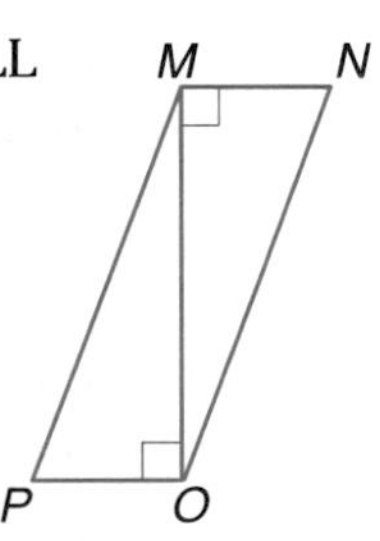

18. HL

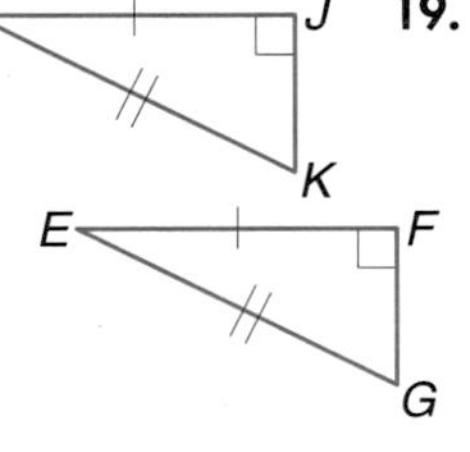

19. LA

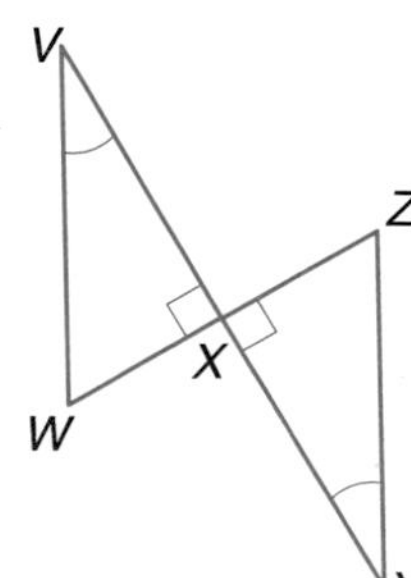

INTEGRATION Algebra

Find the value of *x* so that $\triangle ABC \cong \triangle XYZ$ by the indicated theorem or postulate.

20. $AB = 2x + 6$, $BC = 15$, $AC = 3x + 4$, $XY = 20$, $YZ = x + 8$; LL

21. $m\angle Z = 55°$, $BC = 15x + 2$, $m\angle C = 55°$, $AB = 24$, $YZ = 4x + 13$; LA

22. $YX = 21$, $m\angle X = 9x + 9$, $AB = 21$, $m\angle A = 11x - 3$; LA

23. $AC = 28$, $AB = 7x + 4$, $ZX = 9x + 1$, $YX = 5(x + 2)$; HL

Use the figure at the right to write a two-column proof.

Proof

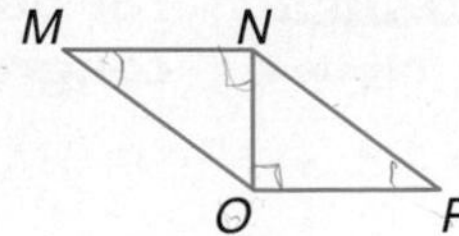

24. **Given:** $\overline{NO} \perp \overline{MN}$
$\overline{NO} \perp \overline{OP}$
$\angle M \cong \angle P$

Prove: $\overline{MO} \cong \overline{NP}$

25. **Given:** $\angle M$ and $\angle P$ are right angles.
$\overline{MN} \parallel \overline{OP}$

Prove: $\overline{MN} \cong \overline{OP}$

Use the figure at the right to write a two-column proof.

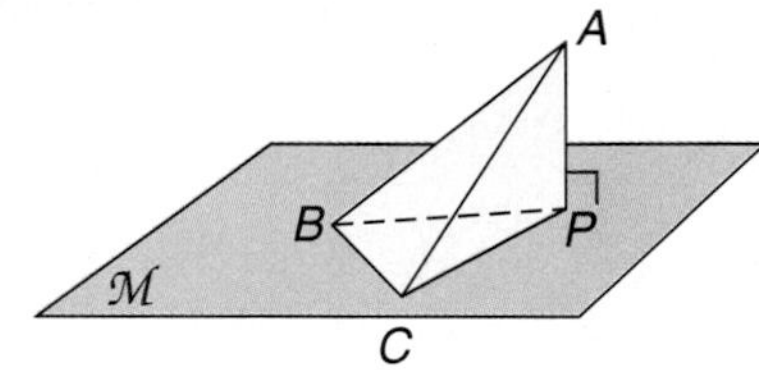

26. **Given:** $\overline{AP} \perp$ plane $\mathcal{M}$
$\overline{AB} \cong \overline{AC}$

Prove: $\triangle BPC$ is isosceles.

27. **Given:** $\angle PBC \cong \angle PCB$
$\overline{AP} \perp$ plane $\mathcal{M}$

Prove: $\angle ABC \cong \angle ACB$

Draw and label a figure for each statement. List the information that is given and the statement to be proved in terms of your figure. Then write a paragraph proof.

28. If the hypotenuse and an acute angle of one right triangle are congruent to the hypotenuse and corresponding acute angle of another right triangle, then the two triangles are congruent. (Theorem 5–6)

29. If one leg and an acute angle of one right triangle are congruent to the corresponding leg and acute angle of another right triangle, then the triangles are congruent. (Theorem 5–7)

30. Corresponding altitudes of congruent triangles are congruent.

31. The two segments that have the midpoint of each leg of an isosceles triangle as one endpoint and are perpendicular to the base of the triangle are congruent.

32. Obtuse triangle RST with obtuse angle at S shares side $\overline{ST}$ with acute triangle UTS. $\overline{RS}$ is congruent to $\overline{UT}$, and $\angle RST$ is supplementary to $\angle UTS$. Prove that the altitude from R to $\overline{ST}$ in $\triangle RST$ is congruent to the altitude from U to $\overline{ST}$ in $\triangle UST$.

Cabri Geometry

33. Use a calculator to explore two right triangles that have congruent corresponding angles. What do you observe?

Critical Thinking

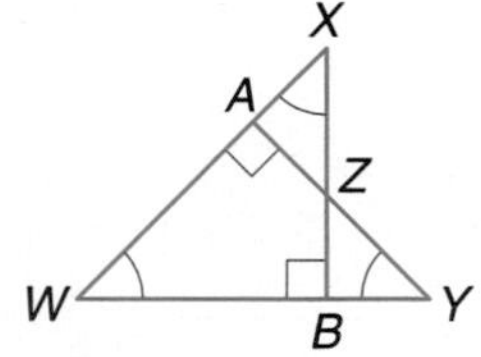

34. In the figure at the right, $m\angle W = m\angle X = m\angle Y = 45$, $\overline{XB} \perp \overline{WY}$, $\overline{YA} \perp \overline{WX}$. If $WZ = 10$, find XY.

Applications and Problem Solving

35. **Transportation** Many tractor-trailer trucks have airfoils on the top of the cab to make the truck more fuel efficient. An airfoil is made from sheets of metal. The side view of an airfoil is shown at the right. List the possible different sets of information that could be used to produce a congruent piece for the other side of the airfoil.

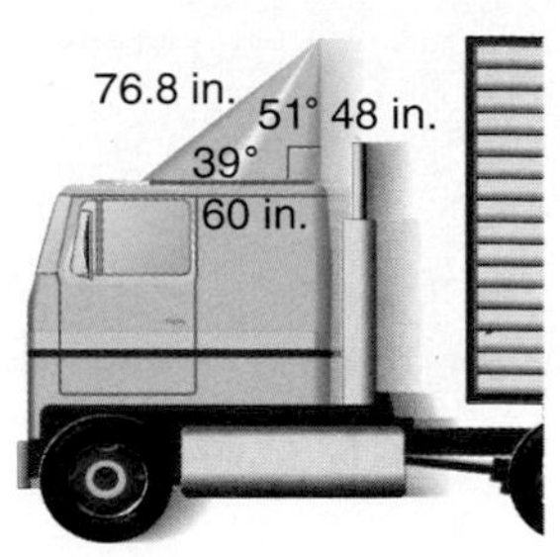

36. **Construction** A wooden bridge is being constructed over a stream in a park. The braces for the support posts came from the lumberyard already cut. The carpenter measures from the top of the support post to a point on the post to find where to attach the brace to the post. Explain why only one measurement must be made to ensure that all of the braces will be in the same relative position.

Mixed Review

37. Draw and label a figure to illustrate that $\overline{PT}$ and $\overline{RS}$ are medians of $\triangle PQR$ and intersect at V. (Lesson 5–1)

38. Given $\triangle ADN \cong \triangle DEO$, $AN = 15$, $ND = 19$, $AD = 27$, and $EO = 4x - 1$, find the value of x. (Lesson 4–3)

39. Draw and label $\triangle LNA$ with $m\angle L > 90$ and $NL = AL$. Then classify the triangle by its angles and sides. (Lesson 4–1)

40. Name five ways to prove that two lines are parallel. (Lesson 3–4)

41. Find the slope of the line that passes through points at $(6, -11)$ and $(4, 9)$. (Lesson 3–3)

42. **SAT Practice** In the figure, $\overline{AB} \parallel \overline{CE}$. If the length of $\overline{DA} = 6$, what is the length of $\overline{DB}$?

A 6 B 7 C 8 D 9

E It cannot be determined from the information given.

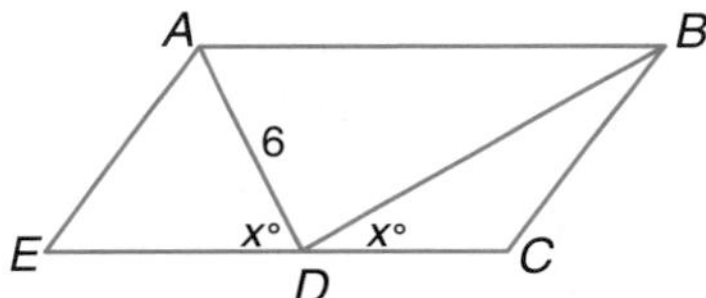

Note: Figure not drawn to scale.

43. Explain how to form the converse, inverse, and contrapositive of a conditional statement. (Lesson 2–2)

Algebra

44. Write $\frac{3}{20}$ as a percent.

45. **Collectibles** The table shows the most valuable editions of Superman comic books in near-mint condition. Express the relation of the year and value shown in the table as a set of ordered pairs. Then state the domain and range of the relation.

Edition	Year	Value($)
Action Comics No. 1	1938	50,000
Superman No. 1	1939	40,000
Action Comics No. 2	1938	5200
Action Comics No. 3	1938	4200
New York World's Fair	1939	4020

Data Update For more information on prices of collectibles, visit: **www.geometry.glencoe.com**

For **Extra Practice**, see page 772.

5-3 Indirect Proof and Inequalities

What **YOU'LL LEARN**

- To use indirect reasoning and indirect proof to reach a conclusion,
- to recognize and apply properties of inequalities to the measures of segments and angles, and
- to solve problems by working backward.

Why **IT'S IMPORTANT**

Indirect reasoning is often used in the legal system and in advertising.

Health

Alzheimer's disease is the fourth leading cause of death in the United States. Unfortunately, doctors have a difficult time diagnosing the disease. When a patient exhibits symptoms that may be Alzheimer's, doctors test for other diseases with similar symptoms. If the tests for the other diseases are negative, the doctor concludes that the patient has Alzheimer's disease. This is an example of **indirect reasoning**.

You have used direct reasoning in the proofs you have encountered up to this point. When using direct reasoning, you start with a true hypothesis and prove that the conclusion is true. With indirect reasoning, you assume that the conclusion is false and then show that this assumption leads to a contradiction of the hypothesis or some other accepted fact, like a postulate, theorem, or corollary. Then, since your assumption has been proved false, the conclusion must be true.

The following steps summarize the process of indirect reasoning. Follow these steps when doing an **indirect proof**.

Steps for Writing an Indirect Proof

1. **Assume that the conclusion is false.**
2. **Show that the assumption leads to a contradiction of the hypothesis or some other fact, such as a postulate, theorem, or corollary.**
3. **Point out that the assumption must be false and, therefore, the conclusion must be true.**

Example 1 **State the assumption you would make to start an indirect proof of each statement. Do *not* write the proofs.**

a. The number 117 is divisible by 13.

b. Akita is the best candidate in the election.

c. $\overline{AB}$ is a median of $\triangle ACD$.

d. $m\angle C < m\angle D$

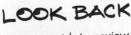

LOOK BACK

You may wish to review direct reasoning and the Law of Detachment in Lesson 2-3.

The first step in writing an indirect proof is to assume that the conclusion is false.

a. The number 117 is not divisible by 13.

b. Akita is not the best candidate in the election.

c. $\overline{AB}$ is not a median of $\triangle ACD$.

d. $m\angle C \geq m\angle D$ *Recall that if $m\angle C$ is not less than $m\angle D$, then it could be greater than or equal to $m\angle D$.*

You have learned about a relationship between exterior angles of a triangle and their remote interior angles. The following theorem is a result of this relationship. This theorem can be proved using an indirect proof.

Theorem 5–8 Exterior Angle Inequality Theorem	**If an angle is an exterior angle of a triangle, then its measure is greater than the measure of either of its corresponding remote interior angles.**

The following statement can be used to define the inequality relationship between two numbers. This definition is often used in proofs.

Definition of Inequality	**For any real numbers *a* and *b*, *a* > *b* if and only if there is a positive number *c* such that *a* = *b* + *c*.**

Proof of Theorem 5–8

Given: $\angle 1$ is an exterior angle of $\triangle MNP$.

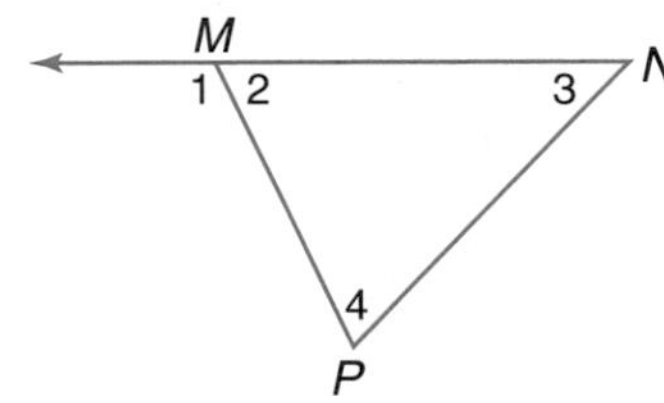

Prove: $m\angle 1 > m\angle 4$

$m\angle 1 > m\angle 3$

Indirect Proof:

Step 1: Make the assumption that $m\angle 1 \ngtr m\angle 3$ and $m\angle 1 \ngtr m\angle 4$. Thus, $m\angle 1 \leq m\angle 3$ and $m\angle 1 \leq m\angle 4$.

Step 2: We will only show that the assumption $m\angle 1 \leq m\angle 3$ leads to a contradiction, since the argument for $m\angle 1 \leq m\angle 4$ uses the same reasoning.

$m\angle 1 \leq m\angle 3$, means that either $m\angle 1 = m\angle 3$ or $m\angle 1 < m\angle 3$. So, we need to consider both cases.

Case 1: $m\angle 1 = m\angle 3$

Since $m\angle 3 + m\angle 4 = m\angle 1$ by the Exterior Angle Theorem, we have $m\angle 3 + m\angle 4 = m\angle 3$ by substitution. Then $m\angle 4 = 0$, which contradicts the fact that the measure of an angle is greater than 0.

Case 2: $m\angle 1 < m\angle 3$

By the Exterior Angle Theorem, $m\angle 3 + m\angle 4 = m\angle 1$. Since angle measures are positive, the definition of inequality implies $m\angle 1 > m\angle 3$ and $m\angle 1 > m\angle 4$. This contradicts the assumption that $m\angle 1 \leq m\angle 3$ and $m\angle 1 \leq m\angle 4$.

Step 3: In both cases, the assumption leads to the contradiction of a known fact. Therefore, the assumption that $m\angle 1 \leq m\angle 3$ must be false, which means that $m\angle 1 > m\angle 3$ must be true. Likewise, $m\angle 1 > m\angle 4$.

Some of the properties of inequalities that you encountered in algebra were used in the proof of Theorem 5–8. For example, in Step 1 of the proof, it was stated that if $m\angle 1 \ngtr m\angle 3$, then $m\angle 1 < m\angle 3$ or $m\angle 1 = m\angle 3$. This statement is an application of the Comparison or Trichotomy Property which states that for any two numbers a and b, either $a > b$, $a < b$, or $a = b$. The following chart gives a list of properties of inequalities you studied in algebra.

Properties of Inequality for Real Numbers	
For all numbers *a, b,* and *c,*	
Comparison Property	$a < b$, $a = b$, or $a > b$.
Transitive Property	**1.** If $a < b$ and $b < c$, then $a < c$. **2.** If $a > b$ and $b > c$, then $a > c$.
Addition and Subtraction Properties	**1.** If $a > b$, then $a + c > b + c$ and $a - c > b - c$. **2.** If $a < b$, then $a + c < b + c$ and $a - c < b - c$.
Multiplication and Division Properties	**1.** If $c > 0$ and $a < b$, then $ac < bc$ and $\frac{a}{c} < \frac{b}{c}$. **2.** If $c > 0$ and $a > b$, then $ac > bc$ and $\frac{a}{c} > \frac{b}{c}$. **3.** If $c < 0$ and $a < b$, then $ac > bc$ and $\frac{a}{c} > \frac{b}{c}$. **4.** If $c < 0$ and $a > b$, then $ac < bc$ and $\frac{a}{c} < \frac{b}{c}$.

Indirect proofs use a problem-solving strategy called **working backward**. After assuming that the conclusion is false, you work backward from the assumption to show that, for the given information, the assumption is false. The strategy of working backward can be used in many other problem-solving situations.

Example 2

Real World
APPLICATION

Banking

Hama's used car loan costs \$207.98 per month. This includes 9% interest paid to the bank for borrowing the money. When Hama bought the car, she was able to negotiate a 10% discount and made a down payment of \$2000. If the loan is for 48 months, what was the asking price of the car?

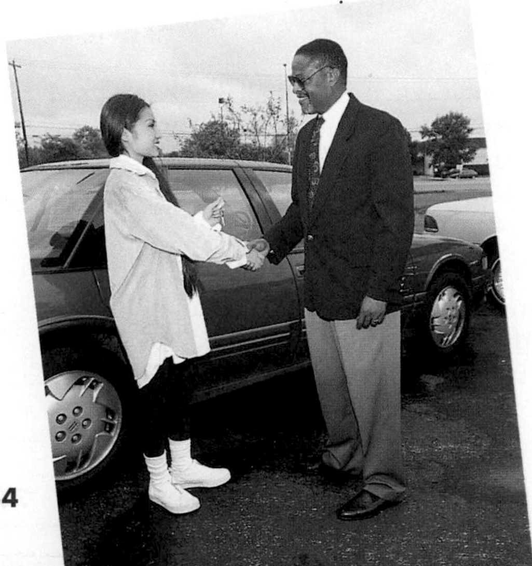

Explore We know the monthly payment, the interest rate, the down payment, the discount, and the length of the loan. We are looking for the asking price of the car.

Plan Work backward through each step to find the asking price.

Solve ***Interest*** The monthly payment is \$207.98 per month including 9% interest. Let p = the price per month before interest is added.

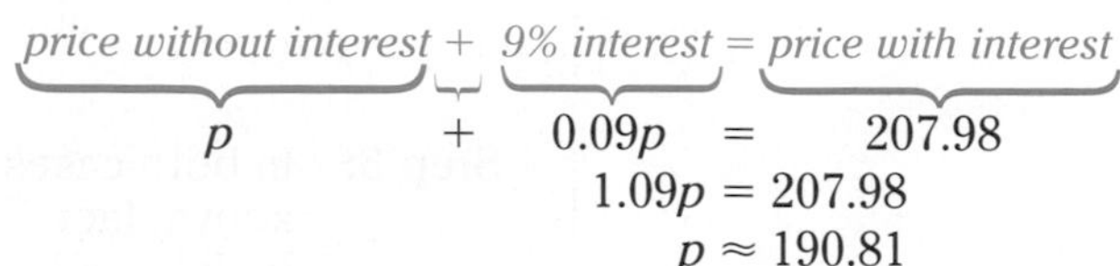
price without interest + 9% interest = price with interest

$$p + 0.09p = 207.98$$
$$1.09p = 207.98$$
$$p \approx 190.81$$

The monthly payment without the interest payment is \$190.81.

Loan amount Let l = the loan amount. The loan amount is the total of the money paid, less the interest.

$$\underbrace{\text{number of payments}}_{48} \underbrace{\times}_{\times} \underbrace{\text{payment amount}}_{190.81} = \underbrace{\text{loan amount}}_{l}$$

$$9158.88 = l$$

The loan amount is $9158.88.

Down payment The down payment was subtracted from the sale price s to determine the loan amount. So to find the sale price, we should add the down payment to the loan amount.

$$\underbrace{\text{loan amount}}_{9158.88} \underbrace{+}_{+} \underbrace{\text{down payment}}_{2000} = \underbrace{\text{sale price}}_{s}$$

$$11{,}158.88 = s$$

The sale price was $11,158.88.

Discount Hama negotiated a 10% discount. Let a represent the asking price.

$$\underbrace{\text{asking price}}_{a} \underbrace{-}_{-} \underbrace{\text{discount}}_{0.10a} = \underbrace{\text{sale price}}_{11{,}158.88}$$

$$0.90a = 11{,}158.88$$

$$a \approx \$12{,}398.76$$

The asking price of the car was $12,399.

Examine Check the solution by starting with the asking price and working forward to find the monthly payment with interest.

CHECK FOR UNDERSTANDING

Communicating Mathematics

Study the lesson. Then complete the following.

1. **Draw** $\triangle ABC$ with an exterior angle at A named $\angle 1$. Which angles have measures less than the measure of $\angle 1$?
2. **Analyze** the argument that given $x > y$ and $z > w$, then $xz > yw$.
3. **Compare and contrast** the properties for equality of real numbers that were reviewed in Lesson 2–4 to the properties of inequality in this lesson.

4. **Assess Yourself** Describe a situation in your life when you have used indirect reasoning or the work-backward, problem-solving strategy.

Guided Practice

State the assumption you would make to start an indirect proof of each statement.

5. Lines ℓ and m intersect at point X.
6. If the alternate interior angles formed by two lines and a transversal are congruent, the lines are parallel.
7. Sabrina ate the leftover pizza.

Use the figure at the right for Exercises 8–10.

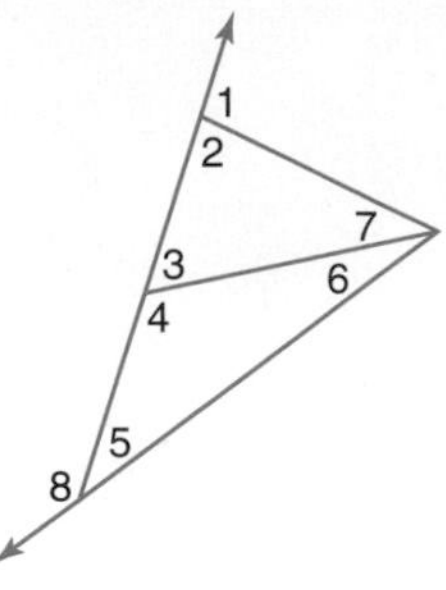

8. List all the angles whose measures are less than $m\angle 1$.
9. List all the angles whose measures are greater than $m\angle 2$.
10. Which is greater, $m\angle 7$ or $m\angle 8$? Justify your answer.

Name the property that justifies each statement.

11. If $-3x < 24$, then $x > -8$.
12. If $AB < CD$, then $AB + RS < CD + RS$.
13. If $m\angle R > m\angle S$ and $m\angle S > m\angle T$ and $m\angle T > m\angle U$, then $m\angle R > m\angle U$.
14. Write a two-column proof.

    Given: B, C, and D lie on $\overleftrightarrow{AE}$.
    $\overline{BD} \cong \overline{AC}$

    Prove: $AC < BE$

15. Write an indirect proof.

    Given: m is not parallel to n.

    Prove: $m\angle 3 \neq m\angle 2$

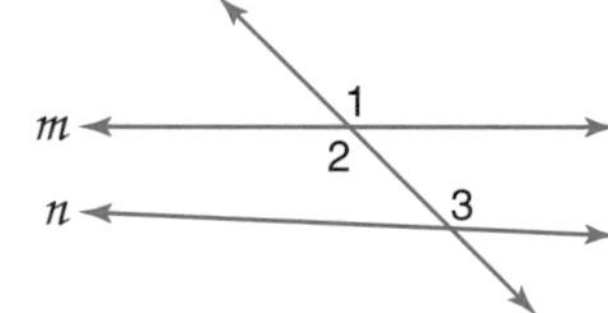

16. **Sales** Mrs. Pinel is offering her students the opportunity to buy discounted graphing calculators. She was able to get the calculators on sale for 20% off at the store, and then a 10% school discount. The cost for each calculator is \$97.05 including the 7% sales tax. What was the original price of each calculator?

EXERCISES

Practice

State the assumption you would make to start an indirect proof of each statement.

17. $\overline{AB} \cong \overline{CD}$
18. An altitude of an isosceles triangle is also a median.
19. The disk is defective.
20. The suspect is guilty.
21. If two altitudes of a triangle are congruent, then the triangle is isosceles.
22. If two lines are cut by a transversal and corresponding angles are congruent, then the lines are parallel.

Use the figure at the right for Exercises 23–28.

23. Which is greater, $m\angle 7$ or $m\angle 5$?
24. Which is greater, $m\angle 4$ or $m\angle 1$?
25. Name an angle whose measure is greater than $m\angle 6$.
26. List all the angles whose measures are greater than $m\angle 2$.
27. List all the angles whose measures are less than $m\angle 7$.
28. If $m\angle 2 > m\angle 6$, which is greater, $m\angle 8$ or $m\angle 1$?

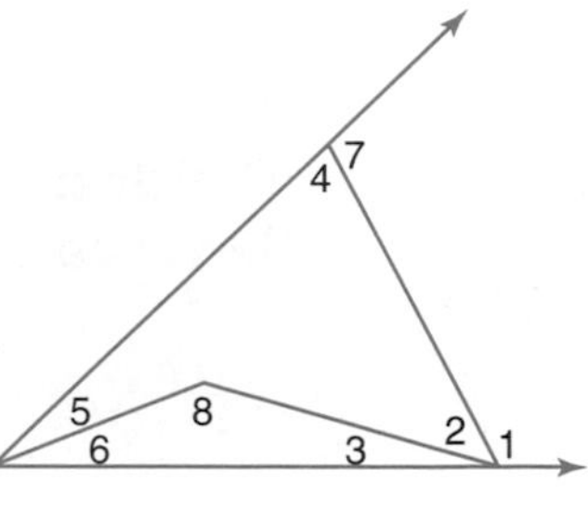

Name the property that justifies each statement.

29. If $5x < 25$, then $x < 5$.
30. Given $m\angle A < m\angle B$ and $m\angle C < m\angle A$, then $m\angle C < m\angle B$.
31. If $RT \neq AB$, then $RT > AB$ or $RT < AB$.
32. Given $x + 5 > 9$, then $x > 4$.
33. If $m\angle X < m\angle Y$, then $\frac{m\angle X}{2} < \frac{m\angle Y}{2}$.
34. If $m\angle X \not> m\angle Y$, then $m\angle X < m\angle Y$ or $m\angle X = m\angle Y$.

Use the figure at the right to write a paragraph proof.

Proof

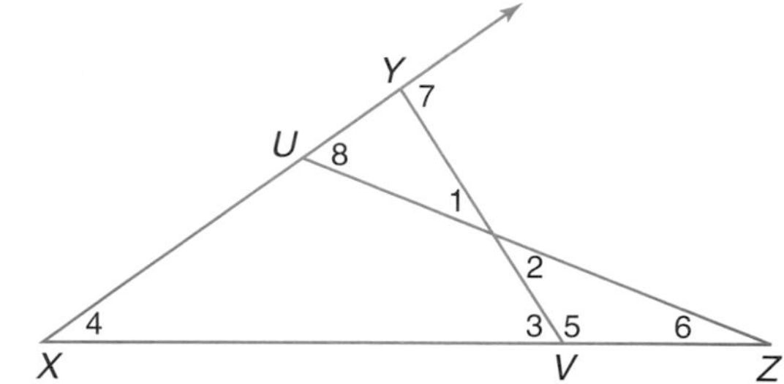

35. **Given:** $\triangle XYV$, $\triangle XZU$

Prove: $m\angle 7 > m\angle 6$

36. **Given:** $\triangle XYV$, $\triangle XZU$
$\angle 4 \cong \angle 1$

Prove: $m\angle 5 > m\angle 2$

Use the figure at the right to write a two-column proof.

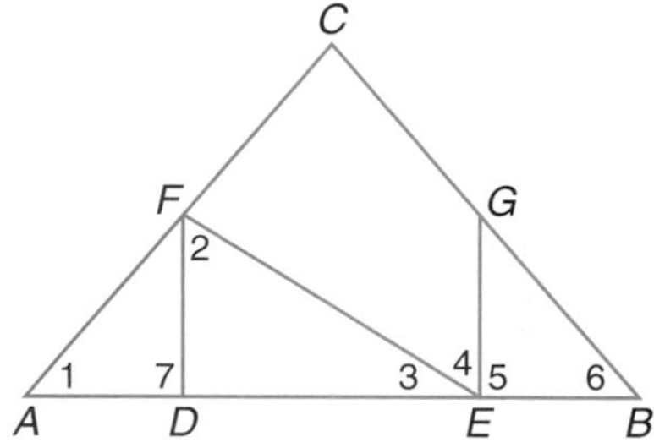

37. **Given:** $\overline{FD} \perp \overline{AB}$
$\overline{GE} \perp \overline{AB}$

Prove: $m\angle 5 > m\angle 2$

38. **Given:** $\overline{AC} \cong \overline{BC}$
$\overline{FD} \perp \overline{AB}$
$\overline{GE} \perp \overline{AB}$

Prove: $m\angle 7 > m\angle 4$

Write an indirect proof.

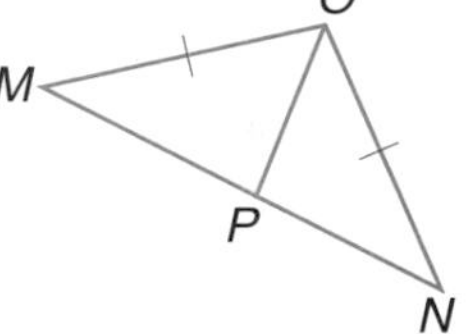

39. **Given:** $\overline{MO} \cong \overline{ON}$
$\overline{MP}$ is not congruent to $\overline{NP}$.

Prove: $\angle MOP$ is not congruent to $\angle NOP$.

40. A triangle can have no more than one acute exterior angle.
41. If no two angles in a triangle are congruent, then no two sides are congruent.

Critical Thinking

42. Mr. Mendez was checking on the date he had attended a four-day conference a year ago. The page in his record book was torn and all that remained of the date for the meeting was *ber 31*. What was the month of the first day of the conference?

Applications and Problem Solving

43. **Work Backward** A guard's prisoner is given the choice of opening one of two doors. Each door leads either to freedom or to the dungeon. A sign on the door on the right reads "This door leads to freedom and the other door leads to the dungeon." The door on the left has a sign that reads "One of these doors leads to freedom and the other leads to the dungeon." A guard tells the prisoner that one of the signs is true and the other is false. Which door should the prisoner choose? Why?

44. **Literature** In Agatha Christie's famous mystery *And Then There Were None*, all ten people on an island were murdered. Because of a storm, no one else had been able to land on or leave the island during the time the murders took place. The murderer thought the police might be able to solve the crime because they knew that the murderer had no background in murder and that, at a certain period of time, there were only four people present. Of those four people, only one person had no background in murder. Explain how the police could use indirect reasoning to determine the identity of the murderer.

45. **Law** The defense attorney said to the jury, "My client is not guilty. According to the police, the crime occurred on July 14 at 6:30 P.M. in Boston. I can prove that at that time my client was attending a business meeting in New York City. A verdict of not guilty is the only possible verdict." Is this an example of indirect reasoning? Explain why or why not.

Mixed Review

46. Prove that if two altitudes of a triangle are congruent, then the triangle is isosceles. (Lesson 5–2)

47. Draw and label $\triangle ABC$ with $\overline{AD}$, which is an altitude and a median. (Lesson 5–1)

48. Draw $\triangle JKL$ with point M on $\overline{KL}$ such that $\overline{JM}$ bisects $\angle KJL$, but $\overline{KM}$ and $\overline{ML}$ are not congruent. (Lesson 4–6)

49. If $\triangle NQD$ has vertices $N(2, -1)$, $Q(-4, -1)$, and $D(-1, -3)$, describe $\triangle NQD$ in terms of its angles and sides. (Lesson 4–1)

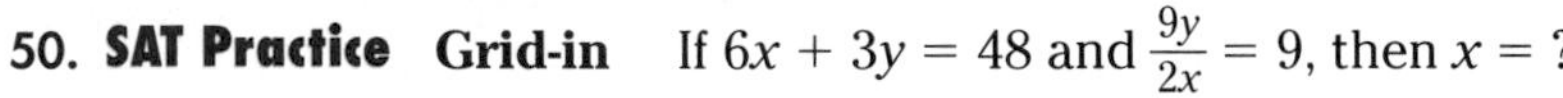

50. **SAT Practice Grid-in** If $6x + 3y = 48$ and $\frac{9y}{2x} = 9$, then $x = ?$

51. Are the pieces of glass in a double-paned window a model of parallel or intersecting planes? (Lesson 3–1)

52. Write the statement *A median of a triangle bisects one side of the triangle* in if-then form. (Lesson 2–2)

53. **Draw a Diagram** Draw a tree diagram to determine the number of lunch specials at Lucky's Restaurant if each special comes with chicken, turkey, or fish; salad or soup; a baked potato or fries; and pie, frozen yogurt, or fruit. (Lesson 1–3)

Algebra

For **Extra Practice**, see page 773.

54. Find the next three terms in the sequence $-2x$, $6x$, $14x$, ...

55. Evaluate $2ab - \frac{c}{d}$ if $a = -9.32$, $b = -0.04$, $c = 0.932$, and $d = 2.5$.

SELF TEST

Refer to the figure at the right for Exercises 1–4.

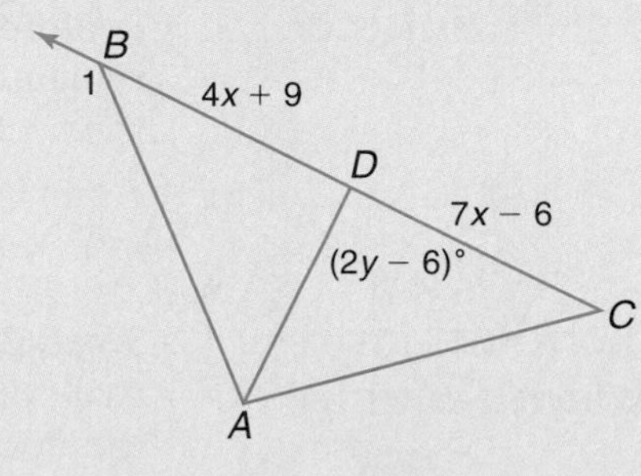

1. What is the value of x if $\overline{AD}$ is a median of $\triangle ABC$? (Lesson 5–1)

2. Find the value of y if $\overline{AD}$ is an altitude of $\triangle ABC$. (Lesson 5–1)

3. Suppose $\overline{AD}$ is the perpendicular bisector of $\overline{BC}$. Is $\triangle ABD \cong \triangle ACD$? Justify your answer. (Lesson 5–2)

4. Given that $\overline{AD}$ is an altitude of $\triangle ABC$, what additional information do you need to prove that $\triangle ABD \cong \triangle ACD$ by HL? (Lesson 5–2)

5. Write an indirect proof for *If $\overleftrightarrow{AB}$ and $\overleftrightarrow{PQ}$ are skew lines, then $\overleftrightarrow{AP}$ and $\overleftrightarrow{BQ}$ are skew lines.* (Lesson 5–3)

5-4 Inequalities for Sides and Angles of a Triangle

What YOU'LL LEARN

- To recognize and apply relationships between sides and angles in a triangle.

Why IT'S IMPORTANT

You can use the inequalities in triangles to solve problems involving transportation and sports.

In Chapter 4, you learned that in a triangle if two sides are congruent, then the angles opposite those sides are congruent. In the following activity, you will investigate the relationship between the measures of two angles of a triangle if the sides opposite those angles are not congruent.

Inequalities for Sides and Angles of Triangles

Materials:  ruler  protractor

- Draw $\triangle ABC$.
- Measure each side of the triangle and record the measures in a table like the one at the right.
- Measure each angle of the triangle. Record each measure in the table.

Side	Measure
$\overline{AB}$ $\overline{BC}$ $\overline{AC}$	
Angle	**Measure**
$\angle A$ $\angle B$ $\angle C$	

Your Turn

a. Describe the measure of the angle opposite the longest side of the triangle in terms of the other angles.

b. Describe the measure of the angle opposite the shortest side of the triangle in terms of the other angles.

c. Repeat the activity for another triangle.

d. Make a conjecture about the relationship between the measures of sides and angles of a triangle.

The Modeling Mathematics activity suggests the following theorem.

Theorem 5-9 **If one side of a triangle is longer than another side, then the angle opposite the longer side has a greater measure than the angle opposite the shorter side.**

This theorem can be abbreviated as "If one side of a $\triangle$ is longer than another, the $\angle$ opp. the longer side > the $\angle$ opp. the shorter side."

The converse of the theorem is also true.

Theorem 5-10 **If one angle of a triangle has a greater measure than another angle, then the side opposite the greater angle is longer than the side opposite the lesser angle.**

This theorem can be abbreviated as "If an $\angle$ of a $\triangle$ > another, the side opp. the greater $\angle$ is longer than the side opp. the lesser $\angle$."

A proof of Theorem 5-9 is shown on the next page. *You will be asked to write a proof of Theorem 5-10 in Exercise 34.*

Proof of Theorem 5–9

Given: $\triangle ABC$
$AC > BA$

Prove: $m\angle C < m\angle ABC$

Paragraph Proof:

Draw auxiliary segment, $\overline{BF}$, shown in red, so that F is between A and C and $\overline{AF} \cong \overline{AB}$.

If two sides of a triangle are congruent, the angles opposite those sides are congruent. Therefore, $\angle 1 \cong \angle 2$.

Notice that $\angle 1$ is an exterior angle of $\triangle BFC$. Since the measure of an exterior angle is greater than the measure of either of its corresponding remote interior angles, $m\angle C < m\angle 1$.

The Angle Addition Postulate allows us to say that $m\angle 2 + m\angle 3 = m\angle ABC$. By the definition of inequality, $m\angle 2 < m\angle ABC$. Thus, $m\angle 1 < m\angle ABC$ by the Substitution Property of Equality.

We can apply the Transitive Property of Inequality to the inequalities $m\angle C < m\angle 1$ and $m\angle 1 < m\angle ABC$ to conclude that $m\angle C < m\angle ABC$.

Theorems 5–9 and 5–10 can be used to prove statements about angle or side relationships in other figures.

Example  1

Proof

Write a two-column proof.

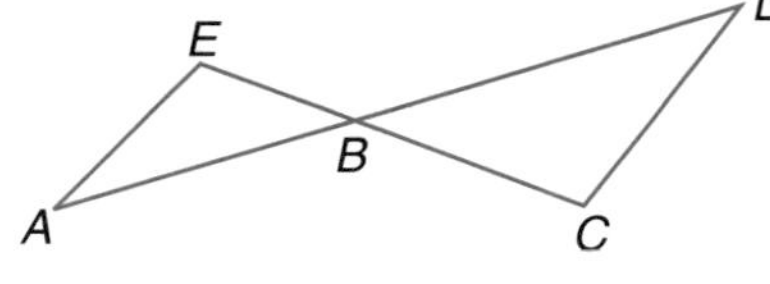

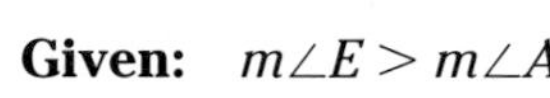

Given: $m\angle E > m\angle A$
$m\angle C > m\angle D$

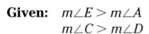

Prove: $AD > EC$

Proof:

Statements	Reasons
1. $m\angle E > m\angle A$, $m\angle C > m\angle D$	**1.** Given
2. $AB > EB$, $BD > BC$	**2.** If one $\angle$ of a $\triangle$ is greater than another, then the side opp. the greater $\angle$ is longer than the side opp. the lesser $\angle$.
3. $AB + BD > EB + BC$	**3.** Addition Property ($\neq$)
4. $AB + BD = AD$, $EB + BC = EC$	**4.** Segment Addition Postulate
5. $AD > EC$	**5.** Substitution Property (=)

Example 2

Algebra

Draw $\triangle RST$ with vertices $R(-2, 4)$, $S(-5, -8)$, and $T(6, 10)$. List the angles in order from the greatest measure to least measure.

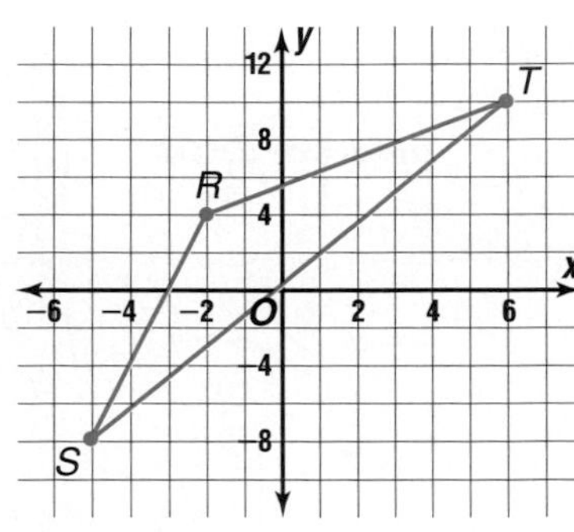

To compare the measures of the angles in $\triangle RST$, you can determine the lengths of the sides and then apply Theorem 5–9. Use the distance formula to find the length of each side of $\triangle RST$.

$$RS = \sqrt{[-2 - (-5)]^2 + [4 - (-8)]^2}$$
$$= \sqrt{(3)^2 + (12)^2}$$
$$= \sqrt{153} \approx 12.37$$

$$ST = \sqrt{(-5 - 6)^2 + (-8 - 10)^2}$$
$$= \sqrt{(-11)^2 + (-18)^2}$$
$$= \sqrt{445} \approx 21.10$$

$$RT = \sqrt{(-2 - 6)^2 + (4 - 10)^2}$$
$$= \sqrt{(-8)^2 + (-6)^2}$$
$$= \sqrt{100} = 10$$

Because $ST > RS > RT$, by Theorem 5–9, $m\angle R > m\angle T > m\angle S$. So, the angles in order from largest to smallest are $\angle R$, $\angle T$, $\angle S$.

LOOK BACK
You can review the distance formula in Lesson 1-4.

In the figure at the right, T is a point not on line m, and R is the point on m such that $\overline{TR} \perp m$. In Lesson 3–7, we defined the distance from T to m as TR. It was stated, without proof, that $\overline{TR}$ was the shortest segment from T to m. Assuming this is true, if L is any point on m other than R, $\overline{TL}$ would have to be longer than $\overline{TR}$. *Why?*

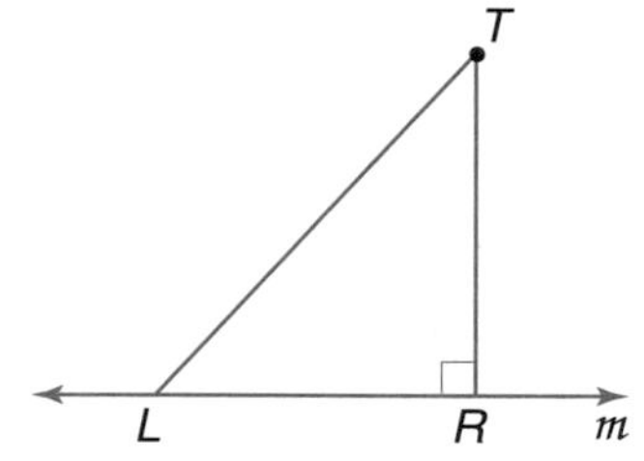

We will restate our assumption about the shortest segment from a point to a line as a theorem and show that it is true as a direct result of Theorem 5–10.

The perpendicular segment from a point to a line is the shortest segment from the point to the line.

Proof of Theorem 5–11

Given: $\overline{TR} \perp m$
$\overline{TL}$ is any segment from T to m that is different from $\overline{TR}$.

Prove: $TL > TR$

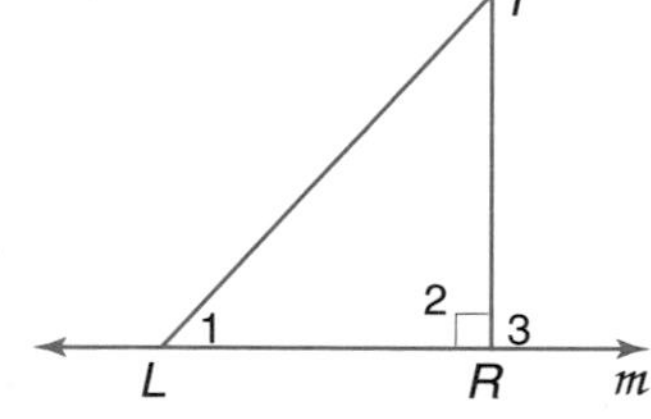

Proof:

Statements	Reasons
1. $\overline{TR} \perp m$	1. Given
2. $\angle 2$ and $\angle 3$ are right angles.	2. $\perp$ lines form four rt. $\angle$s.
3. $\angle 2 \cong \angle 3$	3. All rt. $\angle$s are $\cong$.
4. $m\angle 2 = m\angle 3$	4. Def. $\cong$ $\angle$s
5. $m\angle 3 > m\angle 1$	5. Exterior Angle Inequality Theorem
6. $m\angle 2 > m\angle 1$	6. Substitution Prop. (=)
7. $TL > TR$	7. If an $\angle$ of a $\triangle$ > another, then the side opp. the greater $\angle$ is longer than the side opp. the lesser $\angle$.

Corollary 5–1 follows directly from Theorem 5–11. *You will be asked to prove the corollary in Exercise 35.*

Corollary 5–1	The perpendicular segment from a point to a plane is the shortest segment from the point to the plane.

CHECK FOR UNDERSTANDING

Communicating Mathematics

Study the lesson. Then complete the following.

1. **Draw and label** a triangle that has sides measuring about 5 centimeters, 6.7 centimeters and 7.7 centimeters. Its angle measures are about 40°, 60°, and 80°. Label the angles. Explain how you knew where to place the labels.
2. **Analyze** finding the distance between two lines in a coordinate plane. Could you use the difference between the y-intercepts as the distance? Explain.
3. **Write a convincing argument** for the statement *In an obtuse triangle, the longest side is opposite the obtuse angle.*
4. **State** the conclusions you can draw about $\triangle XYZ$ for each relationship.
   a. $XY = XZ$
   b. $m\angle Y > m\angle Z$
   c. $XY < XZ$

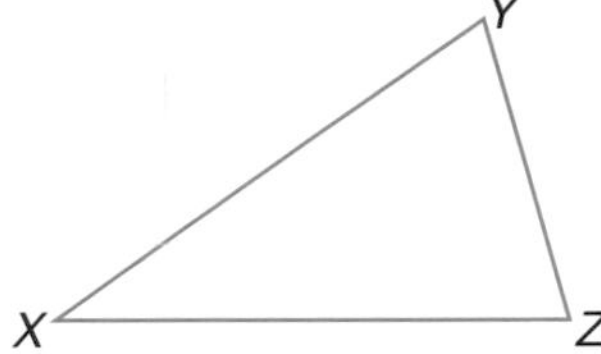

5. Draw obtuse $\triangle ABC$. Construct the distance from A to the line containing $\overline{BC}$ and the distance from B to the line containing $\overline{AC}$. Explain what each distance represents.

Guided Practice

Refer to the figure at the right for Exercises 6–8.

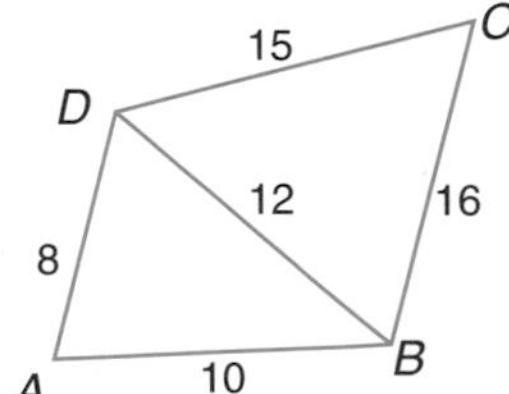

6. Which is greater, $m\angle CBD$ or $m\angle CDB$?
7. Is $m\angle ADB > m\angle DBA$?
8. Which is greater, $m\angle CDA$ or $m\angle CBA$?

Given the angles in the figure at the right, answer each question.

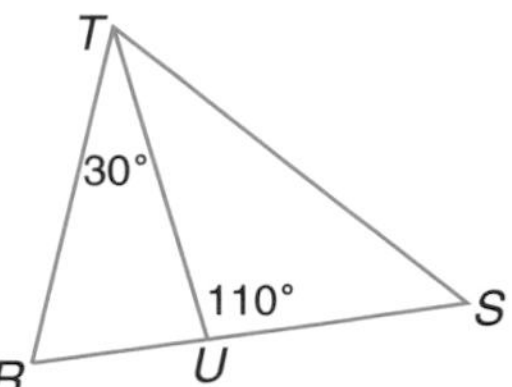

9. Which side of $\triangle RTU$ is the longest?
10. Name the side of $\triangle UST$ that is the longest.
11. If $\overline{TU}$ is an angle bisector, which side of $\triangle RST$ is the longest?
12. Find the value of x and list the sides of $\triangle ABC$ in order from shortest to longest if $m\angle A = 3x + 20$, $m\angle B = 2x + 37$, and $m\angle C = 4x + 15$.

13. Write a paragraph proof for the statement *The altitude to the hypotenuse of a right triangle is shorter than either of the legs.*

14. **Transportation** The line graphed below represents average fuel cost as a function of the speed of a plane. Points A, B, and C represent a DC-10, a DC-9, and a Boeing 737, respectively. For which plane is the prediction the worst? Justify your claim.

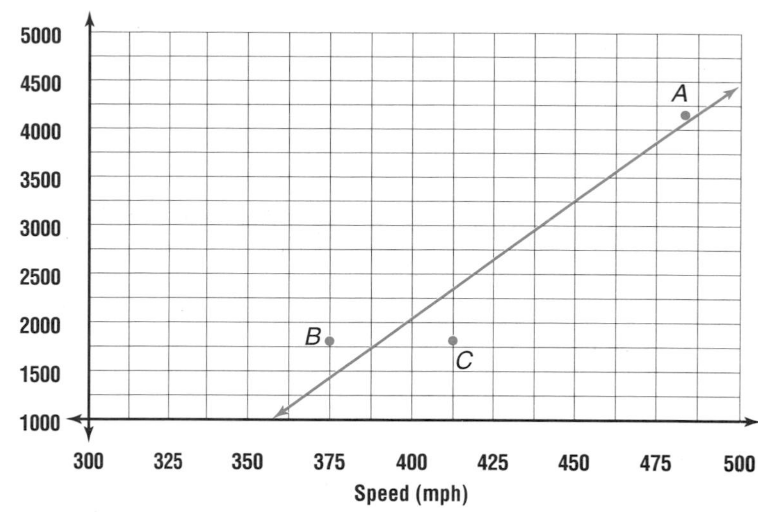

EXERCISES

Practice

Refer to the figure at the right for Exercises 15–17.

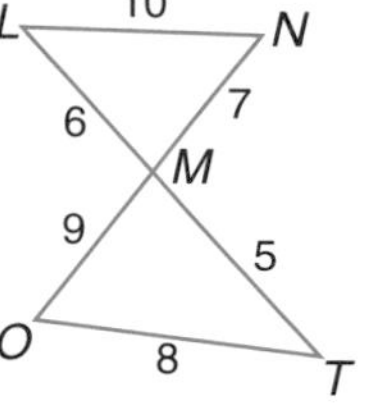

15. Name the angle with the least measure in $\triangle LMN$.
16. Which angle in $\triangle MOT$ has the greatest measure?
17. Name the greatest of the six angles in the two triangles, LMN and MOT.

Refer to the figure at the right for Exercises 18–20.

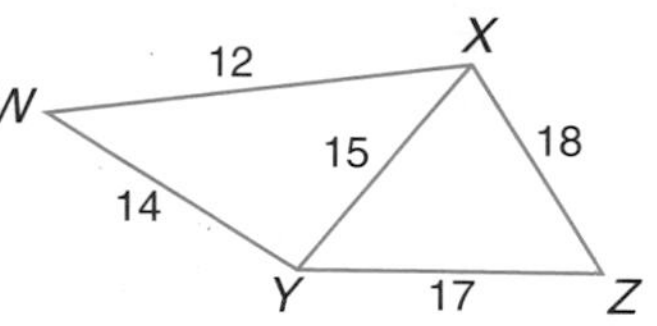

18. Name the angles with the least and the greatest measure in $\triangle XYZ$.
19. What is the angle with the least measure in $\triangle WXY$?
20. Suppose $\angle WXY \cong \angle XYZ$. Which angle of $\triangle XYZ$ or $\triangle WXY$ has the greatest measure?

Refer to the figure at the right for Exercises 21–25.

21. What is the longest segment in $\triangle CED$?
22. Find the longest segment in $\triangle ABE$.
23. Find the longest segment in the figure. Justify your choice.
24. What is the shortest segment in $BCDE$?
25. Is the figure drawn to scale? Explain.

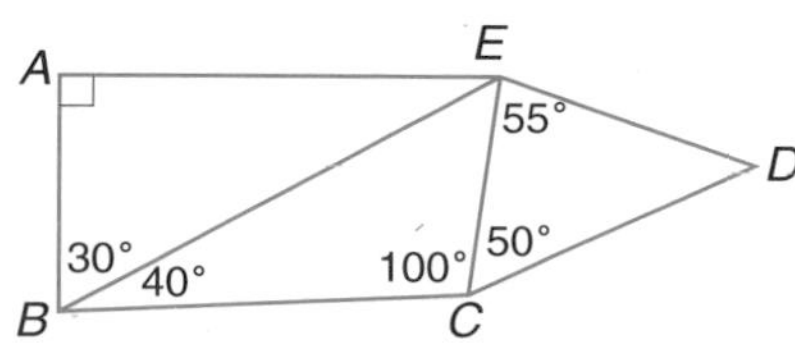

Algebra

Find the value of x and list the sides of $\triangle ABC$ in order from shortest to longest if the angles have the indicated measures.

26. $m\angle A = 9x + 29$, $m\angle B = 93 - 5x$, $m\angle C = 10x + 2$
27. $m\angle A = 9x - 4$, $m\angle B = 4x - 16$, $m\angle C = 68 - 2x$
28. $m\angle A = 12x - 9$, $m\angle B = 62 - 3x$, $m\angle C = 16x + 2$

Write a two-column proof.

29. Given: $\overline{RQ}$ bisects $\angle SRT$.
Prove: $m\angle SQR > m\angle SRQ$

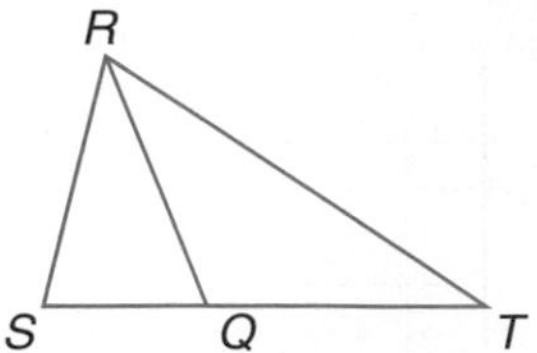

30. Given: $\overline{AB} \perp$ plane BCD
Prove: $AD > BD$

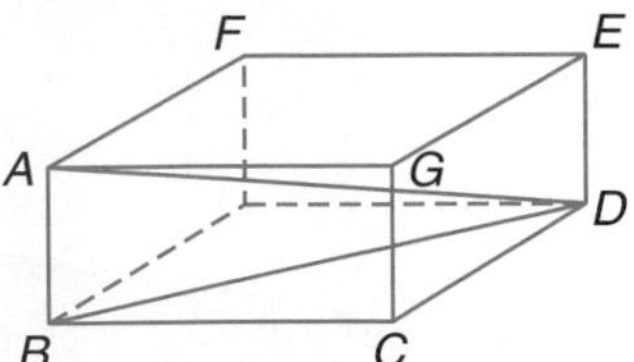

31. Given: D is on $\overline{AC}$.
$\angle BDC$ is acute.
Prove: $BA > BD$

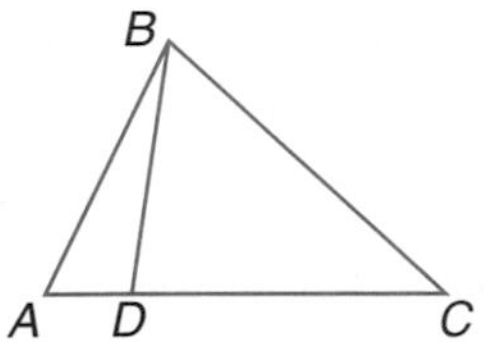

32. Given: $\overline{ON} \cong \overline{OM}$
$\overline{OM} \cong \overline{LM}$
Prove: $m\angle MON > m\angle OMN$

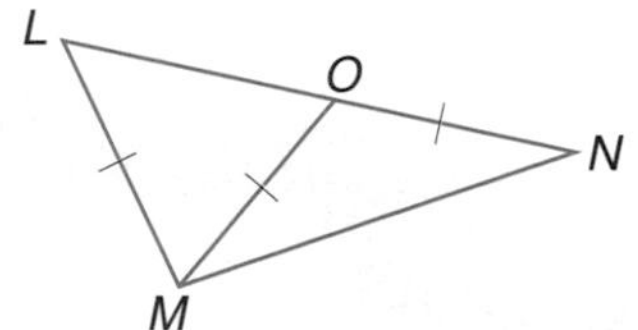

33. Write a paragraph proof for the following statement. *If a triangle is not isosceles, the median to any side of the triangle is greater than the altitude to that side.*

34. Write an indirect proof for Theorem 5–10: *If one angle of a triangle has a greater measure than another angle, then the side opposite the greater angle is longer than the side opposite the lesser angle.*

35. Write a proof for Corollary 5–1: *The perpendicular segment from a point to a plane is the shortest segment from the point to the plane.*

Critical Thinking

36. Find the coordinates of point D on line BC such that $\overline{AD}$ is the shortest distance from A to $\overleftrightarrow{BC}$ for $A(9, 9)$, $B(-1, -11)$, and $C(-2, 2)$.

Applications and Problem Solving

37. Sports Analysts have found that the world record for the 10,000-meter run has been decreasing steadily since 1940. The trend has been described by the equation $y = 30.18 - 0.07x$, where x is the number of years since 1940 and y is the record time in minutes.

a. The table below shows several world record times for the 10,000-meter run. Graph the equation including a point to represent each record.

b. For whose record is the equation the best and worst predictor? Explain.

RECORD HOLDER, COUNTRY	YEAR	TIME
EMIL ZATOPEK, CZECHOSLOVAKIA	1949	29:21.6
VLADIMIR KUTS, SOVIET UNION	1956	28:30.4
RON CLARKE, AUSTRALIA	1965	27:39.4
HENRY RONO, KENYA	1978	27:22.3
ARTURO BARRIOS, MEXICO	1989	27:08.28
HAILE GEBRSELASSIE, ETHIOPIA	1995	26:43.53

38. Recreation Mr. Ramirez is assembling an intricate framework that is part of a jungle gym for his daughter. The instructions say to attach bars to a given bar whose ends are marked as A and B so that $\angle A$ is a 30° angle and $\angle B$ is a 60° angle. The bars available are two different lengths. Which bars should he attach at end A and end B?

Mixed Review

39. **Work Backward** Songan, Lucy, and Tamoko are playing a card game. They have a rule that when a player loses a hand, he or she must subtract enough points from his or her score to double each of the other players' scores. First Songan loses a hand, then Tamoko, then Lucy. If each player now has 8 points, who lost the most points? (Lesson 5–3)

40. Suppose $\triangle ABC$ and $\triangle XYZ$ are right triangles with right angles at $\angle C$ and $\angle Z$. Determine whether $\triangle ABC \cong \triangle XYZ$ if it is given that $\angle A \cong \angle X$ and $\angle B \cong \angle Y$. Explain. (Lesson 5–2)

41. Draw an acute $\triangle DEF$ and construct the angle bisectors of the triangle. (Lesson 5–1)

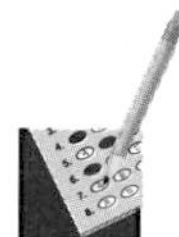

42. **SAT Practice** For all integers $x \neq 2$, let $<x> = \frac{1+x}{x-2}$. Which of the following has the greatest value?

    A $<0>$ **B** $<1>$ **C** $<3>$ **D** $<4>$ **E** $<5>$

43. Determine if the statement *A right triangle is equilateral* is *true* or *false*. If it is false, draw a counterexample. (Lesson 4–1)

44. If possible, write a conclusion. State the law of logic used. (Lesson 2–3)

    (1) $\overline{AX}$ is an altitude of $\triangle ABC$ if $\overline{AX} \perp \overline{BC}$ and X is on $\overline{BC}$.

    (2) $\overline{AX} \perp \overline{BC}$ and X is on $\overline{BC}$.

45. Make a table of values to determine the coordinates of three points that lie on the graph of $y = 2x - 6$. Use the headings x, $2x - 6$, and (x, y). Then use these three points to draw the line in a coordinate plane. (Lesson 1–1)

Algebra

46. Solve $\frac{1}{3} + \frac{5}{x} = \frac{1}{x}$.

47. **Nutrition** One kiwi fruit has 230% of the recommended daily allowance of vitamin C. Write this percent as a decimal.

For **Extra Practice**, see page 773.

WORKING ON THE In·ves·ti·ga·tion

Refer to the Investigation on pages 176–177.

If you read *Alice's Adventures in Wonderland,* you may remember that Alice grows very big and very small. Suppose it was possible to increase the size of a living thing in this way. You would find that the following relationship holds true.

$$\frac{\text{space enclosed by new organism}}{\text{space enclosed by old organism}} > \frac{\text{amount of skin on new organism}}{\text{amount of skin on old organism}}$$

(left: *Volume*; right: *Surface Area*)

1 Some small animals like frogs can respire through their skin, but a larger animal like a human could not support its volume with its surface area. Research human respiration.

2 Most mammals stay cool by releasing heat through their skin and by perspiring. What features of an elephant compensate for its small surface area compared to its volume?

3 Cacti are the trees and shrubs of the hot, dry desert. The surface area of a cactus is minimized to reduce water loss, yet it functions like a tree's branches and leaves. How does the cactus achieve this?

Add the results of your work to your Investigation Folder.

5–5A Using Technology The Sides of a Triangle

A Preview of Lesson 5–5

There is a special relationship among the measures of the sides of a triangle. The graphing calculator program below will allow you to investigate the relationship. The program will generate three random numbers between 0 and 100 and determine if segments with these measures can form a triangle.

```
PROGRAM: TRIANGLE
:int (100rand)→A
:int (100rand)→B
:int (100rand)→C
:Disp "SIDES = ", A, B, C
:If A+B≤C or A+C≤B or B+C≤A
:Then
:Disp "NO TRIANGLE"
:Else
:Disp "TRIANGLE EXISTS"
:Stop
```

Run the program at least ten times. Record the sets of side measures that do or do not form a triangle in a table like the one below. Several sample sets are shown below.

No Triangle	Triangle Exists
63, 8, 97	19, 26, 27
4, 33, 99	51, 40, 73
	28, 79, 95
	54, 85, 97

EXERCISES

Work in groups of three or four.

1. Compare the numbers in each column of your table with the numbers that the other members of your group recorded.

2. Suppose the first number in each set is S1, the second number is S2, and the third number is S3. Find S1 + S2, S2 + S3, and S1 + S3 for the first three sets of numbers in each column.

3. Using the sums from Exercise 2, compare S1 + S2 to S3, S2 + S3 to S1, and S1 + S3 to S2. What do you observe? Compare your results to those of other members of your group.

4. Write a conjecture about the measures of the sides of a triangle based on your observations.

5-5 The Triangle Inequality

What YOU'LL LEARN

- To apply the Triangle Inequality Theorem.

Why IT'S IMPORTANT

You can use the triangle inequality to solve problems involving carpentry and mathematics history.

Design

The Oval is at the heart of the main campus of The Ohio State University in Columbus. The master plan for the campus created in 1893 laid out the open space surrounded by buildings. If you were walking from the Main Library, at A, to Mendenhall Lab, at B, and wanted to get there quickly, would you choose the blue path or the red path? If you chose the blue path, you probably reasoned that a path that is a straight line is the shortest path. You are applying the **Triangle Inequality Theorem**.

Theorem 5-12 *Triangle Inequality Theorem*	**The sum of the lengths of any two sides of a triangle is greater than the length of the third side.**

The Triangle Inequality Theorem shows that some sets of line segments cannot be used to form a triangle because their lengths do not satisfy the inequality. *You will be asked to prove the Triangle Inequality Theorem in Exercise 44.*

Example 1

Probability

If 18, 45, 21, and 52 represent lengths of segments, what is the probability that a triangle can be formed if three of these numbers are chosen at random as lengths of the sides?

Make a list of the possible sets of three measures.

a. 18, 45, 21 **b.** 18, 45, 52 **c.** 18, 21, 52 **d.** 45, 21, 52

Test the triangle inequality for each case.

Inequality test		Sketch	Triangle?
a. Is $18 + 45 > 21$? Is $18 + 21 > 45$? Is $45 + 21 > 18$?	*yes* *no* *yes*	18, 21, 45	no
b. Is $18 + 45 > 52$? Is $18 + 52 > 45$? Is $45 + 52 > 18$?	*yes* *yes* *yes*	18, 45, 52	yes
c. Is $18 + 21 > 52$? Is $18 + 52 > 21$? Is $21 + 52 > 18$?	*no* *yes* *yes*	18, 21, 52	no
d. Is $45 + 21 > 52$? Is $45 + 52 > 21$? Is $21 + 52 > 45$?	*yes* *yes* *yes*	21, 45, 52	yes

A triangle can be formed using two of the four possible combinations. Thus, the probability that a triangle can be formed is $\frac{2}{4}$ or $\frac{1}{2}$.

When you know the lengths of two sides of a triangle, you can determine the range of possible lengths for the third side.

Example 2 INTEGRATION Algebra

If the lengths of two sides of a triangle are 10 centimeters and 15 centimeters, between what two numbers must the measure of the third side fall?

Let t = the length of the third side of the triangle.

By the Triangle Inequality Theorem, each of these inequalities must be true. Graph each on the same number line.

$$10 + 15 > t$$
$$25 > t$$

$$10 + t > 15$$
$$t > 5$$

$$15 + t > 10$$
$$t > -5$$

−10 −5 0 5 10 15 20 25 30

The length of the third side must fall in the range included in all three inequalities.

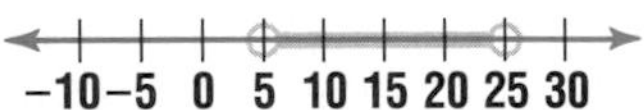

Taking the intersection of these three conditions, t must be greater than 5 and less than 25 or $5 < t < 25$. Any side with length between 5 centimeters and 25 centimeters will form a triangle when the other two sides have lengths 10 centimeters and 15 centimeters.

The Triangle Inequality Theorem can be used to solve problems involving lengths of sides of triangles.

Example 3

The early Egyptians used to make triangles by using a rope with knots tied at equal intervals. Each vertex of the triangle had to occur at a knot. Suppose you had a rope with exactly 13 knots as shown below.

How many different triangles could you make? Sketch each possibility and classify the angles in each case if possible.

Explore The thirteen knots divide the rope into twelve sections of equal length. Thus, the perimeter of each triangle that can be formed will be 12 units. If x, y, and z represent the lengths of the three sides, then $x + y + z = 12$. We also know that $x + y > z$, $x + z > y$, and $y + z > x$ by the Triangle Inequality Theorem.

Plan Make a table of all values of x, y, and z that satisfy $x + y + z = 12$. Then check each possible combination with the Triangle Inequality Theorem.

Solve There are three possible triangles. *How do we know that all the combinations have been checked?*

x	y	z	Triangle?
1	1	10	no
1	2	9	no
1	3	8	no
1	4	7	no
1	5	6	no
2	2	8	no
2	3	7	no
2	4	6	no
2	5	5	yes
3	3	6	no
3	4	5	yes
4	4	4	yes

The triangle with sides 2, 5, and 5 units long is isosceles. The base angles are congruent.

If you sketch a triangle with sides 3, 4, and 5 units long, it appears that the angle opposite the 5-unit side is a right angle. The Pythagorean Theorem confirms this.

The triangle with all three sides 4 units long is equilateral. Each angle measures 60°.

Examine Model the situation with a piece of string to confirm the solution.

CHECK FOR UNDERSTANDING

Communicating Mathematics

Study the lesson. Then complete the following.

1. **Find** three numbers that can be the lengths of the sides of a triangle and three numbers that cannot be the lengths of the sides of a triangle. Justify your reasoning with a drawing.
2. **Determine** if there are any restrictions on the length of the base of a triangle if you are told that it is an isosceles triangle with legs 21 centimeters long. Support your answer with diagrams.
3. **You Decide** A triangle has two sides that are 5 inches and 13 inches long. Jolanda says that the third side could be 8 inches. Brittany says that it could not be 8 inches, but it could be 13 inches. Who is correct and why?

Guided Practice

Determine whether it is possible to draw a triangle with sides of the given measures. Write *yes* or *no*. If yes, draw the triangle.

4. 1, 2, 3
5. 21, 32, 18
6. 11, 6, 2

The measures of two sides of a triangle are given. Between what two numbers must the measure of the third side fall?

7. 21 and 27
8. 5 and 11
9. 30 and 30

Study the figure carefully and indicate whether each statement is *always true, sometimes true,* or *never true*. Justify your answer.

A, B, C, D, 4, 3, 2

10. $DB > 7$
11. $BC < 9$

12. Write a two-column proof.
    Given: $\overline{PO} \cong \overline{OM}$
    Prove: $PO + MN > ON$

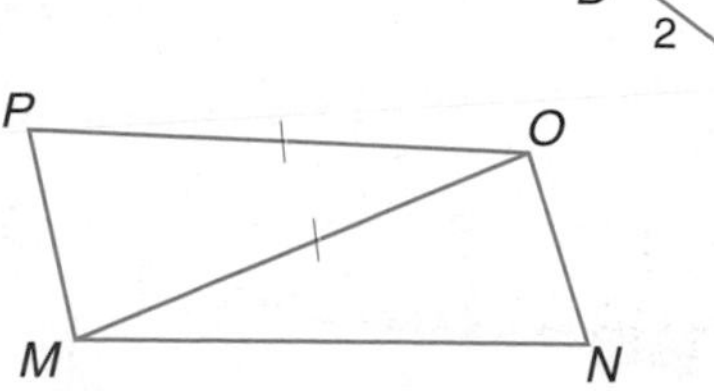

13. **Algebra** Can you make a triangle using $Q(1, 1)$, $R(-2, 5)$, and $S(-5, -4)$ as the vertices? Why or why not?

EXERCISES

Practice

Determine whether it is possible to draw a triangle with sides of the given measures. Write *yes* or *no*. If yes, draw the triangle.

14. 5, 4, 3
15. 5.2, 5.6, 10.1
16. 5, 10, 15
17. 10, 100, 100
18. 301, 8, 310
19. 9, 40, 41
20. 12, 2.2, 14.3
21. 10, 150, 200
22. 84, 7, 115

The measures of two sides of a triangle are given. Between what two numbers must the measure of the third side fall?

23. 15 and 18
24. 14 and 23
25. 22 and 34
26. 21 and 47
27. 64 and 88
28. 99 and 2
29. 47 and 71
30. 104 and 118
31. a and b

Study the figure carefully and indicate whether each statement is *always true, sometimes true,* or *never true.* Justify your answer.

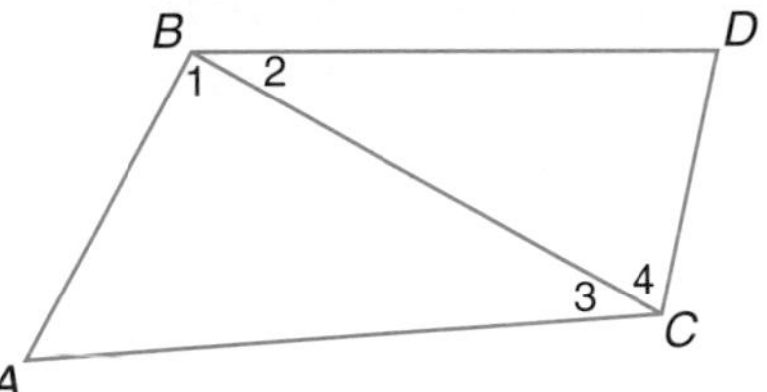

32. If $AB = 5$, $AC = 8$, $DC = 2$, then $BC < 15$.
33. If $DC = 14$, $AC = 18$, $BD = 24$, then $BC = 12$.
34. If $\angle 1 \cong \angle 3$, then $AB > \frac{1}{2}BC$.
35. If $\angle D$ is obtuse, then $AB + AC > BD$.
36. If $\angle 1 \cong \angle 2$, then $AB + AC = BD + DC$.
37. If $\angle A$ is a right angle, then $BC < BA$.

Algebra

Determine whether it is possible to have a triangle with the given vertices. Write *yes* or *no*, and explain your answer.

38. $R(0, 0)$, $S(3, 5)$, $T(5, 3)$
39. $A(2, 3)$, $B(-5, -11)$, $C(-8, 15)$
40. $J(1, -4)$, $K(-3, -20)$, $L(5, 12)$
41. $D(1, 4)$, $E(5, -1)$, $F(1, -4)$

Write a two-column proof.

42. **Given:** $RS = RT$
Prove: $UV + VS > UT$

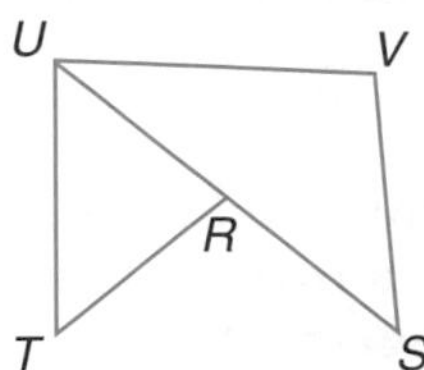

43. **Given:** quadrilateral $ABCD$
Prove: $AD + CD + AB > BC$

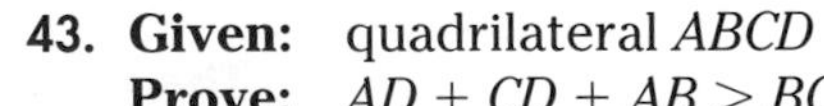

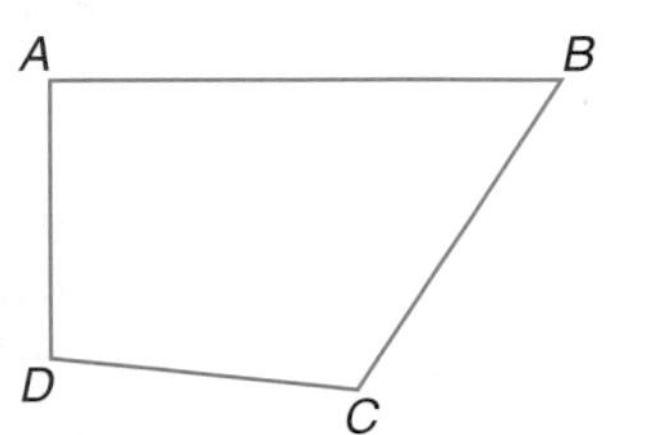

44. Write a paragraph proof for the Triangle Inequality Theorem. (Theorem 5–12)

Given: $\triangle ROS$
Prove: $SO + OR > RS$
(*Hint:* Draw auxiliary segment, $\overline{OT}$, so that O is between R and T and $\overline{OT} \cong \overline{SO}$.)

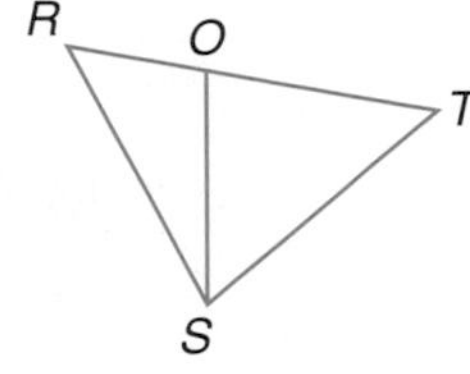

Programming

45. The graphing calculator program tests to see if three measures can form the sides of a triangle.

```
PROGRAM: TESTER
:Disp "ENTER MEASURES"
:Input "SIDE 1:", A
:Input "SIDE 2:", B
:Input "SIDE 3:", C
:If A+B>C and A+C>B and
 B+C>A
:Then
:Disp "TRIANGLE EXISTS"
:Else
:Disp "NO TRIANGLE"
:Stop
```

Use the program to test each set of measures.

a. 5, 7, 12
b. 4.5, 8.85, 6.25
c. 9.87, 12.32, 32.90
d. 112, 89, 75
e. 256, 219, 311

Critical Thinking

46. Is it true that the difference between any two sides of a triangle is less than the third side? Explain your reasoning.

Applications and Problem Solving

47. One side of a triangle is 2 centimeters long. Let a represent the measure of the second side and z represent the measure of the third side. Suppose that $14 < a < 17$ and $13 < z < 17$ and a and z are whole numbers.
   a. **Algebra** List the measures of the sides of the triangles that are possible under those conditions.
   b. **Probability** What is the probability that a randomly chosen triangle that satisfies the conditions will be isosceles?

48. **Carpentry** Salina is building stairs and wants to nail a brace at the base of each stair as shown in the figure. The brace attaches at the bottom of the rise and anywhere along the tread.

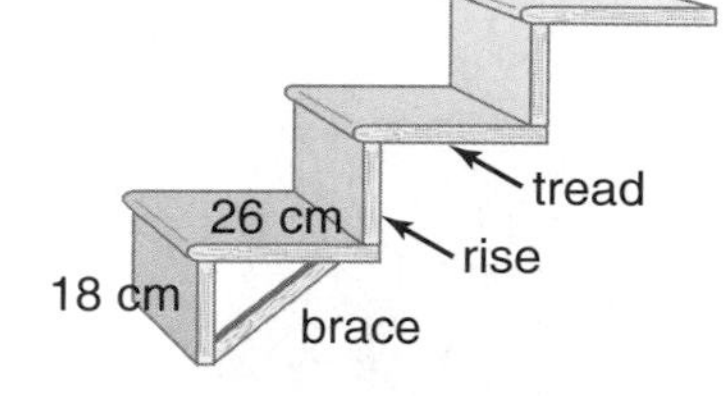

The stairs have an 18-cm rise and a 26-cm tread. There is a pile of braces 5 centimeters, 20 centimeters, 24 centimeters, and 45 centimeters long that Salina can use. Which of the lengths can she use as a brace?

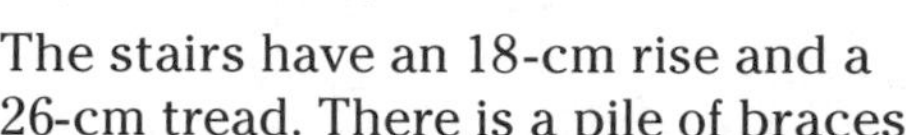

Mixed Review

49. Suppose $m\angle A = 4x + 61$, $m\angle B = 67 - 3x$, and $m\angle C = x + 74$. What is the longest segment in $\triangle ABC$? (Lesson 5–4)

50. State the assumption you would make to start an indirect proof for $\overline{NM}$ is a median of $\triangle NOP$. (Lesson 5–3)

51. If possible, describe a triangle in which the angle bisectors all intersect in a point outside the triangle. If no triangle exists, write *no triangle.* (Lesson 5–1)

52. Refer to the figure at the right. If X is the midpoint of $\overline{AD}$, name the additional parts of $\triangle AXC$ and $\triangle DXB$ that would have to be congruent to prove $\triangle AXC \cong \triangle DXB$ by ASA. (Lesson 4–5)

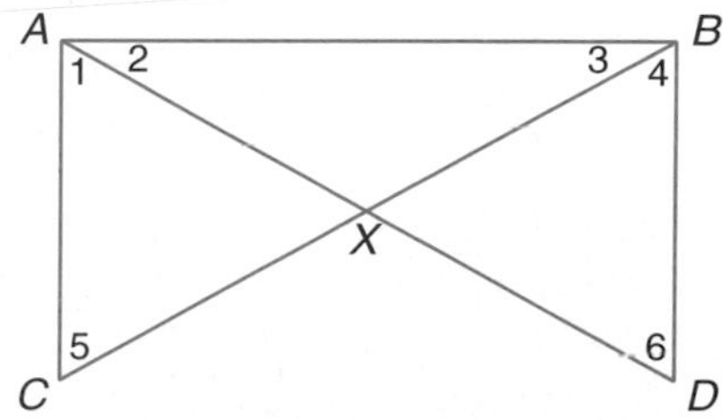

53. The legs of an isosceles triangle are $(6x - 6)$ and $(x + 9)$ units long. Find the length of the legs. (Lesson 4–1)

54. **SAT Practice** The average of x numbers is 15. If the sum of the x numbers is 90, what is the value of x?

A 5 **B** 6 **C** 8 **D** 15 **E** 75

55. Write a two-column proof. (Lesson 3–2)

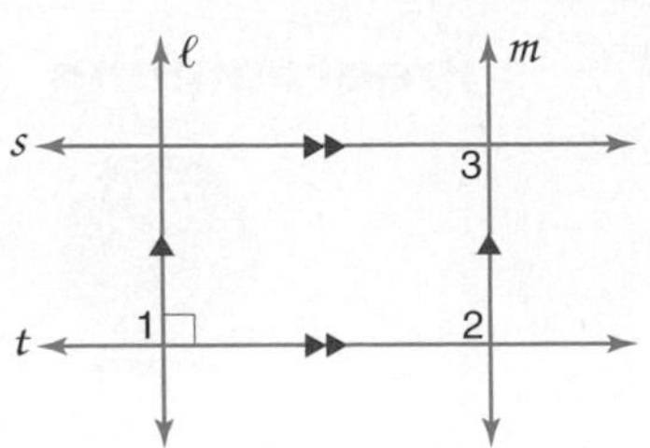

Given: $\ell \parallel m$

$s \parallel t$

$t \perp \ell$

Prove: $\angle 3$ is a right angle.

56. Write the converse of the conditional statement *If three points are collinear, then they lie on a straight line.* Determine if the converse is true or false. If it is false, give a counterexample. (Lesson 2–2)

57. Suppose $m\angle H = 63$. Find the measure of the complement and the supplement of $\angle H$. (Lesson 1–7)

Algebra

58. Simplify $\sqrt{28x^4}$.

59. **Music** Refer to the graph at the right.
   a. Describe the information that the graph shows.
   b. If there is a total of eight million classical music listeners, how many people between 18 and 24 years old listen to classical music?

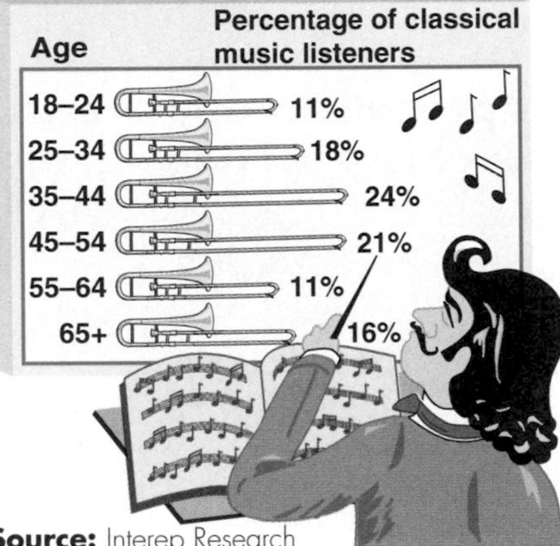

Source: Interep Research

For **Extra Practice**, see page 773.

How Do Turtles Navigate?

The excerpt below appeared in an article in *Earth* in February, 1995.

LOGGERHEAD SEA TURTLES ARE AMONG the world's greatest navigators. They spend much of their lives in the Sargasso Sea, a fertile region in the middle of the Atlantic. But mature loggerheads travel thousands of miles to return to their natal beaches to reproduce.... Recent studies suggest that sea turtles use the changing angle of Earth's magnetic field with respect to the ground, called the angle of inclination, to navigate....At the magnetic poles, the inclination of the field is about 90 degrees....the invisible field lines are nearly perpendicular to the surface of the planet. In contrast, at the equator, the inclination of Earth's magnetic field is zero degrees. That means the field lines flow parallel to the surface. Thus, magnetic inclination is a pretty good stand-in for latitude....subsequent experiments confirmed that the turtles can sense a change as small as three degrees in inclination, and they may well be considerably more sensitive than that. ■

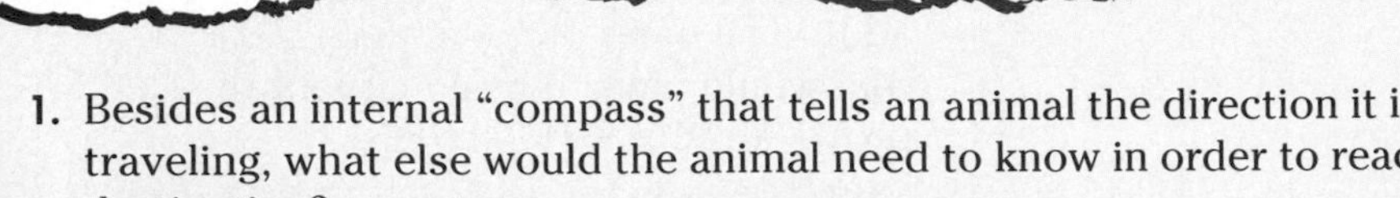

1. Besides an internal "compass" that tells an animal the direction it is traveling, what else would the animal need to know in order to reach its destination?
2. If a town is located at a point that is one-third of the distance from the equator to the North Pole, what would be the angle of inclination of Earth's magnetic field there?
3. How could you find out whether an animal uses the position of the sun to assist it in its migration? How would triangles help in this positioning?

5–6 Inequalities Involving Two Triangles

What YOU'LL LEARN

- To apply the SAS Inequality and the SSS Inequality.

Why IT'S IMPORTANT

You can use the inequalities in triangles to solve problems involving physical therapy and biology.

CAREER CHOICES

If you like to help people, **physical therapy** may be for you!

A career in physical therapy requires a bachelor's degree. Studies include anatomy, physics, and mathematics.

For more information, contact:

American Physical Therapy Assn.
1111 N. Fairfax St.
Alexandria, VA 22314

Real World APPLICATION

Physical Therapy

Physical therapists help people use exercise to regain their strength after an injury or surgery. One exercise a physical therapist may recommend as a part of a rehabilitation program following arm surgery is called a *curl*. To perform a curl, hold a weight in your hand with your arm at your side and your palm facing upward. Raise your forearm by bending at the elbow as far as possible and slowly return to the starting position.

Observe the positions of your arm as you complete a curl. Notice that the measure of the angle between the upper and lower parts of your arm decreases as you raise your hand. What happens to the distance between your hand and your shoulder? A model can help clarify the relationships.

MODELING MATHEMATICS

Inequalities in Two Triangles

Materials:  rubber band ball-bearing compass  centimeter ruler protractor

- Place a rubber band over the tips of a ball-bearing compass as shown. Imagine that the sides of the compass and the rubber band form a triangle, which we will call $\triangle PRT$.
- Measure $\angle R$ and $\overline{PT}$.
- Spread the compass arms farther apart to form a different triangle, $\triangle P'R'T'$.
- Measure $\angle R'$ and $\overline{P'T'}$.

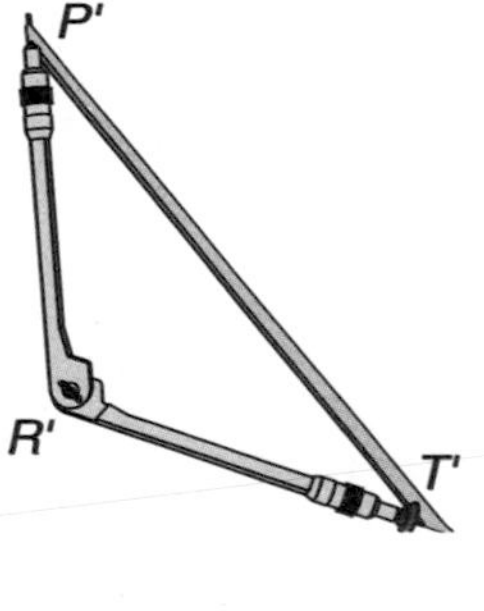

Your Turn

a. Compare $m\angle R$ to $m\angle R'$ and PT to $P'T'$.

b. Change the position of the compass arms again. Make a conjecture about how $m\angle R$ and PT will change. Check your conjecture.

c. Suppose $\overline{AB} \cong \overline{DE}$ and $\overline{BC} \cong \overline{EF}$ in $\triangle ABC$ and $\triangle DEF$. If $m\angle B > m\angle E$, how are AC and DF related?

The Modeling Mathematics activity suggests the following two theorems.

Theorem 5–13 SAS Inequality (Hinge Theorem)	**If two sides of one triangle are congruent to two sides of another triangle and the included angle in one triangle has a greater measure than the included angle in the other, then the third side of the first triangle is longer than the third side in the second triangle.**

Theorem 5–14 SSS Inequality	If two sides of one triangle are congruent to two sides of another triangle and the third side in one triangle is longer than the third side in the other, then the angle between the pair of congruent sides in the first triangle is greater than the corresponding angle in the second triangle.

An indirect proof for the SSS Inequality is given below. *You will be asked to write a proof for the SAS Inequality in Exercise 34.*

Proof of Theorem 5–14

Given: $\overline{RS} \cong \overline{PW}$
$\overline{ST} \cong \overline{WV}$
$RT > PV$

Prove: $m\angle S > m\angle W$

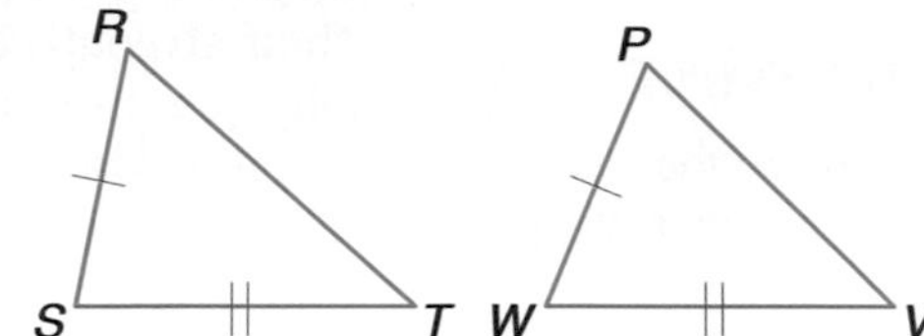

Indirect Proof:

Step 1: Assume $m\angle S \leq m\angle W$.

Step 2: If $m\angle S \leq m\angle W$, then either $m\angle S = m\angle W$ or $m\angle S < m\angle W$.

Case 1: If $m\angle S = m\angle W$, then $\triangle RST \cong \triangle PWV$ by SAS, and $\overline{RT} \cong \overline{PV}$ by CPCTC. Thus, $RT = PV$.

Case 2: If $m\angle S < m\angle W$, then $RT < PV$ by the SAS Inequality.

Step 3: In both cases, our assumptions led to a contradiction of the given information that $RT > PV$. Therefore, the assumption must be false, and the conclusion, $m\angle S > m\angle W$, must be true.

The following examples illustrate geometric and algebraic applications of the SAS Inequality and the SSS Inequality.

Example 1

In the figure at the right, $EF = 2.9$ and $AD = 2.8$. Describe how each pair of angle or segment measures is related.

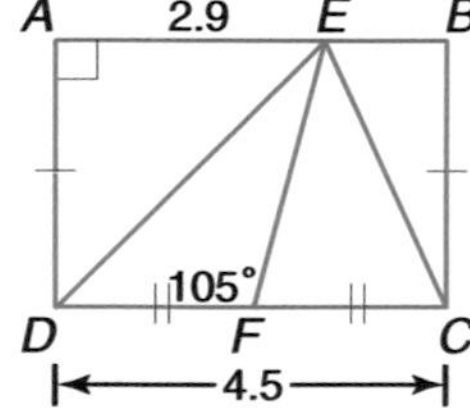

a. $m\angle AED$, $m\angle DEF$

b. DE, EC

a. In $\triangle AED$ and $\triangle DEF$, $\overline{ED} \cong \overline{ED}$, since congruence of segments is reflexive. $EF = 2.9$ and $AE = 2.9$, so $\overline{EF} \cong \overline{AE}$. $DF = 4.5 \div 2$ or 2.25 and $AD = 2.8$. Then the SSS Inequality allows us to conclude that $m\angle AED > m\angle DEF$.

b. Compare $\triangle DEF$ and $\triangle CEF$. $\overline{EF} \cong \overline{EF}$ since congruence of segments is reflexive. The diagram indicates that $\overline{DF} \cong \overline{FC}$. $\angle DFE$ and $\angle EFC$ form a linear pair, so they are supplementary. Then $m\angle DFE + m\angle EFC = 180$. Thus, $105 + m\angle EFC = 180$ or $m\angle EFC = 75$. Therefore, $m\angle DFE > m\angle EFC$. Hence by the SAS Inequality, $DE > EC$.

Example 2 Write a two-column proof.

Given: $\triangle RST$

$\overline{RE} \cong \overline{ST}$

Prove: $RS > TE$

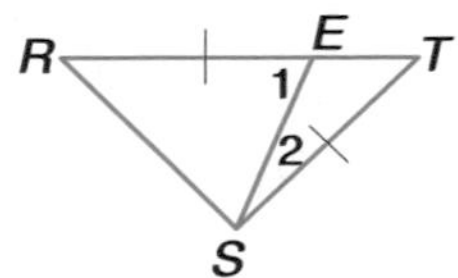

Proof:

Statements	Reasons
1. $\overline{RE} \cong \overline{ST}$	1. Given
2. $\overline{ES} \cong \overline{ES}$	2. Congruence of segments is reflexive.
3. $\angle 1$ is an exterior angle of $\triangle EST$.	3. Def. exterior angle
4. $m\angle 1 > m\angle 2$	4. If an $\angle$ is an ext. angle of a $\triangle$, then its measure > the measure of either of its corr. remote int. $\angle$s.
5. $RS > TE$	5. SAS Inequality

You can use algebra to relate the measures of the angles and sides of two triangles.

Example 3

In the figure, $m\angle SVR = 5x + 15$, $m\angle TVS = 10x - 20$, $RS < ST$, and $\angle VRT \cong \angle VTR$. Write three inequalities to describe possible values for x.

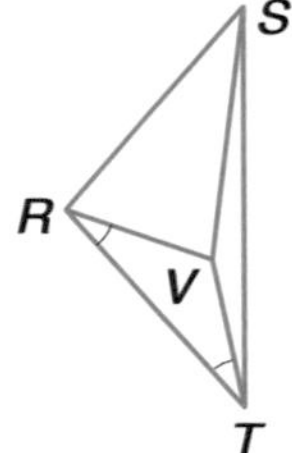

If two angles in a triangle are congruent, then the sides opposite those angles are congruent. Thus, since $\angle VRT \cong \angle VTR$, $\overline{VT} \cong \overline{RV}$. $\overline{SV} \cong \overline{SV}$ because congruence is reflexive. Since $RS < ST$, you know by the SAS Inequality that $m\angle SVR < m\angle TVS$.

Since both $\angle SVR$ and $\angle TVS$ are angles in a triangle, you know that their measures are less than 180. Thus, you can write the following three inequalities.

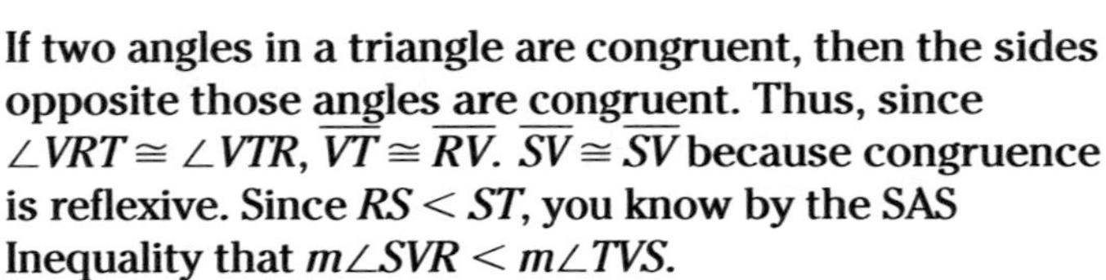

$m\angle SVR < 180$	$m\angle TVS < 180$	$m\angle SVR < m\angle TVS$
$5x + 15 < 180$	$10x - 20 < 180$	$5x + 15 < 10x - 20$
$5x < 165$	$10x < 200$	$35 < 5x$
$x < 35$	$x < 20$	$7 < x$

The range of possible values for x is $7 < x < 20$.

CHECK FOR UNDERSTANDING

Communicating Mathematics

Study the lesson. Then complete the following.

1. **Draw** an isosceles triangle with legs 5 centimeters long and a base 8 centimeters long. Then draw another isosceles triangle with legs 5 centimeters long and a base 2 centimeters long.
   a. Measure the vertex angle of each triangle. Which triangle has the larger vertex angle?
   b. What theorem does this demonstrate?

2. **Compare and contrast** the SAS Inequality Theorem to the SAS Postulate you studied in Chapter 4.

3. **Explain** why you think the SAS Inequality Theorem is subtitled the Hinge Theorem.

4. Tie a loop at one end of a piece of string. Then run the other end of the string through two straws and back through the loop to form a triangle as shown in the figure. Measure the sides of the triangle formed and the angle between the straws. Next, gently pull on the long end of the string to form a different triangle. Measure the sides of the new triangle and the angle between the straws.

   a. Describe what happens to the angle between the straws as the side determined by the string changes.

   b. What theorem does this activity demonstrate?

Guided Practice

Refer to the figure at the right to write an inequality relating the given pair of angle or segment measures.

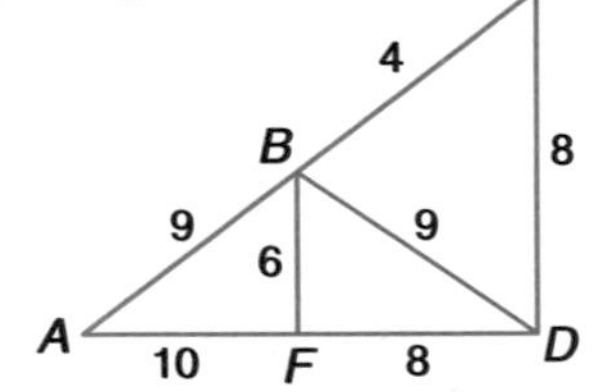

5. $m\angle DBF$, $m\angle BFA$
6. AB, FD
7. $m\angle FDB$, $m\angle BDC$

In the figure, $\overline{SO}$ is a median in $\triangle SLN$, $\overline{OS} \cong \overline{NP}$, $m\angle 1 = 3x - 50$, and $m\angle 2 = x + 30$. Determine if each statement is *always true, sometimes true, or never true.*

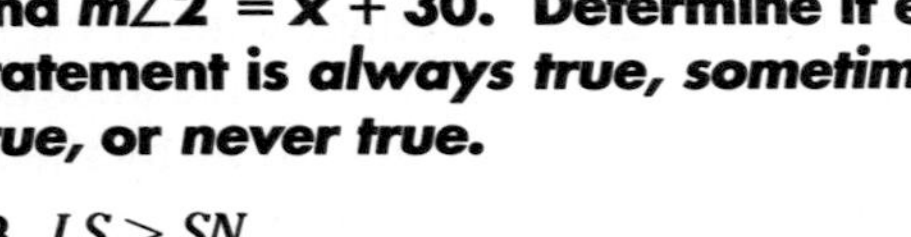

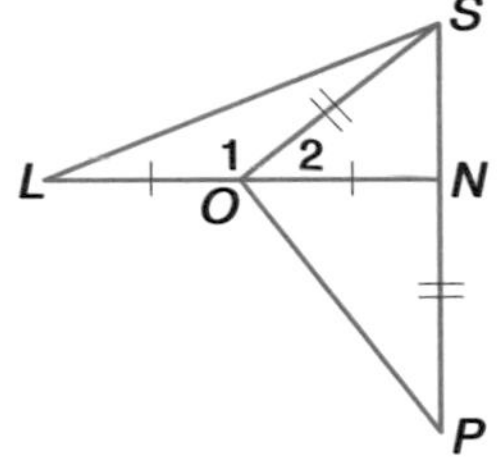

8. $LS > SN$
9. $SN < OP$
10. $x = 45$

11. Write an inequality or pair of inequalities to describe the possible values of x.

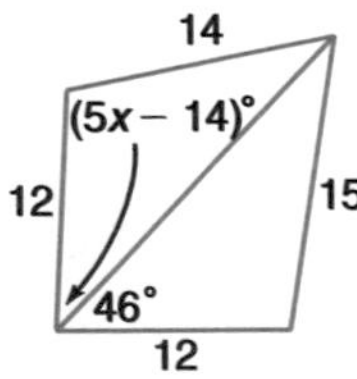

12. Write a two-column proof.

   Given: $\overline{OT} \cong \overline{TV}$

   T is the midpoint of $\overline{SW}$.

   $m\angle STO > m\angle WTV$

   $\overline{RO} \cong \overline{RV}$

   Prove: $RS > RW$

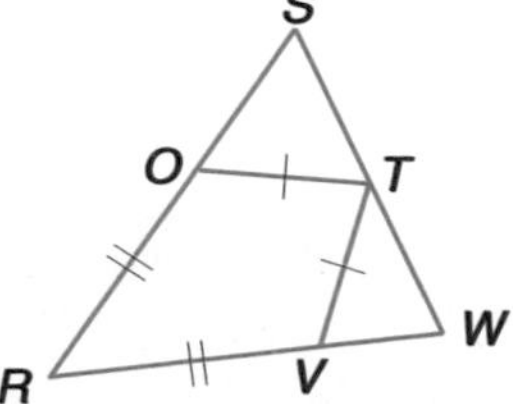

13. **Physical Therapy** Range of motion is the amount that a limb can be moved from the straight position. To determine the range of motion of a patient's arm, find the distance between the palm and the shoulder when the elbow is bent as far as possible. For one patient, the left palm was 8 inches from the shoulder, and the right palm was 3 inches from the shoulder. Which arm has the greater range of motion?

EXERCISES

Practice

Refer to each figure to write an inequality relating the given pair of angle or segment measures.

14. $m\angle 1, m\angle 2$

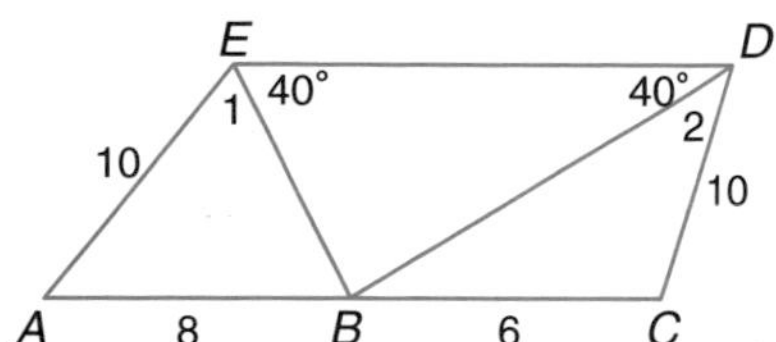

15. TM, RS

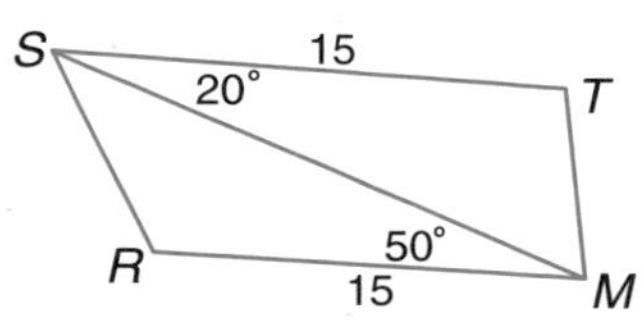

16. $m\angle ACB, m\angle ACD$

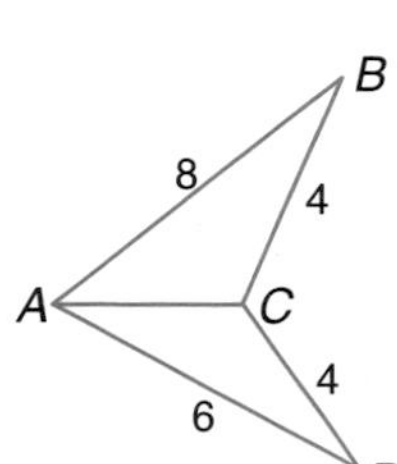

17. FH, GE

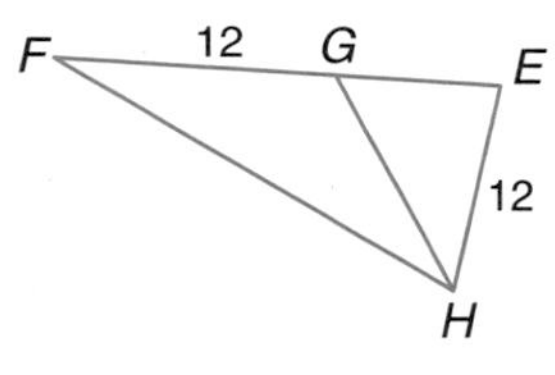

18. OC, AO
19. $m\angle AOD, m\angle AOB$
20. DC, AD

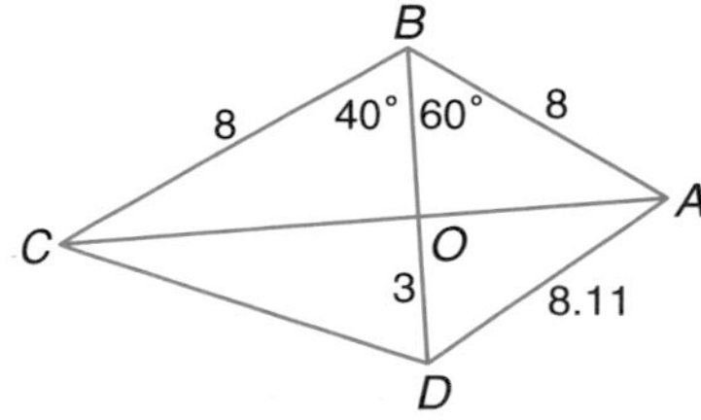

In $\triangle ABC$, $\overline{CM}$ is a median. Determine if each statement is *always true, sometimes true,* or *never true.*

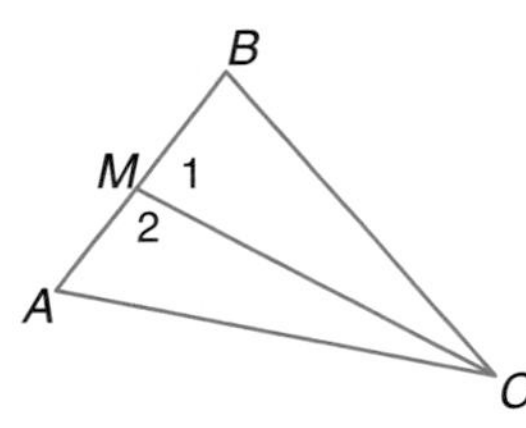

21. If $m\angle 2 < m\angle 1$, then $BC > AC$.
22. If $m\angle B > m\angle A$, then $\angle 1$ is obtuse.
23. If $\angle 2$ is acute, then $m\angle A > m\angle B$.
24. If $m\angle B < m\angle A$, then $m\angle 1 < m\angle 2$.
25. If $m\angle 2 = m\angle 1$, then $BC > AC$.
26. If $m\angle B = 90$, then $AC < BC$.

Algebra

Write an inequality or pair of inequalities to describe the possible values of *x*.

27.

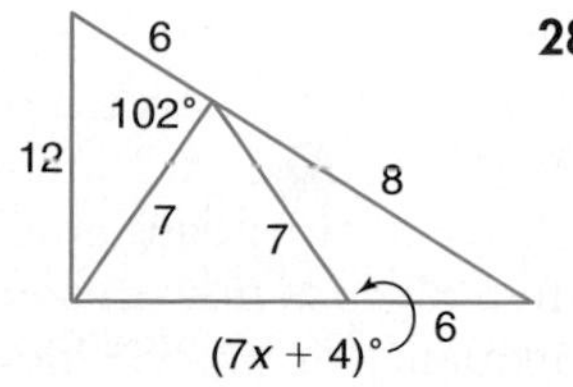

28.

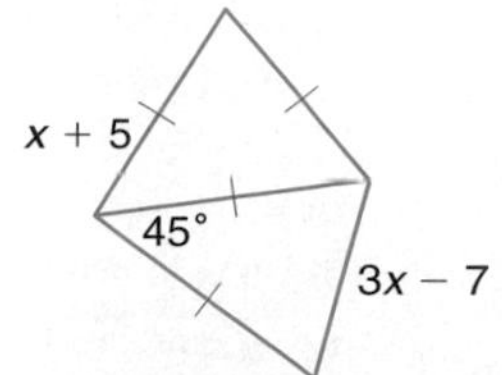

29.

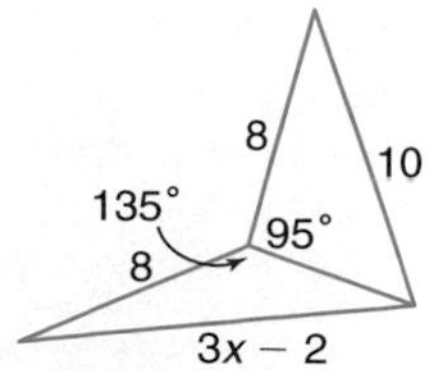

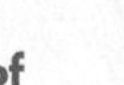

Write a two-column proof.

30. Given: $\overline{DB}$ is a median of $\triangle ABC$.

$m\angle 1 > m\angle 2$

Prove: $m\angle C > m\angle A$

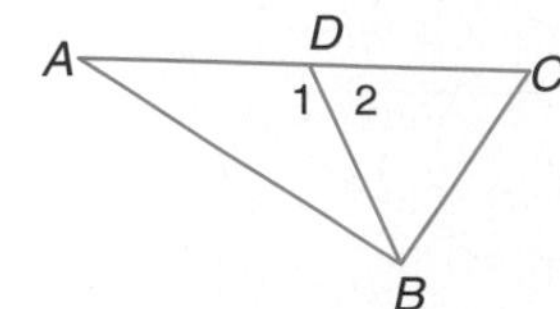

31. Given: $\overline{MN} \cong \overline{QR}$

$\overline{MN} \parallel \overline{QR}$

$m\angle MPQ > m\angle QPR$

Prove: $MQ > QR$

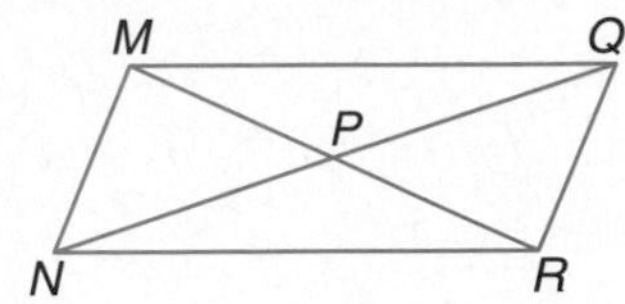

Write a paragraph proof.

32. Given: $\overline{RM} \cong \overline{ST}$

$\overline{RM} \parallel \overline{ST}$

$ST > RS$

Prove: $m\angle ROM > m\angle MOT$

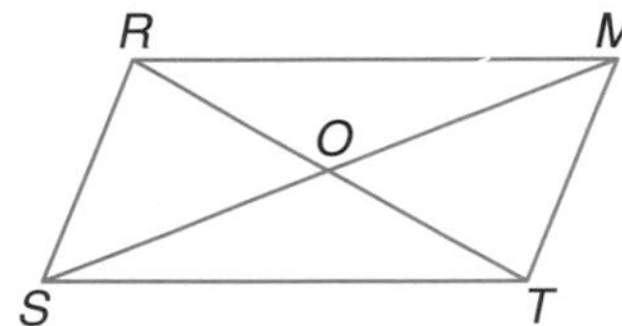

33. Given: $\triangle ABC$ is equilateral.

$\triangle ABD \cong \triangle CBD$

$BD > AD$

Prove: $m\angle BCD > m\angle DAC$

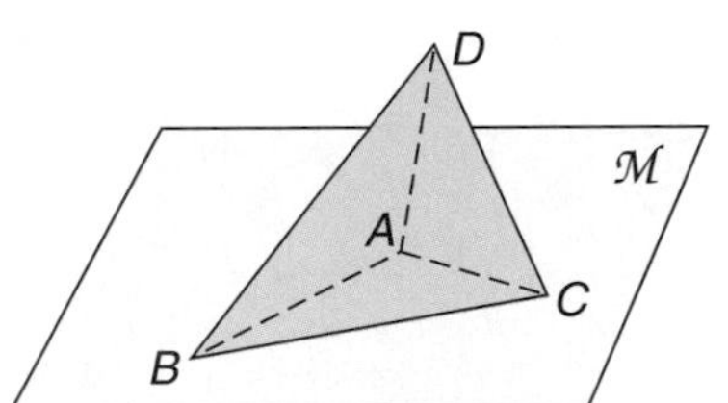

34. Write an indirect proof for the SAS Inequality. (Theorem 5–13)

Given: $\triangle ABC$ and $\triangle DEF$

$\overline{AC} \cong \overline{DF}$

$\overline{BC} \cong \overline{EF}$

$m\angle F > m\angle C$

Prove: $DE > AB$

Critical Thinking

35. Suppose that plane $\mathcal{F}$ bisects $\overline{AC}$ at B and for a point D in $\mathcal{F}$, $DC > DA$. What can you conclude about the relationship between $\overline{AC}$ and plane $\mathcal{F}$?

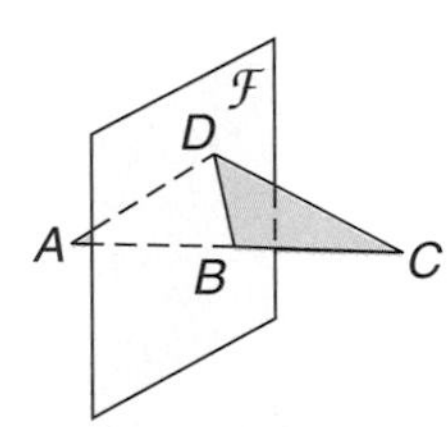

Applications and Problem Solving

36. Biology The fastest human can run about 23 mph for a very short distance, but many animals can run much faster. The formula $v = \frac{0.78s^{1.67}}{h^{1.17}}$, where v is the speed of the animal in meters per second, s is the length of the animal's stride in meters, and h is the height of the animal's hip in meters, can be used to estimate the speed (velocity) of an animal.

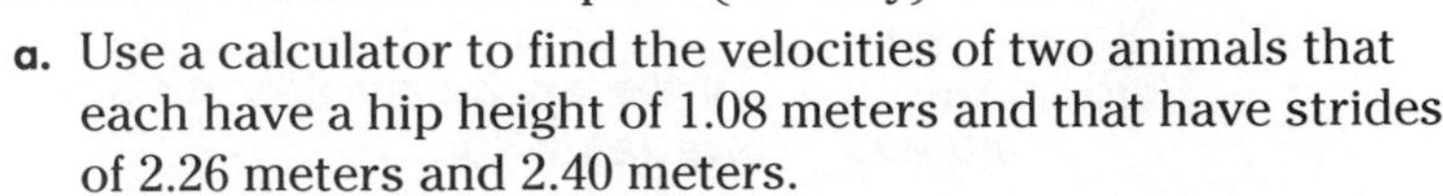

a. Use a calculator to find the velocities of two animals that each have a hip height of 1.08 meters and that have strides of 2.26 meters and 2.40 meters.

b. Draw a mathematical model of the triangles formed by the two animals in part a if one point of the triangle represents the position of each animal's hip and the other two points represent the beginning and end of each animal's stride. Then discuss how this model is related to either the SAS Inequality or the SSS Inequality.

37. **Physics** The discovery of the lever allowed ancient people to accomplish great tasks like building Stonehenge and the Egyptian pyramids. A lever multiplies the force applied to an object. One example of a lever is the nutcracker at the right. Use the SAS or SSS Inequality to explain how to operate the nutcracker.

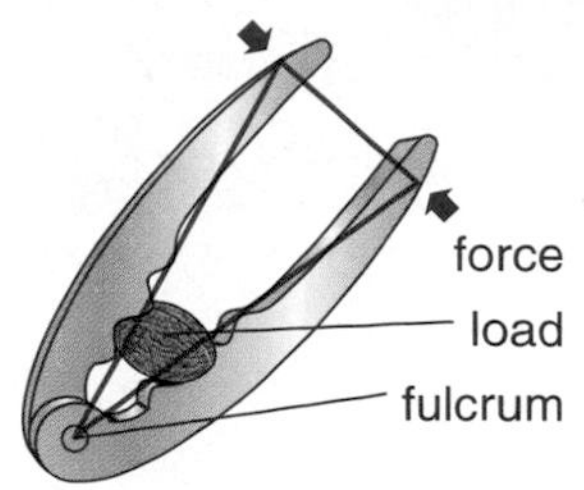

Mixed Review

38. Do points $A(3, 1)$, $B(9, 9)$, and $C(4, 7)$ form a triangle? Explain. (Lesson 5–5)

39. **Algebra** Find the value of x and list the sides of $\triangle FGH$ in order from shortest to longest if $m\angle F = 6x + 25$, $m\angle G = 14x - 18$, and $m\angle H = 65 - 2x$. (Lesson 5–4)

40. **Sailing** A tripping line is often attached to an anchor to allow it to be dislodged from the bottom easily. What ensures that the boat will always be the same distance from the tripping line marker if both the tripping line and the anchor line are taut? (Lesson 5–2)

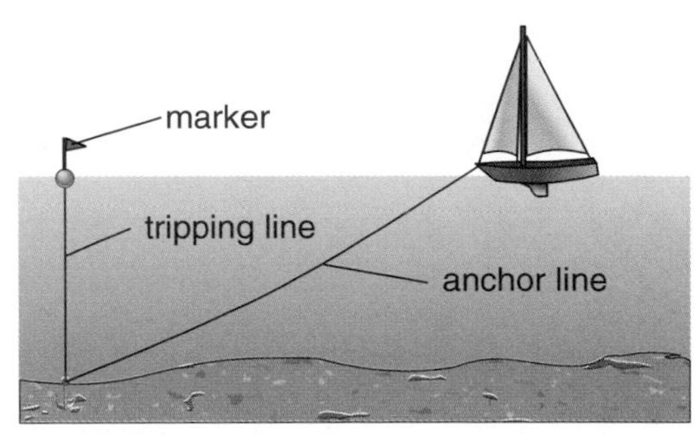

41. **Algebra** $\triangle JKL$ has vertices $J(6, 3)$, $K(-2, 7)$, and $L(9, 0)$. If $\overline{JT}$ is a median of $\triangle JKL$, what are the coordinates of T? (Lesson 5–1)

42. **SAT Practice Grid-in** Set M consists of all multiples of 3 between 13 and 31. Set P consists of all multiples of 4 between 13 and 31. What is one possible number in P but NOT in M?

43. **Algebra** If $\triangle NQD$ has vertices $N(2, -1)$, $Q(-4, -1)$, and $D(-1, 3)$, describe $\triangle NQD$ in terms of its angles and sides. (Lesson 4–1)

44. If possible, write a conclusion. State the law of logic that you use. (Lesson 2–3)
(1) If $2x + 4 = 4x + 8$, then x can be any real number.
(2) $2x + 4 = 4x + 8$

Algebra

For **Extra Practice**, see page 774.

45. Use the distributive property to rewrite $4(8y - 5)$ without parentheses.

46. Solve $|w - 4| = 6$.

WORKING ON THE In•ves•ti•ga•tion

Refer to the Investigation on pages 176–177.

People have long been fascinated with the shape of snowflakes. The oldest recorded statement on snowflakes is found in Chinese writings from the second century B.C. Use reference material or an Internet search to research snowflakes.

1. What is the general structure of a snowflake?
2. How do chemists explain this structure?
3. How is the shape of a snowflake affected by the weather conditions?
4. Compare snowflake structure to the natural patterns described on pages 176–177. Are any of the patterns displayed in snowflakes?

Add the results of your work to your Investigation Folder.

CLOSING THE

In·ves·ti·ga·tion

Refer to the Investigation on pages 166–167.

Once you have recognized a pattern, it is easy to find any irregularities. Scientists have spent centuries studying patterns and then finding irregularities to make discoveries about nature. For example, the planets were discovered because they broke the circular pattern of the movements of the stars. Because of this break from the pattern, the Greeks chose their name, *planetes,* meaning wanderer. Some of the patterns of nature are easy to observe. But we are just learning to recognize some of the patterns that at first appeared to be only random events.

Analyze

You have looked for and researched a pattern found in nature. It is now time to compile your findings and prepare your presentation.

PORTFOLIO ASSESSMENT

You may want to keep your work on this Investigation in your portfolio.

1 Look over the information you have gathered. If you have not already decided, determine whether you present your findings in a poster display, a slide presentation, or a video format.

2 Write a general description of the pattern you observed. Place this description at the beginning of the script for your presentation.

3 Make a list of the places that you observed your pattern. Decide upon the order for each example in your presentation.

4 Write the script to accompany your presentation. Be sure to make the script entertaining as well as informative. If possible, include information on the scientists who have done research in the field.

5 Complete the preparation for the presentation by putting together your poster, compiling your slides, or filming your video.

6 Write a brief summary of your findings to accompany the presentation.

Present

Present your findings to your class.

7 Introduce your presentation by telling how you conducted your research.

8 Explain your poster presentation, show your slides, or play your video.

9 Allow the class members to ask questions or explain how your observations apply to the pattern that they investigated.

Data Collection and Comparison To share and compare your data with other students, visit:
www.geometry.glencoe.com

Chapter Review For additional lesson-by-lesson review, visit: **www.geometry.glencoe.com**

VOCABULARY

After completing this chapter, you should be able to define each term, property, or phrase and give an example or two of each.

Geometry

altitude of a triangle (p. 238)
angle bisector of a triangle (p. 240)
Exterior Angle Inequality Theorem (p. 253)
Hinge Theorem (p. 273)
Hypotenuse-Angle Theorem (p. 246)
Hypotenuse-Leg Postulate (p. 247)
indirect reasoning (p. 252)
indirect proof (p. 252)
Leg-Angle Theorem (p. 247)
Leg-Leg Theorem (p. 245)
median of a triangle (p. 238)
perpendicular bisector of a triangle (p. 238)
SAS Inequality Theorem (p. 273)
SSS Inequality Theorem (p. 274)
Triangle Inequality Theorem (p. 267)

Algebra

Addition Property of Inequality (p. 254)
Comparison Property of Inequality (p. 254)
Division Property of Inequality (p. 254)
inequality (p. 254)
Multiplication Property of Inequality (p. 254)
Subtraction Property of Inequality (p. 254)
Transitive Property of Inequality (p. 254)

Problem Solving

work backward (p. 254)

UNDERSTANDING AND USING THE VOCABULARY

Choose the term from the list that best completes each statement or phrase.

1. The __?__ verifies that if $4x > 28$, then $x > 7$.
2. In $\triangle ABC$, D is the midpoint of $\overline{BC}$. $\overline{AD}$ is a __?__ of $\triangle ABC$.
3. The __?__ (s) and the __?__ (s) of a triangle are perpendicular to the sides of the triangle.
4. In right triangles XYZ and MNO, $\angle X$ and $\angle M$ are the right angles. If $\overline{XY} \cong \overline{MN}$ and $\angle Z \cong \angle O$, the __?__ justifies that the triangles are congruent.
5. According to the __?__, in $\triangle TUV$, $TV + TU > UV$.
6. Eliminating three of four possible answers for a multiple-choice question and then choosing the fourth choice is an example of __?__.
7. The statement *If* $m\angle R > m\angle T$ *and* $m\angle T > m\angle V$, *then* $m\angle R > m\angle V$ is justified by the __?__.
8. The __?__ (s) of a triangle can intersect outside of the triangle.
9. The __?__ can be proved using the Angle-Angle-Side Theorem.
10. In $\triangle MNP$ and $\triangle IJK$, $\overline{MN} \cong \overline{IJ}$, $\overline{PN} \cong \overline{KJ}$, and $MP > IK$. Then according to the __?__, $m\angle N > m\angle J$.

a. Multiplication Property of Inequality
b. altitude
c. Comparison Property of Inequality
d. Division Property of Inequality
e. median
f. Leg-Angle Theorem
g. SSS Inequality
h. Triangle Inequality Theorem
i. work backward
j. indirect reasoning
k. Hypotenuse-Angle Theorem
l. perpendicular bisector
m. Transitive Property of Inequality

STUDY GUIDE AND ASSESSMENT

SKILLS AND CONCEPTS

OBJECTIVES AND EXAMPLES

Upon completing this chapter, you should be able to:

- identify and use medians, altitudes, angle bisectors, and perpendicular bisectors of a triangle (Lesson 5–1)

If $\overline{NP}$ is an altitude of $\triangle NOQ$, then $\overline{NP} \perp \overline{OQ}$.

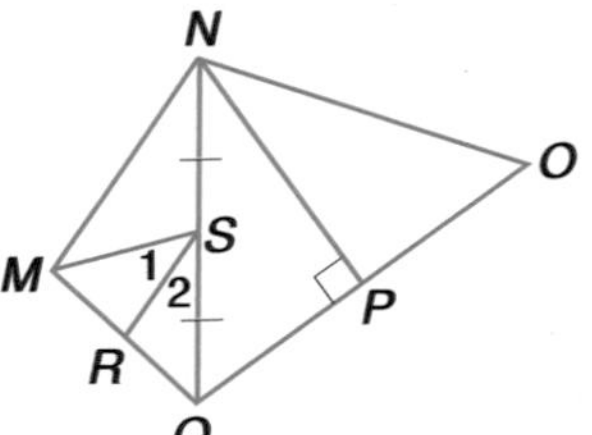

If $\overline{MS}$ is a median of $\triangle MNQ$, then $\overline{SN} \cong \overline{SQ}$.

If $\overline{SR}$ is an angle bisector of $\triangle MSQ$, then $\angle 1 \cong \angle 2$.

REVIEW EXERCISES

Use these exercises to review and prepare for the chapter test.

Refer to the figure at the left for Exercises 11–13.

11. $\overline{NP}$ is an altitude and an angle bisector of $\triangle NOQ$. If $m\angle QNP = 33$, find the measures of the three angles of $\triangle NOQ$.

12. Find the value of x and $m\angle 2$ if $\overline{MS}$ is an altitude of $\triangle MNQ$, $m\angle 1 = 3x + 11$, and $m\angle 2 = 7x + 9$.

13. If $\overline{MS}$ is a median of $\triangle MNQ$, $QS = 3x - 14$, $SN = 2x - 1$, and $m\angle MSQ = 7x + 1$, is $\overline{MS}$ also an altitude of $\triangle MNQ$? Explain.

- recognize and use tests for congruence of right triangles (Lesson 5–2)

Find the values of x and y so that $\triangle ABC \cong \triangle DEF$.

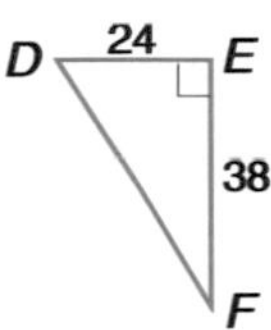

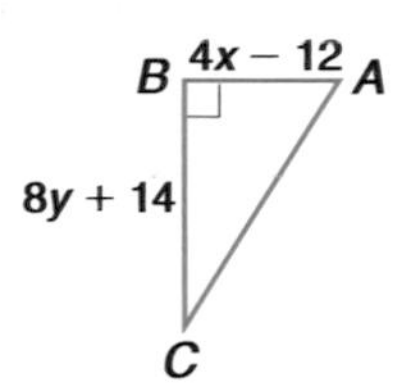

$$AB = DE \qquad BC = EF$$
$$4x - 12 = 24 \qquad 8y + 14 = 38$$
$$4x = 36 \qquad 8y = 24$$
$$x = 9 \qquad y = 3$$

By LL, $\triangle ABC \cong \triangle DEF$ for $x = 9$ and $y = 3$.

Find the values of x and y so that $\triangle ABC \cong \triangle XYZ$ by the indicated theorem or postulate.

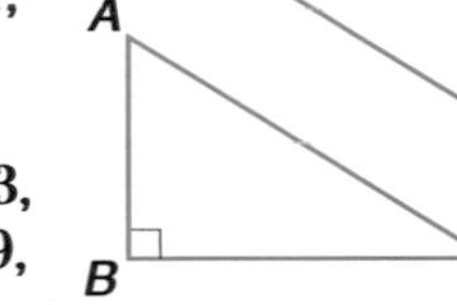

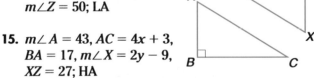

14. $AB = 40$, $XY = 3x + 1$, $AC = 36$, $m\angle C = 3y - 1$, $m\angle Z = 50$; LA

15. $m\angle A = 43$, $AC = 4x + 3$, $BA = 17$, $m\angle X = 2y - 9$, $XZ = 27$; HA

16. $XY = 42$, $m\angle Z = 68$, $AB = 5x - 8$, $m\angle A = 14y - 6$; LA

17. $AC = 12y + 1$, $AB = x + 3$, $XY = 4x - 3$, $YZ = 12$, $XZ = 10y + 3$; HL

- use indirect reasoning and indirect proof to reach a conclusion (Lesson 5–3)

Steps for Writing an Indirect Proof

1. Assume that the conclusion is false.
2. Show that the assumption leads to a contradiction of the hypothesis or some other fact, such as a postulate, theorem, or corollary.
3. Conclude that the assumption must be false, and therefore the conclusion must be true.

Write an indirect proof.

18. **Given:** $\overline{MJ}$ does not bisect $\angle NML$. $\overline{MN} \cong \overline{ML}$
    Prove: $\overline{MJ}$ is not a median of $\triangle NML$.

19. **Given:** $\angle MJN \cong \angle MJL$ $\overline{MJ}$ does not bisect $\angle NML$.
    Prove: $\overline{MJ}$ is an altitude of $\triangle NML$.

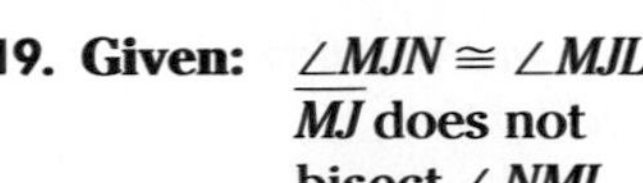

L
M
J
N

OBJECTIVES AND EXAMPLES

- recognize and apply the properties of inequalities to the measures of segments and angles (Lesson 5–3)

Find the value of x if $AC = AB$, $AX = 16$, and $AC = 3x - 2$.

$AC = AB$

$AC = AX + XB$

$AC > AX$

$3x - 2 > 16$

$3x > 18$

$x > 6$

REVIEW EXERCISES

Refer to the figure at the left for Exercises 20–24.

20. Which is greater, $m\angle 3$ or $m\angle ACB$?

21. Name an angle whose measure is greater than $m\angle 10$.

22. Which is greater, $m\angle 6$ or $m\angle 11$?

23. List all the angles whose measures are greater than $m\angle 5$.

24. If $m\angle 7 < m\angle 4$, which is greater, $m\angle 1$ or $m\angle 8$?

- recognize and apply relationships between sides and angles in a triangle (Lesson 5–4)

What is the longest segment in figure *PQRS*? *The figure is not drawn to scale.*

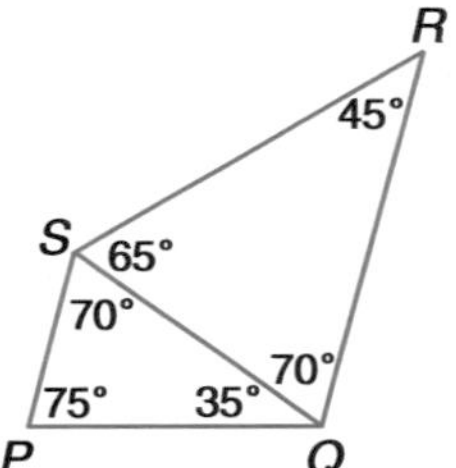

The longest side of $\triangle PQS$ is the side opposite the 75° angle, $\overline{SQ}$. This side is also the shortest side of $\triangle QRS$ since it is opposite the 45° angle. Thus, the longest side is the side opposite the 70° angle in $\triangle QRS$, $\overline{RS}$.

25. Name the shortest segment in the figure at the left.

26. List the sides of $\triangle ABC$ in order from shortest to longest.

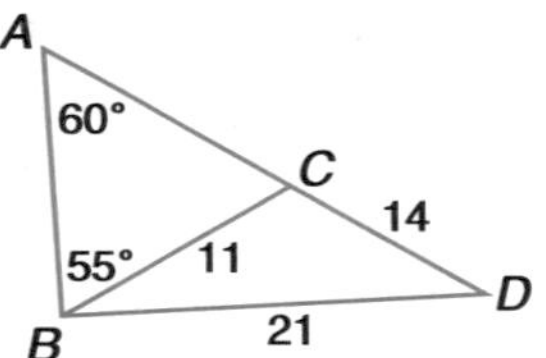

27. Write a two-column proof.

Given: $FG < FH$

Prove: $m\angle 1 > m\angle 2$

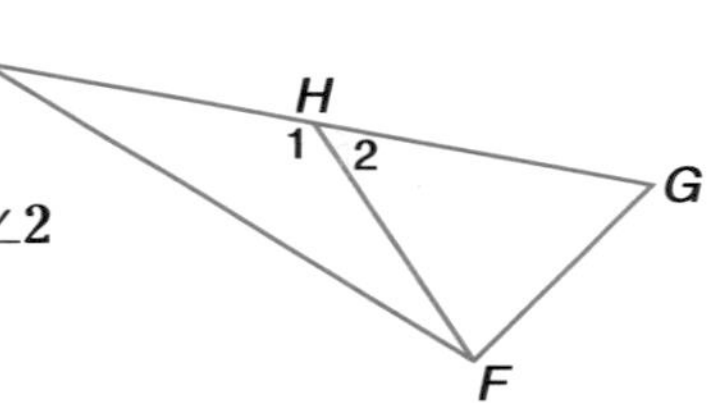

- apply the Triangle Inequality Theorem (Lesson 5–5)

In $\triangle TRI$, $TR = 27$ and $RI = 22$. Between what two numbers is TI?

By the Triangle Inequality Theorem, the following inequalities must be true.

$TI + 27 > 22$	$27 + 22 > TI$	$TI + 22 > 27$
$TI > -5$	$49 > TI$	$TI > 5$

Therefore, TI must be between 5 and 49.

The measures of two sides of a triangle are given. Between what two numbers must the measure of the third side fall?

28. 5 and 11

29. 24 and 7

Is it possible to have a triangle with the given vertices? Write *yes* or *no*, then explain your answer.

30. $A(-4, 13)$, $B(5, -2)$, $C(1, 1)$

31. $D(-5, 2)$, $E(1, -7)$, $F(-3, -1)$

32. How many different scalene triangles are possible if the measures of the three sides must be selected from the measures 2, 3, 4, and 5?

OBJECTIVES AND EXAMPLES

- apply the SAS Inequality and the SSS Inequality (Lesson 5–6)

Refer to the figure below to write an inequality relating AB and AC.

$m\angle AXC = 75 + 18$
$= 93$

$m\angle AXB = 180 - 93$
$= 87$

Since $m\angle AXC > m\angle AXB$, by the SAS Inequality, $AB < AC$.

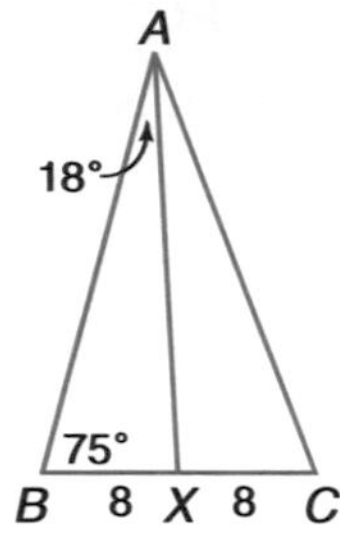

REVIEW EXERCISES

Refer to the figure below to write an inequality relating each pair of angles.

33. $m\angle ALK$, $m\angle ALN$
34. $m\angle ALK$, $m\angle NLO$
35. $m\angle OLK$, $m\angle NLO$
36. $m\angle KLO$, $m\angle ALN$

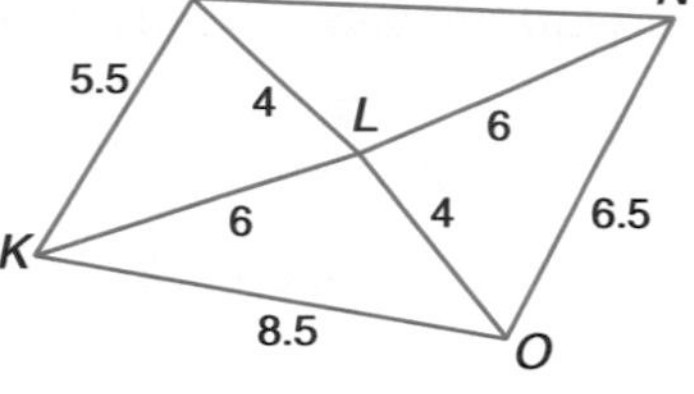

37. Write a two-column proof.
Given: $AD = BC$
Prove: $AC > DB$

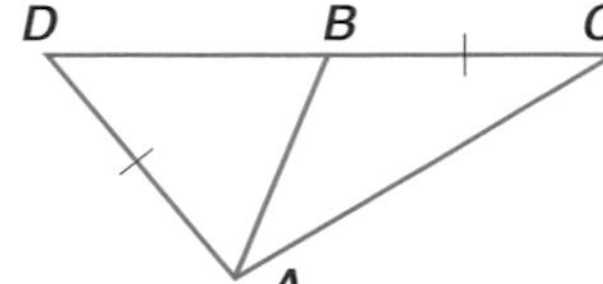

APPLICATIONS AND PROBLEM SOLVING

38. **Construction** Four wood braces of equal length are to be used to support a deck near where the deck meets the wall of a house. One end of each brace will be attached to the bottom of the deck at the same distance from the wall. Explain why the other end of each brace will be attached to the wall at the same distance from the bottom of the deck. (Lesson 5–2)

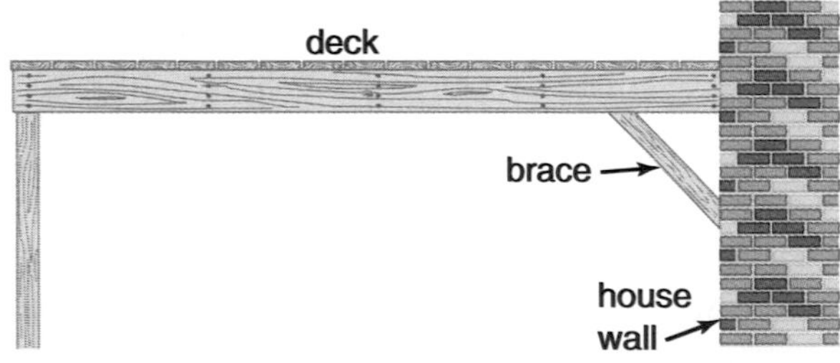

39. **Work Backward** On a game show, all the contestants begin with the same number of points. They are awarded points for questions answered correctly and points are deducted for questions answered incorrectly. Derice answered six 200-point questions correctly. Then he answered a 400-point question and a 500-point question incorrectly. In the final-round question, Derice tripled his score and won with a score of 1500 points. How many points did each player have at the beginning of the game? (Lesson 5–3)

40. **Kitchen Design** When a designer lays out a kitchen, he or she analyzes the work triangle formed by the sink, the stove, and the refrigerator. In the most efficient kitchen designs, the perimeter of the work triangle is less than 26 feet and more than 12 feet and none of the sides of the work triangle are less than 4 feet or more than 9 feet long. Mrs. Alomar would like to move the sink when the kitchen is remodeled. What are the possible distances from the stove to the sink that would meet the recommendations? (Lesson 5–5)

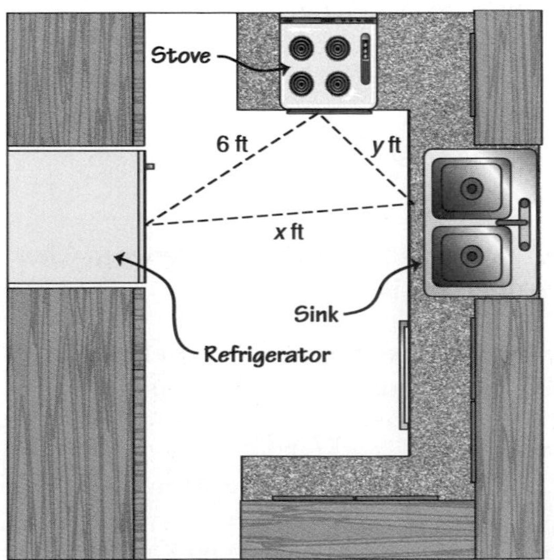

A practice test for Chapter 5 is provided on page 797.

ALTERNATIVE ASSESSMENT

COOPERATIVE LEARNING PROJECT

Projective Geometry In this project, you will apply your knowledge of congruent triangles to projections of triangles.

Follow these steps to explore projections of triangles.

- You will need a projector or similar light source, a flat wall, and construction paper.
- Cut a variety of triangles of different shapes from the construction paper.
- Position the light source so the beam is parallel to the floor.
- Hold two triangles of similar shape in front of the light source, perpendicular to the light source and parallel to the wall. Vary the distance of the triangles from the light source. See if you can find a distance at which the projected images are congruent.

Follow these steps to explore more projections of triangles.

- Now hold two cutouts in front of the projector. Keep one triangle perpendicular to the light source. Try to tilt the other triangle so that its projection is congruent to the projection of the first triangle.
- If you are unable to do this with the first triangle, try the other cutouts.
- If none of them work, note what would have to be different to make it work.
- Try to cut out a triangle that will work.

As a group, discuss what worked and what did not. Try to figure out why it worked and arrive at some general rules to follow. Can you state these rules in terms of numerical relationships between lengths of sides?

Write a report and present your findings to the class. Make your report as attractive as possible, using word-processing and drawing software if available.

THINKING CRITICALLY

Suppose that you proved a theorem from three postulates. Suppose also that you proved the negation of the theorem from the same three postulates. What would you conclude?

PORTFOLIO

Objects of the same height will project shadows of the same length at a given time of day. Have classmates of the same height stand outside in the sunlight. Make measurements of each person's height and the length of each shadow. Draw and label a triangle for each object and its shadow like the one shown below.

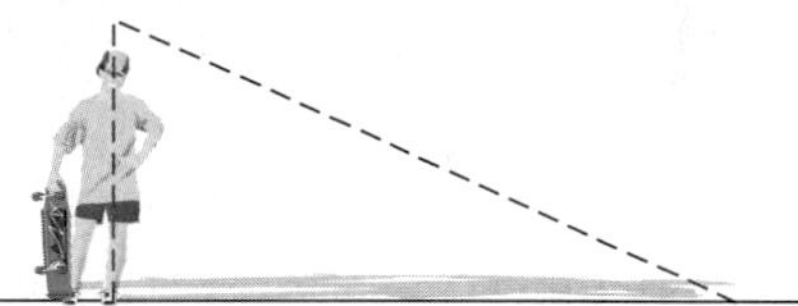

Make measurements at three different times of the day. Compile the data in a table. Determine which triangles are congruent. Keep the results in your portfolio.

LONG-TERM PROJECT

In·ves·ti·ga·tion

This Land Is Your Land

MATERIALS NEEDED

calculator

posterboard

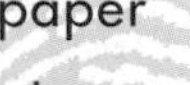
construction paper

glue

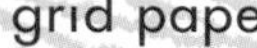
grid paper

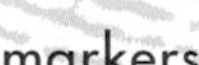
markers

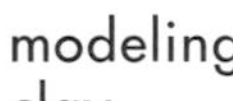
modeling clay

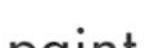
paint

ruler

scissors

masking tape

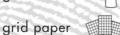

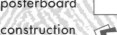

According to historians, the ancient Sumerians of Mesopotamia created the first parks around 2300 B.C. These early parks were private areas set aside by the wealthy for hunting or gardening. The first public parks were probably constructed in Greece. In 1634, the first public park in colonial North America, Boston Common, was established. This park is still a popular recreational area. The National Park Service was founded in 1916 to administer the growing park system in the United States. The system oversees 366 sites, including historic sites, battlefields, lakeshores, memorials, and recreation areas. State and local governments also administer parkland in their area.

Mason County requires that a developer provide one acre of park land for every 100 acres of land that is used for residential or commercial building. The Heflin Development Company is planning a 600-acre residential and business complex on the county's north side, so six acres will be used as park land. The land that has been chosen was previously used as farmland, so it is flat with few trees. The company will allow the designer to choose the shape of the six-acre lot.

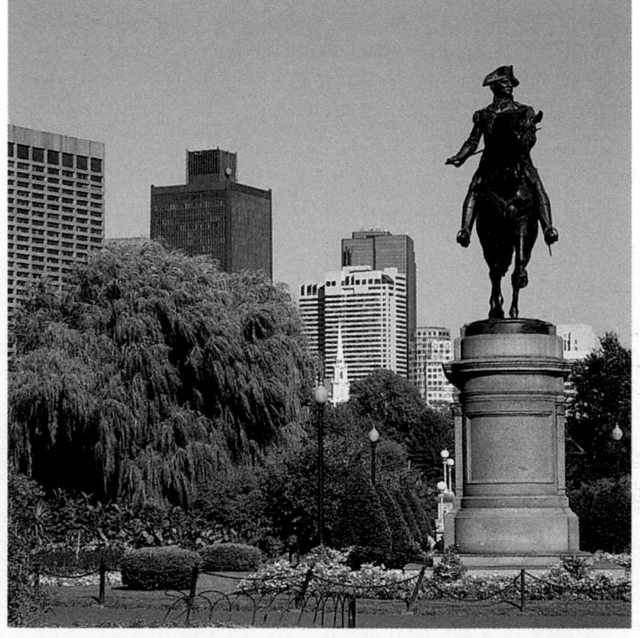

BACKGROUND INFORMATION

The county commissioners have conducted a survey of 1000 residents on possible park features. The results of the survey are given in the chart.

Park Feature	Percent of Respondents Interested
Basketball courts	37%
Bike/jogging path	81%
Community gardens	31%
Fishing pond	6%
Frisbee golf course	4%
Nature trails	74%
Open space	22%
Picnic area	68%
Playground equipment	78%
Racquetball courts	8%
Rose garden	12%
Soccer/football fields	66%
Softball fields	42%
Swimming pool	51%
Tennis courts	48%

Your team of designers has been asked to submit a proposal on how to use the parkland. Your design should incorporate as many of the requested features as possible and meet the desires of as many residents as possible. The budget for the project is limited to $2 million. This budget will pay for design and completion of all elements of the park. If your company's design is chosen, you will supervise the construction of the park.

RESEARCHING THE SITE

Begin by researching the project. Some questions to consider are:

- How much does each park feature cost to construct?
- How much space is available?
- What are the dimensions of each feature?
- Can any park features be combined, such as using the same area for softball and soccer fields during different seasons?
- Will any of the park features require special lighting or safety equipment?
- What features are required that are not on the list, such as restroom facilities and parking space?
- How much maintenance will be required for the park after construction is completed?
- What will be the staff requirements for the park? Will you need rangers, lifeguards, or maintenance personnel? Can any of the positions be filled by volunteers?
- What is a reasonable fee for you to charge for your design and supervisory work?

You will continue working on this Investigation throughout Chapters 6 and 7.

Be sure to keep your research, designs, and other materials in your Investigation Folder.

This Land Is Your Land Investigation

Working on the Investigation
Lesson 6–5, p. 328

Working on the Investigation
Lesson 7–1, p. 345

Working on the Investigation
Lesson 7–5, p. 377

Closing the Investigation
End of Chapter 7, p. 386

CHAPTER

6 Exploring Quadrilaterals

What You'll Learn

In this chapter, you will:

- recognize and define parallelograms, rhombi, rectangles, squares, and trapezoids,
- solve problems by identifying subgoals, and
- use the properties of parallelograms, rhombi, rectangles, squares, and trapezoids to solve problems.

Why It's Important

Real World

Crafts Quiltmaking is a popular art form. Historically, quilting began for practical purposes. In the past, used fabric was often used for quilting, but today, most quilters use new fabrics. Look at the design in the quilt. You can see triangles and quadrilaterals. Other polygons can be used as well. A quilt pattern is similar to a tessellation in that the angles of the figures must meet and leave no gaps in the pattern, In this chapter, you will learn more about different types of quadrilaterals and the relationships between their angle measures and side lengths.

PREREQUISITE SKILLS

To be successful in this chapter, you'll need to understand these concepts and be able to apply them. Refer to the example or to the lesson in parentheses if you need more review before beginning the chapter.

Find the probability of simple events.

Example What is the probability that you will roll an odd number when you roll a die?

$$P(\text{odd}) = \frac{\text{number of odd outcomes}}{\text{number of outcomes}} = \frac{3}{6} \text{ or } \frac{1}{2}$$

A die is rolled. Find each probability.

1. P(even number)
2. P(less than 2)
3. P(greater than 6)
4. P(less than 7)

Find the distance between two points in a coordinate plane.
(Lesson 1–4)

Find the distance between each pair of points.

5. (0, 0), (3, 4)
6. (−8, 5), (10, 5)
7. (4, 3), (4, −1)

Use SSS, SAS, and ASA Postulates to test for triangle congruence.
(Lesson 4–4)

Determine which postulate can be used to prove the triangles are congruent. If it is not possible to prove that they are congruent, write *not possible*.

8. 
9. 
10. 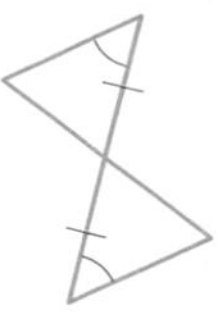

Focus On READING SKILLS

In this chapter, you'll learn about the properties of four-sided polygons, or ***quadrilaterals.*** You will learn the names given to specific quadrilaterals such as *parallelogram*, *rectangle*, *square*, *rhombus*, and *trapezoid*. Some quadrilaterals can be called by several names. For example, a rectangle has both pairs of opposite sides parallel, so it is also a parallelogram. A square is also a rectangle because a rectangle is a quadrilateral with four right angles.

6–1A Using Technology Exploring Parallelograms

A Preview of Lesson 6–1

A **parallelogram** is a four-sided figure with both pairs of opposite sides parallel. You can draw a parallelogram with the geometry program of a calculator by drawing two pairs of intersecting parallel lines.

Step 1 Draw one line anywhere on the screen. Press F2 and choose Line from the menu. Use the arrow key to position the cursor anywhere on the screen and press ENTER. Then move to another place on the screen and press ENTER to position the line.

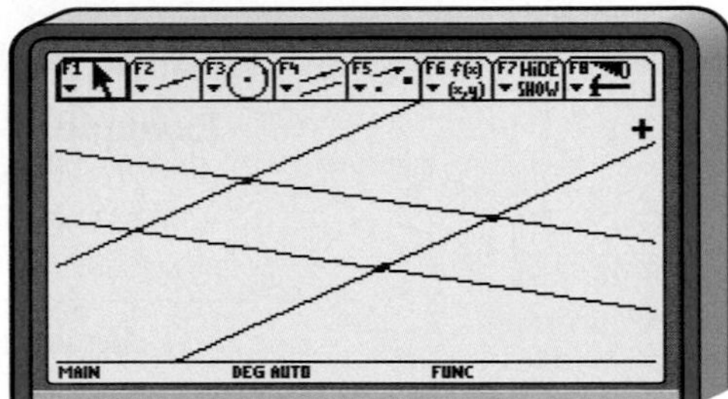

Step 2 Next, use the procedure in Step 1 to draw any line that intersects the first line.

Step 3 To draw a line parallel to the first line, press F4 and choose Parallel Line from the menu. Move the cursor to a point on the first line so that the prompt *parallel to this line* appears. Press ENTER. Then move the cursor to a point through which you would like the line to pass and press ENTER.

Step 4 Use the procedure in step 3 to draw a line parallel to the second line that you drew.

The figure formed by the segments whose endpoints are the points of intersection of the parallel lines is a parallelogram.

EXERCISES

Use the measuring capabilities of the calculator to explore the characteristics of a parallelogram.

1. Measure the opposite sides of the parallelogram that you drew. What do you observe?

2. Make conjectures about the angles in the parallelogram that appear to be congruent, complementary, or supplementary. Measure each angle to check your conjectures.

3. Clear the screen and follow steps 1–4 to draw another parallelogram. Then measure each side and angle. Do the same relationships hold true for this parallelogram?

6–1 Parallelograms

***What* YOU'LL LEARN**

- To recognize and apply the properties of a parallelogram, and
- to find the probability of an event.

***Why* IT'S IMPORTANT**

You can use the properties of parallelograms to solve problems involving transportation and interior design.

Interior Design

Dalia Berlin, ASID, is the owner of an award-winning interior design company in Hollywood, Florida. Ms. Berlin has designed commercial buildings as well as homes in South America and the United States. When Ms. Berlin designs a room, she tries to create an attractive, comfortable, and functional area. She carefully selects each item to suit the room's purpose and overall mood. Important elements she considers are the style of the room, the form of the furniture, color, light, scale, pattern, and texture.

Ceramic tile and wood flooring are often designers' favorite choices for homes. Many styles and colors are available and the floor can be designed in many ways. The wood flooring at the left is laid out in a herringbone pattern.

The herringbone pattern is created using a pattern of **quadrilaterals**. Quadrilaterals are four-sided polygons. Below are some examples and non-examples of quadrilaterals.

CAREER CHOICES

Interior designers need creativity to assemble beautiful and functional rooms. They also need math skills to determine measurements and costs.

For more information, contact:

American Society of Interior Designers (ASID)
608 Massachusetts Ave., NE
Washington, DC 20002

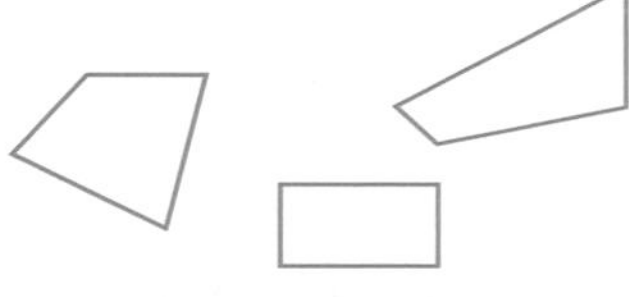

Examples

Non-examples

The special quadrilaterals in the floor shown above are called **parallelograms**. A parallelogram is a quadrilateral with *both pairs of opposite sides parallel.*

The parallelogram at the right has vertices M, N, P, and Q. A symbol for parallelogram $MNPQ$ is $\square MNPQ$. $\overline{MN}$ and $\overline{PQ}$, and $\overline{MQ}$ and $\overline{NP}$ are opposite sides of $\square MNPQ$. The opposite angles are $\angle M$ and $\angle P$, and $\angle N$ and $\angle Q$.

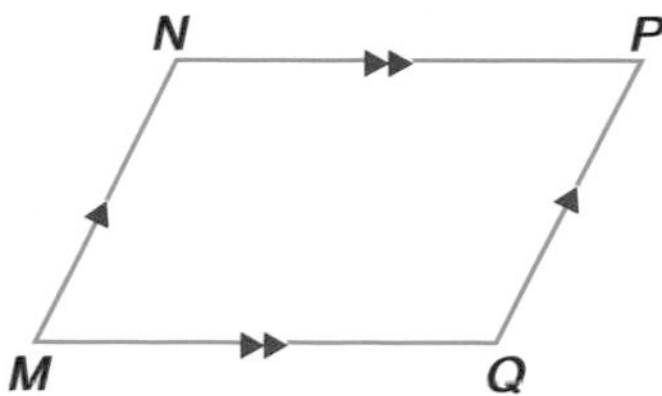

What conjectures can you make about the sides and angles of a parallelogram?

Parallelogram Properties

Materials:  patty paper 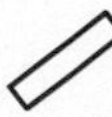 straightedge

- Use a straightedge to draw two sets of intersecting parallel lines on your patty paper like the ones shown below. Label the parallelogram *ABCD*.

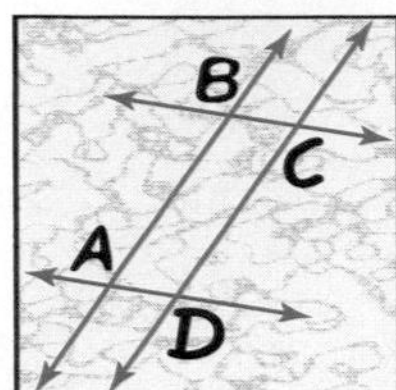

- Place a second patty paper over the first and trace *ABCD*. Label the second parallelogram *PQRS*.

Your Turn

a. Move the second patty paper over the first and compare side $\overline{RS}$ to side $\overline{AB}$. How do the lengths of the opposite sides compare?

b. Rotate and move the second patty paper over the first to compare opposite angles *R* and *A* or angles *S* and *B*. How do the opposite angles compare?

c. Rotate the second patty paper around the first to compare $\angle S$ and $\angle A$. What is their relationship? What conclusion can be drawn?

The Modeling Mathematics activity provides insight into three important properties about parallelograms.

Theorem 6–1	**Opposite sides of a parallelogram are congruent.**
Theorem 6–2	**Opposite angles of a parallelogram are congruent.**
Theorem 6–3	**Consecutive angles in a parallelogram are supplementary.**

You will be asked to prove Theorems 6–1 and 6–2 in Exercises 38 and 39, respectively.

Proof of Theorem 1–1

Write a paragraph proof of Theorem 6–3.

Given: ▱*GEOM*

Prove: $\angle G$ and $\angle E$ are supplementary.

$\angle E$ and $\angle O$ are supplementary.

$\angle O$ and $\angle M$ are supplementary.

$\angle M$ and $\angle G$ are supplementary.

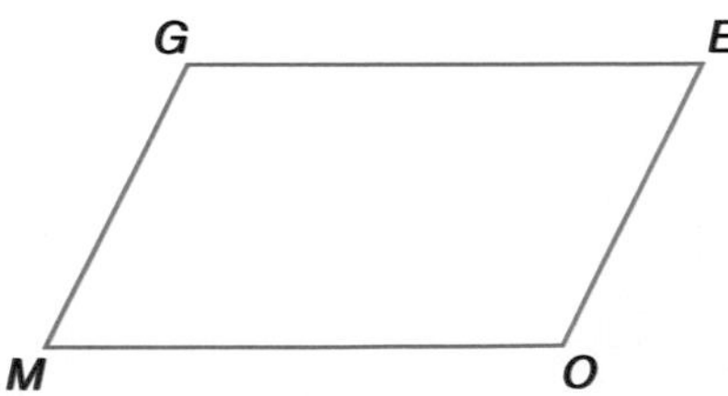

Paragraph Proof:

By the definition of a parallelogram, $\overline{GE} \parallel \overline{MO}$ and $\overline{GM} \parallel \overline{EO}$. For parallels $\overline{GE}$ and $\overline{MO}$, $\overline{GM}$ and $\overline{EO}$ are transversals, and for parallels $\overline{GM}$ and $\overline{EO}$, $\overline{GE}$ and $\overline{MO}$ are transversals. Thus, the consecutive interior angles on the same side of a transversal are supplementary. Therefore, $\angle G$ and $\angle E$, $\angle E$ and $\angle O$, $\angle O$ and $\angle M$, and $\angle M$ and $\angle G$ are supplementary.

Polygons with more than three sides have diagonals. The **diagonals** of a polygon are the segments that connect any two nonconsecutive vertices.

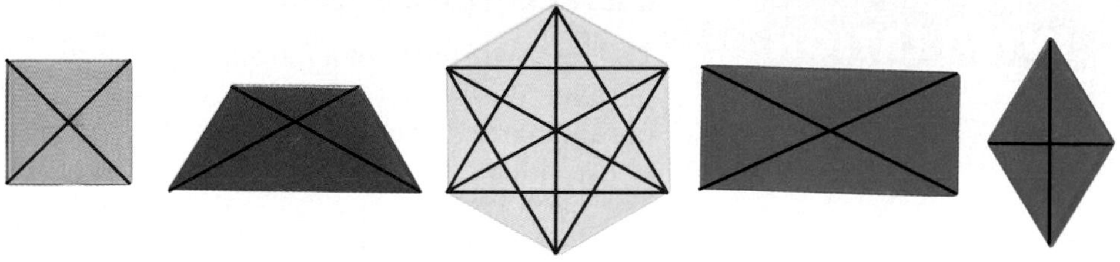

In parallelogram $KLMN$ at the right, the dashed line segments, $\overline{KM}$ and $\overline{LN}$, are diagonals. There is a special relationship between the diagonals of a parallelogram. This relationship is stated in Theorem 6–4. *You will be asked to prove this theorem in Exercise 40.*

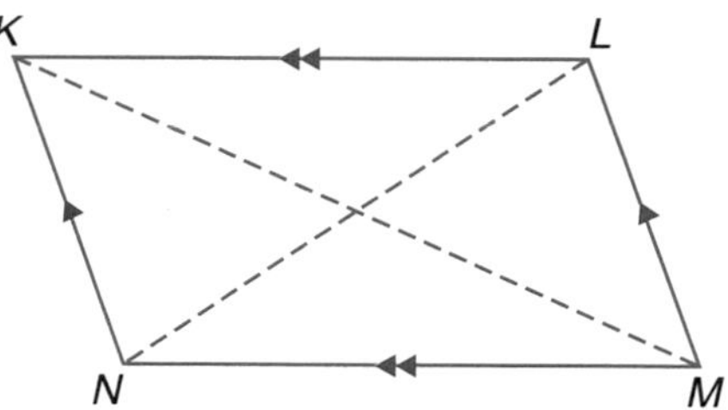

Theorem 6–4	**The diagonals of a parallelogram bisect each other.**

Example **1**

Algebra

***WXYZ* is a parallelogram, $m\angle ZWX = b$, and $m\angle WXY = d$. Find the values of a, b, c, and d.**

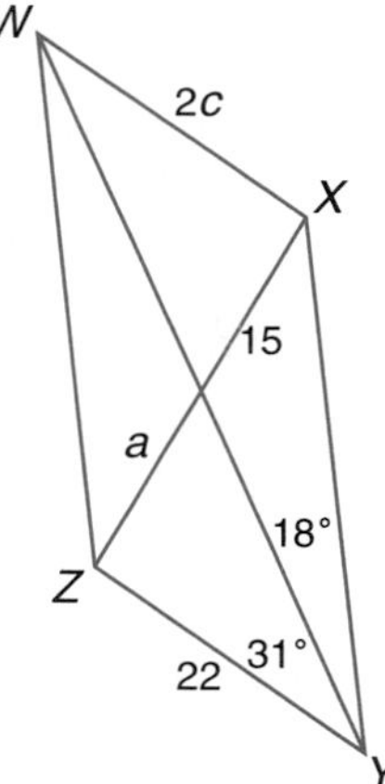

Since the diagonals of a parallelogram bisect each other, $a = 15$.

Opposites angles of a parallelogram are congruent. So $m\angle ZWX = m\angle XYZ$.

$$m\angle ZWX = m\angle XYZ$$
$$b = 31 + 18 \text{ or } 49$$

Opposite sides of a parallelogram are congruent. Therefore, $\overline{WX} \cong \overline{ZY}$.

$$WX = ZY$$
$$2c = 22$$
$$c = 11$$

Since consecutive angles of a parallelogram are supplementary, $\angle WXY$ and $\angle XYZ$ are supplementary.

$$m\angle WXY + m\angle XYZ = 180$$
$$d + (31 + 18) = 180 \quad m\angle WXY = d$$
$$d + 49 = 180$$
$$d = 131$$

You can also use slope to verify properties of a parallelogram.

Example **2**

Algebra

The coordinates of the vertices of *RSTV* are *R*(1, 1), *S*(3, 6), *T*(8, 8), and *V*(6, 3). Determine if *RSTV* is a parallelogram.

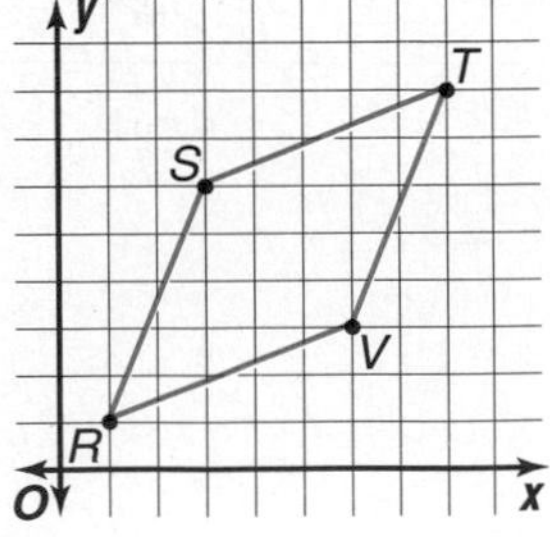

The opposite sides of a parallelogram are parallel. We can determine if *RSTV* is a parallelogram by comparing the slopes of the sides.

slope of $\overline{RS} = \frac{6-1}{3-1}$ or $\frac{5}{2}$

slope of $\overline{TV} = \frac{3-8}{6-8}$ or $\frac{5}{2}$

slope of $\overline{ST} = \frac{8-6}{8-3}$ or $\frac{2}{5}$ slope of $\overline{RV} = \frac{3-1}{6-1}$ or $\frac{2}{5}$

Since the opposite sides have the same slope, $\overline{RS} \parallel \overline{TV}$ and $\overline{ST} \parallel \overline{RV}$. Therefore, *RSTV* is a parallelogram.

The **probability** of an event is the ratio of the number of favorable outcomes to the total number of possible outcomes. For example, the probability of rolling an even number on a die is $\frac{3}{6}$ or $\frac{1}{2}$ because there are 3 even numbers out of 6 possible numbers to roll.

Example **3**

Probability

Two sides of ▱*ABCD* are chosen at random. If $\overline{AB} \ncong \overline{BC}$, what is the probability that the two sides chosen are *not* congruent?

List the possible pairs of sides and determine which are not congruent.

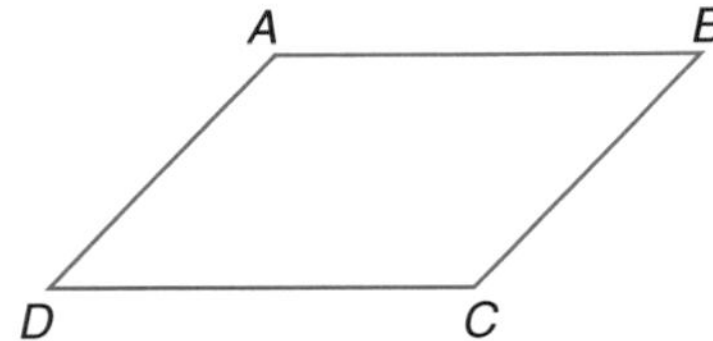

$\overline{AB}, \overline{BC}$	not congruent
$\overline{AB}, \overline{CD}$	congruent
$\overline{AB}, \overline{AD}$	not congruent
$\overline{BC}, \overline{CD}$	not congruent
$\overline{BC}, \overline{AD}$	congruent
$\overline{CD}, \overline{AD}$	not congruent

Four of 6 possibilities are not congruent. So the probability of choosing two sides at random that are not congruent is $\frac{4}{6}$ or $\frac{2}{3}$.

CHECK FOR UNDERSTANDING

Communicating Mathematics

Study the lesson. Then complete the following.

1. **Draw** a quadrilateral and explain why it is a quadrilateral. Then draw a non-example of a quadrilateral.
2. **Compare and contrast** a general quadrilateral and a parallelogram.
3. **Verify** that the diagonals of *RSTV* in Example 2 bisect each other.
4. **Summarize** five properties that hold true for any parallelogram.

5. Draw a parallelogram $QRST$ on a piece of paper. Draw diagonal $\overline{QS}$ with a red marker and diagonal $\overline{TR}$ with a blue marker. Label the intersection of $\overline{QS}$ and $\overline{TR}$, P. Cut along each side and diagonal to form four triangles.
   a. Move and compare the four triangles to determine whether any of the triangles are congruent.
   b. How do the lengths of all the blue segments compare? How do the lengths of all the red segments compare?
   c. What relationship appears to exist for the two diagonals of a parallelogram?

Guided Practice

Complete each statement about ▱*CDFG*. Then name the theorem or definition that justifies your answer.

6. $\overline{CH} \cong$ __?__
7. $\overline{GF} \parallel$ __?__
8. $\angle DCG \cong$ __?__
9. $\overline{DC} \cong$ __?__
10. $\angle DCG$ is supplementary to __?__.
11. $\triangle HGC \cong$ __?__
12. In parallelogram $ABCD$, $AB = 2x + 5$, $m\angle BAC = 2y$, $m\angle B = 120$, $m\angle CAD = 21$, and $CD = 21$. Find the values of x and y.
13. Quadrilateral $WXYZ$ is a parallelogram with $m\angle W = 47$. Find the measure of angles X, Y, and Z.
14. $PQRT$ has vertices $P(-4, 7)$, $Q(3, 0)$, $R(2, -5)$, and $T(-5, 2)$. Determine if $PQRT$ is a parallelogram.

15. Write a two-column proof.

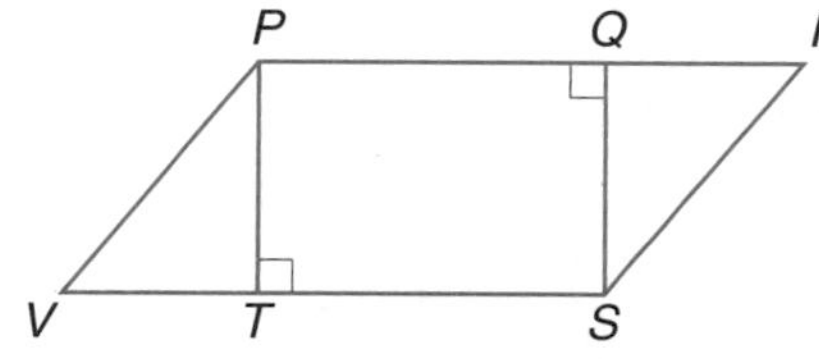

Given: $PRSV$ is a parallelogram.
$\overline{PT} \perp \overline{SV}$
$\overline{QS} \perp \overline{PR}$

Prove: $\triangle PTV \cong \triangle SQR$

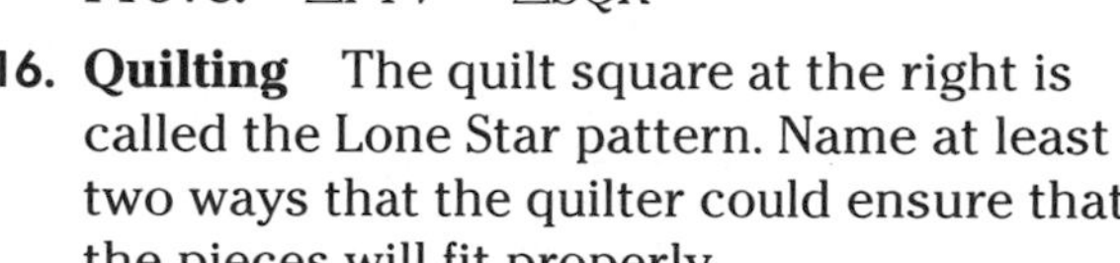

16. **Quilting** The quilt square at the right is called the Lone Star pattern. Name at least two ways that the quilter could ensure that the pieces will fit properly.

EXERCISES

Practice

Complete each statement about ▱*MARK*. Then name the theorem or definition that justifies your answer.

17. $\overline{MK} \parallel$ __?__
18. $\overline{MS} \cong$ __?__
19. $\angle MKR \cong$ __?__
20. $KS = \frac{1}{2}$ __?__
21. $\triangle MAR \cong$ __?__
22. $\overline{AS} \cong$ __?__
23. $\overline{AM} \parallel$ __?__
24. $\angle ARK$ and __?__ are supplementary.
25. $\triangle SAM \cong$ __?__
26. $\angle RKA \cong$ __?__

For each parallelogram, find the values of *x*, *y*, and *z*.

27.

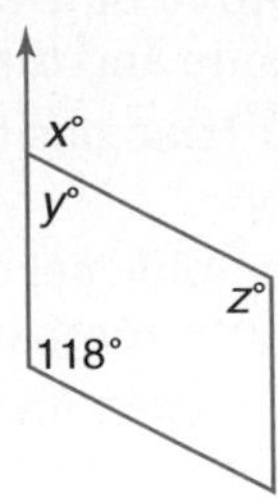

28.

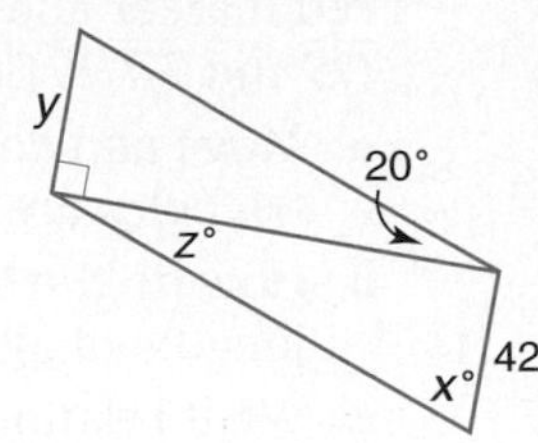

29.

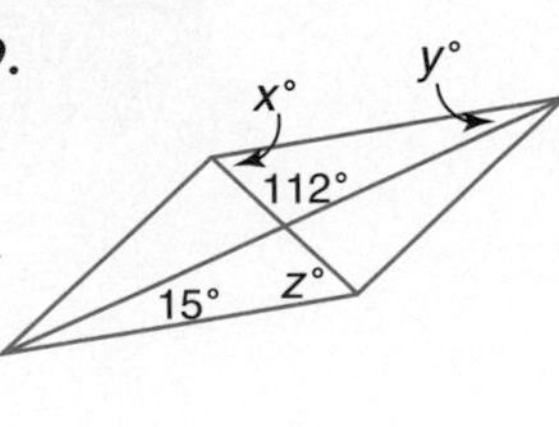

Refer to ▱*CDEF* at the right for Exercises 30–32.

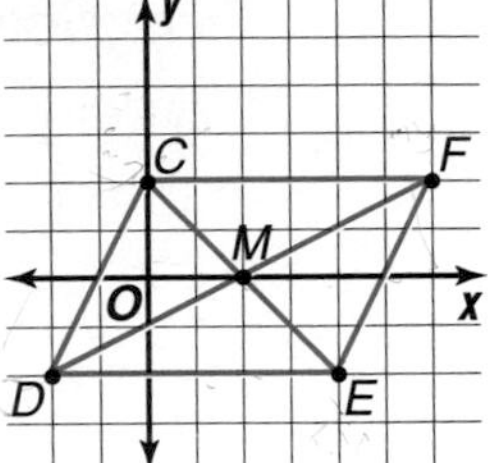

30. Use the distance formula to verify that the diagonals bisect each other.

31. Determine if the diagonals of this parallelogram are congruent.

32. Find the slopes of $\overline{CD}$ and $\overline{CF}$. Are the consecutive sides of *CDEF* perpendicular? Explain.

33. Find all the possible coordinates for the fourth vertex of a parallelogram with vertices $T(4, -1)$, $D(-4, 1)$, and $K(0, 8)$.

34. *NCTM* is a parallelogram with diagonals $\overline{NT}$ and $\overline{MC}$ that intersect at point *Q*. If $NQ = 3a + 18$, $NT = 12a$, $QC = a + 2b$, and $QM = 3b + 1$, find *a*, *b*, and *CM*.

35. If *ABCD* and *PQRS* are parallelograms, find $m\angle APS$.

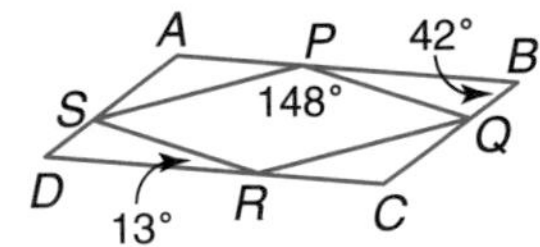

Write a two-column proof.

Proof

36. Given: *SRWV* and *TVXY* are parallelograms.

Prove: $\angle Y \cong \angle R$

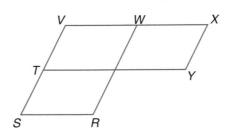

37. Given: ▱*TEAM*
$MS = FS$

Prove: $\angle F \cong \angle TEA$

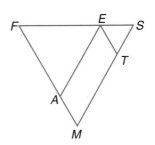

38. Write a paragraph proof of Theorem 6–1.

39. Write a two-column proof of Theorem 6–2.

40. Write a paragraph proof of Theorem 6–4.

41. Draw a parallelogram *MNPQ* with diagonal $\overline{MP}$. Make a conjecture about the sum of the measures of the angles of a parallelogram. Do you think your conjecture holds true for any quadrilateral? Explain.

Critical Thinking

42. Consider parallelogram *RSTV*.

a. As the measure of angle *R* decreases, what must happen to the measure of angle *V*?

b. What is the maximum measure for angle *V*? Explain your reasoning.

Applications and Problem Solving

43. **Language** In a commercial for CompuServe Computer Discount House, the announcer says that he thought a parallelogram was a "telegram for gymnasts at the Olympics." Look up the suffix *gram* in a dictionary. Then make a conjecture about why a parallelogram is named as it is.

44. **Interior Design** Create a unique colorful pattern using parallelograms that could be used to arrange tiles in your kitchen or to create a quilt top.

45. **Transportation** Two tugboats are pulling a ship into harbor. The force exerted by each boat can be represented by a *vector*. A vector is a ray whose length is proportional to a force and whose direction indicates the direction of the force. In the diagram at the right, two forces, A and B, are acting on an object. The net force, or resultant vector R is found by drawing a parallelogram. The resultant is represented by the diagonal that begins at the endpoint of the vectors. Copy vectors X and Y that represent the forces of the tugboats and find the resultant.

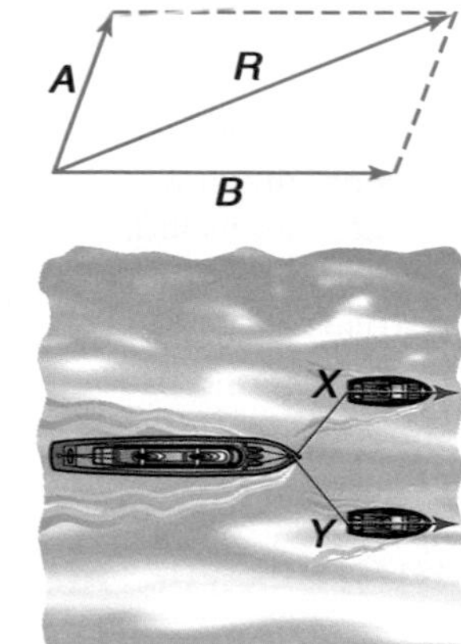

46. **Probability** Suppose $MNOP$ is a parallelogram, but not a rectangle. If you chose two interior angles at random, what is the probability that they would be congruent?

Mixed Review

47. Write an inequality relating the measures of $\angle ADC$ and $\angle ADB$. (Lesson 5–6)

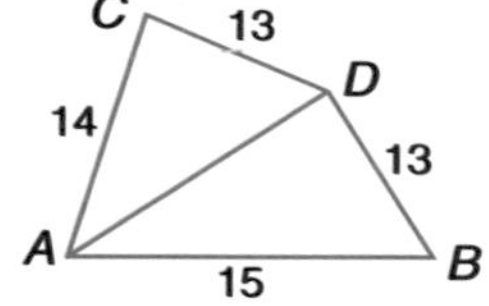

48. Determine whether it is possible to draw a triangle with sides of 6 inches, 9 inches, and 14 inches. Explain. (Lesson 5–5)

49. Given $\triangle KLM$, with $K(-3, 7)$, $L(2, 5)$, and $M(0, -2)$, list the angles in order from least to greatest measure. (Lesson 5–4)

50. If in $\triangle FGH$ and $\triangle JKL$, $\angle H \cong \angle L$ and $GH \cong KL$, what else must be congruent in order to prove that $\triangle FGH \cong \triangle JKL$ by AAS? (Lesson 4–5)

51. **ACT Practice** At a restaurant, a diner uses a coupon for 15% off of the cost of one meal. If the diner orders a meal regularly priced at \$16 and leaves a generous tip of 20% of the discounted meal, how much does she pay in total?

A \$13.60 **B** \$15.64 **C** \$16.32 **D** \$16.80 **E** \$18.72

52. Are the edges of a ruler a model of lines that are *intersecting, parallel,* or *skew*? (Lesson 3–1)

53. Write the conditional *A parallelogram is a quadrilateral with opposite sides parallel* in if-then form. (Lesson 2–2)

54. Refer to the number line to find the coordinate of the midpoint of $\overline{PR}$. (Lesson 1–5)

Algebra

55. **Criminology** The Federal Bureau of Investigation, or FBI, was formed in 1924. In 1980, it had 173.2 million fingerprint cards on file. In 1995, it had 200 million on file. Find the percent of increase from 1980 to 1995. Round to the nearest whole percent.

56. Solve $\frac{d-4}{3} = 5$.

For **Extra Practice**, see page 774.

6–2 Tests for Parallelograms

What YOU'LL LEARN
- To recognize and apply the conditions that ensure a quadrilateral is a parallelogram, and
- to identify and use subgoals in writing proofs.

Why IT'S IMPORTANT

You can use parallelograms to solve problems involving engineering and the arts.

You know that the definition of a parallelogram allows you to prove that a quadrilateral is a parallelogram by proving that opposite sides are parallel. However, there are other tests that can be used.

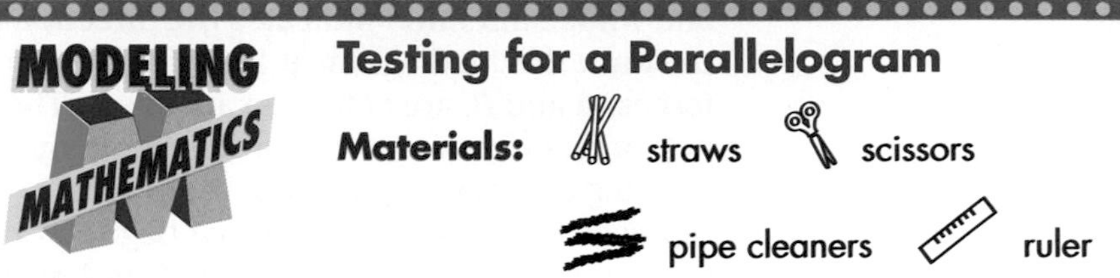

Testing for a Parallelogram

Materials: straws scissors pipe cleaners ruler

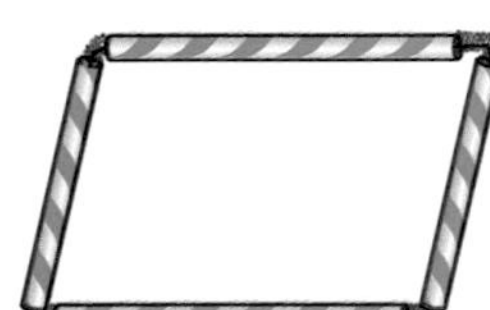

- Cut two straws to one length and two straws to a different length.
- Insert each pipe cleaner in one end of each size of straw. Then form a quadrilateral like the one shown at the right.

Your Turn

a. Shift the position of the sides to form quadrilaterals of different shapes. Measure the distance between opposite sides of the quadrilateral in at least three places. Repeat for several figures. What seems to be true about the opposite sides?

b. What type of quadrilaterals are you forming?

c. How do the measures of pairs of opposite sides compare?

d. What conditions does this activity suggest are sufficient for showing a quadrilateral to be a parallelogram?

The Modeling Mathematics activity leads to Theorem 6–5. Theorems 6–6 and 6–7 are other tests for parallelograms. *You will be asked to prove Theorems 6–6 and 6–7 in Exercises 34 and 35, respectively.*

Theorem 6–5	**If both pairs of opposite sides of a quadrilateral are congruent, then the quadrilateral is a parallelogram.**
Theorem 6–6	**If both pairs of opposite angles of a quadrilateral are congruent, then the quadrilateral is a parallelogram.**
Theorem 6–7	**If the diagonals of a quadrilateral bisect each other, then the quadrilateral is a parallelogram.**

Sometimes when you solve a problem, it is helpful to identify the smaller steps needed to solve the larger problem. **Identifying subgoals** can help you to organize proofs.

Example  **Write a two-column proof for Theorem 6–5.**

Explore We know that both pairs of opposite sides of a quadrilateral are congruent. We need to prove that the quadrilateral must be a parallelogram.

Plan Use quadrilateral *GEOM*.

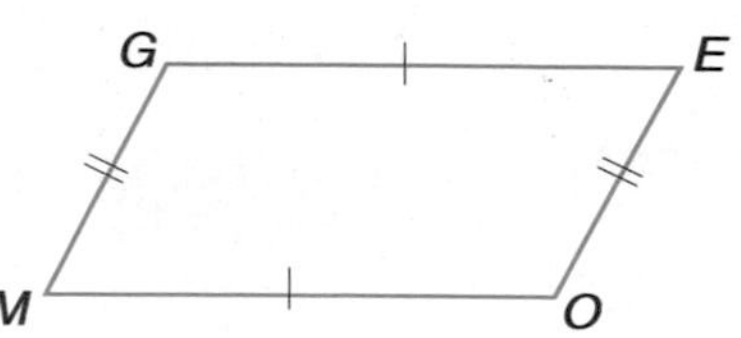

Given: $\overline{GM} \cong \overline{EO}$

$\overline{GE} \cong \overline{MO}$

Prove: *GEOM* is a parallelogram.

Let's set some subgoals before we start the proof. Reasoning backward, we can show that *GEOM* is a parallelogram if we can show that $\overline{GM} \parallel \overline{EO}$ and $\overline{GE} \parallel \overline{MO}$. To show opposite sides parallel, we need a transversal or diagonal that could create alternate interior angles congruent. This can be proved by showing the two triangles formed by the diagonal are congruent and using CPCTC.

Our subgoals are:

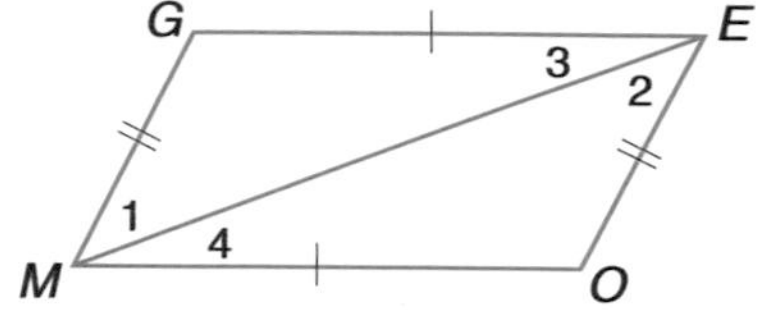

1. Draw diagonal $\overline{EM}$ and prove $\triangle GEM \cong \triangle OME$ by SSS.
2. Use CPCTC to show $\angle 1 \cong \angle 2$ and $\angle 3 \cong \angle 4$.
3. Use alternate interior angles to show that $\overline{GM} \parallel \overline{EO}$ and $\overline{GE} \parallel \overline{MO}$.
4. Show that *GEOM* is a parallelogram by the definition of parallelogram.

Proof:

Solve

Statements	**Reasons**
1. $\overline{GM} \cong \overline{EO}$ $\overline{GE} \cong \overline{MO}$	1. Given
2. Draw $\overline{EM}$.	2. Through any 2 pts. there is 1 line.
3. $\overline{EM} \cong \overline{EM}$	3. Congruence of segments is reflexive.
4. $\triangle GEM \cong \triangle OME$	4. SSS
5. $\angle 1 \cong \angle 2$ $\angle 3 \cong \angle 4$	5. CPCTC
6. $\overline{GM} \parallel \overline{EO}$ $\overline{GE} \parallel \overline{MO}$	6. If ⟷ and alt. int. ∠s are ≅, then the lines are ∥.
7. *GEOM* is a parallelogram.	7. Def. parallelogram

Examine The proof uses a general quadrilateral and shows that if opposite sides are congruent, then the quadrilateral is a parallelogram.

Theorem 6–8 states another test for parallelograms. *You will be asked to prove this theorem in Exercise 36.*

Theorem 6–8 **If one pair of opposite sides of a quadrilateral is both parallel and congruent, then the quadrilateral is a parallelogram.**

We can use the distance formula and the slope formula to determine if a quadrilateral in the coordinate plane is a parallelogram.

Example 2 **Integration: Algebra**

The coordinates of the vertices of quadrilateral *PQRS* are *P*(−5, 3), *Q*(−1, 5), *R*(6, 1), and *S*(2, −1). Determine if quadrilateral *PQRS* is a parallelogram.

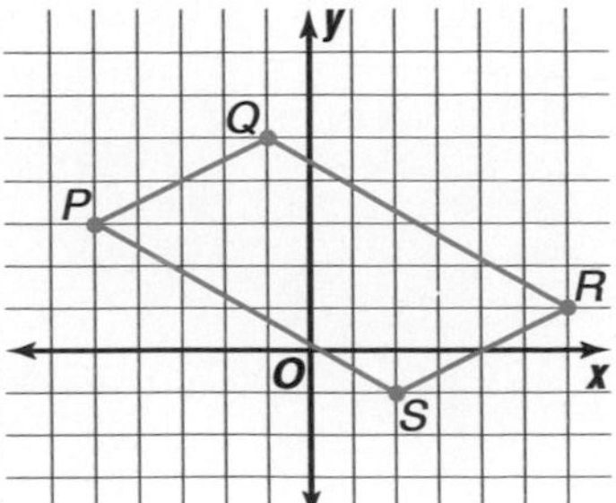

First, find *QR* and *PS* to determine if these opposite sides are congruent.

$QR = \sqrt{(-1-6)^2 + (5-1)^2}$ or $\sqrt{65}$

$PS = \sqrt{(-5-2)^2 + (3-(-1))^2}$ or $\sqrt{65}$

Since $QR = PS$, $\overline{QR} \cong \overline{PS}$.

Next, find the slopes of $\overline{QR}$ and $\overline{PS}$ to determine if $\overline{QR} \parallel \overline{PS}$.

Slope of $\overline{QR} = \frac{5-1}{-1-6}$ or $-\frac{4}{7}$

Slope of $\overline{PS} = \frac{3-(-1)}{-5-2}$ or $-\frac{4}{7}$

Since the slopes are the same, $\overline{QR} \parallel \overline{PS}$.

One pair of opposite sides of *PQRS* are both congruent and parallel, so quadrilateral *PQRS* is a parallelogram.

Here is a summary of the tests to show that a quadrilateral is a parallelogram.

A quadrilateral is a parallelogram if any one of the following is true.
1. Both pairs of opposite sides are parallel. (Definition)
2. Both pairs of opposite sides are congruent. (Theorem 6–5)
3. Both pairs of opposite angles are congruent. (Theorem 6–6)
4. Diagonals bisect each other. (Theorem 6–7)
5. A pair of opposite sides is both parallel and congruent. (Theorem 6–8)

CHECK FOR UNDERSTANDING

Communicating Mathematics

Study the lesson. Then complete the following.

1. **Draw a diagram** that illustrates why only one pair of opposite angles congruent is not enough to prove a quadrilateral is a parallelogram.
2. **Explain** why each quadrilateral is or is not a parallelogram.

a.

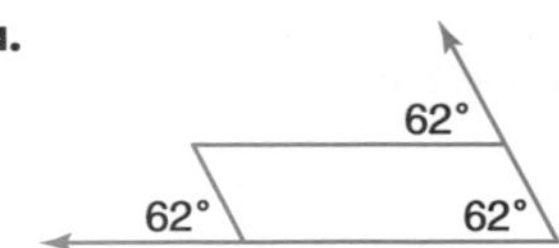

b.

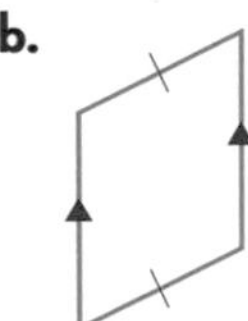

3. **You Decide** Nida says "If a quadrilateral is *not* a parallelogram, then neither pair of opposite sides are parallel." Celina disagrees. Who is correct and why?
4. **Explain** how slope can be used to identify parallelograms in the coordinate plane.

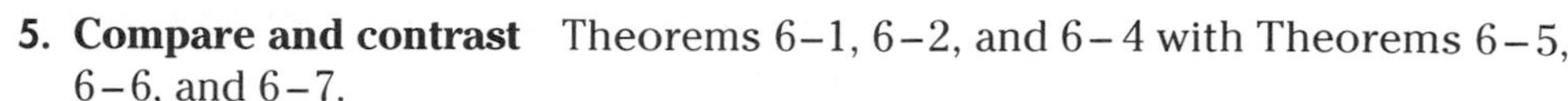

5. **Compare and contrast** Theorems 6–1, 6–2, and 6–4 with Theorems 6–5, 6–6, and 6–7.

6. Write a memory tool you can use to remember the tests for a parallelogram.

Guided Practice

Determine if each quadrilateral is a parallelogram. Justify your answer.

7. 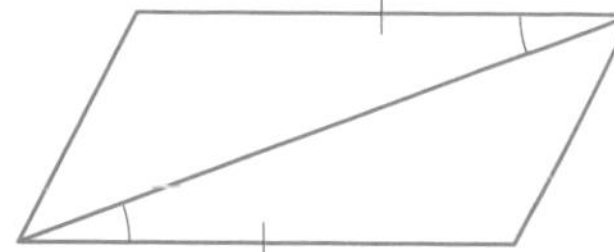

8. 6 cm, 6 cm

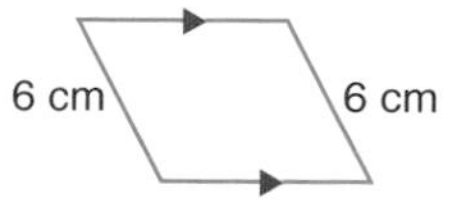

Find the values of *x* and *y* that ensure each quadrilateral is a parallelogram.

9. y, $6x$, $4x + 8$, y^2

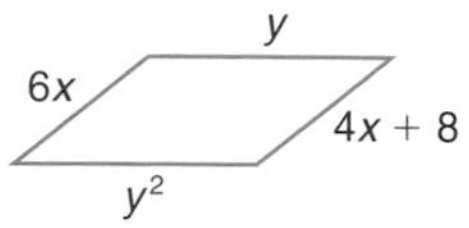

10. $(2x + 8)°$, $120°$, $5y°$

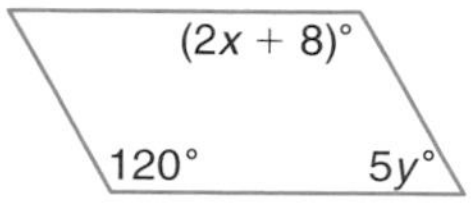

11. Determine if the conditional *If the diagonals of a quadrilateral are congruent, then the quadrilateral is a parallelogram* is true or false. Explain your reasoning.

12. Quadrilateral *GHJK* has vertices $G(-2, 8)$, $H(4, 4)$, $J(6, -3)$, and $K(-1, -7)$. Determine whether *GHJK* is a parallelogram by Theorem 6–8. Justify your answer.

13. Complete a two-column proof.

Given: $\overline{HD} \cong \overline{DN}$
$\angle DHM \cong \angle DNT$

Prove: Quadrilateral *MNTH* is a parallelogram.

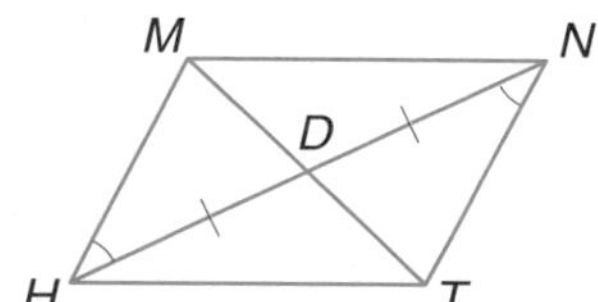

14. **Engineering** Deshon uses an expandable gate to keep his new puppy in the kitchen. As the gate expands or collapses, the shapes that form the gate (such as *ABCD*) always remain parallelograms. Explain why this is true.

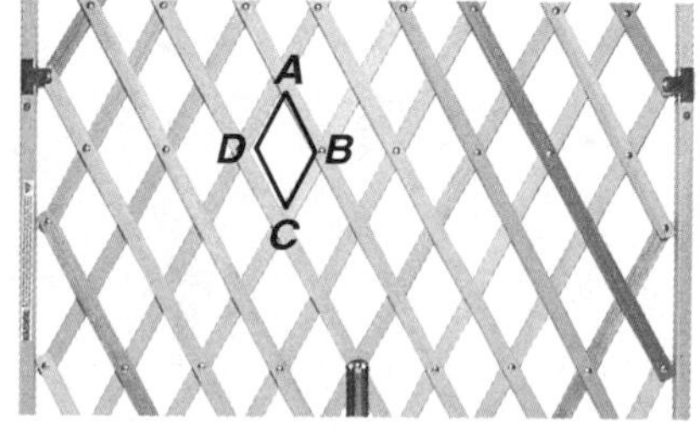

EXERCISES

Practice

Determine if each quadrilateral is a parallelogram. Justify your answer.

15. 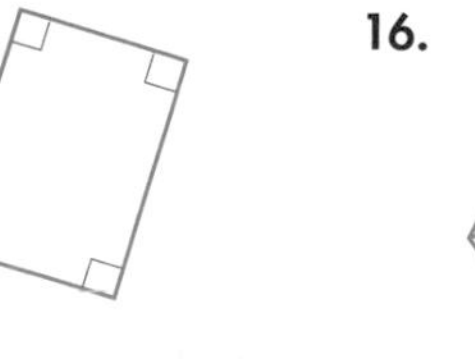

16. 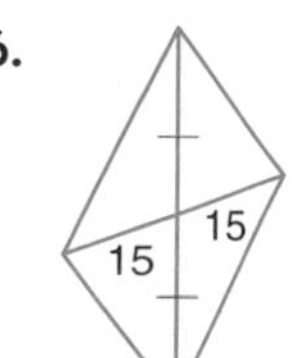

17. 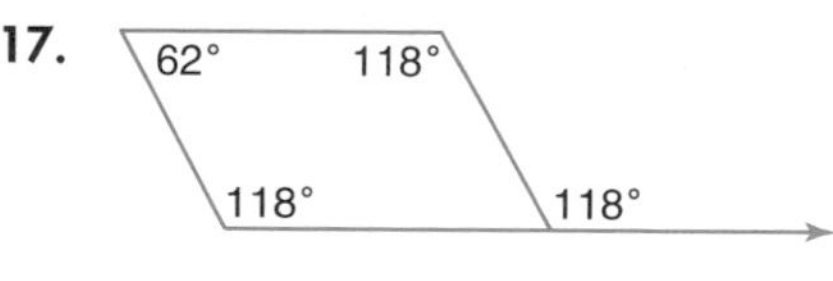

18. 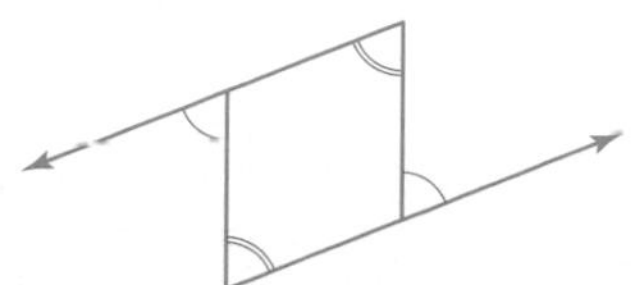

19. 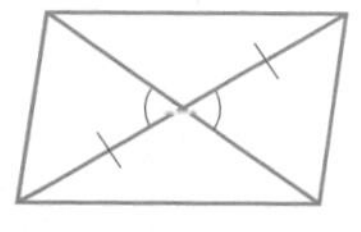

20. 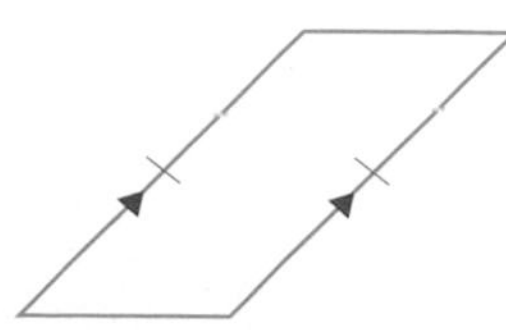

Find the values of x and y that ensure each quadrilateral is a parallelogram.

21. 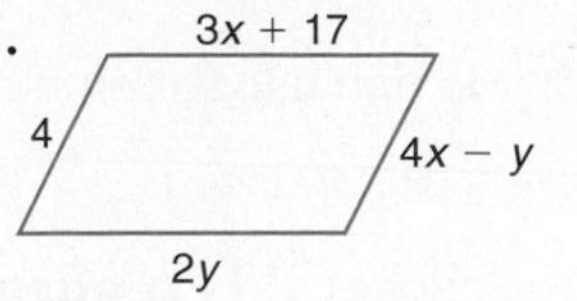

22. 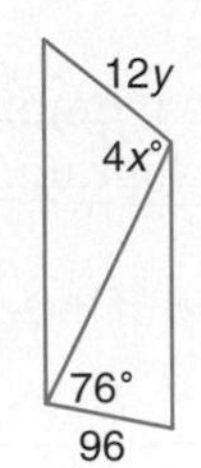

23. 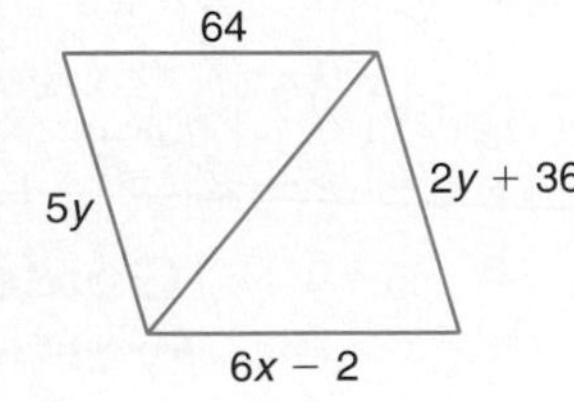

24. 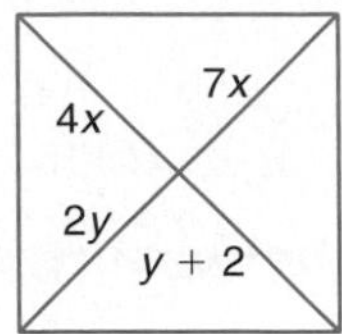

25. 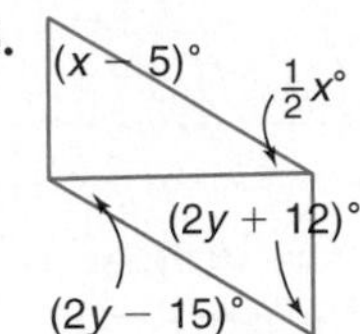

26. 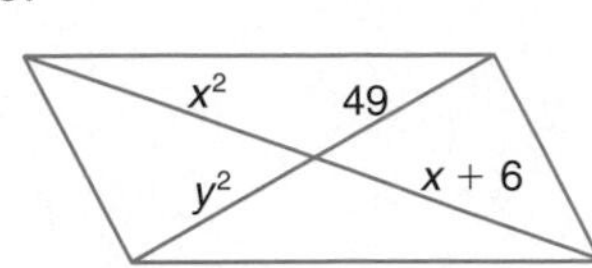

Determine if each conditional is true or false. Explain your reasoning.

27. If two pairs of consecutive sides of a quadrilateral are congruent, then the quadrilateral must be a parallelogram.

28. If all four sides of a quadrilateral are congruent, then the quadrilateral is a parallelogram.

29. If a quadrilateral has one pair of congruent sides and one pair of parallel sides, then the quadrilateral must be a parallelogram.

Determine whether the quadrilateral with the given vertices is a parallelogram by the indicated theorem.

30. $A(5, 6), B(9, 0), C(8, -5), D(3, -2)$; Theorem 6–5

31. $F(-7, 3), G(-3, 2), H(0, -4), J(-4, -3)$; Theorem 6–8

32. $K(-1, 9), L(3, 8), M(6, 2), N(2, 3)$; Theorem 6–7

33. Using a compass and a straightedge, construct a parallelogram *QRST* with one angle congruent to $\angle P$ and sides congruent to $\overline{AB}$ and $\overline{CD}$.

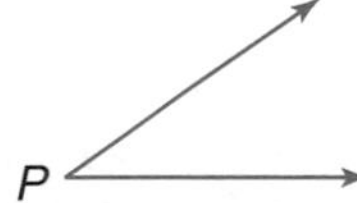

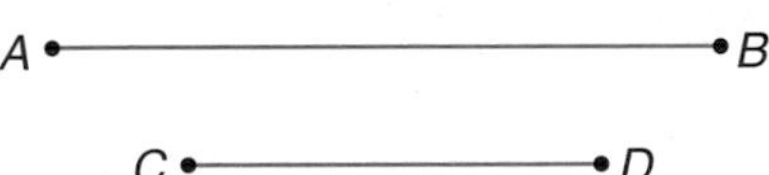

34. Write a paragraph proof of Theorem 6–6.

35. Write a two-column proof of Theorem 6–7.

36. Prove Theorem 6–8.

37. A *regular hexagon* is a six-sided polygon with all sides and all angles congruent. If *ABCDEF* is a regular hexagon, identify the subgoals you would use to prove *FDCA* is a parallelogram. Then write a proof.

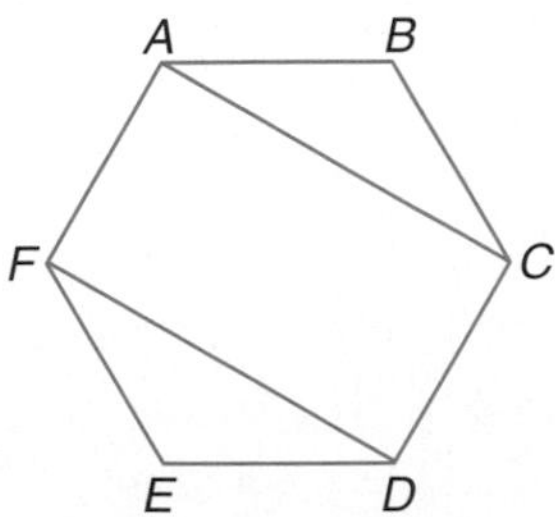

Critical Thinking

38. Ellen claims she has invented a new geometry theorem: *A diagonal of a parallelogram bisects its angles.* She gives the following proof.

Given: $\square MATH$ with diagonal $\overline{MT}$

Prove: $\overline{MT}$ bisects $\angle AMH$ and $\angle ATH$

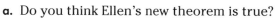

Proof: Since $MATH$ is a parallelogram, $\overline{MH} \cong \overline{AT}$ and $\overline{MA} \cong \overline{HT}$. Since $\overline{MT} \cong \overline{MT}$, $\triangle MHT \cong \triangle MAT$ by SSS. Therefore, $\angle 1 \cong \angle 2$ and $\angle 3 \cong \angle 4$.

a. Do you think Ellen's new theorem is true?

b. Is her proof correct? Explain your reasoning.

Applications and Problem Solving

39. **Construction** Wood lattice panels are made in a configuration of parallelograms.

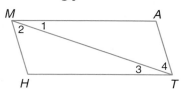

a. Explain how the person who manufactured the panels could verify that the overlapped boards form parallelograms.

b. Look up the word *lattice* in the dictionary. Is the word *crisscross* used in the definition? Why do you think a crisscross pattern results in a series of parallelograms?

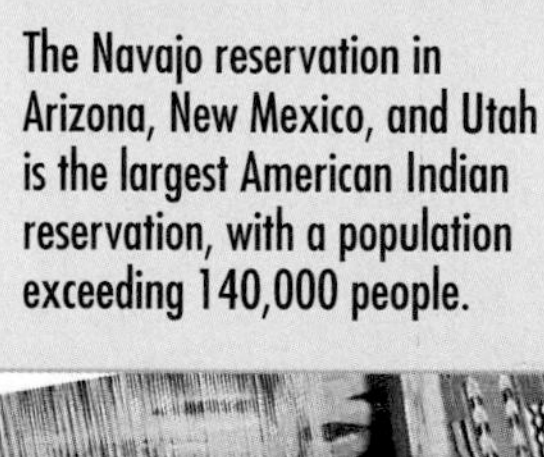

FYI

The Navajo reservation in Arizona, New Mexico, and Utah is the largest American Indian reservation, with a population exceeding 140,000 people.

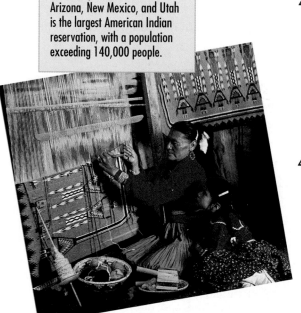

40. **Drafting** Before software became available, blueprints for buildings or mechanical parts were drawn by hand. One tool drafters used, a parallel ruler, is shown at the right. Holding one bar in place and moving the other allowed the drafter to draw a line parallel to the first in many positions. Why does the parallel ruler guarantee that the second line will be parallel to the first?

41. **Arts** The Navajo people are well known for their skill in weaving. The design at the right, known as the Eye-Dazzler, became popular with Navajo weavers in the 1880s. How many parallelograms, not including rectangles, are in the pattern?

Mixed Review

42. In $\square ABCD$, $AB = 2x + 5$, $CD = y + 1$, $AD = y + 5$, and $BC = 3x - 4$. Find the measure of each side. (Lesson 6–1)

43. If the sides of a triangle have measures of $3x + 2$, $8x + 10$, and $5x + 8$, find all possible values of x. (Lesson 5–5)

44. **Work Backward** Juana purchased a jacket for \$190.80. She received a 20% discount off of the original price and was charged a 6% sales tax. What was the original price of the jacket? (Lesson 5–3)

45. If $\triangle DOG \cong \triangle CAT$, what segment in $\triangle CAT$ is congruent to $\overline{GO}$ in $\triangle DOG$? (Lesson 4–3)

46. **SAT Practice** If the sum of two consecutive even integers is 78, then the greater integer equals?

A 33 **B** 35 **C** 38 **D** 40 **E** 42

47. Write a conjecture based on the statement *Both pairs of opposite sides of a quadrilateral are congruent.* Draw a figure to illustrate your conjecture. (Lesson 2–1)

48. If $\angle B \cong \angle L$, $m\angle B = 7x + 29$, and $m\angle L = 9x - 1$, is $\angle B$ acute, right, or obtuse? (Lesson 1–6)

INTEGRATION **Algebra**

49. **Demographics** Do you like to sleep in? If so, you're not alone! A nationwide poll asked people of different ages whether they hit their alarm clock's snooze button. The poll results are shown at the right.

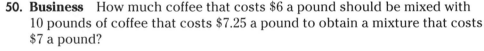

It's Morning Already?!
18–24 52%
25–34 57%
35–44 36%
45–54 30%
55–64 17%
65+ 10%
SNOOZE
TIME 7:00 AM ALARM
Source: Opinion Research for Select Comfort

a. What percent of the people aged 35–44 hit the snooze button?

b. The U.S. Census Bureau projects that there are 25,465,000 people 18 to 24 years old in the United States. How many of those people do you predict hit the snooze button each morning?

50. **Business** How much coffee that costs \$6 a pound should be mixed with 10 pounds of coffee that costs \$7.25 a pound to obtain a mixture that costs \$7 a pound?

For **Extra Practice**, see page 774.

Mathematics and Society

Skyscraper Geometry

The excerpt below appeared in an article in *American Scientist* in July–August, 1996.

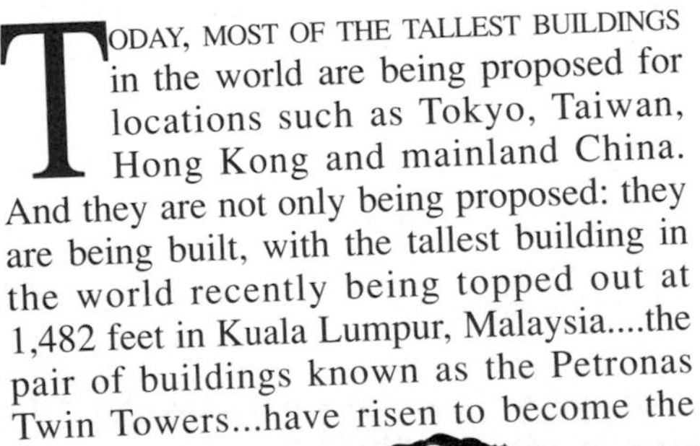

TODAY, MOST OF THE TALLEST BUILDINGS in the world are being proposed for locations such as Tokyo, Taiwan, Hong Kong and mainland China. And they are not only being proposed: they are being built, with the tallest building in the world recently being topped out at 1,482 feet in Kuala Lumpur, Malaysia....the pair of buildings known as the Petronas Twin Towers...have risen to become the world's tallest buildings....The tapering at the top of the building demanded some especially tricky structural engineering, and its geometry necessitated the installation of a wide variety of different-size glass panels....The Twin Towers required about a million and a half square feet of stainless steel cladding and glass, in the form of 32,000 windows, to form a so-called curtain wall. ■

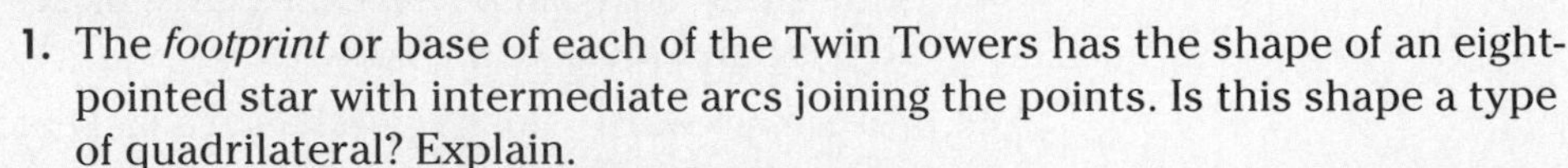

1. The *footprint* or base of each of the Twin Towers has the shape of an eight-pointed star with intermediate arcs joining the points. Is this shape a type of quadrilateral? Explain.
2. If a building's foundation is not secure, the building can settle and begin to tilt, so its walls are no longer vertical. What could happen if the angle of tilt continued to increase each year?
3. If a steel building and a concrete building were built in the shape of identical parallelograms, which would retain its shape better when exposed to high winds? Why?

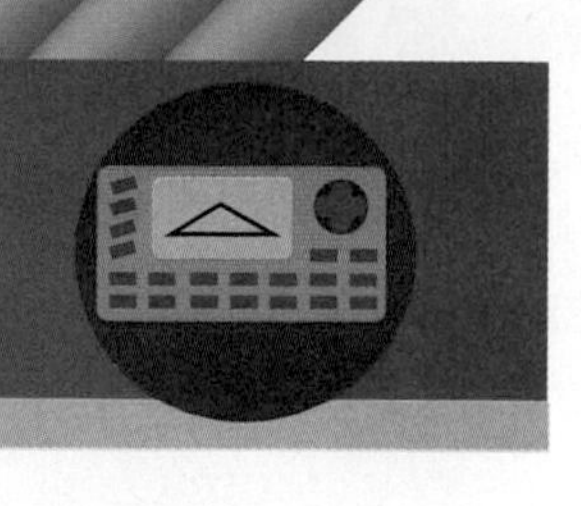

6–3A Using Technology Exploring Rectangles

A Preview of Lesson 6–3

A quadrilateral with four right angles is a **rectangle**. A calculator is a useful tool for exploring some of the characteristics of a rectangle. Use the following steps to draw a rectangle.

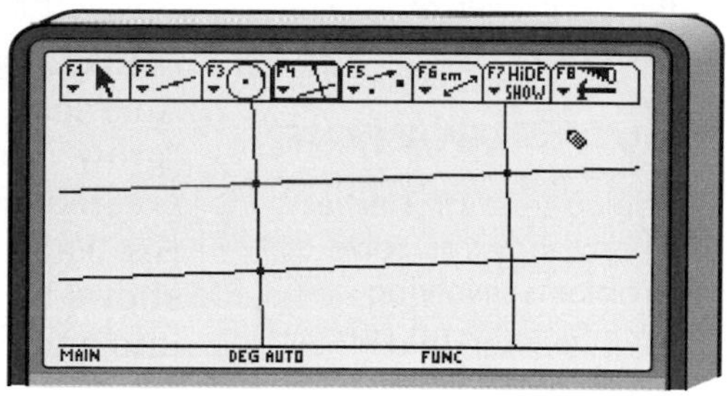

Step 1 Press F2 and choose Line from the menu to draw a line anywhere on the screen. Use the cursor pad to position the cursor anywhere on the screen and press ENTER. Move to another place on the screen and press ENTER to position the line.

Step 2 Next, draw a line perpendicular to the first line. Press F4 and choose Perpendicular Line from the menu. Position the cursor somewhere on the first line so that the prompt *perpendicular to this line* appears. Press ENTER. Now move the cursor to another point you would like the line to pass through and press ENTER.

Step 3 Repeat the procedure in step 2 to draw a line perpendicular to the second line. Then draw a line perpendicular to the third line.

A rectangle is formed by the four segments whose endpoints are the points of intersection of the lines.

EXERCISES

Use the measuring capabilities of the calculator to explore the characteristics of a rectangle.

1. What appears to be true about the opposite sides of the rectangle that you drew? Make a conjecture and then measure each side to check your conjecture.
2. Draw the diagonals of the rectangle by first pressing F2 and choosing Segment from the menu. Then position the cursor over one of the vertices of the rectangle. When the prompt *point at this intersection* appears, press ENTER. Move the cursor to the opposite vertex. Press ENTER when the prompt *point at this intersection* appears. Repeat to draw the other diagonal.
   a. Measure each diagonal. What do you observe?
   b. What is true about the triangles formed by the sides of the rectangle and a diagonal? Justify your conclusion.
3. Clear the screen and follow steps 1–3 to draw another rectangle. Do the relationships you found for the first rectangle you drew hold true for this rectangle also?

6-3

Rectangles

What YOU'LL LEARN

- To recognize and apply the properties of rectangles.

Why IT'S IMPORTANT

You can use properties of rectangles to solve problems involving architecture and sports.

Architecture

Frank Lloyd Wright is one of the most celebrated American architects. He believed that a building should "grow" from its site. The prairie style of homes Wright created emphasizes natural materials and horizontal lines. His "Falling Water" design at Bear Run in Pennsylvania, shows how a structure made up of rectangles can blend into and become a part of a natural setting.

Frank Lloyd Wright

A **rectangle** is a quadrilateral with four right angles. It follows that since both pairs of opposite angles are congruent, a rectangle is a special type of parallelogram. Thus, a rectangle has all the properties of a parallelogram.

However, the diagonals of a rectangle have an additional special relationship. You can use a spreadsheet to explore this relationship.

EXPLORATION — SPREADSHEETS

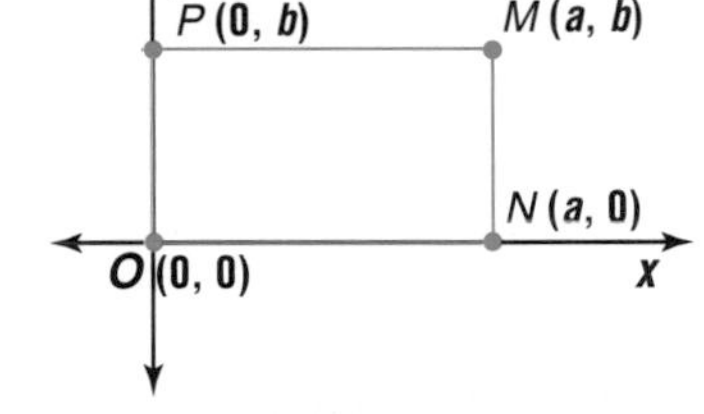

Form a rectangle *MNOP* on a coordinate grid by choosing any point $M(a, b)$ for the first vertex. The other three vertices are $N(a, 0)$, $O(0, 0)$, and $P(0, b)$.

- Using spreadsheet software, enter a in column A and b in column B from the ordered pair (a, b).
- Enter the formula SQR((A1 − 0)^2 + (B1 − 0)^2) in cell C1. This formula will find the distance between M and O. Copy the formula into the other cells in column C.
- Write a similar formula to find the distance between N and P. Enter the formula in the cells in column D.

Your Turn

a. Use the spreadsheet to compare the lengths of the diagonals for at least seven different rectangles.

b. What appears to be true about the diagonals of a rectangle?

This exploration leads us to the following theorem. *You will be asked to prove this Theorem in Exercise 14.*

Theorem 6–9	**If a parallelogram is a rectangle, then its diagonals are congruent.**

You can construct a rectangle using right angles and then verify that the diagonals are congruent.

CONSTRUCTION

Rectangle

Construct a rectangle with a length of 6 centimeters and a width of 3 centimeters.

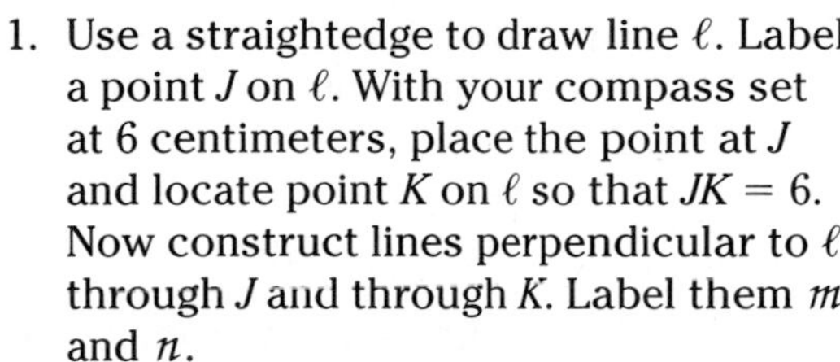

1. Use a straightedge to draw line ℓ. Label a point J on ℓ. With your compass set at 6 centimeters, place the point at J and locate point K on ℓ so that $JK = 6$. Now construct lines perpendicular to ℓ through J and through K. Label them m and n.

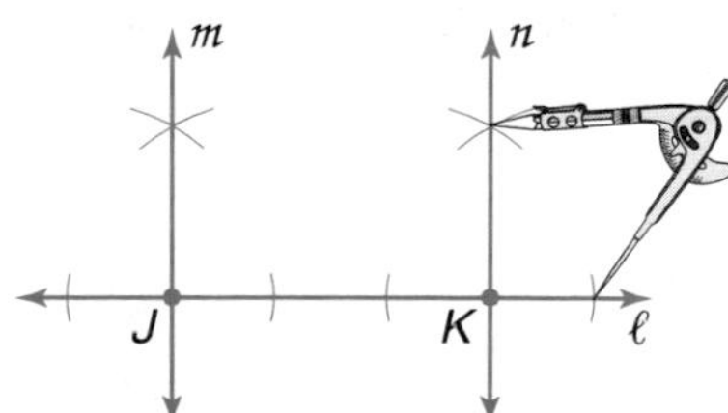

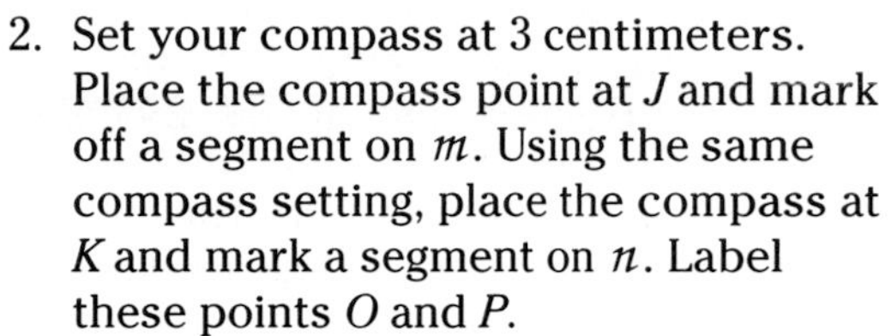

2. Set your compass at 3 centimeters. Place the compass point at J and mark off a segment on m. Using the same compass setting, place the compass at K and mark a segment on n. Label these points O and P.
3. Draw $\overline{OP}$.

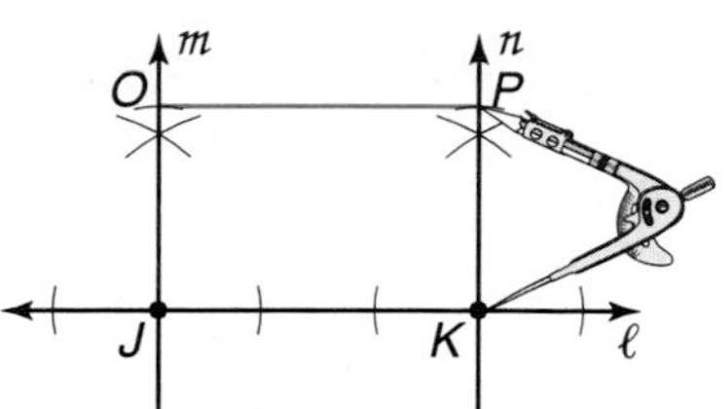

Quadrilateral $JKPO$ is a parallelogram, and all angles are right angles. Therefore, quadrilateral $JKPO$ is a rectangle with a length of 6 centimeters and a width of 3 centimeters.

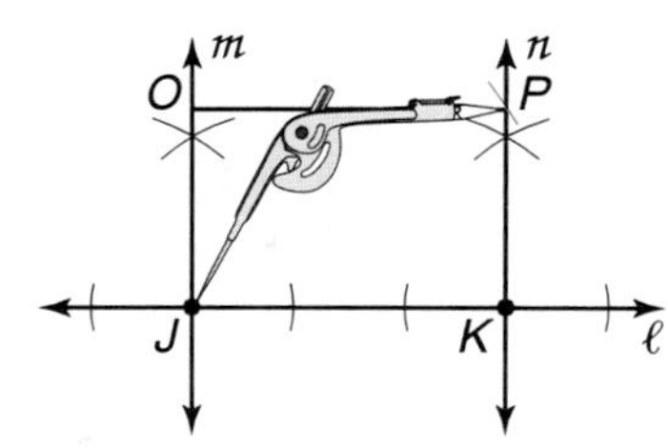

4. Locate the compass setting that represents JP and compare to the setting for OK. The measures should be the same.

Another method of constructing a rectangle often used in the building industry is to use the converse of Theorem 6–9. If the diagonals of a parallelogram are congruent, the parallelogram is a rectangle.

Example 1

Construction

The Mazdrons are building an addition to their house. Mr. Mazdron is cutting an opening for a new window. If he has measured to see that the opposite sides are congruent and that the diagonals are congruent, can Mr. Mazdron be sure that the window opening is a rectangle?

Explore Draw a diagram and label the vertices of the window opening A, B, C, and D. According to Mr. Mazdron's measurements, $\overline{AB} \cong \overline{CD}$, $\overline{BC} \cong \overline{AD}$, and $\overline{AC} \cong \overline{BD}$.

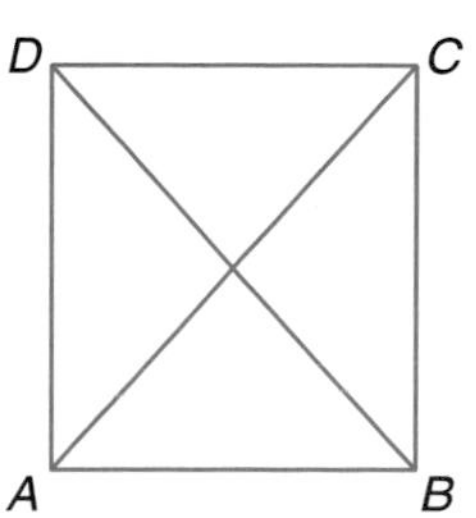

Plan We can show that $ABCD$ is a rectangle by first proving that it is a parallelogram and then proving that it has four right angles.

Solve **Given:** $\overline{AB} \cong \overline{CD}$

$\overline{BC} \cong \overline{AD}$

$\overline{AC} \cong \overline{BD}$

Prove: $ABCD$ is a rectangle.

(continued on the next page)

Paragraph proof:

Since congruence of segments is reflexive, $\overline{AB} \cong \overline{AB}$. So, $\triangle ADB \cong \triangle BCA$ by SSS. Thus, $\angle DAB \cong \angle CBA$ by CPCTC. $ABCD$ is a parallelogram since the opposite sides are congruent. Therefore, since consecutive angles of a parallelogram are supplementary, $\angle DAB$ and $\angle CBA$ are supplementary. So, $m\angle DAB + m\angle CBA = 180$ and $m\angle DAB = m\angle CBA$ by the definitions of supplementary and congruent angles.

By the Substitution Property of Equality, $2m\angle DAB = 180$, so $m\angle DAB = 90$. Therefore, $\angle DAB$ and $\angle CBA$ are right angles. Since the opposite angles of a parallelogram are congruent, $\angle DCB$ and $\angle CDA$ are right angles. By definition, $ABCD$ is a rectangle.

As a result, Mr. Mazdron can be sure that the window opening is a rectangle if the opposite sides are congruent and the diagonals are congruent.

Examine Try to draw a quadrilateral that is not a rectangle, but has opposite sides congruent and diagonals congruent.

Example 1 proves the converse of Theorem 6–9, which is Theorem 6–10.

Theorem 6–10	**If the diagonals of a parallelogram are congruent, then the parallelogram is a rectangle.**

Recall that perpendicular lines have slopes whose product is −1. You can use this fact to verify that a quadrilateral in the coordinate plane is a rectangle.

Example 2

Determine whether parallelogram $ABCD$ is a rectangle, given $A(-6, 9)$, $B(5, 10)$, $C(6, -1)$, and $D(-5, -2)$.

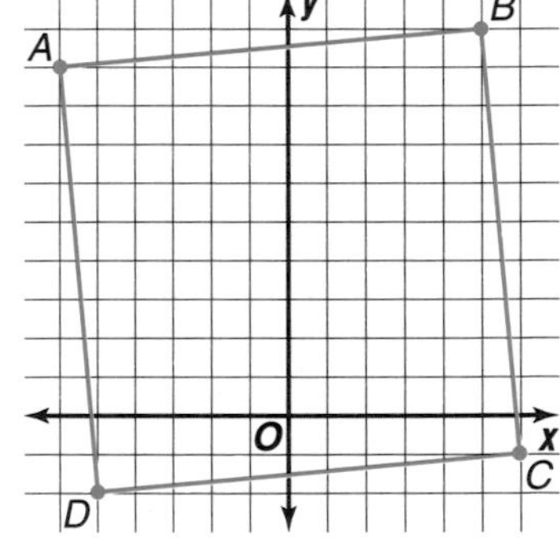

Method 1: Using slopes

slope of $\overline{AB} = \frac{10 - 9}{5 - (-6)}$ or $\frac{1}{11}$

slope of $\overline{CD} = \frac{-1 - (-2)}{6 - (-5)}$ or $\frac{1}{11}$

slope of $\overline{AD} = \frac{9 - (-2)}{-6 - (-5)}$ or -11

slope of $\overline{BC} = \frac{10 - (-1)}{5 - 6}$ or -11

Thus, $\overline{AB} \parallel \overline{CD}$ and $\overline{AD} \parallel \overline{BC}$. In addition, the product of the slopes of consecutive sides is −1. Thus, $\overline{AB} \perp \overline{BC}$, $\overline{BC} \perp \overline{CD}$, $\overline{CD} \perp \overline{AD}$, and $\overline{AD} \perp \overline{AB}$, creating four right angles. Therefore, quadrilateral $ABCD$ is a rectangle.

Method 2: Using diagonals

$AC = \sqrt{(-6 - 6)^2 + (9 - (-1))^2}$ $\quad$ $BD = \sqrt{(5 - (-5))^2 + (10 - (-2))^2}$

$= \sqrt{144 + 100}$ $\quad$ $= \sqrt{100 + 144}$

$= \sqrt{244}$ $\quad$ *$\sqrt{244} \approx 15.62$* $\quad$ $= \sqrt{244}$

Since $ABCD$ is a parallelogram and the diagonals are congruent, $ABCD$ is a rectangle.

Here is a summary of the properties of a rectangle.

If a quadrilateral is a rectangle, then the following properties hold true.
1. Opposite sides are congruent and parallel. **2.** Opposite angles are congruent. **3.** Consecutive angles are supplementary. **4.** Diagonals are congruent and bisect each other. **5.** All four angles are right angles.

CHECK FOR UNDERSTANDING

Communicating Mathematics

Study the lesson. Then complete the following.

1. **Explain** why a rectangle is a special type of parallelogram.
2. **Draw** an example of a quadrilateral with congruent diagonals that is *not* a rectangle.
3. **You Decide** Kalere claims that if the diagonals of a quadrilateral are congruent and bisect each other, then the quadrilateral is a rectangle. Amy says the quadrilateral could be a non-rectangular parallelogram. Who is correct and why?

4. **Assess Yourself** Look for examples of rectangles in the objects around you. Could the objects you see be another shape and still be effective? For example, would a circular window be as useful as a rectangular one? Explain your reasoning.

Guided Practice

Quadrilateral *MNOP* is a rectangle. Find the value of x.

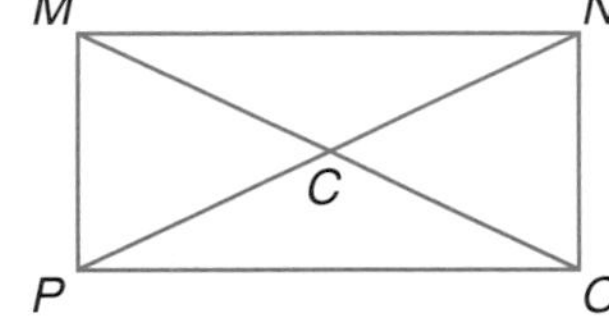

5. $MO = 2x - 8$; $NP = 23$
6. $CN = x^2 + 1$; $CO = 3x + 11$
7. $MO = 4x - 13$; $PC = x + 7$
8. *True* or *false:* If a quadrilateral has opposite sides that are congruent, then it is a rectangle. Justify your answer.
9. *ABCD* has vertices at $A(-3, 1)$, $B(4, 8)$, $C(7, 5)$ and $D(0, -2)$. Use slopes to determine if *ABCD* is a rectangle.

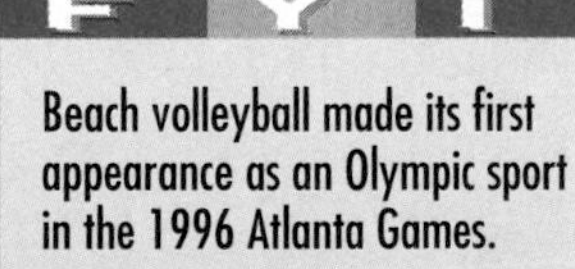

Beach volleyball made its first appearance as an Olympic sport in the 1996 Atlanta Games.

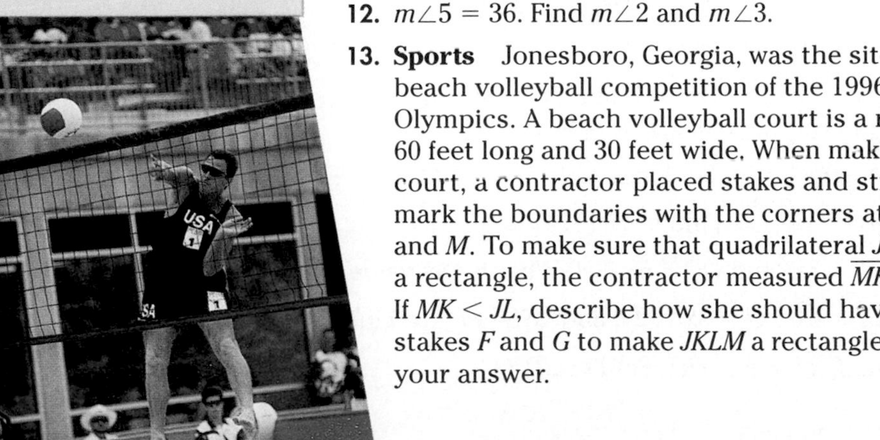

Use rectangle *KLMN* and the given information to solve each problem.

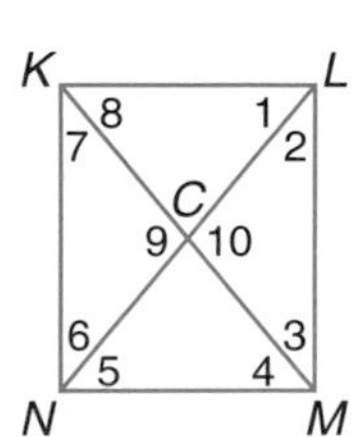

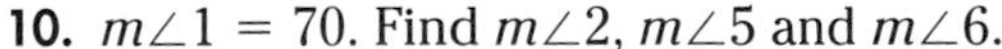

10. $m\angle 1 = 70$. Find $m\angle 2$, $m\angle 5$ and $m\angle 6$.
11. $m\angle 9 = 128$. Find $m\angle 6$, $m\angle 7$ and $m\angle 8$.
12. $m\angle 5 = 36$. Find $m\angle 2$ and $m\angle 3$.
13. **Sports** Jonesboro, Georgia, was the site for the beach volleyball competition of the 1996 Summer Olympics. A beach volleyball court is a rectangle 60 feet long and 30 feet wide. When making the court, a contractor placed stakes and strings to mark the boundaries with the corners at *J*, *K*, *L*, and *M*. To make sure that quadrilateral *JKLM* was a rectangle, the contractor measured $\overline{MK}$ and $\overline{JL}$. If $MK < JL$, describe how she should have moved stakes *F* and *G* to make *JKLM* a rectangle. Explain your answer.

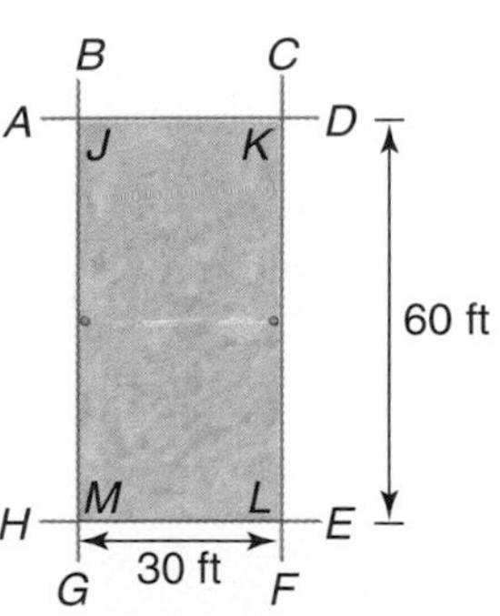

Proof

14. Complete the proof for Theorem 6–9.

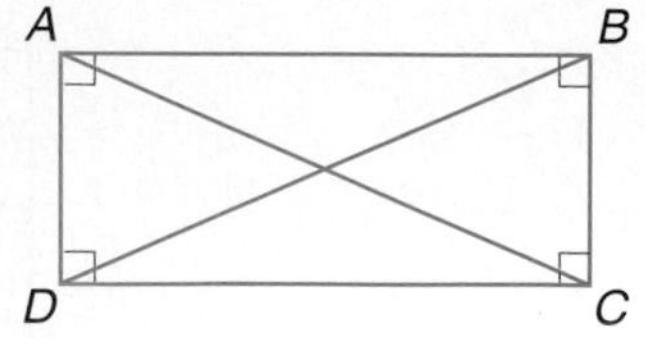

Given: $ABCD$ is a rectangle with diagonals $\overline{AC}$ and $\overline{BD}$.

Prove: $\overline{AC} \cong \overline{BD}$

Proof:

Statements	Reasons
1. $ABCD$ is a rectangle with diagonals $\overline{AC}$ and $\overline{BD}$.	1. Given
2. $\overline{DC} \cong \overline{DC}$	2. ?
3. ?	3. Opp. sides of a ▱ are ≅.
4. $\angle ADC$ and $\angle BCD$ are rt. ∠s.	4. Def. rectangle
5. $\angle ADC \cong \angle BCD$	5. ?
6. ?	6. SAS
7. $\overline{AC} \cong \overline{BD}$	7. ?

EXERCISES

Practice

Quadrilateral *QUAD* is a rectangle. Find the value of *x*.

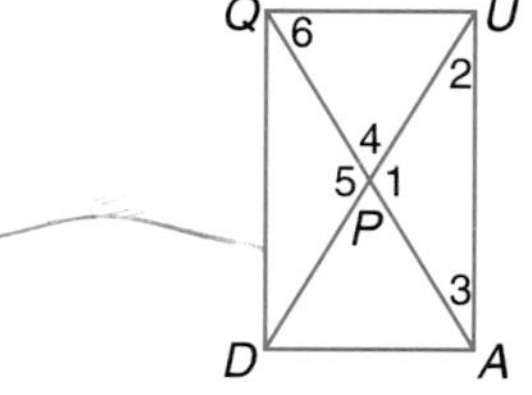

15. $DU = 26$, $QP = 2x + 7$

16. $m\angle 2 = 52$, $m\angle 3 = 16x - 12$

17. $m\angle 4 = 6x - 16$, $m\angle 2 = 2x + 4$

18. $DP = 4x + 1$, $QP = x + 13$

19. $m\angle 3 = 70 - 4x$, $m\angle 6 = 18x - 8$

Determine whether each statement is *true* or *false*. Explain.

20. If a quadrilateral is a rectangle, then it is a parallelogram.

21. If two sides of a quadrilateral are perpendicular, then it is a rectangle.

22. If a parallelogram has a right angle, then it is a rectangle.

23. If all four angles of a quadrilateral are congruent, then it is a rectangle.

Use rectangle *QRST*, parallelogram *QZRC*, and the given information to solve each problem.

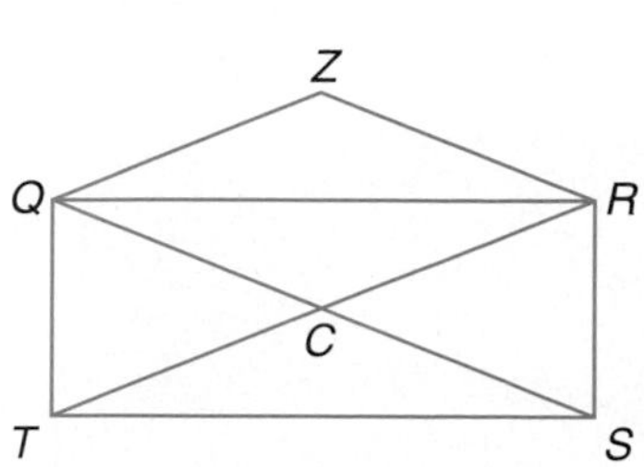

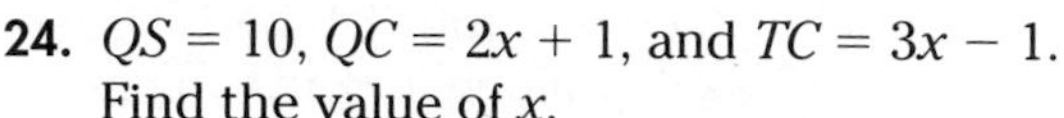

24. $QS = 10$, $QC = 2x + 1$, and $TC = 3x - 1$. Find the value of x.

25. $m\angle TQC = 70$. Find $m\angle QZR$.

26. $m\angle RCS = 35$. Find $m\angle RTS$.

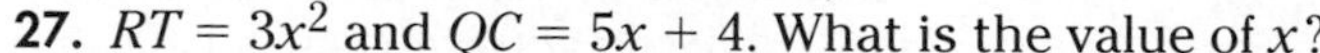

27. $RT = 3x^2$ and $QC = 5x + 4$. What is the value of x?

28. $RZ = 6x$, $ZQ = 3x + 2y$, and $CS = 14 - x$. Find the values of x and y.

29. $m\angle QRT = m\angle TRS$. Find $m\angle TCQ$.

Use rectangle *STUV* and the given information to find each measure.

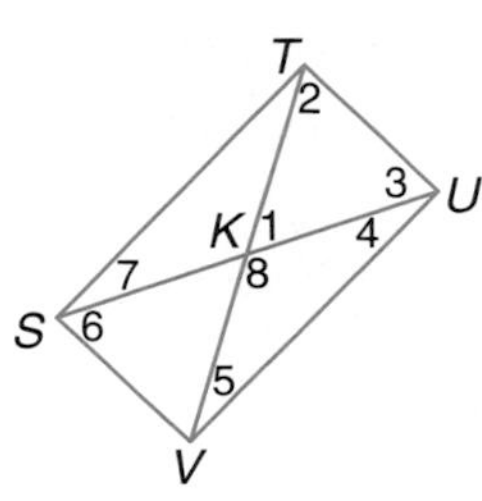

30. If $m\angle 1 = 30$, find $m\angle 2$.

31. If $m\angle 6 = 57$, what is $m\angle 4$?

32. If $m\angle 8 = 133$, find $m\angle 2$.

33. If $m\angle 5 = 16$, what is $m\angle 3$?

Determine if *PQRS* is a rectangle. Justify your answer.

34. $P(9, -1)$, $Q(9, 5)$, $R(-6, 5)$, $S(-6, 1)$

35. $P(6, 2)$, $Q(8, -1)$, $R(10, 6)$, $S(12, 3)$

36. $P(-4, -3)$, $Q(-5, 8)$, $R(6, 9)$, $S(7, -2)$

37. Graph *ABCD* for $A(2, 4)$, $B(-2, 0)$, $C(-1, -7)$, and $D(9, 3)$.

a. Use the distance formula to find *AC* and *BD*.

b. Use the midpoint formula to locate the midpoints of $\overline{AC}$ and $\overline{BD}$.

c. Explain why you cannot conclude that *ABCD* is a rectangle.

38. Graph *WXYZ* with vertices $W(-7, -3)$, $X(0, 4)$, $Y(3, 1)$, and $Z(-4, -7)$.

a. Describe two different methods for determining if *WXYZ* is a rectangle.

b. Is *WXYZ* a rectangle? Justify your reasoning.

Write a two-column proof.

39. Given: *ACDE* is a rectangle.
ABCE is a parallelogram.

Prove: $\triangle ABD$ is isosceles.

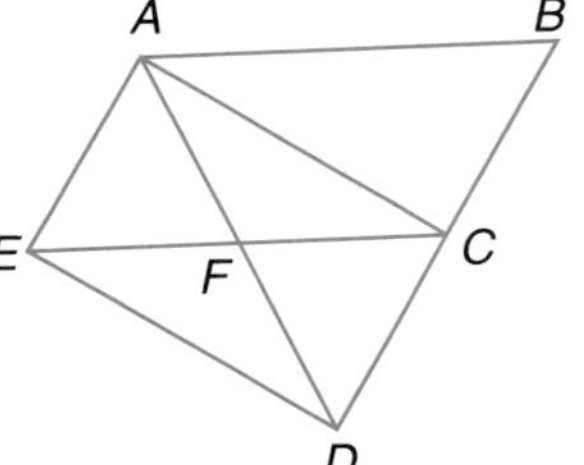

40. Given: *HJLM* is a rectangle.
$\overline{KJ} \cong \overline{NM}$

Prove: $\overline{HK} \cong \overline{LN}$

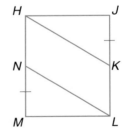

41. Given: *PQMO* and *RQMN* are rectangles.
$\angle SVT \cong \angle UTV$
$\overline{SU}$ and $\overline{TV}$ intersect at *W*.

Prove: *STUV* is a parallelogram.

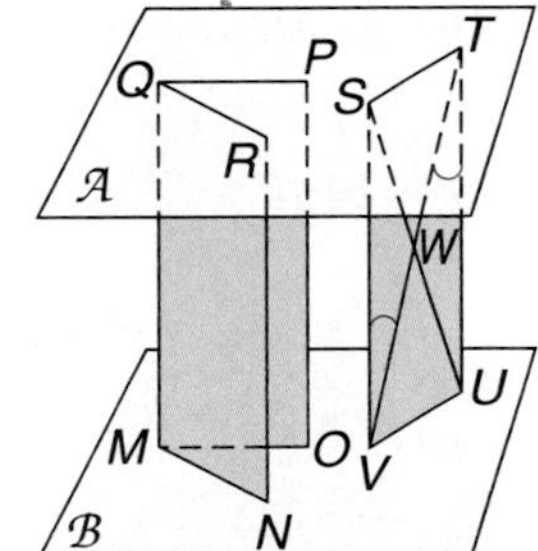

Critical Thinking

42. Spherical Geometry The figure shows a *Saccheri quadrilateral* on a sphere. Notice it has four sides with $\overline{CT} \perp \overline{TR}$ and $\overline{AR} \perp \overline{TR}$ and $\overline{CT} \cong \overline{AR}$. *Recall that all line segments are parts of the great circles.*

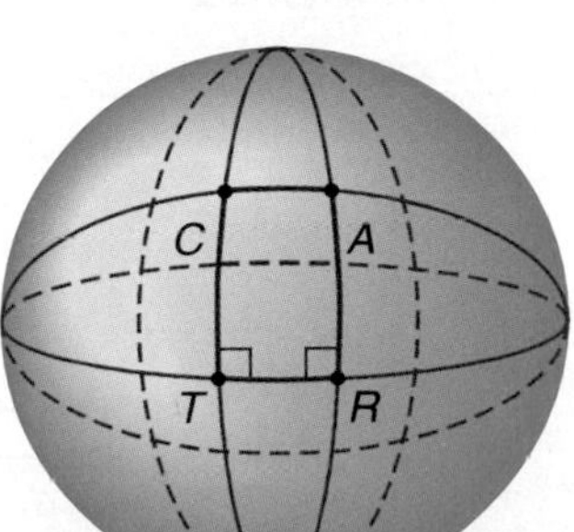

a. What type of angles do *C* and *A* seem to be?

b. Is $\overline{CT}$ parallel to $\overline{AR}$? Explain.

c. How does *AC* compare to *TR*?

d. Can a rectangle exist in spherical geometry? Explain.

LOOK BACK

You can review spherical geometry in Lesson 3-6.

Applications and Problem Solving

43. **Music** Compact discs (CDs) are circular in shape, but packaged in rectangular cases. Why do you think rectangular packaging is used?

44. **Art** *Golden rectangles* have been used by famous artists such as George Seurat, Leonardo DiVinci, and Salvador Dali. In a golden rectangle, the ratio of the length to the width is about 1.618, or the *golden ratio*.
    a. Use your library or search the Internet for information on two other examples of golden rectangles.
    b. Find the length of the diagonal of a golden rectangle whose sides measure 1 unit and 1.618 units.

Mixed Review

45. Quadrilateral $ABCD$ has vertices $A(0, -2)$, $B(-4, 6)$, $C(5, 6)$, and $D(9, 4)$. Is $ABCD$ a parallelogram? (Lesson 6–2)

46. In parallelogram $ABCD$, $AB = 4x + 9$, $m\angle BAC = 5y + 1$, $m\angle D = 75$, $m\angle ACD = 56$, and $CD = 45$. Find the values of x and y. (Lesson 6–1)

47. Name the postulates and theorems that can be used to prove two right triangles congruent. (Lesson 5–2)

48. **Algebra** A median of $\triangle XYZ$ separates side $\overline{XZ}$ into $\overline{XQ}$ and $\overline{QZ}$. If $XQ = r - 3$ and $XZ = 7r - 15$, what is the value of r? (Lesson 5–1)

49. In $\triangle STV$ and $\triangle WXY$, $\angle S \cong \angle W$ and $\overline{ST} \cong \overline{WX}$. What additional information is needed to prove $\triangle STV \cong \triangle WXY$ by ASA? (Lesson 4–5)

50. **SAT Practice** **Grid-in** Five cards are numbered 0 through 4. Two are selected without replacement. What is the probability that the sum of the cards is 3?

51. What algebraic property allows us to say that if $m\angle 1 = m\angle 2$ and $m\angle 2 = m\angle 3$, then $m\angle 1 = m\angle 3$? (Lesson 2–4)

52. Find the measure of the supplement of $\angle Q$ if $m\angle Q = 167$. (Lesson 1–7)

For **Extra Practice**, see page 775.

Algebra

53. Solve $-6 = 5u + 9$.

54. Solve $5 - 9x = 23$.

SELF TEST

Complete each statement about parallelogram *LMNP*. Then name the theorem or definition that justifies your answer. (Lesson 6–1)

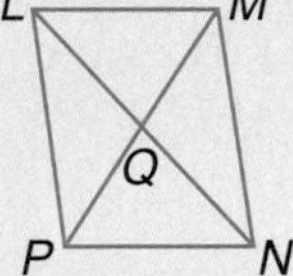

1. $\overline{LM} \parallel$ _?_
2. $\angle LMN \cong$ _?_
3. $\overline{MN} \cong$ _?_
4. $\triangle LMQ \cong$ _?_

Determine if the quadrilateral with the given vertices is a parallelogram. (Lesson 6–2)

5. $A(9, 0)$, $B(2, -5)$, $C(6, -5)$, $D(13, 0)$
6. $E(-1, -1)$, $F(1, 1)$, $G(6, -6)$, $H(-6, 0)$
7. $H(5, 0)$, $I(0, -5)$, $J(-5, 0)$, $K(0, 5)$
8. $L(-2, -1)$, $M(2, 5)$, $N(-10, 13)$, $P(-14, 7)$

Write a proof. (Lesson 6–3)

9. **Given:** ▱$WXZY$
   $\angle 1$ and $\angle 2$ are complementary.
   Prove: $WXZY$ is a rectangle.

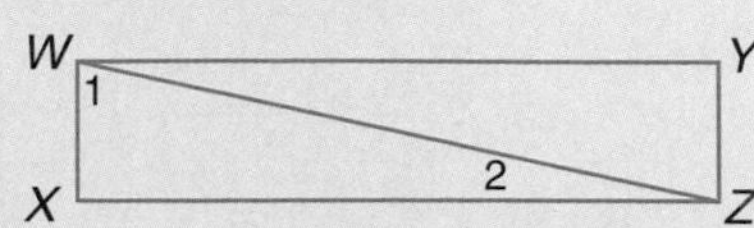

10. **Given:** ▱$KLMN$
    Prove: $PQRS$ is a rectangle.

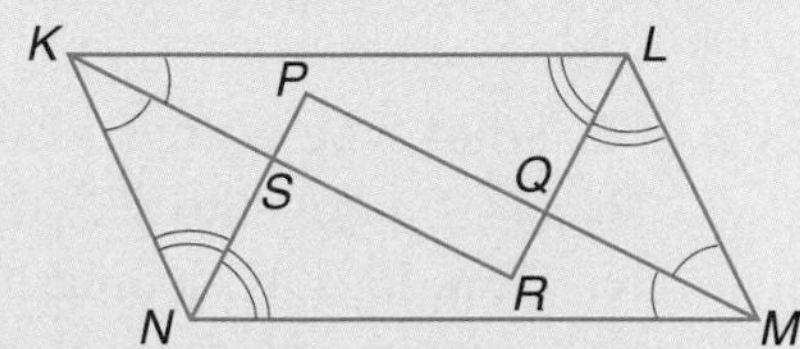

6–4 Squares and Rhombi

What YOU'LL LEARN

- To recognize and apply the properties of squares and rhombi.

Why IT'S IMPORTANT

You can use properties of squares and rhombi to solve problems involving art and construction.

Real World APPLICATION

Optical Illusions

Have you ever looked down a long road and noticed that it seems to grow narrower in the distance? Or have you painted a room a lighter color and found that the room seems bigger than it did before? Then you have experienced an optical illusion. Artists, interior decorators, and clothing designers take advantage of optical illusions to make things appear different than they are.

Study the figure at the right. You may be able to see six cubes or seven cubes depending on which quadrilaterals are considered to be the tops of the cubes. The optical illusion is created using special quadrilaterals called **rhombi** (pronounced rom-bye). A **rhombus** is a quadrilateral with four congruent sides. Since opposite sides of a rhombus are congruent, a rhombus is a parallelogram.

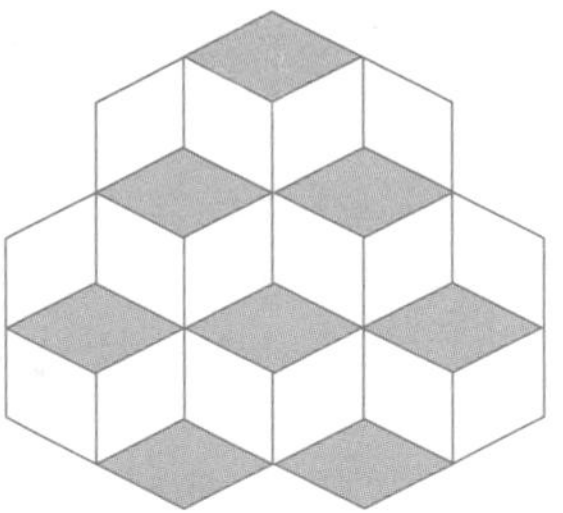

You will be asked to prove Theorems 6–11 and 6–12 in Exercises 19 and 51, respectively.

In addition to all of the properties of a parallelogram, a rhombus has two special relationships that are described in the following theorems. The proof of Theorem 6–13 is given.

Theorem 6–11	**The diagonals of a rhombus are perpendicular.**
Theorem 6–12	**If the diagonals of a parallelogram are perpendicular, then the parallelogram is a rhombus.**
Theorem 6–13	**Each diagonal of a rhombus bisects a pair of opposite angles.**

Proof of Theorem 6–13

Given: *ABCD* is a rhombus.

Prove: Each diagonal bisects a pair of opposite angles.

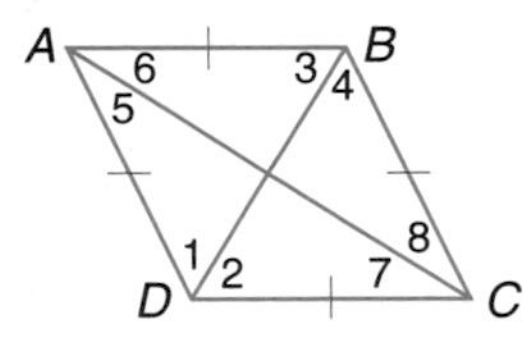

Proof:

Statements	**Reasons**
1. *ABCD* is a rhombus.	**1.** Given
2. *ABCD* is a parallelogram.	**2.** Def. rhombus
3. $\angle ABC \cong \angle ADC$, $\angle BAD \cong \angle BCD$	**3.** Opp. ∠s of a ▱ are ≅.
4. $\overline{AB} \cong \overline{BC} \cong \overline{CD} \cong \overline{DA}$	**4.** Def. rhombus
5. $\triangle ABC \cong \triangle ADC$	**5.** SAS
6. $\angle 5 \cong \angle 6$, $\angle 7 \cong \angle 8$	**6.** CPCTC
7. $\triangle BAD \cong \triangle BCD$	**7.** SAS
8. $\angle 1 \cong \angle 2$, $\angle 3 \cong \angle 4$	**8.** CPCTC
9. Each diagonal bisects a pair of opposite angles.	**9.** Def. ∠ bisector

You can use the properties of a rhombus to solve problems.

Example  **Use rhombus *BCDE* and the given information to find each missing value.**

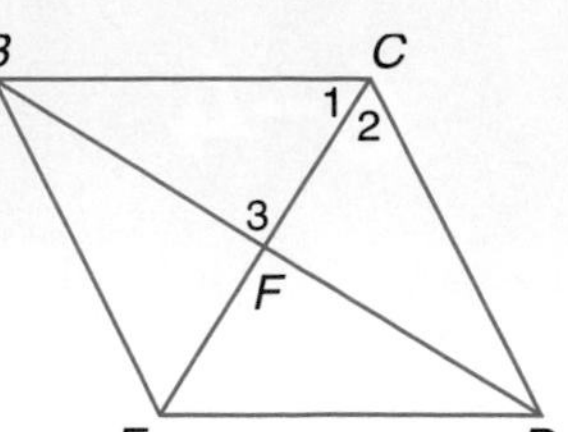

a. If $m\angle 1 = 2x + 20$ and $m\angle 2 = 5x - 4$, find the value of x.

The diagonals of a rhombus bisect a pair of opposite angles. So, $\overline{CE}$ bisects $\angle BCD$ and $m\angle 1 = m\angle 2$.

$$m\angle 1 = m\angle 2$$
$$2x + 20 = 5x - 4$$
$$24 = 3x$$
$$8 = x$$

b. If $BD = 15$, find BF.

Since a rhombus is a parallelogram, its diagonals bisect each other. Thus, if $BD = 15$, $BF = \frac{1}{2}(15)$ or 7.5.

c. If $m\angle 3 = y^2 + 26$, find y.

The diagonals of a rhombus are perpendicular, so $\angle 3$ is a right angle.

$$m\angle 3 = 90$$
$$y^2 + 26 = 90$$
$$y^2 = 64$$
$$y = 8 \text{ or } -8$$

It is possible to construct a rhombus using a compass and a straightedge.

CONSTRUCTION

Rhombus

Construct a rhombus.

1. Draw a segment AD. Set the compass to match the length of $\overline{AD}$. You will use this compass setting for all arcs drawn.

2. Place the compass at point A and draw an arc above $\overline{AD}$. Choose any point on that arc and label it B. Place the compass at point B and draw an arc to the right of B.

3. Then place the compass at point D and draw an arc to intersect the arc drawn from point B. Label the point of intersection C.

4. Use a straightedge to draw $\overline{AB}$, $\overline{BC}$, and $\overline{CD}$.

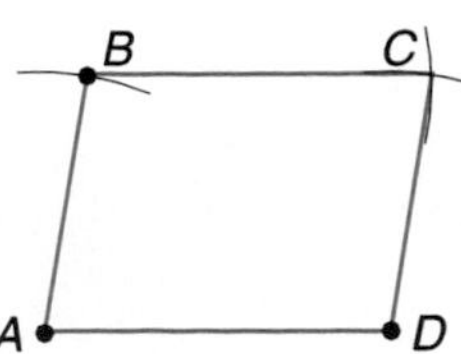

Conclusion: Since all of the sides are congruent, quadrilateral $ABCD$ is a rhombus.

If a quadrilateral is both a rhombus and a rectangle, it is a **square**. A square is a quadrilateral with four right angles and four congruent sides. You can construct a square in the same way that you constructed a rhombus, but a square must have a right angle. *You will be asked to construct a square in Exercise 48.*

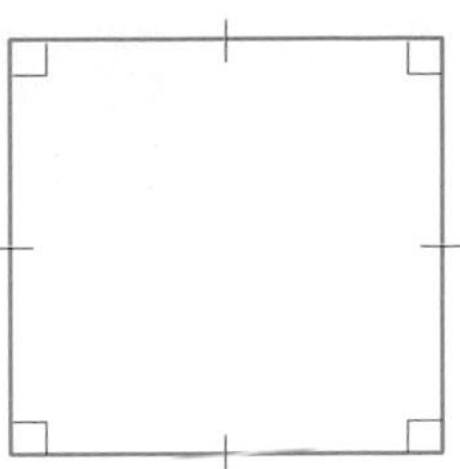

Example 2

Algebra

Determine whether parallelogram *WXYZ* is a rhombus, a rectangle, or a square for *W*(1, 10), *X*(−4, 0), *Y*(7, 2), and *Z*(12, 12).

We can determine if *WXYZ* is a rhombus, a rectangle, or a square by examining its diagonals. The diagonals of a rhombus are perpendicular. So if $\overline{WY} \perp \overline{XZ}$, then *WXYZ* is a rhombus. If the diagonals are congruent, then *WXYZ* is a rectangle. If it is both a rhombus and a rectangle, *WXYZ* is a square.

slope of $\overline{WY} = \frac{10-2}{1-7} = -\frac{8}{6}$ or $-\frac{4}{3}$ slope of $\overline{XZ} = \frac{0-12}{-4-12} = \frac{12}{16}$ or $\frac{3}{4}$

The product of the slopes of the diagonals is $\left(-\frac{4}{3}\right)\left(\frac{3}{4}\right)$ or -1. Therefore, the diagonals are perpendicular, and *WXYZ* is a rhombus.

$$WY = \sqrt{(1-7)^2 + (10-2)^2} = \sqrt{100} \text{ or } 10$$

$$XZ = \sqrt{(-4-12)^2 + (0-12)^2} = \sqrt{400} \text{ or } 20$$

Since the diagonals of *WXYZ* are not congruent, it is not a rectangle. Since a square is a special rectangle, *WXYZ* is not a square.

WXYZ is a rhombus.

Geometric art is one of the ways that geometry is part of everyday life.

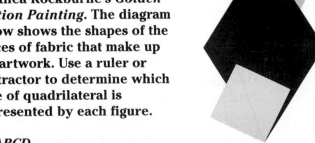

Example 3

Art

The artwork at the right is Dorthea Rockburne's *Golden Section Painting*. The diagram below shows the shapes of the pieces of fabric that make up the artwork. Use a ruler or protractor to determine which type of quadrilateral is represented by each figure.

a. *ABCD*

ABCD is a square because the diagonals are both congruent and perpendicular.

b. *AEFJ*

All four sides of *AEFJ* are congruent. But the consecutive sides are not perpendicular, so *AEFJ* is a rhombus.

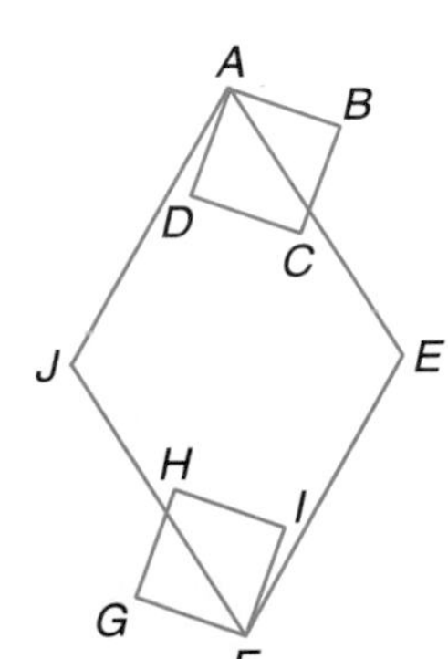

CHECK FOR UNDERSTANDING

Communicating Mathematics

Study the lesson. Then complete the following.

1. **Compare and contrast** the properties of a rhombus and a square. How are they different? How are they the same?
2. One geometry book defines a square as "*an equiangular rhombus.*" Another book defines a square as "*an equilateral rectangle.*" Which definition is correct and why?

3. **Construct** a rhombus with sides that are 6 centimeters long using a compass and straightedge.
4. Copy and complete the summary of the diagonal properties that exist for each of the four types of quadrilaterals in the table below. Write *yes* or *no* in each cell of the table.

Property	Parallelogram	Rectangle	Rhombus	Square
The diagonals bisect each other.				
The diagonals are congruent.				
Each diagonal bisects a pair of opposite angles.				
The diagonals are perpendicular.				

5. Locate two examples of a rhombus or a square in newspaper illustrations. Explain how you know that the shape is a rhombus or a square.

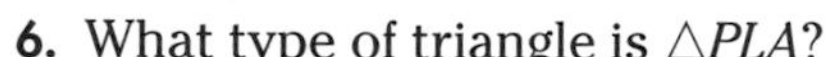

Guided Practice

Use rhombus *PLAN* for Exercises 6–9. Justify your answers.

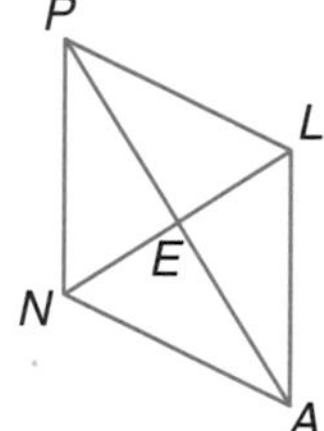

6. What type of triangle is $\triangle PLA$?
7. What type of triangle is $\triangle PEN$?
8. Is $\triangle PEN \cong \triangle AEL$?
9. Is it true that $\overline{PA} \cong \overline{NL}$? Explain.

Use rhombus *RSTV* and the given information to find each value.

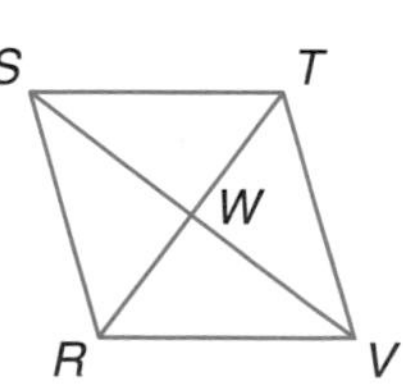

10. If $m\angle RST = 67$, find $m\angle RSW$.
11. Find $m\angle SVT$ if $m\angle STV = 135$.
12. If $m\angle SWT = 2x + 8$, find the value of x.
13. What is the value of x if $m\angle WRV = 5x + 5$ and $m\angle WRS = 7x - 19$?

Determine whether quadrilateral *PARK* is a *parallelogram*, a *rectangle*, a *rhombus*, or a *square* for each set of vertices. List all that apply.

14. $P(-1, 0), A(1, -1), R(2, 1), K(0, 2)$
15. $P(-1, 4), A(-1, 10), R(14, 10), K(14, 4)$
16. $P(2, 11), A(-3, 1), R(8, 3), K(13, 3)$

Name all the quadrilaterals—*parallelogram, rectangle, rhombus,* or *square*—that have each property.

17. All angles are congruent.

18. All sides are congruent.

19. Write a paragraph proof of Theorem 6–11.

Given: $ABCD$ is a rhombus.

Prove: $\overline{AC} \perp \overline{BD}$

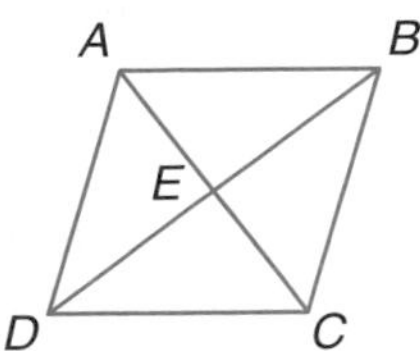

Evidence of Nine men's morris has been found in Ceylon, now Sri Lanka. It was also known to inhabitants of Troy and to the Vikings.

20. **Games** *Nine men's morris* is a game for two players that dates from about 1400 B.C. The game board was one of seven found cut into the roof tiles of the temple in Karnak, Egypt. The game board is made from three squares that share the same center as shown at the right. Use a compass and a straightedge to construct a Nine men's morris board.

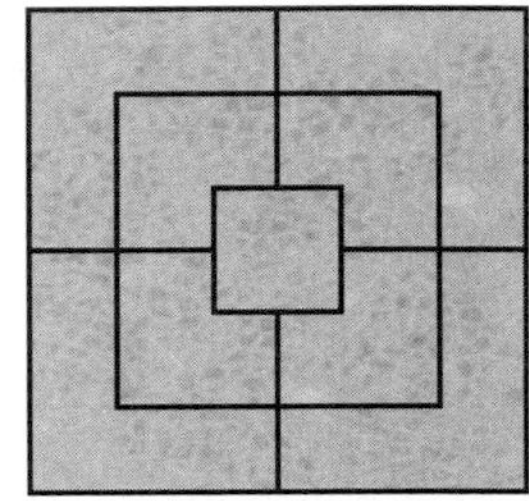

EXERCISES

Practice

Use parallelogram *MNOP* for Exercises 21–27. Justify your answers.

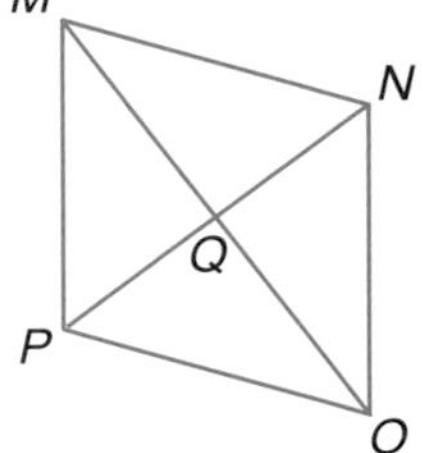

21. If $MNOP$ is not a rhombus, what type of triangle is $\triangle PMN$?
22. If $MNOP$ is a rhombus, what type of triangle is $\triangle PQM$?
23. Is $\triangle PQM \cong \triangle NQM$ if $MNOP$ is a square?
24. Is it true that $\overline{PQ} \cong \overline{NQ}$ if $MNOP$ is a square?
25. If $MNOP$ is a rhombus, must it also be a square?
26. If $\angle NQO$ is right, is $MNOP$ a rhombus?
27. If $\triangle PON$ is isosceles, must $MNOP$ be a square?

Use rhombus *ABCD* and the given information to find each value.

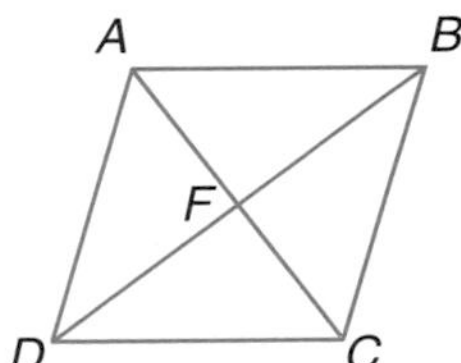

28. If $m\angle BAF = 28$, find $m\angle ACD$.
29. Find the value of x if $m\angle AFB = 16x + 6$.
30. If $m\angle ACD = 34$, find $m\angle ABC$.
31. Find the value of x if $m\angle BFC = 120 - 4x$.
32. What is the value of x if $m\angle BAC = 4x + 6$ and $m\angle ACD = 12x - 18$?
33. If $m\angle DCB = x^2 - 6$ and $m\angle DAC = 5x + 9$, find the value of x.

Determine whether quadrilateral *WXYZ* is a *parallelogram,* a *rectangle,* a *rhombus,* or a *square* for each set of vertices. List all that apply.

34. $W(5, 6), X(7, 5), Y(9, 9), Z(7, 10)$
35. $W(-3, -3), X(1, -6), Y(5, -3), Z(1, 0)$
36. $W(-6, 11), X(-11, -7), Y(-7, -2), Z(-2, -6)$
37. $W(10, 6), X(6, 10), Y(10, 14), Z(14, 10)$

Name all the quadrilaterals—*parallelogram, rectangle, rhombus,* or *square*—that have each property.

38. Diagonals are congruent.
39. One pair of opposite sides are congruent and parallel.
40. All sides congruent *and* all angles are congruent.
41. The diagonals are perpendicular.

Use the Venn diagram at the right to determine whether each statement is *true* or *false*. Explain your reasoning.

42. Every square is a rhombus.
43. Every rhombus is a square.
44. Every rectangle is a square.
45. Every square is a rectangle.
46. All rhombi are parallelograms.
47. Every parallelogram is a rectangle.

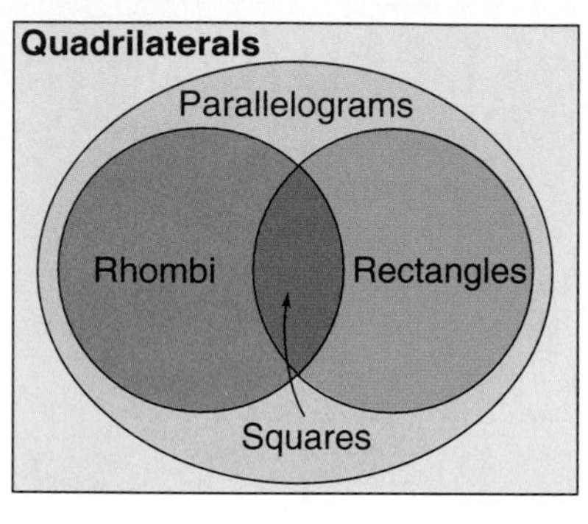

48. Construct square *DEFG* with diagonal congruent to $\overline{AB}$.

Write a two-column proof.

49. **Given:** *JKLM* is a square.
Prove: $\overline{JL} \cong \overline{KM}$

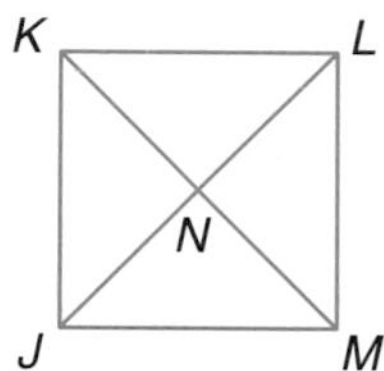

50. **Given:** Rhombus *MTRN*
Prove: $\angle 1$ and $\angle 2$ are complementary.

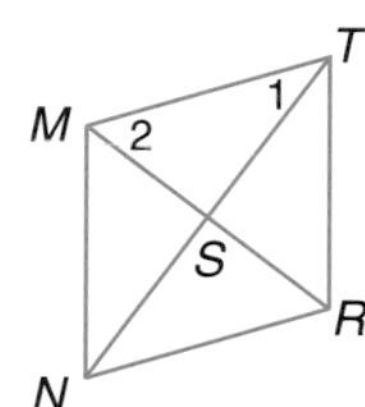

51. Write a paragraph proof of Theorem 6–12.

Critical Thinking

52. Use a ruler to draw a quadrilateral with perpendicular diagonals that is not a rhombus.

Applications and Problem Solving

53. **Flags** Study the flags shown below. Use a ruler and protractor to determine if any of the flags contain parallelograms, rectangles, rhombi, or squares.

Denmark

St. Vincent and The Grenadines

Trinidad and Tobago

54. **Construction** The opening for the reinforced doors of a storage shed is shaped like a square.
  a. Identify the quadrilaterals that make up the doors.
  b. Explain why you think the doors are shaped as they are.

55. **Sports** In the game of racquetball, the server stands between the service line and the short line as he or she delivers the serve. The receiver stands behind the short line. Study the diagram of a racquetball court at the right. If the court is a rectangle, identify each quadrilateral and explain how you determine the shape.

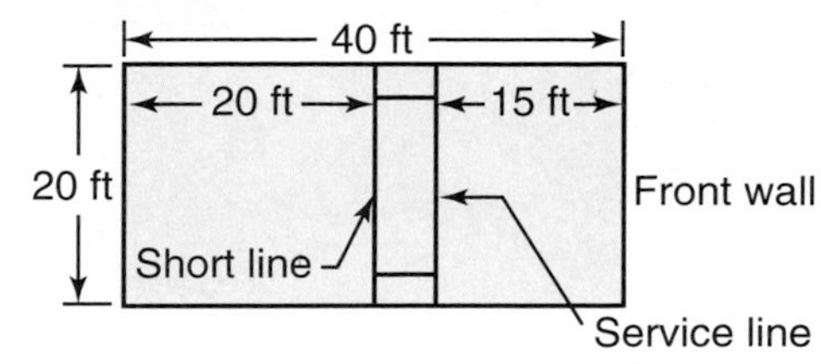

Mixed Review

56. Write a two-column proof. (Lesson 6–3)

Given: $\square$ *WXYZ*

$\angle 1$ and $\angle 2$ are complementary.

Prove: $\square$ *WXYZ* is a rectangle.

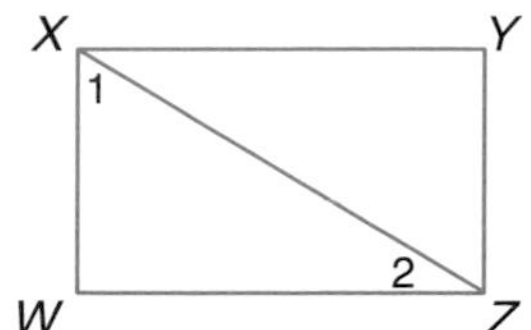

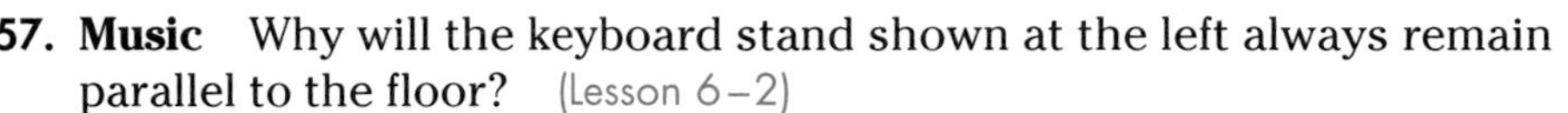

57. **Music** Why will the keyboard stand shown at the left always remain parallel to the floor? (Lesson 6–2)

58. *True* or *false:* The altitude of a triangle can sometimes be located outside the triangle. (Lesson 5–1)

59. $\triangle TUV$ is an isosceles triangle with two vertices $T(7, 10)$ and $U(2, 3)$. If $\angle T$ is the vertex angle, could V be at $(12, 3)$? Explain. (Lesson 4–6)

60. If $\triangle HJK \cong \triangle MNO$, name a segment congruent to $\overline{KH}$. (Lesson 4–3)

61. *True* or *false*: An obtuse triangle can be isosceles. (Lesson 4–1)

62. **ACT Practice** In the figure, *ABCD* is a parallelogram. What must be the coordinates of Point *D*?

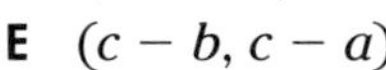

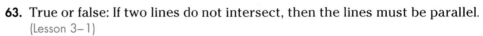

A $(a, c + b)$ **B** $(c + b, a)$
C $(b - c, a)$ **D** $(c - b, a)$
E $(c - b, c - a)$

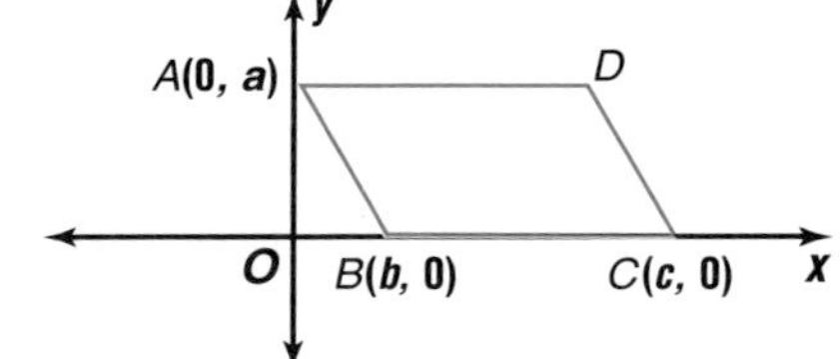

63. True or false: If two lines do not intersect, then the lines must be parallel. (Lesson 3–1)

64. **Civics** Identify the hypothesis and the conclusion of the conditional *If a U.S. citizen is over 18 years old, then he or she may vote.* (Lesson 2–2)

65. Suppose B is between A and C, $AB = 9$, $BC = 4x - 7$, and $AC = 18$. Find the value of x and the measure of $\overline{BC}$. (Lesson 1–4)

Algebra

66. State the domain and range of the relation $\{(8, -2), (7, 6), (-9, -1), (5, 6)\}$.

67. **Money** The graph shows the amount of time in an eight-hour workday that it takes to earn enough money to pay a day's worth of taxes. Find the percent of increase from 1929 to 1999. Round to the nearest whole percent.

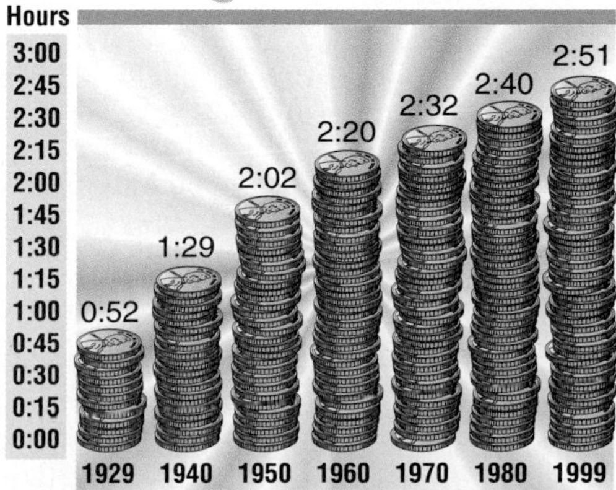

Source: *The Universal Almanac*

Data Update For more information on tax freedom day, visit: **www.geometry.glencoe.com**

For **Extra Practice**, see page 775.

MODELING MATHEMATICS

6–4B Kites

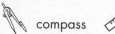

An Extension of Lesson 6–4

Materials:  compass  ruler  protractor

A **kite** is a quadrilateral with exactly two distinct pairs of adjacent congruent sides. You can draw a kite by drawing two noncongruent, nonoverlapping isosceles triangles that share the same base.

Activity Draw a kite *ABCD*.

Step 1 Draw a segment *BD* on a piece of paper.

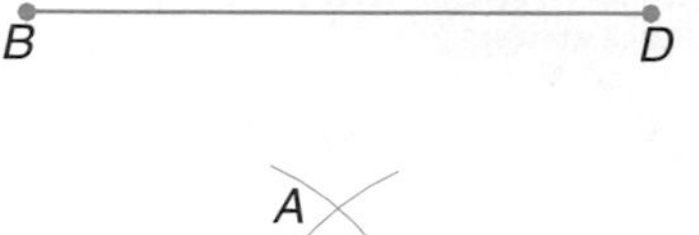

Step 2 Place the tip of your compass at point *B* and draw an arc above $\overline{BD}$. Then without changing the compass setting, move the compass to point *D* and draw an arc to intersect the first one. Label the point of intersection *A*.

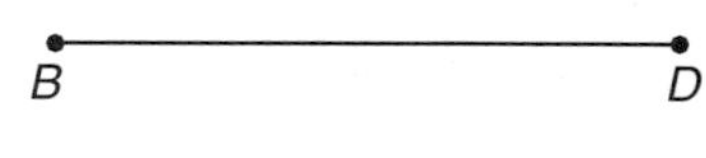

Step 3 Change the compass setting. Place the tip of the compass at *B* and draw an arc below $\overline{BD}$. Then, using the same compass setting, draw an arc from point *D* to intersect the first one. Label the point of intersection *C*.

Step 4 Draw *ABCD*.

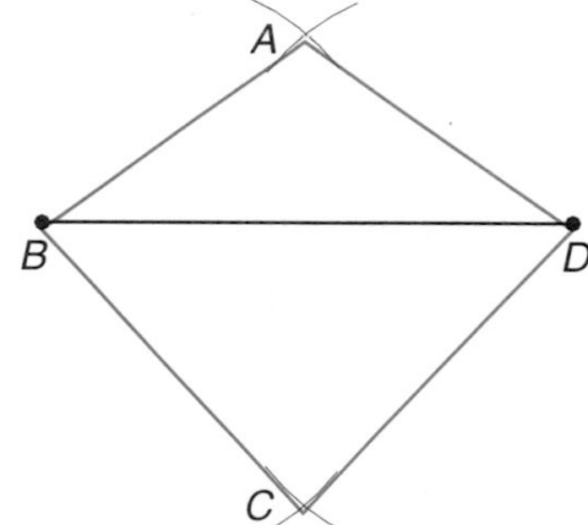

Model

1. Draw $\overline{AC}$ in kite *ABCD*. Use a protractor to measure the angles formed by the intersection of $\overline{AC}$ and $\overline{BD}$.
2. Measure the interior angles of kite *ABCD*. Are any congruent?
3. Label the intersection of $\overline{AC}$ and $\overline{BD}$ as point *E*. Find the lengths of $\overline{AE}$, $\overline{BE}$, $\overline{CE}$, and $\overline{DE}$. How are they related?

Draw

4. Draw another kite *JKLM*. Then repeat Exercises 1–3 using kite *JKLM*.

Write

5. Write as many conjectures about kites as you can. Justify your conjectures.

6-5 Trapezoids

What YOU'LL LEARN

- To recognize and apply the properties of trapezoids.

Why IT'S IMPORTANT

You can use properties of trapezoids to solve problems involving sailing and engineering.

Real World APPLICATION

Civil Engineering

The greatest accomplishment in the field of civil engineering stretches 2150 miles across China. Millions of soldiers and workers spent more than 1800 years creating the Great Wall of China. All of the materials for the wall were carried by human power. The wall is the only human-made object that could be seen by the astronauts walking on the moon.

Terra Cotta Army in tomb of Qin Shi Huangdi

Because it was built over a long period of time, different sections of the Great Wall are made of different materials. Most of the wall that remains is masonry built during the Ming Dynasty. The diagram below shows a cross section of a masonry portion of the wall.

The Great Wall of China
About 25,000 forty-foot watch towers are placed along the Great Wall.
The brick walkways on top of the wall allowed for the movement of troops.
Rubble and dirt provide strength for the wall.
The stone foundation is 22 to 26 feet across and 5 feet deep.

The cross section of the wall is shaped like a **trapezoid**. A trapezoid is a quadrilateral with exactly one pair of parallel sides. The parallel sides are called **bases**. The nonparallel sides are called **legs**. In trapezoid *TRAP* at the right, $\angle T$ and $\angle R$ are one pair of **base angles**. $\angle P$ and $\angle A$ are the other pair.

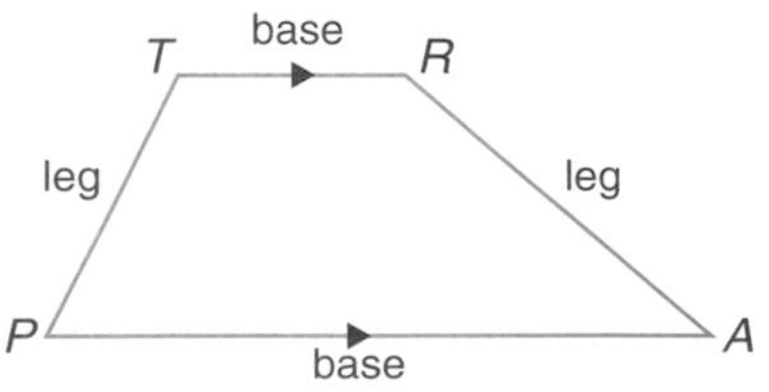

If the legs of a trapezoid are congruent, then the trapezoid is an **isosceles trapezoid**. Isosceles trapezoids have two special relationships that hold true. The proof of Theorem 6–14 is given. *You will be asked to prove Theorem 6–15 in Exercise 38.*

Theorem 6–14	**Both pairs of base angles of an isosceles trapezoid are congruent.**
Theorem 6–15	**The diagonals of an isosceles trapezoid are congruent.**

Proof of Theorem 6–14

Given: *ABCD* is an isosceles trapezoid.
$\overline{BC} \parallel \overline{AD}$
$\overline{AB} \cong \overline{CD}$

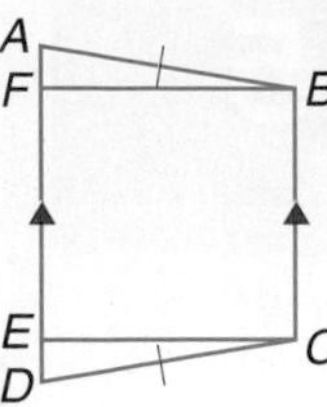

Prove: $\angle A \cong \angle D$
$\angle ABC \cong \angle DCB$

Paragraph Proof:

Draw auxiliary segments so that $\overline{BF} \perp \overline{AD}$ and $\overline{CE} \perp \overline{AD}$. Since $\overline{BC} \parallel \overline{AD}$ and parallel lines are everywhere equidistant, $\overline{BF} \cong \overline{CE}$. Perpendicular lines form right angles, so $\angle BFA$ and $\angle CED$ are right angles. $\triangle BFA$ and $\triangle CED$ are right triangles by definition. Therefore, $\triangle BFA \cong \triangle CED$ by HL. $\angle A \cong \angle D$ by CPCTC.

Since $\angle CBF$ and $\angle BCE$ are right angles and all right angles are congruent, $\angle CBF \cong \angle BCE$. $\angle ABF \cong \angle DCE$ by CPCTC. So, $\angle ABC \cong \angle DCB$ by the Angle Addition Postulate.

You can use algebra to investigate trapezoids on the coordinate plane.

Example 1

Algebra

Verify that isosceles trapezoid *TRAP* with vertices $T(-1, 1)$, $R(-5, -3)$, $A(-4, -10)$, and $P(6, 0)$ has congruent diagonals and congruent base angles.

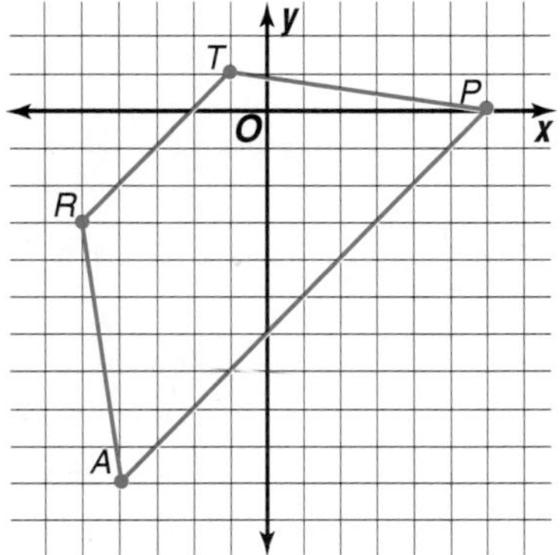

First, show that the diagonals are congruent by using the distance formula to find the lengths of $\overline{RP}$ and $\overline{AT}$.

$$RP = \sqrt{(-5 - 6)^2 + (-3 - 0)^2} = \sqrt{(-11)^2 + (-3)^2} \text{ or } \sqrt{130}$$

$$AT = \sqrt{(-4 - (-1))^2 + (-10 - 1)^2} = \sqrt{(-3)^2 + (-11)^2} \text{ or } \sqrt{130}$$

Since $RP = AT$, the diagonals are congruent.

Next, use triangle congruence to show that the base angles are congruent.

We know that *TRAP* is an isosceles trapezoid, so by the definition of isosceles trapezoid, $\overline{RA} \cong \overline{TP}$. Congruence of segments is reflexive, so $\overline{AP} \cong \overline{PA}$. We have shown that $\overline{RP} \cong \overline{AT}$ using the distance formula. Therefore, $\triangle RAP \cong \triangle TPA$ by SSS. Then we can conclude that $\angle RAP \cong \angle TPA$ by CPCTC.

You can use a similar method with $\triangle TRA$ and $\triangle RTP$ to prove that $\angle R \cong \angle T$.

The **median** of a trapezoid is the segment that joins the midpoints of its legs.

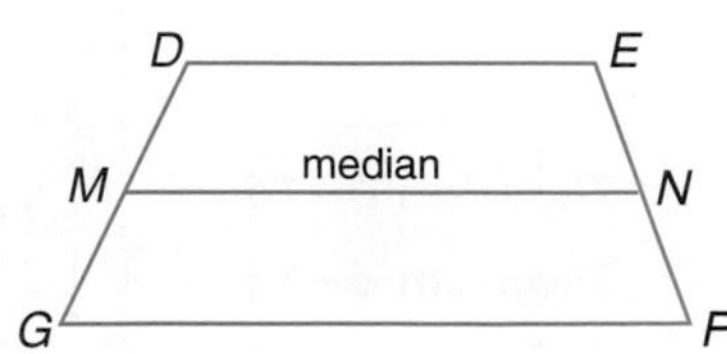

You can construct the median of a trapezoid using a compass and a straightedge.

CONSTRUCTION

Median of a Trapezoid

Construct the median of trapezoid *ZOID*.

Step 1 Draw a trapezoid of any shape or size. Label the vertices *ZOID* with legs $\overline{ZD}$ and $\overline{OI}$.

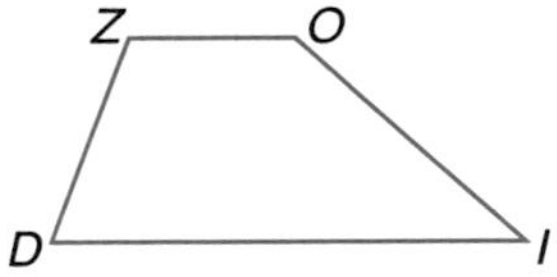

Step 2 Bisect legs $\overline{ZD}$ and $\overline{OI}$. Label the midpoints *M* and *N* as shown.

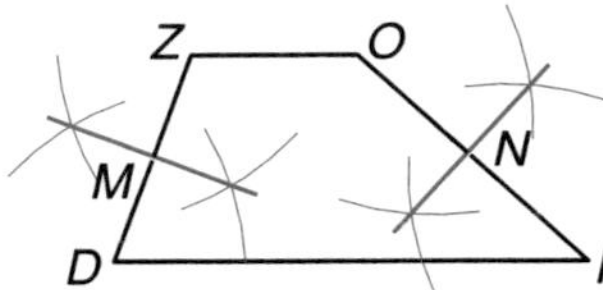

Step 3 Draw line segment *MN*.

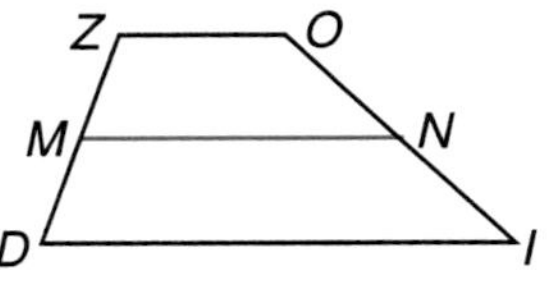

The median has a special relationship to the bases. Use a ruler to measure $\overline{ZO}$, $\overline{ID}$, and $\overline{MN}$ in the figure you constructed. What do you observe about the lengths? This leads to Theorem 6–16.

Theorem 6–16	**The median of a trapezoid is parallel to the bases, and its measure is one-half the sum of the measures of the bases.**

The length of the median can be used to find the length of the bases.

Example 2

Given trapezoid *EZOI* with median $\overline{AB}$, find the value of *x*.

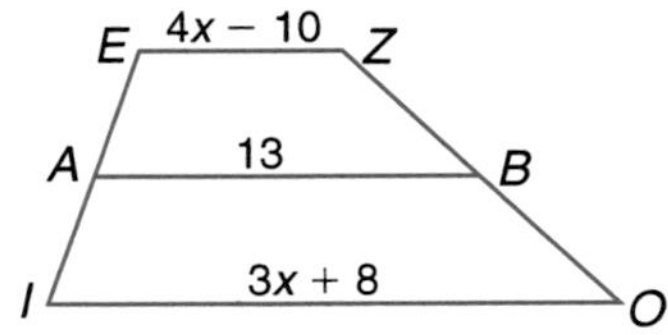

By Theorem 6–16, *AB* equals half the sum of *EZ* and *OI*. We can find the value of *x* by writing and solving an equation.

$AB = \frac{1}{2}(EZ + OI)$

$13 = \frac{1}{2}(4x - 10 + 3x + 8)$ *Substitution Property (=)*

$13 = \frac{1}{2}(7x - 2)$

$26 = 7x - 2$ *Multiplication Property (=)*

$28 = 7x$

$4 = x$

The value of *x* is 4. *Check by substituting 4 for x in the expressions for EZ and OI.*

CHECK FOR UNDERSTANDING

Communicating Mathematics

Read and study the lesson to answer each question.

1. **Draw** an isosceles trapezoid and label the legs and the bases.
2. Refer to Example 1 on page 322. Find the midpoints of $\overline{RA}$ and $\overline{PT}$ and label them M and N, respectively. Use slopes to show that $\overline{MN}$ is parallel to bases $\overline{TR}$ and $\overline{AP}$.
3. Decide whether each statement is *true* or *false*. Explain your reasoning.
   a. A trapezoid is a parallelogram.
   b. The length of the median of a trapezoid is one-half the sum of the lengths of the bases.
   c. The bases of any trapezoid are parallel.
   d. The legs of a trapezoid are always congruent.

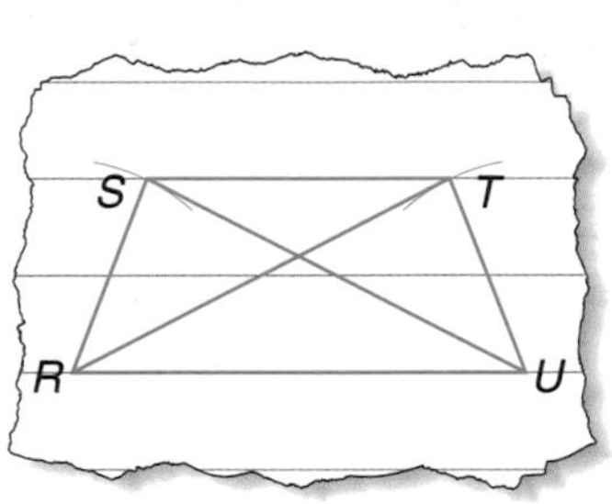

4. Darken two lines on a piece of lined notebook paper to show two parallel lines. Place the point of a compass on a point R on one of the lines, and draw an arc that intersects the second line. Label the point where the arc intersects the line as point S and draw $\overline{RS}$. Place the point of the compass at another point U on the first line, and use the same compass setting to mark a point T. Draw $\overline{TU}$. Make sure that $\overline{SR} \not\parallel \overline{TU}$.
   a. Measure $\angle R$, $\angle S$, $\angle T$, and $\angle U$ of trapezoid $RSTU$. What do you observe about the base angles?
   b. Draw $\overline{RT}$ and $\overline{SU}$. Then use a centimeter ruler to find RT and SU to the nearest millimeter.
   c. What type of trapezoid is $RSTU$?

Guided Practice

***ABCD* is an isosceles trapezoid. Decide whether each statement is *true* or *false*. Explain your reasoning.**

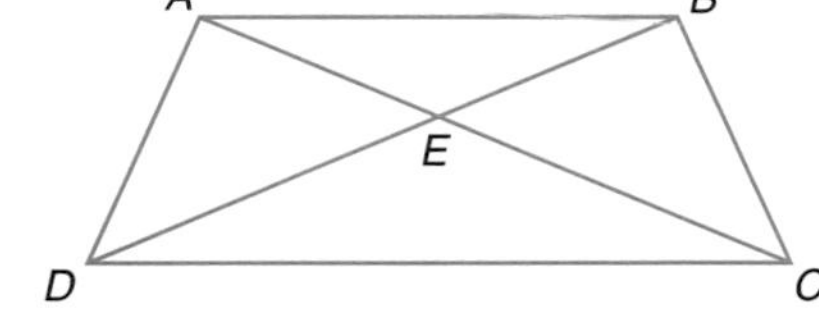

5. $AC = BD$
6. $\overline{AD} \cong \overline{CB}$
7. $\overline{CA}$ and $\overline{BD}$ bisect each other.

***QUAD* is an isosceles trapezoid with bases $\overline{QD}$ and $\overline{AU}$, and median $\overline{EF}$. Use the given information to solve each problem.**

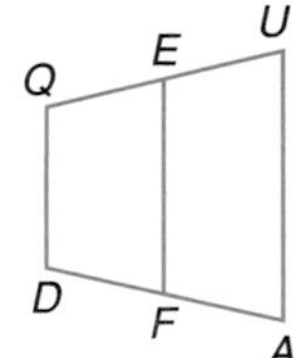

8. If $QU = x^2$ and $AD = 16$, find the value of x.

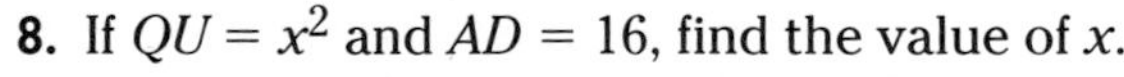

9. If $QE = 17$, find AF.
10. If $m\angle QUA = 62$, find $m\angle ADQ$.

Consider *MNPQ* with vertices at *M*(0, 7), *N*(3, 2), *P*(7, 4) and *Q*(5, 9).

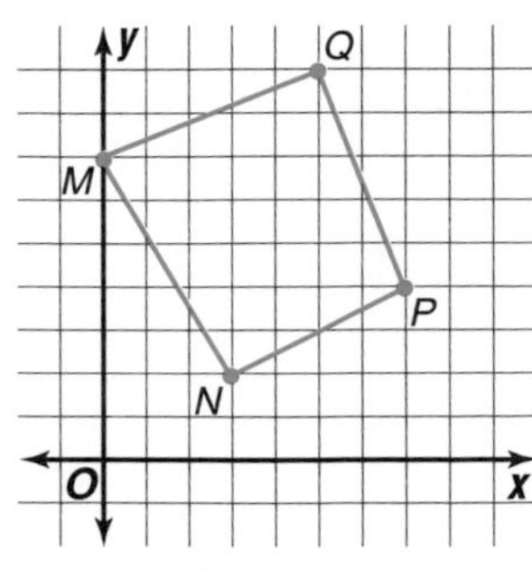

11. Find the coordinates of points R and S if they are the midpoints of $\overline{NP}$ and $\overline{MQ}$.
12. Find the length of $\overline{RS}$.
13. Determine whether $RS = \frac{1}{2}(NP + MQ)$. Explain the result.

Proof **14.** Write a flow proof.

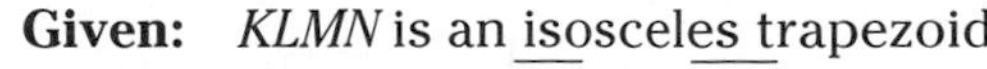

Given: *KLMN* is an isosceles trapezoid with bases $\overline{KL}$ and $\overline{MN}$.

Prove: $\angle 1 \cong \angle 2$

15. Gardening A cold frame is placed over plants on the ground to protect them during cold weather. The top of the frame is glass or plastic to allow the sunlight in and to hold in the heat. The cover slants downward to the south to let in as much sunlight as possible. Identify the shape of each quadrilateral used to construct the cold frame shown at the right.

EXERCISES

Practice

***RSTV* is an isosceles trapezoid. Decide whether each statement is *true* or *false*. Explain your reasoning.**

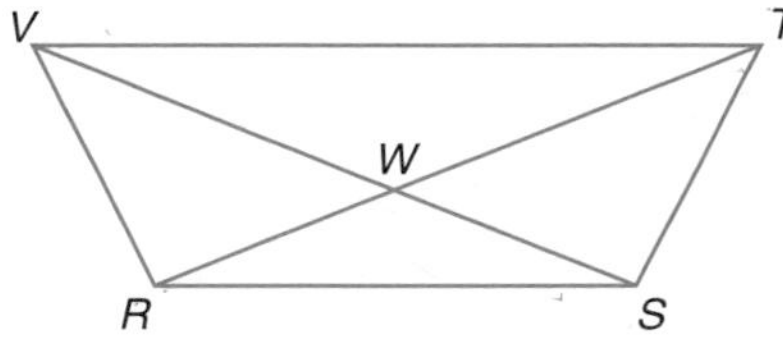

16. $\overline{TR} \perp \overline{SV}$

17. $\angle TRV \cong \angle VST$

18. $\angle RVT \cong \angle STV$

19. $\angle SRV$ and $\angle TVR$ are supplementary.

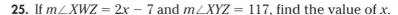

***WXYZ* is an isosceles trapezoid with bases $\overline{WZ}$ and $\overline{XY}$ and median $\overline{MN}$. Use the given information to solve each problem.**

20. Find *MN* if $WZ = 11$ and $XY = 3$.

21. Find $m\angle XMN$ if $m\angle WZN = 78$.

22. If $MN = 10$ and $WZ = 14$, find *XY*.

23. What is the value of *x* if $m\angle MWZ = 15x - 5$ and $m\angle WZN = 90 - 4x$?

24. If $XY = 21.7$ and $ZW = 93.6$, find *MN*.

25. If $m\angle XWZ = 2x - 7$ and $m\angle XYZ = 117$, find the value of *x*.

26. If $MN = 60$, $XY = 4x - 1$, and $WZ = 6x + 11$, find the value of *x*.

27. If $MN = 10x + 3$, $WZ = 11$, and $XY = 8x + 19$, find the value of *x*.

28. If $MN = 2x + 1$, $XY = 8$, and $WZ = 3x - 3$, find the value of *x*.

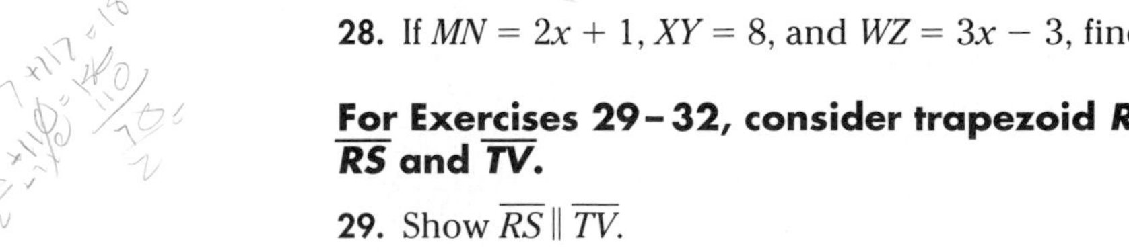

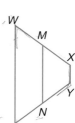

For Exercises 29–32, consider trapezoid *RSTV* with bases $\overline{RS}$ and $\overline{TV}$.

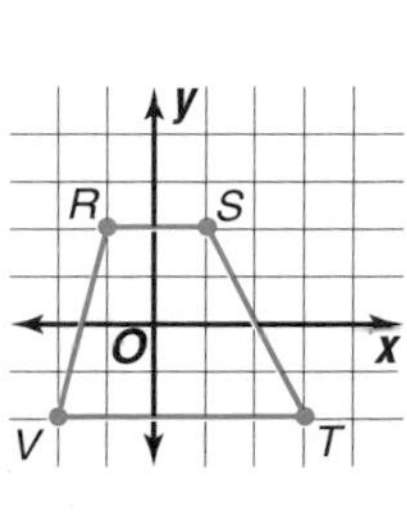

29. Show $\overline{RS} \parallel \overline{TV}$.

30. Is *RSTV* an isosceles trapezoid? Why or why not?

31. Find the coordinates of the endpoints of median $\overline{AB}$ for *RSTV*. Show that $\overline{AB} \parallel \overline{RS}$.

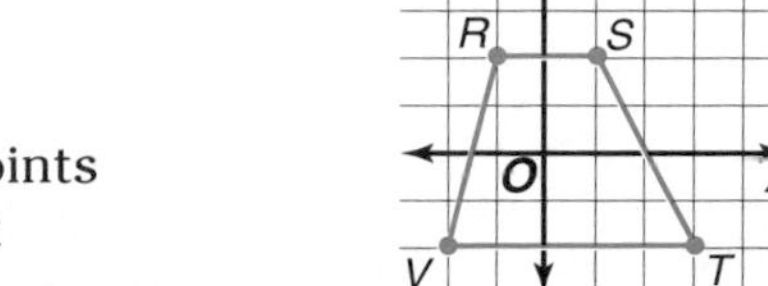

32. Find *AB*, *RS*, and *TV*. Verify that $AB = \frac{1}{2}(RS + TV)$.

Determine whether each figure below is a *trapezoid*, a *parallelogram*, a *rectangle*, or a *quadrilateral*. Choose the most specific term.

33. 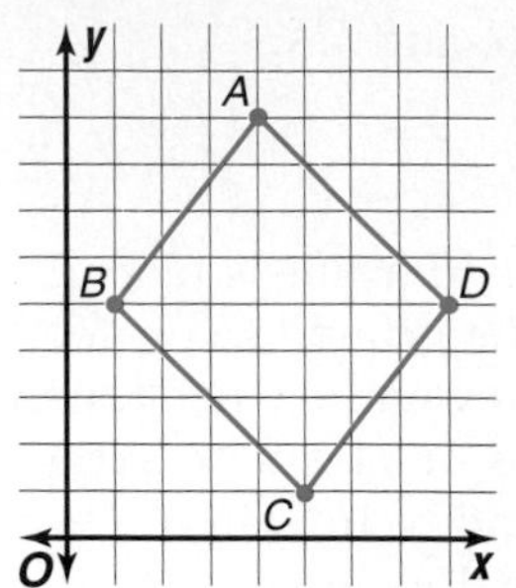

34. 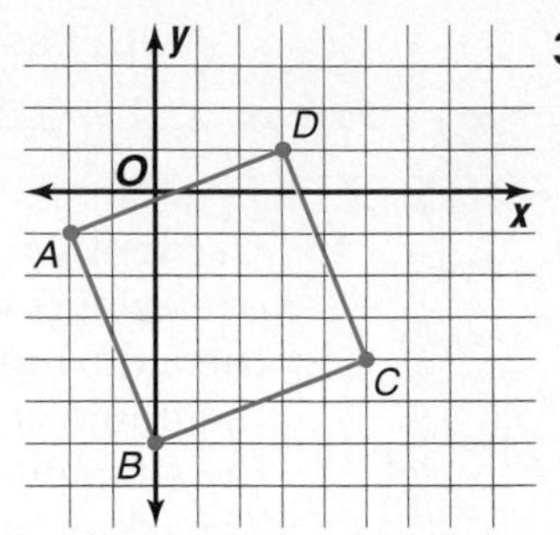

35. 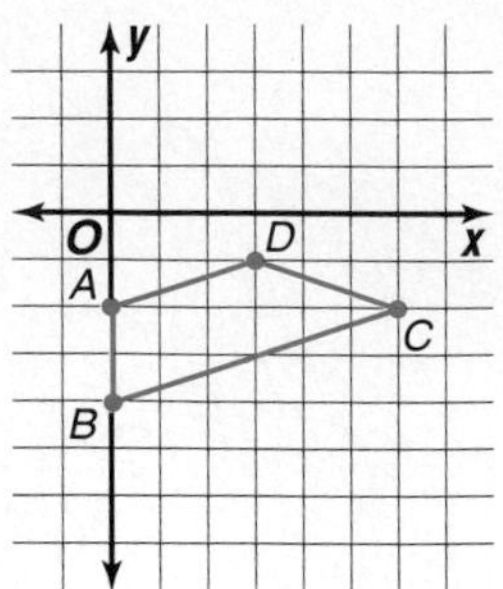

36. In the figure at the right, P is a point not in plane $\mathcal{A}$, and $\triangle XYZ$ is an equilateral triangle in plane $\mathcal{A}$. Plane $\mathcal{B}$ is parallel to plane $\mathcal{A}$ and intersects $\overline{PX}$, $\overline{PY}$, and $\overline{PZ}$ in points J, I, and K, respectively.

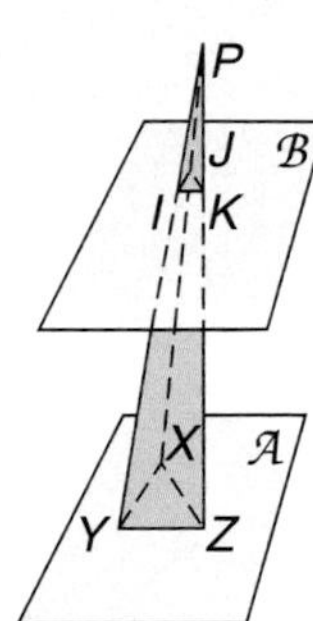

a. How many trapezoids are formed? Name them.

b. If the trapezoids are isosceles, are they congruent? Justify your answer.

c. Where could P be if none of the trapezoids are congruent?

37. In isosceles trapezoid $WXYZ$, $\overline{WZ} \cong \overline{YX}$. If $m\angle W = x$, what is $m\angle Z$?

38. Write a paragraph proof for Theorem 6–15.

Write a two-column proof.

39. Given: $DEFH$ is a trapezoid with bases $\overline{DE}$ and $\overline{FH}$. $\overline{DE} \cong \overline{FG}$

Prove: $DEFG$ is a parallelogram.

40. Given: $JKLM$ is an isosceles trapezoid.

Prove: $\triangle MNL$ is isosceles.

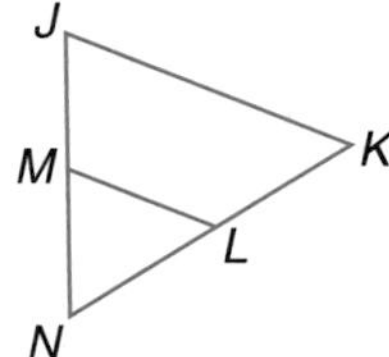

Programming

41. The graphing calculator program will find the measure of the median of a trapezoid $ABCD$ with legs $\overline{AB}$ and $\overline{CD}$.

```
PROGRAM: MEDIAN
: For (A,1,4)
: Disp "ENTER", "COORDINATES:"
: Input "X",X
: X→L1(A)
: Input "Y",Y
: Y→L2(A)
: End
: (L1(1)+L1(2))/2→L1(5)
: (L2(1)+L2(2))/2→L2(5)
: (L1(3)+L1(4))/2→L1(6)
: (L2(3)+L2(4))/2→L2(6)
: √((L1(5)−L1(6))²+(L2(5)−
  L2(6))²→M
: Disp "MEDIAN MEASURE=",M
: Stop
```

Use the program to find the measure of the median of the trapezoid with the given vertices. Round your answers to the nearest hundredth.

a. $A(7, 5)$, $B(7, 6)$, $C(3, 5)$, $D(-1, 3)$

b. $A(2, 10)$, $B(-4, 8)$, $C(-1, 7)$, $D(8, 8)$

c. $A(-5, 7)$, $B(0, 12)$, $C(18, -6)$, $D(2, -22)$

Critical Thinking

42. a. Draw a trapezoid with two right angles. Label the bases and the legs.
    b. Try to draw an isosceles trapezoid with two right angles. Is it possible? Explain why or why not.

43. The median of a trapezoid is 20 millimeters long.
    a. List two possible integral values for the measures of the bases.
    b. How many integral value combinations are possible for the bases? Explain.
    c. How many values are possible for the measures of the bases if the restriction of integral values were removed? Explain.

Applications and Problem Solving

44. **Sailing** Many large sailboats have a *keel* to keep the boat stable in high winds. A keel is shaped like a trapezoid with its top and bottom parallel. The distance across the top of the keel is called the *root chord*, the distance across the bottom of the keel is the *tip chord*, and the distance from the midpoint of the front of the keel to the midpoint of the back of the keel is the *mid-chord*. The chart above shows the root chord and the tip chord for selected types of sailboats. Find the length of the mid-chord for each boat type.

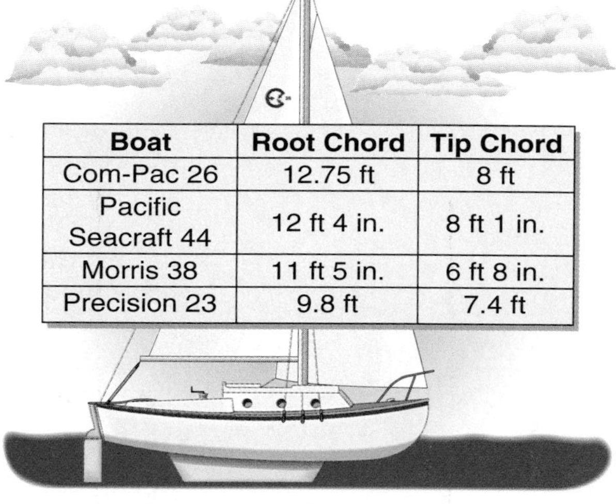

Boat	Root Chord	Tip Chord
Com-Pac 26	12.75 ft	8 ft
Pacific Seacraft 44	12 ft 4 in.	8 ft 1 in.
Morris 38	11 ft 5 in.	6 ft 8 in.
Precision 23	9.8 ft	7.4 ft

45. **Architecture** Tray ceilings make a room appear to be larger than it actually is. The trays are made using wood or drywall panels. What type of quadrilaterals are used to make the tray ceiling shown at the right?

fabulous FIRSTS

Norma Merrick Sklarek (1928–)

Norma Merrick Sklarek was the first African-American woman to become a registered architect.

46. **Make a Table** Copy the table below. Write *yes* or *no* in each table cell according to whether each figure has the given characteristic.

	Parallelogram	Rectangle	Square	Isosceles Trapezoid
diagonals congruent				
two pairs of opposite sides congruent				
one pair of opposite sides congruent				
diagonals perpendicular				
one pair of opposite sides parallel and congruent				

Mixed Review

47. In rhombus *BCDE*, $m\angle B = 68$. Find $m\angle E$. (Lesson 6–4)

48. **Building Materials** Explain why the rectangle is used as the shape of bricks in the construction of walls. (Lesson 6–3)

49. Draw several quadrilaterals to determine if the statement *If the midpoints of the sides of any quadrilateral are connected in clockwise order, they form a parallelogram* is true or false. (Lesson 6–2)

50. Write the assumption you would make to start an indirect proof of the statement *If two lines intersect, then no more than one plane contains them.* (Lesson 5–3)

51. Determine if the figure contains a pair of congruent triangles. Justify your answer. (Lesson 5–2)

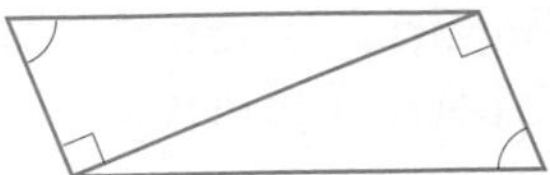

52. Draw and label $\triangle RST$ and $\triangle ABC$ with $\angle S \cong \angle A$ and $\angle R \cong \angle B$. Indicate the additional pair of corresponding parts that would have to be proved congruent in order to prove the triangles congruent by AAS. (Lesson 4–5)

53. **ACT Practice** In the triangle, what is the measure of $\angle A$?

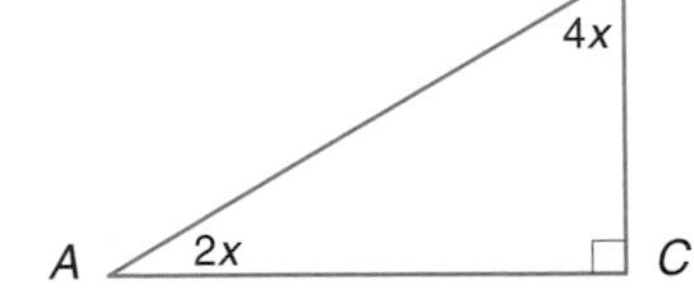

A 15° **B** 30° **C** 60°
D 90° **E** 120°

54. $\angle N$ and $\angle M$ are supplementary. If $\angle N$ is acute, is $\angle M$ acute, right, obtuse, or straight? (Lesson 1–7)

Algebra

55. **Patterns** Use your calculator to find the values of 11^2, 111^2, and 1111^2. Without calculating, use a pattern to determine the value of $11{,}111^2$.

56. Solve $\frac{12}{k} = \frac{2}{9}$.

For **Extra Practice**, see page 775.

WORKING ON THE In·ves·ti·ga·tion

Refer to the Investigation on pages 286–287.

This Land Is Your Land

1 Complete your list of the dimensions of each possible feature of the park.

2 Determine which features you would like to incorporate. You may wish to make two lists: one of features that you feel are necessary and one of features that would be nice if space and money are available. Note the budget for each feature on the lists.

3 Find the area required for each feature you are considering.

Add the results of your work to your Investigation Folder.

HIGHLIGHTS

Chapter Review For additional lesson-by-lesson review, visit: **www.geometry.glencoe.com**

VOCABULARY

After completing this chapter, you should be able to define each term, property, or phrase and give an example or two of each.

Geometry

base of a trapezoid (p. 321)
base angle of a trapezoid (p. 321)
diagonal (p. 293)
isosceles trapezoid (p. 321)
kite (p. 320)
leg of a trapezoid (p. 321)
median of a trapezoid (p. 322)
parallelogram (pp. 290, 291)
quadrilateral (p. 291)
rectangle (pp. 305, 306)
rhombus (p. 313)
rhombi (p. 313)
square (p. 315)
trapezoid (p. 321)

Algebra

probability (p. 294)

Problem Solving

identify subgoals (p. 298)

UNDERSTANDING AND USING THE VOCABULARY

Determine whether each statement is *true* or *false*. If the statement is false, rewrite it to make it true.

1. Every quadrilateral is a parallelogram.
2. If quadrilateral *ABCD* is a parallelogram, then $\overline{AB} \parallel \overline{CD}$.
3. If both pairs of opposite angles in a quadrilateral are congruent, then the quadrilateral is a parallelogram.
4. If *NMOP* is a rectangle, then it is a parallelogram.
5. You can prove that a quadrilateral is a rectangle by proving that the diagonals are congruent.
6. If a quadrilateral is a rhombus or a square, then the diagonals are perpendicular.
7. A square has all of the characteristics of a parallelogram, a rectangle, a rhombus, and a trapezoid.
8. If a quadrilateral has four right angles, then it must be a rectangle.
9. The legs of an isosceles parallelogram are congruent.
10. The median of a trapezoid is parallel to the bases of the trapezoid and its measure is half the sum of the measures of the bases.
11. If *QUAD* is a square, then it is also a parallelogram, a rectangle, a rhombus, a quadrilateral, and a trapezoid.
12. The diagonals of a trapezoid are congruent.

CHAPTER 6 STUDY GUIDE AND ASSESSMENT

SKILLS AND CONCEPTS

OBJECTIVES AND EXAMPLES

Upon completing this chapter, you should be able to:

- recognize and apply the properties of a parallelogram (Lesson 6–1)

If *BCDE* is a parallelogram, then you can make the following statements.

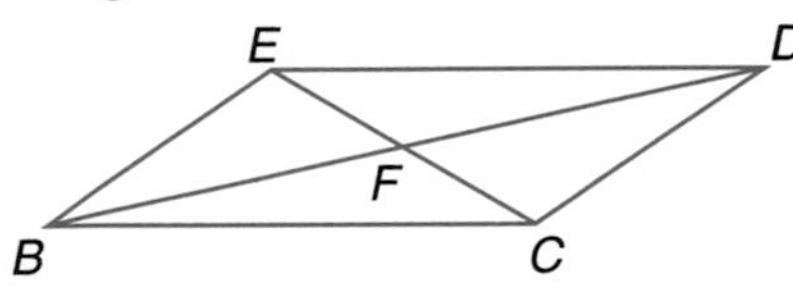

$\overline{BC} \parallel \overline{ED}$ $\overline{BE} \parallel \overline{CD}$

$\overline{BC} \cong \overline{ED}$ $\overline{BE} \cong \overline{CD}$

$\angle EBC \cong \angle CDE$ $\angle BCD \cong \angle DEB$

$\angle EBC$ is supplementary to $\angle BED$.

$\angle EDC$ is supplementary to $\angle DCB$.

$\angle BED$ is supplementary to $\angle EDC$.

$\angle EBC$ is supplementary to $\angle BCD$.

$\overline{BD}$ and $\overline{EC}$ bisect each other.

REVIEW EXERCISES

Use these exercises to review and prepare for the chapter test.

Complete each statement about ▱*WXYZ*. Then name the theorem or definition that justifies your answer.

13. $\angle WZY \cong$ __?__
14. $\overline{WX} \cong$ __?__
15. $\overline{XE} \cong$ __?__
16. $\overline{XY} \cong$ __?__
17. $\triangle XYZ \cong$ __?__
18. $\angle 1 \cong$ __?__
19. $\overline{YE} \cong$ __?__
20. $\angle WZY$ and __?__ are supplementary.

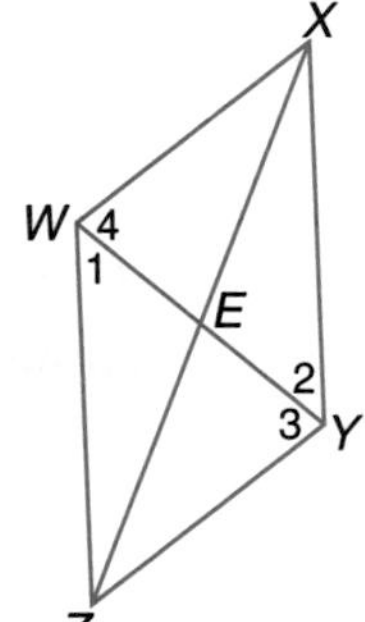

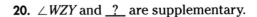

- recognize and apply the conditions that ensure a quadrilateral is a parallelogram (Lesson 6–2)

If $OQ = 20$, $MQ = 4y$, $PQ = x^2$, and $NQ = 36$, find the values of x and y in order for *MNOP* to be a parallelogram.

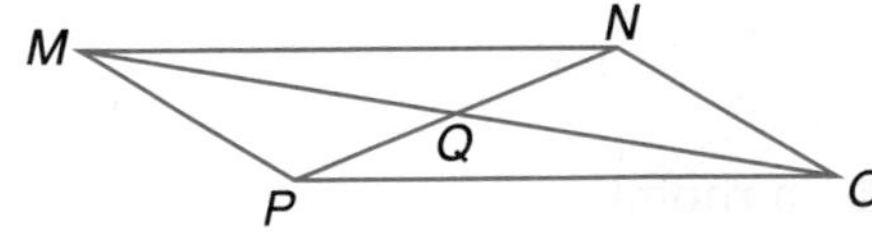

If the diagonals of a quadrilateral bisect each other, then it is a parallelogram. So, $OQ = MQ$ and $PQ = NQ$.

$OQ = MQ$ $PQ = NQ$

$20 = 4y$ $x^2 = 36$

$5 = y$ $x = 6$ or -6

To ensure that *MNOP* is a parallelogram, $x = -6$ or 6 and $y = 5$.

Find the values of *x* and *y* that ensure each quadrilateral is a parallelogram.

21.

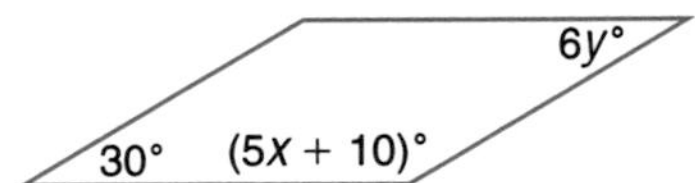

22.

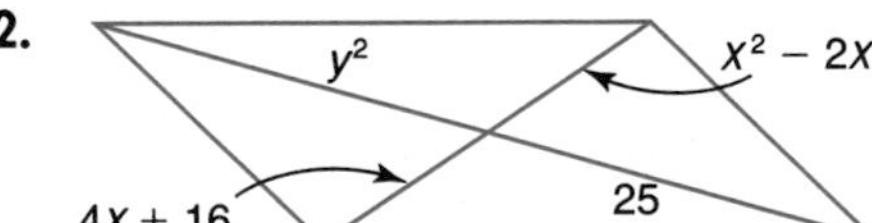

Write a two-column proof.

23. **Given:** ▱*ABCD*
$\overline{AE} \cong \overline{CF}$

Prove: Quadrilateral *EBFD* is a parallelogram.

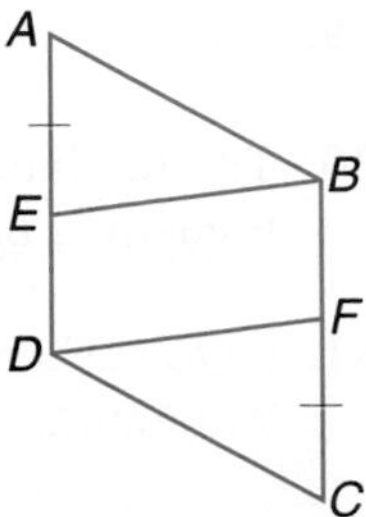

OBJECTIVES AND EXAMPLES

- recognize and apply the properties of rectangles (Lesson 6–3)

If a quadrilateral is a rectangle, then the following properties hold true.

1. Opposite sides are congruent and parallel.
2. Opposite angles are congruent.
3. Consecutive angles are supplementary.
4. Diagonals are congruent and bisect each other.
5. All four angles are right angles.

REVIEW EXERCISES

Quadrilateral *EFGH* is a rectangle. Find the value of *x*.

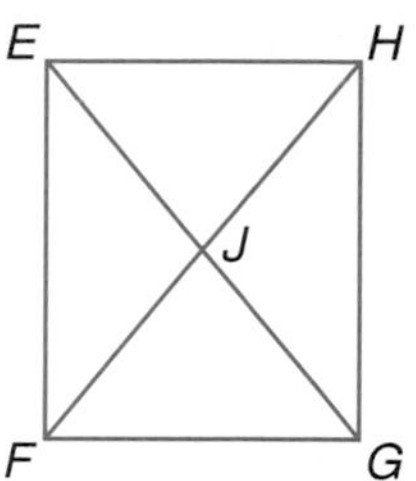

24. $m\angle HEG = 12x + 1$
    $m\angle GEF = 6x - 1$
25. $HF = 5x - 4$
    $EG = 6x - 10$
26. $JF = 8x + 4$
    $EG = 24x - 8$
27. $EF = x^2$
    $HG = 3x - 2$
28. $m\angle FGH = 10x^2$
    $m\angle GHE = 8x^2 + 18$

- recognize and apply the properties of squares and rhombi (Lesson 6–4)

Determine whether quadrilateral *FGHI* with $F(5, 0)$, $G(0, 0)$, $H(2, 3)$, and $I(7, 3)$ is a *parallelogram*, a *rectangle*, a *rhombus*, or a *square*. List all that apply.

$FG = \sqrt{(5-0)^2 + (0-0)^2}$ or 5

$GH = \sqrt{(0-2)^2 + (0-3)^2}$ or $\sqrt{13}$

$HI = \sqrt{(2-7)^2 + (3-3)^2}$ or 5

$IF = \sqrt{(7-5)^2 + (3-0)^2}$ or $\sqrt{13}$

The opposite sides are congruent, so *FGHI* is a parallelogram. All sides are not congruent, so it is not a rhombus or a square. Slopes can determine if *FGHI* is a rectangle.

slope of $\overline{FG} = \frac{0-0}{5-0}$ or 0

slope of $\overline{GH} = \frac{0-3}{0-2}$ or $\frac{3}{2}$

Consecutive sides are not perpendicular, so *FGHI* is a parallelogram, but not a rectangle.

Determine whether quadrilateral *KLMN* is a *parallelogram*, a *rectangle*, a *rhombus*, or a *square* for each set of vertices. List all that apply.

29. $K(4, 8)$, $L(0, 9)$, $M(-2, 1)$, $N(2, 0)$
30. $K(12, 0)$, $L(6, -6)$, $M(0, 0)$, $N(6, 6)$
31. $K(5, 4)$, $L(3, -6)$, $M(0, -10)$, $N(2, 0)$
32. $K(1, 5)$, $L(8, 6)$, $M(15, 5)$, $N(8, 4)$

Use rhombus *PQRS* and the given information to find each value.

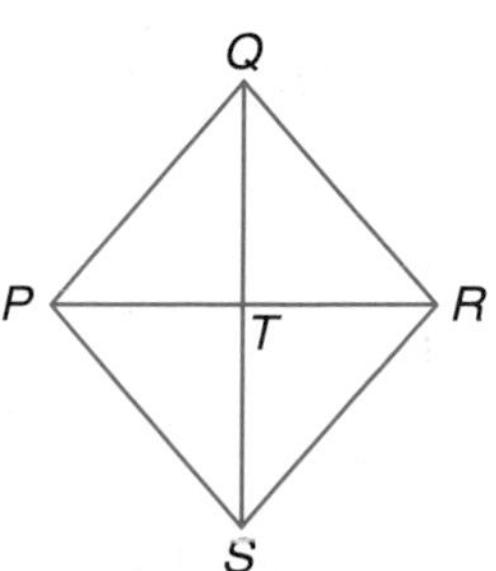

33. If $PT = 14$, find PR.
34. If $m\angle PQT = 34$, find $m\angle PQR$.
35. What is $m\angle STP$?
36. If $RQ = 4x - 1$ and $PQ = 20 + x$, find the value of x.
37. What is $m\angle PQT$ if $m\angle QPT = 52$?
38. Find the values of x and y if $PT = 4x - 8$, $QT = 6y - 9$, $TR = 16 - 2x$, and $TS = 3y + 9$.

OBJECTIVES AND EXAMPLES

- recognize and apply the properties of trapezoids (Lesson 6–5)

If a quadrilateral is a trapezoid, then the following properties hold true.

1. The bases are parallel.
2. The median is parallel to the bases and its measure is half of the sum of the measures of the bases.

If a quadrilateral is an isosceles trapezoid, then the following additional properties hold true.

1. Both pairs of base angles are congruent.
2. The diagonals are congruent.

REVIEW EXERCISES

STUVis a trapezoid with bases $\overline{ST}$ and $\overline{UV}$. Use the figure and the given information to solve each problem.

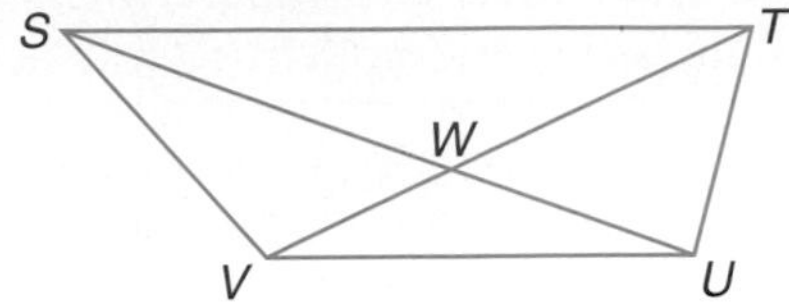

39. If $m\angle STV = 45$, find $m\angle UVT$.
40. If $m\angle TSW = 35$, find $m\angle SUV$.
41. If $m\angle SUV = 47$, find $m\angle TSU$.
42. What is the measure of the median of *STUV* if $ST = 23$ and $UV = 19$?
43. If $m\angle WUV = 23$ and $m\angle TWS = 127$, find $m\angle WVU$.

APPLICATIONS AND PROBLEM SOLVING

44. **Architecture** Oakland-Alameda County Coliseum uses rings of parallel and intersecting concrete columns around the outside of the coliseum to support the roof so that no supports are needed inside. What shape are the quadrilaterals formed by the support columns? (Lesson 6–4)

45. **Kites** Make a kite like those made by children on the island of Molokái, Hawaii.
    a. Fold a sheet of notebook paper in half lengthwise.
    b. Fold again along a diagonal such as $\overline{AB}$.
    c. Tape a stick along $\overline{CD}$.
    d. Turn the kite over and fold the flap back and forth until it stands up straight.
    e. Complete the kite with a tail at *B* and a string to hold about 2 inches from *A* in the flap.

    Identify each quadrilateral formed in construction of the kite. (Lessons 6–1 to 6–5)

A D C B

46. **Art** The painting *Either...Or*, by Josef Albers is shown at the right. Trace the shapes in the painting and label each vertex. Identify each quadrilateral in the painting. (Lessons 6–1 to 6–5)

47. **Civil Engineering** The world's longest single-span suspension bridge is the Humber Bridge in Humberside County, England. The hollow boxes that form the bridge's deck are shaped like trapezoids. Which surfaces of the bridge are parallel? (Lesson 6–5)

A practice test for Chapter 6 is provided on page 798.

ALTERNATIVE ASSESSMENT

COOPERATIVE LEARNING PROJECT

Architecture In this chapter, you learned about a variety of quadrilaterals. Quadrilaterals can be classified into those with right angles and those without right angles. Most buildings have floors, walls, and roofs that are quadrilaterals having four right angles. Can you imagine a building made entirely of quadrilaterals with no right angles?

In this project, you will build a scale model, or make a detailed drawing, of a building constructed almost entirely out of quadrilaterals without right angles.

Follow these steps to complete your building design.

- In the first design, floors and ceilings may be rectangular.
- Design a living room with one wall that is all glass, oriented so it can take in the noonday sun.
- Design a family room with one wall that inclines out over a patio or deck. This provides shelter from the rain for your outdoor enjoyment.
- Design three second floor bedrooms so that each one has two all glass walls, some looking up toward the sky and some looking down for a scenic view of the yard.
- Remember, none of the walls in your house are rectangular.
- As a group, discuss the advantages of your design.

Convention Center, Columbus, Ohio

Follow these steps to complete your next building design.

- In this design, there are no rectangles—no rectangular floors, no rectangular ceilings, or roofs. Even more, not even the steps may use any rectangles.
- As a group, discuss what you will design to walk from one floor to the next.
- Be creative, but make sure everything functions correctly.
- As a group, discuss the advantages and disadvantages of your design.

THINKING CRITICALLY

Find the measure of $\overline{MD}$ in rectangle *ABCD*. (*Hint:* Draw an auxiliary segment.)

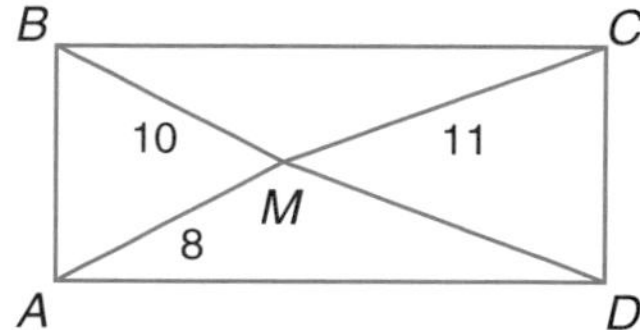

PORTFOLIO

Look through magazines that contain photographs of buildings. Find examples of quadrilaterals in the architecture. Collect these images and organize them by type of quadrilateral. Keep this collection in your portfolio.

STANDARDIZED TEST PRACTICE

CHAPTERS 1–6

Section One: MULTIPLE CHOICE

There are eight multiple-choice questions in this section. After working each problem, write the letter of the correct answer on your paper.

1. If a right triangle has two sides of measures 1 and $\sqrt{2}$, which of the following could be the measure of the third side?

I. 1

II. $\sqrt{2}$

III. $\sqrt{3}$

A. I only
B. II only
C. III only
D. I and III only

2. Find 5% of 6%.

A. 0.11%
B. 0.3%
C. 3%
D. 11%

3. Which of the following is the graph of the solution set of $|x - 3| < 2$?

A. 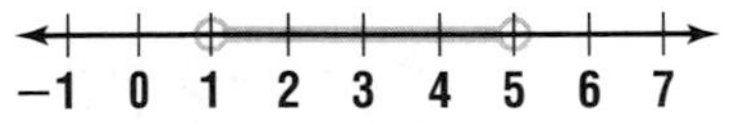 –1 0 1 2 3 4 5 6 7

B. 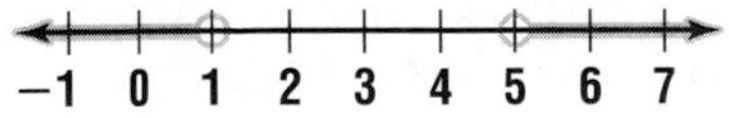 –1 0 1 2 3 4 5 6 7

C. 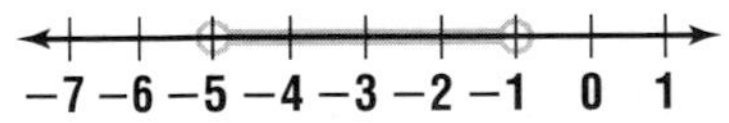 –7 –6 –5 –4 –3 –2 –1 0 1

D. 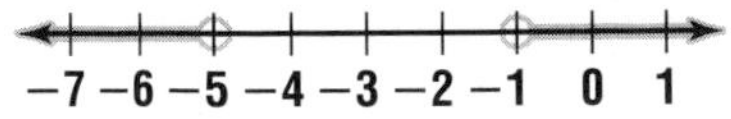 –7 –6 –5 –4 –3 –2 –1 0 1

4. What is the value of 3^{-2}?

A. 9
B. -6
C. $\frac{1}{9}$
D. $-\frac{1}{9}$

5. In the diagram below, which type of segment is $\overline{AD}$?

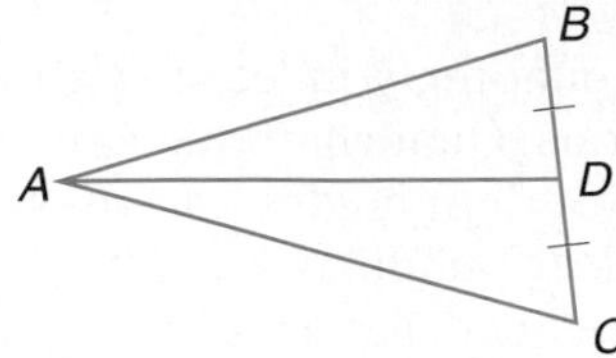

A. angle bisector
B. perpendicular bisector
C. median
D. altitude

6. Which of the following is a false statement?

A. All squares are rectangles.
B. All rhombi are squares.
C. All rectangles are quadrilaterals.
D. All trapezoids are polygons.

7. In right triangle QRS, $m\angle Q = 30$, $m\angle R = 60$, and $QS = 6$. What is the measure of $\overline{QR}$?

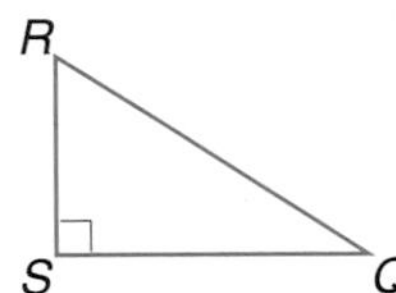

A. $4\sqrt{3}$
B. $6\sqrt{3}$
C. $3\sqrt{3}$
D. $3\sqrt{2}$

8. What is the simplified form of the expression $\frac{a^{-3}bc^2}{a^{-4}b^2c^{-3}}$?

A. $\frac{c^5}{a^7b}$
B. $\frac{ac^5}{b}$
C. $\frac{a^7c^5}{b}$
D. $\frac{a^7}{bc^5}$

Test-Taking Tip When a problem includes a variable, you can use the "plug-in" strategy. That is, replace the variable with numbers that fit the problem.

Section Two: SHORT ANSWER

This section contains seven questions for which you will provide short answers. Write your answer on your paper.

9. What is the lowest common denominator for the fractions $\frac{5}{4x^2y}$, $\frac{7}{6x^2y}$, and $\frac{-4}{15xy}$?

10. Find the sum of the perimeters of a square with sides of length x and an equilateral triangle with sides of length y.

11. If y varies inversely as x and y is 10 when x is $\frac{1}{2}$, find y when x is $\frac{2}{3}$.

12. In the triangle, what is the measure of $\angle BAC$?

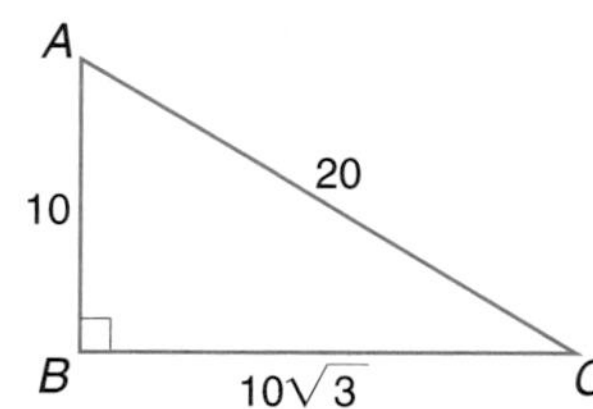

13. Find the solution set for $3 - (x - 5) = 2x - 3(4 - x)$.

14. For all $x \neq 0$, find $\frac{x^6 + x^6 + x^6}{x^2}$.

15. In the figure, $\angle ABC$ is a straight angle, and $\overline{DB}$ is perpendicular to $\overline{BE}$. If $\angle ABD$ measures x degrees, write an expression to represent the measure of $\angle CBE$.

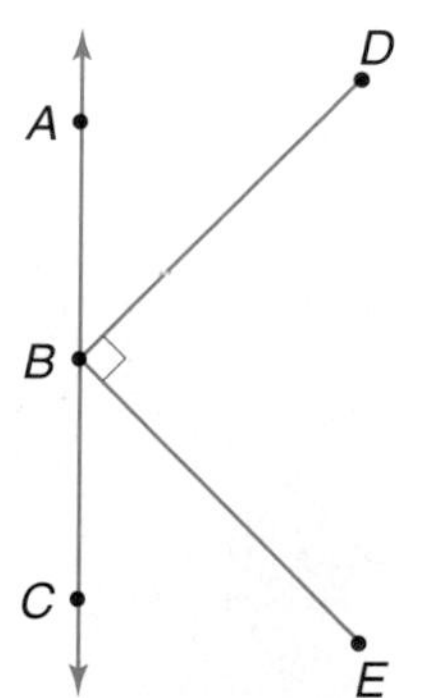

Section Three: COMPARISON

This section contains five comparison problems that involve comparing two quantities, one in column A and one in column B. In certain questions, information related to one or both quantities is centered above them. All variables used represent real numbers.

Compare quantities A and B below.

- Write A if quantity A is greater.
- Write B if quantity B is greater.
- Write C if the two quantities are equal.
- Write D if there is not enough information to determine the relationship.

Column A	Column B
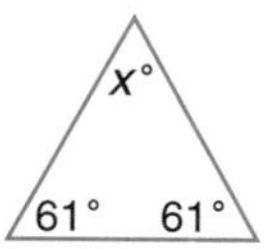	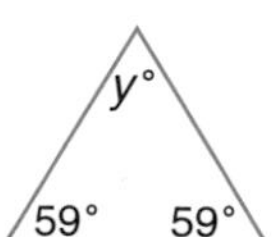
16. x	y
17. $\frac{\frac{1}{2}}{3}$	$\frac{2}{\frac{1}{3}}$
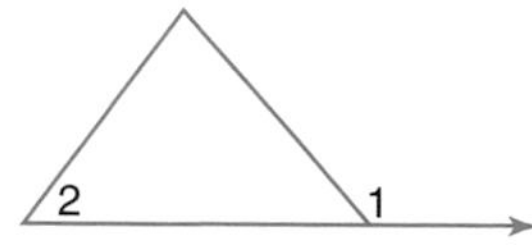	
18. $m\angle 1$	$m\angle 2$
In $\triangle ABC$, $m\angle A > m\angle B$ and $m\angle A > m\angle C$.	
19. AC	AB
$3x - 4y = -2$ $4x + 2y = 12$	
20. x	y

Test Practice For additional test practice questions, visit: **www.geometry.glencoe.com**

CHAPTER 7 Connecting Proportion and Similarity

What You'll Learn

In this chapter, you will:

- recognize and use ratios and proportions,
- identify similar figures and use the properties of similar figures to solve problems,
- use proportional parts of triangles to solve problems,
- solve problems by first solving a simpler problem, and
- recognize and describe characteristics of fractals.

Why It's Important

Real World

Life Science Have you ever looked closely at the leaves on a tree? On a tree, the leaves are not necessarily the same size, but they do have approximately the same shape. This property of having the same shape but not necessarily the same size is called similarity. You see this property in nature, as with tree leaves, and you can see it in other objects. For example, when you take a picture with a camera, the photograph and the scene are similar. In this chapter, you will learn about the relationships between similar polygons

PREREQUISITE SKILLS

To be successful in this chapter, you'll need to understand these concepts and be able to apply them. Refer to the example or to the lesson in parentheses if you need more review before beginning the chapter.

Use the distributive property to simplify expressions. (Lesson 2–4)

Use the distributive property to rewrite each expression without parentheses.

1. $5(2x - 4)$ **2.** $-5(a - 1)$ **3.** $-1(b - 7)$

Use the percent proportion to solve problems involving percents.

Example What is 15% of 30?

$\frac{P}{B} = \frac{r}{100}$ *Use the percent proportion where P is the percentage, B is the base, and $\frac{r}{100}$ is the rate.*

$\frac{P}{30} = \frac{15}{100}$ *Replace B with 30 and r with 15.*

$100 \cdot P = 30 \cdot 15$ *Find the cross products.*

$P = 4.5$ *Divide each side by 100.*

Use the percent proportion to solve each problem.

4. What is 35% of 50?

5. Find 18% of 90.

6. Find 225% of 32.

7. 75 is 125% of what number?

Identify and use medians, altitudes, angle bisectors, and perpendicular bisectors in a triangle. (Lesson 5–1)

Write a short definition for each special segment in a triangle.

8. median **9.** altitude **10.** angle bisector

Focus On

READING SKILLS

In this chapter, you'll be learning about ***similar figures.*** The word *similar* means "having characteristics in common." For two figures to be similar, they must first be the same type of figure, for example, both triangles. Then all the corresponding sides must have the same ratio. The ratio between sides is called the *scale factor.* Recall, that a scale drawing is similar to the actual object. Use the idea of scale drawings to help you think about the term similar. For example, a map is similar to the land being shown, and a house plan is similar to the house that will be built.

7–1

Integration: Algebra Using Proportions

What YOU'LL LEARN

- To recognize and use ratios and proportions, and
- to apply the properties of proportions.

Why IT'S IMPORTANT

You can use proportions to solve problems involving movie props, literature, and meteorology.

Art

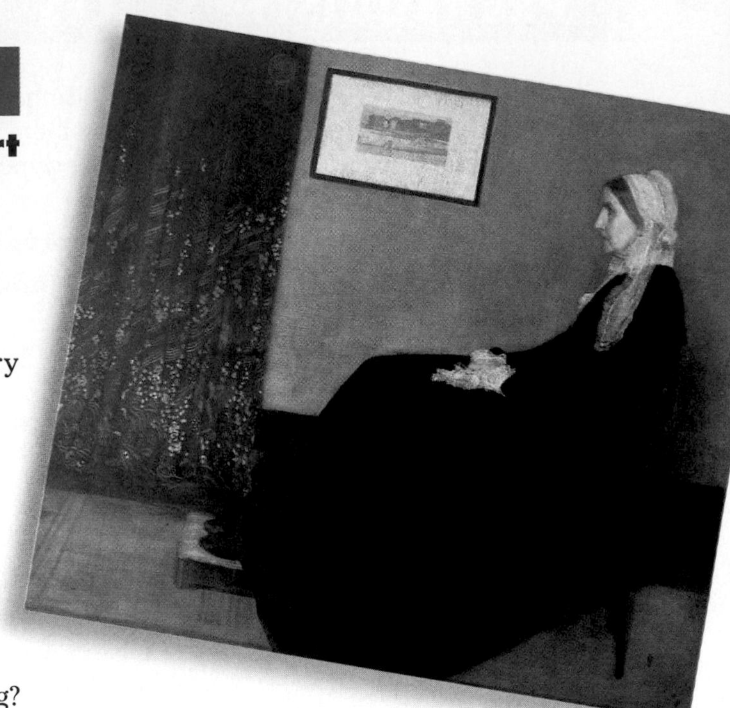

In 1871, American painter James Whistler painted a portrait of his mother called *Arrangement in Black and Gray*. Suppose Whistler drew a preliminary sketch in which his mother's height was 10 inches and the width of the widest part of her gown was $8\frac{1}{3}$ inches. In the actual painting, his mother's height is 50 inches. How wide is the widest part of her gown in the actual painting?

You will be asked to solve this problem in Exercise 16.

You can use **ratios** to solve problems like the one in the application above. A ratio is a comparison of two quantities. The ratio of a to b can be expressed as $\frac{a}{b}$, where b is not zero. The ratio can also be written as $a{:}b$.

Example 1

Entertainment

Special effects in movies are often created using miniature models. In a recent movie, a model jeep 22 inches long was created to look like a real $14\frac{2}{3}$-foot jeep. What is the ratio of the length of the model compared to the length of the real jeep?

$$\frac{\text{length of model}}{\text{length of real jeep}} = \frac{22 \text{ inches}}{176 \text{ inches}} \quad 14\tfrac{2}{3} \text{ feet} = 176 \text{ inches}$$

$$= \frac{1}{8}$$

The ratio comparing the two lengths is $\frac{1}{8}$ or 1:8. The length of the model is $\frac{1}{8}$ the length of a real jeep.

A special ratio often found in nature and used by artists and architects is the *golden ratio*. It is approximately 1:1.618. The *Fibonacci sequence* is related to the golden ratio. You can use a spreadsheet to explore this relation.

EXPLORATION — SPREADSHEETS

Recall that in a spreadsheet each cell is named by the column and row in which it is located (A1 means column A, row 1). Suppose we set up a spreadsheet to evaluate the formula for the Fibonacci sequence and find the ratio between the pairs of consecutive terms of the sequence. Enter the labels in rows 1 and 2 as shown below. Enter 1 in each cell in row 3.

- In cell A4, enter A3 + 1. Copy this formula to cells A5 through A22.
- In cell B5, enter B3 + B4. Copy this formula to cells B6 through B22.
- In cell C3, enter B4/B3. Copy this formula to cells C4 through C22.

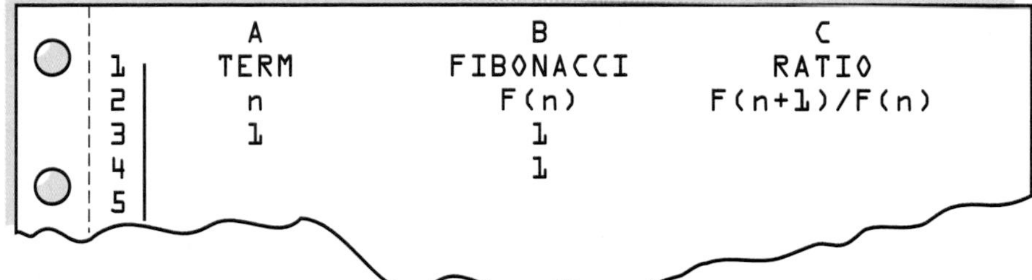

Your Turn

a. Write the first twenty terms of the Fibonacci sequence.

b. Describe the relationship between the terms of the Fibonacci sequence.

c. Find the ratio between each pair of consecutive terms of the sequence. (Round to seven decimal places.) What do you notice about these ratios?

Ratios, like fractions can be simplified. According to Nielsen Media Research, in 1995 there were about 224 televisions for every 100 homes. One radio station reported this as 336 televisions for every 150 homes. How do the two ratios compare? When they are simplified, both ratios are equivalent to $\frac{2.24}{1}$.

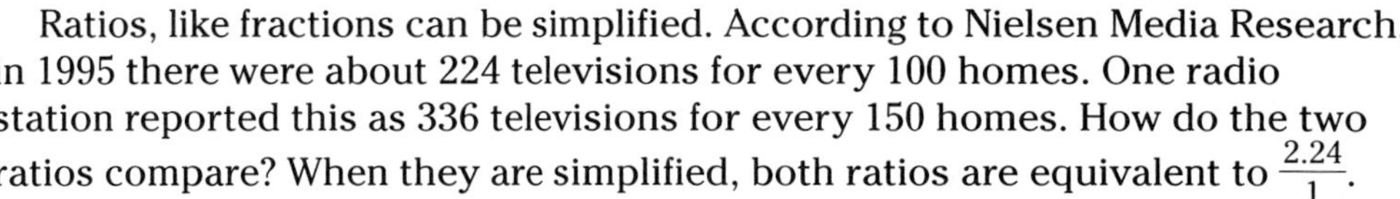

$$\frac{224}{100} = \frac{2.24}{1} \quad (\div 100) \qquad \frac{336}{150} = \frac{2.24}{1} \quad (\div 150)$$

An equation stating that two ratios are equal is a **proportion**. So, $\frac{224}{100} = \frac{336}{150}$ is a proportion. Every proportion has two **cross products**. In the proportion $\frac{224}{100} = \frac{336}{150}$, the cross products are 224 times 150 and 100 times 336. The **extremes** of the proportion are 224 and 150. The **means** are 100 and 336. The cross products of a proportion are equal.

$$\frac{224}{100} = \frac{336}{150}$$
$$224(150) = 100(336)$$
extremes *means*
$$33{,}600 = 33{,}600$$

Consider the general case. $\frac{a}{b} = \frac{c}{d}$

$(bd)\frac{a}{b} = (bd)\frac{c}{d}$ *Multiply each side by bd.*

$da = bc$ *Simplify.*

$ad = bc$ *Commutative Property (×)*

Equality of Cross Products	**For any numbers *a* and *c* and any nonzero numbers *b* and *d*, $\frac{a}{b} = \frac{c}{d}$ if and only if $ad = bc$.**

The product of the means equals the product of the extremes.

Example 2 **Solve $\frac{3t-1}{4} = \frac{7}{8}$ by using cross products.**

$$\frac{3t-1}{4} = \frac{7}{8}$$

$(3t - 1)(8) = (4)\,7$ *Find the cross products*

$24t - 8 = 28$ *Distributive Property*

$24t = 36$ *Add 8 to each side.*

$t = \frac{36}{24}$ or $\frac{3}{2}$ *Divide each side by 24.*

Example 3

Literature

In Chapter 2, you learned that *Alice's Adventures in Wonderland* by Lewis Carroll contains deductive reasoning. It also contains examples of proportions. When Alice ate or drank something, her size changed. ". . . she came suddenly to an open place with a little house in it four feet high . . . . she began nibbling at the right hand bit again, and did not venture to go near the house till she had brought herself down to nine inches high." If Alice was normally about 50 inches tall, how tall would the house have been in Alice's normal world?

You can write a proportion to show the relationship between the measures in the normal world and those in the shrunken world.

$\frac{48}{9} = \frac{x}{50}$ *4 feet = 48 inches*

$(48)(50) = 9x$ *Find the cross products.*

$2400 = 9x$

$266\frac{2}{3} = x$ *Divide each side by 9.*

Note that the numerators are measures associated with the house and the denominators are measures associated with Alice.

The house would have been $266\frac{2}{3}$ inches or about 22 feet high in a normal world.

Ratios can also be used to compare three or more numbers. The expression $a{:}b{:}c$ means that the ratio of the first two numbers is $a{:}b$, the ratio of the last two numbers is $b{:}c$, and the ratio of the first and last numbers is $a{:}c$.

Example 4 **In a triangle, the ratio of the measures of three sides is 8:7:5 and its perimeter is 240 centimeters. Find the measure of each side of the triangle.**

Let $8x$, $7x$, and $5x$ represent the measures of the sides of the triangle. The perimeter of a triangle is the sum of the measures of its sides. Write an equation to represent the perimeter.

$8x + 7x + 5x = 240$

$20x = 240$

$x = 12$ *Divide each side by 20.*

Since $x = 12$, the measures of the sides of the triangle are 8(12) or 96 centimeters, 7(12) or 84 centimeters, and 5(12) or 60 centimeters.

CHECK FOR UNDERSTANDING

Communicating Mathematics

Study the lesson. Then complete the following.

1. In 1994, the median weekly earnings for men was \$522 and for women was \$399. The ratio of the weekly earnings of men to women was $\frac{522}{399} \approx \frac{1.31}{1}$.
   a. **Describe** what the ratio $\frac{1.31}{1}$ represents.
   b. **Identify** the proportion and explain why it is a proportion.
2. **Explain** how you would solve the proportion $\frac{22}{30} = \frac{14}{x}$.
3. **Write** possible proportions if the means of a proportion are 8 and 10, and the extremes are 5 and 16.
4. **Determine** which proportions below are equivalent. Explain your reasoning.
   a. $\frac{7}{8} = \frac{x}{y}$ b. $\frac{y}{x} = \frac{8}{7}$ c. $\frac{y}{7} = \frac{x}{8}$ d. $\frac{7}{x} = \frac{8}{y}$

5. **Assess Yourself** Skim *Alice's Adventures in Wonderland* to find other references to ratio. Use the information in the story to make up several problems about ratio that you think Alice might have used to help her understand her surroundings.

Guided Practice

Express each ratio as a fraction in simplest form.

6. 2 inches on a map represent 150 miles. Find a ratio involving 1 inch.
7. 340 South African rands is equivalent to 18 American dollars. Find the ratio of rands to dollars.

Solve each proportion by using cross products.

8. $\frac{x}{5} = \frac{11}{35}$ 9. $\frac{13}{49} = \frac{26}{7x}$ 10. $\frac{x-2}{x} = \frac{3}{8}$

Use the number line at the right to determine if the given ratios are equal.

A B C D E F G H I
0 20 40 60 80

11. $\frac{CD}{CE} = \frac{EF}{EG}$ 12. $\frac{DE}{CD} = \frac{GI}{EG}$ 13. $\frac{CE}{FG} = \frac{EH}{GH}$

14. In $\triangle ALD$, $DE = ET = TL$.
    a. Find the ratio of DE to EL.
    b. If $\triangle ALD$ is an isosceles triangle with $\overline{AD} \cong \overline{AL}$, find the ratio of $m\angle ADL$ to $m\angle ALD$.

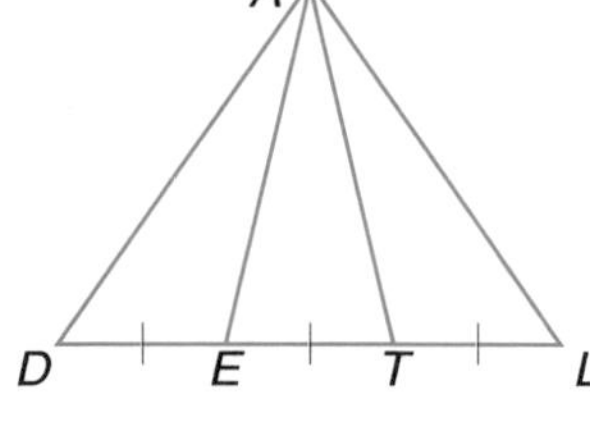

15. The perimeter of a triangle is 72 inches, and the ratio of the measures of the sides is 3:4:5. Find the measure of the sides.

James Whistler

16. **Art** Refer to the application at the beginning of the lesson. In the sketch, the height of Whistler's mother was 10 inches and the width of her gown was $8\frac{1}{3}$ inches. In the actual painting, the height of his mother was 50 inches. Find the width of his mother's gown in the actual painting. Round your answer to the nearest inch.

EXERCISES

Practice **Find each ratio and express it as a fraction in simplest form.**

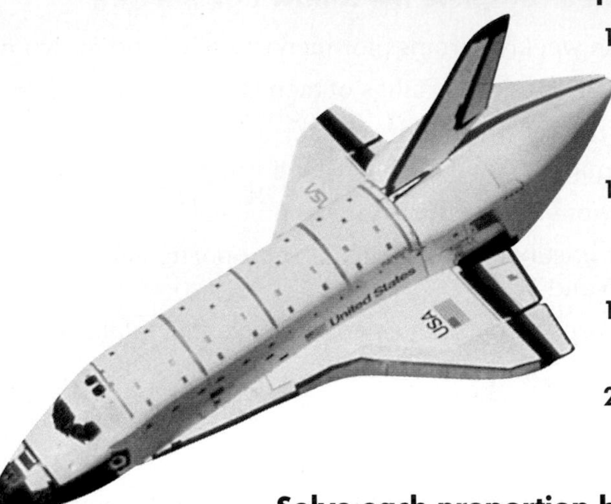

17. By 2010, it may be possible for people to travel in space, at a cost of about \$150,000 for a 150-pound person. Find the ratio of cost to the number of pounds.

18. By 2000, for every 2 new U.S. male workers, there will be 3 new U.S. female workers. Find the ratio of male workers to female workers.

19. A designated hitter made 8 hits in 10 games. Find the ratio of hits to games.

20. There are 76 boys and 89 girls in the sophomore class. Find the ratio of boys to girls.

Solve each proportion by using cross products.

21. $\frac{a}{5.18} = \frac{1}{4}$
22. $\frac{5}{n+3} = \frac{7}{4}$
23. $\frac{7}{11} = \frac{11}{x}$
24. $\frac{3x}{23} = \frac{48}{92}$
25. $\frac{a+1}{a-1} = \frac{5}{6}$
26. $\frac{2}{3x+1} = \frac{1}{x}$

27. A cable that is 42 feet long is divided into lengths in the ratio of 3:4. What are the two lengths into which the cable is divided?

28. In $\triangle ACE$, $\frac{AB}{BC} = \frac{CD}{DE} = \frac{FE}{AF}$. If $AB = 4$, $DE = 9$, $FE = 6$, and $BC = 12$, find the lengths of each side of $\triangle ACE$.

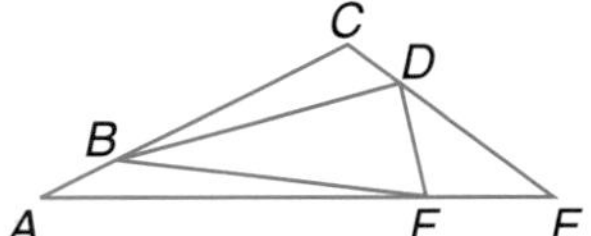

Corresponding sides of polygon *ABCD* are proportional to the sides of polygon *EFGH*.

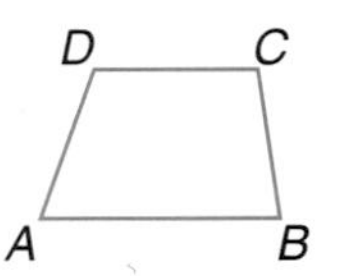

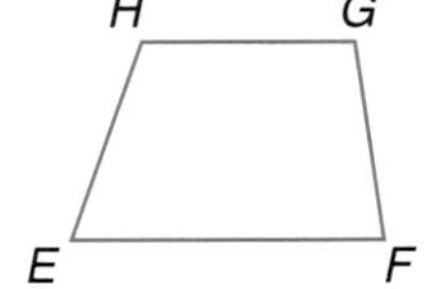

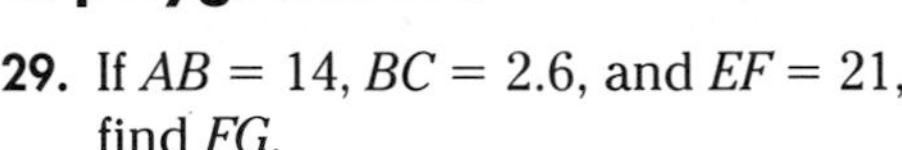

29. If $AB = 14$, $BC = 2.6$, and $EF = 21$, find FG.

30. If $GH = 40$, $FG = 32$, and $DC = 25$, find CB.

31. If $AD = \frac{2}{3}$, $BC = \frac{3}{4}$, and $FG = \frac{1}{2}$, find EH.

32. In the figure at the right, $\frac{AB}{BC} = \frac{AD}{DE}$.

Use proportions to complete the table.

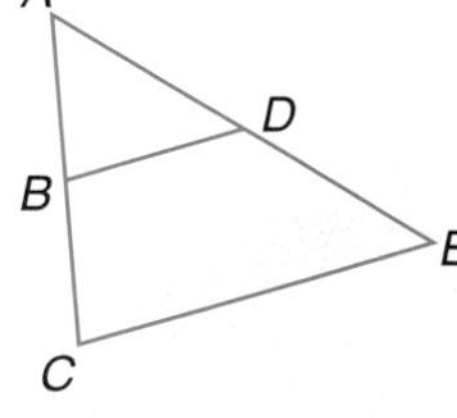

AB	BC	AD	DE	AC	AE
5	8		10		
	14.5		23.2		32
		12		16	20
		17	17	33	

33. The ratio of the measures of three sides of a triangle is 4:6:9, and its perimeter is 190 inches. Find the measure of each side of the triangle.

34. In a triangle, the ratio of the measures of three angles is 2:5:8. Find the measure of each angle in the triangle.

Proportions can be used to find the solutions to percent problems. For example, to find 23% of 15, you can solve the proportion $\frac{23}{100} = \frac{x}{15}$. Use a proportion to solve each problem.

35. Find 32% of 156.

36. Find 175% of 42.

37. What percent of 56 is 14?

38. 32 is 60% of what?

Use cross products to show each of the following is true. Assume that *b* and *d* are not zero.

39. $\frac{a}{b} = \frac{c}{d}$ if $\frac{a+b}{b} = \frac{c+d}{d}$.

40. $\frac{a}{b} = \frac{c}{d}$ if $\frac{a-b}{b} = \frac{c-d}{d}$.

An expression of the form $a = b = c$ means $a = b$, $b = c$, and $a = c$. Solve for *x* and *y*.

41. $\frac{x}{20} = \frac{6}{5} = \frac{y}{10}$

42. $\frac{4}{3} = \frac{x+1}{x+3} = \frac{y+1}{y}$

43. The ratio of the measures of two angles in an isosceles right triangle is 1:2. Explain why this is true.

44. Sandra reduced a rectangle that is 21.3 centimeters by 27.5 centimeters so that it would fit in a 10-centimeter by 10-centimeter area.
    a. Use a proportion to find the maximum dimensions of the reduced rectangle.
    b. What is the percent of reduction of the length?

Critical Thinking

45. **Gas Mileage** At each gas stop on a recent trip, Keshia recorded the distance she traveled since the last stop and the amount of gas she purchased for her midsize car.

Stops	Distance (mi)	Gas (gal)
1	351	13
2	275	11
3	362.5	12.5
4	372	12
5	260	12.8
6	294.4	10

    a. Find the ratios of the distances she drove to the gas she used for each stop.
    b. Can you use the results of part a to find the average number of miles per gallon for her car? Why or why not?

Applications and Problem Solving

CAREER CHOICES

A **meterorologist** studies the atmosphere, including its physical characteristics, and its effects on the environment. A bachelor's degree in meteorology is required, but advanced degrees are often advantageous.

For more information, contact:

American Meteorological Society
45 Beacon Street
Boston, MA 02108-3693

46. **Meteorology** Some people are concerned about the weather when they choose a place to live. The table at the right contains the average number of days per year for precipitation (either rain or snow) and the average number of days that are clear.

City	Days with Precipitation	Clear Days
Albuquerque, NM	59	172
Boston, MA	128	99
Buffalo, NY	168	55
Fresno, CA	44	200
Lubbock, TX	60	164
Montgomery, AL	109	107

Source: *Places Rated Almanac*

    a. Find the ratio of the number of days with precipitation to the number of clear days for each city. Express each ratio as a decimal rounded to the nearest hundredth.
    b. Choose one of the ratios and explain what it means.
    c. Use the information in the table to rank the cities according to precipitation.

47. **Business** Suppose you were the field agent for a company that operates movie theaters around the United States and you were thinking about building a theater in one of the six metropolitan areas listed below. The number of movie screens and the size of the population for each area in a recent year is given in the table.

Metropolitan Area	Population	Number of Movie Screens
Ann Arbor, MI	497,197	45
Kansas City, MO	1,629,241	150
New Haven, CT	538,643	28
Santa Fe, NM	129,841	10
Provo, UT	299,084	22

Source: *Places Rated Almanac*

a. Find the ratio of people per screen for each metropolitan area. Round your answer to next whole number.

b. How could you use this information to help you decide where to build a theater?

48. **Cartography** The scale on a map indicates that 1.5 centimeters represents 200 miles. If the distance on the map between Norfolk, Virginia, and Atlanta, Georgia, measures 2.4 centimeters, how many miles apart are the cities?

Data Update For the latest information on the unemployment rate, visit: **www.geometry.glencoe.com**

49. **Food** There were approximately 255,082,000 people in the United States in a recent year. According to figures from the United States Census, they consumed about 4,183,344,800 pounds of ice cream that year.

a. If there were 406,000 people in the city of Des Moines, Iowa, about how much ice cream might they have been expected to consume?

b. What was the approximate consumption of ice cream per person?

50. **Geography** The area of Texas is 262,017 square miles. The area of Alaska is 570,017 square miles. Find the ratio of Texas area to Alaska area.

51. **Money** The graphic shows how much it costs to make different types of currency.

a. Make a table showing the cost of making each type of currency in cents and the value of each type of currency. Then find the ratio of the cost to the value for each type of currency.

b. Are the ratios equal? Explain.

Money in the Making!

Dollar: 3 cents
Half-dollar: 7.8 cents
Dime: 1.7 cents
Penny: 0.8 cent
Quarter: 3.7 cents
Nickel: 2.9 cents

Source: Treasury Department

c. How much would it cost to make a \$5 bill if the cost were proportional to that of making a nickel? to that of making a quarter?

52. **Economics** Employees with white-collar jobs are usually salaried and have duties that do not require them to wear work clothes or protective clothing, whereas employees with blue-collar jobs are usually paid hourly and require special clothing. As the economy changes, the number of jobs in these areas shifts. The ratio of projected blue-collar jobs to white-collar jobs for a future year is given below. Explain what the ratio indicates for each city.

a. Seattle, WA: 1.75:1

b. Pittsburgh, PA: 20.45:-1

c. New York, NY: -6:-100

Mixed Review

53. Name a quadrilateral that has exactly one pair of parallel sides. (Lesson 6–5)

54. If $JKLM$ is a parallelogram and $m\angle J = 72$, find the measures of $\angle K$, $\angle L$, and $\angle M$. (Lesson 6–1)

55. If two sides of a triangle have measures of 5 and 8, what must be true about the measure of the third side? (Lesson 5–5)

56. **ACT Practice** For the band concert, student tickets cost $2.00 and adult tickets cost $5.00. A total of 200 tickets were sold. If the total sales must be more than $500.00, then what is the minimum number of adult tickets that must be sold?

A 30 **B** 33 **C** 34 **D** 40 **E** 100

57. In a right triangle, one angle measures 38°. Find the measures of the other two angles. (Lesson 4–2)

58. Determine whether the statement *A right triangle is never scalene* is true or false. Explain. (Lesson 4–1)

59. Find the values of a and b in the figure at the right. (Lesson 3–2)

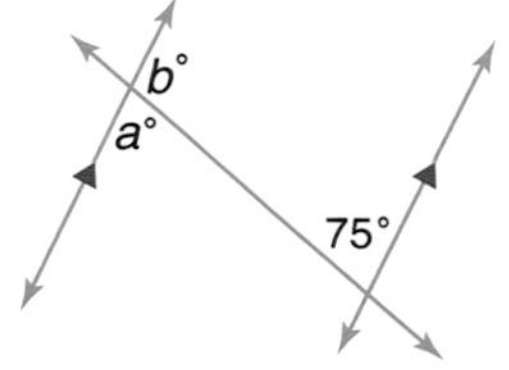

60. **Geology** Determine if a valid conclusion can be reached from the two true statements using the Law of Detachment or the Law of Syllogism. If a valid conclusion is possible, state it and the law of logic that is used. (Lesson 2–3)

(1) If a mineral sample is a sample of quartz, then the sample has a hardness factor of 7.

(2) If a mineral sample has a hardness factor greater than 5, then it can scratch glass.

61. Given that $AP = PB$, is the conjecture that P is the midpoint of $\overline{AB}$ *true* or *false*? Explain. (Lesson 2–1)

For **Extra Practice**, see page 776.

Algebra

62. Solve $2n + 6 < n - 1$.

63. Solve $x^2 + 6x - 16 = 0$.

Refer to the Investigation on pages 286–287.

1 Use grid paper to draw a plan for the park. Begin by drawing the boundary lines of the park.

2 Refer to your list of features. Determine the best location for each feature and add it to the plan.

3 As you add features to the plan, compile a list of costs. Make sure that your project will be in line with the budget. Don't forget to include the costs of roadways in your budget.

Add the results of your work to your Investigation Folder.

7–2 Exploring Similar Polygons

What YOU'LL LEARN

- To identify similar figures, and
- to solve problems involving similar figures.

Why IT'S IMPORTANT

You can use similar polygons to solve problems involving cartography, gardening, and construction work.

Art

In 1928, Charles Demuth used properties of similar polygons to create proportional numeral 5s in his painting *I Saw the Figure 5 in Gold* shown at the right. The numerals are the same shape, but different sizes.

When figures have the same shape but are different sizes, they are called **similar figures**.

Definition of Similar Polygons	Two polygons are similar if and only if their corresponding angles are congruent and the measures of their corresponding sides are proportional.

The definition states that in order for two figures to be similar, the corresponding angles of the two figures must be congruent, and the measures of the corresponding sides must be proportional. The quadrilaterals below are similar.

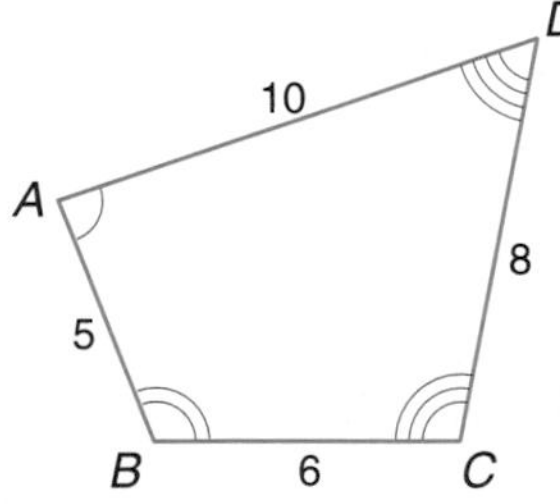

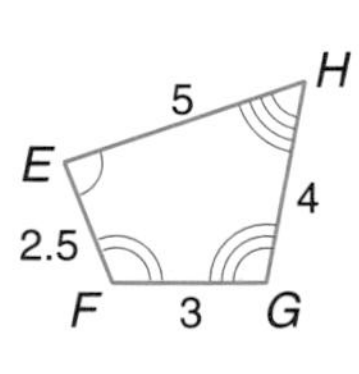

The symbol $\sim$ means *is similar to*. We write quadrilateral $ABCD \sim$ quadrilateral $EFGH$. Just as in congruence, the order of the letters indicates the vertices that correspond. We can make the following statements about quadrilaterals $ABCD$ and $EFGH$.

$$\angle A \cong \angle E \qquad \angle B \cong \angle F \qquad \angle C \cong \angle G \qquad \angle D \cong \angle H$$

$$\frac{AB}{EF} = \frac{BC}{FG} = \frac{CD}{GH} = \frac{DA}{HE} = \frac{2}{1}$$

The ratio of the lengths of two corresponding sides of two similar polygons is called the **scale factor**. The scale factor of quadrilateral $ABCD$ to quadrilateral $EFGH$ is 2. The scale factor of quadrilateral $EFGH$ to quadrilateral $ABCD$ is $\frac{1}{2}$.

You can use an overhead projector and transparencies to investigate characteristics of similar polygons.

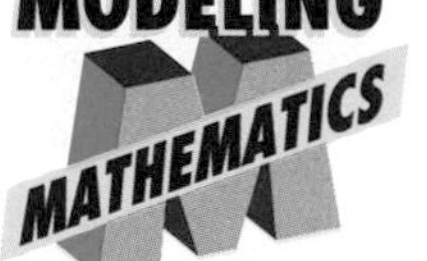

Similar Polygons

Materials: 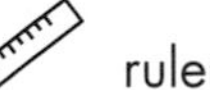 ruler  compass ☐ blank transparency  overhead projector

You can use an overhead projector to create similar polygons.

- Use a ruler and compass to draw a triangle with sides measuring 4 centimeters, 7 centimeters, and 9 centimeters on a blank transparency.
- Project the triangle on a screen by using an overhead projector.
- Measure the sides and the angles of the triangle projected on the screen.

Your Turn

a. What is the ratio between the 4-centimeter side and its projection? between the 7-centimeter side and its projection? between the 9-centimeter side and its projection?

b. What conclusion can you make?

c. What else do you have to check to verify that the original triangle you placed on the screen and its projection are similar?

The properties of proportions can be used to solve problems involving similar polygons.

Example  **Polygon *RSTUV* is similar to polygon *ABCDE*.**

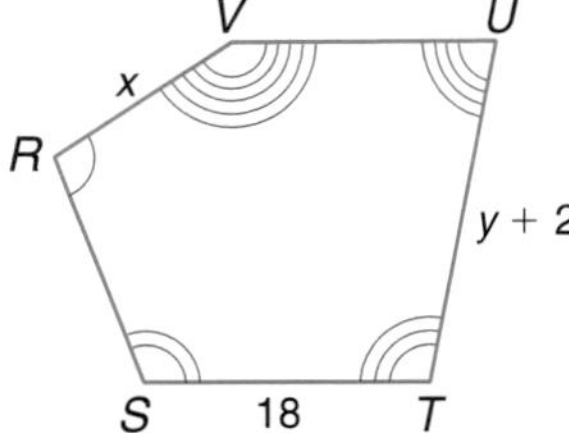

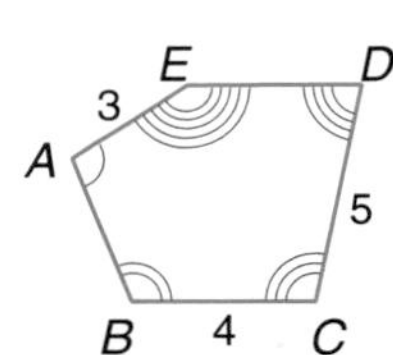

a. Find the scale factor of polygon *RSTUV* to polygon *ABCDE*.

The scale factor is the ratio of the lengths of two corresponding sides.

$$\text{scale factor} = \frac{ST}{BC}$$
$$= \frac{18}{4} \text{ or } \frac{9}{2}$$

b. Find the values of *x* and *y*.

Because the polygons are similar, the corresponding sides are proportional. Thus, we can write proportions to find the values of x and y.

Write a proportion that involves numbers and the variable x.

$\frac{ST}{BC} = \frac{VR}{EA}$

$\frac{18}{4} = \frac{x}{3}$ *$ST = 18, BC = 4, VR = x, EA = 3$*

$54 = 4x$ *Cross products*

$13.5 = x$

Write a proportion that involves numbers and the variable y.

$\frac{ST}{BC} = \frac{UT}{DC}$

$\frac{18}{4} = \frac{y+2}{5}$ *$ST = 18, BC = 4, UT = y + 2, DC = 5$*

$90 = 4y + 8$ *Cross products*

$82 = 4y$

$20.5 = y$

(continued on the next page)

Check: $\frac{VR}{EA} = \frac{x}{3}$ $\qquad$ $\frac{UT}{DC} = \frac{y+2}{5}$

$= \frac{13.5}{3}$ $\qquad$ $= \frac{20.5 + 2}{5}$

$= 4.5 \text{ or } \frac{9}{2}$ $\qquad$ $= 4.5 \text{ or } \frac{9}{2}$

Since both ratios equal the value of the scale factor, the answers are correct.

Because longitude and latitude lines are often used to determine boundaries, many states are shaped like polygons.

Example 2

Cartography

Delaware is a very small state, shaped similarly to a five-sided polygon. The scale on a map of Delaware is approximately 2 centimeters = 50 miles. The distance on the map across Delaware from east to west is 1.4 centimeters. How long would it take to drive across Delaware if you drove at approximately 50 miles per hour?

Create a proportion relating the measurements to the scale to find the distance in miles.

$$\frac{2.0}{50} = \frac{1.4}{x}$$ *centimeters → ← centimeters; miles → ← miles*

$2x = 70$ *Cross products*

$x = 35$ *Divide each side by 2.*

The distance across Delaware is approximately 35 miles. Use the formula $d = rt$ to find the time.

$d = rt$

$35 = 50t$ *d = 35 miles, r = 50 mph*

$\frac{35}{50} = t$

$0.7 = t$

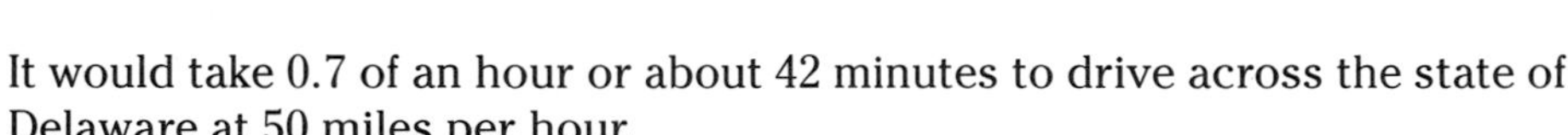

It would take 0.7 of an hour or about 42 minutes to drive across the state of Delaware at 50 miles per hour.

Some transformations will produce similar figures. A **dilation** is a transformation that reduces or enlarges a figure.

Example 3

Algebra

Triangle *ABC* has vertices *A*(0, 0), *B*(8, 0), and *C*(2, 7). If the coordinates of each vertex are multipled by 2, will the new figure be similar to the original?

Explore Find the coordinates of the new triangle. Graph each triangle.

Coordinates	New Coordinates
A(0, 0)	(2(0), 2(0)) or (0, 0)
B(8, 0)	(2(8), 2(0)) or (16, 0)
C(2, 7)	(2(2), 2(7)) or (4, 14)

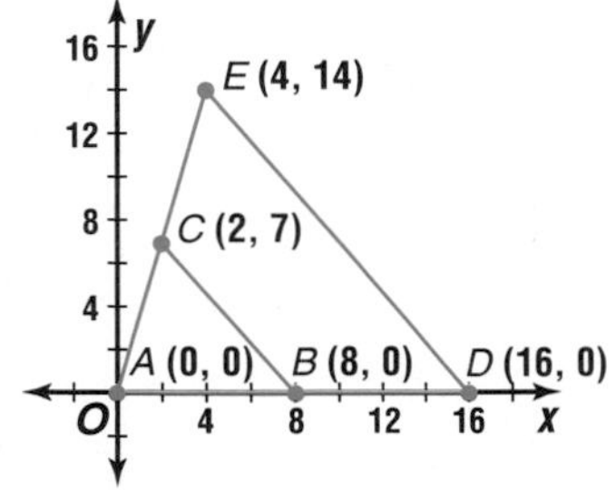

$\triangle ADE$ will be similar to $\triangle ABC$ if the corresponding angles are congruent and the measures of the corresponding sides are proportional.

Plan To determine if the angles are congruent, we can investigate the slopes to see if $\overline{CB} \parallel \overline{ED}$. Then we can use the properties of parallel lines to determine whether the sides are proportional and the distance formula to find the lengths of each side.

Solve The formula for slope is $\frac{y_2 - y_1}{x_2 - x_1}$.

$$\text{slope of } \overline{CB} = \frac{7-0}{2-8} = -\frac{7}{6}$$

$$\text{slope of } \overline{ED} = \frac{14-0}{4-16} = \frac{14}{-12} \text{ or } -\frac{7}{6}$$

Because the slopes are equal and the lines do not contain a common point, $\overline{CB} \parallel \overline{ED}$. If two parallel lines are cut by a transversal, the corresponding angles are congruent, which makes $\angle ABC \cong \angle ADE$ and $\angle ACB \cong \angle AED$. $\angle A \cong \angle A$ because congruence of angles is reflexive. Thus, the corresponding angles in $\triangle ABC$ are congruent to the corresponding angles in $\triangle ADE$.

Now use the distance formula, $d = \sqrt{(x_2 - x_1)^2 + (y_2 - y_1)^2}$, to find the lengths of the sides of each triangle.

In $\triangle ABC$	*In* $\triangle ADE$
$AB = \sqrt{(8-0)^2 + (0-0)^2}$	$AD = \sqrt{(16-0)^2 + (0-0)^2}$
$= \sqrt{64}$ or 8	$= \sqrt{256}$ or 16
$BC = \sqrt{(8-2)^2 + (0-7)^2}$	$DE = \sqrt{(16-4)^2 + (0-14)^2}$
$= \sqrt{85}$	$= \sqrt{340}$ or $2\sqrt{85}$
$AC = \sqrt{(2-0)^2 + (7-0)^2}$	$AE = \sqrt{(4-0)^2 + (14-0)^2}$
$= \sqrt{53}$	$= \sqrt{212}$ or $2\sqrt{53}$

$$\frac{AB}{AD} = \frac{8}{16} = \frac{1}{2} \qquad \frac{BC}{DE} = \frac{\sqrt{85}}{2\sqrt{85}} = \frac{1}{2} \qquad \frac{AC}{AE} = \frac{\sqrt{53}}{2\sqrt{53}} = \frac{1}{2}$$

Therefore, $\frac{AB}{AD} = \frac{BC}{DE} = \frac{AC}{AE}$.

Examine Because the measures of the corresponding sides are proportional and the corresponding angles are congruent, $\triangle ABC$ is similar to $\triangle ADE$. If the coordinates of each vertex are multiplied by 2, the new figure is similar to the original figure.

CHECK FOR UNDERSTANDING

Communicating Mathematics

Study the lesson. Then complete the following.

1. a. **Explain** whether two figures that are congruent are also similar.
   b. **Explain** whether two figures that are similar are also congruent.
2. **You Decide** Yvonne calculated the scale factor between two similar polygons to be $\frac{5}{3}$. Trina found the scale factor to be $\frac{3}{5}$ for the same polygon. Explain how this could happen and who is right.

3. **Draw** the state of Delaware using a scale of 2 centimeters = 30 miles. Is your drawing similar to the one in Example 2? Explain how you know.

4. Every segment in quadrilateral *MNOP* is twice as long as the corresponding segment in quadrilateral *ABCD*. List additional information you would need to determine whether the figures are similar.

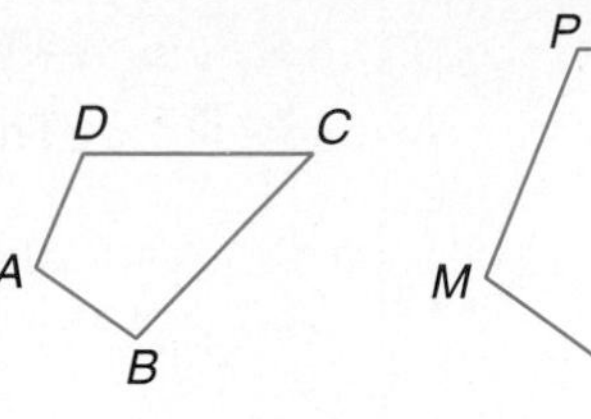

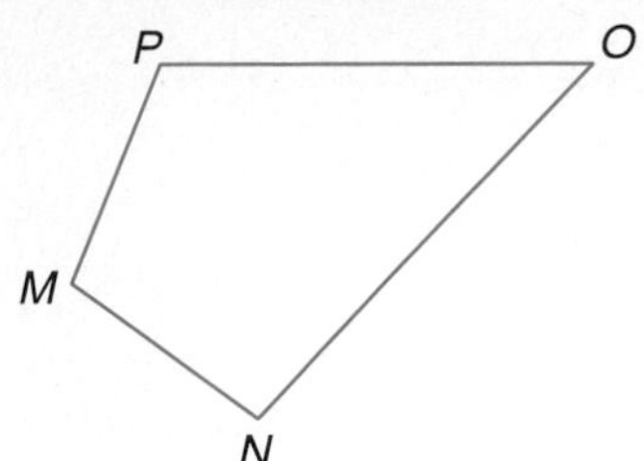

5. A parallelogram with sides 4 centimeters and 8 centimeters long is projected onto a screen. The ratio between a 4-centimeter side and its projection is 2:17.
   a. What is the measurement of the longer side of the parallelogram as it is projected on the screen?
   b. If you move the projector, the measurement of the shorter side of the parallelogram becomes 30 centimeters. What is the new scale factor?

Guided Practice

6. Determine whether the figures below are similar. Justify your answer.

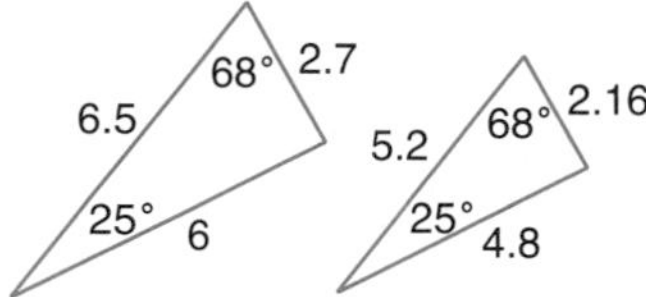

7. The polygons below are similar. Find the values of x and y.

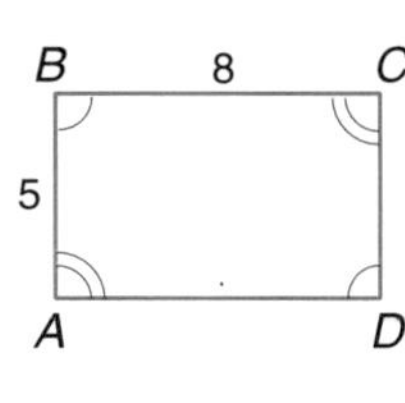

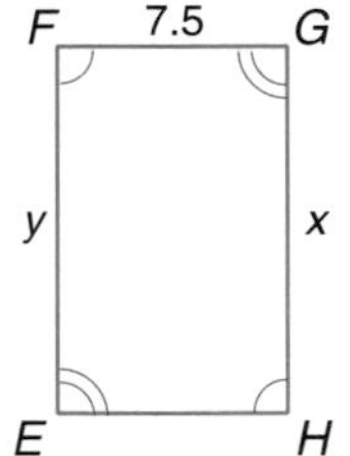

For each statement, write A if the statement is *always* true, S if the statement is *sometimes* true, and N if the statement is *never* true. Draw figures to support your answer.

8. Two right triangles are similar.
9. Two congruent quadrilaterals are similar.
10. Quadrilateral *RSTV* is similar to quadrilateral *LMNO*. The sides of *RSTV* are 6, 10, 12, and 14 inches long. The shortest side of *LMNO* is 9 inches long.
    a. What is the scale factor of *RSTV* to *LMNO*?
    b. Find the two longest sides of *LMNO*.
    c. Find the perimeter of *LMNO*.
    d. What is the ratio of the perimeters of *RSTV* and *LMNO*?

INTEGRATION
Algebra

11. $\triangle ABC$ has vertices $A(0, 0)$, $B(3, 0)$ and $C(0, 4)$. If the coordinates of each vertex are multiplied by 3, will the new triangle be similar to $\triangle ABC$? Why or why not?

12. **Photography** A picture is enlarged by a scale factor of $\frac{5}{4}$ and then enlarged again by the same factor.
    a. If the original picture was 2.5 inches by 4 inches, how large was it after both enlargements?
    b. Write an equation describing the enlargement process.
    c. By what scale factor was the original picture enlarged?

EXERCISES

Practice

Determine whether each pair of figures is similar. Justify your answer.

13.

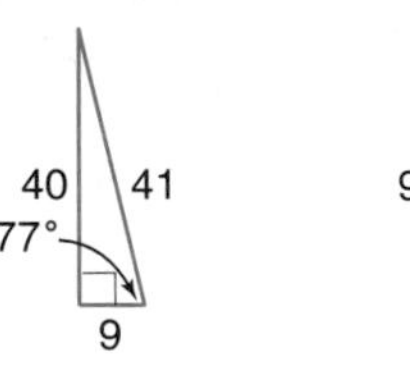

14.

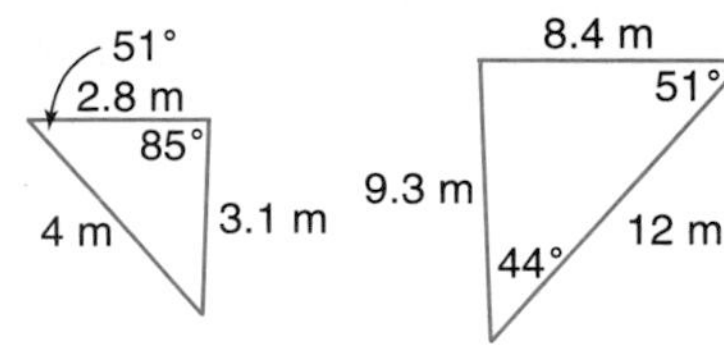

Each pair of polygons is similar. Find the values of *x* and *y*.

15.

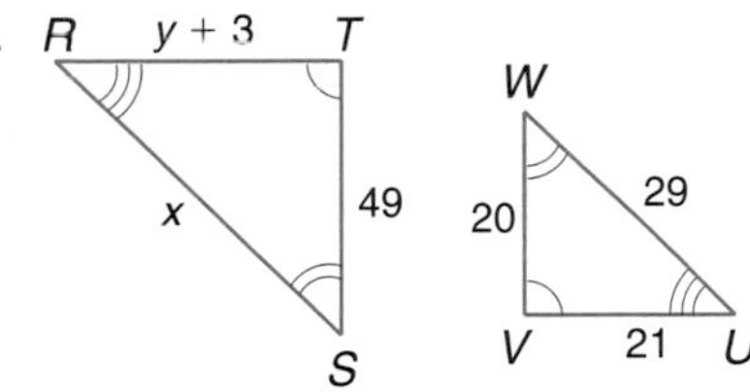

16.

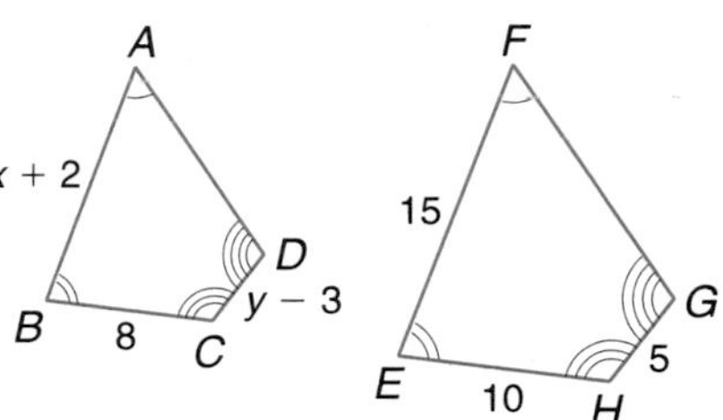

17.

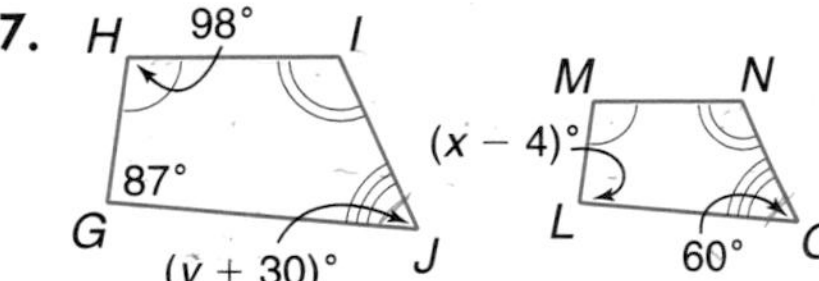

18.

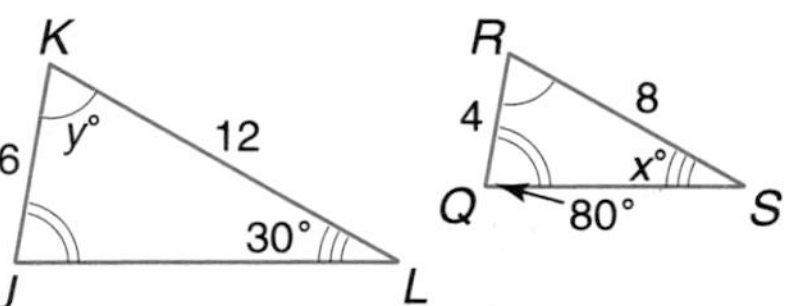

For each statement, write *A* if the statement is *always* true, *S* if the statement is *sometimes* true, and *N* if the statement is *never* true. Draw figures to support your answer.

19. Two rectangles are similar.

20. Two squares are similar.

21. A triangle is similar to a quadrilateral.

22. Two isosceles triangles are similar.

23. Two rhombi are similar.

24. Two obtuse triangles are similar.

25. Two equilateral triangles are similar.

Triangle *RST* is similar to triangle *EGF*.

26. Find the length of the shortest side of $\triangle EFG$.

27. What is the ratio of *RS* to *EG*?

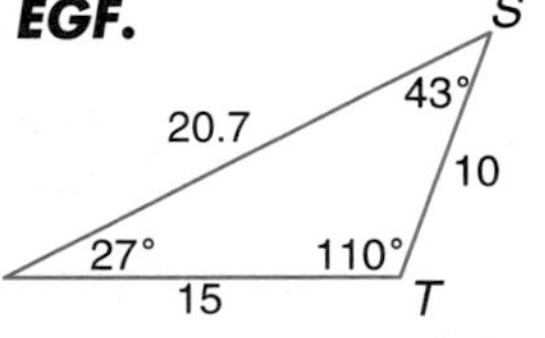

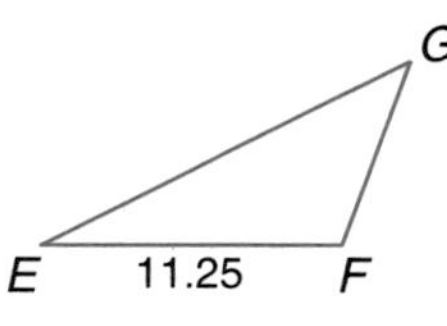

Given trapezoid *ABCD* ~ trapezoid *AEFG*, $m\angle AGF = 108$, *GF* = 14, *AD* = 12, *DG* = 4.5, *EF* = 8, and *AB* = 26, find each of the following.

28. scale factor of *ABCD* to *AEFG*

29. a. *AG* **b.** *DC*

c. $m\angle ADC$ **d.** *BC*

30. a. perimeter of *ABCD*

b. perimeter of *AEFG*

c. ratio of the perimeters of *ABCD* and *AEFG*

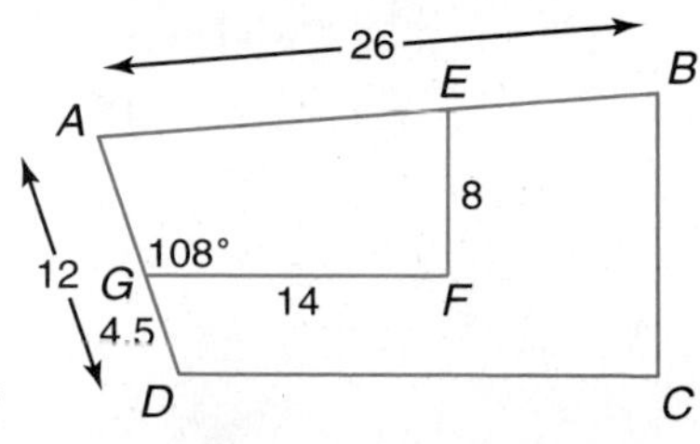

31. Determine which of the following right triangles are similar. Justify your answer.

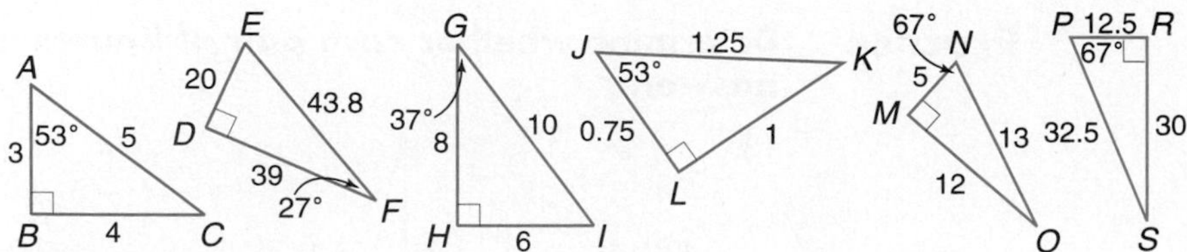

Algebra

32. A triangle has vertices $E(-2, 4)$, $F(6, 4)$, and $H(3, 9)$.

a. If the coordinates of each vertex are increased by 2, describe the new figure. Is it similar to $\triangle EFH$?

b. If the coordinates of each vertex are multiplied by -3, describe the new figure. Is it similar to $\triangle EFH$?

Make a scale drawing of each using the given scale.

33. A basketball court is 84 feet by 50 feet. scale: $\frac{1}{4}$ in. = 4 ft

34. A soccer field is 91 meters by 46 meters. scale: 1 mm = 1 m

35. A tennis court is 36 feet by 78 feet. scale: $\frac{1}{8}$ in. = 1 ft

36. The figure at the right is a block building made out of cubes (each block has the same length, width, and height). The edge of each cube is 1 unit long.

a. Find the total area of the outside surfaces of the building, including the bottom, in square units.

b. Find the perimeter of the base of the building in units.

c. Find the volume of the building in cubes.

d. The builder decides to increase the length, width, and height by 2 times the original measures. What is the ratio of the surface area of the new building to the surface area of the smaller building?

e. What is the ratio of the perimeter of the base of the new building to the base of the smaller building?

f. What is the ratio of the volume of the new building to the volume of the smaller building?

g. Is the new building similar to the first one? Explain why or why not.

Plot the given points on graph paper. Draw *ABCD* and $\overline{MN}$. Find the coordinates for vertices *L* and *O* such that *ABCD* is similar to *NLOM*.

37. $A(2, 0)$, $B(4, 4)$, $C(0, 4)$, $D(-2, 0)$; $M(4, 0)$, $N(12, 0)$

38. $A(0, 0)$, $B(5, 6)$, $C(0, 8)$, $D(-4, 4)$; $M(-2, 14)$, $N(10, 2)$

Critical Thinking

39. Use what you know about slope and distance to show that the two triangles are similar. Triangle *ABC* has vertices $A(0, 0)$, $B(12, 0)$, and $C(6, 9)$. Triangle *DEF* has vertices $D(18, 0)$, $E(26, 0)$, and $F(22, 6)$.

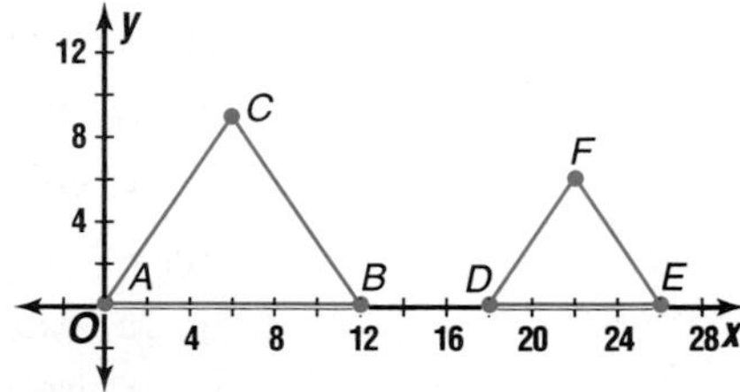

40. Two rectangular solids are similar with a ratio of 4:1 between the corresponding sides.
   a. Suppose you use 16 straws to build the smaller rectangular solid. How many straws would you need to build the larger solid? Explain.
   b. If you were to paint each solid, would the larger solid take four times as much paint? Explain.
   c. Would the larger solid hold exactly four times as much water as the smaller solid? Explain.

Applications and Problem Solving

41. **Construction** A floor plan is given for the first floor of a new house. One inch represents 18 feet. Use the information in the plan to find the dimensions.
   a. living room
   b. deck

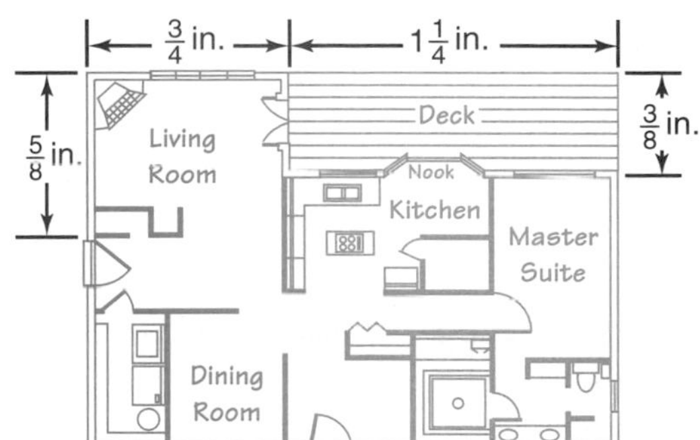

42. **Gardening** The graph at the right compares the average size of a home vegetable garden in 1981 to the average size in 1991. Are these quadrilaterals similar? Explain.

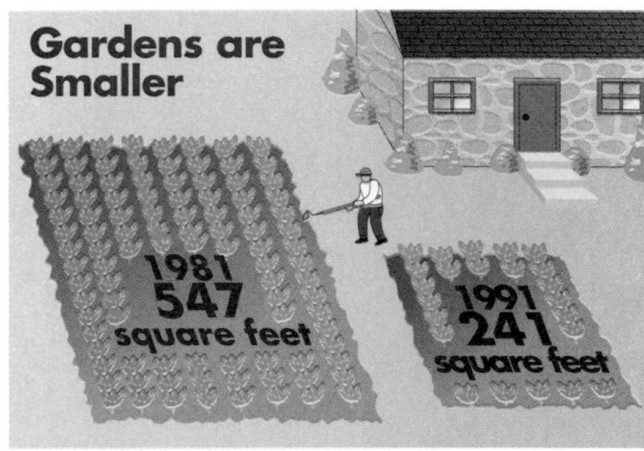

Source: National Garden Association/Gallup Inc.

Mixed Review

43. Solve $\frac{m+3}{12} = \frac{5}{4}$ by using cross products. (Lesson 7–1)

44. Classify the statement *The diagonals of a parallelogram are congruent* as always, sometimes, or never true. (Lesson 6–3)

45. **SAT Practice** A coin was flipped 24 times and came up heads 14 times and tails 10 times. If the first and the last flips were both heads, what is the greatest number of consecutive heads that could have occurred?
   A 7 **B** 9 **C** 10 **D** 13 **E** 14

46. In $\triangle PQR$, $PQ > PR > QR$. List the angles in $\triangle PQR$ in order of measure from greatest to least. (Lesson 5–5)

47. **Work Backward** A number is divided by 2, decreased by 17, multiplied by 3, and subtracted from 400. If the result is 187, what was the original number? (Lesson 5–3)

48. True or false: *AAS is a triangle congruence test.* (Lesson 4–5)

49. Suppose two parallel lines are cut by a transversal and $\angle 1$ and $\angle 2$ are alternate exterior angles. If $m\angle 2 = 84$, find $m\angle 1$. (Lesson 3–2)

50. Write the negation of the statement *Springer spaniels are not dogs.* (Lesson 2–2)

For **Extra Practice**, see page 776.

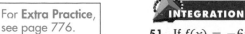

Algebra

51. If $f(x) = -6x - 5$, find $f(2)$.

52. If $f(x) = 5x + 2$, find $f(-1)$.

7-3 Identifying Similar Triangles

What YOU'LL LEARN
- To identify similar triangles, and
- to use similar triangles to solve problems.

Why IT'S IMPORTANT

Identifying similar triangles is the first step in solving problems that apply similar triangles.

History

The Greek mathematician Thales was the first to measure the height of a pyramid by using geometry. He showed that the ratio of a pyramid to a staff was equal to the ratio of one shadow to the other.

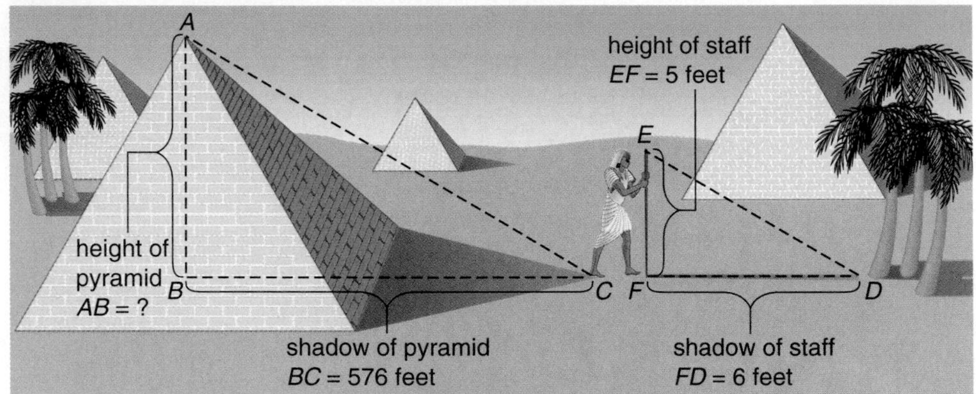

Using the information above, can you find the height of the pyramid?
You will be asked to solve this problem in Exercise 12.

Similar triangles can help you solve problems like the one above. How do you know whether two triangles are similar? In Chapter 4, you learned several tests to determine whether two triangles are congruent. There are also tests to determine whether two triangles are similar.

MODELING MATHEMATICS

Similar Triangles

Materials: ruler  protractor

One way to determine whether triangles are similar is to see if the measures of corresponding sides are proportional.

- Draw $\triangle DEF$ with $m\angle D = 35$, $m\angle F = 80$, and $DF = 4$ centimeters.
- Draw $\triangle RST$ with $m\angle T = 35$, $m\angle S = 80$, and $ST = 7$ centimeters.
- Measure $\overline{EF}$, $\overline{ED}$, $\overline{RS}$, and $\overline{RT}$.
- Calculate the ratios $\frac{FD}{ST}$, $\frac{EF}{RS}$, and $\frac{ED}{RT}$.
- Record your findings in a chart like the one below.

	EF	ED	RS	RT	$\frac{FD}{ST}$	$\frac{EF}{RS}$	$\frac{ED}{RT}$
$\triangle DEF$, $\triangle RST$							

Your Turn

a. What can you conclude about all the ratios?

b. Draw two more triangles with the same angle measures and different side measures. Complete the chart for these two triangles.

c. Repeat the process with two more triangles.

d. Do all of the triangles appear to be similar?

The Modeling Mathematics activity leads to the following postulate.

Postulate 7–1
AA Similarity
(Angle-Angle)

If two angles of one triangle are congruent to two angles of another triangle, then the triangles are similar.

You can use AA Similarity to find missing measurements in triangles.

Example 1

Algebra

Given $\overline{AB} \parallel \overline{CD}$, $AB = 4$, $AE = 3x + 4$, $CD = 8$, and $ED = x + 12$, find AE and DE.

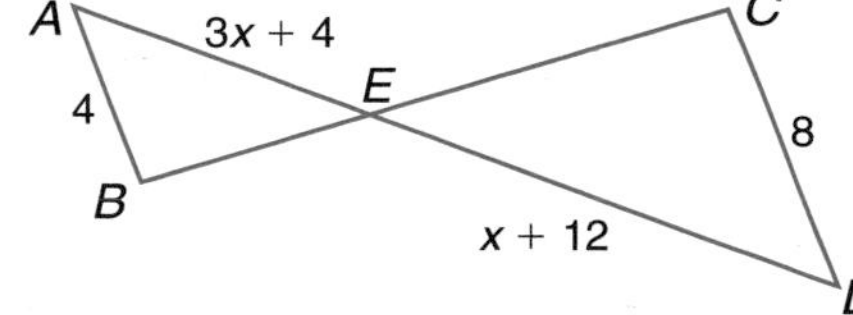

Since $\overline{AB} \parallel \overline{CD}$, $\angle BAE \cong \angle CDE$ and $\angle ABE \cong \angle DCE$ because they are alternate interior angles. By AA Similarity, $\triangle ABE \sim \triangle DCE$. Using the definition of similar polygons, $\frac{AB}{DC} = \frac{AE}{DE}$.

$$\frac{AB}{DC} = \frac{AE}{DE}$$

$\frac{4}{8} = \frac{3x + 4}{x + 12}$ *Substitution Property (=)*

$4(x + 12) = 8(3x + 4)$ *Cross products*

$4x + 48 = 24x + 32$ *Distributive Property (=)*

$-20x = -16$ *Subtract 24x and 48 from each side.*

$x = \frac{16}{20}$ or $\frac{4}{5}$ *Divide each side by −20.*

Now find AE and ED.

$AE = 3x + 4$
$= 3\left(\frac{4}{5}\right) + 4$ or $6\frac{2}{5}$

$ED = x + 12$
$= \frac{4}{5} + 12$ or $12\frac{4}{5}$

It is also possible to prove triangles similar by testing the measures of corresponding sides for proportionality.

Theorem 7–1
SSS Similarity
(Side-Side-Side)

If the measures of the corresponding sides of two triangles are proportional, then the triangles are similar.

Proof of Theorem 7–1

Given: $\frac{PQ}{AB} = \frac{QR}{BC} = \frac{RP}{CA}$

Prove: $\triangle BAC \sim \triangle QPR$

Locate D on $\overline{AB}$ so that $\overline{DB} \cong \overline{PQ}$ and draw $\overline{DE}$ so that $\overline{DE} \parallel \overline{AC}$.

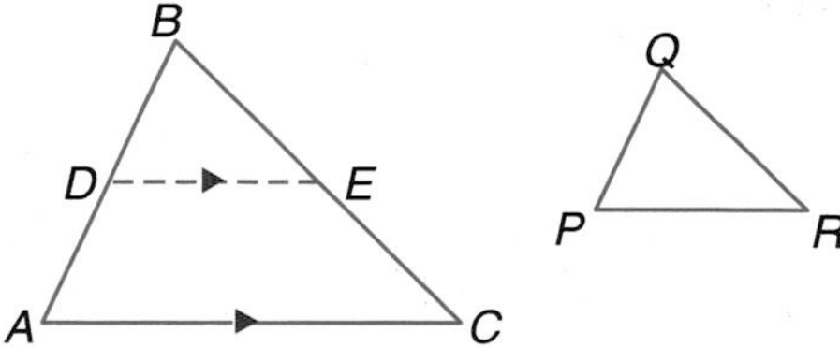

Paragraph Proof:

Since $\overline{DB} \cong \overline{PQ}$, the given proportion will become $\frac{DB}{AB} = \frac{QR}{BC} = \frac{RP}{CA}$. Since $\overline{DE} \parallel \overline{AC}$, $\angle BDE \cong \angle A$ and $\angle BED \cong \angle C$. By AA Similarity, $\triangle BDE \sim \triangle BAC$. By the definition of similar polygons, $\frac{DB}{AB} = \frac{BE}{BC} = \frac{ED}{CA}$. Using the two proportions and substitution, $\frac{QR}{BC} = \frac{BE}{BC}$ and $\frac{RP}{CA} = \frac{ED}{CA}$. This means that $QR = BE$ and $RP = ED$ or $\overline{QR} \cong \overline{BE}$ and $\overline{RP} \cong \overline{ED}$. With these congruences and $\overline{DB} \cong \overline{PQ}$, $\triangle BDE \cong \triangle QPR$ by SSS Postulate. By CPCTC, $\angle B \cong \angle Q$ and $\angle BDE \cong \angle P$. But $\angle BDE \cong \angle A$, so $\angle A \cong \angle P$. By AA Similarity, $\triangle BAC \sim \triangle QPR$.

The next theorem describes another test for similarity of triangles. *You will be asked to prove this theorem in Exercise 33.*

Theorem 7–2* *SAS Similarity* *(Side-Angle-Side)	**If the measures of two sides of a triangle are proportional to the measures of two corresponding sides of another triangle and the included angles are congruent, then the triangles are similar.**

You can use SAS Similarity to determine whether two triangles are similar.

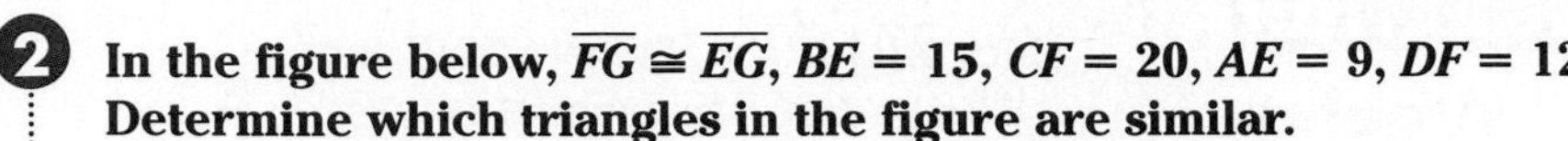

Example 2

In the figure below, $\overline{FG} \cong \overline{EG}$, $BE = 15$, $CF = 20$, $AE = 9$, $DF = 12$. Determine which triangles in the figure are similar.

$\overline{FG} \cong \overline{EG}$ implies $\angle GFE \cong \angle GEF$ because if two sides of a triangle are congruent, the angles opposite those sides are congruent. If the corresponding sides that include the angle are proportional, then the triangles are similar.

$\frac{AE}{DF} = \frac{9}{12}$ or $\frac{3}{4}$ and $\frac{BE}{CF} = \frac{15}{20}$ or $\frac{3}{4}$.

By the Transitive Property of Equality, $\frac{AE}{DF} = \frac{BE}{CF}$.

Thus, we have the measures of two sides of one triangle proportional to the measures of two sides of another and the included angles congruent. So by SAS Similarity, $\triangle ABE$ is similar to $\triangle DCF$.

Similar triangles can also be used to solve problems involving distance.

Example 3

Forestry

FYI

The giant sequoia trees in California reach heights of more than 100 meters.

To estimate the height of a tree, a Girl Scout sights the top of the tree in a mirror that is 34.5 meters from the tree. The mirror is on the ground and faces upward. The scout is 0.75 meters from the mirror, and the distance from her eyes to the ground is about 1.75 meters. How tall is the tree?

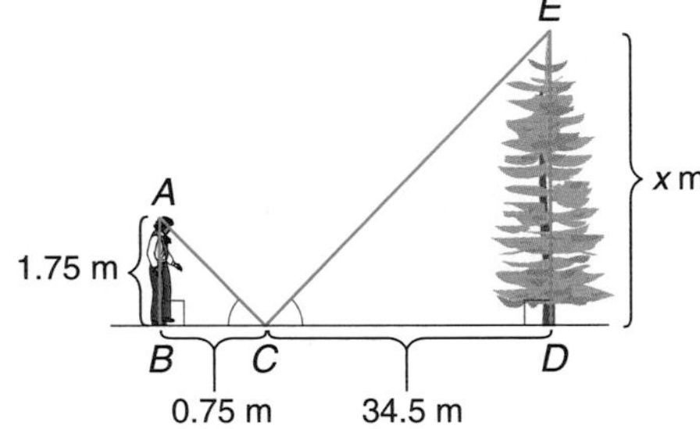

Let x represent the height of the tree. The angle between the ground and the line of sight is congruent to the angle between the ground and the line from the mirror to the top of the tree. If we can show $\triangle ABC$ and $\triangle EDC$ are similar, we can set up a proportion of corresponding sides and solve for x.

Assume $\angle ABC$ and $\angle EDC$ are right angles and, thus, are congruent. Since $\angle ACB \cong \angle ECD$, $\triangle ABC \sim \triangle EDC$ by AA Similarity. Therefore, $\frac{AB}{ED} = \frac{BC}{DC}$.

$\frac{1.75}{x} = \frac{0.75}{34.5}$	*Substitution Property (=)*
$60.375 = 0.75x$	*Cross products*
$80.5 = x$	*Divide each side by 0.75.*

The tree is 80.5 meters high.

Like congruence of triangles, similarity of triangles is reflexive, symmetric, and transitive.

reflexive $\triangle ABC \sim \triangle ABC$

symmetric If $\triangle ABC \sim \triangle DEF$, then $\triangle DEF \sim \triangle ABC$.

transitive If $\triangle ABC \sim \triangle DEF$ and $\triangle DEF \sim \triangle GHI$, then $\triangle ABC \sim \triangle GHI$.

Theorem 7-3	**Similarity of triangles is reflexive, symmetric, and transitive.**

You will be asked to prove this theorem in Exercise 35.

CHECK FOR UNDERSTANDING

Communicating Mathematics

Study the lesson. Then complete the following.

1. **Compare** the tests to prove triangles similar with the tests to prove triangles congruent. How are they alike and how are they different?

2. If you know two triangles are right triangles, how many of the acute angles must be congruent before you know the two triangles are similar? Explain your answer.

3. Use an example to explain what it means to say that similarity of triangles is reflexive, symmetric, and transitive.

4. **You Decide** Lacretia and Song were working with the similar triangles shown at the right. Lacretia used the proportion $\frac{r}{s} = \frac{k}{m}$ for her calculations, and Song used $\frac{r}{k} = \frac{s}{m}$ for hers. Who is right, and why?

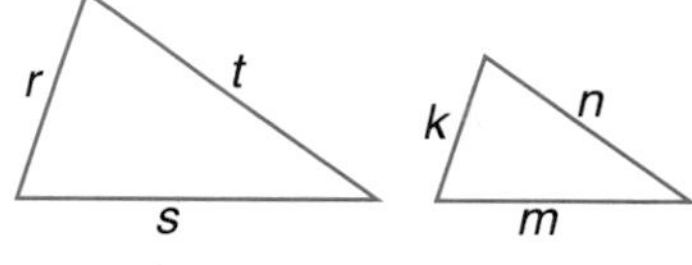

5. Draw $\triangle JKL$ with $m\angle J = 60$, $JK = 3$ centimeters, and $JL = 6$ centimeters. Draw $\triangle QRS$ with $m\angle Q = 60$, $QR = 4.5$ centimeters, and $QS = 9$ centimeters. Measure $\overline{KL}$ and $\overline{RS}$ and calculate the ratios of the corresponding sides. What can you conclude about the ratios?

Guided Practice

Determine whether each pair of triangles is similar. If similarity exists, write a mathematical sentence relating the two triangles. Give a reason for your answer.

6. A, B, C, D, E, F — 32°, 58°

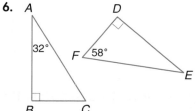

7. G, H, I, J, K, L — 25°, 25°

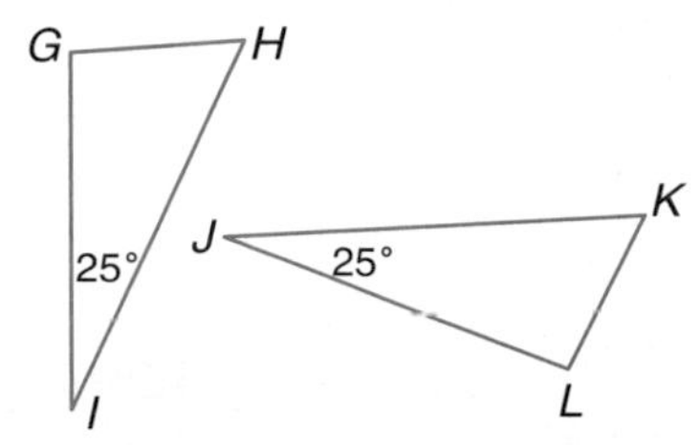

8. Identify the following statement as true or false. If false, state why. *If the measures of the sides of one triangle are three times the measures of the sides of a second triangle, the two triangles are similar.*

9. Identify the similar triangles in the figure at the right. Explain your answer.

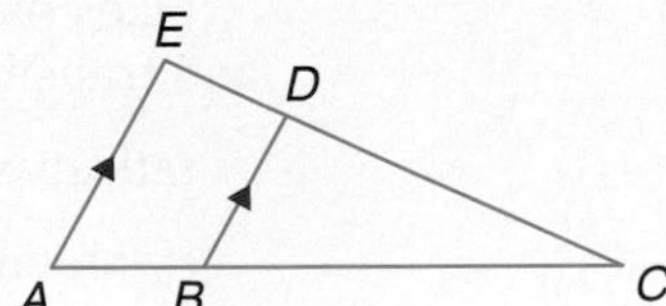

Explain why each pair of triangles is similar and use the given information to find each measure.

10.

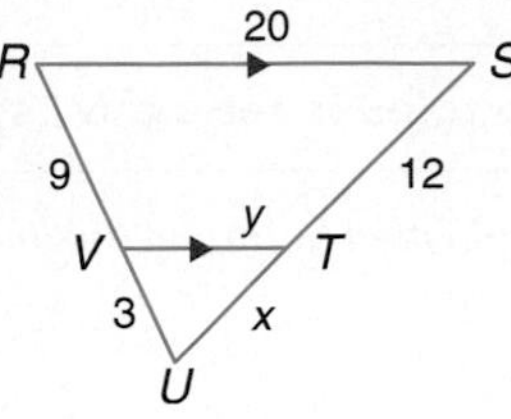

11.

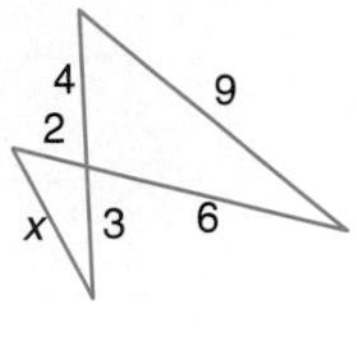

12. **History** Refer to the connection at the beginning of the lesson. If $BC = 576$ feet, $FD = 6$ feet and $EF = 5$ feet, use similar triangles to find the height of the pyramid.

EXERCISES

Practice

Determine whether each pair of triangles is similar. If similarity exists, write a mathematical sentence relating the two triangles. Give a reason for your answer.

13.

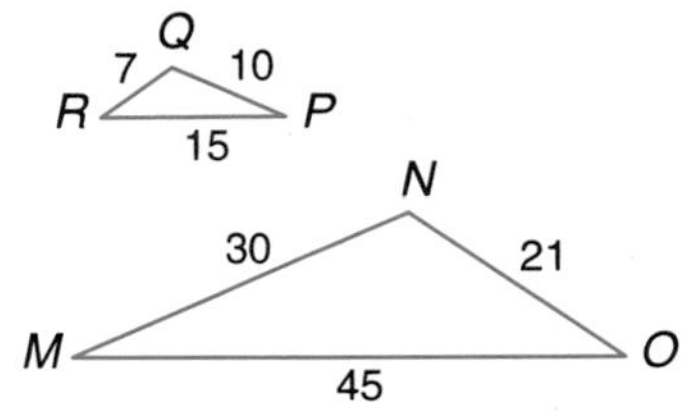

14.

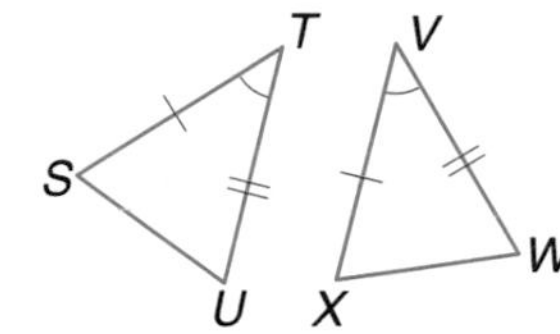

15.

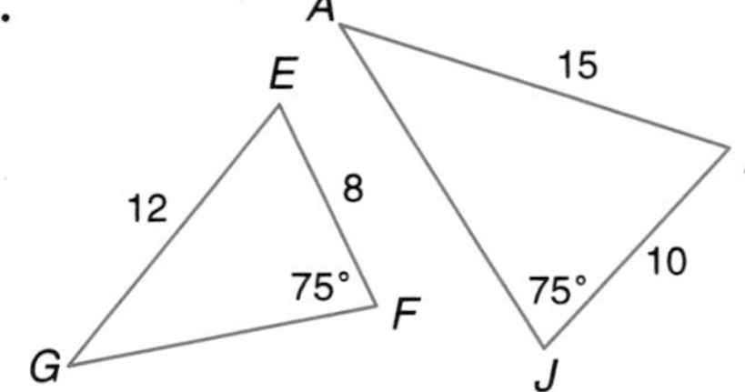

16.

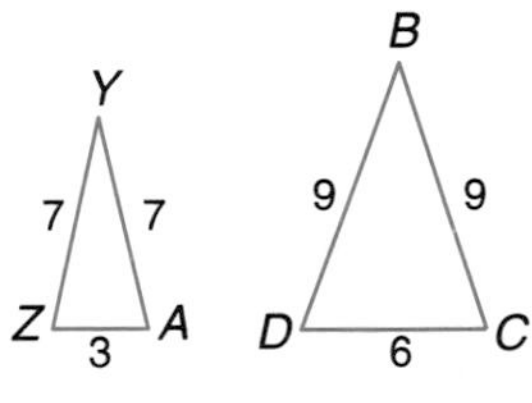

17.

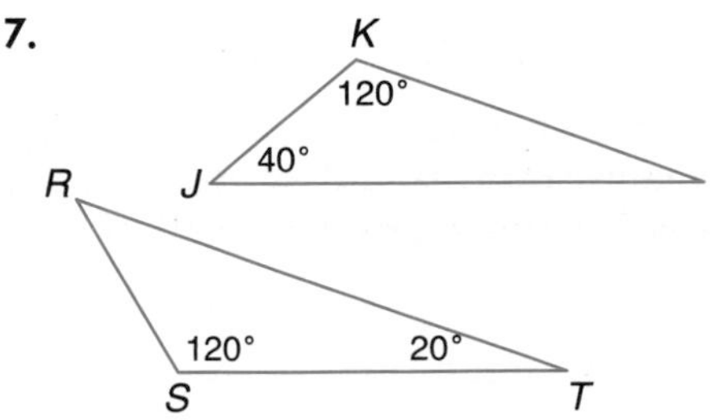

18.

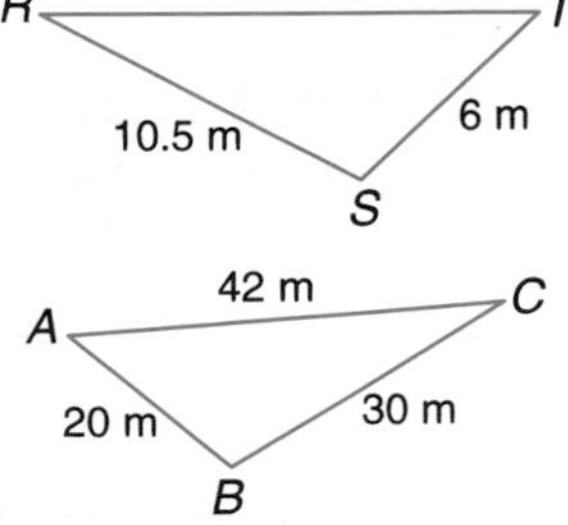

Identify each statement as *true* or *false*. If false, state why.

19. For every pair of similar triangles, there is only one correspondence of vertices that will give you correct angle correspondence and segment proportions.

20. If $\triangle ABC \sim \triangle EFG$ and $\triangle ABC \sim \triangle RST$, then $\triangle EFG \sim \triangle RST$.

21. If the measures of the sides of a triangle are different lengths x, y, and z and the measures of the sides of a second triangle are $x + 1$, $y + 1$ and $z + 1$, the two triangles are similar.

Identify the similar triangles in each figure. Explain your answer.

22.

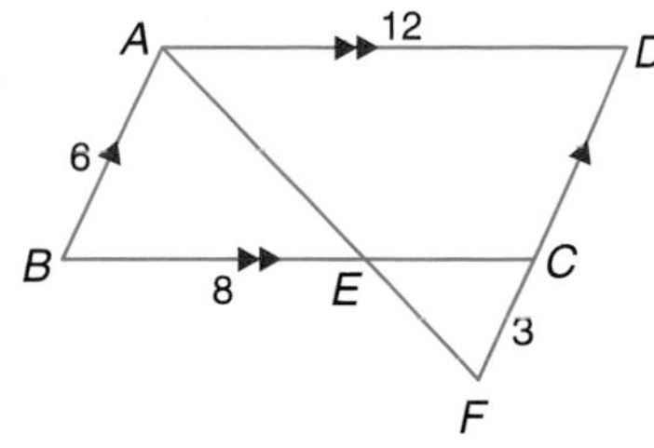

23.

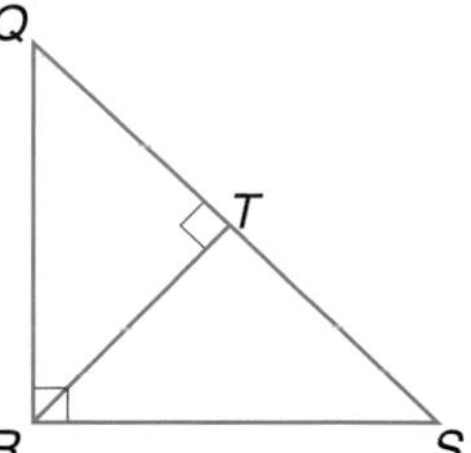

Use the given information to find each measure.

24. If $\overline{ED} \parallel \overline{AB}$, $AB = 10$, $BC = 6$, $AC = 8$, $CD = 5$, and $GE = 3$, find EC, GC, and EF.

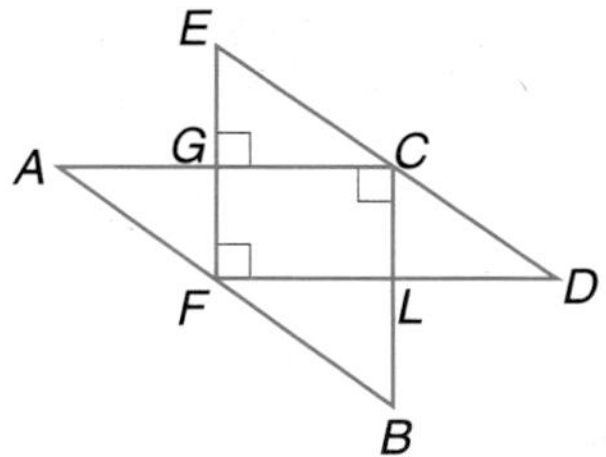

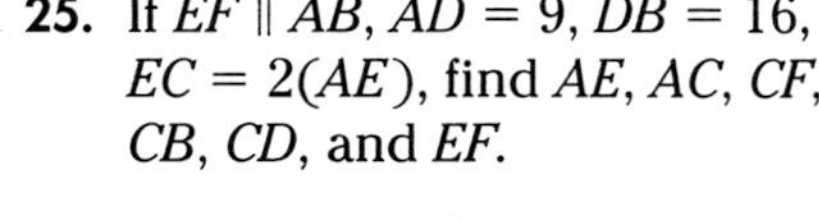

25. If $\overline{EF} \parallel \overline{AB}$, $AD = 9$, $DB = 16$, $EC = 2(AE)$, find AE, AC, CF, CB, CD, and EF.

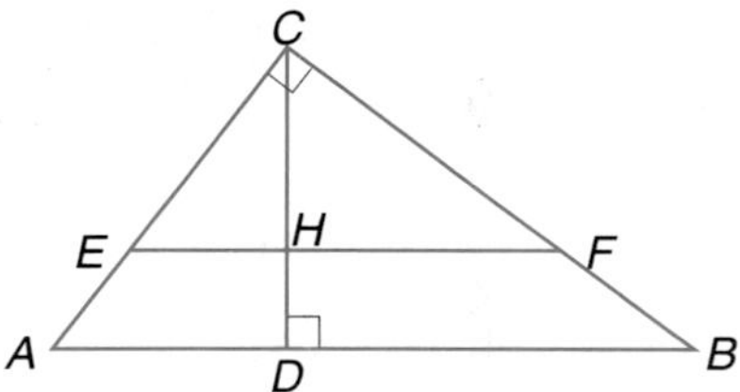

26. If $\frac{FG}{BG} = \frac{KG}{CG}$, $m\angle ABG = 130$, and $m\angle ADJ = 20$, find $m\angle CFD$, $m\angle GKF$, $m\angle GFK$, $m\angle G$, and $m\angle GKJ$.

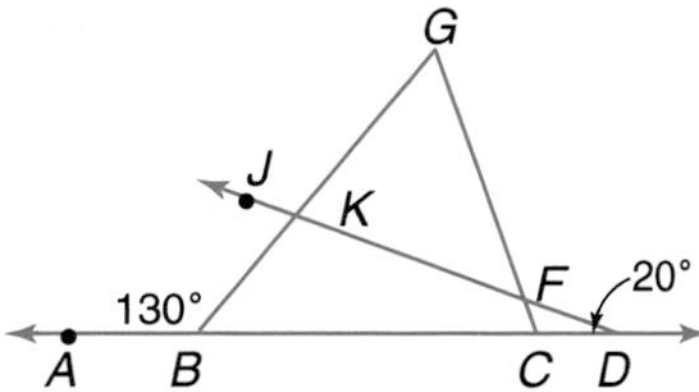

27. If $\angle ABC$ is a right angle, $\overline{BD} \perp \overline{AC}$, $\overline{DE} \perp \overline{BC}$, and $m\angle ACB = 47$, find $m\angle CDE$, $m\angle A$, $m\angle ABD$, and $m\angle BDE$.

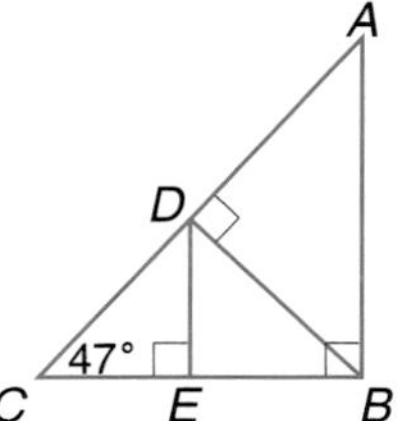

Algebra

28. Graph $\triangle ABC$ and $\triangle TBS$ with vertices $A(-2, -8)$, $B(4, 4)$, $C(-2, 7)$, $T(0, -4)$, and $S(0, 6)$.

a. Show that $\triangle ABC \sim \triangle TBS$.

b. Find the ratio of the perimeters of the two triangles.

Write a two-column proof.

29. **Given:** $\overline{LP} \parallel \overline{MN}$

Prove: $\frac{LJ}{JN} = \frac{PJ}{JM}$

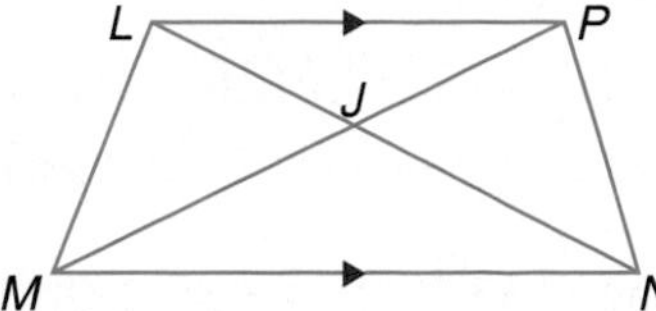

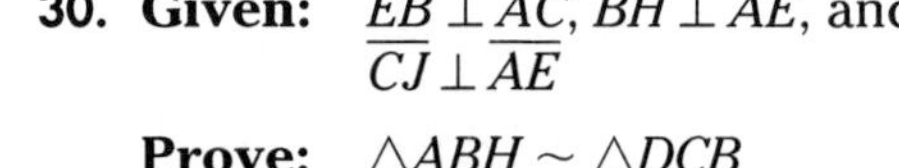

30. **Given:** $\overline{EB} \perp \overline{AC}$, $\overline{BH} \perp \overline{AE}$, and $\overline{CJ} \perp \overline{AE}$

Prove: $\triangle ABH \sim \triangle DCB$

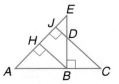

Use the figure below to write a paragraph proof.

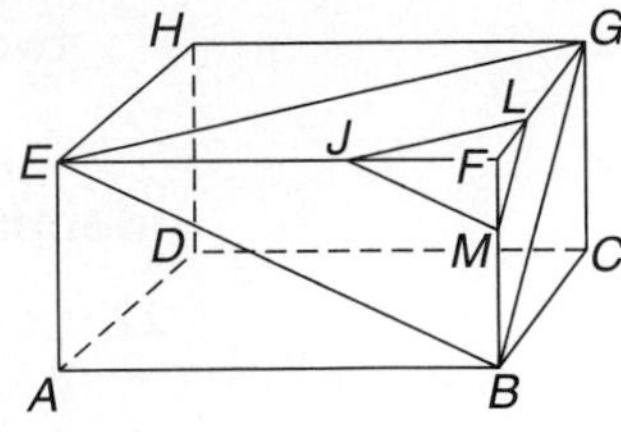

31. **Given:** $\triangle JFM \sim \triangle EFB$
$\triangle LFM \sim \triangle GFB$
Prove: $\triangle JFL \sim \triangle EFG$

32. **Given:** $\overline{JM} \parallel \overline{EB}$
$\overline{LM} \parallel \overline{GB}$
Prove: $\overline{JL} \parallel \overline{EG}$

33. Prove that if the measures of two sides of a triangle are proportional to the measures of two corresponding sides of another triangle and the included angles are congruent, the triangles are similar. (Theorem 7–2)

34. Prove that if the measures of the legs of two right triangles are proportional, the triangles are similar.

35. Prove that similarity of triangles is reflexive, symmetric, and transitive. (Theorem 7–3)

Critical Thinking

36. Is it possible that $\triangle ABC$ is not similar to $\triangle RST$ and that $\triangle RST$ is not similar to $\triangle EFG$, but that $\triangle ABC$ is similar to $\triangle EFG$? Explain.

Applications and Problem Solving

37. **Surveying** Mr. Cardona uses a carpenter's square, an instrument used to draw right angles, to find the distance across a stream. He puts the square on top of a pole that is high enough to sight along $\overline{OL}$ to point P across the river. Then he sights along $\overline{ON}$ to point M. If MK is 2.5 feet and $OK = 5.5$ feet, find the distance KP across the stream.

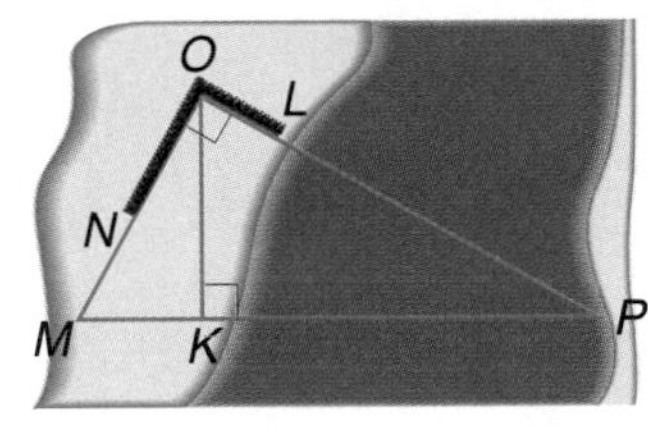

38. **Forestry** A hypsometer as shown below can be used to estimate the height of a tree. Benito looks through the straw to the top of the tree and obtains the readings given below. If Benito's eye was 165 centimeters from the ground and he was 14 meters from the tree, find the height of the tree. (*Hint:* 1 m = 100 cm)

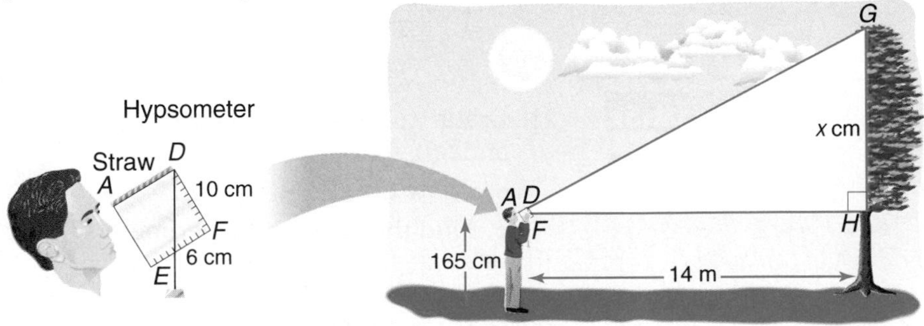

Mixed Review

39. Determine whether the pair of figures at the right is similar. Justify your answer. (Lesson 7–2)

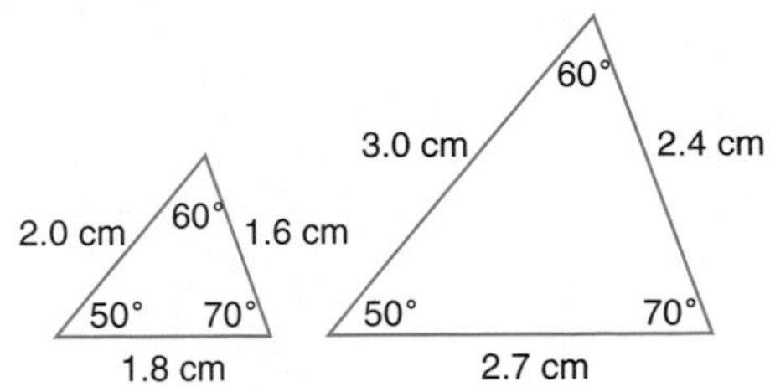

40. **SAT Practice** **Quantitative Comparison**
A rectangle of area 6 has two sides of length a and b, where a and b are integers.

Column A	Column B
$\frac{a}{2}$	$2b$

A if the quantity in Column A is greater
B if the quantity in Column B is greater
C if the quantities are equal
D if the relationship cannot be determined from the information given

41. Determine whether the statement *Consecutive sides of a rhombus are congruent* is true or false. (Lesson 6–4)

42. Two sides of a triangle measure 14 centimeters and 20 centimeters. Determine if the measure of the third side could be 36 centimeters. (Lesson 5–5)

43. Draw and label a figure to illustrate $\triangle FGH$ with altitude $\overline{GR}$, where R does not lie between F and H. (Lesson 5–1)

44. Find the slope of a line that is perpendicular to the line passing through points $G(1, 5)$ and $H(-2, 4)$. (Lesson 3–3)

45. What property justifies the statement $\overline{LM} \cong \overline{LM}$? (Lesson 2–5)

Algebra

46. **Production** In one day, 37,000,000 candies are produced by a company's four plants. Express this number in scientific notation.

For **Extra Practice**, see page 776.

47. Solve $3t - 2 \geq t + 8$.

SELF TEST

1. Solve $\frac{1}{3} = \frac{t}{8-t}$ by using cross products. (Lesson 7–1)
2. Determine whether the pair of figures at the right is similar. Justify your answer. (Lesson 7–2)

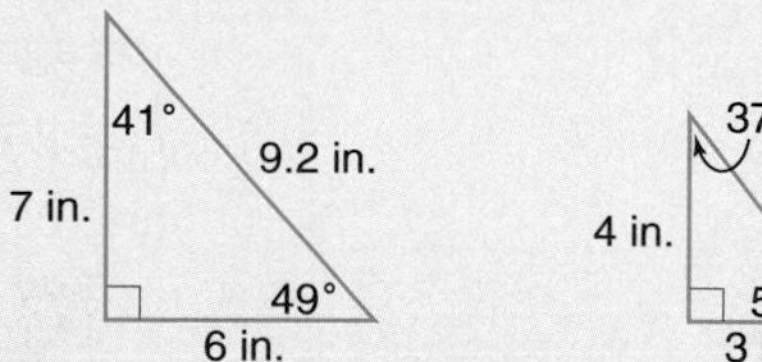

3. **Consumerism** The average American drinks about 4 soft drinks every 3 days. About how many soft drinks will the average person drink in a year? (Lesson 7–1)

Determine whether each pair of triangles is similar using the given information. Explain your answer. (Lesson 7–3)

4. 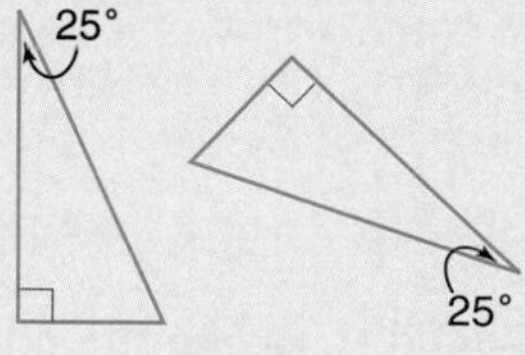

5. 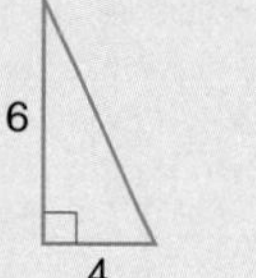

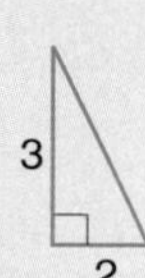

7–4 Parallel Lines and Proportional Parts

What YOU'LL LEARN

- To use proportional parts of triangles to solve problems, and
- to divide a segment into congruent parts.

Why IT'S IMPORTANT

You can use proportions to find the lengths of segments determined by parallel lines.

Music

A harp is a triangular-shaped instrument whose strings can be thought of as parallel lines and whose body and neck as transversals.

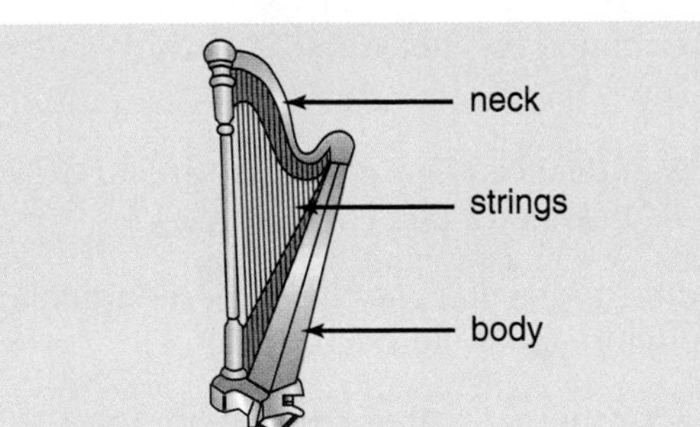

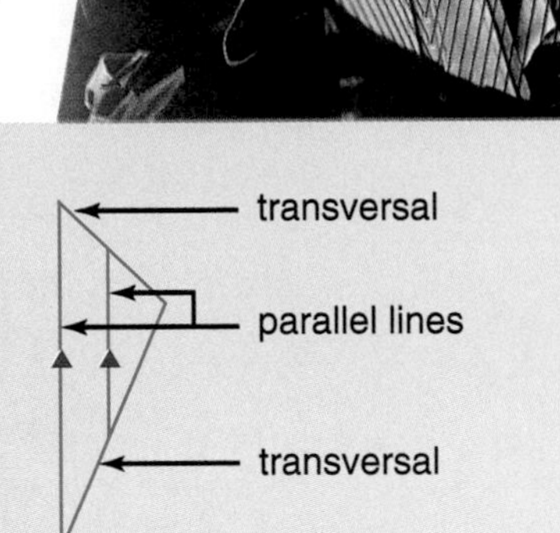

Each string divides the body and the neck proportionally. This is stated in the following theorem.

Theorem 7–4 Triangle Proportionality	If a line is parallel to one side of a triangle and intersects the other two sides in two distinct points, then it separates these sides into segments of proportional lengths.

Proof of Theorem 7–4

Given: $\overline{BD} \parallel \overline{AE}$

Prove: $\frac{BA}{CB} = \frac{DE}{CD}$

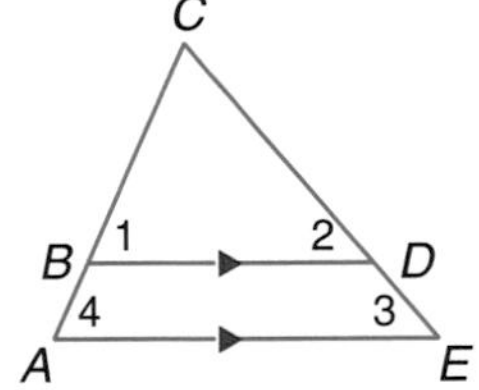

Paragraph Proof:

Since $\overline{BD} \parallel \overline{AE}$, $\angle 4 \cong \angle 1$ and $\angle 3 \cong \angle 2$ because they are corresponding angles. Then, by AA Similarity, $\triangle ACE \sim \triangle BCD$. From the definition of similar polygons, $\frac{CA}{CB} = \frac{CE}{CD}$. Since B is between A and C, $CA = BA + CB$, and since D is between C and E, $CE = DE + CD$. Substituting for CA and CE in the ratio, we get the following proportion.

$$\frac{BA + CB}{CB} = \frac{DE + CD}{CD}$$

$$\frac{BA}{CB} + \frac{CB}{CB} = \frac{DE}{CD} + \frac{CD}{CD}$$

$$\frac{BA}{CB} + 1 = \frac{DE}{CD} + 1 \qquad \textit{Substitution Property(=)}$$

$$\frac{BA}{CB} = \frac{DE}{CD} \qquad \textit{Subtraction Property(=)}$$

Likewise, proportional parts of a triangle can be used to prove the converse of Theorem 7–4. *You will be asked to prove this theorem in Exercise 32.*

Theorem 7–5	**If a line intersects two sides of a triangle and separates the sides into corresponding segments of proportional lengths, then the line is parallel to the third side.**

Example 1

In $\triangle EFG$, $EG = 15$, $EH = 5$, and LG is twice FL. Determine whether $\overline{HL} \parallel \overline{EF}$.

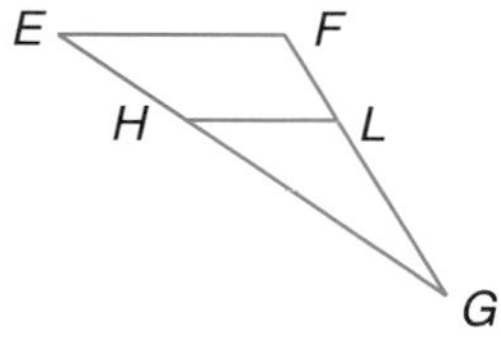

From the Segment Addition Postulate, $EG = EH + HG$. Substitute the known measures.

$$EG = EH + HG$$
$$15 = 5 + HG$$
$$10 = HG$$

Let $x = FL$. Then $2x = LG$. To show $\overline{HL} \parallel \overline{EF}$, we must show that $\frac{EH}{HG} = \frac{FL}{LG}$. Using the information above, $\frac{EH}{HG} = \frac{5}{10}$ or $\frac{1}{2}$ and $\frac{FL}{LG} = \frac{x}{2x}$ or $\frac{1}{2}$. Since the sides have proportional lengths, $\overline{HL} \parallel \overline{EF}$.

The proof of the following theorem is based on Theorems 7–4 and 7–5. *You will be asked to prove this theorem in Exercise 33.*

Theorem 7–6	**A segment whose endpoints are the midpoints of two sides of a triangle is parallel to the third side of the triangle, and its length is one-half the length of the third side.**

In $\triangle RST$, suppose M is the midpoint of $\overline{RS}$ and L is the midpoint of $\overline{RT}$. By Theorem 7–6, $\overline{ML} \parallel \overline{ST}$ and $ML = \frac{1}{2}ST$. This can also be expressed as $2ML = ST$.

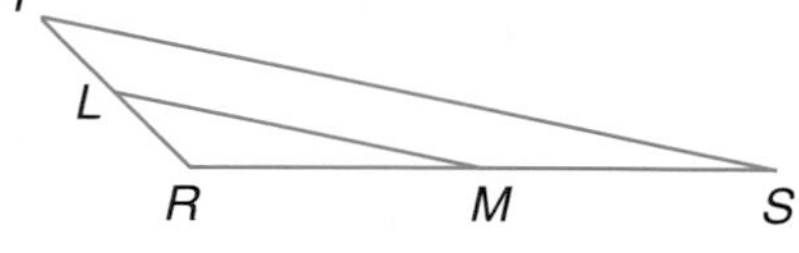

Example 2

Algebra

Triangle ABC has vertices $A(0, 2)$, $B(12, 0)$, and $C(2, 10)$.

a. Find the coordinates of D, the midpoint of $\overline{AB}$, and E, the midpoint of $\overline{CB}$.

b. Show that $\overline{DE}$ is parallel to $\overline{AC}$.

c. Show that $2DE = AC$.

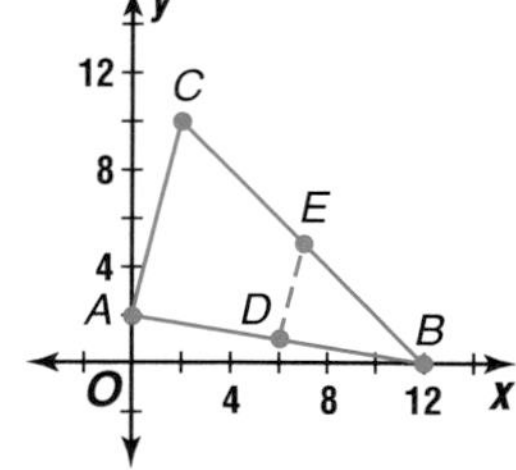

LOOK BACK

Refer to Lesson 1-5 to review the midpoint formula.

a. Use the midpoint formula to find the midpoints of $\overline{AB}$ and $\overline{CB}$.

$$D\left(\frac{x_1 + x_2}{2}, \frac{y_1 + y_2}{2}\right) = D\left(\frac{0 + 12}{2}, \frac{2 + 0}{2}\right) \text{ or } D(6, 1)$$

$$E\left(\frac{x_1 + x_2}{2}, \frac{y_1 + y_2}{2}\right) = E\left(\frac{12 + 2}{2}, \frac{0 + 10}{2}\right) \text{ or } E(7, 5)$$

Thus, the midpoint of $\overline{AB}$ has coordinates (6, 1), and the midpoint of $\overline{CB}$ has coordinates (7, 5).

(continued on the next page)

b. To show that $\overline{AC} \parallel \overline{DE}$, you can show that the slopes of $\overline{AC}$ and $\overline{DE}$ are equal.

slope of $\overline{AC} = \frac{2 - 10}{0 - 2}$ or 4 slope of $\overline{DE} = \frac{1 - 5}{6 - 7}$ or 4

Because the slopes of $\overline{AC}$ and $\overline{DE}$ are equal, $\overline{AC} \parallel \overline{DE}$.

c. To show that $2DE = AC$, we must find the length of $\overline{DE}$ and $\overline{AC}$. Use the distance formula.

$$AC = \sqrt{(0 - 2)^2 + (2 - 10)^2} = \sqrt{4 + 64} = \sqrt{68} \text{ or } 2\sqrt{17}$$

$$DE = \sqrt{(6 - 7)^2 + (1 - 5)^2} = \sqrt{1 + 16} \text{ or } \sqrt{17}$$

Therefore, $2DE = AC$.

We have seen that parallel lines cut the sides of a triangle into proportional parts. Three or more parallel lines separate transversals into proportional parts as stated in the next two corollaries.

Corollary 7–1	**If three or more parallel lines intersect two transversals, then they cut off the transversals proportionally.**
Corollary 7–2	**If three or more parallel lines cut off congruent segments on one transversal, then they cut off congruent segments on every transversal.**

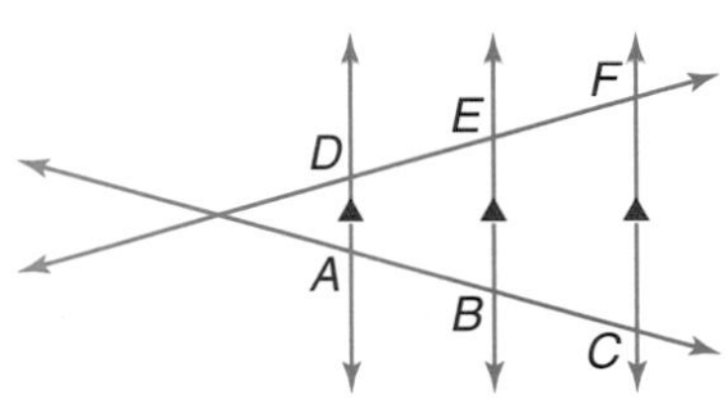

In the figure at the left, $\overleftrightarrow{AD} \parallel \overleftrightarrow{EB} \parallel \overleftrightarrow{FC}$. The transversals $\overleftrightarrow{AC}$ and $\overleftrightarrow{DF}$ have been separated into proportional segments. Sample proportions are listed below.

$$\frac{AB}{BC} = \frac{DE}{EF}, \quad \frac{AC}{DF} = \frac{BC}{EF}, \quad \frac{AC}{BC} = \frac{DF}{EF}$$

The theorems and corollaries about proportionality of triangles can be used to solve practical problems.

Example 3

Real Estate

In Lake Creek, the lots on which houses are to be built are laid out as shown. What is the lake frontage for each of the five lots if the total frontage is 135.6 meters?

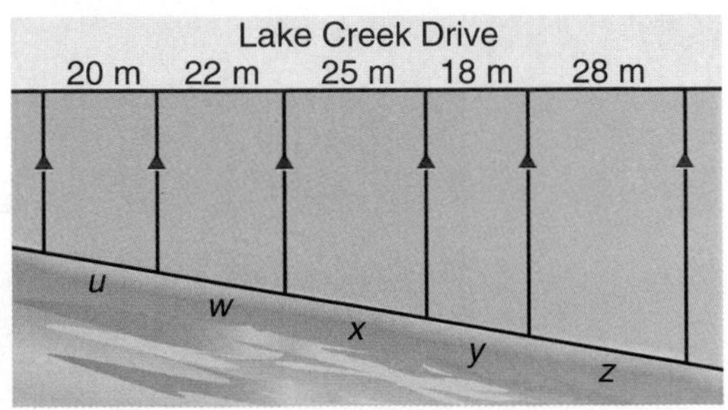

Explore By Corollary 7–1, if three or more parallel lines intersect two transversals, they cut off the transversals proportionally. Thus, $\frac{u}{20} = \frac{w}{22} = \frac{x}{25} = \frac{y}{18} = \frac{z}{28}$.

Plan Use these proportions to find a value for each frontage in terms of one variable, for example, u.

$\frac{u}{20} = \frac{w}{22}$ $\quad$ $\frac{u}{20} = \frac{x}{25}$ $\quad$ $\frac{u}{20} = \frac{y}{18}$ $\quad$ $\frac{u}{20} = \frac{z}{28}$

$w = \frac{22}{20}u$ $\quad$ $x = \frac{25}{20}u$ $\quad$ $y = \frac{18}{20}u$ $\quad$ $z = \frac{28}{20}u$

Solve

$u + w + x + y + z = 135.6$ *Length of the frontage is 135.6 m.*

$u + \frac{22}{20}u + \frac{25}{20}u + \frac{18}{20}u + \frac{28}{20}u = 135.6$ *Substitution Property (=)*

$20u + 22u + 25u + 18u + 28u = 135.6(20)$ *Multiply each side by 20.*

$113u = 2712$ *Simplify.*

$u = 24$ *Divide each side by 113.*

Therefore, $w = \frac{22}{20}u$ $\quad$ $x = \frac{25}{20}u$ $\quad$ $y = \frac{18}{20}u$ $\quad$ $z = \frac{28}{20}u$

$w = \frac{22}{20}(24)$ $\quad$ $x = \frac{25}{20}(24)$ $\quad$ $y = \frac{18}{20}(24)$ $\quad$ $z = \frac{28}{20}(24)$

$w = 26.4$ $\quad$ $x = 30$ $\quad$ $y = 21.6$ $\quad$ $z = 33.6$

The individual lake frontage for each lot is 24 meters, 26.4 meters, 30 meters, 21.6 meters, and 33.6 meters.

Examine By substitution, $\frac{u}{20} = \frac{w}{22} = \frac{x}{25} = \frac{y}{18} = \frac{z}{28} = 1.2$. Since the proportions are equal, the values are correct.

It is possible to separate a segment into two congruent parts by constructing the perpendicular bisector of a segment. However, a segment cannot be separated into three congruent parts by constructing perpendicular bisectors. To do this, you must use parallel lines and the similarity relationships.

CONSTRUCTION

Trisecting a Segment

Separate a segment into three congruent parts.

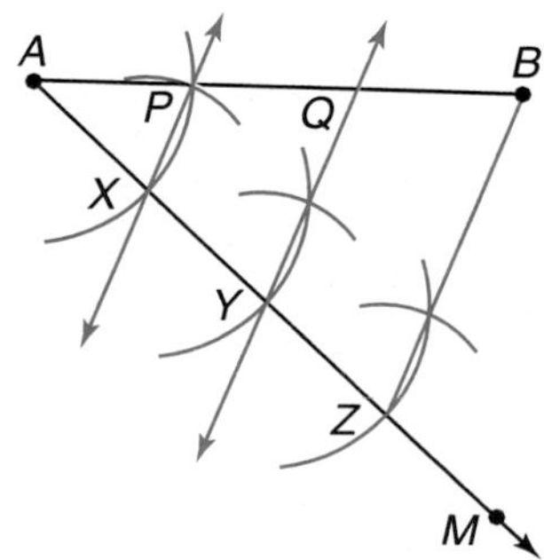

1. Copy $\overline{AB}$ and draw $\overline{AM}$.
2. With a compass point at A, mark off an arc that intersects $\overline{AM}$ at X. Then construct $\overline{XY}$ and $\overline{YZ}$ so that $\overline{AX} \cong \overline{XY} \cong \overline{YZ}$.
3. Draw $\overline{ZB}$. Then construct lines through Y and X that are parallel to $\overline{ZB}$. Call P and Q the intersection points on $\overline{AB}$.

Conclusion: Because parallel lines cut off congruent segments on transversals, $\overline{AP} \cong \overline{PQ} \cong \overline{QB}$. *This process can be used for any number of congruent parts.*

CHECK FOR UNDERSTANDING

Communicating Mathematics

Study the lesson. Then complete the following.

1. **Make a drawing** to show two segments intersected by two lines so that the parts are proportional. Then draw a counterexample.
2. **Explain** the difference between Corollary 7–1 and Corollary 7–2.
3. **Draw** a segment from one side of a triangle to another so that it is parallel to the third side. When will the segment be half the length of the third side of the triangle? Explain why this is the case.
4. **Describe** how to separate a segment into four congruent parts. How should this process change to separate a segment into five congruent parts?

5. **List** at least four methods to prove two lines parallel, including one from this lesson.

Guided Practice

Refer to $\triangle RST$ for Exercises 6 and 7. $\overline{LW} \parallel \overline{TS}$.

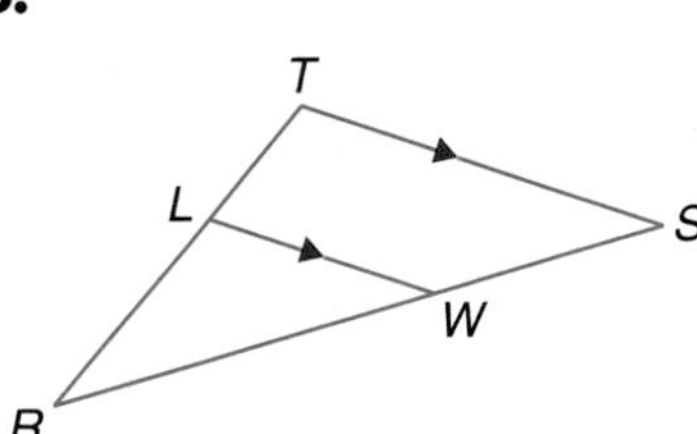

6. Complete each statement.
   a. $\frac{RW}{WS} = \frac{RL}{?}$ b. $\frac{RW}{RS} = \frac{?}{RT}$
7. Determine whether each statement is *true* or *false*. If false, explain why.
   a. If $TL = LR$, then $SW = \frac{1}{2}SR$.
   b. If $TR = 8$, $LR = 3$, and $RW = 6$, then $WS = 16$.

8. Find the values of x and y.

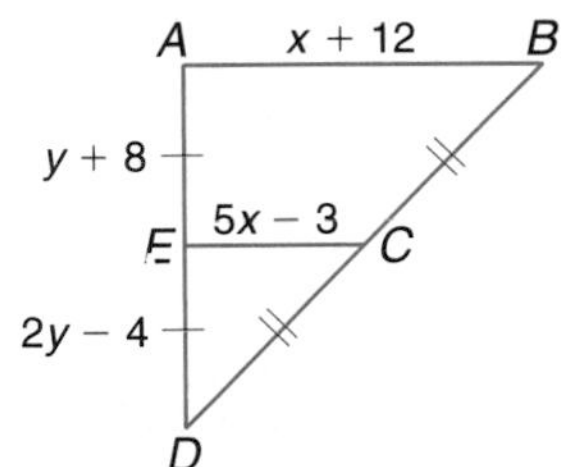

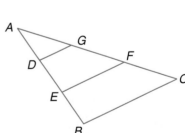

Refer to the figure at the right for Exercises 9–11. Determine whether each conclusion is valid under the given conditions. If the conclusion is valid, give a reason.

9. $\overline{DG} \parallel \overline{EF} \parallel \overline{BC}$, $\overline{AD} \cong \overline{DE} \cong \overline{EB}$. Therefore, $\overline{AG} \cong \overline{FC}$.
10. $\overline{DG} \parallel \overline{BC}$. Therefore, $\frac{AD}{DE} = \frac{AG}{GF}$.

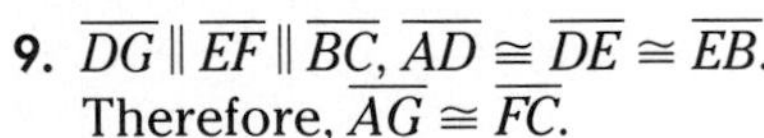

11. $\overline{DG} \parallel \overline{EF} \parallel \overline{BC}$; $DE = x$, $EB = 20$, $GF = x - 5$, $FC = 15$. Therefore, $x = 20$.
12. **Construction** Hai was building a large open stairway and used wood strips as a decoration along the inside wall. If the strips were spaced along the bottom as shown in the diagram at the right, at what distance should he attach the top of the strips if the strips are to be parallel?

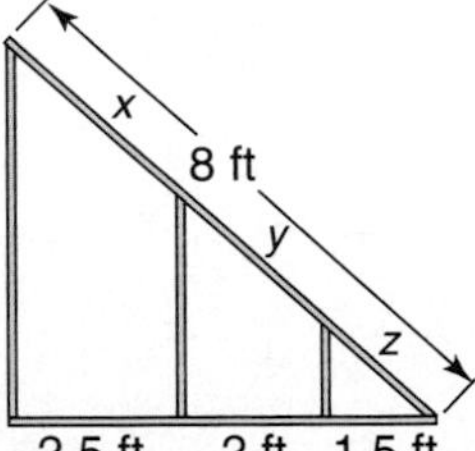

EXERCISES

Practice

In the figure at the right, $\overleftrightarrow{AB} \parallel \overleftrightarrow{CD} \parallel \overleftrightarrow{EF}$. Complete each statement.

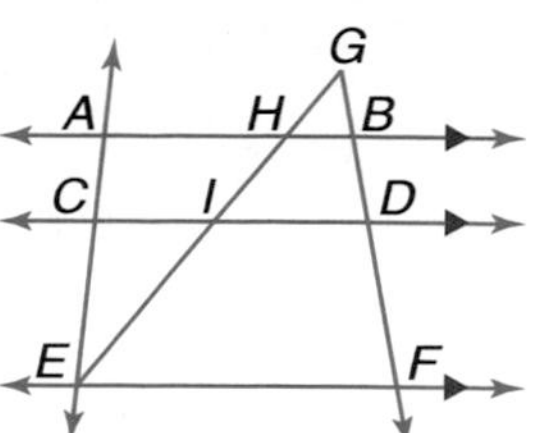

13. $\frac{AC}{BD} = \frac{CE}{?}$

14. $\frac{CE}{IE} = \frac{?}{HI}$

15. $\frac{GH}{GE} = \frac{?}{GF}$

16. $\frac{CE}{?} = \frac{AC}{HI}$

In $\triangle ABC$, $\overline{ED} \parallel \overline{AC}$. Determine whether each statement is *true* or *false*. If false, explain why.

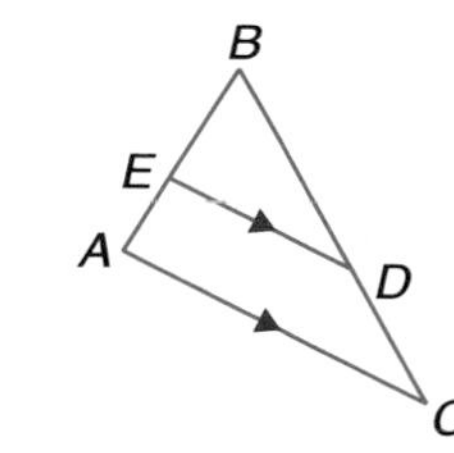

17. If $BE = 4$, $EA = 3$, and $BD = 5$, then $DC = 3\frac{3}{4}$.

18. If $BE = \frac{3}{5}AB$, then $DC = \frac{2}{5}BC$.

19. If $AE = \frac{3}{5}AB$, then $\frac{BD}{DC} = \frac{3}{2}$.

INTEGRATION
Algebra

Find the values of x and y.

20.

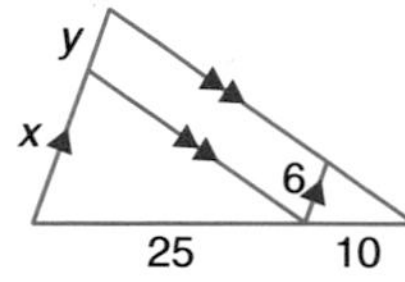

21.

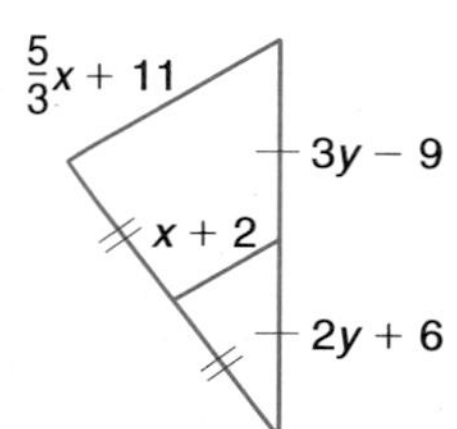

22.

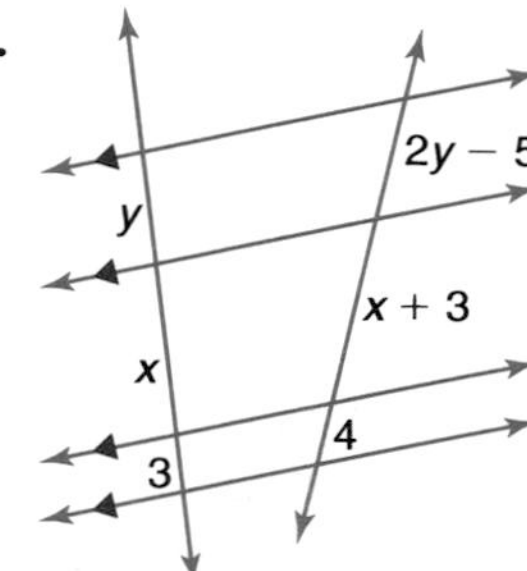

In $\triangle ACE$, find x so that $\overline{DB} \parallel \overline{AE}$.

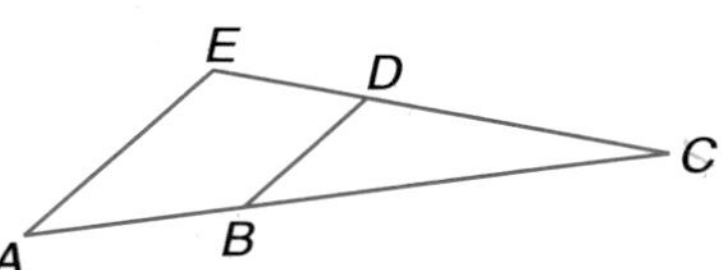

23. $ED = 8$, $DC = 20$, $BC = 25$, $AB = x$

24. $BC = 12$, $AB = 6$, $ED = 8$, $DC = x - 4$

25. $ED = x - 5$, $DC = 15$, $CB = 18$, $AB = x - 4$

INTEGRATION
Algebra

26. Write a paragraph proof to show that $\overline{SM}$ is parallel to $\overline{TW}$ and that $\overline{SM}$ divides the sides $\overline{RT}$ and $\overline{RW}$ in a ratio of 1 to 2.

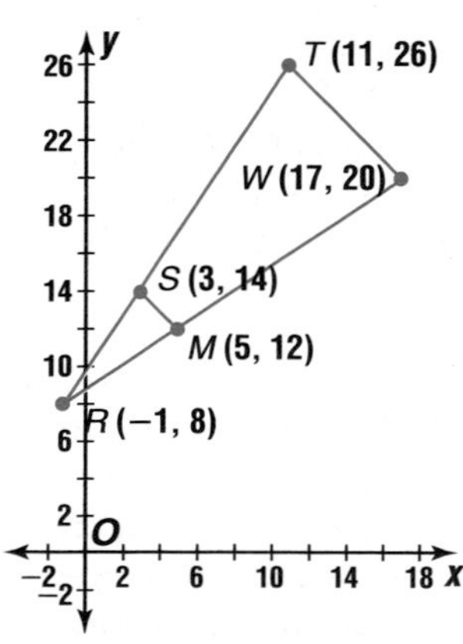

27. If $\overline{BC} \parallel \overline{DE}$ and $2(DE) = BC$, find the coordinates of B and C.

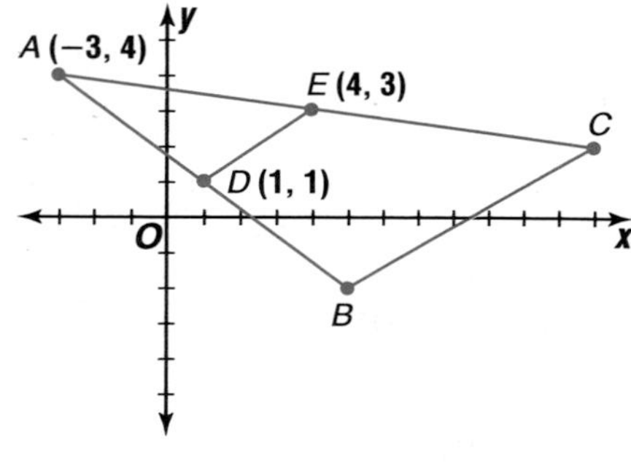

A, B, and C are the midpoints of sides $\overline{DF}$, $\overline{DE}$, and $\overline{FE}$, respectively.

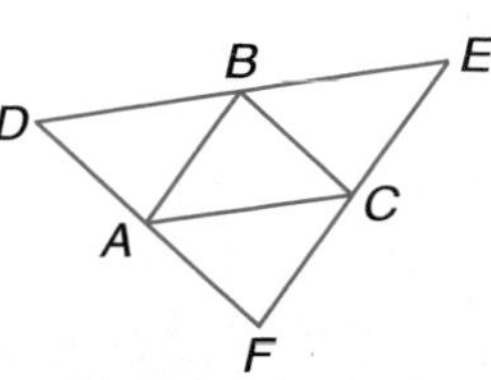

28. If $BC = 11$, $AC = 13$, and $AB = 15$, find the perimeter of $\triangle DEF$.

29. If $DE = 18$, $DA = 10$, and $FC = 7$, find AB, BC, and AC.

30. Draw a segment 8 centimeters long. By construction, separate the segment into three congruent segments.

31. Draw a segment and by construction, separate it into four congruent parts. Draw a segment and separate it into two segments whose ratios are 1 to 4.

Proof

Write a two-column proof for Exercises 32 and 33.

32. If a line intersects two sides of a triangle and separates the sides into corresponding segments of proportional lengths, then the line is parallel to the third side. (Theorem 7–5)

33. A segment whose endpoints are the midpoints of two sides of a triangle is parallel to the third side of the triangle and its length is one-half the length of the third side. (Theorem 7–6)

34. Given $A(2, 12)$ and $B(5, 0)$, find P such that P separates $\overline{AB}$ into two parts with a ratio of 2 to 1.

35. In $\triangle LMN$, $\overline{PR}$ divides sides $\overline{NL}$ and $\overline{MN}$ proportionally. If the points are $N(8, 20)$, $P(11, 16)$, and $R(3, 8)$ and $\frac{LP}{PN} = \frac{2}{1}$, find the coordinates of L and M.

Critical Thinking

36. Draw any quadrilateral $ABCD$ and connect the midpoints E, F, G, H of the sides in order.
    a. Determine what kind of figure $EFGH$ will be. Use the information from this lesson to prove your claim.
    b. Will the same reasoning work with five-sided polygons? Explain why or why not.

37. Draw segment $\overline{AB}$ so it is 7 centimeters long. If $r = 2$ centimeters, $s = 3$ centimeters, and $t = 4$ centimeters, divide $\overline{AB}$ into segments x, y, and z such that $\frac{x}{r} = \frac{y}{s} = \frac{z}{t}$.

Applications and Problem Solving

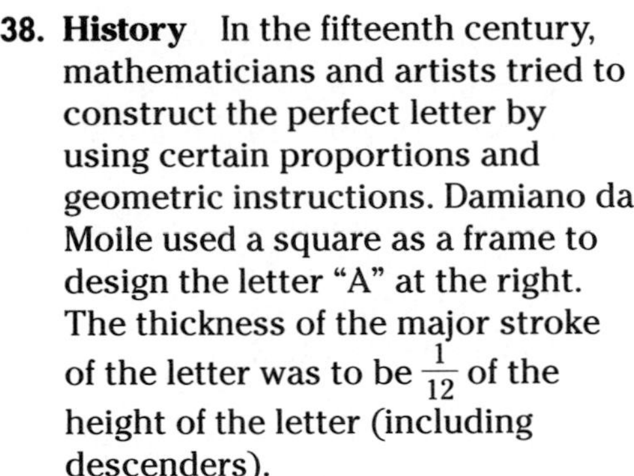

38. **History** In the fifteenth century, mathematicians and artists tried to construct the perfect letter by using certain proportions and geometric instructions. Damiano da Moile used a square as a frame to design the letter "A" at the right. The thickness of the major stroke of the letter was to be $\frac{1}{12}$ of the height of the letter (including descenders).

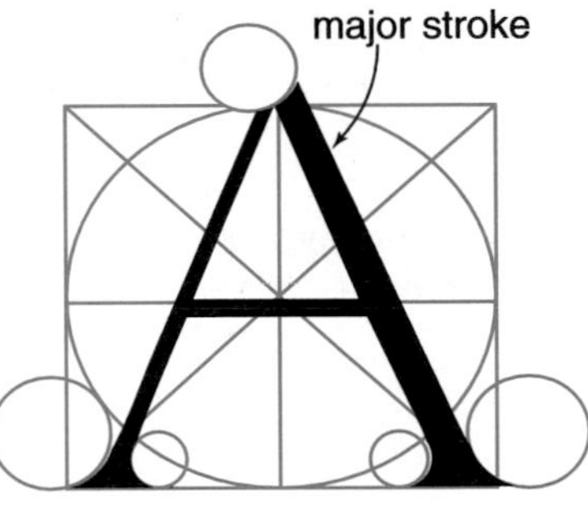

    a. Explain how you can tell that the bar through the middle of the A is half the length between the outside bottom corners of the sides of the letter.
    b. If the letter were 3 centimeters tall, how wide would the major stroke of the A be?

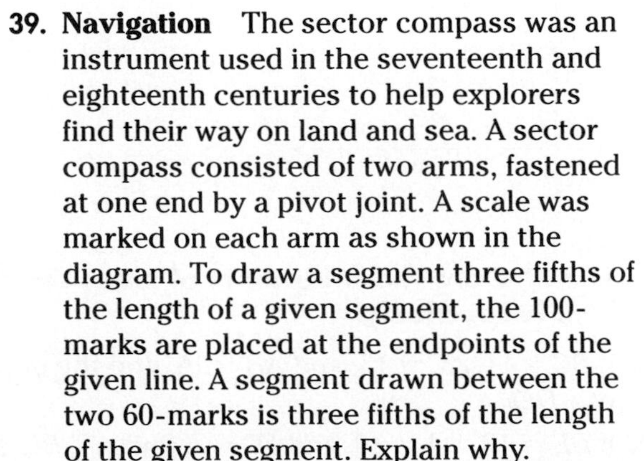

39. **Navigation** The sector compass was an instrument used in the seventeenth and eighteenth centuries to help explorers find their way on land and sea. A sector compass consisted of two arms, fastened at one end by a pivot joint. A scale was marked on each arm as shown in the diagram. To draw a segment three fifths of the length of a given segment, the 100-marks are placed at the endpoints of the given line. A segment drawn between the two 60-marks is three fifths of the length of the given segment. Explain why.

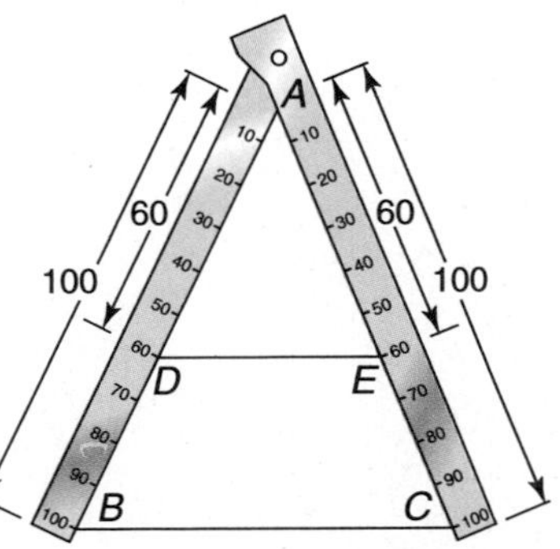

Mixed Review

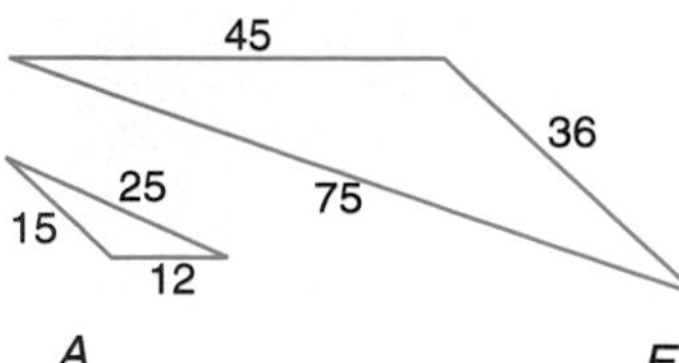

40. Determine if the triangles at the right are similar using the given information. Explain. (Lesson 7–3)

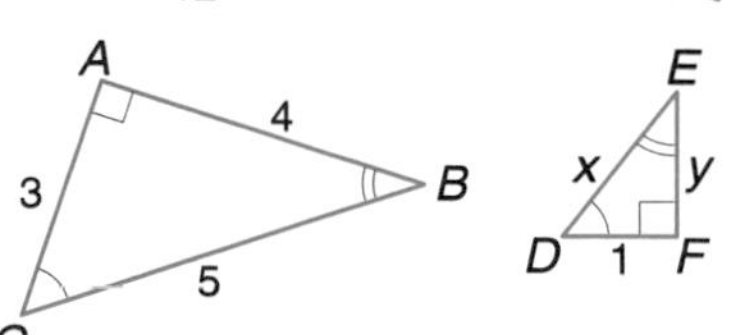

41. Two similar polygons are shown at the right. Find the values of x and y. (Lesson 7–2)

42. Explain whether a rectangle can be a rhombus. (Lesson 6–4)

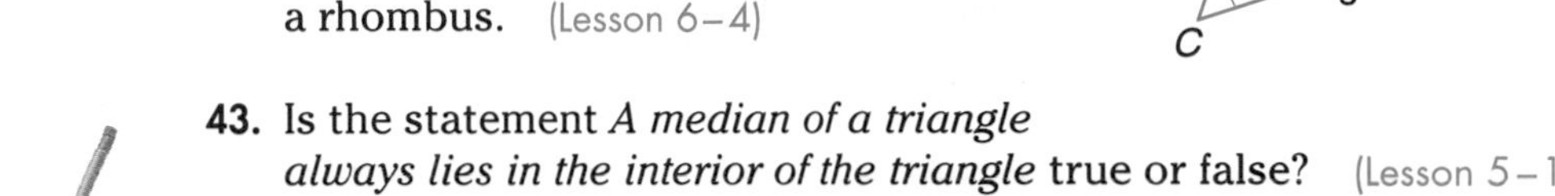

43. Is the statement *A median of a triangle always lies in the interior of the triangle* true or false? (Lesson 5–1)

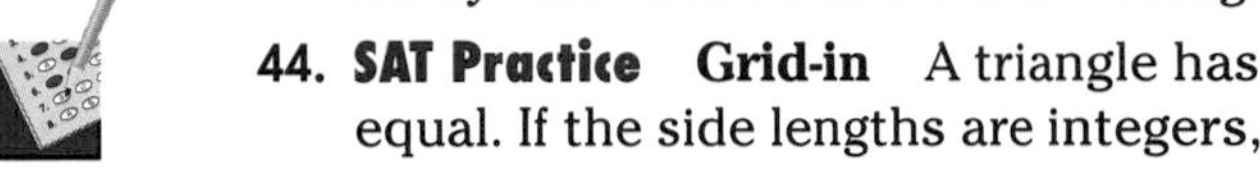

44. **SAT Practice Grid-in** A triangle has base 9, and the other two sides are equal. If the side lengths are integers, what is the shortest possible side?

45. Write a conjecture based on the information that $\angle RST$ and $\angle DSE$ are vertical angles and $\angle DSE \cong \angle ABC$. (Lesson 2–1)

46. Find the coordinates of point B if $M(6, -1)$ is the midpoint of $\overline{AB}$ and the coordinates of A are $(-2, 3)$. (Lesson 1–5)

INTEGRATION **Algebra**

47. Factor $k^2 + 3k - 18$.

For **Extra Practice**, see page 777.

48. **Statistics** The number of grams of sugar contained in a single-serving size of 10 different fruit juices is given below.

33, 23, 7, 29, 23, 27, 40, 29, 34, 22

Find the mean, median, and mode of the data.

Interpreting an Ancient Blueprint

The excerpt below appeared in an article in *Scientific American* in June, 1995.

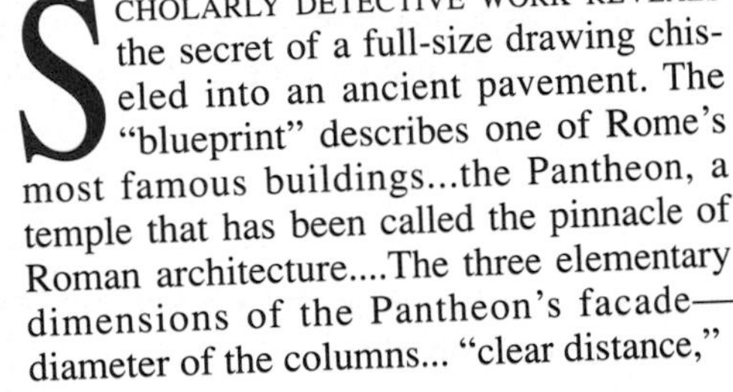

SCHOLARLY DETECTIVE WORK REVEALS the secret of a full-size drawing chiseled into an ancient pavement. The "blueprint" describes one of Rome's most famous buildings...the Pantheon, a temple that has been called the pinnacle of Roman architecture....The three elementary dimensions of the Pantheon's facade—diameter of the columns... "clear distance," or space between the columns...and height of the columns—are in the ratio of 1 to 2 to 9.5. This formula is among those described by Hermogenes, one of the most celebrated architects of the Hellenistic age, as belonging to an ideal facade....For the first time in the exploration of Roman architecture, a numerical recipe for beauty, as known from a textual source, can be tied to an existing monument. ■

1. The chiseled stone blueprint for the Pantheon was drawn at full scale; its dimensions were the same as those of the actual temple. What advantages and disadvantages are there in using this approach?

2. The full-size stone blueprint was identified as the Pantheon by comparing its angles and dimensions. How could you find the size of a large building without drawings or measuring directly?

3. The granite and marble blocks used to build the temple weighed many tons. Why do you think the courtyard, which contained the blueprint and was the place where the stones were measured and cut, was located close to the Tiber River?

7-5 Parts of Similar Triangles

What YOU'LL LEARN

- To recognize and use the proportional relationships of corresponding perimeters, altitudes, angle bisectors, and medians of similar triangles.

Why IT'S IMPORTANT

You can use proportional measures to solve problems involving photography, design, and art.

Games

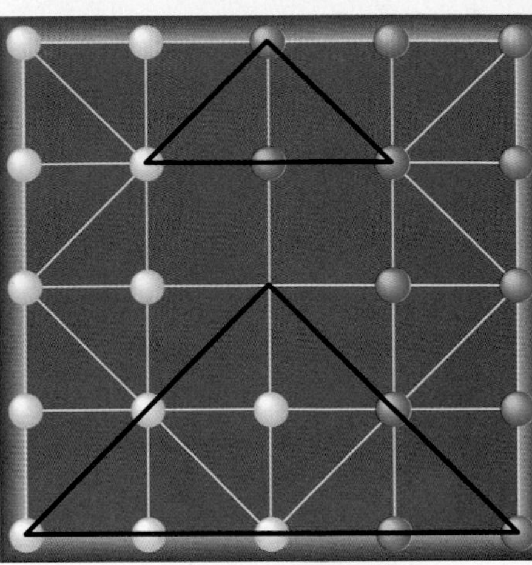

Alquerque is a Spanish game similar to *Checkers*. Two triangular paths are shown on the game board at the right. Notice that the paths are similar in shape and each side of the larger triangle is twice as long as each corresponding side of the smaller triangle. Will a game piece that moves along the path of the larger triangle move twice as far as a game piece that moves along the smaller triangle?

When two triangles are similar, there is a common ratio r such that $\frac{w}{a} = \frac{s}{b} = \frac{t}{c} = r$.

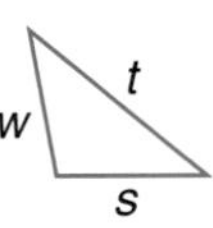

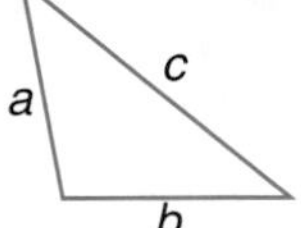

So, $w = ar$, $s = br$, $t = cr$.

The perimeter of the smaller triangle P_1 is $w + s + t$, and the perimeter of the larger triangle P_2 is $a + b + c$.

$$\frac{P_1}{P_2} = \frac{w + s + t}{a + b + c}$$

$$= \frac{ar + br + cr}{a + b + c} \quad \textit{Substitution Property (=)}$$

$$= \frac{r(a + b + c)}{a + b + c} \quad \textit{Distributive Property}$$

$$= r$$

So, the ratio of the perimeters of the two triangles is the same as the ratio between the corresponding sides. Thus, a game piece that moves along the path of the larger triangle will move twice as far as a game piece that moves along the shorter path.

Theorem 7–7 Proportional Perimeters	**If two triangles are similar, then the perimeters are proportional to the measures of corresponding sides.**

Example  **If $\triangle LMN \sim \triangle QRS$, $QR = 40$, $RS = 41$, $SQ = 9$, and $LM = 9$, find the perimeter of $\triangle LMN$.**

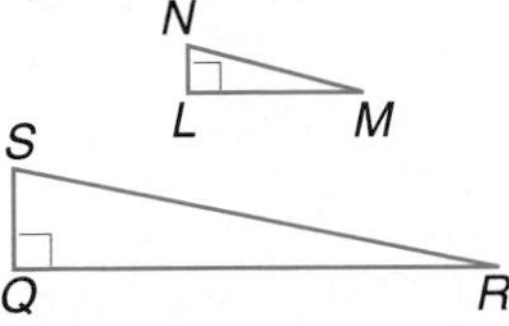

Let x represent the perimeter of $\triangle LMN$. The perimeter of $\triangle QRS = 40 + 41 + 9$ or 90 units.

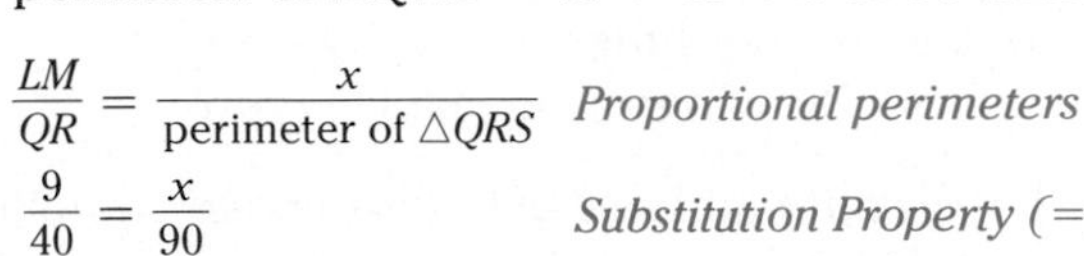

$$\frac{LM}{QR} = \frac{x}{\text{perimeter of } \triangle QRS} \quad \textit{Proportional perimeters}$$

$$\frac{9}{40} = \frac{x}{90} \quad \textit{Substitution Property (=)}$$

$$x = 20.25$$

The perimeter of $\triangle LMN$ is 20.25 units.

When two triangles are similar, the measures of their corresponding sides are proportional. What about the measures of their corresponding altitudes, medians, and angle bisectors? The following theorem states one relationship. *You will be asked to prove this theorem in Exercise 36.*

Theorem 7–8	**If two triangles are similar, then the measures of the corresponding altitudes are proportional to the measures of the corresponding sides.**

Example 2 **In the figure at the right, $\triangle ABC \sim \triangle DEF$. If $\overline{BG}$ is an altitude of $\triangle ABC$, and $\overline{EH}$ is an altitude of $\triangle DEF$, then complete the following.**

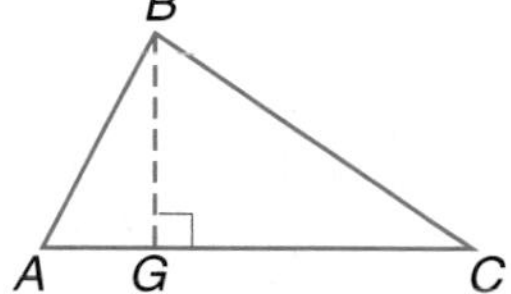

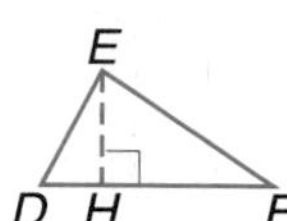

a. $\frac{BG}{EH} = \frac{?}{DE}$ The corresponding measure is AB.

b. $\frac{BG}{EH} = \frac{BC}{?}$ The corresponding measure is EF.

There is a similar theorem involving angle bisectors.

Theorem 7–9	**If two triangles are similar, then the measures of the corresponding angle bisectors of the triangles are proportional to the measures of the corresponding sides.**

Proof of Theorem 7–9

Given: $\triangle RTS \sim \triangle EGF$

$\overline{TA}$ is an angle bisector of $\angle RTS$.

$\overline{GB}$ is an angle bisector of $\angle EGF$.

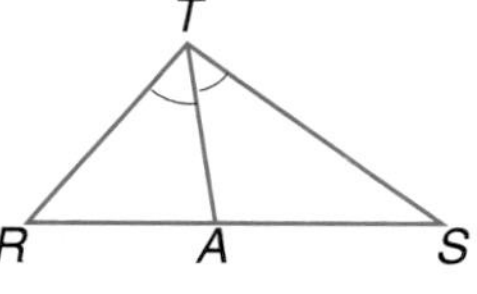

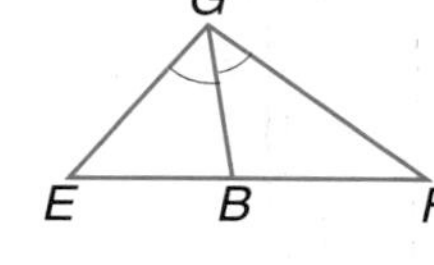

Prove: $\frac{TA}{GB} = \frac{RT}{EG}$

Paragraph Proof:

Since $\triangle RTS \sim \triangle EGF$, $\angle R$ and $\angle E$ are corresponding angles, and $\angle RTS$ and $\angle EGF$ are corresponding angles. Thus, $\angle R \cong \angle E$ and $\angle RTS \cong \angle EGF$. If $\overline{TA}$ and $\overline{GB}$ bisect $\angle RTS$ and $\angle EGF$, respectively, then $\angle RTA \cong \angle EGB$. $\triangle RTA \sim \triangle EGB$ by AA Similarity. Thus, $\frac{TA}{GB} = \frac{RT}{EG}$.

The following theorem states the relationship between the medians of two similar triangles. *You will be asked to prove this theorem in Exercise 39.*

Theorem 7–10	**If two triangles are similar, then the measures of the corresponding medians are proportional to the measures of the corresponding sides.**

The theorems about the relationships of special segments in similar triangles can be used to solve practical problems.

Example 3

Photography

In photography class, Sua made a camera from a box by poking a small hole P in the center of one side of the box. Suppose Sua photographed a flowering bush that was 24 inches tall. How tall will the image of the bush be on the film, if the film is 5 inches from the lens and the camera is 40 inches from the bush?

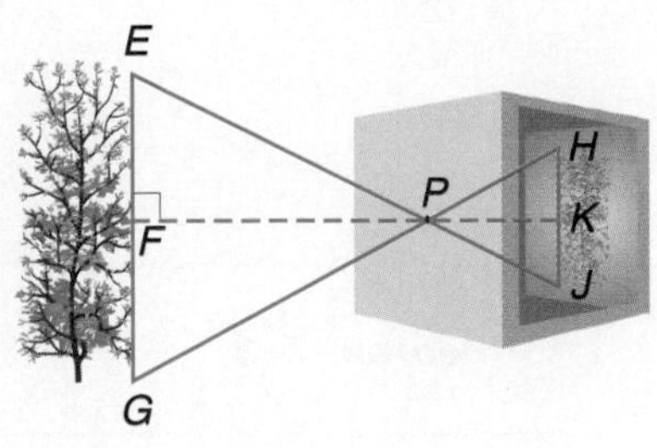

The dashed lines are the altitudes of $\triangle EGP$ and $\triangle JHP$.

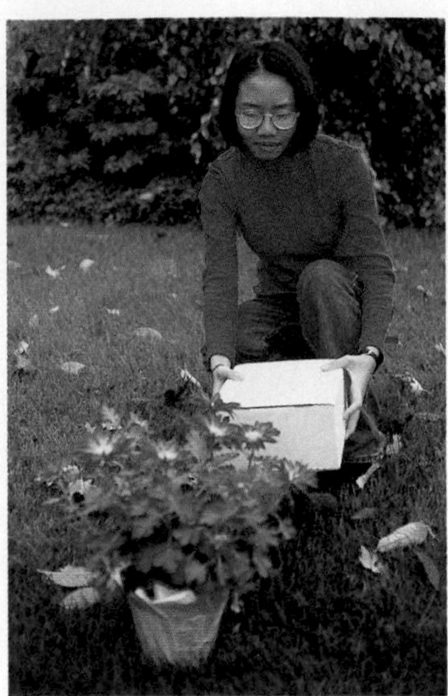

Let b represent the height of the bush on the film. Assume that $\overline{EG} \parallel \overline{HJ}$. $\angle PEF \cong \angle PJH$ and $\angle PGE \cong \angle PHJ$ since they are alternate interior angles. Thus, $\triangle EPG \sim \triangle JPH$ by AA Similarity.

The measures of the corresponding altitudes of similar triangles are proportional to the measures of corresponding sides.

$$\frac{PF}{PK} = \frac{EG}{JH}$$

$$\frac{40}{5} = \frac{24}{b} \quad \text{Substitution Property (=)}$$

$$40b = 120 \quad \text{Cross products}$$

$$b = 3$$

The image of the bush on the film will be 3 inches tall.

There is also a relationship between an angle bisector and the side of the triangle opposite the angle.

Theorem 7–11 Angle Bisector Theorem	**An angle bisector in a triangle separates the opposite side into segments that have the same ratio as the other two sides.**

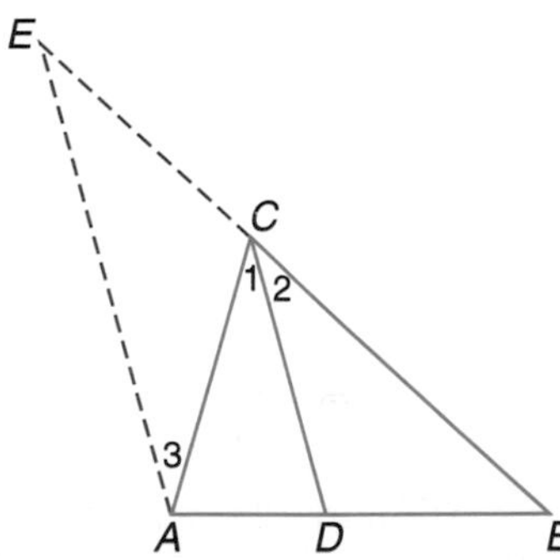

$\overline{CD}$ is the bisector of $\angle ACB$ in $\triangle ABC$. According to Theorem 7–11, $\frac{AD}{DB} = \frac{AC}{BC}$. To prove this, construct a line through point A parallel to $\overline{DC}$ and meeting $\overrightarrow{BC}$ at E. You can prove that $\frac{AD}{DB} = \frac{EC}{BC}$ and $CA = CE$. *You will be asked to prove this theorem in Exercise 40.*

CHECK FOR UNDERSTANDING

Communicating Mathematics

Study the lesson. Then complete the following.

1. **Write** the ratio of the perimeters of two similar triangles if the ratio of their corresponding sides is 3 to 1. Explain how you know.
2. **Explain** what must be true about $\triangle EFG$ and $\triangle RST$ before you can conclude that $\frac{FH}{SW} = \frac{EF}{RS}$.

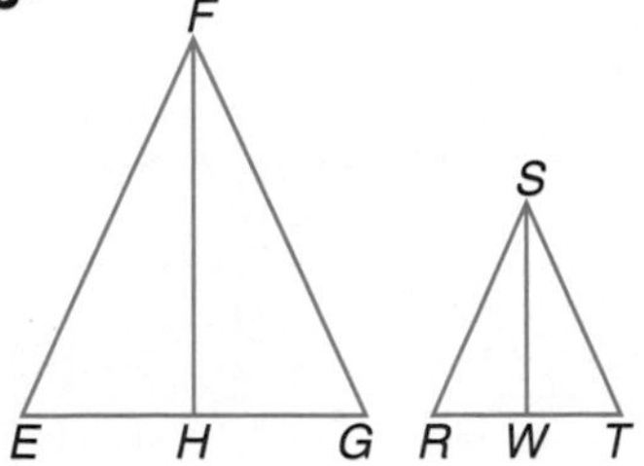

3. **Describe** how Theorem 7–8 applies to the three sets of altitudes in two similar right triangles.

4. **You Decide** Ayashe was given the figure at the right. She concluded that $\frac{AB}{BC} = \frac{DC}{AD}$ by the Angle Bisector Theorem. Was she correct? Why or why not?

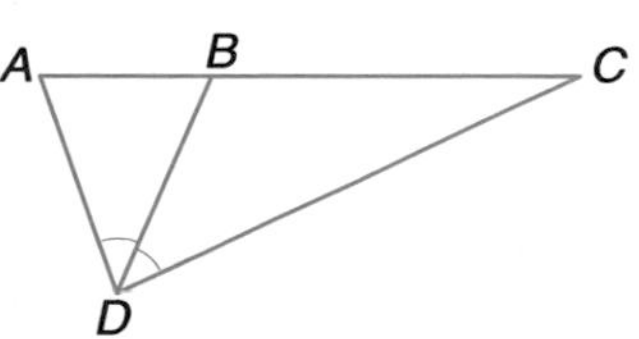

Guided Practice

In the figure at the right, $\triangle ADC \sim \triangle GDE$. Complete each proportion. Give a reason for your answer.

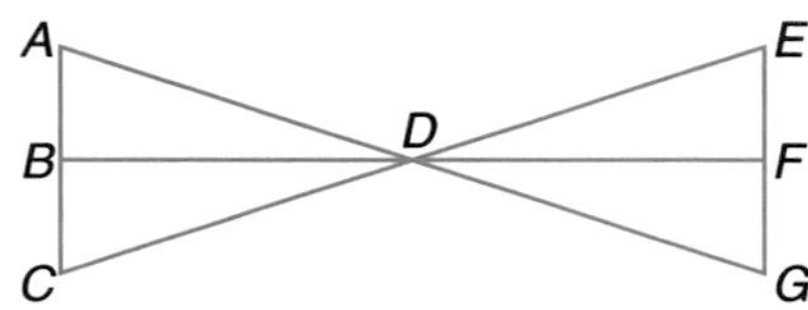

5. If BD bisects $\angle ADC$, $\frac{AD}{DC} = \frac{?}{BC}$.

6. If B and F are midpoints, $\frac{DC}{DE} = \frac{BD}{?}$.

Find the value of x.

7. 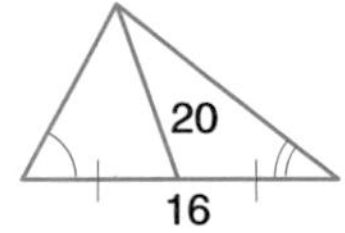

8. 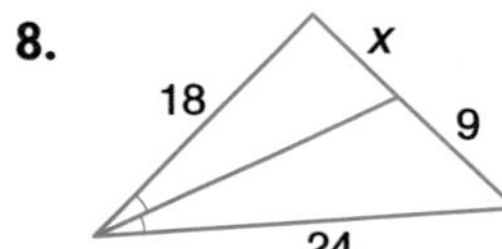

9. In the figure at the right, $\triangle RST \sim \triangle EFG$, and $\overline{SA}$ and $\overline{FB}$ are altitudes. Find FB.

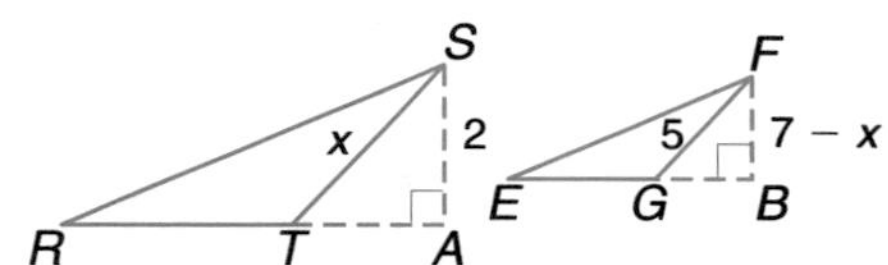

10. Write a two-column proof.

Given: $\overline{VR} \parallel \overline{WS}$

$\overline{WS}$ bisects $\angle RWT$.

Prove: $\frac{RW}{WT} = \frac{RS}{ST}$

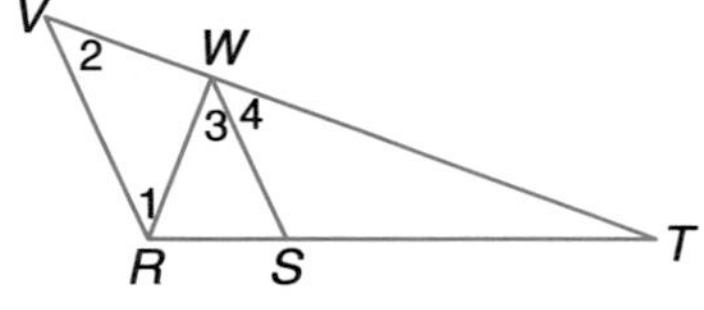

11. **Eliminate the Possibilities** The perimeter of one triangle is 24 centimeters, and the perimeter of a second triangle is 36 centimeters. Find measures for the sides of the two triangles so the triangles will be similar.

EXERCISES

Practice

Refer to $\triangle AEC$ for Exercises 12–17. $\overline{EA} \parallel \overline{DB}$, $EA = 12$, and D and B are midpoints of $\overline{EC}$ and $\overline{AC}$, respectively.

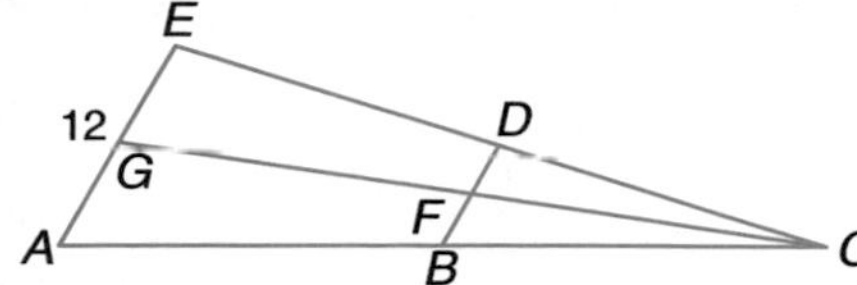

12. If $\angle DCF \cong \angle BCF$, $DF = 2$, and $BC = 8$, $DC = $ __?__.

13. If $\overline{GC} \perp \overline{EA}$ and $FC = 4.3$, $GC = $ __?__.

14. If F and G are midpoints and $GC = 14$, $FC = $ __?__.

15. If $FG = 18$ and $ED = 16$, $FC = \underline{\ ?\ }$.

16. If $\overline{CF} \perp \overline{BD}$ and $\overline{CG} \perp \overline{EA}$, $\frac{CF}{CG} = \frac{?}{CE}$.

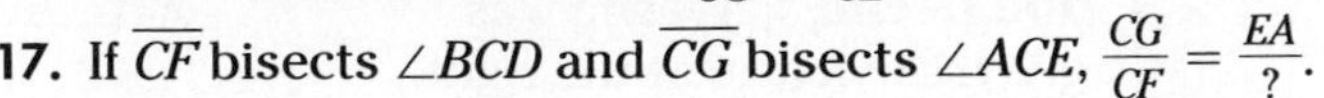

17. If $\overline{CF}$ bisects $\angle BCD$ and $\overline{CG}$ bisects $\angle ACE$, $\frac{CG}{CF} = \frac{EA}{?}$.

Algebra

Find the value of x.

18.

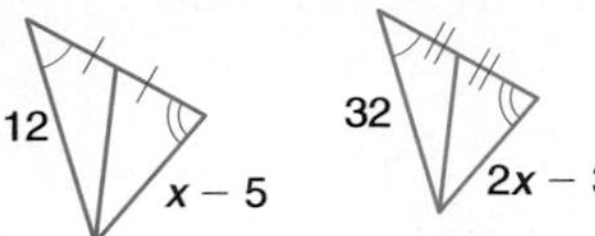

19.

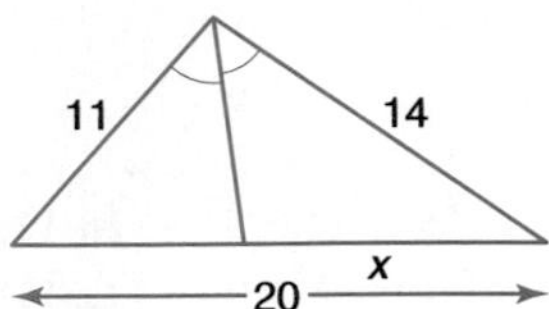

20.

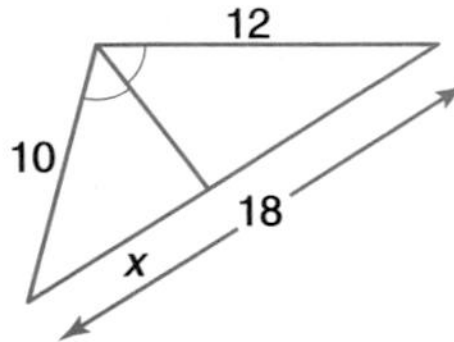

21.

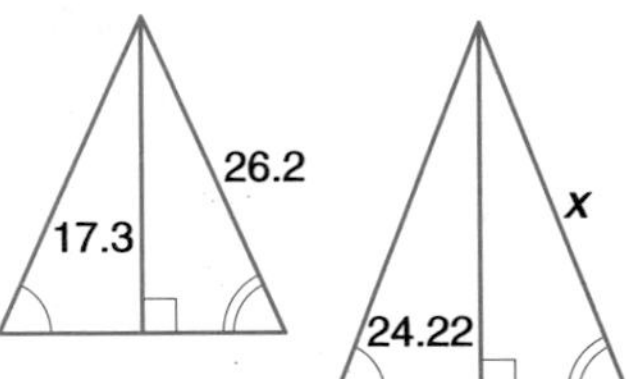

22.

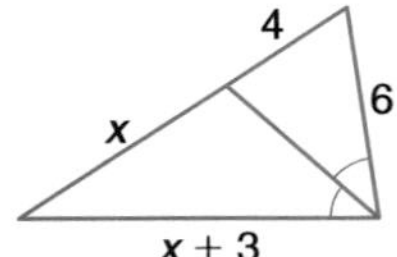

23.

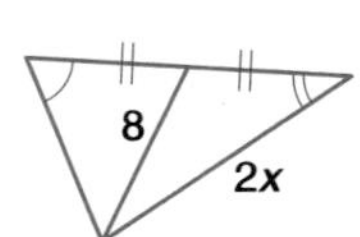

Refer to the figure at the right for Exercises 24–26. $\triangle ABS \sim \triangle RTS$.

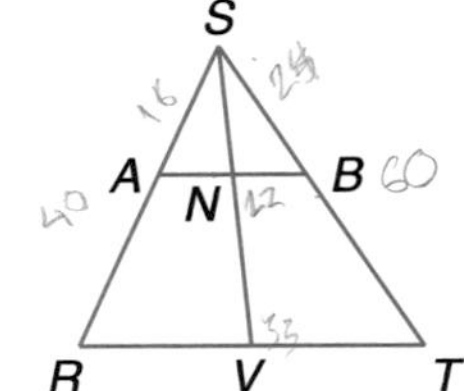

24. $\overline{SV}$ bisects $\angle ASB$. If $AS = 7$ and $AR = 11$, and $SV = 9$, find SN.

25. N is the midpoint of $\overline{AB}$ and V is the midpoint of $\overline{RT}$. If $AB = 16$, $SN = 13$, and $RT = 36$, find SV.

26. If $AB = 22$, $SB = 24$, $AS = 18$, and $RS = 40$, find the perimeter of $\triangle RTS$.

27. In the figure at the right, $\triangle RST \sim \triangle UVW$, $\overline{TA}$ and $\overline{WB}$ are medians. If $TA = 8$, $RA = 3$, $WB = 3x - 6$, and $UB = x + 2$, find UB.

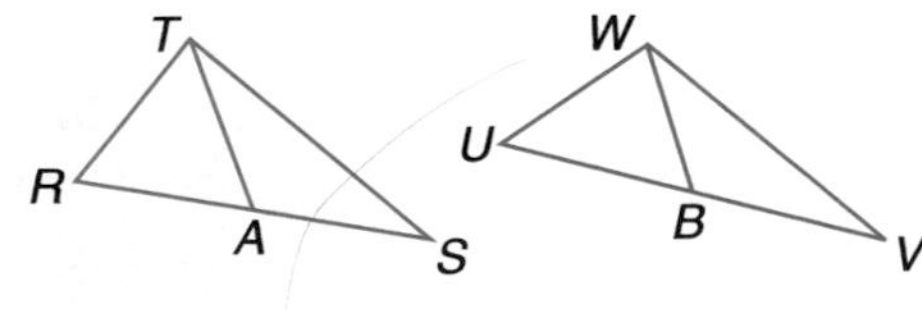

28. The two triangles in the figure at the right are similar. If the perimeter of $\triangle ABC$ is 28, find the value of x.

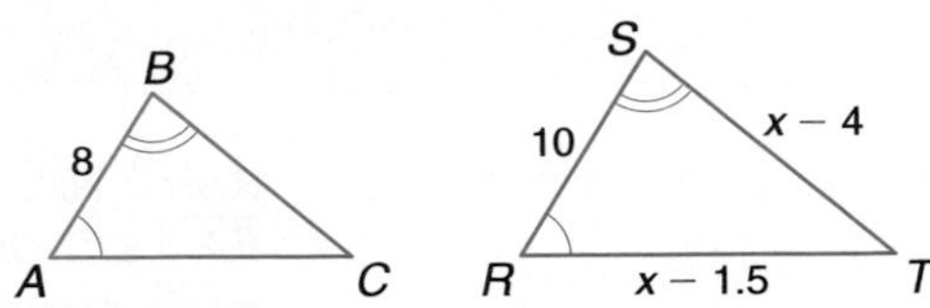

29. In the figure at the right, $\overline{BF}$ bisects $\angle ABC$ and $\overline{AC} \parallel \overline{ED}$. If $BA = 6$, $BC = 7.5$, $AC = 9$, and $DE = 9$, find CF and BD.

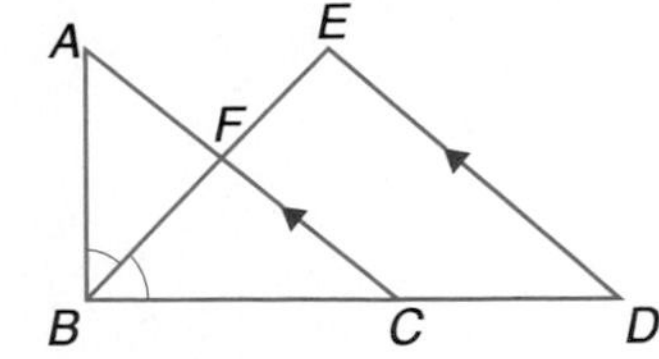

30. In the figure at the right, $\overline{BC} \parallel \overline{EF}$ and $\triangle ABC$ has vertices $A(0, 0)$, $B(6, 0)$ and $C(0, 8)$. Find the perimeter of $\triangle DEF$. Explain your reasoning.

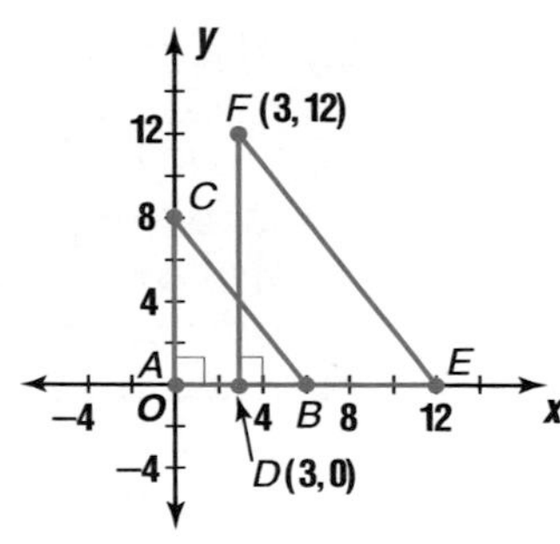

31. The measures of the sides of a triangle are 20, 24, and 30. Find the measures of segments formed where the bisector of the smallest angle meets the opposite side.

32. If the bisector of an angle of a triangle bisects the opposite side, what is the ratio of the measures of the other two sides of the triangle?

33. If $\triangle ABC \sim \triangle RST$, $AB = 10.2$, $RS = 12.24$, and the perimeter of $\triangle RST$ is 32, find the perimeter of $\triangle ABC$.

Write a two-column proof.

34. **Given:** $\overline{RU}$ bisects $\angle SRT$.
$\overline{VU} \parallel \overline{RT}$

Prove: $\frac{SV}{VR} = \frac{SR}{RT}$

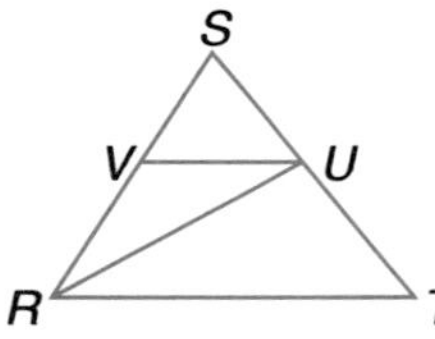

35. **Given:** $\overline{JF}$ bisects $\angle EFG$.
$\overline{EH} \parallel \overline{FG}$, $\overline{EF} \parallel \overline{HG}$

Prove: $\frac{EK}{KF} = \frac{GJ}{JF}$

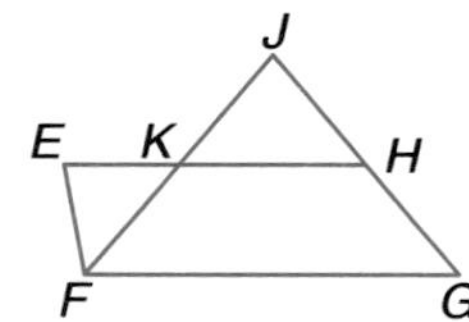

36. **Given:** $\triangle ABC \sim \triangle PQR$
$\overline{BD}$ is an altitude of $\triangle ABC$.
$\overline{QS}$ is an altitude of $\triangle PQR$.

Prove: $\frac{BD}{QS} = \frac{BA}{QP}$

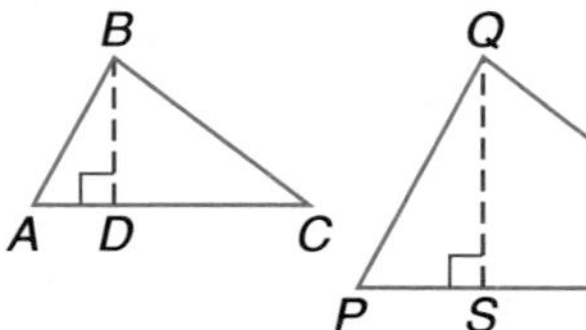

37. Write a flow proof.

Given: $\angle C \cong \angle BDA$

Prove: $\frac{AC}{DA} = \frac{AD}{BA}$

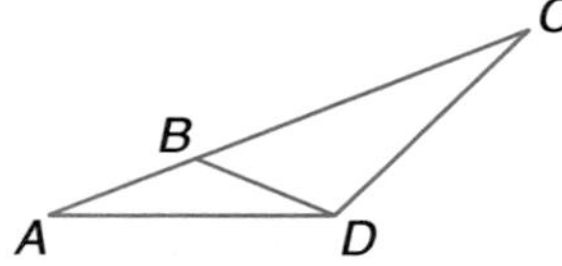

38. Prove using any method.

Given: $\triangle RST \sim \triangle ABC$
W and D are midpoints of $\overline{TS}$ and $\overline{CB}$, respectively.

Prove: $\triangle RWS \sim \triangle ADB$

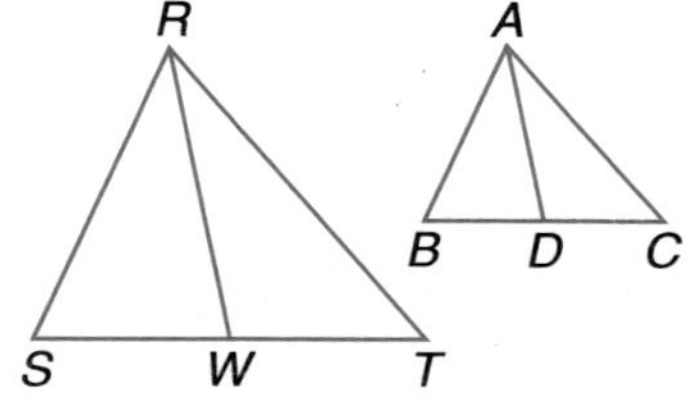

Write a paragraph proof.

39. If two triangles are similar, then the measures of the corresponding medians are proportional to the measures of the corresponding sides. (Theorem 7–10)

40. An angle bisector in a triangle separates the opposite side into segments that have the same ratio as the other two sides. (Theorem 7–11)

Programming

41. The graphing calculator program at the right finds the area of a triangle.

```
PROGRAM: AREA
:Input "SIDE 1 =", A
:Input "SIDE 2 =", B
:Input "SIDE 3 =", C
:If A+B≤C
:Goto 1
:If A+C≤B
:Goto 1
:If B+C≤A
:Goto 1
:(A+B+C)/2→S
:√(S(S−A)(S−B)(S−C))→K
:Disp"AREA=",K
:Stop
:Lbl 1
:Disp "NO TRIANGLE"
:Stop
```

Use the program to find the areas of similar triangles whose sides have the given measures. Then find the ratio of the measures of the sides and the ratio of the areas of the triangles. Write a conjecture about the relationship between the areas of two similar triangles.

a. 3, 4, 5 and 6, 8, 10

b. 8, 11, 17 and 24, 33, 51

c. 5.1, 6.3, 8.1 and 30.6, 37.8, 48.6

Critical Thinking

42. Consider the two rectangular prisms shown at the right.

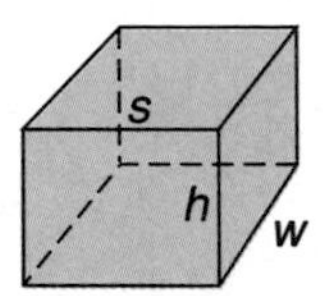

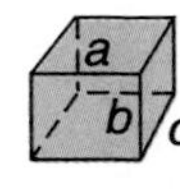

a. What ratios are necessary to determine whether they are similar?

b. If two rectangular prisms are similar, what relationship will exist between the sum of the edges in each figure?

c. If the first prism has dimensions that are three times as large as the second, will the volumes also have a ratio of 1 to 3? Explain why or why not.

Applications and Problem Solving

43. Design Julian had a picture 18 centimeters by 24 centimeters that he wanted enlarged by 30% and then have the inside of the frame edged with navy blue piping. The store only had 110 centimeters of navy blue piping in stock. Will this be enough piping to fit on the inside edge of the frame? Explain.

44. Photography A camera is 10 centimeters long and can have an image no more than 5 centimeters high. Estrella is 165 centimeters tall. How far from the camera should Estrella stand in order to have a full-length picture?

45. Physical Fitness Two triangular jogging paths are laid out in a park as shown. The dimensions of the inner path are 300 meters, 350 meters, and 550 meters. The shortest side of the outer path is 600 meters. Will a jogger on the inner path run half as far as the one on the outer path? Explain.

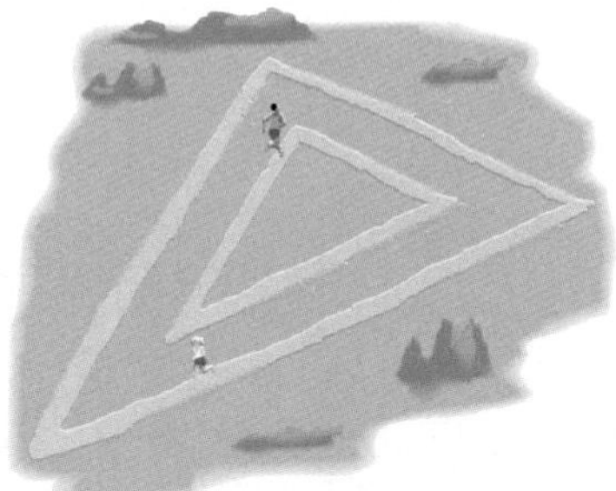

46. Art The design at the right is an abstract pentagonal design that illustrates a sequence of ratios in the relationships among the segments. Copy the figure and measure the segments. Use your measurements to investigate the similar figures in the design. Find any angle bisectors, altitudes or medians and determine if they are illustrations of the theorems in this lesson.

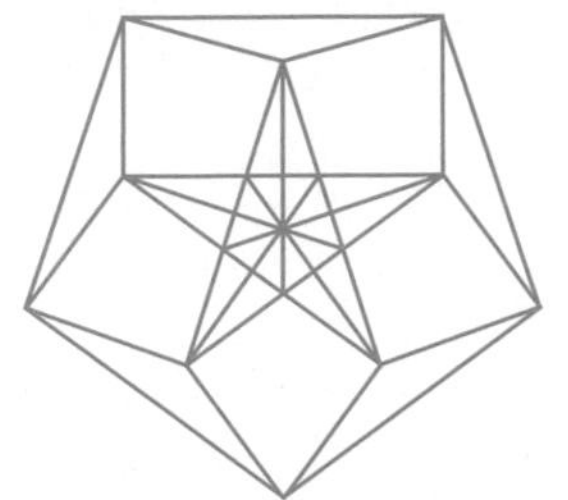

Mixed Review

47. Find the value of x. (Lesson 7–4)

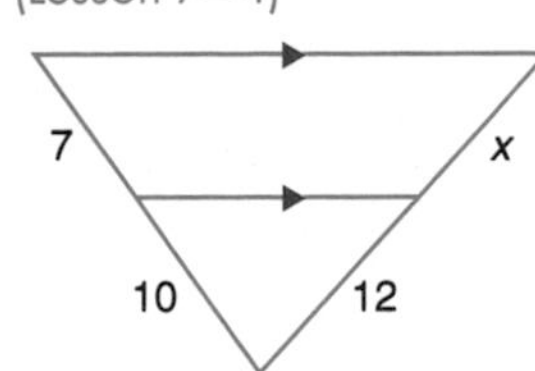

48. Determine whether the pair of triangles below is similar. Explain your answer. (Lesson 7–3)

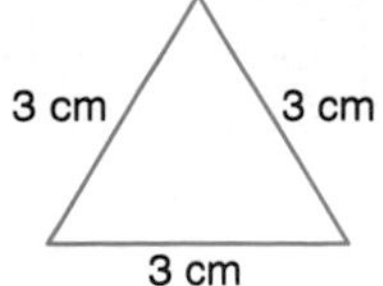

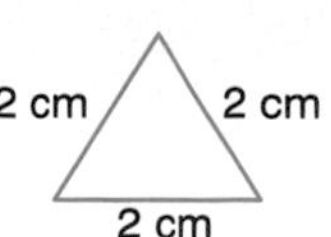

49. Sports The ratio of seniors to juniors on a powder puff football team is 2:3. If there are 21 juniors, how many seniors are on the team? (Lesson 7–1)

50. Determine whether the statement *The diagonals of an isosceles trapezoid are congruent* is true or false. (Lesson 6–5)

51. ACT Practice A rectangular playground is surrounded by an 80-foot long fence. One side of the playground is 10 feet longer than the other. Which of the following equations could be used to find s, the shorter side of the playground?

A $10s + s = 80$ **B** $4s + 10 = 80$ **C** $s(s + 10) = 80$
D $2(s - 10) + 2s = 80$ **E** $2(s + 10) + 2s = 80$

52. Draw and label $\triangle QRS$ with altitudes $\overline{RA}$ and $\overline{QB}$. (Lesson 5–1)

53. If $\triangle ABC$ has sides with measures 4, 8, and 10 and $\triangle RST$ has sides with measures 2, 4, and 5, is $\triangle ABC \cong \triangle RST$? Explain. (Lesson 4–4)

54. Two lines have slopes -4 and $\frac{1}{2}$. State whether or not the lines are perpendicular. Explain. (Lesson 3–3)

55. Write the statement *An 89 is an above average score on the geometry test* in if-then form. (Lesson 2–2)

For **Extra Practice**, see page 777.

INTEGRATION Algebra

56. Solve $|3 - t| \leq 1$.

57. Solve $|3 + 2a| < 11$.

WORKING ON THE In•ves•ti•ga•tion

Refer to the Investigation on pages 286–287.

This Land Is Your Land

1 Review your plan. Make sure that all of the features you wanted to include are placed on the drawing.

2 Check to make sure that your proposal meets the budget requirements. If you are over budget, brainstorm methods for cutting costs. You may wish to cut less popular features, use volunteer labor for construction of less complicated features, or consider building the park in phases when more money is available.

3 Build a three-dimensional model of the park. Draw the boundary lines of the park on a piece of posterboard. Then use modeling clay, construction paper, paint, markers, and whatever else you need to create a three-dimensional model of your design.

Add the results of your work to your Investigation Folder.

7–6

Fractals and Self-Similarity

What YOU'LL LEARN

- To recognize and describe characteristics of fractals, and
- to solve problems by solving a simpler problem.

Why IT'S IMPORTANT

You can observe patterns in nature that are examples of fractal geometry.

One interesting characteristic that can be found in mathematics is that orderly patterns and structure sometimes result from processes that seem to be random and chaotic.

The following activity shows a result from a process called **iteration**. Iteration is a process of repeating the same procedure over and over again.

MODELING MATHEMATICS

Sierpinski Triangle

Materials:  isometric dot paper  ruler

- Draw an equilateral triangle on isometric dot paper where each side is 16 units long (stage 0).
- Connect the midpoints of each side to form another triangle. Shade the center triangle (stage 1).
- Repeat the process using the three nonshaded triangles. Connect the midpoints of each side to form other triangles (stage 2).

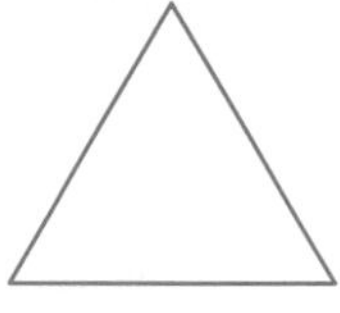
stage 0

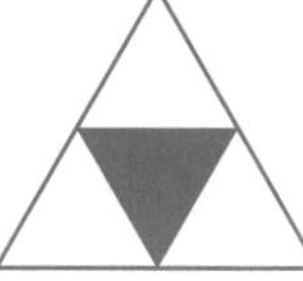
stage 1

stage 2

Your Turn

a. Continue the process through stage 4. How many nonshaded triangles do you have at stage 4?

b. What is the perimeter of a nonshaded triangle in each of the stages?

c. If you continue the process indefinitely, describe what will happen to the perimeter of each nonshaded triangle.

The triangle that will result if you continue to repeat the process in the Modeling Mathematics activity indefinitely is called a **Sierpinski triangle**, named after Polish mathematician Waclaw Sierpinski. A Sierpinski triangle is an example of a mathematical object called a **fractal**. A fractal is a geometric figure that is created using iteration. It is infinite in structure.

Fractal geometry is the geometry of things in nature that are irregular in shape such as clouds, coastlines, or the growth of a tree. Look carefully at the tree in the top picture at the left and observe how the branches have developed. The picture below it shows veins and arteries in a human kidney.

One characteristic of fractals is that they have **self-similar** shapes. That is, the smaller and smaller details of a shape have the same geometrical characteristics as the original, larger form.

Example 1 **Prove that a triangle formed in stage 2 of a Sierpinski triangle is similar to the triangle formed in stage 0.**

The argument will be the same for any triangle in stage 2, so we will use only $\triangle CGJ$ from stage 2.

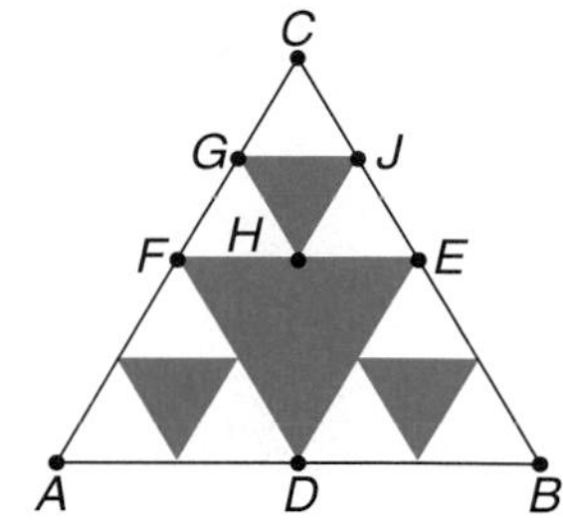

Given: $\triangle ABC$ is equilateral.
D, E, and F are midpoints of sides $\overline{AB}$, $\overline{BC}$, and $\overline{CA}$, respectively.
G, J, and H are midpoints of sides $\overline{FC}$, $\overline{CE}$, and $\overline{FE}$, respectively.

Prove: $\triangle ABC \sim \triangle GJC$

Proof:

Statements	**Reasons**
1. $\triangle ABC$ is equilateral; D, E, F are midpoints of $\overline{AB}$, $\overline{BC}$, $\overline{CA}$; G, J, and H are midpoints of $\overline{FC}$, $\overline{CE}$, $\overline{FE}$	**1.** Given
2. $\overline{FE} \parallel \overline{AB}$	**2.** A segment whose endpoints are the midpoints of 2 sides of a $\triangle$ is $\parallel$ to the third side.
3. $\angle CFH \cong \angle CAB$; $\angle CEF \cong \angle CBA$	**3.** Corresponding $\measuredangle$s Postulate
4. $\triangle FEC \sim \triangle ABC$	**4.** AA Similarity
5. $\overline{GJ} \parallel \overline{FE}$	**5.** A segment whose endpoints are the midpoints of 2 sides of a $\triangle$ is $\parallel$ to the third side.
6. $\angle CGJ \cong \angle CFE$; $\angle CJG \cong \angle CEF$	**6.** Corresponding $\measuredangle$s Postulate
7. $\triangle GJC \sim \triangle FEC$	**7.** AA Similarity
8. $\triangle ABC \sim \triangle GJC$	**8.** Transitive Property ($\sim$)

Thus, using the same reasoning, every triangle in stage 2 is similar to the original triangle in stage 0.

A figure is **strictly self-similar** if any of its parts, no matter where they are located or what size is selected, contain the same figure as the whole. The Sierpinski triangle is strictly self-similar.

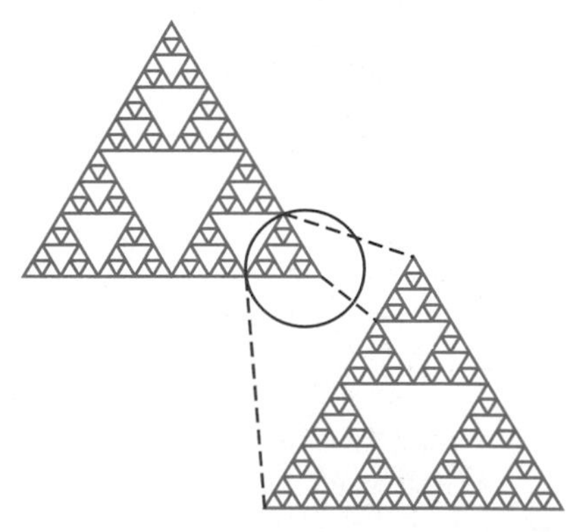

Blaise Pascal

Other repeated patterns are not self-similar. Pascal's triangle is a famous array of numbers that is generated according to a certain pattern. The first number in the initial row is 1. The number in each successive row is the sum of the two numbers above it. How could you find the sum of values in the tenth row?

When you are working on a complicated or unfamiliar problem, it is often helpful to first **solve a simpler problem** that is similar to the original. Then, use the strategy that worked on that simpler problem to solve the original problem. You can use this strategy to find the sum of the values in the tenth row of Pascal's triangle.

Example 2

PROBLEM SOLVING

Solve a Simpler Problem

a. Find a formula in terms of the row number for the sum of the values in any row in the triangle.

b. What is the sum of the values in the tenth row of Pascal's triangle?

a. To find the sum of the values in the tenth row, we can investigate a simpler problem. What is the sum of values in the first four rows of the triangle?

Pascal's Triangle

Row		Sum
1	1	$1 = 2^0$
2	1 1	$2 = 2^1$
3	1 2 1	$4 = 2^2$
4	1 3 3 1	$8 = 2^3$
5	1 4 6 4 1	$16 = 2^4$

It appears that the sum of any row is a power of 2. The formula is 2 to a power that is one less than the row number: $S_n = 2^{n-1}$.

b. The sum of the values in the tenth row will be 2^{10-1} or 512.

There is a relationship between Pascal's triangle and the Sierpinski triangle.

Example 3

Replace each of the even numbers in Pascal's triangle with a 0 and each of the odd numbers with a 1. Then color each 1 and leave the 0s uncolored. Describe the picture you will have if you generate eight rows of Pascal's triangle and color the cells according to the rule.

First, generate eight rows of the triangle and replace each number as even or odd.

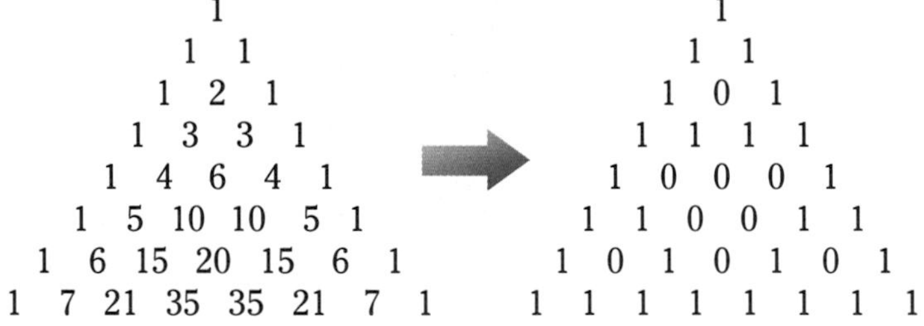

Now color each 1. The result looks like a stage 3 Sierpinski Triangle.

You can generate many other fractal images using an iterative process. For example, trisect a segment of a given length. Replace the middle segment with two segments each the same length as the segment removed, as shown in the diagram at the right. Repeat the process on each of the four segments in stage 2, then continue to repeat the process. The fractal image generated by this process is called a *Koch curve*.

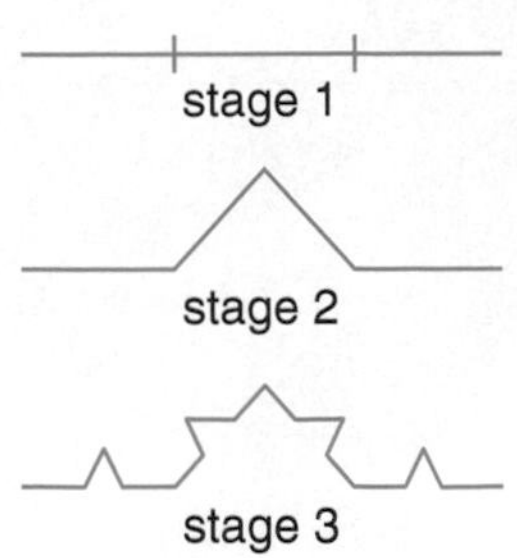

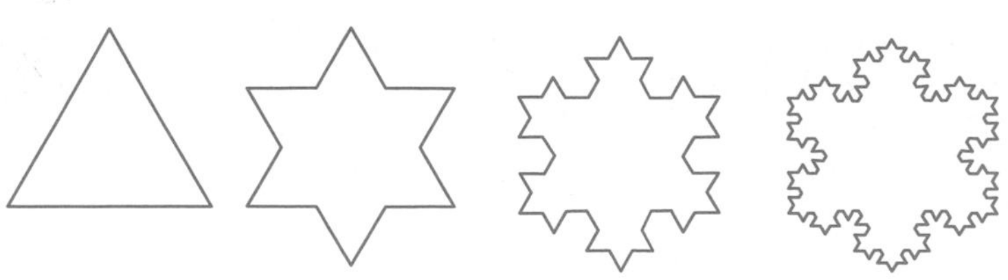

If you generate a Koch curve on each side of an equilateral triangle, you will produce an image called *Koch's snowflake*.

CHECK FOR UNDERSTANDING

Communicating Mathematics

Study the lesson. Then complete the following.

1. **Describe** a specific iterative process. Then draw examples of four stages of that iteration.
2. **Describe** two ways you can find the sum of a row in Pascal's triangle.
3. **State** the number of segments that are in each of the first three stages of Koch's snowflake. What pattern can you use to predict the number of segments in stage 4?

Use grid paper to draw a square that is 27 units long on each side. Connect the trisection points on each side to make nine squares. Then shade the middle square. This is stage 1. Repeat this process on the eight outside squares. The middle square remains a "hole."

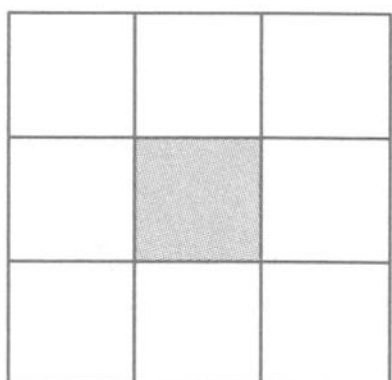

4. Draw stage 2 and stage 3. How many holes are there in each stage?
5. If you continue the process indefinitely, will the figure you obtain be strictly self-similar? Explain.

Guided Practice

Count the number of dots in each arrangement. These numbers are called *triangular numbers*. The second triangular number is 3 because there are three dots in the array.

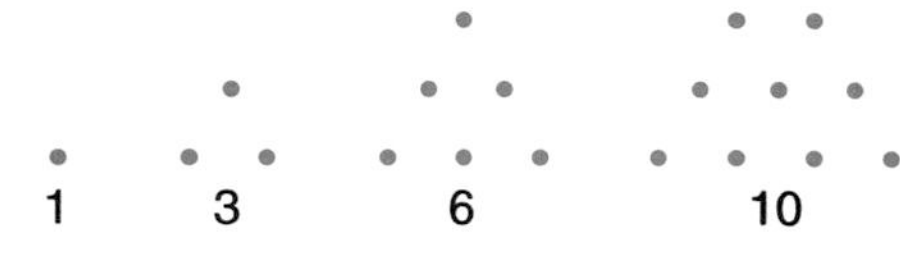

6. How many dots will be in the seventh triangular number?
7. Generate Pascal's triangle and look at the third diagonal from either side. Describe the pattern you see. How does it relate to the triangular numbers?
8. **Science** Find at least two examples in nature excluding those mentioned in the lesson in which a figure appears to be self-similar.

EXERCISES

Practice

Draw an equilateral triangle, trisect the three sides, and connect the points. Shade the three inside triangles as shown at the right.

9. Prove that one of the nonshaded triangles is similar to the original triangle.
10. Using the figure from Exercise 9, trisect the three sides in each of the nonshaded triangles and connect the points as in the original figure. Repeat the process once more on this figure.
    a. Is the new figure strictly self-similar? Explain.
    b. How many nonshaded triangles are in stages 1 and 2?

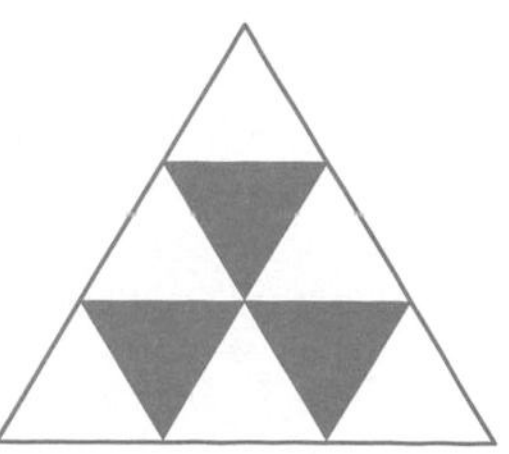

11. Find the value of each expression. Then, use that value as the next x in the expression. Repeat the process until you can make some observations. Describe what happens in each of the iterations.
   a. $\sqrt{x}$, where x initially equals 12
   b. $\frac{1}{x}$, where x initially equals 5
   c. $x^{\frac{1}{3}}$, where x initially equals 0.3

12. Use a calculator to find the square root of 2. Then take the square root of the result. Continue to repeat the process.
   a. What was your result after you repeated the process many times?
   b. Was the process you used an iterative process? Explain.

A "fractal tree" can be drawn by making two new branches from the endpoint of each original branch, each one-third as long as the previous branch.

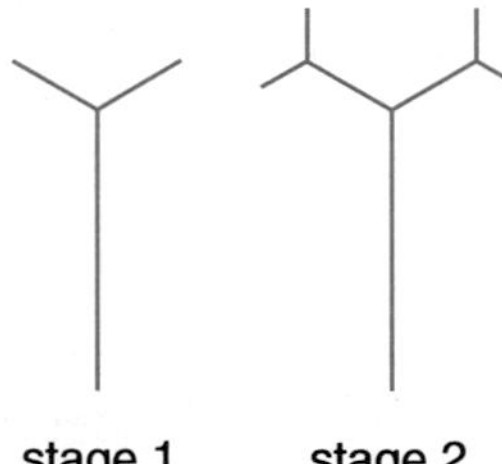

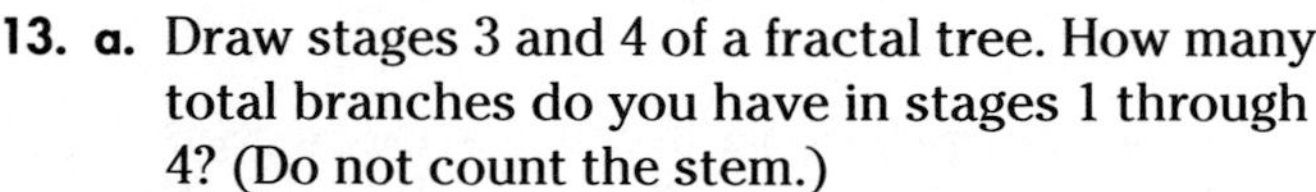

13. a. Draw stages 3 and 4 of a fractal tree. How many total branches do you have in stages 1 through 4? (Do not count the stem.)
   b. Look for a pattern that could be used to predict the number of branches at each stage.

14. Is a fractal tree strictly self-similar? Explain.

15. Generate Pascal's triangle. Divide every entry by 3. If the remainder is 1 or 2, shade the number cell black. If the remainder is 0, leave the cell unshaded. Describe the pattern that emerges.

16. Refer to the Koch curve on page 380.
   a. What is a formula for the number of segments in terms of the stage number? Use your formula to predict the number of segments in stage 8 of a Koch curve.
   b. If the length of the original segment is 1 unit, how long will the segments be in each of the first four stages? What will happen to the length of each segment as the number of stages continues to increase?

17. Refer to the Koch snowflake on page 381. At stage 1, the length of each side is 1 unit.
   a. What is the perimeter at each of the first four stages of a Koch snowflake?
   b. What is a formula for the perimeter in terms of the stage number? Describe the perimeter as the number of stages continues to increase.

18. Consider a Koch snowflake in stage 1. Write a paragraph proof to show that the triangles generated on the sides are similar to the original triangle.

Critical Thinking

19. The fractal in the pictures at the right is the space-filling *Hilbert curve*.
   a. Study the pictures carefully. Then define the iterative process used to generate the curve.
   b. Why do you think it is called "space filling"?

Applications and Problem Solving

20. **Nature** Some of these pictures are of real objects, and others are fractal images of objects. Compare the pictures and identify those you think are of real objects. Describe the characteristics of fractals that are shown in the images.

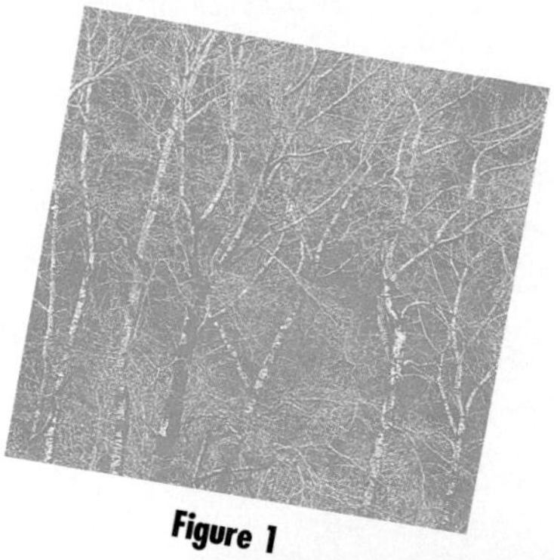

Figure 1

Figure 2

Figure 3

Figure 4

Fractal Web

21. **Solve a Simpler Problem** Look at the diagonals in Pascal's triangle.
    a. Find the sum of the first 25 numbers in the outside diagonal.
    b. Find the sum of the first 50 numbers in the second diagonal.
22. **Art** Describe how artist Edward Berko used iteration and self-similarity in his painting *Fractal Web*, shown at the left.

Mixed Review

23. In the figure, $\triangle RST \sim \triangle WVU$. If $UV = 500$, $VW = 400$, $UW = 300$, and $ST = 1000$, find the perimeter of $\triangle RST$. (Lesson 7–5)

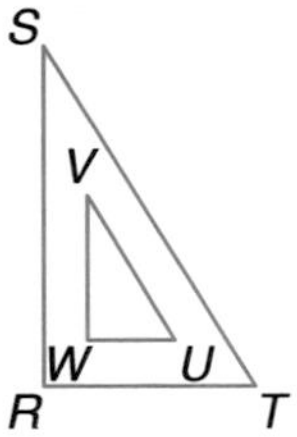

24. Find the values of x and y in the figure below. (Lesson 7–4)

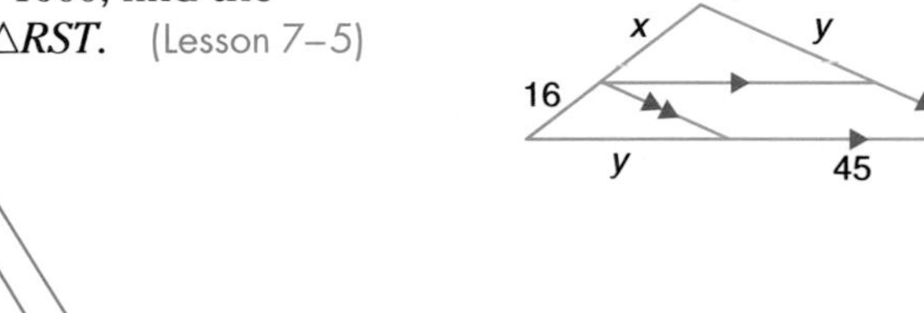

25. Graph $\triangle RST$ with vertices $R(-1, -4)$, $S(1, 3)$, and $T(-6, 2)$. Find the slope of the perpendicular bisector of $\overline{RS}$. (Lesson 5–1)
26. State whether 35, 45, and 55 are possible angle measures of a triangle. Explain. (Lesson 4–2)

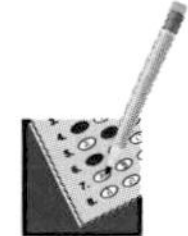

27. **SAT Practice** If the two sides of a triangle measure 12 and 7, then which of the following *cannot* be the perimeter of the triangle?

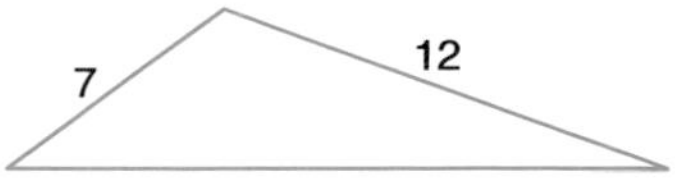

    A 20 **B** 29 **C** 34 **D** 37 **E** 38

Algebra

28. **Biology** Hair grows at a rate of 0.00000001 miles per hour. Express this number in scientific notation.
29. Find the solution set for $y = 3x + 2$ if the domain is $\{-1, 0, 2, 4\}$.

For **Extra Practice**, see page 777.

An Extension of Lesson 7–6

7–6B Chaos Game

Materials: 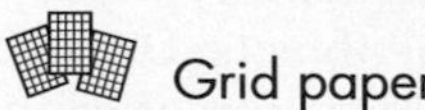 Grid paper 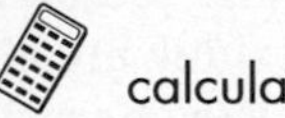 calculator  dice spinner

Fractal designs show that patterns can result from purely random events. The *Chaos game* shows how fractal designs can be created by using random numbers.

Activity 1 **On a coordinate plane, graph the three vertices of an equilateral triangle, *A*(0, 0), *B*(1, 0), and *C*(0.5, 0.9). Let each square represent 0.1 unit.**

- Graph any point *P* inside of the triangle.
- Generate a random number 1, 2, or 3 by using a calculator, tossing a die, or using a spinner that has three congruent parts.
- For a 1, move halfway to *A* and mark the midpoint; for a 2, move halfway to *B* and mark the midpoint, and for a 3, move halfway to *C* and mark the midpoint.

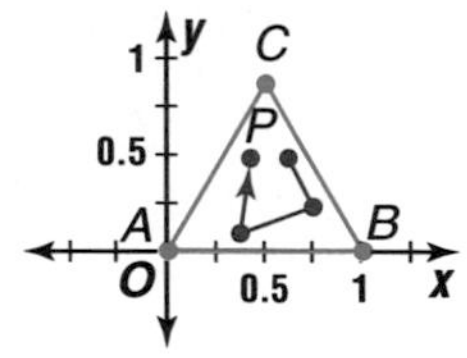

- After you have moved, connect the point to the newly located midpoint. Use that midpoint as your starting point and repeat the process.
- Play the game again with a triangle made on overhead transparency using the same scale as everyone else in class. This time mark only the successive midpoints.
- Play the game to obtain about 50 midpoints. Combine your set of midpoints with others in class by placing your transparencies in a stack. What conclusion can you make?

In Activity 1, you should have discovered that when all of the transparencies are stacked on the overhead projector, a pattern will appear that will be similar to a Sierpinski triangle. You can also use a Sierpinski triangle to determine the probablility of finding certain midpoints.

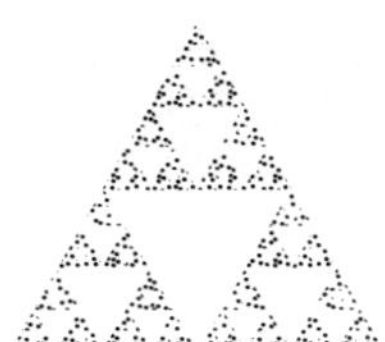

Activity 2 **The triangles in a Sierpinski triangle can be labeled by using *T* for top triangle, *R* for right triangle, and *L* for left triangle as shown in the figure. What is the probability that you will mark a midpoint in triangle *T* after the first random number has been generated? after the second random number has been generated?**

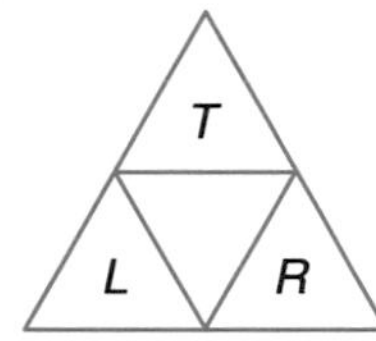

- Make a tree diagram to list the three possible outcomes and probabilities for the first outcome and follow each by the three possible outcomes and probabilities for the second outcome.

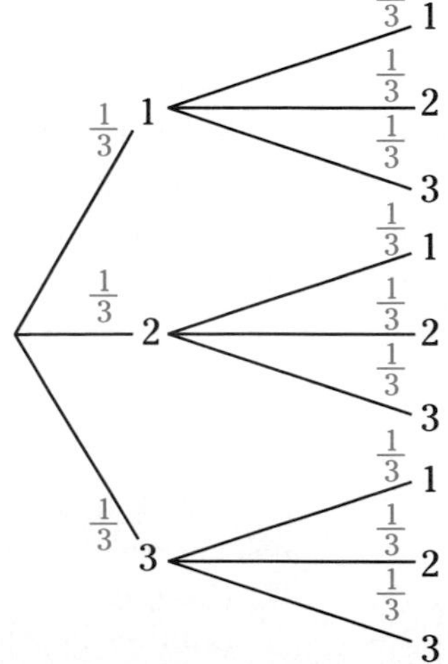

- Look at the pattern in the diagram on the previous page. No matter what initial point you begin with, you have a one-third chance to randomly generate each of the numbers 1, 2, or 3. Suppose 1, 2, and 3 are replaced by *T*, *L*, and *R*, respectively, as shown at the right. If 1 is the number associated with triangle *T*, then you will have a $\frac{1}{3}$ probability of marking a midpoint in triangle *T* after the first random number has been generated. You will have $\frac{1}{3} \cdot \frac{1}{3}$ probability of generating two 1s in a row or *TT*, so you will have a $\frac{1}{9}$ probability of marking a midpoint in triangle *T* after two moves.

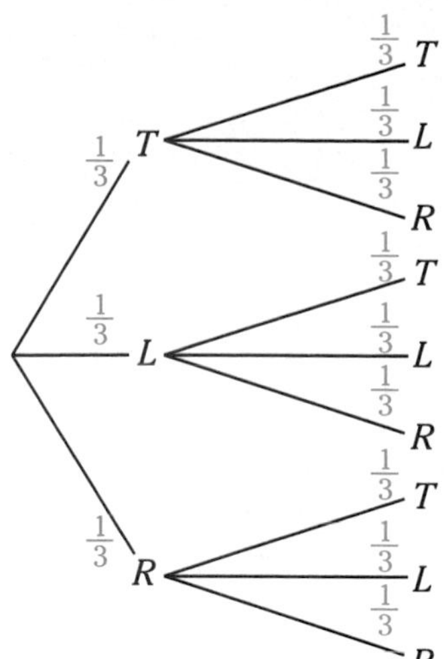

- There are three ways you can end up in triangle *T*. They are *LT*, *RT*, or *TT*. The probability of ending up in *LT* is $\frac{1}{9}$, in *RT*, $\frac{1}{9}$, and in *TT*, $\frac{1}{9}$. Therefore, the probability of ending up in triangle *T* is $\frac{1}{9} + \frac{1}{9} + \frac{1}{9} = \frac{3}{9}$ or $\frac{1}{3}$.

Model

1. Suppose you are playing the Chaos game. Given triangle *ABC* with vertices *A*(0, 0), *B*(1, 0), and *C*(0.5, 0.9), show that when you begin with each of the following points, the midpoint you generate in stage 1 will never end up in the center triangle.
   a. *P*(0.2, 0.4) **b.** *P*(0.5, 0.2)
   c. What is the probability you will end up in triangle *T* after three moves?
   d. What is the probability you will end up in triangle *L* after two moves?
   e. What is the probability you will mark midpoints in triangle *T*, then in triangle *L*?
2. Given $\triangle ABC$ from Exercise 1, begin the game from the point *P*(0.3, 0.5).
   a. Show that in the first move you do not mark a midpoint in the center region.
   b. If you generated the random numbers 1, 1, 2, state the coordinates of each midpoint that you would mark. Indicate whether the midpoint is in $\triangle L$, $\triangle R$, $\triangle T$, or the center triangle.
3. If you are playing the Chaos game, state the probability that you will end up in each location.
   a. *TT* in stage 2 **b.** *T* in stage 3 **c.** *TTT* in stage 3

Write

4. Prove that in playing the Chaos game, if you begin from a point on the perimeter of the original triangle, the midpoint you mark will be on the perimeter of the interior triangle.

CLOSING THE

In·ves·ti·ga·tion

This Land Is Your Land

Refer to the Investigation on pages 276–277.

The competition for a contract to do work for a government agency is regulated by federal, state, and local laws. Each company must submit a bid with a detailed plan for how the work will be completed. The bids and plans are then considered, and the contract is awarded.

Analyze

You have made a master plan and three-dimensional model of your design. A budget has also been completed. Now analyze your design and verify your conclusions.

PORTFOLIO ASSESSMENT

You may want to keep your work on this Investigation in your portfolio.

1. Verify the dimensions and placement of each park feature in your plan.
2. Compare your plan to your three-dimensional model. Make sure that all features are accurately represented on both.
3. Organize your financial bid. It should include the cost of supplies, labor, building materials, and a margin of profit for your company.

Present

Select ten members of your class to represent the county commissioners. The rest of your class represents the members of the general public who have come to speak out on the issues related to the park. Present your proposed park design.

4. Begin your presentation by explaining which park features you chose to incorporate. Explain how you chose the features.
5. Explain how you developed your design. Note any difficulties that you encountered or compromises that had to be made because of space or budget requirements.
6. Discuss the project budget. Explain any overruns and options for reducing the cost of the park.
7. State the advantages of your park design. Then summarize your presentation with a few statements about why your company would be the best one for the job.

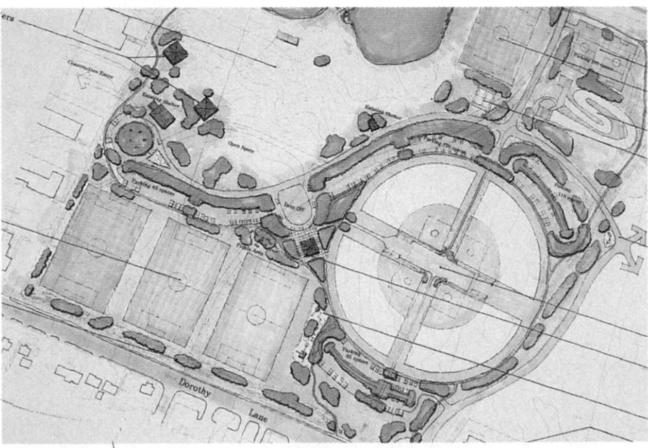

Data Collection and Comparison To share and compare your data with other students, visit: **www.geometry.glencoe.com**

CHAPTER 7 HIGHLIGHTS

Chapter Review For additional lesson-by-lesson review, visit: **www.geometry.glencoe.com**

VOCABULARY

After completing this chapter, you should be able to define each term, property, or phrase and give an example or two of each.

Geometry

dilation (p. 348)
fractal (p. 378)
iteration (p. 378)
scale factor (p. 346)
self-similar (p. 378)
Sierpinski triangle (p. 378)
similar figures (p. 346)
similar polygons (p. 346)
strictly self-similar (p. 379)

Algebra

cross products (p. 339)
equality of cross products (p. 339)
extremes (p. 339)
means (p. 339)
proportion (p. 339)
ratio (p. 338)

Problem Solving

solve a simpler problem (p. 379)

UNDERSTANDING AND USING THE VOCABULARY

State whether each sentence is *true* or *false*. If false, replace the underlined word or expression to make a true sentence.

1. If the measures of the corresponding sides of two triangles are <u>equal</u>, then the triangles are similar.
2. The product of the means <u>equals</u> the product of the extremes.
3. If <u>one</u> angle of a triangle is congruent to <u>one</u> angle of another triangle, then the triangles are similar.
4. A <u>proportion</u> is a comparison of two quantities.
5. If two triangles are similar, then the measures of the corresponding altitudes are proportional to the measures of the corresponding <u>sides</u>.
6. In the proportion $\frac{15}{9} = \frac{5}{3}$, 15 and 3 are called the <u>means</u>.
7. Two polygons are similar if and only if their corresponding sides are <u>proportional</u>.
8. If a line is parallel to one side of a triangle and intersects the other two sides in two distinct points, then it separates these sides into segments of <u>equal</u> lengths.
9. <u>Three</u> or more parallel lines separate transversals into proportional parts.
10. An equation stating that two ratios are equal is a <u>proportion</u>.
11. For any numbers a and c and any nonzero numbers b and d, $\frac{a}{b} = \frac{c}{d}$ if and only if <u>$ab = cd$</u>.
12. A fractal is a geometric figure that has a <u>self-similar</u> shape.

CHAPTER 7 STUDY GUIDE AND ASSESSMENT

SKILLS AND CONCEPTS

OBJECTIVES AND EXAMPLES

Upon completing this chapter, you should be able to:

- recognize and use ratios and proportions (Lesson 7–1)

To solve the equation $\frac{x}{6} = \frac{18}{4}$, find the cross products.

$$\frac{x}{6} = \frac{18}{4}$$
$$4x = 6 \cdot 18$$
$$4x = 108$$
$$x = 27$$

REVIEW EXERCISES

Use these exercises to review and prepare for the chapter test.

Find each ratio and express it as a fraction in simplest form.

13. Loretta's wrist and neck measure 15 centimeters and 30 centimeters, respectively. Find the ratio of her wrist measure to her neck measure.

14. If there are 15 rear sprocket teeth and 55 front sprocket teeth on a bicycle, find the gear ratio, which is the ratio of the number of rear sprocket teeth to the number of front sprocket teeth.

Solve by using cross products.

15. $\frac{x}{8} = \frac{6}{15}$

16. $\frac{7}{x} = \frac{2}{3}$

17. $\frac{1}{a} = \frac{5}{a+5}$

18. $\frac{k+3}{4} = \frac{5k-2}{9}$

- apply and use the properties of proportions (Lesson 7–1)

A telephone pole casts a 33-foot shadow. Nearby a girl 4 feet tall casts a 6-foot shadow. To find the height of the telephone pole, write a proportion.

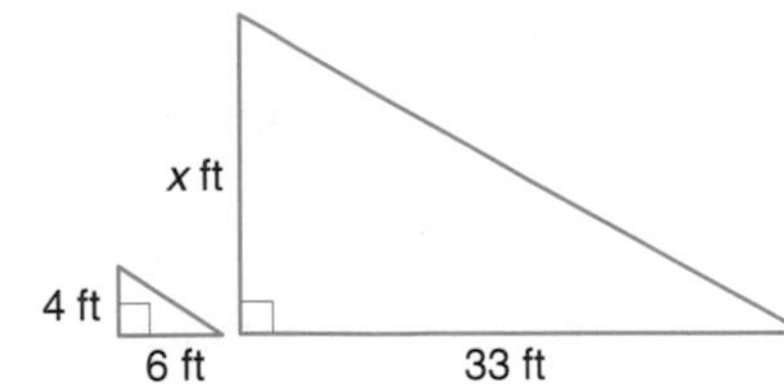

$$\frac{x}{33} = \frac{4}{6}$$
$$6x = 4 \cdot 33$$
$$6x = 132$$
$$x = 22$$

The pole is 22 feet tall.

Corresponding sides of $\triangle ABC$ are proportional to the sides of $\triangle JKL$.

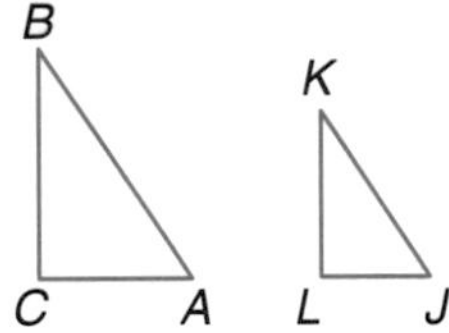

19. If $JK = 7$, $JL = 6$, and $AC = 8$, find AB.

20. If $KL = 7$, $JL = 6$, and $AC = 14$, find BC.

- solve problems involving similar figures (Lesson 7–2)

If $\square WXYZ \sim \square QRST$, find the value of x.

$$\frac{x}{18} = \frac{6}{12}$$
$$x \cdot 12 = 18 \cdot 6$$
$$12x = 108$$
$$x = 9$$

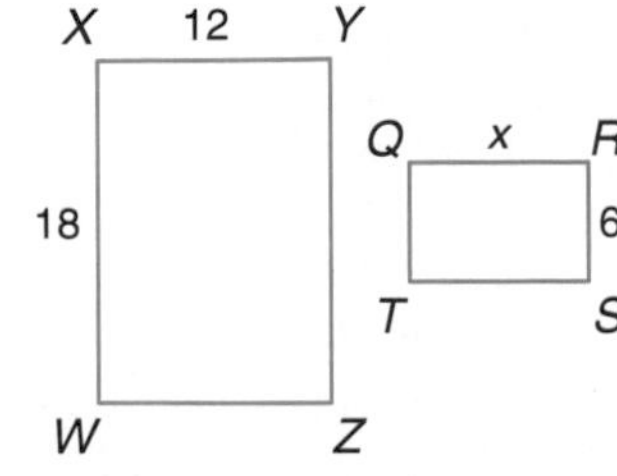

Determine whether each statement is *true* or *false*.

21. All similar triangles are congruent.

22. All congruent triangles are similar.

In the figure below, $\triangle LMN \sim \triangle TUV$.

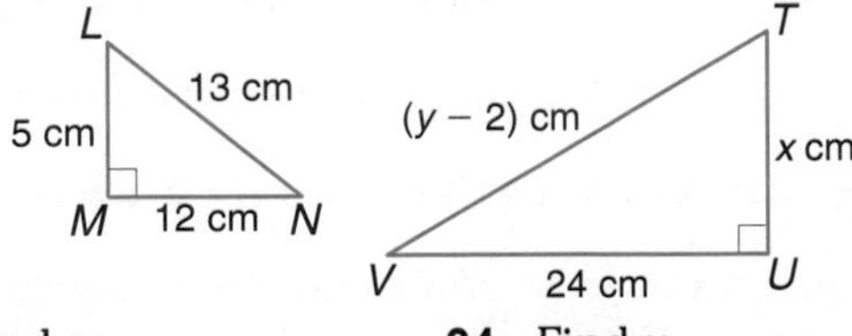

23. Find x.

24. Find y.

OBJECTIVES AND EXAMPLES

- identify similar triangles (Lesson 7–3)

There are three ways to prove that two triangles are similar.

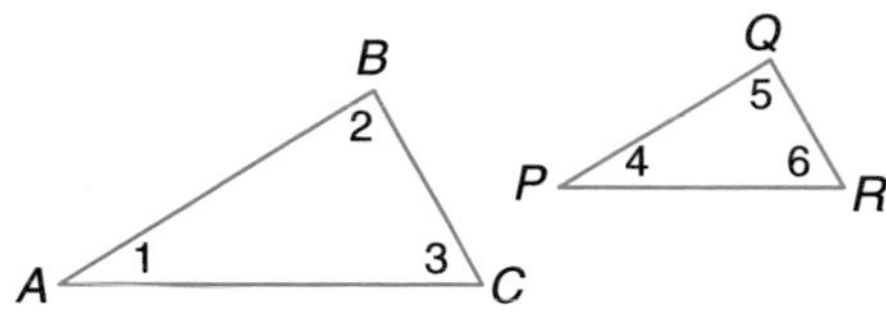

AA Similarity $\angle 1 \cong \angle 4$, $\angle 3 \cong \angle 6$

SSS Similarity $\frac{AB}{PQ} = \frac{BC}{QR} = \frac{CA}{RP}$

SAS Similarity $\frac{AB}{PQ} = \frac{AC}{PR}$, $\angle 1 \cong \angle 4$

REVIEW EXERCISES

Determine whether each pair of triangles is similar. If similarity exists, write a mathematical sentence relating the two triangles. Give a reason for your answer.

25.

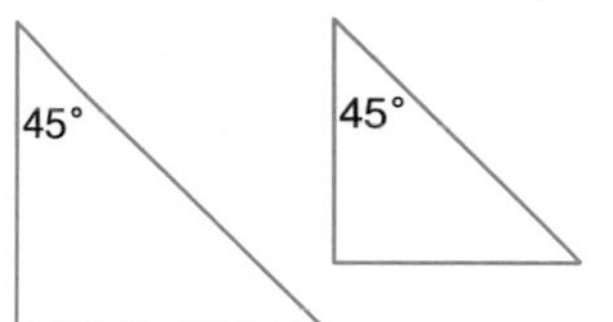

26.

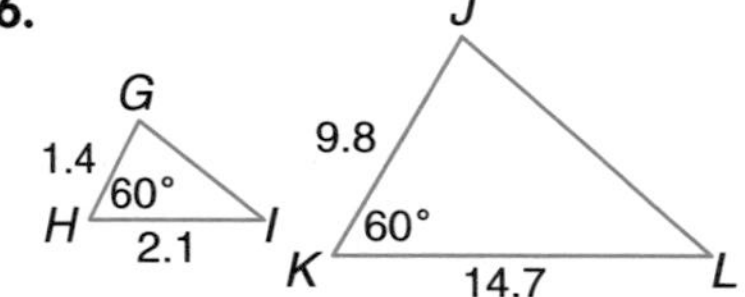

- use proportional parts of triangles to solve problems (Lesson 7–4)

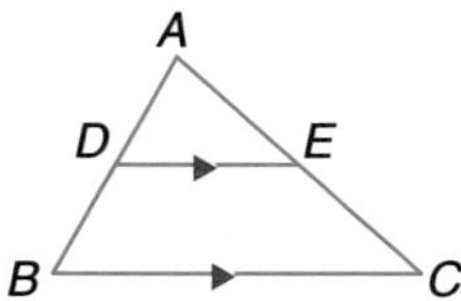

If $\overline{DE} \parallel \overline{BC}$, then $\frac{AD}{DB} = \frac{AE}{EC}$.

Use the figure and the given information to find the value of x.

27. $\overline{PQ} \parallel \overline{DF}$

$EQ = 3$

$DP = 12$

$QF = 8$

$PE = x + 2$

D P E Q F

28. $\overline{SV} \parallel \overline{PR}$

$TS = 5 + x$

$TV = 8 + x$

$VP = 4$

$SR = 3$

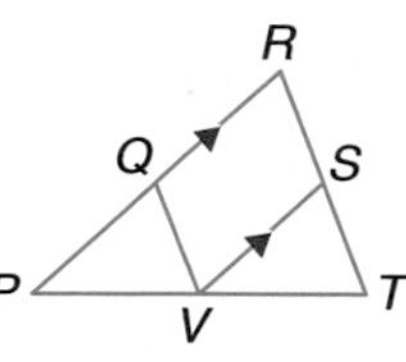

- recognize and use the proportional relationships of corresponding perimeters, altitudes, angles bisectors, and medians of similar triangles (Lesson 7–5)

$\triangle ABC \sim \triangle PQR$

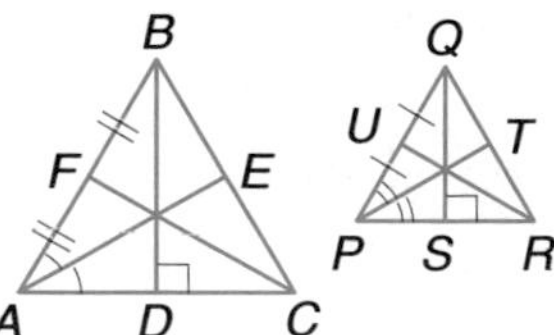

Perimeter $\frac{AB + BC + CA}{PQ + QR + RP} = \frac{CA}{RP}$

Altitude $\frac{BD}{QS} = \frac{BA}{QP}$

Angle bisector $\frac{AE}{PT} = \frac{CA}{RP}$

Median $\frac{CF}{RU} = \frac{CA}{RP}$

29. If $\triangle ABC \sim \triangle XYZ$, find YN.

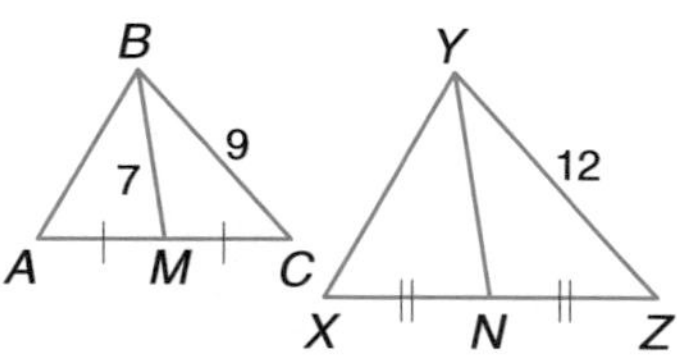

30. If $\triangle STV \sim \triangle PQM$, find the perimeter of $\triangle PQM$.

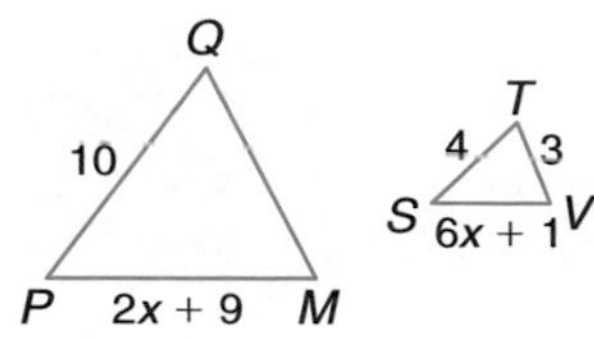

OBJECTIVES AND EXAMPLES

- recognize and describe characteristics of fractals (Lesson 7–6)

The fractal below is formed by trisecting the sides of a square, which forms nine smaller squares, and then removing the middle square. Each stage contains a similar, smaller pattern of the previous stage.

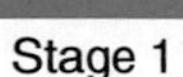

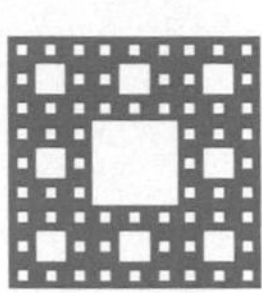

Stage 1 Stage 2 Stage 3

REVIEW EXERCISES

31. Draw stage 3 of the fractal shown below.

32. Explain how self-similarity is evident in a head of cauliflower.

APPLICATIONS AND PROBLEM SOLVING

33. Probability If the probability of rolling a sum of 7 with 2 dice is $\frac{1}{6}$, how many sums of 7 would you expect to get if you rolled the dice 174 times? (Lesson 7–1)

34. Travel A map is scaled so that 1 centimeter represents 15 kilometers. How far apart are two towns if they are 7.9 centimeters apart on the map? (Lesson 7–1)

35. Hobbies A twin-jet airplane especially suited for medium-range flights has a length of 78 meters and a wingspan of 90 meters. If a scale model is made with a wingspan of 36 centimeters, find its length. (Lesson 7–2)

36. Drafting A proportional divider is a drafting instrument that is used to enlarge or decrease a drawing. A screw at T keeps $\overline{AT} \cong \overline{TC}$ and $\overline{DT} \cong \overline{TB}$. $\overline{AB}$ and $\overline{CD}$ can be set so they are divided proportionally at T in any desired ratio. How would you adjust the divider in order to enlarge a design in the ratio 7 to 4? (Lesson 7–3)

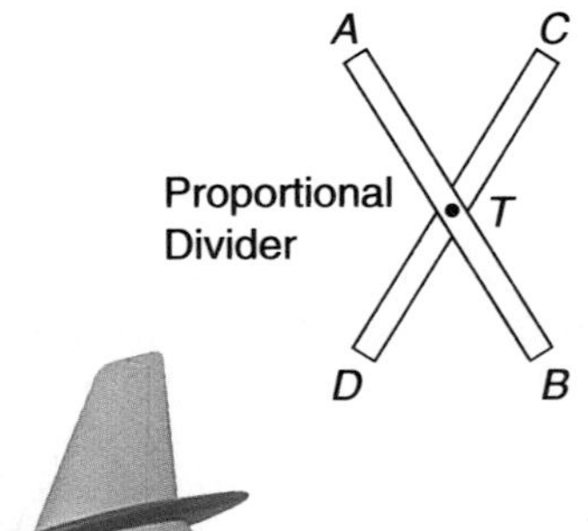

37. Solve a Simpler Problem Find the units digit of 2^{125}. (Lesson 7–6)

A practice test for Chapter 7 is provided on page 799.

ALTERNATIVE ASSESSMENT

COOPERATIVE LEARNING PROJECT

Fractals In this chapter, you learned about fractals. In this project, you will explore fractal patterns in the natural world. You will explore the natural world and discover as many patterns as you can that can be modeled with fractals.

Your goal will be to build a complete collection of natural fractals. If you find more than one of a certain type, share it with others in your group. It may help another student complete his or her collection. You are encouraged to search wherever you can. Photocopies of images contained in books of nature, photography, biology, astronomy, geology, your own photographs or drawings, or even books on fractal geometry can be used.

Follow these steps to complete your collection of natural fractals.

- Collect objects from the natural world: leaves, branches, rocks, and so on. When a real object is not available (cloud, river), find a photograph of it, or photograph or draw it. Find only objects that cannot be represented as conventional geometric objects—quadrilaterals, circles, ellipses, triangles, and so on.
- Refer to a book by Mandelbrot or others to find fractal patterns that look like the objects you found. When available, record the names, or equations, for these patterns.
- Organize all the objects, or images of objects, into the categories defined in the previous step.
- For each category, write one paragraph describing how the pattern is generated. Analyze the basic geometric figure used and the rule for repeating the pattern. Explain whether the repeated geometric figures are similar or congruent.

THINKING CRITICALLY

You are preparing a four-bean salad for lunch. The recipe you followed last time called for 2 cups of beans, and the salad filled a medium-sized bowl. This time you want to make enough salad to fill your largest bowl. The two bowls are shaped similarly, but the diameter of the larger bowl is 30% larger than the diameter of the medium-sized bowl. The volume of a sphere can be found using the formula $V = \frac{4}{3}\pi r^3$, where r is the radius. The volume of each bowl is half of a sphere or $\frac{2}{3}\pi r^3$ cubic units. How many cups of beans do you need so that the recipe fills the larger bowl?

PORTFOLIO

Find a photograph from a book or magazine showing the horizon. Make a photocopy of it. Notice that there is an imaginary point where parallel lines appear to meet. This is a *vanishing point*. There is also an imaginary horizontal line where the sky seems to meet Earth. This is called the *horizon line*. On the photocopy, draw the vanishing point and the horizon line.

Now try to find a photo of the same image with a different vanishing point. Photocopy it, and draw the vanishing point and horizon line on the photocopy. Keep both images in your portfolio.

LONG-TERM PROJECT

In·ves·ti·ga·tion

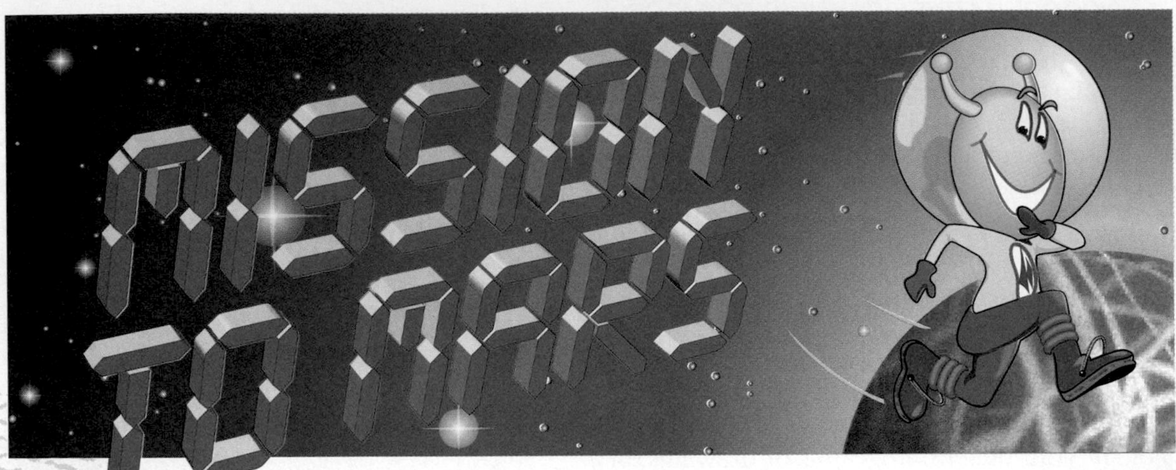

MATERIALS NEEDED

calculator
protractor
ruler
grid paper
markers
modeling clay
paint
scissors
tape

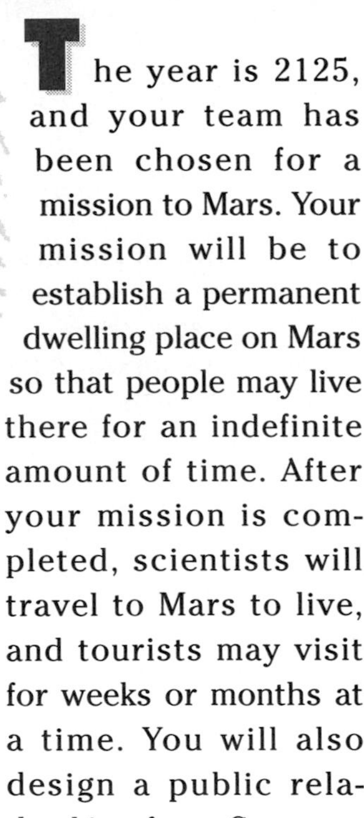

The year is 2125, and your team has been chosen for a mission to Mars. Your mission will be to establish a permanent dwelling place on Mars so that people may live there for an indefinite amount of time. After your mission is completed, scientists will travel to Mars to live, and tourists may visit for weeks or months at a time. You will also design a public relations program to gain backing from Congress and the American public for the mission.

One of the things that the scientists who will follow you to Mars intend to study is the evidence of life on Mars that was discovered in the 1990s. That discovery came from a study of a potato-sized meteorite that was chipped from the surface of Mars by the impact of an asteroid. That meteorite, called ALH84001, contained what is suspected to be microscopic fossils of primitive, bacteria-like organisms. According to their findings, NASA scientists stated that life may have existed on Mars more than 3.6 billion years ago.

NASA has given you the information shown on the next page that has been gathered about Mars to help you in your mission.

Yours will not be the first encounter that the space program has had with Mars. The first spacecraft to approach Mars from Earth was the unoccupied *U.S. Mariner 4.* It traveled within 6118 miles of the planet in 1965. The *Mariner 9* orbited Mars at a distance of about 1000 miles and photographed the entire surface. The *U.S. Viking 1* and *Viking 2* probes landed on Mars in 1976. Both probes transmitted photographs and analyses of the atmosphere and soil back to Earth.

MARS FACTS

- The fourth planet from the Sun, Mars is next beyond Earth.
- The mean distance from Mars to the Sun is 141,600,000 miles, compared with about 93,000,000 miles for Earth.
- At its closest, Mars is 34,600,000 miles from Earth.
- The diameter of Mars is 4223 miles, a little over half the size of Earth.
- Mars orbits the Sun in an elliptical, or oval-shaped, path. The distance from Mars to the Sun varies between about 128,400,000 to 154,800,000 miles.
- Mars takes about 687 Earth-days to make one orbit of the Sun.
- It takes 24 hours, 37 minutes for Mars to rotate once. Earth takes 23 hours, 56 minutes.
- The temperature on the surface of Mars varies from $-225°$ to 63°F.
- The atmosphere of Mars consists of carbon dioxide, nitrogen, argon, oxygen, carbon monoxide, neon, krypton, xenon, and water vapor.
- Mars has two moons, Phobos and Deimos. Phobos is about 5800 miles from the center of Mars, and it travels around Mars once every $7\frac{1}{2}$ hours. Deimos is about 14,600 miles from Mars and completes an orbit every 30 hours.
- The atmospheric pressure on Mars is about 0.1 pound per square inch. That is less than one-hundredth the pressure felt on Earth.

RESEARCH

Brainstorm a list of concerns that need to be addressed by the team. Begin your discussion with the questions below.

- Will it be possible for people to breathe on Mars, or will something have to be done to accommodate respiration?
- What is the best place to locate the space station on Mars?
- How can energy be produced on Mars?
- Where is the best place on Earth to conduct tests of equipment that will be sent to Mars?
- What factors of daily life will be different on Mars? For example, how will the difference in the length of a day and the length of a year affect life?
- How correct are the public perceptions of life on Mars?

You may wish to visit the library or do some research on the NASA Internet home page regarding photographs of Mars and other information that has been gathered about the planet.

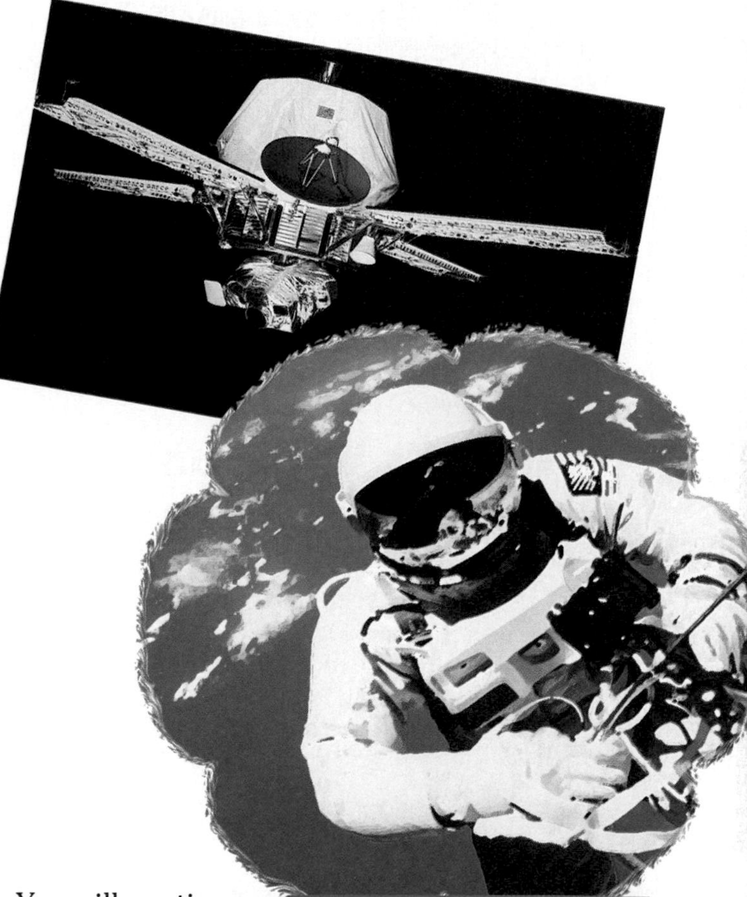

You will continue working on this Investigation throughout Chapters 8 and 9.

Be sure to keep your research, designs, and other materials in your Investigation Folder.

Mission to Mars Investigation

Working on the Investigation
Lesson 8–2, p. 411

Working on the Investigation
Lesson 8–6, p. 436

Working on the Investigation
Lesson 9–3, p. 465

Working on the Investigation
Lesson 9–7, p. 497

Closing the Investigation
End of Chapter 9, p. 504

Applying Right Triangles and Trigonometry

What You'll Learn

In this chapter, you will:

- use the geometric mean to solve problems,
- use the Pythagorean Theorem and its converse,
- use the properties of 45°-45°-90° and 30°-60°-90° triangles,
- use trigonometry to solve triangles, and
- choose the appropriate strategy for solving a problem.

Why It's Important

Real World

Design and Construction Notice the triangular design of the carriage, the forerunner of a modern car, designed in the 1880s. Today, the use of triangles in car designs can be seen most often in the frame, or chassis, of a race car. Of all polygons, the triangle is the most rigid which provides strength to objects like cars and bridges. The triangle is a rigid figure because once the side lengths have been set, the triangle cannot change its shape. In this chapter, you will learn how to use triangle relationships discovered by early mathematicians to find the unknown measures of such triangles.

PREREQUISITE SKILLS

To be successful in this chapter, you'll need to understand these concepts and be able to apply them. Refer to the examples or to the lesson in parentheses if you need more review before beginning the chapter.

Use the Pythagorean Theorem to find the length of the hypotenuse of a right triangle. (Lesson 1–4)

Find the length of the hypotenuse of each right triangle with legs of the lengths given. Round to the nearest hundredth.

1. 3 and 4 **2.** 25 and 35 **3.** 12 and 16 **4.** 9 and 10

Apply the properties of proportions. (Lesson 7–1)

Solve each proportion by using cross products.

5. $\frac{a}{10} = \frac{5}{25}$ **6.** $\frac{16}{63} = \frac{32}{9x}$ **7.** $\frac{b-4}{b} = \frac{2}{3}$ **8.** $25 = \frac{10}{d}$

Simplify radical expressions.

Example 1 Simplify $\sqrt{108x^2y^3z^4}$.

$$\sqrt{108x^2y^3z^4} = \sqrt{2^2 \cdot 3^3 \cdot x^2 \cdot y^3 \cdot z^4}$$ *Prime factorization.*

$$= \sqrt{2^2} \cdot \sqrt{3^2} \cdot \sqrt{3} \cdot \sqrt{x^2} \cdot \sqrt{y^2} \cdot \sqrt{y} \cdot \sqrt{z^4}$$ *Product Property of Square Roots*

$$= 2 \cdot 3 \cdot |x| \cdot |y| \cdot z^2 \cdot \sqrt{3y}$$ *Simplify. (The absolute value of x ensures a nonnegative result.)*

$$= 6|xy|z^2\sqrt{3y}$$

Example 2 Simplify $\frac{\sqrt{6}}{\sqrt{5}}$.

$$\frac{\sqrt{6}}{\sqrt{5}} = \frac{\sqrt{6}}{\sqrt{5}} \cdot \frac{\sqrt{5}}{\sqrt{5}}$$ *Rationalize the denominator.* $\frac{\sqrt{5}}{\sqrt{5}} = 1.$

$$= \frac{\sqrt{30}}{5}$$

Simplify each expression.

9. $\sqrt{12a^2b^3}$ **10.** $\sqrt{2z^2}$ **11.** $\frac{7}{\sqrt{2}}$ **12.** $\frac{2}{\sqrt{2}}$

Focus On READING SKILLS

In this chapter, you'll be learning about special relationships that apply to right triangles. ***Trigonometry*** is the study of triangle measurements. Notice that the words *trigonometry* and *triangle* both begin with the prefix "tri" which means three. In Lesson 8–3, you will find out about the three basic trigonometric ratios, the sine, the cosine, and the tangent.

A Preview of Lesson 8–1

8–1A The Pythagorean Theorem

Materials:  patty paper 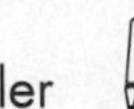 ruler colored pencils

In Chapter 1, you learned that the Pythagorean Theorem relates the measures of the legs and the hypotenuse of a right triangle. Throughout the world, you can find evidence of ancient cultures that used the Pythagorean Theorem before it was officially named in 1909.

You too can discover this relationship among the measures of the sides of a right triangle by using patty paper and algebra.

Activity Use paper folding to develop the Pythagorean Theorem.

Step 1 On a piece of patty paper, make a mark along one side so that the two resulting segments are not congruent. Call one length a and the other length b.

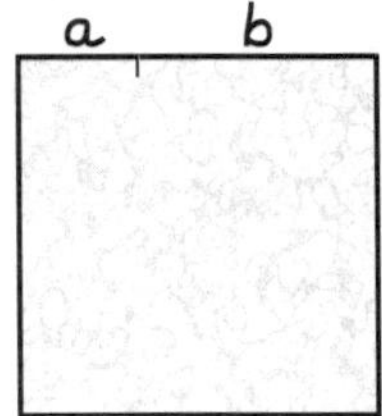

Step 2 Copy these measures on the other sides in the order shown at the right. Fold the paper to divide the square into four sections. Label the area of each section.

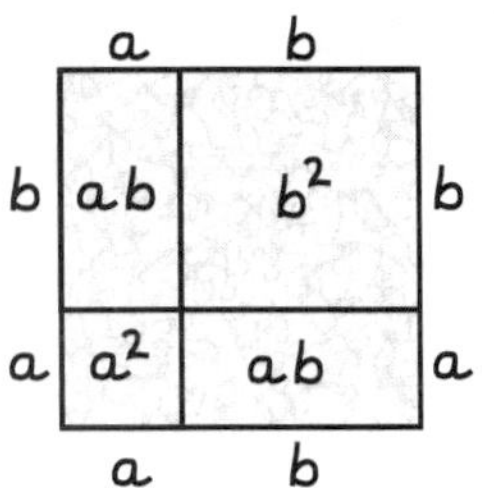

Step 3 On another sheet of patty paper, mark the same lengths a and b on the sides in the different pattern shown at the right.

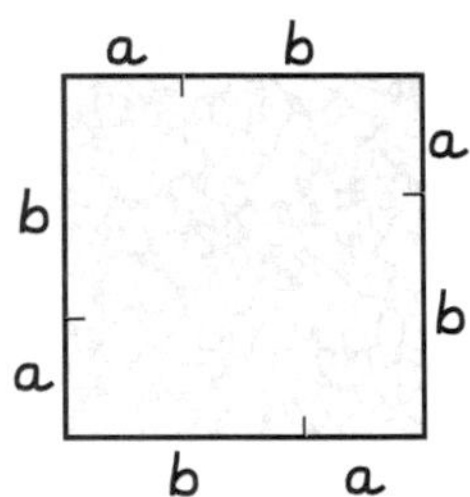

Step 4 Use your straightedge and pencil to connect the marks as shown at the right. Let c represent the length of each hypotenuse.

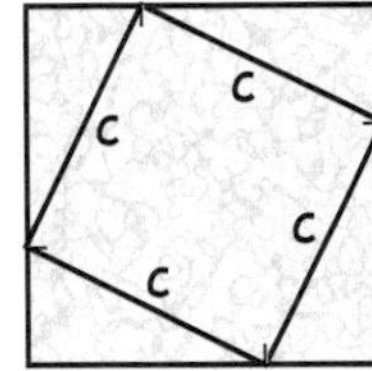

Step 5 Label the area of each section, which is $\frac{1}{2}ab$ for each triangle and c^2 for the square.

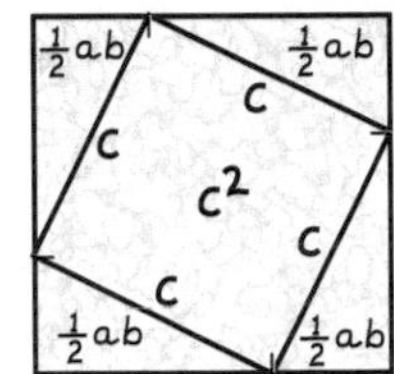

Step 6 Place the squares side by side and color the corresponding regions that have the same area. For example, $ab = \frac{1}{2}ab + \frac{1}{2}ab$.

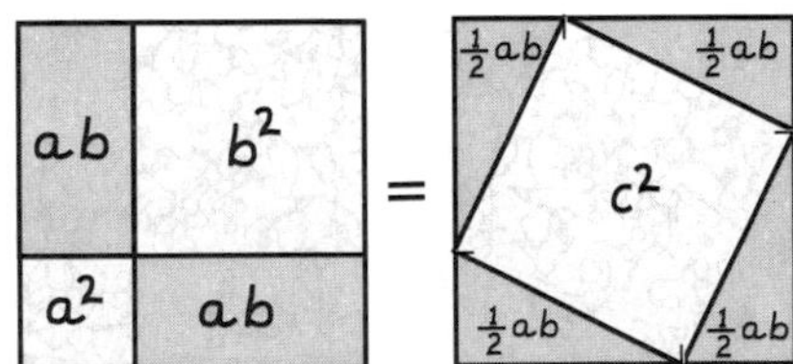

The parts that are not shaded tell us $a^2 + b^2 = c^2$.

Model

1. Use your ruler to find actual measures for a, b, and c. Do these measures confirm that $a^2 + b^2 = c^2$?
2. Repeat the activity with different a and b values. What do you notice?

3. **Explain** why the drawing at the right is an illustration of the Pythagorean Theorem.

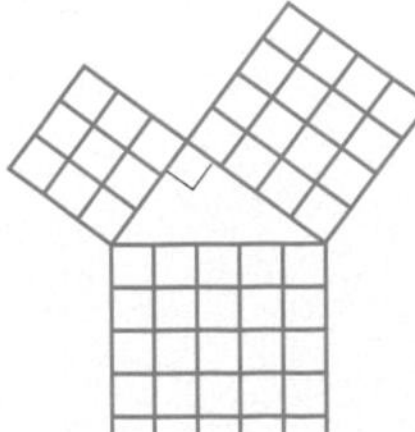

8-1 Geometric Mean and the Pythagorean Theorem

What YOU'LL LEARN

- To find the geometric mean between two numbers,
- to solve problems involving relationships between parts of a triangle and the altitude to its hypotenuse, and
- to use the Pythagorean Theorem and its converse.

Why IT'S IMPORTANT

You can use relationships to find missing measures of parts of right triangles.

Discrete Mathematics

A geometric sequence is one in which each number after the first is found by multiplying the previous number by a given factor. For example, suppose the first number of a sequence is 5 and the factor is 3. You can find the next four numbers in the sequence as shown below.

$$5 \xrightarrow{\times 3} 15 \xrightarrow{\times 3} 45 \xrightarrow{\times 3} 135 \xrightarrow{\times 3} 405$$

Consecutive numbers of the sequence form proportions.

$$\frac{5}{15} = \frac{15}{45} \qquad \frac{15}{45} = \frac{45}{135} \qquad \frac{45}{135} = \frac{135}{405}$$

Note that the denominator of one fraction is the numerator of the next. The **geometric mean** between two positive numbers a and b is the positive number x where $\frac{a}{x} = \frac{x}{b}$. By cross multiplying, we see that $x^2 = ab$ or $x = \sqrt{ab}$. Note that in the proportion, x and x represent the *means* and a and b represent the *extremes*.

Example  **Find the geometric mean between 2 and 10.**

Let x represent the geometric mean.

$\frac{2}{x} = \frac{x}{10}$	*Definition of geometric mean*
$x^2 = 20$	*Cross multiply.*
$x = \sqrt{20}$ or about 4.47	*Take the square root of each side.*

The geometric mean between 2 and 10 is about 4.47.

Consider right triangle ABC with altitude $\overline{CD}$ drawn from the right angle C to the hypotenuse $\overline{AB}$. A special relationship exists for the three right triangles ABC, ACD, and CBD.

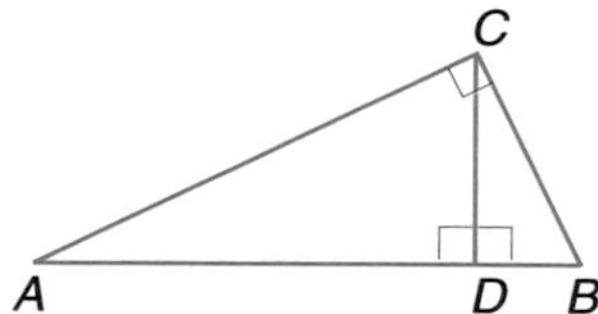

EXPLORATION CABRI GEOMETRY

Use a calculator to draw right triangle ABC shown above. Explore the relationships among the three right triangles by finding the measures of $\angle A$, $\angle ACB$, $\angle B$, $\angle ADC$, $\angle ACD$, $\angle BDC$, and $\angle BCD$.

a. What is the relationship between the measures of $\angle A$ and $\angle BCD$? What is the relationship between the measures of $\angle B$ and $\angle ACD$?

b. Drag point C to another position. Describe the relationship between the measures of $\angle A$ and $\angle BCD$ and between the measures of $\angle B$ and $\angle ACD$.

c. Make a conjecture about $\triangle ABC$, $\triangle ACD$, and $\triangle CBD$.

The results of the Exploration suggest that the following angle relationships are true for right $\triangle ABC$ with right $\angle C$ and altitude $\overline{CD}$.

$\triangle ABC$		$\triangle ACD$		$\triangle CBD$
$\angle A$	$\cong$	$\angle A$	$\cong$	$\angle BCD$
$\angle B$	$\cong$	$\angle ACD$	$\cong$	$\angle B$
$\angle ACB$	$\cong$	$\angle ADC$	$\cong$	$\angle CDB$

This means that all three triangles are similar by the Angle-Angle Postulate.

$$\triangle ABC \sim \triangle ACD \sim \triangle CBD$$

This is more generally stated in Theorem 8–1. *You will be asked to prove this theorem in Exercise 36.*

Theorem 8–1 — **If the altitude is drawn from the vertex of the right angle of a right triangle to its hypotenuse, then the two triangles formed are similar to the given triangle and to each other.**

In the figure at the right, $\triangle ADC \sim \triangle CDB$. So $\frac{AD}{CD} = \frac{CD}{BD}$ because corresponding sides of similar triangles are proportional. Thus, CD is the geometric mean between AD and BD. This is stated in Theorem 8–2. *You will be asked to prove this theorem in Exercise 37.*

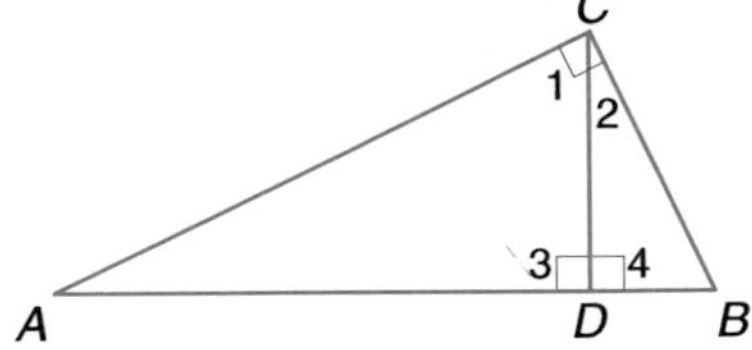

Theorem 8–2 — **The measure of the altitude drawn from the vertex of the right angle of a right triangle to its hypotenuse is the geometric mean between the measures of the two segments of the hypotenuse.**

The geometric mean can be used to estimate hard-to-measure distances.

Example 2

Measurement

To find the height of her school building, Mieko held a book near her eye so that the top and bottom of the building were in line with the edges of the cover. If Mieko's eye level is 5 feet off the ground and she is standing about 10 feet from the building, how tall is the building? ***Assume the building is perpendicular to the ground and the edges of the cover of the book form right angles.***

Draw a diagram of the situation. $\overline{CD}$ is the altitude drawn from the right angle of $\triangle ABC$.

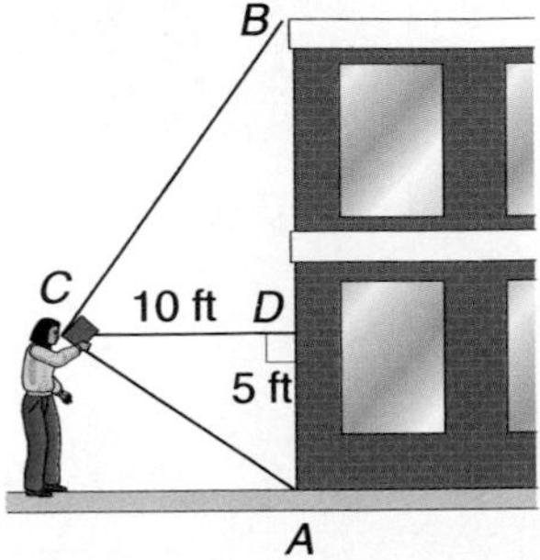

$\frac{AD}{CD} = \frac{CD}{BD}$	*Theorem 8–2*
$\frac{5}{10} = \frac{10}{BD}$	*$AD = 5$, $CD = 10$*
$5BD = 100$	*Cross multiply.*
$BD = 20$	*Division Property (=)*

Mieko estimates that the building is 20 + 5 or 25 feet tall.

The altitude to the hypotenuse of a right triangle determines another relationship between segments.

Theorem 8–3	**If the altitude is drawn to the hypotenuse of a right triangle, then the measure of a leg of the triangle is the geometric mean between the measures of the hypotenuse and the segment of the hypotenuse adjacent to that leg.**

You will be asked to prove this theorem in Exercise 38.

Example 3 **Find a and b in $\triangle TGR$.**

According to Theorem 8–3, we can write the following proportions.

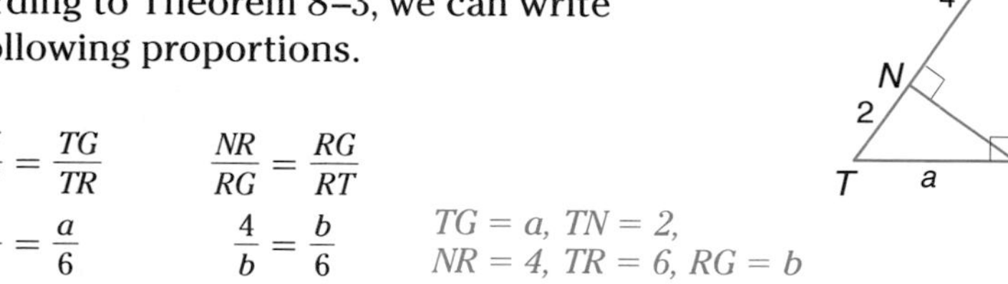

$\frac{TN}{TG} = \frac{TG}{TR}$ $\quad$ $\frac{NR}{RG} = \frac{RG}{RT}$

$\frac{2}{a} = \frac{a}{6}$ $\quad$ $\frac{4}{b} = \frac{b}{6}$ $\quad$ *$TG = a$, $TN = 2$, $NR = 4$, $TR = 6$, $RG = b$*

$a^2 = 12$ $\quad$ $b^2 = 24$ $\quad$ *Cross multiply.*

$a = \sqrt{12}$ $\quad$ $b = \sqrt{24}$ $\quad$ *Take the square root of each side.*

$a \approx 3.46$ $\quad$ $b \approx 4.90$

LOOK BACK

You can refer to Lesson 1–4 for information on the Pythagorean Theorem.

The geometric-mean relationships can be used to prove one of the most important theorems in mathematics, the **Pythagorean Theorem**. Its name comes from Pythagoras, a Greek mathematician from the sixth century B.C. who is said to have been the first to write a proof of the theorem.

Theorem 8–4 Pythagorean Theorem	**In a right triangle, the sum of the squares of the measures of the legs equals the square of the measure of the hypotenuse.**

If c is the measure of the hypotenuse and a and b are the measures of the legs, then $a^2 + b^2 = c^2$.

Proof of Theorem 8–4

Given: right $\triangle ABC$ with right angle at C

Prove: $a^2 + b^2 = c^2$

Proof:

Draw the altitude from C to $\overline{AB}$. Let $AB = c$, $AC = b$, $BC = a$, $AD = x$, $DB = y$, and $CD = h$. Two geometric means now exist.

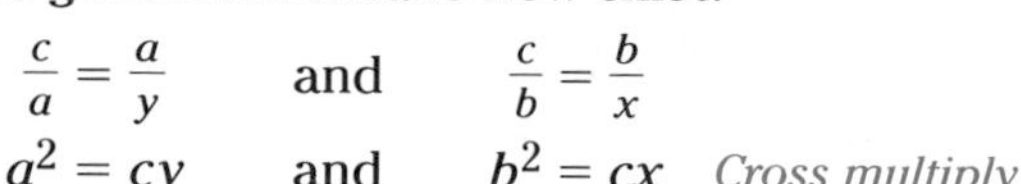

$\frac{c}{a} = \frac{a}{y}$ and $\frac{c}{b} = \frac{b}{x}$

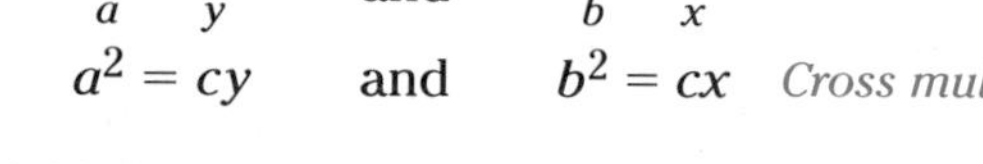

$a^2 = cy$ and $b^2 = cx$ *Cross multiply.*

Add the equations.

$a^2 + b^2 = cy + cx$

$a^2 + b^2 = c(y + x)$ *Factor.*

$a^2 + b^2 = c^2$ *Since $c = y + x$, substitute c for $(y + x)$.*

In Chapter 1, you used the Pythagorean Theorem to find the measures of the sides of a right triangle. The converse of the Pythagorean Theorem is useful to determine if three measures of the sides of a triangle are those of a right triangle. *You will be asked to prove this theorem in Exercise 39.*

Theorem 8–5 Converse of the Pythagorean Theorem	**If the sum of the squares of the measures of two sides of a triangle equals the square of the measure of the longest side, then the triangle is a right triangle.**

A **Pythagorean triple** is a group of three whole numbers that satisfies the equation $a^2 + b^2 = c^2$, where c is the greatest number. One common Pythagorean triple is 3, 4, and 5. If the measures of the sides of a right triangle are whole numbers, the measures are a Pythagorean triple.

Example 

Determine if a triangle with side measures of 74, 91, and 52 forms a right triangle. That is, do 74, 91, and 52 form a Pythagorean triple?

The measure of the longest side is 91. Use the converse of the Pythagorean Theorem.

$$a^2 + b^2 = c^2$$
$$74^2 + 52^2 \stackrel{?}{=} 91^2$$
$$5476 + 2704 \stackrel{?}{=} 8281$$
$$8180 \neq 8281$$

Since $8180 \neq 8281$, the triangle is not a right triangle, and the three numbers do not form a Pythagorean triple.

CHECK FOR UNDERSTANDING

Communicating Mathematics

Study the lesson. Then complete the following.

1. **Describe** how you would find the geometric mean between 5 and 7.
2. **Draw and label** a right triangle with an altitude from the right angle. Explain why the original triangle and the two triangles created by the altitude are similar.
3. **Explain** why a Pythagorean triple can represent the measure of the sides of a right triangle.
4. **Draw and label** a right triangle with an altitude from the right angle. Use your drawing to explain what is meant by "the hypotenuse and the segment of the hypotenuse adjacent to a leg" in Theorem 8–3.
5. **You Decide** Molly draws an equilateral $\triangle RST$ with altitude $\overline{SU}$ as shown at the right. She says that SU is the geometric mean between RU and UT. Sierra disagrees. Who is correct? Explain.

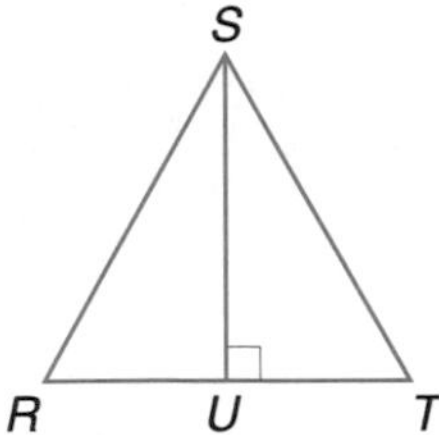

6. On a geoboard or dot paper, form a right triangle with legs one unit long. Build squares on the two legs and the hypotenuse as shown on the right.
   a. What is the area of each square?
   b. What is the relationship between the area of the squares built on the legs and the area of the square built on the hypotenuse?
   c. Form another right triangle and build squares on the two legs and the hypotenuse. How are the areas of the squares related?

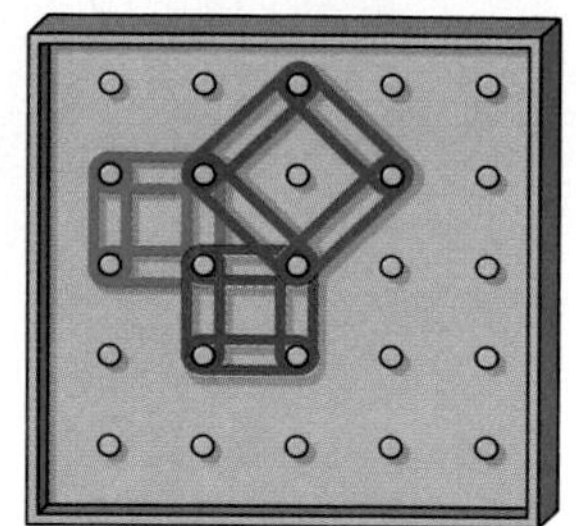

Guided Practice

7. Find the geometric mean between 4 and 25.
8. Refer to right triangle *PTG*.
   a. Name three similar triangles.
   b. Name two angles congruent to $\angle PGT$.

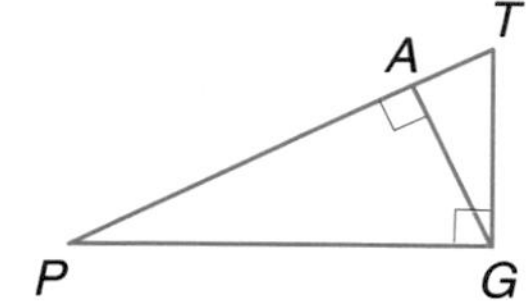

Find the values of *x* and *y*.

9. 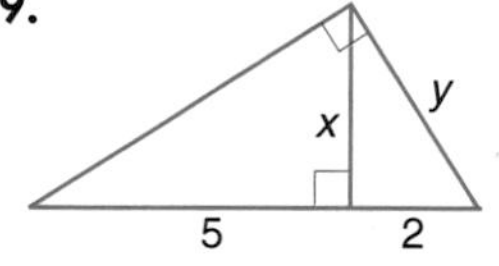

10. 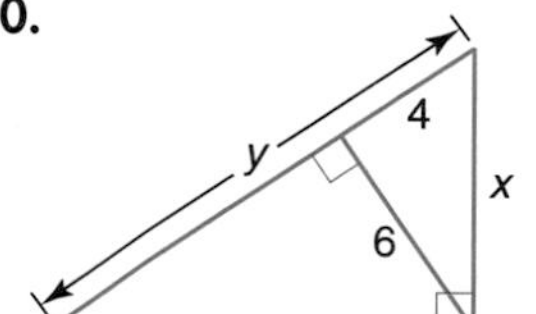

11. 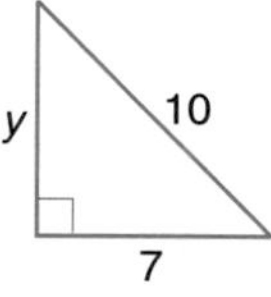

12. Determine if 4, 7.5, and 8.5 are measures of the sides of a right triangle.
13. **Look for a Pattern** Complete the chart of Pythagorean triples.

a	*b*	*c*
3	4	5
6	8	
9	12	
12	16	

   a. A *primitive* Pythagorean triple is set of three numbers with no common factors except 1. Does the chart contain any primitive Pythagorean triples?
   b. Describe the pattern that relates these sets of Pythagorean triples.
   c. Why do you think these Pythagorean triples are called a *family*?
   d. Are the triangles described by a family of Pythagorean triples similar? Explain.
14. **Architecture** The distance between the base of the Leaning Tower of Pisa and the top of the tower is 180 feet. The tower is leaning 16 feet off the perpendicular. Find the distance of the top of the tower from the ground.

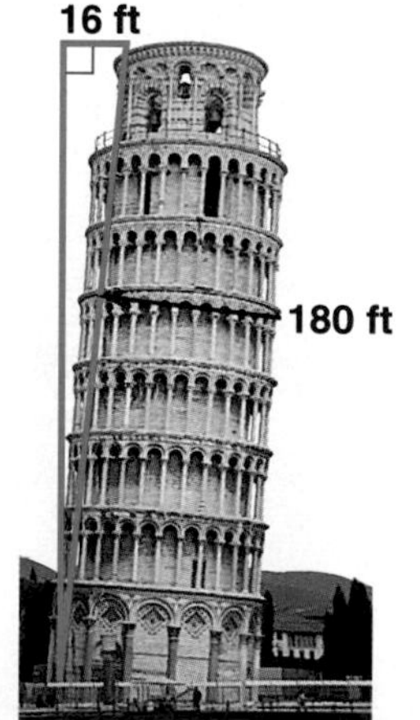

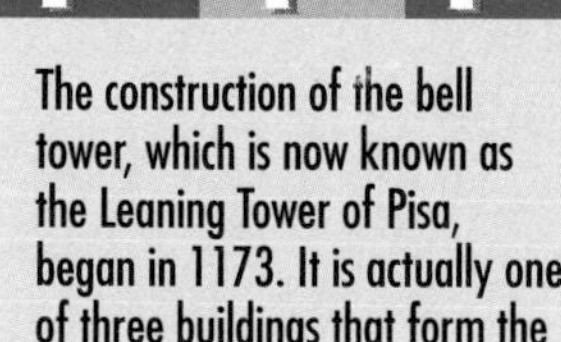

The construction of the bell tower, which is now known as the Leaning Tower of Pisa, began in 1173. It is actually one of three buildings that form the Cathedral of Pisa.

EXERCISES

Practice

Find the geometric mean between each pair of numbers.

15. 4 and 9
16. 4 and $\frac{1}{9}$
17. $\frac{2}{3}$ and $\frac{1}{3}$

Refer to right triangle *ASH* for Exercises 18–20.

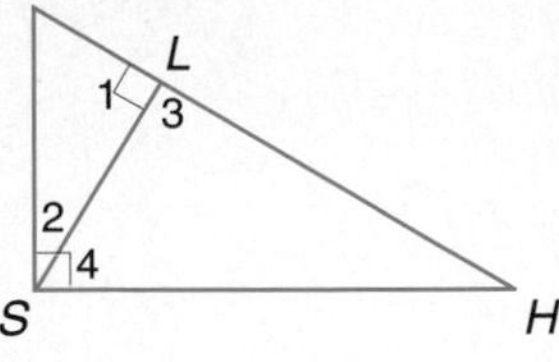

18. Name the three similar triangles.

19. Name two segments for which the given segment is the geometric mean.

a. $\overline{AS}$ **b.** $\overline{LS}$ **c.** $\overline{SH}$

20. Name at least one angle congruent to each angle.

a. $\angle SLH$ **b.** $\angle A$ **c.** $\angle 2$

Find the values of *x* and *y*.

21.

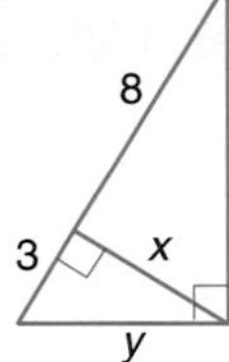

22.

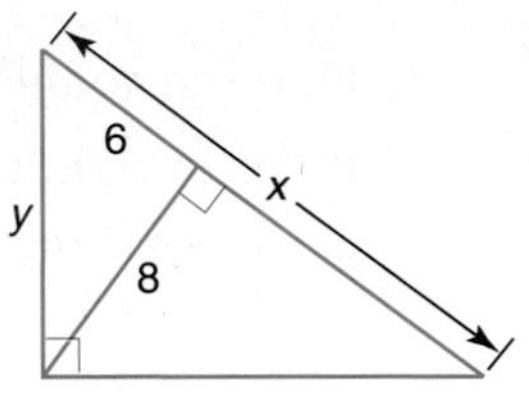

23.

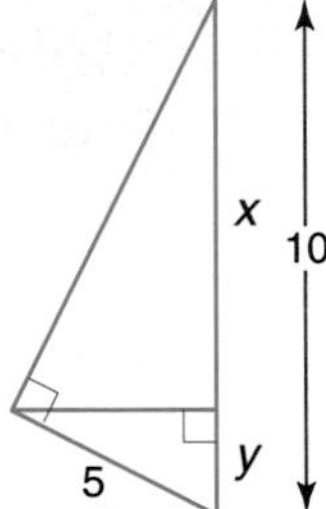

24.

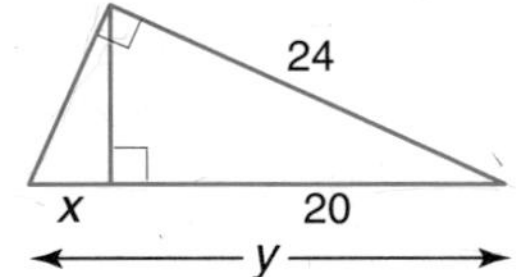

25.

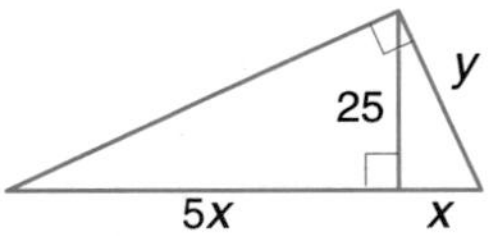

26.

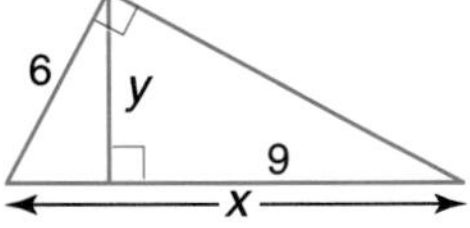

27.

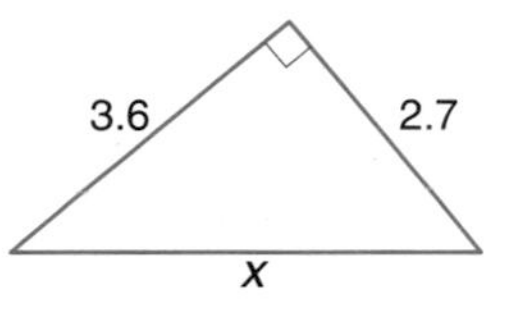

28.

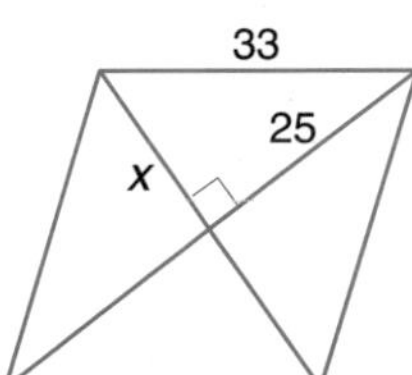

29.

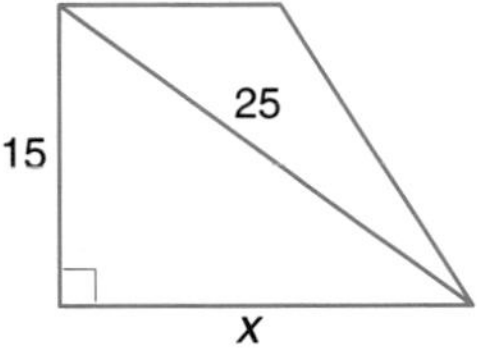

Determine if the given measures are measures of the sides of a right triangle.

30. 5, 12, 13 **31.** 20, 21, 28 **32.** 37, 12, 34

Refer to Exercise 13. For each Pythagorean triple, find two triples in the same family.

33. 8, 15, 17 **34.** 7, 24, 25 **35.** 9, 40, 41

36. Write a two-column proof of Theorem 8–1.

37. Write a paragraph proof of Theorem 8–2.

38. Write a two-column proof of Theorem 8–3.

39. Write a paragraph proof of the converse of the Pythagorean Theorem (Theorem 8–5).

40. Use the Pythagorean Theorem and the figure at the right to show $d = \sqrt{(x_2 - x_1)^2 + (y_2 - y_1)^2}$. That is, show the distance formula is true.

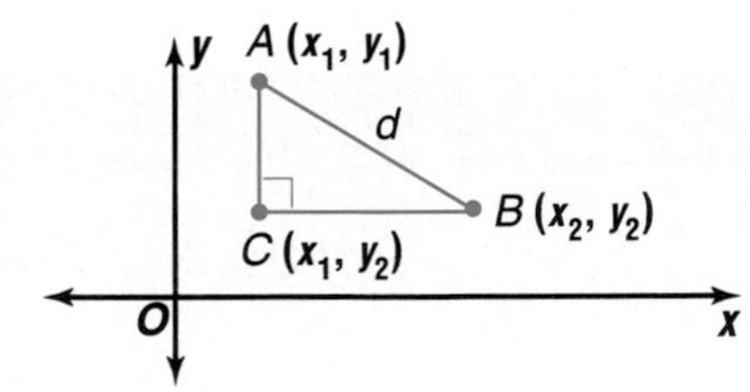

Programming

Euclid

41. The program at the right uses a procedure for finding *Pythagorean triples* that was developed by Euclid around 320 B.C.

Run the program to generate a list of Pythagorean triples.

a. List all the members of the 3-4-5 family that are generated by the program.

b. Ricardo made the conjecture that if three whole numbers are a Pythagorean triple, then their product is divisible by 60. Does Ricardo's conjecture hold true for each triple that is produced by the program?

```
PROGRAM: PYTHTRIP
: For (X,2,6)
: For (Y,1,5)
: If X>Y
: Then
: int(X²−Y²+0.5)→A
: 2XY→B
: int(X²+Y²+0.5)→C
: If A>B
: Then
: Disp B,A,C
: Else
: Disp A,B,C
: End
: End
: Pause
: Disp " "
: End
: End
: Stop
```

Critical Thinking

42. Draw an acute triangle and an obtuse triangle. In the acute triangle, draw an altitude to the longest side. In the obtuse triangle, draw an altitude from the obtuse angle. In either case, are the triangles formed by the altitude similar to the original? Explain.

Applications and Problem Solving

43. **Construction** In the United States, most building codes limit the steepness of the slope of a roof to a 3-4-5 right triangle. A model of this type of roof is shown at the right. A builder wants to put a support brace from point C perpendicular to $\overline{AP}$. Find the length of such a brace.

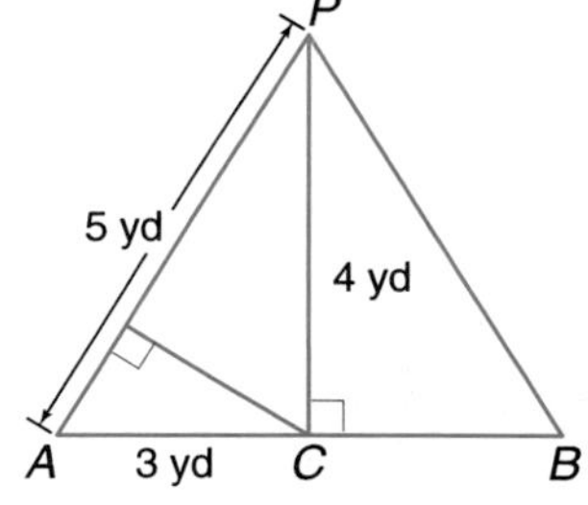

44. **Sailing** The mast of a sailboat is supported by wires called shrouds. In the diagram at the right, the shrouds are shown in red. What is the total length of wire needed to form these shrouds?

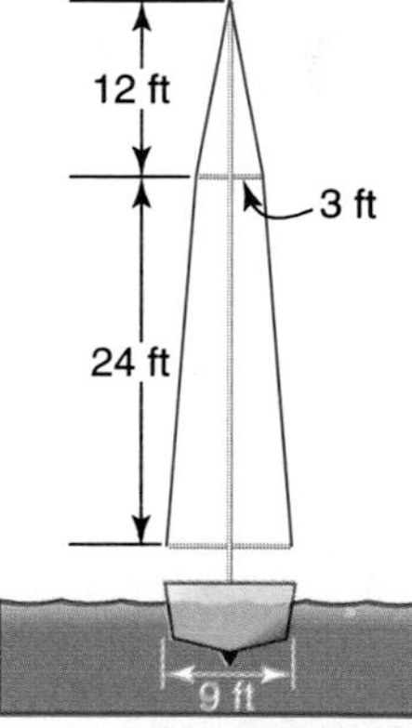

45. **Motion Pictures** In a movie, the Scarecrow is looking for a brain. When the Wizard presents him with a Doctor of Thinkology degree, the Scarecrow immediately announces "The sum of the square roots of any two sides of an isosceles triangle is equal to the square root of the remaining side." Do you agree with the "Scarecrow Theorem"? Explain.

Mixed Review

46. **Solve a Simpler Problem** Donte must finish John Steinbeck's *The Pearl* and write a report by Friday. He is starting to read page 71, and the book ends on page 197. How many pages does Donte have yet to read? (Lesson 7–6)

47. **Travel** A Georgia map is drawn so that 1 centimeter represents 20 miles. If Savannah and Atlanta are 12.7 centimeters apart on the map, how far apart are the cities? (Lesson 7–1)

48. The bases of an isosceles trapezoid measure 10 inches and 22 inches. Find the length of the median of the trapezoid. (Lesson 6–5)

49. In square $LMNP$, $LN = 3x - 2$ and $MP = 2x + 3$. Find LN. (Lesson 6–4)

50. Determine whether the statement *The diagonals of a rectangle bisect the opposite angles* is true or false. Justify your answer. (Lesson 6–3)

51. **SAT Practice** Kendra has at least one quarter, one dime, one nickel, and one penny. If she has three times as many pennies as nickels, the same number of nickels as dimes, and twice as many dimes as quarters, then what is the least amount of money she could have?
A \$0.41 **B** \$0.48 **C** \$0.58 **D** \$0.61 **E** \$0.71

52. If the sides of a triangle have lengths of $(2x + 7)$ meters, $(5x - 3)$ meters, and $(8x + 2)$ meters, find all possible values for x. (Lesson 5–5)

53. The measure of the angles in $\triangle XYZ$ are in the ratio 4:8:12. What are the measures of the angles? (Lesson 4–2)

54. Two vertical angles have measures of $7x + 12$ and $4x + 42$. Find x. (Lesson 2–6)

INTEGRATION **Algebra**

For **Extra Practice**, see page 778.

55. Graph $y = 3x + 1$ by making a table of ordered pairs.

56. Find $(6t + 7a) - (3t + 12a)$.

Mathematics and SOCIETY

Virtual Reality Uses 3D Geometry

The excerpt below appeared in *Popular Mechanics* in July, 1995.

STEP INTO THE "STAR TREK" HOLODECK and you can saunter about a 3-dimensional world, furnished and peopled with lifelike holograms....researchers today are racing toward a similar idea. At the University of North Carolina, computer scientists are developing 3D virtual environments stocked not with computer-wrought scenery, but with real-life images from a remote location. The key: A concept the researchers call a "sea of cameras." At the remote site, video cameras stud the walls and ceiling. Each has a view from a slightly different perspective....At the viewing site, everyone wears virtual-reality visor displays that are fitted with head-tracking sensors. As the participants walk around or turn their heads, the sensors maintain a fix on their positions and viewing directions....Using this technique, researchers have built a 3D image of an operating room. The goal: to let a large audience watch world-class surgeons at work without cramping their style. ■

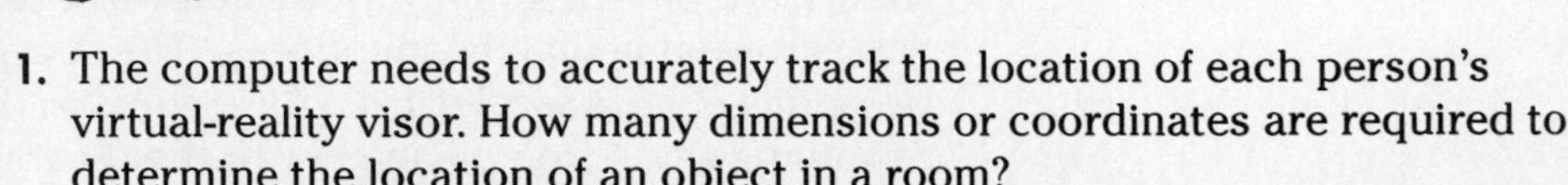

1. The computer needs to accurately track the location of each person's virtual-reality visor. How many dimensions or coordinates are required to determine the location of an object in a room?
2. What advantages would this virtual-reality viewing have over watching the operation in person or on a videotape?
3. Can you think of other situations or applications where this type of virtual-reality viewing could be used?

8-2 Special Right Triangles

What YOU'LL LEARN

- To use the properties of 45°-45°-90° and 30°-60°-90° triangles.

Why IT'S IMPORTANT

You can use properties of special right triangles to find missing measures involved in sports, city planning, and landscaping.

Real World
APPLICATION

Baseball

Years ago, baseball teams often played on asymmetric fields. However, today many teams play in new stadiums with symmetric fields such as the one at Dodger Stadium. A diagram of Dodger Stadium in Los Angeles is shown below.

Suppose a center fielder is standing on the point where a line passing through home plate and second base intersects with a line passing through the foul-ball poles. Is the center fielder closer to dead center field or second base? *This problem will be solved in Example 1.*

DODGER STADIUM (Los Angeles)

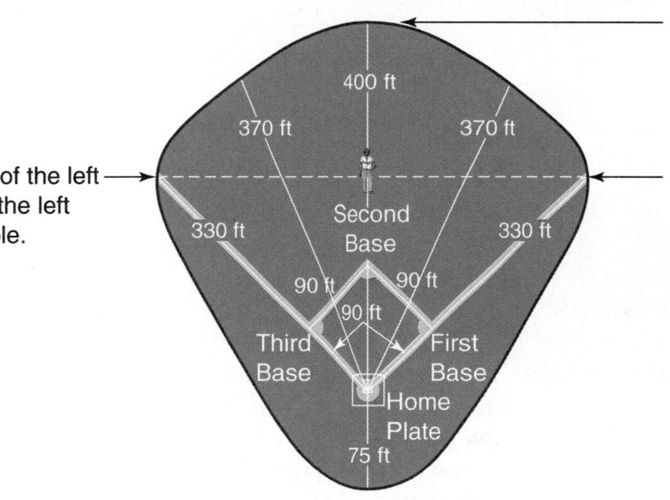

Dead center field is the end of the outfield on a straight line drawn through home plate and second base.
At the end of the left foul line is the left foul-ball pole.
At the end of the right foul line is the right foul-ball pole.
400 ft
370 ft
370 ft
330 ft
330 ft
Second Base
90 ft
90 ft
90 ft
Third Base
First Base
Home Plate
75 ft

In order to solve this problem, we must be able to solve a 45°-45°-90° triangle. The Pythagorean Theorem allows us to discover the special relationship that exists among the sides of a 45°-45°-90° triangle.

Draw a diagonal of a square. The two triangles formed are isosceles right triangles. Let x represent the measure of each side and d represent the measure of the hypotenuse.

$d^2 = x^2 + x^2$ *Pythagorean Theorem*

$d^2 = 2x^2$

$d = \sqrt{2x^2}$ *Take the square root of each side.*

$d = x\sqrt{2}$ *Ignore the negative root, since d is a measure.*

So the length of the hypotenuse of any 45°-45°-90° triangle is $\sqrt{2}$ times the length of a leg.

Theorem 8–6	**In a 45°-45°-90° triangle, the hypotenuse is $\sqrt{2}$ times as long as a leg.**

Example 1

Baseball

Refer to the application at the beginning of the lesson. Is the center fielder closer to dead center field or second base?

First, find the distance from home plate to second base. By Theorem 8–6, this distance d is $90\sqrt{2}$ or about 127.28 feet.

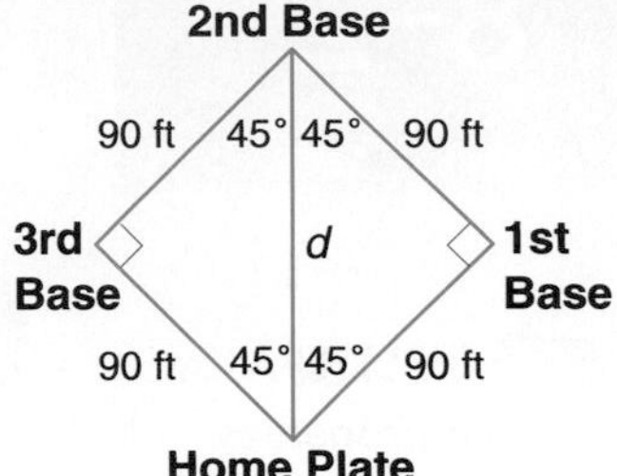

Now, consider the whole field. Draw $\overline{BD}$ from home plate to dead center field and $\overline{AE}$ from the left field foul-ball pole to the right field foul-ball pole. Let C be the intersection point of $\overline{BD}$ and $\overline{AE}$. $\triangle ACB$ is a 45°-45°-90° triangle with an hypotenuse 330 feet long. Let $BC = x$ and $AC = x$ and use the Theorem 8–6 to solve for x, the distance from home plate to the fielder.

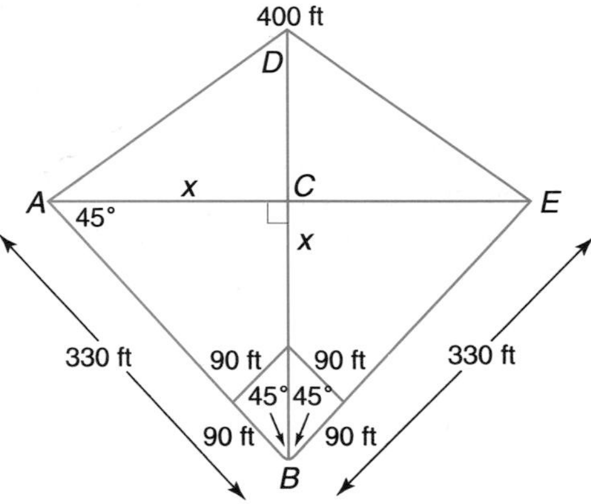

$$x\sqrt{2} = 330$$

$$x = \frac{330}{\sqrt{2}} \quad \textit{Divide each side by } \sqrt{2}.$$

$$x = \frac{330}{\sqrt{2}} \cdot \frac{\sqrt{2}}{\sqrt{2}} \quad \textit{Rationalize the denominator.}$$

$$x = \frac{330\sqrt{2}}{\ }$$

$$x = 165\sqrt{2} \text{ or about } 233.35$$

The distance from home to the center fielder is about 233.35 feet. The distance from the center fielder to second base is about 233.35 − 127.28 or 106.07 feet. The distance from the center fielder to dead center field is about 400 − 233.35 or 166.65 feet. So the center fielder is closer to second base.

There is also a special relationship among the measures of the sides of a 30°-60°-90° triangle.

If an altitude is drawn from any vertex of an equilateral triangle, the triangle is separated into two congruent 30°-60°-90° triangles. Using the Pythagorean Theorem, it is possible to derive a formula relating the lengths of the sides to each other. Let each side of the equilateral triangle have a length of x units. $\overline{LM}$ and $\overline{MN}$ are congruent segments, so each measures $\frac{1}{2}x$ units. Let a represent the measure of the altitude.

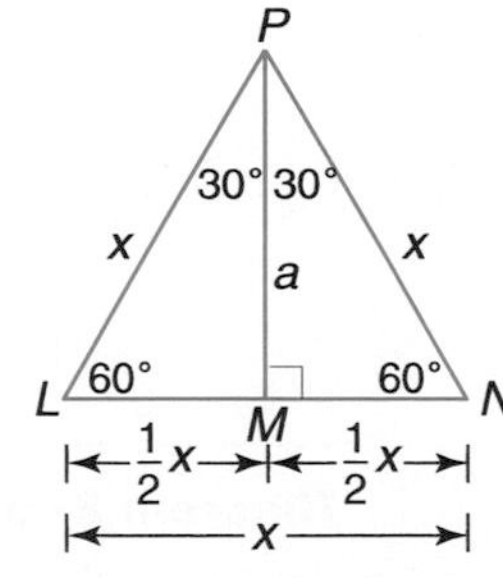

$$(PM)^2 + (LM)^2 = (PL)^2 \quad \textit{Pythagorean Theorem}$$

$$a^2 + \left(\frac{1}{2}x\right)^2 = x^2 \quad PM = a,\ LM = \frac{1}{2}x,\ PL = x$$

$$a^2 + \frac{1}{4}x^2 = x^2$$

$$a^2 = \frac{3}{4}x^2 \quad \textit{Subtraction Property (=)}$$

$$a = \sqrt{\frac{3}{4}x^2} \quad \textit{Take the square root of each side.}$$

$$a = \frac{\sqrt{3}}{2}x \quad \textit{Ignore the negative root since a is a measure.}$$

So, in the 30°-60°-90° triangle, the measures of the sides are x, $\frac{1}{2}x$, and $\frac{\sqrt{3}}{2}x$.

In Lesson 8–1, you learned about families of Pythagorean triples. Multiply the measure of each side by 2 to create a new set of measures for a 30°-60°-90° triangle.

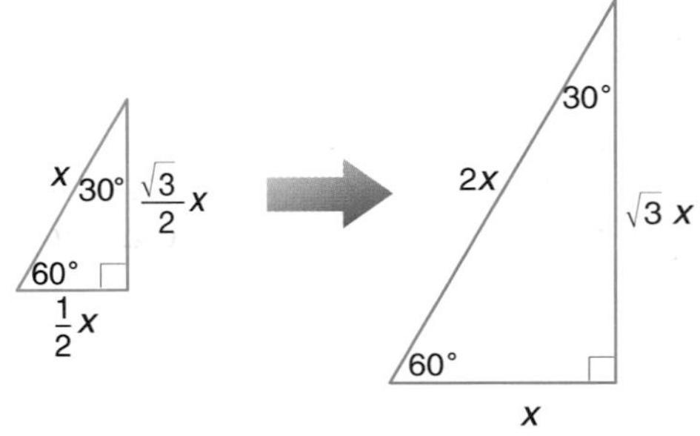

This relationship leads to Theorem 8–7.

Theorem 8–7	**In a 30°-60°-90° triangle, the hypotenuse is twice as long as the shorter leg, and the longer leg is $\sqrt{3}$ times as long as the shorter leg.**

The shorter leg is opposite the 30° angle and the longer leg is opposite the 60° angle.

You can use Theorem 8–7 to quickly find the measures of the sides of a 30°-60°-90° triangle if you know the measure of any side.

Example 2 **A *regular hexagon* is made up of six equilateral triangles. Find AC in the regular hexagon at the right.**

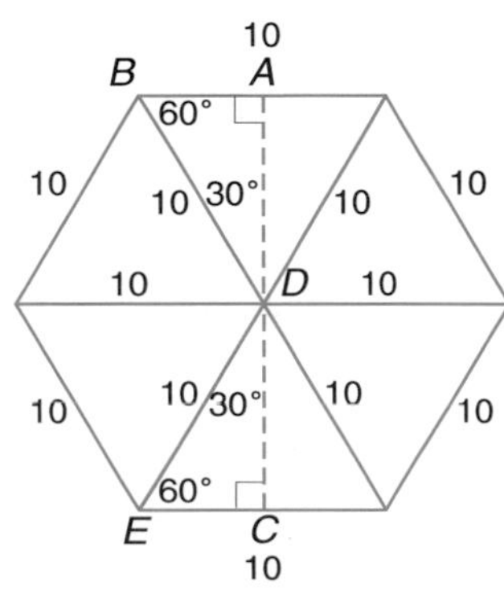

The shorter leg $\overline{AB}$ of $\triangle ABD$ is half as long as the hypotenuse $\overline{BD}$, which is 10 units long. Therefore, $AB = 5$. The longer leg is $\sqrt{3}$ times as long as the shorter leg. So, $AD = 5\sqrt{3}$. Likewise, in $\triangle DEC$, $DC = 5\sqrt{3}$. Since $AC = AD + DC$, $AC = 5\sqrt{3} + 5\sqrt{3}$ or $10\sqrt{3}$.

Special triangles can be graphed in a coordinate plane.

Example **3** **ΔPCD is a 30°-60°-90° triangle with right $\angle C$ and $\overline{CD}$ as the longer leg.**

INTEGRATION
Algebra

a. Find the coordinates of P in Quadrant I for $C(3, 2)$ and $D(9, 2)$.

A graph of the given information can help you visualize the problem.

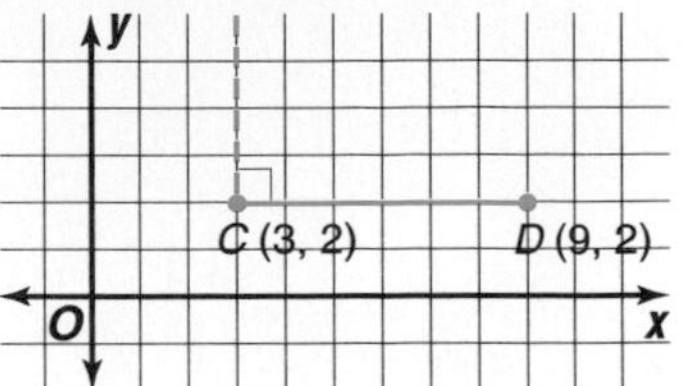

Notice that $\overline{CD}$ lies on a gridline of the coordinate plane and since $\overline{PC}$ will be perpendicular to $\overline{CD}$, it too lies on a gridline. Find the length of $\overline{CD}$.

$$CD = |9 - 3| = 6$$

$\overline{CD}$ is the longer leg, and $\overline{PC}$ is the shorter leg. So, $CD = \sqrt{3}(PC)$ by Theorem 8–7. Use CD to find PC.

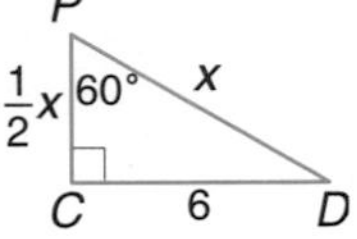

$$CD = \sqrt{3}(PC)$$
$$6 = \sqrt{3}(PC) \quad \textit{CD = 6}$$
$$\frac{6}{\sqrt{3}} = PC \quad \textit{Division Property (=)}$$
$$\frac{6}{\sqrt{3}} \cdot \frac{\sqrt{3}}{\sqrt{3}} = PC \quad \textit{Rationalize the denominator.}$$
$$\frac{6\sqrt{3}}{3} = PC$$
$$2\sqrt{3} = PC$$

The x-coordinate of P is the same as the x-coordinate of C. P is located $2\sqrt{3}$ units above C. So, the coordinates of P are $(3, 2 + 2\sqrt{3})$ or about (3, 5.46).

b. Draw ΔPCD.

Graph points C and D.

Draw $\overline{CD}$.

Graph P, and draw $\overline{PD}$ and $\overline{PC}$.

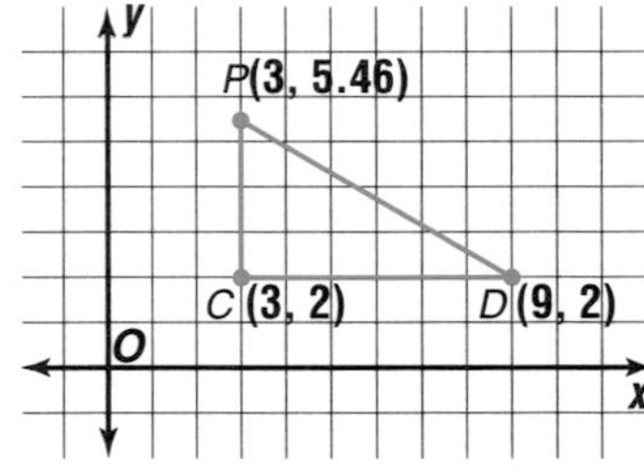

CHECK FOR UNDERSTANDING

Communicating Mathematics

Study the lesson. Then complete the following.

1. **Draw** a 45°-45°-90° triangle. Suppose the measure of one of the legs of the triangle is a. Label the measure of each segment in your figure.
2. **Draw** a 30°-60°-90° triangle. Suppose the measure of the hypotenuse of the triangle is $2x$. Label the measure of each segment in your figure.
3. **You Decide** The students in Ms. Esparza's geometry class have been given the assignment to draw a 30°-60°-90° triangle with $X(0, 0)$, $Y(-5, 0)$, and $\angle X$ as the right angle. Jamal claims there are two triangles that satisfy the conditions. Winona claims there are four such triangles and Mary claims there is only one. Who is correct? Explain.

Guided Practice

Find the values of *x* and *y*.

4\. 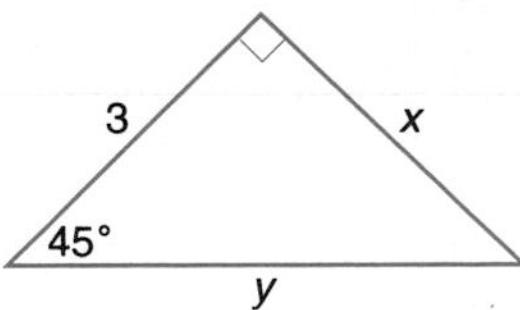

5\. 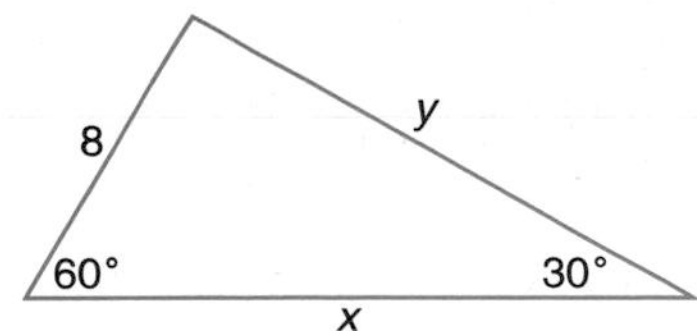

6\. The length of a diagonal of a square is $10\sqrt{2}$ inches. Find the length of one side of the square.

7\. The length of an altitude of an equilateral triangle is $\frac{\sqrt{3}}{2}$ feet. Find the length of a side of the triangle.

8\. The perimeter of a square is 44 meters. Find the length of a diagonal of the square.

Algebra

9\. $\triangle PCD$ is a 30°-60°-90° triangle with right $\angle C$ and $\overline{CD}$ the shorter leg. Find the coordinates of P in Quadrant II for $C(-4, -2)$ and $D(6, -2)$.

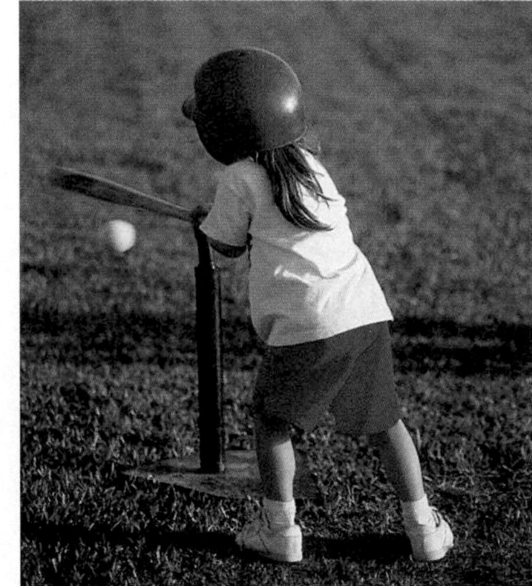

10\. **Tee-Ball** Many younger children like to play a game similar to baseball called tee-ball. Instead of trying to hit a ball thrown by a pitcher, the batter hits the ball off a tee. To accommodate younger children, the bases are only 40 feet apart. Find the distance between home plate and second base in tee-ball.

EXERCISES

Practice

Find the values of *x* and *y*.

11\. 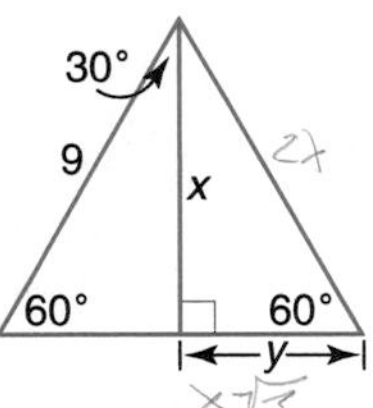

12\. 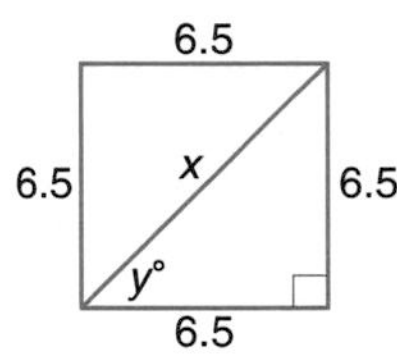

13\. 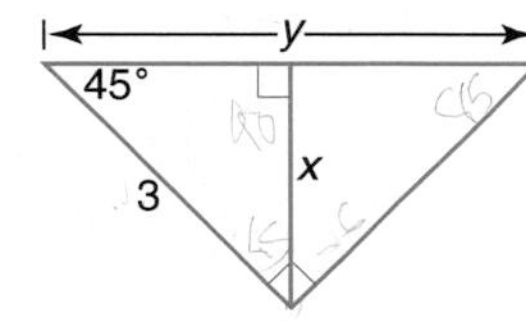

14\. 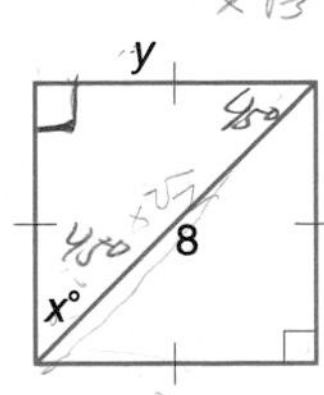

15\. 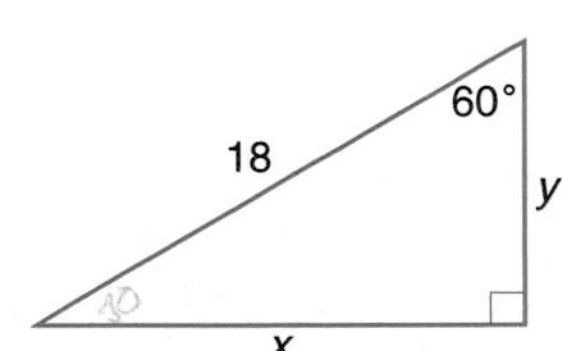

16\. 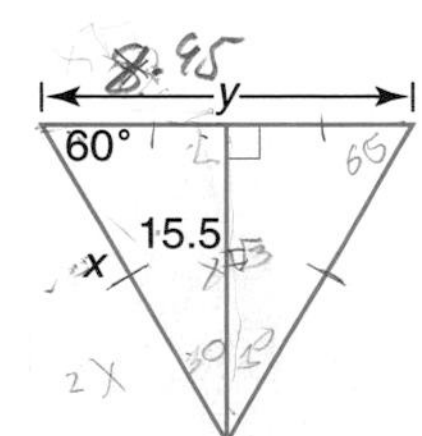

17\. The length of one side of a square is 13.5 centimeters. Find the length of a diagonal of the square.

18\. The length of a diagonal of a square is 10 inches. Find the length of one side of the square.

19\. The length of one side of an equilateral triangle is $6\sqrt{3}$ meters. Find the length of one altitude of the triangle.

20\. The length of an altitude of an equilateral triangle is 12 feet. Find the length of a side of the triangle.

21\. The perimeter of an equilateral triangle is 39 centimeters. Find the length of an altitude of the triangle.

22. The length of a diagonal of a square is $18\sqrt{2}$ millimeters. Find the perimeter of the square.

23. The altitude of an equilateral triangle is 5.2 meters long. Find the perimeter of the triangle.

24. The diagonals of a rectangle are 12 inches long and intersect at an angle of 60°. Find the perimeter of the rectangle.

25. The sum of the squares of the measures of the sides of a square is 196. Find the measure of a diagonal of the square.

Algebra

26. $\triangle PAB$ is a 45°-45°-90° triangle with right $\angle B$. Find the coordinates of P in Quadrant I for $A(-3, 1)$ and $B(4, 1)$.

27. $\triangle PCD$ is a 30°-60°-90° triangle with right $\angle C$ and $\overline{CD}$ the longer leg. Find the coordinates of P in Quadrant III for $C(-3, -6)$ and $D(-3, 7)$.

28. $\triangle PCD$ is a 30°-60°-90° triangle with $m\angle C = 30$ and $\overline{CD}$ the hypotenuse. Find the coordinates of P for $C(2, -5)$ and $D(10, -5)$ if P lies above $\overline{CD}$.

29. Each triangle in the figure at the right is a 30°-60°-90° triangle. Find the value of x.

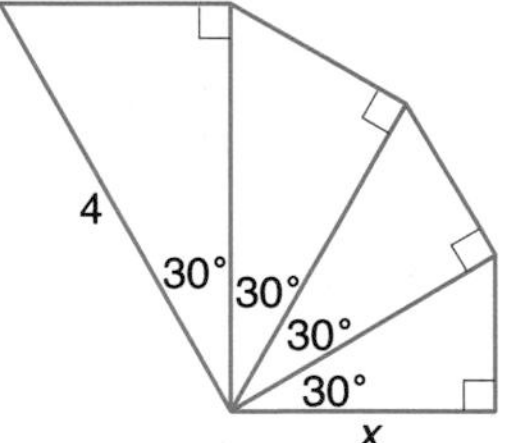

30. Each edge of the cube is s units long.
    a. Write an expression for AF in terms of s.
    b. Find $m\angle BFD$.

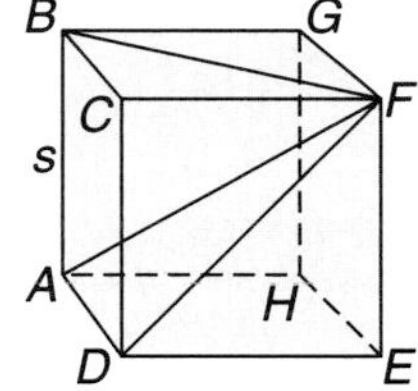

Critical Thinking

31. Two parallel lines are cut by a transversal. The transversal makes a 120° angle with one of the parallel lines. A line bisects the 120° angle. Another line bisects the consecutive interior angle.
    a. Draw the figure.
    b. Show that the triangle formed by the two angle bisectors and the transversal is a 30°-60°-90° triangle.
    c. If the measure of the longer leg is 22, find the measure of the shorter leg of the triangle and the measure of the transversal between the parallel lines.

Applications and Problem Solving

32. **City Planning** Granmichele is a city with a population of about 15,000 people in northern Sicily. The city has a hexagonal design as shown by the aerial view at the right. Consider all sides of the hexagons to be congruent. Suppose the perpendicular distance from the center of the city to each side of the one hexagon is 0.5 mile. Find the perimeter of that hexagon.

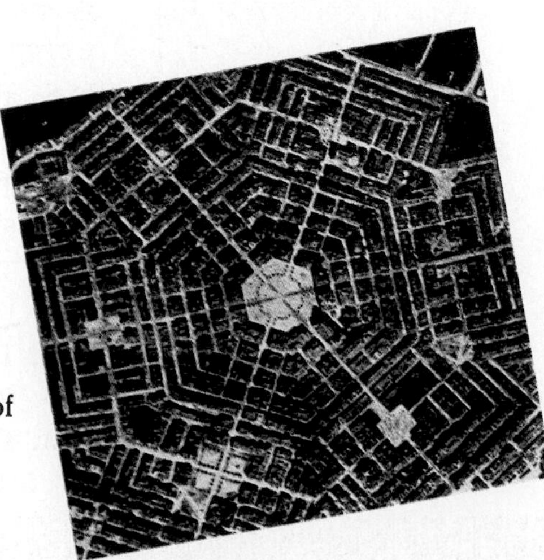

33. **Landscaping** Juana is a landscaper. She wishes to determine the height of a tree. Holding a drafter's 45° triangle so that one leg is horizontal, she sights the top of the tree along the hypotenuse as shown at the right. If she is 6 yards from the tree and her eyes are 5 feet from the ground, find the height of the tree.

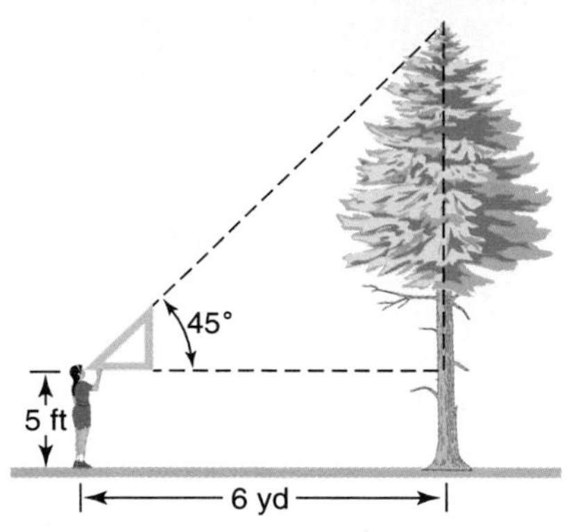

Mixed Review

34. **Sports** Digna is making a ramp to try out her car for a pinewood derby. The ramp support forms a right angle. The base is 12 feet long, and the height is 5 feet. What length of plywood does she need for the ramp? (Lesson 8–1)

35. Name three ways to prove two triangles are similar. (Lesson 7–3)

36. Draw two similar right, scalene triangles. (Lesson 7–2)

37. Is quadrilateral $QRST$ with vertices at $Q(1, 4)$, $R(5, 0)$, $S(-1, -6)$, and $T(-5, -2)$ a rectangle? Justify your answer. (Lesson 6–3)

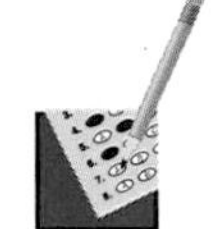

38. **ACT Practice** If $p = -5$, then $5 - p^2 - p =$

    A -15 B -5 C 10 D 30 E 40

39. Graph $x - 2y = 4$ and $J(-3, -1)$. Then construct a perpendicular segment and find the distance from point J to the line. (Lesson 3–5)

For **Extra Practice**, see page 778.

Algebra

40. Simplify $\frac{8k^4}{2k}$.

41. Factor $8g^2 + 32g$.

WORKING ON THE Investigation

Refer to the Investigation on pages 392–393.

On your mission to Mars, you will need to generate power for running equipment in the space station. One option for energy production is solar energy cells. Solar cells convert the energy of sunlight directly into electrical energy.

1 List the equipment that you think will be needed in the space station. Then estimate the energy needed by the space station each day.

2 If solar cells produce about 0.01 watt of electricity per square centimeter, what must be the total area of the solar cells?

3 Many solar cells are hexagonal because congruent hexagons can tessellate, or cover, a flat surface without leaving any gaps. The formula for the area A of a regular hexagon is $A = \frac{1}{2}Pa$, where P is the perimeter. If the solar cell below produces 15 watts, find the length of a side of the cell.

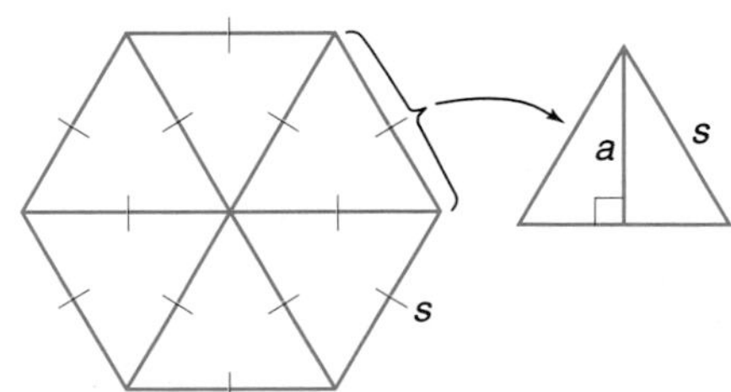

4 How many solar cells like the one above will be needed to power the space station? Is this a reasonable way to produce the energy?

Add the results of your work to your Investigation Folder.

8–3 Integration: Trigonometry Ratios in Right Triangles

What YOU'LL LEARN

- To find trigonometric ratios using right triangles, and
- to solve problems using trigonometric ratios.

Why IT'S IMPORTANT

You can use trigonometric ratios to find missing measures of right triangles involved in aviation, medicine, and astronomy.

Golf

David Leadbetter is rapidly becoming one of the most well-known golf instructors in the world. His students include Nick Price and Debra McHaffie. He invented a golfing aid called the RIGHT ANGLE. The RIGHT ANGLE fits onto the arm and positions the arm correctly on the backswing to add consistency in the swing. The right triangle formed by the arm is modeled by the right triangle below.

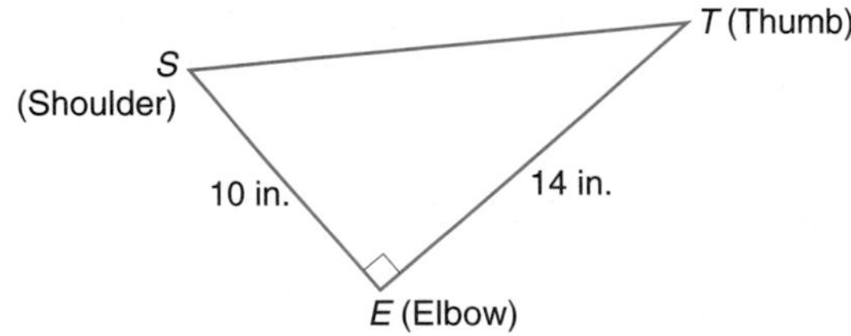

Since $\angle E$ is a right angle, $\triangle SET$ is a right triangle. What is the measure of the angle at the top of the swing represented by $\angle T$? **Trigonometry** will allow us to solve the problem. *You will be asked to find this measure in Exercise 51.*

The word *trigonometry* comes from two Greek terms, *trigon* meaning triangle and *metron* meaning measure. The study of trigonometry involves triangle measurement. A ratio of the lengths of sides of a right triangle is called a **trigonometric ratio**. The three most common ratios are **sine**, **cosine**, and **tangent**. Their abbreviations are *sin*, *cos*, and *tan*, respectively.

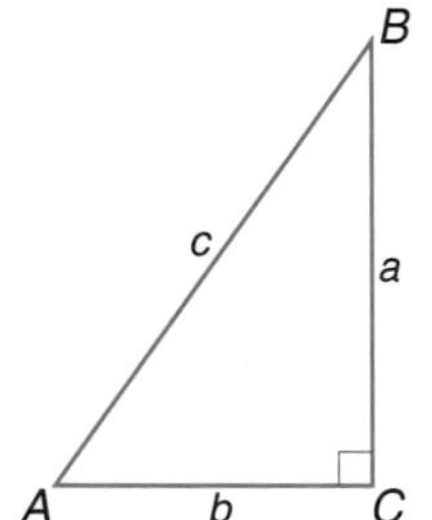

Trigonometric Ratio	Abbreviation	Definition
sine of $\angle A$	sin A	$\frac{\text{leg opposite } \angle A}{\text{hypotenuse}} = \frac{a}{c}$
cosine of $\angle A$	cos A	$\frac{\text{leg adjacent to } \angle A}{\text{hypotenuse}} = \frac{b}{c}$
tangent of $\angle A$	tan A	$\frac{\text{leg opposite } \angle A}{\text{leg adjacent to } \angle A} = \frac{a}{b}$

A represents the measurement of $\angle A$.

Trigonometric ratios are related to the acute angles of a right triangle, *not* the right angle.

You can use paper folding to explore trigonometric ratios.

MODELING MATHEMATICS

Trigonometric Ratios

Materials: paper 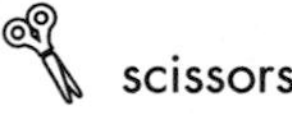 scissors  ruler

- Fold a rectangular piece of paper along a diagonal from A to C. Then cut along the fold to form a right $\triangle ABC$. Write the name of each angle on the inside of the triangle.

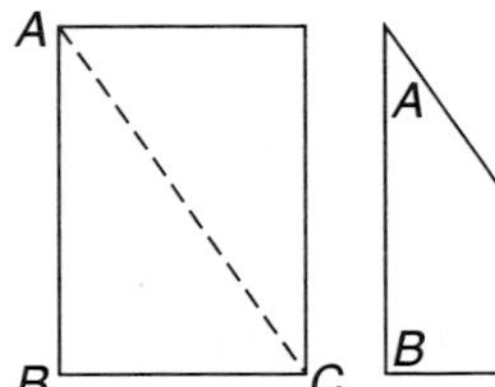

- Fold the triangle so that two segments perpendicular to $\overline{AB}$ result. Label points D, E, F, and G as shown. Use a ruler to find the measures AC, AB, BC, AF, AG, FG, AD, AE, and DE to the nearest millimeter.

Your Turn

a. For $\angle A$, the sine could be expressed as $\frac{DE}{AD}$ for $\triangle ADE$, $\frac{FG}{AF}$ for $\triangle AFG$, and $\frac{BC}{AC}$ for $\triangle ACB$. How do these fractions compare?

b. For $\angle A$, the cosine could be expressed as $\frac{AE}{AD}$ for $\triangle ADE$, $\frac{AG}{AF}$ for $\triangle AFG$, and $\frac{AB}{AC}$ for $\triangle ACB$. How do these fractions compare?

c. For $\angle A$, the tangent could be expressed as $\frac{ED}{AE}$ for $\triangle ADE$, $\frac{GF}{AG}$ for $\triangle AFG$, and $\frac{BC}{AB}$ for $\triangle ACB$. How do these fractions compare?

d. Make a conjecture about the value of a trigonometric function based on your observations.

As the Modeling Mathematics activity suggests, the value of a trigonometric ratio depends *only* on the measure of the angle. It does not depend on the size of the triangle.

Example  **Find the sin *S*, cos *S*, tan *S*, sin *E*, cos *E* and tan *E*. Express each ratio as a fraction and as a decimal.**

$\sin S = \frac{ME}{SE} = \frac{3}{5}$ or 0.6

$\cos S = \frac{SM}{SE} = \frac{4}{5}$ or 0.8

$\tan S = \frac{ME}{SM} = \frac{3}{4}$ or 0.75

$\sin E = \frac{SM}{SE} = \frac{4}{5}$ or 0.8

$\cos E = \frac{ME}{SE} = \frac{3}{5}$ or 0.6

$\tan E = \frac{SM}{ME} = \frac{4}{3}$ or $1.\overline{3}$

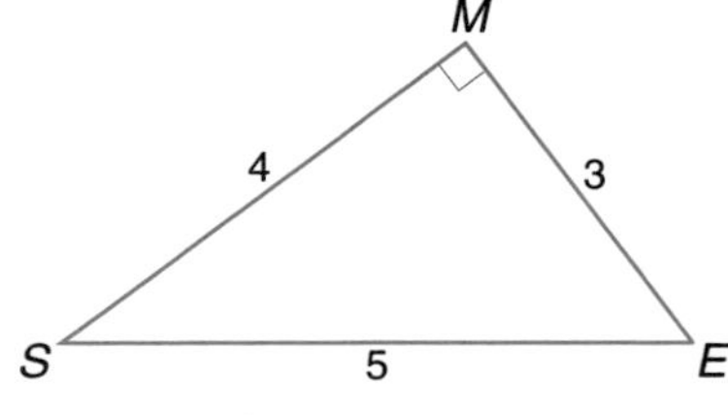

You can use a scientific calculator to evaluate expressions involving trigonometric ratios.

Example  **Find each value using a calculator. Round to the nearest ten thousandths.**

Make sure the calculator is in degree mode.

> **TECHNOLOGY Tip**
>
> If you are using a graphing calculator, the trigonometric ratio name is entered first and then the angle measure.

a. $\cos 41°$

ENTER: 41 [COS] 0.75470958 *Your calculator display may differ slightly from the ones shown.*

$\cos 41° \approx 0.7547$

b. $\sin 78°$

ENTER: 78 [SIN] 0.9781476

$\sin 78° \approx 0.9781$

You can use trigonometric ratios to find missing measures of a right triangle.

Example 3 

Aviation

A plane is one mile above sea level when it begins to climb at a constant angle of 2° for the next 70 ground miles. How far above sea level is the plane after its climb?

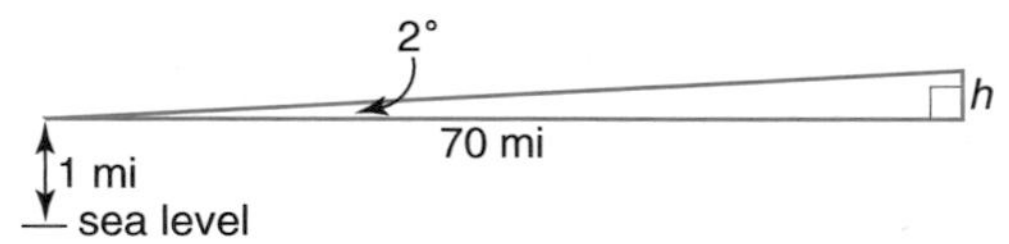

$\tan 2° = \frac{h}{70}$ *$\tan = \frac{\text{opposite}}{\text{adjacent}}$*

$70 \tan 2° = h$ *Multiplication Property (=)*

Use your calculator to find h.

ENTER: 70  2 [TAN] 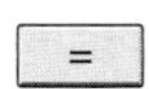 2.444453864

The plane has climbed about 2.4 miles.
Therefore it is about 2.4 + 1 or 3.4 miles above sea level.

You can use trigonometric ratios and a calculator to find the measure of an angle if you know the measures of two sides of the right triangle. The inverse of each trigonometric ratio yields the angle measure. The inverse of the sine, cosine, and tangent ratios are indicated by $\sin^{-1}$, $\cos^{-1}$, and $\tan^{-1}$, respectively.

Example  **Find $m\angle A$ in right triangle ABC for $A(1, 2)$, $B(6, 2)$, and $C(5, 4)$.**

Algebra

Explore The problem gives the coordinates of the vertices of a right triangle and asks for the measure of one of the angles. From the graph, $\angle C$ appears to be the right angle.

Plan Use the distance formula to find the measures of each side. Then find one of the trigonometric ratios for $\angle A$. Use the inverse of that ratio to find $m\angle A$.

CAREER CHOICES

An **airline pilot** must have a commercial pilot's license with an instrument rating issued by the FAA. They must also pass strict physical examinations, including eyesight and hearing. While instruments help pilots to navigate in many situations, they also depend on their visual skills to evaluate angles and distances.

For more information, contact:

Airline Pilots Association
1625 Massachusetts Ave., NW
Washington, DC 20036

Solve

$AB = \sqrt{(6-1)^2 + (2-2)^2}$

$= \sqrt{25 + 0}$ or 5

$BC = \sqrt{(5-6)^2 + (4-2)^2}$

$= \sqrt{1 + 4}$ or $\sqrt{5}$

$AC = \sqrt{(5-1)^2 + (4-2)^2}$

$= \sqrt{16 + 4}$ or $\sqrt{20}$

Use the cosine ratio.

$\cos A = \frac{AC}{AB}$ $\quad \cos = \frac{\text{adjacent}}{\text{hypotenuse}}$

$= \frac{\sqrt{20}}{5}$

Use a scientific calculator to find $m\angle A$.

ENTER: 20 [2nd] [$\sqrt{x}$] [÷] 5 [=] [2nd] [$\cos^{-1}$] 26.56505118

The measure of $\angle A$ is about 26.6.

Examine Use the sine ratio to check the answer.

$\sin A = \frac{BC}{AB}$ $\quad \sin = \frac{\text{opposite}}{\text{hypotenuse}}$

$= \frac{\sqrt{5}}{5}$

ENTER: 5 [2nd] [$\sqrt{x}$] [÷] 5 [=] [2nd] [$\sin^{-1}$] 26.56505118

The answer is correct.

CHECK FOR UNDERSTANDING

Communicating Mathematics

Study the lesson. Then complete the following.

1. **State** the meaning of trigonometry.
2. **Compare and contrast** the sine, cosine, and tangent ratios.
3. **Explain** why trigonometric ratios do not depend on the size of the right triangle.
4. **Distinguish** between cos and $\cos^{-1}$.

5. **Fold** a rectangular piece of paper along a diagonal and cut along the fold to form a right triangle. Fold the triangle so that two more segments perpendicular to one leg of the right triangle result. Label the vertices of the triangles as shown at the right. The *cotangent* is a trigonometric ratio that equals the measure of the adjacent leg divided by the measure of the opposite leg.

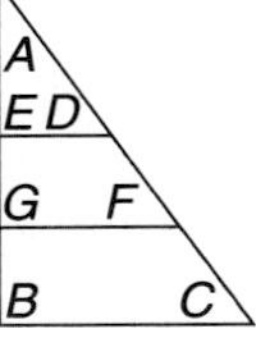

   a. **Write** an expression for the cotangent of A for $\triangle ADE$, $\triangle AFG$, and $\triangle ACB$.
   b. **Measure** the segments to the nearest tenth of a centimeter. Calculate the cotangent of A for each triangle.
   c. Compare the results.

Guided Practice

Find the indicated trigonometic ratio as a fraction and as a decimal rounded to the nearest ten-thousandth.

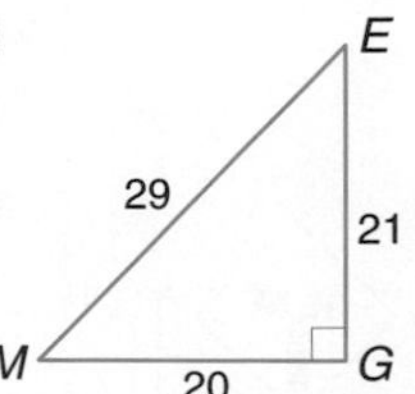

6. $\cos M$ **7.** $\tan E$ **8.** $\sin E$

Find the value of each ratio to the nearest ten-thousandth.

9. $\sin 14°$ **10.** $\cos 51°$

Find the measure of each angle to the nearest tenth of a degree.

11. $\tan A = 2.0035$ **12.** $\cos B = 0.7980$

Find the value of x. Round to the nearest tenth.

13.

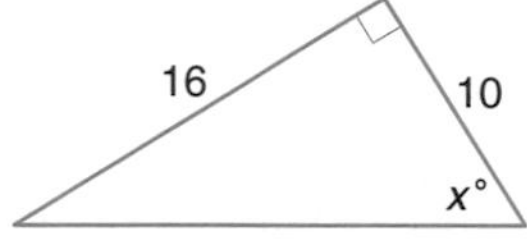

14.

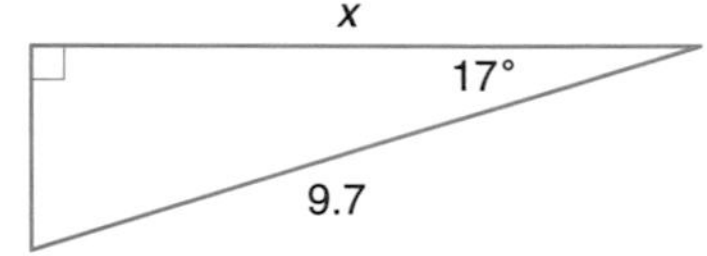

15. Find $m\angle A$ in right $\triangle ABC$ for $A(0, 8)$, $B(3, 0)$, and $C(0, 0)$.

16. Safety To guard against a fall, a ladder should make an angle of 75° or less with the ground. What is the maximum height that a ten-foot ladder can reach safely?

EXERCISES

Practice

Find the indicated trigonometric ratio as a fraction and as a decimal rounded to the nearest ten-thousandth.

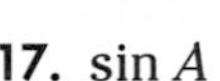

17. $\sin A$ **18.** $\cos x$ **19.** $\cos A$

20. $\sin x$ **21.** $\tan x$ **22.** $\tan A$

23. $\cos B$ **24.** $\sin y$ **25.** $\tan B$

Find the value of each ratio to the nearest ten-thousandth.

26. $\sin 7°$ **27.** $\cos 24°$ **28.** $\tan 54°$

29. $\sin 72°$ **30.** $\cos 42°$ **31.** $\tan 20°$

Find the measure of each angle to the nearest tenth of a degree.

32. $\sin A = 0.7245$ **33.** $\cos B = 0.2493$ **34.** $\tan C = 9.4618$

35. $\sin D = 0.4567$ **36.** $\cos E = 0.1212$ **37.** $\tan R = 0.4279$

Find the value of x. Round to the nearest tenth.

38.

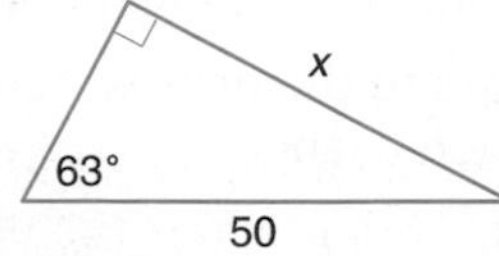

39.

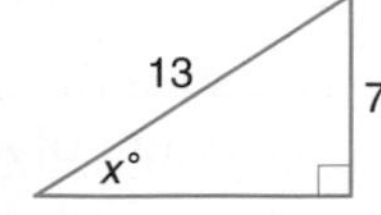

40.

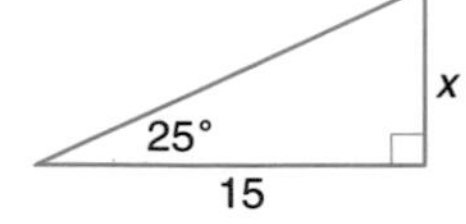

41. 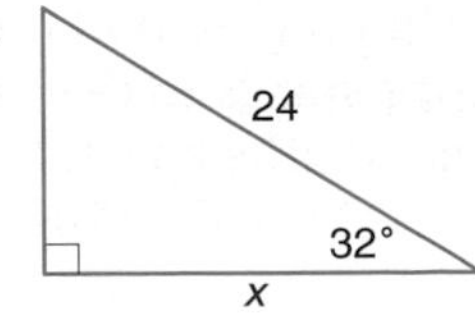

42. 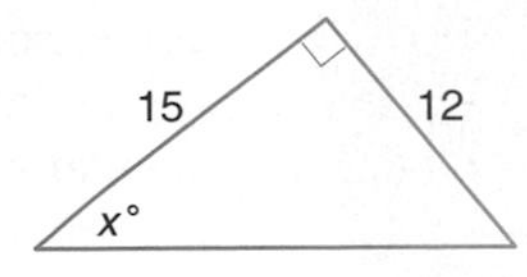

43. 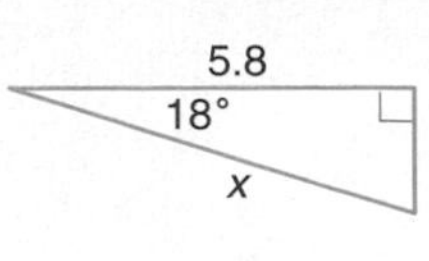

44. Find $m\angle J$ in right $\triangle JCL$ for $J(2, 2)$, $C(2, -2)$, and $L(7, -2)$.

45. Find $m\angle C$ in right $\triangle BCD$ for $B(-1, -5)$, $C(-6, -5)$, and $D(-1, 2)$.

46. Find $m\angle X$ in right $\triangle XYZ$ with vertices $X(-5, 0)$, $Y(7, 0)$, and $Z(0, \sqrt{35})$.

Find the values of *x* and *y*. Round to the nearest tenth.

47. 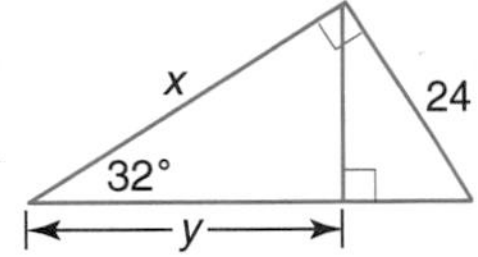

48. 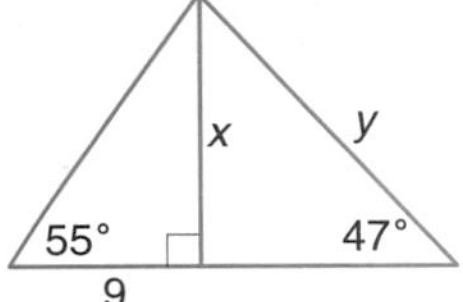

49. 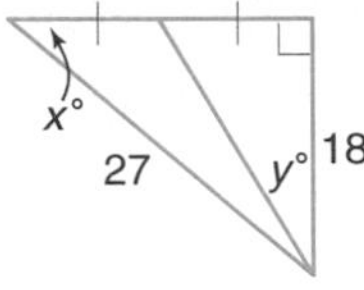

Critical Thinking

50. The figure at the right is one-fourth of a circle drawn in Quadrant I of a coordinate plane. The radius of the circle is 1 unit and each division on the *x*- and *y*-axes is 0.2 unit. Right triangles are formed by drawing a vertical line from the point on the *x*-axis to the circle, then connecting that point with the origin.

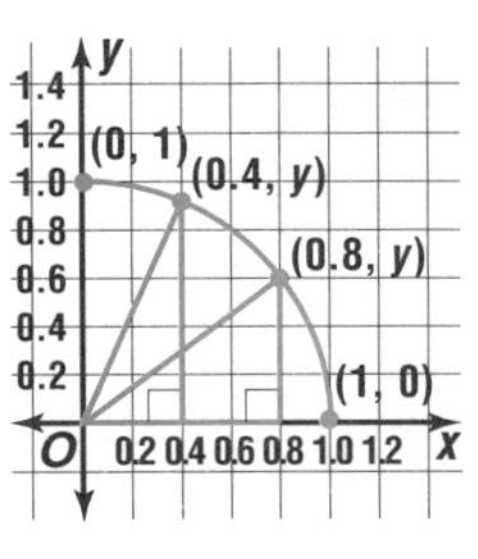

a. Use the Pythagorean Theorem and trigonometry to complete the table of values for each triangle.

Length of Hypotenuse	x value	y value	sin O	cos O
1	0.2			
1	0.4			
1	0.6			
1	0.8			

b. Make a conjecture about the ordered pair (x, y) and the sine and cosine of $\angle O$.

Applications and Problem Solving

51. **Golf** Refer to the application at the beginning of the lesson. Find $m\angle T$.

52. **Medicine** A patient is being treated with radiotherapy for a tumor that is behind a vital organ. In order to prevent damage to the organ, the radiologist must angle the rays to the tumor. If the tumor is 6.3 centimeters below the skin and the rays enter the body 9.8 centimeters to the right of the tumor, find the angle the rays should enter the body to hit the tumor.

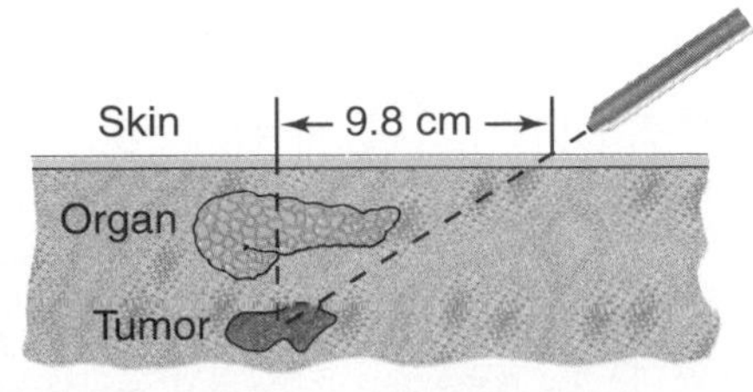

53. Astronomy One way to find the distance between the Sun and a relatively close star is to determine the angles of sight for the star exactly six months apart. Half the measure formed by these two angles of sight is called the *stellar parallax*. Distances in space are sometimes measured in *astronomical units*. An astronomical unit is equal to the average distance between Earth and the Sun. The stellar parallax for the star Alpha Centauri is 0.00021°.

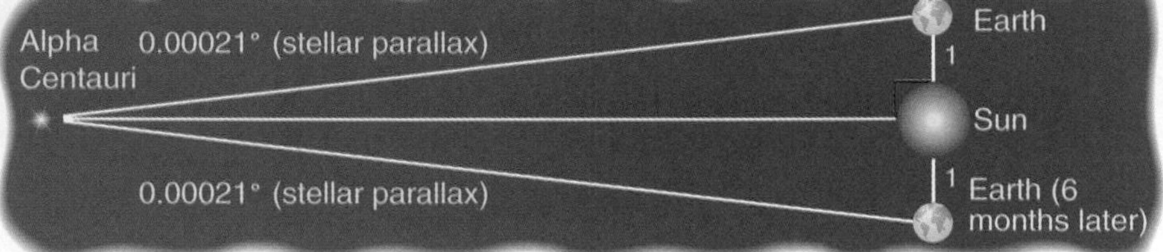

a. Find the distance between Alpha Centauri and the Sun.

b. Make a conjecture as to why this method is used only for relatively close stars.

Mixed Review

54. The measure of the longer leg of a 30°-60°-90° triangle is 14. Write expressions for the measures of the shorter leg and the hypotenuse. (Lesson 8–2)

55. Find the geometric mean between 8 and 9. (Lesson 8–1)

56. In the figure, $\triangle XYZ \sim \triangle XWV$. The perimeter of $\triangle XYZ$ is 34 units, and the perimeter of $\triangle XWV$ is 14 units. If $WV = 4$, find YZ. (Lesson 7–5)

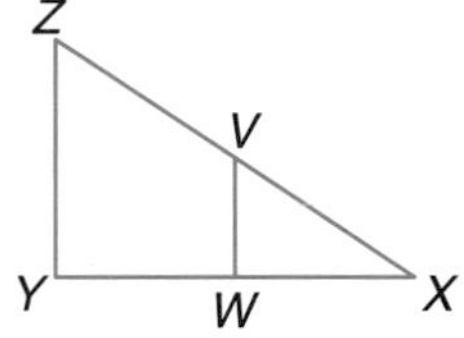

57. Find the value of j. (Lesson 7–2)

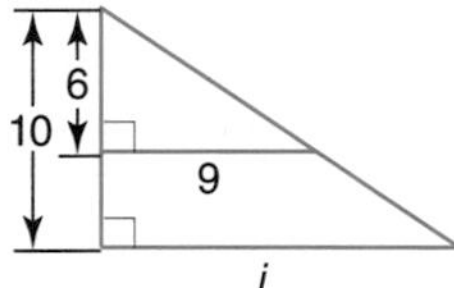

58. If $AC = 6g + 2$ and $AR = 4g - 2$, find RC. (Lesson 6–2)

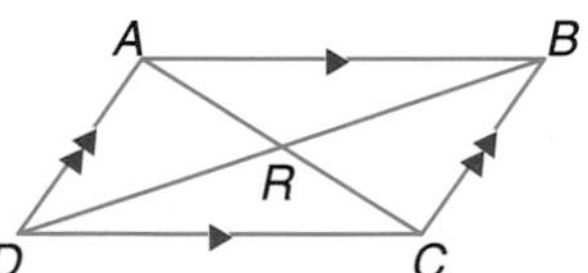

59. SAT Practice Quantitative Comparison

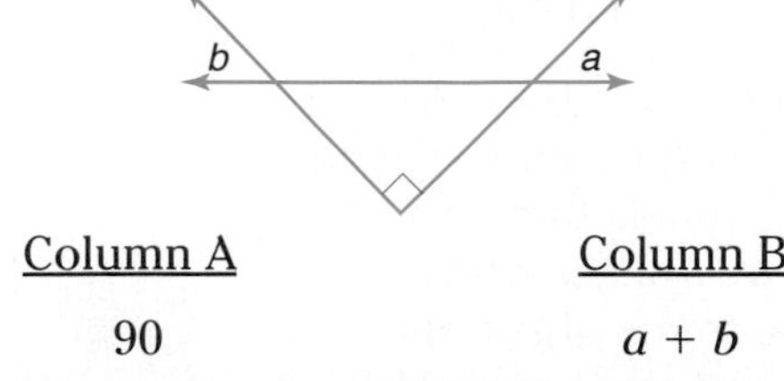

Column A	Column B
90	$a + b$

A if the quantity in Column A is greater
B if the quantity in Column B is greater
C if the two quantities are equal
D if the relationship cannot be determined from the information given

60. Law A man accused of shoplifting claims that he is innocent because he was at a movie with a friend when the crime took place. What needs to be established in order to prove by indirect reasoning that the man is guilty? (Lesson 5–3)

61. Find the values of r and s so that $\triangle RST \cong \triangle XYZ$ by LA. (Lesson 5–2)

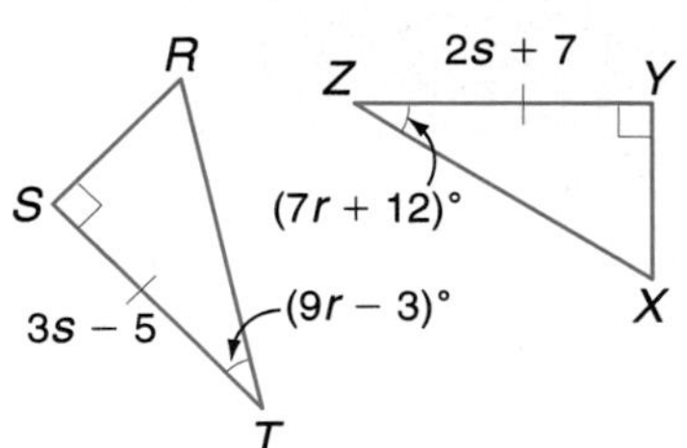

Proof

62. Write a two-column proof. (Lesson 4–5)

Given: $\overline{AB} \perp \overline{BD}$

$\overline{DE} \perp \overline{DB}$

$\overline{DB}$ bisects $\overline{AE}$.

Prove: $\angle A \cong \angle E$

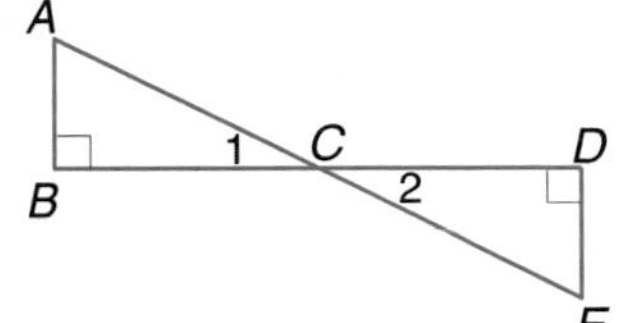

63. If $\triangle GHI \cong \triangle RST$, which angle in $\triangle GHI$ is congruent to $\angle SRT$ in $\triangle RST$? (Lesson 4–3)

Algebra

64. Define a variable, write an inequality, and solve the following problem. *The sum of a number and 11 is at least 25.*

For **Extra Practice**, see page 778.

65. Use substitution to solve the system of equations.

$y = 2x$ and $3x + y = 10$

SELF TEST

1. Find the geometric mean between 15 and 9. (Lesson 8–1)

Find the value of x. (Lesson 8–1)

2.

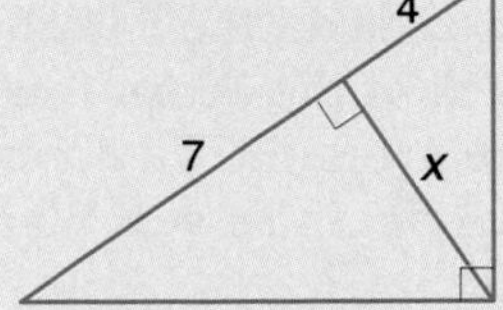

3.

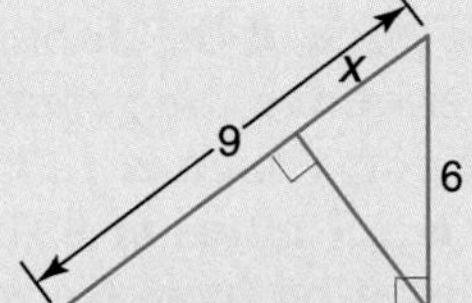

4.

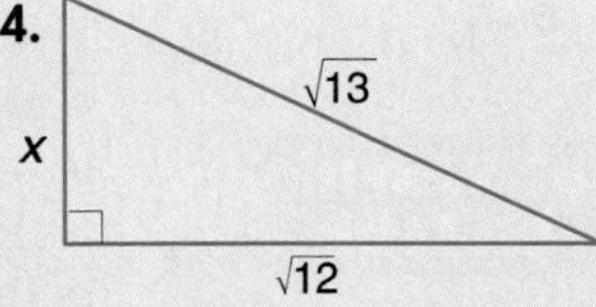

5. The length of one side of a square is 7.5 centimeters. Find the length of a diagonal of the square. (Lesson 8–2)

6. The length of one side of an equilateral triangle is 6 inches. Find the length of one altitude of the triangle. (Lesson 8–2)

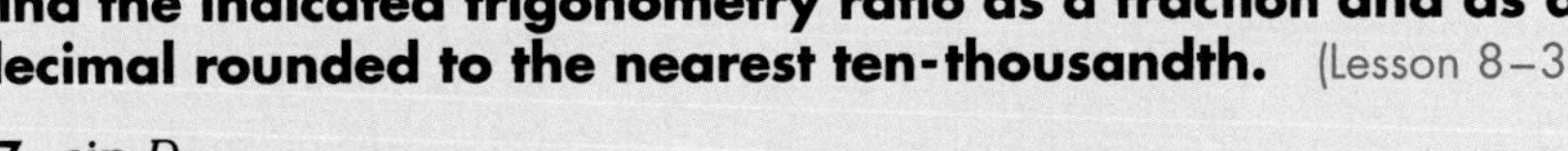

Find the indicated trigonometry ratio as a fraction and as a decimal rounded to the nearest ten-thousandth. (Lesson 8–3)

D, F, E; DF = 2, FE = 3, DE = $\sqrt{13}$

7. $\sin D$

8. $\cos D$

9. $\tan E$

10. Road Construction The 600 block of Powell Street in San Francisco has a steep rise. If it takes 66 feet along the horizontal for the street to rise 10 feet, find the angle the street makes with the horizontal. (Lesson 8–3)

8–4 Angles of Elevation and Depression

What YOU'LL LEARN

- To use trigonometry to solve problems involving angles of elevation or depression.

Why IT'S IMPORTANT

You can use angles of elevation and depression to find missing measures involved in aerospace, architecture, and meteorology.

Tourism

Suppose Arnoldo is on the Skydeck of the Sears Tower looking through a telescope at Hanna who is on the Observation Deck of the John Hancock Center. Hanna is looking through a telescope at Arnoldo. An angle is formed by a horizontal line between the John Hancock Center and the Sears Tower and the line of sight from Hanna to Arnoldo. This angle is called the **angle of elevation**. Another angle is formed by a horizontal line between the Sears Tower and the John Hancock Center and the line of sight from Arnoldo to Hanna. This angle is called the **angle of depression**.

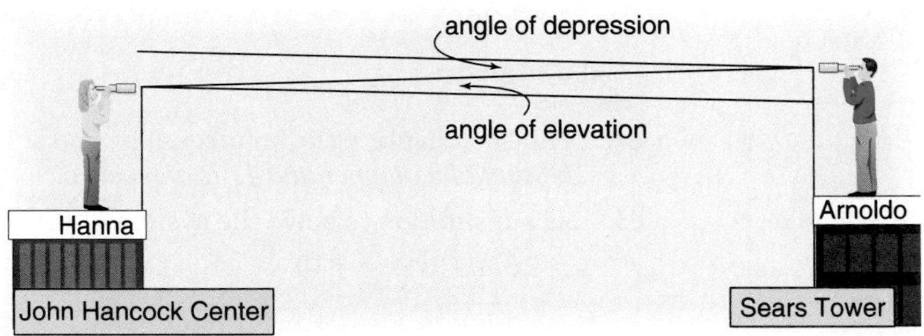

Sometimes drawing a diagram of the situation described in a problem can help you to solve problems involving angles of elevation and depression.

Example 1

Architecture

The Observation Deck of the John Hancock Center is on the 94th floor, which is 1030 feet above the ground. The Skydeck of the Sears Tower is on the 103rd floor, which is 1335 feet from the ground. The John Hancock Center is 1.7 miles or 8976 feet from the Sears Tower. What is the angle of elevation from Hanna to Arnoldo?

Explore The problem gives the height of the Observation Deck and the height of the Skydeck. It also gives the distance between the John Hancock Center and the Sears Tower. It asks for the angle of elevation from Hanna to Arnoldo.

Plan Draw a diagram.

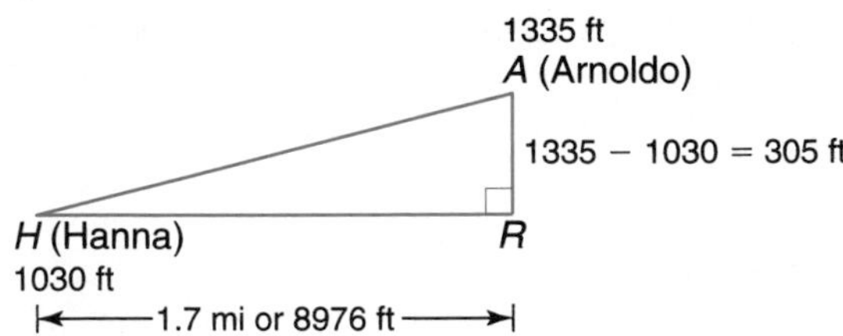

The length of one leg of the right triangle is 1.7 miles or 8976 feet. The length of the other leg is $1335 - 1030$ or 305 feet. The angle of elevation can be found by using $\tan^{-1} H$.

Solve $\tan H = \frac{305}{8976}$

ENTER: 305 [÷] 8976 [=] [2nd] [$\tan^{-1}$] [=]

The angle of elevation is about 1.9°.

Examine Since $\tan A = \frac{8976}{305}$, use a calculator to find $m\angle A$.

ENTER: 8976 [÷] 305 [=] [2nd] [$\tan^{-1}$] [=]

Since $1.9461332 + 88.053867 \approx 90$, the angles are complementary, and the answer is verified.

Angles of elevation or depression to two different objects can be used to find the distance between those objects.

Example 2

Aerospace

On July 20, 1969, Neil Armstrong became the first human to walk on the moon. During this mission, the lunar lander *Eagle* traveled aboard *Apollo 11*. Before sending *Eagle* to the surface of the moon, *Apollo 11* orbited the moon three miles above the surface. At one point in the orbit, the onboard guidance system measured the angles of depression to the far and near edges of a large crater. The angles measured 18° and 25°, respectively. Find the distance across the crater.

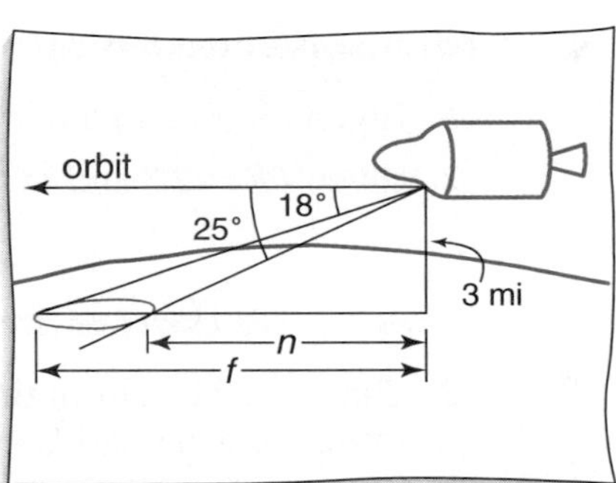

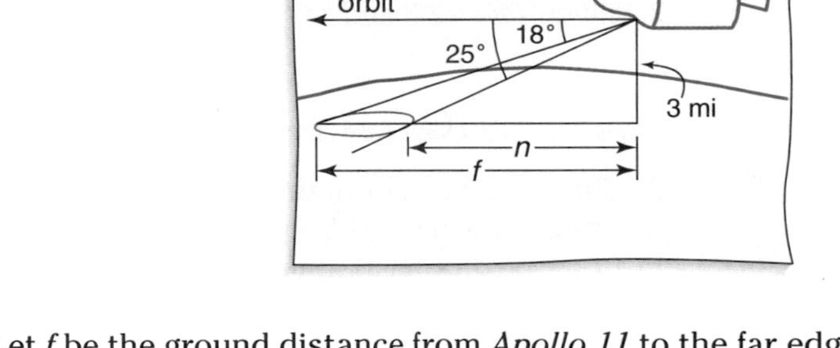

Let f be the ground distance from *Apollo 11* to the far edge of the crater and n be the ground distance to the near edge.

$\tan 18° = \frac{3}{f}$ *$\tan = \frac{\text{opposite}}{\text{adjacent}}$*

$f \tan 18° = 3$ *Cross multiply.*

$f = \frac{3}{\tan 18°}$ *Division Property (=)*

ENTER: 3  [÷] 18 [TAN] 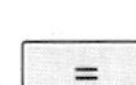 [=]

Make sure your calculator is in degree mode.

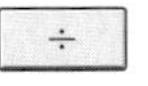
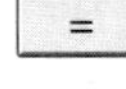

$\tan 25° = \frac{3}{n}$ *$\tan = \frac{\text{opposite}}{\text{adjacent}}$*

$n \tan 25° = 3$ *Cross multiply.*

$n = \frac{3}{\tan 25°}$ *Division Property (=)*

ENTER: 3  [÷] 25 [TAN]  [=]

Since $f \approx 9.2$ and $n \approx 6.4$, the distance across the crater is about $9.2 - 6.4$ or 2.8 miles.

CHECK FOR UNDERSTANDING

Communicating Mathematics

Study the lesson. Then complete the following.

1. **Describe** in your own words the meaning of angle of depression.
2. **Draw** an example showing an angle of elevation. Identify the angle of elevation.
3. **Explain** how you decide whether to use sin, cos, or tan when you are finding the measure of an acute angle in a right triangle.

4. **Assess Yourself** Name three common trigonometric ratios and describe each ratio. What is the meaning of each ratio? When you solve a problem involving trigonometric ratios, do you find a diagram helpful? What suggestions can you make to help your classmates solve these types of problems?

Guided Practice

5. Name the angles of elevation and depression in the figure at the right.

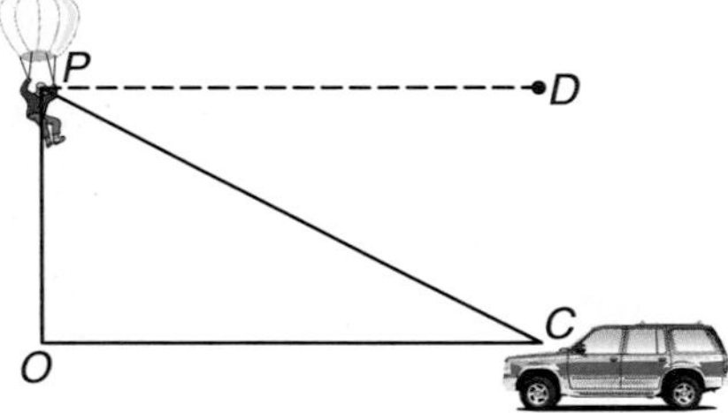

State an equation that would enable you to solve each problem. Then solve. Round answers to the nearest tenth.

6. Given $m\angle P = 15$ and $PQ = 37$, find QR.
7. Given $PR = 2.3$ and $PQ = 5.5$, find $m\angle P$.

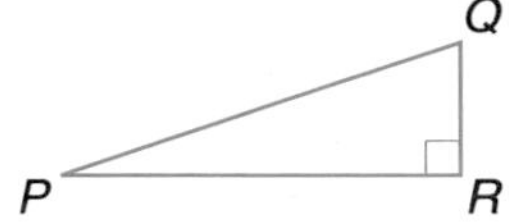

Refer to the chart at the right for Exercises 8–9.

8. Charo is 50 feet from the tallest totem pole. If Charo's eyes are 5 feet from the ground, find the angle of elevation for her line of sight to the top of the totem.
9. Derrick is visiting the San Jacinto State Park outside Houston, Texas. The angle of elevation for his line of sight to the top of the San Jacinto Column is 75°. If his eyes are 6 feet from the ground, how far is he from the base of the column?

Monument Heights

Monument	Height
San Jacinto Column near Houston	570 feet
Gateway to the West Arch St. Louis	630 feet
Washington Monument Washington, D.C.	555 feet
Statue of Liberty New York City	305 feet
Tallest totem pole Alberta Bay, Canada	173 feet

Source: *Comparisons*

Data Update For more information on monument and building heights around the world, visit: **www.geometry.glencoe.com**

10. **Aviation** The cloud ceiling is the lowest altitude at which solid cloud is present. If the cloud ceiling is below a certain level, usually about 61 meters, airplanes are not allowed to take off or land. One way that meteorologists can find the cloud ceiling at night is to shine a searchlight straight up and observe the spot of light on the clouds from a location away from the searchlight.
    a. If the searchlight is located 200 meters from the meteorologist and the angle of elevation to the spot of light on the clouds is 35°, how high is the cloud ceiling?
    b. Can the airplanes land and take off under these conditions? Explain.

EXERCISES

Practice

Name the angles of elevation and depression in each figure.

11. 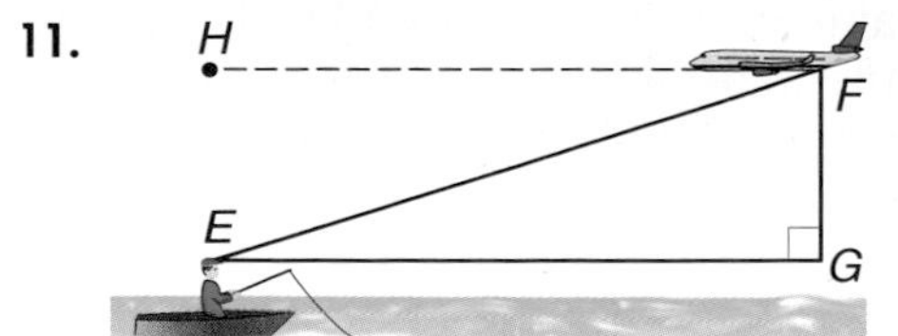

12. 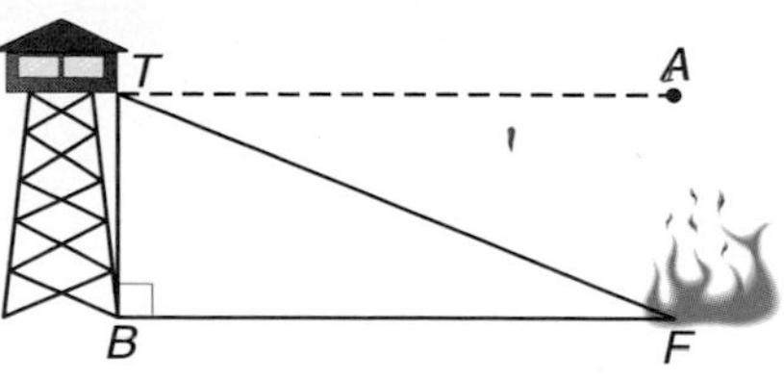

13. 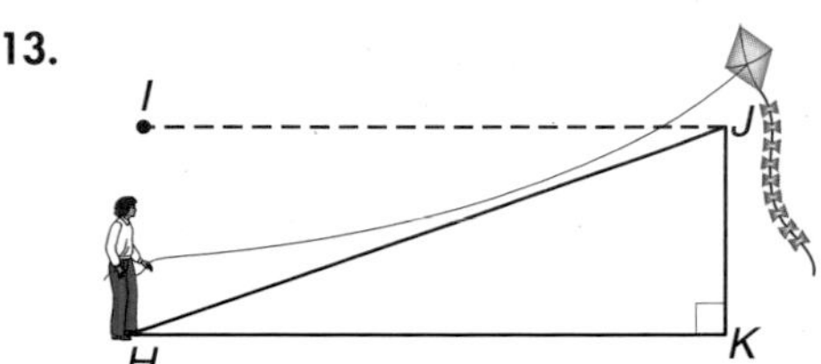

14. 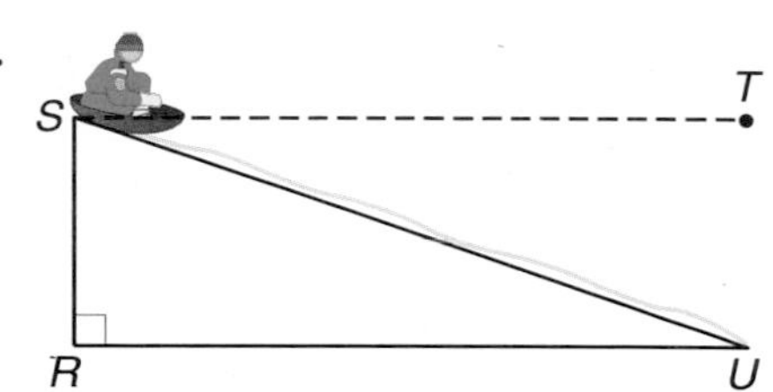

State an equation that would enable you to solve each problem. Then solve. Round answers to the nearest tenth.

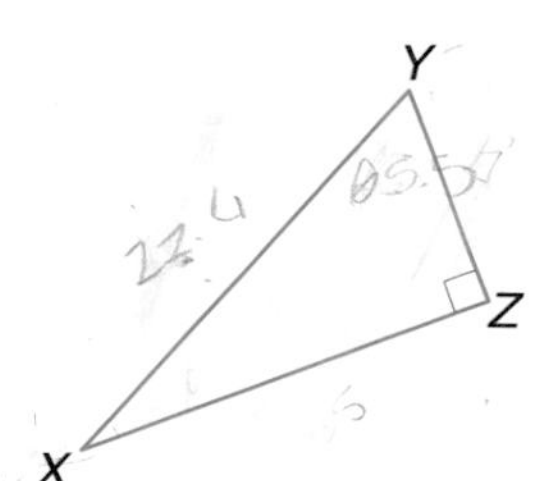

15. Given $YZ = 28$ and $XZ = 54$, find $m\angle Y$.
16. Given $XY = 15$ and $m\angle X = 28$, find YZ.
17. Given $m\angle Y = 66$ and $YZ = 7$, find XY.
18. Given $YZ = 4$ and $XY = 15$, find $m\angle Y$.
19. Given $XZ = 4.5$ and $XY = 6.6$, find $m\angle X$.
20. Given $XY = 22.4$ and $m\angle Y = 65.5$, find XZ.
21. The tallest fountain in the world is located at Fountain Hills, Arizona. If weather conditions are favorable, the water column can reach 625 feet. Suppose Alfonso visits the fountain on a perfect day and his eyes are 5 feet from the ground.
    a. If Alfonso stands 40 feet from the fountain, find the angle of elevation for his line of sight to the top of the spray.
    b. If Alfonso moves so that the angle of elevation for his line of sight to the top of the spray is 75°, how far is he from the base of the spray?

22. After flying at an altitude of 9 kilometers, an airplane starts to descend when its ground distance from the landing field is 175 kilometers. What is the angle of depression for this portion of the flight?
23. A golfer is standing on a tee with the green in a valley below. If the tee is 43 yards higher than the green and the angle of depression from the tee to the hole is 14°, find the distance from the green to the hole.
24. Kierra is flying a kite. She has let out 55 feet of string. If the string makes a 35° angle with the ground, how high above the ground is the kite?
25. A trolley car track rises vertically 40 feet over a horizontal distance of 630 feet. What is the angle of elevation of the track?
26. A ski slope is 550 yards long with a vertical drop of 130 yards. Find the angle of depression of the slope.

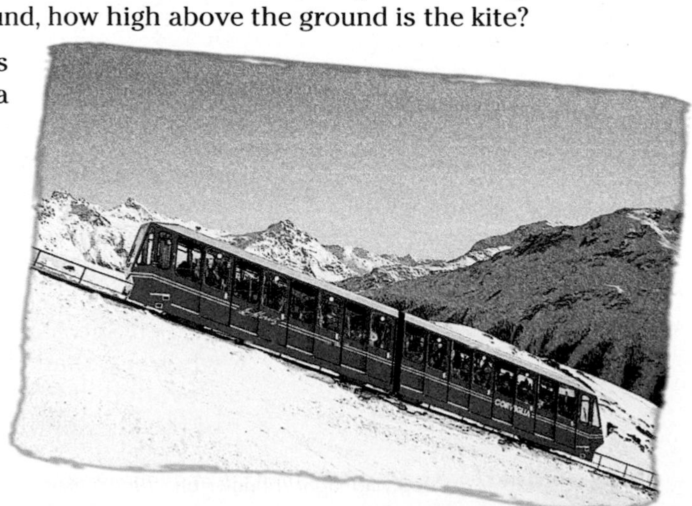

27. The waterway between Lake Huron and Lake Superior separates the United States and Canada at Sault Sainte Marie. The railroad drawbridge located at Sault Saint Marie is normally 13 feet above the water when it is closed. Each section of this drawbridge is 210 feet long. Suppose the angle of elevation of each section is 70°.

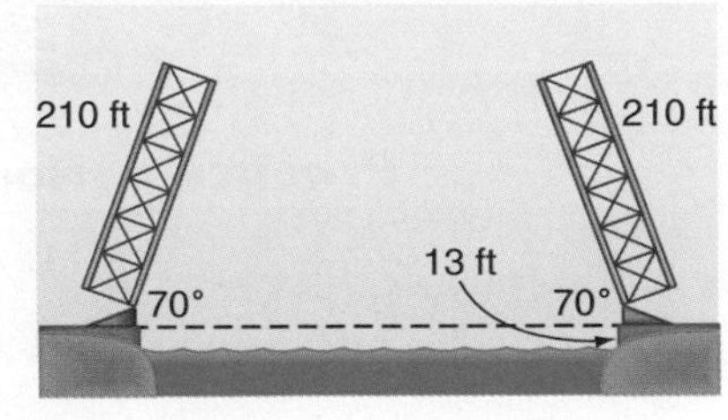

a. Find the distance from the top of a section of the drawbridge to the water.

b. Find the width of the gap created by the two sections of the bridge.

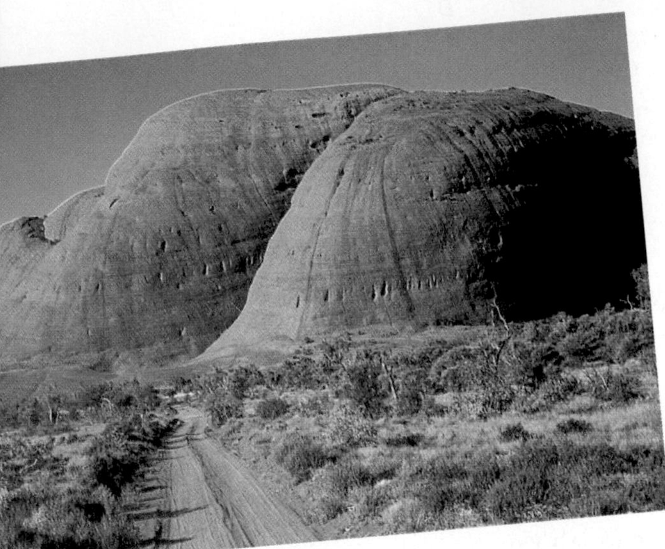

28. Carol is in the Skydeck of the Sears Tower overlooking Lake Michigan. She sights two sailboats going due east from the tower. The angles of depression to the two boats are 42° and 29°. If the Skydeck is 1335 feet high, how far apart are the boats?

29. Ulura or Ayers Rock is a sacred place for Aborigines of the western desert of Australia. Chun-Wei uses a surveying device to measure the angle of elevation to the top of the rock to be 11.5°. He walks half a mile closer and measures the angle of elevation to be 23.9°. How high is Ayers Rock in feet?

Critical Thinking

30. Imagine that a fly and an ant are in one corner of a rectangular box. The end of the box is 4 inches by 6 inches, and the diagonal across the bottom of the box makes an angle of 21.8° with the longer edge of the box. There is food in the corner opposite the insects.

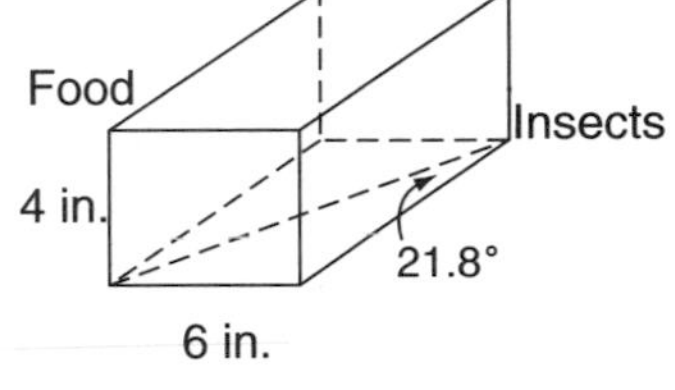

a. What is the shortest distance the fly must fly to get to the food?

b. What is the shortest distance the ant must crawl to get to the food?

31. Given acute $\triangle GME$ with altitude $\overline{EH}$, write a trigonometric expression for the ratio $\frac{GE}{EM}$.

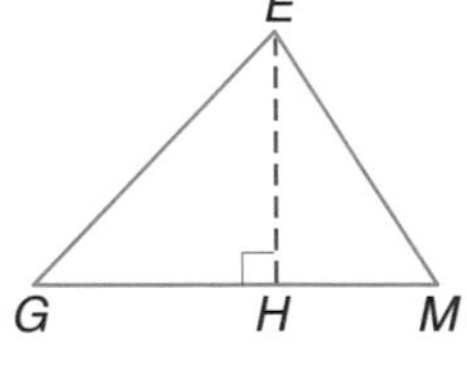

Applications and Problem Solving

32. **Literature** In *The Adventures of Sherlock Holmes: The Adventures of the Musgrave Ritual*, Sherlock Holmes uses trigonometry to solve the mystery. To find a treasure, he must determine where the end of the shadow of an elm tree was located at a certain time of day. Unfortunately, the elm had been cut down, but Mr. Musgrave remembers that his tutor required him to calculate the height of the tree as part of his trigonometry class. Mr. Musgrave tells Sherlock Holmes that the tree was exactly 64 feet. Sherlock needs to find the length of the shadow at a time of day when the shadow from an oak tree is a certain length. The angle of elevation of the sun at this time of day is 33.7°. What was the length of the shadow of the elm?

33. Architecture Diana is an architect who designs houses so that the windows receive minimum sun in the summer and maximum sun in the winter. For Seattle, Washington, the angle of elevation of the sun at noon on the longest day is 66° and on the shortest day is 19°. Suppose a house is designed with a south-facing window that is 6 feet tall. The top of the window is to be installed 1 foot below the overhang.

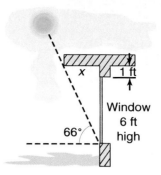

a. How long should Diana make the overhang so that the window gets no direct sunlight at noon on the longest day?

b. Using the overhang from part a, how much of the window will get direct sunlight at noon on the shortest day?

c. To find the angle of elevation of the sun on the longest day of the year where you live, subtract your latitude from 90° and add 23.5°. To find the elevation of the sun on the shortest day, subtract the latitude from 90° and then subtract 23.5°. Draw a solar design for a south-facing window and corresponding overhang for a home in your community.

34. Meteorology Two weather observation stations are 7 miles apart. A weather balloon is located between the stations. From Station 1, the angle of elevation to the weather balloon is 35°. From Station 2, the angle of elevation to the balloon is 54°. Find the altitude of the balloon to the nearest tenth of a mile. (*Hint:* Find the distance from Station 2 to the point directly below the balloon.)

Mixed Review

35. Find the indicated trigonometric ratio as a fraction and as a decimal rounded to the nearest ten-thousandth. (Lesson 8–3)

a. $\sin A$ **b.** $\cos B$ **c.** $\tan A$

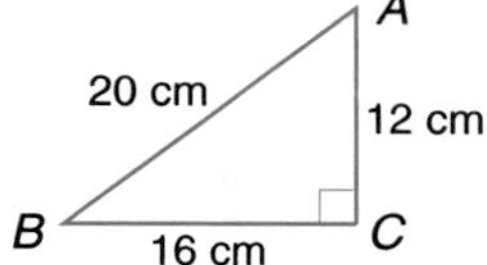

36. The perimeter of an equilateral triangle is 42 centimeters. Find the length of an altitude of the triangle. (Lesson 8–2)

37. Construction Find the length of a diagonal brace needed for a rectangular section of wall that is 6 feet wide and 8 feet high. (Lesson 8–1)

38. ACT Practice In the figure, $\angle A$ is a right angle, AB is 3 units, and BC is 5 units. If the measure of $\angle C$ is θ, what is the value of $\sin \theta$?

A $\frac{3}{5}$ **B** $\frac{3}{4}$ **C** $\frac{4}{5}$ **D** $\frac{5}{4}$ **E** $\frac{5}{3}$

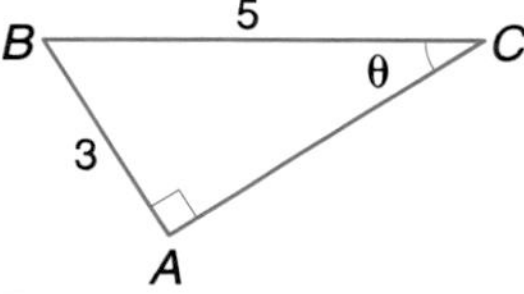

39. Draw a trapezoid with two right angles and one obtuse angle. (Lesson 6–5)

40. Quadrilateral $WXYZ$ is a parallelogram. If $WX = 3g + 7$, $XY = 7h - 1$, $YZ = 6g - 2$, and $WZ = 2h + 9$, find the perimeter of $WXYZ$. (Lesson 6–1)

41. The base of an isosceles triangle is 18 inches long. If the legs are $3y + 21$ and $10y$ inches long, find the perimeter of the triangle. (Lesson 4–6)

Algebra

42. Find the slope and y-intercept of the graph of $2x - y = 16$.

43. Solve the system of inequalities by graphing.
$y < 5$
$y > 2x + 1$

For **Extra Practice**, see page 779.

8–5 Using the Law of Sines

What YOU'LL LEARN

- To use the Law of Sines to solve triangles.

Why IT'S IMPORTANT

You can use the Law of Sines to find missing measures of triangles involved in surveying, aviation, and fire fighting.

Surveying

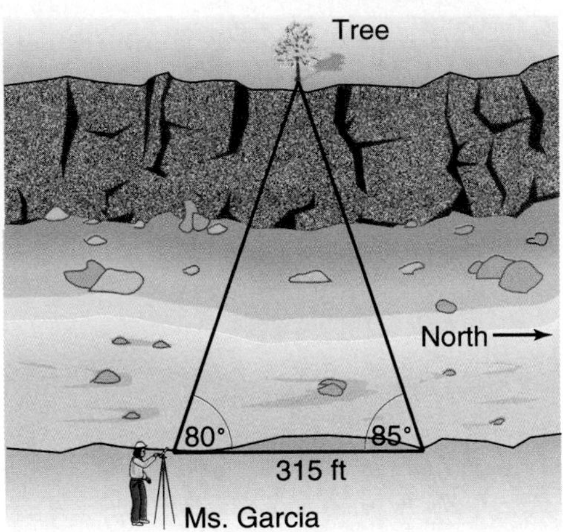

Anna Garcia is a surveyor who has the job of determining the distance across the Rio Grande Gorge in northern New Mexico. Standing at one side of the ridge, she measures the angle formed by the edge of the ridge and the line of sight to a tree on the other side of the ridge. She then walks along the ridge 315 feet north and measures the angle formed by the edge of the ridge and the new line of sight to the same tree. If the first angle is 80° and the second angle is 85°, find the distance across the gorge. *This problem will be solved in Example 2.*

You can use trigonometric functions to solve problems like this that involve triangles that are *not* right triangles. One such method is to use the **Law of Sines**.

Law of Sines	**Let $\triangle ABC$ be any triangle with a, b, and c representing the measures of sides opposite angles with measures A, B, and C, respectively. Then, $\frac{\sin A}{a} = \frac{\sin B}{b} = \frac{\sin C}{c}$.**

Proof of Law of Sines

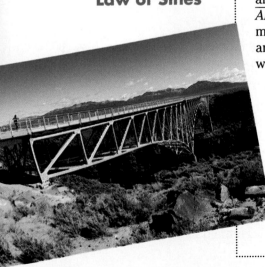

$\triangle ABC$ is a triangle with an altitude from C that intersects $\overline{AB}$ at D. Let h represent the measure of $\overline{CD}$. Since $\triangle ACD$ and $\triangle BCD$ are right triangles, we can find $\sin A$ and $\sin B$.

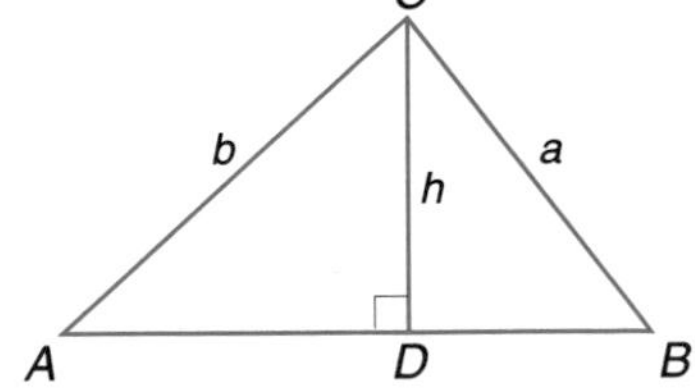

$$\sin A = \frac{h}{b} \qquad \sin B = \frac{h}{a}$$

$$b \sin a = h \qquad a \sin B = h$$

$$b \sin A = a \sin B \quad \textit{Substitution Property (=)}$$

$$\frac{\sin A}{a} = \frac{\sin B}{b} \quad \textit{Divide each side by ab.}$$

The proof can be completed by using a similar technique with another altitude to show that $\frac{\sin A}{a} = \frac{\sin B}{b} = \frac{\sin C}{c}$.

Finding the measures of all the angles and sides of a triangle is called **solving the triangle**. The Law of Sines can be used to solve a triangle in the following cases.

Case 1. You know the measures of two angles and any side of a triangle.
Remember that if you know the measures of two angles of a triangle, you can find the measure of the third angle by using the Angle Sum Theorem.

Case 2. You know the measures of two sides and an angle opposite one of these sides of the triangle.
Examples of this case:
(1) knowing the measures of $\angle A$ and $\angle B$ and length a or b;
(2) knowing the measures of $\angle A$ and $\angle C$ and length a or c;
(3) knowing the measures of $\angle B$ and $\angle C$ and length b or c;

Example 1 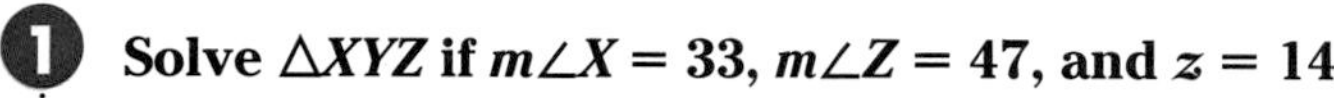 **Solve $\triangle XYZ$ if $m\angle X = 33$, $m\angle Z = 47$, and $z = 14$.**

Since the measures of two angles and a side are known, this is an example of Case 1. Consider the four parts $\angle X$, $\angle Z$, x, and z.

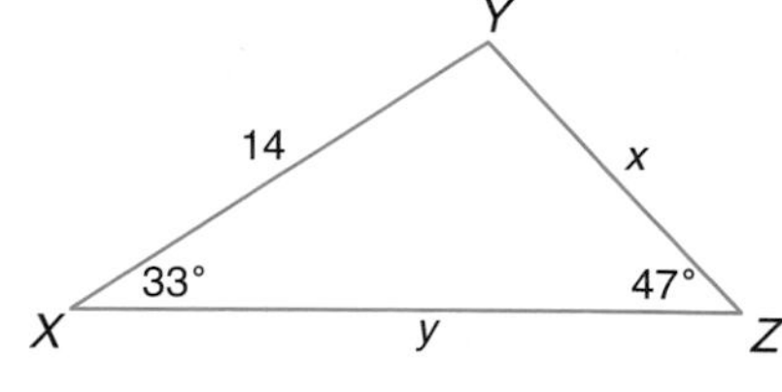

TECHNOLOGY Tip

If you round before finding the final result, your answers will differ slightly from answers in which there was no rounding.

$\frac{\sin Z}{z} = \frac{\sin X}{x}$ *Law of Sines*

$\frac{\sin 47°}{14} = \frac{\sin 33°}{x}$ *$m\angle X = 33$, $m\angle Z = 47$, $z = 14$*

$x \sin 47° = 14 \sin 33°$ *Cross multiply.*

$x = \frac{14 \sin 33°}{\sin 47°}$ *Division Property (=)*

$x \approx 10.4$ *Use a calculator.*

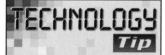

Now we know the measures of two sides and two angles of the triangle. We can find the measure of the third angle by using the Angle Sum Theorem.

$m\angle X + m\angle Y + m\angle Z = 180$ *Angle Sum Theorem*

$33 + m\angle Y + 47 = 180$ *$m\angle X = 33$, $m\angle Z = 47$*

$m\angle Y = 100$ *Subtraction Property (=)*

When finding y, use z instead of x since z is exact and x is approximate.

$\frac{\sin Z}{z} = \frac{\sin Y}{y}$ *Law of Sines*

$\frac{\sin 47°}{14} = \frac{\sin 100°}{y}$ *$m\angle Z = 47$, $m\angle Y = 100$, $z = 14$*

$y \sin 47° = 14 \sin 100°$ *Cross multiply.*

$y = \frac{14 \sin 100°}{\sin 47°}$ *Division Property (=)*

$y \approx 18.9$ *Use a calculator.*

Therefore, $m\angle Y = 100$, $x \approx 10.4$, and $y \approx 18.9$, and the triangle is solved.

In the application at the beginning of the lesson, you know the measures of two angles and the side between the angles.

Example 2 **Refer to the application at the beginning of the lesson. What is the approximate distance across the gorge?**

Surveying

We do not have enough information to solve either right triangle for h using the trigonometric ratios. However, if we can find e using the Law of Sines, we can use sin 80° to find h.

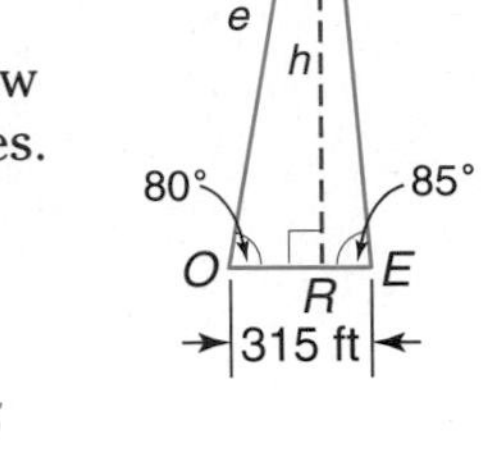

We know the measures of two angles and a side, so this is an example of Case 1. However, we do not know the measure of a side opposite one of the given angles. We must first find $m\angle OGE$.

$m\angle O + m\angle E + m\angle OGE = 180$ *Angle Sum Theorem*

$80 + 85 + m\angle OGE = 180$ *$m\angle O = 80$, $m\angle E = 85$*

$m\angle OGE = 15$ *Subtraction Property (=)*

Choose the proportion that involves e and three known measures.

$\frac{\sin G}{g} = \frac{\sin E}{e}$ *Law of Sines*

$\frac{\sin 15^\circ}{315} = \frac{\sin 85^\circ}{e}$ *$m\angle G = 15$, $m\angle E = 85$, $g = 815$*

$e \sin 15^\circ = 315 \sin 85^\circ$ *Cross multiply.*

$e = \frac{315 \sin 85^\circ}{\sin 15^\circ}$ *Division Property (=)*

$e \approx 1212.4$ *Use a calculator.*

To find h, use right $\triangle GRO$ and sin O.

$\sin O = \frac{h}{e}$ *$\sin = \frac{\text{opposite}}{\text{hypotenuse}}$*

$\sin 80^\circ \approx \frac{h}{1212.4}$ *$m\angle O = 80$, $e \approx 1212.4$*

$1212.4 \sin 80^\circ \approx h$ *Multiplication Property (=)*

$1194.0 \approx h$ *Use a calculator.*

The distance across the gorge is about 1194 feet.

CHECK FOR UNDERSTANDING

Communicating Mathematics

Study the lesson. Then complete the following.

1. **Compare** when you use the Law of Sines rather than the sine ratio as it is defined.
2. **Explain** in your own words the meaning of solving a triangle.
3. **Describe** two cases when the Law of Sines can be used to solve a triangle.

Guided Practice

Draw $\triangle REG$ and mark it with the given information. Write an equation that could be used to find each unknown value. Then find the value to the nearest tenth.

4. If $e = 8.5$, $m\angle G = 31$, and $m\angle E = 68$, find g.
5. If $e = 7$, $g = 11$, and $m\angle G = 37$, find $m\angle E$.

Solve each $\triangle BWY$ described below. Round measures to the nearest tenth.

6. $m\angle Y = 66$, $m\angle W = 59$, $b = 72$
7. $b = 24$, $y = 18$, $m\angle B = 102$
8. $m\angle B = 33$, $m\angle Y = 58$, $w = 22$
9. **Aviation** Two radar stations that are 20 miles apart located an unidentified plane that vanished from their screens at the same time. The first station indicated that the position of the plane made an angle of 43° with the line between the stations. The second station indicated that it made an angle of 48° with the same line.
   a. Draw and label a diagram of this situation.
   b. How far is each station from the point where they lost contact with the plane?

EXERCISES

Practice

Draw $\triangle EPR$ and mark it with the given information. Write an equation that could be used to find each unknown value. Then find the value to the nearest tenth.

10. If $m\angle P = 45$, $m\angle E = 63$, and $p = 22$, find e.
11. If $r = 9$, $e = 13$, and $m\angle E = 47$, find $m\angle R$.
12. If $e = 3.2$, $m\angle P = 52$, and $m\ \angle E = 70$, find p.
13. If $e = 48$, $r = 10$, and $m\angle E = 96$, find $m\angle R$.
14. If $m\angle P = 62$, $m\angle E = 26$, and $r = 2.6$, find p.
15. If $m\angle R = 59$, $p = 8.3$, and $r = 14.8$, find $m\angle P$.

Solve each $\triangle DFR$ described below. Round measures to the nearest tenth.

16. $m\angle R = 71$, $r = 7.4$, $m\angle F = 41$
17. $f = 9.1$, $r = 20.1$, $m\angle R = 107$
18. $m\angle F = 25$, $m\angle D = 52$, $r = 15.6$
19. $m\angle R = 34$, $f = 9.1$, $r = 27$
20. $m\angle D = 38$, $m\angle R = 115$, $d = 8.5$
21. $m\angle D = 43$, $m\angle R = 77$, $d = 0.8$
22. $d = 30$, $r = 9.5$, $m\angle D = 107$
23. $f = 16$, $d = 21$, $m\angle D = 88$
24. $f = 23$, $m\angle F = 45$, $m\angle D = 51$
25. An isosceles triangle has a base of 22 centimeters and a vertex angle of 36°. Find the perimeter.
26. The longest side of a triangle is 34 feet. The measures of two angles of the triangle are 40 and 65. Find the lengths of the other two sides.
27. Given parallelogram $PERA$, find the length of $\overline{PE}$ and $\overline{PR}$.

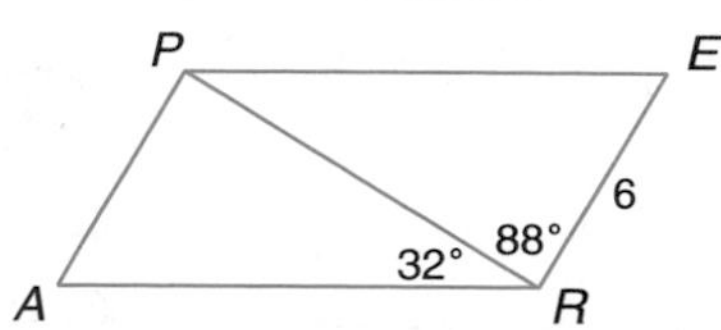

Critical Thinking

28. Refer to the figure at the right. Which segment looks longer, $\overline{AB}$ or $\overline{BC}$? Check your answer by using the Law of Sines to find AB and BC.

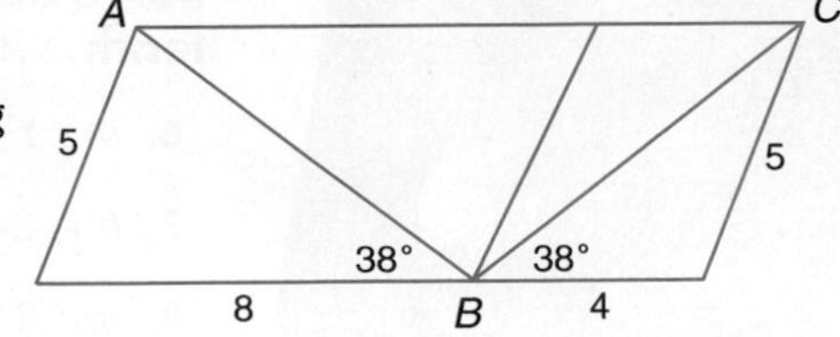

29. Does the Law of Sines hold true for the acute angles of right triangles? Justify your answer.

Applications and Problem Solving

30. **Gardening** Yana is planning a triangular garden. He wants to put a fence around it. The length of one side of the garden is 30 feet. If the angles at each end of this side are 44° and 58°, find the length of the fence needed to enclose the garden.

31. **Fire Fighting** Two ranger stations 5 miles apart spot a forest fire. The Kips station determines that the angle formed by the line of sight to the fire and the line connecting the two stations is 37°. The Rips station determines that the angle formed by the line of sight to the fire and the line connecting the two stations is 52°. How far are the two stations from the fire?

32. **Aviation** Jalisa Thompson is flying an airplane due east. To avoid a severe thunderstorm, she finds it necessary to change her course. She turns the plane 23° toward the north and flies 55 miles. Then she makes another turn of 120° and heads back to her original course.

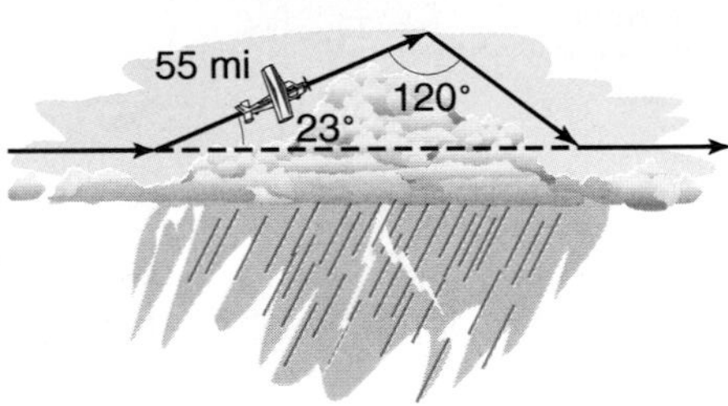

   a. How far must Ms. Thompson fly after her second turn to return to her original course?
   b. What angle must she turn to resume her course?
   c. How many miles did she add to her flight by taking the detour?

Mixed Review

33. **Surveying** A surveyor is 100 meters from a building and finds that the angle of elevation to the top of the building is 23°. If the surveyor's eye level is 1.55 meters above the ground, find the height of the building. (Lesson 8–4)

34. The length of a diagonal of a square is $7\sqrt{2}$. Find the perimeter of the square. (Lesson 8–2)

35. The measures of the sides of a triangle are 5, 9, and 12. If the shortest side of a similar triangle measures 15, what are the measures of its other two sides? (Lesson 7–3)

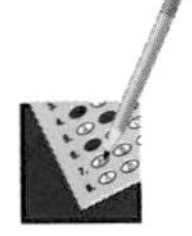

36. **ACT Practice** What is $\cos \theta$ if $\tan \theta = \frac{4}{3}$?

   A $\frac{3}{5}$ **B** $\frac{3}{4}$ **C** $\frac{4}{5}$ **D** $\frac{5}{4}$ **E** $\frac{5}{3}$

37. Use a compass and straightedge to construct a square with sides 6 centimeters long. (Lesson 6–4)

38. The opposite angles of a parallelogram have measures of $9x + 12$ and $15x$. Find the measures of the angles of the parallelogram. (Lesson 6–1)

Algebra

39. Find $(5x^2 + 3xy - 2y) + (4x^2 - xy + 2y)$.

40. Solve $y(y - 12) = 0$.

For **Extra Practice**, see page 779.

8-6 Using the Law of Cosines

What YOU'LL LEARN

- To use the Law of Cosines to solve triangles, and
- to choose the appropriate strategy for solving a problem.

Why IT'S IMPORTANT

You can use the Law of Cosines to find missing measures of triangles involved in technology, astronomy, and sports.

Technology

In 1996, Allen Johnson ran the 110-meter hurdles in 12.92 seconds making him the fourth athlete to run this event in less than thirteen seconds. To help him improve his performance, Johnson used a computer model.

1 Takeoff	2 Hurdle	3 Landing
The distance from his takeoff point to the top of the hurdle was 12.1 feet.	The distance from his takeoff point to his landing point was 19.7 feet.	The distance from the top of the hurdle to his landing point was 8.8 feet.

This problem will be solved in Example 2.

What was the measure of the angle formed by his takeoff point to the top of the hurdle and the top of the hurdle to his landing point?

The **Law of Cosines** allows us to solve a triangle when the Law of Sines cannot be used.

Law of Cosines

Let $\triangle ABC$ be any triangle with a, b, and c representing the measures of sides opposite angles with measures A, B, and C, respectively. Then, the following equations hold true.

$$a^2 = b^2 + c^2 - 2bc \cos A \qquad b^2 = a^2 + c^2 - 2ac \cos B$$
$$c^2 = a^2 + b^2 - 2ab \cos C$$

The Law of Cosines can be used to solve a triangle in the following cases.

Case 1. You know the measures of two sides and the included angle.

Case 2. You know the measures of three sides.

Example 1 **For $\triangle ABC$, find a if $m\angle A = 52$, $c = 7$, and $b = 4$.**

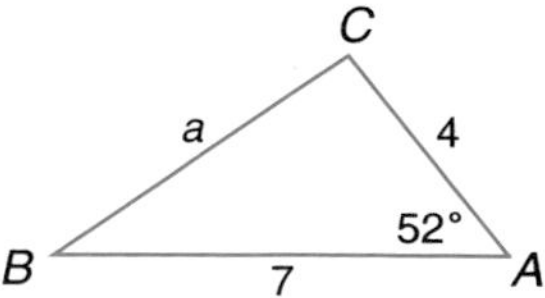

Since the measures of two sides and the included angle are known, this is an example of Case 1 for the Law of Cosines.

$a^2 = b^2 + c^2 - 2bc \cos A$ *Law of Cosines*

$a^2 = 4^2 + 7^2 - 2(4)(7)(\cos 52°)$ *$c = 7$, $b = 4$, $m\angle A = 52$*

$a = \sqrt{4^2 + 7^2 - 2(7)(4)(\cos 52°)}$ *Take the square root of each side.*

$a \approx 5.5$ *Use a calculator.*

In the application at the beginning of the lesson, you know the measures of the three sides and need to find the measure of an angle of the triangle. This is an example of Case 2 of the Law of Cosines.

Example 2

Technology

Refer to the application at the beginning of the lesson. Find the measure of the angle formed by his takeoff point to the top of the hurdle and the top of the hurdle to his landing point.

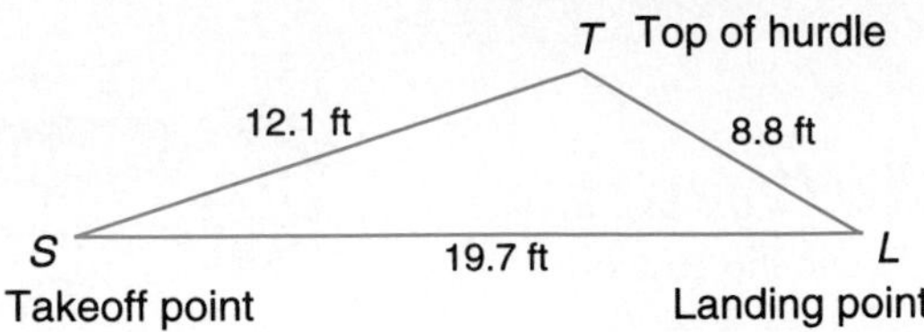

$t^2 = s^2 + \ell^2 - 2s\ell \cos T$	*Law of Cosines*
$19.7^2 = 8.8^2 + 12.1^2 - 2(8.8)(12.1)(\cos T)$	*$t = 19.7$, $s = 8.8$, $\ell = 12.1$*
$19.7^2 - 8.8^2 - 12.1^2 = -2(8.8)(12.1)(\cos T)$	*Subtraction Property*
$\frac{19.7^2 - 8.8^2 - 12.1^2}{-2(8.8)(12.1)} = \cos T$	*Division Property*
$-0.7712 \approx \cos T$	*Use a calculator.*
$140.5 \approx T$	

The measure of the angle is about 140.5.

Most problems can be solved in more than one way. Choosing the most efficient way to solve a problem is sometimes not obvious. **Decision making** is an important problem-solving strategy.

When solving right triangles, you can use the sine, cosine, and/or tangent ratios. When solving other triangles, you can use the Law of Sines and/or the Law of Cosines. You must decide how to solve each problem.

Example 3

Decision Making

Solve $\triangle UVW$ if $m\angle U = 41$, $v = 13$, and $w = 12$.

Since we do not know if this triangle is a right triangle, we must use the Law of Sines or the Law of Cosines. The Law of Cosines must be used to solve a triangle when the measures of two sides and the included angle are given (Case 1).

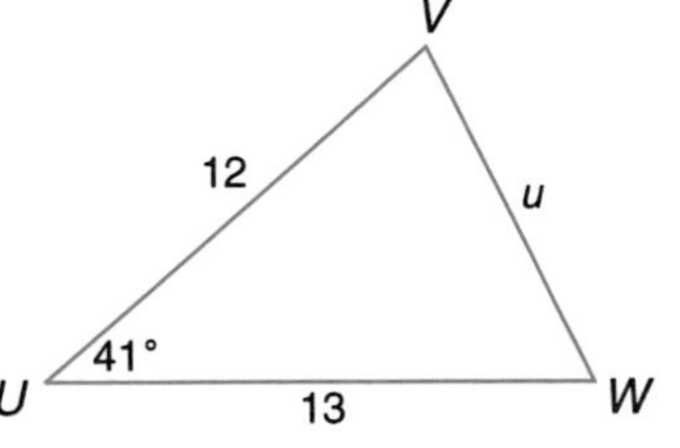

$u^2 = v^2 + w^2 - 2vw \cos U$	*Law of Cosines*
$u^2 = 13^2 + 12^2 - 2(13)(12)(\cos 41°)$	*$v = 13$, $w = 12$, $m\angle U = 41$*
$u = \sqrt{13^2 + 12^2 - 2(13)(12)(\cos 41°)}$	*Take the square root of each side.*
$u \approx 8.8$	*Use a calculator.*

Next, we can find $m\angle V$ or $m\angle W$. Suppose we decide to find $m\angle V$. We can use either the Law of Sines (Case 2) or the Law of Cosines (Case 2) to find this value. In this case, we will use the Law of Sines.

$$\frac{\sin V}{v} = \frac{\sin U}{u}$$ *Law of Sines*

$$\frac{\sin V}{13} \approx \frac{\sin 41°}{8.8}$$ *$v = 13$, $u = 8.8$, $m\angle U = 41$*

$$8.8 \sin V \approx 13 \sin 41°$$ *Cross multiply.*

$$\sin V \approx \frac{13 \sin 41°}{8.8}$$ *Division Property (=)*

$$\sin V \approx 0.9692$$ *Use a calculator.*

$$V \approx 75.7$$ *Use a calculator.*

The last value that we must find is $m\angle W$. We can use the Law of Sines, the Law of Cosines, or the Angle Sum Theorem. We will use the Angle Sum Theorem.

$$m\angle U + m\angle V + m\angle W = 180$$ *Angle Sum Theorem*

$$41 + 75.7 + m\angle W \approx 180$$ *$m\angle U = 41$, $m\angle V \approx 75.5$*

$$m\angle W \approx 63.3$$ *Subtraction Property (=)*

Therefore, $u \approx 8.8$, $m\angle V \approx 75.7$, and $m\angle W \approx 63.3$, and the triangle is solved.

CHECK FOR UNDERSTANDING

Communicating Mathematics

Study the lesson. Then complete the following.

1. **Describe** two cases when the Law of Cosines can be used to solve a triangle.
2. **Compare** the Law of Sines and the Law of Cosines.
3. **Study** Example 3.
   a. Why must you find u first?
   b. Would you prefer to use the Law of Sines or the Law of Cosines to find $m\angle V$? Explain.
   c. Would you prefer to use the Law of Sines, the Law of Cosines, or the Angle Sum Theorem to find $m\angle W$? Explain.

4. a. Draw a right triangle and label its sides and angles. Write an equation for the sine, cosine, and tangent ratios for each acute angle.
   b. Draw an acute triangle and label its sides and angles. State the Law of Sines for the triangle. Write a summary of the cases that use the Law of Sines to solve the triangle.
   c. Draw another acute triangle and label the sides and angles differently than the previous triangle. State the Law of Cosines for the triangle. Write a summary of the cases that use the Law of Cosines to solve the triangle.

Guided Practice

5. Refer to $\triangle DEF$.
   a. Determine whether the Law of Sines or the Law of Cosines should be used first to solve the triangle.
   b. Solve the triangle. Round the measures to the nearest tenth.

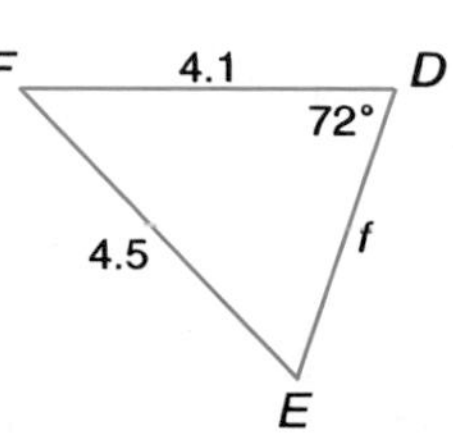

6. In $\triangle WGN$, $w = 19$, $n = 25$, and $m\angle G = 58$.
   a. Sketch $\triangle WGN$.
   b. Determine whether the Law of Sines or the Law of Cosines should be used first to solve the triangle. Explain your choice.
   c. Solve the triangle. Round measures to the nearest tenth.

Solve each $\triangle ABC$ described below. Round measures to the nearest tenth.

7. $m\angle A = 40$, $m\angle B = 59$, $c = 14$
8. $a = 5$, $b = 10$, $c = 13$
9. $a = 51$, $b = 61$, $m\angle B = 19$
10. $a = 20$, $c = 24$, $m\angle B = 47$
11. The measures of the sides of a triangle are 6.8, 8.4, and 4.9. Find the measure of the smallest angle to the nearest tenth.
12. **Golf** In golf, a *slice* is a shot that curves to the right of its intended path (for a right-handed player) and a *hook* curves to the left. Yolanda's shot from the third tee is a 180-yard slice 25° from the path straight to the cup. If the tee is 240 yards from the cup, how far does Yolanda's ball lie from the cup?

EXERCISES

Practice

Determine whether the Law of Sines or the Law of Cosines should be used first to solve each triangle. Then solve each triangle. Round measures to the nearest tenth.

13.
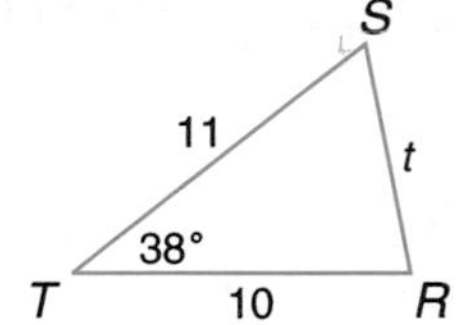

14.
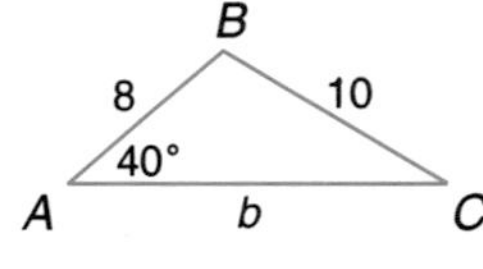

15.
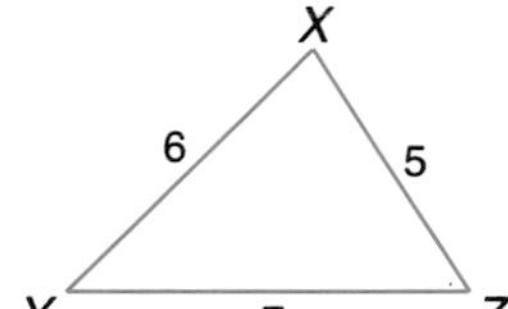

Sketch each $\triangle RGD$. Determine whether the Law of Sines or the Law of Cosines should be used first to solve each triangle. Then solve each triangle. Round measures to the nearest tenth.

16. $m\angle R = 42$, $m\angle D = 77$, $d = 6$
17. $r = 9.1$, $g = 8.3$, $m\angle D = 32$
18. $r = 13$, $g = 16$, $d = 22$
19. $m\angle R = 53$, $m\angle D = 28$, $d = 14.9$

Solve each $\triangle HJK$ described below. Round measures to the nearest tenth.

20. $j = 44$, $h = 54$, $m\angle H = 23$
21. $j = 33$, $h = 56$, $k = 65$
22. $j = 19$, $k = 28$, $m\angle H = 49$
23. $m\angle J = 46$, $m\angle H = 55$, $k = 16$
24. $j = 364$, $h = 669$, $k = 436$
25. $m\angle J = 55$, $h = 6.3$, $k = 6.7$
26. $m\angle J = 27$, $h = 5$, $k = 10$
27. $k = 25$, $j = 27$, $h = 22$
28. $h = 14$, $k = 21$, $m\angle J = 60$
29. $h = 14$, $j = 15$, $k = 16$
30. $m\angle H = 51$, $h = 40$, $k = 35$
31. $h = 5$, $j = 6$, $k = 7$

32. In parallelogram $ABCD$, $AB = 5$, $AD = 6$, and the measure of diagonal $\overline{BD}$ is 7. Find the measure of each angle of the parallelogram.

33. In parallelogram $EFGH$, $EH = 8$, $EF = 11$, and $m\angle E = 110$. Find the measure of each diagonal of the parallelogram.

34. The sides of a triangle are 50 meters, 70 meters, and 85 meters long. Find the measure of the largest angle.

35. The lengths of the diagonals of a rhombus are 26 inches and 18 inches. Find the length of each side of the rhombus and the measure of each angle.

Critical Thinking

36. Explain why each step of the derivation of the Law of Cosines is valid.

a. $c^2 = (a - x)^2 + h^2$

b. $c^2 = a^2 - 2ax + x^2 + h^2$

c. $c^2 = a^2 - 2ax + b^2$

d. $c^2 = a^2 - 2a(b \cos C) + b^2$

e. $c^2 = a^2 + b^2 - 2ab \cos C$

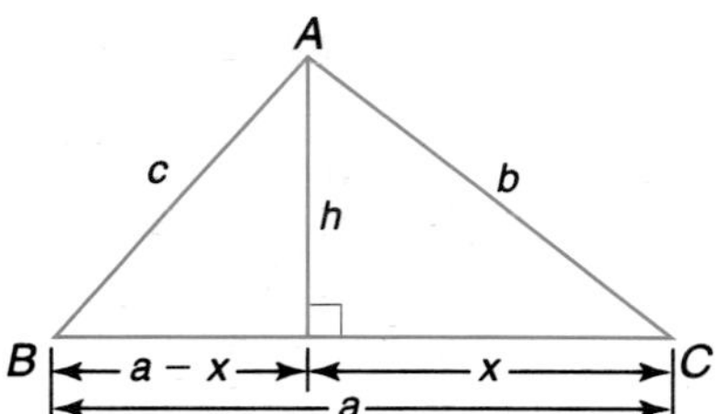

Applications and Problem Solving

37. **Astronomy** The angle formed by the line of sight from Earth to Sirius and the line of sight from Earth to Alpha Centauri is 44°. Find the distance between Sirius and Alpha Centauri.

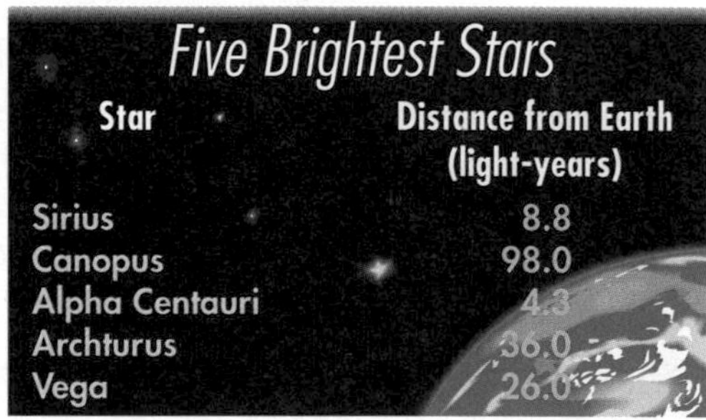

Five Brightest Stars

Star	Distance from Earth (light-years)
Sirius	8.8
Canopus	98.0
Alpha Centauri	4.3
Archturus	36.0
Vega	26.0

38. **Olympic Cycling** The front wheel of an Olympic-type bicycle shown at the left has 14-inch spokes. Each spoke forms a 30° angle with the spoke next to it. Find the distance between the points where the spokes attach to the wheel.

39. **Field Hockey** Juanita and Liz are playing field hockey. Juanita is standing 40 feet from one post of the goal and 45 feet from the other post. Liz is standing 30 feet from one post of the goal and 20 feet from the other post. If the goal is 12 feet wide, which player has a greater angle to make a shot on goal?

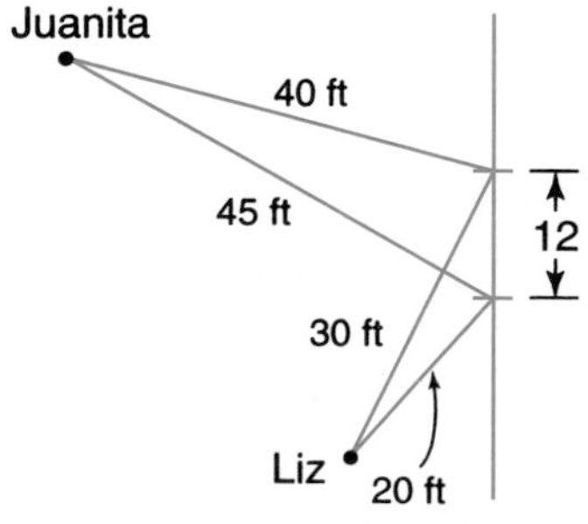

40. **Decision Making** What is the least number of coins you can have and be able to have exact change for any purchase that is less than a dollar?

Mixed Review

41. Solve $\triangle RPQ$ if $m\angle R = 50$, $m\angle P = 75$, and $r = 10$. (Lesson 8–5)

42. **Broadcasting** A guy wire is attached to a tower at a point 100 feet above the ground. If the tower is perpendicular to the ground and the wire makes an angle of 55° with the ground, find the length of the wire. (Lesson 8–3)

43. **SAT Practice** **Grid-in** If L is parallel to M in the figure, what is the value of y?

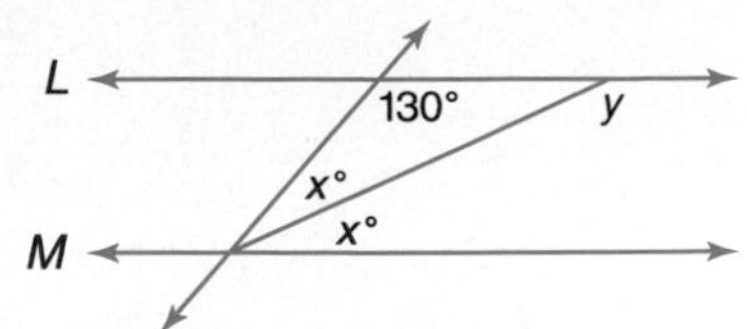

44. Explain the four methods used to determine if a quadrilateral is a parallelogram. (Lesson 6–2)

45. Find all possible ordered pairs for the fourth vertex of a parallelogram with vertices $G(-3, 1)$, $H(1, 5)$, and $I(0, -1)$. (Lesson 6–1)

46. **Quilting** Tim is making a quilt. The pieces of the quilt are made from equilateral triangles. What will he have to check to ensure all the pieces are the right shape and size? (Lesson 4–6)

47. Is the statement *A right triangle can be scalene* true or false? Explain. (Lesson 4–1)

48. Find the slope of the line that passes through points at (3, 0) and (8, −2). (Lesson 3–3)

For **Extra Practice**, see page 779.

Algebra

49. Find $(x + 6)(x + 3)$.

50. Find $\frac{z}{z+4} \div \frac{z+9}{z+4}$.

Refer to the Investigation on pages 392–393.

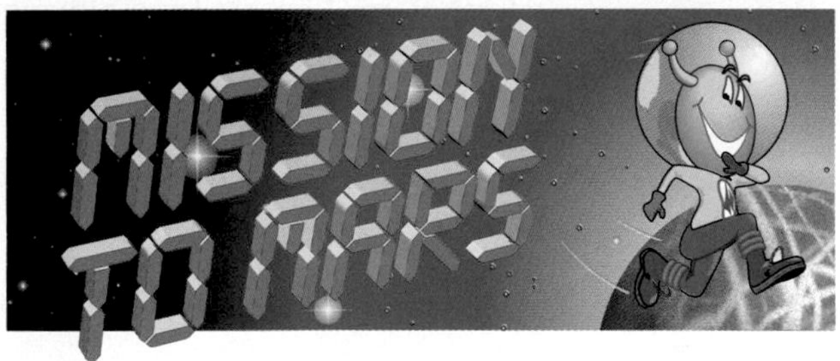

Some of the scientists who will be following you to Mars have expressed concern over the difference in the Martian year. They have asked you to develop a calendar for them to use on Mars.

1. Research the development of the calendar used on Earth. How were the lengths of the months determined? How does the calendar account for the length of time needed for Earth to complete an orbit of the Sun? How were the months named?
2. Refer to the information that you have gathered, along with the information on page 393 of the Investigation. Determine a reasonable number of Martian months and the length of each month.
3. Will you need to have a leap-year system in the Martian calendar? If so, determine how often.
4. Give the months of the Martian calendar names different from those in the Earth calendar. Note your rationale for choosing each name.
5. Complete your calendar and write a clear description of it for the scientists. Include a method for converting the Martian date to the Earth date.

Add the results of your work to your Investigation Folder.

CHAPTER 8 HIGHLIGHTS

Chapter Review For additional lesson-by-lesson review, visit: **www.geometry.glencoe.com**

VOCABULARY

After completing this chapter, you should be able to define each term, property, or phrase and give an example or two of each.

Geometry
Converse of the Pythagorean Theorem (p. 400)
Pythagorean Theorem (p. 399)
Pythagorean triple (p. 400)

Problem Solving
decision making (p. 432)

Trigonometry
angle of depression (p. 420)
angle of elevation (p. 420)
cosine (p. 412)
Law of Cosines (p. 431)
Law of Sines (p. 426)
sine (p. 412)
solving the triangle (p. 427)
tangent (p. 412)
trigonometric ratio (p. 412)
trigonometry (p. 412)

Algebra
extremes (p.397)
geometric mean (p. 397)
means (p.397)

UNDERSTANDING AND USING THE VOCABULARY

Choose the term from the list above that best completes each statement or phrase.

1. In a right triangle, the ratio of the measure of the opposite leg divided by the measure of the hypotenuse is called the __?__ of an angle.
2. In the proportion $\frac{3}{4} = \frac{9}{12}$, 3 and 12 are called the __?__ .
3. The __?__ states that $c^2 = a^2 + b^2 - 2ab \cos C$.
4. The angle formed by a horizontal line and a line of sight down to an object is called the __?__ .
5. The branch of mathematics that involves triangle measurement is called __?__ .
6. In a right triangle, the ratio of the measure of the opposite leg divided by the measure of the adjacent leg is called the __?__ of an angle.
7. A group of three whole numbers that satisfies the equation $a^2 + b^2 = c^2$ is called a(n) __?__ .
8. Six is the __?__ between 4 and 9.
9. The proportion $\frac{\sin A}{a} = \frac{\sin B}{b} = \frac{\sin C}{c}$ is the __?__ .
10. For a right triangle, the __?__ states that the sum of the squares of the measures of the legs equals the square of the measure of the hypotenuse.
11. The sine, cosine, and tangent ratios are three examples of __?__ .
12. In a right triangle, the ratio of the measure of the adjacent leg divided by the measure of the hypotenuse is called the __?__ of an angle.
13. In the proportion $\frac{a}{b} = \frac{c}{d}$, b and c are called the __?__ .
14. The angle formed by a horizontal line and a line of sight up to an object is called the __?__ .

CHAPTER 8 STUDY GUIDE AND ASSESSMENT

SKILLS AND CONCEPTS

OBJECTIVES AND EXAMPLES

Upon completing this chapter, you should be able to:

- find the geometric mean between two numbers (Lesson 8–1)

Find the geometric mean between 16 and 77.

Let x represent the geometric mean.

$\frac{16}{x} = \frac{x}{77}$ *Definition of geometric mean*

$x^2 = 1232$ *Cross multiply.*

$x = \sqrt{1232}$

$x \approx 35.1$

The geometric mean is about 35.1.

REVIEW EXERCISES

Use these exercises to review and prepare for the chapter test.

Find the geometric mean between each pair of numbers.

15. 9 and 27
16. 13 and 39
17. 6 and $\frac{2}{3}$
18. $\frac{3}{4}$ and $\frac{8}{3}$

- solve problems involving relationships between parts of a right triangle and the altitude to its hypotenuse (Lesson 8–1)

If $\triangle ABC$ is a right triangle with altitude $\overline{BD}$, then the following relationships hold true.

$\frac{AD}{BD} = \frac{BD}{DC}$

$\frac{AC}{BC} = \frac{BC}{DC}$

$\frac{AC}{AB} = \frac{AB}{AD}$

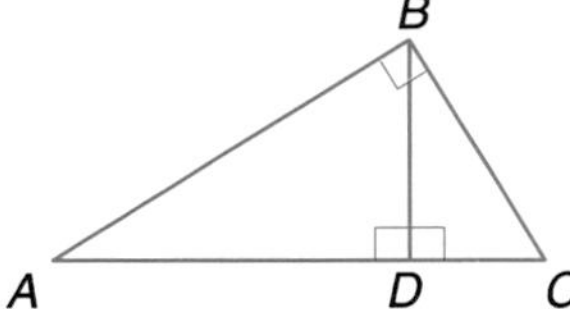

Use right triangle *KLM* and the given information to solve each problem.

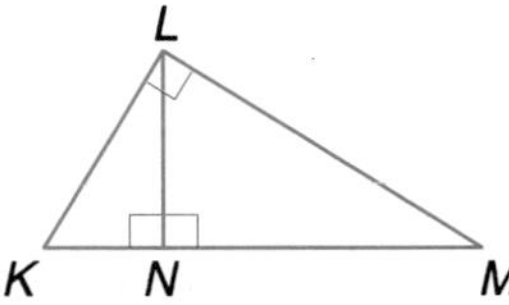

19. Find KL if $KM = 18$ and $KN = 4$.
20. Find LN if $KN = 4$ and $NM = 6$.
21. Find KM if $LM = 19$ and $NM = 14$.
22. Find LK if $KN = \frac{1}{3}$ and $NM = \frac{1}{4}$.
23. Find LM if $KN = 7$ and $NM = 3$.
24. Find KN if $LK = 0.6$ and $KM = 1.5$.

- use the Pythagorean Theorem and its converse (Lesson 8–1)

A right triangle has a hypotenuse 61 inches long and a leg 11 inches long. Find the length of the other leg.

$a^2 + b^2 = c^2$ *Pythagorean Theorem*

$a^2 + 11^2 = 61^2$ *$b = 11$, $c = 61$*

$a^2 + 121 = 3721$

$a^2 = 3600$ *Subtraction Property (=)*

$a = 60$ *Take the square root of each side.*

The other leg of the triangle is 60 inches long.

Find the value of *x*.

25.

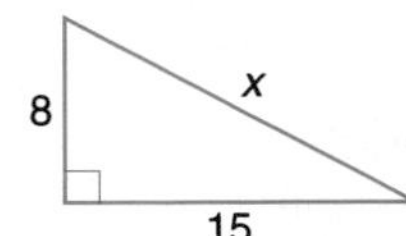

26.

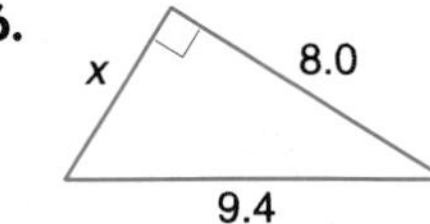

Determine if the given measures are measures of the sides of a right triangle.

27. 19, 24, 30
28. 4, 7.5, 8.5

OBJECTIVES AND EXAMPLES

- use the properties of 45°-45°-90° and 30°-60°-90° triangles. (Lesson 8–2)

In a 45°-45°-90° triangle, the hypotenuse is $\sqrt{2}$ times as long as a leg.

In a 30°-60°-90° triangle, the hypotenuse is twice as long as the shorter leg, and the longer leg is $\sqrt{3}$ times as long as the shorter leg.

REVIEW EXERCISES

Find the value of *x*.

29.

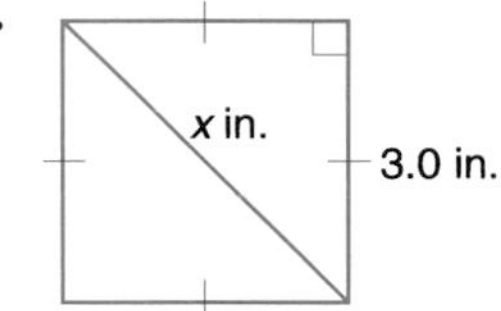

30.

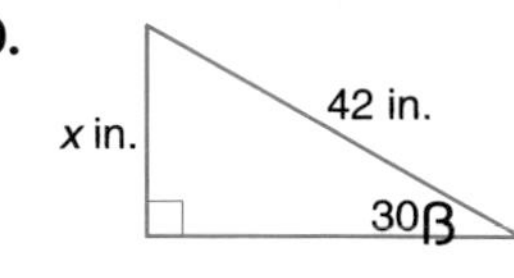

31.

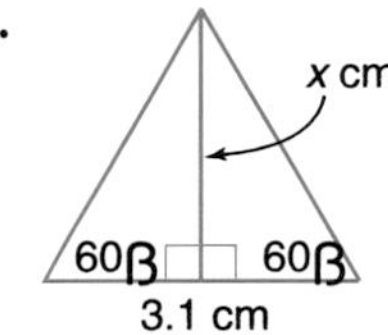

32.

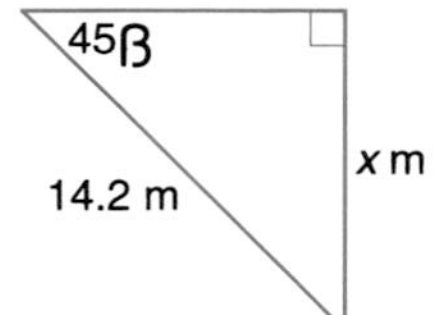

- find trigonometric ratios using right triangles (Lesson 8–3)

$\sin A = \frac{a}{c}$

$\cos A = \frac{b}{c}$

$\tan A = \frac{a}{b}$

A, B, C; sides a, b, c

Find the indicated trigonometric ratio as a fraction and as a decimal rounded to the nearest ten-thousandth.

33. $\sin M$

34. $\tan M$

35. $\cos K$

36. $\tan K$

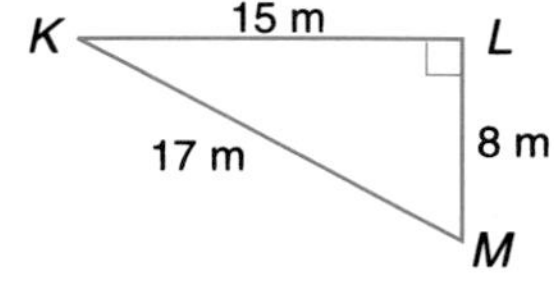

- use trigonometry to solve problems involving angles of elevation and depression (Lesson 8–4)

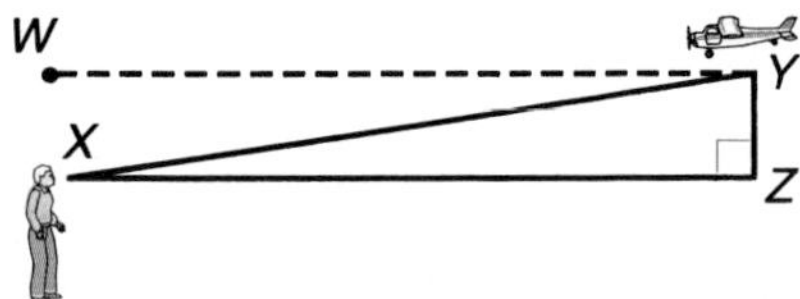

$\angle YXZ$ is the angle of elevation.

$\angle WYX$ is the angle of depression.

Solve each problem. Round answers to the nearest tenth.

37. The top of a lighthouse is 120 meters above sea level. The angle of depression from the top of the lighthouse to the ship is 23°. How far is the ship from the foot of the lighthouse?

38. At a certain time of day, the angle of elevation of the sun is 44°. Find the length of a shadow cast by a building 30 meters high.

39. A railroad track rises 30 feet for every 400 feet of track. What is the measure of the angle of elevation of the track?

OBJECTIVES AND EXAMPLES

- use the Law of Sines to solve triangles (Lesson 8–5)

According to the Law of Sines, $\frac{\sin A}{a} = \frac{\sin B}{b} = \frac{\sin C}{c}$.

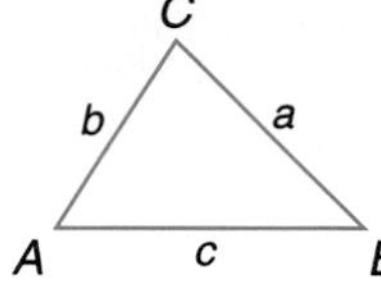

REVIEW EXERCISES

Solve each $\triangle ABC$ described below. Round measures to the nearest tenth.

40. $a = 4.2, b = 6.8, m\angle B = 22$

41. $m\angle B = 46, m\angle C = 83, b = 65$

42. $a = 80, b = 10, m\angle A = 65$

43. $m\angle C = 70, a = 7.5, m\angle A = 30$

- use the Law of Cosines to solve triangles (Lesson 8–6)

According to the Law of Cosines,

$a^2 = b^2 + c^2 - 2bc \cos A$,

$b^2 = a^2 + c^2 - 2ac \cos B$, and

$c^2 = a^2 + b^2 - 2ab \cos C$.

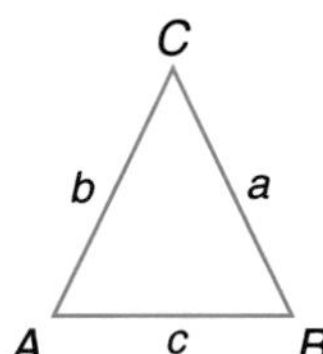

Solve each $\triangle ABC$ described below. Round measures to the nearest tenth.

44. $m\angle C = 55, a = 8, b = 12$

45. $a = 44, c = 32, m\angle B = 44$

46. $m\angle C = 78, a = 4.5, b = 4.9$

47. $a = 6, b = 9, c = 8$

APPLICATIONS AND PROBLEM SOLVING

48. **Recreation** Jeremy is flying a kite whose string makes a 70°-angle with the ground. The kite string is 65 meters long. How far is the kite above the ground? (Lesson 8–3)

49. **Aviation** A jet airplane begins a steady climb of 15° and flies for two ground miles. What was its change in altitude? (Lesson 8–3)

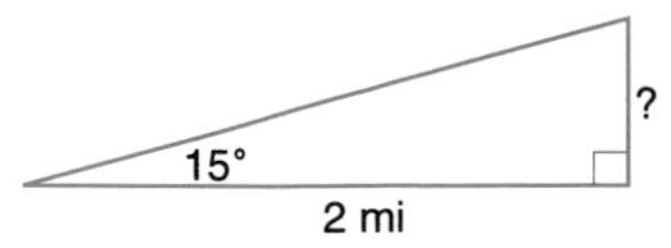

50. **Fire Fighting** Ranger Barojas sights a fire from a fire tower in Wayne National Forest. She finds that the angle of depression to the fire is 22°. If the tower is 75 meters tall, how far is the fire from the base of the tower? (Lesson 8–4)

51. **Architecture** Bianca Jones is an architect designing a new parking garage for the city. The floors of the garage are to be 10 feet apart. The exit ramps between each pair of floors are to be 75 feet long. What is the measure of the angle of elevation of each ramp? (Lesson 8–4)

52. **Aviation** Sonia Ortega flew her airplane 1000 kilometers north before turning 20° clockwise and flying another 700 kilometers. How far is Ms. Ortega from her starting point? (Lesson 8–6)

53. **Decision Making** Which problem-solving strategies might you use to find the remainder for $5^{100} \div 7$? Find the value of the remainder for $5^{100} \div 7$. (Lesson 8–6)

A practice test for Chapter 8 is available on page 800.

ALTERNATIVE ASSESSMENT

COOPERATIVE LEARNING PROJECT

Parallax In this chapter, you explored properties of right triangles and began the study of trigonometry. Your body is one of the best measuring tools available; you never leave home without it. In this project, you will learn to use your eyes and one finger to measure the size of an object if you know its distance, or the distance to an object if you know its size.

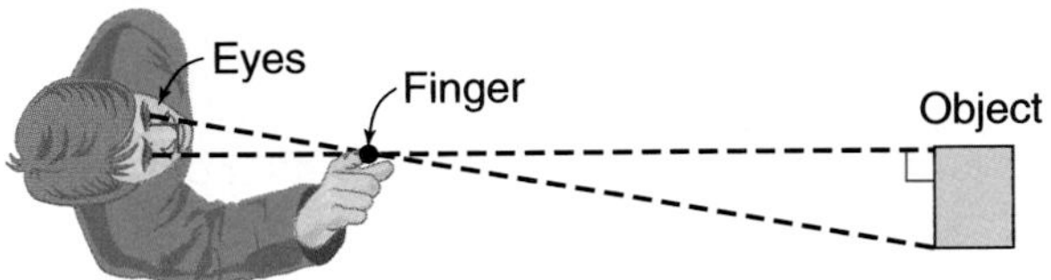

Follow these steps to conduct your measurement experiments.

- Place a chair or desk 10 feet away. As in the diagram above, place your index finger pointing up, about 6 inches in front of your right eye. Make sure your line of sight is perpendicular to the line connecting your eyes. Measure the distance between your eyes. Look with your right eye, then with your left eye.
- Does your finger move from one side of the chair to the other? Move your finger in or out so that through your right eye, your finger is at the left edge of the chair and through your left eye your finger is at the right edge of the chair.
- Use what you know about proportions of similar triangles or trigonometry to determine the size of the chair.
- Now measure the chair to confirm your indirect measurement.
- Now have another student select an object and measure its size. Have the student place the object at a distance from you without telling you the distance.
- Modify the above method to measure the distance to the object.
- As a group, complete the following lists. Make a list of common objects with known sizes in situations in which you could not tell the distance; for example, a car at an unknown distance. Make a list of situations where you would know the distance, but observe an object of unknown size; for example, an object across the street outside your bedroom window.

THINKING CRITICALLY

The angle of the sun's rays varies throughout the year. So, the shadow cast by a sundial at the same time of day will vary throughout the year. Can a sundial be useful all year? Explain.

PORTFOLIO

Trigonometry allows you to measure dimensions you cannot measure directly. Make a list, with photographs or drawings, of ten situations where trigonometry allows you to measure indirectly what cannot be measured directly. Keep this collection in your portfolio.

STANDARDIZED TEST PRACTICE

CHAPTERS 1–8

Section One: MULTIPLE CHOICE

There are nine multiple-choice questions in this section. After working each problem, write the letter of the correct answer on your paper.

1. $(8\sqrt{4})(2\sqrt{9})$

   A. 48

   B. 72

   C. 96

   D. 298

2. If the area of an isosceles right triangle is 36 square units, what is its perimeter?

   A. $(12 + 12\sqrt{2})$ units

   B. $(12 + 6\sqrt{2})$ units

   C. $(6 + 12\sqrt{2})$ units

   D. $(12\sqrt{2})$ units

3. If $af = 6$, $fg = 1$, $ag = 24$, and $a > 0$, find afg.

   A. 4

   B. 6

   C. 12

   D. 144

4. $\sqrt{25 \cdot 64}$

   A. 1600

   B. 400

   C. 80

   D. 40

5. If $x - \frac{2}{x-3} = \frac{x-1}{3-x}$, find the value of x.

   A. 3 or -1

   B. 3 or -3

   C. -1

   D. 3

6. In $\triangle ABC$ below, if $AB = BC$ and if $\angle ABC$ is a right angle, exactly how many 45° angles are formed by pairs of line segments?

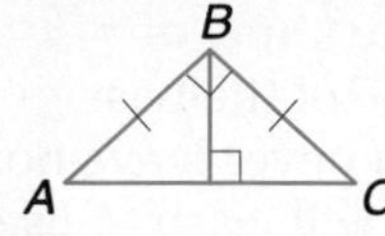

   A. none

   B. two

   C. three

   D. four

7. If a and b are negative numbers, which of the following must be negative?

   A. ab

   B. $(ab)^2$

   C. $a + b$

   D. $a - b$

8. Find the value of x.

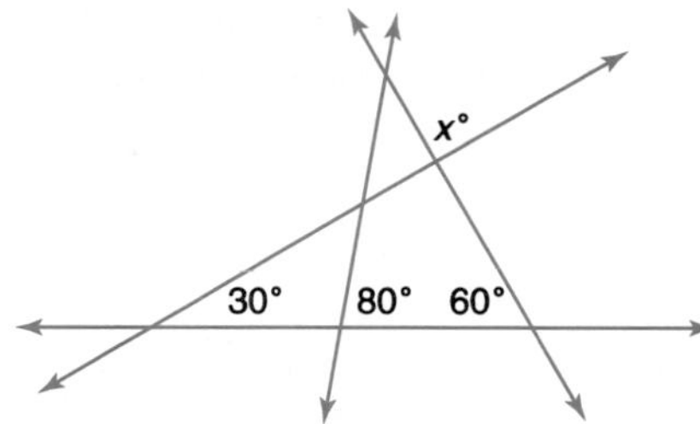

   A. 40

   B. 50

   C. 60

   D. 90

9. Find the value of x if $\overline{AE} \parallel \overline{BD}$, $AB = 10$, $BC = x$, $ED = x + 3$, and $DC = x + 6$.

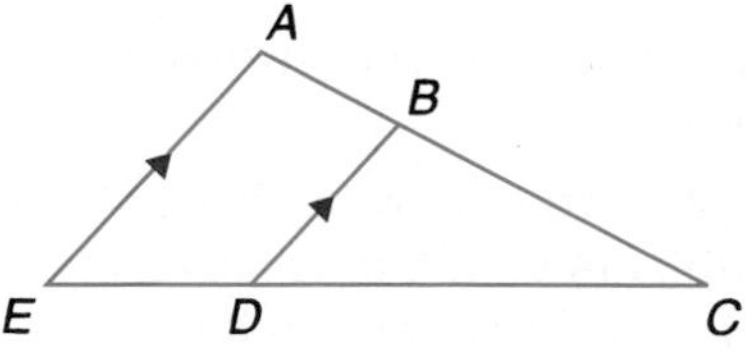

   A. 6

   B. 7

   C. 12

   D. 13

Section Two: SHORT ANSWER

This section contains seven questions for which you will provide short answers. Write your answer on your paper.

10. If the area of $\triangle ABC$ is 30 square units, what is the value of x?

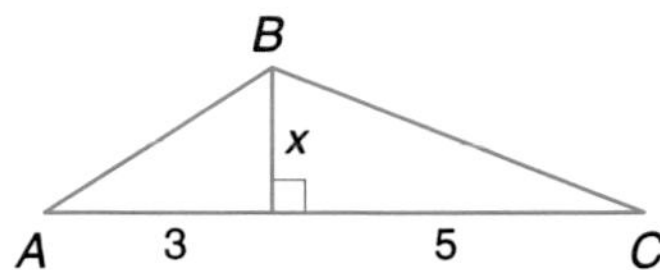

11. Find $\sin \theta$, $\cos \theta$, and $\tan \theta$.

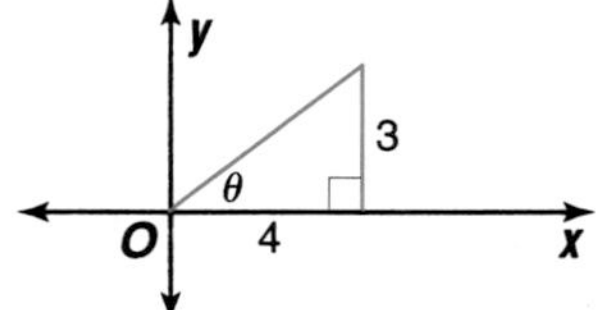

12. In the figure below, $\overline{AB}$ and $\overline{CD}$ are parallel line segments. Find the area of isosceles trapezoid $ABCD$.

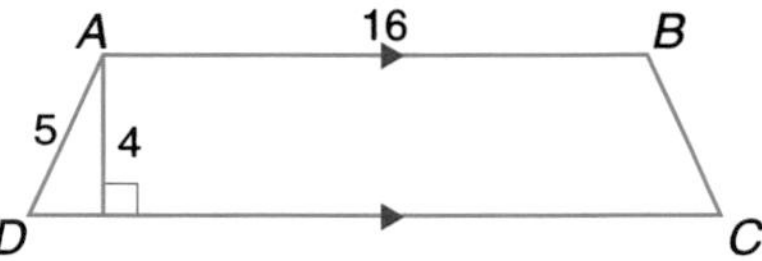

13. For all real numbers a and b, where $b \neq 0$, $a + b = \frac{a^2}{b^2}$, find $(3 + 4)(5 + 3)$.

14. Find the resulting expression when x is subtracted from y and this difference is divided by the sum of x and y.

15. If J, K, and L each represent a different digit in the multiplication problem below, what digit does J represent?

$$\begin{array}{r} 37 \\ \underline{\times 2J} \\ L4K \\ \underline{74} \\ 88K \end{array}$$

16. If the base and height of the triangle are decreased by $2x$ units, what is the area of the resulting triangle?

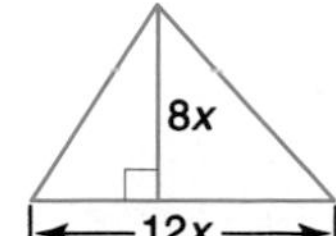

Test-Taking Tip The 3-4-5 right triangle and its multiples like 6-8-10 and 9-12-15 occur frequently on standardized tests. Other commonly used Pythagorean triples include 5-12-13 and 7-24-25.

Section Three: COMPARISON

This section contains five comparison problems that involve comparing two quantities, one in column A and one in column B. In certain questions, information related to one or both quantities is centered above them. All variables used represent real numbers.

Compare quantities A and B below.

- Write A if quantity A is greater.
- Write B if quantity B is greater.
- Write C if the two quantities are equal.
- Write D if there is not enough information to determine the relationship.

Column A	Column B
17. perimeter of $\triangle ABC$	perimeter of $\triangle DEF$

Three angles of a triangle are $x°$, $x°$, and $\frac{x°}{2}$.

Column A	Column B
18. x	60
19. $(3 + 5)^2$	$3^2 + 5^2$
20. the perimeter of a triangle	the perimeter of a rectangle

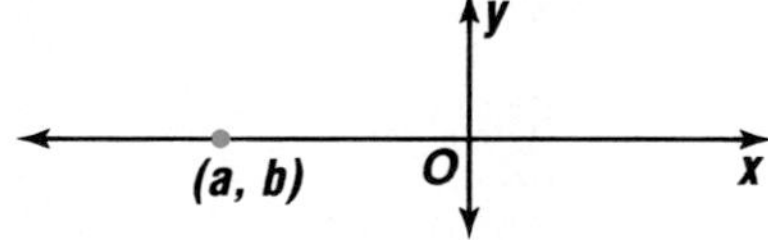

Column A	Column B
21. a	b

Test Practice For additional test practice questions, visit: **www.algebra1.glencoe.com**

CHAPTER

9 Analyzing Circles

What You'll Learn

In this chapter, you will:

- find the degree and linear measures of arcs,
- find the measures of angles in circles,
- solve problems by making circle graphs,
- use properties of chords, tangents, and secants to solve problems, and
- write equations of circles.

Why It's Important

Real World

Sports Did you know that in 1900 more than one million bicycles were being produced per year in the United States? But by about 1910, automobiles and motorcycles had surpassed the bicycle in popularity. All three of these vehicles depend upon the simple geometric figure known as the circle. In this chapter, you will learn about the relationships of special segments and angles relating to the circle.

PREREQUISITE SKILLS

To be successful in this chapter, you'll need to understand these concepts and be able to apply them. Refer to the example or to the lesson in parentheses if you need more review before beginning the chapter.

Use the Pythagorean Theorem. (Lesson 8–1)

Find the value of x. Round to the nearest tenth, if necessary.

1. 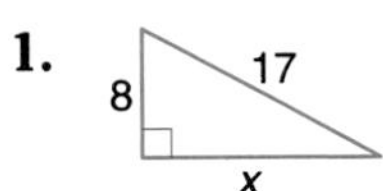

2. 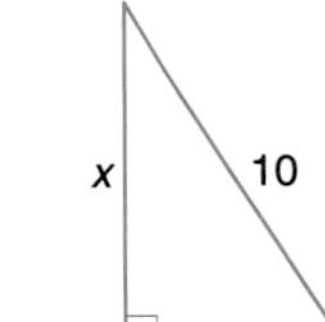

3. 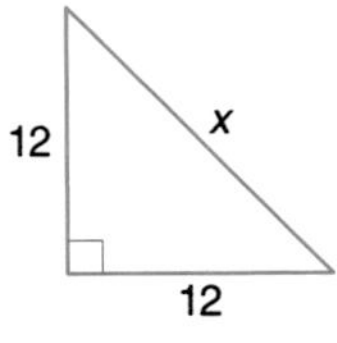

Use proportions to solve percent problems. (Lesson 7–1)

Use a proportion to solve each problem.

4. What percent of 16 is 4?

5. 15 is 12% of what number?

6. Find 32% of 360.

7. Find 55% of 360.

Solve quadratic equations by using the quadratic formula.

Example Solve $x^2 - 2x - 5 = 0$.

$x = \frac{-b \pm \sqrt{b^2 - 4ac}}{2a}$ *Quadratic formula.*

$x = \frac{2 \pm \sqrt{(-2)^2 - 4(1)(-5)}}{2(1)}$ *Substitute the values for a, b, and c into the formula. a = 1, b = −2, c = −5*

$x = \frac{2 \pm 2\sqrt{6}}{2}$ or $1 \pm \sqrt{6}$ *Simplify.*

$x = 1 + \sqrt{6}$ or $x = 1 - \sqrt{6}$

Solve each equation by using the quadratic formula. Approximate irrational roots to the nearest tenth.

8. $x^2 - 4x - 10 = 0$

9. $3x^2 - 2x - 4 = 0$

10. $x^2 = x + 15$

Focus On

READING SKILLS

In this chapter, you'll be learning about ***circles***, ***special segments*** of circles, and ***angles*** relating to circles. An important feature of a circle is its circumference. The word *circumference* begins with "circum" which means "around." Circumference is the perimeter of a circle. Special segments relating to circles can be drawn. Some of these are chords, tangents, and secants.

9–1 Exploring Circles

***What* YOU'LL LEARN**

- To identify and use parts of circles, and
- to solve problems involving the circumference of a circle.

***Why* IT'S IMPORTANT**

You can use properties of circles to solve problems involving surveying, sports, and space travel.

Real World APPLICATION

Surveying

When a house is sold, a land surveyor must construct and draw a *Plat of Survey* that verifies the house and land measurements of the property. In order to measure the dimensions of the property, surveyors use a *trundle wheel.*

Radii is the plural of radius.

A trundle wheel is a model of a **circle**. A circle is the set of all points in a plane that are a given distance from a given point in that plane. The given point is the **center** of the circle. Each spoke of the wheel represents the **radius** of the circle. A radius of a circle is a segment that has one endpoint at the center of the circle and the other endpoint on the circle. Notice that all the spokes of the wheel appear to be congruent. It follows from the definition of a circle that all radii are congruent.

A circle is usually named by its center. The circle at the right is called circle *P*. This is symbolized as $\odot P$.

$\overline{RF}$ and $\overline{KL}$ are **chords** of $\odot P$. A chord of a circle is a segment that has its endpoints on the circle.

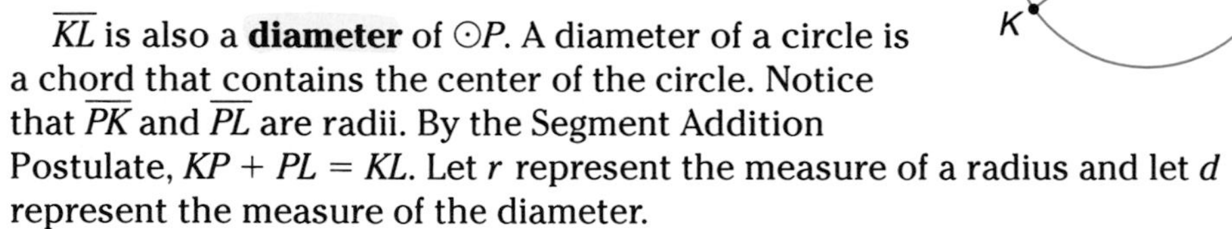

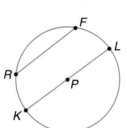

$\overline{KL}$ is also a **diameter** of $\odot P$. A diameter of a circle is a chord that contains the center of the circle. Notice that $\overline{PK}$ and $\overline{PL}$ are radii. By the Segment Addition Postulate, $KP + PL = KL$. Let r represent the measure of a radius and let d represent the measure of the diameter.

$$KP + PL = KL$$
$$r + r = d$$
$$2r = d$$
$$r = \frac{1}{2}d$$

The terms diameter and radius can refer to segments in a circle or the measures of those segments.

Thus, a diameter is twice as long as a radius, or a radius is half as long as a diameter.

Example 1

Real World APPLICATION

Surveying

Refer to the application at the beginning of the lesson. The surveyor's wheel has a diameter of 1.59 feet. What is the length of each spoke of the wheel?

Each spoke is a radius of the wheel.

$$d = 2r$$
$$1.59 = 2r$$
$$0.795 = r$$

The length of each spoke is 0.795 feet.

A trundle wheel is used to measure property lines by counting the number of revolutions. Each time the wheel makes one revolution, the distance it travels is the same as the **circumference** of the wheel. The circumference of a circle is the distance around the circle.

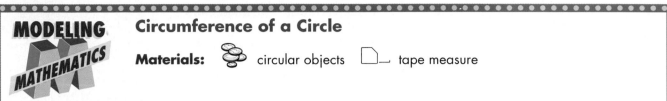

Circumference of a Circle

Materials:  circular objects tape measure

- Choose three or four objects that have a circular shape such as different-sized paper plates, cookies, coins, or old records.

- Use a tape measure to measure the diameter and circumference of each object to the nearest millimeter.

Your Turn

a. Record your results in a table like the one below. In the third column, find the ratio of the circumference C to the diameter d.

Object	d	C	$\frac{C}{d}$
1			
2			
3			

b. What seems to be true about the results for $\frac{C}{d}$ in the third column?

c. Use these results to make a conjecture about the relationship between the circumference and diameter of a circle.

The Modeling Mathematics activity provides insight into the relationship between the radius and the circumference of a circle. By definition, the ratio of the circumference of a circle to its diameter is the irrational number called **π (pi)**. If $\frac{C}{d} = \pi$, then $C = \pi d$. Since $d = 2r$, $C = 2\pi r$.

Circumference of a Circle	**If a circle has a circumference of C units and a radius of r units, then $C = 2\pi r$.**

In this book, we will use a calculator for evaluating expressions involving π. If no calculator is available, 3.14 is a good estimate for π.

Example 2 **Find the circumference of a circle with a radius of 6.8 centimeters.**

$C = 2\pi r$ *Formula for the circumference of a circle*

$C = 2\pi(6.8)$

$C = 13.6\pi$

The *exact* circumference is 13.6π centimeters. To *estimate* the circumference, use a calculator.

Enter: 13.6 [×] [2nd] [π] [=] 42.72566009

The circumference is about 42.7 centimeters.

Circumference can be used to measure distance.

Example 3

Marathons

In 1996, inspectors checked the accuracy of the Olympic marathon course by using the "calibrated bicycle method." The inspectors rode bicycles equipped with counters that tracked the turns of the wheels. If the radius of a bicycle wheel is 13 inches and the counter shows 20,300 revolutions at the end of a course, how long is the course?

The total distance in inches would equal the number of revolutions times the distance per revolution (circumference of the wheel).

$$\begin{aligned} \text{total distance} &= 20{,}300 \cdot (2\pi r) \\ &= 20{,}300 \cdot 2\pi \cdot 13 \\ &= 527{,}800\pi \text{ inches} \end{aligned}$$

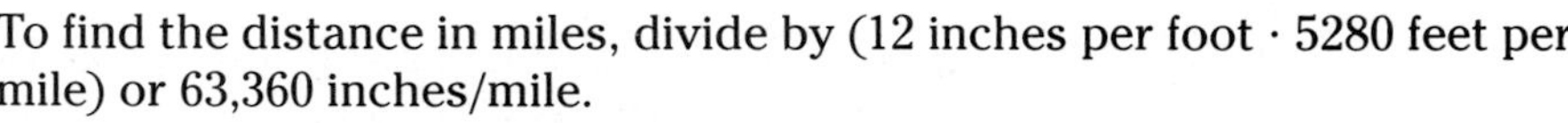

To find the distance in miles, divide by (12 inches per foot · 5280 feet per mile) or 63,360 inches/mile.

$$\begin{aligned} \text{total distance} &= \frac{527{,}800\pi \text{ inches}}{63{,}360 \text{ inches per mile}} \\ &\approx 26.17 \text{ miles} \quad \textit{Use a calculator.} \end{aligned}$$

The course is approximately 26 miles long, which is the length of a marathon.

In Lesson 9–3, you will learn about polygons whose vertices lie on a circle. You can use these kinds of polygons to help find the circumference of circles.

Example 

Find the exact circumference of $\odot P$ shown at the right.

To find the circumference of the circle, we need to know the diameter. Since the triangle is a right triangle, we can use the Pythagorean Theorem to find the measure of the diameter $\overline{JL}$.

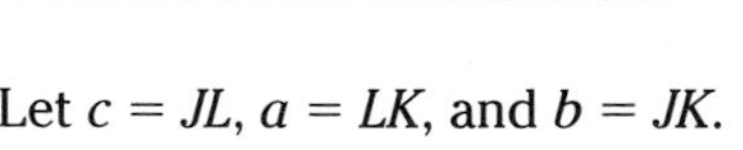

Let $c = JL$, $a = LK$, and $b = JK$.

$$\begin{aligned} a^2 + b^2 &= c^2 \quad \textit{Pythagorean Theorem} \\ 16^2 + 30^2 &= c^2 \quad a = 16,\ b = 30 \\ 1156 &= c^2 \\ 34 &= c \quad \textit{Take the square root of each side.} \end{aligned}$$

The diameter is 34 units. Now use that value to find the circumference.

$$\begin{aligned} C &= \pi d \\ &= \pi \cdot 34 \text{ or } 34\pi \end{aligned}$$

The exact circumference is 34π units.

CHECK FOR UNDERSTANDING

Communicating Mathematics

Study the lesson. Then complete the following.

1. **Explain** why a compass can be used to draw a circle.
2. **Explain** why a diameter is the longest chord of a circle.
3. **You Decide** Fernando claims that "every diameter is a chord of a circle" and Koleka says, "every chord is a diameter of a circle." Who do you think is correct? Explain.
4. **Explain** why $C = \pi d$ and $C = 2\pi r$ are equivalent formulas.

5. Use a dictionary to determine why the word *circle* is used to describe a set of points equidistant from a given point.

Guided Practice

Refer to ⊙*K* for Exercises 6–8.

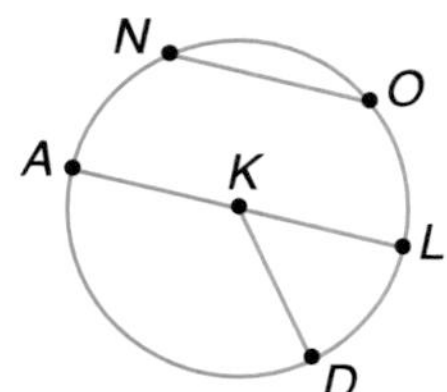

6. Name two chords of ⊙*K*.
7. If $AL = 9.4$, find *KD*.
8. Is $NO < AL$? Explain.

In Exercises 9 and 10, the radius, diameter, or circumference of a circle is given. Find the other measures to the nearest tenth.

9. $r = 7$, $d =$ _?_, $C =$ _?_
10. $C = 76.4$, $d =$ _?_, $r =$ _?_
11. Find the exact circumference of ⊙*T*.

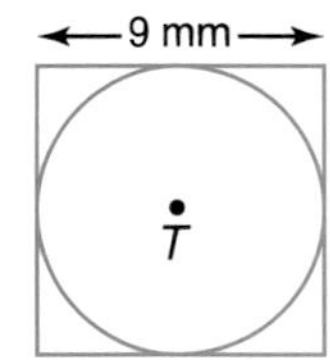

12. Circle *P* has a radius of 6 units, and ⊙*T* has a radius of 4 units.
    a. If $QR = 1$, find *RT*.
    b. If $QR = 1.2$, find *AQ*.

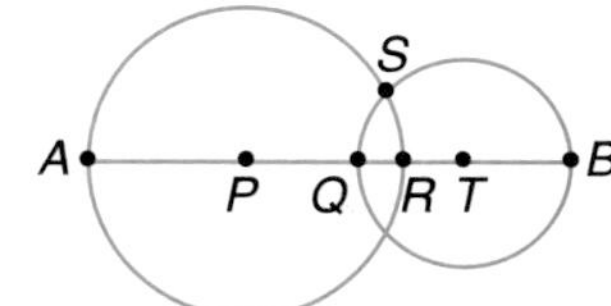

13. **Fencing** The Grabowskis wish to put a fence around their circular pool. The pool has a circumference of 60 feet. If they want the fence to be 6 feet from the pool all the way around, how many feet of fencing will be needed? (*Hint:* Draw a diagram.)

EXERCISES

Practice

Refer to ⊙*M* for Exercises 14–22.

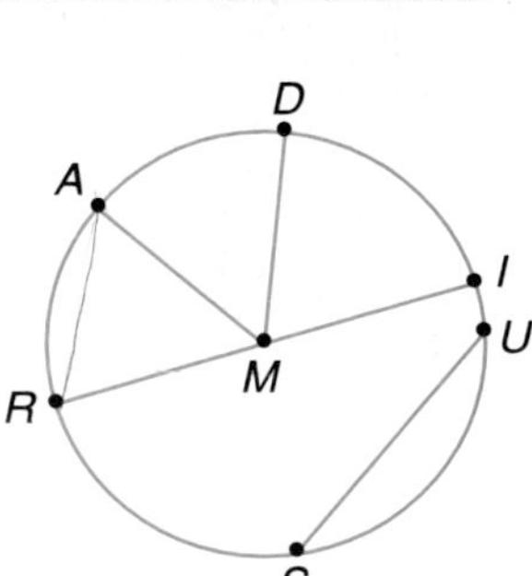

14. Name the center of ⊙*M*.
15. Name a chord that is also a diameter.
16. If $MD = 5$, find *RI*.
17. Is $\overline{MI}$ a chord of ⊙*M*? Explain.
18. Is $\overline{MA} \cong \overline{MI}$? Explain.
19. Name four radii of ⊙*M*.
20. Is $RI > SU$? Explain.
21. If $RI = 11.8$, find *MA*.
22. Draw $\overline{AR}$. What type of triangle is $\triangle MAR$? Explain.

In Exercises 23–28, the radius, diameter, or circumference of a circle is given. Find the other measures to the nearest tenth.

23. $r = 5, d = \underline{\ ?\ }, C = \underline{\ ?\ }$

24. $d = 26.8, r = \underline{\ ?\ }, C = \underline{\ ?\ }$

25. $C = 136.9, d = \underline{\ ?\ }, r = \underline{\ ?\ }$

26. $r = \frac{x}{6}, d = \underline{\ ?\ }, C = \underline{\ ?\ }$

27. $d = 2x, r = \underline{\ ?\ }, C = \underline{\ ?\ }$

28. $C = 2368, d = \underline{\ ?\ }, r = \underline{\ ?\ }$

Find the exact circumference of each circle.

29.

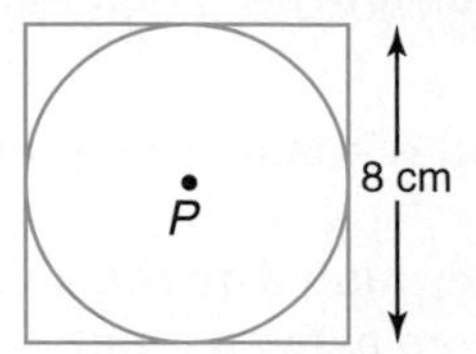

30.

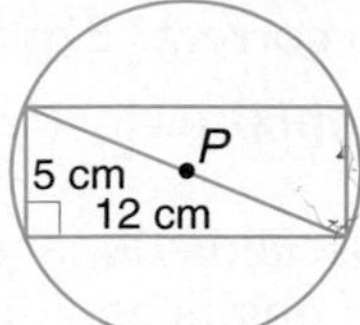

31.

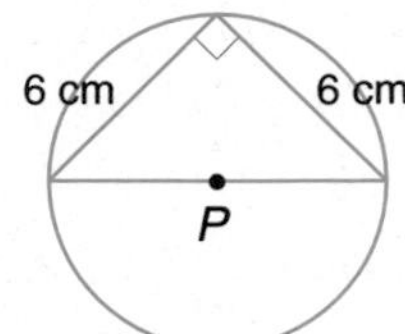

Circle *A* has a radius of 4 units, and ⊙*K* has a radius of 7 units. If *JE* = 2, find each measure.

32. *EK*

33. *IJ*

34. *IC*

35. perimeter of $\triangle ASK$

36. Explain why the perimeter of $\triangle ASK$ equals the length of $\overline{IC}$.

37. The diameter of a circle is 6 feet. If the diameter is doubled, what is the effect on the circumference?

38. The radius of a circle is 18 centimeters. If the radius is divided by 3, what is the effect on the circumference?

Critical Thinking

39. Use the Triangle Inequality Theorem to show diameter $\overline{SA}$ is the longest chord in ⊙*P*. That is, write a paragraph proof that shows $SA > KR$. (*Hint:* Draw $\overline{PK}$ and $\overline{PR}$.)

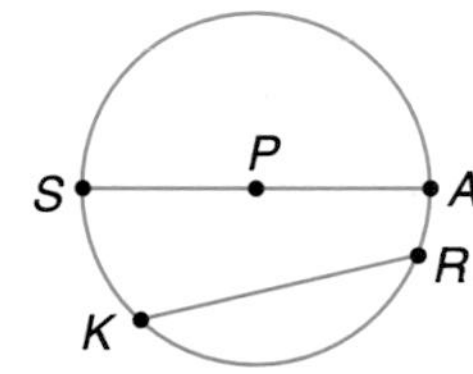

Applications and Problem Solving

40. Sports The new track, soccer, and football field for Prairie Ridge High School is shown below. To the nearest tenth of a yard, find how far a runner will go if she runs around the inside lane of the track four times.

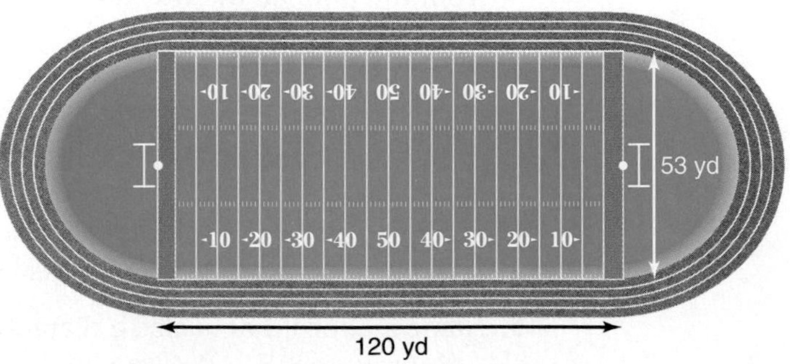

41. Music Using a centimeter tape measure, carefully measure the diameter and circumference of one of your favorite compact discs (CDs). Find the ratio of the circumference and the diameter.

42. Native American Culture About one thousand years ago, wooden poles approximated a gigantic circle 125 meters across in southern Illinois. The structure was a giant solar calendar that kept track of the seasons and the movement of the sun for Native North Americans. What was the circumference of this structure?

43. **Pets** A hamster wheel is 10 centimeters in diameter. If the hamster runs so that the wheel makes 100 revolutions, how far did the hamster run?

44. **Surveying** Refer to Example 3. An inspector uses a 33-centimeter radius bicycle wheel to determine that the total length of a racewalking course is 10 kilometers. How many revolutions did the wheel make?

45. **Crops** The following article appeared in the *Orlando Sentinel* in July, 1996.

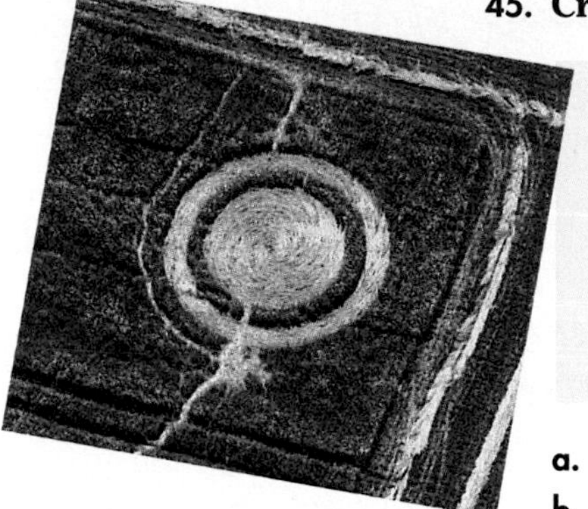

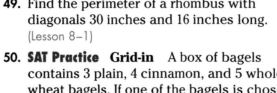

> Authorities were searching Wednesday for the origin of mysterious crop circles that appeared in a nearby barley field this week. "Nothing like this has ever happened around here before," Boyd County Sheriff Duane Pavel said...He said a circle about 25 feet in diameter was surrounded by a wide swath of barley and then another cleared ring...In all, the design was about 42 feet in diameter.

a. Find the circumference of the first circle to the nearest foot.

b. Find the circumference of the entire crop design to the nearest foot.

46. **Space Travel** Bernard A. Harris Jr., M.D. was the first African-American astronaut to walk in space. In 1993, he logged 4,164,183 miles in space on a mission aboard the *STS-55*. If the orbit was 250 miles above Earth and the diameter of Earth is about 8000 miles, find how many orbits around Earth Dr. Harris made on that mission.

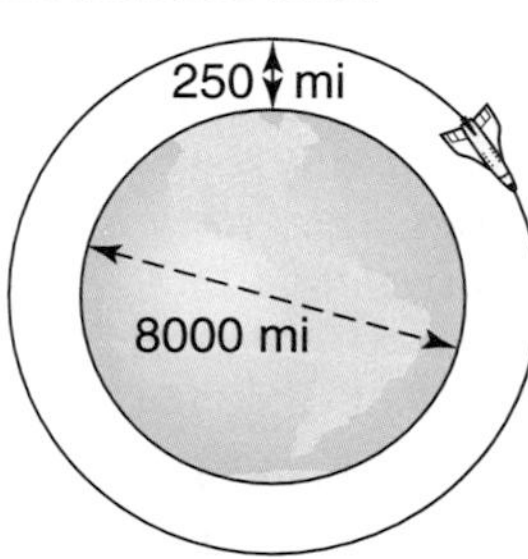

Mixed Review

47. Find the perimeter of $\triangle GEF$ to the nearest tenth of a centimeter. (Lesson 8–6)

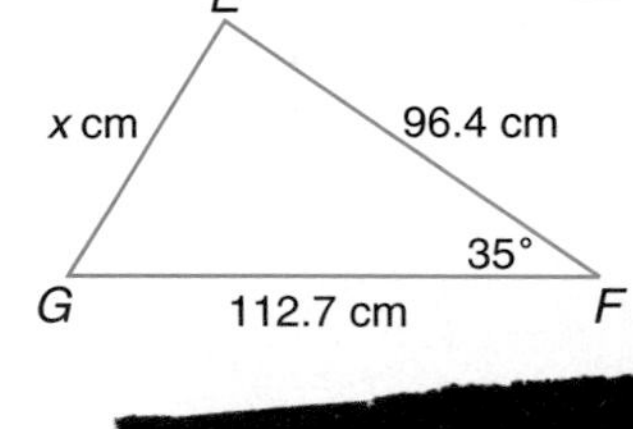

48. **Meteorology** A searchlight 6500 feet from a weather station is turned on. If the angle of elevation to the spot of light on the clouds above the station is 47°, how high is the cloud ceiling? (Lesson 8–4)

49. Find the perimeter of a rhombus with diagonals 30 inches and 16 inches long. (Lesson 8–1)

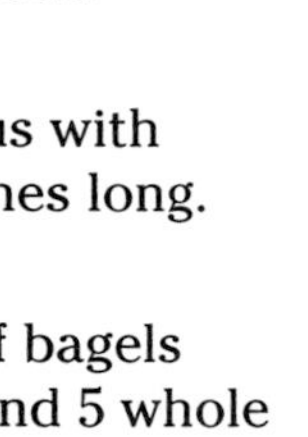

50. **SAT Practice** **Grid-in** A box of bagels contains 3 plain, 4 cinnamon, and 5 whole wheat bagels. If one of the bagels is chosen at random from the box, what is the probability that it will NOT be cinnamon?

51. Solve the proportion $\frac{x}{5} = \frac{7}{2}$. (Lesson 7–1)

52. The median of a trapezoid is 18 inches long. If one base is 29 inches long, what is the length of the other base? (Lesson 6–5)

53. Can 7, 9, and 11 be the measures of the sides of a triangle? Explain. (Lesson 5–5)

54. Find the slope of a line that passes through points $W(2, -9)$ and $C(0, 3)$. (Lesson 3–3)

Algebra

55. Use elimination to solve the system of equations.
$3x - 2y = 8$ and $3x + 4y = -16$

56. Simplify $(a^3b)(a^2b^2)$.

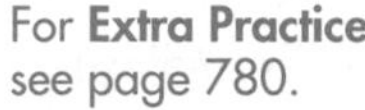

For **Extra Practice**, see page 780.

9–2 Angles and Arcs

What YOU'LL LEARN

- To recognize major arcs, minor arcs, semicircles, and central angles,
- to find measures of arcs and central angles, and
- to solve problems by making circle graphs.

Why IT'S IMPORTANT

Circle graphs are a useful way to display data. They allow you to see important characteristics of the data at a glance.

Statistics

The graph at the right shows how much families planned to spend on their vacations in a recent year. How do you think the artist knew how large to make each section?
This will be solved in Example 1.

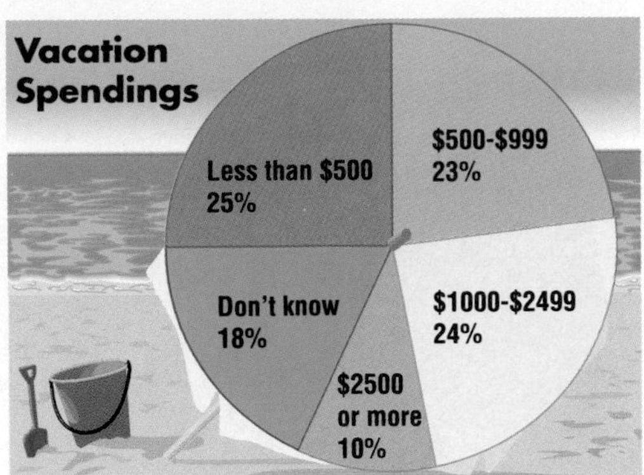

Source: *Economic Analysis of North American Ski Areas*

The graph above is a circle graph. Circle graphs are used to compare parts of a whole, usually as percents. Before drawing a circle graph, the artist needs to know that each section of the graph is a **central angle** of the circle. A central angle is an angle whose vertex is at the center of a circle.

Suppose we draw a diameter of $\odot C$ and name it $\overline{AB}$. We could then construct a diameter perpendicular to $\overline{AB}$ through C and name it $\overline{DE}$. $\angle ACD$, $\angle DCB$, $\angle BCE$, and $\angle ECA$ are central angles with no interior points in common. What is the sum of the measures of these angles?

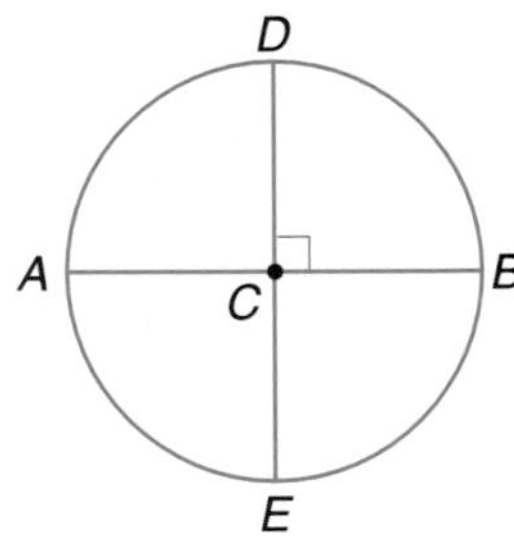

$$90 + 90 + 90 + 90 = 360$$

Sum of Central Angles	**The sum of the measures of the central angles of a circle with no interior points in common is 360.**

The central angles do not have to be right angles for the sum of the measures to be 360.

Example 1

PROBLEM SOLVING

Make a Circle Graph

Refer to the application at the beginning of the lesson. Determine the measure of each central angle used by the artist to draw the circle graph.

The artist knows that in a circle, the sum of the measures of the central angles should be 360. So the central angle that represents each different category of spending amounts for vacation should be proportional to the percent given in the statistics.

Central Angle	Category	%	Number of Degrees
1	Less than $500	25	25% of 360 = 90.0
2	$500 – $999	23	23% of 360 = 82.8
3	$1000 – $2499	24	24% of 360 = 86.4
4	$2500 or more	10	10% of 360 = 36.0
5	Don't know	18	18% of 360 = 64.8
	Total	100	360

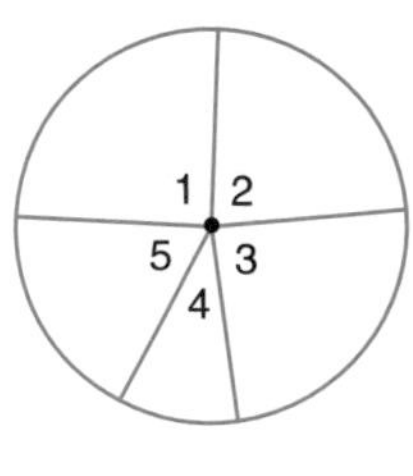

You can use a protractor to verify each angle measure in the graph.

TECHNOLOGY Tip

You can use software that has statistical graphing capabilities to make a circle graph.

From the circle graph, you may have noticed that a central angle separates a circle into two **arcs**.

Two letters are usually used to name a minor arc. However, three letters are often used when the measure of an arc is uncertain.

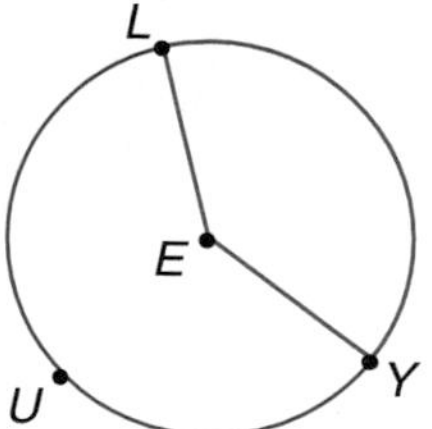

$\widehat{LY}$ is a **minor arc** of $\odot E$.

$\widehat{LUY}$ is a **major arc** of $\odot E$.

Notice that minor arc LY, written $\widehat{LY}$, consists of its endpoints and all points on the circle interior to $\angle LEY$. Major arc LUY, written $\widehat{LUY}$, uses three letters to name the arc and consists of its endpoints and all points on the circle exterior to $\angle LEY$.

A semicircle is usually named by three letters.

Arcs are measured by their corresponding central angles. In $\odot C$ at the right, $m\angle PCM = 110$, so $m\widehat{PM} = 110$. The measure of a minor arc is always less than 180. The measure of a major arc is greater than 180. If the measure of an arc is 180, it is called a **semicircle**. Semicircles are congruent arcs formed when the diameter of a circle separates the circle into two arcs.

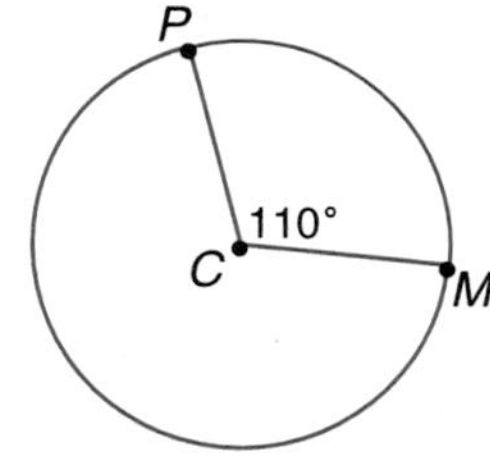

Definition of Arc Measure	**The measure of a minor arc is the measure of its central angle. The measure of a major arc is 360 minus the measure of its central angle. The measure of a semicircle is 180.**

Adjacent arcs are arcs of a circle that have exactly one point in common. As with adjacent angles, the measures of adjacent arcs can be added to find the measure of the arc formed by the adjacent arcs.

Postulate 9–1 Arc Addition Postulate	**The measure of an arc formed by two adjacent arcs is the sum of the measures of the two arcs. That is, if Q is a point on $\widehat{PR}$, then $m\widehat{PQ} + m\widehat{QR} = m\widehat{PQR}$.**

Example 2 **In $\odot E$, $m\angle AEN = 18$, $\overline{JN}$ is a diameter, and $m\angle JES = 90$. Find each measure.**

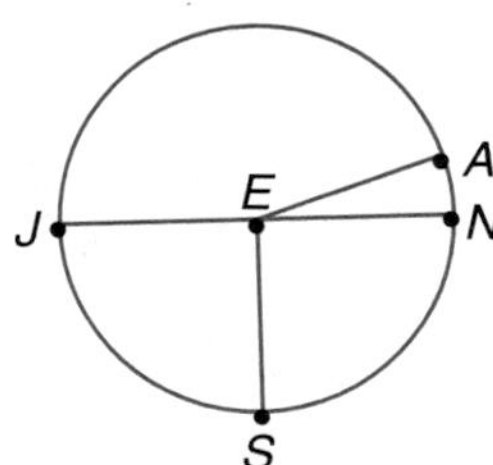

a. $m\widehat{AN}$

Since $\angle AEN$ is a central angle, $m\angle AEN = m\widehat{AN}$. Thus, $m\widehat{AN} = 18$.

b. $m\widehat{JA}$

By the Arc Addition Postulate,

$m\widehat{JAN} = m\widehat{JA} + m\widehat{AN}$

$180 = m\widehat{JA} + 18$ *$\widehat{JAN}$ is a semicircle.*

$m\widehat{JA} = 162$

c. $m\widehat{JAS}$

$\widehat{JAS}$ is the major arc for $\angle JES$. Therefore, $m\widehat{JAS} = 360 - m\widehat{JS}$. Since $m\widehat{JS} = 90$, $m\widehat{JAS} = 360 - 90$ or 270.

You can also use the measure of the central angle to determine the **arc length**. The arc length (or length of an arc) is different from the degree measure of an arc. Suppose a circle was made of string. The length of the arc would be the linear distance of that piece of string representing the arc. The length of the arc is a part of the circumference proportional to the measure of the central angle when compared to the entire circle.

Example 3 **In ⊙*P*, $PR = 9$ and $m\angle QPR = 120$. Find the length of $\widehat{QR}$.**

First, find what part of 360 is represented by $m\angle QPR$.

$$\frac{120}{360} = \frac{1}{3}$$

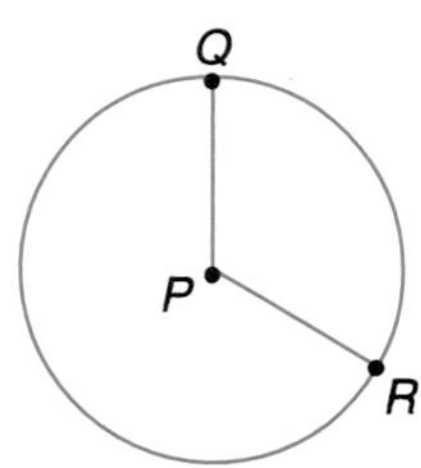

The angle is $\frac{1}{3}$ of the circle, so the length of $\widehat{QR}$ is $\frac{1}{3}$ of the circumference of ⊙*P*.

$$\text{length of } \widehat{QR} = \frac{1}{3}(2\pi r)$$
$$= \frac{1}{3}(2\pi)(9) \quad PR = 9$$
$$= 6\pi \text{ or about 18.8 units}$$

The ripples after a stone is dropped into a pool of water are an example of concentric circles. **Concentric circles** lie in the same plane and have the same center, but have different radii. All circles are **similar circles**, so concentric circles are also similar.

Circles that have the same radius are **congruent circles**. Congruent circles are also similar circles. As with segments and angles, if two arcs of one circle have the same measure, then they are **congruent arcs**. Congruent arcs also have the same arc length.

CHECK FOR UNDERSTANDING

Communicating Mathematics

Study the lesson. Then complete the following.

1. **Explain** why three letters are used to name a major arc of a circle.
2. Refer to ⊙*A*.

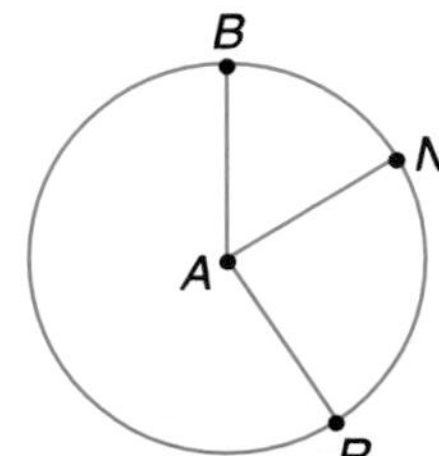

   a. **Explain** why $\widehat{BNR}$ and $\widehat{BRN}$ in ⊙*A* are not the same arc.
   b. **Explain** how to find $m\widehat{BRN}$ in ⊙*A* if $m\widehat{BN} = 25$.
3. **Explain** why a diameter of a circle creates two arcs of degree measure 180.
4. **Discuss** the use of the words *minor* and *major* for arcs. Why are these terms used?

5. Draw a circle A with diameter $\overline{CT}$. Use your protractor to draw radius $\overline{AS}$ so that the measure of central angle SAT is 30.
   a. What is the measure of $\widehat{SCT}$?
   b. Arcs ST and SC have exactly one point in common. Use your protractor to measure $\widehat{SC}$ and $\widehat{TSC}$. Does $m\widehat{TSC} = m\widehat{CS} + m\widehat{ST}$? Compare this observation to the Angle Addition Postulate.

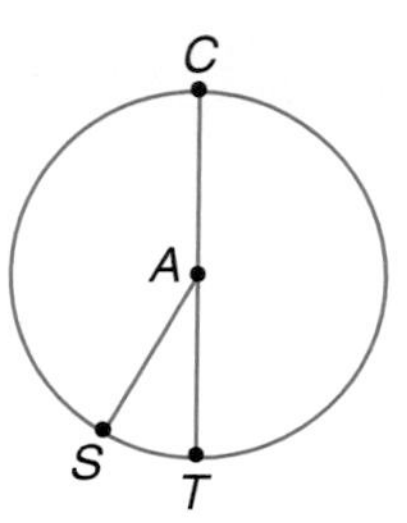

Guided Practice

In ⊙S, $\overline{TE}$ and $\overline{KR}$ are diameters with $m\angle TSR = 42$. Determine whether each arc is a minor arc, a major arc, or a semicircle. Then find the degree measure of each arc.

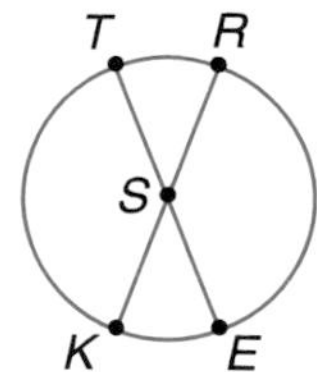

6. $m\widehat{TRE}$ 7. $m\widehat{TK}$ 8. $m\widehat{TRK}$

In ⊙J, $\overline{KM}$ is a diameter with $JL = 18$ and $m\angle KJL = 140$. Find the length of each arc. Round to the nearest tenth.

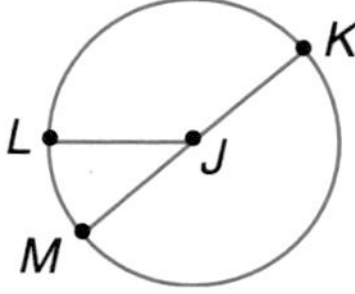

9. $\widehat{LM}$ 10. $\widehat{KL}$ 11. $\widehat{LMK}$

Algebra

In ⊙C, $\overline{IL}$ is a diameter, $m\angle ICR = 3x + 5$, and $m\angle RCL = x - 1$.

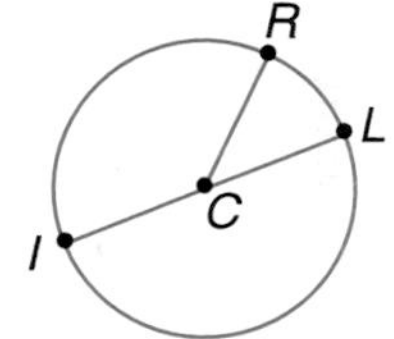

12. Find x.
13. Find $m\angle ICR$.
14. Find $m\widehat{ILR}$.

Determine whether each statement is *true* or *false*. Explain your reasoning.

15. Two concentric circles have the same center.
16. Two concentric circles have the same radius.
17. Two concentric circles never intersect.
18. **Skiing** The graph below shows how people spend their money when they go skiing.
   a. Draw a circle graph by hand or by using computer software. Label each part of the circle graph.
   b. Explain why the total percentage should add up to 100%.

Source: *Economic Analysis of North American Ski Areas*

EXERCISES

Practice

Refer to ⊙E for Exercises 19–36. If $m\angle TEG = 21$ and $\overline{TR}$ is a diameter, determine whether each arc is a minor arc, a major arc, or a semicircle. Then find the degree measure of each arc.

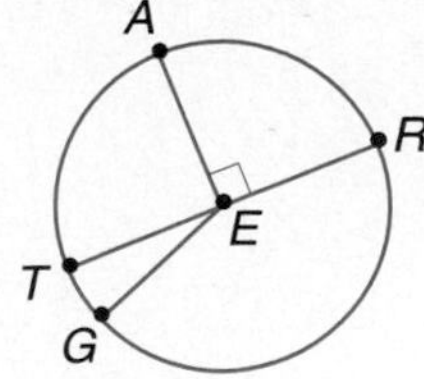

19. $m\overarc{TG}$
20. $m\overarc{ATR}$
21. $m\overarc{AR}$
22. $m\overarc{TAR}$
23. $m\overarc{ATG}$
24. $m\overarc{ARG}$
25. $m\overarc{RAG}$
26. $m\overarc{TAG}$
27. $m\overarc{GR}$

If $TR = 12$, find the length of each arc. Round to the nearest tenth.

28. $\overarc{TG}$
29. $\overarc{ATR}$
30. $\overarc{AR}$
31. $\overarc{TAR}$
32. $\overarc{ATG}$
33. $\overarc{ARG}$
34. $\overarc{RAG}$
35. $\overarc{TAG}$
36. $\overarc{GR}$

Algebra

In ⊙G, $\angle NGE \cong \angle EGT$, $m\angle AGJ = 4x$, $m\angle JGT = 2x + 24$, and $\overline{AT}$ and $\overline{JN}$ are diameters. Find each of the following.

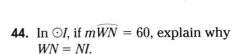

37. x
38. $m\angle AGJ$
39. $m\angle JGT$
40. $m\overarc{NE}$
41. $m\overarc{NJT}$
42. $m\overarc{JNE}$

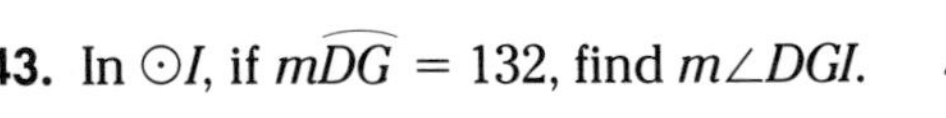

43. In ⊙I, if $m\overarc{DG} = 132$, find $m\angle DGI$.

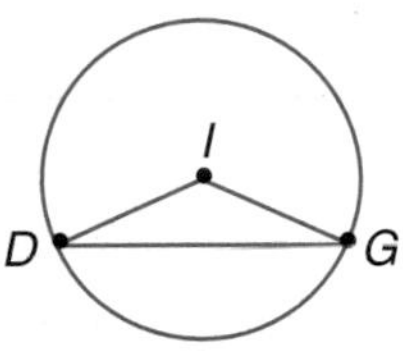

44. In ⊙I, if $m\overarc{WN} = 60$, explain why $WN = NI$.

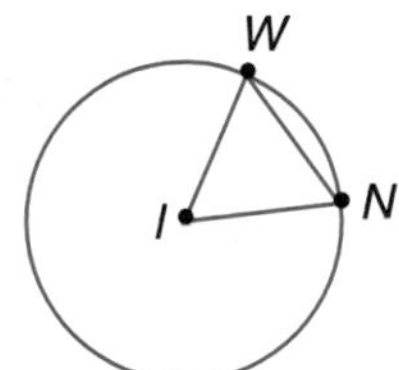

In the figure at the right, P is the center of two concentric circles. Determine whether each statement is *true* or *false*. Explain your reasoning.

45. If $m\overarc{RS} = 25$ and $m\overarc{CD} = 25$, then $\overarc{RS} \cong \overarc{CD}$.

46. If $m\angle CPD = 40$, then $m\overarc{RS}$ and $m\overarc{CD}$ are each 40.
47. If $m\overarc{RS} = m\overarc{CD}$, then $\overarc{RS}$ must be congruent to $\overarc{CD}$.
48. $\overline{PC} \cong \overline{PD}$
49. If $m\overarc{RS} = 42$, then $m\overarc{CKD} = 318$.

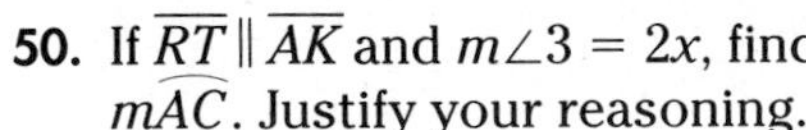

50. If $\overline{RT} \parallel \overline{AK}$ and $m\angle 3 = 2x$, find $m\overarc{AC}$. Justify your reasoning.

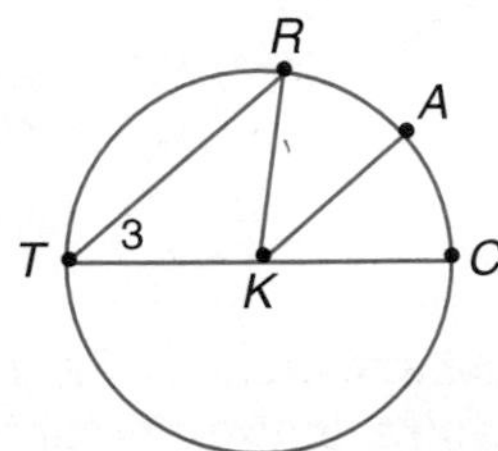

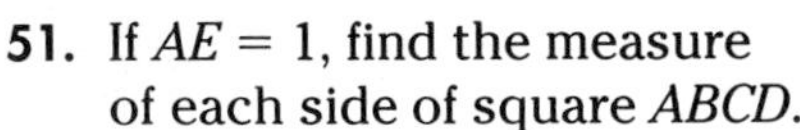

51. If $AE = 1$, find the measure of each side of square $ABCD$.

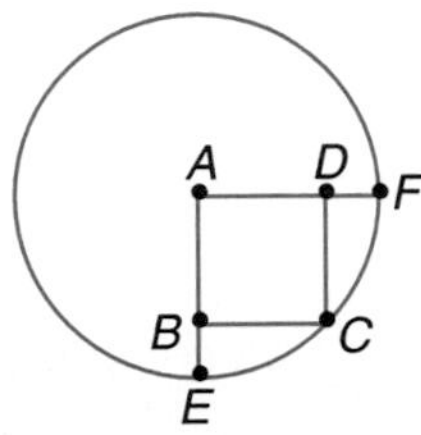

Critical Thinking

52. Draw a diagram to explain how it is possible for two central angles to be congruent, yet their corresponding minor arcs are *not* congruent.

53. Draw a Venn diagram that illustrates the relationship among congruent, similar, and concentric circles.

Applications and Problem Solving

54. Clocks The Floral Clock in Frankfort, Kentucky, has a diameter of 34 feet. The hands on the clock form a central angle with the circular timepiece. Suppose the hour hand is on the 10 and the minute hand is on the 2.

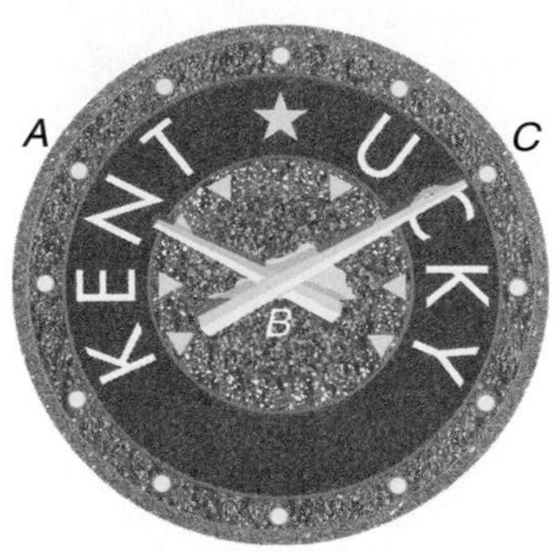

a. Find the measure of central angle *ABC*.

b. Find the arc length of the minor arc.

Where Do All the Apples Go?

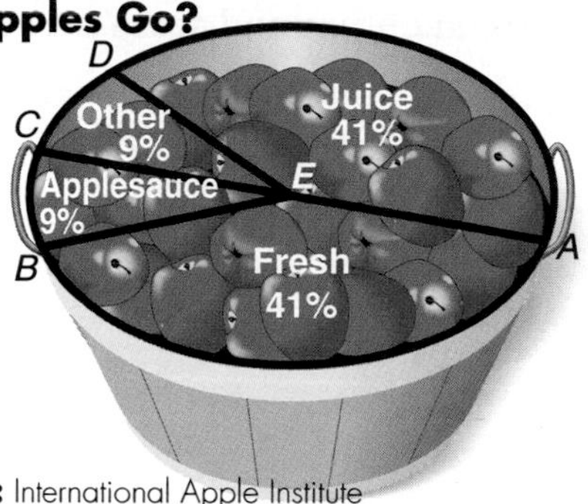

Source: International Apple Institute

55. Produce Each year, the United States produces approximately 11 billion pounds of apples. The graph at the left shows where they go.

a. Since $m\angle AEB = m\angle AED$, then is $m\widehat{AB} = m\widehat{AD}$? Explain.

b. If diameter $AC = 16$, find the length of $\widehat{AB}$ to the nearest tenth.

56. Statistics A survey of students ages 9–18 was taken recently to find with whom students would most like to trade places. They were given five choices and the table at the right shows the results.

Person to Trade Places With	Number of Students
Head of a major company	700
NASA astronaut	350
President of the U.S.	1050
Professional athlete	1850
Teacher	1050

a. Determine the central angle measures to the nearest tenth degree needed to accurately construct a circle graph of the data.

b. Draw and label a circle graph that represents the data. Draw the graph by hand or by using computer software.

c. Explain the advantages of displaying data in a graph rather than in a table. Use the graph you drew in part b to explain your answer.

Mixed Review

57. If the radius of a circle is 22 millimeters, find the circumference. Round to the nearest tenth. (Lesson 9–1)

58. A square has a perimeter of 27 centimeters. Find the length of a diagonal of the square to the nearest tenth. (Lesson 8–2)

59. Find the value of *x* if the triangles are similar. (Lesson 7–5)

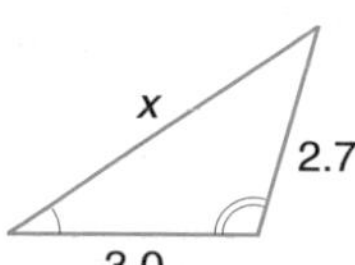

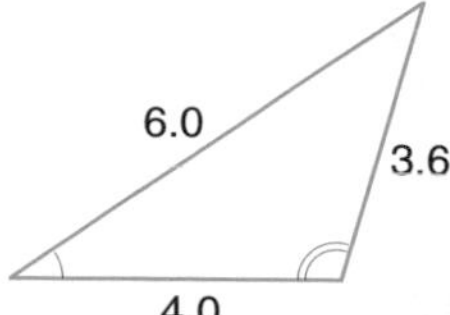

60. Determine whether the statement *If the corresponding sides of two polygons are proportional, the polygons are similar* is true or false. (Lesson 7–2)

61. Given parallelogram $ABCD$ with $AB = 7x - 6$ and $CD = 5x + 14$, find AB. (Lesson 6–1)

62. State the assumption you would make to start an indirect proof of the statement *If two chords are congruent, they are equidistant from the center of the circle.* (Lesson 5–3)

63. **SAT Practice Quantitative Comparison**

20% of x is 25% of 100.

Column A	Column B
x	80

A if the quantity in Column A is greater
B if the quantity in Column B is greater
C if the two quantities are equal
D if the relationship cannot be determined from the information given

64. Describe the difference between obtuse and acute triangles. (Lesson 4–1)

65. Determine whether the statement *If two parallel lines are cut by a transversal, then two consecutive interior angles are congruent* is true or false. (Lesson 3–2)

66. **Sports** Write the converse of the statement *If a student plays a fall sport, then he or she must have a physical exam in the spring.* (Lesson 2–2)

For **Extra Practice**, see page 780.

Algebra

67. Find $(3x^2y^4)(5x^4y)$.

68. Find $\frac{9}{n-3} \cdot \frac{n^2-9}{12}$.

Math and the Sculptor

The excerpt below appeared in *Science News* on February 17, 1996.

From the outside, Helaman Ferguson's modest, two-car garage resembles just about any other garage in his suburban neighborhood....This is Ferguson's studio. Here, he carves stone and other materials to fashion works of art rooted in mathematical ideas.... a graceful, 3-foot-tall figure gradually emerges from a gleaming chunk of white Carrara marble. This sculpture is one of a series based on a type of mathematical form called a minimal surface. It's related to the shape of a soap film stretched across a bent ring.....Sculptors often make a small model...of what they want to carve....Ferguson has pioneered an alternative approach....he can translate geometric forms drawn on the computer screen directly into instructions on how much material the artist should remove at any point from an uncarved stone's surface to reveal the object. ■

1. Which basic arithmetic operation could be used to describe the process of sculpting by stone carving? Which basic arithmetic operation could be used to describe the creation of a piece of art by combining pieces of aluminum, cloth, and wire?
2. If you were viewing a three-dimensional model of a certain sculpture on a computer screen, it might be difficult to clearly see the curvature of all the surfaces. What would you suggest as a way to make the surface curvatures easier to see?

9–3 Arcs and Chords

What YOU'LL LEARN

- To recognize and use relationships among arcs, chords, and diameters.

Why IT'S IMPORTANT

You can use arcs and chords to solve problems involving geology.

Real World APPLICATION

Road Signs

The yellow railroad-crossing sign can be used to help explain special relationships that exist between arcs and chords. The sign is circular and has two perpendicular diameters.

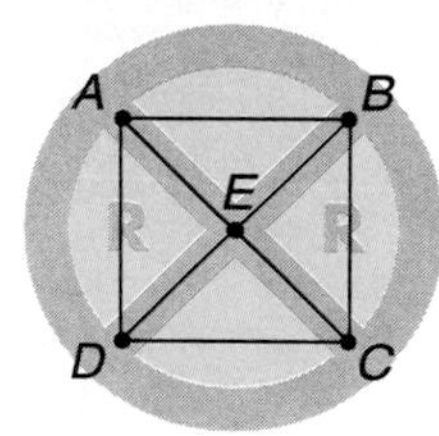

If we label the center of the circle E and endpoints of the diameters A, B, C, and D, we can draw chords $\overline{AB}$, $\overline{BC}$, $\overline{CD}$, and $\overline{DA}$. When a minor arc and a chord share the same endpoints, we call the arc the **arc of the chord**. For example, in the railroad-crossing sign, $\overset{\frown}{AB}$ is the arc of $\overline{AB}$.

Since the diameters in the sign are perpendicular, we know the measure of each central angle is 90. Thus, $m\angle AEB = m\angle BEC = 90$, which implies $m\overset{\frown}{AB} = 90$. Since $m\overset{\frown}{AB} = m\overset{\frown}{BC}$, then $\overset{\frown}{AB} \cong \overset{\frown}{BC}$. How are $\overline{AB}$ and $\overline{BC}$ related? This relation is stated and proved below.

Theorem 9–1 **In a circle or in congruent circles, two minor arcs are congruent if and only if their corresponding chords are congruent.**

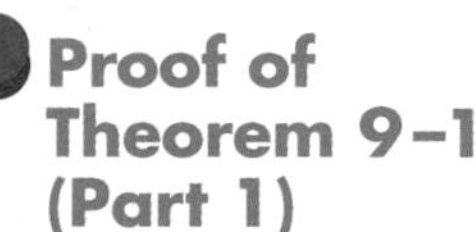

Proof of Theorem 9–1 (Part 1)

Prove that if two minor arcs of a circle are congruent, then their corresponding chords are congruent.

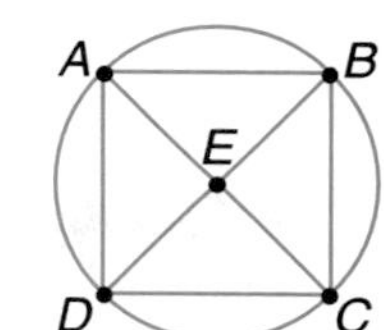

Given: $\odot E$

$\overset{\frown}{AB} \cong \overset{\frown}{DC}$

Prove: $\overline{AB} \cong \overline{DC}$

Proof:

Statements	Reasons
1. $\odot E$, $\overset{\frown}{AB} \cong \overset{\frown}{DC}$	**1.** Given
2. $\overline{AE} \cong \overline{CE}$ $\overline{BE} \cong \overline{DE}$	**2.** All radii of a $\odot$ are $\cong$.
3. $\angle AEB \cong \angle CED$	**3.** Vertical $\angle$s are $\cong$.
4. $\triangle AEB \cong \triangle CED$	**4.** SAS
5. $\overline{AB} \cong \overline{DC}$	**5.** CPCTC

You will be asked to prove the other part of Theorem 9–1 in Exercise 41. That is, prove that if two chords of a circle are congruent, then their corresponding arcs are congruent.

A polygon is an **inscribed polygon** if each of its vertices lies on a circle. The polygon is said to be inscribed in the circle. In the figure at the right, quadrilateral $PQRS$ is inscribed in $\odot C$.

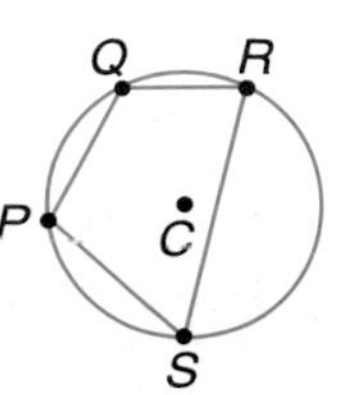

Example 1

Road Signs

A stop sign is an octagon with congruent sides. It can be inscribed in a circle by using the center of the sign and a vertex of the sign as endpoints for the radius of the circle, as in $\overline{QR}$. Find the measure of each of the eight corresponding arcs of a circle around the stop sign shown at the right.

Since all eight chords of the circle (or sides of the sign) are congruent, the eight corresponding arcs are congruent by Theorem 9–1. Therefore, each arc measures $\frac{360}{8}$ or 45.

The following activity provides insight into relationships between arcs and chords.

MODELING MATHEMATICS — Chords and Diameters

Materials:  compass  patty paper  straightedge

- Use a compass to draw a circle on a piece of patty paper. Label the center *K*. Draw a chord that is not a diameter. Name it $\overline{AR}$.
- Fold the paper through *K* so that *A* and *R* coincide. Label this fold as diameter $\overline{EF}$.

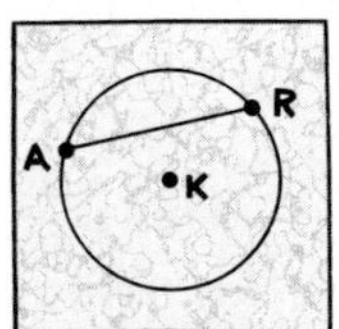

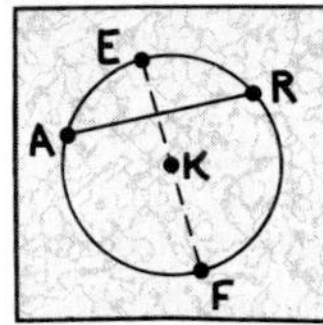

Your Turn

a. When the paper is folded, compare the lengths of $\widehat{AE}$ and $\widehat{ER}$. Then compare the lengths of $\widehat{AF}$ and $\widehat{FR}$.

b. Write a statement about the relationship between a diameter that is perpendicular to a chord and the chord and its arc.

Theorem 9–2	**In a circle, if a diameter is perpendicular to a chord, then it bisects the chord and its arc.**

You will be asked to prove this theorem in Exercise 28.

Example 2

Chords $\overline{CH}$ and $\overline{IR}$ are equidistant from the center of $\odot S$. If $IR = 48$, find CH. SA is the distance from S to $\overline{CH}$, and SB is the distance from S to $\overline{IR}$.

Remember that the distance from a point to a line is measured by a perpendicular segment.

Draw radii $\overline{SH}$ and $\overline{SR}$.

Since $\overline{SB} \perp \overline{IR}$, $BR = 24$ by Theorem 9–2.

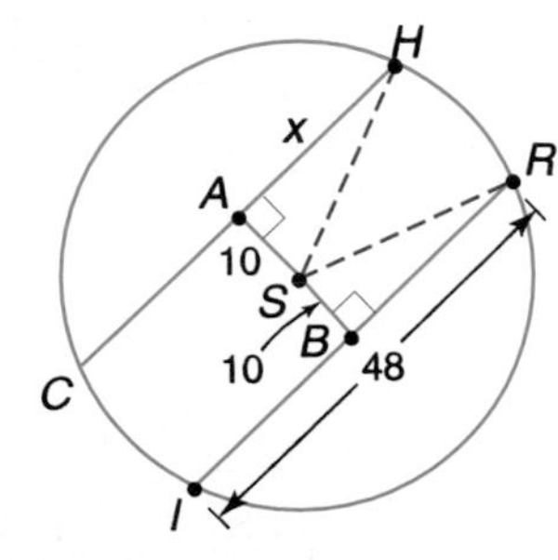

$\triangle SAH$ and $\triangle SBR$ are right triangles.

Since $\overline{SB} \cong \overline{SA}$ (given) and $\overline{SR} \cong \overline{SH}$ (all radii are $\cong$), then $\triangle SAH \cong \triangle SBR$ by HL.

Since $\overline{AH} \cong \overline{BR}$ and $BR = 24$, then $AH = 24$.

Since $\overline{AS}$ is a perpendicular bisector of $\overline{CH}$, by Theorem 9–2, $CH = 48$.

In Example 2, recall that $\overline{CH}$ and $\overline{IR}$ were equidistant from the center S. Also notice that $\overline{CH} \cong \overline{IR}$. This leads to the next theorem.

Theorem 9–3	**In a circle or in congruent circles, two chords are congruent if and only if they are equidistant from the center.**

You will be asked to prove this theorem in Exercises 42 and 43.

Example 3

Suppose a chord of a circle is 10 inches long and 12 inches from the center of the circle. Find the length of the radius.

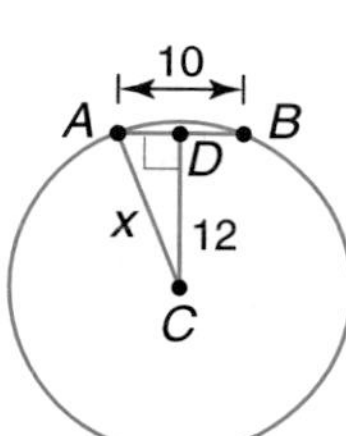

Draw a diagram and then solve.

In $\odot C$, $AB = 10$, $CD = 12$, and $\overline{AC}$ is the radius. Since $\triangle ADC$ is a right triangle, use the Pythagorean Theorem to find x, the length of the radius.

$(AD)^2 + 12^2 = x^2$ *Pythagorean Theorem*

$5^2 + 12^2 = x^2$ *AD = 5 Why?*

$25 + 144 = x^2$

$169 = x^2$

$x = 13$

Therefore, the length of the radius is 13.

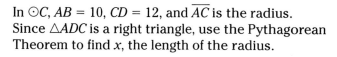

CHECK FOR UNDERSTANDING

Communicating Mathematics

Study the lesson. Then complete the following.

1. **Examine** the labeled railroad-crossing sign at the beginning of the lesson. What type of quadrilateral is *ABCD*? Explain your reasoning.
2. **Extend** Example 1 by connecting every other vertex of the stop sign. What type of figure is formed? Explain how you could verify your conclusion using the theorems from this lesson.
3. **You Decide** Gabriella claims that $\overline{AB}$ bisects $\overline{PQ}$ since $\overline{AB} \perp \overline{PQ}$. Tiarri says $\overline{AB}$ doesn't bisect $\overline{PQ}$ because $\overline{AB}$ is not a diameter of K. Who is right and why?

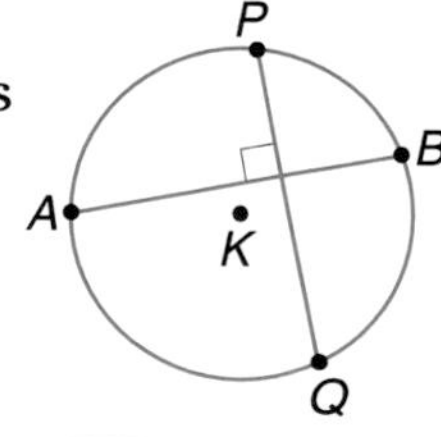

MODELING MATHEMATICS

4. Use a compass to draw a circle on a piece of patty paper. Draw two nonparallel chords and label them $\overline{JK}$ and $\overline{RS}$. Fold the paper so that the endpoints of $\overline{JK}$ coincide. Label this fold as line ℓ. Fold the paper again so that the endpoints of $\overline{RS}$ overlap. Label this fold as line m. Write a conjecture about the intersection of lines ℓ and m. Explain.

Guided Practice

5. State the theorem that justifies the following statement for $\odot H$. If $\overline{HG} \perp \overline{EF}$, then $\overline{EG} \cong \overline{GF}$.

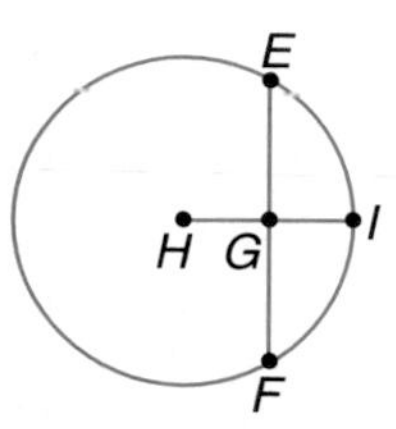

In $\odot P$, $\overline{PQ} \perp \overline{RM}$.

6. Name an arc congruent to $\widehat{QR}$.
7. If $PR = 13$ and $RM = 24$, then find PO.
8. Name a segment congruent to $\overline{PM}$.

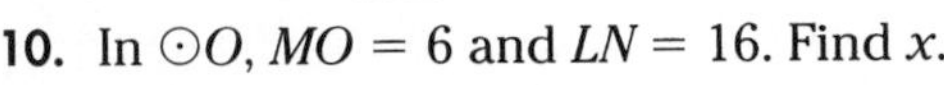

9. In $\odot R$, $TR = 6.4$ and $EN = 10.8$. Find RO to the nearest tenth.

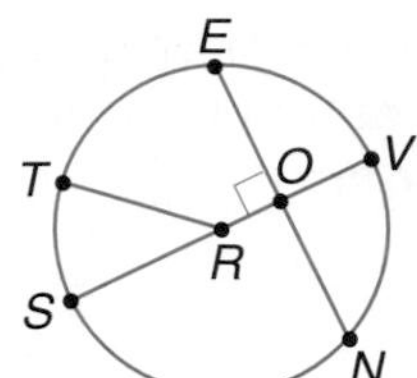

10. In $\odot O$, $MO = 6$ and $LN = 16$. Find x.

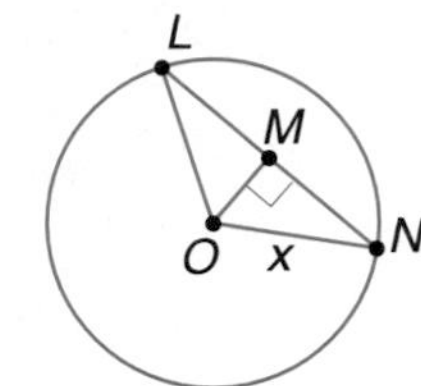

11. Suppose a chord of a circle is 10 inches long and is 12 inches from the center of the circle.
    a. Draw and label a figure.
    b. Find the length of the radius.
12. **Food** Cecilia is barbecuing chicken. The grill on her barbecue is in the shape of a circle with a diameter of 54 centimeters. The horizontal wires are supported by 2 wires that are 12 centimeters apart as shown in the figure at the right. If the grill is symmetrical and the wires are evenly spaced, what is the length of each support wire?

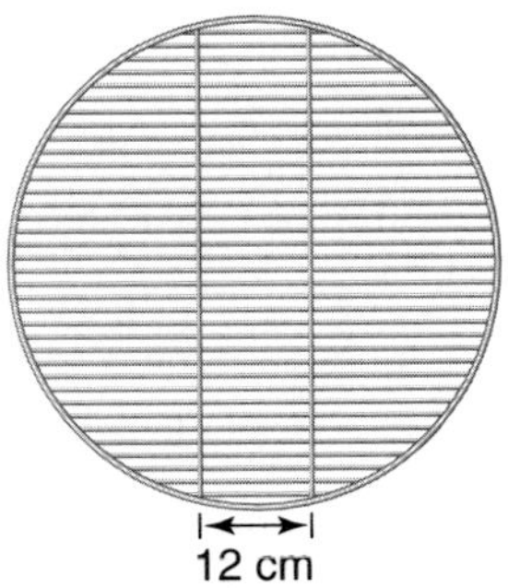

EXERCISES

Practice

State the theorem that justifies each statement.

13. If $\overline{AC} \cong \overline{DG}$, then $\widehat{AC} \cong \widehat{DG}$.
14. If $\overline{JF} \perp \overline{DG}$, then $\overline{DE} \cong \overline{EG}$.
15. If $\overline{KP} \cong \overline{EP}$, then $\overline{AC} \cong \overline{DG}$.

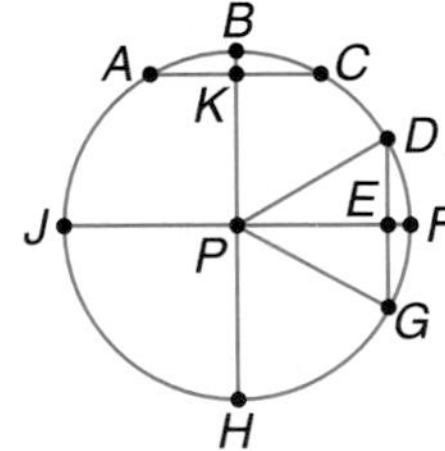

In $\odot D$, $\overline{VR}$ and $\overline{QU}$ are diameters with $\overline{QU} \perp \overline{PR}$ and $\overline{QU} \perp \overline{VW}$.

16. Name the midpoint of $\overline{QU}$.
17. Name the midpoint of $\overline{PR}$.
18. If $VA = 9$, find VW.
19. Name two arcs congruent to $\widehat{QR}$.
20. Name two segments congruent to $\overline{DQ}$.
21. Which chord is longer, $\overline{PR}$ or $\overline{VW}$?
22. If $DS = 14$ and $PR = 32$, find DR to the nearest tenth.
23. Explain why $\overline{VW} \parallel \overline{PR}$.
24. Name a segment congruent to $\overline{QS}$.

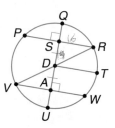

In each circle, *M* is the center. Find each measure.

25. $m\angle CAM$ **26.** $m\widehat{SE}$ **27.** CS

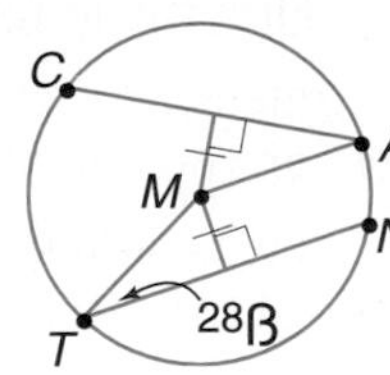

Proof

28. Complete the following proof of Theorem 9–2.

Given: $\odot P$
$\overline{AB} \perp \overline{TK}$

Prove: $\overline{AR} \cong \overline{BR}$
$\widehat{AK} \cong \widehat{BK}$

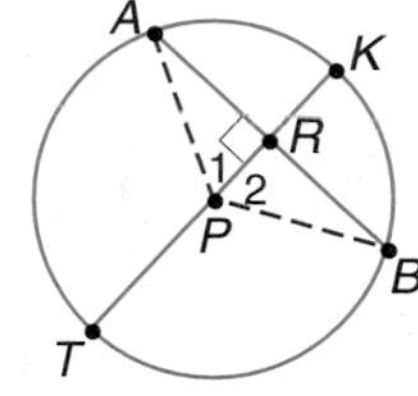

Proof:

Statements	Reasons
a. Draw radii $\overline{PA}$ and $\overline{PB}$.	**a.** ?
b. $\odot P$, $\overline{AB} \perp \overline{TK}$	**b.** ?
c. $\overline{PA} \cong \overline{PB}$	**c.** ?
d. $\overline{PR} \cong \overline{PR}$	**d.** ?
e. ?	**e.** HL
f. $\overline{AR} \cong \overline{BR}$ $\angle 1 \cong \angle 2$	**f.** ?
g. ?	**g.** Def. of $\cong$ arcs

Refer to $\odot O$ with rhombus *RHOM* for Exercises 29 and 30.

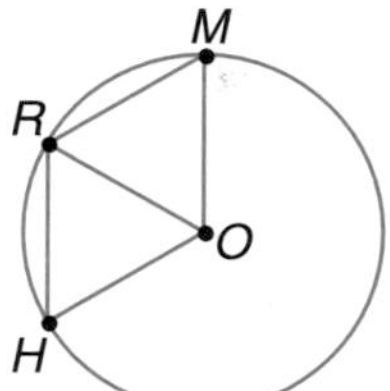

29. Explain why $\widehat{RH} \cong \widehat{MR}$.

30. Explain why $\triangle MRO$ and $\triangle RHO$ are equilateral triangles.

31. In the figures at the right, $\overline{PC} \cong \overline{QT}$ and $\overline{AB} \cong \overline{RS}$. Is $\widehat{AB} \cong \widehat{RS}$? Explain your reasoning.

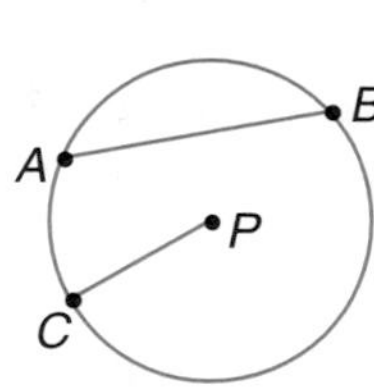

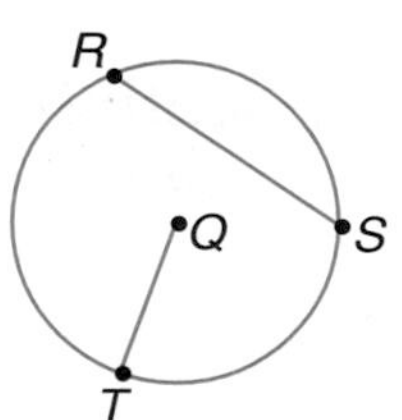

In each figure, *J* is the center of the circle. Find *x* to the nearest tenth.

32. **33.** **34.**

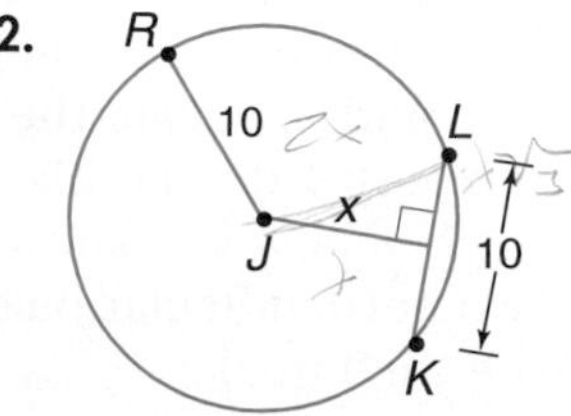

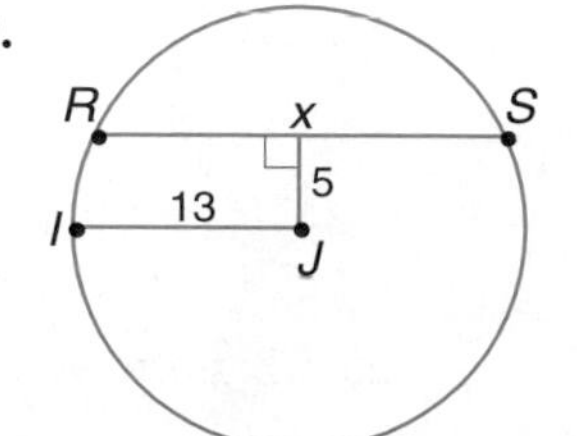

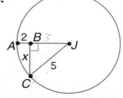

Draw a figure and then solve each problem.

35. Suppose a chord of a circle is 24 centimeters long and is 15 centimeters from the center of the circle. Find the length of the radius.

36. Suppose the diameter of a circle is 34 inches long and a chord is 30 inches long. Find the distance between the chord and the center of the circle.

37. Suppose the diameter of a circle is 50 millimeters long and a chord is 7 millimeters from the center of the circle. Find the length of the chord.

In each figure, O is the center of the circle.

38. $\overline{MA} \cong \overline{TH}$, $MA = 8x + 4$, $TH = 12$, and $OQ = x^2$. Find OA.

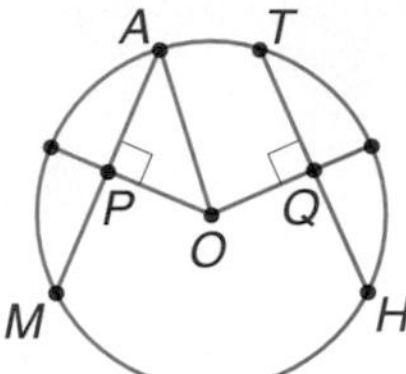

39. $\overline{PA}$ is perpendicular bisector for radius $\overline{OK}$ and $OK = 16$. Find AP.

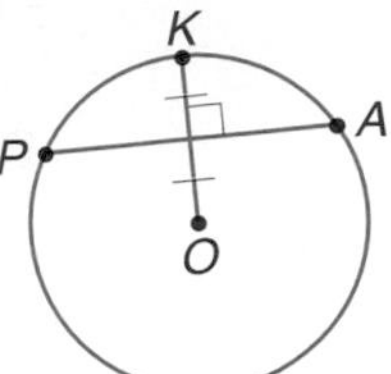

40. Find the length of a chord that is the perpendicular bisector of a radius of length 30 units in a circle.

Proof

Draw a figure and write a proof for each theorem.

41. In a circle, if two chords are congruent, then their corresponding minor arcs are congruent. (second part of Theorem 9–1)

42. In a circle, if two chords are equidistant from the center, then they are congruent. (Theorem 9–3)

43. In a circle, if two chords are congruent, then they are equidistant from the center. (Theorem 9–3)

Critical Thinking

44. Draw a circle R and choose a point P on the circle. Now draw six chords with P as one endpoint. Label the other endpoints A, B, C, D, E, and F. Make a conjecture as to how the lengths of each chord are related to their distances from the center of the circle. Explain your reasoning.

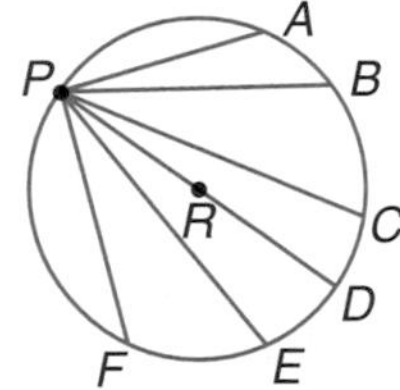

Applications and Problem Solving

45. Geology In order to locate the *epicenter* of an earthquake, geologists need data from three different seismograph stations. During an earthquake, suppose seismograph stations record the same earthquake intensity from Salt Lake City, Utah; Atlanta, Georgia; and Albuquerque, New Mexico.

a. Copy the map at the left or use another map of the United States. Label each station as shown. (Salt Lake City–S, Atlanta–A, Albuquerque–T)

b. Since the stations recorded the same earthquake intensity, they are each equidistant from the epicenter or center of the circle that contains T, S, and A. Locate the epicenter by constructing the perpendicular bisectors of $\overline{ST}$ and $\overline{AT}$. Explain how you know the intersection of the two perpendicular bisectors is the center of the earthquake.

c. Use your map to locate a city near the epicenter.

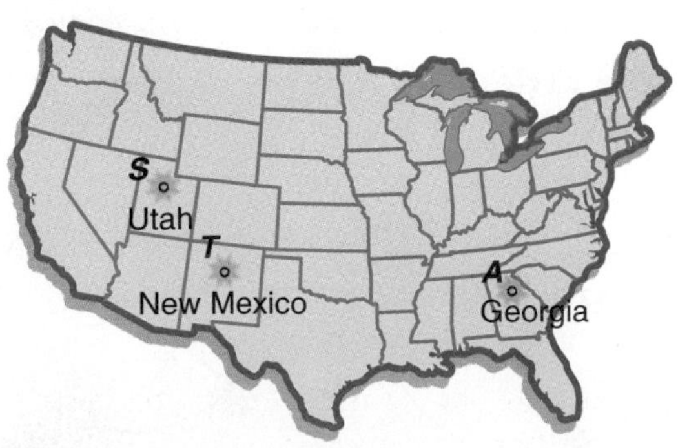

46. **Road Signs** Jodi wants to create a new yield sign that inscribes the yellow isosceles triangle in a circle. Draw an isosceles triangle and explain how Jodi could use the theorems from this lesson to find the center of the circle that contains the yellow yield sign. (*Hint*: See Exercise 4.)

Mixed Review

47. Determine whether the statement *All radii of a circle are congruent* is true or false. Explain your answer. (Lesson 9–2)

48. **Native American Agriculture** The *Hidatsa* are a people of the plains who live in what is now North Dakota. Their gardens, which are sometimes round, include corn, beans, squash, and sunflowers. If a typical round Hidatsa garden is 22 feet across, find the circumference of the garden to the nearest tenth of a foot. (Lesson 9–1)

49. Find the perimeter of $\triangle JKL$. (Lesson 8–2)

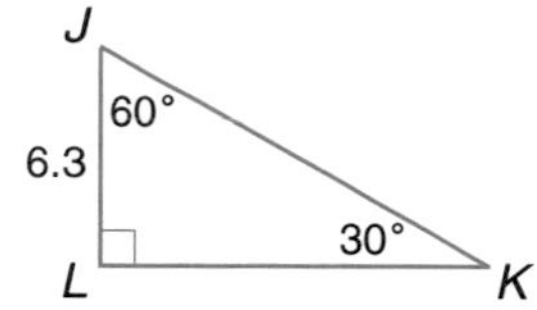

50. **SAT Practice** T is the set of all positive numbers n such that $n < 50$ and $\sqrt{n}$ is an integer. What is the median value of the members of set T?
A 4 **B** 16 **C** 20 **D** 25 **E** 49

51. Determine whether the statement *The diagonals of a parallelogram are always congruent* is true or false. (Lesson 6–1)

52. Can a median of a triangle also be an altitude? If so, what type of triangle is it? (Lesson 5–1)

53. If $\triangle PQR \cong \triangle HLM$, then $\angle R \cong$ __?__. (Lesson 4–3)

54. Identify the hypothesis and conclusion of the conditional *If it is raining, then I will bring an umbrella.* (Lesson 2–2)

For **Extra Practice**, see page 780.

Algebra

55. Find $-6rs(4r^2 + 1) + 8(rs + 2r)$.

56. Find $\frac{21x^3y^5z}{7xy^6z}$.

WORKING ON THE In·ves·ti·ga·tion

Refer to the Investigation on pages 392–393.

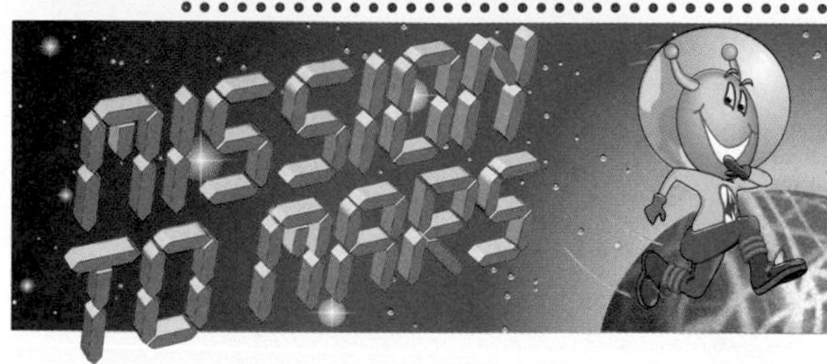

After the space station is established on Mars, a team of scientists will begin studying its moons.

1. Refer to the information on page 393 to find the time it takes Mars to make one complete rotation. Given the amount of time that it takes for the planet to revolve 360°, how many degrees of rotation does Mars make in one hour?
2. Two outposts of the space station will be located so that each moon will pass directly over an outpost on each orbit. Use the information about the orbit times of each moon to determine how often each moon will pass over its outpost.
3. Using the distances from Mars to its moons, find the relative sizes of the moons in Mars' night sky.
4. Determine whether there will there be eclipses of the Sun by either of Mars' moons.

Add the results of your work to your Investigation Folder.

9–4 Inscribed Angles

What YOU'LL LEARN

- To recognize and find measures of inscribed angles, and
- to apply properties of inscribed figures.

Why IT'S IMPORTANT

You can use inscribed angles to solve problems involving civil engineering and carpentry.

Real World APPLICATION

Building Trades

Hardware stores often sell various nuts and bolts that are in a hexagonal shape. These fasteners have numerous uses in the building trades.

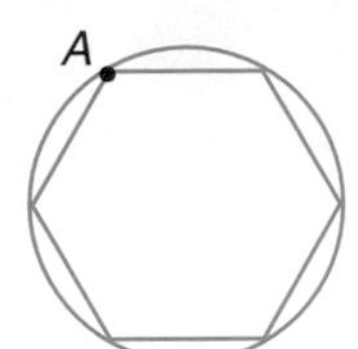

Machinists usually cut hexagonal shapes from round or circular stock. The machinist must trim or cut the stock just right to form a hexagon with all sides congruent. How could the machinist determine the appropriate cuts to make in the stock? *You will be asked to solve this problem in Exercise 16.*

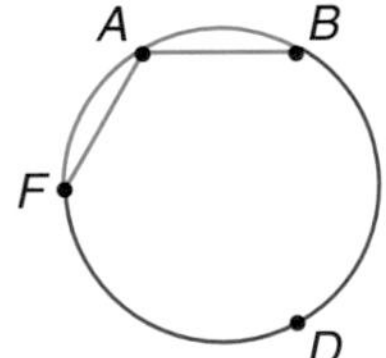

Each angle of the hexagon is an **inscribed angle**. An inscribed angle is an angle whose vertex is on the circle and whose sides each contain chords of the circle. We say that $\angle FAB$ intercepts $\widehat{FDB}$. $\widehat{FDB}$ is called the **intercepted arc** of $\angle FAB$. Notice that $\widehat{FDB}$ lies in the interior of $\angle FAB$.

The following activity explores the relationship between the measures of an inscribed angle and its intercepted arc.

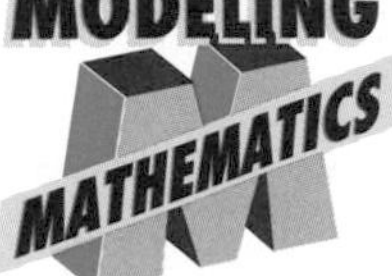

Inscribed Angles

Materials:  protractor  straightedge

- Trace $\odot P$ with inscribed hexagon $ABCDEF$ on a piece of paper. Note that all sides are congruent.

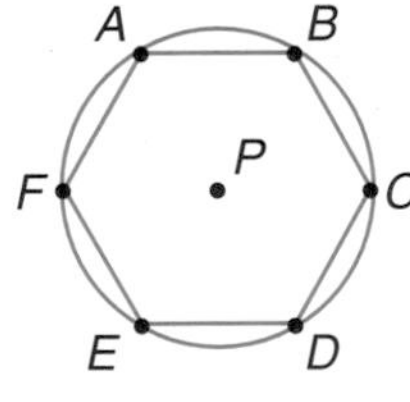

- Draw radii $\overline{PA}$, $\overline{PF}$, and $\overline{PB}$ and label angles 1, 2, 3, and 4.

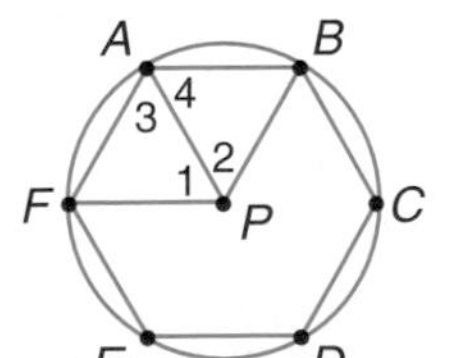

- Use a protractor to measure each numbered angle.

Your Turn

a. Find $m\widehat{FA}$ and $m\widehat{AB}$. Explain your reasoning.
b. Find $m\angle FAB$ and $m\widehat{BDF}$.
c. Make a conjecture regarding the relationship between $m\angle FAB$ and $m\widehat{BDF}$.

The Modeling Math activity suggests the following theorem.

Theorem 9–4	**If an angle is inscribed in a circle, then the measure of the angle equals one-half the measure of its intercepted arc.**

There are three cases to consider when writing a proof of this theorem.

Case 1: The center of the circle lies on one of the sides of the angle.

Case 2: The center of the circle is in the interior of the angle.

Case 3: The center lies in the exterior of the angle.

Proof of Theorem 9–4 (Case 1)

Given: $\angle PRQ$ inscribed in $\odot T$ and $\overline{PR}$ contains T.

Prove: $m\angle PRQ = \frac{1}{2}m\widehat{PQ}$

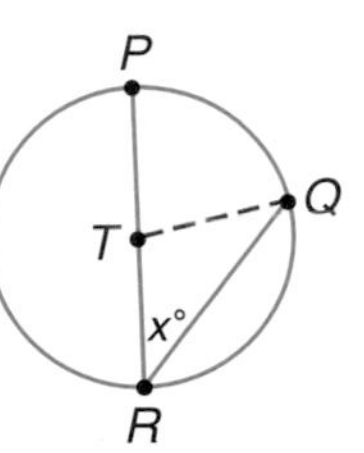

Paragraph Proof:
Draw radius $\overline{TQ}$ and let $m\angle PRQ = x$. Since $\angle PTQ$ is a central angle, $m\widehat{PQ} = m\angle PTQ$. Since $\overline{TQ}$ and $\overline{TR}$ are radii, $\triangle TQR$ is isosceles and $m\angle TQR = x$. By the Exterior Angle Theorem, $m\angle PTQ = 2x$. Therefore, $m\widehat{PQ} = 2x$ and $m\angle PRQ = \frac{1}{2}m\widehat{PQ}$.

You will be asked to prove Cases 2 and 3 of Theorem 9–4 in Exercises 50 and 51, respectively.

Example 1

In the circle at the right, $m\widehat{ST} = 68$. Find $m\angle 1$ and $m\angle 2$.

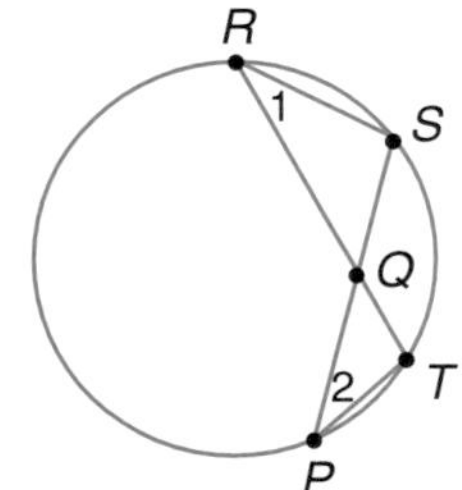

Notice that Q is not the center of the circle.

Since $\widehat{ST}$ is the intercepted arc for both $\angle 1$ and $\angle 2$, apply Theorem 9–4.

$m\angle 1 = \frac{1}{2}m\widehat{ST}$
$\quad = \frac{1}{2}(68)$ or 34

$m\angle 2 = \frac{1}{2}m\widehat{ST}$
$\quad = \frac{1}{2}(68)$ or 34

Therefore, $m\angle 1 = m\angle 2 = 34$.

Example 1 illustrates an important relationship between inscribed angles and intercepted arcs of equal measure.

Theorem 9–5	**If two inscribed angles of a circle or congruent circles intercept congruent arcs or the same arc, then the angles are congruent.**

Suppose an inscribed angle intercepts a semicircle on $\odot P$. The measure of semicircle FHG is 180, so $m\angle FEG = 90$. This is more formally stated in Theorem 9–6.

Notice that $\overline{FG}$ is a diameter of circle P.

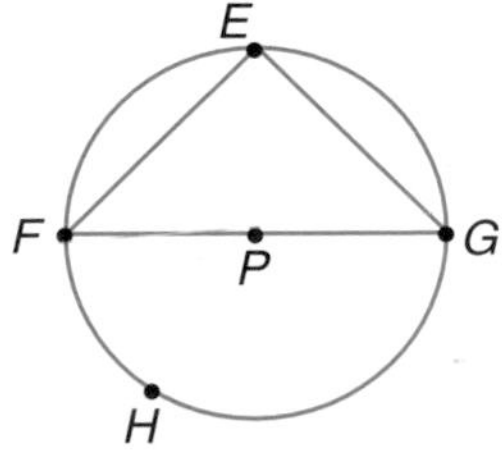

Theorem 9–6	**If an inscribed angle of a circle intercepts a semicircle, then the angle is a right angle.**

You can use this theorem to find the center of a circle given any three points on the circle.

CONSTRUCTION

Finding the Center of a Circle

Given three points on a circle, locate the center of the circle.

1. Draw a circle and locate three points A, B, and C on it.

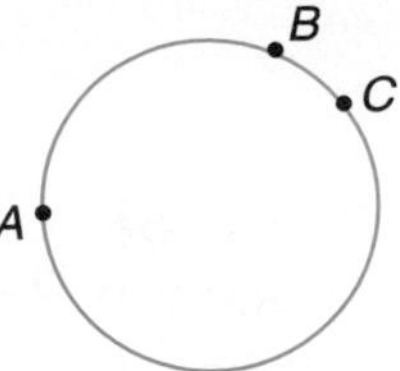

2. Draw $\overline{AB}$. Construct a segment through B perpendicular to $\overline{AB}$ and call it $\overline{BD}$.

   Draw $\overline{AD}$. Since $\overline{BD} \perp \overline{AB}$, $m\angle ABD = 90$ and the arc it intercepts has a measure of 180. This means that $\overline{AD}$ is a diameter.

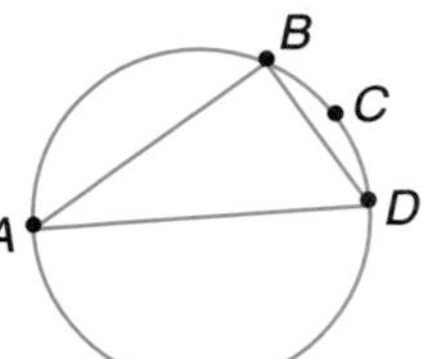

3. Draw $\overline{BC}$ and construct a segment through B perpendicular to $\overline{BC}$ and label it $\overline{BF}$. Draw $\overline{CF}$. Since $m\angle FBC = 90$, $\overline{CF}$ is also a diameter.

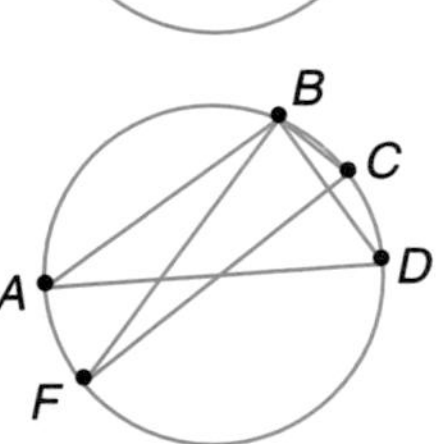

Conclusion: The point that all diameters have in common is the center of the circle. Thus, the intersection of $\overline{AD}$ and $\overline{CF}$ is the center of the circle.

You can also use Theorem 9–6 to find the measures of inscribed angles.

Example 2

Algebra

In $\odot A$, $m\angle 1 = 6x + 11$, $m\angle 2 = 9x + 19$, $m\angle 3 = 4y - 25$, $m\angle 4 = 3y - 9$, and $\widehat{PQ} \cong \widehat{RS}$. Find $m\angle 1$, $m\angle 2$, $m\angle 3$, and $m\angle 4$.

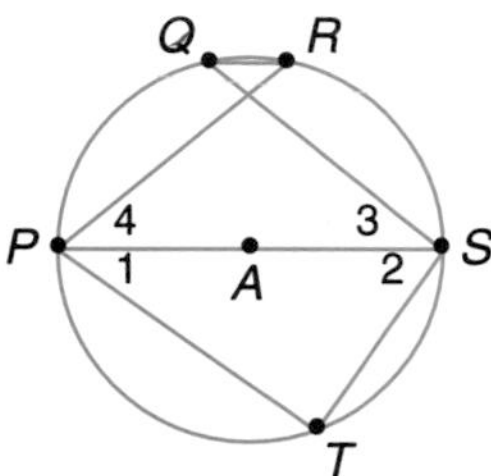

Since $\angle PTS$ is inscribed in a semicircle, by Theorem 9–6, $\angle T$ is a right angle. Therefore, $\triangle PTS$ is a right triangle, and $\angle 1$ and $\angle 2$ must be complementary.

$$m\angle 1 + m\angle 2 = 90 \quad \text{Definition of complementary angles}$$
$$6x + 11 + 9x + 19 = 90$$
$$15x + 30 = 90$$
$$15x = 60$$
$$x = 4$$

Therefore, $m\angle 1 = 6(4) + 11 = 35$ $\qquad$ $m\angle 2 = 9(4) + 19 = 55$

Since $\widehat{PQ} \cong \widehat{RS}$, $\angle 3$ and $\angle 4$ intercept congruent arcs. Thus, $m\angle 3 = m\angle 4$.

$$4y - 25 = 3y - 9$$
$$y = 16$$

Therefore, $m\angle 3 = 4(16) - 25$ or 39

Check: $m\angle 4 = 3(16) - 9$ or 39 ✔

Example 3

Quadrilateral $QRST$ is inscribed in $\odot C$. If $m\angle S = 28$ and $m\angle R = 110$, find $m\angle Q$ and $m\angle T$.

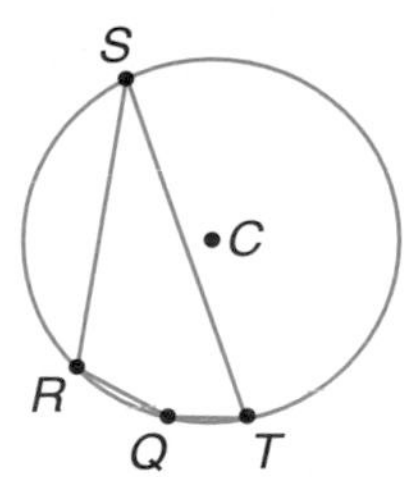

$m\angle S = \frac{1}{2}m\widehat{RQT}$ *Theorem 9–4*

$28 = \frac{1}{2}m\widehat{RQT}$ *$m\angle S = 28$*

$56 = m\widehat{RQT}$

$m\widehat{RST} = 360 - 56$ or 304

Since $\angle Q$ intercepts $\widehat{RST}$,

$m\angle Q = \frac{1}{2}m\widehat{RST}$

$= \frac{1}{2}(304)$ or 152

Use a similar strategy to find $m\angle T$.

$m\angle R = \frac{1}{2}m\widehat{STQ}$ *Theorem 9–4*

$110 = \frac{1}{2}m\widehat{STQ}$ *$m\angle R = 110$*

$220 = m\widehat{STQ}$

$m\widehat{QRS} = 360 - 220$ or 140

Since $\angle T$ intercepts $\widehat{QRS}$,

$m\angle T = \frac{1}{2}m\widehat{QRS}$

$= \frac{1}{2}(140)$ or 70

Notice that $\angle S$ and $\angle Q$ are supplementary and $\angle R$ and $\angle T$ are supplementary.

In Example 3, $\angle R$ and $\angle T$ are opposite angles and supplementary. $\angle Q$ and $\angle S$ are also opposite angles and supplementary. This leads to our next theorem.

Theorem 9–7	**If a quadrilateral is inscribed in a circle, then its opposite angles are supplementary.**

You will be asked to prove this theorem in Exercise 54.

CHECK FOR UNDERSTANDING

Communicating Mathematics

Study the lesson. Then complete the following.

1. **Explain** how an intercepted arc and an inscribed angle are related.
2. In Example 1, $\angle SQT$ appears to be obtuse, yet $m\widehat{ST} = 68$. Should $m\angle SQT$ be 68? Explain why or why not.
3. $\triangle ABC$ is inscribed in a circle so that $\overline{BC}$ is a diameter. What type of triangle is $\triangle ABC$? Explain your reasoning.
4. **Compare and contrast** inscribed angles and central angles of a circle. If they intercept the same arc, how are they related?

5. **Assess Yourself** Write about and describe three uses of hexagonal nuts, bolts or screws that appear at school or home. Why do you think the hexagon shape is used?

Guided Practice

In $\odot P$, $\overline{AC}$ is a diameter.

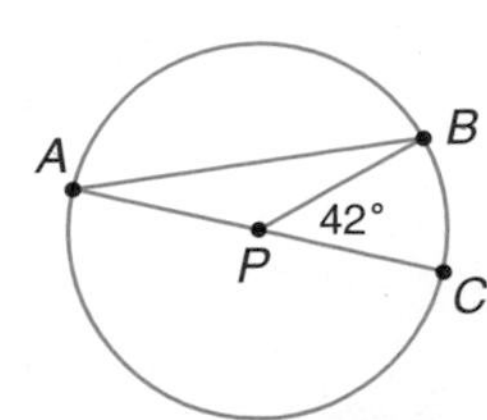

6. Name an inscribed angle.
7. Name the arc intercepted by $\angle BAC$.
8. If $m\angle BPC = 42$, find $m\angle BAC$.

9. Find x in $\odot N$ at the right.

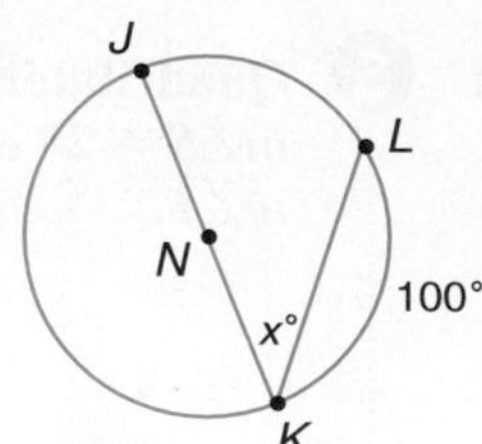

Pentagon *PENTA* is inscribed in $\odot G$. All sides of *PENTA* are congruent. Find each measure.

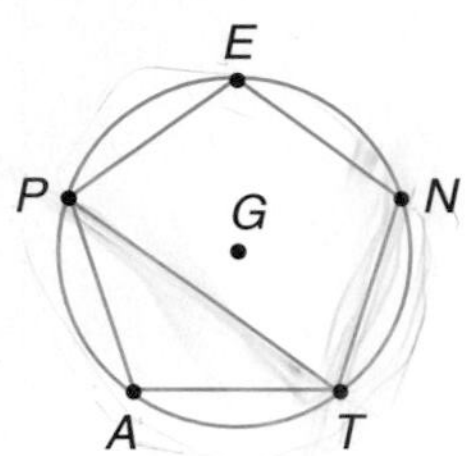

10. $m\widehat{PE}$
11. $m\angle PTN$
12. $m\angle PEN$

In $\odot Z$, $\overline{AB} \parallel \overline{DC}$, $m\widehat{BC} = 94$, and $m\angle AZB = 104$. Find each measure.

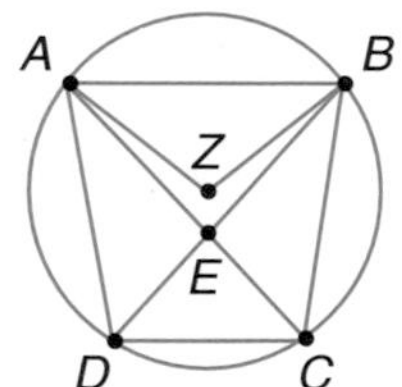

13. $m\widehat{AB}$
14. $m\angle BAC$
15. $m\angle ADB$

16. **Building Trades** Refer to the application at the beginning of the lesson.
    a. Given a round stock, how could a machinist determine the appropriate cuts to make in order to make a hexagonal shape with all sides congruent?
    b. What diameter of round stock should be used to cut a hex nut one-half inch on each side?

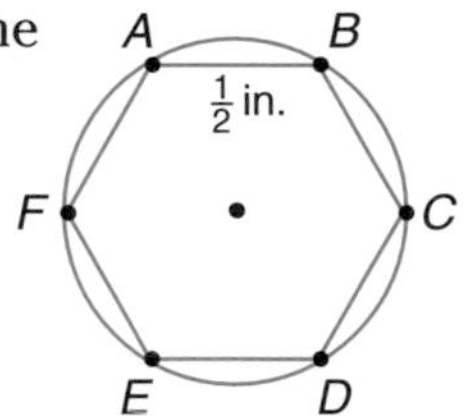

EXERCISES

Practice

In $\odot A$, $\overline{HE}$ is a diameter.

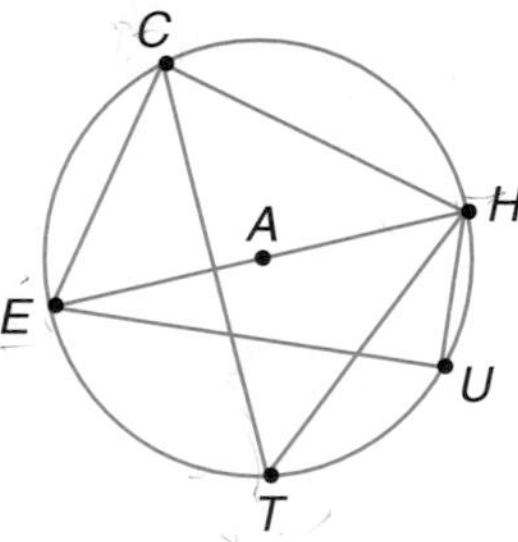

17. Name the intercepted arc for $\angle HTC$.
18. If $m\angle HTC = 52$, find $m\widehat{CH}$.
19. Name the intercepted arc for $\angle TCH$.
20. Find $m\angle ECH$.
21. If $m\angle HTC = 52$, find $m\angle CEH$.
22. Name an inscribed angle.
23. Name a central angle.
24. If $m\angle HTC = 52$, find $m\widehat{CEH}$.

Find the value of x.

25.

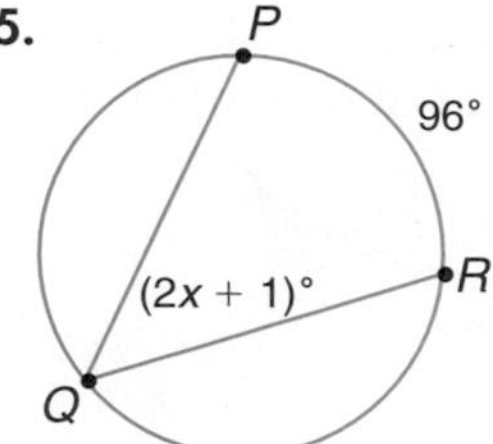

26.

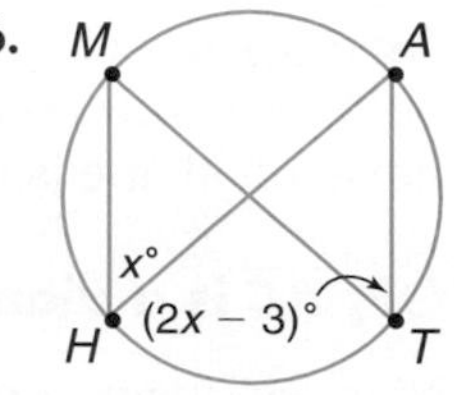

27. $\overline{KR}$ is a diameter o $\odot W$.

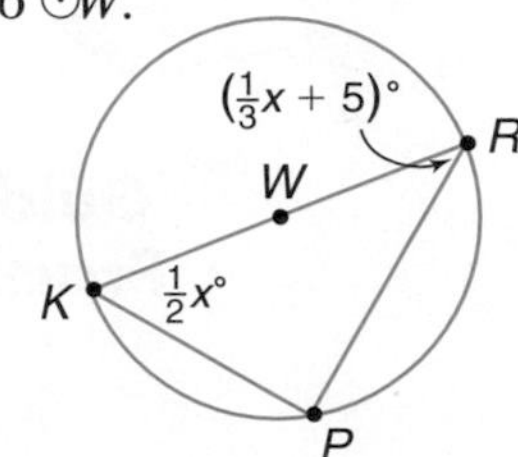

Quadrilateral *SPRT* is inscribed in ⊙*G*, and *PGTS* is a rhombus. If $m\angle PRS = -2x + 42$ and $m\angle SRT = 8x - 18$, find each measure.

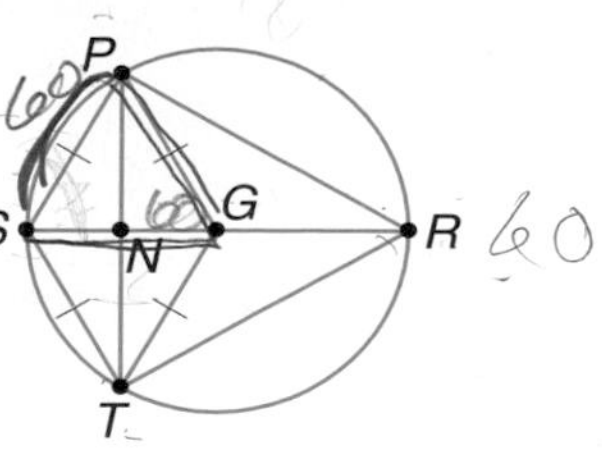

28. $m\overset{\frown}{PS}$
29. $m\overset{\frown}{ST}$
30. $m\angle RST$
31. $m\angle PGR$
32. $m\overset{\frown}{PR}$
33. $m\overset{\frown}{PSR}$
34. $m\angle PNG$
35. $m\overset{\frown}{PRT}$
36. $m\angle PSR$

In ⊙*P*, $\overset{\frown}{EN} = 66$, $m\angle GPM = 89$, and $\overline{GN}$ is a diameter. Find each measure.

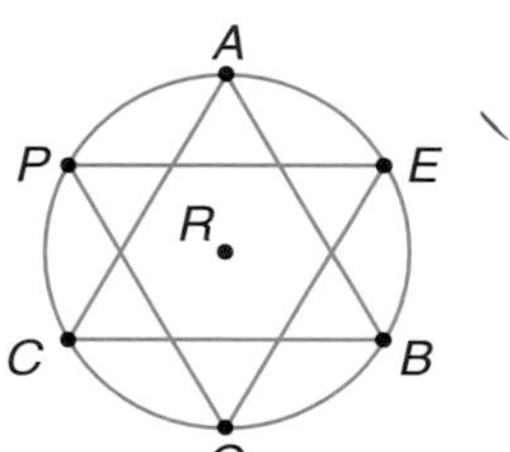

37. $m\overset{\frown}{GM}$
38. $m\overset{\frown}{NM}$
39. $m\overset{\frown}{GE}$
40. $m\angle GEN$
41. $m\angle EGN$
42. $m\angle EMN$
43. $m\angle GNM$
44. $m\angle GME$
45. $m\angle EGM$

46. Equilateral triangle *PEQ* is inscribed in circle *R*. Point *A* bisects $\overset{\frown}{PE}$, point *B* bisects $\overset{\frown}{EQ}$, and point *C* bisects $\overset{\frown}{PQ}$. What must be true about $\triangle ABC$? Explain.

Write a two-column proof.

47. **Given:** $\overline{BR} \parallel \overline{AC}$

Prove: $\overset{\frown}{RA} \cong \overset{\frown}{BC}$

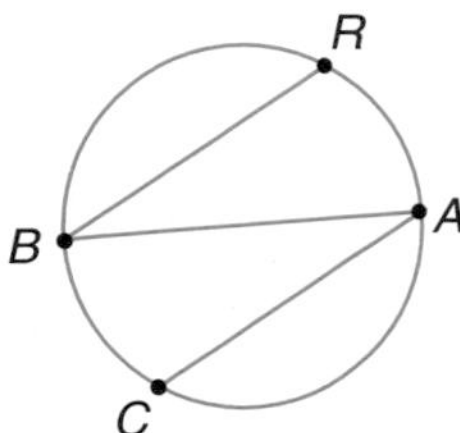

48. **Given:** $\overset{\frown}{MHT}$ is a semicircle.

$\overline{RH} \perp \overline{TM}$

Prove: $\frac{TR}{RH} = \frac{TH}{HM}$

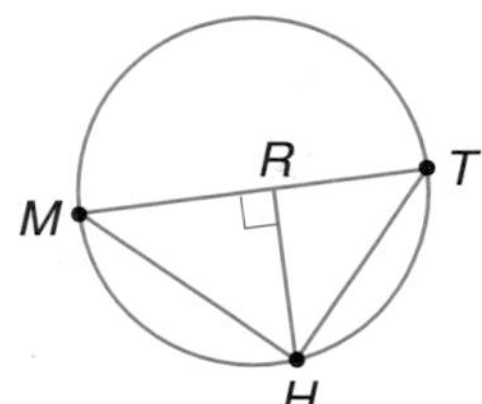

49. Is it possible for a parallelogram to be inscribed in a circle, yet not be a rectangle? Explain your reasoning.

Study the proof for Case 1 of Theorem 9-4. In Case 2 and Case 3, $\overline{PR}$ is not a diameter. To prove Cases 2 and 3, draw diameter $\overline{KR}$ and use Case 1 of Theorem 9-4, the Angle Addition Postulate, and the Arc Addition Postulate.

50. **Case 2**

Given: *T* lies inside $\angle PRQ$.

Prove: $m\angle PRQ = \frac{1}{2}m\overset{\frown}{PQ}$

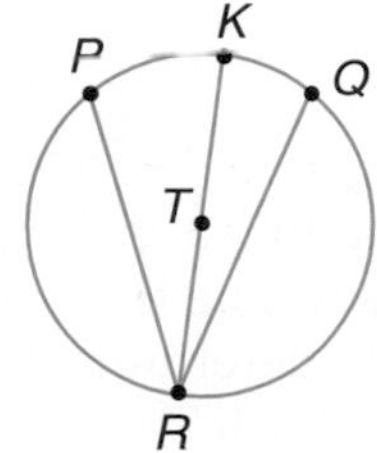

51. **Case 3**

Given: *T* lies outside $\angle PRQ$.

Prove: $m\angle PRQ = \frac{1}{2}m\overset{\frown}{PQ}$

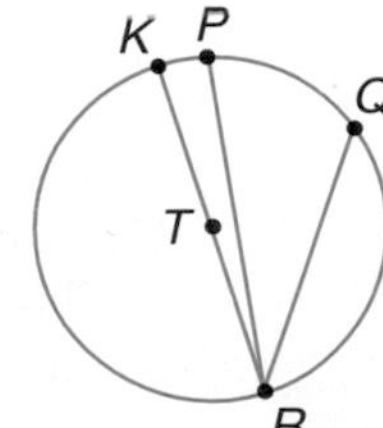

52. Write a two-column proof to show that if two inscribed angles of a circle intercept congruent arcs, then the angles are congruent. (Theorem 9–5)

53. Write a paragraph proof to show that if an inscribed angle of a circle intercepts a semicircle, then the angle is a right angle. (Theorem 9–6)

54. Write a paragraph proof to show that if a quadrilateral is inscribed in a circle, then the opposite angles of the quadrilateral are supplementary. (Theorem 9–7)

Critical Thinking

55. Can an isosceles trapezoid be inscribed in a circle? Write a brief paragraph explaining your reasoning.

Applications and Problem Solving

56. **Civil Engineering** A civil engineer uses the formula $\ell = \frac{2\pi rm}{360}$ to calculate the length ℓ of a curve for a road, given a radius r and a central angle measure m. Find the length of the road from A to B in the diagram at the right. Round to the nearest foot.

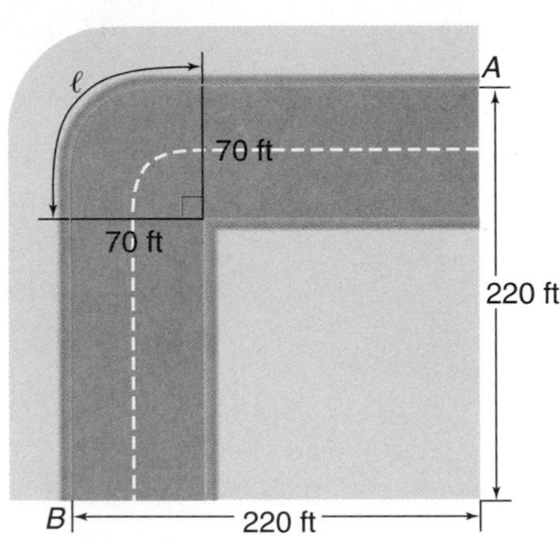

57. **Carpentry** A carpenter needs to find the center of a circular pattern of a parquet floor. How can the carpenter find the center of the circle using only a carpenter's square? Draw a diagram to explain your answer.

Mixed Review

58. **Food** The graph at the right shows where Americans eat take-out food. Find each measure. (Lesson 9–2)
   a. $m\angle ATD$
   b. major arc AB

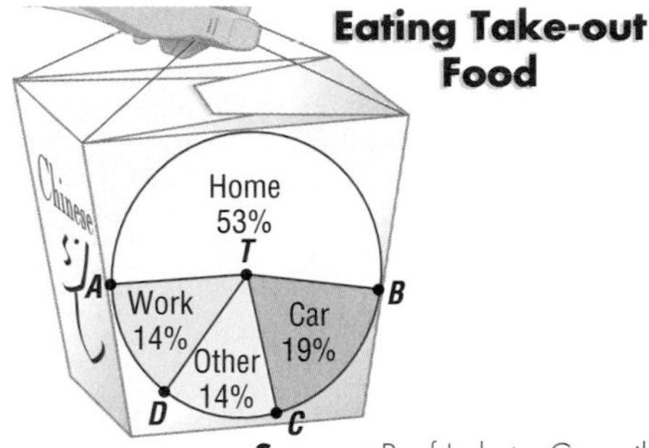

Source: Beef Industry Council

59. **Botany** *Fairy-ring mushrooms* form circles in grassy areas because their rootlike mycelia grow outward from a central point. If a ring of mushrooms is growing 8.5 feet from the central point, find the circumference of the mushroom circle. Round to the nearest tenth of a foot. (Lesson 9–1)

60. Find the geometric mean between 9 and 21. (Lesson 8–1)

61. Draw a segment that is 11 centimeters long. Then by construction, separate the segment into three congruent parts. (Lesson 7–4)

62. $JKLM$ is a rhombus with $m\angle 1 = 37$. Find $m\angle 2$. (Lesson 6–4)

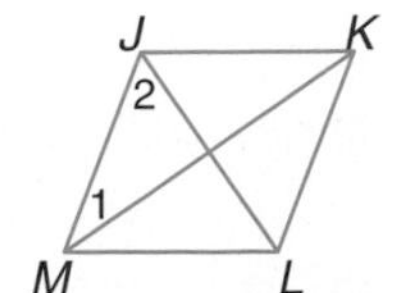

63. **Gardening** Sanequa is planting a rectangular garden. She has placed stakes at what she believes will be the four corners of the rectangle. Before she digs, what can she do to determine if the stakes are in the right places? (Lesson 6–3)

64. SAT Practice **Quantitative Comparison**

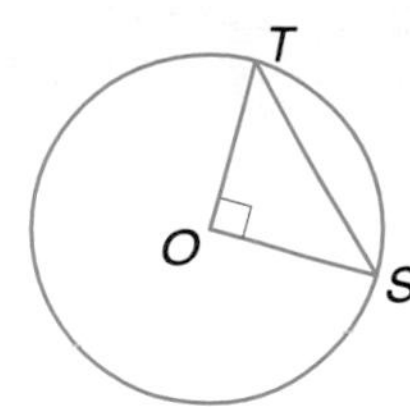

O is the center of a circle.

Column A	Column B
The measure of $\angle OST$	The measure of $\angle OTS$

A if the quantity in Column A is greater
B if the quantity in Column B is greater
C if the two quantities are equal
D if the relationship cannot be determined from the information given

65. Parallel lines ℓ and m are cut by a transversal t, and $\angle 1$ and $\angle 2$ are corresponding angles. If $m\angle 1 = 4x + 3$ and $m\angle 2 = 8x - 7$, what is the value of x? (Lesson 3–4)

66. Write a conjecture given points $A(-1, -3)$, $B(3, 0)$, and $C(3, -3)$. Draw a figure to illustrate your conjecture. (Lesson 2–1)

67. Given points $A(7, 4)$ and $B(-3, 1)$, find AB to the nearest tenth. (Lesson 1–4)

Algebra

68. Electronics The total resistance in ohms R_T of a certain conductor can be given by the equation $\frac{1}{R_T} = \frac{1}{4} + \frac{1}{5}$. Find R_T.

69. Use graphing to solve the system of equations. Check your solution algebraically.

$6a + 2b = 11$

$3a - 2b = 7$

For **Extra Practice**, see page 781.

SELF TEST

Refer to ⊙P. (Lessons 9–1 and 9–2)

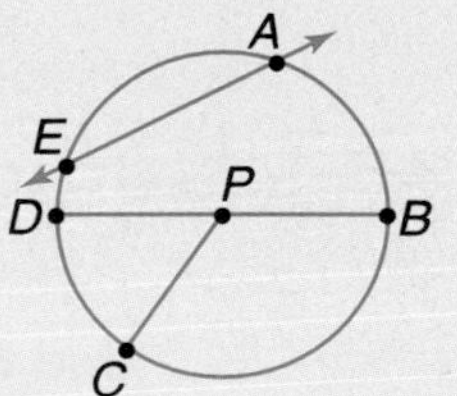

1. Name three radii of ⊙*P*.

2. If $PD = 17$, find the circumference of ⊙*P* to the nearest tenth.

3. If $m\angle CPB = 125$ and $PB = 6$, find the length of $\overset{\frown}{CB}$.

Refer to ⊙O. Find each measure.
(Lessons 9–3 and 9–4)

4. $m\overset{\frown}{ST}$

5. $m\angle TUV$

9–5A Using Technology Exploring Tangents

A Preview of Lesson 9–5

A line that intersects a circle in exactly one point is called a **tangent** to the circle. You can use a calculator to explore some of the characteristics of tangents. Use the following steps to draw two lines that are tangent to a circle.

Step 1 Draw a circle by pressing F3 and choosing 1:Circle from the menu. Use the arrow key to position the cursor anywhere on the screen to locate the center of the circle. Press ENTER. Label the center C. Move to another place on the screen to designate the radius of the circle. Press ENTER to complete the circle.

Step 2 Next, press F2 and choose 1:Point from the menu to draw a point outside the circle. Label the point A.

Step 3 Press F2 and choose Line. Move the cursor to point A so that the prompt THRU THIS POINT occurs. Press ENTER. Move the cursor towards the circle until the line appears to touch the circle at one point. At the prompt ON THIS CIRCLE, press ENTER. Label this point T.

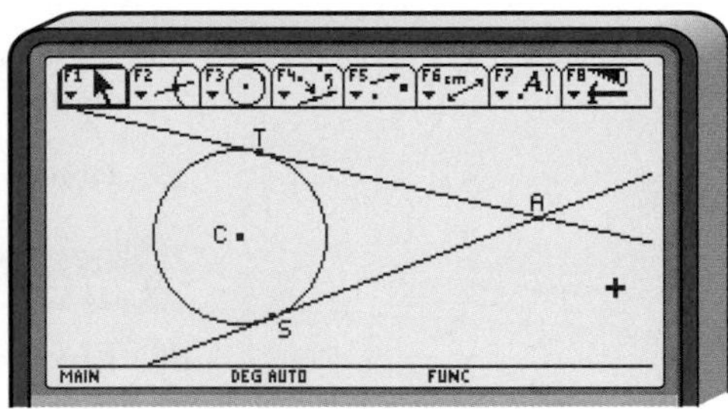

Step 4 Repeat the procedure in Step 3 to draw another line through point A that is tangent to $\odot C$ at point S.

The lines drawn to the circle are tangents to the circle. *Note that these tangents are approximate since it is difficult to find the exact point where the line touches the circle.*

EXERCISES

Use the measuring capabilities of the TI-92 to explore the characteristics of tangents.

LOOK BACK

Refer to Lesson 1-7A to review how to measure angles.

1. Measure $\overline{AT}$ and $\overline{AS}$.
2. Use the hand key to move point A closer to the circle. Make a conjecture about the measurements of $\overline{AT}$ and $\overline{AS}$.
3. Draw radii $\overline{CT}$ and $\overline{CS}$ by pressing F2 and choosing 5:Segment from the menu. Measure $\angle CTA$ and $\angle CSA$.
4. Make a conjecture about the angles formed by a radius and a tangent to a circle.

9–5

Tangents

Astronomy

What **YOU'LL LEARN**

- To recognize tangents and use properties of tangents.

Why **IT'S IMPORTANT**

You can use tangents to solve problems involving astronomy and aerospace.

A *solar eclipse* is an exciting natural phenomenon that occurs when the Moon passes in front of the Sun, blocking it from Earth. Some areas of the world experience a total eclipse, others, a partial eclipse, and some areas experience no eclipse at all.

In the diagram below, the Moon is between the Sun and Earth. The pink region indicates the portion of Earth that will experience a partial eclipse. The narrow blue region indicates a total eclipse. $\overline{AE}$ and $\overline{BF}$ are **tangent** to the circles that represent the Sun and Moon. A line is tangent to a circle if it intersects the circle in *exactly one* point. This point is called the **point of tangency**. *Parts of a tangent line may also be tangent to a circle.*

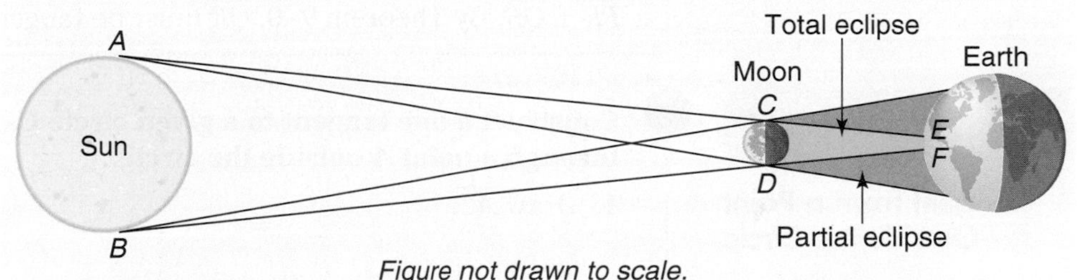

Figure not drawn to scale.

FYI

During the 20th century, 78 total solar eclipses occurred. Only 15 of these eclipses were visible from some part of the United States. The next total eclipse visible from the United States will be in 2017.

A circle separates a plane into three parts. The parts are the **interior**, the **exterior**, and the circle itself. In the figure at the right, T is a point of tangency, X is in the exterior of $\odot P$, and I is in the interior of $\odot P$. $\overline{PT}$ is a radius. Thus, $PX > PT$ and $PI < PT$.

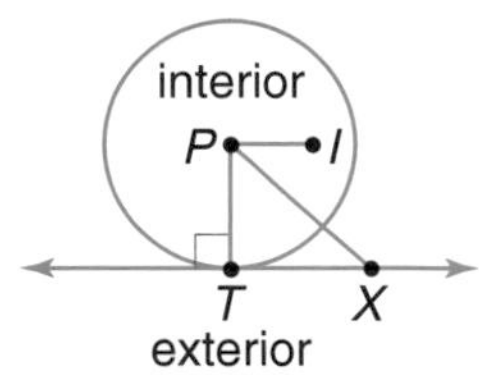

The shortest segment from a point to a line is a perpendicular segment. Thus, $\overline{PT} \perp \overleftrightarrow{TX}$. This leads to Theorem 9–8.

Theorem 9–8	**If a line is tangent to a circle, then it is perpendicular to the radius drawn to the point of tangency.**

You will be asked to prove this theorem in Exercise 47.

Example 1

Algebra

Refer to $\odot C$ with tangent $\overline{AB}$. Find x.

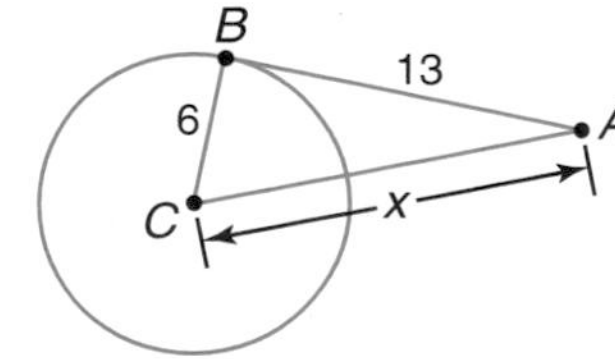

Since $\overline{AB}$ is tangent to $\odot C$, $\overline{AB} \perp \overline{BC}$ by Theorem 9–8. Therefore, $\triangle ABC$ is a right triangle.

By the Pythagorean Theorem,

$$(AB)^2 + (BC)^2 = (AC)^2$$
$$13^2 + 6^2 = x^2$$
$$169 + 36 = x^2$$
$$x^2 = 205$$
$$x = \sqrt{205} \text{ or about } 14.3$$

The converse of Theorem 9–8 is also true and can be used to identify tangents to a circle. *You will prove this theorem in Exercise 46.*

Theorem 9–9 (Converse of Theorem 9–8)	**In a plane, if a line is perpendicular to a radius of a circle at the endpoint on the circle, then the line is a tangent of the circle.**

Example  **Refer to $\odot P$ with radius $\overline{PR}$. Show that $\overline{QR}$ is tangent to $\odot P$.**

Algebra

Since $\overline{PT}$ is also a radius of $\odot P$, then $PT = 5$.

$$PQ = PT + TQ$$
$$= 5 + 8 \text{ or } 13$$

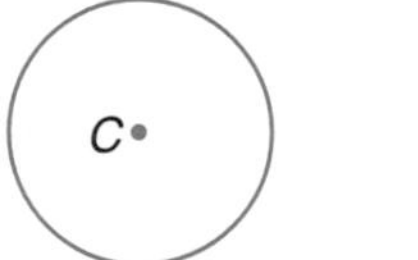

Notice that in $\triangle PQR$, $(PR)^2 + (QR)^2 = (PQ)^2$.

$$5^2 + 12^2 = 13^2$$
$$25 + 144 = 169$$
$$169 = 169$$

By the converse of the Pythagorean Theorem, $\triangle PQR$ is a right triangle with $\overline{PR} \perp \overline{QR}$. By Theorem 9–9, $\overline{QR}$ must be tangent to $\odot P$.

CONSTRUCTION

Tangent from a Point Outside the Circle

Construct a line tangent to a given circle C through a point A outside the circle.

1. Draw $\overline{AC}$.

C
A

2. Construct the perpendicular bisector of $\overline{AC}$. Call this line ℓ. Call X the intersection of ℓ and $\overline{AC}$.
3. Using X as the center, draw a circle with radius measuring XC. Call D and E the intersection points of the two circles.
4. Draw $\overline{AD}$. Then $\overline{AD}$ is tangent to $\odot C$.

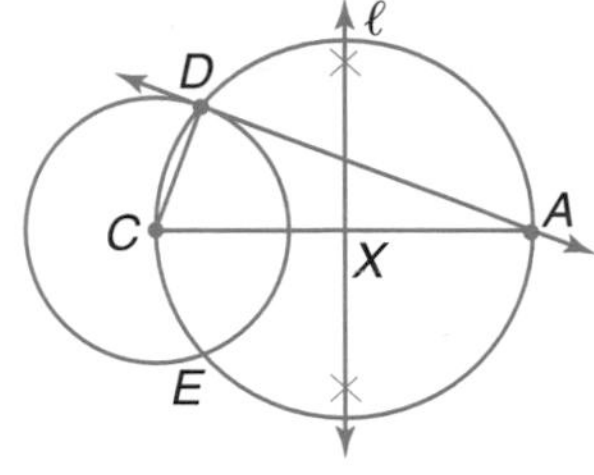

Conclusion: $\overline{AD}$ is a tangent to $\odot C$ if $\overline{AD} \perp \overline{DC}$. How do you know $\angle CDA$ is a right angle?

A line or line segment that is tangent to two circles in the same plane is called a **common tangent** of the two circles. There are two types of common tangents.

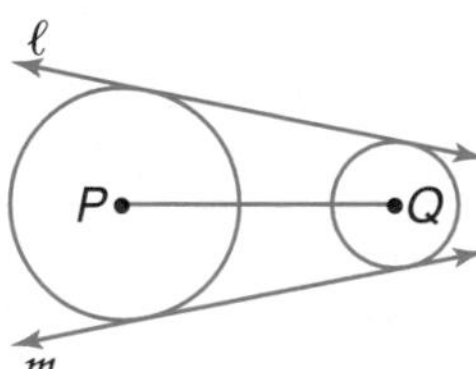

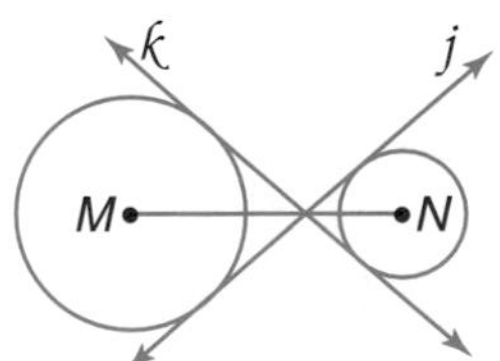

Common external tangents do not intersect the segment whose endpoints are the centers of the circles. In the figure above, lines ℓ and m are common external tangents to $\odot P$ and $\odot Q$.

Common internal tangents intersect the segment whose endpoints are the centers of the circles. In the figure above, lines j and k are common internal tangents to $\odot M$ and $\odot N$.

In the solar eclipse model, $\overline{EC}$ and $\overline{ED}$ are examples of two **tangent segments** drawn from a common point E outside the circle (the Moon). An important relationship exists between $\overline{EC}$ and $\overline{ED}$.

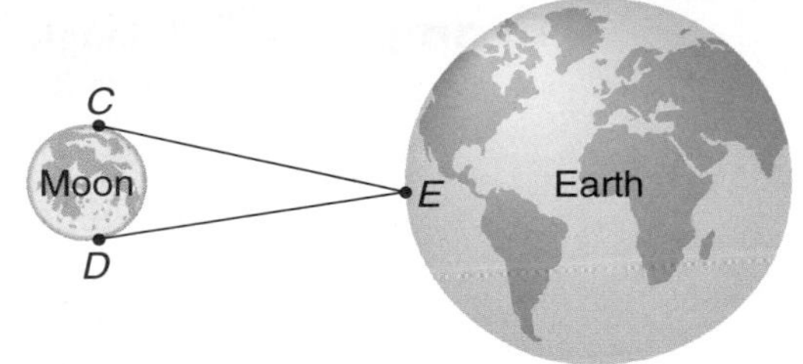

In the figure at the right, note that $\overrightarrow{PQ}$ and $\overrightarrow{PR}$ are both tangent to $\odot C$. Also, $\overline{PQ}$ and $\overline{PR}$ are tangent segments. By drawing $\overline{PC}$, $\overline{CR}$, and $\overline{CQ}$, two right triangles are formed. These triangles are congruent by HL, which leads us to the following theorem. *You will be asked to prove this theorem in Exercise 48.*

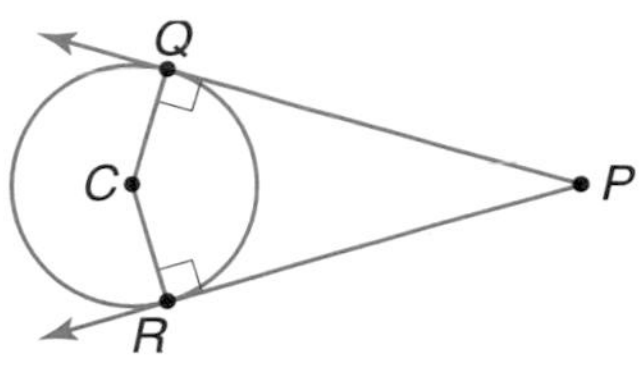

Theorem 9–10	**If two segments from the same exterior point are tangent to a circle, then they are congruent.**

We have worked with problems involving polygons inscribed in circles. We can use Theorem 9–10 to solve problems involving **circumscribed polygons**. A polygon is circumscribed *about* a circle if each side of the polygon is tangent to the circle. The following two statements are equivalent.

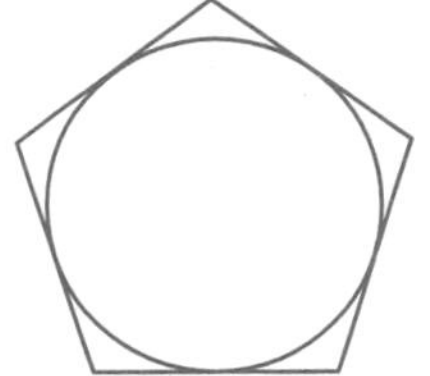

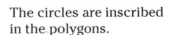

The polygons are circumscribed about the circles.

The circles are inscribed in the polygons.

CONSTRUCTION

Circle Inscribed in a Polygon

Construct a circle inscribed in a given triangle ABC.

1. Construct the angle bisectors of $\angle A$ and $\angle C$. Extend the bisectors to meet at point X.
2. Construct a line from X perpendicular to $\overline{AC}$. Label the intersection of the perpendicular line and $\overline{AC}$, Y.
3. Setting the compass length equal to XY, draw $\odot X$.

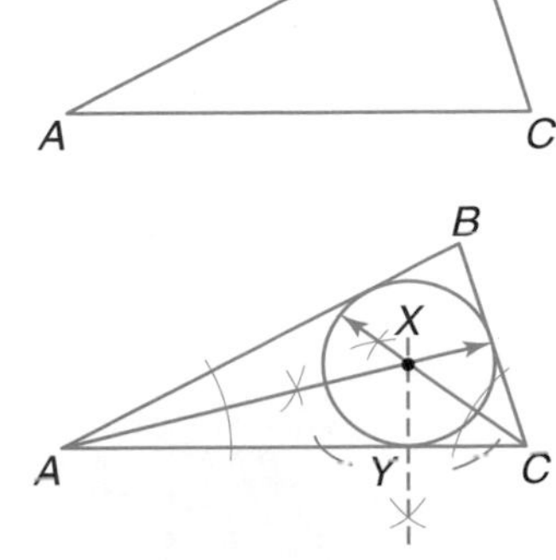

Conclusion: $\odot X$ is inscribed in $\triangle ABC$. $\overline{AB}$, $\overline{BC}$, and $\overline{AC}$ are tangent to $\odot X$.

You can use Theorem 9–10 to solve problems involving circumscribed polygons.

Example 3 **Triangle TRW is circumscribed about $\odot A$. If the perimeter of $\triangle TRW$ is 50, $TK = 3$, and $WM = 9.5$, find TR.**

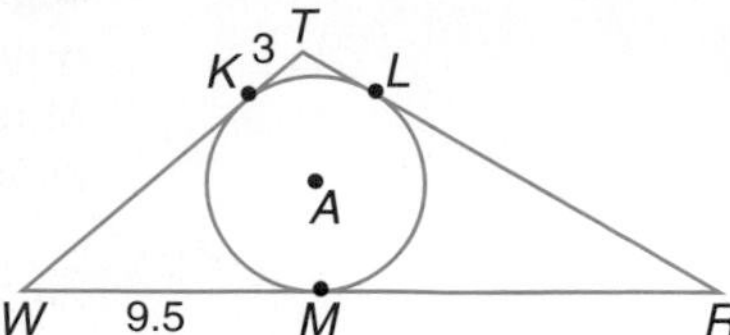

Since $\odot A$ is inscribed in $\triangle TRW$ or circumscribed by $\triangle TRW$, K, L, and M are points of tangency. Thus, by Theorem 9–10, $WK = WM = 9.5$, $TK = TL = 3$, and $LR = RM$.

To find TR, let $\overline{LR}$ and $\overline{RM}$ have length x and use the known perimeter.

$$P = WK + KT + TL + LR + RM + WM \quad \textit{Perimeter of } \triangle TRW$$
$$50 = 9.5 + 3 + 3 + x + x + 9.5$$
$$50 = 25 + 2x$$
$$25 = 2x$$
$$12.5 = x$$

The measures of $\overline{LR}$ and $\overline{RM}$ are each 12.5.

$$TR = TL + LR \quad \textit{Segment Addition Postulate}$$
$$= 3 + 12.5$$
$$= 15.5$$

Therefore, the measure of $\overline{TR}$ is 15.5.

CHECK FOR UNDERSTANDING

Communicating Mathematics

Study the lesson. Then complete the following.

1. Refer to the diagram of a solar eclipse in the application at the beginning of the lesson. Name a pair of common external tangents and a pair of common internal tangents.
2. **Draw** two circles having two common external tangents and one common internal tangent.
3. **State** the number of tangents that can be drawn to a circle for each case. Explain your answers.
   a. through a point outside the circle
   b. through a point inside the circle
   c. through a point on the circle
4. **You Decide** Sang Hee claims that two lines tangent to a circle must always intersect. Anjula says that she thinks it's possible for two lines tangent to a circle to be parallel. Who do you think is right? Explain. Draw a diagram to support your answer.

5. Draw circle P with radius $\overline{PQ}$ on patty paper. Fold the paper so the crease goes through Q and no other point of the circle. Put two points, R and S, on the line. Fold the paper along $\overline{PQ}$.
   a. What is the relationship of $\overline{PQ}$ to $\overleftrightarrow{RS}$?
   b. Draw a different radius and repeat the activity. What is true of any radius and the tangent through the endpoint of the radius?

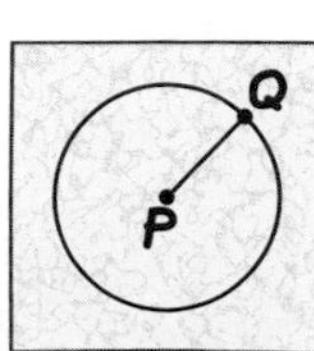

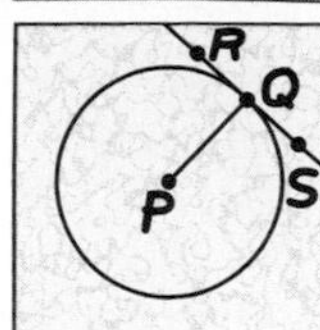

Guided Practice

Refer to ⊙M for Exercises 6–13. $\overline{LK}$ and $\overline{LE}$ are tangent to ⊙M, m∠EML = 66, KM = 15, and LK = 36. Find each measure.

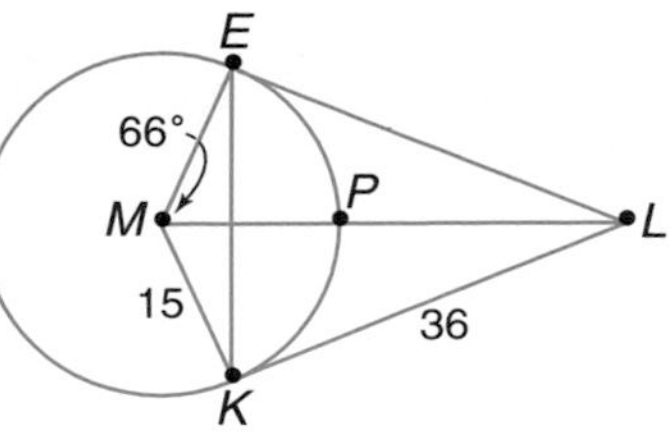

6. $m\angle MKL$
7. $m\angle ELM$
8. EL
9. $m\widehat{KPE}$

Complete each statement.

10. $\triangle KML \cong$ __?__
11. $\overline{KM} \perp$ __?__
12. $\triangle KLE$ is a(n) __?__ triangle.
13. Explain why $\angle EMK$ and $\angle ELK$ are supplementary.

For each ⊙C, find the value of *x*. Assume that segments that appear to be tangent are tangent.

14.

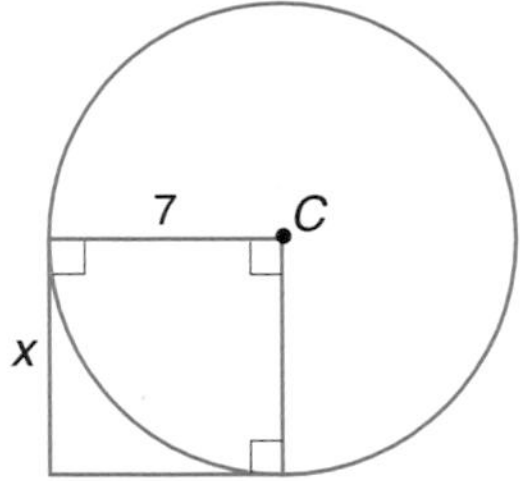

15.

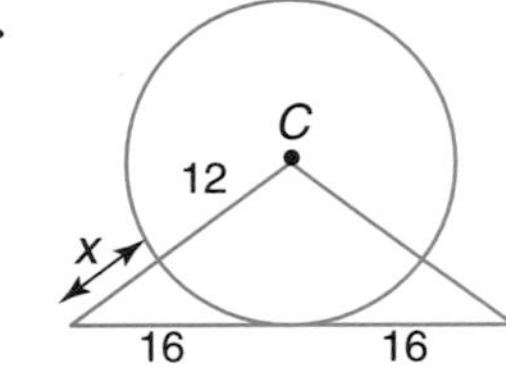

16. **Aerospace** A spacecraft is 4000 kilometers above Earth's surface. If the radius of Earth is about 6400 kilometers, find the distance between the spacecraft and the horizon. Round to the nearest tenth of a kilometer.

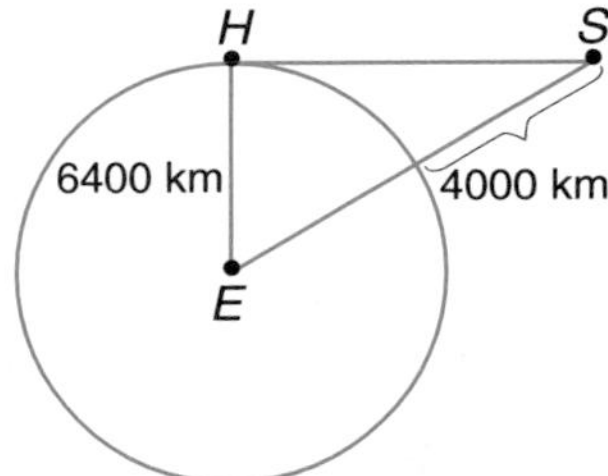

EXERCISES

Practice

Refer to the figure below for Exercises 17–30. $\overline{HL}$, $\overline{GF}$, and $\overline{AJ}$ are tangent to both ⊙P and ⊙Q, m∠APC = 60, PC = 5, and QE = 2. Find each measure.

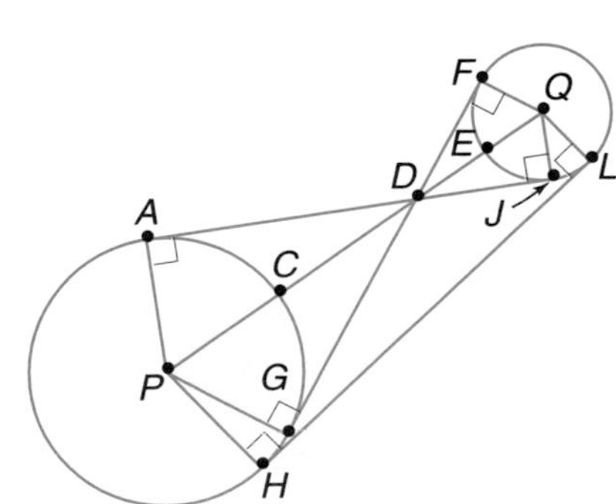

17. AD
18. $m\angle DAP$
19. PD
20. DF
21. $m\widehat{CA}$
22. $m\angle ADP$
23. $m\angle QJD$
24. DQ
25. DJ
26. $m\widehat{FEJ}$
27. $m\angle QDJ$
28. AJ
29. Name two common internal tangents to ⊙P and ⊙Q.
30. Explain why $\overline{GF} \cong \overline{AJ}$.

For each $\odot Q$, find the value of x. Assume that segments that appear to be tangent are tangent.

31.

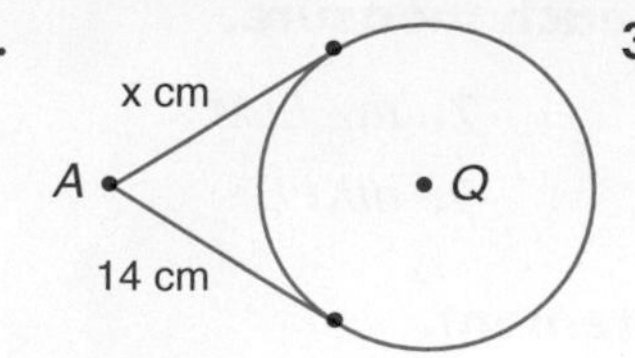

32.

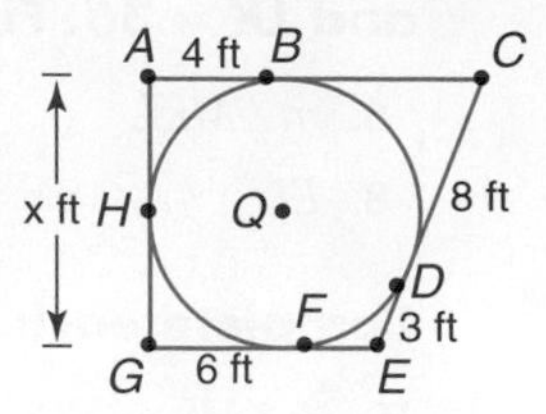

33.

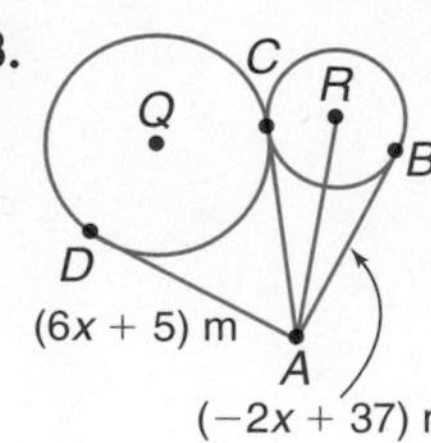

34.

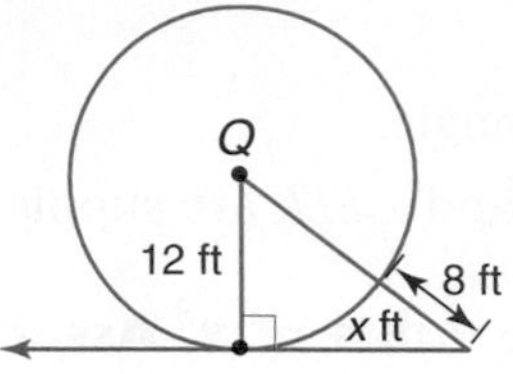

35.

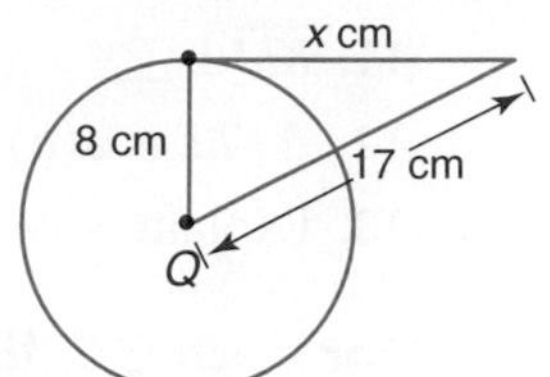

36.

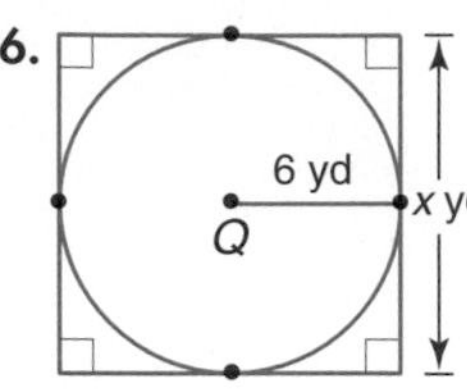

Proof

37. Complete the following proof.

Given: $\overline{AB}$ and $\overline{AR}$ are tangent to $\odot K$.

Prove: $\angle BAK \cong \angle RAK$

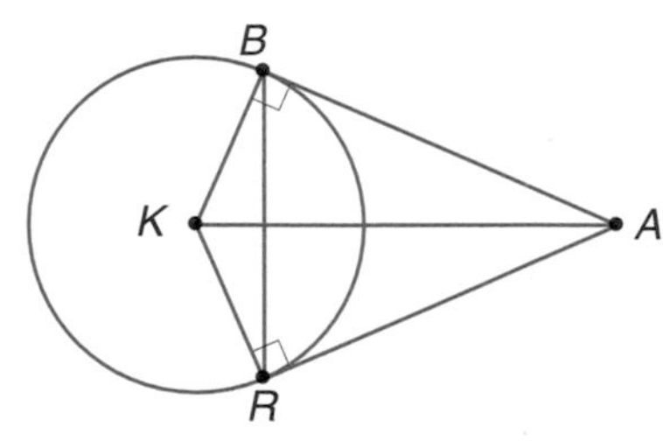

Proof:

Statements	Reasons
a. $\overline{AB}$ and $\overline{AR}$ are tangent to $\odot K$.	a. ?
b. $\overline{AB} \cong \overline{AR}$	b. ?
c. $\overline{BK} \cong \overline{RK}$	c. ?
d. $\overline{AK} \cong \overline{AK}$	d. ?
e. $\triangle BAK \cong \triangle RAK$	e. ?
f. $\angle BAK \cong \angle RAK$	f. ?

Algebra

Line ℓ is tangent to $\odot A$ at B.

38. Find the radius of $\odot A$.

39. Find the slope of line ℓ.

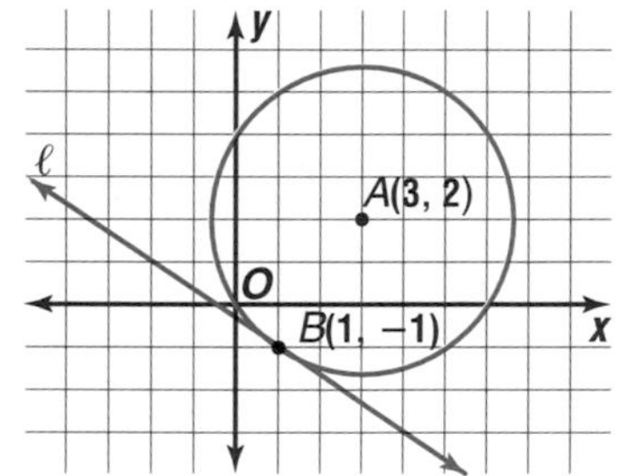

40. $\odot B$ is inscribed in quadrilateral $TANG$, $GN = 10$, $TG = 12$, and $AT = 18$. If the point of tangency bisects $\overline{AT}$, find AN.

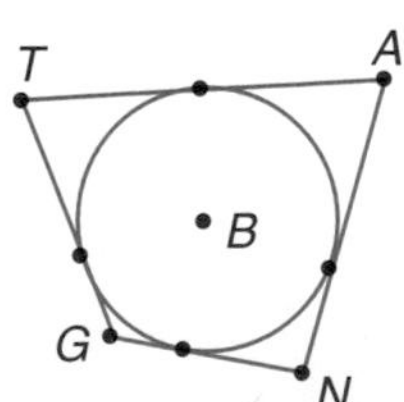

41. Draw a circle. Label the center P. Locate a point on the circle. Label it A. Construct a tangent to $\odot P$ at A.

42. Draw a circle. Label the center Q. Draw a point exterior to the circle. Label it B. Construct a tangent to $\odot Q$ containing B.

43. $\odot P$ is inscribed in hexagon $QRSTUV$. If the radius of $\odot P$ is 8 and all sides of $QRSTUV$ are congruent, find QN. Explain your reasoning.

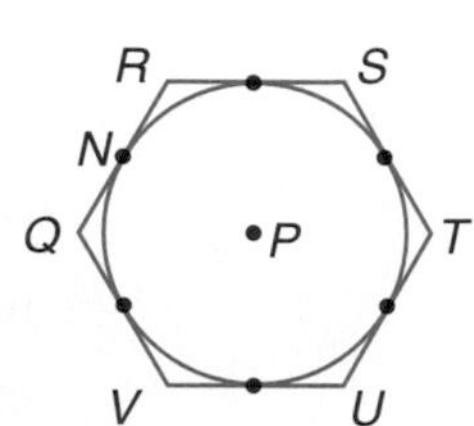

44. $\odot P$ is inscribed in right $\triangle CTA$. Find the perimeter of $\triangle CTA$ if the radius of $\odot P$ is 5 and $CT = 18$.

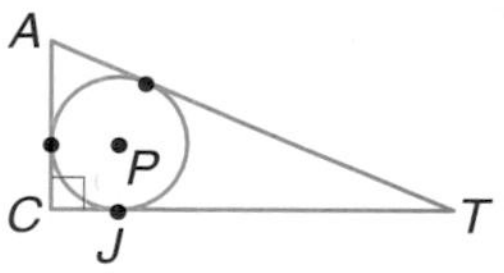

Proof

45. $\overline{GR}$ is tangent to $\odot D$ at G, and $\overline{AG} \cong \overline{DG}$. Write a paragraph proof to show that $\overline{GA}$ bisects $\overline{RD}$.

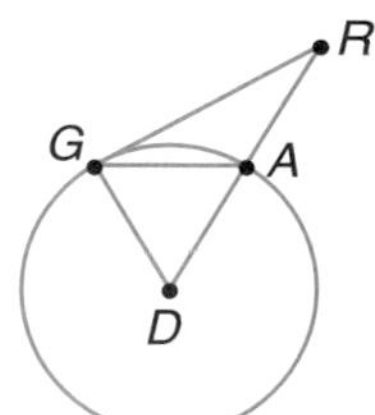

46. Write an indirect proof.

Given: $\ell \perp \overline{AB}$
$\overline{AB}$ is a radius of $\odot A$.

Prove: ℓ is tangent to $\odot A$. (Theorem 9–9)

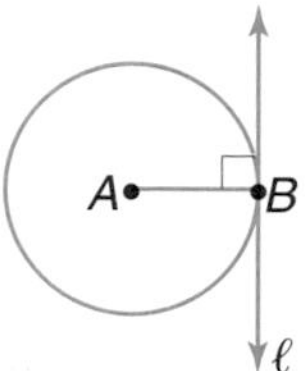

Write a two-column proof.

47. If a line is tangent to a circle, then it is perpendicular to the radius drawn to the point of tangency. (Theorem 9–8)

48. If two segments from the same exterior point are tangent to a circle, then they are congruent. (Theorem 9–10)

Critical Thinking

49. A *unit circle* is a circle with a radius of 1. In the figure, $\odot O$ is a unit circle with $\overline{QR}$ tangent to $\odot O$ at R.

a. Use the right triangle trigonometry definition of tangent on page 412 to find $\tan \theta$.

b. How do these two different definitions of tangent compare?

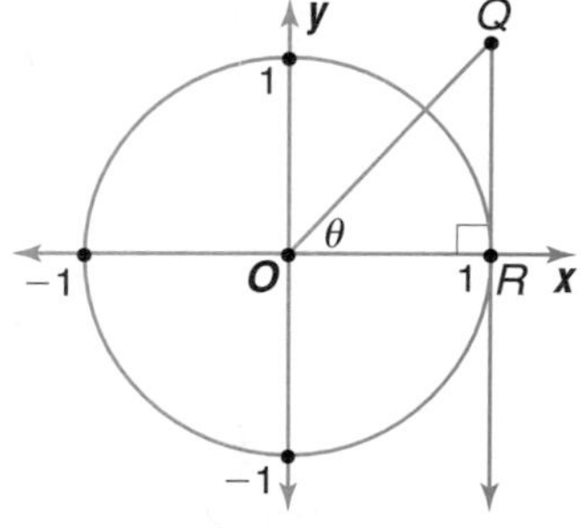

Applications and Problem Solving

50. **Astronomy** During an eclipse, the darkest portion of the shadow cast by the Moon is called the *umbra*. A *total* eclipse occurs when the umbra hits Earth's surface. The distance from S to E is about 93,000,000 miles. Round each answer to the nearest mile.

Celestial Body	Radius (miles)
Earth	3964
Moon	1080
Sun	432,000

a. Find RT.

b. Use similar triangles to find TN.

c. Use $\overline{TN}$ and right $\triangle MNT$ to find MT.

d. Find OE.

e. Use your answer in part d to approximate how far apart Earth and the Moon must be during an eclipse in order to experience a total eclipse with the umbra casting a shadow on part of Earth.

51. **Olympics** At the 1996 Summer Olympics, Anthony Washington of Aurora, Colorado, finished fourth in the discus event with a throw of 65.42 meters. Suppose he wound up in a circular pattern to throw the discus along a tangent line (path) to the circle. Use the figure to find the radius of the circle.

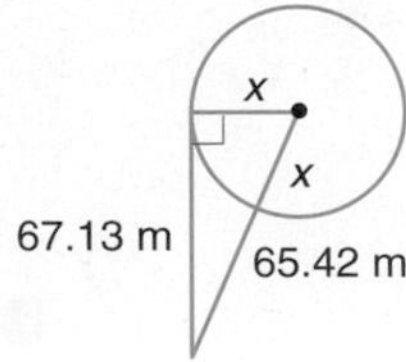

Mixed Review

52. Use the figure below to find $m\widehat{TZ}$. (Lesson 9–4)

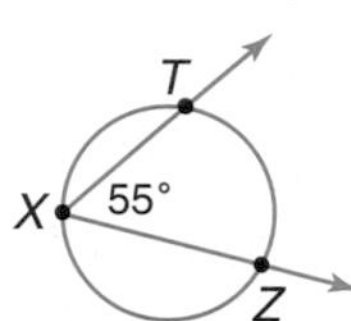

53. In ⊙O, find AC. (Lesson 9–3)

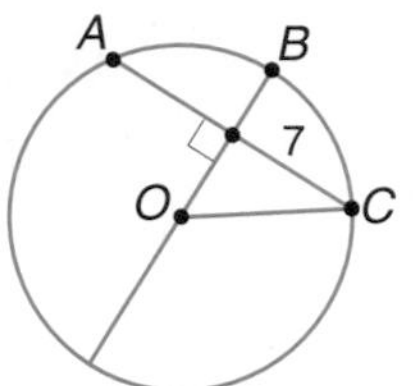

54. 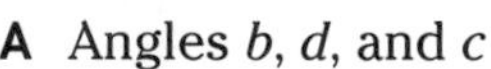 In the figure, line L is parallel to line M. Line N intersects both L and M, with angles a, b, c, d, e, f, g, and h, as shown. Which of the following lists includes all of the angles that are supplementary to $\angle a$?

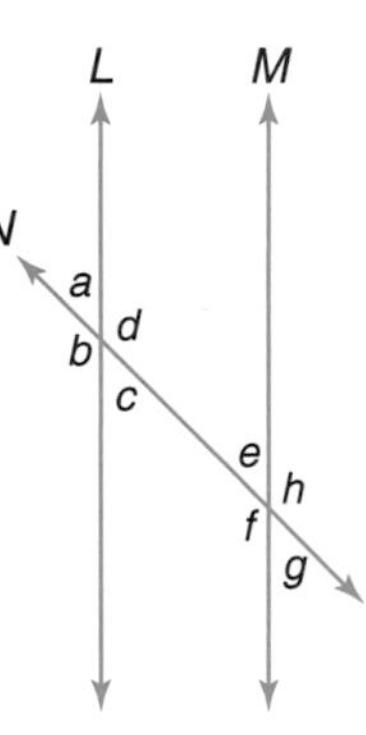

A Angles b, d, and c
B Angles b, d, f, and h
C Angles c, e, and g
D Angles d, c, h, and g
E Angles e, f, g, and h

55. Two triangles are similar, and their corresponding sides are in a ratio of 3:5. What is the measure of a side in the second triangle that corresponds to a side that measures 15 centimeters in the first triangle? (Lesson 7–3)

56. Construct a rectangle with length of 6 centimeters and width of 5 centimeters. (Lesson 6–3)

57. Write an inequality to describe the possible values of x. (Lesson 5–6)

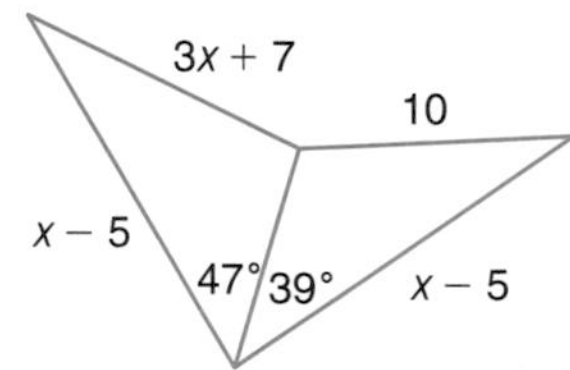

58. Can a triangle have two obtuse angles? Justify your answer. (Lesson 4–2)

59. **Biology** Determine if a valid conclusion can be reached from the two true statements using the Law of Detachment or the Law of Syllogism. If a valid conclusion is possible, state it and the law that is used. If a valid conclusion does not follow, write *no conclusion*. (Lesson 2–3)

(1) All species in the plant family *Maalvaceae* have five petals.
(2) A wild rose has five petals.

60. Draw opposite rays $\overrightarrow{QS}$ and $\overrightarrow{QR}$. (Lesson 1–6)

Algebra

Factor each polynomial.

61. $x^2 + 16x + 64$

62. $9p^2 - 56p - 49$

For **Extra Practice**, see page 781.

9–6 Secants, Tangents, and Angle Measures

What YOU'LL LEARN

- To find the measures of angles formed by intersecting secants and tangents in relation to intercepted arcs.

Why IT'S IMPORTANT

A knowledge of secants, tangents, and angle measures will help you solve problems involving civil engineering, astronomy, and meteorology.

Civil Engineering

The *Luis I* bridge in Oporto, Portugal, was designed by T. Seyrig and completed in 1885. He worked closely with Gustav Eiffel (Eiffel tower) who built a railroad bridge of similar design in Gazabit, France.

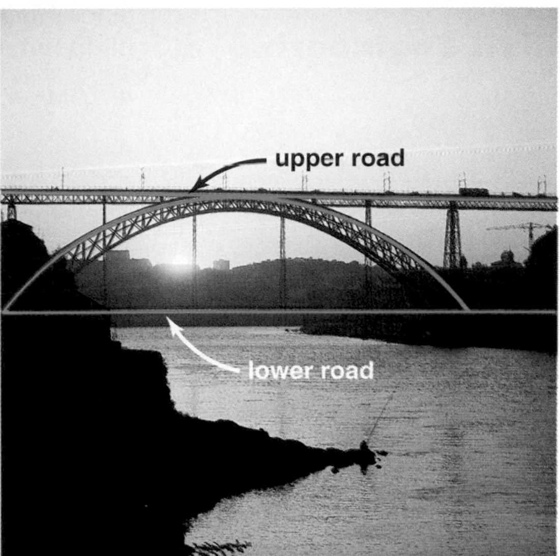

The roads represent the two different ways that lines or segments can intersect a circle.

- The upper road appears to touch the circular support arch.
- The lower road appears to intersect the circular support arch in two places.

In Lesson 9–5, you learned that a line that touches or intersects a circle in exactly one point is a tangent. A line that intersects a circle in exactly *two* points is called a **secant** of the circle. A secant of a circle contains a chord of the circle.

You have calculated the measures of angles in circles whose vertices are either on the circle or are the center of the circle. You can also find the measures of angles formed by secants and tangents.

The *Luis I* bridge provides an example of an angle whose vertex is located *on* the circle at the point where tangent $\overleftrightarrow{TC}$ intersects the circle. $\overrightarrow{CA}$ is a secant ray with $\overline{CA}$ representing the chord from the point of tangency of the arch to the lower foot of the arch. Angle TCA intercepts $\overset{\frown}{CA}$ like the inscribed angle relationship studied in Lesson 9–4. The measure of $\angle TCA$ is one-half the measure of the intercepted arc or $\frac{1}{2}m\overset{\frown}{CA}$.

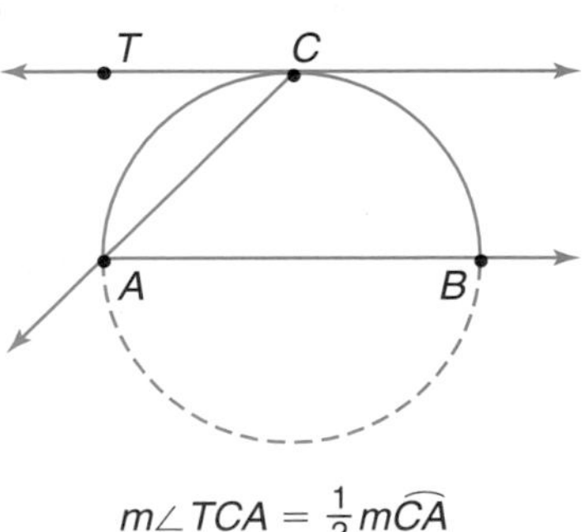

$m\angle TCA = \frac{1}{2}m\overset{\frown}{CA}$

Theorem 9–11	If a secant and a tangent intersect at the point of tangency, then the measure of each angle formed is one-half the measure of its intercepted arc.

You will be asked to prove this theorem in Exercise 47.

Example 1

Civil Engineering

The Darby bridge was built from 1777–1779 near Coalbrookdale, London. If the arch of the bridge is a semicircle and *M*, the midpoint of the semicircle, is the point of tangency, find $m\angle AMP$.

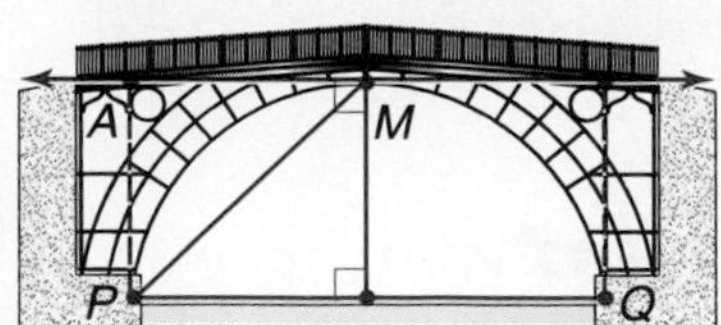

Since $\widehat{PMQ}$ is a semicircle with M as the midpoint, $m\widehat{PM} = 90$. By Theorem 9–11,

$$m\angle AMP = \frac{1}{2}m\widehat{PM}$$
$$= \frac{1}{2}(90) \text{ or } 45$$

Therefore, $m\angle AMP$ is 45.

There is also a special relationship between the angles formed by two secants and the arcs they intercept.

MODELING MATHEMATICS

Angle Vertex in the Interior of a Circle

Materials: compass  straightedge

- Draw a circle and choose point V in the interior of the circle.

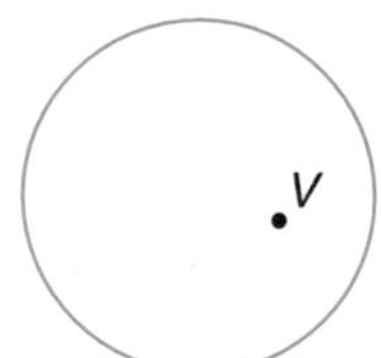

- Draw two secants intersecting at V. Label W, X, Y, Z, and $\angle 1$ as shown.

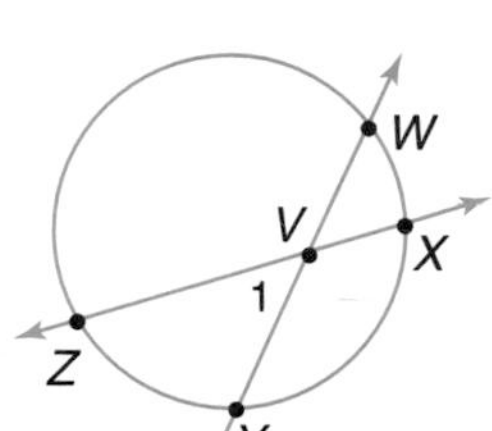

- Draw $\overline{XY}$ and label $\angle 2$ and $\angle 3$. Note that $m\angle 2 = \frac{1}{2}m\widehat{WX}$ and $m\angle 3 = \frac{1}{2}m\widehat{ZY}$, since $\angle 2$ and $\angle 3$ are inscribed angles.

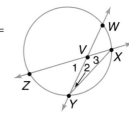

Your Turn

a. Explain why $m\angle 1 = m\angle 2 + m\angle 3$.

b. Use part a to write an equation for $m\angle 1$ in terms of the measures of $\widehat{WX}$ and $\widehat{ZY}$.

c. Make a conjecture about the measure of an angle formed by two secants in relation to the arcs of the circle.

The Modeling Mathematics activity illustrates Theorem 9–12. *You will be asked to prove this theorem in Exercise 45.*

Theorem 9–12	**If two secants intersect in the interior of a circle, then the measure of an angle formed is one-half the sum of the measures of the arcs intercepted by the angle and its vertical angle.**

In the circle at the right, two secants intersect inside the circle. From Theorem 9–12, we can conclude $m\angle 1 = \frac{1}{2}(m\widehat{RU} + m\widehat{ST})$ and $m\angle 2 = \frac{1}{2}(m\widehat{TU} + m\widehat{RS})$. Since $m\angle 1 = m\angle 3$, $m\angle 3 = \frac{1}{2}(m\widehat{RU} + m\widehat{ST})$ and, likewise, $m\angle 4 = \frac{1}{2}(m\widehat{TU} + m\widehat{RS})$.

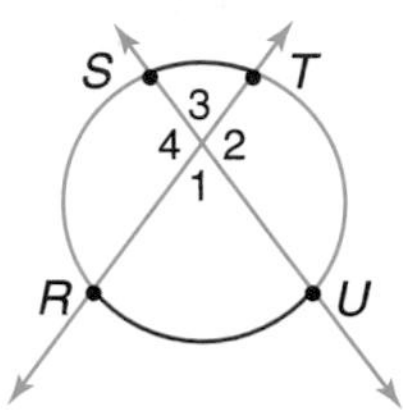

You can also use inscribed angles to find the measure of an angle formed by two secants that intersect outside the circle

Example  **If $m\widehat{BE} = 100$ and $m\widehat{AR} = 25$, find $m\angle S$.**

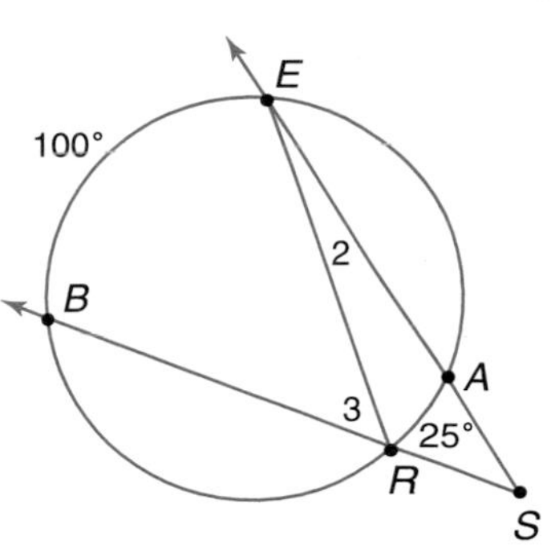

$\angle 3$ and $\angle 2$ are inscribed angles, so $m\angle 3 = \frac{1}{2}m\widehat{BE}$ and $m\angle 2 = \frac{1}{2}m\widehat{AR}$.

$m\angle 3 = m\angle S + m\angle 2$	*Exterior ∠ Th.*
$\frac{1}{2}m\widehat{BE} = m\angle S + \frac{1}{2}m\widehat{AR}$	*Substitution Prop. (=)*
$\frac{1}{2}(100) = m\angle S + \frac{1}{2}(25)$	*Substitution Prop. (=)*
$50 = m\angle S + 12.5$	
$37.5 = m\angle S$	*Substitution Prop. (=)*

In Example 2, notice that $m\angle 1 = \frac{1}{2}(m\widehat{BE} - m\widehat{AR})$. This illustrates Theorem 9–13.

Theorem 9–13 **If two secants, a secant and a tangent, or two tangents intersect in the exterior of a circle, then the measure of the angle formed is one-half the positive difference of the measures of the intercepted arcs.**

There are three possible cases to illustrate Theorem 9–13.

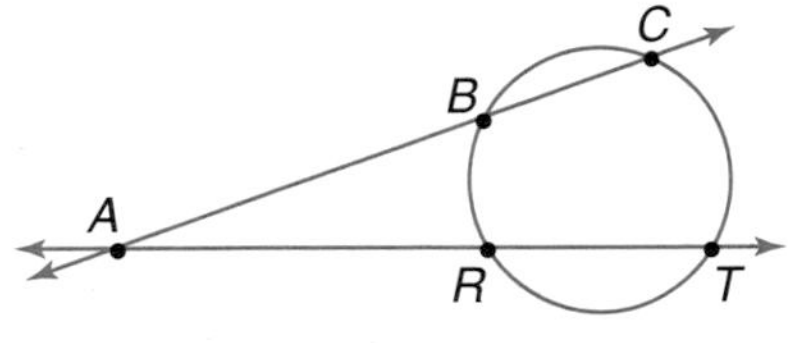

Case 1: two secants

$m\angle CAT = \frac{1}{2}(m\widehat{CT} - m\widehat{BR})$

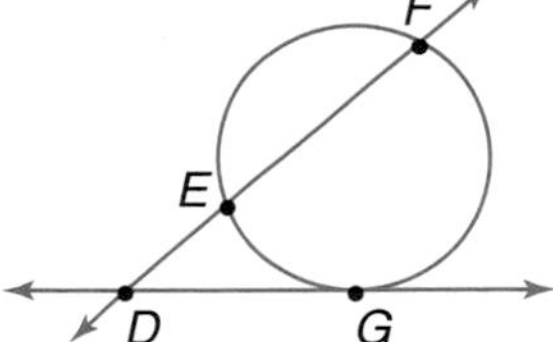

Case 2: a secant and a tangent

$m\angle FDG = \frac{1}{2}(\widehat{FG} - \widehat{EG})$

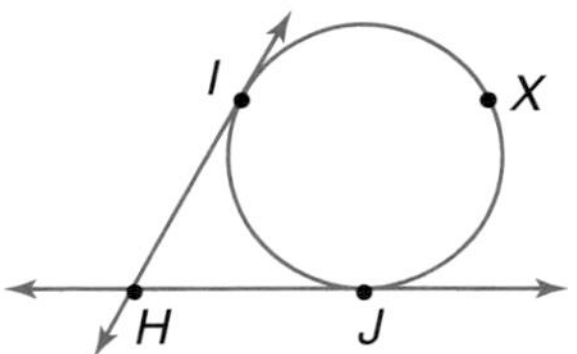

Case 3: two tangents

$m\angle IHJ = \frac{1}{2}(\widehat{IXJ} - \widehat{IJ})$

You will be asked to prove each of these cases in Exercise 46.

Example 3

a. Use ⊙K to find the value of x.

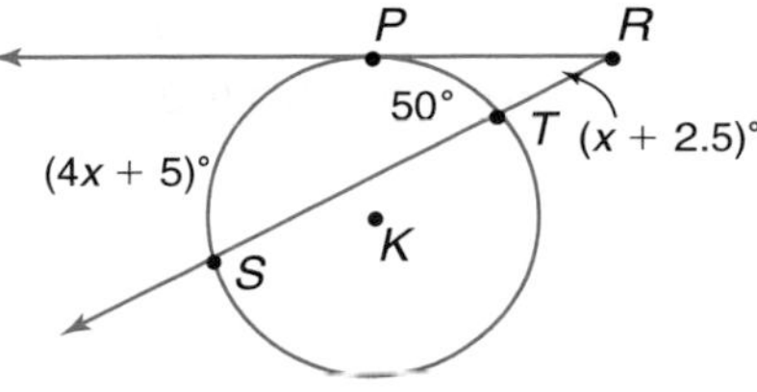

$\angle R$ is formed by a secant and a tangent.

$$m\angle R = \frac{1}{2}(m\widehat{PS} - m\widehat{PT})$$

$$x + 2.5 = \frac{1}{2}[(4x + 5) - 50]$$

$$x + 2.5 = \frac{1}{2}(4x - 45)$$

$$x + 2.5 = 2x - 22.5$$

$$25 = x$$

(continued on the next page)

b. Use $\odot S$ to find the value of y.

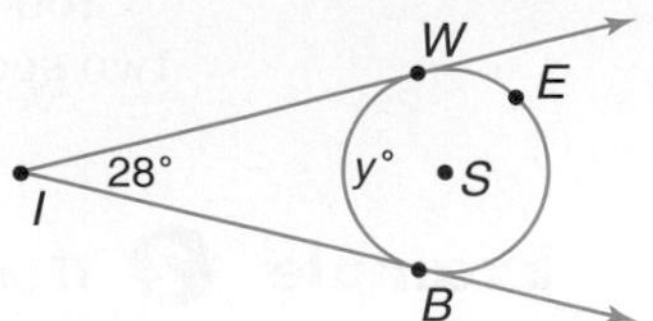

$\angle I$ is formed by two tangents. If $m\overset{\frown}{WB} = y$, then $m\overset{\frown}{WEB} = 360 - y$. *Why?*

$$m\angle I = \frac{1}{2}(m\overset{\frown}{WEB} - m\overset{\frown}{WB})$$

$$28 = \frac{1}{2}[(360 - y) - y]$$

$$28 = \frac{1}{2}(360 - 2y)$$

$$28 = 180 - y$$

$$y = 152$$

CHECK FOR UNDERSTANDING

Communicating Mathematics

Study the lesson. Then complete the following

1. **Explain** how to find the measure of an angle formed by two secants whose vertex is in the *exterior* of a circle.
2. **You Decide** Using $\odot P$, Angelica argues that $m\angle 1 = \frac{1}{2}(m\overset{\frown}{JK} + m\overset{\frown}{ML})$ by Theorem 9–12. Sommer says $m\angle 1 = m\overset{\frown}{JK}$ since $\angle 1$ is a central angle. Who is right and why?

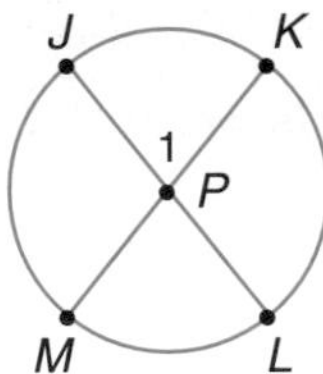

3. **Examine** $\odot S$ in part b of Example 3.
   a. What would be the value of y if $m\angle I$ were 100?
   b. Make a conjecture about the relationship between $m\angle I$ and the measure of the intercepted minor arc WB.
4. $\overline{BD}$ is a diameter of $\odot P$ with tangent ray $\overrightarrow{DE}$. Provide two different reasons why $m\angle BDE$ is a right angle.

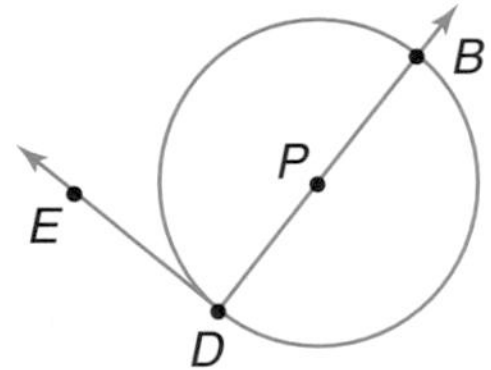

5. The word secant comes from the Latin word *secare*, meaning "to cut." Explain why secant is the word used for a line that intersects a circle in exactly two points. How is a secant different from a tangent?

Guided Practice

6. Find the measure of $\angle 3$.

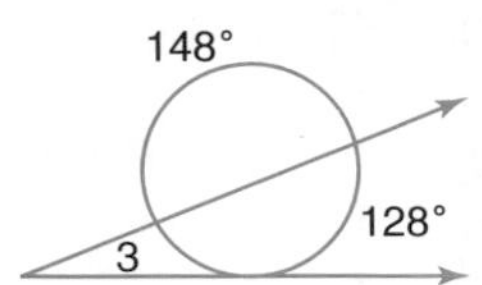

In $\odot K$, $m\overset{\frown}{OB} = 98$, $m\overset{\frown}{OY} = 28$, $m\overset{\frown}{YD} = 62$, and $m\overset{\frown}{DA} = 38$. Find each measure.

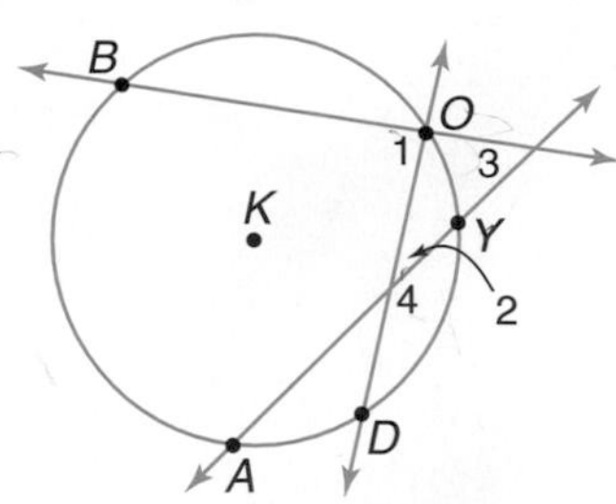

7. $m\overset{\frown}{AB}$
8. $m\angle 1$
9. $m\angle 2$
10. $m\angle 3$

11. Given $\odot L$, find the value of x.

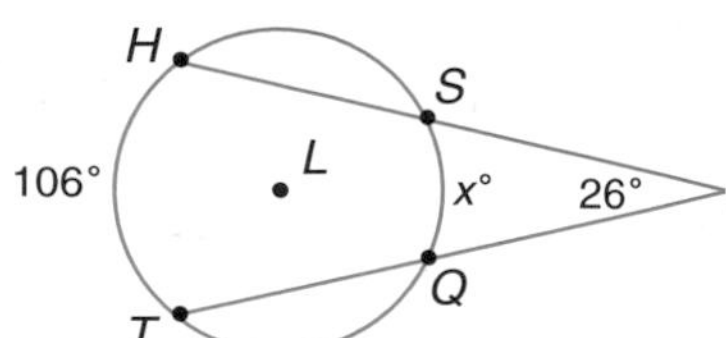

12. Refer to $\odot K$.

a. Find the value of x.

b. Find $m\angle AET$.

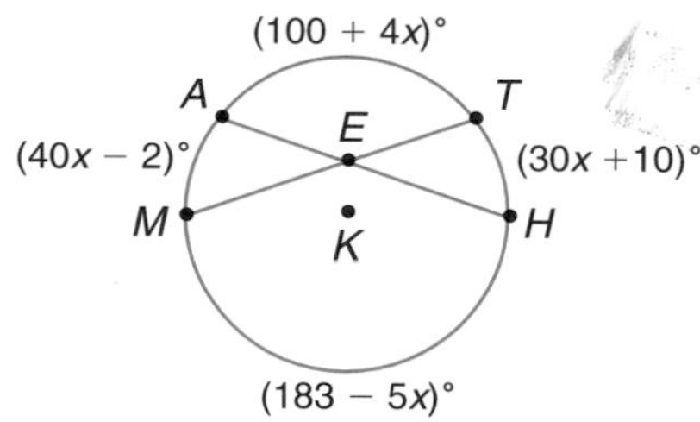

13. Astronomy A geostationary satellite that orbits about 33,000 miles above Earth rotates in the same direction as Earth and, as a result, appears to hover directly over a fixed point on the equator. Use the figure below to measure the arc on the equator of Earth visible to a geostationary satellite.

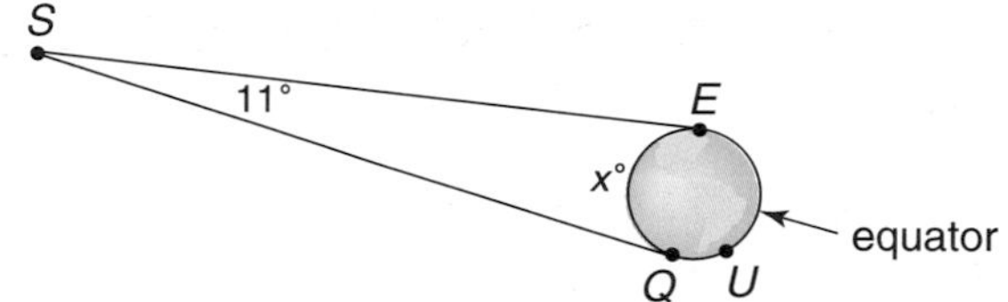

EXERCISES

Practice

Find the measure of each numbered angle.

14.

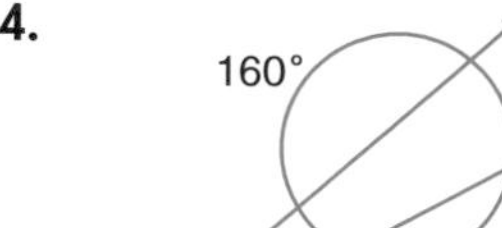
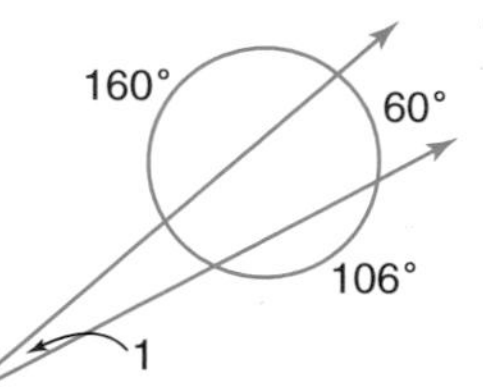

15.

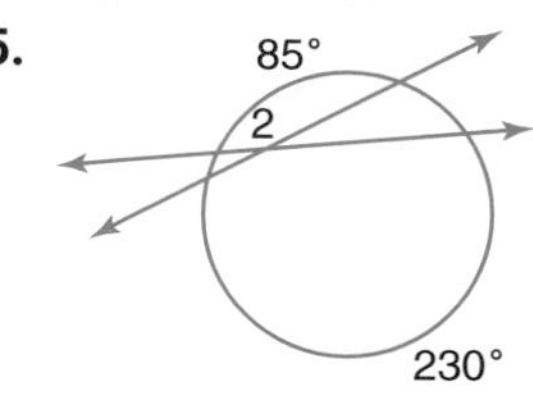

16.

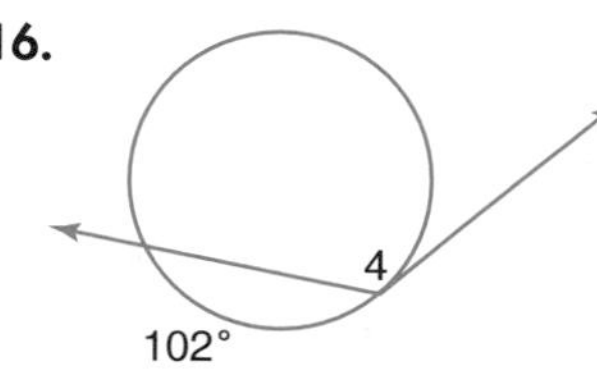

$\overleftrightarrow{CE}$ is tangent to $\odot S$ at E, and $m\widehat{IG} = 80$, $m\widehat{GE} = 140$, and $m\widehat{ER} = 130$. Find each measure.

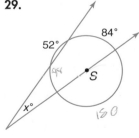

17. $m\angle GRE$

18. $m\widehat{IR}$

19. $m\angle T$

20. $m\angle GAE$

21. $m\angle IEC$

22. $m\angle IED$

23. $m\angle IGA$

24. $m\widehat{GIE}$

25. $m\angle IAG$

26. $m\angle TIE$

27. $m\angle CER$

28. $m\angle DEF$

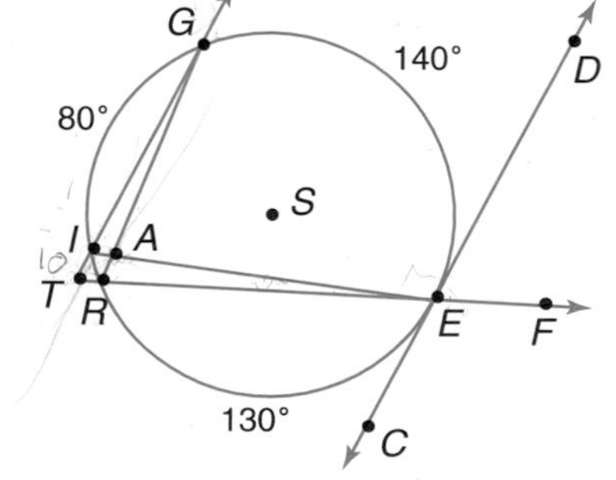

Given $\odot S$, find the value of x.

29.

84°
52°
S
x°

30.

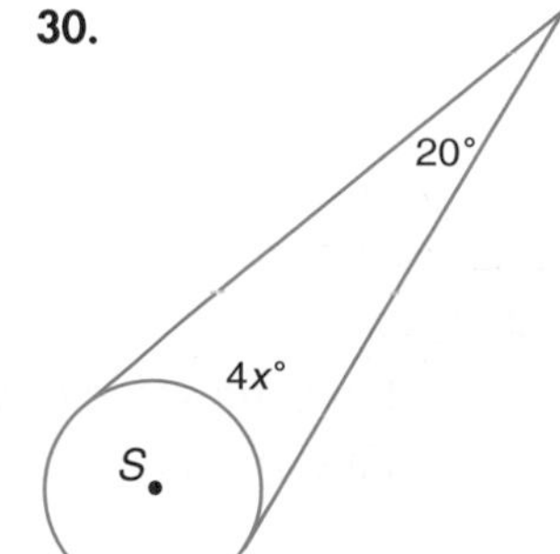

31.

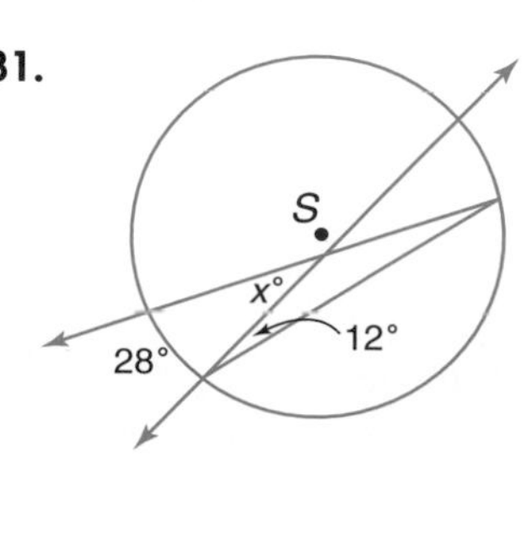

32. In the figure at the right, $m\overparen{CD} = 116$, $m\overparen{BE} = 38$, and $m\overparen{ATF} = 219$. Find $m\overparen{AF}$.

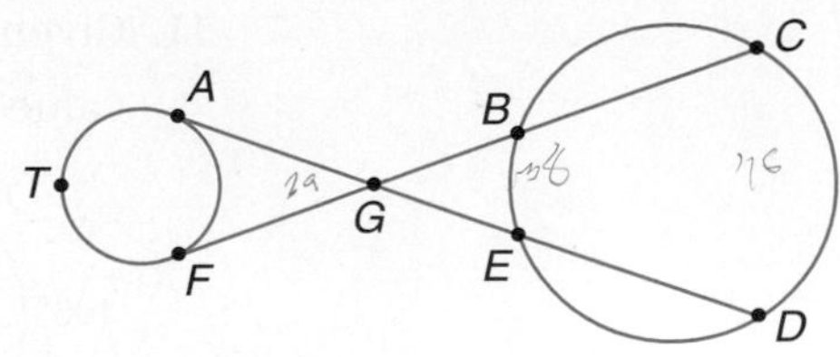

33. In $\odot X$, $m\overparen{AK} = 108$, $m\overparen{RE} = 118$, $m\angle KRE = 30$, and $m\angle KME = 52$. Find each measure.

a. $m\overparen{SR}$

b. $m\overparen{AS}$

c. $m\angle KPE$

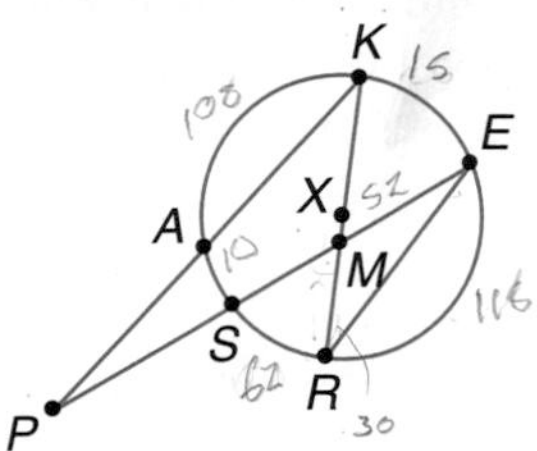

34. In the figure, $\overline{TA}$ and $\overline{TG}$ are tangent to $\odot P$. Find the values of x, y, and z.

35. In $\odot T$, $m\overparen{UD} = 106$, $m\overparen{QU} = 96$, and $m\angle Q = 92$. Find $m\angle A$.

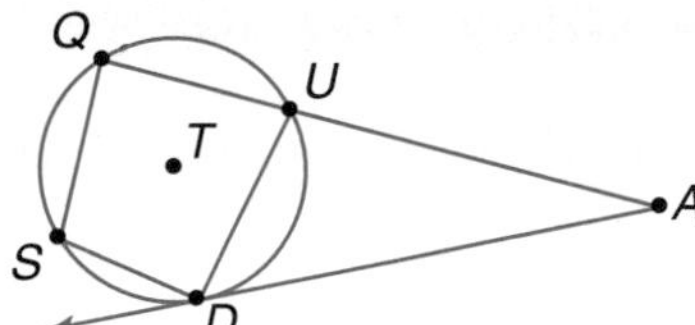

36. In the figure, $\overline{MR}$ and $\overline{MT}$ are tangent to $\odot B$ and $HTMR$ is a rhombus. Find $m\overparen{RT}$.

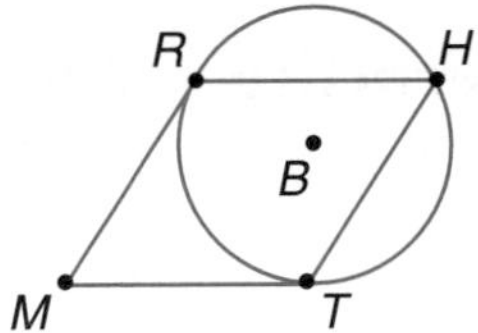

In the figure, quadrilateral *GERA* is inscribed in $\odot P$, with $\overleftrightarrow{TA}$ tangent to $\odot P$ at *A*, $m\angle REG = 78$, $m\overparen{AR} = 46$, and $\overline{AR} \parallel \overline{GE}$. Find each measure.

37. $m\angle GAR$

38. $m\overparen{AG}$

39. $m\angle TAR$

40. $m\angle GAN$

41. $m\overparen{GE}$

42. $m\overparen{RE}$

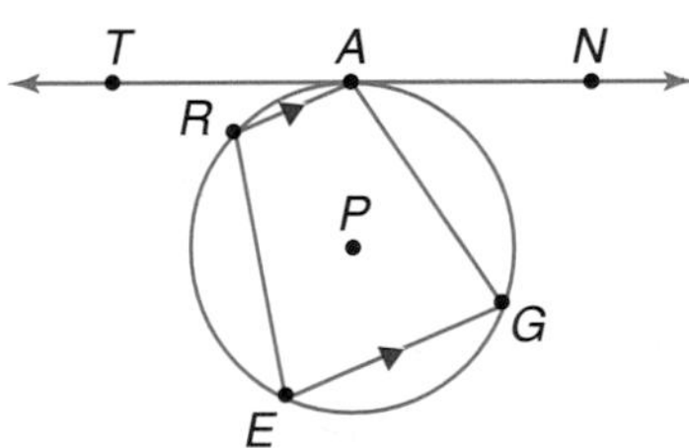

43. Refer to $\odot U$.

a. Find the value of x.

b. Find $m\angle P$.

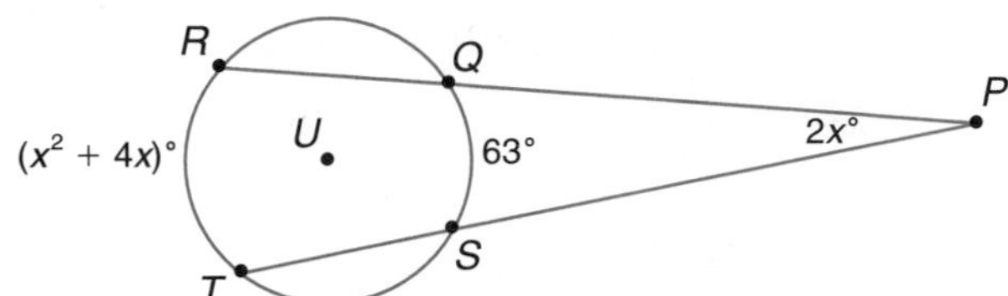

44. Refer to $\odot D$.

a. Find the value of x.

b. Find $m\overparen{ABT}$.

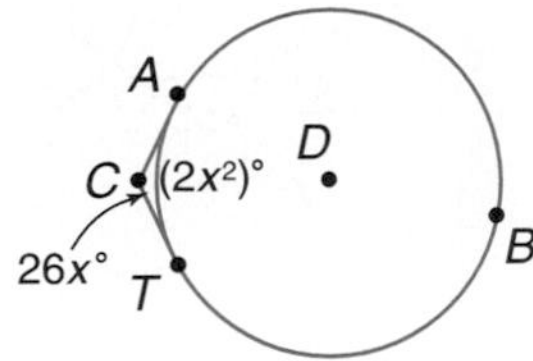

Write a proof.

45. If two secants intersect in the interior of a circle, then the measure of an angle formed is one-half the sum of the measures of the arcs intercepted by the angle and its vertical angle. (Theorem 9–12)

46. If two secants, a secant and a tangent, or two tangents intersect in the exterior of a circle, then the measure of the angle formed is one-half the positive difference of the measures of the intercepted arcs. (Theorem 9–13)

a. Case 1

Given: $\overleftrightarrow{AC}$ and $\overleftrightarrow{AT}$ are secants to the circle.

Prove: $m\angle CAT = \frac{1}{2}(m\overset{\frown}{CT} - m\overset{\frown}{BR})$

(*Hint:* Draw $\overline{CR}$.)

b. Case 2

Given: $\overleftrightarrow{DG}$ is a tangent to the circle.
$\overleftrightarrow{DF}$ is a secant to the circle.

Prove: $m\angle FDG = \frac{1}{2}(m\overset{\frown}{FG} - m\overset{\frown}{GE})$

(*Hint:* Draw $\overline{FG}$.)

c. Case 3

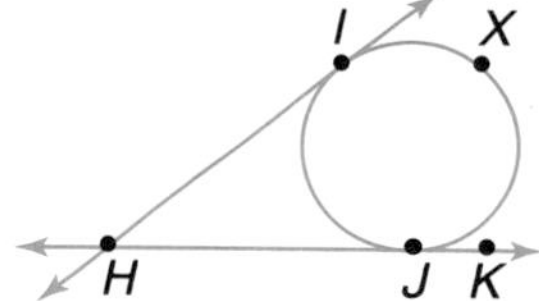

Given: $\overleftrightarrow{HI}$ and $\overleftrightarrow{HJ}$ are tangents to the circle.

Prove: $m\angle IHJ = \frac{1}{2}(m\overset{\frown}{IXJ} - m\overset{\frown}{IJ})$

(*Hint:* Draw $\overline{IJ}$.)

47. Write a paragraph proof for Theorem 9–11.

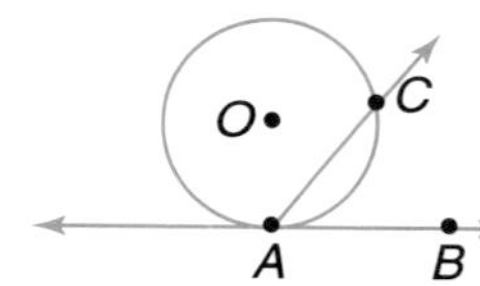

a. Given: $\overleftrightarrow{AB}$ is a tangent of $\odot O$.
$\overleftrightarrow{AC}$ is a secant of $\odot O$.
$\angle CAB$ is acute.

Prove: $m\angle CAB = \frac{1}{2}m\overset{\frown}{CA}$

(*Hint:* Draw diameter $\overline{AD}$.)

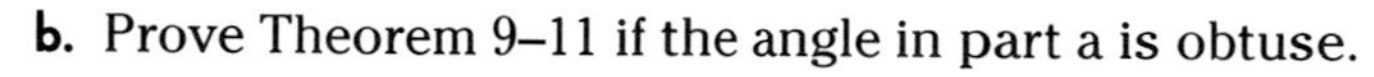

b. Prove Theorem 9–11 if the angle in part a is obtuse.

Critical Thinking

48. A set of points are *concyclic* if they all lie on the same circle. Explain why the vertices of any rectangle must always be concyclic.

Applications and Problem Solving

49. Meteorology Every rainbow is really a full circle whose center is at a point in the sky directly opposite the Sun. The position of a rainbow in the sky varies according to the viewer's position, but its angular size, $\angle ABC$, is always 42°, as shown in the diagram below. If $m\overset{\frown}{CD} = 160$, find the measure of the visible part of the rainbow, $m\overset{\frown}{AC}$.

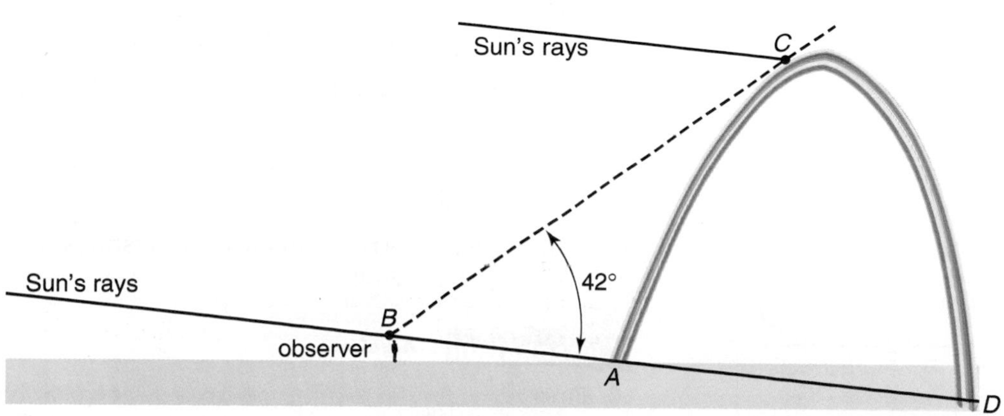

F Y I

It has been calculated that it would have taken 1500 men ten years to move all the stones at Stonehenge, some of which weighed as much as 50 tons.

50. Ancient Monuments Stonehenge, built around 2100 B.C., is one of the most extraordinary and mysterious artifacts in Great Britain. The stones are arranged in a circular pattern according to the movements of Earth and the moon. If $m\overset{\frown}{AB} = 71$ and $m\overset{\frown}{BC} = 118$, find $m\angle AMB$, the angle that the north/south axis makes with the axis of the farthest north moonrise.

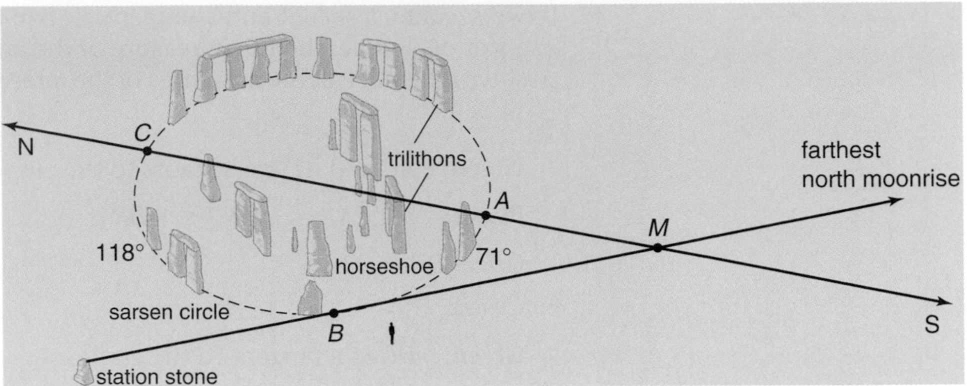

Mixed Review

51. Find the value of x. (Lesson 9–5)

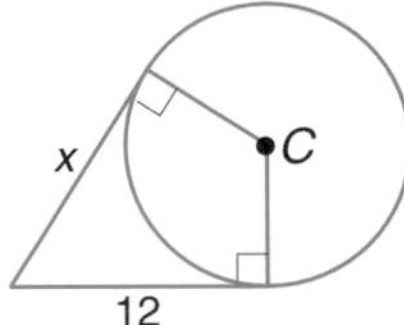

52. Determine whether $\angle A$ is an inscribed angle. Explain. (Lesson 9–4)

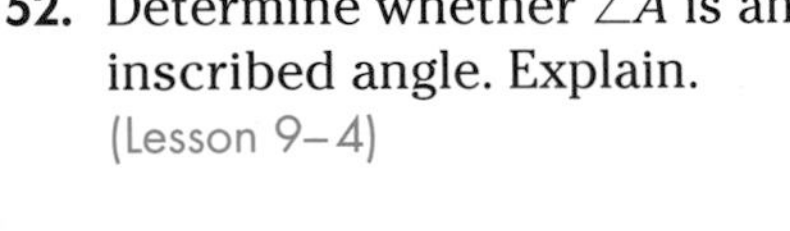

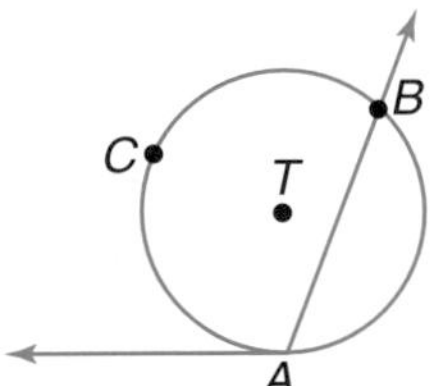

53. **ACT Practice** In the figure, if $AE = 10$, what is the value of h?

A 6 B 8 C 10
D 12 E 18

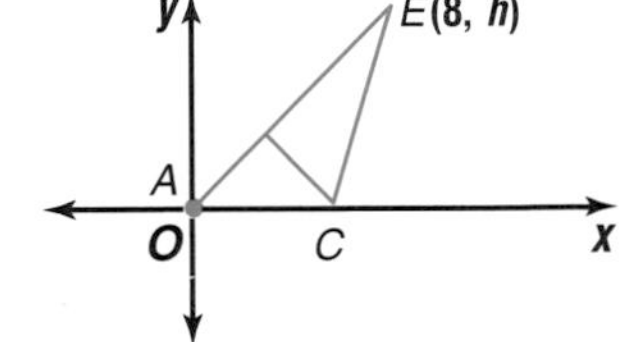

54. State the trigonometric ratio used to find the measure of $\angle Q$. Then find the measure to the nearest degree. (Lesson 8–3)

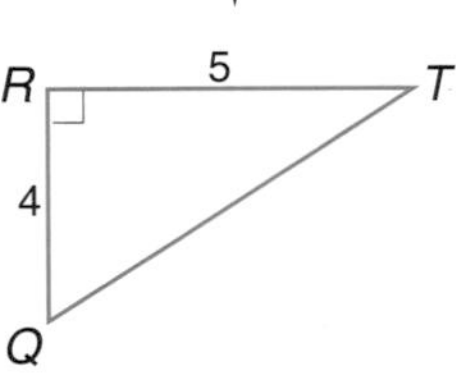

55. **Travel** A tour book costs 450 Taiwanese dollars. Solve the proportion $\frac{25.74}{1} = \frac{450}{x}$ to find the cost in U.S. dollars x. (Lesson 7–1)

56. In parallelogram $PQRS$, $m\angle P = 110$. Find $m\angle Q$. (Lesson 6–1)

57. Determine whether it is possible to draw a triangle with sides measuring 17 centimeters, 10 centimeters, and 29 centimeters. (Lesson 5–5)

58. Determine whether the statement *Skew lines intersect in at least one point* is true or false. (Lesson 3–1)

59. Name the property of equality that justifies the statement *If $3x = 12$, then $x = 4$.* (Lesson 2–4)

Algebra

60. Find $\frac{6w^2}{5} \cdot \frac{20}{12w}$.

61. Solve $x^2 + 2x - 3 = 0$.

For **Extra Practice**, see page 781.

9–7 Special Segments in a Circle

What YOU'LL LEARN

- To use properties of chords, secants, and tangents to solve segment measure problems.

Why IT'S IMPORTANT

A knowledge of special segments in a circle is helpful in solving problems involving carpentry and architecture.

Engineering

The indoor track being designed for the new Forest View Athletic Club must fit on an area 214 feet by 48 feet. The track design can be described as a rectangle with an arc at each end. The distance from the top of the arc to the side of the rectangle cannot exceed 12 feet.

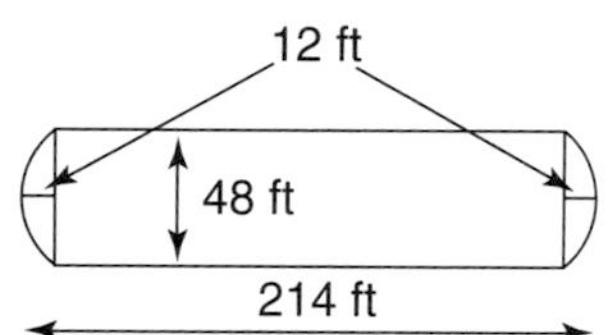

In order to draw the plans correctly, the architect needs to know the radius of the circle containing the arc. How could the architect find this radius? *This problem will be solved in Example 1.*

The side of the rectangle in the diagram above is a chord of the circle containing the arc. There are special relationships among the measures of chords in a circle. The following Exploration investigates such a relationship.

EXPLORATION — CABRI GEOMETRY

- Use a calculator or geometry software to construct $\odot P$ and draw two chords, $\overline{AB}$ and $\overline{CD}$, that intersect in the interior of $\odot P$.
- Label the intersection of the chords E.
- Use the calculator or software to measure $\overline{AE}$, $\overline{EB}$, $\overline{EC}$ and $\overline{ED}$.
- Find the products $AE \cdot EB$ and $EC \cdot ED$.

Your Turn

a. Repeat the activity for two other circles.

b. In each case, compare $AE \cdot EB$ with $EC \cdot ED$. Make a conjecture about these two products.

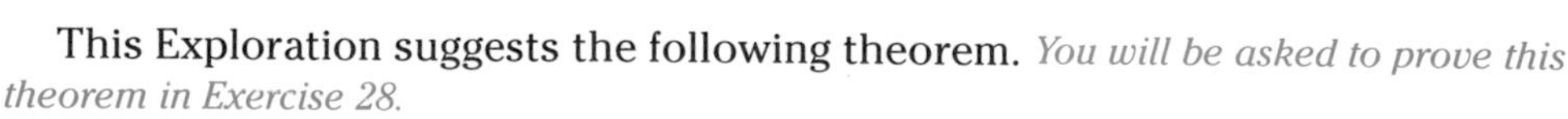

This Exploration suggests the following theorem. *You will be asked to prove this theorem in Exercise 28.*

Theorem 9–14	**If two chords intersect in a circle, then the products of the measures of the segments of the chords are equal.**

We can use Theorem 9–14 to solve the problem at the beginning of the lesson.

Example 1 **Refer to the application at the beginning of the lesson. How can the engineer find the radius to accurately draw the arc for the track?**

Engineering

Explore To accurately draw the arc for the track, the engineer needs to know the radius of the circle.

Plan First, sketch the circle that contains the arc of the track for easier reference. Label the chords, which intersect at E, as $\overline{AB}$ and $\overline{CD}$. $\overline{DE}$ is 12 feet long and is part of the diameter of the circle because it is perpendicular to the chord and meets the chord at its center.

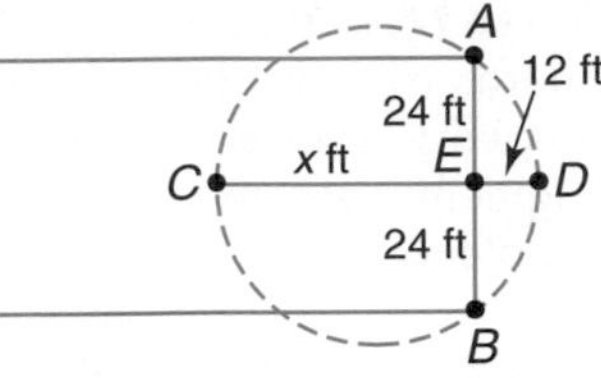

Solve We now have two intersecting chords, and we know the measures of three of the four segments. Let x represent the unknown measure.

$$DE \cdot EC = AE \cdot EB$$
$$12 \cdot x = 24 \cdot 24$$
$$12x = 576$$
$$x = 48$$

The radius of the circle is half the diameter.

$$DC = DE + EC$$
$$= 12 + 48 \text{ or } 60$$

The radius of the circle is $\frac{1}{2}(60)$ or 30 feet.

Examine Use the Pythagorean Theorem to check the triangle with vertices at A, E, and the center of the circle.

$$24^2 + (30 - 12)^2 \stackrel{?}{=} 30^2$$
$$576 + 324 \stackrel{?}{=} 900$$
$$900 = 900 \checkmark$$

The measures check.

In the figure at the right, $\overline{SC}$ and $\overline{SA}$ are **secant segments** and contain chords $\overline{EC}$ and $\overline{AN}$ of $\odot T$. The parts of $\overline{SC}$ and $\overline{SA}$ that are exterior to the circle are called **external secant segments**. $\overline{SE}$ and $\overline{SN}$ are examples of external secant segments.

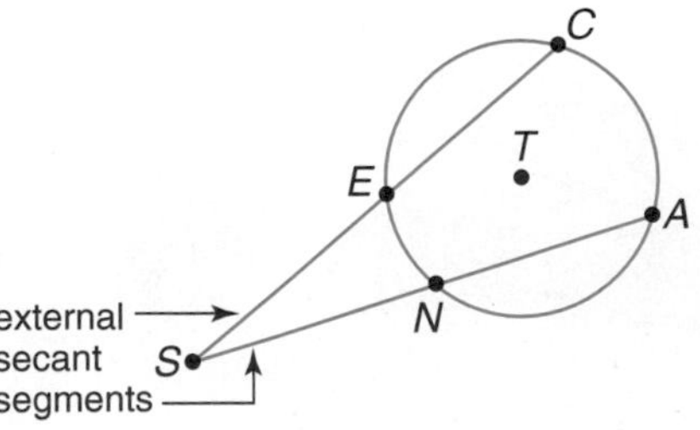

The segments formed by two secant segments also have a special relationship. Recall from Theorem 9–14, that for the two chords intersecting in $\odot E$, $AE \cdot EC = DE \cdot EB$. It is interesting that this relationship also holds true when the intersection of $\overline{AC}$ and $\overline{BD}$, E, occurs outside the circle.

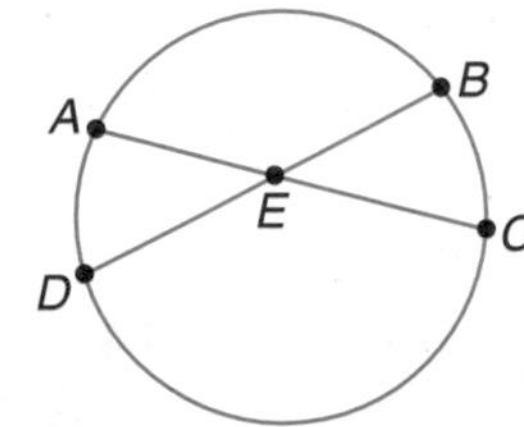

In the figure at the right, $\overline{AC}$ and $\overline{BD}$ are extended outside the circle to intersect at E as secant segments. It is true that $EA \cdot EC = ED \cdot EB$. Notice that $\overline{EA}$ and $\overline{ED}$ are external secant segments.

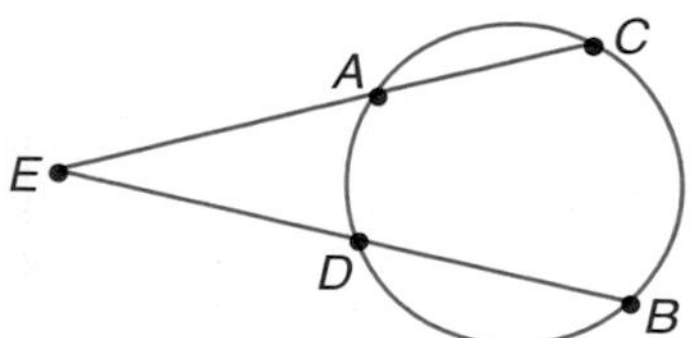

Theorem 9–15	**If two secant segments are drawn to a circle from an exterior point, then the product of the measures of one secant segment and its external secant segment is equal to the product of the measures of the other secant segment and its external secant segment.**

You will be asked to prove this theorem in Exercise 33.

Example 2 **Use the figure at the right to find OG.**

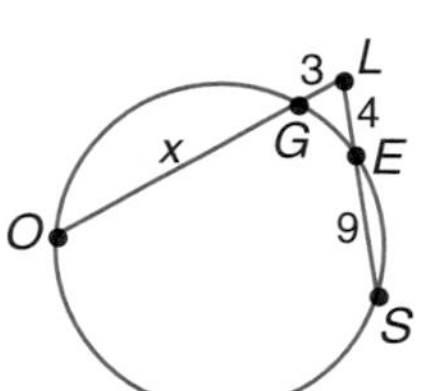

Let x represent OG.

$$LG \cdot LO = LE \cdot LS$$
$$3(x + 3) = 4 \cdot 13$$
$$3x + 9 = 52$$
$$3x = 43$$
$$x = 14\frac{1}{3}$$

Therefore, $OG = 14\frac{1}{3}$.

Suppose one of the secant segments from the exterior point is moved to become a tangent segment. In the figure at the right, $\overline{EC}$ is a tangent segment to the circle. $\overline{EC}$ now represents both the entire segment and the portion of the segment external to the circle. Thus, $EC \cdot EC = ED \cdot EB$. This suggests Theorem 9–16.

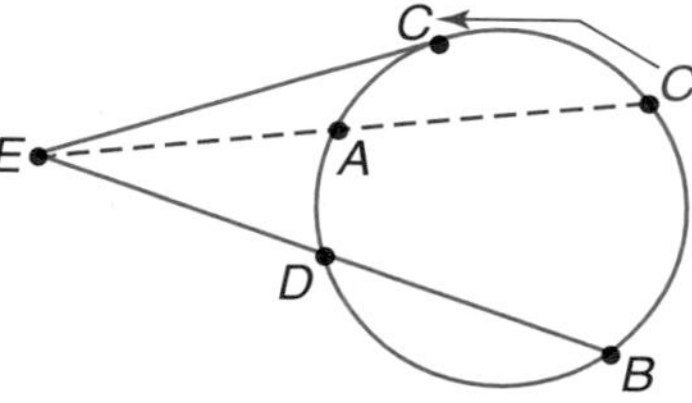

Theorem 9–16	**If a tangent segment and a secant segment are drawn to a circle from an exterior point, then the square of the measure of the tangent segment is equal to the product of the measures of the secant segment and its external secant segment.**

You will be asked to prove this theorem in Exercise 35.

Example 3 **In the figure at the right, $\overline{MA}$ is tangent to $\odot P$. Find the value of x.**

Algebra

$(AM)^2 = MB \cdot MC$ *Theorem 9–16*

$$10^2 = x(x + 6)$$
$$100 = x^2 + 6x$$
$$0 = x^2 + 6x - 100$$

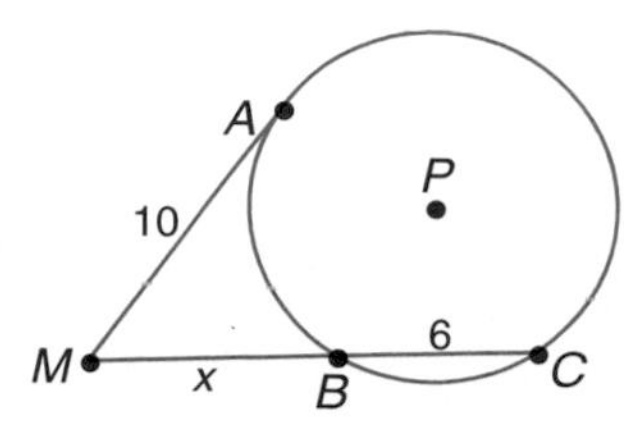

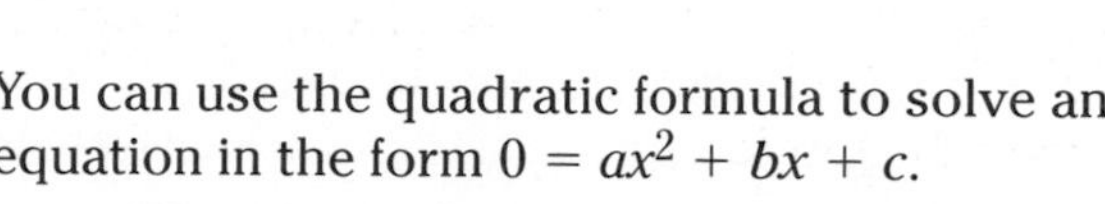

You can use the quadratic formula to solve an equation in the form $0 = ax^2 + bx + c$.

(continued on the next page)

$$x = \frac{-b \pm \sqrt{b^2 - 4ac}}{2a}$$

$$x = \frac{-6 \pm \sqrt{6^2 - 4(1)(-100)}}{2(1)} \quad a = 1, b = 6, c = -100$$

$$x = \frac{-6 \pm \sqrt{436}}{2}$$

$$x = \frac{-6 + \sqrt{436}}{2} \qquad x = \frac{-6 - \sqrt{436}}{2}$$

$$x \approx 7.4 \qquad x \approx -13.4$$

Disregard the negative value. Thus, $x \approx 7.4$.

CHECK FOR UNDERSTANDING

Communicating Mathematics

Study the lesson. Then complete the following.

1. **Identify** an external secant segment in the figure at the right. Explain why you believe your choice is correct.

2. **Explain** why $CO \cdot OG = EO \cdot OM$.

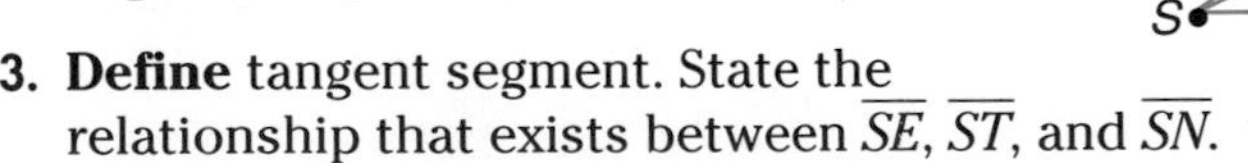

3. **Define** tangent segment. State the relationship that exists between $\overline{SE}$, $\overline{ST}$, and $\overline{SN}$.

4. **You Decide** Ayashe uses the figure at the right to claim that $KA \cdot AN = KL \cdot LO$ by Theorem 9–15. Is she right? Explain why or why not.

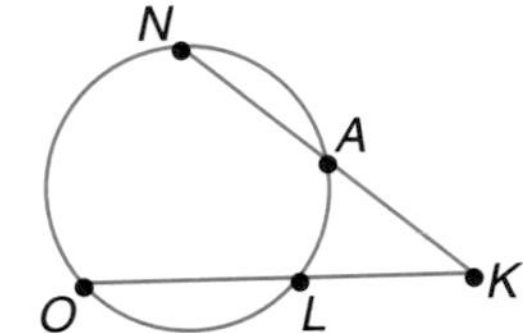

5. **Examine** Example 1. The radius of the circle containing an arc 12 feet from the side of the rectangle was 30 feet, so the side was 30 − 12 or 18 feet from the center. Suppose the arc was to have been a semicircle. Would the "distance from the center" have been greater or less than 12 feet? Explain.

Guided Practice

Find the value of *x* to the nearest tenth. Assume that segments that appear to be tangent are tangent.

6.

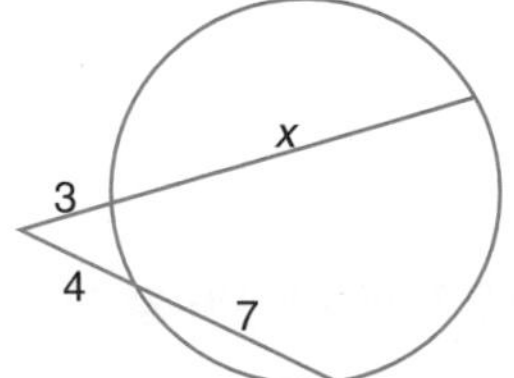

7.

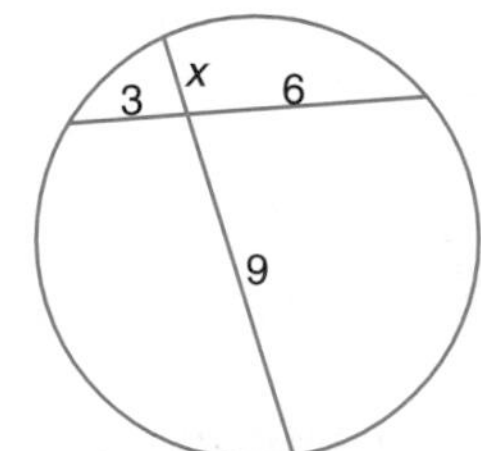

8.

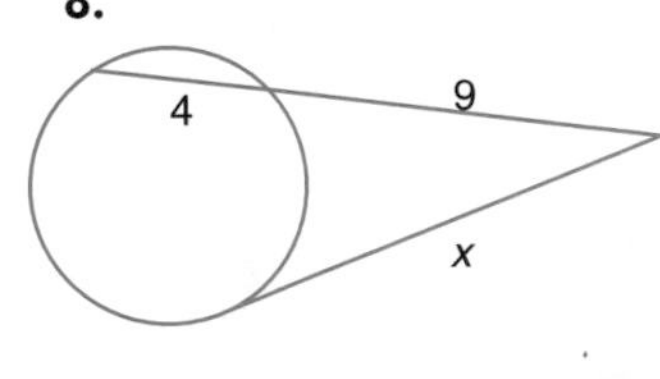

In ⊙*A*, $\overline{TS} \perp \overline{RE}$ with *TS* = 10 and *RE* = 3. Find each measure.

9. *TE* and *ES*
10. *PE*
11. *PR*
12. radius of ⊙*A*

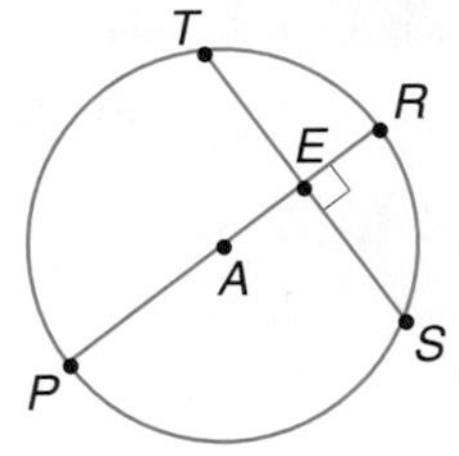

13. **Carpentry** An arch over a double-door entrance is 200 centimeters wide and 60 centimeters high. Find the radius of the circle that contains the arch. (*Hint:* Use the steps in Exercises 9–12.)

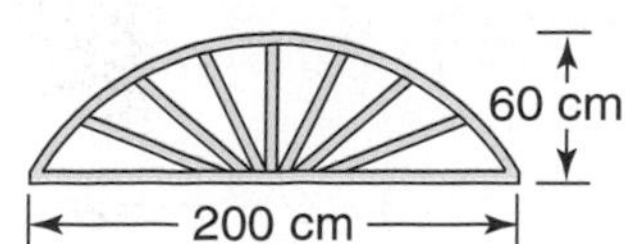

EXERCISES

Practice **Find the value of x to the nearest tenth. Assume that segments that appear to be tangent are tangent.**

14.
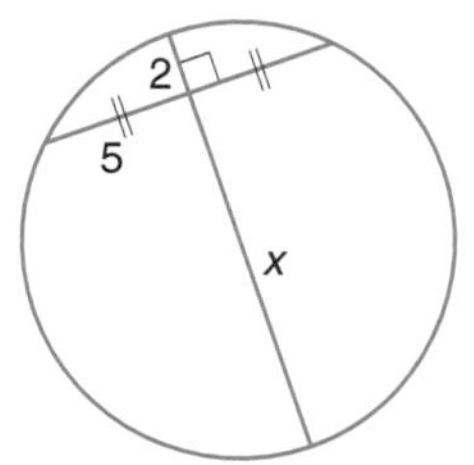

15.
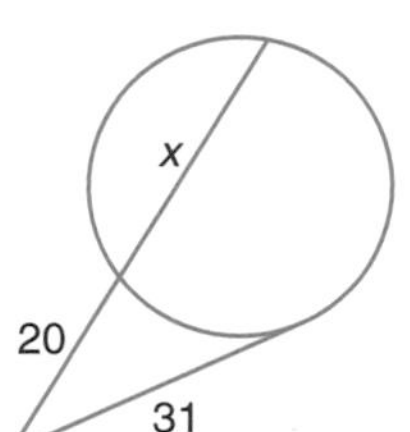

16.
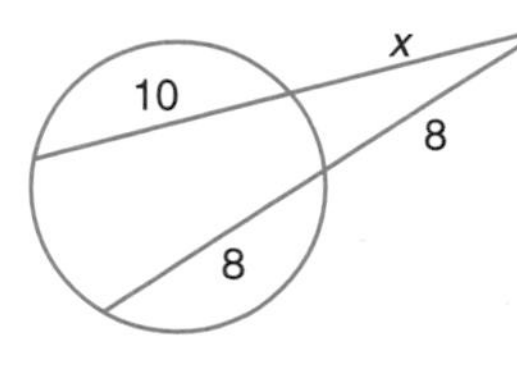

17.
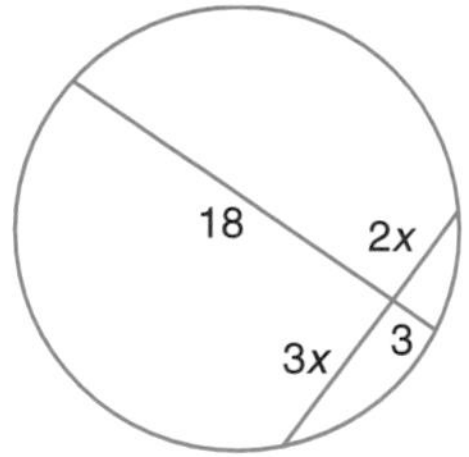

18.
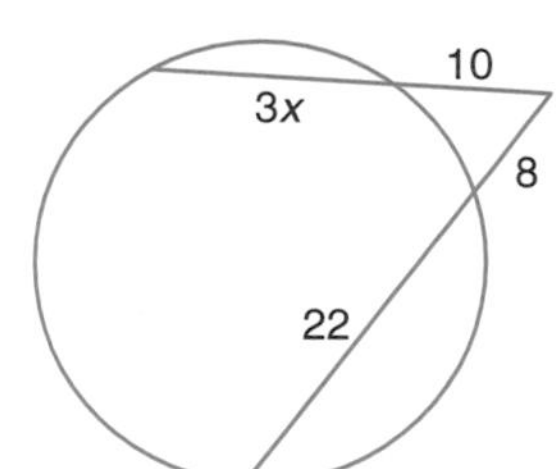

19.
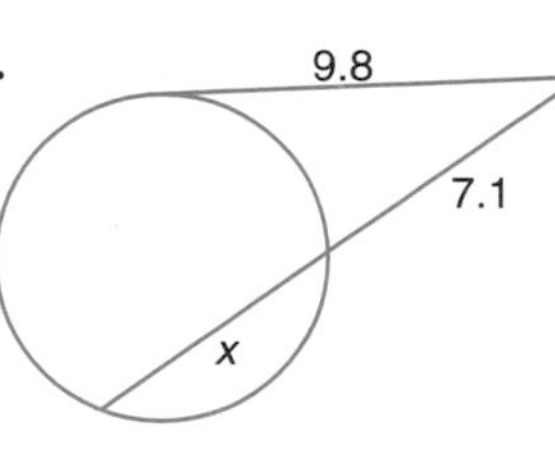

20.
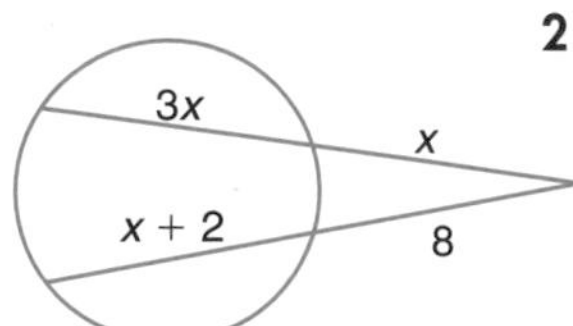

21.
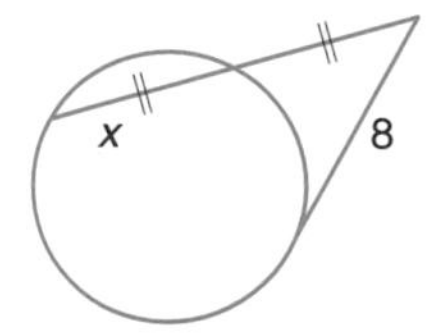

22.
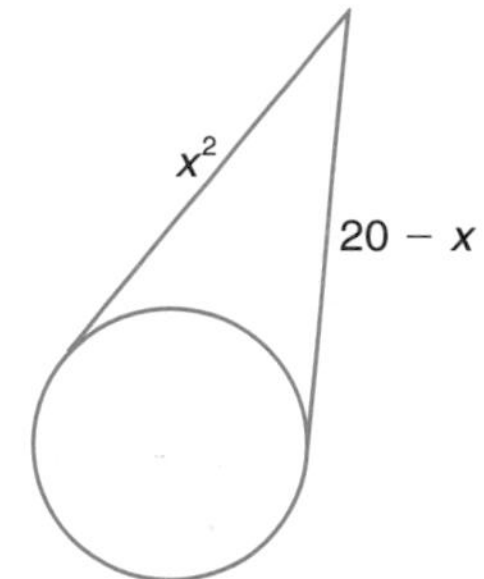

In $\odot D$, $EX = 24$, $DE = 7$, $\overline{XT}$ is a secant segment, and $\overline{EX}$ and $\overline{AX}$ are tangent segments. Find each measure.

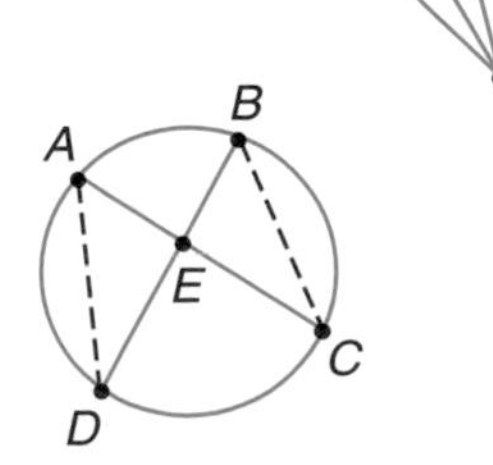

23. AX

24. DX

25. QX

26. TX

27. Show two different techniques for finding QX.

Proof

28. Complete the reasons for each step in the proof of Theorem 9–14.

Given: $\overline{AC}$ and $\overline{BD}$ intersect at E.

Prove: $AE \cdot EC = BE \cdot ED$

A B E C D

Statements	Reasons
a. Draw $\overline{AD}$ and $\overline{BC}$ forming $\triangle DAE$ and $\triangle CBE$.	a. ?
b. $\angle A \cong \angle B$, $\angle D \cong \angle C$	b. ?
c. $\triangle DAE \sim \triangle CBE$	c. ?
d. $\frac{AE}{BE} = \frac{ED}{EC}$	d. ?
e. $AE \cdot EC = BE \cdot ED$	e. ?

29. In $\odot K$ find KR.

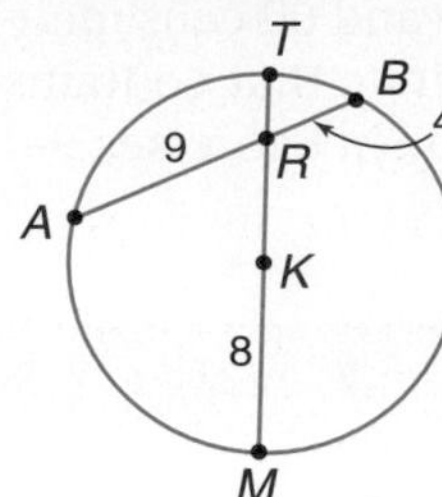

30. $\odot P$ and $\odot Q$ are tangent at S. If $AB = 8$, $CD = 9$, and $BC = 10$, find DE.

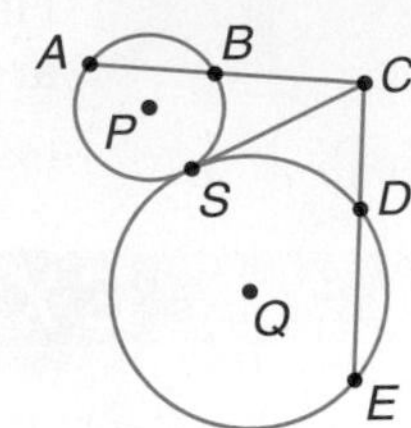

31. Find the values of x and y.

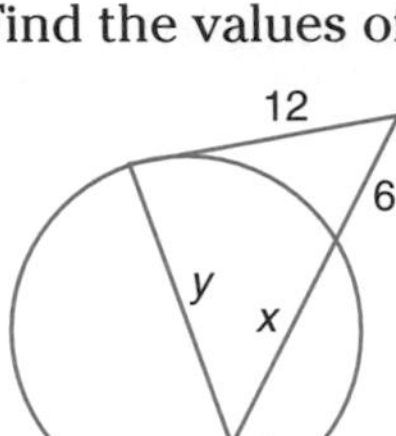

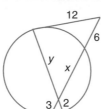

32. Find the values of x and y.

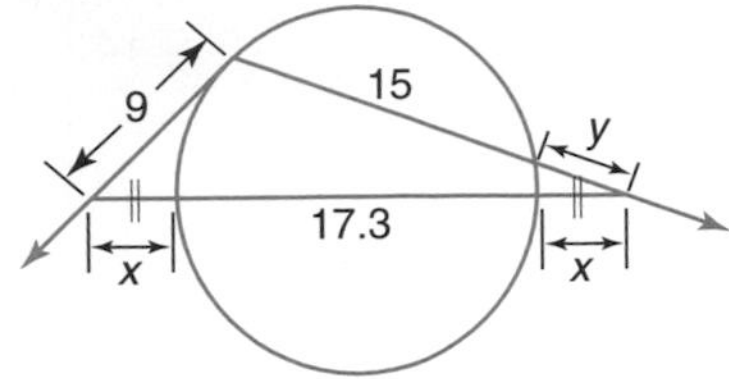

Proof

33. Prove Theorem 9–15. Start with a figure like the one shown at the right. Draw $\overline{AB}$ and $\overline{CD}$.

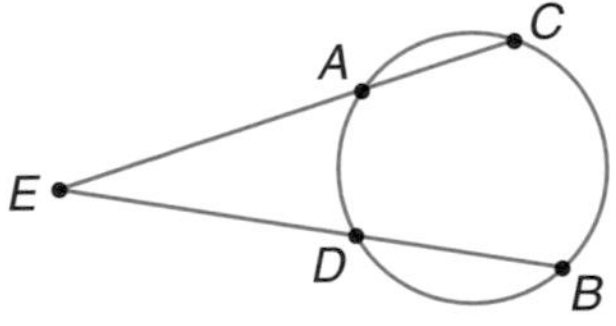

Write a two-column proof.

34. Given: $\odot H$

$\overline{AO} \perp \overline{DM}$

Prove: $OT \cdot TA = (TM)^2$

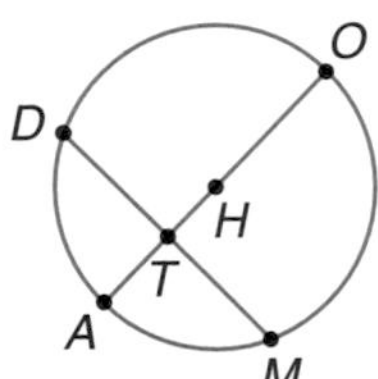

35. Prove Theorem 9–16.

Given: $\overline{XY}$ is tangent to $\odot A$.

$\overline{WY}$ is a secant segment.

Prove: $(XY)^2 = WY \cdot ZY$

(*Hint:* Draw $\overline{WX}$ and $\overline{XZ}$.)

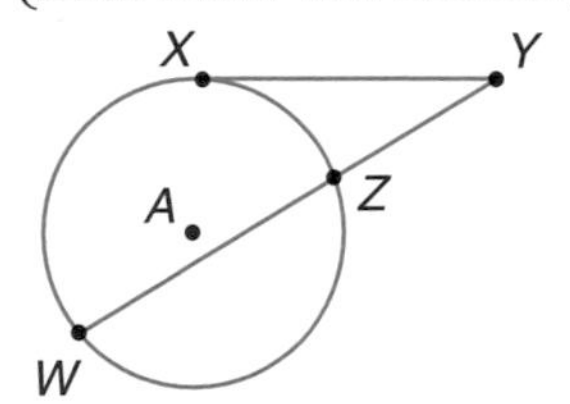

Programming

36. The graphing calculator program finds the measure of a segment in a circle. If you know the measures of $\overline{AE}$, $\overline{EB}$, and $\overline{CE}$, the program will find the measure of $\overline{ED}$.

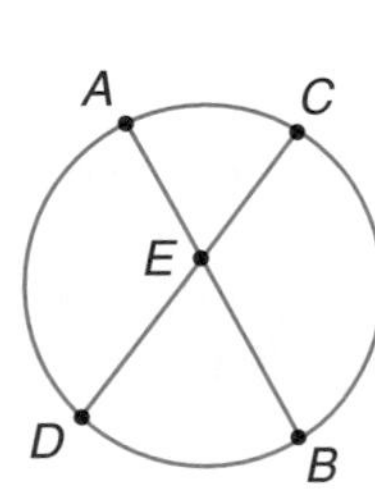


```
PROGRAM:SEGMENTS
:Disp "ENTER THE",
 "MEASURES"
:Input "AE = ", A
:Input "EB = ", E
:Input "CE =", C
:(A*E)/C→D
:Disp "ED = ", D
:Stop
```

Find the measure of $\overline{ED}$ for each of the following.

a. $AE = 8$, $EB = 6$, $CE = 3$

b. $AE = 9$, $EB = 2$, $CE = 7$

c. $AE = 10$, $EB = 30$, $CE = 18$

Critical Thinking

37. In $\odot P$, $\overline{BC} \cong \overline{CD}$. Show that $AB = \sqrt{2} \cdot BC$.

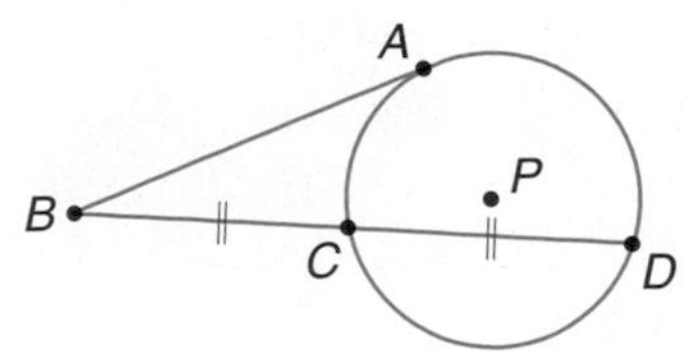

Applications and Problem Solving

Data Update For more information on domed stadiums in the U.S., visit: **www.geometry.glencoe.com**

38. **Architecture** The Louisiana Super Dome has a diameter that measures 680 feet. If the center of the dome is 113 feet higher than the sides of the stadium, how long is the radius of the sphere that forms the dome?

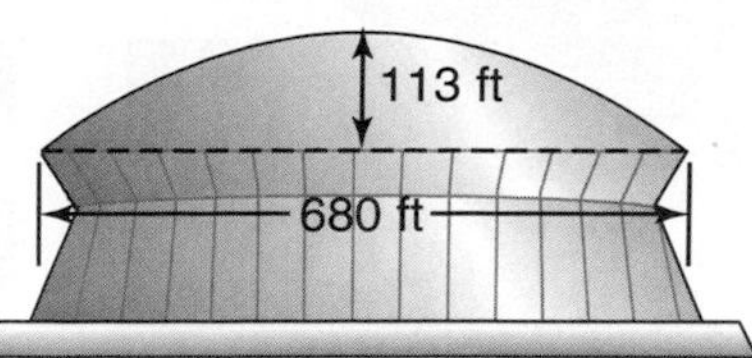

39. **Roman Architecture** The Roman Coliseum has many "entrances" in the shape of a door with an arch top. The ratio of the arch width to the arch height is 7:3. Find the ratio of the arch width to the radius of the circle that contains the arch.

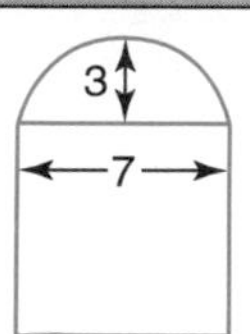

Mixed Review

40. Find $m\angle 1$. (Lesson 9–6)

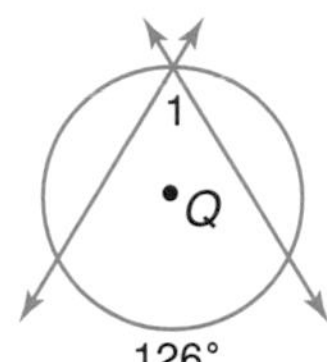

41. Find the value of x. (Lesson 9–5)

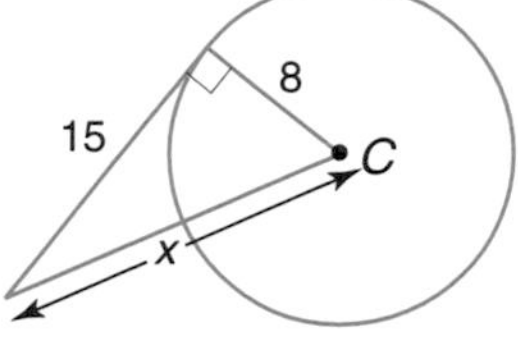

42. If the length of a side of a square is 30.4 yards, find the length of the diagonal to the nearest tenth of a yard. (Lesson 8–2)

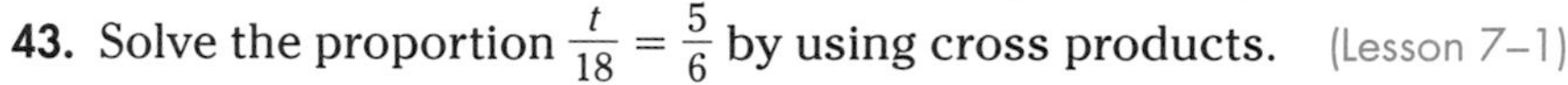

43. Solve the proportion $\frac{t}{18} = \frac{5}{6}$ by using cross products. (Lesson 7–1)

44. **SAT Practice** If 4 more than x is 5 less than y, what is x in terms of y?
A $y - 1$ **B** $y - 9$ **C** $y + 9$ **D** $y - 5$ **E** $y + 1$

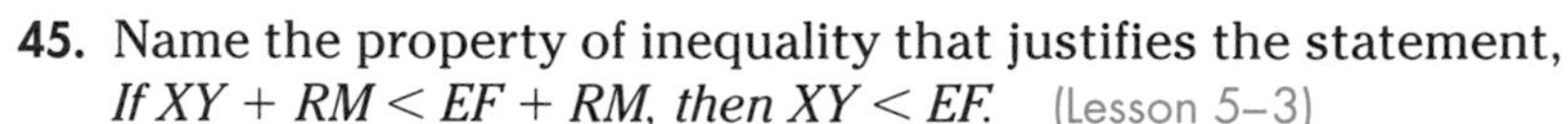

45. Name the property of inequality that justifies the statement, *If $XY + RM < EF + RM$, then $XY < EF$.* (Lesson 5–3)

46. Complete the statement with *always*, *sometimes*, or *never*. "Complementary angles are __?__ congruent." (Lesson 2–6)

Algebra

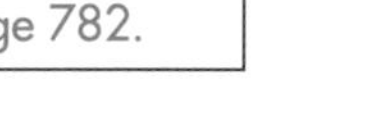

For **Extra Practice**, see page 782.

47. Find $(2c - 1)(c + 7)$.

48. Factor $25a^2 - 4$.

WORKING ON THE

In·ves·ti·ga·tion

Refer to the Investigation on pages 392–393.

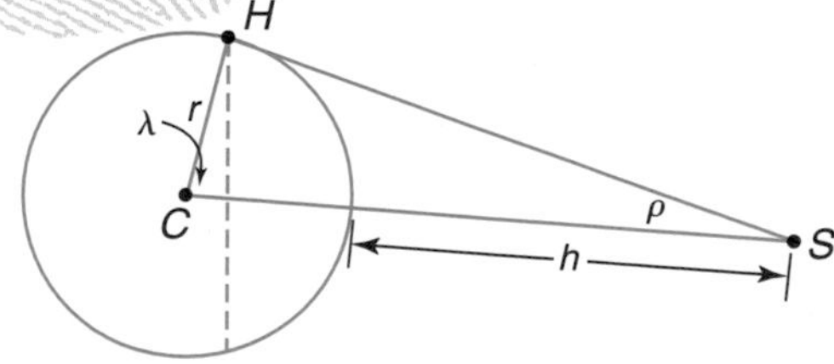

When a spacecraft is in orbit around a planet, only a portion of the surface of the planet can be seen at any one time. The circle that is the boundary of the visible area is called the *horizon circle.*

1 A satellite will be placed in orbit around Mars to relay information from the space station to Earth. The following diagram shows the position of the satellite *S* in relationship to Mars. *C* is the center of Mars, and *H* is a point on the horizon circle. Find the relationships among ρ (rho), λ (lambda), h, and r.

2 If the satellite is in an orbit 160 miles above the surface of Mars, find the angular radius λ of the horizon circle seen by the satellite.

3 Approximately what percent of the surface of Mars will be visible by the satellite at any given moment?

Add the results of your work to your Investigation Folder.

9–8 Integration: Algebra Equations of Circles

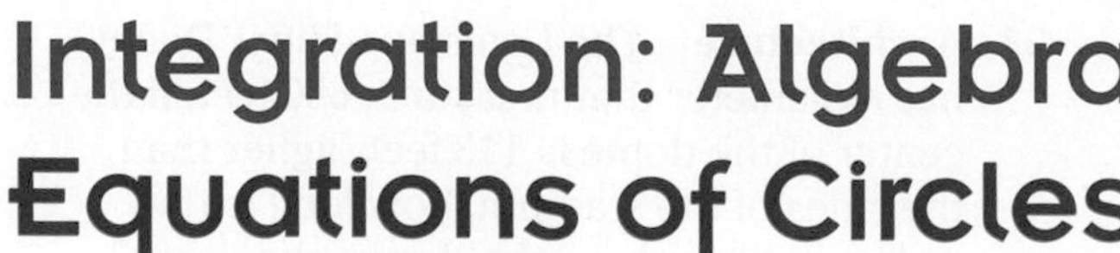

What YOU'LL LEARN

- To write and use the equation of a circle in the coordinate plane.

Why IT'S IMPORTANT

You can use equations of circles to solve problems involving aerodynamics and meteorology.

Real World APPLICATION

Aerodynamics

When a supersonic airplane's speed exceeds the speed of sound, cone-shaped shockwaves are produced as a result of overlapping waves of sound. The graph below shows the spherical sound waves produced by a supersonic airplane.

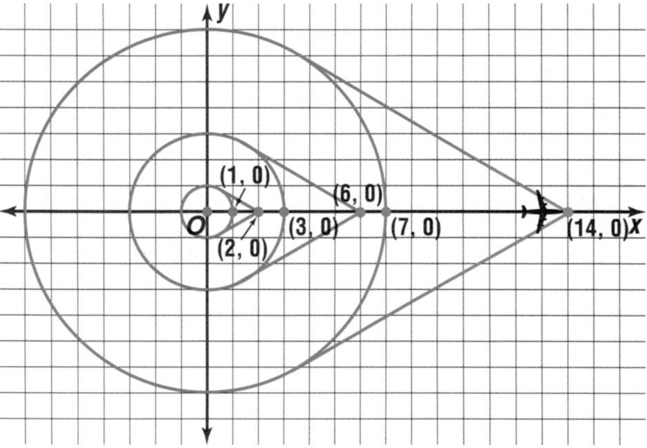

The wave is shown as an expanding circle in a two-dimensional coordinate plane. Notice that when the radius of the shockwave is 1 unit, the airplane is 2 units from the origin. A shockwave radius of 3 shows the airplane 6 units from the origin, and a shockwave radius of 7 shows the airplane 14 units from the origin. So, as sound travels from the center of each circle, the airplane is traveling twice as fast.

FYI

When Chuck Yeager broke the sound barrier in 1947 by flying at 700 mph, no one knew that he had two broken ribs and an immobile right arm from a riding accident the day before.

As supersonic aircraft travel through the sound barrier, it is sometimes useful to track the impact of the shockwaves being produced. To do this, it helps to know the equation for a circle that represents the size of the shockwave.

You can use the distance formula to write the equation for a circle given its center and radius. The circle at the right has its center at $C(-1, 4)$ and a radius of 5 units. Let $P(x, y)$ be any point on C. Then $\overline{CP}$ is a radius of the circle. The distance between $P(x, y)$ and $C(-1, 4)$ is 5 units.

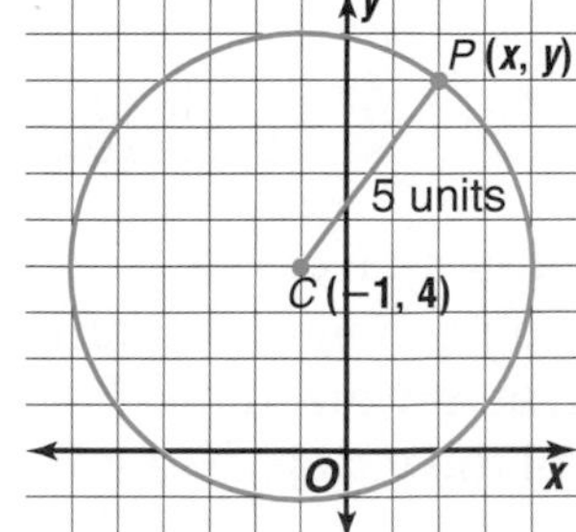

$$PC = 5$$

$$\sqrt{(x - (-1)^2 + (y - 4)^2} = 5 \quad \textit{Distance formula}$$

$$(x + 1)^2 + (y - 4)^2 = 25 \quad \textit{Square each side.}$$

Thus, an equation for the circle with center at $C(-1, 4)$ and radius of 5 units is $(x + 1)^2 + (y - 4)^2 = 25$. Study the pattern in the chart to find the equation of a circle if the center of the circle is (h, k) and the radius is r.

Center	Radius	Equation of Circle
(1, 1)	$\sqrt{2}$	$(x-1)^2+(y-1)^2=2$
(2, 4)	5	$(x-2)^2+(y-4)^2=25$
(h, k)	r	$(x-h)^2+(y-k)^2=r^2$
(0, 0)	r	$x^2+y^2=r^2$

Standard Equation of a Circle	**In general, an equation for a circle with center at (h, k) and a radius of r units is $(x-h)^2+(y-k)^2=r^2$.**

The values of h and k tell how the circle is translated from the origin. That is, $x^2 + y^2 = r^2$ is moved h units horizontally and k units vertically.

Example 1 **Write an equation for a circle with center $C(-3, 6)$ and a diameter of 6 units.**

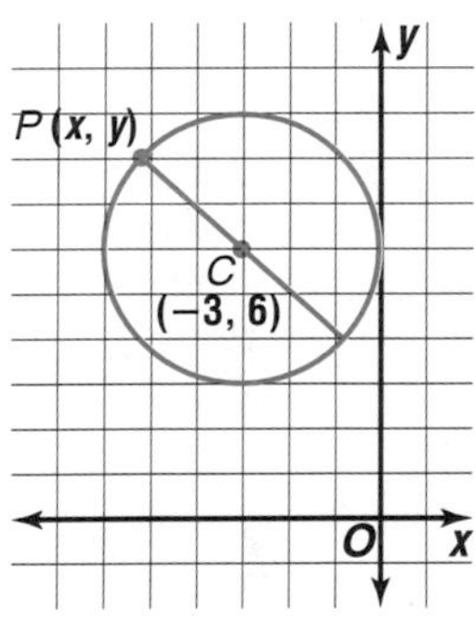

Since the diameter is 6, it follows that the radius must be 3.

Use the equation of a circle.

$$(x-h)^2+(y-k)^2=r^2$$
$$(x-(-3))^2+(y-6)^2=3^2 \quad (h, k) = (-3, 6),\ r = 3$$
$$(x+3)^2+(y-6)^2=9$$

An equation for the circle is $(x+3)^2+(y-6)^2=9$.

You can use the equation of a circle to draw its graph.

Example 2 **Graph the circle whose equation is $(x+2)^2+(y-3)^2=10$.**

Notice that the circle is translated 2 units to the left and 3 units up from the origin.

First, rewrite the equation in the form $(x-h)^2+(y-k)^2=r^2$.

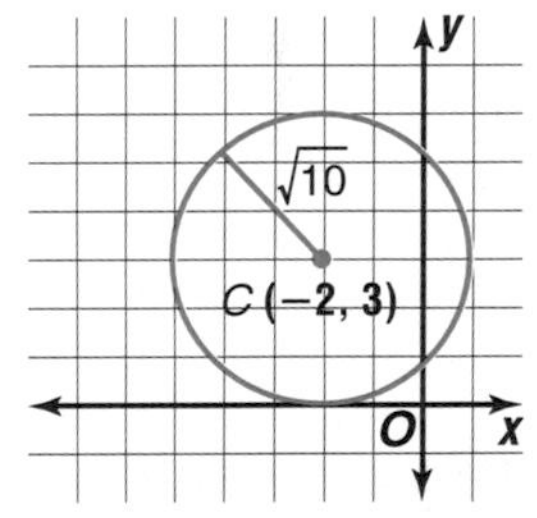

$$(x+2)^2+(y-3)^2=10$$
$$(x-(-2))^2+(y-3)^2=(\sqrt{10})^2$$

Therefore, $h=-2$, $k=3$, and $r=\sqrt{10}\approx 3.1$.

The center is at $(-2, 3)$ and the radius is approximately 3.1 units long.

Examples 1 and 2 explain how to write the equation of a circle and how to graph a circle given the center and radius. However, any three noncollinear points determine a circle. You can use the following steps to write the equation of a circle determined by three noncollinear points.

Step 1 Draw the triangle formed by the three points.

Step 2 Construct the perpendicular bisectors of two of the sides. The center of the circle is their point of intersection.

Step 3 Find the distance between the center and any of the three given points. This is the radius of the circle.

Step 4 Use the center and radius to write an equation of the circle.

Example 3 **A radio station needs to locate a broadcasting tower so that it reaches three cities. Find the equation of a circle that passes through the cities if their coordinates are $A(-2, 4)$, $B(10, 4)$, and $C(8, -3)$.**

Broadcasting

Step 1 First, graph the points and draw $\triangle ABC$.

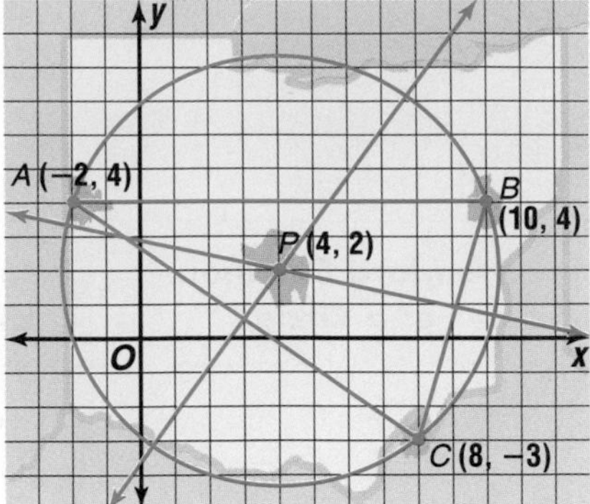

Step 2 Construct the perpendicular bisectors of $\overline{BC}$ and $\overline{AC}$. The coordinates of their intersection at P appear to be (4, 2). Draw the circle by using $\overline{PA}$ as the radius.

Step 3 The radius of the circle is the distance between the center $P(4, 2)$ and any one of the three vertices.

$$d = \sqrt{(x_2 - x_1)^2 + (y_2 - y_1)^2}$$

$$= \sqrt{(10 - 4)^2 + (4 - 2)^2} \quad (x_1, y_1) = (4,2),\ (x_2, y_2) = (10, 4)$$

$$= \sqrt{6^2 + 2^2}$$

$$= \sqrt{40} \quad \text{The radius of } \odot P \text{ is } \sqrt{40}.$$

Step 4 Find an equation of the circle.

$$(x - h)^2 + (y - k)^2 = r^2$$

$$(x - 4)^2 + (y - 2)^2 = (\sqrt{40})^2 \quad (h, k) = (4, 2),\ r = \sqrt{40}$$

$$(x - 4)^2 + (y - 2)^2 = 40$$

An equation for the circle is $(x - 4)^2 + (y - 2)^2 = 40$.

CAREER CHOICES

A **radio announcer** uses special skills in communication to relate news, weather, and entertainment to the listening audience. Formal training in broadcast journalism from a college or technical school is valuable; well-done audition tapes often get announcers their jobs.

For more information, contact:

National Association of Broadcasters
1771 N St., NW
Washington, DC 20036

CHECK FOR UNDERSTANDING

Communicating Mathematics

Study the lesson. Then complete the following.

1. **Explain** how the distance formula and the equation of a circle relate to each other.
2. In Example 3, the center of the circle $P(4, 2)$ was conveniently located at a vertex of the graph paper. Explain how you could use the equations of the two perpendicular lines to determine the location of the center.

3. Aircraft traveling at approximately the speed of sound are assigned a *Mach number* of 1.
   a. What does the Mach number represent?
   b. Research to find the name and nationality of the physicist who developed the concept of the Mach number.

Guided Practice

4. Determine the coordinates of the center and the measure of the radius for the circle whose equation is $(x + 2)^2 + (y + 7)^2 = 81$.

Graph each circle whose equation is given. Label the center and measure of the radius on each graph.

5 $x^2 + y^2 = 4$

6. $x^2 + (y - 3)^2 = 5$

Write an equation of circle *P* based on the given information.

7. center: $P(-2, 3)$
   radius: $\sqrt{11}$

8. 

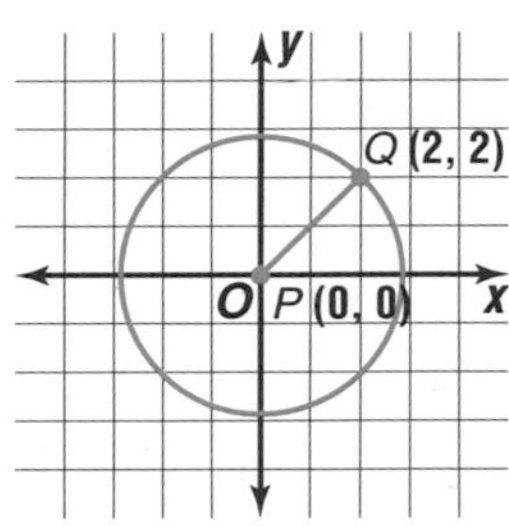

9. $\odot P$ contains the three points $A(4, 0)$, $B(0, 4)$, and $C(-4, 0)$.

10. **Supersonics** Refer to Example 1. Then copy the diagram shown below on graph paper.

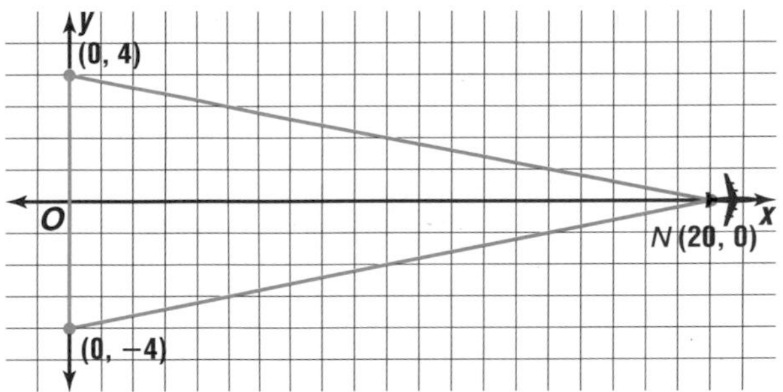

Draw your own circle tangent to the two segments and estimate the speed of the aircraft to produce the shockwave. The speed is about how many times the speed of sound?

EXERCISES

Practice

Determine the coordinates of the center and the measure of the radius for each circle whose equation is given.

11. $\left(x - \frac{3}{4}\right)^2 + (y + 3)^2 = \frac{81}{4}$

12. $(x + 4)^2 + y^2 - 121 = 0$

13. $(x - 0.5)^2 + (y + 3.1)^2 = 17.64$

Graph each circle whose equation is given. Label the center and measure of the radius on each graph.

14. $x^2 + y^2 = 16$

15. $(x + 3)^2 + y^2 = 9$

16. $(x + 2)^2 + (y - 3)^2 = 49$

17. $x^2 + (y - 1)^2 = 8$

18. $\left(x - \frac{2}{5}\right)^2 + \left(y + \frac{1}{2}\right)^2 = \frac{1}{4}$

19. $(x + 5)^2 + (y - 9)^2 = 20$

Write an equation of circle *P* based on the given information.

20. center: $P(0, 0)$
    radius: 5

21. center: $P(-1, 4)$
    radius: $\sqrt{15}$

22. center: $P\left(0, -\frac{3}{2}\right)$
    radius: $\frac{4}{3}$

23. 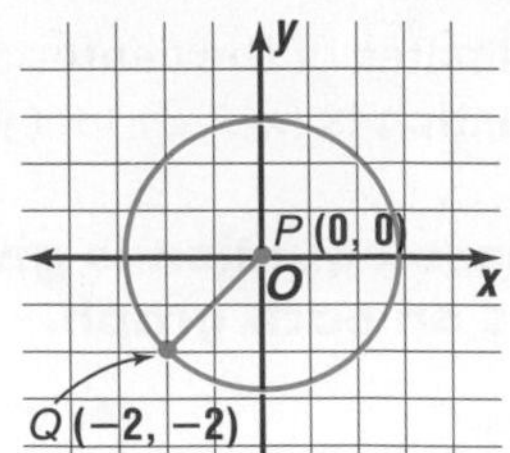

24. 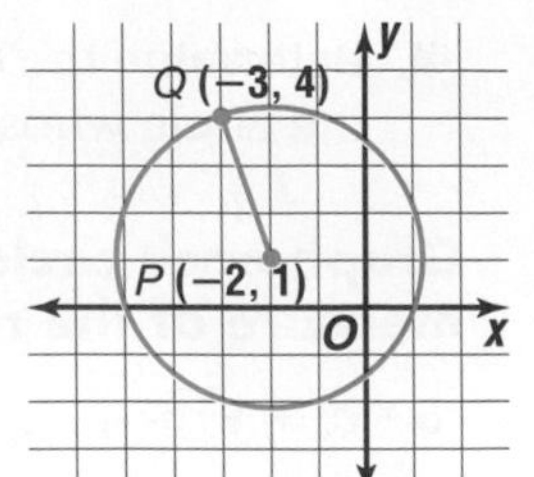

25. 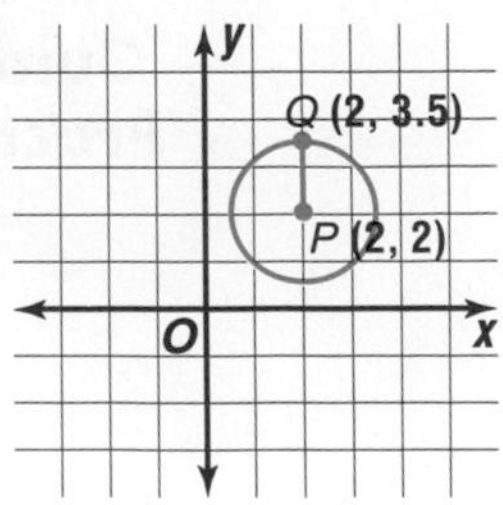

26. Write an equation of the circle that has a diameter of 12 units and whose center is translated 18 units to the left and 7 units down from the origin.

27. The graphs of $x = 4$ and $y = -1$ are tangent to a circle that has its center in the third quadrant and a diameter of 14. Write an equation of the circle.

28. Write an equation of the circle that has a diameter whose endpoints are at (2, 7) and (−6, 15).

Graph the circle $(x - 6)^2 + (y + 2)^2 = 36$ and the line having the given equation. Determine whether the line is a secant or tangent of the circle. Explain your reasoning.

29. $y = 2x - 2$ 30. $x = 0$ 31. $y = -\frac{1}{6}x + 5$ 32. $y = -x$

33. Use points $A(4, 4)$, $B(0, -12)$, and $C(-4, 6)$ to find each.

a. Draw $\triangle ABC$ and construct the perpendicular bisectors of two sides.

b. Determine the intersection of the two bisectors, P, which is the center of the circle.

c. Draw the circle and use either A, B, or C with center P to find the radius.

d. Write an equation of the circle passing through points A, B, and C.

34. Repeat the steps in Exercise 33 to write an equation of a circle passing through points $D(-2, 4)$, $E(4, 8)$, and $F(6, -8)$.

35. a. Write an equation of the circle passing through points $A(0, 6)$, $B(6, 0)$, and $C(6, 6)$.

b. What type of triangle is $\triangle ABC$? Explain.

c. What type of segment is $\overline{AB}$?

36. Choose three noncollinear points that form an obtuse triangle in the coordinate plane. Would the center of the circle containing those points lie in the interior or exterior of the triangle? Justify your response.

Cabri Geometry

37. You can use a calculator to generate and display equations and coordinates of circles.

a. Display the x- and y-axes by pressing F8 and selecting 9:Format. Then select 2:RECTANGULAR from the Coordinate Axes option. Use F3 to draw a circle.

b. Press F6 and select 5:Equation & Coordinates. Select the circle to find its equation.

c. Select the center point of the circle to find the coordinates of the point.

d. Move or change the circle and find the new equation and the new coordinates of the center point.

Critical Thinking

38. a. Draw the graphs of $x^2 + y^2 = 20$ and $x - 2y = 0$.
b. Locate the two points of intersection and label them A and B.
c. Find the distance $\overline{AB}$ between the two points.
d. State the relationship between $\overline{AB}$ and the circle.

Applications and Problem Solving

39. **Aerodynamics** The diagram at the right shows a supersonic jet that travels at four times the speed of sound. Use the measurements to determine $m\angle BCD$. (*Hint*: Use trigonometry to find angle measures.)

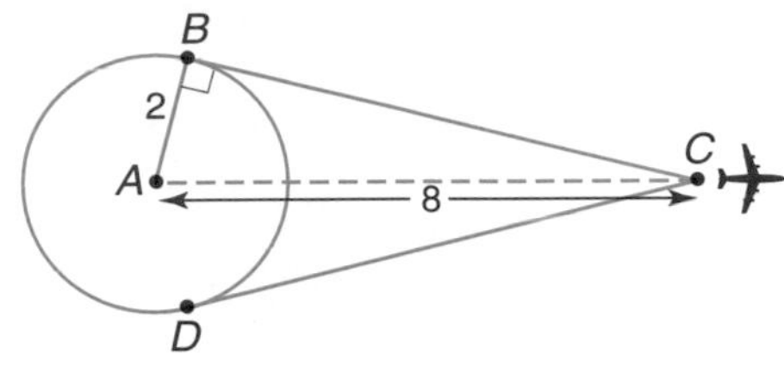

40. **Meteorology** Meteorologists often track the path of severe storms using radar. If the center of a storm is at $C(0, 0)$ and each concentric ring of the radar image has a width of 1 unit, determine the equation of the third concentric circle that encompasses the major part of the storm.

Mixed Review

41. Use the figure below to find the value of x. (Lesson 9–7)

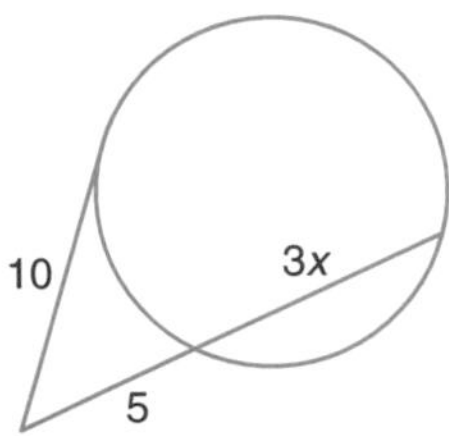

42. If the measure of inscribed angle ABC is 42, what is the measure of the intercepted arc AC? (Lesson 9–4)
43. In a circle with a radius of 7 inches, a chord is 5 inches from the center of the circle. What is the length of the chord? (Lesson 9–3)
44. Solve $\triangle DHK$ if $m\angle K = 37$, $k = 8.1$, and $m\angle H = 62$. Round measures to the nearest tenth. (Lesson 8–5)

45. **ACT Practice** For all $a \neq 4$, $\frac{a^2 - 16}{4a - 16} = ?$

A $a + 16$ B $a + 1$ C $\frac{a-4}{4}$ D $\frac{a}{4}$ E $\frac{a+4}{4}$

46. In a right triangle, one angle measures 28°. Find the measures of the other two angles. (Lesson 4–2)
47. Justify the statement $QT = QT$ with a property from algebra or a property of congruent segments. (Lesson 2–5)

Algebra

48. Solve the system of equations.
$y = x - 3$ and $2x + y = -3$
49. **Savings** Jaclyn is investing \$8000 in a savings certificate that matures in 5 years. The annual interest rate is 7% compounded quarterly. The equation $y = 8000(1 + 0.0175)^{4t}$ represents the balance after t years. Find the amount of Jaclyn's investment after 5 years.

For **Extra Practice**, see page 782.

Refer to the Investigation on pages 392–393.

Because it is Earth's closest neighbor, people have long been fascinated by Mars. Hundreds of movies have been made about fictional Martians, and many real-life explorations have been made there.

Analyze

You have prepared research and plans for a mission to Mars. It is now time to analyze your work and prepare your report.

PORTFOLIO ASSESSMENT

You may want to keep your work on this Investigation in your portfolio.

1 Look over your calculations regarding the energy needs of the space station. Verify your calculations. Then write a short description of the energy system you recommend for the station.

2 Review the Martian calendar you prepared. Make sure it is concise and complete. Then place the calendar aside to add to your report later.

3 Evaluate your calculations regarding the positions of Mars's moons. Rewrite your solutions and conclusions neatly to add to your report.

Write

Your report to NASA should explain your process for researching each of your recommendations. Also include materials that they could use to launch a campaign for presenting the project to the public and to Congress.

4 Include a neat, detailed description of the research you conducted to solve each problem that the scientists raised.

5 Review the list of questions that your team generated in your brainstorming session at the beginning of the Investigation. Are there any questions you have not answered that you could research and add to your report?

6 Research the terrain of Mars. Include any maps or photographs that you find in your report.

7 Investigate the history of the study of Mars. Add a summary to your report.

8 Write a proposal for ways that NASA can promote the Mars space station project to Congress and the public. Include descriptions of the benefits of the project. You may wish to investigate products and innovations that were results of previous NASA projects.

9 Demonstrate how your research could be adapted to future expeditions to other planets.

Data Collection and Comparison To share and compare your data with other students, visit:
www.geometry.glencoe.com

HIGHLIGHTS

Chapter Review For additional lesson-by-lesson review, visit: **www.geometry.glencoe.com**

VOCABULARY

After completing this chapter, you should be able to define each term, property, or phrase and give an example or two of each.

Geometry

adjacent arc (p. 453)
arc (p. 453)
arc length (p. 454)
arc measure (p. 453)
arc of the chord (p. 459)
center (p. 446)
central angle (p. 452)
chord (p. 446)
circle (p. 446)
circumference (p. 447)
circumscribed polygon (p. 477)
common external tangent (p. 476)
common internal tangent (p. 476)
common tangent (p. 476)
concentric circles (p. 454)
congruent arcs (p. 454)
congruent circles (p. 454)
diameter (p. 446)
exterior of a circle (p. 475)
external secant segment (p. 492)
inscribed angle (p. 466)
inscribed polygon (p. 459)
intercepted arc (p. 466)
interior of a circle (p. 475)
major arc (p. 453)
minor arc (p. 453)
pi (p. 447)
point of tangency (p. 475)
radius (p. 446)
secant (p. 483)
secant segment (p. 492)
semicircle (p. 453)
similar circles (p. 454)
standard equation of a circle (p. 499)
sum of central angles (p. 452)
tangent (pp. 474–475)
tangent segment (p. 477)

Problem Solving

make a circle graph (p. 452)

UNDERSTANDING AND USING THE VOCABULARY

Choose the letter of the term that best matches each phrase.

1. an angle whose vertex is at the center of the circle
2. circles lying in the same plane that have the same center but different radii
3. a line that intersects a circle in exactly two points
4. a segment that has its endpoints on the circle
5. an angle whose vertex is on the circle and whose sides contain chords of the circle
6. circles that have the same radius
7. a line that intersects a circle in exactly one point
8. the distance around a circle
9. arcs of a circle that have exactly one point in common
10. arcs of congruent circles that have the same measure

a. chord
b. congruent circles
c. inscribed angle
d. congruent arcs
e. adjacent arcs
f. tangent
g. circumference
h. secant
i. central angle
j. concentric circles

SKILLS AND CONCEPTS

OBJECTIVES AND EXAMPLES	REVIEW EXERCISES

Upon completing this chapter, you should be able to:

- identify and use parts of circles, and solve problems involving the circumference of circles (Lesson 9–1)

$\overline{JK}$, $\overline{LM}$, and $\overline{HI}$ are chords of $\odot C$. $\overline{LM}$ is the diameter, and $\overline{CQ}$, $\overline{CL}$, and $\overline{CM}$ are radii. If $CM = 9$, find the circumference.

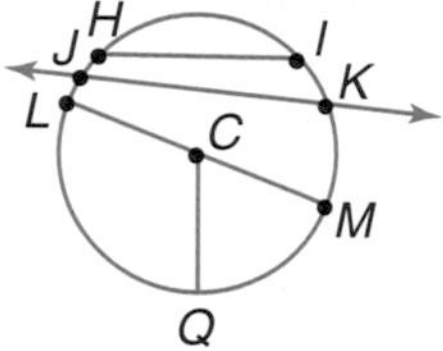

$$C = 2\pi r$$
$$= 2\pi(9)$$
$$= 18\pi$$ The circumference of circle P is 18π units.

Use these exercises to review and prepare for the chapter test.

Refer to $\odot T$ for Exercises 11–13.

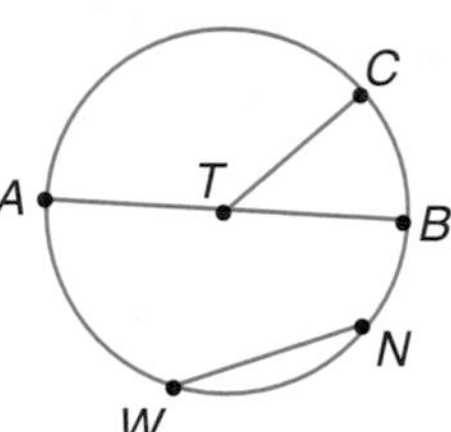

11. Name a chord that is also a diameter.

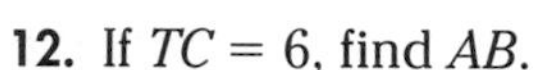

12. If $TC = 6$, find AB.
13. If $AT = 14$, find the circumference to the nearest tenth.

- find the measures of arcs and central angles (Lesson 9–2)

$m\angle FOG = 36$

$m\widehat{FG} = 36$

If $FO = 4$, find the length of $\widehat{FG}$.

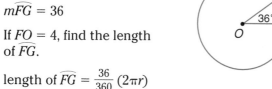

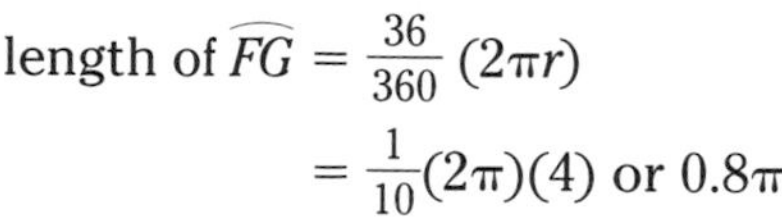

$$\text{length of } \widehat{FG} = \frac{36}{360}(2\pi r)$$
$$= \frac{1}{10}(2\pi)(4) \text{ or } 0.8\pi$$

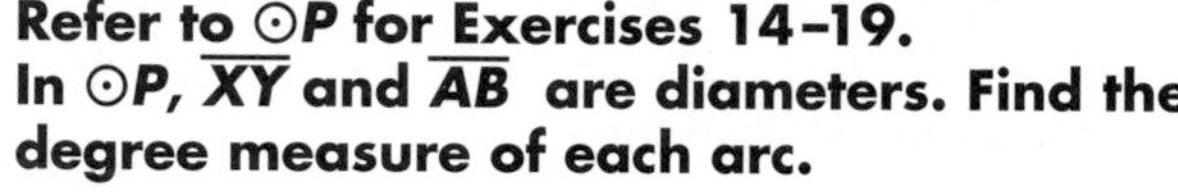

Refer to $\odot P$ for Exercises 14–19. In $\odot P$, $\overline{XY}$ and $\overline{AB}$ are diameters. Find the degree measure of each arc.

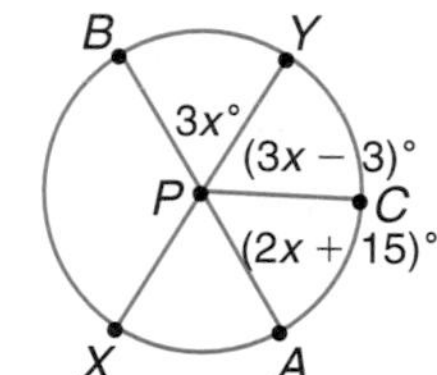

14. $m\widehat{YC}$
15. $m\widehat{BC}$
16. $m\widehat{BX}$

Find the length of each arc if $BA = 9$. Round to the nearest tenth.

17. $\widehat{CA}$
18. $\widehat{YAX}$
19. $\widehat{ABY}$

- recognize and use relationships between arcs, chords, and diameters (Lesson 9–3)

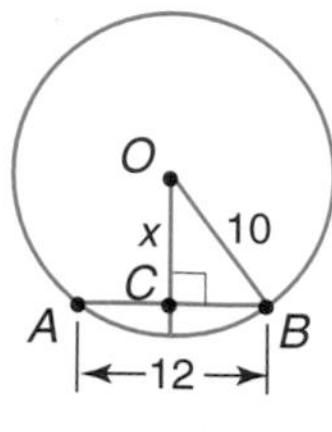

Find the value of x.

$$CB = \frac{1}{2}AB$$
$$= \frac{1}{2}(12) \text{ or } 6$$
$$(OC)^2 + (CB)^2 = (OB)^2$$
$$x^2 + 6^2 = 10^2$$
$$x^2 + 36 = 100$$
$$x^2 = 64$$
$$x = 8$$

20. A chord is 5 centimeters from the center of a circle with a radius of 13 centimeters. Find the length of the chord.
21. Suppose a 24-centimeter chord of a circle is 32 centimeters from the center of the circle. Find the length of the radius.
22. Suppose the diameter of a circle is 20 centimeters long and a chord is 16 centimeters long. Find the distance between the chord and the center of the circle.
23. Suppose the diameter of a circle is 10 inches long and a chord is 6 inches long. Find the distance between the chord and the center of the circle.

OBJECTIVES AND EXAMPLES

- recognize and find measures of inscribed angles (Lesson 9–4)

$m\angle XYZ = \frac{1}{2}m\widehat{XZ}$

$= \frac{1}{2}(78)$

$= 39$

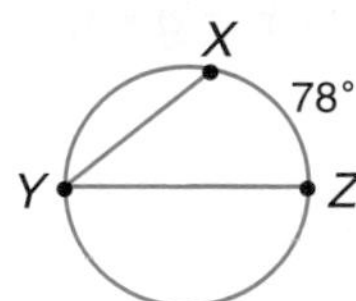

REVIEW EXERCISES

In ⊙P, $\overline{AB} \parallel \overline{CD}$, $m\widehat{BD} = 72$, and $m\angle CPD = 144$. Find each measure.

24. $m\angle DAB$
25. $m\widehat{CD}$
26. $m\widehat{CA}$
27. $m\angle CDA$
28. $m\widehat{AB}$

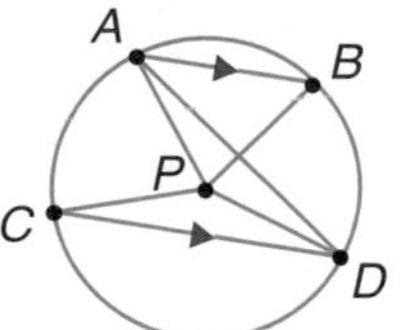

- recognize and use properties of tangents (Lesson 9–5)

Find the values of x and y.

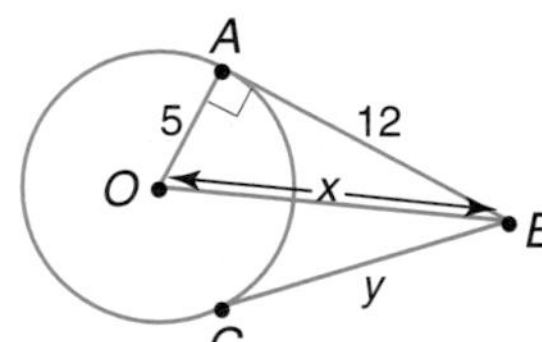

$5^2 + 12^2 = x^2$

$25 + 144 = x^2$

$169 = x^2$

$13 = x$

$CB = AB$

$y = 12$

For each ⊙C, find the value of x. Assume that segments that appear to be tangent are tangent.

29.

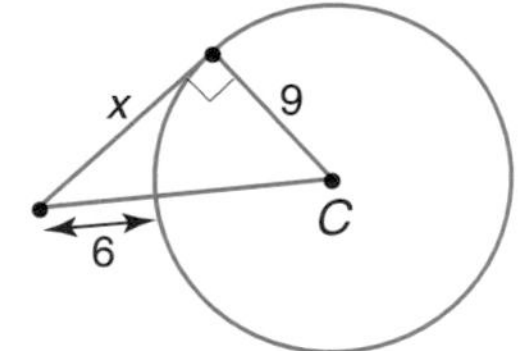

30.

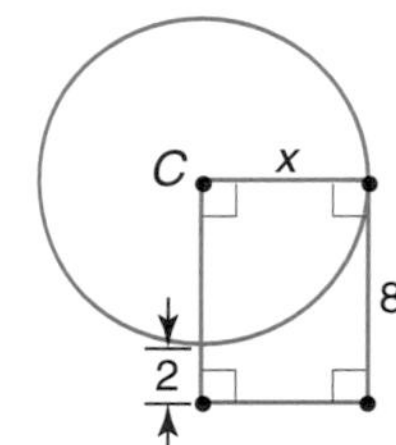

- find the measures of angles formed by intersecting secants and tangents in relation to intercepted arcs (Lesson 9–6)

$m\angle DXC = \frac{1}{2}(m\widehat{DC} + m\widehat{AB})$

$= \frac{1}{2}(60 + 20)$

$= \frac{1}{2}(80)$ or 40

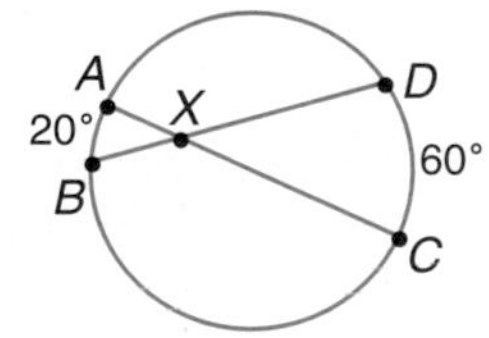

In ⊙P, $m\widehat{AB} = 29$, $m\angle AEB = 42$, $m\widehat{BG} = 18$, and $\overline{AC}$ is a diameter. Find each measure.

31. $m\angle DEC$
32. $m\widehat{CD}$
33. $m\angle GFD$
34. $m\widehat{AD}$
35. $m\angle AED$
36. $m\widehat{GC}$

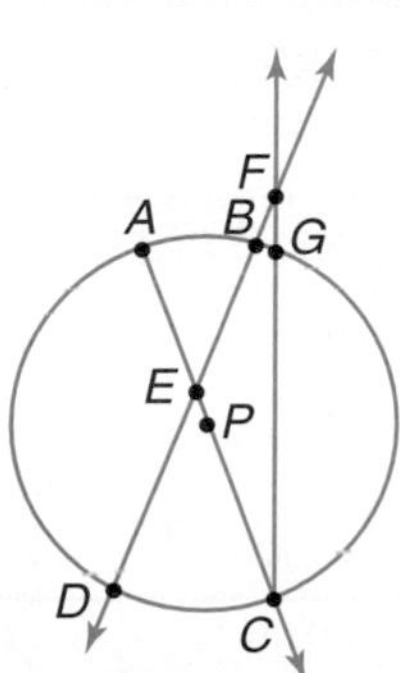

OBJECTIVES AND EXAMPLES

- use properties of chords, secants, and tangents to solve segment measure problems (Lesson 9–7)

Find the value of x.

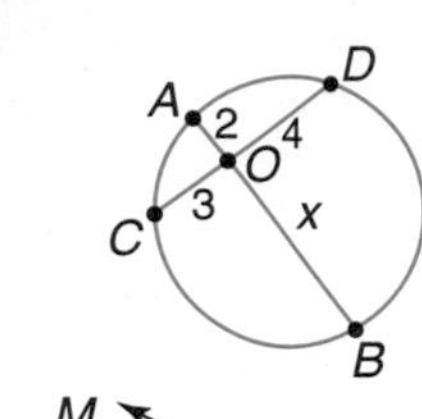

$$(AO)(OB) = (CO)(OD)$$
$$2x = 3 \cdot 4$$
$$2x = 12$$
$$x = 6$$

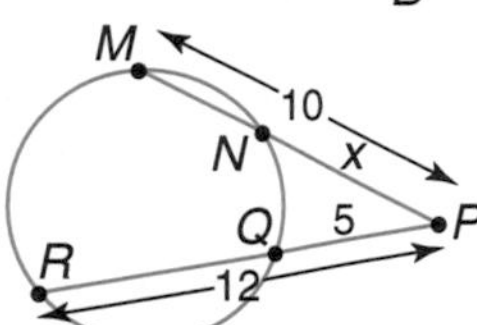

$$(MP)(NP) = (QP)(RP)$$
$$10x = 5 \cdot 12$$
$$10x = 60$$
$$x = 6$$

REVIEW EXERCISES

Find the value of *x* to the nearest tenth. Assume that segments that appear to be tangent are tangent.

37.

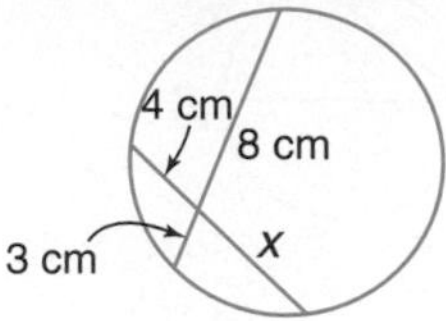

38.

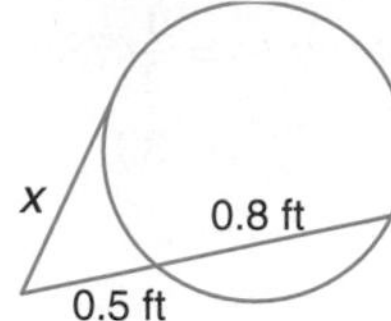

39.

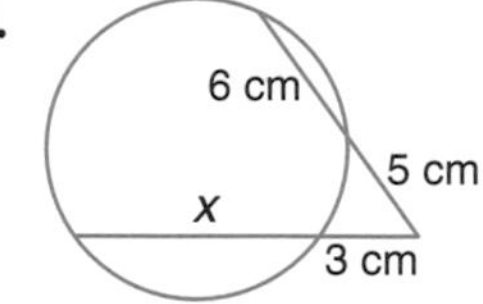

40.

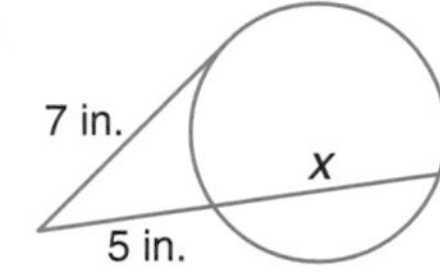

- write and use the equation of a circle in the coordinate plane (Lesson 9–8)

Write the equation of a circle with center at $(-1, 4)$ and radius 3.

$$(x - h)^2 + (y - k)^2 = r^2$$
$$[x - (-1)]^2 + (y - 4)^2 = 3^2$$
$$(x + 1)^2 + (y - 4)^2 = 9$$

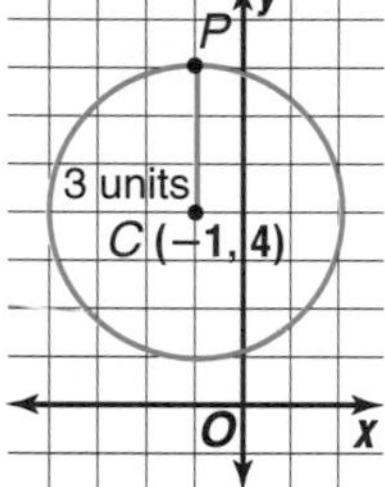

Write the equation of each circle based on the given information.

41. center: $P(-4, 3)$
radius: 6

42.

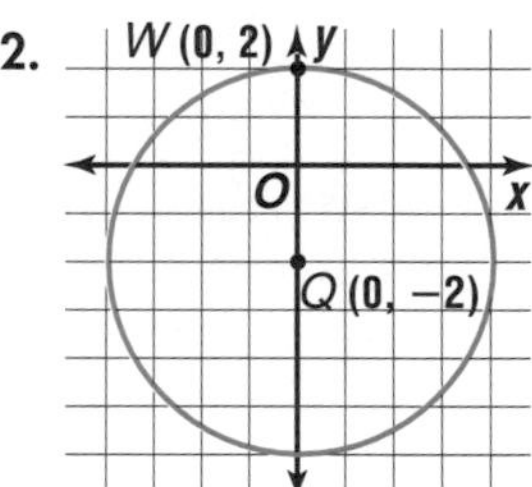

APPLICATIONS AND PROBLEM SOLVING

43. Make a Circle Graph Ms. Perez is opening a music store. She gathers information to help her decide what to sell. Use the information in the table below to make a circle graph showing the portion of sales for each type of recording. Draw the graph by hand or use graphing software. (Lesson 9–2)

Music Sales	
Compact discs	52%
Cassettes	40%
45s	6%
12-inch singles	2%

44. Construction The support of a bridge is in the shape of an arc. The span of the bridge (the length of the chord connecting the endpoints of the arc) is 28 meters. The highest point of the arc is 5 meters above the imaginary chord connecting the endpoints of the arc. What is the radius of the circle that forms the arc? (Lesson 9–3)

45. Crafts Sara uses wooden spheres to make paperweights to sell at craft shows. She cuts off a flat surface for each base. If the original sphere has a radius of 4 centimeters and the diameter of the flat surface is 6 centimeters, what is the height of the paperweight? (Lesson 9–3)

A practice test for Chapter 9 is provided on page 801.

ALTERNATIVE ASSESSMENT

COOPERATIVE LEARNING PROJECT

Tilted circles Circular and spherical objects are everywhere throughout the universe. In this project, you will apply your knowledge of circles to planetary rings and spiral galaxies. Planetary rings and spiral galaxies are circular in shape, but because they tilt toward or away from us, they appear elliptical. The greater the tilt, the less circular they look.

An ellipse has a major and minor axis. You will measure these and use this information to determine the degree of tilt.

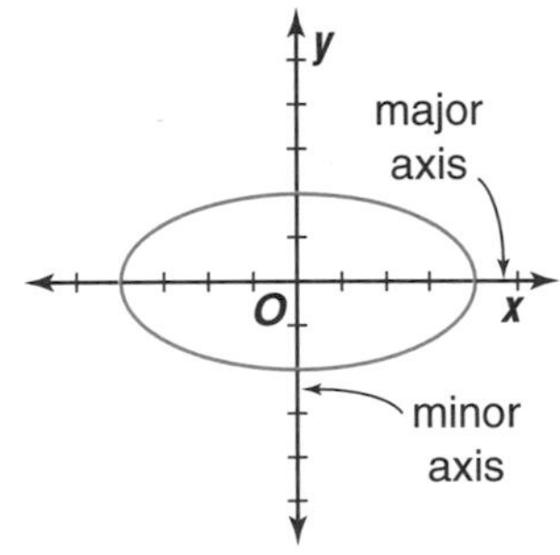

Follow these steps to determine the degree of tilt.

- Find five photographs of the planet Saturn with its rings at different degrees of tilt. Find five photographs of spiral galaxies with varying degrees of tilt.
- For the first photograph, measure the major and minor axes. Find the ratio of minor to major axis.
- Now cut out a 10-inch diameter circle from construction paper. Have one student hold it while another student views it edgewise, so that it appears paper thin. Slowly tilt it toward the observer until the proportions look the same as the photograph.
- Use a protractor to measure the angle of tilt of the circle. Use a ruler to measure the ratio of the minor to major axis as they appear to the student observing the tilted circle.
- Repeat the same procedure for each photograph and label each angle of tilt.

Follow these steps to discover how to use trigonometry to confirm the degree of tilt.

- Look at your tilted 10-inch diameter circle again. Do you see a right triangle that will allow you to apply trigonometry? Look at it from the side. Look at it from the observer's point of view.
- What information do you have? What information do you need to compute? Do you need to use sine, cosine, or tangent to draw your conclusion?
- Apply your trigonometric method to the measurements of major and minor axes from the photographs to compute the angle of tilt.

THINKING CRITICALLY

Circle graphs are geometric models used to represent nongeometric information. Think of other geometric models used to represent non-geometric information. Why do you think geometric models are useful? Are there ways in which they can be misleading?

PORTFOLIO

Before it was discovered that Earth revolved around the Sun, it was believed that the sun and all the planets revolved around Earth in perfect circular orbits. This model was known as the Ptolemaic system. Read about this model and draw a diagram of the solar system according to Ptolemy. Keep this diagram in your portfolio.

LONG-TERM PROJECT

In•ves•ti•ga•tion

Just For Kicks

MATERIALS NEEDED

- ruler
- protractor
- calculator
- posterboard
- markers

Football, or soccer as it is called in the United States, is the most popular game in the world. It is enjoyed by millions of people in more than 150 countries. A game similar to soccer was played in China as early as 400 B.C. The rules that we use in a game of soccer today were established in England in the 1800s. Every four years, the best soccer teams in the world compete for the sport's top prize, the World Cup. The Fédération Internationale de Football Association (FIFA) is the world soccer authority. It establishes the rules for international play, including professional, semiprofessional, and interclub play.

Imagine that your group is a design team for a sporting goods manufacturer. Seeing how popular soccer is becoming in the United States, your company has decided to get involved in this growing market by producing a soccer ball. It is your team's job to design the ball that your company will produce.

CHECKING OUT THE COMPETITION

The marketing research division has produced design prints of the two best-selling soccer balls on the market. The top-selling soccer ball is made by Everkick. The design of this ball is comprised of pentagons and hexagons.

Below are three views of the Everkick soccer ball. Analyze the design pattern and determine the number of pentagons and hexagons it takes to make a ball. Explain your conclusion using mathematical reasoning.

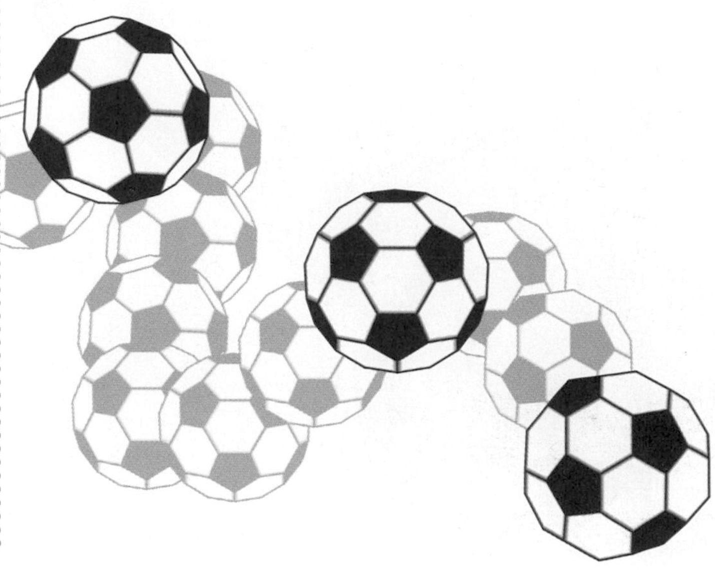

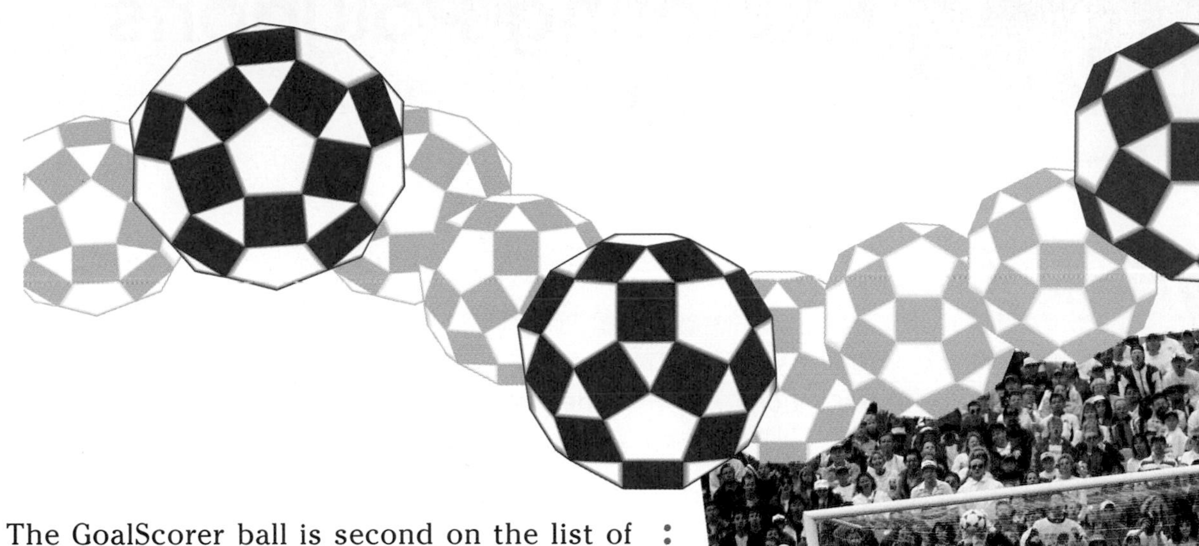

The GoalScorer ball is second on the list of best-selling soccer balls. It has a different design, which is comprised of pentagons, triangles and squares. Above are three views of the GoalScorer ball. Analyze the design pattern and determine the number of pentagons, triangles and squares it takes to make a ball. Explain your conclusion using mathematical reasoning.

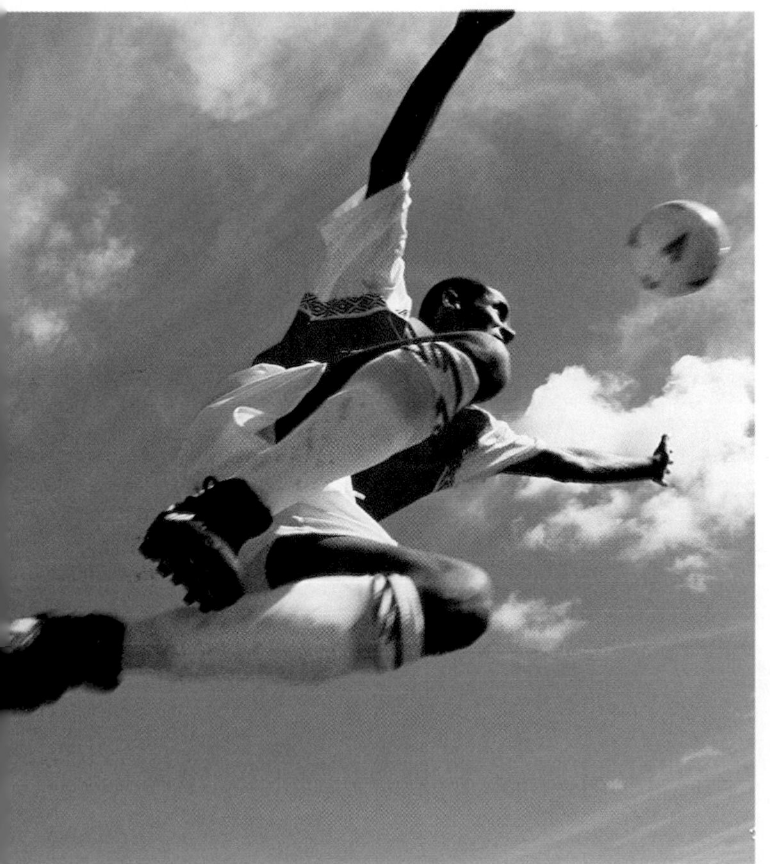

Discuss the advantages and disadvantages of each design. Brainstorm a list of ideas for your soccer ball design.

You will continue working on this Investigation throughout Chapters 10 and 11.

Be sure to keep your list of observations and materials in your Investigation Folder.

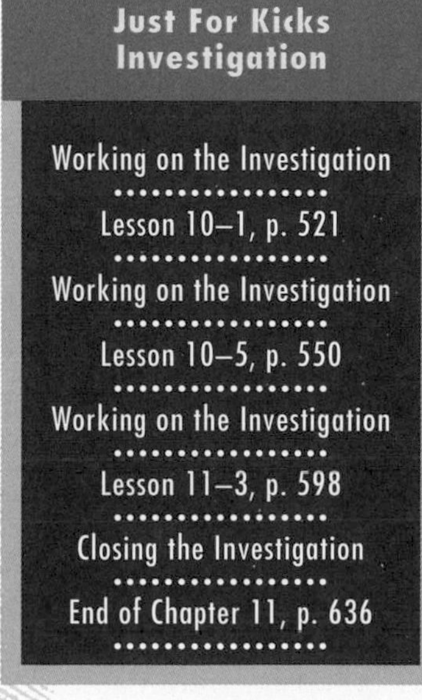

Just For Kicks Investigation
Working on the Investigation
Lesson 10–1, p. 521
Working on the Investigation
Lesson 10–5, p. 550
Working on the Investigation
Lesson 11–3, p. 598
Closing the Investigation
End of Chapter 11, p. 636

CHAPTER

10 Exploring Polygons and Area

What You'll Learn

In this chapter, you will:

- find measures of interior and exterior angles of polygons,
- find areas of polygons and circles,
- solve problems involving geometric probability,
- determine characteristics of networks, and
- solve problems by using guess and check.

Why It's Important

Real World

Entertainment Interlocking polygons are often used in structures you see everyday. In the framework for the roller coaster, you can see parallelograms and triangles. You can see polygons also in the picture of Spaceship Earth near Orlando, Florida. It is often necessary to find the area of polygons. Area formulas of some polygons are related. For example, the formulas for the area of a parallelogram and a triangle are closely related to the formula for the area of a rectangle. You will learn and use formulas for areas of various polygons in this chapter.

PREREQUISITE SKILLS

To be successful in this chapter, you'll need to understand these concepts and be able to apply them. Refer to the lesson in parentheses if you need more review before beginning the chapter.

Use the properties of 45°–45°–90° triangles. (Lesson 8–2)

Find the values of x and y.

1. 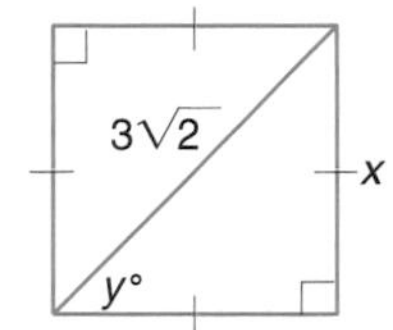

2. 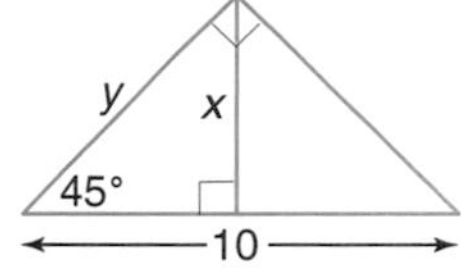

3. 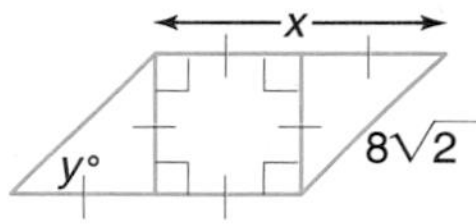

Find the slopes of lines. (Lesson 3–3)

Find the slope of the line passing through the given points.

4. $M(-6, 6), N(3, 10)$

5. $C(-4, 2), D(5, 2)$

6. $A(9, -1), B(5, 0)$

7. $X(0, 7), Y(-2, 4)$

Solve problems involving the circumference of a circle. (Lesson 9–1)

The radius, diameter, or circumference of a circle is given. Find the other measures to the nearest tenth.

8. $r = 4, d = ?, C = ?$

9. $r = ?, d = 10, C = ?$

10. $r = ?, d = ?, C = 42$

READING SKILLS

In this chapter, you'll be learning about ***polygons*** and formulas for finding areas. In Lesson 10–1, you will learn the names for various polygons. Sometimes you can use prefixes to determine the number of sides of a polygon. For example, a triangle has three sides, since the prefix "tri" means three. A hexagon has six sides, since the prefix "hex" means six. An octagon has eight sides, since the prefix "octa" means eight. (Remember, an octopus has eight legs.) Believe it or not, the prefix "dodeca" means twelve.

10-1 Polygons

What YOU'LL LEARN

- To identify and name polygons,
- to find the sum of the measures of interior and exterior angles of convex polygons and measures of interior and exterior angles of regular polygons, and
- to solve problems involving angle measures of polygons.

Why IT'S IMPORTANT

You can use polygons to solve problems involving architecture, landscaping, and recreation.

Real World APPLICATION

Architecture

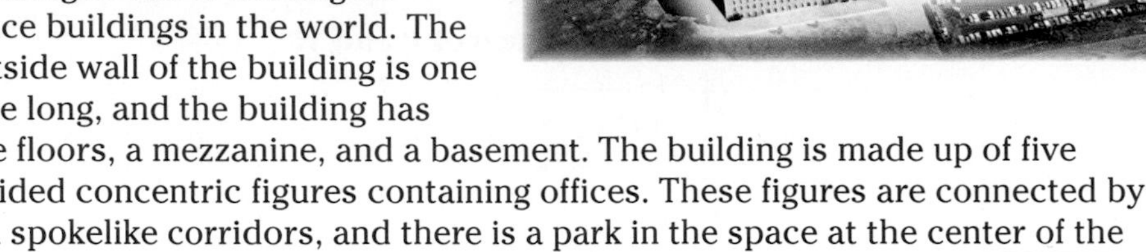

Completed in 1943, the Pentagon building is one of the largest office buildings in the world. The outside wall of the building is one mile long, and the building has five floors, a mezzanine, and a basement. The building is made up of five 5-sided concentric figures containing offices. These figures are connected by ten spokelike corridors, and there is a park in the space at the center of the innermost 5-sided figure. Each figure in the Pentagon building is a **polygon**.

The term *polygon* is derived from the Greek word meaning "many-angled." Look at the figures below. The figures on the left are polygons, and the figures on the right are not.

polygons

not polygons

You can determine the properties of polygons using these figures.

Definition of Polygon	**A polygon is a closed figure formed by a finite number of coplanar segments such that** **1. the sides that have a common endpoint are noncollinear, and** **2. each side intersects exactly two other sides, but only at their endpoints.**

A **convex** polygon is a polygon such that no line containing a side of the polygon contains a point in the interior of the polygon. A polygon that is not convex is *nonconvex* or **concave**.

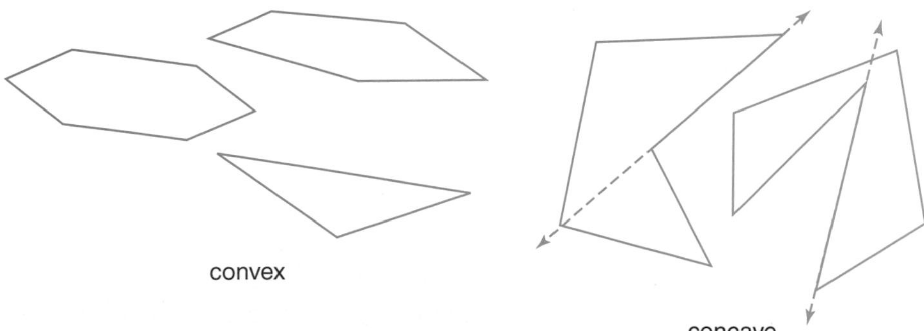

convex

concave

If a line segment in each of the concave polygons is extended, then the polygon fails to satisfy the definition for convex because a line containing a side of the polygon has a point on the interior of the polygon.

Polygons may be classified by the number of sides they have. The chart at the right gives some common names for polygons. In general, a polygon with n sides is called an ***n*-gon**. This means the *nonagon* can also be called a 9-gon.

Number of Sides	Polygon
3	triangle
4	quadrilateral
5	pentagon
6	hexagon
7	heptagon
8	octagon
9	nonagon
10	decagon
12	dodecagon
n	n-gon

When referring to a polygon, we use its name and list the vertices in consecutive order. Pentagon *RSTUV* and pentagon *TUVRS* are two possible correct names for the polygon at the right. In pentagon *RSTUV*, all the sides are congruent, and all the angles are congruent. When a polygon has these characteristics, it is called a **regular polygon**.

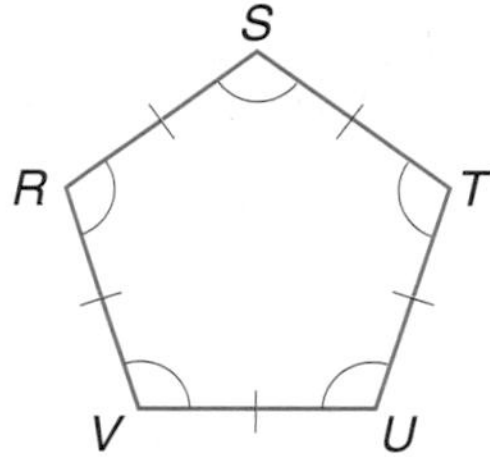

Definition of Regular Polygon	**A regular polygon is a convex polygon with all sides congruent and all angles congruent.**

An equilateral triangle is a regular polygon.

Example 1

Name and classify each polygon

a. by the number of sides,

b. as convex or concave, and

c. as regular or not regular.

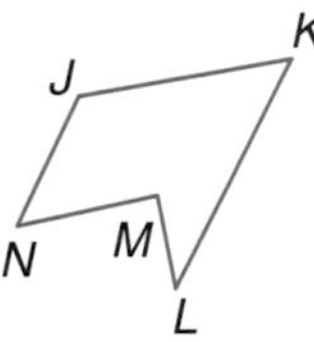

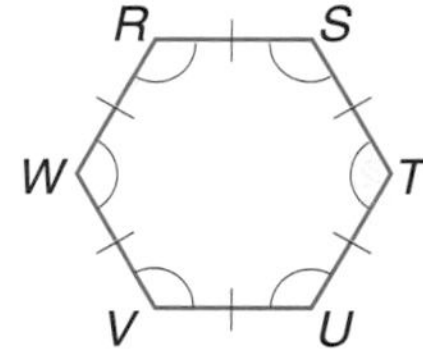

a. Polygon *JKLMN* has five sides. It is a pentagon.

b. If $\overline{ML}$ is extended, it passes through the interior of the pentagon. So, the pentagon is concave.

c. Since it is concave, pentagon *JKLMN* cannot be regular.

a. Polygon *RSTUVW* has six sides, so it is a hexagon.

b. No lines containing sides of the hexagon pass through the interior. Therefore, the polygon is convex.

c. Since all sides and all angles are congruent, hexagon *RSTUVW* is regular.

Consider each convex polygon below with all possible diagonals drawn from one vertex.

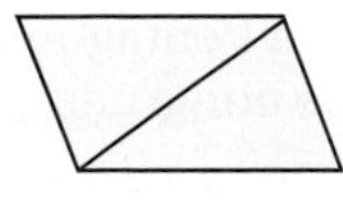
quadrilateral

pentagon

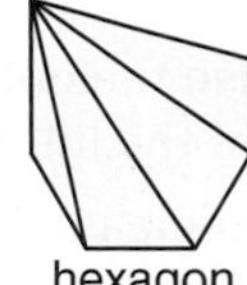
hexagon

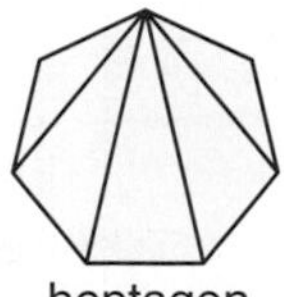
heptagon

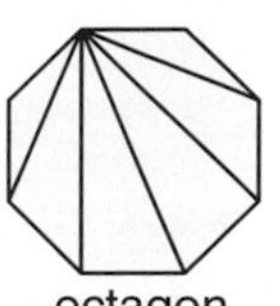
octagon

Notice that in each case, the polygon is separated into triangles. The sum of the measures of the angles of each polygon can be found by adding the measures of the angles of the triangles. Since the sum of the measures of the angles in a triangle is 180, we can easily find this sum. Make a chart to find the sum of the angle measures for several convex polygons.

TECHNOLOGY Tip

A spreadsheet can be used to calculate each sum. If cell B2 contains the number of sides, cell B3 can contain the formula, B2 − 2, and the formula for B4 is 180 * B3.

Convex Polygon	Number of Sides	Number of Triangles	Sum of Angle Measures
triangle	3	1	(1 · 180) or 180
quadrilateral	4	2	(2 · 180) or 360
pentagon	5	3	(3 · 180) or 540
hexagon	6	4	(4 · 180) or 720
heptagon	7	5	(5 · 180) or 900
octagon	8	6	(6 · 180) or 1080

Look for a pattern in the angle measures. In each case, the sum of the angle measures is 2 less than the number of sides in the polygon times 180. So in an n-gon, the sum of the angle measures will be $(n - 2)180$ or $180(n - 2)$. This conclusion, based on inductive reasoning, is stated formally in Theorem 10–1.

Theorem 10–1 Interior Angle Sum Theorem

If a convex polygon has *n* sides and *S* is the sum of the measures of its interior angles, then $S = 180(n - 2)$.

The Pentagon building, described in the application at the beginning of the lesson, has congruent sides and angles. Do you think that the angles formed in each of the five concentric pentagon-shaped buildings have the same measure, or are their measures different because the pentagons are of different sizes? Use Example 2 below to help you think about your answer. *You will be asked to write about this in Exercise 4.*

Example 

INTEGRATION Algebra

Find the measure of each interior angle of the largest pentagon-shaped section of the Pentagon building.

Use the Interior Angle Sum Theorem to find the sum of the angle measures in a convex pentagon.

$S = 180(n - 2)$

$S = 180(5 - 2)$ *For a pentagon, $n = 5$.*

$S = 180(3)$ or 540

Because the outer pentagon of the Pentagon building is a regular pentagon, each of the interior angles will be congruent. Therefore, the measure of each angle is $\frac{540}{5}$ or 108.

Example 3 **Find the measure of each interior angle in quadrilateral *RSTU* if the measure of each consecutive angle is a consecutive multiple of *x*.**

Algebra

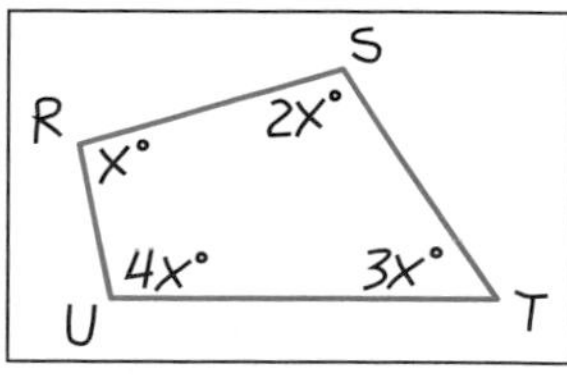

Draw a quadrilateral *RSTU* and label the measure of each angle. Let $m\angle R = x$, $m\angle S = 2x$, $m\angle T = 3x$, and $m\angle U = 4x$.

Since $n = 4$, the sum of the measures of the interior angles is $180(4 - 2)$ or 360. Write an equation to express the sum of the measures of the interior angles of the quadrilateral.

$360 = m\angle R + m\angle S + m\angle T + m\angle U$

$360 = x + 2x + 3x + 4x$ *Substitution Property (=)*

$360 = 10x$ *Combine like terms.*

$36 = x$ *Divide each side by 10.*

Use the value of *x* to find the measure of each angle.

$m\angle R = 36$, $m\angle S = 2 \cdot 36$ or 72, $m\angle T = 3 \cdot 36$ or 108, and $m\angle U = 4 \cdot 36$ or 144.

The Interior Angle Sum Theorem identifies a relationship among the interior angles of a convex polygon. Is there a relationship among the exterior angles of a convex polygon?

MODELING MATHEMATICS

Sum of the Exterior Angles of a Polygon

Materials:  straightedge  protractor

- Draw a triangle, a convex quadrilateral, a convex pentagon, a convex hexagon, and a convex heptagon.
- Extend the sides of each polygon to form exactly one exterior angle at each vertex.
- Use a protractor to measure each exterior angle of each polygon and record it on your drawing.

Your Turn

a. Copy the table below and find the sum of the exterior angles for each of your polygons.

b. What conjecture can you make?

Polygon	triangle	quadrilateral	pentagon	hexagon	heptagon
number of exterior angles					
sum of measures of exterior angles					

From the Modeling Mathematics activity, you should have discovered that the sum of the measures of the exterior angles of any convex polygon is 360. This is known as the Exterior Angle Sum Theorem.

Theorem 10–2 Exterior Angle Sum Theorem

If a polygon is convex, then the sum of the measures of the exterior angles, one at each vertex, is 360.

You will be asked to prove this theorem in Exercise 53.

Example 4 **Use the Exterior Angle Sum Theorem to find the measure of each interior angle and exterior angle of a regular octagon.**

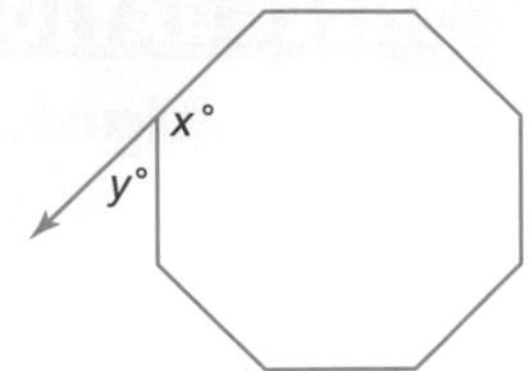

A regular octagon has eight congruent interior angles. Because each exterior angle is supplementary to an interior angle and all interior angles are congruent, all the exterior angles are also congruent. So the measure of each exterior angle y is $\frac{360}{8}$ or 45.

The measure of each interior angle x is $180 - 45$ or 135.

CHECK FOR UNDERSTANDING

Communicating Mathematics

Study the lesson. Then complete the following.

1. Determine if each figure below is a polygon. If the figure is not a polygon, explain why it is not.

   a.  **b.** 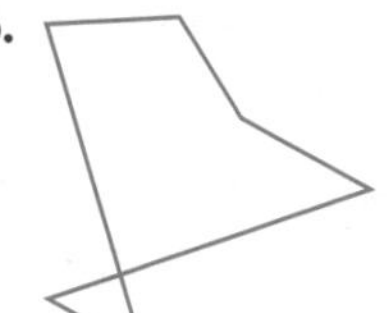 **c.**  **d.** 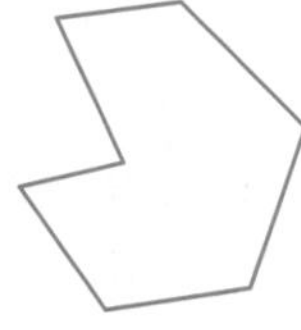

2. **You Decide** Mariana thinks that the measure of an exterior angle of a regular 15-gon is 156. Juanita says that the measure is 24. Who is correct and why?
3. **Draw** a concave pentagon and explain why it is concave.
4. **Describe** the relationship between the angles in each of the five concentric pentagon-shaped buildings in the Pentagon building described at the beginning of the lesson.
5. **Explain** how you would find the number of sides in a regular polygon if you know the measure of one of its interior angles.

FYI

The Pentagon building has one of the world's largest private telephone systems. It handles more than 200,000 calls per day and utilizes about 100,000 miles of cable.

6. Demonstrate the Exterior Angle Sum Theorem with exterior angles of any polygon.
   a. Draw a polygon on a sheet of paper. Extend the sides of the polygon to form one exterior angle at each vertex. Label the angles by number and cut them out.
   b. Arrange the angles so that all the vertices are at the same point and no angles share interior points. What do you notice?
   c. Will a different polygon have the same angle sum?

Guided Practice

Refer to polygon *ABCDEF* for Exercises 7 and 8.

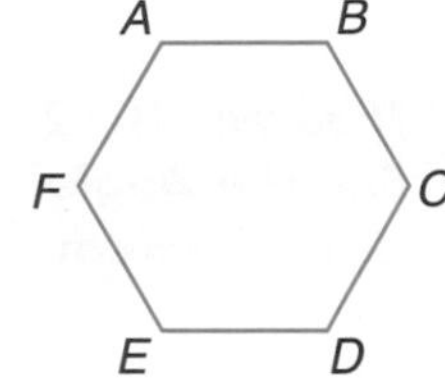

7. Name the sides of the polygon.
8. Is the polygon concave or convex? Explain.

Find the sum of the measures of the interior angles of each convex polygon

9. decagon

10. 21-gon

The measure of an exterior angle of a regular polygon is given. Find the number of sides of the polygon.

11. 30

12. 8

The number of sides of a regular polygon is given. Find the measures of an interior angle and an exterior angle of the polygon.

13. 8

14. a

The measure of an interior angle of a regular polygon is given. Find the number of sides in each polygon.

15. 120

16. 150

17. The measures of the interior angles of a pentagon are x, $3x$, $2x - 1$, $6x - 5$, and $4x + 2$. Find the measure of each angle.

18. **Landscaping** Anita Cruz is placing a large pentagon-shaped gazebo in her yard. The flower beds around the gazebo will be rectangular. What is the measure of the angle between two of the flower beds?

EXERCISES

Practice

Refer to polygon *ABCDEF* for Exercises 19–21.

19. Name the vertices of the polygon.

20. Is the polygon regular?

21. Name the polygon in two other ways.

Find the sum of the measures of the interior angles of each convex polygon.

22. 11-gon

23. 26-gon

24. 90-gon

25. 46-gon

26. x-gon

27. $3m$-gon

The measure of an exterior angle of a regular polygon is given. Find the number of sides of the polygon.

28. 72

29. 45

30. 18

31. 14.4

32. 20

33. x

The number of sides of a regular polygon is given. Find the measures of an interior angle and an exterior angle for each polygon. Round to the nearest hundredth.

34. 30

35. 16

36. 22

37. 14

38. $3m$

39. $x + 2y$

The measure of an interior angle of a regular polygon is given. Find the number of sides in each polygon.

40. 135

41. 144

42. 157.5

43. 176.4

44. 165.6

45. $\frac{180(s - 2)}{s}$

Find the measure of each angle.

46. 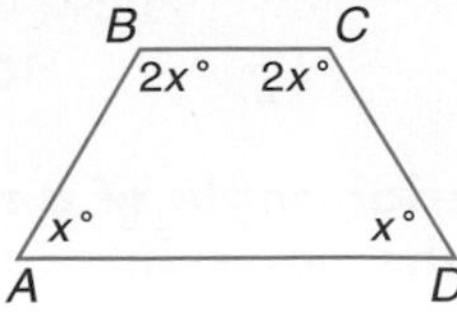

47. 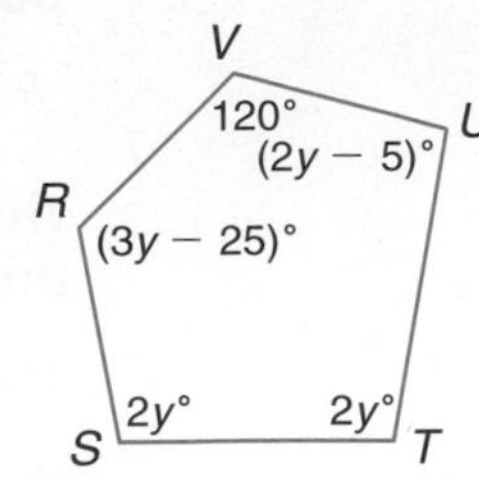

48. If the exterior angle of a regular polygon measures 36°, find the sum of the measures of the interior angles.

49. If the exterior angle of a regular polygon measures 60°, find the sum of the measures of the interior angles.

50. Use the star at the right to find each of the following.

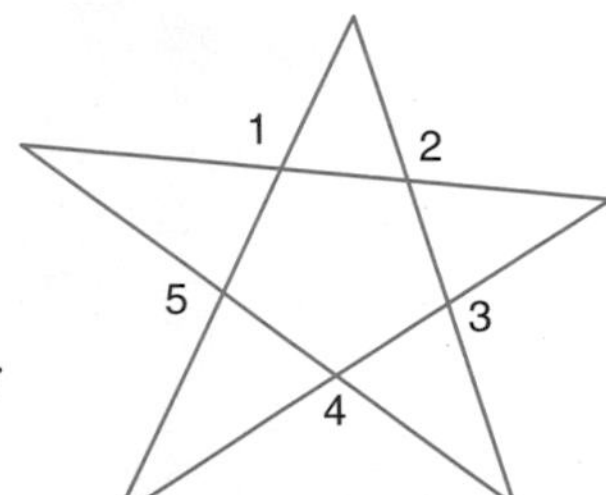

   a. the sum of the measures of the numbered angles in the exterior of the star
   b. the sum of the angles at each point of the star

51. If the measure of each of the interior angles of a regular polygon is 100 more than the measure of each of the exterior angles, name the polygon.

52. Each of the measures of the interior and exterior angles of a regular polygon are in a ratio of 5:1. Find the measures of an interior and an exterior angle and the name of the polygon.

Critical Thinking

53. Use algebra to show that if an *n*-gon is convex, then the sum of the measures of the exterior angles, one at each vertex, is 360 (Exterior Angle Sum Theorem).

54. Let *x*, *y*, and *z* represent the measures of an exterior angle of a regular 20-gon, a regular decagon, and a regular dodecagon, respectively. Without calculating exact measures, order *x*, *y*, and *z* from least to greatest. Explain your reasoning.

Applications and Problem Solving

55. **Recreation** The soccer ball design at the right is made up of two types of regular polygons. Pentagons are surrounded and interlocked by regular hexagons. Could these polygons be interlocked in a plane? Explain.

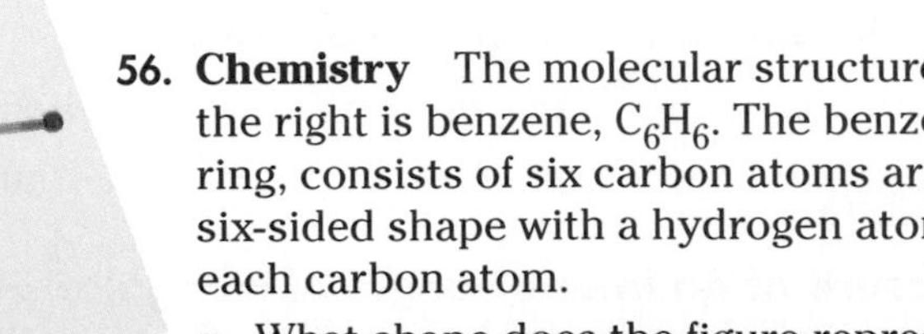

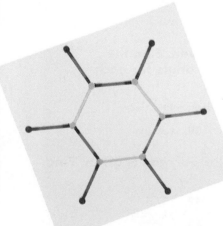

56. **Chemistry** The molecular structure represented at the right is benzene, C_6H_6. The benzene molecule, or ring, consists of six carbon atoms arranged in a flat, six-sided shape with a hydrogen atom attached to each carbon atom.

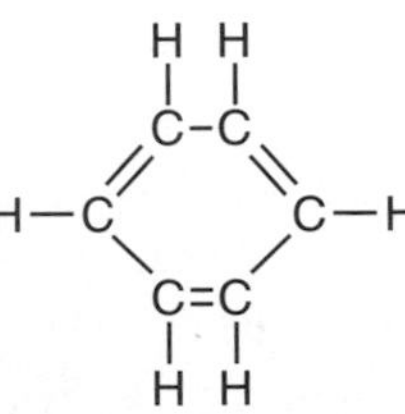

   a. What shape does the figure represent?
   b. Find the measure of each interior angle.
   c. Label all the measures of the exterior angles that are formed between the carbon and hydrogen atoms in the given representation.

Mixed Review

57. Sketch the graph of the circle whose equation is $(x + 5)^2 + (y + 2)^2 = 16$. Label the center C and the radius measure on the graph. (Lesson 9–8)

58. If the $m\widehat{QS} = 140$ and $m\widehat{QTS} = 220$, find $m\angle DRS$. (Lesson 9–6)

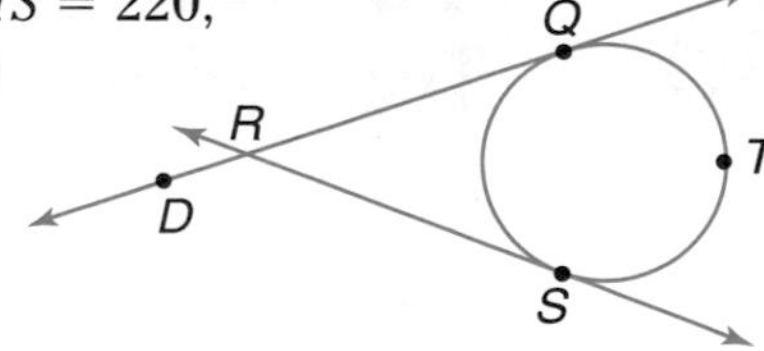

59. **Cycling** A bicycle tire has a diameter of 16 inches. How far does the bike travel during one rotation of the wheel? (Lesson 9–1)

60. **SAT Practice Grid-in** In the figure, segment DA bisects $\angle BAC$, and segment DC bisects $\angle BCA$. If the measure of $\angle ADC = 120°$, then what is the measure of $\angle B$?

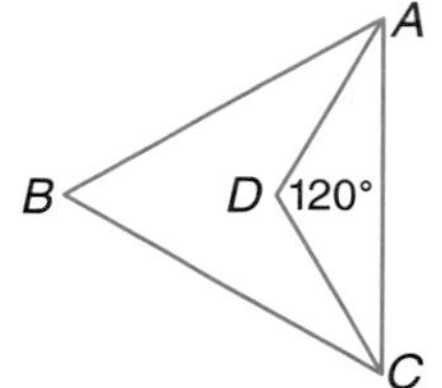

61. The measures of the legs of a right triangle are 4.0 and 5.6 centimeters. Find the measure of the hypotenuse. Round to the nearest tenth. (Lesson 8–1)

62. Use a compass and straightedge to construct a rhombus with sides equal to 8 centimeters. (Lesson 6–4)

63. List the sides of $\triangle STV$ from largest to smallest if $m\angle S = 2x + 7$, $m\angle T = 5x - 4$, and $m\angle V = 4x + 12$. (Lesson 5–4)

64. Classify the statement *An isosceles triangle has three congruent sides* as *always*, *sometimes*, or *never* true. (Lesson 4–1)

65. What property justifies the statement $m\angle K = m\angle K$? (Lesson 2–4)

Algebra

For **Extra Practice**, see page 782.

66. Simplify $12\sqrt{15} + 7\sqrt{15}$.

67. Solve $4^y = 4^{2y-3}$.

WORKING ON THE In·ves·ti·ga·tion

Refer to the Investigation on pages 510–511.

Study the patterns on the soccer balls shown on pages 510 and 511. Assume that all the polygons are regular.

1 Investigate the sum of the measures of the different combinations of angles that meet at a common vertex in each pattern.

2 Do the patterns fill the plane totally?

3 Explain your findings. How does this affect the design your team will propose for your company's soccer ball?

Add the results of your work to your Investigation Folder.

A Preview of Lesson 10–2

10–2A Tessellations and Transformations

Materials:  plain paper pencil

A **tessellation** is the result of covering a plane surface with a repeated pattern of figures without any spaces between the figures. Tessellations can be formed by transformations such as translations (slides), reflections (flips), and rotations (turns).

Activity 1 Make a tessellation using a translation.

Step 1 Start by drawing a square. Then draw a triangle inside the top and a curved figure with a dot inside the left of the square as shown.

Step 2 Translate the triangle on the top side to the bottom side.

Step 3 Translate the figure on the left side to the right side to complete the pattern unit.

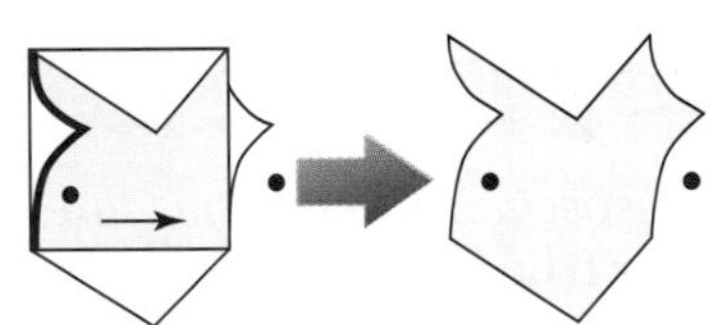

Step 4 Repeat this pattern unit on a tessellation of squares. *It is sometimes helpful to complete one pattern unit, cut it out, and trace it for the other units.*

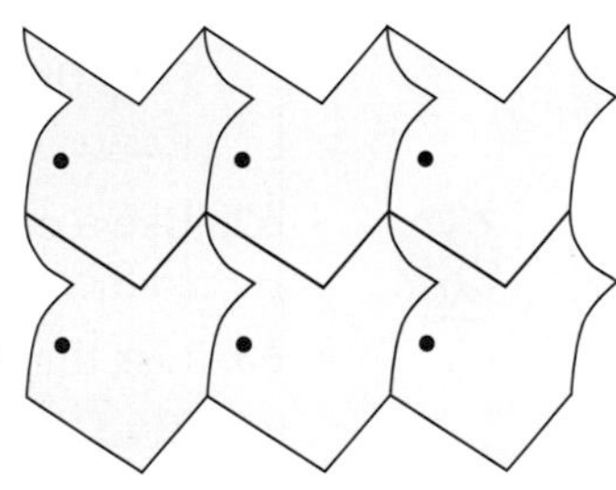

Activity 2 Make a tessellation using a rotation.

Step 1 Start by drawing an equilateral triangle. Then draw another triangle inside the right side of the triangle as shown below.

Step 2 Rotate the triangle so you can copy the change on the side indicated.

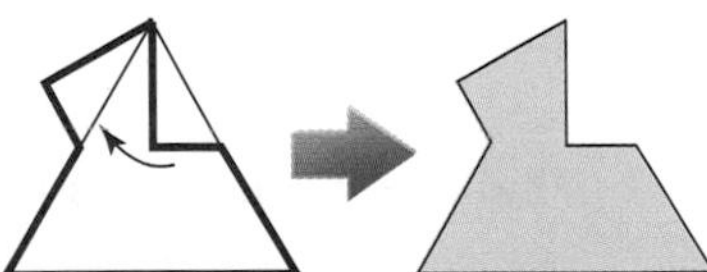

Step 3 Repeat this pattern unit on a tessellation of equilateral triangles. Alternating colors are used to best show the tessellation.

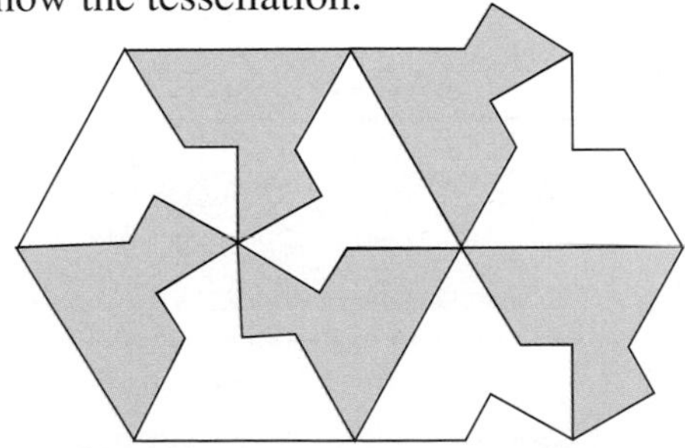

Write

1. Is the area of the original square in Step 1 of Activity 1 the same as the area of the new shape in Step 2? Explain.
2. Describe how you would create the unit for the pattern shown at the right.

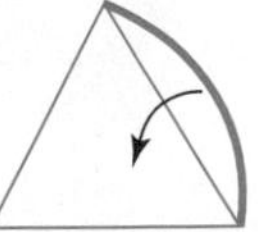

Draw **Make a tessellation for each pattern described. Use a tessellation of two rows of three squares as your base.**

3. 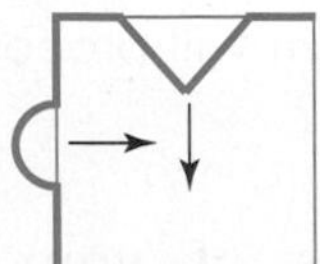
4. 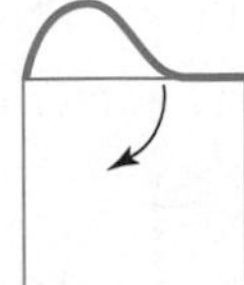
5. 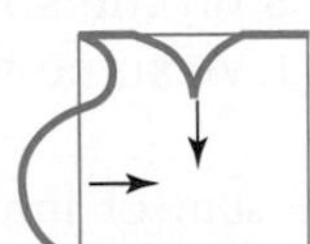

10-2 Tessellations

What YOU'LL LEARN

- To identify regular and uniform (semi-regular) tessellations,
- to create tessellations with specific attributes, and
- to solve problems by using guess and check.

Why IT'S IMPORTANT

You can use tessellations to solve problems involving interior design and quilting.

Patterns that cover a plane with repeating figures so there is no overlapping or empty spaces are called **tessellations**. A **regular tessellation** uses only one type of regular polygon. In a tessellation, the sum of the measures of the angles of the polygons surrounding a point (at a vertex) is 360.

You can use what you know about angle measures in regular polygons to help determine which polygons tessellate.

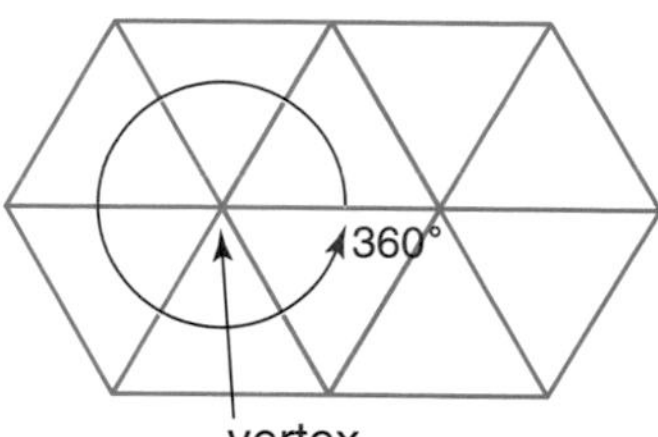

MODELING MATHEMATICS

Tessellations of Regular Polygons

Materials:  pattern blocks

- Use the pattern blocks to form regular tessellations of triangles, squares, and hexagons.
- Find the measure of one interior angle for each polygon.

Your Turn

a. Copy and complete the table below.
b. Which regular polygons tessellate?
c. What conjecture can you make?

Polygon	triangle	square	pentagon	hexagon	heptagon	octagon
Does it tessellate the plane?						
measure of one interior angle						

From the Modeling Mathematics activity, you found that if a regular polygon has an interior angle with a measure that is a factor of 360, then the polygon will tessellate the plane.

Example 1 **Determine if a regular 20-gon will tessellate the plane.**

Find the measure of each interior angle of the regular 20-gon.

$$\frac{180\,(20-2)}{20} = 162 \qquad n = 20$$

Since 162 is not a factor of 360, the 20-gon will not tessellate the plane.

A tessellation pattern can contain any number of figures. Tessellations containing the same combination of shapes and angles at each vertex are called **uniform**. The pattern below at the left will create a uniform tessellation. The pattern below at the right will form a tessellation, but it will not be uniform.

uniform

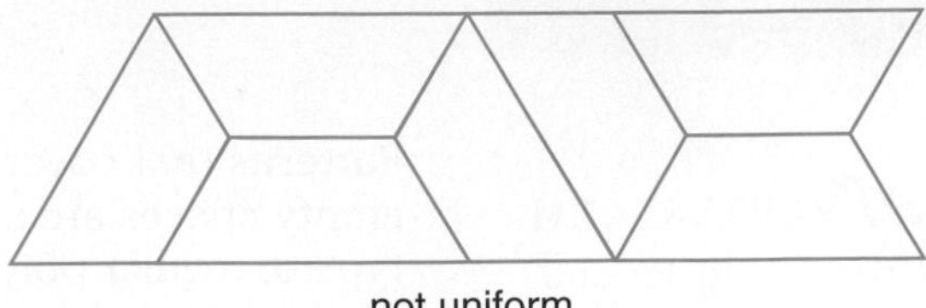
not uniform

Uniform tessellations containing two or more regular polygons are called **semi-regular**.

Example  **Determine whether a semi-regular tessellation can be created from regular octagons and squares, all having sides 1 unit long.**

There are two methods to solve this problem.

Method 1: Make a model

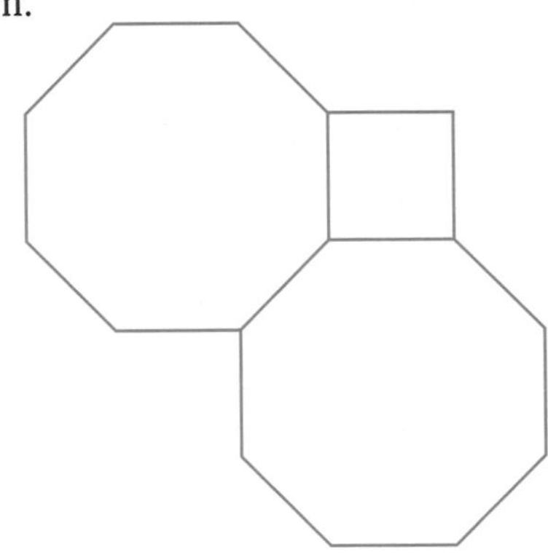

Either trace or place octagon pattern blocks side by side. You will notice that the space at each vertex could be filled in by a square. The uniform pattern at each vertex is a square and two octagons.

Method 2: Solve algebraically

Calculate the measure of each interior angle of a regular octagon.

$$\frac{180(n-2)}{n} = \frac{180(8-2)}{8} \text{ or } 135 \quad \textit{n = 8}$$

The measure of each interior angle is 135.

In the pattern, two octagons are adjacent, so there will be two angles with measures of 135 each.

$135 + 135 = 270$ *Find the sum.*

$360 - 270 = 90$ *Subtract the sum from 360.*

At each vertex, there is a 90° angle remaining. The sides will be congruent because the octagon is regular. Therefore, the figure could be a square.

The problem-solving strategy **guess and check**, or trial and error, is a powerful strategy. To use this strategy, first make an educated guess to answer a problem. Then use the conditions given in the problem to check if the answer is correct. If the first guess is not correct, use the information gathered from that guess to continue guessing until you find the correct answer.

Example **3** **Use the following three shapes and determine whether they can tessellate a row.**

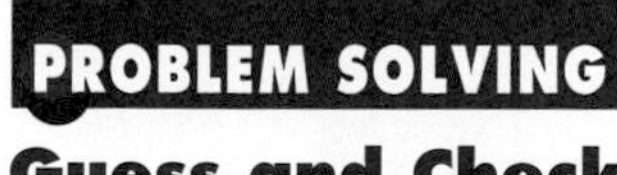

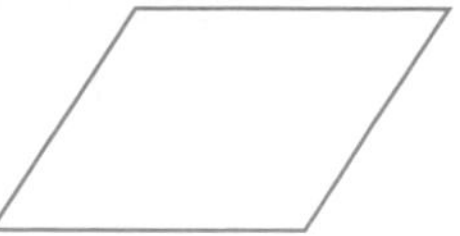

Use a model and the guess-and-check method to arrive at a possible answer.

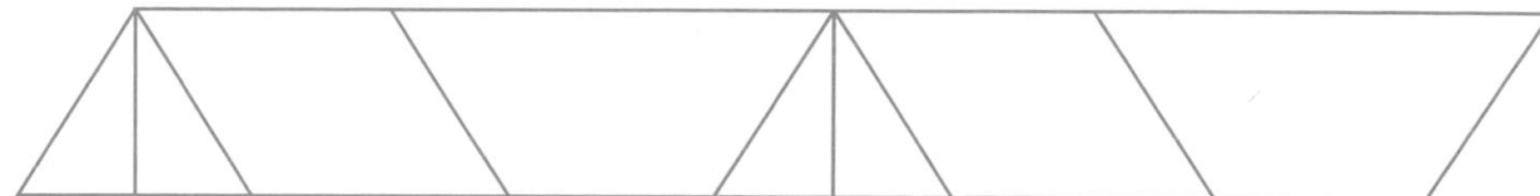

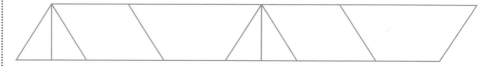

CHECK FOR UNDERSTANDING

Communicating Mathematics

Study the lesson. Then complete the following.

1. **Explain** why a rectangle will tessellate in a plane.
2. What is the difference between a regular and a semi-regular tessellation?
3. **Describe** the only three regular tessellations. Why are these the only ones?

4. Use the following blocks to create a tessellation.

5. Draw a tessellation that is semi-regular.

Guided Practice

Determine whether each regular polygon tessellates in a plane. If so, draw a sample figure.

6. decagon
7. equilateral triangle
8. Determine whether the pattern at the right will tesselate.

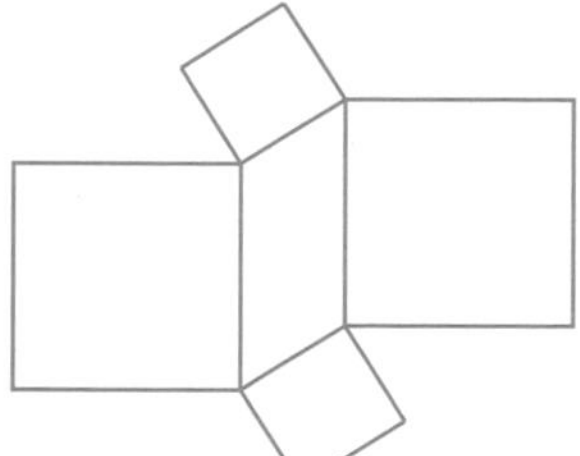

Determine whether each tessellation is *uniform, regular,* or *semi-regular.* Name all possibilities.

9. 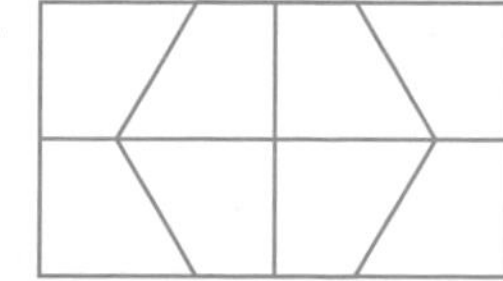

10. 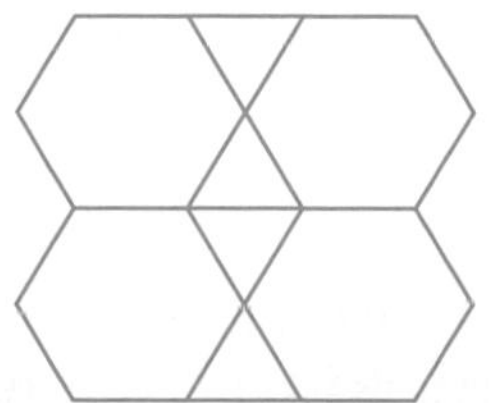

11. **Guess and Check** Replace each ? with an operation symbol and add parentheses in the following to make it true.

$$8 \ ? \ 4 \ ? \ 7 \ ? \ 3 = 31$$

EXERCISES

Practice

Determine whether each figure tessellates a plane. If so, draw a sample figure.

12. rhombus with no right angles
13. regular pentagon
14. regular14-gon
15. nonequilateral triangle

Determine whether each pattern will tessellate a plane.

16. trapezoid and triangle

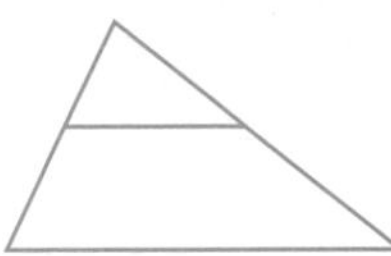

17. two rhombi

18. regular pentagon and triangle

Determine whether each tessellation is *regular, uniform,* or *semi-regular*. Name all possibilities.

19. hexagon and triangle

20. hexagon

21. square and triangle

22. hexagon, square, and triangle

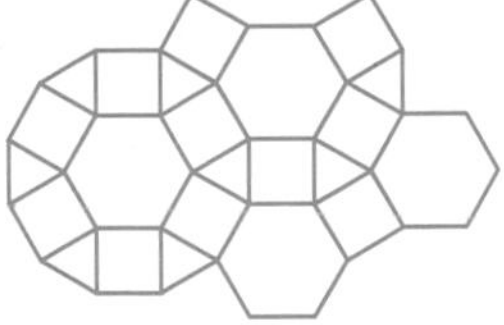

23. Draw a regular tessellation in a plane.
24. Draw a uniform tessellation in a plane.

Cabri Geometry

25. Use the reflection tool on a calculator to create a tessellation by reflecting a polygon over each of its sides. Draw your tessellation.

Critical Thinking

26. What could be the measures of the interior angles in a pentagon that tessellate a plane? Is this tessellation regular? Is it uniform?
27. Can a tessellation be uniform but not semi-regular? Explain.
28. Can a tessellation be semi-regular but not uniform? Explain.

Applications and Problem Solving

29. Guess and Check Copy the figure at the right. Find the numbers that go in the squares whose sum on each side are the circled numbers between each pair of squares.

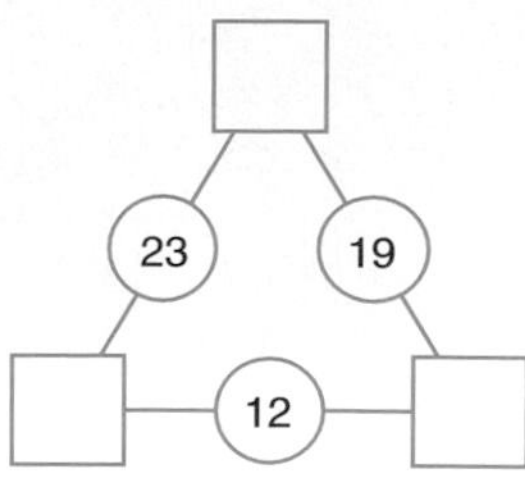

30. Interior Design Aiyana's family is tiling their rectangular recreation room in the basement with square tiles arranged so that the diagonal of the square is parallel to the wall.

a. Draw a sketch of the pattern.

b. Describe the shape of the figure needed so that the entire floor is covered with tiles.

31. Guess and Check Use the number equation below to answer each of the following.

a. If D = 5, find the unique digit represented by each letter.

b. How does the problem change if more than one letter represents the same digit?

```
 DYNALD
+SERALD
 RYCERT
```

32. Quilting In the quilt at the right, what tessellation is shown? Describe the tessellation using the terms presented in this lesson.

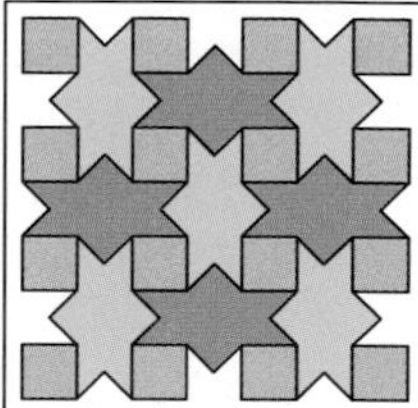

Mixed Review

33. The measure of an interior angle of a regular polygon is 160. Determine the number of sides of the polygon. (Lesson 10–1)

34. State the equation you would use to find the value of x if $\overline{RS}$ is tangent to $\odot Q$. Then find the value of x. (Lesson 9–7)

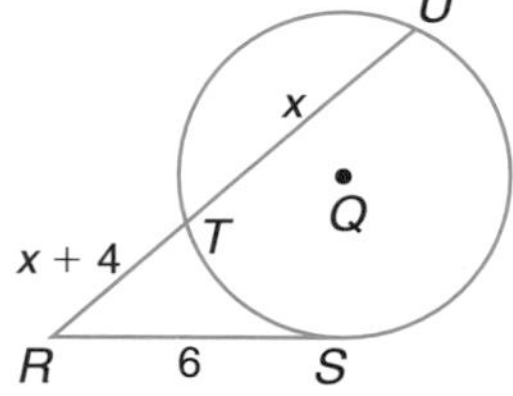

35. A chord of a circle is 10 inches long and is 12 inches from the center of the circle. What is the radius of the circle? (Lesson 9–3)

36. Make a Graph The national parks in the United States are used for many recreational activities. In a recent year, they were used for driving off-road vehicles 65,808 times, for camping 173,597 times, for hunting 46,760 times, for fishing 23,392 times, for boating 18,491 times, and for winter sports 3119 times. Make a circle graph of this data. (Lesson 9–2)

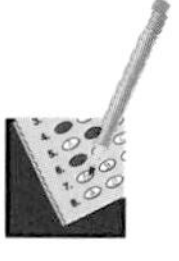

37. ACT Practice For all x, $(8x^4 - 2x^2 + 3x - 5) - (2x^4 + x^3 + 3x + 5) = ?$

A $6x^4 - x^3 - 2x^2 - 10$ **B** $6x^4 - 3x^2$

C $6x^4 + x^3 - 2x^2 + 6x$ **D** $6x^4 - 3x^2 + 6x - 10$

E $0x^4 - x^3 - 2x^2 + 6x$

Algebra

38. Determine the slope of the line that passes through points at (1, 8) and (−4, 7).

39. Express the relation shown in the mapping at the right as a set of ordered pairs. Then state the domain, range, and inverse of the relation.

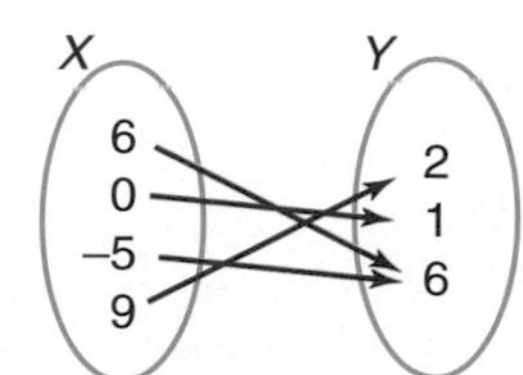

For **Extra Practice**, see page 783.

10–3A Using Technology Area of Parallelograms

A Preview of Lesson 10–3

You can use a calculator to draw a rectangle and find its area. Then using the same parallel lines and base length as the rectangle, you can draw a parallelogram and find its area. Compare the area of the rectangle to the area of the parallelogram.

Use these steps to draw a rectangle.

- Draw and label a horizontal line segment AB.
- At point A, draw a perpendicular line and label a point C on that line above $\overline{AB}$.
- Draw a line through C parallel to $\overline{AB}$.
- Construct point D on that line such that $CD = AB$.
- Use 4:Polygon on the F3 menu to draw a rectangle over $ABDC$.
- Then use 2:Area on the F6 menu to find the area of the polygon (rectangle $ABDC$).

Use the following steps to construct a parallelogram that is not a rectangle but has a base congruent to $\overline{AB}$ and is within the same parallel lines, $\overleftrightarrow{AB}$ and $\overleftrightarrow{CD}$, as rectangle $ABDC$.

- Using rectangle $ABDC$ from above, label a point E between C and D that is on $\overleftrightarrow{CD}$.
- Label point F on $\overleftrightarrow{CD}$ such that $EF = AB$.
- Use 4:Polygon on the F3 menu to trace parallelogram $ABFE$.
- Then use 2:Area on the F6 menu to find the area of the polygon (parallelogram $ABFE$).

EXERCISES

Analyze your drawing.

1. What is the length of $\overline{AB}$?
2. What is the height of $\overline{AC}$?
3. What is the area of rectangle $ABDC$?
4. What is the height of parallelogram $ABFE$?
5. What is the area of parallelogram $ABFE$?
6. Compare the areas of the two figures. Explain what you found.
7. Use the rule for finding the area of a rectangle to make a conjecture about the rule for finding the area of a parallelogram.

Test your conjecture.

Use a calculator to draw and label each pair of figures so that they have the same area.

8. a square and a parallelogram that is not a square
9. a square and a rectangle that is not a square
10. two noncongruent rectangles

10–3 Area of Parallelograms

What YOU'LL LEARN

- To find areas of parallelograms.

Why IT'S IMPORTANT

You can use the area of parallelograms to solve problems involving housing, carpeting, and paving.

Real World APPLICATION

Architecture

Both old and new buildings in New York City have geometric figures in their designs. One of the most prevalent figures in construction is a *parallelogram.*

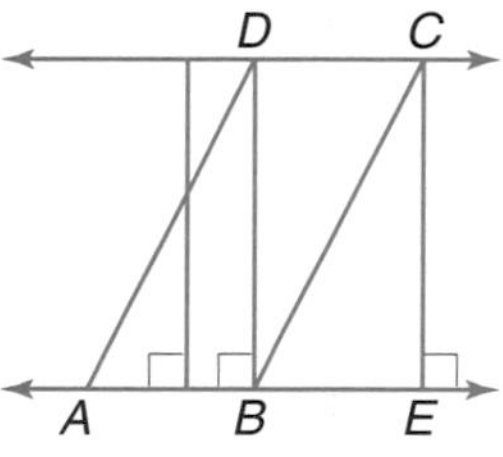

Any side of a parallelogram can be called a **base**. For each base, there is a corresponding **altitude**. In ▱$ABCD$, $\overline{AB}$ is considered to be the base. A corresponding altitude is any perpendicular segment between the parallel lines $\overleftrightarrow{AB}$ and $\overleftrightarrow{DC}$. So in ▱$ABCD$, $\overline{CE}$ and $\overline{DB}$ are altitudes. The length of an altitude is called the **height** of the parallelogram.

Tallest Buildings in New York City (feet, stories)

1. World Trade Center (1368, 110)
2. Empire State Building (1250, 102)
3. Chrysler Building (1046, 77)
4. American International (950, 67)
5. 40 Wall Tower (927, 71)

In Chapter 6, we learned that parallelograms have special properties. These properties will be used to investigate the **area** of a parallelogram. The *area* of a figure is the number of square units contained in the interior of the figure.

As you recall, the area of a rectangle, A square units, can be found by using the formula $A = \ell w$, where ℓ units is the length of the rectangle and w units is the width. The formula for the area of a parallelogram is closely related to the formula for the area of a rectangle.

MODELING MATHEMATICS

Area of a Parallelogram

Materials: plain paper pencil

- Draw and cut out ▱$ABCD$.
- Fold ▱$ABCD$ so that A lies on B and C lies on D. A rectangle results.

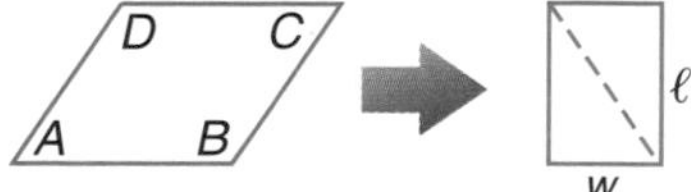

- This rectangle has an area of ℓw. Since there are 2 congruent layers of this rectangle, the area of the parallelogram is $2\ell w$.
- Unfolding the parallelogram, you see that $2w$ equals the base b of the parallelogram and that ℓ is its height h.

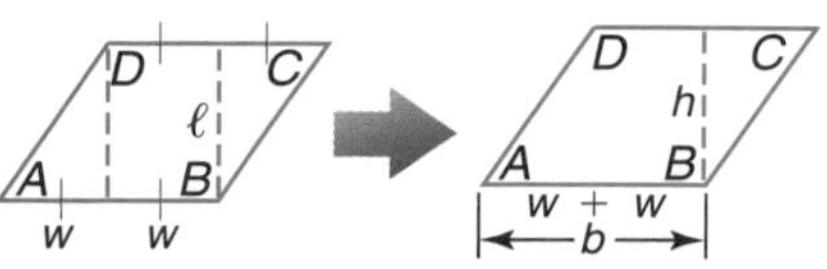

Your Turn

a. Use a ruler to measure $\overline{AB}$ or b, w, and ℓ or h.

b. Compare these measures.

c. Summarize your findings in a formula.

Area of a Parallelogram	If a parallelogram has an area of A square units, a base of b units, and a height of h units, then $A = bh$.

Example 1 **Find the area of parallelogram *RSTU*.**

Explore Study the diagram to determine what information is given, what information you will use, and what information you need.

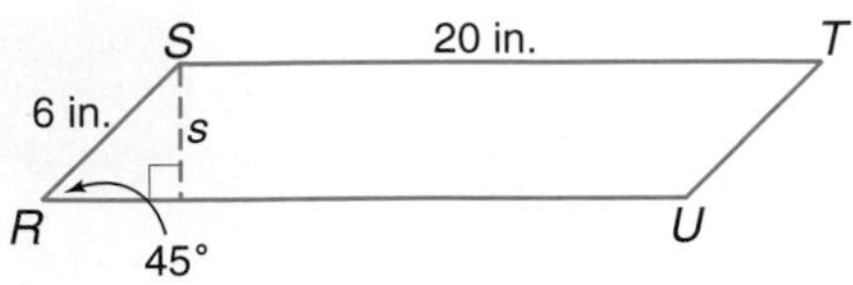

LOOK BACK

Refer to Lesson 8-2 to review relationships in 45°-45°-90° triangles.

Plan To find the area of the parallelogram, we must know the base and the height. The base, given in the figure, is 20 inches long. Since $m\angle R = 45$, the height can be found by using the 45°-45°-90° triangle that is formed when the altitude is drawn from S.

Solve Find the height by using the length of the hypotenuse. Recall that the length of the hypotenuse is $s\sqrt{2}$, where s is the length of the leg.

$$6 = s\sqrt{2}$$ *Substitute 6 for the hypotenuse.*

$$\frac{6}{\sqrt{2}} = \frac{s\sqrt{2}}{\sqrt{2}}$$ *Solve for s.*

$$\frac{6\sqrt{2}}{\sqrt{2}\sqrt{2}} = s$$ *Rationalize the denominator.*

$$3\sqrt{2} = s$$

The formula for the area of the parallelogram is $A = bh$, so $A = (20)(3\sqrt{2})$ or $60\sqrt{2}$. Therefore, the area of parallelogram *RSTU* is $60\sqrt{2}$ or about 84.85 square inches.

Examine Find the area of a 20-inch by 6-inch rectangle. The area of the parallelogram should be less than the area of the rectangle.

Example 2 **The coordinates of the vertices of a quadrilateral are at $A(3, 7)$, $B(7, 1)$, $C(1, -3)$, and $D(-3, 3)$.**

Algebra

a. Graph each point and draw the quadrilateral. Is it a *square*, a *rectangle*, or a *parallelogram*?

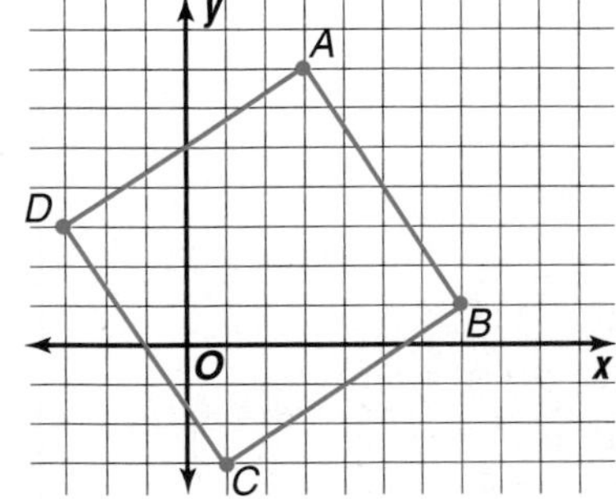

First, determine the slopes of each side.

slope of $\overline{AB} = \frac{7-1}{3-7}$ or $\frac{3}{-2}$

slope of $\overline{BC} = \frac{1-(-3)}{7-1}$ or $\frac{2}{3}$

slope of $\overline{CD} = \frac{-3-3}{1-(-3)}$ or $\frac{-3}{2}$

slope of $\overline{AD} = \frac{7-3}{3-(-3)}$ or $\frac{2}{3}$

The slopes of opposite sides are the same, so the opposite sides are parallel. The slopes of consecutive sides are negative reciprocals of each other, so those sides are perpendicular.

Since consecutive sides are perpendicular, opposite sides are parallel, and all sides are congruent, the quadrilateral is a rectangle.

Next, we can use the distance formula to find the length of two adjacent sides.

$AB = \sqrt{(3-7)^2 + (7-1)^2}$ or $\sqrt{52}$ $\qquad s = \sqrt{(x_1 - x_2)2 + (y_1 - y_2)^2}$

$BC = \sqrt{(7-1)^2 + (1-(-3))^2}$ or $\sqrt{52}$

Since two adjacent sides are congruent, the rectangle is a square.

b. Find the area of quadrilateral *ABCD*.

Use the formula for the area of a square.

$A = s^2$

$= (\sqrt{52})^2$ *Substitute side length.*

$= 52$

The area of the square is 52 square units.

Sometimes a figure can be divided into several figures. To find its area, you must find the area of each figure and then find the sum of those areas. This is stated in the postulate below.

Postulate 10–1 **The area of a region is the sum of the areas of all of its nonoverlapping parts.**

We can use this postulate to solve the following problem.

Example 3

Carpeting

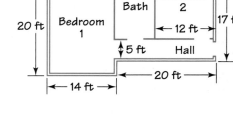

Tyrone's grandparents are moving into a retirement home. Before they move in, they want to recarpet the two bedrooms and hall. Find the amount of carpeting in square yards that is needed to cover this floor space.

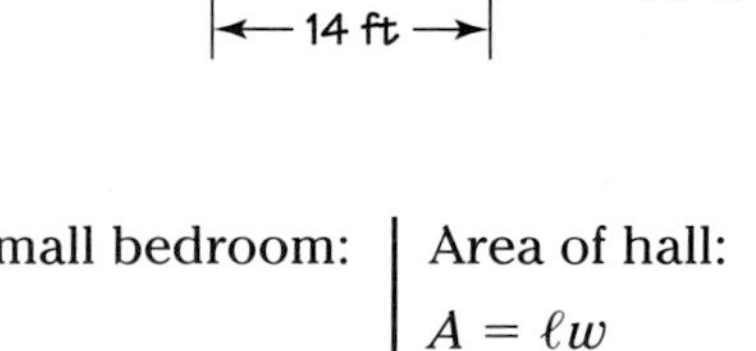

First, divide the carpeted area into regions whose areas you can find.

Area of large bedroom:	Area of small bedroom:	Area of hall:
$A = \ell w$	$A = \ell w$	$A = \ell w$
$A = (20)(14)$	$A = 12(17 - 5)$	$A = (20)(5)$
$A = 280$	$A = 144$	$A = 100$

The total area of the bedrooms and the hall is 280 + 144 + 100 or 524 square feet.

There are 9 square feet in one square yard, so divide the total number of square feet by 9 to find the number of square yards of carpeting needed.

$\frac{524}{9} \approx 58.2$

Therefore, Tyrone's grandparents will need 59 square yards of carpeting.

CHECK FOR UNDERSTANDING

Communicating Mathematics

Study the lesson. Then complete the following.

1. **Describe** the difference between 4 meters and 4 square meters.
2. **Compare** the areas of rectangle *ABCD* and parallelogram *CDEF*.

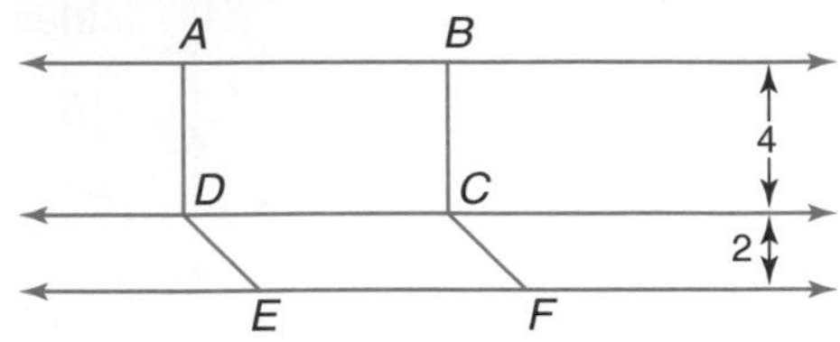

3. How can you add two squares to increase the area of the figure at the right by two square units, but not change its perimeter?

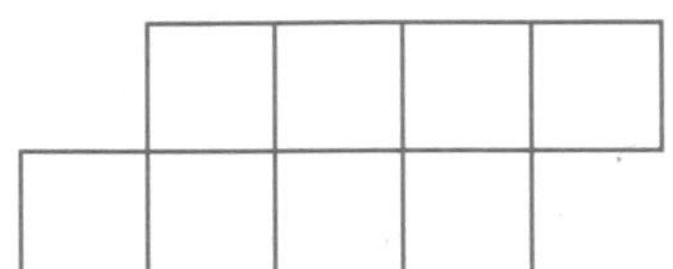

4. Draw and cut out a rectangle. Make two cuts on the rectangle to make it into a parallelogram. These two cuts must form two congruent shapes. Describe these two cuts and the shapes. Is there more than one way to make the cuts? Explain.

Guided Practice

Find the area of each figure. Assume the angles that appear to be right are right angles.

5. 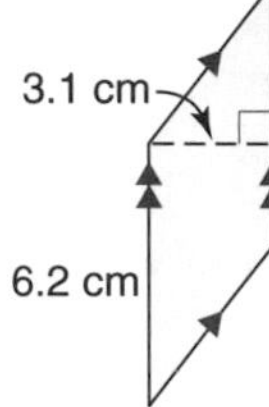

6. 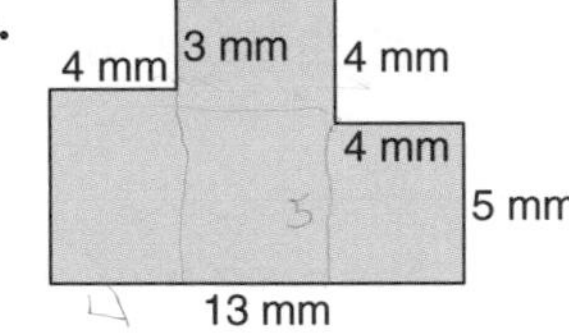

7. 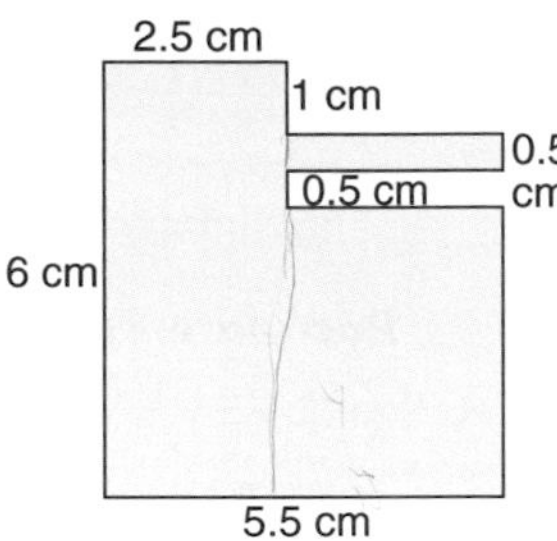

8. The coordinates of the vertices of a quadrilateral are $(-1, -1)$, $(1, 2)$, $(6, 2)$, and $(4, -1)$. Graph the points and draw the quadrilateral and an altitude. Identify the quadrilateral as a *square*, *rectangle*, or a *parallelogram*, and find its area.

9. **Maintenance** The University of North Carolina in Chapel Hill is resealing the parking lot around Peabody Hall. It costs \$0.52 per square yard to seal a parking lot. How much will it cost the University for their parking lot project?

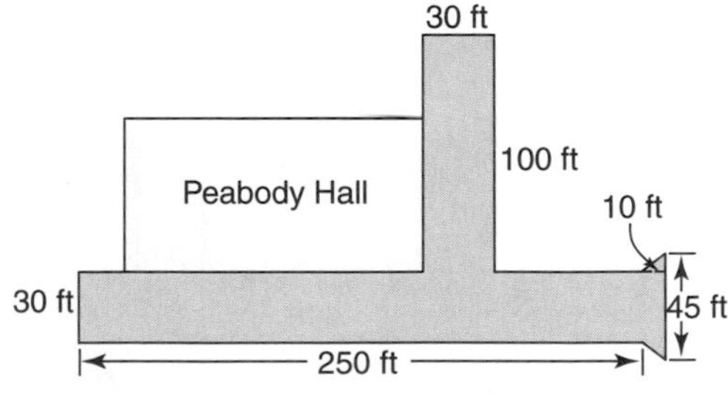

EXERCISES

Practice

Find the area of each figure or shaded region. Assume the angles that appear to be right are right angles.

10. 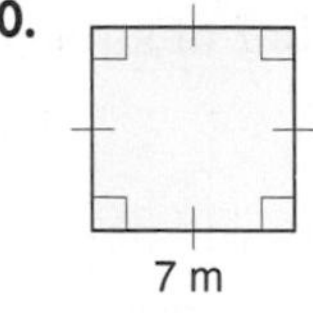

11. 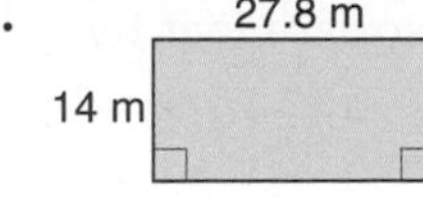

12. 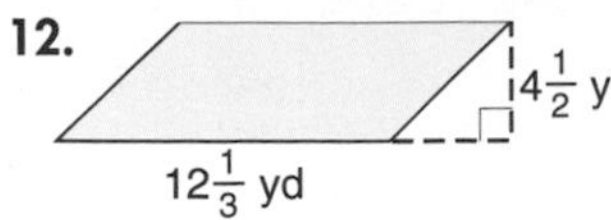

13. 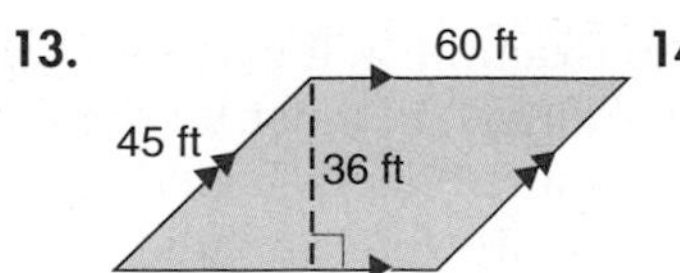

14. 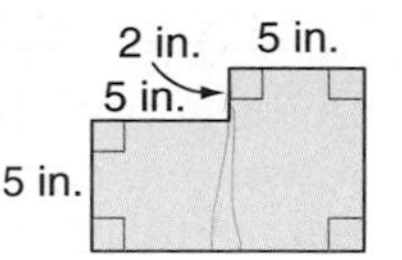

15. 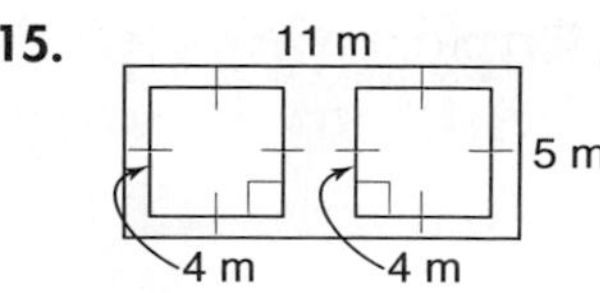

16. 10 mm, 5 mm 5 mm, $5\sqrt{2}$ mm, 5 mm 5 mm

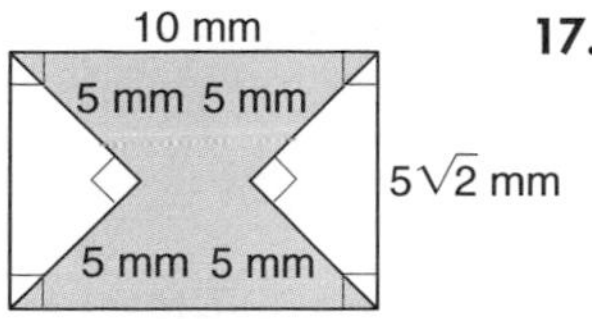

17. 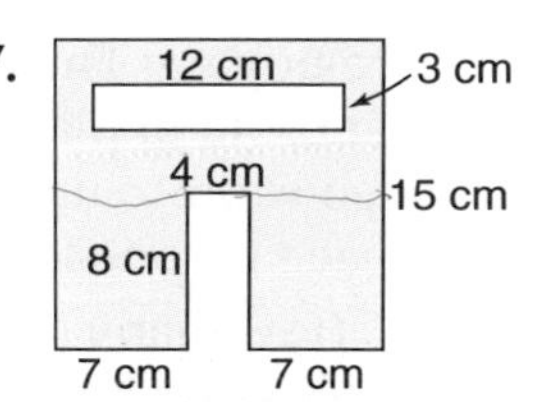

18. 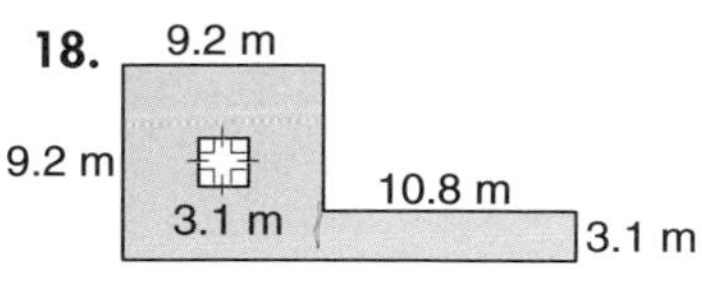

The coordinates of the vertices of a quadrilateral are given. Graph the points and draw the quadrilateral and an altitude.

a. Identify the quadrilateral as a *square*, a *rectangle*, or a *parallelogram*.

b. Find its area.

19. (4, −2), (10, −4), (10, −2), (4, −4)

20. (−4, −5), (2, −5), (4, −8), (−2, −8)

21. (1, 10), (−1, 7), (2, 5), (4, 8)

22. How many 6-feet-by-8-feet rectangles can fit into a rectangle 18 feet long by 48 feet wide? Draw a figure to justify your answer.

23. Find the area of the parallelogram at the right.

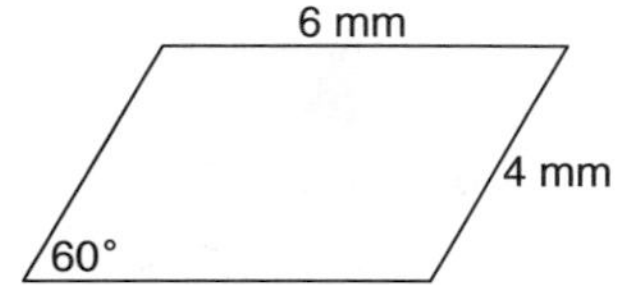

24. When the length of each side of a square is increased by 5 inches, the area of the resulting square is 2.25 times the area of the original square. What is the area of the original square?

25. Find the area of the shaded region of the congruent overlapping squares if *E* is the intersection of the diagonals of the square *ABDC*. Explain your solution.

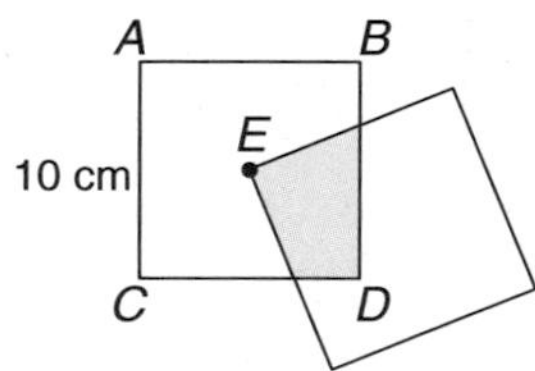

Cabri Geometry

26. Use a calculator to draw several parallelograms that have bases of the same length, but different heights. Compare the areas of the parallelograms to find the shape of the parallelogram that has a maximum area and the shape of the parallelogram that has a minimum area.

Critical Thinking

27. Given that the area of parallelogram *MNOP* is 48 square units and $MX = 3$, find the length of the diagonal $\overline{NP}$.

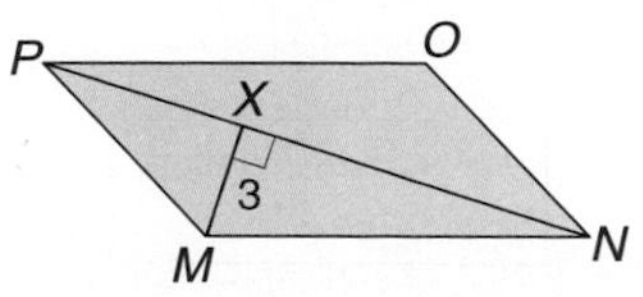

Applications and Problem Solving

28. Carpeting Refer to Example 3. If carpeting comes in widths of 4 yards and the direction the carpet is laid must be the same throughout the house, how much carpeting must be purchased to cover the bedrooms and the hall in the diagram? Justify your calculations with a drawing.

29. Housing At the right is the floor plan of the home that the Summers are buying. They want to add a sunroom in place of the patio. They also want to increase their living space by one-fourth.

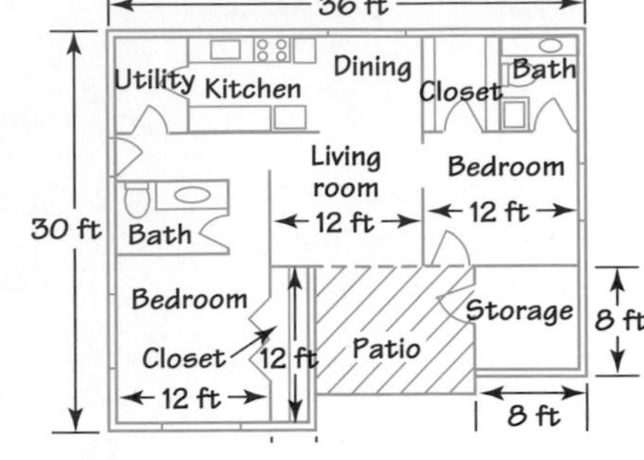

a. How many square feet are in the existing house?

b. What is the area that will be added to the existing house?

c. What could be the dimensions of the new sunroom?

d. Make a sketch of your design for the addition.

Mixed Review

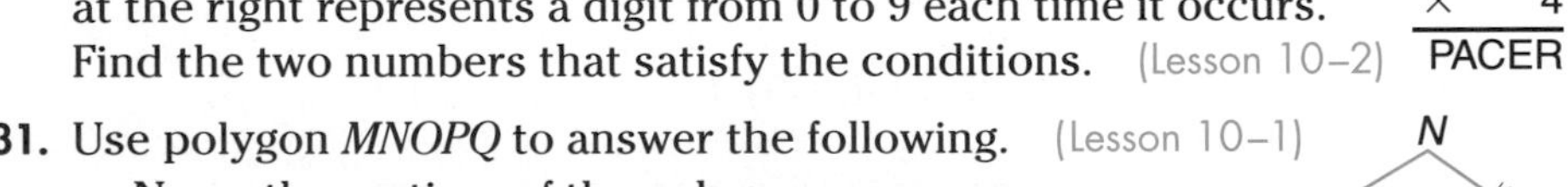

30. Guess and Check Each letter in the multiplication problem at the right represents a digit from 0 to 9 each time it occurs. Find the two numbers that satisfy the conditions. (Lesson 10–2)

$$\begin{array}{r} \text{RECAP} \\ \times \quad 4 \\ \hline \text{PACER} \end{array}$$

31. Use polygon *MNOPQ* to answer the following. (Lesson 10–1)

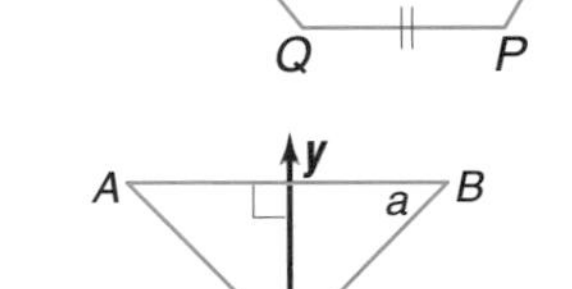

a. Name the vertices of the polygon.

b. Is the polygon convex or concave?

c. Is the polygon regular? Explain.

d. Name the polygon according to its sides.

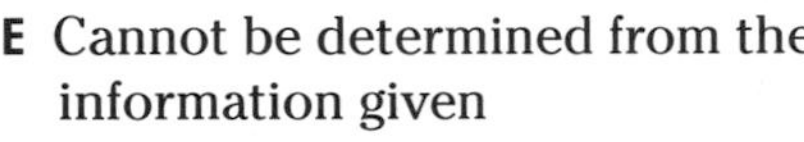

32. ACT Practice If $\tan a = 1$, then what is the sum of the slopes of each side of $\triangle ABC$?

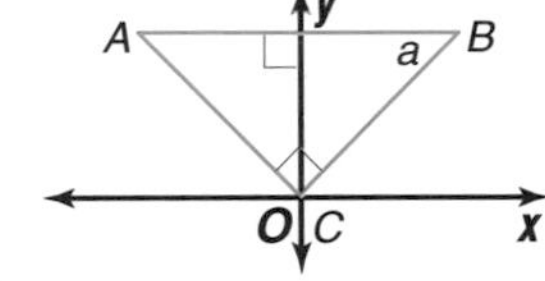

A -1 B 0 C $\frac{1}{2}$ D 1

E Cannot be determined from the information given

33. Write a two-column proof. (Lesson 6–5)

Given: trapezoid *ABCD*, $\overline{AB} \parallel \overline{DC}$

Prove: $\angle A$ and $\angle D$ are supplementary.

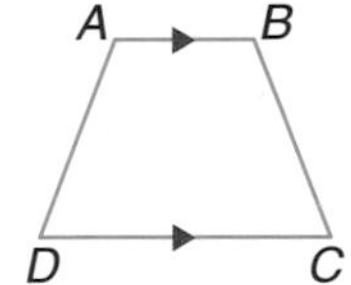

34. In the figure, $\overline{EF} \perp \overline{EH}$, $\overline{EH} \perp \overline{HG}$, and $\overline{HF} \perp \overline{JG}$. Name the segment whose length represents the distance between the following points and lines. (Lesson 3–5)

a. G to $\overleftrightarrow{HF}$

b. E to $\overleftrightarrow{HG}$

c. F to $\overleftrightarrow{EH}$

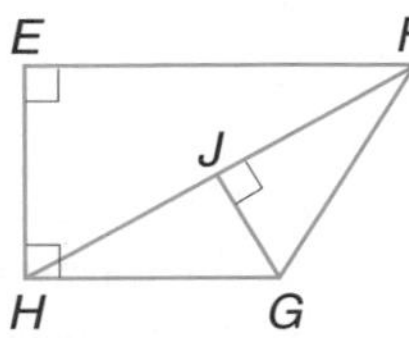

35. Draw and label a figure showing planes $\mathcal{M}$, $\mathcal{N}$, and $\mathcal{L}$ that do not intersect. (Lesson 1–2)

Algebra

36. Solve $a^2 - 5a + 6 = 0$

37. Determine which ordered pairs are solutions to $y > 4x - 3$.

a. $(0, 2)$ b. $(3, 0)$ c. $(1, -1)$ d. $(-4, 2)$

For **Extra Practice**, see page 783.

10–4 Area of Triangles, Rhombi, and Trapezoids

What YOU'LL LEARN

- To find areas of triangles, rhombi, and trapezoids.

Why IT'S IMPORTANT

You can use area of triangles, rhombi, and trapezoids to solve problems involving real estate, home repair, and design.

Real Estate

When a new development is planned, it is divided into lot parcels. These lots sell for varying amounts and there are several factors that determine the price: area, location (corner, cul-de-sac, etc.), trees and terrain, and so on. The map at the right shows the layout of the streets and lots in the Linworth Village development in Columbus, Ohio.

The developers made the lots into shapes that were not just rectangular or square. Since the area of a lot is a major factor in its cost, the developer must use geometry to find the area and understand the property of area described in Postulate 10–2.

Postulate 10–2	**Congruent figures have equal areas.**

You know formulas for the areas of squares and rectangles. Now, use the following activity to discover the formula for the area of a triangle.

MODELING MATHEMATICS

Area of a Triangle

Materials: 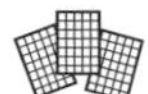 grid paper  scissors  straightedge

You can determine a formula for the area of a triangle by using the area of a rectangle.

- Draw a triangle on grid paper so that one edge is along a horizontal line as shown at the right. Label the vertices of the triangle *A*, *B*, and *C*.
- Draw a line perpendicular to $\overline{AC}$ through *A*.
- Draw a line perpendicular to $\overline{AC}$ through *C*.
- Draw a line parallel to $\overline{AC}$ through *B*.
- Label the points of intersection of the lines drawn as *D* and *E* as shown.
- Find the area of rectangle *ACDE* in square units.
- Cut out rectangle *ACDE*. Then cut out $\triangle ABC$. Place the two smaller pieces over $\triangle ABC$ to completely cover the triangle.

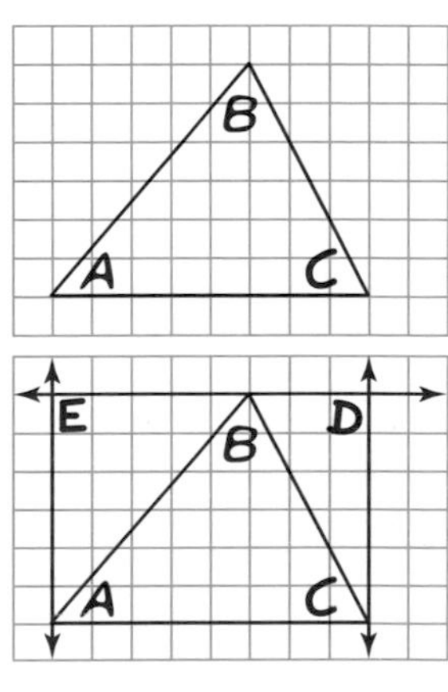

Your Turn

a. What do you observe about the two smaller triangles and $\triangle ABC$?

b. What portion of rectangle *ACDE* is $\triangle ABC$?

c. What is a formula that could be used to find the area of $\triangle ABC$?

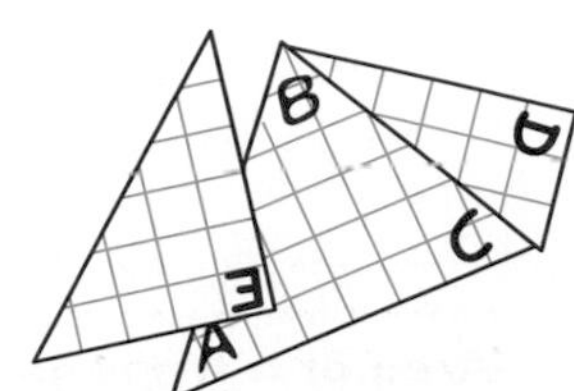

Since the combined area of the two smaller triangles *ABE* and *CBD* is the same as the area of the larger triangle *ABC*, we can see that the area of *ABC* is one-half the area of the rectangle. In Lesson 10–3, we found the area of a parallelogram by multiplying the measures of the base and the height. This leads us to the formula for the area of a triangle.

Area of a Triangle	**If a triangle has an area of *A* square units, a base of *b* units, and a corresponding height of *h* units, then $A = \frac{1}{2}bh$.**

You can find the areas of some shapes by finding the total areas of the figures that make up that shape.

Example  **Find the area of quadrilateral *ABCD* if *BE* = 12, *ED* = 5, and *AC* = 20.**

area of quadrilateral *ABCD* = area of $\triangle ABC$ + area of $\triangle ADC$

Find the areas of $\triangle ABC$ and $\triangle ACD$.

area of $\triangle ABC$:

$$A = \frac{1}{2}bh$$
$$= \frac{1}{2}(20)12$$
$$= 120$$

area of $\triangle ACD$:

$$A = \frac{1}{2}bh$$
$$= \frac{1}{2}(20)(5)$$
$$= 50$$

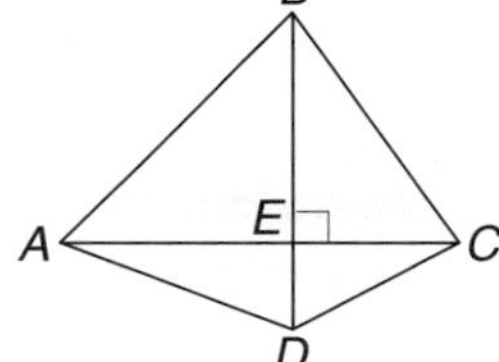

The area of quadrilateral *ABCD* is 120 + 50 or 170 square units.

Now, let's use the formula for the area of a triangle to derive the formula for the area of a trapezoid.

Suppose we are given trapezoid *RSTU* with diagonal $\overline{SU}$. The two bases of the trapezoid are parallel. Therefore, the altitude *h* from vertex *U* to the extension of base $\overline{RS}$ is the same length as the altitude from vertex *S* to the base $\overline{UT}$. Since the area of the trapezoid is the area of the two nonoverlapping parts, we can write the equation below.

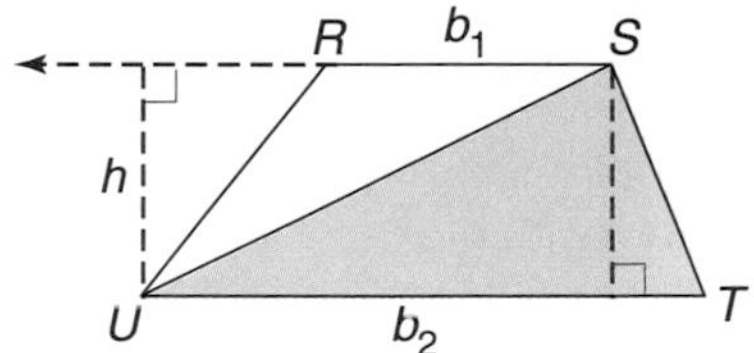

area of trapezoid *RSTU* = area of $\triangle RSU$ + area of $\triangle STU$

area of trapezoid *RSTU* = $\frac{1}{2}(b_1)h + \frac{1}{2}(b_2)h$ *Let RS be b_1 and UT be b_2.*

$= \frac{1}{2}(b_1 + b_2)h$ *Factor.*

$= \frac{1}{2}h(b_1 + b_2)$ *Commutative Property (=)*

This leads us to the formula for the area of a trapezoid.

Area of a Trapezoid	**If a trapezoid has an area of *A* square units, bases of b_1 units and b_2 units, and height of *h* units, then $A = \frac{1}{2}h(b_1 + b_2)$.**

If you know the area of a figure and some of its measures, you can find missing measures.

Example 2 **Find the height of a trapezoid that has an area of 287 square inches and bases of 38 inches and 44 inches.**

Use the formula for the area of a trapezoid and solve for h.

$$A = \frac{1}{2}h(b_1 + b_2)$$

$$287 = \frac{1}{2}h(38 + 44) \quad \textit{A = 287, } b_1 = 38,\ b_2 = 44$$

$$287 = 41h$$

$$7 = h \quad \textit{Divide each side by 41.}$$

The height of the trapezoid is 7 inches.

The formula for the area of a triangle can also be used to derive the formula for the area of a rhombus. *You will be asked to derive this formula in Exercise 3.*

Area of a Rhombus	**If a rhombus has an area of A square units and diagonals of d_1 and d_2 units, then $A = \frac{1}{2}d_1d_2$.**

Example 3

Real World APPLICATION

Landscaping

Sonja Washington wants to place decorative brick edging around a flower garden that is in the shape of a rhombus. One diagonal is 12 feet long, and the area is 264 square feet. How much brick must she purchase?

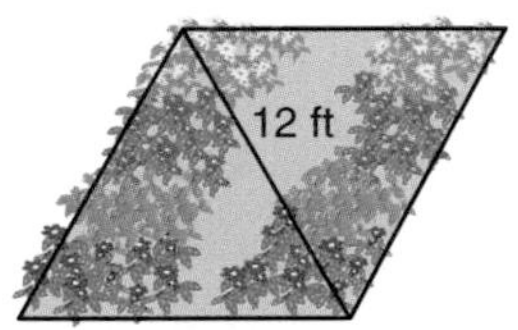

To solve the problem, we need to find the perimeter of the rhombus. First, use the area formula to find the length of the other diagonal.

$$A = \frac{1}{2}d_1d_2$$

$$264 = \frac{1}{2}(12)d_2$$

$$264 = 6d_2$$

$$44 = d_2$$

The diagonals are 12 feet and 44 feet long.

The diagonals of a rhombus are perpendicular and they bisect each other. We can use the Pythagorean Theorem to find the length of a side.

$$s^2 = 6^2 + 22^2$$

$$s^2 = 36 + 484$$

$$s^2 = 520$$

$$s = \sqrt{520} \approx 22.8$$

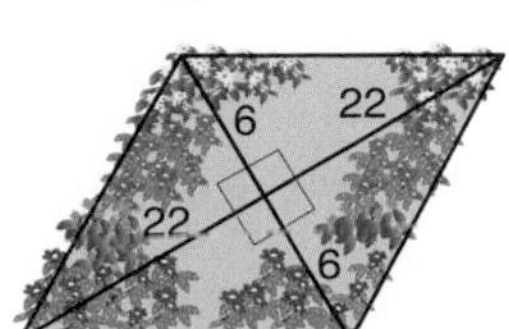

Since each side is about 22.8 feet long, the perimeter is about 4(22.8) or 91.2 feet.

Therefore, Ms. Washington needs to buy 92 feet of brick edging.

The following chart summarizes the area formulas you have learned thus far.

Summary of Area Formulas			
Square	$A = s^2$	Triangle	$A = \frac{1}{2}bh$
Rectangle	$A = \ell w$	Trapezoid	$A = \frac{1}{2}h(b_1 + b_2)$
Parallelogram	$A = bh$	Rhombus	$A = \frac{1}{2}d_1 d_2$

CHECK FOR UNDERSTANDING

Communicating Mathematics

Study the lesson. Then complete the following.

1. **Explain** whether $\triangle ABC$ and $\triangle DEF$ have the same area.

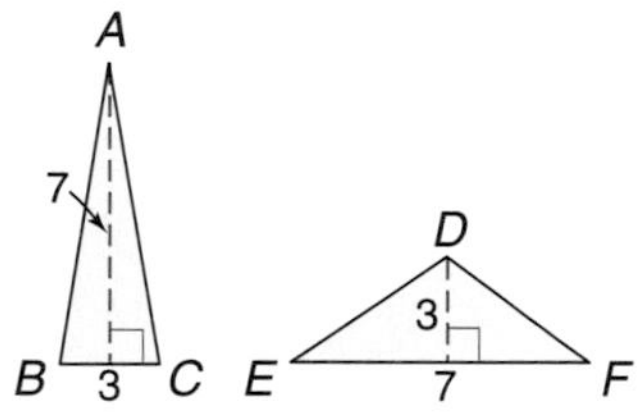

2. **Analyze** the following quilting patterns and determine the formula(s) you would use to find the area of each section in each pattern.

a. 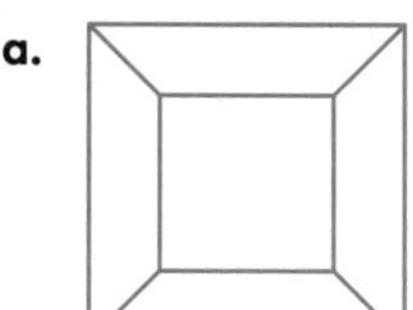 **b.** 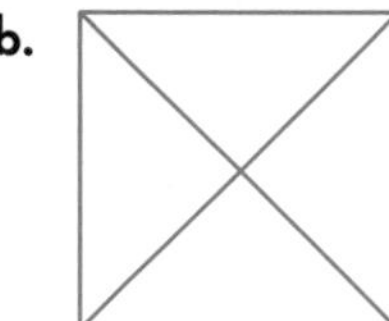 **c.** 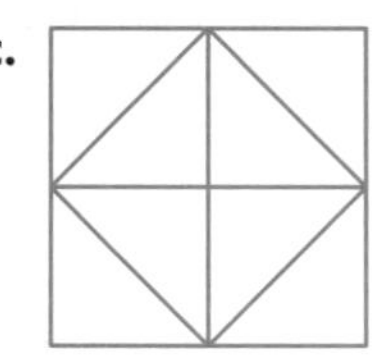 **d.** 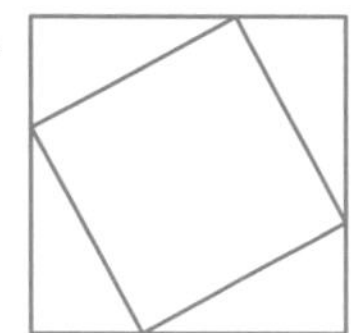

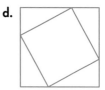

3. **Describe** how the formula for the area of a rhombus is related to the formula for the area of a triangle.

4. **Draw** a trapezoid and measure the height and each base. Determine three different ways to find the area of the trapezoid. Show and explain your work.

MODELING MATHEMATICS

5. Draw a right triangle, an acute triangle, and an obtuse triangle on a piece of paper. Cut out each of them. Cut one line through each of the triangles and position the two pieces to form another figure.
   a. Describe the figures formed by each pair of pieces.
   b. What is true of the areas of the original triangles and the new figures?

Guided Practice

Find the area of each figure.

6. 

7. 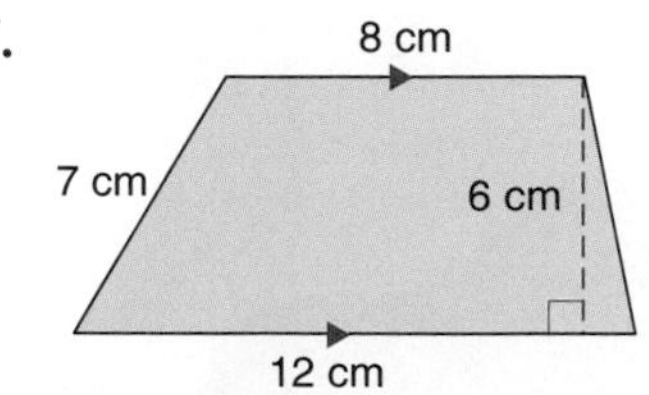

8. If the area of $\triangle ABC$ is 24 square units, find the value of x.

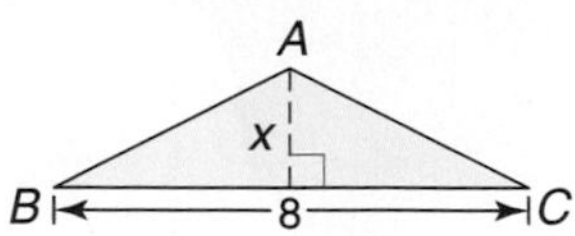

9. A rhombus has side lengths of 12 inches each and an angle measure of 120. Find the area of the rhombus.

10. **Cleaning** City Carpeting will clean up to 325 square feet of carpeting for $78. Yolanda Alvarez has three rectangular bedrooms which all have carpeting. Their measures are 8 feet by 13 feet, 8 feet by 10 feet, and 12 feet by 12 feet. Will Mrs. Alvarez have to pay more than $78 to have the three bedrooms cleaned? Explain.

EXERCISES

Practice

Find the area of each figure.

11. 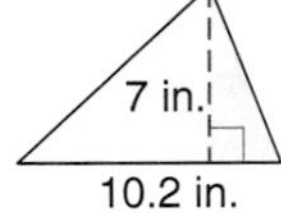

12. 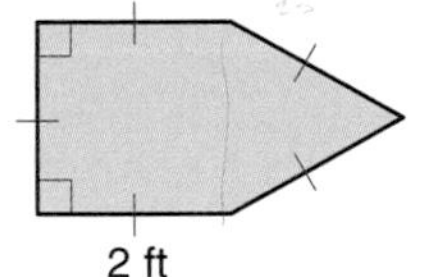

13. 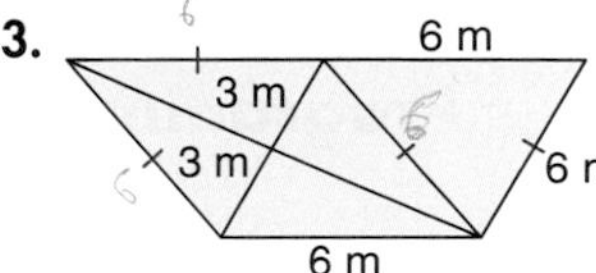

14. 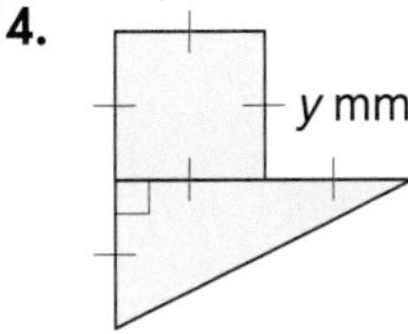

15. 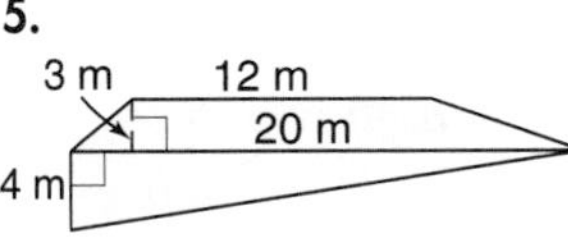

16. 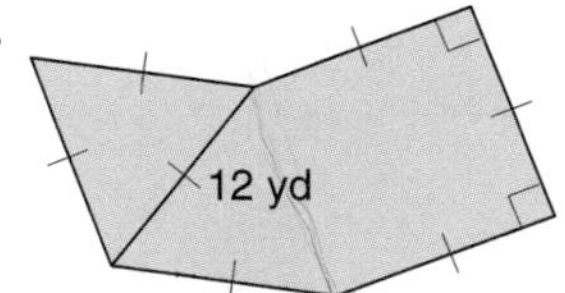

Find the value of *x* in each figure.

17. $A = 12$

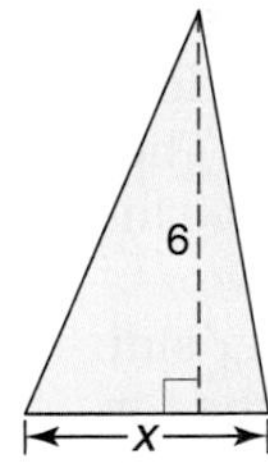

18. $A = 56$

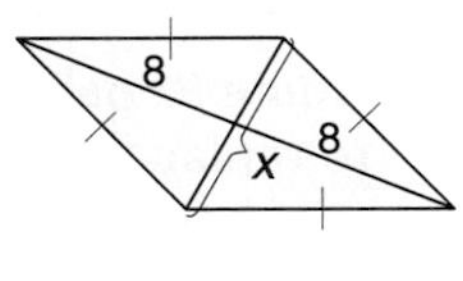

19. $A = 95$

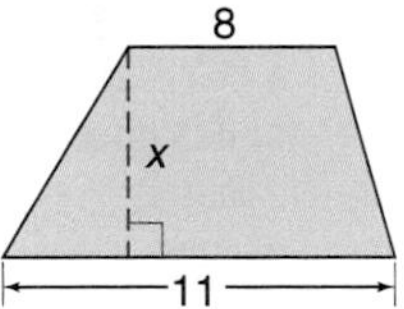

20. A trapezoid has an area of 75 square inches, and its two bases are 8 and 17 inches long. Find the height of the trapezoid.

21. A rhombus has an angle measure of 120, and its longer diagonal has a length of 10 inches. Find the area of the rhombus.

22. The shorter diagonal of a rhombus is 10 inches long, and the angle the diagonal forms with the side measures 60°. Find the area of the rhombus.

23. Find the area of an equilateral triangle whose perimeter is 24 centimeters.

24. A trapezoid has an area of 126 square feet, a height of 9 feet, and one base of 13 feet. Find the length of the second base.

25. The area of an isosceles trapezoid is 77 square inches, the height is 4 inches, and the congruent sides of the trapezoid are each 5 inches long. Find the lengths of the bases.

26. What is the relationship between the base of a triangle and the base of a parallelogram that both have an area of 96 square centimeters and a height of 12 centimeters?

Critical Thinking

27. Use $\triangle BCD$ and $\square ABDE$ at the right to derive the formula for the area of a trapezoid.

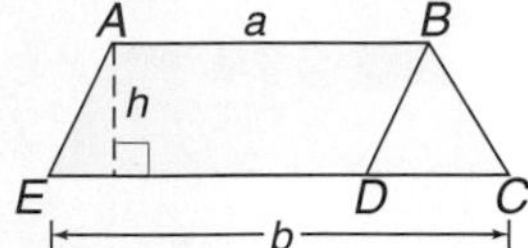

28. In the figure, the vertices of quadrilateral *HBDF* intersect the square *ACEG* and divides its sides into segments whose measures have a ratio of 1:2.

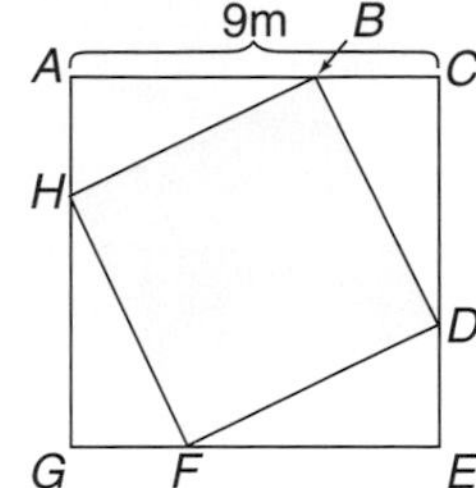

a. Find the area of quadrilateral *HBDF*.

b. What type of figure is *HBDF*?

c. What is the relationship between the areas of quadrilateral *HBDF* and square *ACEG*?

Programming

29. The graphing calculator program finds the area of a trapezoid given the height and the measures of the bases.

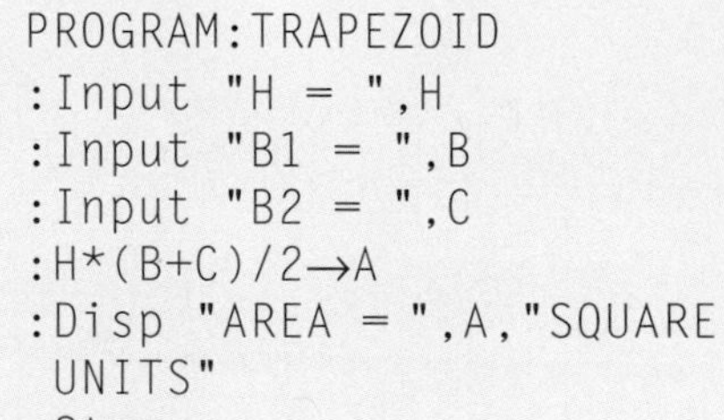

```
PROGRAM:TRAPEZOID
:Input "H = ",H
:Input "B1 = ",B
:Input "B2 = ",C
:H*(B+C)/2→A
:Disp "AREA = ",A,"SQUARE
 UNITS"
:Stop
```

Use the program to find the area of each trapezoid.

a. $h = 5, b_1 = 8, b_2 = 6$

b. $h = 3.5, b_1 = 7.1, b_2 = 8.4$

Applications and Problem Solving

30. **Real Estate** Kamali Narula planned to sell a trapezoid-shaped portion of her property that has parallel sides measuring 147.08 feet and 57.62 feet. The other sides measure 144.34 feet and 90.54 feet while the depth of the property is 50.74 feet.

a. What is the area of this property?

b. If an acre is 43,560 square feet, what percent of an acre is this property?

31. **Home Repair** Rosa Llarena needed to replace the roof on her home. The roof had two large trapezoids, one in the front and one in the back, two smaller trapezoids, one on either side of the roof, and a rectangle on the top. If each of the trapezoids has the same height and a package of shingles covers 100 square feet, how many packages of shingles should Ms. Llarena buy?

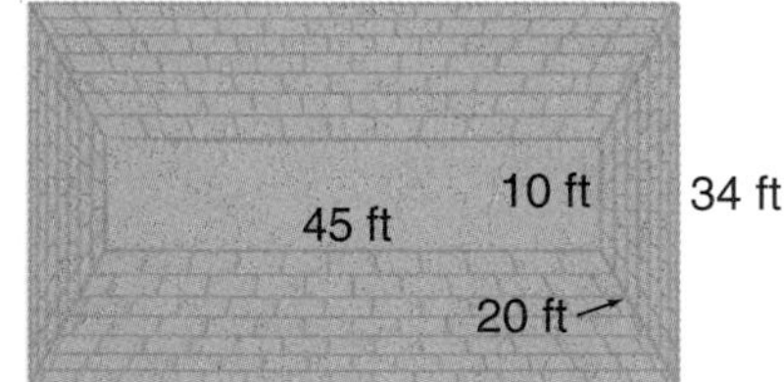

32. **Design** Zephia Watson was planning a meeting and was not satisfied with the usual place marker that could only be seen from two directions. She wanted to develop place markers that could be seen from three directions. Use what you know about triangles, rhombi, and trapezoids to design a creative pattern for a place marker that Zephia could use.

> **CAREER CHOICES**
>
> An **industrial designer** develops and designs manufactured products and materials. Creativity, persistence, problem-solving skills, and communication of ideas, both visually and verbally, are required for this career.
>
> For more information, contact:
>
> The Industrial Designers Society of America
> 1142-E Walker Road
> Great Falls, VA 22066

Mixed Review

33. The area of a rectangle is 520 square units, and its perimeter is 106 units. What are its dimensions? (Lesson 10–3)

34. Is the tessellation shown at the right *uniform*, *regular*, or *semi-regular*? Name all possibilities. (Lesson 10–2)

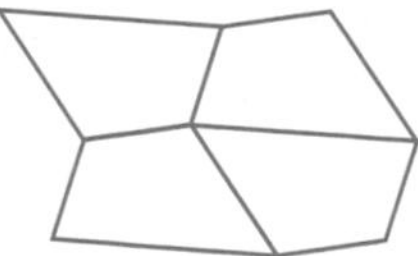

35. Find the value of h and j in the two concentric circles below. (Lesson 9–7)

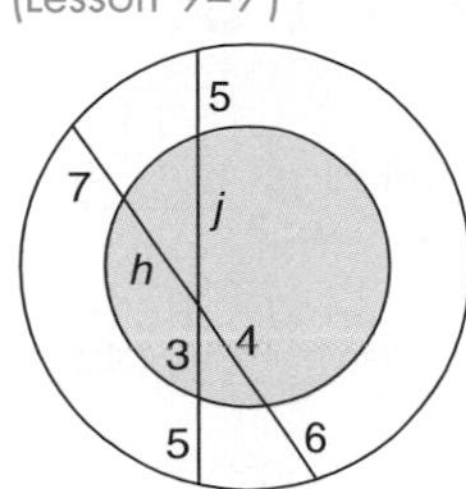

36. Find the value of r to the nearest tenth. (Lesson 8–3)

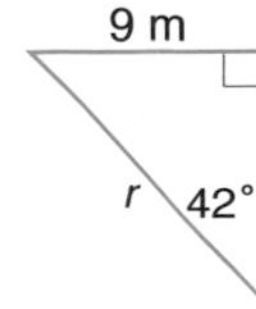

37. In quadrilateral $JULY$, $m\angle U = 120$ and $m\angle L = 50$. Is $JULY$ a parallelogram? Justify your answer. (Lesson 6–1)

38. **SAT Practice** **Grid-in** The average of a and b is 18, and the ratio of a to b is 5 to 4. What is the value of $a - b$?

Algebra

For **Extra Practice**, see page 783.

39. Find $\frac{k^3\ell m^2}{x^2y^2} \div \frac{k^3m}{x^3y^2}$.

40. Find the supplement of a 64° angle.

SELF TEST

1. **Refer to polygon *ABCDEFG*.** (Lesson 10–1)
   a. Identify the type of polygon.
   b. Is it regular?
   c. Is it convex or concave?

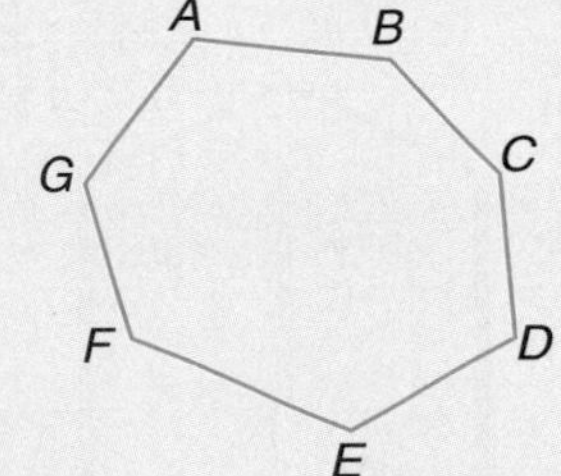

Find the measure of an interior angle of each regular polygon. (Lesson 10–1)

2. 20-gon

3. dodecagon

The measure of an interior angle of a regular polygon is given. Find the measure of an exterior angle and the number of sides of each regular polygon. (Lesson 10–1)

4. 160

5. $168\frac{3}{4}$

6. Is the tessellation shown below *regular*, *uniform*, or *semi-regular*? Name all possibilities. (Lesson 10–2)

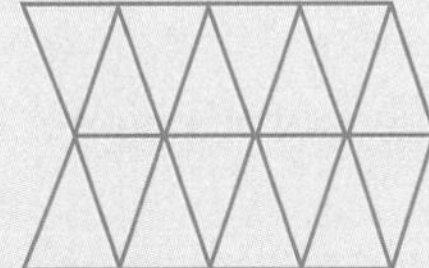

7. Find the area of the green region below. (Lesson 10–3)

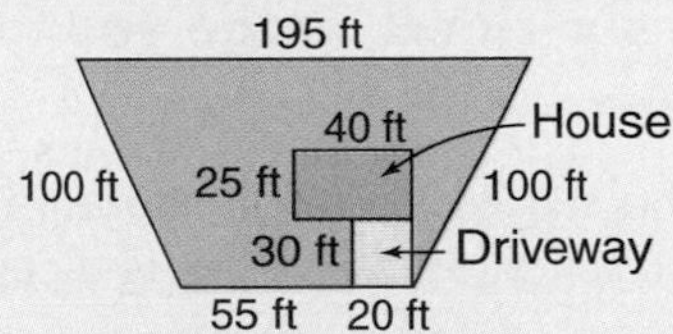

Find the area of each region. (Lesson 10–4)

8.

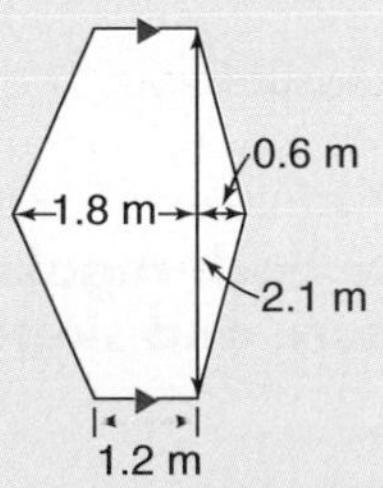

9.

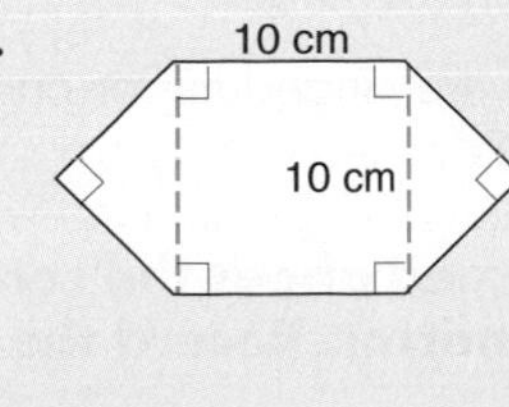

10. **Guess and Check** Place two addition symbols and two subtraction symbols in the expression below to make it a true equation. Note that digits can be put together to form larger numbers, but not rearranged. (Lesson 10–2)

$$3\ 5\ 6\ 4\ 7\ 5\ 2\ 6\ 1\ 7 = 712$$

MODELING MATHEMATICS

10–5A Regular Polygons

A Preview of Lesson 10–5

Materials:  straightedge  hinged mirror protractor plain paper

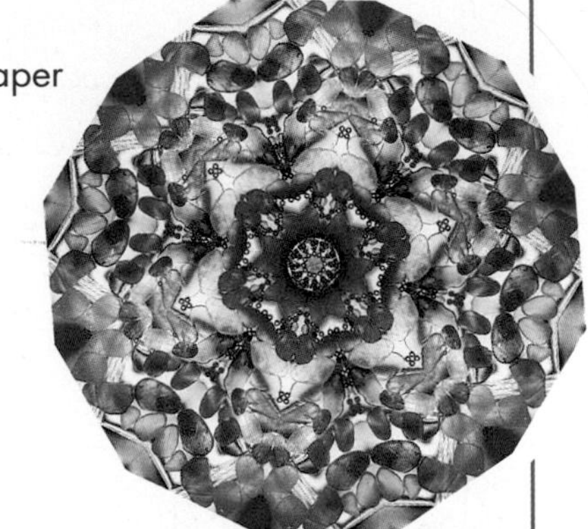

Mirrors are often used to create images in kaleidoscopes and telescopes as well as in interior design. You can use a hinged mirror to create any *n*-gon that is regular. The mirror uses multiple reflections to create the number of sides you desire.

Activity Use a hinged mirror to create a regular hexagon.

Step 1 Use a straightedge to draw a line on a piece of plain paper. Position the hinged mirror so that the sides of the mirror intersect the line at approximately the same distance from the hinge of the mirror.

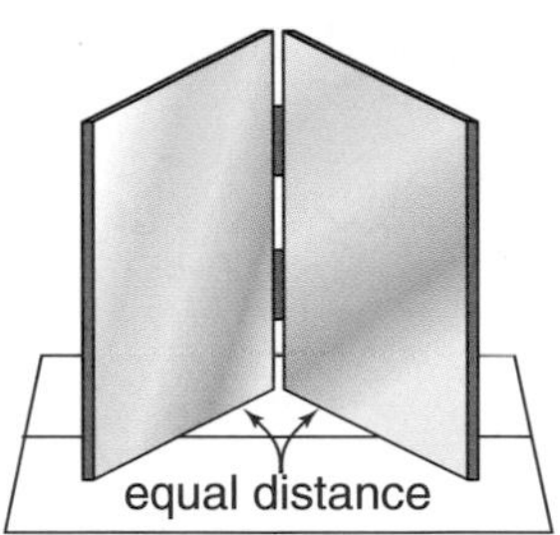

Step 2 Sketch what you see. Place a protractor on top of the mirror so that you can measure the angle of the mirror. Record this measurement.

Step 3 Open the mirror wider. Make sure the line still intersects the mirror's edges at equal distances from the hinge. What changes occur in the figure you see? Sketch the figure and record the angle of the mirror.

Step 4 Open or close the mirror until the image you see has six sides. Sketch the figure and record the angle of the mirror.

Write **Use the six-sided image you formed in Step 4 above to answer these questions.**

1. Find the sum of the angle measures for those angles whose vertex is the mirror's hinge. (*Hint:* Each of these angles is a central angle.)
2. a. How many angles are there in the image for each number of sides?
   b. What is the measure of each angle?
3. How does the measure of the hinged mirror angle compare with the measure you calculated in Exercise 2b?

Draw **Use what you have learned about the central angles to create each regular polygon with a hinged mirror. Record the central angle measure and sketch the polygon.**

4. triangle
5. square
6. pentagon
7. octagon
8. decagon
9. dodecagon

10–5 Area of Regular Polygons and Circles

What YOU'LL LEARN

- To find areas of regular polygons, and
- to find areas of circles.

Why IT'S IMPORTANT

You can use the area of polygons and circles to solve problems involving landscape design and architecture.

Real World APPLICATION

Gardening

The central feature of Elizabeth Park in Hartford, Connecticut, is the Rose Garden. Over 15,000 rosebushes of approximately 800 varieties are planted in the garden. In 1896, Theodore Wirth designed the garden with a rustic summer house encircled with concentric circles of rosebushes. Wide paths radiate from the center and divide the garden into square and circular sections.

The polygon and the circle are closely related. Both the circumference and the area of circles can be approximated using the perimeter and area of regular polygons. All regular polygons can be inscribed in a circle.

CONSTRUCTION

Regular Hexagon

Use a compass and straightedge to construct a regular hexagon.

1. Use your compass to draw a $\odot P$. Point P will also be the **center** of the hexagon. The radius of $\odot P$ is the **radius** of the hexagon and is congruent to a side of the hexagon.

2. Using the same compass setting, place the compass point on the circle and draw an arc, labeling the point of intersection with the circle, A.

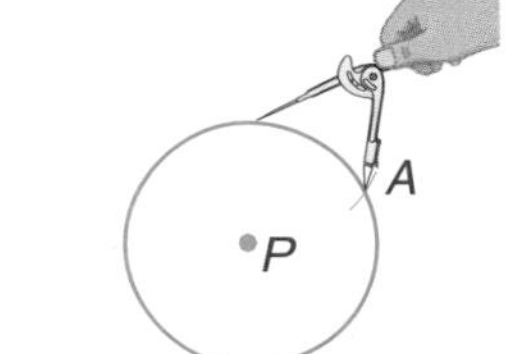

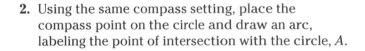

3. Place the compass point on A and draw another arc. Label the point of intersection with the circle, B.

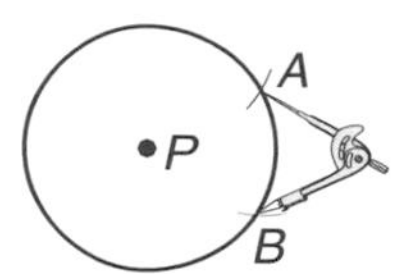

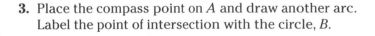

4. Continue this process, labeling points C, D, E, and F, on the circle. Point F should be where the point of the compass was placed to locate point A.

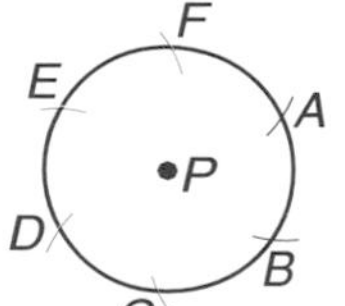

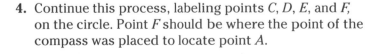

5. Use a straightedge to connect A, B, C, D, E, and F, consecutively.

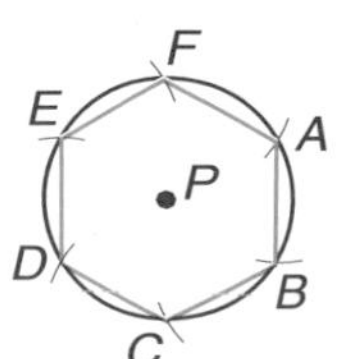

Conclusion: $ABCDEF$ is a regular hexagon inscribed in $\odot P$ because in a circle, if minor arcs are congruent, then their corresponding chords are congruent.

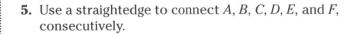

GLOBAL CONNECTIONS

Circles or regular multisided polygons have often been used as plans for cities or monuments. Stonehenge, built about 1800–1400 B.C., is a circular monument in England. Baghdad, the capital of Iraq, began as a circular village in the early 600s.

In regular hexagon $ABCDEF$, suppose we draw $\overline{PE}$, $\overline{PD}$, and $\overline{PT}$ so that $\overline{PT}$ is perpendicular to $\overline{ED}$. A segment like $\overline{PT}$ that is drawn from the center of a regular polygon perpendicular to a side of the polygon is called an **apothem**.

$\triangle PED$ is an isosceles triangle, since sides $\overline{PE}$ and $\overline{PD}$ are radii of $\odot P$. If all of the radii of hexagon $ABCDEF$ were drawn, they would separate the hexagon into six triangles all congruent to $\triangle PED$.

Now, since the area of a region is the sum of the areas of its nonoverlapping parts, you can find the area of the hexagon by adding the areas of the triangles. Since $\overline{PT}$ is perpendicular to $\overline{ED}$, it is an altitude of $\triangle PED$ as well as an apothem of hexagon $ABCDEF$. Let a represent the measure of $\overline{PT}$ and s represent the length of a side of the hexagon.

$$\text{area of } \triangle PED = \frac{1}{2}bh \quad \textit{Formula for area of a triangle}$$
$$= \frac{1}{2}sa$$

The area of one of the six triangles is $\frac{1}{2}sa$ units^2. So the area of the hexagon is $6\left(\frac{1}{2}sa\right)$ units^2. Notice that the perimeter of hexagon $ABCDEF$ is $6s$ units. Therefore, if the perimeter of the hexagon is P units, the area will be $\frac{1}{2}Pa$ square units. This area formula can be used for *any* regular polygon.

Area of a Regular Polygon	**If a regular polygon has an area of A square units, a perimeter of P units, and an apothem of a units, then $A = \frac{1}{2}Pa$.**

Example 1 **Find the area of a regular octagon that has a perimeter of 72 inches.**

To use the formula for the area of a regular polygon, we must find the length of the apothem $\overline{OP}$.

The central angles of $ABCDEFGH$ are all congruent. Therefore, the measure of each angle is $\frac{360}{8}$ or 45. Since $\overline{OP}$ is an apothem of the octagon, it is perpendicular to and bisects $\overline{AB}$. It also bisects central angle AOB. Therefore, $m\angle BPO = 90$ and $m\angle BOP = 22.5$. Since the perimeter is 72 inches, each side of the octagon is $\frac{72}{8}$ or 9 inches. Therefore, $PB = \frac{1}{2}(9)$ or 4.5.

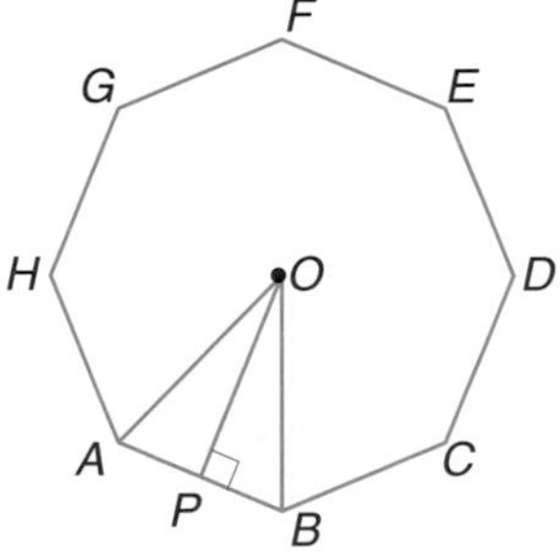

Use a trigonometric ratio to find the length of $\overline{OP}$.

$$\tan \angle BOP = \frac{PB}{OP}$$
$$\tan 22.5 = \frac{4.5}{OP} \quad m\angle BOP = 22.5,\ PB = 4.5$$
$$OP(\tan 22.5) = 4.5 \quad \textit{Multiply each side by OP.}$$
$$OP = \frac{4.5}{\tan 22.5} \quad \textit{Divide each side by tan 22.5.}$$
$$OP \approx 10.9 \quad \textit{Use a calculator.}$$

Now use the formula for the area of a regular polygon.

$$A = \frac{1}{2}Pa$$
$$A \approx \frac{1}{2}(72)(10.9)$$
$$A \approx 392.4$$

The area of the regular octagon is about 392.4 square inches.

You can use a calculator to help derive the formulas for the area and circumference of a circle from the areas and perimeters of regular polygons.

EXPLORATION — SCIENTIFIC CALCULATOR

Suppose each regular polygon is inscribed in a circle of radius r. Copy and complete the following chart. Round to the nearest hundredth.

	(triangle)	(pentagon)	(octagon)	(decagon)	(20-gon)	(50-gon)
Number of Sides	3	5	8	10	20	50
Measure of a Side	$1.73r$	$1.18r$	$0.77r$	$0.62r$	$0.31r$	$0.126r$
Perimeter						
Measure of Apothem	$0.5r$	$0.81r$	$0.92r$	$0.95r$	$0.99r$	$0.998r$
Area						

a. What happens to the perimeters as the number of sides increases?

b. What happens to the areas as the number of sides increases?

LOOK BACK

Refer to Lesson 9-1 to review circumference of a circle.

From this Exploration, you found that as the number of sides increases, the perimeter approaches $6.28r$, and the area approaches $3.14r^2$. Notice that $6.28 = 2 \times 3.14$ and 3.14 is an approximation of the irrational number called π (pi). So $6.28r \approx 2\pi r$ and $3.14r^2 \approx \pi r^2$. Therefore, the perimeters approach the circumference of the circle, and the areas approach the area of the circle.

Area of a Circle	**If a circle has an area of A square units and a radius of r units, then $A = \pi r^2$.**

Example 2

Design

The area of a circular pool is approximately 7850 square feet. The owner wants to replace the tiling at the edge of the pool.

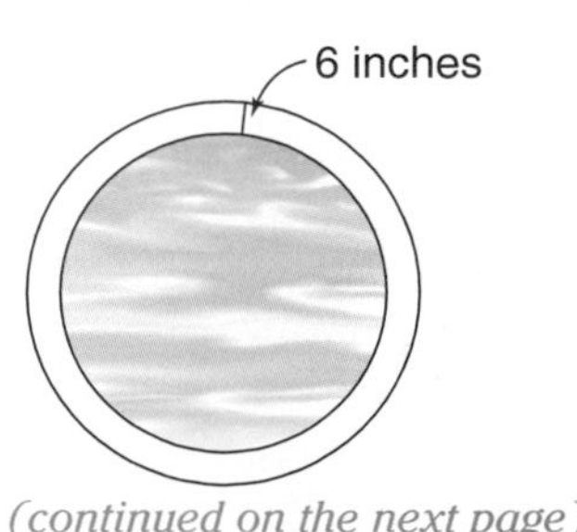

a. He plans to use tiles that are each 6-inch squares since the edging is 6 inches wide. How many tiles should he purchase?

(continued on the next page)

Use the area formula to find the radius of the pool.

$$A = \pi r^2 \quad \textit{Formula for area of circle}$$
$$7850 = \pi r^2$$
$$\frac{7850}{\pi} = r^2 \quad \textit{Divide each side by } \pi.$$
$$50.0 \approx r \quad \textit{Use a calculator.}$$

Now use the circumference formula to find the distance around the pool.

$$C = 2\pi r \quad \textit{Formula for circumference}$$
$$C \approx 2\pi(50)$$
$$C \approx 100\pi$$
$$C \approx 314.2 \quad \textit{Use a calculator.}$$

Divide the circumference by the length of each tile to find the number of tiles to go around the edge of the pool.

$$\frac{314.2}{6} \approx 52.4$$

The owner should purchase 52 tiles and space evenly.

b. Once the square tiles are placed around the circular pool, there will be extra space between each square. What shape of tile will fill this space and how many tiles of this shape will he need to go around the pool?

The best shape would be a triangle. The owner will need the same number of triangular tiles as 6-inch square tiles.

In Lesson 10–3, you learned how to find the area of a specific region by adding the areas of various shapes.

Example 3 **Find the area of the shaded region if $m\angle BAC = m\angle BCA = 60$.**

The area of the shaded region is the area of $\odot O$ minus the area of $\triangle ABC$.

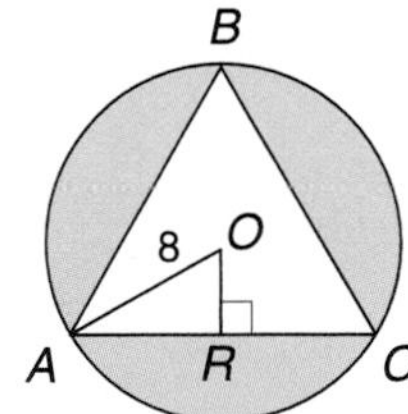

Find the area of the circle.

$$A = \pi r^2$$
$$A = \pi(8^2) \quad r = 8$$
$$A = 64\pi$$
$$A \approx 201.1 \quad \textit{Use a calculator.}$$

Now find the area of $\triangle ABC$. $\triangle AOR$ is a 30°-60°-90° triangle, so $OR = 4$ and $AR = 4\sqrt{3}$. Since $AC = 2(AR)$, $AC = 2(4\sqrt{3})$ or $8\sqrt{3}$. The perimeter of $\triangle ABC$ is $3(8\sqrt{3})$ or $24\sqrt{3}$. Now use the formula for the area of a regular polygon.

$$A = \frac{1}{2}Pa$$
$$= \frac{1}{2}(24\sqrt{3})(4)$$
$$A \approx 83.1$$

The area of the shaded region is about 201.1 − 83.1 or 118.0 square units.

CHECK FOR UNDERSTANDING

Communicating Mathematics

Study the lesson. Then complete the following.

1. **Explain** how you would find the apothem of a regular hexagon if you know its perimeter.
2. **Describe** the relationship between the radius and the apothem of a regular polygon.
3. The circumference of a circle is sometimes expressed as $C = \pi d$, where d is the measure of the diameter of the circle. Why is this formula equivalent to $C = 2\pi r$?

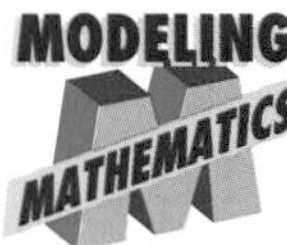

4. Using graph paper, draw several regular polygons. Inscribe a circle in each polygon and then circumscribe a circle about each polygon. Describe when a radius and an apothem will be the same and when they will be different.

Guided Practice

For Exercises 5–7, use the figure at the right.

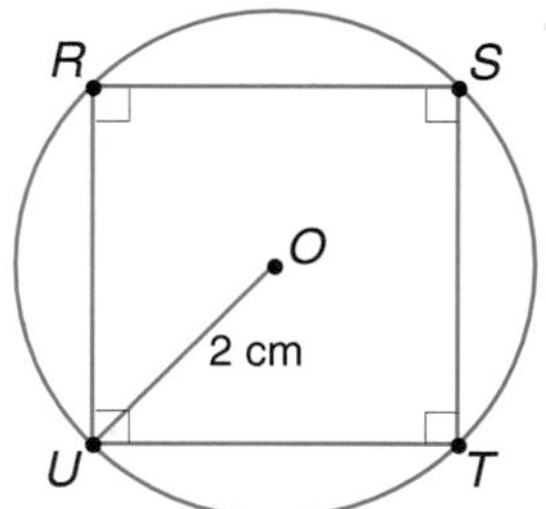

5. Find the perimeter of *RSTU*.
6. Find the area and the circumference of $\odot O$.
7. How do the perimeter of *RSTU* and the circumference of $\odot O$ compare?

Find the apothem, the area, and the perimeter of each regular polygon. Round to the nearest tenth.

8. 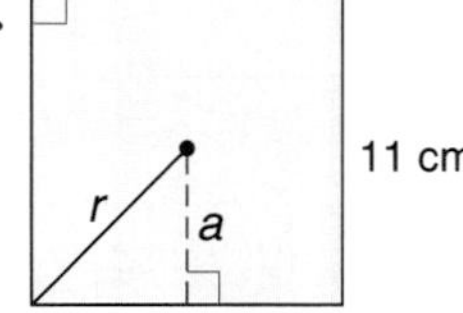

9. 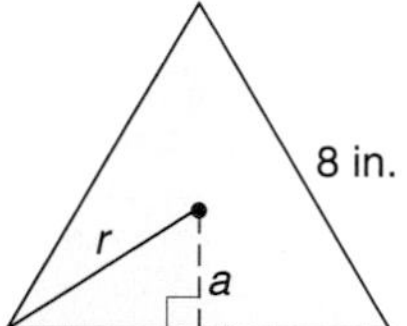

10. Find the circumference of a circle with an area of 10π square feet. Round to the nearest tenth.

Find the area of each shaded region. Assume that all polygons are regular. Round to the nearest tenth.

11. 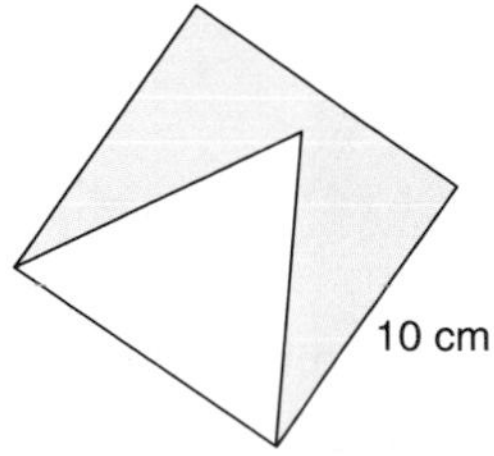

12. 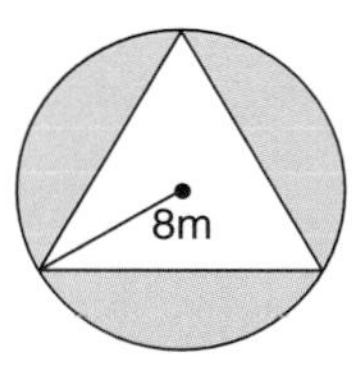

13. **Sports** Shorem High School colors are blue and white. On the basketball court, the entire free throw areas and the entire large center circle of the basketball court are painted blue. What is the area of the court that is painted blue?

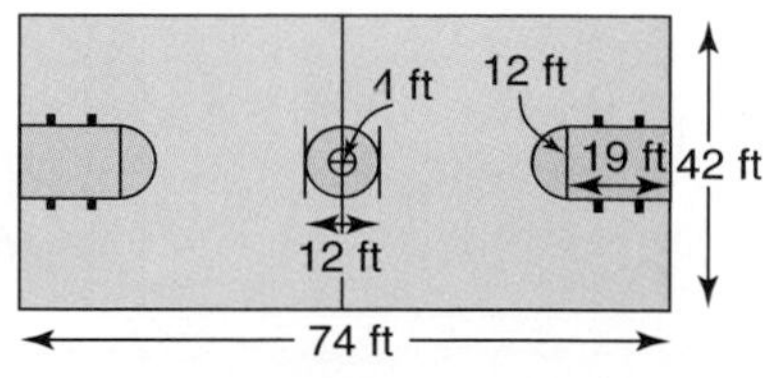

EXERCISES

Practice **Find the area of each regular polygon described. Round to the nearest tenth.**

14. square with apothem length of 12 centimeters

15. triangle with side length of 15.5 inches

16. square with perimeter of $84\sqrt{2}$ meters

17. hexagon with perimeter of 60 feet

18. octagon with side length of 10 kilometers

19. hexagon with apothem length 24 inches

Find the circumference and the area of circles with the given radius. Round to the nearest tenth.

20. 34 meters **21.** 8.5 centimeters **22.** 15 millimeters

23. 10.25 inches **24.** $5\frac{1}{3}$ feet **25.** $21\frac{3}{4}$ centimeters

Find the area of each shaded region. Assume that all polygons are regular. Round to the nearest tenth.

26. 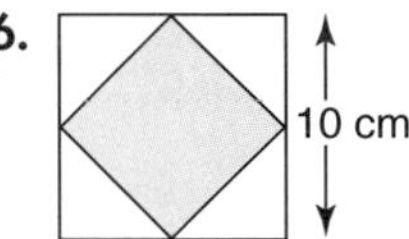

27. 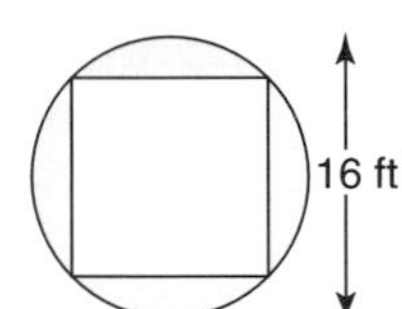

28. 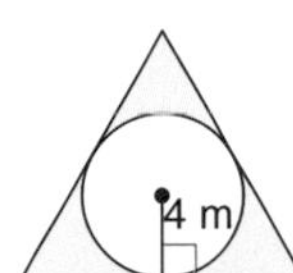

29. 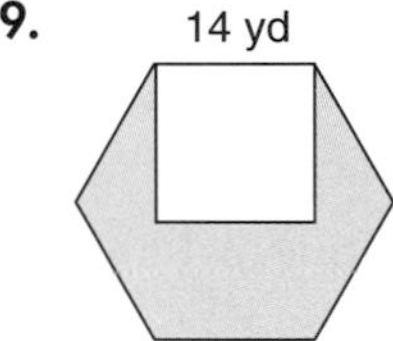

30. 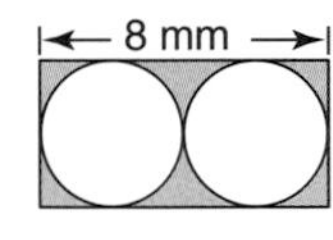

31. 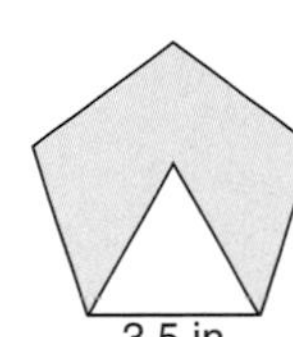

32. 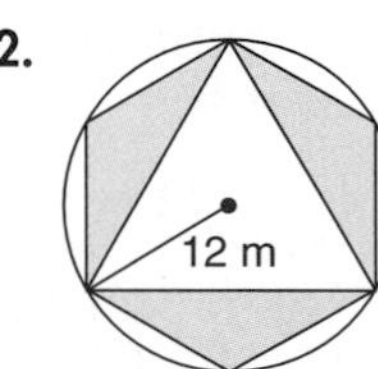

33. 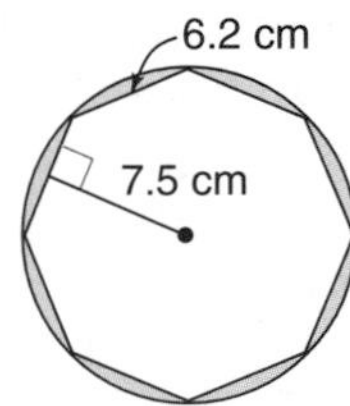

34. 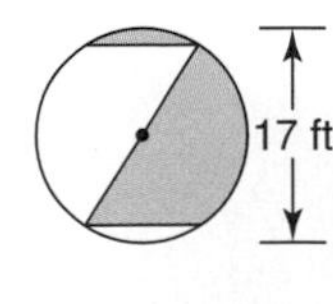

35. A circle is inscribed in a rhombus whose diagonals are 24 and 32 feet long.
  a. Find the area of the circle.
  b. What is the area of the region that is inside the rhombus but outside the circle?

36. A circle inscribes a regular hexagon and circumscribes another. If the radius of the circle is 10 units long, find the ratio of the area of the smaller hexagon to the area of the larger hexagon.

Critical Thinking

37. The number 3.14 is often used to approximate π. Use the findings in the Exploration on page 545 to explain why 3.14 is a good approximation of π.

38. Find the ratio of the area of $\triangle ABC$ to square $BCDE$.

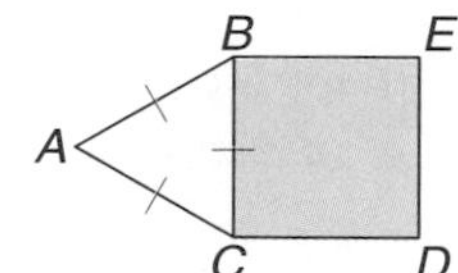

Applications and Problem Solving

39. **Shopping** The Pizza Shoppe sells a 12-inch cheese pizza for $6.98 and a 16-inch cheese pizza for $9.98.
    a. Which will give Kwam the most pizza, buying two 12-inch pizzas or one 16-inch pizza?
    b. Will Kwam get the best deal if he chooses to buy the option that will give him the most pizza? Explain.

40. **Gardening** Refer to the application about Elizabeth Park at the beginning of the lesson. Use the diameters of the gazebo and the two concentric rose garden plots to answer the following questions.

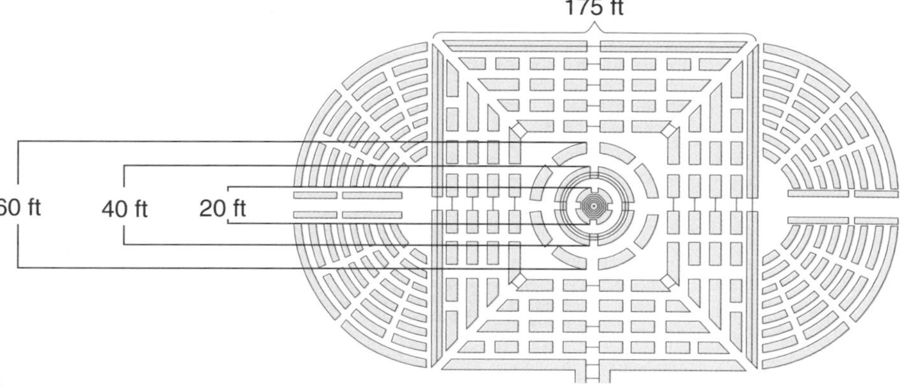

a. Find the area and the perimeter of the entire Rose Garden. The rectangular section in the center is a square.
b. What is the total of the circumferences of the three concentric circles formed by the gazebo and the two circular rose garden plots?
c. Each rose plot has a width of 5 feet. What is the area of the path between the outer two complete circles of rose garden plots?

FYI

Elizabeth Park began in 1894 when a tract of land was bequeathed to the City of Hartford, Connecticut, by the will of Charles H. Pond to be used as a park and named for his wife, Elizabeth.

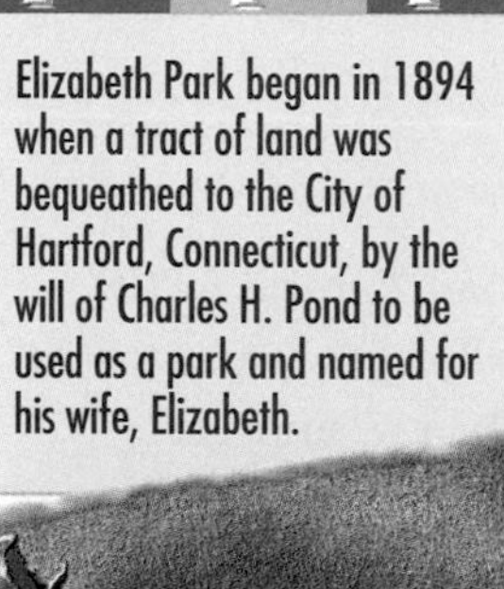

41. **Architecture** The Anraku Temple in Japan is composed of four octagonal floors of different sizes that are separated by four octagonal roofs of different sizes. If the sides of the octagonal floors are in a ratio of 1:3:5:7, are the areas of each of the four octagonal floors in the same ratio? Explain your answer.

Mixed Review

42. The perimeter of the triangle at the right is 18 units. Find the area of the triangle. (Lesson 10–4)

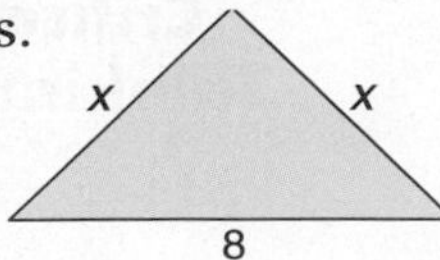

43. Find the area of the parallelogram at the right. (Lesson 10–3)

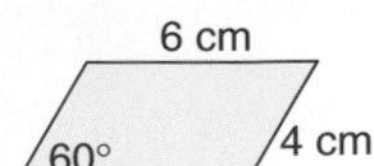

44. Will a regular dodecagon tessellate in a plane? If so, draw a sample figure. (Lesson 10–2)

45. Write an equation for the circle whose diameter has endpoints at $(-1, 6)$ and $(5, 2)$. (Lesson 9–8)

46. **SAT Practice** If 2 packages contain a total of 16 cookies, how many cookies are there in 7 packages?

A 16 **B** 23 **C** 40 **D** 56 **E** 112

47. Identify the similar triangles in the figure at the right. Explain your answer. (Lesson 7–3)

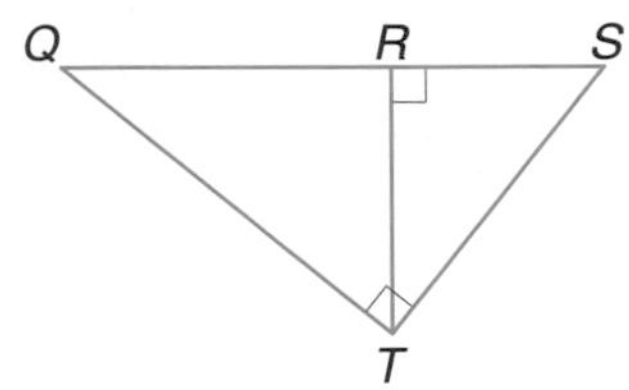

48. Graph $A(7, 6)$, $B(3, 7)$, $C(5, 11)$, and $D(9, 10)$. Draw quadrilateral $ABCD$. Determine if $ABCD$ is a parallelogram. Justify your answer. (Lesson 6–2)

49. The measures of the angles in a triangle are $x + 16$, $8x + 7$, and $11x - 3$. Is the triangle *acute*, *obtuse*, *right*, or *equiangular*? (Lesson 4–2)

For **Extra Practice**, see page 784.

Algebra

50. Factor $21xy^3 + 3y^5$.

51. Simplify $(2a^3)^2$.

Refer to the Investigation on pages 510–511.

Most adult soccer teams, including Major League and World Cup teams, use a size 5 soccer ball. A size 5 ball must have a circumference of 27 to 28 inches and weigh 14 to 16 ounces.

1 About how many polygons of each type appear on each soccer ball design shown on pages 510 and 511?

2 Estimate the length of an edge of a pentagon in each design. Then use that estimate to find the area of each type of polygon.

3 What is the total area of the figures on each ball design?

4 Do you think one of the two designs would be less expensive to manufacture than the other? Explain.

5 Use your results to describe some things you will consider when you determine a proposed design for your company's soccer ball.

Add the results of your work to your Investigation Folder.

10–6 Integration: Probability
Geometric Probability

What YOU'LL LEARN

- To use area to solve problems involving geometric probability.

Why IT'S IMPORTANT

You can use geometric probability to solve problems involving navigation and entertainment.

Real World APPLICATION

Radio

WGLN radio station is having a "Listen and Call" contest. A song of the day is announced and played at 7:00 A.M. each day. Then, once during each hour of the day, that song is played. Since WGLN is at 91.1 on the radio dial, the ninety-first person to call the station when the song of the day begins to play wins $100. If you turn on your radio at 2:35 P.M., what is the probability that you have missed the start of the song of the day when it is played in the 2:00 P.M. to 3:00 P.M. hour?

This problem can be solved using **geometric probability**. Geometric probability involves using principles of length and area to find the probability of an event. One of the principles of geometric probability is stated in Postulate 10–3.

Postulate 10–3 Length Probability Postulate	If a point on $\overline{AB}$ is chosen at random and C is between A and B, then the probability that the point is on $\overline{AC}$ is $\frac{\text{length of } \overline{AC}}{\text{length of } \overline{AB}}$.

To find the probability that you have missed the start of the song of the day, draw a line segment to represent the time from 2:00 to 3:00. Let point C represent the time that you started listening.

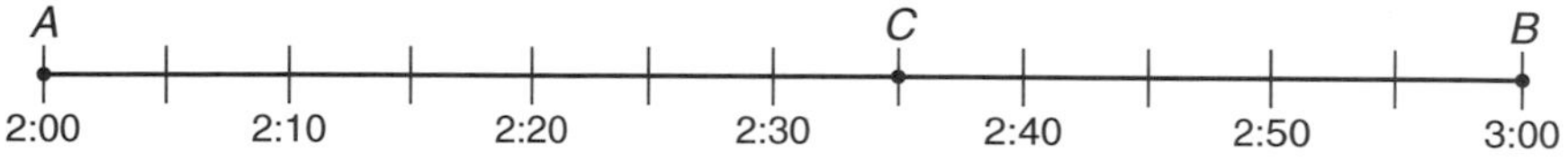

The entire hour is represented by the length of $\overline{AB}$. The length of $\overline{AC}$ represents the time from 2:00 to 2:35, and the length of $\overline{CB}$ represents the time from 2:35 to 3:00.

You have missed the song of the day if the station does not play the song after 2:35. So, the probability of having missed the song is $\frac{\text{length of } \overline{AC}}{\text{length of } \overline{AB}}$.

$$P(\text{missed the song}) = \frac{\text{length of } \overline{AC}}{\text{length of } \overline{AB}}$$

$$= \frac{35 \text{ minutes}}{60 \text{ minutes}}$$

$$= \frac{7}{12}$$

The probability that you missed the song is $\frac{7}{12}$ or about 58%.

Postulate 10–4 states a second principle of geometric probability.

Postulate 10–4 Area Probability Postulate	If a point in region *A* is chosen at random, then the probability that the point is in region *B*, which is in the interior of region *A*, is $\frac{\text{area of region } B}{\text{area of region } A}$.

Example 1

Real World APPLICATION

Festivals

Every summer the Hayti Heritage Center in Durham, North Carolina, has a festival at Eno River Park. The Center invites entertainers and vendors to celebrate the diversity of the community.

The rectangular area of the park where vendors are located has shaded and sunny areas as shown in the diagram. Seventy-five percent of the park is used by vendors. The other 25% is for walkways. Each vendor space has exactly the same area.

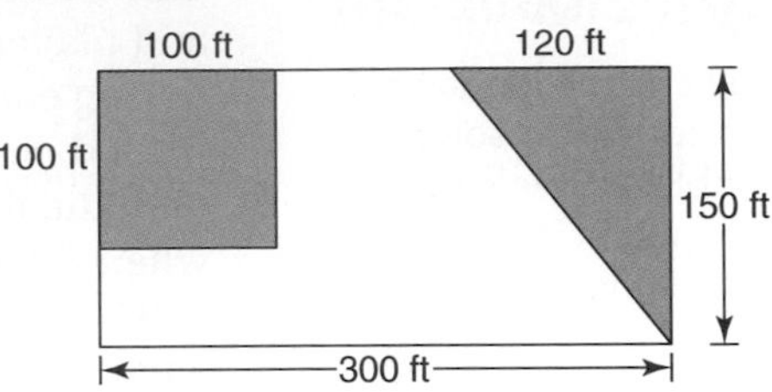

Because it is usually very hot during the festival, the shaded spots are most desired. If vendors are randomly placed in the park, what is the probability that a vendor will be placed in a shaded spot?

Explore The dimensions of the entire park and also the dimensions of the shaded areas are given. We need to subtract the area of the walkways away from these areas. Then we can use the Area Probability Postulate to find the probability.

Plan To find the probability that a vendor will be in a shaded area, we must find the area of the entire region without the walkways and the area of the shaded regions without the walkways.

Solve

- Find the area of the entire region without the walkways.

  area of entire region = (150)(300) or 45,000
  area of entire region without walkways = 0.75(45,000) or 33,750

  The vendor area of the entire region is 33,750 square feet.

- Find the area of the shaded regions without the walkways.

  area of square region = (100)(100) or 10,000
  area of square region without walkways = 0.75(10,000) or 7500

  area of triangular region = $\frac{1}{2}$(120)(150) or 9000

  area of triangular region without walkways = 0.75(9000) or 6750

  The area of the shaded regions without the walkways is 7500 + 6750 or 14,250 square feet.

- Find the probability of being in a shaded spot.

$$P(\text{shaded region}) = \frac{\text{shaded area}}{\text{total area}}$$
$$= \frac{14{,}250}{33{,}750} \text{ or } \frac{19}{45}$$

The probability that the vendor would be placed in a shaded region is $\frac{19}{45}$ or about 42%.

Examine Look at the diagram. Is about 40% of the total area shaded?

Sometimes when you are finding a geometric probability, you need to find the area of a **sector of a circle**. A sector of a circle is a region of a circle bounded by a central angle and its intercepted arc.

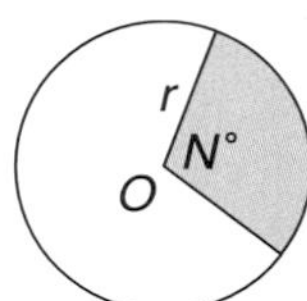

Area of a Sector of a Circle	If a sector of a circle has an area of A square units, a central angle measuring $N°$, and a radius of r units, then $A = \frac{N}{360}\pi r^2$.

Example 2 **Nuela has to spin either a 3 or a 5 to stay in the game. What is the probability that she will spin either one of these two numbers?**

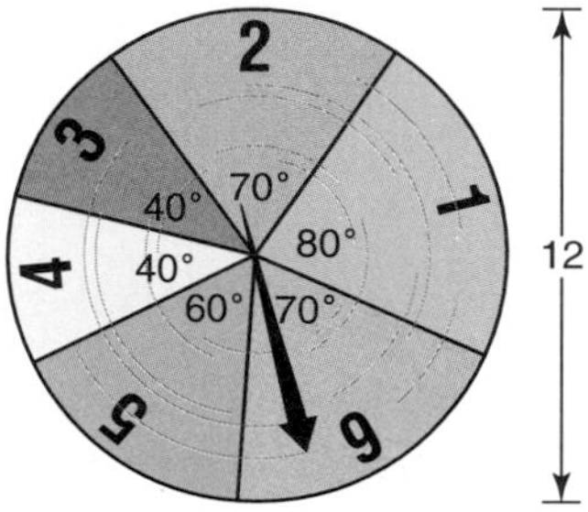

Find the area of the circle and the areas of the appropriate sectors of the circle.

area of the 3 sector	area of the 5 sector	area of circle
$A = \frac{N}{360}\pi r^2$	$A = \frac{N}{360}\pi r^2$	$A = \pi r^2$
$= \frac{40}{360}\pi(6)^2$	$= \frac{60}{360}\pi(6)^2$	$= \pi(6)^2$
$= \frac{1440\pi}{360}$	$= \frac{2160\pi}{360}$	$= 36\pi$
$= 4\pi$	$= 6\pi$	

The total area of the two sectors is 10π.

Now find the geometric probability.

$$P(3 \text{ or } 5) = \frac{\text{area of sectors}}{\text{area of circle}}$$
$$= \frac{4\pi + 6\pi}{36\pi}$$
$$= \frac{10\pi}{36\pi} \text{ or } \frac{5}{18}$$

The probability that Nuela will spin a 3 or a 5 is $\frac{5}{18}$ or about 28%.

CHECK FOR UNDERSTANDING

Communicating Mathematics

Study the lesson. Then complete the following.

1. **Compare and contrast** the Length Probability Postulate with the Area Probability Postulate.

2. **You Decide** Terrence needs to find the area of a region that has an irregular shape. Adina said that he could place the region in a square dartboard, throw darts randomly at the square, and then use the number of times the darts land in that region to find the area of the region. Terrence said this was impossible. Who is right? Can the probability of an event occurring be used to estimate the area of a region?

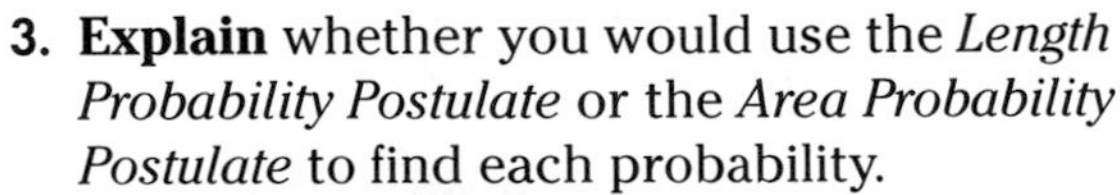

3. **Explain** whether you would use the *Length Probability Postulate* or the *Area Probability Postulate* to find each probability.
   a. the probability that a point is in a particular region
   b. the probability that a point is on a particular segment

4. **Draw** a target made up of three concentric circles so that the probability of a dart landing in each ring, as the circumferences get larger, increases by a factor of 24. Justify your drawing.

Guided Practice

Find the probability that a point chosen at random on $\overline{DI}$ is also a part of each of the following segments.

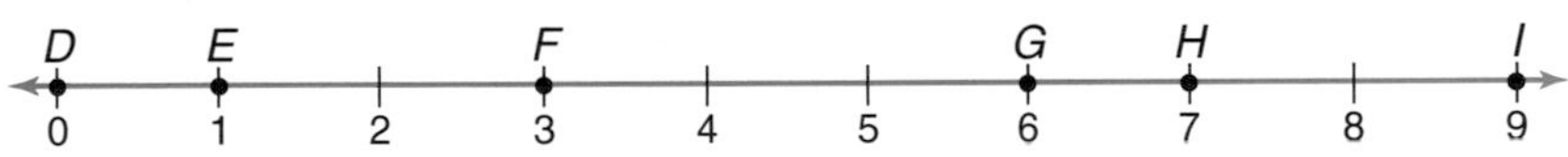

5. $\overline{EF}$

6. $\overline{FI}$

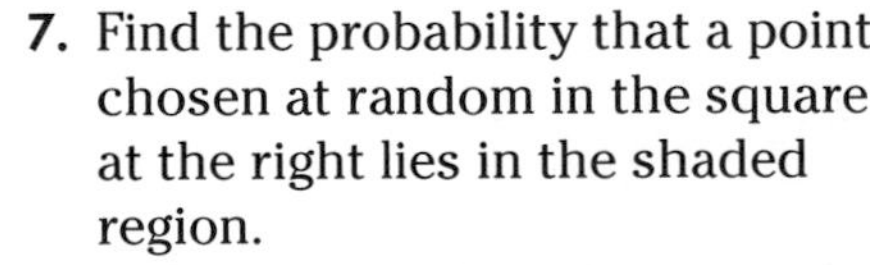

7. Find the probability that a point chosen at random in the square at the right lies in the shaded region.

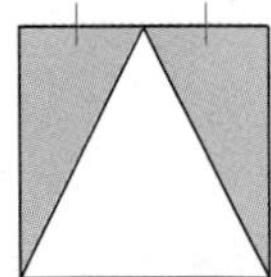

8. To win at a carnival game, you must throw a dart at a board that is 6-feet by 3-feet and hit one of the 25 playing cards on the board. The playing cards are each $2\frac{1}{2}$-inches by $3\frac{1}{2}$-inches.
   a. Draw a diagram of the dartboard.
   b. What is the probability that a randomly thrown dart that hits the board hits a playing card? Round to the nearest hundredth.
   c. Does the arrangement of the cards on the board affect the probability? Explain.

9. **Entertainment** The children at Jasmine's birthday party are playing a game for prizes. Each child tosses a beanbag at the target on the floor. Depending on where one marked corner of the beanbag lands, the following prizes are given:

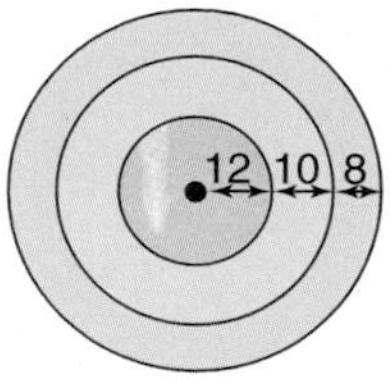

red: jumbo squirt gun
blue: yo-yo
green: candy bar

Measurements are in inches.

If the marked corner of the beanbag lands on the target, find the probability of a child winning a jumbo squirt gun. Round to the nearest hundredth.

EXERCISES

Practice

10. What is the probability that a random point on $\overline{AB}$ will be closer to point A than to point X?

11. A point is chosen at random on $\overline{XY}$. If Z is the midpoint of $\overline{XY}$, W is the midpoint of $\overline{XZ}$, and V is the midpoint of $\overline{WY}$, what is the probability that the point is on $\overline{XV}$?

Find the probability that a point chosen at random in each figure lies in the shaded region.

12. 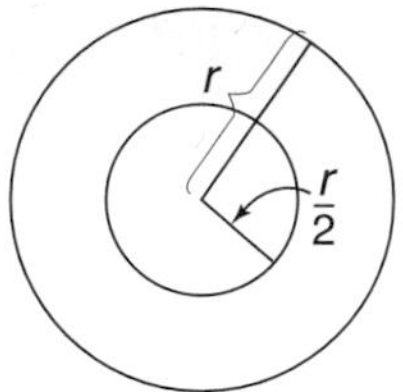

13. 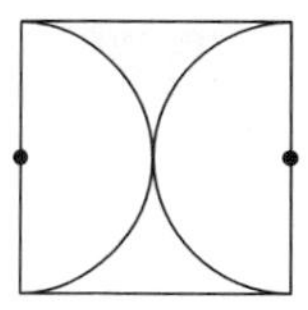

14. 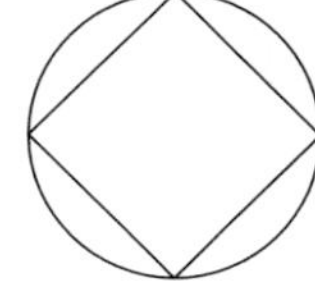

Find the probability for each outcome on the game spinner shown at the right.

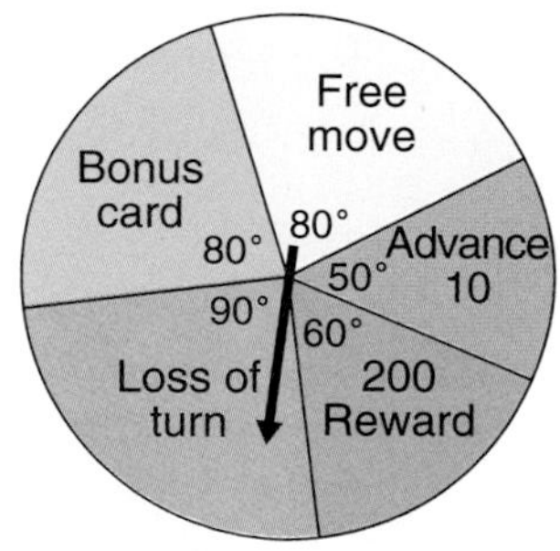

15. free move
16. loss of turn
17. draw bonus card
18. get $200 reward or a free move

Find the probability for each outcome on the target shown at the right. The center ring has a radius of 1 unit. Each successive ring has a radius 1 unit greater than the previous one.

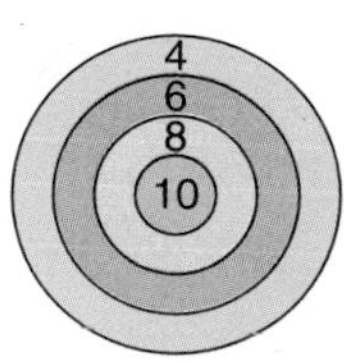

19. 10 points
20. 8 points
21. 6 points 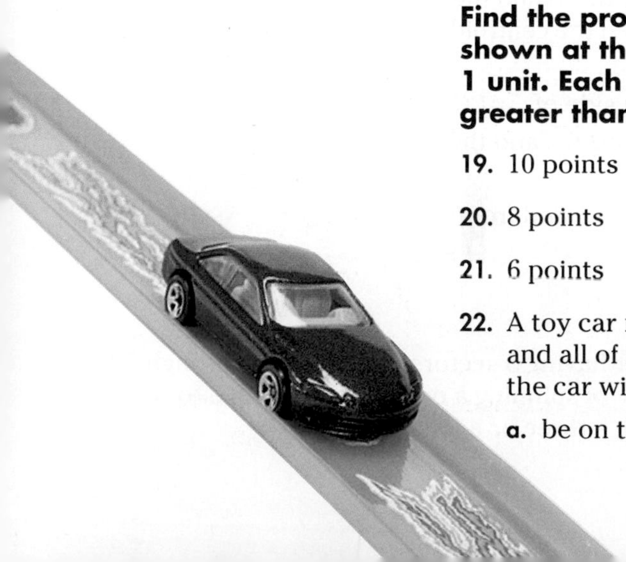
22. A toy car moves on tracks that consist of the sides of a regular hexagon and all of its diagonals. What is the probability that at a random moment the car will:
   a. be on the longest diagonal?
   b. not be on the perimeter?

23. A, B, C, D, E, and F are consecutive collinear points such that $AD = BF$, $BC = CD$, B is the midpoint of $\overline{AD}$, and E is the midpoint of $\overline{DF}$.
  a. Draw the figure.
  b. What is the probability that a random point will fall on $\overline{AB}$?
  c. What is the probability that a random point will fall on $\overline{DE}$?
  d. What is the probability that a random point will fall on $\overline{BE}$?

24. Ms. Perez often gives a short one-problem quiz during her 48-minute geometry class. If a student is out of the room at the time she gives the quiz, he or she will miss that grade. Natalia was fifteen minutes late for class. What is the probability that she missed the start of the quiz?

LOOK BACK

Refer to Lesson 5-1 to review median of a triangle.

25. What is the probability that a random point on a median of a triangle will fall between the centroid of a triangle (the intersection point of the medians) and the side to which the median is connected?

26. You tell a friend that you will arrive at the mall sometime between 12:00 and 12:30. Since she is not sure she will be able to come, your friend tells you not to wait for her any longer than 10 minutes. If your friend comes at 12:25, what is the probability that she missed you?

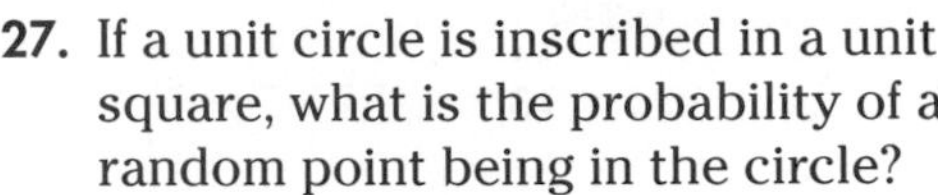

27. If a unit circle is inscribed in a unit square, what is the probability of a random point being in the circle?

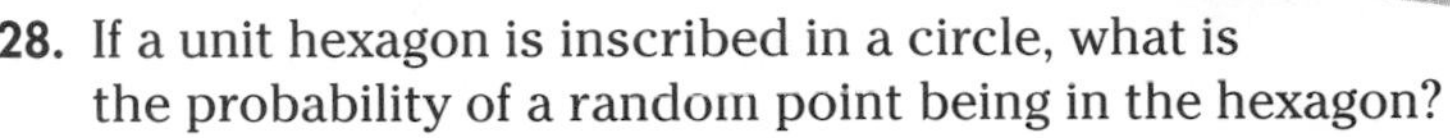

28. If a unit hexagon is inscribed in a circle, what is the probability of a random point being in the hexagon?

29. The figure at the right shows two squares. Determine the value of s if the shaded part is 75% of the area of the figure.

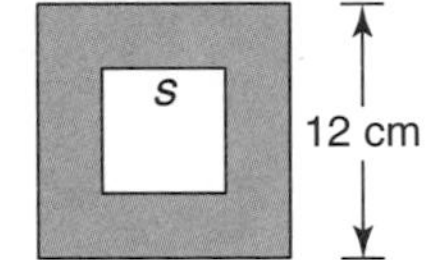

30. A circle is circumscribed about a square with side lengths of 7 centimeters. If a dart is thrown at random into the circle, what is the probability that it lands in the circle, but outside the square?

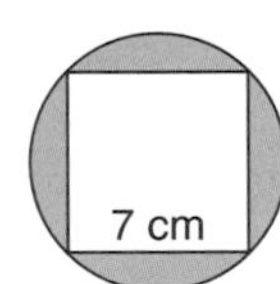

Cabri Geometry

31. Use a calculator and the steps below to find the probability that a point on the median of a triangle is between the centroid and the midpoint of the side.
  a. Construct $\triangle ABC$.
  b. Construct midpoints on each side of $\triangle ABC$.
  c. Construct two medians, $\overline{AD}$ and $\overline{BF}$, and their point of intersection G.
  d. Construct the third median.
  e. Measure the distance from B to F and from F to G.
  f. Find the ratio of $\overline{FG}$ to $\overline{BF}$.

Critical Thinking

32. How would you design a spinner having 5 sectors, numbered 1 through 5, if you want to have a probability of spinning a one, a two, or a five to be twice the probability of spinning a three or a four?

Applications and Problem Solving

33. **Accommodations** The convention center in Washington, D.C., lies in the northwest sector of the city between New York and Massachusetts Avenues. These streets intersect at a 130° angle. If the amount of hotel space is evenly distributed over an area with that intersection as the center and a radius of 1.5 miles, what is the probability that a visitor, randomly assigned to a hotel in the city, will be housed in the sector that contains the convention center?

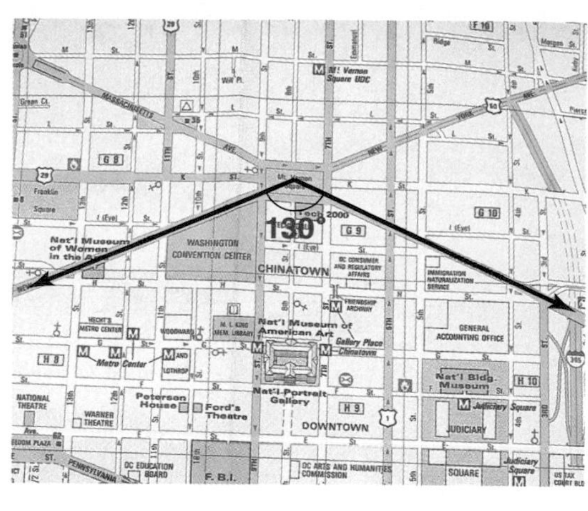

34. **Conventions** The Annual Meeting of the National Council of Teachers of Mathematics will be held in San Antonio, Texas, in 2003. If there are 1100 sessions evenly distributed over the three full days of the convention, what is the probability that a particular session will fall on the first day of the meeting?

35. **Navigation** As part of a scuba diving exercise, a 12-foot by 3-foot rectangular-shaped rowboat was sunk in a quarry. A motorboat takes a scuba diver to a random spot in the triangular section of the quarry and anchors there so that the diver can search for the rowboat.

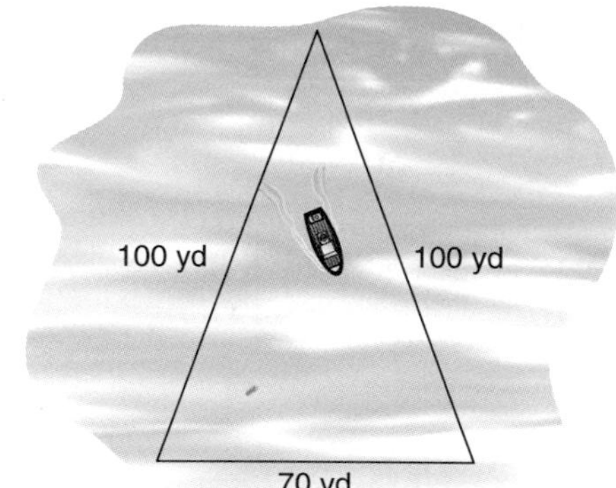

   a. What is the approximate area of the triangular section of the quarry?
   b. What is the area of the rowboat?
   c. What is the probability that the motorboat will anchor over the sunken rowboat?

Mixed Review

36. Find the area of an octagon with an apothem 7.5 feet long and a side 6.2 feet long. (Lesson 10–5)

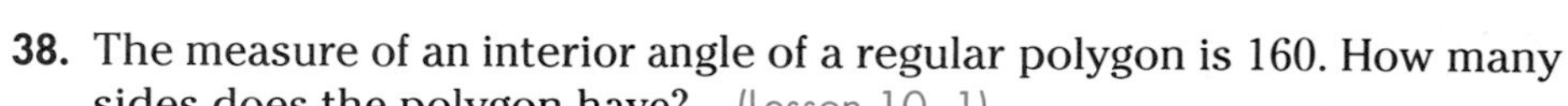

37. If the lengths of the sides of a triangle are multiplied by 3, what is the ratio of the area of the new triangle to the area of the old triangle? (Lesson 10–4)

38. The measure of an interior angle of a regular polygon is 160. How many sides does the polygon have? (Lesson 10–1)

39. Find $m\widehat{IJK}$. (Lesson 9–4)

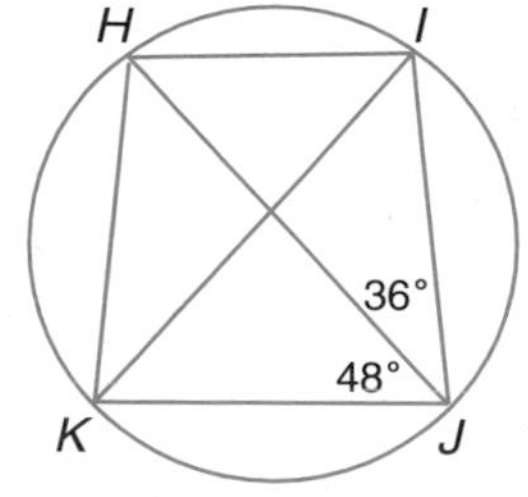

40. **Gardening** A flower bed is in the shape of an obtuse triangle with the shortest side measuring 7.5 feet. One angle measures 103°, and the side opposite that angle measures 14 feet. Find the measures of the remaining sides and angles. (Lesson 8–5)

41. **SAT Practice** **Quantitative Comparison**

Set A: $\{-4, 2, 7, 8, 10, 12\}$
Set B: $\{15, 30, -4, 3, 8, -7\}$

Column A	Column B
The median of Set A	The average of Set B

A if the quantity in Column A is greater
B if the quantity in Column B is greater
C if the two quantities are equal
D if the relationship cannot be determined from the information given

INTEGRATION **Algebra**

42. Solve $2x^2 - 3x + 1 = 0$.

43. **Percent** In general, the more education a person has, the higher the earnings for that person. The graphic shows the mean annual income in a recent year for people 18 years or older, by level of education. Use the graph to find the percent of increase in each level of education.

Data Update For more information on education and earnings, visit: **www.geometry.glencoe.com**

For **Extra Practice**, see page 784.

Higher Education, Higher Earnings

Mean annual income for people 18 or older, by level of education:

Level of education	Mean annual income
Professional	$74,560
Doctorate	$54,904
Master	$40,368
Bachelor	$32,629
Associate	$24,398
Some college, no degree	$19,666
High school graduate	$18,737
Not high school graduate	$12,809

Seeing Double with Telescopes

The excerpt below appeared in *Science News* in June, 1996.

The domes of eight telescopes dot the barren landscape atop...an extinct Hawaiian volcano. Reigning supreme among them...has been the 10-meter W.M. Keck Telescope. Now, the world's largest optical telescope has a partner....Like the original, Keck II uses a mirror composed of 36 [hexagonal] glass tiles to form a parabolic reflecting surface equivalent to that of a single 10-meter mirror....By adjusting individual tiles of the segmented mirror 100 times a second, the telescope is expected to produce images with a resolution as sharp as 0.04 arc second....That's like distinguishing between two barely touching pennies viewed from a distance of 600 kilometers. Spaced 85 meters apart, Keck I and Keck II can make simultaneous observations of the same heavenly body. In that way, the paired instruments would act as a single, 85-meter telescope. ■

1. Why are optical telescopes usually located on mountains or volcanoes in remote areas?
2. If two 5-meter telescopes were paired 40 meters apart, they would have the same capability as what size single-mirror telescope?
3. If a parabolic mirror is a curved surface, are the hexagons regular? Explain your answer.

10–7 Integration: Graph Theory Polygons as Networks

What YOU'LL LEARN

- To recognize nodes and edges as used in graph theory,
- to determine if a network is traceable, and
- to determine if a network is complete.

Why IT'S IMPORTANT

You can use graph theory to solve games and problems involving routing.

Real World APPLICATION

Internet

The international computer information system, Internet, was called the information highway in the early 1990s. It is the networking of computers, connections, and information. Internet is composed of nodes that allow for information to be transferred from one computer site to another.

The diagram at the right represents a **network**. Such a diagram illustrates a branch of mathematics called **graph theory**. In networking, the points are called **nodes**, and the paths connecting the nodes are called **edges**. Edges can be straight or curved.

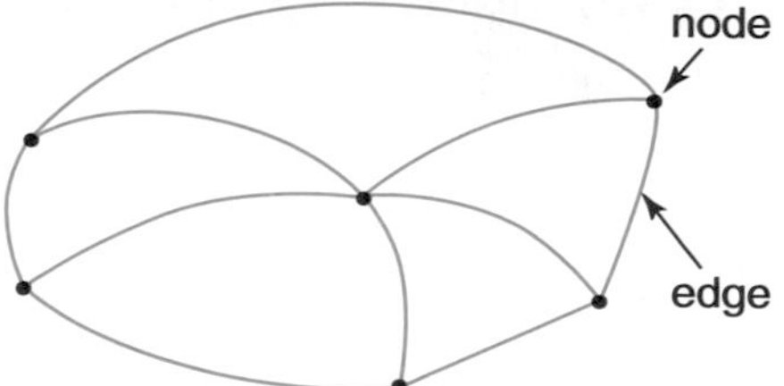

Straight edges can be used to form *closed* or *open* graphs. If each edge of a closed graph intersects exactly two other edges only at the endpoints, then the graph forms a polygon.

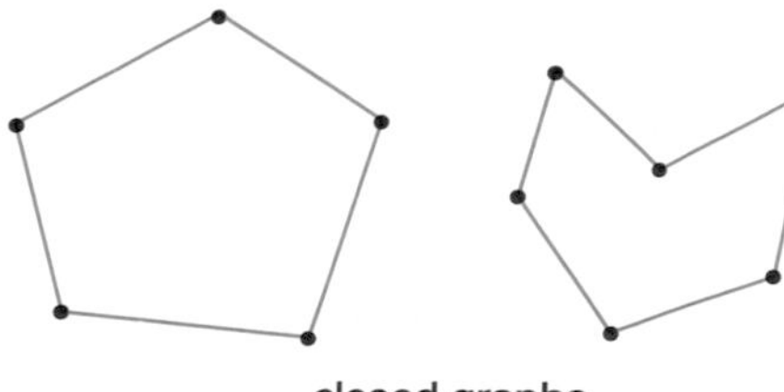

closed graphs

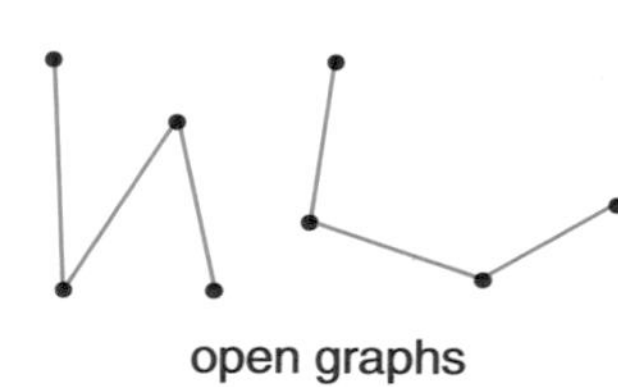

open graphs

Intersection points of edges are not nodes unless the network is shown with a solid vertex point.

However, not all pairs of nodes are connected by an edge in some networks. A network like this is called **incomplete**. A **complete network** has exactly one path between each pair of nodes.

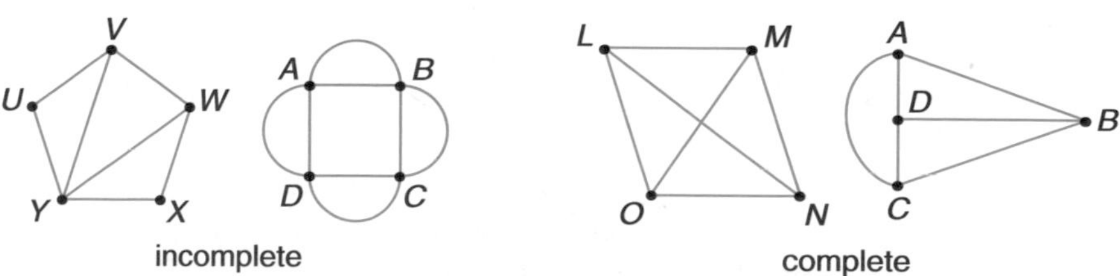

incomplete complete

You have probably seen puzzles that ask you to trace over a figure without lifting your pencil and without tracing any lines more than once. In graph theory, if all nodes can be connected and each edge of a network can be covered exactly once, then a network is said to be **traceable**.

Example 1

Routing

County highway inspectors need to visually inspect roads that connect several small towns. They must travel a route that takes them over each section of road exactly once. Use the networks of towns shown below to determine if this is possible for each set of towns. That is, determine if each network is traceable.

a. three towns

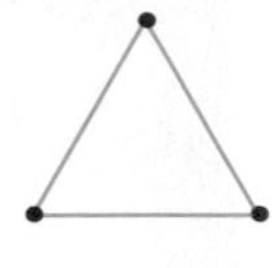

yes

b. three towns

yes

c. four towns

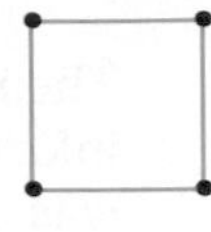

yes

d. four towns

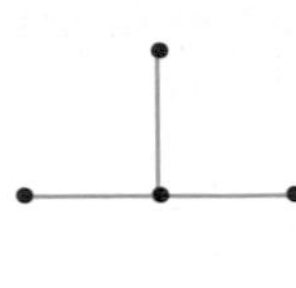

no

e. five towns

no

f. five towns

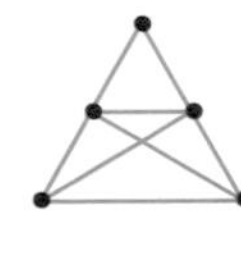

yes

g. six towns

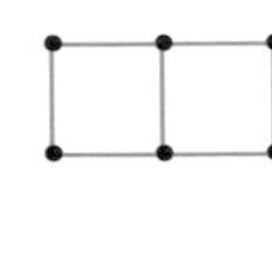

yes

h. six towns

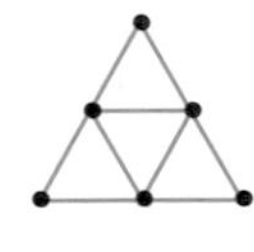

yes

The **degree of a node** is the number of edges that are connected to that node. The traceability of a network is related to the degrees of the nodes in the network. What conjecture can you make about the design of the networks in Example 1 and their traceability?

Network Traceability Tests

A network is traceable if and only if one of the following is true.
1. All of the nodes in the network have even degrees.
2. Exactly two nodes in the network have odd degrees.

Example 

Determine if each network is traceable and complete. If not complete, name the edges that need to be added to make it complete.

a.

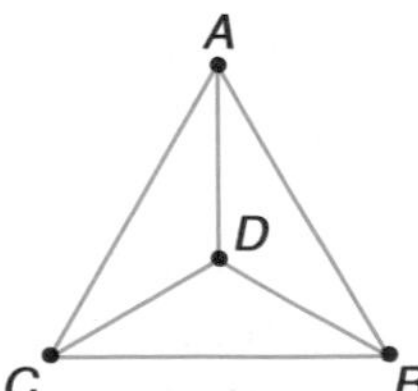

The network is not traceable because all of the nodes have odd degrees.

Since all nodes are connected, the network is complete.

b.

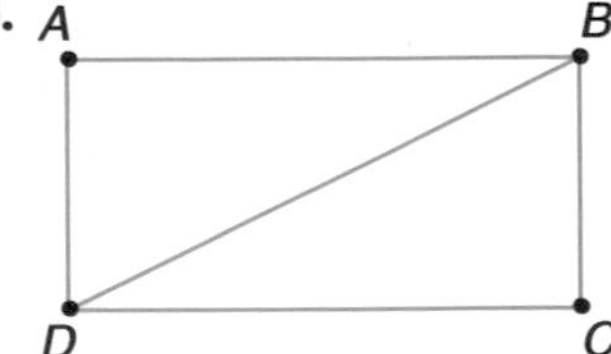

Exactly two nodes, B and D, have odd degrees. So the network is traceable.

The network is incomplete because nodes A and C are not connected.

Example 

a. Find the degree of each node in the network shown at the right.

b. Is the network traceable?

c. If so, list a sequence of segments that demonstrates traceability.

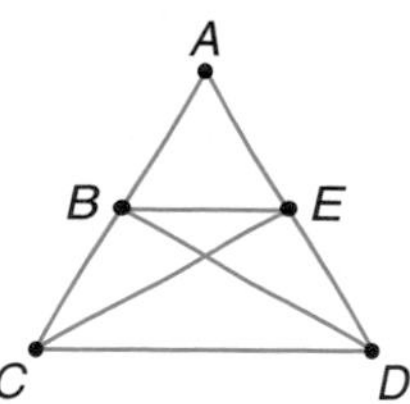

a. A: degree 2, B: degree 4, C: degree 3, D: degree 3, E: degree 4

b. Exactly two nodes, C and D, have odd degrees. The network is traceable.

c. $\overline{CB}$ to $\overline{BA}$ to $\overline{AE}$ to $\overline{ED}$ to $\overline{DB}$ to $\overline{BE}$ to $\overline{EC}$ to $\overline{CD}$

Look at the network in Example 3. Did it matter where you started in order to determine the sequence of segments to show its traceability? When a path goes through a node, it uses two edges. When a traceable network has an odd node, it must be a starting or finishing point of the traceability path. Therefore, you could start at vertex C or D and end at the one in which you didn't start. What would happen if there were three vertices with odd degree? *It would not be traceable.*

When you determined if a network was traceable, you probably had to try different strategies on some of the figures. The following patterns may have become apparent as you worked.

Starting Points for Network Traceability	**If a node has an odd degree, then the tracing must start or end at that node.** **If there are more than two nodes of odd degree, then the network is not traceable.** **If the network has no node of odd degree, then the tracing can start at any node and will end at the starting point.**

CHECK FOR UNDERSTANDING

Communicating Mathematics

Study the lesson. Then complete the following.

1. **Compare** a network to a convex polygon.
2. **You Decide** Amiri thought that the traceability of a network was related to whether the network was open or closed. Terri did not think that open or closed had anything to do with traceability. Who was correct? Explain.
3. **Draw** an example of a network that is traceable and incomplete.
4. **Explain** why only two nodes can have an odd degree.

5. **Assess Yourself** Describe in your journal why this network is or is not traceable. Can you determine visually whether a network is traceable? Draw a traceable and a not traceable network and show how the pattern works.

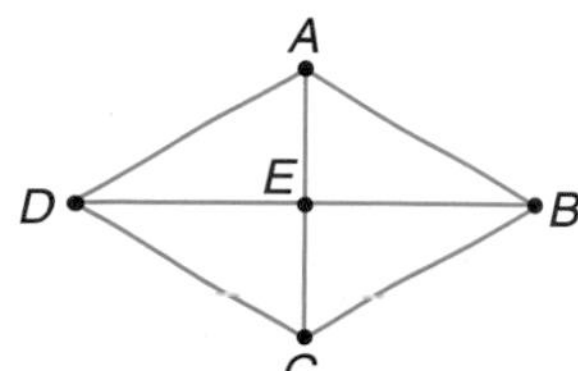

Guided Practice

6. Find the degree of each node in the network at the right.

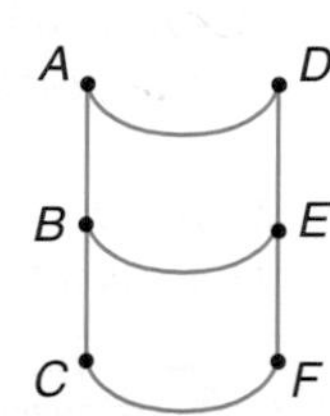

Consider each network.

a. Determine if it is traceable and complete.

b. If traceable, list a sequence of segments that demonstrates traceability.

c. If not complete, name the edges that need to be added to make it complete.

7. 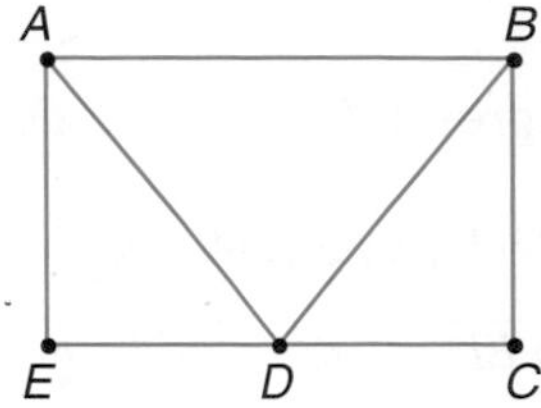

8. 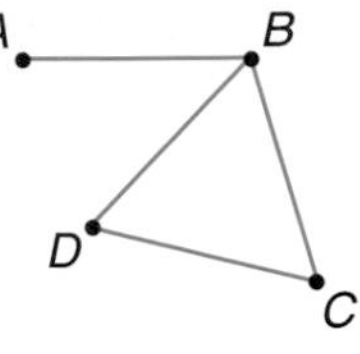

9. **Neighborhoods** Three neighbors decided to share the cost of groundskeeping equipment for the summer and the winter. One purchased a lawn mower, another purchased a snowblower, and the third purchased various gardening, pruning, and trimming tools. All the equipment purchased was stored in three separate sheds at the back of the three properties.

   a. Draw a network that shows possible paths for each of the neighbors.

   b. Do the paths have to cross? Explain.

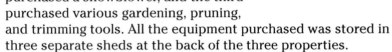

EXERCISES

Practice

Find the degree of each node in each network.

10. 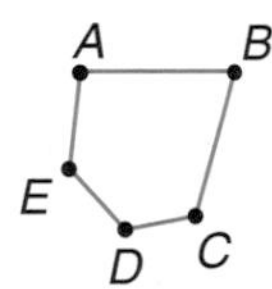

11. 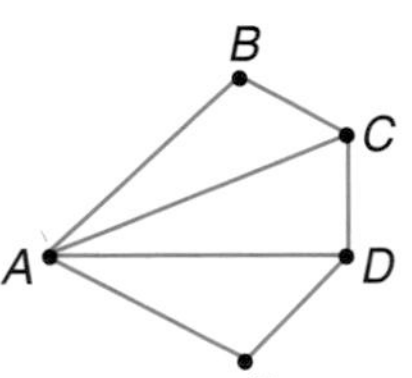

12. 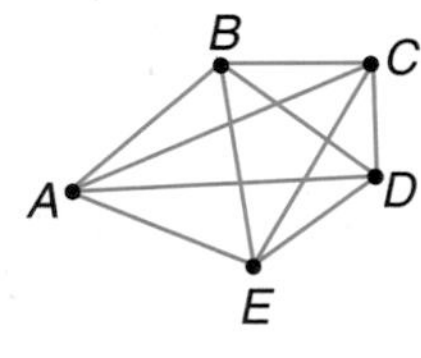

Consider each network.

a. Determine if it is traceable and complete.

b. If traceable, list a sequence of segments that demonstrates traceability.

c. If not complete, name the edges that need to be added to make it complete.

13. 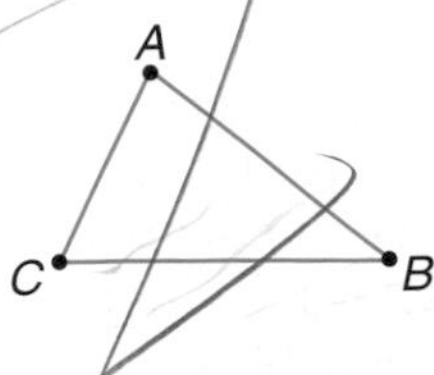

14. 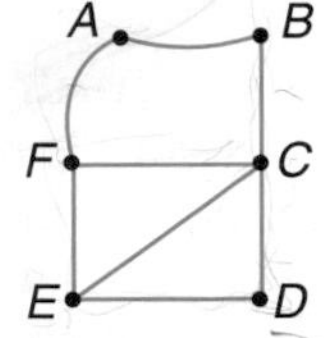

15. 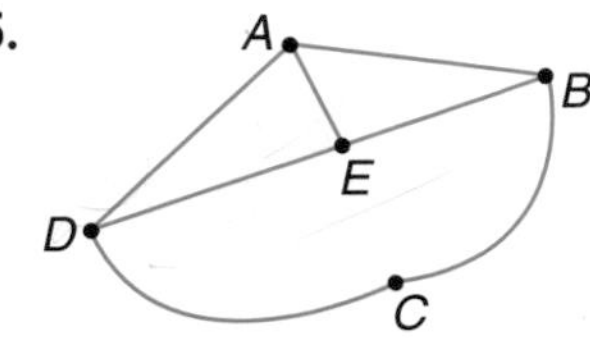

16. 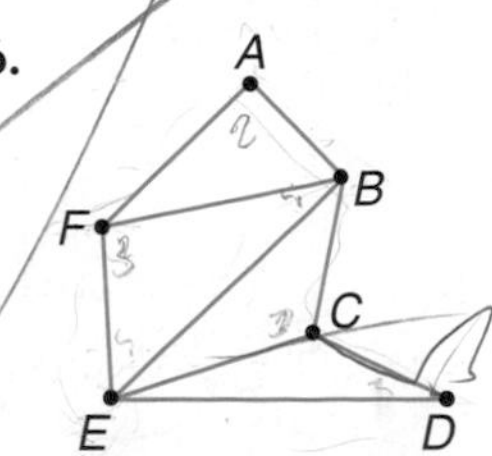

17. 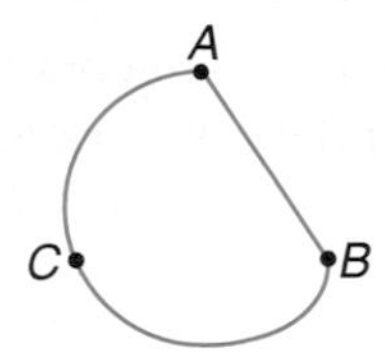

18. 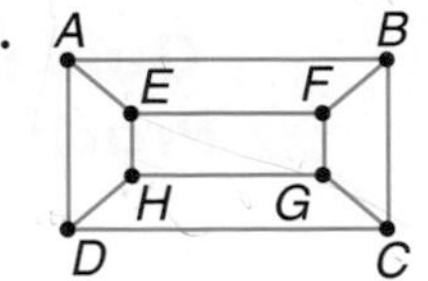

Determine whether the network puzzles below *can't be traced, can be traced and started at one odd vertex and finished at the other,* or *can be traced and started and finished at the same vertex.* If it can be traced, show how.

19. 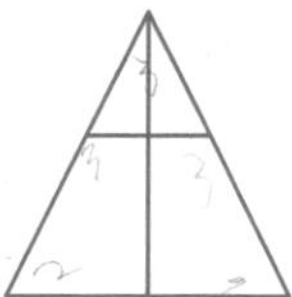

20. 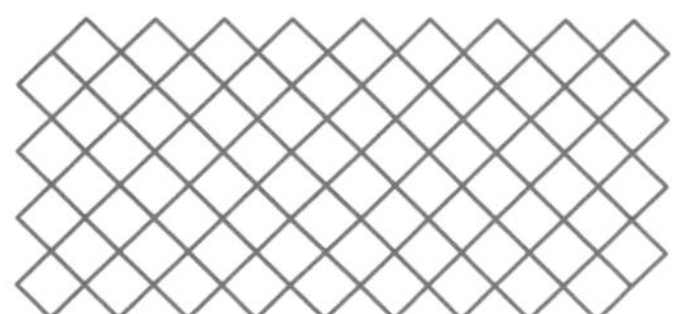

If it is possible, draw a network that satisfies the following conditions.

21. both traceable and complete

22. complete, but not traceable

23. not traceable and incomplete

24. If you can use your pencil to draw a path through a network so that the path starts and ends at two different nodes and no node is passed more than once, the path is called an *Euler* (pronounced OI-ler) *path*. If the path can be drawn so that the path starts and ends at the same node and no node is passed more than once, the path is called an *Euler circuit*.

  a. Does the network at the right have an Euler path?
  b. Does the network contain an Euler circuit?
  c. Do you think a network could contain both an Euler path and an Euler circuit? Explain.

Critical Thinking

25. Show why this statement is true: *If a network has no node of odd degree, then the tracing can start at any node and will end at the starting point.*

26. a. Find the degree of each node in the network at the right.

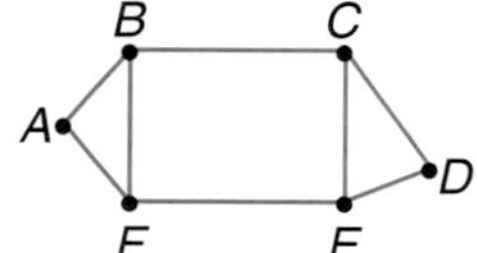

  b. Find the sum of the degrees of all the nodes.
  c. Compare this sum to the number of edges in the network. Is there a relationship between the number of edges and the sum of the degrees of the nodes?
  d. Draw several networks to verify your answer.

Applications and Problem Solving

27. **Math History** In 1735, Leonhard Euler used networks to solve a famous problem about the bridges in the city of Köenigsberg, which is now Kaliningrad. Köenigsberg had seven bridges connecting both sides of the Predgel River to two islands in the river. Try to find a path that will take you over all seven bridges without crossing the same bridge twice.

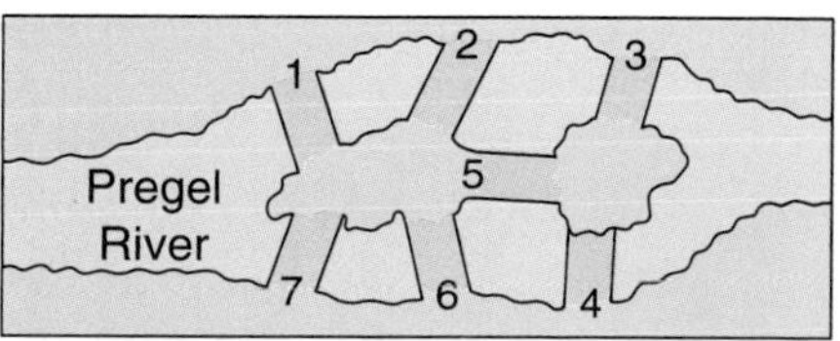

  a. Draw the network.
  b. Is the network traceable? Explain your answer.
  c. If the bridge is not traceable, could you add one bridge to make it traceable?

28. Safety The fire department suggests that families establish exit paths in the case of an emergency. Use the floor plan at the right to establish a path that can be used to check to make sure that everyone is out of the house. Because children often hide behind doors in an emergency, the path should go through each door once.

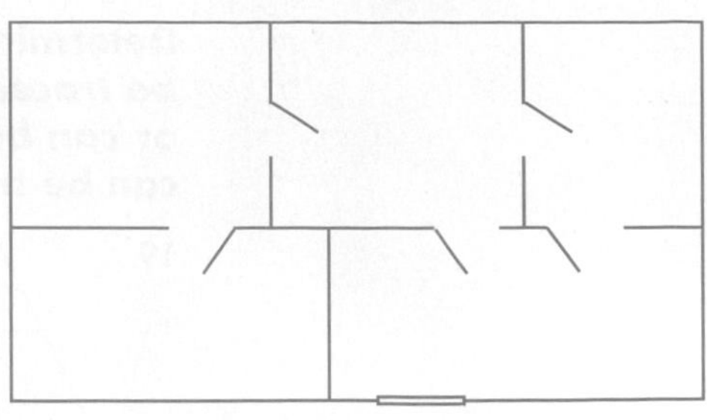

a. Draw a network that represents a plan for checking the rooms.

b. Determine if the floor plan is traceable and if so, show how.

29. Games Shongo children from Zaire, a country in central Africa, spend time tracing patterns in the sand near their homes. The designs often look like those found on the clothing and woodcarvings their parents make. These patterns are actually traceable networks. Design your own network and, like the Shongo children, pattern it after a design on your clothes, the wallpaper in your bedroom, or another design. Exchange with a classmate when completed to see if they can trace your design.

Mixed Review

30. If a point is chosen at random in the interior of $\odot O$ with radius 5, what is the probability that the point is in the interior of right triangle AOB where A and B lie on $\odot O$ and the right angle is at O? (Lesson 10–6)

31. If $QRSTV$ is a regular pentagon and $ABVT$ is a square, find the area of the shaded region. (Lesson 10–5)

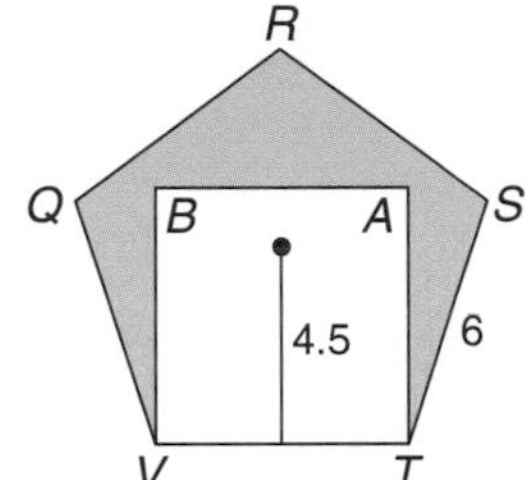

32. ACT Practice What is the area, in square feet, of a circle that has a circumference of 10 feet?

A $\frac{10}{\pi^2}$ **B** $\frac{25}{\pi^2}$ **C** $\frac{10}{\pi}$ **D** $\frac{25}{\pi}$ **E** 25

33. In $\triangle RST$, X lies on $\overline{RT}$ and Y lies on $\overline{RS}$. Determine the value of g that would make $\overline{YX} \parallel \overline{ST}$ if $RS = 12$, $RY = 8$, $RX = 10$, and $XT = g$. (Lesson 7–4)

34. Write the given statement and draw the figure for a proof of the statement *If two lines intersect, then at least one plane contains both lines.* Then write the assumption you would make to write an indirect proof. (Lesson 5–3)

35. Given $\triangle UVW \cong \triangle HJK$, $UV = 9x - 12$, $VW = 12x - 30$, $UW = 2x + 6$, and $JK = 7x - 5$, find the length of $\overline{JK}$. (Lesson 4–3)

Algebra

36. Simplify $6\frac{1}{4} \div 1\frac{1}{4}$.

37. Use substitution to solve the system of equations. If the system does not have exactly one solution, state whether it has *no* solution, or *infinitely many* solutions.

$2s = 10r + 6$ and $s = 5r - 1$

For **Extra Practice**, see page 784.

Chapter Review For additional lesson-by-lesson review, visit: **www.geometry.glencoe.com**

VOCABULARY

After completing this chapter, you should be able to define each term, property, or phrase and give an example or two of each.

Geometry

altitude (p. 529)
apothem (p. 544)
area (p. 529)
base (p. 529)
center (p. 543)
complete network (p. 559)
concave (p. 514)
convex (p. 514)
degree of a node (p. 560)
edge (p. 559)
geometric probability (p. 551)
graph theory (p. 559)
height (p. 529)
incomplete network (p. 559)
n-gon (p. 515)
network (p. 559)
nodes (p. 559)
polygon (p. 514)
radius (p. 543)
regular polygon (p. 515)
regular tessellation (p. 523)
sector of a circle (p. 553)
semi-regular (p. 524)
tessellation (p. 522, 523)
traceable (p. 559)
uniform (p. 524)

Problem Solving

guess and check (p. 524)

UNDERSTANDING AND USING THE VOCABULARY

Determine whether each statement is *true* or *false*. If it is false, change the underlined word to make it true.

1. An <u>apothem</u> connects the center of a polygon with one of its vertices.
2. An <u>equilateral triangle</u> is a regular polygon.
3. Figure 3 is <u>convex</u>.
4. Figure 2 is a <u>regular polygon</u>.
5. The degree of node *A* in Figure 1 is <u>4</u>.
6. Figure 3 is a regular <u>apothem</u>.
7. Figure 1 is <u>traceable</u>.
8. Figure 2 is a <u>complete</u> network.
9. The segment drawn in Figure 3 is an <u>altitude</u>.
10. Figure 4 is a <u>uniform</u> tessellation.

Figure 1

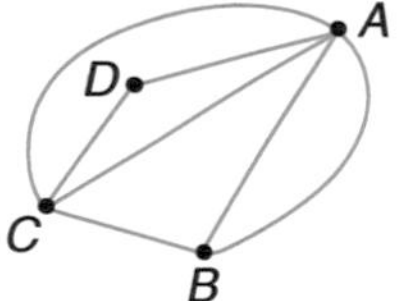

Figure 2

Figure 3

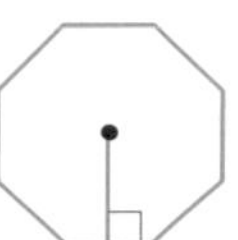

Figure 4

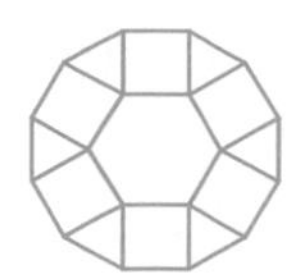

CHAPTER 10 STUDY GUIDE AND ASSESSMENT

SKILLS AND CONCEPTS

OBJECTIVES AND EXAMPLES

Upon completing this chapter, you should be able to:

- find measures of angles in polygons (Lesson 10–1)

Interior Angle Sum Theorem (Theorem 10–1)

If a convex polygon has n sides and S is the sum of the measures of its interior angles, then $S = 180(n - 2)$.

Exterior Angle Sum Theorem (Theorem 10–2)

If a polygon is convex, then the sum of the measures of the exterior angles, one at each vertex, is 360.

REVIEW EXERCISES

Use these exercises to review and prepare for the chapter test.

11. Find the measure of an interior angle of a regular decagon.

12. The sum of the measures of the interior angles of a convex polygon is 1980. How many sides does the polygon have?

13. A regular polygon has 20 sides. Find the measure of an interior angle and an exterior angle of the polygon.

14. The measure of each interior angle of a regular polygon is eleven times that of an exterior angle. How many sides are in the polygon?

- identify regular and uniform (semi-regular) tessellations (Lesson 10–2)

A *regular tessellation* uses only one type of regular polygon.

A *uniform tessellation* contains the same combination of shapes and angles at each vertex.

Uniform tessellations that contain two or more regular polygons are *semi-regular*.

Determine whether the following tessellations are *regular, uniform,* or *semi-regular.* Name all possibilites.

15.

16.

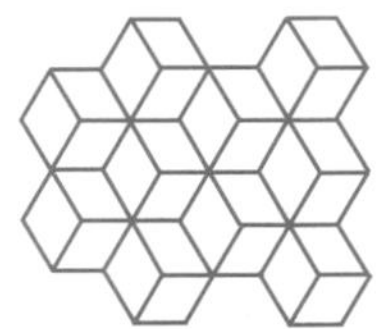

- find the areas of parallelograms (Lesson 10–3)

Find the area of the parallelogram.

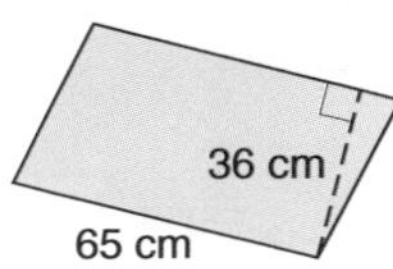

$A = bh$ *Formula for area of a parallelogram*

$A = (65)(36)$ $b = 65, h = 36$

$A = 2340$ square centimeters

Find the area of each shaded region.

17.

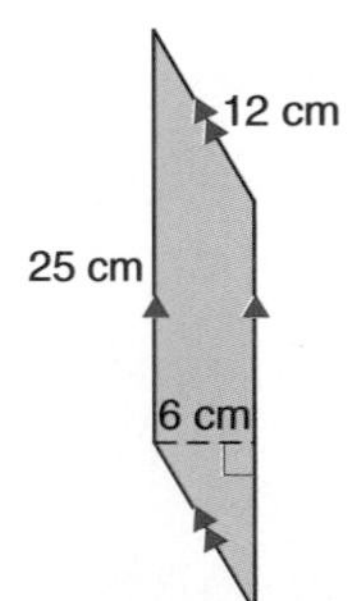

18.

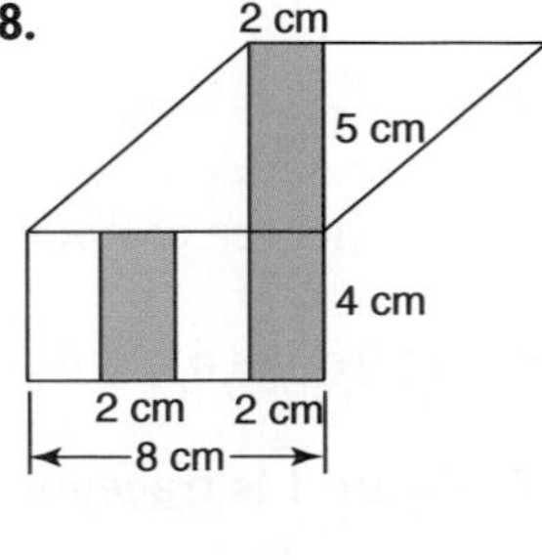

19. The area of parallelogram *ABCD* is 134.19 square inches. If the base is 18.9 inches long, find the height.

OBJECTIVES AND EXAMPLES

- find the areas of triangles, rhombi, and trapezoids (Lesson 10–4)

Area Formulas	
Triangle	$A = \frac{1}{2}bh$
Rhombus	$A = \frac{1}{2}d_1d_2$
Trapezoid	$A = \frac{1}{2}h(b_1 + b_2)$

Find the area of each figure.

20. 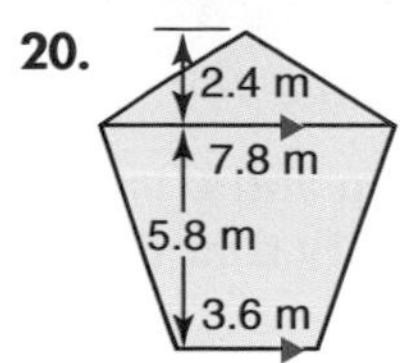

21. 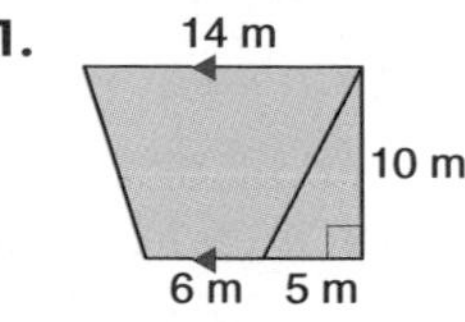

22. A rhombus has a diagonal 8.6 centimeters long and an area of 54.18 square centimeters. What is the length of each side?

23. The area of an isosceles trapezoid is 7.2 square feet. The perimeter is 6.8 feet. If a leg is 1.9 feet long, find the height.

- find the areas of regular polygons (Lesson 10–5)

Find the area of a regular hexagon with an apothem 12 centimeters long.

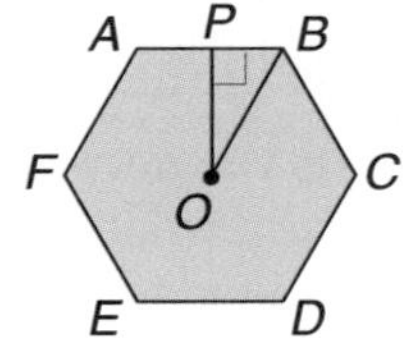

Since $\triangle POB$ is a 30°-60°-90° triangle, $PB = \frac{12}{\sqrt{3}}$ or $4\sqrt{3}$.

So, $AB = 8\sqrt{3}$ and $P = 48\sqrt{3}$.

$A = \frac{1}{2}Pa$

$A = \frac{1}{2}(48\sqrt{3})(12)$

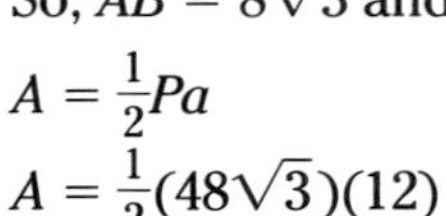

$A = 288\sqrt{3}$ or about 498.8 cm^2

Find the area of each regular polygon. Round to the nearest tenth.

24. an equilateral triangle with an apothem 8.9 inches long

25. a pentagon with an apothem 0.4 feet long

26. a hexagon with sides 64 millimeters long

27. a square with an apothem n centimeters long

28. Find the area of the regular pentagon shown at the right with a perimeter of 45 inches.

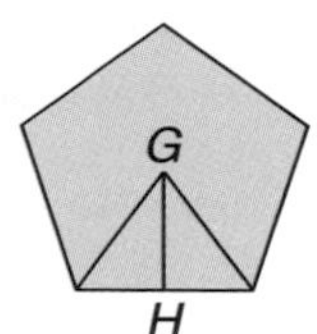

- find the area of a circle (Lesson 10–5)

If a circle has a radius of r units, then its area is πr^2.

Find the area of a circle with a diameter of 14 centimeters.

$A = \pi r^2$

$A = \pi(7)^2$ *d = 14, so r = 7.*

$A = 49\pi$

The area is 49π or about 153.9 square centimeters.

Find the circumference and area of a circle with the given radius. Round to the nearest tenth.

29. 7 mm

30. 19 in.

31. 0.9 ft

32. $3\frac{1}{3}$ cm

33. The diagonal of a square is 8 feet. Find the circumference and the area of a circle inscribed in the square.

OBJECTIVES AND EXAMPLES

- use area to solve problems involving geometric probability (Lesson 10–6)

Length Probability Postulate (*Postulate 10–3*)

If a point on $\overline{AB}$ is chosen at random and C is between A and B, then the probability that the point is on $\overline{AC}$ is $\frac{\text{length of } \overline{AC}}{\text{length of } \overline{AB}}$.

Area Probability Postulate (*Postulate 10–4*)

If a point in region A is chosen at random, then the probability that the point is in region B, which is in region A, is $\frac{\text{area of region } B}{\text{area of region } A}$.

OBJECTIVES AND EXAMPLES

34. During the morning rush hour, a bus arrives at the bus stop at Indianola and Morse Roads every seven minutes. A bus will wait for thirty seconds before departing. If you arrive at a random time, what is the probability that there will be a bus waiting?

35. Circle O has a radius of 4.5 centimeters. What is the probability that a point chosen at random will lie in the shaded region?

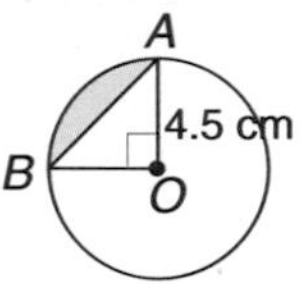

- determine if a network is traceable or complete (Lesson 10–7)

The network at the right is traceable, since exactly two of the nodes have odd degrees. However, it is not complete since there is no edge between M and A.

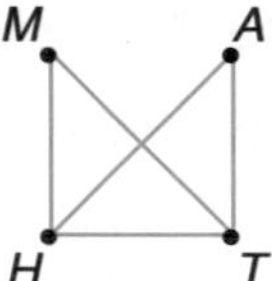

Use the network at the right to answer each question.

36. Name the nodes and edges in the network.

37. Is the network traceable?

38. Is the network complete? If not, what edges need to be added for the network to be complete?

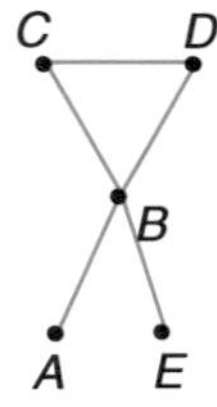

APPLICATIONS AND PROBLEM SOLVING

39. Guess and Check Using the digits 1, 2, 3, 4, 5, and 6 only once, find two whole numbers whose product is as great as possible. (Lesson 10–2)

40. Interior Design John and Cynthia are buying paint for the walls of their kitchen. Three walls are each 8 feet long, and one is 10 feet long. The walls are all 8 feet high. If a gallon of paint covers 400 square feet, how many gallons of paint will they need for two coats of paint? (Lesson 10–3)

41. Manufacturing A wooden planter has a square base and sides that are shaped like trapezoids. An edge of the base is 8 inches long, and the top edge of each side is 10 inches long. If the height of a side is 12 inches, how much wood does it take to make a planter? (Lesson 10–4)

A practice test for Chapter 10 is provided on page 802.

ALTERNATIVE ASSESSMENT

COOPERATIVE LEARNING PROJECT

Networks In this chapter, you saw that polygons can be described by networks. Networks have many applications. In this project, you will apply networks to several areas.

Follow these steps to explore applications of networks.

- Each step of a proof follows from a previous step in the proof or a proved theorem or postulate. Select a proof from any chapter in this text and draw a network for the proof.
- Your daily schedule can be represented by a network. Each activity occurs in a sequence. Draw a network for your schedule for tomorrow.
- Making a decision can be difficult because there are so many choices. Each choice can be represented as a branch on a tree. Each choice can lead to several possible outcomes. Each outcome can be represented by a branch from the choice branch. Select a decision you will be faced with in the near future and draw a network for the possible choices and their outcomes.
- As a group, discuss other applications of networks.

THINKING CRITICALLY

It is possible to find the area of a region by adding the areas of its nonoverlapping parts. How would you determine the area of a region if you know the areas of regions that do overlap?

PORTFOLIO

Make plans for a gazebo you might build for your yard or a nearby park. Its base should be a regular polygon. Its sides should consist of rectangles, and the roof should consist of triangles joined together. How might you use what you've learned in this chapter to determine how much paint would be needed for the gazebo? Include a drawing of the gazebo in your portfolio.

STANDARDIZED TEST PRACTICE

CHAPTERS 1–10

Section One: MULTIPLE CHOICE

There are eight multiple-choice questions in this section. After working each problem, write the letter of the correct answer on your paper.

1. The circle below with center P is divided into six equal sectors. If the area of the circle is 144π, find the perimeter of the shaded sector.

A. $12 + 3\pi$
B. $24 + 4\pi$
C. $24 + 24\pi$
D. $12 + 4\pi$

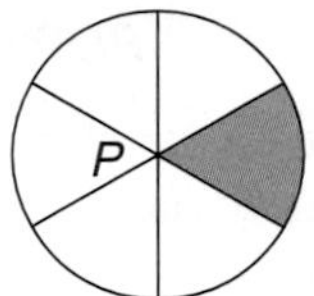

2. If $3x = \pm\sqrt{25} - \sqrt{16}$, then find the solution set for x.

A. $\left\{\frac{1}{3}, -3\right\}$ **B.** $\left\{-\frac{1}{3}, 9\right\}$
C. $\left\{\frac{1}{3}, \frac{1}{9}\right\}$ **D.** $\left\{-\frac{1}{3}, -\frac{1}{9}\right\}$

3. If $2x + 4 = 9$, then find $x - \frac{1}{2}$.

A. 1
B. 2
C. 12
D. 14

4. In the circle below, C is the center of the circle with radius r, and $ABCD$ is a rectangle. Find the length of diagonal $\overline{BD}$.

A. r
B. $r\frac{\sqrt{2}}{2}$
C. $r\sqrt{3}$
D. $r\frac{\sqrt{3}}{2}$

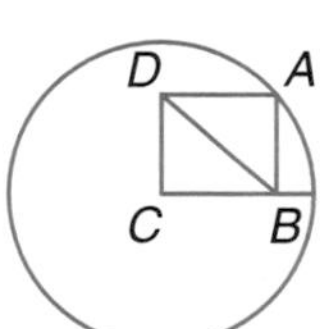

5. In the circle below, the shaded region is what fractional part of the area of the circle?

A. $\frac{1}{40}$
B. $\frac{1}{60}$
C. $\frac{1}{10}$
D. $\frac{1}{20}$

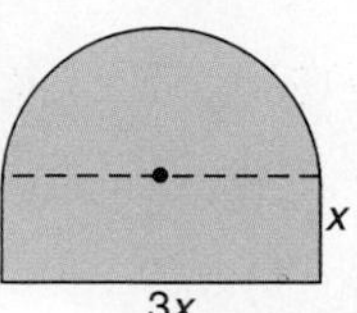

6. In the figure below, isosceles right triangle BEF overlaps square $ABCD$. If $AB = 2$ and $EB = 4$, find the area of $CDEF$.

A. 2
B. $\sqrt{2}$
C. $\frac{\sqrt{2}}{2}$
D. 1

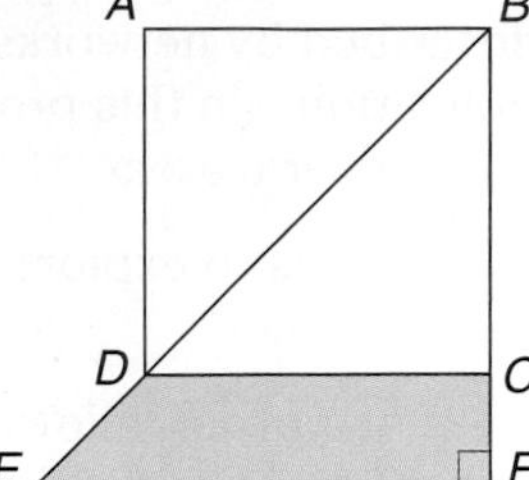

7. What are the values of x for which $\frac{x(x+1)}{(x-4)} \cdot \frac{1}{(x+3)}$ is undefined?

A. $-3, 4$
B. $3, -4$
C. $1, 3, -4$
D. $-1, -3, 4$

8. The graph of $y = 4x + 8$ is which of the following?

A. a horizontal line
B. a vertical line
C. a line that rises to the right
D. a line that falls to the right

Section Two: SHORT ANSWER

This section contains seven questions for which you will provide short answers. Write your answer on your paper.

9. If $x^2 - 11 < x^2 + 2x - 5 < x^2 + 25$, find an inequality that represents x.

10. The figure below is formed from a semicircle and a rectangle. The diameter of the semicircle is the length of the rectangle. Write a formula for the area of the figure if the length of the rectangle is $3x$ and the width is x.

Test-Taking Tip On standarized tests, *average* is the same as *mean.*

11. The average of six numbers is 10, and the average of ten other numbers is 6. Find the average of all sixteen numbers.

12. In the figure, P and Q lie on $\odot O$. $\overline{OP}$ is 8 units long, and $\angle POQ$ measures 80°. Find the length of $\widehat{PQ}$.

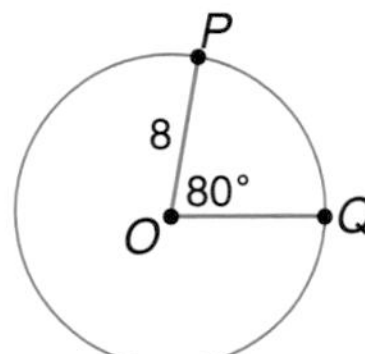

13. Simplify $\frac{8x^5y^{-2}z}{16x^{-2}yz^2}$.

14. Find the value of x in the figure at the right.

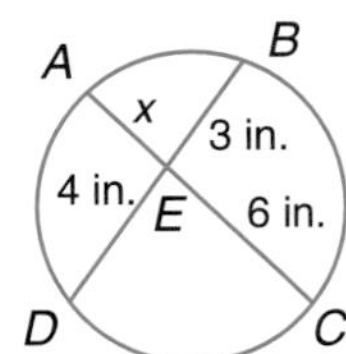

15. The figure below is composed of two squares with areas shown. What is the value of x?

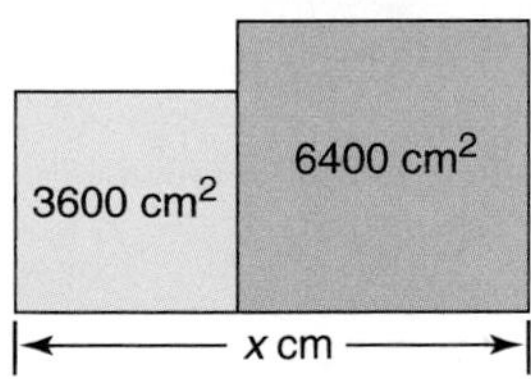

Test Practice For additional test practice questions, visit: **www.geometry.glencoe.com**

Section Three: COMPARISON

This section contains five comparison problems that involve comparing two quantities, one in column A and one in column B. In certain questions, information related to one or both quantities is centered above them. All variables used represent real numbers.

Compare quantities A and B below.

- Write A if quantity A is greater.
- Write B if quantity B is greater.
- Write C if the two quantities are equal.
- Write D if there is not enough information to determine the relationship.

Column A	Column B
Parallelogram P and rectangle R have equal bases and equal areas. P is not a rectangle.	
16. perimeter of R	perimeter of P
17. the length of the hypotenuse of right triangle MNP	the length of side $\overline{PQ}$ of equilateral triangle PQR
18. the largest prime factor of 858	the largest prime factor of 2310

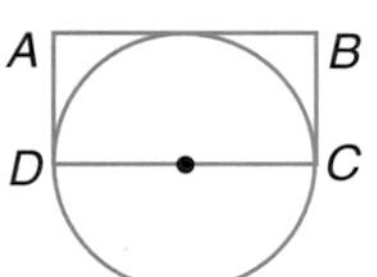

Column A	Column B
19. perimeter of $ABCD$	circumference of the circle with diameter $\overline{DC}$

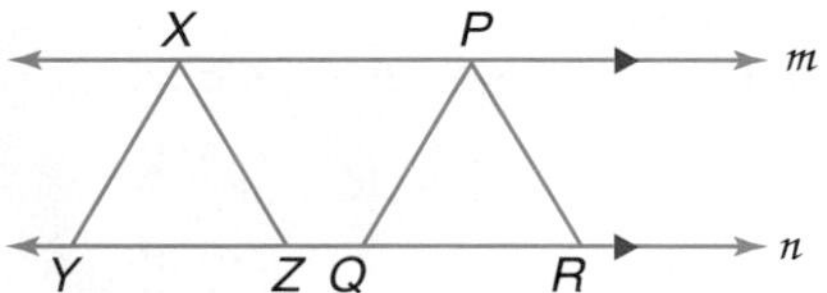

Column A	Column B
20. the length of the altitude of $\triangle XYZ$ drawn to side $\overline{YZ}$	the length of the altitude of $\triangle PQR$ drawn to side $\overline{QR}$

CHAPTER

11

Investigating Surface Area and Volume

What You'll Learn

In this chapter, you will:

- describe and draw cross sections and other slices of three-dimensional figures,
- draw three-dimensional figures,
- make two-dimensional nets for three-dimensional solids,
- find lateral areas, surface areas, and volumes of solids, and
- identify and state properties of congruent or similar solids.

Why It's Important

Real World

Parades The floats in the picture are from the annual two-hour Tournament of Roses Parade held every January 1 in Pasadena, California. This parade features floats entirely decorated with flowers, leaves, bark, mosses, seeds, or other living or dried plant parts. The people who design and decorate these floats are concerned with both the surface area and volume of floats. The surface area of the float is the surface that must be totally covered with plant parts. The volume is the amount of space that is occupied by the plant parts used on the float. In this chapter, you will learn methods for finding the surface area and volume of three-dimensional figures.

Everything's
Coming up Roses

PREREQUISITE SKILLS

To be successful in this chapter, you'll need to understand these concepts and be able to apply them. Refer to the lesson in parentheses if you need more review before beginning the chapter.

Find areas of triangles, rhombi, and trapezoids. (Lesson 10–4)

Find the area of each figure.

1. 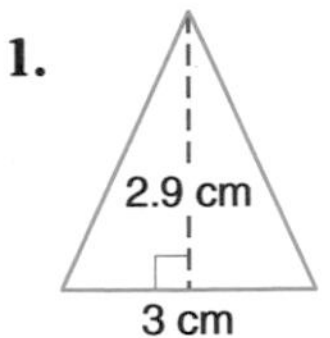

2. 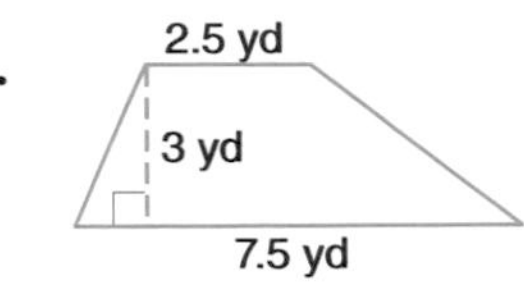

3. 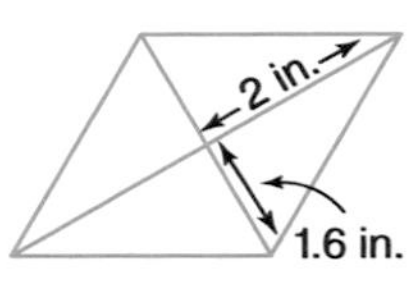

Find areas of regular polygons and circles. (Lesson 10–5)

Find the area of each regular polygon or circle described. Round to the nearest tenth.

4. hexagon with a side length of 7 cm

5. circle with a diameter of 13 ft

6. octagon with perimeter of 80 in.

7. an isosceles right triangle with a hypotenuse of $3\sqrt{2}$ m

Solve problems using trigonometric ratios. (Lesson 8–3)

Find the values of x and y. Round to the nearest tenth.

8. 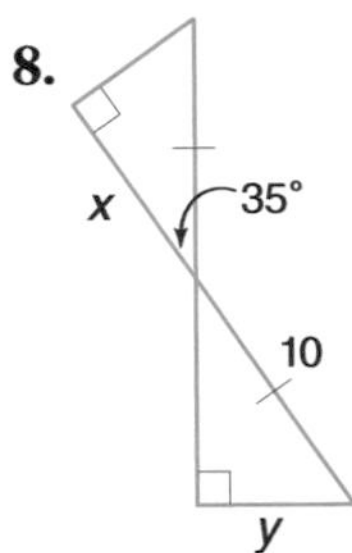

9. 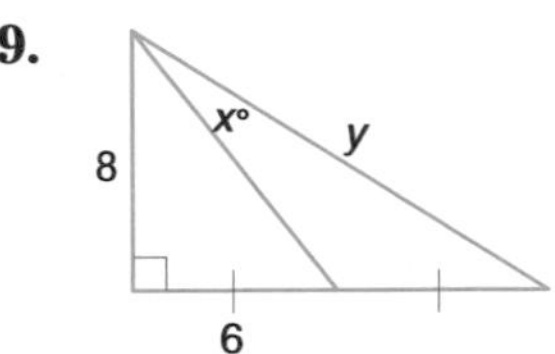

10. 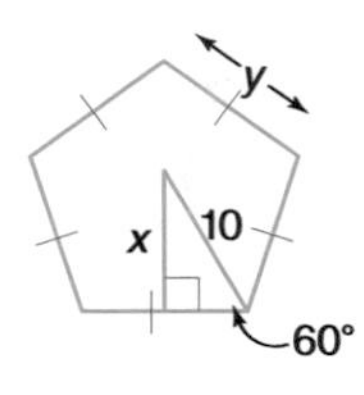

Focus On READING SKILLS

In this chapter, you'll learn about ***surface area*** and ***volume*** of polygons. It is helpful to connect figures having formulas that are related. For example, a cylinder and a cone have volume formulas that are related. If a cylinder and a cone have the same size circular base and the same height, then the volume of the cone is one-third the volume of the cylinder.

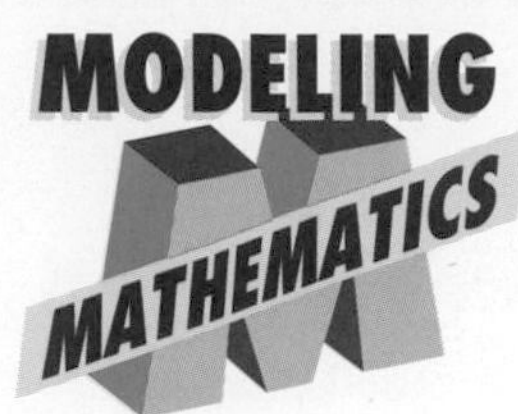

11–1A Cross Sections and Slices of Solids

Materials:  modeling clay ∿ dental floss

A Preview of Lesson 11–1

Every day you deal with three-dimensional objects, or **solids**. A **slice** of a solid is the plane figure formed by making a straight cut across the solid. We can investigate solids and their slices using modeling clay and dental floss.

Activity Make a cylinder.

Step 1 Roll a piece of modeling clay on your desk to form a thick tube. Then use dental floss to cut off the ends. Make your cuts perpendicular to the sides of the tube. The figure you have formed is a **cylinder**. The ends of your cylinder should be circles. Sketch a diagram of the cylinder.

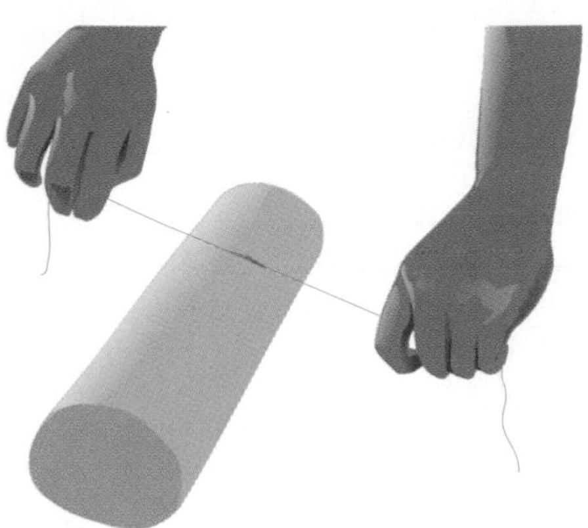

Step 2 Set your cylinder on the table on one of its flat surfaces. Make a cut parallel to the table. What shape is the new surface that is exposed by making this slice?

Step 3 Make a cut along a diameter of a base of your cylinder and perpendicular to the table. A rectangular surface will be exposed. What are the dimensions of the rectangle?

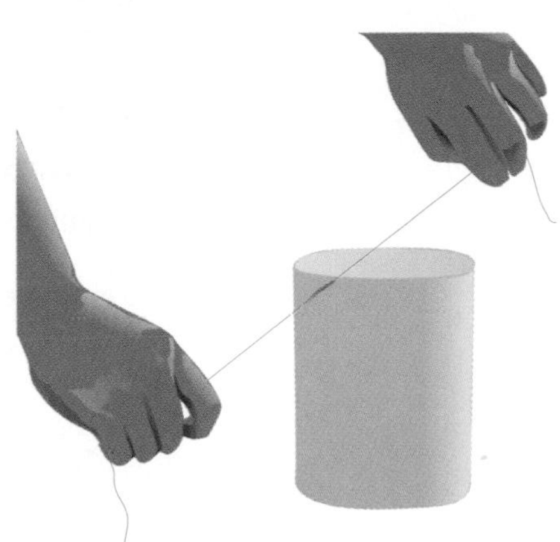

Model

1. Make a **triangular prism**. First roll a tube of modeling clay. Then flatten one side of the tube by pressing it against the table. Turn the tube and flatten another side, then flatten the rest to form a tube that is triangular instead of round. Cut off the ends with dental floss to form a solid like the one shown at the right.

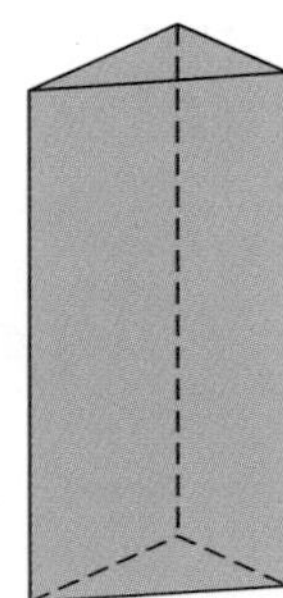

Write

2. A slice made parallel to the two parallel bases of a prism is called a **cross section**. What shape is a cross section of a triangular prism? Use dental floss to make a cut to test your conjecture.
3. What shape would be revealed if you made a cut along one of the medians of the triangular end of the prism? Use dental floss to make a cut to test your conjecture.

11-1

Exploring Three-Dimensional Figures

What YOU'LL LEARN

- To use top, front, side, and corner views of three-dimensional solids to make models, and
- to describe and draw cross sections and other slices of three-dimensional figures.

Why IT'S IMPORTANT

Different perspectives of three-dimensional figures are used in art and architecture.

History

The ancient Egyptians left many remnants of their civilization for us to appreciate. One of these is Cleopatra's Needles, two giant granite obelisks, or pillars, that once stood at the entrance to the Temple of the Sun just north of Cairo. One of Cleopatra's Needles was given to the United Kingdom by the Turkish people in gratitude for their help in resisting Napoleon. The other obelisk is displayed in Central Park in New York City.

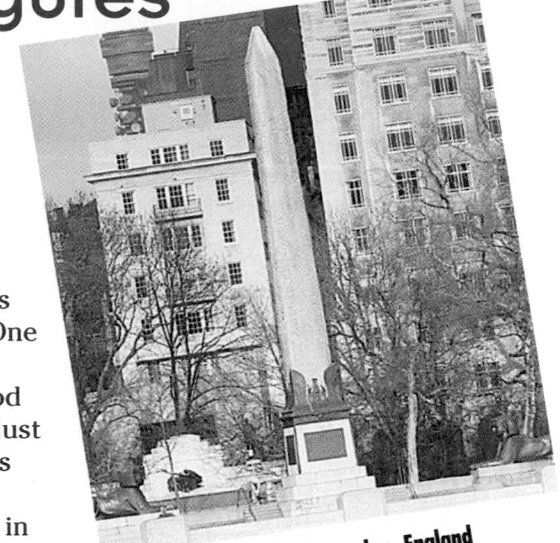

Cleopatra's Needle, London, England

Looking at a photograph of just one view of Cleopatra's Needle does not allow you to determine its three-dimensional shape. Front, top, and side views give you more information.

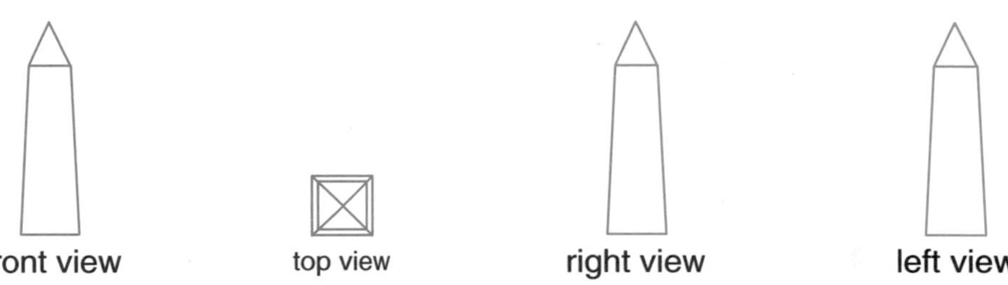

If you are given top, front, and side views of a three-dimensional object, or **solid,** you can **make a model** of the object. Making a model is often helpful in solving problems.

Example 1

PROBLEM SOLVING

Make a Model

Various views of a solid figure are shown below. The edge of one block represents one unit of length. A dark segment indicates a break in the surface.

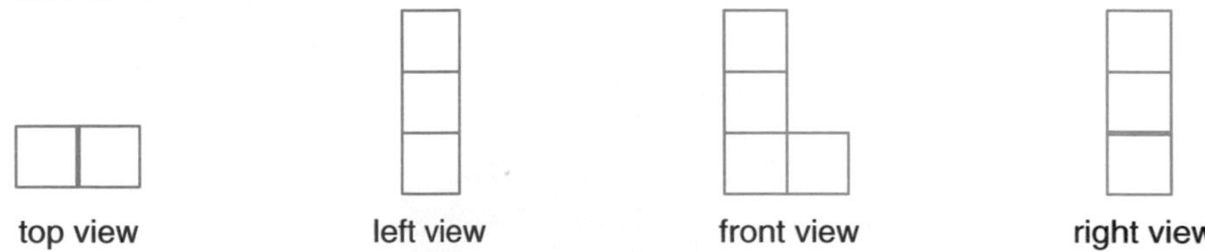

a. Make a model of the figure.

b. Draw the back view of the figure.

a. Use each view to determine the number of blocks needed.

- The top view tells us that there are two columns of blocks. The heavy segment tells us that one column is taller than the other. Start with two blocks as the base of the figure.
- The left view tells us that the figure is 3 blocks high. Since there are no heavy segments in this view, you know that the left column is 3 blocks high. Add blocks to match this view. *(continued on the next page)*

CAREER CHOICES

Graphic artists use various perspectives to portray ideas using a variety of media. They may use hand drawn artwork, computer generated art, or film to create designs. Art classes in drawing and painting in high school will help prepare a student for a BA degree in fine arts, visual communications, or graphic design.

For more information, contact:

The American Institute of Graphic Arts
164 Fifth Ave.
New York, NY 10010

- The front view shows the figure has 3 blocks in the left column and 1 in the right. Our figure meets this requirement.
- The right view says that the figure is 3 blocks high, but there is a heavy segment at the 1-block level. This means that the right column is exactly 1 block high. Our model still works!
- After completing your model, check to see that all views correspond to the model.

b. Now turn the figure to the back view and draw what you see. No heavy segments are needed since all of the surfaces are flush at the back.

back view

A **corner view** or **perspective view** is the view from a corner of the figure. You can use isometric dot paper to draw a corner view. The figure at the right shows a corner view of a cube.

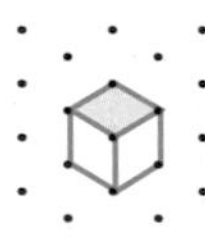

The portion of a drawing that represents the top surface of a three-dimensional figure is often shaded to give a three-dimensional look.

Example  **From the views given below, make a model and then draw a corner view.**

top view

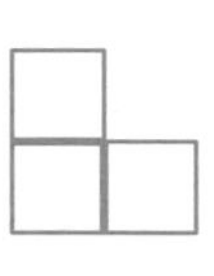

left view

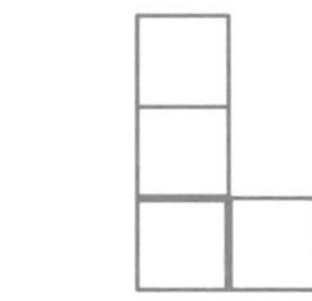

front view

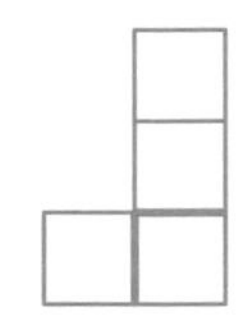

right view

Use cubes to make a model that satisfies each view.

- The top view tells us that the base has 3 blocks and that the figure has columns of different heights.
- The left view shows blocks flush with the surface. Add blocks to the left back column so that the figure is 3 blocks tall.
- The front view and right view show us that the other columns are 1 block tall. Your model should look like the one shown at the left.

To make a corner drawing, turn your model so that you are looking at the edge of the highest column and the other columns as well.

Connect dots on the isometric dot paper to represent the edges you see. Shade the tops of each column.

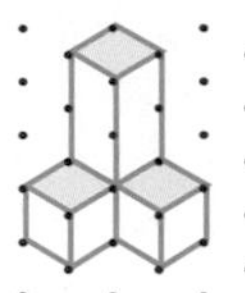

All of the surfaces of the solids in Examples 1 and 2 are flat surfaces or **faces**. Solids with all flat surfaces that enclose a single region of space are called **polyhedrons** or **polyhedra**. All of the faces are polygons, and the line segments where the faces intersect are called **edges**.

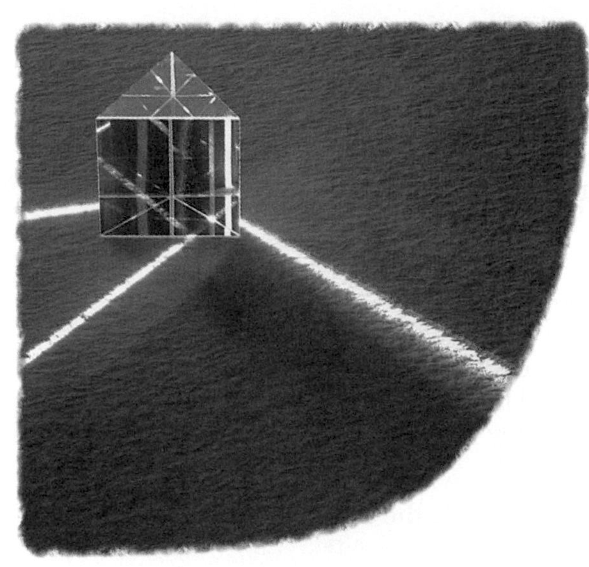

A **prism** is a polyhedron with two congruent faces that are polygons contained in parallel planes. These two faces are called the **bases** of the prism. The other faces, called **lateral faces**, are shaped like parallelograms. In the prism at the right, pentagons *ABCDE* and *FGHJK* are the bases, and ▱*AEKF* and ▱*DJHC* are two of the lateral faces.

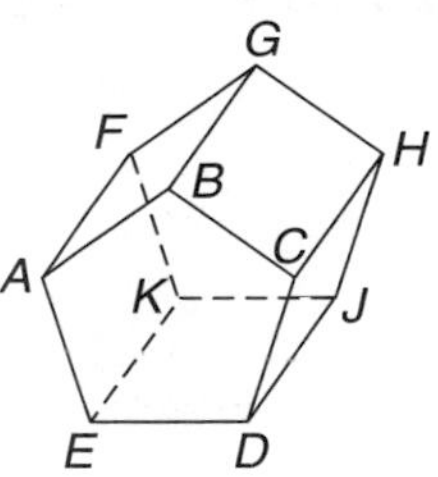

regular prism

Prisms are named by the shape of their bases. A **regular prism** is a prism whose bases are regular polygons. A **cube** is a prism in which all the faces are squares.

A polyhedron that has all faces except one intersecting at one point is a **pyramid**. A **cylinder** is a solid with congruent bases in a pair of parallel planes. However, since its bases are circular regions, it is not a polyhedron. A **cone** has a circular base and a vertex. A **sphere** is a set of points in space that are a given distance from a given point.

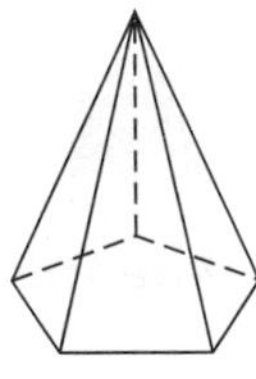
pyramid

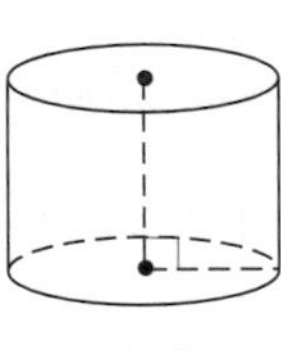
cylinder

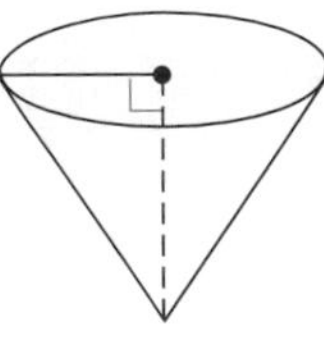
cone

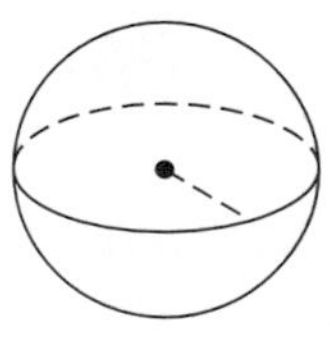
sphere

A polyhedron is **regular** if all of its faces are shaped like congruent regular polygons. Since all of the faces of a regular polyhedron are regular and congruent, all of the edges of a regular polyhedron are congruent. There are exactly five types of regular polyhedra. These are called the **Platonic solids**, because Plato described them so fully in his writings.

Plato with his students

4 faces
tetrahedron

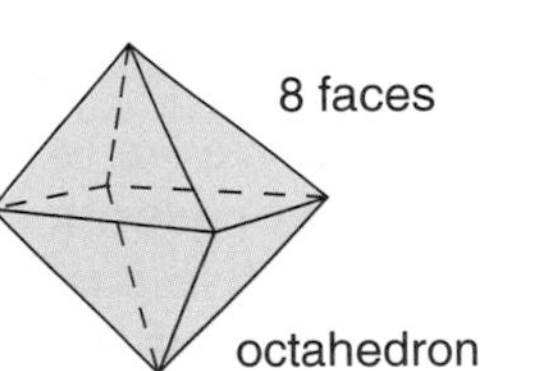
8 faces
octahedron

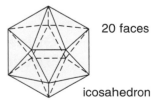

20 faces
icosahedron

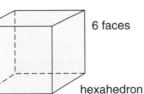

6 faces
hexahedron

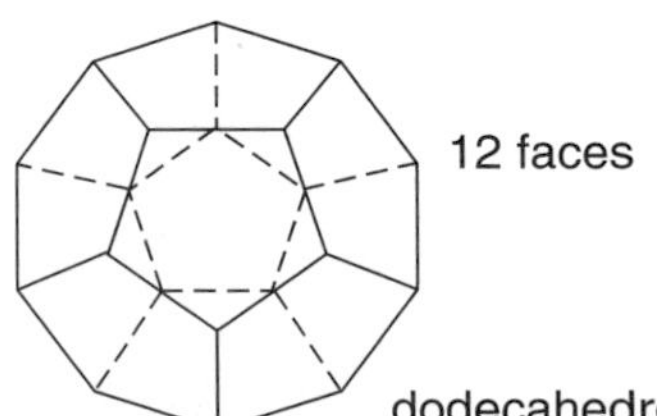
12 faces
dodecahedron

Interesting shapes occur when a plane intersects, or **slices**, a solid figure. If the plane is parallel to the base or bases of the solid, then the intersection of the plane and the solid is called a **cross section** of the solid.

Example 3 At Mike's Delicatessen, a slicer is used to slice whole pieces of meat and cheese for sandwiches. Suppose Dan had a cylindrical piece of cheddar cheese.

Foods

a. What type of slices would you get if the cheese is placed in the slicer so that the long edge of the cheese is perpendicular to the blade?

b. Suppose Dan placed the cheese with the long edge parallel to the blade. Describe the first ten slices of cheese he would cut.

c. Which slice is a cross section of the cylinder?

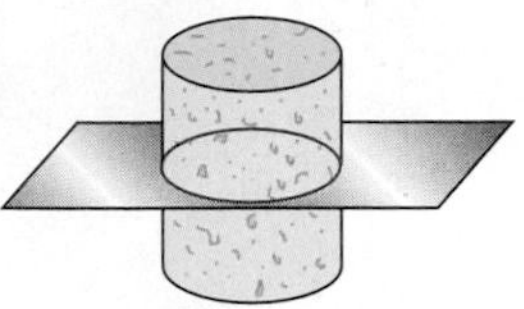

a. If the cylinder is placed so that the long edge is perpendicular to the blade, then the slices would be circles congruent to the bases of the cylinder.

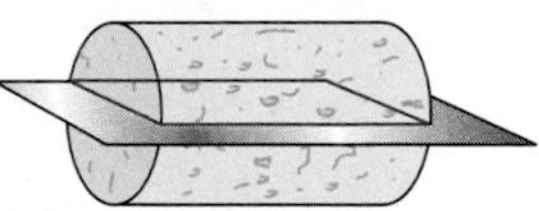

b. Placing the long edge of the cylinder parallel to the blade would produce differently shaped slices of cheese. Each slice would be rectangular and as long as the cylinder. Each successive slice would be wider until you reached the middle of the original cylinder. Then they would be less wide with each slice. *Why?*

c. The circular slice is the cross section because the slice was made parallel to the bases.

CHECK FOR UNDERSTANDING

Communicating Mathematics

Study the lesson. Then complete the following.

1. Refer to the application at the beginning of the lesson. Is an obelisk a polyhedron? Explain.
2. **Compare and contrast** a cylinder with a prism and a pyramid with a cone. How are they alike and how are they different?
3. **You Decide** Evita says that a sphere is a polyhedron. Rachel disagrees. Who is correct and why?

MATH JOURNAL

4. Study the types of solids shown on page 7. Identify a real-life example of an object of each shape.

Guided Practice

Various views of a solid figure are shown below. The edge of one block represents one unit of length. A dark segment indicates a break in the surface. Make a model of each figure. Then draw the back view of the figure.

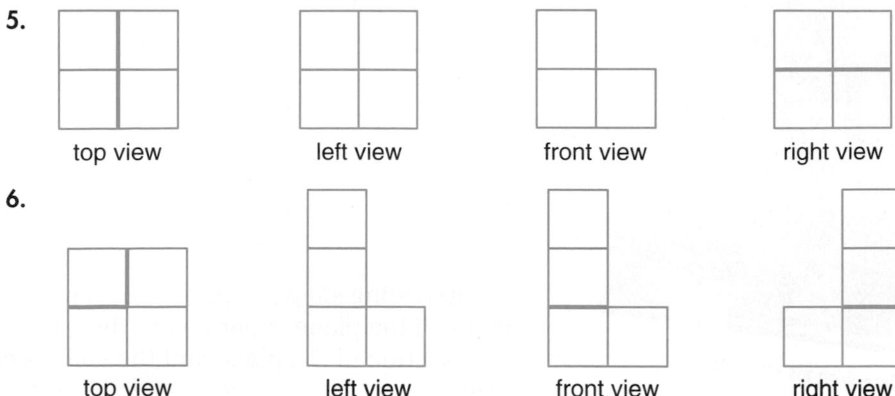

From the views of a solid figure given below, draw a corner view.

7.  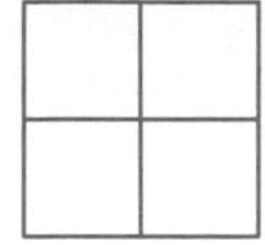 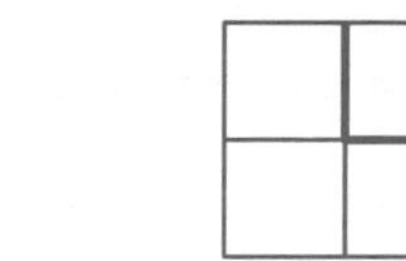

8. 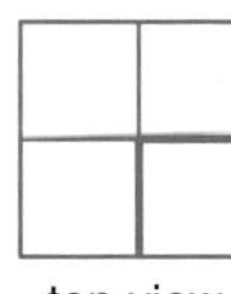

9. The corner view of a figure is given at the right. Draw the top, left, front, right, and back views of the figure.

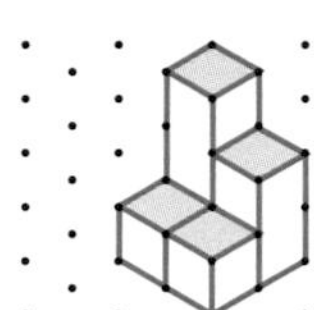

Determine if each figure is a polyhedron. Name the edges, faces, and vertices of each polyhedron.

10. 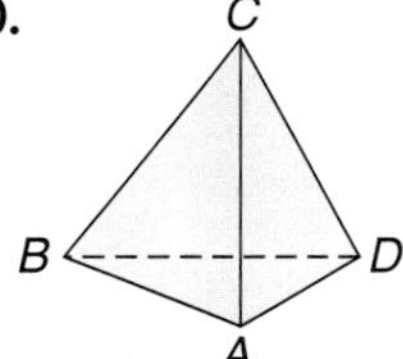

11. 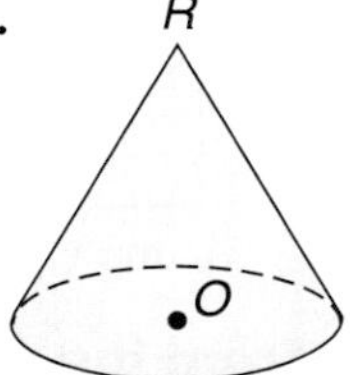

12. Suppose pyramid *ABCDE* is sliced by a plane. What is the shape of each slice?
    a. The plane is parallel to the base and does not contain *E*. 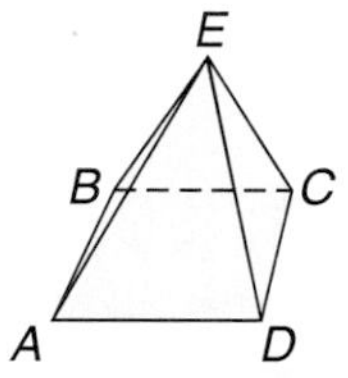
    b. The plane is perpendicular to the base and parallel to $\overline{AB}$ and $\overline{CD}$ and does not contain *E*.
    c. The plane is perpendicular to the base and passes through *E*.

13. **Art** The photograph at the left shows a front view of *Cubi XIX* by David Smith. Choose one element of the sculpture and draw a left, right, and back view.

EXERCISES

Practice **Various views of a solid figure are shown below. The edge of one block represents one unit of length. A dark segment indicates a break in the surface. Make a model of each figure. Then draw the back view of the figure.**

14. 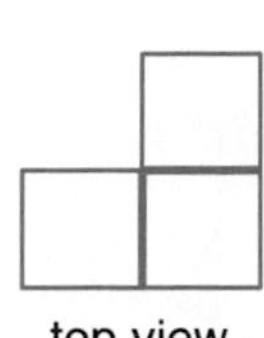 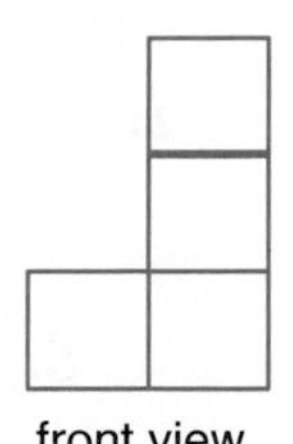 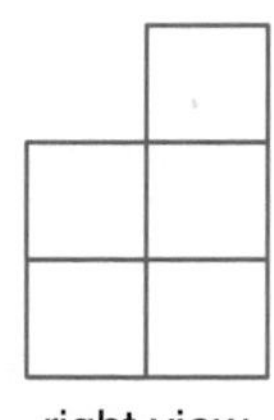

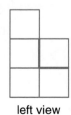

Various views of a solid figure are shown below. The edge of one block represents one unit of length. A dark segment indicates a break in the surface. Make a model of each figure. Then draw the back view of the figure.

15.

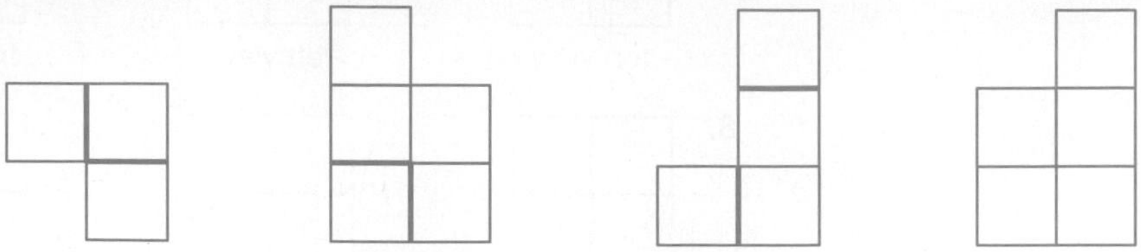

top view left view front view right view

16.

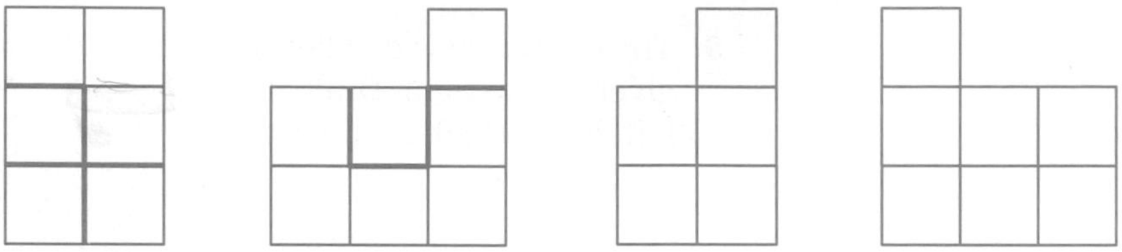

top view left view front view right view

17.

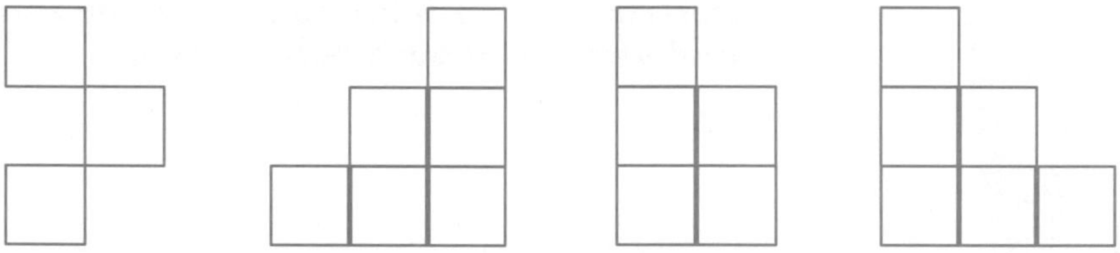

top view left view front view right view

From the views of a solid figure given below, draw a corner view.

18.

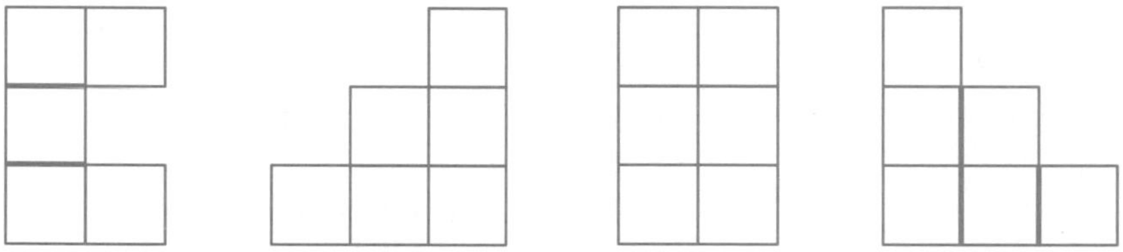

top view left view front view right view

19.

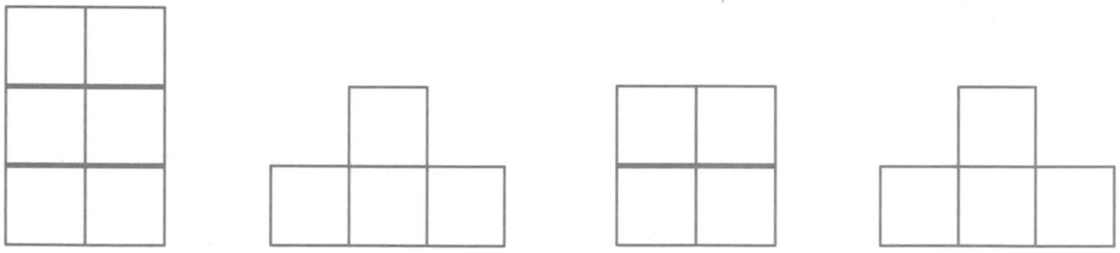

top view left view front view right view

20.

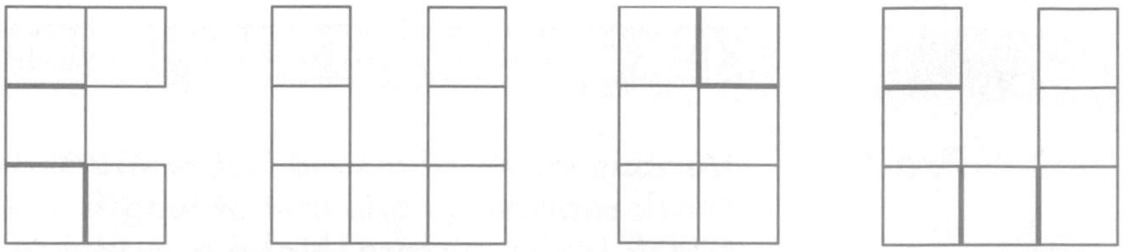

top view left view front view right view

Describe the slice resulting from each cut of the cone.

21.

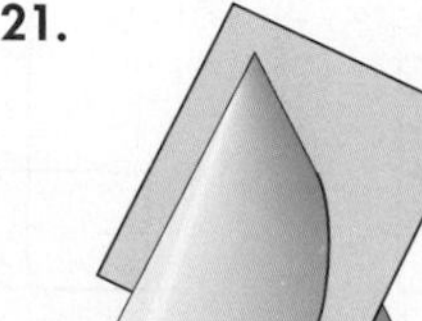

22.

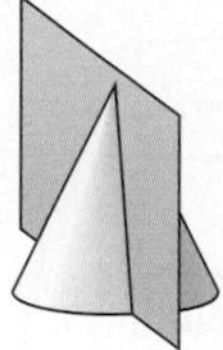

23.

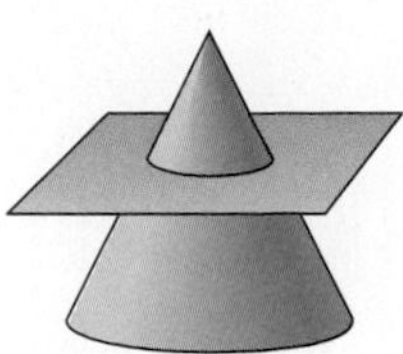

The corner view of a figure is given. Draw the top, left, front, right, and back views of the figure.

24. 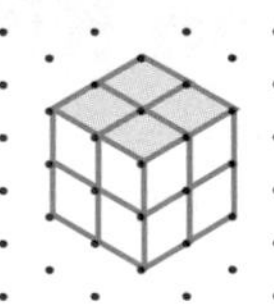

25. 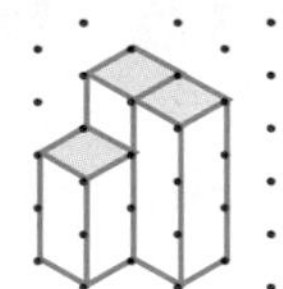

26. 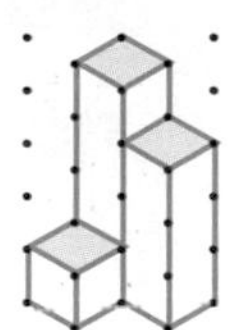

Determine if each figure is a polyhedron. Name the edges, faces, and vertices of each polyhedron.

27. 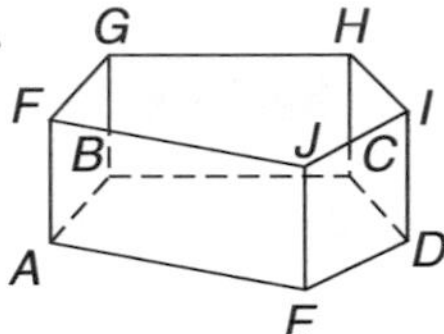

28. 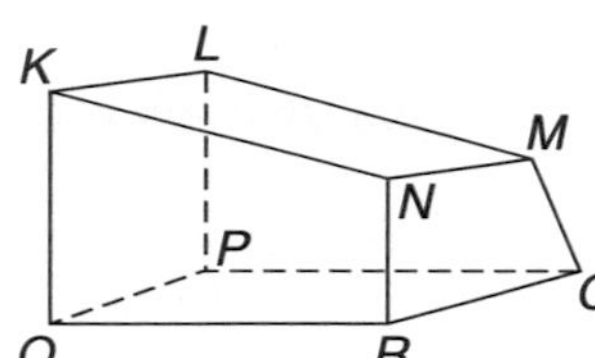

Describe the shape of the intersection in each situation.

29. A cube is intersected by a plane that is parallel to the base.
30. A sphere is intersected by a plane that contains the center of the sphere.
31. The plane perpendicular to the base of a cone and containing the vertex intersects the cone.

Critical Thinking

Determine whether each statement is *true* or *false*. Explain.

32. A plane can intersect a cube in a trapezoid.
33. The intersection of a plane and a cylinder can be a triangle.
34. The cross section of a prism is always congruent to the bases.

Applications and Problem Solving

35. **Architecture** Architects make elevation drawings to show how buildings appear from each side. Draw a set of elevation drawings for your home or school.
36. **Medicine** Physicians use magnetic resonance imaging, called MRI scans, to examine their patients for injuries or disease. Research MRI scans in the library or on the Internet. How does an MRI scanner use the concepts discussed in this lesson?

Mixed Review

37. Find the degree of each node in the network at the right. Is the network traceable? (Lesson 10–7)
38. Find the sum of the measures of the interior angles of a 22-gon. (Lesson 10–1)

39. **ACT Practice** Point $B(3, 4)$ is the midpoint of line segment AC. If point A has coordinates $(1, 7)$, what are the coordinates of point C?

    A $(-1, 10)$ **B** $\left(2, \frac{11}{2}\right)$ **C** $(4, 11)$ **D** $(5, 1)$ **E** $(7, 1)$

40. **Engineering** A diagram of a right triangular brace has side measures of 2.7, 3.0, and 5.3. Is the diagram correct? Explain. (Lesson 8–1)
41. Determine whether $ABCD$ is a parallelogram for $A(-3, 3)$, $B(3, 4)$, $C(1, -1)$, and $D(-4, -2)$. Explain your answer. (Lesson 6–2)
42. Graph the line that passes through the point at $(2, -3)$ and is parallel to the line through points at $(-2, 6)$ and $(4, -6)$. (Lesson 3–3)

Algebra

For **Extra Practice**, see page 785.

43. Evaluate $\sqrt{t}$ if $t = 324$.
44. State whether $\{(7, 7), (8, 12), (12, 1), (0, 7)\}$ is a function.

An Extension of Lesson 11–1

11–1B Tetrahedron Kites

Materials: straws string glue colored tissue paper

You can use straws and string to make a tetrahedron. Then you can combine several tetrahedra to form a type of box kite that really flies.

Activity 1 Use strings and straws to make a tetrahedron.

Step 1 For each tetrahedron, you will need 6 straws, a 40-inch piece of string, and a 20-inch piece of string.

Step 2 Thread the 40-inch string through three of the straws. Tie the string so the straws form a triangle with one end of the string very short and the other end very long.

Step 3 Place two straws on the 20-inch string and tie them to the vertices of the triangle that do not contain the original knot.

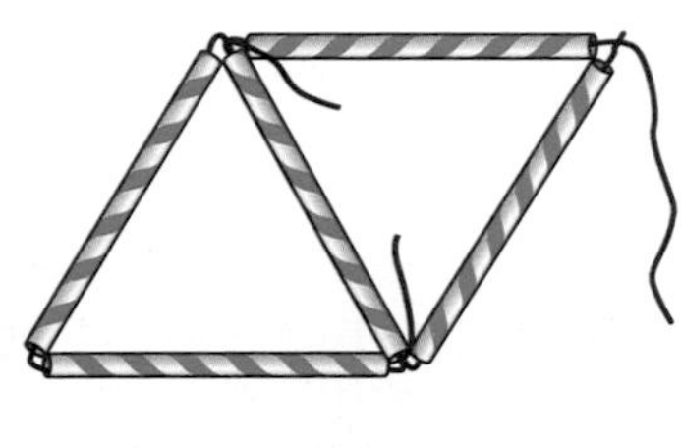

Step 4 Place one straw on the long end of the original 40-inch string and tie it (*A*) to the free vertex (*B*) of the equilateral triangle to form the tetrahedron.

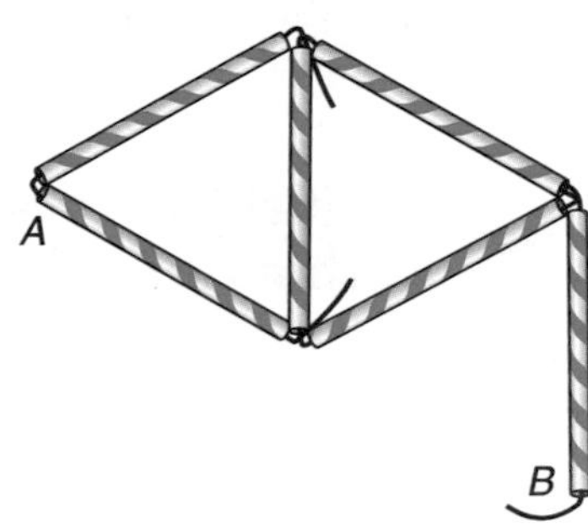

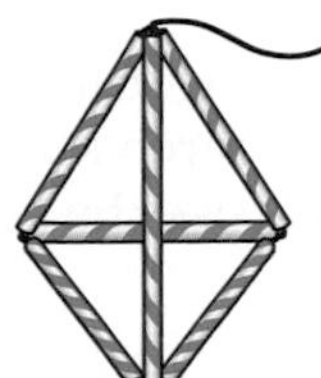

To make a tetrahedron kite, you will need at least four tetrahedra. In order to make the kite fly, you will have to add material to each tetrahedron to catch the wind. The following activity tells how to make the pattern for each tetrahedron's surface and how to assemble the tetrahedra to form a kite.

Activity 2 Make a tetrahedron kite.

Step 1 Draw a template similar to the one below as a pattern for cutting tissue paper.

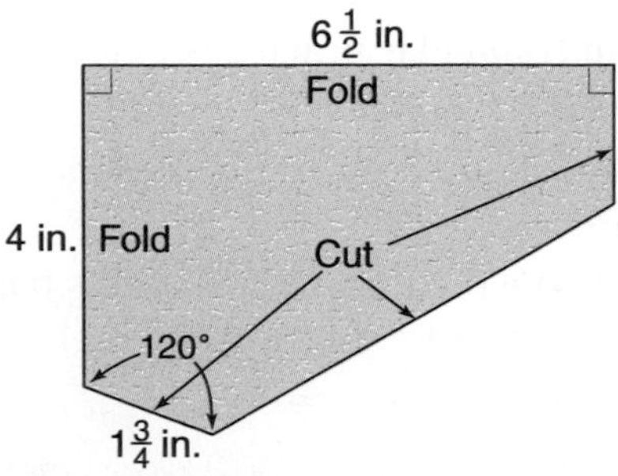

Note which sides say "Fold" and which ones say "Cut."

Step 2 Fold a 20-inch by 26-inch piece of tissue paper into fourths. Then fold the resulting 10-inch by 13-inch paper into fourths. The end result is a rectangle that is 5-inches by 6.5-inches.

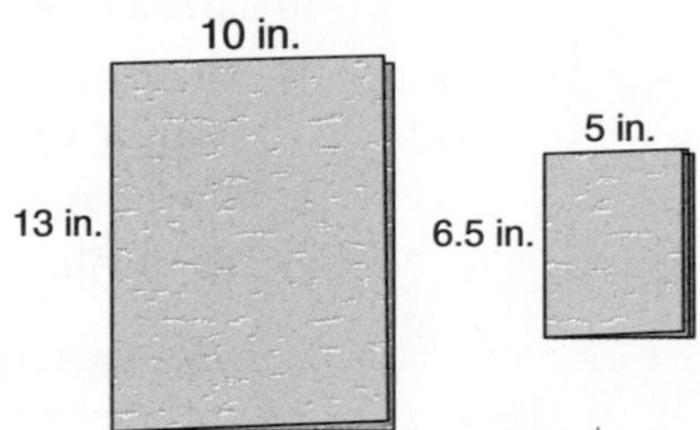

Step 3 Place the pattern template on the paper matching the pattern edges with the folds of the paper. Cut along the three lines. Separate the pieces of paper. You should have four shapes like the one shown below.

Step 4 Lay one edge of the tetrahedron on the horizontal crease of the tissue paper. Glue the other edges around the four straws that attach to that edge.

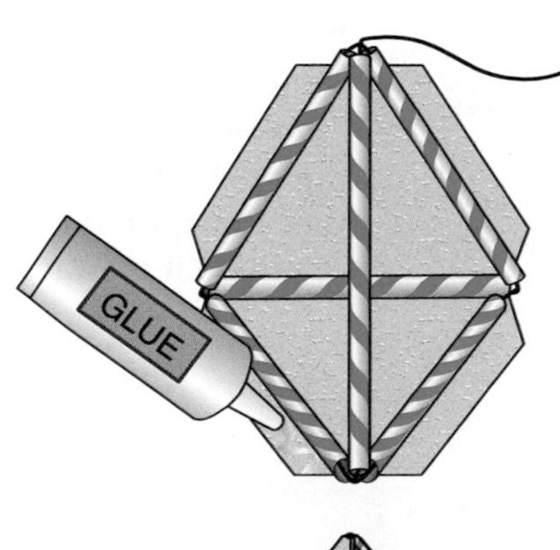

Step 5 Use the long ends of the strings to connect four tetrahedra to form a larger tetrahedron. This is your box kite.

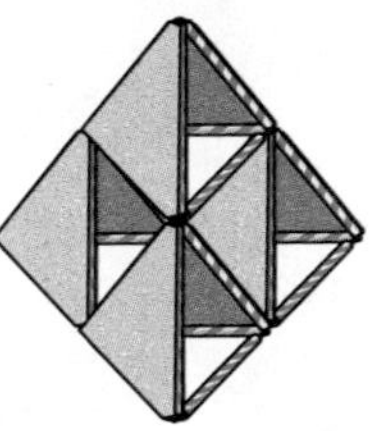

Write

1. Describe the types of triangles used to form the tetrahedron.
2. Explain how you determined how each tetrahedron should be connected to the others in order to form a tetrahedron that would catch the wind.
3. Estimate what percent of the outer surface of the tetrahedron is covered with tissue paper. Explain how you arrived at your estimate.
4. What is the ratio of the size of the smaller tetrahedron to the larger one?

Model

5. a. Attach a 16-inch string to the top (point *A*) and bottom (point *B*) of the front edge of the top tetrahedron. Put a loop in the middle of the string.

   b. Attach a 24-inch string to the back bottom vertex (point *C*) of the tetrahedron, then thread it through the loop, and tie it to the other back bottom vertex (point *D*) of the tetrahedron.

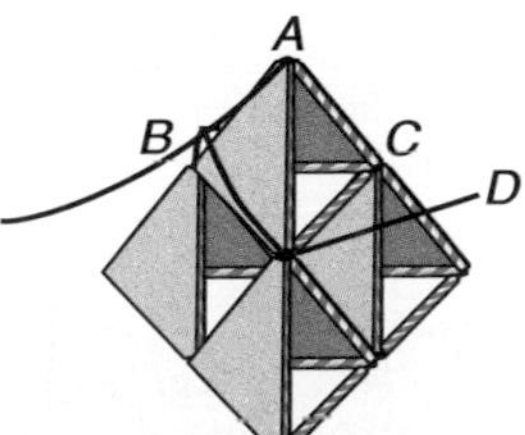

   c. Attach the flying string to the loop and find a gentle breeze to test your kite.

6. Suppose you wanted to build a larger tetrahedron kite. How many small tetrahedra would you need for a kite that is three layers high? four layers high?

11-2 Nets and Surface Area

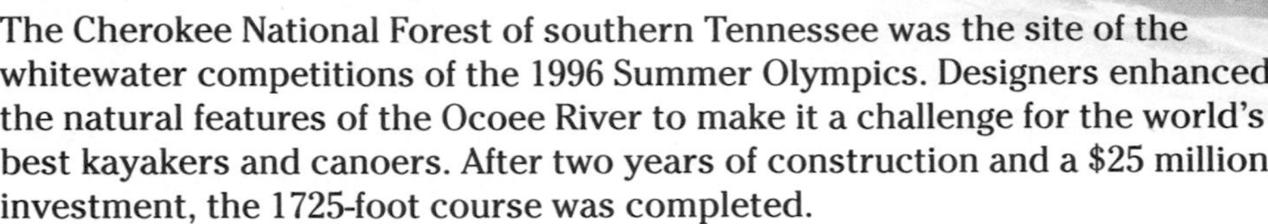

What YOU'LL LEARN

- To draw three-dimensional figures on isometric dot paper,
- to make two-dimensional nets for three-dimensional solids, and
- to find surface areas.

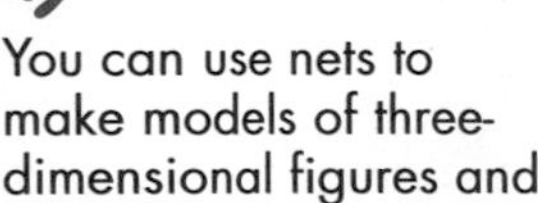

Why IT'S IMPORTANT

You can use nets to make models of three-dimensional figures and find their surface area.

Real World APPLICATION

Sports

The Cherokee National Forest of southern Tennessee was the site of the whitewater competitions of the 1996 Summer Olympics. Designers enhanced the natural features of the Ocoee River to make it a challenge for the world's best kayakers and canoers. After two years of construction and a $25 million investment, the 1725-foot course was completed.

In sports such as kayaking where speed is important, designers are conscious of the **surface area** of the equipment. The surface area of a three-dimensional object or solid is the sum of the areas of its outer surfaces. If a piece of sporting equipment has less surface area, it will have less friction with anything it contacts. The reduced friction allows it to travel faster.

In this lesson, we will consider the surfaces of polyhedrons. Recall that the surfaces that make up a polyhedron are polygons. Some common polyhedra are shown below.

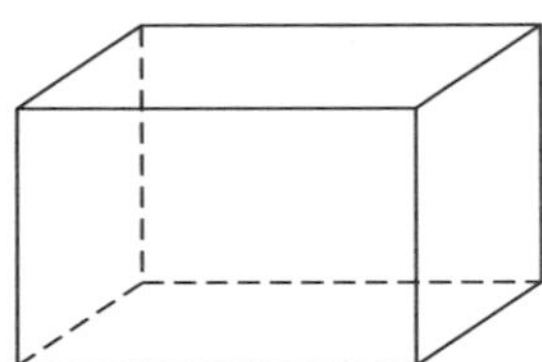

rectangular prism

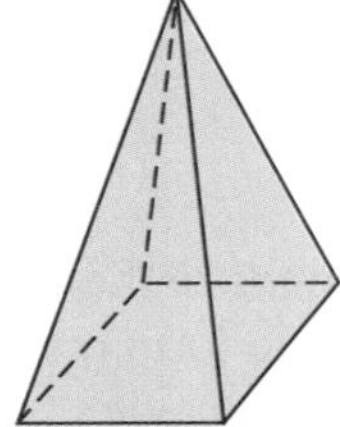

square pyramid

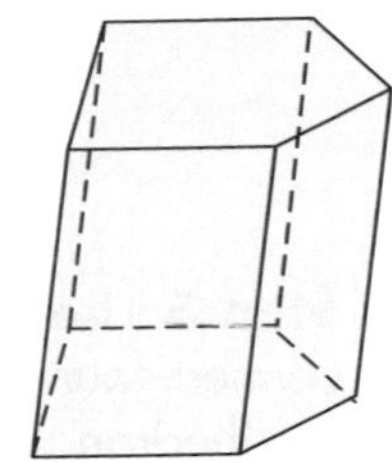

pentagonal prism

In Lesson 11–1, you drew corner views of three-dimensional figures using isometric dot paper. You can also use isometric dot paper to draw geometric solids.

Example  **Sketch a rectangular solid 3 units high, 4 units long, and 5 units wide using isometric dot paper.**

- Draw the top of the solid 4 by 5 units.
- Draw a segment 3 units down from each vertex. Hidden edges are shown by dashed lines.
- Connect the corresponding vertices.

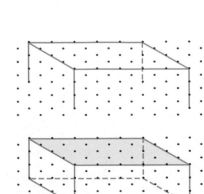

Imagine that you cut a cardboard box along its edges and laid it out flat. The result is a two-dimensional figure known as a **net**. Nets can be made for any solid figure. A net for triangular pyramid $ABCD$ is shown at the right. Nets are very useful in seeing the polygons whose areas must be computed to find the surface area of a polyhedron. For the pyramid, the surface area is the sum of the areas of $\triangle ABC$, $\triangle ACD$, $\triangle ABD$, and $\triangle BCD$.

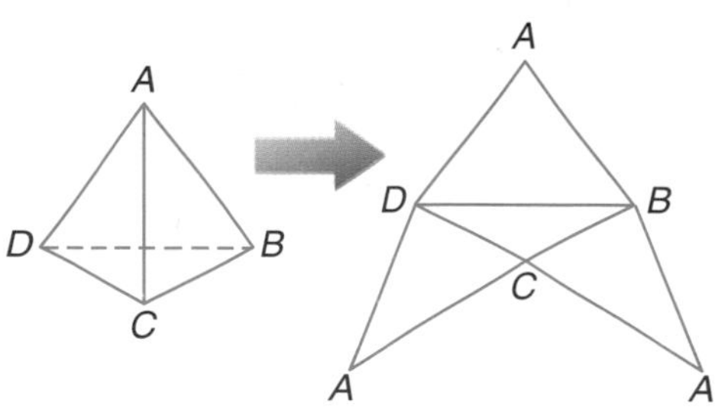

Example 2 **Refer to the rectangular prism in Example 1.**

a. Use rectangular dot paper to draw a net for the prism.

b. Find the surface area of the prism.

a. Choose one surface to begin your drawing. Then visualize unfolding the solid along the edges so that it is a two-dimensional figure.

This is only one of several possible nets that you could draw for the prism. However, the surface area should be the same regardless of which net is used.

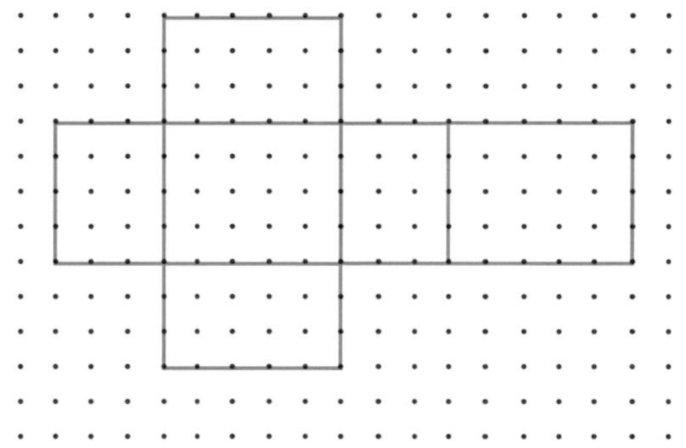

b. Find the area of each rectangle in the net. The surface area is the sum of these areas.

$(3 \cdot 5) + (5 \cdot 4) + (3 \cdot 4) + (5 \cdot 4) + (3 \cdot 4) + (3 \cdot 5)$

$= 15 + 20 + 12 + 20 + 12 + 15$

$= 94$ square units

Often times there are several different nets that could be used to produce the same solid. Could you design a different one for the rectangular prism drawn in Example 1?

Example 3 **Real World APPLICATION** **Set Design**

Alegria is covering a box 40 centimeters high, 30 centimeters long, and 20 centimeters wide with expensive fabric to use as a prop in a play. She wants to cover the box using the smallest amount of fabric possible. What size of fabric is needed for a net to cover the box?

Explore We know the dimensions of the box and need to find the size of the smallest piece of fabric that could be used to cover the box.

Plan First, we can sketch the box using the given dimensions. Next, we can find the dimensions of each face of the box. Then we can draw nets for the box to find the smallest piece of fabric that can be used.

Solve Draw a corner view of the box using isometric dot paper. Let each dot represent 10 centimeters.

(continued on the next page)

Now, determine the dimensions of each face of the box.

top and bottom	3×2
small ends	4×2
sides	3×4

Use grid paper and these dimensions to draw each face of the box. Then arrange the rectangles to form a net so that the total width and the total length of the net uses the minimum amount of fabric. Two of the nets are shown below.

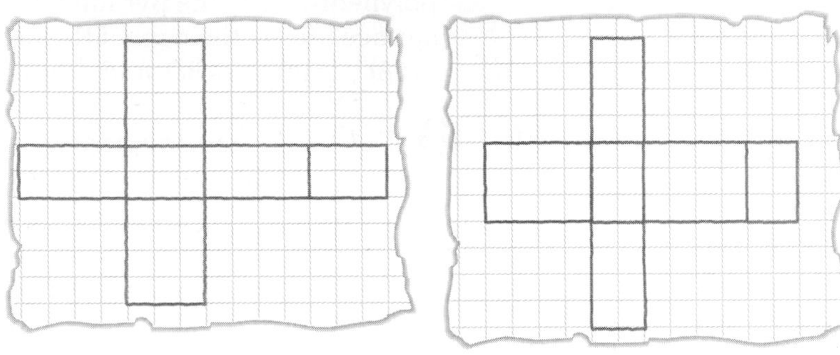

total length = 140 cm
total width = 100 cm
area of fabric = 14,000 cm^2

total length = 110 cm
total width = 120 cm
area of fabric = 13,200 cm^2

The net on the right uses less fabric. Alegria needs an 110-cm by 120-cm piece of fabric.

Examine Copy one of the nets on rectangular grid paper and fold it into a box. Try covering the box in different-sized pieces of paper to verify the solution.

CHECK FOR UNDERSTANDING

Communicating Mathematics

Study the lesson. Then complete the following.

1. **Compare and contrast** isometric dot paper and rectangular dot paper. When is each type of paper useful?
2. What is the relationship between a net and surface area?
3. Refer to the application at the beginning of the lesson. Explain how kayak design is related to the geometric concept of surface area.
4. **Determine** whether the figure at the right produces a cube when folded. Why or why not? Copy the figure on rectangular dot paper, cut it out, and fold to check your answer.

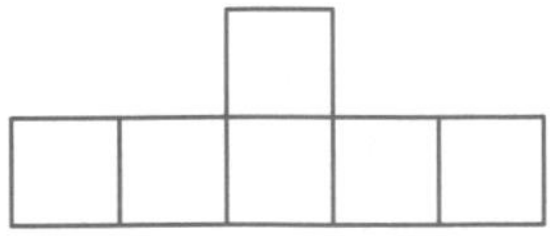

Guided Practice

Use isometric dot paper to draw each polyhedron.

5. a rectangular solid 1 unit high, 5 units long, and 3 units wide
6. a cube 4 units on each edge
7. Given the polyhedron at the right, copy its net and label the remaining vertices.

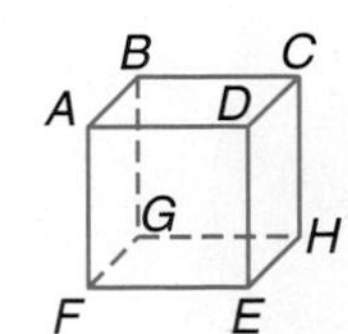

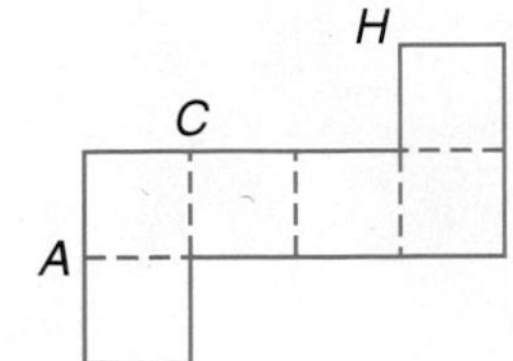

Use rectangular dot paper to draw a net for each solid. Then find the surface area of the solid.

8. 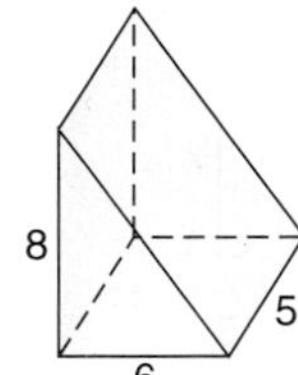

9. 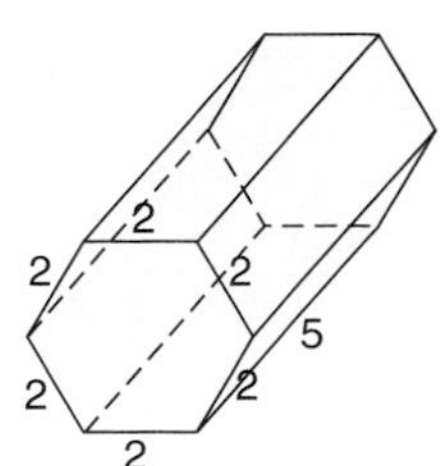

Wright Brothers' plane at Kittyhawk, 1903

10. **Aeronautical Engineering** The surface area of the wing on an aircraft is used to determine a design factor known as wing loading. If the total weight of the aircraft and its load is w and the total surface area of its wings is s, then the formula for the wing-loading factor l is $l = \frac{w}{s}$. If the wing-loading factor is exceeded, the pilot must either reduce the fuel load or remove passengers or cargo. Find the wing-loading factor for the Wright Brothers' plane if it had a maximum take-off weight of 750 pounds and the surface area of the wings was 532 square feet.

EXERCISES

Practice

Use isometric dot paper to draw each polyhedron.

11. a rectangular prism 3 units high, 6 units long, and 4 units wide
12. a cube 6 units on each edge
13. a rectangular prism 9 units high, 3 units long, and 2 units wide
14. a triangular prism 2 units high, whose bases are equilateral triangles with sides 5 units long
15. a cube with a surface area of 150 square units
16. a triangular prism 5 units high, whose bases are triangles with sides 3, 4, and 5 units long

Given each polyhedron, copy its net and label the remaining vertices.

17. 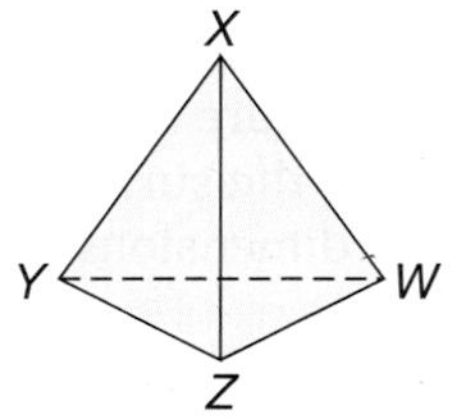

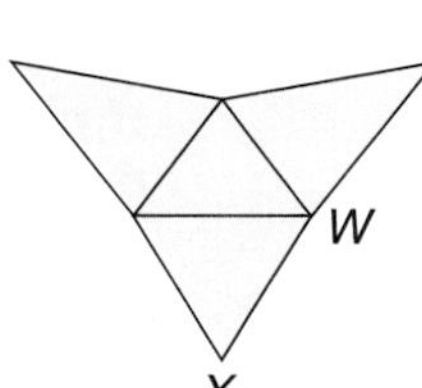

18. 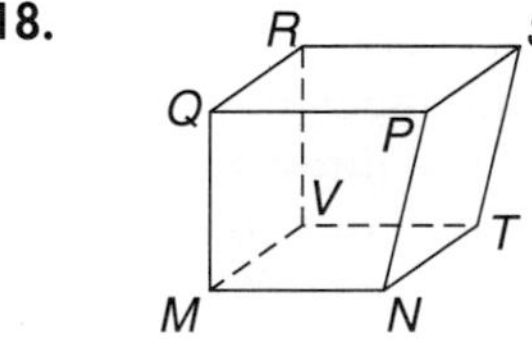

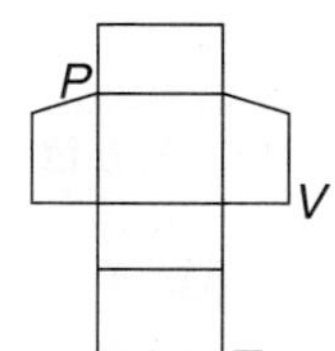

19. 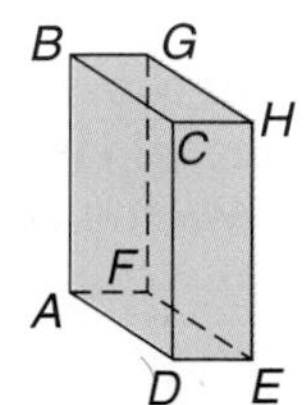

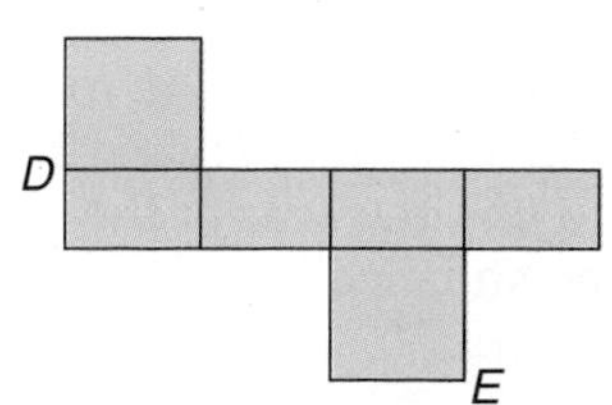

Use rectangular dot paper to draw a net for each solid. Then find the surface area of the solid.

20. 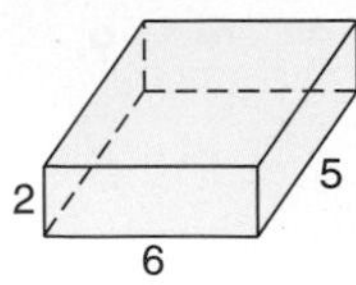

21. 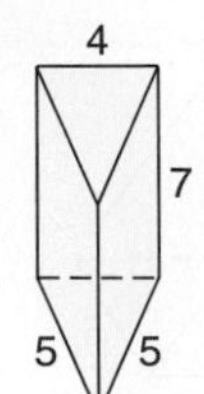

22. 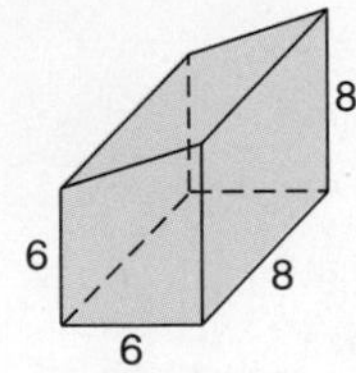

23. 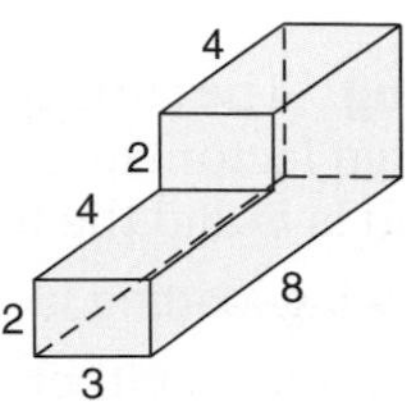

24. 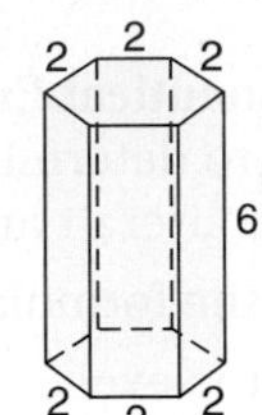

25. 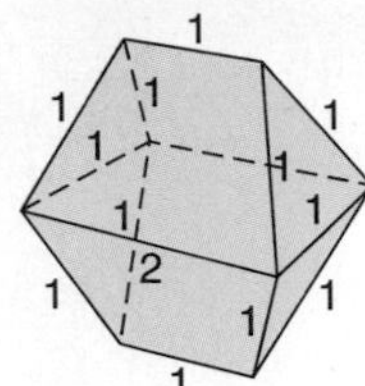

Use isometric dot paper to draw the polyhedron represented by each net.

26. 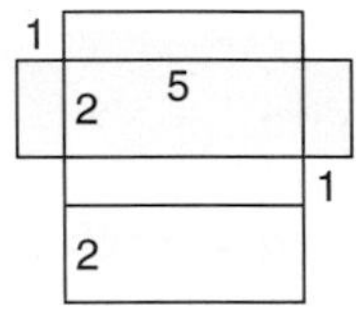

27. 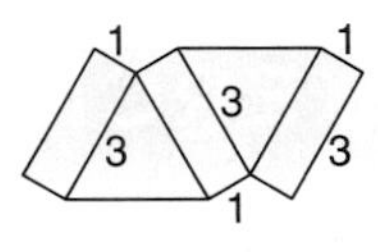

28. 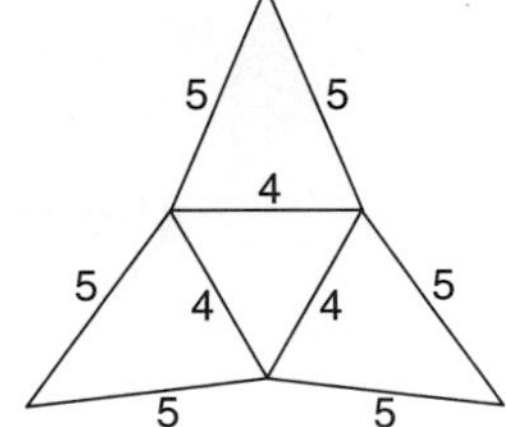

29. Investigate the change in surface area that results from a change in the dimensions of a solid.

a. Draw a net for each solid and find its surface area.

b. Double each dimension of each solid. Then draw a net for each new solid. Find the surface area of the new solid.

c. Study the results of part b. Write a conjecture about the surface area of a solid whose dimensions have been tripled. Check your conjecture by drawing a net and finding the surface area of one solid whose dimensions have been tripled.

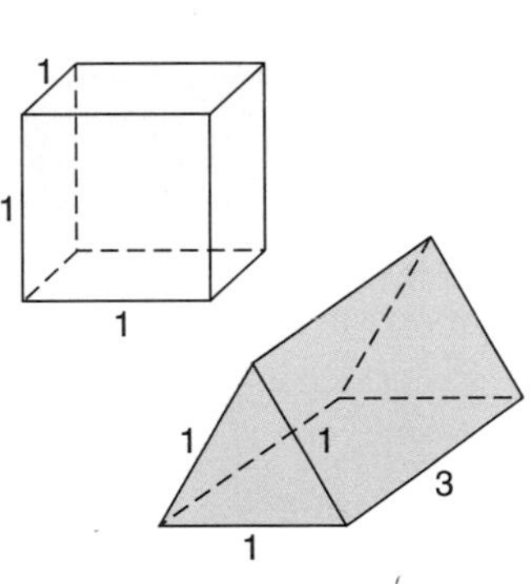

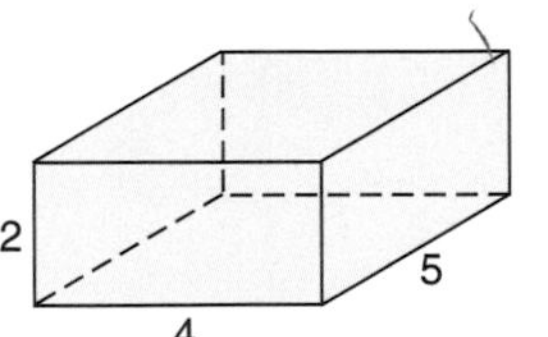

Critical Thinking

30. Suppose you wrote a different integer on each face of cube *ABCDEFGH*. Then a is the sum of the integers on the faces that meet at vertex A, b is the sum of the integers on the faces that meet at vertex B, and so on. Suppose $N = a + b + c + d + e + f + g + h$. What is the greatest number by which N must be divisible?

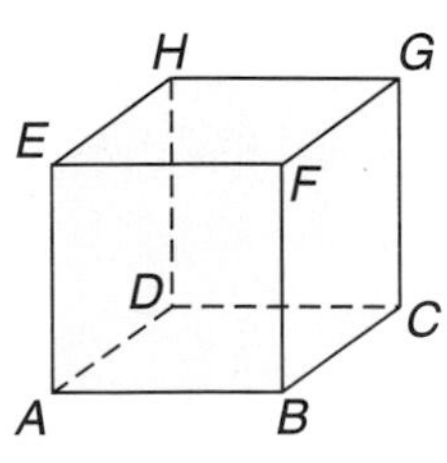

Applications and Problem Solving

31. Construction The roof shown at the right is a hip-and-valley style. Use the dimensions given to find the area of the roof that would need to be shingled.

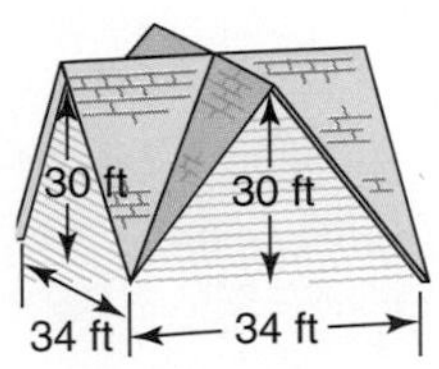

32. **Chemistry** Mineral crystals form as solids with flat faces. The forms of three common minerals are shown below. Draw a net of each crystal.

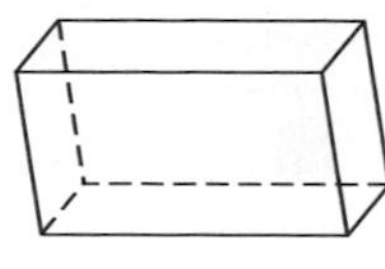

borax

quartz

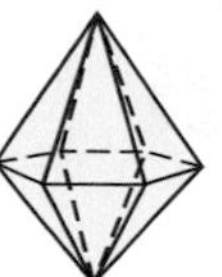

calcite

33. **Games** Many board games use a standard die like the one shown at the right. The opposite faces of a standard die are marked with numbers of dots so that their sum is 7. Determine whether each numbered net can be folded to result in a standard die.

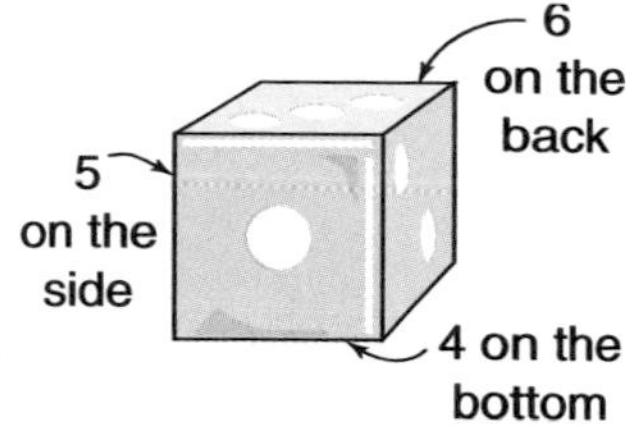

a.

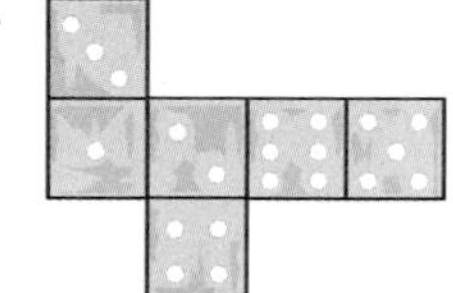

b.

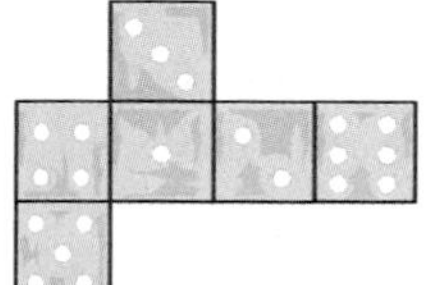

c.

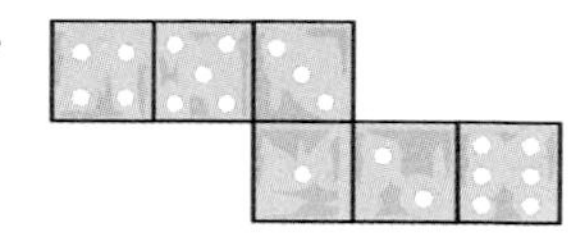

Mixed Review

34. The corner view of a figure is given at the right. Draw the top, left, front, right, and back views of the figure. (Lesson 11–1)

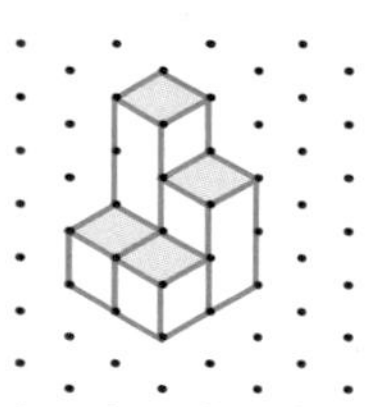

35. Find the area of a trapezoid whose median is 8.5 feet long and whose altitude is 7.1 feet long. (Lesson 10–4)

36. **SAT Practice** **Quantitative Comparison**
O is the center of a circle with $CA = CB$. Segment AB passes through point O.

Column A	Column B
x	30

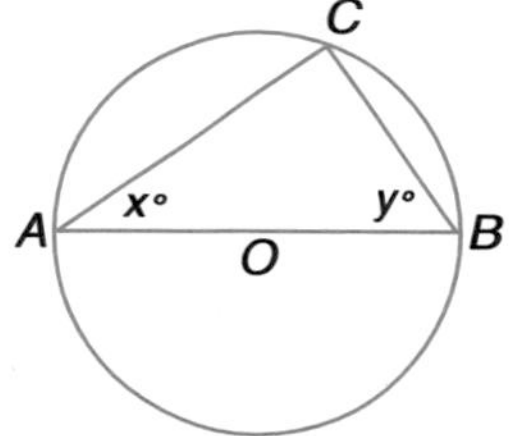

Note: Figure not drawn to scale.

A if the quantity in Column A is greater
B if the quantity in Column B is greater
C if the two quantities are equal
D if the relationship cannot be determined from the information given

37. If $b = 6.4$, $m\angle A = 38$, and $m\angle B = 79$ in $\triangle ABC$, find a. (Lesson 8–5)

38. What is the length of a diagonal of a square that has sides 53.7 meters long? (Lesson 8–2)

39. If $\overline{AE} \parallel \overline{BD}$ and $\overline{AE} \perp \overline{ED}$, find AB. (Lesson 7–4)

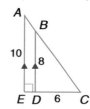

For **Extra Practice**, see page 785.

Algebra

40. Solve $d - (-3) < 13$.

41. Solve $-0.15 \geq y + (-0.03)$.

MODELING MATHEMATICS

An Extension of Lesson 11–2

11–2B Plateau's Problem

Materials:  toothpicks gumdrops thread  water dishwashing detergent deep bowl

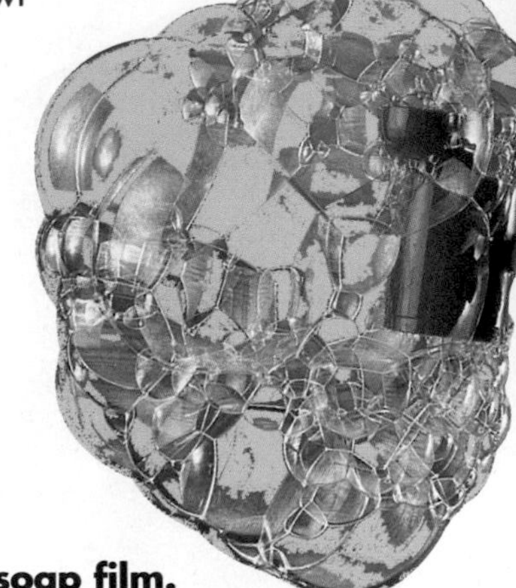

Children and adults alike have long been fascinated with bubbles. Joseph Antoine Ferdinand Plateau (1801–1883) spent much of his life studying the surface properties of fluids. The problem of finding the surfaces of three-dimensional figures that have the least surface area is now called Plateau's problem in his honor. You can investigate Plateau's problem with soap film even though it was not solved mathematically until 1976.

Activity **Make a model and find the surface area of the model using soap film.**

Step 1 Make a soap mixture from a gallon of water and 6 ounces of dishwashing detergent.

Step 2 Build a triangular pyramid, a cube, and a triangular prism. Use toothpicks to represent the edges of the solids and gumdrops to represent the vertices of the solids.

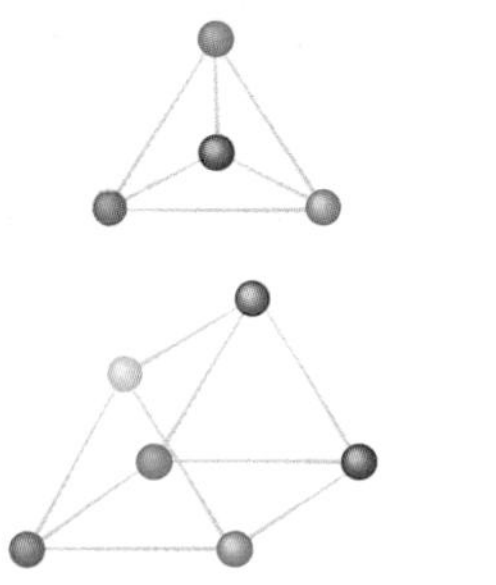

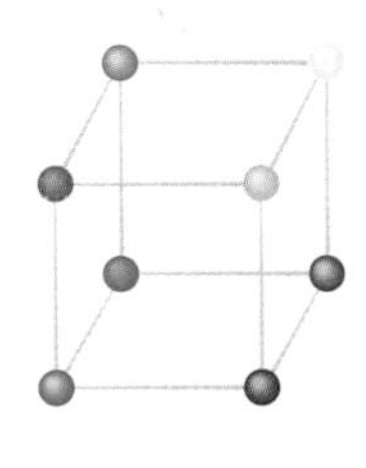

Step 3 Cut a piece of thread and loop it around an edge of your pyramid. Hold onto both ends of the thread and dip the pyramid into the soap mixture. Remove the pyramid from the soap, and observe the shape made by the soap bubble. Dip the cube and the prism in the soap and observe the shapes made by the bubbles.

Write

1. Make a conjecture about the way that the soap will cling to the toothpicks to make the minimum surface area on a cube.

Model

2. Make solids of other shapes using toothpicks and gumdrops. Dip in the soap mixture and observe the results.
3. Does a soap bubble cling to the sides of a solid? Explain.
4. Measure the angle made by three soap surfaces that meet in a line. What does the angle made by four soap surfaces that meet in a line measure?
5. Why do you think the soap behaves as it does? Was the surface made by the soap what you expected? Explain.

11-3 Surface Area of Prisms and Cylinders

What YOU'LL LEARN

- To find the lateral area and surface area of a right prism, and
- to find the lateral area and surface area of a right cylinder.

Why IT'S IMPORTANT

You can solve surface area problems involving manufacturing, gardening, and camping.

Real World APPLICATION

Architecture

You have probably seen buildings covered in brick, stucco, aluminum siding, and wood. But have you ever seen a building covered in corn? In 1892, the citizens of Mitchell, South Dakota, erected the first Corn Palace. The Corn Palace that stands in Mitchell now was built in 1919 and is covered in colored corn and grain each September. The locally grown decorations now cost about $40,000 annually and feed the birds through the winter.

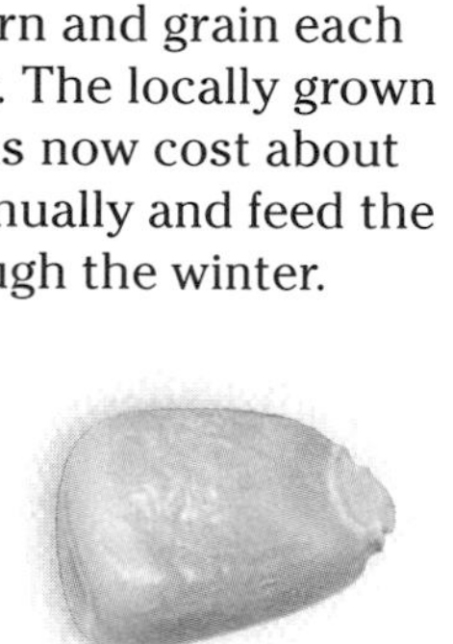

How do you think the citizens of Mitchell go about covering the Corn Palace each year? The first step is to determine the amount of grain to purchase. The amount of grain needed is a function of the surface area of the building. The surface area of the Corn Palace can be estimated by finding the surface area of a rectangular prism. Prisms have the following characteristics.

- Two faces, called **bases**, are formed by congruent polygons that lie in parallel planes.
- The **lateral faces**, the faces that are not bases, are formed by parallelograms.

altitude

base

lateral edge

base

lateral face

- The intersections of two adjacent lateral faces are called **lateral edges** and are parallel segments.

A segment perpendicular to the planes containing the two bases, with an endpoint in each plane is called an **altitude** of the prism. The length of an altitude of a prism is called the **height** of the prism. A prism whose lateral edges are also altitudes is called a **right prism**. If a prism is not right, then it is an **oblique prism**.

You can classify a prism by the shape of its bases.

The lateral faces of right prisms are rectangles.

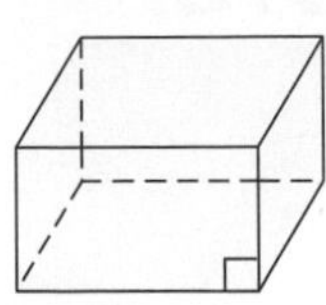
Right rectangular prism

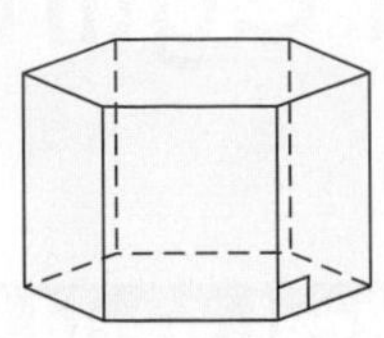
Right hexagonal prism

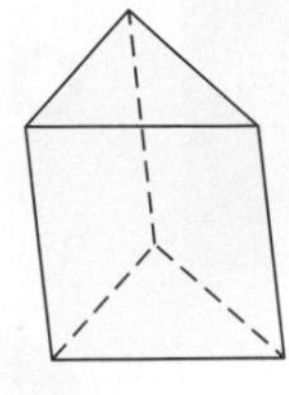
Oblique triangular prism

The lateral edges of oblique prisms are not altitudes.

You can find the surface area T of a prism by adding the areas of each of the faces. However, if you can use a formula, the process is much faster. The first step in finding a formula for the surface area of a prism is to find a formula for the area of the lateral faces.

The area of all the lateral faces of a prism is called the **lateral area** L. As a result of the Distributive Property, the lateral area of a right prism can be found by multiplying the height by the perimeter P of the base as shown in its net. For example in the hexagonal prism at the right, a, b, c, d, e, and f are the measures of the sides of the base.

$$L = ah + bh + ch + dh + eh + fh$$

$$= (a + b + c + d + e + f)h$$

$$= Ph \qquad P = a + b + c + d + e + f$$

This formula, $L = Ph$, can be used to find the lateral area of any right prism.

Lateral Area of a Right Prism	**If a right prism has a lateral area of L square units, a height of h units, and each base has a perimeter of P units, then $L = Ph$.**

The bases of a right prism are congruent, so they have the same area B. Thus, the surface area T of a right prism is found by adding the lateral area to the area of both bases $2B$.

Surface Area of a Right Prism	**If the total surface area of a right prism is T square units, its height is h units, and each base has an area of B square units and a perimeter of P units, then $T = Ph + 2B$.**

Example 

Find the lateral area and the surface area of a right triangular prism with a height of 20 inches and a right triangular base with legs of 8 and 6 inches.

First, use the Pythagorean Theorem to find the measure of the hypotenuse, c.

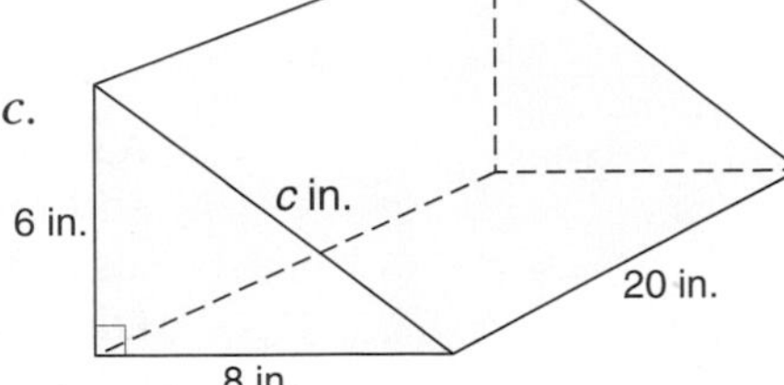

$c^2 = 6^2 + 8^2$

$c^2 = 100$

$c = 10$

Next, use the value of c to find the perimeter.

$P = 6 + 8 + 10$ or 24

Now find the area of a base.

$$B = \frac{1}{2}bh$$

$$B = \frac{1}{2}(6)(8) \text{ or } 24 \text{ in}^2 \quad b = 6, h = 8$$

Finally, find the surface area.

$$T = Ph + 2B$$

$$T = 24 \cdot 20 + 2 \cdot 24 \text{ or } 528 \text{ in}^2 \quad P = 24, h = 20, B = 24$$

The formulas for the lateral area and the surface area of a right prism can be used to estimate areas of other three-dimensional objects.

Example 2 **Refer to the application at the beginning of the lesson.**

Architecture

a. Estimate the area of the Corn Palace to be covered if its base is 152 by 196 feet and it is 40 feet tall, not including the turrets.

b. Suppose a bushel of grain can cover 15 square feet. How many bushels of grain will it take to cover the Corn Palace?

c. Will the actual amount of grain needed be higher or lower than the estimate?

a. Assume that the Corn Palace is a right rectangular prism. Then since neither the floor nor roof of the Palace will be covered, the area to consider is the lateral area.

$$\begin{aligned} L &= Ph \\ &= [2(152 + 196)][40] \\ &= 27{,}840 \text{ ft}^2 \end{aligned}$$

b. It will take $27{,}840 \div 15$ or about 1856 bushels of grain to cover the Corn Palace.

c. The actual amount of grain needed will be higher because the area of the turrets was not accounted for in the estimate.

FYI

Actually, the walls of the Corn Palace are not covered entirely by grain. Fourteen huge grain murals cover a large portion of the surface area.

A **cylinder** is another common type of solid. Like a prism, a cylinder has parallel, congruent **bases**. However, the bases of a cylinder are circles. The **axis** of the cylinder is the segment whose endpoints are centers of these circles.

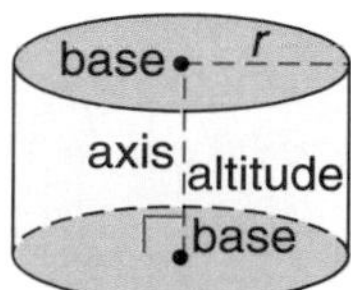

right cylinder

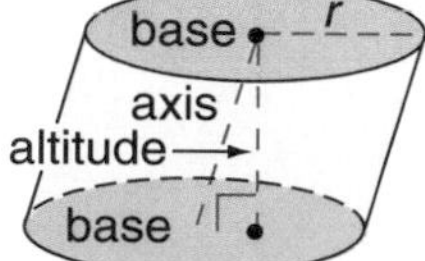

oblique cylinder

An **altitude** of a cylinder is a segment perpendicular to the planes containing the bases with an endpoint in each plane. If the axis of a cylinder is also an altitude of the cylinder, then the cylinder is called a **right cylinder**. Otherwise, the cylinder is an **oblique cylinder**.

The figure below is a net for a right cylinder.

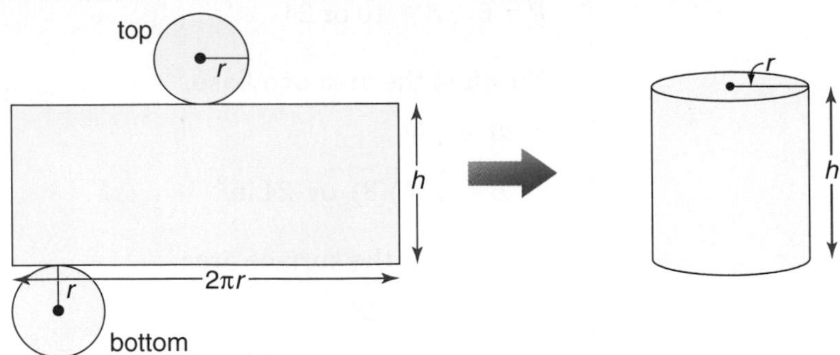

The lateral area of a cylinder is the area of the curved surface. Like a prism, the lateral area of a cylinder can be found by using the formula $L = Ph$. However, since the base is a circle, the perimeter is the circumference ($2\pi r$) of the circle. Thus, the curved surface is a rectangle whose width is the height of the cylinder, h units, and whose length is the circumference of one of its bases, $2\pi r$ units.

Lateral Area of a Right Cylinder	**If a right cylinder has a lateral area of *L* square units, a height of *h* units, and the bases have radii of *r* units, then $L = 2\pi rh$.**

The surface area of a cylinder is the sum of the lateral area and the areas of the bases. Each base is a circle whose area is πr^2.

Surface Area of a Right Cylinder	**If a right cylinder has a total surface area of *T* square units, a height of *h* units, and the bases have radii of *r* units, then $T = 2\pi rh + 2\pi r^2$.**

Example 3

Manufacturing

Find the number of square feet of aluminum used to make 50,000 cylindrical vegetable cans if each can is 5 inches tall and 2.5 inches in diameter.

If the diameter of the base is 2.5 inches, then the radius is 1.25 inches.

$T = 2\pi rh + 2\pi r^2$

$T = 2\pi(1.25)(5) + 2\pi(1.25)^2$

$T = 12.5\pi + 3.125\pi$

$T = 15.625\pi$

$T \approx 49.1 \text{ in}^2$

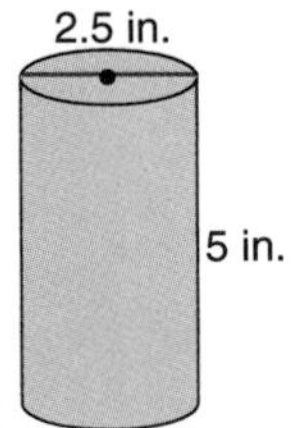

The surface area of each can is about 49.1 square inches, so the area of aluminum used for 50,000 cans is about $50{,}000 \times 49.1$ or about 2,455,000 square inches. There are 144 square inches in a square foot. Therefore, about $2{,}455{,}000 \div 144$ or about 17,049 square feet of aluminum are used.

You can use the formulas for lateral and surface area to find missing measures.

Example 4 **The surface area of a right cylinder is 301.6 cm². If the height is 8 centimeters, find the radius of the base.**

*Remember that for $ax^2 + bx + c = 0$, the value of x can be found using $x = \frac{-b \pm \sqrt{b^2 - 4ac}}{2a}$. This is the **quadratic formula**.*

$$T = 2\pi rh + 2\pi r^2$$
$$301.6 = 2\pi r(8) + 2\pi r^2$$
$$301.6 = 50.3r + 6.3r^2$$
$$0 = 6.3r^2 + 50.3r - 301.6$$

Now use the quadratic formula to solve this equation. Let r replace x.

$$r = \frac{-b \pm \sqrt{b^2 - 4ac}}{2a}$$
$$= \frac{-50.3 \pm \sqrt{50.3^2 - 4(6.3)(-301.6)}}{2(6.3)} \quad a = 6.3,\ b = 50.3, \text{ and } c = -301.6.$$
$$\approx 4.0 \text{ or } -12.0 \quad \textit{Use a calculator.}$$

Since a circle cannot have a negative radius, the radius of the base must be 4.0 centimeters.

CHECK FOR UNDERSTANDING

Communicating Mathematics

Read and study the lesson to answer these questions.

1. Refer to the triangular prism in Example 1.
   a. What are the shapes of the polygonal faces that make up the solid? Be specific.
   b. Draw a net of the solid on rectangular dot paper.
   c. Draw a corner view of the solid on isometric dot paper.
2. What dimensions of a soft-drink can would be important for you to know if you are designing a new label for the can? Explain.
3. **Compare and contrast** lateral area and surface area.

4. **Describe** in your own words the difference between a right prism and an oblique prism. Draw an example of each.

Guided Practice

Refer to the prism at the right to answer each of the following.

5. Is the prism a right prism or an oblique prism? Explain.
6. What is the shape of its bases and lateral faces?
7. If the bases are regular polygons with sides 6 units long, find the perimeter of the base.
8. If the perimeter of the base is 60 units and the length of a lateral edge is 15 units, find the lateral area of the prism.

Find the surface area of the right rectangular prism for each set of measures.

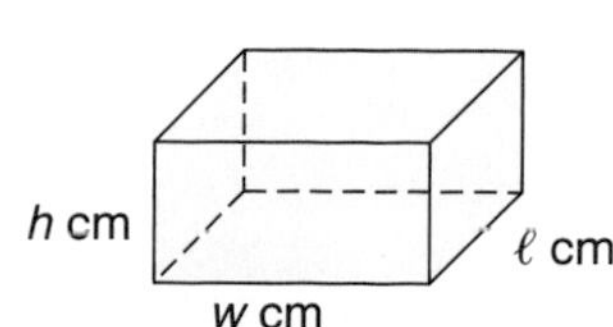

9. $\ell = 8$, $w = 4$, and $h = 2$
10. $\ell = 6.5$, $w = 6.5$, and $h = 6.5$

Find the surface area of a cylinder with each set of measures. Round to the nearest tenth.

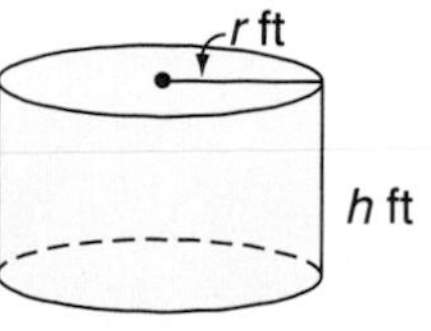

11. $r = 4$ and $h = 6$
12. $r = 8.3$ and $h = 6.6$

13. The height of a right cylinder is 28 inches, and its surface area is 1977.7 square inches. Find the radius of the base of the cylinder.
14. **Manufacturing** Many baking pans are given a special coating to make food stick to the surface less.
    a. A rectangular cake pan is 9 inches by 13 inches and 2 inches deep. What is the area of the surface to be coated?
    b. Round cake pans have a diameter of 9 inches and a height of 2 inches. Find the area of the surface to be coated.

EXERCISES

Practice

Use the rectangular prism at the right to answer each of the following.

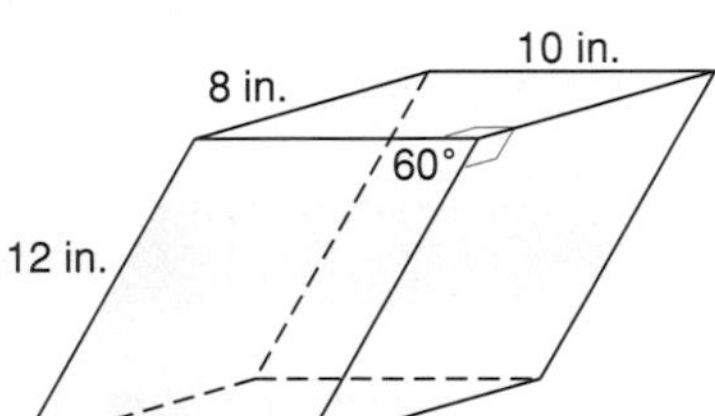

15. Is the prism a right prism or an oblique prism?
16. What shape are its bases and lateral faces?
17. Find the perimeter of the base.
18. What is the lateral area of the prism?
19. Find the surface area of the prism.

Find the surface area of the right triangular prism for each set of measures. Round to the nearest tenth.

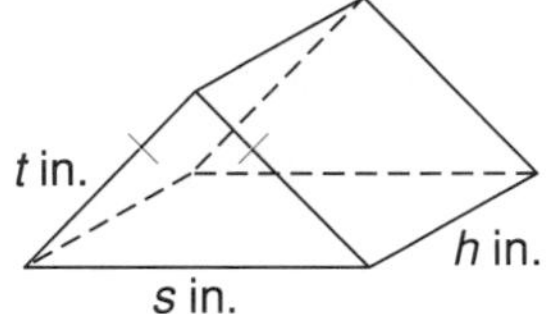

20. $t = 8$, $s = 14$, and $h = 7$
21. $t = 10$, $s = 8$, and $h = 20.4$
22. $t = 14$, $s = 18$, and $h = 30.5$

Find the lateral area and the surface area of each right prism. Round to the nearest tenth.

23. 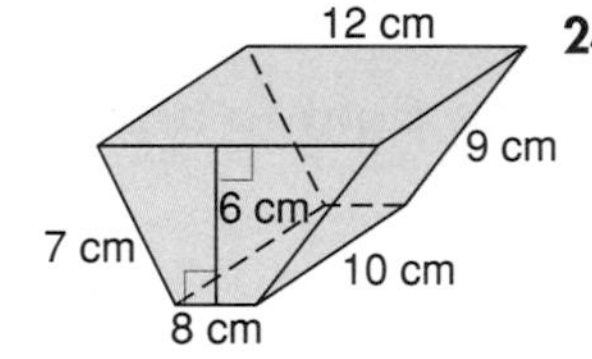

24. 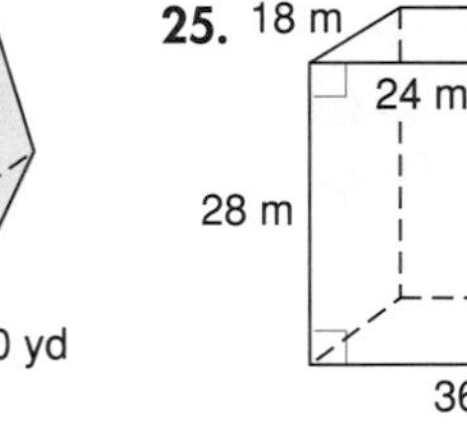

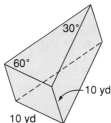

25. 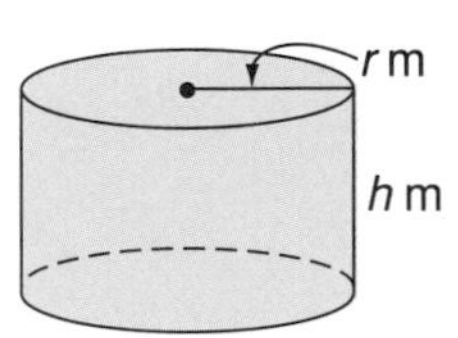

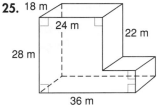

Find the surface area of a cylinder with each set of measures. Round to the nearest tenth.

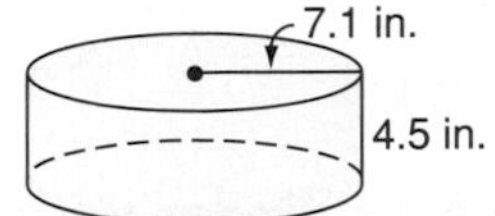

26. $r = 11$ and $h = 11$
27. $r = 13$ and $h = 15.8$
28. $r = 6.8$ and $h = 1.9$

Find the lateral area and the surface area of each right cylinder. Round to the nearest tenth.

29.

30. 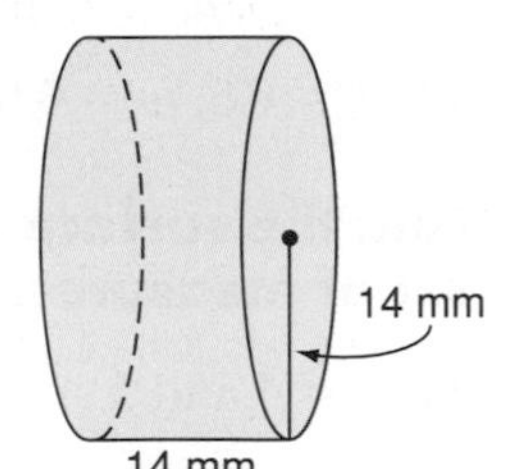

31. 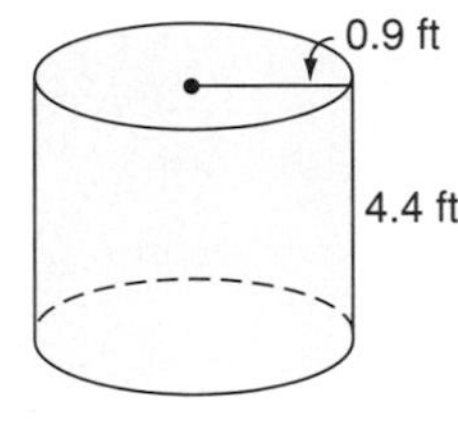

32. The surface area of a cube is 864 square units. What is the length of a lateral edge of the cube?

33. Suppose the lateral area of a right rectangular prism is 144 square centimeters. If its length is three times its width and its height is twice its width, find its surface area.

34. A cylinder has a surface area of 301.6 square meters. Find the diameter of the base if the cylinder is 8 meters tall.

Critical Thinking

35. Suppose the height of a right prism is doubled. What effect does it have on the surface area?

36. Suppose the height of a right cylinder is tripled. Is the surface area tripled? Explain.

Applications and Problem Solving

37. **Gardening** A greenhouse is designed to allow gardeners to grow plants in areas where they would not normally grow. The surface of a greenhouse is covered with plastic or glass. Find the amount of plastic that would be needed to construct a greenhouse like the one shown at the right.

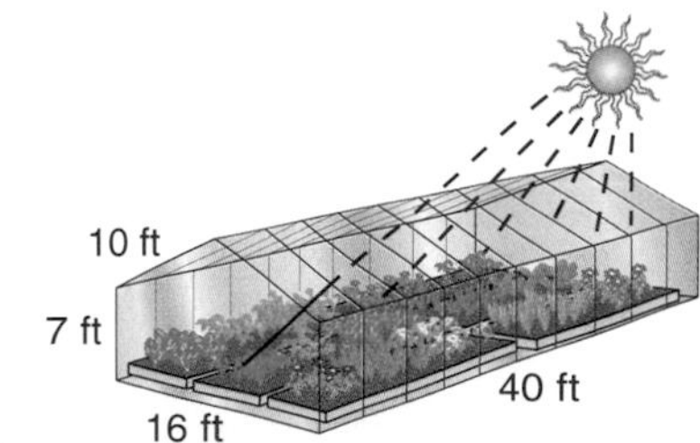

38. **Agriculture** The acid associated with filling a silo can weaken the cement walls and seriously damage the silo's structure. So the inside of the silo must occasionally be resurfaced. The cost of the resurfacing is a function of the lateral area of the inside of the silo. Find the lateral area of a silo 13 meters tall with an interior diameter of 5 meters.

39. **Camping** Campers can use a solar cooker to use the energy of the Sun to prepare food. You can make a solar cooker from supplies you have on hand. The reflector in the cooker shown at the right is half of a cardboard cylinder covered with aluminum foil. The reflector is 18 inches long and has a diameter of $5\frac{1}{2}$ inches. How much aluminum foil was needed to cover the inside of the reflector?

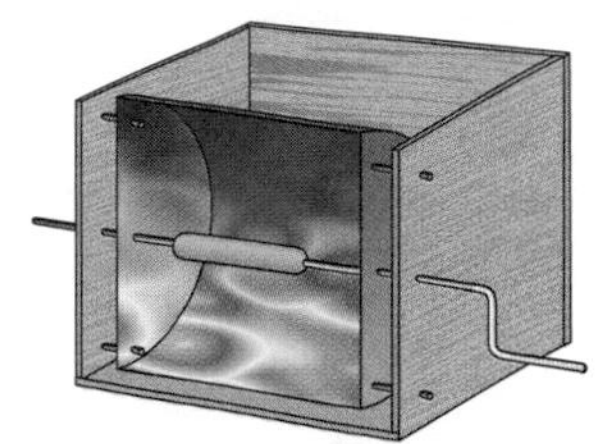

Mixed Review

40. Draw a net for the solid shown at the right. (Lesson 11–2)

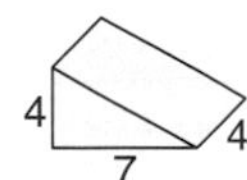

41. **SHOE**

a. What view of the Washington Monument is shown in the photograph in the comic—corner, top, left, right, front, or back? (Lesson 11–1)

b. Find a photograph of the Washington Monument in a reference book or on the Internet. Draw the top, left, right, front, back, and corner views of the monument. (Lesson 11–1)

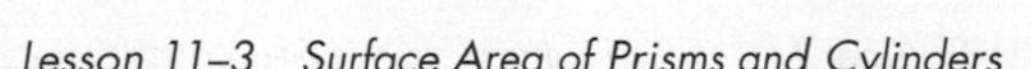

42. **Probability** Find the probability that a point chosen at random in the figure is in the shaded region. Round to the nearest hundredth. (Lesson 10–6)

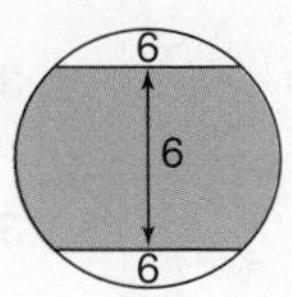

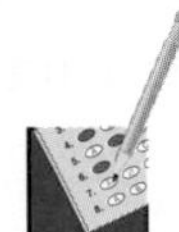

43. One diagonal of a rhombus with an area of 1664 square centimeters is 52 centimeters long. Find the length of the other diagonal. (Lesson 10–4)

44. **SAT Practice** In parallelogram $ABCD$, $BD = 6$ and $AD = 10$. What is the area of $ABCD$?

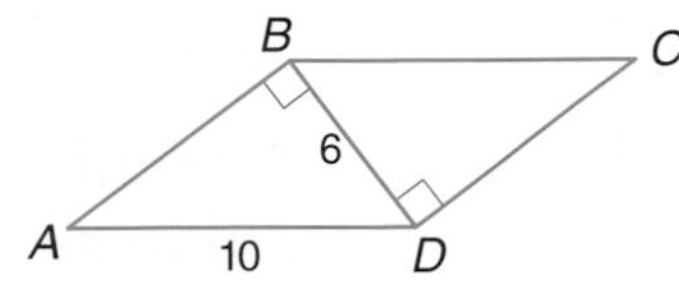

A 24 **B** 30
C 48 **D** 60
E It cannot be determined from the information given.

45. The vertices of $\triangle ABC$ are $A(9, 5)$, $B(6, 2)$, and $C(7, -3)$. Find the slope of the altitude to $\overline{BC}$. (Lesson 5–1)

46. In $\triangle JKL$, $m\angle J = 10x - 7$, $m\angle K = 63$, and $m\angle L = 7x + 5$. Classify $\triangle JKL$ by its angles and sides. Justify your answer. (Lesson 4–6)

47. If possible, write a valid conclusion. State the law of logic that you used. (Lesson 2–3)
    (1) A triangle is equiangular if it is equilateral.
    (2) Each angle of an equilateral triangle measures 60°.

Algebra

48. Determine whether $\frac{8}{5}x + \frac{4}{y} = 10$ is a linear equation. If it is linear, rewrite it in standard form, $Ax + By = C$.

49. **Statistics** Explain whether a scatter plot showing cities' annual per capita consumption of coffee and the population of cities would probably show a *positive*, *negative*, or *no* correlation.

For **Extra Practice**, see page 785.

WORKING ON THE **In·ves·ti·ga·tion**

Refer to the Investigation on pages 510–511.

Just For Kicks

Soccer balls that are made of leather or vinyl are stitched together by following a pattern. The pattern can be created by first drawing a net of the surface of the ball.

1 Choose a length for the edge of a pentagon in each design shown on pages 510 and 511. Then draw a net for each design.

2 Which design will require more stitching to complete?

3 Which ball would be easier to manufacture?

4 Note your observations to consider when completing your soccer ball design.

Add the results of your work to your Investigation Folder.

11-4 Surface Area of Pyramids and Cones

What YOU'LL LEARN

- To find the lateral area and surface area of a regular pyramid, and
- to find the lateral area and surface area of a right circular cone.

Why IT'S IMPORTANT

Pyramids and cones have been used in structures since ancient times.

You can use an envelope to make a *triangular pyramid.*

MODELING MATHEMATICS **Triangular Pyramids**

Materials: 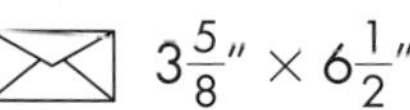 $3\frac{5}{8}'' \times 6\frac{1}{2}''$ envelope

 straightedge  scissors

- Seal a $3\frac{5}{8}'' \times 6\frac{1}{2}''$ envelope. Then draw both diagonals of the envelope.

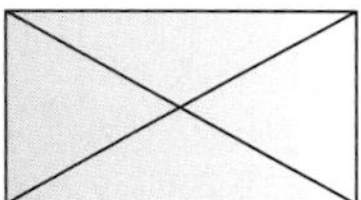

- Fold and crease along each diagonal. Then fold and crease along the perpendicular bisector of the long edge of the envelope. Label the point of intersection of the diagonals as point *A*.

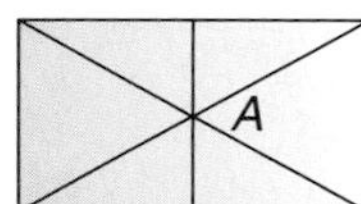

- Carefully cut from the top of the envelope along each diagonal to point *A*. Remove the triangular piece.

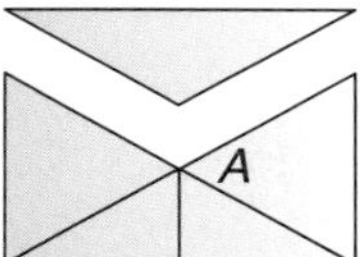

- Open the envelope. Fold along the perpendicular bisector so that the top corners come together. Tuck one corner inside the other and push until the corner meets the bottom edge. The solid formed is a triangular pyramid.

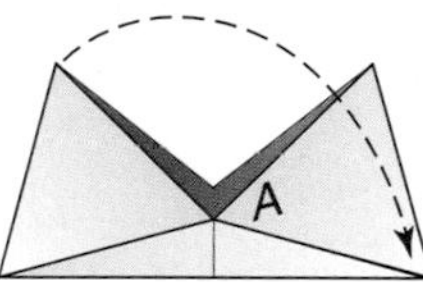

Your Turn

a. What shape is each face of the pyramid?

b. Is each face of the pyramid congruent? Explain.

c. Suppose that the face with the letter *A* on it is the base of the pyramid. What do you observe about the other faces?

d. Repeat the activity using a $4\frac{1}{8}'' \times 9\frac{1}{2}''$ envelope and describe your results.

A triangular pyramid is just one type of pyramid. A pyramid has the following characteristics.

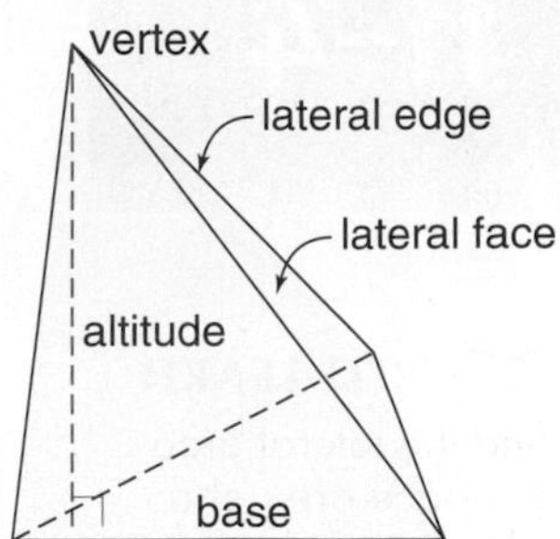

- All the faces except one intersect at a point called the **vertex**.
- The **base** is the face that does not intersect the other faces at the vertex. The base is always a polygon.
- The faces that intersect at the vertex are called **lateral faces** and form triangles. The edges of the lateral faces that have the vertex as an endpoint are called **lateral edges**.
- The **altitude** is the segment from the vertex perpendicular to the base.

If the base of a pyramid is a regular polygon and the segment whose endpoints are the center of the base and the vertex is perpendicular to the base, then the pyramid is called a **regular pyramid**. In a regular pyramid, the segment whose endpoints are the center of the base and the vertex is also the altitude.

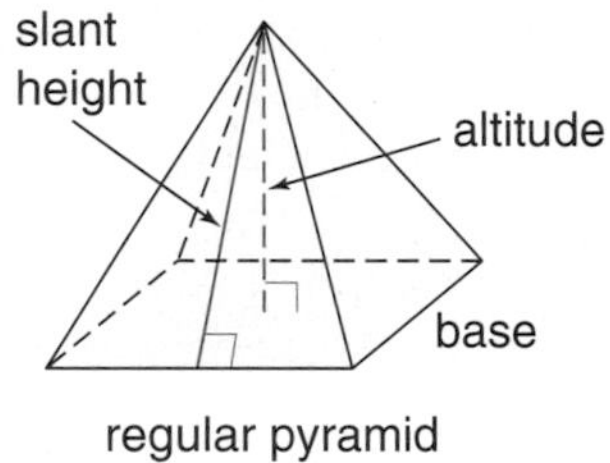

regular pyramid

In a regular pyramid, all of the lateral faces are congruent isosceles triangles. The height of each lateral face is called the **slant height** ℓ of the pyramid.

The figure below is a regular hexagonal pyramid. Its lateral area L can be found by adding the areas of all its congruent triangular faces as shown in its net.

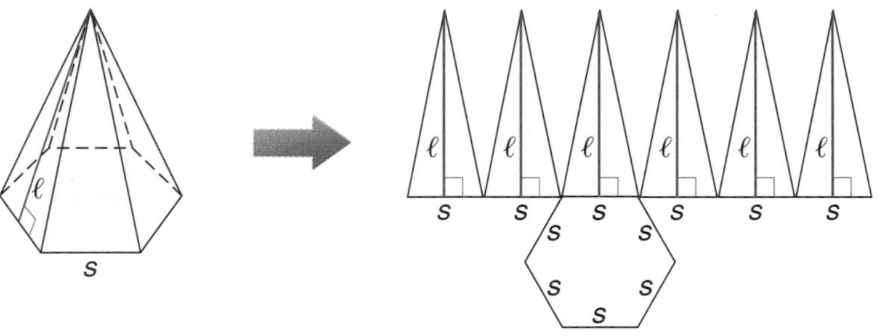

$$L = \frac{1}{2}s\ell + \frac{1}{2}s\ell + \frac{1}{2}s\ell + \frac{1}{2}s\ell + \frac{1}{2}s\ell + \frac{1}{2}s\ell$$
$$= \frac{1}{2}(s + s + s + s + s + s)\ell$$
$$= \frac{1}{2}P\ell \quad P = (s + s + s + s + s + s)$$

This suggests the following formula.

Lateral Area of a Regular Pyramid	If a regular pyramid has a lateral area of L square units, a slant height of ℓ units, and its base has a perimeter of P units, then $L = \frac{1}{2}P\ell$.

Example 1

Architecture

The Luxor Hotel in Las Vegas, Nevada, is shaped like a gigantic black glass pyramid. The base of the pyramid is a square with edges 646 feet long. The hotel is 350 feet tall. Find the area of the glass on the Luxor.

The area of the glass is the lateral area of the pyramid.

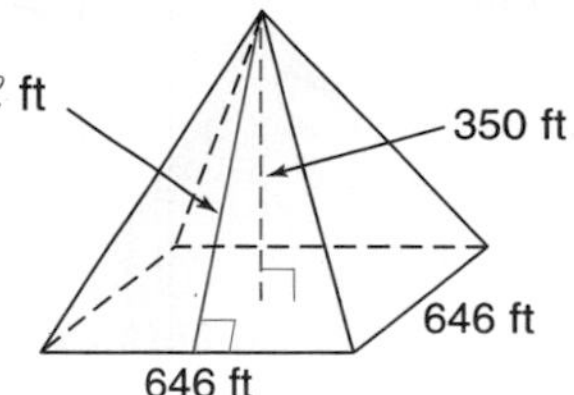

The altitude and the slant height are a leg and the hypotenuse of a right triangle. The other leg is half of the measure of a side of the base. Use the Pythagorean Theorem, $c^2 = a^2 + b^2$, to find the slant height of the pyramid.

$$\ell^2 = 350^2 + \left(\frac{1}{2} \cdot 646\right)^2$$
$$\ell^2 = 226{,}829$$
$$\ell = \sqrt{226{,}829} \text{ or about } 476.3 \text{ ft}$$

$$L = \frac{1}{2}P\ell$$
$$\approx \frac{1}{2}(4 \cdot 646)(476.3)$$
$$\approx 615{,}379.6$$

The area of the glass on the Luxor is about 615,380 ft^2.

The surface area of a regular pyramid is the sum of its lateral area and the area of its base.

Surface Area of a Regular Pyramid	**If a regular pyramid has a surface area of *T* square units, a slant height of ℓ units, and its base has a perimeter of *P* units and an area of *B* square units, then $T = \frac{1}{2}P\ell + B$.**

Example 2

A regular pyramid has a slant height of 13 centimeters and a height of 12 centimeters. If the base is a regular pentagon, find the surface area of the pyramid.

The slant height, the altitude, and the apothem form a right triangle. Use the Pythagorean Theorem to find the length of the apothem.

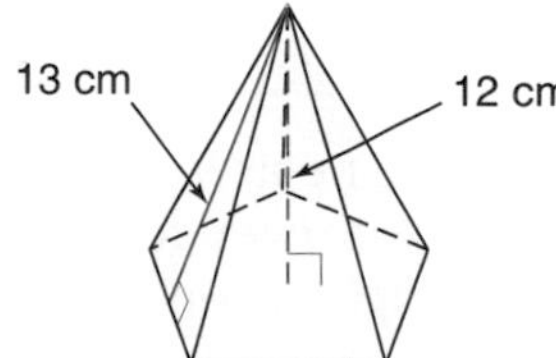

$$13^2 = 12^2 + a^2$$
$$25 = a^2$$
$$5 = a$$

(continued on the next page)

LOOK BACK

You can review the trigonometric functions in Lesson 8-3.

Now find the length of the sides of the base of the pyramid. The central angle of the pentagon measures $\frac{360°}{5}$ or 72°. Let a represent the measure of the angle formed by a radius and the apothem. Thus, $a = \frac{72}{2}$ or 36.

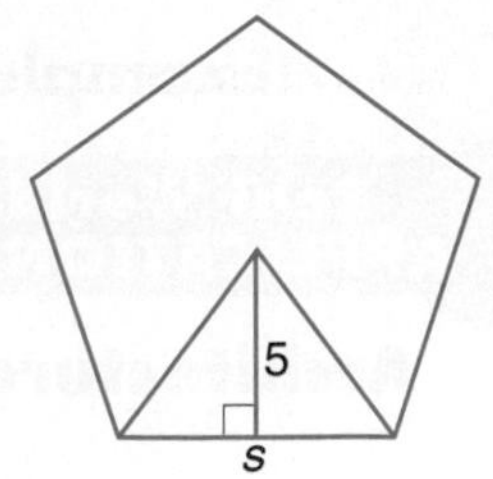

Use trigonometry to find the length of the sides.

$$\tan 36° = \frac{\frac{1}{2}s}{5}$$

$$5 \tan 36° = \frac{1}{2}s$$

$$10 \tan 36° = s$$

$$7.3 \approx s$$

Now use the formula to find the surface area of the pyramid.

$$T = \frac{1}{2}P\ell + B$$

$$\approx \frac{1}{2}(5 \cdot 7.3)(13) + \frac{1}{2}(5 \cdot 7.3)(5) \quad B = \frac{1}{2}Pa$$

$$\approx 328.5$$

The surface area of the pyramid is about 328.5 cm^2.

The figure at the right is a **circular cone**. Its **base** is a circle, and its **vertex** is at V. Its **axis**, $\overline{VX}$, is the segment whose endpoints are the vertex and the center of the base. The segment that has the vertex as one endpoint and is perpendicular to the base is called the **altitude** of the cone.

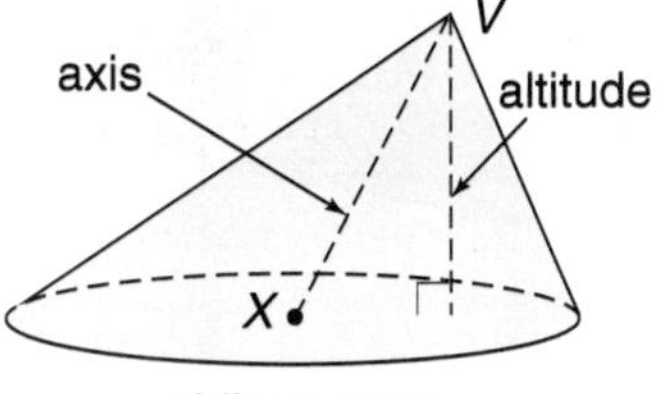

oblique cone

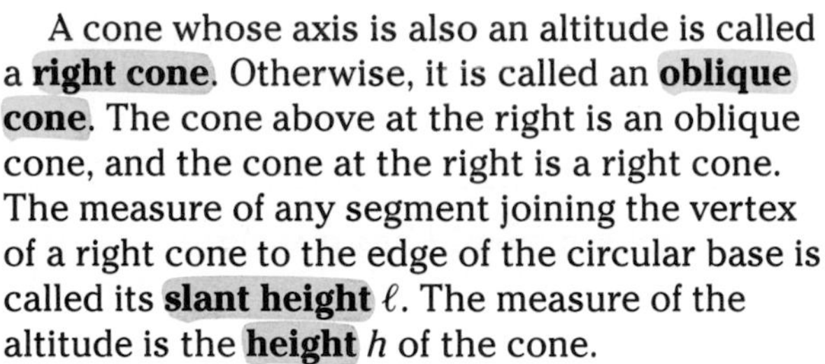

A cone whose axis is also an altitude is called a **right cone**. Otherwise, it is called an **oblique cone**. The cone above at the right is an oblique cone, and the cone at the right is a right cone. The measure of any segment joining the vertex of a right cone to the edge of the circular base is called its **slant height** ℓ. The measure of the altitude is the **height** h of the cone.

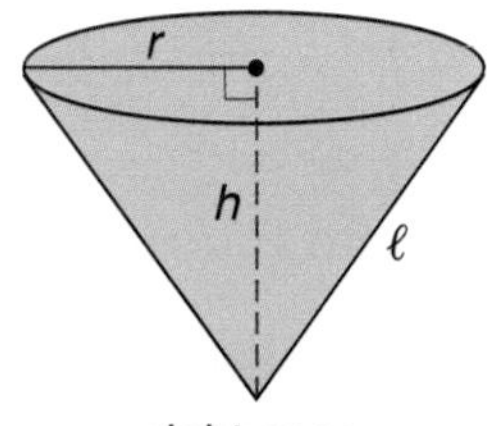

right cone

You will use formulas similar to the formulas for finding the lateral area and the surface area of a regular pyramid to find those same measures for a right cone. In the net for a cone, the region of the cone that is not the base is a sector of a circle whose radius is the slant height ℓ.

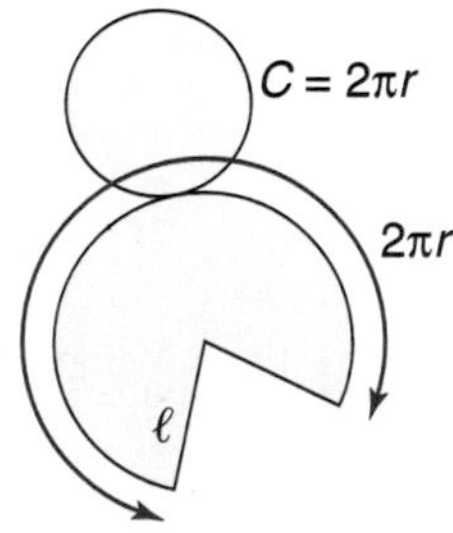

The area of the sector is proportional to the area of the circle. Notice that the arc length of this sector is equal to the circumference of the base of the original cone, $2\pi r$.

$$\frac{\text{area of sector}}{\text{area of circle}} = \frac{\text{measure of arc}}{\text{circumference of circle}}$$

$$\frac{\text{area of sector}}{\pi\ell^2} = \frac{2\pi r}{2\pi\ell}$$

$$\text{area of sector} = \pi r\ell \quad \textit{Multiply each side by } \pi\ell^2.$$

Lateral Area and Surface Area of a Right Circular Cone	**If a right circular cone has a lateral area of L square units, a surface area of T square units, a slant height of ℓ units, and the radius of the base is r units, then $L = \pi r\ell$ and $T = \pi r\ell + \pi r^2$.**

Example 3 **Find the lateral area and the surface area of a cone if the slant height is 13 feet and the diameter of the base is 10 feet.**

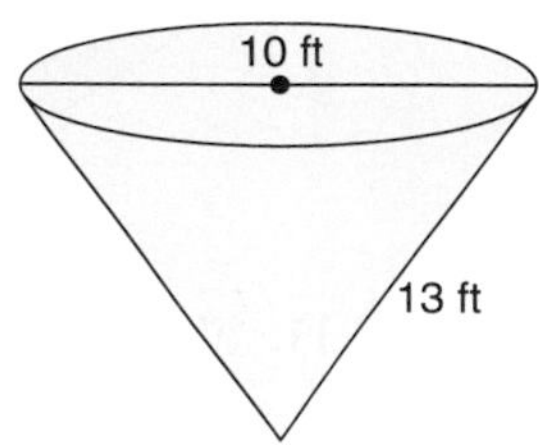

If the diameter is 10 feet, then the radius is $\frac{1}{2} \times 10$ or 5 feet.

$$L = \pi r\ell$$
$$= \pi(5)(13)$$
$$= 65\pi$$
$$\approx 204.2 \text{ ft}^2$$

$$T = \pi r\ell + \pi r^2$$
$$= \pi(5)(13) + \pi(5)^2$$
$$= 90\pi$$
$$\approx 282.7 \text{ ft}^2$$

CHECK FOR UNDERSTANDING

Communicating Mathematics

Study the lesson. Then complete the following.

1. **Compare and contrast** the lateral edges of a pyramid and those of a prism.
2. **Analyze** the change in the shape of the base of a regular pyramid as the number of sides increases.
   a. What shape is it?
   b. As this transformation takes place, what happens to the shape of the pyramid?
3. **Sketch** a regular pyramid. Which is longer, a lateral edge or the slant height? Explain.
4. **Describe** a cone in which the axis is not also the altitude of the cone.

5. Cut the side of a cone-shaped drinking cup from the brim to the vertex. Flatten out the cup.
   a. What is the shape of the surface?
   b. Measure the radius and the angle formed by the cut edges.
   c. Find the area of the surface. How does the area relate to the formula for the lateral area of a cone?

Guided Practice

Determine whether the condition given is characteristic of a *pyramid* or *prism, both,* or *neither.*

6. The lateral faces are parallelograms.
7. It can have as few as five faces.

Find the lateral area and the surface area of each regular pyramid or right cone. Round to the nearest tenth.

8.
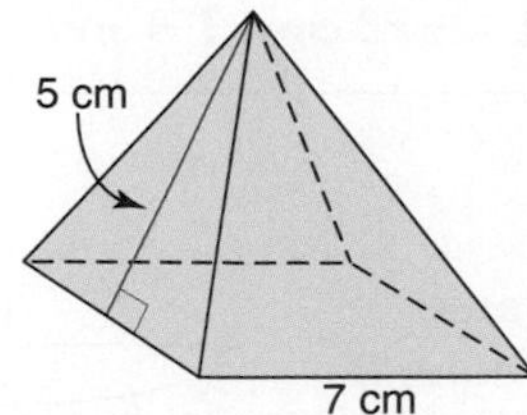

9.
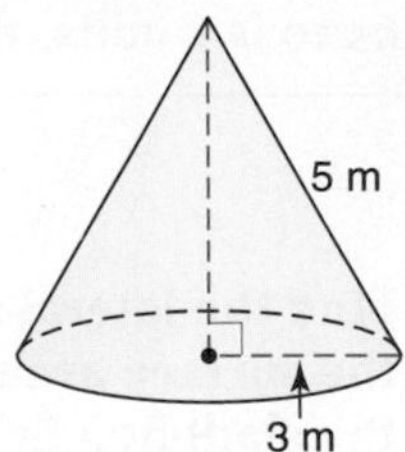

10.
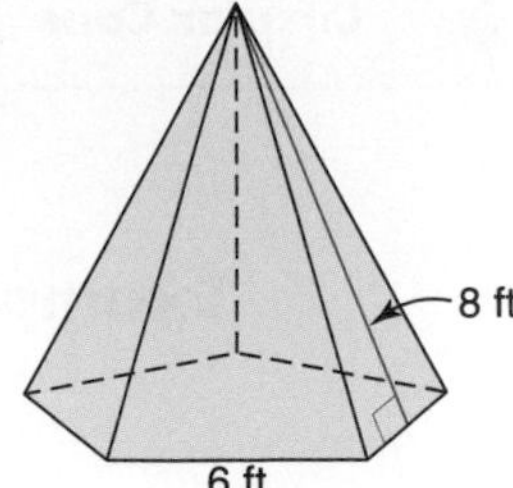

11. Find the surface area of the solid at the right. Round to the nearest tenth.

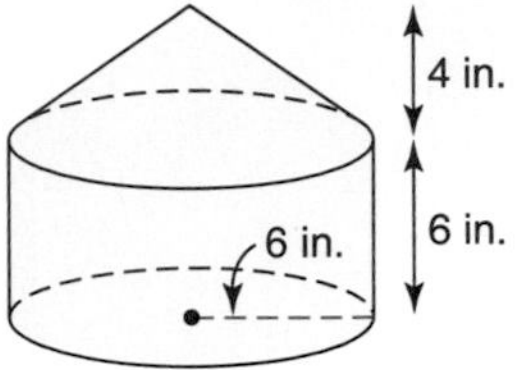

12. The base of a rectangular pyramid is 15 inches long and 8 inches wide. The height is 4 inches. Find the surface area of the pyramid if all of the lateral edges are congruent.

13. **History** Historians believe that the Great Pyramids of Egypt were once covered with gold or white stones that have worn away or have been removed to be used for other purposes. The diagram at the right shows the approximate dimensions of the Great Pyramid of Khufu. Find the surface area that would have to have been covered.

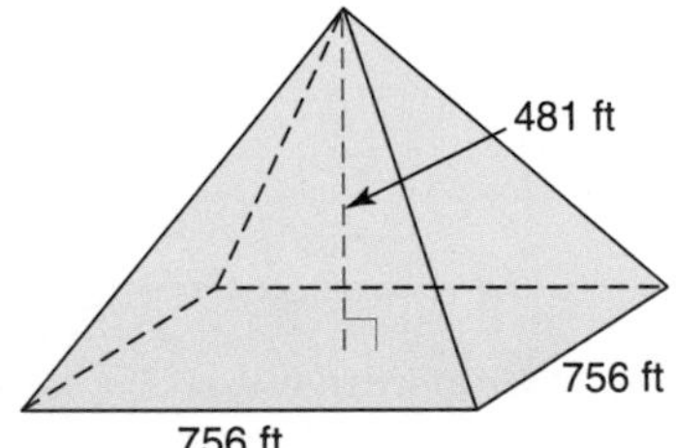

EXERCISES

Practice

Determine whether the condition given is characteristic of a *pyramid* or *prism, both,* or *neither.*

14. There is exactly one base.
15. It always has an even number of faces.
16. There are two bases.
17. The lateral faces are triangles.
18. It has the same number of lateral faces as vertices.
19. There can be as few as four faces.
20. The number of edges is always even.

Find the lateral area and the surface area of each regular pyramid or right cone. Round to the nearest tenth.

21.
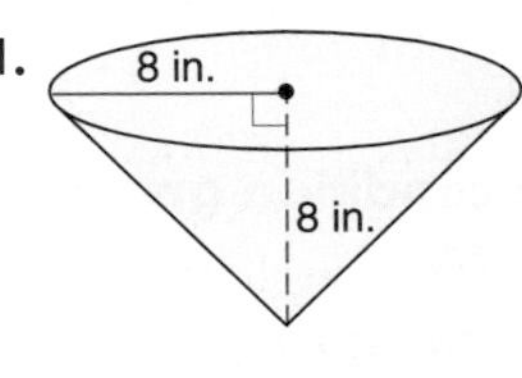

22.
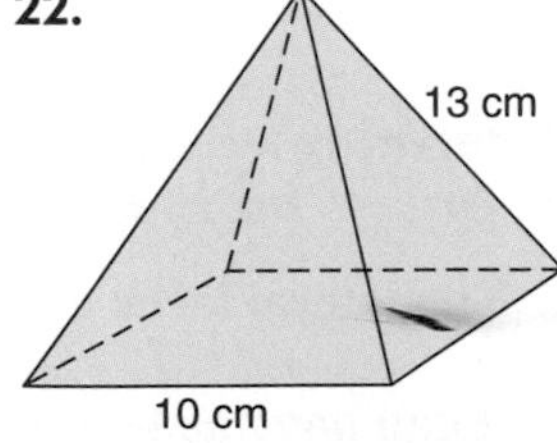

23.
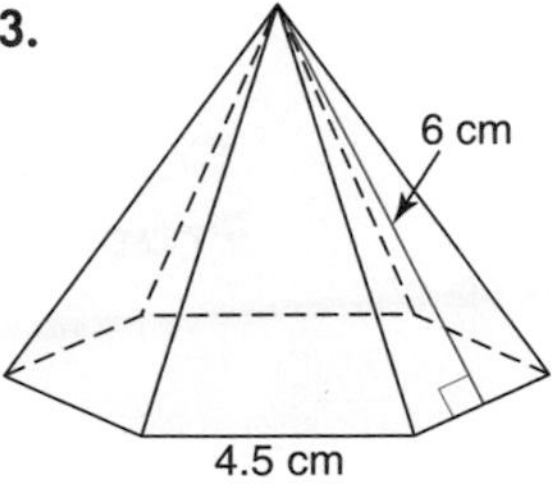

24. 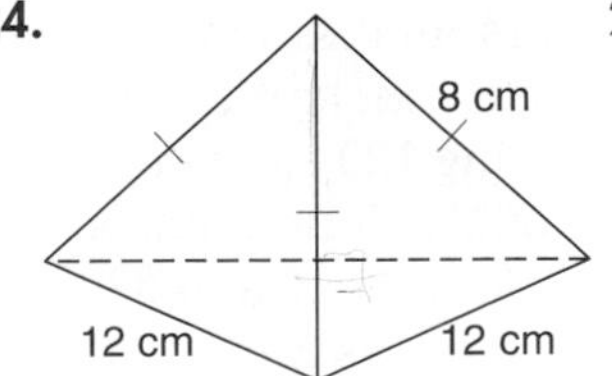

25. 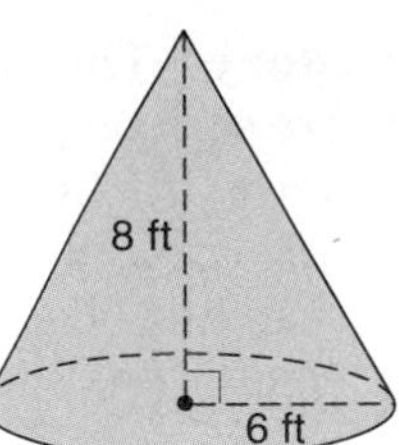

26. 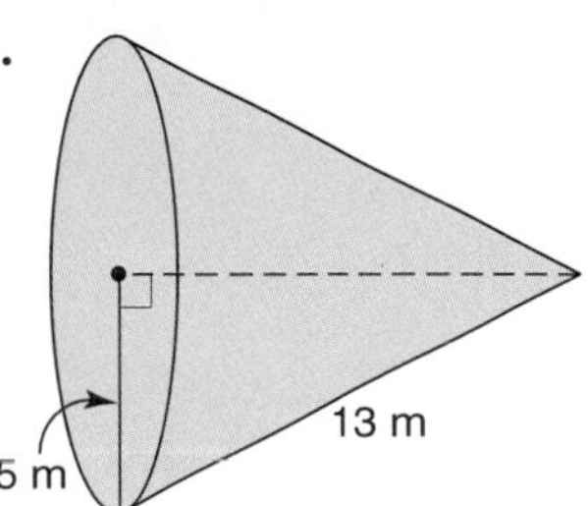

Find the surface area of each solid. Round to the nearest tenth.

27. 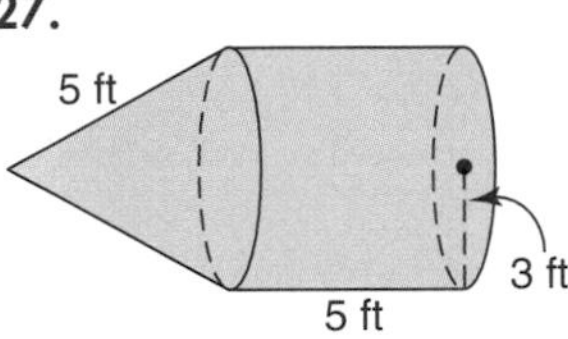

28. 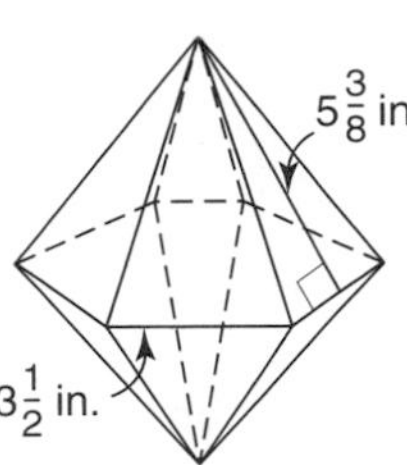

29. 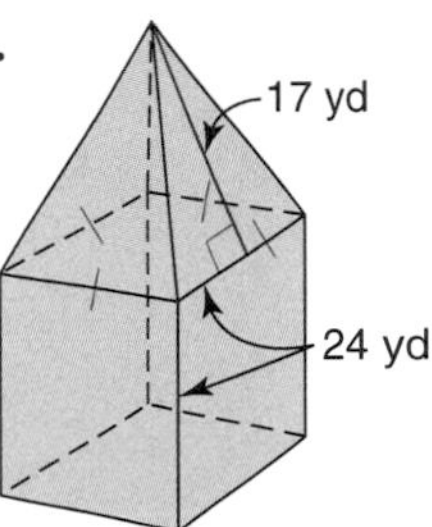

30. A regular pyramid has a slant height of 13 feet. The area of its square base is 100 square feet. Find its surface area.

31. In the given cube, A, B, and C are the vertices of the base of the pyramid with vertex D. If the edge of the cube is 8 units long, find the lateral area and the surface area of the pyramid.

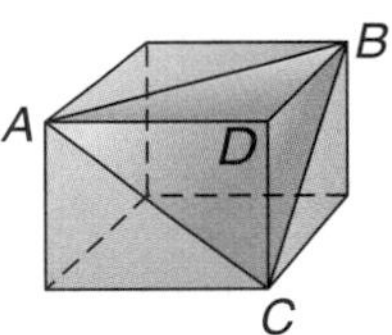

32. A *frustum* is the part of a solid that remains after the top portion of the solid has been cut off by a plane parallel to the base.

a. The figure below is a frustum of a regular pyramid. Find its lateral area.

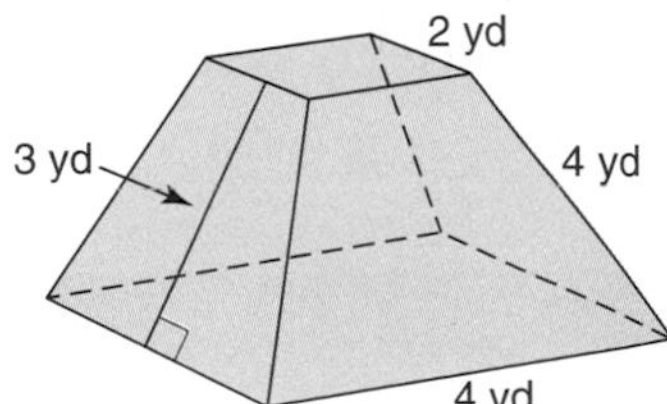

b. Find the surface area of the frustum of a cone shown below.

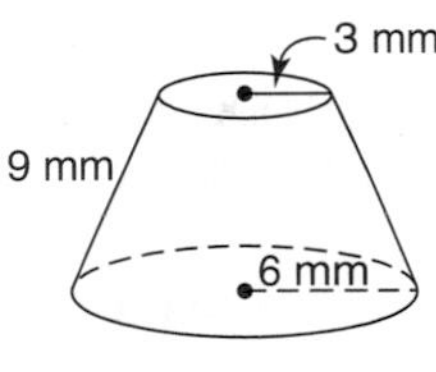

Critical Thinking

33. If you were to move the vertex of a right cone down the axis toward the center of the base, what would happen to the lateral area of the cone? Be as specific as possible and demonstrate the validity of your answer with a series of diagrams.

Applications and Problem Solving

34. **Dwellings** The largest tepee in the United States belongs to Dr. Michael Doss of Washington, D.C. Dr. Doss is a member of Montana's Crow Tribe. The tepee is a right cone with a diameter of 42 feet and a slant height of about 47.9 feet. How much canvas was used to cover the tepee?

35. **Art** In 1921, Italian immigrant Simon Rodia bought a home in Los Angeles, California, and began building conical towers in his backyard. The structures are made of steel mesh and cement mortar with no rivets, bolts, or welds. The first tower completed was the East Tower, which stands 55 feet high. The diameter of the base of the East Tower is $8\frac{1}{2}$ feet. Find the lateral area of the tower.

36. **History** The Cahokia Mounds stand close to East St. Louis, Illinois, where there was once the largest ancient city in America, north of Mexico. The inhabitants began building 120 earthen pyramids there around A.D. 600. The largest mound, Monk's Mound, has a height of 30.5 meters and a rectangular base with sides 216.6 and 329.4 meters long. Find the lateral area of Monk's Mound. (*Hint:* Is Monk's Mound a regular pyramid?)

Mixed Review

37. If the lateral area of a right rectangular prism is 784 square centimeters, its length is three times its width, and its height is twice its width, find its surface area. (Lesson 11–3)

38. From the views given below, draw a corner view. (Lesson 11–1)

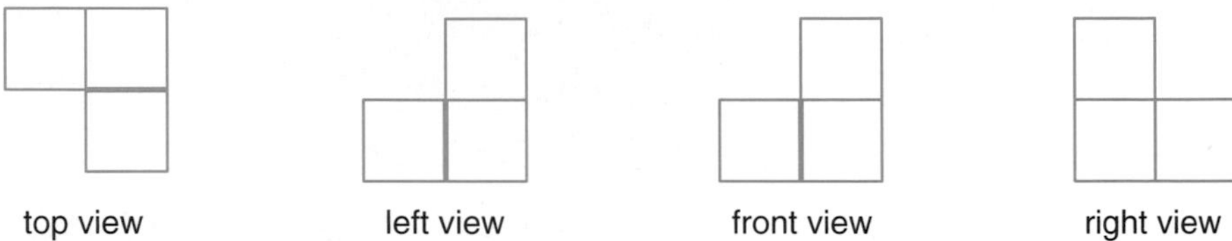

39. Find the area of a regular nonagon with an apothem 12.2 centimeters long and a side 10.4 centimeters long. (Lesson 10–5)

40. **SAT Practice** In a history class with 32 students, the ratio of girls to boys is 5 to 3. How many more girls are there than boys?

A 2 **B** 8 **C** 12 **D** 15 **E** 20

41. **Photography** Kandhi is taking pictures at the zoo. If a gnu is 4.3 feet tall, the film is 1 inch from the camera lens, and the camera lens is standing 5 feet from the gnu, how tall will the gnu's image be on the film? (Lesson 7–5)

42. Graph $(-6, -3)$ and $(1, 8)$. Then draw the lipe that passes through the points. Through which quadrants does the line pass? (Lesson 1–1)

Algebra

43. Translate the sentence *The quantity x is equal to 18 more than the square of b* into an equation.

44. Use elimination to solve the system of equations.

$2x + 4y = -14$ and $3x - 5y = 23$

For **Extra Practice**, see page 786.

SELF TEST

1. The corner view of a figure is given at the right. Draw the top, left, front, right, and back views of the figure. (Lesson 11–1)

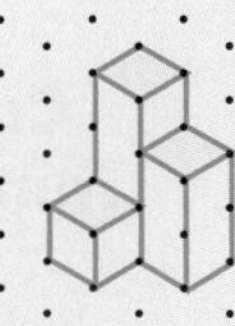

2. Use isometric dot paper to draw a corner view of a rectangular solid 2 units high, 4 units long, and 2 units wide. Then draw a net of the solid on rectangular dot paper. (Lesson 11–2)

Find the lateral area and the surface area of each solid below. Round to the nearest tenth. (Lesson 11–3)

3.

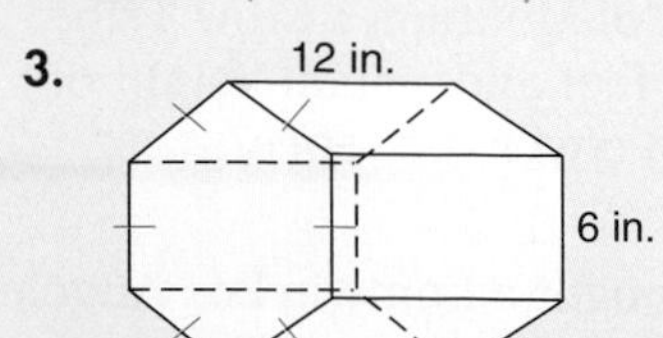

4.

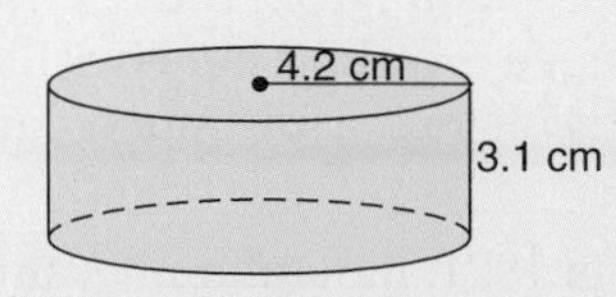

5. **Architecture** The Transamerica Tower in San Francisco is a regular pyramid with a square base that is 149 feet on each side and a height of 853 feet. Find its lateral area. (Lesson 11–4)

11–5 Volume of Prisms and Cylinders

What **YOU'LL LEARN**

- To find the volume of a right prism, and
- to find the volume of a right cylinder.

Why **IT'S IMPORTANT**

Prisms and cylinders are common shapes used for containers.

The measure of the amount of space that a figure encloses is the **volume** of the figure. Volume is measured in cubic units. You can use the skills you learned about creating solid figures from different views of the figure to investigate the volume of a right rectangular prism.

MODELING MATHEMATICS

Volume of a Right Rectangular Prism

Materials:  small cubes

- Make a model of a prism for the left, front, and top views given below.

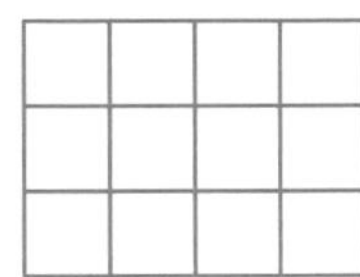
left view

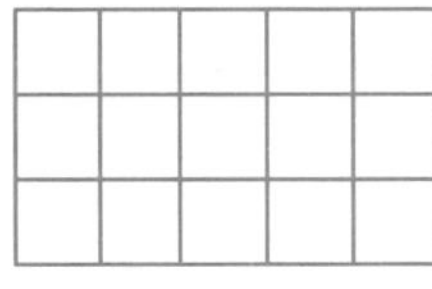
front view

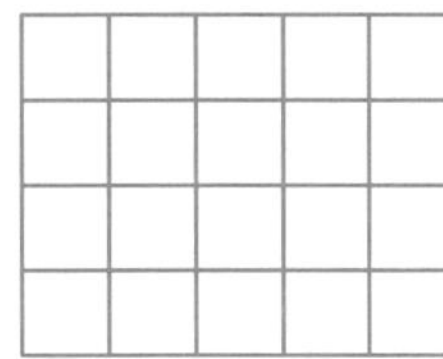
top view

- Disassemble the prism and count the number of cubes that make up the prism.
- Multiply the length, width, and height of the prism.

Your Turn

a. How do the number of cubes used to make the prism and the product of the length, width, and height of the prism compare?

b. Repeat the activity using a different prism and describe your results.

c. Write a formula for the volume of a right rectangular prism.

The Modeling Mathematics activity leads us to the following formula for the volume of a right prism.

Volume of a Right Prism	**If a right prism has a volume of *V* cubic units, a base with an area of *B* square units, and a height of *h* units, then *V* = *Bh*.**

Example  **Find the volume of the right triangular prism shown at the right.**

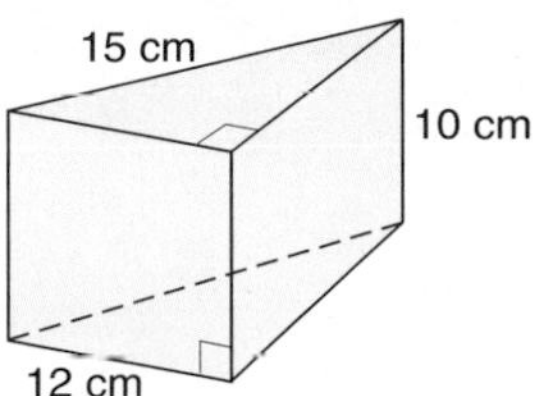

First, use the Pythagorean Theorem to find the height of the base of the prism.

$$a^2 + 12^2 = 15^2$$
$$a^2 = 15^2 - 12^2$$
$$a^2 = 81$$
$$a = 9$$

(continued on the next page)

Now find the volume of the prism.

$V = Bh$

$V = \left(\frac{1}{2} \cdot 12 \cdot 9\right) \cdot 10$

$V = 540 \text{ cm}^3$

The volume formula can be used to solve real-life problems that involve objects shaped like prisms.

Example 2

Sports

Before the athletes could hit the water for the 1996 Olympic Games, the new main competition pool at the Georgia Tech Aquatic Center had to be filled. The pool is 50 meters long and 25 meters wide. The adjustable bottom of the pool can be up to 3 meters deep for competition and as shallow as 0.3 meter deep for recreation. A liter of water has a volume of 0.001 cubic meter. Suppose the pool was filled to the recreational level and then the floor was lowered to the competition level. How much water had to be added to fill the pool?

Explore We know the dimensions of the pool when the floor is set for competition and when it is set for recreation. We also know the amount of water needed for a cubic meter of volume. We need to find the amount of water needed to fill the pool.

Plan First find the volume of water needed to fill the pool for competition and the volume of water needed to fill the pool for recreation. Find the difference in the two volumes. Then determine the number of liters of water needed to fill the difference in volume.

Solve When the pool is set for competition, it is a right rectangular prism 50 meters long, 25 meters wide, and 3 meters deep. The pool is 50 meters long, 25 meters wide, and 0.3 meters deep when it is set for recreation.

Competition level

$V = Bh$

$V = (50 \cdot 25)(3)$

$V = 3750 \text{ m}^3$

Recreation level

$V = Bh$

$V = (50 \cdot 25)(0.3)$

$V = 375 \text{ m}^3$

The difference in volumes is $3750 - 375$ or 3375 m^3.

A liter of water occupies 0.001 cubic meter of volume. Thus, $3375 \div 0.001$ or 3,375,000 liters of water were needed to fill the pool.

Examine Draw a model of the prism of water that needs to be added. Then find its volume and amount of water needed to confirm the solution.

Changing the dimensions of a prism affects the surface area and the volume of the prism. You can investigate the changes by using a spreadsheet.

EXPLORATION SPREADSHEETS

Remember that a spreadsheet is like a table in which you enter data and formulas to calculate desired quantities.

- Use column A to enter the length, column B to enter the width, and column C to enter the height of the prism.
- Enter the formula (2A1+ 2B1)*C1 + 2*(A1*B1) in cell D1. This formula will find the surface area of the prism. Copy the formula into the other cells in column D.
- Write a formula to find the volume of the prism. Enter the formula in the cells in column E.
- Use a spreadsheet to find the surface areas and volumes of prisms with the dimensions given below in columns A, B, and C.

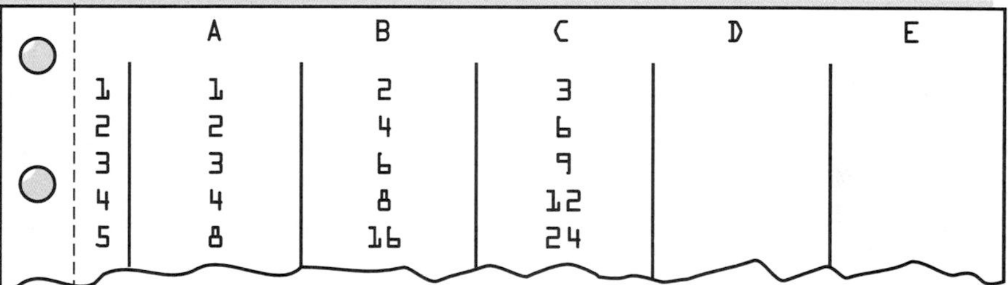

	A	B	C	D	E
1	1	2	3		
2	2	4	6		
3	3	6	9		
4	4	8	12		
5	8	16	24		

- Print the spreadsheet. If a printer is not available, copy the information into a table.

Your Turn

a. Compare the dimensions of prisms 1 and 2, prisms 2 and 4, and prisms 4 and 5.

b. Compare the surface areas of prisms 1 and 2, prisms 2 and 4, and prisms 4 and 5.

c. Compare the volumes of prisms 1 and 2, prisms 2 and 4, and prisms 4 and 5.

d. Write a statement about the change in the surface area and volume of a prism when the dimensions are doubled.

Like a right prism, the formula for the volume of a right cylinder is $V = Bh$. The area of the base is the area of a circle, πr^2.

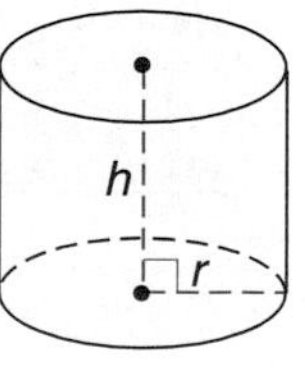

$$\text{volume} = \text{area of base} \times \text{height}$$
$$V = Bh$$
$$V = \pi r^2 \cdot h$$
$$\text{or } \pi r^2 h$$

Volume of a Right Cylinder	**If a right cylinder has a volume of *V* cubic units, a height of *h* units, and a radius of *r* units, then $V = \pi r^2 h$.**

Example 3 **Find the volume of each right cylinder. Round to the nearest tenth.**

a.

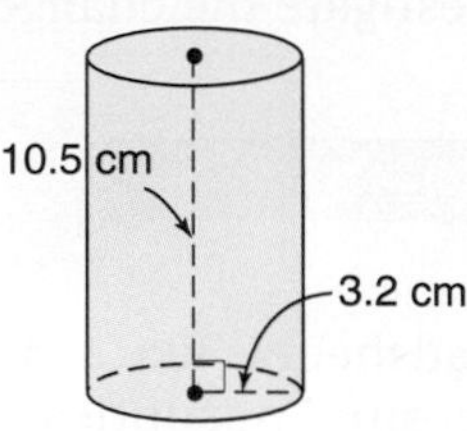

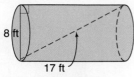

In this cylinder, the height h is 10.5 cm and the radius r is 3.2 cm.

$$V = \pi r^2 h$$
$$= \pi(3.2)^2(10.5)$$
$$= 107.52\pi$$
$$\approx 337.8 \text{ cm}^3$$

b.

8 ft

17 ft

The diameter of the base, the diagonal, and the lateral edge of the cylinder form a right triangle. Use the Pythagorean Theorem to find the height.

$$h^2 + 8^2 = 17^2$$
$$h^2 = 17^2 - 8^2$$
$$h^2 = 225$$
$$h = 15$$

Now find the volume.

$$V = \pi r^2 h$$
$$= \pi(4)^2(15)$$
$$= 240\pi$$
$$\approx 754.0 \text{ ft}^3$$

CHECK FOR UNDERSTANDING

Communicating Mathematics

Study the lesson. Then complete the following.

1. **Research** to find three items that are sold by volume. State the units of measure used in their sale.
2. **Describe** the relationship between the volumes of two geometric solids that are the same size and shape.
3. **Compare and contrast** the procedures for finding the volume of a prism and finding the volume of a cylinder.
4. **You Decide** Shalena says there are 3 cubic feet in a cubic yard. Kiki thinks there are 27 cubic feet in a cubic yard. Who is correct and why?

5. Make a model of a prism for the top, front, and right views given below.

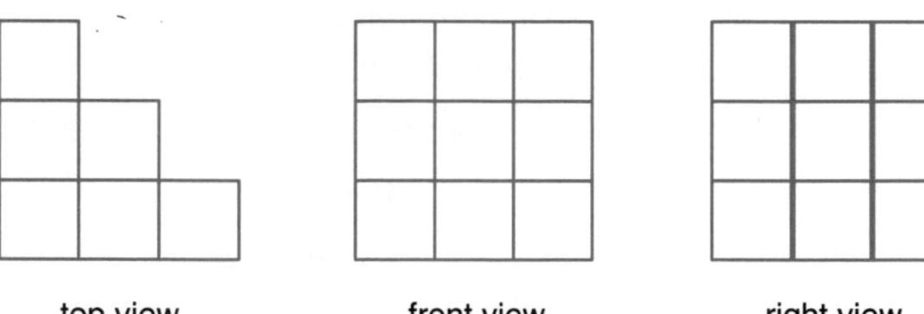

   a. Count the number of cubes used to build the figure.
   b. Compute the volume using the formula for the volume of a prism.
   c. Does the number of cubes you counted verify the formula? If not, why not?

Guided Practice **Find the volume of each right prism or right cylinder. Round to the nearest tenth.**

6. 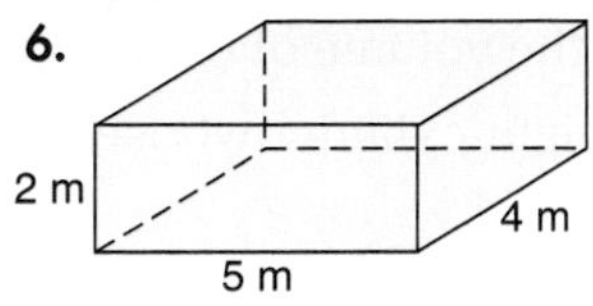

7. 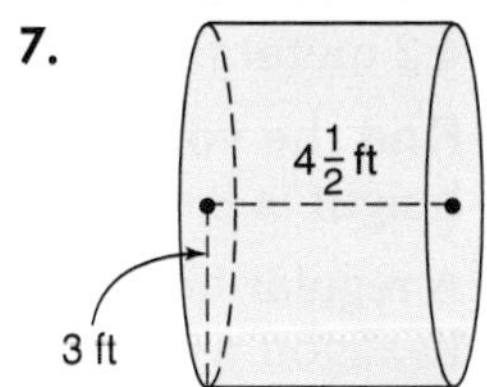

8. 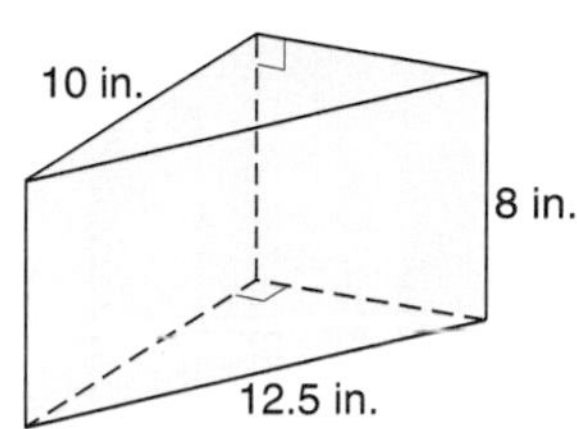

9. What is the volume of a right cylinder whose diameter is 12 yards long and has a height of 15 yards? Round to the nearest tenth.

10. Find the volume of a right hexagonal prism that has a height of 10 centimeters and whose base is a regular hexagon with sides of 6 centimeters. Round to the nearest cubic centimeter.

11. Find the volume of the partial right cylinder shown below. Round to the nearest tenth.

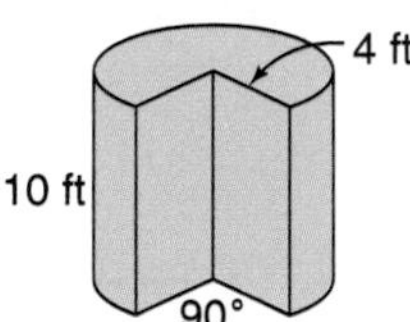

12. A right prism is formed by folding this net. Find its volume.

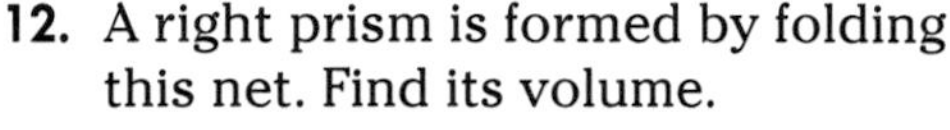

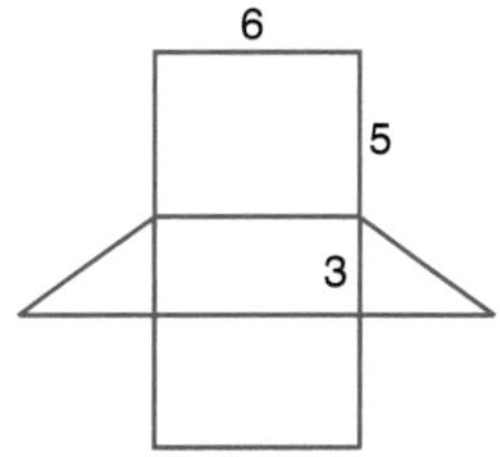

13. **Meteorology** On April 1, 1960, the world's first weather satellite, *TIROS 1,* was launched. *TIROS 1* orbited Earth taking pictures as it passed over sites. The weather satellites in service today are often placed in orbits so that they are over the same area all of the time. One of these satellites is cylindrical with a diameter of 7 feet and a height of 12 feet. What is the volume available for carrying weather instruments and other hardware?

EXERCISES

Practice **Find the volume of each right prism or right cylinder. Round to the nearest tenth.**

14. 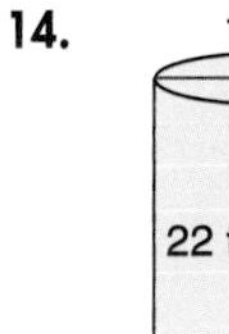

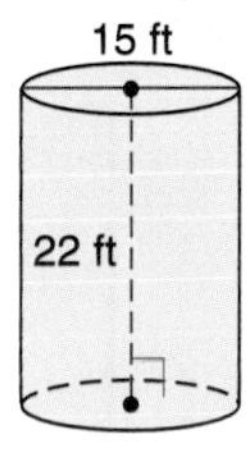

15. 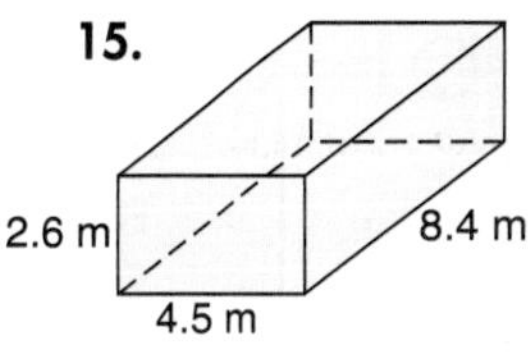

16. 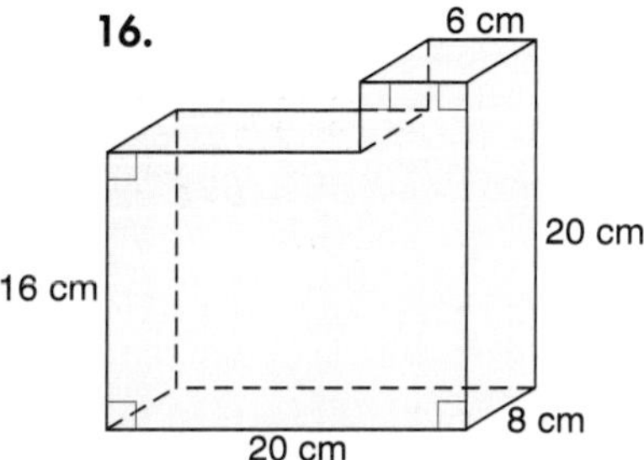

17. 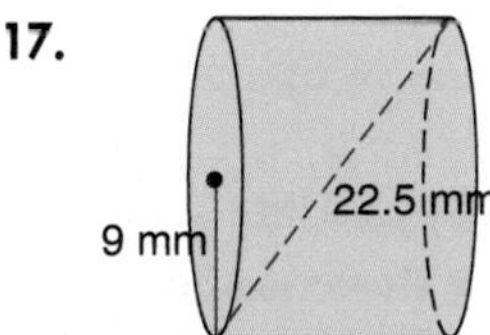

18. 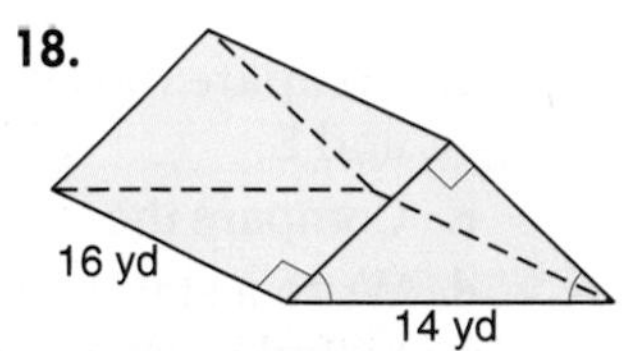

19. 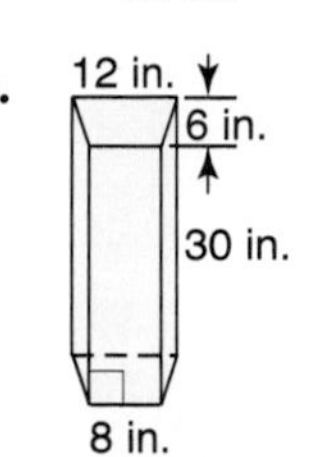

20. What is the volume of a cube that has an edge of 6 meters long?

21. The base of a right prism has an area of 25 square meters. The prism is 4.2 meters high. Find the volume of the prism.

22. Find the volume of a right cylinder whose diameter is 12 centimeters and height is 8 centimeters.

23. A regular hexagonal prism has a length of 40 feet and a base that is a regular hexagon with sides 5 feet long. Find the volume of the prism.

24. A cylinder has a height of 4 meters and a volume of about 452.4 cubic meters. What is the diameter of the base of the cylinder?

25. Find the length of a lateral edge of a right prism with a volume of 648 cubic inches and a base whose area is 36 square inches.

Find the volume of each solid.

26.

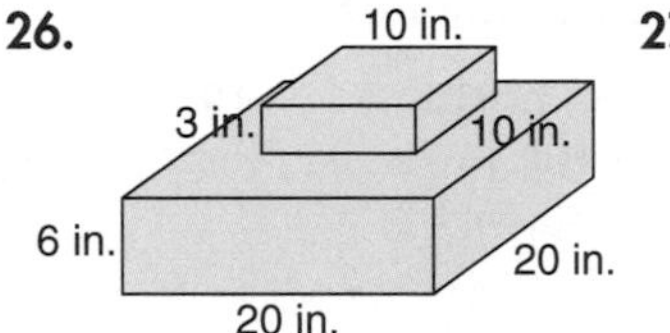

27.

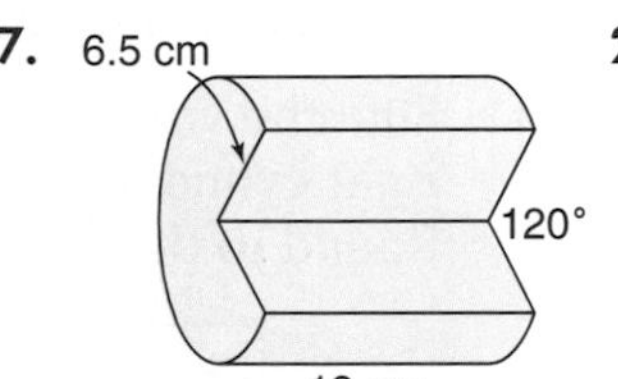

28.

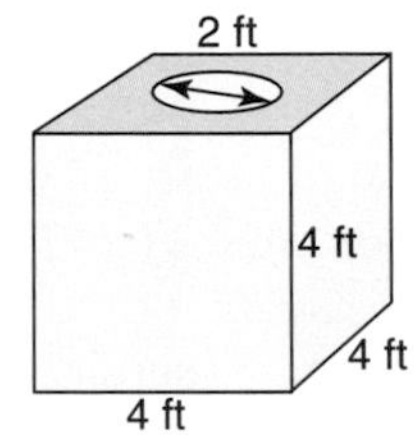

Find the volume of the solid formed by each net.

29.

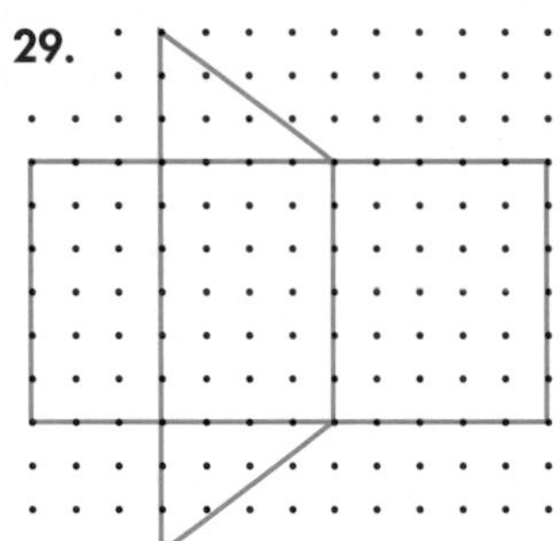

30.

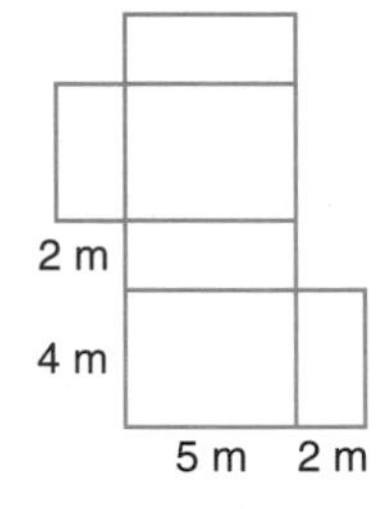

31.

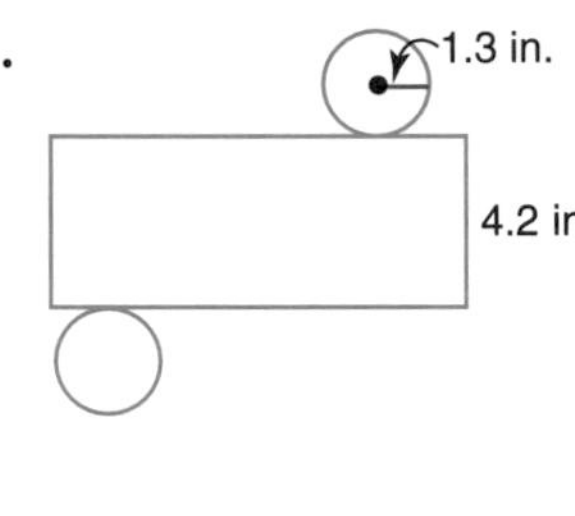

32. **Technology** Use a spreadsheet to investigate the change in the surface area and volume of a right cylinder when the dimensions are changed. Use column A to name cylinders A, B, C, D, and E. Use column B to enter the radius of the base, column C for the height, column D for the lateral area, and column E for the volume. Complete the spreadsheet for the values given below.

Cylinder	Radius	Height	Lateral Area	Volume
A	1	2		
B	1	4		
C	1	8		
D	2	2		
E	4	2		

a. How do the dimensions of cylinders A and B, B and C, A and D, and D and E compare?

b. Compare the lateral areas of cylinders A and B, B and C, A and D, and D and E.

c. Compare the volumes of cylinders A and B, B and C, A and D, and D and E.

d. Write a statement about the change in the lateral area and volume of a cylinder when its height is doubled, and a statement about the change in the lateral area and volume of a cylinder when its radius is doubled.

Critical Thinking

33. *True* or *false*: The volume of every oblique prism or oblique cylinder is equal to the area of its base times it height. Justify your answer.

Applications and Problem Solving

34. **Engineering** Machinists make parts for complicated pieces of equipment for manufacturing. While making a part for a weaving machine, a machinist drilled a hole in a block of copper as shown at the right.

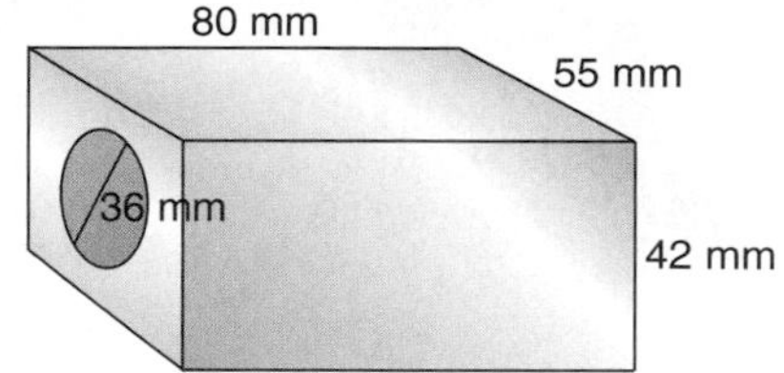

   a. Find the volume of the resulting solid.
   b. The *density* of a substance is its mass per unit of volume. At room temperature, the density of copper is 8.9 grams per cubic centimeter. What is the mass of this block?

35. **Music** To play a concertina, you push and pull the end plates and press the keys. Then the air pressure causes vibrations of the metal reeds that make the notes. When fully expanded, the concertina at the right is 36 inches from one endplate to the other. If the concertina is compressed, it is 7 inches from one endplate to the other. Find the volume of air in the instrument when it is fully expanded and when it is compressed. (*Hint*: There is no air in the endplates.)

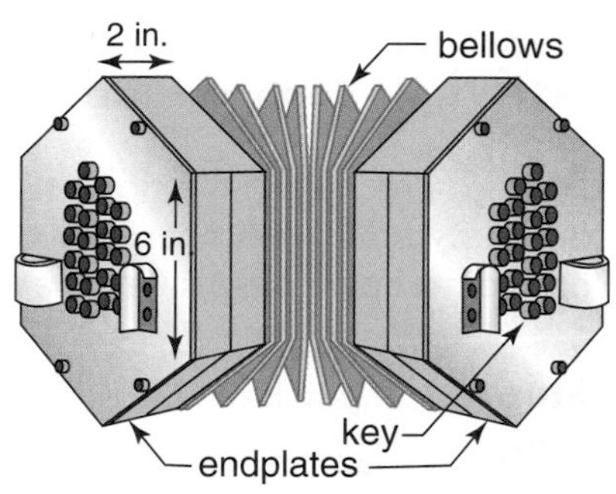

36. **Sports** The diving pool at the Georgia Tech Aquatic Center was used for the springboard and platform diving competitions of the 1996 Olympic Games. The pool measures 78 feet wide, 78 feet long, and 17 feet deep. If it takes about $7\frac{1}{2}$ gallons of water to fill one cubic foot of space, approximately how many gallons of water are needed to fill the diving pool?

37. **Automotive Engineering** A car muffler reduces the noise that the exhaust makes when it is released. Study the diagram of the muffler at the right. If each of the louver tubes has a diameter of 3 inches, find the total volume of the louver tubes.

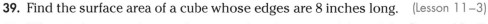

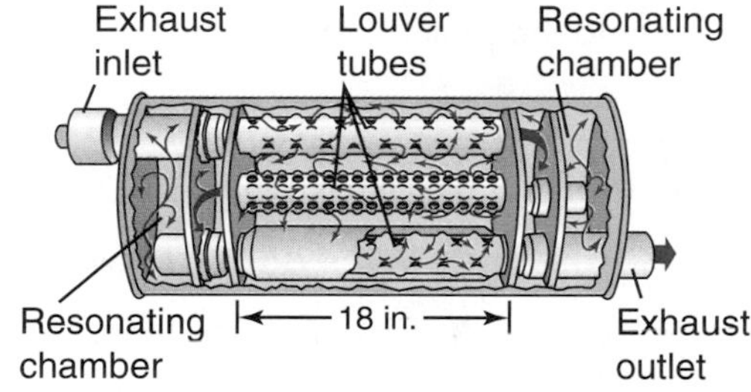

Mixed Review

38. **Highway Management** A building used to store salt is pyramid-shaped with height 32 feet and slant height 56 feet. If the salt pile is cone-shaped, what is the greatest radius of the base of the pile? (Lesson 11–4)

39. Find the surface area of a cube whose edges are 8 inches long. (Lesson 11–3)

40. What is the area of a circular rug with a diameter of 4 yards? (Lesson 10–5)

41. **SAT Practice** **Grid-in** Forty-eight trees were planted in a park. If $\frac{1}{4}$ of them are evergreens and $\frac{1}{2}$ of the remaining trees are oaks, how many oaks were planted?

42. *True* or *false*: All diameters of a circle are congruent. (Lesson 9–1)

43. If $\triangle RST \sim \triangle MNO$, name the proportional parts of the triangles. (Lesson 7–3)

44. Determine if $\overleftrightarrow{AB} \parallel \overleftrightarrow{CD}$ for $A(6, 0)$, $B(8, 4)$, $C(0, -3)$, and $D(2, 1)$. (Lesson 3–3)

Algebra

For **Extra Practice**, see page 786.

45. Factor $54c^2d$ completely. Do not use exponents.

46. Find the degree of $x^3y + 5x^4y^2$.

11–6A Investigating Volumes of Pyramids and Cones

A Preview of Lesson 11–6

Materials: 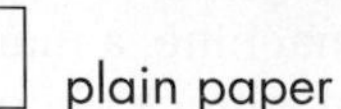 plain paper 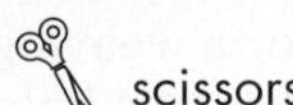 scissors 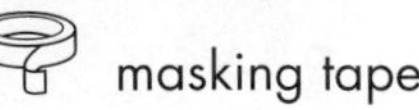 masking tape

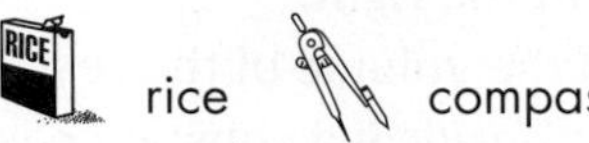 rice compass 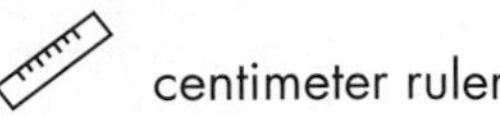 centimeter ruler

You can compare the volumes of prisms and pyramids and the volumes of cylinders and cones by making a model of each solid and filling them with rice.

Activity **Compare the volumes of a prism and pyramid with the same height and base.**

Step 1 The nets at the right will fold into a prism and a pyramid with open tops. Draw the two nets on plain paper using a centimeter ruler.

Step 2 Cut out the nets and fold on the dashed lines.

Step 3 Fully tape the edges together to form models of the solids.

Step 4 Estimate how much larger the volume of the prism is than the volume of the pyramid.

Step 5 Fill the pyramid with rice. Then pour this rice into the prism. Repeat until the prism is filled.

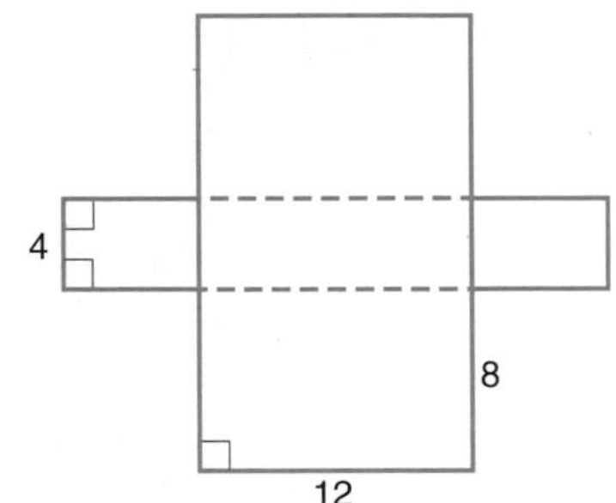

All measures are centimeters.

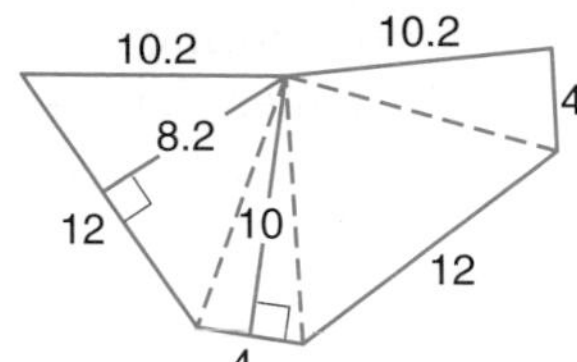

Write

1. How many pyramids of rice did it take to fill the prism?
2. What do you know about the areas of the bases of the prism and the pyramid?
3. Compare the heights of the prism and the pyramid.
4. Write a formula for the volume of a pyramid.

Model

5. The nets below fold into a cylinder and a cone. Repeat the activity above using these nets.

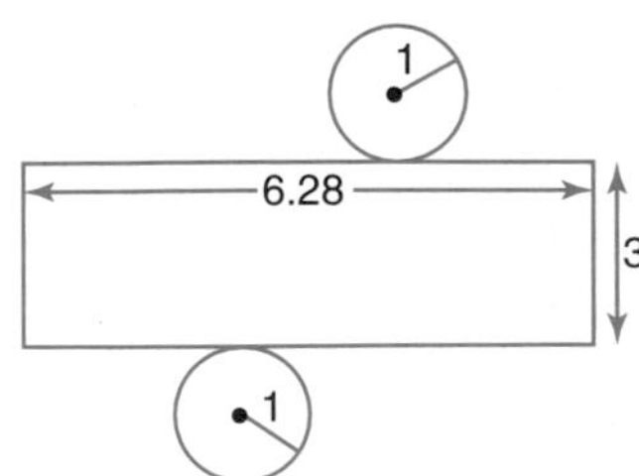

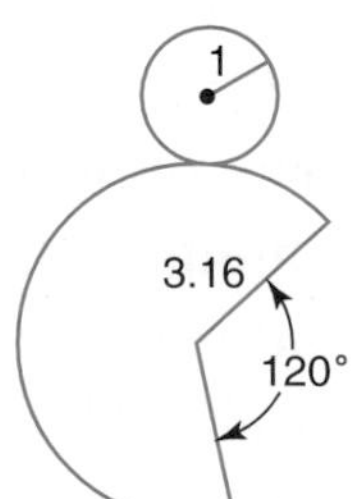

Write

6. How many cones of rice did it take to fill the cylinder?
7. What do you know about the areas of the bases of the cone and the cylinder?
8. Compare the heights of the cone and the cylinder.
9. Write a formula for the volume of a cone.

11-6

Volume of Pyramids and Cones

What YOU'LL LEARN

- To find the volume of a pyramid, and
- to find the volume of a circular cone.

Why IT'S IMPORTANT

You can use pyramids and cones in solving problems about history, geology, and engineering.

Real World APPLICATION

History

The American Indians who lived in the plains followed herds of buffalo, which they hunted for food and hides for making shelters and clothes. When hunting, they lived in tepees made by stretching buffalo skins over poles. The tepees were many-sided pyramids that resembled cones. The American Heritage Center at the University of Wyoming was designed with a conical shape to represent a celebration of the culture of the American Indians.

The heating and cooling systems chosen for the American Heritage Center had to be able to handle the amount of space that had to be heated or cooled. A formula for the volume of a cone was used since the building is shaped like a cone. *You will find the volume of the American Heritage Center in Example 1.*

Study the figures at the right. The cone and cylinder have the same base and height, and the pyramid and prism have the same base and height. As you can see, the volume of the cone is less than the volume of the cylinder, and the volume of the pyramid is less than the volume of the prism. In the Modeling Mathematics on page 614, you discovered that the ratio of the volumes in each case is 1:3. This relationship is stated in the formulas below.

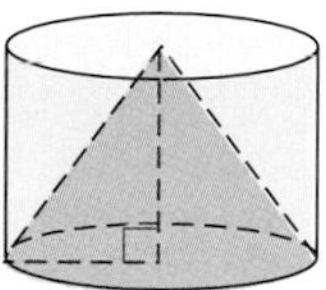

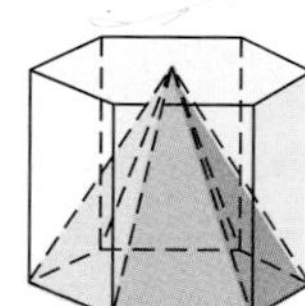

Volume of a Right Circular Cone	**If a right circular cone has a volume of V cubic units, a height of h units, and the area of the base is B square units, then $V = \frac{1}{3}Bh$.**

Volume of a Right Pyramid	**If a right pyramid has a volume of V cubic units, a height of h units, and the area of the base is B square units, then $V = \frac{1}{3}Bh$.**

Example 1

Mechanical Engineering

Refer to the application at the beginning of the lesson. The American Heritage Center has a height of 77 feet. The area of the base is about 38,000 square feet. Find the volume of air that the heating and cooling systems would have to be able to accommodate. Round to the nearest tenth.

$$\begin{aligned} V &= \frac{1}{3}Bh \\ &\approx \frac{1}{3}(38{,}000)(77) \\ &\approx 975{,}333.3 \text{ ft}^3 \end{aligned}$$

The heating and cooling systems would have to accommodate about 975,333 cubic feet of air.

You can use trigonometry to help solve many problems involving the volume of solids.

Example 

INTEGRATION

Trigonometry

Find the volume of each solid. Round to the nearest tenth.

a.

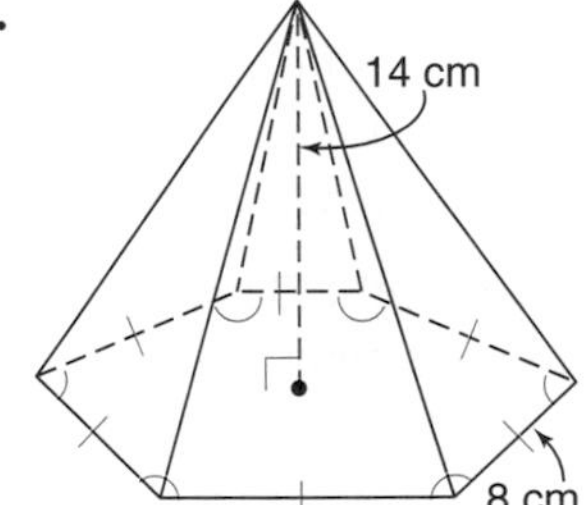

$$\begin{aligned} V &= \frac{1}{3}Bh \\ &= \frac{1}{3}\left(\frac{1}{2}Pa\right)h \\ &= \frac{1}{3}\left(\frac{1}{2} \cdot 48 \cdot 4\sqrt{3}\right)14 \\ &= 448\sqrt{3} \\ &\approx 776.0 \text{ cm}^3 \end{aligned}$$

b.

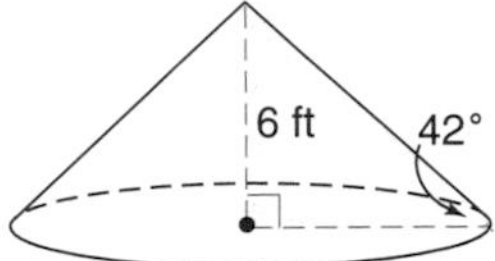

Use trigonometry to find the radius of the base.

$$\begin{aligned} \tan 42^\circ &= \frac{6}{r} \\ r &= \frac{6}{\tan 42^\circ} \\ r &\approx 6.7 \end{aligned}$$

Now find the volume.

$$\begin{aligned} V &= \frac{1}{3}Bh \\ &\approx \frac{1}{3}\pi(6.7)^2(6) \\ &\approx 89.78\pi \\ &\approx 282.1 \text{ ft}^3 \end{aligned}$$

If you solve problems using calculators without rounding, your answers may vary slightly.

In this chapter so far, only the formulas for the volume of right prisms, cylinders, pyramids, and cones have been presented. Do you think the same formulas can be applied to all oblique solids?

Study the photograph at the right. It shows two matching stacks of index cards. The stack on the left represents a right prism and the stack on the right represents an oblique prism. Since the stacks have the same number of cards, with all cards the same size and shape, the two prisms represented by the stacks have the same volume. This observation was first made by Cavalieri, an Italian mathematician of the seventeenth century. It is known as **Cavalieri's Principle**.

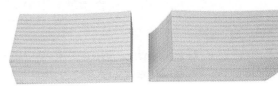

Cavalieri's Principle	**If two solids have the same height and the same cross-sectional area at every level, then they have the same volume.**

As a result of Cavalieri's Principle, if a prism has a base with an area of B square units and a height of h units, then its volume is Bh cubic units, whether it is right or oblique. The volume formulas for cylinders, cones, and pyramids hold whether they are right or oblique.

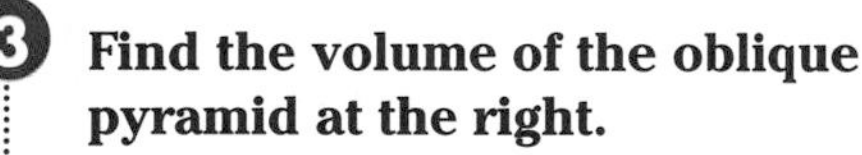

Example 3 **Find the volume of the oblique pyramid at the right.**

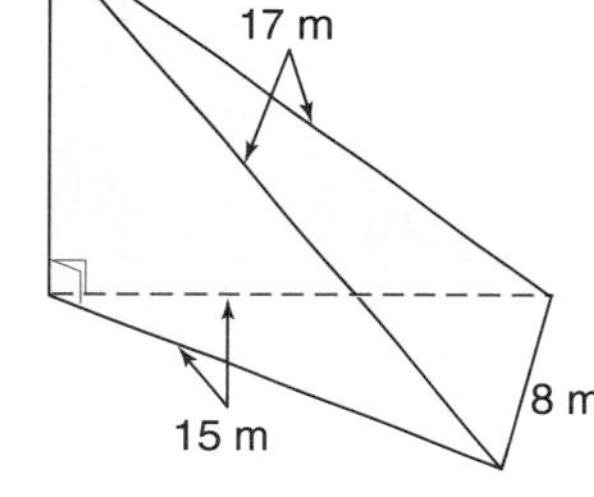

First find the height of the pyramid.

$$h^2 + 15^2 = 17^2$$
$$h^2 = 64$$
$$h = 8$$

Now find the area of the base.

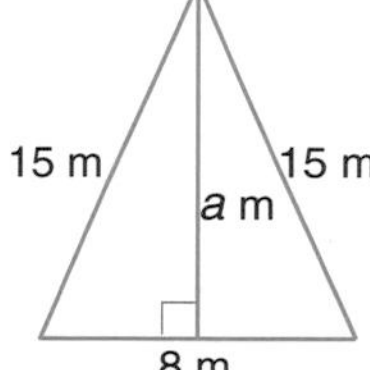

$$a^2 + 4^2 = 15^2$$
$$a^2 = 209$$
$$a = \sqrt{209}$$

Finally, find the volume.

$$V = \frac{1}{3}Bh$$
$$= \frac{1}{3}\left(\frac{1}{2} \cdot 8 \cdot \sqrt{209}\right)8$$
$$\approx 154.2$$

The volume is about 154.2 cubic meters.

CHECK FOR UNDERSTANDING

Communicating Mathematics

Study the lesson. Then complete the following.

1. **Explain** how the volume of a cone is related to that of a cylinder with the same altitude and a base congruent to that of the cone.
2. Suppose you found the cross section of an oblique prism and a right prism that had the same length, width, and height. How would the cross sections compare? Explain.

3. **Describe** how you would find the volume of an empty ice-cream cone.

Guided Practice

Find the volume of each pyramid or cone. Round your answer to the nearest tenth.

4.
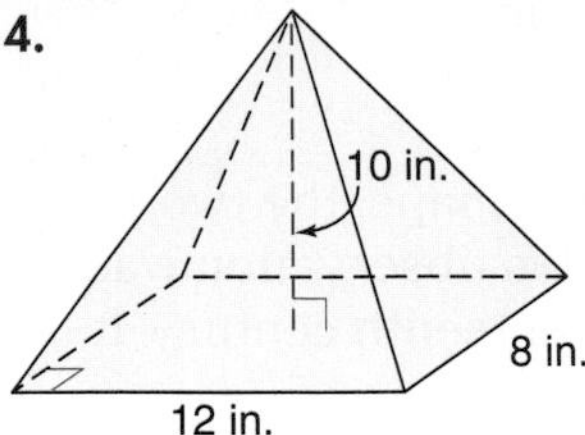

5. 6.
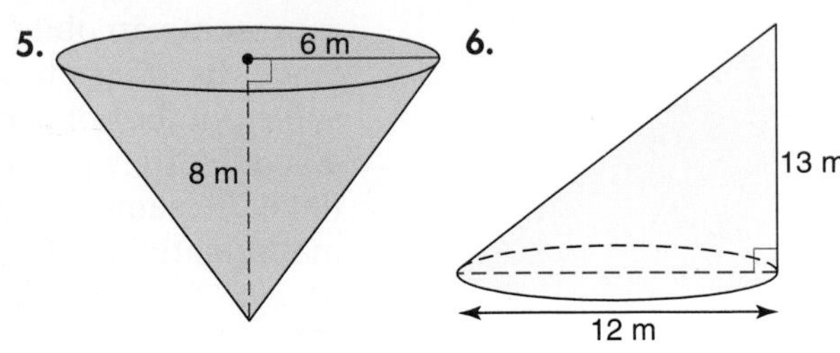

7. Find the volume of the solid shown at the right.

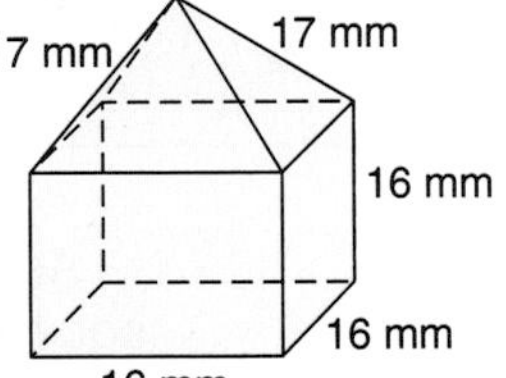

8. What is the volume of a right cone with a slant height of 20 feet and a 60° angle at the vertex of the cone?

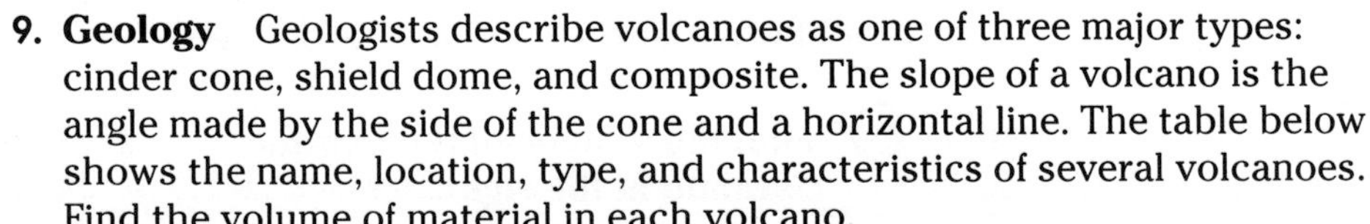

9. **Geology** Geologists describe volcanoes as one of three major types: cinder cone, shield dome, and composite. The slope of a volcano is the angle made by the side of the cone and a horizontal line. The table below shows the name, location, type, and characteristics of several volcanoes. Find the volume of material in each volcano.

Data Update For more information on volcanoes, visit: **www.geometry.glencoe.com**

Volcano	Location	Type	Characteristics
Mauna Loa	Hawaii, U.S.A.	shield dome	9100 m tall, 97 km across at base
Fuji	Honshu, Japan	composite	3776 m tall, slope of 30°
Paricutín	Michoacán, Mexico	cinder cone	410 m tall, 33° slope
Vesuvius	Campania, Italy	composite	120 across at base, 1303 m tall

EXERCISES

Practice

Find the volume of each pyramid or cone. Round to the nearest tenth.

10.
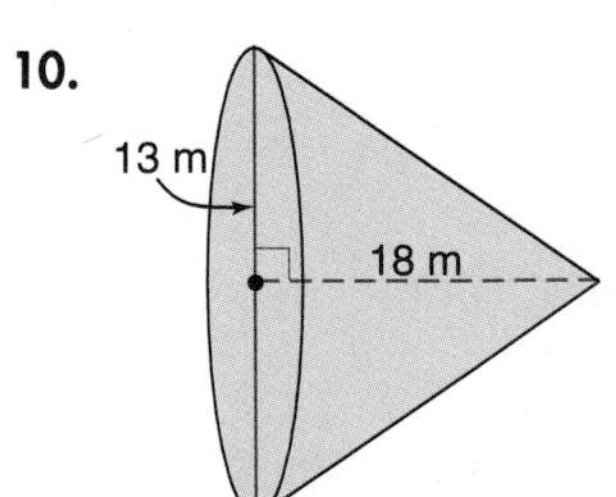

11.
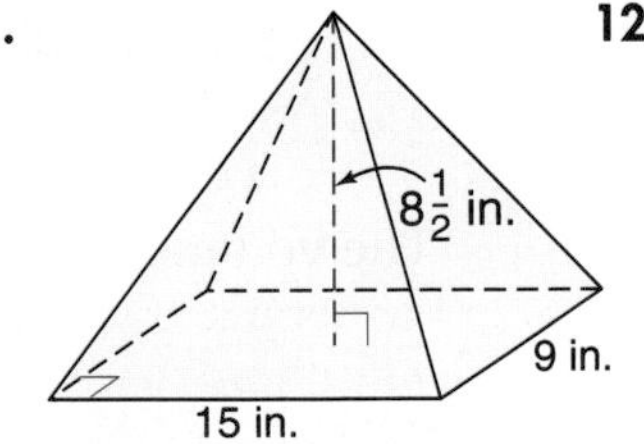

12.
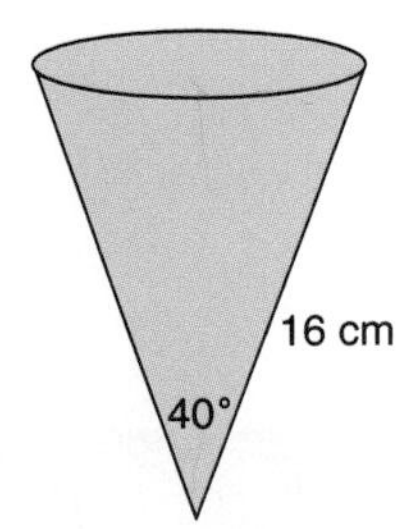

13. 14 ft, 22 ft

14.
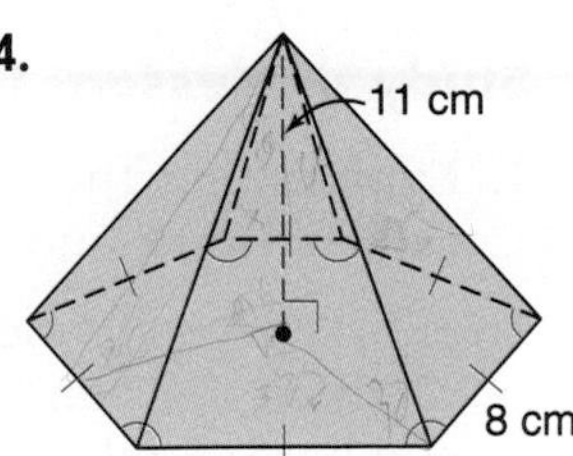

15.
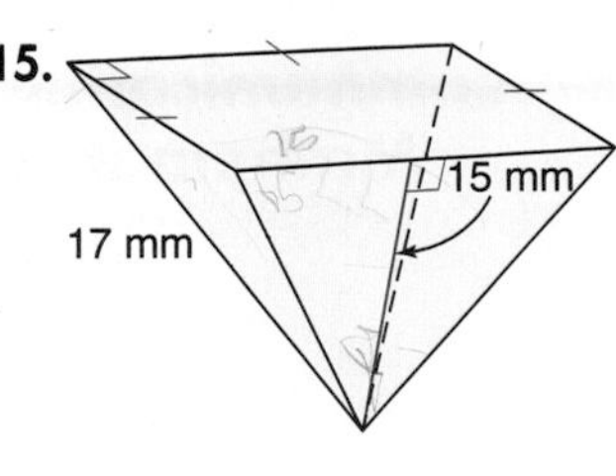

16.

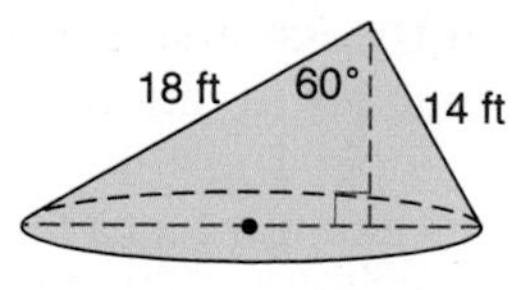

17.

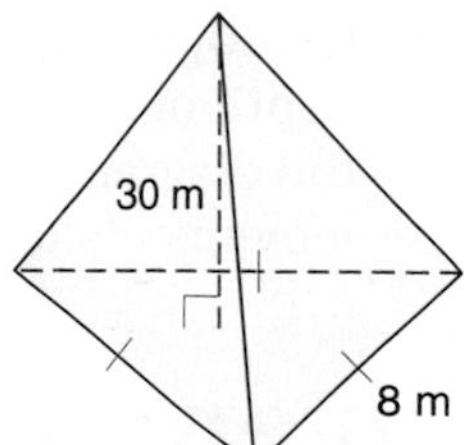

18.

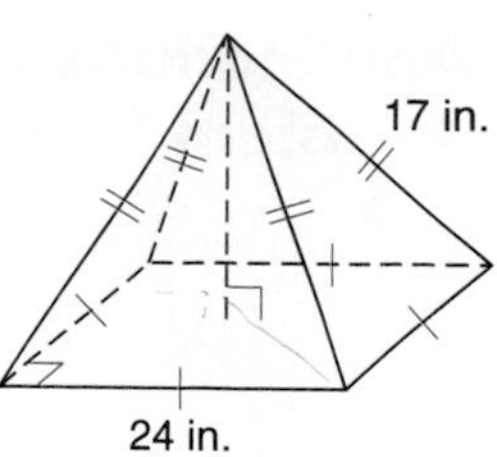

Find the volume of each solid. Round to the nearest tenth.

19.

12 units
10 units
18 units

20.

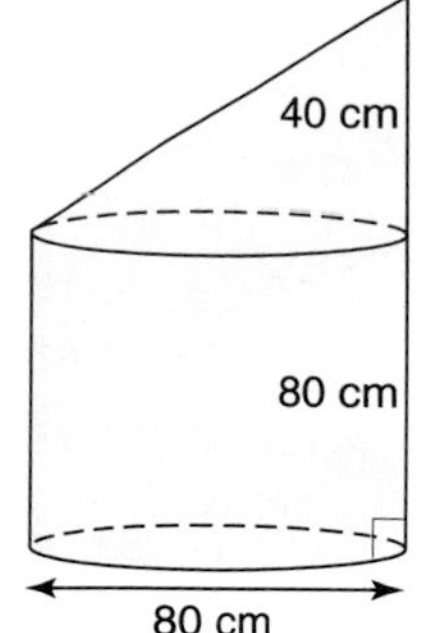

21.

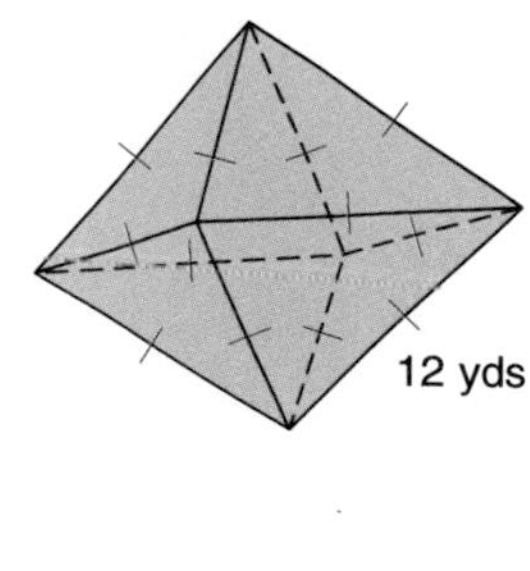

22. A pyramid has a volume of 729 cubic units. If the area of the base is 243 square units, what is the height of the pyramid?

23. One right circular cone is inside a larger right circular cone. The two cones have the same axis, the same vertex, and the same height. Find the volume of the space between the cones if the diameter of one cone is 6 inches, the diameter of the other is 9 inches, and the height of both is 5 inches.

24. A regular tetrahedron is a pyramid with four congruent equilateral triangles as its only faces. If the length of one edge of a tetrahedron measures 12 units, find the volume of the tetrahedron.

Programming

25. The graphing calculator program finds the volume of a rectangular pyramid or of a circular cone rounded to the nearest cubic unit.

```
PROGRAM:VOLUME
:Disp "CHOOSE:"
:Disp "1: PYRAMID"
:Input "2: CONE", C
:If C=2
:Goto 2
:Input "LENGTH = ", L
:Input "WIDTH = ", W
:Input "HEIGHT = ", H
:round(L*W*H/3,0)→V
:Disp "VOLUME = ", V
:Stop
:Lbl 2
:Input "RADIUS = ", R
:Input "HEIGHT = ", H
:round(π*R²*H/3,0)→V
:Disp "VOLUME = ", V
:Stop
```

Use the program to find the volume of each solid.

a. a pyramid that is 20 feet tall with a base 16 feet long and 14 feet wide

b. a cone with a diameter of 44 inches and a height of 54 inches

c. a pyramid with a base 3 meters by 5 meters and a height of 7 meters

d. How could you change the program to find the volume of a prism or a cylinder?

e. Would the program have to be altered to find the volume of an oblique pyramid or cone? Explain.

Critical Thinking

26. Write a ratio comparing the volumes of the solids shown at the right. Justify your answer.

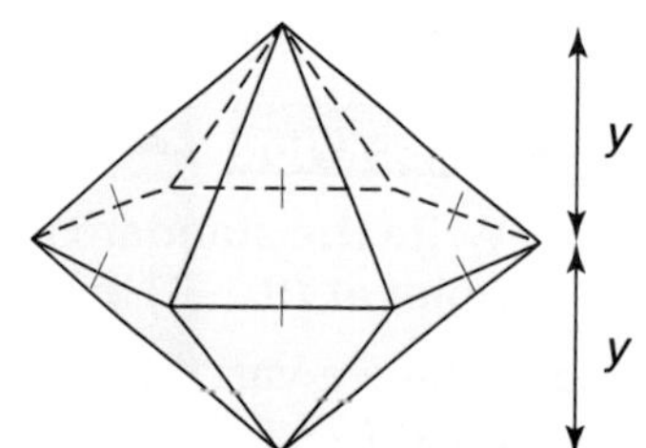

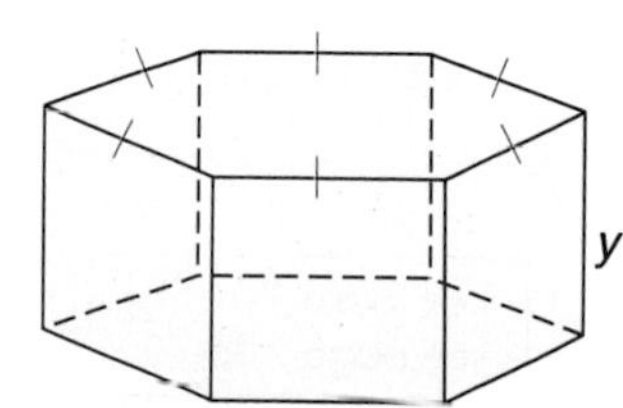

Applications and Problem Solving

27. **Geology** The way that the ice inside glaciers moves depends on the depth of the ice. The changes in different areas of a glacier are shown in the diagram below. How does the volume of the ice change as each of the movements takes place? Explain.

Glacier Movement

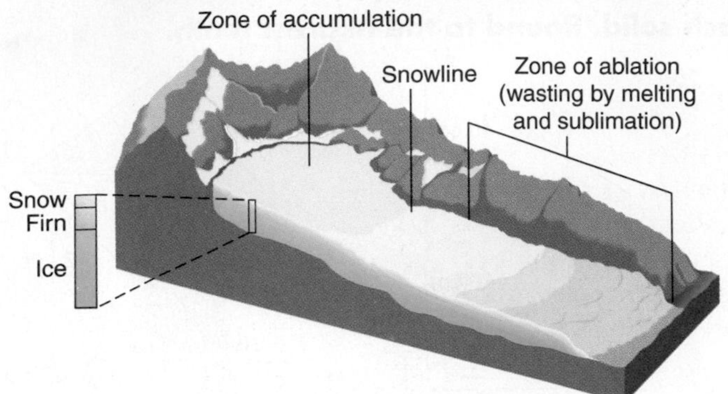

In the top layers of a glacier, ice slides along the base.

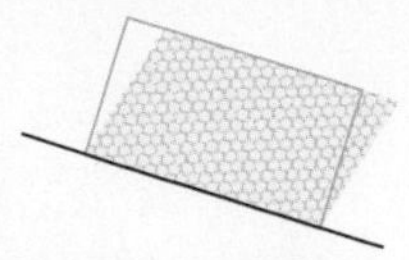

At lower levels, grains of snow and ice move within the glacier.

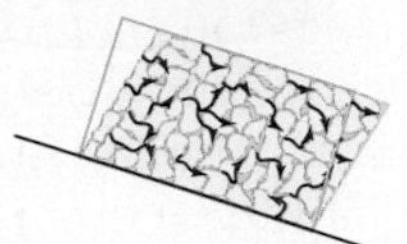

Pressure can cause ice to move by melting and refreezing.

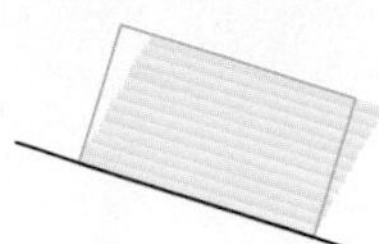

Sheets of ice can slip in planes to move a glacier.

28. **Architecture** In an attempt to rid Florida's Lower Sugarloaf Key of mosquitoes, Richter Perky built a tower to attract bats. The Perky Bat Tower is a frustum of a pyramid with a square base. Each side of the base of the tower is 15 feet long, the top is a square with sides 8 feet long, and the tower is 35 feet tall. How many cubic feet of space does the tower supply for bats?

Mixed Review

29. **Engineering** The oil drilling platform called the Statfjord B is located off the coast of Norway in the North Sea. The base of the platform is made up of 24 concrete cylinders or cells. Twenty of the cells are used for oil storage. The pillars that support the platform deck rest on the four other cells. Find the total volume of the storage cells. (Lesson 11–5)

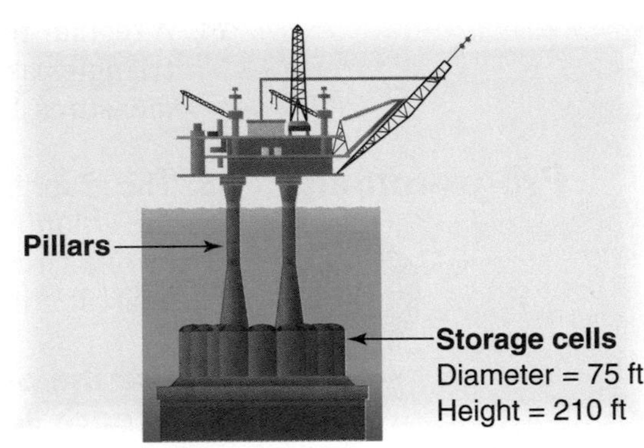

30. Find the surface area of a cylindrical water tank that is 8 meters tall and has a diameter of 8 meters. (Lesson 11–3)

31. A rhombus with a 56-inch diagonal has an area of 1344 in^2. Find the length of a side of the rhombus. (Lesson 10–4)

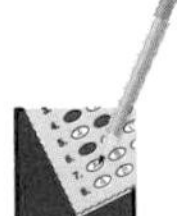

32. **ACT Practice** What is the complete factorization of $16a^3 - 54b^3$?
    - **A** $(2a - 3b)(4a^2 + 6ab + 9b^2)$
    - **B** $2(2a - 3b)(4a^2 + 6ab + 9b^2)$
    - **C** $2(2a - 3b)(4a^2 - 6ab + 9b^2)$
    - **D** $2(2a + 3b)(4a^2 + 6ab + 9b^2)$
    - **E** $(2a - 3b)(4a^2 - 6ab + 9b^2)$

33. The sides of a right triangle have measures of $x + 8$, $x + 1$, and $x + 9$ units. Find the value of x. (Lesson 8–1)

Algebra

34. Write the standard form of an equation of the line that passes through the point at $(9, -3)$ and has a slope of -1.

35. Define a variable, write an inequality, and solve *if nine times a number is at most 108.*

For **Extra Practice**, see page 786.

11-7 Surface Area and Volume of Spheres

What **YOU'LL LEARN**

- To recognize and define basic properties of spheres,
- to find the surface area and the volume of a sphere.

Why **IT'S IMPORTANT**

You can use spheres to solve problems involving geography, astronomy, and housing.

History

According to legend, Christopher Columbus set out to prove that Earth was round when he discovered America. But in reality, Greek mathematicians not only knew that Earth was round sixteen centuries earlier, they had calculated its circumference!

Eratosthenes of Cyrene (275–194 B.C.) was director of the Alexandrian Library, which was the academic center of the ancient world. Eratosthenes used the position of the Sun and the distance between two cities to calculate that the circumference of Earth was 250,000 stades or about 39,375 km. This is remarkably close to the current accepted value of 40,075 km.

Earth is roughly a sphere. To visualize a sphere, consider infinitely many congruent circles in space, all with the same point for their center. Considered together, all these circles form a sphere. In space, a sphere is the set of all points that are a given distance from a given point called its **center**.

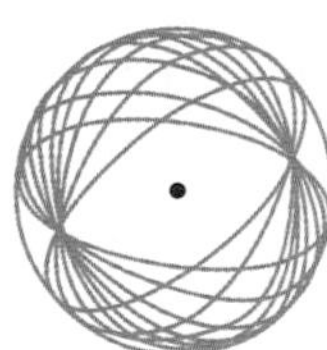

There are several special segments and lines related to spheres.

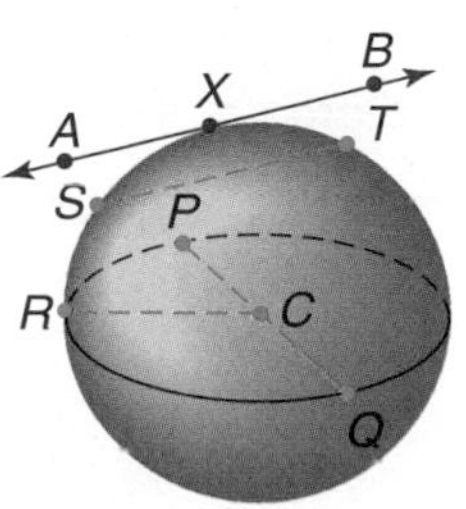

- A segment whose endpoints are the center of the sphere and a point on the sphere is a **radius** of the sphere. In the figure, $\overline{CR}$, $\overline{CP}$, and $\overline{CQ}$ are radii.
- A **chord** of a sphere is a segment whose endpoints are points on the sphere. In the figure, $\overline{TS}$ and $\overline{PQ}$ are chords.
- A chord that contains the sphere's center is a **diameter** of the sphere. In the figure, $\overline{PQ}$ is a diameter.
- A **tangent** to a sphere is a line that intersects the sphere in exactly one point. In the figure, $\overleftrightarrow{AB}$ is tangent to the sphere at X.

A plane can intersect a sphere in a point or in a circle.

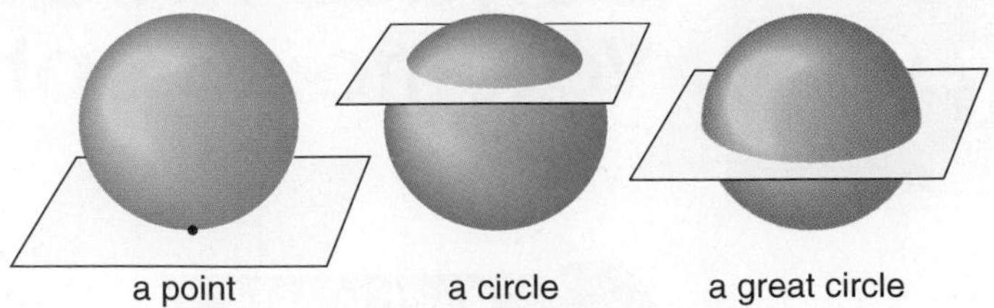

LOOK BACK

You can review great circles of spheres in Lesson 3-6.

When a plane intersects a sphere so that it contains the center of the sphere, the intersection is called a **great circle**. A great circle has the same center as the sphere, and its radii are also radii of the sphere. On the surface of a sphere, the shortest distance between any two points is the length of the arc of a great circle passing through those two points. Each great circle separates a sphere into two congruent halves called **hemispheres**.

You can use a model to investigate the surface area of a sphere.

MODELING MATHEMATICS

Surface Area of a Sphere

Materials: polystyrene ball  scissors

 tape  straight pins

- Cut the ball along a great circle. Trace around the edge to draw a circle. Then cut out the paper circle.

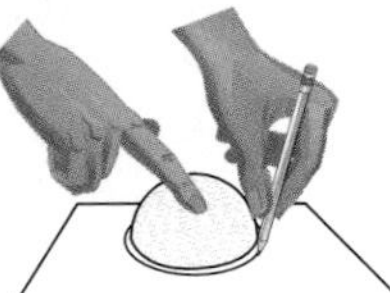

- Fold the circle into eighths. Then unfold and cut the eight pieces apart. Tape the pieces back together in the arrangement shown at the right.

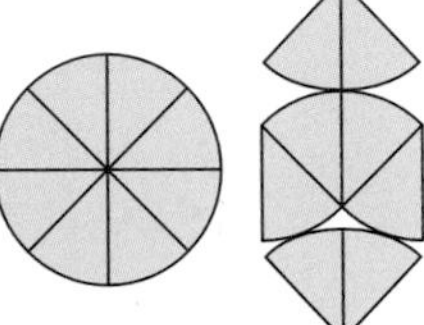

- Use tape or a straight pin to put the two pieces of the ball back together. Then use pins to attach the pattern to the sphere.

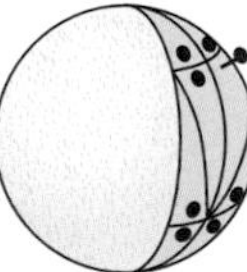

Your Turn

a. Estimate how much of the total surface of the sphere was covered by the pattern.

b. What is the area of the pattern in terms of r, the radius of the sphere?

c. Write a formula for the surface area of the sphere.

The Modeling Mathematics activity leads us to the formula for the surface area of a sphere.

Surface Area of a Sphere	**If a sphere has a surface area of T square units and a radius of r units, then $T = 4\pi r^2$.**

Many sports use balls that are shaped like spheres. When the balls are made, the manufacturer needs to find its surface area to determine the amount of material needed.

Example 1

Sports

Find the surface area of each sports ball described.

a. An NCAA basketball has a radius of $4\frac{3}{4}$ inches.

b. An Olympic-sized volleyball has a circumference of 27 inches.

a. $T = 4\pi r^2$

$= 4\pi\left(4\frac{3}{4}\right)^2$

$\approx 283.5 \text{ in}^2$ *Use a calculator.*

The surface area of an NCAA basketball is about 283.5 square inches.

b. Use the formula for circumference to find the radius of a volleyball.

$C = 2\pi r$

$27 = 2\pi r$

$4.30 \approx r$ *Use a calculator.*

Now find the surface area.

$T = 4\pi r^2$

$\approx 4\pi(4.30)^2$

≈ 232.4 *Use a calculator.*

The surface area of an Olympic-sized volleyball is about 232.4 square inches.

The *Associated Press'* Most Outstanding Players in Women's College Basketball

1999, Chamique Holdsclaw (Tennessee)

1998, Chamique Holdsclaw (Tennessee)

1997, Kara Wolters (Connecticut)

1996, Jennifer Rizzotti (Connecticut)

1995, Rebecca Lobo (Connecticut)

You can relate finding a formula for the volume of a sphere to finding the volume of a right pyramid and the surface area of a sphere.

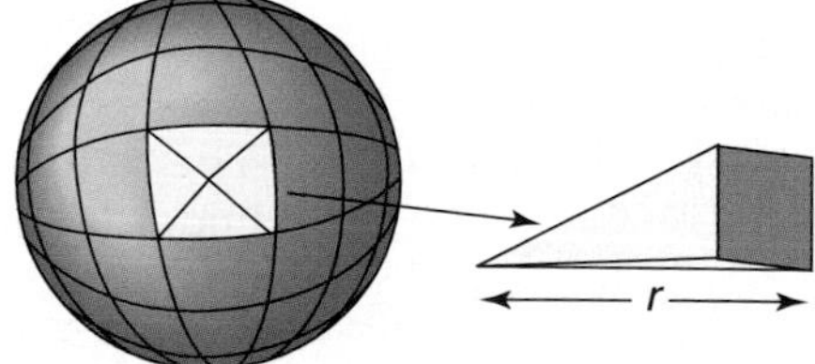

Picture separating the space inside a sphere into infinitely many small pyramids, all with their vertices located at the center of the sphere. This is shown above. Observe that the height of these very small pyramids is equal to the radius r of the sphere. The sum of the areas of all the pyramid bases equals the surface of the sphere.

Each pyramid has a volume of $\frac{1}{3}Bh$, where B is the area of its base and h is its height. The volume of the sphere is equal to the sum of the volumes of all the infinitely many small pyramids. Thus, the volume V of the sphere can be represented as follows.

$$V = \frac{1}{3}B_1h_1 + \frac{1}{3}B_2h_2 + \frac{1}{3}B_3h_3 + \ldots + \frac{1}{3}B_nh_n$$

$$= \frac{1}{3}B_1r + \frac{1}{3}B_2r + \frac{1}{3}B_3r + \ldots + \frac{1}{3}B_nr$$

$$= \frac{1}{3}r(B_1 + B_2 + B_3 + \ldots + B_n)$$

$$= \frac{1}{3}r(4\pi r^2)$$ *Replace $(B_1 + B_2 + B_3 + \ldots + B_n)$, which is the surface area of the sphere, with $4\pi r^2$.*

$$= \frac{4}{3}\pi r^3$$

Volume of a Sphere	**If a sphere has a volume of V cubic units and a radius of r units, then $V = \frac{4}{3}\pi r^3$.**

Example 2 **Find the volume of each sphere. Round to the nearest tenth.**

a.

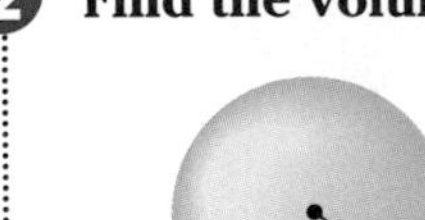

$$V = \frac{4}{3}\pi r^3$$
$$= \frac{4}{3}\pi(18)^3$$
$$\approx 24{,}429.0 \text{ cm}^3$$

b.

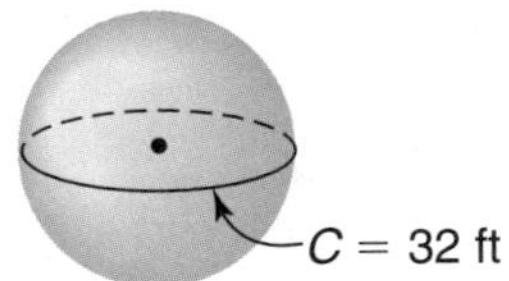

Find the radius of the sphere.

$$C = 2\pi r$$
$$32 = 2\pi r$$
$$\frac{16}{\pi} \text{ ft} = r$$

Now find the volume.

$$V = \frac{4}{3}\pi r^3$$
$$= \frac{4}{3}\pi\left(\frac{16}{\pi}\right)^3$$
$$\approx 553.3 \text{ ft}^3$$

FYI

Archimedes thought this relationship was so important that he asked to have the figure shown engraved on his tombstone.

Over 2200 years ago, the mathematician Archimedes discovered a relationship between the volume of a cylinder and an inscribed sphere.

The radius of the sphere is r units. So the radius of the cylinder is also r units. The height of the cylinder is the diameter of the sphere, $2r$.

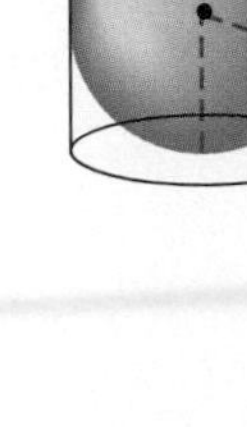

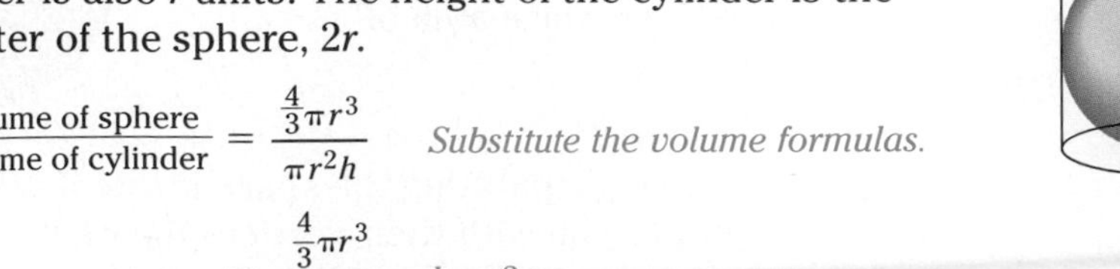

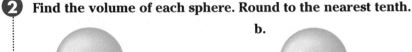

$$\frac{\text{volume of sphere}}{\text{volume of cylinder}} = \frac{\frac{4}{3}\pi r^3}{\pi r^2 h}$$ *Substitute the volume formulas.*

$$= \frac{\frac{4}{3}\pi r^3}{\pi r^2(2r)}$$ *$h = 2r$*

$$= \frac{\frac{4}{3}\pi r^3}{2\pi r^3}$$

$$= \frac{\frac{4}{3}}{2} \text{ or } \frac{2}{3}$$ *Simplify.*

The ratio of the volume of a sphere to that of a cylinder in which it is inscribed is $\frac{2}{3}$.

CHECK FOR UNDERSTANDING

Communicating Mathematics

Study the lesson. Then complete the following.

1. **Draw and label a diagram** of a sphere with a great circle.
2. Is a sphere a polyhedron? Justify your answer.
3. **Explain** how the formula for the volume of a sphere was developed in this lesson.
4. **Compare and contrast** squares and cubes, and circles and spheres.

5. **Assess Yourself** Describe some models of spheres that you encounter in your life.

Guided Practice

Describe each object as a model of a *circle*, a *sphere*, or *neither*.

6. baseball
7. jelly jar

Determine whether each statement is *true* or *false*.

8. All chords of a sphere are diameters.
9. All diameters of a sphere are chords.
10. If a great circle of a sphere is congruent to a great circle of another sphere, then the spheres are congruent.

In the figure, *C* is the center of the sphere, and plane $\mathcal{B}$ intersects the sphere in circle *R*.

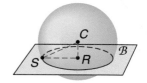

11. Suppose $CR = 4$ and $SR = 3$. What is the length of a radius of the sphere?
12. If the radius of the sphere is 13 units and the radius of $\odot R$ is 12 units, find CR.

13. **Meteorology** The figure below shows the differences meteorologists say there are between rain and drizzle. Assume that rain and drizzle fall in spherical drops. Write a statement or two about the surface area and volume of drops of rain and drizzle.

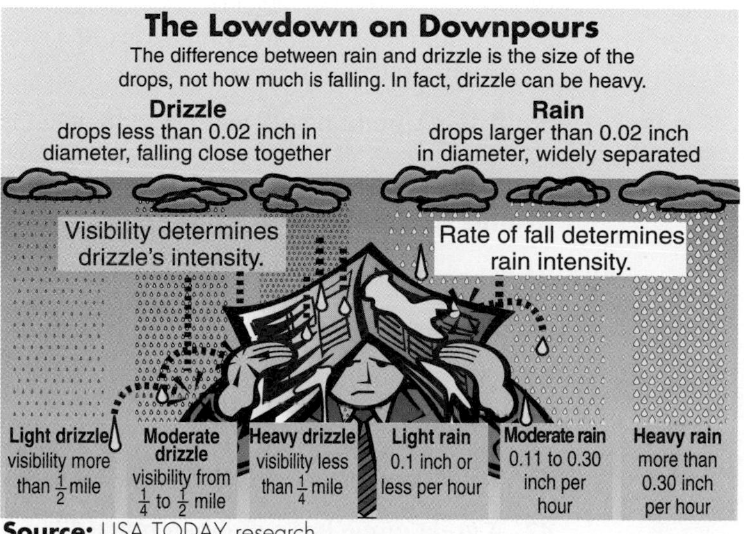

The Lowdown on Downpours
The difference between rain and drizzle is the size of the drops, not how much is falling. In fact, drizzle can be heavy.
Drizzle
drops less than 0.02 inch in diameter, falling close together
Rain
drops larger than 0.02 inch in diameter, widely separated
Visibility determines drizzle's intensity.
Rate of fall determines rain intensity.
Light drizzle visibility more than $\frac{1}{2}$ mile
Moderate drizzle visibility from $\frac{1}{4}$ to $\frac{1}{2}$ mile
Heavy drizzle visibility less than $\frac{1}{4}$ mile
Light rain 0.1 inch or less per hour
Moderate rain 0.11 to 0.30 inch per hour
Heavy rain more than 0.30 inch per hour

Source: USA TODAY research

Find the surface area and volume of each sphere described below. Round to the nearest tenth.

14. A radius is 12 centimeters long.

15. One of its great circles has an area of 113.04 square feet.

16. The volume of a sphere is $\frac{32}{3}\pi$ cubic meters. Find the surface area of the sphere.

EXERCISES

Practice

Describe each object as a model of a *circle*, a *sphere*, or *neither*.

17. compact disc
18. the Moon
19. shot put
20. football
21. hot-air balloon
22. soda can

Determine whether each statement is *true* or *false*.

23. In a sphere, all the radii are congruent.

24. A sphere is contained in a plane.

25. A sphere's longest chord will always pass through the sphere's center.

26. A radius of a great circle of a sphere is also a radius of the sphere.

27. All of the chords of a sphere are congruent.

28. Two spheres may intersect in exactly one point.

29. In a sphere, two different great circles may intersect in exactly one point.

30. If two spheres intersect, their intersection may be a circle.

31. The intersection of two spheres with congruent radii may be a great circle.

In the figure, *O* is the center of the sphere, and plane *C* intersects the sphere in ⊙*R*.

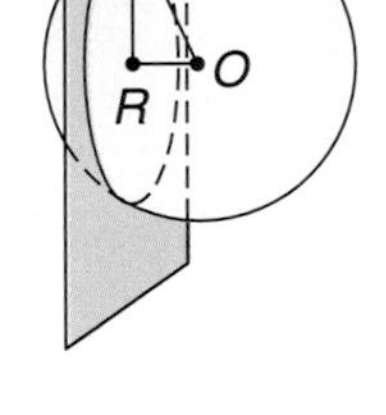

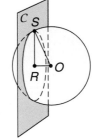

32. If $OR = 9$ and $SR = 12$, find OS.

33. Given $OS = 16$ and $RS = 12.8$, find OR.

34. If the radius of the sphere is 15 units and the radius of the circle is 10 units, what is OR?

35. If O and R are distinct points, is ⊙R a great circle?

36. If M is a point on ⊙R and $OS = 18$, what is OM?

Find the surface area and volume of each sphere described below. Round to the nearest tenth.

37. The radius is 25 inches long.

38. The radius of a great circle is 14.5 centimeters.

39. The diameter is 450 meters.

40. A radius is $6\frac{1}{2}$ inches long.

41. The diameter of a great circle is 3.4 meters.

42. A great circle has a circumference 43.96 centimeters.

43. What is the volume of a sphere if its surface area is 16π cm^2?

44. A sphere is circumscribed about a cube with a volume of 1728 cubic centimeters. What is the volume of the sphere?

45. Find the ratio of the radii of two spheres if the surface area of one is 4 times the surface area of the other.

Critical Thinking

46. Desta, a pilot, is leaving for a flight. Her friend Mei needs to fly to another city. Desta offers to take her. Mei says that she doesn't want to make Desta go out of her way. Desta replies that no matter where Mei is going it will not take her out of her way. Explain how this could be true.

47. A plane slices a sphere 5 centimeters from its center. The sphere has a radius of 13 centimeters. What is the area of the slice to the nearest hundredth?

Applications and Problem Solving

48. **Astronomy** Imagine that the lines of longitude and latitude are projected on the sky as if Earth was surrounded by a giant sphere with the stars on it. This sphere is called the *celestial sphere*.
    a. Is a longitude line in the celestial sphere a great circle?
    b. Is a latitude line in the celestial sphere a great circle?
    c. Can you see an entire hemisphere of the celestial sphere?

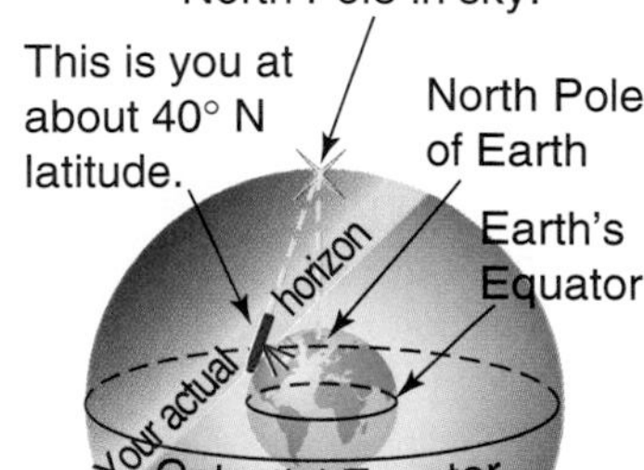

49. **Food** The first ice-cream cone was made at the World's Fair in St. Louis in 1904 when an ice-cream seller ran out of cups. Suppose a sugar cone for ice cream is 10 centimeters deep and has a diameter of 4 centimeters. A scoop of ice cream with a diameter of 4 centimeters rests on the top of the cone.
    a. If all the ice cream melts into the cone, will the cone overflow?
    b. If the cone does not overflow, what percent of the cone will be filled?

50. **Dwellings** The traditional shelter of the Algonquian people of northern Canada, Greenland, Alaska, and eastern Siberia was the igloo. Igloos were made of hard-packed snow cut into blocks 2 to 3 feet long and 1 to 2 feet wide. The blocks were put together in a spiral that grew smaller at the top to form a hemispherical dome. An air vent at the top allowed fresh air to enter and the long entrance trapped cold air to keep the interior warm. Find the surface area and volume of the living area in the igloo shown above.

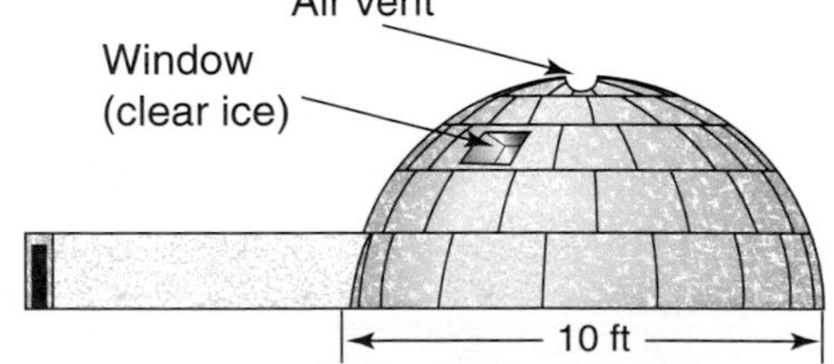

51. **Architecture** Buckminster Fuller was an engineer and inventor who worked to create new home designs that maximized space with a minimum amount of materials. Research Fuller's designs in a reference book or on the Internet. How does his work relate to the information in this lesson?

Mixed Review

52. Refer to the figure at the right. (Lesson 11–6)

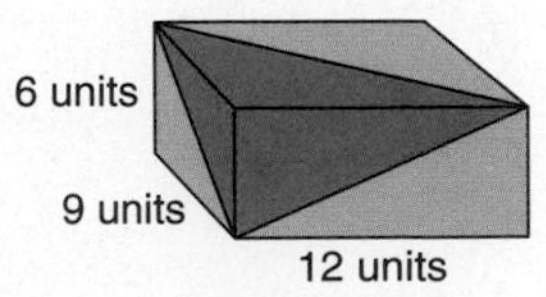

   a. Find the volume of the pyramid cut from the rectangular solid shown.
   b. What is the ratio of the volume of the pyramid to the volume of the rectangular solid?

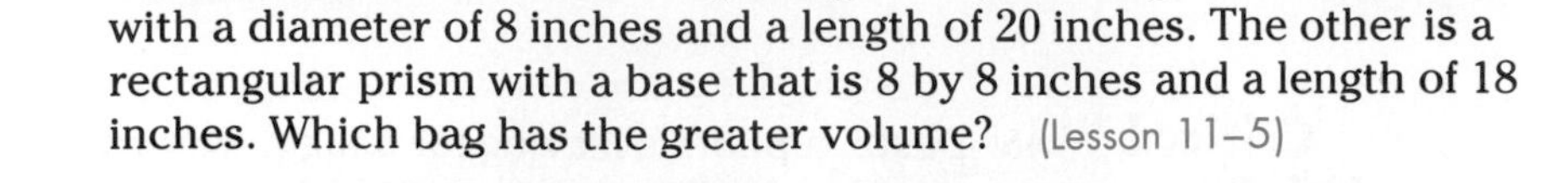

53. **Consumerism** Lucita is comparing two sports bags. One is cylindrical with a diameter of 8 inches and a length of 20 inches. The other is a rectangular prism with a base that is 8 by 8 inches and a length of 18 inches. Which bag has the greater volume? (Lesson 11–5)

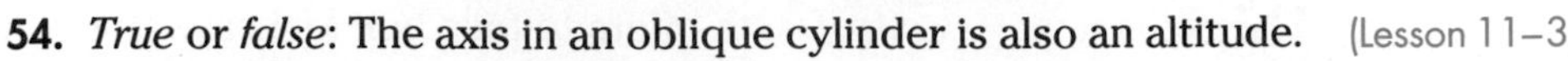

54. *True* or *false*: The axis in an oblique cylinder is also an altitude. (Lesson 11–3)

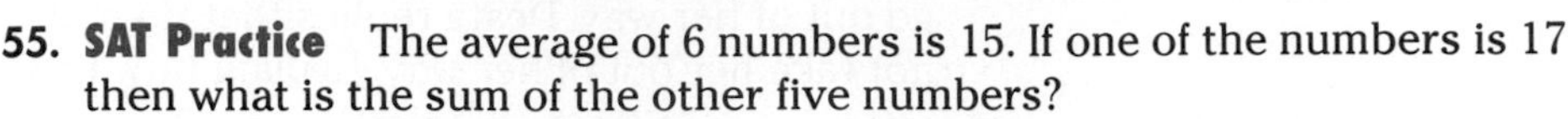

55. **SAT Practice** The average of 6 numbers is 15. If one of the numbers is 17, then what is the sum of the other five numbers?
   A 73 **B** 75 **C** 85 **D** 90 **E** 102

56. What are the measures of an interior angle and an exterior angle of a regular 24-gon? (Lesson 10–1)

57. Quadrilateral $ABCD$ is inscribed in a circle. If $m\angle A = 63$, find $m\angle C$. (Lesson 9–4)

58. *True* or *false*: A radius of a circle is a chord of the circle. (Lesson 9–3)

59. The bases of a trapezoid are $3x + 4$ and $5x - 2$ units long. If the median measures 21 units, what is the value of x? (Lesson 6–5)

For **Extra Practice**, see page 787.

60. **Work Backward** Decrease a number by 52, multiply by 12, add 20, divide by 4, and the result is 32. What was the original number? (Lesson 5–3)

Algebra

61. Solve $4n + 7 < 15$.

62. Solve $8(1 - 2x) \geq 28$.

Dimples for Distance

The excerpt below appeared in an article in *Popular Science* in February, 1995.

A GOLF BALL'S PERFORMANCE IS LARGELY determined by its dimple pattern....Dimple patterns work with the ball's spin to move the ball through the air....A ball with no dimples will travel 130 yards and behave much like a bullet, with a straight trajectory. A dimpled ball, on the other hand, can travel 280 yards, rising through the air because of its lift. Dimple design has become quite sophisticated, and patterns are now complex geometric constructions, such as icosahedrons, octahelixes, and cuboctahedrons. Symmetry is a key element because it ensures that no matter how the ball is spinning, it will fly straight. Designers have become obsessed with fitting more and more of these impressions on the spheroids. Ball designers...hit on the concept that one way to fit more dimples was to increase the surface area....the solution was obvious: make a bigger ball. ■

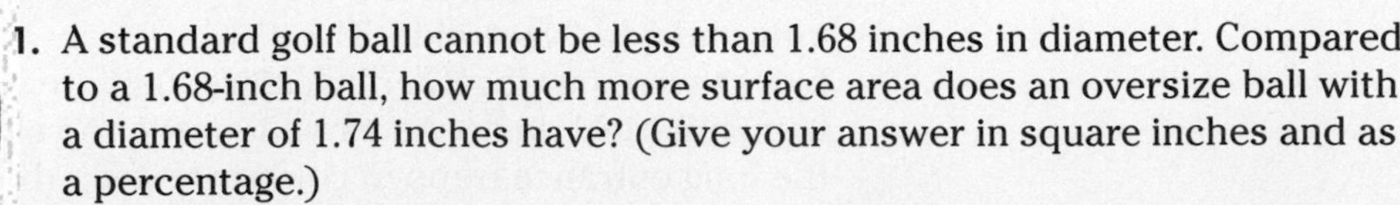

1. A standard golf ball cannot be less than 1.68 inches in diameter. Compared to a 1.68-inch ball, how much more surface area does an oversize ball with a diameter of 1.74 inches have? (Give your answer in square inches and as a percentage.)
2. Why might a larger ball not travel as far as designers expect? Explain.
3. Think of another sport or game that uses a ball. Explain why the geometric characteristics of the ball, such as size, shape, and surface features, are necessary or helpful.

11-8

Congruent And Similar Solids

What YOU'LL LEARN

- To identify congruent or similar solids, and
- to state properties of congruent solids.

Why IT'S IMPORTANT

You can apply what you learned about similar and congruent polygons to similar and congruent solids.

Folk Art

Artists have been making traditional matryoshka dolls in Russia and the Ukraine for centuries. Each doll in a set opens to reveal a smaller doll of the same shape. Sets of dolls usually contain 3, 6, or 12 dolls and are often painted to allow telling a story as the dolls are opened. Old dolls were hand carved, but now the dolls are made using a wood lathe. Matryoshka dolls are **similar solids**.

Similar solids are solids that have exactly the same shape but not necessarily the same size. You can determine if two solids are similar by comparing the ratios of corresponding linear measurements. For example, in the similar solids at the right, $\frac{1}{4} = \frac{3}{12} = \frac{5}{20}$. The ratio of the measures is called the **scale factor**. In two similar polyhedra, all of the corresponding faces are similar and all of the corresponding edges are proportional. Below are examples of pairs of similar and non-similar solids.

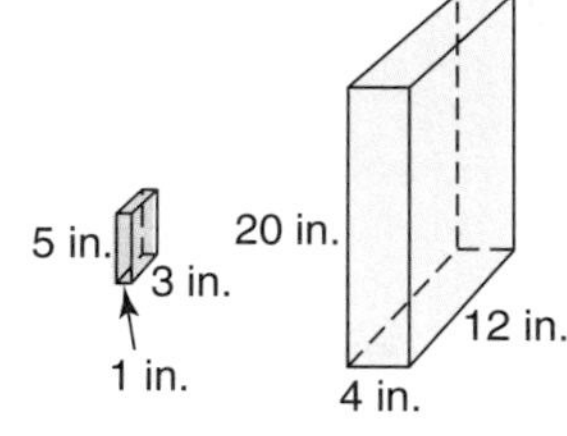

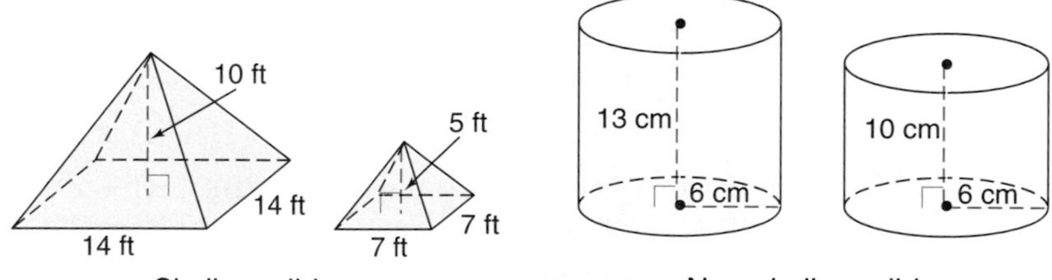

Similar solids

Non-similar solids

If the ratio of corresponding measurements of two solids is 1:1, then the solids are **congruent**. For two solids to be congruent, all of the following conditions must be met.

Two solids are congruent if:
• the corresponding angles are congruent, • corresponding edges are congruent, • areas of corresponding faces are equal, and • the volumes are equal.

Congruent polyhedra are exactly the same shape and exactly the same size.

Example 1 **Determine if each pair of solids are *similar*, *congruent*, or *neither*.**

a.

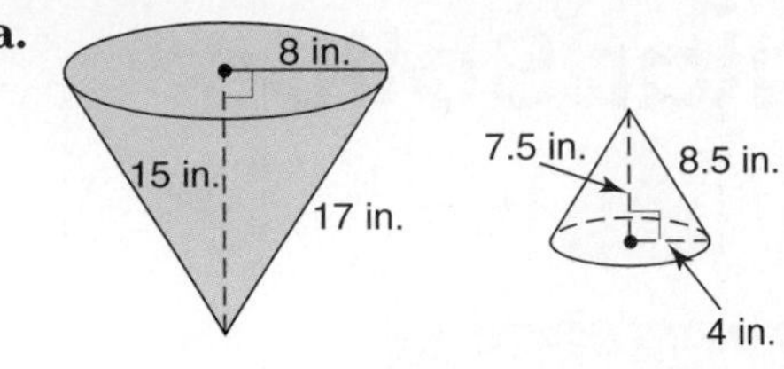

Find the ratios between corresponding parts of the cones.

$$\frac{\text{radius of larger cone}}{\text{radius of smaller cone}} = \frac{8}{4} \text{ or } 2$$

$$\frac{\text{height of larger cone}}{\text{height of smaller cone}} = \frac{15}{7.5} \text{ or } 2$$

$$\frac{\text{slant height of larger cone}}{\text{slant height of smaller cone}} = \frac{17}{8.5} \text{ or } 2$$

The ratios of the measures are equal, so we can conclude that the cones are similar. Since the scale factor is not 1, the solids are not congruent.

b.

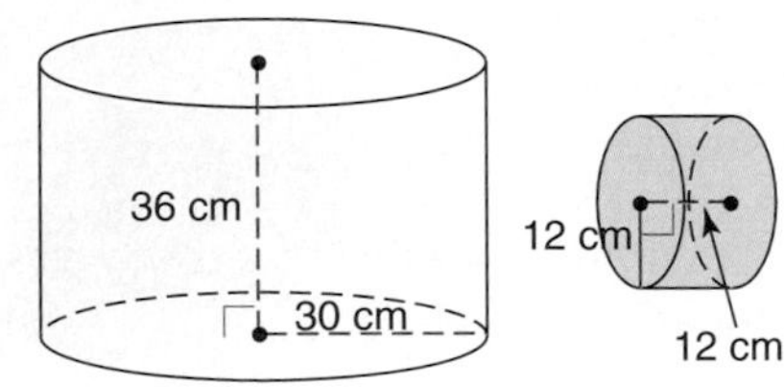

Compare the ratios between corresponding parts of the cylinders.

$$\frac{\text{radius of larger cylinder}}{\text{radius of smaller cylinder}} = \frac{36}{12} \text{ or } \frac{3}{1}$$

$$\frac{\text{height of larger cylinder}}{\text{height of smaller cylinder}} = \frac{30}{12} \text{ or } \frac{5}{2}$$

Since the ratios are not the same, the cylinders are not similar.

You can investigate the relationships between similar solids using spreadsheets.

EXPLORATION — SPREADSHEETS

- In column A, enter the labels *length*, *width*, *height*, *surface area*, *volume*, *scale factor*, *ratio of surface areas*, and *ratio of volumes* in the first eight cells.
- You will be using columns B, C, D, and E for four similar prisms.
- Type the formula (2B1+ 2B2)*B3 + 2*(B1*B2) in cell B4. This formula will find the surface area of a prism B. Copy the formula into the other cells in row 4.
- Write a similar formula to find the volume of a prism. Enter the formula in the cells in row 5.
- Enter the formula C1/B1 in cell C6, enter D1/B1 in cell D6, and so on. These formulas find the scale factor of prism B and each other solid.
- Type the formula C4/B4 in cell C7, type D4/B4 in cell D7, and so on. This formula will find the ratio of the surface area of prism B to the surface area of each other prism.
- Write a formula for the ratio of the volume of prism C to the volume of prism B. Enter the formula in cell C8. Enter similar formulas in cell 8 of each other column.
- Use the spreadsheet to find the surface areas, volumes, and ratios for prisms with the dimensions given on the next page.

	A	B	C	D	E
1	length	1	2	3	4
2	width	4	8	12	16
3	height	6	12	18	24
4	surface area				
5	volume				
6	scale factor				
7	ratio of surface areas				
8	ratio of volumes				

Your Turn

a. Compare the ratios in cells 6, 7, and 8 of columns C, D, and E. What do you observe?

b. Write a statement about the ratio between the surface areas of two solids if the scale factor is $a{:}b$.

c. Write a statement about the ratio between the volumes of two solids if the scale factor is $a{:}b$.

The Exploration leads us to the following theorem.

Theorem 11–1	**If two solids are similar with a scale factor of $a{:}b$, then the surface areas have a ratio of $a^2{:}b^2$ and the volumes have a ratio of $a^3{:}b^3$.**

Example 2

Tourism

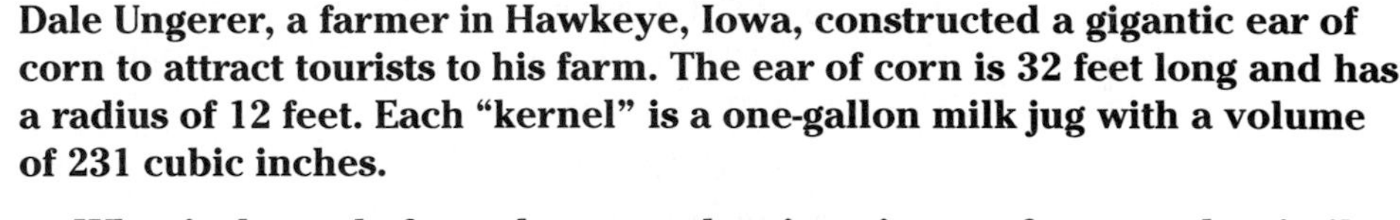

Dale Ungerer, a farmer in Hawkeye, Iowa, constructed a gigantic ear of corn to attract tourists to his farm. The ear of corn is 32 feet long and has a radius of 12 feet. Each "kernel" is a one-gallon milk jug with a volume of 231 cubic inches.

a. What is the scale factor between the gigantic ear of corn and a similar real ear of corn that is 14 inches long?

b. Estimate the volume of a kernel of the 14-inch real ear of corn.

a. Write the ratio between corresponding measures of the ears of corn.

$$\frac{\text{length of gigantic ear}}{\text{length of real ear}} = \frac{384}{14} \quad \textit{32 feet is } 32 \cdot 12 \textit{ or 384 inches.}$$
$$= \frac{192}{7}$$

The scale factor is 192:7.

b. According to Theorem 11–1 if the scale factor is $a{:}b$, then the ratio of the volumes is $a^3{:}b^3$.

$$\frac{\text{volume of gigantic ear}}{\text{volume of real ear}} = \frac{a^3}{b^3}$$
$$= \frac{192^3}{7^3} \quad a = 192,\ b = 7$$
$$= \frac{7{,}077{,}888}{343}$$

(continued on the next page)

Use a proportion to find the volume of a kernel on the real ear of corn.

$$\frac{7,077,888}{343} = \frac{231}{x}$$

$7,077,888x = 79,233$ *Multiply cross products.*

$x \approx 0.011$

The volume of a kernel on the real ear of corn is about 0.011 in^3.

CHECK FOR UNDERSTANDING

Communicating Mathematics

Study the lesson. Then complete the following.

1. **Describe** how you can determine if two polyhedra are similar.
2. **Explain** how the surface areas and volumes of similar solids are related.
3. **Compare and contrast** similar solids and congruent solids.

Guided Practice

Determine if each pair of solids is *similar, congruent,* or *neither.*

4.

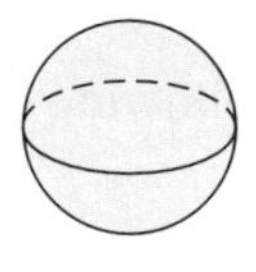
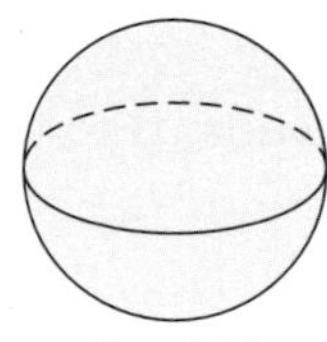

$C = 32$ ft $\quad C = 45$ ft

5.

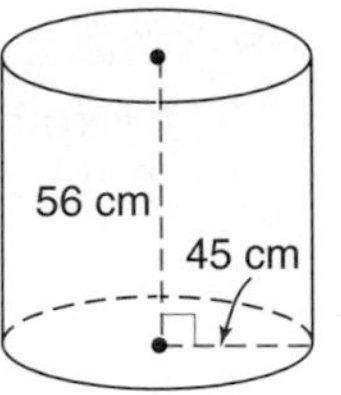

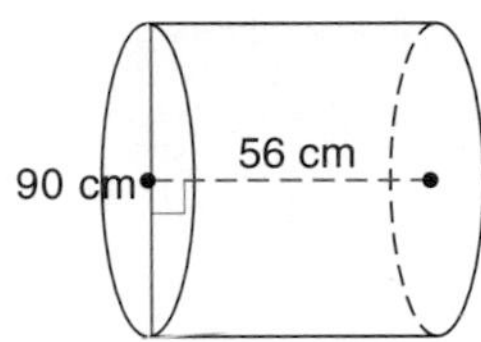

Refer to the cones at the right.

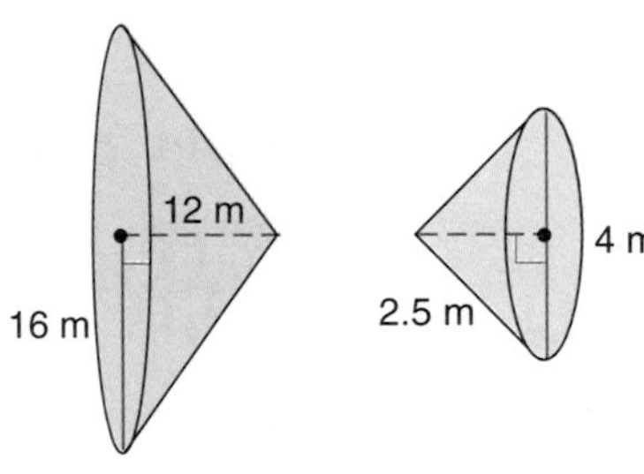

6. What is the ratio of the height of the larger cone to the height of the smaller cone?
7. Find the ratio of the surface areas.
8. What is the ratio of the circumferences?
9. If the volume of the larger cone is x cubic meters, what is the volume of the smaller cone?

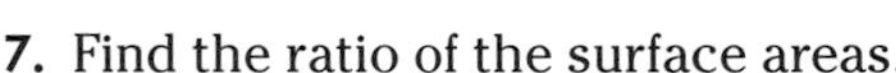

Determine if each statement is *true* or *false.* If the statement is false, rewrite it so that it is true.

10. All spheres are similar.
11. If two pyramids have square bases, then they must be similar.
12. If the edge of one cube is twice that of another cube, then its surface area is twice that of the smaller cube.

13. **World Records** According to the Guinness Book of World Records, the world's smallest car is a miniature version of a 1936 Model AA sedan. The scale factor between the full-sized car and the miniature car is 1000:1.
    a. If the door handle of the full-sized car is 15 centimeters long, how long is the door handle on the miniature car?
    b. If the surface area of the miniature car is x square centimeters, what is the surface area of the full-sized car?

EXERCISES

Practice **Determine if each pair of solids is *similar, congruent,* or *neither.***

14.

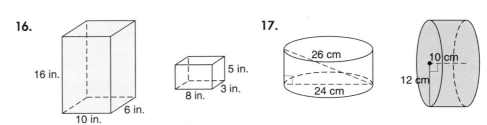

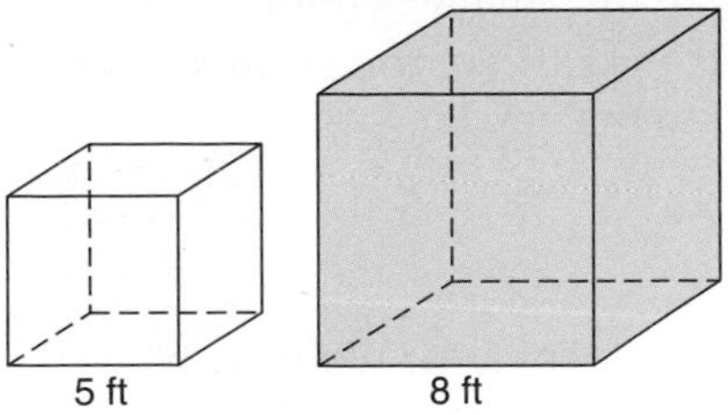

5 ft

8 ft

15.

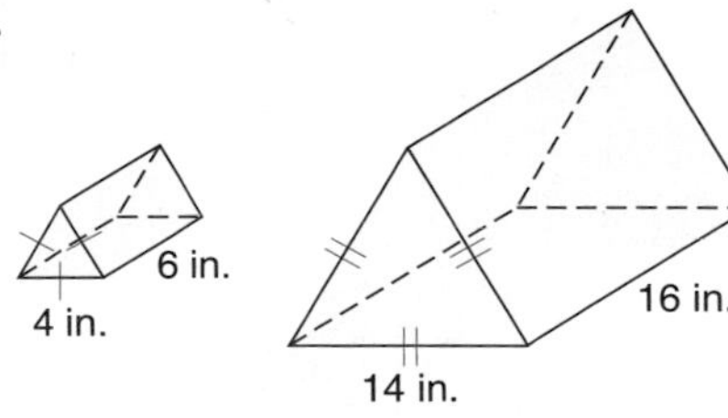

16.

16 in.

10 in.

6 in.

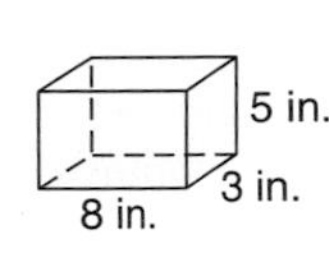

17.

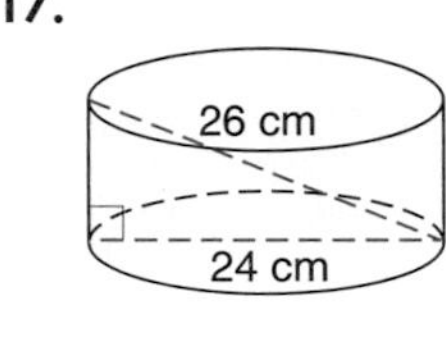

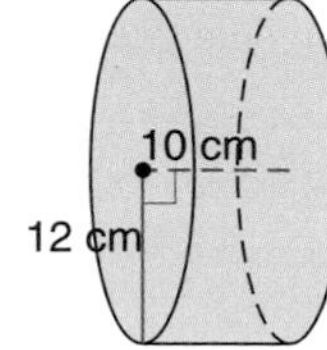

18.

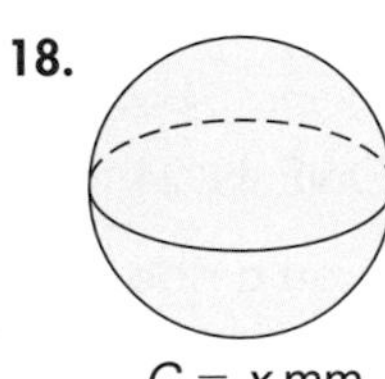

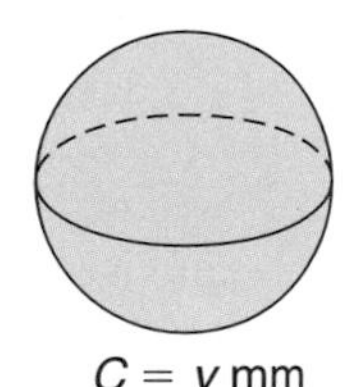

19.

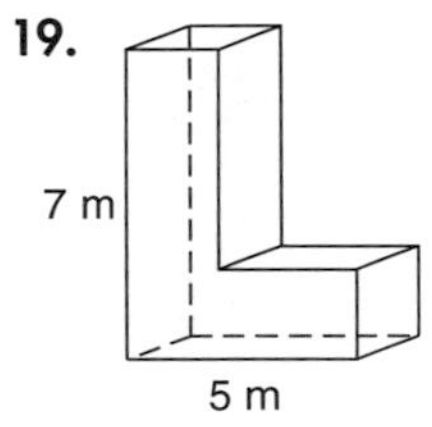

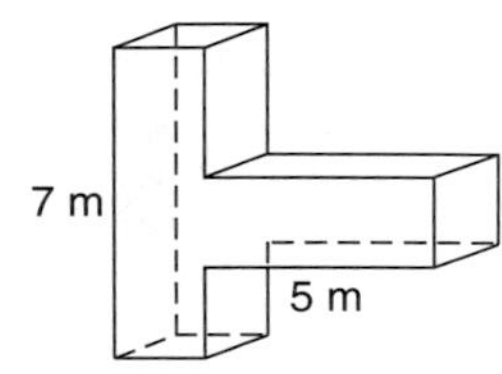

The two right rectangular prisms shown at the right are similar.

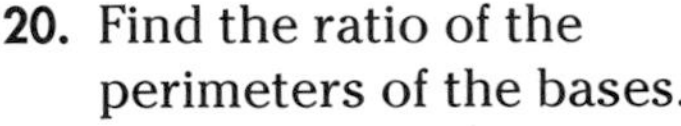

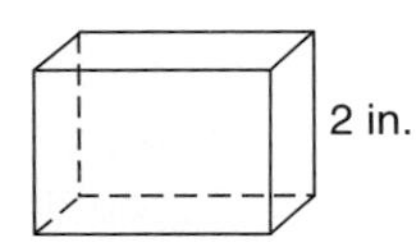

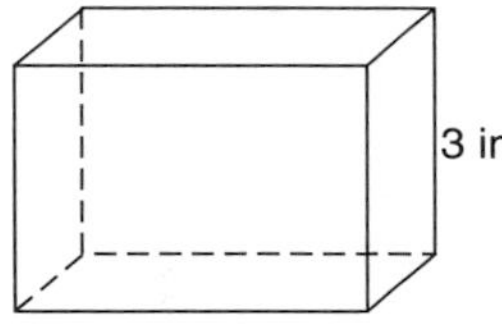

20. Find the ratio of the perimeters of the bases.

21. What is the ratio of the volumes?

22. Suppose the volume of the larger prism is 54 cm^3. Find the volume of the smaller prism.

23. The diameters of two similar cylinders are in the ratio of 4 to 5. If the volume of the smaller cylinder is 48π cubic units and the diameter of the larger cylinder is 10 units, what is the height of the larger cylinder?

Determine if each statement is *true* or *false*. If the statement is false, rewrite it so that it is true.

24. All cubes are similar.

25. If an edge length of a cube is tripled, then its volume is nine times greater.

26. If the surface area of one of two similar pyramids is one-fourth the surface area of the other, then the volume of the smaller is $\frac{1}{8}$ of the volume of the larger.

27. Doubling the radius of a sphere doubles the surface area.

28. If two solids are congruent, then their volumes are equal.

29. When a cone is cut by a plane parallel to its base, a cone similar to the original is formed. Suppose the slant height of the smaller cone is half that of the original.

a. What is the ratio of the volume of the frustum to that of the original cone? to the smaller cone?

b. What is the ratio of the surface area of the frustum to that of the original cone? to the smaller cone?

Critical Thinking

30. Is there a value for x for which the rectangular prisms at the right would be similar? Explain.

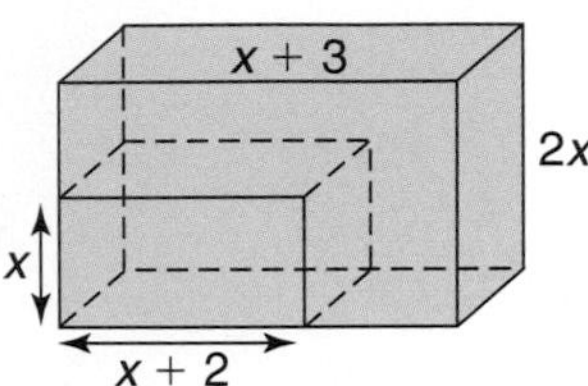

31. Write a convincing argument to show that all spheres are similar.

Applications and Problem Solving

32. Literature Read the poem *One Inch Tall* by Shel Silverstein.

ONE INCH TALL

If you were only one inch tall, you'd ride a worm to school.
The teardrop of a crying ant would be your swimming pool.
A crumb of cake would be a feast
And last you seven days at least,
A flea would be a frightening beast
If you were one inch tall.

If you were one inch tall, you'd walk beneath the door,
And it would take about a month to get down to the store.
A bit of fluff would be your bed,
You'd swing upon a spider's thread,
And wear a thimble on your head
If you were one inch tall.

You'd surf across the kitchen sink upon a stick of gum.
You couldn't hug your mama, you'd just have to hug her thumb.
You'd run from people's feet in fright,
To move a pen would take all night,
(This poem took fourteen years to write —
'Cause I'm just one inch tall.)

a. Choose one of the statements in the poem and determine whether it is correct. For example, could you swing upon a spider's thread if you were one inch tall?

b. A person who is 6 feet tall has lungs that have a surface area of 810 ft^2 and a volume of $\frac{1}{2}$ liter. Estimate the surface area and volume of your lungs if you were one inch tall.

33. Architecture Citizens of Padova, Italy built a model of the Basilica di Sant' Antonio di Padova from empty beverage cans. The model was in a scale of 1:4. The model measured 96 feet by 75 feet by 56 feet. Find the dimensions of the actual Basilica di Sant' Antonio di Padova.

34. **Food** The world's largest cherry pie was made by the Oliver Rotary Club of Oliver, British Columbia, Canada. It measured 20 feet in diameter and was completed on July 14, 1990. Most pies are 8 inches in diameter. If the largest pie was similar to a standard pie, how much greater was the volume of the largest pie?

Mixed Review

35. **World Records** The world's largest ball of string is 13 feet $2\frac{1}{2}$ inches in diameter. It took 2 years to complete and is now on display in Valley View, Texas. Find the surface area and volume of the ball of string. (Lesson 11–7)

36. Find the volume of the solid formed by the net shown at the right. (Lesson 11–5)

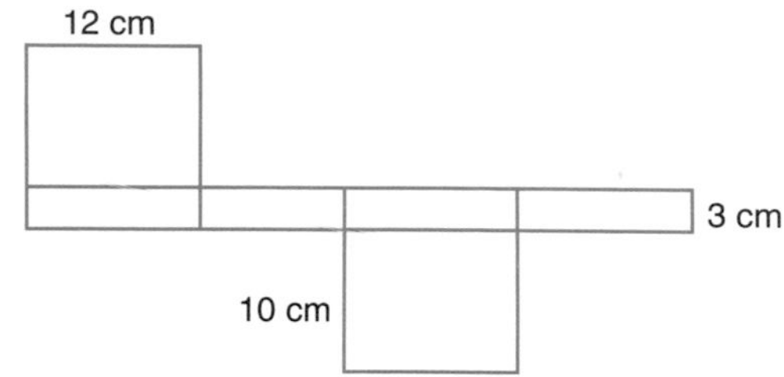

37. **ACT Practice** In the standard (x, y) coordinate plane, which of the following is the equation of a circle if the endpoints of its diameter are at $(-3, 4)$ and $(3, -4)$?
   A $x^2 + y^2 = 25$ **B** $x^2 + y^2 = 100$
   C $(x + 3)^2 + (y - 4)^2 = 10$ **D** $(x + 3)^2 + (y - 4)^2 = 100$
   E $(x - 3)^2 + (y + 4)^2 = 100$

38. The coordinates of the vertices of a quadrilateral are $(-2, 3)$, $(3, 5)$, $(-2, -6)$, and $(3, -4)$. Graph the quadrilateral and determine whether it is a *square*, a *rectangle*, or a *parallelogram*. Then find its area. (Lesson 10–3)

39. Solve $\triangle DFG$ if $m\angle D = 45$, $m\angle G = 37$, and $DG = 15$. (Lesson 8–5)

40. **Given:** $\angle S \cong \angle W$
   $\overline{SY} \cong \overline{WY}$
   Prove: $\overline{ST} \cong \overline{WV}$ (Lesson 4–4)

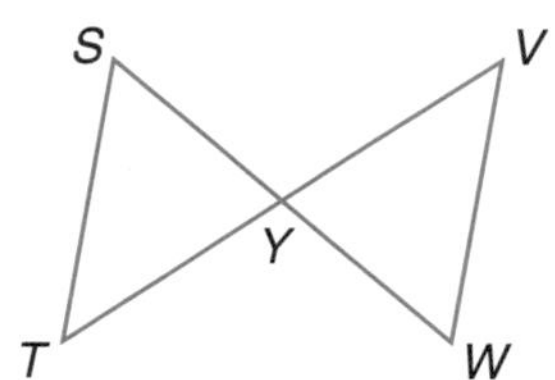

Algebra

41. Determine which ordered pairs are solutions of $y \geq 8x + 2$.
   a. (0, 2) **b.** (3, 24) **c.** (1, −1) **d.** (−4, −30)

42. **Probability** The graph at the right shows the chances of an amateur bowling a perfect 300 game.
   a. What is the probability of a man bowling a perfect game?
   b. What are the odds of a woman bowling a perfect game?

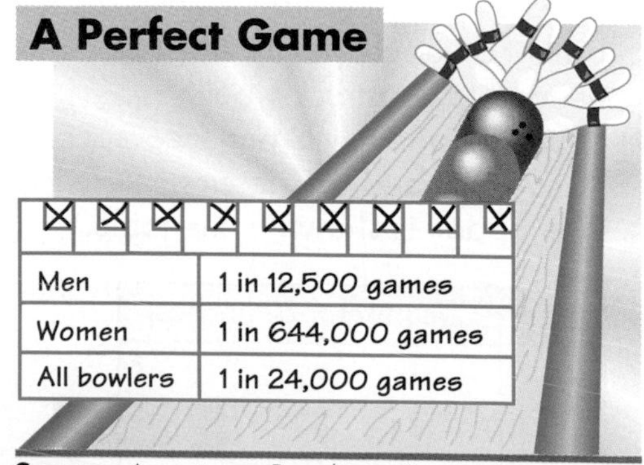

Men	1 in 12,500 games
Women	1 in 644,000 games
All bowlers	1 in 24,000 games

Source: American Bowling Congress

For **Extra Practice**, see page 787.

CLOSING THE

In·ves·ti·ga·tion

Just For Kicks

Refer to the Investigation on pages 510–511.

Soccer players must master many skills to be effective. They include *kicking* the ball, *passing* the ball to another player, *heading* or hitting the ball with the head, *dribbling* or moving the ball while running, and *tackling* or stealing the ball.

Analyze

It is now time to analyze your findings and complete your soccer ball design.

PORTFOLIO ASSESSMENT

You may want to keep your work on this Investigation in your portfolio.

1 Note the characteristics that are the same and those that are different for the two most popular soccer ball designs. List the characteristics you will incorporate in your design.

2 Use various polygons to create a spherical-shaped object. Build a design model using construction paper and tape. Draw a blueprint design of the ball showing at least three different views of the soccer ball. For each polygon, state the quantity needed to construct a ball.

3 Your company will produce soccer balls in sizes 3, 4, and 5. The circumference of each type of ball is shown in the table.

Size	Circumference (inches)
3	23
4	25
5	27–28

Prepare a design specification report for all three sizes of soccer balls. For each size of ball, state the volume, diameter, circumference, and surface area. Also, list the dimensions of each polygon used in your design, including the length of each side, the measure of an interior angle, the perimeter, and the area of each polygon.

4 Research the cost of leather and vinyl. Then estimate the cost of materials for a ball of each size. If possible, interview a sporting goods manufacturer regarding the cost of labor for constructing a product and the selling price. Complete a cost analysis and suggest a retail price for each size of soccer ball.

Present

The class represents the board of directors of your company. Present your proposed design model and a written report.

5 Summarize the process you used for creating your design and write a justification to the executive board explaining the advantages of your design.

Data Collection and Comparison To share and compare your data with other students, visit: **www.geometry.glencoe.com**

HIGHLIGHTS

Chapter Review For additional lesson-by-lesson review, visit: **www.geometry.glencoe.com**

VOCABULARY

After completing this chapter, you should be able to define each term, property, or phrase and give an example or two of each.

Geometry

altitude (pp. 591, 593, 600, 602)
axis (pp. 593, 602)
base (pp. 577, 591, 593, 600, 602)
Cavalieri's Principle (p. 617)
center (p. 621)
chord (p. 621)
circular cone (p. 602)
cone (p. 577)
congruent solids (p. 629)
corner view (p. 576)
cross section (pp. 574, 577)
cube (p. 577)
cylinder (pp. 574, 577, 593)
diameter (p. 621)
edge (p. 577)
face (p. 577)
great circle (p. 622)
height (pp. 591, 602)
hemisphere (p. 622)
lateral area (p. 592)
lateral edge (pp. 591, 600)
lateral face (pp. 577, 591, 600)
net (p. 585)
oblique cone (p. 602)
oblique cylinder (p. 593)
oblique prism (p. 591)
perspective view (p. 576)
Platonic solids (p. 577)
polyhedron (p. 577)
prism (p. 577)
pyramid (p. 577)
radius (p. 621)
regular polyhedron (p. 577)
regular prism (p. 577)
regular pyramid (p. 600)
right cone (p. 602)
right cylinder (p. 593)
right prism (p. 591)
scale factor (p. 629)
similar solids (p. 629)
slant height (pp. 600, 602)
slice (pp. 574, 577)
solid (p. 574)
sphere (p. 577)
surface area (p. 584)
tangent (p. 621)
triangular prism (p. 574)
vertex (pp. 600, 602)
volume (p. 607)

Problem Solving

make a model (p. 575)

UNDERSTANDING AND USING THE VOCABULARY

Choose the correct best term to complete each statement.

1. If a plane that is parallel to the base or bases of a solid slices the solid, then the intersection of the plane and the solid is a (cross section, corner view).
2. A (cylinder, prism) is a polyhedron with two congruent faces that are polygons in parallel planes.
3. If the axis of a cone is perpendicular to the base of the cone, then the cone is a(n) (right cone, oblique cone).
4. A (net, slice) is a two-dimensional figure that shows the shapes that could be folded to form a three-dimensional figure.
5. The lateral area of a solid is the area of the (lateral faces, lateral edges).
6. In a regular pyramid, the height of a lateral face is the (altitude, slant height) of the pyramid.
7. In a (right, regular) prism, the bases are regular polygons, and the lateral faces are perpendicular to the bases.
8. A ball bearing is a model of a (sphere, great circle).
9. If two solids are (similar, congruent), then they have the same shape but may have different sizes.
10. The height of one solid is a units and the height of a similar solid is b units. Then the scale factor is (a:b, a^2:b^2).

SKILLS AND CONCEPTS

OBJECTIVES AND EXAMPLES

Upon completing this chapter, you should be able to:

- use top, front, side, and corner views of three-dimensional solids to make models (Lesson 11–1)

From the views given, draw a perspective view.

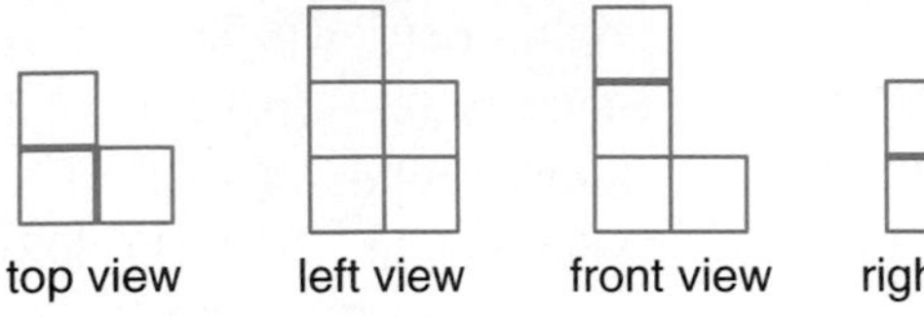

The top view tells us that the base has 3 blocks and that the columns have different heights. The other views allow us to determine the height of each column.

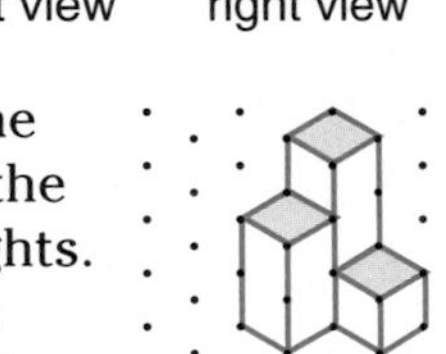

REVIEW EXERCISES

Use these exercises to review and prepare for the chapter test.

11. From the views of a solid figure given, draw a corner view.

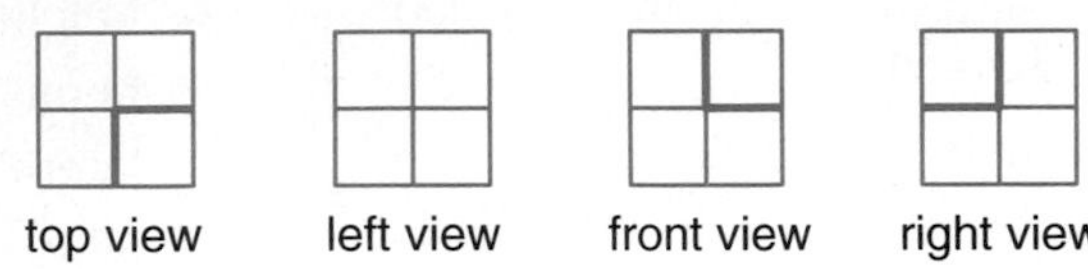

12. The corner view of a figure is given. Draw the top, left, front, right, and back views of the solid.

- make two-dimensional nets for three-dimensional solids (Lesson 11–2)

A polyhedron and one of its two-dimensional nets are shown below.

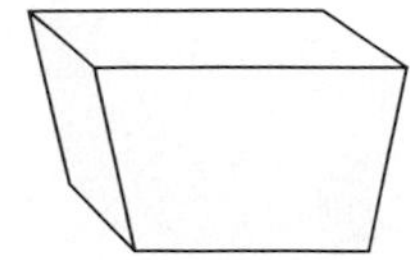

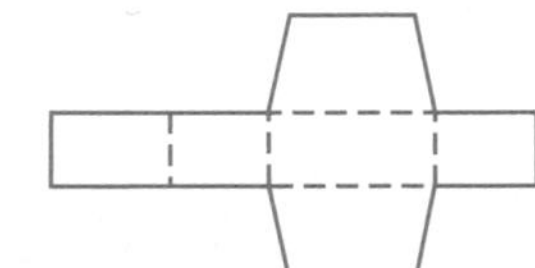

13. Use isometric dot paper to draw a rectangular prism that is 4 units high, 7 units long, and 2 units wide.

14. Use rectangular dot paper to draw a net for the solid shown at the right. Then find the surface area of the solid.

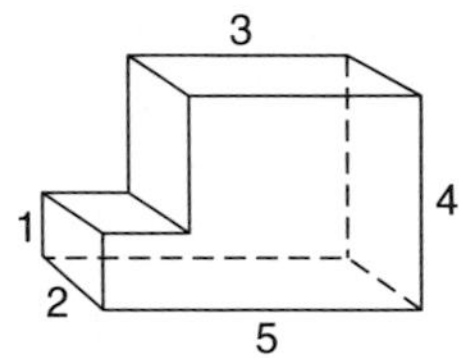

- find the lateral areas and surface areas of right prisms and right cylinders (Lesson 11–3)

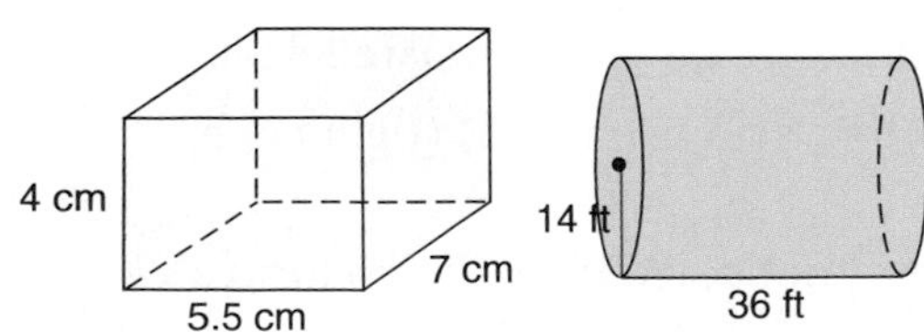

$L = Ph$

$= (25)(4)$

$= 100 \text{ cm}^2$

$T = Ph + 2B$

$= 100 + 2(5.5 \cdot 7)$

$= 177 \text{ cm}^2$

$L = 2\pi rh$

$= 2\pi(14)(36)$

$\approx 3166.7 \text{ ft}^2$

$T = 2\pi rh + 2\pi r^2$

$\approx 3166.7 + 2\pi(14)^2$

$\approx 4398.2 \text{ ft}^2$

Find the lateral area and the surface area of each right prism or right cylinder. Round to the nearest tenth.

15.

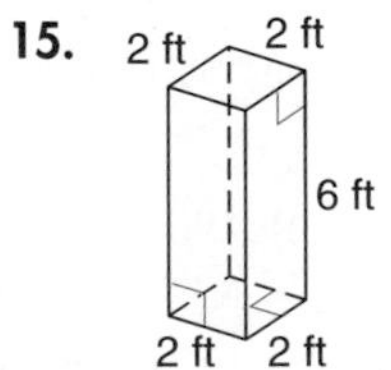

16.

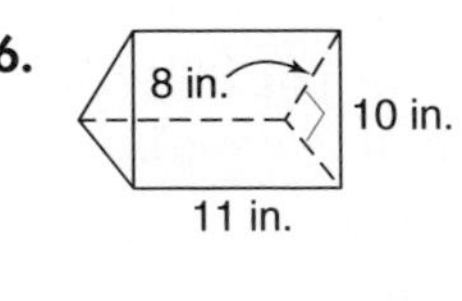

17.

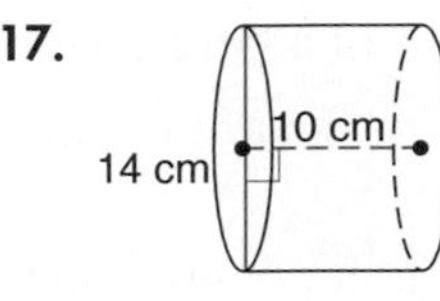

18.

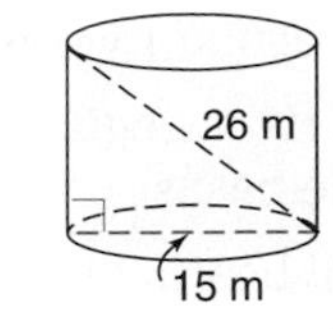

OBJECTIVES AND EXAMPLES

- find the lateral areas and surface areas of regular pyramids and right circular cones (Lesson 11–4)

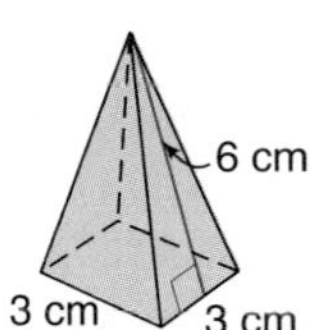

$L = \frac{1}{2}P\ell$

$= \frac{1}{2}(12)(6)$ or 36 cm^2

$T = L + B$

$= 36 + 9$

$= 45 \text{ cm}^2$

$L = \pi r\ell$

$= \pi(5)(13)$

≈ 204.2 ft

$T = \pi r\ell + \pi r^2$

$= 204.2 + \pi(5^2)$

$\approx 282.7 \text{ ft}^2$

REVIEW EXERCISES

Find the lateral area and the surface area of each regular prism or right cone. Round to the nearest tenth.

19.

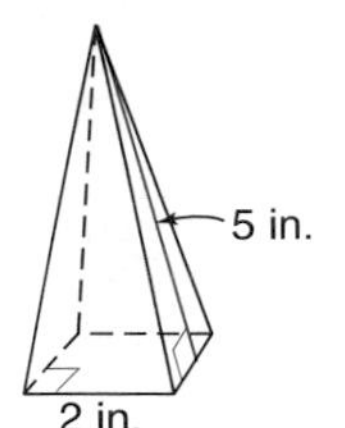

20.

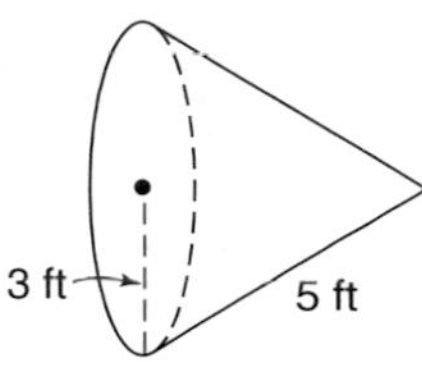

21.

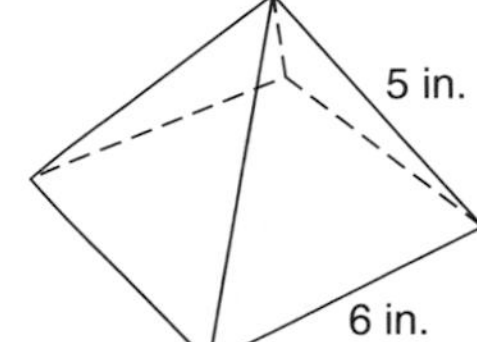

22.

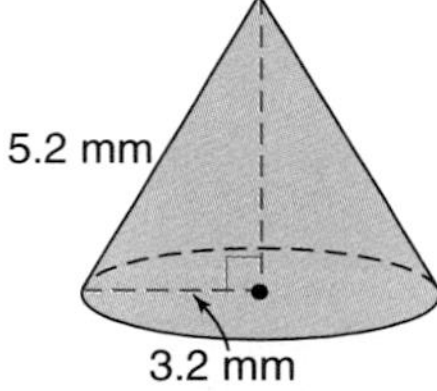

OBJECTIVES AND EXAMPLES

- find the volumes of right prisms and right cylinders (Lesson 11–5)

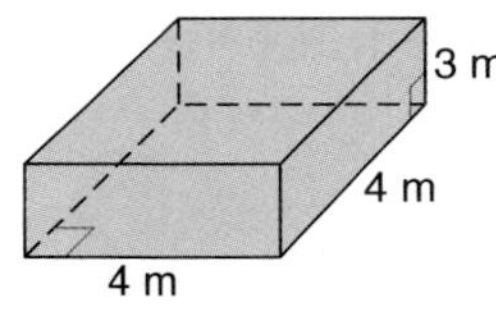

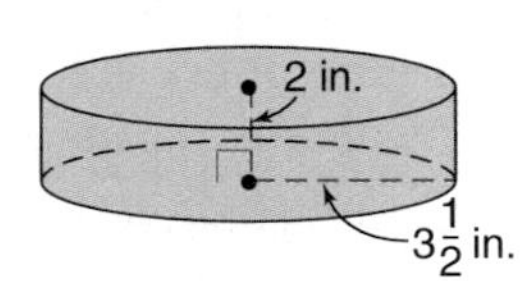

$V = Bh$

$= (16)(3)$

$= 48 \text{ m}^3$

$V = \pi r^2 h$

$= \pi\left(3\frac{1}{2}\right)^2 2$

$\approx 77.0 \text{ in}^3$

REVIEW EXERCISES

Find each of the following. Round to the nearest tenth.

23. What is the volume of a hexagonal prism if its radius is 10 centimeters and its height is 20 centimeters?

24. A right cylinder has a radius of 10 centimeters and a height of 20 centimeters. Find the volume of the cylinder.

25. Find the volume of a right cylinder if its diameter is 10 feet and its height is 13 feet.

OBJECTIVES AND EXAMPLES

- find the volumes of circular cones and pyramids (Lesson 11–6)

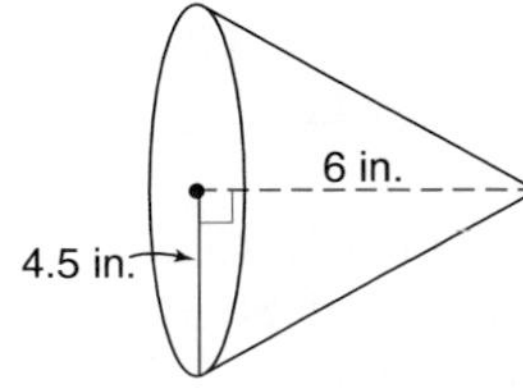

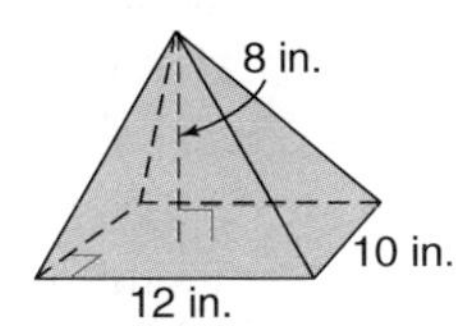

$V = \frac{1}{3}Bh$

$= \frac{1}{3}\pi(4.5^2)(6)$

$\approx 127.2 \text{ in}^3$

$V = \frac{1}{3}Bh$

$= \frac{1}{3}(12 \cdot 10)(8)$

$= 320 \text{ in}^3$

REVIEW EXERCISES

Find each of the following. Round to the nearest tenth.

26. The base of a triangular pyramid is an equilateral triangle with sides 9 centimeters long. The pyramid's height is 15 centimeters. Find the volume of the pyramid.

27. What is the volume of a right circular cone if its height is 22 centimeters and its radius is 11 centimeters?

28. The circumference of the base of a right circular cone is 62.8 millimeters. The cone has a height of 15 millimeters. Find the volume of the cone.

OBJECTIVES AND EXAMPLES

- find the surface area and volume of a sphere (Lesson 11–7)

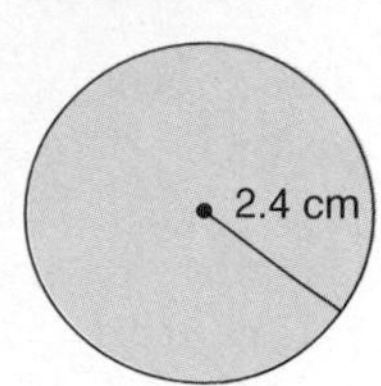

$T = 4\pi r^2$ $\quad V = \frac{4}{3}\pi r^3$

$= 4\pi(2.4)^2$ $\quad = \frac{4}{3}\pi(2.4)^3$

$\approx 72.4 \text{ cm}^2$ $\quad \approx 57.9 \text{ cm}^3$

REVIEW EXERCISES

Answer each of the following.

29. *True* or *false*: In a sphere, all great circles are congruent.
30. The radius of the moon is approximately 1080 miles. Find its surface area.
31. The area of a great circle of a sphere is 50.24 square centimeters. Find the volume of the sphere.

- identify congruent or similar solids (Lesson 11–8)

Determine if the cylinders are *congruent*, *similar*, or *neither*.

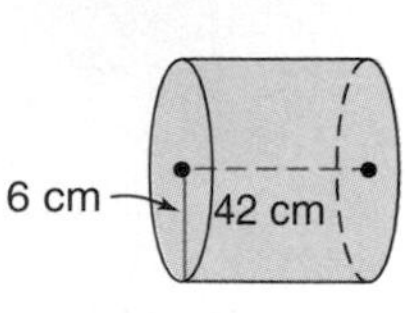

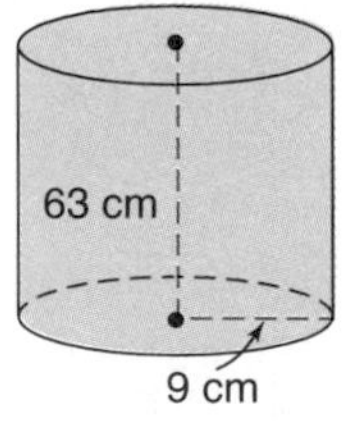

$\frac{\text{radius of larger cylinder}}{\text{radius of smaller cylinder}} = \frac{9}{6} \text{ or } \frac{3}{2}$

$\frac{\text{height of larger cylinder}}{\text{height of smaller cylinder}} = \frac{63}{42} \text{ or } \frac{3}{2}$

Since the ratios are the same, the cylinders are similar. However since the ratio is not 1, the cylinders are not congruent.

32. Determine if the solids below are *similar*, *congruent*, or *neither*.

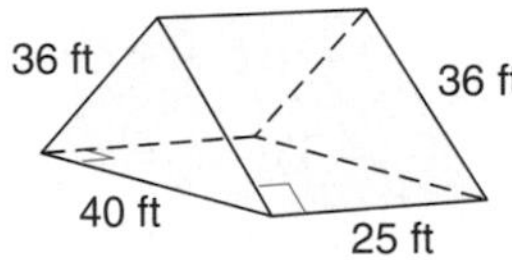

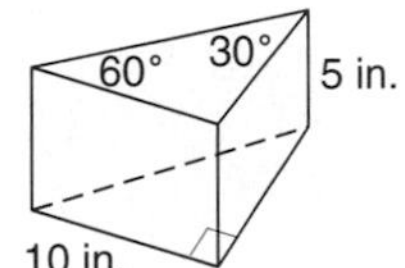

Determine if each statement is *true* or *false*. If the statement is false, rewrite it so that it is true.

33. Two congruent prisms are also similar.
34. The radius of a sphere is twice the radius of another sphere. Thus, the surface area of the larger sphere is four times the surface area of the smaller sphere.
35. A solid is never similar to itself.

APPLICATIONS AND PROBLEM SOLVING

36. **Home Maintenance** Different types of paint are used for different purposes. A wall with a rough texture will diffuse more light than a smooth-surfaced one. A smooth wall can withstand more traffic and cleaning. When minute differences are considered, a wall with a rough texture will have a greater surface area than the same wall with a smooth texture. Study the magnified paint surfaces below. Which finish has the most surface area? (Lesson 11–2)

Ceiling, Flat, Eggshell, Semi-Gloss, Gloss

37. **Engineering** In 1986, the Water Spheroid was completed in Edmond, Oklahoma. This water tank is the largest in the world, with a diameter of 218 feet. Find its surface area and volume. (Lesson 11–7)

38. **Communications** Coaxial cable is used to transmit long-distance telephone calls, cable television programming, and other communications. A typical coaxial cable contains 22 coaxials, or copper tubes, with copper wire and plastic insulation. A 22-coaxial cable is about 3 inches in diameter. What is the lateral area and the volume of a coaxial cable that is 500 feet long? (Lessons 11–3, 11–5)

39. **Music** The largest guitar in the world was built by students in the Shakamak High School in Jasonville, Indiana. If a folk guitar is 18 inches wide and the largest guitar is 38 feet 2 inches wide, find the scale factor between the guitars. (Lesson 11–8)

A practice test for Chapter 11 is provided on page 803.

ALTERNATIVE ASSESSMENT

COOPERATIVE LEARNING PROJECT

Chartography In this chapter, you learned about the surface area of figures that have smooth surfaces. An acre of land has an area of 43,560 square feet, but this applies to a perfectly flat surface, which is rare on Earth. In this project, you will estimate the surface area of a plot of land with a contour that is not flat.

The U.S.G.S. (United States Geological Survey) provides contour maps for every region of the United States. You can obtain these from the library, federal government office, or some bookstores. Obtain the map for your region, or the nearest region which contains hills and valleys. You will select a rectangular region of the map and estimate its surface area.

WEST KENYA: Contour Map

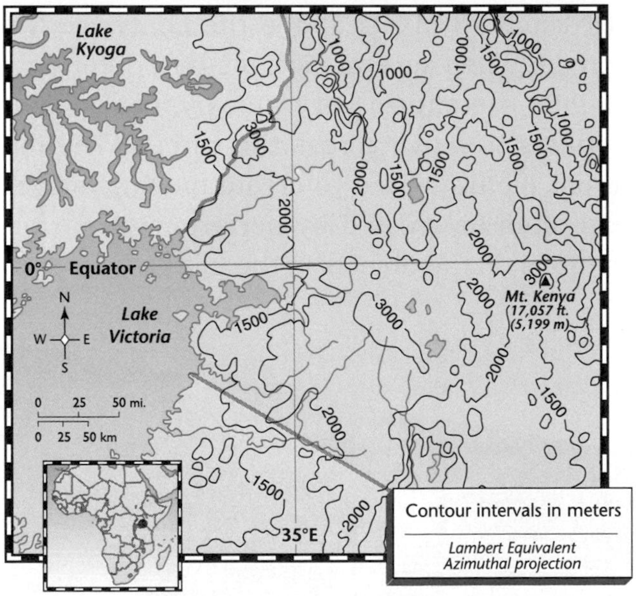

Contour intervals in meters

Lambert Equivalent Azimuthal projection

Follow these steps to estimate the surface area of the selected region.

- Draw a border around the rectangular region you will measure.
- Subdivide the rectangle into smaller regions so that the boundary of each region is a polygon.
- The vertices of a polygonal region may be at different elevations. Select regions where the vertices all lie in the same plane.
- Use formulas for area to determine the area of each polygon.
- Sum up the areas to obtain an estimate of the surface area for the entire plot of land.

Now imagine that the area is to be flooded to create a recreational lake. You need to know how much water is needed to fill the area so that the deepest point is six feet. As a group, discuss the method you would use to estimate the volume of water needed. Compute the volume.

THINKING CRITICALLY

Discuss the sentence *Length is to area as area is to volume*. Include examples or drawings to suggest your conclusions.

PORTFOLIO

Select a problem from this chapter that you found challenging. Write a paragraph about how you solved it, and why it was so challenging. Keep this problem and your essay in your portfolio.

LONG-TERM PROJECT

Investigation

MATERIALS NEEDED

- grid paper
- ruler
- markers
- calculator

Were you amazed when Forrest Gump shook hands with President Kennedy? Or astounded when dinosaurs came to life in Jurassic Park? Then you have experienced the magic that digital computers can create at the movies.

The field of computer special effects for movies began when George Lucas couldn't find a company to create the effects he envisioned for his 1976 space movie. At the time, filmmakers considered Lucas' ideas experimental and not commercially viable. But by 1996, roughly half of the movies released used digital visuals of some kind!

Computer graphics are created using *coordinate geometry*. The computer screen is comprised of thousands of little lights, called pixels, arranged in rows and columns across and down the screen. An image is created by assigning each pixel a color. Pixels are identified by a coordinate position on the screen. For example, (34, 78) may refer to a pixel that is 34 pixels to the right and 78 pixels above the center of the screen. When several pictures that differ only slightly are shown very quickly, the subject appears to move across the screen. The different pictures used to create moving images can often be created by performing transformations on the objects in the picture.

THE SCREEN

To develop your skill in creating computer graphics, begin by creating a simple design on a "screen." Use graph paper to create a model of a computer screen that measures 50 pixels horizontally and 30 pixels vertically. Let the center of the screen be at (0, 0). What are the coordinates of each corner of the screen? Lay the model screen over an existing picture or draw a picture on the screen. Your picture could be a geometric figure or a simple illustration. What are the coordinates of ten different points used to create the picture? What are the colors of the pixels you identified? Could you give someone else the coordinates and instructions so that they could recreate your picture without seeing the finished picture?

THE MOVIE

Work with the members of your group to choose a subject and plot for a "movie." You will be drawing the images to make the objects or characters move. Then the movie images can be bound as a small book. When the pages of the book are flipped quickly, the images of your movie will appear to come to life.

You will continue working on this Investigation throughout Chapters 12 and 13.

Be sure to keep your computer screen model, movie ideas, and other materials in your Investigation Folder.

Movie Magic Investigation

Working on the Investigation
Lesson 12–1, p. 651

Working on the Investigation
Lesson 12–6, p. 686

Working on the Investigation
Lesson 13–6, p. 737

Working on the Investigation
Lesson 13–8, p. 753

Closing the Investigation
End of Chapter 13, p. 754

CHAPTER

12 Continuing Coordinate Geometry

What You'll Learn

In this chapter, you will:

- write and graph linear equations,
- relate statistics and equations of lines to geometric concepts,
- prove theorems using coordinate proofs,
- perform operations with vectors, and
- locate points in space.

Why It's Important

Real World

Media The first modern comic book, Funnies on Parade, was printed in 1933. Within weeks, all 10,000 copies were sold, and a new industry was born. Now that computers can create a virtual reality, comic book characters have entered a new dimension. Three-dimensional displays and data-input sensors allow interaction with characters that appear to be there in the flesh. In some types of computer software, each point in a particular scene, or frame, is given coordinates in three dimensions. Matrices, which you will learn about in Lesson 12–5, can be used to show the coordinates of points.

PREREQUISITE SKILLS

To be successful in this chapter, you'll need to understand these concepts and be able to apply them. Refer to the lesson in parentheses if you need more review before beginning the chapter.

Use slope to identify parallel and perpendicular lines. (Lesson 3–3)

Determine the slope of each line described.

1. a line parallel to the line that passes through $A(0, 0)$, $M(1, -2)$
2. a line perpendicular to the line that passes through $Y(1, 4)$, $C(-3, -4)$
3. a line parallel to the line that passes through $D(0, 4)$, $G(-3, 0)$
4. a line perpendicular to the line that passes through $E(10, 10)$, $Z(5, 1)$

Find the midpoint of a segment. (Lesson 1–5)

N is the midpoint of $\overline{AY}$. For each pair of points given, find the coordinates of the third point.

5. $A(0, -1)$, $Y(-5, 4)$
6. $N(0, 0)$, $Y(-2, 2)$
7. $A(1, -1)$, $N(10, 10)$
8. $A(5, 0)$, $Y(-3, 6)$

Write the equation of a circle in the coordinate plane. (Lesson 9–8)

Write an equation of circle P based on the given information.

9. center: $P(0, 0)$; radius: 4
10. center: $P(-1, 1)$; radius: 10
11. endpoints of the diameter are $(4, -3)$ and $(2, 1)$
12. endpoints of the diameter are $(0, 0)$ and $(6, 6)$

Focus On READING SKILLS

In this chapter, you'll be learning about ***coordinate geometry.*** This topic includes the coordinate plane, graphing linear equations, and vectors. The coordinate plane, linear equations, and vectors are closely related. Recall that the coordinate plane is a grid with an x– and y–axis. The word *linear* contains the word *line*. When you graph a linear equation on the coordinate plane, the graph is a line. Vectors are like line segments whose direction and length are important features. Vectors can be graphed on a coordinate plane similar to linear equations.

12-1 Integration: Algebra Graphing Linear Equations

What YOU'LL LEARN

- To graph linear equations using the intercepts method and the slope-intercept method.

Why IT'S IMPORTANT

You can use linear equations to model relationships and solve real-world problems.

Many relationships can be represented using linear equations.

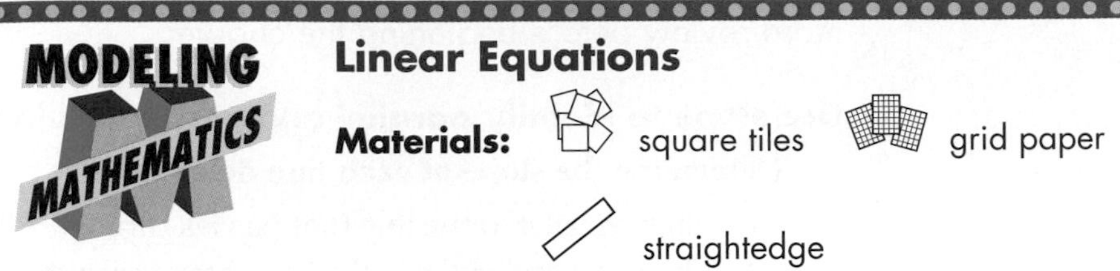

Linear Equations

Materials: square tiles, grid paper, straightedge

- The figures below represent the first three figures in a sequence. The perimeter of each figure is noted in the table below at the right.

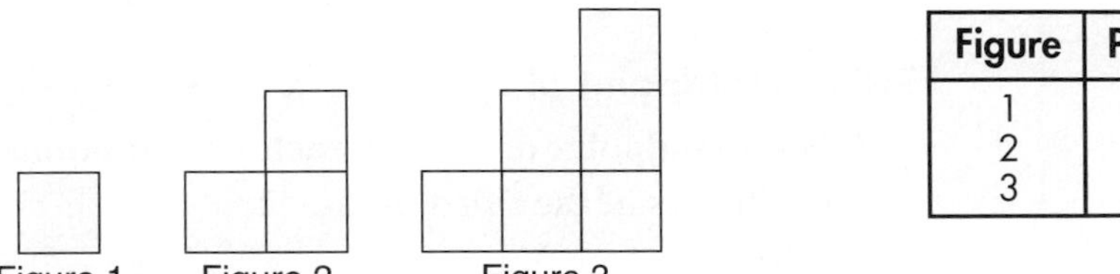

Figure 1 Figure 2 Figure 3

Figure	Perimeter
1	4
2	8
3	12

- Arrange square tiles in the patterns shown. Then extend the pattern to make the fourth, fifth, and sixth figures.
- Copy the table shown above. Then find the perimeter of figures 4, 5, and 6 and record them in your table.

Your Turn

a. Suppose that the number of the figure is x and its perimeter is y. Verify that for each figure you drew, $y = 4x$.

b. Graph each ordered pair (x, y) on a coordinate plane.

c. Draw a line through the points.

d. Use the graph to estimate the perimeter of figure 40 in this sequence.

The line that you graphed in the Modeling Mathematics activity has the equation $y = 4x$. Every line drawn in a coordinate plane has a corresponding **linear equation** that describes the line algebraically. A linear equation can be written in the form $Ax + By = C$, where A, B, and C are any real numbers and A and B are not both zero. A linear equation in the form $Ax + By = C$ is said to be in **standard form**.

The equations $2x + y = 10$, $3x = 5y - 8$, $y = 2$, and $\frac{1}{2}x + \frac{3}{4}y = 0$ are all linear equations. Each can be written in the form $Ax + By = C$. The equations $y = 2x^2 + 3$ and $x + \frac{1}{y} = 0$ are not linear equations. *Why not?*

The graph of a linear equation, $Ax + By = C$, is the set of all points with coordinates (x, y) that satisfy the equation. One technique for graphing a linear equation is called the **intercepts method**. This method uses two special values, the *x-intercept* and the *y-intercept*. The x-intercept is the value of x when y equals zero. Likewise, the y-intercept is the value of y when x equals zero. This method can be used as long as neither A nor B is zero.

Example 1

Finance

In this application, for any ordered pair, (x, y), x represents the number of months, and y represents the amount owed to Aunt Guillerma after x months.

Maita is buying a used car from her Aunt Guillerma for \$4500. She agreed to pay her \$150 each month. The equation $y = 4500 - 150x$ represents the amount owed, y, after x payments. Graph the equation by finding the intercepts.

First write the equation in standard form.

$$y = 4500 - 150x$$
$$150x + y = 4500$$

Let $x = 0$ to find the y-intercept.

$$150(0) + y = 4500$$
$$y = 4500$$

Let $y = 0$ to find the x-intercept.

$$150x + 0 = 4500$$
$$150x = 4500$$
$$x = 30$$

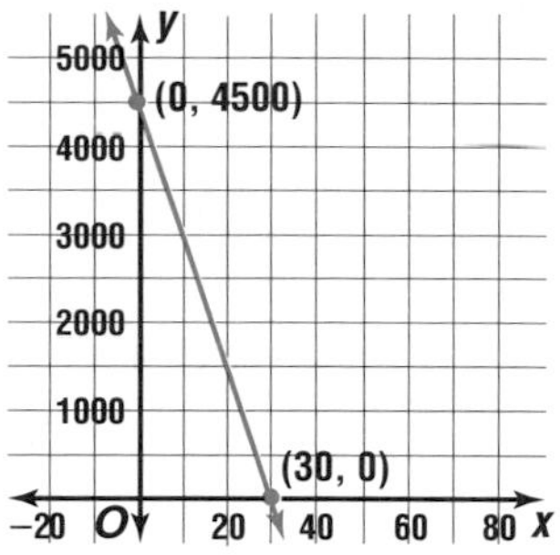

The x-intercept is 30, and the y-intercept is 4500. To graph $150x + y = 4500$, plot (30, 0) and (0, 4500) and draw a line through the points.

In the example above, the amount owed is said to be a **function** of the number of payments. A function is a relationship between input and output. In a function, one or more operations are performed on the input to get the output, and there is exactly one output for each input.

When an equation is rewritten so that one side has y by itself, another method for graphing is more convenient to use. Use a graphing calculator, graphing software, or graph paper to graph the *family of equations* that appears below.

$$y = \frac{2}{3}x + 2 \qquad y = \frac{2}{3}x + 0 \qquad y = \frac{2}{3}x - 3$$

The graphs are parallel lines, and the slope of each line is $\frac{2}{3}$. How does the slope of each line compare to its equation? In each case, the slope of each line is equal to the coefficient of x in its equation. Also the y-intercept of each line is equal to the constant term of the equation. A linear equation written in the form $y = mx + b$ is called the **slope-intercept form** of the equation.

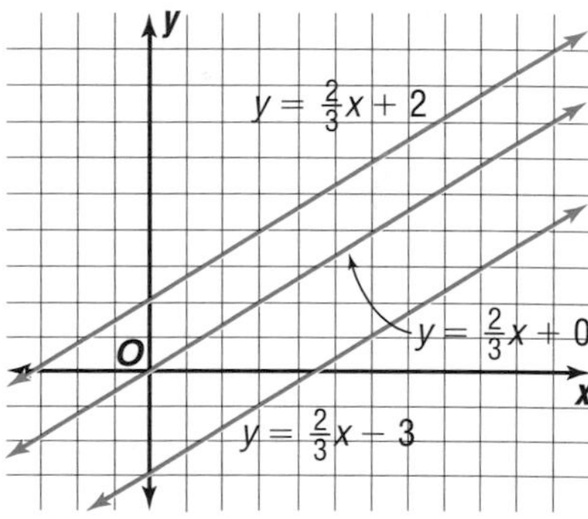

LOOK BACK

Refer to Lesson 3-3 to review how to find the slope of a line.

Theorem 12–1 Slope-Intercept Form	If the equation of a line is written in the form $y = mx + b$, m is the slope of the line, and b is the y-intercept.

In Theorem 12–1, notice that if $x = 0$, $y = b$. Therefore, b is the y-intercept. If $x = 1$, then $y = m + b$. Since the ordered pairs for two points on the line have coordinates $(0, b)$ and $(1, m + b)$, the slope is $\frac{(m + b) - b}{1 - 0}$ or m.

Horizontal and vertical lines are special cases. The graph of an equation of the form $x = a$ is a vertical line and has an undefined slope. If the equation of a vertical line is written in the standard form, $Ax + By = C$, then $B = 0$. The graph of an equation of the form $y = b$ is a horizontal line and has a slope of 0. When the equation of a horizontal line is written in the standard form, $A = 0$.

You can graph a line if you know its slope and y-intercept.

Example 2 **Graph $y = -\frac{1}{3}x + 4$ using the slope and the y-intercept.**

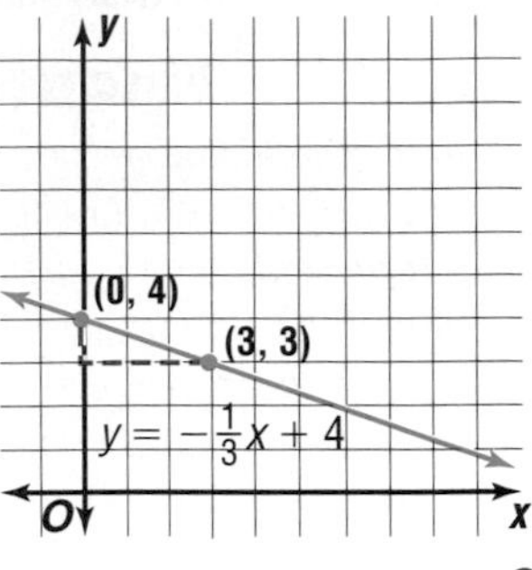

Since the equation is in slope-intercept form, the slope is $-\frac{1}{3}$, and the y-intercept is 4.

The ordered pair for the y-intercept, 4, is (0, 4). Graph (0, 4). From this point, move 1 unit down and 3 units right. This point, which has coordinates (3, 3), must also lie on the line. Draw the line containing the two points.

The slope-intercept form of a linear equation can also be used to determine if two lines are parallel or perpendicular without graphing the equations. Remember that if two nonvertical lines are parallel, they have the same slope. If two lines are perpendicular and neither is vertical, the product of their slopes is -1.

Example 3 **Use the description to graph each line.**

a. the line parallel to the graph of $y = 3x + 8$ through the point at $(1, -2)$

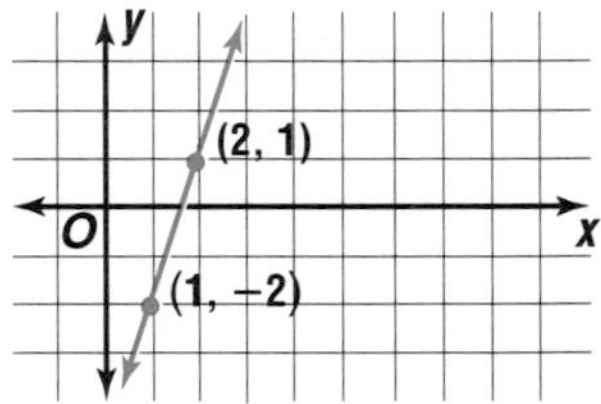

Graph $(1, -2)$. By Postulate 3–2, the line must have the same slope as the graph of $y = 3x + 8$. Its slope is 3. So, from the point at $(1, -2)$, move up 3 units and right 1 unit. This point, which has coordinates (2, 1), must also lie on the line. Draw the line containing the two points.

b. the line perpendicular to the graph of $y = 2x$ through the point at (2, 4)

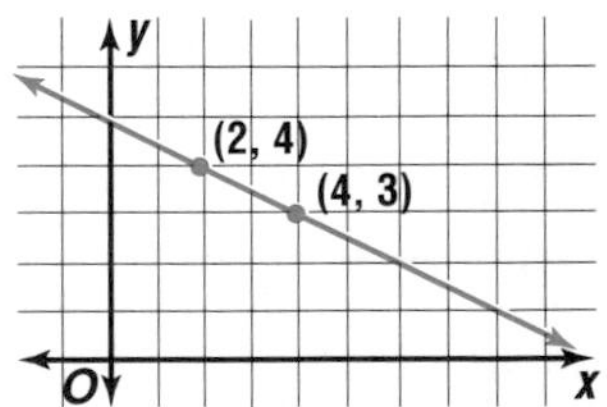

Graph (2, 4). By Postulate 3–3, the product of the slopes of two perpendicular lines is -1. Since the slope of the line for $y = 2x$ is 2, the slope of the perpendicular must be $-\frac{1}{2}$. From the point at (2, 4), move down 1 unit and right 2 units. This point, which has coordinates (4, 3), must also lie on the line. Draw the line containing the two points.

CHECK FOR UNDERSTANDING

Communicating Mathematics

Study the lesson. Then complete the following.

1. **Describe** how you would determine if a given point is on a given line.
2. **Explain** why the slope of a vertical line is undefined.
3. **Describe** two methods of graphing a linear equation. Which method is more convenient for graphing $5x - 7y = 16$? Explain.

4. **Determine** whether $y = 3x + 4$ is a linear equation. Explain your reasoning.

5. The figures at the right form a sequence. Figure number x has a perimeter of y units.
   a. Construct the first eight figures of the sequence.
   b. Verify that the equation $y = 2x + 2$ holds true for the first eight figures.
   c. Then graph the equation on a coordinate grid.
   d. What is the perimeter of figure 40?

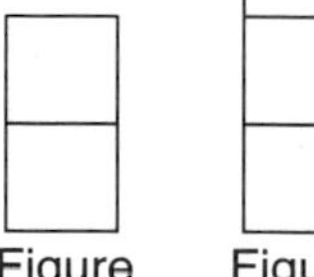

Guided Practice

Find the x- and y-intercepts and slope of the graph of each equation.

6. $y = x$
7. $10x + 35y = 280$
8. Graph $10x - 15y = 120$ by using the intercepts method.
9. Graph $y = -0.5x + 20$ by using the slope and the y-intercept.

Use the description to graph each line.

10. $m = -1$; passes through $A(0,1)$
11. line parallel to the line $y = 2x - 5$; passes through $B(0, -4)$
12. **Physics** The velocity of an object v in feet per second, that is dropped from a given height can be expressed by the equation $v = 32t$ where t is the time in seconds since the object was dropped.
    a. Graph $v = 32t$.
    b. What does the slope of the line represent?
    c. Why is the v-intercept 0?

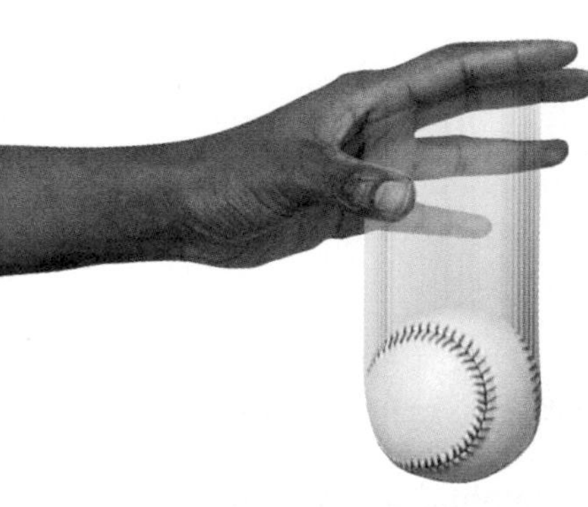

EXERCISES

Practice

Find the x- and y-intercepts and slope of the graph of each equation.

13. $y = -x$
14. $x + 3y = 0$
15. $x = 2$
16. $y = 0$
17. $x = 0$
18. $18x - 42y = 210$

Graph each equation. Explain the method you used.

19. $y = -4x + 3$
20. $x + y = 8$
21. $3x + 7y = 0$
22. $y = \frac{1}{4}x + 6$
23. $10x + 25y = 100$
24. $21x - 7y = 14$

Use the description to graph each line.

25. $m = 0$; $b = 2$
26. line perpendicular to the line $y = 2x - 5$; passes through $C(0, -4)$
27. line parallel to the y-axis through $D(-8, 15)$
28. line perpendicular to the y-axis through $E(-8, 15)$
29. $m = \frac{3}{10}$; passes through $F(0, 32)$

30. Graph several lines having a slope of -4. Describe how the equations of these lines are alike and how they are different.

31. Graph several lines passing through the point $G(0, -1)$. Describe how the equations of these lines are alike and how they are different.

32. Describe the relationship among the graphs of the three lines whose equations are $3x - y = 10$, $3y + x = 6$, and $y = 3x - 2$.

33. What is the slope of any line that is parallel to the y-axis? Explain why this is true.

34. The equations for two lines are $3x + 4y = 12$ and $Ax + By = 10$. Find an A and B that will satisfy each condition.
   a. The two lines are parallel.
   b. The two lines are perpendicular.

35. The equation of a line is $y = -2x + 6$. Suppose you added 3 to the x value and 2 to the y value of every ordered pair of points on the line.
   a. How would the graph of the line change?
   b. How would the equation of the line change?

Graphing Calculator

36. Enter the equation $y = -2x + 4$ into the Y= list of a TI-82/83 graphing calculator and then press ZOOM 6. Determine the equation of a line that is perpendicular to it, enter the equation in the Y= list, and press ZOOM 5 to visually check your answer.

Critical Thinking

37. The graphs of $5x + 2y = 12$ and $5x + 2y = 2$ are parallel lines. Find the equation of the line that is parallel to both lines and lies midway between them. Explain why your answer is correct.

Applications and Problem Solving

38. **Transportation** According to research by *Travel & Leisure* magazine, the equations $y = 1.80x + 1.70$, $y = 1.80x + 1.80$, and $y = 1.40x + 1.50$ represent the costs of a taxi in San Francisco, Philadelphia, and Atlanta, respectively. The slope is the cost per mile, and the constant term is the initial fee. Graph the equations on the same coordinate axes and compare the cost of taking a taxi in these three cities.

39. **Nutrition** The equation $C = 12f + 180$ describes the relationship between the amount of fat f in grams, and the number of Calories C in some food items.
   a. What is the slope of the line? What does it indicate about the relationship between fat and Calories in the food items?
   b. What is the y-intercept?
   c. Graph the equation.
   d. If a food item contains 30 grams of fat, how many Calories would you expect to be in that item?

40. **Civil Engineering** The maximum road grade recommended by the Federal Highway Commission is 12%. This means that the slope of a road should be no more than $\frac{12}{100}$ or 0.12. At the maximum grade, a road would change 12 feet vertically for every 100 feet horizontally.
   a. How many horizontal feet will it take for a road with the maximum grade to drop 240 feet?
   b. If the maximum grade for railroad tracks is 2%, how many more miles will it take a train to climb 500 feet than it will take a car on a road that has the maximum grade?

FYI

According to *The Guiness Book of World Records*, the steepest street in the world is Baldwin Street, Dunedin, New Zealand. It has a maximum road grade of 79%.

41. **Transportation** A small van can carry 10 people and a large van can carry 15 people. Let x represent the number of small vans and y the number of large vans. If 120 people need transportation to a camp in the mountains, an equation describing the transportation of the people is $10x + 15y = 120$.
    a. Explain what $10x$ represents.
    b. What are the intercepts? What does each represent?
    c. Graph the equation using the intercepts.
    d. Is it possible to have an arrangement of 6 small vans and 4 large vans? Explain why or why not.
    e. Does the line representing the equation pass through the point at (3, 6)? Explain why or why not.

Mixed Review

42. **World Records** The world's smallest model train was built by Jean Damery of Paris, France. Built in a 1:1000 scale, the engine on Damery's train measures $\frac{5}{16}$ inch. How long is the engine on a life-sized train similar to Damery's? (Lesson 11–8)

43. If the height of a parallelogram is 3 times the base, and the area of the parallelogram is 108 square meters, find the height and the base. (Lesson 10–3)

44. **ACT Practice** In the figure, what is the area of ΔABC in terms of x?

    A $60 \cos x$ B $120 \cos x$
    C $24 \tan x$ D $60 \sin x$
    E $120 \sin x$

B
10
$x°$
A 12 C

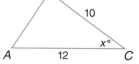

45. **Aviation** A jet takes off from Columbus International Airport at an angle of 22° with the ground. The jet covers 6.9 miles of ground distance before leveling off. At what altitude does the jet level off? (Lesson 8–3)

46. Determine whether the statement *The opposite angles of a parallelogram are supplementary* is true or false. (Lesson 6–1)

47. Graph $\triangle ABC$ and list the angles from least to greatest measure for $A(1, 3)$, $B(5, -2)$, and $C(-2, -4)$. (Lesson 5–4)

48. Find the measure of $\overline{AB}$ for $A(5, -2)$, and $B(-3, -7)$. (Lesson 3–5)

For **Extra Practice**, see page 787.

Algebra

49. Find the factors of 22.

50. Find $5y(8y^3 + 7y^2 - 3y)$.

WORKING ON THE
In•ves•ti•ga•tion

Refer to the Investigation on pages 642–643.

1 Begin creating your movie by drawing a central object or character.

2 Draw the image on a computer screen model.

3 Then create a list of pixels to be colored in different colors. You may wish to use lines to enclose the figure and list the colors as regions instead of individual pixels.

Add the results of your work to your Investigation Folder.

12–2A Using Technology Writing Equations

A Preview of Lesson 12–2

You can use a graphing calculator to plot points and determine an equation of the line that passes through the points.

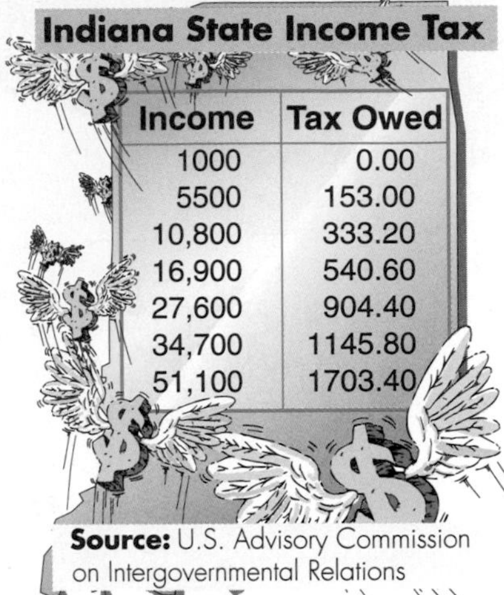

Indiana State Income Tax

Income	Tax Owed
1000	0.00
5500	153.00
10,800	333.20
16,900	540.60
27,600	904.40
34,700	1145.80
51,100	1703.40

Source: U.S. Advisory Commission on Intergovernmental Relations

The table at the right shows the amount of Indiana state income tax owed by several single taxpayers with no dependents. Find an equation for the amount of tax owed, y, if a taxpayer has an income of x dollars.

Step 1 Clear the statistical lists by pressing [STAT] 4 [2nd] [L1] [,] [2nd] [L2] [ENTER].

Step 2 Next, enter the data into the statistical lists.To access the lists, press [STAT] [ENTER]. Enter the income amounts into list L1. Then press the right arrow button and enter the tax amounts into list L2.

Step 3 To have the calculator find the equation of the line, press [STAT] [▶] then choose LinReg($ax + b$) and press [ENTER].

The calculator will display values for a and b for the equation $y = ax + b$. In this case, $a = 0.034$ and $b = -34$. So an equation for the amount of Indiana state tax owed by a single taxpayer with no dependents is $y = 0.034x - 34$.

EXERCISES

Study the results of the activity.

1. Choose any two ordered pairs from the list and find the slope between them.

2. Use the equation $y = mx + b$. Substitute the x and y values of one data point and the slope you found in Exercise 1 for x, y, and m, respectively. Then solve for b.

3. Write the equation for the relationship in slope-intercept form. How does the equation you found compare to the equation that the calculator found?

4. Use the calculator to find an equation that represents the data in the table at the right. Then follow the steps in Exercises 1–3 to find the equation. Compare your results.

x	y
−5	4.75
−3	5.25
0	6.0
2	6.5
6	7.5

12-2 Integration: Algebra Writing Equations of Lines

What YOU'LL LEARN

- To write an equation of a line given information about its graph, and
- to solve problems by using equations.

Why IT'S IMPORTANT

You can write equations to solve problems involving sports and computers.

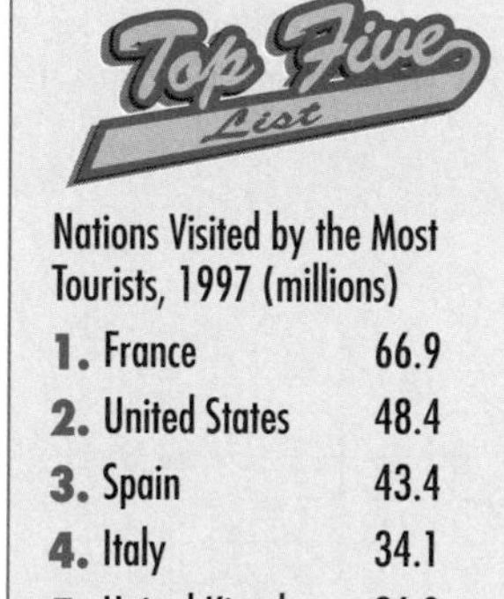

Top Five List

Nations Visited by the Most Tourists, 1997 (millions)

1. France	66.9
2. United States	48.4
3. Spain	43.4
4. Italy	34.1
5. United Kingdom	26.0

Real World APPLICATION

Tourism

Do you dream of visiting far away places? One of the world's most popular vacation spots is Rio de Janeiro, Brazil. The harbor there has been called one of the seven wonders of the natural world because of its beautiful, low mountain ranges.

Like most countries other than the United States, Brazil uses the Celsius scale to report their temperatures. In the United States, the Fahrenheit scale is used. According to U.S. Department of Commerce data, the high temperature in Rio de Janeiro averages 25° Celsius during the month of January. What is the equivalent Fahrenheit temperature?

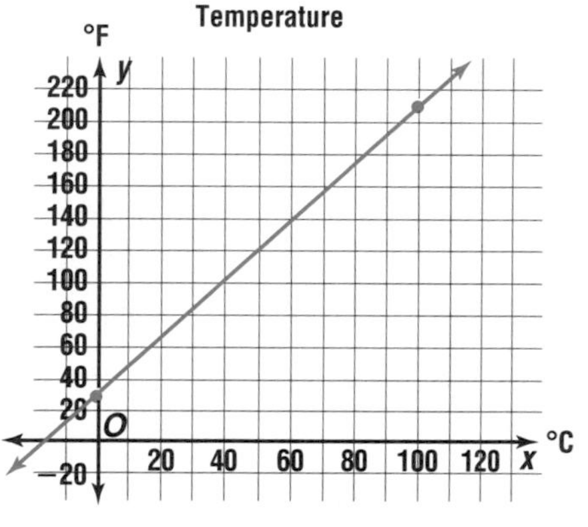

The graph at the left shows the relationship between temperatures in Celsius and temperatures in Fahrenheit. The y-intercept, 32, represents the temperature where water freezes. Let $A(100, 212)$ represent the temperature at which water boils. We can use the ordered pairs (0, 32) and (100, 212) to find the slope of the line.

$$\frac{y_2 - y_1}{x_2 - x_1} = \frac{212 - 32}{100 - 0}$$

$$= \frac{180}{100} \text{ or } \frac{9}{5}$$

Since the graph of this relationship is a line, it can be represented by a linear equation. If we substitute $\frac{9}{5}$ for m and 32 for b in $y = mx + b$, we can find the slope-intercept form of the equation for this line. Thus, the equation is $y = \frac{9}{5}x + 32$, where y represents the temperature in degrees Fahrenheit and x represents the temperature in degrees Celsius.

In general, you can write an equation of a line if you are given:

- **Case 1:** the slope and the y-intercept,
- **Case 2:** the slope and the coordinates of a point on the line, or
- **Case 3:** the coordinates of two points on the line.

The application above illustrates Case 1.

Example 1 illustrates Case 2. You can find an equation of a line given the slope and the coordinates of a point on the line using the **point-slope form** of a linear equation. The point-slope form is $y - y_1 = m(x - x_1)$, where (x_1, y_1) are the coordinates of a point on the line and m is the slope of the line.

Example 1

Write an equation of the line whose slope is −2 and passes through the point at (4, 10).

Method 1: Point-Slope Form

$y - y_1 = m(x - x_1)$ *Use point-slope form.*

$y - 10 = -2(x - 4)$ *Substitute −2 for m and (4, 10) for (x_1, y_1).*

The point-slope form of the equation is $y - 10 = -2(x - 4)$.

Method 2: Slope-Intercept Form

Since we know that the slope is −2, we can substitute −2 for m in $y = mx + b$.

$$y = -2x + b$$

The point at (4, 10) is on the line. Substitute the coordinates of this point into the equation to find the value of b.

$y = -2x + b$

$10 = -2(4) + b$ *y = 10 and x = 4*

$10 = -8 + b$

$18 = b$ *Add 8 to each side.*

The slope-intercept form of the equation of the line is $y = -2x + 18$.

The first step in Case 3 is to find the slope by using the two ordered pairs. Then you can write the equation in point-slope form by using either of the two ordered pairs and simplifying.

Example 2

Write an equation for each line.

a. line ℓ

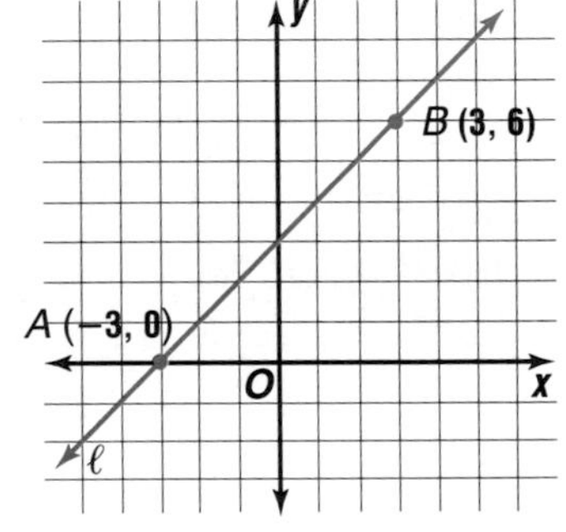

Find the slope of line ℓ by using the points $A(-3, 0)$ and $B(3, 6)$.

$$m = \frac{y_2 - y_1}{x_2 - x_1} = \frac{6 - 0}{3 - (-3)} = \frac{6}{6} \text{ or } 1$$

Now use the point-slope form to write the linear equation.

$y - y_1 = m(x - x_1)$ *Use point-slope form.*

$y - 0 = 1(x - (-3))$ *The slope is 1, and the point with coordinates (−3, 0) is on the line.*

$y = x + 3$

The slope-intercept form of the equation of the line is $y = x + 3$.

b. a line that is parallel to line ℓ and passes through the point at (4, 3)

Since the slope of line ℓ is 1, the slope of a line parallel to it is 1. *Why?*

$y - y_1 = m(x - x_1)$ *Use point-slope form.*

$y - 3 = 1(x - 4)$ *The slope is 1 and the point with coordinates (4, 3) is on the line.*

$y - 3 = x - 4$

$y = x - 1$

The slope-intercept form of the equation of the line is $y = x - 1$.

One strategy for solving problems is to **write and solve an equation.**

Example 3 **In 1998, the first year of operation, the publisher of a computer magazine had revenues of \$2.9 million. By 2000, revenues were \$25.2 million. If the increase in earnings remains constant, how much will the company earn in its eighth year of operation?**

PROBLEM SOLVING

Write an Equation

Explore You know how much the company made in the third year and how much it made in the first year.

Plan Define the variables and write a linear equation that describes the company's earnings.

Let d represent dollars and let t represent time in years.

Solve The line passes through the points at (1, 2.9) and (3, 25.2). Find the slope of the line.

$$m = \frac{d_2 - d_1}{t_2 - t_1}$$

$$= \frac{25.2 - 2.9}{3 - 1}$$

$$= \frac{22.3}{2} \text{ or } 11.15$$

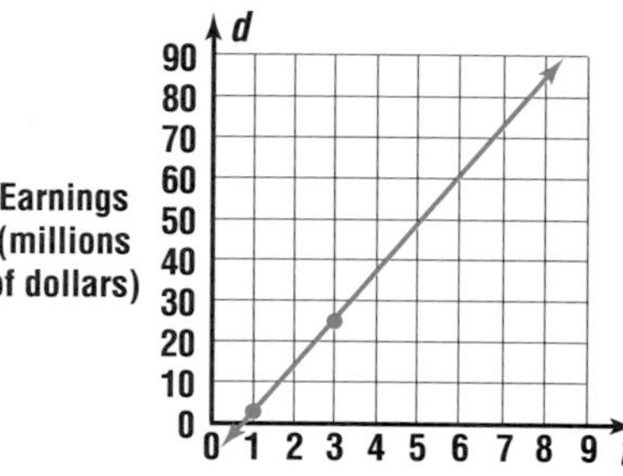

Now use the point-slope form to write the linear equation.

$d - d_1 = m(t - t_1)$ *Use point-slope form.*

$d - 25.2 = 11.15(t - 3)$ *Use the coordinates of either point for (t_1, d_1).*

$d - 25.2 = 11.15t - 33.45$ *We chose (3, 25.2).*

$d = 11.15t - 8.25$

Use the equation to find how much money the company is predicted to earn in its eighth year.

$$d = 11.15t - 8.25$$

$$d = 11.15(8) - 8.25$$

$$d = 89.2 - 8.25$$

$$d = 80.95$$

If the rate of increase remains the same, the company will earn \$80.95 million in its eighth year. This is represented by the ordered pair (8, 80.95).

Examine Check by looking at the x-coordinate of the graph of the line when y is about 80.95.

CHECK FOR UNDERSTANDING

Communicating Mathematics

Study the lesson. Then complete the following.

1. **Explain** how you would write the equation of a line whose slope is 3 and y-intercept is -10.
2. **Write** equations for two different lines that pass through the point at (7, 2).
3. **Explain** why it is important to define a variable before you write an equation to solve a problem.

4. **State** the slope and y-intercept of line a. Then write its equation in slope-intercept form.
5. **Explain** why the slope of line b is 1 if line a is perpendicular to line b.

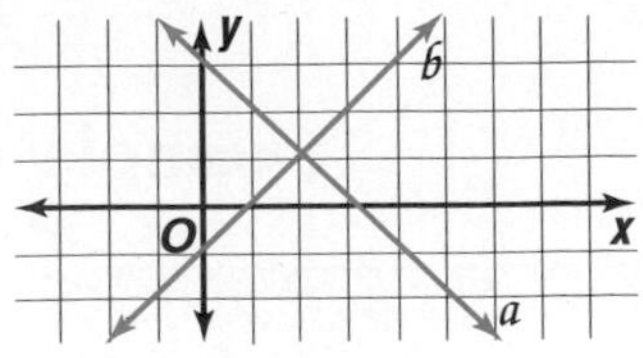

Guided Practice

Write the equation in slope-intercept form of the line having the given slope and passing through the point with the given coordinates.

6. 4, (1, 5)
7. $-\frac{1}{3}$, (−2, 4)

Write the equation in slope-intercept form of the line that satisfies the given conditions.

8. $m = -3$; y-intercept $= 5$
9. $m = \frac{1}{3}$; passes through the point at (6, −1)
10. $m = 5$; x-intercept is −2
11. perpendicular to $y = \frac{1}{2}x - 5$; through the point at (4, −1)
12. **Business** A news retrieval service allows a customer to have news delivered to his or her computer via the Internet. The Dow Jones News Retrieval Service, which specializes in stock market news, advertises a \$29.95 start-up fee and a \$19.95 yearly fee with the first year's fee waived.
    a. Write an equation that will show the cost of using the service over the years. What assumptions do you have to make?
    b. Explain how you would know if the point at (3, 89.8) is on the graph of the equation.

EXERCISES

Practice

Refer to the graph for Exercises 13–15. State the slope and y-intercept for each line. Then write the equation of the line in slope-intercept form.

13. a
14. b
15. c

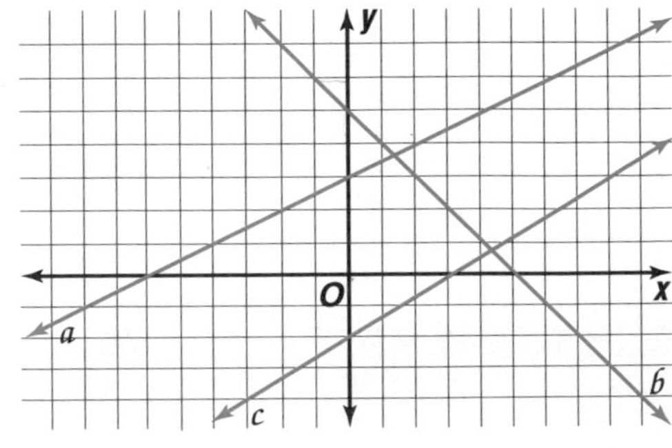

Write the equation in slope-intercept form of the line having the given slope and passing through the point with the given coordinates.

16. 0.3, (0, −6)
17. 0.5, (40, 19)
18. $\frac{1}{3}$, (−3, −15)
19. 1, (2, 4)
20. $-\frac{3}{4}$, (24, 3)
21. 0, (2.5, 8)

Write the equation in slope-intercept form of the line that satisfies the given conditions.

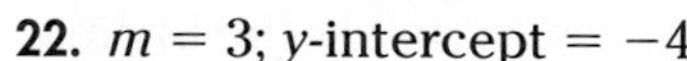

22. $m = 3$; y-intercept $= -4$
23. $m = 0$; y-intercept $= 6$
24. x-intercept is 5; y-intercept is 3
25. passes through points at $(2, -3)$ and $(-1, -9)$
26. parallel to $y = \frac{1}{2}x - 5$; passes through the point at $(4, -1)$
27. $m = -0.1$; passes through the point at $(20, 17)$
28. $m = 5$; passes through the point at $(-8, -7)$
29. parallel to $y = 5x - 2$; passes through the point at $(4, 0)$

30. parallel to the y-axis; passes through the point at $(4, -3)$
31. perpendicular to the y-axis; passes through the point at $(-5, 10)$
32. x-intercept is 5; y-intercept is -1
33. passes through points at $(4, -1)$ and $(-2, -1)$
34. passes through points at $(5, 7)$ and $(-3, 11)$

Use the information to write an equation representing the cost in relation to time.

35. A lawn mower costs \$340, and the price has increased an average of \$20 per year.
36. The cost of a local call from a pay phone was 25¢ in 1985. The cost has not changed since then.
37. Line a has slope $-\frac{3}{5}$ and passes through the point at $(6, 3)$. Line b is perpendicular to line a. Write an equation for line b in slope-intercept form if lines a and b have the same x-intercept.
38. Write the equation in slope-intercept form of the line that is the perpendicular bisector of the segment whose endpoints have the coordinates $(5, -3)$ and $(11, 7)$.
39. $\angle ABC$ is a right angle. If the coordinates of points A and B are $(4, 6)$ and $(5, 9)$, respectively, write an equation for line BC.
40. Verify that $\triangle DEF$ is a right triangle for $D(0, -4)$, $E(2, -3)$, and $F(-4, 9)$.

Programming

41. The graphing calculator program below finds the equation of a line containing two points at (x_1, y_1) and (x_2, y_2). Then it displays the equation in slope-intercept form.

```
PROGRAM:WRITEQN
:ClrDraw:AxesOff
:Input "X1=", Q
:Input "Y1=", R
:Input "X2=", S
:Input "Y2=", T
:Text (8, 5, "EQUATION IS")
:If S-Q = 0
:Goto 1
:(T−R)/(S−Q)→M
:(R-M*Q)→B
:If M=0
:Then
:Text (14, 5, "Y=", R)
:Else
:If B<0
:Then
:abs(B)→B
:Text (14, 5, "Y= ", M,
 "X−", B)
:Else
:Text (14, 5, "Y= ", M,
 "X+", B)
:Stop
:Lbl 1
:Text (14, 5, "X= ", S)
```

a. Use the program to find the equation of the line that passes through $A(6, -3)$ and $B(3, -2)$.

b. Use the program to check your equations in Exercises 33–34.

Critical Thinking

42. The inverse of any equation is found by interchanging the variables and solving the equation for y.
  a. Determine the equation that is the inverse of $y = 0.125x + 11$.
  b. Write an equation that is its own inverse.
  c. Compare the graphs of any two inverse equations with the graph of the equation that is its own inverse. Describe the relationship.

43. Study the design at the right.
  a. Create a similar design by writing four equations and graphing them.
  b. If you moved the center of the design from (0, 0) to (1, 3) how would your equations change?

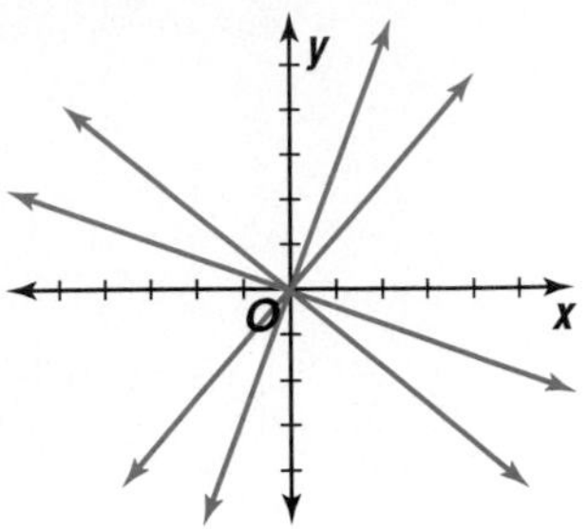

Applications and Problem Solving

44. **Write an Equation** A long distance phone company advertises that with its new program, calls during the prime daytime hours cost $0.22 per minute and calls during evening hours are $0.10 per minute. Zina Harrison's bill for long distance calls in September was $28.52 before taxes. If she was billed for eight more than twice as many minutes during the evening as during the day, how many minutes was she billed for use during the day?

45. **Computers** An Integrated Services Digital Network, or ISDN, uses telephone lines to transmit digital signals. The charges for an ISDN depend on the distance from the company to your home. A telephone company charges one ISDN customer $192 for 6 months of service and $576 for 18 months of service. The bill for a second customer was $456 for one year and $570 for 15 months of service.
  a. Write an equation that describes the cost of using each ISDN in terms of months of service.
  b. Describe the difference between the bills for the two customers based on what you know from the equations.

46. **Sports** It costs about $1200 to equip a typical Olympic runner in 1996. The costs are expected to increase $82 per year. Write the linear equation in slope-intercept form that represents the approximate cost of equipping a runner in the future. Assume that the rate of increase remains constant.

Mixed Review

47. Graph $6x - 4y = 3$. State the slope and y-intercept of the line. (Lesson 12–1)

48. Find the volume of a cylinder with a radius of 5 meters and a height of 2 meters. (Lesson 11–5)

49. **Entertainment** One midway game at the Virginia State Fair is a dart game involving a board that is 6 feet by 8 feet with forty 8-inch-square targets attached. If you throw a dart at random and it hits the board, what is the probability that you will hit a target? (Lesson 10–6)

Refer to the graph at the right for Exercises 50 and 51.

50. Write the equation of circle R. (Lesson 9–8)

51. Describe a circle S that has no common tangents with circle R. (Lesson 9–1)

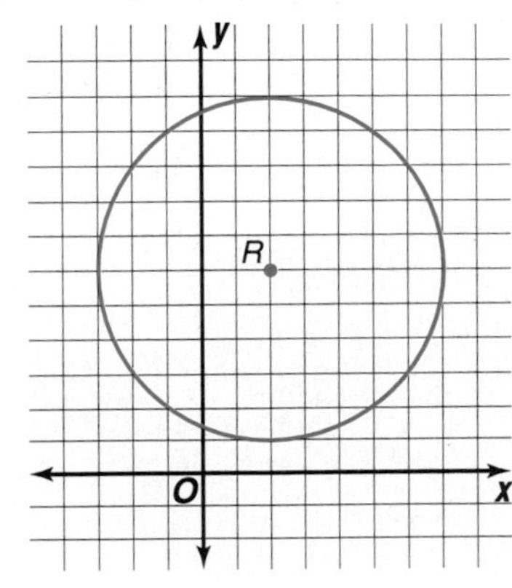

52. **ACT Practice** If $ax = bx + c$, then what is the value of x in terms of a, b, and c?

A $\frac{c}{a+b}$ **B** $\frac{b}{a+c}$ **C** $\frac{c}{a-b}$ **D** $\frac{b}{c-a}$ **E** $\frac{b+c}{a}$

53. Suppose $\triangle EFG \sim \triangle JKL$. If $EG = 3.6$, $JL = 2.4$, and the perimeter of $\triangle JKL$ is 6.8, find the perimeter of $\triangle EFG$. (Lesson 7–5)

Proof

54. **Given:** $\overline{MN} \cong \overline{QP}$

$\overline{MQ} \cong \overline{NP}$

Prove: $\overline{MN} \parallel \overline{QP}$

(Lesson 4–4)

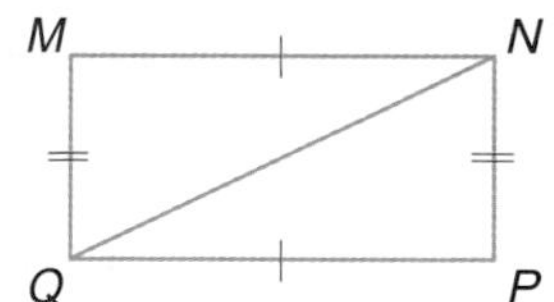

INTEGRATION **Algebra**

55. Simplify $\frac{8a^2bc}{28abc^2}$. State the excluded values of a, b, and c.

56. **Cosmetics** The graph shows the percent of teenage girls who use each different type of cosmetic.

a. Do you think there are girls who use more than one type of cosmetic listed? Explain.

b. Would a circle graph be appropriate for displaying this data? Why or why not?

Source: Simmons Market Research Bureau

For **Extra Practice**, see page 788.

Mapping the Undersea Landscape

The excerpt below appeared in an article in *Earth* in June, 1996.

TOP-SECRET DATA RECENTLY declassified by the Navy has enabled scientists to view the seafloor almost as if the oceans had been drained completely of water....scientists knew...that much of the seafloor's topography is mirrored on the surface by slight variations in sea level. This is due to the subtle gravitational pull exerted by seamounts and other hidden features of the deep....a satellite called Geosat...carried out its mission for the Navy, spinning a dense web of tracks over the oceans while conducting a detailed global survey of the height of the sea surface....The gravity data reveals all discrete seafloor features larger than about 3,000 feet high and six miles across....a vast improvement over prior global maps. ■

1. Who might benefit from having these improved maps of the ocean floor?
2. Which type of orbit did the Geosat satellite have to follow in order to map Earth's oceans, north-to-south between the poles or east-to-west along the equator? Explain.
3. Ships using sonar could provide even more detailed maps of the ocean floor. What might be the main advantage of using a satellite for this task?

12-3 Integration: Algebra and Statistics Scatter Plots and Slopes

***What* YOU'LL LEARN**

- To relate statistics and equations of lines to geometric concepts.

***Why* IT'S IMPORTANT**

Statistics are used to find relationships between quantities in fields such as education and sports.

Real World APPLICATION

Aviation

Many proponents of the 55 mile per hour speed limit on U.S. highways say that it is less expensive to run a car at lower speeds. The average speed at which a car is operated is just one of many factors that affect its operating costs. Do you think that the speed of an airplane affects its cost of operation? The table at the right shows the average airborne speed and average operating cost for the most commonly used models.

Plane	Speed (mph)	Cost per hr
B747-100	478	$5571
DC-10-10	439	$4504
B747-400	507	$6592
DC-10-40	444	$3904
B727-200	357	$2241
L-1011-100/200	429	$3783
DC-10-30	479	$4349
A300-600	393	$3648
MD11	484	$5024
L-1011-500	486	$3958
B757-200	397	$2362

Source: *The World Almanac*

Because an airplane must maintain a minimum speed to remain airborne, the reasonable values of the domain and range are restricted.

A **scatter plot**, like the one below, shows the relationship between speed and cost by plotting a set of data as ordered pairs in the coordinate plane. In the graph below, we can fit a line as a way to summarize the data, and find an equation that will express the approximate cost in terms of speed.

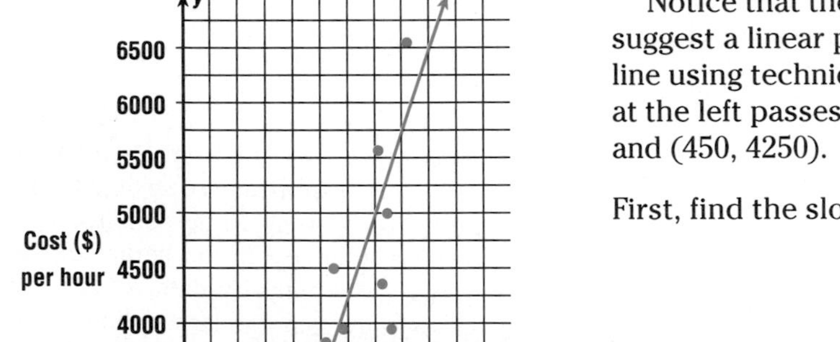

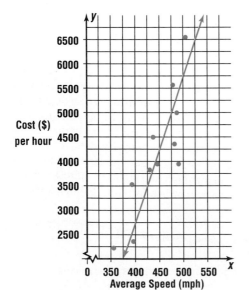

Notice that the points do not lie in a straight line, but they do suggest a linear pattern. You can determine the equation for this line using techniques from algebra. The line drawn in the graph at the left passes through points with coordinates (525, 6500) and (450, 4250).

First, find the slope of the line.

$$m = \frac{y_2 - y_1}{x_2 - x_1}$$

$$= \frac{6500 - 4250}{525 - 450}$$

$$= \frac{2250}{75} \text{ or } 30$$

Substitute the slope and the coordinates of one of the points into the point-slope form to find the equation.

$$y - y_1 = m(x - x_1)$$

$$y - 4250 = 30(x - 450)$$ *The slope is 30, and one point has coordinates (450, 4250).*

$$y - 4250 = 30x - 13{,}500$$

$$y = 30x - 9250$$

An equation that relates the airborne speed and the cost is $y = 30x - 9250$. *Since the scatter plot shows a pattern, it is reasonable to say that the average speed at which an airplane flies may be a factor in its cost of operation.*

You can use the equation to predict the cost per hour for a given speed or find the speed that predicts a given cost per hour. Find the approximate cost of traveling at a speed of 500 miles per hour by substituting 500 for x in the equation.

$$y = 30x - 9250$$
$$y = 30(500) - 9250$$
$$y = 15{,}000 - 9250 \text{ or } 5750$$

The cost per hour will be approximately $5750.

Whenever you work with coordinates, it is a good idea to begin by graphing the information you are given.

Example  **Determine whether $K(-8, 12)$, $E(4, -6)$, and $B(6, -9)$ are collinear.**

One way to approach the problem is to find the equation of $\overleftrightarrow{KE}$ and see if the coordinates of B satisfy the equation. That is, does B lie on $\overleftrightarrow{KE}$? First find the slope of $\overleftrightarrow{KE}$.

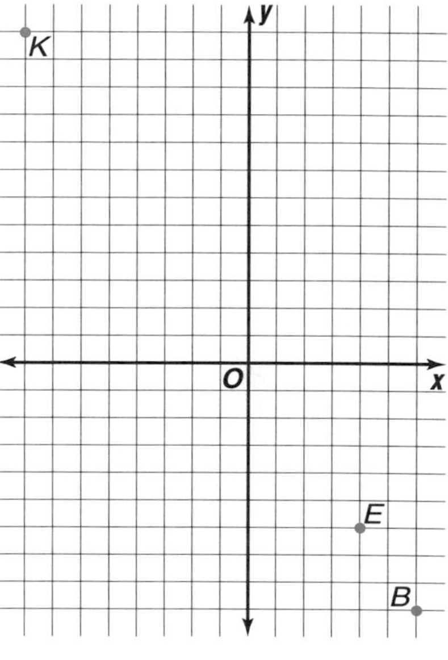

$$m = \frac{y_2 - y_1}{x_2 - x_1}$$
$$= \frac{12 - (-6)}{-8 - 4}$$
$$= \frac{18}{-12} \text{ or } -\frac{3}{2}$$

$$y - y_1 = m(x - x_1)$$
$$y - 12 = -\frac{3}{2}(x - (-8))$$
$$y - 12 = -\frac{3}{2}x - 12$$
$$y = -\frac{3}{2}x$$

The equation of $\overleftrightarrow{KE}$ is $y = -\frac{3}{2}x$. Since $-9 = -\frac{3}{2}(6)$, $B(6, -9)$ satisfies the equation, and the points are collinear.

You can also use algebraic techniques to find the coordinates of the point of intersection of two lines.

Example 2

Algebra

Find the coordinates of the point of intersection of the medians $\overline{CD}$ and $\overline{BE}$ for $\triangle ABC$ if the vertices are $A(4, -5)$, $B(8, 11)$, and $C(0, 15)$.

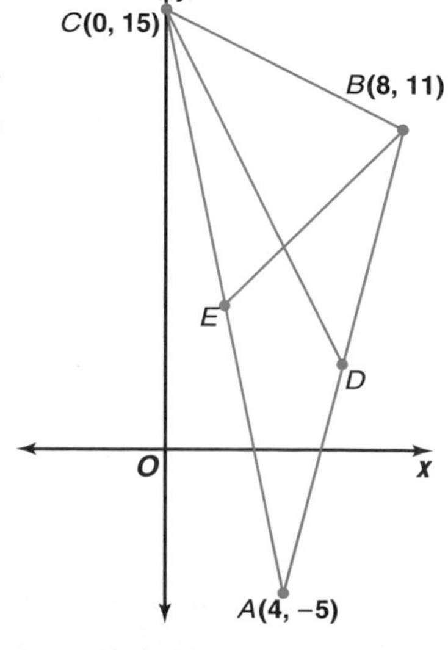

The median is a segment whose endpoints are the vertex of an angle and the midpoint of the side opposite that angle. To find the coordinates of D and E, first find the midpoints of $\overline{AB}$ and $\overline{AC}$. Use the midpoint formula.

(continued on the next page)

midpoint of $\overline{AB}$

$$D(x, y) = \left(\frac{x_1 + x_2}{2}, \frac{y_1 + y_2}{2}\right) = \left(\frac{4 + 8}{2}, \frac{-5 + 11}{2}\right) = (6, 3)$$

midpoint of $\overline{AC}$

$$E(x, y) = \left(\frac{x_1 + x_2}{2}, \frac{y_1 + y_2}{2}\right) = \left(\frac{4 + 0}{2}, \frac{-5 + 15}{2}\right) = (2, 5)$$

Now find the equation of the line containing each median. Find the slope of each segment.

slope of $\overline{CD}$

$$m = \frac{y_2 - y_1}{x_2 - x_1} = \frac{15 - 3}{0 - 6} = \frac{12}{-6} \text{ or } -2$$

Since the point at (0, 15) is the y-intercept, use the slope-intercept form.

$y = mx + b$

$y = -2x + 15$

slope of $\overline{BE}$

$$m = \frac{y_2 - y_1}{x_2 - x_1} = \frac{11 - 5}{8 - 2} = \frac{6}{6} \text{ or } 1$$

Use the point-slope form.

$y - y_1 = m(x - x_1)$

$y - 5 = 1(x - 2)$

$y - 5 = x - 2$

$y = x + 3$

Find the coordinates of the intersection of the lines containing the medians, $y = -2x + 15$ and $y = x + 3$.

$y = x + 3$

$-2x + 15 = x + 3$ *Substitution Property of Equality*

$-3x = -12$

$x = 4$

Substitute 4 into one of the equations to find y.

$y = x + 3$

$y = 4 + 3$

$y = 7$

The two medians of the triangle intersect at (4, 7).

CHECK FOR UNDERSTANDING

Communicating Mathematics

Study the lesson. Then complete the following.

1. **Explain** a way in which a scatter plot would be useful.
2. **Describe** two ways to determine whether three points are collinear.
3. **You Decide** Conchita and Dalila were analyzing a scatter plot of data. Conchita decided to draw a line to summarize the data by connecting the first and the last data point. Dalila thought she had a better procedure because she drew a line that had the same number of data points on either side. Who is correct? Explain your reasoning.

Guided Practice

Determine whether the three points listed are collinear.

4. $B(4, 9)$, $C(7, 18)$, $D(-2, -9)$
5. $R(-6, 10)$, $S(-1, -1)$, $T(5, 1)$

Side $\overline{RS}$ of $\triangle RST$ lies on line a. The equation of line a is $y = -\frac{2}{5}x + 8$. The coordinates of T are $(-10, 15)$. Identify each of the following as *true* or *false*. If the statement is false, explain why.

6. The altitude from T to $\overline{RS}$ lies on the line $y = -\frac{5}{2}x - 4$.
7. Point R could have coordinates $(10, 4)$.
8. Point T could be at the intersection of line a and the line whose equation is $y = 4x - 20$.
9. Point S could be at the intersection of line a and the line whose equation is $x = 5$.
10. **Sports** The table below contains the winning times in selected Olympics for the men's 100-meter backstroke.

YEAR	WINNER, COUNTRY	TIME
1912	Arno Bieberstein, Ger	84.6
1920	Warren Kealoha, USA	75.2
1928	George Koiac, USA	68.2
1936	Adolph Kiefer, USA	65.9
1952	Yoshinobu Oyakawa, USA	65.4
1960	David Thiele, Australia	61.9
1972	Roland Matthes, E Ger	56.58
1980	Bengt Baron, Sweden	56.33
1988	Daichi Suzuki, Japan	55.05
1996	Jeff Rouse, USA	54.10

Source: *The World Almanac*, 1996

   a. Draw a scatter plot to show how year x and time y are related.
   b. Write an equation that relates the year to the approximate winning time.
   c. Predict the winning time for the 2000 Olympics Games.
   d. If you used this equation to predict for the year 2020, what time would you get? Is this a reasonable prediction? Explain.

Jeff Rouse

EXERCISES

Practice

Determine whether the three points listed are collinear.

11. $A(0, 0), B(6, 4), C(-3, -2)$
12. $D(19, -13), E(14, 18), F(10, 10)$
13. $M(6, 10), N(5, 10), P(-2, -6)$
14. $G(0, 0), H(4, 12), J(7, 21)$
15. $R(1, -3), S(4, -18), T(-3, 17)$
16. $U(4, 3), V(-2, -9), W(-7, 9)$

17. The table at the right lists United States birthrates. The rates are per 1000 people living in the United States.
   a. Draw a scatter plot to show how year x and birthrate y are related.
   b. Write an equation that relates the year to the approximate birthrate.
   c. What does the slope tell you about the birthrate every ten years?
   d. Predict the birthrate for 2000. How reliable do you think your prediction is? Explain your reasoning.

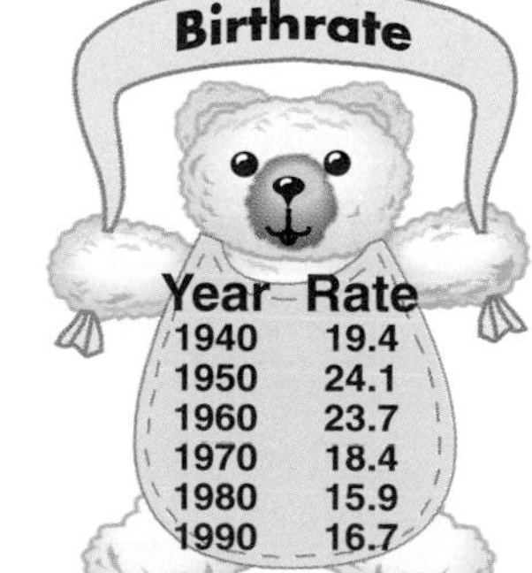

Birthrate

Year	Rate
1940	19.4
1950	24.1
1960	23.7
1970	18.4
1980	15.9
1990	16.7

Source: *National Center for Health Statistics*

18. Find the equations of the lines containing the legs of an isosceles triangle if they meet at $(3, 8)$ and a vertex of a base angle is at the x-intercept of the graph of $-x + 7y = 3$, which contains the base.

The vertices of $\triangle ABC$ are $A(2, 18)$, $B(-2, -4)$, and $C(6, 12)$.

19. Write the equation of the line containing the altitude to $\overline{BC}$.

20. Write the equation of the line containing the perpendicular bisector of $\overline{BC}$.

21. Write the equation of the line containing the median from A to $\overline{BC}$.

22. The vertices of $\triangle RST$ are $R(-10, 2)$, $S(-2, -4)$, and $T(-3, 3)$. Find the equation of the altitude to $\overline{RS}$ and of the median to $\overline{RS}$. What conclusion can you make?

23. The vertices of $\triangle ABC$ are $A(-4, -6)$, $B(-2, 8)$, and $C(6, 4)$.
    a. Write the equation of the line through D and E, the midpoints of $\overline{AB}$ and $\overline{AC}$, respectively.
    b. What is the relationship between the line through D and E and the line containing $\overline{BC}$? Explain your reasoning.
    c. Find the distance between $\overline{BC}$ and $\overline{DE}$.

24. Find the coordinates of D, E, and F in $\triangle DEF$ if the vertices lie on the lines $y = 4x + 7$, $y = x - 2$, and $y = \frac{2}{5}x + 3\frac{2}{5}$.

An equation of a circle is $(x - 3)^2 + (y - 5)^2 = 100$.

25. Show that the point with coordinates $(-3, 13)$ is on the circle. Find another point you know will be on the circle.

26. Is the point at $(-2, 5)$ inside the circle or outside the circle? Explain.

27. Write the equation of the tangent to the circle at $(-3, 13)$.

28. Consider the points $A(4, 15)$, $B(-2, 7)$, and $C(13, 2)$.
    a. Find D such that D is $\frac{1}{3}$ the distance from $\overline{AB}$ to C.
    b. The median fit line is the line that passes through D and is parallel to $\overline{AB}$. Write the equation of the median fit line.
    c. Point D could be considered an "average" point for the triangle. Explain.

Critical Thinking

29. Plot the points $A(2, 14)$, $B(0, 10)$, $C(12, 34)$ and $D(6, 8)$. Which of the two equations, $y = 1.8x + 7.3$ or $y = 2x + 10$, seems like a better fit for summarizing the linear pattern? Explain your reasoning.

Applications and Problem Solving

Real World

30. **Education** The SAT is a college entrance examination given in two parts, mathematics and verbal. The 1998 mean math and verbal scores for the twelve most populous states are shown in the table below.

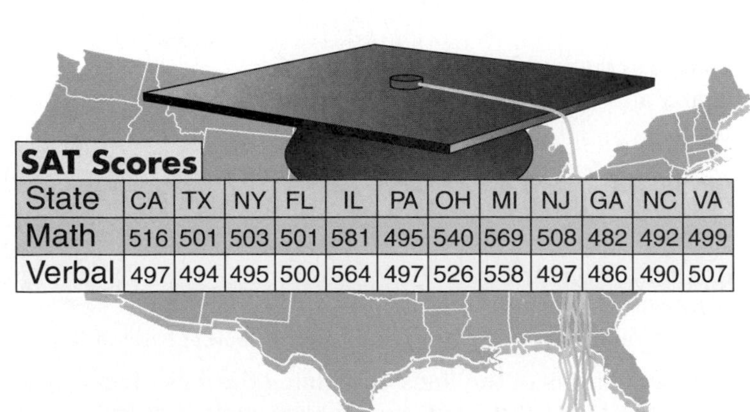

SAT Scores

State	CA	TX	NY	FL	IL	PA	OH	MI	NJ	GA	NC	VA
Math	516	501	503	501	581	495	540	569	508	482	492	499
Verbal	497	494	495	500	564	497	526	558	497	486	490	507

Source: The College Entrance Examination Board

a. Draw a scatter plot to show how the scores are related.
b. Does it make a difference if you plot (verbal, math) or (math, verbal)? Explain why or why not.
c. Write an equation that relates a state's mean math score to its mean verbal score.
d. The mean SAT score on the math portion in the state of Maine was 501 in 1998. What prediction might you make for its mean verbal score?

Car Costs

Car	Original price	Operating cost per year	Fixed cost per year
A	\$24,087	\$2771	\$7978
B	\$42,638	\$2910	\$11,383
C	\$21,047	\$2585	\$7183
D	\$12,543	\$2006	\$5210
E	\$20,286	\$2597	\$7430
F	\$10,402	\$1913	\$4902
G	\$16,228	\$2365	\$6499
H	\$15,141	\$2099	\$5734
I	\$32,895	\$2899	\$11,830
J	\$20,288	\$2214	\$7872
K	\$15,470	\$1948	\$6246

31. **Statistics** The base price for buying a new car in a recent year is given in the table along with the annual operating costs and fixed costs. Operating costs include items such as gas, oil, maintenance, and repairs. Fixed costs include insurance, license, taxes, and finance charges.
    a. Make a scatter plot using any two columns containing dollar amounts.
    b. Draw a line that represents the data. Find the equation of the line.

Mixed Review

32. *True* or *false:* To find an equation for the line that passes through the points at (21, 4) and (5, 0), first find the y-intercept and then find the slope. (Lesson 12–2)
33. Find the x- and y-intercepts and the slope of the graph of $3x - 6y = 18$. (Lesson 12–1)
34. Find the volume of a sphere with a radius of 5 inches. (Lesson 11–7)

35. **SAT Practice Grid-in** Segment AB is perpendicular to segment BD. Segment AB and segment CD bisect each other at point x. If $AB = 16$ and $CD = 20$, what is the length of BD?
36. **Food** A round pizza is cut into 12 congruent pieces. What is the measure of the central angle of each piece? (Lesson 9–2)
37. *True* or *false:* In $\odot C$, if D is a point on the circle, then $\overline{CD}$ is a radius of the circle. (Lesson 9–1)

For Extra Practice, see page 788.

Algebra

38. Find $(m + 3n)^2$.
39. Find $(2a^2 + 3a - 7) \div (2a + 1)$.

SELF TEST

Find the x- and y-intercepts and slope of the graph of each equation. (Lesson 12–1)

1. $y = 12$
2. $3x - 4y = 24$

Use the description to graph each line. (Lesson 12–1)

3. $m = -2$; $b = 3$
4. perpendicular to the line $y = 3x + 4$; passes through $A(1, 2)$

Write the equation in slope-intercept form of the line that satisfies the given conditions. (Lesson 12–2)

5. $m = 5$, y-intercept = 15
6. $m = 0.35$; passes through the point at (4, 1.9)
7. passes through points at $(-10, -1)$ and $(8, 1)$
8. **Sports** Nascha is lining a soccer field. The length should be 75 yards shorter than 3 times the width. If the perimeter is 370 yards, what are the length and the width of the field? (Lesson 12–2)

Determine whether the three points listed are collinear. (Lesson 12–3)

9. $A(9, 0)$, $B(4, 2)$, $C(2, -1)$
10. $L(0, 4)$, $M(2, 3)$, $N(-4, 6)$

12-4 Coordinate Proof

What YOU'LL LEARN
- To prove theorems using coordinate proofs.

Why IT'S IMPORTANT

You can use coordinate proofs to prove theorems in geometry and algebra.

You can use paper folding to discover geometric relationships that can be proved using a coordinate proof.

Linear Equations

Materials:  scissors  ruler 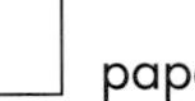 paper

- Cut a quadrilateral *ABCD* with no two sides parallel or congruent from a piece of paper.
- Fold to find the midpoint of $\overline{AB}$ and pinch to make a crease at the point. Mark the point with a pencil.
- Fold to find the midpoints of $\overline{BC}$, $\overline{CD}$, and $\overline{DA}$. Mark the points.
- Draw the quadrilateral formed by the midpoints of the segments.

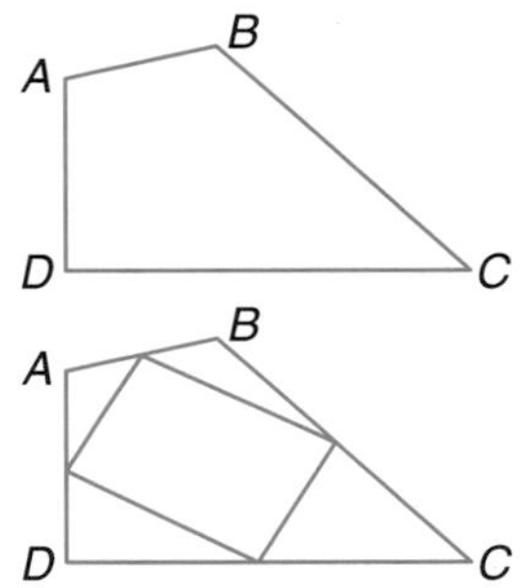

Your Turn

a. Measure each side of the quadrilateral determined by the midpoints. Record your measurements on the sides.

b. What type of quadrilateral is formed by the midpoints? Justify your answer.

In the Modeling Mathematics activity, you discovered that the quadrilateral formed by connecting the midpoints of the sides of any quadrilateral is a parallelogram. You can prove this using a **coordinate proof**. In geometry, we can assign coordinates to figures and use the coordinates to prove theorems. An important part of planning a coordinate proof is the placement of the figure on the coordinate plane. *This proof will be completed in Example 2.*

Guidelines for Placing Figures on a Coordinate Plane	**1. Use the origin as a vertex or center.** **2. Place at least one side of a polygon on an axis.** **3. Keep the figure within the first quadrant if possible.** **4. Use coordinates that make computations as simple as possible.**

Example  **Position and label a parallelogram with one side *a* units long on the coordinate plane.**

- Use the origin as vertex *A* of the parallelogram.
- Place the side of length *a* on the positive *x*-axis. Label the vertices *A*, *B*, *C*, and *D*.
- Since *B* is on the *x*-axis, its *y*-coordinate is 0. Its *x*-coordinate is *a* because $\overline{AB}$ is *a* units.

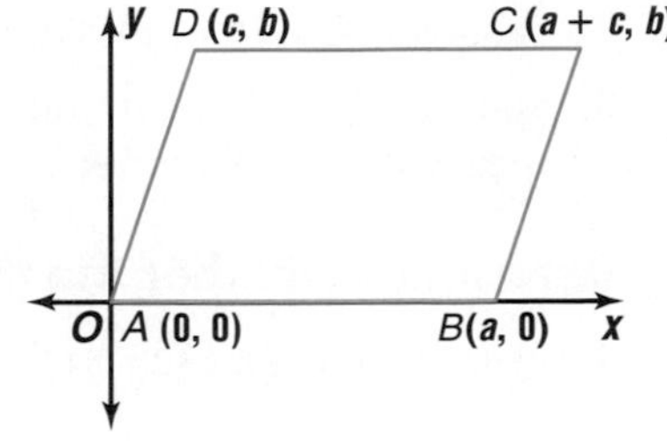

- Let D have coordinates (c, b).
- Since the altitude in a parallelogram from D is equal to the altitude from C, the y-coordinate at C will be the same as the y-coordinate at D. The x-coordinate at C will be the sum of the x-coordinates at B and D, so C will have coordinates $(a + c, b)$.

Some examples of figures placed on the coordinate plane are given below. The figures have been placed so that the coordinates of the vertices are as simple to use as possible.

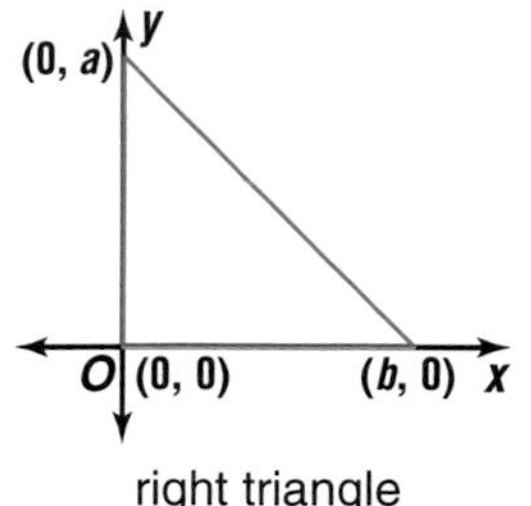

right triangle

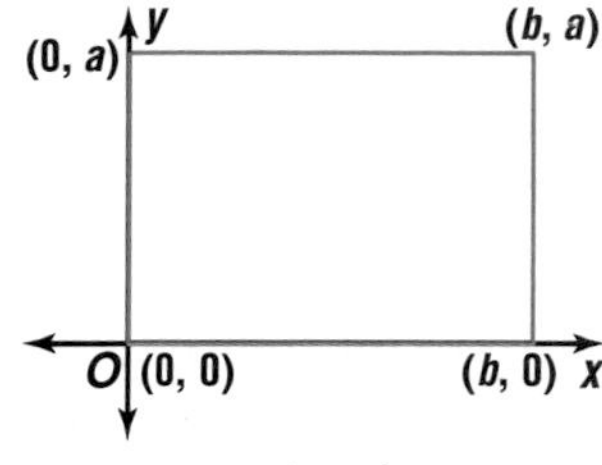

rectangle

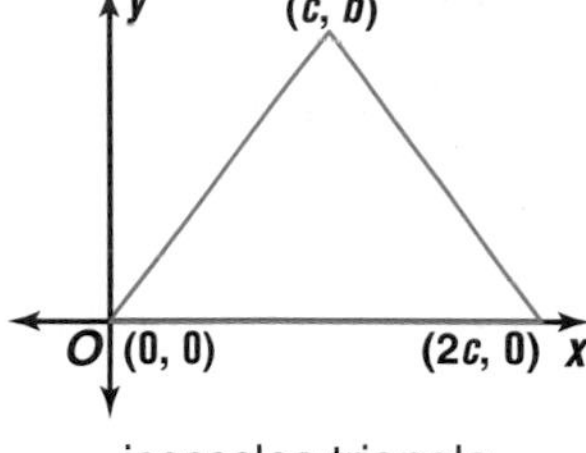

isosceles triangle

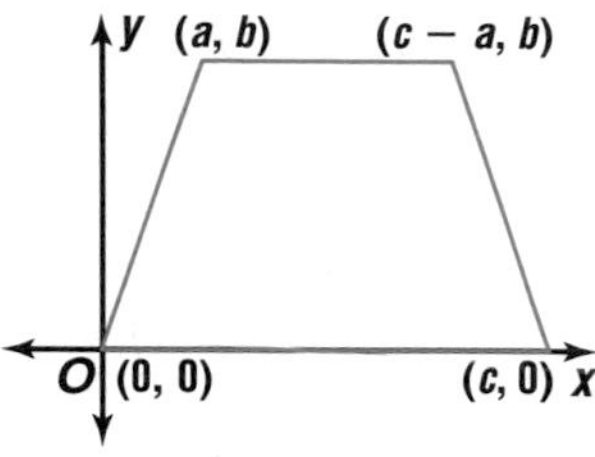

isosceles trapezoid

Example 2

Algebra

Use a coordinate proof to prove that the segments joining the midpoints of the sides of a quadrilateral form a parallelogram.

Given: $RSTV$ is a quadrilateral.
A, B, C, and D are midpoints of sides $\overline{RS}$, $\overline{ST}$, $\overline{TV}$, and $\overline{VR}$, respectively.

Prove: $ABCD$ is a parallelogram.

Proof:

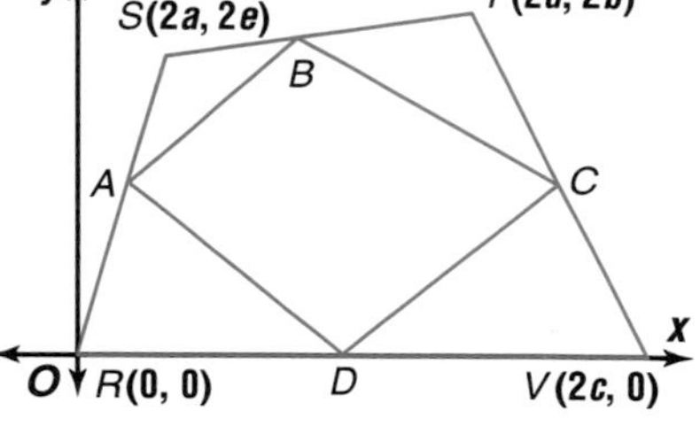

Place quadrilateral $RSTV$ on the coordinate plane and label coordinates as shown. (Using coordinates that are multiples of 2 will make the computation easier.) By the midpoint formula, the coordinates of A, B, C, and D are $A\left(\frac{2a}{2}, \frac{2e}{2}\right) = (a, e)$; $B\left(\frac{2d + 2a}{2}, \frac{2e + 2b}{2}\right) = (d + a, e + b)$; $C\left(\frac{2d + 2c}{2}, \frac{2b}{2}\right) = (d + c, b)$; and $D\left(\frac{2c}{2}, \frac{0}{2}\right) = (c, 0)$.

Find the slopes of $\overline{AB}$ and $\overline{DC}$.

slope of $\overline{AB}$	slope of $\overline{DC}$
$m = \frac{y_2 - y_1}{x_2 - x_1}$	$m = \frac{y_2 - y_1}{x_2 - x_1}$
$= \frac{(e + b) - e}{(d + a) - a}$	$= \frac{0 - b}{c - (d + c)}$
$= \frac{b}{d}$	$= \frac{-b}{-d}$ or $\frac{b}{d}$

The slopes of $\overline{AB}$ and $\overline{DC}$ are the same so the segments are parallel.

Use the distance formula to find AB and DC.

(continued on the next page)

$$AB = \sqrt{((d + a) - a)^2 + ((e + b) - e)^2} \qquad DC = \sqrt{((d + c) - c)^2 \ (b - 0)^2}$$
$$= \sqrt{d^2 + b^2} \qquad\qquad = \sqrt{d^2 + b^2}$$

Thus, $AB = DC$. Therefore, $ABCD$ is a parallelogram because if one pair of opposite sides of a quadrilateral are both parallel and congruent, then the quadrilateral is a parallelogram.

CHECK FOR UNDERSTANDING

Communicating Mathematics

Study the lesson. Then complete the following.

1. **Explain** why it is often helpful to place as many vertices of a polygon as possible on an axis when planning a coordinate proof.
2. **Write** a sample problem in which it would be helpful to place a figure on a coordinate grid so that the coordinates are multiples of 2.
3. **You Decide** Lorenza positioned a triangle on a coordinate plane shown at the right. She claimed it was an equilateral triangle. Dashiki disagreed. Who is correct? Explain your reasoning.

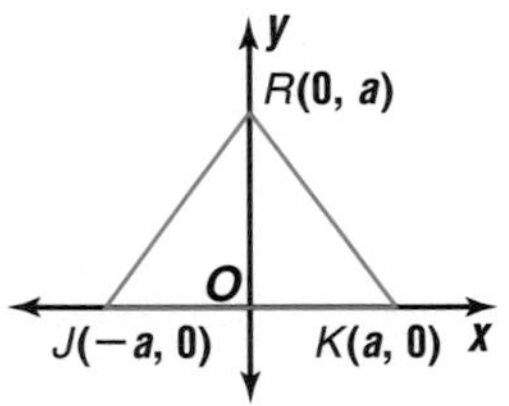

4. Select an area from your community or region and place it on a coordinate plane. Identify the coordinates of at least one point of interest in the area. Write a paragraph explaining what you did and how you selected the coordinates.

Guided Practice

Name the missing coordinates in terms of the given variables.

5. CBD is an isosceles right triangle.
6. $PQRS$ is a rectangle.

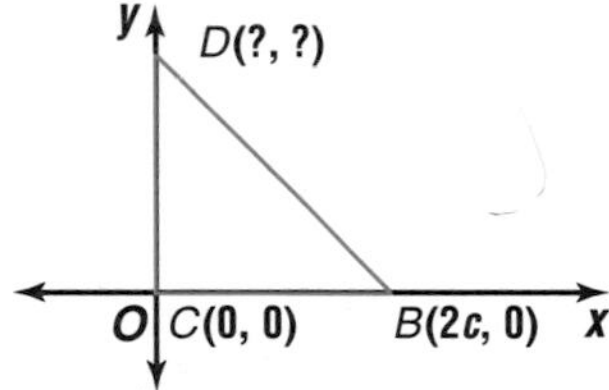

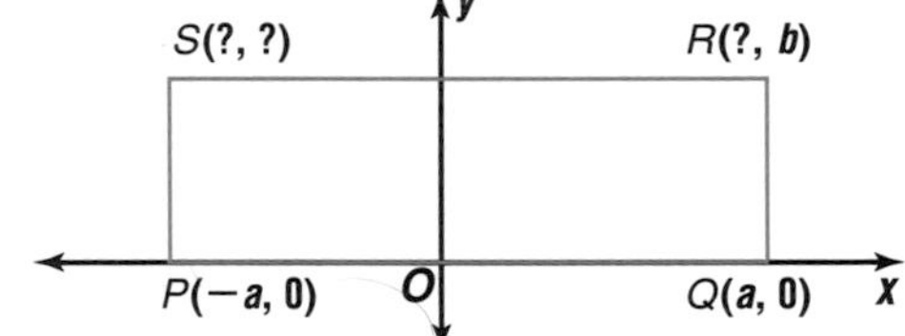

7. Use quadrilateral $ABCD$ with coordinates as indicated in the figure to answer each of the following.
   a. What kind of quadrilateral is $ABCD$? Explain your answer.
   b. Find the midpoints of $\overline{AC}$ and $\overline{DB}$. What conclusion can you make?
   c. Find the slope of $\overline{AC}$.
   d. Find slope of $\overline{DB}$.
   e. What conclusion can you make about $\overline{AC}$ and $\overline{DB}$?

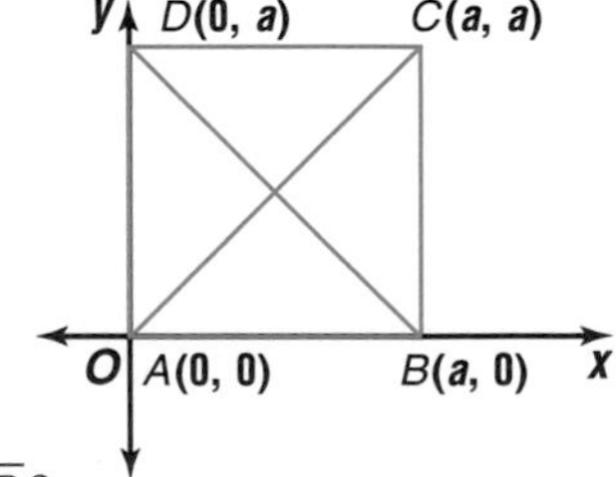

Use the triangle at the right to prove each of the following.

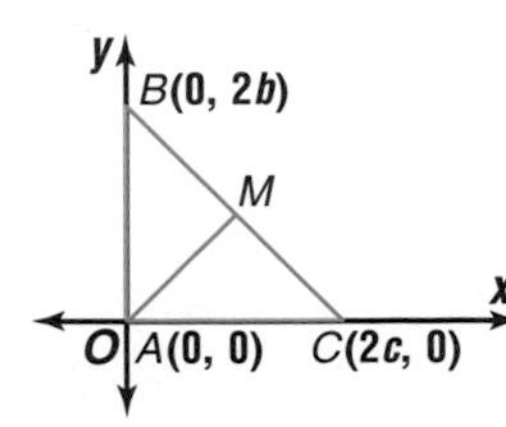

8. The midpoint of the hypotenuse is equidistant from each of the vertices.
9. The measure of the median to the hypotenuse is one-half the measure of the hypotenuse.

10. Position and label a parallelogram on the coordinate plane. Then write a coordinate proof to prove that the diagonals of a parallelogram bisect each other.

11. **Recreation** Refer to the map of Yellowstone National Park shown below. If a helicopter flew directly from Morning Glory Pool to Inspiration Point, what distance would it fly?

Yellowstone National Park

Old Faithful, the most famous geyser in the park, thrills visitors with streams of water shot 100 feet into the air.

Another popular attraction is *Morning Glory Pool*. The hot water of this pool brings minerals to the surface that make it resemble a morning glory flower in shape and color.

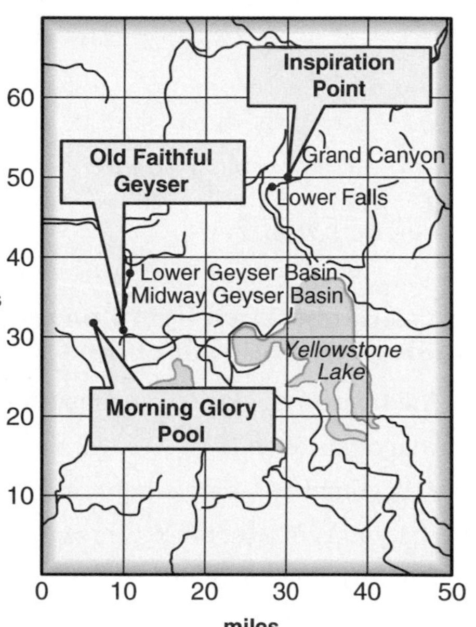

Two waterfalls grace the Yellowstone River that flows through the park. *Inspiration Point*, near the Lower Falls, offers an especially beautiful view.

Yellowstone is home to more than 275 species of birds and nearly 50 other kinds of animals. Trumpeter swans, bald eagles, elk, grizzly bears, moose, bighorn sheep, and cougars are just some of the residents.

EXERCISES

Practice **Name the missing coordinates in terms of the given variables.**

12. *KLMN* is an isosceles trapezoid.

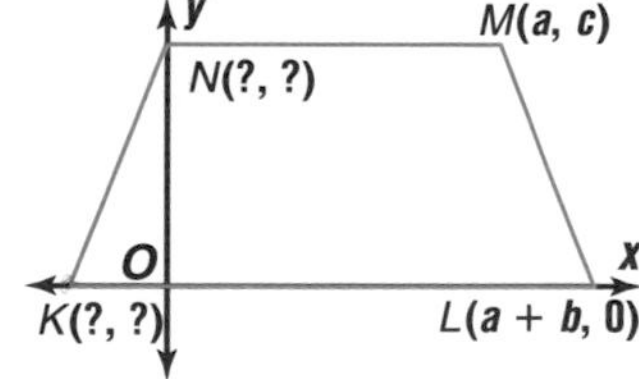

13. $\triangle DEF$ is isosceles.

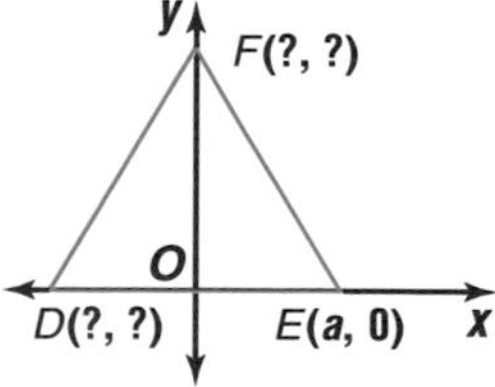

14. *PQRS* is a parallelogram.

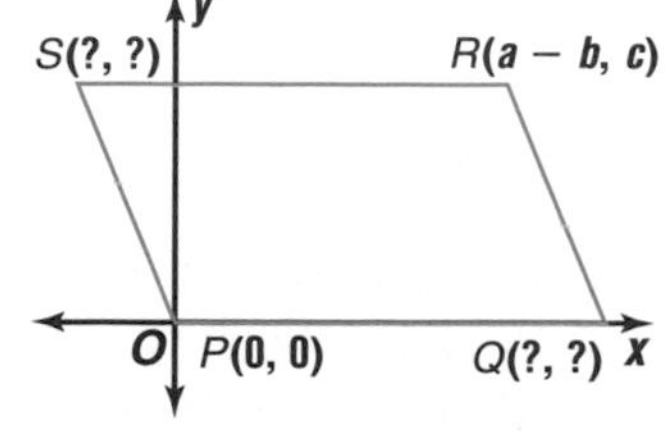

15. *ABCDEF* is a regular hexagon.

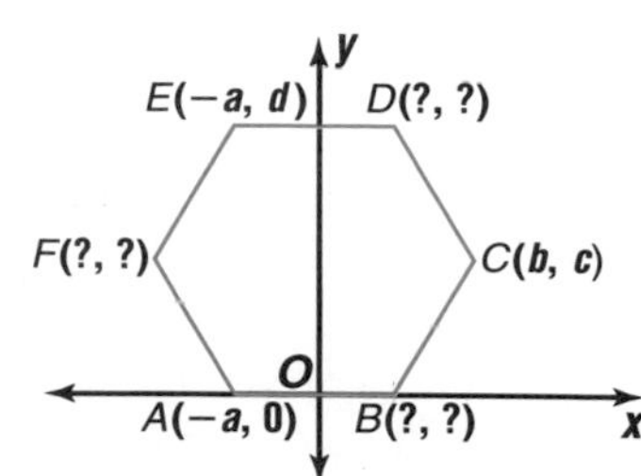

Proof **Prove using a coordinate proof.**

16. *JFHK* is an isosceles trapezoid.

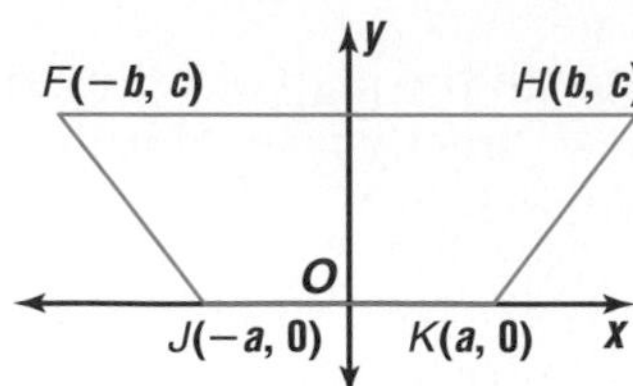

17. $\triangle LMN$ is a right triangle.

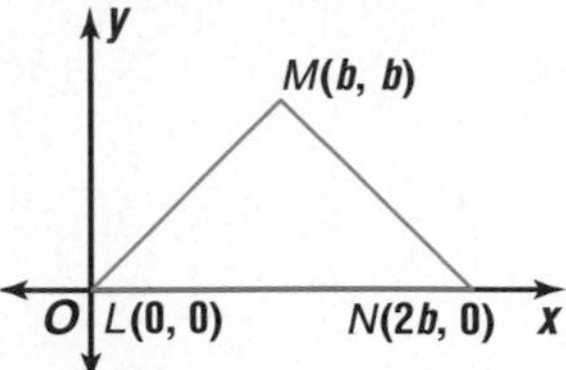

18. $\triangle RST$ is equilateral.

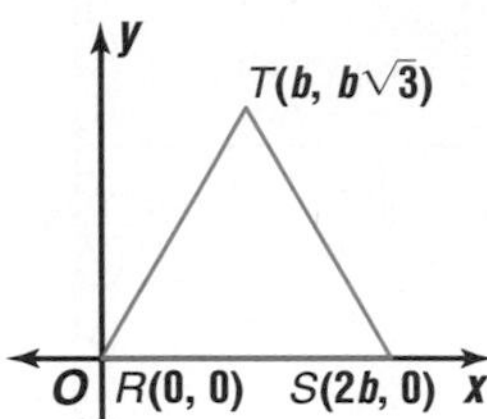

19. *EFGH* is a rhombus.

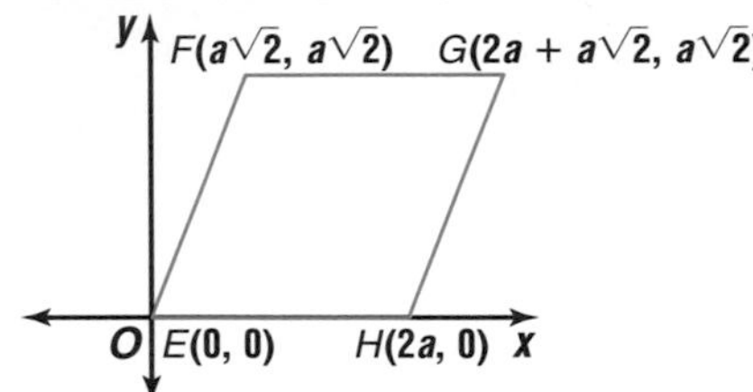

Position and label each figure on the coordinate plane. Then write a coordinate proof for each of the following.

20. The medians to the legs of an isosceles triangle are congruent.
21. The diagonals of an isosceles trapezoid are congruent.
22. If the diagonals of a parallelogram are congruent, then it is a rectangle.
23. The diagonals of a rectangle are congruent.
24. The three segments joining the midpoints of the sides of an isosceles triangle form another isosceles triangle.
25. The segments joining the midpoints of the opposite sides of a quadrilateral bisect each other.
26. If the diagonals of a parallelogram are perpendicular, then the parallelogram is a rhombus.
27. The segments joining the midpoints of the sides of an isosceles trapezoid form a rhombus.
28. If a line segment joins the midpoints of two sides of a triangle, then it is parallel to the third side.
29. If a line segment joins the midpoints of two sides of a triangle, then its length is equal to one-half the length of the third side.
30. The line segments joining the midpoints of the sides of a rectangle form a rhombus.

Critical Thinking

31. Point *A* has coordinates (0, 0), and *B* has coordinates (a, b).
    a. Find the coordinates of point *C* so $\triangle ABC$ is a right triangle.
    b. Find the coordinates of point *C* so $\triangle ABC$ is isosceles.
    c. Find the coordinates of *C* and *D* so *ABCD* is a rectangle.

Applications and Problem Solving

32. **Theater** An orchestra pit is in the shape of a semicircle with a 44-foot diameter. If a scale drawing of the orchestra pit is assigned coordinates as shown at the right, find the equation of the line that bisects the orchestra pit into two congruent parts.

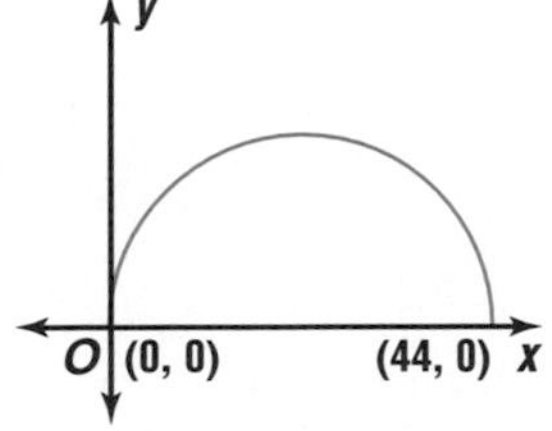

33. Air Traffic Control An airplane is 4 kilometers east and 5 kilometers north of the airport while a second airplane is 3 kilometers west and 8 kilometers south. Assuming the planes are flying at the same altitude, find the distance between the airplanes.

34. Statistics Consider scalene triangle ABC.

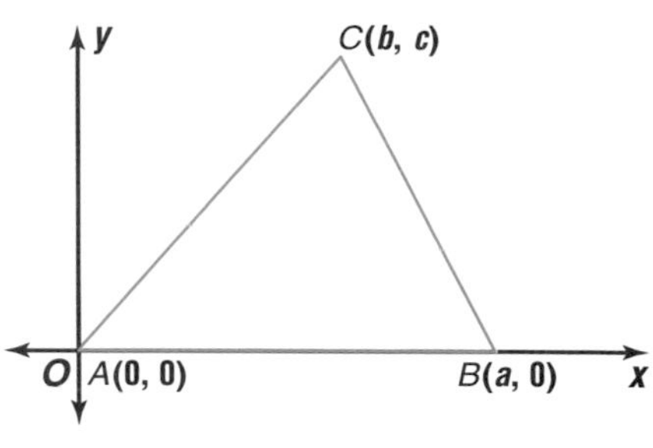

a. Find the coordinates of point D if the x-coordinate of D is the mean of the x-coordinates of the vertices of $\triangle ABC$ and the y-coordinate is the mean of the y-coordinates of the vertices of $\triangle ABC$.

b. Prove that D is the intersection of the medians of $\triangle ABC$.

Mixed Review

FYI

The Census Bureau has been taking a census of the United States every ten years since 1790. The primary purpose is to distribute seats for the U.S. House of Representatives.

35. Politics The table shows the number of millions of Hispanic-American citizens of voting age in several congressional election years. (Lesson 12–3)

a. Draw a scatter plot to show how the year and the number of voters are related.

b. Write an equation that relates the year and the number of millions of Hispanic-Americans that are eligible voters.

c. Predict how many Hispanic-Americans will be eligible to vote in the congressional elections of 2010.

Hispanic-American Voters

Year	Voters
1980	8.2
1982	8.8
1984	9.5
1986	11.8
1988	12.9
1990	13.8
1992	14.7
1994	17.5
1996	18.4

Source: U.S. Bureau of the Census

36. ACT Practice What is the slope of a line perpendicular to the line represented by the equation $2x - 8y = 16$?

A -4 B -2 C $\frac{-1}{4}$ D $\frac{1}{4}$ E 4

Refer to the regular square pyramid for Exercises 37–38.

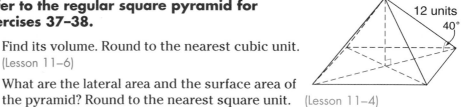

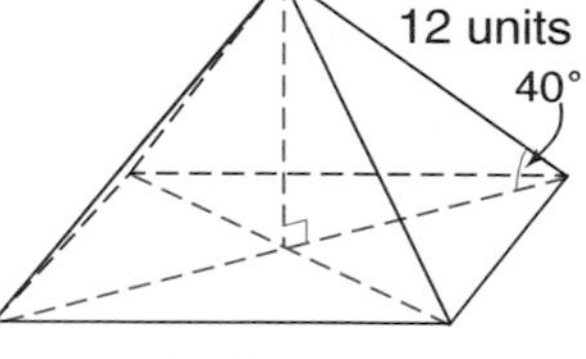

37. Find its volume. Round to the nearest cubic unit. (Lesson 11–6)

38. What are the lateral area and the surface area of the pyramid? Round to the nearest square unit. (Lesson 11–4)

39. A car tire has a radius of 8 inches. How far does the car travel in one revolution of the tire? (Lesson 9–1)

40. The sides of a rhombus are each 25 units long. If a diagonal makes an angle of 30° with a side, what is the length of each diagonal to the nearest tenth? (Lesson 8–3)

INTEGRATION **Algebra**

41. State whether $7b^3 - 4ab$ is a polynomial. If it is a polynomial, identify it as a *monomial*, a *binomial*, or a *trinomial*.

42. Factor $b^2 + 6b + 12$, if possible. If the trinomial cannot be factored using integers, write *prime*.

For **Extra Practice**, see page 788.

A Preview of Lesson 12–5

12–5A Representing Motion

Materials:  battery-powered car  meterstick, butcher paper, graph paper, straightedge, masking tape

When a power boat travels up or down a river, the speed of the river current affects how far the boat travels in a given amount of time. How do you think the current affects the motion of a boat that is traveling across the river? You can use a model car to investigate this situation.

Activity 1 Investigate the motion of the boat if there is no current in the river.

Step 1 The butcher paper represents the river and the car represents the boat. Place a piece of masking tape on the floor and lay the butcher paper over it so that the tape shows past each edge of the paper.

Step 2 Set your car at the edge of the paper so that its left wheels align with the tape. Make a mark on the edge of the paper where the left rear tire of the car is placed. Then allow the car to travel across the width of the paper and stop it when its front wheels reach the edge.

Step 3 Suppose that the starting position of the car is the origin and the tape is the y-axis. Measure the distance that the car traveled vertically and horizontally from the origin.

Step 4 Graph the ending position of the car as point A on a coordinate plane. Then draw segment OA.

Activity 2 Now investigate the motion of the boat in a river with a current.

Step 1 Place the car at the origin again. When the car is released, have one of your group members pull the paper at a constant speed to simulate the current. Stop the river motion and the car when the car reaches the opposite side of the river. Mark the ending position of the car on the floor with a piece of tape.

Step 2 Measure the distance that the car traveled vertically and horizontally from the origin.

Step 3 Graph the ending position of the car as point B on the coordinate plane. Then draw segment OB.

Step 4 Write an ordered pair to represent the ending position of the mark from which the car was released. Graph this as point C on the coordinate plane. Then draw segment OC.

Segment $\overline{OA}$ represents the displacement of the boat without the effect of the current, $\overline{OC}$ represents the current, and $\overline{OB}$ represents the displacement of the boat with the current.

Draw 1. Draw $\overline{BC}$ and $\overline{AB}$ as dashed segments on your graph.

Write 2. What type of figure is quadrilateral $OABC$?

3. How is $\overline{OB}$ related to quadrilateral $OABC$?

4. How does the current affect the motion of the boat?

12–5 Vectors

What YOU'LL LEARN

- To find the magnitude and direction of a vector,
- to determine if two vectors are equal, and
- to perform operations with vectors.

Why IT'S IMPORTANT

You can use vectors to solve problems in physics and transportation.

fabulous FIRSTS

Benjamin Oliver Davis, Jr. (1912–)

The leader of the first squadron to graduate from the Tuskegee Institute was Benjamin Oliver Davis, Jr. He went on to become the first Air Force general and the first African-American to command an air base.

Real World APPLICATION

Aviation

In 1942, the United States Army established the first and only training facility for African-American pilots in Tuskegee, Alabama. About six hundred men were trained at this facility during World War II. Known as The Tuskegee Airmen, these pilots distinguished themselves in battles in Europe. Their efforts opened the door for the integration of the United States armed forces.

In addition to flying skills, pilots at the Tuskegee Institute were taught to plot a course. Even today, pilots must file a flight plan with the Federal Aviation Agency before taking off. Often a pilot plans a course for a trip involving flying to one airport, refueling, and then flying to another airport. A course like this may be represented by the diagram at the right. In this case, the course shows that the pilot travels 200 miles at a direction of 22° east of north and then travels 100 miles at a direction of 65° east of north.

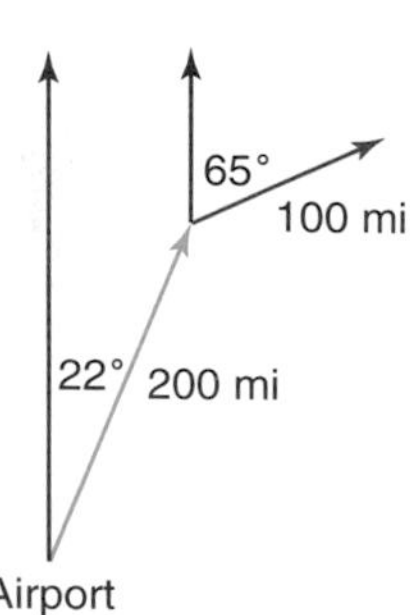

The distance and direction of the flight can be represented by a directed segment called a **vector**. A vector is any quantity that has both **magnitude** (length) and **direction**. In this case, the length of the segment represents the distance flown by the plane. Aviators use the angle that the vector makes with north to indicate direction.

In symbols, a vector is written as $\vec{v}$ or $\overrightarrow{AB}$. A vector can be either in **standard position**, with the initial point at the origin, or it can be drawn anywhere in the coordinate plane. A vector can be represented by an ordered pair (change in x, change in y). In the diagram below, vector OB ($\overrightarrow{OB}$) can be represented by the ordered pair (3, 4). To represent $\overrightarrow{CD}$ as an ordered pair, find the change in x and the corresponding change in y and write it as an ordered pair.

$$\overrightarrow{CD} = (x_2 - x_1, y_2 - y_1)$$
$$= (8 - 4, -2 - 6)$$
$$= (4, -8)$$

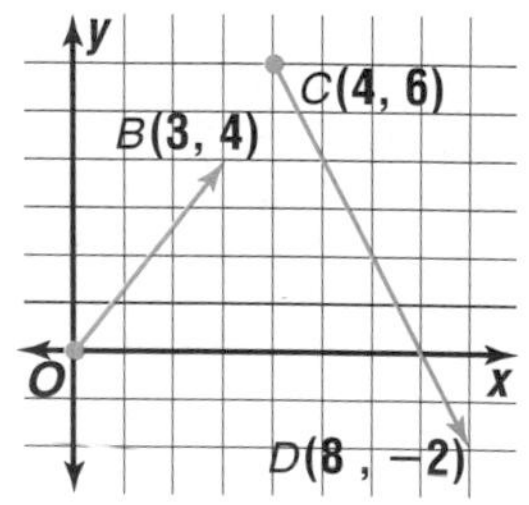

Because the magnitude and direction are not changed by moving, or translating, the vector (4, −8) represents the same vector as $\overrightarrow{CD}$.

You can use the distance formula to find the magnitude, or length, of a vector. The symbol for the magnitude of $\overrightarrow{AB}$ is $|\overrightarrow{AB}|$. In geometry, the direction of a vector is the measure of the angle that the vector, or an equal vector in standard position, forms with the positive x-axis. You can use the trigonometric ratios to find the direction of a vector.

Example 1 **Given $C(8, -2)$ and $D(4, 6)$, find the magnitude and direction of $\overrightarrow{CD}$.**

magnitude

$$|\overrightarrow{CD}| = \sqrt{(8-4)^2 + (-2-6)^2}$$
$$= \sqrt{80} \text{ or about 8.9 units}$$

direction

$$\tan C = \frac{8}{4}$$
$$m\angle C \approx 63.4 \quad \textit{Use a calculator.}$$

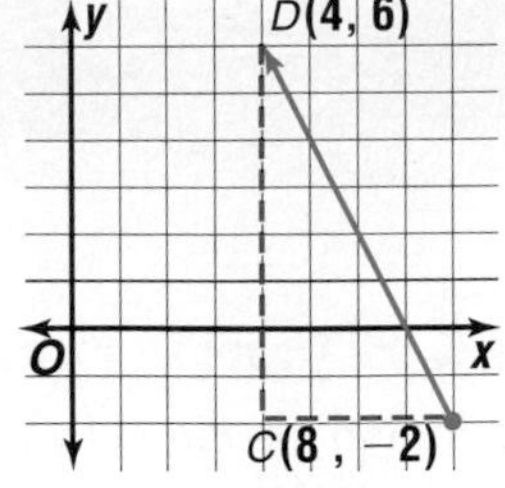

A vector in standard position that is equal to $\overrightarrow{CD}$ forms a 63.4° angle with the negative x-axis, so it forms a 180 − 63.4 or 116.6° angle with the positive x-axis. Thus, $\overrightarrow{CD}$ has a magnitude of about 8.9 units and a direction of about 116.6°.

Two vectors are equal only if they have the same magnitude and direction. They are parallel if they have the same direction or slope.

$\overrightarrow{RS} \parallel \overrightarrow{AB}$	Both have a slope of $\frac{4}{3}$, but have different lengths.
$\overrightarrow{RS} = \overrightarrow{KL}$	Both have a length of 5 and a slope of $\frac{4}{3}$.
$\overrightarrow{RS} \neq \overrightarrow{EC}$	Both have a length of 5, but have different slopes.

$\overrightarrow{RS}$ is *not* considered to be parallel to $\overrightarrow{ET}$ because they have opposite directions.

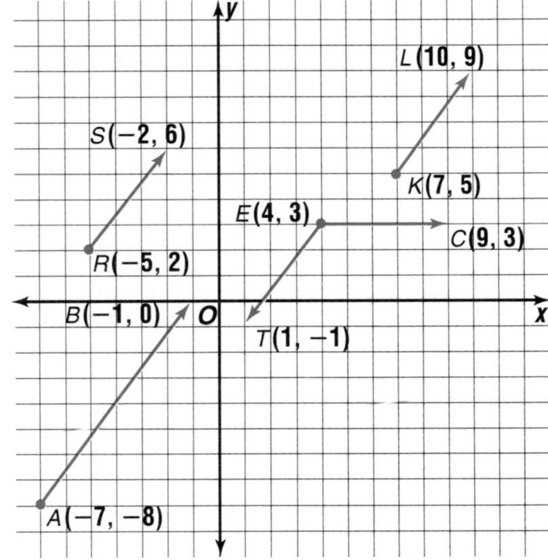

A vector can be multiplied by a constant that will change the magnitude of the vector but not affect the direction. If $\vec{v} = (1, 5)$, then $3\vec{v} = (3 \times 1, 3 \times 5)$ or $(3, 15)$. Now compare the magnitudes of $\vec{v}$ and $3\vec{v}$.

$$|\vec{v}| = \sqrt{1^2 + 5^2} = \sqrt{1 + 25} = \sqrt{26}$$

$$|3\vec{v}| = \sqrt{3^2 + 15^2} = \sqrt{9 + 225} = \sqrt{234} = 3\sqrt{26}$$

Notice that multiplying the vector by 3 tripled its magnitude. Multiplying a vector by a constant is called **scalar multiplication**.

It is also possible to add vectors. Suppose a plane flew from Minneapolis (M) to St. Louis (S) and then from St. Louis to New York City (N). This has the same result as flying directly from Minneapolis to New York City. In terms of vectors, $\overrightarrow{MS} + \overrightarrow{SN} = \overrightarrow{MN}$.

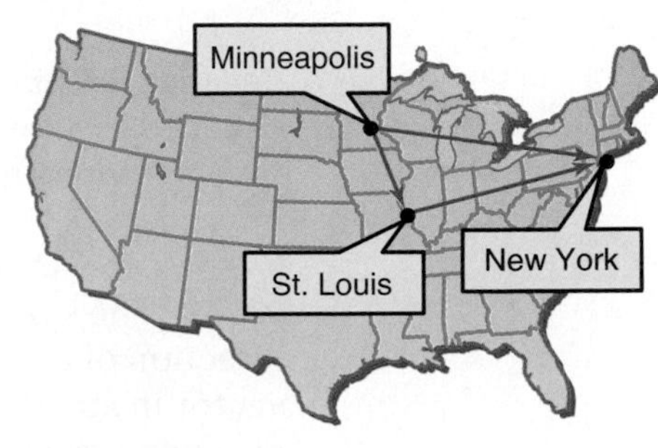

To add vectors, you can use the **parallelogram law**. The **resultant** or sum of two vectors is the diagonal of the parallelogram made by using the given vectors as sides. To add $\overrightarrow{QR}$ and $\overrightarrow{QT}$, draw parallelogram $QRST$ using the magnitude and direction of $\overrightarrow{QR}$ for side $\overline{TS}$ and that of $\overrightarrow{QT}$ for side $\overline{RS}$. The sum of the vectors is the diagonal of the parallelogram from Q to S.

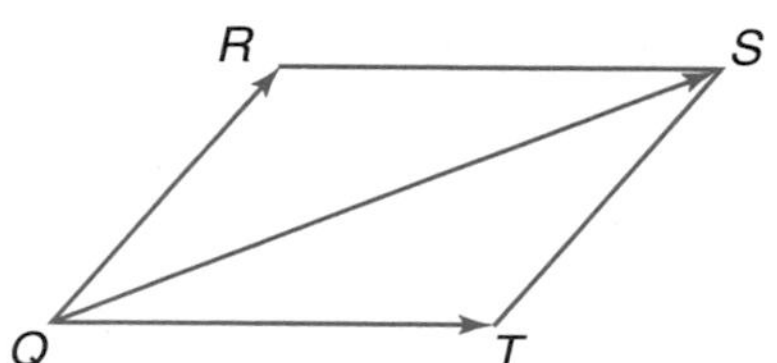

EXPLORATION — CABRI GEOMETRY

A certain calculator can create a resultant vector that is the sum of two vectors.

- Use 9: Format on [F8] to turn on the coordinate axes of the calculator.
- Create a vector with its initial point at the origin.
- Create a vector with a different magnitude and direction having its initial point at the origin.
- Create the resultant vector that is the sum of the two vectors using the Vector Sum tool in the Construction menu [F4].

Your Turn

a. Use 5:Equation & Coordinates on the [F6] menu to find the coordinates of the endpoints of all three vectors.

b. How do the coordinates of the three endpoints compare?

c. Write a conjecture about the coordinates of two vectors and the coordinates of their sum.

Vectors can also be added by adding their coordinates.

$$(a, b) + (c, d) = (a + c, b + d)$$
$$\overrightarrow{AB} + \overrightarrow{BC} = \overrightarrow{AC}$$
$$(6, 2) + (1, 2) = (6 + 1, 2 + 2)$$
$$= (7, 4)$$

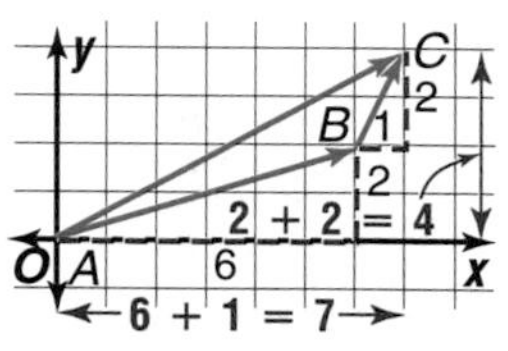

Vectors are used in physics to represent motion or forces acting upon objects.

Example 2 CONNECTION Physics

A river has a current of 2 kilometers per hour. A swimmer can swim at a rate of 3.5 kilometers per hour. How does the current affect the speed and direction of the swimmer as she swims across the river?

Use coordinates to make a model. Let each unit represent 0.5 kilometer. $\overrightarrow{OS}$ is the vector that represents the speed and direction of the swimmer. $\overrightarrow{OC}$ is the vector that represents the speed and direction of the current.

$\overrightarrow{OS} = (3.5, 0)$ $\quad$ $\overrightarrow{OC} = (0, 2)$

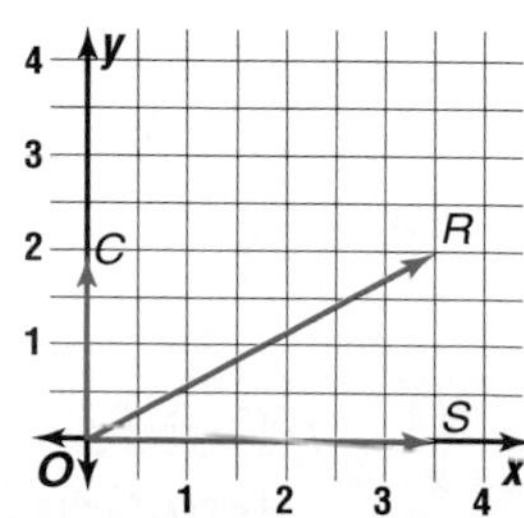

(continued on the next page)

$\overrightarrow{OR}$ is the resultant of $\overrightarrow{OS} + \overrightarrow{OC}$.

$$\begin{aligned}\overrightarrow{OR} &= \overrightarrow{OS} + \overrightarrow{OC}\\ &= (3.5, 0) + (0, 2)\\ &= (3.5 + 0, 0 + 2) \text{ or } (3.5, 2)\end{aligned}$$

Find the magnitude of $\overrightarrow{OR}$.

$$\begin{aligned}|\overrightarrow{OR}| &= \sqrt{3.5^2 + 2^2}\\ &= \sqrt{12.25 + 4}\\ &= \sqrt{16.25}\\ &\approx 4.03\end{aligned}$$

Find the direction of $\overrightarrow{OR}$.

$$\begin{aligned}\tan x &= \frac{2}{3.5} \qquad \tan x = \frac{RS}{OS}\\ \tan x &\approx 0.57\\ x &\approx 29.7\end{aligned}$$

The current pushes the swimmer off course by about 29.7° and increases her speed to about 4.03 kilometers per hour.

Vectors can also be represented in an array called a *matrix*. The vector (4, 2) can be written as a **column matrix** $\begin{bmatrix}4\\2\end{bmatrix}$. You can add column matrices by adding corresponding entries.

Example 3 Given $\vec{v} = \begin{bmatrix}3\\7\end{bmatrix}$ and $\vec{u} = \begin{bmatrix}4\\-1\end{bmatrix}$, find each sum.

Algebra

a. $\vec{v} + \vec{u}$

$$\vec{v} + \vec{u} = \begin{bmatrix}3\\7\end{bmatrix} + \begin{bmatrix}4\\-1\end{bmatrix} = \begin{bmatrix}3+4\\7+(-1)\end{bmatrix} = \begin{bmatrix}7\\6\end{bmatrix}$$

b. $\vec{v} + 2\vec{u}$

$$\vec{v} + 2\vec{u} = \begin{bmatrix}3\\7\end{bmatrix} + 2\begin{bmatrix}4\\-1\end{bmatrix} = \begin{bmatrix}3+2(4)\\7+2(-1)\end{bmatrix} = \begin{bmatrix}3+8\\7+(-2)\end{bmatrix} = \begin{bmatrix}11\\5\end{bmatrix}$$

CHECK FOR UNDERSTANDING

Communicating Mathematics

Study the lesson. Then complete the following.

1. **Describe** the difference between two parallel vectors and two equal vectors.
2. **Explain** how two vectors can have the same slope but are not considered parallel vectors.
3. **Draw** two vectors that meet the conditions.
   a. They have the same magnitude but different directions.
   b. They have the same direction but different magnitudes.

MATH JOURNAL

4. **Assess Yourself** List three different methods for adding vectors. Which of these methods do you prefer and why?

Guided Practice

Sketch each vector. Then find the magnitude to the nearest tenth and the direction to the nearest degree.

5. $\overrightarrow{AB} = (2, 11)$
6. $\overrightarrow{RS}$ if $R(-1, 3)$ and $S(4, 8)$

Use the coordinate plane to answer each question. Explain your answers.

7. Which vectors are parallel?
8. Which vectors are equal?
9. Name a pair of vectors that have the same direction but different magnitudes.

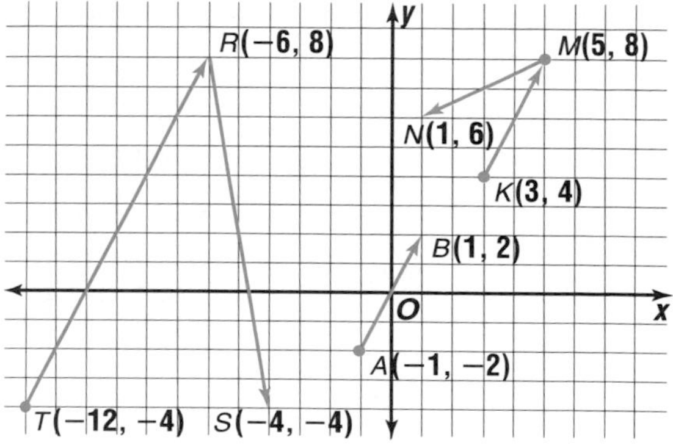

Given $\vec{u} = (-4, 10)$, $\vec{v} = (0, 5)$, and $\vec{w} = (6, 2)$, represent each of the following as an ordered pair.

10. $\vec{u} + 2\vec{v}$

11. $3\vec{w} - (\vec{u} + \vec{v})$

Find each sum or difference.

12. $\begin{bmatrix} 1 \\ 3 \end{bmatrix} + \begin{bmatrix} -5 \\ 2 \end{bmatrix}$

13. $\begin{bmatrix} 2 \\ -6 \end{bmatrix} + \begin{bmatrix} 0 \\ 3 \end{bmatrix} - \begin{bmatrix} -4 \\ -1 \end{bmatrix}$

Copy each pair of vectors and draw a resultant vector.

14. 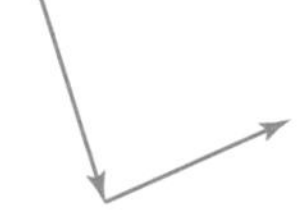

15. 

16. **Physics** Suppose the river in Example 2 is 1 kilometer wide.
   a. How long will it take the swimmer to cross the river?
   b. How far downstream from the point perpendicular across from her starting point will she land?

EXERCISES

Practice

Sketch each vector. Then find the magnitude to the nearest tenth and the direction to the nearest degree.

17. $\overrightarrow{AB} = (1, 4)$

18. $\vec{v} = (4, -9)$

19. $\overrightarrow{AB}$ if $A(4, 2)$ and $B(7, 22)$

20. $\overrightarrow{CD}$ if $C(0, -20)$ and $D(40, 0)$

21. $\overrightarrow{EF}$ if $E(0, 6)$ and $F(-6, 0)$

22. $\overrightarrow{GH}$ if $G(12, -4)$ and $H(19, 1)$

Given a path from *A* south 25 units to *B*, then east 10 units to *C*, answer each question.

23. What is the total length of the path?
24. What is the magnitude of $\overrightarrow{AC}$?
25. What is the direction from A to C?

Find each sum or difference.

26. $\begin{bmatrix} 9 \\ 21 \end{bmatrix} + \begin{bmatrix} 8 \\ -2 \end{bmatrix}$

27. $\begin{bmatrix} 2 \\ 4 \end{bmatrix} + \begin{bmatrix} -2 \\ -3 \end{bmatrix}$

28. $\begin{bmatrix} -5 \\ -8 \end{bmatrix} + \begin{bmatrix} 6 \\ -1 \end{bmatrix}$

29. $\begin{bmatrix} 12 \\ 4 \end{bmatrix} - \begin{bmatrix} 4 \\ -1 \end{bmatrix}$

30. $\begin{bmatrix} 5 \\ 4 \end{bmatrix} - \begin{bmatrix} 7 \\ -3 \end{bmatrix}$

31. $\begin{bmatrix} 3 \\ -3 \end{bmatrix} + \begin{bmatrix} -3 \\ 4 \end{bmatrix} - \begin{bmatrix} 2 \\ -5 \end{bmatrix}$

Given $A(2, 4)$, $B(1, 7)$, $C(-1, 0)$, $D(-3, 6)$, $E(4, 6)$, $F(7, 7)$, $G(-3, -1)$, $H(-2, -4)$, and $I(1, -3)$, draw $\overrightarrow{AB}$, $\overrightarrow{CD}$, $\overrightarrow{EF}$, $\overrightarrow{HG}$, and $\overrightarrow{HI}$. Use your diagram to answer each question. Explain your answers.

32. Which vectors are parallel?

33. Which vectors are equal?

34. Which vectors have the same magnitude?

Given the quadrilateral *ACDE*, complete each statement.

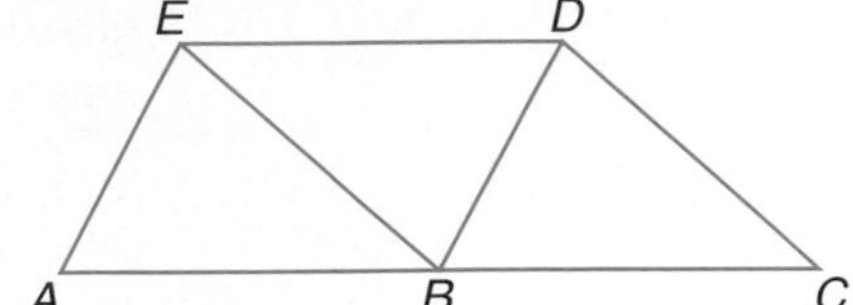

35. $\overrightarrow{AE} + \overrightarrow{AB} = \underline{\ ?\ }$

36. $\overrightarrow{BC} + \overrightarrow{CD} = \underline{\ ?\ }$

37. $\overrightarrow{BA} + \overrightarrow{AE} = \underline{\ ?\ }$

38. $\overrightarrow{EB} + \overrightarrow{ED} = \underline{\ ?\ }$

Given $\vec{v} = (3, 1)$ and $\vec{u} = (5, 8)$, represent $\vec{w}$ as an ordered pair.

39. $\vec{u} + 2\vec{v} = \vec{w}$

40. $\vec{u} + 2\vec{w} = \vec{v}$

41. $\vec{w} = 4\vec{u}$

42. $\vec{w} = 2\vec{u} + 3\vec{u}$

43. $\vec{w} = \vec{v} + \vec{u}$

44. $3\vec{v} - \vec{w} = \vec{u}$

45. Complete the proof.

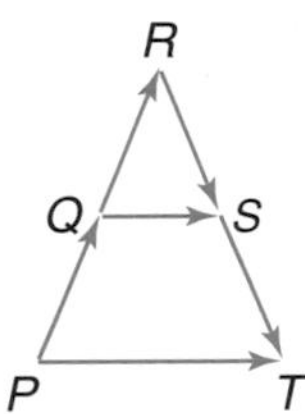

Given: $\overrightarrow{PQ} = \overrightarrow{QR}$ and $\overrightarrow{RS} = \overrightarrow{ST}$

Prove: $\overrightarrow{QS} = \frac{1}{2}\overrightarrow{PT}$

Proof:

Statements	Reasons
1. $\overrightarrow{QR} + \overrightarrow{RS} = \overrightarrow{QS}$	1. ?
2. $\overrightarrow{PR} + \overrightarrow{RT} = \overrightarrow{PT}$	2. ?
3. $\overrightarrow{PQ} = \overrightarrow{QR}$, $\overrightarrow{RS} = \overrightarrow{ST}$	3. ?
4. $2\overrightarrow{QR} + 2\overrightarrow{RS} = \overrightarrow{PT}$	4. ?
5. $2(\overrightarrow{QR} + \overrightarrow{RS}) = \overrightarrow{PT}$	5. ?
6. $2\overrightarrow{QS} = \overrightarrow{PT}$	6. ?
7. $\overrightarrow{QS} = \frac{1}{2}\overrightarrow{PT}$	7. ?

46. Prove that addition of vectors is associative.

47. How would you define $a \cdot \begin{bmatrix} c \\ b \end{bmatrix}$? Explain your reasoning.

48. If $R(4, 12)$, $S(9, 2)$, $T(x, 8)$, and $M(2, 7)$, find x if $\overrightarrow{RS} \parallel \overrightarrow{TM}$.

49. Vector (k, m) has its initial point at the origin and is on a line that passes through the origin. Prove that if (r, s) is a scalar multiple of (k, m), then (r, s) lies on that line.

If two vectors are represented by a single ordered pair, it is possible to determine whether they are perpendicular by using the dot product test. If $\vec{a} = (x_1, y_1)$ and $\vec{b} = (x_2, y_2)$, their dot product, $\vec{a} \cdot \vec{b}$, is $x_1 \cdot x_2 + y_1 \cdot y_2$. If the dot product is 0, the vectors are perpendicular. Find each dot product and determine whether the vectors are perpendicular. Write *yes* or *no*.

50. $(3, -5) \cdot (-6, 10)$

51. $(3, -5) \cdot (5, 3)$

52. $(3, -5) \cdot (-5, 3)$

Critical Thinking

53. Points C, D, E, F, G, and H are noncollinear, and $\overrightarrow{CD} + \overrightarrow{FG} = 0$.

a. What is the relationship between $\overrightarrow{CD}$ and $\overrightarrow{FG}$?

b. If $\overrightarrow{DE} + \overrightarrow{GH} = 0$ and $\overrightarrow{EF} + \overrightarrow{HC} = 0$, what is true about the polygon with vertices C, D, E, F, G, and H?

Applications and Problem Solving

54. **Transportation** A truck driver drives 40 miles due north and then 75 miles due west. What is the direction and distance of the truck from the starting point?

55. **Sailing** A boat is traveling north at 20 miles per hour. A wind from the west is blowing the boat eastward at 5 miles per hour. Find the speed of the boat and the direction in which it is moving.

56. **Physics** A force of 40 pounds is acting on an object at a 30° angle and a force of 20 pounds is acting on the same object at a 60° angle. In what direction and with what force will the object move?

57. **Sports** Two soccer players kick the ball at the same time. One exerts a force of 72 newtons east. The other exerts a force of 45 newtons north. What is the magnitude and direction of the resultant force on the ball?

Mixed Review

58. $\triangle ABC$ is a right isosceles triangle. M is the midpoint of $\overline{AB}$. Use a coordinate proof to show that $\overline{CM}$ is perpendicular to $\overline{AB}$. (Lesson 12–4)

59. **Transportation** The graphic shows the number of thousands of people who travel to work alone or in car pools in various states. (Lesson 12–3)

interNET CONNECTION

Data Update For more statistics on carpooling, visit: **www.geometry.glencoe.com**

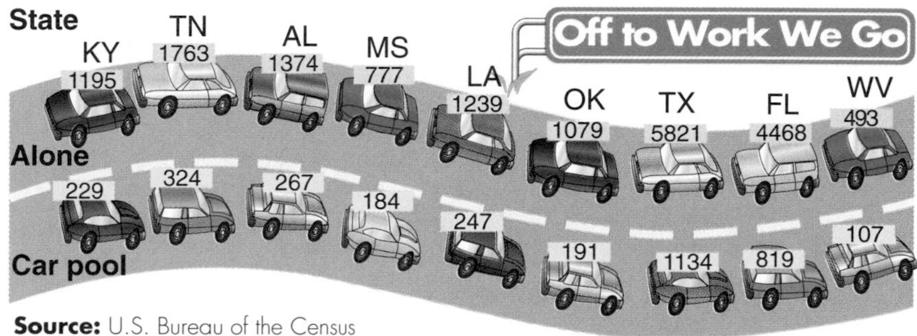

Source: U.S. Bureau of the Census

a. Write an equation that relates the number of people who drive alone to the number of people who carpool. You may wish to use a scatter plot.

b. The Census Bureau found that there are 2281 thousand workers who drove to work alone in the state of Virginia. Predict how many people carpool in Virginia.

60. Find the volume of a regular square pyramid if its lateral edges are 12 centimeters and the base is 16 centimeters on a side. (Lesson 11–6)

61. Triangle ABC has an area of 88 square meters. If the height is 8 meters, what is the length of the base? (Lesson 10–4)

62. Find the measure of an interior angle of a regular polygon with 36 sides. (Lesson 10–1)

63. $\angle NMP$ is inscribed in a circle. If $m\angle NMP = 62$, what is the measure of the arc intercepted by $\angle NMP$? (Lesson 9–4)

64. **SAT Practice** A rectangular swimming pool has a volume of 16,320 cubic feet, a depth of 8 feet, and a length of 85 feet. What is the width of the pool, in feet?

A 24 **B** 48 **C** 74 **D** 192 **E** 2040

65. The perimeter of an equilateral triangle is 72 inches. What is the length of an altitude of the triangle? (Lesson 8–2)

66. The number 10 has exactly four factors, 1, 2, 5, and 10. Find the least number with exactly five factors. (Lesson 7–6)

67. Find the length of the median of a trapezoid whose vertices are at (2, 1), (4, 0), (7, 3), and (8, 7). (Lesson 6–5)

For **Extra Practice**, see page 789.

Algebra

68. Solve $n^2 + 36n = 0$.

69. Find $(3u^2 - 4) \div 2u$.

12-6 Coordinates in Space

What YOU'LL LEARN

- To locate a point in space,
- to use the distance and midpoint formulas for points in space, and
- to determine the center and radius of a sphere.

Why IT'S IMPORTANT

A three-dimensional coordinate system is a good way to represent locations in space.

Real World APPLICATION

Marine Biology

Hal Whitehead, his wife Lindy Weilgart, and their three children have an unusual home away from home. Every two years, the family spends months sailing the waters near the Galápagos Islands in search of sperm whales. Hal, Lindy, and the other scientists and students who join the expeditions study the social habits and communications of the whales.

Sperm whales make clicking noises that biologists can use to locate them. A special microphone, called a hydrophone, is used to listen underwater. When whales are located, their location is recorded in a chart. One way they can record an underwater location is by using the latitude, longitude, and depth. They would write an ordered triple like (2°N, 150°W, −1500 ft) to record the location.

In a coordinate plane, the ordered pair for each point has two numbers, or coordinates, to describe its location because a plane has two dimensions. In space, each point requires three numbers, or coordinates, to describe its location because space has three dimensions. In space, the *x*-, *y*-, and *z*-axes are perpendicular to each other.

A point in space is represented by an **ordered triple** of real numbers (x, y, z). In the figure at the right, the ordered triple $(-1, 2, 4)$ locates point *P*. Notice that a parallelogram is used to help show perspective and convey the idea of the third dimension.

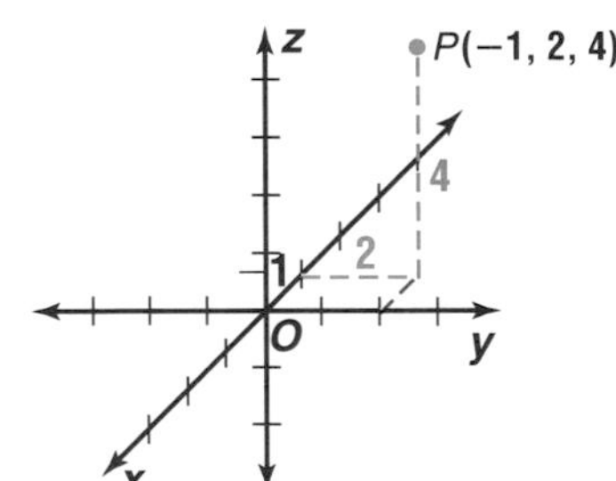

CAREER CHOICES

The oceans are the largest unexplored territories left on Earth. Each year **marine biologists** learn more about the plants and animals that live there.

For more information, contact:

National Ocean Industries Association
1050 17th St., NW, Suite 700
Washington, DC 20036

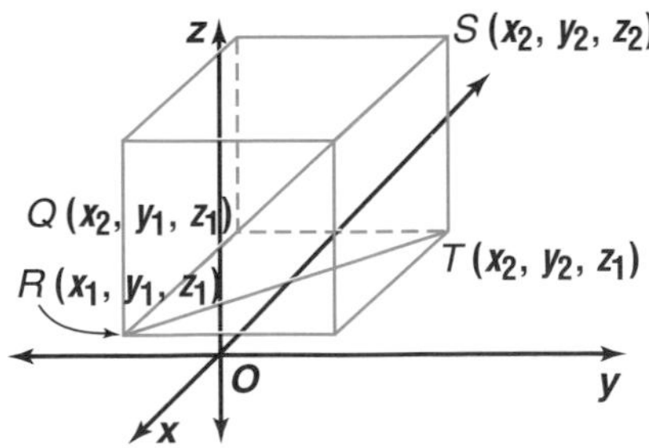

Just as the Pythagorean Theorem can be used to find the formula for the distance between two points in a plane, it can also be used to find the formula for the distance between two points in space.

Since *R* and *T* are both in the same plane that is parallel to the *xy*-plane, the *z*-coordinates for both points are the same. Since $\triangle QRT$ is a right triangle, the distance between points *R* and *T* can be found as follows.

$RT = \sqrt{(x_2 - x_1)^2 + (y_2 - y_1)^2 + (z_1 - z_1)^2}$ or $\sqrt{(x_2 - x_1)^2 + (y_2 - y_1)^2}$

Therefore, $(RT)^2 = (x_2 - x_1)^2 + (y_2 - y_1)^2$.

Likewise, *S* and *T* are in the same plane that is parallel to the *yz*-plane, so the *x*-coordinates for both points are the same. Since $\triangle RTS$ is a right triangle, the distance between *S* and *T* can be found as follows.

$ST = \sqrt{(x_2 - x_2)^2 + (y_2 - y_2)^2 + (z_2 - z_1)^2}$ or $\sqrt{(z_2 - z_1)^2}$

Therefore, $(ST)^2 = (z_2 - z_1)^2$.

$(RS)^2 = (RT)^2 + (TS)^2$	*Pythagorean Theorem*
$(RS)^2 = ((x_2 - x_1)^2 + (y_2 - y_1)^2) + ((z_2 - z_1)^2)$	*Substitution Property (=)*
$RS = \sqrt{(x_2 - x_1)^2 + (y_2 - y_1)^2 + (z_2 - z_1)^2}$	*Take the square root of each side.*

Theorem 12-2

Given two points $A(x_1, y_1, z_1)$ and $B(x_2, y_2, z_2)$ in space, the distance between A and B is given by the following equation.

$$AB = \sqrt{(x_2 - x_1)^2 + (y_2 - y_1)^2 + (z_2 - z_1)^2}$$

This formula is an extension of the distance formula in the two-dimensional coordinate system. The midpoint formula can also be extended to the three-dimensional system.

Suppose M is the midpoint of $\overline{AB}$, a segment in space. The midpoint has the following coordinates.

$$\left(\frac{x_1 + x_2}{2}, \frac{y_1 + y_2}{2}, \frac{z_1 + z_2}{2}\right)$$

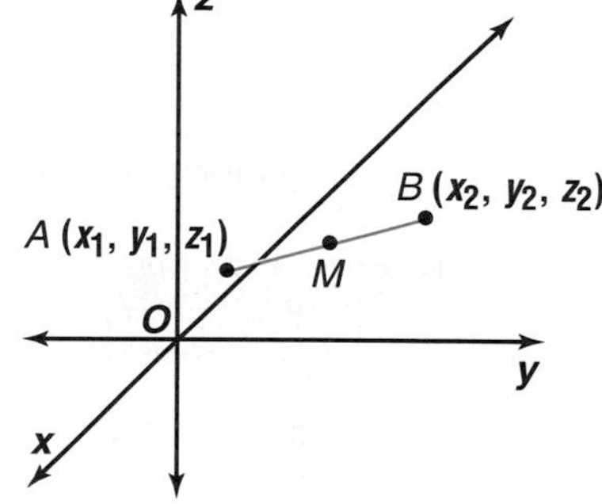

Example 1 **Consider $F(4, 14, 8)$, $G(1, 2, -1)$, and $H(2, 6, 2)$.**

a. Use the distance formula to prove that the points are collinear.

b. Determine whether H is the midpoint of $\overline{FG}$.

a. Use the distance formula to find the lengths of $\overline{GH}$, $\overline{FH}$, and $\overline{GF}$.

$$GH = \sqrt{(1 - 2)^2 + (2 - 6)^2 + (-1 - 2)^2} = \sqrt{26}$$

$$HF = \sqrt{(4 - 2)^2 + (14 - 6)^2 + (8 - 2)^2} = \sqrt{104} \text{ or } 2\sqrt{26}$$

$$GF = \sqrt{(1 - 4)^2 + (2 - 14)^2 + (-1 - 8)^2} = \sqrt{234} \text{ or } 3\sqrt{26}$$

Since $3\sqrt{26} = \sqrt{26} + 2\sqrt{26}$, $GF = GH + HF$.

By the Segment Addition Postulate, if $GF = GH + HF$, then H is between G and F. Thus, G, F, and H are collinear points.

b. Use the midpoint formula to find the coordinates of the midpoint of $\overline{FG}$. Let $(4, 14, 8)$ be (x_1, y_1, z_1) and $(1, 2, -1)$ be (x_2, y_2, z_2).

$$\left(\frac{4+1}{2}, \frac{14+2}{2}, \frac{8+(-1)}{2}\right) = \left(\frac{5}{2}, \frac{16}{2}, \frac{7}{2}\right) = (2.5, 8, 3.5)$$

The coordinates of the midpoint are $(2.5, 8, 3.5)$. Thus, H is not the midpoint of $\overline{FG}$.

The formula for the equation of a sphere is an extension of the formula for the equation of a circle. The equation of a sphere whose center is at (0, 0, 0) and whose radius is r units long is as follows.

$$x^2 + y^2 + z^2 = r^2$$

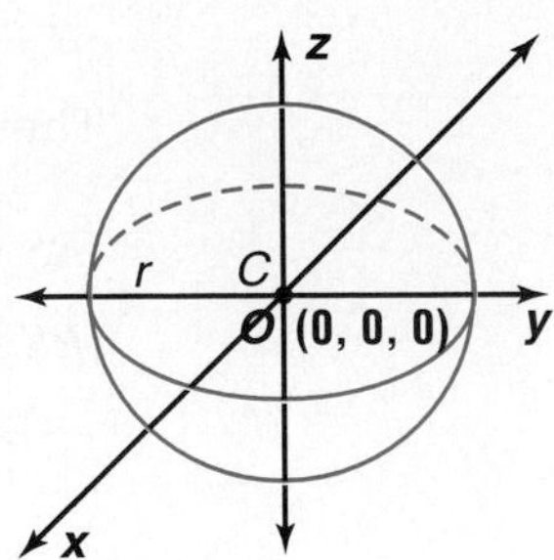

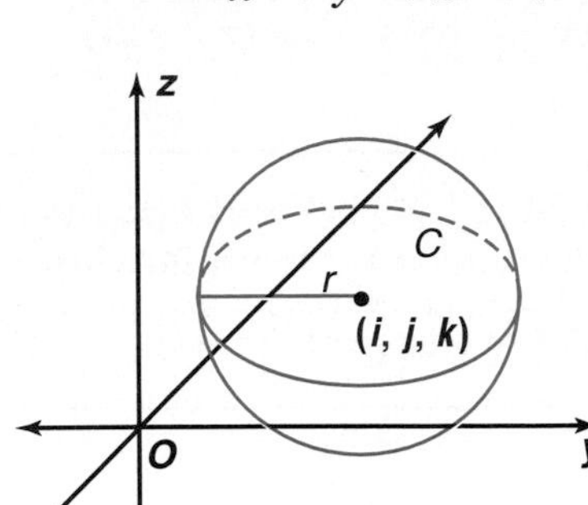

The equation of a sphere whose center is at (i, j, k) and whose radius is r units long is as follows.

$$(x - i)^2 + (y - j)^2 + (z - k)^2 = r^2$$

Example 2

Algebra

Write an equation for the sphere that has a diameter with endpoints at (−5, 10, −2) and (11, 22, 14).

The center of the sphere is the midpoint of the diameter.

First, find the midpoint of the diameter.

Let (−5, 10, −2) be (x_1, y_1, z_1) and (11, 22, 14) be (x_2, y_2, z_2).

$$\left(\frac{-5 + 11}{2}, \frac{10 + 22}{2}, \frac{-2 + 14}{2}\right) = \left(\frac{6}{2}, \frac{32}{2}, \frac{12}{2}\right)$$

$$= (3, 16, 6) \quad \textit{Center of sphere}$$

Next, find the radius.

Let $i = 3$, $j = 16$, and $k = 6$.

$r^2 = (x - i)^2 + (y - j)^2 + (z - k)^2$

$r^2 = (-5 - 3)^2 + (10 - 16)^2 + (-2 - 6)^2$ *Substitute 3 for i, 16 for j, and 6 for k. Substitute −5 for x, 10 for y, and −2 for z.*

$r^2 = (-8)^2 + (-6)^2 + (-8)^2$

$r^2 = 164$

The equation of the sphere is $(x - 3)^2 + (y - 16)^2 + (z - 6)^2 = 164$.

You can solve problems using a three-dimensional coordinate system.

Example 3

Aviation

An airplane at an elevation of 2 miles is 50 miles east and 100 miles north of an airport. Another airplane at an elevation of 2.5 miles is 240 miles west and 140 miles north of the airport. Find the distance between the two planes.

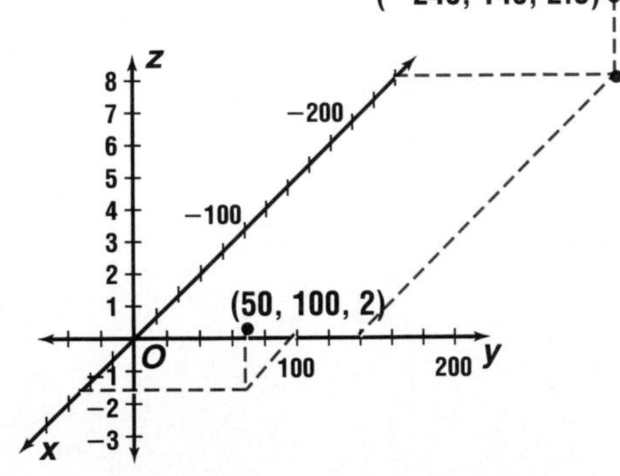

Explore Since the distances are relative to the same airport, the airport can be represented by the origin. The coordinates of the first airplane are (50, 100, 2). The other plane has coordinates (−240, 140, 2.5).

Plan Use Theorem 12–2 to find the distance between the two airplanes.

Solve
$$\sqrt{(x_2 - x_1)^2 + (y_2 - y_1)^2 + (z_2 - z_1)^2}$$
$$= \sqrt{(50 - (-240))^2 + (100 - 140)^2 + (2 - 2.5)^2}$$
$$= \sqrt{290^2 + (-40)^2 + (-0.5)^2}$$
$$= \sqrt{85{,}700.25} \approx 292.75$$

The airplanes are about 292.75 miles apart.

Examine Model the problem on a three-dimensional coordinate system. Draw a rectangular solid having the coordinates of the two planes as vertices. Check the solution by using the Pythagorean Theorem.

CHECK FOR UNDERSTANDING

Communicating Mathematics

Study the lesson. Then complete the following.

1. **Explain** how to locate the point at $(-3, 1, -2)$ in space. Then sketch the point on a three-dimensional coordinate system.
2. **Compare** the midpoint and distance formulas in space to the midpoint and distance formulas in a plane.
3. **Describe** the sphere whose equation is $(x + 2)^2 + (y + 4)^2 + (z - 4)^2 = 25$.

Guided Practice

Plot each point in a three-dimensional coordinate system.

4. $J(2, 5, 6)$
5. $L(4, 2, -3)$

Determine the distance between each pair of points and determine the coordinates of the midpoint of the segment containing them.

6. $R(0, 0, 4)$ and $S(0, -1, 6)$
7. $A(19, -8, 40)$ and $B(20, -2, 18)$

Identify each of the following as *true* or *false*. If the statement is false, explain why.

8. The point at $(0, 0, 5)$ lies on the z-axis.
9. The distance between the points at $(0, 0, 0)$ and (x, y, z) can be expressed as $\sqrt{x^2 + y^2 + z^2}$.
10. The point at $(2, 1, 5)$ lies on the sphere whose equation is $(x + 2)^2 + (y - 4)^2 + (z - 1)^2 = 25$.
11. Determine the coordinates of the center and the measure of the radius for the sphere with equation $(x - 5)^2 + (y + 2)^2 + z^2 = 36$.

Write the equation of the sphere using the given information.

12. center at $(-1, 2, 5)$, radius of 10
13. $\overline{AB}$ is a diameter where $A(3, -5, 18)$ and $B(-7, 11, 32)$

14. **Marine Biology** A group of sperm whales was found feeding at a location of (3°N, 152°W, −1500 ft). The next day a group is found at (2°N, 154°W, −1300 ft). The distance between latitude lines that differ by one degree is about 60 nautical miles. Near the equator, the distance between longitude lines that differ by one degree is about 69 nautical miles. A nautical mile is 6076.115 feet.
    a. Write an ordered pair for each group sighted in terms of nautical miles from (0, 0, 0).
    b. Find the distance between the locations of the groups.

EXERCISES

Practice

Plot each point in a three-dimensional coordinate system.

15. $A(2, 1, 5)$
16. $B(3, 5, 4)$
17. $C(-6, 0, 0)$
18. $D(-1, 2, -2)$
19. $E(4, -2, 8)$
20. $F(-3, -4, -5)$

Determine the distance between each pair of points, and determine the coordinates of the midpoint of the segment connecting them.

21. $J(3, -7, 0)$ and $K(5, 1, 7)$
22. $L(17, -22, -41)$ and $M(-19, 34, -53)$
23. $A(6, 1, 1)$ and $B(9, 0, 1)$
24. $C(4, -8, 12)$ and $D(7, 20, 18)$
25. $E(3, 7, -1)$ and $F(5, 7, 2)$
26. $G(2, 2, 2)$ and $H(-25, 4, 18)$

Identify each of the following as *true* or *false*. If the statement is false, explain why.

27. Every point on the yz-plane has coordinates (c, y, z) for any real number c.
28. The point at $(1, 8, -12)$ is inside the sphere $(x - 3)^2 + (y - 5)^2 + (z + 2)^2 = 9$.
29. The intersection of the xy-plane, the yz-plane, and the xz-plane is the point at $(0, 0, 0)$.
30. The set of points in space 5 units from the point at $(1, -1, 3)$ can be described by the equation $(x - 1)^2 + (y + 1)^2 + (z - 3)^2 = 25$.
31. The set of points equidistant from $A(2, 5, 8)$ and $B(-3, 4, 7)$ is a line that is the perpendicular bisector of $\overline{AB}$.

Determine the coordinates of the center and the measure of the radius for each sphere whose equation is given.

32. $x^2 + (y - 3)^2 + (z + 8)^2 = 81$
33. $(x - 5)^2 + (y + 4)^2 + (z - 10)^2 = 9$
34. $x^2 + y^2 + (z - 3)^2 = 49$
35. $(x + 4)^2 + (y - 2)^2 + (z + 12)^2 = 18$

Write the equation of the sphere using the given information.

36. The center is at $(-5, 11, -3)$, and the radius is 4.
37. The center is at $(-2, 3, -4)$, and it contains the point at $(5, -1, -1)$.
38. The diameter has endpoints at $(14, -8, 32)$ and $(-12, 10, 12)$.
39. It is concentric with the sphere with equation $(x + 5)^2 + (y - 4)^2 + (z - 19)^2 = 9$, and it has a radius of 6 units.
40. It is inscribed in a cube determined by the points at $(0, 0, 0)$, $(4, 0, 0)$, $(0, 4, 0)$, and $(4, 4, 4)$.
41. Find the perimeter of a triangle with vertices $A(-1, 3, 2)$, $B(0, 2, 4)$, and $C(-2, 0, 3)$.
42. Show that $A(2, -1, 3)$, $B(-3, 5, -6)$, and $C(7, -7, 12)$ are collinear.

43. Show that $\triangle ABC$ is an isosceles right triangle if the vertices are $A(3, 2, -3)$, $B(5, 8, 6)$, and $C(-3, -5, 3)$.

44. Consider $R(6, 1, 3)$, $S(4, 5, 5)$, and $T(2, 3, 1)$.
 a. Determine the measures of $\overline{RS}$, $\overline{ST}$, and $\overline{RT}$.
 b. If $\overline{RS}$, $\overline{ST}$, and $\overline{RT}$ are the sides of a triangle, what type of triangle is $\triangle RST$?

45. Find the surface area and volume of the rectangular prism at the right.

46. Find z if the distance between $R(5, 4, -1)$ and $S(3, -2, z)$ is 7.

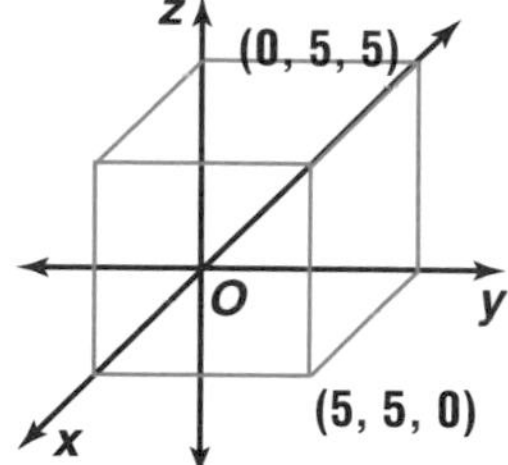

Proof

47. Write a coordinate proof to prove that the diagonals of a rectangular prism are congruent and bisect each other.

48. The center of a sphere is at $(4, -2, 6)$, and the endpoint of a diameter is at $(8, 10, -2)$. What are the coordinates of the other endpoint of the diameter?

Critical Thinking

49. The sphere whose equation is $(x - 1)^2 + (y - 4)^2 + (z - 8)^2 = 25$ is inscribed in a cube. Determine the coordinates of the cube and make a drawing of the figure.

Applications and Problem Solving

50. **Statistics** Each year *Money* magazine collects data on cities and uses that data to rate the cities in terms of the "best" places to live. The data collected on the unemployment rate, the property taxes, and the number of robberies per 100,000 people for four cities are given in the table. Suppose the "best" city was the city that was the closest to 0 in each of the categories.

City	Unemployment Rate	Average Property Tax	Violent Crimes per 100,000 people
Austin, TX	3.5%	2800	539
Gainesville, FL	3.8%	1875	1377
Rochester, MN	2.5%	1300	176
Seattle, WA	4.7%	2250	542

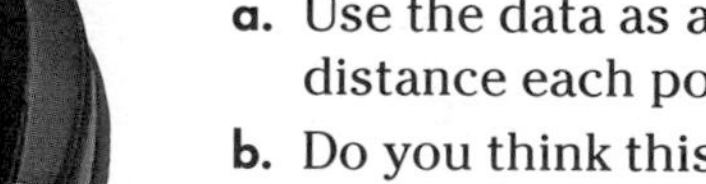

Source: *Money*, September 1996

 a. Use the data as an ordered triple and rank the cities in terms of the distance each point is from (0, 0, 0).
 b. Do you think this method of ranking cities is fair? Explain your reasoning.

51. **Recreation** Two children are playing a three-dimensional tic-tac-toe game. Three *X*s or three *O*s in any row wins the game. The positions of the *X*s are at (2, 1, 1) and (2, 3, 3), and the position of the *O* is at (1, 2, 3). Where should the next *O* be placed? Explain your answer.

52. **Sports** Two hot-air balloons take off from the same location. One hot-air balloon is 8 miles west and 10 miles south of the take-off point and 0.3 mile above the ground. The other hot-air balloon is 4 miles west and 8 miles south of the take-off point and 0.4 mile above the ground. Find the distance between the hot-air balloons.

Mixed Review

53. Sketch $\overrightarrow{LN}$ if L is at $(-3, 5)$ and N is at $(6, 7)$. Then find the magnitude of $\overrightarrow{LN}$ to the nearest tenth and the direction to the nearest degree. (Lesson 12–5)

54. Position and label a right triangle on the coordinate plane. Then write a coordinate proof to show that the measure of the median to the hypotenuse of a right triangle is half of the measure of the hypotenuse. (Lesson 12–4)

55. **Communication** There are currently 349 public television stations in the United States. About 11 new stations begin broadcasting each year. (Lesson 12–2)
   a. Write an equation that represents the total number of public television stations x years from now if the rate of increase stays the same.
   b. Approximately how many public television stations will there be in fifteen years if the rate of increase stays the same?

56. If the volume of a sphere is 972π cubic centimeters, what is the diameter of the sphere? (Lesson 11–7)

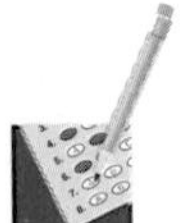

57. **ACT Practice** If $x \neq 0, \pm 1$, then $(x^{-2} - x^2)(1 - x^2)^{-1} =$

   A -1 B 1 C $\frac{1 - x^2}{x^2}$ D $\frac{x^2 - 1}{x^2}$ E $\frac{1 + x^2}{x^2}$

58. A circle is inscribed in a square whose diagonal is 18 feet long. Find the circumference and the area of the circle. (Lesson 10–5)

59. The sides of a parallelogram are 18 and 32 meters long. One angle of the parallelogram is 135°. What is the area of the parallelogram? (Lesson 10–4)

60. The sum of the measures of eight interior angles of a convex nonagon is 1210. What is the measure of the ninth angle? (Lesson 10–1)

61. Suppose $\triangle ABC \sim \triangle DEF$. If $AC = 30$, $DF = 18$, and $AB = 7.5$, find DE. (Lesson 7–3)

For **Extra Practice**, see page 789.

Algebra

62. Factor $v^2 - 16$.

63. Solve $y + 3 = -15$.

WORKING ON THE

In•ves•ti•ga•tion

Refer to the Investigation on pages 642–643.

Making believable movies with computer graphics requires drawing three-dimensional objects on a two-dimensional surface. Computers begin perspective drawings by finding the three-dimensional coordinates of the object being drawn. Then algebra is used to transform these to two-dimensional coordinates. The image created is called a projection. The formulas $X = x(-\cos a°) + y$ and $Y = x(-\sin a°) + z$ will draw a projection in which the x-axis is horizontal, the y-axis is vertical, and the z-axis is drawn at an angle of $a°$ with the x-axis. If the three-dimensional coordinates of a point are (x, y, z), then the projected coordinates are (X, Y).

A cube has vertices $A(5, 0, 5)$, $B(5, 5, 5)$, $C(5, 5, 0)$, $D(5, 0, 0)$, $E(0, 0, 5)$, $F(0, 5, 5)$, $G(0, 5, 0)$, and $H(0, 0, 0)$.

1. Draw the cube using a projection with $a = 45°$.
2. Then create a drawing of an image for your movie using a projection.

Add the results of your work to your Investigation Folder.

CHAPTER 12 HIGHLIGHTS

Chapter Review For additional lesson-by-lesson review, visit: **www.geometry.glencoe.com**

VOCABULARY

After completing this chapter, you should be able to define each term, property, or phrase and give an example or two of each.

Geometry

coordinate proof (p. 666)
direction of a vector (p. 673)
magnitude of a vector (p. 673)
ordered triple (p. 680)
parallelogram law (p. 675)
resultant (p. 675)
standard position (p. 673)
vector (p. 673)

Algebra

column matrix (p. 676)
function (p. 647)
intercepts method (p. 646)
linear equation (p. 646)
point-slope form (p. 653)
scalar multiplication (p. 674)
slope-intercept form (p. 647)
standard form (p. 646)

Statistics

scatter plot (p. 660)

Problem Solving

write and solve an equation (p. 655)

UNDERSTANDING AND USING THE VOCABULARY

Choose the term from the list above that best completes each statement.

1. An important part of _?_ is the placement of the figure on the coordinate plane.
2. Multiplying a vector by a constant is called _?_ .
3. The equation $y = -3x + 5$ is written in _?_ .
4. The _?_ can be found by using the distance formula.
5. The equation $4x + 5y = -5$ is written in _?_ .
6. The _?_ can be used to add vectors.
7. The equation of the line suggested by the points on a(n) _?_ can be used to make predictions.
8. A point in space is represented by a(n) _?_ of real numbers.
9. An equation whose graph is a straight line is called a(n) _?_ .
10. The equation $y + 6 = -3(x - 1)$ is written in _?_ .

SKILLS AND CONCEPTS

OBJECTIVES AND EXAMPLES

Upon completing this chapter, you should be able to:

- graph linear equations using the intercepts method (Lesson 12–1)

$x + 4y = 4$

x-intercept

$x + 4(0) = 4$

$x = 4$

$(4, 0)$

y-intercept

$0 + 4y = 4$

$y = 1$

$(0, 1)$

REVIEW EXERCISES

Use these exercises to review and prepare for the chapter test.

Graph each equation using the intercepts method.

11. $y = 3x + 3$
12. $5x - y = 10$

- graph linear equations using the slope and y-intercept (Lesson 12–1)

$y = 4x - 1$

slope

$m = 4$

y-intercept

$y = 4(0) - 1$

$y = -1$

$(0, -1)$

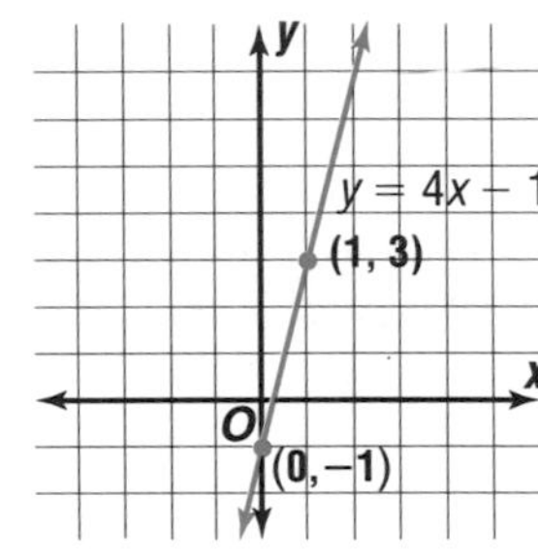

Graph each equation using the slope and y-intercept.

13. $y = 5x - 3$
14. $x + 2y = 4$

- write an equation of a line given information about its graph (Lesson 12–2)

The slope of the line that passes through points at $(4, -3)$ and $(2, 1)$ is $\frac{1-(-3)}{2-4}$ or -2. To find an equation of the line, use the point-slope form.

$y - y_1 = m(x - x_1)$ *Point-slope form*

$y - 1 = -2(x - 2)$ *$m = -2, (x_1, y_1) = (2, 1)$*

$y - 1 = -2x + 4$

$y = -2x + 5$

Write the equation in slope-intercept form of the line that satisfies the given conditions.

15. $m = 4$; y-intercept $= 2$
16. $m = 5$; passes through the point at $(-1, 3)$
17. passes through points at $(-7, 4)$ and $(-5, -6)$
18. x-intercept $= 5$; y-intercept $= 3$
19. line parallel to the graph of $y = x - 5$; passes through the point at $(0, 8)$
20. line perpendicular to the graph of $y = 3x - 1$; passes through the point at $(6, 0)$

OBJECTIVES AND EXAMPLES

- relate equations of lines to geometric concepts (Lesson 12–3)

To find the equation of the perpendicular bisector of a segment whose endpoints have coordinates (5, −3) and (−1, 1), first find the coordinates of the midpoint.

$$\left(\frac{5-1}{2}, \frac{-3+1}{2}\right) = \left(\frac{4}{2}, \frac{-2}{2}\right) \text{ or } (2, -1)$$

Then find the slope of the segment.

$$\frac{-3-1}{5-(-1)} = \frac{-4}{6} \text{ or } -\frac{2}{3}$$

The perpendicular bisector will pass through (2, −1) and have a slope of $\frac{3}{2}$.

$$y - y_1 = m(x - x_1)$$

$$y - (-1) = \frac{3}{2}(x - 2)$$

$$y = \frac{3}{2}x - 4$$

The equation is $y = \frac{3}{2}x - 4$.

REVIEW EXERCISES

The vertices of △*XYZ* are at *X*(2, −1), *Y*(6, 1), and *Z*(0, −3).

21. Find the equations of the lines containing the sides of △*XYZ*.
22. Find the equations of the lines containing the medians of △*XYZ*.
23. Find the equations of the lines containing the altitudes of △*XYZ*.
24. Find the equations of the lines containing perpendicular bisectors of the sides of △*XYZ*.

- prove theorems using coordinate proofs (Lesson 12–4)

When planning a coordinate proof, use the following guidelines to position the figure on a coordinate plane.

- Use the origin as a vertex or center.
- Position at least one side of a polygon on a coordinate axis.
- Keep the figure in the first quadrant if possible.
- Use coordinates that make computations simple.

Prove using a coordinate proof.

25. The segment through the midpoints of the nonparallel sides of a trapezoid is parallel to the bases.
26. The length of the segment through the midpoints of the nonparallel sides of a trapezoid is one-half the sum of the lengths of the bases.

- find the magnitude and direction of a vector (Lesson 12–5)

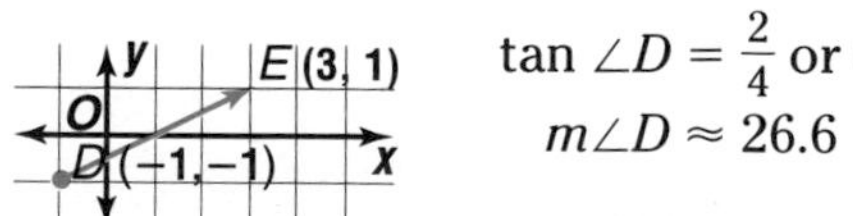

$\tan \angle D = \frac{2}{4}$ or 0.5

$m\angle D \approx 26.6$

$$|\overrightarrow{DE}| = \sqrt{(3-(-1))^2 + (1-(-1))^2}$$

$$= \sqrt{4^2 + 2^2}$$

$$= 20 \text{ or about } 4.5$$

$\overrightarrow{DE}$ has a magnitude of about 4.5 units and a direction of about 26.6°.

Find the magnitude and direction of each vector. Round to the nearest tenth.

27. $\vec{v} = (7, 1)$
28. $\overrightarrow{AB}$ with $A(1, 0)$ and $B(7, 5)$

OBJECTIVES AND EXAMPLES

- perform operations with vectors (Lesson 12–5)

Find the sum of $\vec{a}$ and $\vec{b}$.

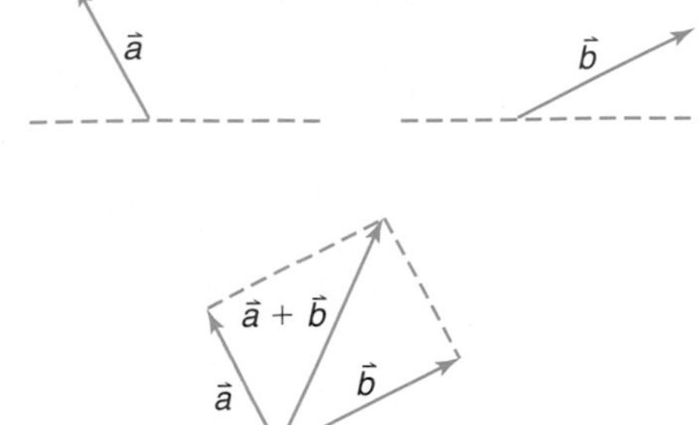

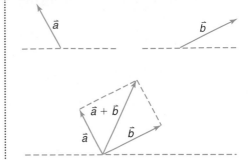

REVIEW EXERCISES

Given $\vec{v} = (0, 8)$ and $\vec{s} = (4, 0)$, represent each of the following as an ordered pair.

29. $\vec{v} + \vec{s}$

30. $3\vec{v} + \vec{s}$

Find each sum.

31. $\begin{bmatrix} 4 \\ 6 \end{bmatrix} + \begin{bmatrix} 3 \\ 2 \end{bmatrix}$

32. $\begin{bmatrix} 2 \\ -3 \end{bmatrix} + \begin{bmatrix} 0 \\ 9 \end{bmatrix}$

- use the distance and midpoint formulas for points in space (Lesson 12–6)

The distance between the points at (2, 2, 2) and (−6, 0, 5) is $\sqrt{(-6-2)^2 + (0-2)^2 + (5-2)^2}$ or $\sqrt{77}$ units. The coordinates of the midpoint of a line segment whose endpoints are (2, 2, 2) and (−6, 0, 5) are $\left(\frac{2-6}{2}, \frac{2+0}{2}, \frac{2+5}{2}\right)$ or (−2, 1, 3.5).

Determine the distance between each pair of points and the coordinates of the midpoint of the segment whose endpoints are given.

33. $A(3, -3, 1), B(7, -3, 5)$

34. $C(2, 4, 6), D(0, 2, 4)$

Write the equations of the sphere given the coordinates of the center and measure of the radius.

35. (0, 0, 0), 5

36. (−1, 2, −3), 4

APPLICATIONS AND PROBLEM SOLVING

37. **Business** Just Like Grandma's Bake Shop spends \$1400 a month for rent and utilities. For each day of operation, they spend \$500 for employees' wages, benefits, and baking supplies. If x represents the number of days of operation in a month, $y = 500x + 1400$ represents the cost of operations for the month. Draw a graph that represents the cost of operation for one week. What does the y-intercept represent? (Lesson 12–1)

38. **Economics** In 1960, a candy bar cost 10¢. In 1996, the same candy bar cost 60¢. Write an equation describing the cost of the candy bar over time. If the rate of change in price has been constant, how much will the candy bar cost in the year 2010? (Lesson 12–2)

39. **Architecture** The table below shows the height and number of stories of several notable structures around the world. (Lesson 12–3)
    a. Draw a scatter plot to show how height x and number of stories y are related.
    b. Write an equation that relates the height to the number of stories.
    c. The Governor Philip Tower in Sydney, Australia, is 745 feet tall. How many stories do you think the tower has?

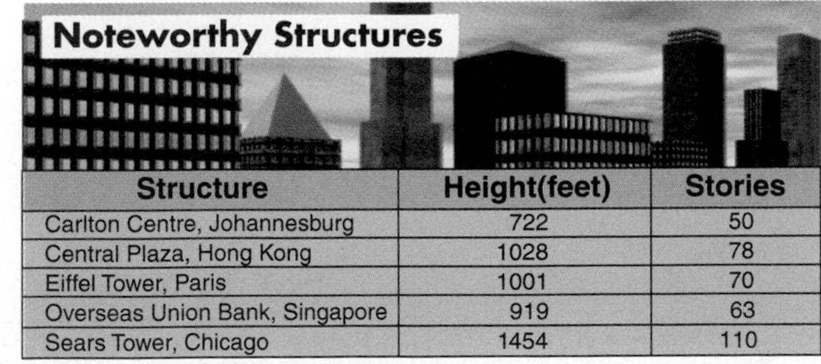

Noteworthy Structures

Structure	Height(feet)	Stories
Carlton Centre, Johannesburg	722	50
Central Plaza, Hong Kong	1028	78
Eiffel Tower, Paris	1001	70
Overseas Union Bank, Singapore	919	63
Sears Tower, Chicago	1454	110

Source: *World Almanac*, 1996

A practice test for Chapter 12 is provided on page 804.

ALTERNATIVE ASSESSMENT

COOPERATIVE LEARNING PROJECT

Architectural Design In this chapter, you learned how to write coordinates for points in space. In this project, you will apply this know-how to describe a simple building you plan to construct.

Design a simple building with surfaces that are polygons. Follow these steps to describe your building mathematically.

- Construct a scale model of your design. Only the frame of the building is needed.
- Remove the top and two adjoining sides of a cardboard box. Mark the edges of the remaining sides, x, y, and z.
- Position the model so that one of its corners is at the origin.
- Using a ruler, determine the coordinates of each point on the model that is the intersection of two line segments.
- Record the ordered triples for these points in a table.
- Use three pieces of graph paper to model the positive quadrants in the plane. Label the edges of each plane with appropriate axes.
- Plot the points from the table on each graph.
- Connect the appropriate points. The result is a view of your building from three perspectives.

THINKING CRITICALLY

Is vector addition commutative? Justify your answer with a proof.

PORTFOLIO

Find a photograph of a building with simple geometric structure. The photograph should view the building from one side. Position the lower left corner of the building at the origin of a two-dimensional coordinate system. Determine values of points that are the intersection of line segments. From this, determine linear equations for each line. Keep the photograph and the equations in your portfolio.

STANDARDIZED TEST PRACTICE

CHAPTERS 1–12

Section One: MULTIPLE CHOICE

There are nine multiple-choice questions in this section. After working each problem, write the letter of the correct answer on your paper.

1. If $4^a = 16$, then find $2^a \times 2$.
   - A. 4
   - B. 8
   - C. 16
   - D. 32

2. If the volume of a cube is 125 cm^3, find its surface area.
   - A. 250 cm^2
   - B. 150 cm^2
   - C. 100 cm^2
   - D. 125 cm^2

3. If the operation # is defined by the equation $x \# y = 2x + y$, what is the value of a in the equation $2 \# a = a \# 3$?
   - A. -1
   - B. 0
   - C. 1
   - D. 4

4. If x is between 0 and 1, which of the following increases as x increases?

   I. $x + 1$

   II. $1 - x^2$

   III. $\frac{1}{x}$

   - A. I only
   - B. III only
   - C. I and II only
   - D. I, II, and III

5. If $x + 2 = 0$, then find x^3.
   - A. 4
   - B. -4
   - C. 8
   - D. -8

6. In the figure below, if $\overline{PQ} \perp \overline{QT}$, $\overline{QR} \perp \overline{RT}$, and $\overline{QS} \perp \overline{SV}$, what is the value of $y - x$?
   - A. 60
   - B. 45
   - C. 30
   - D. 20

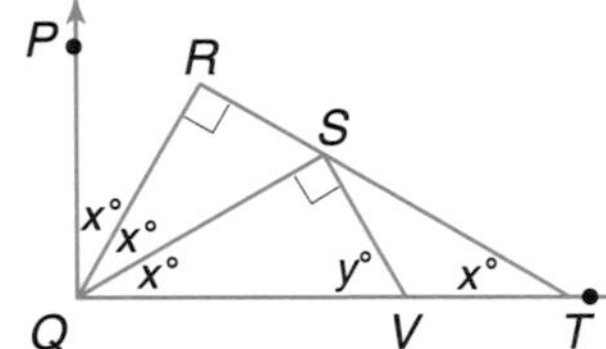

7. If $\frac{x}{3}$ is the measure of the perimeter of a square, then find the measure of the area of the square.
   - A. $\frac{x^2}{144}$
   - B. $\frac{x^2}{36}$
   - C. $\frac{x^2}{16}$
   - D. $\frac{x^2}{9}$

8. A cylindrical can has a diameter of 12 inches and a height of 8 inches. If one gallon of liquid occupies 231 cubic inches, what is the capacity of the can in gallons rounded to the nearest tenth?
   - A. 3.9 gallons
   - B. 4.2 gallons
   - C. 2.3 gallons
   - D. 4.4 gallons

9. What is the mean of $a + 5$, $2a - 4$, and $3a + 8$?
   - A. $2a$
   - B. $6a + 9$
   - C. $2a + 3$
   - D. $3a + 3$

Section Two: SHORT ANSWER

This section contains seven questions for which you will provide short answers. Write your answer on your paper.

10. Right triangles *ABC* and *DEF* are similar. If $BC = 9$, $AC = 21$, and $EF = 24$, find *DF*.

11. Find the volume of a right cylinder with a radius of $3x$ meters and a height of $6x$ meters.

12. In the figure, find x if *O* is the center of the circle.

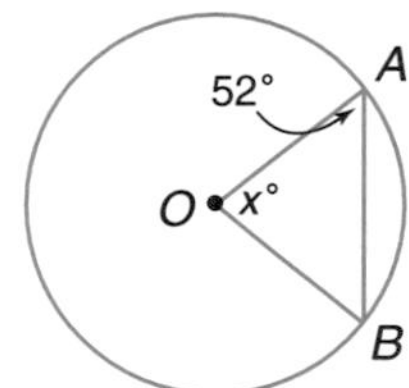

13. The volume of a sphere *S* is $\frac{4}{3}\pi r^3$, and the volume of sphere *Q* is $2\pi r^3$. The volume of *S* is what percent of the volume of *Q*?

14. Find the distance in three-dimensional space between points at $(-1, 2, -3)$ and $(2, 4, -1)$

15. A cone and a cylinder both have a height of 48 units and a radius of 15 units. Find the ratios of their volumes.

16. The corner is cut from a rectangular piece of cardboard as shown below. Find the area of the remaining cardboard.

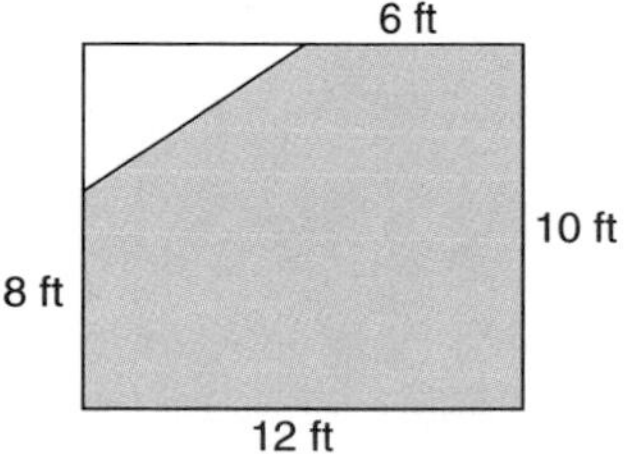

Test Practice For additional test practice questions, visit: **www.geometry.glencoe.com**

Test-Taking Tip Memorize the formula for the volume of a rectangular solid, $V = Bh$. When an SAT question requires finding the volume of another type of three-dimensional figure, the formula will be given in the question or in the instructions for the section of questions.

Section Three: COMPARISON

This section contains five comparison problems that involve comparing two quantities, one in column A and one in column B. In certain questions, information related to one or both quantities is centered above them. All variables used represent real numbers.

Compare quantities A and B below.

- Write A if quantity A is greater.
- Write B if quantity B is greater.
- Write C if the two quantities are equal.
- Write D if there is not enough information to determine the relationship.

	Column A	Column B
	$x > 0$	
17.	the perimeter of a square with side of length $9x$	the perimeter of an equilateral triangle with side of length $12x$
	$3y - x = 8$ $y + x = 2$	
18.	x	y
19.	the volume of a sphere with a radius of 2	the volume of a hemisphere with a radius of 4
	$0 < x < 1$	
20.	$2x$	x^2
21.	the volume of a cylinder with a radius of 2	the volume of a cylinder with a radius of 4

CHAPTER 13

Investigating Loci and Coordinate Transformations

What You'll Learn

In this chapter, you will:

- draw, locate, or describe a locus in a plane or in space,
- determine line symmetry and point symmetry,
- draw reflections, translations, rotations, and dilations on the coordinate plane, and
- solve problems by making a table.

Why It's Important

Real World

Life Science Beautifully-colored butterflies have fascinated people for centuries. Ancient Egyptians, Chinese, Japanese, and Native Americans have been inspired by the beauty of butterflies to use their images in art and pottery. Today, over 170,000 butterflies and moths have been identified. Look carefully at the shape of a butterfly. In Lesson 13–5, you will learn about a type of geometric transformation called a reflection. A butterfly is an example of a reflection. The butterfly has one line of symmetry, the body. If the butterfly was folded along the line of symmetry, one wing would correspond almost exactly to the other wing. Other geometric transformations that you will learn about in this chapter include translations, rotations, and dilations.

PREREQUISITE SKILLS

To be successful in this chapter, you'll need to understand these concepts and be able to apply them. Refer to the lesson in parentheses if you need more review before beginning the chapter.

Find areas of circles. (Lesson 10–5)

Find the area of each circle. Round to the nearest hundredth.

1. circle with radius of 3.2 m
2. circle with diameter of 1 mi
3. circle with radius of 5 in.
4. circle with circumference of 62.83 mm

Find areas of regions. (Lesson 10–3)

Find the area of each shaded region. Round to the nearest hundredth.

5. 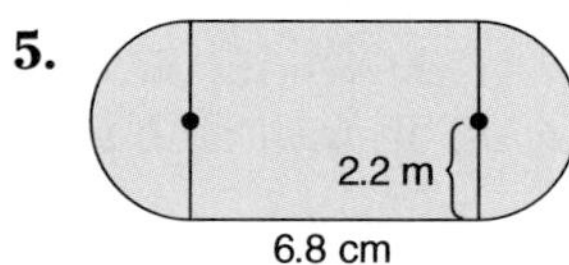

6. 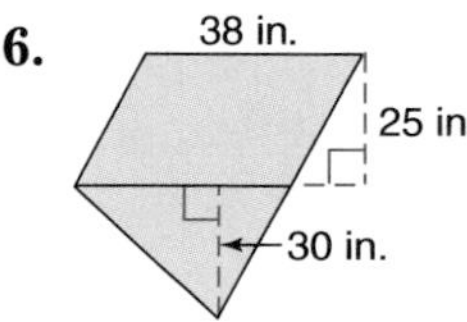

7. 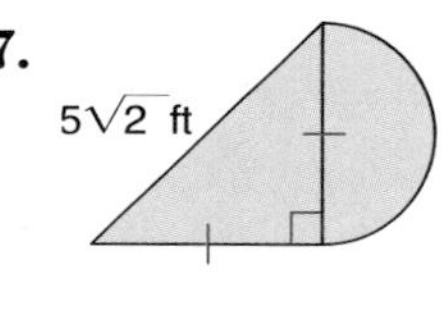

Graph linear equations. (Lesson 12–1)

Graph each equation.

8. $2x + 3y = 6$
9. $y = x$
10. $x - 5y = 10$

Focus On

READING SKILLS

In this chapter, you'll be learning about ***transformations*** in geometry. The terms used for the transformations give you clues to their meaning. A *reflection* is like seeing your reflection in a mirror. A *rotation* involves rotating a geometric figure. A *dilation* makes the figure a different size while maintaining the shape of the object. The ophthalmologist sometimes dilates your pupils, or makes them larger, to see features of you eye more clearly. In geometry, however, a dilation can make the figure larger or smaller. A *translation* is sometimes called a glide because it just slides the figure to a new location.

13–1

What Is Locus?

What YOU'LL LEARN

- To locate, draw, and describe a locus in a plane or in space.

Why IT'S IMPORTANT

Using loci is a way to describe many sets of points that have unique characteristics.

Astronomy

In our solar system, nine planets orbit the Sun. The path of each planet is different, and each path satisfies a set of conditions. The path of each planet can be defined and predicted from these conditions. A path can be thought of as a **locus** of points. The word locus comes from a Latin word meaning "location" or "place." The plural of locus is *loci* (pronounced *LOW-sigh*).

FYI

The Solar system consists of the Sun, which is a star, and all of the objects that travel around it. These include nine planets and their satellites (moons), asteroids, meteoroids, comets, interplanetary dust, and electrically charged gas called plasma.

In geometry, a figure is a locus if it is the set of all the points that satisfy a given condition or set of conditions. A locus may also be defined as the path of a moving point satisfying a given set of conditions. A locus may be one or more points, lines, planes, surfaces, or any combination of these. A locus can occur in a plane or in space.

In order to describe a locus for a given problem, you should follow these steps.

Problem Find the locus of all points in a plane that are 5 centimeters from a given point in that plane.

Step 1 Draw the given figure. — •C — The given figure is point C.

Step 2 Locate points that satisfy the conditions. — (5 cm, •C) — Draw points that are 5 centimeters from point C. Locate enough points to suggest the shape of the locus.

Step 3 Draw a curve or line. — (•C) — The points suggest a circle.

Step 4 Describe the locus. — The locus of the points in the plane that are 5 centimeters from given point C is a circle with a radius of 5 centimeters.

The steps below summarize the procedure for determining a locus.

Procedure for Determining Locus	**After reading the problem carefully, follow these steps.** **1. Draw the given figure.** **2. Locate the points that satisfy the given conditions.** **3. Draw a smooth geometric figure.** **4. Describe the locus.**

A locus in a plane may be different from a locus in space that has the same description.

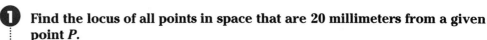

Locus of All Points Equidistant from Two Parallel Lines	
In a Plane	**In Space**
a, *ℓ*, *b* Line *ℓ* is the locus.	*a*, *M*, *b* Plane *M* is the locus.
The locus of points equidistant from two parallel lines in a plane is a line located halfway between the two lines.	The locus of points equidistant from two parallel lines in space is a plane located halfway between the two lines.

Example  **Find the locus of all points in space that are 20 millimeters from a given point *P*.**

Step 1 *Draw the given figure.*

• *P*

The given figure is point *P*.

Step 2 *Locate the points that satisfy the given conditions.*

20 mm • *P*

Draw enough points to determine the shape.

Step 3 *Draw a smooth geometric figure.*

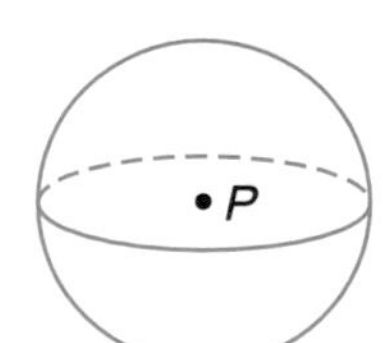

The points describe a sphere.

Step 4 *Describe the locus.* The locus of the points in space that are 20 millimeters from given point *P* is a sphere with a radius of 20 millimeters.

Many loci can be determined by theorems you have learned in previous chapters.

Example **2** **Determine the locus of all points in a plane that are equidistant from the sides of a given angle and are in the interior of that angle.**

Step 1 *Draw the given figure.*

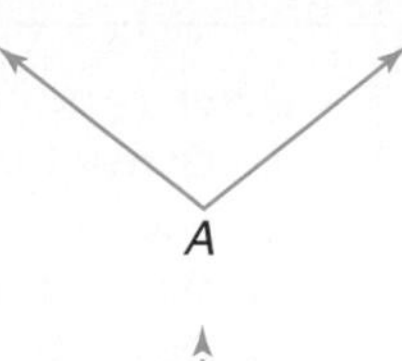

The given figure is $\angle A$.

Step 2 *Locate the points that satisfy the given conditions.*

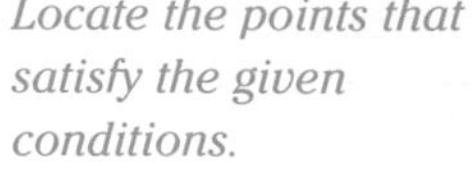

A point in the interior of an angle equidistant from its sides lies on the bisector of the angle. (Theorem 5–4)

Step 3 *Draw a smooth geometric figure.*

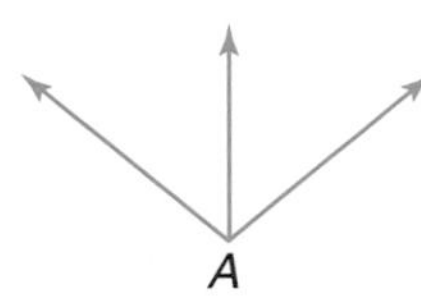

Draw the angle bisector.

Step 4 *Describe the locus.* The locus of the points in the plane that are equidistant from the sides of $\angle A$ and are in the interior of $\angle A$ is the angle bisector of $\angle A$.

A locus can be composed of more than one figure.

Example  **Draw a diagram and describe the locus of points in space that are 0.5 inch from the endpoints of a given line segment.**

The locus of all points in space at a specific distance from a given point is a sphere. Thus, for this problem, the locus of points is two spheres each with a radius of 0.5 inch whose centers are the endpoints of the given line segment.

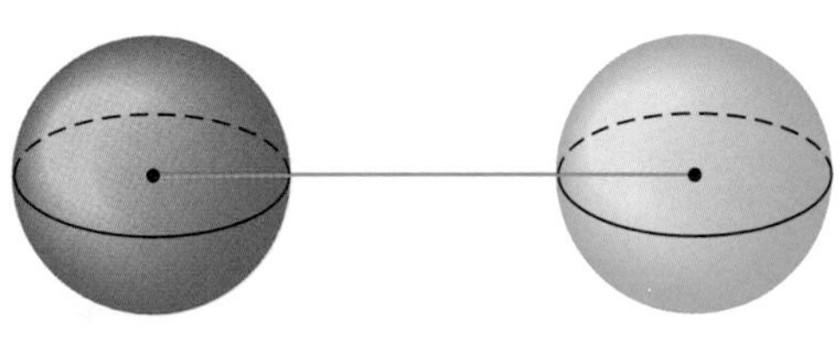

A locus can also describe a region in a real-life situation.

Example  **Paloma designed a type of sprinkler that looks like a small tractor. It moves along a straight line as it sprays water in a circular motion. Suppose this sprinkler can cast a spray 30 feet in diameter and travels 10 feet in an hour.**

Real World APPLICATION

Engineering

a. Draw a figure showing the boundary of the sprinkler's spray in an hour.

When not moving, the sprinkler makes a circular spray. As it moves, the circle moves leaving a wet path.

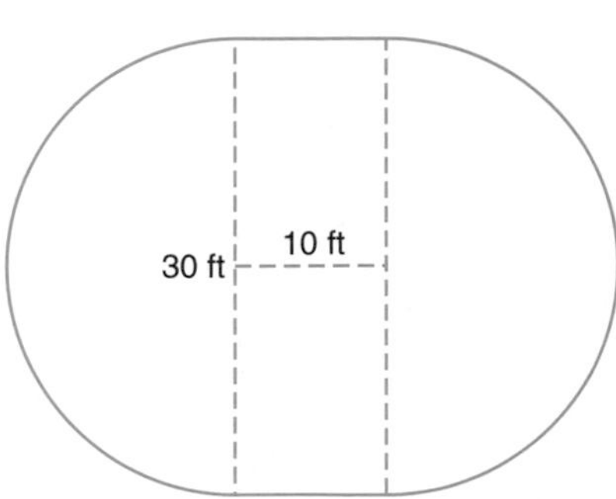

b. Describe the locus.

The locus of points is a rectangle having dimensions 10 feet by 30 feet with a semicircle of radius 15 feet attached to each end.

c. How large is the area that this sprinkler will water?

LOOK BACK

Refer to Lesson 10-3 to review area of several regions and Lesson 10-5 to review area of circles.

$$\begin{aligned}\text{total area} &= \text{area of rectangle} + \text{area of circle}\\ &= \ell w + \pi r^2\\ &= (10)(30) + \pi(15)^2\\ &= 300 + 225\pi\\ &\approx 1006.86 \quad \textit{Use a calculator.}\end{aligned}$$

The sprinkler will water about 1007 square feet.

CHECK FOR UNDERSTANDING

Communicating Mathematics

Study the lesson. Then complete the following.

1. **Explain** why a locus is a geometric figure.
2. **Describe** the locus defined by *a circle with a radius r.*
3. **Explain** the difference between a locus in a plane and a locus in space.

Guided Practice

Draw a figure and describe the locus of points in a plane that satisfy each set of conditions.

4. all points on or in the interior of a right angle and equidistant from the rays that form the angle
5. all points equidistant from two given points

Draw a figure and describe the locus of points in space that satisfy each set of conditions.

6. all points 10 millimeters from a given line m
7. all points equidistant from the vertices of a given square
8. Write a locus description of the drawing at the right.

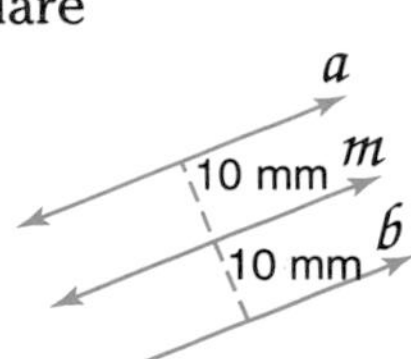

9. Use isosceles trapezoid *ABCD* in plane $\mathcal{R}$ to describe the locus of points in plane $\mathcal{R}$ equidistant from the vertices of *ABCD*.

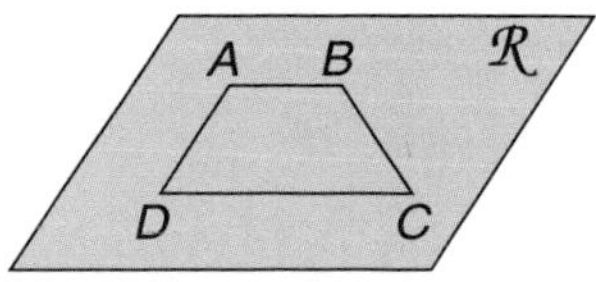

10. **Sports** In high school basketball, three points are awarded for a shot that is made from a distance greater than 19 feet 9 inches from the basket.
    a. Draw a figure showing the locus of points on a basketball court from which a 3-point shot can be made.
    b. Describe the locus.

EXERCISES

Practice

Draw a figure and describe the locus of points in a plane that satisfy each set of conditions.

11. all points equidistant from the centers of two given intersecting circles of the same radius
12. all points on the midpoint of the radii of given circle C whose radius is 6 centimeters
13. all points r units from a given line ℓ
14. all points that are the third vertices of triangles having a given base with length b and a given altitude length a
15. all points equidistant from the vertices of a given regular pentagon
16. all points on the perpendicular bisector of all chords in a given circle
17. all points that are equidistant from two concentric circles with larger radius r and smaller radius s

Draw a figure and describe the locus of points in space that satisfy each set of conditions.

18. all points 3 feet from the floor in your classroom
19. all points 4 meters from a given plane $\mathcal{M}$
20. all points that are the centers of the wheels of a car moving in a straight line
21. all points that are third vertices of equilateral triangle ABC with side length s and with $\overline{BC}$ in plane $\mathcal{M}$
22. all points equidistant to three noncollinear points in a given plane $\mathcal{P}$
23. all points equidistant from two opposite vertices of a cube whose edge is one unit long
24. all points that are the centers of spheres with radius 7 units and tangent to a sphere of radius r units with center C
25. Write a locus description of the figure at the right.

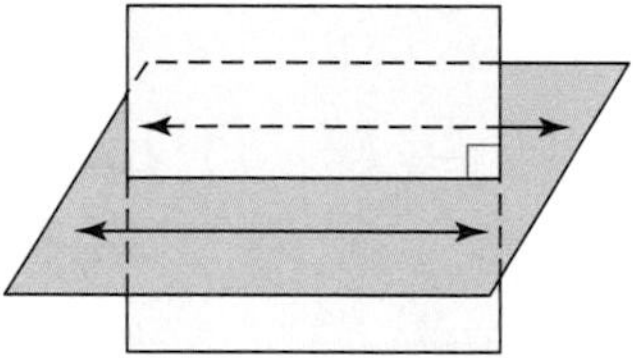

Refer to $\odot O$ and inscribed $\triangle ABC$ in plane $\mathcal{R}$. Draw a figure and describe each locus in Exercises 26–28.

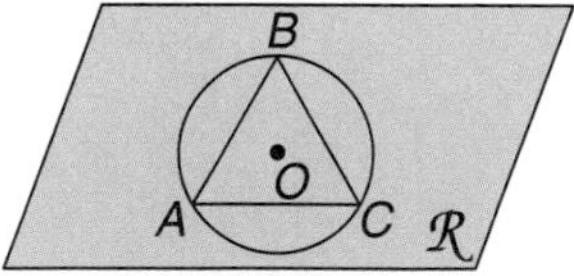

26. all points within r units of point O
27. all points in plane $\mathcal{R}$ on the intersecting lines tangent to $\odot O$ at the points A, B, and C
28. all points that are the midpoints of all chords of $\odot O$ of a given length
29. Given $\angle ABC$ with bisector $\overline{BD}$ in a plane, describe the locus of the centers of circles that are tangent to $\overline{BD}$ and the outside ray of the angle.
30. Describe the locus of points in space that are all the centers of circles that contain two given points.
31. Given a hypotenuse of length h and an altitude of length a from the right angle of a right triangle, what is the locus of points in space formed by the right angle vertex of all possible right triangles?

32. Suppose a semicircle has center $(a, 0)$. What is the result of rotating the semicircle in space about the x-axis?

Graphing Calculator

33. Graph the functions $y = 0$ and $y = x^2$ on your calculator.
   a. What is the locus of points in the plane equidistant from the two graphs?
   b. Write an equation that represents this locus.

Critical Thinking

34. Draw two perpendicular lines a and b. Describe the locus of points in a plane equidistant from these two lines.
35. Mrs. Richter is part of a team selected to plan the placement of new fire stations in a town where a fire station is required to be within two miles of each building. Some of the members of the team thought that some buildings would be within two miles of more than one fire station; others thought this would never happen. Who was correct. Draw a diagram to support your answer.

Applications and Problem Solving

36. **Sports** Two friends who live on opposite sides of town usually ride bicycles together three days a week. Use the map to determine the locus of points where they can meet that are equidistant from both homes.

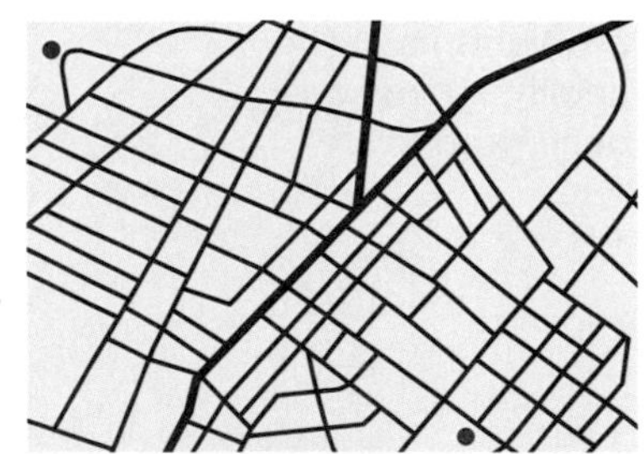

37. **Copper Mining** What is the locus of points that describes how particles will disperse in an explosion above ground if the expected distance a particle could travel is 300 feet?

Mixed Review

38. Draw a network to represent a regular octagon and all of its diagonals. (Lesson 10–7)
39. **Space** A satellite is orbiting 220 miles above Earth. If the diameter of Earth is about 8000 miles, what is the length of the longest line of sight from the satellite to Earth? (Lesson 9–7)
40. **SAT Practice** A video store stocks 250 different movie titles. If 26% of the titles are action movies and 14% are comedies, how many are *not* action movies or comedies?

   A 75 **B** 150 **C** 185 **D** 210 **E** 215
41. The shortest side of a triangle is 18 centimeters long. If two angles of the triangle measure 108° and 56°, find the lengths of the other two sides. (Lesson 8–5)
42. Classify the statement *The diagonals of a rhombus are congruent* as always, sometimes, or never true. (Lesson 6–4)
43. **Algebra** Given $\triangle YLO \cong \triangle GRN$, $m\angle Y = 60$, $m\angle L = 78$, $m\angle O = 42$, and $m\angle N = \frac{2}{3}x + 6$, find x. (Lesson 4–3)
44. If $\angle 1$ and $\angle 2$ are congruent alternate interior angles and are supplementary, describe the lines that form them. (Lesson 3–4)

Algebra

45. Graph $y = 2x - 4$ by using the x- and y-intercepts.
46. Determine whether $9x = 5 - 3y$ is a linear equation. If it is linear, rewrite it in standard form, $Ax + By = C$.

For **Extra Practice**, see page 789.

13-2 Integration: Algebra Locus and Systems of Linear Equations

What YOU'LL LEARN

- To find the locus of points that are solutions of a system of equations by graphing, substitution, or elimination.

Why IT'S IMPORTANT

You can use systems of equations to solve problems involving health, sports, and demographics.

Real World APPLICATION

Physical Fitness

Millions of people exercise regularly to maintain or improve their cardiovascular health. During exercise many people monitor their heart rate to make sure that they reach a pulse rate that exercises but does not overtax their heart. The recommended maximum heart rate is a function of the age of the person and can be represented in a graph. This graph represents a locus of points that satisfy a certain set of conditions that can also be given in a table and represented by a linear equation.

Graph **Table**

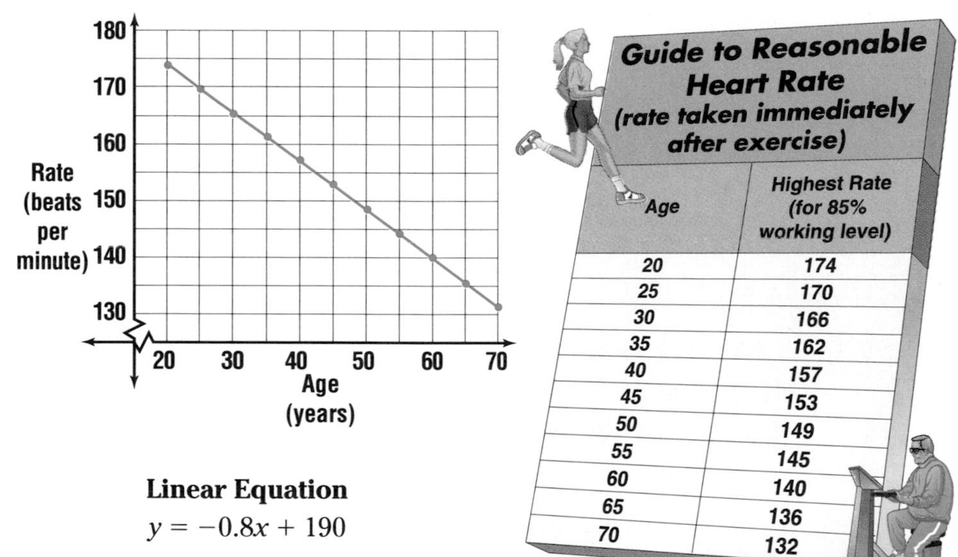

Guide to Reasonable Heart Rate *(rate taken immediately after exercise)*

Age	Highest Rate (for 85% working level)
20	174
25	170
30	166
35	162
40	157
45	153
50	149
55	145
60	140
65	136
70	132

Source: *Longevity* by Kenneth R. Pelletier

Example 1

Travel

A to Z Travel offered a vacation package plan to the Krivaneks that was based on a plane fare of \$350 per person and a rate of \$125 per day per person and included a hotel room and a rental car.

a. Write an equation that represents the total cost of this plan per person based on the number of days that the Krivaneks stay.

Let x = the number of days stayed and let y = the total cost of the plan.

$$\underbrace{\text{total cost}}_{y} \; \underbrace{=}_{=} \; \underbrace{\text{(rate)}}_{125} \; \underbrace{\times}_{\cdot} \; \underbrace{\text{(number of days)}}_{x} \; \underbrace{+}_{+} \; \underbrace{\text{plane fare}}_{350}$$

$$y = 125x + 350$$

b. Draw a graph of the locus of points that represents this plan.

$y = 125x + 350$		
x	y	(x, y)
2	600	(2, 600)
6	1100	(6, 1100)
10	1600	(10, 1600)

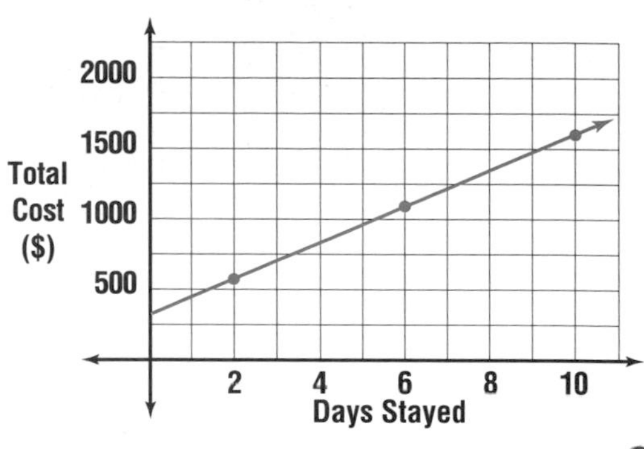

Relationships exist between pairs of lines. To find the locus of points that are common to two lines, you can use graphing to solve the system of equations.

Example 2

Consumerism

The Krivaneks then went to Northwest Travel to see what plan they could offer. Northwest Travel presented a plan that had a plane fare of $470 per person and a rate of $95 per day per person and included a hotel room and a rental car. Describe the locus of points in which the total cost of this plan and the total cost of the plan described in Example 1 are the same.

Explore You have two travel-plan options. One plan has a plane fare of $350 per person and a rate of $125 per day per person. The other plan has a plane fare of $470 per person and a rate of $95 per day per person.

Plan Write an equation for each of these plans and graph them on the same coordinate grid. Construct a table of values for these two equations. Then describe the locus of points in which the total cost of each plan is equal.

Solve Let x = the number of days stayed and y = the total cost of the plan.

$$\underbrace{\text{total cost}}_{y} = \underbrace{\text{(rate)}}_{95} \underbrace{\times}_{\cdot} \underbrace{\text{(number of days)}}_{x} \underbrace{+}_{+} \underbrace{\text{plane fare}}_{470}$$

$$y = 95x + 470$$

The equation for the A to Z plan in Example 1 is $y = 125x + 350$.

The graph of these two equations is shown at the right. The graphs intersect at $P(4, 850)$. Therefore, P is the locus of points that satisfy the graph of both equations, $y = 95x + 470$ and $y = 125x + 350$. Therefore, if the Krivaneks stay for 4 days, the total cost of each plan will be $850.

(continued on the next page)

Examine Study the table at the right. The table confirms the point of intersection is (4, 850).

x	95x + 470	125x + 350
1	565	475
2	660	600
3	755	725
4	**850**	**850**
5	945	975
6	1040	1100
7	1135	1225
8	1230	1350
9	1325	1475
10	1420	1600

Notice that the graphs in Example 2 intersect at the point with coordinates (4, 850). Since this point lies on both graphs, its coordinates satisfy both $y = 95x + 470$ and $y = 125x + 350$. The equations $y = 95x + 470$ and $y = 125x + 350$ are called a **system of equations**. The solution of this system is (4, 850).

You may remember from algebra that a system of equations can be solved by algebraic methods as well as by graphing. Two algebraic methods are the *substitution method* and the *elimination method*. The substitution method involves solving one equation for one of the variables and then substituting that value for the variable in the second equation. The elimination method involves combining two equations by addition or subtraction to eliminate one of the variables.

Example 3

Algebra

Find the locus of points that satisfy the graphs of both equations $5x - 2y = 11$ and $y = x - 1$.

Method 1: Substitution

The second equation states that $y = x - 1$. Therefore, $x - 1$ can be substituted for y in the first equation.

$$5x - 2y = 11$$

$$5x - 2(x - 1) = 11 \quad \textit{Substitute } x - 1 \textit{ for } y \textit{ and solve.}$$

$$5x - 2x + 2 = 11$$

$$3x + 2 = 11$$

$$3x = 9$$

$$x = 3 \quad \textit{The x-coordinate is 3.}$$

Find y by substituting 3 for x in the second equation.

$$y = x - 1$$

$$y = 3 - 1$$

$$y = 2 \quad \textit{The y-coordinate is 2.}$$

The point with coordinates (3, 2) is the locus of points that satisfy the graphs of both equations $5x - 2y = 11$ and $y = x - 1$.

Method 2: Elimination

Rewrite the second equation in standard form, $Ax + By = C$.

$$y = x - 1$$
$$-x + y = -1 \quad \textit{Subtract x from each side.}$$

Sometimes adding or subtracting two equations will eliminate a variable. In this case, adding or subtracting the two equations will not eliminate a variable. If both sides of the second equation are multiplied by 5, the system can be solved by adding the two equations.

$5x - 2y = 11$

$-x + y = -1$ **Multiply by 5.** →

$$5x - 2y = 11$$
$$\underline{-5x + 5y = -5}$$
$$3y = 6 \quad \textit{Add to eliminate x.}$$
$$y = 2 \quad \textit{The y-coordinate is 2.}$$

Substitute 2 for y in the second equation. Then solve for x.

$$y = x - 1$$
$$2 = x - 1$$
$$3 = x \quad \textit{The x-coordinate is 3.}$$

The point with coordinates (3, 2) is the locus of points that satisfy the graphs of both equations $5x - 2y = 11$ and $y = x - 1$.

Note in Example 3 that the solution to the system of equations was the same regardless of the method used. However, one method of solution may be simpler than another.

A graphing calculator can also help you find the locus of points that satisfy the graphs of two linear equations.

EXPLORATION
GRAPHING CALCULATOR

Use a graphing calculator to find the locus of points that satisfy the graphs of $y = 3.4x + 2.1$ and $y = -5.1x + 8.3$.

- Graph both functions.
- Use the CALC menu and select 5:Intersect from the menu.
- As the calculator prompts you, press ENTER to select each line and determine the intersection point.

The approximate coordinates of the locus will appear at the bottom of the screen.

Your Turn

a. According to your exploration, what are the coordinates of the point of intersection to the nearest hundredth?

b. Solve the system of equations algebraically. Is your solution close to the results found on the graphing calculator?

c. When do you think finding a solution to a system of equations with a graphing calculator is appropriate? Does this method of solution always give an accurate solution?

CHECK FOR UNDERSTANDING

Communicating Mathematics

Study the lesson. Then complete the following.

1. **Explain** whether a point could possibly satisfy all of a certain set of conditions in the definition of a locus if the point is not in the locus.
2. **Describe** the possible loci of points that satisfy the graphs of two linear equations.
3. **Health** The suggested maximum pulse rate for a person 15 years old is 160 beats per minute. The suggested maximum pulse rate for a person 40 years old is 140 beats per minute. If the pulse rate decreases at a constant rate, what is the equation whose graph models the locus of points that represent the maximum heart rate for any age a?

4. Determine whether the most efficient way to find the locus of points that satisfy the graphs of $3x + 2y = 9$ and $-x + 3y = 8$ is graphing, substitution, or elimination. Explain your answer.

Guided Practice

Graph each pair of equations to find the locus of points that satisfy the graphs of both equations.

5. $y = 2x - 2$
   $x + y = 4$
6. $y = 3x + 1$
   $y = -3x + 1$

Cleveland

Oakland

Use either substitution or elimination to find the locus of points that satisfy the graphs of both equations.

7. $2x - y = 4$
   $x - y = 5$
8. $y = \frac{1}{3}x - 4$
   $y = -x + 1$
9. $y = 4x$
   $x + y = 10$
10. $12 - 3y = 4x$
    $40 + 4x = 10y$
11. **Population** In 1990, Cleveland, Ohio, had a population of 505,616, which was decreasing at the rate of about 68,000 people in 10 years. In 1990, Oakland, California, had a population of 372,242, which was increasing at a rate of 33,000 people in 10 years. If the rates of decrease and increase remain constant, describe the locus of points that represent the time at which the populations of Cleveland and Oakland will be equal.

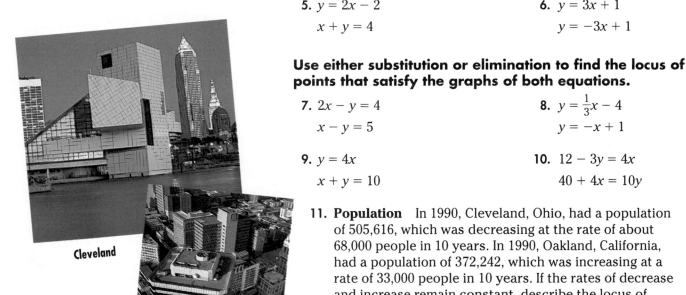

EXERCISES

Practice

Graph each pair of equations to find the locus of points that satisfy the graphs of both equations.

12. $x + 3y = 5$
    $y = 2x - 10$
13. $y = 4x + 7$
    $y = -x - 3$
14. $4x - 3y = 6$
    $2x - 3y = 12$
15. $y = -3x$
    $6y - x = -38$
16. $2x + y = 4$
    $x - y = 2$
17. $x - 2y = 0$
    $y = 2x - 3$

Use either substitution or elimination to find the locus of points that satisfy the graphs of both equations.

18. $x + 2y = 10$
$y = x - 4$

19. $3x + y = 8$
$y = x$

20. $x + y = 4$
$y = x + 6$

21. $2x + y = 9$
$x = 4y$

22. $y = -x + 1$
$y = 6$

23. $y = -2x + 2$
$y = \frac{x}{2} + 2$

24. $3x - y = 11$
$x + y = 5$

25. $x - 4y = 9$
$2x - 3y = 8$

26. $x - 2y = 7$
$2x + 3y = 0$

27. $x + 4y = 27$
$x + 2y = 21$

28. $y = x + 7$
$3x + y = 8$

29. $y = \frac{1}{3}x$
$x + 2y = -3$

30. The graphs of $y = 3$, $y = 7$, $y = 2x$, and $y = 2x - 13$ intersect to form a parallelogram.

a. Find the coordinates of the vertices.

b. Find the area of the parallelogram.

c. Find the locus of points that satisfy the graphs of all four equations.

31. Find the locus of points that satisfy the graphs of $(x + 3)^2 - (y + 4)^2 = 9$ and $x = 2$.

32. Given $\overline{AG}$ with endpoints $A(-1, -3)$ and $G(9, 0)$, find the equation that represents the locus of points that is the perpendicular bisector of $\overline{AG}$.

33. Find the locus of points that satisfy the graphs of $y = (x - 1)^2$ and $y = x + 1$.

Critical Thinking

34. Write three pairs of linear equations so that one pair would best be solved by using substitution, another by using elimination, and the third by graphing. Explain your reasoning for each pair of equations.

Applications and Problem Solving

35. Sports For an opening-game promotion, the Orange Hill Volleyball Boosters purchased caps for all children under 12 years at \$1.50 each and computer mouse pads for everyone 12 years and over at \$2.50 each. The ticket price for the game for children under 12 was \$1.00 and \$4.00 for everyone else. What is the locus of points that represents the distribution of tickets sold if merchandise costs were \$405 and ticket sales were \$550?

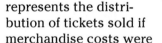

Data Update For more information on state populations, visit: **www.geometry.glencoe.com**

36. Population From 1993 to 2020, the number of people age 65 or older will surge as baby boomers head into old age. Pennsylvania's population of persons age 65 or older was 1,908,000 in 1993 and is expected to be 2,303,000 by 2020. In 1993, the population in Texas of persons age 65 or older was 1,835,000, and in 2020, it will be 3,640,000. Will there be a time when Pennsylvania and Texas have the same population of people age 65 or older? If so, describe the locus of points that will represent that time.

Mixed Review

37. Draw and describe the locus of points in a plane formed by the vertex of the right triangles with hypotenuse $\overline{DC}$. (Lesson 13–1)

38. The measures of the exterior angles of a convex pentagon are g, $2g$, $1.5g$, $0.5g$, and g. Find the value of g and the measure of each exterior angle. (Lesson 10–1)

39. If the equation of $\odot O$ is $(x - 2)^2 + (y - 3)^2 = 25$, find the value of j so that the point at $(6, j)$ lies on the circle. (Lesson 9–8)

40. **SAT Practice** A bag contains 6 blue balls, 8 red balls, and 2 white balls. If three balls are removed at random and no ball is returned to the bag after removal, what is the probability that all three balls will be red?

A $\frac{1}{10}$ **B** $\frac{1}{8}$ **C** $\frac{3}{8}$ **D** $\frac{3}{7}$ **E** $\frac{1}{2}$

41. Draw a quadrilateral with exactly one obtuse angle and one pair of parallel sides. What kind of quadrilateral is it? (Lesson 6–5)

42. Find $m\angle 1$ and $m\angle 2$. Then classify the triangle by its angles and by its sides. (Lesson 4–2)

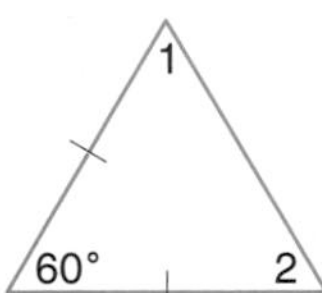

43. Parallel lines m and n are cut by transversal p forming angles 1 through 8. If all the even-numbered angles have measures of 35, draw a figure labeling the lines and the angles and determine the measures of all odd-numbered angles. (Lesson 3–2)

Algebra

44. The sum of three consecutive odd integers is -39. What are the integers?

For **Extra Practice**, see page 790.

45. Determine the value of r so the line that passes through points at $(2, r)$ and $(0, 3)$ has a slope of $\frac{3}{2}$.

Mathematics and SOCIETY

Fractal Models for City Growth

The excerpt below appeared in *Science News* on January 6, 1996.

DURING THE LAST DECADE, RESEARCHERS have begun to examine the possibility of using mathematical forms called fractals to capture the irregular shapes of developing cities....A magnified portion of a fractal looks very similar, if not identical, to the overall structure. Further magnification reveals details that again resemble the full pattern. Hence, fractal objects look the same whatever the magnification. Zooming in for a closer view doesn't smooth out the irregularities or end the branching....The recent focus on fractal models of urban development represents one aspect of a renewed interest in the importance of local actions, individual decisions, and self-organization in shaping cities....realistic models of how local activities lead to large-scale patterns and order may eventually serve as important tools for planning and predicting urban development. ■

1. If you took a long-range photo of a fractal river from a satellite and then took a much closer photo of just a portion of the river, what would you notice about the images in the two photos?
2. How does the concept of locus play a part in city planning? What factors might be considered to predict and plan growth?

13-3 Intersection of Loci

What YOU'LL LEARN

- To solve locus problems that satisfy more than one condition.

Why IT'S IMPORTANT

You can use the intersection of loci to solve problems involving agriculture and sports.

You can use tracing paper to create a triangle that has certain specifications.

MODELING MATHEMATICS

Creating a Triangle

Materials:  tracing paper  ruler  protractor 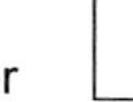 plain paper

Use tracing paper to create a triangle ABC that contains $\angle A$, a side AC units long, and an altitude BD units long.

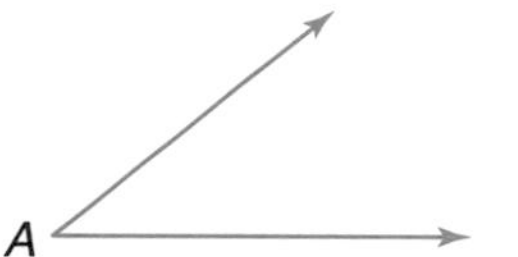

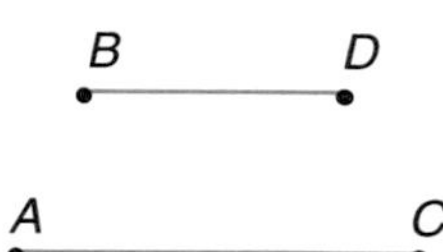

- Trace $\angle A$ on a piece of tracing paper. Extend the rays to the edges of the paper.
- Lay the paper over $\overline{AC}$, aligning both As and a ray of $\angle A$ with $\overline{AC}$. Mark the position of C on a ray of $\angle A$.

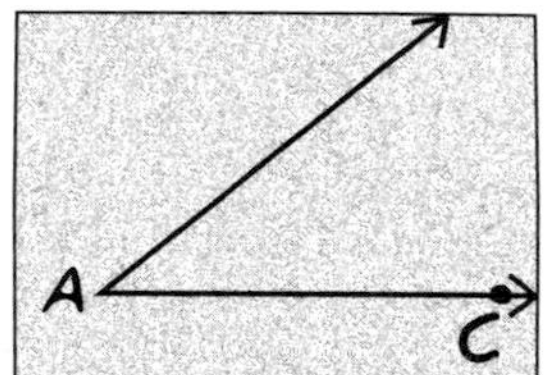

- On another piece of paper, draw a line ℓ and draw a segment perpendicular to ℓ with an endpoint on ℓ that is the same length as $\overline{BD}$.

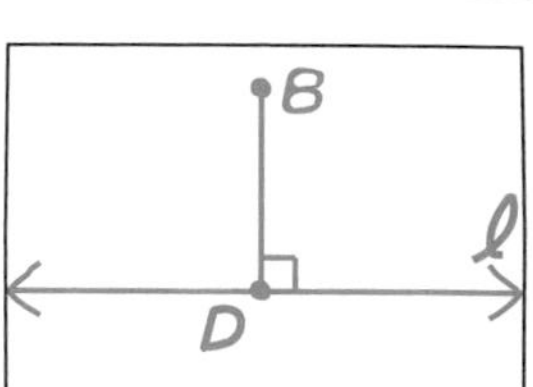

- Place the tracing paper over your drawing of ℓ and $\overline{BD}$ so that ℓ and $\overline{AC}$ coincide. Move the tracing paper right or left, keeping the two lines aligned, until point B appears to lie on the other ray of $\angle A$.

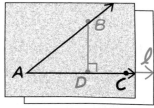

- Mark the position of B on your angle and draw $\triangle ABC$.

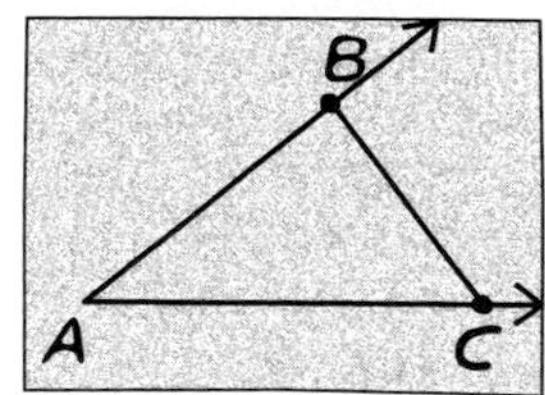

Your Turn

a. Measure $\angle BAC$, $\overline{AC}$, and $\overline{BD}$ in $\triangle ABC$. Are they the same measures as the original measures of $\angle A$, $\overline{AC}$, and $\overline{BD}$?

b. Repeat the activity putting C on the other ray. How does your result compare to $\triangle ABC$ in the activity?

c. Are there other possible ways to draw $\triangle ABC$?

In Chapters 4 and 5, you learned that different sets of angles and sides could define unique triangles. Let's investigate the relationship between one given angle, one given altitude, and one given side of a triangle and how they can define one or more triangles.

Example 1 **Determine the ways in which the locus of points that satisfy the conditions in the Modeling Mathematics activity on the previous page can be drawn.**

Draw line ℓ. Construct $\overline{AC}$ on line ℓ.

The locus of points for B, BD units from $\overline{AC}$, is two parallel lines on either side of line ℓ. Construct the locus of points BD units from ℓ.

Construct $\angle A$ on the upper and lower sides of the base and mark the points where the rays intersect the parallel lines.

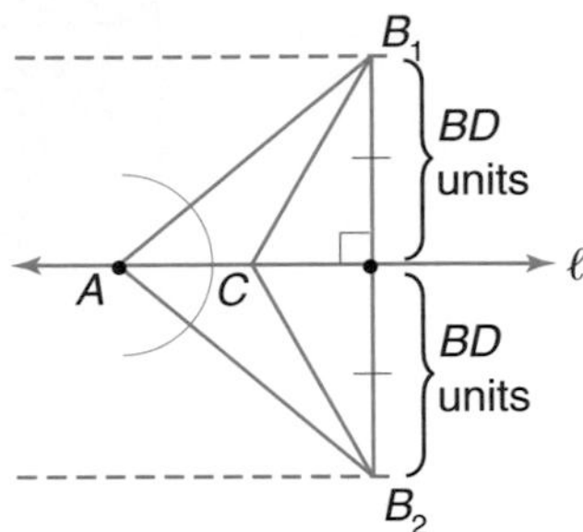

Note: $\triangle AB_1C$ and $\triangle AB_2C$ are congruent.

Label the points B_1 and B_2. Draw $\overline{B_1C}$ and $\overline{B_2C}$.

There are two possible ways to draw $\triangle ABC$.

Often loci satisfy several conditions. Such loci can usually be determined by finding the intersection of the loci that meet each separate condition.

Example 2 **An above-ground irrigation system with a radius of 100 meters is positioned at the intersection of four fields as shown at the right.**

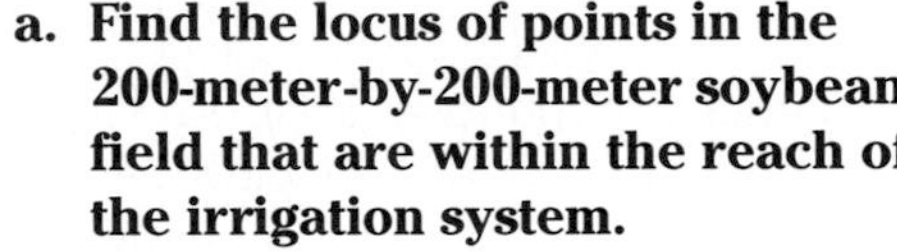

Agriculture

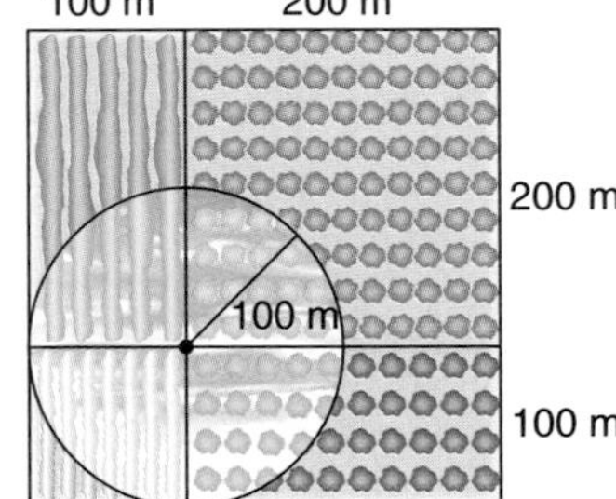

a. Find the locus of points in the 200-meter-by-200-meter soybean field that are within the reach of the irrigation system.

b. What percent of the soybean field is in the locus?

FYI

The soybean is one of the world's most inexpensive and useful sources of protein. Soybeans grown on an acre of land can provide about 10 times as much protein as beef cattle raised on the same land.

a. The locus of the irrigation system arm is the set of points on and within a circle with a radius of 100 meters. Only one fourth of this locus is in the 200-meter-by-200-meter soy bean field. Therefore, the locus of points in the soybean field is a sector that covers one fourth of the circle.

b. First, find the area of the sector that is irrigated.

$A = \frac{1}{4}\pi r^2$

$= \frac{1}{4}\pi(100)^2$ or about 7854 square meters *Use a calculator.*

The area of the soybean field is 200^2 or 40,000 square meters.

To find the percent of the soybean field that is in the locus, divide the area of the sector by the area of the field and write as a percent.

$\frac{7854}{40,000} = 0.19635$ or about 19.6%

About 19.6% of the field is in the locus.

Example 3 **Find the locus of points in the coordinate plane that satisfy the graphs of $y = (x - 1)^2$ and $y = 5x - 11$.**

Algebra

The locus of points that satisfy $y = (x - 1)^2$ can be drawn by first making a table of some of the values and then sketching the graph.

x	$(x - 1)^2$	y	(x, y)
-2	$(-2 - 1)^2$	9	$(-2, 9)$
-1	$(-1 - 1)^2$	4	$(-1, 4)$
0	$(0 - 1)^2$	1	$(0, 1)$
1	$(1 - 1)^2$	0	$(1, 0)$
2	$(2 - 1)^2$	1	$(2, 1)$

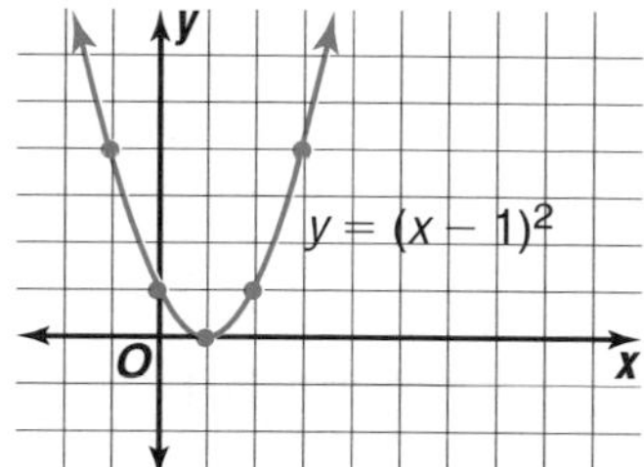

Therefore, from the graph, the locus of points that satisfy $y = (x - 1)^2$ is a parabola with the vertex at (1, 0).

The locus of points that satisfy the graph of $y = 5x - 11$ is a line with slope 5 that passes through $(2, -1)$.

Method 1: Solve this system of equations graphically.

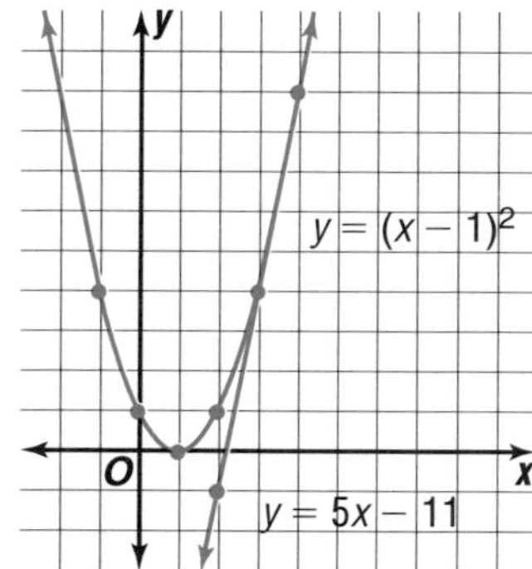

The graphs intersect at (3, 4) and (4, 9).

Method 2: Solve this system of equations algebraically. Substitute $5x - 11$ for y in the first equation.

$$
\begin{aligned}
y &= (x - 1)^2 \\
5x - 11 &= (x - 1)^2 \\
5x - 11 &= x^2 - 2x + 1 \\
0 &= x^2 - 7x + 12 \\
0 &= (x - 3)(x - 4) \\
x &= 3 \text{ or } x = 4
\end{aligned}
$$

Using substitution, $y = 4$ for $x = 3$ and $y = 9$ for $x = 4$. The solutions are (3, 4) and (4, 9).

The locus of points that satisfy both conditions are points at (3, 4) and (4, 9).

The conditions of loci can also be in space. These loci conditions will be drawn and described differently than those in a plane.

Example 4 **Describe the locus of points in space that satisfy the graph of $y = 2x - 10$.**

The locus of points is a plane perpendicular to the xy-plane whose intersection with the xy-plane is the graph of $y = 2x - 10$.

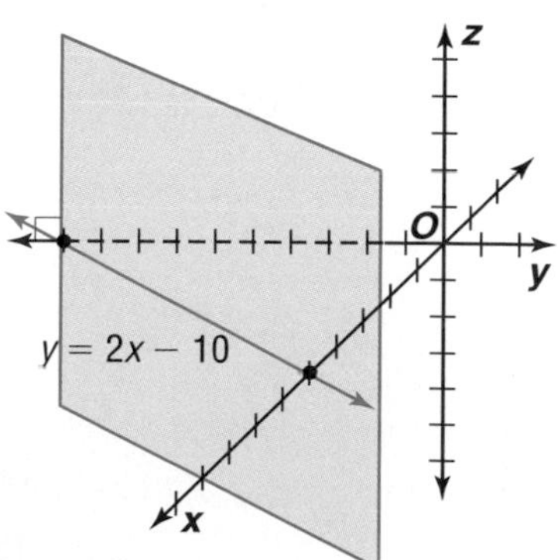

CHECK FOR UNDERSTANDING

Communicating Mathematics

Study the lesson. Then complete the following.

1. **Explain** two ways to determine the locus of points that satisfy the graphs of two equations in a coordinate plane.
2. **Draw** and describe all possible loci involving a line and a parabola in a coordinate plane.
3. How many $\triangle ABC$s can be drawn with a given $\angle A$ and given sides of length AB and BC?

4 Use the same $\overline{AC}$ and $\overline{BD}$ from the Modeling Mathematics activity on page 709, but draw $\angle A$ so that it is obtuse. Determine $\triangle ABC$ with these new specifications. What did you find? Did you get the same or different results? Explain.

Guided Practice

5. Describe the different possible intersections of two circles.

Draw a diagram to find the locus of points that satisfy the conditions. Then describe the locus.

6. all the points in the coordinate plane 5 units from the origin and 3 units from the x-axis
7. all points in a plane 7 inches from the vertex of an angle and equidistant to the sides of the angle
8. the graphs of $y = (x + 4)^2$ and $y = -x - 2$

Describe the locus of points in space that satisfy the graph of each equation.

9. $x = 5$
10. $(x - 2)^2 + (y - 5)^2 = 16$

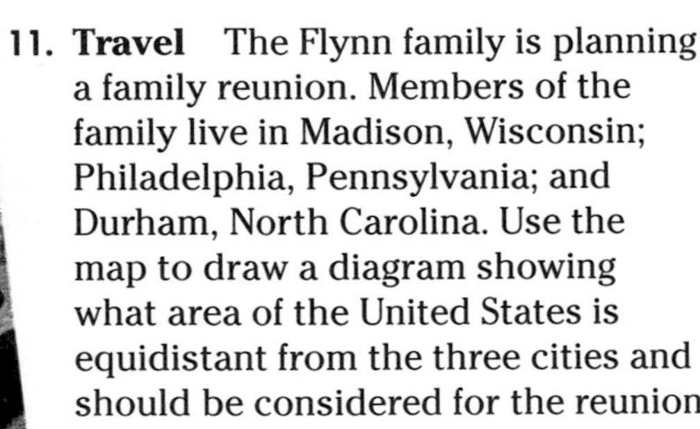

11. **Travel** The Flynn family is planning a family reunion. Members of the family live in Madison, Wisconsin; Philadelphia, Pennsylvania; and Durham, North Carolina. Use the map to draw a diagram showing what area of the United States is equidistant from the three cities and should be considered for the reunion.

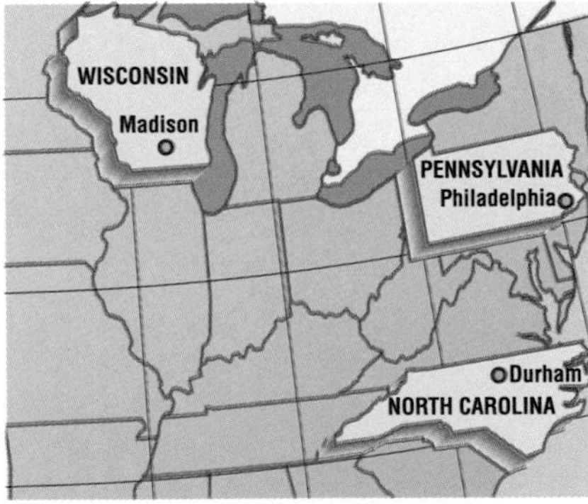

EXERCISES

Practice

12. Determine the least and the greatest number of points in the intersection of a circle and a line MN that satisfy these conditions:
    a. M and N are in the exterior of the circle.
    b. M and N are in the interior of the circle.
    c. M and N are on the circle.
    d. M is in the exterior and N is on the circle.

Describe the different possible intersections of each pair of figures.

13. a point and a line
14. a line and a plane
15. a cube and a line
16. two spheres

Draw a diagram to find the locus of points that satisfy the conditions. Then describe the locus.

17. all the points in the coordinate plane 6 units from the origin and equidistant from the x-and y-axes
18. all the points in a plane equidistant from two given points and equidistant from two parallel lines
19. all the points in space equidistant from a cylinder
20. all points in a plane that are 2 centimeters from a given line and 3 centimeters from a given point on the line
21. all points in space 5 centimeters from a line perpendicular to two parallel planes that lie in the two planes
22. all points in the plane that are equidistant from the rays of an angle and equidistant from two points on one side of the angle
23. all points in space equidistant from two intersecting planes

Describe the geometric figure whose locus in space satisfies the graph of each equation.

24. $y = -3x + 11$
25. $y = (x - 9)^2$
26. $y + x = 0$
27. $(x - 4)^2 + (y - 1)^2 = 25$
28. $(x + 6)^2 + (y - 3)^2 + (z - 1)^2 = 25$
29. $(x - 3)^2 + (y - 4)^2 + (z - 5)^2 = 16$

30. Construct an isosceles triangle with base $\overline{RT}$ and vertex angle S. How many triangles can be drawn that satisfy these conditions?

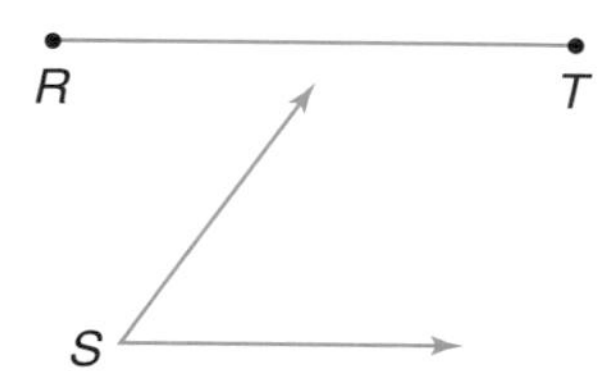

Find the locus of points in the coordinate plane that satisfy the graphs of both equations or inequalities.

31. $(x + 3)^2 + (y - 1)^2 = 25$ and $y = -3x - 13$
32. $x + y = 6$ and $(x - 1)^2 + y^2 = 17$
33. $x^2 + y^2 = 4$ and $y = x + 6$
34. $y > 0$ and $x^2 + y^2 \leq 9$

35. Circumscribe an equilateral triangle around a circle.
36. Point M is on line ℓ and point N is not on the line. Construct a circle that is tangent to line ℓ, and contains both points M and N.

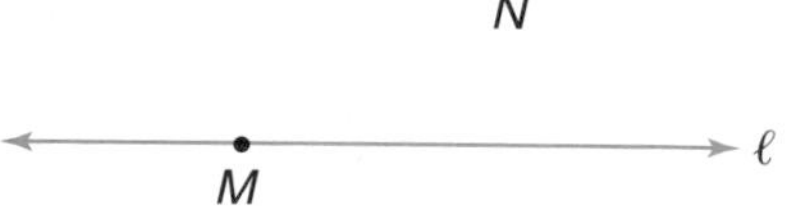

Critical Thinking

37. How would the construction in Example 1 be different if the side given were $\overline{BC}$?
38. Place a point in a plane with two intersecting lines so that the locus of points equidistant from the two lines and a distance t from the point are at a maximum.

Applications and Problem Solving

39. **Management** The Mathematics Club from Northwood High School was planning a trip to Washington, D.C. On one day, three groups of students wanted to visit three attractions, the Capitol, Union Station, and the Air and Space Museum. Use the map to locate a meeting point equidistant from these three buildings.

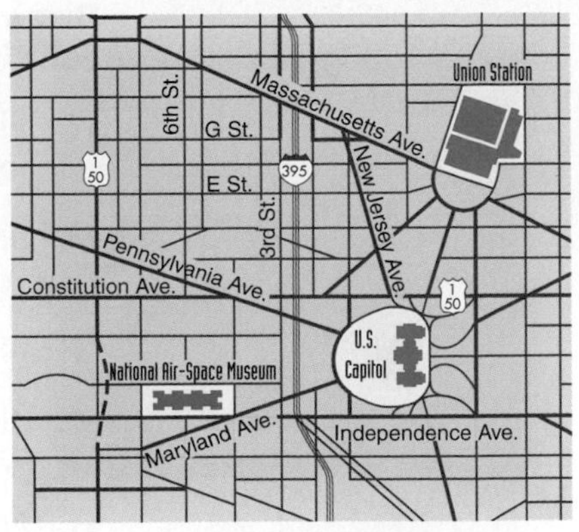

40. **Agriculture** On the field at the right, where would you place a circular irrigation system with a radius of 50 meters, so that the maximum area of the field is irrigated?

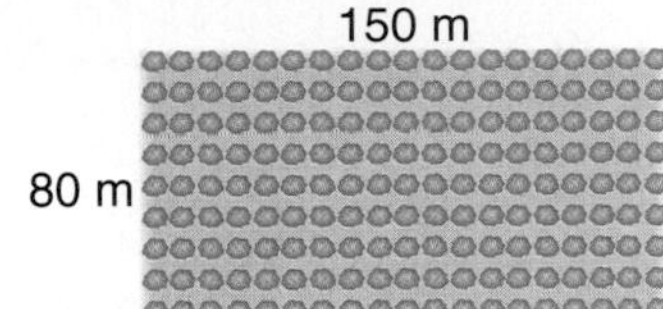

41. **Sports** Josh is the goalie for a 16-year-and-under soccer team. His kicks from the goal area average 160 feet in length before they hit the ground. What is the locus of points representing the area in which most of his kicks from the goal area land?

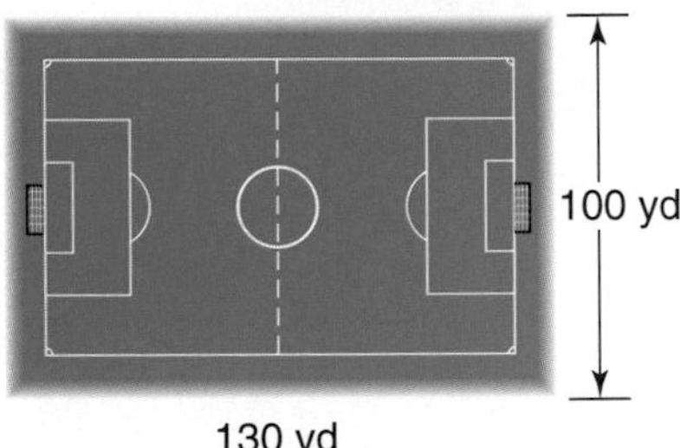

Mixed Review

42. Use an algebraic method to find the locus of points that satisfy the graphs of $9x + y = 20$ and $3x + 3y = 12$. (Lesson 13–2)

43. Draw and describe the locus of points in a plane that are equidistant from the points of intersection of two given circles. (Lesson 13–1)

44. Write the equation of the sphere with center at $(4, -3, -2)$ and the point at $(-2, -5, 7)$ on the sphere. (Lesson 12–6)

45. Find the surface area and volume of a sphere with a great circle that has a circumference of 15.7 meters. (Lesson 11–7)

46. A trapezoid has a median 17.5 centimeters long. If its height is 18 centimeters, find the area of the trapezoid. (Lesson 10–4)

47. **Sports** Two cross-country skiers left from a ski cabin at 8:00 A.M. The first skier traveled in a path 17° east of north, and the second skier traveled 22° west of north. By 9:00 A.M., the first skier had gone 15 miles, and the second skier had gone 17 miles. How far apart are the skiers? (Lesson 8–6)

48. **ACT Practice** What are all the values of y for which $y^2 < 1$?

A $y < -1$ **B** $-1 < y < 1$ **C** $y > -1$
D $y < 1$ **E** $y > 1$

49. If $\triangle ABC \cong \triangle CDE$, name the corresponding congruent parts of the two triangles. (Lesson 4–3)

Algebra

50. Solve $x^2 + 8x - 5 = 0$ by completing the square. Leave irrational roots in simplest radical form.

51. Solve $x^2 + 5x - 3 = 0$ by using the quadratic formula. Approximate irrational roots to the nearest hundredth.

For **Extra Practice**, see page 790.

13-4 Mappings

What YOU'LL LEARN

- To recognize an isometry or congruence transformation, and
- to solve problems by making a table.

Why IT'S IMPORTANT

Mappings can be used to describe the movement of a figure in a plane.

Real World APPLICATION

Engineering

The Golden Gate Bridge in San Francisco is tied down to an anchor block with many cables that support the bridge. The cables are arranged like the roots of a tree causing the braces on the cables to be different sizes.

Each cable is bound by each of the four braces labeled A, B, C, and D. Every contact point for a cable in one brace can be associated with a contact point in another corresponding brace. These correspondences are called **mappings** or **transformations**. A transformation maps a **preimage** onto an **image**.

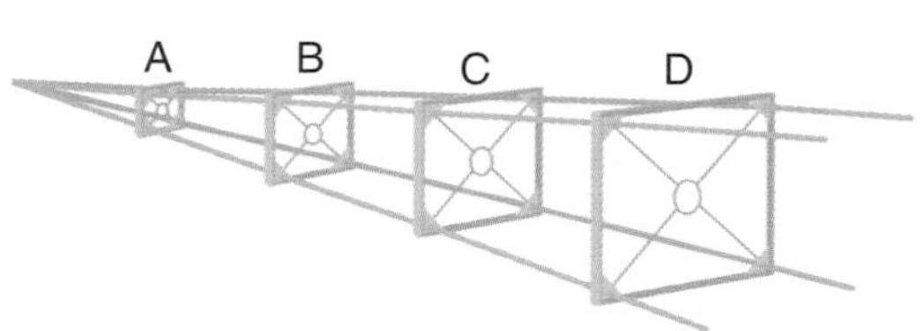

Definition of Transformation	**In a plane, a mapping is a transformation if each preimage point has exactly one image point, and each image point has exactly one preimage point.**

Below are some of the types of transformations.

dilation

A figure can be enlarged or reduced.

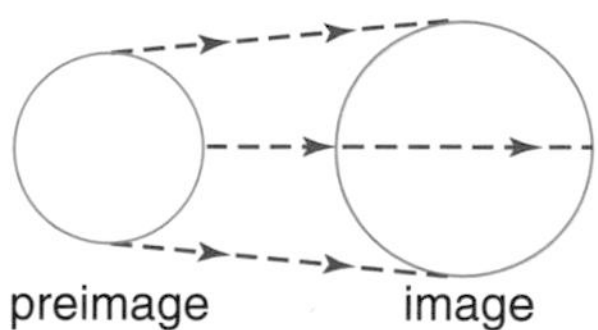

rotation

A figure can be turned.

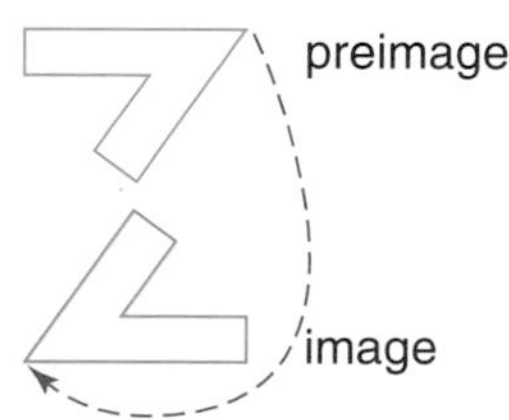

translation

A figure can be slid.

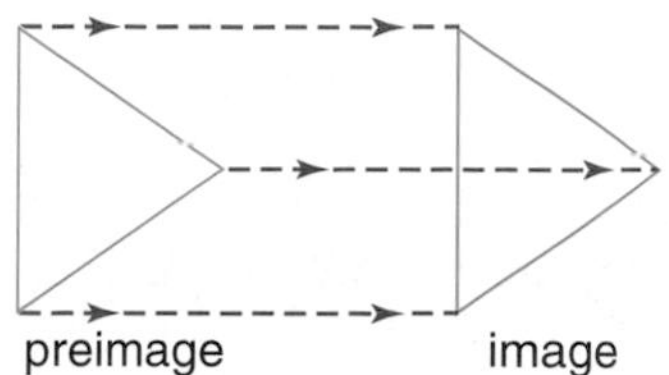

reflection

A figure can be flipped.

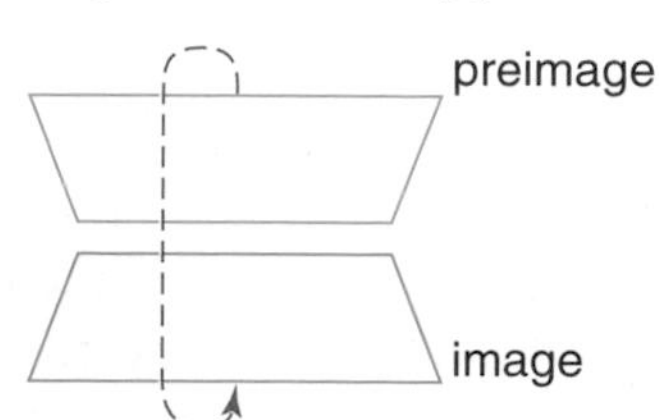

Example  **Refer to the application at the beginning of the lesson.**

a. What type of transformation is used in the braces of the Golden Gate Bridge?

The brace transformation on the Golden Gate Bridge is a dilation.

b. Determine whether *A*, *B*, and *C* are preimages of *D*.

Yes, each previous image is a preimage of the enlargement.

The symbol $\rightarrow$ is used to indicate a mapping. For example, $\triangle ABC \rightarrow \triangle PQR$ means each point in $\triangle ABC$ is mapped onto exactly one point of $\triangle PQR$. $\triangle ABC$ is the preimage, and $\triangle PQR$ is the image. The order of the letters indicates the correspondence of the preimage to the image; that is, *A* and *P* are corresponding vertices, *B* and *Q* are corresponding vertices, and *C* and *R* are corresponding vertices.

When a geometric figure and its transformation image are congruent, the mapping is called an **isometry** or a **congruence transformation**.

Definition of Isometry (Congruence Transformation)	**In a plane, an isometry is a transformation that maps every segment to a congruent segment.**

Transformations can occur in the coordinate plane.

Example  **Show that $\triangle WNR \rightarrow \triangle KAP$ in the coordinate plane at the right is an isometry.**

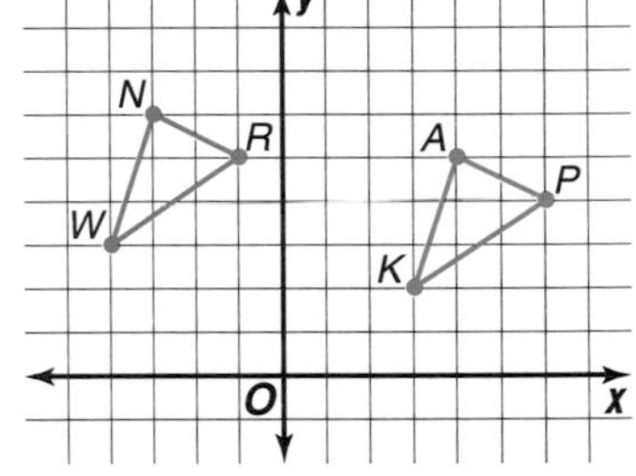

Use the distance formula to show that the sides of $\triangle WNR$ are congruent to the sides of $\triangle KAP$.

The coordinates are $W(-4, 3)$, $N(-3, 6)$, $R(-1, 5)$, $K(3, 2)$, $A(4, 5)$, and $P(6, 4)$.

Use the distance formula to find the measure of each side.

$WN = \sqrt{(-4 + 3)^2 + (3 - 6)^2}$ or $\sqrt{10}$

$NR = \sqrt{(-3 + 1)^2 + (6 - 5)^2}$ or $\sqrt{5}$

$WR = \sqrt{(-4 + 1)^2 + (3 - 5)^2}$ or $\sqrt{13}$

$KA = \sqrt{(3 - 4)^2 + (2 - 5)^2}$ or $\sqrt{10}$

$AP = \sqrt{(4 - 6)^2 + (5 - 4)^2}$ or $\sqrt{5}$

$KP = \sqrt{(3 - 6)^2 + (2 - 4)^2}$ or $\sqrt{13}$

LOOK BACK

You can review the distance formula in Lesson 1-4.

Since the measures of the corresponding sides are equal, $WN = KA$, $NR = AP$, and $WR = KP$. So, the corresponding sides are congruent, and $\triangle WNR \cong \triangle KAP$ by SSS.

Therefore, the mapping $\triangle WNR \rightarrow \triangle KAP$ is a congruence transformation or an isometry.

When a figure and its transformation image are similar, the mapping is called a **similarity transformation**.

Example 3 **In the figure at the right, $\overline{AC} \rightarrow \overline{DF}$.**

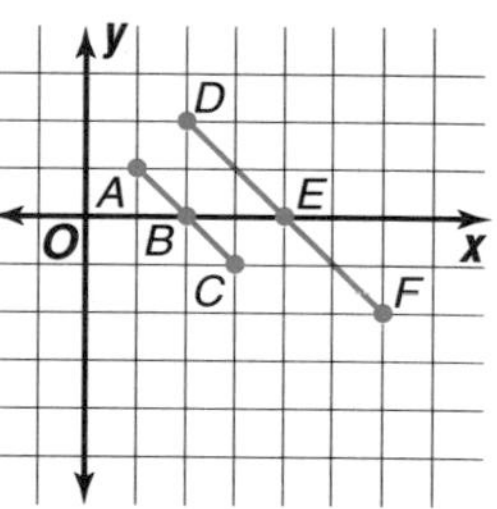

a. Is this a transformation?

b. If so, what type?

a. A and D are corresponding points, C and F are corresponding points, and B and E are corresponding points. Every point on $\overline{AC}$ corresponds to a point on $\overline{DF}$, and every point on $\overline{DF}$ corresponds to a point on $\overline{AC}$. Thus, the mapping is a transformation.

b. Use the Pythagorean Theorem to determine the length of each segment.

$AB = \sqrt{1^2 + 1^2}$ or $\sqrt{2}$ $\qquad DE = \sqrt{2^2 + 2^2}$ or $2\sqrt{2}$

$BC = \sqrt{1^2 + 1^2}$ or $\sqrt{2}$ $\qquad EF = \sqrt{2^2 + 2^2}$ or $2\sqrt{2}$

$AC = \sqrt{2} + \sqrt{2}$ or $2\sqrt{2}$ $\qquad DF = 2\sqrt{2} + 2\sqrt{2}$ or $4\sqrt{2}$

LOOK BACK

You can review the Pythagorean Theorem by referring to Lesson 1-4.

Since $\overline{AC} \not\cong \overline{DF}$, this is *not* a congruence transformation. But, since $\frac{AB}{DE} = \frac{BC}{EF} = \frac{\sqrt{2}}{2\sqrt{2}}$ or $\frac{1}{2}$ and $\frac{AC}{DF} = \frac{2\sqrt{2}}{4\sqrt{2}}$ or $\frac{1}{2}$, then $\overline{AC}$ and $\overline{DF}$ are proportional in every respect. Therefore, $\overline{AC} \rightarrow \overline{DF}$ is a similarity transformation.

Sometimes you can observe patterns in mappings by **making a table**.

Example 4 **The figure at the right shows a reflection of $\triangle ABC$ over the x-axis.**

PROBLEM SOLVING

Make a Table

a. Make a table to compare the coordinates of the vertices of the preimage and the image.

$\triangle ABC$		$\triangle DEF$	
x	y	x	y
−3	2	−3	−2
2	5	2	−5
7	3	7	−3

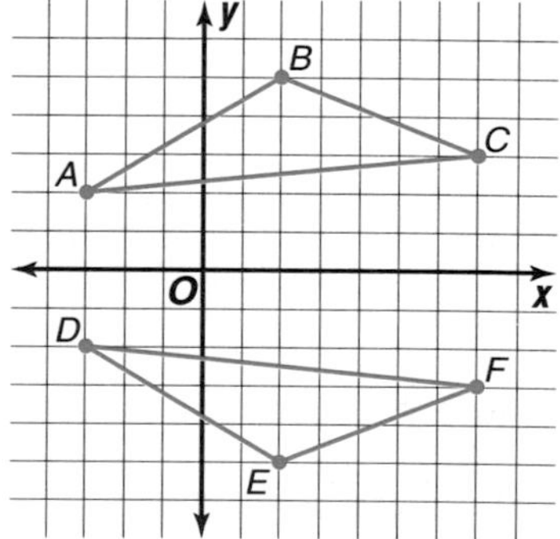

b. Describe any pattern shown in the table.

The x-coordinates are exactly the same, and the y-coordinates are the opposite of each other.

c. Use the pattern to predict the image of (a, b) reflected over the x-axis.

For (a, b), the x-coordinate of the image would be a, and the y-coordinate would be $-b$. The image of (a, b) would be $(a, -b)$.

CHECK FOR UNDERSTANDING

Communicating Mathematics

Study the lesson. Then complete the following.

1. **Explain** why in Example 1 there is more than one preimage.
2. Compare transformations that have isometry to those that are enlargements. How are they similar and how are they different?
3. **Describe** the four different transformations in your own words.
4. **Explain** how tables are helpful in problem solving.

Guided Practice

Suppose trapezoid *RSTU* → trapezoid *ABCD*.

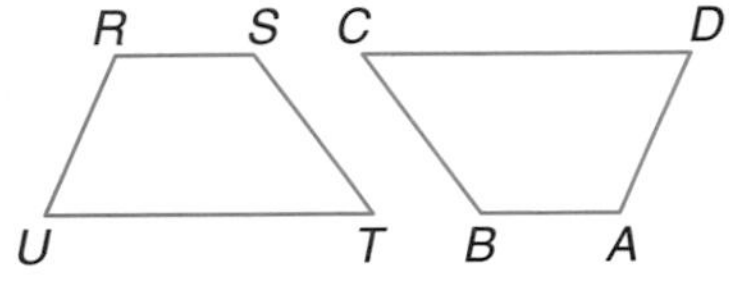

5. Name the image of $\overline{TU}$.
6. Name the preimage of $\angle ADC$.
7. Name the preimage of $\angle BCD$.
8. Name the image of $\overline{ST}$.
9. What transformations were necessary in the mapping?
10. Describe the transformation(s) that occurs in the mapping of $\triangle ABC \rightarrow \triangle DEC$.

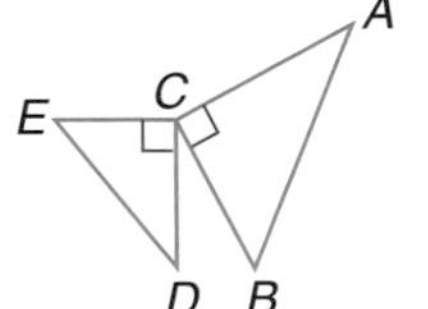

11. Determine if pentagon *QRSTU* → pentagon *ABCDE* is an isometry.

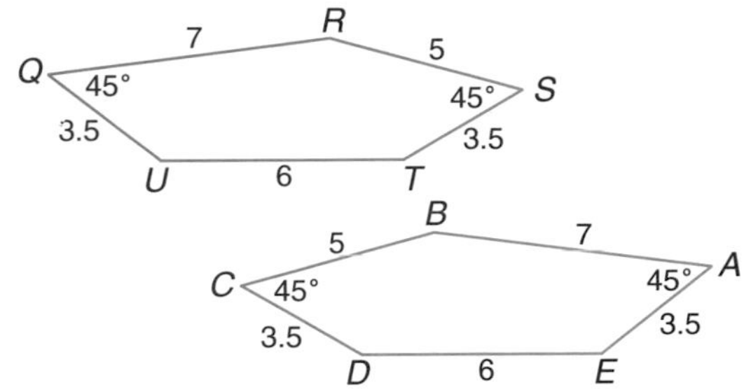

12. **Make a Table** Refer to the arrangement of circles at the right. Use a table to find how to place the digits from 1 to 6 in the circles so that every side of the triangle has the same sum.

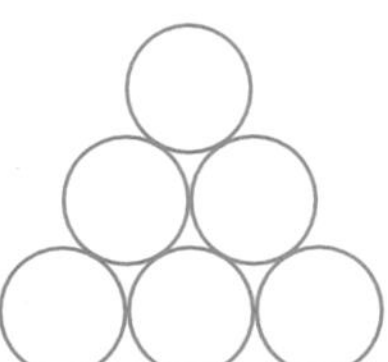

EXERCISES

Practice

Suppose $\triangle ABC \rightarrow \triangle EFD$.

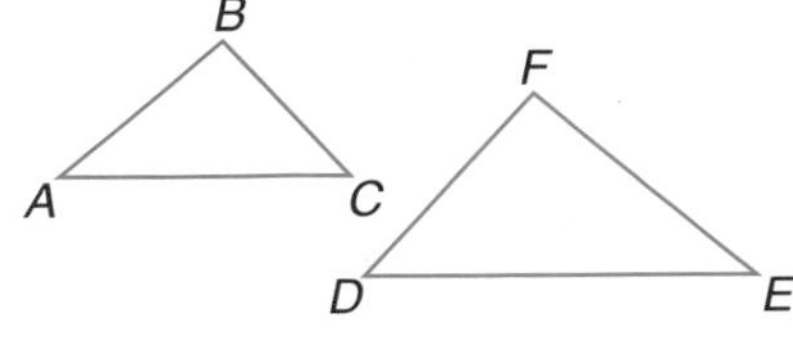

13. Name the preimage of $\angle FED$.
14. Name the image of $\angle C$.
15. Name the image of $\overline{AC}$.
16. Name the preimage of $\overline{DE}$.
17. What transformations were necessary in the mapping?

Describe the transformations that occurred in the mappings.

18. 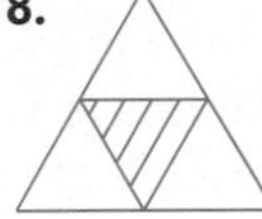

19. 

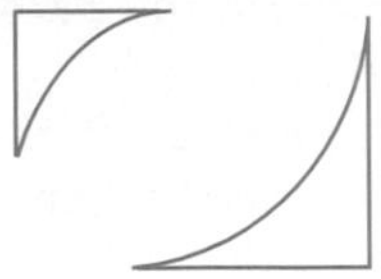

20. 

Determine if each transformation is an isometry.

21. $\triangle RST \rightarrow \triangle WUV$

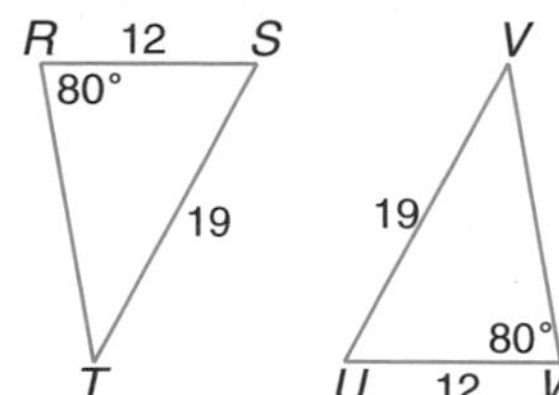

22. hexagon $ABCDEF \rightarrow$ hexagon $GLKJIH$

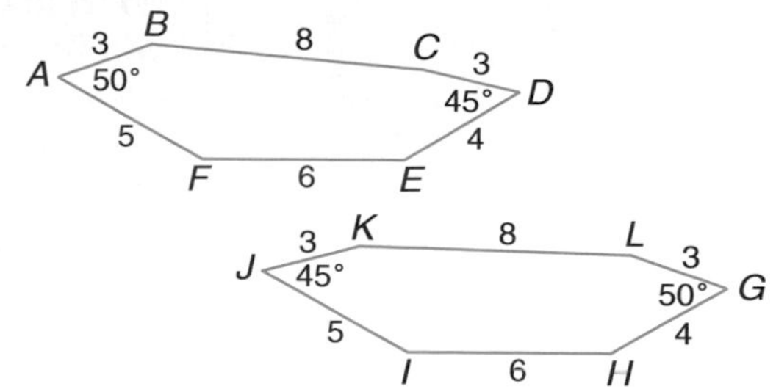

23. In a coordinate plane, quadrilateral $ABCD \rightarrow$ quadrilateral $EFGH$ for $A(-4, 0)$, $B(0, 3)$, $C(1, -2)$, $D(-3, -4)$, $E(0, -4)$, $F(-3, 0)$, $G(2, 1)$, and $H(4, -3)$.
   a. Graph the quadrilaterals and describe the transformation.
   b. Is the transformation an isometry?

Each figure below has a preimage that is an isometry. Write the image of each given preimage listed in Exercises 24–29.

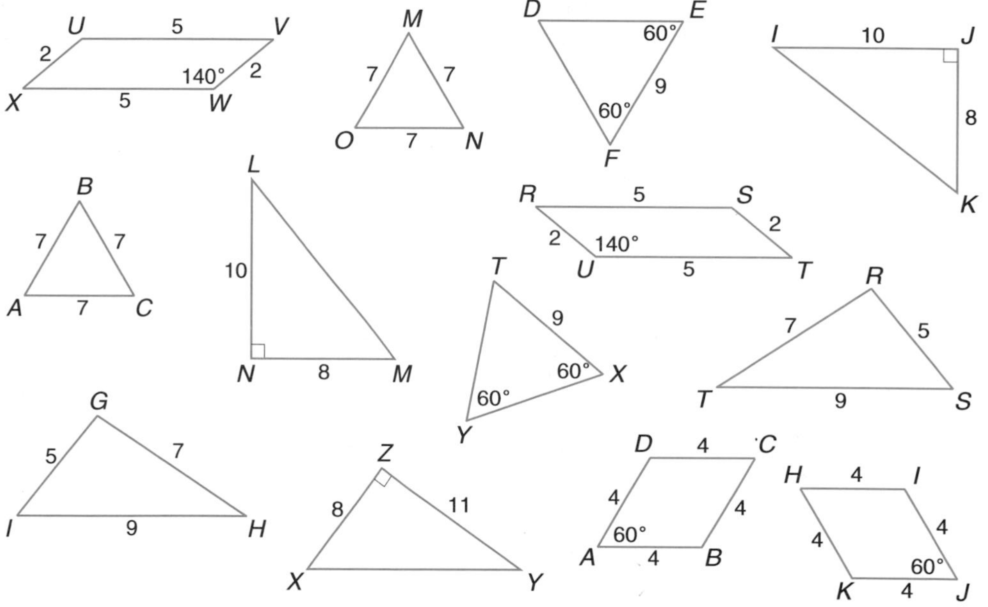

24. $\triangle ABC$

25. quadrilateral $UVWX$

26. $\triangle DEF$

27. $\triangle RST$

28. $\triangle LMN$

29. quadrilateral $ABCD$

30. **Make a Table** Use the figure at the right and a table of coordinates to determine the coordinates of the image after a reflection of the given preimage over the x-axis.

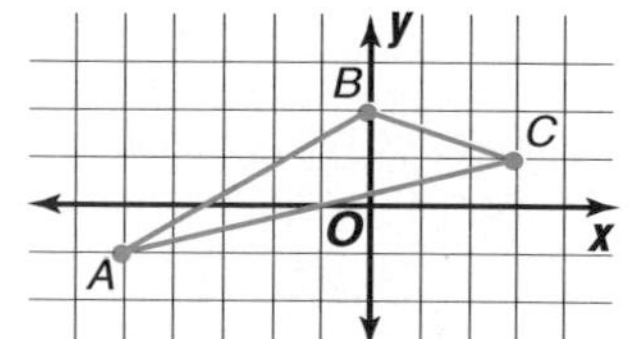

31. **Make a Table** Use a table of coordinates and the given figure to determine the coordinates of the image after a slide 4 units down and a reflection across the y-axis.

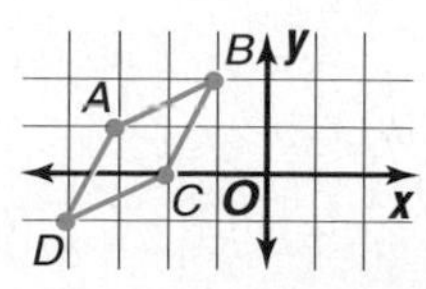

Programming

32. The graphing calculator program below draws a triangle and its image when it slides a certain horizontal and vertical distance. You enter the coordinates of its vertices and the distances to slide. Then the mapping will be drawn.

```
PROGRAM: MAPPING
:Disp "ENTER VERTICES"
 For (N, 1, 3)
:Input "X", X
:X→L1(N)
:Input "Y", Y
:Y→L2(N)
:End
:ClrDraw
:ZStandard
:Line(L1(1), L2(1), L1(2),
 L2(2))
:Line(L1(2), L2(2), L1(3),
 L2(3))
:Line(L1(1), L2(1), L1(3),
 L2(3))
:Pause
:Input "HORIZONTAL MOVE ", H
:Input "VERTICAL MOVE ", V
:Line(L1(1)+H, L2(1)+V,
 L1(2)+H, L2(2)+V)
:Line(L1(2)+H, L2(2)+V,
 L1(3)+H, L2(3)+V)
:Line(L1(1)+H, L2(1)+V,
 L1(3)+H, L2(3)+V)
:Stop
```

TECHNOLOGY Tip

After the graph of $\triangle ABC$ is shown, press ENTER to continue with the program.

Use the program to draw $\triangle ABC$ and its translated image.

a. $A(2, 3)$, $B(5, 9)$, $C(0, 4)$; horizontal move: 2; vertical move: 1

b. $A(-4, 3)$, $B(1, 7)$, $C(3, -2)$; horizontal move: 4; vertical move: -3

c. $A(3, 6)$, $B(5, 2)$, $C(-2, 8)$; horizontal move: -4; vertical move: -5

d. How are the triangles and their images related?

e. Is the slide performed by the program an isometry? Justify your answer.

Critical Thinking

Separate each figure into two congruent figures. Explain what transformation is used on one of the two congruent figures to form the given original figure.

33.

34.

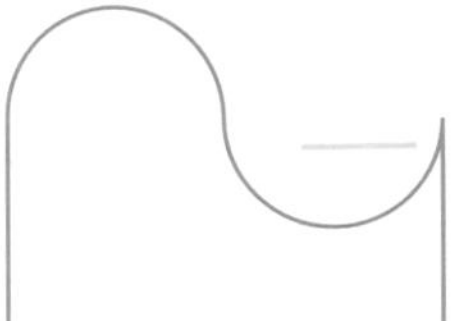

Applications and Problem Solving

35. **Writing** Small children often write letters backward. What transformations did a child use in writing her name?

AIJUL

36. **Art** What transformations were used in the pattern on the Incan tunic shown at the left?

37. **Fractals** What transformations are apparent in the development of the fractal below called the Networked-Growing Twig?

Step 1 | Step 2 | Step 3 | Final Image

Mixed Review

38. Describe and draw a diagram of the locus of all points in a plane equidistant from two given parallel lines and a given distance from an intersecting line. (Lesson 13–3)

39. Graph the equations $y = x + 3$ and $3y + x = 5$ to find the locus of points that satisfy the graphs of both equations. (Lesson 13–2)

40. **SAT Practice** A student bought four college textbooks that cost \$29.50, \$18.95, \$25.90, and \$32.45. She paid one half of the total amount herself and borrowed the rest from her mother. She repaid her mother in 4 equal monthly payments. How much was each of the 4 monthly payments?
A \$6.68 **B** \$13.35 **C** \$21.36 **D** \$26.70 **E** \$53.40

41. Write the equation in slope-intercept form of the line that passes through points at (7, 6) and (3, −2). (Lesson 12–2)

42. Classify the statement *Chords that are equidistant from the center of a circle are congruent* as always, sometimes, or never true. (Lesson 9–3)

43. In right triangle XYZ, $m\angle X = 30$ and $m\angle Y = 60$. If $XY = 10$, find YZ and XZ. (Lesson 8–2)

Algebra

44. Write an equation in functional notation for the relation shown in the table at the right.

x	1	2	3	4	5
$f(x)$	4	7	10	13	16

For **Extra Practice**, see page 790.

45. Fourteen is 17.5% of what number?

SELF TEST

1. Point R is on line ℓ. What is the locus of points in space that are 8 centimeters from ℓ and 8 centimeters from R? (Lesson 13–1)

2. Points P and Q are 6 centimeters apart. What is the locus of points in a plane that are equidistant from P and Q and are 8 centimeters from P? Sketch the locus. (Lesson 13–1)

Use either substitution or elimination to find the locus of points that satisfy the graphs of both equations. (Lesson 13–2)

3. $x - 7y = 13$
$3x - 5y = 23$

4. $x = 8 + 3y$
$2x - 5y = 8$

5. Sketch the circle for $x^2 + y^2 = 10$ and the line for $y = 3x + 10$. Describe the locus of points that satisfy both graphs. (Lesson 13–3)

Describe the different possible intersections of each pair of figures. (Lesson 13–3)

6. a plane and a sphere in space

7. two perpendicular lines and a circle in a plane

8. Describe and draw the locus of points in the plane of $\angle DEF$, equidistant from the sides of the angle, and 4 centimeters from $\overline{EF}$. (Lesson 13–3)

Suppose $\triangle ABC \rightarrow \triangle EBD$. (Lesson 13–4)

9. Name the image of $\overline{CB}$.

10. Name the preimage of $\angle BDE$.

13-5 Reflections

What YOU'LL LEARN

- To name, recognize, and draw reflected images, lines of symmetry, and points of symmetry.

Why IT'S IMPORTANT

Reflections can be used to solve problems involving mirrors, billiards, and golf.

Real World APPLICATION

Mirrors

As light strikes the full-length mirror in the photo at the right, it reflects an image that appears to be behind the mirror, but it is reversed from left to right. The image is the same size as the woman it reflects. It also appears to be the same distance from the mirror as the woman.

A reflection is sometimes called a flip of the figure.

A **reflection** is a type of transformation. Just as the mirror is a plane of reflection for the woman and her image, line ℓ is a **line of reflection** for P and its image R. If this page were folded along line ℓ, P and R would coincide. This means that ℓ is the perpendicular bisector of $\overline{PR}$. Therefore, point Q is the midpoint of $\overline{PR}$. Since point Q is already on the line of reflection, Q is its own image.

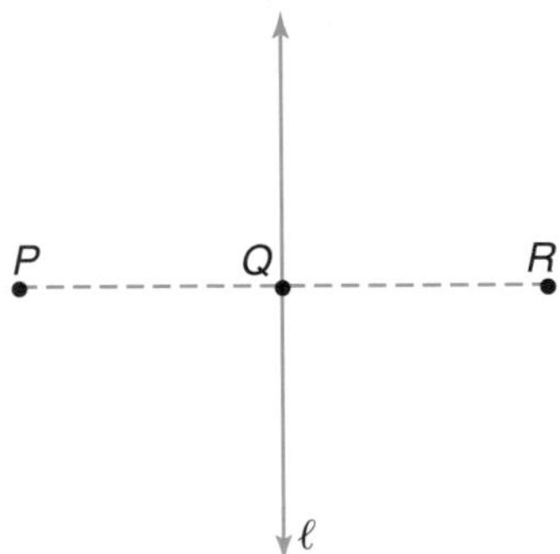

It is also possible to reflect an image with respect to a point. Consider the figure above without line ℓ. Point P is the reflection of R with respect to Q. Therefore, Q is a **point of reflection**.

P Q R

What happens with the reflected images of collinear points S, T, and U? The reflected points X, Y, and Z are also collinear. Thus, it is said that reflections preserve collinearity.

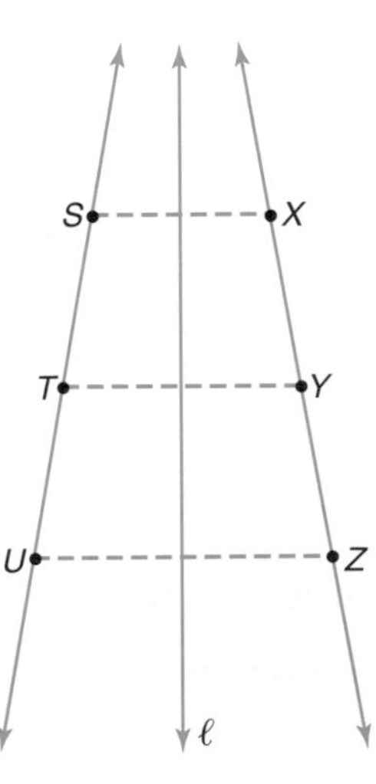

Reflections also preserve betweenness of points. Point T is between points S and U. Likewise, the reflection of T, point Y, is between reflected points X and Z.

In the figure below, $\triangle ABC$ is reflected over the y-axis. $\triangle XYZ$ is the reflected image of $\triangle ABC$. Are there any other relationships between the two triangles? By measuring the corresponding parts of the triangles, it appears that the two triangles are also congruent.

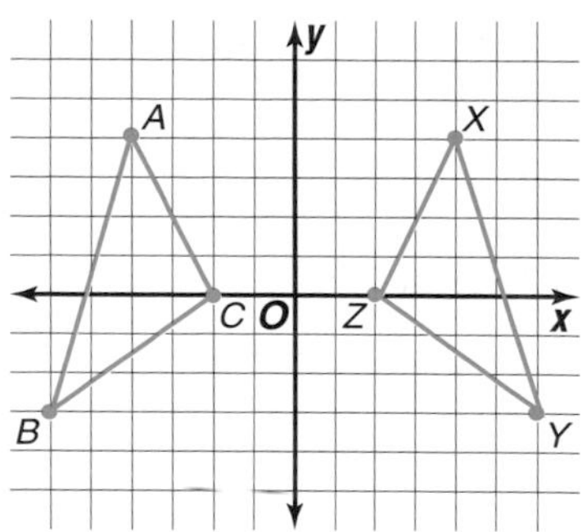

In addition to preserving collinearity and betweenness of points, reflections also preserve angle measure and distance measure.

Points A, B, and C can be read in a counterclockwise order. $\triangle ABC$ is said to have a counterclockwise orientation.

Corresponding points X, Y, and Z are then in a clockwise order. $\triangle XYZ$ is said to have a clockwise orientation.

CONSTRUCTION

Reflecting a Triangle

Construct the reflected image of $\triangle ABC$ over line ℓ.

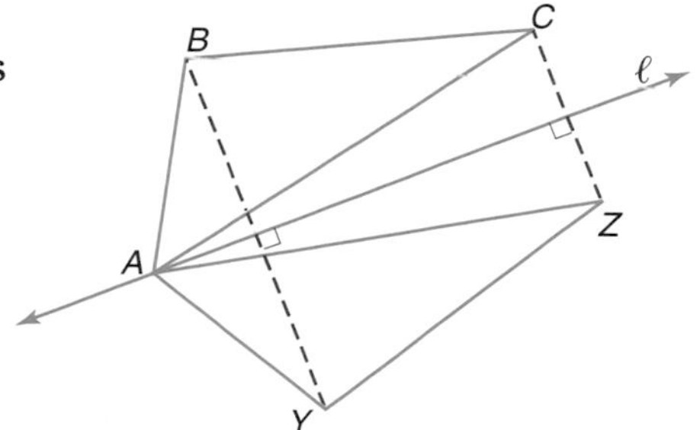

1. Since A is on line ℓ, A is its own reflection. Construct perpendiculars from B and C through line ℓ.
2. Locate Y and Z so that line ℓ is the perpendicular bisector of $\overline{BY}$ and $\overline{CZ}$. Points Y and Z are the reflected points of B and C.
3. Connect vertices A, Y, and Z.

Conclusion: Since points A, Y, and Z are reflected images of points A, B, and C, $\triangle AYZ$ is the reflection of $\triangle ABC$.

Reflections are also used in playing many sports.

Example 1

Recreation

A top or bottom spin of the cue ball will affect the angle at which the cue ball rebounds off another ball, while a side spin of the cue ball will affect the angle at which the cue ball rebounds off a cushion.

Jenelle is playing billiards. She wants to use the white cue ball to hit the black 8 ball into the lower right pocket. How can she use a reflection to make the shot?

She can mentally reflect the pocket over the upper side of the pool table. Then she can align the eight ball and the reflected pocket to find point P on the side of the table. If the ball is hit without a spin, she can aim the cue ball to hit the 8 ball so that it will hit point P on the table and then bounce off the table side at an angle that will take it to the pocket.

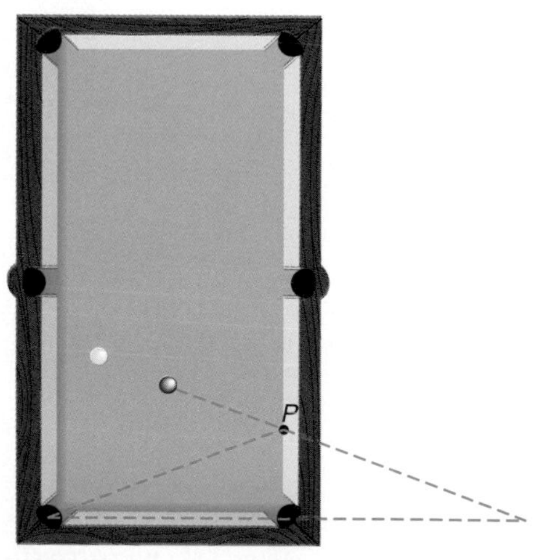

A line of reflection can be drawn through some plane figures so that one side is a reflected image of the other side. This line of reflection is called a **line of symmetry**. This line of symmetry can be determined and drawn in certain figures.

The union of any figure and its reflected image is always a figure that has a line of symmetry.

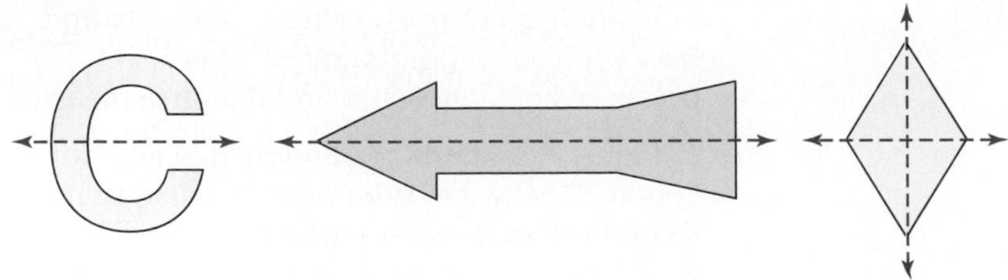

Example  **Determine and show all the lines of symmetry for each figure below.**

a. equilateral triangle

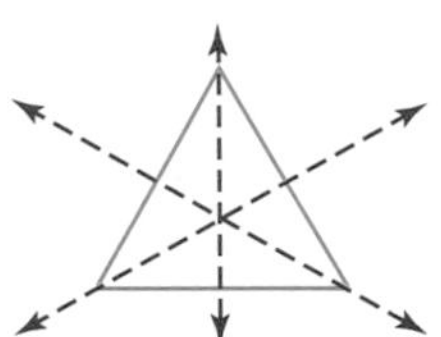

3 lines of symmetry

b. line segment

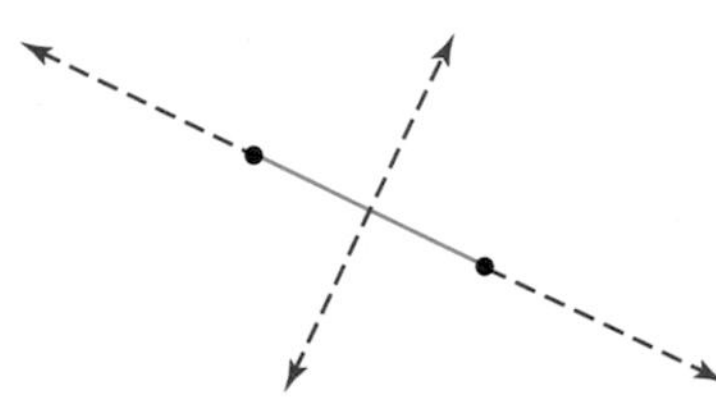

2 lines of symmetry

In part b of Example 2, note that one of the lines of symmetry is a perpendicular bisector of the segment. You can use perpendicular bisectors to find images of reflections.

Example  **a. Draw the reflected image of $\triangle PQR$ over the x-axis.**

Reflect each point over the x-axis so that the x-axis is the perpendicular bisector of the segment formed by each point and its image.

Connect the reflected image points to form $\triangle P'Q'R'$.

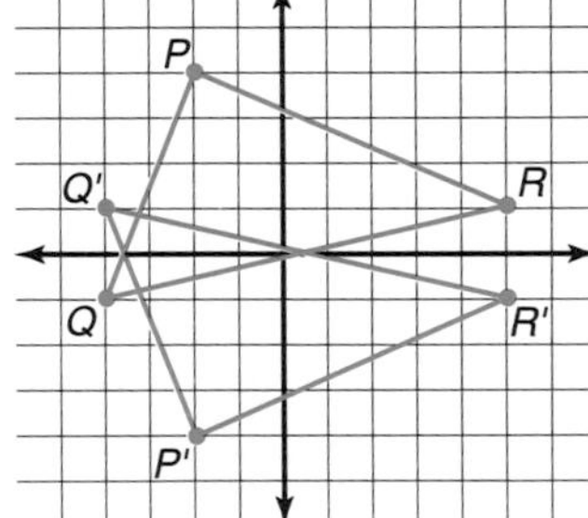

b. Describe any pattern you see in the coordinates of the triangle and its reflected image.

$Q(-4, -1) \rightarrow Q'(-4, 1)$

$P(-2, 4) \rightarrow P'(-2, -4)$

$R(5, 1) \rightarrow R'(5, -1)$

In a reflection over the x-axis, the x-coordinate remains the same, and the y-coordinate is the opposite of the y-coordinate of the preimage.

A pattern also exists when you reflect a figure over the y-axis.

Example 4 **The vertices of quadrilateral $PQRS$ are $P(-5, 4)$, $Q(-1, -1)$, $R(-3, -6)$, and $S(-7, -3)$, and the vertices of quadrilateral $P'Q'R'S'$ are $P'(5, 4)$, $Q'(1, -1)$, $R'(3, -6)$, and $S'(7, -3)$.**

a. Construct a table of these vertices and look for a pattern in the x- and y-coordinates. What type of transformation occurred?

The x-coordinates of quadrilateral $P'Q'R'S'$ are the opposites of the x-coordinates of quadrilateral $PQRS$. The y-coordinates stayed the same. Thus, the pattern indicates that the vertices must have been reflected over the y-axis.

quadrilateral PQRS	quadrilateral P'Q'R'S'
$P(-5, 4)$,	$P'(5, 4)$
$Q(-1, -1)$	$Q'(1, -1)$
$R(-3, -6)$	$R'(3, -6)$
$S(-7, -3)$	$S'(7, -3)$

b. Graph the preimage, the image, and the line of reflection to verify the transformation.

Each point is an equal distance from the y-axis, which is the line of reflection.

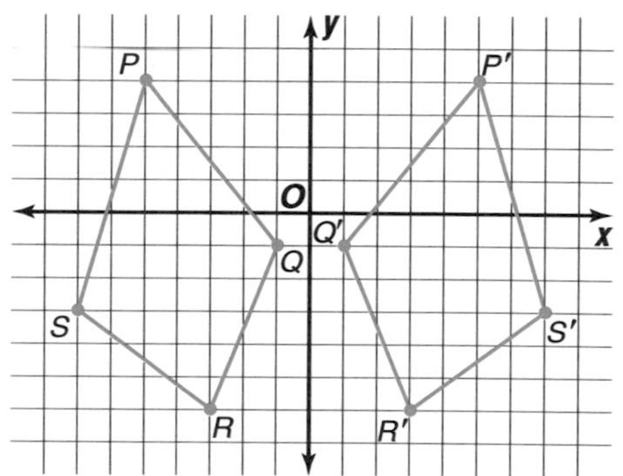

For many figures, a point can be found that is a point of reflection for all points on the figure. This point of reflection is called a **point of symmetry**. A point of symmetry must be a midpoint for all segments that pass through it and have endpoints on the figure. In the two figures below, P and Q are midpoints of the segments drawn. In the third figure, R is not a point of symmetry because R is not the midpoint of $\overline{XZ}$.

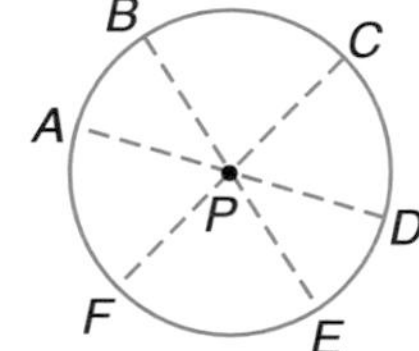

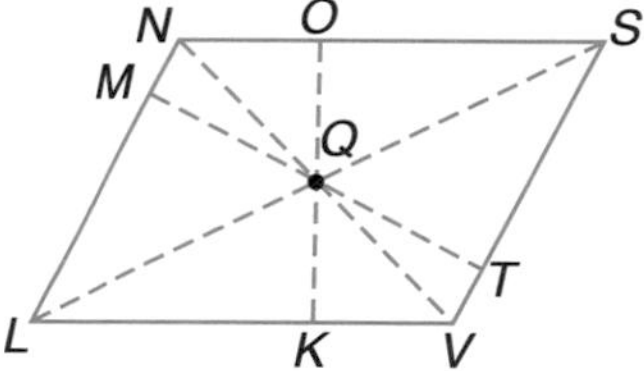

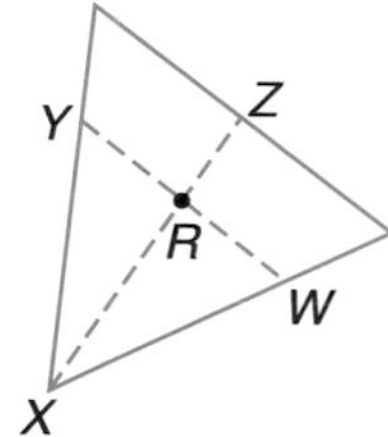

P and Q are points of symmetry.

R is not a point of symmetry.

CHECK FOR UNDERSTANDING

Communicating Mathematics

Study the lesson. Then complete the following.

1. **Describe** the relationship between the line of reflection and the line segment that joins a point and its reflected image.
2. **Name** four properties that are preserved in reflections.
3. **You Decide** Damaris was asked to name the geometric plane figure that has the most lines of symmetry. Damaris thinks it is the square. Sarah disagrees. She thinks there is another geometric plane figure that has more lines of symmetry than a square. Who is right?

Guided Practice

For the figure at the right, name the reflected image of each of the following over line ℓ.

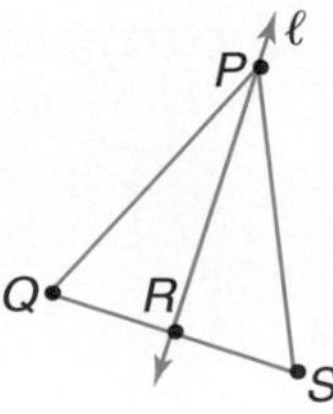

4. P
5. $\overline{PQ}$
6. $\angle PQS$
7. $\triangle QSP$

Copy each figure. Draw the reflected image of each figure over line p.

8.

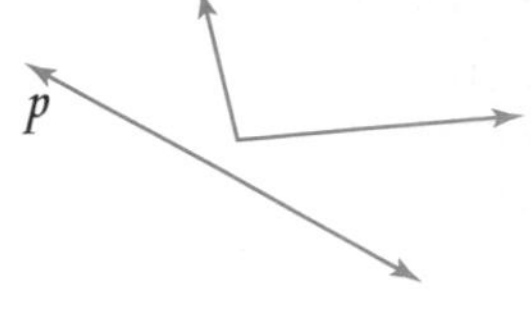

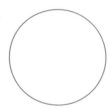

9.

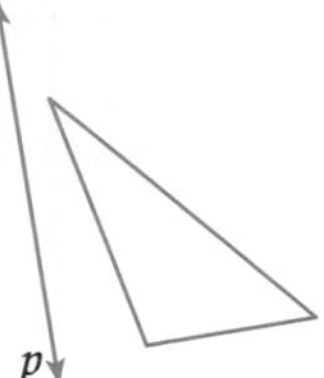

Determine how many lines of symmetry each figure has. Then, identify those figures that have point symmetry.

10.

11.

12.

13.

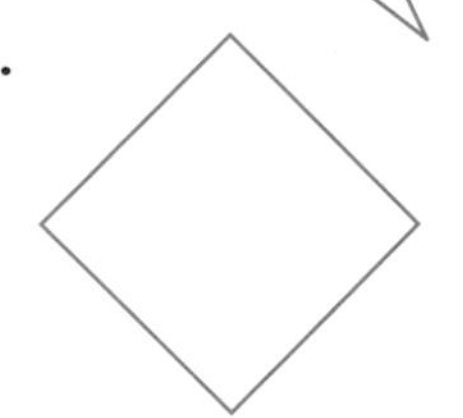

14. Graph $\triangle JKM$ with vertices $J(3, -2)$, $K(2, 3)$, $M(-4, 0)$ on a coordinate plane. Then draw its reflected image over the line that is the graph of $y = 2$.

15. **Grooming** Jamal uses two mirrors to check the appearance of his hair from the back. Using the measurements as indicated in the figure, how far away does the image of the back of his head appear to Jamal?

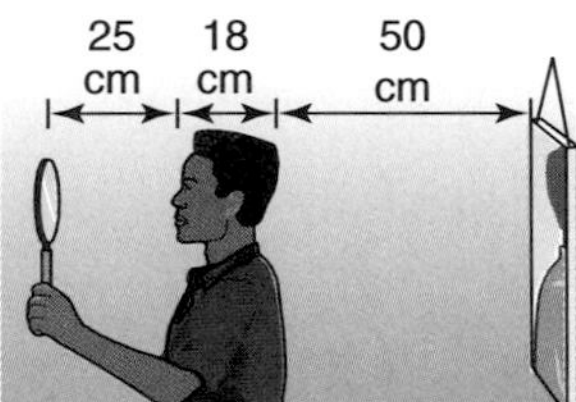

EXERCISES

Practice

For the figure below, name the reflected image of each figure over line m.

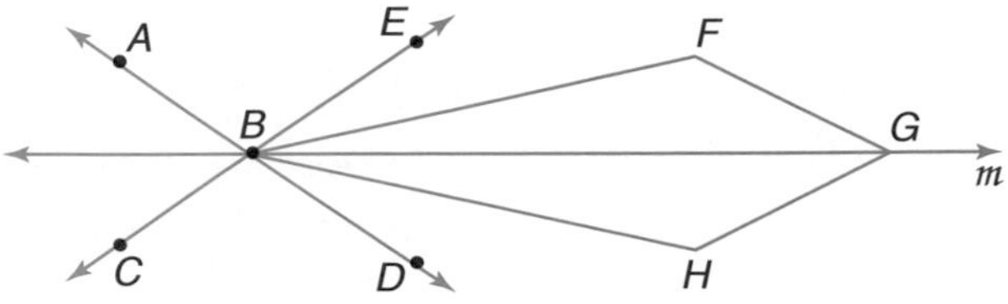

16. A
17. $\overline{FG}$
18. $\angle ABE$
19. $\angle ABC$
20. $\overline{BG}$
21. $\triangle BFG$
22. $\angle GBF$
23. quadrilateral $FGHB$

Copy each figure. Draw all possible lines of symmetry. If none exist, write *none*.

24. N

25. T

26. X

27. D

Copy each figure. Use a straightedge to draw the reflected image of each figure over line p.

28.

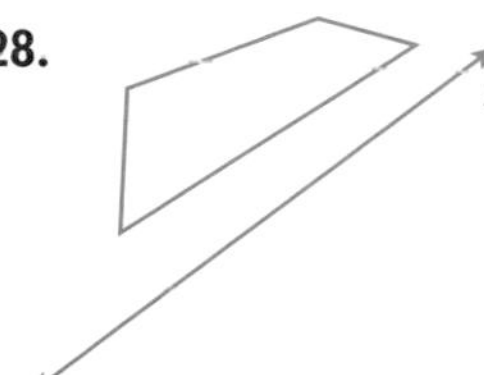

29.

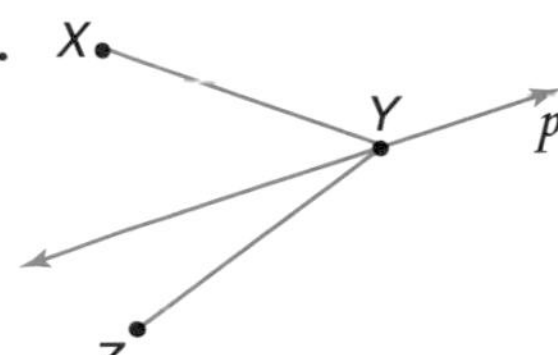

30.

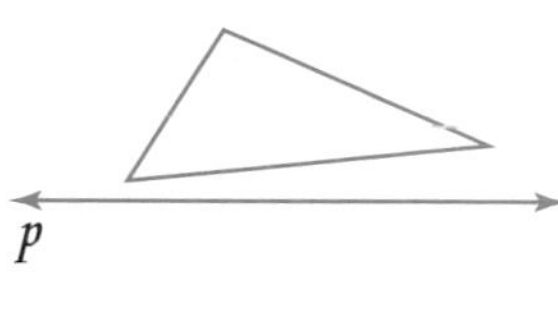

31.

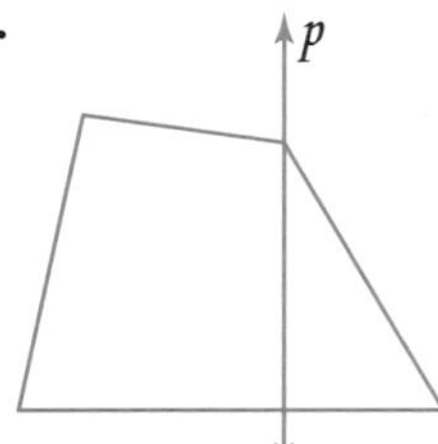

32.

33.

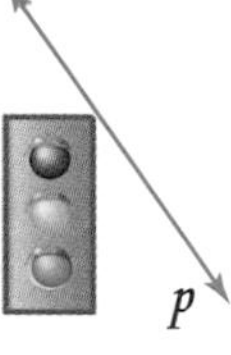

For each figure, determine whether ℓ is a line of symmetry. Write *yes* or *no*. Explain your answer.

34.

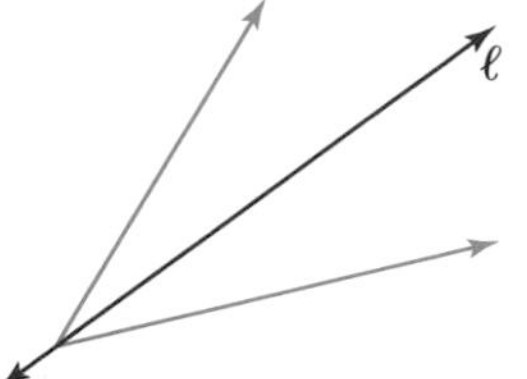

35.

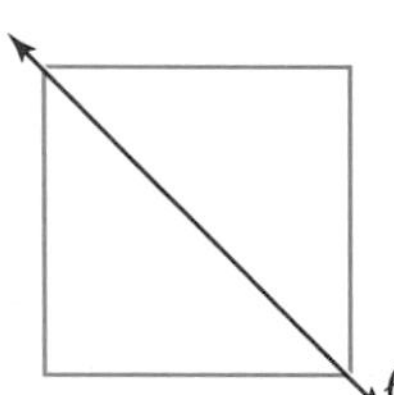

36.

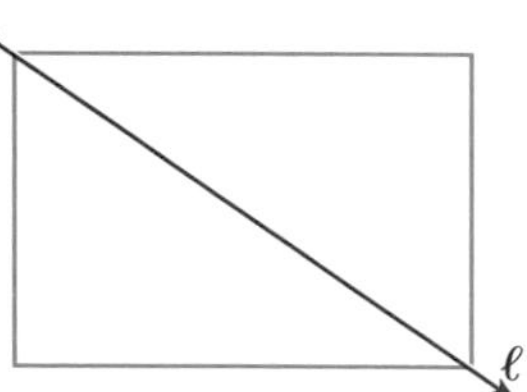

37.

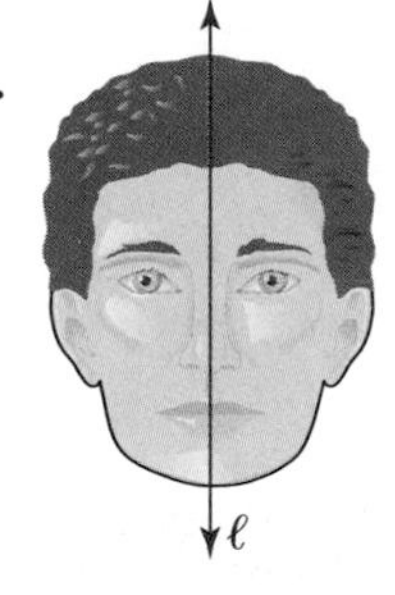

38.

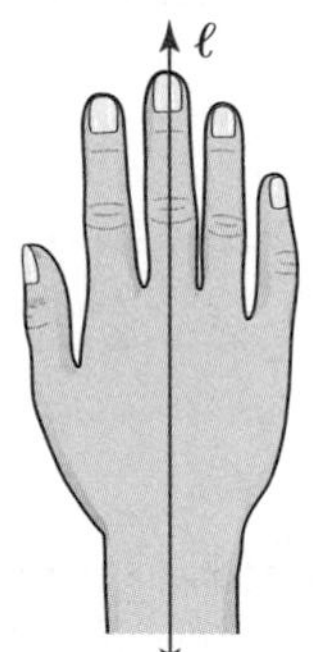

39.

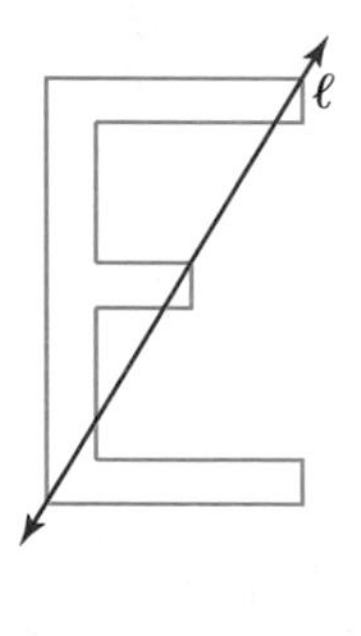

40. In Example 3, what pattern do you notice when there is a reflection over the x-axis?

Graph each figure on a coordinate plane. Then draw its reflected image over the indicated line of reflection.

41. $\overline{AB}$ for $A(4, 3)$ and $B(0, 3)$ over the x-axis.

42. Trapezoid $ABCD$ for $A(2, 2)$, $B(2, -1)$, $C(-1, -2)$, $D(-1, 3)$ over the graph of $y = 3$.

43. $\triangle PQR$ for $P(0, 6)$, $Q(-4, 0)$, $R(-4, -6)$ over the graph of $y = x$.

Copy each figure below. Indicate any points of symmetry, If none exist, write *none*.

44. 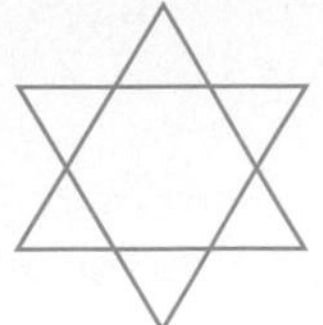

45. 

46. 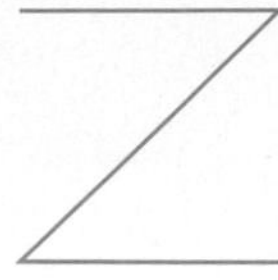

47. 

48. 

49. 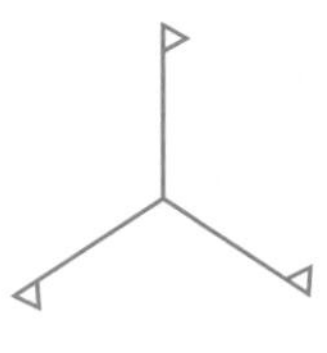

For each figure, indicate if the figure has *line symmetry, point symmetry*, or *both*.

50. 

51. 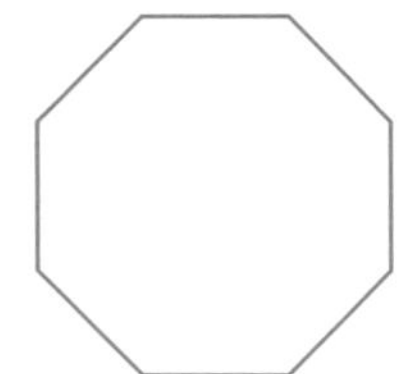

52. 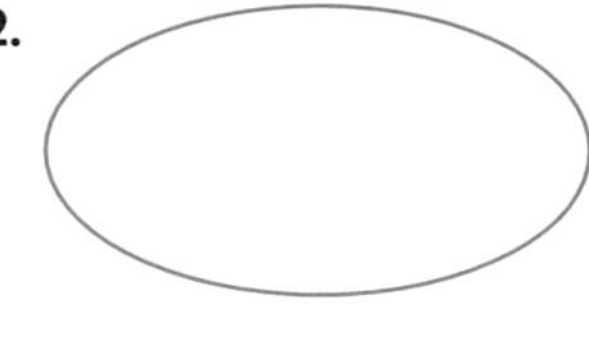

Critical Thinking

53. Study Example 3, Example 4, and Exercise 40. Make a conjecture for reflecting over each of the following.
    a. y-axis
    b. x-axis
    c. graph of $y = x$

54. Copy the figure at the right. Draw the reflected image of pentagon $ABCDE$ over line ℓ. Then reflect that image over line m. Label the vertices A', B', C', D', and E' to correspond to A, B, C, D, and E, respectively.

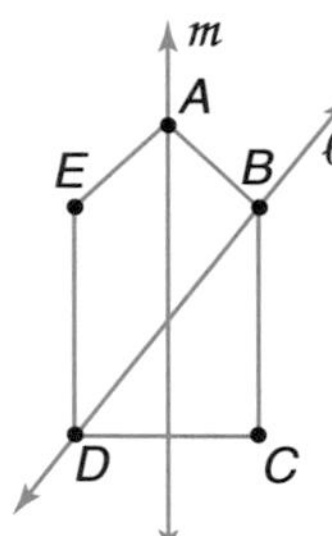

Applications and Problem Solving

55. **Algebra** For the parabola shown, give an example of a line of reflection and a point of reflection.

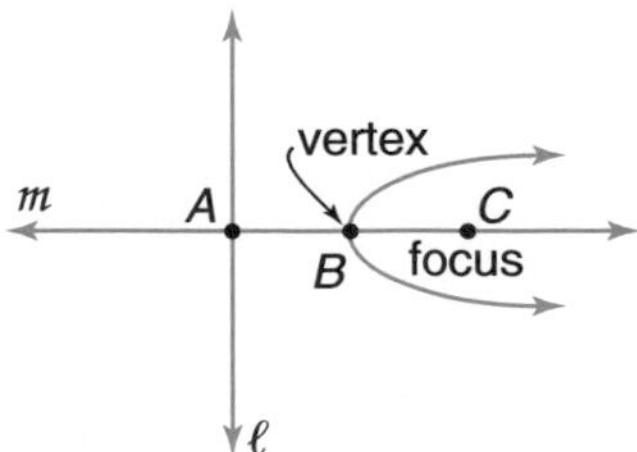

56. **Tessellations** Draw a regular hexagon and reflect it over each of its sides.
    a. Describe the pattern that you made. Do any of the areas of the figures overlap?
    b. After the first set of reflections, suppose you perform the same reflections on the resulting hexagons. Describe the pattern you see.
    c. What would happen if you reflected a regular octagon in the same way?

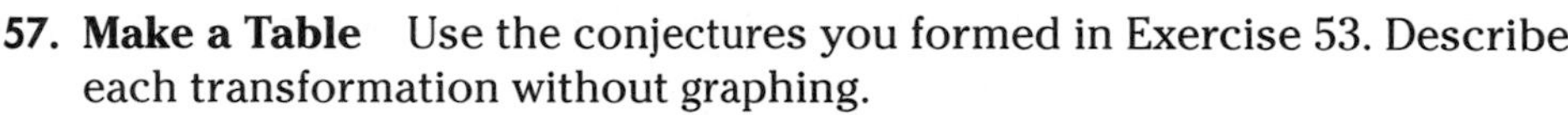

57. Make a Table Use the conjectures you formed in Exercise 53. Describe each transformation without graphing.

a. The vertices of $\triangle XYZ$ on a coordinate plane are $X(-2, 1)$, $Y(0, 5)$, and $Z(6, 2)$, and the vertices of $\triangle X'Y'Z'$ are $X'(-2, -1)$, $Y'(0, -5)$, and $Z'(6, -2)$.

b. The vertices of $\triangle ABC$ on a coordinate plane are $A(3, -2)$, $B(4, 1)$, and $C(-3, 2)$, and the vertices of $\triangle A'B'C'$ are $A'(-3, -2)$, $B'(-4, 1)$, and $C'(3, 2)$.

58. Recreation Moira is playing miniature golf. If the ball is at point B, how can she make a hole-in-one at H for the situation shown at the right?

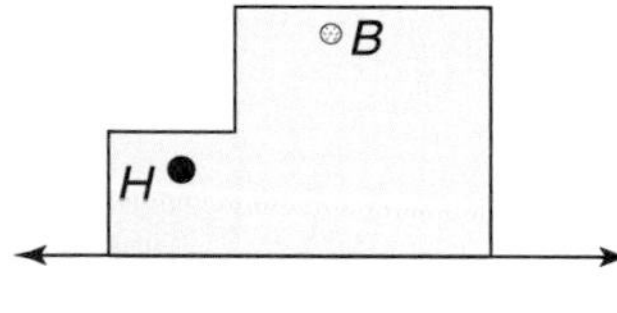

Mixed Review

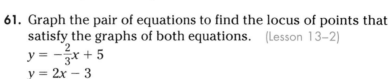

59. Suppose quadrilateral $ABCD \rightarrow$ quadrilateral $RSTU$. (Lesson 13–4)

a. Name the image of $\angle B$.

b. Name the preimage of $\overline{TU}$.

60. **Emergency Services** Two emergency weather sirens, that are 10 miles apart, are located in Shelbyville. One has a radius of 8 miles, and the other has a radius of 5 miles. Draw the locus of points where both sirens can be heard. (Lesson 13–3)

61. Graph the pair of equations to find the locus of points that satisfy the graphs of both equations. (Lesson 13–2)

$y = -\frac{2}{3}x + 5$

$y = 2x - 3$

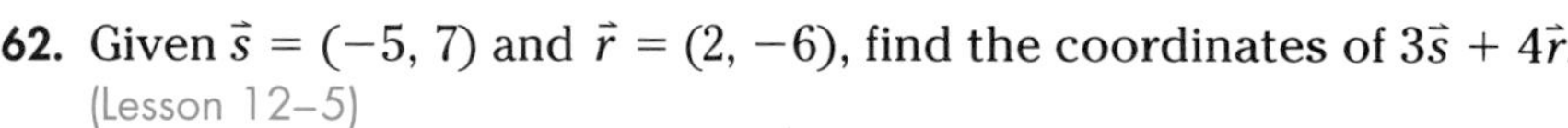

62. Given $\vec{s} = (-5, 7)$ and $\vec{r} = (2, -6)$, find the coordinates of $3\vec{s} + 4\vec{r}$. (Lesson 12–5)

63. Find the equation of the line containing the perpendicular bisector of the segment with endpoints at $(-2, -3)$ and $(4, 5)$. (Lesson 12–3)

64. Complete: In the linear equation $y = mx + b$, m represents the __?__ of the line. (Lesson 12–1)

65. **ACT Practice** How many points of intersection exist if the equations $(x - 5)^2 + (y - 5)^2 = 4$ and $y = -x$ are graphed on the same coordinate plane?

A none **B** one **C** two **D** three **E** four

66. Find the area of the parallelogram. (Lesson 10–3)

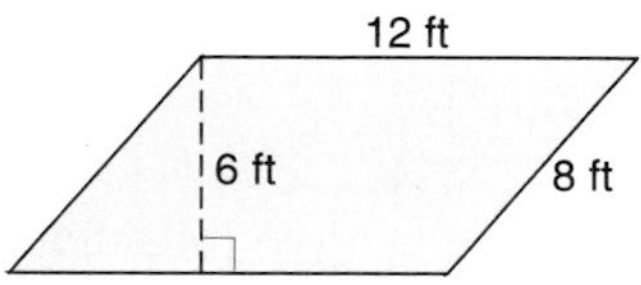

67. Complete: The longest chords of a circle are __?__. (Lesson 9–1)

68. Use the figure to complete each statement with < or >. (Lesson 5–3)

a. $m\angle 1$ __?__ $m\angle 4$

b. $m\angle 2$ __?__ $m\angle 6$

c. $m\angle 7$ __?__ $m\angle 3$

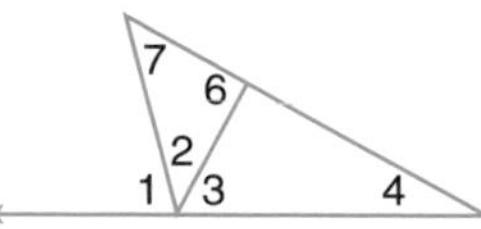

Algebra

69. Solve $|8n - 5| < 0$.

70. Solve $8b \geq 64$.

For **Extra Practice**, see page 791.

MODELING MATHEMATICS

A Preview of Lesson 13–6

13–6A Reflections and Translations

Materials: paper  straightedge geomirror

A geomirror is a construction instrument that allows you to find the reflection image of a figure. Use a geomirror with the following activity to investigate translations.

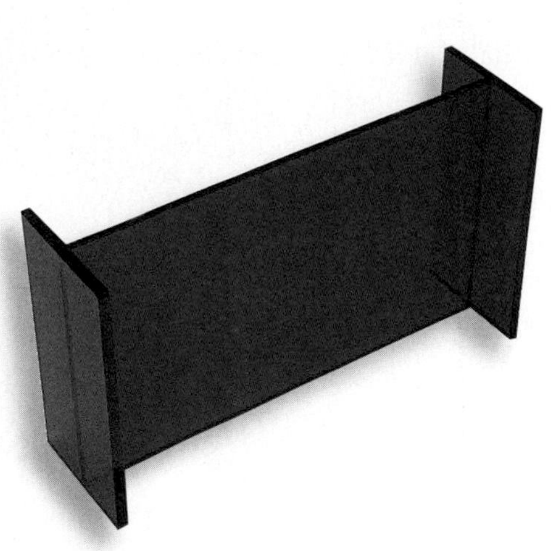

- Draw two parallel lines ℓ and m and a triangle ABC.
- Place the geomirror so that the edge is aligned with line ℓ. Look into the geomirror to see the reflection image of $\triangle ABC$.
- Use a straightedge to draw the image of $\triangle ABC$. Label the vertices A', B', and C'.
- Align the edge of the geomirror with line m. Look into the geomirror to see the reflection image of $\triangle A'B'C'$.
- Draw the image of $\triangle A'B'C'$. Label the vertices A'', B'', and C''.

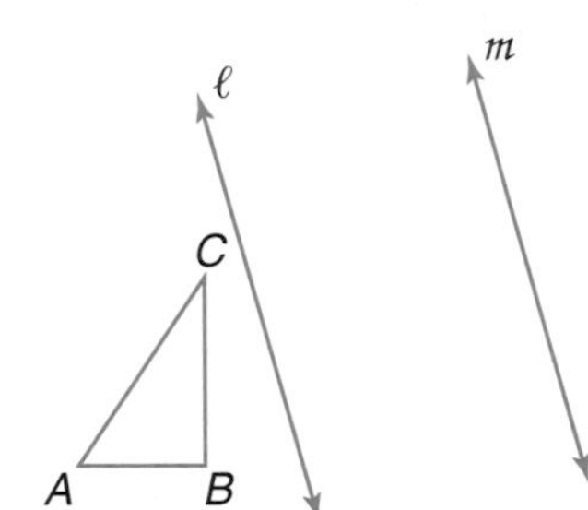

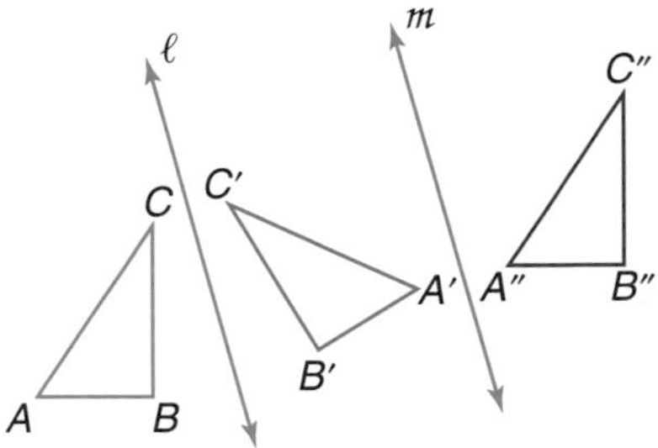

Write **Examine your drawings.**

1. You know that a reflection changes the orientation of its image. What happens when a figure is reflected twice?
2. Are the points A, A',and A'' collinear? How about B, B',and B'' and C, C',and C''? Use a straightedge to verify your answer.
3. Compare the lengths of $\overline{AA''}$, $\overline{BB''}$, and $\overline{CC''}$, and the distance between ℓ and m.
4. Describe how you could map $\triangle ABC$ onto $\triangle A''B''C''$ in one motion instead of two reflections.

Model

5. **Use the geomirror to test whether two reflections will translate a regular polygon.**
   a. Draw two parallel lines and a regular polygon.
   b. Reflect the regular polygon twice and compare the preimage and the image.
   c. Does your conjecture in Exercise 4 about mapping in one motion instead of two reflections hold true?

13–6

Translations

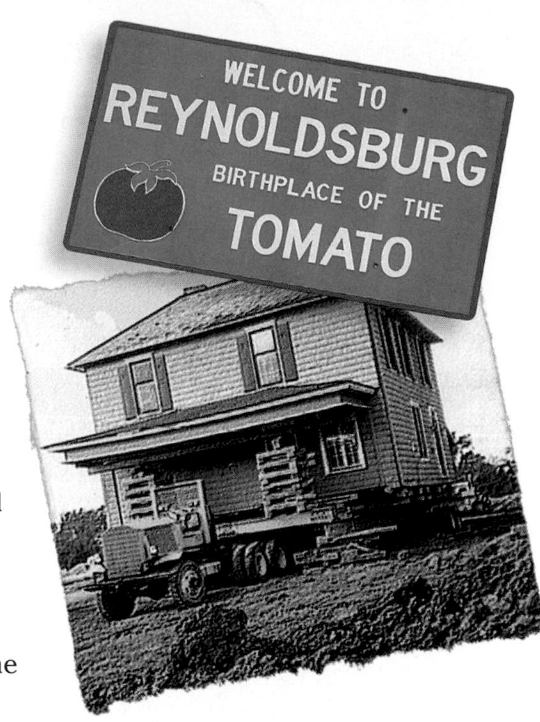

What YOU'LL LEARN

- To name and draw translation images of figures with respect to parallel lines.

Why IT'S IMPORTANT

Translations are used to describe linear movement in the real-world as well as on a coordinate plane.

Transportation

In order to preserve an old building that was going to be demolished to widen a road, the Reynoldsburg (Ohio) Historical Society had the building moved to another location. The result of a movement in one direction is a transformation called a **translation**. In a translation, all points are moved the same distance in the same direction.

One way to find a translation image is to perform one reflection after another with respect to two parallel lines. For example, the leftmost mitten is first reflected over line ℓ. Then that image is reflected over line m. Two successive reflections such as this are called a **composite of reflections**.

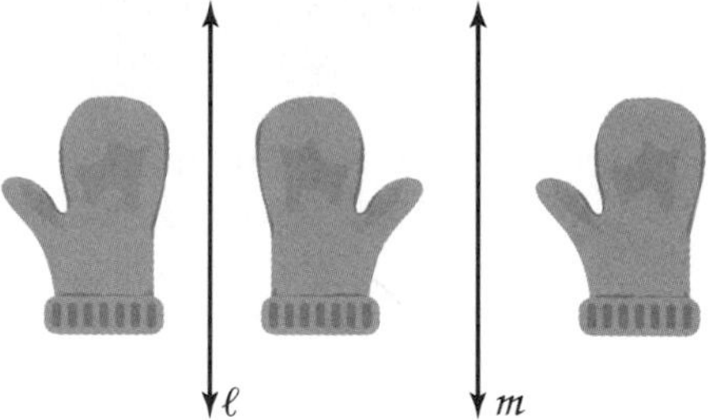

Do you see a relationship between the first and the last images? A translation is often referred to as a glide.

Since translations are composites of two reflections, all translations are isometries. As a result, all properties preserved by reflections are preserved by translations. These properties include collinearity, betweenness of points, and angle and distance measure.

Example 1

Algebra

Find the translation image of ▱PQRS with respect to the parallel lines s and t.

Using the coordinate plane, reflect ▱PQRS over line s. Then draw the reflection image of ▱P′Q′R′S′ with respect to line t.

The image reflected over line t is the translation image of ▱PQRS.

Is there a way that ▱PQRS can be directly translated to ▱P″Q″R″S″?

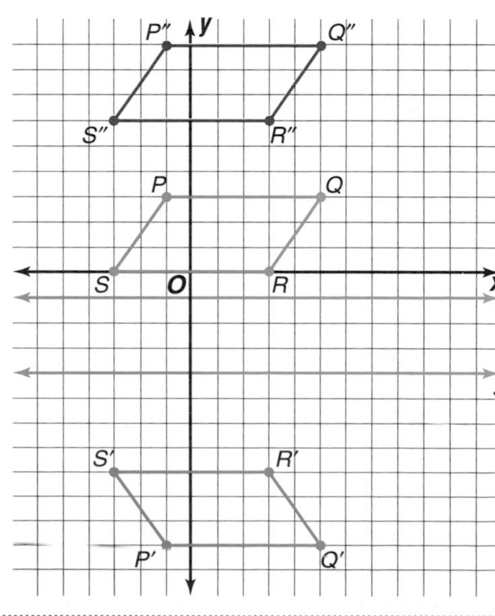

Since a translation is an isometry, congruence of shapes is preserved. You can use a translation to draw a prism by drawing the base and translating it to form a congruent base.

MODELING MATHEMATICS

Drawing Prisms Using Translations

Materials:  thin cardboard, plain paper  straightedge

Use a translation to draw a triangular prism.

- Cut a triangle out of a piece of thin cardboard. Label its angles 1, 2, and 3.

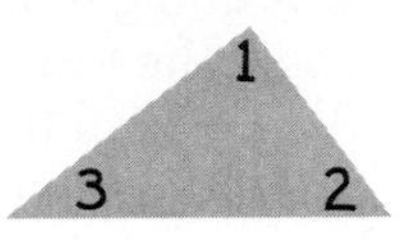

- Place the triangle on a piece of notebook paper, aligning the common side of ∠2 and ∠3 with a horizontal rule. Trace the triangle on the paper. Label the vertices so that the vertex of ∠1 is *X*, the vertex of ∠2 is *Y*, and the third vertex is *Z*.
- Now slide the cutout triangle to another place on the paper, making sure that the common side of ∠2 and ∠3 is still aligned with a horizontal rule. Trace the cutout triangle again. Label the vertices of this triangle *X'*, *Y'*, and *Z'* so that they correspond to the vertices of the first triangle you drew.

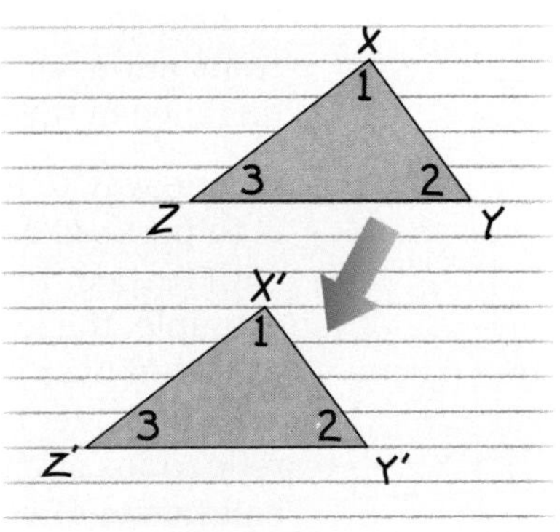

- Use a straightedge to draw $\overline{XX'}$, $\overline{YY'}$, and $\overline{ZZ'}$.

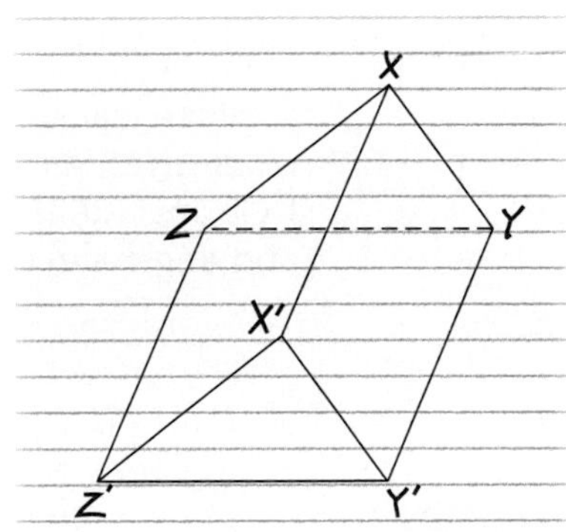

Your Turn

a. Why is $\overline{ZY}$ a dashed segment?

b. When you started with two triangular bases, how many more faces evolved when the prism was completed?

c. Describe the shapes of the faces that were formed from this translation. How many of each shape are there?

Translations on a coordinate plane are easily drawn if you know the direction in which the figure is moving horizontally and vertically. Let's use a table to look at the three images from Example 1.

▱PQRS		▱P′Q′R′S′		▱P″Q″R″S″
$P(-1, 3)$	→	$P'(-1, -11)$	→	$P''(-1, 9)$
$Q(5, 3)$	→	$Q'(5, -11)$	→	$Q''(5, 9)$
$R(3, 0)$	→	$R'(3, -8)$	→	$R''(3, 6)$
$S(-3, 0)$	→	$S'(-3, -8)$	→	$S''(-3, 6)$

From ▱*PQRS* to ▱*P″Q″R″S″*, the *x*-coordinates stay the same, but the *y*-coordinates have been translated 6 units up. So, (x, y) maps to $(x, y + 6)$. In general, if (a, b) describes the translation horizontally a units and vertically b units, then the image of (x, y) is $(x + a, y + b)$.

Example 2 **Translate $\triangle ABC$ with vertices $A(5, 4)$, $B(3, -1)$, and $C(0, 2)$, so that A' is located at $(3, 1)$. Graph both triangles, and state the coordinates of B' and C'.**

Graph $\triangle ABC$ and A'.

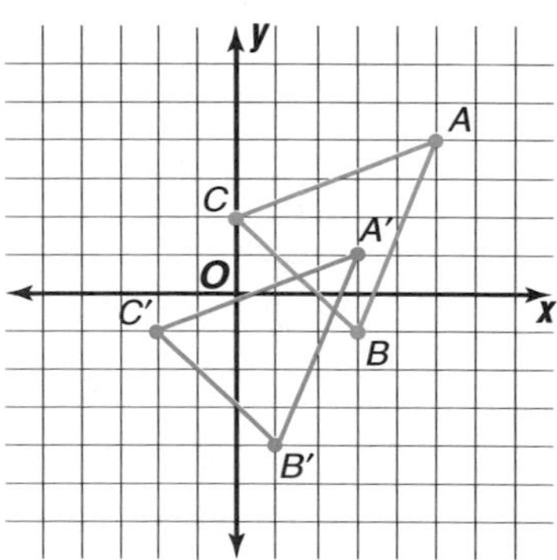

Point A was translated 2 units left and 3 units down to become A'. Perform this same translation on B and C to determine the coordinates of B' and C'.

B' has coordinates of $(1, -4)$, and C' has coordinates of $(-2, -1)$.

CHECK FOR UNDERSTANDING

Communicating Mathematics

Study the lesson. Then complete the following.

1. **Name** the properties that are preserved in a reflection. Since a translation consists of two successful reflections, what properties will be preserved in a translation?

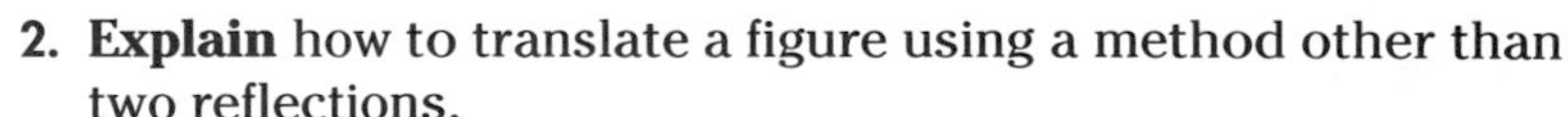

2. **Explain** how to translate a figure using a method other than two reflections.
3. **You Decide** Johnna says that the designs of wallpaper are usually designs of translations. Do you agree with her? Why or why not? You may use drawings to support your answer.
4. Suppose $\triangle BCD$ is slid along the *y*-axis until D has coordinates $(0, -8)$. What are the new coordinates of point B?

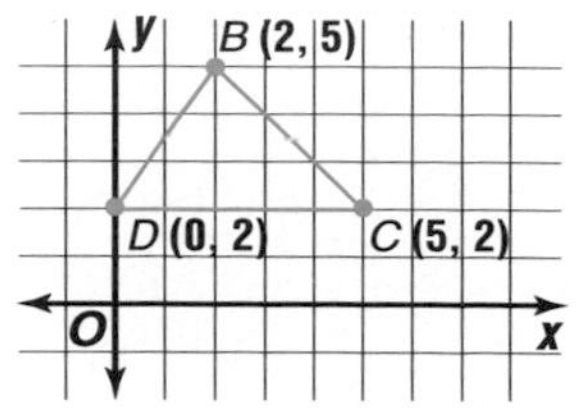

5. Refer to the Modeling Mathematics activity. Use a translation to draw a pentagonal prism.
   a. How many additional faces were formed from these bases in the translation?
   b. Write a generalization about the number of faces and their shapes when a prism is formed from a translation.

Guided Practice

For each of the following, lines ℓ and m are parallel. Determine whether each red figure is a translation image of the blue figure. Write *yes* or *no*. Explain your answer.

6. 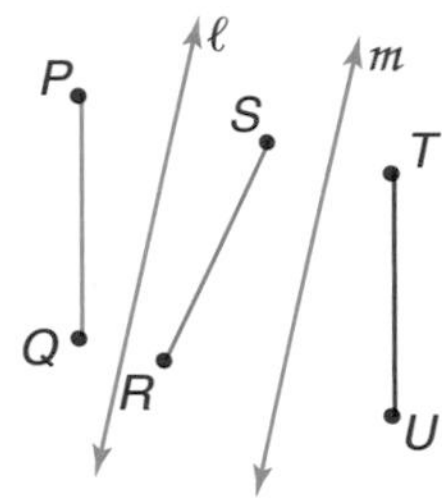

7. 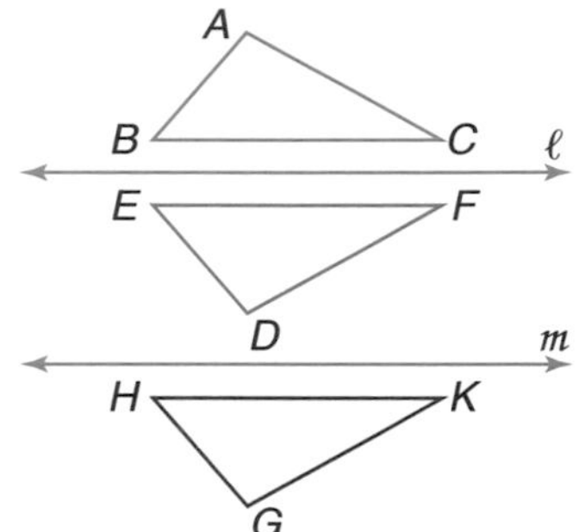

Graph each preimage on a coordinate plane. Then graph the image after performing the translation given, and list the coordinates of that image.

8. $S(-3, 5)$; slide 3 down, 4 right
9. $\triangle EFG$, $E(-4, 1)$, $F(-1, 3)$, $G(-1, 1)$; reflection over the graph of $x = -2$ and then the graph of $x = 2$
10. $\square HIJK$, $H(-2, 7)$, $I(2, 9)$, $J(2, 7)$, $K(-2, 5)$; slide 3 units right, 6 units down
11. **Design** The diagram at the right shows the floor plan of Eleanor's kitchen. She recently had her kitchen remodeled and her refrigerator (square A) was moved. The new position is at square B. Describe the move. *Each square on the diagram represents 3×3 or 9 square feet.*

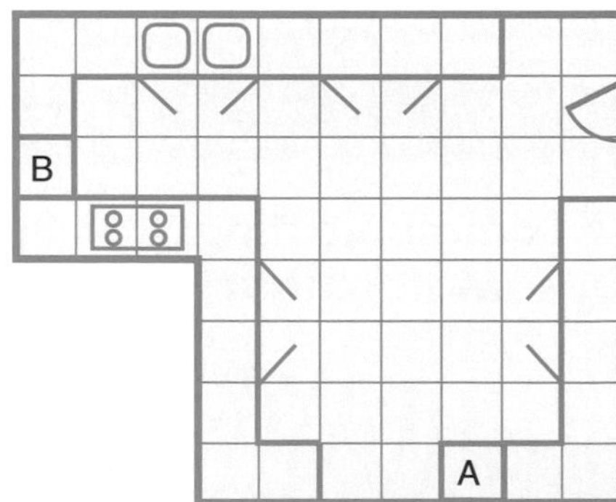

EXERCISES

Practice

In the figure at the right, $a \parallel b$. Name the translation image for each point with respect to line a, then line b.

12. K	13. M
14. J	15. P
16. L	17. N

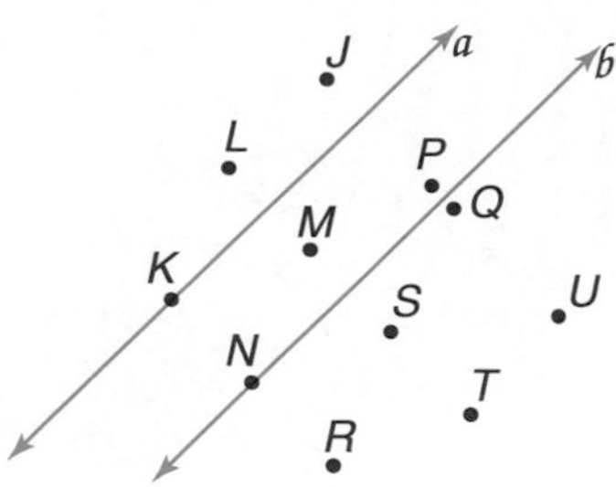

In each figure for Exercises 18–21, $\ell \parallel m$. Determine whether each red figure is a translation image of the blue figure. Write *yes* or *no*. Explain your answer.

18.

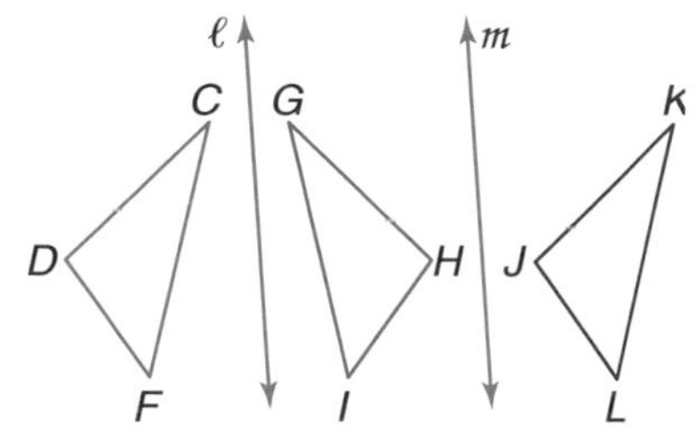

19.

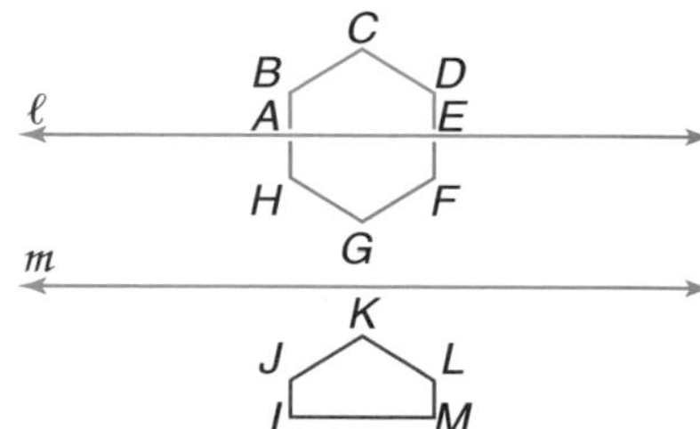

20.

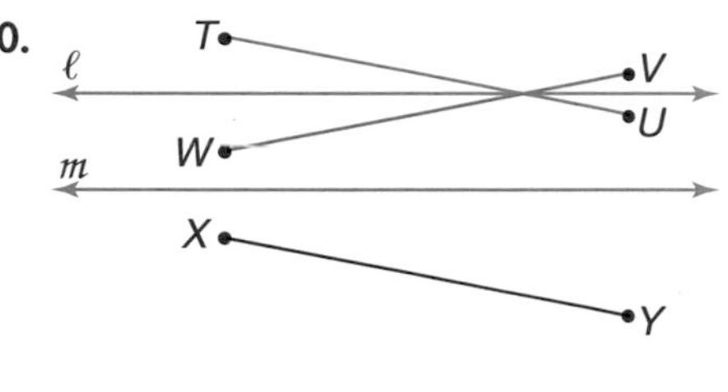

21.

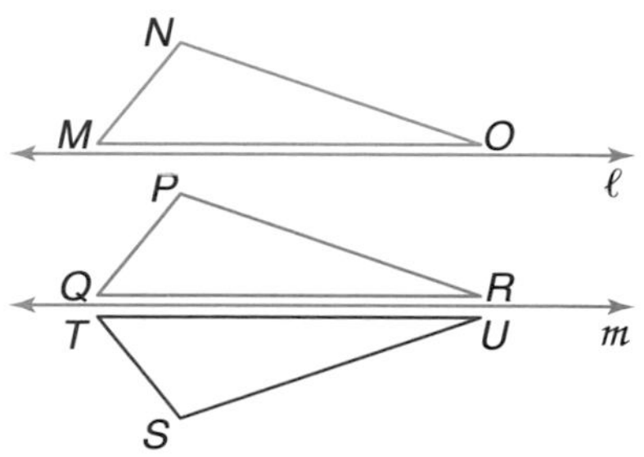

Graph each preimage on a coordinate plane. Then graph the image after performing the translation given, and list the coordinates of that image.

22. $\overline{WX}$, $W(4, 8)$, $X(7, 5)$; slide 5 left, 3 down

23. Quadrilateral $ABCD$, $A(-8, 4)$, $B(-3, 6)$, $C(-4, 3)$, $D(-7, 2)$; reflect over the graph of $y = 2$ and then the graph of $y = -2$

24. $\triangle CAT$, $C(-2, -1)$, $A(0, 2)$, $T(2, -2)$; reflect over the graph of $x = 1$ and then the graph of $x = -3$

25. Trapezoid $NICK$, $N(2, 3)$, $I(6, 2)$, $C(6, -1)$, $K(-2, 1)$; slide N to $N'(-4, 1)$

Copy each figure on a coordinate plane. Then find the translation image of each geometric figure with respect to the parallel lines n and ℓ.

26.

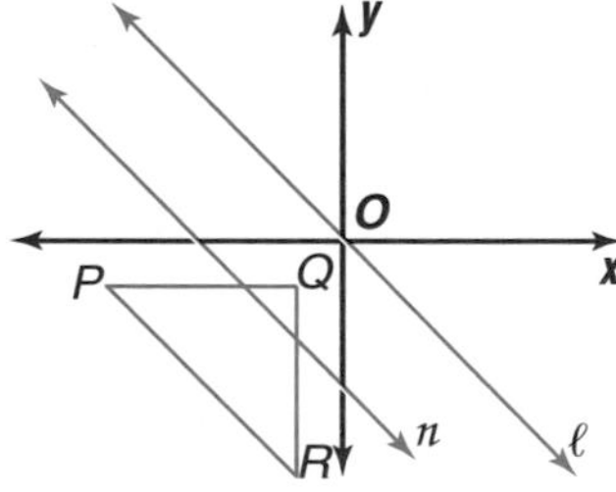

27.

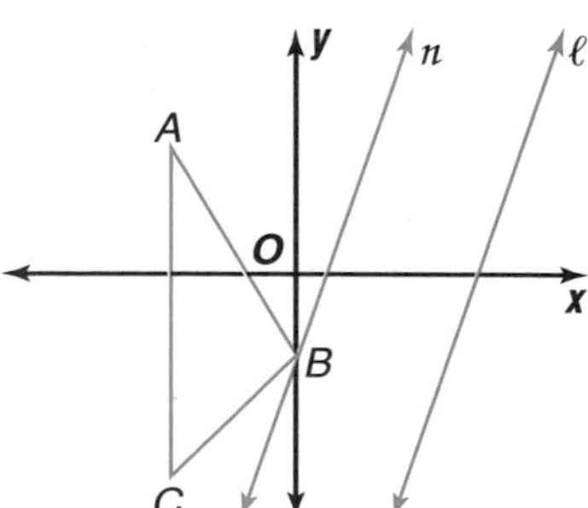

28.

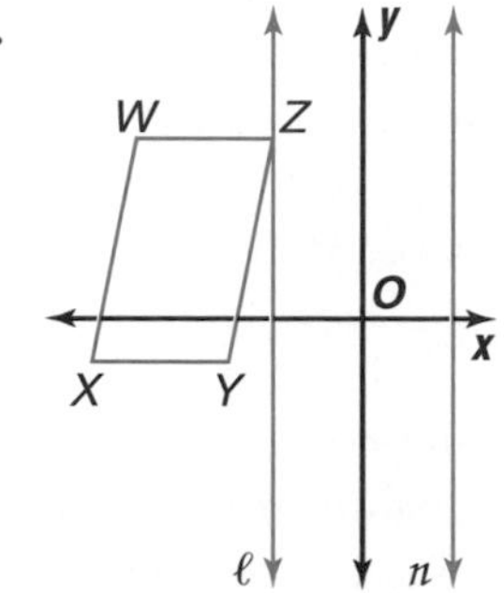

29.

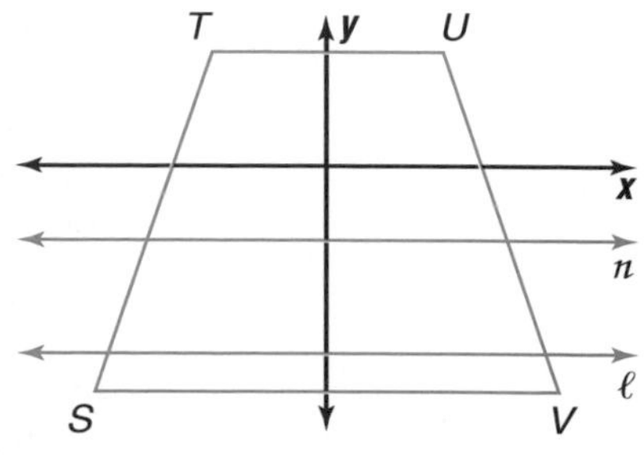

Use the figures to name each triangle. Assume $n \parallel t$.

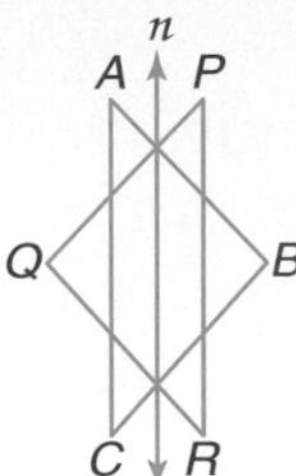

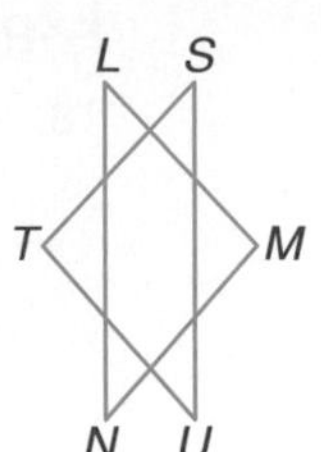

30. reflection image of $\triangle ABC$ with respect to n and t
31. reflection image of $\triangle PQR$ with respect to n and t
32. Plan a proof to show that the translation image of $\triangle ABC$ with respect to parallel lines ℓ and m preserves angle measure and distance measure.

Critical Thinking

33. Triangle RST has vertices $R(3, -7)$, $S(7, -4)$, and $T(9, -8)$. Triangle ABC has vertices $A(3, 3)$, $B(7, 6)$, and $C(9, 2)$. If $\triangle ABC$ is the translation image of $\triangle RST$ with respect to two parallel lines, find the equations that represent two possible parallel lines.

Applications and Problem Solving

34. **Art** Mbwane is an illustrator working on a gemology book. One section is about quartz crystals that are six-sided prisms. Using a translation, draw a prism with hexagonal bases that could be used as a sketch for the quartz crystals.
35. **Environment** In the photograph, a cloud of dense gas and dust pour out of Surtsey, a volcanic island off the south coast of Iceland. If the cloud blows 40 miles north and then 30 miles east, make a sketch to show the translation of the smoke particles. Then indicate the shortest path that would take the particles to the same position.

Mixed Review

36. Copy quadrilateral $QRST$. Draw the reflection image of quadrilateral $QRST$ with respect to line ℓ. (Lesson 13–5)

Q R T S ℓ

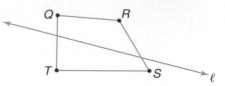

37. Describe the locus of all points in space that are a given distance from a line segment and equidistant from the endpoints of the line segment. (Lesson 13–3)

38. Use an algebraic method to find the locus of points that satisfy the graphs of both equations. (Lesson 13–2)

$2x - 6y = 8$
$3x + 4y = -4$

39. Given ▱$QRST$, complete each statement. (Lesson 12–5)
 a. $\overrightarrow{SR} + \overrightarrow{RQ} = \underline{\ ?\ }$
 b. $\overrightarrow{QR} + \overrightarrow{RT} = \underline{\ ?\ }$
 c. $\overrightarrow{TQ} + \overrightarrow{TS} = \underline{\ ?\ }$

40. Graph the line with slope equal to $\frac{3}{4}$ and that passes through the point at $(0, -3)$. Then write the equation for the line. (Lesson 12–2)

41. **SAT Practice** **Quanitative Comparison**

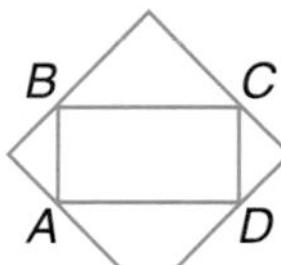

The area of the square is 36.
Points A, B, C, and D are on the square.

Column A	Column B
Perimeter of the rectangle *ABCD*	24

A if the quantity in Column A is greater
B if the quantity in Column B is greater
C if the two quantities are equal
D if the relationship cannot be determined from the information given

42. Describe how an equilateral triangle is also an isosceles triangle. (Lesson 4–1)

Algebra

For **Extra Practice**, see page 791.

43. Factor $8x^2 + 48xy + 72y^2$.

44. Factor $16a^2 - 72a + 81$.

Refer to the Investigation on pages 642–643.

Computer movie images can use reflections to create a variety of effects. For example, a bouncing ball, a reflection in a mirror, and any type of symmetric object could be made by reflecting an image.

1 Brainstorm with your group about ways that reflections can be used to make your movie.

2 Incorporate one or more reflections into your movie.

3 Write a sentence describing how each reflection affects the coordinates of the object being reflected in your movie.

Add the results of your work to your Investigation Folder.

13–6B Using Technology Translating Polygons with Vectors

An Extension of Lesson 13–6

LOOK BACK

Refer to Lesson 12-5 to review vectors.

TECHNOLOGY Tip

To scroll the whole screen so that a figure is visible, hold the 2nd key down and press the direction pad to move the screen in the desired direction.

You can use a calculator to translate a triangle in the direction and magnitude (length) of a given vector.

- Draw a triangle by pressing F3 and selecting 3:Triangle.
- A vector is needed beside the triangle in order to determine how you want the triangle translated—direction and magnitude of the translation. To draw a vector on your screen, press F2 and then select 7:Vector.
- Now translate the triangle by placing the cursor on the triangle (The message will say, "THIS TRIANGLE.") and press F5 and select 1:Translation.

  The message will say, "TRANSLATE THIS TRIANGLE." Press ENTER. Then move the cursor to the vector. The message will say, "BY THIS VECTOR." Press ENTER. The translated triangle will appear.
- Hide the vector by pressing F7 and selecting 1:Hide/Show. Then place the cursor on the vector and press ENTER.
- Use the button F6 and select 5:Equation & Coordinates to find the coordinates of the six vertices. A sample screen is shown at the right.

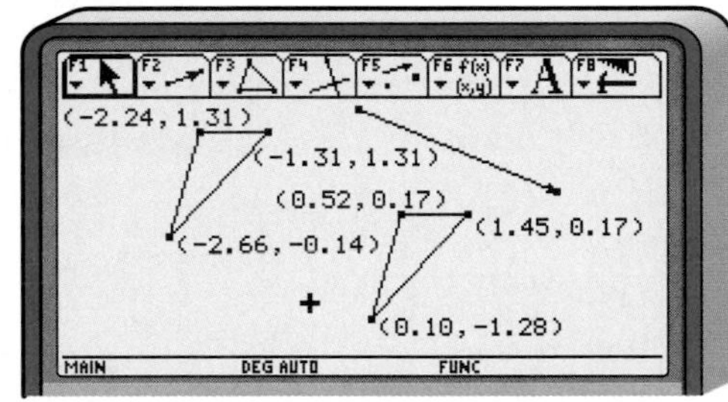

EXERCISES

Examine your drawings.

1. Were you able to see all the parts of the translated triangle?
2. Compare the coordinates of the corresponding vertices. What do you notice?
3. Describe the translation in terms of horizontal and vertical movement. Make a conjecture.

Test your conjecture.

4. Use the TI-92 to test a translation of a regular polygon.
   a. Draw a regular polygon.
   b. Draw a vector in a different direction.
   c. Translate the regular polygon and compare the preimage and the image coordinates.
   d. Does your conjecture in Exercise 3 about describing translations hold true?

13–7 Rotations

***What* YOU'LL LEARN**

- To name and draw rotation images of figures with respect to intersecting lines.

***Why* IT'S IMPORTANT**

Rotations are an important part of motion, which affects every aspect of your life.

Real World APPLICATION

Industry

Modern windmills called wind turbines are used to generate electric current. These turbines are made with two or three blades that rotate on a shaft. As a windmill rotates, each blade moves in a circular motion to a new position. This is a **rotation**, which is another type of transformation.

Turning motions such as a doorknob being turned or the movement of the paddles on a paddle wheel involve rotation around a fixed point called the **center of rotation**.

Remember, a translation is a composite of two reflections over parallel lines.

MODELING MATHEMATICS

Rotating a Triangle

Materials:  grid paper  tracing paper

- Use grid paper to draw two intersecting lines p and q on a coordinate plane.
- Draw $\triangle ABC$ so that it does not intersect either line.
- Reflect $\triangle ABC$ over line p. Label the image $A'B'C'$.
- Reflect $A'B'C'$ over line q. Label this reflection $A''B''C''$. This is the rotated triangle.
- Trace the original triangle on a separate sheet of paper.
- Place the traced triangle over $\triangle ABC$ on the grid paper. Using your pencil as the center of rotation, place it on the point of intersection for lines p and q. Turn your paper until the traced triangle coincides with the image of $\triangle A''B''C''$.

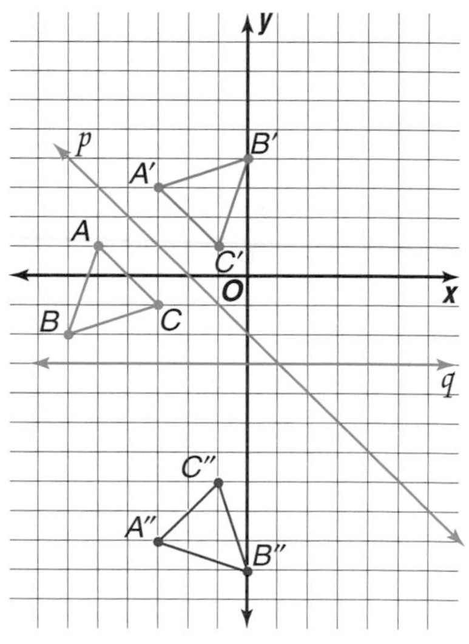

Your Turn

a. This rotation consists of two reflections. What properties were preserved in this rotation?

b. Is a rotation an isometry? Explain.

This Modeling Mathematics activity shows that a rotation is a composite of two reflections with respect to two intersecting lines.

Example 1 **Find the rotation image of rhombus *ABCD* over lines *m* and *n* using reflections.**

First reflect rhombus *ABCD* over line *m*. Label the image *A′B′C′D′*.

Next, reflect the image over line *n*.

Label this reflection *A′B″C″D″*.

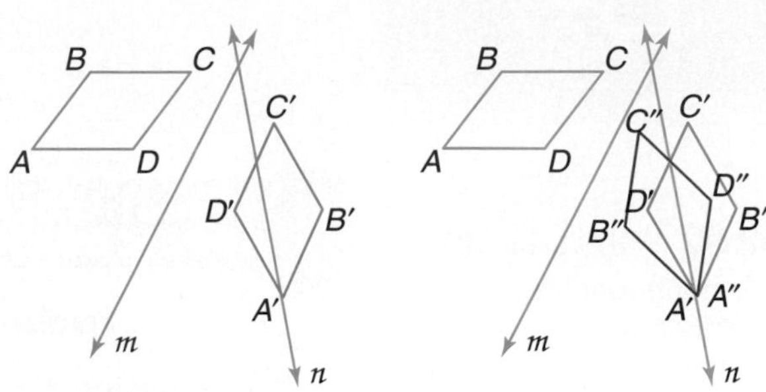

Rhombus *A′B″C″D″* is a rotation image of rhombus *ABCD*.

There are two methods by which a rotation can be performed.

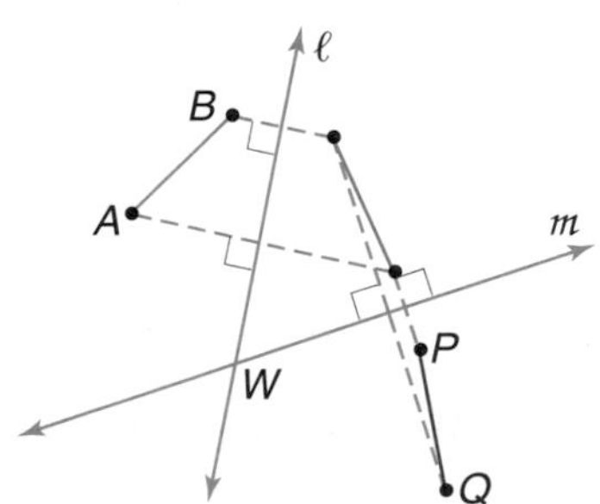

A rotation image can be determined by a composite of two successive reflections over two intersecting lines. $\overline{AB}$ is reflected over ℓ and then *m*. The rotation image of $\overline{AB}$ is $\overline{PQ}$.

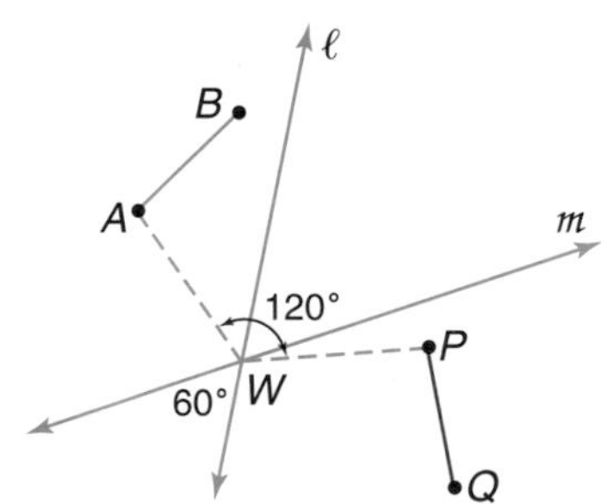

A rotation image can also be determined by using the angle formed by the intersecting lines. This is called the **angle of rotation**. The point at which the two lines intersect is the center of rotation.

There is a relationship between the angle formed by the intersecting lines and the angle of rotation.

Postulate 13–1	**In a given rotation, if *A* is the preimage, *P* is the image, and *W* is the center of rotation, then the measure of the angle of rotation ∠*AWP* is twice the measure of the angle formed by the intersecting lines of reflection.**

Example  **Find the rotation image of $\overline{XY}$ over intersecting lines ℓ and *m*, which form a 40° angle, by using the angle of rotation.**

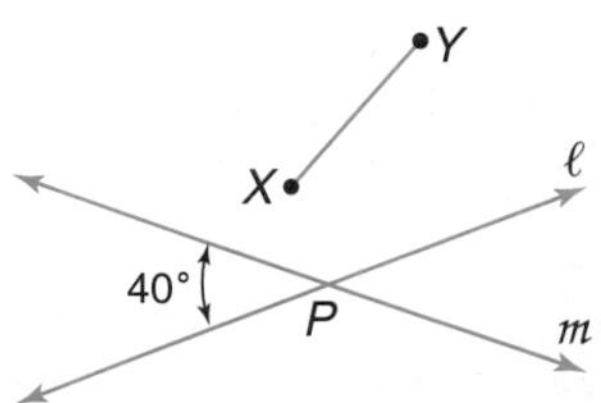

Explore Since the angle formed by lines ℓ and *m* measures 40°, the angle of rotation measures 2(40°) or 80°.

Plan Draw the angles of rotation and rotate each point.

Solve Draw $\angle XPR$ so that its measure is 80 and $\overline{XP} \cong \overline{PR}$. Draw $\angle YPQ$ so that its measure is 80 and $\overline{YP} \cong \overline{PQ}$. Connect R and Q to form the rotation image of $\overline{XY}$.

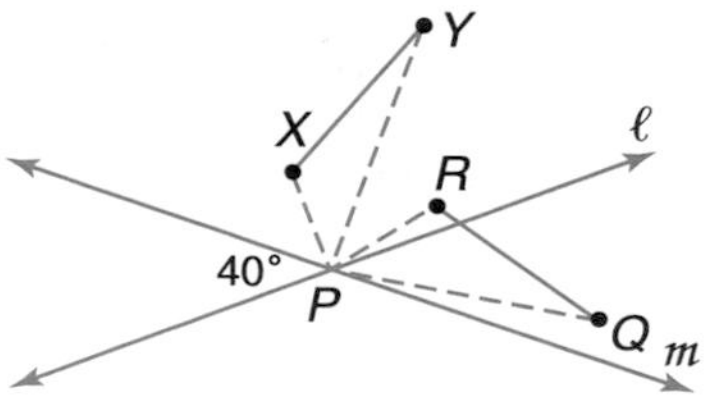

Examine Reflect $\overline{XY}$ over the two lines to verify that $\overline{RQ}$ is the rotation image of $\overline{XY}$.

Example 3

PROBLEM SOLVING

Look for a Pattern

$\triangle XYZ$ is rotated 180° around the origin on the coordinate plane to form $\triangle X'Y'Z'$.

a. Make a table of the coordinates of the vertices of $\triangle XYZ$ and $\triangle X'Y'Z'$.

$\triangle XYZ$	$\triangle X'Y'Z'$
$X(-2, 1)$	$X'(2, -1)$
$Y(2, -3)$	$Y'(-2, 3)$
$Z(3, 5)$	$Z'(-3, -5)$

b. Use the table to find a pattern in the coordinates of the vertices of $\triangle XYZ$ and $\triangle X'Y'Z'$. Make a conjecture about what happens in a coordinate plane for a 180° rotation.

The x- and y-coordinates of the image are the opposites of the preimage coordinates.

$P(x, y) \rightarrow P(-x, -y)$ for a rotation of 180° around the origin.

Example 4

Real World APPLICATION

Design

One of the two screws securing a stop sign has come out, and the sign is hanging upside down. Assuming the remaining screw can be considered the center of rotation, find the angle of rotation needed to straighten the sign.

A stop sign is the shape of a regular octagon. From the center of rotation, imagine lines to each of the corners of the sign. The angles formed by the intersecting lines form 45° angles, because 360 ÷ 8 is 45. The measure of the angle of rotation is 2(45) or 90. The bottom side of the sign would have to be rotated two times or 180°.

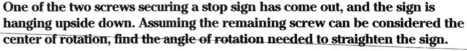

CHECK FOR UNDERSTANDING

Communicating Mathematics

Study the lesson. Then complete the following.

1. **Compare and contrast** a translation and a rotation.
2. **Describe** two techniques that can be used to locate a rotation image with respect to two intersecting lines.
3. **Determine** if a regular hexagon has 60° rotational symmetry. Explain.

4. Point Y is the image of A under a rotation.
   a. Copy the figure and construct a point that could be the center of the rotation.
   b. Construct a second point that could be the center of rotation.
   c. Generalize your conclusions for parts a and b.

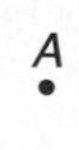

5. Two lines intersect to form an angle of 37°. What is the angle of rotation?

Guided Practice

Refer to the figure at the right for Exercises 6–11.

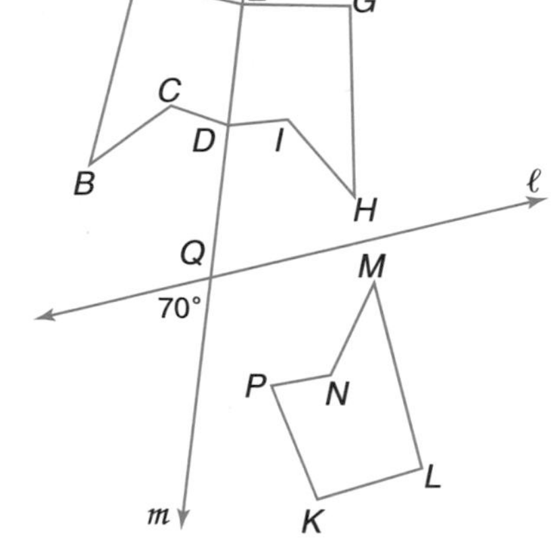

6. Find the reflection image of pentagon $ABCDE$ with respect to line m.
7. Find the reflection image of pentagon $DEGHI$ with respect to line ℓ.
8. Find the rotation image of pentagon $ABCDE$ with respect to lines ℓ and m.
9. Name the angle of rotation used to rotate C.
10. Find the angle measure of $\angle AQL$.
11. Find the rotation image of $\overline{CE}$ with respect to lines m and ℓ.
12. Copy the figure below. Then use a composite of reflections to find the rotation image with respect to lines n and t.

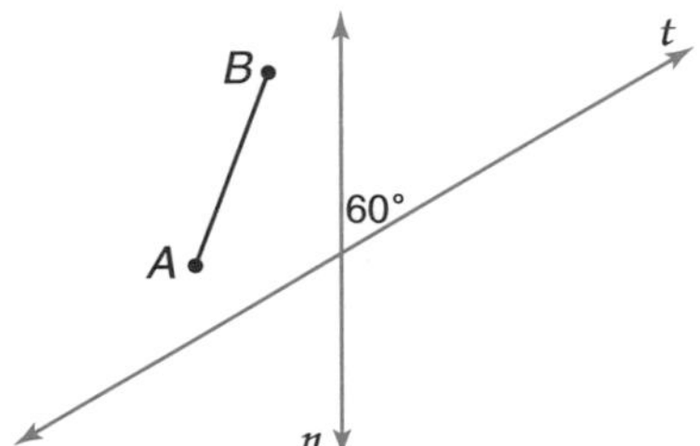

13. Copy the figure below. Use the angle of rotation to find the rotation image with respect to lines n and t.

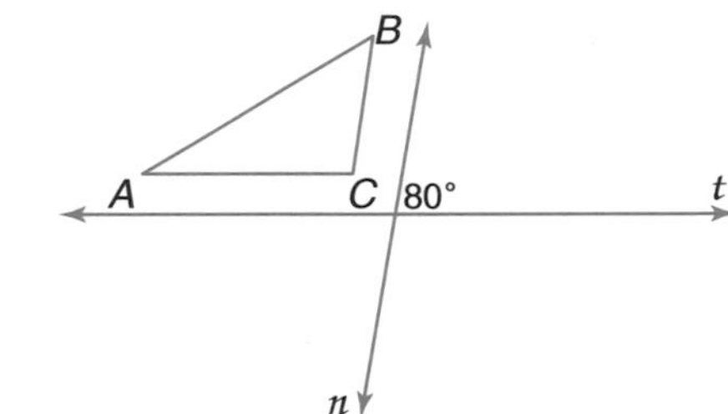

14. **Gymnastics** The high bar is a men's gymnastics event. If the drawing at the right is put on a coordinate grid with the hand grips at (0, 0), then the upper left body figure has toe coordinates of (−5, 6), and the lower right body figure has toe coordinates of (5, −6).
    a. What is the change that occurred in the coordinates?
    b. What kind of a rotation is this?

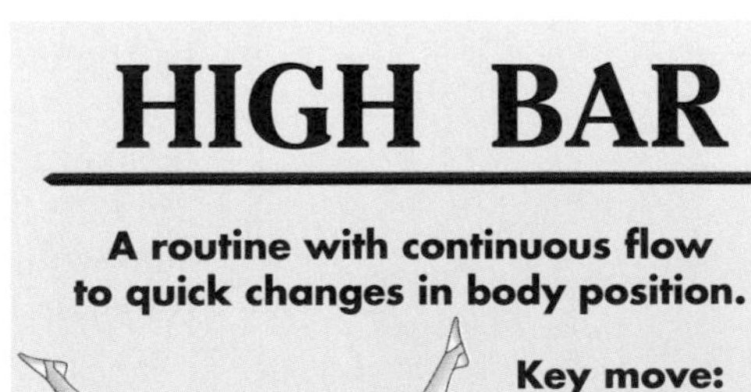

HIGH BAR
A routine with continuous flow to quick changes in body position.
Key move:
Giant swing. As the body swings around the bar the body should be straight with a slight hollow to the chest.
Height: 8½ feet
Length: 8 feet

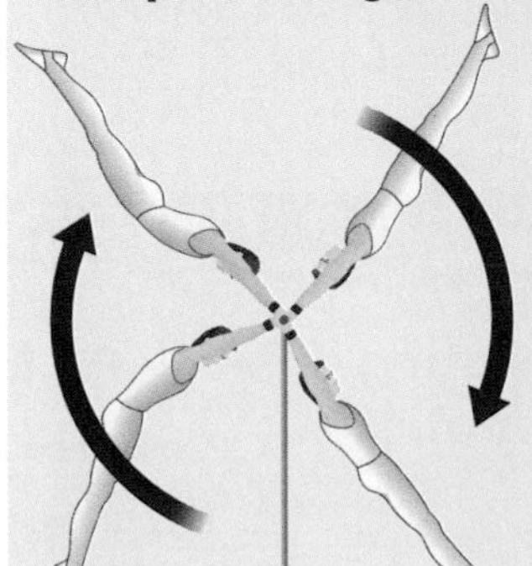

EXERCISES

Practice

For each of the following, determine whether the indicated composition of reflections is a rotation. Explain your answer.

15.

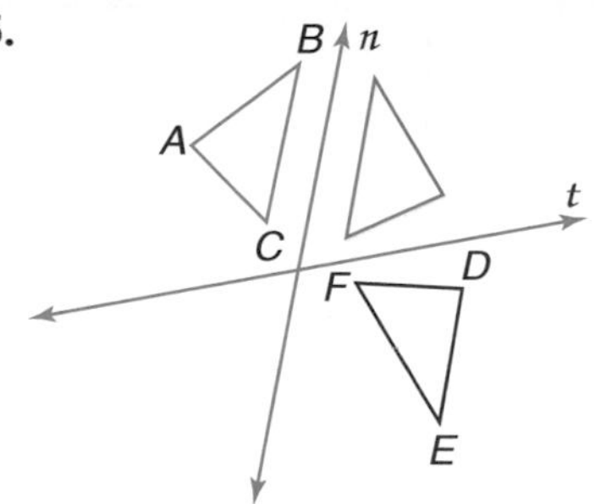

16.

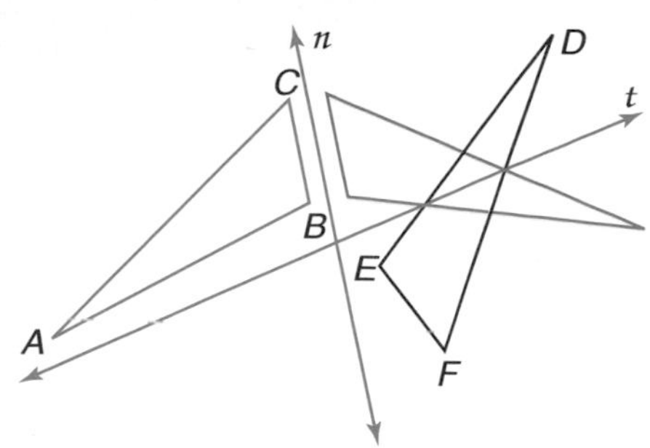

Two lines intersect to form an angle with the following measure. Find the angle of rotation for each.

17. $55°$ **18.** $28.5°$ **19.** $74°$

Copy each figure. Then use a composite of reflections to find the rotation image with respect to lines ℓ and t.

20.

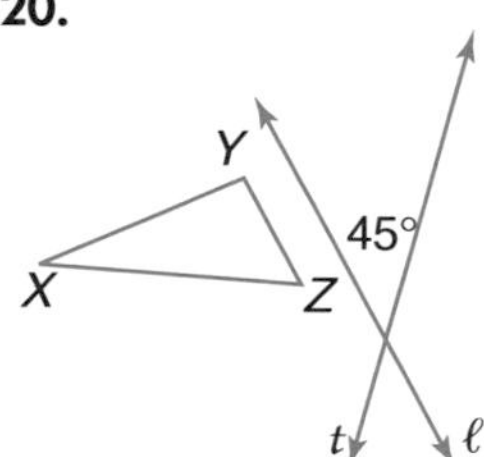

21.

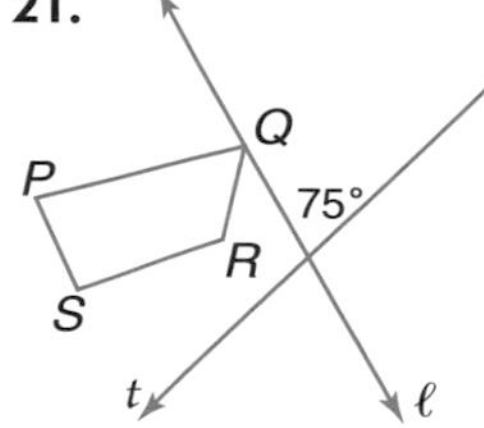

22.

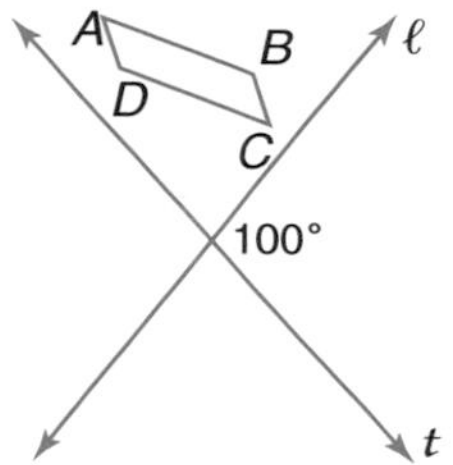

Copy each figure. Then use the angle of rotation to find the rotation image with respect to lines ℓ and t.

23.

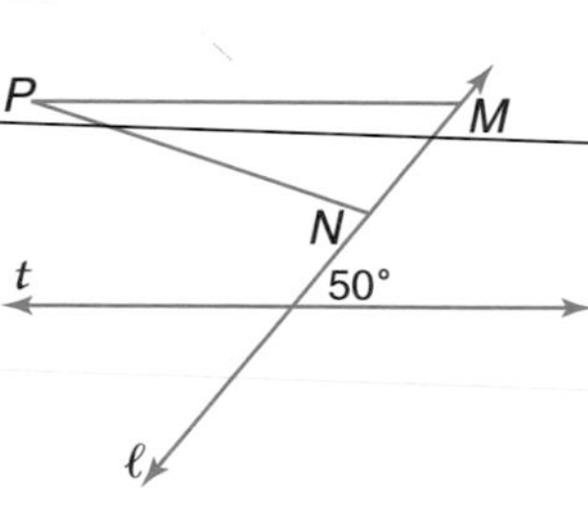

24.

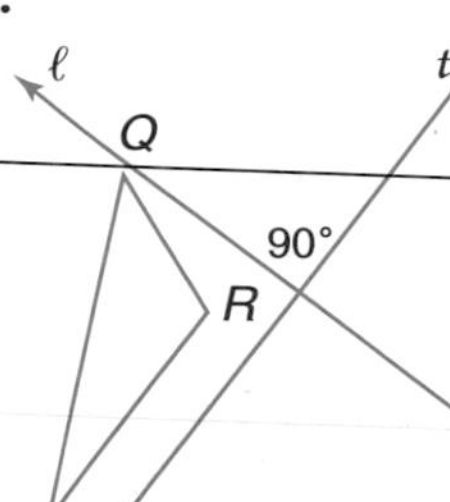

25.

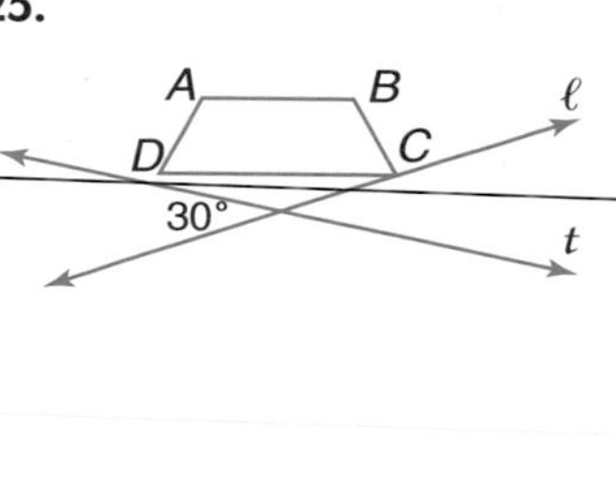

In the following figures, a rotation of $\triangle ABC$ has been performed with respect to lines p and q. Use a protractor to find the angle of rotation.

26.

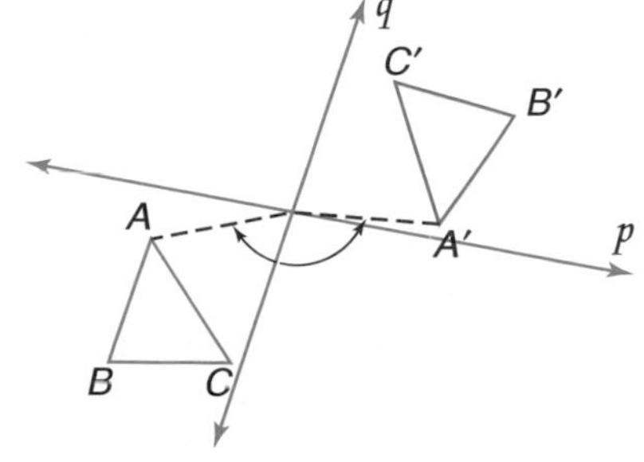

27.

Copy and complete the table below. Determine whether each translation preserves the given property. Write *yes* or *no*.

		angle measure	betweenness of points	orientation	collinearity	distance measure
28.	reflection					
29.	translation					
30.	rotation					

31. Draw a segment and two intersecting lines. Find the rotation image of the segment with respect to the two intersecting lines.

32. Draw an isosceles triangle and two intersecting lines. Find the rotation image of the triangle with respect to the two intersecting lines.

33. Kite *ABCD* in the figure at the right is soaring upward. If kite *ABCD* goes through a 180° rotation, it will form kite *STUV*. Draw kite *STUV* on the coordinate grid.

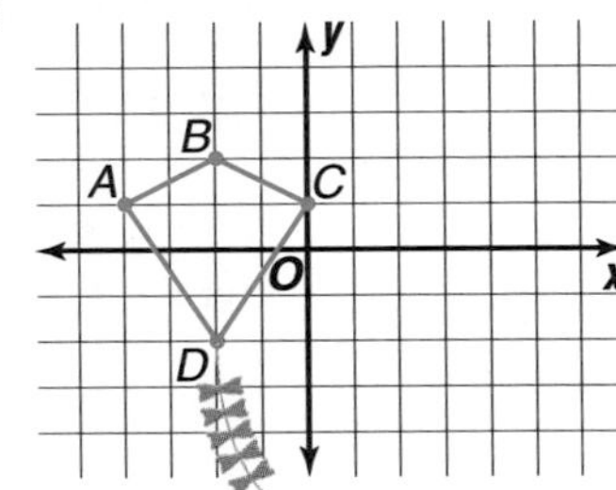

Hexagon *ABCDEF* is a regular hexagon with diagonals as drawn. Point *C* is the rotation image of point *A*.

A F B E C D

34. What is the measure of the angle formed by each set of intersecting lines? Explain your answer.

35. What is the measure of the angle of rotation?

36. Determine the number of rotations needed to rotate *C* to *A*'s original position if the rotation is counterclockwise.

In the figure at the right, $\ell \parallel m$. Name the type of transformation represented by the mappings shown.

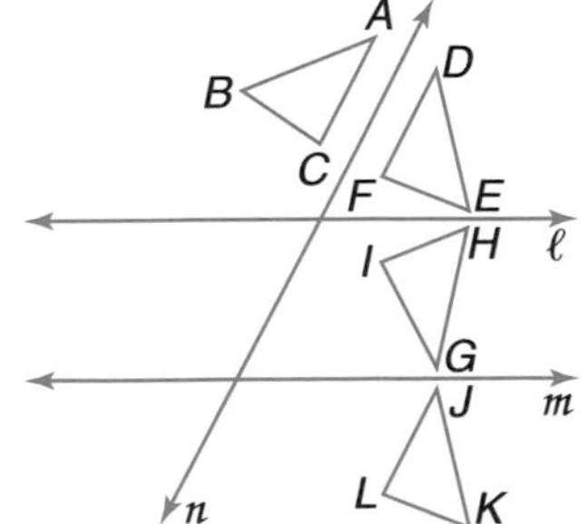

37. $\triangle GHI \to \triangle DEF$

38. $\triangle ABC \to \triangle GHI$

39. $\triangle DEF \to \triangle JKL$

40. $\triangle ABC \to \triangle DEF$

Critical Thinking

41. If a rotation is performed on a coordinate plane, what angles of rotation would make the rotations easier? Explain your answer.

42. What capital letters produce the same letter after being rotated 180°?

Applications and Problem Solving

43. Recreation A Ferris wheel's motion is an example of a rotation.

a. What is the measure of the angle of rotation if seat 1 of a 16-seat Ferris wheel is moved to the seat 5 position?

b. If seat 1 of a 16-seat Ferris wheel is rotated 135°, find the seat whose position it now occupies.

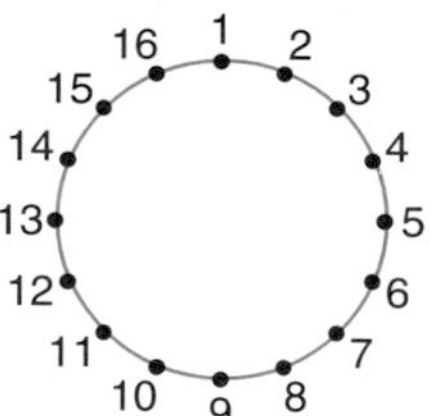

44. Technology A computer designer can form various solids by rotating a 2-dimensional geometric figure about a line.

a. Describe the solid that would be formed by a rotation of $\triangle PQR$ about the y-axis.

b. Describe the geometric figure that should be rotated about the y-axis to form a cube.

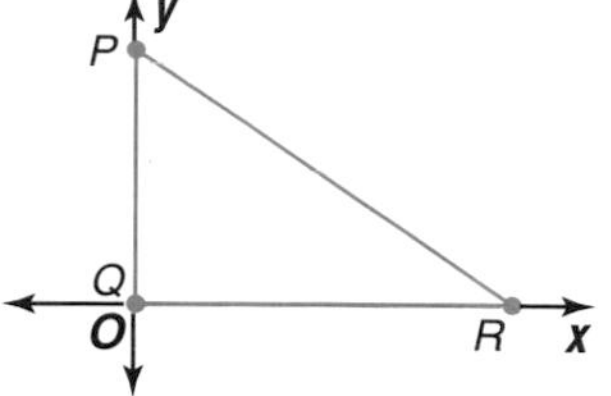

Mixed Review

45. In the tessellation shown at the right, it is possible to slide, or glide, the lower left quadrilateral so it fits exactly on top of the lower right quadrilateral. Is it possible to glide the lower left quadrilateral so it fits exactly on any of the other quadrilaterals? (Lesson 13–6)

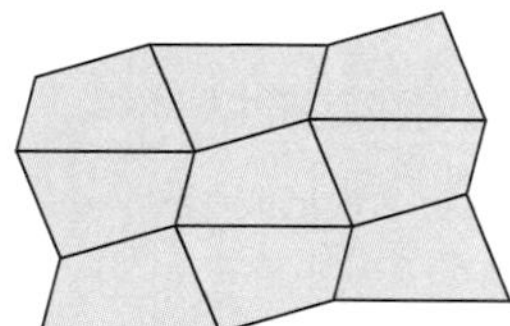

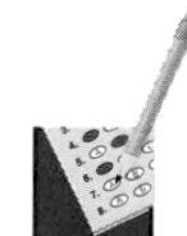

46. **SAT Practice Grid-in** A refreshment stand sells a large tub of popcorn for twice the price of a box of popcorn. If 60 tubs were sold for a total of \$150 and the total popcorn sales were \$275, how many boxes of popcorn were sold?

47. Describe the locus of points in a plane that are 3 inches from a circle with a radius measuring 5 inches. (Lesson 13–1)

48. Prove the statement *If both pairs of opposite sides of a quadrilateral are congruent, then the quadrilateral is a parallelogram* using a coordinate proof. (Lesson 12–4)

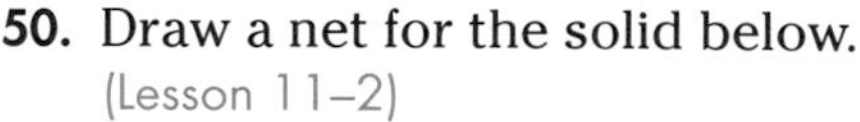

49. Find the volume of the cone below. (Lesson 11–6)

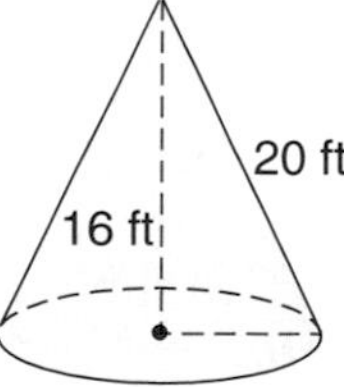

50. Draw a net for the solid below. (Lesson 11–2)

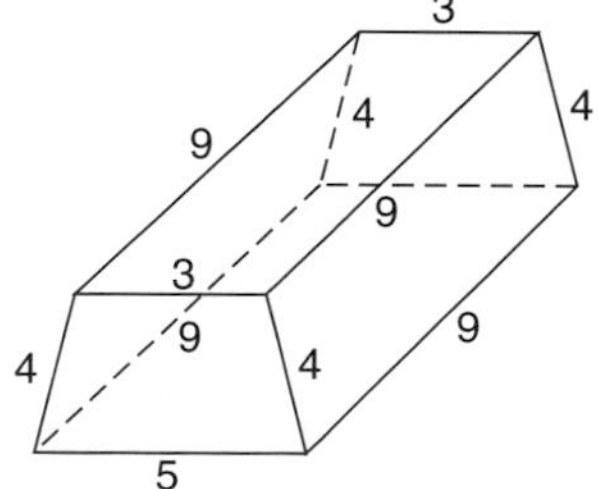

51. Find the value of x in $\odot P$. (Lesson 9–4)

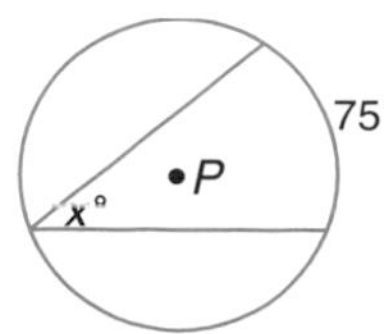

52. Is $\triangle QRS \sim \triangle CAB$? If so, find the missing measures. (Lesson 7–3)

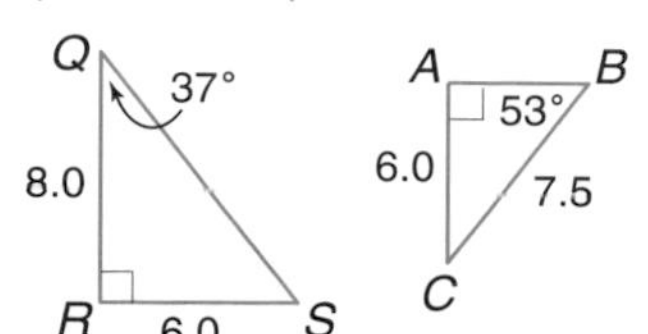

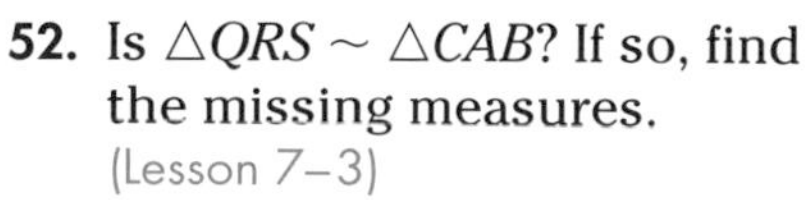

Algebra

53. Find $\frac{4}{x+9} - \frac{3}{x+9}$.

54. Factor $3z^2 - 75$.

For **Extra Practice**, see page 792.

13-8

Dilations

What YOU'LL LEARN

- To use scale factors to determine if a dilation is an enlargement, a reduction, or a congruence transformation, and
- to find the center and scale factor for a given dilation, and vice versa.

Why IT'S IMPORTANT

You can use dilations to solve problems in photography, drafting, and art.

Video Cameras

One night, while experimenting with his video camera, Mick hooked up the camera to play on the television. His son Troy was very amused to see himself on television. He also noticed that he looked smaller on television than in real life. This is an example of yet another transformation called a **dilation**.

We have already studied how translations, reflections, and rotations of geometric shapes produce figures that are congruent to each other. These transformations are all isometries.

A dilation is a transformation that alters the size of the geometric figure, but does not change its shape. So, a dilation is a *similarity transformation*.

CAREER CHOICES

A **cinematographer** is a photographer of motion pictures. An artistic eye and hands-on experiences are the best qualifications for this job. Courses in cinematography are often offered by photography institutes and universities or through special seminars.

For more information, contact:

American Society of Media Photographers
Washington Rd., Suite 502
Princeton Junction, NJ 08550

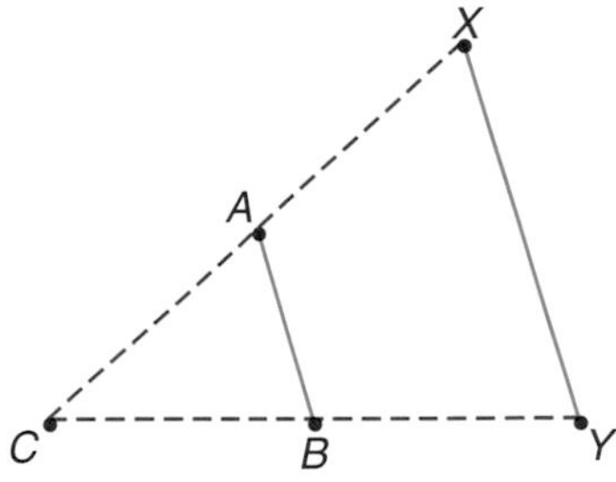

In the figure above, the measure of the distance from C to a point on $\overline{XY}$ is twice the distance from C to the corresponding point on $\overline{AB}$.

$$CX = 2(CA)$$

$$CY = 2(CB)$$

In this transformation, $\overline{AB}$ with **center** C and a **scale factor** of 2 is enlarged to $\overline{XY}$.

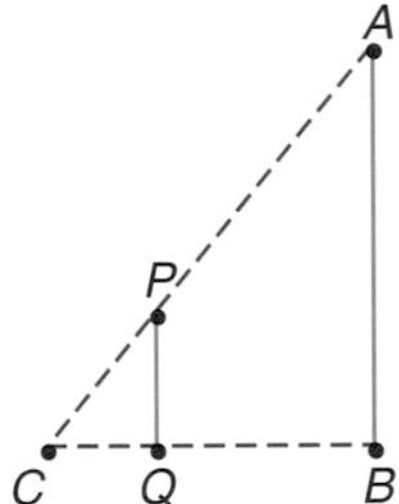

In the figure above, the measure of the distance from C to a point on $\overline{PQ}$ is one-third the distance from C to the corresponding point on $\overline{AB}$.

$$CP = \frac{1}{3}(CA)$$

$$CQ = \frac{1}{3}(CB)$$

In this transformation, $\overline{AB}$ with **center** C and a **scale factor** of $\frac{1}{3}$ is reduced to $\overline{PQ}$.

Theorem 13–1	**If a dilation with center C and a scale factor k maps A onto E and B onto D, then $ED = k(AB)$.**

You will be asked to prove this theorem in Exercise 46.

Notice that when the scale factor is 2, the figure is enlarged. When the scale factor is $\frac{1}{3}$, the figure is reduced.

In general, if k is the scale factor for a dilation with center C, then the following is true.

> **If $k > 0$, P', the image of point P, lies on $\overrightarrow{CP}$, and $CP' = k \cdot CP$.**
> **If $k < 0$, P', the image of point P, lies on the ray opposite $\overrightarrow{CP}$, and $CP' = |k| \cdot CP$.** *The center of a dilation is always its own image.*
>
> **If $|k| > 1$, the dilation is an enlargement.**
> **If $0 < |k| < 1$, the dilation is a reduction.**
> **If $|k| = 1$, the dilation is a congruence transformation.**

Example 1 **Given center C and each scale factor k, find the dilation image of $\overline{XY}$.**

a. $k = \frac{3}{4}$

Since $k < 1$, the dilation is a reduction.

Draw $\overline{CX}$ and $\overline{CY}$.

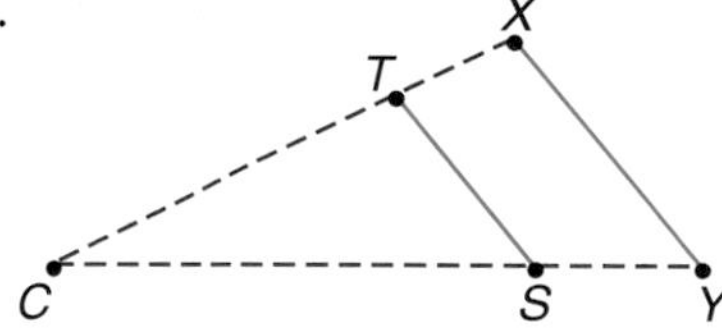

$CT = \frac{3}{4}(CX)$

$CS = \frac{3}{4}(CY)$

$\overline{TS}$ is the dilation of $\overline{XY}$ with scale factor $\frac{3}{4}$.

b. $k = 5$

Since $k > 1$, the dilation is an enlargement.

Draw $\overline{CX}$ and $\overline{CY}$.

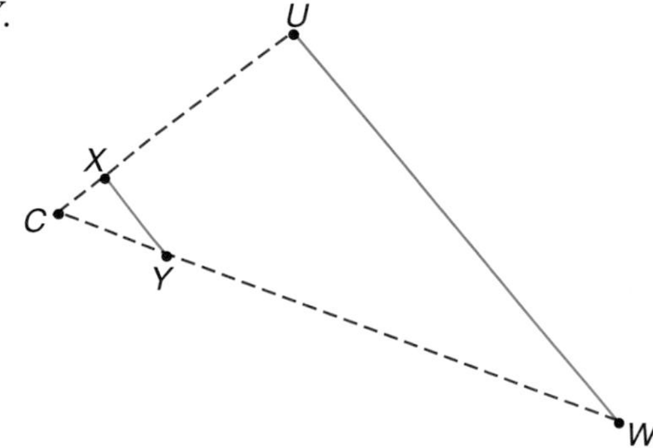

$CU = 5(CX)$

$CW = 5(CY)$

$\overline{UW}$ is the dilation of $\overline{XY}$ with scale factor 5.

In Chapter 7, you learned about scale factors of similar figures. Dilations form similar figures. If you know the measurements of an image and its dilation, you can determine the scale factor.

Example 2 **Determine the scale factor used for each dilation with center *C*. The dilation image in the figure is red.**

a. 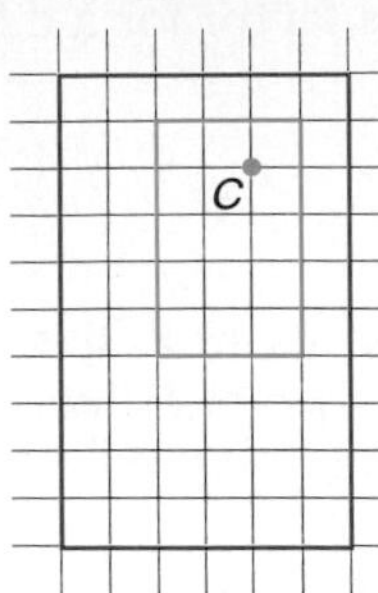

b. 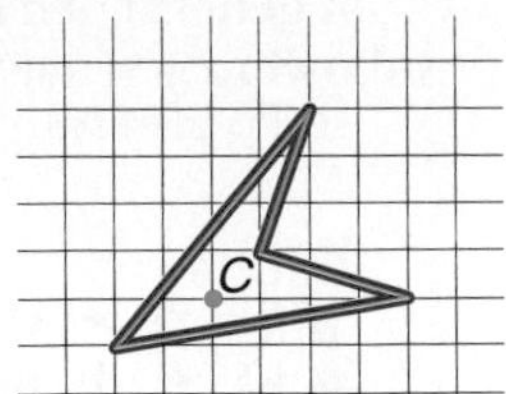

c. 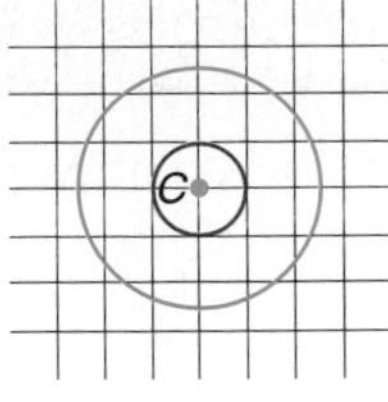

Count the grids to determine sizes of each figure and its dilation image. Compare these values to obtain a scale factor.

a. The dilation is an enlargement of the preimage. The ratio of the measures of the dilation to the original is 6:3 so the scale factor is 2.

b. The dilation is the same size as the preimage so it is a congruence transformation. The scale factor is 1.

c. The dilation is a reduction of the preimage. The ratio of the diameters of the dilation to the original is 1:2.5. The scale factor is $\frac{1}{2.5}$ or $\frac{2}{5}$.

You can also use algebraic skills to find the scale factor of a dilation.

Example 3 

Algebra

The dilation image for trapezoid *ABCD* with center *G* is shown in the coordinate plane below. Compare the length of $\overline{AB}$ with the length of its image to find the scale factor of the dilation.

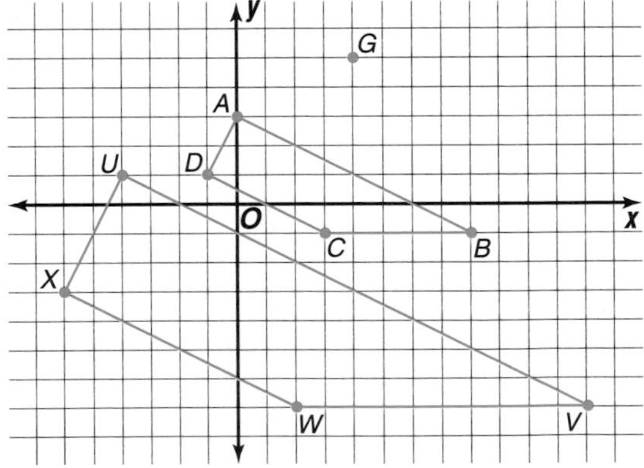

Use the distance formula to find the lengths.

$$AB = \sqrt{(3 + 1)^2 + (0 - 8)^2} \text{ or } 4\sqrt{5}$$

$$UV = \sqrt{(1 + 7)^2 + (-4 - 12)^2} \text{ or } 8\sqrt{5}$$

$\overline{UV}$, the image, is 2 times as long as the preimage $\overline{AB}$. The scale factor is $\frac{8\sqrt{5}}{4\sqrt{5}}$ or 2.

Example 4

Real World APPLICATION

Scale Drawing

Maria had to make a scale drawing of her bedroom for a class project. If her bedroom is 16 feet long and 12 feet wide, what scale factor can she use to produce her drawing on an $8\frac{1}{2}$-inch by 11-inch piece of construction paper? What would be the size of this scale drawing?

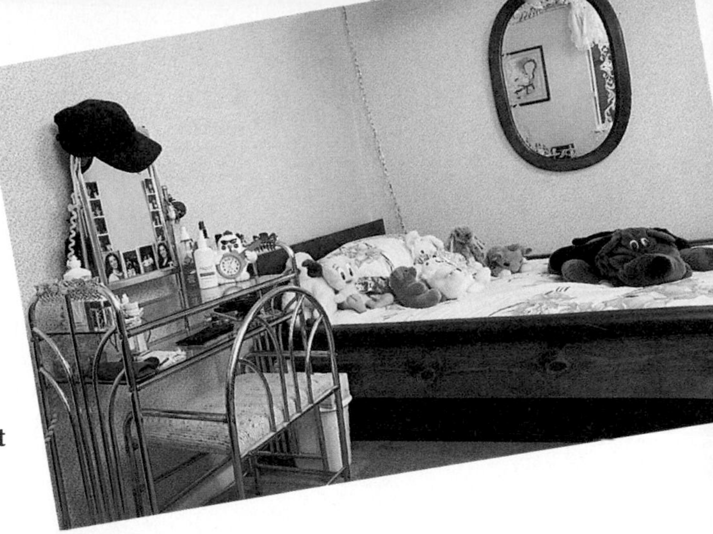

The scale factor Maria uses for the reduction must allow the drawing to fit on the paper. Even though there are many scales she could use, Maria decides to let 1 inch represent 2 feet. So, the scale factor is $\frac{1 \text{ inch}}{2 \text{ feet}}$. The size of her drawing will be 16 feet $\times \frac{1 \text{ inch}}{2 \text{ feet}}$ or 8 inches in length and 12 feet $\times \frac{1 \text{ inch}}{2 \text{ feet}}$ or 6 inches in width.

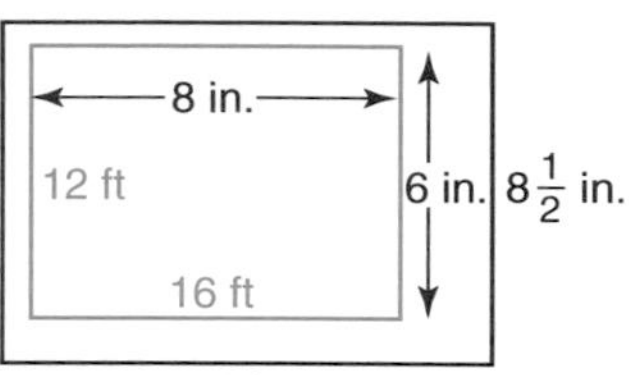

CHECK FOR UNDERSTANDING

Communicating Mathematics

Study the lesson. Then complete the following.

1. **Explain** the difference between a dilation and other transformations that you have studied.
2. **Identify** how you can determine if a dilation is a reduction or an enlargement.
3. **Explain** how a negative value for k affects the dilation image.
4. **Discuss** why a dilation is also called a similarity transformation.

5. **Assess Yourself** Which of the transformations studied do you find the easiest to perform? Which do you find are the most difficult? Explain why.

Guided Practice

In the figure, $\triangle XYZ$ is a dilation image of $\triangle ABC$ with a scale factor of 8. Complete.

6. If $QB = 6$, then $QY =$ __?__.
7. BC __?__ YZ
8. If $XY = 32$, then $AB =$ __?__.
9. If $m\angle BCA = 62$, then $m\angle YZX =$ __?__.
10. $\triangle ABC$ is __?__ to $\triangle XYZ$.

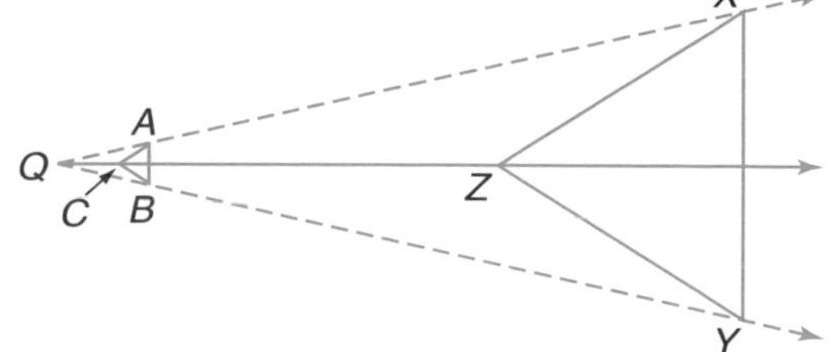

For each scale factor, determine if the dilation is an *enlargement*, a *reduction*, or a *congruence transformation*.

11. 1.00
12. $5\frac{2}{3}$
13. -0.75
14. $\frac{3}{5}$

On a coordinate plane, graph the segment whose endpoints are given. Using the origin as the center of dilation and a scale factor of 2, draw the dilation image. Repeat using a scale factor of $\frac{1}{2}$.

15. $A(3, -3), B(-2, -2)$

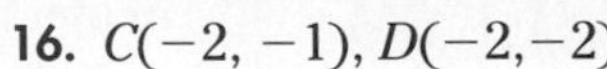

16. $C(-2, -1), D(-2, -2)$

17. For the given figure at the right, a dilation with center C produced the figure in red. What is the scale factor for this transformation?

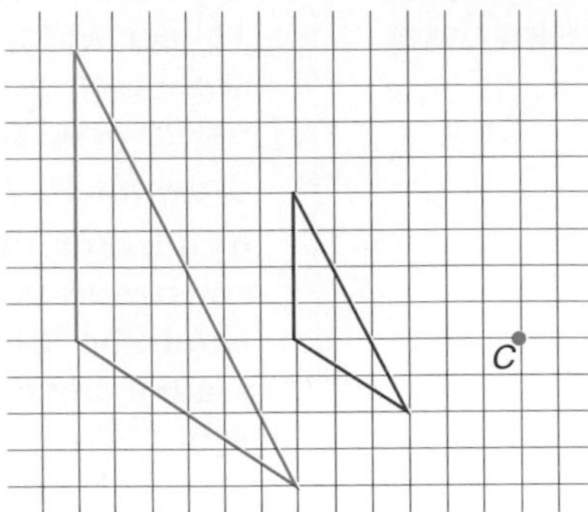

18. **Publishing** Tara is doing a layout for a new book. She has a drawing that is 12 centimeters by 9 centimeters, but the maximum space available for the drawing is 8 centimeters by 6 centimeters. If she wants the drawing to be as large as possible in the book, what scale factor should she use to reduce the original drawing?

EXERCISES

Practice

For each scale factor, find the image of *A* with respect to a dilation with center *P*.

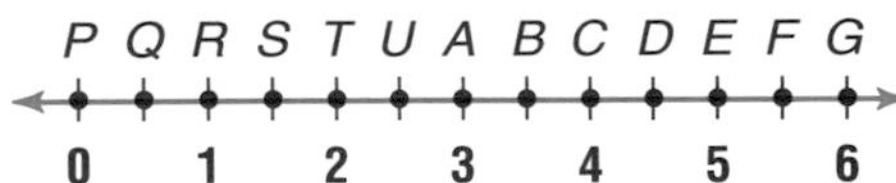

19. $1\frac{1}{6}$ 20. $\frac{2}{3}$ 21. $\frac{1}{2}$ 22. $1\frac{5}{6}$

Find the measure of the dilation image of $\overline{AB}$ with the given scale factor.

23. $AB = 5, k = -6$
24. $AB = \frac{2}{3}, k = \frac{1}{2}$
25. $AB = 16, k = 1.5$
26. $AB = 12, k = \frac{1}{4}$
27. $AB = 3, k = -1$
28. $AB = 3.1, k = -5$

A dilation with center *C* and a scale factor of *k* maps *A* onto *D* and *B* onto *E*. Find $|k|$ for each dilation. Then determine whether each dilation is an *enlargement*, a *reduction*, or a *congruence transformation*.

29. $CD = 10, CA = 5$
30. $CB = 6, CE = 4$
31. $AB = 7, DE = 7$
32. $DE = 12, AB = 4$
33. $CE = 7, CB = 28$
34. $CE = 18, CB = 27$

For each figure, a dilation with center *C* produced the figure in red. What is the scale factor for each transformation?

35.

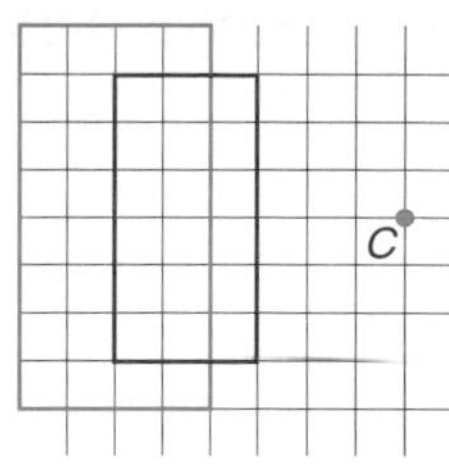

36.

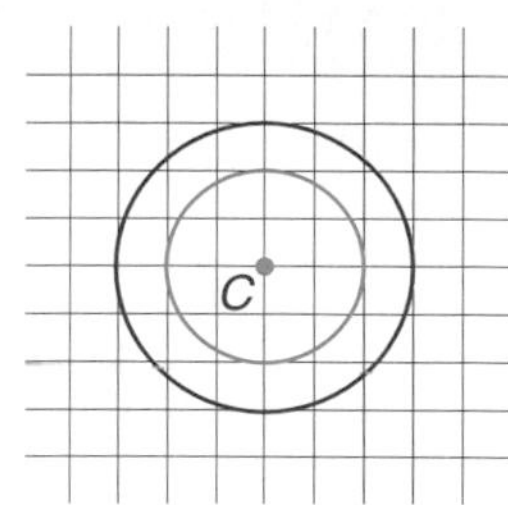

37.

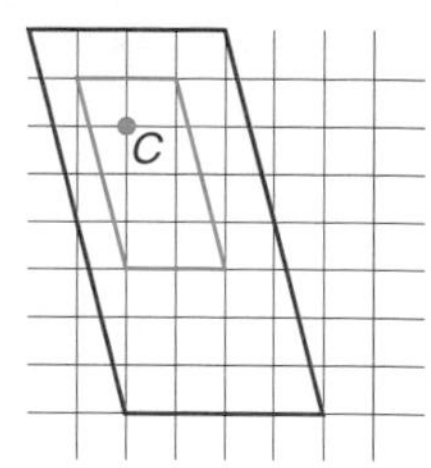

On a coordinate plane, graph the polygons whose vertices are given. Using the origin as the center of dilation and a scale factor of 2, draw the dilation image. Repeat using a scale factor of $\frac{1}{2}$.

38. $A(3, 4)$, $B(6, 10)$, $C(-3, 5)$

39. $D(6, 5)$, $E(4, 5)$, $F(3, 7)$

40. $G(-1, 4)$, $H(0, 1)$, $I(2, 3)$

41. $J(1, -2)$, $K(4, -3)$, $L(6, -1)$

42. $M(1, 2)$, $N(3, 3)$, $O(3, 5)$, $P(1, 4)$

43. $Q(4, 2)$, $R(-4, 6)$, $S(-6, -8)$, $T(6, -10)$

44. A dilation of a rectangle has a scale factor of 4.

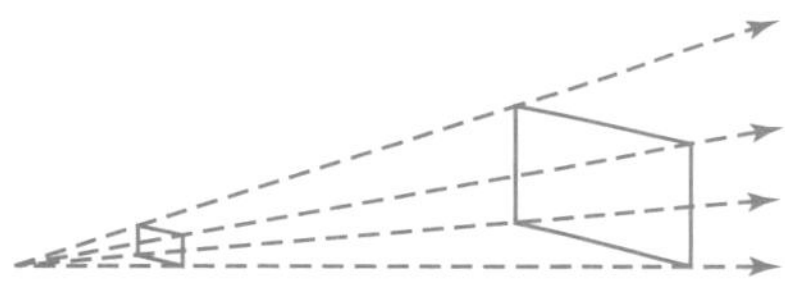

a. What is the effect of the dilation on the perimeter of the rectangle?
b. What is the effect of the dilation on the area of the rectangle?

45. A dilation of a cube has a scale factor of 3.

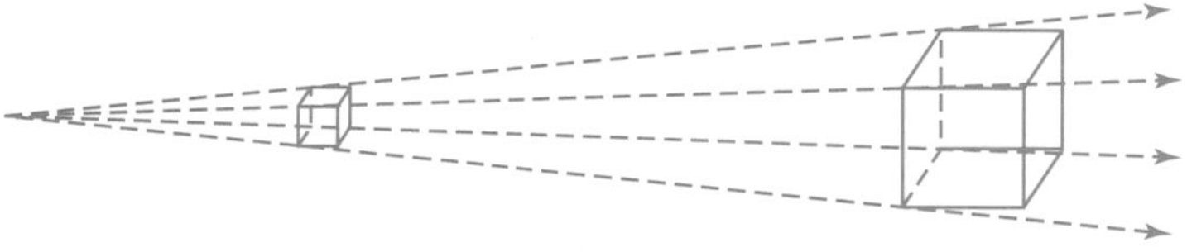

a. What is the effect of the dilation on the surface area of the cube?
b. What is the effect of the dilation on the volume of the cube?

Proof

46. Write a paragraph proof for *If a dilation with center C and a scale factor k maps A onto E and B onto D, then ED = k(AB).* (Theorem 13–1)

Critical Thinking

47. Graph $\triangle ABC$ with vertices $A(3, 4)$, $B(4, 3)$, and $C(2, 1)$ and $\triangle RST$ with vertices $R(7.5, 2.5)$, $S(10, 0)$, and $T(5, -5)$. If $\triangle RST$ is the dilation image of $\triangle ABC$, find the coordinates of the center and the scale factor.

Applications and Problem Solving

48. Photography An 8-inch by 10-inch photograph is being reduced by a scale factor of $\frac{3}{4}$. What are the dimensions of the new photograph?

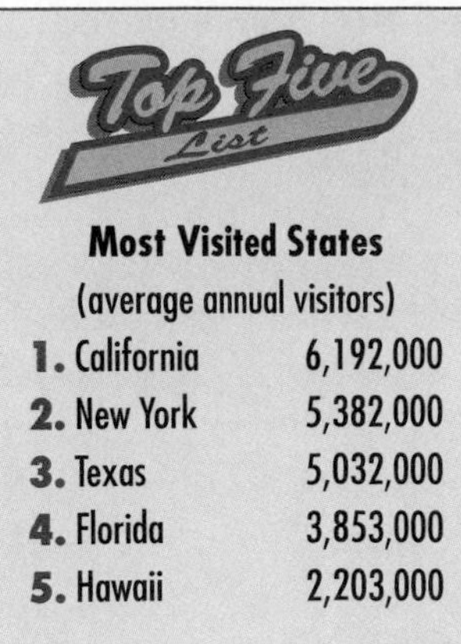

49. Travel In trying to calculate how far she must travel for an appointment, Gunja measured the distance between Richmond, Virginia, and Charlotte, North Carolina, on a map. The distance on the map was 2.25 inches, and the scale factor was 1 inch equals 150 miles. How far must she travel?

50. Art The picture below is Mauritz C. Escher's *Fish and Scales*.

a. Describe how Escher has used dilations in this piece of art.

b. How are tessellation images used in this piece of art?

Mixed Review

51. Draw and label $\triangle EFG$. Draw the dilation image of $\triangle EFG$ with a scale factor of $\frac{1}{2}$ and center C. (Lesson 13–8)

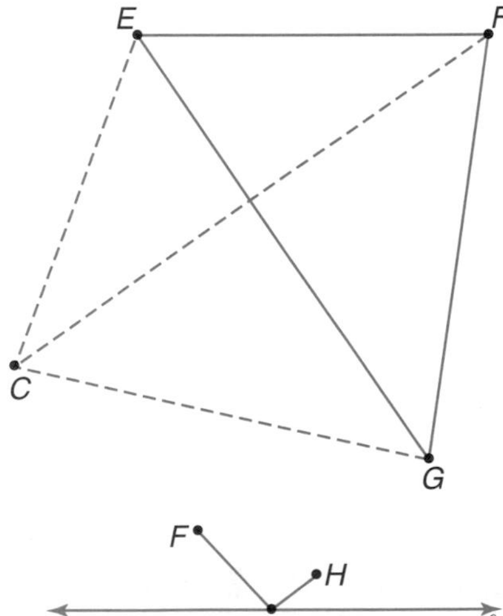

52. What is the measure of the angle of rotation if two lines intersect to form a $63°$ angle? (Lesson 13–7)

53. Copy the figure at the right. Then find the translation image of the geometric figure with respect to parallel lines ℓ and m. (Lesson 13–6)

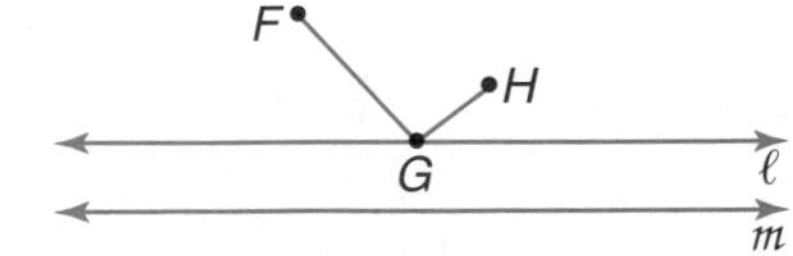

54. Describe the procedure for determining locus. (Lesson 13–1)

55. Determine the distance between points $L(-2, 3, 7)$ and $M(5, 6, -4)$. (Lesson 12–6)

56. **Water Management** A water tank holds 17,650 gallons of water. (Lesson 12–2)

a. If it is losing 895 gallons per hour, write an equation to describe the number of gallons left after x hours.

b. After how many hours will the tank be empty?

57. ACT Practice In the figure, if point B lies on the perpendicular bisector of $\overline{AC}$, what is the area of $\triangle ABC$?

A 15 **B** 30 **C** 50 **D** 1602

E Cannot be determined from the information given

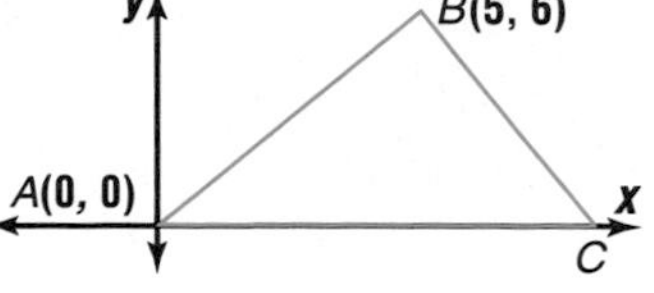

58. What is the measure of $\angle B$? Justify your answer. (Lesson 8–2)

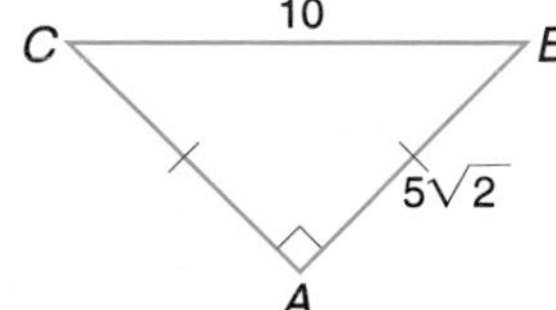

59. If $\overline{GJ}$ bisects $\angle HGK$, $GH = 9$, $GK = 12$, and $JK = 10$, find HJ. (Lesson 7–5)

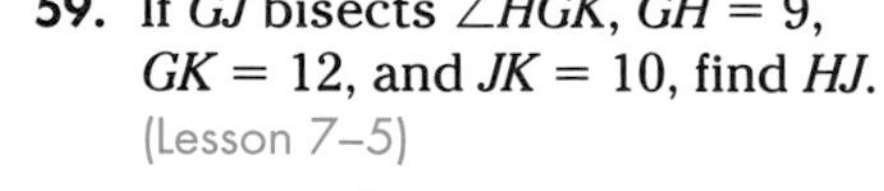
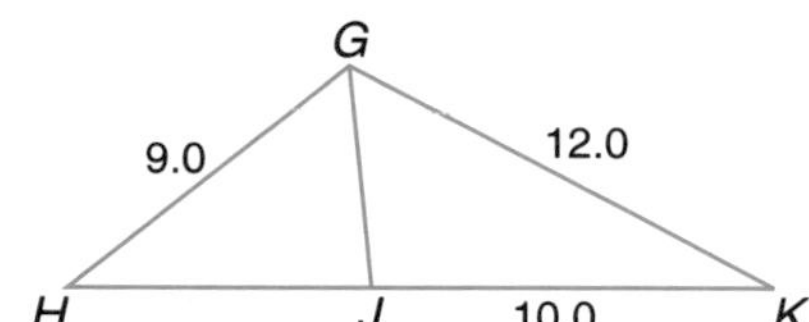

Algebra

60. Use elimination to solve the system of equations.

$3p + 4q = -5$
$5p - 12q = -27$

61. Probability For the dinner special at the Silver Inn, you can select one item from each of the categories at the right.

a. Draw a tree diagram showing all of the dinner special combinations.

b. What is the probability that a customer will select fish with a baked potato?

Entree	Potato	Vegetable
chicken fish	baked sweet fried	broccoli peas

For **Extra Practice**, see page 792.

WORKING ON THE

In·ves·ti·ga·tion

Refer to the Investigation on pages 642–643.

In the making of a movie about dinosaurs, small models of dinosaurs were used to produce the images of dinosaurs. Then, those images were enlarged and added to the scenes with the actors. Dilations can be used to create similar effects with graphic images.

1 Use a dilation to change the size and position of a drawing you created for your movie.

2 Write a sentence describing how the dilation affects the coordinates of the object being dilated in your movie.

3 Is there any part of the story that could be accomplished by using a dilation?

Add the results of your work to your Investigation Folder.

CLOSING THE

In·ves·ti·ga·tion

Refer to the Investigation on pages 642–643.

In 1995, the movie about a friendly ghost became the first movie to feature a character that was completely a digital image. In the near future, filmmakers believe we may see a movie produced with a digital human character. As the power of computers grows, making amazing special effects that were once impossible will become more and more simple.

Production

PORTFOLIO ASSESSMENT

You may want to keep your work on this Investigation in your portfolio.

1 Complete the story line for your movie.

2 If your movie will need sound, write a script or choose music.

3 Make a list of characters and background sets that must be designed. Then complete each design.

4 Use the images of the sets and characters to create at least fifteen images to make the movie. Bind the images into a book.

5 Flip the movie book to test the flow of the movie. Add or change images as needed to complete the production.

Screening

6 Choose another group of students in your class to represent the executives of your production company.

7 Write an introduction to the movie, choose a title, and create promotional posters and previews.

8 Show your movie to the executives. Present the promotional materials and previews.

9 Explain your technique for making the movie. Include any sketches or story research.

Data Collection and Comparison To share and compare your data with other students, visit: **www.geometry.glencoe.com**

CHAPTER 13 HIGHLIGHTS

interNET CONNECTION

Chapter Review For additional lesson-by-lesson review, visit: **www.geometry.glencoe.com**

VOCABULARY

After completing this chapter, you should be able to define each term, property, or phrase and give an example or two of each.

Geometry

angle of rotation (p. 740)
center of dilation (p. 746)
center of rotation (p. 739)
composite of reflections (p. 731)
congruence transformation (p. 716)
dilation (p. 746)
image (p. 715)
isometry (p. 716)
line of reflection (p. 722)
line of symmetry (p. 724)
locus (p. 696)
mapping (p. 715)
point of reflection (p. 722)
point of symmetry (p. 725)
preimage (p. 715)
reflection (p. 722)
rotation (p. 739)
scale factor (p. 746)
similarity transformation (p. 717)
transformation (p. 715)
translation (p. 731)

Algebra

system of equations (p. 704)

Problem Solving

look for a pattern (p. 741)
make a table (p. 717)

UNDERSTANDING AND USING THE VOCABULARY

State whether each sentence is *true* or *false*. If false, replace the underlined word to make a true sentence.

1. The <u>scale factor</u> in figure e is 2.
2. An <u>image</u> is a set of all points that satisfy a given set of conditions.
3. In figure c, $\triangle JKL$ demonstrates a <u>translation</u> image of $\triangle ABC$.
4. Figure b is an example of an <u>isometry</u>.
5. If quadrilateral $KLMN \to$ quadrilateral $UVWX$, the <u>preimage</u> of $\overline{LM}$ is $\overline{VW}$.
6. Figure d represents a <u>rotation</u>.
7. A <u>transformation</u> maps a preimage onto an image.
8. Figure a represents a <u>dilation</u>.
9. In figure c, $\triangle ABC$ is the <u>image</u> of $\triangle JKL$.
10. In figure c, $\triangle XYZ$ is a reflection over line ℓ of $\triangle ABC$.

a.

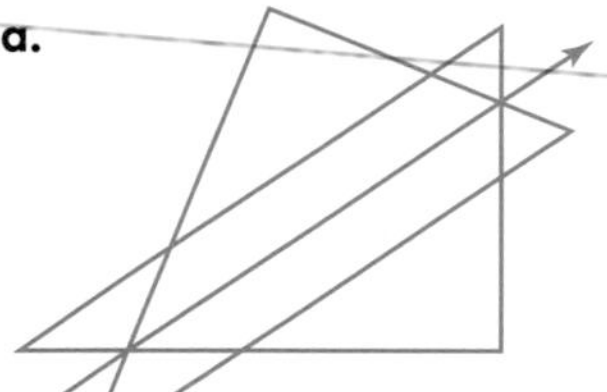

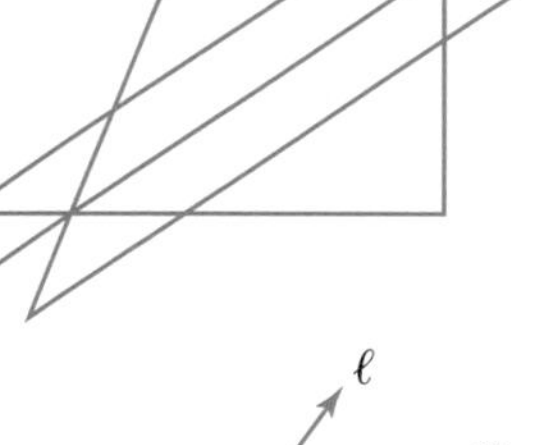

b.

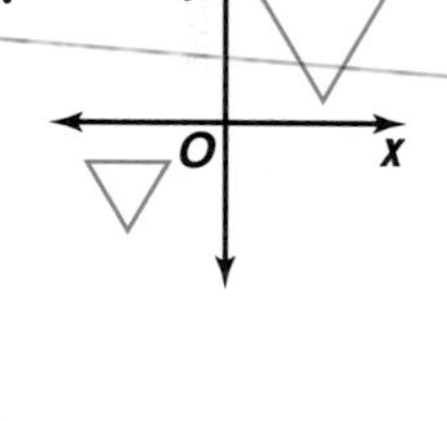

c.

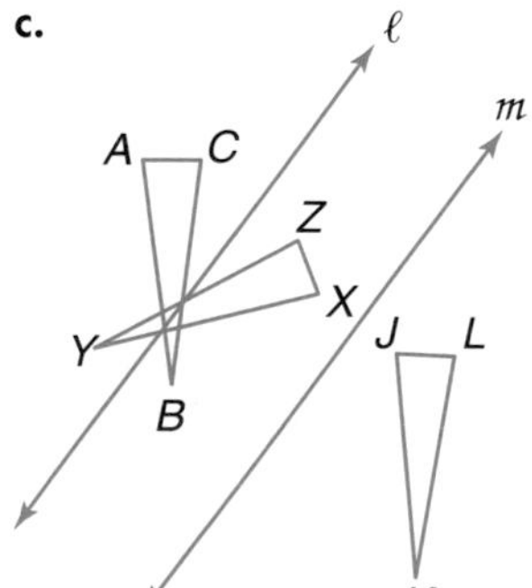

d.

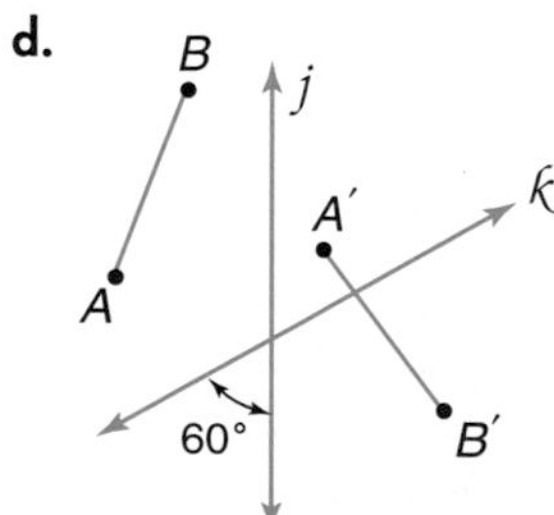

e.

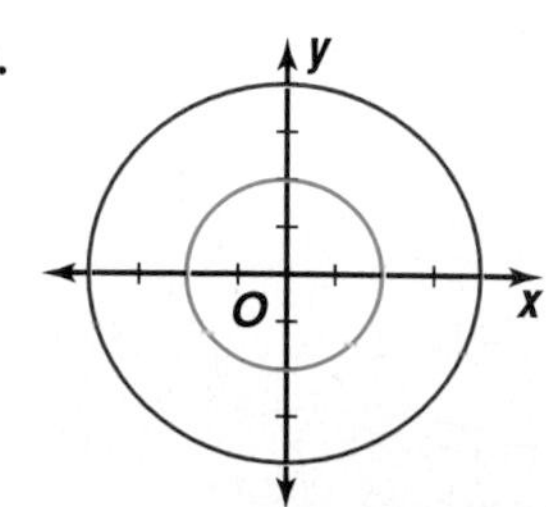

CHAPTER 13 STUDY GUIDE AND ASSESSMENT

SKILLS AND CONCEPTS

OBJECTIVES AND EXAMPLES

Upon completing this chapter, you should be able to:

- locate, draw, and describe a locus in a plane or in space (Lesson 13–1)

The locus of all points in a plane that are equidistant from the endpoints of a given line segment is the perpendicular bisector of the line segment.

The locus of all points in space that are equidistant from the endpoints of a given line segment is the plane that is the perpendicular bisector of the line segment.

REVIEW EXERCISES

Use these exercises to review and prepare for the chapter test.

Describe the locus of points for each set of conditions.

11. all points in a plane that are less than 6 centimeters from a given point
12. all points in space that are equidistant from two given points
13. all points in a plane that are equidistant from three noncollinear points A, B, and C

- find the locus of points that are solutions of a system of equations by graphing, by substitution, or by elimination (Lesson 13–2)

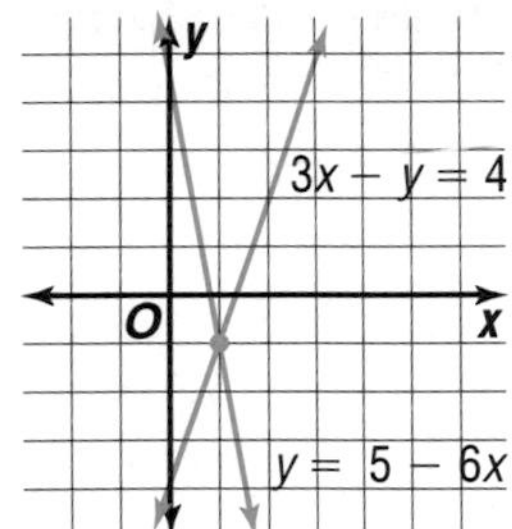

$3x - y = 4$

$y = 5 - 6x$

The locus of points that satisfy both graphs is $(1, -1)$.

Graph each pair of equations to find the locus of points that satisfy the graphs of both equations.

14. $y = x - 2$
    $2x + y = 13$

15. $3x - 4y = -1$
    $-2x + y = -1$

Use either substitution or elimination to find the locus of points that satisfy the graphs of both equations.

16. $y = 2x - 1$
    $x + y = 7$

17. $3x + y = 5$
    $2x + 3y = 8$

- solve locus problems that satisfy more than one condition (Lesson 13–3)

The locus of points in a plane 3 inches from a line and 4 inches from a point on the line is the four points where the parallel lines of one locus intersect the circle of the other.

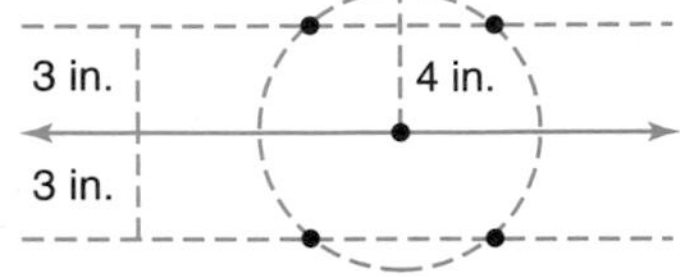

Draw a diagram to find the locus of points that satisfy each set of conditions. Then describe the locus.

18. all points in a plane equidistant from lines j and k, which intersect at point P, and are 2 centimeters from point P
19. all points in space that are 3 centimeters from the given points P and Q, which are 5 centimeters apart
20. all points in a plane equidistant from two parallel lines and also equidistant from two points on one of the lines

OBJECTIVES AND EXAMPLES

- name the image and preimage of a mapping (Lesson 13–4)

$\triangle RST \rightarrow \triangle R'S'T'$

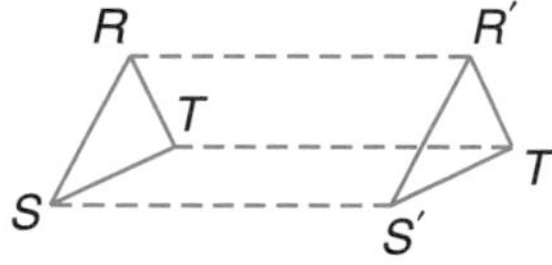

The image of $\angle S$ is $\angle S'$. The preimage of $\overline{R'T'}$ is $\overline{RT}$.

REVIEW EXERCISES

Suppose $\triangle ABE \rightarrow \triangle CBD$.

21. Name the preimage of D.
22. Name the image of $\angle ABE$.
23. Name the image of $\overline{AE}$.
24. Name the preimage of $\angle D$.
25. Describe the transformation that occurred in the mapping.

- draw reflection images, lines of symmetry, and points of symmetry (Lesson 13–5)

M is the reflection of X over line ℓ.

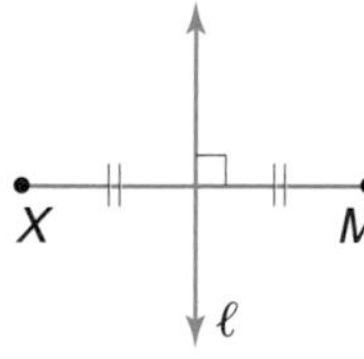

Copy each figure. Then draw the reflection image over line ℓ.

26.

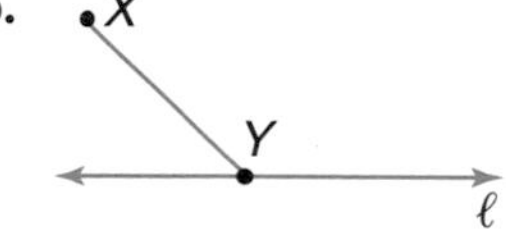

27.

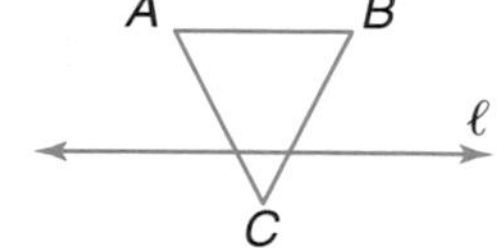

- name and draw translation images of figures with respect to parallel lines (Lesson 13–6)

$\triangle XYZ$ is the translation image of $\triangle ABC$ with respect to parallel lines ℓ and m.

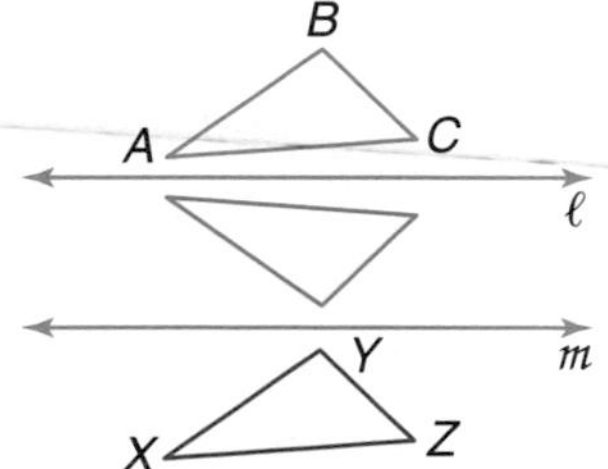

Copy each figure. Then find the translation image of each geometric figure with respect to parallel lines ℓ and m.

28.

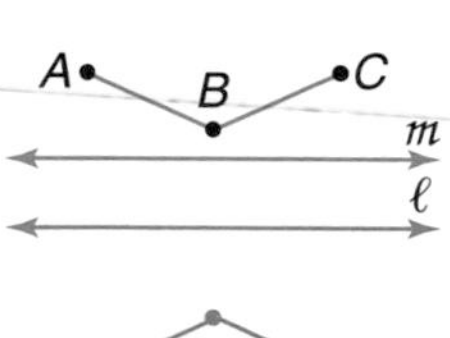

29.

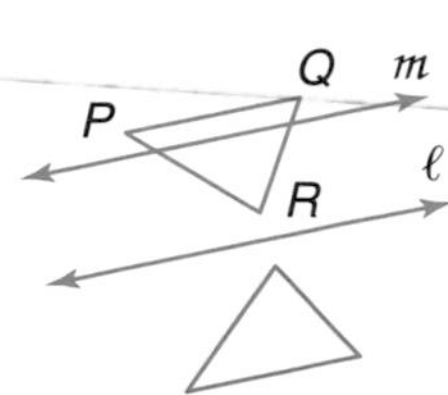

- name and draw rotation images of figures with respect to intersecting lines (Lesson 13–7)

$\triangle RST$ is the rotation image of $\triangle DEF$ with respect to lines r and t.

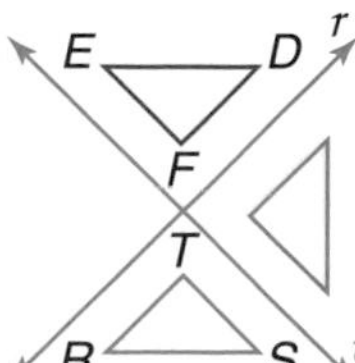

Refer to the figure below for Exercises 30–31.

30. Find the measure of the angle of rotation of the. rotation image with respect to lines t and m.
31. Copy the figure at the right. Then draw the rotation image with respect to lines t and m.

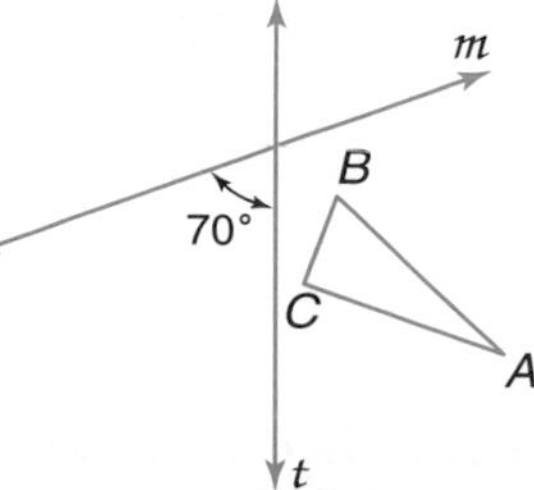

OBJECTIVES AND EXAMPLES

- find the dilation image for a given center and scale factor (Lesson 13–8)

$\triangle XYZ$ is the dilation image of $\triangle PQR$ with center A and scale factor of $\frac{1}{2}$. Since the absolute value of the scale factor is less than 1, the dilation is a reduction.

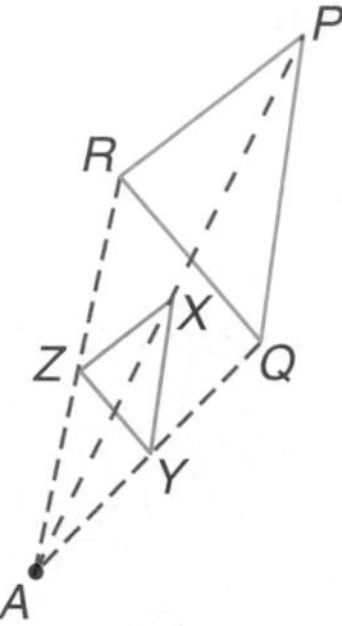

REVIEW EXERCISES

Refer to the figure at the right for Exercises 32 and 33.

32. Copy the figure at the right. Then draw the dilation image with a scale factor of 3 and center at P.

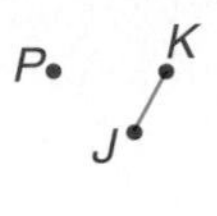

33. Is the image an *enlargement*, a *reduction*, or a *congruence transformation*? Explain your answer.

APPLICATIONS AND PROBLEM SOLVING

34. **Pet care** Perez has 2 dogs, Toby and Gretel. Part of each day, he ties the dogs in the backyard. Toby's 5-foot leash is tied to a long horizontal pole 4 feet above the ground so that it can slide along the pole. Gretel is tied to a stake with a 15-foot leash that can pivot around the stake. Explain how Perez should place the stake so that Toby and Gretel cannot tangle their leashes or get into a fight. (Lesson 13–3)

35. **Recreation** Dominique is playing billiards. How can he hit the 5 ball with the cue ball (all white) if the cue ball must first hit a side of the billiard table? Draw a diagram. (Lesson 13–5)

36. **Photography** Use the photo below to describe the transformation that it depicts. (Lesson 13–5)

37. **Manufacturing** Mr. Rex designs paddle fans for Functional Fans Inc. He designs one fan with 5 paddles. What is the measure of the angle of rotation if paddle A moves to the position of paddle B? (Lesson 13–6)

A practice test for Chapter 13 is provided on page 805.

ALTERNATIVE ASSESSMENT

COOPERATIVE LEARNING PROJECT

Motion pictures In this chapter, you became familiar with several types of transformations. Imagine you are out for a walk. With your eyes looking directly ahead, you notice that each point in your visual field is moving. In fact, each point is moving from the vanishing point straight ahead on the road, toward the periphery of your visual field. You take a turn to the right. What just happened to the points in your visual field? In this project, you will write transformations for this situation.

Follow these steps to write the transformations.

- Take a walk down the hallway. Take a photograph of the view ahead at intervals of five steps. (Drawings can be used instead of photos.) Walk in a straight line until you reach a turn in the hallway. Walk through the turn slowly, taking another photograph every two steps.
- Have the photographs developed and printed. On each photograph, place the origin of the coordinates at the vanishing point.
- For the sequence of photographs prior to reaching the turn, select one point in each quadrant and assign coordinates to the points. Write an equation for the transformation that describes the motion of the points away from the origin and toward the periphery.
- For the sequence of photographs of the walk through the turn, select a few points in the field of view. Write transformations for the turning sequence.

THINKING CRITICALLY

How many points are contained in a locus?

PORTFOLIO

Rent one of your favorite movies and view it. Find a sequence in the film where there are interesting camera angles and changes of camera angles. Write a short essay describing the sequence and the types of transformations for each camera movement. Keep this essay in your portfolio.

STANDARDIZED TEST PRACTICE

CHAPTERS 1–13

Section One: MULTIPLE CHOICE

There are eight multiple-choice questions in this section. After working each problem, write the letter of the correct answer on your paper.

1. Ten segments of equal length form the figure below. If the perimeter of $\triangle XYZ$ is eight inches, what is the perimeter of octagon *RSTUVWYZ*?

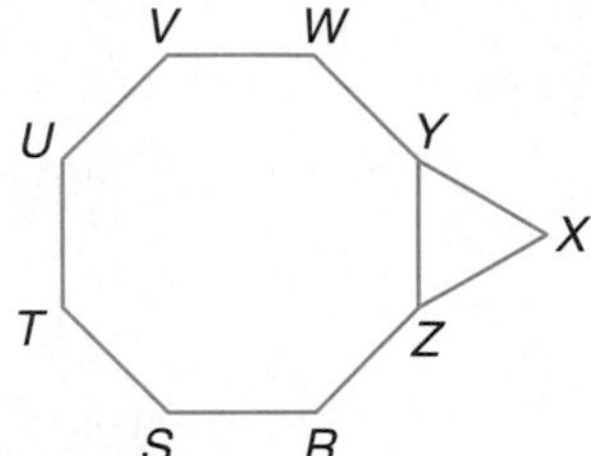

A. $16\frac{2}{3}$ inches

B. $21\frac{1}{3}$ inches

C. 16 inches

D. $18\frac{2}{3}$ inches

2. How many hours equal 282 minutes?

A. $4\frac{1}{4}$ hours

B. $4\frac{2}{3}$ hours

C. $4\frac{4}{5}$ hours

D. $4\frac{7}{10}$ hours

3. Three vertices of a parallelogram are at (2, 1), (−1, −3), and (6, 4). Which of the following could be the coordinates of the remaining vertex?

A. (0, 3) **B.** (3, 0)

C. (−1, 4) **D.** (6, 1)

4. Find the value of p if the ratio of p to q is 2 to 3 and the ratio of q to r is 6 to 11.

A. 4 **B.** −4

C. 8 **D.** −8

5. 8(624) + 624 is equivalent to which of the following?

A. 5(624) + 3(624) **B.** 5(624) + 4(624)

C. 4(624) + 6(624) **D.** 3(624) + 9(624)

6. Which of the following represents the solution set of the inequality $-3 - x < 2x < 9 - x$?

A. $\{x \mid -1 < x < 3\}$

B. $\{x \mid 1 < x < 3\}$

C. $\{x \mid -3 < x < 1\}$

D. $\{x \mid -3 < x < -1\}$

7. Solve $\frac{a}{3a + 6} - \frac{a}{5a + 10} = \frac{2}{5}$.

A. −2 **B.** −3

C. −3 and −2 **D.** 3 and 2

8. Find the decimal form of $\frac{2}{1000} + \frac{3}{10} + \frac{4}{100}$.

A. 0.342 **B.** 0.432

C. 0.234 **D.** 0.243

Section Two: SHORT ANSWER

This section contains seven questions for which you will provide short answers. Write your answer on your paper.

9. One angle of a triangle measures 68°. The measures of the other two angles are in a 1 to 3 ratio. Find the measures of the other two angles.

10. If the center of square *ABCD* is at the origin and the sides of the square are parallel to the *x*- and *y*-axes, find the position of the four vertices if the square is rotated 180° about the *y*-axis and then rotated 180° about the *x*-axis.

Test-Taking Tip There are only four answer choices, A, B, C, and D, on a quantitative comparison question. But an SAT answer sheet has a bubble for choice E. Make sure that you don't mark choice E when you mean to choose D.

11. Simplify $\frac{3}{4 - x} \div \frac{x}{x^2 - 16}$.

12. Find the maximum number of common tangents that can be drawn to any two circles.

13. Find the solution of $\sqrt{y^2 - 4y + 9} = y - 1$.

14. Find x if m is parallel to n.

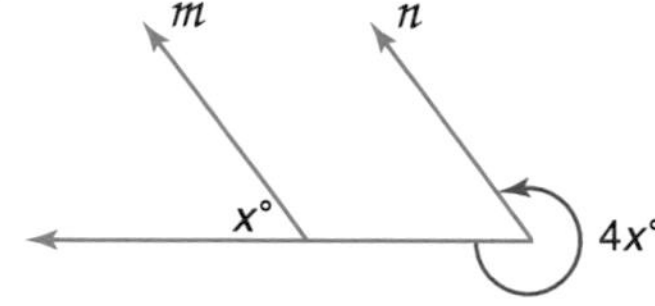

15. In the following diagram, $MN = NO = OP = PQ = QR = 1$. Find the length of $\overline{MR}$.

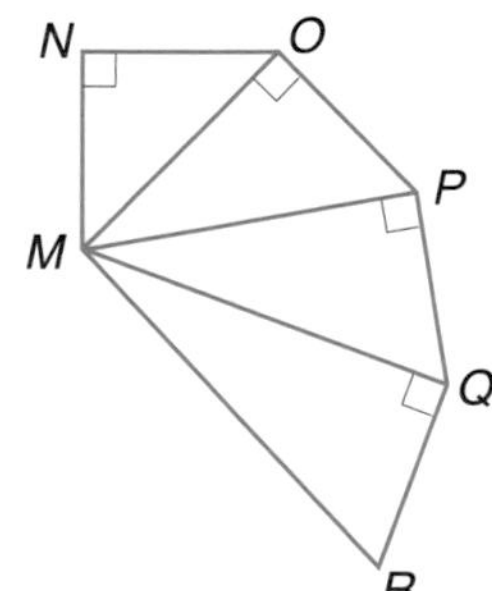

Test Practice For additional test practice questions, visit: **www.geometry.glencoe.com**

Section Three: COMPARISON

This section contains five comparison problems that involve comparing two quantities, one in column A and one in column B. In certain questions, information related to one or both quantities is centered above them. All variables used represent real numbers.

Compare quantities A and B below.

- Write A if quantity A is greater.
- Write B if quantity B is greater.
- Write C if the two quantities are equal.
- Write D if there is not enough information to determine the relationship.

	Column A	Column B
16.	$\sqrt{3^4}$	9
17.	a given chord in a given circle	the radius of the same circle

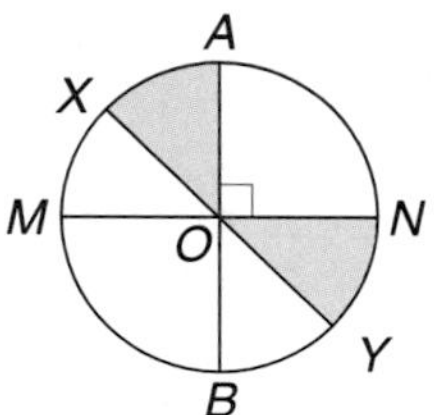

O is the center of a circle with $\overline{MN}$, $\overline{AB}$, and $\overline{XY}$ as diameters,

	Column A	Column B
18.	30% of the circle's area	shaded part of the circle

$a @ b$ is defined to be $\frac{b}{a - b}$ and $a \neq b$.

	Column A	Column B
19.	$a @ b$	$b @ a$

$Q = \{1, 2, 3, 4, 5, 6, 7, 8, 9\}$

	Column A	Column B
20.	the product of the odd integers in Q	the product of the even integers in Q

STUDENT HANDBOOK

For Additional Assessment

For Reference

EXTRA PRACTICE

EXTRA PRACTICE

Lesson 1-1 **Write the ordered pair for each point shown at the right.**

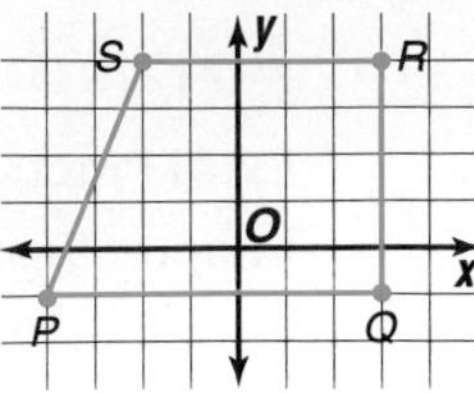

1. P
2. Q
3. R
4. S

Graph each point on the same coordinate plane.

5. $D(-4, 3)$
6. $E(-3, -3)$
7. $F(0, 4)$

Points $R(-1, -5)$ and $S(3, 3)$ lie on the graph of $y = 2x - 3$. Determine whether each point is collinear with R and S.

8. $T(0, -3)$
9. $U(-0.5, -4)$
10. $V(2, -1)$

Lesson 1-2 **Refer to the figure at the right to name each of the following.**

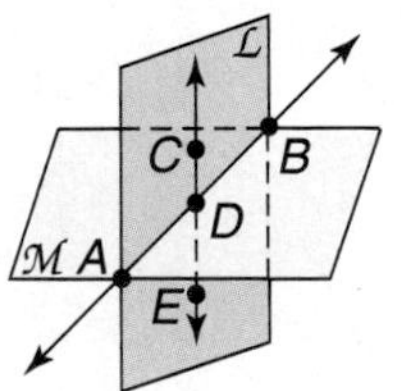

1. the intersection of planes $\mathcal{M}$ and $\mathcal{L}$
2. a point collinear with points C and E
3. a plane containing line $\overleftrightarrow{CE}$

Draw and label a figure for each relationship.

4. Lines $\overleftrightarrow{AB}$ and ℓ both contain point X.
5. Lines $\overleftrightarrow{PQ}$ and m, and plane $\mathcal{K}$ intersect at T.

Refer to figure at the right to answer each question.

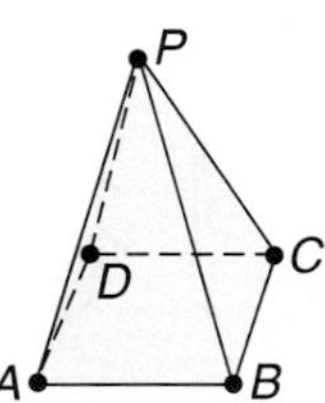

6. The flat surfaces of the figure are called *faces*. Name the five planes that contain the faces of the pyramid.
7. Name three lines that intersect at B.

Lesson 1-3 **Find the perimeter and area of each rectangle.**

1. 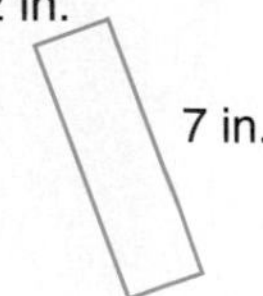

2. 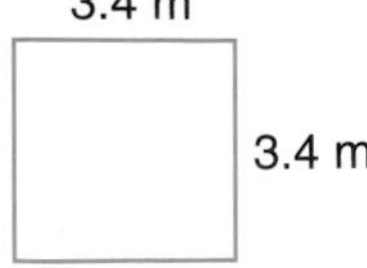

3. 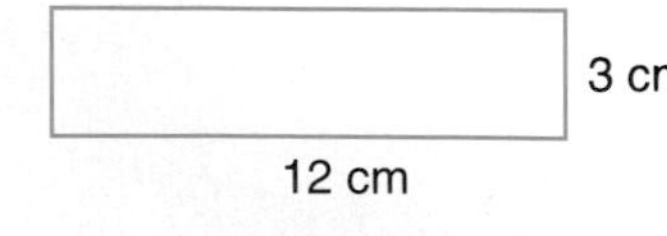

Find the missing measure in each formula if $P = 2\ell + 2w$ and $A = \ell w$.

4. $\ell = 6, w = 4.5, A = ?$
5. $\ell = 12, w = 8, P = ?$
6. $P = 17, \ell = 5.5, w = ?$
7. $A = 91, w = 7, \ell = ?$

Find the maximum area for the given perimeter of a rectangle. State the length and width of the rectangle.

8. 36 cm
9. 68 in.
10. 118 yd

Lesson 1-4 Refer to the number line below to find each measure.

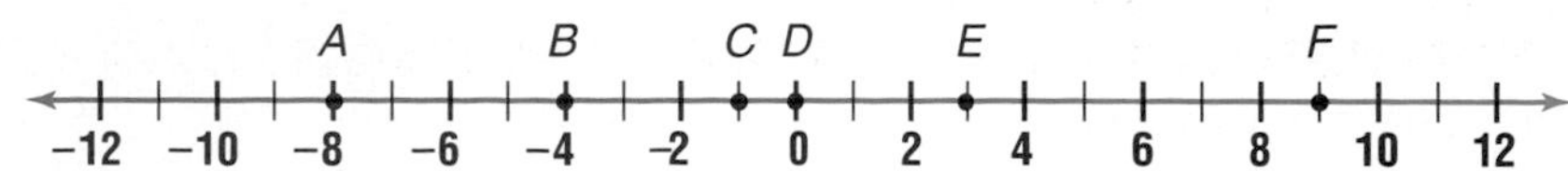

1. CD
2. BF
3. CF
4. EB

Refer to the coordinate plane at the right to find each measure. Round your answers to the nearest hundredth.

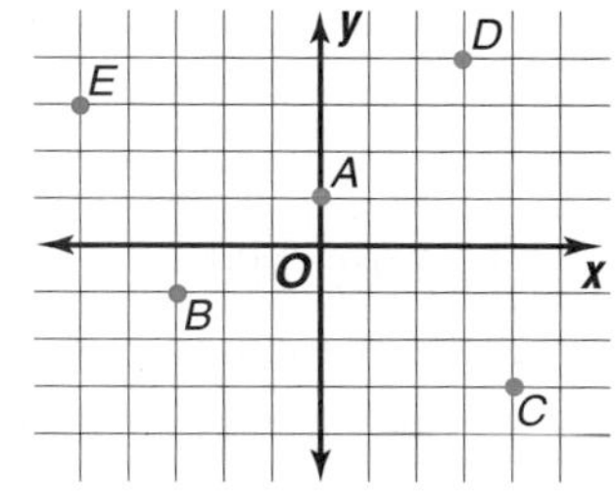

5. BA
6. ED
7. AC
8. CD

Use the Pythagorean Theorem to find the missing length x in each right triangle.

9.  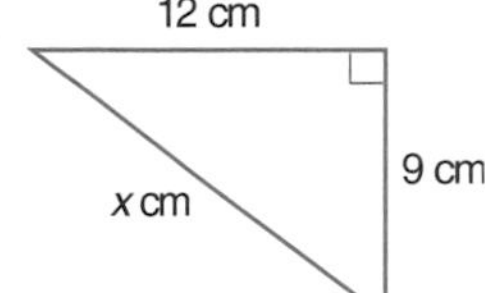

10. 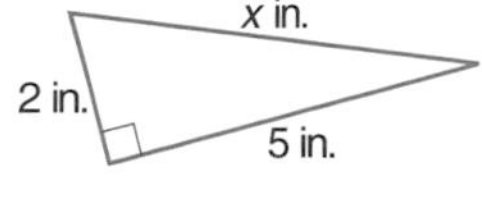

11. 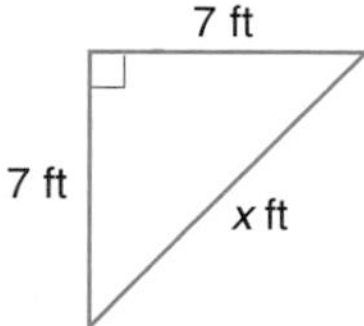

Lesson 1-5 Refer to the number line at the right to find the coordinate of the midpoint of each segment.

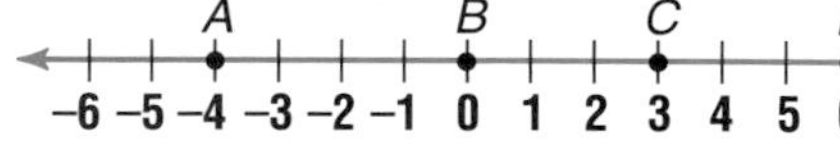

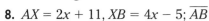

1. $\overline{BD}$
2. $\overline{AB}$
3. $\overline{BC}$

Given three points A, B, and M, and M is the midpoint of $\overline{AB}$, find the coordinates of the third point.

4. $A(-9, 4), B(3, -2)$
5. $M(-1, 8), B(2, 6)$
6. $A(-5, -4), M(1, -5)$
7. $B(7, 3), A(2, 4)$

In the figure at the right, $\overline{CX}$ bisects $\overline{AB}$ at X, and $\overleftrightarrow{CD}$ bisects $\overline{XB}$ at Y. For each of the following, find the value of x and the measure of the indicated segment.

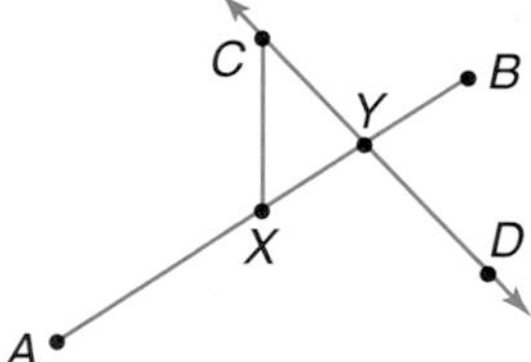

8. $AX = 2x + 11, XB = 4x - 5; \overline{AB}$
9. $YB = 23 - 2x, XY = 2x + 3; \overline{AB}$
10. $AB = 5x - 4, XY = x + 1; \overline{AX}$

Lesson 1-6 In the figure at the right, $\overrightarrow{RQ}$ and $\overrightarrow{RS}$ are opposite rays, and $\overrightarrow{RU}$ bisects $\angle QRT$.

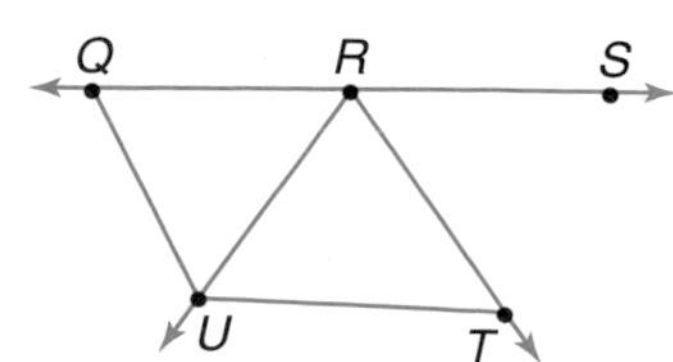

1. Name two congruent angles.
2. Name three angles that have U as a vertex.
3. List all the angles that have $\overrightarrow{RT}$ as a side.
4. If $m\angle SRT = 17$ and $m\angle TRU = 80$, what is the measure of $\angle SRU$?
5. If $m\angle QRU = 4x - 3$ and $m\angle TRU = 2x + 23$, find $m\angle QRU$.
6. Given that $m\angle SRU = 3y + 8$ and $m\angle URQ = 2y + 7$, find $m\angle URQ$.
7. Find $m\angle URT$ if $m\angle URQ = 3n + 11$ and $m\angle QRT = 9n - 14$.

EXTRA PRACTICE

Lesson 1-7 **Refer to the figure at the right for Exercises 1–6.**

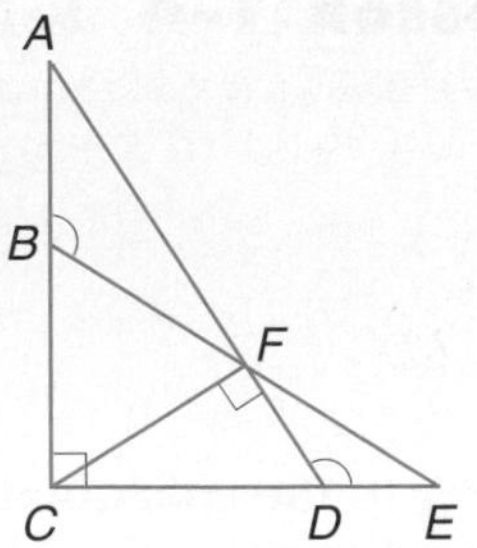

1. Name a pair of vertical angles.
2. Which angle is complementary to $\angle ECF$?
3. Name a pair of angles that are congruent and supplementary.
4. Identify $\angle ABF$ and $\angle CBF$ as adjacent, vertical, complementary, supplementary, or as a linear pair. List all possibilities.
5. Can you assume that $\angle ABF \cong \angle EDF$? Explain.
6. Does $\angle EDF$ appear to be obtuse, acute, or right?
7. The measure of an angle is one-third the measure of its supplement. Find the measure of the angle.
8. Suppose $\angle C$ and $\angle D$ are complementary. Find $m\angle C$ and $m\angle D$ if $m\angle C = 3x + 5$ and $m\angle D = 4x - 6$.

Lesson 2-1 **Determine if each conjecture is *true* or *false* based on the given information. Explain your answer.**

1. **Given:** $\overline{AB}$, $\overline{BC}$, and $\overline{CD}$

   Conjecture: A, B, C, and D are collinear.
2. **Given:** x is a real number.

   Conjecture: x^2 is a nonnegative number.
3. **Given:** $W(-2, 3)$, $X(1, 7)$, $Y(5, 4)$, and $Z(2, 0)$

   Conjecture: $WXYZ$ is a square.

Write a conjecture based on the given information. If appropriate, draw a figure to illustrate your conjecture.

4. **Given:** $\angle ABC$ and $\angle DBE$ are vertical angles.
5. **Given:** Lines ℓ and m intersect to form right angles.
6. **Given:** $\angle QRS$ and $\angle QRT$ form a linear pair.

Lesson 2-2 **Identify the hypothesis and conclusion of each conditional statement.**

1. If a container holds 32 ounces, then it holds a quart.
2. If a candy bar is a Caramel Crunch, then it contains caramel.

Write each conditional in if-then form.

3. Right angles are congruent.
4. A car has four wheels.
5. A triangle contains exactly three angles.

Write the converse, inverse, and contrapositive of each conditional. Determine if the converse, inverse, and contrapositive are *true* or *false*. If false, give a counterexample.

6. If the distance of a race is 10 kilometers, then it is about 6.2 miles.
7. An apple is a fruit.

Lesson 2-3 **Determine if statement (3) follows from statements (1) and (2) by the Law of Detachment or the Law of Syllogism. If it does, state which law was used. If it does not, write *invalid*.**

1. (1) All pilots must pass a physical examination.
   (2) Kris Thomas must pass a physical examination.
   (3) Kris Thomas is a pilot.
2. (1) If a student is enrolled at Lyons High, then the student has an ID number.
   (2) Joel Nathan is enrolled at Lyons High.
   (3) Joel Nathan has an ID number.

Determine if a valid conclusion can be reached from the two true statements using the Law of Detachment or the Law of Syllogism. If a valid conclusion is possible, state it and the law that is used. If a valid conclusion does not follow, write *no conclusion*.

3. (1) Basalt is an igneous rock.
   (2) Igneous rocks were formed by volcanoes.
4. (1) If a quadrilateral is a rectangle, then it has four right angles.
   (2) A rectangle has diagonals that are congruent.

Lesson 2-4 **Name the property of equality that justifies each statement.**

1. If $AB + BC = AC$ and $AC = EF + GH$, then $AB + BC = EF + GH$.
2. If $x + y = 9$ and $x - y = 12$, then $2x = 21$.

Copy the proof. Then name the property that justifies each statement.

3. **Given:** $x - 1 = \frac{x - 10}{-2}$

   Prove: $x = 4$

Statements	Reasons
1. $x - 1 = \frac{x - 10}{-2}$	1. ?
2. $-2(x - 1) = x - 10$	2. ?
3. $-2x + 2 = x - 10$	3. ?
4. $12 = 3x$	4. ?
5. $4 = x$	5. ?
6. $x = 4$	6. ?

Lesson 2-5 **Justify each statement with a property from algebra or a property of congruent segments.**

1. If $2MN = TS$, then $MN = \frac{1}{2}TS$.
2. If $AN - 8 = IN - 8$, then $AN = IN$.
3. If $EF = GH$ and $GH = JK$, then $EF = JK$.

Write the given and prove statements you would use to prove the theorem. Draw a figure if applicable.

4. If two angles are vertical, then they are congruent.
5. The diagonals of a rectangle are congruent.

Lesson 2-6 **Complete each statement with *always*, *sometimes*, or *never*.**

1. Congruent angles are __?__ vertical angles.
2. If two angles form a linear pair, then they are __?__ supplementary.
3. If two lines are perpendicular, they __?__ form acute angles.

Refer to the figure at the right to answer each question.

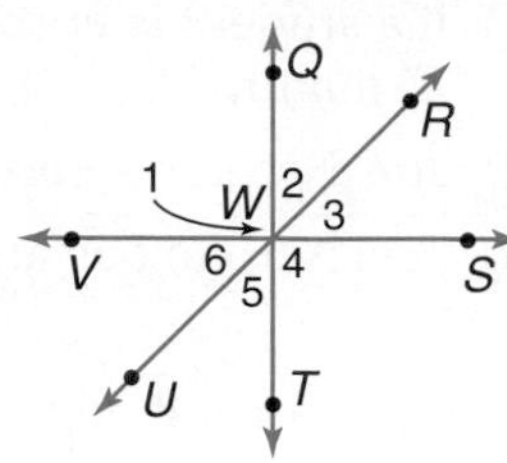

4. Name three pairs of vertical angles.
5. Name eight pairs of angles that form a linear pair.
6. If $m\angle 1 = 4x + 14$ and $m\angle 4 = 3x + 33$, find the value of x and $m\angle 4$.
7. $\angle 2$ and $\angle 3$ are complementary. If $m\angle 2 = 2x - 14$ and $m\angle 3 = x + 17$, find $m\angle 3$.
8. If $m\angle 5 = 3x - 7$ and $m\angle UWQ = 8x$, find $m\angle UWQ$.

Lesson 3-1 **Describe each of the following as *intersecting*, *parallel*, or *skew*.**

1. a floor and a ceiling
2. plaid fabric
3. flag pole in a football stadium and the 50-yard line on the field
4. train tracks

Determine whether each statement is *true* or *false*, Explain Your reasoning

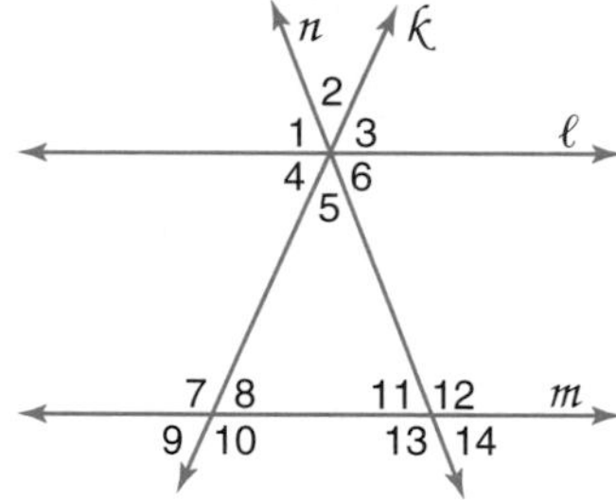

5. $\angle 6$ and $\angle 11$ are alternate interior angles.
6. $\angle 4$ and $\angle 9$ are alternate exterior angles.
7. $\angle 7$ and $\angle 11$ are corresponding angles.

State the transversal that forms each pair of angles. Then identify the special angle name for each pair of angles.

8. $\angle 3$ and $\angle 8$
9. $\angle 14$ and $\angle 7$
10. $\angle 8$ and $\angle 5$

Lesson 3-2 **In the figure, $m\angle 2 = 62$, $m\angle 1 = 41$, $\overline{XS} \parallel \overline{YT}$, and $\overline{SY} \parallel \overline{TZ}$. Find the measure of each angle.**

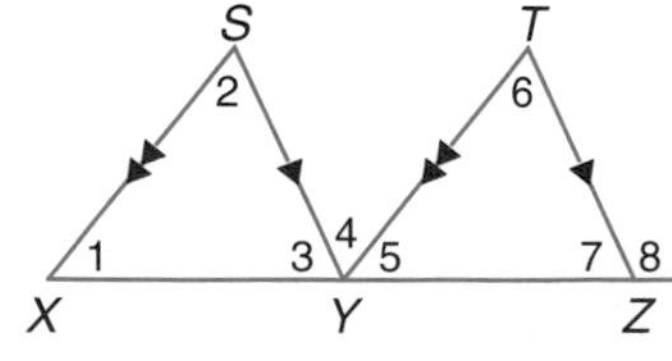

1. $\angle 4$
2. $\angle 3$
3. $\angle 5$
4. $\angle 6$
5. $\angle 7$
6. $\angle 8$

Find the values of x, y, and z in each figure.

7.

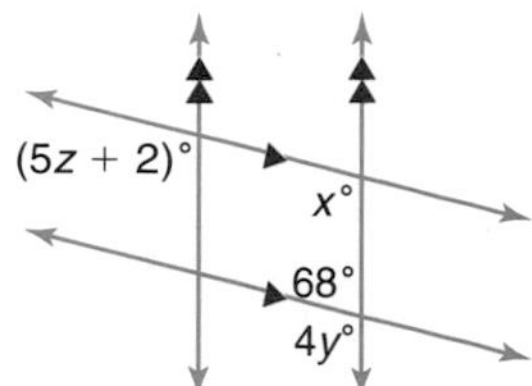

8.

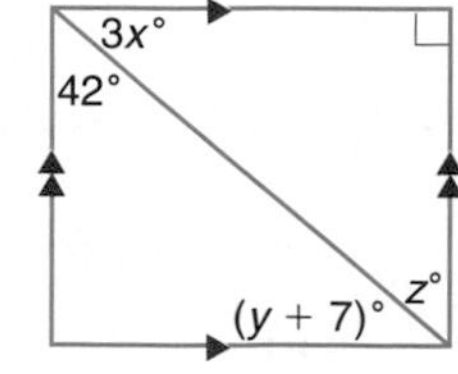

9.

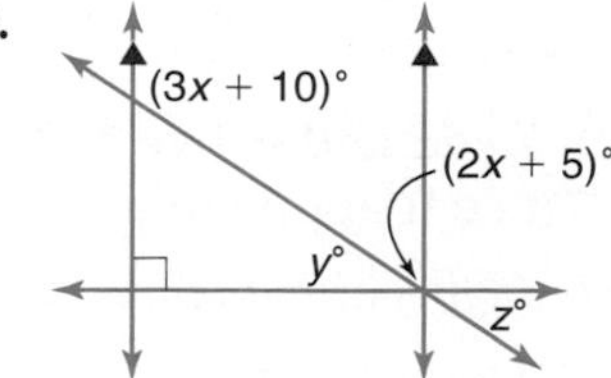

Lesson 3-3 **Determine the slope of each line named below.**

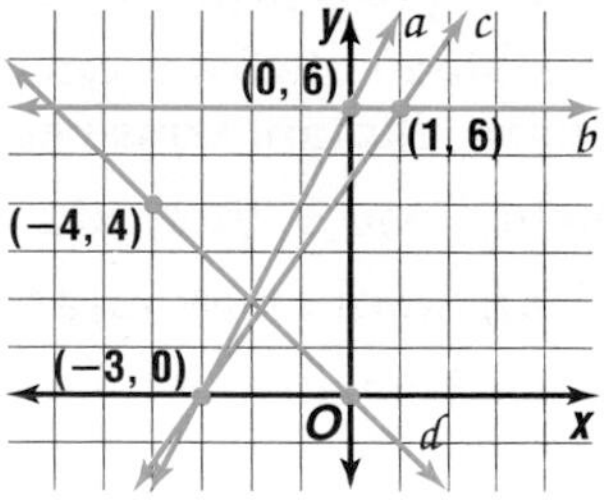

1. a
2. b
3. c
4. d
5. any line perpendicular to b
6. any line parallel to a
7. any line perpendicular to c

Graph the line that satisfies each description.

8. slope = 0, passes through $P(2, 6)$
9. slope = $\frac{2}{5}$, passes through $P(0, -2)$
10. passes through $P(2, 1)$ and is parallel to $\overline{AB}$ with $A(-2, 5)$ and $B(1, 8)$

Lesson 3-4 **Given the following information, determine which lines, if any, are parallel. State the postulate or theorem that justifies your answer.**

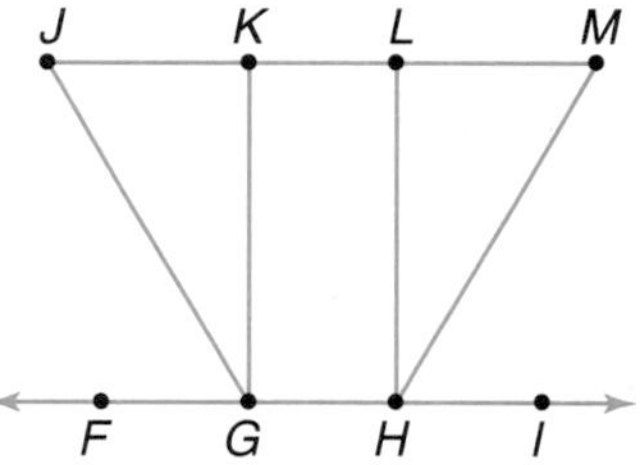

1. $\angle HLK \cong \angle GKJ$
2. $\angle IHL \cong \angle HLK$
3. $\angle FGJ \cong \angle KJG$
4. $m\angle GJK + m\angle HLK = 180$

Find the value of x so that $\ell \parallel m$.

5.

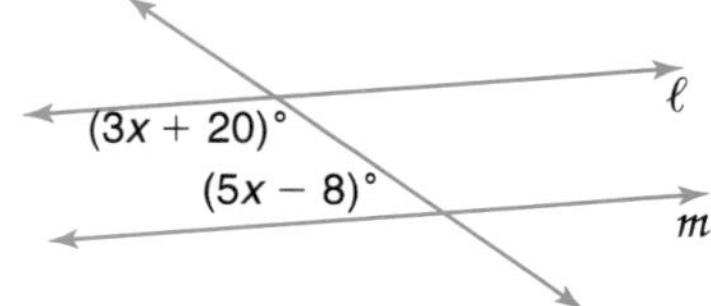

6.

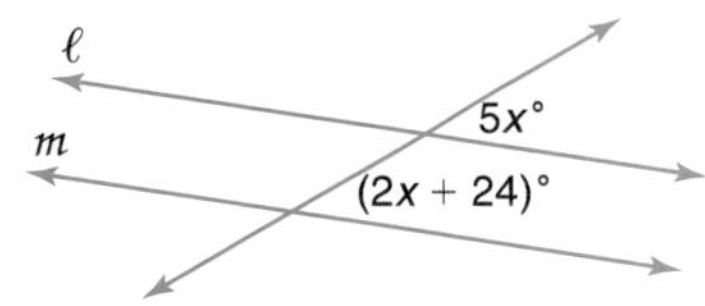

7.

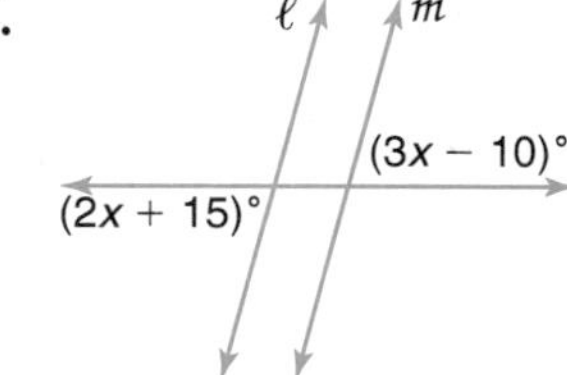

Lesson 3-5 **Copy each figure and draw the segment that represents the distance indicated.**

1. C to $\overline{DE}$

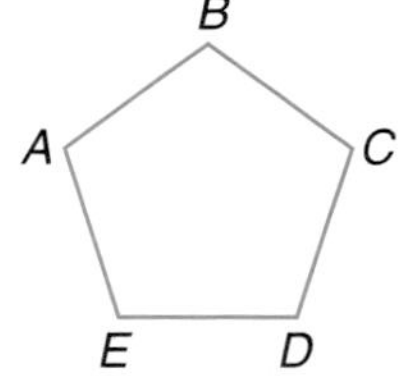

2. N to $\overline{MP}$

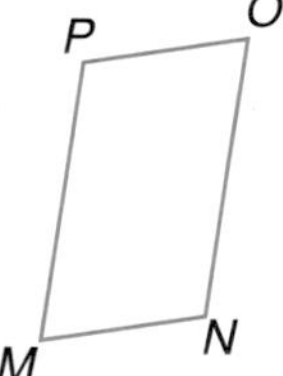

3. R to $\overline{ST}$

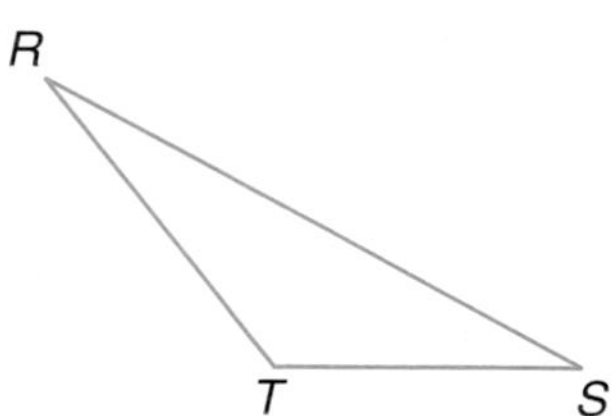

Graph each equation and plot the given ordered pair. Then construct a perpendicular segment and find the distance from the point to the line.

4. $x = 1$, $(4, 5)$
5. $y = -3$, $(-2, 4)$
6. $y = -2x + 3$, $(-2, -3)$

In the figure at the right, $\overline{QR} \perp \overline{RT}$, $\overline{QR} \perp \overline{QU}$, and $\overline{VT} \perp \overline{RU}$. Name the segment whose length represents the distance between the following points and lines.

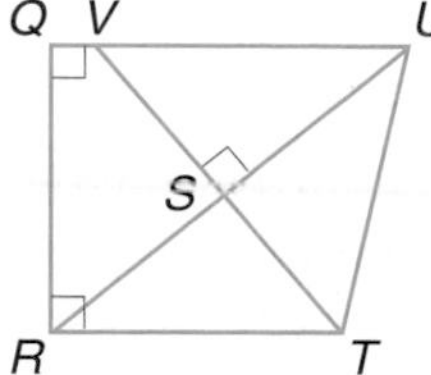

7. U to $\overline{QR}$
8. Q to $\overline{RT}$
9. V to $\overline{RU}$
10. T to $\overline{US}$

EXTRA PRACTICE

Lesson 3-6 **If spherical points are restricted to be nonpolar points, decide which statements from Euclidean geometry are true in spherical geometry. If false, explain your reasoning.**

1. Given points A and B, there is a unique straight line containing A and B.
2. Perpendicular lines divide a plane into four infinite regions.
3. Parallel lines have no point of intersection.

For each property listed from plane Euclidean geometry, write a corresponding statement for non-Euclidean spherical geometry.

4. A line segment is the shortest path between two points.
5. A line has infinite length.
6. The intersecting lines intersect at one point.
7. If three points are collinear, exactly one is between the other two.

Lesson 4-1 **Refer to the figure for Exercises 1–5.**

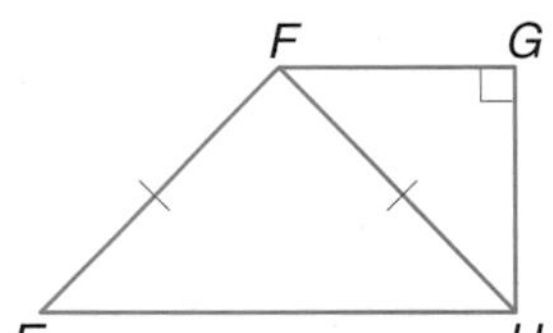

1. Name the hypotenuse.
2. Name the base angles.
3. Name the side opposite $\angle EFH$.
4. Name the vertex angle.
5. Name the angle opposite $\overline{FG}$.

Use the distance formula to classify each triangle by the measures of its sides.

6. $\triangle ABC$ with vertices $A(6, 4)$, $B(-2, 4)$, and $C(2, 7)$.
7. $\triangle PQR$ with vertices $P(-3, 4)$, $Q(0, 1)$, and $R(2, 3)$.

Copy each sentence. Fill in the blank with *sometimes*, *always*, or *never*.

8. Scalene triangles are _?_ obtuse.
9. Equilateral triangles are _?_ acute.
10. Right triangles are _?_ isosceles.

Lesson 4-2 **Find the value of x.**

1. 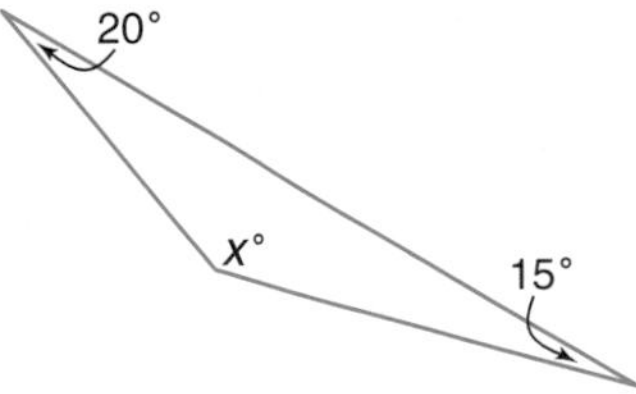

2. 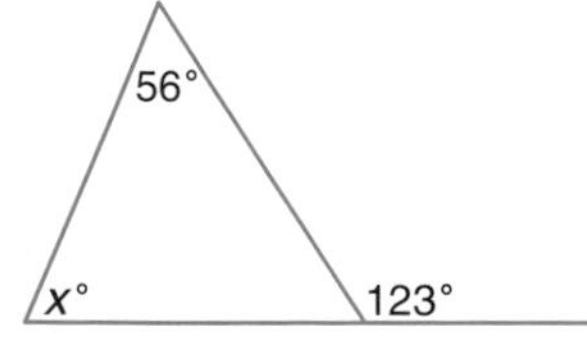

3. 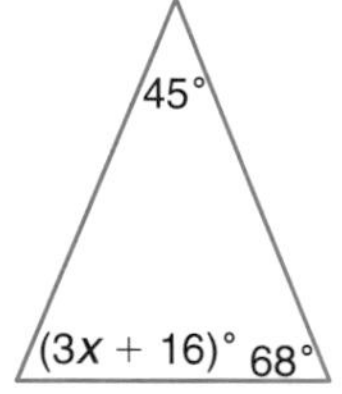

If $\overline{PQ}$ is perpendicular to $\overline{QR}$, find the measure of each angle in the figure at the right.

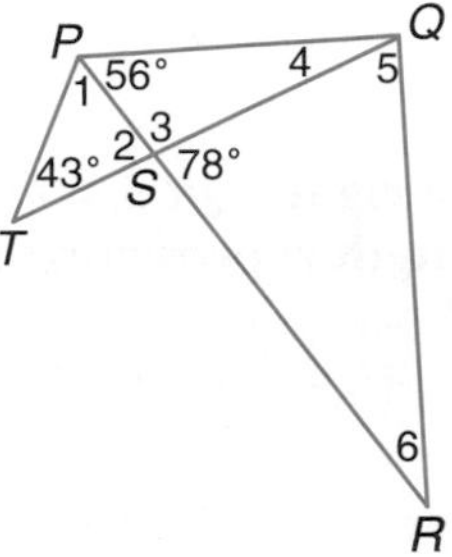

4. $m\angle 2$
5. $m\angle 1$
6. $m\angle 3$
7. $m\angle 4$
8. $m\angle 5$
9. $m\angle 6$

Lesson 4-3 **For each pair of congruent triangles, name all the corresponding sides and angles. Use ↔ to indicate each correspondence. Draw a figure for each pair of triangles, and mark the corresponding parts.**

1. $\triangle MNO \cong \triangle JKL$
2. $\triangle XYZ \cong \triangle ZPR$
3. $\triangle DEF \cong \triangle ABC$

Complete each congruence statement.

4. $\triangle MST \cong \triangle$ _?_

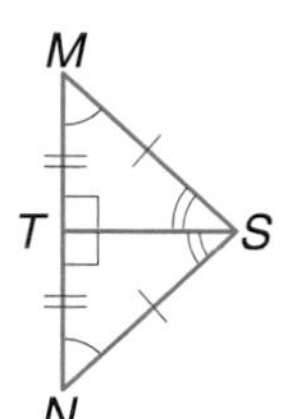

5. $\triangle ABX \cong \triangle$ _?_

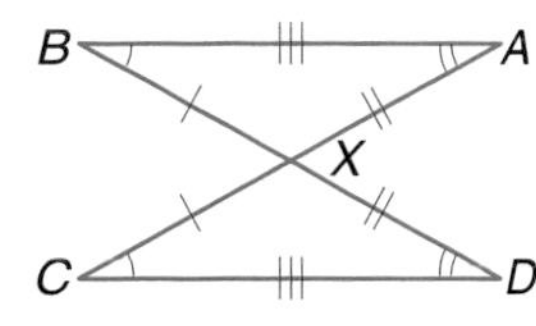

6. $\triangle WYX \cong \triangle$ _?_

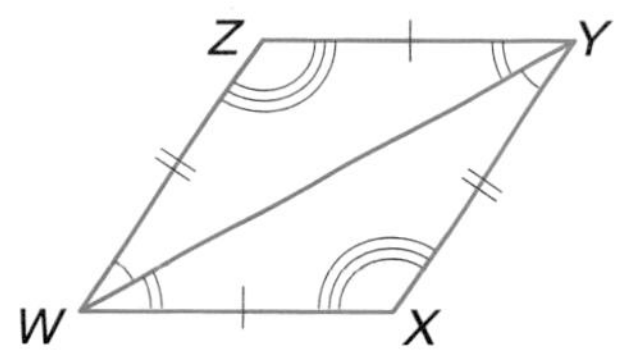

7. Given $\triangle QRS \cong \triangle LMN$, $RS = 8$, $QR = 14$, $QS = 10$, and $MN = 3x - 25$.
   a. Draw and label a figure to show the congruent triangles.
   b. Find the value of x.

Lesson 4-4 **Determine which postulate can be used to prove the triangles congruent. If it is not possible to prove them congruent, write *not possible*.**

1.

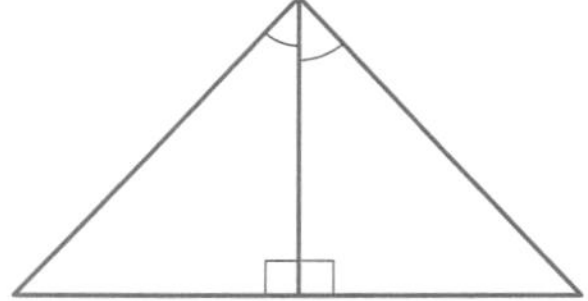

2.

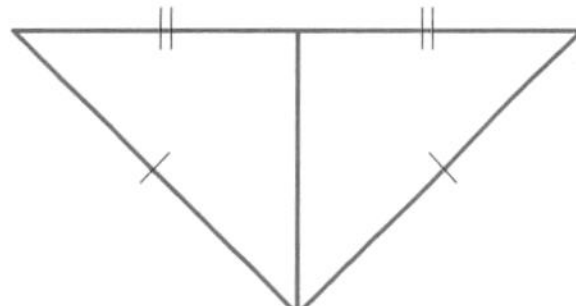

3.

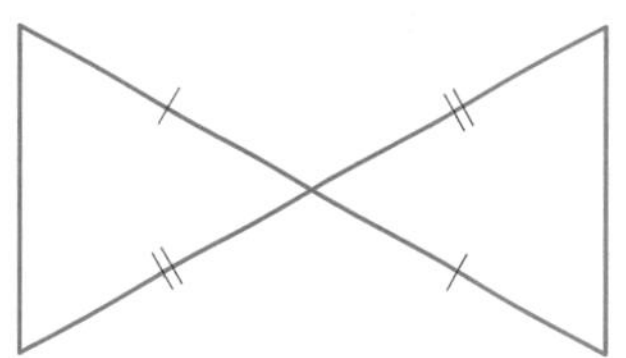

Determine if each pair of triangles is congruent. Justify your answer.

4. The vertices of $\triangle JKL$ are $J(-3, 1)$, $K(-8, 5)$, and $L(-1, 8)$, and the vertices of $\triangle STU$ are $S(0, 1)$, $T(4, 6)$, and $U(7, -1)$.

5. The vertices of $\triangle BCD$ are $B(-4, 1)$, $C(-1, 2)$, and $D(-1, 4)$, and the vertices of $\triangle NMR$ are $N(1, -1)$, $M(0, -3)$, and $R(4, -4)$.

6. Write a proof.

   Given: $\angle A \cong \angle D$
   $\overline{AO} \cong \overline{OD}$

   Prove: $\triangle AOB \cong \triangle DOC$

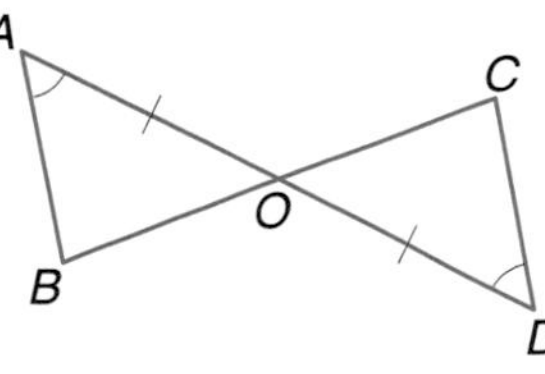

Lesson 4-5 **Refer to the figure at the right for Exercises 1–5.**

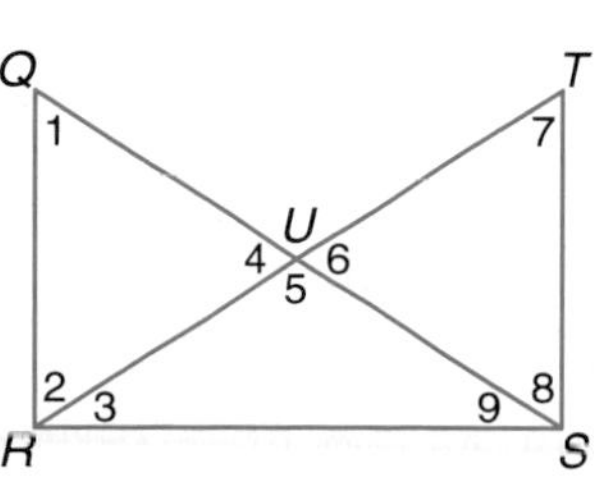

1. Name the included side for $\angle 2$ and $\angle 4$.

2. Name a nonincluded side for $\angle 6$ and $\angle 8$.

3. $\overline{TS}$ is included between what two angles?

4. Name a pair of angles so that $\overline{RS}$ is not included.

5. If $\angle 1 \cong \angle 8$, $\angle 4 \cong \angle 6$, and $\overline{QR} \cong \overline{ST}$, then $\triangle QUR \cong \triangle$ _?_ by _?_ .

EXTRA PRACTICE

Lesson 4-6 **Refer to the figure at the right for Exercises 1–5.**

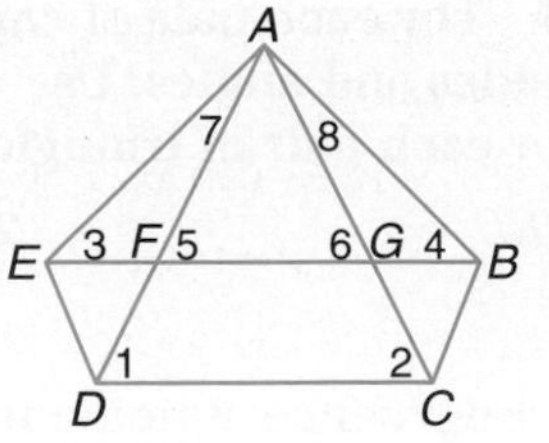

1. If $\overline{FA} \cong \overline{GA}$, name two congruent angles.
2. If $\overline{AE} \cong \overline{AB}$, name two congruent angles.
3. If $\overline{AD} \cong \overline{AC}$, name two congruent angles.
4. If $\angle 6 \cong \angle 3$, name two congruent segments.
5. If $\angle 5 \cong \angle 4$, name two congruent segments.

Find the value of *x*.

6.

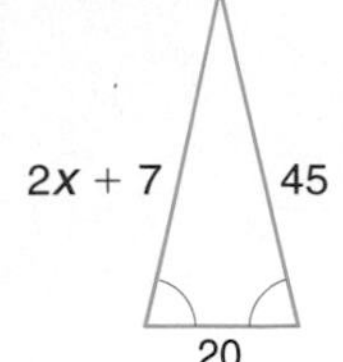

7.

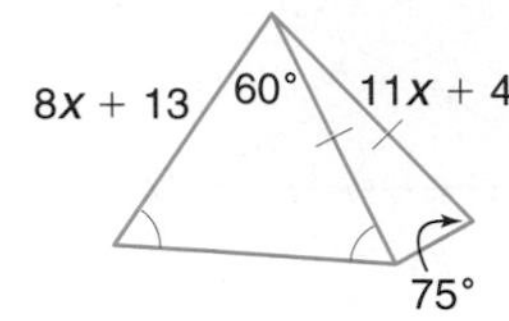

8.

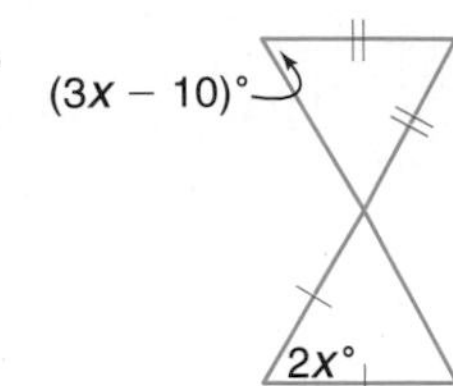

Lesson 5-1 **Draw and label a figure to illustrate each situation.**

1. $\overline{PT}$ and $\overline{RS}$ are medians of $\triangle PQR$ and intersect at V.
2. $\overline{AD}$ is a median and an altitude of $\triangle ABC$.
3. $\overline{EG}$ and $\overline{FG}$ are altitudes of $\triangle EFG$.

Refer to $\triangle EFG$. Write at least one conclusion you can make from each statement.

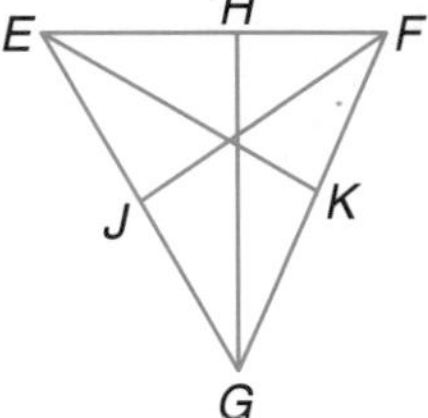

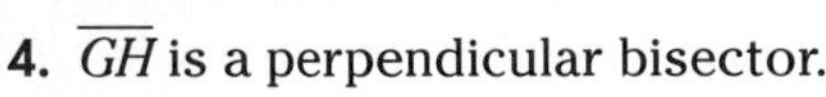

4. $\overline{GH}$ is a perpendicular bisector.
5. $\overline{FJ}$ is a median.
6. $\angle FEK \cong \angle GEK$

If possible, describe a triangle for which each statement is true. If no triangle exists, write *no such triangle*.

7. The three medians of a triangle intersect at a point inside the triangle.
8. The three angle bisectors of a triangle intersect at a point outside the triangle.
9. The three altitudes of a triangle intersect at a vertex of the triangle.

Lesson 5-2 **State the additional information needed to prove each pair of triangles congruent by the given theorem or postulate.**

1. LL

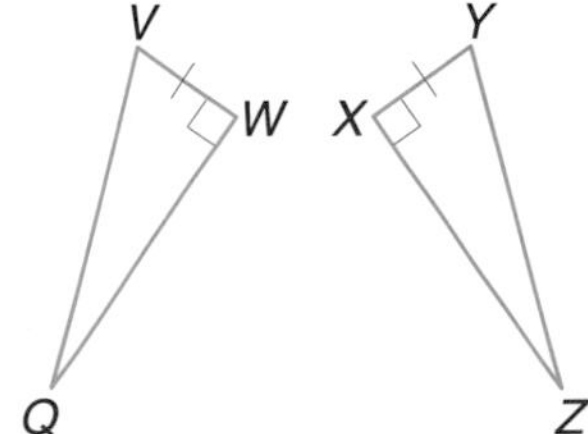

2. HA

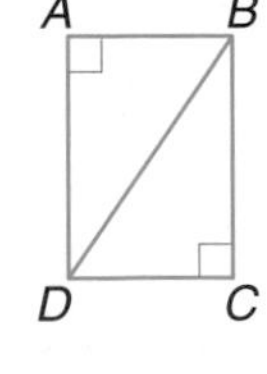

3. HL

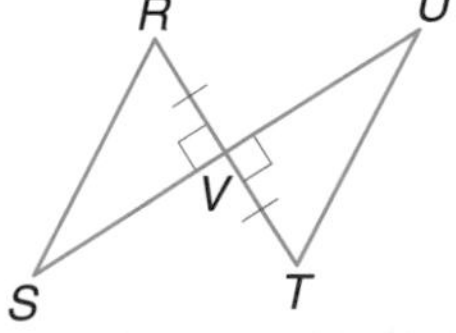

Find the value of *x* so that $\triangle DEF \cong \triangle PQR$ by the indicated theorem or postulate.

4. $DF = 6x + 1$, $FE = 8$, $PR = 10x - 19$, $RQ = 8$; HL
5. $FE = 9$, $m\angle F = 8x - 3$, $RQ = 9$, $m\angle R = 7x + 4$; LA

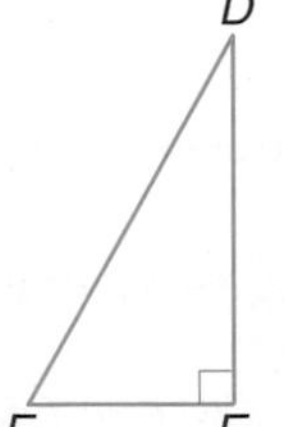

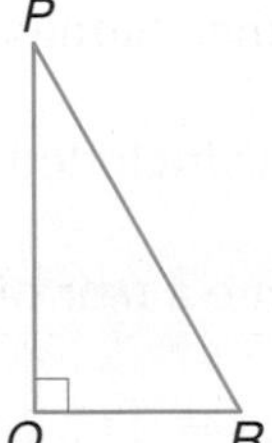

Lesson 5–3 **State the assumption you would make to start an indirect proof of each statement.**

1. Points M, N, and P are collinear
2. Triangle ABC is an acute triangle.
3. The angle bisector of the vertex angle of an isosceles triangle is also an altitude of the triangle.

Refer to the figure at the right for Exercises 4–6.

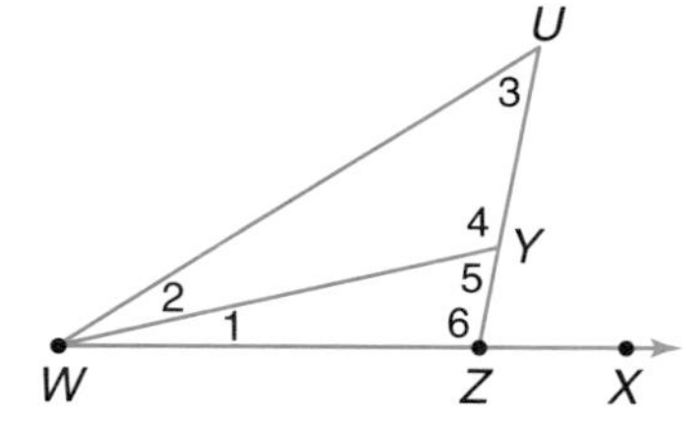

4. Which is greater, $m\angle 4$ or $m\angle 5$?
5. If $m\angle 1 < m\angle 3$, which is greater $m\angle 5$ or $m\angle 1$?
6. Name an angle whose measure is less than $m\angle UWZ$.
7. Write an indirect proof.

   Given: $\overline{PQ} \cong \overline{PR}$

   $m\angle 1 \neq m\angle 2$

   Prove: $\overline{PZ}$ is not a median of $\triangle PQR$.

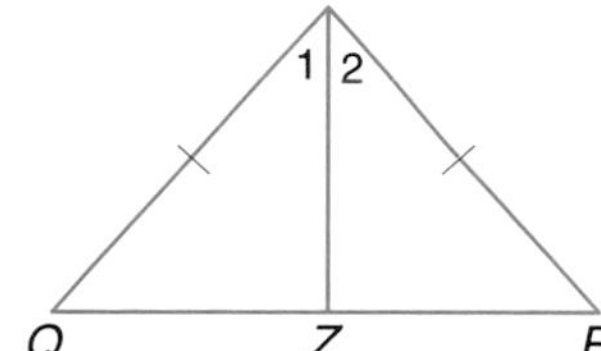

Lesson 5–4 **Refer to the figure at the right for Exercises 1–2.**

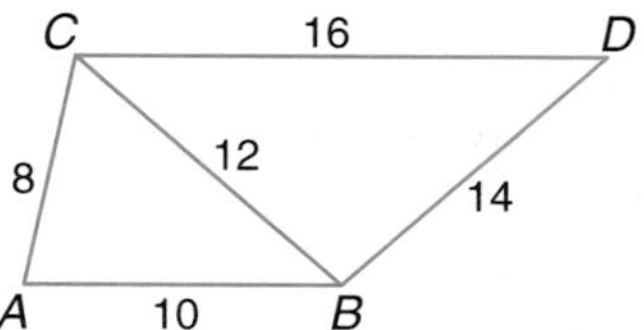

1. Name the angles with the least and greatest measures in $\triangle ABC$.
2. Name the angles with the least and greatest measures in $\triangle BCD$.

Find the value of x and list the sides of $\triangle PQR$ in order from shortest to longest if the angles have the indicated measures.

3. $m\angle P = 7x + 8$, $m\angle Q = 8x - 10$, $m\angle R = 7x + 6$
4. $m\angle P = 3x + 44$, $m\angle Q = 68 - 3x$, $m\angle R = x + 61$
5. Write a two-column proof.

   Given: $QR > QP$

   $\overline{PR} \cong \overline{PQ}$

   Prove: $m\angle P > m\angle Q$

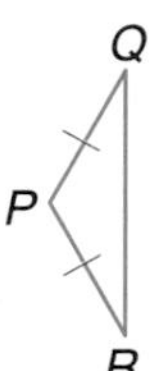

Lesson 5–5 **Determine whether it is possible to draw a triangle with sides of the given measures. Write *yes* or *no*. If yes, draw the triangle.**

1. 12, 11, 17
2. 5, 100, 100
3. 4.7, 9, 4.1
4. 2.3, 12, 12.2

The measures of two sides of a triangle are given. Between what two numbers must the measure of the third side fall?

5. 12 and 15
6. 4 and 13
7. 21 and 17

Determine whether it is possible to have a triangle with the given vertices. Write *yes* or *no*, and then explain your answer.

8. $A(4, -3)$, $B(0, 0)$, $C(-4, 3)$
9. $G(-2, 4)$, $H(-6, 5)$, $I(-3, 3)$

EXTRA PRACTICE

Lesson 5-6 Refer to the figure at the right to write an inequality relating the given pair of angle or segment measures.

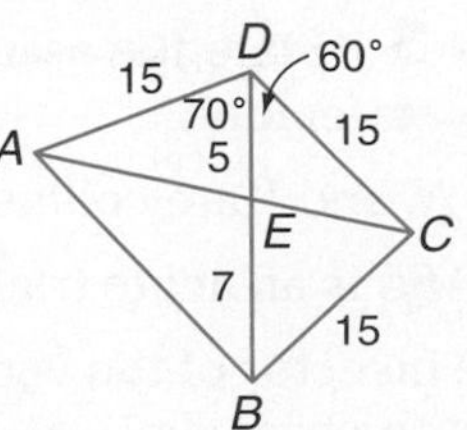

1. AE, CE
2. $m\angle DCE$, $m\angle ECB$
3. AB, BC

Write an inequality or pair of inequalities to describe the possible values of x.

4. 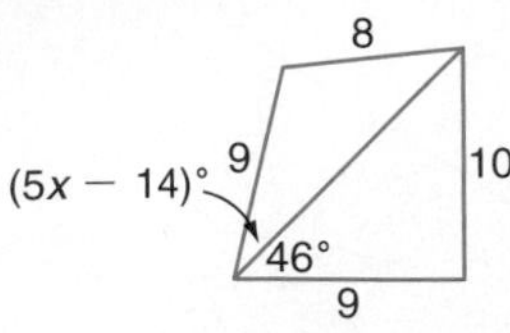

5. 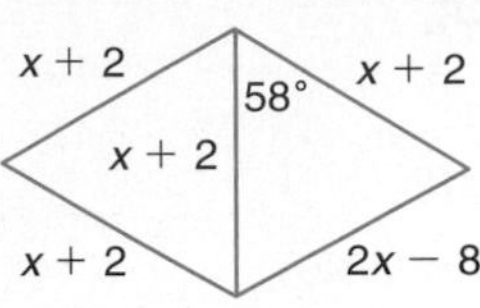

6. 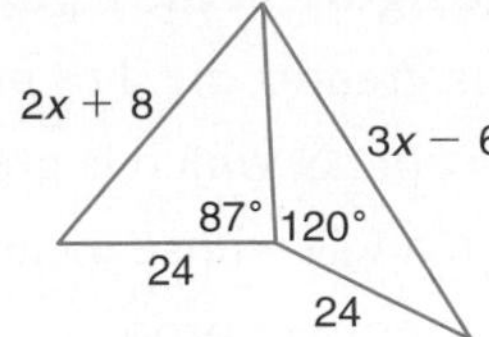

7. Write a proof.
Given: $\overline{TR} \cong \overline{EU}$
Prove: $TE > RU$

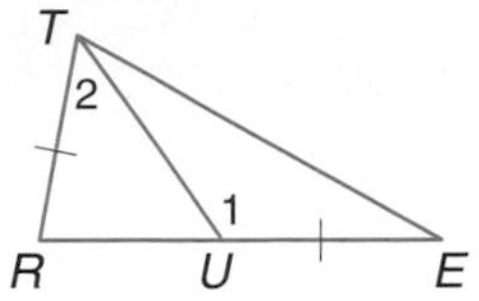

Lesson 6-1 Complete each statement about $\square ABCD$. Then name the theorem or definition that justifies your answer.

1. $\overline{AB} \parallel$?
2. $\overline{DA} \cong$?
3. $\triangle ADC \cong$?
4. $\angle CDA \cong$?
5. $\overline{DE} \cong$?
6. $\angle BAC \cong$?

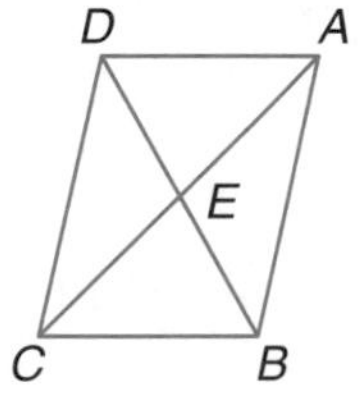

For each parallelogram, find the values of x, y, and z.

7. 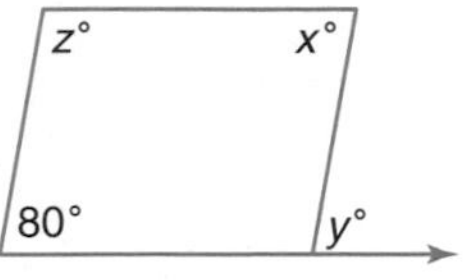

8. 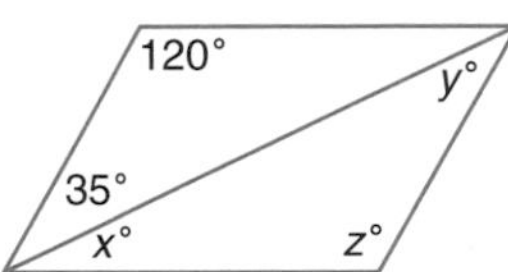

9. 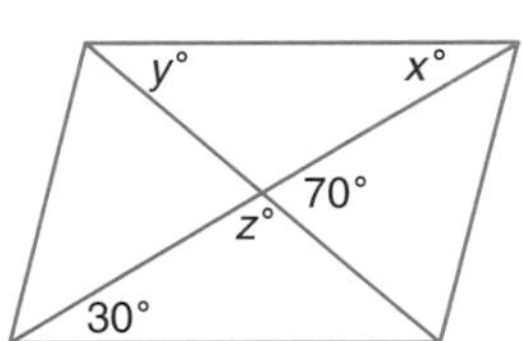

Lesson 6-2 Determine if each quadrilateral is a parallelogram. Justify your answer.

1. 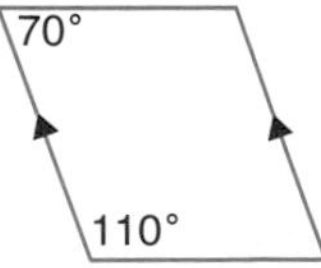

2. 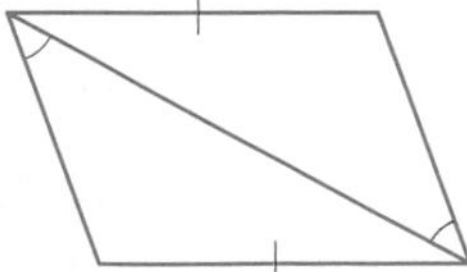

3. 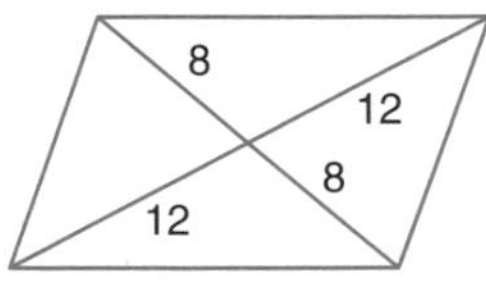

Find the values of x and y that ensure each quadrilateral is a parallelogram.

4. 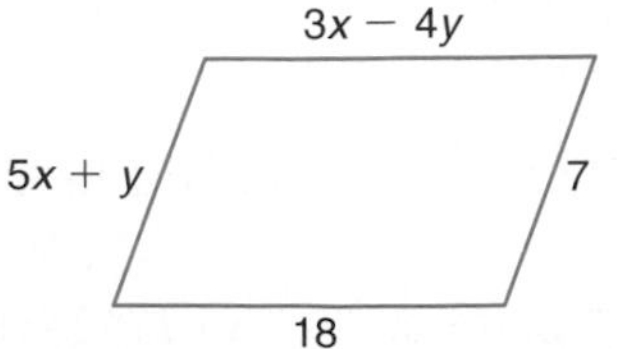

5. 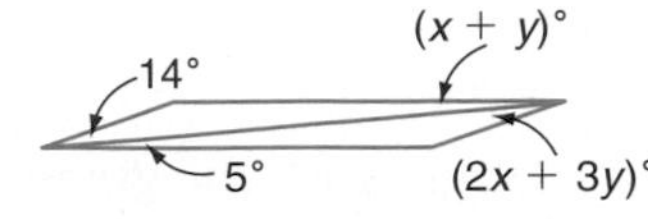

6. 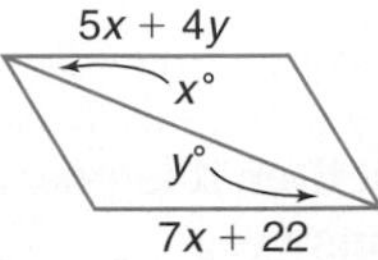

Lesson 6-3 **Use rectangle *QRST* and the given information to solve each problem.**

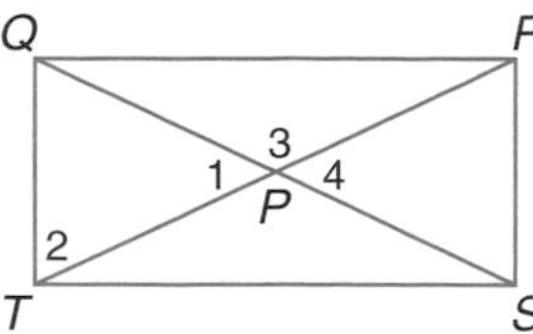

1. $QP = 6$, find RT.
2. $QT = 8$, find RS.
3. $PT = 3x$ and $PS = 18$, find x.
4. $m\angle 1 = 55$, find $m\angle 2$.
5. $m\angle 3 = 110$, find $m\angle 4$.

Determine whether each statement is *true* or *false*. Justify your answer.

6. If a parallelogram has congruent diagonals, then it is a rectangle.
7. If the diagonals of a quadrilateral bisect each other, then it is a rectangle.
8. If a quadrilateral is a rectangle, then it has four right angles.

Determine whether *PQRS* is a rectangle. Justify your answer.

9. $P(12, 2)$, $Q(12, 8)$, $R(-3, 8)$, $S(-3, 2)$
10. $P(0, -3)$, $Q(4, 8)$, $R(11, 7)$, $S(7, -4)$

Lesson 6-4 **Use rhombus *IJKL* and the given information to find each value.**

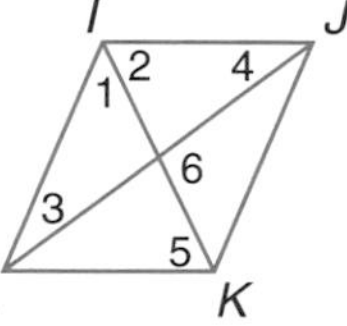

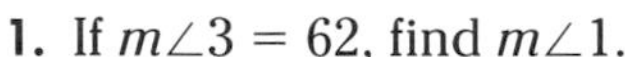
1. If $m\angle 3 = 62$, find $m\angle 1$.
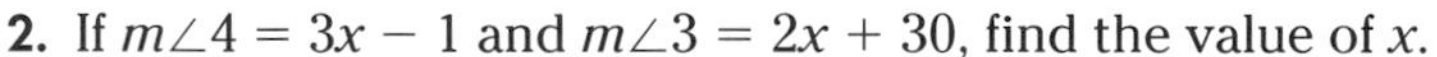
2. If $m\angle 4 = 3x - 1$ and $m\angle 3 = 2x + 30$, find the value of x.
3. If $m\angle 5 = 2(x + 1)$ and $m\angle 3 = 4(x + 1)$, find the value of x.
4. If $m\angle 6 = 7x + 13$, find the value of x.
5. If $m\angle LKJ = x^2 - 17$ and $m\angle 2 = x + 23$, find the value of x.

Name all the quadrilaterals—*parallelogram*, *rectangle*, *rhombus*, or *square*—that have each property.

6. The opposite sides are parallel.
7. The opposite sides are congruent.
8. All angles congruent.

Lesson 6-5 ***PQRS* is an isosceles trapezoid with bases $\overline{PS}$ and $\overline{QR}$ and median $\overline{TV}$. Use the given information to solve each problem.**

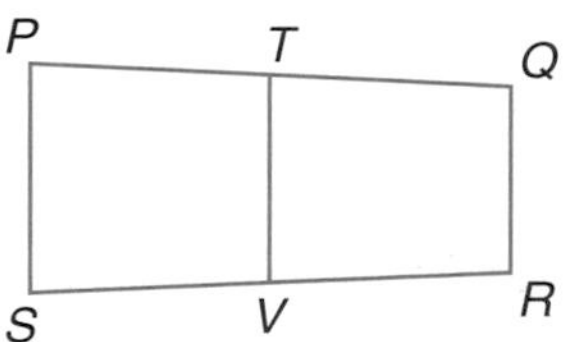

1. If $PS = 20$ and $QR = 14$, find TV.

2. If $QR = 14.3$ and $TV = 23.2$, find PS.
3. If $TV = x + 7$ and $PS + QR = 5x + 2$, find x.
4. If $m\angle RVT = 57$, find $m\angle QTV$.
5. If $m\angle VTP = a$, find $m\angle TPS$ in terms of a.

Determine whether each figure is a *trapezoid*, a *parallelogram*, a *rectangle*, or a *quadrilateral*.

6. 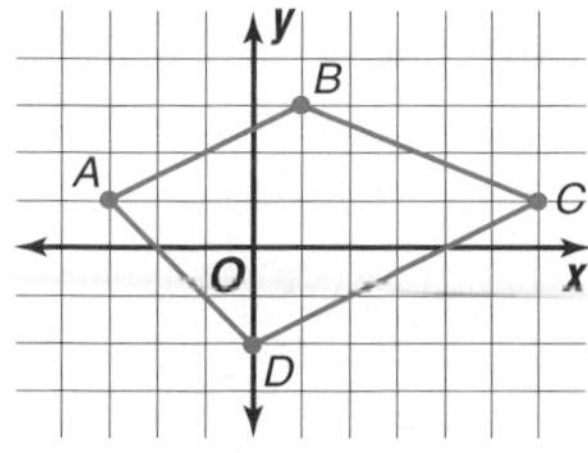

7. 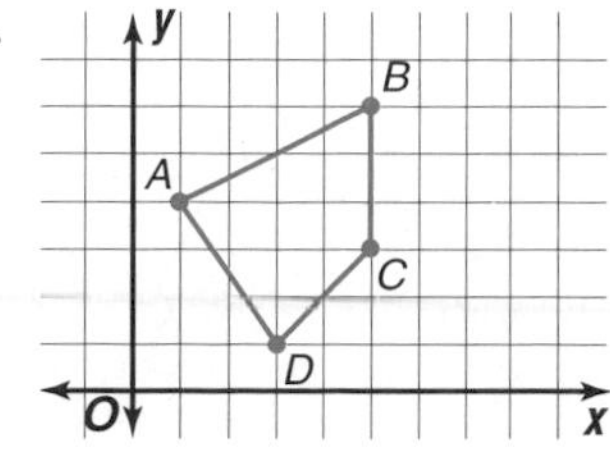

8. 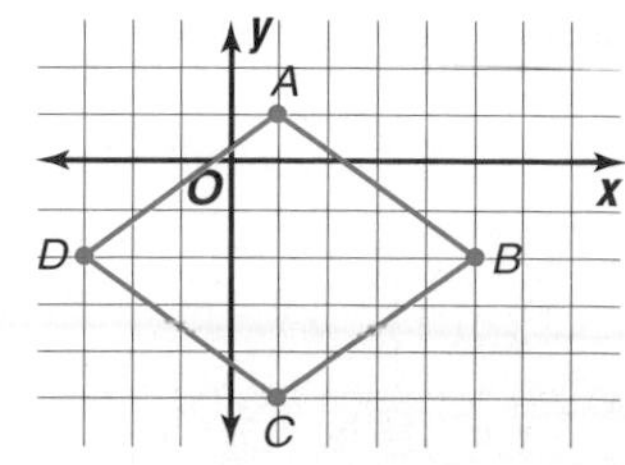

Lesson 7-1 **Find each ratio and express it as a fraction in simplest form.**

1. There are 96 girls and 74 boys in the senior class. Find the ratio of girls to boys.
2. A 5-pound can of cashews cost \$15.00. Find the ratio of pounds to cost.
3. A designated hitter made 6 hits in 9 games. Find the ratio of hits to games.

Solve each proportion by using cross products.

4. $\frac{4}{n} = \frac{7}{8}$
5. $\frac{x+1}{x} = \frac{7}{2}$
6. $\frac{5}{17} = \frac{2x}{51}$

Corresponding sides of polygon *PQRS* are proportional to the sides of polygon *ABCD*.

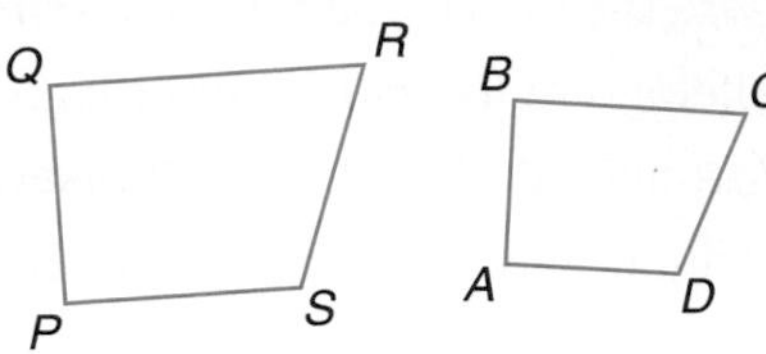

7. If $AB = 4$, $AD = 8$, and $PQ = 6$, find PS.
8. If $RS = 4.5$, $CD = 6.3$, and $BC = 7$, find QR.
9. If $CD = 21.8$, $DA = 43.6$, and $SR = 33$, find SP.

Lesson 7-2 **Each pair of figures is similar. Find the values of *x* and *y*.**

1. 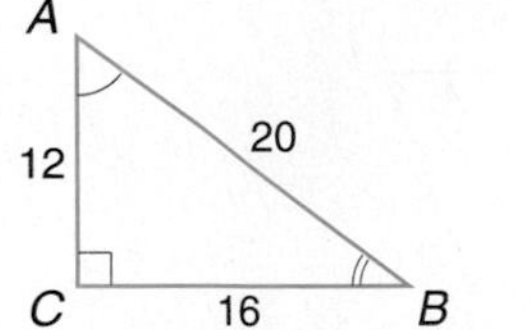

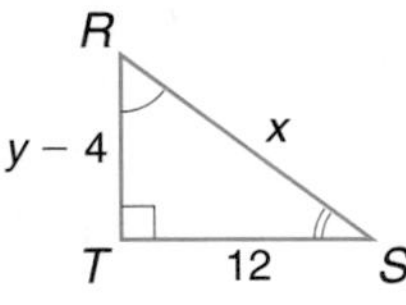

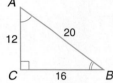

2. 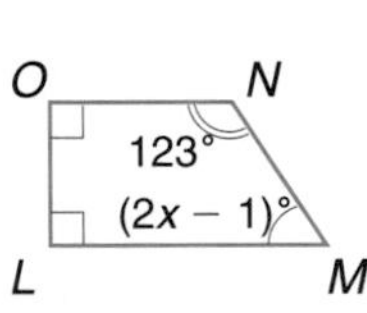

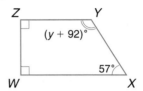

Given quadrilateral *ABCD* ~ quadrilateral *EFGH*, find each of the following.

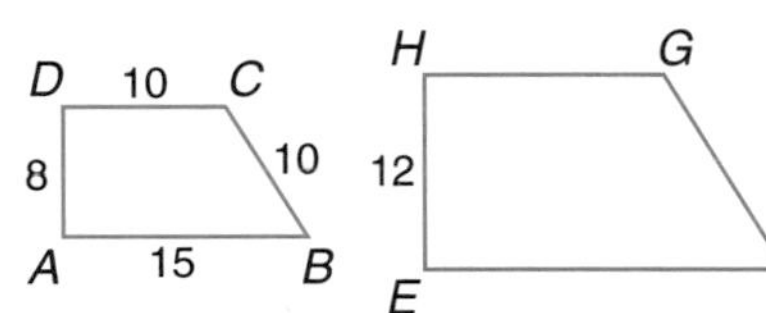

3. scale factor of *ABCD* to *EFGH*
4. **a.** *GH* **b.** *FG* **c.** *EF*
5. **a.** perimeter of *ABCD*
   b. perimeter of *EFGH*
   c. ratio of the perimeters of *ABCD* and *EFGH*

Lesson 7-3 **Determine whether each pair of triangles is similar. Give a reason for your answer. If similarity exists, write a mathematical sentence relating the two triangles.**

1. 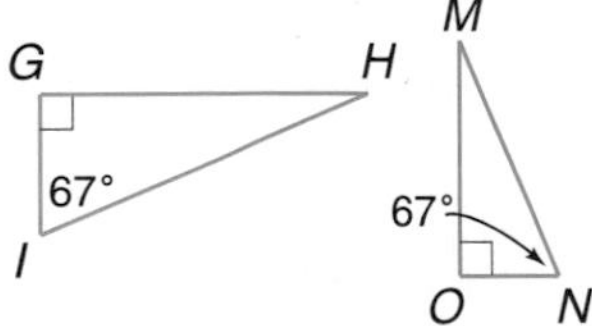

2. 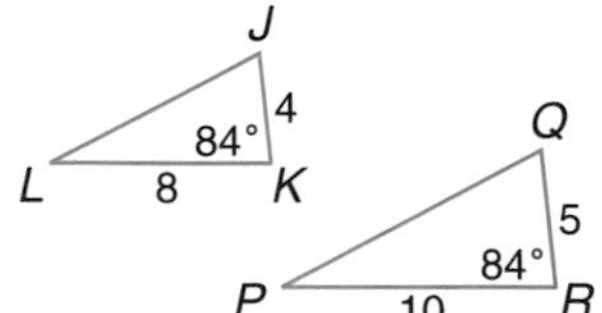

3. 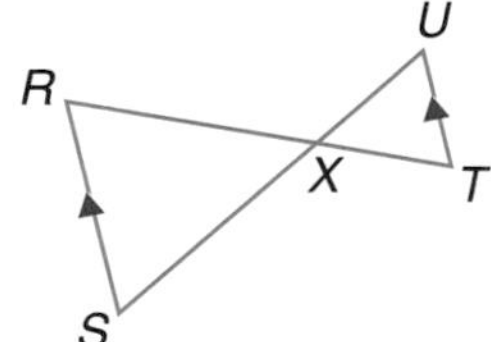

Use the given information to find each measure.

4. $\overline{BE} \parallel \overline{CD}$, find *CD*, *AC*, and *BC*.

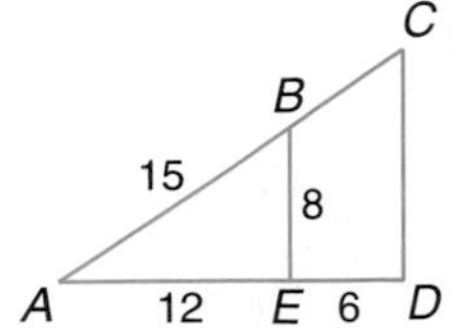

5. If *RSTU* is a parallelogram, find *SV*, *WS*, and *RW*.

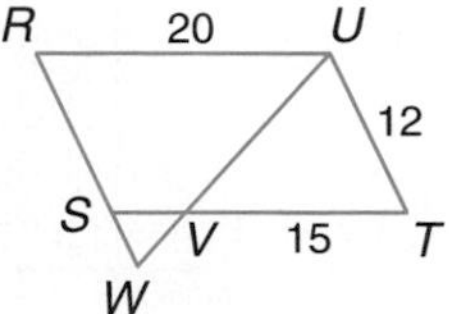

Lesson 7-4 In the figure at the right, $AB \parallel CD \parallel EF$. Complete each statement.

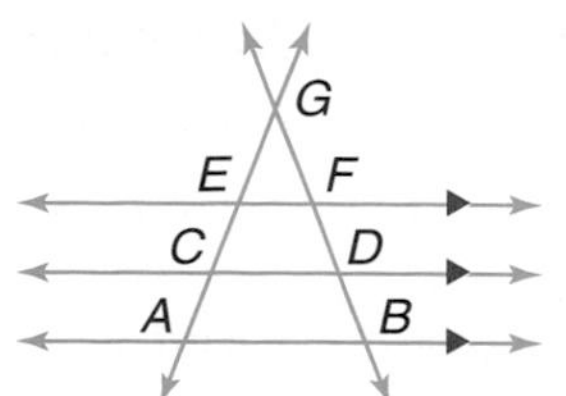

1. $\frac{CE}{AC} = \frac{DF}{?}$
2. $\frac{AE}{AC} = \frac{?}{BD}$
3. $\frac{?}{DF} = \frac{AE}{CE}$
4. $\frac{EG}{CE} = \frac{FG}{?}$

In $\triangle ACE$, $\overline{BD} \parallel \overline{AE}$. Determine whether each statement is *true* or *false*. If false, explain why.

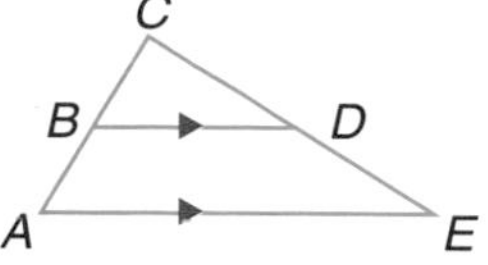

5. $\frac{BC}{ED} = \frac{AB}{CD}$
6. $\frac{AB}{BC} = \frac{DE}{CD}$

In $\triangle XYZ$, find x so that $\overline{YZ} \parallel \overline{MN}$.

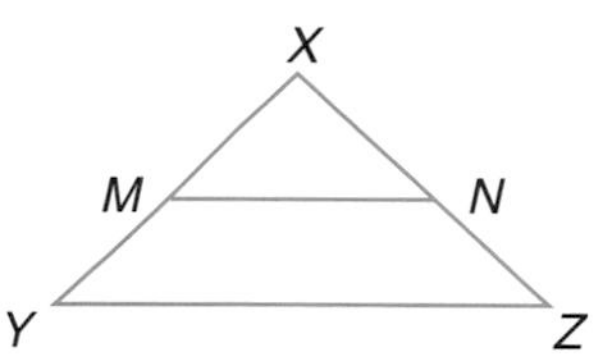

7. $YM = 6, MX = 9, NX = 12, ZN = x.$
8. $XN = 20, NZ = 16, XM = x - 6, MY = 20.$
9. $MX = 5, MY = x - 2, NX = 9, NZ = x + 6.$

Lesson 7-5 In the figure at the right, $\triangle ABC \sim \triangle DEF$.

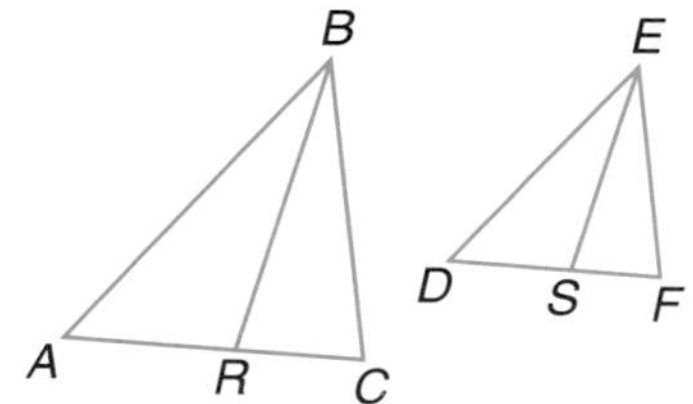

1. If $\overline{AR} \cong \overline{RC}$, $\overline{DS} \cong \overline{SF}$, $AC = 20$, $DF = 12$, and $ES = 5$, $BR = \underline{\ ?\ }$.
2. If $\overline{BR} \perp \overline{AC}$ and $\overline{ES} \perp \overline{DF}$, $\frac{BR}{ES} = \frac{?}{EF}$.
3. If $\overline{BR}$ and $\overline{ES}$ are angle bisectors, $AB = 12$, $DE = 5$, and $ES = 4$, $BR = \underline{\ ?\ }$.

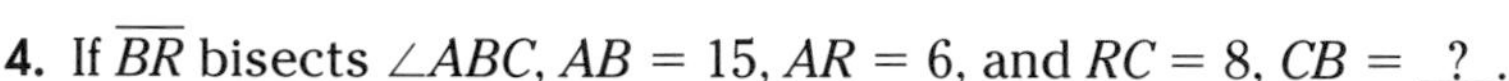

4. If $\overline{BR}$ bisects $\angle ABC$, $AB = 15$, $AR = 6$, and $RC = 8$, $CB = \underline{\ ?\ }$.

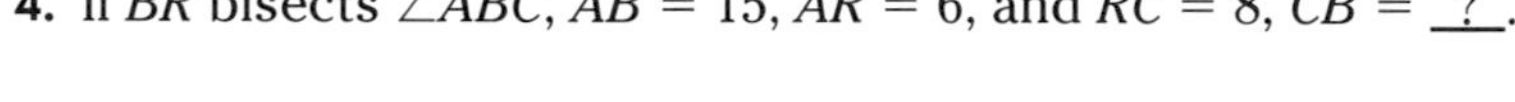

5. If the perimeter of $\triangle DEF$ is 30, $AC = 12$, and $DF = 8$, find the perimeter of $\triangle ABC$.

Lesson 7-6 Find the value of each expression. Then, use that value as the next x in the expression. Repeat the process until you can make some observations. Describe what happens in each of the iterations.

1. $\sqrt[3]{x}$, where x initially equals 28
2. $\frac{1}{x^2}$, where x initially equals 2
3. $x^{\frac{1}{2}}$, where x initially equals 0.4

Use grid paper to draw a square that is 20 units long on each side. Connect the midpoints on each side to make four squares. Shade the bottom right square. This is stage 1. Repeat this process in the the unshaded squares.

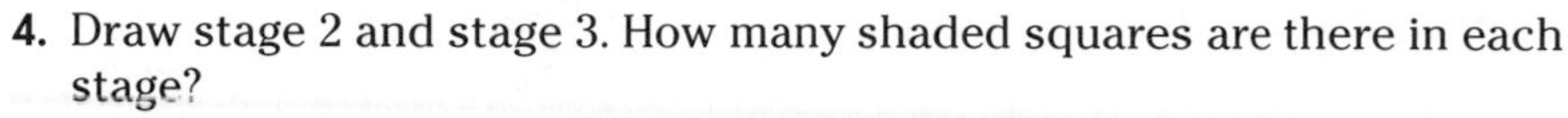

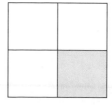

4. Draw stage 2 and stage 3. How many shaded squares are there in each stage?
5. If your continue the process indefinitely, will the figure you obtain be strictly self similar? Explain.

Lesson 8-1 Find the geometric mean between each pair of numbers.

1. 3 and 5
2. $\frac{1}{4}$ and 9
3. $\frac{3}{8}$ and $\frac{8}{3}$

Find the values of x and y.

4. 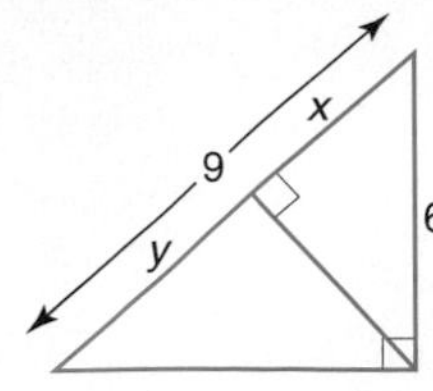

5. 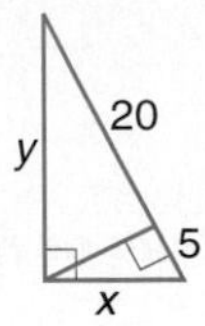

6. 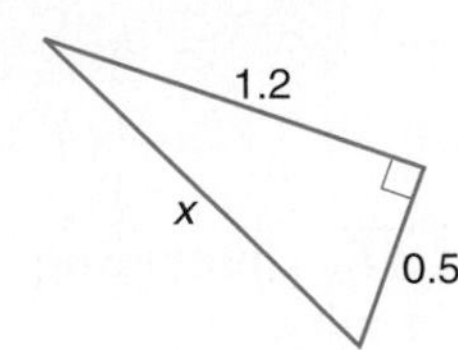

Determine if the given measures are measures of the sides of a right triangle.

7. 25, 20, 15
8. 1.6, 3.0, 3.4
9. 18, 34, 39

Lesson 8-2 Find the values of x and y.

1. 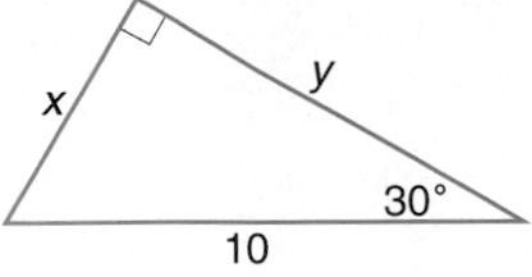

2. 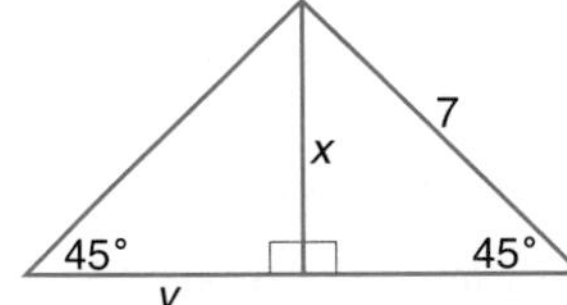

3. 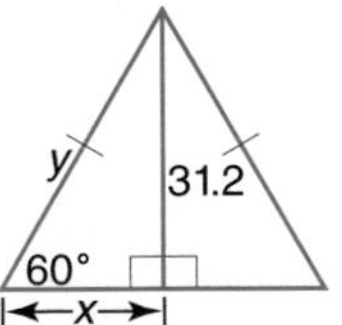

4. The length of one side of a square is 17 meters. Find the length of a diagonal of the square.
5. The length of an altitude of an equilateral triangle is $\frac{\sqrt{3}}{3}$ yards. Find the length of one side of the triangle.
6. The length of a diagonal of a square is 6 inches. Find the length of one side of the square.
7. The perimeter of an equilateral triangle is 42 centimeters. Find the length of an altitude of the triangle.

Lesson 8-3 Find the indicated trigonometric ratio as a fraction and as a decimal rounded to the nearest ten-thousandth.

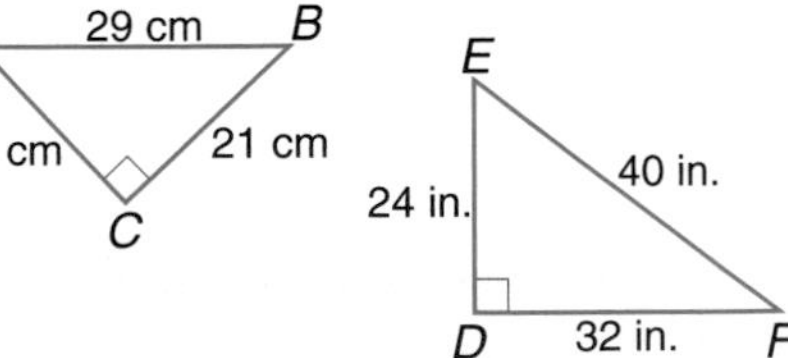

1. sin A

2. sin B

3. tan A

4. cos B

5. cos E

6. tan F

Find the value of x. Round to the nearest tenth.

7. 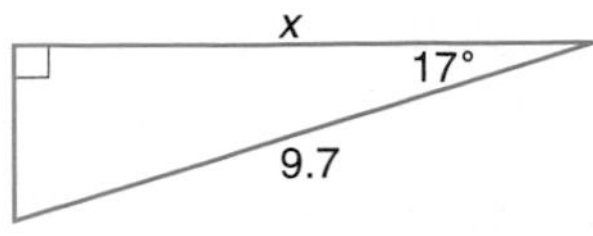

8. 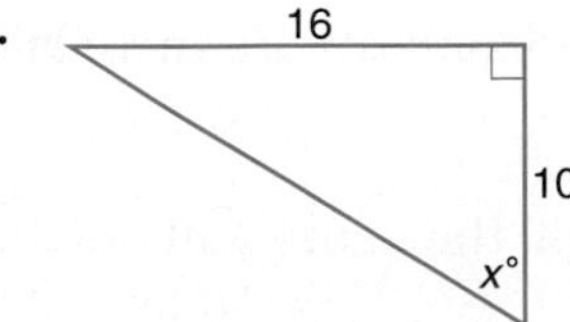

9. 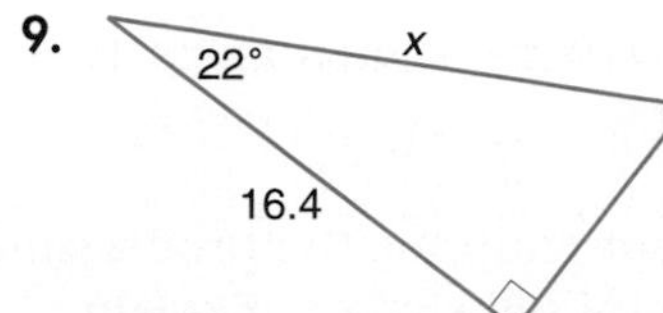

Lesson 8-4 **State an equation that would enable you to solve each problem. Then solve. Round answers to the nearest tenth.**

1. Given $m\angle A = 15$ and $AL = 37$, find LN.
2. Given $m\angle A = 47$ and $LN = 10$, find AL.
3. Given $AN = 13.4$ and $m\angle A = 16$, find AL.
4. Given $m\angle L = 72$ and $AN = 13$, find LN.
5. Given $LN = 33.6$ and $m\angle L = 74$, find AN.
6. A surveyor is standing 100 meters from a bridge. She determines that the angle of elevation to the top of the bridge is 35°. The surveyor's eye level is 1.5 meters above the ground. Find the height of the bridge.
7. A ladder leaning against the side of a house forms an angle of 65° with the ground. The foot of the ladder is 8 feet from the building. Find the length of the ladder.
8. From the top of a lighthouse, the angle of depression to a buoy is 25°. If the top of the lighthouse is 150 feet above sea level, find the distance from the buoy to the foot of the lighthouse.
9. Bill Owens is an architect designing a new parking garage for the city. The floors of the garage are to be 10 feet apart. The exit ramps between the pair of floors are to be 75 feet long. What is the measurement of the angle of elevation of each ramp?
10. A jet takes off and rises at an angle of 19° with the ground until it hits 28,000 feet. How much ground distance is covered in miles? (*Hint*: 5280 feet = 1 mile)

Lesson 8-5 **Draw $\triangle RST$ and mark it with the given information. Write an equation that could be used to find each unknown value. Then find the value to the nearest tenth.**

1. If $s = 2.8$, $m\angle R = 53$, and $m\angle S = 61$, find r.
2. If $s = 36$, $t = 12$, and $m\angle S = 98$, find $m\angle T$.
3. If $m\angle R = 70$, $m\angle S = 23$, and $t = 2.2$, find r.
4. If $m\angle T = 55$, $r = 9$, and $t = 11$, find $m\angle R$.

Solve each $\triangle EFG$ described below. Round measures to the nearest tenth.

5. $m\angle G = 70$, $g = 8$, $m\angle E = 30$
6. $e = 10$, $g = 25$, $m\angle G = 124$
7. $m\angle E = 29$, $m\angle F = 62$, $g = 11.5$
8. $m\angle G = 35$, $e = 7.5$, $g = 24$
9. $m\angle F = 36$, $m\angle G = 119$, $f = 8$
10. $m\angle F = 47$, $m\angle G = 73$, $e = 0.9$

Lesson 8-6 **Sketch each $\triangle ERQ$. Determine whether the Law of Sines or the Law of Cosines should be used first to solve each triangle. Then solve each triangle. Round measures to the nearest tenth.**

1. $m\angle E = 40$, $m\angle Q = 70$, $q = 4$
2. $e = 11$, $r = 10.5$, $m\angle Q = 35$
3. $e = 11$, $r = 17$, $m\angle R = 42$
4. $m\angle E = 56$, $m\angle Q = 26$, $q = 12.2$

Solve each $\triangle PSV$ described below. Round measures to the nearest tenth.

5. $p = 51$, $s = 61$, $m\angle S = 19$
6. $p = 5$, $s = 12$, $v = 13$
7. $p = 20$, $v = 24$, $m\angle S = 47$
8. $m\angle P = 40$, $m\angle V = 59$, $s = 14$
9. $p = 345$, $v = 648$, $s = 442$
10. $m\angle P = 29$, $v = 5$, $s = 4.9$

EXTRA PRACTICE

Lesson 9-1 **Refer to the circle at the right.**

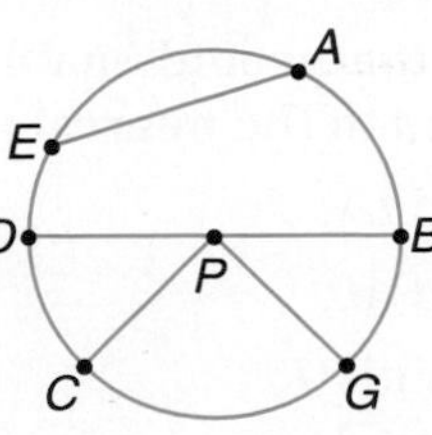

1. Name the center of ⊙P.
2. Is $\overline{PD}$ a chord of ⊙P? Explain.
3. If $PC = 6$, find DB.
4. Name a chord that is not a diameter.
5. Is $\overline{PC} \cong \overline{PB}$? Explain.
6. If $DB = 17$, find PG.

In Exercises 7–9, the radius, diameter, or circumference of a circle is given. Find the other measurements to the nearest tenth.

7. $r = 3.8$, $d = \underline{?}$, $C = \underline{?}$
8. $C = 11$, $d = \underline{?}$, $r = \underline{?}$
9. $d = x$, $r = \underline{?}$, $C = \underline{?}$

Find the exact circumference of each circle.

10.

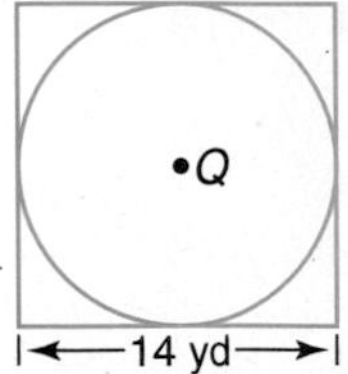

11.

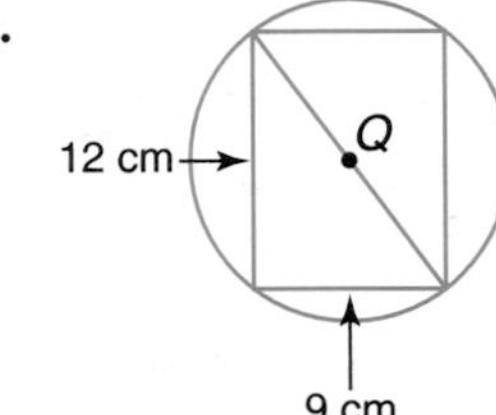

12.

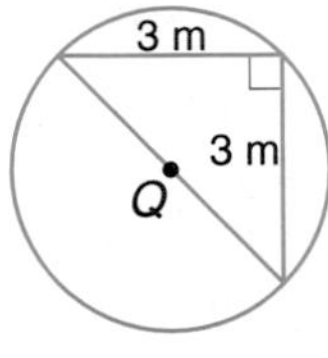

Lesson 9-2 **Refer to ⊙R for Exercises 1–8. In ⊙R, $m\angle QRS = 40$ and $m\angle SRT = 90$ with diameters $\overline{XS}$ and $\overline{QV}$. Determine whether each arc is a minor arc, a major arc, or a semicircle. Then find the degree measure of each arc.**

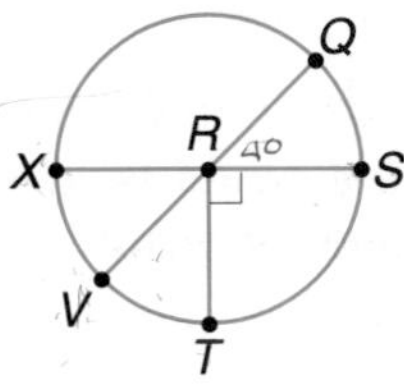

1. $m\overset{\frown}{XQ}$
2. $m\overset{\frown}{VSX}$
3. $m\overset{\frown}{QXV}$
4. $m\overset{\frown}{TSV}$

If $XS = 16$, find the length of each arc. Round to the nearest tenth.

5. $\overset{\frown}{XQ}$
6. $\overset{\frown}{VSX}$
7. $\overset{\frown}{QXV}$
8. $\overset{\frown}{TSV}$

Lesson 9-3 **In ⊙A, $\overline{YS}$ and $\overline{ZR}$ are diameters, and $\overline{SY} \perp \overline{QT}$.**

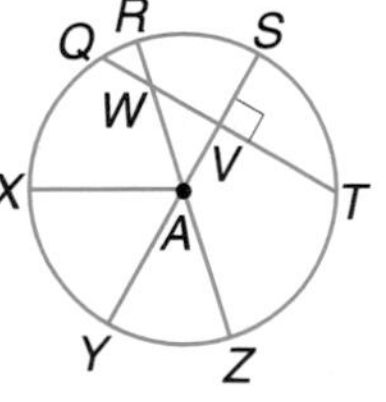

1. Name the midpoint of $\overline{RZ}$.
2. Name an arc congruent to $\overset{\frown}{ST}$.
3. Name a segment congruent to $\overline{VT}$.
4. Which segment is longer, $\overline{WA}$ or $\overline{VA}$?
5. Name a segment congruent to $\overline{RZ}$.

In each circle, O is the center. Find each measure.

6. AC

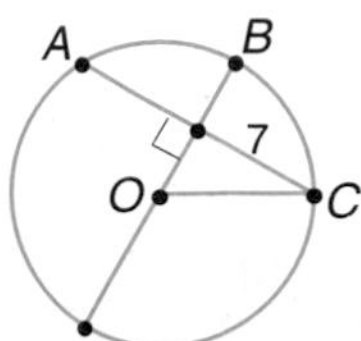

7. $m\overset{\frown}{JK}$

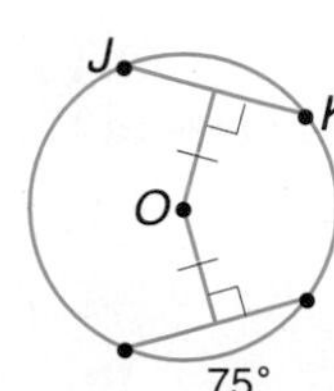

8. ON

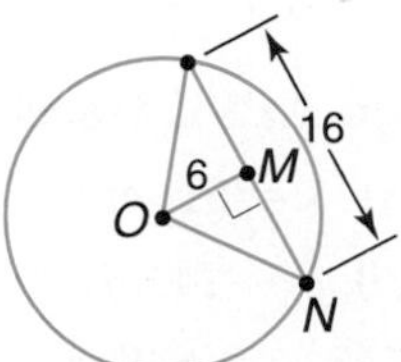

Lesson 9-4 In $\odot O$, $\overline{GM}$ is a diameter.

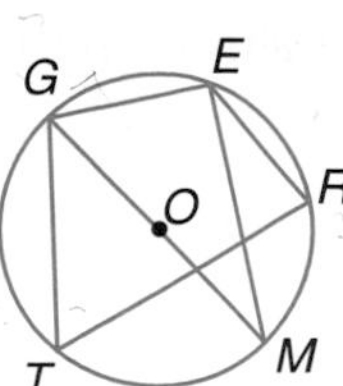

1. Name the intercepted arc for $\angle GTR$.
2. If $m\angle GTR = 62$, find $m\widehat{GR}$.
3. Find $m\angle MEG$.
4. If $m\angle GTR = 62$, find $m\widehat{GTR}$.
5. Name an inscribed angle.

Quadrilateral $GHIJ$ is inscribed in $\odot Q$. $m\angle GHI = 2x$, $m\angle HGJ = 2x - 10$, and $m\angle IJG = 2x + 10$. Find each measure.

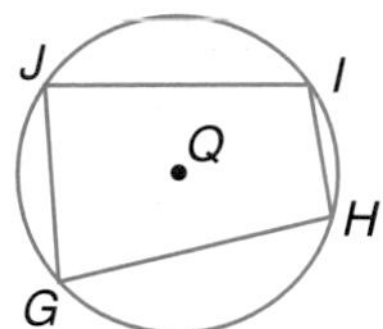

6. $m\angle GHI$
7. $m\widehat{GJI}$
8. $m\widehat{JIH}$
9. $m\widehat{JGH}$

Lesson 9-5 $\overline{AB}$ and $\overline{CD}$ are both tangent to $\odot P$ and $\odot Q$, $AP = 8$, $BQ = 5$, and $m\angle CPE = 45$. Find each measure.

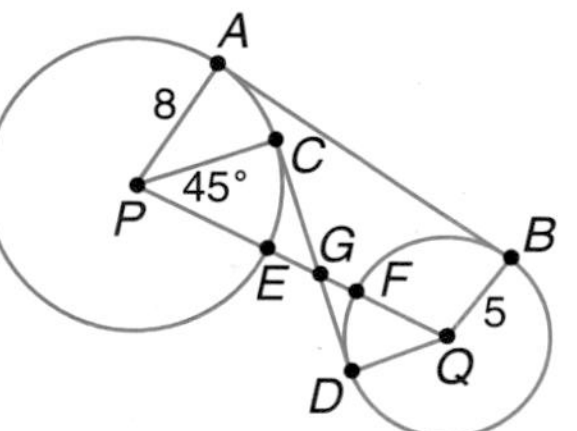

1. $m\widehat{CE}$
2. $m\angle PCG$
3. $m\angle CGP$
4. CG
5. $m\angle FQD$
6. $m\widehat{DF}$
7. DQ
8. DG

For each circle C, find the value of x. Assume that segments that appear to be tangent are tangent.

9.

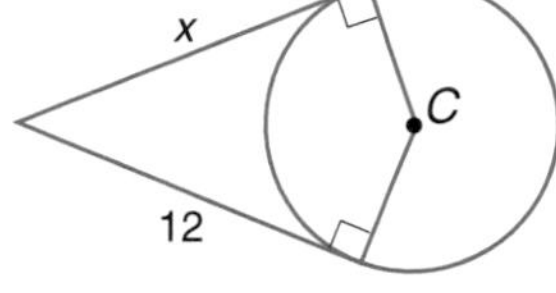

10.

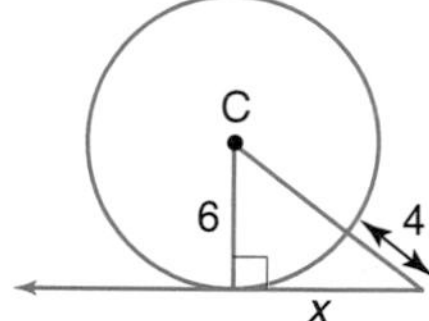

11.

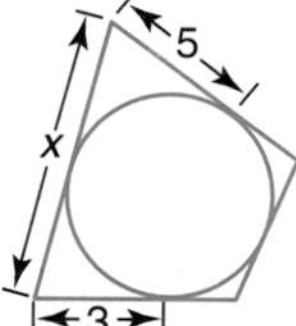

Lesson 9-6 $\overline{AF}$ and $\overline{AB}$ are tangent to $\odot Y$, and $m\widehat{BC} = 84$, $m\widehat{CD} = 38$, $m\widehat{DE} = 64$, and $m\widehat{EF} = 60$. Find each measure.

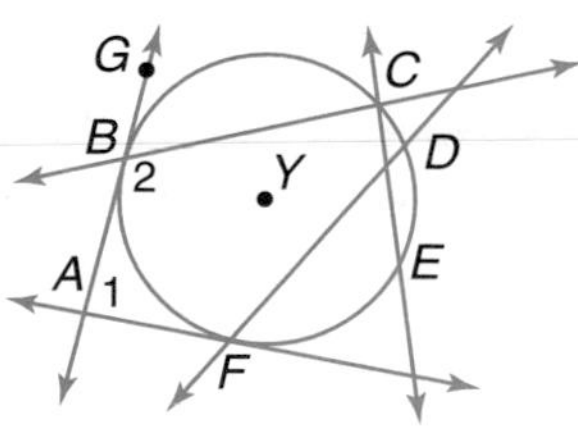

1. $m\widehat{BF}$
2. $m\widehat{BDF}$
3. $m\angle 1$
4. $m\widehat{BFC}$
5. $m\angle 2$
6. $m\angle GBC$

Find the value of x.

7.

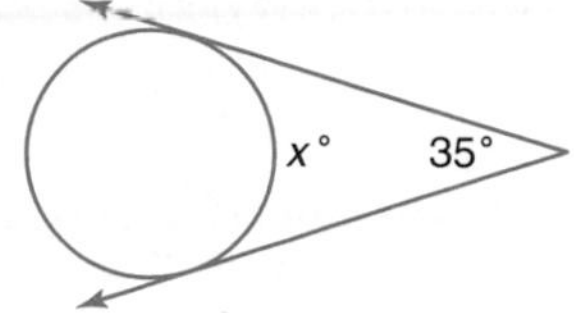

8.

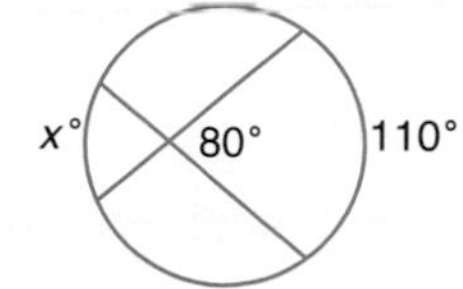

9.

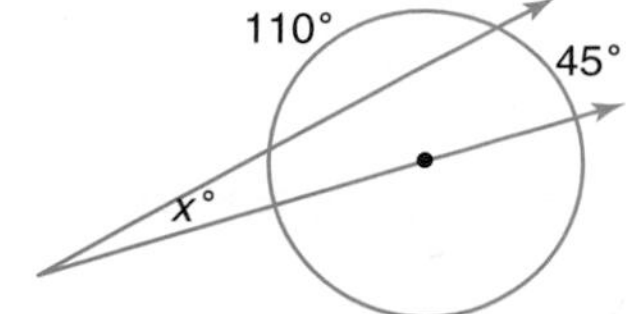

Lesson 9-7 Find the value of x to the nearest tenth. Assume segments that appear tangent to be tangent.

1. 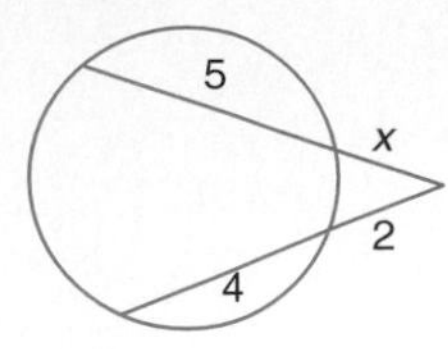

2. 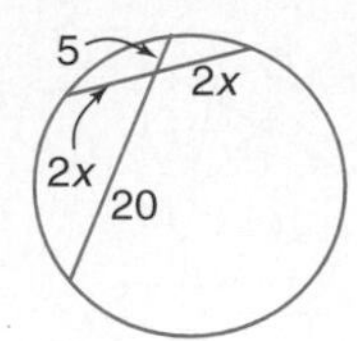

3. 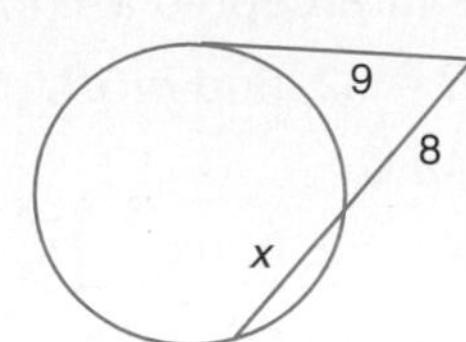

In $\odot L$, $\overline{NM}$ and $\overline{NO}$ are tangent segments, $LO = 8$, $NM = 13$, and $\overline{QN}$ is a secant segment. Find each measure.

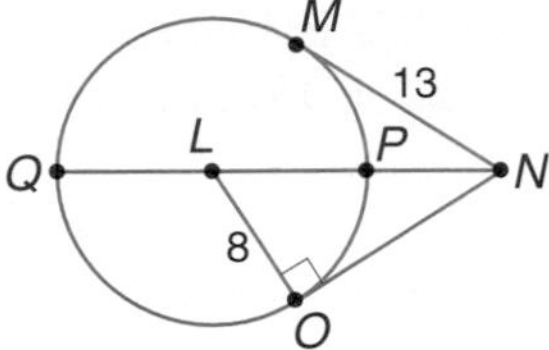

4. NO

5. NL

6. PN

Lesson 9-8 Determine the coordinates of the center and the measure of the radius for each circle whose equation is given.

1. $(x + 2)^2 + (y - 7)^2 = 81$

2. $\left(x - \frac{2}{3}\right)^2 + (y + 4)^2 - 169 = 0$

Graph each circle whose equation is given. Label the center and measure of the radius on each graph.

3. $(x - 7)^2 + (y + 5)^2 = 4$

4. $(x + 3)^2 + (y + 6)^2 = 49$

Write an equation of circle P based on the given information.

5. center: $P(3, 4)$
radius: 6

6. center: $P(-2, 7)$
radius: $\sqrt{17}$

Lesson 10-1 Find the sum of the measures of the interior angles of each convex polygon.

1. 15-gon

2. 59-gon

3. $2t$-gon

The measure of an exterior angle of a regular polygon is given. Find the number of sides of the polygon.

4. 24

5. 60

6. $51\frac{3}{7}$

The number of sides of a regular polygon is given. Find the measures of an interior angle and an exterior angle for each polygon.

7. 4

8. 16

9. s

The measure of an interior angle of a regular polygon is given. Find the number of sides in each polygon.

10. 160

11. 177.6

12. 120

Lesson 10-2 Determine whether each figure tessellates in a plane. If so, draw a sample figure.

1. isosceles triangle
2. quadrilateral with no two sides congruent

Determine if each pattern will tessellate.

3. regular octagon and equilateral triangle

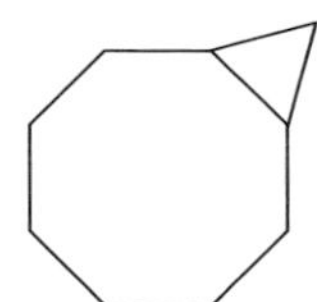

4. regular pentagon and square

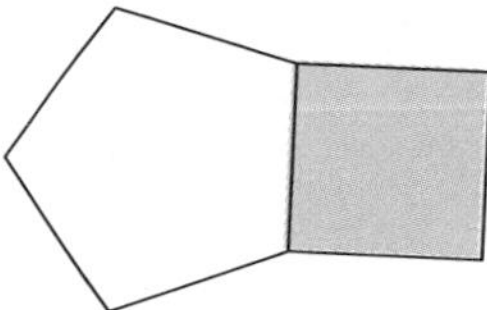

Determine whether each tessellation is *regular*, *uniform*, or *semi-regular*. Name all possibilities.

5. quadrilateral

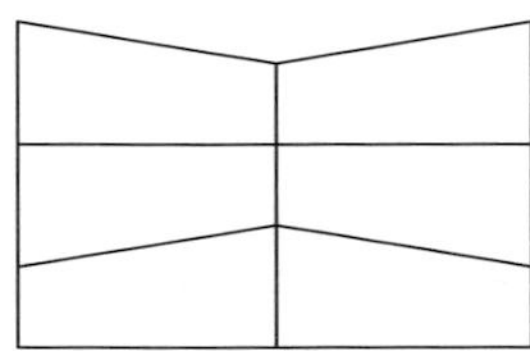

6. rhombus

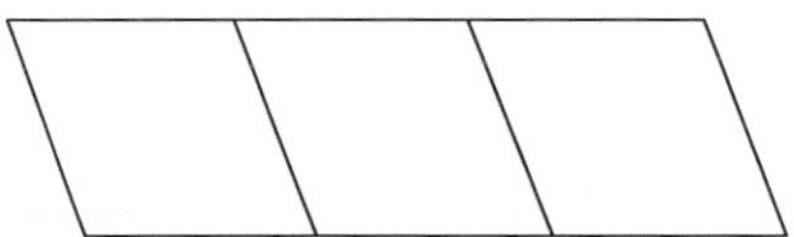

Lesson 10-3 Find the area of each figure. Assume that angles that appear to be right are right angles.

1.

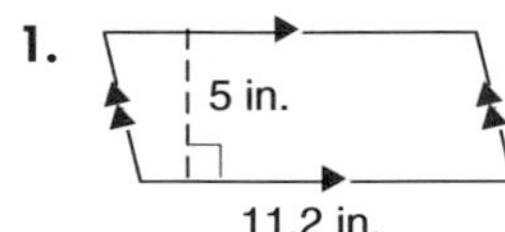

2.

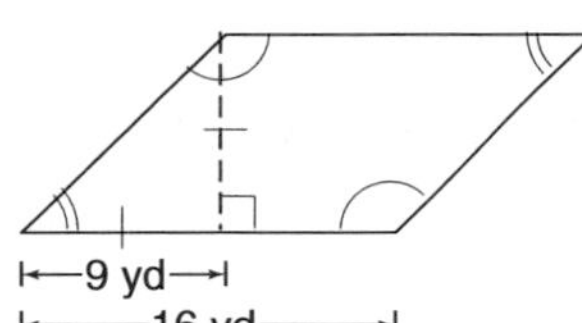

3.

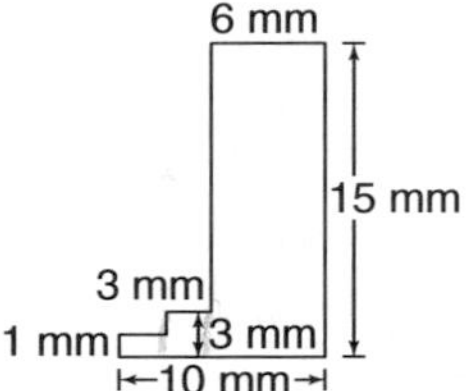

The coordinates of the vertices of a quadrilateral are given. Graph the points and draw the quadrilateral and an altitude. Identify the quadrilateral as a *square, rectangle*, or a *parallelogram*, and find its area.

4. (3, 4), (2, 1), (8, 4), (9, 7)
5. (0, −5), (3, −4), (1, 2), (−2, 1)

Lesson 10-4 Find the area of each figure.

1.

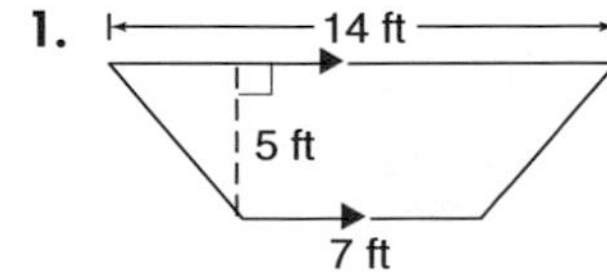

2.

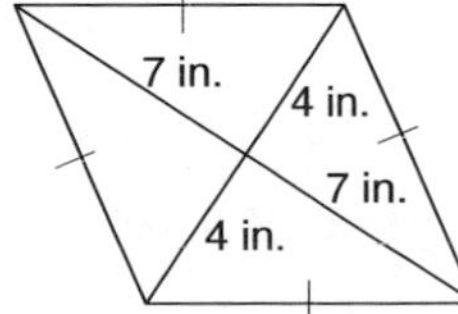

3.

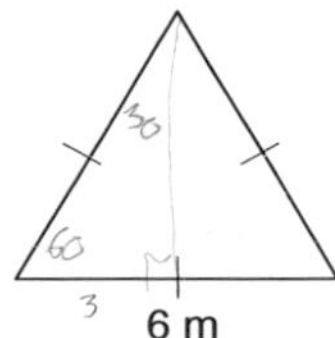

4. The perimeter of a trapezoid is 29 inches. Its nonparallel sides are 4 inches and 5 inches long. If the height of the trapezoid is 3 inches, find its area.
5. A rhombus has a perimeter of 52 units and a diagonal 24 units long. Find the area of the rhombus.
6. The area of an isosceles trapezoid is 36 square centimeters. The perimeter is 28 centimeters. If a leg is 5 centimeters long, find the height of the trapezoid.

Lesson 10-5 **Find the area of each regular polygon described. Round to the nearest tenth.**

1. triangle with an apothem length of 5.8 centimeters
2. square with apothem length of 8 inches
3. a hexagon with a side length of 19.1 millimeters
4. a pentagon with side length of 13.0 miles
5. Find the circumference and the area of a circle with radius of 18 inches. Round to the nearest tenth.

Find the area of each shaded region. Assume that all polygons are regular. Round to the nearest tenth.

6. 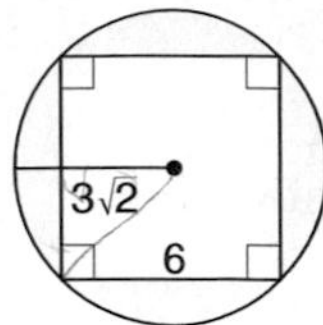

7. 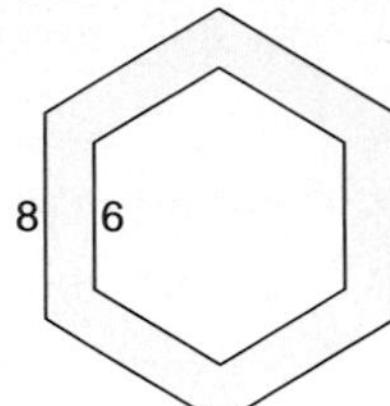

8. 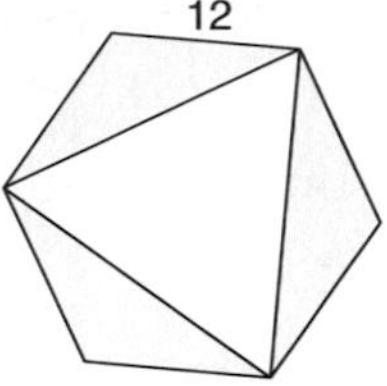

Lesson 10-6 **Find the probability that a point chosen at random in each figure lies in the shaded region.**

1. 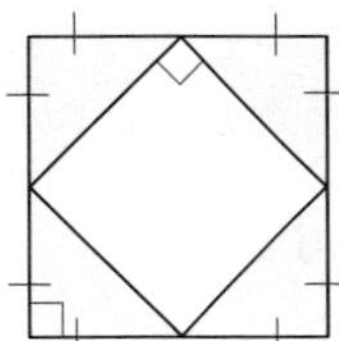

2. 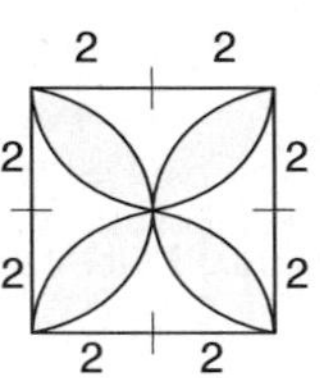

3. 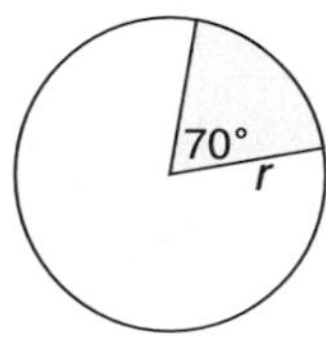

4. A, B, C, D, and E are consecutive collinear points such that $\overline{AB} \cong \overline{BD}$, $\overline{BC} \cong \overline{CD}$, and $\overline{AB} \cong \overline{DE}$.
   a. Draw a diagram of the figure.
   b. What is the probability that a random point will fall on $\overline{AC}$?
   c. What is the probability that a random point will fall on $\overline{AD}$?

Lesson 10-7 **Find the degree of each node in each network.**

1. 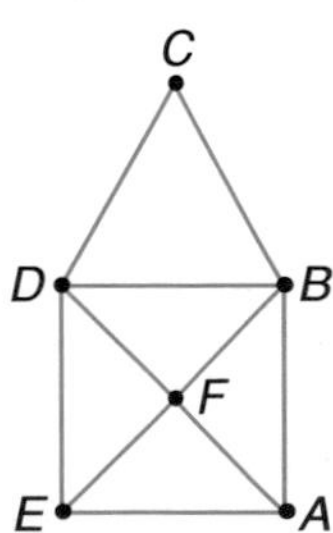

2. 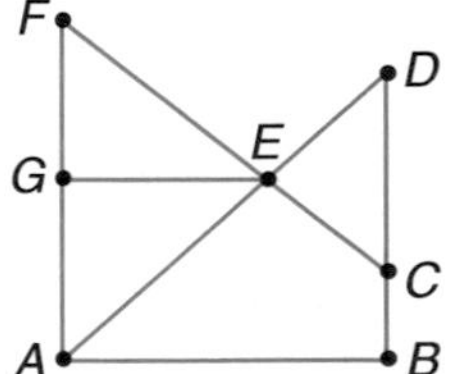

3. 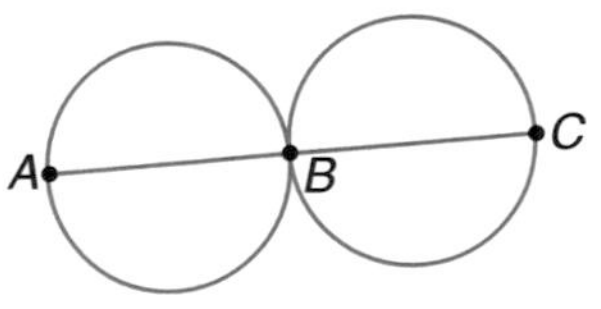

Determine if each network is traceable and complete. If a network is not complete, name the edges that need to be added to make it complete.

4. 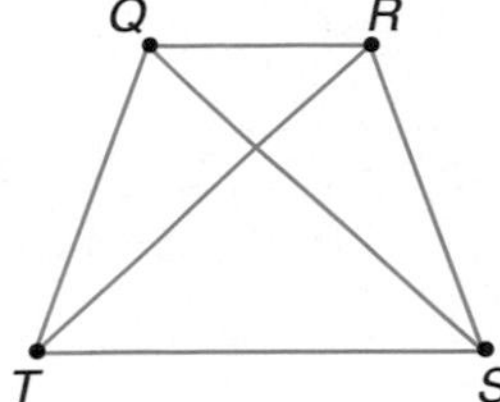

5. 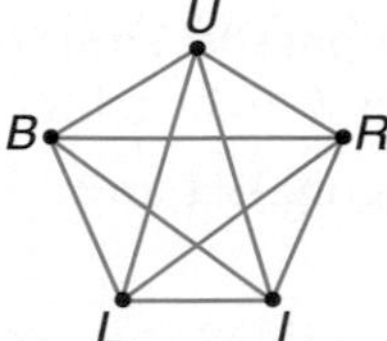

6. 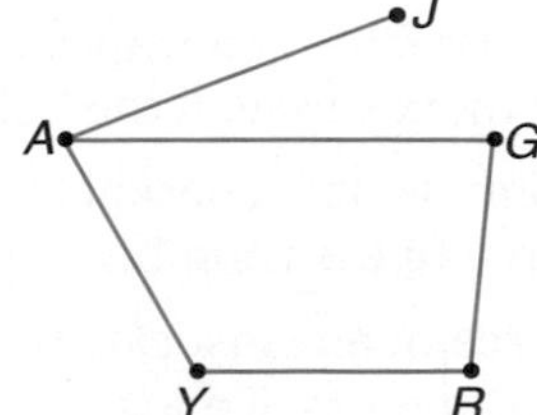

Lesson 11–1 Various views of a solid figure are given. The edge of one block represents one unit of length. A dark segment indicates a break in the surface. Make a model of each figure. Then draw the back view of the figure.

1. 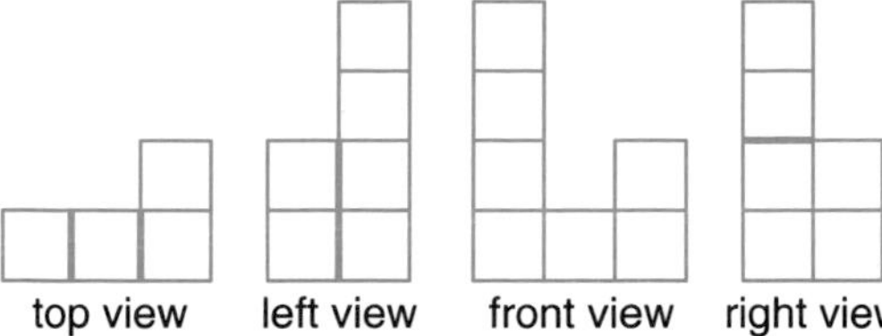

2. 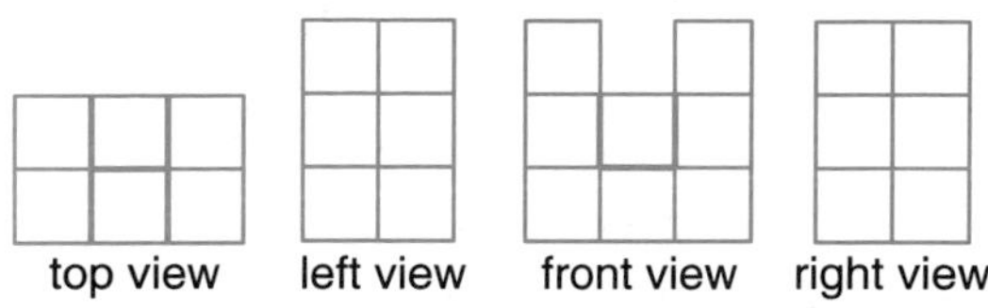

The corner view of a figure is given. Draw the top, left, front, right, and back views of the figure.

3. 

4. 

5. 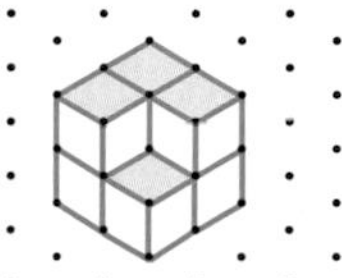

Lesson 11–2 Use isometric dot paper to draw each polyhedron.

1. a rectangular prism 4 units high, 7 units long, and 2 units wide
2. a cube 3 units on each edge
3. a triangular prism 4 units high, whose bases are equilateral triangles with sides 2 units long

Use rectangular dot paper to draw a net for each solid. Then find the surface area of the solid.

4. 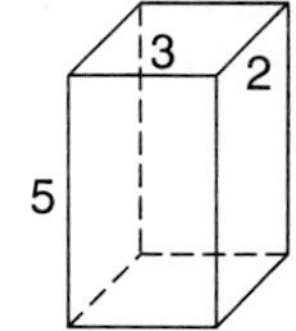

5. 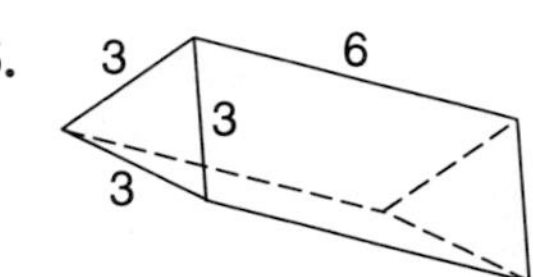

6. 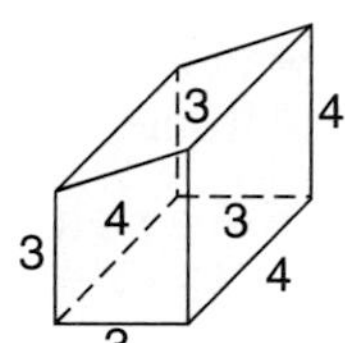

Lesson 11–3 Find the surface area of each right prism. Round to the nearest tenth.

1. 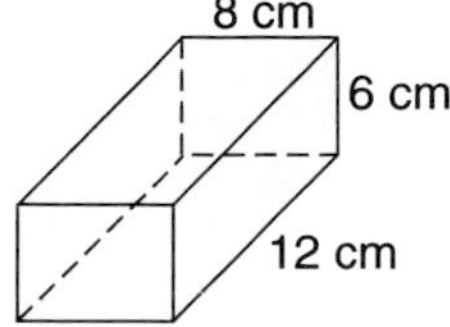

2. 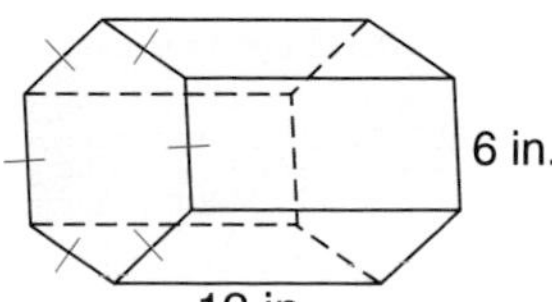

3. 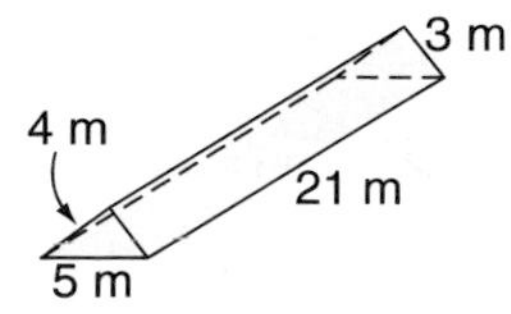

Find the lateral area and surface area of each right cylinder. Round to the nearest tenth.

4. 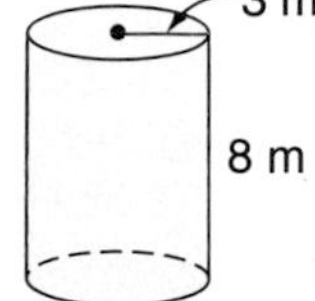

5. 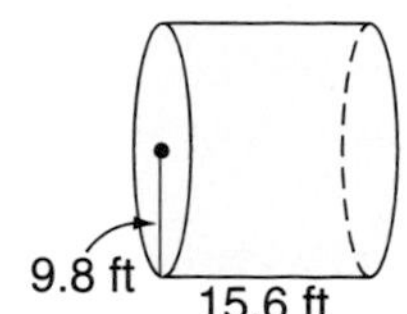

6. 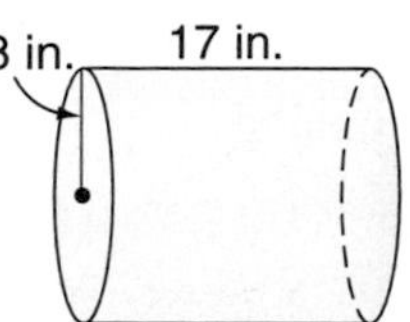

EXTRA PRACTICE

EXTRA PRACTICE

Lesson 11-4 **Find the surface area of each solid. Round to the nearest tenth.**

1. 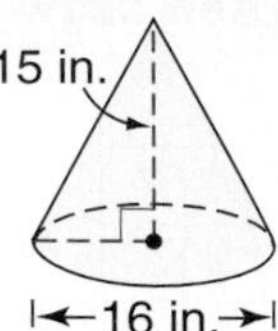

2. 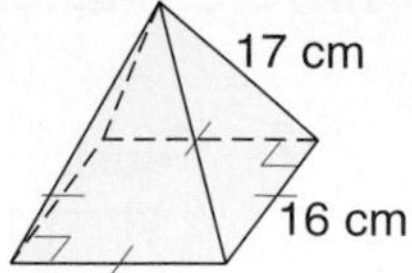

3. 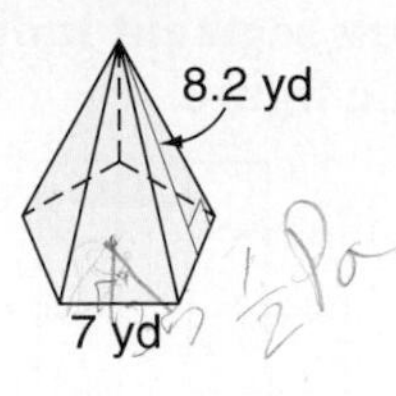

4. 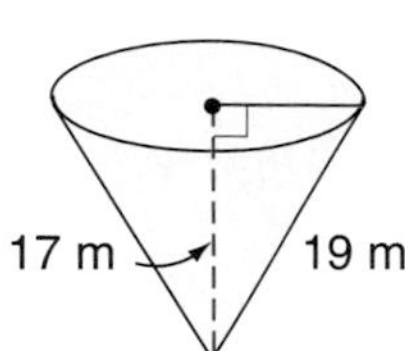

5. 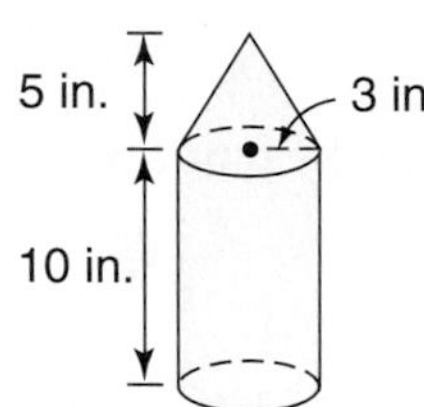

6. 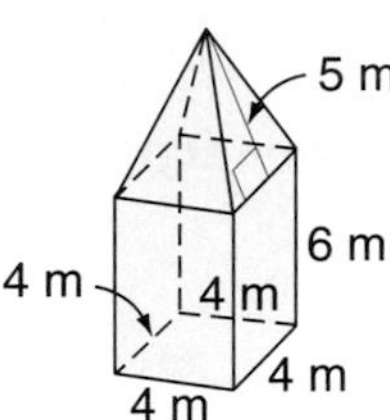

Lesson 11-5 **Find the volume of each prism or right cylinder. Round to the nearest tenth.**

1. 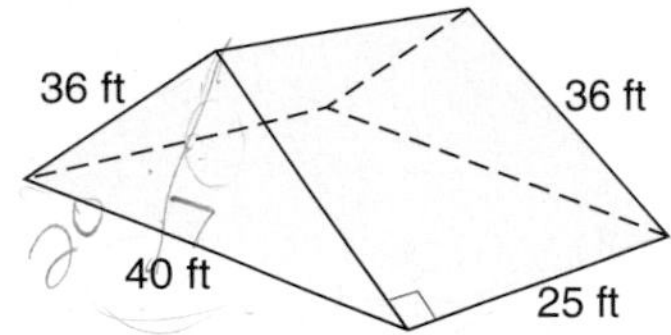

2. 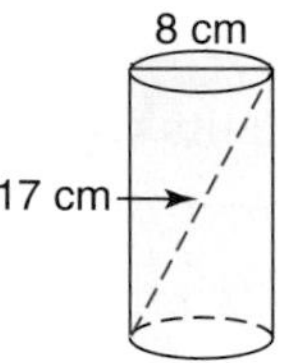

3. 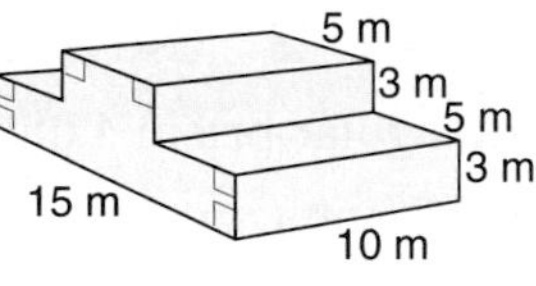

4. The base of a right prism has an area of 17.5 square centimeters. The prism is 14 centimeters high. Find the volume of the prism.
5. Find the volume of a right cylinder whose radius is 3.2 centimeters and height is 10.5 centimeters.
6. Find the volume of a right prism whose base has an area of 16 square feet and a height of 4.2 feet.
7. Find the volume of a cube for which a diagonal of one of its faces measures 12 meters.

Lesson 11-6 **Find the volume of each pyramid or cone. Round to the nearest tenth.**

1. 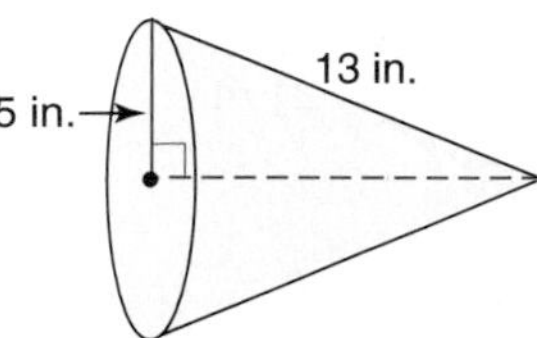

2. 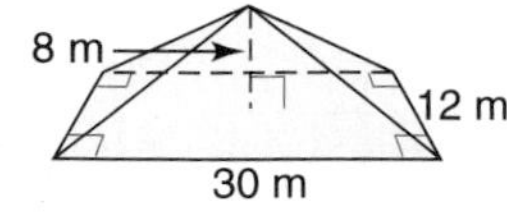

3. 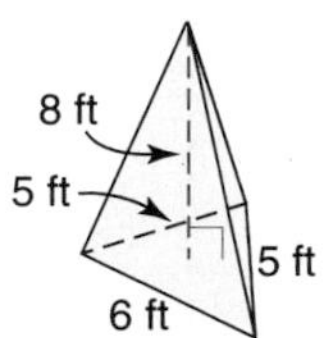

4. 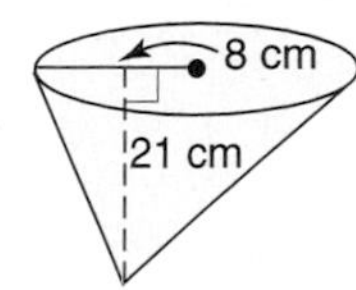

5. 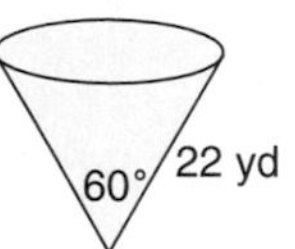

6. 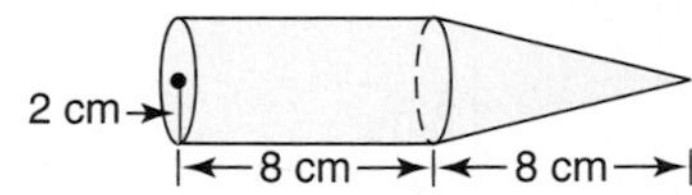

Lesson 11-7 **Determine whether each statement is *true* or *false*.**

1. All diameters of a sphere are congruent.
2. A plane and a sphere may intersect in exactly two points.
3. A diameter of a great circle is a diameter of the sphere.
4. All great circles of the same sphere are congruent.

In the figure, P is the center of the sphere, and plane $\mathcal{B}$ intersects the sphere in $\odot R$.

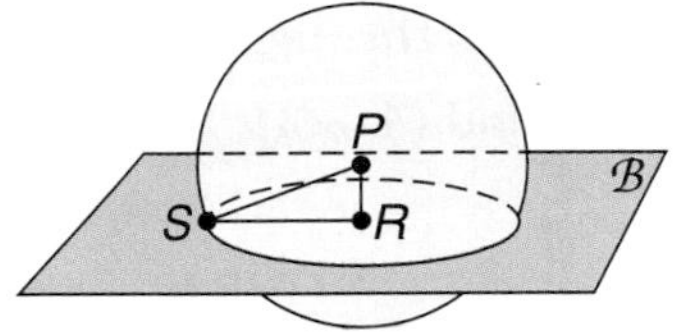

5. Suppose $PS = 15$ and $PR = 9$, find RS.
6. Suppose $PS = 26$ and $RS = 24$, find PR.

Find the surface area and volume of each sphere described below. Round to the nearest tenth.

7. The diameter is 400 feet long.
8. A great circle has a circumference of 18.84 meters.

Lesson 11-8 **Determine if each pair of solids is *similar*, *congruent*, or *neither*.**

1.
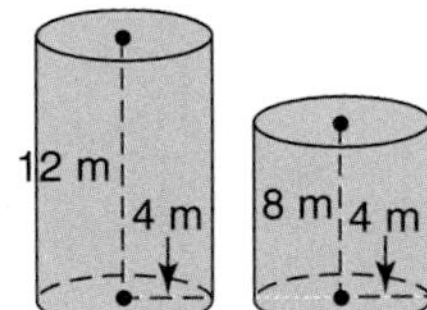

2.
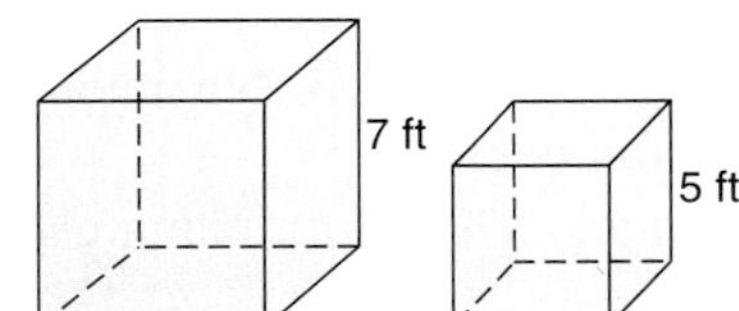

3.
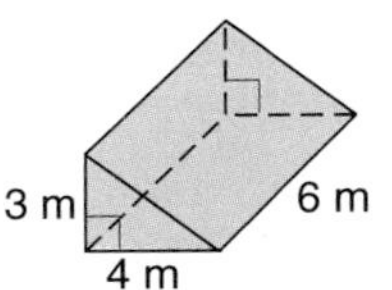

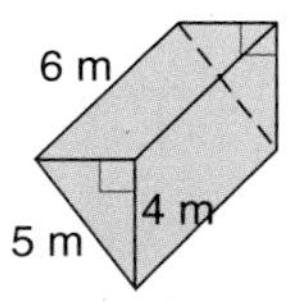

The two right rectangular prisms shown at the right are similar.

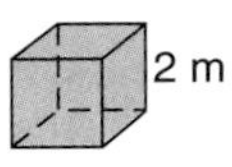

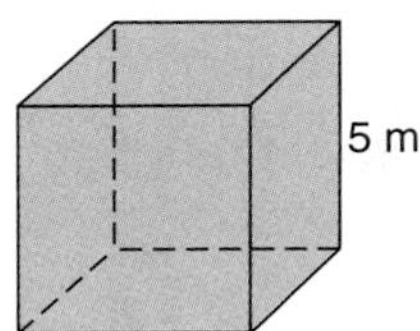

4. Find the ratio of the surface areas.
5. Suppose the volume of the smaller prism is 18 cubic meters. Find the volume of the larger prism.

Lesson 12-1 **Find the x- and y-intercepts and slope of the graph of each equation.**

1. $4x - y = 4$
2. $x = 4$
3. $x + 2y = 6$
4. $3x - 6y = 6$

Graph each equation. Explain the method you used to draw the graph.

5. $y = 4x - 2$
6. $5x + 2y = 10$
7. $y = 2x - 10$
8. $3x - y = 6$

Use the description to graph each line.

9. $m = -\frac{1}{3}$; $b = -2$
10. parallel to y-axis through $A(2, -1)$
11. perpendicular to the line $y = 2x - 4$ through $G(0, 3)$
12. $m = \frac{2}{3}$; passes through $D(-2, 1)$

EXTRA PRACTICE

EXTRA PRACTICE

Lesson 12-2 **Write the equation in slope-intercept form of the line having the given slope and passing through the point with the given coordinates.**

1. $-4, (-3, -2)$
2. $\frac{1}{6}, (12, -3)$
3. $0, (6, 7)$

Write the equation in slope-intercept form of the line that satisfies the given conditions.

4. $m = \frac{3}{4}$; y-intercept $= 8$
5. parallel to $y = -4x + 1$; passes through the point at $(-3, 1)$
6. perpendicular to the y-axis; passes through the point at $(-8, 2)$
7. passes through points at $(0, 3)$ and $(4, -3)$

Lesson 12-3 **Determine whether the three points listed are collinear.**

1. $A(9, 0), B(4, 2), C(2, -1)$
2. $X(6, 9), Y(3, -1), Z(4, 0)$
3. $L(0, 4), M(2, 3), N(-4, 6)$
4. The table below lists the Federal Income Tax due from a single person with the given taxable income for 1995.

Taxable Income	11,905	7412	22,898	19,054	10,995	3268	18,753
Tax Due	1789	1114	3491	2861	1646	491	2816

   a. Draw a scatter plot to show how taxable income x and Federal Income Tax due y are related.
   b. Write an equation that relates a singe person's taxable income and their approximate Federal Income Tax due.
   c. Angela's taxable income for 1995 was $12,982. Approximately how much did she owe in Federal Income Tax?

The vertices of $\triangle RST$ are $R(-6, -8)$, $S(6, 4)$, and $T(-6, 10)$.

5. Write the equations of the lines containing the medians of $\triangle RST$.
6. Write the equations of the lines containing the altitudes of $\triangle RST$.

Lesson 12-4 **Name the missing coordinates in terms of the given variables.**

1. $\triangle RST$ is isosceles and right.

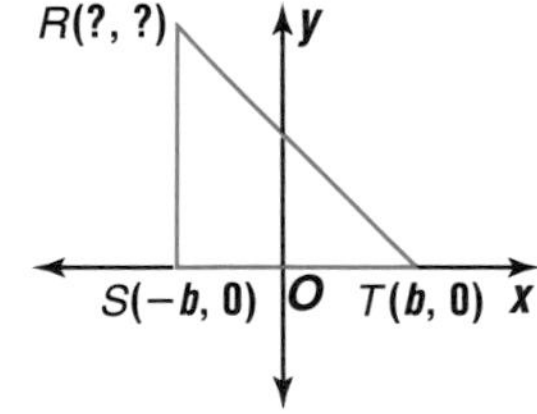

2. $ABCD$ is an isosceles trapezoid.

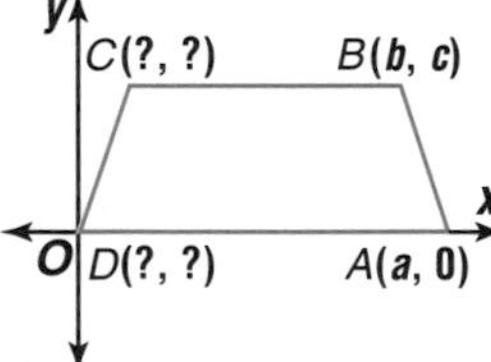

Prove using a coordinate proof.

3. $\triangle ABC$ is an isosceles triangle.

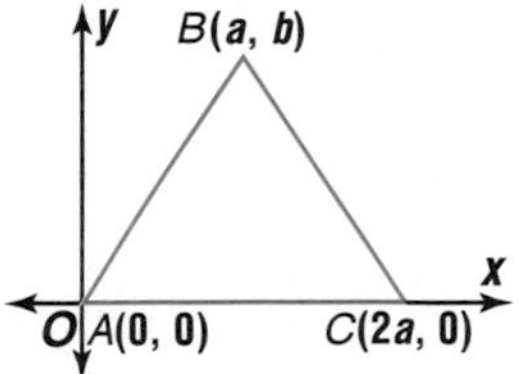

4. $\triangle PQR$ is a right triangle.

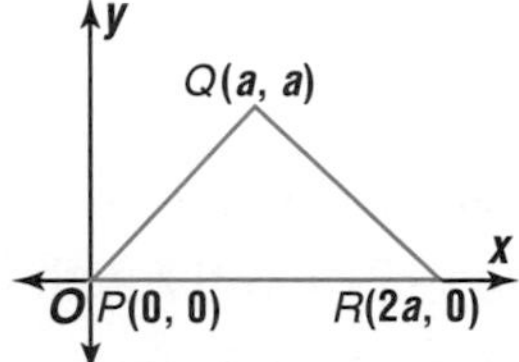

Lesson 12-5 **Sketch each vector. Then find the magnitude to the nearest tenth and the direction to the nearest degree.**

1. $\overrightarrow{AB}$ if $A(4, 2)$ and $B(8, 6)$
2. $\overrightarrow{RS}$ if $R(-2, 4)$ and $S(5, 10)$

Find the sum or difference of the given vectors.

3. $\begin{bmatrix} 2 \\ -1 \end{bmatrix} + \begin{bmatrix} -6 \\ 3 \end{bmatrix}$
4. $\begin{bmatrix} 3 \\ 5 \end{bmatrix} - \begin{bmatrix} 1 \\ 7 \end{bmatrix}$
5. $\begin{bmatrix} 7 \\ 9 \end{bmatrix} + \begin{bmatrix} 6 \\ -4 \end{bmatrix} - \begin{bmatrix} 12 \\ 3 \end{bmatrix}$

Given the quadrilateral *RPTQ*, complete each statement.

6. $\overrightarrow{RP} + \overrightarrow{PT} = \underline{\ ?\ }$
7. $\overrightarrow{PH} + \overrightarrow{HR} = \underline{\ ?\ }$

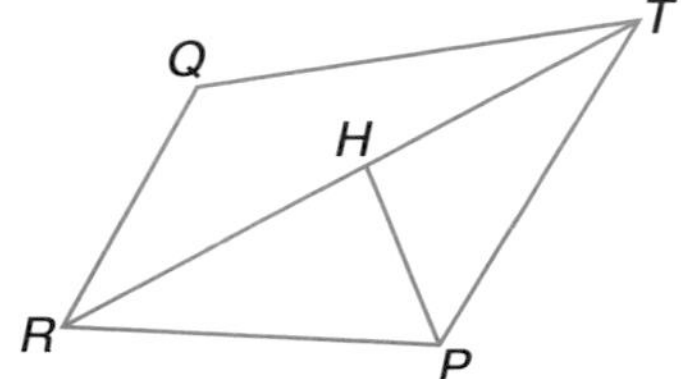

Given $\vec{v} = (2, 5)$ and $\vec{u} = (7, 1)$, represent each of the following as an ordered pair.

8. $\vec{v} + \vec{u}$
9. $\vec{v} + 2\vec{u}$
10. $2\vec{u} + 3\vec{v}$

Lesson 12-6 **Plot each point in a three-dimensional coordinate system.**

1. $(2, -1, 4)$
2. $(3, 1, -4)$

Determine the distance between each pair of points, and determine the coordinates of the midpoint of the segment connecting them.

3. $A(0, -2, 5)$ and $B(-3, 4, -2)$
4. $J(9, 1, 0)$ and $K(5, -7, 4)$

Determine the coordinates of the center and the measure of the radius for each sphere whose equation is given.

5. $(x - 6)^2 + (y + 5)^2 + (z - 1)^2 = 81$
6. $x^2 + (y - 2)^2 + (z - 4)^2 = 4$

Write the equation of the sphere using the given information.

7. The center is at $(6, -1, 3)$, and the radius is 12.
8. A diameter has endpoints at $(-3, 5, 7)$ and $(5, -1, 5)$.
9. The center is at $(0, -2, 1)$, and the diameter is 16.
10. The center is at $(2, -2, -1)$, and one endpoint of a radius is located at $(5, -5, 2)$.

Lesson 13-1 **Draw a figure and describe the locus of points in a plane that satisfy each set of conditions.**

1. all points that are 3 inches from a circle with a radius of 6 inches
2. all points equidistant from the endpoints of a given line segment
3. all points that are equidistant from two intersecting lines
4. all points that are the midpoints of parallel chords of a given circle

Draw a figure and describe the locus of points in space that satisfy each set of conditions.

5. all points 4 meters from a given line m
6. all points that are equidistant from two intersecting planes
7. all points that are 6 inches from a plane
8. all points equidistant from all points on a circle

Lesson 13–2 Graph each pair of equations to find the locus of points that satisfy the graphs of both equations.

1. $3x - 2y = 10$
   $x + y = 0$
2. $x + 2y = 7$
   $y = 2x + 1$
3. $x + y = 6$
   $x - y = 2$
4. $y = x - 1$
   $x + y = 11$

Use either substitution or elimination to find the locus of points that satisfy the graphs of both equations.

5. $y = 3x$
   $x + 2y = -21$
6. $x - y = 5$
   $x + y = 25$
7. $x - y = 6$
   $x + y = 5$
8. $y = x - 1$
   $4x - y = 19$
9. $x - 2y = 5$
   $3x - 5y = 8$
10. $9x + 7y = 4$
    $6x - 3y = 18$

Lesson 13–3 Describe the different possible intersections of the figures.

1. a sphere and a plane
2. two concentric circles and a line
3. a line and a sphere
4. a circle and a plane

Draw a diagram to find the locus of points that statisfy the conditions. Then describe the locus of points.

5. all points in a coordinate plane that are 5 units from the graph of $x = 6$ and equidistant from the graphs of $y = 1$ and $y = 7$
6. all points in a plane that are 2 inches from a given line and 5 inches from a point on the line
7. all points in a plane that are equidistant from two intersecting lines and 3 units from the point of intersection

Describe the geometric figure whose locus in space satisfies the graph of each equation.

8. $(x - 2)^2 + (y + 4)^2 = 36$
9. $(x - 3)^2 + (y - 4)^2 + (z - 5)^2 = 16$

Lesson 13–4 Describe the transformations that occurred in the mappings.

1. 

2. 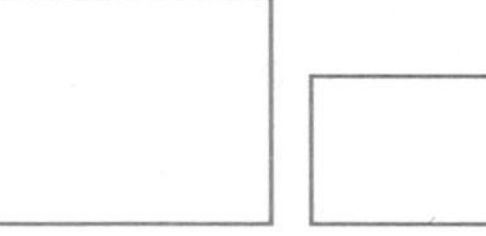

Each figure below has a preimage that is an isometry. Write the image of each preimage given in Exercises 3–8.

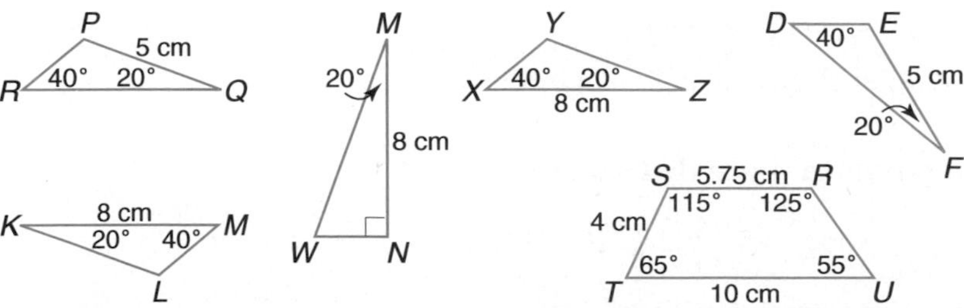

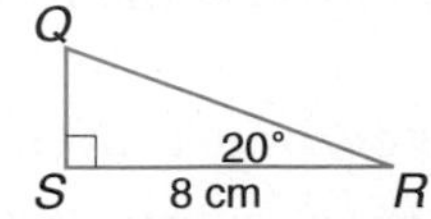

3. $\triangle MWN$
4. $\triangle PQR$
5. $\triangle SRQ$
6. $\triangle LMK$
7. quadrilateral $RSTU$
8. $\triangle ZYX$

Lesson 13-5 For the figure at the right, name the reflection image of each of the following over line ℓ.

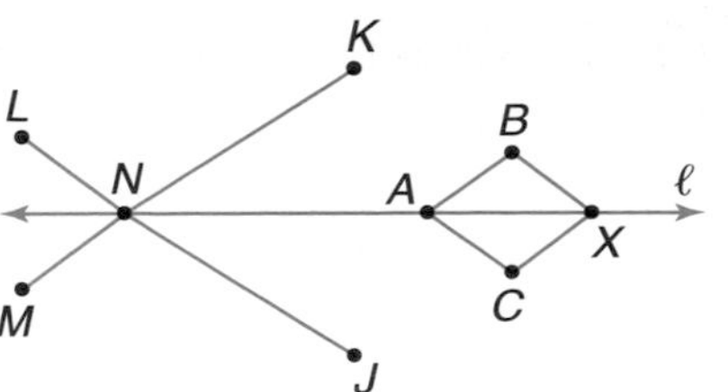

1. M
2. K
3. $\overline{KM}$
4. $\triangle BXA$
5. N
6. $\angle LNK$

Copy each figure. Use a straightedge to draw the reflection image of each figure over line m.

7. 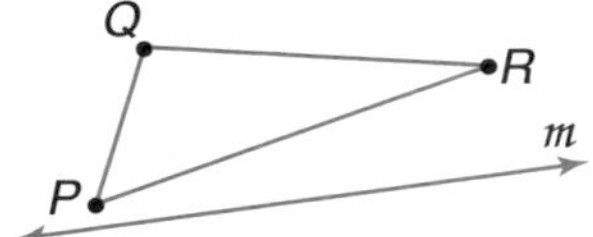

8. 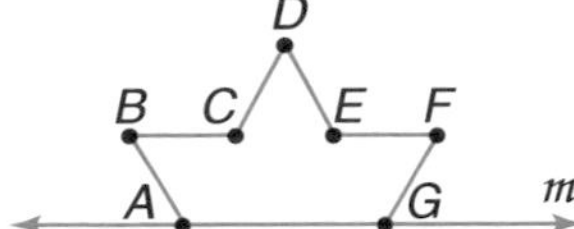

9. 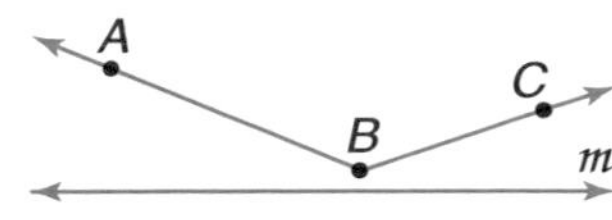

For each figure, determine whether p is a line of symmetry. Write *yes* or *no*. Explain your answer.

10. 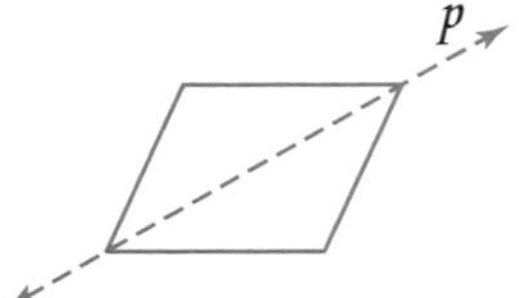

11. 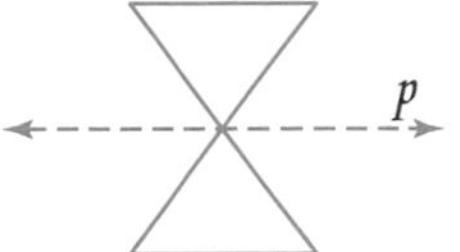

12. 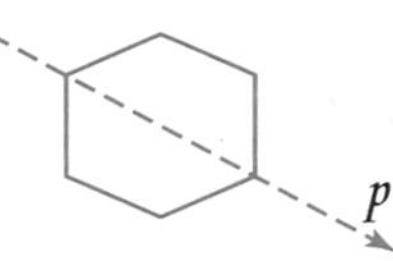

Graph each figure on a coordinate plane. Draw the reflection image for each over the indicated line of reflection.

13. $\overline{ST}$: $S(-1, 2)$, $T(3, 4)$
    line of reflection: y-axis
14. $\triangle XYZ$: $X(1, 5)$, $Y(0, -2)$, $Z(-1, 1)$
    line of reflection: $x = 1$

Lesson 13-6 For each of the following, lines ℓ and m are parallel. Determine whether each red figure is a translation image of the blue figure. Write *yes* or *no*. Explain your answer.

1. 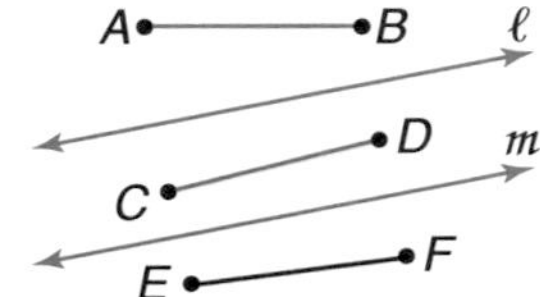

2. 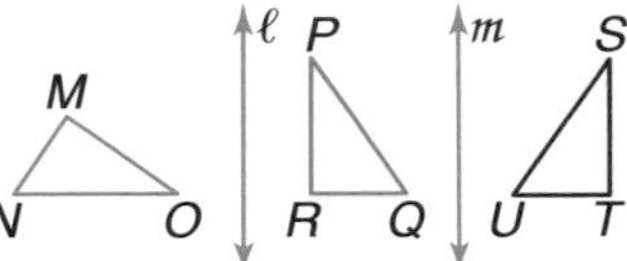

3. 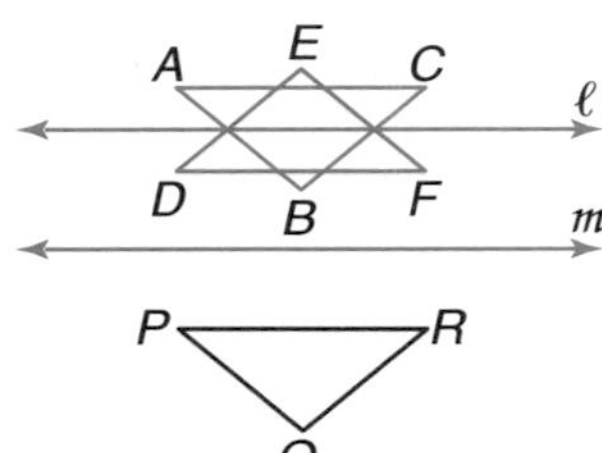

Graph each of the following preimages on a coordinate plane. Then graph the image after performing the translation given, and list the coordinates of the vertices of that image.

4. $\overline{RS}$, $R(2, 5)$, $S(-4, -2)$; slide 3 right, 4 down
5. $\overline{AB}$, $A(2, -1)$, $B(0, 2)$; reflect over the graph of $y = 1$ and then over the graph of $y = 4$
6. $\triangle DFG$, $D(-2, 5)$, $F(3, 6)$, $G(2, 1)$; slide D to $D'(0, 2)$
7. $\triangle XYZ$, $X(-1, -2)$, $Y(0, 3)$, $Z(1, 1)$; reflect over the graph of $y = -2$ and then over the graph of $y = 0$

Copy each figure on a coordinate plane. Then find the translation image of each geometric figure with respect to the parallel lines m and ℓ.

8. 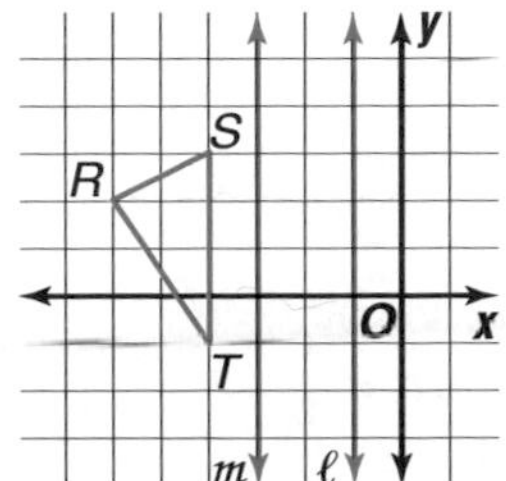

9. 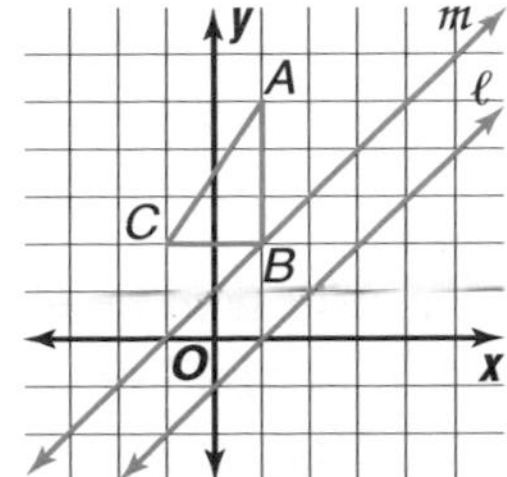

10. 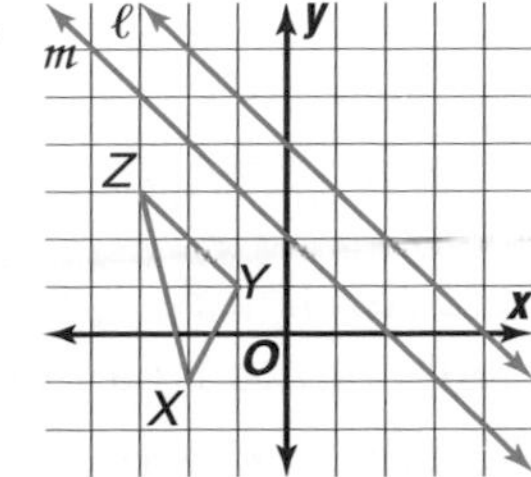

Lesson 13-7 For each of the following, determine whether the indicated composition of reflections is a rotation. Explain your answer.

1. 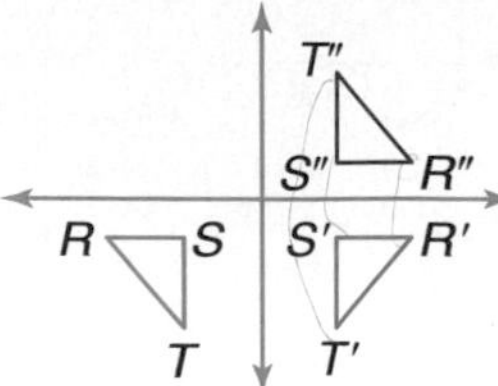

2. 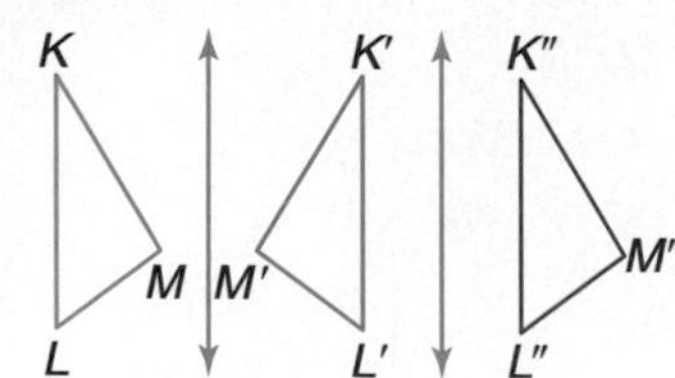

3. 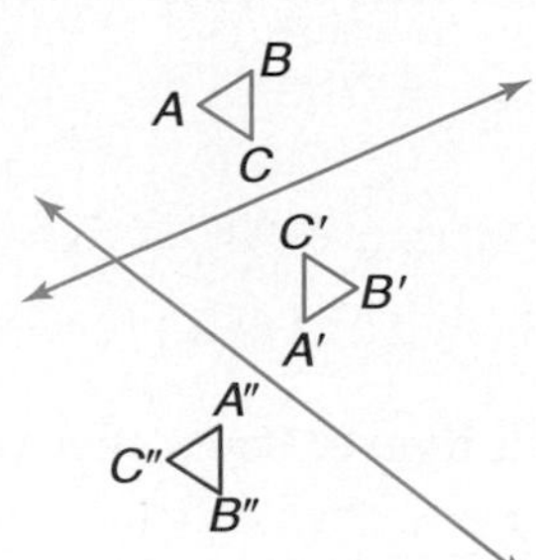

Copy each figure. Then use a composite of reflections to find the rotation image with respect to lines ℓ and t.

4. 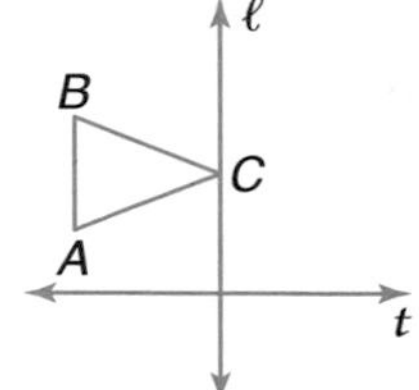

5. 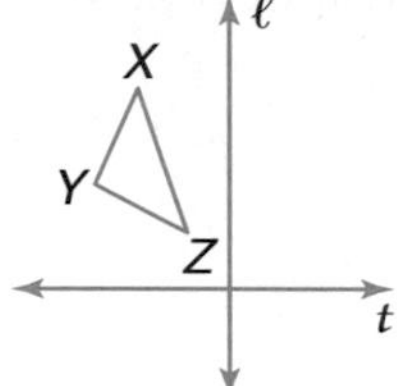

6. 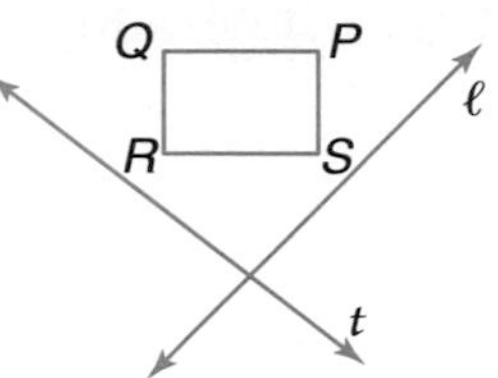

Copy each figure. Then use the angle of rotation to find the image with respect to lines ℓ and t.

7. 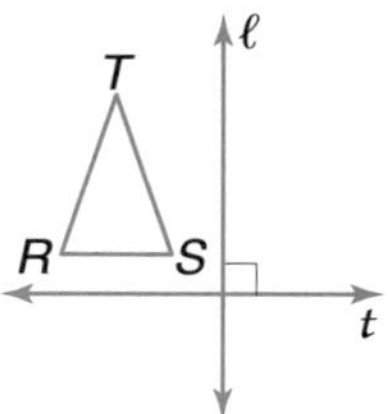

8. 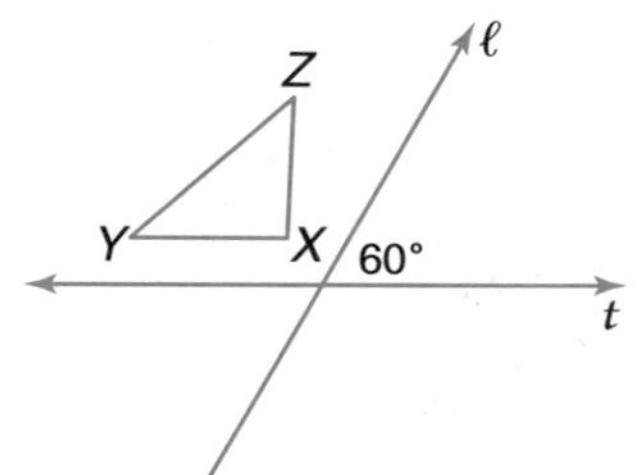

9. 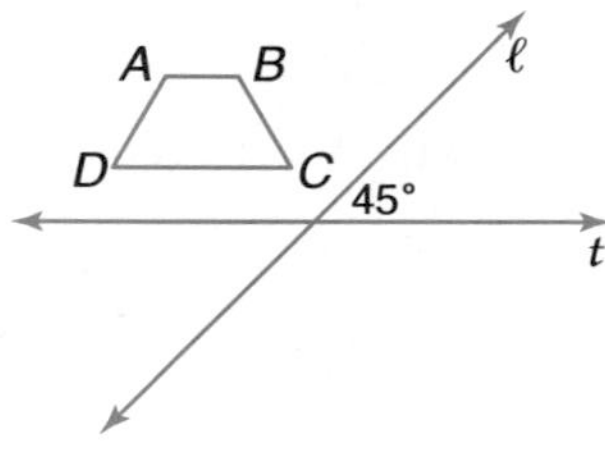

Lesson 13-8 Find the measure of the dilation image of $\overline{AB}$ with the given scale factor.

1. $AB = 4, k = -3$
2. $AB = \frac{3}{5}, k = 25$
3. $AB = 18, k = \frac{2}{3}$

A dilation with center C and a scale factor of k maps A onto D and B onto E. Find $|k|$ for each dilation. Then determine whether each dilation is an enlargement, a reduction, or a congruence transformation.

4. $CE = 18, CB = 9$
5. $AB = 3, DE = 1$
6. $AB = 3, DE = 4$

For each figure, a dilation with center C produced the figure in red. What is the scale factor for each transformation?

7. 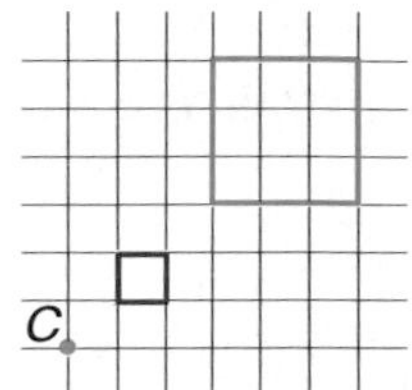

8. 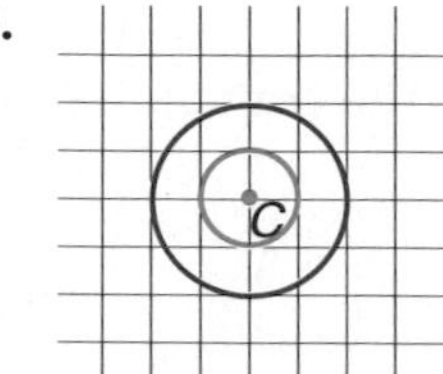

9. 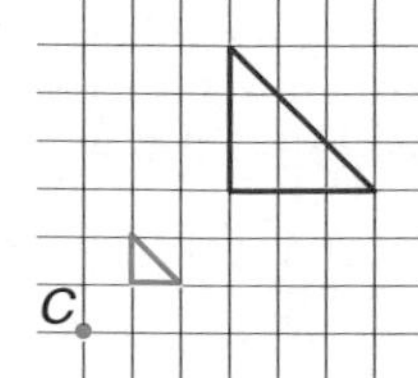

Graph each set of ordered pairs. Then connect the points in order. Using the origin as the center of dilation and a scale factor of 2, draw the dilation image. Repeat using a scale factor of $\frac{1}{2}$.

10. $(4, 2), (7, -2), (-1, -3)$
11. $(2, 0), (3, 4), (-1, 4), (-4, 2)$

Draw and label a figure for each relationship.

1. Lines ℓ, m, and n all intersect at point X.
2. Planes $\mathcal{Q}$ and $\mathcal{R}$ do not intersect.
3. Plane $\mathcal{P}$ contains point A but does not contain $\overline{BC}$.
4. $A(4, 3)$, $B(-2, -4)$, $C(-3, 4)$, and $D(0, -1)$ lie on a coordinate plane.

Refer to the coordinate grid at the right to answer each question.

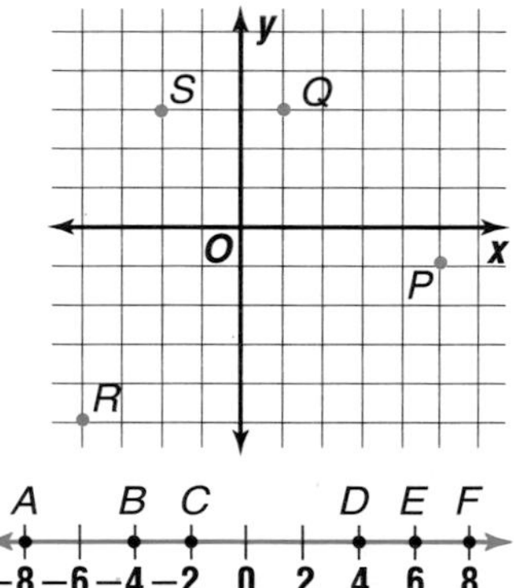

5. What ordered pair names point S?
6. What is the length of $\overline{PQ}$?
7. What are the coordinates of the midpoint of $\overline{QR}$?
8. What is the y-coordinate of any point collinear to Q and S?

Refer to the number line to answer each question.

A B C D E F
−8 −6 −4 −2 0 2 4 6 8

9. What is the measure of $\overline{AC}$?
10. What is the coordinate of the midpoint of $\overline{CF}$?
11. What segment is congruent to $\overline{BF}$?
12. What is the coordinate of G if C is between D and G and $DG = 14$?

Refer to the figure at the right to answer each question.

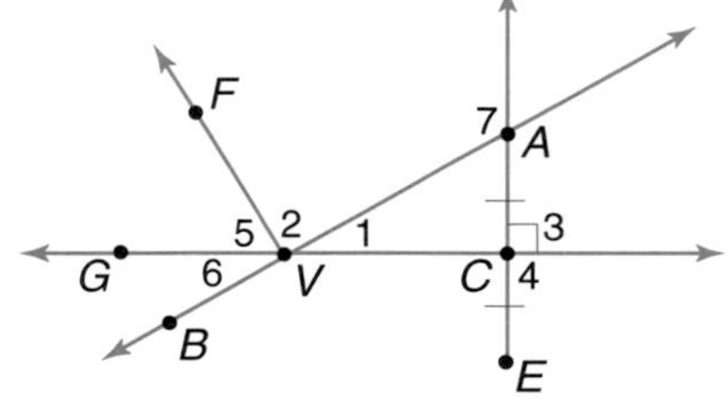

13. Which of the numbered angles appears to be obtuse?
14. Name a pair of congruent supplementary angles.
15. Name a pair of adjacent angles that do not form a linear pair.
16. If $\angle 1 \cong \angle 5$, does $\overrightarrow{VG}$ bisect $\angle BVF$?
17. *True* or *false*: $\overleftrightarrow{CG}$ bisects $\overline{AE}$.
18. If $AC = 4x + 1$ and $CE = 16 - x$, find AE.
19. If $m\angle BVF = 7x - 1$ and $m\angle FVA = 6x + 12$, is $\overleftrightarrow{AB} \perp \overrightarrow{VF}$?
20. Which two angles must be complementary if $\angle AVF$ is a right angle?
21. If $m\angle 5 = 3x + 14$, $m\angle 6 = x + 30$, and $m\angle FVB = 9x - 11$, find $m\angle FVB$.

Use the city block at the right to answer each question.

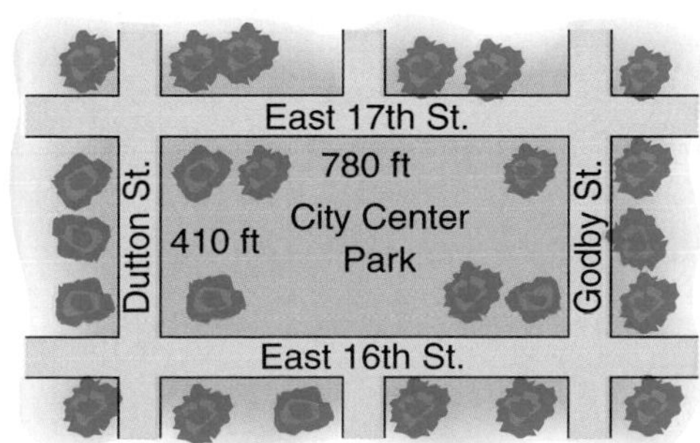

22. How far will Kelli walk her dog if they walk around the block?
23. The entire block is a park. What is the area of the park?
24. The Department of Parks and Recreation wants to pave a path that goes through the park at a diagonal from corner to corner. How long will the path be?
25. **Sports** Three darts are thrown at the target shown at the right. If we assume that each dart lands within a ring, how many different point totals are possible? List them.

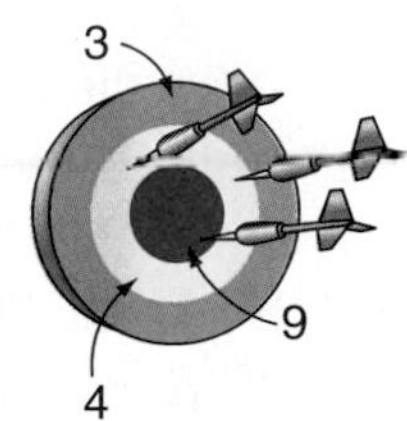

CHAPTER 2 TEST

Determine if each conjecture is *true* or *false*. Explain your answer and give a counterexample for any false conjecture.

1. **Given:** x is a real number.
   Conjecture: $-x < 0$
2. **Given:** $\angle 1 \cong \angle 2$
   Conjecture: $\angle 2 \cong \angle 1$
3. **Given:** $\angle 1$ and $\angle 2$ form a linear pair.
   Conjecture: $m\angle 1 + m\angle 2 = 180$
4. **Given:** $3x^2 = 48$
   Conjecture: $x = 4$

Write each conditional statement in if-then form. Identify the hypothesis and conclusion of each conditional. Then write the converse, inverse, and contrapositive of each conditional.

5. A rolling stone gathers no moss.
6. An apple a day keeps the doctor away.
7. Two parallel planes do not intersect.
8. Two points make a line.
9. Through any two points there is exactly one line.

Determine if a valid conclusion can be reached from the two true statements using the Law of Detachment or the Law of Syllogism. If a valid conclusion is possible, state it and the law that is used. If a valid conclusion does not follow, write *no conclusion*.

10. (1) Wise investments with Petty-Bates pay off.
    (2) Investments that pay off build for the future.
11. (1) Perpendicular lines intersect.
    (2) Lines ℓ and m are perpendicular.
12. (1) Vertical angles are congruent.
    (2) $\angle 1$ is congruent to $\angle 2$.
13. (1) All integers are real numbers.
    (2) 7 is an integer.

Name the property of equality that justifies each statement.

14. If $m\angle A = m\angle B$, then $m\angle B = m\angle A$.
15. If $x + 9 = 12$, then $x = 3$.
16. If $2ST = 4UV$, then $ST = 2UV$.
17. If $AB = 7$ and $CD = 7$, then $AB = CD$.
18. **Transportation** An empty commuter train picked up passengers on Monday morning at a rate of one passenger at the first stop, three at the second stop, five at the third stop, seven at the fourth, and so on.
    a. How many passengers got on the train at the 15th stop?
    b. What was the total number of passengers on the train after 10 stops?

Copy and complete the following proofs.

19. **Given:** $\overline{AC} \cong \overline{BD}$
    Prove: $\overline{AB} \cong \overline{CD}$

Statements	Reasons
a. $\overline{AC} \cong \overline{BD}$	a. ?
b. $AC = BD$	b. ?
c. ? ?	c. Segment Addition Postulate
d. $AB + BC = BC + CD$	d. ?
e. $AB = CD$	e. ?
f. ?	f. Def. $\cong$ seg.

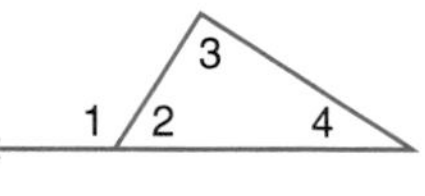

20. **Given:** $m\angle 1 = m\angle 3 + m\angle 4$
    Prove: $m\angle 3 + m\angle 4 + m\angle 2 = 180$

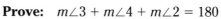

Statements	Reasons
a. ?	a. Given
b. $m\angle 2 = m\angle 2$	b. ?
c. $m\angle 1 + m\angle 2 = m\angle 3 + m\angle 4 + m\angle 2$	c. ?
d. $\angle 1$ and $\angle 2$ form a linear pair.	d. ?
e. $\angle 1$ and $\angle 2$ are supplementary.	e. ?
f. ?	f. Def. suppl. $\angle$s
g. $m\angle 3 + m\angle 4 + m\angle 2 = 180$	g. ?

CHAPTER 3 TEST

In the figure, $\ell \parallel m$. Determine whether each statement is *true* or *false*. Justify your answer.

1. $\angle 1$ and $\angle 14$ are alternate exterior angles.

2. $\angle 5$ and $\angle 11$ are consecutive interior angles.
3. $\angle 2$ and $\angle 6$ are vertical angles.
4. $\angle 6$ and $\angle 12$ are supplementary angles.
5. $\angle 3 \cong \angle 8$
6. $m\angle 7 + m\angle 10 = 180$
7. $m\angle 4 + m\angle 5 + m\angle 11 = 180$

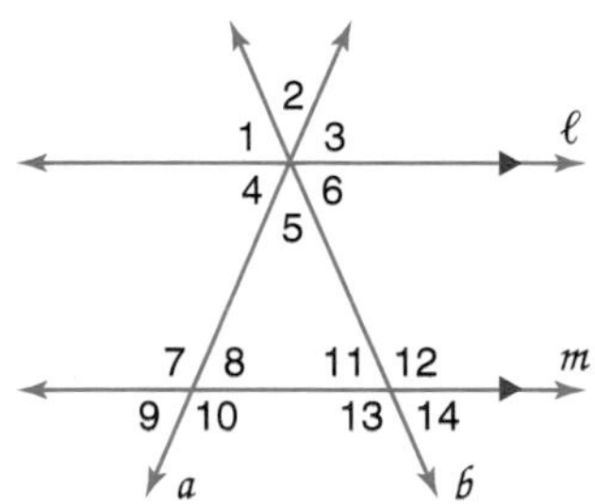

Use the figure at the right to answer each question.

8. If $\angle 5 \cong \angle 4$, which lines are parallel and why?
9. If $\angle 3 \cong \angle 9$, which lines are parallel and why?
10. If $m\angle 4 + m\angle 9 + m\angle 6 = 180$, which lines are parallel and why?

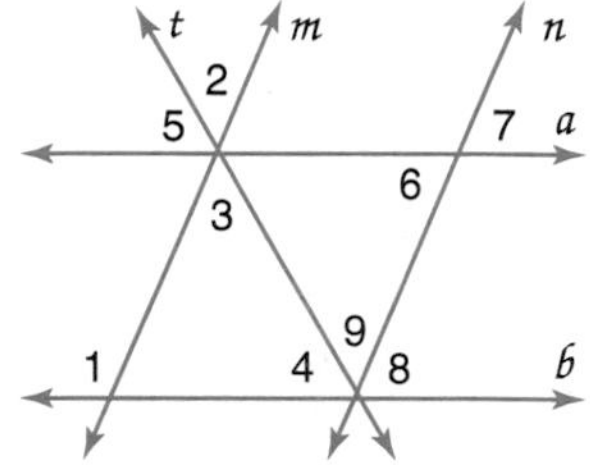

Determine the slope of each line.

11. $\overleftrightarrow{EF}$
12. $\overleftrightarrow{AB}$
13. any line parallel to $\overleftrightarrow{CD}$
14. any line perpendicular to $\overleftrightarrow{EF}$

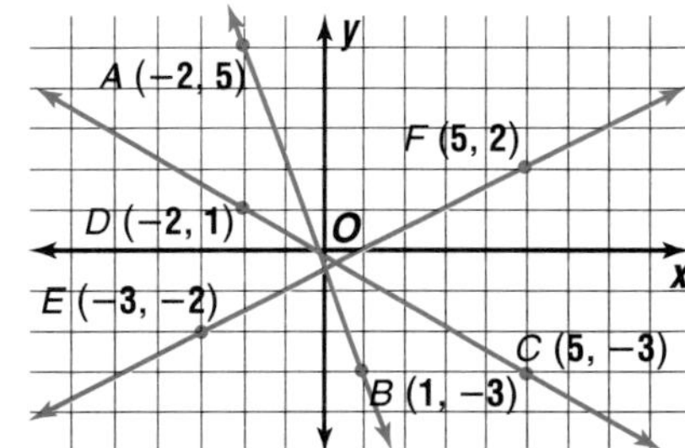

Name the segment in the figure at the right whose length represents the distance between the following points and lines.

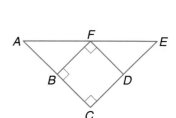

15. from F to $\overleftrightarrow{CE}$
16. from A to $\overleftrightarrow{CE}$
17. from D to $\overleftrightarrow{BC}$
18. Write a two-column proof.

    Given: $\angle 1 \cong \angle 2$

    $\overline{ST} \parallel \overline{PR}$

    Prove: $\angle P \cong \angle R$

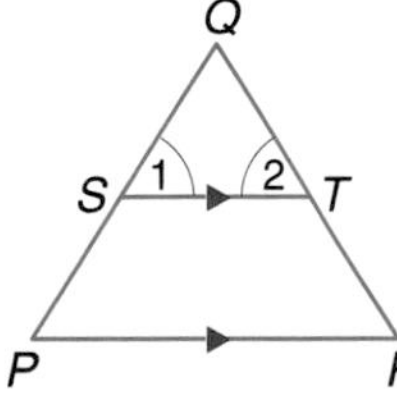

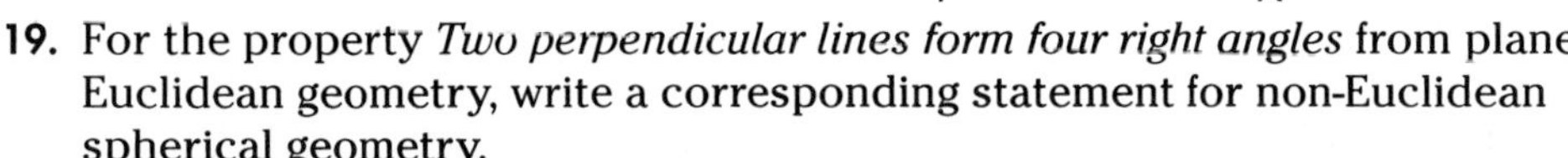

19. For the property *Two perpendicular lines form four right angles* from plane Euclidean geometry, write a corresponding statement for non-Euclidean spherical geometry.
20. **Business** The Carpet Experts cleaning team can clean the carpet in a room that is 10 feet by 10 feet in 20 minutes. Draw a diagram and find how long it would take them to clean a walk-in closet that is 5 feet by 5 feet.

CHAPTER 4 TEST

Use figure *PQRST* to answer each of the following.

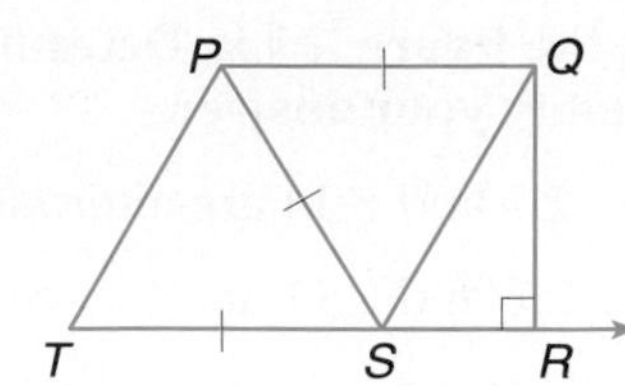

1. Name the isosceles triangle(s).
2. Which triangle is a right triangle?
3. Which side of $\triangle PTS$ is opposite $\angle T$?
4. Name the triangle(s) that appear to be scalene.

In the figure, $\overline{GH} \cong \overline{GL}$, $\overline{GI} \cong \overline{GK}$, $\overline{GJ} \perp \overline{HL}$, $m\angle 3 = 30$, and $m\angle 4 = 20$. Find each measure.

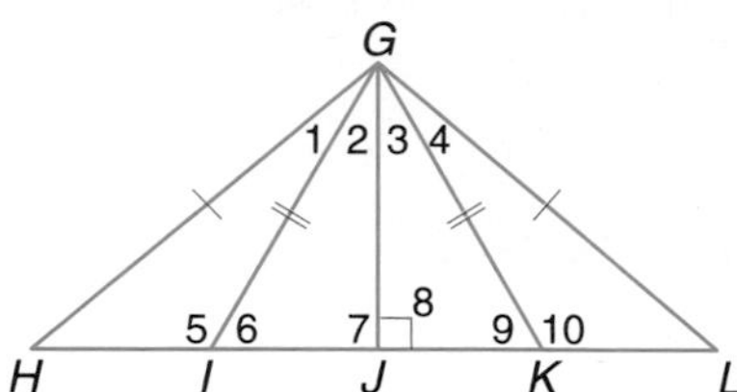

5. $m\angle 8$
6. $m\angle 6$
7. $m\angle 2$
8. $m\angle 10$
9. $m\angle H$
10. $m\angle 1$

Use the figure at the right to answer each of the following.

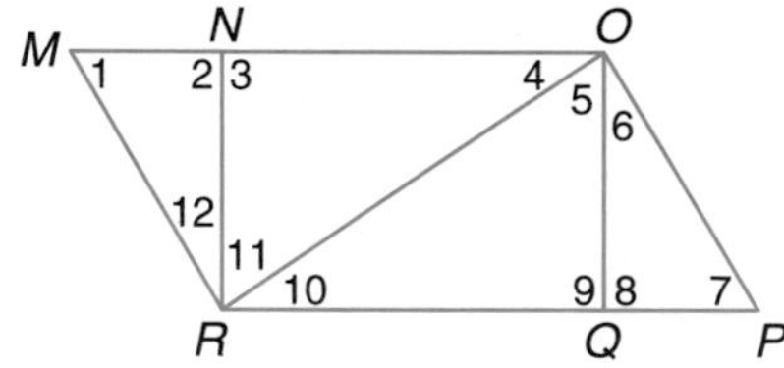

11. $\overline{RO}$ is included between what two angles?
12. If $\angle 1 \cong \angle 7$ and $\overline{MO} \parallel \overline{RP}$, then $\triangle MOR \cong$ _?_ by _?_ .
13. If $\overline{NO} \cong \overline{RQ}$ and $\overline{NR} \cong \overline{OQ}$, then $\triangle ONR \cong$ _?_ by _?_ .
14. If $\overline{MO} \parallel \overline{RP}$ and $\overline{QO} \parallel \overline{RN}$, then $\triangle NRO \cong$ _?_ by _?_ .
15. If $\overline{NR} \perp \overline{MO}$, $\overline{RP} \perp \overline{OQ}$, $\overline{MN} \cong \overline{QP}$, and $\overline{NR} \cong \overline{OQ}$, then $\triangle MNR \cong$ _?_ by _?_ .

For each triangle, find the value of *x*.

16.

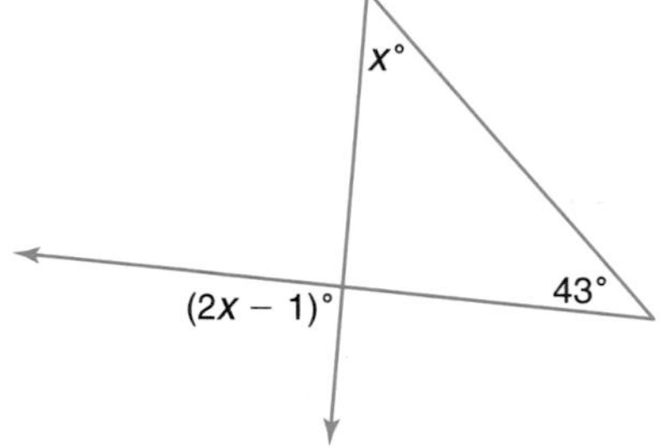

17.

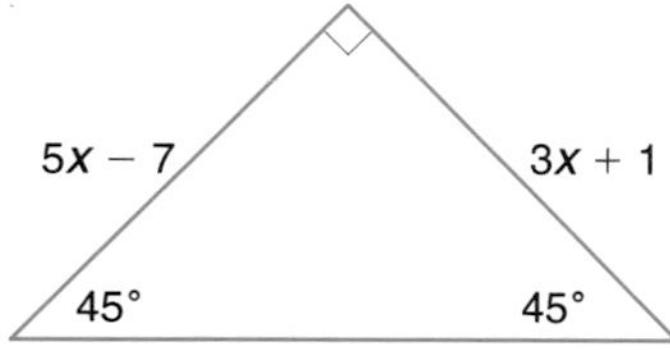

18. **Eliminate the Possibilities** Umeko, Jim, and Gwen each participate in an extracurricular activity and have an after-school job. One of them is in the Spanish Club, one is in the Drama Club, and one is in the marching band. One of them is a pizza delivery person, one is a math tutor, and one is a lifeguard at the community pool. Jim is tutoring the Drama Club member's brother in Algebra. The Spanish Club member cannot swim or drive a car. The lifeguard is teaching Umeko to read music. Who does what?

Write a proof.

19. **Given:** $\overline{AC} \perp \overline{BD}$

$\angle B \cong \angle D$

Prove: C is the midpoint of $\overline{BD}$.

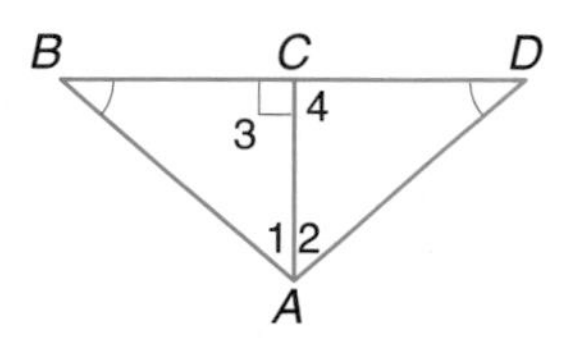

20. **Given:** $m\angle ACB = 110$

$m\angle DAC = 40$

Prove: $\triangle ACD$ is isosceles.

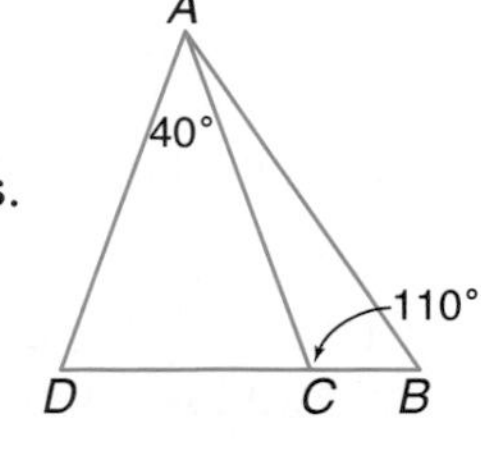

CHAPTER 5 TEST

In $\triangle AHW$, $m\angle A = 64$ and $m\angle AWH = 36$. If $\overline{WP}$ is an angle bisector and $\overline{HQ}$ is an altitude, find each measure.

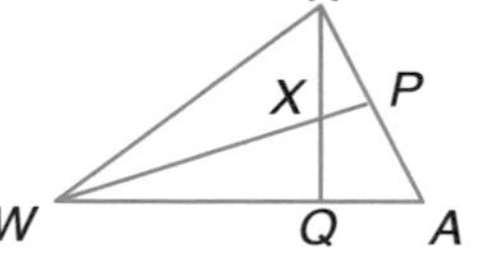

1. $m\angle AQH$
2. $m\angle AHQ$

3. $m\angle APW$
4. $m\angle HXW$
5. If $\overline{WP}$ is a median, $AP = 3y + 11$, and $PH = 7y - 5$, find AH.

Refer to the figure at the right. Complete each statement with < or >.

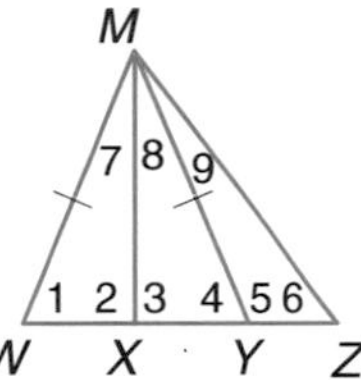

6. $m\angle 2$ __?__ $m\angle 9$
7. $m\angle 6$ __?__ $m\angle 1$
8. If $m\angle 8 < m\angle 7$, then WX __?__ XY.
9. If $MX < MZ$, then $m\angle 4$ __?__ $m\angle 5$.

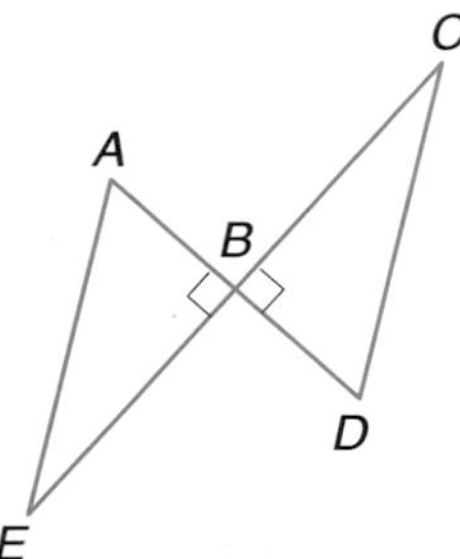

Find the value of x so that $\triangle ABE \cong \triangle DBC$ by the indicated theorem or postulate.

10. $AB = 2x$, $BE = 3x - 1$, $BC = 23$, $BD = 5x - 24$; LL
11. $AB = x + 6$, $AE = 2x + 13$, $BD = 15$, $CD = 4x - 5$; HL

12. List the least and the greatest angle measures in $\triangle LAN$.

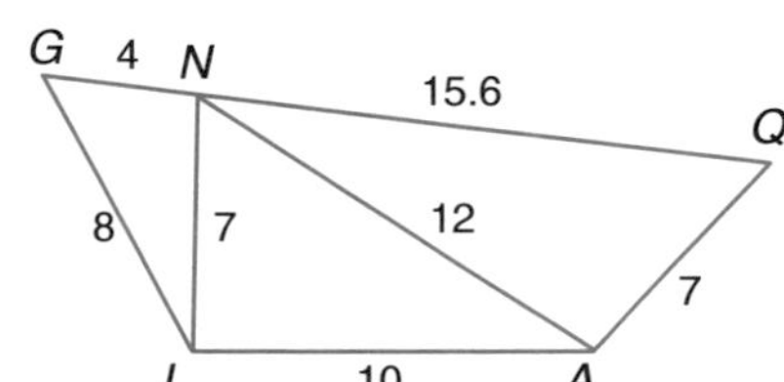

13. List all the angles whose measures are greater than $m\angle 6$.

14. Find the longest segment in $\triangle ABC$ if $m\angle A = 5x + 31$, $m\angle B = 74 - 3x$, and $m\angle C = 4x + 9$.

15. Determine whether it is possible to have a triangle with vertices $A(1, -1)$, $B(7, 7)$, and $C(2, -5)$. Write *yes* or *no*, and explain.

16. Write an inequality to describe the possible values of x.

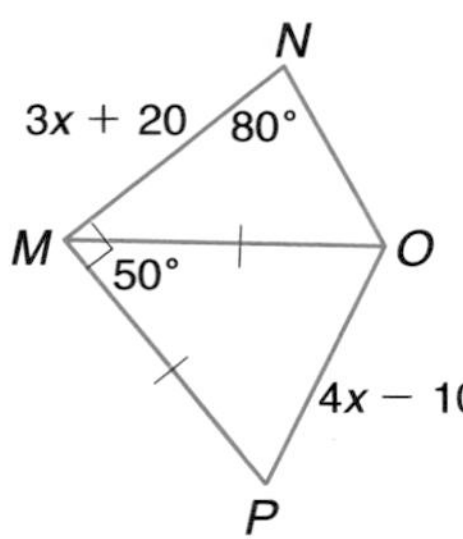

Refer to the figure at the right for Questions 17 and 18.

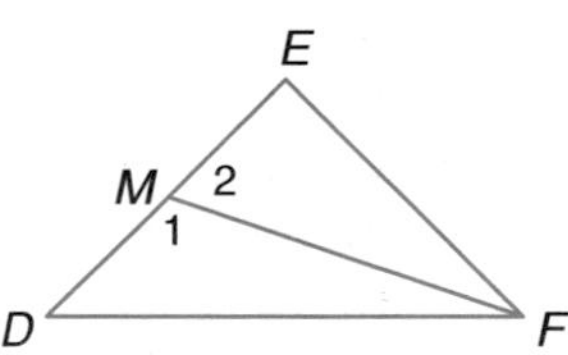

17. If M is the midpoint of $\overline{DE}$, $m\angle 1 > m\angle 2$, $DF = 13x - 5$, and $EF = 7x + 25$, find all possible values of x.

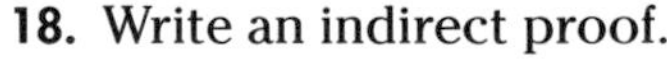

18. Write an indirect proof.

Given: $\overline{FM}$ is a median of $\triangle DEF$.
$m\angle 1 < m\angle 2$

Prove: $DF \neq EF$

Write a two-column proof.

19. **Given:** $NO = QP$
$PN > OQ$

Prove: $MP > MO$

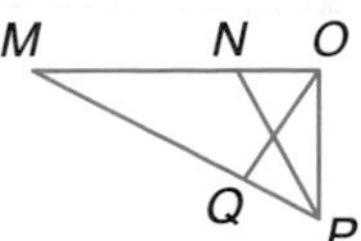

20. **Given:** $\overline{AD} \perp \overline{DC}$
$\overline{AB} \perp \overline{BC}$
$AB = DC$

Prove: $\overline{DC} \perp \overline{BC}$

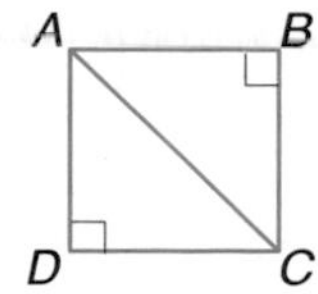

CHAPTER 6 TEST

Complete each statement about ▱*DHGF*. Then name the theorem or definition that justifies your answer.

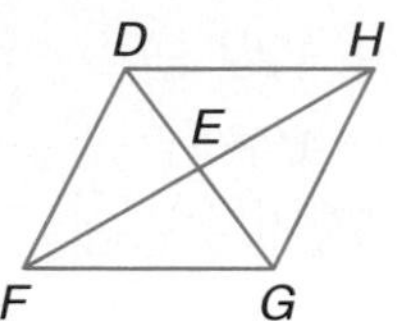

1. $\overline{DE} \cong$ __?__
2. $\angle FDH \cong$ __?__
3. $\overline{FD} \parallel$ __?__
4. $\triangle FDG \cong$ __?__

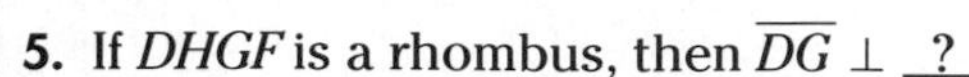

5. If *DHGF* is a rhombus, then $\overline{DG} \perp$ __?__.
6. If *DHGF* is a rhombus, then $\overline{DG}$ bisects __?__ and __?__.

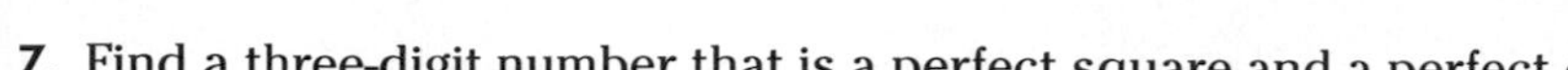

7. Find a three-digit number that is a perfect square and a perfect cube.

Determine whether *ABCD* is a parallelogram for each set of vertices. Justify your answer.

8. $A(-2, 6)$, $B(2, 11)$, $C(3, 8)$, $D(-1, 3)$
9. $A(7, -3)$, $B(4, -2)$, $C(6, 4)$, $D(12, 2)$

Determine whether each conditional is *true* or false. If false, draw a counterexample.

10. If a quadrilateral has four right angles, then it is a rectangle.
11. If the diagonals of a quadrilateral are perpendicular, then it is a rhombus.
12. If a quadrilateral has all four sides congruent, then it is a square.
13. If a quadrilateral has opposite sides congruent and one right angle, then it is a rectangle.

Find the values of *x* and *y* that ensure each quadrilateral is a parallelogram.

14.

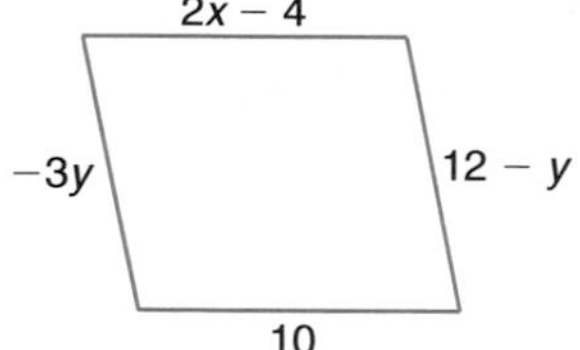

15.

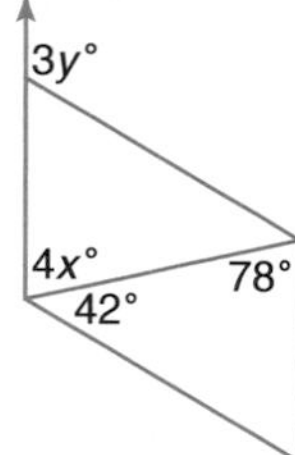

***PQRS* is an isosceles trapezoid with bases $\overline{QR}$ and $\overline{PS}$. Use the figure and the given information to find each measure.**

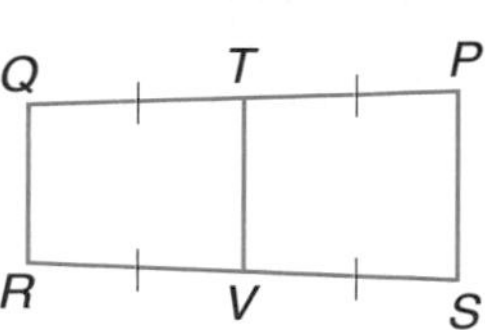

16. If $PS = 32$ and $TV = 26$, find QR.
17. If $m\angle QTV = 79$, find $m\angle TVS$.
18. If $QR = 9x$ and $PS = 13x$, find TV.
19. If $QR = x - 3$, $PS = 2x + 4$, and $TV = 3x - 10$, find QR, PS, and TV.
20. Use rhombus *MNOP* to find the value of *x* if $m\angle MPS = 4x + 7$ and $m\angle OPS = 7x - 38$.

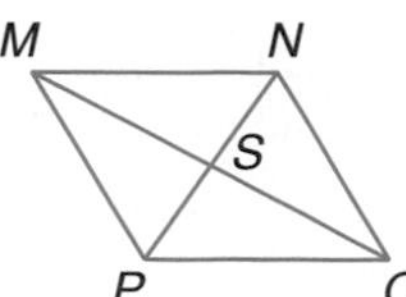

Solve each proportion.

1. $\frac{x}{28} = \frac{60}{16}$

2. $\frac{21}{1 - x} = \frac{7}{x}$

3. A recent survey showed that 14 out of 25 teens prefer orange juice over apple juice. How many of the 1600 students at Dublin High School would you expect to prefer orange juice?

Identify each statement as *true* or *false*. If false, state why.

4. All equilateral triangles are similar.

5. All isosceles triangles are similar.

Determine whether each pair of triangles is similar. Justify with a reason. If similarity exists, write a mathematical sentence relating the two triangles.

6.

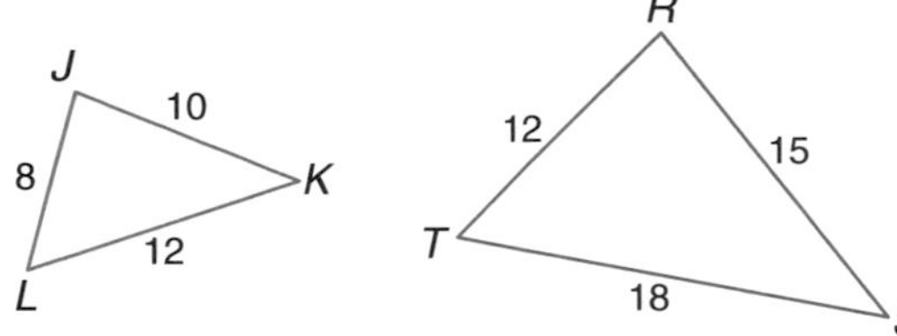

7.

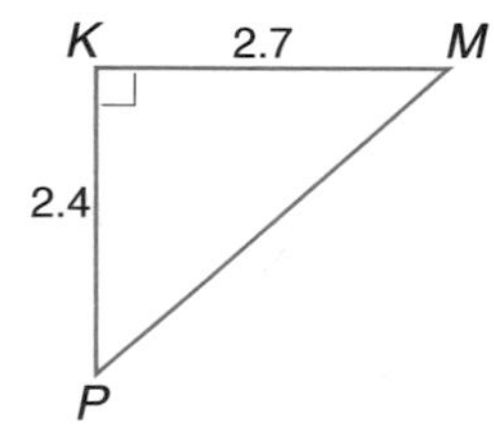

Use $\triangle ACE$ to find the value of x.

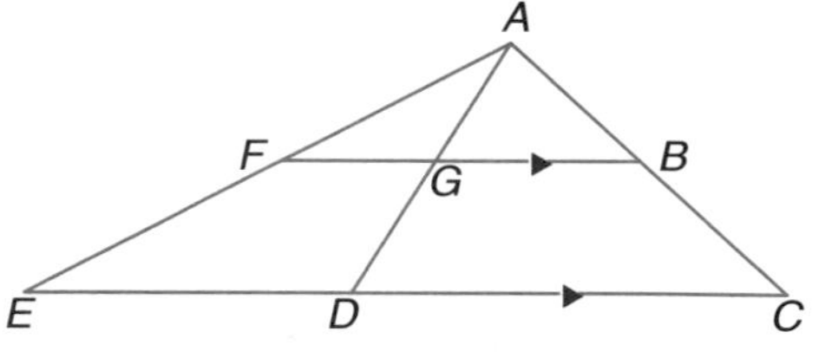

8. $AF = 3x + 1$, $FE = 18$, $AG = 8$, $GD = 9$

9. $AB = BC = AF = FE$, $ED = 2x + 5$, $DC = 4x - 7$, $FB = 17$

10. Find the value of $a^{\frac{1}{2}}$, where a initially equals 15. Then, use that value as the next a in the expression. Repeat the process until you can make some observations. Describe what happens in the iteration.

11. If $\overline{BD} \parallel \overline{AE}$ and $3(BD) = AE$, find the coordinates of A and E.

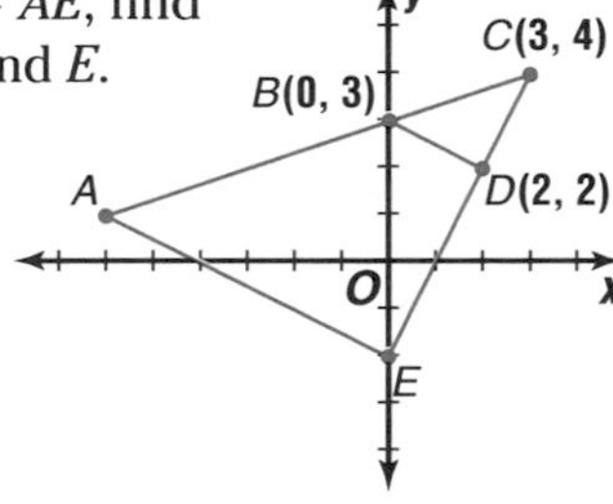

12. $\triangle JKL \sim \triangle BKA$, and the perimeter of $\triangle BKA$ is 42. Find the value of x.

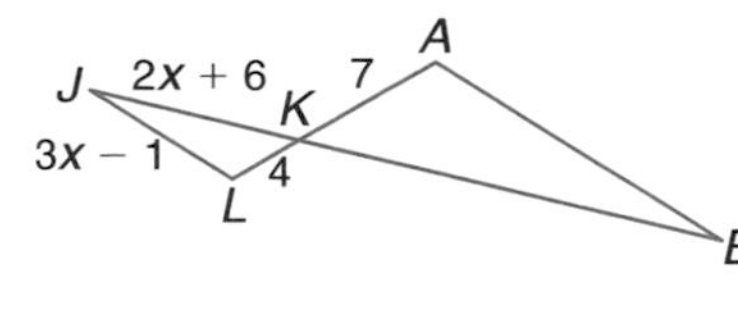

Use the figure at the right and the given information to find the value of x.

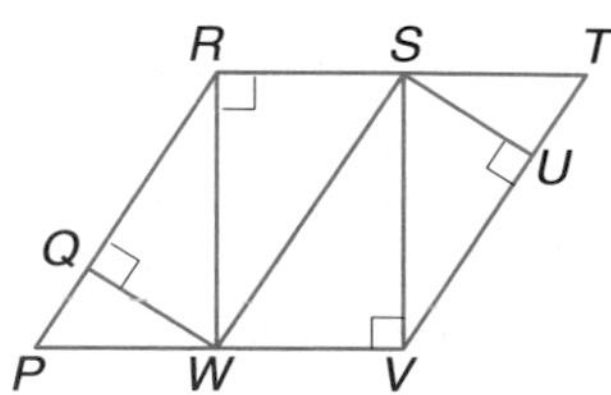

13. $\triangle RWP \sim \triangle VST$,
$PW = x$, $SU = 5$,
$QW = 1$, $ST = x + 5$

14. $\triangle WRS \sim \triangle VUS$,
$SW = 3$, $SV = 2$,
$RS = 1\frac{1}{3}$, $SU = x + \frac{2}{3}$

Complete each of the following.

15. **Given:** $\triangle ABC \sim \triangle RSP$
D is the midpoint of $\overline{AC}$.
Q is the midpoint of $\overline{PR}$.

Prove: $\triangle SPQ \sim \triangle BCD$

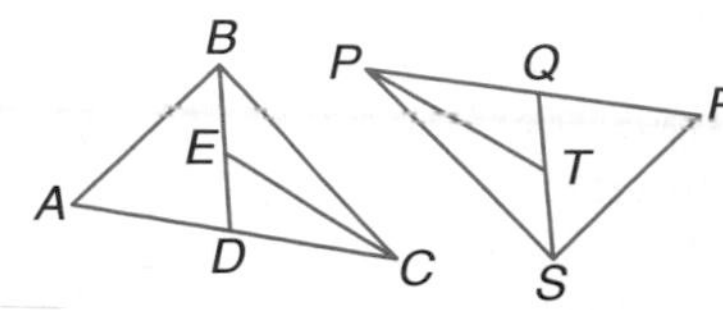

CHAPTER 8 TEST

Find the geometric mean between each pair of numbers.

1. 3 and 12
2. 5 and 4
3. 28 and 56

Use the figure below and the given information to find each measure. Round answers to the nearest tenth.

4. Find QS if $PS = 8$ and $SR = 5$.
5. Find QP if $SP = 9.5$ and $SR = 3$.

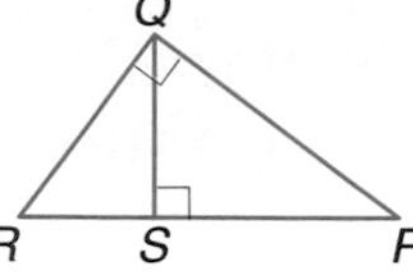

Determine if the given measures are measures of the sides of a right triangle.

6. 10, 26, 24
7. 12, 27, 19
8. 39, 80, 89

Find the value of *x*.

9. 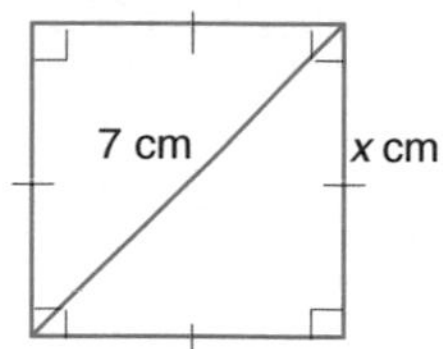

10. 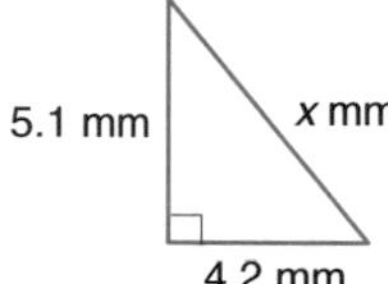

11. 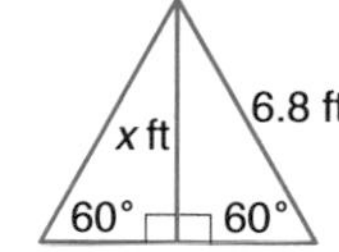

12. The perimeter of an equilateral triangle is 51 inches. Find the length of an altitude of the triangle.

Find the indicated trigonometric ratio as a fraction and as a decimal rounded to the nearest ten-thousandth.

13. $\sin d°$
14. $\tan c°$
15. $\cos a°$
16. $\tan b°$

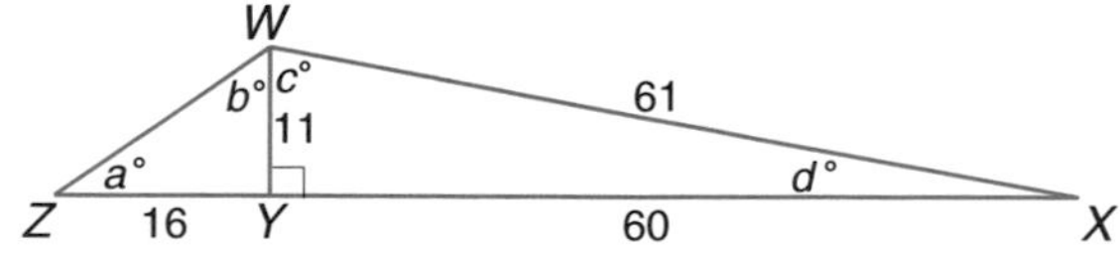

17. **Fire Safety** A firefighter's 36-foot ladder leans against a building. The top of the ladder touches the building 28 feet above the ground. What is the measure of the angle the ladder forms with the ground?

18. The longest side of a triangle is 30 centimeters long. Two of the angles have measures of 45 and 79. Find the measures of the other two sides and the remaining angle.

19. **Gemology** A jeweler is making a sapphire earring in the shape of an isosceles triangle with sides 22, 22, and 28 millimeters long. What is the measure of the vertex angle?

20. **Algebra** Which problem-solving strategy might you use to find the fractional part of the odd whole numbers less than 100 that are perfect squares? Find the fraction.

Refer to ⊙N for Questions 1–6.

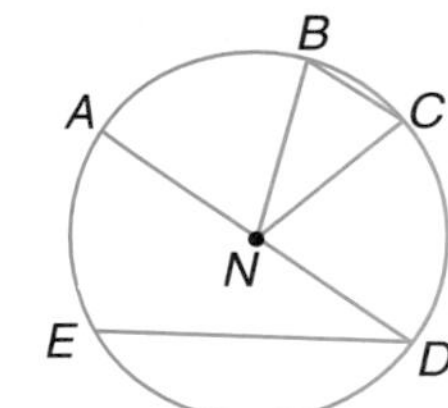

1. Name a radius of ⊙N.
2. What kind of a triangle is $\triangle BNC$?
3. Is $ED > AD$? Explain.
4. If $AD = 21$, what is the radius?
5. If $NC = 5$, what is the circumference?
6. If $C = 126.5$, what is the diameter?

Find the exact circumference of each circle.

7. 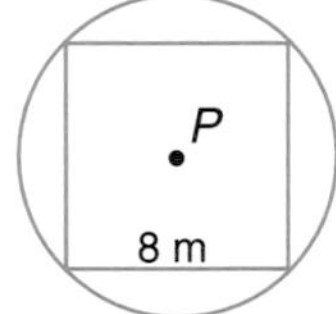

8. 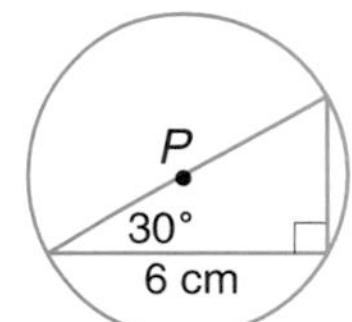

For each ⊙C, find the value of x. Assume that segments that appear to be tangent are tangent.

9. 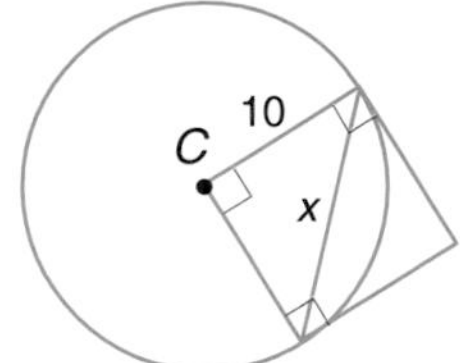

10. 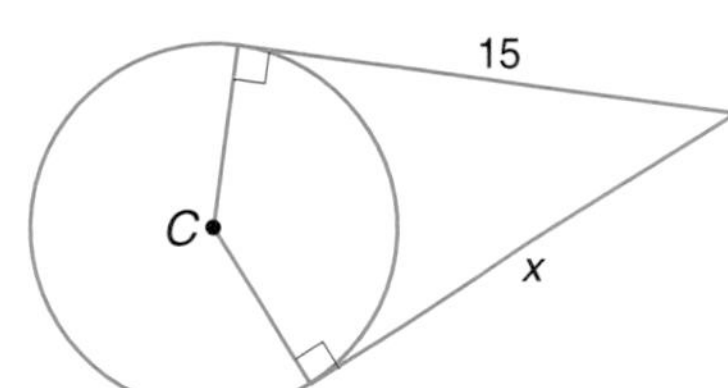

11. 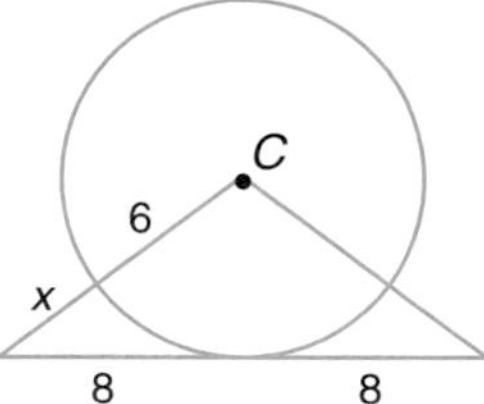

12. 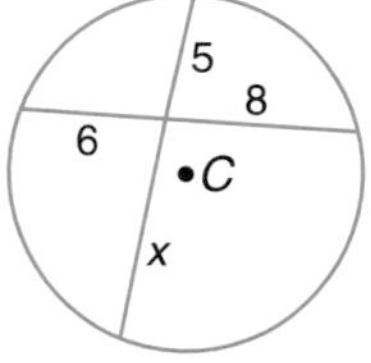

13. 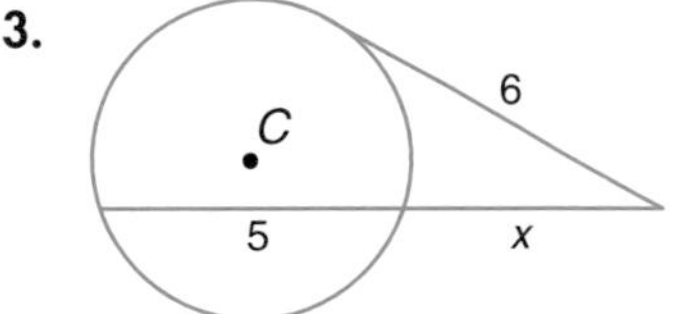

14. 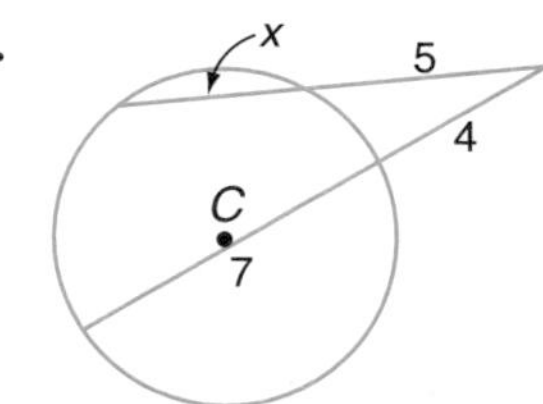

In ⊙P, $\overline{AB} \parallel \overline{CD}$, $m\widehat{BD} = 42$, $m\widehat{BE} = 12$, and $\overline{CF}$ and $\overline{AB}$ are diameters. Find each measure.

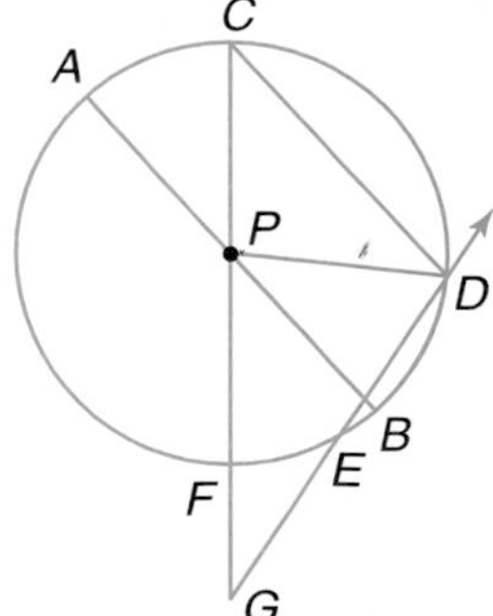

15. $m\widehat{AC}$
16. $m\widehat{CD}$
17. $m\angle BPF$
18. $m\angle CPD$
19. $m\widehat{AF}$
20. $m\angle G$
21. $m\angle FCD$
22. $m\angle EDC$
23. Suppose the diameter of a circle is 10 inches long and a chord is 6 inches long. Find the distance between the chord and the center of the circle.
24. **Make a Graph** In 1995, a survey of parents of children under the age of 18 was taken to find how parents wake up their children. Of the parents surveyed, 64 said their kids wake on their own, 88 said their kids wake with an alarm, 172 said their kids have them call to wake them, and 76 said their kids have other ways of waking. Draw and label a circle graph that represents the data.
25. **Crafts** Teri plans to paint a jack-o-lantern face on a circular piece of wood. She wants it to be able to stand freely on a shelf so she cuts along a chord of the circle to form a flat bottom. If the radius of the circle is 4 inches and the chord is 5 inches, find the height of the final product.

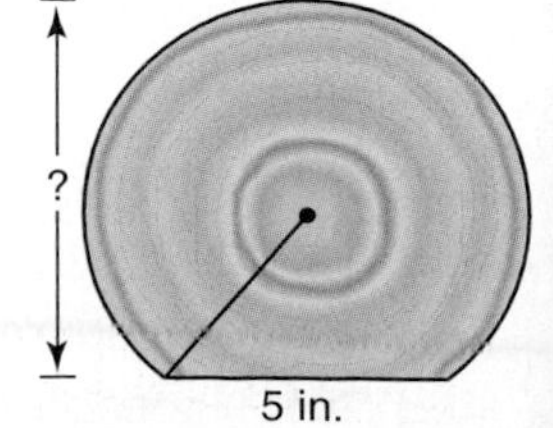

CHAPTER 10 TEST

Refer to the polygon at the right for Questions 1–3.

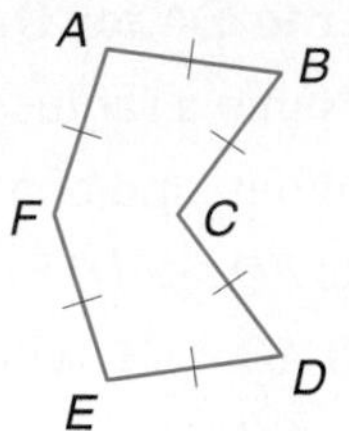

1. Classify the polygon by the number of sides.
2. Classify the polygon as convex or concave.
3. Classify the polygon as regular or not regular. Explain.
4. Find the sum of the measures of the interior angles of a convex dodecagon.
5. Find the measure of one interior angle and one exterior angle of a regular 15-gon.
6. Draw a regular tessellation.
7. Determine whether the tessellation at the right is *regular, uniform,* or *semi-regular.* Name all possibilities.

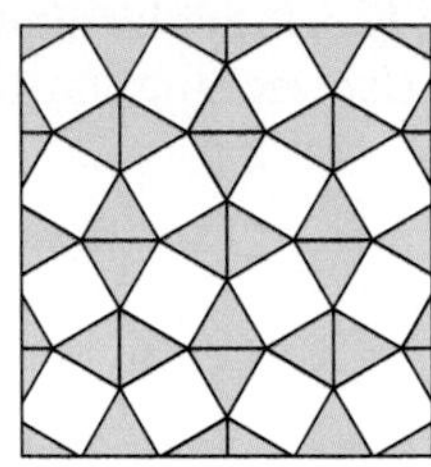

8. Use the strategy of guess and check to find the least prime number greater than 720.
9. Find the area of a parallelogram with a base of 4.95 feet and a height of 0.51 feet.
10. A square has a perimeter of 258 inches. Find the length of one side and the area of the square.
11. A triangle has a base of 16 feet and a height of 30.6 feet. Find its area.
12. The area of $\triangle ABC$ is $(3x^2 + 6x)$ square units. If the height is $3x + 6$ units, find the measure of the base.
13. The median of a trapezoid is 13 feet 6 inches long. If the height is 10 feet, what is the area of the trapezoid?
14. A regular hexagon has sides 10 centimeters long. What is its area?

$\odot O$ has a diameter of 4.5 inches. Find the missing measures. Round to the nearest tenth.

15. What is the area of $\odot O$?
16. What is the circumference of $\odot O$?
17. A sector of $\odot O$ has a central angle measuring 30°. Find the area of the sector.

A circular dartboard has a diameter of 18 inches. The bull's-eye has a diameter of 3 inches, and the blue ring around the bull's-eye is 4 inches wide.

18. What is the probability that a randomly-thrown dart that hits the dartboard will land in the bull's-eye?
19. What is the probability that a randomly-thrown dart that hits the dartboard will land in the blue ring?
20. Draw a complete network with 6 nodes.
    a. How many edges does it have?
    b. Find the degree of each node.
    c. Is it traceable? If so, list a sequence of segments for traceability.

1. From the views of the given solid figure, draw a corner view.

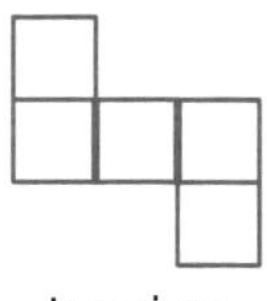
top view

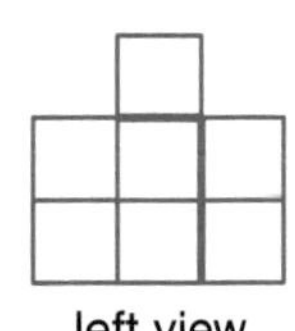
left view

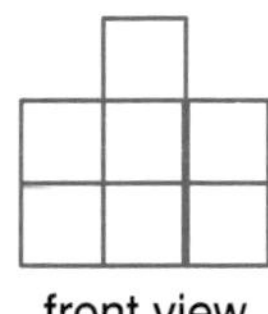
front view

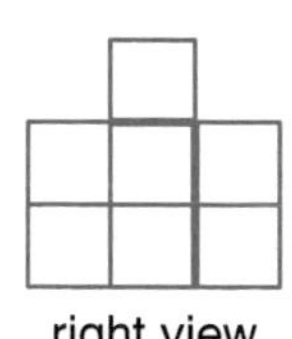
right view

2. Use isometric dot paper to draw a rectangular prism with a square base that is 5 units on each side and a height that is 10 units.
3. Use rectangular dot paper to sketch a net of a right square pyramid.

Find the surface area of each solid figure. Round to the nearest tenth.

4. 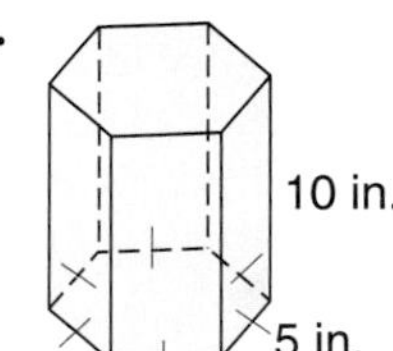

5. 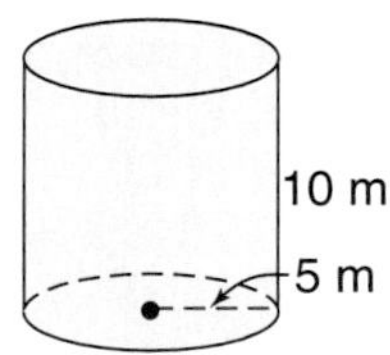

6. 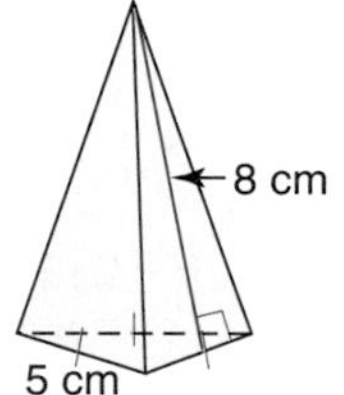

7. a right circular cone with a radius of 2.7 millimeters and a height of 30 millimeters
8. a sphere with a diameter of 6 inches

Find the volume of each solid figure. Round to the nearest tenth.

9. 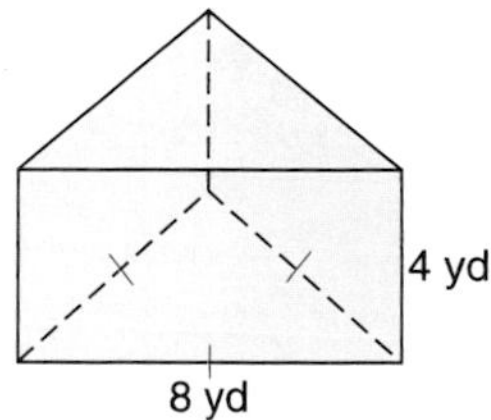

10. 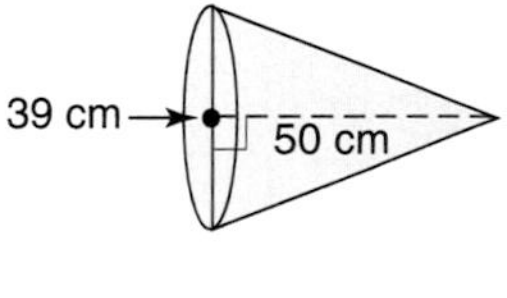

11. 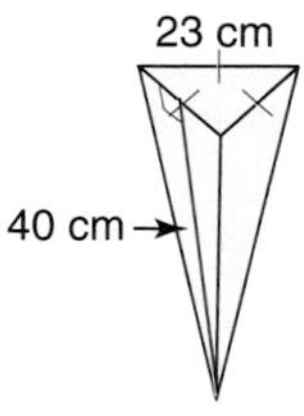

12. a right cylinder with a diameter of 39 centimeters and a height of 50 centimeters
13. a sphere with a radius of 18 millimeters
14. **Sports** A rectangular swimming pool is 4 meters wide and 10 meters long. A concrete walkway is poured around the pool. The walkway is 1 meter wide and 0.1 meter deep. What is the volume of the concrete?
15. The base of a right prism is a right triangle with legs 6 inches and 8 inches. If the height of the prism is 16 inches, find the lateral area of the prism.

CHAPTER 12 TEST

Graph each equation. State the slope and *y*-intercept.

1. $x + 2y = 6$
2. $x = 3$
3. $y = -5x$

Write the equation in slope-intercept form of the line that satisfies the given conditions.

4. $m = -4$; passes through the point at $(3, -2)$
5. passes through points at $(-4, 11)$ and $(-6, 3)$
6. parallel to the graph of $y = 2x - 5$; passes through the point at $(-1, -4)$
7. parallel to the y-axis; passes through the point at $(-4, -2)$

Given $\vec{a} = (-3, 5)$ and $\vec{b} = (0, 7)$, represent each of the following as an ordered pair.

8. $\vec{a} + \vec{b}$
9. $\vec{b} - 3\vec{a}$

Determine the coordinates of the center and the measure of the radius for each sphere whose equation is given.

10. $(x - 4)^2 + (y - 5)^2 + (z + 2)^2 = 81$
11. $x^2 + y^2 + z^2 = 7$
12. Given $\vec{v} = (-5, -3)$, find the magnitude of $\vec{v}$.
13. Given $A(3, 7)$ and $B(-2, 5)$, find the magnitude of $\overline{AB}$.
14. Determine the distance between points at $(2, 4, 5)$ and $(2, 4, 7)$.
15. Determine the midpoint of the segment whose endpoints are $X(0, -4, 2)$ and $Y(3, 0, 2)$.
16. Write the equation of the sphere whose diameter has endpoints at $P(-3, 5, 7)$ and $Q(5, -1, 5)$.
17. Write an equation for the perpendicular bisector of the segment whose endpoints are at $(5, 2)$ and $(1, -4)$.
18. Name the missing coordinates for the parallelogram in terms of the given variables.

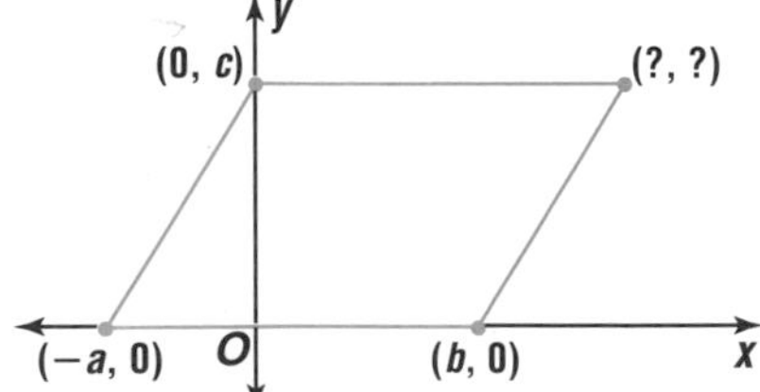

19. Use a coordinate proof to show that $\overline{NK}$ is a perpendicular bisector of a side of $\triangle LOM$.

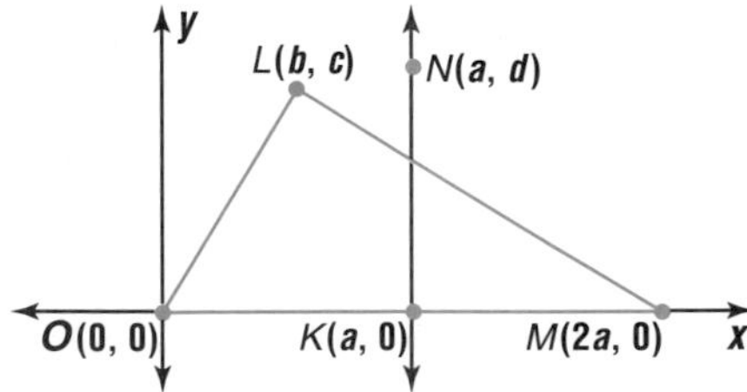

20. The table below lists radio advertising revenue in billions of dollars.

Year	1980	1982	1984	1986	1988	1990	1992	1994
Revenue	4	4.8	5.9	7	7.8	8.7	8.5	10.3

a. Draw a scatter plot to show how year x and advertising revenue y are related.
b. Write an equation that relates the year to the dollar revenue.
c. Graph the equation.

Describe the locus of points that satisfy each set of conditions.

1. all the points in a plane that are 5 inches from a given line ℓ
2. all the points in a plane that are 3.5 centimeters from a vertex of a given rhombus
3. all the points in space that are 4 meters from a given corner of a room
4. all the points in space that are 10 feet from a given line n

Describe all the possible ways the figures can intersect.

5. two concentric spheres and a line
6. two parallel planes and a sphere

Graph the following pair of equations to find the locus of points that satisfy the graphs of both equations.

7. $y = 4x$
   $x + y = 5$

Use either substitution or elimination to find the locus of points that satisfy the graphs of both equations.

8. $3x - 2y = 10$
   $x + y = 0$
9. $x - 4y = 7$
   $2x + y = -4$

Determine if each of the following is an isometry. Write *yes* or *no*.

10. reflection
11. translation
12. rotation
13. dilation

Describe each of the following as a *reflection*, a *rotation*, a *translation*, or a *dilation*.

14. 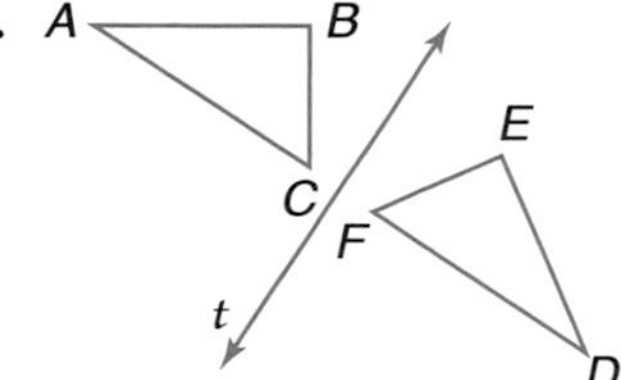

15. 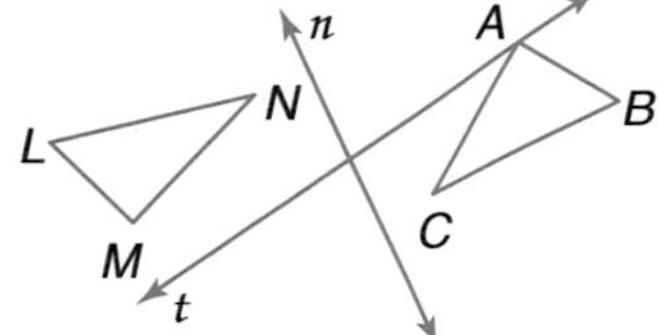

16. 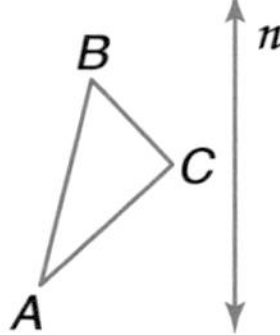

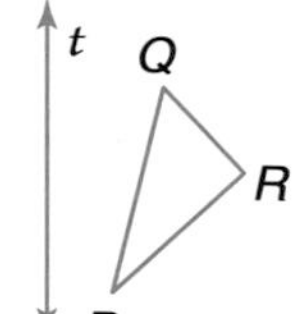

17. Copy the figure at the right. Then draw all the possible lines of symmetry.

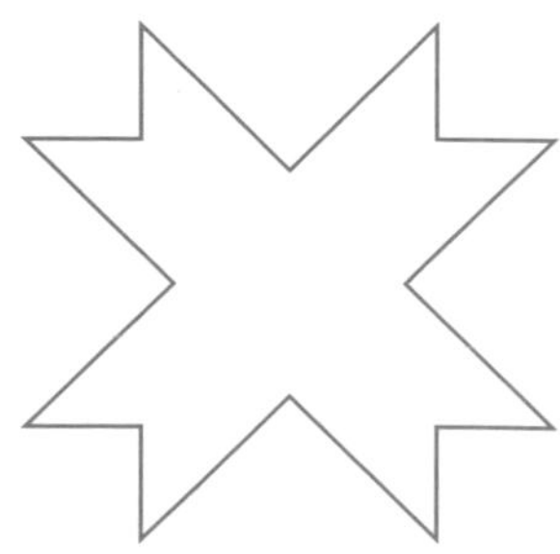

For each scale factor, determine whether the image is an *enlargement*, a *reduction*, or a *congruence transformation*.

18. $k = \frac{2}{3}$
19. $k = 1$

20. **Entertainment** On a TV game show called "The Big 24", a contestant is paid \$7 for each correct answer, but must pay back \$5 for each wrong answer. After answering 24 questions on the show, Sam broke even. How many questions did he answer correctly?

POSTULATES, THEOREMS, AND COROLLARIES

Chapter 1 Discovering Points, Lines, Planes, and Angles

Postulate 1–1 Ruler Postulate	The points on any line can be paired with the real numbers so that, given any two points P and Q on the line, P corresponds to zero, and Q corresponds to a positive number. (29)
Postulate 1–2 Segment Addition Postulate	If Q is between P and R, then $PQ + QR = PR$. If $PQ + QR = PR$, then Q is between P and R. (29)
Postulate 1–3 Protractor Postulate	Given $\overrightarrow{AB}$ and a number r between 0 and 180, there is exactly one ray with endpoint A, extending on each side of $\overleftrightarrow{AB}$, such that the measure of the angle formed is r. (46)
Postulate 1–4 Angle Addition Postulate	If R is in the interior of $\angle PQS$, then $m\angle PQR + m\angle RQS = m\angle PQS$. If $m\angle PQR + m\angle RQS = m\angle PQS$, then R is in the interior of $\angle PQS$. (46)
Theorem 1–1 Midpoint Theorem	If M is the midpoint of $\overline{AB}$, then $\overline{AM} \cong \overline{MB}$. (39)

Chapter 2 Connecting Reasoning and Proof

Postulate 2–1	Through any two points there is exactly one line. (79) *(Through any 2 pts. there is 1 line.)*
Postulate 2–2	Through any three points not on the same line there is exactly one plane. (79) *(Through any 3 noncollinear pts. there is 1 plane.)*
Postulate 2–3	A line contains at least two points. (79) *(A line contains at least 2 pts.)*
Postulate 2–4	A plane contains at least three points not on the same line. (79) *(A plane contains at least 3 noncollinear pts.)*
Postulate 2–5	If two points lie in a plane, then the entire line containing those two points lies in that plane. (79) *(If 2 pts. are in a plane, then the line that contains them is in the plane.)*
Postulate 2–6	If two planes intersect, then their intersection is a line. (79) *(The intersection of 2 planes is a line.)*
Theorem 2–1	Congruence of segments is reflexive, symmetric, and transitive. (101) *(Reflexive Prop. of ≅ Segments, Symmetric Prop. of ≅ Segments, Transitive Prop. of ≅ Segments.)*
Theorem 2–2 Supplement Theorem	If two angles form a linear pair, then they are supplementary angles. (107)
Theorem 2–3	Congruence of angles is reflexive, symmetric, and transitive. (107) *(Reflexive Prop. of ≅ ∠s, Symmetric Prop. of ≅ ∠s, Transitive Prop. of ≅ ∠s.)*
Theorem 2–4	Angles supplementary to the same angle or to congruent angles are congruent. (108) *(∠s supp. to same ∠ or ≅ ∠s are ≅.)*
Theorem 2–5	Angles complementary to the same angle or to congruent angles are congruent. (109) *(∠s compl. to same ∠ or ≅ ∠s are ≅.)*

Theorem 2–6 All right angles are congruent. (109) *(All rt. $\angle$s are $\cong$.)*

Theorem 2–7 Vertical angles are congruent. (109) *(Vert. $\angle$s are $\cong$.)*

Theorem 2–8 Perpendicular lines intersect to form four right angles. (110) *($\perp$ lines form 4 rt. $\angle$s.)*

Chapter 3 Using Perpendicular and Parallel Lines

Postulate 3–1 Corresponding Angles Postulate It two parallel lines are cut by a transversal, then each pair of corresponding angles is congruent. (132)

Theorem 3–1 Alternate Interior Angles Theorem If two parallel lines are cut by a transversal, then each pair of alternate interior angles is congruent. (132)

Theorem 3–2 Consecutive Interior Angles Theorem If two parallel lines are cut by a transversal, then each pair of consecutive interior angles is supplementary. (132)

Theorem 3–3 Alternate Exterior Angles Theorem If two parallel lines are cut by a transversal, then each pair of alternate exterior angles is congruent. (132)

Theorem 3–4 Perpendicular Transversal Theorem In a plane, if a line is perpendicular to one of two parallel lines, then it is perpendicular to the other. (133)

Postulate 3–2 Two nonvertical lines have the same slope if and only if they are parallel. (139)

Postulate 3–3 Two nonvertical lines are perpendicular if and only if the product of their slopes is -1. (139)

Postulate 3–4 If two lines in a plane are cut by a transversal so that the corresponding angles are congruent, then the lines are parallel. (147) *(If ⇄ and corr. $\angle$s are $\cong$, then the lines are $\parallel$.)*

Postulate 3–5 Parallel Postulate If there is a line and a point not on the line, then there exists exactly one line through the point that is parallel to the given line. (147)

Theorem 3–5 If two lines in a plane are cut by a transversal so that a pair of alternate exterior angles is congruent, then the two lines are parallel. (147) *(If ⇄ and alt. ext. $\angle$s are $\cong$, then the lines are $\parallel$.)*

Theorem 3–6 If two lines in a plane are cut by a transversal so that a pair of consecutive interior angles is supplementary, then the lines are parallel. (147) *(If ⇄ and consec. int. $\angle$s are supp., then the lines are $\parallel$.)*

Theorem 3–7 If two lines in a plane are cut by a transversal so that a pair of alternate interior angles is congruent, then the lines are parallel. (147) *(If ⇄ and alt. int. $\angle$s are $\cong$, then the lines are $\parallel$.)*

Theorem 3–8 In a plane, if two lines are perpendicular to the same line, then they are parallel. (147) *(In a plane, if 2 lines are $\perp$ to the same line, they are $\parallel$.)*

Chapter 4 Identifying Congruent Triangles

Theorem 4–1 ***Angle Sum Theorem***	The sum of the measures of the angles of a triangle is 180. (189)
Theorem 4–2 ***Third Angle Theorem***	If two angles of one triangle are congruent to two angles of a second triangle, then the third angles of the triangles are congruent. (190)
Theorem 4–3 ***Exterior Angle Theorem***	The measure of an exterior angle of a triangle is equal to the sum of the measures of the two remote interior angles. (190)
Corollary 4–1	The acute angles of a right triangle are complementary. (192) *(The acute $\angle$s of a rt. $\triangle$ are comp.)*
Corollary 4–2	There can be at most one right or obtuse angle in a triangle. (192) *(There can be at most 1 rt. or obtuse $\angle$ in a $\triangle$.)*
Theorem 4–4	Congruence of triangles is reflexive, symmetric, and transitive. (198) *(Congruence of triangles is (reflexive/symmetric/transitive).)*
Postulate 4–1 ***SSS Postulate***	If the sides of one triangle are congruent to the sides of a second triangle, then the triangles are congruent. (206)
Postulate 4–2 ***SAS Postulate***	If two sides and the included angle of one triangle are congruent to two sides and the included angle of another triangle, then the triangles are congruent. (207)
Postulate 4–3 ***ASA Postulate***	If two angles and the included side of one triangle are congruent to two angles and the included side of another triangle, the triangles are congruent. (207)
Theorem 4–5 ***AAS***	If two angles and a nonincluded side of one triangle are congruent to the corresponding two angles and side of a second triangle, the two triangles are congruent. (214)
Theorem 4–6 ***Isosceles Triangle Theorem***	If two sides of a triangle are congruent, then the angles opposite those sides are congruent. (222)
Theorem 4–7	If two angles of a triangle are congruent, then the sides opposite those angles are congruent. (223) *(If 2 $\angle$s of a $\triangle$ are $\cong$, the sides opp. the $\angle$s are $\cong$.)*
Corollary 4–3	A triangle is equilateral if and only if it is equiangular. (224) *(An (equilateral/equiangular) $\triangle$ is (equiangular/equilateral).)*
Corollary 4–4	Each angle of an equilateral triangle measures 60°. (224) *(Each $\angle$ of an equilateral $\triangle$ measures 60°.)*

Chapter 5 Applying Congruent Triangles

Theorem 5–1	A point on the perpendicular bisector of a segment is equidistant from the endpoints of the segment. (238) *(A pt. on the $\perp$ bisector of a segment is equidistant from the endpts. of the segment.)*
Theorem 5–2	A point equidistant from the endpoints of a segment lies on the perpendicular bisector of the segment. (238) *(A pt. equidistant from the endpts. of a segment lies on the $\perp$ bisector of the segment.)*
Theorem 5–3	A point on the bisector of an angle is equidistant from the sides of the angle. (240) *(A pt. on the bisector of an $\angle$ is equidistant from the sides of the $\angle$.)*

Theorem 5–4	A point on or in the interior of an angle and equidistant from the sides of an angle lies on the bisector of the angle. (240) *(A pt. on or in the int. of an ∠ and equidistant from the sides of an ∠ lies on the bisector of the ∠.)*
Theorem 5–5 ***LL***	If the legs of one right triangle are congruent to the corresponding legs of another right triangle, then the triangles are congruent. (245)
Theorem 5–6 ***HA***	If the hypotenuse and an acute angle of one right triangle are congruent to the hypotenuse and corresponding acute angle of another right triangle, then the two triangles are congruent. (246)
Theorem 5–7 ***LA***	If one leg and an acute angle of one right triangle are congruent to the corresponding leg and acute angle of another right triangle, then the triangles are congruent. (247)
Postulate 5–1 ***HL***	If the hypotenuse and a leg of one right triangle are congruent to the hypotenuse and corresponding leg of another right triangle, then the triangles are congruent. (247)
Theorem 5–8 ***Exterior Angle Inequality Theorem***	If an angle is an exterior angle of a triangle, then its measure is greater than the measure of either of its corresponding remote interior angles. (253)
Theorem 5–9	If one side of a triangle is longer than another side, then the angle opposite the longer side has a greater measure than the angle opposite the shorter side. (259) *(If one side of a △ is longer than another, the ∠ opp. the longer side > the ∠ opp. the shorter side.)*
Theorem 5–10	If one angle of a triangle has a greater measure than another angle, then the side opposite the greater angle is longer than the side opposite the lesser angle. (259) *(If an ∠ of a △ > another, the side opp. the greater ∠ is longer than the side opp. the lesser ∠.)*
Theorem 5–11	The perpendicular segment from a point to a line is the shortest segment from the point to the line. (261) *(The ⊥ segment from a pt. to a line is the shortest segment from the pt. to the line.)*
Corollary 5–1	The perpendicular segment from a point to a plane is the shortest segment from the point to the plane. (262) *(The ⊥ segment from a pt. to a plane is the shortest segment from the pt. to the plane.)*
Theorem 5–12 ***Triangle Inequality Theorem***	The sum of the lengths of any two sides of a triangle is greater than the length of the third side. (267)
Theorem 5–13 ***SAS Inequality (Hinge Theorem)***	If two sides of one triangle are congruent to two sides of another triangle, and the included angle in one triangle is greater than the included angle in the other, then the third side of the first triangle is longer than the third side in the second triangle. (273)
Theorem 5–14 ***SSS Inequality***	If two sides of one triangle are congruent to two sides of another triangle and the third side in one triangle is longer than the third side in the other, then the angle between the pair of congruent sides in the first triangle is greater than the corresponding angle in the second triangle. (274)

Chapter 6 Exploring Quadrilaterals

Theorem 6–1	Opposite sides of a parallelogram are congruent. (292) *(Opp. sides of a ▱ are ≅.)*

Theorem 6–2 Opposite angles of a parallelogram are congruent. (292) *(Opp. $\angle$s of a ▱ are $\cong$.)*

Theorem 6–3 Consecutive angles in a parallelogram are supplementary. (292) *(Consec. $\angle$s of a ▱ are supp.)*

Theorem 6–4 The diagonals of a parallelogram bisect each other. (293) *(Diagonals of a ▱ bisect each other.)*

Theorem 6–5 If both pairs of opposite sides of a quadrilateral are congruent, then the quadrilateral is a parallelogram. (298) *(If opp. sides of a quad. are $\cong$, it is a ▱.)*

Theorem 6–6 If both pairs of opposite angles of a quadrilateral are congruent, then the quadrilateral is a parallelogram. (298) *(If both pairs of opp. $\angle$s of a quad. are $\cong$, it is a ▱.)*

Theorem 6–7 If the diagonals of a quadrilateral bisect each other, then the quadrilateral is a parallelogram. (298) *(If the diagonals of quad. bisect, it is a ▱.)*

Theorem 6–8 If one pair of opposite sides of a quadrilateral are both parallel and congruent, then the quadrilateral is a parallelogram. (299) *(If a pair of opp. sides of a quad. are $\cong$ and $\parallel$, it is a ▱.)*

Theorem 6–9 If a parallelogram is a rectangle, then its diagonals are congruent. (306) *(If a ▱ is a rect., then its diagonals are $\cong$.)*

Theorem 6–10 If the diagonals of a parallelogram are congruent, then the parallelogram is a rectangle. (308) *(If the diagonals of a ▱ are $\cong$, then the ▱ is a rect.)*

Theorem 6–11 The diagonals of a rhombus are perpendicular. (313) *(Diagonals of a rhom. are $\perp$.)*

Theorem 6–12 If the diagonals of a parallelogram are perpendicular, then the parallelogram is a rhombus. (313) *(If the diagonals of a ▱ are $\perp$, then the ▱ is a rhombus.)*

Theorem 6–13 Each diagonal of a rhombus bisects a pair of opposite angles. (313) *(Each diagonal of a rhom. bisects opp. $\angle$s.)*

Theorem 6–14 Both pairs of base angles of an isosceles trapezoid are congruent. (321) *(Base $\angle$s of an iso. trap. are $\cong$.)*

Theorem 6–15 The diagonals of an isosceles trapezoid are congruent. (321) *(The diagonals of an isos. trap. are $\cong$.)*

Theorem 6–16 The median of a trapezoid is parallel to the bases, and its measure is one-half the sum of the measures of the bases. (323) *(Median of a trap. is $\parallel$ to the bases. Length of median of a trap. = $\frac{1}{2}$(sum of the lengths of bases).)*

Chapter 7 Connecting Proportion and Similarity

Postulate 7–1
AA Similarity If two angles of one triangle are congruent to two angles of another triangle, then the triangles are similar. (355)

Theorem 7–1
SSS Similarity If the measures of the corresponding sides of two triangles are proportional, then the triangles are similar. (355)

Theorem 7–2
SAS Similarity If the measures of two sides of a triangle are proportional to the measures of two corresponding sides of another triangle and the included angles are congruent, then the triangles are similar. (356)

Theorem 7–3	Similarity of triangles is reflexive, symmetric, and transitive. (357) *(Reflexive Prop. of ~ △s, Symmetric Prop. of ~ △s, Transitive Prop. of ~ △s.)*
Theorem 7–4 Triangle Proportionality	If a line is parallel to one side of a triangle and intersects the other two sides in two distinct points, then it separates these sides into segments of proportional lengths. (362)
Theorem 7–5	If a line intersects two sides of a triangle and separates the sides into corresponding segments of proportional lengths, then the line is parallel to the third side. (363)
Theorem 7–6	A segment whose endpoints are the midpoints of two sides of a triangle is parallel to the third side of the triangle and its length is one-half the length of the third side. (363)
Corollary 7–1	If three or more parallel lines intersect two transversals, then they cut off the transversals proportionally. (364)
Corollary 7–2	If three or more parallel lines cut off congruent segments on one transversal, then they cut off congruent segments on every transversal. (364)
Theorem 7–7 Proportional Perimeters	If two triangles are similar, then the perimeters are proportional to the measures of corresponding sides. (370)
Theorem 7–8	If two triangles are similar, then the measures of the corresponding altitudes are proportional to the measures of the corresponding sides. (371)
Theorem 7–9	If two triangles are similar, then the measures of the corresponding angle bisectors are proportional to the measures of the corresponding sides. (371)
Theorem 7–10	If two triangles are similar, then the measures of the corresponding medians are proportional to the measures of the corresponding sides. (371)
Theorem 7–11 Angle Bisector Theorem	An angle bisector in a triangle separates the opposite side into segments that have the same ratio as the other two sides. (372)

Chapter 8 Applying Right Triangles and Trigonometry

Theorem 8–1	If the altitude is drawn from the vertex of the right angle of a right triangle to its hypotenuse, then the two triangles formed are similar to the given triangle and to each other. (398)
Theorem 8–2	The measure of the altitude drawn from the vertex of the right angle of a right triangle to its hypotenuse is the geometric mean between the measures of the two segments of the hypotenuse. (398)
Theorem 8–3	If the altitude is drawn to the hypotenuse of a right triangle, then the measure of a leg of the triangle is the geometric mean between the measures of the hypotenuse and the segment of the hypotenuse adjacent to the leg. (399)
Theorem 8–4 Pythagorean Theorem	In a right triangle, the sum of the squares of the measures of the legs equals the square of the measure of the hypotenuse. (399)
Theorem 8–5 Converse of the Pythagorean Theorem	If the sum of the squares of the measures of two sides of a triangle equals the square of the measure of the longest side, then the triangle is a right triangle. (395)

Theorem 8–6	In a 45°-45°-90° triangle, the hypotenuse is $\sqrt{2}$ times as long as a leg. (405)
Theorem 8–7	In a 30°-60°-90° triangle, the hypotenuse is twice as long as the shorter leg, and the longer leg is $\sqrt{3}$ times as long as the shorter leg. (407)

Chapter 9 Analyzing Circles

Postulate 9–1 Arc Addition Postulate	The measure of an arc formed by two adjacent arcs is the sum of the measures of the two arcs. That is, if Q is a point on $\overparen{PR}$, then $m\overparen{PQ} + m\overparen{QR} = m\overparen{PQR}$. (453)
Theorem 9–1	In a circle or in congruent circles, two minor arcs are congruent if and only if their corresponding chords are congruent. (459) *(In a ⊙ or in ≅ ⊙s, 2 minor arcs are ≅ if and only if their corr. chords are ≅.)*
Theorem 9–2	In a circle, if a diameter is perpendicular to a chord, then it bisects the chord and its arc. (460) *(In a ⊙, if a diameter is ⊥ to a chord, then it bisects the chord and its arc.)*
Theorem 9–3	In a circle or in congruent circles, two chords are congruent if and only if they are equidistant from the center. (461) *(In a ⊙ or in ≅ ⊙s, 2 chords are ≅ if and only if they are equidistant from the center.)*
Theorem 9–4	If an angle is inscribed in a circle, then the measure of the angle equals one-half the measure of the intercepted arc. (466) *(If an ∠ is inscribed in a ⊙, then the measure of the ∠ = $\frac{1}{2}$ the measure of the intercepted arc.)*
Theorem 9–5	If two inscribed angles of a circle or congruent circles intercept congruent arcs or the same arc, then the angles are congruent. (467) *(If 2 inscribed ∠s of a ⊙ or ≅ ⊙s intercept ≅ arcs or the same arc, then the ∠s are ≅.)*
Theorem 9–6	If an inscribed angle of a circle intercepts a semicircle, then the angle is a right angle. (467) *(If an inscribed ∠ of a ⊙ intercepts a semicircle, then the ∠ is a rt. ∠.)*
Theorem 9–7	If a quadrilateral is inscribed in a circle, then its opposite angles are supplementary. (469) *(If a quad. is inscribed in a ⊙, then its opp. ∠s are supp.)*
Theorem 9–8	If a line is tangent to a circle, then it is perpendicular to the radius drawn to the point of tangency. (475)
Theorem 9–9	In a plane, if a line is perpendicular to a radius of a circle at the endpoint on the circle, then the line is a tangent of the circle. (476)
Theorem 9–10	If two segments from the same exterior point are tangent to a circle, then they are congruent. (477)
Theorem 9–11	If a secant and a tangent intersect at the point of tangency, then the measure of each angle formed is one-half the measure of its intercepted arc. (483)
Theorem 9–12	If two secants intersect in the interior of a circle, then the measure of an angle formed is one-half the sum of the measure of the arcs intercepted by the angle and its vertical angle. (484)
Theorem 9–13	If two secants, a secant and a tangent, or two tangents intersect in the exterior of a circle, then the measure of the angle formed is one-half the positive difference of the measures of the intercepted arcs. (485)
Theorem 9–14	If two chords intersect in a circle, then the products of the measures of the segments of the chords are equal. (491)

Theorem 9–15	If two secant segments are drawn to a circle from an exterior point, then the product of the measures of one secant segment and its external secant segment is equal to the product of the measures of the other secant segment and its external secant segment. (493)
Theorem 9–16	If a tangent segment and a secant segment are drawn to a circle from an exterior point, then the square of the measure of the tangent segment is equal to the product of the measures of the secant segment and its external secant segment. (493)

Chapter 10 Exploring Polygons and Area

Theorem 10–1 Interior Angle Sum Theorem	If a convex polygon has n sides and S is the sum of the measures of its interior angles, then $S = 180(n - 2)$. (516)
Theorem 10–2 Exterior Angle Sum Theorem	If a polygon is convex, then the sum of the measures of the exterior angles, one at each vertex, is 360. (517)
Postulate 10–1	The area of a region is the sum of the areas of all of its nonoverlapping parts. (531)
Postulate 10–2	Congruent figures have equal areas. (535)
Postulate 10–3 Length Probability Postulate	If a point on $\overline{AB}$ is chosen at random and C is between A and B, then the probability that the point is on $\overline{AC}$ is $\frac{\text{length of } \overline{AC}}{\text{length of } \overline{AB}}$. (551)
Postulate 10–4 Area Probability Postulate	If a point in region A is chosen at random, then the probability that the point is in region B, which is in the interior of region A, is $\frac{\text{area of region } B}{\text{area of region } A}$. (552)

Chapter 11 Investigating Surface Area and Volume

Theorem 11–1	If two solids are similar with a scale factor of $a:b$, then the surface areas have a ratio of $a^2:b^2$ and the volumes have a ratio of $a^3:b^3$. (631)

Chapter 12 Continuing Coordinate Geometry

Theorem 12–1 Slope-Intercept Form	If the equation of a line is written in the form $y = mx + b$, m is the slope of the line and b is the y-intercept. (647)
Theorem 12–2	Given two points $A(x_1, y_1, z_1)$ and $B(x_2, y_2, z_2)$ in space, the distance between A and B is given by the following equation. (681) $AB = \sqrt{(x_2 - x_1)^2 + (y_2 - y_1)^2 + (z_2 - z_1)^2}$

Chapter 13 Investigating Loci and Coordinate Transformations

Postulate 13–1	In a given rotation, if A is the preimage, P is the image, and W is the center of rotation, then the measure of the angle of rotation, $\angle AWP$, equals twice the measure of the angle formed by intersecting lines of reflection. (740)
Theorem 13–1	If a dilation with center C and a scale factor k maps A onto E and B onto D, then $ED = k(AB)$. (746)

GLOSSARY

A

absolute value (29) For any real number, the number of units the number is from zero on the number line.

acute angle (46) An angle whose degree measure is less than 90.

acute triangle (180) A triangle all of whose angles are acute angles.

adjacent angles (53) Two angles in the same plane that have a common vertex and a common side, but no common interior points.

adjacent arcs (453) Two arcs of a circle that have exactly one point in common.

alternate exterior angles (126) In the figure, transversal t intersects lines ℓ and m. $\angle 5$ and $\angle 3$, and $\angle 6$ and $\angle 4$ are alternate exterior angles.

alternate interior angles (126) In the figure above, transversal t intersects lines ℓ and m. $\angle 1$ and $\angle 7$, and $\angle 2$ and $\angle 8$ are alternate interior angles.

altitude of a cone (602) The segment from the vertex to the plane of the base and perpendicular to the plane of the base.

altitude of a cylinder (593) A segment perpendicular to the base planes and having an endpoint in each plane.

altitude of a parallelogram (529) Any perpendicular segment between the lines containing two of the parallel sides.

altitude of a prism (591) A segment perpendicular to the base planes of the prism, with an endpoint in each plane. The length of an altitude is called the height of the prism.

altitude of a pyramid (600) The segment from the vertex to the plane of the base and perpendicular to the plane of the base.

altitude of a triangle (238) A segment from a vertex of the triangle to the line containing the opposite side and perpendicular to the line containing that side.

angle (44) A figure consisting of two noncollinear rays with a common endpoint. The rays are the sides of the angle. The common endpoint of the rays is the vertex of the angle. An angle separates a plane into three parts, the interior of the angle, the exterior of the angle, and the angle itself.

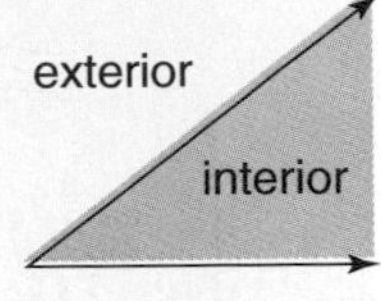

angle bisector (48) The ray, QS, is the bisector of $\angle PQR$ if S is in the interior of the angle and $\angle PQS \cong \angle RQS$.

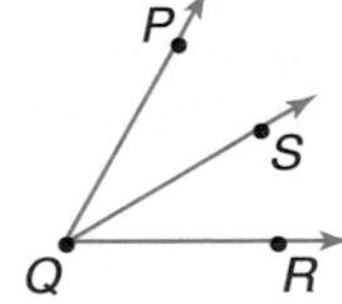

angle bisector of a triangle (240) A segment that bisects an angle of the triangle and has one endpoint at a vertex of the triangle and the other endpoint at another point on the triangle.

angle of depression (420) An angle formed by a horizontal line and the line of sight to an object below the level of the horizontal.

angle of elevation (400) An angle formed by a horizontal line and the line of sight to an object above the level of the horizontal.

angle of rotation (740) The angle of rotation, $\angle ABC$, is determined by the preimage A, the center of rotation B, and the rotation image C.

apothem (544) A segment that is drawn from the center of a regular polygon perpendicular to a side of the polygon.

arc (453) An unbroken part of a circle.

arc length (454) The linear distance representing the arc. The length of the arc is a part of the circumference proportional to the measure of the central angle when compared to the entire circle.

arc measure (453) The degree measure of a minor arc is the degree measure of its central angle. The degree measure of a major arc is 360 minus the degree measure of its central angle. The degree measure of a semicircle is 180.

arc of a chord (459) A minor arc that has the same endpoints as a chord is called an arc of the chord.

area (19, 529) The number of square units contained in the interior of a figure.

auxiliary line (189) A line or line segment added to a given figure to help in proving a result.

axis **1.** (6) In a coordinate plane, the *x*-axis is the horizontal number line and the *y*-axis is the vertical number line. **2.** (593) The axis of a cylinder is the segment whose endpoints are the centers of the bases. **3.** (602) The axis of a cone is the segment whose endpoints are the vertex and the center of the base.

B

base **1.** (182) In an isosceles triangle, the side opposite the vertex angle is called the base. **2.** (321) In a trapezoid, the parallel sides are called bases. **3.** (529) Any side of a parallelogram can be called a base. **4.** (577, 591) In a prism, the bases are the two faces formed by congruent polygons that lie in parallel planes, all of the other faces being parallelograms. **5.** (593) In a cylinder, the bases are the two congruent and parallel circular regions that form the ends of the cylinder. **6.** (600) In a pyramid, the base is the face that does not intersect the other faces at the vertex. The base is a polygonal region. **7.** (602) In a cone, the base is the flat, circular portion of the cone.

base angle **1.** (182) In an isosceles triangle, either angle formed by the base and one of the legs is called a base angle. **2.** (321) In the trapezoid at the right, $\angle A$ and $\angle D$, and $\angle B$ and $\angle C$ are pairs of base angles.

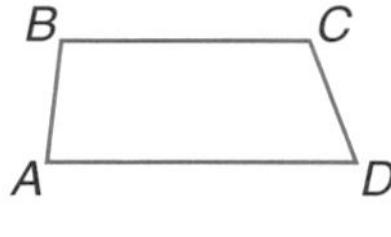

between (28) In general, *B* is between *A* and *C* if and only if *A*, *B*, and *C* are collinear and $AB + BC = AC$.

A B C

C

Cavalieri's Principle (617) If two solids have the same height and the same cross-sectional area at every level, then they have the same volume.

center of a circle (446) A point from which all given points in a plane are equidistant.

center of a regular polygon (543) The common center of the inscribed and circumscribed circles of the regular polygon.

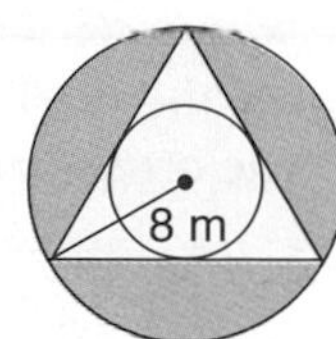

center of a sphere (621) The point from which a set of all points in space are a given distance.

center of dilation (747) A fixed point used for measurement when altering the size of a geometric figure without changing its shape.

center of rotation (739) A fixed point around which shapes move in a circular motion to a new position.

central angle **1.** (452) For a given circle, an angle that intersects the circle in two points and has its vertex at the center of the circle. **2.** (544) An angle formed by two segments drawn to consecutive vertices of a regular polygon from its center.

chord **1.** (446) For a given circle, a segment whose endpoints are points on the circle. **2.** (621) For a given sphere, a segment whose endpoints are on the sphere.

circle (446) A set of points that consists of all points in a plane that are a given distance from a given point in the plane, called the center.

circumference (447) The limit of the perimeters of the inscribed regular polygons as the number of sides increases.

circumscribed polygon (477) A polygon is circumscribed about a circle if each side of the polygon is tangent to the circle.

collinear points (8) Points that lie on the same line.

column matrix (676) A matrix that has only one column.

common external tangent (476) A common tangent that does not intersect the segment whose endpoints are the centers of the two circles.

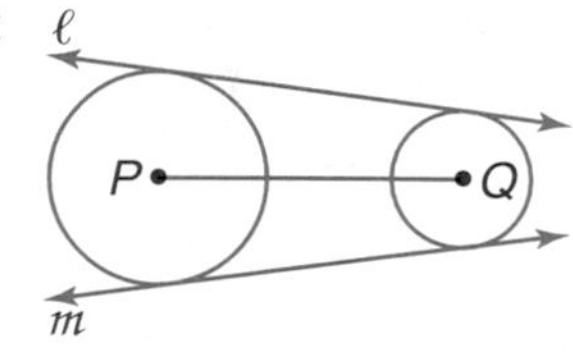

common internal tangent (476) A common tangent that intersects the segment whose endpoints are the centers of the two circles.

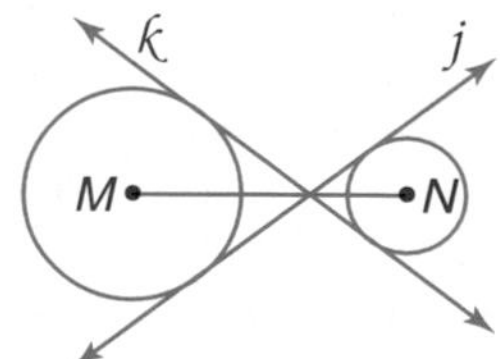

common tangent (476) A line that is tangent to two circles that are in the same plane.

compass (38) An instrument used to draw circles and arcs of circles.

complementary angles (55) Two angles whose degree measures have a sum of 90.

complete network (559) In graph theory, a network that has at least one path between each pair of nodes.

composite of reflections (731) Two successive reflections.

concave polygon (514) A polygon for which there is a line containing a side of the polygon and a point in the interior of the polygon.

concentric circles (454) Circles that lie in the same plane and have the same center but have different radii.

conclusion (76) In a conditional statement, the statement that immediately follows the word *then*.

conditional statement (76) A statement of the form "If A, then B." The part following *if* is called the hypothesis. The part following *then* is called the conclusion.

cone (577, 602) A solid with a circular base, a vertex *V* not contained in the same plane as the base, and a lateral surface area composed of all points in the segments connecting the vertex to the edge of the base.

congruence transformation (196, 716) When a geometric figure and its transformation image are congruent, the mapping is called a congruence transformation or isometry.

congruent angles (47) Angles that have the same measure.

congruent arcs (454) Two arcs of the same or congruent circles that have the same measure.

congruent circles (454) Circles that have the same radius.

congruent segments (31) Segments that have the same length.

congruent solids (629) Two solids are congruent if all of the following conditions are met.

1. The corresponding angles are congruent.
2. Corresponding edges are congruent.
3. Areas of corresponding faces are equal.
4. The volumes are equal.

congruent triangles (196) Triangles that have their corresponding parts congruent.

conjecture (70) An educated guess.

consecutive interior angles (126) In the figure, transversal t intersects lines ℓ and m. There are two pairs of consecutive interior angles: $\angle 8$ and $\angle 1$, and $\angle 7$ and $\angle 2$.

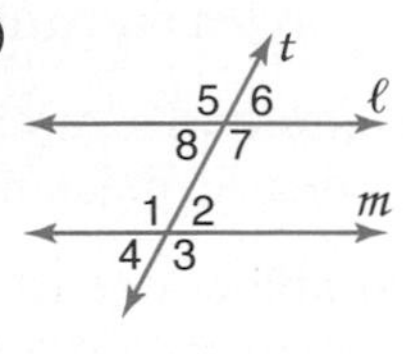

contrapositive (78) The statement formed by negating both the hypothesis and conclusion of the converse of a conditional statement.

converse (77) The statement formed by interchanging the hypothesis and conclusion of a conditional statement.

convex polygon (514) A polygon for which there is no line that contains both a side of the polygon and a point in the interior of the polygon.

coordinate (7) The numbers in an ordered pair of numbers. The first number is called the *x*-coordinate and the second number is called the *y*-coordinate.

coordinate plane (6) A plane for which two perpendicular number lines that intersect at their zero points have been used to match the points of the plane one-to-one with ordered pairs of numbers.

coordinate proof (666) A proof that uses figures in a coordinate plane so that geometric results can be proved by means of algebra.

coplanar points (13) Points that lie in the same plane.

corner view (576) The view from a corner of a figure (also called perspective view).

corollary (192) A statement that can be easily proven using a theorem is called a corollary of that theorem.

corresponding angles (126) In the figure, transversal t intersects lines ℓ and m. There are four pairs of corresponding angles: $\angle 5$ and $\angle 1$, $\angle 8$ and $\angle 4$, $\angle 6$ and $\angle 2$, and $\angle 7$ and $\angle 3$.

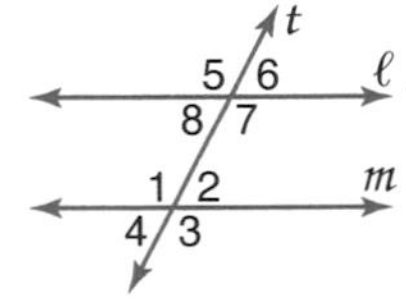

cosine (412) For an acute angle of a right triangle, the ratio of the length of the leg adjacent to the acute angle to the length of the hypotenuse.

counterexample (72) An example used to show that a given general statement is not always true.

cross products (339) In the proportion $\frac{a}{b} = \frac{c}{d}$, where $b \neq 0$ and $d \neq 0$, the cross products are ad and bc. The proportion is true if and only if the cross products are equal.

cross section (574, 577) The intersection of a plane parallel to the base or bases of a solid.

cube (577) A prism in which all the faces are squares.

cylinder (577) A figure whose bases are formed by congruent circles in parallel planes.

deductive reasoning (86) A system of reasoning used to reach conclusions that must be true whenever the assumptions on which the reasoning is based are true.

degree (45) A unit of measure used in measuring angles and arcs of circles. An arc of a circle with a measure of $1°$ is $\frac{1}{360}$ of the entire circle.

degree of a node (560) In a network, the number of edges meeting at a given node.

diagonal (293) In a polygon, a segment joining nonconsecutive vertices of the polygon.

diameter **1.** (446) In a circle, a chord that contains the center of the circle. **2.** (621) In a sphere, a segment that contains the center of the sphere, and whose endpoints are on the sphere.

dilation (348, 746) A transformation determined by a center point C and a scale factor $k > 0$. For any point P in the plane, the image P' of P is the point on $\overrightarrow{CP}$ such that $CP' = k \cdot CP$.

direction of a vector (673) The measure of the angle that the vector forms with the positive x-axis or any other horizontal line.

distance between a point and a line (154) For a point not on a given line, the length of the segment perpendicular to the line from the point. If the point is on the line, then the distance between the point and the line is zero.

distance between two parallel lines (156) The distance between one of the lines and any point on the other line.

edge **1.** (577) In a polyhedron, a line segment in which a pair of faces intersect. **2.** (559) In graph theory, a path connecting two nodes.

equiangular triangle (181) A triangle with all angles congruent.

equidistant (156) The distance between two lines measured along a perpendicular line to the line is always the same.

equilateral triangle (181) A triangle with all sides congruent.

exterior angle of a polygon (190) An angle that forms a linear pair with one of the angles of the polygon. In the figure, $\angle 2$ is an exterior angle.

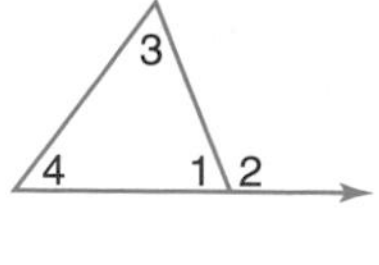

exterior angles (126) In the figure, transversal t intersects lines ℓ and m. The exterior angles are $\angle 3$, $\angle 4$, $\angle 5$, and $\angle 6$.

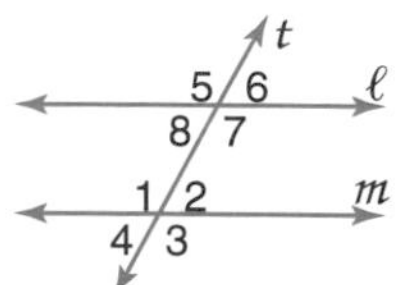

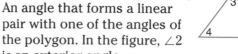

exterior point **1.** (45) For a given angle, a point that is neither on the angle or in the interior of the angle. **2.** (475) For a circle, a point whose distance from the center of the circle is greater than the radius of the circle.

external secant segment (492) The part of a secant segment that is exterior to the circle. *See secant segment.*

extremes (339, 397) In the proportion, $\frac{a}{b} = \frac{c}{d}$, the numbers a and d.

face (577) In a polyhedron, the flat polygonal surfaces that intersect to form the edges of the polyhedron.

flow proof (191) A proof which organizes a series of statements in logical order, starting with the given statements. Each statement along with its reasons is written in a box. Arrows are used to show how each statement leads to another.

fractals (378) A figure generated by repeating a special sequence of steps infinitely often. Fractals often exhibit self-similarity.

function (647) A relationship between input and output. In a function, one or more operations are performed on the input to get the output, and there is exactly one output for each input.

geometric mean (397) For any positive numbers a and b, the positive number x such that $\frac{a}{x} = \frac{x}{b}$.

geometric probability (551) Involves using the principles of length and area to find the probability of an event.

graph theory (559) The study of properties of figures consisting of points, called nodes, and paths that connect the nodes in various ways.

great circle (622) For a given sphere, the intersection of the sphere and a plane that contains the center of the sphere.

H

height (529, 591, 602) The length of an altitude of a figure.

hemisphere (622) One of the two congruent parts into which a great circle separates a given sphere.

hypotenuse (181) In a right triangle, the side opposite the right angle.

hypothesis (76) In a conditional statement, the statement that immediately follows the word *if*.

I

if and only if (139) When both a conditional and its converse are true.

if-then statement (76) A compound statement of the form "if A, then B", where A and B are statements.

image (715) The result of a transformation. If A is mapped onto A', then A' is called the image of A. The preimage of A' is A.

included angle (207) In a triangle, the angle formed by two sides is the included angle for these two sides.

included side (207) The side of a triangle that forms a side of two given angles.

incomplete network (559) In graph theory, a network with at least one pair of nodes not connected by an edge.

indirect proof (252) Proof by contradiction. In an indirect proof, one assumes that the statement to be proved is false. One then uses logical reasoning to deduce a statement that contradicts a postulate, theorem, or one of the assumptions. Once a contradiction is obtained, one concludes that the statement assumed false must in fact be true.

indirect reasoning (252) Reasoning that assumes the conclusion is false and then shows that this assumption leads to a contradiction of the hypothesis or some other accepted fact, like a postulate, theorem, or corollary. Then, since the assumption has been proved false, the conclusion must be true.

inductive reasoning (70) Reasoning that uses a number of specific examples to arrive at a plausible generalization or prediction. Conclusions arrived at by inductive reasoning lack the logical certainty of those arrived at by deductive reasoning.

inequality (254) For any real numbers a and b, $a > b$ if there is a positive number c such that $a = b + c$.

inscribed angle (466) An angle having its vertex lie on a given circle and containing two chords of the circle.

inscribed polygon (459) A polygon is inscribed in a circle if each of its vertices lies on the circle.

intercepted arc (466) An angle intercepts an arc if and only if each of the following conditions holds.

1. The endpoints of the arc lie on the angle.
2. All points of the arc, except the endpoints, are in the interior of the circle.
3. Each side of the angle contains an endpoint of the arc.

intercepts method (646) A technique for graphing a linear equation by locating the x- and y-intercepts.

interior **1.** (45) A point is in the interior of an angle if it does not lie on the angle itself and it lies on a segment whose endpoints are on the sides of the angle. **2.** (475) A point is in the interior of a circle if the measure of the segment joining the point to the center of the circle is less that the measure of the radius.

interior angles (126) In the figure, transversal t intersects lines ℓ and m. The interior angles are $\angle 1$, $\angle 2$, $\angle 7$, and $\angle 8$.

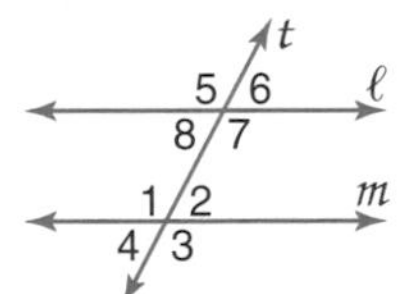

intersection (14) The intersection of two figures is the set of points that are in both figures.

inverse (78) The denial of a statement.

isometry (716) A mapping for which the original figure and its image are congruent.

isosceles trapezoid (321) A trapezoid in which the legs are congruent. Both pairs of base angles are congruent and the diagonals are congruent.

isosceles triangle (181) A triangle with at least two sides congruent. The congruent sides are called legs. The angles opposite the legs are base angles. The angle formed by two legs is the vertex angle. The third side is the base.

iteration (378) A process of repeating the same procedure over and over again.

kite (320) A quadrilateral with exactly two distinct pairs of adjacent congruent sides.

lateral area (592) For prisms, pyramids, cylinders, and cones, the area of the figure not including the bases.

lateral edge **1.** (591) In a prism, the intersection of two adjacent lateral faces. Lateral edges are parallel segments that join corresponding vertices of the bases. **2.** (600) In a pyramid, lateral edges are the edges of the lateral faces that join the vertex to vertices of the base.

lateral faces **1.** (591) In a prism, a face that is not a base of the figure. **2.** (600) In a pyramid, faces that intersect at the vertex.

Law of Cosines (431) Let $\triangle ABC$ be any triangle with a, b, and c representing the measures of sides opposite the angles with measures A, B, and C, respectively. Then the following equations hold true.

$$a^2 = b^2 + c^2 - 2bc \cos A$$
$$b^2 = a^2 + c^2 - 2ac \cos B$$
$$c^2 = a^2 + b^2 - 2ab \cos C$$

Law of Detachment (86) If $p \rightarrow q$ is a true conditional and p is true, then q is true.

Law of Sines (426) Let $\triangle ABC$ be any triangle with a, b, and c representing the measures of sides opposite the angles with measures A, B, and C, respectively. Then, $\frac{\sin A}{a} = \frac{\sin B}{b} = \frac{\sin C}{c}$.

Law of Syllogism (87) If $p \rightarrow q$ and $q \rightarrow r$ are true conditionals, then $p \rightarrow r$ is also true.

leg **1.** (181) In a right triangle, the sides opposite the acute angles. **2.** (182) In an isosceles triangle, the congruent sides. **3.** (321) In a trapezoid, the nonparallel sides.

line (12) A basic undefined term of geometry. Lines extend indefinitely and have no thickness or width. In a figure, a line is shown with arrows at each end. Lines are usually named by lower case script letters or by writing capital letters for two points on the line, with a double arrow over the pair of letters.

linear equation (646) An equation that can be written in the form $Ax + By = C$, where A, B, and C are real numbers, with A and B not both 0.

linear pair (53) A pair of adjacent angles whose noncommon sides are opposite rays.

line of reflection (722) Line ℓ is a line of reflection for a figure if, for every point A of the figure not on ℓ, there is a point A' of the figure such that ℓ is the perpendicular bisector of $\overline{AA'}$. The points A and A' are reflection images of each other.

line of symmetry (724) A line that can be drawn through a plane figure so that the figure on one side is the reflection image of the figure on the opposite side.

locus (696) In geometry, a figure is a locus if it is the set of all points and only those points that satisfy a given condition.

magnitude of a vector (673) The length of a vector.

major arc (453) If $\angle APB$ is a central angle of circle P, and C is any point on the circle and in the exterior of the angle, then points A and B and all points of the circle exterior to $\angle APB$ form a major arc called $\overset{\frown}{ACB}$ Three letters are needed to name a major arc.

mapping (715) A one-to-one correspondence between points of two figures.

means (339, 397) In the proportion, $\frac{a}{b} = \frac{c}{d}$, the numbers b and c.

measure **1.** (28) The length of $\overline{AB}$, written AB, is the distance between A and B. **2.** (45) A protractor can be used to find the measure of an angle.

median **1.** (238) In a triangle, a segment that joins a vertex of the triangle and the midpoint of the opposite side. **2.** (322) In a trapezoid, the segment joining the midpoints of the legs.

midpoint (36) A point M is the midpoint of segment PQ if M is between P and Q, and $PM = MQ$.

minor arc (453) If $\angle APB$ is a central angle of circle P, then points A and B and all points on the circle interior to the angle form a minor arc called $\overset{\frown}{AB}$.

negation (78) The denial of a statement.

net (585) A two-dimensional figure that, when folded, forms the surfaces of a three-dimensional object.

network (559) A figure consisting of points, called nodes, and edges that join various nodes to one another.

***n*-gon** (515) A polygon with *n* sides.

node (559) In graph theory, the points of a network.

noncollinear points (8) Points that do not lie on the same line.

non-Euclidean geometry (164) The study of geometrical systems which are not in accordance with the Parallel Postulate of Euclidean geometry.

oblique cone (602) A cone that is not a right cone.

oblique cylinder (593) A cylinder that is not a right cylinder.

oblique prism (591) A prism that is not a right prism.

obtuse angle (46) An angle with degree measure greater than 90 and less than 180.

obtuse triangle (180) A triangle with an obtuse angle.

opposite rays (44) Two rays $\overrightarrow{BA}$ and $\overrightarrow{BC}$ such that *B* is between *A* and *C*.

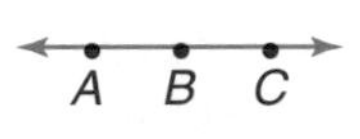

ordered pair (6) A pair of numbers given in a specific order. Ordered pairs are used to locate points in a plane.

ordered triple (680) Three numbers given in a specific order. Ordered triples are used to locate points in space.

origin (6) In a coordinate plane or a three-dimensional coordinate system, the point where the coordinate axes intersect (usually designated by *O*).

paragraph proof (39) A proof written in the form of a paragraph (as opposed to a two-column proof).

parallel lines (124) Lines in the same plane that do not intersect.

parallelogram (290, 291) A quadrilateral in which both pairs of opposite sides are parallel. Any side of a parallelogram may be called a base. For each base, a segment called an altitude is a segment perpendicular to the base and having its endpoints on the lines containing the base and the opposite side.

parallelogram law (675) The parallelogram law is the basis for a method for adding two vectors. The two vectors with the same initial point form part of a parallelogram. The resultant or sum of the two vectors is the diagonal of the parallelogram.

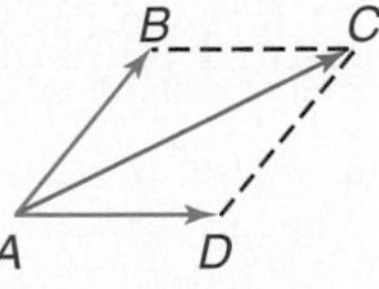

perpendicular bisector (238) A bisector of a segment that is perpendicular to the segment.

perpendicular bisector of a triangle (238) A line or line segment that passes through the midpoint of a side of a triangle and is perpendicular to that side.

perpendicular lines (53) Two lines that intersect to form a right angle.

perspective view (576) The view from a corner of a figure (also called corner view).

pi (447) The ratio of the circumference of a circle to its diameter.

plane (12) A basic undefined term of geometry. Planes can be thought of as flat surfaces that extend indefinitely in all directions and have no thickness. In a figure, a plane is often represented by a parallelogram. Planes are usually named by a capital script letter or by three noncollinear points on the plane.

plane Euclidean geometry (164) Geometry based on Euclid's axioms dealing with a system of points, lines, and planes.

Platonic solid (577) Any one of the five regular polyhedrons: tetrahedron, hexahedron, octahedron, dodecahedron, or icosohedron.

point (12) A basic undefined term of geometry. Points have no size. In a figure, a point is represented by a dot. Points are named by capital letters.

point of reflection (722) Point *S* is the reflection of point *R* with respect to point *Q*, the point of reflection, if *Q* is the midpoint of the segment drawn from *R* to *S*.

point of symmetry (725) The point of reflection for all points of a figure.

point of tangency (475) For a line that intersects a circle in only one point, the point in which they intersect.

point-slope form (654) An equation of the form $y - y_1 = m(x - x_1)$ for the line passing through a point whose coordinates are (x_1, y_1) and having a slope of *m*.

polygon (180, 514) A figure in a plane that meets the following conditions.

1. It is a closed figure formed by three or more coplanar segments called sides.
2. Sides that have a common endpoint are noncollinear.
3. Each side intersects exactly two other sides, but only at their endpoints.

polyhedron (577) A closed three-dimensional figure made up of flat polygonal regions. The flat regions formed by the polygons and their interiors are called faces. Pairs of faces intersect in line segments called edges. Points where three or more edges intersect are called vertices.

postulate (28) A statement that describes a fundamental relationship between the basic terms of geometry. Postulates are accepted as true without proof.

preimage (715) For a transformation, if A is mapped onto A', then A is the preimage of A'.

prism (577) A solid with the following characteristics.

1. Two faces, called bases, are formed by congruent polygons that lie in parallel planes.
2. The faces that are not bases, called lateral faces, are formed by parallelograms.
3. The intersections of two adjacent lateral faces are called lateral edges and are parallel segments.

probability (294) The ratio that tells how likely it is that an event will occur.

$$P(\text{event}) = \frac{\text{number of successful outcomes}}{\text{total number of possible outcomes}}$$

proof (39, 94) A logical argument showing that the truth of a hypothesis guarantees the truth of the conclusion.

proportion (339) An equation of the form $\frac{a}{b} = \frac{c}{d}$ that states that two ratios are equivalent.

protractor (45) A tool used to find the degree measure of angles.

pyramid (577) A solid with the following characteristics.

1. All the faces, except one face, intersect at a point called the vertex.
2. The face that does not contain the vertex is called the base and is a polygonal region.
3. The faces meeting at the vertex are called lateral faces and are triangular regions.

Pythagorean Theorem (30, 399) In a right triangle, the sum of the squares of the measures of the legs equals the sum of the square of the measure of the hypotenuse.

Pythagorean triple (400) A set of numbers, a, b, and c, that satisfy the equation $a^2 + b^2 = c^2$.

quadrant (6) One of the four regions into which the two perpendicular axes of a coordinate plane divide the plane.

quadrilateral (291) A four-sided polygon.

radius **1.** (446) A radius of a circle is a segment whose endpoints are the center of the circle and a point on the circle. **2.** (543) A segment is a radius of a regular polygon if it is a radius of a circle circumscribed about the polygon. **3.** (621) A radius of a sphere is a segment whose endpoints are the center and a point on the sphere.

ratio (338) A comparison of two numbers using division.

ray (44) $\overrightarrow{PQ}$ is a ray if it is the set of points consisting of $\overline{PQ}$ and all points S for which Q is between P and S.

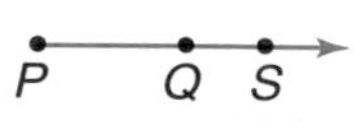

rectangle (305, 306) A quadrilateral with four right angles.

reflection (722) A transformation that flips a figure over a line called the line of reflection. *Also see line of reflection and point of reflection.*

regular polygon (515) A convex polygon with all sides congruent and all angles congruent.

regular polyhedron (577) A polyhedron in which all faces are congruent regular polygons.

regular prism (577) A right prism whose bases are regular polygons.

regular pyramid (600) A pyramid whose base is a regular polygon and in which the segment from the vertex to the center of the base is perpendicular to the base. This segment is called the altitude of the pyramid.

regular tessellation (523) A tessellation consisting entirely of regular polygons.

remote interior angles (190) The angles of a triangle that are not adjacent to a given exterior angle.

resultant (675) The sum of two or more vectors.

rhombus (313) A quadrilateral with all four sides congruent.

right angle (46) An angle whose degree measure is 90.

right circular cone (602) A cone that has a circular base and whose axis (the segment from the vertex to the center of the base) is perpendicular to the base. The axis is also the altitude of the cone.

right cylinder (593) A cylinder whose axis is also an altitude.

right prism (591) A prism in which the lateral edges are also altitudes.

right triangle (180) A triangle with a right angle. The side opposite the right angle is called the hypotenuse. The other two sides are called legs.

rotation (739) A transformation that is the composite of two reflections with respect to two intersecting lines. The intersection of the two lines is called the center of rotation.

Ruler Postulate (28) The points on any line can be paired with real numbers so that, given any two points *P* and *Q* on the line, *P* corresponds to zero, and *Q* corresponds to a positive number.

S

scalar multiplication (674) Multiplication of a vector by a real number.

scale factor **1.** (346) The ratio of the lengths of two corresponding sides of two similar polygons. **2.** (629) The ratio of the lengths of two corresponding sides of two similar solids. **3.** (746) For a dilation transformation with center *C*, the number *k* such that $ED = k(AB)$, where *E* is the image of *A* and *D* is the image of *B*.

scalene triangle (181) A triangle with no two sides congruent.

scatter plot (660) Two sets of data plotted as ordered pairs in a coordinate plane.

secant (483) For a circle, a line that intersects the circle in exactly two points.

secant segment (492) A segment from a point exterior to a circle to a point on the circle and containing a chord of the circle. The part of a secant segment that is exterior to the circle is called an external secant segment.

sector (553) A region bounded by a central angle and the intercepted arc.

segment (12) A part of a line that consists of two points, called endpoints, and all the points between them.

segment bisector (38, 48) A segment, line, or plane that intersects a segment at its midpoint.

self-similar (378) If any parts of a fractal image are replicas of the entire image, the image is self-similar.

semicircle (453) Either of the two parts into which a circle is separated by a line containing a diameter of the circle.

semi-regular (524) Uniform tessellations containing two or more regular polygons.

side **1.** (44) For an angle, the two rays that form the sides of the angle. **2.** (180) For a polygon, a segment joining two consecutive vertices of the polygon.

Sierpinski triangle (378) A fractal triangle created by connecting midpoints of an equilateral triangle. The resulting fractal is named after the Polish mathematician, Waclaw Sierpinski.

similar circles (454) Circles that have different radii.

similar figures (346) Figures that have the same shape but that may differ in size.

similar polygons (346) Two polygons are similar if there is a correspondence between their vertices such that corresponding angles are congruent and the measures of corresponding sides are proportional.

similar solids (629) Solids that have exactly the same shape but not necessarily the same size.

similarity transformation (717) When a figure and its transformation image are similar.

sine (412) For an acute angle of a right triangle, the ratio of the length of the leg opposite the acute angle to the length of the hypotenuse.

skew lines (125) Lines that do not intersect and are not in the same plane.

slant height **1.** (600) For a regular pyramid, the height of lateral face. **2.** (602) For a right circular cone, the length of any segment joining the vertex to the edge of the circular base.

slice of a solid (574, 577) The figure formed by making a straight cut across the solid.

slope (138) For a (nonvertical) line containing two points (x_1, y_1) and (x_2, y_2), the number *m* given by $m = \frac{y_2 - y_1}{x_2 - x_1}$ where $x_2 \neq x_1$.

slope-intercept form (647) A linear equation of the form $y = mx + b$. The graph of such an equation has slope m and y-intercept b.

solid (574) A three-dimensional figure consisting of all of its surface points and all of its interior points.

solving the triangle (427) Finding the measures of all the angles and sides of a triangle.

space (12) A boundless three-dimensional set of all points.

sphere (577, 621) In space, the set of all points that are a given distance from a given point, called the center.

spherical geometry (164) The branch of geometry which deals with a system of points, great circles (lines), and spheres (planes) (also known as Riemannian geometry).

spreadsheets (21) Computer programs designed especially for creating charts involving many calculations.

square (315) A quadrilateral with four right angles and four congruent sides.

standard form (646) For linear equations, an equation of the form $Ax + By = C$, where A and B are not both zero.

straight angle (44) A figure formed by two opposite rays.

straightedge (31) An instrument used to draw lines.

strictly self-similar (379) A figure is strictly self-similar if any of its parts, no matter where they are located or what size is selected, contain the same figure as the whole.

supplementary angles (55) Two angles whose degree measures have a sum of 180.

surface area (584) The sum of the areas of all faces and side surfaces of a three-dimensional figure.

system of equations (704) A group of two or more equations with the same variables.

T

tangent **1.** (412) For an acute angle of a right triangle, the ratio of the length of the leg opposite the acute angle to the length of the leg adjacent to the acute angle. **2.** (474) A tangent to a circle is a line in the plane of the circle that intersects the circle in exactly one point. The point of intersection is called the point of tangency. **3.** (621) A tangent to a sphere is a line or plane that intersects the sphere in exactly one point.

tangent segment (477) A segment AB such that one endpoint is on a circle, the other is outside the circle, and the line AB is tangent to the circle.

tessellation (522, 523) Tile-like patterns formed by repeating shapes to fill a plane without gaps or overlaps.

theorem (39, 101) A statement, usually of a general nature, that can be proved by appeal to postulates, definitions, algebraic properties, and rules of logic.

traceable network (559) A network that can be traced in one continuous path without retracing any edge. A network is traceable if its nodes all have even degrees or if exactly two nodes have odd degrees.

transformation (715) In a plane, a mapping for which each point has exactly one image point and each image point has exactly one preimage point.

translation (731) A composite of two reflections over two parallel lines. A translation slides figures the same distance in the same direction.

transversal (126) A line that intersects two or more lines in a plane at different points.

trapezoid (321) A quadrilateral that has exactly one pair of parallel sides. The parallel sides of a trapezoid are called bases. The nonparallel sides are called legs. The pairs of angles with their vertices at the endpoints of the same base are called base angles. The line segment joining the midpoints of the legs of a trapezoid is called the median. An altitude is a segment perpendicular to the lines containing the bases and having its endpoints on these lines.

triangle (180) A figure formed by the segments determined by three noncollinear points. The three segments are called sides. The endpoints are called the vertices of the triangle. A triangle separates a plane into three parts, the triangle, its interior, and its exterior.

trigonometric ratio (412) A ratio of the measures of two sides of a right triangle is called a trigonometric ratio.

trigonometry (412) The study of the properties of triangles and trigonometric functions and their applications.

two-column proof (94) A formal proof in which statements are listed in one column and the reasons for each statement are listed in a second column.

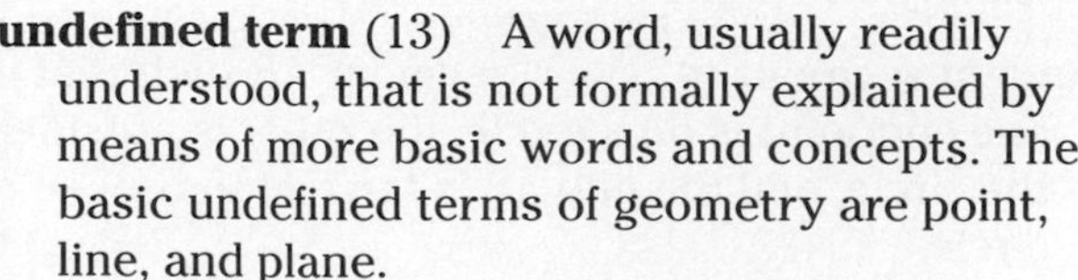

undefined term (13) A word, usually readily understood, that is not formally explained by means of more basic words and concepts. The basic undefined terms of geometry are point, line, and plane.

uniform (524) Tessellations containing the same combination of shapes and angles at each vertex.

vector (673) A directed segment. Vectors possess both magnitude (length) and direction.

vertex **1.** (44) For an angle, the common endpoint of the two rays that form the angle. **2.** (180) In a polygon, the endpoints of the sides is called a vertex. **3.** (577) In a polyhedron, where three or more edges intersect. **4.** (600) In a pyramid, the vertex that is not contained in the base of the pyramid. **5.** (602) In the figure, the vertex of the cone is point *V*.

vertex angle (182) In an isosceles triangle, the angle formed by the congruent sides (legs).

vertical angles (53) Two nonadjacent angles formed by two intersecting lines.

volume (607) The measure of the amount of space enclosed by a three dimensional figure.

***x*-axis** (6) The horizontal number line in a coordinate plane.

***x*-coordinate** (6) The first number in an ordered pair of numbers.

***y*-axis** (6) The vertical number line in a coordinate plane.

***y*-coordinate** (6) The second number in an ordered pair of numbers.

SPANISH GLOSSARY

absolute value/valor absoluto (29) El número de unidades que un número real dista de cero en una recta numérica.

acute angle/acutángulo (46) Ángulo que mide menos de 90 grados.

acute triangle/triángulo acutángulo (180) Triángulo en el que todos los ángulos son agudos.

adjacent angles/ángulos adyacentes (53) Dos ángulos en el mismo plano que tienen el vértice y un lado en común, pero sin puntos interiores en común.

adjacent arcs/arcos adyacentes (453) Dos arcos de un círculo que tienen solo un punto en común.

alternate exterior angles/ ángulos alternos externos (126) En la figura, la transversal t interseca las rectas ℓ y m. $\angle 5$ y $\angle 3$, y $\angle 6$ y $\angle 4$ se conocen como ángulos alternos externos.

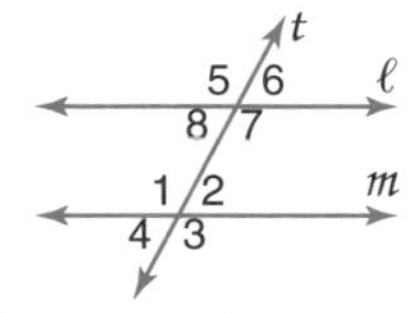

alternate interior angles/ángulos alternos internos (126) En la figura anterior, la transversal t interseca las rectas ℓ y m. $\angle 1$ y $\angle 7$, y $\angle 2$ y $\angle 8$ son ángulos alternos internos.

altitude of a cone/altitud de un cono (602) Segmento perpendicular trazado desde el vértice del cono al plano que contiene la base.

altitude of a cylinder/altitud de un cilindro (593) Segmento perpendicular a los planos que contienen las bases del cilindro y que tiene un extremo en cada uno de los planos.

altitude of a parallelogram/altitud de un paralelogramo (529) Cualquier segmento perpendicular entre las rectas que contienen dos de los lados paralelos del paralelogramo.

altitude of a prism/altitud de un prisma (591) Segmento perpendicular a los planos que contienen las bases del prisma y con un extremo en cada uno de ellos. A la medida de la altitud se le llama la altura del prisma.

altitude of a pyramid/altitud de una pirámide (600) Segmento perpendicular a la base trazado desde el vértice de la pirámide hasta el plano que contiene la base.

altitude of a triangle/altitud de un triángulo (238) Segmento perpendicular trazado del vértice de un triángulo a la recta que contiene el lado opuesto.

angle/ángulo (44) Figura que con-consiste de dos rayos no colineales con un extremo común. Los rayos son los lados del ángulo y el extremo común es su vértice. Un ángulo separa el plano en tres partes: el interior y el exterior del ángulo y el ángulo mismo.

exterior

interior

angle bisector/bisectriz de un ángulo (48) El rayo, QS, es la bisectriz del $\angle PQR$ si S está en el interior del ángulo y $\angle PQS \cong \angle RQS$.

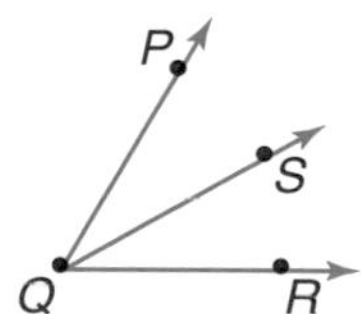

angle bisector of a triangle/bisectriz del ángulo de un triángulo (240) Segmento que biseca el ángulo de un triángulo y que tiene un extremo en un vértice del triángulo y el otro extremo en otro punto del triángulo.

angle of depression/ángulo de depresión (420) Ángulo formado por una recta horizontal y la recta de visión a un objeto, debajo del nivel de la recta horizontal.

angle of elevation/ángulo de elevación (400) Ángulo formado por una recta horizontal y la recta de visión a un objeto, encima del nivel de la recta horizontal.

angle of rotation/ángulo de rotación (740) El ángulo de rotación, $\angle ABC$, está determinado por la preimagen A, el centro de rotación B y la imagen de rotación C.

apothem/apotema (544) Segmento perpendicular a un lado trazado desde el centro de un polígono regular hasta uno de sus lados.

arc/arco (453) Parte continua de un círculo.

arc length/longitud de arco (454) Distancia lineal que representa el arco. La longitud de un arco es una parte de la circunferencia proporcional a la medida del ángulo central correspondiente cuando se compara con el círculo completo.

arc measure/medida de arco (453) La medida en grados de un arco menor es la medida de su ángulo central. La de un arco mayor es igual a 360°, menos la medida de su ángulo central. La de un semicírculo es 180°.

arc of a chord/arco de cuerda (459) Un arco menor que tiene los mismos extremos que una cuerda dada.

area/área (19, 529) El número de unidades cuadradas contenidas en el interior de una figura.

auxiliary line/recta auxiliar (189) Recta o segmento de recta que se agrega a una figura para ayudar a demostrar un resultado.

axis/eje **1.** (6) En un plano de coordenadas, el eje x es la recta numérica horizontal y el eje y es es la recta numérica vertical. **2.** (593) El eje de un cilindro es el segmento cuyos extremos son los centros de las bases. **3.** (602) El eje de un cono es el segmento cuyos extremos son el vértice y el centro de la base.

B

base/base **1.** (182) En un triángulo isósceles, el lado opuesto al ángulo del vértice. **2.** (321) En un trapecio, las bases son los lados paralelos. **3.** (529) Cualquier lado de un paralelogramo recibe el nombre de base. **4.** (577, 591) En un prisma, las bases son las dos caras formadas por polígonos congruentes que yacen en planos paralelos, mientras que todas las otras caras son paralelogramos. **5.** (593) En un cilindro, las bases son las regiones circulares, congruentes y paralelas que forman los extremos del cilindro. **6.** (600) En una pirámide, la base es la cara que no interseca las otras caras en el vértice. La base es una región poligonal. **7.** (602) En un cono, la base es la porción plana y circular del cono.

base angle/ángulo basal **1.** (182) Cualquiera de los ángulos de un triángulo isósceles formado por la base y uno de los catetos. **2.** (321) En el trapecio de la derecha, $\angle A$ y $\angle D$, y $\angle B$ y $\angle C$ are son pares de ángulos basales.

between/estar entre (28) B está y entre A y C si y solo si A, B y C son colineales y $AB + BC = AC$.

C

Cavalieri's Principle/Principio de Cavalieri (617) Si dos sólidos tienen la misma altura y la misma área transversal en cada nivel, entonces tienen el mismo volumen.

center of a circle/centro de un círculo (446) Punto desde el cual equidistan todos los puntos en un plano.

center of a regular polygon/centro de un polígono regular (543) El centro común de los círculos inscritos y circunscritos del polígono regular.

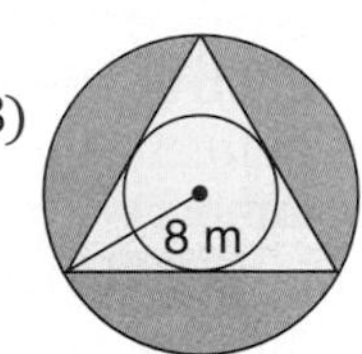

center of a sphere/centro de una esfera (621) Un punto en el espacio del cual equidistan todos los puntos de una esfera.

center of dilation/centro de dilatación (747) Un punto fijo que se usa para la medición cuando se altera el tamaño de una figura geométrica sin cambiar su forma.

center of rotation/centro de rotación (739) Un punto fijo alrededor del cual las figuras se mueven circularmente a una nueva posición.

central angle/ángulo central **1.** (452) Ángulo que interseca un círculo en dos puntos y que tiene su vértice en el centro del círculo. **2.** (544) Ángulo formado por dos segmentos trazados desde el centro de un polígono regular hasta los vértices consecutivos del mismo.

chord/cuerda **1.** (446) Segmento de recta cuyos extremos están en un círculo. **2.** (621) Segmento de recta cuyos extremos están en una esfera.

circle/círculo (446) Conjunto de puntos del plano que están a una distancia dada de un punto dado del plano, llamado centro.

circumference/circunferencia (447) El límite de los perímetros de los polígonos regulares inscritos a medida que aumenta el número de lados.

circumscribed polygon/polígono circunscrito (477) Un polígono está circunscrito a un círculo si cada lado del polígono es tangente al círculo.

collinear points/puntos colineales (8) Puntos que yacen sobre la misma recta.

column matrix/matriz columna (676) Matriz que solo tiene una columna.

common external tangent/ tangente externa común (476) Tangente común que no interseca el segmento cuyos extremos son los centros de los dos círculos.

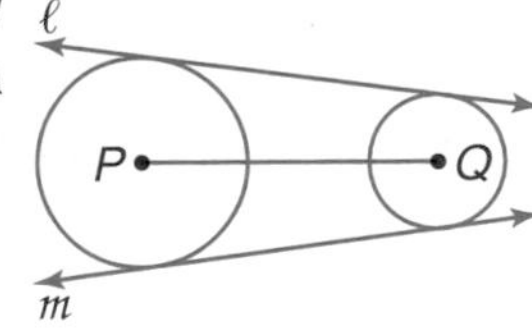

common internal tangent/ tangente interna común (476) Tangente común que interseca el segmento cuyos extremos son los centros de dos círculos.

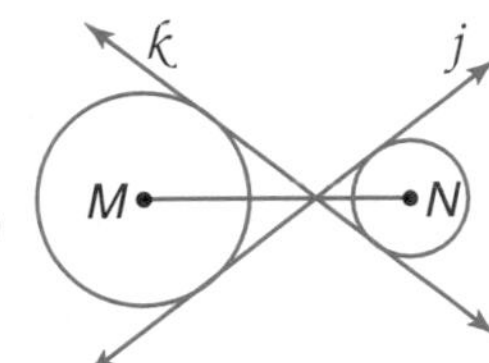

common tangent/tangente común (476) Recta tangente a dos círculos que están en el mismo plano.

compass/compás (38) Instrumento usado para trazar círculos y arcos de círculos.

complementary angles/ángulos complementarios (55) Dos ángulos cuyas medidas suman 90°.

complete network/red completa (559) En teoría de grafos, un circuito que tiene al menos un camino entre cada par de nodos.

composite of reflections/composición de reflexiones (731) Dos reflexiones sucesivas.

concave polygon/polígono cóncavo (514) Polígono para el cual existe una recta que contiene un lado del polígono y un punto interior del polígono.

concentric circles/círculos concéntricos (454) Círculos que yacen sobre el mismo plano y que tienen el mismo centro pero distintos radios.

conclusion/conclusión (76) La afirmación que le sigue a *entonces* en un enunciado condicional.

conditional statement/enunciado condicional (76) Enunciado de la forma “Si A, entonces B”, que puede escribirse en la forma *si-entonces*. La parte que sigue a *si* se llama hipótesis y la que sigue a *entonces* se llama conclusión.

cone/cono (577, 602) Sólido de base circular, un vértice *V* fuera del plano de la base y un área de superficie lateral compuesta de todos los puntos de los segmentos que conectan al vértice con las aristas de la base.

congruence transformation/transformación de congruencia (196, 716) Cuando una figura geométrica y su imagen son congruentes, la transformación se llama transformación de congruencia o isometría.

congruent angles/ángulos congruentes (47) Ángulos que tienen la misma medida.

congruent arcs/arcos congruentes (454) Dos arcos que tienen la misma medida.

congruent circles/círculos congruentes (454) Círculos que tienen el mismo radio.

congruent segments/segmentos congruentes (31) Segmentos que tienen la misma longitud.

congruent solids/sólidos congruentes (629) Dos sólidos son congruentes si se cumplen todas las siguientes condiciones:

1. Los ángulos correspondientes son congruentes.
2. Las aristas correspondientes son congruentes.
3. Las caras correspondientes son congruentes.
4. Los volúmenes son congruentes.

congruent triangles/triángulos congruentes (196) Triángulos cuyas partes correspondientes son congruentes.

conjecture/conjetura (70) Suposición informada.

consecutive interior angles/ángulos consecutivos internos (126) En la figura, la transversal t interseca las rectas ℓ y m. Hay dos pares de ángulos consecutivos internos: $\angle 8$ y $\angle 1$, y $\angle 7$ y $\angle 2$.

contrapositive/antítesis (78) Negación de la hipótesis y de la conclusión del recíproco de un enunciado condicional dado.

converse/recíproco (77) En un enunciado condicional dado, el enunciado formado al intercambiar la hipótesis y la conclusión del enunciado original.

convex polygon/polígono complejo (514) Polígono para el que no existe recta alguna que contenga un lado del polígono y un punto interior del polígono.

coordinate/coordenada (7) Los números en un par ordenado de números. El primer número se llama la coordenada *x* y el segundo, la coordenada *y*.

coordinate plane/plano de coordenadas (6) Plano en el que se han usado dos rectas numéricas perpendiculares que se intersecan en sus puntos cero para aparear los puntos del plano con pares ordenados de números.

coordinate proof/prueba de coordenada (666) Demostración que usa figuras en un plano coordenado para demostrar resultados algebraicos.

coplanar points/puntos coplanares (13) Puntos que están en el mismo plano.

corner view/vista de esquina (576) También llamada vista de perspectiva, es la vista desde una esquina de una figura.

corollary/corolario (192) La afirmación que puede demostrarse fácilmente mediante un teorema es el corolario de dicho teorema.

corresponding angles/ángulos correspondientes (126) En la figura, la transversal t interseca las rectas ℓ y m. Hay cuatro pares de ángulos correspondientes: $\angle 5$ y $\angle 1$; $\angle 8$ y $\angle 4$; $\angle 6$ y $\angle 2$; y $\angle 7$ y $\angle 3$.

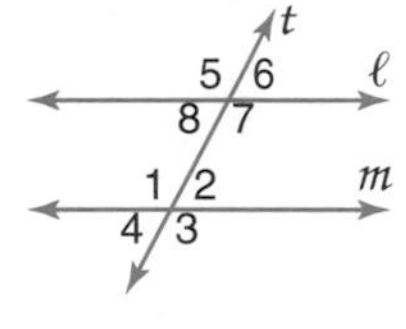

cosine/coseno (412) La razón de la longitud del cateto adyacente al ángulo agudo, a la longitud de la hipotenusa, en un ángulo agudo de un triángulo rectángulo.

counterexample/contraejemplo (72) Un ejemplo que se usa para demostrar que un enunciado dado no es siempre cierto.

cross products/productos cruzados (339) En la proporción $\frac{a}{b} = \frac{c}{d}$, donde $b \neq 0$ y $d \neq 0$, los productos cruzados son ad bc. La proporción es verdadera si y solo si los productos cruzados son iguales.

cross section/corte transversal (574, 577) La intersección de un sólido con un plano paralelo a la base o bases del sólido.

cube/cubo (577) Prisma en el que todas las caras son cuadradas.

cylinder/cilidro (577) Figura cuyas bases están formadas por círculos congruentes que yacen en planos paralelos.

D

deductive reasoning/razonamiento deductivo (86) Sistema de razonamiento usado para llegar a conclusiones que deben ser verdaderas cada vez que las suposiciones en las que está basado el razonamiento son verdaderas.

degree/grado (45) Unidad de medida para medir ángulos y arcos de círculo. El arco de un círculo con una medida de 1° es $\frac{1}{360}$ del círculo completo.

degree of a node/grado de un nodo (560) Número de aristas que convergen en un nodo de una red.

diagonal/diagonal (293) Segmento que une vértices no consecutivos de un polígono.

diameter/diámetro **1.** (446) Cuerda que contiene el centro de un círculo. **2.** (621) Segmento que contiene el centro de una esfera y cuyos extremos están en la esfera.

dilation/dilatación (348, 746) Transformación determinada por un punto central C y un factor de escala $k > 0$. Para cada punto P del plano, la imagen P' de P es el punto en $\overrightarrow{CP}$ que satisface $CP' = k \cdot CP$.

direction of a vector/dirección de un vector (673) Medida del ángulo entre un vector y el eje x positivo o cualquier otra recta horizontal.

distance between a point and a line/distancia entre un punto y una recta (154) Para un punto fuera de una recta, la longitud del segmento perpendicular trazado desde el punto a la recta. Si el punto está sobre la recta, entonces la distancia entre el punto y la recta es cero.

distance between two parallel lines/distancia entre dos paralelas (156) Distancia entre una de las rectas y cualquier punto en la otra recta.

E

edge/arista **1.** (577) Segmento de recta en el que se intersecan dos caras de un poliedro. **2.** (559) En teoría de grafos, un camino que conecta dos nodos.

equiangular triangle/triángulo equiangular (181) Aquél en que todos los ángulos son congruentes.

equidistant/equidistante (156) La distancia entre dos rectas medida a lo largo de una perpendicular a la recta es siempre la misma.

equilateral triangle/triángulo equilátero (181) Aquél en que todos los lados son congruentes.

exterior angle of a polygon/ángulo exterior de un polígono (190) Ángulo que forma un par lineal con uno de los ángulos del polígono. En la figura, $\angle 2$ es un ángulo exterior.

exterior angles/ángulos exteriores (126) En la figura, la transversal t interseca las rectas ℓ y m. Los ángulos exteriores son $\angle 3$, $\angle 4$, $\angle 5$ y $\angle 6$.

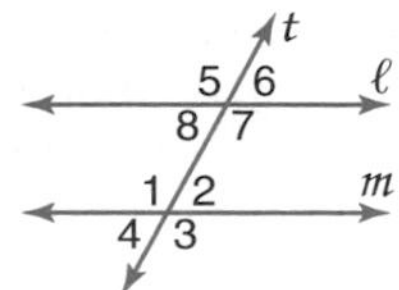

exterior point/punto exterior **1.** (45) Para un ángulo dado, un punto que no está ni en el ángulo ni en el interior del ángulo. **2.** (475) En un círculo, un punto cuya distancia al centro del círculo es mayor que el radio del mismo.

external secant segment/segmento externo de la secante (492) La parte de un segmento de secante exterior al círculo. *Ver segmento de secante.*

extremes/extremos (339, 397) Los números a y d, en la proporción, $\frac{a}{b} = \frac{c}{d}$.

F

face/cara (577) Superficies poligonales planas de un poliedro que al intersecarse forman las aristas del poliedro.

flow proof/demostración de flujo (191) La que organiza una serie de enunciados en orden lógico, comenzando con los enunciados dados. Cada enunciado junto con sus razones se escribe en un rectángulo. Se usan flechas para demostrar cómo cada enunciado conduce a otro.

fractals/fractales (378) Figura generada por la repetición infinita de una sucesión especial de pasos. A menudo exhiben autosemejanza.

function/función (647) Relación entre las entradas y las salidas. Una o más operaciones son ejecutadas en las entradas para obtener las salidas y hay una única salida para cada entrada.

G

geometric mean/media geométrica (397) Para números positivos a y b cualesquiera, el número positivo x tal que $\frac{a}{x} = \frac{x}{b}$.

geometric probability/probabilidad geométrica (551) El uso de los principios de longitud y área para calcular la probabilidad de un evento.

graph theory/teoría de grafos (559) Estudio de las propiedades de figuras que consisten de puntos llamados nodos y de caminos que conectan los nodos de varias formas.

great circle/círculo máximo (622) Intersección de una esfera con un plano que contiene el centro de la esfera.

H

height/altura (529, 591, 602) La longitud de la altitud de una figura.

hemisphere/semiesfera (622) Una de las dos partes congruentes en las que un círculo máximo divide una esfera.

hypotenuse/hipotenusa (181) Lado opuesto al ángulo recto en un triángulo rectángulo.

hypothesis/hipótesis (76) La afirmación que sigue inmediatamente a la palabra *si* en un enunciado condicional.

I

if and only if/si y solo si (139) Cuando un enunciado condicional y su recíproco son verdaderos.

if-then statement/enunciado si-entonces (76) Enunciado compuesto que tiene la forma "si A, entonces B", donde A y B son enunciados.

image/imagen (715) El resultado de una transformación. Si A es mapeado en A', A' se llama la imagen de A. La preimagen de A' es A.

included angle/ángulo incluido (207) El ángulo formado por dos lados de un triángulo recibe el nombre de ángulo incluido de esos lados.

included side/lado incluido (207) Lado de un triángulo, común a dos de sus ángulos.

incomplete network/red incompleta (559) En teoría de grafos, una red que tiene al menos un par de nodos no conectados por una arista.

indirect proof/demostración indirecta (252) Demostración por contradicción. Uno asume que el enunciado por demostrar es falso, en una demostración indirecta. Usando razonamiento lógico, deduce un enunciado que contradice un postulado, un teorema o una de las suposiciones. Al obtener una contradicción, concluye que el enunciado que se supuso falso debe ser cierto.

indirect reasoning/razonamiento indirecto (252) Razonamiento que supone que la conclusión es falsa y demuestra que esa suposición contradice la hipótesis o algún hecho aceptado, como un postulado, teorema o corolario. Luego, como se ha demostrado que la suposición es falsa, la conclusión debe ser verdadera.

inductive reasoning/razonamiento inductivo (70) Razonamiento que usa ejemplos específicos para llegar a una generalización o predicción plausible. Las conclusiones obtenidas mediante este tipo de razonamiento carecen de la certeza lógica de aquellas a las que se ha llegado mediante el razonamiento deductivo.

inequality/desigualdad (254) Para números reales a y b cualesquiera, $a > b$ si existe un número positivo c tal que $a = b + c$.

inscribed angle/ángulo inscrito (466) Un ángulo que tiene su vértice en un círculo dado y cuyos lados son dos cuerdas del mismo.

inscribed polygon/polígono inscrito (459) Un polígono está inscrito en un círculo, si y solo si, todos sus vértices están en el círculo.

intercepted arc/arco interceptado (466) Un ángulo interseca un arco si se cumplen las condiciones siguientes:

1. Los extremos del arco yacen en el ángulo.
2. Todos los puntos del arco, excepto los extremos, están en el interior del ángulo.
3. Cada lado del ángulo contiene un extremo del arco.

intercepts method/método de las intersecciones axiales (646) Técnica para trazar la gráfica de una ecuación lineal usando las intersecciones de esta con los ejes de coordenadas.

interior/interior **1.** (45) Un punto está en el interior de un ángulo si no está en el ángulo mismo, pero está en un segmento cuyos extremos yacen en los lados del ángulo. **2.** (475) Un punto está en el interior de un círculo si la medida del segmento que une el punto con el centro del círculo es menor que la medida del radio del círculo.

interior angles/ángulos interiores (126) En la figura, la transversal t interseca las rectas ℓ y m. Los ángulos interiores son $\angle 1$, $\angle 2$, $\angle 7$ y $\angle 8$.

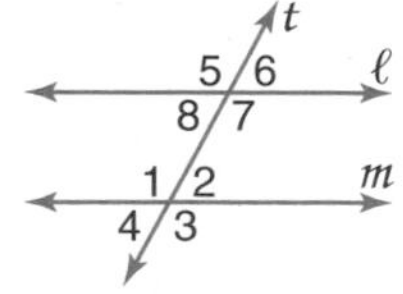

intersection/intersección (14) La intersección de dos figuras es el conjunto de puntos que están en ambas figuras.

inverse/inversa (78) Negación de un enunciado.

isometry/isometría (716) Una relación en que la figura original y su imagen son congruentes.

isosceles trapezoid/trapecio isósceles (321) Trapecio en que los catetos son congruentes. Ambas bases son congruentes y las diagonales también son congruentes.

isosceles triangle/triángulo isósceles (181) El que tiene al menos dos lados congruentes. Los lados congruentes se llaman catetos. Los ángulos opuestos a los catetos se llaman ángulos basales. El ángulo formado por los catetos es el ángulo del vértice y el tercer lado es la base.

iteration/iteración (378) Proceso en que se repite el mismo procedimiento una y otra vez.

K

kite/cometa (320) Cuadrilátero que tiene exactamente dos pares distintos de lados congruentes adyacentes.

L

lateral area/área lateral (592) En prismas, pirámides, cilindros y conos, el área de la figura excluyendo las bases.

lateral edge/arista lateral **1.** (591) La intersección de dos caras laterales adyacentes de un prisma. Las aristas laterales son segmentos paralelos que unen los vértices correspondientes de las bases. **2.** (600) Las aristas laterales de una pirámide son las aristas de las caras laterales que unen el vértice a los vértices de la base.

lateral faces/caras laterales **1.** (591) En un prisma, una cara que no es la base del prisma. **2.** (600) En una pirámide, las caras que se intersecan en el vértice de la pirámide.

Law of Cosines/Ley de los cosenos (431) Sea $\triangle ABC$ un triángulo cualquiera y sean a, b y c los lados opuestos a los ángulos con medidas A, B y C, respectivamente. Entonces las siguientes ecuaciones son verdaderas:

$$a^2 = b^2 + c^2 - 2bc \cos A$$
$$b^2 = a^2 + c^2 - 2ac \cos B$$
$$c^2 = a^2 + b^2 - 2ab \cos C$$

Law of Detachment/Ley de indiferencia (86) Si $p \rightarrow q$ es un enunciado condicional verdadero y p es verdadero, entonces q es verdadero.

Law of Sines/Ley de los senos (426) Sea $\triangle ABC$ un triángulo cualquiera y sean a, b y c los lados opuestos a los ángulos con medidas A, B y C, respectivamente. Entonces, $\frac{\sin A}{a} = \frac{\sin B}{b} = \frac{\sin C}{c}$.

Law of Syllogism/Ley del silogismo (87) Si $p \rightarrow q$ y $q \rightarrow r$ son enunciados condicionales verdaderos, entonces $p \rightarrow r$ es también verdadero.

leg/cateto **1.** (181) Los lados opuestos a los ángulos agudos en un triángulo rectángulo. **2.** (182) Los lados congruentes en un triángulo isósceles. **3.** (321) Los lados no paralelos en un trapecio.

line/recta (12) Término primitivo en geometría. Las rectas se extienden indefinidamente en ambos sentidos y no tienen grosor ni ancho. Las rectas se muestran con flechas en cada extremo y se designan mediante letras caligráficas minúsculas o mediante letras mayúsculas que designan dos puntos en la recta, con una flecha doble trazada encima del par de letras.

linear equation/ecuación lineal (646) Ecuación que tiene la forma $Ax + By = C$, donde A, B y C son números reales, con al menos uno de A o B distinto de 0.

linear pair/par lineal (53) Par de ángulos adyacentes cuyos lados no comunes son rayos opuestos.

line of reflection/línea de reflexión (722) La recta ℓ es una línea de reflexión de una figura si para cada punto A de la figura que no está en ℓ, existe un punto A' de la figura de modo que ℓ es la bisectriz perpendicular de $\overline{AA'}$. Los puntos A y A' son imagenes reflexivas uno del otro.

line of symmetry/línea de simetría (724) Línea que puede trazarse a través de una figura plana de modo que la figura en un lado de la recta sea la imagen reflexiva de la figura en el otro lado.

locus/lugar geométrico (696) Una figura es un lugar geométrico si es el conjunto de todos los puntos y, solo esos puntos que satisfacen una condición dada.

magnitude of a vector/magnitud de un vector (673) La longitud del vector.

major arc/arco mayor (453) Si $\angle APB$ es un ángulo central del círculo P y C es un punto cualquiera en el círculo y en el exterior del ángulo, entonces los puntos A y B y todos los puntos del círculo exteriores al $\angle APB$ forman el arco mayor $\widehat{ACB}$. Se necesitan tres letras para identificar un arco mayor.

mapping/relación (715) Correspondencia uno-a-uno entre los puntos de dos figuras.

means/medios (339, 397) Los números b y c en la proporción, $\frac{a}{b} = \frac{c}{d}$.

measure/medida **1.** (28) La longitud de $\overline{AB}$, designada por AB, es la distancia entre A y B. **2.** (45) Para calcular la medida de un ángulo puede usarse un transportador.

median/mediana **1.** (238) Segmento que une un vértice de un triángulo con el punto medio del lado opuesto. **2.** (322) El segmento que une los puntos medios de los catetos de un trapecio.

midpoint/punto medio (36) El punto M es el punto medio del segmento PQ si M está entre P y Q y $PM = MQ$.

minor arc/arco menor (453) Si $\angle APB$ es un ángulo central del círculo P, entonces los puntos A y B y todos los puntos del círculo, interiores al $\angle APB$, forman un arco que recibe el nombre de arco menor y que se designa por $\widehat{AB}$.

negation/negación (78) Negación de un enunciado.

net/red (585) Figura bidimensional que una vez plegada forma la superficie de un objeto tridimensional.

network/red (559) Figura que consiste de puntos llamados nodos y aristas que unen los nodos.

***n*-gon/*en*ágono** (515) Polígono de n lados.

node/nodo (559) En teoría de grafos, los puntos de una red.

noncollinear points/puntos no colineales (8) Puntos que no están en la misma recta.

non-Euclidean geometry/geometría no Euclidiana (164) El estudio de los sistemas geométricos que no satisfacen el Postulado de las Paralelas de la geometría Euclidiana.

O

oblique cone/cono oblicuo (602) Cono que no es un cono recto.

oblique cylinder/cilindro oblicuo (593) Cilindro que no es un cilindro recto.

oblique prism/prisma oblicuo (591) Prisma que no es un prisma recto.

obtuse angle (46) Ángulo que mide más de 90° y menos de 180°.

obtuse triangle/triángulo obtuso (180) Triángulo que tiene un ángulo obtuso.

opposite rays/rayos opuestos (44) Dos rayos $\overrightarrow{BA}$ y $\overrightarrow{BC}$ tales que B está entre A y C.

A B C

ordered pair/par ordenado (6) Un par de números dados en un orden específico. Se s se usan para ubicar puntos en un plano.

ordered triple/triple ordenado (680) Tres números dados en un orden específico. Se usan para ubicar puntos en el espacio.

origin/origen (6) En un plano de coordenadas o en un sistema de coordenadas tridimensional, el punto en el que se intersecan los ejes (designado generalmente con *O*).

P

paragraph proof/demostración de párrafo (39) Demostración escrita en forma de párrafo (en contraste con la demostración de dos columnas).

parallel lines/rectas paralelas (124) Rectas en el mismo plano que no se intersecan.

parallelogram/paralelogramo (290, 291) Cuadrilátero en el que ambos pares de lados opuestos son paralelos. Cualquier lado de un paralelogramo puede llamarse base. Para cada base, el segmento denominado altitud es un segmento perpendicular a la base. Dicho segmento tiene sus extremos en las rectas que contienen la base y el lado opuesto.

parallelogram law/ley del paralelogramo (675) La ley del paralelogramo se usa para sumar dos vectores. Los dos vectores con el mismo punto inicial forman parte de un paralelogramo. La resultante o suma de los dos vectores es la diagonal del paralelogramo.

perpendicular bisector/bisectriz perpendicular (238) Bisectriz de un segmento que es perpendicular al segmento.

perpendicular bisector of a triangle/bisectriz perpendicular de un triángulo (238) Recta o segmento de recta que pasa por el punto medio del lado de un triángulo y que es perpendicular al lado.

perpendicular lines/rectas perpendiculares (53) Dos rectas que se intersecan para formar un ángulo recto.

perspective view/vista de perspectiva (576) Vista desde una esquina de una figura, también llamada vista de esquina.

pi/pi (447) La razón de la circunferencia de un círculo a su diámetro.

plane/plano (12) Término primitivo en geometría. Los planos pueden visualizarse como superficies planas que se extienden indefinidamente en todas direcciones y que no tienen grosor. En una figura, los planos se representan generalmente mediante paralelogramos y se designan, por lo general, con una letra caligráfica mayúscula o mediante tres puntos no colineales del plano.

plane Euclidean geometry/geometría del plano euclidiano (164) Geometría basada en los axiomas de Euclides, un sistema de puntos, rectas y planos.

Platonic solid/sólido platónico (577) Cualquiera de los cinco poliedros regulares: tetraedro, hexaedro, octaedro, dodecaedro o icosaedro.

point/punto (12) Término primitivo indefinido en geometría. Los puntos no tienen tamaño. En una figura, los puntos se representan con puntos dibujados y son designados mediante letras mayúsculas.

point of reflection/punto de reflexión (722) El punto *S* es la reflexión del punto *R* con respecto al punto *Q*, si *Q* es el punto medio del segmento que une *R* con *S*.

point of symmetry/punto de simetría (725) El punto de reflexión de todos los puntos de una figura.

point of tangency/punto de tangencia (475) Para una recta que interseca un círculo en un único punto, el punto en el que se intersecan.

point-slope form/forma punto-pendiente (654) Ecuación de la forma $y - y_1 = m(x - x_1)$ de la recta que pasa por el punto (x_1, y_1) y que tiene pendiente m.

polygon/polígono (180, 514) Una figura en el plano que cumple con las siguientes condiciones:

1. Es una figura cerrada formada por tres o más segmentos coplanares llamados lados.
2. Los lados que tienen un extremo común no son colineales.
3. Cada lado interseca exactamente dos lados, pero solo en sus extremos.

polyhedron/poliedro (577) Figura cerrada tridimensional que consiste de regiones poligonales planas. Las regiones planas formadas por los polígonos y sus interiores se llaman caras. Dos caras se intersecan en segmentos llamados aristas. Los puntos en que se intersecan tres o más aristas se llaman vértices.

postulate/postulado (28) Enunciado que describe una relación fundamental entre los términos geométricos primitivos. Los postulados se aceptan como verdaderos sin necesidad de demostración.

preimage/preimagen (715) Para una transformación, si A está relacionado con A', entonces A es la preimagen de A'.

prism/prisma (577) Sólido con las siguientes características:

1. Dos caras, llamadas bases, formadas por polígonos congruentes que yacen en planos paralelos.
2. Las caras que no son bases, llamadas caras laterales, están formadas por paralelogramos.
3. Las intersecciones de dos caras adyacentes laterales se llaman aristas laterales y son segmentos paralelos.

probability/probabilidad (294) Razón que indica el grado de certeza de que un evento ocurra.

$$P\,(\text{evento}) = \frac{\text{número de resultados exitosos}}{\text{número total de posibles resultados}}$$

proof/demostración (39, 94) Argumento lógico que muestra que la verdad de una hipótesis garantiza la verdad de la conclusión.

proportion/proporción (339) Ecuación de la forma $\frac{a}{b} = \frac{c}{d}$ que afirma que dos razones son equivalentes.

protractor/transportador (45) Instrumento que se usa para calcular la medida de los ángulos.

pyramid/pirámide (577) Sólido con las siguientes características:

1. Todas las caras, excepto una, se intersecan en un punto llamado vértice.
2. La cara que no contiene el vértice se llama base y es una región poligonal.
3. Las caras que se encuentran en el vértice se llaman caras laterales y son regiones triangulares.

Pythagorean Theorem/teorema de Pitágoras (30, 399) En un triángulo rectángulo, la suma de los cuadrados de las medidas de los catetos es igual al cuadrado de la medida de la hipotenusa.

Pythagorean triple/triplete de Pitágoras (400) Tres números, a, b y c que satisfacen la ecuación $a^2 + b^2 = c^2$.

Q

quadrant/cuadrante (6) Una de las cuatro regiones en que los ejes perpendiculares de un plano de coordenadas dividen dicho plano.

quadrilateral/cuadrilátero (291) Polígono de cuatro lados.

R

radius/radio **1.** (446) El radio de un círculo es un segmento cuyos extremos son el centro del círculo y un punto en el mismo. **2.** (543) El radio de un polígono regular es el radio del círculo circunscrito al polígono. **3.** (621) El radio de una esfera es un segmento cuyos extremos son el centro de la esfera y un punto en la misma.

ratio/razón (338) Comparación de dos números usando división.

ray/rayo (44) $\overrightarrow{PQ}$ es un rayo si es el conjunto de puntos que consisten de P, Q y todos los puntos S para los cuales Q se encuentra entre P y S.

P Q S

rectangle/rectángulo (305, 306) Cuadrilátero que tiene cuatro ángulos rectos.

reflection/reflexión (722) Transformación que relaciona simétricamente una figura a través de una línea llamada línea de reflexión. *Véase también línea de reflexión y punto de reflexión.*

regular polygon/polígono regular (515) Polígono convexo cuyos lados son todos congruentes y cuyos ángulos son también congruentes.

regular polyhedron/poliedro regular (577) Poliedro en el que todas las caras son polígonos regulares congruentes.

regular prism/prisma regular (577) Prisma recto cuyas bases son polígonos regulares.

regular pyramid/pirámide regular (600) Pirámide cuya base es un polígono regular y en la cual el segmento desde el vértice al centro de la base es perpendicular a la base. Este segmento se llama la altitud de la pirámide.

regular tessellation/teselado regular (523) Teselado que consiste únicamente de polígonos regulares.

remote interior angles/ángulos internos no adyacentes (190) Los ángulos de un triángulo que no son adyacentes a un ángulo exterior dado.

resultant/resultante (675) La suma de dos o más vectores.

rhombus/rombo (313) Cuadrilátero cuyos lados son todos congruentes.

right angle/ángulo rectángulo (46) Un ángulo que mide 90°.

right circular cone/cono circular recto (602) Cono cuyo eje (el segmento trazado desde el vértice al centro de la base) es perpendicular a la base. El eje es asimismo la altitud del cono.

right cylinder/cilindro recto (593) Cilindro cuyo eje es también una altitud.

right prism/prisma recto (591) Prisma en el que las aristas laterales son también altitudes.

right triangle/triángulo rectángulo (180) Triángulo que tiene un ángulo recto. El lado opuesto al ángulo recto se llama hipotenusa. Los otros dos lados se llaman catetos.

rotation/rotación (739) Transformación que es la composición de dos reflexiones con respecto a dos rectas que se intersecan. La intersección de las rectas se llama el centro de rotación.

Ruler Postulate/postulado de la regla (28) Establece que los puntos de una recta cualquiera pueden aparearse con números reales de modo que, dados dos puntos cualesquiera *P* y *Q* en la recta, *P* corresponda a cero y *Q* corresponda a un número positivo.

S

scalar multiplication/multiplicación escalar (674) Multiplicación de un vector por un número real.

scale factor/factor de escala **1.** (346) La razón de las longitudes de dos lados correspondientes de dos polígonos similares. **2.** (629) La razón de las longitudes de dos lados correspondientes de dos sólidos similares. **3.** (746) Para una transformación de dilatación de centro *C*, el número *k* tal que $ED = k(AB)$, donde *E* es la imagen de *A* y *D* es la imagen de *B*.

scalene triangle/triángulo escaleno (181) Aquél que no tiene ningún par de lados congruentes.

scatter plot/diagrama de dispersión (660) Dos conjuntos de datos graficados como pares ordenados en un plano de coordenadas.

secant/secante (483) Recta que interseca un círculo exactamente en dos puntos.

secant segment/segmento de secante (492) Segmento trazado desde un punto exterior a un círculo hasta un punto en el círculo y que contiene una cuerda del círculo. La parte de un segmento de secante que es exterior al círculo se llama segmento externo de la secante.

sector/sector (553) Región acotada por un ángulo central y el arco interceptado.

segment/segmento (12) Parte de una recta que consiste de dos puntos, llamados extremos, y de todos los puntos que están entre ellos.

segment bisector/bisectriz de segmento (38, 48) Segmento, recta o plano que interseca un segmento en su punto medio.

self-similar/autosemejanza (378) Si cualquier parte de una imagen fractal es una réplica de la imagen total, la imagen es autosemejante.

semicircle/semicírculo (453) Cualquiera de las dos partes en que un círculo es dividido por una recta que contiene un diámetro del círculo.

semi-regular/semi-regular (524) Teselados uniformes que contienen dos o más polígonos regulares.

side/lado **1.** (44) Los dos rayos que forman los lados de un ángulo. **2.** (180) Segmento que une dos vértices consecutivos de un polígono.

Sierpinski triangle/triángulo de Sierpinski (378) Triángulo fractal creado al conectar los puntos medios de los lados de un triángulo equilátero. El fractal resultante lleva el nombre del matemático polaco Waclau Sierpinsk.

similar circles/círculos semejantes (454) Círculos que tienen distintos radios.

similar figures/figuras semejantes (346) Figuras que tienen la misma forma, pero que pueden diferir en tamaño.

similar polygons/polígonos semejantes (346) Dos polígonos son semejantes si existe una correspondencia entre sus vértices de modo que los ángulos correspondientes sean congruentes y las medidas de los lados correspondientes sean proporcionales.

similar solids/sólidos semejantes (629) Sólidos que tienen exactamente la misma forma, pero no necesariamente el mismo tamaño.

similarity transformation/transformación de semejanza (717) Cuando una figura y su imagen transformada son semejantes.

sine/seno (412) Para un ángulo agudo de un triángulo rectángulo, la razón de la longitud del cateto opuesto al ángulo a la longitud de la hipotenusa.

skew lines/rectas alabeadas (125) Rectas que no se intersecan y que no están en el mismo plano.

slant height/altura oblicua **1.** (600) La longitud de la altitud de una cara lateral en una pirámide. **2.** (602) La longitud de cualquier segmento que une el vértice de un cono recto circular con la arista de la base circular.

slice of a solid/corte de un sólido (574, 577) Figura que se forma al hacer un corte recto a lo largo de un sólido.

slope/pendiente (138) Para una recta no vertical que contiene los puntos (x_1, y_1) y (x_2, y_2), el número m dado por $m = \frac{y_2 - y_1}{x_2 - y_1}$ donde $x_2 \neq x_1$.

slope-intercept form/forma pendiente-intersección (647) Ecuación lineal de la forma $y = mx + b$. La gráfica de tal ecuación tiene pendiente m e intersección y igual a b.

solid/sólido (574) Figura tridimensional que consiste de todos los puntos de su superficie más todos sus puntos interiores.

solving the triangle/resolviendo un triángulo (427) Hallar todas las medidas de los ángulos y lados de un triángulo.

space/espacio (12) Conjunto tridimensional no acotado de todos los puntos.

sphere/esfera (577, 621) Conjunto de puntos en el espacio que están a una distancia dada de un punto dado llamado centro.

spherical geometry/geometría esférica (164) También conocida con el nombre de geometría riemaniana, es la rama de la geometría que estudia los sistemas de puntos, círculos máximos (rectas) y esferas (planos).

spreadsheets/hojas de cálculo (21) Programas de computación diseñados para crear tablas que involucran cálculos.

square/cuadrado (315) Cuadrilátero que tiene cuatro ángulos rectos y cuyos lados son todos congruentes entre sí.

standard form/forma estándar (646) Para las ecuaciones lineales, una ecuación de la forma $Ax + By = C$, donde A y B no son ambos cero.

straight angle/ángulo extendido (44) Figura formada por dos rayos opuestos.

straightedge/regla (31) Instrumento usado para trazar rectas.

strictly self-similar/estrictamente autosemejanza (379) Una figura es estrictamente autosemejante si cualquiera de sus partes, sin importar su ubicación o tamaño, contiene la figura completa.

supplementary angles/ángulos suplementarios (55) Dos ángulos cuyas medidas suman 180°.

surface area/área de superficie (584) La suma de las áreas de todas las caras y superficies laterales de una figura tridimensional.

system of equations/sistema de ecuaciones (704) Un grupo de dos o más ecuaciones con las mismas variables.

tangent/tangente **1.** (412) Para un ángulo agudo de un triángulo rectángulo, la razón de la longitud del cateto opuesto al ángulo a la longitud del cateto adyacente al ángulo. **2.** (474) Una tangente a un círculo es una recta en el plano del círculo que lo interseca en un único punto. El punto de intersección se llama punto de tangencia. **3.** (621) Una tangente a una esfera es una recta o plano que interseca la esfera en un único punto.

tangent segment/segmento de tangente (477) Un segmento AB tal que un extremo está en un círculo, el otro está fuera de él y la recta AB es tangente al círculo.

tessellation/teselado (522, 523) Patrones que semejan mosaicos que se forman repitiendo figuras para así cubrir completamente un plano sin traslapación.

theorem/teorema (39, 101) Un enunciado, a menudo de carácter general, que puede ser demostrado apelando a postulados, definiciones, propiedades algebraicas y reglas de lógica.

traceable network/red trazable (559) Red que puede recorrerse completamente sin pasar dos veces por una misma arista. Una red es trazable si todos sus nodos tienen grado par o si hay exactamente dos nodos de grado impar.

transformation/transformación (715) En un plano, una relación en que cada punto tiene un único punto imagen y cada punto imagen tiene un único punto preimagen.

translation/traslación (731) Composición de dos reflexiones a través de rectas paralelas. Una traslación desliza las figuras la misma distancia y en la misma dirección.

transversal/transversal (126) Recta que interseca dos o más rectas en el plano en puntos distintos.

trapezoid/trapecio (321) Cuadrilátero con sólo un par de lados paralelos llamados bases. Los lados no paralelos son los catetos. Los pares de ángulos con sus vértices en los extremos de la misma base se llaman ángulos basales. El segmento de recta que une los puntos medios de los catetos es la mediana. Una altitud es un segmento perpendicular a las rectas que contienen las bases y que tiene sus extremos en estas rectas.

triangle/triángulo (180) Figura formada por los segmentos determinados por tres puntos no colineales. Los tres segmentos se llaman lados. Los extremos de los segmentos se llaman vértices. Un triángulo divide el plano en tres partes: el triángulo, su interior y su exterior.

trigonometric ratio/razón trigonométrica (412) Razón de las medidas de dos lados de un triángulo rectángulo.

trigonometry/trigonometría (412) Estudio de las propiedades de los triángulos y de las funciones trigonométricas y sus aplicaciones.

two-column proof/demostración a dos columnas (94) Demostración formal en la que los enunciados aparecen en una columna y las razones de cada enunciado en la otra columna.

undefined term/término primitivo (13) Una palabra, habitualmente fácil de entender, que no se puede explicar mediante palabras y conceptos más básicos. Los términos primitivos de la geometría son el punto, la recta y el plano.

uniform/uniforme (524) Teselados que contienen la misma combinación de formas y ángulos en cada vértice.

vector/vector (673) Segmento dirigido. Los vectores tienen magnitud (longitud) y dirección.

vertex/vértice **1.** (44) El extremo común de los dos rayos que forman un ángulo. **2.** (180) El extremo de un lado cualquiera en un polígono. **3.** (577) El lugar de intersección de tres o más aristas de un poliedro. **4.** (600) En una pirámide, el vértice que no está contenido en la base de la pirámide. **5.** (602) En la figura, el vértice del cono es el punto *V*.

vertex angle/ángulo del vértice (182) El ángulo formado por los lados congruentes (catetos) de un triángulo isósceles.

vertical angles/ángulos opuestos por el vértice (53) Dos ángulos no adyacentes formados por dos rectas que se intersecan.

volume/volumen (607) Medida de la cantidad de espacio encerrado por una figura tridimensional.

x*-axis/eje *x (6) La recta numérica horizontal de un plano de coordenadas.

x*-coordinate/coordenada *x (6) El primer número en un par ordenado de números.

y*-axis/eje *y (6) La recta numérica vertical de un plano de coordenadas.

y*-coordinate/coordenada *y (6) El segundo número en un par ordenado de números.

SELECTED ANSWERS

CHAPTER 1 DISCOVERING POINTS, LINES, PLANES, AND ANGLES

Pages 9–11 Lesson 1-1

5. $(-4, -3)$ **6–7.**

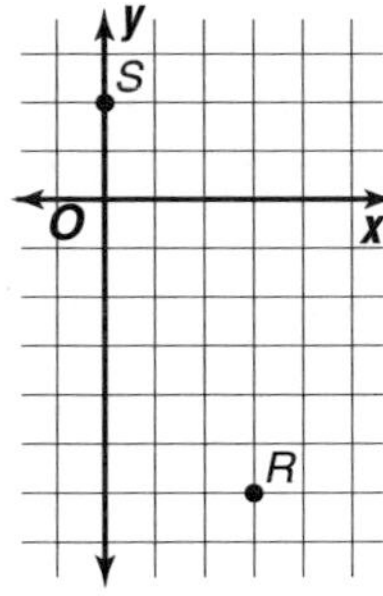

9. noncollinear **11.** $(4, 3)$ **13.** $(-3, 2)$ **15.** $(1, -4)$

17–22.

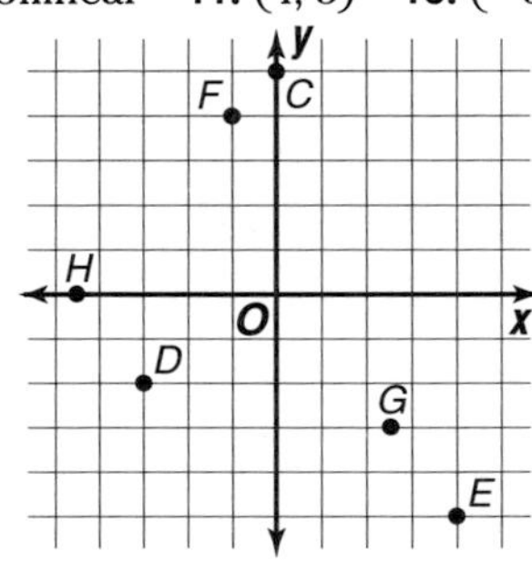

23. collinear **25.** noncollinear **27.** $(0, 4)$ **29.** $(3, -2)$ **31.** $(-5, 0)$ **33.** Sample points: $(0, -9)$, $(1, -3)$, $(2, 3)$

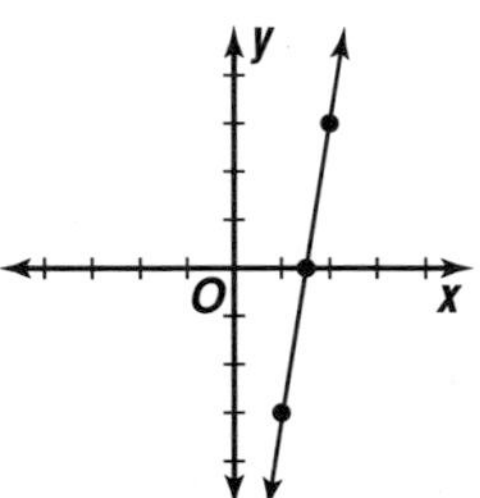

sample point not on line: $(0, 0)$ **35a.** 2; I, IV **35b.** $(2, 0)$; $(6, 0)$ **35c.** Sample answer: $(4, -3)$ **35d.** -3 **35e.** 2 **37.** $(-3, -4)$; Substitute the -3 for x and -4 for y in each of the equations to make sure these coordinates satisfy both equations. Since it satisfies both equations, these are the coordinates of a point that lies on the graphs of both equations. In order for the point to lie on two distinct lines, it must be the point they have in common, or their intersection.

39.

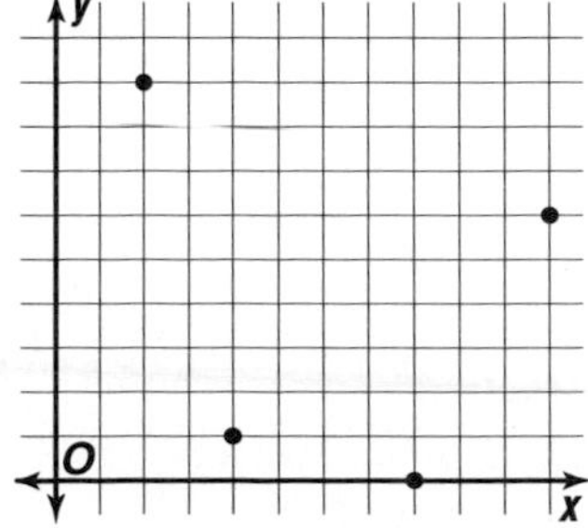

41. 1 **43.** 4 **45.** 6 **47a.** 5 or more hours **47b.** 1920 parents **47c.** 1–4 hours.

Pages 16–18 Lesson 1-2

7. point **9.** line **11.** Sample answer: plane $\mathcal{A}$

13.

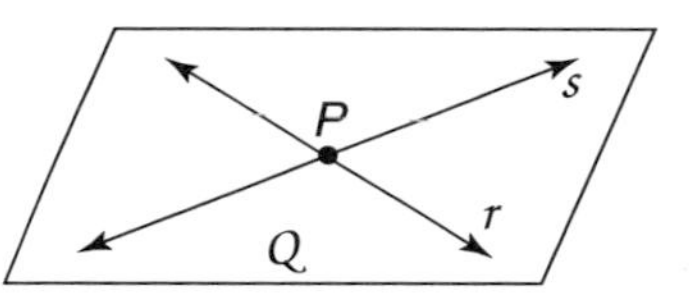

15. $\overleftrightarrow{PR}$ **17.** $\overleftrightarrow{PR}$

19. line **21.** plane **23.** lines **25.** point **27.** line **29–33.** Sample names for figures are given. **29.** $\overleftrightarrow{BF}$ **31.** F, C, E, or D **33.** $\mathcal{R}$

35. sample answer:

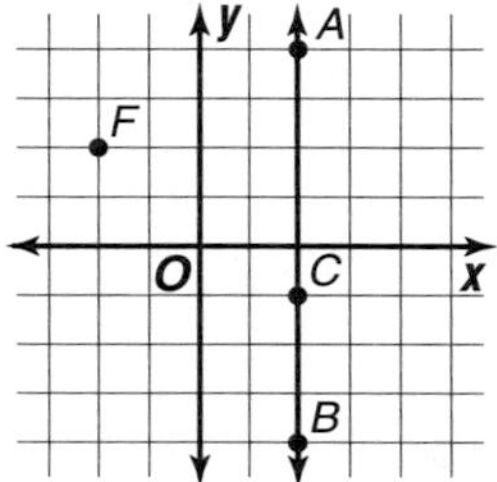

37. sample answer:

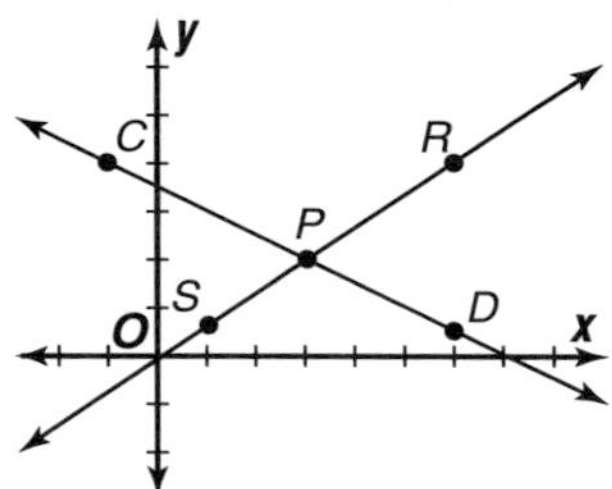

39.

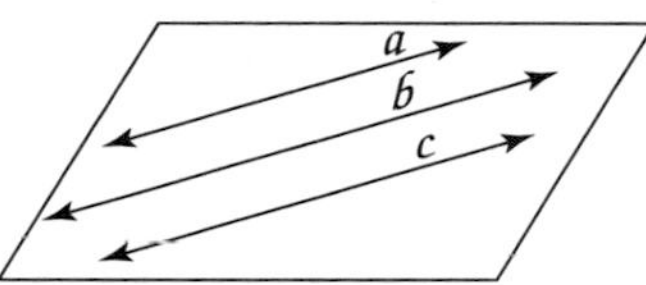

41.

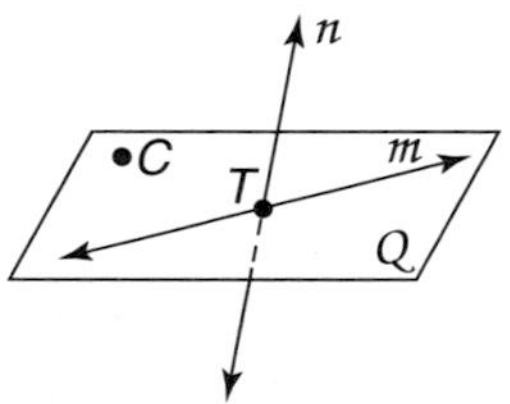

43–51. Sample names for figures are given. **43.** AFH, BHD, CDE, EGF, AGC, FED **45.** $\overleftrightarrow{GE}$, $\overleftrightarrow{DE}$, $\overleftrightarrow{FE}$ **47.** $\overleftrightarrow{GE}$, $\overleftrightarrow{GC}$ **49.** G **51.** Yes; you can draw $\overleftrightarrow{BG}$ and $\overleftrightarrow{HE}$ to form a plane that fits diagonally through the figure.

53.

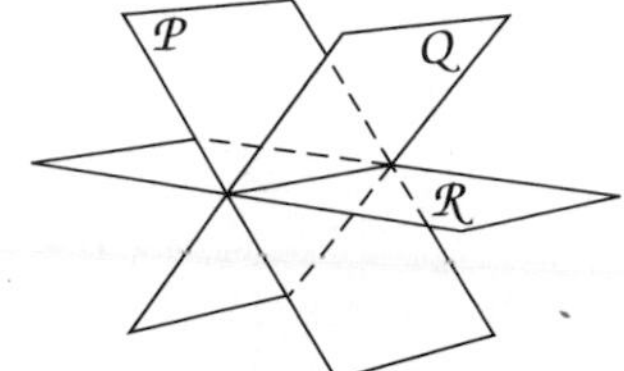

55.

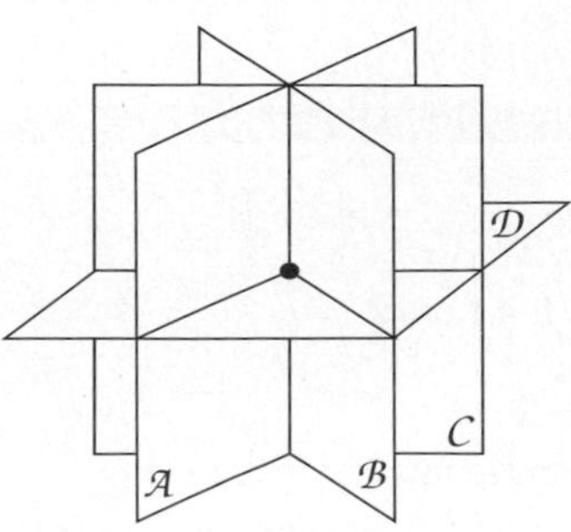

57. maximum: 6, minimum: 0

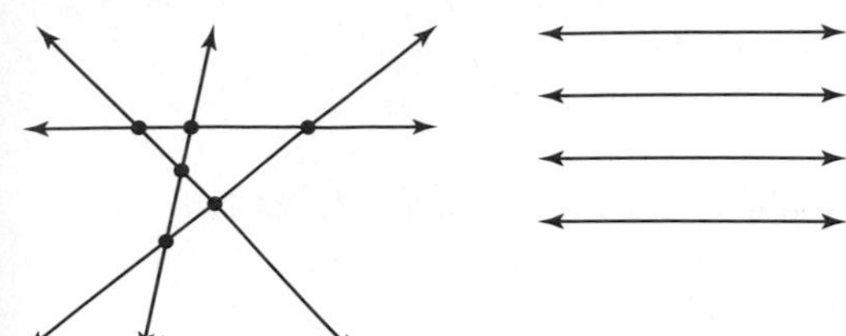

61.

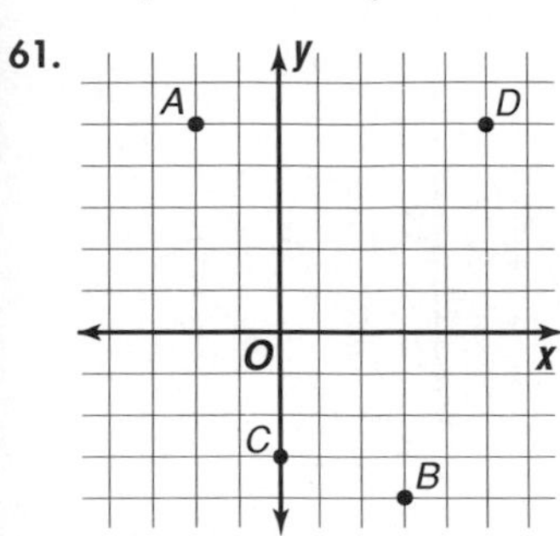

63. B **65.** −11 **67.** −4 **69.** −4 **71.** 7

Pages 22–27 Lesson 1–3

5. 28 cm, 40 cm^2 **7.** 28 **9.** 2 **11.** $T = 5d + t$ **13.** 81 mi^2; $\ell = w = 9$ mi **15.** 26 m, 30 m^2 **17.** 10 cm, 6.25 cm^2 **19.** 14 mi, 8.25 mi^2 **21.** 20 **23.** 32 **25.** 6 **27.** 17 **29.** 30 **31.** 49 in^2; $\ell = w = 7$ in. **33.** 64 ft^2; $\ell = w = 8$ ft **35.** $14\frac{1}{16}$ yd^2; $\ell = w = 3\frac{3}{4}$ yd **37.** 60 units^2 **39.** 17.5 units^3 **41.** 4753.125 ft^2 **43a.** B **43b.** C **45.** Sample answers: $\overline{ST}$, $\overline{SV}$, $\overline{SR}$ **47.** B **49.** $7x^2 + 7x$ **51.** $8c - 3d$

Page 26 Self Test

1a. (4, 0) **1b.** (1, −3) **1c.** (−2, 1) **3.** $\overleftrightarrow{MN}$, $\overleftrightarrow{NM}$, $\overleftrightarrow{OM}$, $\overleftrightarrow{MO}$, $\overleftrightarrow{ON}$, $\overleftrightarrow{NO}$, line q **5.** No; any three of the points are coplanar, but the fourth does not lie in the same plane as the other three.

7.

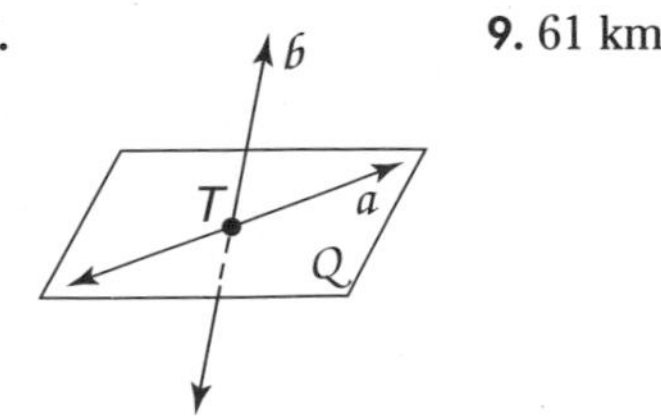

9. 61 km

Page 27 Lesson 1–4A

1. any number **3.** The calculator shows the measurement of the entire segment containing $\overline{EA}$, not just $\overline{EA}$.

5. Sample answer:

7. Sample answer:

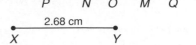

9. Sample answer:

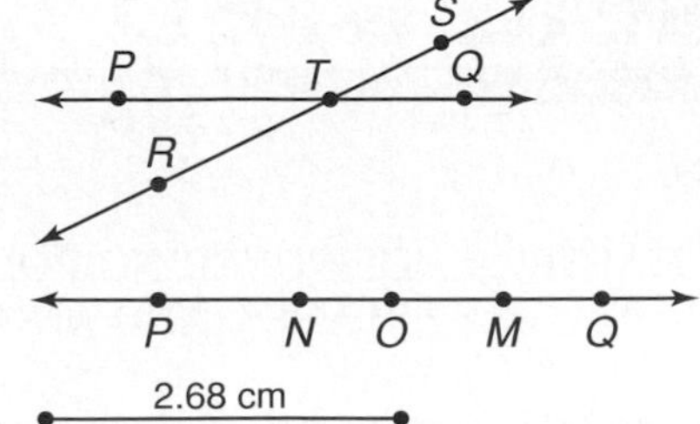

Pages 32–35 Lesson 1–4

5. 8 units **7.** 23 **9.** 3.61 **11.** $NP < QR$ **13.** 3, 20

15a.

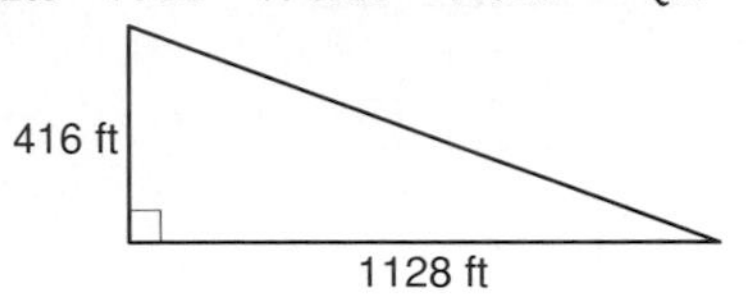

15b. about 1202 feet **17.** 4 **19.** 5 **21.** 4 **23.** $5\frac{1}{3}$ **25.** 8.35 **27.** 9 **29.** 1.41 **31.** 5 **33.** $FJ = JC$ **35.** 10 **37.** $\sqrt{2} \approx 1.414$ **39.** 2, 7

41.

43.

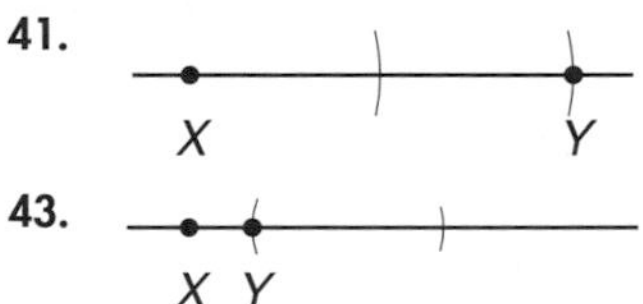

45a.

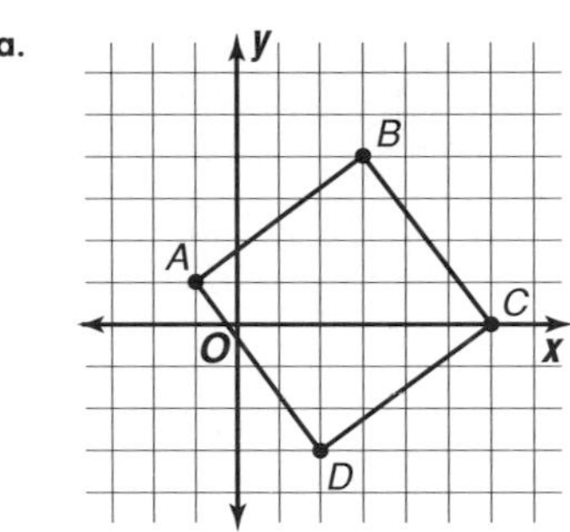

45b. $A = 25$ units^2, $P = 20$ units **47a.** $E(6, 4)$, $F(4, 6)$ **47b.** The x-coordinate will be the same as the x-coordinate of F and the y-coordinate will be the same as the y-coordinate of E. Thus, $G(4, 4)$. **47c.** $DG = GB$; use the distance formula to find DG and GB, which yields about 2.83. **49.** The distance is measured along a number line. The distance from 6 to 9 is 3 units. **51.** 6 cm

53.

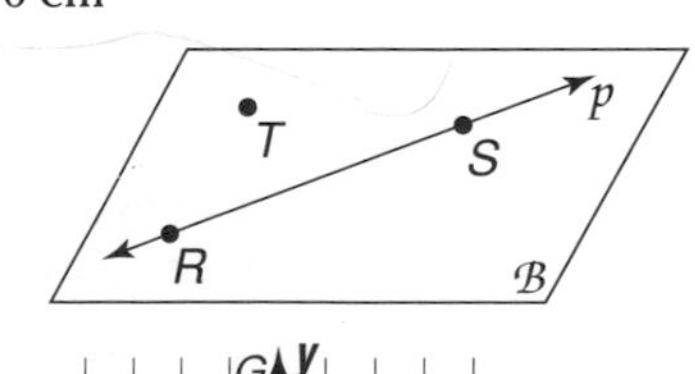

55.

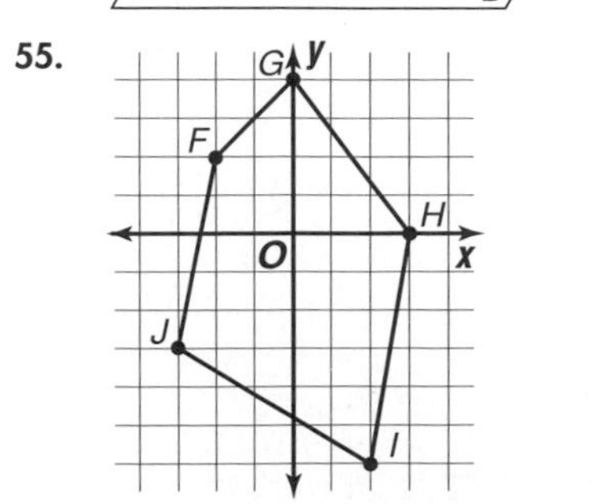

57. $3x^2 + 18x$ **59.** $x^2 - x - 12$

Pages 40–43 Lesson 1–5

7. true **9.** (1.5, −2) **13a.** 2; 18 **13b.** 1; 2 **15.** −2 **17.** −3.5 **19.** False; sample answer: E is the midpoint of $\overline{HJ}$. **21.** true **23.** true **25.** $S = -18$ **27.** $Y(2, 5)$ **29.** $X(-6, -4)$ **31.** $Z\left(\frac{8}{3}, 11\right)$ **33.** 8; 60 **35.** 1; 1 **37.** 2; 6 **39.** (scale 2:5), $AB = 5$ in., $AC = \frac{1}{8}AB$

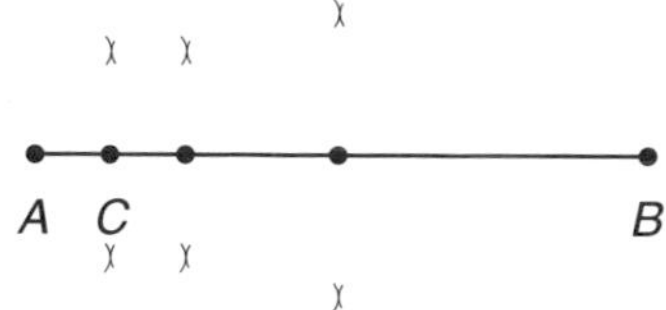

41. $R(4, -1), S(6, -5)$

43. S T E P

By the Segment Addition Property, we know that $SP = SE + EP$ and $SE = ST + TE$. By substituting for SE in the first equation, $SP = ST + TE + EP$.

45a–b.

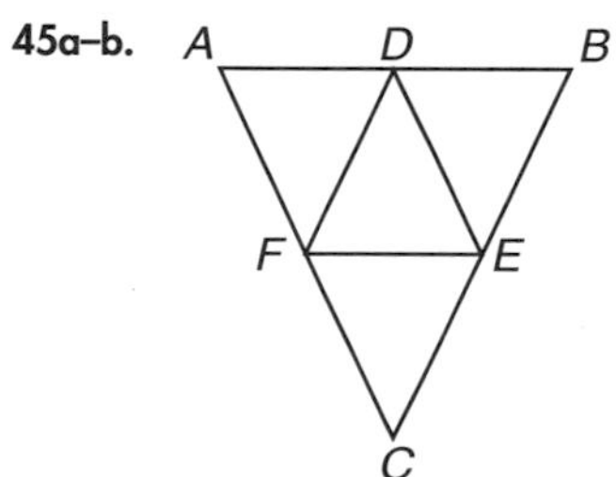

45d. The perimeter of the larger triangle is twice that of the smaller triangle. **45e.** The area of the larger triangle is 4 times that of the smaller triangle. Sample answer: If you make four copies of the small triangle, you can fit them on the surface of the large rectangles. **47.** 159 **49a.** path 1: 8 units; path 2: 8 units; path 3: 5.83 units **49b.** path 3, because you can't drive diagonally through city blocks.

51a.

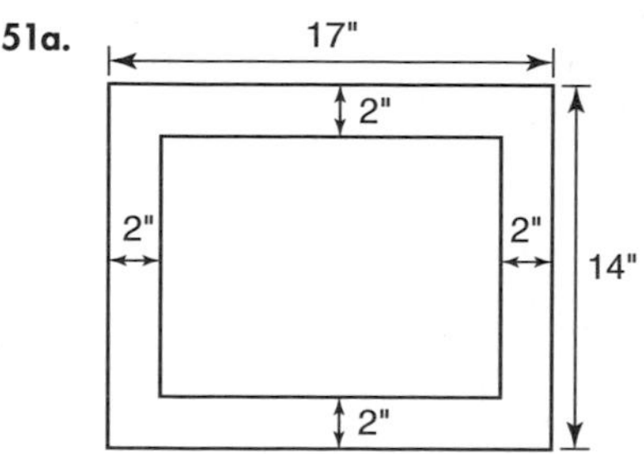

51b. 130 in^2

53.

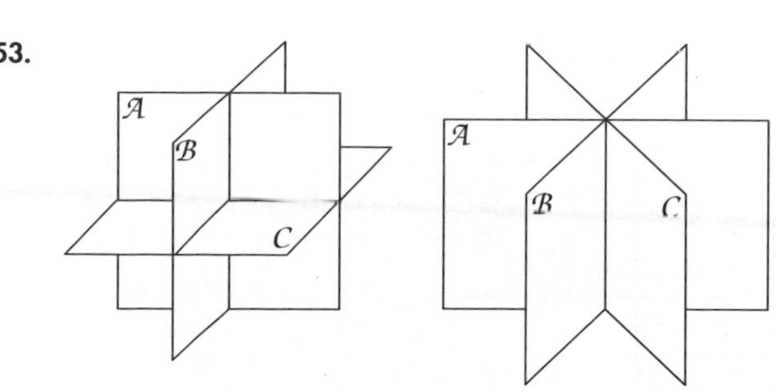

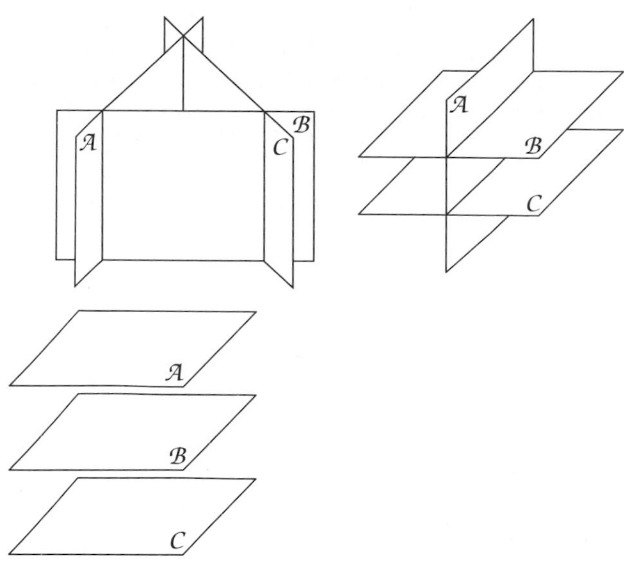

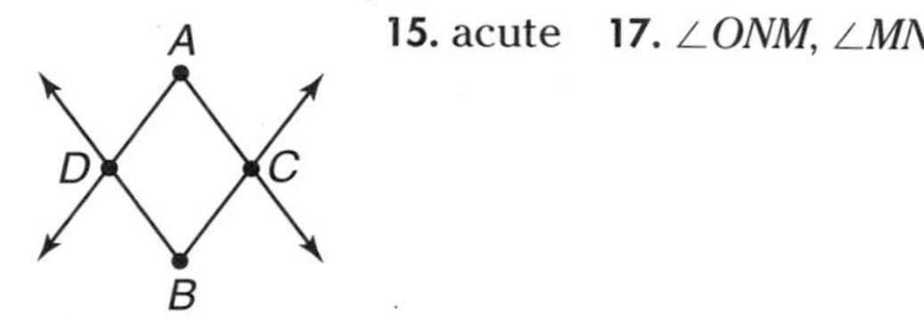

55. $x = 4$

Pages 49–51 Lesson 1–6

7. $\angle AFB$, $\angle BFA$ **9.** $\overrightarrow{FD}$; $\angle EFC$ **11.** $m\angle 3$ **13.** A, D, C, B **15.** acute **17.** $\angle ONM$, $\angle MNR$ **19.** I or P **21.** obtuse **23.** $\angle NML$, $\angle NMK$, $\angle NMJ$, $\angle NMP$, $\angle NMO$ **25.** 120 **27.** $\angle JMQ$ and $\angle NMK$ **29.** **31.**

33. 50 **35.** 22 **37.** 2.8 **39a.** 1, 3, 6, 10, 15 **39b.** For 4 rays, there are $(4 \cdot 3) \div 2$ or 6 angles. For 5 rays, there are $(5 \cdot 4) \div 2$ or 10 angles. For 6 rays, there are $(6 \cdot 5) \div 2$ or 15 angles. **39c.** 21, 45 **39d.** $a = \frac{n(n-1)}{2}$, for a = number of angles and n = number of rays **41a.** acute **41b.** 15 **43.** 10 **45.** 5.5 mm **47.**

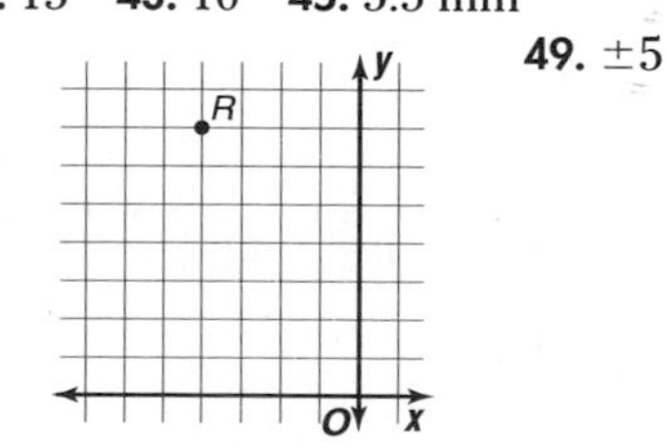

49. ± 5

Page 52 Lesson 1–7A

1a. $\angle CBA$ and $\angle DBE$; $\angle CBD$ and $\angle EBA$ **1b.** The measures of the vertical angles are equal. **5.** Regardless of the position of the lines, vertical angles have equal measures.

Pages 58–60 Lesson 1–7

7. $\overline{AD}$ **9.** $\angle AFB$ and $\angle BFC$, $\angle BFC$ and $\angle CFD$, or $\angle CFD$ and $\angle DFE$ **11.** No, because there are no markings or measures given to indicate that the segments are congruent. **13.** 148 **15.** $\angle YUW$ and $\angle XUV$ or $\angle YUX$ and $\angle VUW$ **17.** $\angle TWU$ **19.** No, there are no measures or markings to indicate that right angles are present.

21. congruent, adjacent supplementary, linear pair **23.** No, there is no indication that $\angle SRT$ is a 90°angle. **25.** right **27a.** perpendicular **27c.** The angle bisectors are perpendicular, because the angles formed by the two lines measure 45° + 45° or 90°. **29.** 112, 68 **31.** 67.8; 22.2 **33.** 36, 17, 75, 15, 55, 35

35. Given: $\angle PQR$ and $\angle RQS$ are complementary angles; $m\angle PQR = 45$.

Prove: $\angle PQR \cong \angle RQS$

Sample paragraph proof: Since $\angle PQR$ and $\angle RQS$ are complementary, $m\angle PQR + m\angle RQS = 90$. Substitute 45 for $m\angle PQR$ in the equation. This results in, $45 + m\angle RQS = 90$. By subtracting 45 from each side, $m\angle PQR = 45$. Since $\angle PQR$ and $\angle QRS$ have the same measure, $\angle PQR \cong \angle QRS$.

37. Sample answer: Complementary is defined as something that completes, and complementary angles complete a right angle. Supplementary is defined as something that completes or makes an addition, and supplementary angles add up to be a straight angle.

39. obtuse **41.** about 14 feet 2 inches

43.

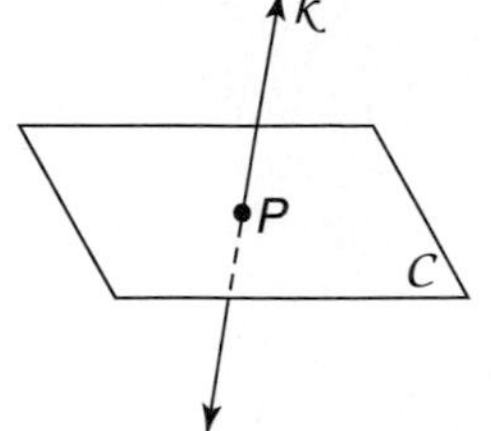

45. $4ab$

Page 61 Chapter 1 Highlights

1. quadrants **3.** supplementary **5.** line **7.** origin **9.** Vertical

Pages 62–64 Chapter 1 Study Guide and Assessment

11–14.

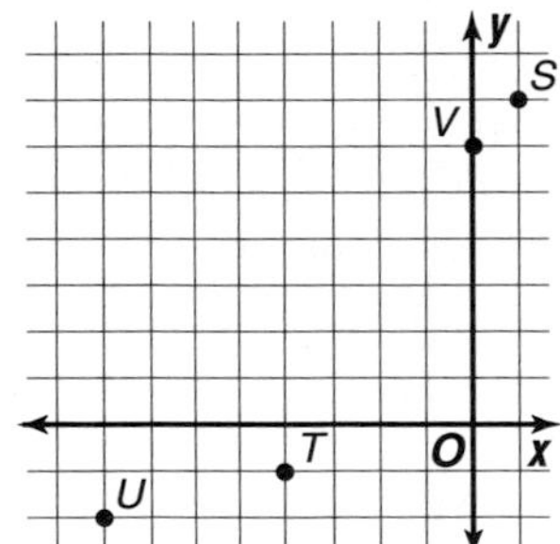

15. Quadrant III **17.** point A **19.** 45 mph **21.** 56.25 cm^2 **23.** $3\frac{1}{2}$ **25.** 5 **27.** 12.21 **29.** (5.5, −0.5) **31.** 12 **33.** yes **35.** 148 **37.** 40 **39.** $\angle PTN$ **41.** 120, 60

43.

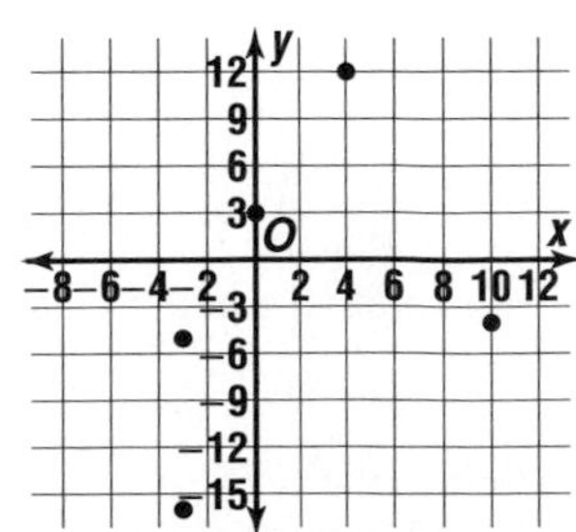

45a. As the loft of the club increases, the ball will travel higher in the air and for a shorter distance, as long as the ball is struck with the same amount of force.

45b.

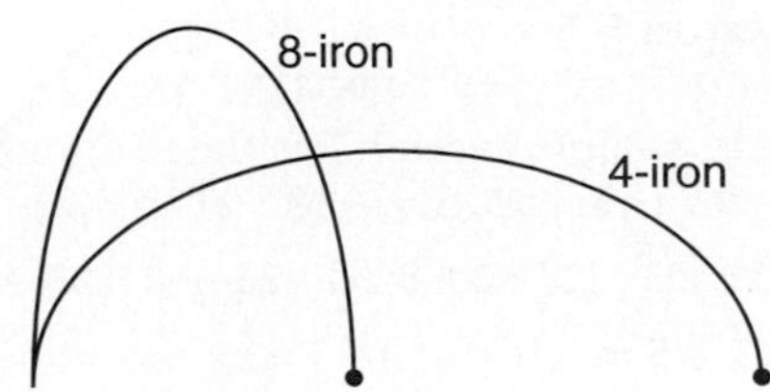

CHAPTER 2 CONNECTING REASONING AND PROOF

Pages 72–75 Lesson 2–1

7. If ℓ and m are perpendicular, then they form a right angle. **9.** Points H, I, and J are noncollinear. **11.** Sample answer: Earth is flat, Earth is the center of the universe. **13.** False; $XZ + YZ = XY$ by the Segment Addition Post.

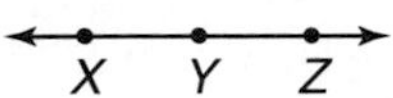

15. Points A, B, and C do not lie on a line.

A •

• C

B •

17. X, Y, Z, and W are noncollinear

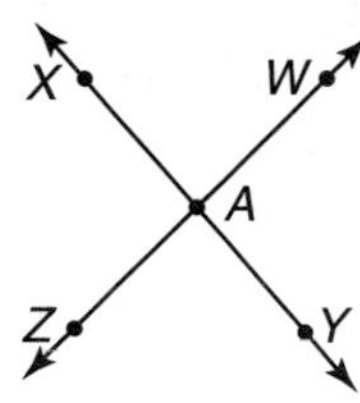

19. Points R, S, and T are collinear.

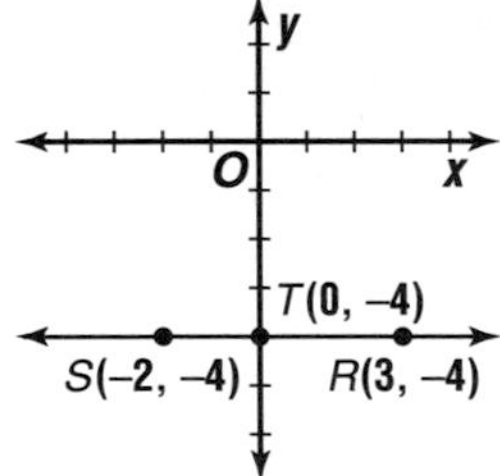

21. $m\angle ABD = m\angle CBD$

23. false; counterexample:

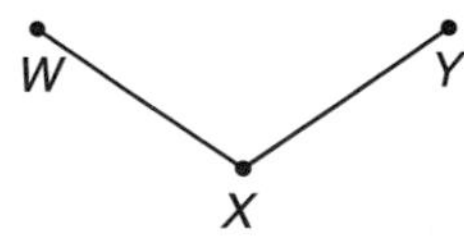

25. False; K, L, and M are collinear.

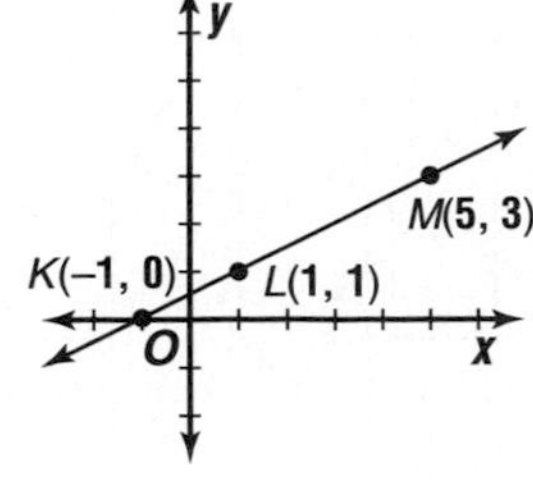

27. *PQRST* is a pentagon.

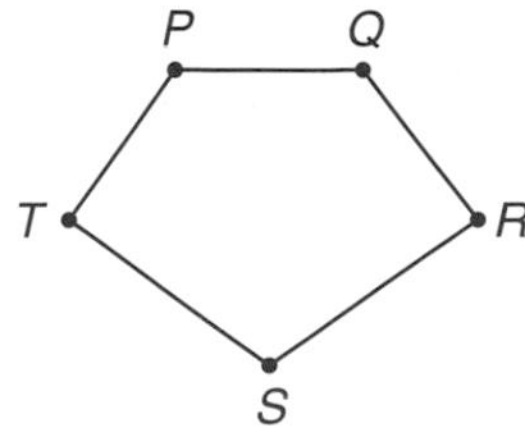

29a. They are congruent. **29b.** Sample answer: Use the Angle option on the F6 menu to find the measures of $\angle PQS$ and $\angle QSR$. **31.** The ball will strike two rails and then continue on a path to the corner on the opposite side of the table at the opposite end. **33.** 20, no **35.** (10, 12) **37.** 3; 13 cm by 9 cm **39.** $\frac{3}{20}$

Pages 80–83 Lesson 2–2

7. Hypothesis: three points lie on a line; Conclusion: they are collinear **9.** If angles have the same measure, then they are congruent. **11.** Four points are coplanar. **13. Converse:** If an angle is a right angle, then it measures 90°; true. **Inverse:** If an angle does not measure 90°, then it is not a right angle; true. **Contrapositive:** If an angle is not a right angle, then it does not measure 90°; true. **15.** false **17.** false **19.** Hypothesis: a man hasn't discovered something that he will die for; Conclusion: he isn't fit to live **21.** Hypothesis: we would have new knowledge; Conclusion: we must get a whole world of new questions **23.** Hypothesis: $3x - 5 = -11$; Conclusion: $x = -2$ **25.** If people are happy, then they rarely correct their faults. **27.** If angles are adjacent, then they have a common vertex. **29.** If angles have measures between 90 and 180, then they are obtuse. **31.** A book is not a mirror. **33.** Rectangles are squares. **35.** You do not live in Dallas. **37. Converse:** If there are three points that are noncollinear, then they are not on the same line; true. **Inverse:** If there are three points on the same line, then they are collinear; true. **Contrapositive:** If there are three points that are collinear, then they are on the same line; true. **39. Converse:** If an angle has a measure less than 90, then it is acute; true. **Inverse:** If an angle is not acute, then it does not have a measure less than 90; true. **Contrapositive:** If an angle does not have a measure less than 90, then it is not acute; true. **41. Converse:** If you don't live in Illinois, then you don't live in Chicago; true. **Inverse:** If you live in Chicago, then you live in Illinois; true. **Contrapositive:** If you live in Illinois, then you live in Chicago; false. Counterexample: you may live in another city in Illinois. **43.** true **45.** true **47.** three

49. one **51a.** doubles **51b.** quadruples **51c.** is multiplied by 8 **53a.** The Hatter is right; Alice exchanged the hypothesis and conclusion. **53b.** These statements are converses of each other. **55a.** If you try Casa Fiesta, then you're looking for a fast, easy way to add some fun to your family's menu. **55b.** They are a fast, easy way to add fun to your family's menu. **55c.** No; the conclusion is implied. **57.** Sample answer: $\angle C \cong \angle D$ **59.** E

61. acute **63.** 12 in. by 8 in., 15 in. by 10 in., 18 in. by 12 in., 21 in. by 14 in. **65.** no **67.** false

Page 84 Lesson 2–2B

1. The result is the same, $x < -3$. **3.** true **5.** False; if $12 - 3x > 23 - 14x$, then $x > 1$.

Pages 88–91 Lesson 2–3

7. yes; syllogism **9.** Patricia Gorman should get 8 hours of sleep each day; detachment. **11.** If the measure of an angle is less than 90, then it is not obtuse; syllogism. **13a.** You'll become a true Beatles fan. **13b.** no conclusion **13c.** no conclusion **13d.** You'll love this album. **15.** yes; detachment **17.** invalid **19.** invalid **21.** invalid **23.** Odina lives to eat; detachment. **25.** If *M* is the midpoint of $\overline{AB}$, then $\overline{AM} \cong \overline{MB}$; syllogism. **27.** no conclusion **29.** Planes $\mathcal{M}$ and $\mathcal{N}$ intersect in a line; detachment. **31.** Line *t* is perpendicular to line *q*; detachment. **33.** $\overleftrightarrow{XY}$ lies in plane $\mathcal{P}$; detachment. **35a.** (2) I like pizza with everything. (3) I'll like Jimmy's pizza. **35b.** (2) If you like Jimmy's pizza, then you are a pizza connoisseur. (3) If you like pizza with everything, then you are a pizza connoisseur. **37.** (1) If a person is a baby, then the person is not logical. (statement 1) (2) If a person is not logical, then the person is despised. (statement 3) (3) If a person is a baby, then the person is despised. (Law of Syllogism) (4) If a person is despised, then the person cannot manage a crocodile. (contrapositive of statement 2) (5) If a person is a baby, then the person cannot manage a crocodile. (Law of Syllogism) **39.** 1

41. If two lines intersect, then they are perpendicular; false. Counterexample:

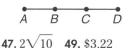

43. Sample answer: *A*, *B*, *C*, and *D* are collinear.

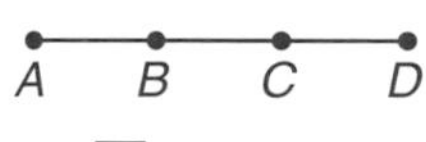

47. $2\sqrt{10}$ **49.** \$3.22

Page 91 Self Test

1. False; counterexample:

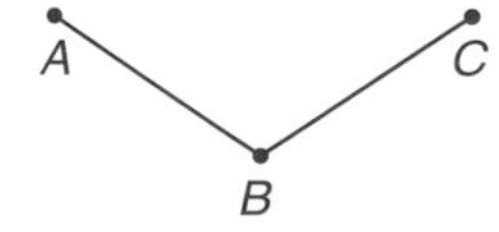

3a. Sample answer: insufficient light or water **3b.** Yes; the fungus is killing the plants. **3c.** Introduce the fungus to some healthy plants and see if they droop. **5.** If there are clouds, then it is raining. **7.** If a figure does not have four sides, then it is not a square. **9.** invalid

Pages 95–99 Lesson 2–4

7. Addition Property (=) **9a.** 1 **9b.** 3 **9c.** 4 or 5 **9d.** 5 or 4 **9e.** 2

11. Given: $k = \frac{\Delta\ell}{\ell(T - t)}$

Prove: $T = \frac{\Delta\ell}{k\ell} + t$

Proof:

Statements	Reasons
1. $k = \frac{\Delta\ell}{\ell(T - t)}$	1. Given
2. $k(T - t) = \frac{\Delta\ell}{\ell}$	2. Multiplication Property (=)
3. $T - t = \frac{\Delta\ell}{k\ell}$	3. Division Property (=)
4. $T = \frac{\Delta\ell}{k\ell} + t$	4. Add. Property (=)

13. Symmetric Property (=) **15.** Multiplication Property (=) or Division Property (=) **17.** Division Property (=) **19.** Transitive Property (=) **21.** Subtraction Property (=) **23a.** 2 **23b.** 1 **23c.** 4 **23d.** 3 **23e.** 5 **23f.** 6 **25a.** Given **25b.** Multiplication Property (=) **25c.** Distributive Property **25d.** Subtraction Property (=) **25e.** Division Property (=) **27a.** $m\angle TUV = 90$, $m\angle XWV = 90$, $m\angle 1 = m\angle 3$ **27b.** Substitution Property (=) **27c.** Angle Addition Postulate **27d.** $m\angle 1 + m\angle 2 = m\angle 3 + m\angle 4$ **27e.** Substitution Property (=) **27f.** $m\angle 2 = m\angle 4$

29. Given: $2x + 6 = 3 + \frac{5}{3}x$

Prove: $x = -9$

Proof:

Statements	Reasons
1. $2x + 6 = 3 + \frac{5}{3}x$	1. Given
2. $3(2x + 6) = 3(3 + \frac{5}{3}x)$	2. Multiplication Property (=)
3. $6x + 18 = 9 + 5x$	3. Distributive Property
4. $x + 18 = 9$	4. Subtraction Property (=)
5. $x = -9$	5. Subtraction Property (=)

31. Sample answer: Both properties are transitive; equality relates numbers, congruence relates sets of points.

33. Given: $AC = BD$

Prove: $AB = CD$

Proof:

Statements	Reasons
1. $AC = BD$	1. Given
2. $AB + BC = AC$ $BC + CD = BD$	2. Segment Addition Postulate
3. $AB + BC = BC + CD$	3. Substitution Property (=)
4. $AB = CD$	4. Subtraction Property (=)

35. If $m\angle 1 \neq 27$, then $\angle 1$ is not acute; false, if $m\angle 1 = 32$, then $\angle 1$ is acute. **37.** Complement does not exist; 21.

41. $5m - n$

Pages 103–106 Lesson 2–5

5. Symmetric Property of ≅ Segments **7.** Addition Property (=) **9.** Given: x is a whole number. Prove: x is an integer. **11a.** 3 **11b.** 1 or 4 **11c.** 2 **11d.** 1 or 4

13. Given: $WX = XY$

Prove: $WY = 2XY$

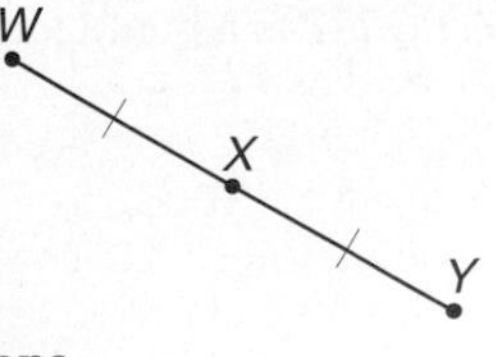

Proof:

Statements	Reasons
1. $WX = XY$	1. Given
2. $WY = WX + XY$	2. Segment Addition Post.
3. $WY = XY + XY$	3. Substitution Property (=)
4. $WY = 2XY$	4. Substitution Property (=)

15. Reflexive Property of ≅ Segments **17.** Addition Property (=) **19.** Transitive Property of ≅ Segments **21.** Distributive Property **23.** Given: $x > 2$ and x is prime. Prove: x is odd.

25. Given: $AB = CD$, $EF = CD$

Prove: $AB = EF$

27. Given: A, B are on ℓ.

Prove: $\overline{AB}$ is on ℓ.

29. Given: x is a rational number.

Prove: x is a real number.

31a. Given **31b.** Definition ≅ Segments **31c.** $PM = MS$, $RM = MQ$ **31d.** Segment Addition Postulate **31e.** Substitution Property (=) **31f.** Substitution Property (=) **31h.** Division Property (=) **31i.** Definition ≅ Segments

33. Given: $NL = NM$
$AL = BM$

Prove: $NA = NB$

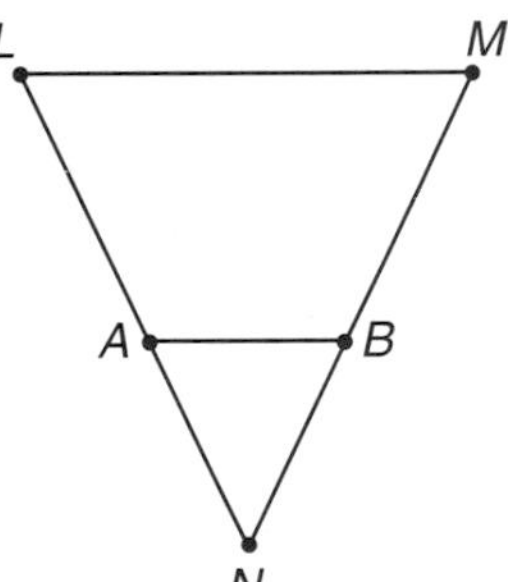

Proof:

Statements	Reasons
1. $NL = NM$ $AL = BM$	1. Given
2. $NL = NA + AL$ $NM = NB + BM$	2. Segment Addition Postulate
3. $NA + AL = NB + BM$	3. Substitution Property (=)
4. $NA + BM = NB + BM$	4. Substitution Property (=)
5. $NA = NB$	5. Subtraction Property (=)

35. Sample answers: $\overline{LN} \cong \overline{QO}$ and $\overline{LM} \cong \overline{MN} \cong \overline{RS} \cong \overline{ST} \cong \overline{QP} \cong \overline{PO}$ **37a.** about 15.2 cm **37b.** Yes; the scales are different. **39.** Law of Syllogism; if lines are parallel, then they have no point in common. **41.** D **43.** 136

45. $2\frac{1}{2}$ yd

Pages 111–114 Lesson 2-6

11. $m\angle 1 = 121$, $m\angle 2 = 59$ **13d.** $m\angle XYZ = m\angle 1 + m\angle 2$ **13e.** Substitution Property (=) **13f.** Definition of supp. $\angle$s **15.** always **17.** always **19.** sometimes **21.** sometimes **23.** never **25.** sometimes **27.** $m\angle 5 = m\angle 6 = 58$ **29.** $m\angle 1 = 124$, $m\angle 2 = 56$ **31.** $m\angle 5 = m\angle 6 = 62$ **33a.** $\angle 1$ and $\angle 8$, $\angle 2$ and $\angle 6$, $\angle 3$ and $\angle 5$, $\angle 4$ and $\angle 7$ **33b.** $\angle AXB$ and $\angle BXE$, $\angle BXC$ and $\angle CXF$, $\angle CXD$ and $\angle DXH$, $\angle DXE$ and $\angle EXG$, $\angle EXF$ and $\angle FXA$, $\angle FXH$ and $\angle HXB$, $\angle HXG$ and $\angle GXC$, $\angle GXA$ and $\angle AXD$

35. Given: $\angle ABD$ and $\angle CBD$ form a linear pair. $\angle YXZ$ and $\angle WXZ$ form a linear pair. $\angle ABD \cong \angle YXZ$

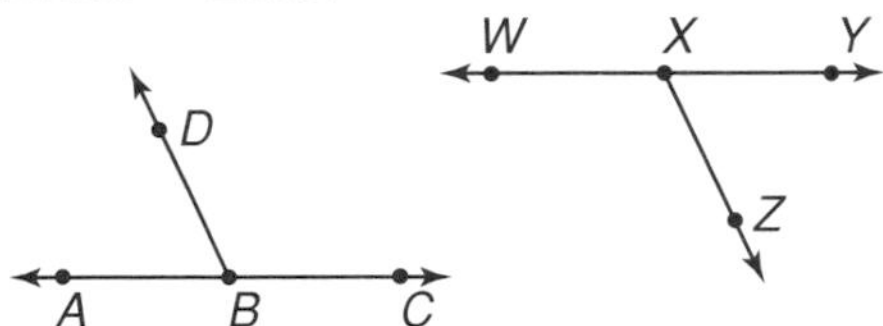

Prove: $\angle CBD \cong \angle WXZ$

Proof:

Statements	Reasons
1. $\angle ABD$ and $\angle CBD$ form a linear pair, $\angle YXZ$ and $\angle WXZ$ form a linear pair, $\angle ABD \cong \angle YXZ$.	1. Given
2. $\angle ABD$ and $\angle CBD$ are supplementary, $\angle YXZ$ and $\angle WXZ$ are supplementary.	2. Supp. Theorem
3. $\angle CBD \cong \angle WXZ$	3. $\angle$s supp. to $\cong$ $\angle$s are $\cong$.

37. Given: $\angle A \cong \angle B$

Prove: $\angle B \cong \angle A$

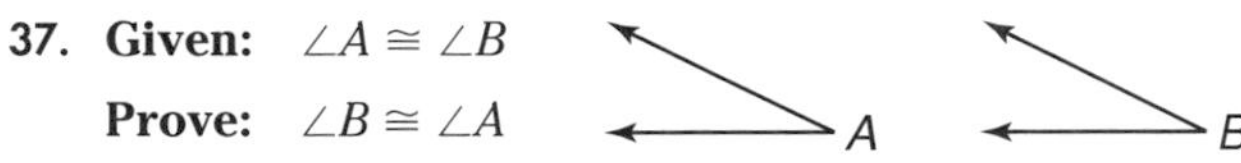

Proof:

Statements	Reasons
1. $\angle A \cong \angle B$	1. Given
2. $m\angle A = m\angle B$	2. Definition of $\cong$ $\angle$s
3. $m\angle B = m\angle A$	3. Symmetric Property (=)
4. $\angle B \cong \angle A$	4. Definition of $\cong$ $\angle$s

39. Given: $\angle 1$ and $\angle 2$ form a linear pair. $\angle 1$ is a right angle.

Prove: $\angle 2$ is a right angle.

Proof:

Statements	Reasons
1. $\angle 1$ and $\angle 2$ form a linear pair. $\angle 1$ is a right angle.	1. Given
2. $m\angle 1 + m\angle 2 = 180$	2. Definition of linear pair
3. $m\angle 1 = 90$	3. Definition of rt. $\angle$
4. $90 + m\angle 2 = 180$	4. Substitution Property (=)
5. $m\angle 2 = 90$	5. Subtraction Property (=)
6. $\angle 2$ is a right angle.	6. Definition of rt. $\angle$

41. No, they are parallel; hold at an angle and look at it. **43a.** The first pair of vertical lines appear to curve inward, the second pair appear to curve outward. **43b.** They appear parallel.

43c. Given: $\angle 4 \cong \angle 2$

Prove: $\angle 3 \cong \angle 1$

Proof:

Statements	Reasons
1. $\angle 4 \cong \angle 2$	1. Given
2. $\angle 4$ and $\angle 3$ form a linear pair. $\angle 2$ and $\angle 1$ form a linear pair.	2. Definition of linear pair
3. $\angle 4$ and $\angle 3$ are supplementary. $\angle 2$ and $\angle 1$ are supplementary.	3. If 2 $\angle$s form a linear pair, then they are supplementary.
4. $\angle 3 \cong \angle 1$	4. $\angle$s supplementary to $\cong$ $\angle$s are $\cong$.

45. Substitution Property (=) **47.** 5, 15 **49.** 15 ft by 5 ft, 13 ft by 6 ft, 11 ft by 7 ft, 9 ft by 8 ft **51.** Thursday; the largest audience is available.

Page 115 Chapter 2 Highlights

1. k **3.** b **5.** c **7.** a **9.** h **11.** j

Pages 116–118 Chapter 2 Study Guide and Assessment

13. true; definition of midpoint **15.** If something is a cloud, then it has a silver lining. **17.** If a rock is obsidian, then it is a glassy rock produced by a volcano. **19. Converse:** If a rectangle is a square, then it has four congruent sides. **Inverse:** If a rectangle does not have four congruent sides, then it is not a square. **Contrapositive:** If a rectangle is not a square, then it does not have four congruent sides. **21. Converse:** If a month has 31 days, then it is January. **Inverse:** If the month is not January, then it does not have 31 days. **Contrapositive:** If a month does not have 31 days, then it is not January. **23.** $\angle A$ and $\angle B$ have measures with a sum of 180; Law of Detachment. **25.** The sun is in constant motion; Law of Syllogism. **27.** Reflexive Property (=) **29.** Subtraction Property (=) **31.** Substitution Property (=) **33.** never **35.** never **37.** 56 pairs **39.** Substitution Property (=) or Transitive Property (=)

CHAPTER 3 USING PERPENDICULAR AND PARALLEL LINES

Pages 127–129 Lesson 3-1

5. intersecting **7.** False; a transversal intersects 2 lines in a plane. **9.** $\overline{AH}$; alternate interior **11.** plane *ABC* and plane *DEF*

13.

15. Skew lines; the planes are flying in different directions and at different altitudes. **17.** intersecting **19.** parallel **21.** parallel **23.** False; the angles are not formed by 2 lines and a transversal. **25.** True; the angles are in corresponding positions when the transversal m intersects r and s. **27.** False; the angles are formed by lines ℓ and m and transversal s. **29.** ℓ; alternate exterior **31.** ℓ; alternate interior **33.** q; alternate exterior **35.** $\overline{AE}$ and $\overline{DR}$, $\overline{AD}$ and $\overline{ER}$ **37.** none **39.** plane ADR, plane DRM, plane ERM, plane AEM

41. **43.**

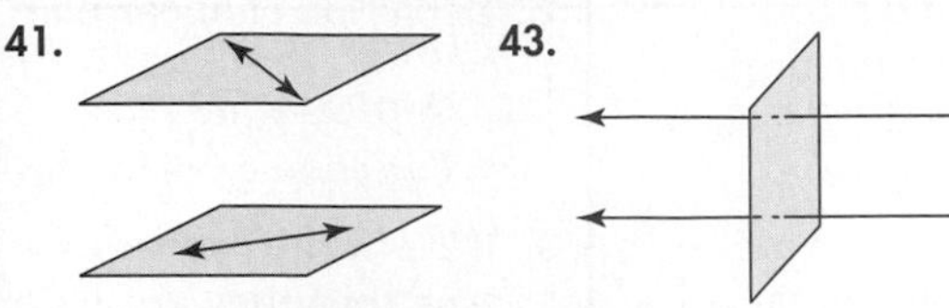

45a. $m\angle 1 = m\angle 2 = 160$ **45b.** The measures of alternate interior angles are equal. **47.** Sample answer: parallel circuits in electronics, parallel story lines in literature

49. 24 handshakes

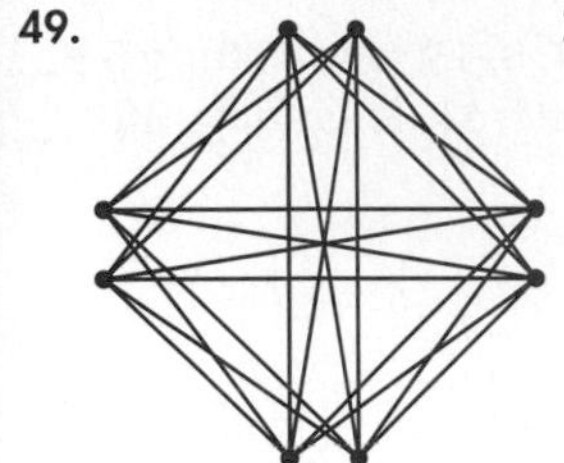

51. Given: $PARL$ is a parallelogram.

Prove: $\angle 1 \cong \angle 2$
$\angle 3 \cong \angle 4$

53. If something is a cloud, then it is composed of millions of water droplets. **55.** B **57.** Explore the problem; plan the solution; solve the problem; and examine the solution. **59.** $\frac{2}{5}$

Page 130 Lesson 3–2A

1. consecutive interior angles: $\angle CEF$ and $\angle AFE$, $\angle DEF$ and $\angle BFE$; alternate exterior angles: $\angle CEG$ and $\angle BFH$, $\angle DEG$ and $\angle AFH$; alternate interior angles: $\angle CEF$ and $\angle BFE$, $\angle DEF$ and $\angle AFE$; corresponding angles: $\angle CEG$ and $\angle AFE$, $\angle DEG$ and $\angle BFE$, $\angle CEF$ and $\angle AFH$, $\angle DEF$ and $\angle BFH$ **3a.** Corresponding angles are congruent. **3b.** Alternate interior angles are congruent. **3c.** Alternate exterior angles are congruent. **5a.** The measure of each of the eight angles is 90. **5b.** A transversal that is perpendicular to one of two parallel lines is perpendicular to the other.

Pages 134–137 Lesson 3–2

7. Alternate Interior Angles Theorem **9.** 107 **11.** 48 **13.** 59 **15.** $x = 12$, $y = 10$ **17.** Alternate Interior Angles Theorem **19.** Corresponding Angles Postulate **21.** 49 **23.** 49 **25.** 131 **27.** 47 **29.** 107 **31.** 49 **33.** 42 **35.** 42 **37.** 120 **39.** $x = 90$, $y = 15$, $z = 13.5$ **41.** Since $\angle 4$ and $\angle 8$ are alternate exterior angles formed when 2 parallel lines are cut by a transversal, $m\angle 4 = m\angle 8$ or $2x - 25 = x + 26$. Solving this equation, $x = 51$. Therefore, $m\angle 8 = 51 + 26$ or 77. Since $\angle 2$ and $\angle 8$ are vertical angles, $m\angle 2 = m\angle 8$ or $m\angle 2 = 77$.

43.
1. Given
2. Perpendicular lines form 4 right angles.
3. Definition of right angle
4. Corresponding Angles Postulate
5. Definition of congruent angles
6. Substitution Property (=)
7. Definition of right angle
8. Definition of $\perp$ lines

45. Given: $\ell \parallel m$

Prove: $\angle 3$ and $\angle 5$ are supplementary. $\angle 4$ and $\angle 6$ are supplementary.

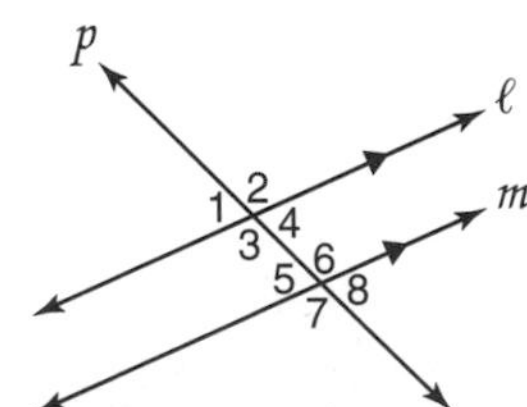

Proof:
We are given that $\ell \parallel m$. If two parallel lines are cut by a transversal, corresponding angles are congruent. So, $\angle 1 \cong \angle 5$ and $\angle 2 \cong \angle 6$ or $m\angle 1 = m\angle 5$ and $m\angle 2 = m\angle 6$. Since $\angle 1$ and $\angle 3$ form a linear pair and $\angle 2$ and $\angle 4$ form a linear pair, $\angle 1$ and $\angle 3$ are supplementary and $\angle 2$ and $\angle 4$ are supplementary. By the definition of supplementary angles, $m\angle 1 + m\angle 3 = 180$ and $m\angle 2 + m\angle 4 = 180$. By the Substitution Property (=), $m\angle 5 + m\angle 3 = 180$ and $m\angle 6 + m\angle 4 = 180$. Therefore, $\angle 3$ and $\angle 5$ are supplementary and $\angle 4$ and $\angle 6$ are supplementary by the definition of supplementary.

47. Given: $\overline{MQ} \parallel \overline{NP}$
$\angle 4 \cong \angle 3$

Prove: $\angle 1 \cong \angle 5$

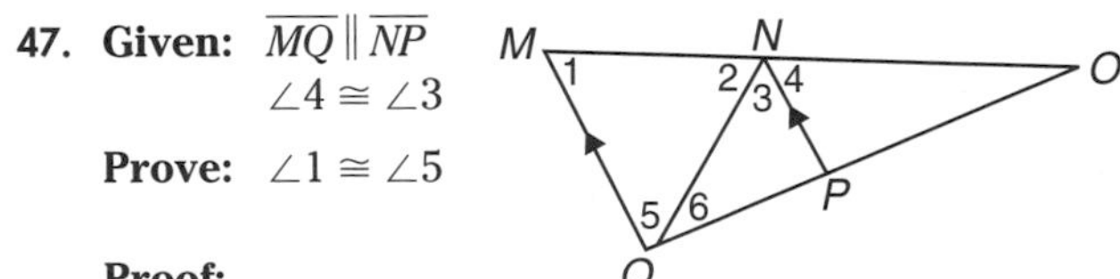

Proof:

Statements	Reasons
1. $\overline{MQ} \parallel \overline{NP}$ $\angle 4 \cong \angle 3$	1. Given
2. $\angle 3 \cong \angle 5$	2. Alternate Interior Angles Theorem
3. $\angle 4 \cong \angle 5$	3. Congruence of angles is transitive.
4. $\angle 4 \cong \angle 1$	4. Corresponding Angles Postulate
5. $\angle 1 \cong \angle 5$	5. Congruence of angles is transitive.

49. Since $\overline{AB} \parallel \overline{DC}$, we know $\angle 1 \cong \angle 4$ by the Alternate Interior Angle Theorem. However, $\angle 3$ and $\angle 2$ are not alternate interior angles for a transversal and sides $\overline{AB}$ and $\overline{DC}$. They are alternate interior angles for a transversal and sides $\overline{AD}$ and $\overline{BC}$ which may or may not be parallel. **51a.** Exclude; the transversal was used to show the number of teams not making the playoffs. **51b.** Sample answers: to make money, to create excitement during the season

53. yes; Transitive Property **55.** 43 **57a.** yes **57b.** Law of Detachment **59.** right angles **61.** 12 **63.** 20

Pages 141–145 Lesson 3–3

7. −5; falling **9.** $\frac{3}{4}$ **11.** 0 **13.**

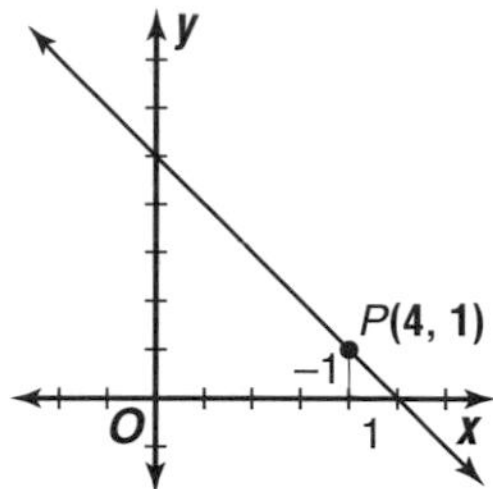

15. 8 **17.** $-\frac{3}{2}$; falling **19.** 0; horizontal **21.** $\frac{1}{2}$; rising **23.** $-\frac{5}{4}$ **25.** 1 **27.** undefined **29.** $\frac{4}{5}$ **31.** 0

33.

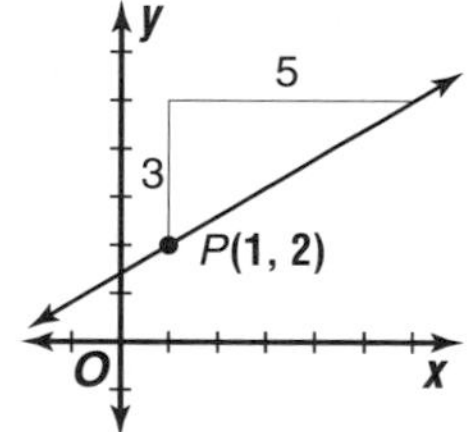

35.

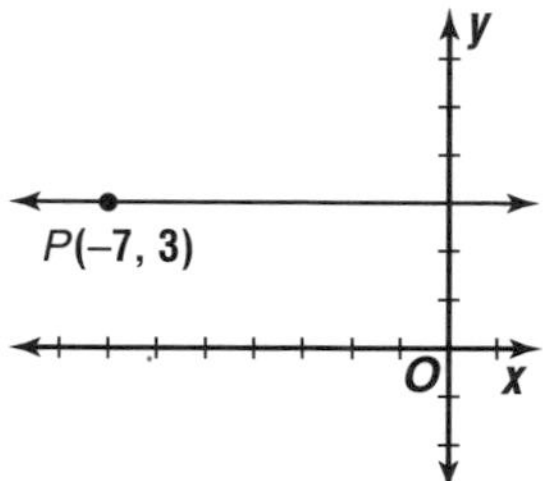

37.

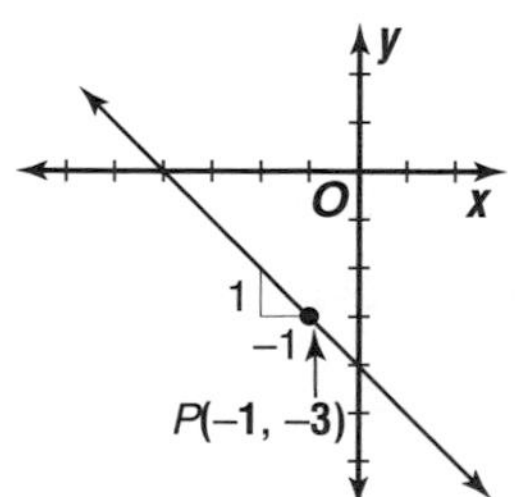

39. yes; slope of $\overleftrightarrow{MA} = -\frac{3}{5}$, slope of $\overleftrightarrow{TH} = -\frac{3}{5}$ **41.** yes; slope of $\overleftrightarrow{PQ} = \frac{1}{2}$, slope of $\overleftrightarrow{RS} = -2$, $\frac{1}{2}(-2) = -1$

43. 13

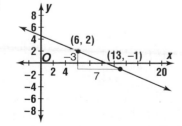

45a. slope of $\overline{PQ} = -1$, slope of $\overline{RS} = -1$, slope of $\overline{QR} = 1$, slope of $\overline{PS} = 1$ **45b.** $-1(1) = -1$ and $1(-1) = -1$

45c.
$$PQ = \sqrt{(5-1)^2 + (2-6)^2} = \sqrt{32}$$
$$QR = \sqrt{(1-(-3))^2 + (6-2)^2} = \sqrt{32}$$
$$RS = \sqrt{(-3-1)^2 + (2-(-2))^2} = \sqrt{32}$$
$$PS = \sqrt{(5-1)^2 + (2-(-2))^2} = \sqrt{32}$$

45d. square **47a.** $-\frac{5}{6}$ **47b.** $-\frac{5}{2}$ **47c.** $\frac{5}{2}$ **47d.** $-\frac{3}{4}$ **49.** yes; $0.64 < 0.88$ **51a.** $\frac{1}{18}$ **51b.** 1.5 feet **53.** $x = 16$, $y = 11$

55. Given: $\angle 1 \cong \angle 2$

Prove: $\angle 1 \cong \angle 3$

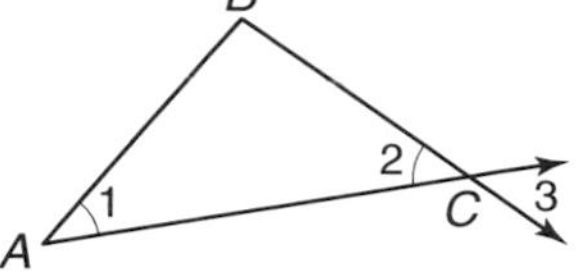

Proof:

Statements	Reasons
1. $\angle 1 \cong \angle 2$	1. Given
2. $\angle 2 \cong \angle 3$	2. Vertical ∠s are ≅.
3. $\angle 1 \cong \angle 3$	3. Congruence of angles is transitive.

57.

$A = 2\pi r^2 + 2\pi rh$	*Given*
$A - 2\pi r^2 = 2\pi rh$	*Subtraction Property (=)*
$\frac{A - 2\pi r^2}{2\pi r} = h$	*Division Property (=)*

59. Sample answer: $\left(\frac{1}{2}\right)^2 = \frac{1}{4}, \frac{1}{2} > \frac{1}{4}$ **61.** A **63.** 22

Page 145 Self Test

1. $\overline{AE}$, $\overline{BD}$ **3.** plane *ABC* and plane *EDF* **5.** 40 **7.** $-\frac{1}{3}$ **9.** $-\frac{4}{3}$

Pages 149–153 Lesson 3–4

7. $a \parallel b$; If ⇄, and a pair of alt. int. ∠s is ≅, then the lines are ∥. **9.** $\ell \parallel m$; If ⇄, and corr. ∠s are ≅, then the lines are ∥. **11.** 7

13. 1. Given
2. Vertical ∠s are ≅.
3. Congruence of ∠s is transitive.
4. If ⇄, and corr. ∠s are ≅, then the lines are ∥.

15. $\overleftrightarrow{EC} \parallel \overleftrightarrow{HF}$; If ⇄, and corr. ∠s are ≅, then the lines are ∥.

17. $\overleftrightarrow{JK} \parallel \overleftrightarrow{BL}$; If ⇄, and a pair of consecutive int. ∠s is supplementary, then the lines are ∥. **19.** $p \parallel q$; If ⇄, and a pair of alt. ext. ∠s is congruent, then the lines are ∥. **21.** $p \parallel q$; If ⇄, and a pair of consecutive int. ∠s is supplementary, then the lines are ∥. **23.** $\ell \parallel m$; If ⇄, and a pair of consecutive int. ∠s is supplementary, then the lines are ∥. **25.** 13 **27.** 9 **29.** $x = 10, y = 3$ **31.** $\ell \parallel m$; If ⇄, and a pair of consecutive int. ∠s is supplementary, then the lines are ∥.

33. 1. Given
2. Definition of linear pair
3. ∠2 and ∠3 are supplementary.
4. 2 ∠s supplementary to the same ∠ are ≅.
5. If ⇄, and corr. ∠s are ≅, then the lines are ∥.

35. Given: $\angle 2 \cong \angle 1$
$\angle 1 \cong \angle 3$
Prove: $\overline{ST} \parallel \overline{YZ}$

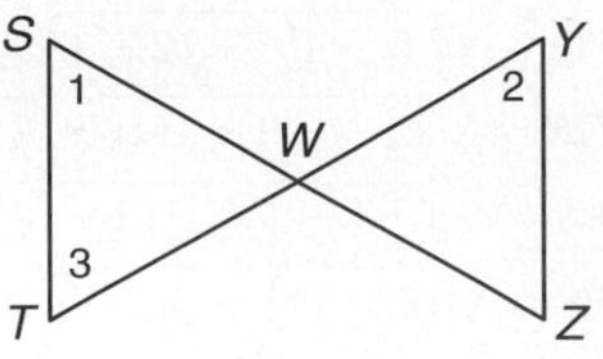

Proof:

Statements	Reasons
1. $\angle 2 \cong \angle 1$, $\angle 1 \cong \angle 3$	1. Given
2. $\angle 2 \cong \angle 3$	2. Congruence of ∠s is transitive.
3. $\overline{ST} \parallel \overline{YZ}$	3. If 2 lines are cut by a transversal, and a pair of alt. int. ∠s are ≅, then the lines are ∥.

37. Given: $\overline{AU} \perp \overline{QU}$
$\angle 1 \cong \angle 2$
Prove: $\overline{DQ} \perp \overline{QU}$

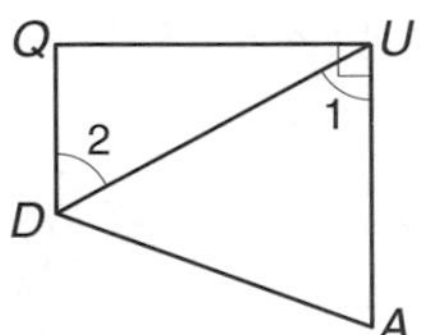

Proof:

Statements	Reasons
1. $\overline{AU} \perp \overline{QU}$, $\angle 1 \cong \angle 2$	1. Given
2. $\overline{AU} \parallel \overline{DQ}$	2. If 2 lines are cut by a transversal, and a pair of alt. int. ∠s is ≅, then the lines are ∥.
3. $\overline{DQ} \perp \overline{QU}$	3. Perpendicular Transversal Theorem

39a. Sample drawing:

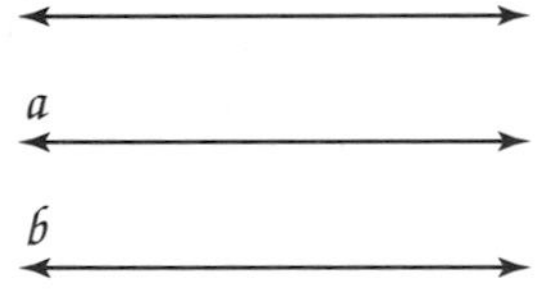

39b. Draw a transversal intersecting all three lines and number the angles formed.

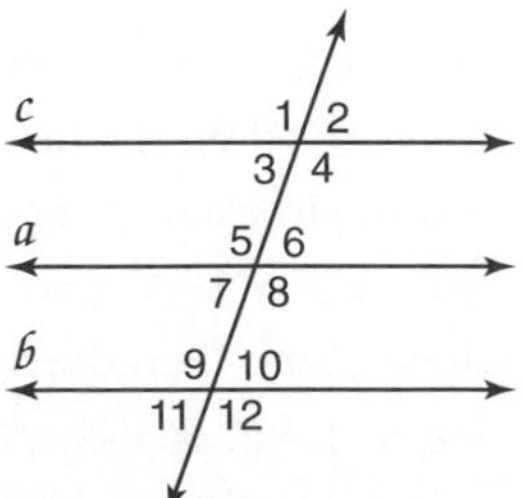

Sample plan for proof: Use corresponding angles to show $\angle 6 \cong \angle 10$ and $\angle 2 \cong \angle 6$. Then $\angle 2 \cong \angle 10$ which is enough to show that $b \parallel c$. **41.** yes **43.** $\frac{4}{5}$; rising
45. Intersecting; they all meet at the hub.
47. Lines p and m never meet; Law of Detachment.
49. (2, 9) **51.** $(2x - 3y)(8x + y)$

Pages 158–161 Lesson 3–5

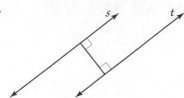

7.

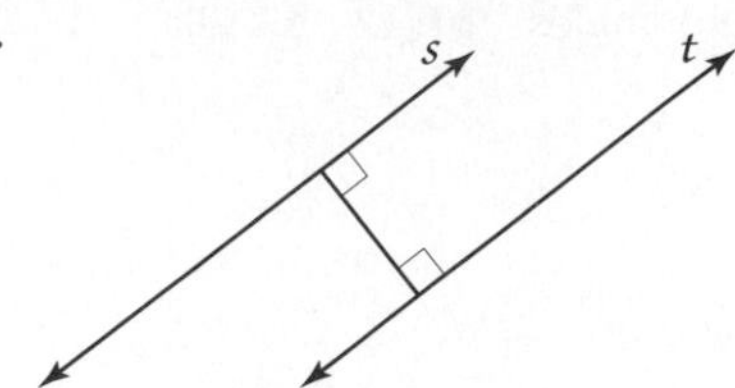

9. 1

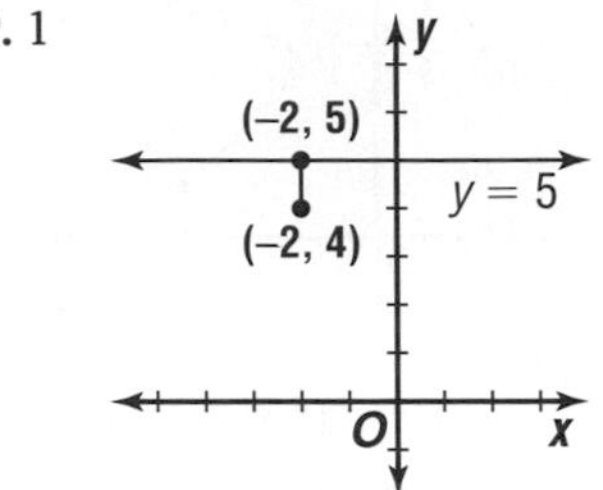

11. $\overline{PS}$ **13.** It is everywhere equidistant.

15.

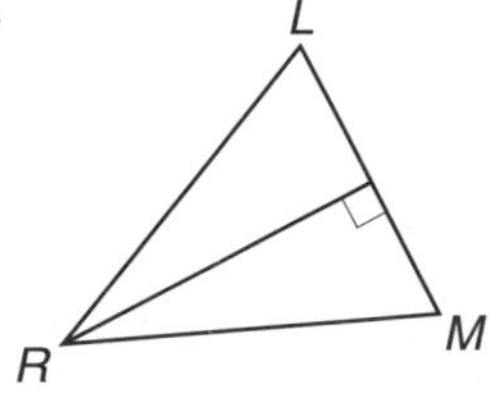

17.

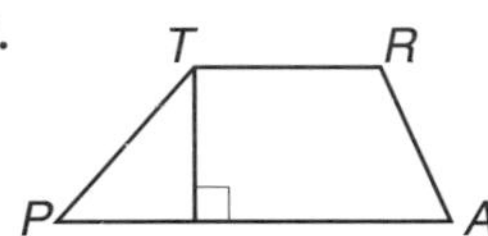

19.

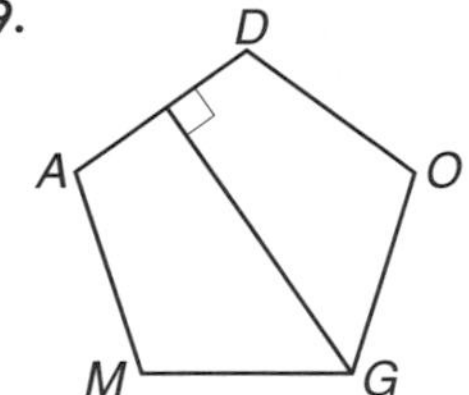

21. Yes; the lines are everywhere equidistant.

23. 5

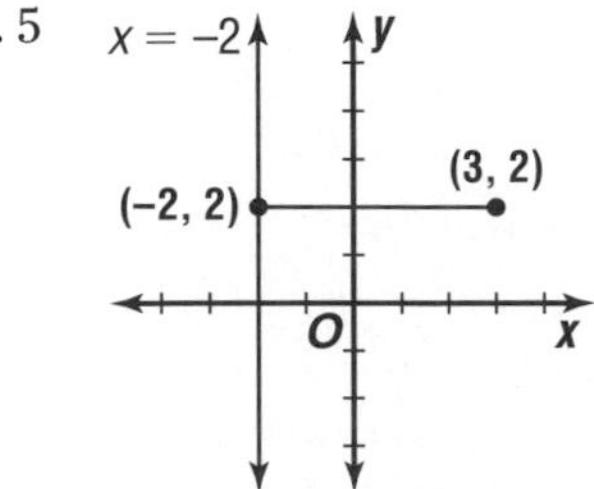

25. $\sqrt{5} \approx 2.24$

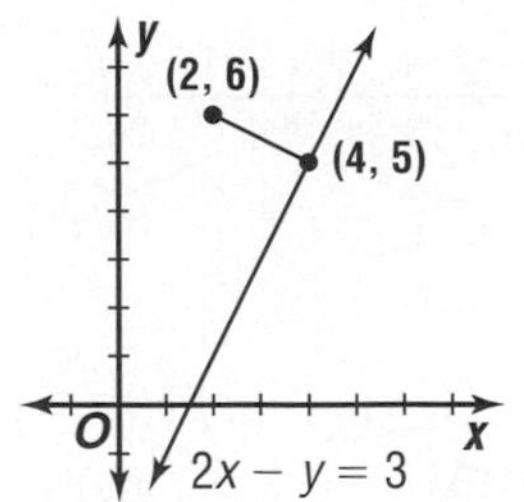

27. $\sqrt{13} \approx 3.61$

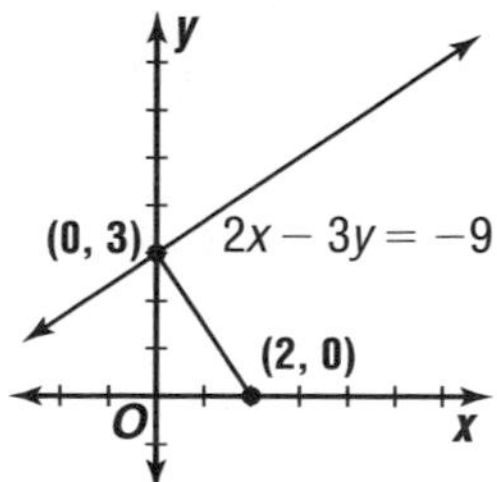

29. $\overline{MT}$ 31. $\overline{GM}$ 33. $\overline{HM}$ 35. $\sqrt{24.5} \approx 4.95$

37. $d = \frac{|3 \cdot 2 + 4 \cdot 5 - 1|}{\sqrt{3^2 + 4^2}} = 5$; Yes; both equal 5. 39a. yes

39b. no 41. $p \nparallel q$; cons. int. ∠s are not supp. 43. $m\angle 2 = m\angle 4$; They are corresponding angles. 45. They are right angles.

47.

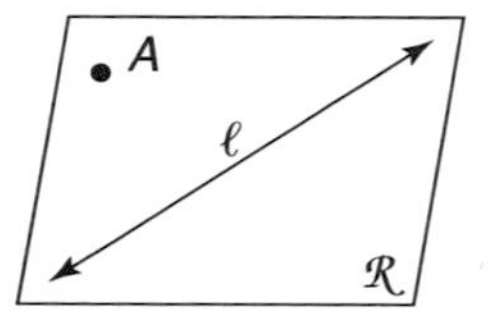

49. 8.1

Page 162 Lesson 3–5B

1. 1.41 3. 3.16

Pages 167–169 Lesson 3–6

5. true 7. 41° N, 29° E 9. The great circle is finite.
13. False; in spherical geometry, a line has finite length.
15. true 17. true 19. 30° N, 85° W 21. 19° N, 99° W
23. Reno, Nevada 25. There exists no parallel lines.
27. A pair of perpendicular great circles divides the sphere into 4 finite congruent regions. 29. There exists no parallel lines.
31. 1

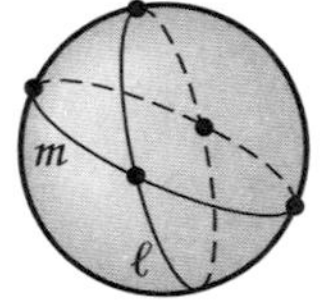

33.

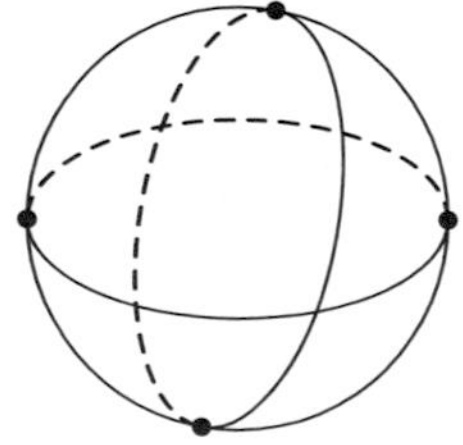

33a. They are each perpendicular to each other.
33b. They lie at the intersection of the other 2 circles.
35a. In a plane, if 2 lines are perpendicular to the same line, then they are parallel. 35b. Yes, 2 intersecting great circles can both be perpendicular to another great circle.

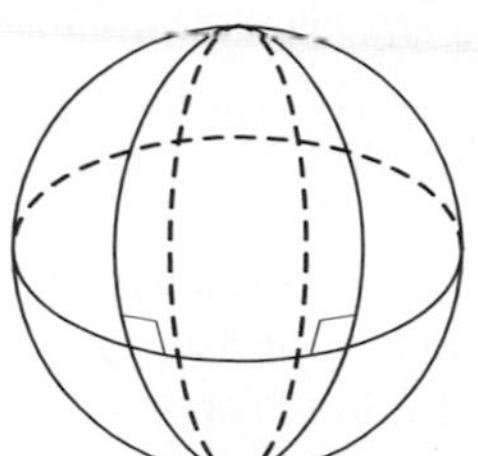

37. Sample answer: The spaceship will eventually return to its starting point. 39. The 2 lines must be perpendicular to the transversal.

41. 21

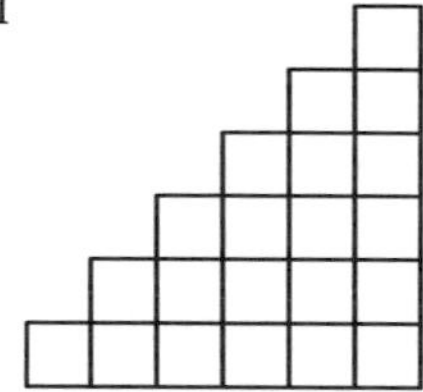

43. Sample answer: If the measure of an angle is 90, the angle is a right angle. Hypothesis: if the measure of an angle is 90; Conclusion: the angle is a right angle
45. Commutative (+)

Page 171 Chapter 3 Highlights

1. f 3. d 5. g 7. c 9. b 11. j

Pages 172–174 Chapter 3 Study Guide and Assessment

13. t and m 15. Sample answer: ℓ and n 17. $\angle 3 \cong \angle 5$, $\angle 3 \cong \angle 6$, $\angle 1 \cong \angle 4$, $\angle 5 \cong \angle 6$, $\angle GAB \cong \angle 10$, $\angle GAB \cong \angle ABH$, $\angle FAE \cong \angle 9$ 19. $\angle 3$, $\angle 8$ 21. 6; rising 23. $-\frac{1}{3}$; falling 25. $-\frac{9}{7}$; $\frac{7}{9}$ 27. $\frac{1}{6}$; -6 29. $\overline{AD} \parallel \overline{EF}$; $\angle 1 \cong \angle 4$ since vertical angles are congruent. So, $\angle 4$ and $\angle 2$ are supplementary. The lines are parallel because consecutive interior angles are supplementary.
31. $\overline{AD} \parallel \overline{EF}$; corr. ∠s are ≅. 33. $\overline{RQ}$ 35. $\overline{PQ}$ or $\overline{MR}$
37. The intersection of 2 great circles creates 8 angles.
39. The shortest path between two points is an arc on the great circle passing through the points.
41. 36 students

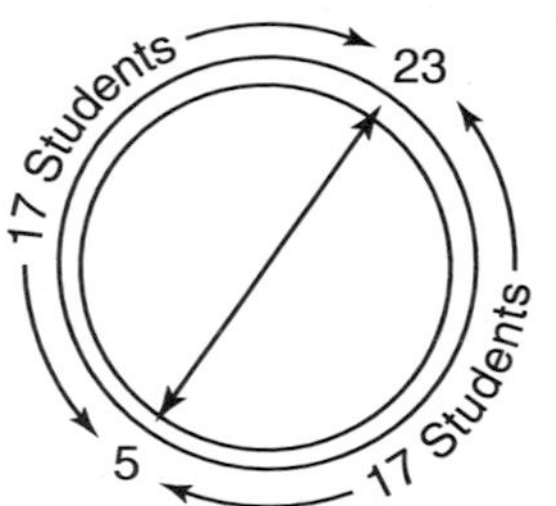

43. 65 mph

CHAPTER 4 IDENTIFYING CONGRUENT TRIANGLES

Pages 183–187 Lesson 4–1

7. $\overline{SM}$ 9. Sample answer: right, isosceles

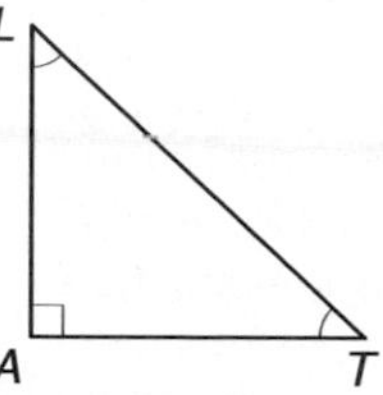

11. $x = 7$; $EQ = 25$, $QU = 25$, $EU = 25$ 13. sometimes
15a. right, scalene; obtuse, scalene

15b. Sample answer:

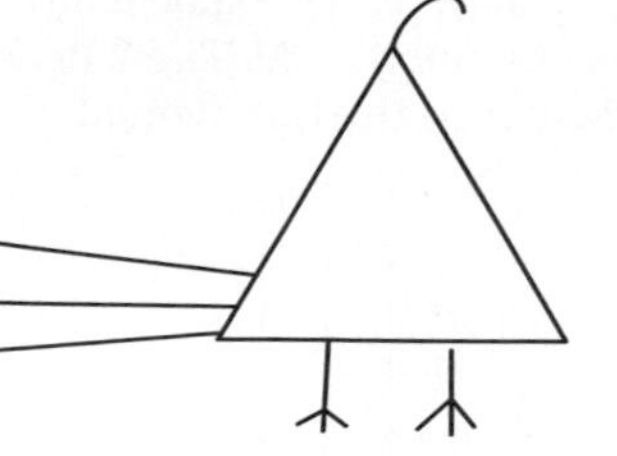

15c. Sample answer:

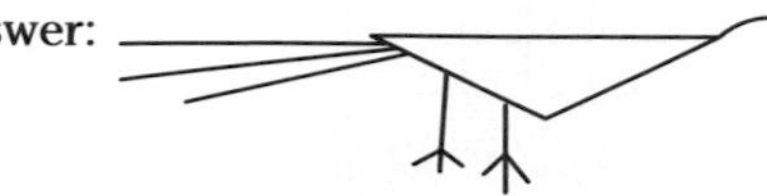

17. $\overline{LM}$ **19.** $\overline{LM}$ **21.** $\angle BLM$, $\angle BML$ **23.** $\overline{BL}$, $\overline{BM}$
25. Sample answer: right, isosceles

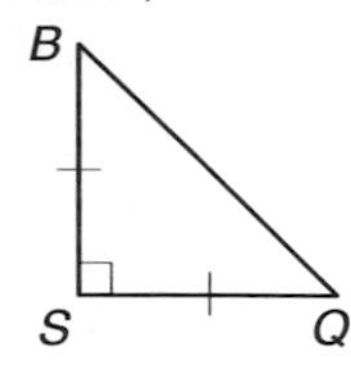

27. Sample answer: scalene

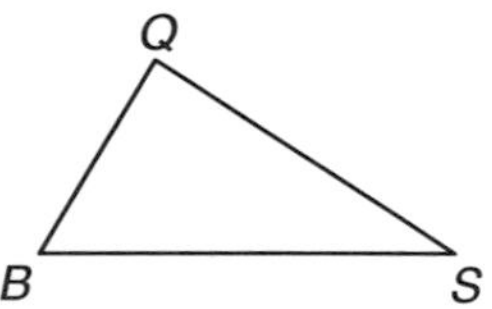

29. Sample answer: obtuse, isosceles

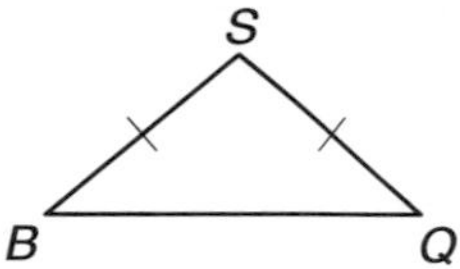

31. $x = 5$; $HK = 12$, $HT = 12$, $KT = 12$

33. $PQ = \sqrt{(3-0)^2 + (6-6)^2} = \sqrt{9+0} = \sqrt{9} = 3$
$QR = \sqrt{(3-3)^2 + (0-6)^2} = \sqrt{0+36} = \sqrt{36} = 6$
$PR = \sqrt{(3-0)^2 + (0-6)^2} = \sqrt{9+36} = \sqrt{45} = 3\sqrt{5}$; scalene

35. $KL = \sqrt{(-2-4)^2 + (0-0)^2} = \sqrt{36+0} = \sqrt{36} = 6$
$LM = \sqrt{(1-(-2))^2 + (5-0)^2} = \sqrt{9+25} = \sqrt{34}$
$KM = \sqrt{(1-4)^2 + (5-0)^2} = \sqrt{9+25} = \sqrt{34}$
isosceles

37. never **39.** sometimes **41.** always

43. 2. Definition of perpendicular
3. Perpendicular Transversal Theorem
4. $\angle RST$ is a right angle.
5. $\triangle RST$ is a right triangle.

45. $\angle E$; Since $23 < 2x + 2 + 10 + x + 4 < 32$, $\frac{7}{3} < x < \frac{16}{3}$. If $2x + 2 = 10$, $x = 4$. If $2x + 2 = x + 4$, $x = 2$. If $x + 4 = 10$, $x = 6$. The only value of x that satisfies the conditions is $x = 4$, so $\overline{EF}$ and $\overline{ED}$ are the legs of the isosceles triangle and $\angle E$ is the vertex. **47.** If a triangle is an acute triangle, the square of the length of the longest side is less than the sum of the squares of the lengths of the other two sides. If a triangle is an obtuse triangle, the square of the length of the longest side is greater than the sum of the squares of the lengths of the other two sides. **49a.** $\triangle ABC$, $\triangle ADG$, $\triangle AHM$, $\triangle ANS$ **49b.** none **49c.** $\triangle BED$, $\triangle CFG$, $\triangle BJH$, $\triangle CKM$, $\triangle DIH$, $\triangle GLM$, $\triangle BPN$, $\triangle CQS$, $\triangle DON$, $\triangle GRS$ **51.** infinite number **53.** (1) If two lines in a plane are cut by a transversal so that corresponding angles are congruent, then the lines are parallel. (2) If two lines in a plane are cut by a transversal so that a pair of alternate interior angles is congruent, then the lines are parallel. (3) If two lines in a plane are cut by a transversal so that a pair of consecutive interior angles is supplementary, then the lines are parallel. (4) If two lines in a plane are cut by a transversal so that a pair of alternate exterior angles is congruent, then the lines are parallel. (5) In a plane, if two lines are perpendicular to the same line, then they are parallel. **55.** Sample answer: the line formed by the intersection of the floor and a wall of a room and the line formed by two walls on the opposite side of the room **57.** Transitive Property of Equality; Sample answer: The Law of Syllogism states if $p \rightarrow q$ and $q \rightarrow r$, then $p \rightarrow r$. The Transitive Property of Equality states if $p = q$ and $q = r$, then $p = r$. **59.** 45° **61.** −3

Page 188 Lesson 4–2A
3. The sum of the measures of the angles of a triangle is 180.

Pages 192–195 Lesson 4–2
7. 139 **9.** 54 **11a.** 45 **11b.** 90 **11c.** right triangle **13.** 30 **15.** 91 **17.** 115 **19.** 75 **21.** 68 **23.** 40 **25.** 26 **27.** 14 **29.** 51 **31.** scalene; none of the angles are congruent so none of the sides are congruent.

33.

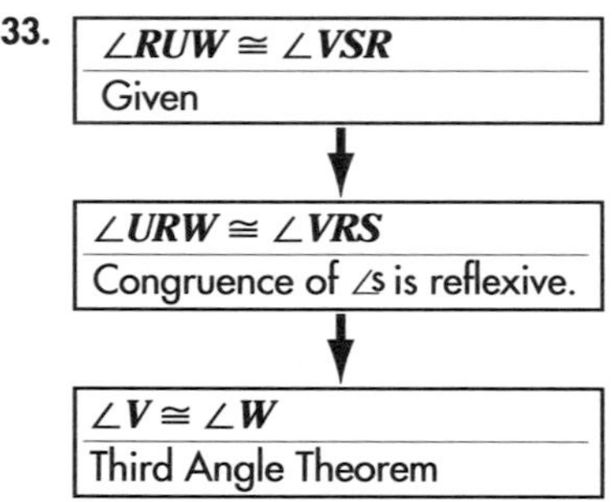

35. Given: $\triangle RED$ is equiangular.
Prove: $m\angle R = m\angle E = m\angle D = 60$

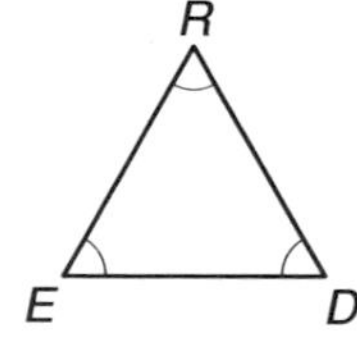

Proof:

Statements	Reasons
1. $\triangle RED$ is equiangular.	1. Given
2. $\angle R \cong \angle E \cong \angle D$	2. Definition of equiangular triangle
3. $m\angle R = m\angle E = m\angle D$	3. Definition of congruent angles
4. $m\angle R + m\angle E + m\angle D = 180$	4. Angle Sum Theorem
5. $m\angle R + m\angle R + m\angle R = 180$	5. Substitution Property (=)
6. $3m\angle R = 180$	6. Substitution Property (=)
7. $m\angle R = 60$	7. Division Property (=)
8. $m\angle R = m\angle E = m\angle D = 60$	8. Substitution Property (=)

37. Given: $\triangle RST$
$\angle R$ is a right angle.

Prove: $\angle S$ and $\angle T$ are complementary.

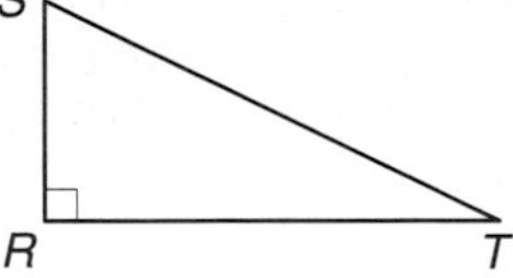

Proof:

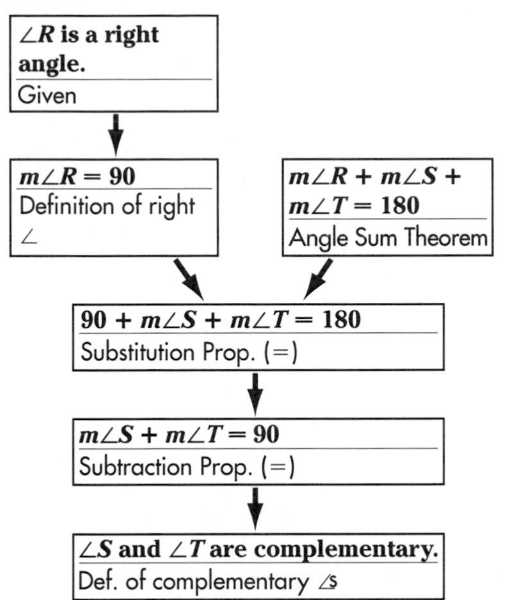

39. 360 **41a.** No; two of the angles are right angles, so the sum of their measures is 180. **41b.** The sum of the measures is greater than 180 and less than or equal to 360. The sum of the measures of two angles is 180. The measure of the third angle is greater than 0 and less than or equal to 180. **41c.** If the measure of the angle at the pole is less than 90, the triangle is an acute triangle. **41d.** If the measure of the angle at the pole is greater than 90, the triangle is an obtuse triangle. **41e.** If the measure of the angle at the pole is 90, the triangle is a right triangle. **43.** 130 **45.** No; If they are parallel, $m\angle 1 = m\angle 5$.

47. 28 games

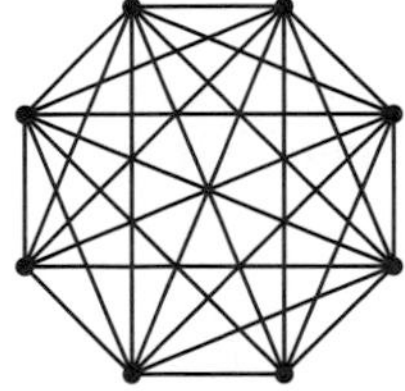

49. A **51.** III **53a.** 260 women **53b.** 395 women

Pages 199–203 Lesson 4–3

7. $\angle A \longleftrightarrow \angle X$, $\angle B \longleftrightarrow \angle Y$, $\angle C \longleftrightarrow \angle Z$; $\overline{AB} \longleftrightarrow \overline{XY}$, $\overline{AC} \longleftrightarrow \overline{XZ}$, $\overline{BC} \longleftrightarrow \overline{YZ}$

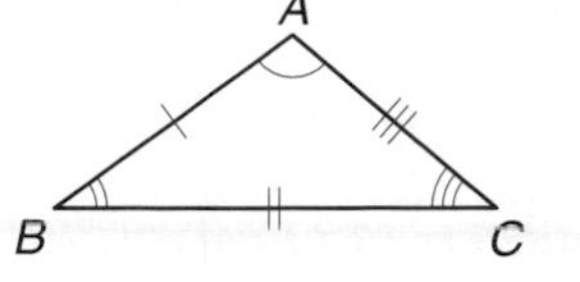

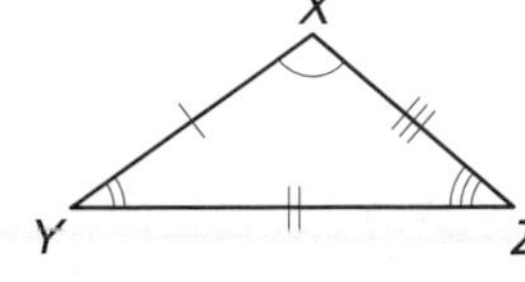

9. 1. Given
2. Congruence of segments is reflexive.
3. Congruence of angles is reflexive.
4. Definition of congruent triangles

11a.

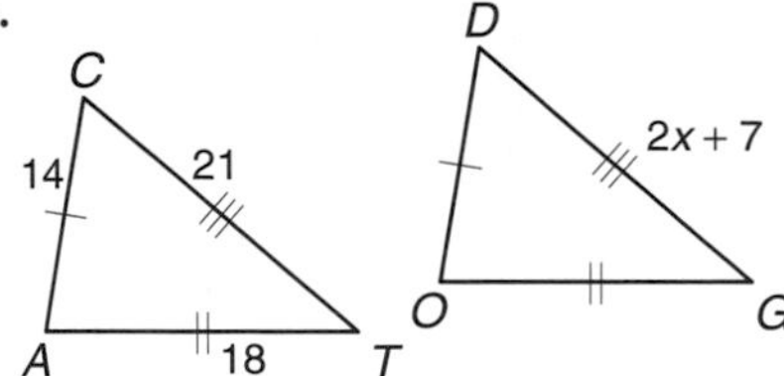

11b. 7

13. Given: $\triangle MXR$ is a right isosceles triangle with $\angle X$ the vertex angle.
$\overline{XY} \perp \overline{MR}$
Y is the midpoint of $\overline{MR}$.
$\angle M \cong \angle R$
$\overline{YX}$ bisects $\angle MXR$

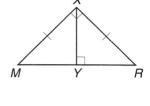

Prove: $\triangle MXY \cong \triangle RXY$

Proof:

Statements	Reasons
1. $\triangle MXR$ is isosceles with vertex $\angle X$.	1. Given
2. $\overline{XM} \cong \overline{XR}$	2. Definition of isosceles $\triangle$
3. $\overline{XY} \perp \overline{MR}$	3. Given
4. $\angle XYR$ and $\angle XYM$ are right ∡s.	4. Definition of $\perp$ lines
5. $\angle XYR \cong \angle XYM$	5. All right ∡s are $\cong$.
6. Y is the midpoint of $\overline{MR}$.	6. Given
7. $\overline{MY} \cong \overline{RY}$	7. Definition of midpoint
8. $\angle M \cong \angle R$	8. Given
9. $\overline{YX}$ bisects $\angle MXR$.	9. Given
10. $\angle MXY \cong \angle RXY$	10. Definition of angle bisector
11. $\overline{XY} \cong \overline{XY}$	11. Congruence of segments is reflexive.
12. $\triangle MXY \cong \triangle RXY$	12. Definition of $\cong$ triangles

15. $\triangle HJK$, $\triangle HJI$, $\triangle IKJ$, $\triangle IKH$; $\triangle IAJ$, $\triangle JAK$, $\triangle KAH$, $\triangle HAI$; $\triangle HGA$, $\triangle KLA$, $\triangle JBA$, $\triangle IEA$, $\triangle HEA$, $\triangle KGA$, $\triangle JLA$, $\triangle IBA$; $\triangle HDE$, $\triangle KFG$, $\triangle JML$, $\triangle ICB$; $\triangle EDA$, $\triangle GFA$, $\triangle LMA$, $\triangle BCA$ **17.** $\overline{PQ} \longleftrightarrow \overline{RS}$, $\overline{PR} \longleftrightarrow \overline{RT}$, $\overline{QR} \longleftrightarrow \overline{ST}$; $\angle P \longleftrightarrow \angle R$, $\angle Q \longleftrightarrow \angle S$, $\angle R \longleftrightarrow \angle T$

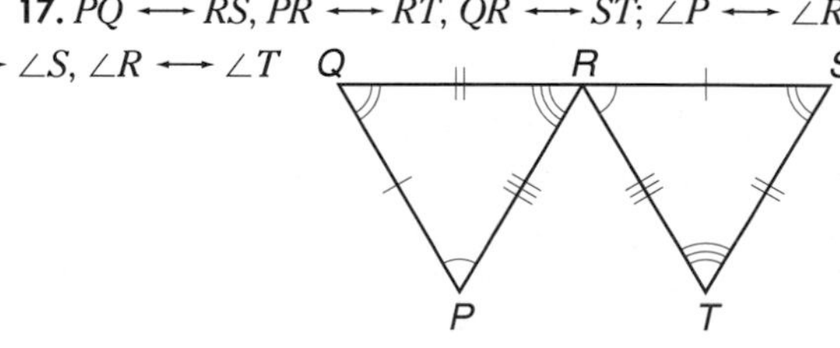

19. WXY **21.** ERG **23a.** Given **23b.** Given **23c.** Congruence of segments is reflexive. **23d.** Given **23e.** Definition of $\perp$ lines **23f.** Given **23g.** Definition of $\perp$ lines **23h.** All right ∡s are $\cong$. **23i.** Given **23j.** Alternate Interior Angle Theorem **23k.** Given **23l.** Alternate Interior Angle Theorem **23m.** Definition of congruent triangles

25a.

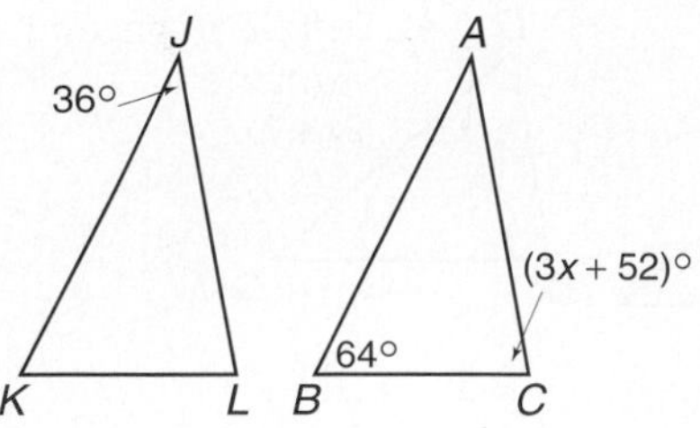

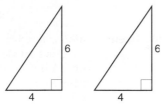

25b. $\frac{28}{3}$ **27.** Sample answer:

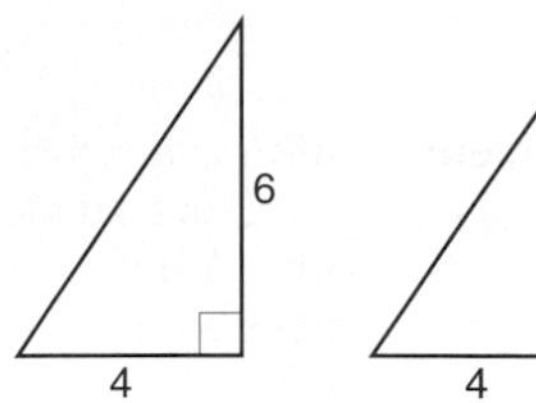

29. Sample answer:

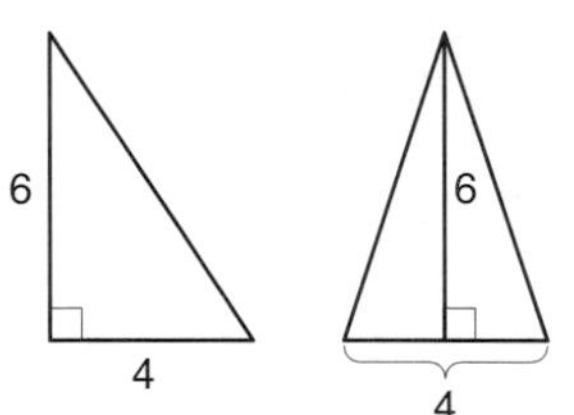

31. true; Since $AD = AB + BD$ and $AD = AC + CD$, $AB + BD = AC + CD$. By CPCTC $\overline{BD} \cong \overline{AC}$, so $BD = AC$. By Subtraction Property (=), $AB = CD$ and $\overline{AB} \cong \overline{CD}$ or $\overline{DC} \cong \overline{AB}$. **33.** true; CPCTC **35.** true; Since $m\angle AND = m\angle ANB + m\angle BND$ and $m\angle AND = m\angle ANC + m\angle CND$, $m\angle ANB + m\angle BND = m\angle ANC + m\angle CND$. By CPCTC, $\angle BND \cong \angle ANC$, so $m\angle BND = m\angle ANC$. By Subtraction Property (=), $m\angle ANB = m\angle CND$ and $\angle ANB \cong \angle CND$.

37. Given: $\triangle SIO \cong \triangle OIS$

Prove: $\triangle SIO$ is an isosceles triangle.

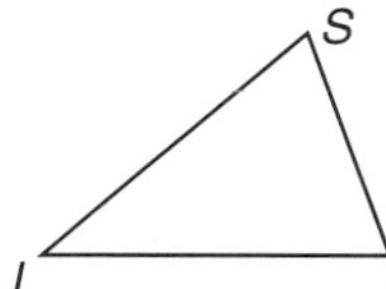

Proof:

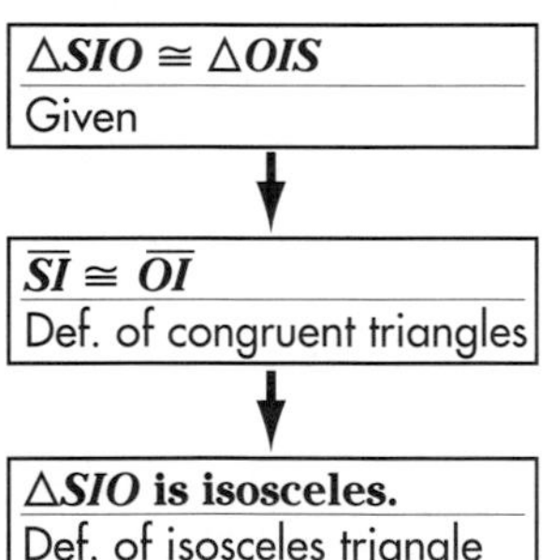

39a. 4 **39b.** 0 **41.** $\triangle RQU \cong \triangle RSU$, $\triangle QUP \cong \triangle SUT$, $\triangle RUP \cong \triangle RUT$ **43.** 33 **45.** $\angle 1 \cong \angle 7 \cong \angle 9$, $\angle 10 \cong \angle 8$, $\angle 3 \cong \angle 5 \cong \angle 6$, $\angle 2 \cong \angle 4$; $\angle 1$ and $\angle 10$, $\angle 7$ and $\angle 10$, $\angle 7$ and $\angle 8$, $\angle 8$ and $\angle 9$, $\angle 9$ and $\angle 10$, $\angle 1$ and $\angle 8$, $\angle 2$ and $\angle 6$, $\angle 2$ and $\angle 5$, $\angle 5$ and $\angle 4$, $\angle 4$ and $\angle 3$, $\angle 3$ and $\angle 2$, $\angle 6$ and $\angle 4$ are supplementary. **47.** Substitution Property (=) **49.** a straight angle **51.** $\{x \mid -3 < x \leq 2\}$

Page 203 Self Test

1. $\triangle DBC$, $\triangle ABD$ **3.** $\overline{AD}$, $\overline{AC}$ **5.** 107 **7.** 125 **9.** $\triangle YZX$

Page 205 Lesson 4–4A

1. It is congruent to $\triangle ABC$. **3.** It is congruent to $\triangle ABC$. **5.** If the sides of one triangle are congruent to the sides of another triangle, the triangles are congruent. **7.** If two angles and the included side of one triangle are congruent to two angles and the included side of another triangle, the triangles are congruent.

Pages 209–213 Lesson 4–4

7. SSS

9. **1.** Given
2. Definition of segment bisector
3. Given
4. Alternate Interior Angle Theorem
5. Vertical angles are ≅.
6. ASA

11. Given: $\overline{MO} \cong \overline{PO}$
$\overline{NO}$ bisects $\overline{MP}$.

Prove: $\triangle MNO \cong \triangle PNO$

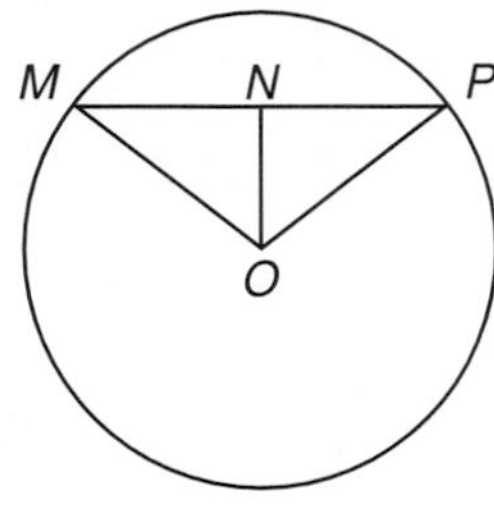

Proof:

Statements	Reasons
1. $\overline{MO} \cong \overline{PO}$, $\overline{NO}$ bisects $\overline{MP}$	1. Given
2. $\overline{MN} \cong \overline{PN}$	2. Definition of segment bisector
3. $\overline{NO} \cong \overline{NO}$	3. Congruence of segments is reflexive.
4. $\triangle MNO \cong \triangle PNO$	4. SSS

13a. the two middle frameworks **15.** SAS **17.** SAS **19.** SSS

21. $PQ = \sqrt{(0-(-1))^2 + (6-(-1))^2} = \sqrt{50} = 5\sqrt{2}$

$QR = \sqrt{(2-0)^2 + (3-6)^2} = \sqrt{13}$

$PR = \sqrt{(2-(-1))^2 + (3-(-1))^2} = \sqrt{25} = 5$

$XY = \sqrt{(5-3)^2 + (3-1)^2} = \sqrt{8} = 2\sqrt{2}$

$YZ = \sqrt{(8-5)^2 + (1-3)^2} = \sqrt{13}$

$XZ = \sqrt{(8-3)^2 + (1-1)^2} = \sqrt{25} = 5$

No; $\overline{QR} \cong \overline{YZ}$ and $\overline{PR} \cong \overline{XZ}$, but $\overline{PQ}$ is not congruent to $\overline{XY}$.

23. $\angle SRU \cong \angle TRU$

25. Given: $\angle J \cong \angle L$
B is the midpoint of $\overline{JL}$.

Prove: $\triangle JHB \cong \triangle LCB$

Proof:

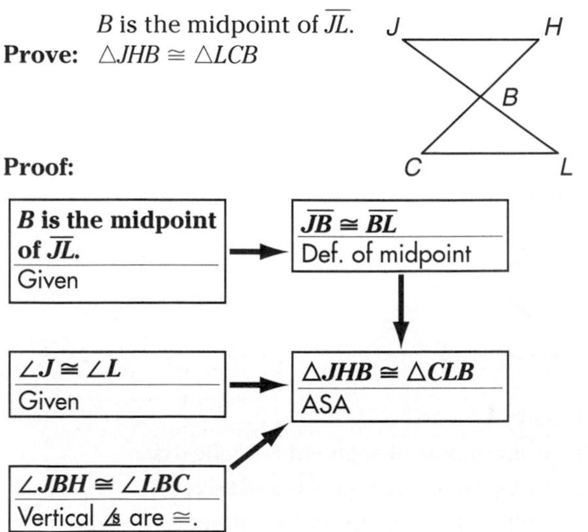

27. Given: $\triangle MGR$ is an isosceles triangle with vertex $\angle MGR$.
K is the midpoint of $\overline{MR}$.

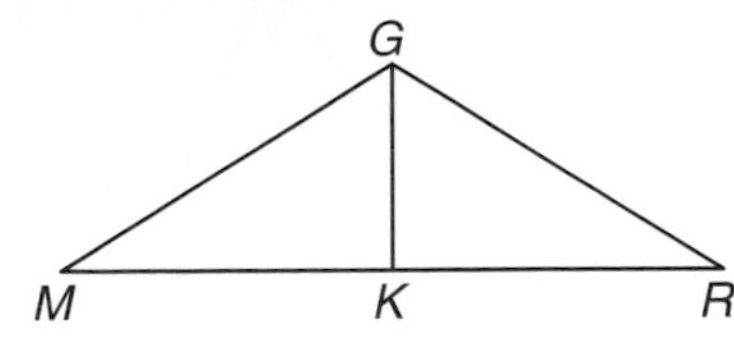

Prove: $\triangle MGK \cong \triangle RGK$

Proof:

Statements	Reasons
1. $\triangle MGR$ is an isosceles triangle with vertex $\angle MGR$.	1. Given
2. $\overline{GM} \cong \overline{GR}$	2. Definition of isosceles triangle
3. K is the midpoint of $\overline{MR}$.	3. Given
4. $\overline{MK} \cong \overline{RK}$	4. Definition of midpoint
5. $\overline{GK} \cong \overline{GK}$	5. Congruence of segments is reflexive.
6. $\triangle MGK \cong \triangle RGK$	6. SSS

29. Given: $\overline{RL} \parallel \overline{DC}$
$\overline{LC} \parallel \overline{RD}$

Prove: $\angle R \cong \angle C$

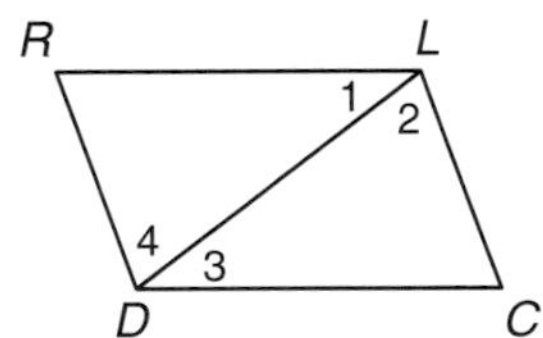

Proof:

Statements	Reasons
1. $RL \parallel \overline{DC}$	1. Given
2. $\angle 1 \cong \angle 3$	2. Alternate Interior Angle Theorem
3. $\overline{LC} \parallel \overline{RD}$	3. Given
4. $\angle 2 \cong \angle 4$	4. Alternate Interior Angle Theorem
5. $\overline{LD} \cong \overline{DL}$	5. Congruence of segments is reflexive.
6. $\triangle RLD \cong \triangle CDL$	6. ASA
7. $\angle R \cong \angle C$	7. CPCTC

31. Given: $\angle 1 \cong \angle 2$
$\angle 3 \cong \angle 4$
$\overline{LA} \cong \overline{RU}$

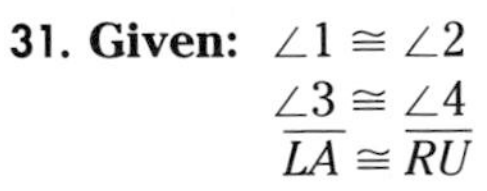

Prove: $\triangle WLU \cong \triangle WRA$

Proof:

Statements	Reasons
1. $\angle 1 \cong \angle 2$, $\angle 3 \cong \angle 4$, $\overline{LA} \cong \overline{RU}$	1. Given
2. $LA = RU$	2. Definition of congruent segments
3. $LU = LA + AU$, $RA = RU + AU$	3. Segment Addition Postulate
4. $RA = LA + AU$	4. Substitution Property (=)
5. $LU = RA$	5. Substitution Property (=)
6. $\overline{LU} \cong \overline{RA}$	6. Definition of congruent segments
7. $\triangle WLU \cong \triangle WRA$	7. ASA

33. Given: $\angle 5 \cong \angle 7$
$\angle 3 \cong \angle 4$
$\overline{LW} \cong \overline{RW}$

Prove: $\overline{LU} \cong \overline{RA}$

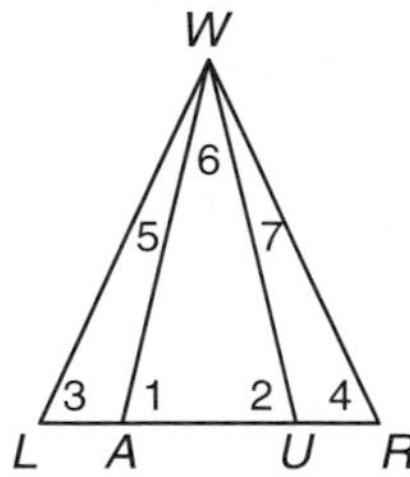

Proof:

Statements	Reasons
1. $\angle 5 \cong \angle 7$	1. Given
2. $m\angle 5 = m\angle 7$	2. Definition of congruent angles
3. $m\angle LWU = m\angle 5 + m\angle 6$, $m\angle RWA = m\angle 7 + m\angle 6$	3. Angle addition
4. $m\angle RWA = m\angle 5 + m\angle 6$	4. Substitution Property (=)
5. $m\angle LWU = m\angle RWA$	5. Substitution Property (=)
6. $\angle LWU \cong \angle RWA$	6. Definition of congruent angles
7. $\angle 3 \cong \angle 4$, $\overline{LW} \cong \overline{RW}$	7. Given
8. $\triangle LWU \cong \triangle RWA$	8. ASA
9. $\overline{LU} \cong \overline{RA}$	9. CPCTC

35. $\angle A$ is not the included angle. **37.** Since Jamal is perpendicular to the ground, two right triangles are formed and the two right angles are congruent. The angles of sight are the same and his height is the same, so the triangles are congruent by ASA. By CPCTC, the distances are the same and the method is valid. **39.** 30 **41.** $\frac{1}{10}$ **43.** Sample answer: The plants need more sunlight to live. Place them somewhere with more sun and see if they survive. **45.** 1.5 hours

Pages 217–221 Lesson 4–5

7. Sample answer: $\angle 1$ and $\angle 5$ **9.** Yes; Since $\angle 7 \cong \angle 11$, $\overline{AD} \parallel \overline{EV}$. Then $\angle 6 \cong \angle 10$, by the Alt. Interior Angle Theorem. Then you have two sides and the included angle congruent.

11. 1. Given
2. Vertical ∠s are ≅.
3. AAS
4. CPCTC

13. Given: $\overline{NM}$ bisects $\overline{RD}$.
$\angle 7 \cong \angle 8$

Prove: $\overline{MD} \cong \overline{NR}$

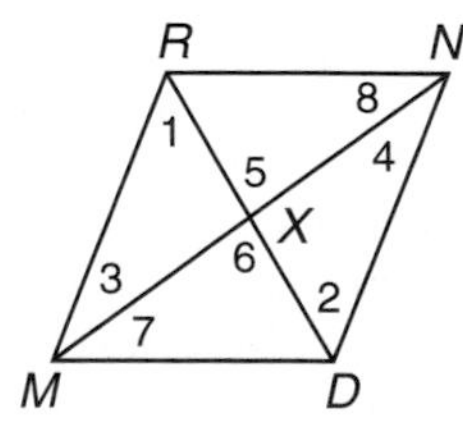

Proof:

Statements	Reasons
1. $\overline{NM}$ bisects $\overline{RD}$	1. Given
2. $\overline{RX} \cong \overline{XD}$	2. Definition of bisector
3. $\angle 7 \cong \angle 8$	3. Given
4. $\angle 6 \cong \angle 5$	4. Vertical ∠s are ≅.
5. $\triangle DXM \cong \triangle RXN$	5. AAS
6. $\overline{MD} \cong \overline{NR}$	6. CPCTC

15. $\overline{FD}$ **17.** $\overline{DR}$ or $\overline{RS}$ **19.** $\angle 10$ and $\angle 11$ **21.** $\angle 1$ and $\angle 4$ or $\angle 2$ and $\angle 4$ **23.** *SRD*; AAS **25.** $\overline{FD} \cong \overline{CD}$ and $\overline{DR} \cong \overline{DT}$ **27a.** Given **27b.** Given **27c.** Given **27d.** $\perp$ lines form 4 rt. $\triangle$s. **27e.** $\perp$ lines form 4 rt. $\triangle$s. **27f.** All rt. $\triangle$s are $\cong$. **27g.** Given **27h.** AAS **27i.** CPCTC

29. Given: $\angle A \cong \angle D$
$\angle EBC \cong \angle ECB$
$\overline{AE} \cong \overline{DE}$

Prove: $\triangle ABE \cong \triangle DCE$

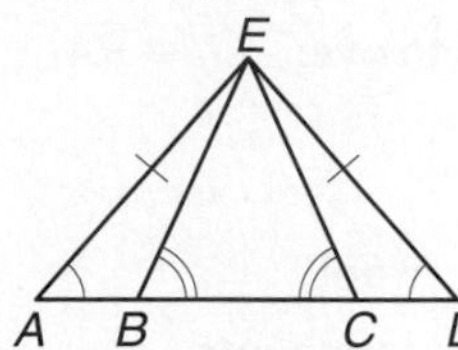

Proof:

Since $\angle EBC \cong \angle ECB$, by definition of congruent angles $m\angle EBC = m\angle ECB$. Since $\angle ABE$ and $\angle EBC$ are linear pairs, the angles are supplementary and $m\angle ABE + m\angle EBC = 180$. Likewise, $m\angle DCE + m\angle ECB = 180$. By substitution, $m\angle ABE + m\angle EBC = m\angle DCE + m\angle ECB$ and $m\angle ABE + m\angle ECB = m\angle DCE + m\angle ECB$. Using the Subtraction Property of Equality, $m\angle ABE = m\angle DCE$. By the definition of congruent angles, $\angle ABE \cong \angle DCE$. Since we are given that $\angle A \cong \angle D$ and $\overline{AE} \cong \overline{DE}$, $\triangle ABE \cong \triangle DCE$ by AAS.

31. Given: $\angle 3 \cong \angle 2$
$\angle T \cong \angle N$
$\overline{TC} \cong \overline{NA}$

Prove: $\triangle TEA \cong \triangle NEC$

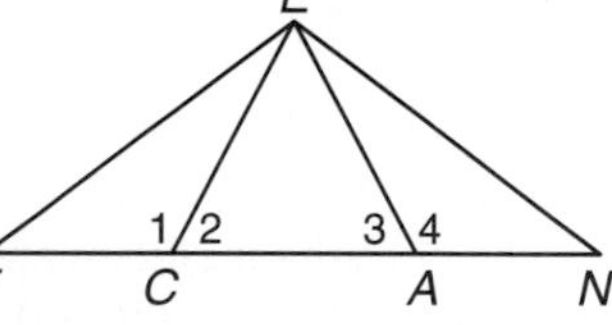

Proof:

Statements	Reasons
1. $\angle 3 \cong \angle 2$, $\angle T \cong \angle N$, $\overline{TC} \cong \overline{NA}$	1. Given
2. $TC = NA$	2. Definition of $\cong$ segments
3. $TA = TC + CA$, $NC = NA + AC$	3. Segment Addition Postulate
4. $TA = NA + AC$	4. Substitution Property (=)
5. $TA = NC$	5. Substitution Property (=)
6. $\overline{TA} \cong \overline{NC}$	6. Definition of $\cong$ segments
7. $\triangle TEA \cong \triangle NEC$	7. ASA

33. Given: $\overline{FP} \parallel \overline{ML}$
$\overline{FL} \parallel \overline{MP}$

Prove: $\overline{PM} \cong \overline{LF}$

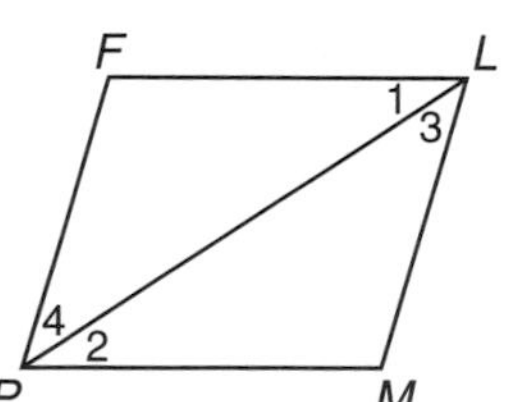

Proof:

Statements	Reasons
1. $\overline{FP} \parallel \overline{ML}$	1. Given
2. $\angle 3 \cong \angle 4$	2. Alternate Interior Angle Theorem
3. $\overline{FL} \parallel \overline{MP}$	3. Given
4. $\angle 1 \cong \angle 2$	4. Alternate Interior Angle Theorem
5. $\overline{PL} \cong \overline{LP}$	5. Congruence of segments is reflexive.
6. $\triangle FLP \cong \triangle MPL$	6. ASA
7. $\overline{PM} \cong \overline{LF}$	7. CPCTC

35. Given: $\triangle GVR$ is an isosceles triangle with base $\overline{GR}$.
$\triangle TVS$ is an isosceles triangle with base $\overline{TS}$.
$\overline{GV} \cong \overline{TV}$
$\angle 5 \cong \angle 6$

Prove: $\overline{GR} \cong \overline{TS}$

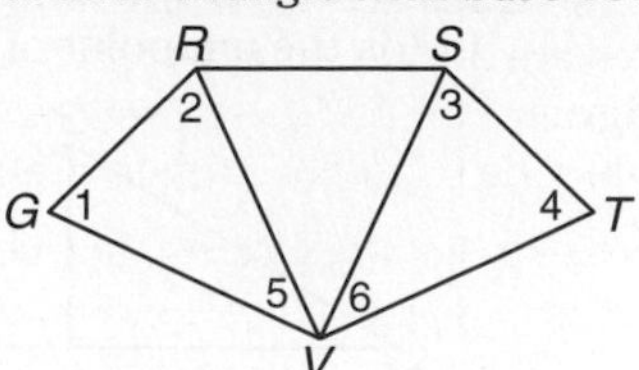

Proof:

Statements	Reasons
1. $\triangle GVR$ is an isosceles triangle with base $\overline{GR}$.	1. Given
2. $\overline{RV} \cong \overline{GV}$	2. Definition of isosceles triangle
3. $\triangle TVS$ is an isosceles triangle with base $\overline{TS}$.	3. Given
4. $\overline{TV} \cong \overline{SV}$	4. Definition of isosceles triangle
5. $\overline{GV} \cong \overline{TV}$	5. Given
6. $\overline{RV} \cong \overline{TV}$	6. Congruence of segments is transitive.
7. $\overline{RV} \cong \overline{SV}$	7. Congruence of segments is transitive.
8. $\angle 5 \cong \angle 6$	8. Given
9. $\triangle GRV \cong \triangle TSV$	9. SAS
10. $\overline{GR} \cong \overline{TS}$	10. CPCTC

37. Given: $\overline{PX} \cong \overline{LT}$
$\triangle PRL$ is an isosceles triangle with base $\overline{PL}$.

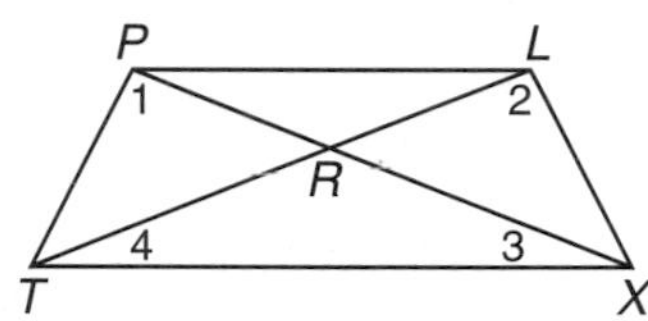

Prove: $\triangle TRX$ is an isosceles triangle.

Proof:

Statements	Reasons
1. $\overline{PX} \cong \overline{LT}$	1. Given
2. $PX = LT$	2. Definition of $\cong$ segments
3. $PX = PR + RX$, $LT = LR + RT$	3. Segment Addition Postulate
4. $PR + RX = LR + RT$	4. Substitution Property (=)
5. $\triangle PRL$ is an isosceles triangle with base $\overline{PL}$	5. Given
6. $\overline{PR} \cong \overline{LR}$	6. Definition of isosceles triangle
7. $PR = LR$	7. Definition of $\cong$ segments
8. $PR + RX = PR + RT$	8. Substitution Property (=)
9. $RX = RT$	9. Subtraction Property (=)
10. $\overline{RX} \cong \overline{RT}$	10. Definition of $\cong$ segments
11. $\triangle TRX$ is an isosceles triangle.	11. Definition of isosceles triangle

39. No; two equiangular triangles are an example of AAA, but the sides are not necessarily congruent. **41.** The guy wires are all the same length because the triangles are congruent by AAS. **43.** Neither player has a greater angle. The triangles formed are congruent by SSS and therefore the angles are congruent. **45.** no **47.** $-\frac{1}{9}$ **49.** a straight angle **51.** inverse; 18

Pages 224–228 Lesson 4–6

5. $\angle 5 \cong \angle 11$ **7.** $\overline{MT} \cong \overline{MR}$ **9.** 56

11a. $AB = \sqrt{(5-2)^2 + (2-5)^2} = \sqrt{18} = 3\sqrt{2}$

$BC = \sqrt{(2-5)^2 + (-1-2)^2} = \sqrt{18} = 3\sqrt{2}$

$AC = \sqrt{(2-2)^2 + (-1-5)^2} = \sqrt{36} = 6$

$\overline{AB} \cong \overline{BC}$

11b. $\angle A \cong \angle C$

13. Given: $\triangle ABC$
$\angle A \cong \angle C$
Prove: $\overline{AB} \cong \overline{CB}$

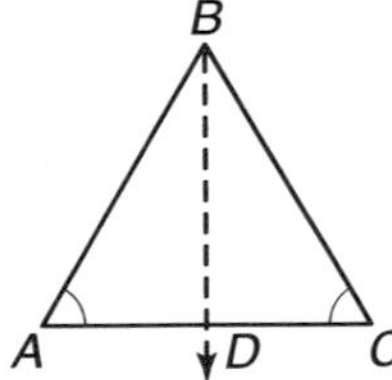

Proof:

Statements	Reasons
1. Let $\overrightarrow{BD}$ bisect $\angle ABC$.	1. Protractor Postulate
2. $\angle ABD \cong \angle CBD$	2. Definition of angle bisector
3. $\angle A \cong \angle C$	3. Given
4. $\overline{BD} \cong \overline{BD}$	4. Congruence of segments is reflexive.
5. $\triangle ABD \cong \triangle CBD$	5. AAS
6. $\overline{AB} \cong \overline{CB}$	6. CPCTC

15. $\angle 10 \cong \angle 6$ **17.** $\angle 11 \cong \angle 9$ **19.** $\angle OEF \cong \angle OFE$
21. $\overline{EF} \cong \overline{EO}$ **23.** $\overline{FB} \cong \overline{FA}$

25. Sample answer:

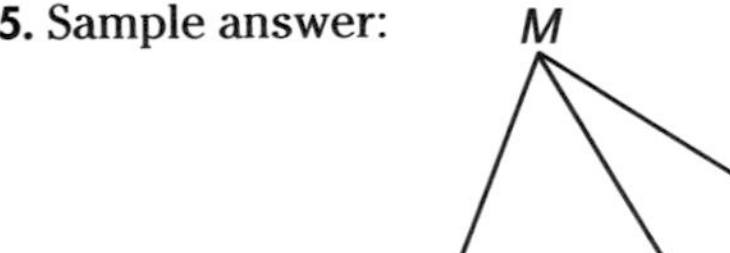

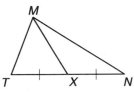

27. 60 **29.** 45 **31.** 18

33. $DE = \sqrt{(-2-(-4))^2 + (-1-(-3))^2} = \sqrt{8} = 2\sqrt{2}$

$EF = \sqrt{(0-(-2))^2 + (-3-(-1))^2} = \sqrt{8} = 2\sqrt{2}$

$DF = \sqrt{(0-(-4))^2 + (-3-(-3))^2} = \sqrt{16} = 4$

$\overline{DE} \cong \overline{EF}$, so the triangle is isosceles.

slope of $\overline{DE} = \frac{-1-(-3)}{-2-(-4)} = \frac{2}{2} = 1$

slope of $\overline{EF} = \frac{-3-(-1)}{0-(-2)} = \frac{-2}{2} = -1$

Since $1(-1) = -1$, $\overline{DE} \perp \overline{EF}$ and $\triangle DEF$ is right isosceles.

35. $m\angle 2 = 120$, $m\angle 3 = 30$, $m\angle 4 = 60$, $m\angle 5 = 60$, $m\angle 6 = 60$, $m\angle 7 = 120$, $m\angle 8 = 90$

37. Given: $\angle 5 \cong \angle 6$
$\overline{FR} \cong \overline{GS}$
Prove: $\angle 4 \cong \angle 3$

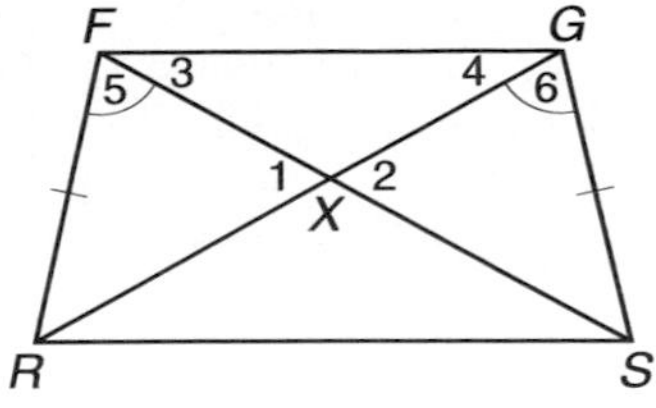

Proof:

Statements	Reasons
1. $\angle 5 \cong \angle 6$, $\overline{FR} \cong \overline{GS}$	1. Given
2. $\angle 1 \cong \angle 2$	2. Vertical ∠s are ≅.
3. $\triangle FXR \cong \triangle GXS$	3. AAS
4. $\overline{FX} \cong \overline{GX}$	4. CPCTC
5. $\angle 4 \cong \angle 3$	5. Isosceles Triangle Theorem

39. Given: $\triangle CAN$ is an isosceles triangle with vertex $\angle N$.
$\overline{CA} \parallel \overline{BE}$
Prove: $\triangle NEB$ is an isosceles triangle.

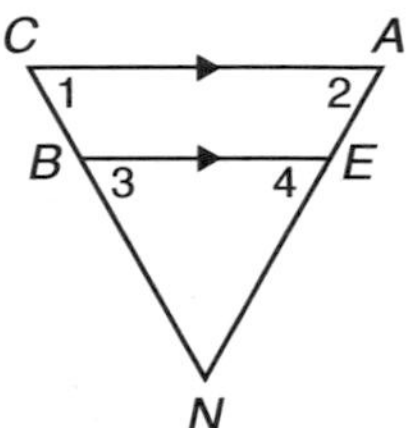

Proof:

Statements	Reasons
1. $\triangle CAN$ is an isosceles triangle with vertex $\angle N$.	1. Given
2. $\overline{NC} \cong \overline{NA}$	2. Definition of isosceles triangle
3. $\angle 2 \cong \angle 1$	3. Isosceles Triangle Theorem
4. $\overline{CA} \parallel \overline{BE}$	4. Given
5. $\angle 1 \cong \angle 3$, $\angle 4 \cong \angle 2$	5. Corresponding Angles Postulate
6. $\angle 2 \cong \angle 3$	6. Congruence of angles is transitive.
7. $\angle 4 \cong \angle 3$	7. Congruence of angles is transitive.
8. $\overline{BN} \cong \overline{NE}$	8. If 2 ∠s of a △ are ≅, then the sides opposite those ∠s are ≅.
9. $\triangle NEB$ is an isosceles triangle.	9. Definition of isosceles triangle

41. Given: $\triangle IOE$ is an isosceles triangle with base $\overline{OE}$.
$\overline{AO}$ bisects $\angle IOE$.
$\overline{AE}$ bisects $\angle IEO$.
Prove: $\triangle EAO$ is an isosceles triangle.

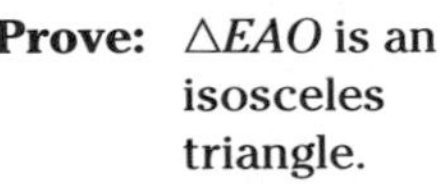

Proof:

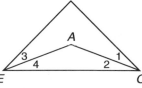

Statements	Reasons
1. $\overline{AO}$ bisects $\angle IOE$. $\overline{AE}$ bisects $\angle IEO$.	1. Given
2. $\angle 1 \cong \angle 2$, $\angle 3 \cong \angle 4$	2. Definition of angle bisector
3. $m\angle 1 = m\angle 2$, $m\angle 3 = m\angle 4$	3. Definition of congruent angles

Statements	Reasons
4. $m\angle IOE = m\angle 1 + m\angle 2$ $m\angle IEO = m\angle 3 + m\angle 4$	**4.** Angle Addition Postulate
5. $m\angle IOE = 2m\angle 2$ $m\angle IEO = 2m\angle 4$	**5.** Substitution Property (=)
6. $\triangle IOE$ is an isosceles triangle with base $\overline{OE}$.	**6.** Given
7. $\overline{IE} \cong \overline{IO}$	**7.** Definition of isosceles triangle
8. $\angle IOE \cong \angle IEO$	**8.** Isosceles Triangle Theorem
9. $m\angle IOE = m\angle IEO$	**9.** Definition of congruent angles
10. $2m\angle 2 = 2m\angle 4$	**10.** Substitution Property (=)
11. $m\angle 2 = m\angle 4$	**11.** Division Property (=)
12. $\angle 2 \cong \angle 4$	**12.** Definition of $\cong$ angles
13. $\overline{AE} \cong \overline{AO}$	**13.** If 2 ∠s of a △ are ≅, then the sides opposite those ∠s are ≅.
14. $\triangle AEO$ is an isosceles triangle.	**14.** Definition of isosceles triangle

43. Given: $\triangle MNO$ is an equilateral triangle.

Prove: $m\angle M = m\angle N = m\angle O = 60$

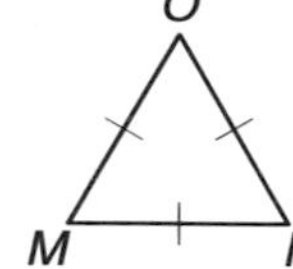

Proof:

Statements	Reasons
1. $\triangle MNO$ is an equilateral triangle.	**1.** Given
2. $\overline{MN} \cong \overline{MO} \cong \overline{NO}$	**2.** Definition of equilateral triangle
3. $\angle M \cong \angle N \cong \angle O$	**3.** Isosceles Triangle Theorem
4. $m\angle M = m\angle N = m\angle O$	**4.** Definition of $\cong$ angles
5. $m\angle M + m\angle N + m\angle O = 180$	**5.** Angle Sum Theorem
6. $3m\angle M = 180$	**6.** Substitution Property (=)
7. $m\angle M = 60$	**7.** Division Property (=)
8. $m\angle M = m\angle N = m\angle O = 60$	**8.** Substitution Property (=)

45.

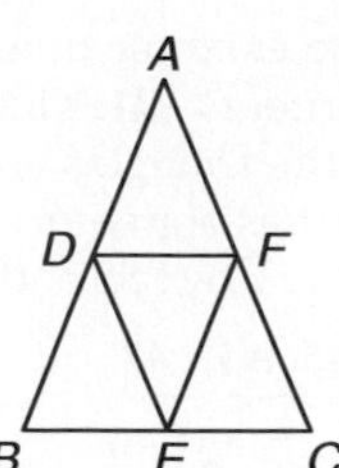

45a. $\triangle DEB \cong \triangle FEC$; Since E is the midpoint of $\overline{BC}$, $\overline{BE} \cong \overline{CE}$. Since $\overline{AC} \cong \overline{AB}$, $\angle B \cong \angle C$. Since D is the midpoint of $\overline{AB}$, F is the midpoint of $\overline{AC}$, and $\overline{AB} \cong \overline{AC}$; $\overline{AD} \cong \overline{AF}$. Therefore, $\triangle DEB \cong \triangle FEC$ by SAS. **45b.** $\triangle ABC$, $\triangle ADF$, $\triangle DEF$; We are given $\triangle ABC$ is an isosceles triangle. Since D is the midpoint of $\overline{AB}$, F is the midpoint of $\overline{AC}$, and $\overline{AB} \cong \overline{AC}$; $\overline{AD} \cong \overline{AF}$ and $\triangle ADF$ is an isosceles triangle. Since $\triangle DEB \cong \triangle FEC$, $\overline{DE} \cong \overline{FE}$ and $\triangle DEF$ is an isosceles triangle. **47.** point M; If the surface is level then it is horizontal. The plumb line will be vertical so it is perpendicular to the surface. A perpendicular dropped from the vertex angle of an isosceles triangle will pass through the midpoint of the base. **49.** SSA means 2 sides and a nonincluded angle are congruent to the corresponding sides and angle. It cannot be used as a proof for congruent triangles. SAS means 2 sides and the included angle are congruent to 2 sides and the included angle. It can be used as a proof for congruent triangles. **51.** $\angle C \leftrightarrow \angle P$, $\angle D \leftrightarrow \angle Q$, $\angle E \leftrightarrow \angle R$, $\overline{CD} \leftrightarrow \overline{PQ}$, $\overline{DE} \leftrightarrow \overline{QR}$, $\overline{CE} \leftrightarrow \overline{PR}$ **53.** 12 units **55.** C **57.** $2\sqrt{29} \approx 10.77$; (2, 6) **59.** 2; −9

Page 229 Chapter 4 Highlights

1. true **3.** false; base angles **5.** false; nonincluded **7.** true **9.** false; flow proof **11.** true **13.** false; base **15.** true

Pages 230–232 Chapter 4 Study Guide and Assessment

17. $\triangle SUV$, $\triangle STU$ **19.** $\triangle SWT$, $\triangle VWU$ **21.** $\overline{SV}$ **23.** 90 **25.** 65 **27.** 40 **29.** 55 **31.** 25 **33.** 55 **35.** 65 **37.** $\overline{RT}$ **39.** $\angle ONM$ **41.** $\angle STR$

43. Given: $\overline{AM} \parallel \overline{CR}$
B is the midpoint of $\overline{AR}$.

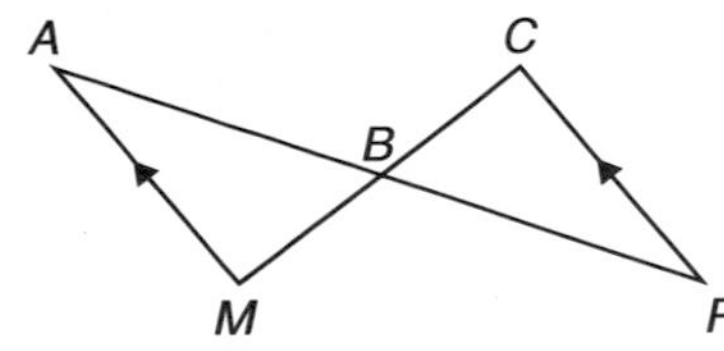

Prove: $\overline{AM} \cong \overline{RC}$

Proof:

Statements	Reasons
1. $\overline{AM} \parallel \overline{CR}$	1. Given
2. $\angle A \cong \angle R$	2. Alternate Interior Angle Theorem
3. B is the midpoint of $\overline{AR}$.	3. Given
4. $\overline{AB} \cong \overline{RB}$	4. Definition of midpoint
5. $\angle ABM \cong \angle RBC$	5. Vertical ∠s are ≅.
6. $\triangle ABM \cong \triangle RBC$	6. ASA
7. $\overline{AM} \cong \overline{RC}$	7. CPCTC

45. Given: $\overline{AC} \cong \overline{EC}$
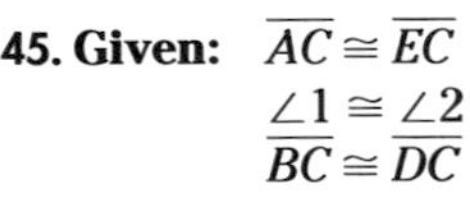
$\angle 1 \cong \angle 2$
$\overline{BC} \cong \overline{DC}$

Prove: $\angle B \cong \angle D$

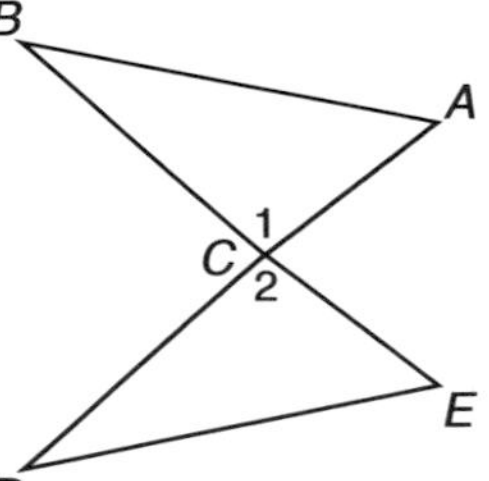

Proof:

Statements	Reasons
1. $\overline{AC} \cong \overline{EC}$ $\angle 1 \cong \angle 2$ $\overline{BC} \cong \overline{DC}$	1. Given
2. $\triangle ABC \cong \triangle EDC$	2. SAS
3. $\angle B \cong \angle D$	3. CPCTC

47. 32 **49.** 36 **51.** 135

CHAPTER 5 APPLYING CONGRUENT TRIANGLES

Pages 242–244 Lesson 5–1

7.

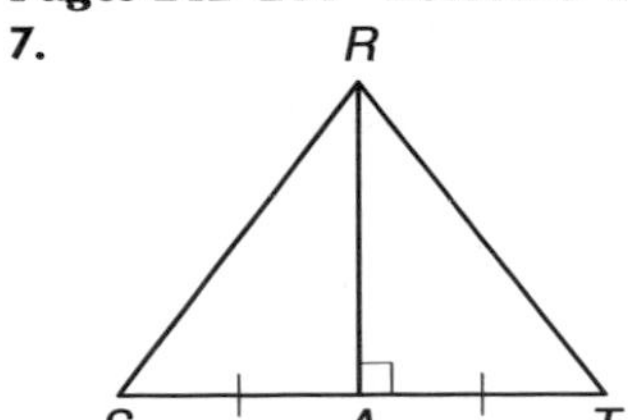

9.

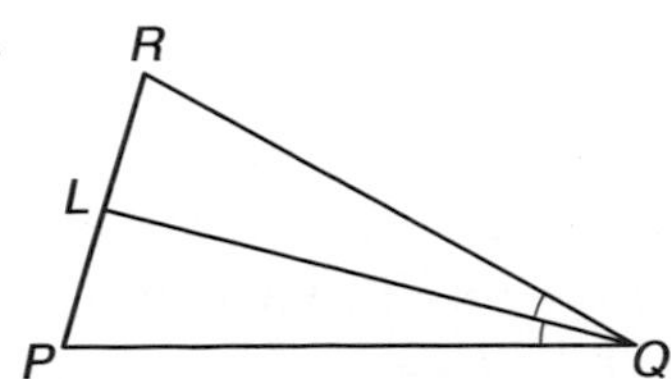

Sample answers given for Exercises 11–13. **11.** N is the midpoint of $\overline{SE}$; $\overline{RN}$ is a median of $\triangle RES$ **13.** $\overline{NR}$ is the angle bisector of $\angle SRE$ **15a.** (7, 5) **15b.** $\frac{7}{5}$ **15c.** No, because $\frac{7}{5} \cdot -3 \neq -1$. **15d.** No, $AT \approx 17.2$ and $AB \approx 21.5$ by using the distance formula. **17.** The flag is located at the intersection of the angle bisector between the west and shore roads and the perpendicular bisector of the segment joining the lookout tower to the entrance.

19.

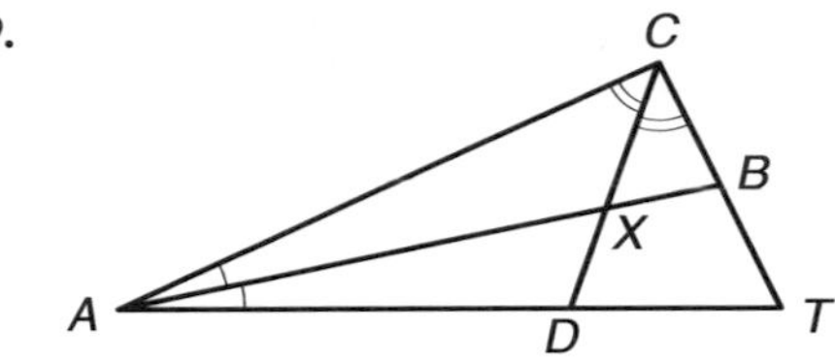

21.

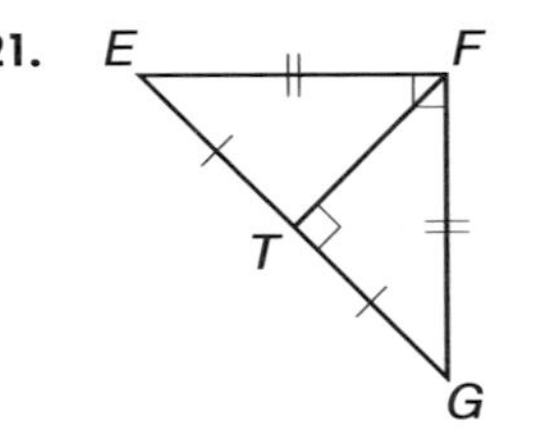

23.

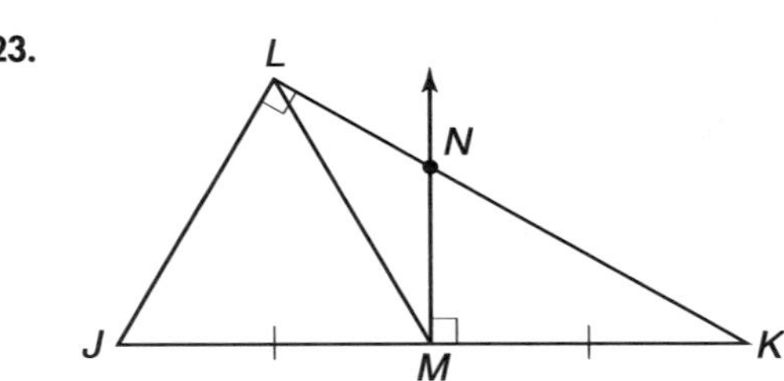

25. $\overline{AE} \cong \overline{EC}$; $\overline{BE} \perp \overline{AC}$; $\overline{AB} \cong \overline{BC}$ **27.** $\angle CAD \cong \angle DAB$ **29.** In a right triangle the altitudes intersect at the vertex of the right angle. **31.** an obtuse triangle **33a.** 40 **33b.** 35 **33c.** 18

35. Given: $\overline{UW}$ is the perpendicular bisector of $\overline{XZ}$.

Prove: For any point V on $\overline{UW}$, $VX = VZ$.

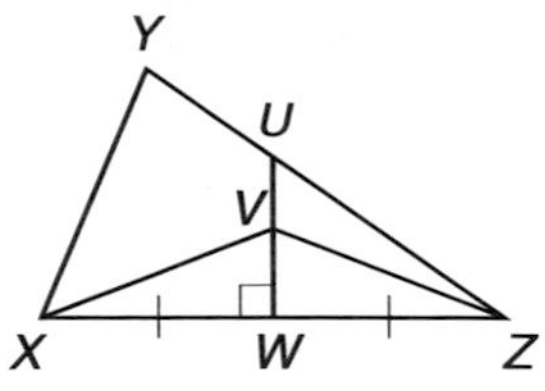

Proof:

Statements	Reasons
1. $\overline{UW}$ is the ⊥ bisector of $\overline{XZ}$.	1. Given
2. W is the midpoint of $\overline{XZ}$.	2. Def. ⊥ bisector
3. $\overline{XW} \cong \overline{WZ}$	3. Def. midpoint
4. $\overline{UW} \perp \overline{XZ}$	4. Def. ⊥ bisector
5. $\angle XWV$, $\angle ZWV$ are right ∠s.	5. ⊥ lines form four rt. ∠s.
6. $\angle XWV \cong \angle ZWV$	6. All rt. ∠s are ≅.
7. $\overline{VW} \cong \overline{VW}$	7. Congruence of segments is reflexive.
8. $\triangle XWV \cong \triangle ZWV$	8. SAS
9. $\overline{VX} \cong \overline{VZ}$	9. CPCTC
10. $VX = VZ$	10. Def. ≅ segments

37. Given: $\overline{BD}$ bisects $\angle ABC$.

Prove: $DE = DF$

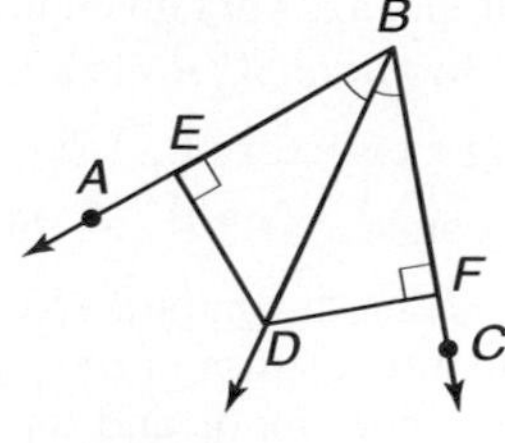

Proof:

Statements	Reasons
1. $\overline{BD}$ bisects $\angle ABC$.	1. Given
2. $\angle ABD \cong \angle CBD$	2. Def. $\angle$ bisector
3. Let DE = distance from D to $\overline{AB}$, and DF = distance from D to $\overline{BC}$.	3. Def. distance from a point to a line
4. $\overline{DE} \perp \overline{AB}$, $\overline{DF} \perp \overline{BC}$	4. Def. distance from a point to a line
5. $\angle DEB$, $\angle DFB$ are rt. $\angle$s.	5. $\perp$ lines form four rt. $\angle$s.
6. $\angle DEB \cong \angle DFB$	6. All rt. $\angle$s are $\cong$.
7. $\overline{BD} \cong \overline{BD}$	7. Congruence of segments is reflexive.
8. $\triangle DEB \cong \triangle DFB$	8. AAS
9. $\overline{DE} \cong \overline{DF}$	9. CPCTC
10. $DE = DF$	10. Def. $\cong$ segments

39. Given: $\overline{LT}$ is a median. $\triangle RLS$ is isosceles with base $\overline{RS}$.

Prove: $\overline{LT}$ bisects $\angle SLR$.

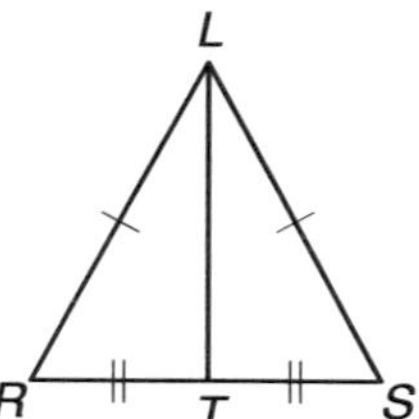

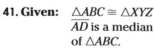

Proof:

Statements	Reasons
1. $\overline{LT}$ is a median. $\triangle RLS$ is isosceles with base $\overline{RS}$.	1. Given
2. $\overline{RT} \cong \overline{TS}$	2. Def. median
3. $\overline{RL} \cong \overline{LS}$	3. Def. isosceles $\triangle$
4. $\overline{LT} \cong \overline{LT}$	4. Congruence of segments is reflexive.
5. $\triangle RLT \cong \triangle SLT$	5. SSS
6. $\angle TLR \cong \angle TLS$	6. CPCTC
7. $\overline{LT}$ bisects $\angle SLR$.	7. Def. $\angle$ bisector

41. Given: $\triangle ABC \cong \triangle XYZ$
$\overline{AD}$ is a median of $\triangle ABC$.
$\overline{XW}$ is a median of $\triangle XYZ$.

Prove: $\overline{AD} \cong \overline{XW}$

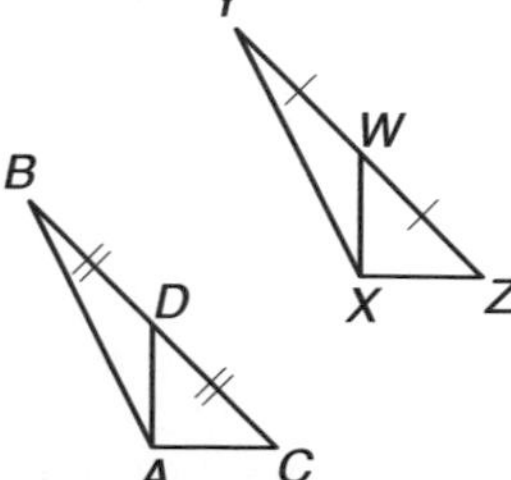

Proof:

Statements	Reasons
1. $\triangle ABC \cong \triangle XYZ$ $\overline{AD}$ is a median of $\triangle ABC$. $\overline{XW}$ is a median of $\triangle XYZ$.	1. Given
2. $\overline{AB} \cong \overline{XY}$, $\angle B \cong \angle Y$, $\overline{BC} \cong \overline{YZ}$	2. CPCTC
3. $BC = YZ$	3. Def. $\cong$ segments
4. D is the midpoint of $\overline{BC}$. W is the midpoint of $\overline{YZ}$.	4. Def. median
5. $\overline{BD} \cong \overline{DC}$, $\overline{YW} \cong \overline{WZ}$	5. Def. midpoint
6. $BD = DC$, $YW = WZ$	6. Def. $\cong$ segments
7. $BC = BD + DC$, $YZ = YW + WZ$	7. Seg. Addition Post.
8. $BC = BD + BD$, $YZ = YW + YW$	8. Substitution Prop. (=)
9. $2BD = 2YW$	9. Substitution Prop. (=)
10. $BD = YW$	10. Division Prop. (=)
11. $\overline{BD} \cong \overline{YW}$	11. Def. $\cong$ segments
12. $\triangle ABD \cong \triangle XYW$	12. SAS
13. $\overline{AD} \cong \overline{XW}$	13. CPCTC

43. Acute triangle: all segments meet inside the triangle; right triangle: altitudes meet at the vertex of the right angle, other segments meet inside the triangle; obtuse triangle: altitudes meet outside the triangle, rest meet inside **45a.** 4 **45b.** 8

45c.

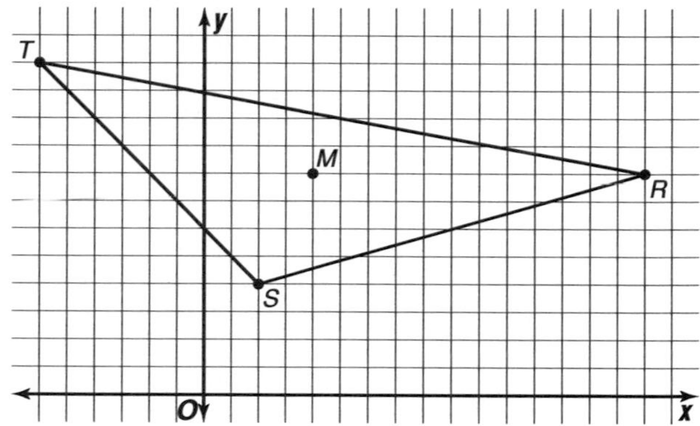

45d. The medians intersect at M. **47.** $BA = 5$, $AY = 8$, $BY = 5$ **49.** yes; both have slope $-\frac{2}{3}$

51. Given: $\overline{PQ}$ bisects $\overline{AB}$ at point M.

Prove: $\overline{AM} \cong \overline{MB}$

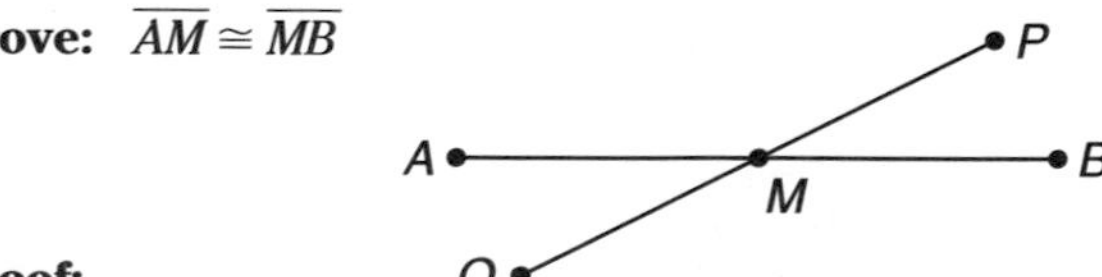

Proof:

Statements	Reasons
1. $\overline{PQ}$ bisects $\overline{AB}$ at point M.	1. Given
2. M is the midpoint of $\overline{AB}$.	2. Def. bisector
3. $AM = MB$	3. Def. midpoint
4. $\overline{AM} \cong \overline{MB}$	4. Def. $\cong$ segments

53. S; $RS + ST = RT$ **55.** 45 meters

Pages 248–51 Lesson 5–2

7. $\overline{ST} \cong \overline{TU}$ **9.** 6 **11.** 9 **13.** yes, LL **15.** $\overline{JT} \cong \overline{MR}$ and $\angle J \cong \angle M$ or $\overline{JT} \cong \overline{MR}$ and $\angle T \cong \angle R$ **17.** $\overline{MN} \cong \overline{OP}$ **19.** $\overline{VX} \cong \overline{XY}$ or $\overline{WX} \cong \overline{XZ}$ **21.** 1 **23.** 3

25. Given: $\angle M$ and $\angle P$ are right angles.
$\overline{MN} \parallel \overline{OP}$
Prove: $\overline{MN} \cong \overline{OP}$

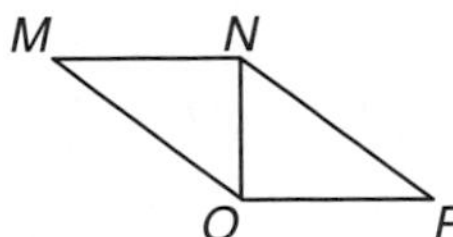

Proof:

Statements	Reasons
1. $\angle M$ and $\angle P$ are right angles. $\overline{MN} \parallel \overline{OP}$	1. Given
2. $\triangle MNO$ and $\triangle PON$ are rt. $\triangle$s.	2. Def. rt. $\triangle$
3. $\angle MNO \cong \angle PON$	3. If ∥, alt. int. ∠s are ≅.
4. $\overline{NO} \cong \overline{NO}$	4. Congruence of segments is reflexive.
5. $\triangle MNO \cong \triangle PON$	5. HA
6. $\overline{MN} \cong \overline{OP}$	6. CPCTC

27. Given: $\angle PBC \cong \angle PCB$
$\overline{AP} \perp$ plane $\mathcal{M}$
Prove: $\angle ABC \cong \angle ACB$

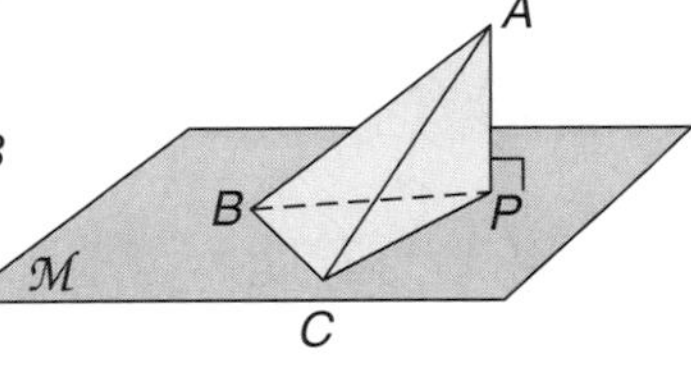

Proof:

Statements	Reasons
1. $\angle PBC \cong \angle PCB$ $\overline{AP} \perp$ plane $\mathcal{M}$	1. Given
2. $\angle APB$ and $\angle APC$ are rt. ∠s.	2. $\perp$ lines form four rt. ∠s.
3. $\triangle APB$ and $\triangle APC$ are rt. $\triangle$s.	3. Def. rt. $\triangle$
4. $\overline{BP} \cong \overline{CP}$	4. Isosceles Triangle Th.
5. $\overline{AP} \cong \overline{AP}$	5. Congruence of segments is reflexive.
6. $\triangle APC \cong \triangle APB$	6. LL
7. $\overline{AC} \cong \overline{AB}$	7. CPCTC
8. $\angle ABC \cong \angle ACB$	8. Isosceles Triangle Th.

29. Case 1:

Given: $\triangle ABC$ and $\triangle DEF$ are right triangles.
$\overline{AC} \cong \overline{DF}$
$\angle C \cong \angle F$

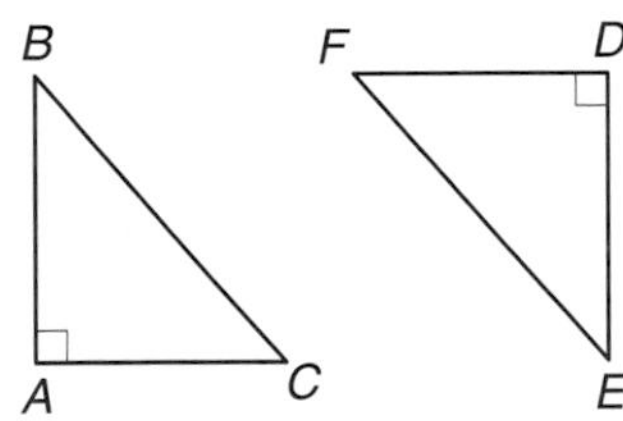

Prove: $\triangle ABC \cong \triangle DEF$

Proof:
It is given that $\triangle ABC$ and $\triangle DEF$ are right triangles, $\overline{AC} \cong \overline{DF}$, and $\angle C \cong \angle F$. By the definition of right triangle, $\angle A$ and $\angle D$ are right angles. Thus, $\angle A \cong \angle D$ since all right angles are congruent. Therefore, $\triangle ABC \cong \triangle DEF$ by Angle-Side-Angle.

Case 2:

Given: $\triangle ABC \cong \triangle DEF$ are right triangles.
$\overline{AC} \cong \overline{DF}$
$\angle B \cong \angle E$

Prove: $\triangle ABC \cong \triangle DEF$

Proof:
It is given that $\triangle ABC$ and $\triangle DEF$ are right triangles, $\overline{AC} \cong \overline{DF}$, and $\angle B \cong \angle E$. By the definition of right triangle, $\angle A$ and $\angle D$ are right angles. Thus, $\angle A \cong \angle D$ since all right angles are congruent. Therefore, $\triangle ABC$ and $\triangle DEF$ by Angle-Angle-Side.

31. Given: $\triangle LMN$ is isosceles with base $\overline{LN}$.
O is the midpoint of $\overline{LM}$.
P is the midpoint of $\overline{NM}$.
$\overline{OQ} \perp \overline{LN}$,
$\overline{PR} \perp \overline{LN}$

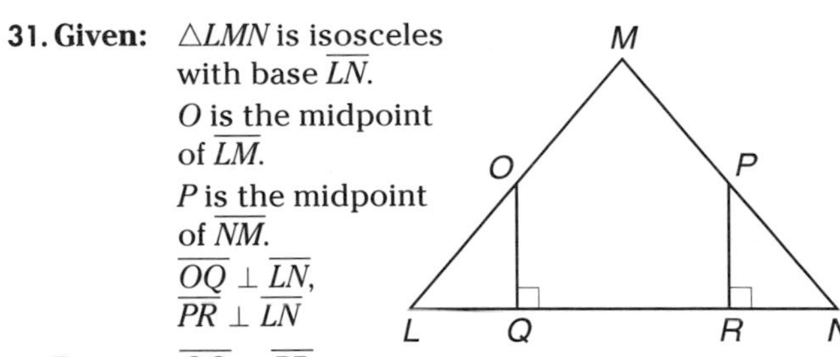

Prove: $\overline{OQ} \cong \overline{PR}$

Proof:
It is given that $\triangle LMN$ is isosceles. By the definition of isosceles triangle, $\overline{LM} \cong \overline{NM}$. Thus, $LM = NM$ by the definition of congruent segments. $\angle L \cong \angle N$ because if two sides of a triangle are congruent, then the angles opposite the sides are congruent also. We were given that O is the midpoint of $\overline{LM}$ and P is the midpoint of $\overline{NM}$. The definition of midpoint lets us say that $\overline{LO} \cong \overline{OM}$ and $\overline{NP} \cong \overline{PM}$. Thus by the definition of congruent segments, $LO = OM$ and $NP = PM$. By the segment addition postulate, $LM = LO + OM$ and $NM = NP + PM$. Thus, $LO + OM = NP + PM$ by substitution. Then $2LO = 2NP$ by substitution. By the division property of equality, $LO = NP$. Then $\overline{LO} \cong \overline{NP}$ by the definition of congruent segments. It was given that $\overline{OQ} \perp \overline{LN}$ and $\overline{PR} \perp \overline{LN}$. Then $\angle OQL$ and $\angle PRN$ are right angles since perpendicular lines form four right angles. Then $\triangle OQL$ and $\triangle PRN$ are right triangles by the definition of right triangle. By Hypotenuse-Angle, $\triangle OQL \cong \triangle PRN$. Therefore, $\overline{OQ} \cong \overline{PR}$ by CPCTC.

33. The corresponding sides are proportional, but not necessarily congruent. **35.** legs: 48 in. and 60 in.; hypotenuse and an angle: 76.8 in. and 51° or 76.8 in. and 39°; leg and an angle: 60 in. and 51°, or 60 in. and 39°, 48 in. and 51°, or 48 in. and 39°; hypotenuse and a leg: 76.8 in. and 60 in. or 78.6 in. and 48 in.

37.

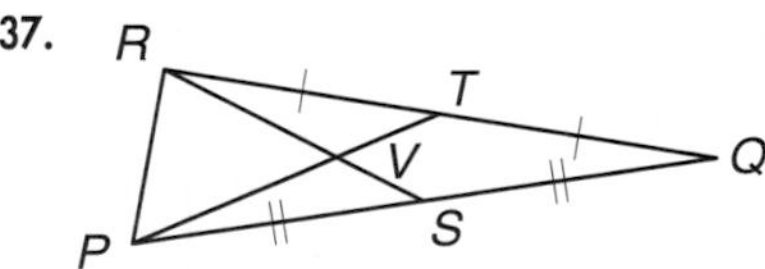

39. obtuse, isosceles

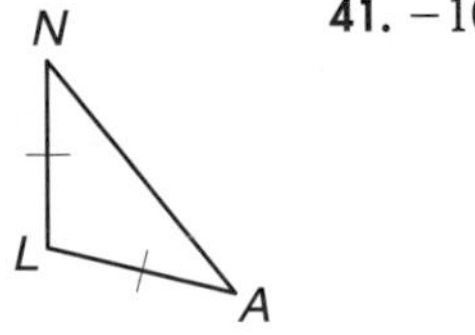

41. -10

43. converse: interchange the hypothesis and conclusion; **inverse:** negate the hypothesis and the conclusion; **contrapositive:** negate and interchange the hypothesis and conclusion **45.** {(1938, 50,000), (1939, 40,000), (1938, 5200), (1938, 4200), (1939, 4020)}; D = {1938, 1939}; R = {50,000, 40,000, 5200, 4200, 4020}

Pages 255–258 Lesson 5-3

5. Lines ℓ and m do not intersect at point X. **7.** Sabrina did not eat the leftover pizza. **9.** $\angle 1, \angle 4, \angle 8$ **11.** Division Property of Inequality **13.** Transitive Property of Inequality

15. Given: m is not parallel to n.

Prove: $m\angle 3 \neq m\angle 2$

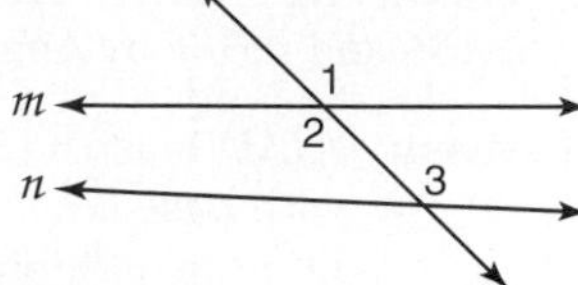

Proof:

Step 1: Assume $m\angle 3 = m\angle 2$.

Step 2: $\angle 3$ and $\angle 2$ are alternate interior angles. If 2 lines are cut by a transversal so that alternate interior angles are congruent, then the lines are parallel. This means that $m \parallel n$. However, that contradicts the given statement.

Step 3: Therefore, since the assumption leads to a contradiction, the assumption must be false. Thus, $m\angle 3 \neq m\angle 2$.

17. $\overline{AB}$ is not congruent to $\overline{CD}$. **19.** The disk is not defective. **21.** If two altitudes of a triangle are congruent, then the triangle is not isosceles. **23.** $m\angle 7$ **25.** Sample answer: $\angle 1$ **27.** $\angle 2, \angle 3, \angle 4, \angle 5, \angle 6$ **29.** Division Property of Inequality **31.** Comparison Property of Inequality **33.** Division Property of Inequality

35. Given: $\triangle XYV, \triangle XZV$

Prove: $m\angle 7 > m\angle 6$

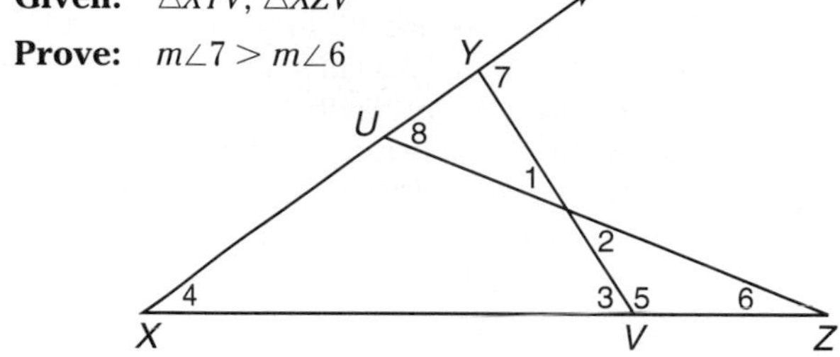

Proof:
Since if an angle is an exterior angle of a triangle its measure is greater than the measure of either corresponding interior angle, we can say that $m\angle 7 > m\angle 8$ and $m\angle 8 > m\angle 6$. Thus, by the Transitive Property of Inequality, $m\angle 7 > m\angle 6$.

37. Given: $\overline{FD} \perp \overline{AB}$, $\overline{GE} \perp \overline{AB}$

Prove: $m\angle 5 > m\angle 2$

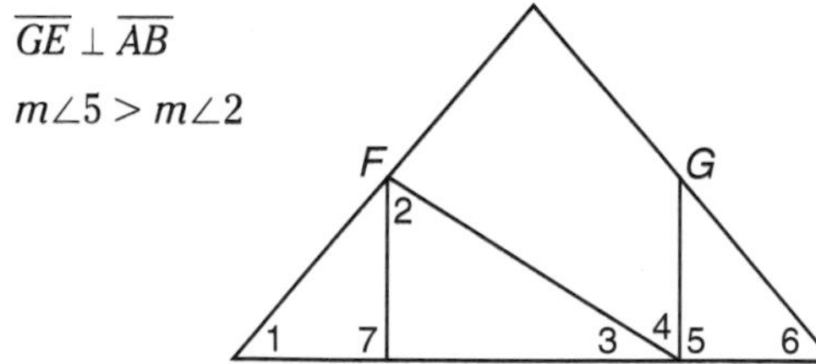

Proof:

Statements	Reasons
1. $\overline{FD} \perp \overline{AB}$, $\overline{GE} \perp \overline{AB}$	1. Given
2. $\angle 7$ and $\angle 5$ are rt. ∠s.	2. $\perp$ lines form four rt. ∠s.
3. $m\angle 7 > m\angle 2$	3. Exterior Angle Inequality Th.
4. $\angle 7 \cong \angle 5$	4. All rt. ∠s are $\cong$.
5. $m\angle 7 = m\angle 5$	5. Def. $\cong$ ∠s
6. $m\angle 5 > m\angle 2$	6. Substitution Prop. of Inequality

39. Given: $\overline{MO} \cong \overline{ON}$, $\overline{MP}$ is not congruent to $\overline{NP}$.

Prove: $\angle MOP$ is not congruent to $\angle NOP$.

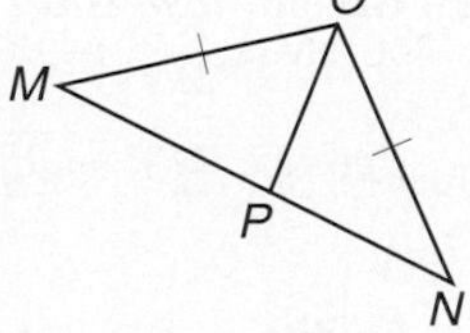

Proof:
Assume that $\angle MOP \cong \angle NOP$. We are given that $\overline{MO} \cong \overline{ON}$. If two sides of a triangle are congruent, then the angles opposite the sides are congruent also. Thus, $\angle M \cong \angle N$. Therefore, $\triangle MOP \cong \triangle NOP$ by ASA. We can then conclude that $\overline{MP} \cong \overline{NP}$. But this contradicts the given information. Therefore, the assumption is incorrect and $\angle MOP$ is not congruent to $\angle NOP$.

41. Given: $\triangle XYZ$, $m\angle X \neq m\angle Y \neq m\angle Z$

Prove: $XY \neq YZ \neq XZ$

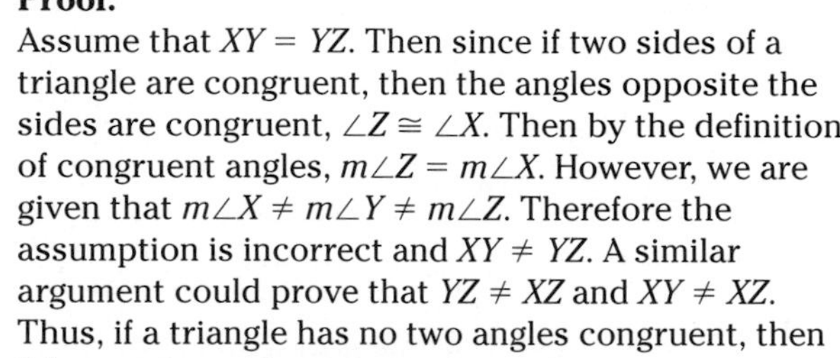

Proof:
Assume that $XY = YZ$. Then since if two sides of a triangle are congruent, then the angles opposite the sides are congruent, $\angle Z \cong \angle X$. Then by the definition of congruent angles, $m\angle Z = m\angle X$. However, we are given that $m\angle X \neq m\angle Y \neq m\angle Z$. Therefore the assumption is incorrect and $XY \neq YZ$. A similar argument could prove that $YZ \neq XZ$ and $XY \neq XZ$. Thus, if a triangle has no two angles congruent, then it has no two sides congruent.

43. The door on the left. If the sign on the door on the right were true, then both signs would be true. But one sign is false, so the sign the on the door on the right must be false. **45.** Yes. If you assume the client was at the scene of the crime, it is contradicted by his presence at the meeting in New York City. Thus, the assumption he was present at the crime is wrong.

47.

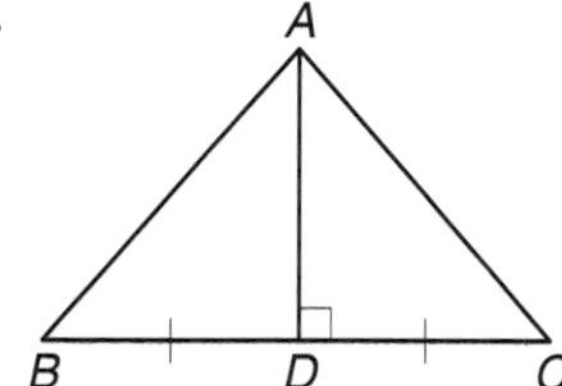

49. obtuse, isosceles
51. parallel **53.** 36 lunch specials **55.** 0.3728

Page 258 Self Test

1. 5 **3.** yes, by LL or SAS

5. Given: $\overleftrightarrow{AB}$ and $\overleftrightarrow{PQ}$ are skew.

Prove: $\overleftrightarrow{AP}$ and $\overleftrightarrow{BQ}$ are skew.

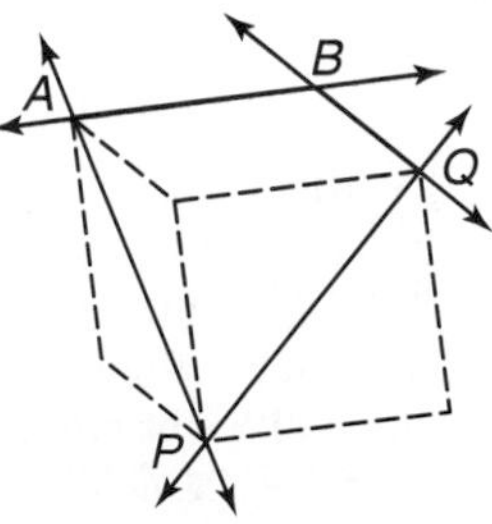

Proof:

Assume that $\overleftrightarrow{AP}$ and $\overleftrightarrow{BQ}$ are not skew. Then either $\overleftrightarrow{AP} \parallel \overleftrightarrow{BQ}$ or $\overleftrightarrow{AP}$ and $\overleftrightarrow{BQ}$ intersect. In either case, A, B, P, and Q are coplanar. It is given that $\overleftrightarrow{AB}$ and $\overleftrightarrow{PQ}$ are skew, so $\overleftrightarrow{AB}$ and $\overleftrightarrow{PQ}$ do not lie in the same plane, and A and B

are not coplanar with P and Q. This is a contradiction. Therefore, $\overleftrightarrow{AP}$ and $\overleftrightarrow{BQ}$ must be skew.

Lesson 262–265 Lesson 5–4

7. yes **9.** $\overline{TU}$ **11.** $\overline{TS}$

13. Given: $\angle B$ is a right angle.
$\overline{BD}$ is an altitude.

Prove: $BD < AB$ and $BD < BC$

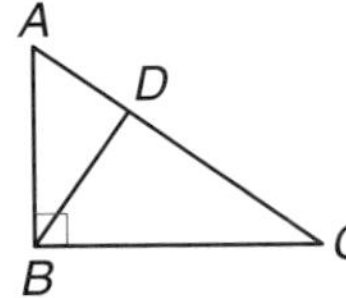

Proof:

We are given that $\overline{BD}$ is an altitude. Therefore by the definition of altitude, $\overline{BD} \perp \overline{AC}$. Then since perpendicular lines form four right angles, $\angle ADB$ and $\angle CDB$ are right angles. Then $\triangle ADB$ and $\triangle CDB$ are right triangles by the definition of right triangle. $\angle BAD$ and $\angle DBA$, and $\angle BCD$ and $\angle DBC$ are complementary since acute angles of right triangles are complementary. By the definition of complementary angles, $m\angle BAD + m\angle DBA = 90$ and $m\angle BCD + m\angle DBC = 90$. The definition of inequality allows us to say that $m\angle BAD < 90$ and $m\angle BCD < 90$. $m\angle ADB = 90$ and $m\angle CDB = 90$ by the definition of a right angle. Thus by substitution, $m\angle BAD < m\angle ADB$ and $m\angle BCD < m\angle CDB$. Therefore in $\triangle BCD$, $BD < AB$; and in $\triangle ABD$, $BD < BC$.

15. $\angle N$ **17.** $\angle T$ **19.** $\angle WYX$ **21.** $\overline{CE}$ **23.** $\overline{BE}$, In $\triangle EDC$, $\overline{EC}$ is longest. In $\triangle BEC$, $\overline{BE}$ is longest. In $\triangle ABE$, $\overline{BE}$ is longest. Thus, $\overline{BE}$ is the longest segment. **25.** No, the segments do not have the correct relative lengths. **27.** $x = 12$; $m\angle A = 104$, $m\angle B = 32$, $m\angle C = 44$; so $AC < AB < BC$.

29. Given: $\overline{RQ}$ bisects $\angle SRT$.

Prove: $m\angle SQR > m\angle SRQ$

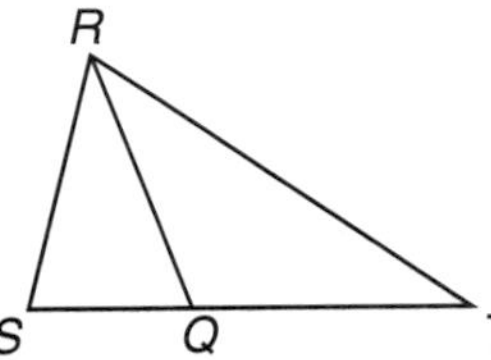

Proof:

Statements	Reasons
1. $\overline{RQ}$ bisects $\angle SRT$.	1. Given
2. $\angle SRQ \cong \angle QRT$	2. Def. $\angle$ bisector
3. $m\angle SRQ = m\angle QRT$	3. Def. $\cong$ $\angle$s
4. $m\angle SQR > m\angle QRT$	4. Exterior Angle Inequality Th.
5. $m\angle SQR > m\angle SRQ$	5. Substitution Prop. (=)

31. Given: D is on $\overline{AC}$.
$\angle BDC$ is acute.

Prove: $BA > BD$

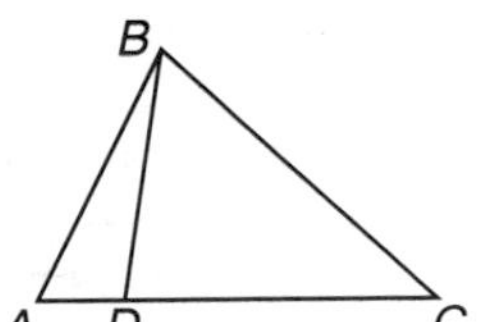

Proof:

Statements	Reasons
1. D is on $\overline{AC}$. $\angle BDC$ is acute.	1. Given
2. $m\angle BDC < 90$	2. Def. acute $\angle$
3. $m\angle BDC + x = 90$	3. Def. inequality
4. $m\angle BDC = 90 - x$	4. Subtraction Prop. (=)
5. $\angle BDC$ and $\angle BDA$ are a linear pair.	5. Def. linear pair
6. $m\angle BDC + m\angle BDA = 180$	6. Supplement Th.
7. $(90 - x) + m\angle BDA = 180$	7. Substitution Prop. (=)
8. $m\angle BDA = 90 + x$	8. Subtraction Prop. (=)
9. $m\angle BDA > 90$	9. Def. inequality
10. $\angle BDA$ is obtuse.	10. Def. obtuse
11. $\angle BAD$ and $\angle ABD$ are acute angles.	11. There can be at most 1 rt. or obtuse $\angle$ in a $\triangle$.
12. $m\angle BAD < 90$	12. Def. acute $\angle$
13. $m\angle BDA > m\angle BAD$	13. Transitive Prop. of Inequality
14. $BA > BD$	14. If an $\angle$ of a $\triangle$ > another $\angle$, the side opp. the greater $\angle$ is longer than the side opp. the lesser $\angle$.

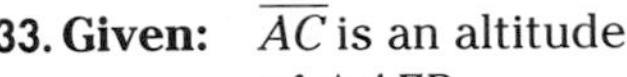

33. Given: $\overline{AC}$ is an altitude of $\triangle AEB$.
$\overline{AF}$ is a median.
$\overline{AE} \not\cong \overline{AB}$

Prove: $AF > AC$

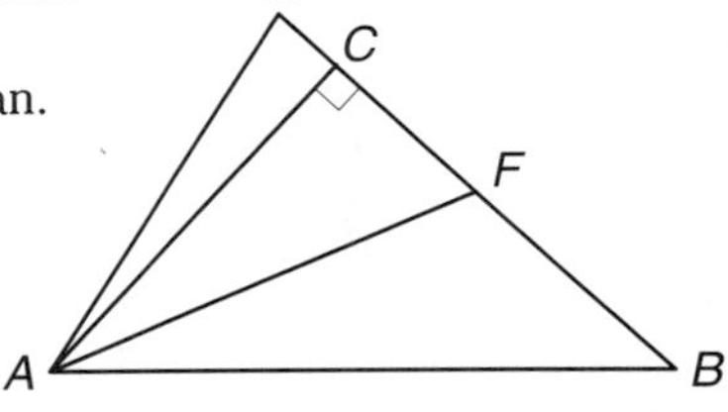

Proof:

We are given that $\overline{AC}$ is an altitude of $\triangle AEB$. $\overline{AC} \perp \overline{EB}$ by the definition of an altitude. The perpendicular segment from a point to a line is the shortest segment from the point to the line, so $\overline{AC}$ is the shortest segment from A to $\overline{EB}$. The only way that $\overline{AF}$ is not longer than $\overline{AC}$ is if they are the same segment. If they are the same segment, then C and F are the same point and $EC = CB$ by the definition of median. In that case, $\angle ACE$ and $\angle ACB$ are right angles since perpendicular lines form four right angles. Then $\triangle ACE$ and $\triangle ACB$ are right triangles by the definition of right triangle. $\overline{AC} \cong \overline{AC}$ because congruence of segments is reflexive. Then $\triangle ACE \cong \triangle ACB$ by HL. Then $\overline{AE} \cong \overline{AB}$ by CPCTC. But this contradicts the given information. Thus, C and F do not correspond and $AF > AC$.

35. Given: $\overline{PQ} \perp$ plane $\mathcal{M}$

Prove: $\overline{PQ}$ is the shortest segment from P to plane $\mathcal{M}$.

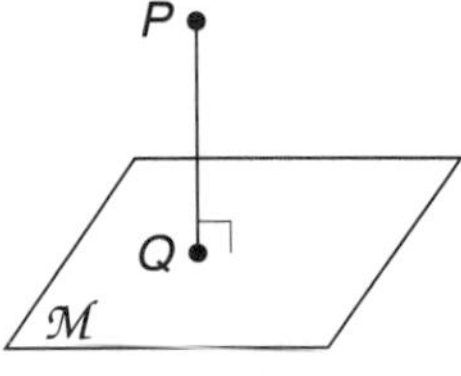

Proof:

By definition, $\overline{PQ}$ is perpendicular to plane $\mathcal{M}$ if it is perpendicular to every line in $\mathcal{M}$ that intersects it. But since the perpendicular segment from a point to a line is the shortest segment from the point to the line, that perpendicular segment is the shortest segment from the point to each of these lines. Therefore, $\overline{PQ}$ is the shortest segment from P to $\mathcal{M}$.

37a.

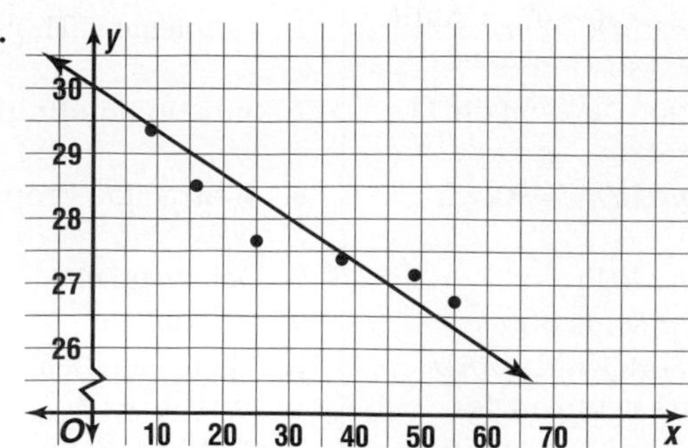

37b. The equation is the best predictor for Henry Rono's record in 1978. It is the worst predictor for Ron Clarke's record in 1965. **39.** Songan **43.** False; the hypotenuse must be longer than the other two sides.

45.

x	2x – 6	(x, y)
0	–6	(0, –6)
2	–2	(2, –2)
9	2	(4, 2)

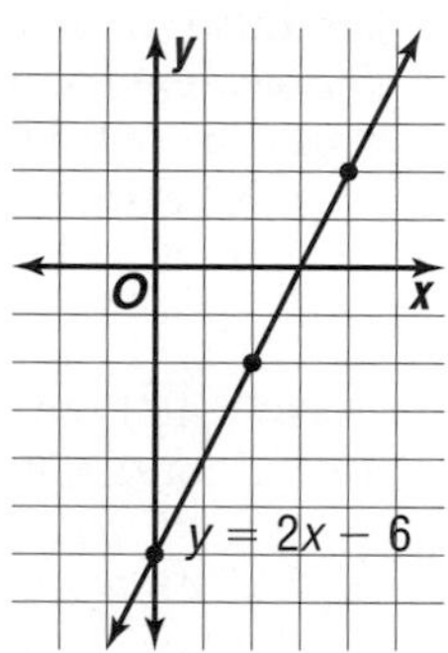

47. 2.3

Page 266 Lesson 5–5A

3. In the column of sets that do not make a triangle, either $S1 + S2 \le S3$, $S2 + S3 \le S1$, or $S1 + S3 \le S2$. In the column of sets that do make a triangle, $S1 + S2 > S3$, $S2 + S3 > S1$, and $S1 + S3 > S2$.

Pages 269–272 Lesson 5–5

5. yes

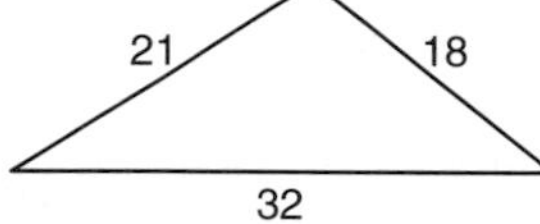

7. 6 and 48 **9.** 0 and 60 **11.** always true; $DB < 7$ and $BC < 2 + DB$ **13.** Yes. The measures of the sides are 5, $\sqrt{51}$ and $\sqrt{90}$ and the Triangle Inequality Theorem applies. **15.** yes

17. yes **19.** yes

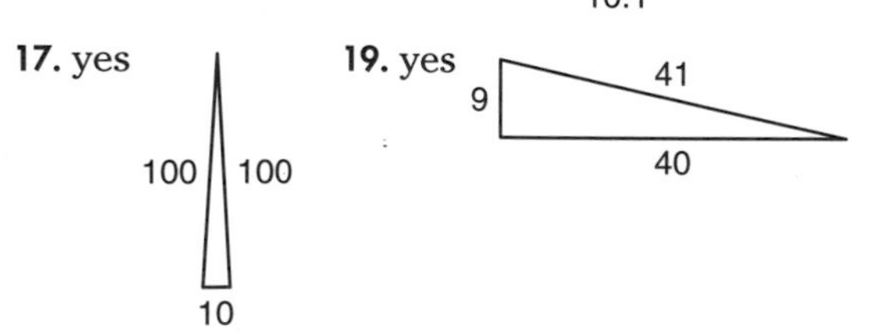

21. no **23.** 3 and 33 **25.** 12 and 56 **27.** 24 and 152 **29.** 24 and 118 **31.** $|a - b|$ and $a + b$ **33.** sometimes true; $BD + DC < BC$ and $BC + DC > BC$, so $10 < BC < 38$. 12 is one of the acceptable values for BC. **35.** always true; $AB + AC > BC$ and $BC + DC > BD$, so $AB + AC > BD - DC$, then $AB + AC > BD$. **37.** never true; $\overline{BA}$ is the perpendicular, so it is the shortest distance from B to $\overline{AC}$. **39.** yes; $AB = \sqrt{245} \approx 15.7$, $AC = \sqrt{244} \approx 15.6$, $BC = \sqrt{685} \approx 26.2$ These measures satisfy the Triangle Inequality Theorem. **41.** yes; $DE = \sqrt{41} \approx 6.4$, $DF = 8$, $EF = 5$ These measures satisfy the Triangle Inequality Theorem.

43. Given: quadrilateral $ABCD$

Prove: $AD + CD + AB > BC$

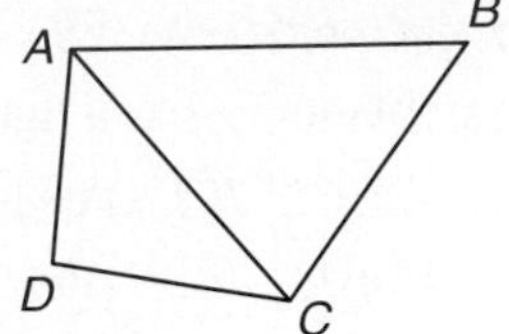

Proof:

Statements	Reasons
1. quadrilateral $ABCD$	1. Given
2. Draw $\overline{AC}$.	2. Through any 2 pts. there is 1 line.
3. $AD + CD > AC$ $AB + AC > BC$	3. Triangle Inequality
4. $AC > BC - AB$	4. Subtraction Prop. of Inequality
5. $AD + CD > BC - AB$	5. Transitive Prop. of Inequality
6. $AD + CD + AB > BC$	6. Addition Prop. of Inequality

45a. no triangle **45b.** Triangle exists. **45c.** no triangle **45d.** Triangle exists. **45e.** Triangle exists. **47a.** a is either 15 cm, or 16 cm; z is 14 cm, 15 cm, or 16 cm. The possible triangles that can be made from sides with those measures are (2 cm, 15 cm, 14 cm), (2 cm, 15 cm, 15 cm), (2 cm, 15 cm, 16 cm), (2 cm, 16 cm, 15 cm), (2 cm, 16cm, 16 cm) **47b.** $\frac{2}{5}$ **49.** $\overline{AC}$ **51.** no triangle **53.** 12 units

55. Given: $\ell \parallel m$, $s \parallel t$, $t \perp \ell$

Prove: $\angle 3$ is a right angle.

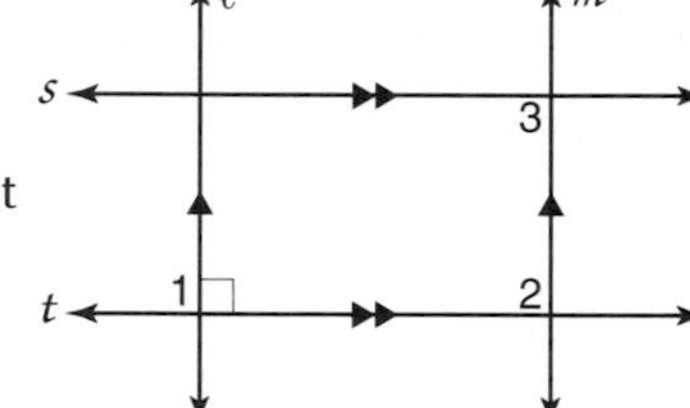

Proof:

Statements	Reasons
1. $\ell \parallel m$, $s \parallel t$, $t \perp \ell$	1. Given
2. $\angle 1$ is a right angle.	2. $\perp$ lines form four rt. $\angle$s.
3. $m\angle 1 = 90$	3. Def. rt. $\angle$
4. $\angle 1 \cong \angle 2$	4. Corresponding Angles Postulate
5. $m\angle 2 = 90$	5. Substitution Prop. (=)
6. $\angle 2$ and $\angle 3$ are supplementary.	6. Consecutive Interior Angle Theorem
7. $m\angle 2 + m\angle 3 = 180$	7. Def. supp.
8. $90 + m\angle 3 = 180$	8. Substitution Prop. (=)
9. $m\angle 3 = 90$	9. Subtraction Prop. (=)
10. $\angle 3$ is a right angle.	10. Def. rt. $\angle$

57. 27; 117 **59a.** the percentage of classical music listeners who are in different age groups **59b.** 880,000 people

Pages 276–279 Lesson 5–6

5. $m\angle DBE < m\angle BFA$ **7.** $m\angle FDB > m\angle BDC$ **9.** never **11.** $2.8 < x$ and $x < 12$ **13.** right arm **15.** $TM < RS$ **17.** $FH > GE$ **19.** $m\angle AOD > m\angle AOB$ **21.** always **23.** always **25.** never **27.** $x < 14$ **29.** $x > 4$

31. Given: $\overline{MN} \cong \overline{QR}$
$\overline{MN} \parallel \overline{QR}$
$m\angle MPQ >$ $m\angle QPR$

Prove: $MQ > QR$

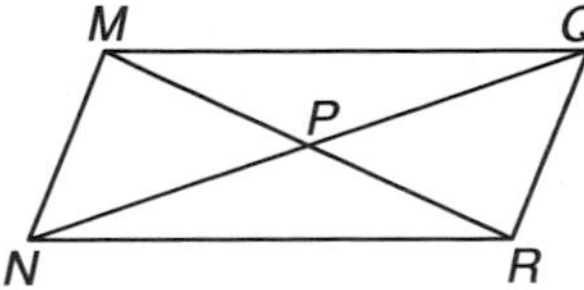

Proof:

Statements	Reasons
1. $\overline{MN} \cong \overline{QR}$, $\overline{MN} \parallel \overline{QR}$, $m\angle MPQ > m\angle QPR$	1. Given
2. $\angle MNQ \cong \angle NQR$	2. If $\parallel$, alt. int. $\angle$s are $\cong$.
3. $\angle MPN \cong \angle QPR$	3. Vertical $\angle$s are $\cong$.
4. $\triangle MPN \cong \triangle RPQ$	4. AAS
5. $\overline{MP} \cong \overline{RP}$	5. CPCTC
6. $\overline{PQ} \cong \overline{PQ}$	6. Congruence of segments is reflexive.
7. $MQ > QR$	7. SAS Inequality

33. Given: $\triangle ABC$ is equilateral.
$\triangle ABD \cong \triangle CBD$
$BD > AD$

Prove: $m\angle BCD >$ $m\angle DAC$

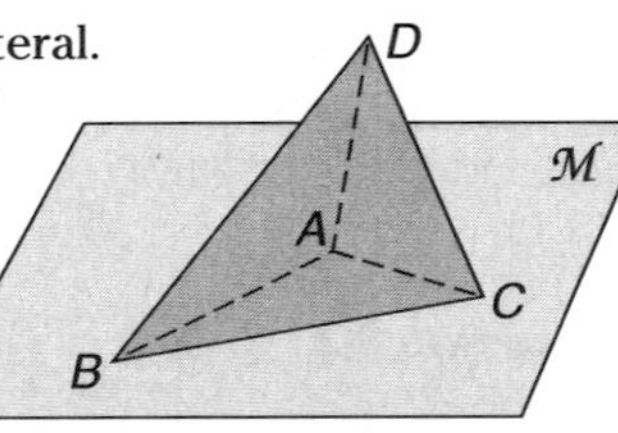

Proof:
It is given that $\triangle ABC$ is equilateral, $\triangle ABD \cong \triangle CBD$, and $BD > AD$. By the definition of an equilateral triangle, $\overline{AC} \cong \overline{BC}$. $\overline{DC} \cong \overline{DC}$ because congruence of segments is reflexive. The SSS Inequality allows us to say that $m\angle BCD > m\angle DCA$. $\overline{BC} \cong \overline{BA}$ by the definition of an equilateral triangle. $\overline{DB} \cong \overline{DB}$ because congruence of segments is reflexive. Thus, $\triangle DBC \cong \triangle DBA$ by SAS. Then by CPCTC, $\overline{DC} \cong \overline{DA}$. Therefore since if two sides of a triangle are congruent then the angles opposite those sides are congruent, $\angle DAC \cong \angle DCA$. $m\angle DAC = m\angle DCA$ by the definition of congruent angles. Then by substitution, $m\angle BCD > m\angle DAC$.

35. $\overline{AC}$ is not perpendicular to plane $\mathcal{F}$. **37.** The nutcracker is an example of an application of the SAS inequality. As force is applied to the arms of the lever, the distance between the ends of the arms of the lever is decreased. According to the SSS inequality, this makes the angle between the arms of the lever get smaller. This is the same angle as the one in the triangle formed by the arms of the lever and the segment between the points where the nut meets the arms of the lever. As this angle get smaller, the segment between the points where the nut meets the arms of the lever gets shorter, thereby crushing the nut. **39.** 6; $\overline{FG}, \overline{GH}, \overline{FH}$ **41.** $\left(3\frac{1}{2}, 3\frac{1}{2}\right)$ **43.** acute, isosceles **45.** $32y - 20$

Page 281 Chapter 5 Highlights
1. d **3.** b and l **5.** h **7.** m **9.** k

Pages 282–284 Chapter 5 Study Guide and Assessment
11. $m\angle NQO = m\angle NOQ = 57$, $m\angle ONQ = 66$ **13.** No; $m\angle MSQ \neq 90$. **15.** $x = 6, y = 26$ **17.** $x = 2, y = 1$

19. Given: $\triangle MJN \cong \triangle MJL$.
$\overline{MJ}$ does not bisect $\angle NML$.

Prove: $\overline{MJ}$ is an altitude of $\triangle NML$.

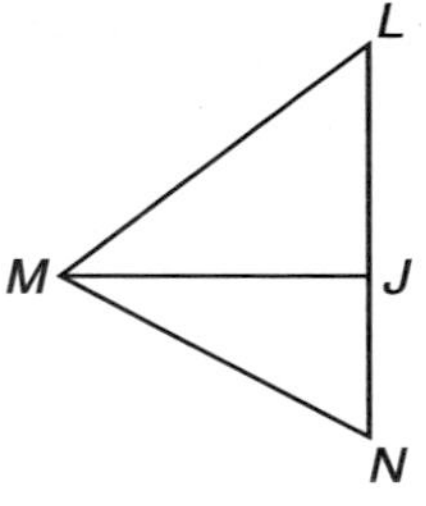

Proof:
We assume that $\overline{MJ}$ is not an altitude of $\triangle LMN$. This means that $\overline{MJ}$ is not perpendicular to $\overline{LN}$ by the definition of altitude. So, by the definition of perpendicular, this means $\angle MJN$ is not a right angle. Therefore, $\angle MJN$ is either acute or obtuse. Since $\angle MJL$ and $\angle MJN$ form a linear pair, these two angles are supplementary. Hence, if $\angle MJL$ is acute, $\angle MJN$ is obtuse (or vice versa), by the definition of supplementary. In either case, $\angle MJL \not\cong \angle MJN$. This is a contradiction, and hence our assumption must be false. Therefore $\overline{MJ}$ is an altitude of $\triangle LMN$.

21. Sample answer: $\angle 7$ **23.** $\angle YXB, \angle ACB, \angle 9, \angle 4, \angle 11$
25. $\overline{SP}$

27. Given: $FG < FH$

Prove: $m\angle 1 <$ $m\angle 2$

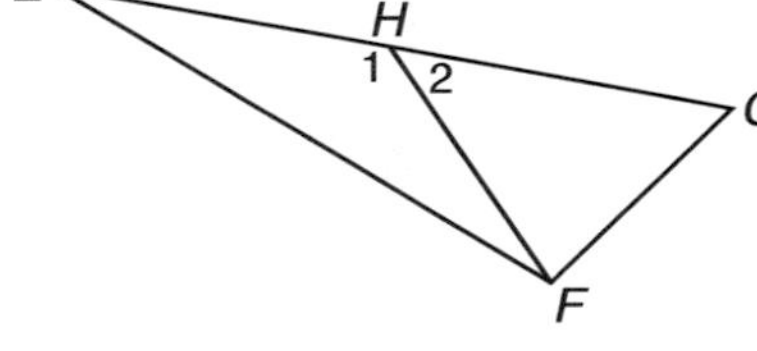

Proof:

Statements	Reasons
1. $FG < FH$	1. Given
2. $m\angle FGH > m\angle 2$	2. If one side of a $\triangle$ is longer than another, the $\angle$ opp. the longer side $>$ the $\angle$ opp. the shorter side.
3. $m\angle 1 > m\angle FGH$	3. Exterior Angle Inequality Th.
4. $m\angle 1 > m\angle 2$	4. Transitive Property ($\neq$)

29. 17 and 31 **31.** No, the distances between the points do not satisfy the triangle inequality. **33.** $m\angle ALK < m\angle ALN$ **35.** $m\angle OLK > m\angle NLO$

37. Given: $AD = BC$

Prove: $AC > DB$

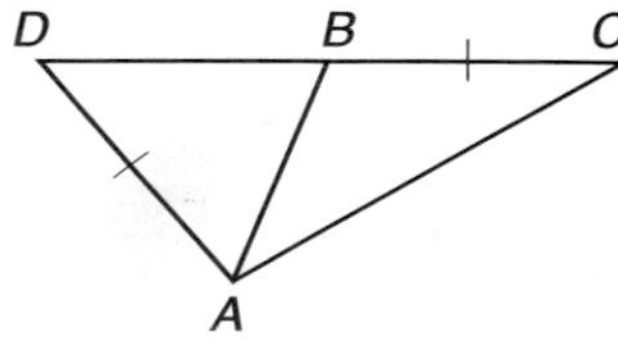

Proof:

Statements	Reasons
1. $AD = BC$	1. Given
2. $\overline{AD} \cong \overline{BC}$	2. Def. $\cong$ segments
3. $\overline{BA} \cong \overline{BA}$	3. Congruence of segments is reflexive.
4. $m\angle CBA > m\angle DAB$	4. Exterior Angle Inequality Th.
5. $AC > DB$	5. SAS Inequality Th.

39. 200 points

CHAPTER 6 EXPLORING QUADRILATERALS

Page 290 Lesson 6–1A

1. Sample answer: The opposite sides are congruent.
3. Sample answer: Yes, the same relationships hold true.

Pages 295–297 Lesson 6–1

7. $\overline{DC}$; opposite sides of a parallelogram are parallel.
9. $\overline{GF}$; opposite sides of a parallelogram are congruent.
11. $\triangle HDF$; SSS. Since the diagonals of a parallelogram bisect each other, $\overline{HC} \cong \overline{HF}$ and $\overline{HG} \cong \overline{HD}$ and since the opposite sides of a parallelogram are congruent, $\overline{GC} \cong \overline{DF}$. **13.** $m\angle Y = 47$, $m\angle X = 133 = m\angle Z$

15. Given: *PRSV* is a parallelogram. $\overline{PT} \perp \overline{SV}$, $\overline{QS} \perp \overline{PR}$

Prove: $\triangle PTV \cong \triangle SQR$

Proof:

Statements	Reasons
1. *PRSV* is a parallelogram. $\overline{PT} \perp \overline{SV}$ $\overline{QS} \perp \overline{PR}$	1. Given
2. $\angle V \cong \angle R$	2. Opp. ∠s of a ▱ are ≅.
3. $\overline{PV} \cong \overline{RS}$	3. Opp. sides of a ▱ are ≅.
4. $\angle PTV$ and $\angle SQR$ are rt. ∠s.	4. ⊥ lines form four rt. ∠s.
5. $\triangle PTV$ and $\triangle SQR$ are rt. △s.	5. Def. rt. △
6. $\triangle PTV$ and $\triangle SQR$	6. HA

17. $\overline{AR}$; opposite sides of a parallelogram are parallel.
19. $\angle MAR$; opposite angles of a parallelogram are congruent. **21.** $\triangle RKM$; SSS. $\overline{MA} \cong \overline{RK}$ and $\overline{KM} \cong \overline{RA}$ because opposite sides of a parallelogram are congruent. $\overline{RM} \cong \overline{RM}$ because congruence of segments is reflexive.
23. $\overline{RK}$; opposite sides of a parallelogram are parallel.
25. $\triangle SKR$; SAS. Diagonals bisect each other, so $\overline{SA} \cong \overline{SK}$ and $\overline{SM} \cong \overline{SR}$. $\angle ASM \cong \angle KSR$ because they are vertical angles. **27.** $x = 118$; $y = 62$; $z = 118$ **29.** $x = 53$; $y = 15$; $z = 53$ **31.** The diagonals are not congruent.
33. (0, −8), (8, 6), or (−8, 10) **35.** 19

37. Given: ▱*TEAM*, $MS = FS$

Prove: $\angle F \cong \angle TEA$

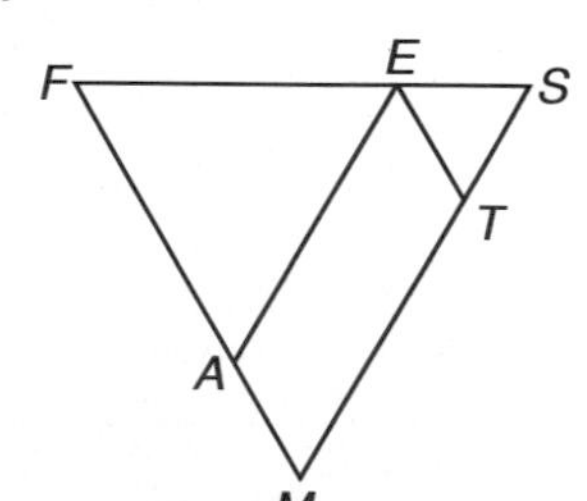

Proof:

Statements	Reasons
1. ▱*TEAM* $MS = FS$	1. Given
2. $\angle F \cong \angle M$	2. Isosceles Triangle Th.
3. $\angle M \cong \angle TEA$	3. Opp. ∠s of a ▱ are ≅.
4. $\angle F \cong \angle TEA$	4. Congruence of angles is transitive.

39. Given: *ABCD* is a parallelogram.

Prove: $\angle BAD \cong \angle DCB$, $\angle ABC \cong \angle CDA$

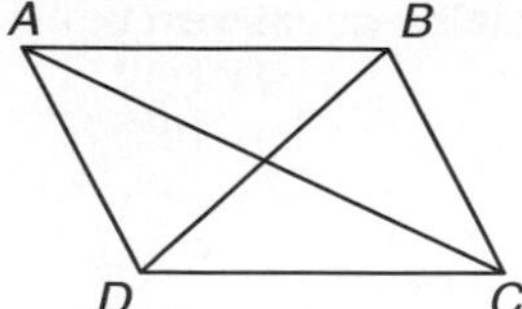

Proof:

We are given that *ABCD* is a parallelogram. Since the opposite sides of a parallelogram are congruent, $\overline{AD} \cong \overline{BC}$, $\overline{AB} \cong \overline{CD}$. $\overline{BD} \cong \overline{BD}$ and $\overline{AC} \cong \overline{AC}$ because congruence of segments is reflexive. By Side-Side-Side, $\triangle BAD \cong \triangle DCB$ and $\triangle ABC \cong \triangle CDA$. Therefore we can conclude that $\angle BAD \cong \angle DCB$ and $\angle ABC \cong \angle CDA$ by CPCTC.

41. Sum = 360°; yes, two triangles are always formed by any quadrilateral and a diagonal.

45.

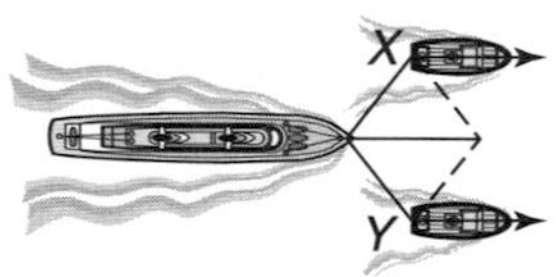

47. $m\angle ADC < m\angle ADB$ **49.** $\angle M$, $\angle K$, $\angle L$ **51.** C **53.** If a quadrilateral has opposite sides parallel, then it is a parallelogram. **55.** 15%

Pages 301–304 Lesson 6–2

7. Yes; the triangles are congruent by SAS, so both pairs of opposite sides are congruent. **9.** $x = 4$, $y = 1$
11. False; the diagonals being congruent does not ensure that opposite sides will be congruent or parallel.

13. Given: $\overline{HD} \cong \overline{DN}$, $\angle DHM \cong \angle DNT$

Prove: Quadrilateral *MNTH* is a parallelogram.

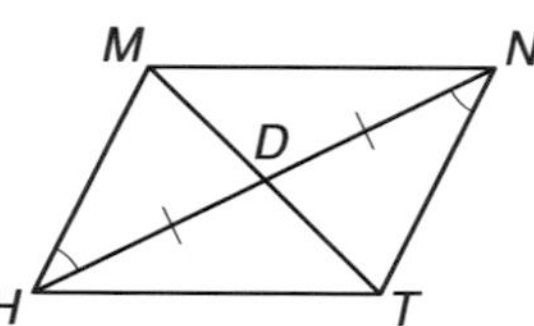

Proof:

Statements	Reasons
1. $\overline{HD} \cong \overline{DN}$ $\angle DHM \cong \angle DNT$	1. Given
2. $\angle MDH \cong \angle TDN$	2. Vert ∠s are ≅.
3. $\triangle MDH \cong \triangle TDN$	3. ASA
4. $\overline{MH} \cong \overline{TN}$	4. CPCTC
5. $\overline{MH} \parallel \overline{TN}$	5. If 2 lines are cut by a transversal, and alt. int. ∠s are ≅, then the lines are ∥.
6. *MNTH* is a parallelogram.	6. If one pair of opp. sides of a quad. are both ∥ and ≅, then the quad. is a ▱.

15. Yes, both pairs of opposite angles are congruent.
17. Yes, both pairs of opposite angles are congruent.
19. No, none of the tests for parallelograms is met.
21. $x = 5$, $y = 16$ **23.** $x = 11$, $y = 12$ **25.** $x = 64$, $y = 23.5$
27. false, counterexample:

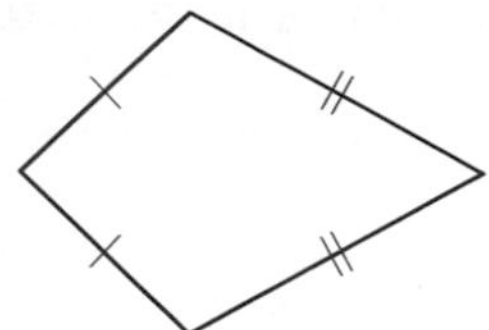

29. false, counterexample:

31. a parallelogram; slope of $\overline{FG} = -\frac{1}{4}$, slope of $\overline{HJ} = -\frac{1}{4}$, $FG = \sqrt{17}$, $HJ = \sqrt{17}$ since one pair of opposite sides is both parallel and congruent, *FGHJ* is a parallelogram.

33.

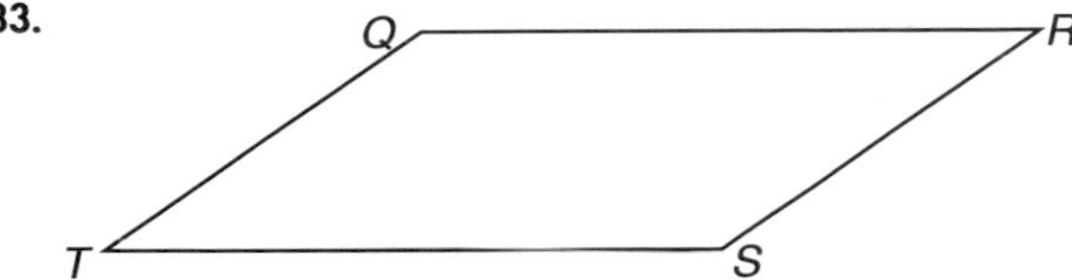

35. Given: $\overline{BD}$ bisects $\overline{AC}$.
$\overline{AC}$ bisects $\overline{BD}$.

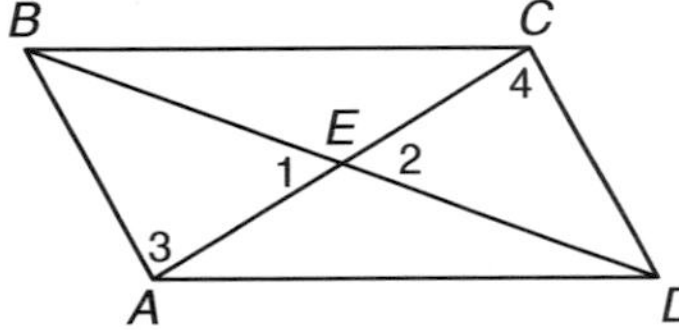

Prove: *ABCD* is a parallelogram.

Proof:

Statements	Reasons
1. $\overline{BD}$ bisects $\overline{AC}$. $\overline{AC}$ bisects $\overline{BD}$.	1. Given
2. $\overline{AE} \cong \overline{CE}$, $\overline{BE} \cong \overline{DE}$	2. Def. segment bisector
3. $\angle 1 \cong \angle 2$	3. Vert. ∠s are ≅.
4. $\triangle BEA \cong \triangle DEC$	4. SAS
5. $\angle 3 \cong \angle 4$, $\overline{AB} \cong \overline{CD}$	5. CPCTC
6. $\overline{AB} \parallel \overline{CD}$	6. If ⇸ , and alt. int. ∠s are ≅, then the lines are ∥.
7. *ABCD* is a parallelogram.	7. If a pair of opp. sides of a quad. are ≅ and ∥, it is a ▱.

37. Subgoals:
1. Prove that $\triangle ABC \cong \triangle DEF$.
2. Use CPCTC to say that $\overline{AC} \cong \overline{FD}$.
3. State that $\overline{FA} \cong \overline{DC}$ by the definition of a regular hexagon.
4. Use both pair of opposite sides congruent to show *FDCA* is a parallelogram.

Given: *ABCDEF* is a regular hexagon.
Prove: *FDCA* is a ▱.

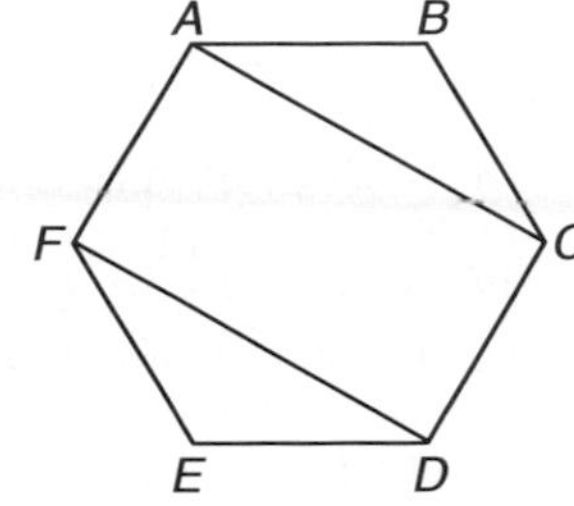

Proof:

Statements	Reasons
1. *ABCDEF* is a regular hexagon.	1. Given
2. $\overline{AB} \cong \overline{DE}$, $\overline{BC} \cong \overline{EF}$, $\angle E \cong \angle B$	2. Def. reg. hexagon
3. $\triangle ABC \cong \triangle DEF$	3. SAS
4. $\overline{AC} \cong \overline{DF}$	4. CPCTC
5. $\overline{FA} \cong \overline{CD}$	5. Def. reg. hexagon
6. *FDCA* is a ▱.	6. If opp. sides of a quad. are ≅, it is a ▱.

39a. Sample answer: Measure each pair of opposite segments. **39b.** Sample answer: Yes. The pattern yields quadrilaterals with guaranteed parallel opposite sides. **41.** 42 **43.** no solution **45.** $\overline{TA}$ **47.** The quadrilateral is a parallelogram. **49a.** 36% **49b.** 13,241,800

Page 305 Lesson 6–3A

1. The opposite sides of a rectangle are congruent.
3. Yes; the opposite sides are congruent and the diagonals are congruent.

Pages 309–312 Lesson 6–3

5. $x = 15.5$ **7.** $x = 13.5$ **9.** *ABCD* is a rectangle. slope of $\overline{AB} = 1$; slope of $\overline{BC} = -1$; slope of $\overline{CD} = 1$; slope of $\overline{DA} = -1$ **11.** 26; 26; 64 **13.** Left; When stakes *F* and *G* are moved left an appropriate distance, $\overline{MK}$ lengthens and its midpoints will coincide with the midpoint of $\overline{JL}$, which becomes shorter after the move. **15.** 3 **17.** 12 **19.** 2
21. false; A counterexample:

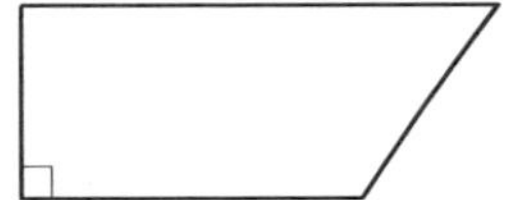

23. True; The sum of the measures of the angles of a quadrilateral is 360. If all four angles are congruent, then the measure of each angle is $\frac{360}{4}$ or 90. Thus, all four angles are right and the definition of a rectangle is a quadrilateral with four right angles. **25.** 140 **27.** $-\frac{2}{3}$ or 4 **29.** 90 **31.** 33 **33.** 74 **35.** no; not a quadrilateral

37.

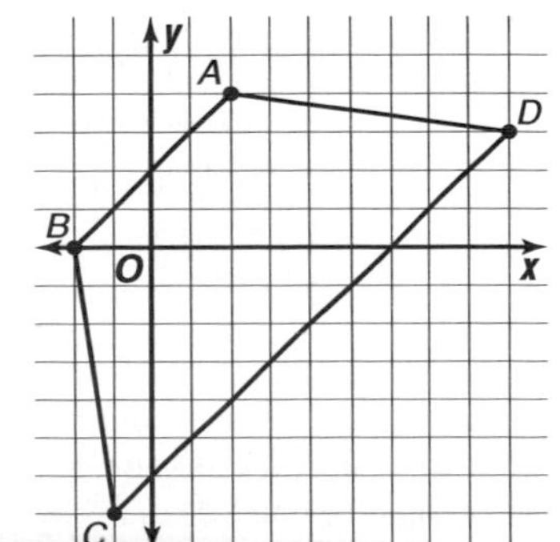

37a. $AC = \sqrt{130}$; $BD = \sqrt{130}$ **37b.** midpoint of $\overline{AC} = \left(\frac{1}{2}, -1\frac{1}{2}\right)$; midpoint of $\overline{BD} = \left(3\frac{1}{2}, 1\frac{1}{2}\right)$ **37c.** *ABCD* is not a rectangle because it is not a parallelogram.

39. Given: $ACDE$ is a rectangle.
$ABCE$ is a parallelogram.

Prove: $\triangle ABD$ is isosceles.

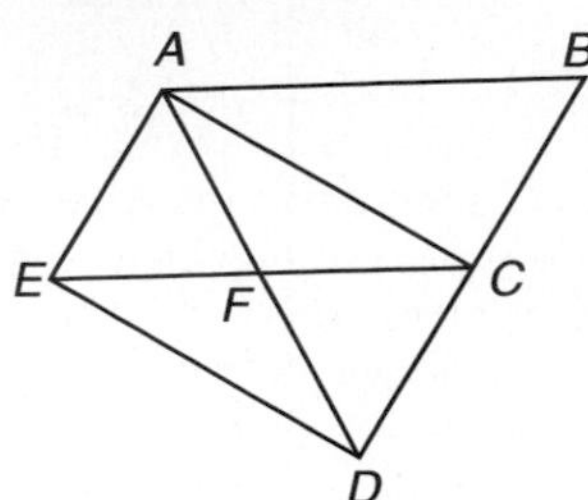

Proof:

Statements	Reasons
1. $ACDE$ is a rectangle. $ABCE$ is a parallelogram.	1. Given
2. $\overline{AD} \cong \overline{CE}$	2. If a ▱ is a rect. then its diagonals are ≅.
3. $\overline{AB} \cong \overline{CE}$	3. Opp. sides of a ▱ are ≅.
4. $\overline{AD} \cong \overline{AB}$	4. Congruence of segments is transitive.
5. $\triangle ABD$ is isosceles.	5. Def. isosceles △

41. Given: $PQMO$ and $RQMN$ are rectangles.
$\angle SVT \cong \angle UTV$

Prove: $STUV$ is a parallelogram.

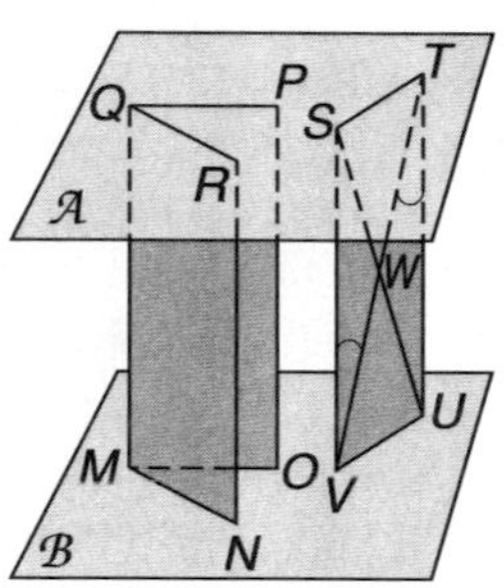

Proof:

Statements	Reasons
1. $PQMO$ and $RQMN$ are rectangles. $\angle SVT \cong \angle UTV$ $\overline{SU}$ and $\overline{TV}$ intersect at W.	1. Given
2. $\overline{PQ} \parallel \overline{MO}$ and $\overline{RQ} \parallel \overline{MN}$	2. Def. ▱
3. plane $\mathcal{N} \parallel$ plane $\mathcal{M}$	3. Def. ∥ planes
4. $S, U\ T, V$ and W are in the same plane.	4. Def. intersecting lines
5. $\overline{ST} \parallel \overline{VU}$	5. Def. ∥ lines
6. $\overline{SV} \parallel \overline{TU}$	6. If ⇹, and alt. int. ∠s are ≅, then the lines are ∥.
7. $STUV$ is a ▱.	7. Def. ▱

43. Sample answer: A rectangular package fits better in a box or display unit than a circular package would. **45.** no **45a.** Answers will vary. Sample answers: In architecture the golden rectangle is used in many well-known structures such as the Parthenon. In marketing: Many consumer products such as credit cards are approximately golden rectangles. **45b.** The diagonal is about 1.96 units long. **47.** HA, LL, and HL **49.** $\angle T \cong \angle X$ **51.** Transitive Property (=) **53.** −3

Page 312 Self Test

1. $\overline{PN}$; opposite sides of a parallelogram are parallel by definition. **3.** $\overline{LP}$; opposite sides of a parallelogram are congruent. **5.** yes **7.** yes

9. Given: ▱$WXZY$
$\angle 1$ and $\angle 2$ are complementary.

Prove: $WXZY$ is a rectangle.

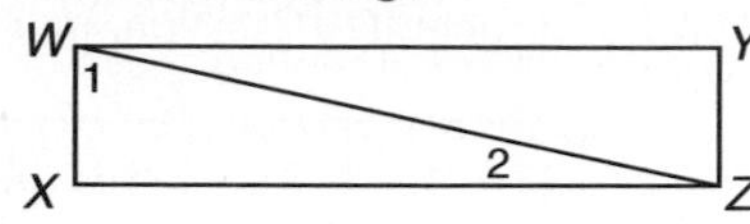

Proof:

Statements	Reasons
1. ▱$WXZY$ $\angle 1$ and $\angle 2$ are complementary.	1. Given
2. $m\angle 1 + m\angle 2 = 90$	2. Def. complementary ∠s
3. $m\angle 1 + m\angle 2 + m\angle X = 180$	3. Angle Sum Theorem
4. $90 + m\angle X = 180$	4. Substitution Prop. (=)
5. $m\angle X = 90$	5. Subtraction Prop. (=)
6. $\angle X \cong \angle Y$	6. Opp. ∠s of a ▱ are ≅.
7. $m\angle Y = 90$	7. Substitution Prop. (=)
8. $\angle X$ and $\angle XWY$ are supp. $\angle X$ and $\angle XZY$ are supp.	8. Consec. ∠s of a ▱ are supp.
9. $m\angle X + m\angle XWY = 180$ $m\angle X + m\angle XZY = 180$	9. Def. Supp. ∠s
10. $90 + m\angle XWY = 180$ $90 + m\angle XZY = 180$	10. Substitution Prop. (=)
11. $m\angle XWY = 90$ $m\angle XZY = 90$	11. Subtraction Prop. (=)
12. $\angle Y$, $\angle XWY$, and $\angle XZY$ are rt. ∠s.	12. Def. rt. ∠s
13. $WXZY$ is a rect.	13. Def. rect.

Pages 316–319 Lesson 6–4

7. Right; the diagonals of a rhombus are perpendicular, so $\angle PEN$ is right. **9.** No, the diagonals of a rhombus are not congruent unless the rhombus is a square. **11.** 22.5 **13.** 12 **15.** parallelogram, a rectangle **17.** rectangle, square

19. Given: $ABCD$ is a rhombus.

Prove: $\overline{AC} \perp \overline{BD}$

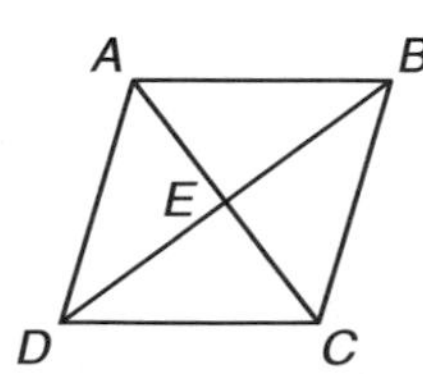

Proof:
The definition of a rhombus states that the four sides of the quadrilateral are congruent. Therefore, $\overline{AB} \cong \overline{BC} \cong \overline{CD} \cong \overline{AD}$. A rhombus is a parallelogram and the diagonals of a parallelogram bisect each other, so $\overline{BD}$ bisects $\overline{AC}$ at E. $\overline{AE} \cong \overline{CE}$ by the definition of a bisector. $\overline{BE} \cong \overline{BE}$ because congruence of segments is reflexive. Thus, $\triangle ABE \cong \triangle CBE$ by SSS. $\angle BEA \cong \angle BEC$ by CPCTC. $\angle BEA$ and $\angle BEC$ form a linear pair and if

two angles form a linear pair they are supplementary. Thus, $\angle BEA$ and $\angle BEC$ are supplementary. By the definition of supplementary angles, $m\angle BEA + m\angle BEC = 180$. Also, $m\angle BEA = m\angle BEC$ by the definition of congruent angles. So by substitution, $2m\angle BEA = 180$. Then $m\angle BEA = 90$ by the division property of equality. $\angle BEA$ is right by the definition of a right angle. Therefore, $\overline{AC} \perp \overline{BD}$ by the definition of perpendicular lines.

21. Scalene; the consecutive sides of *MNOP* are not congruent. **23.** Yes; the sides of *MNOP* are congruent and the diagonals are perpendicular, so the triangles are congruent by SAS. **25.** No; all squares are rhombi, but not all rhombi are squares. **27.** No; *MNOP* could be a rhombus if $\overline{PO} \cong \overline{ON}$, but the conditions are not sufficient for it to be a square. **29.** 5.25 **31.** 7.5 **33.** 12 **35.** parallelogram, rhombus **37.** parallelogram, rectangle, rhombus, square **39.** parallelogram, rectangle, rhombus, square **41.** rhombus, square **43.** False; some rhombi are not squares. **45.** True; the set of squares is a subset of the set of rectangles. **47.** False; some parallelograms are not rectangles.

49. Given: *JKLM* is a square.
Prove: $\overline{JL} \cong \overline{KM}$

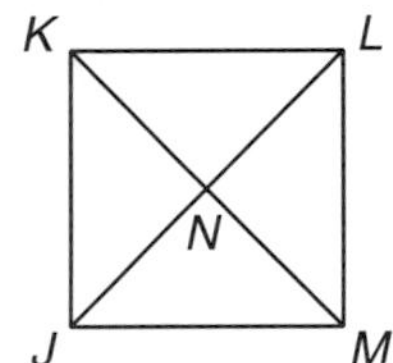

Proof:

Statements	Reasons
1. *JKLM* is a square.	1. Given
2. *JKLM* is a rectangle.	2. Def. square
3. $\overline{JL} \cong \overline{KM}$	3. If a ▱ is a rect. then its diagonals are ≅.

51. Given: *ABCD* is a parallelogram. $\overline{AC} \perp \overline{BD}$
Prove: *ABCD* is a rhombus.

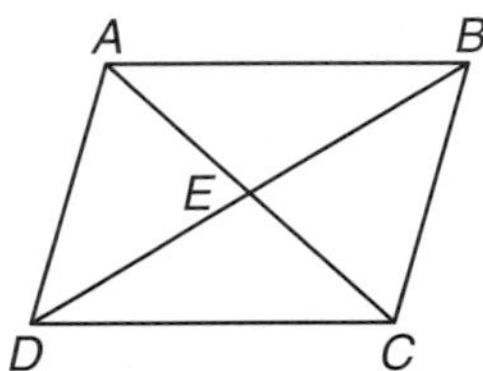

Proof:

We are given that *ABCD* is a parallelogram. The diagonals of a parallelogram bisect each other, so $\overline{AE} \cong \overline{EC}$. $\overline{BE} \cong \overline{BE}$ because congruence of segments is reflexive. We are also given that $\overline{AC} \perp \overline{BD}$. Thus, $\angle AEB$ and $\angle BEC$ are right angles by the definition of perpendicular lines. Then $\angle AEB \cong \angle BEC$ because all right angles are congruent. Therefore, $\triangle AEB \cong \triangle BEC$ by SAS. $\overline{AB} \cong \overline{BC}$ by CPCTC. Opposite sides of parallelograms are congruent, so $\overline{AB} \cong \overline{CD}$ and $\overline{BC} \cong \overline{AD}$. Then since congruence of segments is transitive, $\overline{AB} \cong \overline{CD} \cong \overline{BC} \cong \overline{AD}$. All four sides of *ABCD* are congruent, so *ABCD* is a rhombus by definition.

53. The flag of Denmark contains two red squares and two rectangles. The flag of St. Vincent and The Grenadines contains a blue rectangle, a green rectangle, a yellow rectangle, a blue and yellow rectangle, a yellow and green rectangle, and three green rhombi. The flag of Trinidad and Tobago contains two white parallelograms, and one black parallelogram. **55.** The quadrilateral formed by the back wall, the two side walls, and the short line is a square because all the sides are congruent and the angles are right. The quadrilateral formed by the front wall, the two side walls, and the short line is a square because all the sides are congruent and the angles are right. The back wall, the two side walls and the service line form a rectangle because the angles are all right, but the sides are not all congruent. The front wall, the two side walls and the service line form a rectangle because the angles are all right, but the sides are not all congruent. The two side walls, the short line, and the service line form a rectangle because the angles are all right, but the sides are not all congruent.
57. The legs are made so that they will bisect each other, so the quadrilateral formed by the ends of the legs is always a parallelogram. Therefore, the table top is parallel to the floor. **59.** Yes; If *V* is (12, 3), $TU = UV$. **61.** true **63.** false **65.** 4; 9 **67.** 229%

Page 320 Lesson 6–4B

5. Sample answer: For kite *ABCD*, $\angle B \cong \angle D$; $\overline{AC} \perp \overline{BD}$; $\overline{AC}$ bisects $\overline{BD}$, $\overline{AB} \cong \overline{AD}$; $\overline{BC} \cong \overline{CD}$; $\triangle ABC \cong \triangle ADC$.

Pages 324–328 Lesson 6–5

5. True; the diagonals of an isosceles trapezoid are congruent. **7.** False; if the diagonals bisected each other, *ABCD* would be a parallelogram not a trapezoid. **9.** 17 **11.** (5, 3), $\left(2\frac{1}{2}, 8\right)$ **13.** RS $\neq \frac{1}{2}(NP + MQ)$ because *MNPQ* is not a trapezoid. **15.** Sides are trapezoids; front, back & tops are rectangles. **17.** True; $\overline{TV} \cong \overline{TV}$ because congruence of segments is reflexive, $\angle TVR \cong \angle VTS$ because base angles of an isosceles trapezoid are congruent, and $\overline{TR} \cong \overline{VS}$ because diagonals of an isosceles trapezoid are congruent; thus $\triangle TRV \cong \triangle VST$ by SAS and $\angle TRV \cong \angle VST$ by CPCTC. **19.** True; $\overline{TV} \parallel \overline{SR}$ and consecutive interior angles are supplementary. **21.** 78 **23.** 5 **25.** 35 **27.** 2 **29.** slope of $\overline{RS} = 0$, slope of $\overline{TV} = 0$ **31.** midpoint of $\overline{RV} = \left(\frac{-1 + (-2)}{2}, \frac{2 + (-2)}{2}\right)$ or $\left(-1\frac{1}{2}, 0\right)$; midpoint of $\overline{ST} = \left(\frac{1 + 3}{2}, \frac{2 + (-2)}{2}\right)$ or (2, 0); slope of $\overline{AB} = 0$; slope of $\overline{RS} = 0$ **33.** parallelogram **35.** trapezoid **37.** $180 - x$

39. Given: *DEFH* is a trapezoid with bases $\overline{DE}$ and $\overline{FH}$. $\overline{DE} \cong \overline{FG}$
Prove: *DEFG* is a parallelogram.

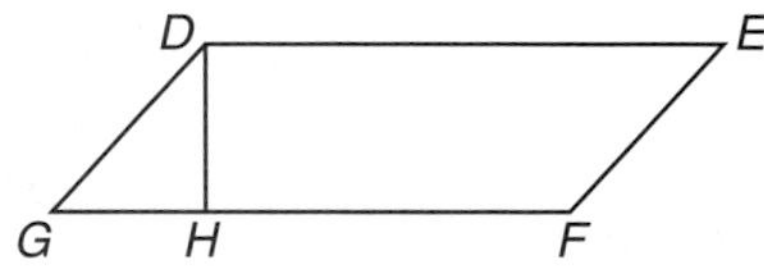

Proof:

Statements	Reasons
1. *DEFH* is a trapezoid with bases $\overline{DE}$ and $\overline{FH}$. $\overline{DE} \cong \overline{FG}$	1. Given
2. $\overline{DE} \parallel \overline{FH}$	2. Def. base of a trapezoid
3. *DEFG* is a parallelogram.	3. If a pair of opp. sides of a quad. are ≅ and ∥, it is a ▱.

41a. 6.18 **41b.** 4.74 **41c.** 26.6 **43a.** Sample answer: 10 and 30 **43b.** A measure cannot be negative and the average of the measures of the bases is 20. So the possible measures of the bases are 1 and 39, 2 and 38, ... and 19 and 21. The measures of the bases cannot be the same or the figure would be a parallelogram. Thus, there are 19 possible combinations. **43c.** There are an infinite number of combinations of possible measures if the integer restriction is removed because there is an infinite number of positive values with an average of 20. **45.** Top is a square, sides are isosceles trapezoids. **47.** 112 **49.** true **51.** yes; LA or AAS **53.** B **55.** 123,454,321

Page 329 Chapter 6 Highlights

1. False; Every parallelogram is a quadrilateral. **3.** true **5.** False; You can prove that a parallelogram is a rectangle by proving that the diagonals are congruent. **7.** False; A square has all of the characteristics of a parallelogram, a rectangle, and a rhombus, but not the characteristics of a trapezoid. **9.** False; The legs of an isosceles trapezoid are congruent. **11.** If *QUAD* is a square, then it is also a parallelogram, a rectangle, a rhombus, and a quadrilateral. But *QUAD* is not a trapezoid.

Pages 330–332 Chapter 6 Study Guide and Assessment

13. $\angle WXY$; Opposite ∠s in a ▱ are ≅. **15.** $\overline{EZ}$; Diagonals of a ▱ bisect each other. **17.** $\triangle ZWX$; SAS, or SSS **19.** $\overline{EW}$; Diagonals of a ▱ bisect each other. **21.** $x = 28$, $y = 5$

23. Given: ▱*ABCD*
$\overline{AE} \cong \overline{CF}$

Prove: Quadrilateral *EBFD* is a parallelogram.

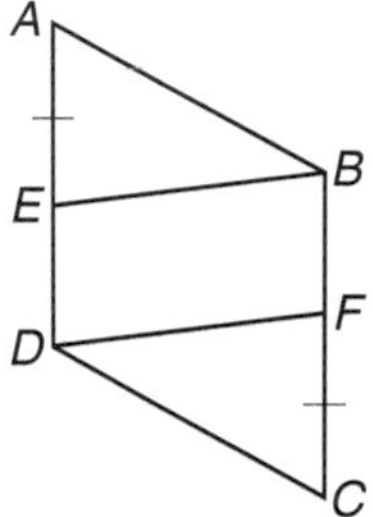

Proof:

Statements	**Reasons**
1. ▱*ABCD*, $\overline{AE} \cong \overline{CF}$	1. Given
2. $\overline{BA} \cong \overline{DC}$	2. Opp. sides of a ▱ are ≅.
3. $\angle A \cong \angle C$	3. Opp. ∠s of a ▱ are ≅.
4. $\triangle BAE \cong \triangle DCF$	4. SAS
5. $\overline{BE} \cong \overline{DF}$, $\angle BEA \cong \angle DFC$	5. CPCTC
6. $\overline{BC} \parallel \overline{AD}$	6. Def. ▱
7. $\angle DFC \cong \angle FDE$	7. Alt. Int. ∠ Th.
8. $\angle BEA \cong \angle FDE$	8. Congruence of angles is transitive.
9. $\overline{BE} \parallel \overline{DF}$	9. If ⇹, and corr. ∠s are ≅, the lines are ∥.
10. Quadrilateral *EBFD* is a parallelogram.	10. If a pair of opp. sides of a quad. are ≅ and ∥, it is a ▱.

25. 6 **27.** 1 or 2 **29.** rectangle, parallelogram **31.** parallelogram **33.** 28 **35.** 90 **37.** 38 **39.** 45 **41.** 47 **43.** 30 **45.** the kite flaps and the back are all trapezoids **47.** the deck top and the road surface

CHAPTER 7 CONNECTING PROPORTION AND SIMILARITY

Pages 341–345 Lesson 7–1

7. $\frac{170}{9}$ **9.** 14 **11.** yes **13.** no **15.** 18 in., 24 in., 30 in. **17.** $\frac{1000}{1}$ **19.** $\frac{4}{5}$ **21.** 1.295 **23.** $\frac{121}{7} \approx 17.29$ **25.** -11 **27.** 18 ft, 24 ft **29.** 3.9 **31.** $\frac{4}{9}$ **33.** 40 in., 60 in., 90 in. **35.** 49.92 **37.** 25%

39.
$$\begin{aligned} \frac{a+b}{b} &= \frac{c+d}{d} \\ (a+b)d &= (c+d)b \\ ad + bd &= cb + db \\ ad + bd &= cb + bd \\ ad + bd &= cb + bd \\ ad + bd - bd &= cb + bd - bd \\ ad &= cb \\ \frac{a}{b} &= \frac{c}{d} \end{aligned}$$

41. $x = 24$, $y = 12$ **43.** This is true because one of the angles measures 90° and the other two measure 45°. The ratio of 45:90 is 1:2. **45a.** The ratios (miles per gallon) are 27, 25, 29, 31, 20.3125, 29.44. **45b.** You cannot find the average number of miles per gallon by adding the number of miles per gallon at each stop and dividing by 6 because Keshia drove a different distance between each stop. To find the average, you must divide the total distance by the total amount of gas for all six stops. The average is about 26.86 miles per gallon. **47a.** 11,049; 10,862; 19,238; 12,985; 13,595 **47b.** You could find the ratio of people per movie screen to see which is smallest; Kansas City is the lowest at 10,862 people per screen, followed by Ann Arbor at 11,049 people per screen. **49a.** 6,658,400 lb **49b.** 16.4 lb

51a.

Currency	Cost of Making (cents)	Value (cents)	Ratio (cost/value)
penny	0.8	1	$\frac{0.8}{1} = 0.8$
nickel	2.9	5	$\frac{2.9}{5} = 0.58$
dime	1.7	10	$\frac{1.7}{10} = 0.17$
quarter	3.7	25	$\frac{3.7}{25} = 0.148$
half dollar	7.8	50	$\frac{7.8}{50} = 0.156$
dollar bill	3	100	$\frac{3}{100} = 0.03$

51b. Sample answer: No, it costs the most to make a penny since the ratio is the greatest at 0.8 and it costs the least to make a dollar bill since the ratio is the smallest at 0.03. **51c.** \$2.90, \$0.74 **53.** trapezoid **55.** It is less than 13 and greater than 3. **57.** 90, 52 **59.** 75, 105 **61.** False; *A*, *B*, and *P* are not necessarily collinear. **63.** -8, 2

Pages 350–353 Lesson 7–2

7. 12, 12 **9.** A **11.** Yes; the corresponding angles are congruent and the corresponding sides are proportional with a scale factor of $\frac{1}{3}$. **13.** No; the side on the second triangle measuring 18.4 should be 21.6. **15.** 71.05, 48.45 **17.** 91, 30 **19.** S **21.** N **23.** S **25.** A **27.** $\frac{4}{3}$ **29a.** 7.5 **29b.** 22.4 **29c.** 108 **29d.** 12.8 **31.** $\triangle ABC \sim \triangle IHG \sim \triangle JLK$ and $\triangle NMO \sim \triangle RPS$ **33.** Students should draw a rectangle $5\frac{1}{4}$ in. by $\frac{31}{8}$ in. **35.** Students should draw a rectangle $4\frac{1}{2}$ in. by $9\frac{3}{4}$ in. **37.** $L(16, 8)$ and $O(8, 8)$ or $L(16, -8)$ and $O(8, -8)$

39. In triangles ABC and DEF, $\overline{AC}$ and $\overline{DF}$ both have slope $\frac{3}{2}$ and $\overline{BC}$ and $\overline{EF}$ both have slope $-\frac{3}{2}$. Thus, $\overline{AC} \parallel \overline{DF}$ and $\overline{BC} \parallel \overline{EF}$. The horizontal axis is a transversal for both pairs of parallel lines. Thus, corresponding angles BAC and EDF are $\cong$. In the same way, corresponding angles B and E are $\cong$. $\angle C \cong \angle F$ because if two ∠s in a △ are ≅ to two ∠s in another △, the third ∠s are also ≅. Using the distance formula, $\frac{AC}{DF} = \frac{AB}{DE} = \frac{BC}{EF} = \frac{3}{2}$. Corresponding sides have the same ratio, and the angles are congruent, so the figures are similar. **41a.** 13.5 ft by 11.25 ft **41b.** 22.5 ft by 6.75 ft **43.** 12 **45.** D **47.** 176 **49.** 84 **51.** −17

Pages 357–361 Lesson 7–3

7. no **9.** $\triangle AEC \sim \triangle BDC$; AA Similarity **11.** SAS Similarity; $x = 4.5$ **13.** yes; $\triangle MNO \sim \triangle PQR$; SSS Similarity **15.** no **17.** yes; $\triangle RST \sim \triangle JKL$; AA Similarity **19.** False; this is not true for equilateral or isosceles triangles. **21.** False; the proportions are not the same. **23.** $\triangle QRS \sim \triangle QTR$ (AA Similarity); $\triangle QRS \sim \triangle RTS$ (AA Similarity); $\triangle QTR \sim \triangle RTS$ (Transitive Prop. of ~ △s) **25.** $AE = 5$, $AC = 15$, $CF = 13\frac{1}{3}$, $CB = 20$, $CD = 12$, $EF = 16\frac{2}{3}$ **27.** $m\angle CDE = 43$, $m\angle A = 43$, $m\angle ABD = 47$, $m\angle DBE = 43$, $m\angle BDE = 47$

29. Given: $\overline{LP} \parallel \overline{MN}$

Prove: $\frac{LJ}{JN} = \frac{PJ}{JM}$

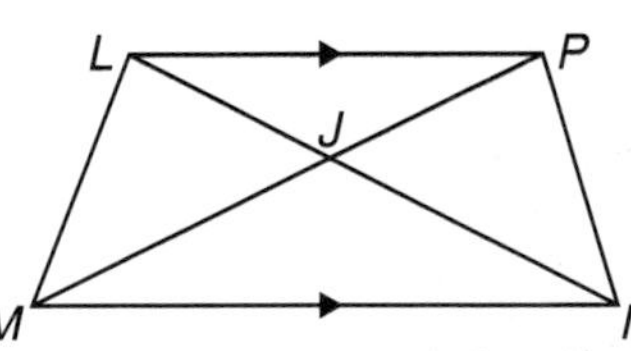

Proof:

Statements	Reasons
1. $\overline{LP} \parallel \overline{MN}$	1. Given
2. $\angle PLN \cong \angle LNM$, $\angle LPM \cong \angle PMN$	2. Alternate Interior ∠ Theorem
3. $\triangle LPJ \sim \triangle NMJ$	3. AA Similarity
4. $\frac{LJ}{JN} = \frac{PJ}{JM}$	4. Corr. sides of ~ △s are proportional.

31. Given: $\triangle JFM \sim \triangle EFB$, $\triangle LFM \sim \triangle GFB$

Prove: $\triangle JFL \sim \triangle EFG$

Proof:

Since $\triangle JFM \sim \triangle EFB$ and $\triangle LFM \sim \triangle GFB$, then by the definition of similar triangles, $\frac{JF}{EF} = \frac{MF}{BF}$ and $\frac{MF}{BF} = \frac{LF}{GF}$. By the Transitive Property of Equality, $\frac{JF}{EF} = \frac{LF}{GF}$. $\angle F \cong \angle F$, by the Reflexive Property of congruent angles. Then, by SAS Similarity, $\triangle JFL \sim \triangle EFG$.

33. Given: $\angle B \cong \angle E$, $\frac{AB}{DE} = \frac{BC}{EF}$

Prove: $\triangle ABC \sim \triangle DEF$

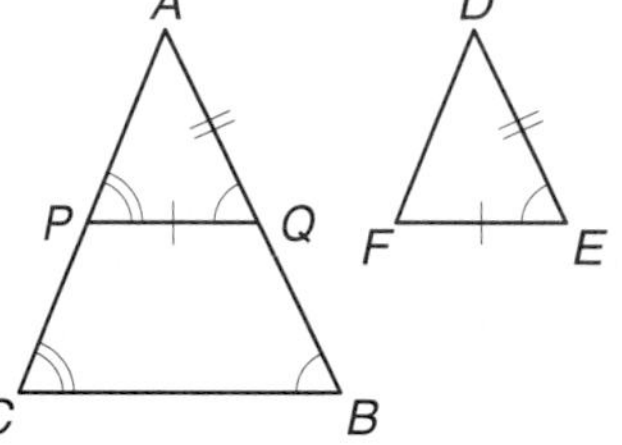

Proof:

Statements	Reasons
1. Draw $\overline{QP} \parallel \overline{BC}$ so that $\overline{QP} \cong \overline{EF}$	1. Parallel Postulate
2. $\angle APQ \cong \angle C$, $\angle AQP \cong \angle B$	2. Corresponding ∠s Postulate
3. $\angle B \cong \angle E$	3. Given
4. $\angle AQP \cong \angle E$	4. Transitive Prop. of ≅ ∠s
5. $\triangle ABC \sim \triangle AQP$	5. AA Similarity
6. $\frac{AB}{AQ} = \frac{BC}{QP}$	6. Def. of ~ △s
7. $\frac{AB}{DE} = \frac{BC}{EF}$	7. Given
8. $AB \cdot QP = AQ \cdot BC$; $AB \cdot EF = DE \cdot BC$	8. Equality of cross products
9. $QP = EF$	9. Def. of ≅ segments
10. $AB \cdot EF = AQ \cdot BC$	10. Substitution Prop. (=)
11. $AQ \cdot BC = DE \cdot BC$	11. Substitution Prop. (=)
12. $AQ = DE$	12. Div. Prop. (=)
13. $\overline{AQ} \cong \overline{DE}$	13. Def. of ≅ segments
14. $\triangle AQP \cong \triangle DEF$	14. SAS
15. $\angle APQ \cong \angle F$	15. CPCTC
16. $\angle C \cong \angle F$	16. Transitive Prop. of ≅ ∠s
17. $\triangle ABC \sim \triangle DEF$	17. AA Similarity

35. Proof of Reflexive Property

Given: $\triangle ABC$

Prove: $\triangle ABC \sim \triangle ABC$

Proof:

Statements	Reasons
1. $\triangle ABC$	1. Given
2. $\angle A \cong \angle A$, $\angle B \cong \angle B$	2. Reflexive Prop. of ≅ ∠s
3. $\triangle ABC \sim \triangle ABC$	3. AA Similarity

Proof of Symmetric Property

Given: $\triangle ABC \sim \triangle DEF$

Prove: $\triangle DEF \sim \triangle ABC$

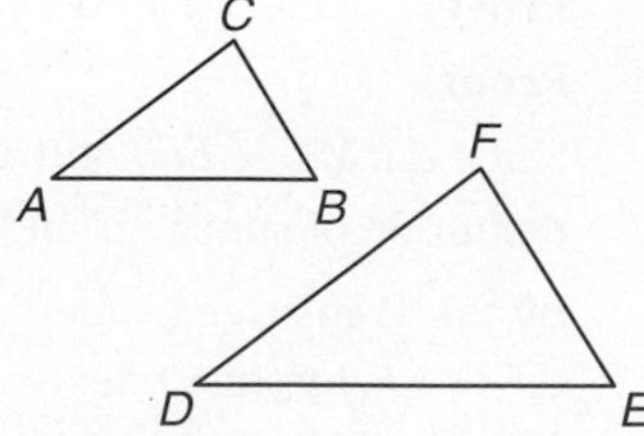

Proof:

Statements	Reasons
1. $\triangle ABC \sim \triangle DEF$	1. Given
2. $\angle A \cong \angle D$ $\angle B \cong \angle E$	2. Def. of $\sim$ polygons
3. $\angle D \cong \angle A$ $\angle E \cong \angle B$	3. Symmetric Prop. of $\cong$ ∡
4. $\triangle DEF \sim \triangle ABC$	4. AA Similarity

Proof of Transitive Property

Given: $\triangle ABC \sim \triangle DEF$
$\triangle DEF \sim \triangle GHI$

Prove: $\triangle ABC \sim \triangle GHI$

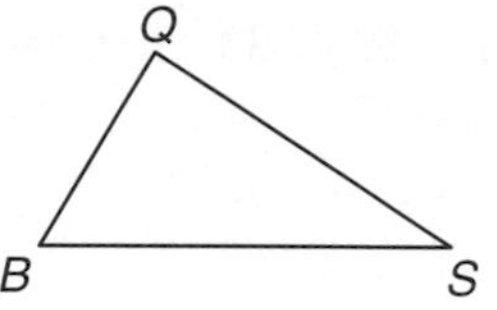

Proof:

Statements	Reasons
1. $\triangle ABC \sim \triangle DEF$ $\triangle DEF \sim \triangle GHI$	1. Given
2. $\angle A \cong \angle D$ $\angle B \cong \angle E$ $\angle D \cong \angle G$ $\angle E \cong \angle H$	2. Def. of $\sim$ polygons
3. $\angle A \cong \angle G$ $\angle B \cong \angle H$	3. Transitive Prop. of $\cong$ ∡
4. $\triangle ABC \sim \triangle GHI$	4. AA Similarity

37. 12.1 feet **39.** Yes; corresp. ∡ are $\cong$ and $\frac{2.0}{3.0} = \frac{1.6}{2.4} = \frac{1.8}{2.7}$. **41.** true

43.

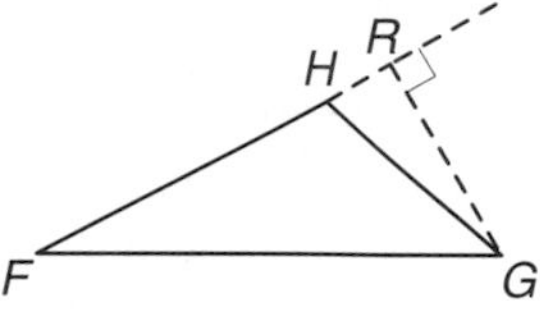

45. Reflexive Prop. of $\cong$ Segments **47.** $\{t \mid t \geq 5\}$

Page 361 Self Test

1. 2 **3.** about 487 soft drinks **5.** yes; SAS Similarity

Pages 366–369 Lesson 7–4

7a. true **7b.** false; $RS = 16$ **9.** Yes; if 3 or more $\parallel$ lines cut off $\cong$ segments on one transversal, then they cut off $\cong$ segments on every transversal. **11.** Yes; if 3 or more $\parallel$ lines intersect 2 transversals, then they cut off the transversals proportionately. **13.** DF **15.** GB **17.** true **19.** false; $\frac{BD}{DC} = \frac{2}{3}$ **21.** 21, 15 **23.** 10 **25.** 10 **27.** $B(5, -2)$; $C(11, 2)$ **29.** $AB = 7$, $BC = 10$, $AC = 9$

31.

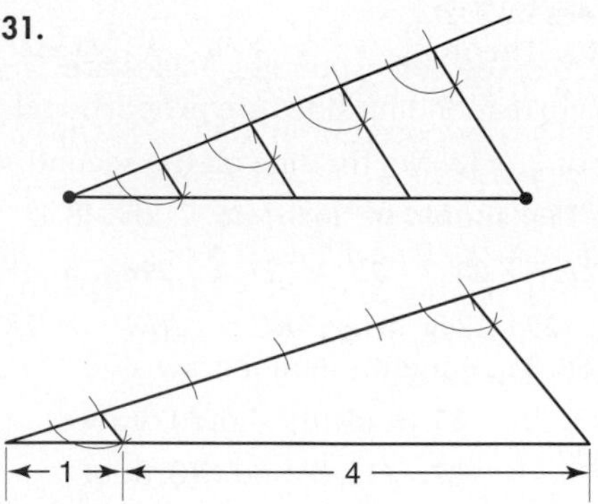

33. Given: D is the midpoint of $\overline{AB}$.
E is the midpoint of $\overline{AC}$.

Prove: $\overline{DE} \parallel \overline{BC}$
$DE = \frac{1}{2}BC$

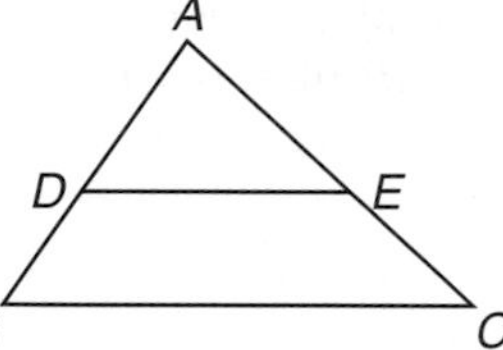

Proof:

Statements	Reasons
1. D is the midpoint of $\overline{AB}$. E is the midpoint of $\overline{AC}$.	1. Given
2. $\overline{AD} \cong \overline{DB}$ $\overline{AE} \cong \overline{EC}$	2. Def.of midpoint
3. $AD = DB$ $AE = EC$	3. Def. of $\cong$ segments
4. $AB = AD + DB$ $AC = AE + EC$	4. Segment Addition Postulate
5. $AB = AD + AD$ $AC = AE + AE$	5. Substitution Prop. (=)
6. $AB = 2AD$ $AC = 2AE$	6. Substitution Prop. (=)
7. $\frac{AB}{AD} = 2$ $\frac{AC}{AE} = 2$	7. Division Prop. (=)
8. $\frac{AB}{AD} = \frac{AC}{AE}$	8. Transitive Prop. (=)
9. $\angle A \cong \angle A$	9. Reflexive prop. of $\cong$ ∡
10. $\triangle ADE \sim \triangle ABC$	10. SAS Similarity
11. $\angle ADE \cong \angle ABC$	11. Def. of $\sim$ polygons
12. $\overline{DE} \parallel \overline{BC}$	12. If ⇸, and corr. ∡ are $\cong$, the lines are $\parallel$.
13. $\frac{BC}{DE} = \frac{AB}{AD}$	13. Def. of $\sim$ polygons
14. $\frac{BC}{DE} = 2$	14. Substitution Prop. (=)
15. $2DE = BC$	15. Mult. Prop. (=)
16. $DE = \frac{1}{2}BC$	16. Division Prop. (=)

35. $L(17, 8)$; $M(-7, -16)$ **37.** $x = \frac{14}{9}$ cm, $y = \frac{7}{3}$ cm, $z = \frac{35}{9}$ cm **39.** $\triangle ABC \sim \triangle ADE$ by SAS Similarity, so $\frac{AD}{AB} = \frac{DE}{BC}$. $AD = 60$ and $AB = 100$ so $\frac{60}{100} = \frac{DE}{BC}$. $\frac{3}{5} = \frac{DE}{BC}$ and $\frac{3}{5}BC = DE$. **41.** $1\frac{2}{3}, 1\frac{1}{3}$ **43.** true **45.** $\angle RST \cong \angle ABC$ **47.** $(k - 3)(k + 6)$

Pages 373–377 Lesson 7-5

5. AB; Angle Bisector Theorem **7.** 15 **9.** When solved algebraically, $x = 5$ or 2, which means $FB = 2$ or 5. Geometrically, $FB = 5$ is not possible. In ΔFBG, which is a right triangle, the hypotenuse is 5 and since the hypotenuse is the longest side of the triangle, leg FB cannot also be 5 units long. Thus FB must equal 2.
11. Sample answer: 6, 8, 10 and 9, 12, 15 **13.** 8.6 **15.** 18
17. DB **19.** $11\frac{1}{5}$ **21.** 36.68 **23.** 6 **25.** 29.25 **27.** 36
29. 5, 13.5 **31.** $8\frac{8}{9}$, $11\frac{1}{9}$ **33.** 26.14

35. Given: $\overline{JF}$ bisects $\angle EFG$.
$\overline{EH} \parallel \overline{FG}$; $\overline{EF} \parallel \overline{HG}$

Prove: $\frac{EK}{KF} = \frac{GJ}{JF}$

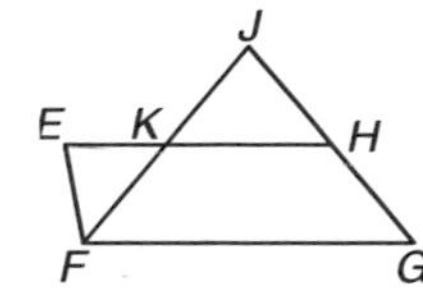

Proof:

Statement	Reasons
1. $\overline{EH} \parallel \overline{FG}$; $\overline{EF} \parallel \overline{HG}$ $\overline{JF}$ bisects $\angle EFG$.	1. Given
2. $\angle EFK \cong \angle KFG$	2. Def. of $\angle$ bisector
3. $\angle KFG \cong \angle JKH$	3. Corresponding $\angle$ Postulate
4. $\angle JKH \cong \angle EKF$	4. Vertical $\angle$s are $\cong$.
5. $\angle EFK \cong \angle EKF$	5. Transitive Prop. (=)
6. $\angle EFK \cong \angle FJH$	6. Alternate Interior $\angle$ Theorem
7. $\angle EKF \cong \angle FJH$	7. Transitive Prop. (=)
8. $\Delta EKF \sim \Delta GJF$	8. AA Similarity
9. $\frac{EK}{KF} = \frac{GJ}{JF}$	9. Def. of $\sim \Delta$s

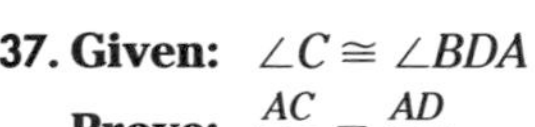

37. Given: $\angle C \cong \angle BDA$

Prove: $\frac{AC}{DA} = \frac{AD}{BA}$

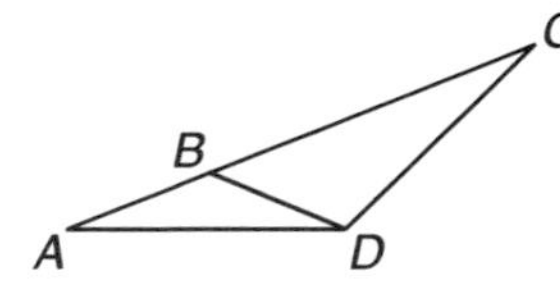

Proof:

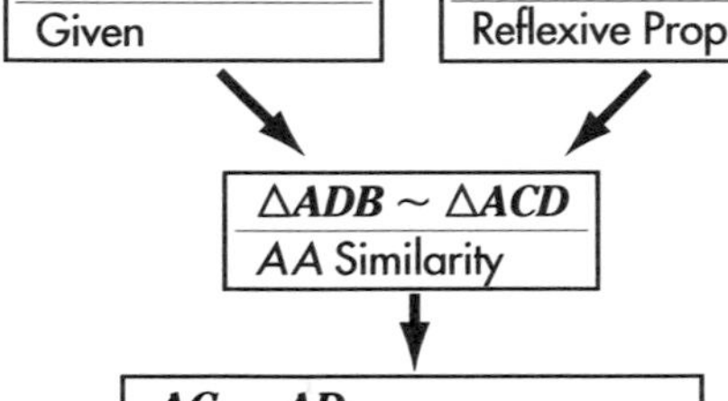

$\angle C \cong \angle BDA$ — Given

$\angle A \cong \angle A$ — Reflexive Prop. of $\cong$ $\angle$s

$\Delta ADB \sim \Delta ACD$ — AA Similarity

$\frac{AC}{DA} = \frac{AD}{BA}$ — Definition of similar polygons

39. Given: $\Delta ABC \sim \Delta RST$
$\overline{AD}$ is a median of ΔABC.
$\overline{RU}$ is a median of ΔRST.

Prove: $\frac{AD}{RU} = \frac{AB}{RS}$

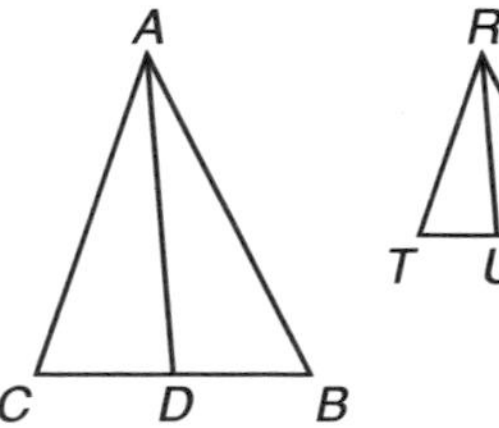

Proof:

We are given that $\Delta ABC \sim \Delta RST$, $\overline{AD}$ is median of ΔABC, and $\overline{RU}$ is a median of ΔRST. So, by the definition of median, $CD = DB$ and $TU = US$. According to the definition of similar polygons, $\frac{AB}{RS} = \frac{CB}{TS}$. $CB = CD + DB$ and $TS = TU + US$ by the Segment Addition Postulate.

Substituting, $\frac{AB}{RS} = \frac{CD + DB}{TU + US}$

$\frac{AB}{RS} = \frac{DB + DB}{US + US}$

$\frac{AB}{RS} = \frac{DB}{US}$

$\angle B \cong \angle S$ by the definition of similar polygons and $\Delta ABD \sim \Delta RSU$ by SAS Similarity. Therefore, $\frac{AD}{RU} = \frac{AB}{RS}$ by the defintion of similar polygons.

41a. 6; 24; 1:2; 1:4 **41b.** 35.5; 319.5; 1:3; 1:9 **41c.** 16.1; 578.3; 1:6; 1:36; Sample answer: If the ratio of the measures of the sides of two similar triangles is $a{:}b$, then the ratio of their areas is $a^2{:}b^2$. **43.** Yes; the enlarged picture will take approximately 109.2 cm of piping. **45.** yes; proportional perimeters **47.** 8.4 **49.** 14 seniors **51.** E **53.** No; the measures of corresponding sides are not equal. **55.** If a geometry test score is 89, then it is above average.
57. $\{a \mid -7 < a < 4\}$

Pages 381–383 Lesson 7-6

7. 1, 3, 6, 10, 15, ...; each difference is 1 more than the preceding difference. The triangular numbers are the numbers in the diagonal.

9. Given: ΔABC is equilateral.
$CD = \frac{1}{3}CB$
$CE = \frac{1}{3}CA$

Prove: $\Delta CED \sim \Delta CAB$

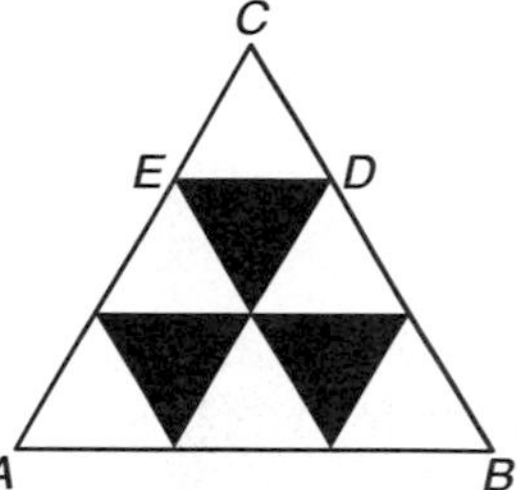

Proof:

Statements	Reasons
1. ΔABC is equilateral. $CD = \frac{1}{3}CB$; $CE = \frac{1}{3}CA$	1. Given
2. $\overline{AC} \cong \overline{BC}$	2. Def. of equilateral Δ
3. $AC = BC$	3. Def of $\cong$ segments
4. $\frac{1}{3}AC = \frac{1}{3}CB$	4. Mult. Prop. (=)
5. $CD = CE$	5. Substitution Prop. (=)
6. $\frac{CD}{CB} = \frac{CE}{CB}$	6. Division Prop. (=)
7. $\frac{CD}{CB} = \frac{CE}{CA}$	7. Substitution Prop. (=)
8. $\angle C \cong \angle C$	8. Reflexive Prop. of $\cong$ $\angle$s
9. $\Delta CED \sim \Delta CAB$	9. AA Similarity

11a. converges to 1 **11b.** alternates between 0.2 and 5.0
11c. converges to 1

13a.

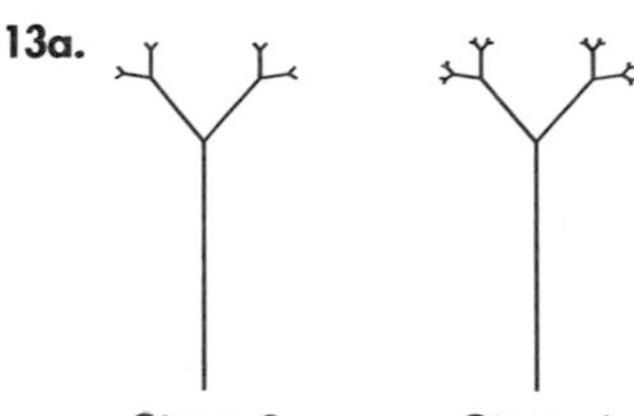

stage 1: 2, stage 2: 6, stage 3: 14, stage 4: 30 **13b.** Sample answer: at stage n, the number of branches would be $S_n = 2(1 - 2^n)$. **15.** Similar to stage 1 of the Sierpinski triangle variation from Exercise 10.

17a. stage 1: 3 units
stage 2: $3 \cdot \frac{4}{3}$ or 4 units
stage 3: $3 \cdot \frac{4}{3} \cdot \frac{4}{3} = 3\left(\frac{4}{3}\right)^2$ or $5\frac{1}{3}$ units
stage 4: $3 \cdot \frac{4}{3} \cdot \frac{4}{3} \cdot \frac{4}{3} = 3\left(\frac{4}{3}\right)^3$ or $7\frac{1}{9}$ units
17b. $P = 3\left(\frac{4}{3}\right)^{n-1}$; as the stages increase, the perimeter continues to increase. The perimeter will approach infinity. **19a.** Trisect each of the three edges; replace the middle section on the center edge with three segments of length equal to the length removed; replace the first section on each of the outside edges with three segments of length equal to that removed. Repeat. **21a.** 25 **21b.** 1275 **23.** 2400 units **25.** $-\frac{2}{7}$ **27.** E **29.** $\{(-1, -1), (0, 2), (2, 8), (4, 14)\}$

Page 385 Lesson 7–6B

1a. The midpoint of $\overline{PA}$ is (0.1, 0.2) in $\triangle L$; the midpoint of $\overline{PB}$ is (0.6, 0.2) in $\triangle R$; the midpoint of $\overline{PC}$ is (0.35, 0.65) in $\triangle T$. **1b.** The midpoint of $\overline{PA}$ is (0.25, 0.1) in $\triangle L$; the midpoint of $\overline{PB}$ is (0.75, 0.1) in $\triangle R$; the midpoint of $\overline{PC}$ is (0.5, 0.55) in $\triangle T$. **1c.** $\frac{1}{27}$ **1d.** $\frac{1}{9}$ **1e.** $\frac{1}{9}$ **3a.** $\frac{1}{9}$ **3b.** $\frac{1}{3}$ **3c.** $\frac{1}{27}$

Page 387 Chapter 7 Highlights

1. false; proportional **3.** false; two, two **5.** true **7.** true **9.** true **11.** false; $ad = bc$

Pages 388 390 Chapter 7 Study Guide and Assessment

13. $\frac{1}{2}$ **15.** 3.2 **17.** 1.25 **19.** $9\frac{1}{3}$ **21.** false **23.** 10 **25.** no; not enough information **27.** 2.5 **29.** $9\frac{1}{3}$
31.

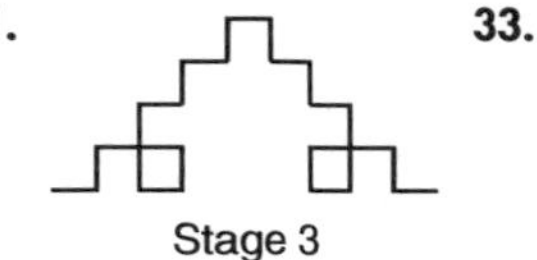

Stage 3

33. 29 **35.** 31.2 cm **37.** 2

CHAPTER 8 APPLYING RIGHT TRIANGLES AND TRIGONOMETRY

Page 396 Lesson 8–1A

1. yes. **3.** The sum of the areas of the two smaller squares $(a^2 + b^2)$ is equal to the area of the larger square (c^2).

Pages 401–404 Lesson 8–1

7. 10 **9.** $\sqrt{10} \approx 3.16$; $\sqrt{14} \approx 3.74$ **11.** $\sqrt{51} \approx 7.14$
13.

a	b	c
3	4	5
6	8	10
9	12	15
12	16	20

13a. yes; 3, 4, 5 **13b.** Each triple is a multiple of the triple 3, 4, and 5. **13c.** Sample answer: The triples are all multiples of the triple 3, 4, and 5. **13d.** Yes; the measures of the sides are always multiples of 3, 4, and 5. **15.** 6 **17.** $\frac{\sqrt{2}}{3} \approx 0.47$ **19a.** $\overline{AH}, \overline{AL}$ **19b.** $\overline{AL}, \overline{LH}$ **19c.** $\overline{LH}, \overline{AH}$ **21.** $2\sqrt{6} \approx 4.90$; $\sqrt{33} \approx 5.74$ **23.** $7\frac{1}{2}$; $2\frac{1}{2}$ **25.** $5\sqrt{5} \approx 11.18$; $5\sqrt{30} \approx 27.39$ **27.** 4.5 **29.** 20 **31.** no

33. Sample answers: 16, 30, 34; 24, 45, 51 **35.** Sample answers: 18, 80, 82; 27, 120, 123

37. Given: $\angle ADC$ is a right angle. $\overline{DB}$ is an altitude of $\triangle ADC$.

Prove: $\frac{AB}{DB} = \frac{DB}{CB}$

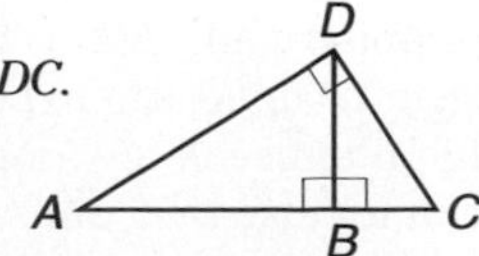

Proof:
It is given that $\triangle ADC$ is a right triangle and $\overline{DB}$ is an altitude of $\triangle ADC$. $\triangle ADC$ is a right triangle by the definition of a right triangle. Therefore, $\triangle ADB \sim \triangle DCB$, because if the altitude is drawn from the vertex of the right angle to the hypotenuse of a right triangle, then the two triangles formed are similar to the given triangle and to each other. So $\frac{AB}{DB} = \frac{DB}{CB}$ by definition of similar polygons.

39. Given: $\triangle ABC$ with sides of measure a, b, and c, where $c^2 = a^2 + b^2$

Prove: $\triangle ABC$ is a right triangle.

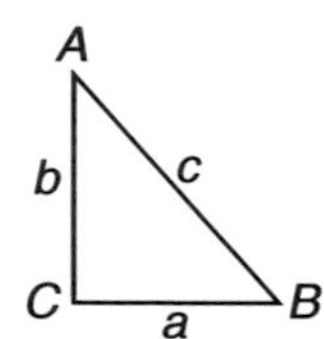

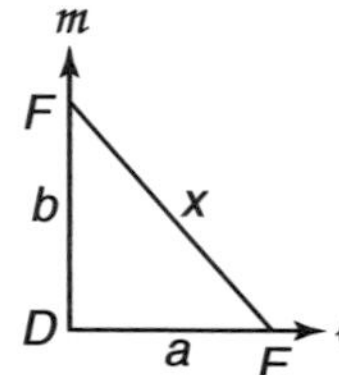

Proof:

Draw $\overline{DE}$ on line ℓ with measure equal to a. At D, draw line $m \perp \overline{DE}$. Locate point F on m so that $DF = b$. Draw $\overline{FE}$ and call its measure x. Because $\triangle FED$ is a right triangle, $a^2 + b^2 = x^2$. But $a^2 + b^2 = c^2$, so $x^2 = c^2$ or $x = c$. Thus, $\triangle ABC \cong \triangle FED$ by SSS. This means $\angle C \cong \angle D$. Therefore, $\angle C$ must be a right angle, making $\triangle ABC$ a right triangle. **41a.** 3, 4, 5; 6, 8, 10; 12, 16, 20; 24, 32, 40; 27, 36, 45 **41b.** yes **43.** 2.4 yd **45.** no; Sample counterexample: isosceles triangle with sides measuring 4, 4, and 7

$$\sqrt{4} + \sqrt{4} \stackrel{?}{=} \sqrt{7}$$
$$2 + 2 \stackrel{?}{=} 2.65$$
$$4 \neq 2.65$$

47. 254 mi **49.** 13

51. D **53.** 30, 60, 90

55.

x	y
0	1
1	4
−1	−2

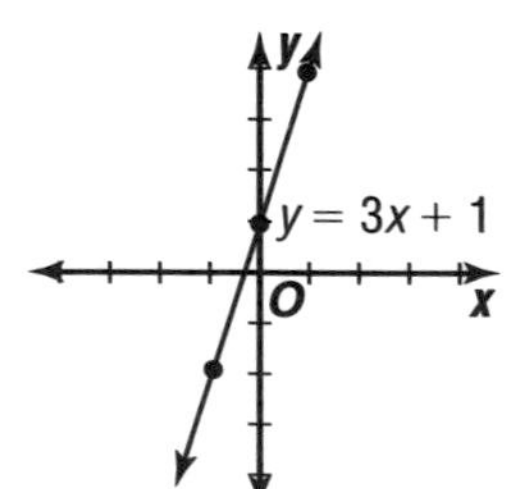

Pages 409–411 Lesson 8–2

5. 16; $8\sqrt{3} \approx 13.86$ **7.** 1 ft **9.** $(-4, -2 + 10\sqrt{3})$ or about $(-4, 15.32)$ **11.** $4.5\sqrt{3} \approx 7.79$; 4.5 **13.** $\frac{3\sqrt{2}}{2} \approx 2.12$; $3\sqrt{2} \approx 4.24$ **15.** $9\sqrt{3} \approx 15.59$; 9 **17.** $13.5\sqrt{2}$ or about 19.09 cm **19.** 9 m **21.** $6.5\sqrt{3}$ or about 11.26 cm **23.** $10.4\sqrt{3}$ or about 18.01 m **25.** $7\sqrt{2} \approx 9.90$ **27.** $\left(-3 - \frac{13\sqrt{3}}{3}, -6\right)$ or about $(-10.51, -6)$ **29.** 2.25

31a.

31b. $m\angle BCD = 120$, $m\angle BCE = 60$, $m\angle ABC = 180 - 120 = 60$, $m\angle EBC = 30$, $m\angle BEC = 180 - 60 - 30 = 90$; $\triangle BEC$ is a 30°-60°-90° triangle. **31c.** $\frac{22\sqrt{3}}{3} \approx 12.70$; $\frac{44\sqrt{3}}{3} \approx 25.40$ **33.** 23 ft **35.** AA Similarity, SSS Similarity, SAS Similarity **37.** Yes; since $QS = 2\sqrt{26}$ and $RT = 2\sqrt{26}$, the diagonals are congruent.

39. $\sqrt{5}$

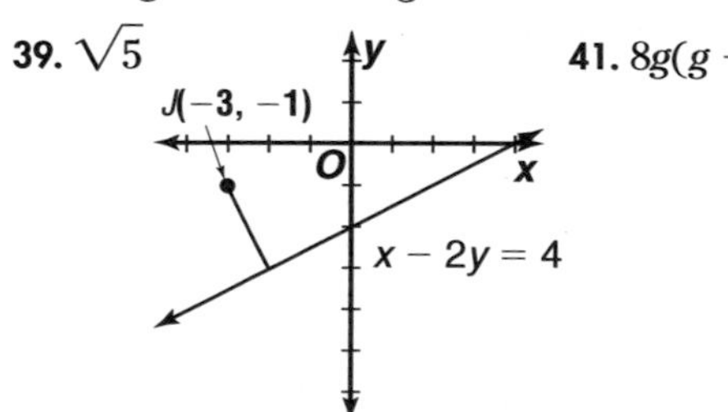

41. $8g(g + 4)$

Pages 416–420 Lesson 8–3

7. $\frac{20}{21} \approx 0.9524$ **9.** 0.2419 **11.** 63.5 **13.** 58.0 **15.** about 20.6 **17.** $\frac{\sqrt{26}}{26} \approx 0.1961$ **19.** $\frac{5\sqrt{26}}{26} \approx 0.9806$ **21.** $\frac{5}{1} = 5.0000$ **23.** $\frac{\sqrt{26}}{26} \approx 0.1961$ **25.** $\frac{5}{1} = 5.0000$ **27.** 0.9135 **29.** 0.9511 **31.** 0.3640 **33.** 75.6 **35.** 27.2 **37.** 23.2 **39.** 32.6 **41.** 20.4 **43.** 6.1 **45.** about 54.5 **47.** 38.4; 32.6 **49.** 41.8; 29.2 **51.** about 35.5 **53a.** about 272,837 astronomical units **53b.** The stellar parallax would be too small. **55.** $6\sqrt{2} \approx 8.49$ **57.** 15 **59.** C **61.** 7.5; 12 **63.** $\angle HGI$ **65.** (2, 4)

Page 420 Self Test

1. $3\sqrt{15} \approx 11.62$ **3.** 4 **5.** $7.5\sqrt{2} \approx 10.61$ **7.** $\frac{3\sqrt{13}}{13} \approx 0.8321$ **9.** $\frac{2}{3} \approx 0.6667$

Pages 422–425 Lesson 8–4

5. $\angle OCP$; $\angle DPC$ **7.** $\cos P = \frac{2.3}{5.5}$; 65.3 **9.** about 151.1 ft **11.** $\angle GEF$; $\angle HFE$ **13.** $\angle KHJ$; $\angle IJH$ **15.** $\tan Y = \frac{54}{28}$; 62.6 **17.** $\cos 66° = \frac{7}{XY}$; 17.2 **19.** $\cos X = \frac{4.5}{6.6}$; 47.0 **21a.** about 86.3° **21b.** about 166.1 ft **23.** about 177.7 yd **25.** about 3.6° **27a.** about 210.3 ft **27b.** about 276.4 ft **29.** about 993.0 ft **31.** $\frac{\sin M}{\sin G}$ **33a.** about 3.12 ft **33b.** the bottom 5.93 ft of the window **35a.** $\frac{4}{5} = 0.8000$ **35b.** $\frac{4}{5} = 0.8000$ **35c.** $\frac{4}{3} \approx 1.3333$ **37.** 10 ft

39.

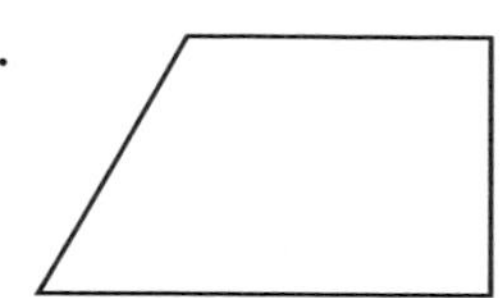

41. 78 in.

43.

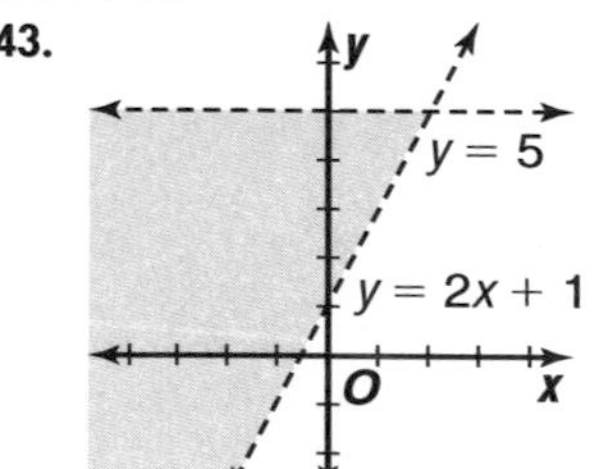

Pages 429–430 Lesson 8–5

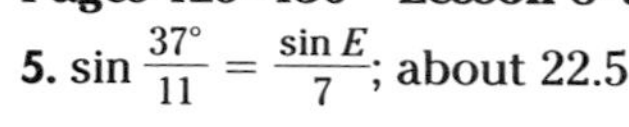

5. $\sin \frac{37°}{11} = \frac{\sin E}{7}$; about 22.5

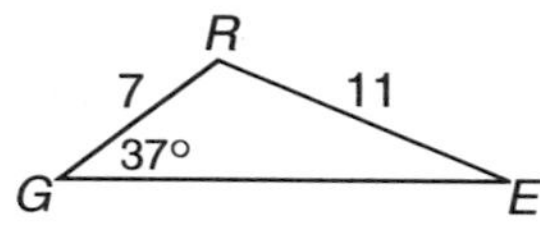

7. $m\angle Y \approx 47.2$, $m\angle W \approx 30.8$, $w = 12.6$

9a.

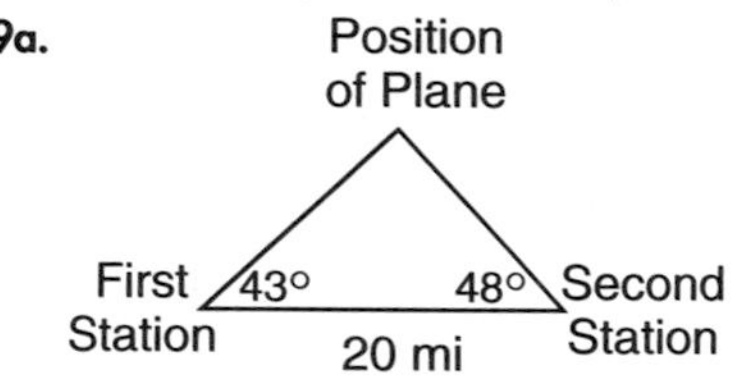

9b. about 14.9 mi, about 13.6 mi

11. $\sin \frac{47°}{13} = \frac{\sin R}{9}$; about 30.4

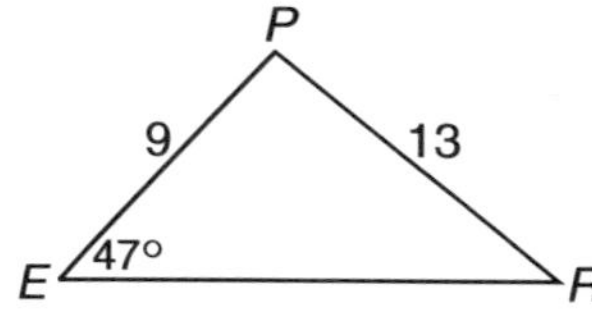

13. $\sin \frac{96°}{48} = \frac{\sin R}{10}$; about 12.0

15. $\frac{\sin 59°}{14.8} = \frac{\sin P}{8.3}$; about 28.7

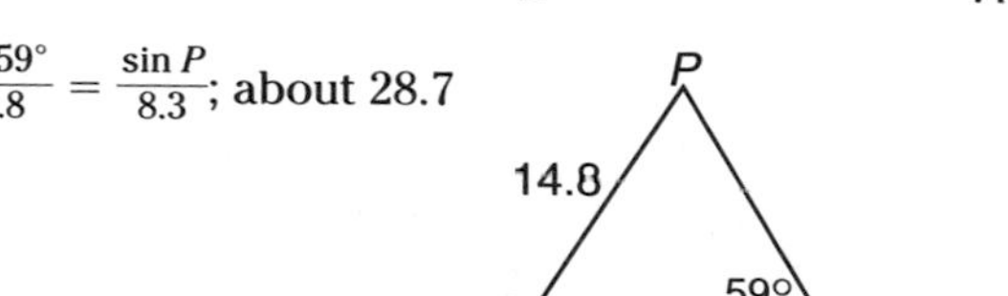

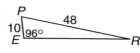

17. $m\angle F \approx 25.7$, $m\angle D \approx 47.3$, $d \approx 15.4$ **19.** $m\angle F \approx 10.9$, $m\angle D \approx 135.1$, $d \approx 34.1$ **21.** $m\angle F = 60$, $f \approx 1.0$, $r \approx 1.1$ **23.** $m\angle F \approx 49.6$, $m\angle R \approx 42.4$, $r \approx 14.2$ **25.** about 93.2 cm **27.** about 11.3; about 9.8 **29.** Yes; by definition, $\sin A = \frac{a}{c}$ and $\sin B = \frac{b}{c}$. According to the Law of Sines, $\frac{\sin A}{a} = \frac{\sin B}{b}$.

$$\frac{\frac{a}{c}}{a} \stackrel{?}{=} \frac{\frac{b}{c}}{b}$$

$$\frac{1}{c} = \frac{1}{c}$$

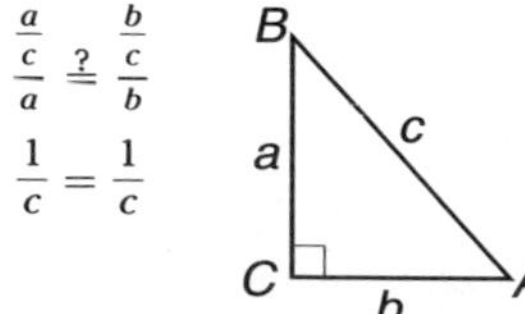

So the Law of Sines holds true for the acute angles of a right triangle. **31.** about 3.0 mi and 3.9 mi **33.** about 44.0 ft **35.** 27, 36 **39.** $9x^2 + 2xy$

Pages 433–436 Lesson 8–6

5a. Law of Sines **5b.** $m\angle E \approx 60.1$, $m\angle F \approx 47.9$, $f \approx 3.5$ **7.** $m\angle C = 81$, $a \approx 9.1$, $b \approx 12.1$ **9.** $m\angle A \approx 15.8$, $m\angle C \approx 145.2$, $c \approx 106.9$ **11.** about 35.7 **13.** Law of Cosines; $t \approx 6.9$, $m\angle R \approx 79.0$, $m\angle S \approx 63.0$ **15.** Law of Cosines; $m\angle X \approx 78.5$, $m\angle Y \approx 44.4$, $m\angle Z \approx 57.1$

17.

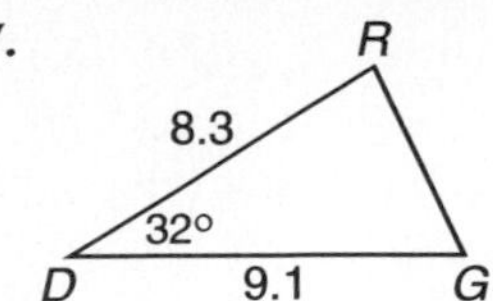

Law of Cosines; $d \approx 4.9$, $m\angle G \approx 63.8$, $m\angle R \approx 84.2$

19.

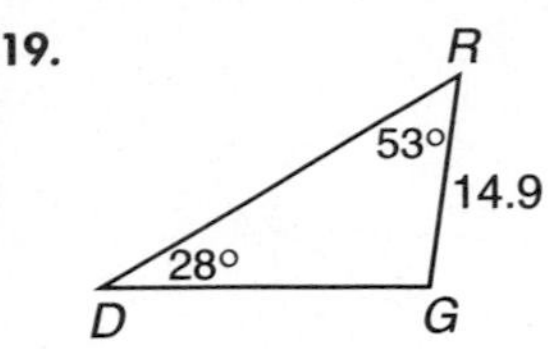

Law of Sines; $m\angle G = 99$, $g \approx 31.3$, $r \approx 25.3$ **21.** $m\angle H \approx 59.5$, $m\angle J \approx 30.5$, $m\angle K \approx 90.0$ **23.** $m\angle K = 79$, $h \approx 13.4$, $j \approx 11.7$ **25.** $j \approx 6.0$, $m\angle H \approx 59.3$, $m\angle K \approx 65.7$ **27.** $m\angle H \approx 49.9$, $m\angle J \approx 69.8$, $m\angle K \approx 60.3$ **29.** $m\angle H \approx 53.6$, $m\angle J \approx 59.6$, $m\angle K \approx 66.8$ **31.** $m\angle H \approx 44.4$, $m\angle J \approx 57.1$, $m\angle K \approx 78.5$ **33.** $HF \approx 15.7$, $EG \approx 11.2$ **35.** 15.8, 15.8, 15.8, 15.8; 110.7, 69.3, 110.7, 69.3 **37.** about 6.4 light-years **39.** Liz **41.** $m\angle Q = 55$, $p \approx 12.6$, $q \approx 10.7$ **43.** 155 **45.** (4, 3), (−2, 7), (−4, −5) **47.** True; Sample explanation: A 3-4-5 triangle is a right triangle with sides of different lengths. **49.** $x^2 + 9x + 18$

Page 437 Chapter 8 Highlights

1. sine **3.** Law of Cosines **5.** trigonometry **7.** Pythagorean triple **9.** Law of Sines **11.** trigonometric ratios **13.** means

Pages 438–440 Chapter 8 Study Guide and Assessment

15. $9\sqrt{3} \approx 15.59$ **17.** 2 **19.** $6\sqrt{2} \approx 8.49$ **21.** about 25.79 **23.** $\sqrt{30} \approx 5.48$ **25.** 17 **27.** no **29.** $3\sqrt{2} \approx 4.24$ **31.** $1.55\sqrt{3} \approx 2.68$ **33.** $\frac{15}{17} \approx 0.8824$ **35.** $\frac{15}{17} \approx 0.8824$ **37.** about 282.7 m **39.** about 4.3° **41.** $m\angle A = 51$, $a \approx 70.2$, $c \approx 89.7$ **43.** $m\angle B = 80$, $b \approx 14.8$, $c \approx 14.1$ **45.** $b \approx 30.6$, $m\angle A \approx 87.3$, $m\angle C \approx 48.7$ **47.** $m\angle A \approx 40.8$, $m\angle B \approx 78.6$, $m\angle C \approx 60.6$ **49.** about 0.5 mi **51.** about 7.7° **53.** Sample answer: Look for a pattern; 2.

CHAPTER 9 ANALYZING CIRCLES

Pages 449–451 Lesson 9–1

7. 4.7 **9.** $d = 14$, $C = 44.0$ **11.** 9π **13.** 97.7 feet **15.** $\overline{RI}$ **17.** No; it is a radius. **19.** $\overline{RM}$, $\overline{AM}$, $\overline{DM}$, $\overline{IM}$ **21.** 5.9 **23.** $d = 10$, $C = 31.4$ **25.** $d = 43.6$, $r = 21.8$ **27.** $r = x$, $C = 6.3x$ **29.** 8π cm **31.** $6\sqrt{2}\pi$ cm **33.** 6 **35.** 20 **37.** The circumference is doubled.

39. Given: $\odot P$ with diameter $\overline{SA}$ and chord $\overline{KR}$

Prove: $SA > KR$

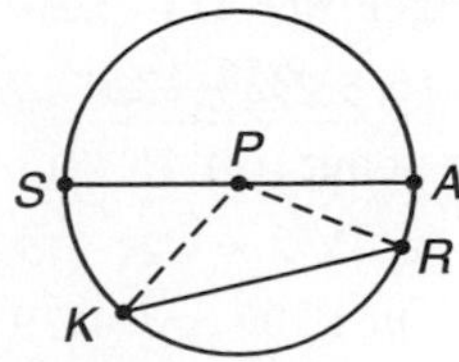

Proof:

By the Segment Addition Postulate, $SA = SP + PA$. Draw $\overline{PK}$ and $\overline{PR}$ since through any two points there is one line. Since all radii of a circle are congruent, $\overline{SP} \cong \overline{PK}$ and $\overline{PA} \cong \overline{PR}$. By substitution, $SA = PK + PR$. By the Triangle Inequality Theorem, $PK + PR > KR$. By substitution, $SA > KR$.

41. $d \approx 11.8$ cm, $C \approx 37.1$ cm, $\frac{C}{d} \approx 3.144$ or π **43.** 3141.6 cm **45a.** 79 ft **45b.** 132 ft **47.** 273.9 cm **49.** 68 in. **51.** 17.5 **53.** Yes; the sum of any two sides is greater than the length of the third side. **55.** (0, −4)

Pages 455–458 Lesson 9–2

7. minor; 138 **9.** 12.6 **11.** 69.1 **13.** 137 **15.** true; def. of concentric circles **17.** True; concentric circles have the same center. **19.** minor, 21 **21.** minor, 90 **23.** minor, 111 **25.** major, 201 **27.** minor, 159 **29.** 28.3 **31.** 18.8 **33.** 26.1 **35.** 35.5 **37.** 26 **39.** 76 **41.** 256 **43.** 24 **45.** False; arcs are not of the same circle. **47.** False; arcs are not of same circle. **49.** true **51.** $\frac{\sqrt{2}}{2}$

53. Sample answer:

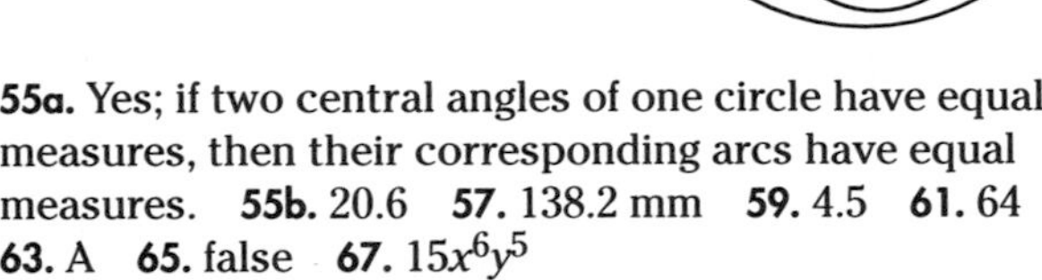

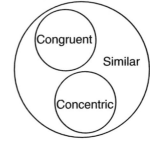

55a. Yes; if two central angles of one circle have equal measures, then their corresponding arcs have equal measures. **55b.** 20.6 **57.** 138.2 mm **59.** 4.5 **61.** 64 **63.** A **65.** false **67.** $15x^6y^5$

Pages 461–465 Lesson 9–3

5. In a circle, if a diameter is perpendicular to a chord, then it bisects the chord and its arc. **7.** 5 **9.** 3.4

11a.

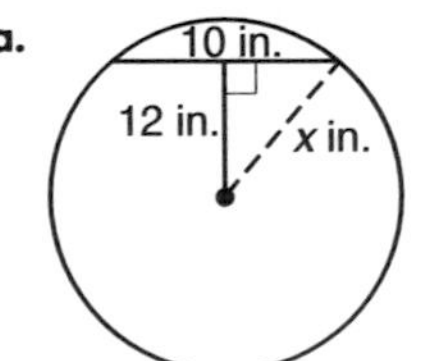

11b. 13 in.

13. In a circle or in congruent circles, two minor arcs are congruent if and only if their corresponding chords are congruent. **15.** In a circle or in congruent circles, two chords are congruent if and only if they are equidistant from the center. **17.** S **19.** $\widehat{PQ}$, $\widehat{VU}$, and $\widehat{UW}$ **21.** Neither; they are $\cong$. **23.** They are perpendicular to the same line. **25.** 28 **27.** 21 **29.** $\overline{RH} \cong \overline{MR}$ **31.** Yes; since radii are congruent, $\odot P \cong \odot Q$. Since the corresponding chords are congruent, by Theorem 9–1, $\widehat{AB} \cong \widehat{RS}$. **33.** 24

35. 19.2 cm

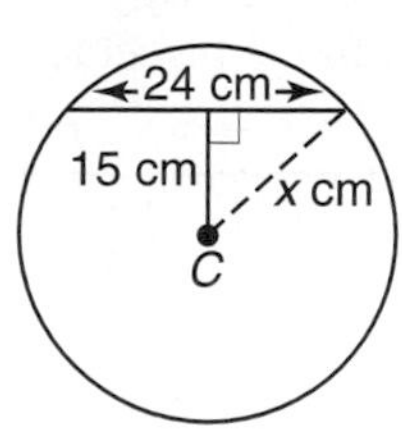

37. 48 mm

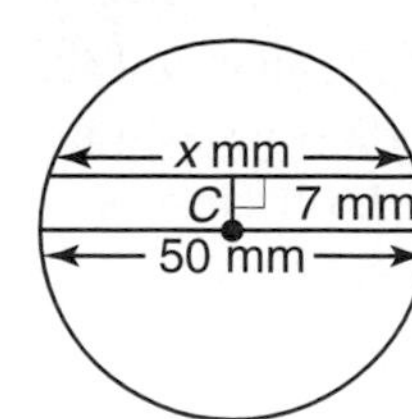

39. 27.7

41. Given: $\odot O$
$\overline{AB} \cong \overline{CD}$

Prove: $\widehat{AB} \cong \widehat{CD}$

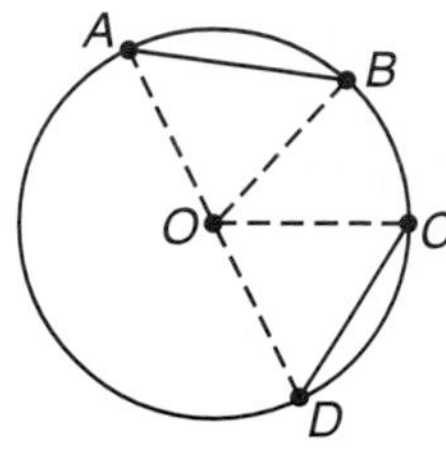

Proof:

Statements	Reasons
1. Draw radii $\overline{OA}$, $\overline{OB}$, $\overline{OC}$, and $\overline{OD}$.	1. Through any 2 pts. there is 1 line.
2. $\overline{OA} \cong \overline{OC}$ $\overline{OB} \cong \overline{OD}$	2. All radii of a $\odot$ are $\cong$.
3. $\overline{AB} \cong \overline{CD}$	3. Given
4. $\triangle ABO \cong \triangle CDO$	4. SSS
5. $\angle AOB \cong \angle COD$	5. CPCTC
6. $\widehat{AB} \cong \widehat{CD}$	6. In a $\odot$, 2 minor arcs are $\cong$ if and only if their corr. central $\angle$s are $\cong$.

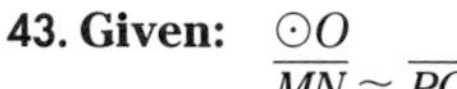

43. Given: $\odot O$
$\overline{MN} \cong \overline{PQ}$

Prove: $\overline{OA} \cong \overline{OB}$

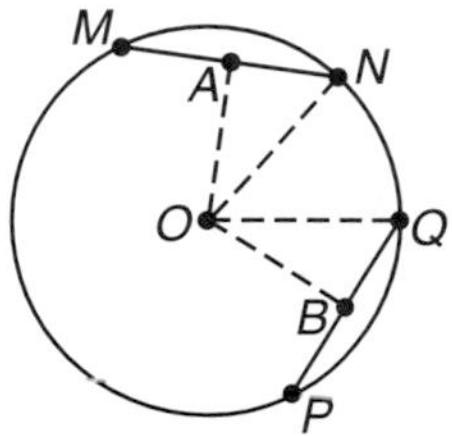

Proof:

Statements	Reasons
1. Draw radii $\overline{ON}$ and $\overline{OQ}$. Draw $\overline{OA}$ so that $\overline{OA} \perp \overline{MN}$ and draw $\overline{OB}$ so that $\overline{OB} \perp \overline{PQ}$.	1. Through any 2 pts. there is 1 line.
2. $\overline{OA}$ bisects $\overline{MN}$. $\overline{OB}$ bisects $\overline{PQ}$.	2. $\overline{OA}$ and $\overline{OB}$ can be extended to form radii, and a radius $\perp$ to a chord bisects the chord.
3. $AN = \frac{1}{2}MN$ $BQ = \frac{1}{2}PQ$	3. Def. of $\perp$ bisector
4. $\overline{MN} \cong \overline{PQ}$	4. Given
5. $MN = PQ$ $AN = BQ$	5. Def. of $\cong$ segments
6. $\overline{ON} \cong \overline{OQ}$	6. All radii of a $\odot$ are $\cong$.
7. $\triangle AON \cong \triangle BOQ$	7. HL
8. $\overline{OA} \cong \overline{OB}$	8. CPCTC

45b. It is the only point equidistant from T, S, and A. **45c.** Minneapolis, Minnesota **47.** True; all points on a circle are equidistant from the center. **49.** 29.8 **51.** false **53.** $\angle M$ **55.** $-24r^3s + 2rs + 16r$

Pages 469–473 Lesson 9–4

7. $\widehat{BC}$ **9.** 40 **11.** 72 **13.** 104 **15.** 52 **17.** $\widehat{HC}$ **19.** $\widehat{TCH}$ **21.** 52 **23.** There are none shown in the figure. **25.** 23.5 **27.** 102 **29.** 60 **31.** 120 **33.** 240 **35.** 240 **37.** 89 **39.** 114 **41.** 33 **43.** 44.5 **45.** 78.5

47. Given: $\overline{BR} \parallel \overline{AC}$

Prove: $\widehat{RA} \cong \widehat{BC}$

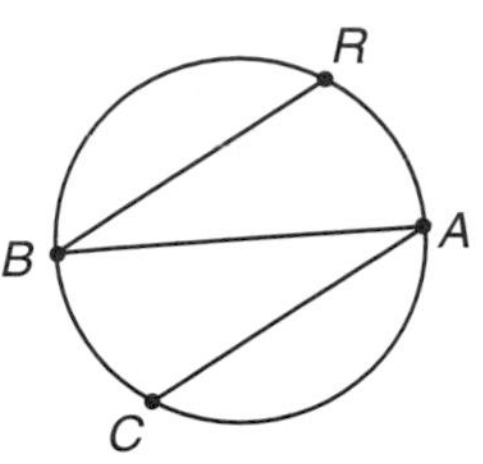

Proof:

Statements	Reasons
1. $\overline{BR} \parallel \overline{AC}$	1. Given
2. $\angle RBA \cong \angle BAC$	2. Alt. Int. $\angle$ Theorem
3. $m\angle RBA = m\angle BAC$	3. $\cong$ $\angle$s have $=$ measures.
4. $m\angle RBA = \frac{1}{2}m\widehat{RA}$ $m\angle BAC = \frac{1}{2}m\widehat{BC}$	4. If an $\angle$ is inscribe in a $\odot$, the measure of the $\angle$ = $\frac{1}{2}$ the measure of its intercepted arc.
5. $\frac{1}{2}m\widehat{RA} = \frac{1}{2}m\widehat{BC}$	5. Substitution Prop. (=)
6. $m\widehat{RA} = m\widehat{BC}$	6. Mult. Prop. (=)
7. $\widehat{RA} \cong \widehat{BC}$	7. Arcs that have = measures and are in the same $\odot$ are $\cong$.

49. No; opposite angles must be congruent and supplementary. Therefore, they each must be 90° or right angles.

51.
$$\begin{aligned} m\angle PRQ &= m\angle KRQ - m\angle KRP \\ &= \tfrac{1}{2}m\widehat{KQ} - \tfrac{1}{2}m\widehat{KP} \\ &= \tfrac{1}{2}(m\widehat{KQ} - m\widehat{KP}) \\ &= \tfrac{1}{2}\, m\widehat{PQ} \end{aligned}$$

53. Given: $\widehat{PQR}$ is a semicircle of $\odot C$.

Prove: $\angle PQR$ is a right $\angle$.

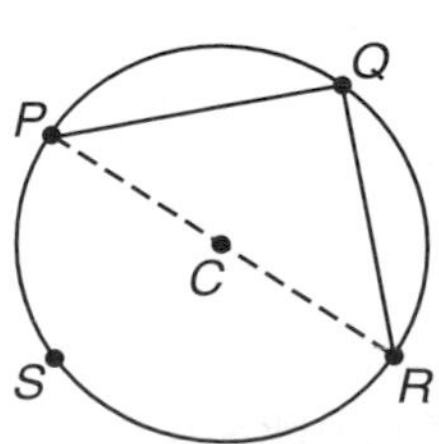

Proof:

Since $\widehat{PQR}$ is a semicircle, $\widehat{PSR}$ is also a semicircle and has a degree measure of 180. From the diagram, $\angle PQR$ is an inscribed angle, and $m\angle PQR = \frac{1}{2}(m\widehat{PSR})$ or 90. As a result, $\angle PQR$ is a right angle.

55. yes **57.** Position the carpenter's square so that the vertex of the right angle of the square is on the circle and the square forms an inscribed angle. Since the inscribed angle is a right angle, the measure of the intercepted arc is 180. So, the points where the square crosses the circle are the ends of a diameter (Theorem 9–6). Mark these points and draw the diameter. Draw another diameter using the same method. The point where the two diameters intersect is the center of the circle.

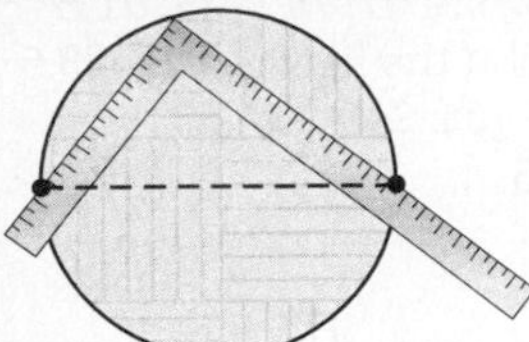

59. 53.4 ft **63.** Sample answer: measure the diagonals to see if they are congruent. **65.** 2.5 **67.** 10.4 **69.** $\left(2, -\frac{1}{2}\right)$

Page 473 Self Test

1. $\overline{PD}, \overline{PB}, \overline{PC}$ **3.** 13.1 **5.** 115

Pages 479–482 Lesson 9–5

7. 24 **9.** 132 **11.** $\overline{KL}$ **13.** In quadrilateral *KLEM*, $\angle K$ and $\angle E$ are right angles and the sum of their measures equals 180. Therefore, the sum of $m\angle L$ and $m\angle M$ must equal 180 and the angles must be supplementary. **15.** 8 **17.** $5\sqrt{3} \approx 8.7$ **19.** 10 **21.** 60 **23.** 90 **25.** $2\sqrt{3} \approx 3.5$ **27.** 30 **29.** $\overline{GF}$ and $\overline{AJ}$ **31.** 14 **33.** 4 **35.** 15 **37a.** Given **37b.** If 2 segments from the same exterior point are tangent to a ⊙, then they are ≅. **37c.** Radii of the same circle are ≅. **37d.** Reflexive Property (=) **37e.** SSS **37f.** CPCTC **39.** $-\frac{2}{3}$

41.

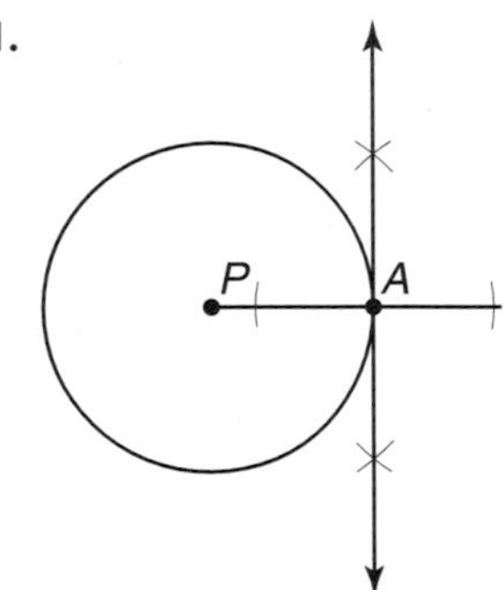

43. 4; $\triangle PQR$ is equilateral and $\overline{QN} \cong \overline{NR}$.

45. Given: $\overline{GR}$ is tangent to $\odot D$ at G. $\overline{AG} \cong \overline{DG}$

Prove: $\overline{GA}$ bisects $\overline{RD}$.

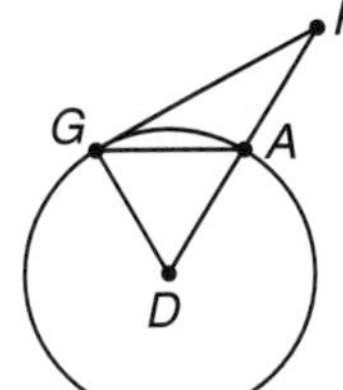

Proof:

Since $\overline{DA}$ is a radius, $\overline{DG} \cong \overline{DA}$. Since $\overline{AG} \cong \overline{DG} \cong \overline{DA}$, $\triangle GDA$ is equilateral. Therefore, each angle has a measure of 60. Since $\overline{GR}$ is tangent to $\odot D$, $m\angle RGD = 90$. Since $m\angle AGD = 60$, then by Angle Addition Post., $m\angle RGA = 30$. If $m\angle DAG = 60$, then $m\angle RAG = 120$. Then $m\angle R = 30$. Thus, $\triangle RAG$ is isosceles. By Transitive Prop. (=), $\overline{RA} \cong \overline{DA}$. Thus, $\overline{GA}$ bisects $\overline{RD}$.

47. Given: $\overleftrightarrow{CA}$ is tangent to the circle at A.

Prove: $\overline{XA} \perp \overleftrightarrow{CA}$

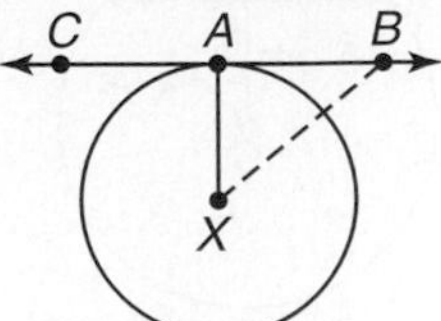

Proof:

Statements	Reasons
1. $\overleftrightarrow{CA}$ is tangent to the circle at A.	1. Given
2. Pick any point on $\overleftrightarrow{CA}$ other than A and call it B. Draw $\overline{XB}$.	2. Through any 2 pts. there is 1 line.
3. $\overleftrightarrow{CA}$ intersects $\odot X$ at exactly one point, A, and B lies in the exterior of $\odot X$.	3. Def. of tangent
4. $XA < XB$	4. The measure of a segment joining an exterior pt. to the center of a ⊙ is greater than the measure of a radius.
5. $\overline{XA} \perp \overleftrightarrow{CA}$	5. $\overline{XA}$ is the shortest segment from X to $\overleftrightarrow{CA}$.

49a. $\overline{QR}$ **49b.** One definition is a ratio of the measure of the leg opposite the acute angle to the measure of the leg adjacent to the acute angle. Another definition is a line in a plane that intersects a circle in the plane in exactly one point. **51.** 1.73 meters **53.** 14 **55.** 25 cm **57.** $x > 5$ **59.** no conclusion **61.** $(x + 8)^2$

Pages 486–490 Lesson 9–6

7. 134 **9.** 33 **11.** 54 **13.** 169 **15.** 157.5 **17.** 70 **19.** 65 **21.** 70 **23.** 5 **25.** 105 **27.** 65 **29.** 20 **31.** 26 **33a.** 44 **33b.** 30 **33c.** 15 **35.** 26 **37.** 102 **39.** 23 **41.** 94 **43a.** $\sqrt{63} \approx 7.9$ **43b.** $2\sqrt{63} \approx 15.9$

45. Given: Secants $\overleftrightarrow{AC}$ and $\overleftrightarrow{BD}$ intersect at X inside $\odot P$.

Prove: $m\angle AXB = \frac{1}{2}(m\overset{\frown}{AB} + m\overset{\frown}{CD})$

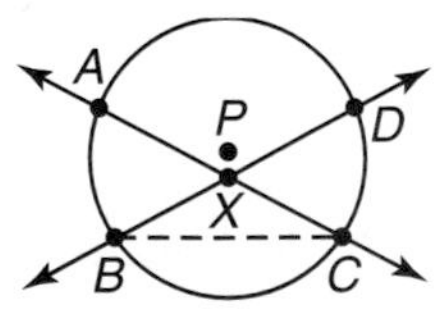

Proof:

Statements	Reasons
1. Secants $\overleftrightarrow{AC}$ and $\overleftrightarrow{BD}$. intersect at X inside $\odot P$	1. Given
2. Draw $\overline{BC}$.	2. Through any 2 pts. there is 1 line.
3. $m\angle XBC = \frac{1}{2}m\overset{\frown}{CD}$ $m\angle XCB = \frac{1}{2}m\overset{\frown}{AB}$	3. An ∠ inscribed in a ⊙ has the measure of $\frac{1}{2}$ the measure of its intercepted arc.

4. $m\angle AXB = m\angle XCB + m\angle XBC$	4. Exterior Angle Theorem
5. $m\angle AXB = \frac{1}{2}m\overset{\frown}{AB} + \frac{1}{2}m\overset{\frown}{CD}$	5. Substitution Prop. (=)
6. $m\angle AXB = \frac{1}{2}(m\overset{\frown}{AB} + m\overset{\frown}{CD})$	6. Distributive Prop.

47a. Given: $\overleftrightarrow{AB}$ is a tangent to $\odot O$. $\overrightarrow{AC}$ is a secant to $\odot O$. $\angle CAB$ is acute.

Prove: $m\angle CAB = \frac{1}{2}m\overset{\frown}{CA}$

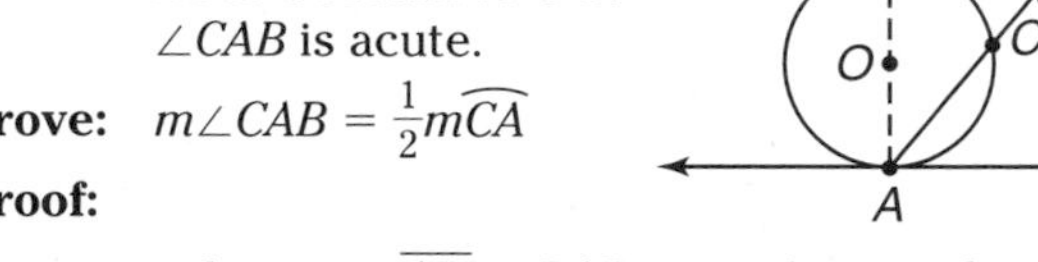

Proof:

Construct diameter $\overline{AD}$. $\angle DAB$ is a right $\angle$ with measure 90, and $\overset{\frown}{DCA}$ is a semicircle with measure 180, because a line is $\perp$ to the radius at the point of tangency if it is tangent to a $\odot$. Since $\angle CAB$ is acute, C is in the interior of $\angle DAB$, so by the Angle and Arc Addition Postulates, $m\angle DAB = m\angle DAC + m\angle CAB$ and $m\overset{\frown}{DCA} = m\overset{\frown}{DC} + m\overset{\frown}{CA}$. By substitution, $90 = m\angle DAC + m\angle CAB$ and $180 = m\overset{\frown}{DC} + m\overset{\frown}{CA}$. So, $90 = \frac{1}{2}m\overset{\frown}{DC} + \frac{1}{2}m\overset{\frown}{CA}$ by Division Prop. (=), and $m\angle DAC + m\angle CAB = \frac{1}{2}m\overset{\frown}{DC} + \frac{1}{2}m\overset{\frown}{CA}$ by substitution. $m\angle DAC = \frac{1}{2}m\overset{\frown}{DC}$ since $\angle DAC$ is inscribed, so substitution yields $\frac{1}{2}m\overset{\frown}{DC} + m\angle CAB = \frac{1}{2}m\overset{\frown}{DC} + \frac{1}{2}m\overset{\frown}{CA}$. By Subtraction Prop. (=), $m\angle CAB = \frac{1}{2}m\overset{\frown}{CA}$.

47b. Given: $\overleftrightarrow{AB}$ is a tangent to $\odot O$. $\overrightarrow{AC}$ is a secant to $\odot O$.

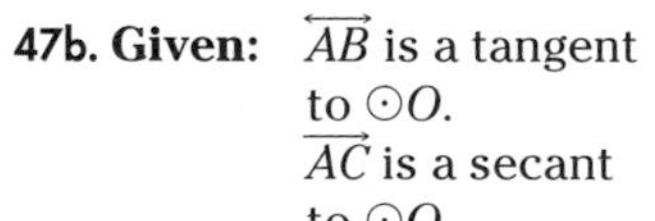

Prove: $m\angle CAB = \frac{1}{2}m\overset{\frown}{CDA}$

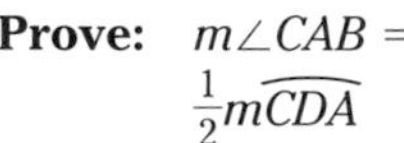

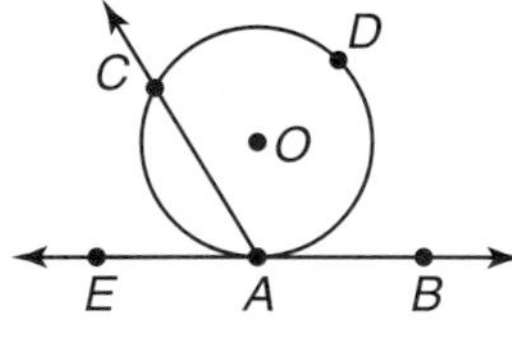

Proof:

$\angle CAB$ and $\angle CAE$ form a linear pair, so $m\angle CAB + m\angle CAE = 180$. Since $\angle CAB$ is obtuse, $\angle CAE$ is acute and Case 1 applies, so $m\angle CAE = \frac{1}{2}m\overset{\frown}{CA}$. $m\overset{\frown}{CA} + m\overset{\frown}{CDA} = 360$, so $\frac{1}{2}m\overset{\frown}{CA} + \frac{1}{2}m\overset{\frown}{CDA} = 180$ by Division Prop. (=), and $m\angle CAE + \frac{1}{2}m\overset{\frown}{CDA} = 180$ by substitution. By the Transitive Prop. (=), $m\angle CAB + m\angle CAE = m\angle CAE + \frac{1}{2}m\overset{\frown}{CDA}$, so by Subtraction Prop. (=), $m\angle CAB = \frac{1}{2}m\overset{\frown}{CDA}$.

49. 76 **51.** 12 **53.** A **55.** \$17.48 **57.** no **59.** Division Prop. (=) or Mult. Prop. (=) **61.** −3, 1

Pages 494–497 Lesson 9–7

7. 2 **9.** 5, 5 **11.** $11\frac{1}{3}$ **13.** $113\frac{1}{3}$ cm **15.** 28.1 **17.** 3 **19.** 6.4 **21.** 5.7 **23.** 24 **25.** 18

27. $DQ + QX = DX$ *Segment Add. Post.*
$QX = DX - DQ$
$= 25 - 7$ *DQ = DE = 7*
$= 18$

$(EX)^2 = QX \cdot TX$ *Theorem 9–16*
$24^2 = x(x + 14)$
$0 = x^2 + 14x - 576$
$x = 18$ *Quadratic formula*

29. 5.3 **31.** 15; 22.5

33. Given: $\overline{EC}$ and $\overline{EB}$ are secant segments.

Prove: $EA \cdot EC = ED \cdot EB$

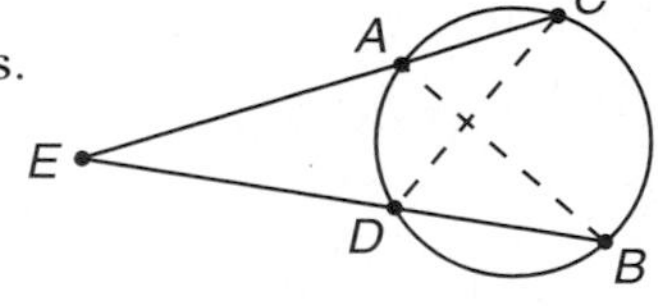

Proof:

Statements	Reasons
1. $\overline{EC}$ and $\overline{EB}$ are secant segments.	1. Given
2. Draw $\overline{AB}$ and $\overline{CD}$.	2. Through any 2 pts. there is 1 line.
3. $\angle DEC \cong \angle AEB$	3. Reflexive Prop. of $\cong$ ∠s
4. $\angle ECD \cong \angle EBA$	4. If 2 inscribed ∠s of a $\odot$ or $\cong$ $\odot$s intercept $\cong$ arcs or the same arc, then the ∠s are $\cong$.
5. $\triangle ABE \sim \triangle DCE$	5. AA Similarity
6. $\frac{EA}{ED} = \frac{EB}{EC}$	6. Def. of ~ polygons
7. $EA \cdot EC = ED \cdot EB$	7. Cross products

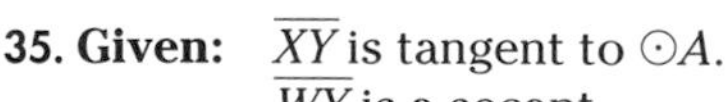

35. Given: $\overline{XY}$ is tangent to $\odot A$. $\overline{WY}$ is a secant segment to $\odot A$.

Prove: $(XY)^2 = WY \cdot ZY$

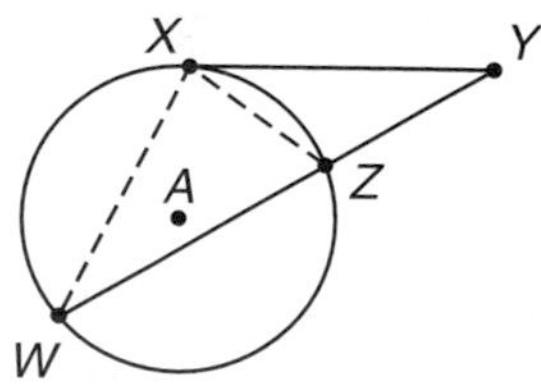

Proof:

Statements	Reasons
1. $\overline{XY}$ is tangent to $\odot A$. $\overline{WY}$ is a secant segment to $\odot A$.	1. Given
2. Draw $\overline{XZ}$ and $\overline{XW}$.	2. Through any 2 pts. there is 1 line.
3. $m\angle XWZ = \frac{1}{2}m\overset{\frown}{XZ}$	3. If an $\angle$ is inscribed in a $\odot$, the measure of the $\angle = \frac{1}{2}$ the measure of its intercepted arc.
4. $m\angle YXZ = \frac{1}{2}m\overset{\frown}{XZ}$	4. If a secant and a tangent intersect at the pt. of tangency, then the measure of the $\angle$ formed $= \frac{1}{2}$ the measure of its intercepted arc.
5. $m\angle YXZ = m\angle XWZ$	5. Substitution Prop. (=)
6. $\angle YXZ \cong \angle XWZ$	6. Def. of $\cong$ ∠s
7. $\angle Y \cong \angle Y$	7. Reflexive Prop. of $\cong$ ∠s
8. $\triangle YXZ \sim \triangle YWX$	8. AA Similarity
9. $\frac{XY}{WY} = \frac{ZY}{XY}$	9. Def. of ~ polygons
10. $(XY)^2 = WY \cdot ZY$	10. Cross products

37. $(AB)^2 = BC(BD)$ *Th. 9–16*
$= BC(BC + CD)$ *Seg. Add. Post.*
$= BC(BC + BC)$ *BC = CD*
$= BC(2BC)$
$= 2(BC)^2$
$AB = \sqrt{2}BC$

39. 7:3.5 **41.** 17 **43.** 15 **45.** Subtraction Prop. (≠)
47. $2c^2 + 13c - 7$

Pages 501–503 Lesson 9–8

5.

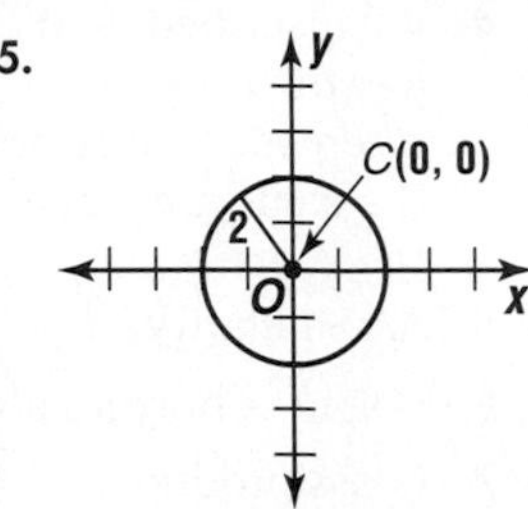

7. $(x + 2)^2 + (y - 3)^2 = 11$
9. $x^2 + y^2 = 16$ **11.** $\left(\frac{3}{4}, -3\right), \frac{9}{2}$
13. (0.5, −3.1), 4.2

15.

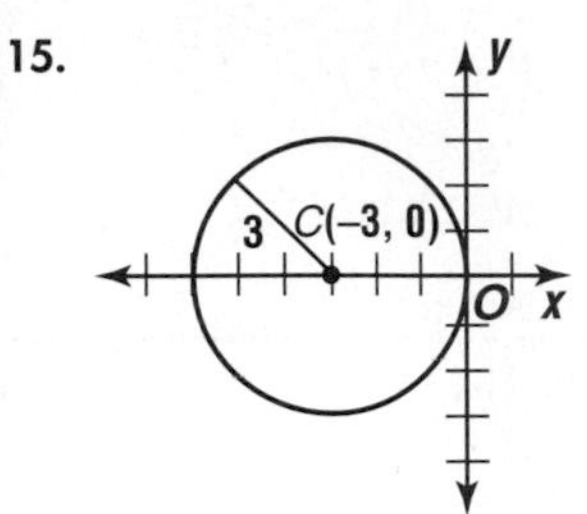

17.

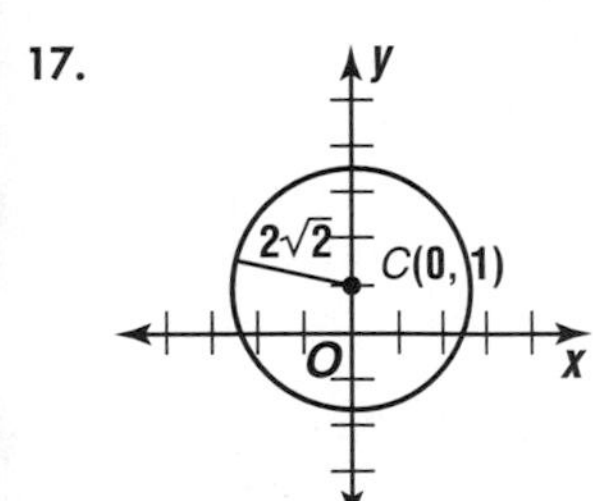

19.

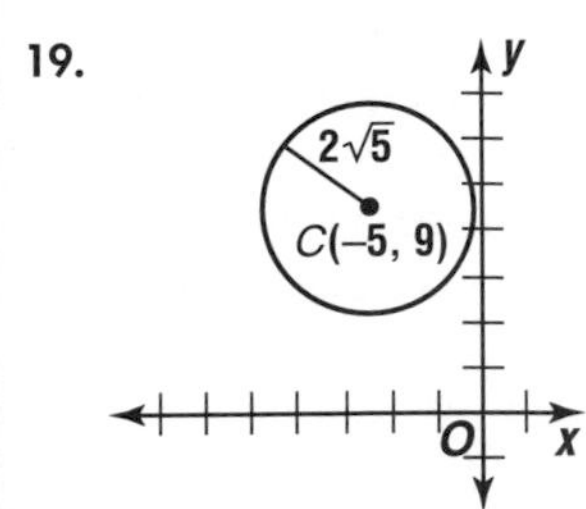

21. $(x + 1)^2 + (y - 4)^2 = 15$
23. $x^2 + y^2 = 8$
25. $(x - 2)^2 + (y - 2)^2 = 2.25$
27. $(x + 3)^2 + (y + 8)^2 = 49$

29–32.

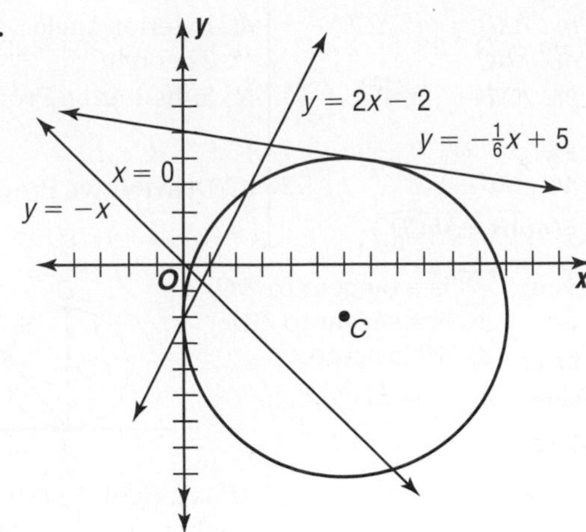

29. secant **31.** tangent

33a, c.

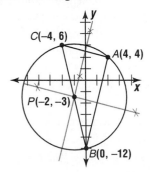

33b. $P(-2, -3)$ **33c.** $\sqrt{85} \approx 9.2$ **33d.** $(x + 2)^2 + (y + 3)^2 = 85$ **35a.** $(x - 3)^2 + (y - 3)^2 = 18$ **35b.** right triangle **35c.** diameter **39.** 29 **41.** 5 **43.** 9.8 in. **45.** E **47.** Reflexive Prop. (=) **49.** $10,377.82

Page 505 Chapter 9 Highlights

1. i **3.** h **5.** c **7.** f **9.** e

Pages 506–508 Chapter 9 Study Guide and Assessment

11. $\overline{AB}$ **13.** 88.0 **15.** 123 **17.** 4.5 **19.** 19.1 **21.** about 34.2 cm **23.** 4 in. **25.** 144 **27.** 36 **29.** 12 **31.** 42 **33.** 18.5 **35.** 138 **37.** 6.0 cm **39.** 15.3 cm **41.** $(x + 4)^2 + (y - 3)^2 = 36$

43. Portion of Sales per Recording Type

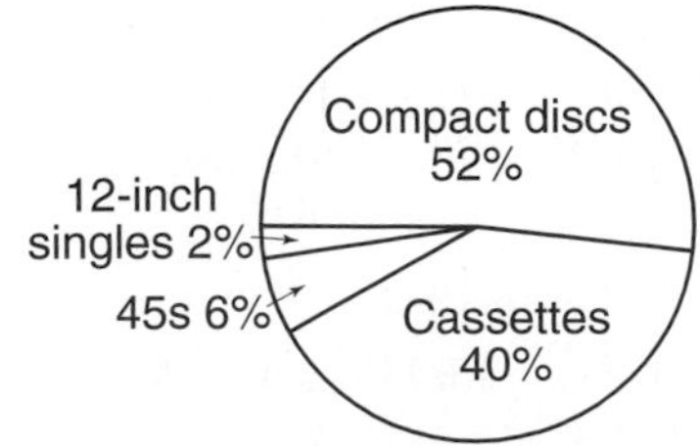

45. about 6.6 cm

CHAPTER 10 EXPLORING POLYGONS AND AREA

Pages 518–521 Lesson 10–1

7. $\overline{AB}, \overline{BC}, \overline{CD}, \overline{DE}, \overline{EF}, \overline{FA}$ **9.** 1440 **11.** 12 **13.** 135, 45 **15.** 6 **17.** $x = 34, 3x = 102, 2x - 1 = 67, 6x - 5 = 199, 4x + 2 = 138$ **19.** *A, B, C, D, E, F* **21.** Sample answers: *BCDEFA* and *FABCDE* **23.** 4320 **25.** 7920

27. $180(3m - 2)$ **29.** 8 **31.** 25 **33.** $\frac{360}{x}$ **35.** 157.5, 22.5 **37.** 154.29, 25.71 **39.** $\frac{180(x + 2y - 2)}{x + 2y}, \frac{360}{x + 2y}$ **41.** 10 **43.** 100 **45.** s **47.** $m\angle R = 125$, $m\angle S = 100$, $m\angle T = 100$, $m\angle U = 95$, $m\angle V = 120$ **49.** 720 **51.** nonagon

53. Consider the sum of the measures of the exterior angles for an n-gon.

sum of measures of exterior angles	=	sum of measures of linear pairs	−	sum of measures of interior angles
	=	$n \cdot 180$	−	$180(n - 2)$
	=	$180n$	−	$180n + 360$
	=	360		

So, the sum of the exterior angle measures is 360 for *any* convex polygon. **55.** No; the sum of the measures of the angles at any vertex is less than 360, so the polygons would not interlock in a plane.

57.

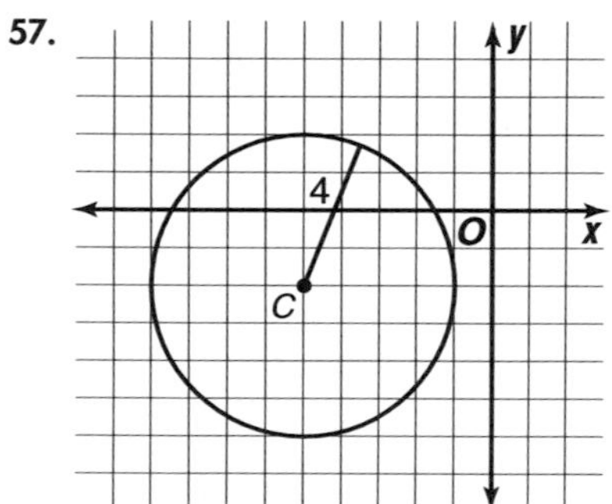

59. $16\pi \approx 50.27$ in. **61.** 6.9 cm **63.** $\overline{ST}, \overline{SV}, \overline{TV}$ **65.** Reflexive Property (=) **67.** 3

Page 522 Lesson 10–2A

1. Yes; whatever space is taken out of the square is then added onto the outside of the square. The area does not change, only the shape changes.

3.

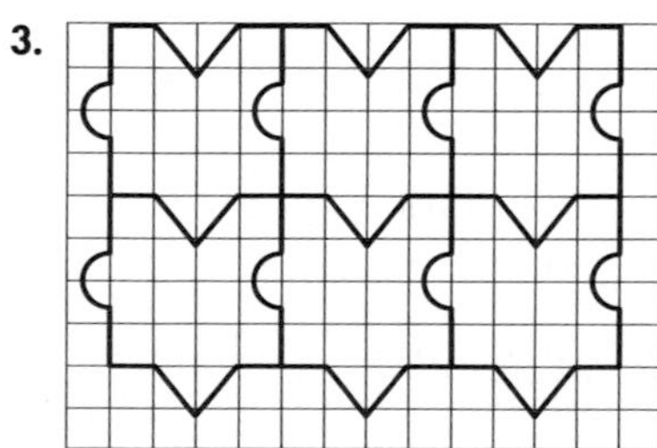

5.

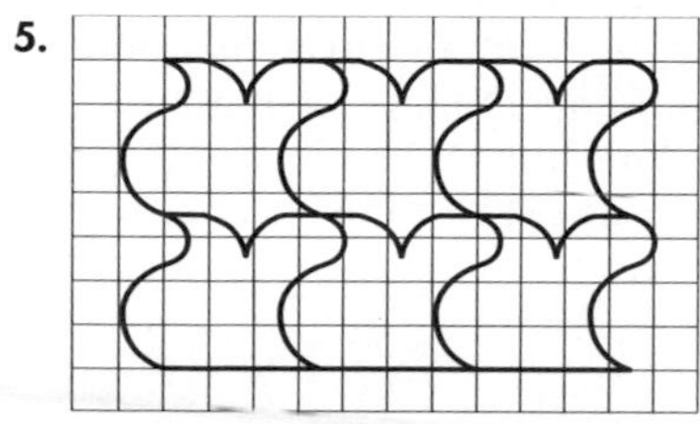

Pages 525–527 Lesson 10–2

7. yes; interior angle = 60 **9.** regular **11.** $(8 - 4) \times 7 + 3 = 31$ **13.** no; interior angle = 108

15. yes; sample answer:

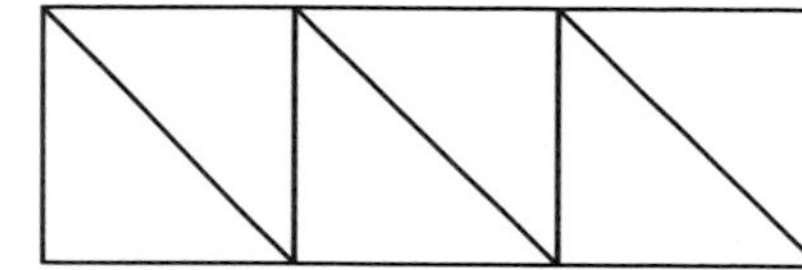

17. yes **19.** none **21.** uniform, semi-regular

23. Sample answer:

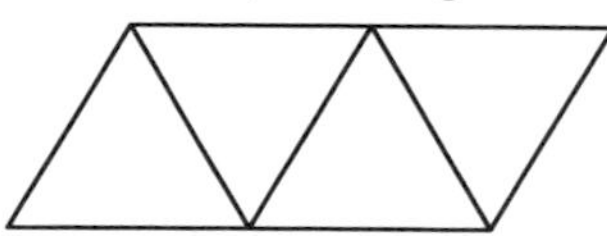

27. Yes; a tessellation can be uniform with only one regular polygon. Therefore, it would not be semi-regular.

29.

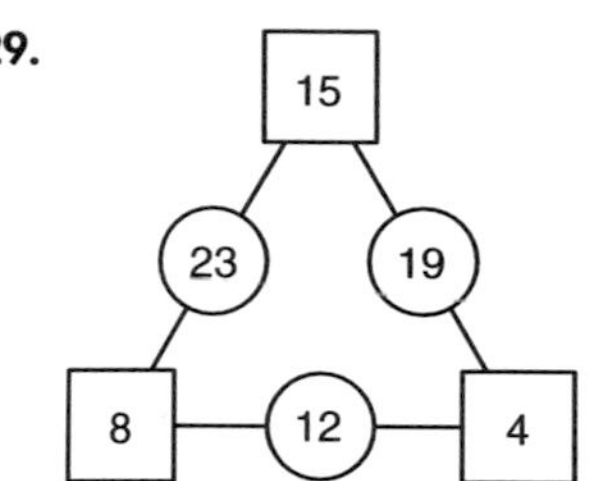

31a. $D = 5$, $Y = 2$, $N = 6$, $E = 9$, $A = 4$, $L = 8$, $S = 1$, $R = 7$, $C = 3$, and $T = 0$ **31b.** More answers will result; sample answers: 561,535 + 207,535 = 769,070. **33.** 18 **35.** 13 in. **37.** A **39.** {(6, 6), (0, 1), (−5, 6), (9, 2)}; $D = \{6, 0, -5, 9\}$; $R = \{2, 1, 6\}$; inverse = {(6, 6), (1, 0), (6, −5), (2, 9)}

Page 528 Lesson 10–3A

7. $A = bh$

9. Sample answer:

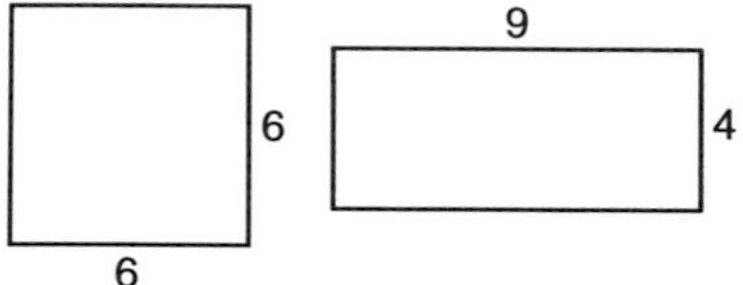

Pages 532–534 Lesson 10–3

5. 19.22 cm^2 **7.** 28.5 cm^2 **9.** \$628.33 **11.** 389.2 m^2 **13.** 2160 ft^2 **15.** 23 m^2 **17.** 202 cm^2 **19.** rectangle; 12 $units^2$ **21.** square; 13 $units^2$ **23.** $12\sqrt{3}$ mm^2 **25.** 25 cm^2; $\frac{1}{4}$ area of square **27.** $NP = 16$ units **29a.** 828 ft^2 **29b.** 207 ft^2 **29c.** Sample answer: 12 ft by 17.25 ft **31a.** M, N, O, P, Q **31b.** convex **31c.** No; its sides are not all congruent. **31d.** pentagon

33. Given: trapezoid $ABCD$, $\overline{AB} \parallel \overline{DC}$

Prove: $\angle A$ and $\angle D$ are supplementary.

Proof:

Statements	Reasons
1. trapezoid $ABCD$, $\overline{AB} \parallel \overline{DC}$	1. Given
2. $\angle A$ and $\angle D$ are supplementary.	2. Consecutive Interior Angle Theorem

35.

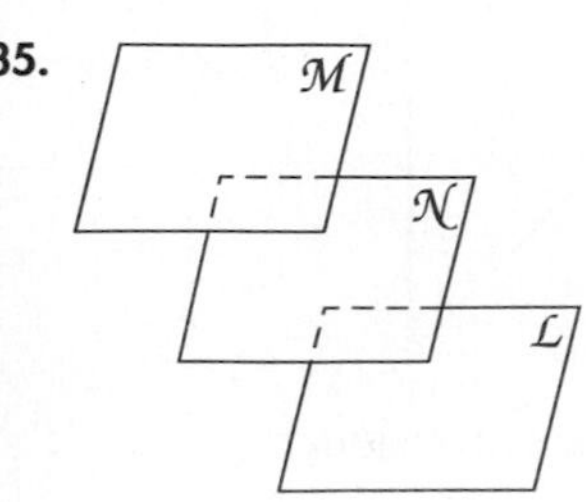

37. a, d

Pages 538–541 Lesson 10–4

7. 60 cm^2 **9.** $72\sqrt{3} \approx 124.7$ in^2 **11.** 35.7 in^2 **13.** $27\sqrt{3} \approx 46.8$ m^2 **15.** 88 m^2 **17.** 4 **19.** 10 **21.** about 28.9 in^2 **23.** $16\sqrt{3} \approx 27.7$ cm^2 **25.** 16.25 in., 22.25 in. **27.** total area = area of parallelogram + area of triangle

$$\begin{aligned} &= bh + \tfrac{1}{2}bh \\ &= ah + \tfrac{1}{2}(b - a)h \\ &= ah + \left(\tfrac{1}{2}b - \tfrac{1}{2}a\right)h \\ &= ah + \tfrac{1}{2}bh - \tfrac{1}{2}ah \\ &= \tfrac{1}{2}ah + \tfrac{1}{2}bh \\ &= \tfrac{1}{2}h(a + b) \end{aligned}$$

29a. 35 units^2 **29b.** 27.125 units^2 **31.** 30 packages **33.** 13 units by 40 units **35.** $h = 45, j = 60$ **37.** No; consecutive angles in a parallelogram must be supplementary. **39.** ℓmx

Page 541 Self Test

1a. heptagon **1b.** no **1c.** convex **3.** 150 **5.** $11\frac{1}{4}$, 32 **7.** 9200 ft^2 **9.** 150 cm^2

Page 542 Lesson 10–5A

1. 360 **3.** They are the same. **5.** 90 **7.** 45 **9.** 30

Pages 547–550 Lesson 10–5

5. $8\sqrt{2} \approx 11.31$ cm **7.** They are close in length. **9.** 2.3 in., 27.7 in^2, 24 in. **11.** 56.7 cm^2 **13.** $(456 + 72\pi) \approx$ 682.19 ft^2 **15.** 104.0 in^2 **17.** 259.8 ft^2 **19.** 1995.3 in^2 **21.** 53.4 cm, 227.0 cm^2 **23.** 64.4 in., 330.1 in^2 **25.** 136.7 cm, 1486.2 cm^2 **27.** 73.1 ft^2 **29.** 313.2 yd^2 **31.** 15.8 in^2 **33.** 20.9 cm^2 **35a.** 92.16π ft^2 **35b.** $(384 - 92.16\pi)$ ft^2 **37.** As the polygon increases in the number of sides, the length of the apothem has a limit of the radius of the circle, and the measure of the side of a polygon becomes increasingly small and has as factors of the limit (2)(3.14). **39a.** two 12-inch pizzas **39b.** No; the unit cost of two 12-inch pizzas is more expensive than the unit cost of one 16-inch pizza. **41.** No; they will increase by the squares of 1, 3, 5, and 7, or in other words, 1, 9, 25, and 49. This is because the apothem has a factor of the side and so does the perimeter. **43.** $12\sqrt{3}$ cm^2 **45.** $(x - 2)^2 + (y - 4)^2 = 13$ **47.** $\triangle QRT \sim \triangle QTS$ by AA, $\triangle QTS \sim \triangle TRS$ by AA, $\triangle QRT \sim \triangle TRS$ by Transitive Property **49.** acute **51.** $4a^6$

Pages 554–558 Lesson 10–6

5. $\frac{2}{9}$ **7.** $\frac{1}{2}$ **9.** $\frac{4}{25}$ or 0.16 **11.** $\frac{5}{8}$ or 0.625 **13.** $\frac{4 - \pi}{4} \approx 21.5\%$ **15.** $\frac{2}{9}$ **17.** $\frac{2}{9}$ **19.** $\frac{1}{16}$ **21.** $\frac{5}{16}$

23a. A B C D E F

23b. $\frac{1}{3}$ **23c.** $\frac{1}{6}$ **23d.** $\frac{1}{2}$ **25.** $\frac{1}{3}$ **27.** $\frac{\pi}{4} \approx 78.5\%$ **29.** 6 cm **31f.** $\frac{1}{3}$ **33.** $\frac{13}{36} \approx 36.1\%$ **35a.** 3279 yd^2 **35b.** 36 ft^2 **35c.** $\frac{36}{29{,}508} \approx 0.12\%$ **37.** 9:1 **39.** 192 **41.** C **43.** 46.3%, 5.0%, 24.1%, 33.7%, 23.7%, 36.0%, 35.8%

Pages 562–564 Lesson 10–7

7a. traceable, incomplete **7b.** $\overline{AE}, \overline{ED}, \overline{DC}, \overline{CB}, \overline{BD}, \overline{DA}, \overline{AB}$ **7c.** $\overline{AC}, \overline{BE}, \overline{CE}$

9a.

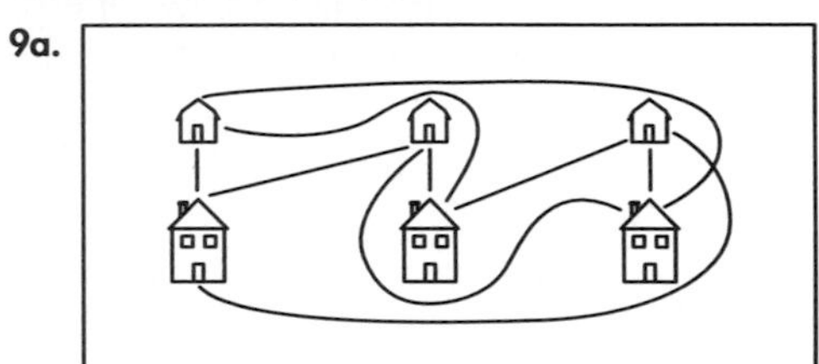

9b. yes; at least once **11.** A: degree 4, B: degree 2, C: degree 3, D: degree 3, E: degree 2 **13a.** traceable, complete **13b.** $\overline{AB}, \overline{BC}, \overline{CA}$ **13c.** none **15a.** not traceable, incomplete **15b.** none **15c.** $\overline{AC}, \overline{BD}, \overline{CE}$ **17a.** traceable, complete **17b.** $\overline{AB}, \overline{BC}, \overline{CA}$, **17c.** none **19.** can't be traced

21. Sample answer:

23. Sample answer:

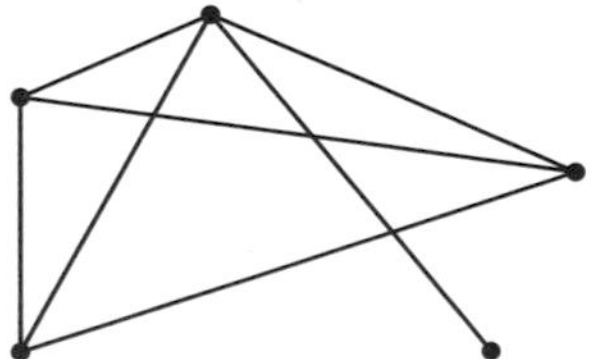

25. When a path goes through a node, it uses two edges, so wherever you start you use one edge from that node and the edges of every subsequent node are both used, one for entering and one for exiting. Therefore, the starting node is the only node left to go to because only one edge was used to start.

27a.

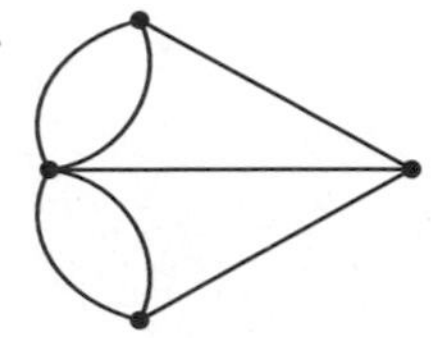

27b. No; all four nodes have odd degree.

27c.

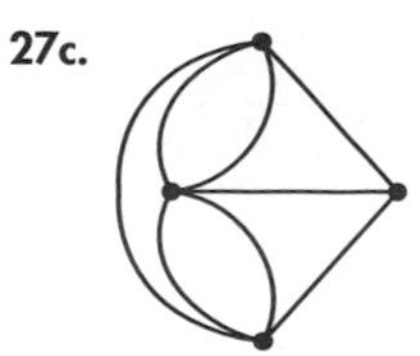

31. 31.5 units^2 **33.** 5 **35.** 30 units **37.** no solution

Page 565 Chapter 10 Highlights

1. false; radius **3.** true **5.** false; 5 **7.** true **9.** false; apothem

Pages 566–568 Chapter 10 Study Guide and Assessment

11. 144 **13.** 162; 18 **15.** regular, uniform **17.** 150 cm^2 **19.** 7.1 in. **21.** 125 m^2 **23.** 4.8 ft **25.** 0.6 ft^2 **27.** $4n^2$ cm^2 **29.** 44.0 mm; 153.9 mm^2 **31.** 5.7 ft; 2.5 ft^2 **33.** $4\pi\sqrt{2} \approx$ 17.8 ft, $8\pi \approx 25.1$ ft^2 **35.** 9.08% **37.** yes **39.** 631 and 542 **41.** 496 in^2

CHAPTER 11 INVESTIGATING SURFACE AREA AND VOLUME

Page 574 Lesson 11-1A

3. a rectangle whose length is the length of the prism

Pages 578–581 Lesson 11-1

5. **7.**

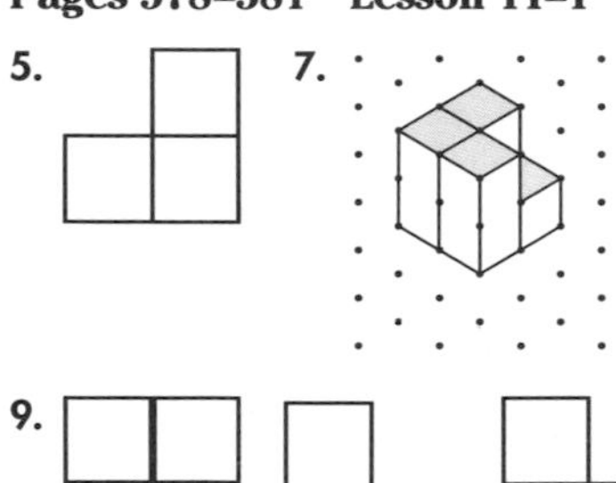

9.

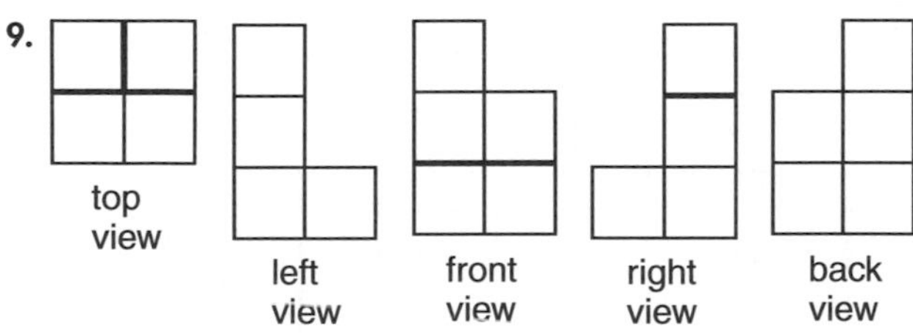

11. no

15. **17.**

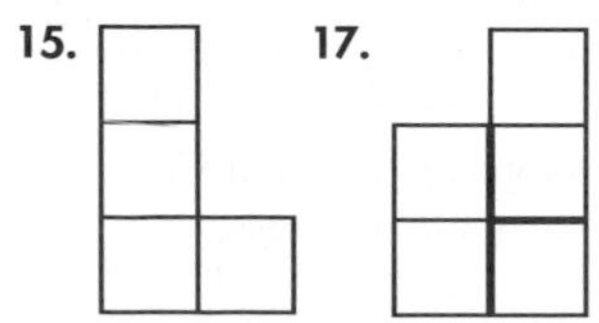

19.

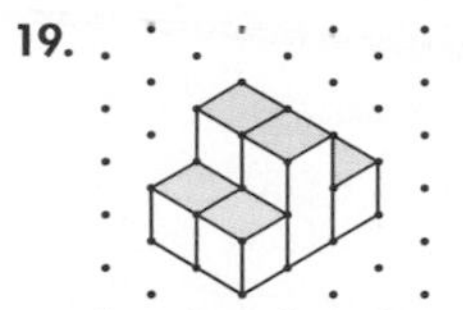

21. upside down U **23.** circle

25.

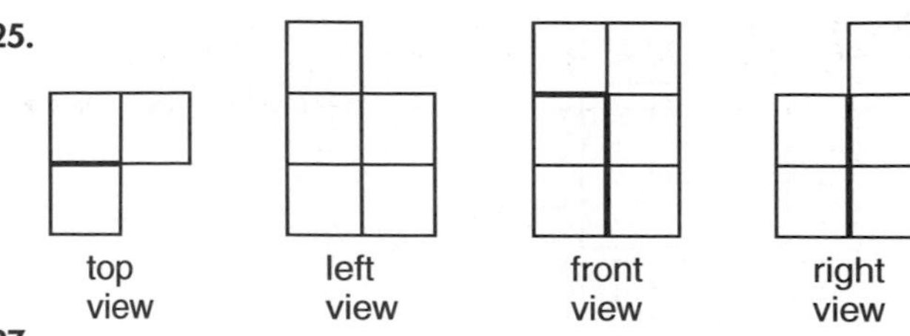

27. yes; edges: $\overline{AB}, \overline{BC}, \overline{CD}, \overline{DE}, \overline{EA}, \overline{AF}, \overline{BG}, \overline{CH}, \overline{DI}, \overline{EJ}, \overline{FG}, \overline{GH}, \overline{HI}, \overline{IJ}, \overline{JF}$; faces: $ABCDE$, $ABGF$, $BCHG$, $CDIH$, $DEJI$, $EAFJ$, $FGHIJ$; vertices: $A, B, C, D, E, F, G, H, I, J$ **29.** square **31.** triangle **33.** False; The only polygon that can be formed by the intersection of a plane and a cylinder is a rectangle. **37.** $A = 5, B = 5, C = 5, D = 5$, not traceable **39.** D **41.** No; the slope of $\overline{AB} = \frac{1}{6}$, but the slope of $\overline{CD} = \frac{1}{5}$ and $\overline{AB} \parallel \overline{CD}$ if $ABCD$ is a parallelogram. **43.** 18

Pages 582–583 Lesson 11-1B

1. equilateral

Pages 586–589 Lesson 11-2

5.

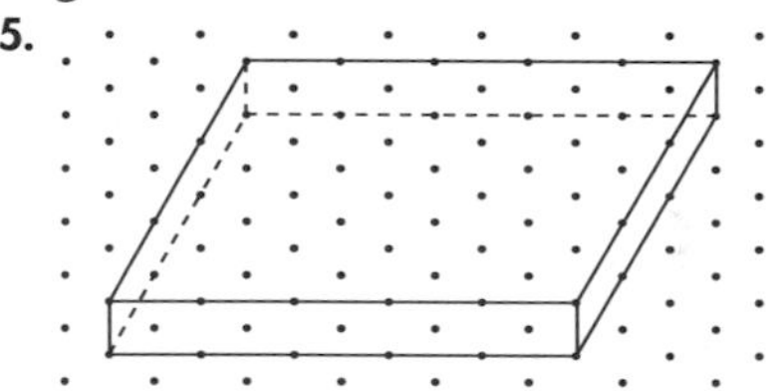

7.

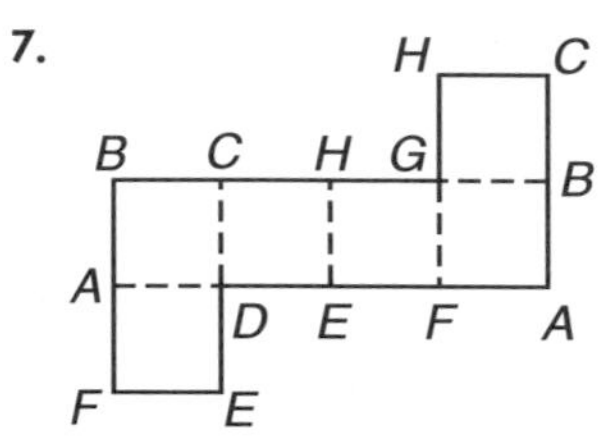

9. surface area = 80.8 units^2; sample net:

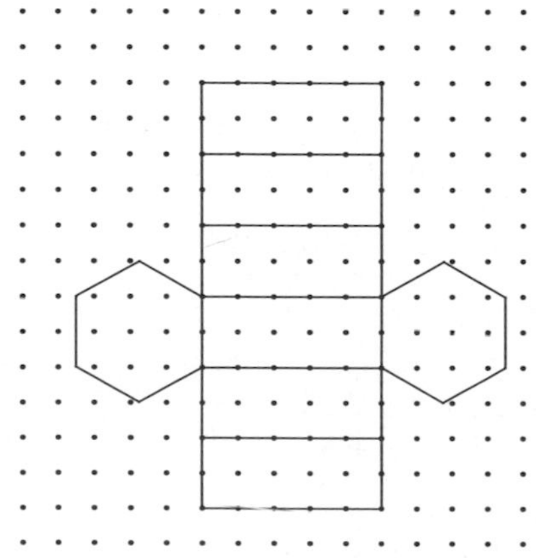

11.

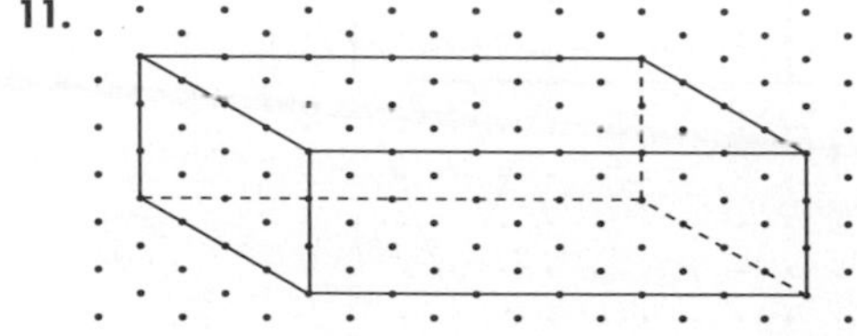

13.

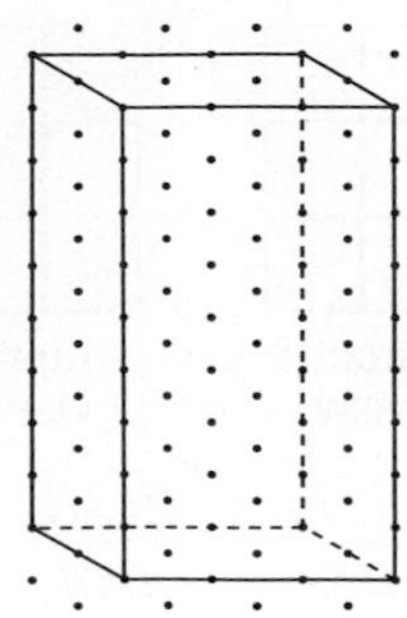

15.

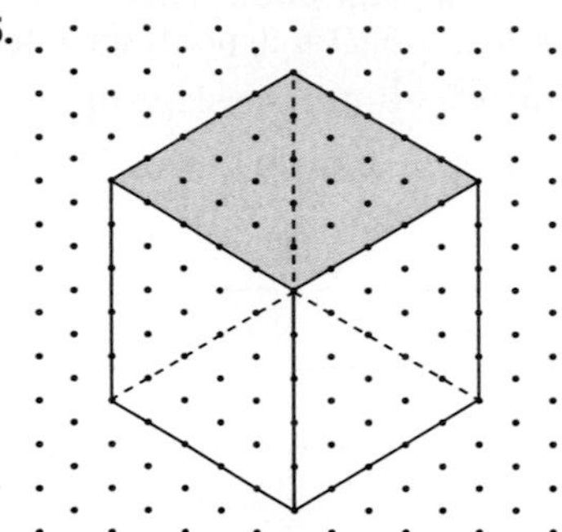

17.

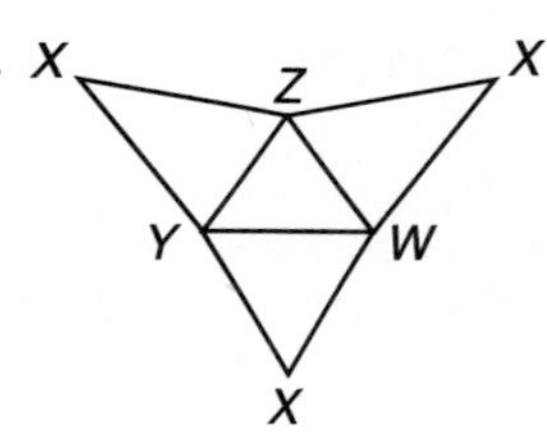

19.

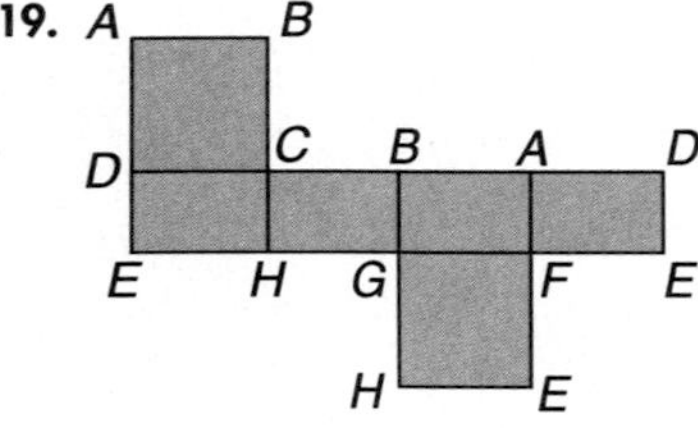

21. surface area ≈ 116.3 units^2;
sample net:

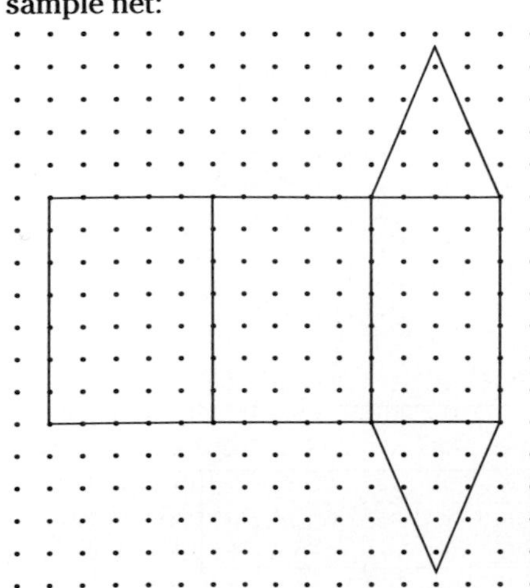

23. surface area = 120 units^2;
sample net:

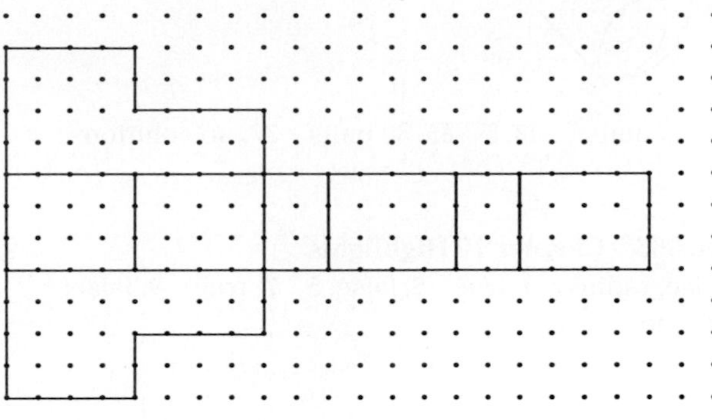

25. surface area = 6.9 units^2;
sample net:

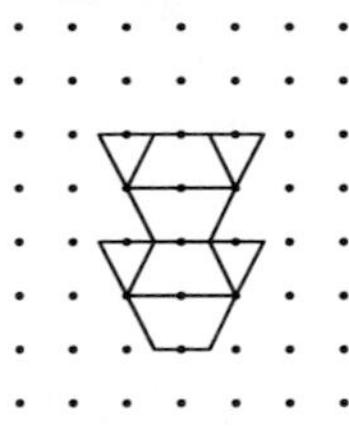

27.

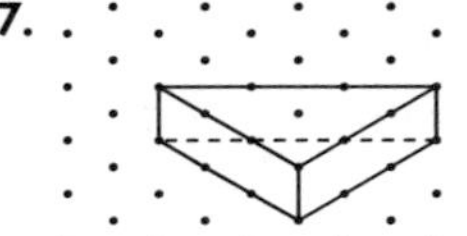

29a.

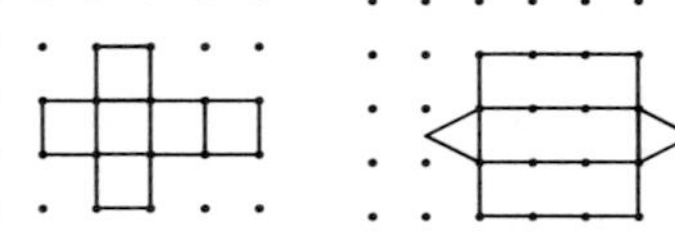

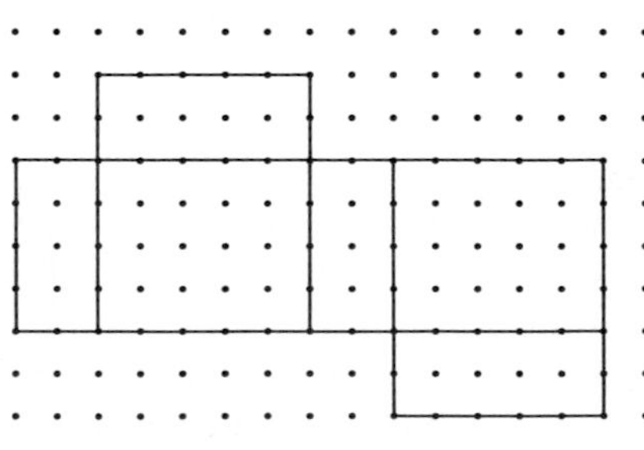

cube surface area = 6 units^2; triangular prism surface area = 9.9 units^2; rectangular prism surface area = 76 units^2

29b.

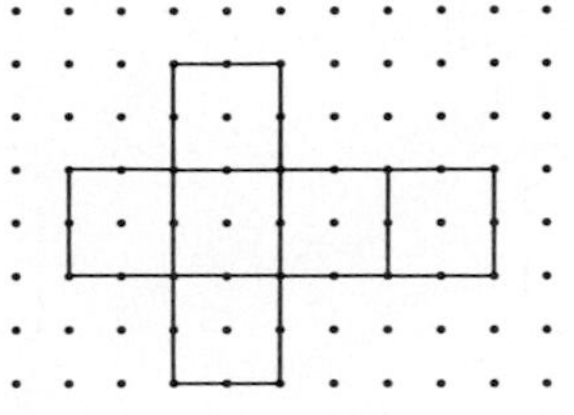

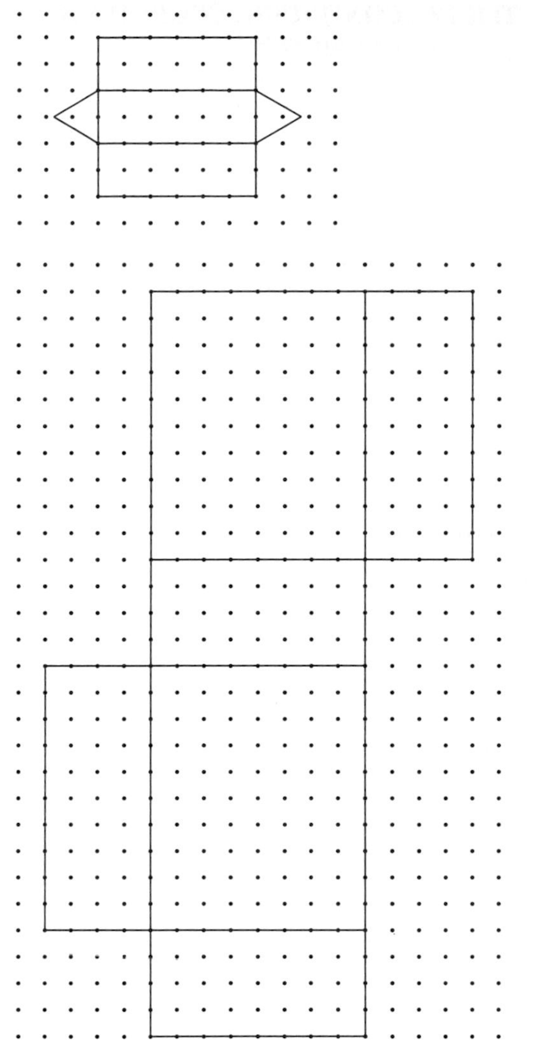

cube surface area = 24 units^2; triangular prism surface area ≈ 39.5 units^2; rectangular prism surface area = 304 units^2 **29c.** The surface area of a solid whose dimensions have been tripled is nine times the surface area of the original solid. **31.** 2344.8 ft^2 **33a.** yes **33b.** no **33c.** yes **35.** 60.35 ft^2 **37.** ≈4.0 **39.** 2.5 **41.** $\{y \mid y \leq -0.12\}$

Page 590 Lesson 11–2B
3. No, it forms a shape inside of the frame.

Pages 595–598 Lesson 11–3
5. Right prism; the lateral edges are perpendicular to the bases. **7.** 30 units **9.** 112 cm^2 **11.** 251.3 ft^2 **13.** 8.6 in. **15.** oblique prism **17.** 36 in. **19.** 559.8 in^2 **21.** 644.5 in^2 **23.** 360 cm^2; 480 cm^2 **25.** 2304 m^2; 3792 m^2 **27.** 2352.4 m^2 **29.** 200.7 in^2; 517.5 in^2 **31.** 24.9 ft^2; 30.0 ft^2 **33.** 198 cm^2 **35.** If the height of the prism is doubled, the *lateral area* is doubled. The surface area is increased, but it is not doubled. **37.** 1680 ft^2 **39.** 179.3 in^2 **41a.** top
41b.

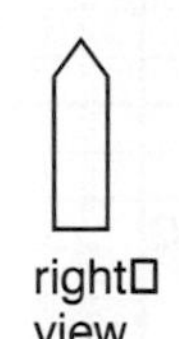

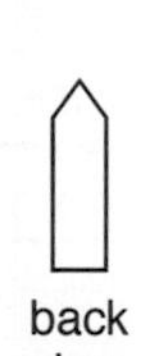

43. 64 cm **45.** $\frac{1}{5}$ **47.** A triangle is equiangular if each angle measures 60°; Law of Syllogism **49.** positive

Pages 603–606 Lesson 11–4
7. both **9.** 47.1 m^2; 75.4 m^2 **11.** 475.2 in^2 **13.** 924,974.6 ft^2 **15.** neither **17.** pyramid **19.** pyramid **21.** 284.3 in^2; 485.2 in^2 **23.** 81 cm^2; 133.6 cm^2 **25.** 188.5 ft^2; 301.6 ft^2 **27.** 169.6 ft^2 **29.** 3696 yd^2 **31.** L = 96 units^2; T = 151.4 units^2 **33.** The lateral area approaches the area of the base. This can be seen by showing a series of cones cut on their slant heights and folded out into sectors. As the altitude approaches zero, the slant height approaches the radius of the base, and the sector narrows, approaching a complete circle. **35.** 736.5 ft^2 **37.** 1078 cm^2 **39.** 570.96 cm^2 **41.** 0.86 inches **43.** $x = b^2 + 18$

Page 606 Self Test
1.

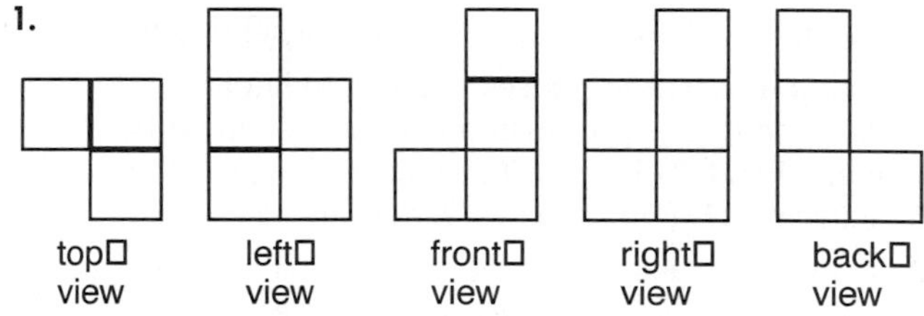

3. L = 432 in^2; T = 619.1 in^2 **5.** about 255,161.7 ft^2

Pages 610–613 Lesson 11–5
7. 127.2 ft^3 **9.** about 1696.5 yd^3 **11.** about 377.0 ft^3 **13.** 461.8 ft^3 **15.** 98.3 m^3 **17.** 3435.3 mm^3 **19.** 1800 in^3 **21.** 105 m^3 **23.** 2598.1 ft^3 **25.** 18 in. **27.** 1592.8 cm^3 **29.** 36 units^3 **31.** 22.3 in^3 **33.** Sample answer: true; Imagine you sliced the oblique cylinder or prism parallel to its base to form a large number of pieces. Then you could slide the pieces so that they stack as a right cylinder or prism. This sliding will not affect the volume or the height. Since the volume of the right cylinder or prism is the product of the area of the base and the height, then the volume of the oblique cylinder or prism is the product of the area of the base and the height. **35.** fully expanded: 2993.0 in^3; fully compressed: 280.6 in^3 **37.** 381.7 in^3 **39.** 384 in^2 **41.** 18 **43.** $\overline{RS}$, $\overline{MN}$; $\overline{RT}$, $\overline{MO}$; $\overline{ST}$, $\overline{NO}$ **45.** $2 \cdot 3 \cdot 3 \cdot 3 \cdot c \cdot c \cdot d$

Page 614 Lesson 11–6A
1. 3 **3.** The heights of the prism and the pyramid are the same. **7.** The areas of the bases of the cone and the cylinder are the same. **9.** $V = \frac{1}{3}Bh$

Pages 617–620 Lesson 11–6
5. 301.6 m^3 **7.** about 5178.8 mm^3 **9.** Mauna Loa: 22,415.8 km^3; Fuji: 169.1 km^3; Paricutín: 171,137,610 m^3; Vesuvius: 4,912,194 m^3 **11.** 382.5 in^3 **13.** 4515.5 ft^3 **15.** 1082.8 m^3 **17.** 277.1 m^3 **19.** 5730.3 units^3 **21.** 814.6 yd^3 **23.** 58.9 in^3 **25a.** 1493 ft^3 **25b.** 27,370 in^3 **25c.** 35 m^3 **25d.** Delete the "/3" in each of the lines that contains the volume formulas. **25d.** No, the formulas are the same for a right or oblique solid. **27.** According to Cavalieri's Principle, the volume of each solid stays the same. **29.** 18,555,031.6 ft^3 **31.** 36.9 in. **33.** 4 **35.** Let x = the number; $9x \leq 108$; $\{x \mid x \leq 12\}$

SELECTED ANSWERS

Pages 625–628 Lesson 11–7
7. neither **9.** true **11.** 5 **13.** drizzle: $T < 0.005$ in^2, $V < 0.000034$ in^3; rain: $T > 0.005$ in^2, $V > 0.000034$ in^3 **15.** $T = 452.4$ ft^2; $V = 904.8$ ft^3 **17.** circle **19.** sphere **21.** neither **23.** true **25.** true **27.** false **29.** false **31.** false **33.** 9.6 **35.** no **37.** $T = 7854.0$ in^2; $V = 65{,}449.8$ in^3 **39.** $T = 636{,}172.5$ m^2; $V = 47{,}712{,}938.4$ m^3 **41.** $T = 36.3$ m^2; $V = 20.6$ m^3 **43.** $\frac{32}{3}\pi$ cm^3 or about 33.5 cm^3 **45.** 2:1 **47.** 452.39 cm^2 **49a.** no **49b.** 80% **51.** Sample answer: Buckminster Fuller designed geodesic domes. The domes are portions of spheres. **53.** the bag shaped like a rectangular prism **55.** A **57.** 117 **59.** 5 **61.** $\{n \mid n < 2\}$

Pages 632–635 Lesson 11–8
5. congruent **7.** 64:1 **9.** $\frac{x}{512}$ m^3 **11.** False; if two pyramids have square bases and all the linear measurements are proportional, then they must be similar. **13a.** 0.015 cm or 0.15 mm **13b.** 1,000,000x cm^2 **15.** neither **17.** congruent **19.** neither **21.** 8:27 **23.** $3\frac{3}{4}$ in. **25.** False; if an edge length of a cube is tripled, then its volume is twenty-seven times greater. **27.** False; doubling the radius of a sphere quadruples the surface area. **29a.** 7:8; 7:1 **29b.** 3:4; 3:1 **31.** Since the only linear measure involved in a sphere is the radius, the linear measures of two spheres are always proportional. **33.** 384 ft × 300 ft × 224 ft **35.** $T = 548.08$ ft^2; $V = 1206.5$ ft^3 **37.** A **39.** $m\angle F = 98$, $DF \approx 9.1$, $FG \approx 10.7$ **41.** a and d

Page 637 Chapter 11 Highlights
1. cross section **3.** right cone **5.** lateral faces **7.** regular **9.** similar

Pages 638–640 Chapter 11 Study Guide and Assessment

11.
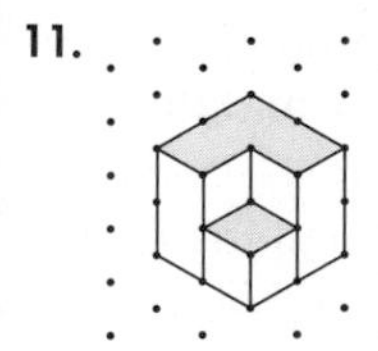

13.
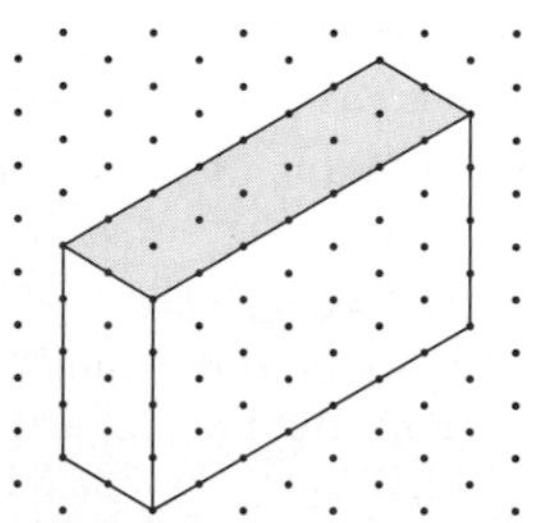

15. $L = 48$ ft^2; $T = 56$ ft^2 **17.** $L = 439.8$ cm^2; $T = 747.7$ cm^2 **19.** $L = 20$ in^2; $T = 24$ in^2 **21.** $L = 48$ in^2; $T = 84$ in^2 **23.** 5196.2 cm^3 **25.** 1021.0 ft^3 **27.** 2787.6 cm^3 **29.** true **31.** 267.9 cm^3 **33.** true **35.** False; a solid is always similar to itself. **37.** 149,301.0 ft^2; 5,424,604.8 ft^3 **39.** 18:458 or about 1:25.4

CHAPTER 12 CONTINUING COORDINATE GEOMETRY

Pages 649–651 Lesson 12–1
7. 28; 8; $-\frac{2}{7}$

9.
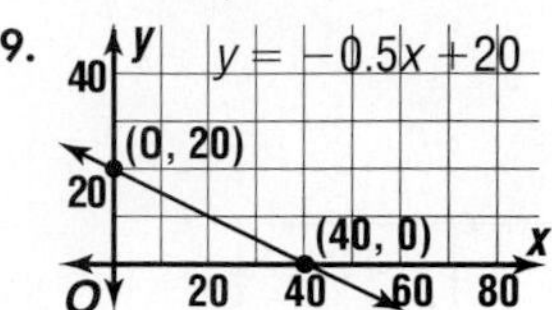

11.
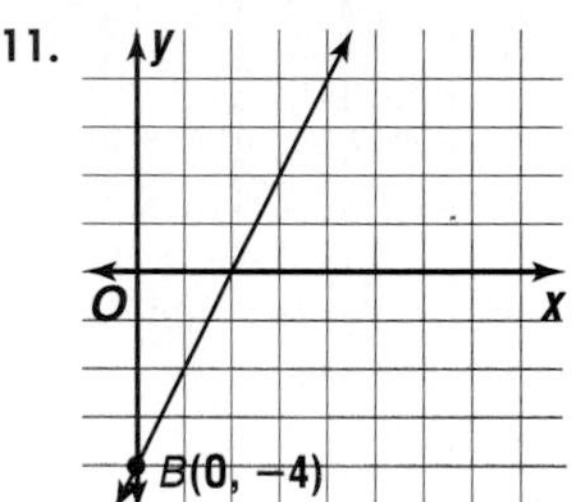

13. 0; 0; −1 **15.** 2; none; undefined **17.** 0; none; undefined

19.
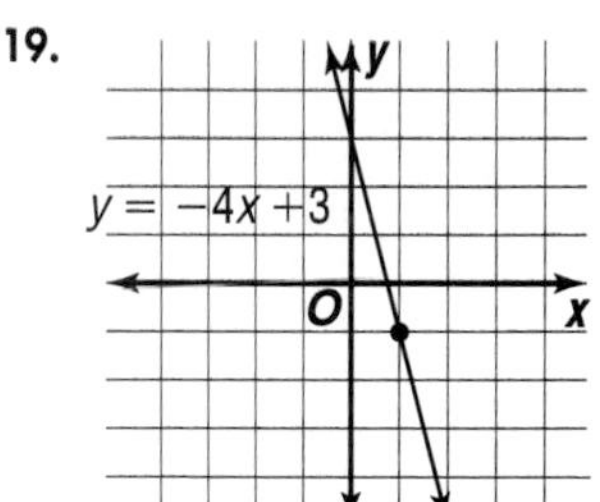

21.
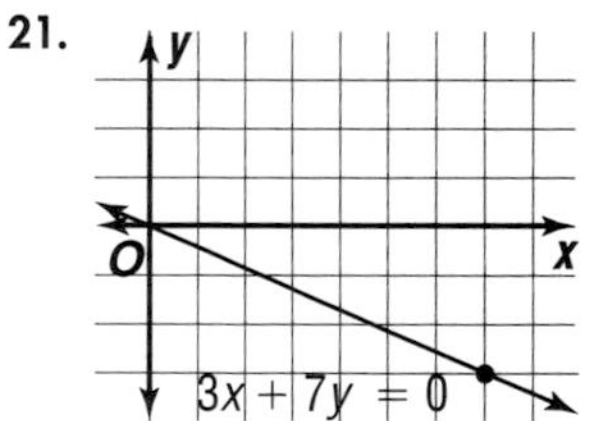

23.
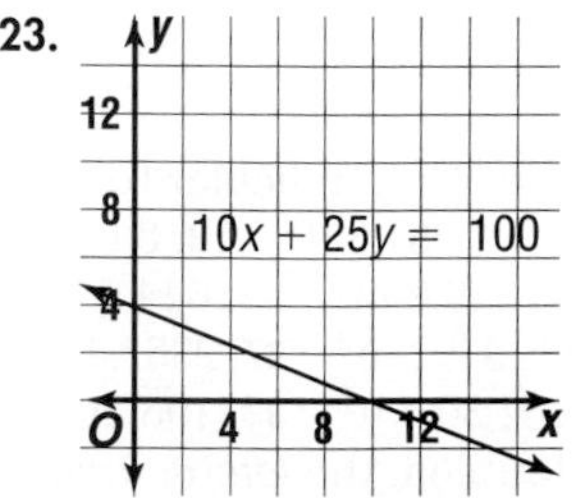

25.
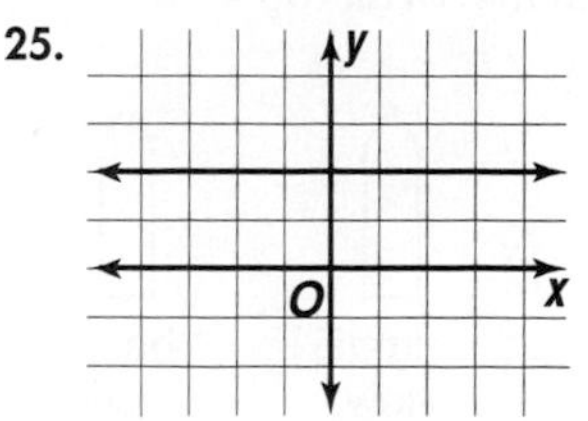

27. 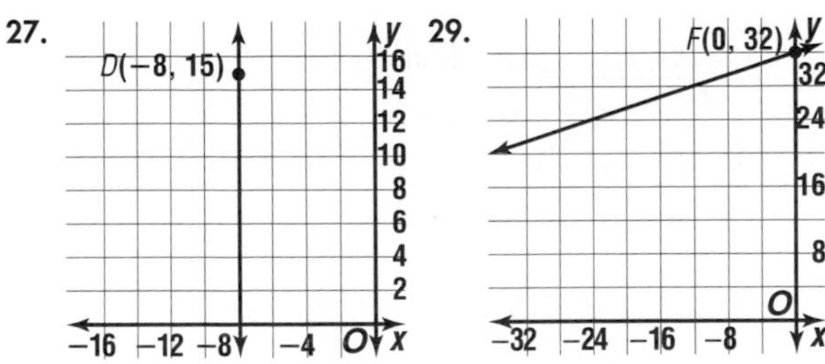

29.

31. Sample answer: All are of the form $y = mx - 1$, but all have a different value for m. **33.** The slope is undefined. Sample answer: The x-coordinates of points on a vertical line are the same. When finding the slope, the denominator (the change in x) is zero and division by zero is undefined. **35a.** The new line would be parallel to the first line with y-intercept 14. **35b.** m would still be -2, b would change to 14 **37.** $y = -2.5x + 3.5$; Sample answer: The distance between two parallel lines is the length of any perpendicular segment that connects points on the two lines. The distance between the y-intercepts is 5. The y–intercept midway between the two lines is 3.5. Since the line is parallel to the other two, the slope must be the same. **39a.** 12; for every gram of fat in the food, there are 12 times more Calories **39b.** 180

39c. 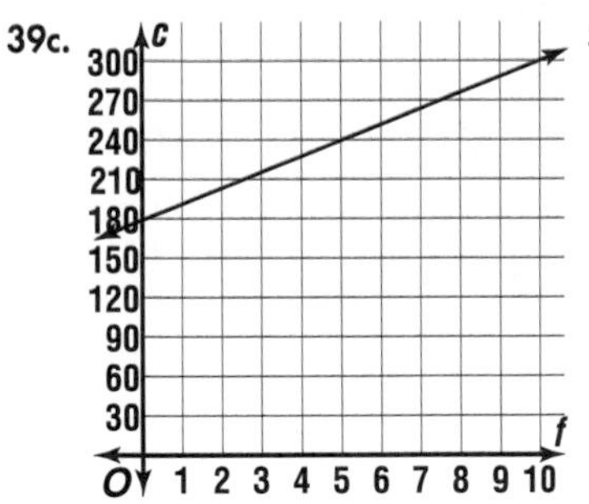

39d. 540 Calories

41a. the number of people that can be carried in x small vans **41b.** x: 12; the number of small vans if all of the people were to be transported in small vans and none in the large vans; y: 8, the number of large vans if all of the people were to be transported in the large vans and none in the small vans

41c. 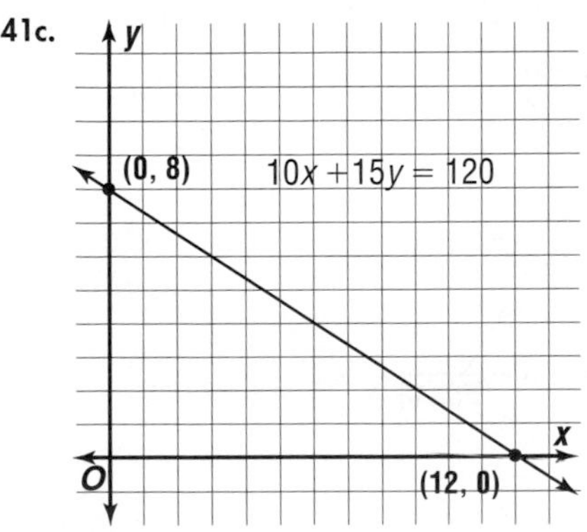

41d. yes; $10(6) + 15(4) = 120$ **41e.** yes; $10(3) + 15(6) = 120$ **43.** 18 m, 6 m **45.** 2.8 miles **47.** $\angle C$, $\angle A$, $\angle B$ **49.** 1, 2, 11, 22

Page 652 Lesson 12–2A

1. The slope between any two points is 0.034.
3. $y = 0.034x - 34$

Pages 656–659 Lesson 12–2

7. $y = -\frac{1}{3}x + \frac{10}{3}$ **9.** $y = \frac{1}{3}x - 3$ **11.** $y = -2x + 7$ **13.** $\frac{1}{2}, 3; y = \frac{1}{2}x + 3$ **15.** $\frac{2}{3}, -2; y = \frac{2}{3}x - 2$ **17.** $y = 0.5x - 1$ **19.** $y = x + 2$ **21.** $y = 8$ **23.** $y = 6$ **25.** $y = 2x - 7$ **27.** $y = -0.1x + 19$ **29.** $y = 5x - 20$ **31.** $y = 10$ **33.** $y = -1$ **35.** $c = 20t + 340$ **37.** $y = \frac{5}{3}x - \frac{55}{3}$ **39.** $y = -\frac{1}{3}x + \frac{32}{3}$ **41a.** $y = -\frac{1}{3}x - 1$ **43b.** The y-intercepts will all change. **45a.** $y = 32x$; $y = 38x$ **45b.** Sample answer: Ameritech ISDN service is always cheaper for the first customer.

47. 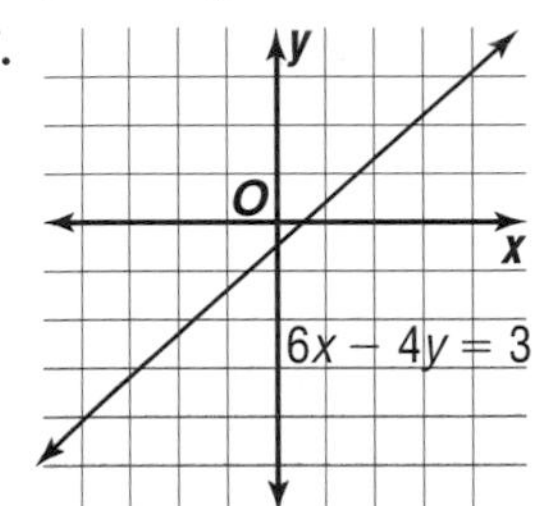

$m = \frac{3}{2}; b = -\frac{3}{4}$

49. 0.37 **51.** Sample answer: A circle S that has all of its points in the interior of circle R would have no common tangents with circle R. **53.** 10.2 **55.** $\frac{2a}{7c}$; $a \neq 0$, $b \neq 0$, $c \neq 0$

Pages 662–665 Lesson 12–3

5. no **7.** False; $4 \neq -10 + 8$ **9.** true **11.** yes **13.** no **15.** yes **17a.**

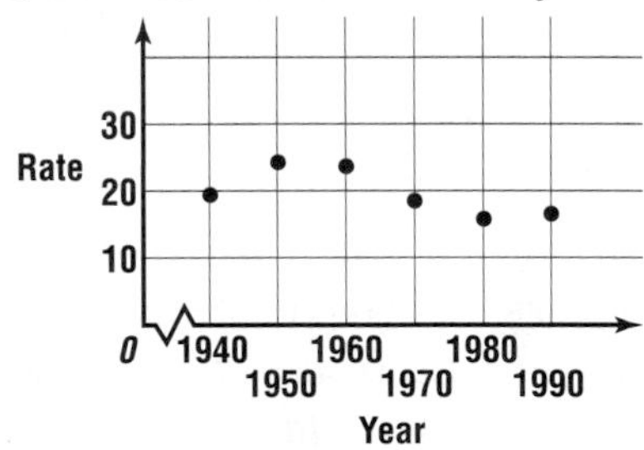

17b. Answers will vary. The points (1950, 24.1) and (1990, 16.7) produce slope of -0.185 and the equation $y = -0.185x + 384.85$. **17c.** Sample answer: Each year the birthrate decreases by 0.185 so every ten years about 2 fewer children are born per 1000 people. **17d.** Sample answer: 14.85; It will be reliable only if the trend continues as it has in the past 50 years. **19.** $y = -\frac{1}{2}x + 19$ **21.** $x = 2$ **23a.** $y = -\frac{1}{2}x - \frac{1}{2}$ **23b.** The line through the midpoint of the two sides of a triangle is parallel to the third side and equal to half of the length of the third side. Point D will have coordinates $(-3, 1)$ and $E(1, -1)$. The slope of the line containing $\overline{BC}$ is $-\frac{1}{2}$. The slope of $\overline{DE} = -\frac{1}{2}$. The lines do not share a common point but have the same slope so they are parallel. $BC = \sqrt{(-2-6)^2 + (8-4)^2} = \sqrt{80}$ or $4\sqrt{5}$. $DE = \sqrt{(-3-1)^2 + (1-(-1))^2} = 2\sqrt{5}$. Thus, the length of the segment connecting the midpoints is one half the length of the third side. **23c.** (0, 7) is a point on $\overline{BC}$ and on the perpendicular from $\overline{BC}$ to $D(-3, 1)$. The distance from $\overline{DE}$ to $\overline{BC}$ is $\sqrt{45} = 3\sqrt{5}$ or 6.71. **25.** Sample answer: $(-3-3)^2 + (13-5)^2 =$

$(-6)^2 + (8)^2 = 36 + 64$ or 100; (9, −3), (11, 11), (11, −1) are some of the possible points. **27.** $y = 0.75x + 5.25$ **29.** Sample answer: One approach might be to use the vertical distance as a measure of error. A, B, and C are on the line $y = 2x + 10$, but D is not. The error in predicting y for $x = 6$ at point D is 14. The error in predicting using $y = 1.8x + 7.3$ for point A is 3.1, B is 2.7, C is 5.1 and D is 10.1 for a total of 21. The better line is $y = 2x + 10$. **31b.** Sample answer: $y = 0.264x + 4339.5$ where x is the original cost and y is the sum of the operating and fixed costs or the total cost of operating the car after it has been purchased. **33.** 6, −3; $\frac{1}{2}$ **35.** 6 **37.** true **39.** $a + 1 - \frac{8}{2a + 1}$

Page 665 Self Test

1. no x-intercept; 12; 0

3.

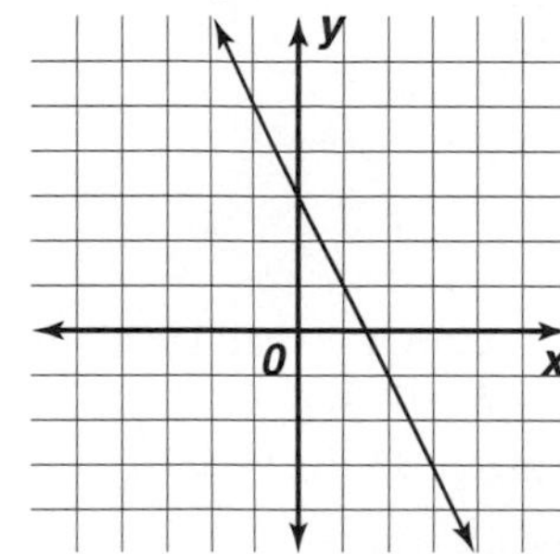

5. $y = 5x + 15$ **7.** $y = \frac{1}{9}x + \frac{1}{9}$ **9.** no

Pages 668–671 Lesson 12–4

5. $D(0, 2c)$ **7a.** square; $AB = BC = CD = DA$; and $\angle DAB$ is right. **7b.** Both midpoints are $\left(\frac{a}{2}, \frac{a}{2}\right)$; $\overline{AC}$ and $\overline{BD}$ bisect each other. **7c.** The slope of $\overline{AC}$ is 1. **7d.** The slope of $\overline{DB}$ is −1. **7e.** $\overline{AC}$ and $\overline{DB}$ are perpendicular. **9.** Let M be the midpoint of $\overline{BC}$. The coordinates of M will be $\left(\frac{2c}{2}, \frac{2b}{2}\right) = (c, b)$. $MA = \sqrt{(c-0)^2 + (b-0)^2} = \sqrt{c^2 + b^2}$. $BC = \sqrt{(2c-0)^2 + (0-2b)^2} = 2\sqrt{c^2 + b^2}$. Since $2MA = BC$, $MA = \frac{1}{2}BC$. **11.** about 29.2 miles **13.** $D(-a, 0)$, $F(0, b)$ **15.** $B(a, 0)$, $D(a, d)$; $F(-b, c)$ **17.** Slope of $\overline{LM}$ is $\frac{b-0}{b-0} = \frac{b}{b}$ or 1. Slope of $\overline{MN}$ is $\frac{b-0}{b-2b} = \frac{b}{-b}$ or −1. $\overline{LM}$ and $\overline{MN}$ are perpendicular and $\triangle LMN$ is a right triangle.

19. $EF = \sqrt{(a\sqrt{2} - 0)^2 + (a\sqrt{2} - 0)^2} = \sqrt{2a^2 + 2a^2}$
$= \sqrt{4a^2}$

$FG = \sqrt{((2a + a\sqrt{2}) - a\sqrt{2})^2 + (a\sqrt{2} - a\sqrt{2})^2}$
$= \sqrt{(2a)^2 + 0^2} = \sqrt{4a^2}$

$GH = \sqrt{((2a + a\sqrt{2}) - 2a)^2 + (a\sqrt{2} - 0)^2}$
$= \sqrt{2a^2 + 2a^2} = \sqrt{4a^2}$

$EH = \sqrt{(0-0)^2 + (2a-0)^2} = \sqrt{0^2 + (2a)^2} = \sqrt{4a^2}$

$EF = FG = GH = EH$

$\overline{EF} \cong \overline{FG} \cong \overline{GH} \cong \overline{EH}$ $EFGH$ is a rhombus.

21.

y
D(b, c) C(a − b, c)
A(0, 0) B(a, 0) x

$DB = \sqrt{(a-b)^2 + (0-c)^2} = \sqrt{(a-b)^2 + c^2}$

$AC = \sqrt{((a-b)-0)^2 + (c-0)^2} = \sqrt{(a-b)^2 + c^2}$

$DB = AC$ and $\overline{DB} \cong \overline{AC}$

$AC = \sqrt{(a+b-0)^2 + (c-0)^2}$

$BD = \sqrt{(b-a)^2 + (c-0)^2}$

But $AC = BD$ and $\overline{DB} \cong \overline{AC}$.

23.

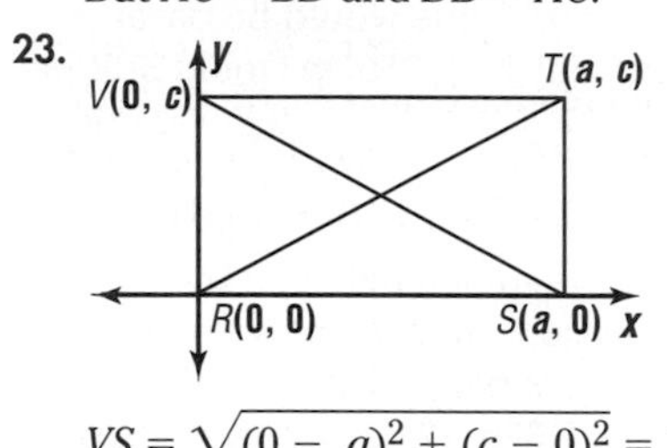

$VS = \sqrt{(0-a)^2 + (c-0)^2} = \sqrt{a^2 + c^2}$

$RT = \sqrt{(a-0)^2 + (c-0)^2} = \sqrt{a^2 + c^2}$

$VS = RT$ and $\overline{VS} \cong \overline{RT}$

25.

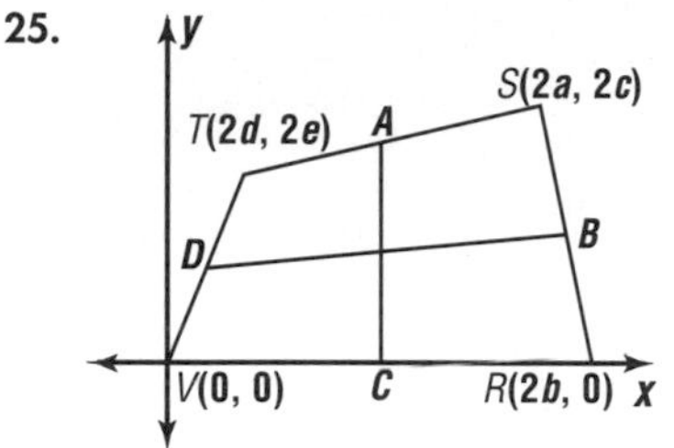

Midpoint A of $\overline{TS}$ is $\left(\frac{2d + 2a}{2}, \frac{2e + 2c}{2}\right)$ or $(d + a, e + c)$. Midpoint B of $\overline{SR}$ is $\left(\frac{2a + 2b}{2}, \frac{2c + 0}{2}\right)$ or $(a + b, c)$. Midpoint C of $\overline{VR}$ is $\left(\frac{0 + 2b}{2}, \frac{0 + 0}{2}\right)$ or $(b, 0)$. Midpoint D of $\overline{TV}$ is $\left(\frac{0 + 2d}{2}, \frac{0 + 2e}{2}\right)$ or (d, e). Midpoint of $\overline{AC}$ is $\left(\frac{d + a + b}{2}, \frac{e + c + 0}{2}\right)$ or $\left(\frac{a + b + d}{2}, \frac{c + e}{2}\right)$. Midpoint of $\overline{DB}$ is $\left(\frac{d + a + b}{2}, \frac{e + c}{2}\right)$ or $\left(\frac{a + b + d}{2}, \frac{c + e}{2}\right)$. $\overline{AC}$ and $\overline{DB}$ bisect each other.

27.

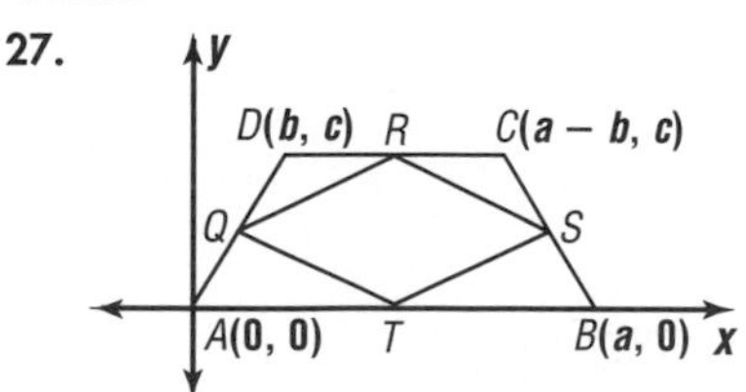

$QR = \sqrt{\left(\frac{b}{2} - \frac{a}{2}\right)^2 + \left(\frac{c}{2} - c\right)^2} = \frac{\sqrt{b^2 - 2ab + a^2 + c^2}}{2}$;

$TS = \left(\frac{2a - b}{2} - \frac{a}{2}\right)^2 + \left(\frac{c}{2} - 0\right)^2 = \frac{\sqrt{b^2 - 2ab + a^2 + c^2}}{2}$;

$QT = \left(\frac{b}{2} - \frac{a}{2}\right)^2 + \left(\frac{c}{2} - 0\right)^2 = \frac{\sqrt{b^2 - 2ab + a^2 + c^2}}{2}$;

$RS = \sqrt{\left(\frac{a}{2} - \frac{2a - b}{2}\right)^2 + \left(c - \frac{c}{2}\right)^2} = \frac{\sqrt{b^2 - 2ab + a^2 + c^2}}{2}$;

$QR = TS = QT = RS$, so $\overline{QR} \cong \overline{TS} \cong \overline{QT} \cong \overline{RS}$ and $QRST$ is a rhombus.

29. 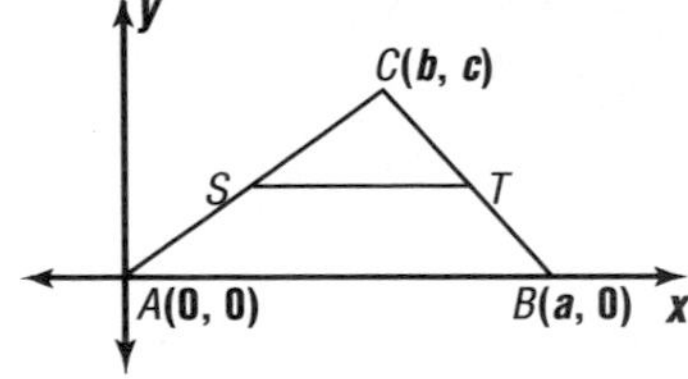

$$ST = \sqrt{\left(\frac{a+b}{2} - \frac{b}{2}\right)^2 + \left(\frac{c}{2} - \frac{c}{2}\right)^2} = \sqrt{\left(\frac{a}{2}\right)^2 + 0^2}$$
$$= \sqrt{\left(\frac{a}{2}\right)^2} = \frac{a}{2}$$
$$AB = \sqrt{(a-0)^2 + (0-0)^2} = \sqrt{a^2 + 0^2} = \sqrt{a^2} = a$$
$$ST = \frac{1}{2}AB$$

31a. Two possible coordinates for C are $(a, 0)$ or $(0, b)$. **31b.** Two possible coordinates for C are $(2a, 0)$ or $(0, 2b)$. **31c.** Two possible solutions are $C(a - b, a + b)$, $D(-b, a)$ and $C(a + b, b - a)$, $D(b, -a)$. **33.** $\sqrt{218} \approx 14.8$ km
35a.

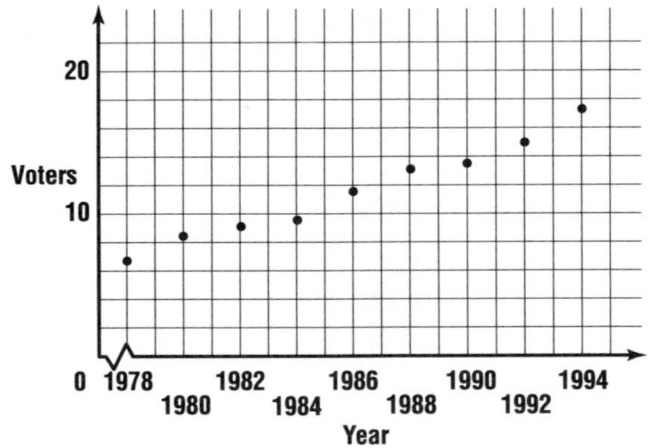

35b. Sample answer: $y = 0.66875x - 1315.9875$
35c. Sample answer: 28.2 million **37.** 435 units^3
39. about 50.3 inches **41.** yes; binomial

Page 672 Lesson 12–5A
3. It is a diagonal.

Pages 676–679 Lesson 12–5
5. $\sqrt{125} \approx 11.2$; 80° **7.** $\overrightarrow{AB} \parallel \overrightarrow{KM} \parallel \overrightarrow{TR}$ **9.** $\overrightarrow{TR}$ and $\overrightarrow{AB}$ or $\overrightarrow{TR}$ and $\overrightarrow{KM}$ **11.** $(22, -9)$ **13.** $\begin{bmatrix} 6 \\ -2 \end{bmatrix}$

15.

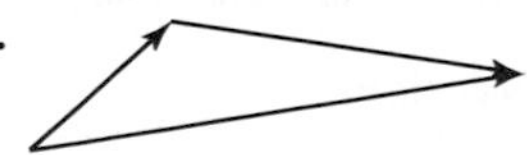

17. $\sqrt{17} \approx 4.1$ units, 76°
19. $\sqrt{409} \approx 20.2$ units, 81° **21.** $\sqrt{72} \approx 8.5$ units, $-135°$
23. 35 units **25.** $-68°$ **27.** $\begin{bmatrix} 0 \\ 1 \end{bmatrix}$ **29.** $\begin{bmatrix} 8 \\ 5 \end{bmatrix}$ **31.** $\begin{bmatrix} -2 \\ 6 \end{bmatrix}$
33. $\overrightarrow{AB} = \overrightarrow{HG}$ **35.** $\overrightarrow{AD}$ **37.** $\overrightarrow{BE}$ **39.** $(11, 10)$ **41.** $(20, 32)$; $\overrightarrow{EF} = \overrightarrow{HI}$ **43.** $(8, 9)$
45. **1.** Definition of vector addition
2. Definition of vector addition
3. Given
4. Substitution Property (=)
5. Distributive Property (=)
6. Substitution Property (=)
7. Multiplication Property (=)

47. $\begin{bmatrix} ac \\ ab \end{bmatrix}$; The process has to model the way scalar multiplication of vectors written as ordered pairs is written. **49.** A line through the origin will have equation $y = cx$ for some constant c. If (k, m) lies on the line, then $m = ck$. If (r, s) is a scalar multiple of (k, m), then $(r, s) = c(k, m) = (ck, cm)$. Substituting into $y = cx$, $cm = c(ck)$. Dividing by c, $m = ck$ which is true because (k, m) lies on the line. Therefore, (r, s) lies on the line also. **51.** yes **53a.** They have the same magnitude but opposite directions. **53b.** It is a hexagon. **55.** $\sqrt{425} \approx 20.62$ miles per hour at a direction of 76° **57.** 84.9 newtons, 32° northeast **59a.** Sample answer: Using points (1195, 229) and (5821, 1134) the equation is $y = 0.2x - 4$. **59b.** 441 thousand **61.** 22 m **63.** 124
65. $12\sqrt{3} \approx 20.8$ in. **67.** $\frac{9\sqrt{2}}{2} \approx 6.364$ units **69.** $\frac{3}{2}u - \frac{2}{u}$

Pages 683–686 Lesson 12–6
5.

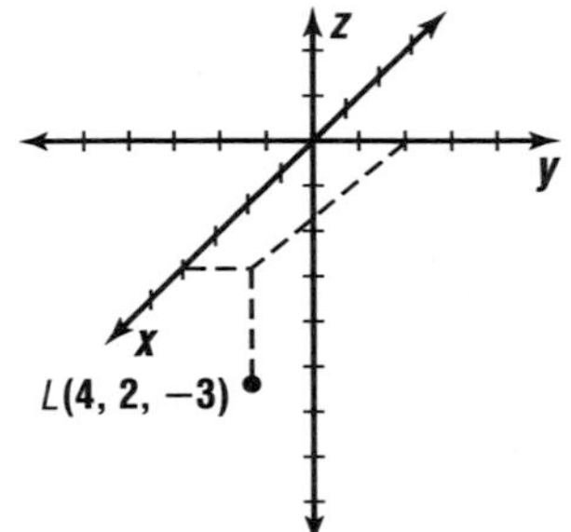

7. $\sqrt{521} \approx 22.83$; $(19.5, -5, 29)$ **9.** true
11. $(5, -2, 0)$, 6 **13.** $(x + 2)^2 + (y - 3)^2 + (z - 25)^2 = 138$

15.

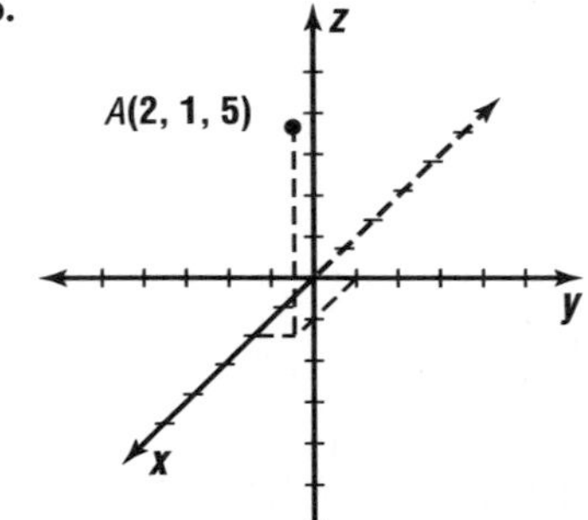

17.

19.

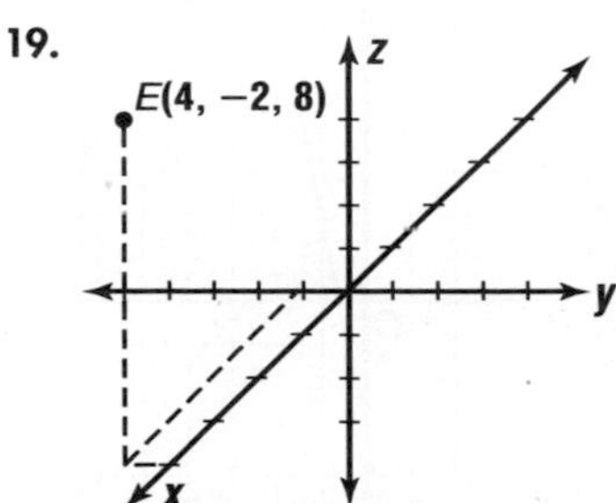

21. $\sqrt{117} \approx 10.8$ units; (4, –3, 3.5) **23.** $\sqrt{10} \approx 3.16$ units; (7.5, 0.5, 1) **25.** $\sqrt{13} \approx 3.6$ units; (4, 7, 0.5) **27.** False; $c = 0$; the points will be of the form $(0, y, z)$. **29.** true **31.** False; it is a plane that contains that line. **33.** (5, –4, 10), 3 **35.** (–4, 2, –12), $\sqrt{18} \approx 4.24$ **37.** $(x + 2)^2 + (y - 3)^2 + (z + 4)^2 = 74$ **39.** $(x + 5)^2 + (y - 4)^2 + (z - 19)^2 = 36$ **41.** $\sqrt{6} + \sqrt{11} + 3$ or about 8.77 units **43.** $AB = 11$, $AC = 11$ so $\triangle ABC$ is isosceles. If $AB^2 + AC^2 = BC^2$, $\triangle ABC$ will be a right triangle. $BC = \sqrt{242}$ and $11^2 + 11^2 = 242$, so $\triangle ABC$ is a right triangle. **45.** 150 units^2, 125 units^3

47.

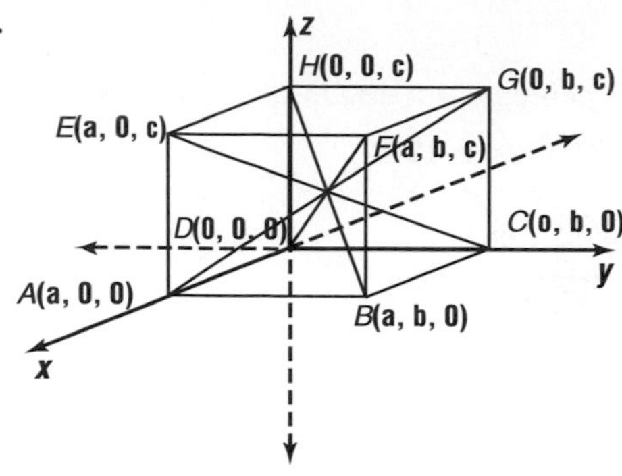

$AG = \sqrt{(a - 0)^2 + (0 - b)^2 + (0 - c)^2} = \sqrt{a^2 + b^2 + c^2}$
$BH = \sqrt{(a - 0)^2 + (b - 0)^2 + (0 - c)^2} = \sqrt{a^2 + b^2 + c^2}$
$CE = \sqrt{(0 - a)^2 + (b - 0)^2 + (0 - c)^2} = \sqrt{a^2 + b^2 + c^2}$
$DF = \sqrt{(0 - a)^2 + (0 - b)^2 + (0 - c)^2} = \sqrt{a^2 + b^2 + c^2}$
$AG = BH = CE = DF$ and $\overline{AG} \cong \overline{BH} \cong \overline{CE} \cong \overline{DF}$
Midpoint of $\overline{AG}$ is $\left(\frac{a+0}{2}, \frac{0+b}{2}, \frac{0+c}{2}\right)$ or $\left(\frac{a}{2}, \frac{b}{2}, \frac{c}{2}\right)$.
Midpoint of $\overline{CE}$ is $\left(\frac{0+a}{2}, \frac{b+0}{2}, \frac{0+c}{2}\right)$ or $\left(\frac{a}{2}, \frac{b}{2}, \frac{c}{2}\right)$.
Midpoint of $\overline{BH}$ is $\left(\frac{a+0}{2}, \frac{b+0}{2}, \frac{0+c}{2}\right)$ or $\left(\frac{a}{2}, \frac{b}{2}, \frac{c}{2}\right)$.
Midpoint of $\overline{DF}$ is $\left(\frac{0+a}{2}, \frac{0+b}{2}, \frac{0+c}{2}\right)$ or $\left(\frac{a}{2}, \frac{b}{2}, \frac{c}{2}\right)$.
$\overline{AG}$, $\overline{CE}$, $\overline{BH}$, and $\overline{DF}$ all bisect each other.
49. (6, –1, 3), (6, 9, 3), (–4, 9, 3), (–4, –1, 3), (6, –1, 13), (6, 9, 13), (–4, 9, 13), (–4, –1, 13) **51.** (2, 2, 2) to block a row of Xs **53.** $|\overrightarrow{LN}| = \sqrt{85} \approx 9.2$; 13° **55a.** $y = 11x + 349$ **55b.** 514 **57.** E **59.** $288\sqrt{2}$ or about 407.3 m^2 **61.** 4.5 **63.** –18

Page 687 Chapter 12 Highlights
1. coordinate proof **3.** slope-intercept form **5.** standard form **7.** scatter plot **9.** linear equation

Page 688–690 Chapter 12 Study Guide and Assessment
11.

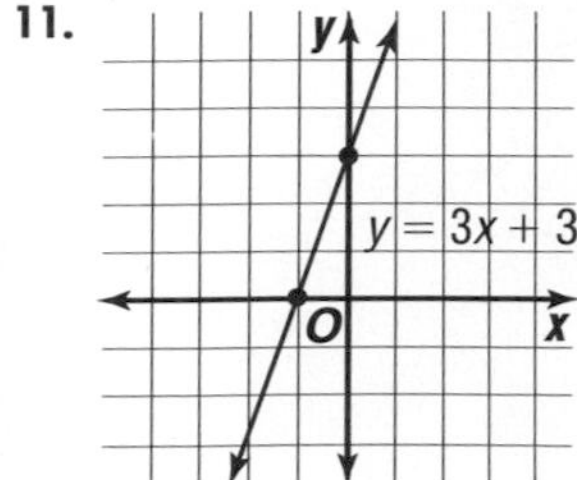

13.

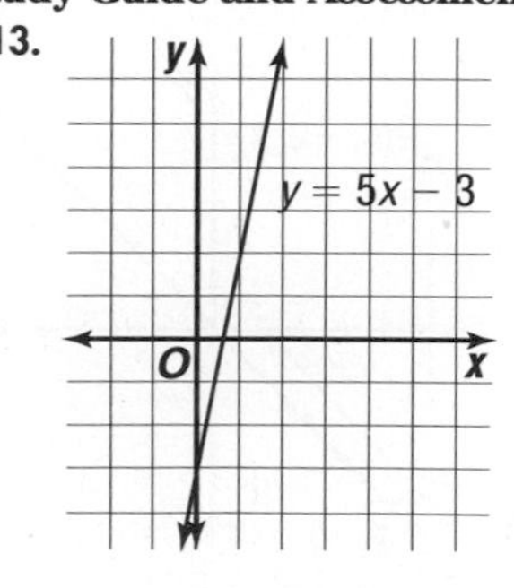

15. $y = 4x + 2$ **17.** $y = -5x - 31$ **19.** $y = x + 8$
21. $y = \frac{1}{2}x - 2$, $y = \frac{2}{3}x - 3$, $y = x - 3$
23. $y = -2x - 3$, $y = -\frac{3}{2}x + 2$, $y = -x + 7$
25.

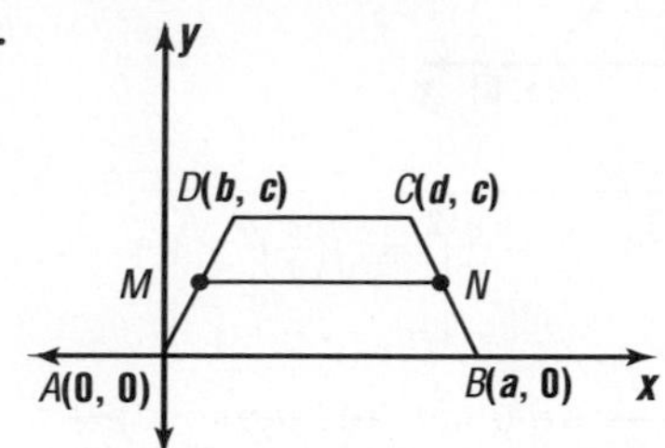

Midpoint M is $\left(\frac{b+0}{2}, \frac{c+0}{2}\right)$ or $\left(\frac{b}{2}, \frac{c}{2}\right)$.
Midpoint N is $\left(\frac{a+d}{2}, \frac{c+0}{2}\right)$ or $\left(\frac{a+d}{2}, \frac{c}{2}\right)$.
Slope of $\overline{DC}$ is $\frac{c-c}{d-b}$ or $\frac{0}{d-b}$ or 0.
Slope of $\overline{MN}$ is $\frac{\frac{c}{2} - \frac{c}{2}}{\frac{a+d}{2} - \frac{b}{2}}$ or $\frac{0}{\frac{a+d-b}{2}}$ or 0.
Slope of $\overline{AB}$ is $\frac{0-0}{a-0}$ or $\frac{0}{a}$ or 0.
$\overline{DC} \parallel \overline{MN} \parallel \overline{AB}$
27. $\sqrt{50} \approx 7.1$ units; about 8.1° **29.** (4, 8) **31.** $\begin{bmatrix} 7 \\ 8 \end{bmatrix}$
33. $4\sqrt{2} \approx 5.7$ units; (5, –3, 3) **35.** $x^2 + y^2 + z^2 = 25$
37. \$1400 a month for rent and utilities

39a.

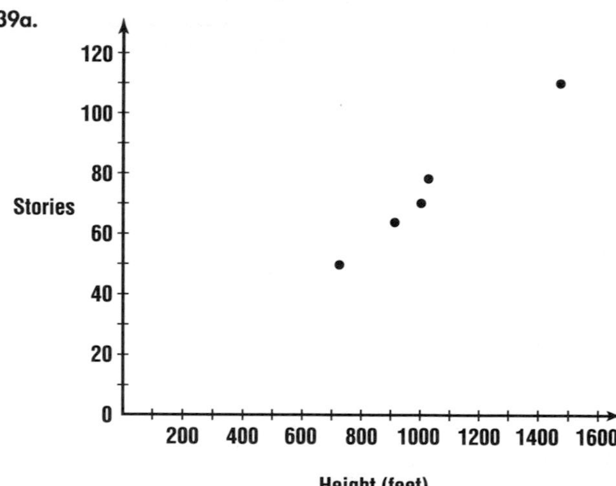

39b. Sample answer: $y = 0.09x - 15$ **39c.** Sample answer: 52 stories. The tower actually has 54 stories.

CHAPTER 13 INVESTIGATING LOCI AND COORDINATE TRANSFORMATIONS

Pages 699–701 Lesson 13–1
5. the bisector of the segment that joins the two points;

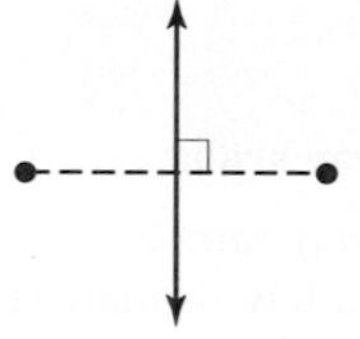

7. a line perpendicular to the plane of the square through the point of intersection of the diagonals;

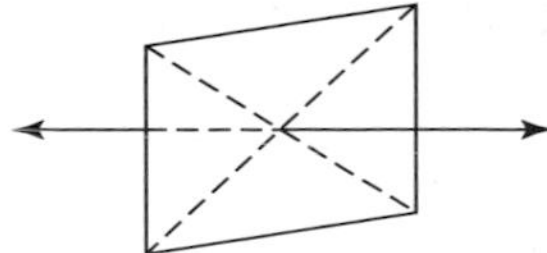

9. the perpendicular bisector of $\overline{AB}$ and $\overline{DC}$ in plane $\mathcal{R}$;
11. the line that joins the two points of intersection;

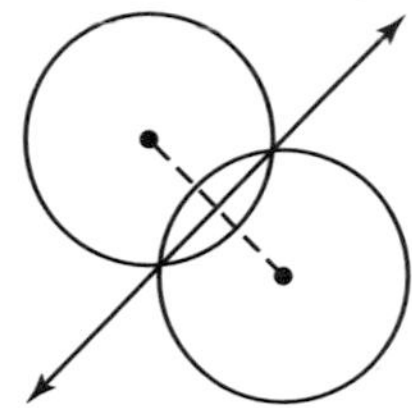

13. a pair of parallel lines, one on each side of ℓ, and each r units from ℓ;

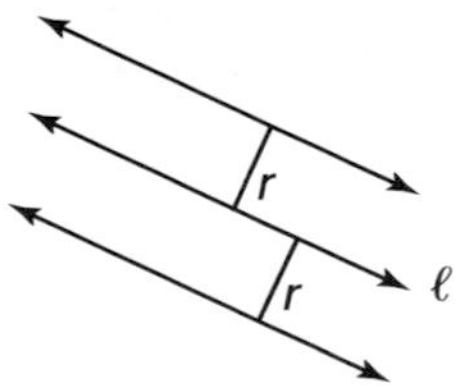

15. center of the pentagon;

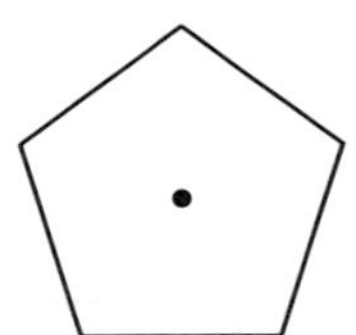

17. concentric circle of radius $s + \frac{(r-s)}{2}$;

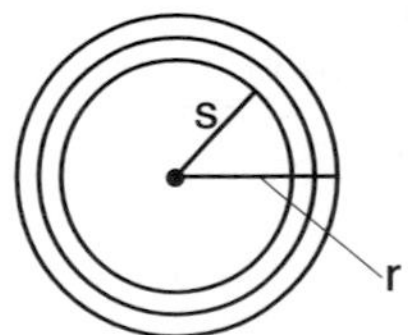

19. a pair of parallel planes, one on each side of plane $\mathcal{M}$, and each 4 m from $\mathcal{M}$;

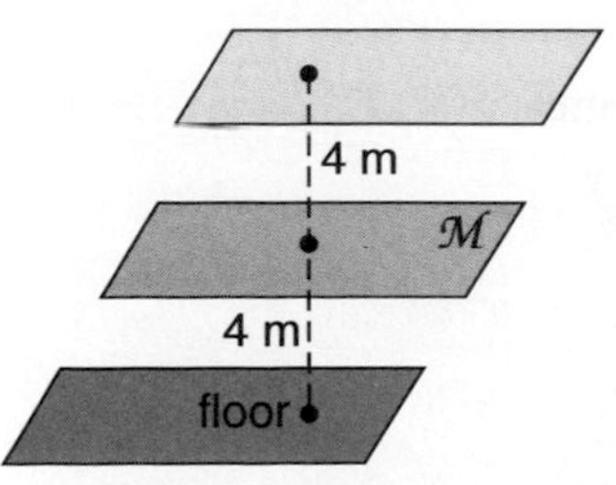

21. a circle with radius $\frac{\sqrt{3}s}{2}$ units in a plane perpendicular to the base with center on the midpoint of the base;

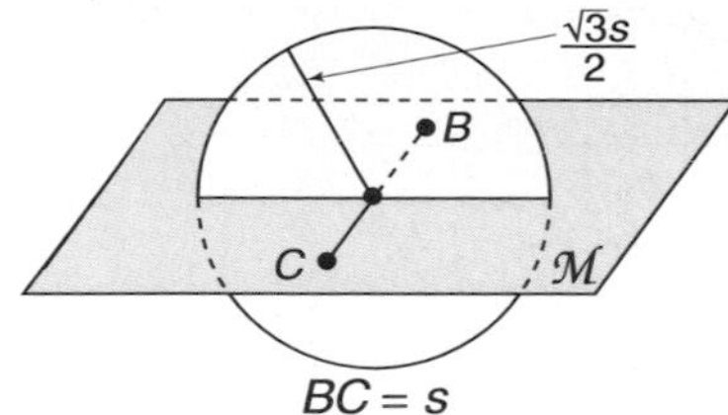

23. a plane perpendicular to the line containing the opposite vertices of the cube;

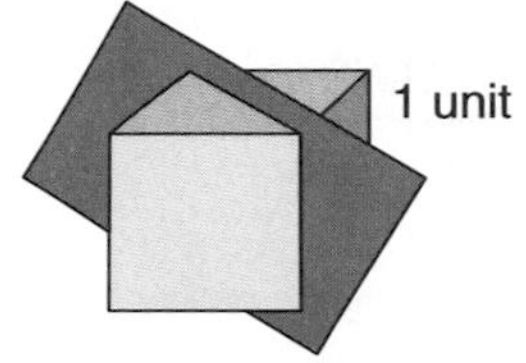

25. a plane midway between two parallel lines, perpendicular to the plane containing the two parallel lines **27.** a triangle similar to $\triangle ABC$ containing points A, B, and C;

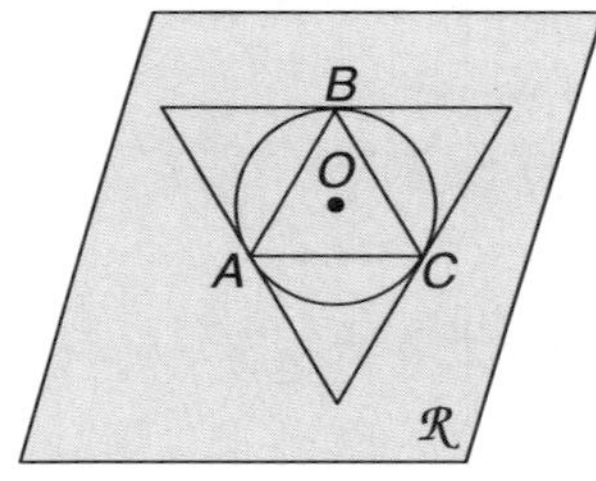

29. the rays that bisect $\angle ABD$ and $\angle BDC$ **31.** a pair of circles with radius a and center on h in two parallel planes perpendicular to the hypotenuse **33a.** a curve, parabola with ordinate one-half that of $y = x^2$
33b. $y = \frac{x^2}{2}$ **35.** When three or more circles are tangent there is an area among them that is not covered, therefore the circles must overlap in order to cover all of the area. This would then put some of the buildings within two miles of more than one fire station;

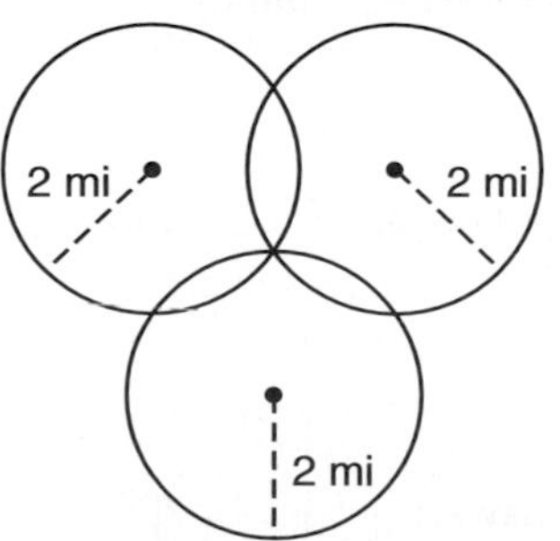

37. half of a sphere with a radius of 300 ft **39.** 1345 miles
41. about 54.1 cm and 62.1 cm **43.** $x = 54$

45.

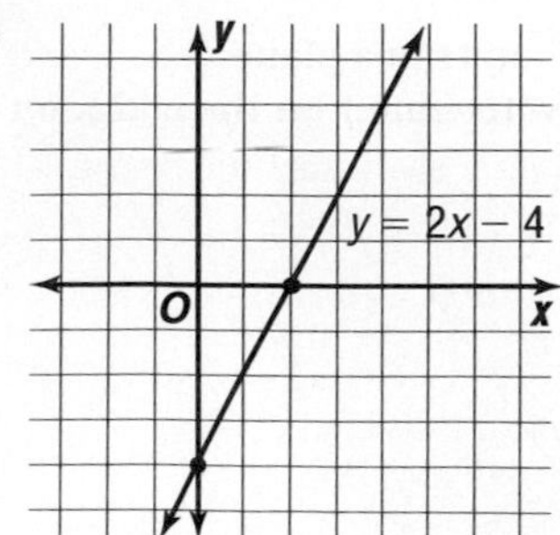

Pages 706–708 Lesson 13–2

5. (2, 2) **7.** (−1, −6) **9.** (2, 8) **11.** in about 13 years, 2003 **13.** (−2, −1);

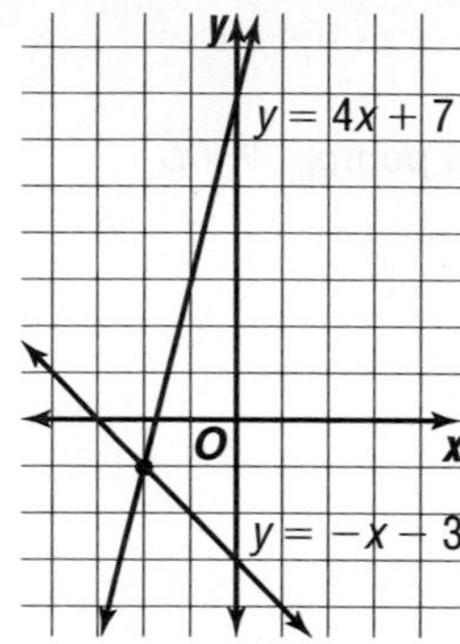

15. (2, −6);

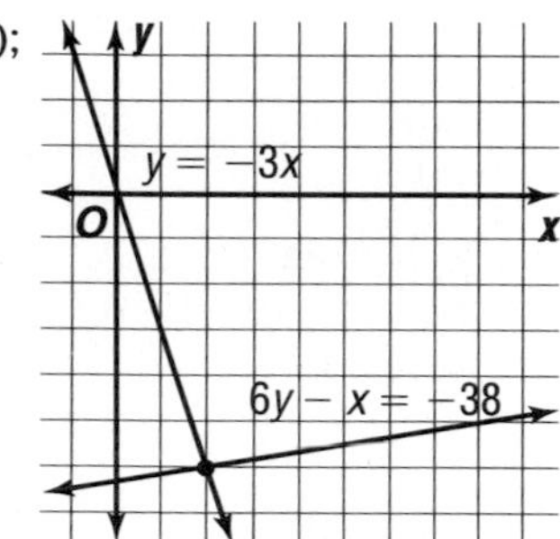

17. (2, 1);

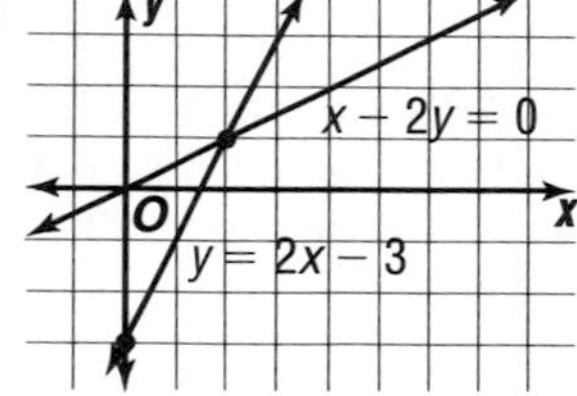

19. (2, 2) **21.** (4, 1) **23.** (0, 2) **25.** (1, −2) **27.** (15, 3) **29.** $\left(-\frac{9}{5}, -\frac{3}{5}\right)$ **31.** (2, 0) and (2, 8) **33.** (0, 1) and (3, 4) **35.** (70, 120) **37.** a circle with radius $\frac{DC}{2}$ and the midpoint of $\overline{DC}$ as its center;

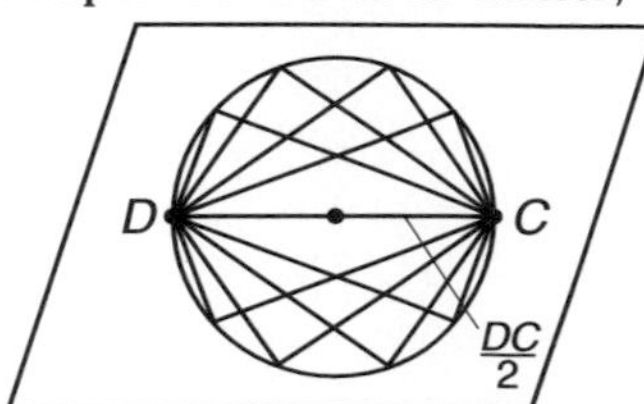

39. 0 **41.** trapezoid; Sample answer:

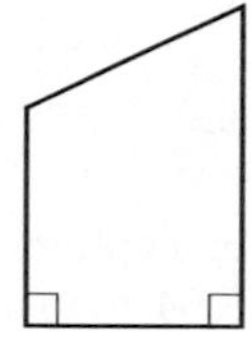

43. Sample answer:

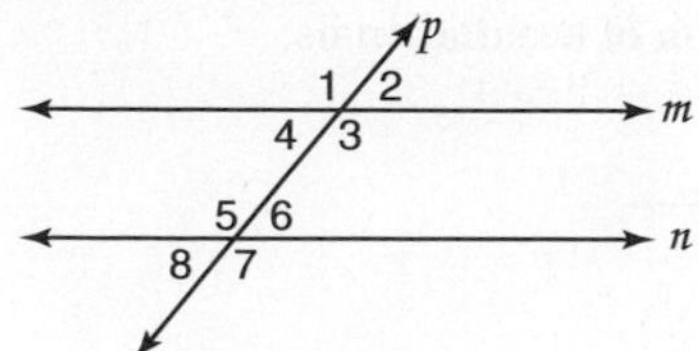

all odd-numbered angles have measure 145 **45.** 6

Pages 712–714 Lesson 13–3

5. none, a point, two points, or a circle
7. point on the angle bisector 7 in. from the vertex;

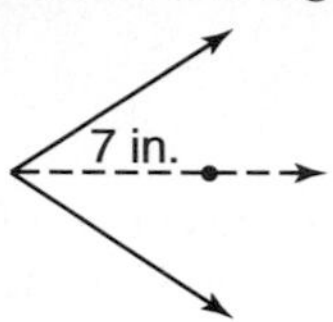

9. a plane that is perpendicular to the *xy*-plane intersecting the *xy*-plane at $x = 5$ **11.** Construct the perpendicular bisectors of each side of the triangle formed by these three cities. This intersection area is the area that should be considered; Columbus, Ohio, area. **13.** none or a point **15.** none, a point, two points, or a line segment

17. four points whose coordinates are $(\pm 3\sqrt{2}, \pm 3\sqrt{2})$

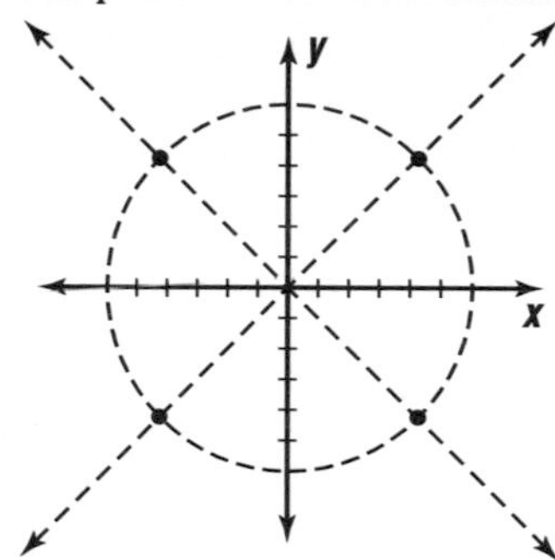

19. the axis of the cylinder

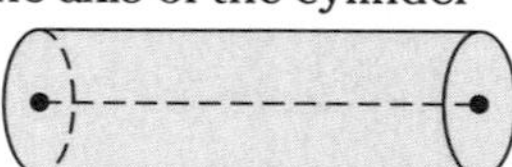

21. two circles with a radius of 5 cm that each lie in the parallel planes and whose centers are the intersection of the perpendicular line with each plane

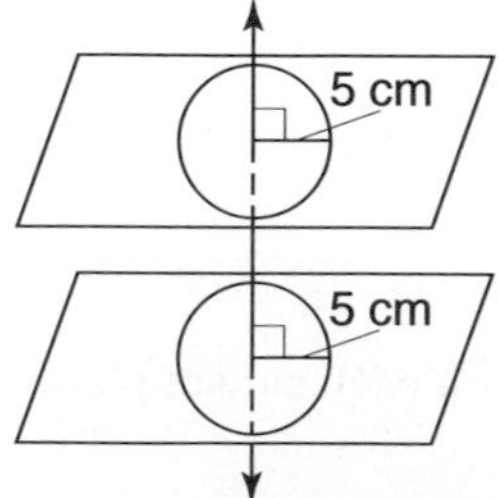

23. set of two intersecting planes;

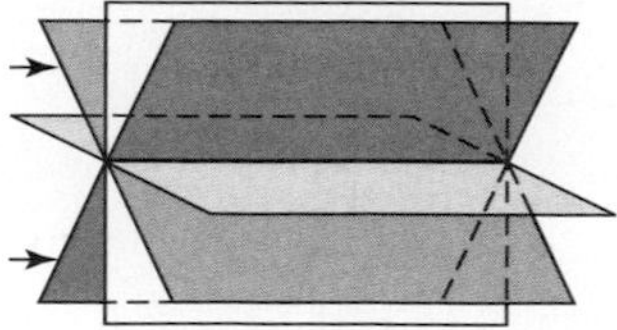

25. a parabola that opens up with vertex at (9, 0) **27.** circle with center (4, 1) and radius 5 **29.** a sphere with center (3, 4, 5) and radius 4 **31.** (−6, 5), (−3, −4) **33.** no points
35.

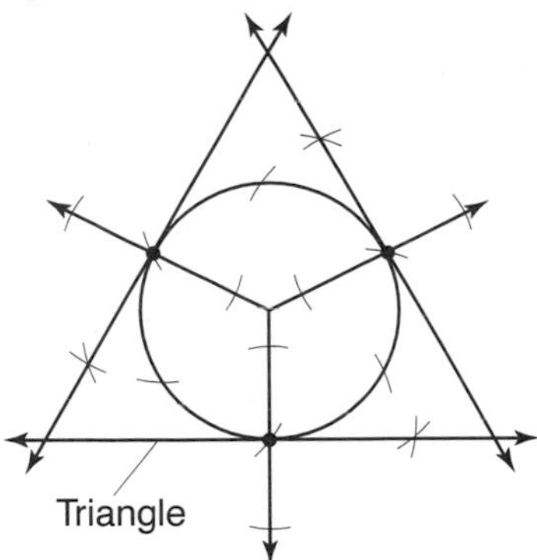

37. There would be many solutions. **39.** the corner of E St. and 3rd St. **41.** a semicircle with radius 80 ft and center at the center of the goal
43. ⊥ bisector of the segment connecting the two points of intersection;

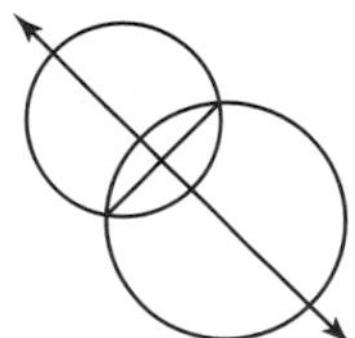

45. $T \approx 78.5\text{ m}^2$; $V \approx 65.4\text{ m}^3$ **47.** 10.8 miles
49. $\angle A \cong \angle C$, $\angle B \cong \angle D$, $\angle C \cong \angle E$, $\overline{AB} \cong \overline{CD}$, $\overline{BC} \cong \overline{DE}$, $\overline{AC} \cong \overline{CE}$ **51.** −5.54, 0.54

Pages 718–721 Lesson 13–4

5. $\overline{CD}$ **7.** $\angle STU$ **9.** two reflections or a rotation **11.** Yes; all angles and sides are congruent. **13.** $\angle BAC$ **15.** $\overline{ED}$ **17.** enlargement and reflection **19.** enlargement, reflection **21.** Yes; all angles and sides are congruent.
23a. rotation;

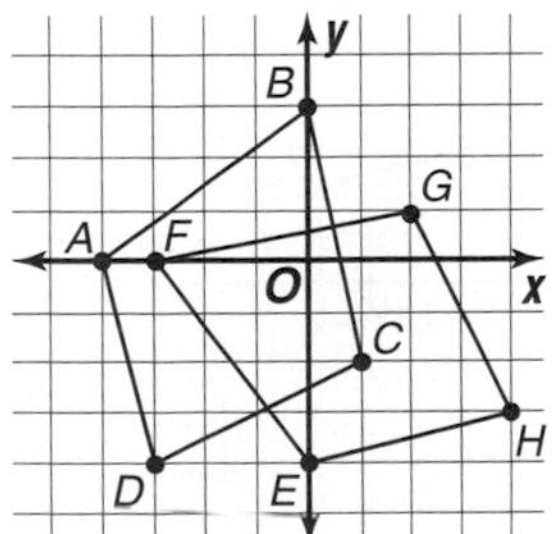

23b. yes **25.** quadrilateral *SRUT* **27.** $\triangle GIH$ **29.** quadrilateral *HIJK* **31.** $A'(3, -3)$, $B'(1, -2)$, $C'(2, -4)$, $D'(4, -5)$ **33.** two reflections or a rotation;

35. reflection of order and refection of L and J
37. reflections **39.** (−1, 2) **41.** $y = 2x - 8$ **43.** 5, $5\sqrt{3}$
45. B

Page 721 Self Test

1. the circle that is the intersection of the cylindrical surface with axis ℓ and radius 8 cm, and the sphere with center *R* and radius 8 m **3.** (6, −1)
5. one point;(−3, 1);

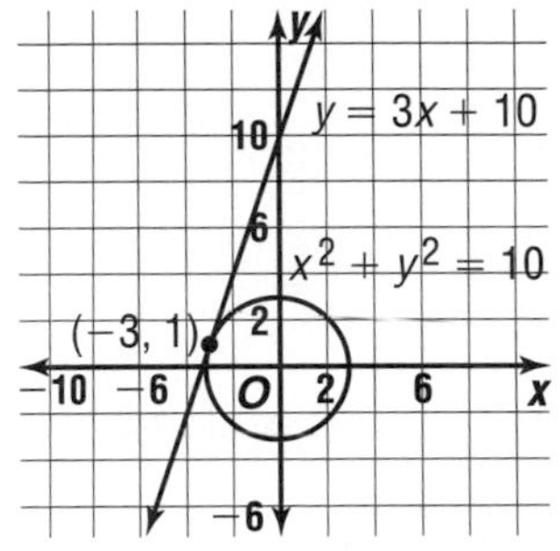

7. 0, 1, 2, 3, or 4 points **9.** $\overline{DB}$

Pages 726–729 Lesson 13–5

5. $\overline{PS}$ **7.** $\triangle SQP$
9.

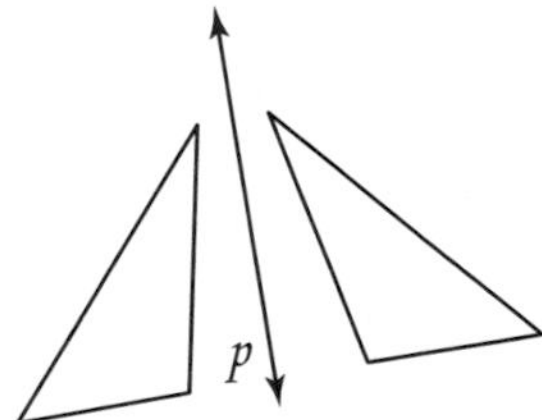

11. no lines of symmetry; no point symmetry **13.** four lines of symmetry; point symmetry **15.** 168 cm **17.** $\overline{HG}$
19. $\angle CBA$ **21.** $\triangle BHG$ **23.** quadrilateral *HGFB*

25. **27.**

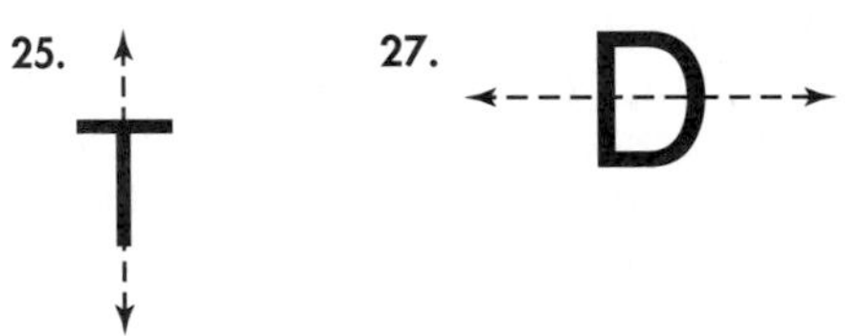

29.

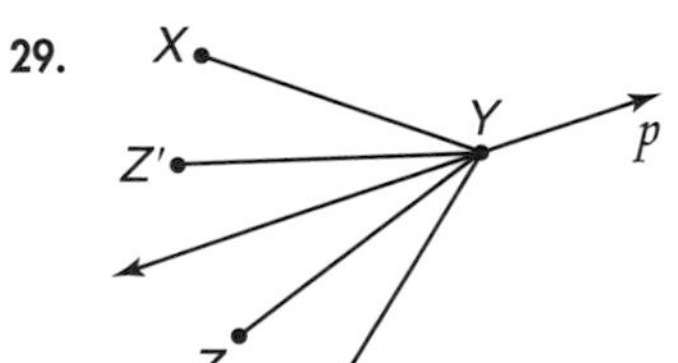

31.

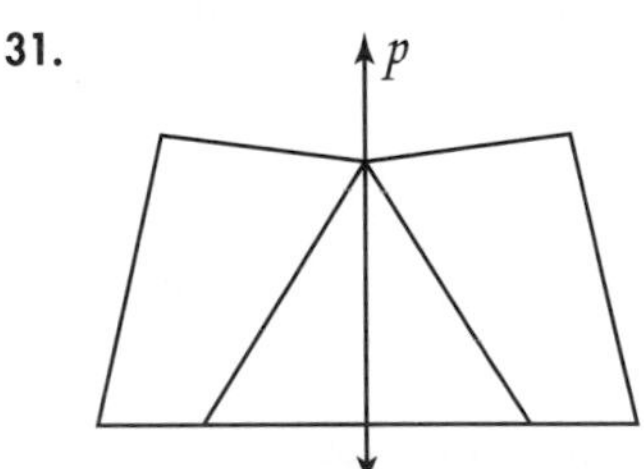

33.

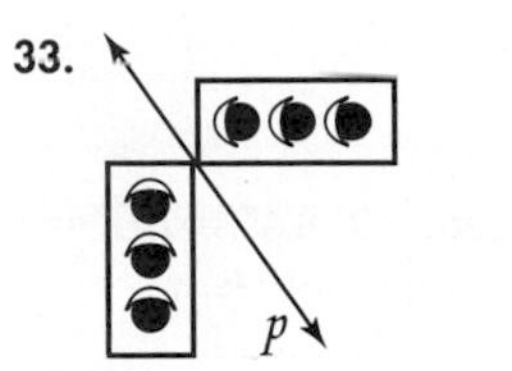

35. Yes; ℓ is the line of symmetry because it is possible to find, for every point A, another point B so that ℓ is the perpendicular bisector of $\overline{AB}$. **37.** Yes; ℓ is the line of symmetry because it is possible to find, for every point A, another point B so that ℓ is the perpendicular bisector of $\overline{AB}$. **39.** No; not all points are the same distance from ℓ.

41.

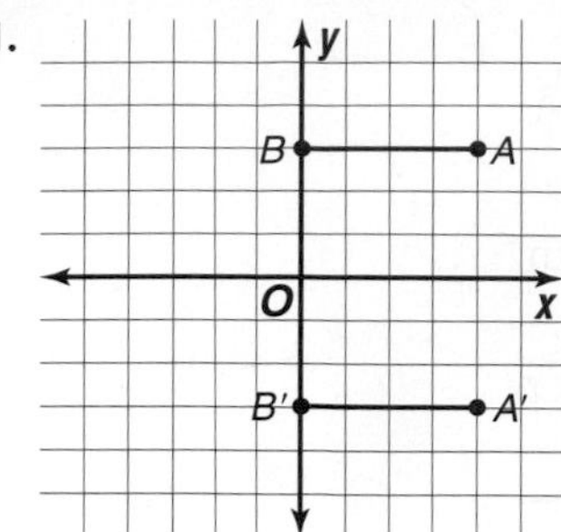

43.

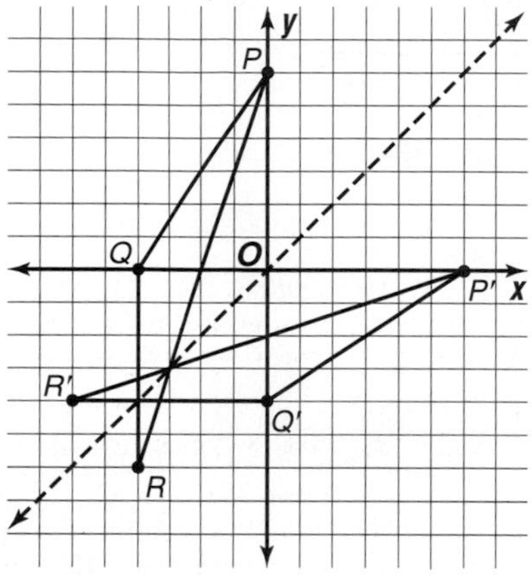

45. none **47.** none **49.** none **51.** both
53a. $(x, y) \rightarrow (-x, y)$ when reflected over the y-axis
53b. $(x, y) \rightarrow (x, -y)$ when reflected over the x-axis
53c. $(x, y) \rightarrow (y, x)$ when reflected over line $y = x$
55. Line m is a line of reflection, the vertex is a point of reflection. **57a.** The y-coordinates are negated and the x-coordinates are unchanged. It is reflected over the x-axis. **57b.** The x-coordinates are negated and the y-coordinates are unchanged. It is reflected over the y-axis. **59a.** $\angle S$ **59b.** $\overline{CD}$

61. (3, 3);

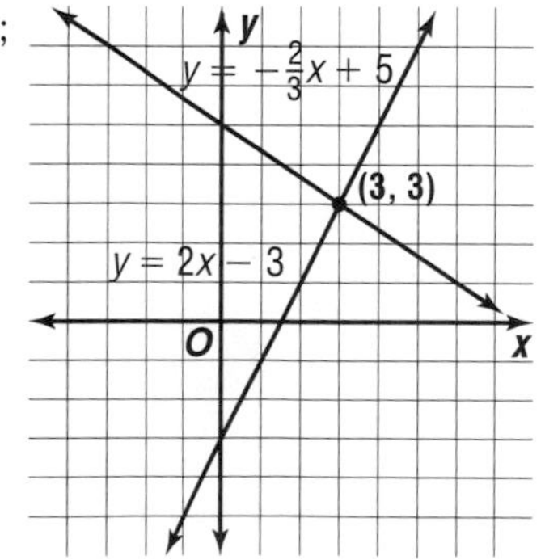

63. $y = -\frac{3}{4}x + \frac{7}{4}$ **65.** A **67.** diameters **69.** $\varnothing$

Page 730 Lesson 13–6A

1. The orientation remains unchanged. **3.** $AA'' = BB'' = CC'' = 2$(distance from ℓ to m) **5b.** The image is a translation of the preimage. **5c.** yes

Pages 734–737 Lesson 13–6

7. No; $\triangle HGK$ is not turned correctly.
9. $E'(4, 1)$, $F'(7, 3)$, $G'(7, 1)$;

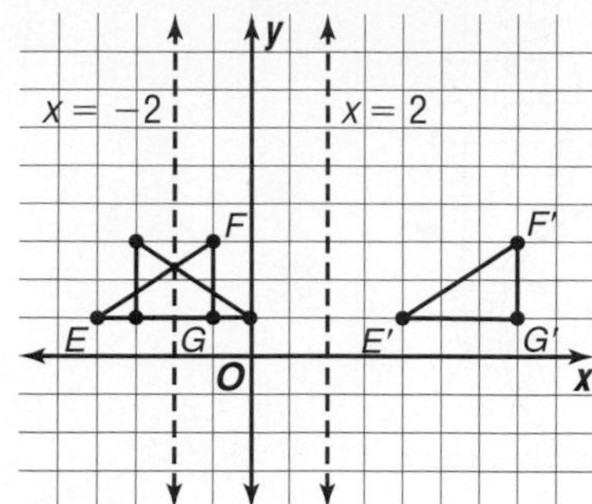

11. It is moved 15 ft from the wall and 21 ft to the left, then rotated. **13.** T **15.** U **17.** none **19.** Yes; it is one reflection after another with respect to two parallel lines. **21.** No; it is a translation and then a reflection with respect to a line.

23. $A'(-8, -4)$, $B'(-3, -2)$, $C'(-4, -5)$, $D'(-7, -6)$;

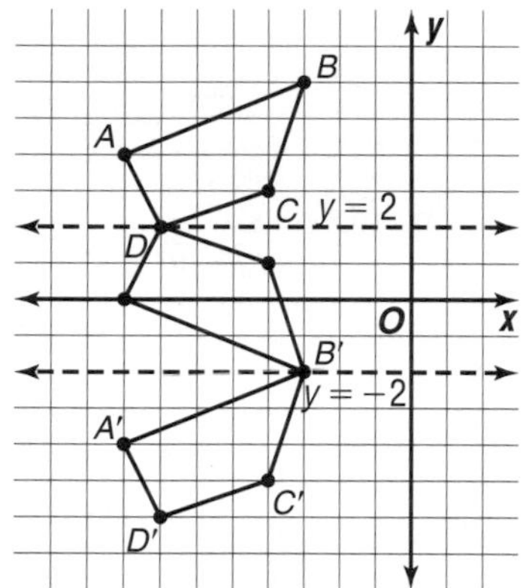

25. $N'(-4, 1)$, $I'(0, 0)$, $C'(0, -3)$, $K'(-8, -1)$;

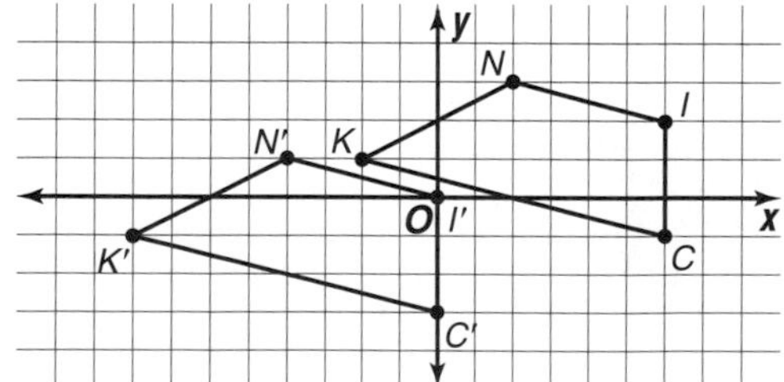

27.

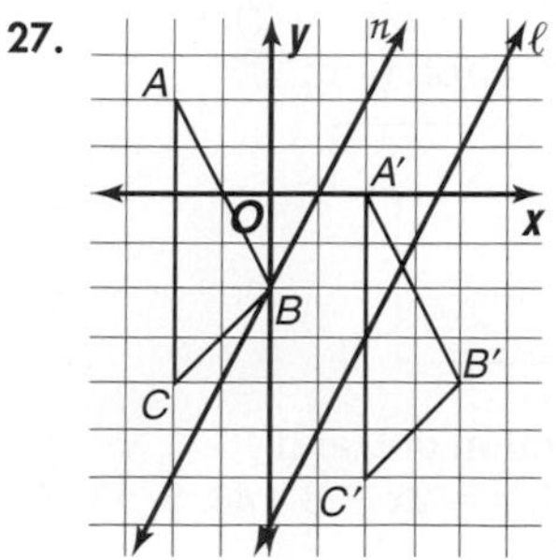

29.

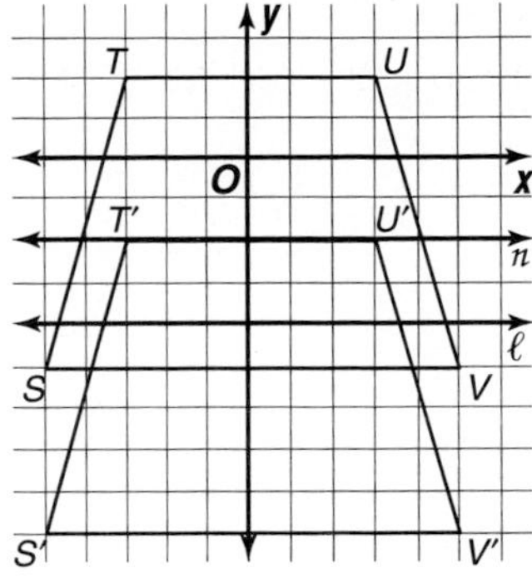

31. $\triangle STU$ **33.** Sample answer: $y = 1$, $y = -4$

35.

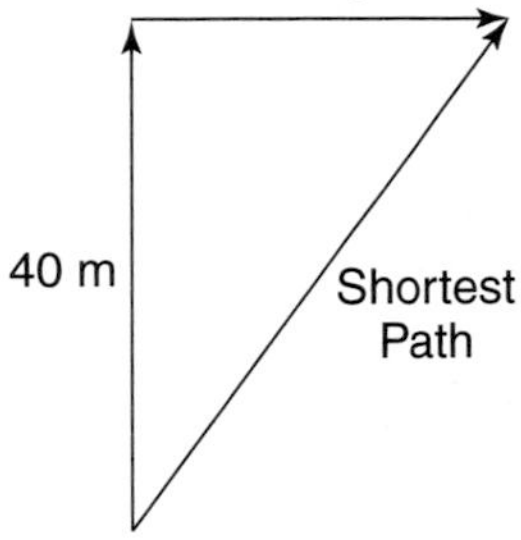

40 m

Shortest Path

37. a circle **39a.** $\overrightarrow{SQ}$ **39b.** $\overrightarrow{QT}$ **39c.** $\overrightarrow{TR}$ **41.** B

43. $8(x + 3y)^2$

Pages 742–745 Lesson 13–7

5. 74° **7.** pentagon *PKLMN* **9.** $\angle CQN$ **11.** $\overline{NK}$

13.

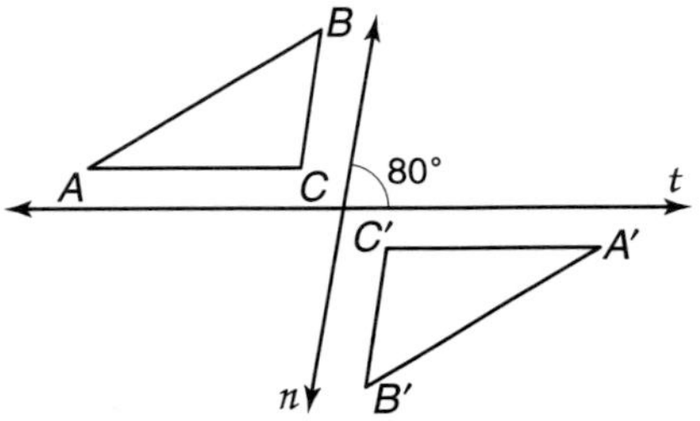

A B C 80° t C′ A′ n B′

15. Yes; it is a proper successive reflection with respect to two intersecting lines. **17.** 110 **19.** 148

21.

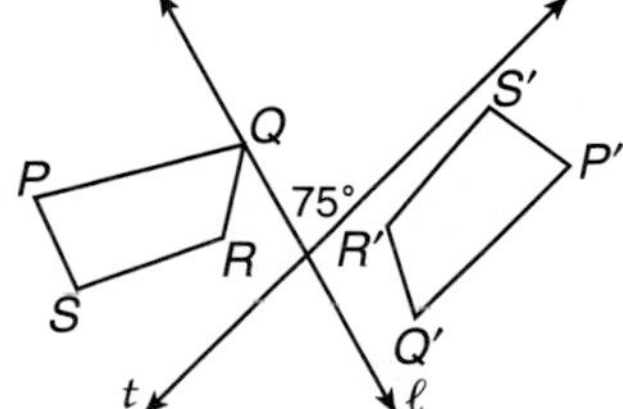

23.

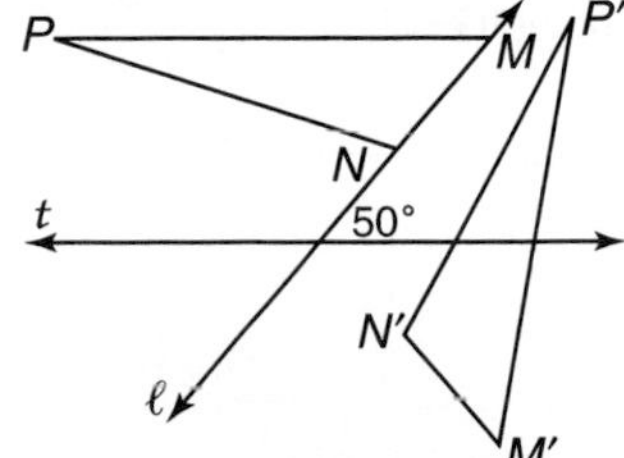

25.

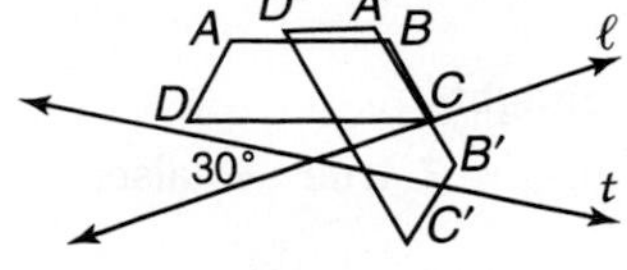

27.

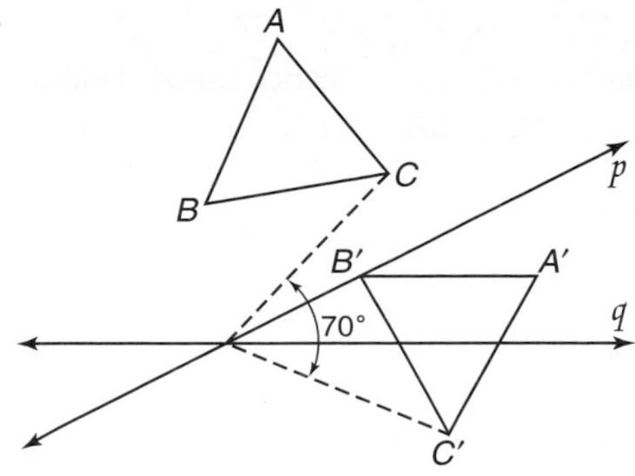

29. yes, yes, yes, yes, yes

33.

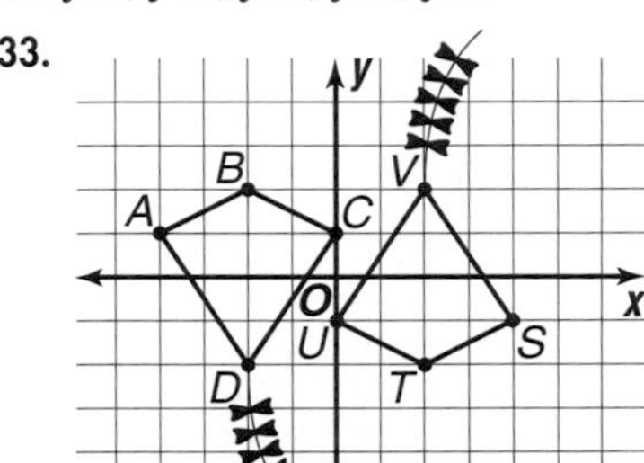

35. 120 **37.** reflection **39.** translation **41.** Angles of rotation with measures of 90 or 180 would be easier on a coordinate plane because of the grids used in graphing. **43a.** 90 **43b.** seat 7 **45.** yes; the upper left, upper right, and center quadrilaterals **47.** two circles concentric to the given circle, one with a radius of 2 in. and the other with a radius of 8 in. **49.** $768\pi \approx 2412.7 \text{ ft}^3$ **51.** 37.5

53. $\frac{1}{x + 9}$

Pages 749–753 Lesson 13–8

7. < **9.** 62 **11.** congruence transformation

13. reduction

15. Scale factor of 2:

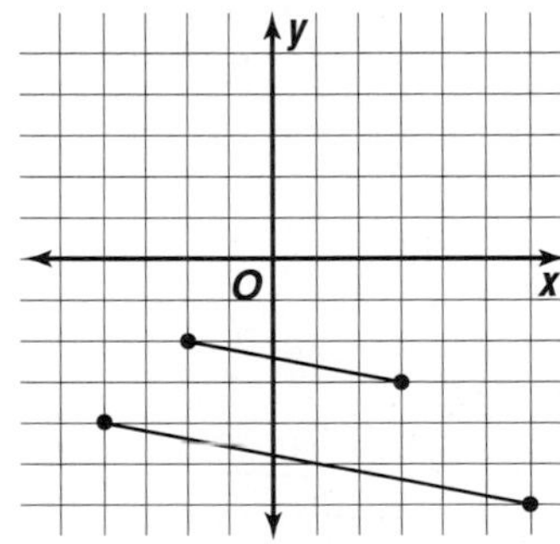

Scale factor of $\frac{1}{2}$:

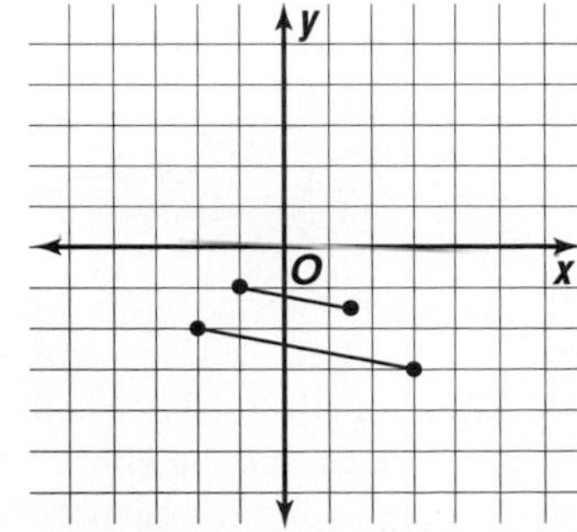

17. $\frac{1}{2}$ **19.** *B* **21.** *S* **23.** 30 **25.** 24 **27.** 3
29. 2, enlargement **31.** 1, congruence transformation
33. $\frac{1}{4}$, reduction **35.** $\frac{3}{4}$ **37.** 2

39. Scale factor of 2:

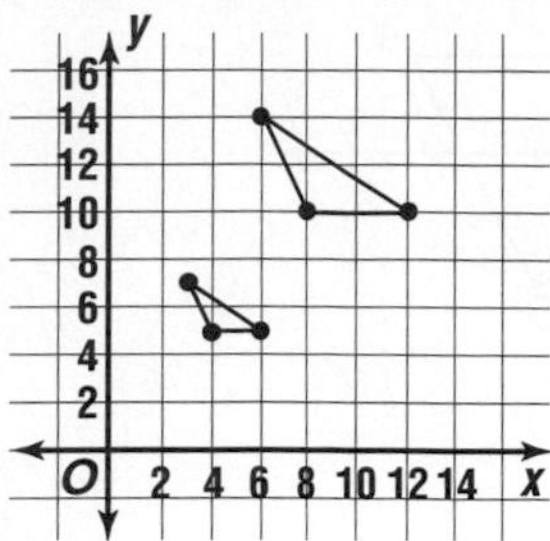

Scale factor of $\frac{1}{2}$:

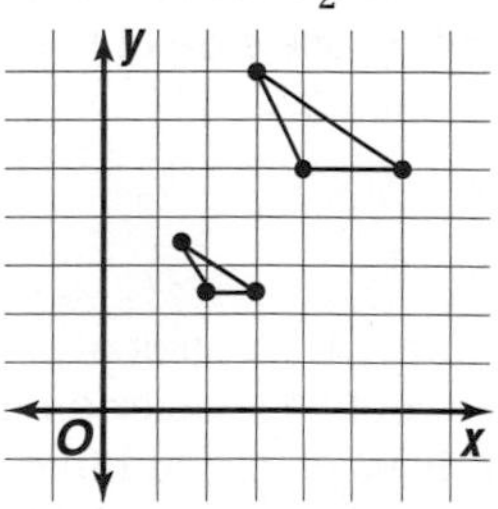

41. Scale factor of 2:

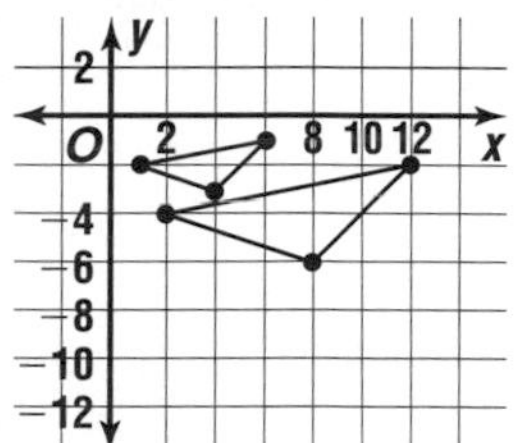

Scale factor of $\frac{1}{2}$:

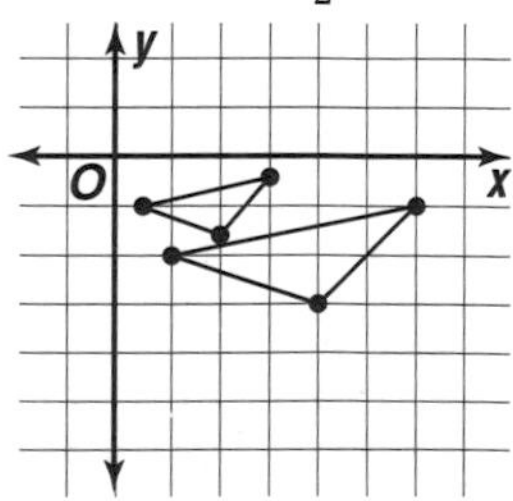

43. Scale factor of 2:

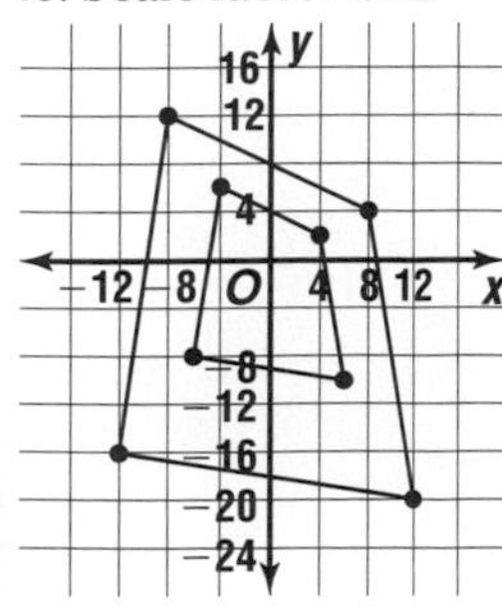

Scale factor of $\frac{1}{2}$:

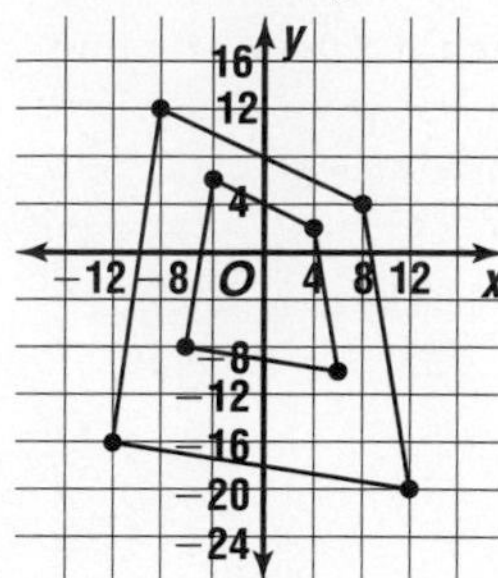

45a. The surface area of the image will be nine times the surface area of the preimage. **45b.** The volume of the image will be 27 times the volume of the preimage.
47. (0, 5); $\frac{5}{2}$ **49.** 337.5 miles

51.

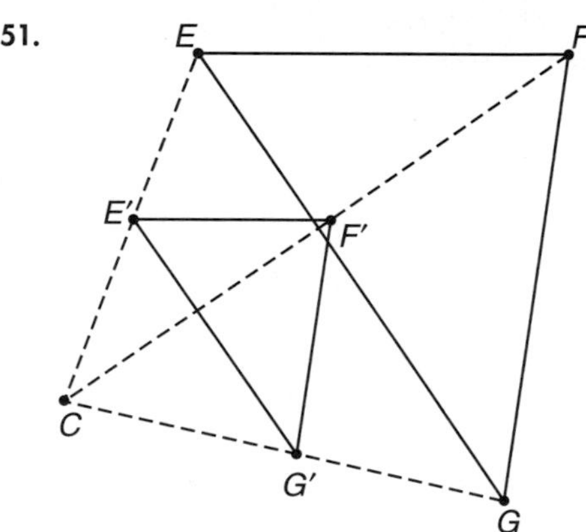

53.

F, H, F″, G, ℓ, H″, m, G″

55. $\sqrt{179} \approx 13.4$ units **57.** B **59.** 7.5

61a.

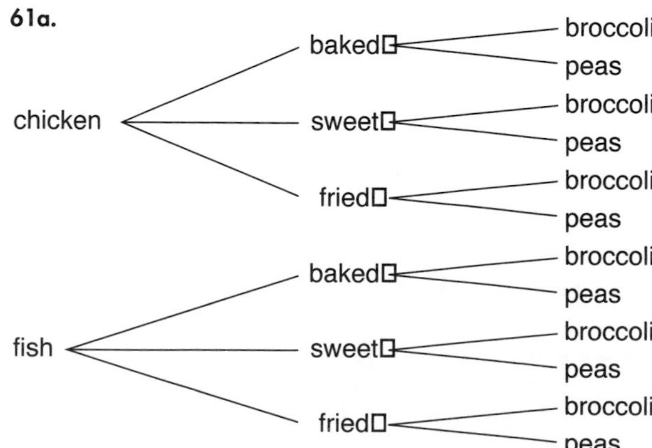

61b. $\frac{2}{12}$ or $\frac{1}{6}$

Page 755 Chapter 13 Highlights

1. true **3.** true **5.** false; image **7.** true **9.** false; preimage

Pages 756–758 Chapter 13 Study Guide and Assessment

11. the interior of a circle with center at the given point and radius 6 cm **13.** the circumcenter; the point where the three perpendicular bisectors of the segments joining *A*, *B*, and *C* intersect **15.** (1, 1) **17.** (1, 2)

19. the circle that is the intersection of spheres *P* and *Q* with radii 3 cm;

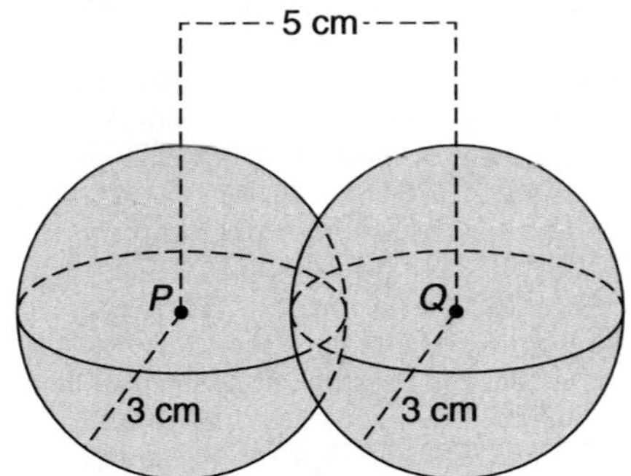

21. *E* **23.** $\overline{CD}$ **25.** reflection

27. **29.**

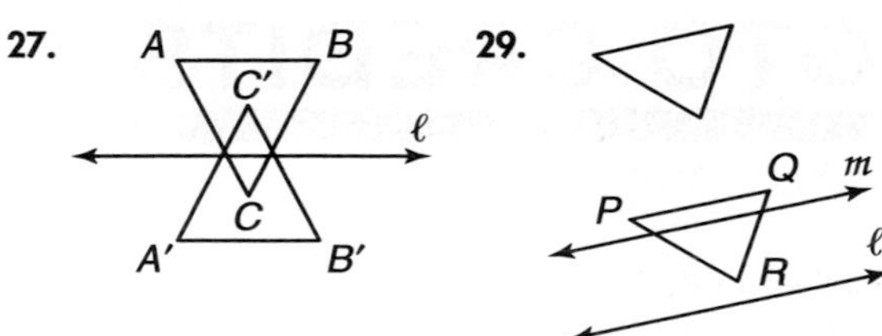

31.

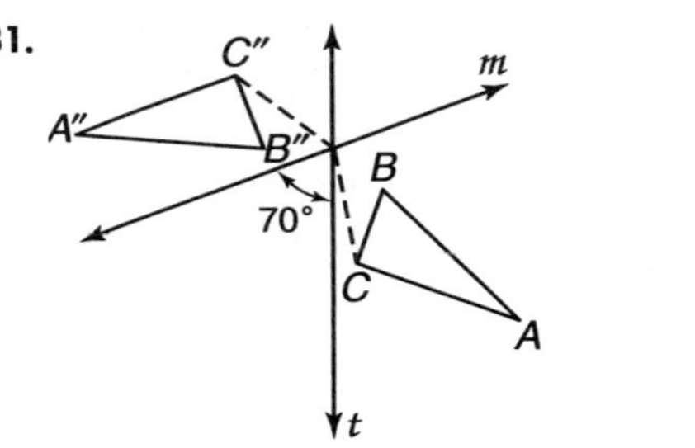

33. enlargement; $|3| > 1$ **35.** Imagine the reflection of the 5 ball with respect to the line formed by one side of the billiard table, and aim for the reflection. **37.** 144

PHOTO CREDITS

PHOTO CREDITS

COVER (background) Paul Raftery/Arcaid, (center) Joanne Lotter/Tom Stack & Associates, (bottom) Mark Tomalty/Masterfile; **iii** (t)Aaron Haupt Photography, (b)SuperStock; **iv** (t)courtesy Cindy Boyd, (c)courtesy Gail Burrill, (b)courtesy Jerry Cummins; **v** (t)courtesy Tim Kanold, (c)courtesy Carol Malloy, (b)courtesy Mrs. Marie Yunker; **viii** (t)Doug Martin, (b)B. Graham/FPG; **ix** Mark Burnett; **x** (t)Mark Burnett, (b)North Wind Picture Archives; **xi** (r)John Evans Photography, (bl)Steve Lissau; **xii** David Ball/The Stock Market; **xiii** (t)courtesy of Norma Merrick Sklarek, (b)SHOE © 2-22-89 Tribune Media Services, Inc. All rights reserved; **xiv** John Evans Photography; **xv** courtesy Mark Clingan; **xvi** (l)David Hamilton/The Image Bank, (r)David Noble/FPG; **xvii** Aaron Haupt Photography; **4** (l)Nissan Motor Corporation, USA, (r)British Museum, Ref:1904-6-21-34; **6** (t)(br)Doug Martin, (bl)Giboux/Gamma-Liaison, International; **9** Kaku Kurita/ Gamma-Liaison, International; **10** Arslanian-Liaison/ Gamma-Liaison, International; **11** (l)Loviny Chrix/Gamma-Liaison, International, (r)John Veltri/Photo Researchers; **13** Doug Martin; **16** Mark Steinmetz; **18** Doug Martin; **19** Ric Ergenbright Photography; **20** (t)courtesy Johnson Publishing Company, (b)Mark Steinmetz; **22** Doug Martin; **24** Steve Lissau; **28** (tr)Eric Sander/Gamma-Liaison, International, (c)Doug Martin, (bl)THE FAR SIDE, 5-25-93 © 1993 FarWorks, Inc./ Dist. by Universal Press Syndicate; **30** Mark Steinmetz; **32** North Wind Picture Archives; **33** Martyn Goddard/ Colorific; **34** PEANUTS ® reprinted by permission of United Feature Syndicate, Inc; **35** Jeremy Scott/International Stock; **38** Mark Burnett; **43** (t)Doug Martin, (b)Mark Burnett; **44** Mark Burnett; **49** (t)B. Graham/FPG, (b)Dick Luria/ Science Source/Photo Researchers; **58** Richard Martin/Agence Vandystadt/ Duomo; **64** Mark Steinmetz; **65** The Museum of Modern Art, New York. Acquired through the Lillie P. Bliss Bequest; **66** Aaron Haupt Photography; **67** (t)Rich Brommer, (bl)Cincinnati Art Museum, John J. Emery Fund, (br)file photo; **68** (l)Bob Mullenix, (r)Mary Evans Picture Library; **70** SHOE, 2-22-89. © 1989 Tribune Media Services, Inc. All rights reserved.; **71** Aaron Haupt Photography; **73** courtesy JPL; **74** Ancient Art & Architecture Collection, Ltd.; **75** Cyberware; **76** Stock Montage; **77** Mary Evans Picture Library; **79** PEOPLE Weekly © 1996/John Zich; **80** (t)John Madere/The Stock Market, (b)Stock Montage/Charles Walker Collection; **82** SuperStock; **83** Aaron Haupt Photography; **86** CBS Entertainment; **87** Ancient Art & Architecture Collection, Ltd; **88** (t)Larry Hamill, (b)LGI Photo Agency; **89** Mark Burnett; **90** (t)Ken Frick, (b)Aaron Haupt Photography; **92** George Anderson; **95** Bob Daemmrich Photo, Inc; **99** Aaron Haupt Photography; **100** David Frazier Photolibrary; **106** Gerard Photography; **108** Boltin Picture Library; **113** Stock Montage; **118** David Frazier Photolibrary; **119** Hess/The Image Bank; **122** (l)DUOMO/William R. Sallaz, (r)North Wind Picture Archives; **124** (l)Grant Heilman/Grant Heilman Photography, (r)David Ball/The Stock Market; **125** Charles O'Rear; **126** (t)file photo, (b)Charles O'Rear; **128** Guido Rossi/The Image Bank; **137** Mark Scott/FPG; **138** Salt Lake City Convention & Visitors Bureau; **140** Doug Martin; **142** Eliot Cohen; **144** Aaron Haupt Photography; **146** (t)file photo, (b)Roger K. Burnard; **150** Doug Martin; **157, 158, 160, 161** (l)Architectural Association Photo Library, London/John Edward Linden, (r)Architectural Association Photo Library, London/V. Bennett; **163** (t)Aaron Haupt Photography, (b)Doug Martin; **165** (l)TRIP/Eric Smith, (r)David Noble/FPG; **167** Aaron Haupt Photography; **168** (t)Mark E. Gibson, (ct)Travelpix/FPG, (cb)Alan Schein/The Stock Market, (b)Rafael Macia/Photo Researchers; **169** Mary Evans Picture Library/Explorer; **170** Ken Van Dyne; **174** Aaron Haupt Photography; **175** David Frazier Photolibrary; **176** William J. Weber; **177** Gail Shumway/FPG; **178** (l)Tim Courlas, (r)SuperStock; **180** (t)Sharon Kurgis, (c)Aaron Haupt Photography, (b)Yousuf Karsh/Woodfin Camp & Associates; **181** Mark E. Gibson; **189** courtesy Carolyn Shoemaker; **196** courtesy Diane Leighton; **200** Museum of Modern Art; **202** Doug Martin; **204** Mark Steinmetz; **206** Doug Martin; **212** SuperStock; **213** David Park/Science Photo Library/ Photo Researchers; **221, 222** John Evans; **227** (t)Mark Burnett, (b)SuperStock; **228** Mark Burnett; **232** Doug Martin; **233** Patty Jedick; **236** (l)Robert Frerck/The Stock Market, (r)M. C. Escher © 1996 Cordon Art, Baarn, Holland. All rights reserved; **241** courtesy Jack Friedman/Park Board of Parkville, MO; **242** Aaron Haupt Photography; **245** David Brownell; **246** Aaron Haupt Photography; **249** Robert Frerck/ Odyssey; **252, 254** Aaron Haupt Photography; **263** file photo; **267** Doug Martin; **268** Jim Brown/The Stock Market; **272** Lynn Stone; **273** Aaron Haupt Photography; **278** Lynn M. Stone; **280** Tracy Borland; **285** Aaron Haupt Photography; **286** David Noble/FPG; **287** Aaron Haupt Photography; **288** (l)Doug Martin, (r)Kent State University Museum, gift of the Silverman/Rodgers Collection, 1983. 1.2158; **291**(t)courtesy Dalia Berlin, ASID, (b)Mark Burnett; **293** Mark Burnett; **301** (l)Margaret Kois/The Stock Market, (r)Mark Burnett; **303** (l)Jerry Jacka Photography, (r)Mark Burnett; **304** Neil Troiano/Arcaid; **306** (t)Jerry Irwin/SuperStock, (b)Hans Namuth/Photo Researchers; **308** Doug Martin; **309** DUOMO/ Mitchell Layton; **313** Mark Burnett; **315** High Museum of Art, © 1997 Dorothea Rockburne/Artists Rights Society (ARS), New York; **319** Aaron Haupt Photography; **320** SuperStock; **321** Telegraph Colour Library/FPG International; **325** Gil Ullom; **327** (l)courtesy Norma Merrick Sklarek, (r)Doug Martin; **332** (tl)Mark E. Gibson, (tr)Yale University Art Gallery, gift of Anni Albers and The Josef Albers Foundation, Inc., (b)Colin Molyneux/ Molyneux Associates/The Image Bank; **333** Doug Martin; **336** (l)Runk/Schoenberger from Grant Heilman, (r)Mark Steinmetz; **338** Scala/Art Resource, NY; **339** John Evans; **340** SuperStock; **341** (t)Stock Montage/Newberry Library, (b)Hulton Getty Collection/Woodfin Camp & Associates; **342** NASA; **345** Smithsonian Institution; **346** The Metropolitan Museum of Art, Alfred Stieglitz Collection, 1949(49.59.1); **348** John Evans; **350, 352** Mark Burnett; **353** Elaine Shay; **356** David W. Hamilton/The Image Bank; **361** Mark Burnett; **362** The Kobal Collection; **368** North Wind Picture Archives; **369** Stock Montage, Charles Walker Collection; **372** Doug Martin; **376** Tony Demin/International Stock; **377** Mark Burnett; **378** Michael D. McGuire; **379** North Wind Picture Archives; **383** (l)Edward Berko, oil on wood 48" × 36" ©1991, courtesy private collection, NYC, fig.1 © 92 Lawrence Hudetz, fig.2 Loren Carpenter, fig.3 Peitgen, Jurgens, Saupe, FRACTALS FOR THE CLASSROOM, Springer-Verlag, NY; fig.4 © 1987 Lawrence Hudetz; **390, 391** Mark Burnett; **394** (l)courtesy Mercedes-Benz of North America, (r)Philip Wallick/International Stock; **398** Doug Martin; **401** Ed Wheeler/The Stock Market; **403** Hulton Getty Collection/Woodfin Camp & Associates; **404** (t)Kunio Owaki/The Stock Market, (b)The Kobal Collection; **405** (t)Mark Steinmetz, (b)David Madison/Tony Stone Images; **406** David Madison/Tony Stone Images; **409** Grant V. Faint/The Image Bank; **410** from THE NATURE OF CITIES © 1955 Paul Theobald, Chicago; **412** (t)AP/Wide World Photos, (b)Doug Martin; **414** (t)Joe Towers/The Stock Market, (b)Bob Daemmrich/ Stock Boston; **416** Doug Martin; **418** (t)NASA, (b)file photo; **421** NASA; **423** (l)Peter

Menzel/Stock Boston, (r)ITTC Productions/The Image Bank; **424** (t)G. R. Roberts, (b)Jake Baker/The Image Bank; **425** (tl)file photo, (bl)David Frazier Photolibrary, (br)KS Studios; **426** Thomas Braise/The Stock Market; **428** Mastrorillo/The Stock Market; **429** (t)Philip Jon Bailey/Stock Boston, (b)Jim Dwyer/Stock Boston; **430** Aaron Haupt/Photo Researchers, Inc.; **431** Paul J. Sutton/DUOMO; **434, 440** Doug Martin; **441** Tom McGuire; **444** (l)Mike Yamashita/Westlight, (r)Chuck Mason/ International Stock; **446** Doug Martin; **447** John Evans Photography; **448** (t)Neal Lauron/Sipa Press/Woodfin Camp & Associates, (b)Bob Daemmrich/Stock Boston; **451** (t)AP Photo/Journal Star, Randy Hampton/Wide World Photos, (b)Aaron Haupt Photography; **454** Bavaria/Viesti Associates; **458** Claire Ferguson; **459** Larry Hamill; **462** Rick Weber; **466** Mark Steinmetz; **472** Hans Reinhard/Okapia/ Photo Researchers; **475** file photo; **482** William J. Weber; **483** Guy Marche/FPG; **487** NASA/Science Photo Library/ Photo Researchers; **491** Doug Martin; **498** (l)UPI/Bettmann, (r)Benjamin Rondel/The Stock Market; **503** Edi Ann Otto; **508** Rick Weber; **509** JPL; **511** (t)Bob Daemmrich/Stock Boston, (b)Chuck Mason/ International Stock; **512** (l)J. Messerschmidt/The Stock Market, (r)Paul L. Ruben; **514** Photo Researchers, Inc.; **516** Mark Burnett; **519** Elaine Shay; **520** (l)Aaron Haupt Photography, (r)Tim Courlas; **523** Aaron Haupt Photography; **529** Joseph DiChello, Jr.; **531** Aaron Haupt Photography; **532** Tim Barnwell/Stock Boston; **537** Aaron Haupt Photography; **539** Doug Martin; **542** Steve Grohe/The Stock Market; **543** (t)courtesy Knox Parks Foundation, (b)Jack McConnell; **547** Mark Burnett; **549** (l)Ruth Dixon, (r)American Photo Library, Tokyo; **551** Lou Jones/The Image Bank, (inset)Aaron Haupt Photography; **552** courtesy St. Joseph's Historic Foundation; **553** Doug Martin; **554** (t)Mark Burnett, (b)Mark Steinmetz; **555** Mark Burnett; **556** Jon Feingersh/Stock Boston; **557** (t)Mark Burnett, (b)file photo; **558** Philip Rosenberg/Pacific Stock; **560** Joseph DiChello, Jr.; **562, 568** Mark Steinmetz; **569** Rudy VonBriel; **572** (l)AP Photo/ Damian Dovarganes/ Wide World Photos, (r)(br)North Wind Picture Archives; **574** Mark Burnett; **575** Ancient Art & Architecture Collection; **576** Mark Burnett; **577** (t)David Parker/Science Photo Library/Photo Researchers, (b)Laurie Platt Winfrey; **578** Mark Burnett; **579** Tate Gallery, London/Art Resource, NY; **584** (t)Jeff Gnass/The Stock Market, (b)Woodfin Camp & Associates; **585** Mark Burnett; **587** Mercury Archives/The Image Bank; **589** Mark Burnett; **590** Aaron Haupt Photography; **591** (t)Jeff Gnass/ The Stock Market, (l)(b)Mark Burnett; **593** Mark E. Gibson; **596** Mark Steinmetz; **597** (l)file photo, (r) © Tribune Media Services. All rights reserved. Reprinted with permission.; **601** Grant V. Faint/The Image Bank; **608** Bob Daemmrich/ Stock Boston; **615** (t)Library of Congress, (b)John Elk/ Tony Stone Images; **616** Timothy Hursley; **617** Mark Steinmetz; **621** (t)Library of Congress, (b)Egyptian National Museum, Cairo, Egypt/SuperStock; **623** Mark Steinmetz; **627** Doug Martin; **628** Mark Steinmetz; **629** Matt Meadows; **631** AP Photo/Des Moines Register, Harry Baumert; **632** Nippondenso Company, Ltd; **635** Mark Burnett; **642, 643** The Kobal Collection; **644** (l)Mark Burnett, (r)John E. Evans Photography; **649** John E. Evans Photography; **651** Les Losego/Crown Studio; **653** Telegraph Colour Library/FPG; **657** John E. Evans Photography; **659** The Living Earth, Inc; **663** (t)DUOMO/ Ben Van Hook, (b)Fiona Pragoff/The Image Bank; **671** Peter Gridley/FPG; **673** (t)Corbis-Bettmann, (b)UPI/Corbis/ Bettmann; **680** Flip Nicklin/Minden Pictures; **684** Kim Westerskov/ Tony Stone Images; **685** Aaron Haupt Photography; **691** file photo; **694** (l)Art Trade, Bonhams, London/ Bridgeman Art Library, London/SuperStock, (r)Mark Burnett; **696** Tom McHugh/ Science Source/Photo Researchers; **698, 699, 703, 704** Aaron Haupt Photography; **706** (t)Brownie Harris/The Stock Market, (b)Mark Downey Photographer; **707** Mark Burnett; **708** Doug Martin; **710** Mark Burnett; **712** Rafael Macia/ Photo Researchers; **715** Greg Probst/Tony Stone Images; **720** Dumbarton Oaks; **722, 729** (l)Aaron Haupt Photography, (r)Tim Courlas; **730, 731** (t)Mark Burnett, (b)courtesy Alcy Haden/ Reynoldsburg, OH, Chamber of Commerce; **733** Mark Burnett; **734** Richard Gross Photo; **736** SuperStock; **739** A. & L. Sinibaldi/Tony Stone Images; **741** Mark Burnett; **746, 749, 750** Aaron Haupt Photography; **752** M. C. Escher © 1996 Cordon Art, Baarn, Holland. All rights reserved.; **754** The Kobal Collection; **758** (t)David Ball/The Stock Market, (b)Mark Burnett; **759** Brooks/Brown/Photo Researchers.

APPLICATIONS & CONNECTIONS INDEX

P

INDEX

A

B

D

E

F

G

H

I

INDEX

P

Q

R

S

FORMULAS

Perimeter and Circumference

Square	$P = 4s$
Rectangle	$P = 2\ell + 2w$
Circumference of a circle	$C = 2\pi r$ or πd

Area

Circle	$A = \pi r^2$
Parallelogram	$A = bh$
Rectangle	$A = \ell w$ or $A = bh$
Regular Polygon	$A = \frac{1}{2}Pa$
Rhombus	$A = \frac{1}{2}d_1 d_2$
Sector of a circle	$A = \frac{N}{360}\pi r^2$
Square	$A = s^2$
Trapezoid	$A = \frac{1}{2}h(b_1 + b_2)$
Triangle	$A = \frac{1}{2}bh$

Surface Area of Solid

Lateral Area

Regular pyramid	$L = \frac{1}{2}P\ell$
Right circular cone	$L = \pi r\ell$
Right cylinder	$L = 2\pi rh$
Right prism	$L = Ph$

Total Surface Area

Regular pyramid	$T = \frac{1}{2}P\ell + B$
Right circular cone	$T = \pi r\ell + \pi r^2$
Right cylinder	$T = 2\pi rh + 2\pi r^2$
Right prism	$T = Ph + 2B$
Sphere	$T = 4\pi r^2$

Volume

Cube	$V = s^3$
Right circular cone	$V = \frac{1}{3}Bh$
Right cylinder	$V = \pi r^2 h$
Right prism	$V = Bh$
Right pyramid	$V = \frac{1}{3}Bh$
Right rectangular prism	$V = \ell wh$
Sphere	$V = \frac{4}{3}\pi r^3$

Pythagorean Theorem

In a right triangle, for side lengths a, b, and c, $a^2 + b^2 = c^2$.

Distance

Between two points on a coordinate plane

$$d = \sqrt{(x_2 - x_1)^2 + (y_2 - y_1)^2}$$

Between two points A and B on a number line

$$d = |a - b|$$

Arc length $\ell = \frac{N}{360} \cdot 2\pi r$

From a point to a line with equation

$Ax + By + C = 0$ $\quad d = \dfrac{|Ax_1 + By_1 + C|}{\sqrt{A^2 + B^2}}$

Midpoint

Between two points on a coordinate plane

$$\frac{x_1 + x_2}{2}, \frac{y_1 + y_2}{2}$$

Between two points A and B on a number line

$$\frac{a + b}{2}$$

Slope

Slope of a line $\quad m = \dfrac{y_2 - y_1}{x_2 - x_1}$

Equations for Figures on a Coordinate Plane

Circle	$(x - h)^2 + (y - k)^2 = r^2$
Slope-intercept form of a line	$y = mx + b$
Point-slope form of a line	$y - y_1 = m(x - x_1)$
Standard form of a line	$Ax + By = C$

Sphere with center (i, j, k)

$$(x - i)^2 + (y - j)^2 + (z - k)^2 = r^2$$

Trigonometry

Law of Cosines

$$a^2 = b^2 + c^2 - 2bc \cos A$$
$$b^2 = a^2 + c^2 - 2ac \cos B$$
$$c^2 = a^2 + b^2 - 2ab \cos C$$

Law of Sines

$$\frac{\sin A}{a} = \frac{\sin B}{b} = \frac{\sin C}{c}$$

SQUARES AND APPROXIMATE SQUARE ROOTS

n	n^2	$\sqrt{n}$	n	n^2	$\sqrt{n}$
1	1	1.000	51	2601	7.141
2	4	1.414	52	2704	7.211
3	9	1.732	53	2809	7.280
4	16	2.000	54	2916	7.348
5	25	2.236	55	3025	7.416
6	36	2.449	56	3136	7.483
7	49	2.646	57	3249	7.550
8	64	2.828	58	3364	7.616
9	81	3.000	59	3481	7.681
10	100	3.162	60	3600	7.746
11	121	3.317	61	3721	7.810
12	144	3.464	62	3844	7.874
13	169	3.606	63	3969	7.937
14	196	3.742	64	4096	8.000
15	225	3.873	65	4225	8.062
16	256	4.000	66	4356	8.124
17	289	4.123	67	4489	8.185
18	324	4.243	68	4624	8.246
19	361	4.359	69	4761	8.307
20	400	4.472	70	4900	8.367
21	441	4.583	71	5041	8.426
22	484	4.690	72	5184	8.485
23	529	4.796	73	5329	8.544
24	576	4.899	74	5476	8.602
25	625	5.000	75	5625	8.660
26	676	5.099	76	5776	8.718
27	729	5.196	77	5929	8.775
28	784	5.292	78	6084	8.832
29	841	5.385	79	6241	8.888
30	900	5.477	80	6400	8.944
31	961	5.568	81	6561	9.000
32	1024	5.657	82	6724	9.055
33	1089	5.745	83	6889	9.110
34	1156	5.831	84	7056	9.165
35	1225	5.916	85	7225	9.220
36	1296	6.000	86	7396	9.274
37	1369	6.083	87	7569	9.327
38	1444	6.164	88	7744	9.381
39	1521	6.245	89	7921	9.434
40	1600	6.325	90	8100	9.487
41	1681	6.403	91	8281	9.539
42	1764	6.481	92	8464	9.592
43	1849	6.557	93	8649	9.644
44	1936	6.633	94	8836	9.695
45	2025	6.708	95	9025	9.747
46	2116	6.782	96	9216	9.798
47	2209	6.856	97	9409	9.849
48	2304	6.928	98	9604	9.899
49	2401	7.000	99	9801	9.950
50	2500	7.071	100	10000	10.000

TRIGONOMETRIC RATIOS

Angle	sin	cos	tan	Angle	sin	cos	tan
0°	0.0000	1.0000	0.0000	45°	0.7071	0.7071	1.0000
1°	0.0175	0.9998	0.0175	46°	0.7193	0.6947	1.0355
2°	0.0349	0.9994	0.0349	47°	0.7314	0.6820	1.0724
3°	0.0523	0.9986	0.0524	48°	0.7431	0.6691	1.1106
4°	0.0698	0.9976	0.0699	49°	0.7547	0.6561	1.1504
5°	0.0872	0.9962	0.0875	50°	0.7660	0.6428	1.1918
6°	0.1045	0.9945	0.1051	51°	0.7771	0.6293	1.2349
7°	0.1219	0.9925	0.1228	52°	0.7880	0.6157	1.2799
8°	0.1392	0.9903	0.1405	53°	0.7986	0.6018	1.3270
9°	0.1564	0.9877	0.1584	54°	0.8090	0.5878	1.3764
10°	0.1736	0.9848	0.1763	55°	0.8192	0.5736	1.4281
11°	0.1908	0.9816	0.1944	56°	0.8290	0.5592	1.4826
12°	0.2079	0.9781	0.2126	57°	0.8387	0.5446	1.5399
13°	0.2250	0.9744	0.2309	58°	0.8480	0.5299	1.6003
14°	0.2419	0.9703	0.2493	59°	0.8572	0.5150	1.6643
15°	0.2588	0.9659	0.2679	60°	0.8660	0.5000	1.7321
16°	0.2756	0.9613	0.2867	61°	0.8746	0.4848	1.8040
17°	0.2924	0.9563	0.3057	62°	0.8829	0.4695	1.8807
18°	0.3090	0.9511	0.3249	63°	0.8910	0.4540	1.9626
19°	0.3256	0.9455	0.3443	64°	0.8988	0.4384	2.0503
20°	0.3420	0.9397	0.3640	65°	0.9063	0.4226	2.1445
21°	0.3584	0.9336	0.3839	66°	0.9135	0.4067	2.2460
22°	0.3746	0.9272	0.4040	67°	0.9205	0.3907	2.3559
23°	0.3907	0.9205	0.4245	68°	0.9272	0.3746	2.4751
24°	0.4067	0.9135	0.4452	69°	0.9336	0.3584	2.6051
25°	0.4226	0.9063	0.4663	70°	0.9397	0.3420	2.7475
26°	0.4384	0.8988	0.4877	71°	0.9455	0.3256	2.9042
27°	0.4540	0.8910	0.5095	72°	0.9511	0.3090	3.0777
28°	0.4695	0.8829	0.5317	73°	0.9563	0.2924	3.2709
29°	0.4848	0.8746	0.5543	74°	0.9613	0.2756	3.4874
30°	0.5000	0.8660	0.5774	75°	0.9659	0.2588	3.7321
31°	0.5150	0.8572	0.6009	76°	0.9703	0.2419	4.0108
32°	0.5299	0.8480	0.6249	77°	0.9744	0.2250	4.3315
33°	0.5446	0.8387	0.6494	78°	0.9781	0.2079	4.7046
34°	0.5592	0.8290	0.6745	79°	0.9816	0.1908	5.1446
35°	0.5736	0.8192	0.7002	80°	0.9848	0.1736	5.6713
36°	0.5878	0.8090	0.7265	81°	0.9877	0.1564	6.3138
37°	0.6018	0.7986	0.7536	82°	0.9903	0.1392	7.1154
38°	0.6157	0.7880	0.7813	83°	0.9925	0.1219	8.1443
39°	0.6293	0.7771	0.8098	84°	0.9945	0.1045	9.5144
40°	0.6428	0.7660	0.8391	85°	0.9962	0.0872	11.4301
41°	0.6561	0.7547	0.8693	86°	0.9976	0.0698	14.3007
42°	0.6691	0.7431	0.9004	87°	0.9986	0.0523	19.0811
43°	0.6820	0.7314	0.9325	88°	0.9994	0.0349	28.6363
44°	0.6947	0.7193	0.9657	89°	0.9998	0.0175	57.2900
45°	0.7071	0.7071	1.0000	90°	1.0000	0.0000	∞